MW00913933

DIAGNOSTIC IMAGING
ABDOMEN

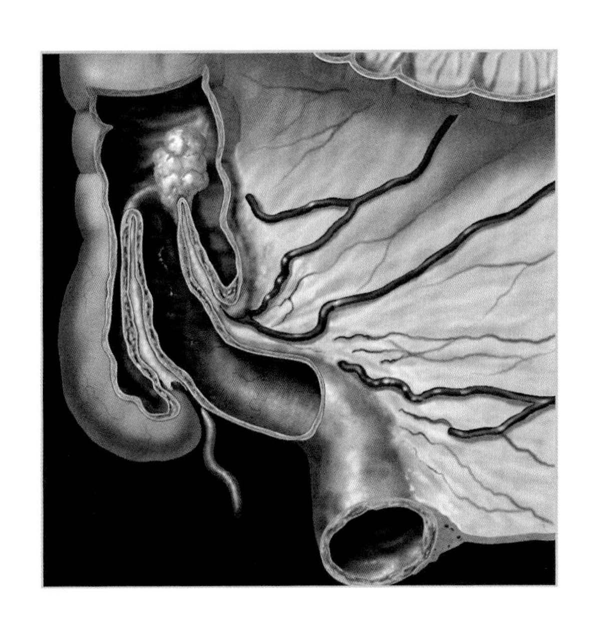

DIAGNOSTIC IMAGING
ABDOMEN

Michael P. Federle, MD, FACR
Professor of Radiology
Chief Abdominal Imaging
University of Pittsburgh Medical Center

R. Brooke Jeffrey, MD
Professor of Radiology
Chief of Abdominal Imaging
Stanford University Medical Center

Terry S. Desser, MD
Assistant Professor of Radiology
Stanford University School of Medicine

Venkata Sridhar Anne, MD
Clinical Research Fellow
Radiology Department
Abdominal Imaging Division
University of Pittsburgh Medical Center

Andres Eraso, MD
Assistant Clinical Professor of Radiology
Louisiana State University Health Sciences Center
Radiologist
Veterans Affairs Medical Center
New Orleans, Louisiana

Joseph Jen-Sho Chen, BA
Medical Student; Research Fellow
Radiology Department
Abdominal Imaging Division
University of Pittsburgh Medical Center

Shalini Guliani-Chabra, MD
Radiology Resident
Radiology Department – Abdominal Imaging Division
University of Pittsburgh Medical Center

Karen M. Pealer, BA, CCRC
Clinical Research Coordinator
Radiology Department – Abdominal Imaging Division
University of Pittsburgh Medical Center

AMIRSYS®
Names you know, content you trust®

AMIRSYS®

Names you know, content you trust®

First Edition

Composition by Amirsys Inc, Salt Lake City, Utah

Printed by Friesens, Altona, Manitoba, Canada

ISBN: 1-4160-2541-3
ISBN: 0-8089-2316-1 (International English Edition)

Notice and Disclaimer

Library of Congress Cataloging-in-Publication Data

Diagnostic imaging. Abdomen / Michael P. Federle ... [et al.].— 1st ed.
p. ; cm.
Includes bibliographical references and index.
ISBN 1-4160-2541-3
1. Abdomen—Imaging.
[DNLM: 1. Radiography, Abdominal—methods. 2. Diagnostic Imaging—methods. 3. Digestive System Diseases—radiography. WI 900 D536 2004] I. Title: Abdomen. II. Federle, Michael P.

RC944.D526 2004
617.5'50754—dc22

2004022701

This book is dedicated to my wife, Lynne, who graciously endured many months of late dinners during its writing, and to my colleagues and trainees at the University of Pittsburgh Medical Center, who enthusiastically joined in my quest for the "great cases" to include in the book.

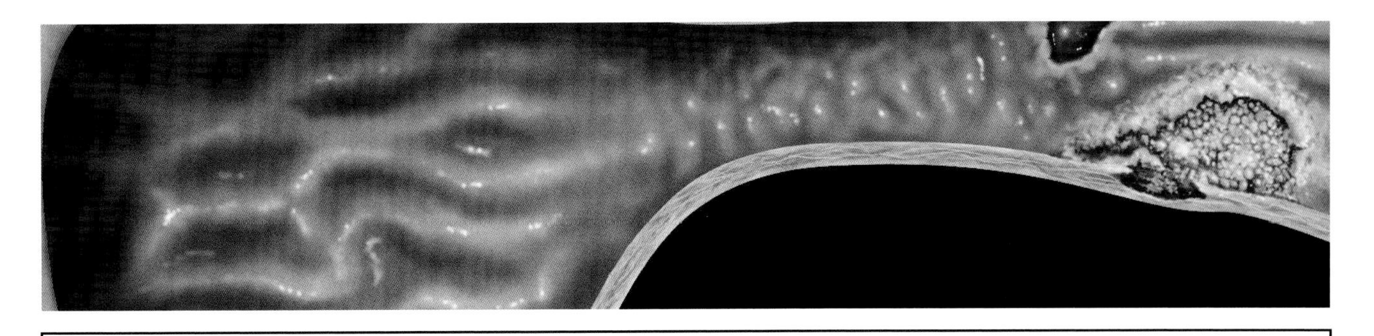

DIAGNOSTIC IMAGING: ABDOMEN

We at Amirsys and Elsevier are proud to present Abdomen, the fifth volume in our acclaimed *Diagnostic Imaging* series. This precedent-setting, image- and graphic-packed series began with David Stoller's Orthopaedics. The next three books, Brain, Head and Neck, and Spine are now joined by the first of three volumes that will focus on abdominal imaging issues. Mike Federle and his team lead off with Abdomen. Subsequent topics will include Obstetrics and Gynecology.

The unique bulleted format of the *Diagnostic Imaging* series allows our authors to present approximately twice the information and four times the images per diagnosis compared to the old-fashioned traditional prose textbook. All the *DI* books follow the same format, which means the same information is in the same place: Every time! In every organ system. The innovative visual differential diagnosis "thumbnail" that provides an at-a-glance look at entities that can mimic the diagnosis in question has been highly popular. "Key Facts" boxes provide a succinct summary for quick, easy review.

In summary, *Diagnostic Imaging* is a product designed with you, the reader, in mind. Today's typical practice settings demand efficiency in both image interpretation and learning. We think you'll find the new Abdomen volume a highly efficient and wonderfully rich resource that will be the core of your reference collection in abdominal imaging. Enjoy!

Anne G. Osborn, MD
Executive Vice President and Editor-in-Chief, Amirsys Inc.

H. Ric Harnsberger, MD
CEO, Amirsys Inc.

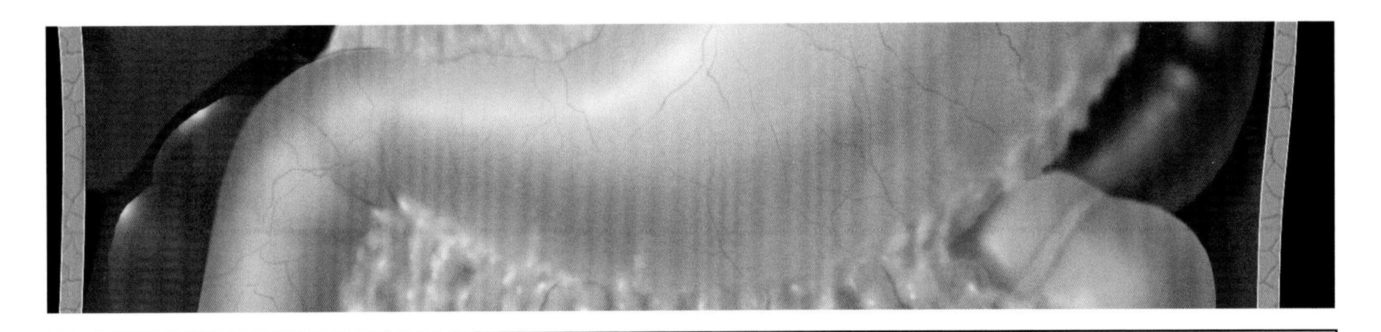

FOREWORD

Those of us who are abdominal imagers have been eagerly awaiting the Diagnostic Imaging: Abdomen book, having admired the prior volumes in this remarkable series of imaging texts. We are also pleased to see that the editor of this volume is Michael Federle, MD, who is Professor of Radiology and Chief of Abdominal Imaging at the University of Pittsburgh Medical Center.

I have been privileged to know Michael for over 20 years. I share his passion for abdominal imaging, but I have always been in awe of his clinical and teaching skills. He has used all his experience, judgment and expertise to produce a teaching tool that will be useful at multiple levels of training, from medical student to practicing radiologist. The format of the Diagnostic Imaging series makes it easy to learn the relevant imaging and clinical aspects of every major disease or lesion confronting the radiologist, and should facilitate meaningful communication with referring physicians.

A unique feature of these texts is the abundance of beautiful medical illustrations that virtually animate the disease process and the accompanying radiographs. The imaging studies are remarkably excellent, carefully selected and cropped to highlight every major manifestation of every Diagnosis. The illustrated Differential Diagnosis series and the accompanying text are other unique features, making it easy to distinguish among several diseases that may have similar clinical or imaging features.

This text covers all of the major gastrointestinal and genitourinary organs and disease processes from congenital through traumatic, inflammatory and neoplastic. Between the individual "Diagnoses" (chapters) and the Differential Diagnoses listed for each, you will find a comprehensive coverage of the factual material covered in the multi-volume encyclopedic prose textbooks. The Anatomic Overview and Imaging Issues for each section are written by Dr. Federle to provide the "glue" to pull the individual Diagnoses together, to suggest optimal imaging protocols, and to provide extensive tables of Custom Differential Diagnoses that the reader will be referring to on a regular basis.

I have often instructed my residents that the successful radiologist understands three aspects of a disease process: *What does the lesion look like on imaging studies? What does the lesion do on a histopathological and clinical level? What do these pathological results look like on imaging?* Apparently, Michael takes the same approach, and this book will provide the understanding of these aspects of every GI and GU process likely to be encountered in clinical practice.

Modern imaging is very powerful in revealing the source of a patient's symptoms or signs. This book, as with the others in the Diagnostic Imaging series, succeeds magnificently in conveying the understanding necessary to make optimal use of our imaging tools.

Alec J. Megibow, MD, MPH, FACR
Professor and Vice Chairman for Education
New York University School of Medicine

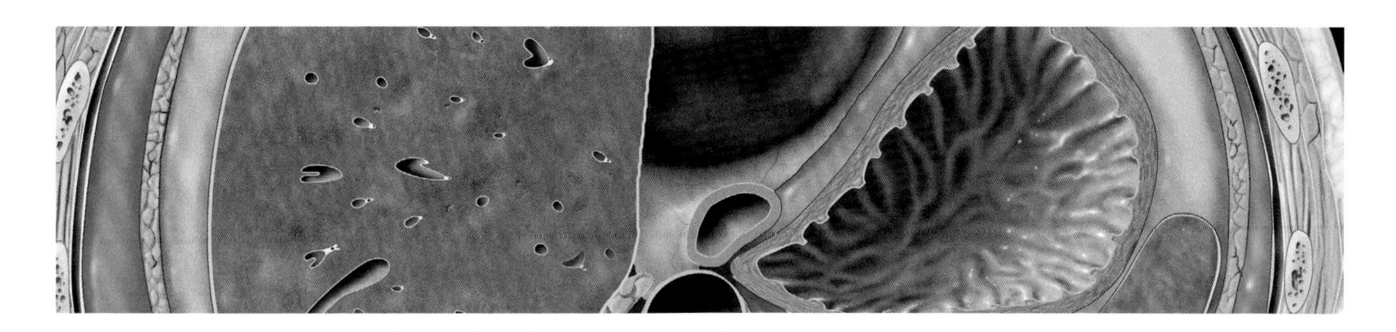

PREFACE

This is a book that almost did not happen. For many years I have wanted to write a comprehensive text of abdominal imaging that would serve as a "readable reference" for radiology residents and practicing physicians. Increasing clinical demands and the frustrations inherent in writing and editing a traditional multivolume-multiauthor text seemed to preclude this. Ric Harnsberger approached me several years ago with his vision of an entirely new type of textbook and method of writing, with senior authors in all fields of radiology being the driving force. The text and images would be entered into a proprietary computer program (the Amirsys authoring tool) that would help organize the data into easily accessible, bulleted text rather than traditional prose, saving an estimated 50% of unnecessary verbiage. The reader would find the key facts, not just in imaging, but in clinical, pathological and treatment options of every important diagnosis, and he would find it in the same place for every diagnosis and every book in the series. Moreover, we would ignore the usual publishers' discouragement of the use of color images, and would use color images and original artwork lavishly. I signed on with some trepidation.

The use of the Amirsys "authoring tool" provided and additional benefit of allowing for a certain uniformity of style and depth of coverage that is usually impossible to achieve in a multiauthor text. At the University of Pittsburgh, we assembled a team of bright young people who performed much of the literature review and data entry, freeing me to concentrate on editing, overview and selecting the best images and illustrations for maximum teaching value. Our proprietary authoring tool even facilitated the updating and entry of references; you will note many references from within the past year, previously impossible to achieve with the lag time to publication of a traditional textbook.

The Pittsburgh team of contributors includes Venkata Anne, MD, who also assisted with the research and writing of our first Amirsys book, the Pocket Radiologist, Top 100 Diagnoses in Abdominal Imaging. Dr. Anne has augmented his residency in radiology with a thorough review of available texts and journal articles to acquire an encyclopedic knowledge of abdominal imaging, and he has contributed substantially to the majority of diagnoses included in our Pocket Radiologist and current text.

This comprehensive work would not have been accomplished without the dedication and efforts of many people:

Andres Eraso, MD, a former resident and fellow at Pitt, was my main assistant in reviewing our teaching files and medial records to select the most compelling and informative images for the book. Joe Chen is a medical student and future radiologist who was part of the research and writing team, along with Shalini Guliani, MD a radiologist trained in India who is gaining additional training in our radiology residency. Their co-authorship of diagnoses is noted in the table of contents, and their excellent work is both acknowledged and appreciated.

Karen Pealer has been my research assistant for many years and was absolutely instrumental in the success of this project. Karen used all of her computer and communication skills to help me organize, edit, enter and illustrate each of the diagnostic "chapters" and introductory sections. She was also the conduit of information and problem solving between Pittsburgh and Amirsys headquarters in Salt Lake City. I suppose it would have been possible for an old dog like me to learn all the new computer tricks necessary to complete this project, but it would have been much more difficult and time-consuming.

The Stanford team was led by my career-long colleague and friend, Brooke Jeffrey, MD, himself the author of several excellent textbooks and innumerable articles and other contributions. Brooke has always been one of the top radiologists and educators in the world, and we are proud to have him as part of this project. Terry Desser, MD is the other Stanford faculty radiologist who contributed dozens of diagnoses that are all beautifully researched, written and illustrated.

The Amirsys team all contributed their own expertise. I particularly want to acknowledge Ric Harnsberger whose unfailing vision and enthusiasm have seen this through some difficult times. The medical illustrations, so key to understanding and such an Amirsys "trademark", are the work of Rich Coombs, MS and James Cooper, MD I absolutely love what they have done and you will be seeing them in my lectures for many years!

Michael P. Federle, MD, FACR

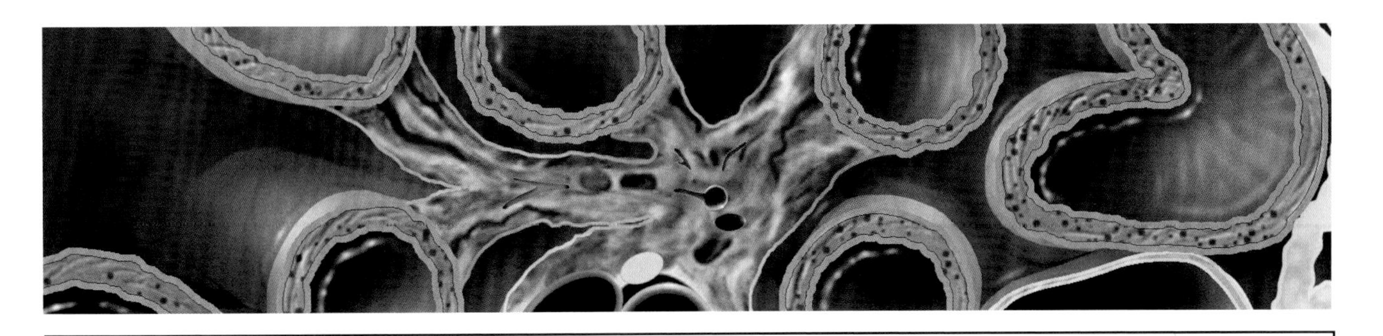

ACKNOWLEDGMENTS

Illustrations

Richard Coombs, MS
James Cooper, MD

Art Direction and Design

Lane R. Bennion, MS
Richard Coombs, MS

Image/Text Editing

Angie D. Mascarenaz
Cassie L. Dearth
Kaerli Main
Roth LaFleur
David Harnsberger

Medical Text Editing

Andre Macdonald, MD
Doug Green, MD
David Avrin, MD
Akram Shaaban, MD

Production Lead

Melissa A. Morris

SECTIONS

PART I
GI Tract and Abdominal Cavity

PART II
Hepatobiliary and Pancreas

PART III
Genitourinary and Retroperitoneum

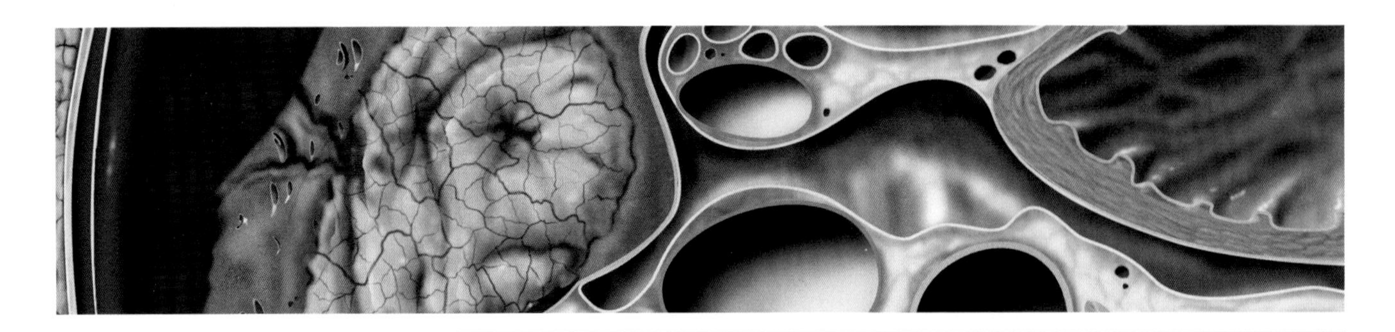

TABLE OF CONTENTS

PART I
GI Tract and Abdominal Cavity

SECTION 1
Peritoneum, Mesentery, and Abdominal Wall

Introduction and Overview

Infection

Inflammation

External Hernias

Internal Hernias

Trauma

Neoplasm, Benign

Neoplasm, Malignant

SECTION 2
Esophagus

Introduction and Overview

Congenital

Infection

Inflammation

Degenerative

Esophageal Diverticula

Trauma

Neoplasm, Benign

Neoplasm, Malignant

SECTION 3
Gastroduodenal

Introduction and Overview

Congenital

Inflammation

Trauma

Neoplasm, Benign

Neoplasm, Malignant

Treatment Related

Miscellaneous

SECTION 4
Small Intestine

Introduction and Overview

Congenital

Infection

Inflammation

Vascular Disease

Trauma

Neoplasm, Benign

Neoplasm, Malignant

Miscellaneous

SECTION 5
Colon

Introduction and Overview

Infection

Inflammation and Ischemia

Vascular Disorder

Metabolic - Inherited

Trauma

Neoplasm, Benign

Neoplasm, Malignant

Treatment Related

SECTION 2
Biliary System

Introduction and Overview

Congenital

Infection

Inflammation

Neoplasm, Primary

Treatment Related

SECTION 3
Pancreas

Introduction and Overview

Normal Variants and Congenital

Inflammation

Trauma

Neoplasm, Benign

Neoplasm, Malignant

PART III
Genitourinary and Retroperitoneum

SECTION 1
Retroperitoneum

Introduction and Overview

Normal Variants

Inflammation

Trauma

Neoplasm

Treatment Related

SECTION 2
Adrenal

Introduction and Overview

Infection

Vascular - Traumatic

Metabolic

Neoplasm, Benign

Neoplasm, Malignant

SECTION 3
Kidney and Urinary Tract

Introduction and Overview

Normal Variants and Pseudolesions

Congenital

Infection

Inflammation

Vascular

Trauma

Neoplasm, Benign

Neoplasm, Malignant

SECTION 4
Ureter

Congenital

Infection

SECTION 5
Bladder

Introduction and Overview

Congenital

Infection

Inflammation

Trauma

Neoplasm

SECTION 6
Genital Tract (Male)

Introduction and Overview

Congenital

Infection

Inflammation

Trauma

Neoplasm

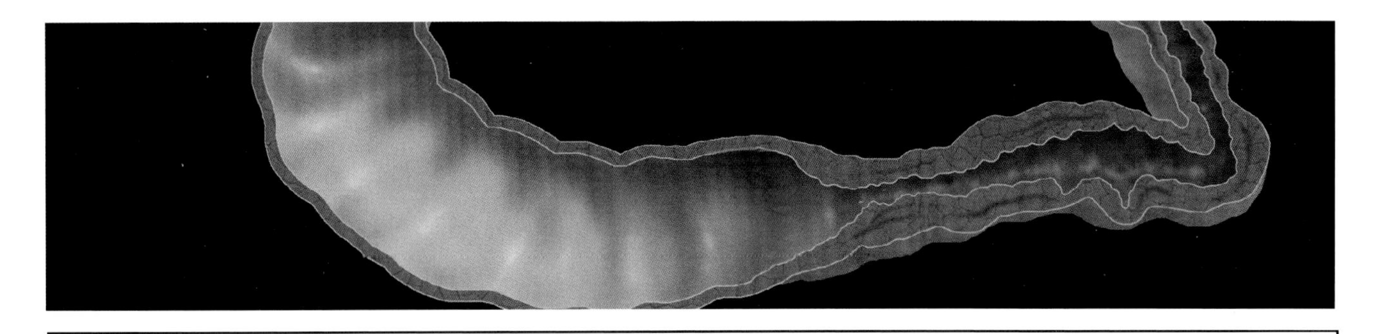

ABBREVIATIONS

ACKD: Acquired Cystic Kidney Disease
ADPKD: Autosomal Dominant Polycystic Kidney Disease
ADPLD: Autosomal Dominant Polycystic Liver Disease
AFB: Acid Fast Bacilli
AML: Angiomyolipoma
AML: Renal Angiomyolipoma
ANA: Antinuclear Antibodies
APUD: Amine Precursor Uptake & Decarboxylation
ARPKD: Autosomal Recessive Polycystic Kidney Disease
ATN: Acute Tubular Necrosis
AVM: Arterio-Venous Malformation
BE: Barium Enema
BPH: Benign Prostatic Hypertrophy
CA: Cancer
CBD: Common Bile Duct
CEA: Carcinomembryonic Antigen
CECT: Contrast Enhanced Computed Tomography
CMV: Cytomegalovirus
DIC: Disseminated Intravascular Coagulation
DPL: Diagnostic Peritoneal Lavage
DVT: Deep Venous Thrombosis
EBV: Ebstein-Barr Virus
EHE: Epitheloid Hemangioendothelioma
ERCP: Endoscopic Retrograde Cholangiapancreatogram
ESR: Erythrocyte Sedimentation Rate
EUS: Endoscopic Ultrasound
FAPS: Familial Adenomatous Polyposis Syndrome
FDG: 18F Flourodeoxyglucose
FNA: Fine Needle Aspiration
FUDR: Floxuridine
GB: Gallbladder
GBM: Glomerular Basement Membrane
GE: Gastroesophageal
GERD: Gastroesophageal Reflux Disease
GIST: Gastrointestinal Stromal Tumor
GU: Genitourinary
HAP: Hepatic Arterial Phase
HAV: Hepatis A
HBV: Hepatis B
HCC: Hepatocellular Carcinoma
HCV: Hepatis C
HELLP: Hemolysis, Elevated Liver Enzymes, Low Platelets
HHT: Hereditary Hemorrhagic Telangiectasia
HIV: Human Immunodeficiency Virus
HLA: Human Leukocyte Antigen
HNPCC: Hereditary Nonpolyposis Colorectal Cancer
HPV: Human Papilloma Virus
HSV: Herpes Simplex Virus
HU: Hounsfield Units
IHBD: Intrahepatic Bile Ducts
IMA: Inferior Mesenteric Artery
IMV: Inferior Mesenteric Vein
IPMT: Intraductal Papillary Mucinous Tumor
ITP: Idiopathic Thrombocytopenic Purpura
IVC: Inferior Vena Cava
IVU: Excretory Urography
Lap. Port: Laproscopy Port
LES: Lower Esophageal Sphincter
LLQ: Left Lower Quadrant
LUQ: Left Upper Quadrant
MAI: Atypical Myobacterial Infection
MALT: Mucosa-Asociated Lymphoid Tissue
MEN: Multiple Endocrine Neoplasia
Mets: Metastasis
MIP: Maximum Intensity Projection
MPD: Main Pancreatic Duct
MRCP: MR Cholangiopancreatography
MRS: MR Spectroscopy
MVA: Motor Vehicle Accidents
NASH: Nonalcoholic Steatohepatis
NECT: Non-enhanced Computed Tomography
NET: Neuroendocrine Tumor
NG: Nasogastric
NHL: Non-Hodgkin Lymphoma
NSAID: Non-steroidal Anti-inflammatory Drug
O-W-R: Osler-Weber-Rendu Disease
PBC: Primary Biliary Cirrhosis
PD: Pancreatic Duct
PET: Positron Emission Tomograpghy
PJS: Peutz-Jeghers Syndrome
PMN: Polymorphonuclear Leukocytes
PMP: Psuedomyxoma Peritonei
PPD: Positive Purified Protein Derivative
PSC: Primary Sclerosing Cholangitis
PSS: Progressive Systematic Sclerosis
PTC: Percutaneous Transhepatic Cholangiography
PUD: Peptic Ulcer Disease
PVP: Portal Venous Phase
RAO: Right Anterior Oblique
RARE: Rapid Acquisition with Relaxation Enhancement
RCC: Renal Cell Carcinoma
RES: Reticuloendothelial System
RLQ: Right Lower Quadrant
RUQ: Right Upper Quadrant
RYGB: Roux-enY Gastric Bypass
SB: Small Bowel
SBFT: Small Bowel Follow Through
SBO: Small Bowel Obstruction, Closed Loop Obstruction
SMA: Superior Mesenteric Artery
SMV: Superior Mesenteric Vein
SSFSE: Single Shot Fast Spin Echo
SVC: Superior Vena Cava
TB: Tuberculosis
TCC: Transitional Cell Carcinoma
TI: Terminal Ileum
TNM: Tumor, Nodal, Metastasis
TURP: Transurethral Prostatectomy
UC: Ulcerative Colitis
UGI: Upper GI Series
UPJ: Ureteropelvic Junction
US: Ultrasound
UTI: Urinary Tract Infection
UVJ: Ureterovesical Junction
VZV: Varicella-Zoster Virus
WBC Scan: White Blood Cell Scan
XPGN: Xanthogranulomatous Pyelonephritis
ZES: Zollinger-Ellison Syndrome

DIAGNOSTIC IMAGING
ABDOMEN

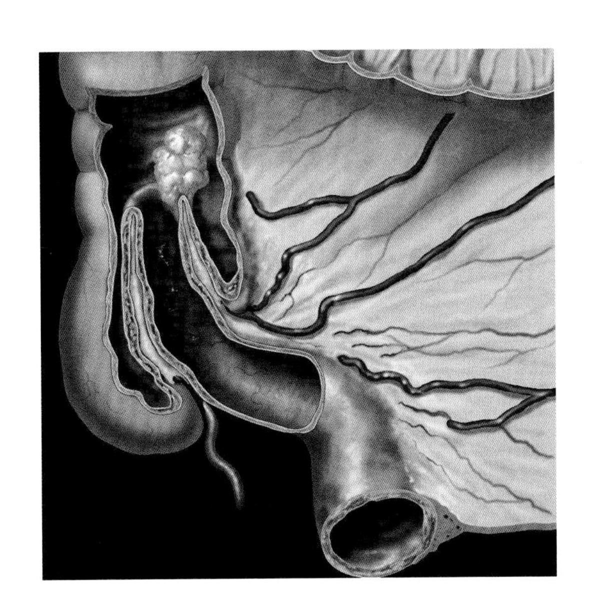

PART I

GI Tract and Abdominal Cavity

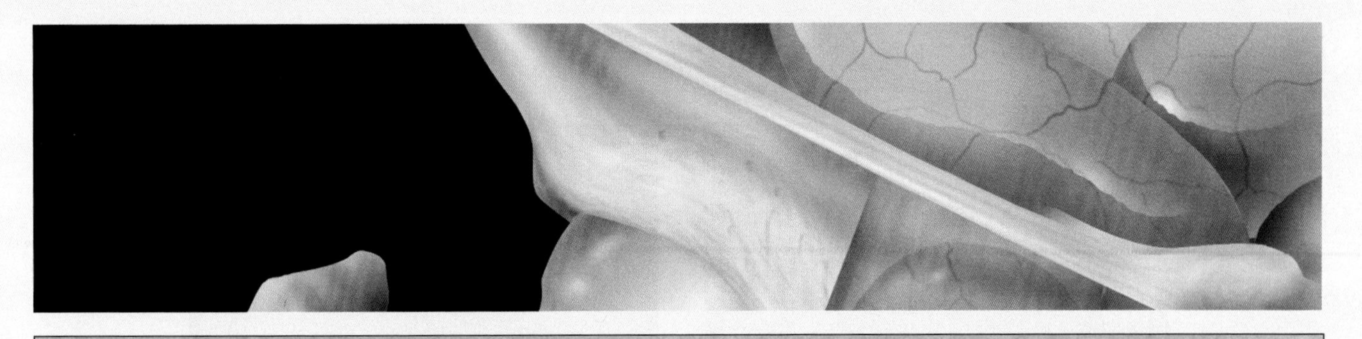

SECTION 1: Peritoneum, Mesentery, and Abdominal Wall

Introduction and Overview

Infection

Inflammation

External Hernias

Internal Hernias

Trauma

Neoplasm, Benign

Neoplasm, Malignant

PERITONEUM, MESENTERY, AND ABDOMINAL WALL

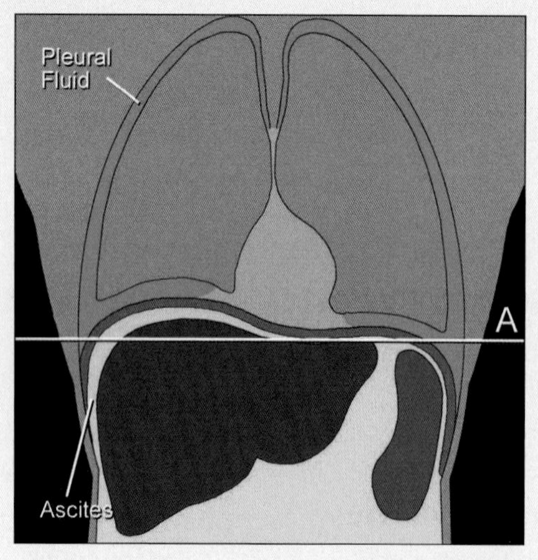

Graphic shows pleural fluid (green) outside the confines of the diaphragm, and ascites (yellow) within.

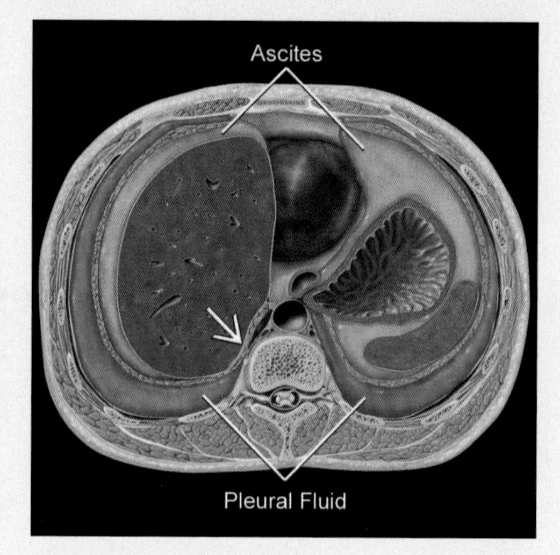

Graphic shows pleural fluid & ascites relative to the diaphragm. Pleural fluid extends to "touch" the spine. Ascites is held medially by diaphragm; excluded from bare area of liver (arrow).

TERMINOLOGY

Definitions

- Peritoneal cavity
  - The potential space within the abdomen and pelvis that is lined with peritoneum
- Peritoneum
  - Single layer of mesothelium that covers a thin layer of connective tissue
  - Parietal peritoneum lines the abdominal and pelvic muscular walls
  - Visceral peritoneum covers the bowel (serosa) and abdominal viscera
- Greater omentum
  - Four-layered fold of peritoneum that includes fat, blood vessels, nerves, and lymph nodes
  - Drapes over bowel, separating it from anterior abdominal wall
  - Joins stomach and transverse colon (as the gastro-colic ligament), then continues inferiorly
- Lesser omentum
  - Connects the lesser curvature of stomach and duodenum to undersurface of liver
- Omental bursa
  - Synonymous with lesser sac
- Mesentery
  - Two layers of peritoneum (plus vessels, nerves, etc.) connecting bowel to posterior abdominal wall
  - Covers and suspends jejunum and ileum
  - Transverse and sigmoid mesocolon are analogous in structure and function (root of transverse mesocolon separates upper and lower compartments of the peritoneal cavity
- Ligament
  - Fold of peritoneum that connects and supports structures within peritoneal cavity (e.g., gastro-colic ligament)
- Hernias
  - Abnormal opening in abdominal wall due to trauma, surgery, or weakness of supporting structures
  - May be inapparent on clinical exam (obesity, deep location) or mistaken for an abdominal mass
  - Easily demonstrated on CT, including complications (e.g., bowel obstruction)
  - Ventral hernia - through linea alba
  - Spigelian hernia - along lateral margin of rectus muscle, through defect in aponeuroses of internal oblique and transverse muscles
  - Umbilical
  - Lumbar - through posterolateral abdominal wall
    - Inferior lumbar triangle, just above iliac crest
    - Superior lumbar (Grynfelt triangle), between 12th rib and erector spinae muscles
  - Inguinal
    - Indirect through deep inguinal ring
  - Femoral - adjacent to femoral vessels
  - Obturator - between pectineus and obturator muscles
  - Diaphragmatic
    - Esophageal - hiatal
    - Traumatic
    - Morgagni - congential, anterior
    - Bochdalek - congential, posterior

ANATOMY-BASED IMAGING ISSUES

Key Concepts or Questions

- Peritoneum and pleura are almost identical in structure and function
  - Are subject to the same inflammatory and neoplastic processes
    - E.g., empyema and abdominal abscess; mesothelium (pleural and peritoneal)
  - Imaging manifestations are similar
    - E.g., smooth thickening with infection; nodular thickening with tumor

PERITONEUM, MESENTERY, AND ABDOMINAL WALL

DIFFERENTIAL DIAGNOSIS

Mesenteric or omental tissue mass

Common
- Hematoma
- Lymphoma
- ⇒ (Especially non-Hodgkin lymphoma)
- Lymphadenopathy
- Pancreatitis
- Carcinomatosis, metastases

Uncommon
- Mesothelioma
- Desmoid
- Mesenchymal (benign and malignant)
- ⇒ (E.g., lipoma, liposarcoma)
- Carcinoid

Cystic mesenteric or omental mass

Common
- Loculated ascites
- Abscess
- Metastases
- ⇒ (E.g., ovarian, cystadenocarcinoma
- Pseudocyst (pancreatitis)

Uncommon
- Pseudomyxoma peritonei
- Mesenteric cyst, lymphangioma
- Cystic mesothelioma
- Urachal cyst
- "Cystic" (caseated) lymph nodes
- ⇒ Mycobacterial (TB, MAI); Whipple disease

        - Mesothelioma and peritoneal (or pleural) metastases are indistinguishable by imaging
- How do you determine on CT or MR whether a peridiaphragmatic fluid collection is intrathoracic (i.e., pleural) or intra-abdominal (i.e., ascites, abscess)?
    - Relationship to diaphragm
        - Outside the confines of the diaphragm = intrathoracic
        - Inside (medial to) the diaphragm = abdominal
    - Pleural fluid "touches" the body wall and paraspinal region
        - Ascites is held medially by the diaphragm
    - Pleural fluid has a "fuzzy" (indistinct) margin as it abuts the top of liver and spleen
        - Ascites has "sharp" interface
    - Do not mistake atelectatic lung for diaphragm
        - Atelectatic lower lobe appears as a tapering curvilinear structure that is wider medially and on more cephalic sections
        - Diaphragm is a continuous curvilinear "line" of uniform width that moves toward the body wall on more caudal sections
    - Pleural fluid displaces lung from body wall
        - Ascites displaces abdominal viscera medially
- Fluid in peritoneal cavity accumulates and spreads predictably
    - First, near site of origin
        - E.g., cirrhosis with ascites in right subphrenic and subhepatic spaces)
    - Influence of diaphragmatic motion
        - Creates "suction", favoring collection in subphrenic spaces
    - Influence of gravity
        - Pelvis is most dependent recess
        - Morison pouch (posterior subhepatic space); most dependent in upper abdomen
        - Right paracolic gutter wider and more dependent than left
        - Paracolic gutters are a common pathway for spread of fluid between upper abdomen and pelvis
    - Fluid between mesenteric leaves
        - Usually of bowel/mesenteric origin
        - Has triangular, sharp wedge shape due to peritoneal reflections
    - Lesser sac
        - Communicates with greater peritoneal cavity through an opening in the hepatoduodenal ligament (foramen of Winslow)
        - Most patients with benign transudative ascites will not have fluid in the lesser sac (unless ascites is massive)
        - Extensive fluid in lesser sac suggests a local source (such as pancreatitis) or disseminated peritoneal tumor or infection
    - Pouch of Douglas
        - = Rectovaginal space (women), rectovesical space (men)
        - Most dependent portion of peritoneal cavity
        - Common site for pooling of infected or malignant ascites (= abscess or peritoneal pelvic implant or tumor)

Imaging Approaches
- CT is best overall
- Ultrasound can detect small peritoneal implants in the presence of extensive ascites
- Laparoscopy detects many seed-like tumor implants missed by imaging

CLINICAL IMPLICATIONS

Clinical Importance
- Greater omentum
    - Functions as "nature's bandage" (adheres to, and limits the spread of hemorrhage and inflammation
    - Usually prevents bowel perforations (e.g., diverticulitis, appendicitis) from leading to generalized peritonitis
    - Common site for metastatic disease
        - Gastrointestinal tract and ovarian tumors most common
        - Usually accompanied by ascites
    - Used by surgeons to cover "raw" surface of cut or traumatized liver

PERITONEUM, MESENTERY, AND ABDOMINAL WALL

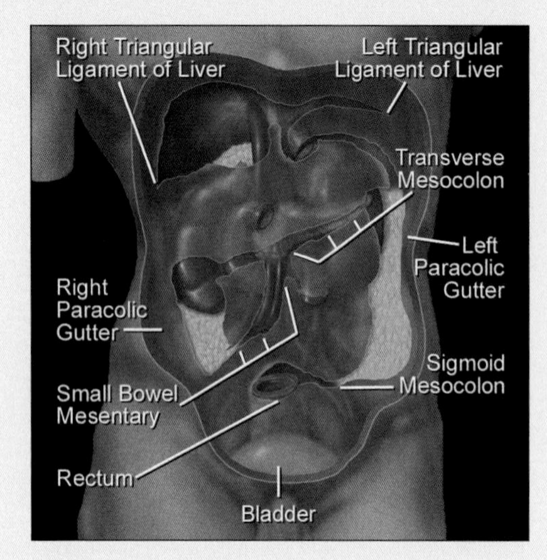

Graphic shows eviscerated abdomen and the major peritoneal reflections. The root of the transverse mesocolon separates the upper & lower portions of the peritoneal cavity.

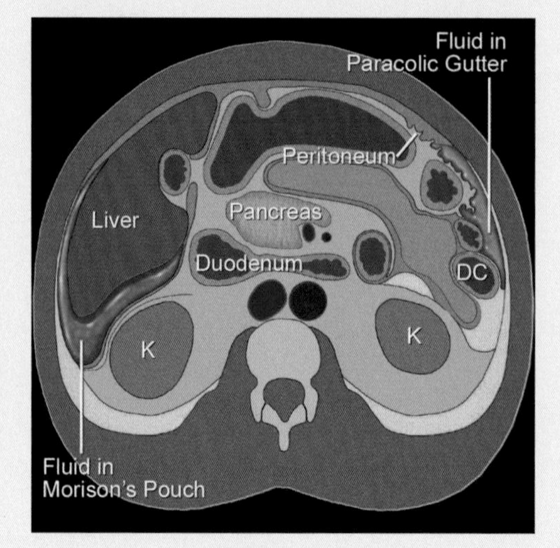

Graphic shows fluid in paracolic gutters. Morison pouch is the cephalic continuation of the right paracolic gutter & the most dependent recess in the the upper abdomen.

- Ascites
  - Accumulation of fluid in the peritoneal cavity, usually due to increased production
  - Common etiologies: Heart failure, cirrhosis, venous or lymphatic obstruction, peritoneal infection or malignancy
  - Attenuation (density) of ascites not very helpful in determining etiology
- Abscess (intraperitoneal)
  - Approximately 1/3 contain gas, usually as small bubbles mixed with fluid
  - Enhancing wall, rounded margins, and mass effect are usually seen
  - Common etiologies: Diverticulitis, Crohn disease, surgery (especially involving colon, stomach, biliary tree)
- Hemorrhage
  - Common etiologies: Trauma, anticoagulation
  - Appearance varies by age and location
  - Acute hemorrhage with active extravasation = extravasation of blood isodense to opacified vessels
  - Clotted blood: Usually 45 to 70 HU, heterogeneous, first accumulates near site of hemorrhage (= "sentinel clot" sign)
  - Lysed clot, free blood in abdomen: Usually 20 to 45 HU
  - Hematocrit effect: Settling of cellular elements; fluid level; sign of anticoagulation
- Tumor
  - Common etiologies: Non-Hodgkin lymphoma (NHL), carcinomas of the GI tract (including pancreas and biliary), ovary and uterus
  - Appearance
    - Infiltration and thickening of mesentery or omentum; mesentery may appear "pleated" and stiff, like a fan
    - Nodular or strand-like soft tissue densities
    - Omental "cake" (soft tissue mass separating bowel from anterior abdominal wall)
    - Calcified foci (usually ovarian carcinoma or malignant teratoma)
    - Rounded mass (usually in mesentery, due to NHL)
    - "Cystic" masses (usually ovarian cystadenocarcinoma)
    - Pseudomyxoma peritonei (massive, septated collections of mucinous fluid; scalloped surface of liver and spleen); appendiceal or ovarian cancer

CUSTOM DIFFERENTIAL DIAGNOSIS

Misty (infiltrated) mesentery

- Mesenteric edema
  - Portal hypertension, heart failure, hypoalbuminemia
- Lymphedema
  - Surgery, congenital, radiation therapy
- Inflammation
  - Pancreatitis, inflammatory and infectious bowel disease
  - Fibrosing (sclerosing) mesenteritis
- Hemorrhage
  - Trauma, anticoagulation, bowel ischemia
- Neoplasms
  - Lymphoma, carcinoid, mesothelioma
  - Ovarian, colon, pancreatic carcinoma

SELECTED REFERENCES

1. Dähnert W: Radiology Review Manual (4th ed), Philadelphia, Lippincott, Williams and Wilkins. 615-721, 2000
2. Ghahremani GG: Abdominal and Pelvic HHernias: In Gore RM, Levine MS (eds) Textbook of Gastrointestinal Radiology (2nd ed), Philadelphia, WB Saunders. 1993-2009, 2000
3. Heiken JP, et al: Peritoneal Cavity and Retroperitoneum: Normal Anatomy and Examination Techniques. In Gore RM, Levine MS (eds) Textbook of Gastrointestinal Radiology (2nd ed), Philadelphia, WB Saunders. 1930-1947, 2000

IMAGE GALLERY

Typical

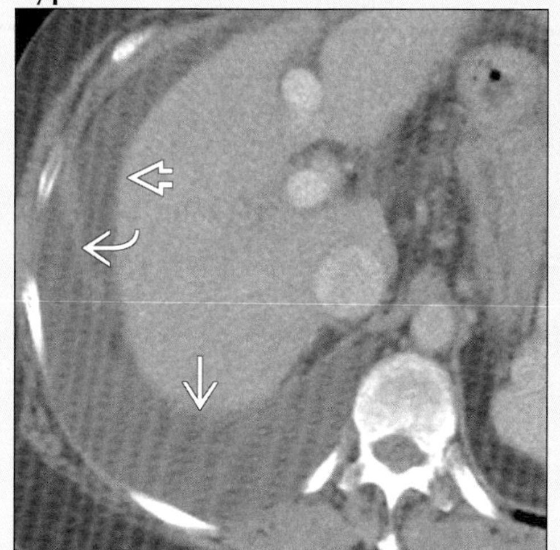

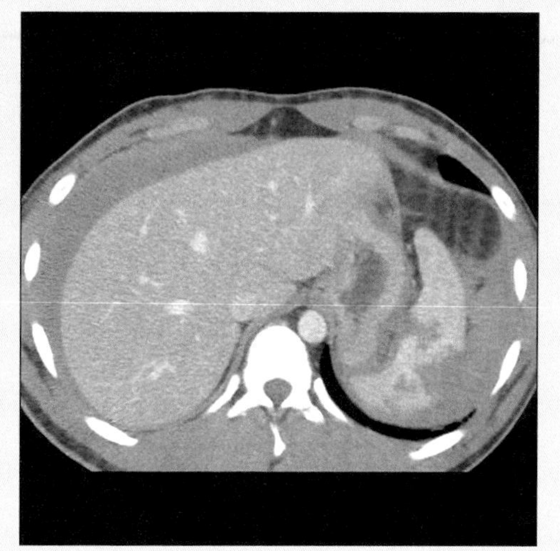

(Left) *Axial CECT shows pleural fluid (arrow) and ascites (open arrow); note interface with surface of liver, spine, and diaphragm (curved arrow).* ***(Right)*** *Axial CECT shows "sentinel clot", high density blood near splenic laceration, lower density blood adjacent to liver.*

Typical

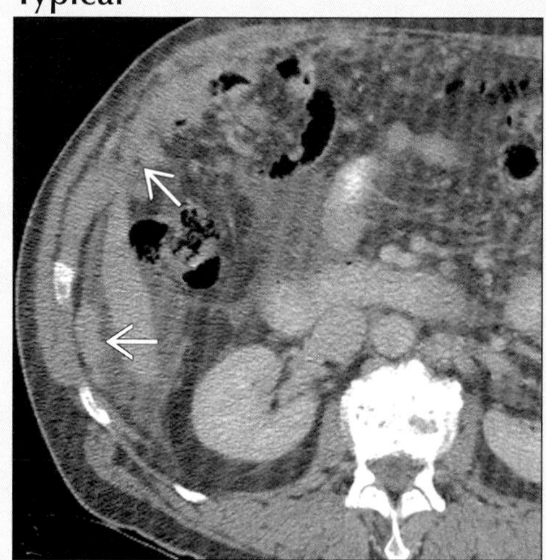

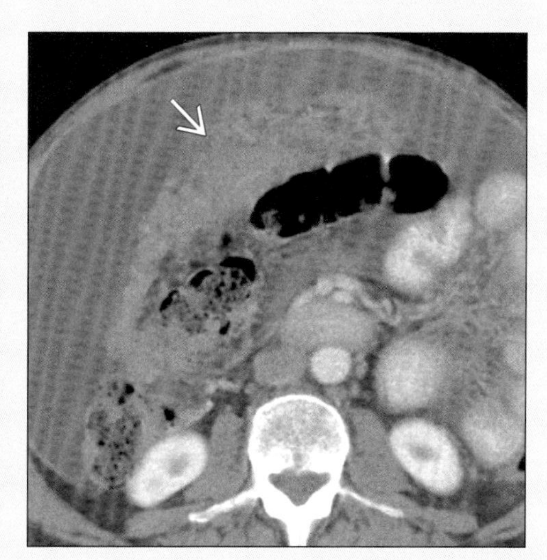

(Left) *Axial CECT shows peritoneal and omental metastases (arrows) from gallbladder carcinoma.* ***(Right)*** *Axial CECT shows "omental cake" (arrow) between transverse colon and abdominal wall; ovarian carcinoma.*

Typical

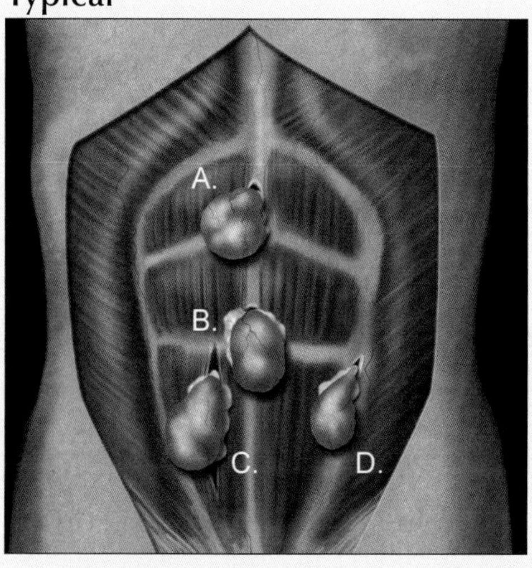

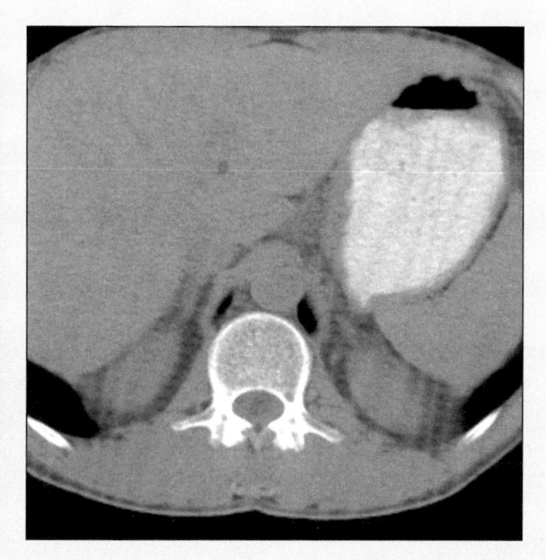

(Left) *Graphic shows typical anterior abdominal wall hernias; A = ventral, epigastric; B = umbilical; C = incisional; D = Spigelian.* ***(Right)*** *Axial NECT shows bilateral Bochdalek hernias with cephalic displacement of kidneys toward the thorax.*

ABDOMINAL ABSCESS

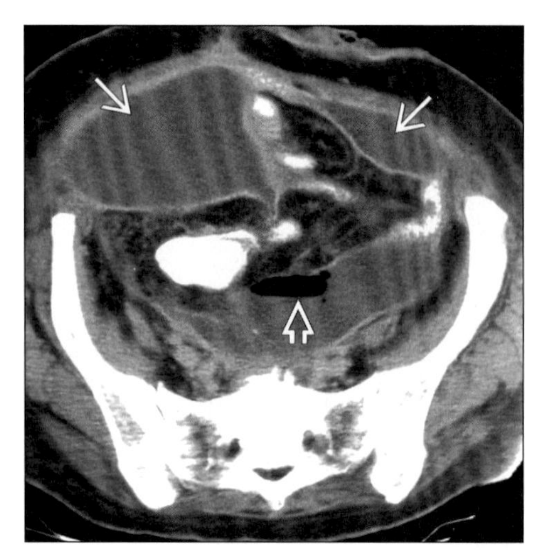

Axial CECT shows pyogenic post-op abscess (arrows) after bowel resection. Note multiple fluid collections with enhancing rims; gas is seen only in pelvic abscess (open arrow).

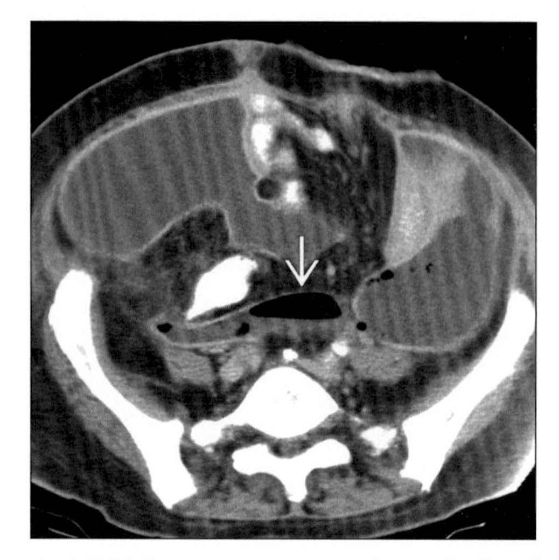

Axial CECT shows pyogenic post-op abscess after bowel resection. Note multiple fluid collections with enhancing rims. Gas noted in pelvic abscess (arrow), but not in other collections.

TERMINOLOGY

Definitions

- Localized abdominal collection of pus

IMAGING FINDINGS

General Features

- Best diagnostic clue: Fluid collection with mass effect & enhancing rim with or without gas bubbles or air-fluid level on CECT
- Location: Anywhere within abdominal cavity; intraparenchyma; within intra- or extraperitoneal spaces
- Size: Highly variable; 2-15 cm in diameter; microabscesses < 2 cm
- Morphology: Low density fluid collection with peripheral enhancing rim

Radiographic Findings

- Radiography
  - Ectopic gas (50% of cases)
  - Air-fluid level
  - Soft tissue "mass"
  - Focal ileus
  - Loss of soft tissue-fat interface
  - Subphrenic abscess: Pleural effusion and lower lobe atelectasis
- Fluoroscopy
  - Abscess sinogram
    - Useful after percutaneous drainage
    - Defines catheter position in dependent portion of abscess
    - Detection of fistulas to bowel, pancreas or biliary duct

CT Findings

- NECT: Low attenuation fluid collection, mass effect, gas in 50% of cases
- CECT: Peripheral rim-enhancement

MR Findings

- T1WI: Low signal
- T2WI: Intermediate to high signal fluid collection
- T1 C+
  - Similar to CECT
  - Low signal fluid collection with enhancing rim

Ultrasonographic Findings

- Real Time
  - Complex fluid collection with internal low level echoes, membranes or septations on US
  - Dependent echoes representing debris
  - Fluid-fluid level

DDx: Spectrum of Cystic Abdominal Lesions Mimicking Abscess

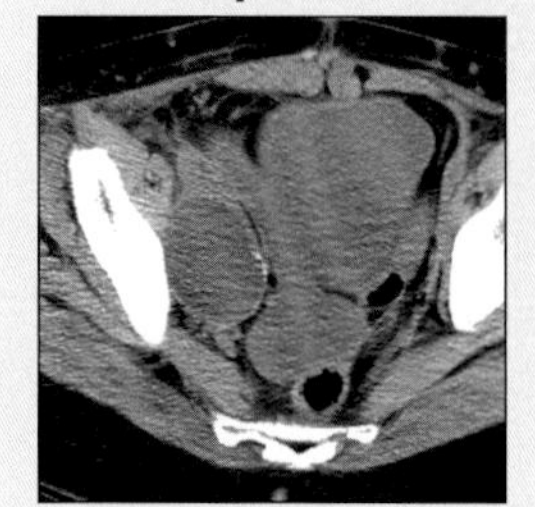

Lymphocele

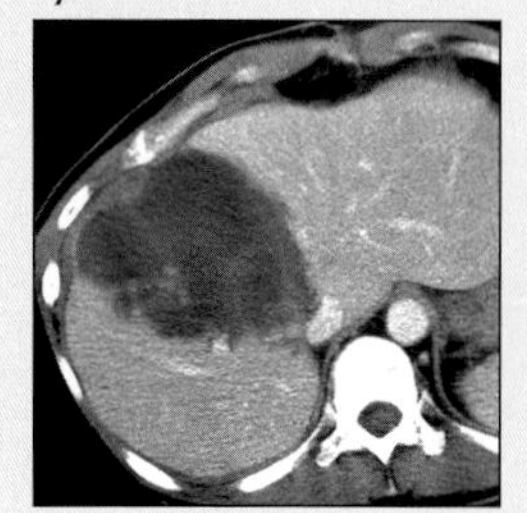

Biloma

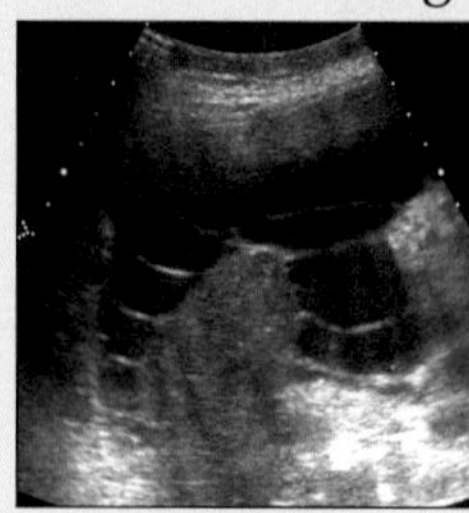

Loculated Ascites

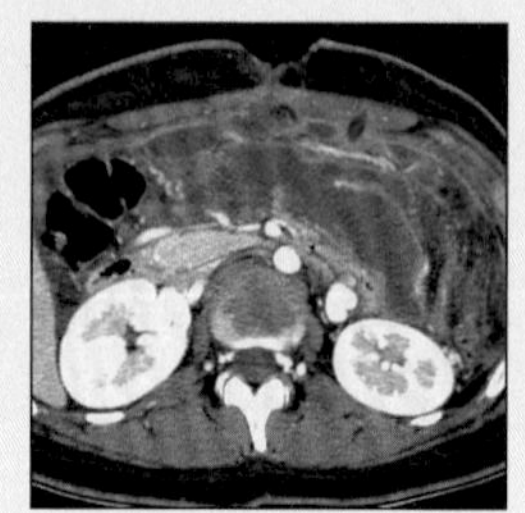

Panc Pseudocyst

ABDOMINAL ABSCESS

Key Facts

Terminology

- Localized abdominal collection of pus

Imaging Findings

- Best diagnostic clue: Fluid collection with mass effect & enhancing rim with or without gas bubbles or air-fluid level on CECT
- Location: Anywhere within abdominal cavity; intraparenchyma; within intra- or extraperitoneal spaces
- NECT: Low attenuation fluid collection, mass effect, gas in 50% of cases
- CECT: Peripheral rim-enhancement
- Complex fluid collection with internal low level echoes, membranes or septations on US
- Best imaging tool: CECT

Pathology

- General path comments: Pus collection; peripheral fibrocapillary "capsule"; often polymicrobial from enteric organisms
- Enteric perforation

Clinical Issues

- Most common signs/symptoms: Fever, chills; abdomen pain; increased heart rate, decreased blood pressure if septic
- Variable depending on extent of abscess, patient's immune system status; excellent prognosis
- Percutaneous abscess drainage (PAD)

Diagnostic Checklist

- Diagnostic mimics: Biloma, lymphocele, pseudocyst

- High amplitude linear echoes with reverberation artifacts representing gas bubbles
- Inflamed fat adjacent to abscess: Echogenic mass

- Color Doppler
  - Hypervascular periphery
  - Avascular center of abscess
  - Hyperemic inflamed fat

Nuclear Medicine Findings

- Gallium scan
  - Useful for fever of unknown origin
  - Nonspecific: Positive with tumor such as lymphoma and granulomatous lesions
- WBC scan
  - 73-83% sensitivity
  - False positives with bowel infarct or hematoma
- Newer agents
  - Indium-labeled polyclonal IgG
  - Tc99m-labeled monoclonal antibody

Imaging Recommendations

- Best imaging tool: CECT
- Protocol advice: Oral & I.V. contrast, 150 ml I.V. contrast at 2.5 ml/sec

DIFFERENTIAL DIAGNOSIS

Lymphocele

- History of lymph node dissection
- Fluid collections with mass (often bilateral) along lymphatic drainage
- Attenuation values -10 HU to +10 HU

Biloma

- History of biliary or hepatic surgery
- Perihepatic fluid collection commonly in gallbladder fossa or Morison pouch
- Attenuation value 0-15 HU

Loculated ascites

- Evidence for cirrhosis or chronic liver disease
- Minimal or no mass effect
- Often passively conforms to peritoneal space
- May contain septations on US

Pancreatic fluid collection/pseudocyst

- History of pancreatitis
- Associated pancreatic necrosis on CECT
- Location: Highly variable but most often within pancreatic parenchyma, lesser sac, anterior pararenal space, transverse mesocolon
- Pseudocyst requires several weeks to develop peripheral pseudocapsule

PATHOLOGY

General Features

- General path comments: Pus collection; peripheral fibrocapillary "capsule"; often polymicrobial from enteric organisms
- Genetics
  - Increased risk if genetically altered immune response
  - Diabetics have increased incidence of gas-forming abscesses
- Etiology
  - Enteric perforation
    - Appendicitis
    - Diverticulitis
    - Crohn disease
  - Post-operative
    - Typically intraperitoneal spaces such as cul-de-sac, Morison pouch and subphrenic spaces
  - Bacteremia
  - Trauma
- Epidemiology
  - Most commonly due to post-operative complication
  - Microabscesses due to fungal infections in immunocompromised patients
  - Higher incidence in diabetics, immunocompromised patients and posto-perative patients

Gross Pathologic & Surgical Features

- Often adherent omentum or bowel loops; pus collection
- May or may not have "capsule"

ABDOMINAL ABSCESS

Microscopic Features

- PMN and white cell debris
- Bacteria, fungi detected

Staging, Grading or Classification Criteria

- Organism: Bacterial, fungal amebic
- Related to organ of origin (i.e., liver abscess)
- Intraperitoneal
- Extraperitoneal
- Communicating
  - Underlying fistula to GI tract
  - Connection to biliary tract or pancreatic duct

CLINICAL ISSUES

Presentation

- Most common signs/symptoms: Fever, chills; abdomen pain; increased heart rate, decreased blood pressure if septic
- Clinical profile: Leukocytosis, + blood cultures and elevated ESR

Demographics

- Age: Any age
- Gender: M = F

Natural History & Prognosis

- Variable depending on extent of abscess, patient's immune system status; excellent prognosis

Treatment

- Options, risks, complications
  - Percutaneous abscess drainage (PAD)
    - 80% success rate of percutaneous drainage
    - Patient selection critical for success
    - Best candidates for PAD have well-localized, fluid-filled abscesses > 3 cm with safe catheter access route
  - Contraindications for PAD related to patient
    - Coagulopathy with prothrombin time > 3 sec
    - International normalized ratio > 1.5
    - Platelets < 50,000 uL
  - Contraindications for PAD related to abscess
    - Infected necrosis (i.e., pancreatic abscess)
    - Gas-forming infection such as emphysematous pancreatitis
    - Soft tissue infection (i.e., phlegmon)
    - No safe access route for catheter insertion
  - Surgery indications
    - Extensive intraperitoneal abscesses
    - Debridement of necrotic infected tissue
    - Failed PAD
  - Antibiotic therapy
    - Abscesses < 3 cm

DIAGNOSTIC CHECKLIST

Consider

- Diagnostic mimics: Biloma, lymphocele, pseudocyst

Image Interpretation Pearls

- Half of abscesses don't contain gas or air-fluid levels; mass effect & enhancing rim highly suggestive in appropriate clinical context

SELECTED REFERENCES

1. Men S et al: Percutaneous drainage of abdominal abcess. Eur J Radiol. 43(3):204-18, 2002
2. Benoist S et al: Can failure of percutaneous drainage of postoperative abdominal abscesses be predicted? Am J Surg. 184(2):148-53, 2002
3. Cinat ME et al: Determinants for successful percutaneous image-guided drainage of intra-abdominal abscess. Arch Surg. 137(7):845-9, 2002
4. Betsch A et al: CT-guided percutaneous drainage of intra-abdominal abscesses: APACHE III score stratification of 1-year results. Acute Physiology, Age, Chronic Health Evaluation. Eur Radiol. 12(12):2883-9, 2002
5. Ralls PW: Inflammatory disease of the liver. Clin Liver Dis. 6(1):203-25, 2002
6. Harisinghani MG et al: CT-guided transgluteal drainage of deep pelvic abscesses: indications, technique, procedure-related complications, and clinical outcome. Radiographics. 22(6):1353-67, 2002
7. Lohela P: Ultrasound-guided drainages and sclerotherapy. Eur Radiol. 12(2):288-95, 2002
8. Maggard MA et al: Surgical diverticulitis: treatment options. Am Surg. 67(12):1185-9, 2001
9. vanSonnenberg E et al: Percutaneous abscess drainage: update. World J Surg. 25(3):362-9; discussion 370-2, 2001
10. Green BT: Splenic abscess: report of six cases and review of the literature. Am Surg. 67(1):80-5, 2001
11. Deck AJ et al: Perinephric abscesses in the neurologically impaired. Spinal Cord. 39(9):477-81, 2001
12. Jacobs JE et al: Computed tomography evaluation of acute pancreatitis. Semin Roentgenol. 36(2):92-8, 2001
13. Krige JE et al: ABC of diseases of liver, pancreas, and biliary system. BMJ. 322(7285):537-40, 2001
14. Zibari GB et al: Pyogenic liver abscess. Surg Infect (Larchmt). 1(1):15-21, 2000
15. Sirinek KR: Diagnosis and treatment of intra-abdominal abscesses. Surg Infect (Larchmt). 1(1):31-8, 2000
16. Barakate MS et al: Pyogenic liver abscess: a review of 10 years' experience in management. Aust N Z J Surg. 69(3):205-9, 1999
17. Barakate MS et al: Pyogenic liver abscess: a review of 10 years' experience in management. Aust N Z J Surg. 69(3):205-9, 1999
18. Kimura K et al: Amebiasis: modern diagnostic imaging with pathological and clinical correlation. Semin Roentgenol. 32(4):250-75, 1997
19. Montgomery RS et al: Intraabdominal abscesses: image-guided diagnosis and therapy. Clin Infect Dis. 23(1):28-36, 1996
20. Snyder SK et al: Diagnosis and treatment of intra-abdominal abscess in critically ill patients. Surg Clin North Am. 62(2):229-39, 1982

ABDOMINAL ABSCESS

IMAGE GALLERY

Typical

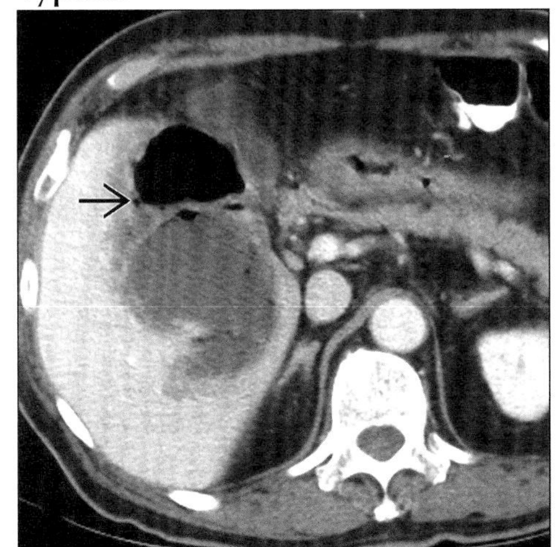

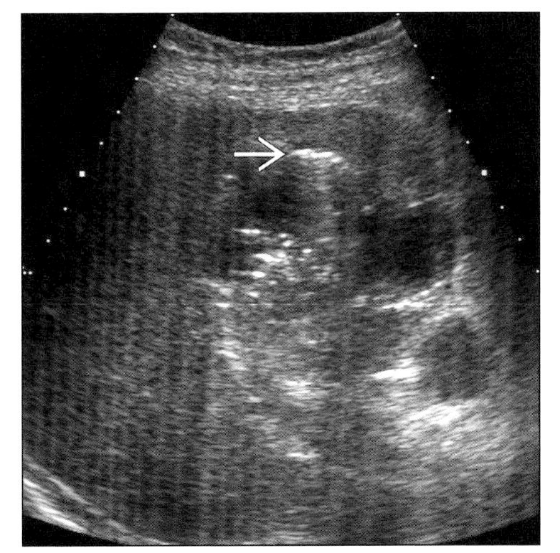

(Left) *Axial CECT demonstrates gas-forming pyogenic liver abscess in diabetic patient. Note air-fluid level within abscess cavity (arrow).* ***(Right)*** *Axial US shows liver abscess demonstrating linear high amplitude echoes with "dirty" distal acoustic shadowing representing gas (arrow).*

Typical

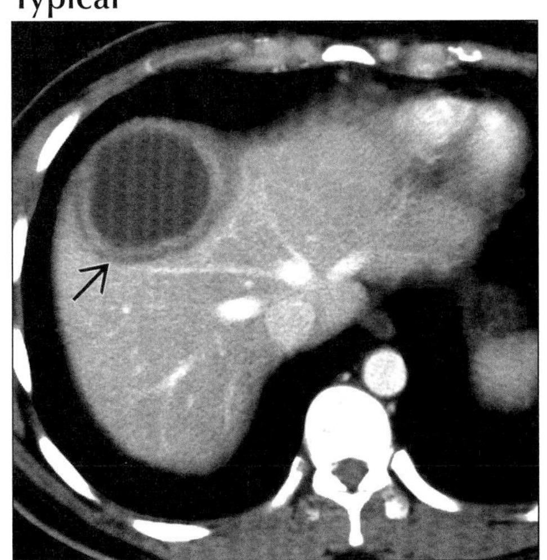

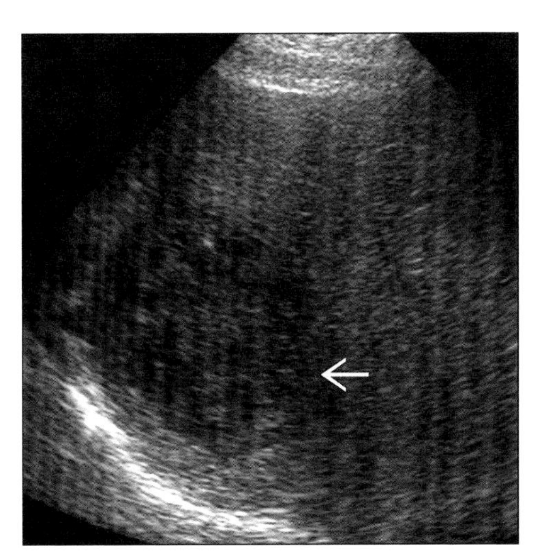

(Left) *Axial CECT of amebic abscess. Note peripheral low attenuation zone of edema (arrow) surrounding abscess.* ***(Right)*** *Sagittal US of amebic abscess. Note low level echoes (arrow) within hypoechoic mass and lack of distal acoustic enhancement.*

Typical

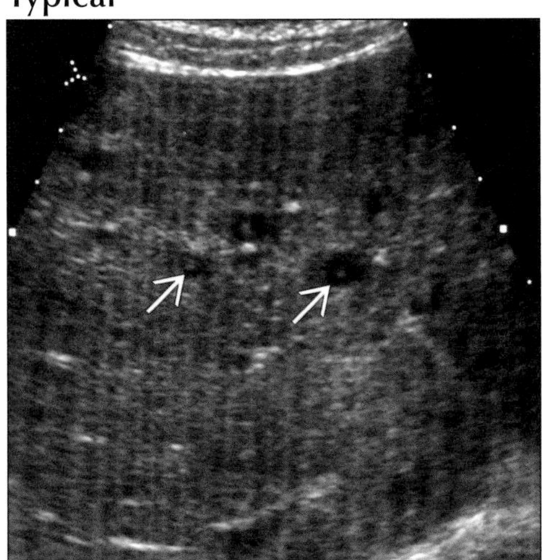

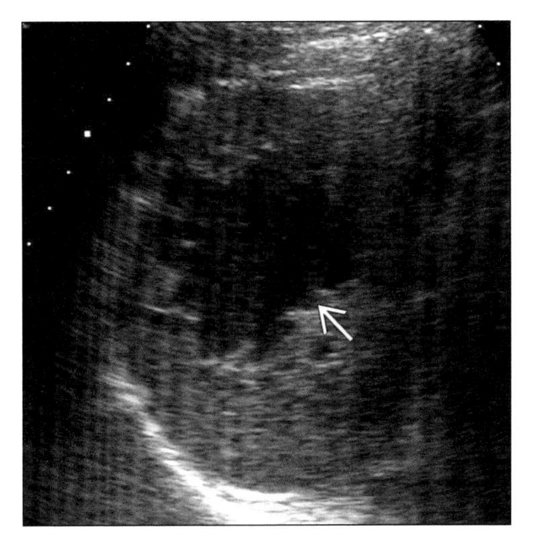

(Left) *Sagittal US of fungal microabscesses due to systemic Candidiasis. Note multiple "target" lesions (arrows).* ***(Right)*** *Sagittal US of spleen demonstrates abscess with low-level echoes (arrow).*

PERITONITIS

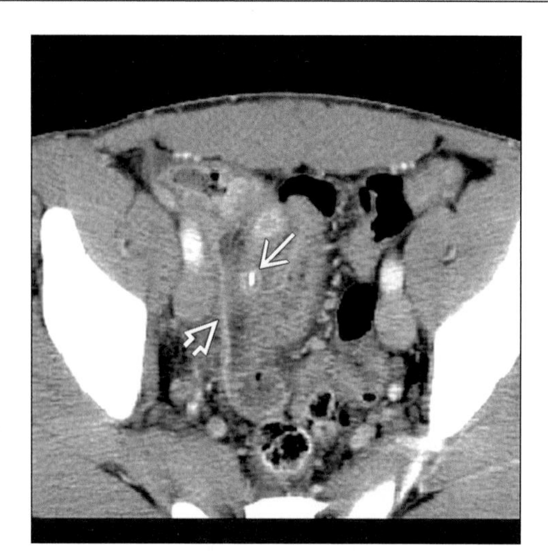

Axial CECT of perforated appendicitis with peritonitis. Note linear appendicolith (arrow), symmetric thickening of the peritoneum (open arrow) and adjacent low attenuation pus.

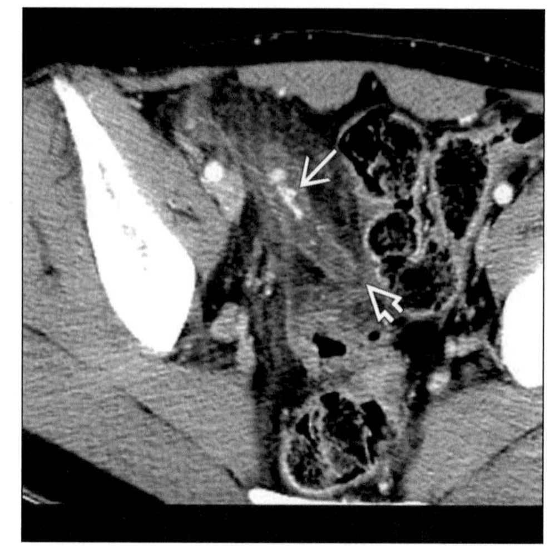

Axial CECT of perforated appendicitis with peritonitis shows multiple appendicoliths (arrow) with nonenhancing necrotic tip of appendix (open arrow) & surrounding soft tissue infiltration.

TERMINOLOGY

Definitions

- Infectious or inflammatory process involving peritoneum or peritoneal cavity

IMAGING FINDINGS

General Features

- Best diagnostic clue: Ascites, symmetric enhancement of peritoneum with fat stranding of abdominal fat
- Location: Peritoneal surface, mesentery, omentum
- Size: Variable, may be focal or diffuse
- Morphology: Symmetric thickening of peritoneum

Radiographic Findings

- Radiography
  - Evidence of ascites: More than 500 ml of fluid required for plain film diagnosis
    - Bulging of flanks
    - Indistinct psoas margin
    - Small bowel loops floating centrally
    - Lateral edge of liver displaced medially (Hellmer sign); visible in 80% of patients with significant ascites
    - Pelvic "dog's ear"; present in 90% of patients with significant ascites
    - Medial displacement of cecum and ascending colon; present in 90% of patients with significant ascites
  - +/- Free air
  - Hydropneumoperitoneum
  - Air in lesser sac with perforated gastric ulcer
- Contrast studies
  - Perforated ulcer with contrast leak on UGI
  - Perforation of diverticulum in diverticulitis on contrast enema

CT Findings

- CECT
  - Ascites, enhancing peritoneum with smooth thickening, infiltration and soft tissue stranding of fat within mesentery on CECT
  - +/- Gas bubbles, low attenuation nodes in TB peritonitis

MR Findings

- T1WI: Low attenuation peritoneal fluid
- T2WI: High attenuation peritoneal fluid
- T1 C+
  - Thickened enhancing peritoneum
  - Low attenuation peritoneal fluid

Ultrasonographic Findings

- Real Time

DDx: Spectrum of Intraperitoneal Fluid

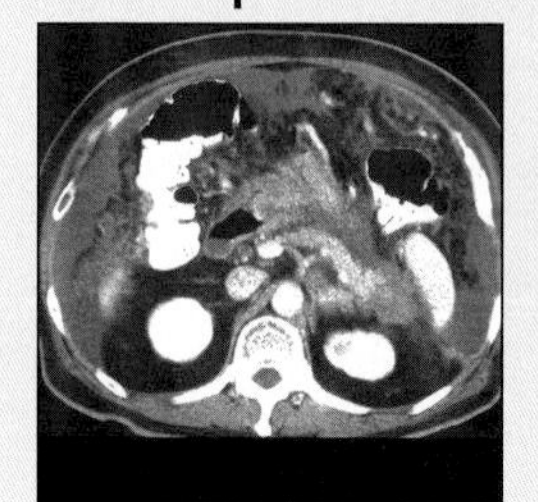

Carcinomatosis

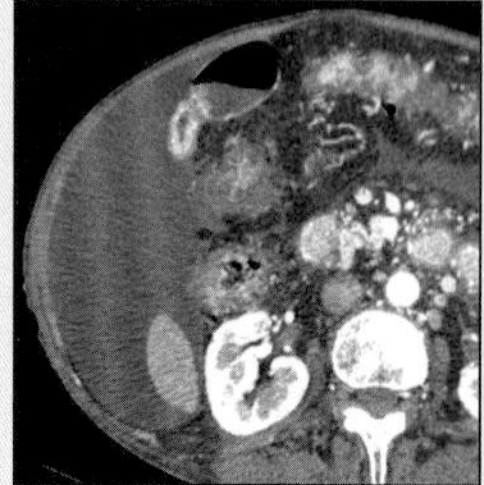

Benign Ascites

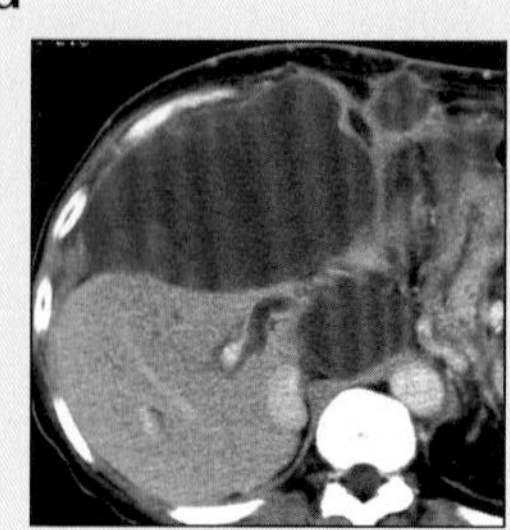

Pseudomyxoma

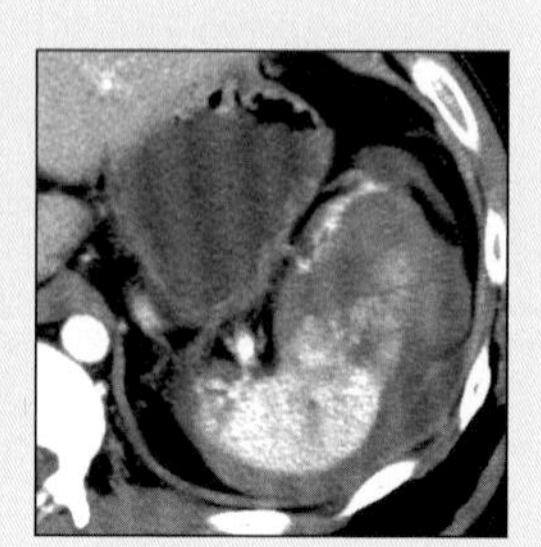

Hemoperitoneum

PERITONITIS

Key Facts

Terminology

- Infectious or inflammatory process involving peritoneum or peritoneal cavity

Imaging Findings

- Best diagnostic clue: Ascites, symmetric enhancement of peritoneum with fat stranding of abdominal fat
- Ascites, enhancing peritoneum with smooth thickening, infiltration and soft tissue stranding of fat within mesentery on CECT
- T1WI: Low attenuation peritoneal fluid
- T2WI: High attenuation peritoneal fluid
- Peritoneal fluid, septations, thickened echogenic mesentery on US
- Best imaging tool: CECT

Top Differential Diagnoses

- Peritoneal carcinomatosis
- Benign ascites
- Pseudomyxoma peritonei
- Hemoperitoneum

Pathology

- General path comments: Pus in peritoneal cavity, thickened peritoneum or mesentery

Clinical Issues

- Most common signs/symptoms: Fever, abdominal pain and distension

Diagnostic Checklist

- Peritoneal carcinomatosis
- Symmetric enhancement of thickened peritoneum

  - Peritoneal fluid, septations, thickened echogenic mesentery on US
  - Dilated fallopian tube with fluid-debris level (pyosalpinx) in pelvic inflammatory disease (PID)
  - Complex adnexal cystic masses in PID
    - Tubo-ovarian abscesses (TOA)
- Color Doppler
  - Hyperemic thickened echogenic fat
    - Associated with gastrointestinal source of inflammation

Imaging Recommendations

- Best imaging tool: CECT
- Protocol advice
  - Oral and IV contrast (150 ml injected at 2.5ml/sec)
  - Rectal contrast to distinguish colon from pelvic infection
  - 5 mm collimation with 5 mm reconstruction interval

DIFFERENTIAL DIAGNOSIS

Peritoneal carcinomatosis

- Nodular implants on peritoneum
- Omental caking
- Ascites
- Mesenteric nodules and adenopathy

Benign ascites

- Cirrhosis
- Bile leak
  - Caused by trauma, surgery, liver biopsy, biliary drainage
- Pancreatic ascites
  - Caused by pancreatic duct leakage
- Chylous ascites
- Urine ascites
  - Caused by bladder perforation
- Congestive heart failure (CHF), fluid overload

Pseudomyxoma peritonei

- Massive accumulation of gelatinous ascites in peritoneal cavity
- Scalloping of liver and spleen contour
- Rupture of mucinous tumor of appendix
- Calcified cystic implants
- Cystic masses attached to ligaments such as falciform or gastrohepatic ligament

Hemoperitoneum

- High attenuation intraperitoneal fluid
- Free lysed blood measuring 30-45 Hounsfield units (HU)
- Clotted blood measuring 60 HU
- Active arterial extravasation
  - Isodense with adjacent major arterial structures
  - Large surrounding hematoma

PATHOLOGY

General Features

- General path comments: Pus in peritoneal cavity, thickened peritoneum or mesentery
- Etiology
  - Spontaneous
    - Secondary bacterial infection of chronic ascites
    - Younger patients have higher incidence of pneumococcal or hemolytic streptococcal infection
  - Bacterial
    - Bowel perforation
    - Pelvic inflammatory disease (PID)
    - Infected intrauterine device (IUD)
    - Ruptured tubo-ovarian abscess
    - Gastric or duodenal ulcer
    - Ruptured appendicitis
    - Ruptured diverticulitis
  - TB
    - Ingestion of tuberculous sputum with development of TB peritonitis
  - Traumatic
    - Duodenum, jejunum, distal ileum most common sites
    - Small bowel injury may present 4-6 weeks post-trauma

PERITONITIS

- Bowel injury from deceleration injury
- Colonic injuries rare, but have rapid clinical onset of peritonitis
- Iatrogenic
  - Inadvertent bowel perforation during laparotomy or diagnostic/therapeutic paracentesis
  - Post-operative anastomotic leak
  - Retained foreign body during surgery
  - Dropped gallstones during laparoscopic cholecystectomy
- Epidemiology
  - Increased incidence in patients with chronic ascites
    - Cirrhosis
    - Peritoneal dialysis
  - Increased incidence in patients with risk factors for PID
    - IUD
    - Multiple sexual partners

Gross Pathologic & Surgical Features

- Pus in peritoneal cavity
- Inflammatory changes in mesentery
- Inflammatory adhesions
- Hyperemia of adherent omentum or mesentery

Microscopic Features

- > 500 leukocytes per mm^3 indicates infected ascites

Staging, Grading or Classification Criteria

- Localized
  - Walled off infection
- Diffuse
  - Multiple peritoneal compartments involved

CLINICAL ISSUES

Presentation

- Most common signs/symptoms: Fever, abdominal pain and distension

Demographics

- Age: All ages
- Gender: No predilection for male or female

Natural History & Prognosis

- Sepsis if not treated promptly
- Prognosis determined by primary etiology
  - Excellent if localized and no evidence of septicemia
  - Poor if generalized peritonitis and gram-negative septicemia

Treatment

- Options, risks, complications
  - Etiology of peritonitis determines treatment
  - Correct underlying cause (i.e., perforated ulcer)
  - Antibiotic therapy
    - Early PID
    - Soft tissue inflammation (phlegmon) from appendicitis or diverticulitis
  - Surgery for failed antibiotic therapy
  - Surgery for perforated viscus
    - Appendicitis
    - Duodenal ulcer
    - Diverticulitis

DIAGNOSTIC CHECKLIST

Consider

- Peritoneal carcinomatosis
- Causes of water-attenuation ascitic fluid
  - Transudate (e.g., cirrhosis)
  - Urine
  - Chyle
  - Bile
  - Pancreatic juice

Image Interpretation Pearls

- Symmetric enhancement of thickened peritoneum
- Inflammatory changes with adjacent fat of mesentery and omentum
- Enlarged fallopian tube (pyosalpinx) with fluid-fluid level & complex adnexal mass (TOA) in PID

SELECTED REFERENCES

1. Shetty H et al: Treatment of infections in peritoneal dialysis. Contrib Nephrol. (140):187-94, 2003
2. Alberti LE et al: Spontaneous bacterial peritonitis in a patient with myxedema ascites. Digestion. 68(2-3):91-3, 2003
3. Cheadle WG et al: The continuing challenge of intra-abdominal infection. Am J Surg. 186(5A):15S-22S; discussion 31S-34S, 2003
4. Troidle L et al: Continuous peritoneal dialysis-associated peritonitis: a review and current concepts. Semin Dial. 16(6):428-37, 2003
5. Yao V et al: Role of peritoneal mesothelial cells in peritonitis. Br J Surg. 90(10):1187-94, 2003
6. Marshall JC et al: Intensive care unit management of intra-abdominal infection. Crit Care Med. 31(8):2228-37, 2003
7. Witte MB et al: Repair of full-thickness bowel injury. Crit Care Med. 31(8 Suppl):S538-46, 2003
8. Malangoni MA: Current concepts in peritonitis. Curr Gastroenterol Rep. 5(4):295-301, 2003
9. Chow KM et al: Indication for peritoneal biopsy in tuberculous peritonitis. Am J Surg. 185(6):567-73, 2003
10. Sabri M et al: Pathophysiology and management of pediatric ascites. Curr Gastroenterol Rep. 5(3):240-6, 2003
11. Veroux M et al: A rare surgical complication of Crohn's diseases: free peritoneal perforation. Minerva Chir. 58(3):351-4, 2003
12. Reijnen MM et al: Pathophysiology of intra-abdominal adhesion and abscess formation, and the effect of hyaluronan. Br J Surg. 90(5):533-41, 2003
13. Hanbidge AE et al: US of the peritoneum. Radiographics. 23(3):663-84; discussion 684-5, 2003
14. Runyon BA: Strips and tubes: improving the diagnosis of spontaneous bacterial peritonitis. Hepatology. 37(4):745-7, 2003
15. Brook I: Microbiology and management of intra-abdominal infections in children. Pediatr Int. 45(2):123-9, 2003
16. Sivit CJ et al: Imaging of acute appendicitis in children. Semin Ultrasound CT MR. 24(2):74-82, 2003
17. Nishie A et al: Fitz-Hugh-Curtis syndrome. Radiologic manifestation. J Comput Assist Tomogr. 27(5):786-91, 2003

PERITONITIS

IMAGE GALLERY

Typical

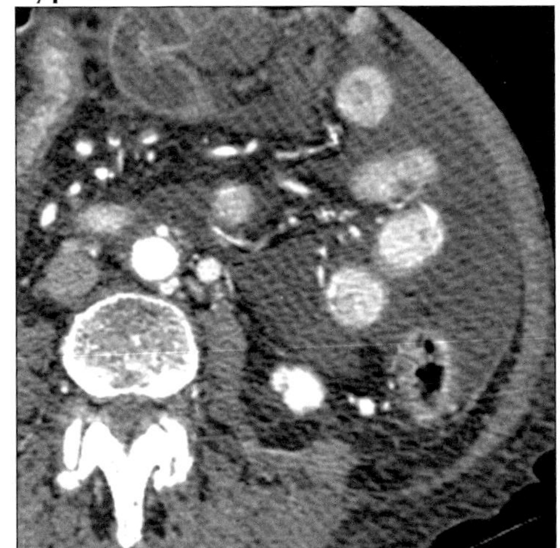

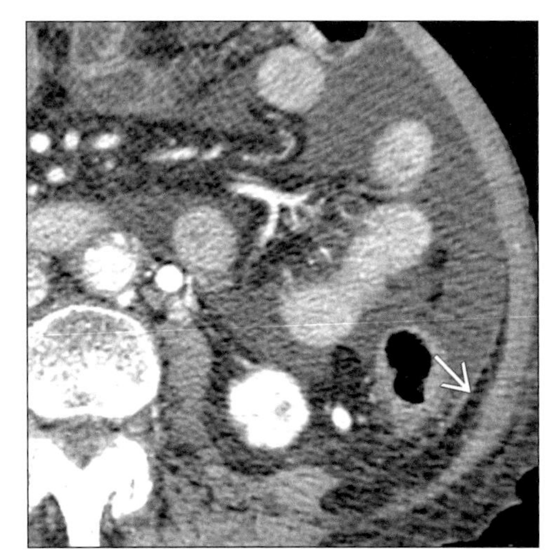

(Left) *Axial CECT of spontaneous bacterial peritonitis. Note marked ascites.* ***(Right)*** *Axial CECT of spontaneous bacterial peritonitis. Note symmetric thickening of parietal peritoneum in left flank (arrow).*

Typical

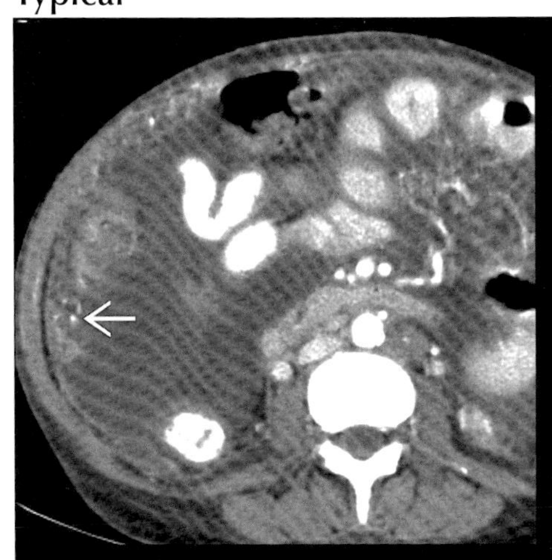

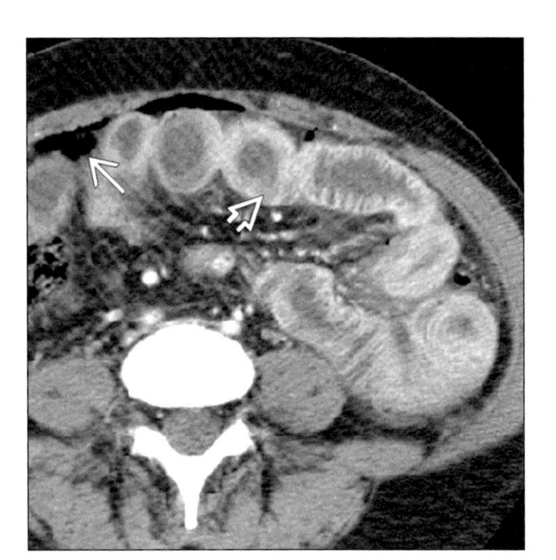

(Left) *Axial CECT of TB peritonitis demonstrates ascites, nodular thickening of omentum (arrow) and peritoneum.* ***(Right)*** *Axial CECT of traumatic bowel perforation with serosal inflammation and peritonitis. Note pneumoperitoneum (arrow) and hyperemic thickened small bowel (open arrow).*

Typical

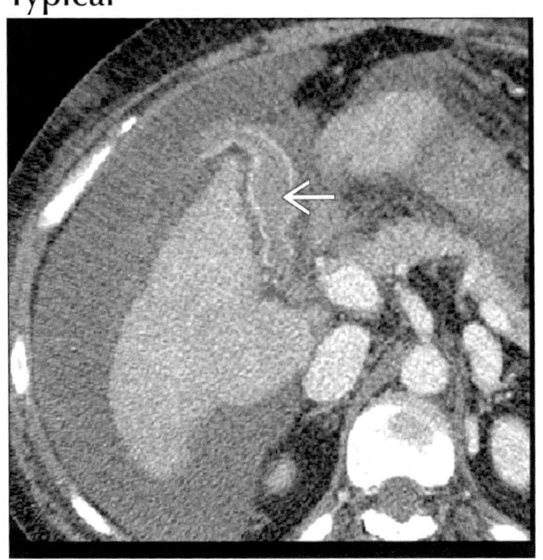

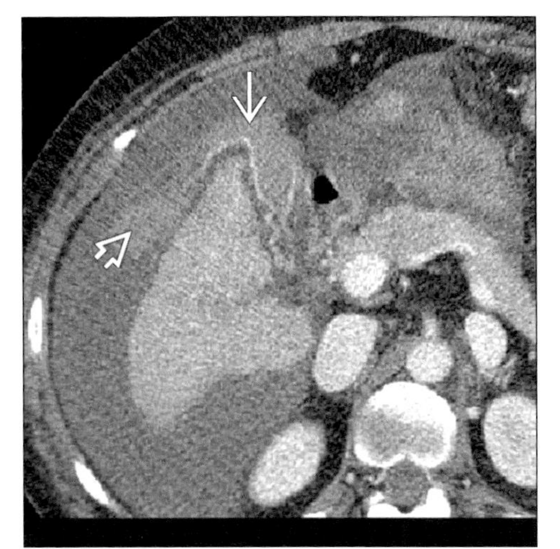

(Left) *Axial CECT of gallbladder perforation with bile peritonitis and hemoperitoneum. Note clot in gallbladder (arrow) and massive perihepatic low density fluid.* ***(Right)*** *Axial CECT of gallbladder perforation w/bile peritonitis & hemoperitoneum. Note interruption of gallbladder wall (arrow) w/adjacent clot (open arrow), and large amount of bile in peritoneal cavity.*

SCLEROSING MESENTERITIS

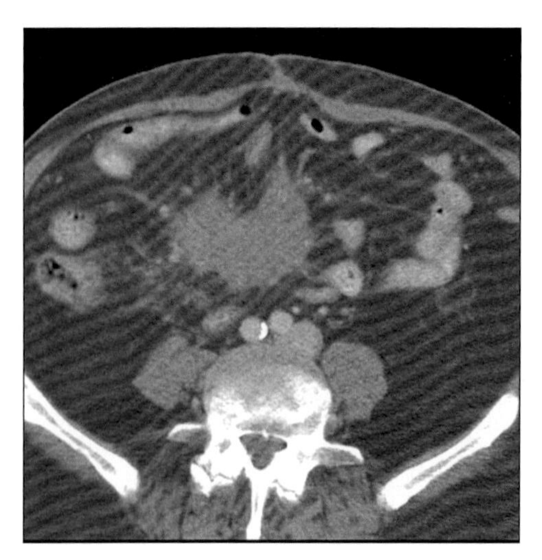

Axial CECT shows infiltrative mesenteric mass that encases blood vessels. Note engorgement of mesenteric veins.

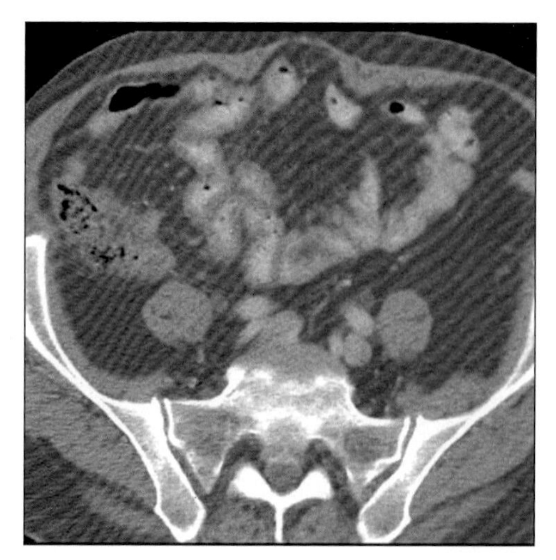

Axial CECT shows mildly thickened bowel wall and mesenteric venous congestion due to fibrosing mediastinitis.

TERMINOLOGY

Abbreviations and Synonyms

- Retractile mesenteritis, fibrosing mesenteritis, mesenteric panniculitis, mesenteric lipodystrophy, liposclerotic mesenteritis, systemic nodular panniculitis

Definitions

- Complex mesenteric inflammatory disorder of unknown etiology

IMAGING FINDINGS

General Features

- Best diagnostic clue: Fibrofatty mesenteric mass that encases mesenteric vessels but preserves a fat halo around vessels
- Location
  - Most common site: Root of small bowel mesentery
  - Occasionally: Colon (transverse/rectosigmoid)
  - Rarely: Peripancreatic, omentum & retroperitoneum
- Morphology
  - Mostly characterized by a mixture of mesenteric of:
    - Chronic inflammation
    - Fat necrosis & fibrosis
  - Mass that is often obscured in leaves of mesentery
- Key concepts
  - Rare, benign, nonspecific process involving mesenteric fat
  - Histologically: Classified into three types based on predominant tissue type in the mass
    - Mesenteric panniculitis: Chronic/acute inflammation & fat necrosis more than fibrosis
    - Mesenteric lipodystrophy: Fat necrosis more than inflammation & fibrosis
    - Retractile mesenteritis: Fibrosis/retraction more than inflammation/fat necrosis
  - Retractile mesenteritis
    - Considered as final, more invasive/chronic form
  - Often associated with other idiopathic inflammatory disorders (more than one condition may be present)
  - May coexist with malignancy (e.g., lymphoma, breast/lung/colon cancer & melanoma)

Radiographic Findings

- Fluoroscopic guided barium study
  - Involved bowel loops
    - Dilated/displaced/fixed/narrowed/tethering
  - Fold thickening: Submucosal infiltration/edema
  - Thumbprinting of bowel wall
    - Submucosal edema due to ischemia/lymphedema
  - Preservation of mucosal pattern
    - Clue in differentiating from carcinoma
  - Luminal narrowing
    - Common finding in retractile mesenteritis

DDx: Infiltrated Mesentery +/- Mass

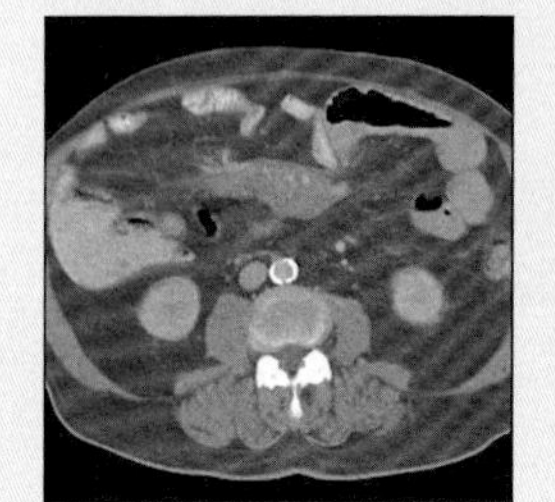

Treated Lymphoma

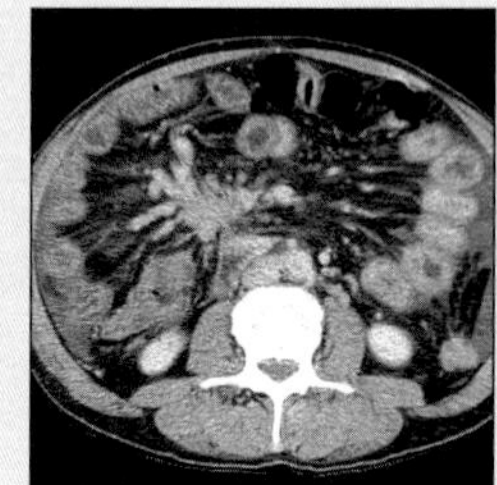

Carcinoid

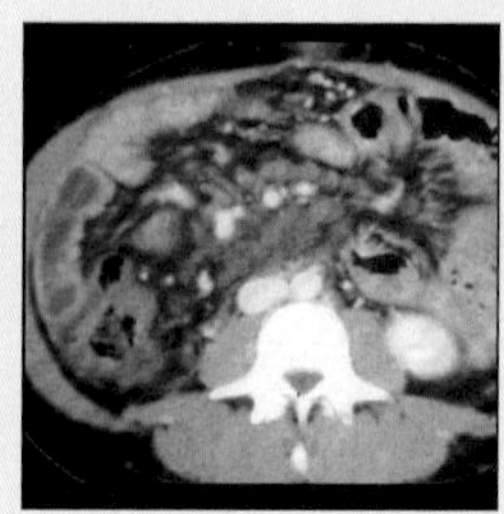

Cirrhosis

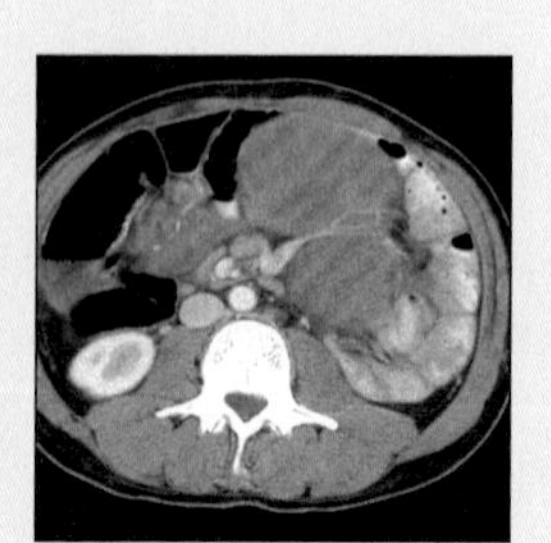

Desmoid Tumor

SCLEROSING MESENTERITIS

Key Facts

Terminology
- Retractile mesenteritis, fibrosing mesenteritis, mesenteric panniculitis, mesenteric lipodystrophy, liposclerotic mesenteritis, systemic nodular panniculitis
- Complex mesenteric inflammatory disorder of unknown etiology

Imaging Findings
- Best diagnostic clue: Fibrofatty mesenteric mass that encases mesenteric vessels but preserves a fat halo around vessels
- Area of subtle increased attenuation in mesentery (inflamed fat)
- Calcification; enlarged mesenteric lymph nodes
- Fatty necrotic cystic mass may be seen
- "Fat ring" sign: Preservation of fat around vessels
- Pseudocapsule: Peripheral band of soft tissue attenuation that limits normal mesentery from inflammatory process

Top Differential Diagnoses
- Non-Hodgkin lymphoma
- Carcinoid tumor
- Mesenteric edema
- Desmoid tumor (fibromatosis)
- Carcinomatosis (mesenteric metastases)

Diagnostic Checklist
- "Fat ring" sign: Differentiates sclerosing mesenteritis from lymphoma, carcinoid & mesenteric metastases
- Check for other idiopathic inflammatory disorders & malignancies (breast/lung/colon cancers; melanoma)

CT Findings
- Findings vary depending on predominant tissue
  - Area of subtle increased attenuation in mesentery (inflamed fat)
    - May be solitary/multiple, well-/ill-defined
  - Calcification; enlarged mesenteric lymph nodes
  - Fatty necrotic cystic mass may be seen
  - May show Infiltration of pancreas or portahepatis
  - Encasement of mesenteric vessels & collateral vessels
    - Narrowing/occlusion seen on contrast study
  - "Fat ring" sign: Preservation of fat around vessels
    - Hypodense fatty halo surrounding mesenteric vessels & nodules
    - Predominantly seen in mesenteric panniculitis
    - Differentiates sclerosing mesenteritis from other mesenteric processes (e.g., lymphoma, carcinoid tumor, carcinomatosis)
  - Pseudocapsule: Peripheral band of soft tissue attenuation that limits normal mesentery from inflammatory process
    - Seen in mesenteric panniculitis phase
    - Enhancement of pseudocapsule may be seen
  - "Misty mesentery": Nonspecific sign
    - Increased attenuation of mesentery
    - Evidence of small mesenteric nodes seen
    - No discrete soft tissue mass
    - Seen in any pathology that infiltrates mesentery
  - Thickening/infiltration/displacement/narrowing of bowel loops
  - Solid soft tissue mass usually in root of small bowel mesentery (fibrous tissue)
    - Single/large/lobulated or ill-defined increased density mass with linear radiating strands (fibroma-rare)
    - Small mesenteric soft tissue nodules of increased density (fibromatosis)
- CTA
  - Delineates relationship of mass to mesenteric vasculature

MR Findings
- Variable signal intensity: Due to inflammation/fat/fibrosis/vascular/Ca++
- Mesenteric panniculitis & lipodystrophy
  - T1WI: Mixed signal intensity
  - T2WI: Mixed signal intensity
- Retractile mesenteritis: In mature fibrotic reaction
  - T1WI: Decreased signal intensity
  - T2WI: Very low signal intensity
  - Gradient-echo MR image with flip angle 30°
    - Narrowing/occlusion of flow in mesenteric vessels
    - Collateral vessels are seen

Imaging Recommendations
- CT with 3D volume rendering is optimal study

DIFFERENTIAL DIAGNOSIS

Non-Hodgkin lymphoma
- Large discrete/confluent mesenteric nodes
- Lymphoma, no calcification unless treated
- Nodal mass in root of mesentery may mimic sclerosing mesenteritis
- Preservation of fat ring sign favors a diagnosis of sclerosing mesenteritis

Carcinoid tumor
- Discrete enhancing mass in bowel wall
- Hypervascular liver metastases
- Increased urinary 5-HIAA
- Mimic sclerosing mesenteritis due to
  - Ill-defined, infiltrating soft tissue mass in root of mesentery with associated calcification & desmoplastic reaction
  - Preservation of fat ring sign favors a diagnosis of sclerosing mesenteritis

Mesenteric edema
- Fluid infiltrates mesentery & increases attenuation of mesenteric fat, simulating sclerosing mesenteritis
- Seen in conditions like
  - Cirrhosis, hypoalbuminemia, heart failure

SCLEROSING MESENTERITIS

- Portal or mesenteric vein thrombosis, vasculitis
- Pancreatitis

Desmoid tumor (fibromatosis)

- Usually discrete solid mass
- Associated with Gardner syndrome
- Occur in traumatized sites
- Diagnosis: Biopsy & histologic analysis

Carcinomatosis (mesenteric metastases)

- Mesenteric implants can mimic CT appearance of sclerosing mesenteritis
- Implants: Not only root of mesentery, but also seen in omentum, surface of liver, spleen or bowel
- Calcification: In case of mucinous adenocarcinoma
  - Example: Ovarian or colon cancer
- Ascites is common in carcinomatosis

PATHOLOGY

General Features

- General path comments
  - Excessive fat deposition → inflammation → fatty degeneration/necrosis
  - Fibrosis → thickened mesentery contracts → adhesions → nodular changes
- Etiology
  - Exact etiology remains unknown
  - Possible causative factors
    - Autoimmune, infection, trauma, ischemia
    - Prior abdominal surgery, neoplastic
- Epidemiology
  - Prevalence: 0.6%
  - Peak incidence in 6th & 7th decades
- Associated abnormalities
  - Idiopathic inflammatory disorders
    - Retroperitoneal fibrosis, sclerosing cholangitis
    - Riedel thyroiditis, orbital pseudotumor

Gross Pathologic & Surgical Features

- Encapsulated firm/hard masses
- Nodules of fat, areas of necrosis & fibrosis
- Thickened mesentery/adhesions/displaced bowel loops

Microscopic Features

- Fat with lipid-laden macrophages & fibrous septa
- Lymphocytes/plasma cells/eosinophils; calcifications
- Invasion of bowel muscle/submucosa
- Mucosa is spared

CLINICAL ISSUES

Presentation

- Most common signs/symptoms
  - Abdominal pain, fever, nausea, vomiting, weight loss, diarrhea
  - Abdominal tenderness & palpable mass
  - Incidental finding in an asymptomatic patient
- Lab data
  - Increased ESR & decreased hematocrit (anemia)
  - PAS positive (histological DDx-Whipple)
- Diagnosis
  - Percutaneous or surgical excisional biopsy

Demographics

- Age: 2nd-8th decade of life; average age 60-70 years
- Gender: M:F = 1.8:1

Natural History & Prognosis

- Complications
  - Bowel or ureteral obstruction
  - Ischemia due to mesenteric vessels narrowing
- Prognosis
  - Partial or complete resolution
  - Nonprogressive course or aggressive course

Treatment

- Steroids, colchicine, immunosuppressive agents
  - Effective before fibrotic change
- Surgical excision
  - Fibrosis & retraction with obstructive symptoms

DIAGNOSTIC CHECKLIST

Consider

- Rule out other pathologies which can mimic sclerosing mesenteritis

Image Interpretation Pearls

- CT appearances vary depending on predominant tissue component (fat, inflammation, or fibrosis)
- "Fat ring" sign: Differentiates sclerosing mesenteritis from lymphoma, carcinoid & mesenteric metastases
- Check for other idiopathic inflammatory disorders & malignancies (breast/lung/colon cancers; melanoma)

SELECTED REFERENCES

1. Horton KM et al: CT findings in sclerosing mesenteritis (panniculitis): spectrum of disease. Radiographics. 23(6): 1561-7, 2003
2. Seo BK et al: Segmental misty mesentery: analysis of CT features and primary causes. Radiology. 226(1): 86-94, 2003
3. Lawler LP et al: Sclerosing mesenteritis: depiction by multidetector CT and three-dimensional volume rendering. AJR Am J Roentgenol. 178(1): 97-9, 2002
4. Horton KM et al: Volume-rendered 3D CT of the mesenteric vasculature: normal anatomy, anatomic variants, and pathologic conditions. Radiographics. 22(1): 161-72, 2002
5. Daskalogiannaki M et al: CT evaluation of mesenteric panniculitis: prevalence and associated diseases. AJR Am J Roentgenol. 174(2): 427-31, 2000
6. Sabate JM et al: Sclerosing mesenteritis: imaging findings in 17 patients. AJR Am J Roentgenol. 172(3): 625-9, 1999
7. Fujiyoshi F et al: Retractile mesenteritis: small-bowel radiography, CT, and MR imaging. AJR Am J Roentgenol. 169(3): 791-3, 1997
8. Mindelzun RE et al: The misty mesentery on CT: differential diagnosis. AJR Am J Roentgenol. 167(1): 61-5, 1996
9. Kronthal AJ et al: MR imaging in sclerosing mesenteritis. AJR Am J Roentgenol. 156(3): 517-9, 1991

IMAGE GALLERY

Typical

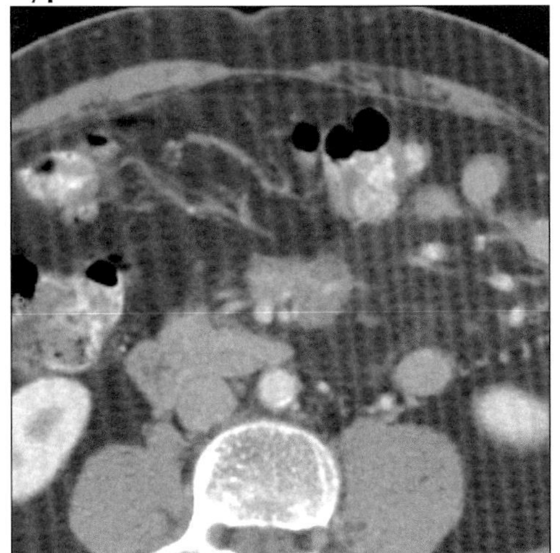

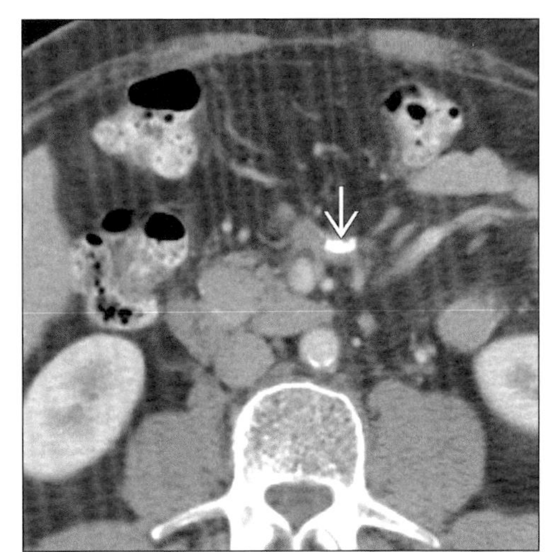

(Left) *Axial CECT shows mesenteric mass encasing vessels, causing mesenteric venous distention.* ***(Right)*** *Axial CECT shows focal calcification within mesenteric mass (arrow) and vascular engorgement.*

Typical

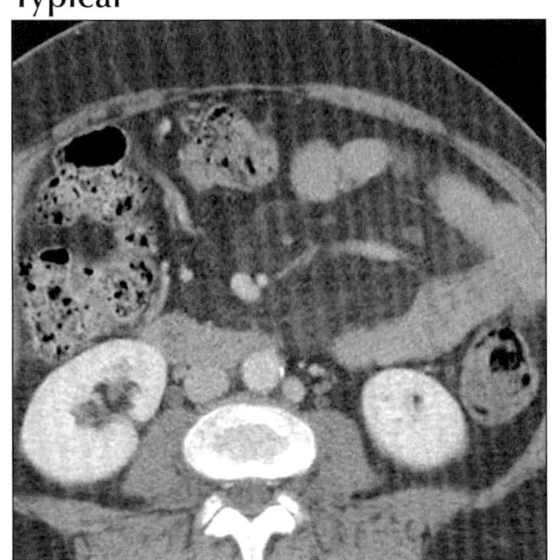

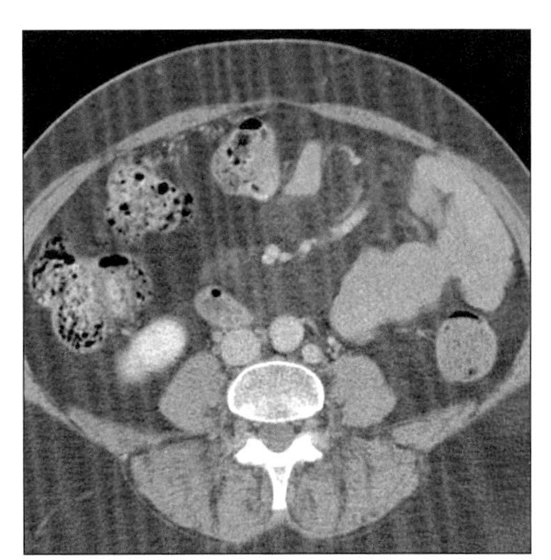

(Left) *Axial CECT in a 59 year old woman with "misty mesentery" and mesenteric adenopathy due to mesenteritis.* ***(Right)*** *Axial CECT in a 59 year old woman with mesenteritis and subtle infiltration of mesenteric fat.*

Typical

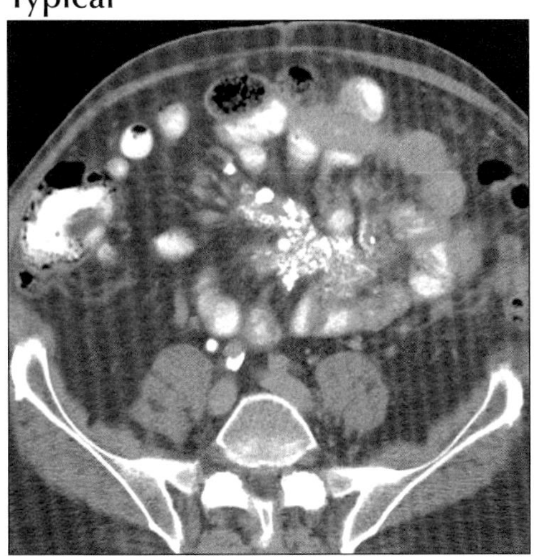

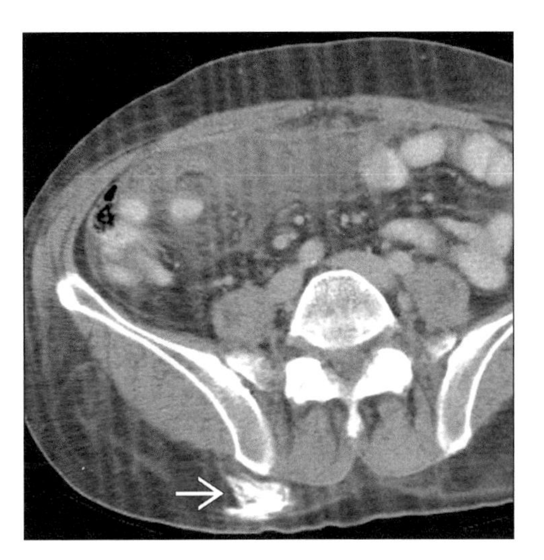

(Left) *Axial CECT of an 84 year old patient with retractile mesenteritis. Note calcified, spiculated mass that encases vessels and distorts bowel.* ***(Right)*** *Axial CECT in a patient with dermatomyositis (note subcutaneous calcification, arrow) along with sclerosing mesenteritis.*

ASCITES

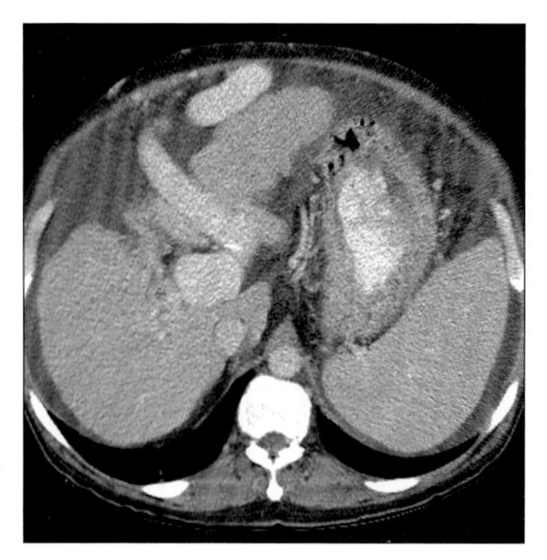

Axial CECT shows ascites due to hepatic cirrhosis, with large varices and splenomegaly.

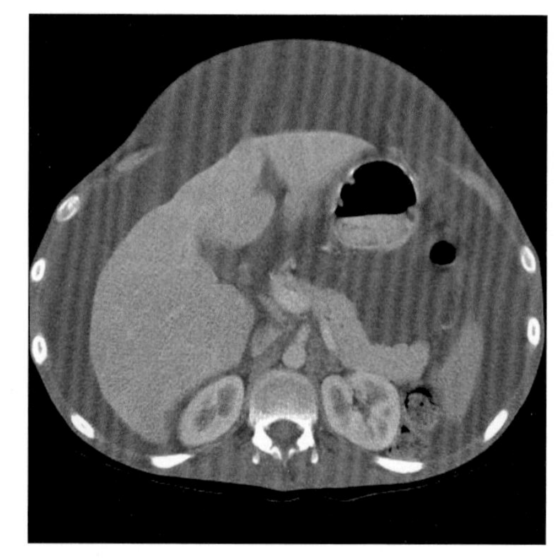

Axial CECT shows massive ascites due to right heart failure.

TERMINOLOGY

Abbreviations and Synonyms

- Intraperitoneal fluid collection

Definitions

- Pathologic accumulation of fluid within peritoneal cavity

IMAGING FINDINGS

General Features

- Best diagnostic clue: Diagnostic paracentesis
- Location
  - In uncomplicated cases, fluid flows to most dependent position
  - Morison pouch (hepatorenal fossa)
    - Most dependent upper abdominal recess
  - Pelvis
    - Most dependent space
  - Paracolic gutters
    - Lateral to ascending & descending colon
  - Subphrenic spaces
    - Not dependent, but fill due to suction effect of diaphragmatic motion
  - Lesser sac
    - Usually does not fill with ascites
    - Exceptions: Tense ascites, local source (gastric ulcer or pancreatitis)
    - Otherwise, usually due to carcinomatosis or infected ascites
- Morphology
  - Free-flowing fluid: Shaped by surrounding structures & does not deform normal shape of adjacent organs
    - Fluid insinuates itself between organs
  - Loculated: Rounded, bulging contour, encapsulated
    - Does not conform to organ margins
- Key concepts
  - Transudate, exudate, hemorrhagic, pus
  - Chylous, bile, pancreatic, urine, cerebrospinal fluid
  - Pseudomyxoma peritonei, neonatal ascites

Radiographic Findings

- Plain abdominal film: Insensitive for diagnosis
  - Hellmer sign; lateral edge of liver medially displaced from adjacent thoracoabdominal wall
  - Obliteration of hepatic & splenic angle
  - Symmetric densities on sides of bladder ("dog's ear")
  - Medial displacement of ascending & descending colon; lateral displacement of properitoneal fat line
  - Indirect signs: Diffuse abdominal haziness; bulging of flanks; poor visualization of psoas & renal outline
    - Separation of small bowel loops; centralization of floating gas-containing small bowel
  - Chest film: Elevation of diaphragm, sympathetic pleural effusion; with massive ascites

DDx: Fluid in Peritoneal Cavity

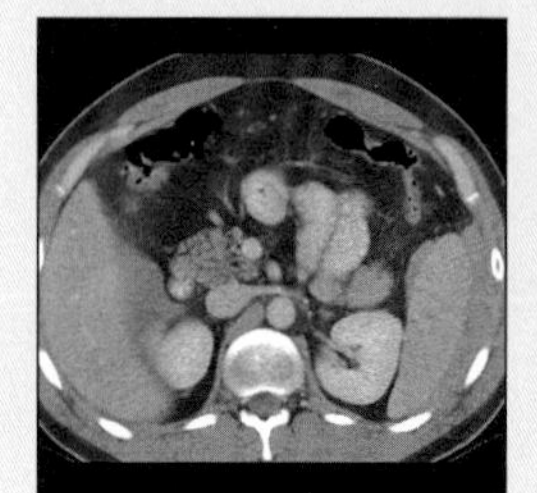

Hemoperitoneum

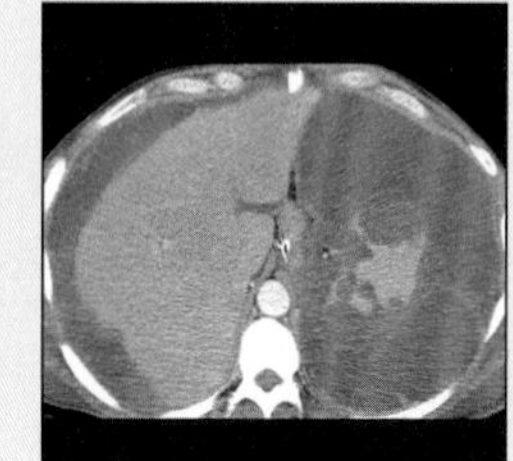

Pseudonmyx. Perit.

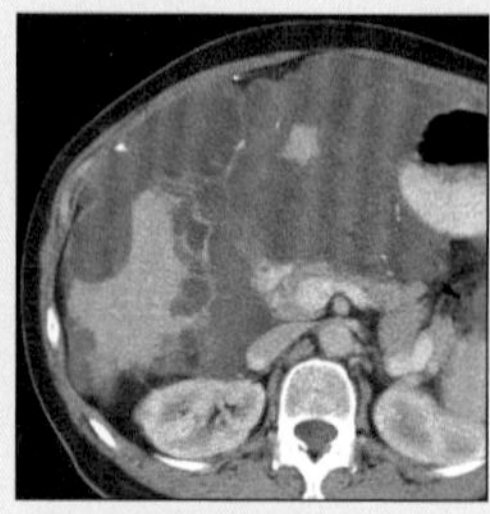

Ovarian Mets

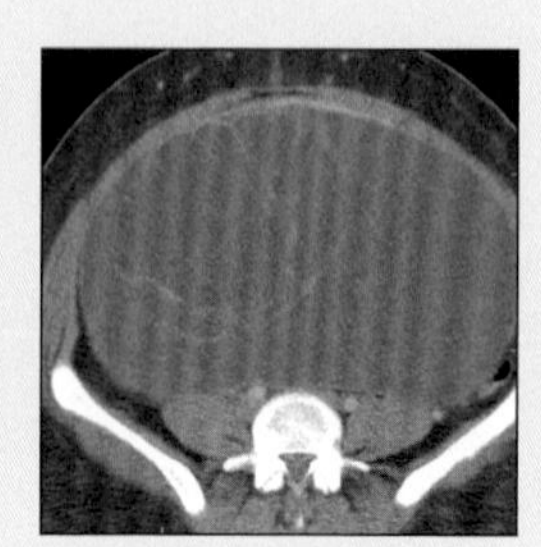

Ovarian Tumor

ASCITES

Key Facts

Terminology
- Pathologic accumulation of fluid within peritoneal cavity

Imaging Findings
- Simple ascites: Low density free fluid collection
- 0-30 Hounsfield units (HU); transudate fluid
- Small amounts seen in right perihepatic space, Morison pouch, pouch of Douglas
- Larger amounts of fluid in paracolic gutters
- Massive ascites; distends peritoneal spaces
- Associated evidence of liver, heart, kidney failure
- Complicated ascites: Exudates; infection, inflammation, malignancy
- Loculated ascites: Adhesions, malignancy, or infection

Top Differential Diagnoses
- Hemoperitoneum
- Malignant ascites
- Infectious ascites
- Cystic peritoneal metastases

Clinical Issues
- Diagnosis: Paracentesis; US guidance or blind tap
- Complication: Spontaneous bacterial peritonitis, respiratory compromise, anorexia

Diagnostic Checklist
- Difficult to characterize nature & underlying cause of peritoneal fluid collections on basis of imaging alone

CT Findings
- Simple ascites: Low density free fluid collection
  - 0-30 Hounsfield units (HU); transudate fluid
  - Small amounts seen in right perihepatic space, Morison pouch, pouch of Douglas
  - Larger amounts of fluid in paracolic gutters
    - Centralization of bowel loops; triangular configuration within leaves of mesentery
  - Massive ascites; distends peritoneal spaces
  - Associated evidence of liver, heart, kidney failure
    - Liver diseases; around liver (92%), pelvis (77%), paracolic gutters (69%), Morison pouch (63%)
- CT findings in other intraperitoneal collections
  - Exudates: Density of ascitic fluid increases with increasing protein content
  - Pseudomyxoma peritonei: Large low attenuation collection; multiseptate, loculations
    - Multiple cystic-appearing masses; calcification
    - Thickening of peritoneal & omental surface
    - Scalloping of liver & spleen contour
  - Chylous ascites: Less than 0 HU; intraperitoneal & extraperitoneal water density fluid (in trauma)
  - Bile ascites: Less than 20 HU; typically in right or left supramesocolic spaces
    - Located adjacent to liver or biliary structures
    - Bilomas have sharp margins
  - Urine ascites: Nonspecific CT appearance
    - Intravenously administered contrast material accumulates after renal concentration & excretion
  - Cerebrospinal fluid ascites: Small amounts of free fluid normal with ventriculoperitoneal shunt
    - Localized collection in association with tip of shunt tube; pathologic, implies malfunction
  - Pancreatic ascites: Peripancreatic, lesser sac, anterior pararenal space
    - Disruption of pancreatic duct or severe pancreatitis

MR Findings
- Transudate: Hypointense on T1WI
  - Hyperintense on T2WI
- Exudate: Intermediate to short T1 & long T2 values
  - T1 relaxation time decreases with increased protein
- Ascites hyperintense on gradient-echo images (fluid motion; transmitted pulsations)

Ultrasonographic Findings
- Uncomplicated ascites: Homogeneous, freely mobile, anechoic collection; deep acoustic enhancement
  - Free fluid: Acute angles where fluid borders organs
    - Shifts with change in patient position
    - Compresses with increased transducer pressure
- Complicated ascites: Exudates; infection, inflammation, malignancy
  - Internal echoes: Coarse (blood); fine (chyle)
  - Loculation; atypical fluid distribution
  - Multiple septa (tuberculous peritonitis, pseudomyxoma peritonei)
  - Matted or clumped; infiltrated bowel loops
  - Thickened interfaces between fluid & adjacent structures; peritoneal lining, omental thickening
- Loculated ascites: Adhesions, malignancy, or infection
  - No movement with change in patient position
  - Non compressible with increased probe pressure
- Thickening of gall bladder wall; more than 3 mm in benign ascites; in carcinomatosis less than 3 mm thick
- Small free fluid in cul-de-sac; physiologic in women
- Sonolucent band; small amounts of fluid in Morison pouch, around liver
- Polycyclic, "lollipop", arcuate appearance
  - Small bowel loops arrayed on either side of vertically floating mesentry; seen with massive ascites
- Transverse & sigmoid colon usually float on top of fluid (nondependent gas content when patient supine)
  - Ascending & descending colon do not float
- Right kidney may be displaced anteriorly & laterally
- Triangular fluid cap; overdistended bladder obscures
  - Fluid displaced to peritoneal reflection adjacent to uterine fundus

Imaging Recommendations
- US: Simplest, sensitive, small volume (5-10 ml) visualized, cost-effective
- CT: If US equivocal; as part of diagnostic work-up for cause, nature of ascites

ASCITES

DIFFERENTIAL DIAGNOSIS

Hemoperitoneum

- High attenuation fluid; > 30 HU (30-60 HU)
  - Unless present for > 48 hours or diluted by ascites, urine, bile, bowel contents
  - Active hemorrhage: Fluid collection isodense to contrast-enhanced blood vessels
- Sentinal clot: Focal heterogeneous collection of high density (> 60 HU); accumulates near site of bleed

Malignant ascites

- Loculated collections; fluid in greater & lesser sac
- Bowel loops tethered along posterior abdominal wall
- Thickening of peritoneum; peritoneal seeding
- Hepatic, splenic, lymph node lesions; mass arising from ovary, gut, pancreas

Infectious ascites

- Higher attenuation fluid (20-30 HU)
- Partial loculations; multiple septa
- Peritoneal enhancement; frank abscess
- Tuberculosis, acquired immunodeficiency syndrome, fungal infections

Cystic peritoneal metastases

- Massive ascites; loculated fluid collections
- Thickening of greater omentum
- Small nodules along peritoneal surface
- Apparent thickening of mesenteric vessels (fluid within leaves of mesentery)
- Increased density of linear network in mesenteric fat
- Adnexal mass of cystic density (ovarian, Krukenberg)

PATHOLOGY

General Features

- General path comments: Diminished effective volume (hydrostatic vs. colloid osmotic pressure); overflow
- Etiology
  - Hepatic: Cirrhosis, portal hypertension
    - Budd-Chiari, portal vein thrombosis, alcoholic hepatitis, fulminant hepatic failure
  - Cardiac: Congestive heart failure, constrictive pericarditis, cardiac tamponade
  - Renal: Nephrotic syndrome, chronic renal failure
  - Neoplasm: Colon, gastric, pancreatic, hepatic, ovarian; metastatic disease (breast/lung etc.)
    - Meig syndrome, mesothelioma
  - Infections: Tuberculosis
    - Acquired immunodeficiency syndrome
    - Bacterial, fungal or parasitic infections
  - Trauma: Blunt, penetrating or iatrogenic
    - Diagnostic/therapeutic peritoneal lavage
    - Bile ascites: Trauma, cholecystectomy, biliary or hepatic surgery, biopsy, percutaneous drainage
    - Urine ascites: Direct or seat belt trauma to bladder or collecting system; instrumentation
    - Cerebrospinal fluid: Ventriculoperitoneal shunts
    - Chylous: Blunt, penetrating, surgical trauma
  - Hypoalbuminemia; protein-losing enteropathy
  - Miscellaneous: Myxedema, marked fluid overload

Gross Pathologic & Surgical Features

- Transudate; clear, colorless or straw colored
- Exudate; yellowish or hemorrhagic
- Neoplasm; bloody, clear or chylous
- Pyogenic; turbid, chylous; yellowish white, milky
- Pseudomyxoma peritonei; gelatinous, mucinous

Microscopic Features

- Ascitic fluid may contain blood cells, colloids, protein molecules, or crystalloids (such as glucose) & water
  - Protein content: Less than 2.5 g/dl transudate; greater than 2.5 g/dl exudate
  - Polymorphonuclear leukocyte count greater than 500/cu. mm suggests infection or pancreatic ascites

CLINICAL ISSUES

Presentation

- Most common signs/symptoms: Asymptomatic, abdominal discomfort & distension, weight gain
- Physical examination: Bulging flanks, flank dullness, fluid wave, umbilical hernia, penile or scrotal edema
- Diagnosis: Paracentesis; US guidance or blind tap
  - Indications: All patients with new onset ascites
    - Chronic ascites with fever, abdominal pain, renal insufficiency, or encephalopathy
  - Fluid analysis: Protein, lactate dehydrogenase
    - Amylase, blood cell count with differential bacteriology, cytology, pH, triglycerides
    - Serum-ascites albumin gradient; greater than 1.1 g/dL indicates portal hypertension

Natural History & Prognosis

- Complication: Spontaneous bacterial peritonitis, respiratory compromise, anorexia

Treatment

- Sodium restriction & diuretics
- Therapeutic paracentesis
- Refractory cases: Large volume paracentesis
  - Peritoneovenous shunting; LeVeen, Denver
  - Transjugular intrahepatic portosystemic shunting
  - Liver transplantation

DIAGNOSTIC CHECKLIST

Consider

- Difficult to characterize nature & underlying cause of peritoneal fluid collections on basis of imaging alone

SELECTED REFERENCES

1. Hanbidge AE et al: US of the peritoneum. Radiographics. 23(3):663-84; discussion 684-5, 2003
2. Jeffery J et al: Ascitic fluid analysis: the role of biochemistry and haematology. Hosp Med. 62(5): 282-6, 2001
3. Heneghan MA et al: Pathogenesis of ascites in cirrhosis and portal hypertension. Med Sci Monit. 6(4):807-16, 2000
4. Habeeb KS et al: Management of ascites. Paracentesis as a guide. Postgrad Med. 101(1): 191-2, 195-200, 1997
5. Henriksen JH et al: Ascites formation in liver cirrhosis: the how and the why. Dig Dis. 8(3):152-62, 1990

IMAGE GALLERY

Typical

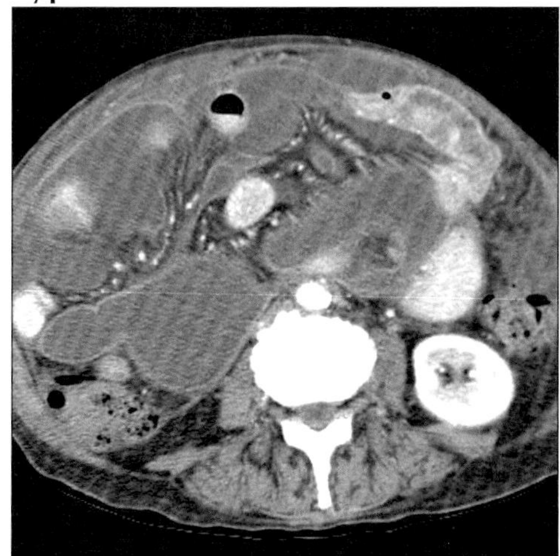

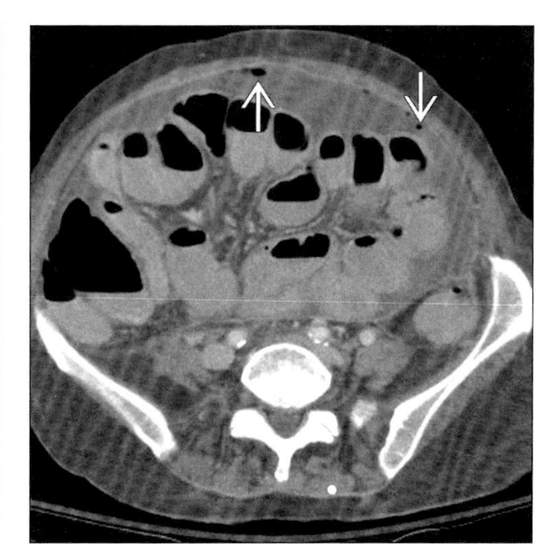

(Left) *Axial CECT shows loculated ascites due to peritoneal dialysis. Note mass effect and contrast-enhancing wall.* ***(Right)*** *Axial CECT in a cirrhotic patient with spontaneous bacterial peritonitis, loculation, enhancing rim and gas (arrows).*

Typical

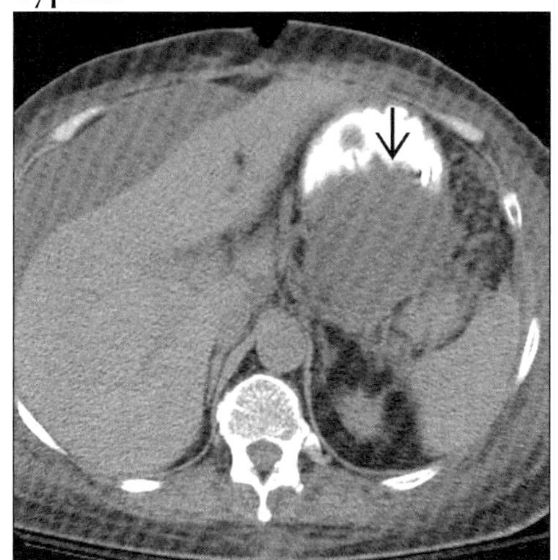

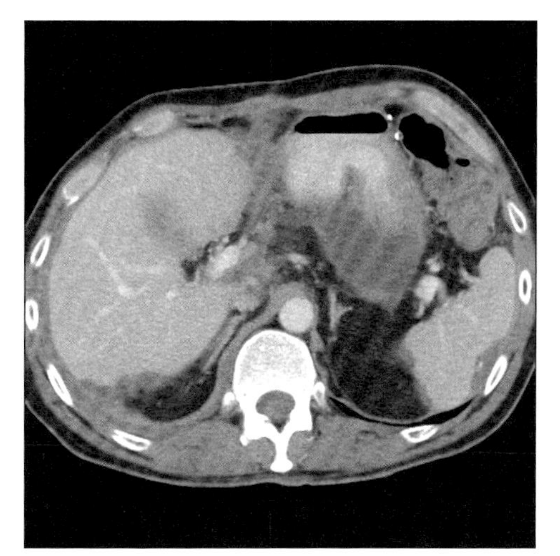

(Left) *Axial NECT in patient with pancreatitis. Note prominent lesser sac collection of fluid (arrow).* ***(Right)*** *Axial CECT shows fluid in lesser sac and "scalloped" surface of liver and spleen. Peritoneal carcinomatosis.*

Typical

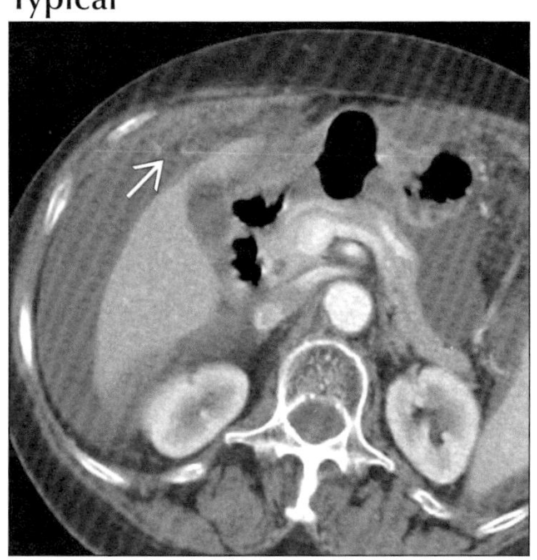

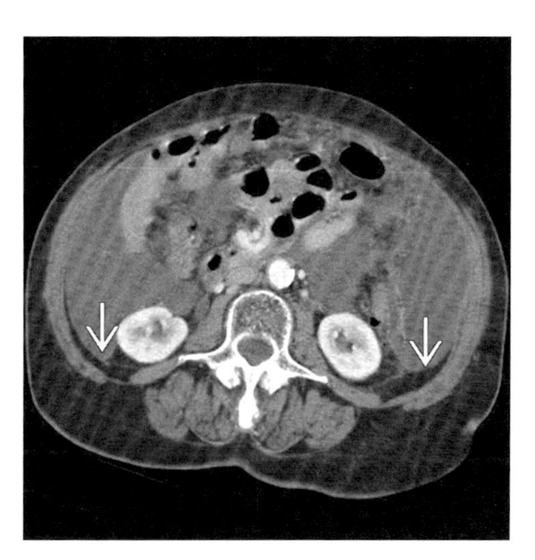

(Left) *Axial CECT shows ascites, including lesser sac, along with thickened, "smudged" omentum (arrow). Malignant ascites.* ***(Right)*** *Axial CECT shows loculated ascites, peritoneal thickening (arrows), nodular omentum. Ovarian carcinoma metastases.*

OMENTAL INFARCT

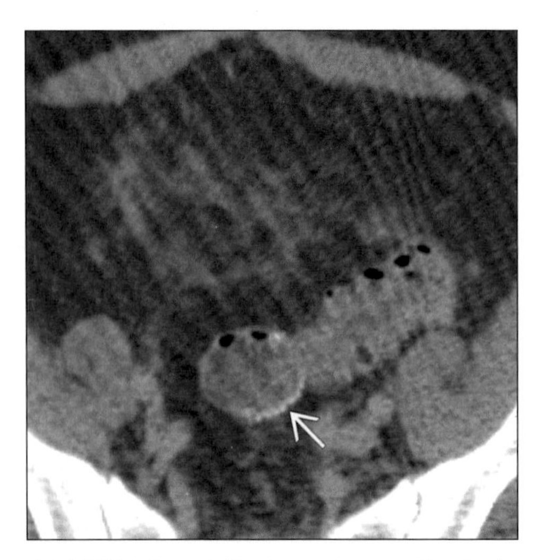

Axial CECT shows infiltration and mass effect in the omentum adjacent to the sigmoid colon. Note staple line (arrow) from prior sigmoid resection 2 months previously.

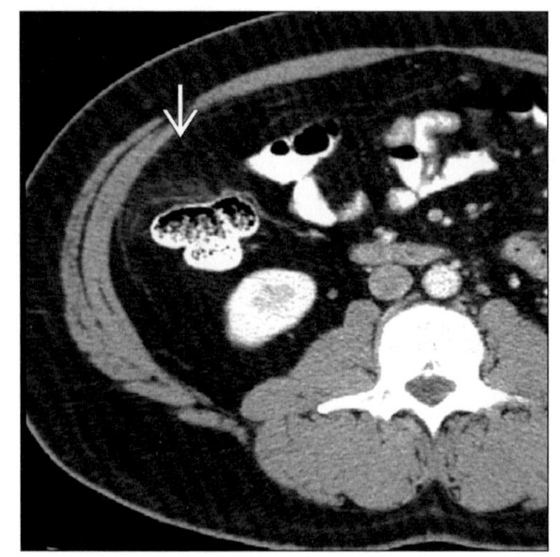

Axial CECT shows increased attenuation stranding in the omental fat (arrow) adjacent to the ascending colon (Courtesy D. Green, MD).

TERMINOLOGY

Abbreviations and Synonyms

- Omental infarction

Definitions

- Necrosis caused by interruption of arterial blood supply to omentum

IMAGING FINDINGS

General Features

- Best diagnostic clue: Focal mass of heterogeneous omental fat ± a capsule in right lower quadrant
- Size: Varies from 3.5-15.0 cm
- Morphology
  - Inflamed ± hemorrhagic omental fat
  - Triangular, ovoid or cake-like in shape

CT Findings

- Heterogeneous fatty mass + hyperattenuated streaks
- Located between anterior abdominal wall & colon
- Pericolonic inflammatory changes
- Adherence to colon or parietal peritoneum
- Thickening of overlying abdominal wall (rare)
- Free fluid may be visualized

Ultrasonographic Findings

- Real Time
  - Hyperechoic, non-mobile/compressible, fixed mass
  - Free fluid may be seen

Imaging Recommendations

- Helical CECT with oral contrast

DIFFERENTIAL DIAGNOSIS

Acute appendicitis

- Appendicolith may be seen
- Periappendiceal soft tissue stranding
- Wall thickening of cecum or terminal ileum
- Right lower quadrant fluid or abscess

Epiploic appendagitis

- Paracolic fatty mass + hyperattenuating ring
- Location: More common in LLQ (rectosigmoid)
- Pericolonic fat stranding; thickened peritoneum/bowel

Omental torsion

- Due to omental cysts, hernias, tumors, adhesions
- Fibrous + fatty folds converging towards torsion

Pancreatitis

- Focal or diffuse enlargement of pancreas
- Fluid collections, infiltration of peripancreatic fat

DDx: Heterogeneous Fat in Omentum or Mesentery

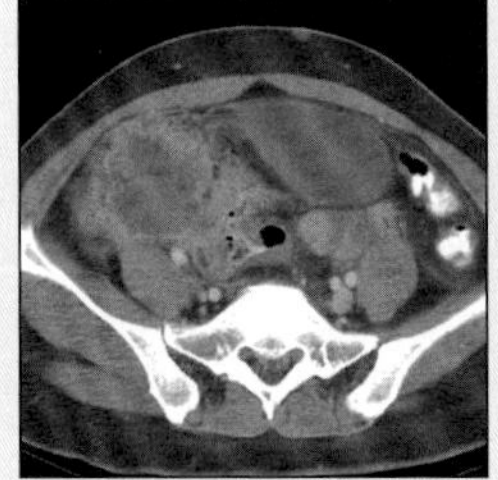

Appy. Abscess

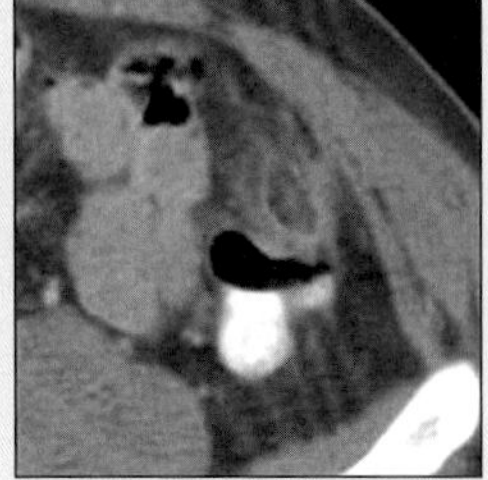

Epiploic Append.

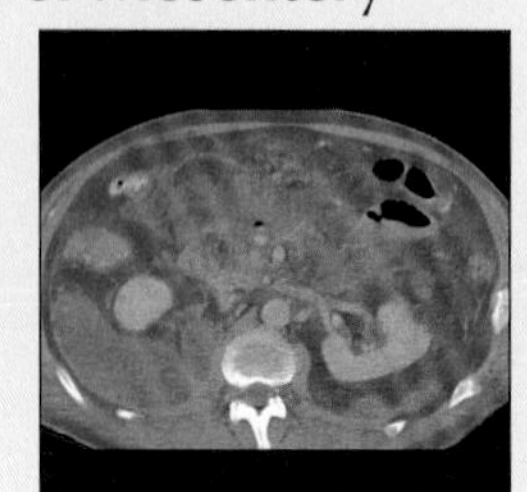

Pancreatitis

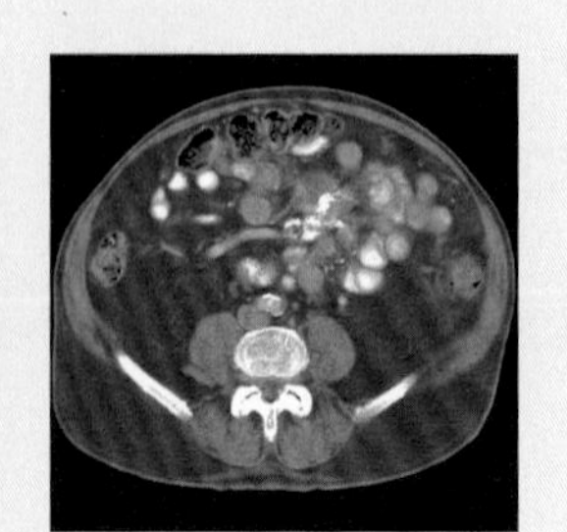

Fibrosing Mesent.

OMENTAL INFARCT

Key Facts

Imaging Findings
- Best diagnostic clue: Focal mass of heterogeneous omental fat ± a capsule in right lower quadrant
- Located between anterior abdominal wall & colon
- Pericolonic inflammatory changes
- Adherence to colon or parietal peritoneum

Top Differential Diagnoses
- Acute appendicitis
- Epiploic appendagitis
- Omental torsion
- Pancreatitis
- Fibrosing sclerosing mesenteritis

Diagnostic Checklist
- Omental infarct in case of acute abdominal pain with absence of constitutional symptoms or ↑ WBC
- CT: Fatty mass with hyperattenuated streaks between abdominal wall & colon on right side

- Abscess, pseudocyst, gallstones
- Chest: Pleural effusion & basal atelectasis

Fibrosing sclerosing mesenteritis
- Soft tissue mass usually in root of small bowel mesentery (fibrous tissue)
- Thickening, infiltration, displacement, narrowing of bowel loops

PATHOLOGY

General Features
- General path comments
  - Hemorrhagic infarction with fat necrosis
  - Followed by inflammatory infiltrate
  - Fibrosis & retraction → healing or autoamputation
- Etiology
  - Unclear, abnormal omental vascular system
    - Right epiploic vessels involved in 90% of cases
  - Precipitating factors: Obesity, vascular congestion, kinking of vessels, ↑ in intra-abdominal pressure
- Epidemiology: Adults (85%), children (15%)

Gross Pathologic & Surgical Features
- Infarcted omentum often adherent to parietal peritoneum or colon
- Serosanguineous fluid in peritoneal cavity

Microscopic Features
- Inflammatory infiltrate: Predominantly plasmocytic, lymphocytic and histiocytic cells

CLINICAL ISSUES

Presentation
- Most common signs/symptoms
  - Acute abdominal pain, right-sided in 90% of cases
  - Fever; ± palpable mass in RLQ
- Lab-data: WBC & ESR normal or mildly elevated

Demographics
- Age
  - Elderly obese people (85% of cases)
  - Less common in children (15% of cases)
- Gender: M > F

Natural History & Prognosis
- Complications: Abscess, adhesions, bowel obstruction
- Prognosis
  - Usually self limiting and will resolve spontaneously

Treatment
- Conservative management
- Laparoscopic excision

DIAGNOSTIC CHECKLIST

Consider
- Omental infarct in case of acute abdominal pain with absence of constitutional symptoms or ↑ WBC

Image Interpretation Pearls
- CT: Fatty mass with hyperattenuated streaks between abdominal wall & colon on right side

SELECTED REFERENCES

1. Macari M et al: The acute right lower quadrant: CT evaluation. Radiologic clinics of North America. 14(6):1117-36, 2003
2. Grattan-Smith JD et al: Omental infarction in pediatric patients: sonographic and CT findings. AJR Am J Roentgenol. 178(6):1537-9, 2002
3. McClure MJ et al: Radiological features of epiploic appendagitis and segmental omental infarction. Clin Radiol. 56(10):819-27, 2001
4. Puylaert JB: Right-sided segmental infarction of the omentum: clinical, US, and CT findings. Radiology. 185(1):169-72, 1992

IMAGE GALLERY

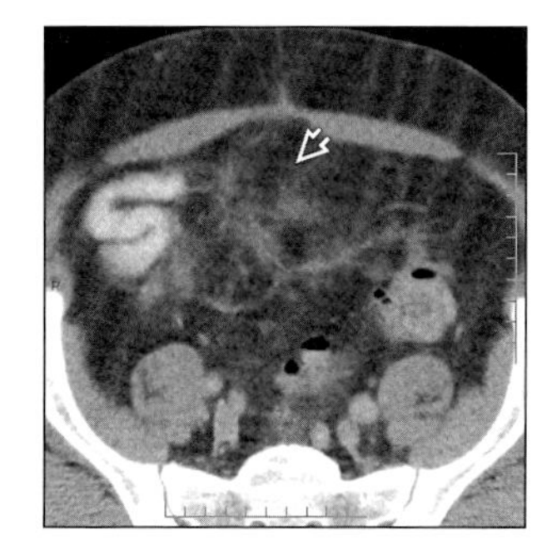

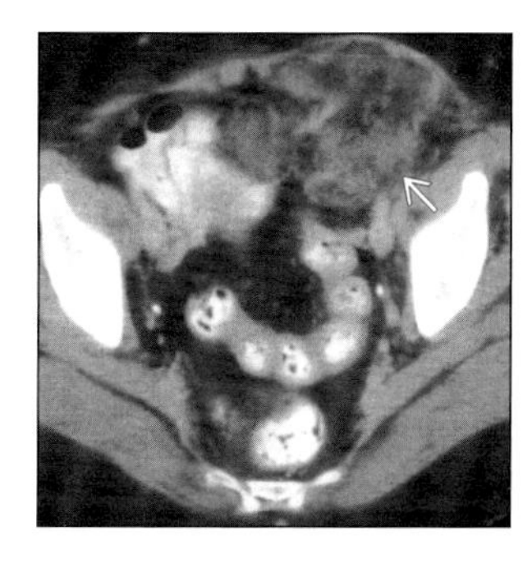

(Left) *Axial NECT shows infiltration of omental fat (arrow) (Courtesy D. Avrin, MD).* ***(Right)*** *Axial CECT shows heterogeneous "mass" of mixed fat density in the omentum (arrow), displacing the anterior abdominal wall.*

INGUINAL HERNIA

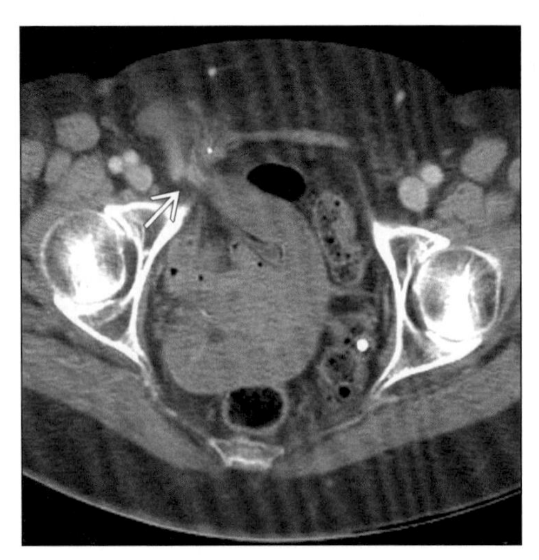
Axial CECT shows dilated small bowel entering inguinal canal, collapsed bowel leaving it (arrow).

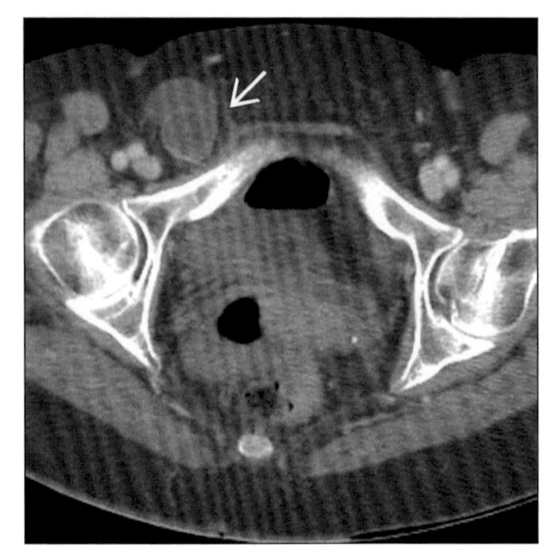
Axial CECT shows "knuckle" of fluid-filled small intestine strangulated within right inguinal hernia (arrow).

TERMINOLOGY

Abbreviations and Synonyms

- Inguinal hernia (IH)
- Pelvic & groin hernia

Definitions

- External hernia: Abnormal protrusion of intra-abdominal tissue
  - Through defect in abdominal or pelvic wall & extending outside abdominal cavity
  - IH: Inguinal location of hernia orifice

IMAGING FINDINGS

General Features

- Location
  - Indirect IH: Passes through internal inguinal ring, down the inguinal canal & emerges at external ring
    - Can extend along spermatic cord into scrotum; complete hernia
    - In females, hernia follows course of round ligament of uterus into labium majus
    - Passes lateral to epigastric vessels (lateral umbilical fold) & is also known as lateral IH
    - Juxtafunicular: Indirect hernia passes outside spermatic cord
  - Direct IH: Occurs in floor of inguinal canal, through Hesselbach triangle
    - Protrudes medial to inferior epigastric vessels (IEV)
    - It is not contained in spermatic cord & it generally does not pass into scrotum
    - Medial umbilical fold divides Hesselbach triangle into medial & lateral parts
    - Medial & lateral direct IH
- Morphology
  - Indirect IH in males within spermatic cord has smooth contour & elongated oblique course
    - Juxtafunicular hernias: Have a more irregular contour; do not protrude into a preformed sac
    - Dissect through subcutaneous fat & fibrous tissue
  - Direct IH: Broad & dome-shaped; appears as a small bulge in groin; short & blunt aperture

Radiographic Findings

- Radiography
  - Films of abdomen with patient supine can indicate incarceration or strangulation
    - Convergence of distended intestinal loops toward inguinal region
    - Soft tissue density or gas-containing mass overlying obturator foramen on affected side
  - Barium examination of small or large bowel
    - Tapered narrowing or obstruction of intestinal segments as it enters hernia orifice

DDx: Mass Near Inguinal Ligament

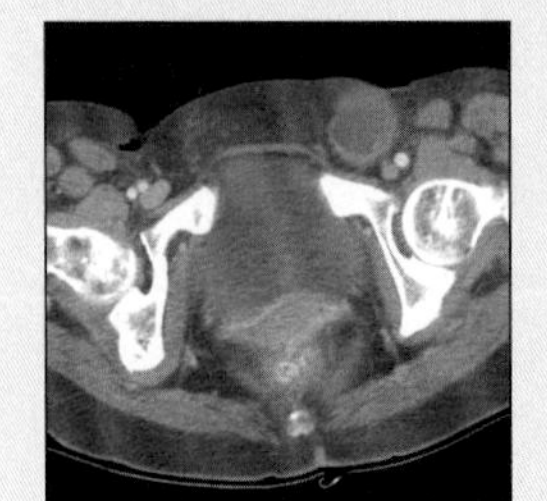
Femoral Hernia

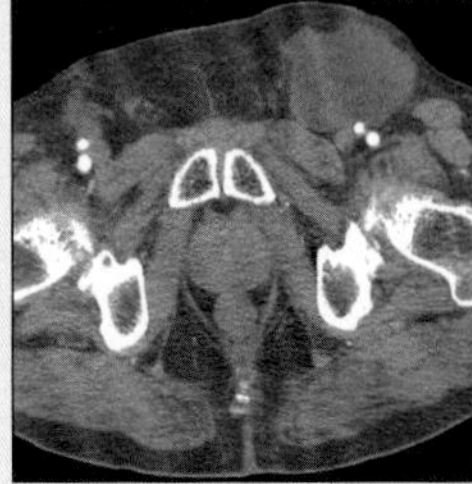
Hematoma

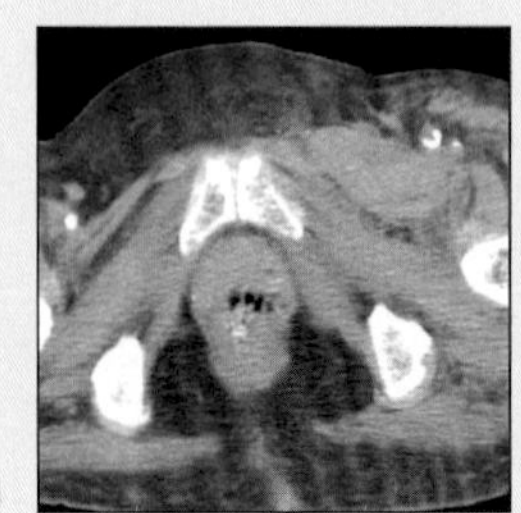
Hematoma

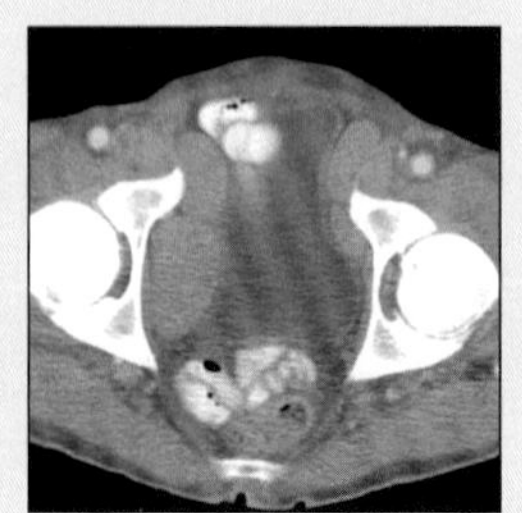
Lymphadenopathy

INGUINAL HERNIA

Key Facts

Imaging Findings
- Indirect IH: Passes through internal inguinal ring, down the inguinal canal & emerges at external ring
- Direct IH: Occurs in floor of inguinal canal, through Hesselbach triangle
- Indirect IH in males within spermatic cord has smooth contour & elongated oblique course
- Direct IH: Broad & dome-shaped; appears as a small bulge in groin; short & blunt aperture
- Indirect IH: May see well-defined ovoid mass in groin
- Collapsed bowel loops & mesenteric fat in hernia sac
- Neck of indirect IH can be demonstrated at deep inguinal ring lateral to IEV
- Whereas direct IH remain medial to IEV throughout

Top Differential Diagnoses
- Femoral hernia
- Iatrogenic
- Lymphadenopathy

Pathology
- 75-80% of all hernias occur in inguinal region
- Indirect IH are 5 times more common than direct IH
- Contents: Include small bowel loops or mobile colon segments such as sigmoid, cecum & appendix

Clinical Issues
- Diagnosis: History & physical examination
- Indirect IH five to ten times more common in men
- Complications: Incarceration; strangulation

- ○ Attempt should be made to reduce hernia manually under fluoroscopy
  - ○ Visualize afferent & efferent loops of protruding intestine
- Herniography: Indirect IH; emerges from lateral inguinal fossa & protrudes medially
  - ○ Roughly parallel to superior pubic ramus
  - ○ Persistent processus vaginalis has a width of 1 to 2 mm; may extend into scrotum
    - ▪ Communicating hydrocele
  - ○ When length of sac exceeds 4 cm, it usually widens abruptly beyond external opening of inguinal canal
  - ○ Widened internal opening of inguinal canal is seen as triangular outpouching of lateral inguinal fossa
    - ▪ Its acute apex is directed inferomedially & its medial border is concave
  - ○ No plicae or indentations visualized lateral to an indirect hernia
  - ○ Open Nuck's canal in women; same herniographic appearance as patent processus vaginalis in men
- More lateral direct hernia, protruding from medial inguinal fossa; usually dome-shaped with wide neck
- More medially located direct IH; protrude from supravesical fossa & are usually smaller

CT Findings
- Indirect IH: May see well-defined ovoid mass in groin
  - ○ Collapsed bowel loops & mesenteric fat in hernia sac
- Neck of indirect IH can be demonstrated at deep inguinal ring lateral to IEV
  - ○ Whereas direct IH remain medial to IEV throughout
- CT useful when there is suspicion that another disease process is either mimicking or precipitating hernia

MR Findings
- In obese patients; physical examination difficult
- Dynamic evaluation in multiple imaging planes may have advantages

Ultrasonographic Findings
- Real Time
  - ○ May see bowel loops peristalse within hernia
  - ○ Useful when patient presents non-urgently with history suggesting reducible hernia
    - ▪ US real time examination allows patient to stand upright & perform Valsalva maneuver
  - ○ Valsalva maneuver: In direct herniation distended pampiniform plexus displaced by hernia sac
    - ▪ In indirect hernia impaired swelling of pampiniform plexus seen
- Color Doppler
  - ○ To distinguish among types of groin hernias
    - ▪ Demonstrate inferior epigastric artery (origin &/or trunk segment) & its relationship with hernia sac

Imaging Recommendations
- Best imaging tool
  - ○ US; CT; MRI for demonstrating acutely strangulated hernia in obese patients
    - ▪ In problem cases where there is clinical diagnostic uncertainty
- Protocol advice
  - ○ CT: Oral + intravenous CECT; axial plane; thinner slice collimation (e.g., 5 mm)
    - ▪ Frequent image reconstructions to show IEV

DIFFERENTIAL DIAGNOSIS

Femoral hernia
- Medial position within femoral canal posterior to line of inguinal ligament; caudal & posterior to IH
- Frequently has a narrow neck; neck remains below inguinal ligament & lateral to pubic tubercle
- More common in women

Iatrogenic
- Arterial puncture following arteriography; needle biopsy or aspiration
  - ○ Hematoma formed may extend into rectus muscle or lateral abdominal wall muscles
  - ○ Blood can track directly from groin, along transversalis fascia & transversus abdominis muscle
  - ○ CT, US, MR: Appearance of blood; extent of lesion; changes over time
  - ○ Pseudoaneurysm: Perivascular, rounded mass; neck + track connecting it with injured artery

INGUINAL HERNIA

Lymphadenopathy

- Appears as mass near inguinal ligament
- CT, US: Help differentiate hernia contents from other masses involving groin & scrotum
  - Such as hydrocele, varix, lipoma of spermatic cord, undescended testicle, abscess, tumor

PATHOLOGY

General Features

- Etiology
  - Indirect IH considered to be congenital defect
    - Patency of processus vaginalis; weakness of crus lateralis at lateral aspect of inguinal canal
  - Direct IH considered acquired lesion
    - Weakness in transversalis fascia of posterior wall of inguinal canal in Hesselbach triangle
- Epidemiology
  - 75-80% of all hernias occur in inguinal region
  - Indirect IH are 5 times more common than direct IH
  - Incidence: IH occurs in 1-3% of all children
    - In premature infants, incidence is one-half to two times greater
  - Approximately 5% of men develop IH during their lifetime & require an operation
  - Bilateral patent processus vaginalis occurs in up to 10% of patients with indirect IH

Gross Pathologic & Surgical Features

- Contents: Include small bowel loops or mobile colon segments such as sigmoid, cecum & appendix
- Sliding IH: Partially retroperitoneal organs
  - Urinary bladder, distal ureters or ascending or descending colon, included in herniation
  - Retroperitoneal structures constitute wall of sac
  - Blood vessels supplying herniated segments, may be injured during surgical repair or trauma
- Littre hernia: Meckel diverticulum in hernia sac
- Richter hernia: Only portion of bowel circumference in sac (antimesenteric)
- Incomplete IH: Sac not extended through external inguinal ring
- Diverticular direct IH: Protrudes from either medial inguinal or supravesical fossa
  - Small opening in otherwise normal transverse fascia
  - Distinct circumscribed neck & usually protrudes more in anterior than inferior direction
- Potential indirect hernias are associated with an undescended testis; a testis in inguinal canal
  - Testicular or spermatic cord hydrocele

CLINICAL ISSUES

Presentation

- Asymptomatic; sudden appearance of lump in groin; intermittently present; ± groin pain; palpable bulge
- Physical examination: Recumbent & upright position; may be reducible; bowel sounds audible; ± tender
  - Indirect hernia lightly touches tip of finger
    - Examining finger placed along spermatic cord at scrotum & passed into external ring along canal
    - During maneuvers that ↑ intra-abdominal pressure
  - Direct hernia causes bulge forward low in canal
- Incarcerated or strangulated hernia: Bowel distension; painful & often tense swelling in groin or scrotum
- Diagnosis: History & physical examination

Demographics

- Age
  - Indirect IH may occur from infancy to old age but generally occur by fifth decade of life
  - Direct IH increases in occurrence with age
- Gender
  - Indirect IH five to ten times more common in men
  - Direct IH occurs mostly in men & seldom in women

Natural History & Prognosis

- Pediatric IH: Almost always indirect; higher risk of incarceration
  - Usually on right (60-75%); often bilateral (10-15%)
- Recurrent hernia: Groin hernias recur after herniorrhaphy in up to 20% of patients
  - Direct IH may develop after repair of an indirect hernia
  - Diverticular hernias are a form of recurrence
- Multiple hernias: One is usually a direct type
  - May obscure smaller hernias that can be clinically significant
- Saddlebag, pantaloon, combined IH: Simultaneous occurrence of direct & indirect IH in same groin
  - Separation of two adjacent hernia sacs by IEV creates bilocular appearance
- Indirect IH accounts for 15% of intestinal obstructions
- Diverticulitis; appendicitis; primary or metastatic tumor may occur within hernia sac
- Complications: Incarceration; strangulation
  - Direct IH rarely becomes incarcerated & less often associated with strangulation

Treatment

- Laparoscopic or open hernia repair

DIAGNOSTIC CHECKLIST

Consider

- Hernias that protrude from lateral inguinal fossa are indirect
- Those from medial & supravesical fossae are direct

SELECTED REFERENCES

1. van den Berg JC: Inguinal hernias: MRI and ultrasound. Semin Ultrasound CT MR. 23(2): 156-73, 2002
2. Zhang GQ et al: Groin hernias in adults: value of color Doppler sonography in their classification. J Clin Ultrasound. 29(8): 429-34, 2001
3. Shadbolt CL et al: Imaging of groin masses: inguinal anatomy and pathologic conditions revisited. Radiographics. 21 Spec No: S261-71, 2001
4. Toms AP et al: Illustrated review of new imaging techniques in the diagnosis of abdominal wall hernias. Br J Surg. 86(10): 1243-9, 1999
5. Hahn-Pedersen J et al: Evaluation of direct and indirect inguinal hernia by computed tomography. Br J Surg. 81(4): 569-72, 1994

INGUINAL HERNIA

IMAGE GALLERY

Typical

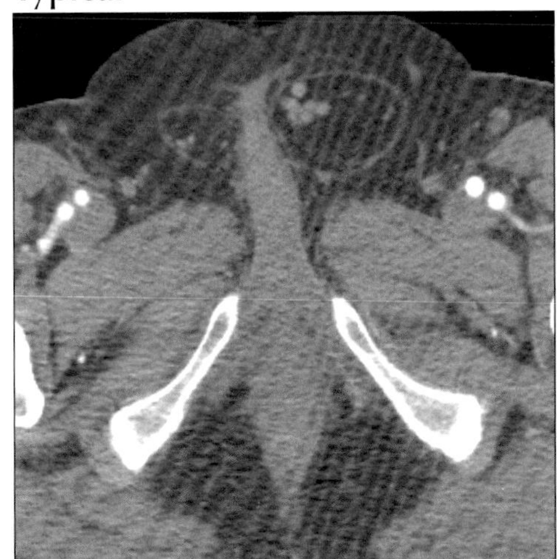

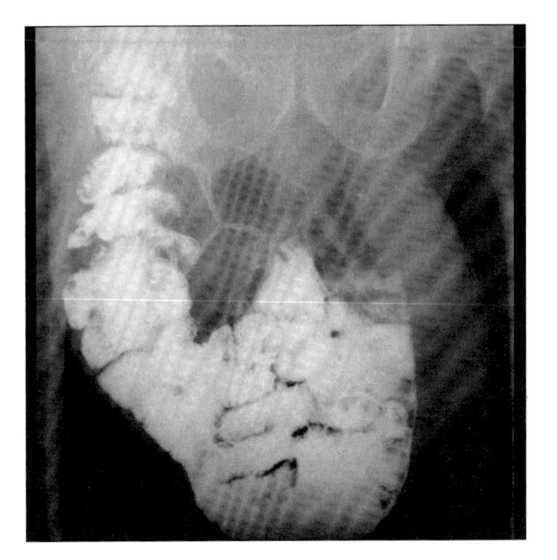

(Left) *Axial CECT shows left inguinal hernia containing only fat and spermatic cord.* ***(Right)*** *Small bowel follow through (SBFT) shows almost entire small intestine within scrotum due to right inguinal hernia.*

Typical

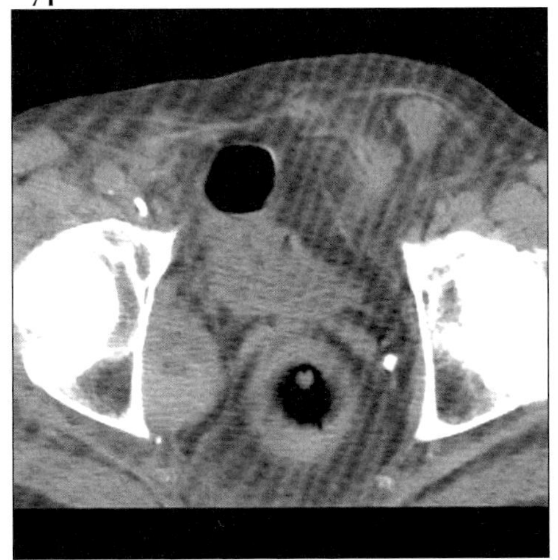

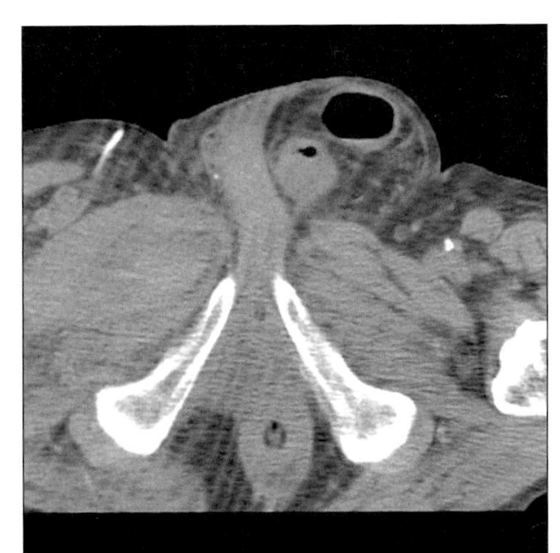

(Left) *Axial CECT shows left inguinal hernia at upper end of inguinal canal. There is mass effect due to herniated fat and bowel.* ***(Right)*** *Axial CECT shows left inguinal hernia containing sigmoid colon. Also note right thigh hematoma.*

Typical

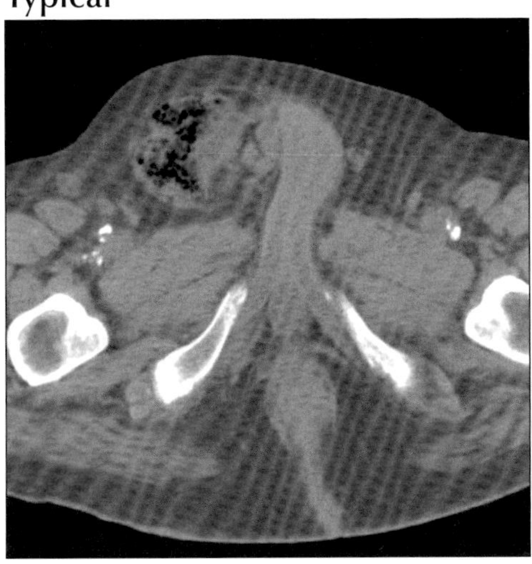

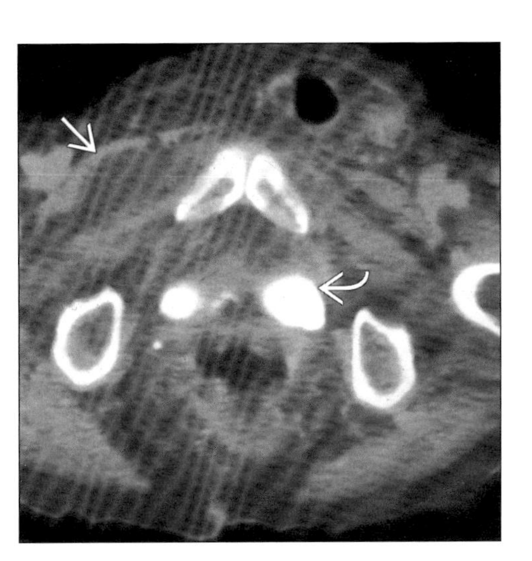

(Left) *Axial CECT shows right inguinal hernia with colon in upper scrotum.* ***(Right)*** *Axial CECT in elderly woman shows left inguinal hernia, right obturator hernia (arrow) and pessary (curved arrow).*

FEMORAL HERNIA

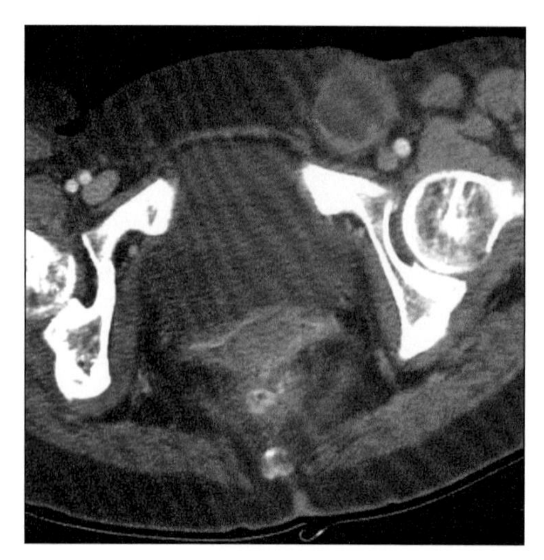

Axial CECT shows left femoral hernia containing "knuckle" of strangulated bowel.

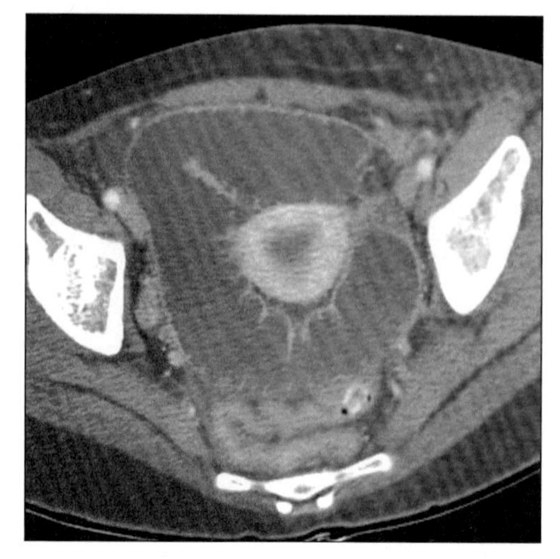

Axial CECT shows small bowel obstruction due to strangulated femoral hernia.

TERMINOLOGY

Abbreviations and Synonyms

- Femoral hernia (FH)
- Crural hernia; Enteromerocele; Femorocele

Definitions

- FH occurs when intra-abdominal contents protrude along femoral sheath in femoral canal

IMAGING FINDINGS

General Features

- Best diagnostic clue: History + clinical examination
- Location
  - Hernia contents protrude posteriorly to inguinal ligament
    - Anteriorly to pubic ramus periosteum (Cooper ligament), medially to femoral vessels
  - FH traverses femoral canal & presents as mass at level of foramen ovale
  - Neck of FH always remains below inguinal ligament & lateral to pubic tubercle
- Morphology
  - FH protrudes at right angle to inguinal canal
  - Has a narrow neck & characteristic pear shape

Radiographic Findings

- Herniography: Characteristic posterior "inbulging"
  - Due to proximity of symphysis pubis to posterior & medial aspect of hernia

CT Findings

- Anatomic localization; relationship to femoral triangle; medial to femoral vein
- Identify contents of sac; differentiating hernia from other causes of groin swelling

Ultrasonographic Findings

- In patients with undiagnosed pain; mass in groin
- Questionable inguinal or femoral region hernias
- Post-operative complaints after herniorrhaphy
- Possible femoral aneurysms or pseudoaneurysms

DIFFERENTIAL DIAGNOSIS

Inguinal hernia

- FH differentiated from inguinal hernia by its medial position within femoral canal
  - Posterior to line of inguinal ligament
- Like direct inguinal hernia, FH arises from medial inguinal fossa
  - But FH protrudes in caudal rather than anterior direction

DDx: Mass Near Femoral Artery

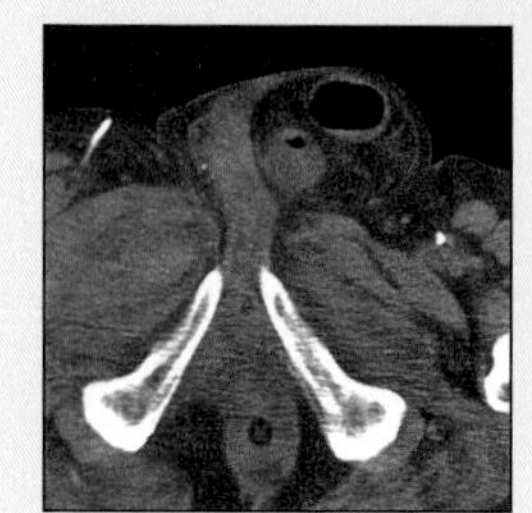

Inguinal Hernia

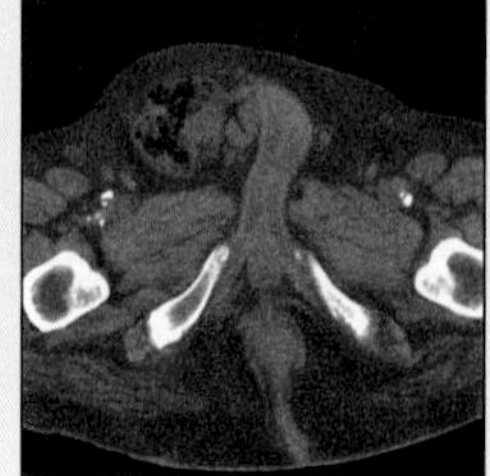

Inguinal Hernia

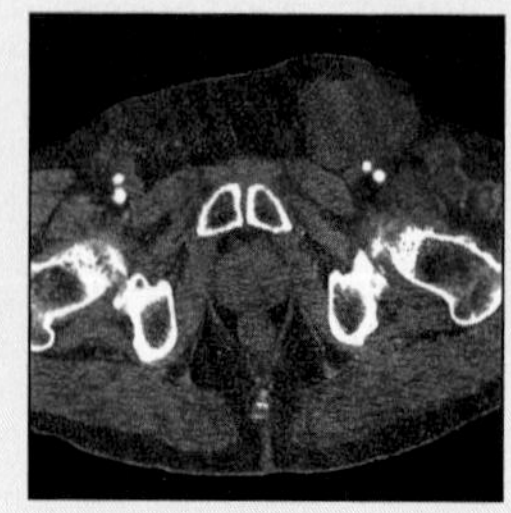

Hematoma

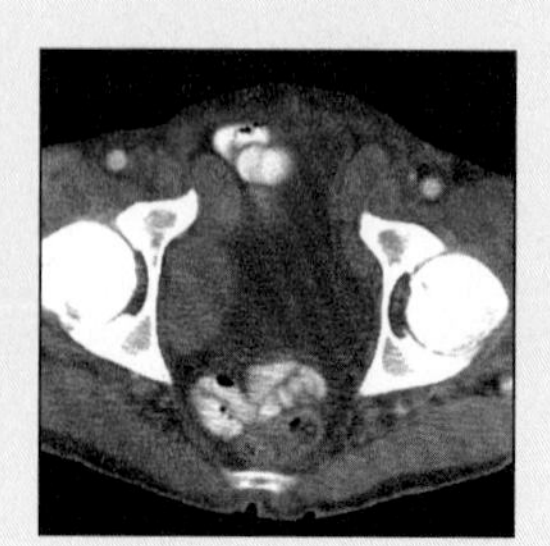

Lymphadenopathy

FEMORAL HERNIA

Key Facts

Terminology

- FH occurs when intra-abdominal contents protrude along femoral sheath in femoral canal

Imaging Findings

- FH traverses femoral canal & presents as mass at level of foramen ovale
- Neck of FH always remains below inguinal ligament & lateral to pubic tubercle
- Has a narrow neck & characteristic pear shape

Top Differential Diagnoses

- Inguinal hernia
- Iatrogenic
- Lymphadenopathy

Clinical Issues

- Gender: Occurs predominantly in women (M:F = 1:3)
- 25-40% are incarcerated; because of a narrow neck

Iatrogenic

- Hematoma; following arterial puncture in femoral sheath during arteriography; needle biopsy etc.
- CT; US: Appearance of blood + extent of lesion

Lymphadenopathy

PATHOLOGY

General Features

- Etiology
  - There may be congenital defect in insertion of transversalis fascia to ileopubic tract
    - Forms superior & anterior limit of lacuna vasorum
    - If it is weakened, femoral canal dilates, permitting formation of FH
  - Associated with ↑ intra-abdominal pressure
- Epidemiology
  - FH accounts for less than 1% of all groin hernias in children
  - In adults accounts for approximately 5-10% of groin hernias
  - Accounts for about one third of groin hernias in women

Gross Pathologic & Surgical Features

- Hernia contents: Usually properitoneal fat, edge of omentum or loop of small bowel
- Richter hernia: Usually in older women with FH

CLINICAL ISSUES

Presentation

- Swelling; ± pain; dragging sensation in groin
- Lump usually felt in top of thigh, below groin crease
  - Large FH may bulge over inguinal ligament
- Neck palpable lateral & inferior to pubic tubercle
- Nausea, vomiting, severe abdominal pain may occur with strangulated hernia
- Often difficult to diagnose clinically; deep location of femoral canal & abundance of overlying adipose tissue

Demographics

- Age
  - Incidence: 36% in over 80 years age group
    - 16% among patients in their seventies
  - Children are rarely affected
- Gender: Occurs predominantly in women (M:F = 1:3)

Natural History & Prognosis

- FH may displace or narrow femoral vein; may descend along saphenous vein
- Complications: Incarceration; strangulation
  - 25-40% are incarcerated; because of a narrow neck
  - FH is 8-12 times more prone to incarceration & strangulation than inguinal hernia
  - Due to firm & unyielding margins of femoral ring
- Morbidity: Significantly related to intestinal obstruction
- Mortality: 1% in 70-79 age group, 5% in 80-90 years age group; associated with intestinal obstruction

Treatment

- Femoral sheath defect closed by apposing Cooper ligament & posterior reflection of inguinal ligament
- Repair using a mini-incision, "tension free" technique, utilizing mesh
- Often floor of inguinal canal is also reinforced, using transversalis fascia

SELECTED REFERENCES

1. Toms AP et al: Illustrated review of new imaging techniques in the diagnosis of abdominal wall hernias. Br J Surg. 86(10): 1243-9, 1999
2. Radcliffe G et al: Reappraisal of femoral hernia in children. Br J Surg. 84(1): 58-60, 1997
3. Chamary VL: Femoral hernia: intestinal obstruction is an unrecognized source of morbidity and mortality. Br J Surg. 80(2): 230-2, 1993

IMAGE GALLERY

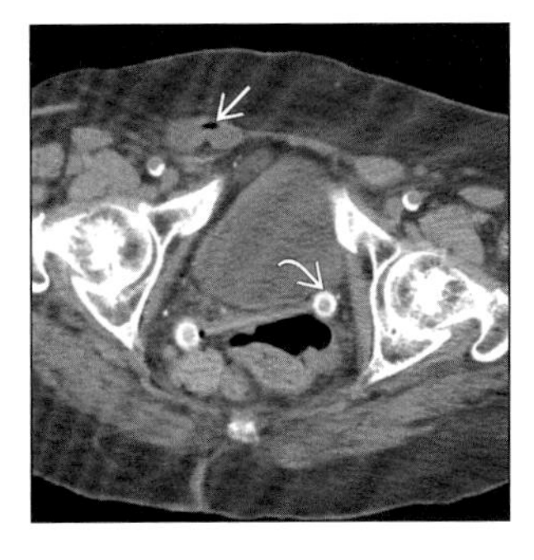

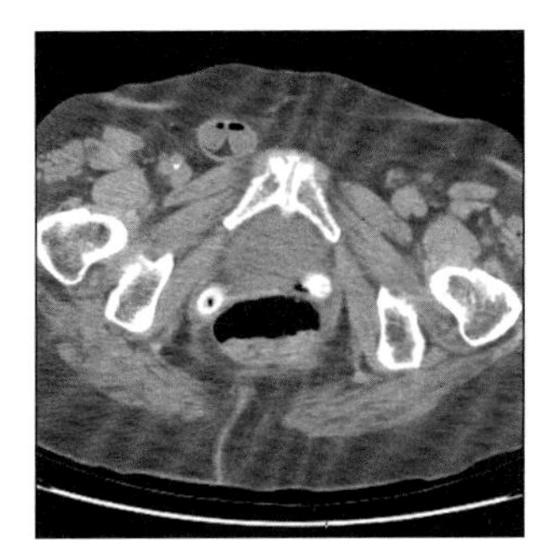

(Left) *Axial CECT in elderly woman shows right femoral hernia (arrow) and pessary (curved arrow).* ***(Right)*** *Axial CECT shows right femoral hernia containing small bowel that caused obstruction.*

VENTRAL HERNIA

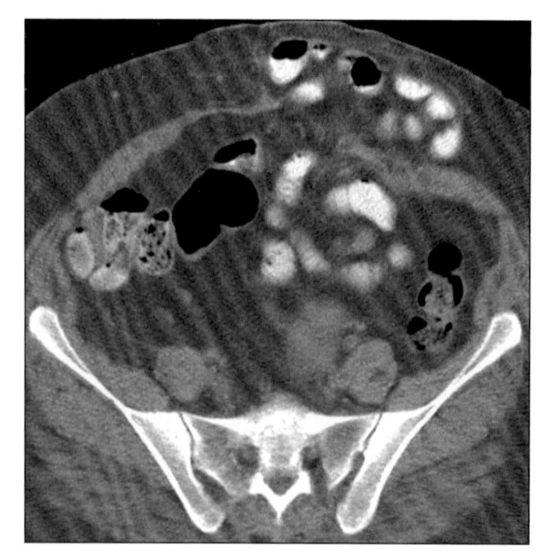

Axial CECT shows herniation of small bowel through wide ventral hernia at site of prior paramedian surgical incision.

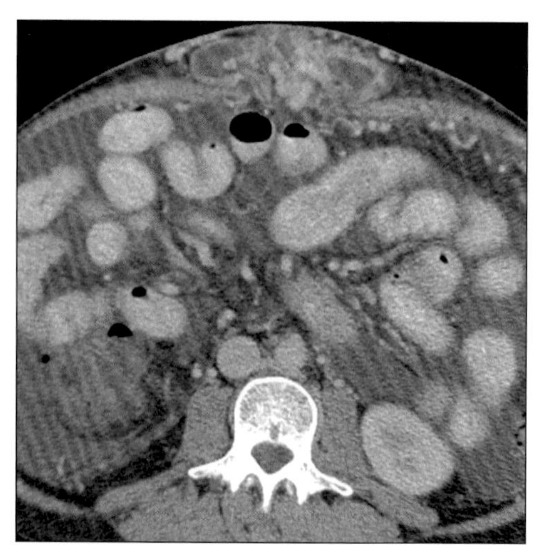

Axial CECT shows ventral hernia containing ascites and varices in a patient with cirrhosis (Umbilicus at a lower section).

TERMINOLOGY

Definitions

- Term "ventral hernia" (VH) encompasses herniations through anterior & lateral aspects of abdominal wall
  - Incisional Hernia (IH): Hernia resulting from abdominal wall incisions

IMAGING FINDINGS

General Features

- Location
  - Majority occur in midline; emerge through aponeurosis forming linea alba
  - Epigastric hernia: Above umbilicus
  - Hypogastric hernia: Below umbilicus
  - IH: Along midline, paramedian incisions; although any surgical scar may be a potential site

Radiographic Findings

- Fluoroscopic evaluation of gastrointestinal tract with barium; may reveal clinically occult, unsuspected IH
  - Areas of old surgical scars may be viewed in profile while patient strains; detect reducible hernias
- Distended bowel loops proximal to obstruction in relationship to locally tender part of abdominal wall

CT Findings

- Defect in peritoneal & fascial layers of ventral abdominal wall
  - Through which omentum or knuckle of intestines protrude into subcutaneous fat
- Early dehiscence of muscle layer in anterior abdominal wall closure; without disruption of overlying skin
- Useful when contents & orifice not palpable; interstitial or interparietal location
  - Herniated segments dissect & "hide" between muscular & fascial layers of abdominal wall
- Differentiate incarcerated hernia & post-operative hematoma

Ultrasonographic Findings

- Actual defect & herniation in ventral wall of abdomen

DIFFERENTIAL DIAGNOSIS

Parastomal hernia

- Following stoma formation of an ileostomy or colostomy

Spigelian hernia

- Anterolateral; defect involving lateral border of rectus sheath

DDx: Anterior Abdominal Wall Defect or Mass

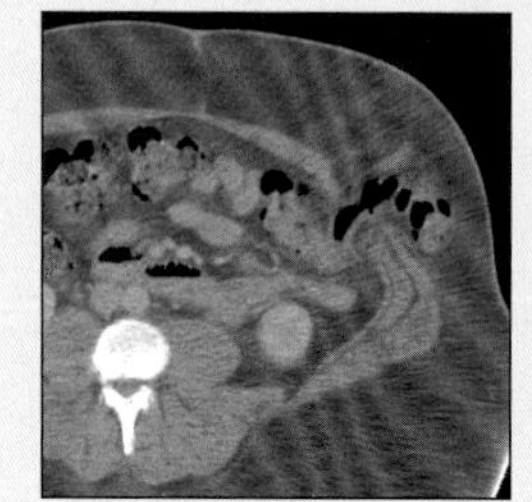

Parastomal Hernia

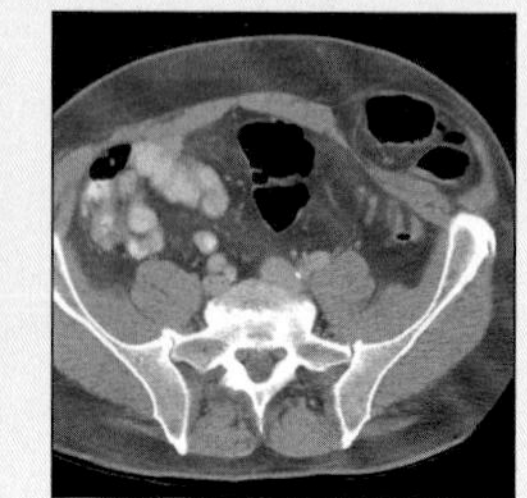

Spigelian Hernia

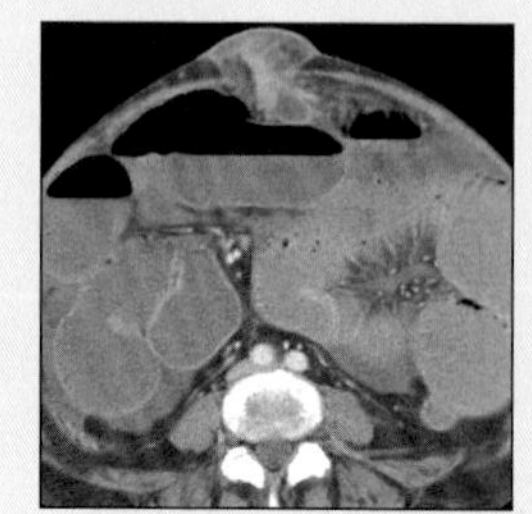

Umbilical Hernia

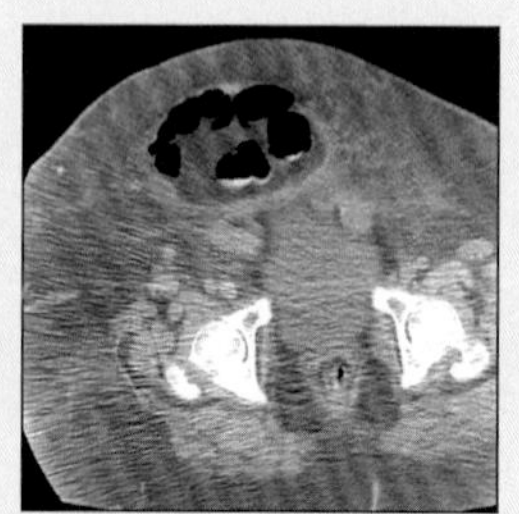

Pannus

VENTRAL HERNIA

Key Facts

Terminology
- Term "ventral hernia" (VH) encompasses herniations through anterior & lateral aspects of abdominal wall
- Incisional Hernia (IH): Hernia resulting from abdominal wall incisions

Imaging Findings
- Majority occur in midline; emerge through aponeurosis forming linea alba
- Defect in peritoneal & fascial layers of ventral abdominal wall

Top Differential Diagnoses
- Parastomal hernia
- Spigelian hernia
- Umbilical hernia
- Hematoma or abscess
- Pannus

Umbilical hernia
- Umbilical defect; patent umbilical ring

Hematoma or abscess
- No defect in abdominal wall
- Distinguish gas in abscess from gas in bowel

Pannus
- Mid-abdomen may protrude over lower, simulating hernia

PATHOLOGY

General Features
- Etiology
  - Congenital; spontaneous; primary defect
  - Acquired: Latrogenic; previous abdominal surgery, laparoscopy, peritoneal dialysis, stab wound etc.
  - Factors which ↑ likelihood of hernia occurrence
    - Patient-related: Collagen biochemistry, obesity, age over 65, pulmonary disease, uremia, diabetes, steroids, malignancy, trauma
    - Technical factors: Wound infection, suture material, types of incisions & closures
- Epidemiology
  - More than 80% of VH result from prior surgery
    - IH reported to occur after 0.2-26% of abdominal procedures
  - Majority of incisional hernias are ventral

Gross Pathologic & Surgical Features
- Portions of greater omentum, properitoneal fat, or a bowel loop protrude anteriorly

CLINICAL ISSUES

Presentation
- Bulge or swelling on abdominal wall; can become larger & pain aggravated with exertion
- IH: Tend to occur during first 4 months after surgery
  - Progressive enlargement usually manifested within first post-operative year
  - 5-10% remain clinically silent for several years
  - Most IH are incidental findings at imaging
  - Initially, vague abdominal discomfort; localized tenderness of healed scar
  - Advanced stage: Persistent bulging mass resulting from incarcerated bowel loops may be seen
- Symptoms out of proportion to objective findings if incarceration or strangulation occurs
- 10% of IH not detected on physical examination
  - Diagnosis difficult in obese patients, severe pain or distension, keloid, thick panniculus

Natural History & Prognosis
- Complications: Incarceration & strangulation of contents; may occur frequently

Treatment
- Repair techniques: Open suture; open mesh; laparoscopic mesh
- Laparoscopic repair: Lower recurrence rate (0-9%)
- With development of prosthetic mesh safe to place intraperitoneally, recurrence rate has ↓ to under 5%

DIAGNOSTIC CHECKLIST

Consider
- Accurate demonstration of size & site of hernial orifice, along with contents of sac
  - May be useful in assessing potential risk of strangulation or likely success of hernia repair

SELECTED REFERENCES

1. Millikan KW: Incisional hernia repair. Surg Clin North Am. 83(5):1223-34, 2003
2. Thoman DS et al: Current status of laparoscopic ventral hernia repair. Surg Endosc. 16(6):939-42, 2002
3. Yahchouchy-Chouillard E et al: Incisional hernias. I. Related risk factors. Dig Surg. 20(1):3-9, 2003;20(1):3-9.

IMAGE GALLERY

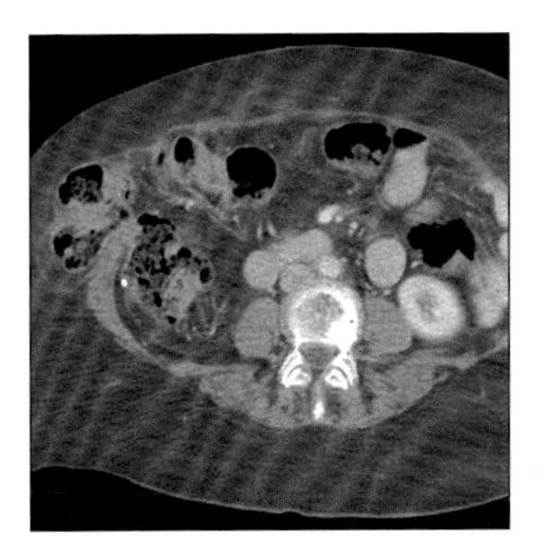
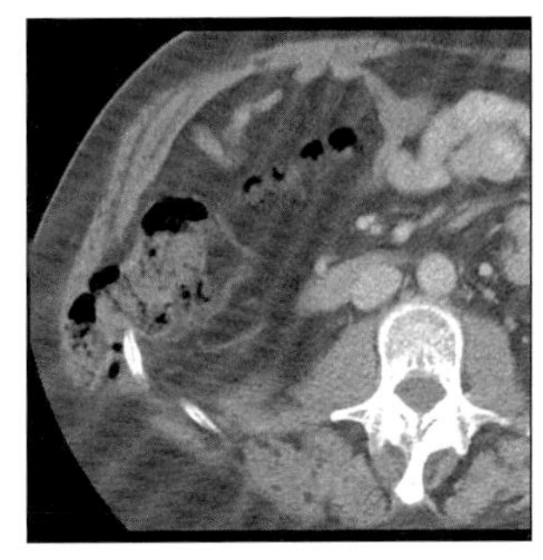

*(**Left**) Axial CECT shows a lateral ventral hernia. (**Right**) Axial CECT shows a lateral incisional hernia containing colon.*

SPIGELIAN HERNIA

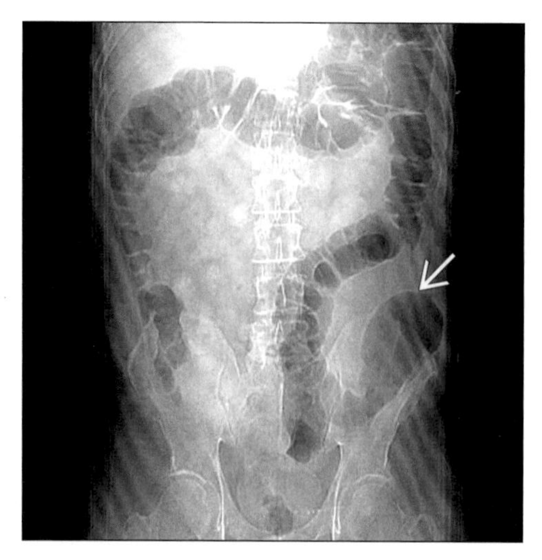

Radiograph shows dilated colon due to left Spigelian hernia containing descending colon (arrow).

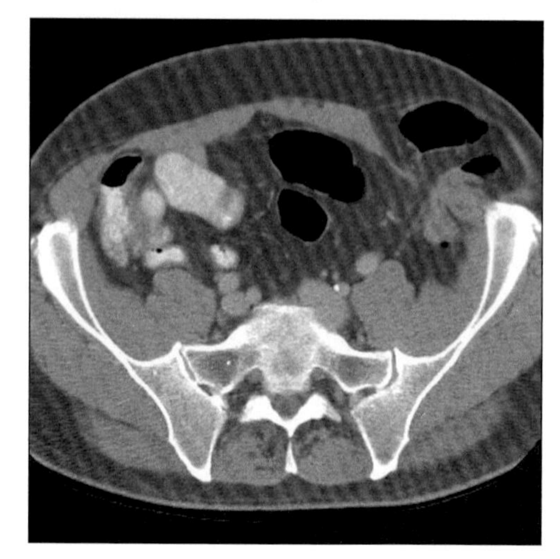

Axial CECT shows Spigelian hernia just lateral to rectus muscle, with obstructed, herniated descending colon.

TERMINOLOGY

Abbreviations and Synonyms

- Spigelian hernia (SH)

Definitions

- Spigelian hernia is a protrusion of intra-abdominal fat or bowel through a defect in '"Spigelian aponeurosis"

IMAGING FINDINGS

General Features

- Location
  - Spigelian aponeurosis: Part of anterior abdominal wall aponeurosis
    - Lies between linea semilunaris laterally & lateral edge of rectus muscle medially
  - SH is classically found at position cranial to junction of inferior epigastric vessels & Spigelian aponeurosis
    - Caudal to this point it is termed "low SH"
  - Anatomically: Hernia sac extends through defect
    - In transversalis fascia, transversus abdominis, & internal oblique aponeurosis
    - Lies deep to external oblique aponeurosis
    - Classic SH: External oblique aponeurosis remains intact & hernial sac is intermuscular
  - Clinically: In fascia below level of umbilicus, lateral to junction of semilunar line & arcuate lines
- Morphology
  - Defect size range: 1-7.5 cm diameter
    - Most commonly; 1-3 cm

Radiographic Findings

- Barium enema or small bowel examination: Position & contents of hernia
  - May show distended bowel loops proximal to a narrowing or obstruction

CT Findings

- Hernia defect involving lateral border of rectus sheath
- Interstitial protrusion of omentum or intestinal loops

DIFFERENTIAL DIAGNOSIS

Ventral hernia

- Majority occur in midline; emerge through aponeurosis forming linea alba
- Epigastric or hypogastric: Above or below umbilicus
  - Most SH are caudal & lateral to umbilical level

Umbilical hernia

- CT; US: "Knuckle" of bowel, fat, ascites protruding through umbilical defect in midline

DDx: Anterior Abdominal Wall Defect

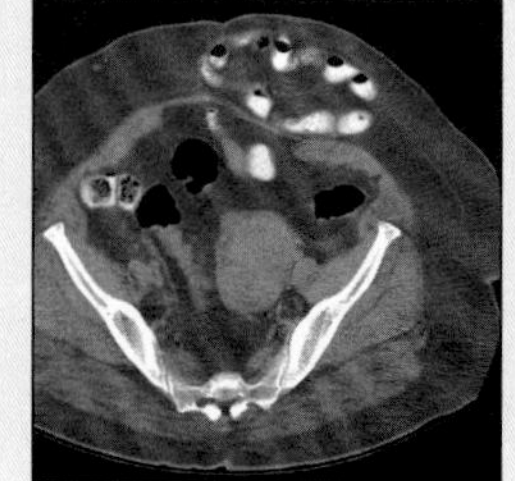

Ventral Hernia

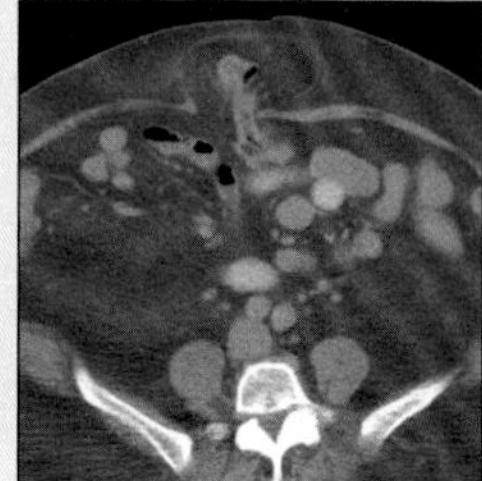

Umbilical Hernia

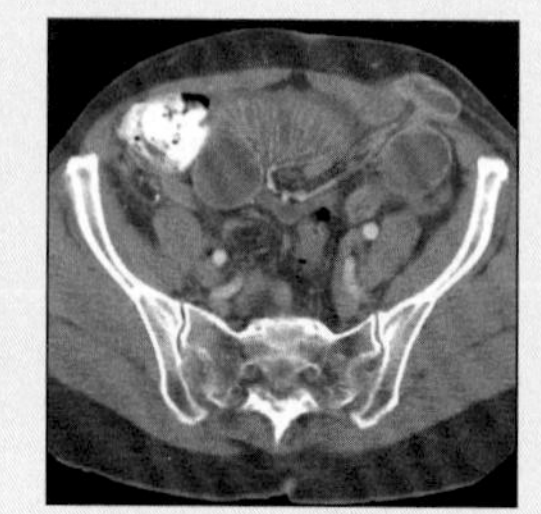

Lap. Port Hernia

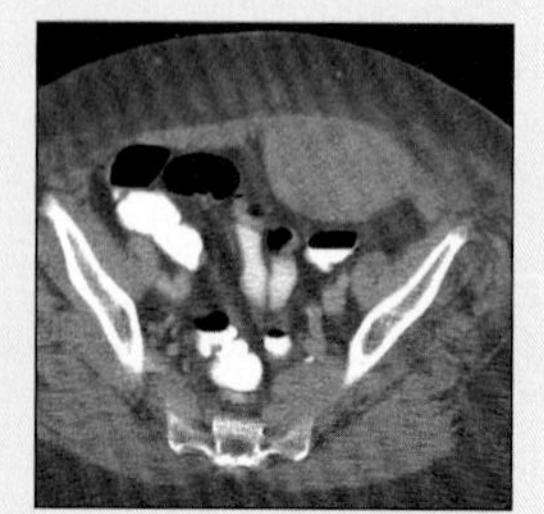

Rectus Hematoma

SPIGELIAN HERNIA

Key Facts

Terminology

- Spigelian hernia is a protrusion of intra-abdominal fat or bowel through a defect in '"Spigelian aponeurosis"

Imaging Findings

- Classic SH: External oblique aponeurosis remains intact & hernial sac is intermuscular
- Hernia defect involving lateral border of rectus sheath
- Interstitial protrusion of omentum or intestinal loops

Top Differential Diagnoses

- Ventral hernia
- Umbilical hernia
- Hernia through laparoscopy port (lap. port)
- Rectus sheath hematoma

Pathology

- Frequency: 2% of anterior abdominal hernias

Hernia through laparoscopy port (lap. port)

- Several potential sites
- Usually medial or lateral to Spigelian site

Rectus sheath hematoma

- Cylindrical, heterogeneous, encapsulated mass

PATHOLOGY

General Features

- Etiology: Variations in SHs site, size, age suggest multifactorial etiology; including congenital
- Epidemiology
  - Frequency: 2% of anterior abdominal hernias
  - SH is considered rare; but more likely underreported
  - 37 reported cases in children (newborn-17 y)
    - Bilateral hernia: 15% incidence in children
- Associated abnormalities
  - Undescended testis (17%) in children; ipsilateral
  - Anterior wall defects: Gastroschisis, omphalocele, bladder or cloacal extrophy, prune belly syndrome
  - Coexisting ventral, inguinal, umbilical hernia

Gross Pathologic & Surgical Features

- Interparietal or interstitial herniation
- Contents: Omentum & short segment of small or large intestine (may contain empty sac, testis, ovary)

CLINICAL ISSUES

Presentation

- Asymptomatic; intestinal obstruction
- Prolonged history of intermittent pain
- Slight swelling or vanishing anterolateral mass
- Difficult to diagnose clinically; due to deep anatomic location & insidious development
- Patients tend to be obese; making firm diagnosis on physical examination difficult
  - May result in surgical exploration undertaken on suspicion raised by history alone

Demographics

- Gender
  - In adult: M = F
  - In children: M:F = 2.1:1

Natural History & Prognosis

- Omentum within SH may infarct & cause symptoms
- Because hernial ring is usually small, irreducibility & strangulation are not uncommon
- Presence of congenital SH predisposes to development of ipsilateral undescended testis

Treatment

- Tension free repair using a prosthetic mesh
- Repair of pediatric SH utilizes endogenous tissue

DIAGNOSTIC CHECKLIST

Consider

- Intra-operative US is a valid option for accurate localization of SH, especially in obese patients
  - Extensive intra-operative dissection, distortion of tissue planes, & morbidity risks may be avoided

SELECTED REFERENCES

1. Losanoff JE et al: Spigelian hernia in a child: case report and review of the literature. Hernia. 6(4): 191-3, 2002
2. Losanoff JE et al: Recurrent Spigelian hernia: a rare cause of colonic obstruction. Hernia. 5(2): 101-4, 2001
3. Losanoff JE et al: Incarcerated Spigelian hernia in morbidly obese patients: the role of intraoperative ultrasonography for hernia localization. Obes Surg. 7(3): 211-4, 1997

IMAGE GALLERY

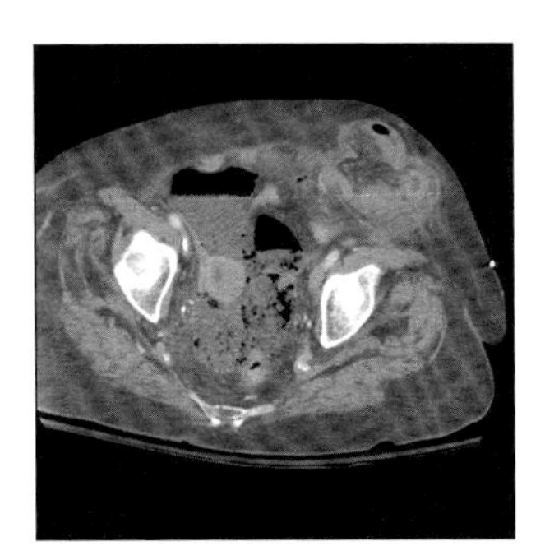

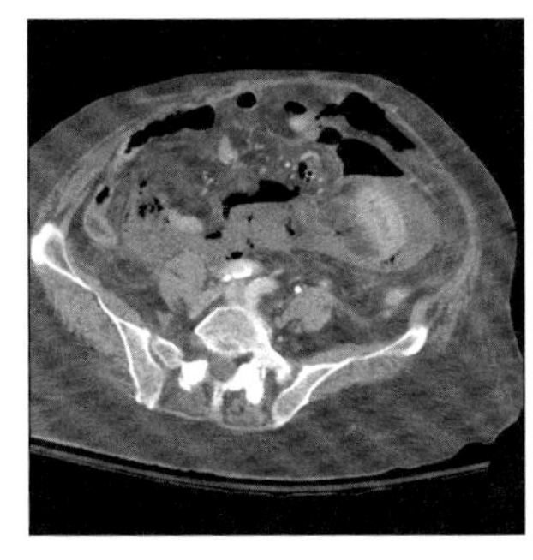

*(**Left**) Axial CECT shows left Spigelian hernia. (**Right**) Axial CECT shows partial small bowel (SB) obstruction due to Spigelian hernia.*

OBTURATOR HERNIA

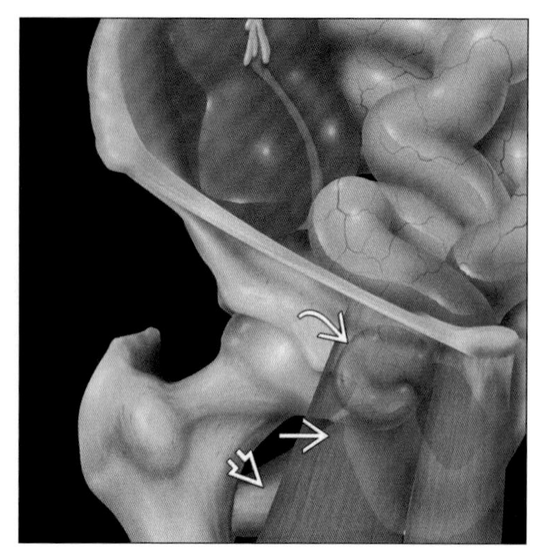

Graphic shows bowel obstruction due to obturator hernia (curved arrow) with strangulated bowel lying deep to pectineus muscle (arrow) and superficial to obturator externus (open arrow).

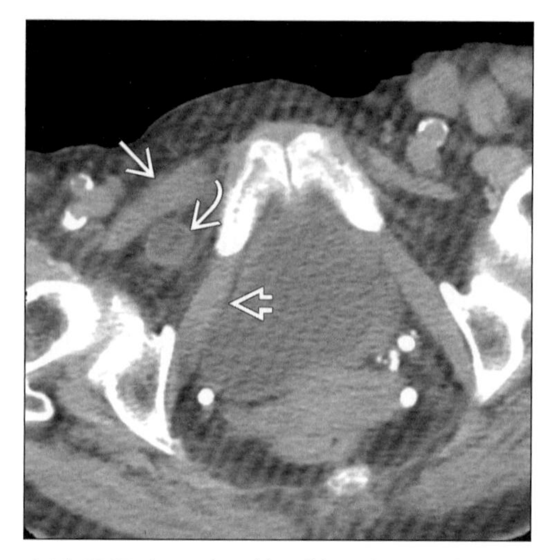

Axial CECT shows knuckle of bowel (curved arrow) lying between pectineus (arrow) and obturator muscles (open arrow).

TERMINOLOGY

Abbreviations and Synonyms

- Obturator hernia (OH)

Definitions

- Pelvic hernia; protruding through obturator foramen

IMAGING FINDINGS

General Features

- Best diagnostic clue: CT evidence of herniated bowel lying between pectineus & obturator muscles in elderly woman
- Location
  - Site of herniation is obturator canal in superolateral aspect of obturator foramen
  - More common on right
- Morphology
  - Hernia between: Pectineus & obturator externus muscles (externus, most common)
    - Superior & middle fasciculi of obturator externus muscle; external & internal obturator membranes

Radiographic Findings

- Abdominal radiographs or barium studies
  - Small bowel obstruction; fixed loop containing gas or contrast medium in obturator region

CT Findings

- Soft tissue mass or opacified loop that protrudes through obturator foramen
  - Extends between pectineus & obturator muscles
  - Sac exits pelvis near obturator vessels & nerve

Imaging Recommendations

- Best imaging tool: CT; pelvis & upper aspect of thigh

DIFFERENTIAL DIAGNOSIS

Inguinal hernia

- Indirect: Through inguinal canal → external ring
  - Females; course of round ligament into labium majus
  - Males; along spermatic cord → scrotum
- Direct: Weak area medial to inferior epigastric vessels

Sciatic hernia

- Through greater sciatic foramen → laterally into subgluteal region

Perineal hernia

- Anterior: Through urogenital diaphragm
- Posterior: Between levator ani & coccygeus muscle

DDx: Defect in Pelvic Floor

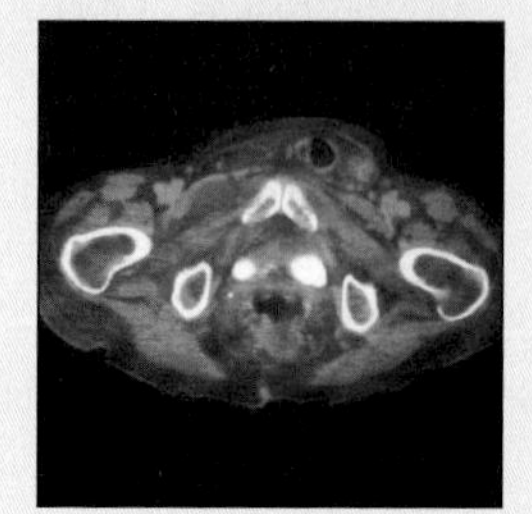

Inguinal + Obturator

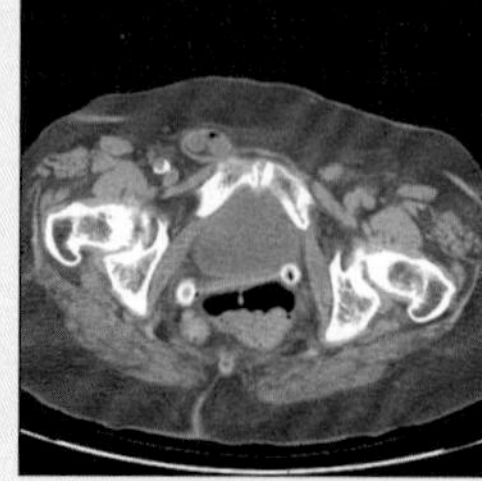

Femoral Hernia

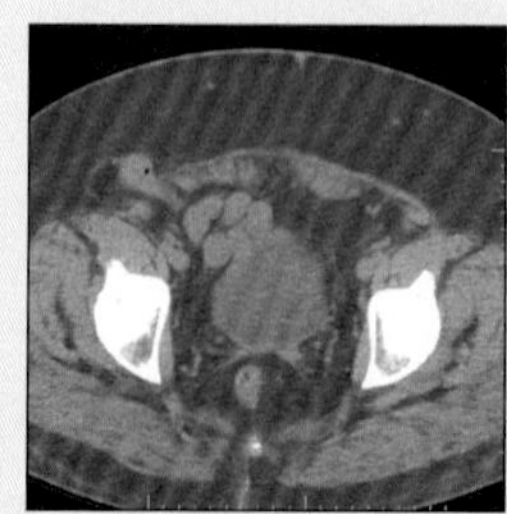

Femoral Hernia

OBTURATOR HERNIA

Key Facts

Imaging Findings

- Site of herniation is obturator canal in superolateral aspect of obturator foramen
- Extends between pectineus & obturator muscles
- Sac exits pelvis near obturator vessels & nerve

Top Differential Diagnoses

- Inguinal hernia
- Sciatic hernia
- Perineal hernia

Pathology

- Accounts for 0.05-0.14% of all hernias

Clinical Issues

- Most patients present with acute or recurrent small bowel obstruction; partial > complete
- Elderly emaciated women with chronic disease
- Mortality rate: 13-40%

PATHOLOGY

General Features

- General path comments
  - Anatomy: Obturator canal; obliquely oriented fibro-osseous tunnel; 2-3 cm long & 1 cm in diameter
    - Obturator nerves & vessels course through it
- Etiology
  - Defect in pelvic floor or laxity
  - Chronic lung disease, constipation, kyphoscoliosis, pregnancy; predispose by ↑ intra-abdominal pressure
- Epidemiology
  - Accounts for 0.05-0.14% of all hernias
    - 0.2-1.6% of all small bowel obstruction
  - Bilateral OH rare; 6% incidence
  - May have inguinal and femoral hernia also

Gross Pathologic & Surgical Features

- Usually contains an ileal loop
  - May involve other viscera & pelvic adnexa (large bowel, omentum, fallopian tube, appendix)

CLINICAL ISSUES

Presentation

- Most patients present with acute or recurrent small bowel obstruction; partial > complete
- May present as tender mass in obturator region on rectal or vaginal examination
- Howship-Romberg sign; pain in medial aspect of thigh with abduction, extension, internal rotation of knee
  - Positive in 20-50% of cases
  - Obturator nerve irritation; compression by hernia
- Absent adductor reflex in thigh
- Rarity & non-specific signs contribute to late diagnosis; & is made correctly in 10-30% of cases

Demographics

- Age: Mean age: 82; range: 65-95 years
- Gender
  - M:F = 1:6-9
  - 80-90% of OH occur in elderly women
    - Owing to enlargement of obturator canal after pregnancies & aging; broader pelvis in women
    - Elderly emaciated women with chronic disease

Natural History & Prognosis

- Protruding structures often incarcerated in canal or space between pectineus & obturator muscles
- Morbidity & mortality rates significantly high
  - Group of debilitated patients with chronic disease; undergo late operation for this elusive diagnosis
  - Mortality rate: 13-40%

Treatment

- Majority require resection of strangulated small bowel
- Abdominal/inguinal approach for reduction & repair
- Contralateral side exploration is recommended

DIAGNOSTIC CHECKLIST

Consider

- In any elderly, debilitated, chronically ill woman
  - Symptoms & signs of recurrent small-bowel obstruction (no history of surgery or hernias)
  - With pain along ipsilateral thigh & knee
  - High index of suspicion must be maintained

SELECTED REFERENCES

1. Lo CY et al: Obturator hernia presenting as small bowel obstruction. Am J Surg. 167(4): 396-8, 1994
2. Chan MY et al: Obturator hernia--case reports. Ann Acad Med Singapore. 23(6): 911-3, 1994
3. Zerbey AL 3rd et al: Bilateral obturator hernias: case report, radiographic characteristics, and brief review of literature. Comput Med Imaging Graph. 17(6): 465-8, 1993

IMAGE GALLERY

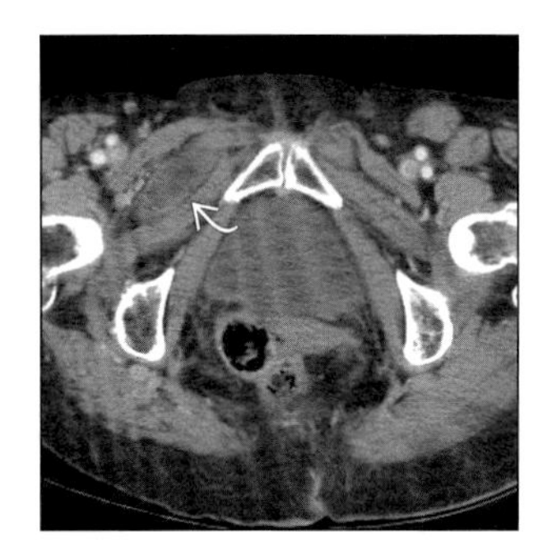

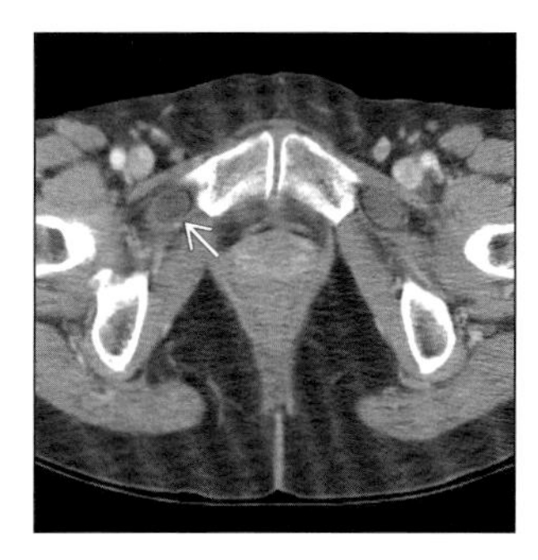

(Left) *Axial CECT shows obturator hernia with small bowel strangulated (curved arrow) between pectineus + obturator externus muscles.* ***(Right)*** *Axial CECT shows right obturator hernia (arrow).*

PARADUODENAL HERNIA

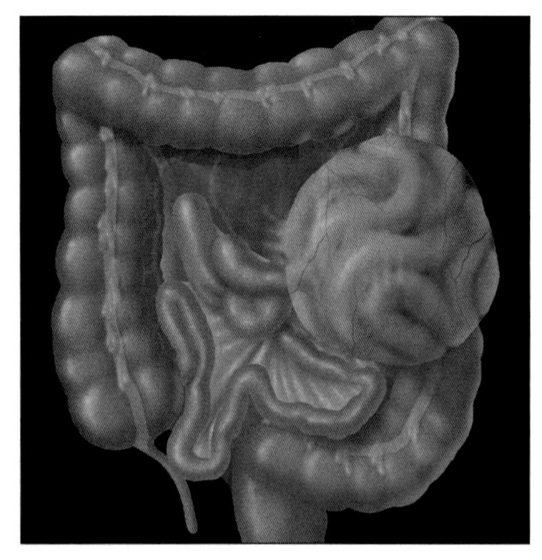

Graphic shows left paraduodenal hernia containing dilated, proximal jejunal loops in a peritoneal "sac".

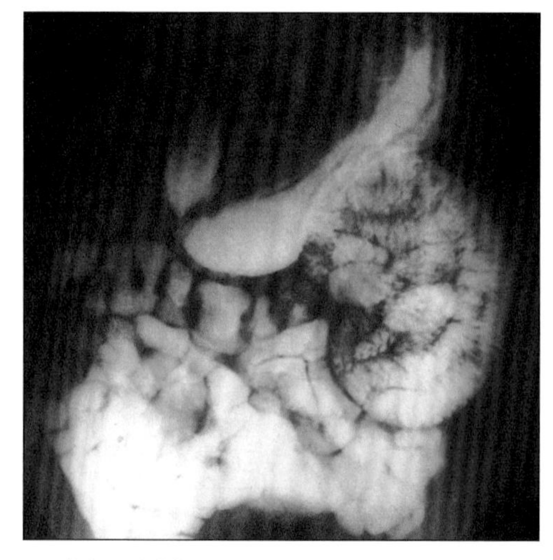

Small bowel follow through (SBFT) shows cluster of jejunal segments that seem to lie within a confining sac.

TERMINOLOGY

Definitions

- Protrusion of bowel loops through a congenital or acquired defect of mesentery within abdominal cavity

IMAGING FINDINGS

General Features

- Best diagnostic clue: Cluster of dilated bowel loops with distorted mesenteric vessels on CECT
- Location
  - Left paraduodenal hernia (75%)
    - Via paraduodenal (lateral to 4th part) mesenteric fossa of Landzert close to ligament of Treitz
  - Right paraduodenal hernia (25%)
    - Via jejunal mesentericoparietal fossa of Waldeyer
- Key concepts
  - Classification of hernias based on anatomic location
    - Internal or intra-abdominal: Herniation of bowel loops via defect within abdominal cavity
    - External: Prolapse of bowel loops via defect in wall of abdomen or pelvis
    - Diaphragmatic: Protrusion of bowel loops via hiatus or congenital defect
  - Subclassification of internal hernias
    - Paraduodenal hernia
    - Transmesenteric post-operative hernia
    - Foramen of Winslow, pericecal hernias
    - Intersigmoid & transomental hernias
  - Subclassification of paraduodenal based on location
    - Left paraduodenal hernia (75%)
    - Right paraduodenal hernia (25%)
  - Second most common subtype of internal hernia after transmesenteric post-operative hernia
  - Usually congenital or rarely acquired
  - Rare cause of small-bowel obstruction
  - 0.6-5.8% of small-bowel obstructions (SBO) are due to internal hernias
    - Paraduodenal accounts 1% or less of all SBO cases

Radiographic Findings

- Radiography
  - Plain x-ray abdomen (supine)
    - "Closed loop": Markedly distended segment of small bowel (indicates SBO)
- Fluoroscopic guided small-bowel follow through
  - Crowding of bowel loops in an abnormal location
    - Location: Displaced to left or right side of colon
    - Small-bowel is often absent from pelvis
  - Left paraduodenal hernia
    - A circumscribed ovoid mass of jejunal loops in left upper quadrant lateral to ascending duodenum
  - Right paraduodenal hernia
    - Ovoid mass of small-bowel loops lateral & inferior to descending duodenum

DDx: Cluster of Dilated Bowel Loops

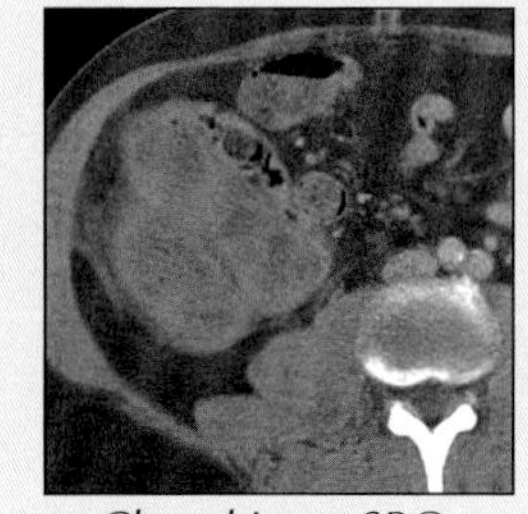

Closed Loop SBO

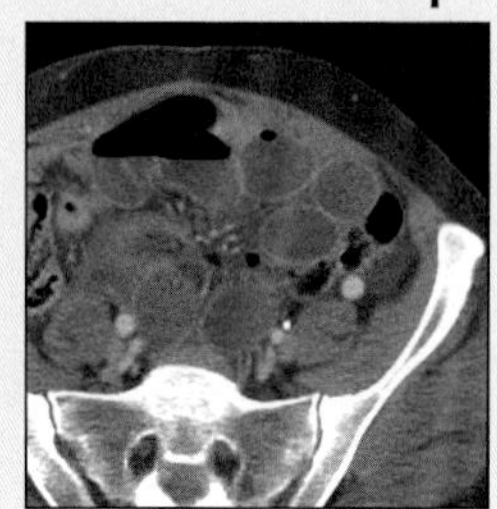

Closed Loop SBO

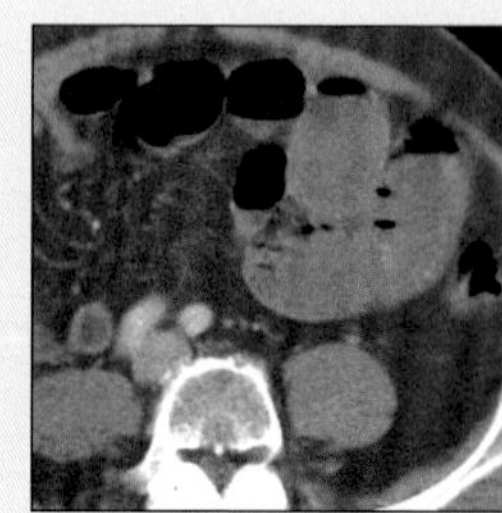

SBO

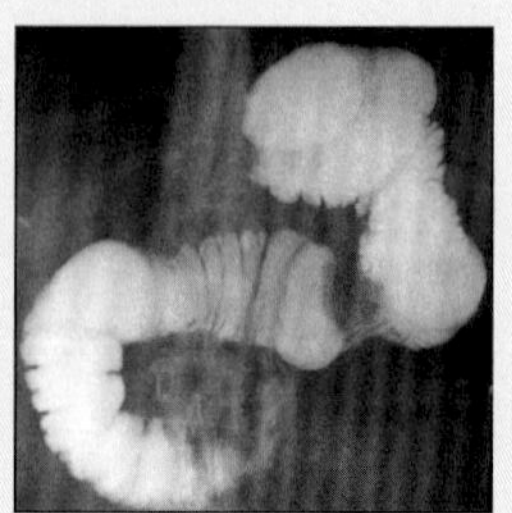

TMI Hernia

PARADUODENAL HERNIA

Key Facts

Terminology

- Protrusion of bowel loops through a congenital or acquired defect of mesentery within abdominal cavity

Imaging Findings

- Best diagnostic clue: Cluster of dilated bowel loops with distorted mesenteric vessels on CECT
- Left paraduodenal hernia (75%)
- Right paraduodenal hernia (25%)
- Crowding of bowel loops in an abnormal location
- A circumscribed ovoid mass of jejunal loops in left upper quadrant lateral to ascending duodenum
- Ovoid mass of small-bowel loops lateral & inferior to descending duodenum
- Varying degrees of small bowel obstruction
- Mass effect: Displacement of bowel loops

Top Differential Diagnoses

- Closed loop obstruction (SBO)
- Transmesenteric (internal) hernia (TMI hernia)

Pathology

- Congenital or developmental mesenteric anomalies
- Acquired: As a complication of surgery or trauma
- Herniation via abnormal mesenteric fossa of Landzert (seen in 2% of autopsies)
- Herniation via abnormal mesentericoparietal fossa of Waldeyer (seen in 1% of autopsies)

Diagnostic Checklist

- Rule out other causes of dilated small-bowel loops
- Cluster of dilated small-bowel loops lateral to ascending or descending duodenum with crowded/twisted mesenteric vessels

---

  - Appear as they are contained in a sac or confining border
  - Varying degrees of small bowel obstruction
  - Point of transition between dilated & nondilated bowel (common)
  - Some degree of fixation, stasis & delayed flow of contrast seen in herniated bowel
  - Inability to displace clustered loops by manual palpation or change in position
    - Right side herniated loops more fixed than left
  - Lateral films: Useful in demonstrating retroperitoneal displacement of herniated bowel loops

CT Findings

- Left-sided paraduodenal hernia
  - Evidence of small bowel obstruction (SBO)
  - Encapsulated, cluster or sac-like mass of small bowel loops
    - Location: Between pancreatic body/tail & stomach, to left of ligament of Treitz
  - Mass effect: Displacement of bowel loops
    - Posterior stomach wall & duodeno-jejunal junction (inferior & medially)
    - Transverse colon inferiorly
  - Mesenteric vessels: Crowded & engorged
- Right-sided paraduodenal hernia
  - Cluster or encapsulated small bowel loops
    - Location: Lateral & inferior to descending duodenum
  - Superior mesenteric vein: Rotated anteriorly & to left side
  - Twisting of vascular jejunal branches behind SMA & into hernial sac
  - Ascending colon: Lies lateral to hernia sac
  - Cecum: Normal site
  - Right ureter: Laterally displaced

Ultrasonographic Findings

- Real Time: Dilated small-bowel loops

Angiographic Findings

- Conventional
  - Superior mesenteric arteriogram
    - Normal jejunal branches arise from left margin of main trunk & abruptly turn to right & pass behind it to supply herniated loops

Imaging Recommendations

- Helical CECT; small-bowel follow through

DIFFERENTIAL DIAGNOSIS

Closed loop obstruction (SBO)

- Obstruction of small-bowel at two points
- Tends to involve mesentery & prone to produce a volvulus
  - Represents most common cause of strangulation
- Etiology
  - Most often caused by an adhesive band
  - Occasionally by an internal or external hernia
- Imaging features
  - Markedly distended segment of fluid-filled small-bowel
  - Volvulus: C- or U-shaped or "coffee bean" configuration of bowel loop
  - Stretched mesenteric vessels converging toward site of torsion
  - Adjacent collapsed loops at site of obstruction
    - Round, oval or triangular in shape
  - "Beak sign"
    - Fusiform tapering at point of torsion/obstruction
  - "Whirl sign"
    - Due to tightly twisted mesentery with volvulus
- May be indistinguishable from paraduodenal hernia
  - Especially if associated with volvulus

Transmesenteric (internal) hernia (TMI hernia)

- Iatrogenic (post-operative)
  - Abdominal surgery: Roux-en-Y gastric bypass (73%)
  - Most common subtype of internal hernia in adults
- Congenital (mesenteric defect)
  - Most common subtype in pediatric age group
  - Location: Close to ligament of Treitz/ileocecal valve
  - Mesenteric defect size: Usually 2 to 5 cm in diameter

PARADUODENAL HERNIA

- Herniating loops: Small-bowel (jejunum/ileum)
- Imaging findings
  - Cluster of dilated small-bowel loops
    - Right side abdomen more common
    - Small-bowel obstruction (100%)
  - Crowding or twisting of mesenteric vessels
  - Hepatic flexure displacement: Inferiorly/posteriorly
  - May be indistinguishable from paraduodenal hernia without surgical history

PATHOLOGY

General Features

- Etiology
  - Congenital or developmental mesenteric anomalies
  - Acquired: As a complication of surgery or trauma
  - Pathogenesis & mechanism
    - Congenital: Anomalies in mesenteric fixation of ascending or descending colon lead to abnormal openings through which hernia may occur
    - Acquired: Abnormal mesenteric defects are created
    - Abnormal mesenteric fixation may lead to abnormal mobility of small bowel & right colon, which facilitates herniation
  - Left paraduodenal hernia
    - Approximately 75% occur on left
    - Herniation via abnormal mesenteric fossa of Landzert (seen in 2% of autopsies)
    - Location: Lateral to ascending part of duodenum
    - Herniation of bowel loops into a pocket of distal transverse & descending mesocolon, posterior to superior mesenteric artery (SMA)
    - Herniating segment: Proximal loops of jejunum
  - Right paraduodenal hernia
    - Approximately 25% occur on right
    - Herniation via abnormal mesentericoparietal fossa of Waldeyer (seen in 1% of autopsies)
    - Location: Jejunal mesentery, immediately behind SMA & inferior to transverse part of duodenum
    - Herniation of bowel loops into a pocket of ascending mesocolon
    - Herniating segment: Proximal loops of jejunum
- Epidemiology
  - Autopsy incidence
    - Internal hernias: 0.2-0.9%
    - Paraduodenal: One half of internal hernias

Gross Pathologic & Surgical Features

- Dilated bowel loops herniating via a mesenteric defect

CLINICAL ISSUES

Presentation

- Most common signs/symptoms
  - Smaller ones: Easily reducible & clinically silent
  - Larger ones
    - Vague discomfort, abdominal distension
    - Periumbilical colicky pain, bowel obstruction
    - Palpable mass, localized tenderness
  - Small-bowel obstruction
    - Low grade, chronic & recurrent
    - May be high grade & acute

Demographics

- Age
  - Both children & adult age group
  - Typically present between 4th & 6th decade
- Gender: M:F = 3:1

Natural History & Prognosis

- Complications
  - Volvulus, ischemia, strangulation
  - Bowel gangrene, shock & death
- Prognosis
  - Early surgical correction: Good
  - Delayed surgical correction & complications: Poor

Treatment

- Laparotomy, incise enclosing mesentery
- Bowel decompression
- Avoid injury to superior & inferior mesenteric vessels
- Surgical correction of mesenteric defect

DIAGNOSTIC CHECKLIST

Consider

- Rule out other causes of dilated small-bowel loops

Image Interpretation Pearls

- Cluster of dilated small-bowel loops lateral to ascending or descending duodenum with crowded/twisted mesenteric vessels

SELECTED REFERENCES

1. Catalano OA et al: Internal hernia with volvulus and intussusception: case report. Abdom Imaging, 2004
2. Blachar A et al: Gastrointestinal complications of laparoscopic Roux-en-Y gastric bypass surgery: clinical and imaging findings. Radiology. 223(3): 625-32, 2002
3. Blachar A et al: Internal hernia: an increasingly common cause of small bowel obstruction. Semin Ultrasound CT MR. 23(2): 174-83, 2002
4. Blachar A et al: Radiologist performance in the diagnosis of internal hernia by using specific CT findings with emphasis on transmesenteric hernia. Radiology. 221(2): 422-8, 2001
5. Blachar A et al: Bowel obstruction following liver transplantation: clinical and ct findings in 48 cases with emphasis on internal hernia. Radiology. 218(2): 384-8, 2001
6. Blachar A et al: Internal hernia: clinical and imaging findings in 17 patients with emphasis on CT criteria. Radiology. 218(1): 68-74, 2001
7. Khan MA et al: Paraduodenal hernia. Am Surg. 64(12): 1218-22, 1998
8. Suchato C et al: CT findings in symptomatic left paraduodenal hernia. Abdom Imaging. 21(2): 148-9, 1996
9. Zarvan NP et al: Abdominal hernias: CT findings. AJR Am J Roentgenol. 164:1391-1395, 1995
10. Hamy A et al: Left-sided paraduodenal internal hernia containing sigmoid colon: diagnosis based on findings on barium examinations. AJR Am J Roentgenol. 162:1500-1501, 1994
11. Lee GH et al: CT imaging of abdominal hernias. AJR Am J Roentgenol. 161:1209-1213, 1993
12. Meyers MA: Paraduodenal hernias. Radiologic and arteriographic diagnosis. Radiology. 95: 29-37, 1970

IMAGE GALLERY

Typical

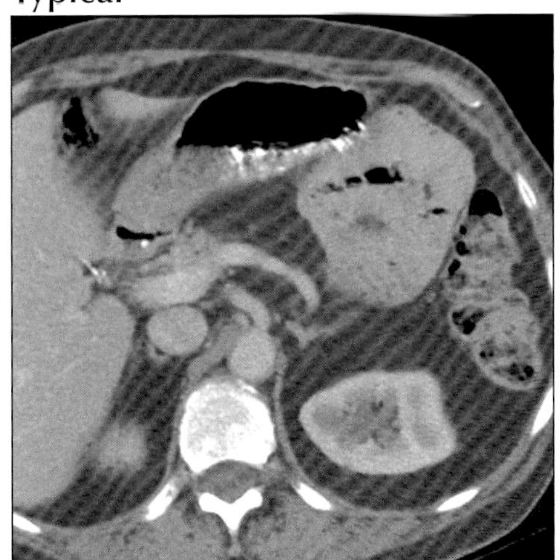

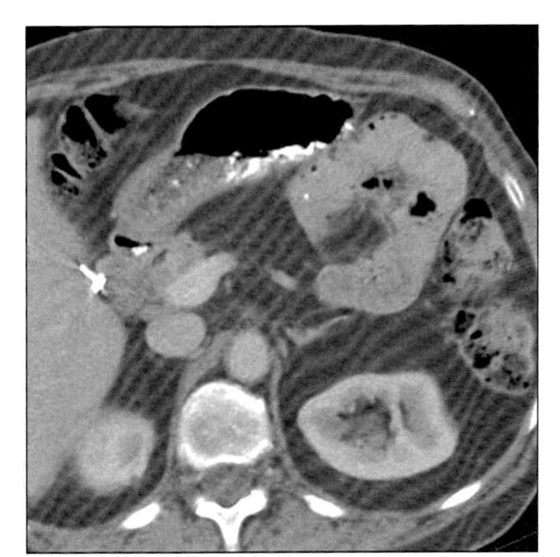

(Left) *Axial CECT shows left paraduodenal hernia with cluster of bowel between pancreas and stomach.* ***(Right)*** *Axial CECT shows left paraduodenal hernia with inward-directed mesenteric vessels.*

Typical

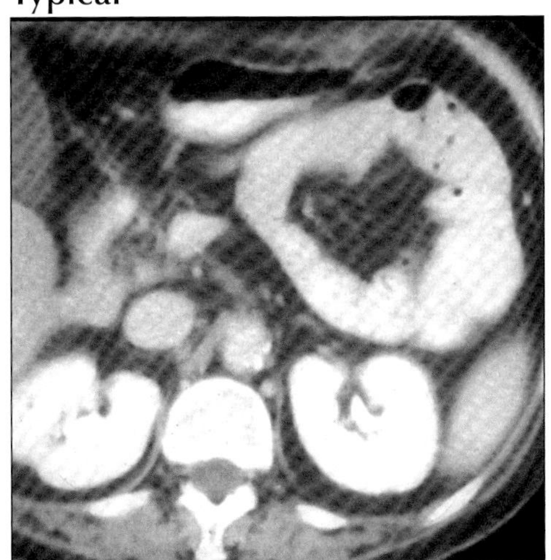

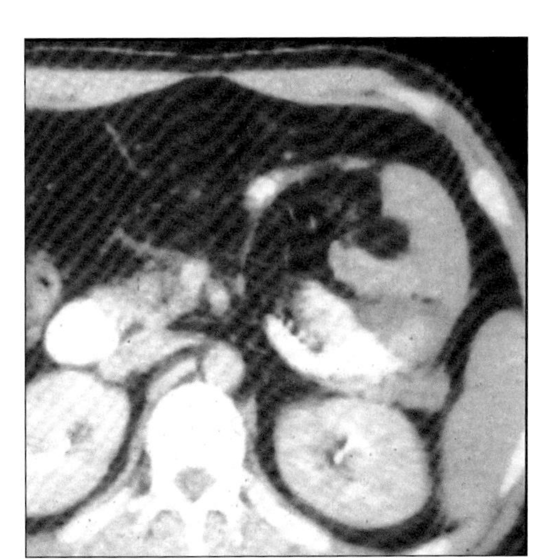

(Left) *Axial CECT shows left paraduodenal hernia with sac of bowel behind stomach and inward-directed, engorged mesenteric vessels.* ***(Right)*** *Axial CECT shows sac of dilated bowel with dilated, distorted mesenteric vessels.*

Typical

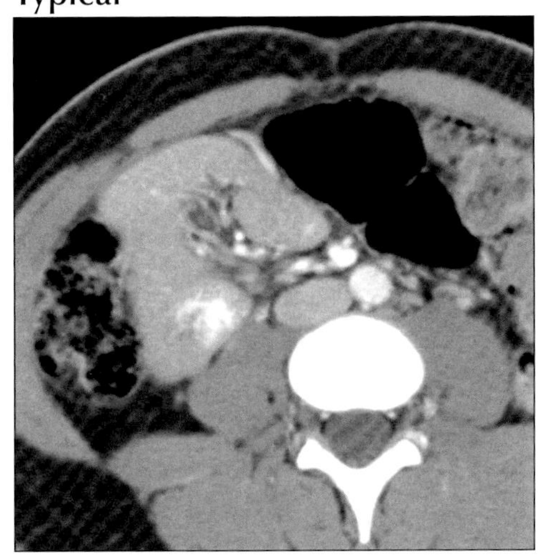

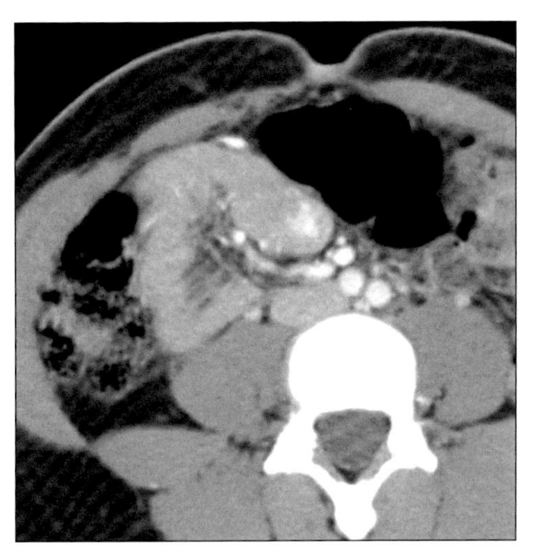

(Left) *Axial CECT shows cluster of dilated bowel media to ascending colon with distorted, engorged vessels (Courtesy D. Meyers, MD).* ***(Right)*** *Axial CECT shows right paraduodenal hernia with dilated jejunum and its mesenteric vessels twisted and displaced (Courtesy D. Meyers, MD).*

TRANSMESENTERIC POST-OPERATIVE HERNIA

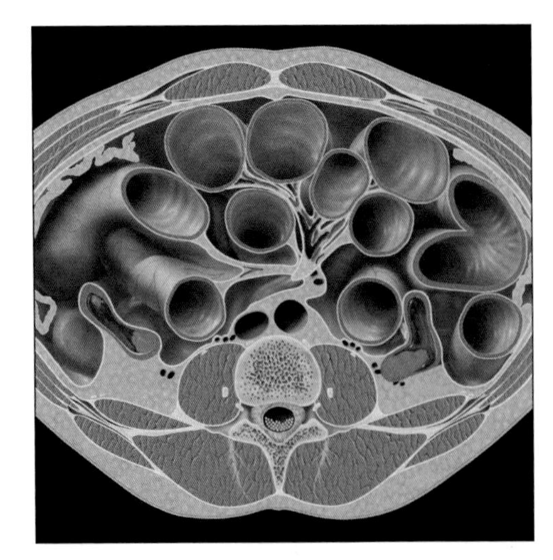

Graphic shows dilated small bowel (SB) that has herniated through a mesenteric defect. Note peripheral position of SB and medial displacement of colon, displaced mesenteric vessels.

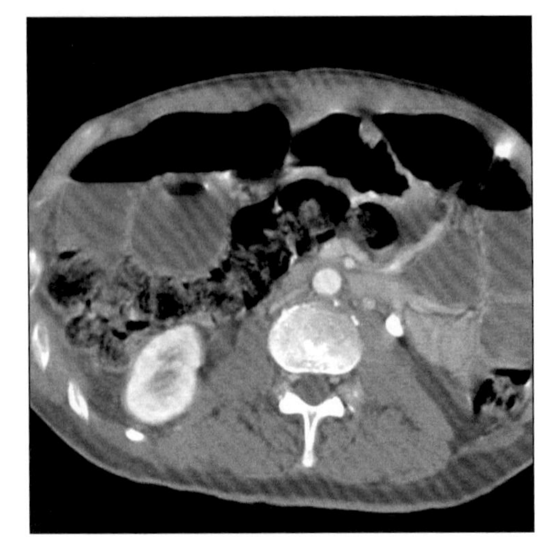

Axial CECT shows dilated SB with no overlying omental fat, displacing transverse colon medially and inferiorly.

TERMINOLOGY

Definitions

- Protrusion of bowel loops through an acquired defect of mesentery within abdominal cavity

IMAGING FINDINGS

General Features

- Best diagnostic clue: Cluster of dilated small bowel loops with distorted mesenteric vessels
- Location
  - Abnormal opening in mesentery of small bowel or colon
  - Hernia post Roux-en-Y gastric bypass surgery via
    - Transverse mesocolon (80%)
    - Small bowel mesentery (14%)
    - Behind Roux loop (6%): Peterson type hernia
  - Hernia post liver transplant via
    - Transverse mesocolon (more common)
    - Small bowel mesentery
- Size: Mesenteric defect varies from few mm to few cm
- Key concepts
  - Classification of hernias based on anatomic location
    - Internal or intra-abdominal: Herniation of bowel loops via defect within abdominal cavity
    - External: Prolapse of bowel loops via defect in wall of abdomen or pelvis
    - Diaphragmatic: Protrusion of bowel loops via hiatus, congenital or acquired defect
  - Subclassification of internal hernias
    - Transmesenteric hernia
    - Paraduodenal hernia
    - Foramen of Winslow, pericecal hernias
    - Intersigmoid & transomental hernias
  - Transmesenteric hernia two types based on etiology
    - Transmesenteric post-operative hernia (most common subtype of all internal hernias in adults)
    - Transmesenteric congenital hernia (most common subtype of internal hernias in pediatric age group)
  - 0.6-5.8% cases of small bowel obstruction (SBO) are due to internal hernias in selected populations
    - Transmesenteric post-operative hernia accounts more than half of these cases of SBO
    - Examples: Roux-en-Y gastric bypass surgery, liver transplantation, small & large bowel surgery

Radiographic Findings

- Radiography
  - Plain x-ray abdomen (supine)
    - "Closed loop": Markedly distended segment of small bowel (indicates SBO)
    - Crowded & dilated small bowel loops in an abnormal location
    - Multiple air fluid levels

DDx: Cluster of Dilated Bowel Loops

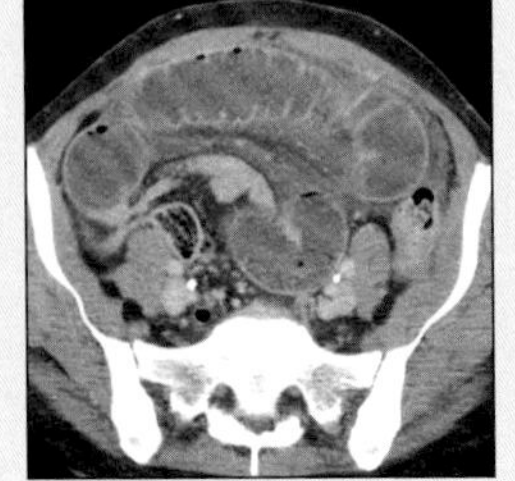

SBO Adhesions

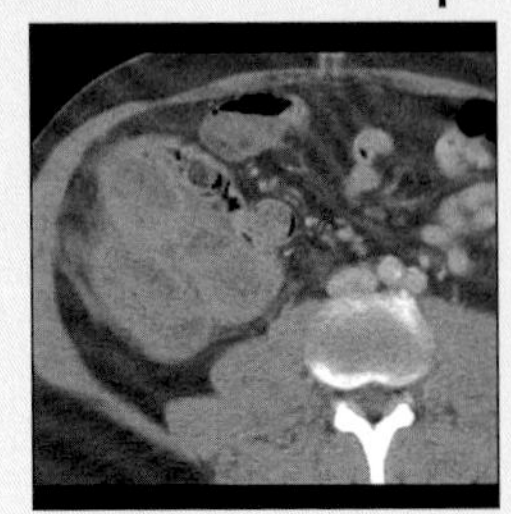

Closed Loop SBO

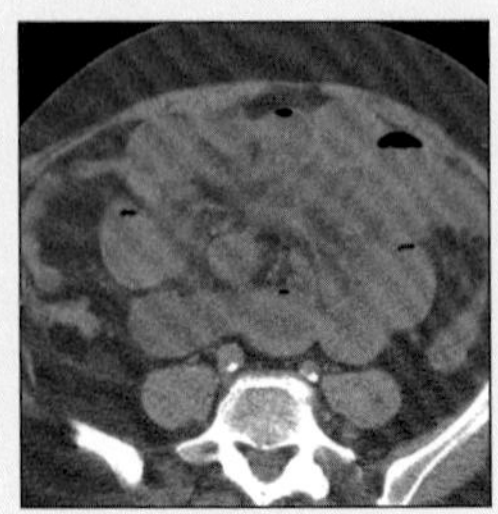

SB Volvulus

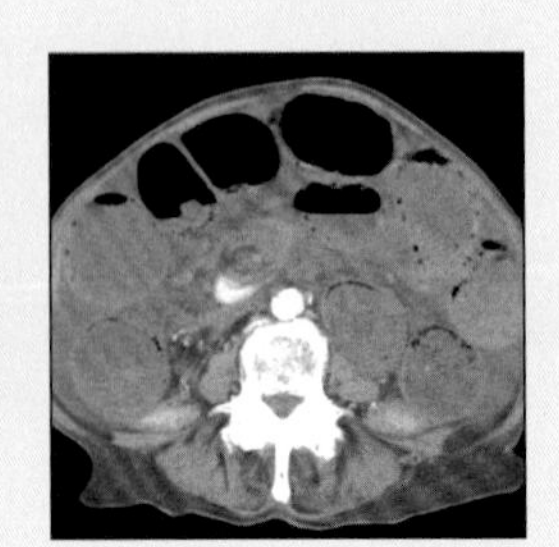

SB Volvulus

TRANSMESENTERIC POST-OPERATIVE HERNIA

Key Facts

Terminology

- Protrusion of bowel loops through an acquired defect of mesentery within abdominal cavity

Imaging Findings

- Best diagnostic clue: Cluster of dilated small bowel loops with distorted mesenteric vessels
- Evidence of small bowel obstruction (100%)
- Transition point dilated-nondilated (100%)
- Location: Usually adjacent to abdominal wall
- Hernia usually not encapsulated or enveloped in a sac
- Mesenteric vessels: Engorged, crowded or twisted
- "Whirl sign": Small bowel volvulus with twisted mesenteric vessels

Top Differential Diagnoses

- Closed loop obstruction (SBO)

Pathology

- Roux-en-Y gastric bypass
- Liver transplantation, small or large bowel surgery
- Post-operative: Due to prior abdominal surgery abnormal spaces or mesenteric defects are created

Clinical Issues

- Smaller: Easily reducible & clinically silent
- Periumbilical pain, symptoms of bowel obstruction
- Gender: Females more than males

Diagnostic Checklist

- Differentiate from other types of internal hernias
- Cluster of dilated small bowel loops which are not encapsulated, not enveloped in a sac located adjacent to abdominal wall usually on right side with crowded or twisted mesenteric vessels

- Fluoroscopic guided small bowel follow through
  - Crowding of bowel loops in an abnormal location
    - Location: Right side abdomen (more common)
  - Do not appear as they are contained in a sac or confining border
  - Varying degrees of small bowel obstruction (SBO)
  - Point of transition between dilated & nondilated bowel (common)
  - Some degree of fixation, stasis & delayed flow of contrast seen in herniated bowel
  - Inability to displace clustered loops by manual palpation or change in position
  - Lateral films: Useful in demonstrating displacement of herniated bowel loops

CT Findings

- Cluster of dilated small bowel loops
  - Evidence of small bowel obstruction (100%)
  - Transition point dilated-nondilated (100%)
  - Location: Usually adjacent to abdominal wall
- Hernia usually not encapsulated or enveloped in a sac
- Displacement of overlying omental fat of herniated bowel loop (74%)
- Mesenteric vessels: Engorged, crowded or twisted
- Main mesenteric trunk: Right or left displacement
- Colon displacement
  - Hepatic flexure displaced inferiorly & posteriorly (more common)
  - Ascending & descending colon displaced medially (less common)
- Thick bowel wall & ascites (more common in cases with bowel ischemia)
- "Whirl sign": Small bowel volvulus with twisted mesenteric vessels
- Transmesenteric smaller hernias after Roux-en-Y surgery (via transverse mesocolon)
  - Small retrogastric cluster of small bowel loops
  - Mass effect on posterior stomach wall
  - Redundant dilated Roux loop
  - No colon or fat displacement

Ultrasonographic Findings

- Real Time: Dilated small bowel loops in an abnormal location

Angiographic Findings

- Conventional
  - Superior mesenteric arteriogram
    - Abrupt angulation & displacement of visceral branches as they pass through mesenteric defect to supply herniated loops

Imaging Recommendations

- Helical CECT; small bowel follow through

DIFFERENTIAL DIAGNOSIS

Closed loop obstruction (SBO)

- Obstruction of small bowel at two points
- Tends to involve mesentery & prone to produce a volvulus
  - Represents most common cause of strangulation
- Etiology
  - Most often caused by an adhesive band
  - Occasionally by an internal or external hernia
- Imaging features
  - Markedly distended segment of fluid-filled small bowel
  - Volvulus
    - C or U-shaped or "coffee bean" configuration of bowel loop
  - Stretched mesenteric vessels converging toward site of torsion
  - Adjacent collapsed loops at site of obstruction
    - Round, oval or triangular in shape
  - "Beak sign"
    - Fusiform tapering at point of torsion or obstruction
  - "Whirl sign"
    - Due to tightly twisted mesentery with volvulus
  - May be indistinguishable from transmesenteric hernia especially if associated with volvulus

TRANSMESENTERIC POST-OPERATIVE HERNIA

PATHOLOGY

General Features

- Etiology
  - Transmesenteric post-operative type
    - Roux-en-Y gastric bypass
    - Liver transplantation, small or large bowel surgery
  - Transmesenteric congenital type
    - Mesenteric defect located close to ligament of Treitz or ileocecal valve
    - Size of mesenteric defect: Usually 2-5 cm in diameter
  - Pathogenesis & mechanism
    - Post-operative: Due to prior abdominal surgery abnormal spaces or mesenteric defects are created
    - Congenital: Developmental mesenteric anomalies
    - Abnormal mesenteric fixation or defects may lead to abnormal mobility of small bowel loops which facilitates herniation
    - Herniating segment: Small bowel loops (jejunum/ileum)
    - Herniation: Transient or intermittent
- Epidemiology
  - Autopsy incidence
    - Internal hernias: 0.2-0.9%
  - Transmesenteric post-operative hernia
    - Accounts more than 50% of internal hernias
- Associated abnormalities
  - Roux-en-Y gastric bypass loop
  - Liver transplant
  - Evidence of small or large bowel surgery

Gross Pathologic & Surgical Features

- Abnormal opening in mesentery of small bowel/colon
- Dilated small bowel loops herniating via a mesenteric defect
- Distorted or twisted mesenteric vessels

CLINICAL ISSUES

Presentation

- Most common signs/symptoms
  - Smaller: Easily reducible & clinically silent
  - Larger
    - Vague discomfort, abdominal distension
    - Periumbilical pain, symptoms of bowel obstruction
    - Palpable mass, localized tenderness
  - Small bowel obstruction
    - Low grade, chronic & recurrent
    - May be high grade & acute
  - Onset usually months after original surgery

Demographics

- Age
  - Transmesenteric post-operative
    - Usually obese adult age group who have undergone Roux-en-Y gastric bypass surgery
    - Typically between 4th & 6th decade
- Gender: Females more than males

Natural History & Prognosis

- Complications
  - Volvulus, ischemia, strangulation
  - Bowel gangrene, shock & death
- Prognosis
  - Early surgical correction: Good
  - Delayed surgical correction & complications: Poor

Treatment

- Laparotomy, bowel decompression
- Avoid injury to superior & inferior mesenteric vessels
- Surgical correction of mesenteric defect

DIAGNOSTIC CHECKLIST

Consider

- Differentiate from other types of internal hernias

Image Interpretation Pearls

- Cluster of dilated small bowel loops which are not encapsulated, not enveloped in a sac located adjacent to abdominal wall usually on right side with crowded or twisted mesenteric vessels

SELECTED REFERENCES

1. Catalano OA et al: Internal hernia with volvulus and intussusception: case report. Abdom Imaging, 2004
2. Filip JE et al: Internal hernia formation after laparoscopic Roux-en-Y gastric bypass for morbid obesity. Am Surg. 68(7): 640-3, 2002
3. Blachar A et al: Gastrointestinal complications of laparoscopic Roux-en-Y gastric bypass surgery: clinical and imaging findings. Radiology. 223(3): 625-32, 2002
4. Blachar A et al: Internal hernia: an increasingly common cause of small bowel obstruction. Semin Ultrasound CT MR. 23(2): 174-83, 2002
5. Blachar A et al: Radiologist performance in the diagnosis of internal hernia by using specific CT findings with emphasis on transmesenteric hernia. Radiology. 221(2): 422-8, 2001
6. Huang YC et al: Left paraduodenal hernia presenting as intestinal obstruction: report of one case. Acta Paediatr Taiwan. 42(3): 172-4, 2001
7. Blachar A et al: Bowel obstruction following liver transplantation: clinical and ct findings in 48 cases with emphasis on internal hernia. Radiology. 218(2): 384-8, 2001
8. Delabrousse E et al: Strangulated transomental hernia: CT findings. Abdom Imaging. 26(1): 86-8, 2001
9. Blachar A et al: Internal hernia: clinical and imaging findings in 17 patients with emphasis on CT criteria. Radiology. 218(1): 68-74, 2001
10. Rha SE et al: CT and MR imaging findings of bowel ischemia from various primary causes. Radiographics. 20(1): 29-42, 2000
11. Ha HK et al: Usefulness of CT in patients with intestinal obstruction who have undergone abdominal surgery for malignancy. AJR Am J Roentgenol. 171(6): 1587-93, 1998
12. Khanna A et al: Internal hernia and volvulus of the small bowel following liver transplantation. Transpl Int. 10(2): 133-6, 1997

TRANSMESENTERIC POST-OPERATIVE HERNIA

IMAGE GALLERY

Typical

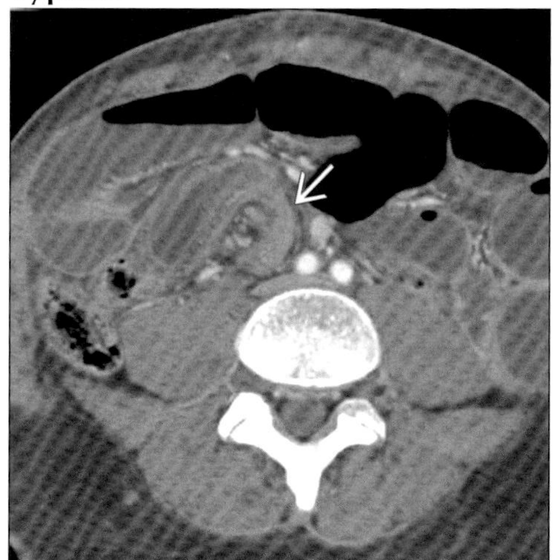

(Left) *Axial CECT shows herniation of bowel through mesenteric defect, causing twisting of mesenteric vessels (arrow) and SB obstruction.* ***(Right)*** *Surgical photograph shows shows mesenteric defect through which all the visualized bowel had herniated and obstructed.*

Typical

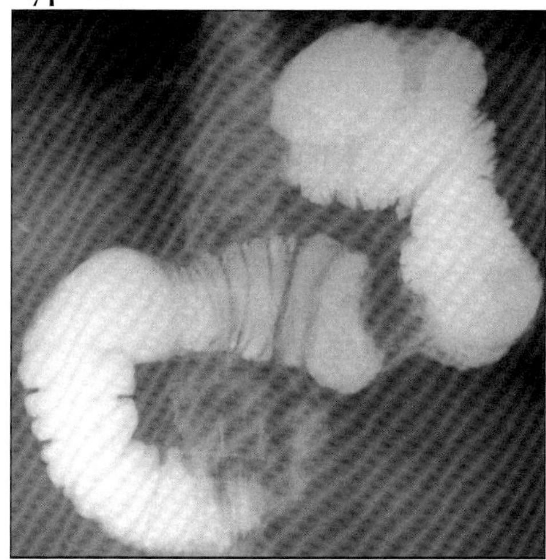

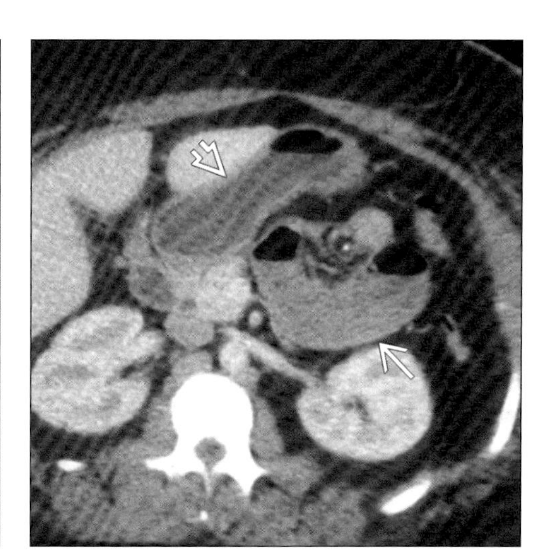

(Left) *SBFT following Roux-en-Y gastric bypass surgery (RYGB) shows herniation, dilation and twisting of Roux limb due to internal hernia.* ***(Right)*** *Axial CECT following RYGB shows dilated, twisted Roux limb (arrow) lying behind bypassed stomach (open arrow).*

Typical

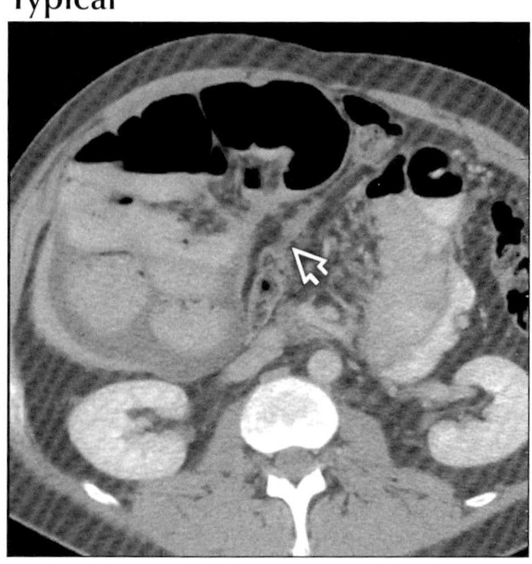

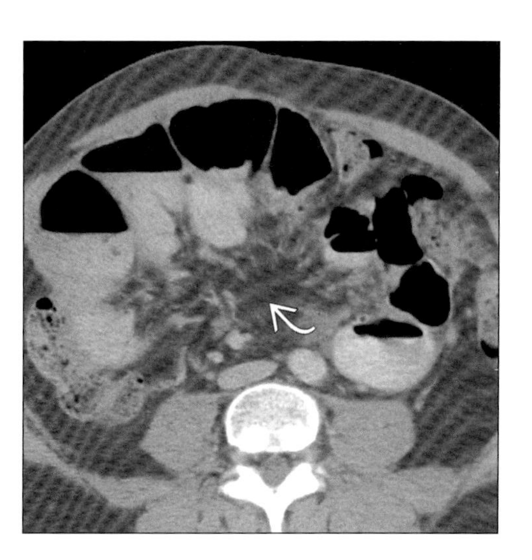

(Left) *Axial CECT shows cluster of dilated SB lateral to and displacing colon (open arrow). Focal ascites and engorged vessels.* ***(Right)*** *Axial CECT shows transmesenteric internal hernia with displaced, crowded mesenteric vessels converging at the mesenteric defect (curved arrow).*

TRAUMATIC ABDOMINAL WALL HERNIA

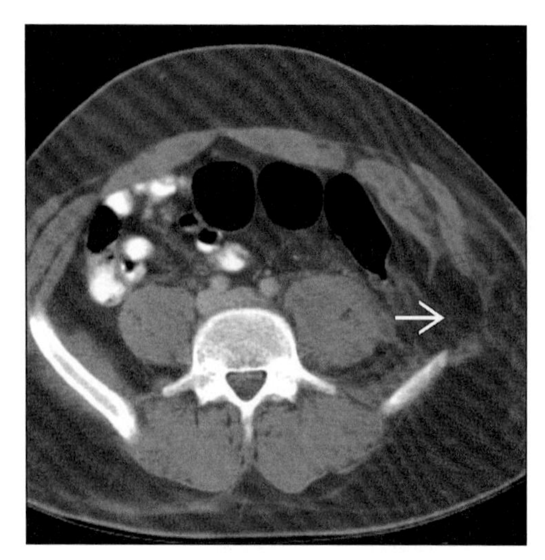

Axial CECT shows traumatic avulsion of abdominal wall muscles from pelvic insertion (arrow).

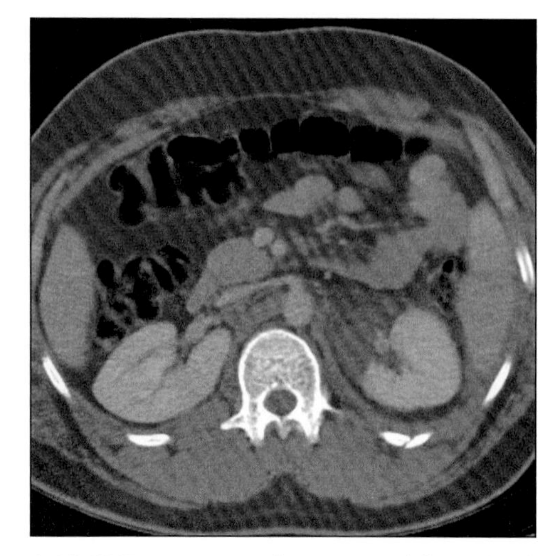

Axial CECT in patient with traumatic abdominal wall hernia shows renal laceration and fractured ribs.

TERMINOLOGY

Abbreviations and Synonyms

- Traumatic abdominal wall hernia (TAWH)
- Handlebar hernia

Definitions

- Traumatic disruption of musculature & fascia of anterior abdominal wall
  - With herniation of intestinal loops into subcutaneous space
- Handlebar hernia (HH) is a localized abdominal wall hernia caused by handlebar (or similar) injury

IMAGING FINDINGS

General Features

- Best diagnostic clue
  - History of recent trauma, without skin penetration
    - No evidence of previous hernia defect at site of injury
- Location
  - Focal; lower abdominal quadrant defect; lateral to rectus sheath; inguinal region; in blunt trauma
  - Larger, diffuse abdominal wall defects; sustained in motor vehicle accidents
  - Region of iliac crest; in seat belt injury (site of junction of lap & shoulder strap)
  - Through rent in retroperitoneum (rare)
- Size: Anatomical defects vary from small tears to large disruptions
- Morphology
  - Differing patterns of muscular & fascial disruption
    - Due to different types of forces involved & tensile properties of various areas in abdominal wall
    - Most elastic structure, skin, remains intact

CT Findings

- Size & site of defect
- Identify nature of herniated contents
- Rent in abdominal wall through which intestinal loops protrude into subcutaneous space
- Avulsion of abdominal wall muscles; all layers of abdominal wall may be disrupted
- Associated visceral injury

Imaging Recommendations

- Best imaging tool: CT

DIFFERENTIAL DIAGNOSIS

Spigelian hernia

- Through semilunar line at lateral border of rectus abdominis muscle

DDx: Anterior Abdominal Wall Defect or Mass

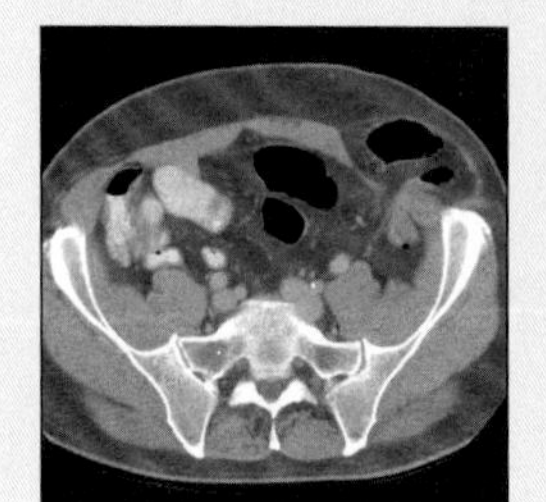

Spigelian Hernia

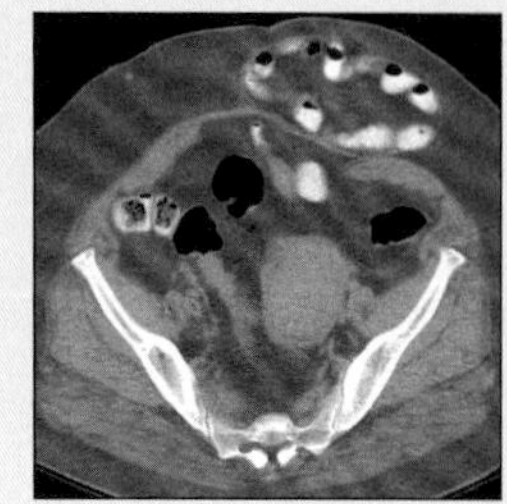

Ventral Hernia

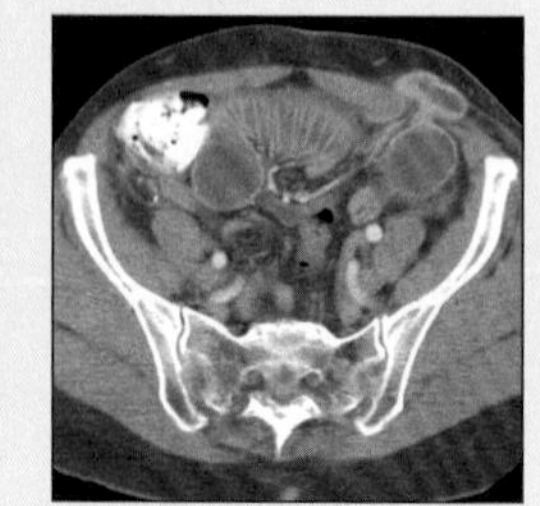

Lap. Port Hernia

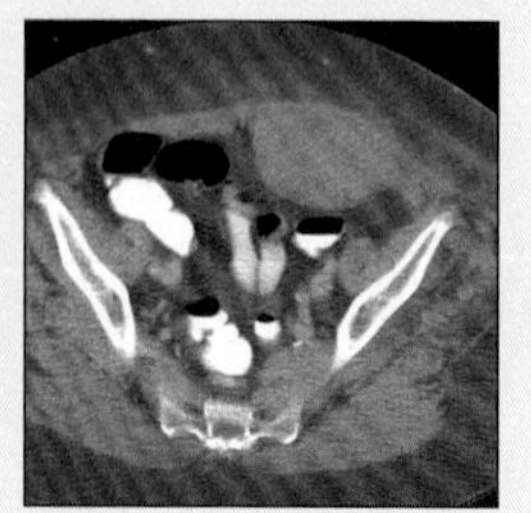

Rectus Hematoma

TRAUMATIC ABDOMINAL WALL HERNIA

Key Facts

Terminology
- Traumatic disruption of musculature & fascia of anterior abdominal wall
- With herniation of intestinal loops into subcutaneous space
- Handlebar hernia (HH) is a localized abdominal wall hernia caused by handlebar (or similar) injury

Imaging Findings
- Most elastic structure, skin, remains intact

Top Differential Diagnoses
- Spigelian hernia
- Ventral hernia
- Rectus hematoma

- Traumatic hernia may sometimes occur at typical Spigelian hernia site

Ventral hernia
- Site of previous surgery, laparoscopy (lap port), or stab wounds
- Midline; paramedian; small hernial aperture

Rectus hematoma

PATHOLOGY

General Features
- General path comments
  - Produced by sudden application of blunt force
    - Insufficient to penetrate skin but strong enough to disrupt muscle & fascia
  - Sudden ↑ in intra-abdominal pressure in combination with direct impact
  - Persistent & severe cough ("internal trauma")
    - Acting over a weak abdominal wall subjected to stress due to repetitive muscular contractions
  - Shearing force applied across bony prominences
- Etiology
  - Low energy injuries: Impact on small blunt object
    - HH: Produced by impaction of a bicycle handlebar on abdominal wall
  - High energy injuries: Motor vehicle accidents
  - Even after relatively minor trauma; in children
- Epidemiology
  - TAWH remains a rare clinical entity
  - Diffuse defects are 1.4 to 1.6 times more frequent than localized defects
- Associated abnormalities
  - Duodenal or mesenteric hematoma, pancreatic injuries with HH
  - Intra-abdominal injuries with high energy TAWH
    - Both bowel & solid viscera are at high risk

CLINICAL ISSUES

Presentation
- Varying degrees of abdominal tenderness
- Abdominal skin ecchymosis or abrasions ("seat belt ecchymosis")
- Reducible swelling or a cough impulse

Natural History & Prognosis
- May not be diagnosed at initial presentation; present as delayed consequence of trauma
- Complications: Bowel strangulation; ischemia

Treatment
- Low energy injuries: Local exploration; incision overlying defect; laparoscopic; mesh technique
- High energy injuries: Immediate exploratory laparotomy; midline incision; primary repair

DIAGNOSTIC CHECKLIST

Consider
- Poor correlation between site of impact & abdominal wall defect
- Can be easily missed in presence of major pelvic & abdominal lesions
- Careful review of associated injuries to bowel & mesentery should be undertaken
  - CT findings may correlate poorly with severity of injury in these areas

SELECTED REFERENCES

1. Losanoff JE et al: Handlebar hernia: ultrasonography-aided diagnosis. Hernia. 6(1):36-8, 2002
2. Vasquez JC et al: Traumatic abdominal wall hernia caused by persistent cough. South Med J. 92(9): 907-8, 1999
3. Damschen DD et al: Acute traumatic abdominal hernia: case reports. J Trauma. 36(2): 273-6, 1994

IMAGE GALLERY

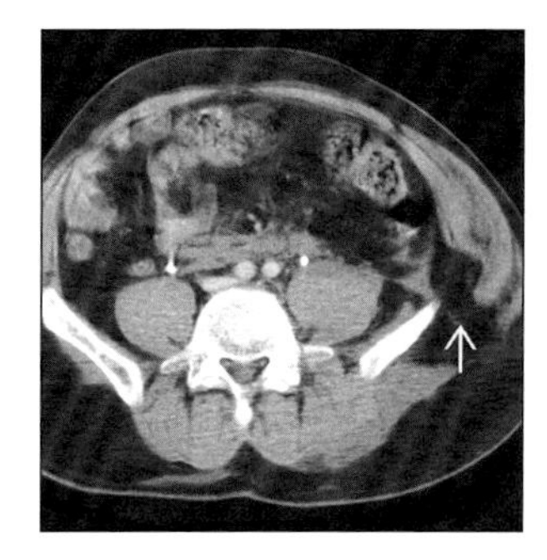

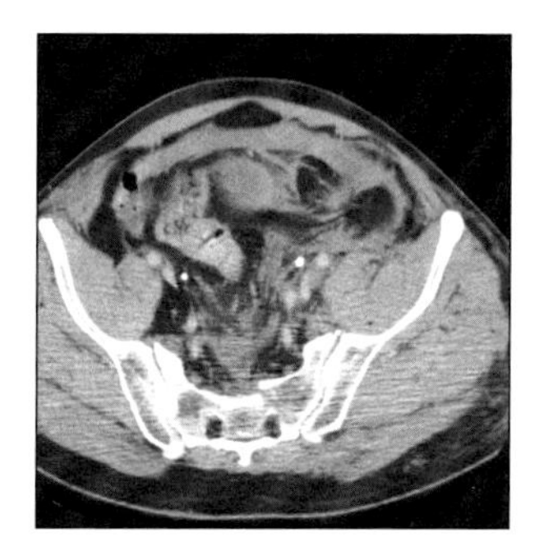

*(**Left**) Axial CECT shows traumatic avulsion of muscles from pelvis (arrow). (**Right**) Axial CECT shows pelvic fractures and mesenteric bleeding in patient with traumatic abdominal wall hernia.*

MESENTERIC TRAUMA

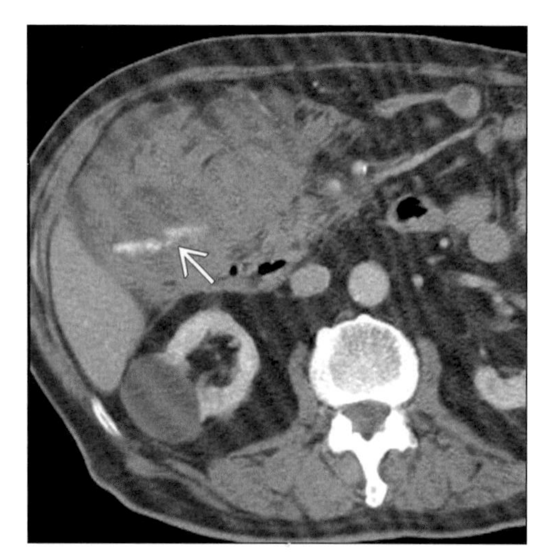

Axial CECT shows large omental/mesenteric hematoma with active bleeding (arrow).

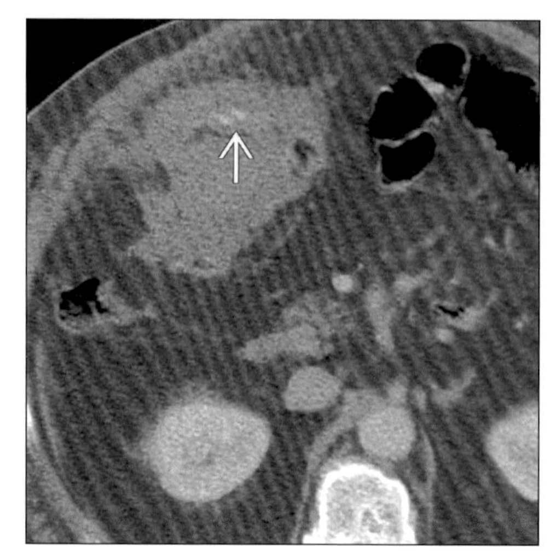

Axial CECT shows large omental/mesenteric hematoma and active bleeding (arrow).

TERMINOLOGY

Definitions

- Injury to mesentery; often with bowel injury

IMAGING FINDINGS

General Features

- Best diagnostic clue: CT; mesenteric hemorrhage, focal bowel wall thickening, free peritoneal fluid
- Location: Mesentery of small bowel injured five times more frequently than colonic mesentery

CT Findings

- Free intraperitoneal fluid: Hemoperitoneum in absence of detectable solid visceral injury
  - Fluid that is less dense than blood may represent bowel contents; ominous finding
  - Interloop fluid or trapped between mesenteric leaves
- Mesenteric stranding: heterogeneous attenuation
  - Hemorrhage or edema; diffuse or focal
- Active hemorrhage: Mesenteric fluid collection that is isodense to contrast-enhanced blood vessels
- "Sentinel clot sign": Focal heterogeneous high density (> 60 HU) collection in mesentery
  - Indicates mesenteric source of bleed
- Mesenteric pseudoaneurysm: Rare; indicates substantial mesenteric + vascular injury
  - High likelihood of continued, renewed hemorrhage
- CT signs of coexisting bowel injury
  - Bowel wall thickening: Focal; > 3 mm thick or disproportionate to normal bowel wall segments
    - Perforation or bowel ischemia or infarction
    - Enhance more than normal bowel wall segment
    - Early after injury, focal + limited to area of injury; CT finding can guide surgeon's exploration
    - Delayed detection → inflammation generalized, bowel wall thickening may become diffuse
  - Free intraperitoneal air: Tiny bubbles of extraluminal gas
    - Amount of air may be substantial or minimal
    - Often, only minimal air seen; anteriorly near liver
    - Trapped within or between folds of mesentery
  - Free retroperitoneal air: Traumatic disruption of duodenum or colon
  - Extraluminal enteric contrast: Indicative of bowel perforation; most dense near site of perforation
- Associated injuries to other organs; spleen, liver etc.
- CT helps distinguish operable from nonoperable cases
  - Incidence of specific findings
    - Free peritoneal fluid (96%) + mesenteric infiltration (86%) + focal bowel-wall thickening (61%)

DDx: Mesenteric Edema/Blood after Trauma

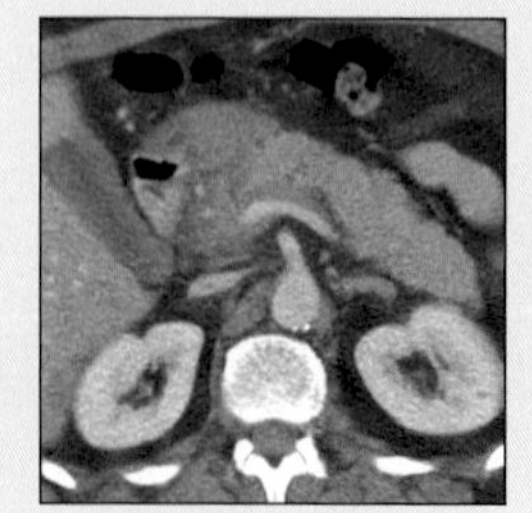

Pancreatic Trauma

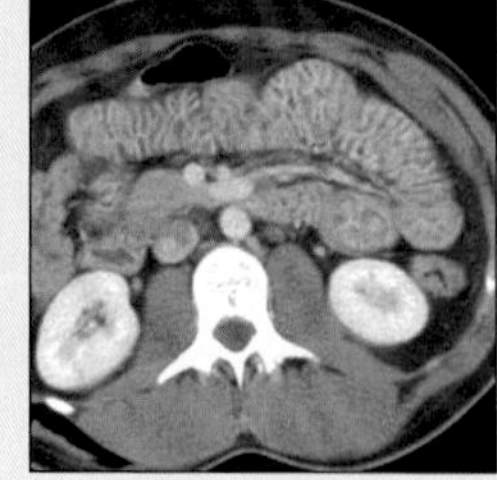

Shock Bowel

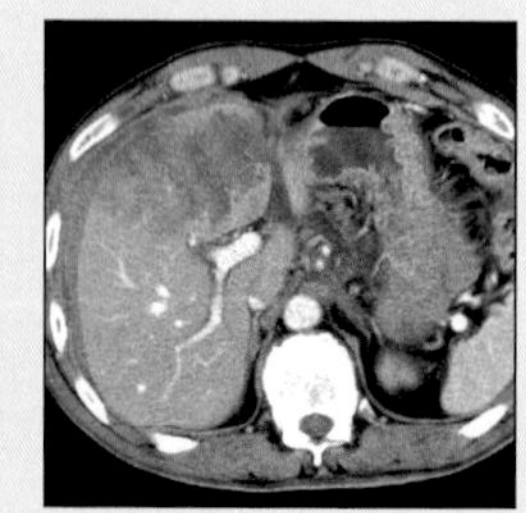

Liver Laceration

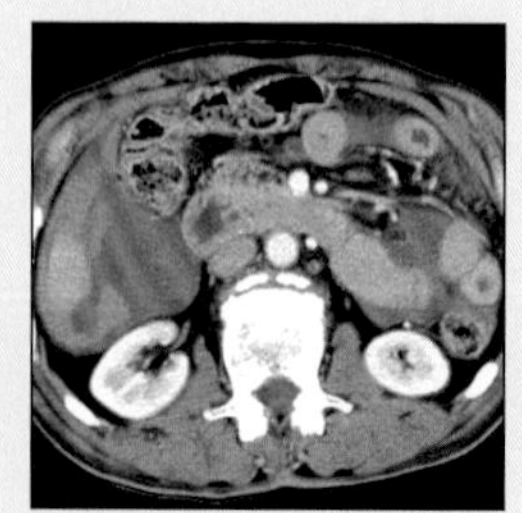

Liver Laceration

MESENTERIC TRAUMA

Key Facts

Imaging Findings
- Best diagnostic clue: CT; mesenteric hemorrhage, focal bowel wall thickening, free peritoneal fluid

Top Differential Diagnoses
- Traumatic pancreatitis
- Shock bowel
- Hemoperitoneum

Pathology
- Blunt abdominal trauma: Motor vehicle accidents
- Assault; fall from height; lap belt; child abuse
- Isolated injuries of mesentery are rare

Clinical Issues
- Delay in diagnosis: ↑ Morbidity & mortality (5-65%)

    - + Associated abdominal injuries (43%) + free air (32%) in surgical candidates
  - In nonoperable cases: Bowel thickening (84%)
    - Less frequently peritoneal fluid (21%), mesenteric infiltration (26%) or associated injuries (5%)

Imaging Recommendations
- Best imaging tool: CT; high negative predictive value

DIFFERENTIAL DIAGNOSIS

Traumatic pancreatitis
- Peripancreatic infiltration; edema, fluid collections, peripancreatic hematoma, loss of fat planes
- Thickening of anterior perirenal fascia

Shock bowel
- Extensive infiltration of mesenteric fat planes
- Diffuse bowel wall thickening + abnormally intense contrast enhancement of bowel mucosa
- Resolves soon after adequate fluid resuscitation

Hemoperitoneum
- Hyperattenuating fluid in peritoneum; typically not as hyperattenuating as extravasated contrast material
- 30-50 HU; less dense if ascites or bowel contents
- Source of bleed: Visceral laceration/"sentinel clot" sign

PATHOLOGY

General Features
- Etiology
  - Blunt abdominal trauma: Motor vehicle accidents
  - Assault; fall from height; lap belt; child abuse
- Epidemiology
  - Isolated injuries of mesentery are rare
  - Incidence: 5% following blunt abdominal trauma

CLINICAL ISSUES

Presentation
- Triad of abdominal tenderness, rigidity, absent bowel sounds (present in 1/3 of patients)
- Serial abdominal examinations; may be unreliable
- Diagnostic peritoneal lavage: Insensitive to retroperitoneal injuries; ↑ nontherapeutic laparotomies
- Fluid or gas from lavage makes CT diagnosis difficult

Natural History & Prognosis
- Delay in diagnosis: ↑ Morbidity & mortality (5-65%)

Treatment
- Surgery: Active bleeding; most bowel injuries
- Non-operative: Isolated mesenteric or bowel wall hematoma

DIAGNOSTIC CHECKLIST

Consider
- High index of suspicion required
  - Findings can be subtle
- Active mesenteric bleeding or combination of mesenteric hematoma + bowel wall thickening
  - Usually require surgery

SELECTED REFERENCES

1. Hanks PW et al: Blunt injury to mesentery and small bowel: CT evaluation. Radiol Clin North Am. 41(6):1171-82, 2003
2. Strouse PJ et al: CT of bowel and mesenteric trauma in children. Radiographics. 19(5):1237-50, 1999
3. Rizzo MJ et al: Bowel and mesenteric injury following blunt abdominal trauma: evaluation with CT. Radiology. 173(1): 143-8, 1989

IMAGE GALLERY

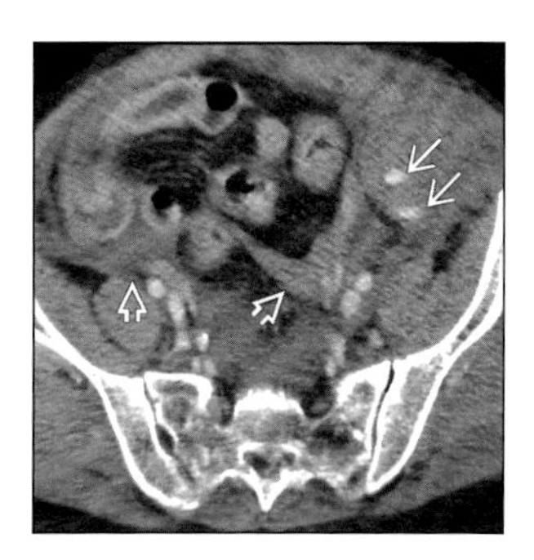
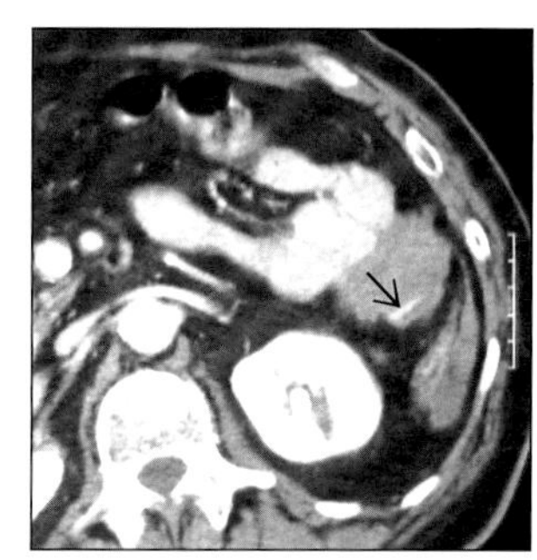

(Left) *Axial CECT shows mesenteric hematoma + active bleeding (arrows). Also note "triangular" and interloop collections of blood (open arrows).* ***(Right)*** *Axial CECT shows sentinel clot and active bleeding (arrow) in mesentery and blood in left paracolic gutter.*

TRAUMATIC DIAPHRAGMATIC RUPTURE

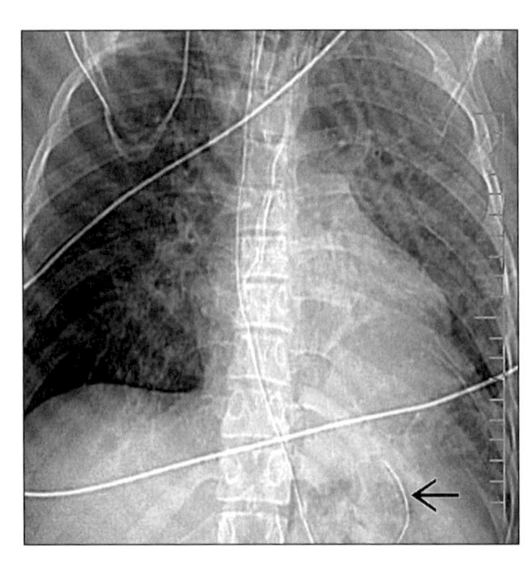

Chest radiograph shows "elevated" and indistinct left hemidiaphragm, tip of NG tube pointed up (arrow).

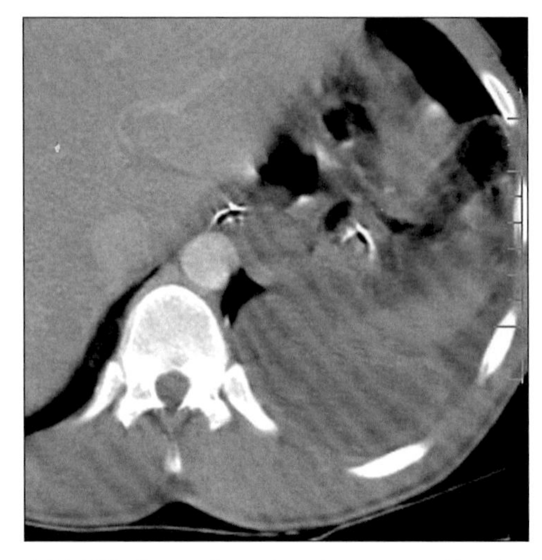

Axial CECT shows "dependent viscera" sign with spleen and bowel abutting posterior ribs.

TERMINOLOGY

Definitions

- Diaphragmatic rupture with or without herniation of abdominal contents into thorax

IMAGING FINDINGS

General Features

- Best diagnostic clue: Discontinuity of hemidiaphragm with air-filled bowel loops in thorax
- Location
  - 90-98% on left side in location
    - Posterolateral part of diaphragm medial to spleen
  - 2-4% on right side
- Size
  - Blunt trauma: Most tears > 10 cm in length
  - Penetrating trauma
    - Gun shot wounds (large blast injuries)
    - Stab wounds (smaller injuries)
- Key concepts
  - Due to blunt & penetrating trauma
  - Occur in 0.8-5.0% of all trauma patients
  - Accounts for 5% of all diaphragmatic hernias
  - 90% of all strangulated diaphragmatic hernias
  - Herniated organs: Stomach, colon, small bowel, omentum, spleen, liver
  - Associated injuries (52-100%): Rib fractures; bowel, splenic, liver & renal injuries
  - Posterolateral defects diagnosed at CT in 6% of nontraumatic, asymptomatic adults
    - May mimic diaphragmatic tears
    - Seen more commonly on left side
    - Represent congenital asymptomatic Bochdalek hernias

Radiographic Findings

- Radiography
  - Nonvisualization of diaphragmatic contour
  - Abnormally elevated hemidiaphragm contour
  - Contralateral shift of mediastinum, rib fractures
  - Lower lobe soft tissue density mass: Herniated solid organ, omentum or airless bowel loop
  - Intrathoracic herniation of a hollow viscus (stomach, colon, small bowel with air-fluid levels)
  - Site of diaphragmatic tear: Focal constricted gas filled bowel loop (collar sign)
  - Hydropneumothorax: Strangulation & lung injury
  - Visualization of nasogastric (NG) tube above left hemidiaphragm
  - Lung effusion, contusion, atelectasis & phrenic nerve palsy can mask diaphragmatic injury
- Fluoroscopic guided contrast studies
  - "Collar sign": Focal constricted (site of tear) contrast filled bowel loop partly in thorax & abdomen

DDx: Elevate or Deformed Diaphragm

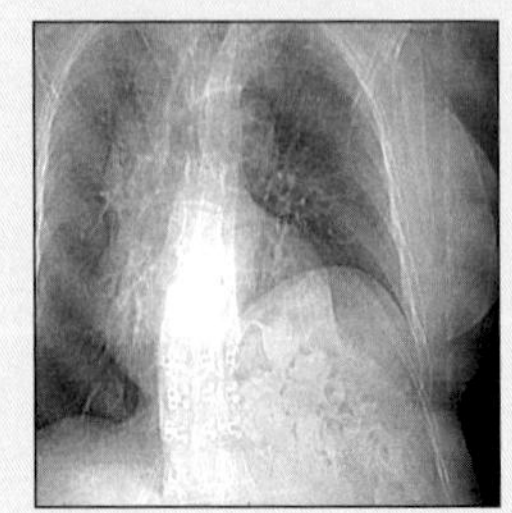

Paralyzed Diaphragm

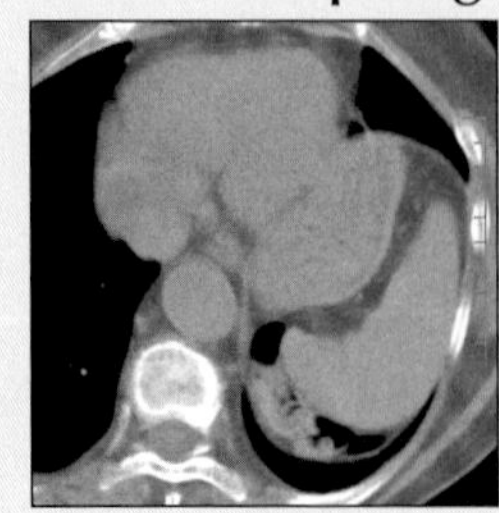

Paralyzed Diaphragm

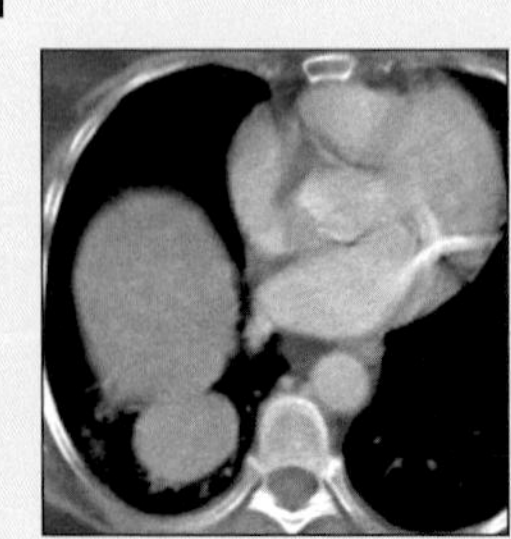

Eventration

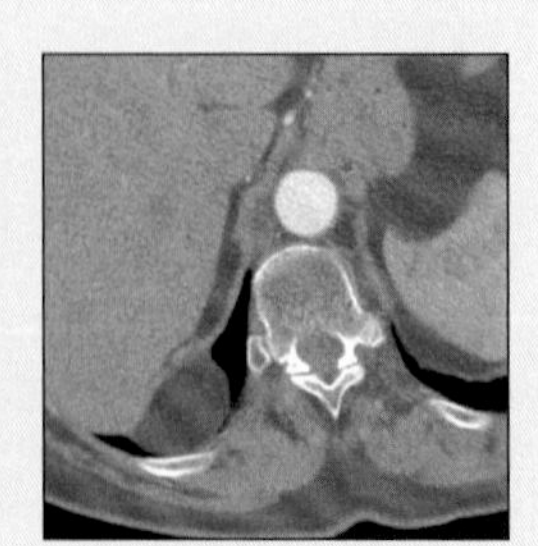

Bochdalek Hernia

TRAUMATIC DIAPHRAGMATIC RUPTURE

Key Facts

Imaging Findings

- Best diagnostic clue: Discontinuity of hemidiaphragm with air-filled bowel loops in thorax
- 90-98% on left side in location
- Discontinuity of hemidiaphragm
- Intrathoracic herniation of abdominal contents
- Left side (collar sign): Waist-like constriction of hollow viscus at site of diaphragmatic tear
- Right side (collar sign): Can appear as a focal indentation of liver (usually in posterolateral aspect)
- "Dependent viscera" sign: Seen in acute rupture or large tear of diaphragm due to blunt trauma
- Active extravasation of contrast near diaphragm
- "Thick crus" sign: Thickened crus of hemidiaphragm
- Penetrating diaphragmatic injury: Depiction of trajectory of a missile or puncturing instrument

Top Differential Diagnoses

- Eventration
- Diaphragmatic paralysis
- Pleural effusion & pulmonary or extrapleural mass
- Congenital foramen (hernia) of Bochdalek
- Congenital foramen (hernia) of Morgagni

Diagnostic Checklist

- Consider helical NE + CECT with multiplanar reformation images
- Intrathoracic herniation of abdominal contents
- "Dependent viscera sign": Herniated viscera are no longer supported posteriorly by injured diaphragm & fall to a dependent position against posterior ribs
- Lung contusion, effusion, atelectasis & phrenic nerve palsy can mask underlying diaphragmatic injury

CT Findings

- Discontinuity of hemidiaphragm
- Intrathoracic herniation of abdominal contents
  - Left side: Stomach & colon (most common viscera)
  - Right side: Liver (most common viscus)
- "Collar sign"
  - Left side (collar sign): Waist-like constriction of hollow viscus at site of diaphragmatic tear
  - Right side (collar sign): Can appear as a focal indentation of liver (usually in posterolateral aspect)
  - Abdominal contents lateral to diaphragm
- "Dependent viscera" sign: Seen in acute rupture or large tear of diaphragm due to blunt trauma
  - Right: Upper one-third of liver abuts posterior ribs
  - Left: Stomach, spleen or bowel abuts posterior ribs
  - Large pleural effusion may mask this CT sign
- Active extravasation of contrast near diaphragm
- "Thick crus" sign: Thickened crus of hemidiaphragm
  - Low sensitivity & specificity
- Penetrating diaphragmatic injury: Depiction of trajectory of a missile or puncturing instrument
  - Localized soft tissue swelling
  - Subcutaneous emphysema
  - Focal extravasation of I.V. contrast material
  - Injury to solid or hollow organ

MR Findings

- Normal diaphragm on T1 & T2WI
  - Continuous hypointense band
- Traumatic diaphragmatic hernia
  - Abrupt disruption of diaphragmatic contour
  - Intrathoracic herniation of abdominal fat or viscera

Imaging Recommendations

- Best imaging tool
  - Helical NE + CECT (multiplanar reformation)
  - MR most accurate, but not practical for unstable patient

DIFFERENTIAL DIAGNOSIS

Eventration

- Eventration (thin hypoplastic diaphragm)
  - Diaphragmatic muscular developmental defect
  - Persistence of a thin aponeurotic sheet of tissue (hypoplastic diaphragm)
  - More common on right side & affects anterior part of hemidiaphragm
  - Chest radiography
    - Elevation of anteromedial aspect of diaphragm
    - Hump on hemidiaphragm (typical of eventration)
  - CT: Abdominal contents bulge into thoracic cavity
    - Mimic & difficult to differentiate from true hernia
  - Unilateral: May be associated with Beckwith-Wiedemann syndrome, trisomy 13, 15, 18
  - Bilateral: May be associated with toxoplasmosis & cytomegalovirus (CMV)

Diaphragmatic paralysis

- Bilateral or unilateral
- Etiology: Unilateral diaphragmatic paralysis
  - Phrenic nerve compression by a tumor (> common)
  - Trauma: Injury to phrenic nerve or brachial plexus
  - Infection, diabetes mellitus, idiopathic
- Imaging (unilateral diaphragmatic paralysis)
  - Chest x-ray
    - Elevated hemidiaphragm
    - Small lung volumes & atelectasis
  - Fluoroscopy: Paradoxical movement may be seen
    - Upward movement of hemidiaphragm on inspiration or sniffing
  - CT: Mediastinal pathology (tumor) may be detected
  - MRI of cervical spine: Tumor or infective pathology

Pleural effusion & pulmonary or extrapleural mass

- Can mask & mimic traumatic diaphragmatic hernia
- Clinical history & imaging helps in differentiation

Congenital foramen (hernia) of Bochdalek

- Failure of closure of pleuroperitoneal membrane by 9th week of gestational age

TRAUMATIC DIAPHRAGMATIC RUPTURE

- Accounts for 85-90% of all non-hiatal hernias
- Epidemiology: Approximately 1 per 3000 live births
- Location: Left posterolateral (80%), right (15%), bilateral (5%)
- Size: Usually large compared to Morgagni hernia
- Left: Herniated organs (omental fat, bowel, spleen, left lobe of liver, stomach, kidney, pancreas)
- Right: Herniated organs (part of liver, gallbladder, small bowel, kidney)
- Associated with other congenital anomalies, IUGR
- Chest x-ray
  - Thorax: Air-filled bowel loops or soft tissue density
  - Mediastinal shift to right ± pulmonary hypoplasia
- Fluoroscopic guided contrast studies
  - Opacified stomach or bowel loops in left thorax

Congenital foramen (hernia) of Morgagni

- Anteromedial parasternal defect of diaphragm due to maldevelopment of central tendon
- Accounts for 2-4% of all non-hiatal hernias
- Epidemiology: 1 in 100, 000 live births
- Location: Usually right side, have a covering sac
- Size: Usually small compared to Bochdalek hernia
- Herniated organs: Fat, transverse colon, liver
- Associated with other congenital anomalies
- Chest X-ray
  - AP view: Soft tissue or air density along right cardiophrenic angle
  - Lateral view: Anterior location of herniated bowel
- Fluoroscopic guided contrast studies
  - Opacified bowel loops in right anterior thorax

PATHOLOGY

General Features

- General path comments
  - Anatomy of diaphragm
    - Dome-shaped, musculotendinous structure
    - Location: Bottom of pleural cavity & top of abdominal cavity
    - Consists of a central tendon, with right & left leaflets composed of striated muscles
    - Three large openings: Aortic, esophageal & inferior vena caval apertures
    - Covered by parietal pleura & peritoneum except for bare area of liver
    - Composed of two parts: Lumbar or posterior, costal or anterior diaphragm
- Etiology
  - Blunt trauma
    - Motor vehicle accidents (MVA)
    - Fall from a height
  - Penetrating: Gun shot & stab wounds
  - Mechanism of trauma
    - Blunt trauma: Increase intraabdominal pressure
    - Penetrating: Related to missile or stabbing weapon
- Epidemiology: 0.8-5.0% of all trauma patients
- Associated abnormalities: Solid & hollow visceral injuries, pelvic & rib fractures

CLINICAL ISSUES

Presentation

- Most common signs/symptoms
  - Asymmetry of chest Wall, respiratory distress
  - Pain, tightness, diaphoresis & discomfort
  - Hypotension, tachycardia
  - Decreased breath sounds on affected side
  - Bowel sounds heard in the chest

Demographics

- Age: More common in young age group
- Gender: Males more than females

Natural History & Prognosis

- Complications
  - Malrotation or malfixation
  - Bowel obstruction & strangulation
  - Torsion & devascularization
  - Lung: Hemopneumothorax
- Prognosis
  - Early diagnosis & repair: Good prognosis
  - Delayed diagnosis & repair: Poor prognosis
  - Morbidity & mortality rate: Up to 50% in visceral herniation & strangulation

Treatment

- Surgical correction & repair

DIAGNOSTIC CHECKLIST

Image Interpretation Pearls

- Consider helical NE + CECT with multiplanar reformation images
- Intrathoracic herniation of abdominal contents
- "Dependent viscera sign": Herniated viscera are no longer supported posteriorly by injured diaphragm & fall to a dependent position against posterior ribs
- Lung contusion, effusion, atelectasis & phrenic nerve palsy can mask underlying diaphragmatic injury

SELECTED REFERENCES

1. Yao DC et al: Using contrast-enhanced helical CT to visualize arterial extravasation after blunt abdominal trauma: incidence and organ distribution. AJR Am J Roentgenol. 178(1):17-20, 2002
2. Larici AR et al: Helical CT with sagittal and coronal reconstructions: accuracy for detection of diaphragmatic injury. AJR Am J Roentgenol. 179(2):451-7, 2002
3. Iochum S et al: Imaging of diaphragmatic injury: a diagnostic challenge? Radiographics. 22 Spec No:S103-16; discussion S116-8, 2002
4. Bergin D et al: The "dependent viscera" sign in CT diagnosis of blunt traumatic diaphragmatic rupture. AJR Am J Roentgenol. 177(5):1137-40, 2001
5. Killeen KL et al: Helical CT of diaphragmatic rupture caused by blunt trauma. AJR Am J Roentgenol. 173(6):1611-6, 1999
6. Shackleton KL et al: Traumatic diaphragmatic injuries: spectrum of radiographic findings. Radiographics. 18(1):49-59, 1998

TRAUMATIC DIAPHRAGMATIC RUPTURE

IMAGE GALLERY

Typical

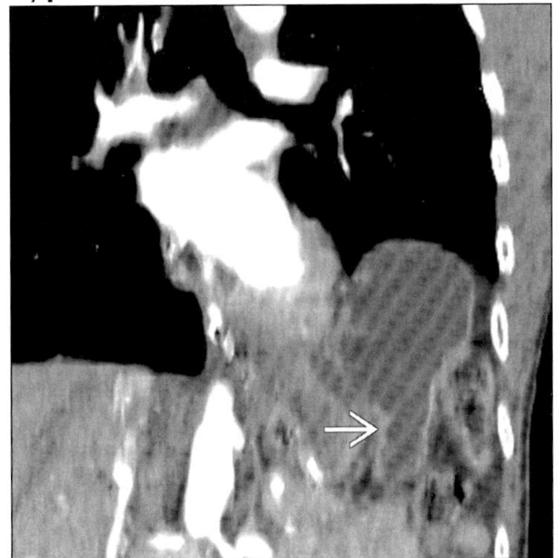

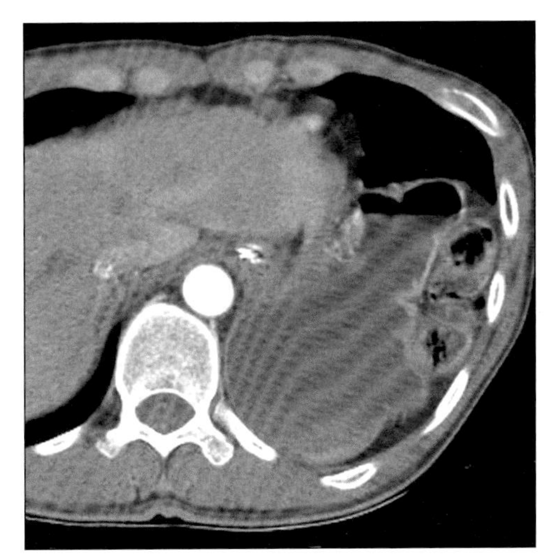

(Left) *Coronal reformation from axial CECT shows herniation of abdominal contents into left thorax. Note constriction of stomach (arrow) at site of tear.* ***(Right)*** *Axial CECT shows dependent viscus sign as stomach abuts posterior ribs.*

Typical

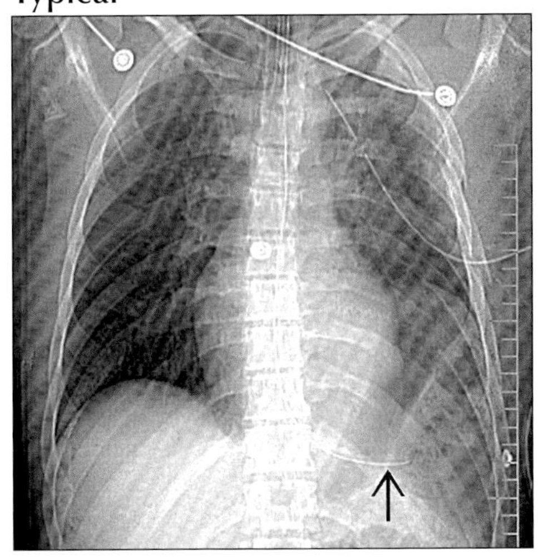

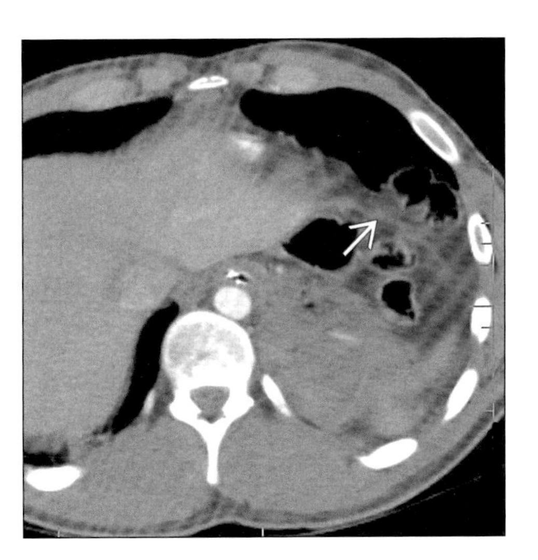

(Left) *Chest radiograph shows "elevated", distorted diaphragm and high position of NG tube (arrow).* ***(Right)*** *Axial CECT shows dependent viscera, plus colon and abdominal fat lying lateral to diaphragm (arrow), indicating thoracic position.*

Typical

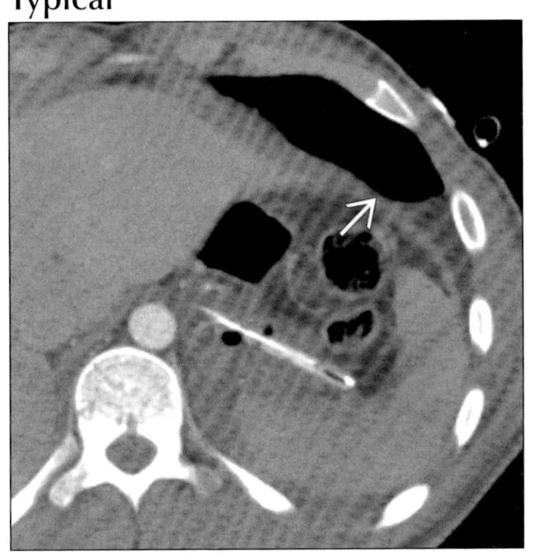

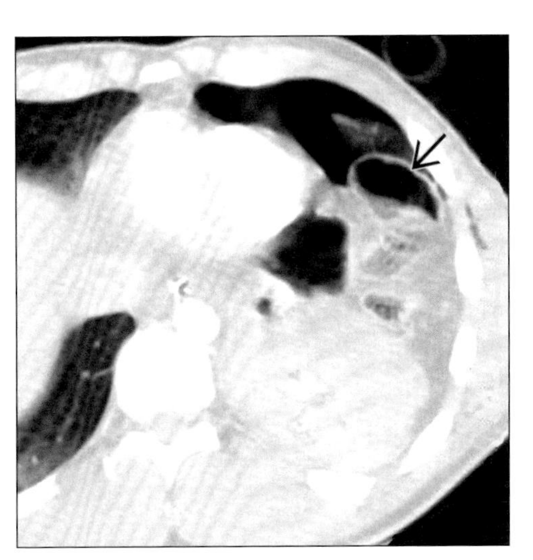

(Left) *Axial CECT shows spleen in dependent position and abdominal fat lateral to diaphragm (arrow).* ***(Right)*** *Axial CECT at "lung windows" shows splenic flexure of colon (arrow) abutting lung and pneumothorax.*

MESENTERIC CYST

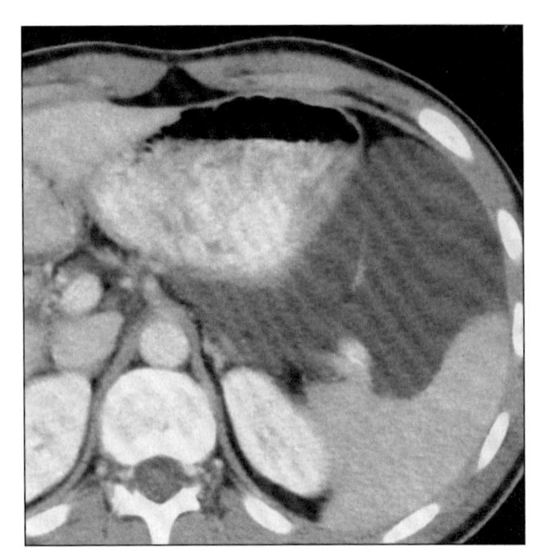

Axial CECT shows water density mass with very thin wall, "indented" by a mesenteric vessel. Cystic lymphangioma.

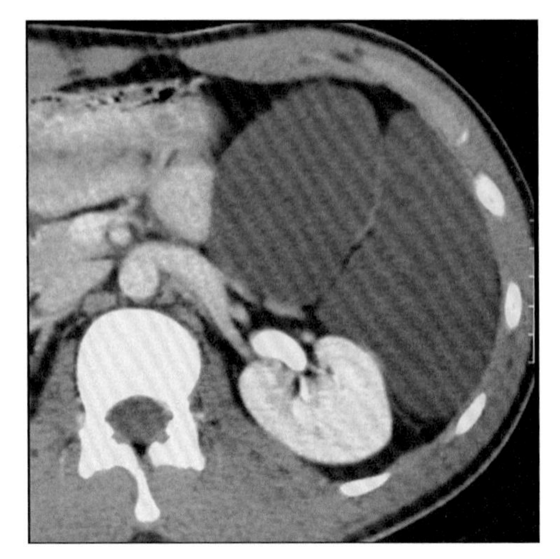

Axial CECT shows water density, thin-walled mass, "indented" by a mesenteric vessel. Cystic lymphangioma.

TERMINOLOGY

Definitions

- Generic descriptive term for a cystic mass arising in the mesentery or omentum, not from an abdominopelvic viscus
- In this and other reviews, may refer to cystic lymphangioma unless otherwise specified

IMAGING FINDINGS

General Features

- Best diagnostic clue: Fluid-filled mass in the mesentery
- Location: Occur anywhere in the mesentery or omentum
- Size: Few mm to 40 cm in diameter
- Morphology
  - May extend to retroperitoneum
  - Simple or multiple, unilocular or multilocular

CT Findings

- Well-circumscribed mass with varying attenuation, depends on the fluid (usually water density, (serous) or less (chylous), rarely hemorrhagic)
- ± Fine calcifications along cyst wall

Ultrasonographic Findings

- Real Time
  - Fluid-filled cystic structure with thin internal septa
  - ± Internal echoes (debris, hemorrhage or infection)

Imaging Recommendations

- Best imaging tool: CECT or CEMR

DIFFERENTIAL DIAGNOSIS

Cystic lymphangioma

- Most common type of mesenteric cyst
- Endothelial cell lining the cyst with foam cells and thin walls that contain lymphatic spaces, lymphoid tissue and smooth muscle
- Frequent attachment to bowel
- Predominantly in male children

Loculated ascites

- Significant size collection may be indistinguishable from mesenteric cyst
- Evaluate other causes (e.g., neoplasm, cirrhosis)

Enteric duplication cyst

- Unilocular or multilocular mass with thick walls (has enteric lining and muscular wall)
- Contrast-enhancement of cystic walls

Enteric cyst

- Hypoechoic mass with few, thin septa and without a visible wall

DDx: Mesenteric Mass - Cystic

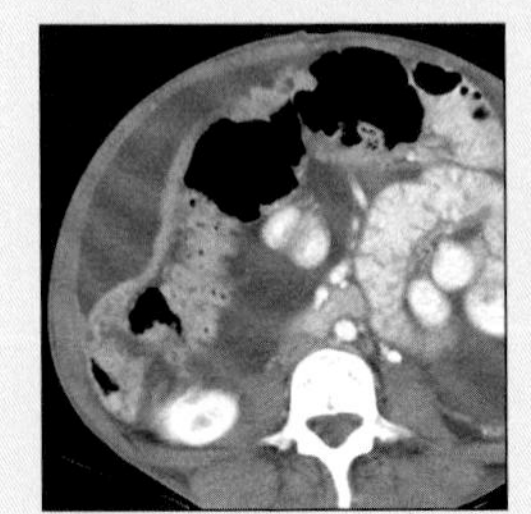

Loculated Ascites

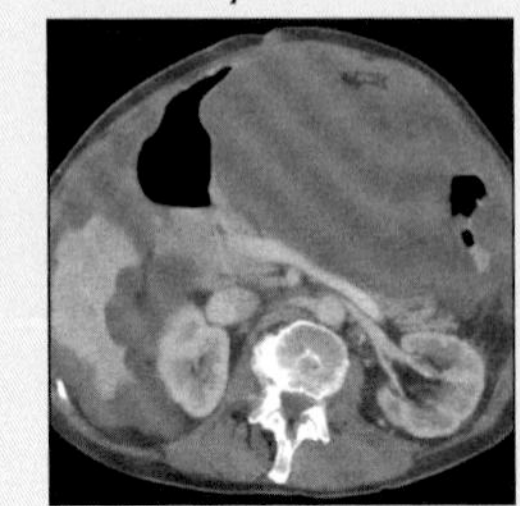

Pseudomyxoma

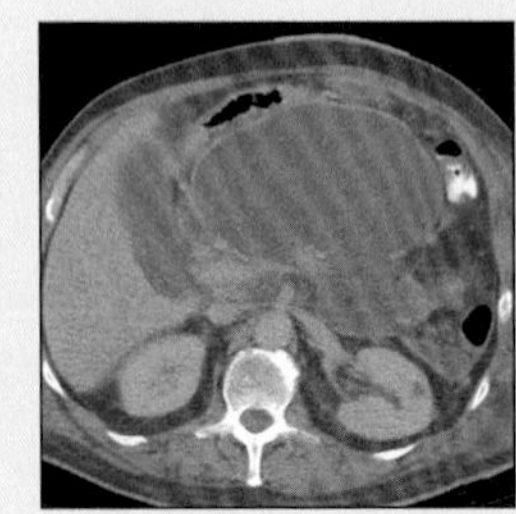

Pseudocyst

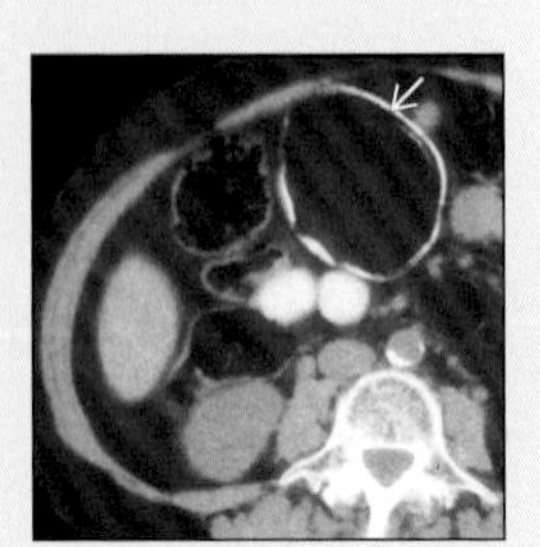

Cystic Teratoma

MESENTERIC CYST

Key Facts

Imaging Findings
- Best diagnostic clue: Fluid-filled mass in the mesentery
- ± Internal echoes (debris, hemorrhage or infection)
- Best imaging tool: CECT or CEMR

Top Differential Diagnoses
- Cystic lymphangioma
- Loculated ascites
- Enteric duplication cyst
- Enteric cyst
- Pancreatic or non-pancreatic pseudocyst
- Cystic mesothelioma
- Cystic teratoma

Diagnostic Checklist
- First exclude visceral origin of "cyst"
- Histologic analysis is necessary to establish diagnosis
- Fluid-filled cyst; thin internal septa

Pancreatic or non-pancreatic pseudocyst
- Unilocular or multilocular, visible wall
- Sequela of pancreatitis or mesenteric hematoma

Cystic mesothelioma
- Unilocular, anechoic thin-walled mass with acoustic enhancement; rare, benign tumor

Cystic teratoma
- Cystic mass with septa and calcification, contains fat-density fluid

PATHOLOGY

General Features
- Epidemiology
  - Cystic lymphangioma
    - Rare; 1/140,000 in general admission, 1/20,000 in pediatric admission

Gross Pathologic & Surgical Features
- Thin-walled, multiseptated, serous or chylous fluid contents
- Often attached to bowel wall

Microscopic Features
- Cuboidal or columnar cell lining the cyst without smooth muscle or lymphatic spaces within the walls

CLINICAL ISSUES

Presentation
- Most common signs/symptoms
  - Abdominal distention, vague abdominal pain
  - Vomiting, palpable abdominal mass
  - Acute abdomen (10% of cases)
  - Small bowel obstruction (most common in children)

Demographics
- Age
  - Cystic lymphangioma
    - Children and young adults; 33% < 15 years of age

Natural History & Prognosis
- Complications
  - Intestinal obstruction, volvulus, hemorrhage, rupture, infection, sepsis, cystic torsion and obstruction of the urinary and biliary tract
- Prognosis
  - Good after surgery, 0-13.6% recurrence rate

Treatment
- Enucleation of cyst ± bowel resection

DIAGNOSTIC CHECKLIST

Consider
- First exclude visceral origin of "cyst"
- Histologic analysis is necessary to establish diagnosis

Image Interpretation Pearls
- Fluid-filled cyst; thin internal septa

SELECTED REFERENCES

1. de Perrot M et al: Mesenteric cysts. Toward less confusion? Dig Surg. 17(4):323-8, 2000
2. Stoupis C et al: Bubbles in the belly: imaging of cystic mesenteric or omental masses. Radiographics. 14(4):729-37, 1994
3. Ros PR et al: Mesenteric and omental cysts: histologic classification with imaging correlation. Radiology. 164(2):327-32, 1987
4. Vanek VW et al: Retroperitoneal, mesenteric, and omental cysts. Arch Surg. 119(7):838-42, 1984

IMAGE GALLERY

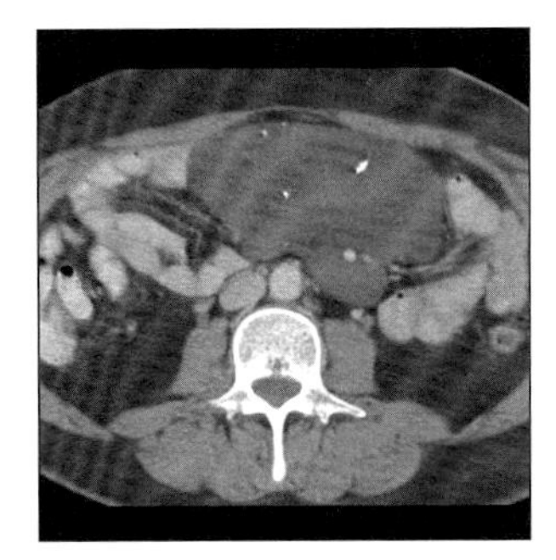

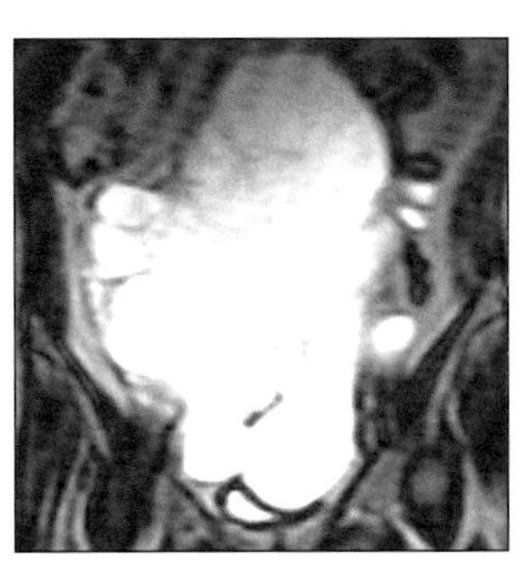

(Left) *Axial CECT shows cystic lymphangioma, a thin-walled water density mesenteric mass, with scattered calcifications in septa.* ***(Right)*** *Coronal T2WI MR shows large, multiloculated water intensity cystic lymphangioma.*

DESMOID

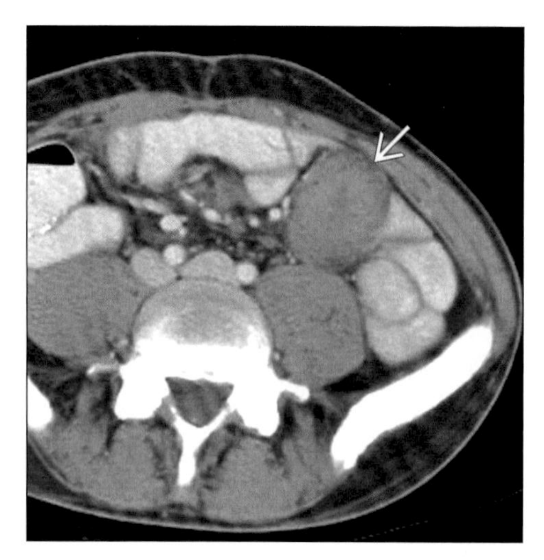

Axial CECT in a 36 year old man 12 months following colectomy for Gardner syndrome shows solid mesenteric desmoid (arrow).

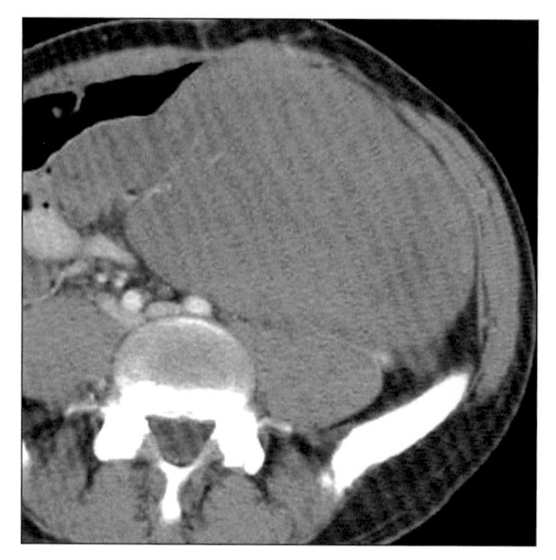

Axial CECT 20 months following colectomy for Gardner syndrome shows rapid growth of mesenteric mass; desmoid tumor.

TERMINOLOGY

Abbreviations and Synonyms

- Aggressive fibromatosis

Definitions

- Rare, benign, locally aggressive, nonencapsulated tumor of connective or fibrous tissue

IMAGING FINDINGS

General Features

- Best diagnostic clue: Small bowel mesentery or abdominal wall mass arising from scar of prior surgery
- Location
  - Abdominal
    - Mesentery: Small bowel (most common)
    - Musculature: Rectus, internal/external oblique, psoas, pelvic (rare)
    - Retroperitoneum
  - Extra-abdominal
    - Bladder, ribs & pelvic bones
- Size: Mass may range from 5-20 cm
- Morphology: Well-/ill-defined, tan or white, hard fibrous tissue mass
- Key concepts
  - Abdominal desmoids can be solitary or multiple
  - Locally aggressive mesenteric primary tumor
  - Tend to arise in musculoaponeurotic planes
    - Desmoid tumor of abdominal wall
  - Tendency to invade locally & recur + grow very rapidly, especially in Gardner syndrome
  - May involve bowel loops, bladder, ribs, pelvic bones
  - Sometimes classified as a low grade fibrosarcoma, or as a subgroup of fibromatosis
  - Usually associated with Gardner syndrome
    - Familial polyposis coli, osteomas, dental defects, congenital pigmented lesions of retina
    - Desmoid tumor, mesenteric fibromatosis
    - Epidermoid (sebaceous) cyst & fibromas of skin
    - Periampullary, adrenal, thyroid & liver carcinomas
  - 75% of desmoid tumors: Previous abdominal surgery
  - 18-20% Gardner syndrome cases develop desmoids
  - Accounts for 45% of fibrous lesions in Gardner

Radiographic Findings

- When associated with Gardner syndrome
  - Fluoroscopic guided double contrast studies
    - Familial polyposis: Innumerable varied sized radiolucent filling defects

CT Findings

- NECT
  - Abdominal desmoids can be solitary or multiple
  - Mesenteric desmoids
    - Soft tissue mass with well or ill-defined margins

DDx: Mesenteric Mass - Solid

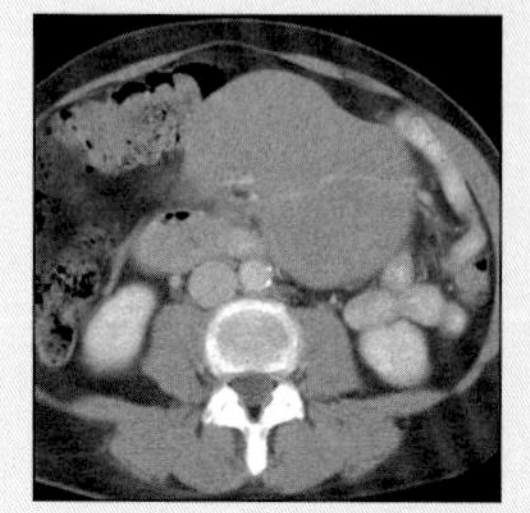

Lymphoma

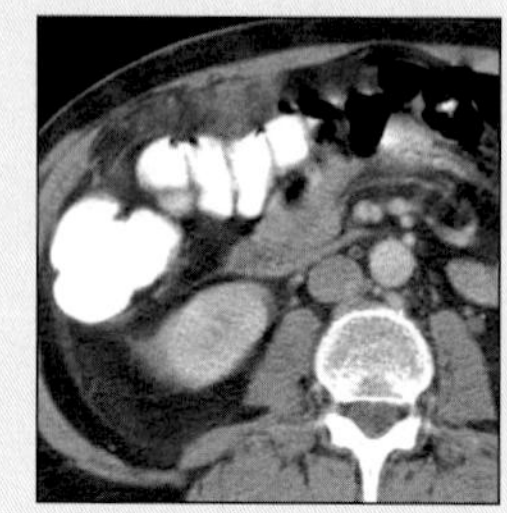

Omental Metastases

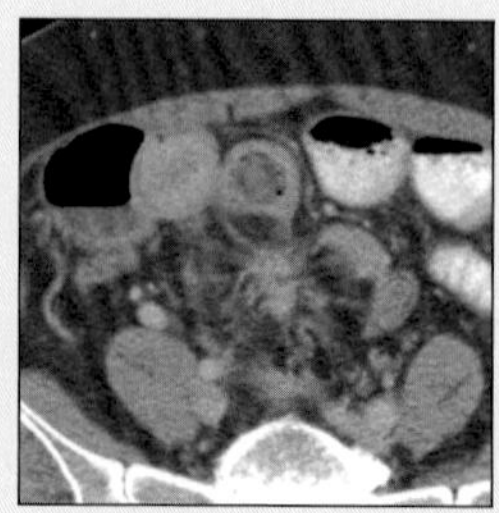

Carcinoid

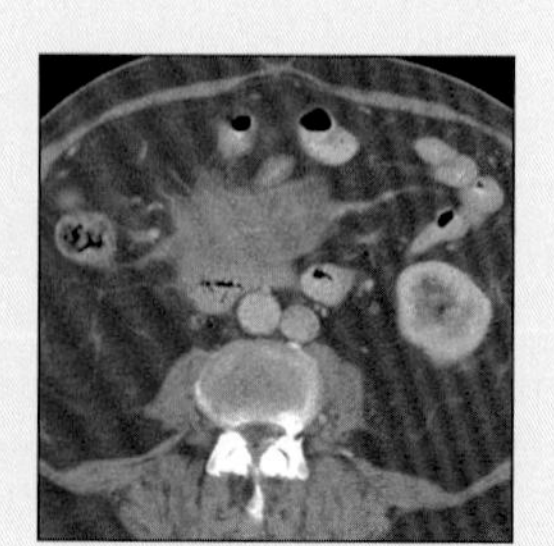

Fibros. Mesenteritis

DESMOID

Key Facts

Terminology
- Rare, benign, locally aggressive, nonencapsulated tumor of connective or fibrous tissue

Imaging Findings
- Best diagnostic clue: Small bowel mesentery or abdominal wall mass arising from scar of prior surgery
- Tendency to invade locally & recur + grow very rapidly, especially in Gardner syndrome
- Soft tissue mass with well or ill-defined margins
- "Whorled appearance": Radiating fibrotic strands into adjacent mesenteric fat
- ± Displacement, retraction, compression: Bowel loops

Top Differential Diagnoses
- Lymphoma
- Omental or Mesenteric metastases
- Carcinoid
- Mesothelioma
- Fibrosing mesenteritis
- Hematoma

Pathology
- Previous abdominal surgery (75% of cases)
- 18-20% of patients with Gardner syndrome develop desmoid tumor
- May be "rock hard", resistant to percutaneous biopsy

Diagnostic Checklist
- Check history of abdominal surgery & colonic polyps
- Rule out other solid mesenteric masses
- Soft tissue density mesenteric mass ± invasion, displacement or encasement of bowel loops & vessels

---

        - Isodense relative to muscle
        - Usually involves small bowel mesentery
        - "Whorled appearance": Radiating fibrotic strands into adjacent mesenteric fat
        - ± Displacement, retraction, compression: Bowel loops
        - ± Infiltration into adjacent organs & musculature
        - ± Small bowel obstruction
    - Abdominal wall desmoids
        - Usually solid with well or partially well-defined margins
        - Homo-/heterogeneous density (higher HU than muscle)
        - Involves rectus or oblique muscles
    - Large tumors with necrosis: Hypodense
- CECT
    - Mesenteric desmoid: Minimal enhancement
    - Abdominal wall desmoid: May enhance
    - Show extent of mesenteric & small bowel invasion
    - ± Encased or displaced mesenteric vasculature
- CT colonography after colonic air insufflation (endoluminal images)
    - Small or large, sessile or pedunculated polyps

MR Findings
- T1WI
    - Poorly or well-circumscribed mass
    - Hypointense to muscle
- T2WI: Hypointense or variable signal intensity
- T1 C+: Variable enhancement
- Multiplanar MR imaging
    - Show origin, extent & invasion of tumor

Ultrasonographic Findings
- Real Time: Well-defined mesenteric mass with variable echogenicity

Imaging Recommendations
- Helical NE + CECT
- Multiplanar MR imaging & T1 C+

DIFFERENTIAL DIAGNOSIS

Lymphoma
- Median age 60 years, predominantly non-Hodgkin
- Imaging
    - Tiny, round soft tissue densities to large, bulky masses
    - "Sandwich" sign: Lobulated, confluent mesenteric masses surrounding superior mesenteric artery/vein
    - Associated retroperitoneal adenopathy confirms

Omental or Mesenteric metastases
- Usually multiple; less well-defined

Carcinoid
- Desmoplastic reaction (not found in desmoids)
- Most common primary tumor of small bowel
    - Usually seen in ileum
- Soft tissue mass in mesentery + radiating strands
    - Strands represent thickened neurovascular bundles
- Segmental thickening of adjacent bowel loops
- 50-85% of cases have nodal metastases
- Liver metastases seen in 80% of lesions > 2 cm
- Encasement of mesenteric vessels → bowel ischemia

Pseudocyst (pancreatic)
- Cystic mass with infiltration of peripancreatic fat
- More common in body & tail + changes of pancreatitis
- NECT
    - Round or oval; hypodense (near water HU)
    - Lobulated, mixed HU lesion (hemorrhagic/infected)
    - Acute pancreatitis: Enlarged pancreas
    - Chronic pancreatitis: Gland atrophy, dilated MPD & intraductal calculi
- CECT: Enhancement of wall, not contents
- MRCP: Hyperintense pseudocyst

Mesothelioma
- Arises from serosal lining of pleural & peritoneal cavity
- Mostly affects males exposed to asbestos
- Peritoneal cavity is involved alone or in association with pleural disease
- Imaging
    - Malignant peritoneal mesothelioma

DESMOID

- Peritoneal thickening, irregularity & nodularity
- ± Omental & mesenteric involvement; ascites
- Adhesions & fixation of bowel loops
- Rarely present as large solitary mesenteric mass
  - Benign cystic peritoneal mesothelioma
    - Multiloculated cystic mass lesion
    - Usually seen in women of reproductive age
    - Strong predilection for pelvic viscera

Fibrosing mesenteritis

- Usually less mass-like than desmoid
- Infiltrative, can be desmoplastic mimicking carcinoid

Hematoma

- Cause: Blunt trauma, excessive anticoagulation, thrombocytopenia
- Acute hematoma
  - Typically quite dense (50-60 HU)
  - Focal/dispersed between leaves of mesentery
  - ↓ In density attaining of water HU by 2 weeks

PATHOLOGY

General Features

- General path comments
  - Fibroproliferative lesion resembling scar tissue
  - Desmoid tumors may be an intermediate step between a reparative process and a true malignancy
- Genetics: May be due to somatic mutation in APC gene 5q
- Etiology
  - Exact cause is unknown
  - Previous abdominal surgery (75% of cases)
    - Pre-existing surgical scar
  - Most often in women of childbearing age
  - Gardner syndrome patients
    - Due to mutation in APC gene (5q)
    - After surgery in mesentery or abdominal wall
  - May be sporadic
- Epidemiology
  - Usually rare
  - 18-20% of patients with Gardner syndrome develop desmoid tumor
    - Mesenteric more than abdominal wall desmoid
  - Increased incidence in child bearing age women
- Associated abnormalities: Components of Gardner syndrome

Gross Pathologic & Surgical Features

- Tan/white, well or poorly defined, firm mass
- May be "rock hard", resistant to percutaneous biopsy

Microscopic Features

- Well-differentiated fibroblasts invading surrounding tissues
- Elongated spindle-shaped cells of uniform appearance with dense bands of collagen

CLINICAL ISSUES

Presentation

- Most common signs/symptoms
  - Asymptomatic; abdominal pain, palpable mass
  - Acute abdominal findings
    - Due to bowel obstruction, ischemia
- Most common sign: Mass in abdominal wall at site of prior surgery

Demographics

- Age: 70% of cases seen between 20-40 years old
- Gender: M:F = 1:3

Natural History & Prognosis

- Complications
  - Locally aggressive growth pattern
  - Necrosis, cystic degeneration → abscess
  - Rarely metastasize
  - High recurrence rate: 25-65%
- Prognosis
  - Poor prognostic signs
    - Size of tumor (> 10 cm), multiplicity
    - Extensive infiltration & tethering of bowel loops
    - Encasement of mesenteric vessels & ureters

Treatment

- Wide surgical resection is treatment of choice
- Radiation therapy: Successful in abdominal wall rather than mesentery
- Steroids, NSAIDs (e.g., sulindac)
- Antiestrogen (e.g., tamoxifen) therapy
- Low-dose chemotherapy & interferon therapy
- Complete resection may require small bowel transplantation

DIAGNOSTIC CHECKLIST

Consider

- Check history of abdominal surgery & colonic polyps
- Look for other components of Gardner syndrome
- Rule out other solid mesenteric masses

Image Interpretation Pearls

- Soft tissue density mesenteric mass ± invasion, displacement or encasement of bowel loops & vessels

SELECTED REFERENCES

1. Sheth S et al: Mesenteric neoplasms: CT appearances of primary and secondary tumors and differential diagnosis. Radiographics. 23(2):457-73; quiz 535-6, 2003
2. Healy JC et al: MR appearances of desmoid tumors in familial adenomatous polyposis. AJR Am J Roentgenol. 169(2):465-72, 1997
3. Mindelzun RE et al: The misty mesentery on CT: differential diagnosis. AJR Am J Roentgenol. 167(1):61-5, 1996
4. Kawashima A et al: CT of intraabdominal desmoid tumors: is the tumor different in patients with Gardner's disease? AJR Am J Roentgenol. 162(2):339-42, 1994
5. Ichikawa T et al: Abdominal wall desmoid mimicking intra-abdominal mass: MR features. Magn Reson Imaging. 12(3):541-4, 1994
6. Einstein DM et al: Abdominal desmoids: CT findings in 25 patients. AJR Am J Roentgenol. 157(2):275-9, 1991
7. Casillas J et al: Imaging of intra- and extraabdominal desmoid tumors. Radiographics. 11(6):959-68, 1991

DESMOID

IMAGE GALLERY

Typical

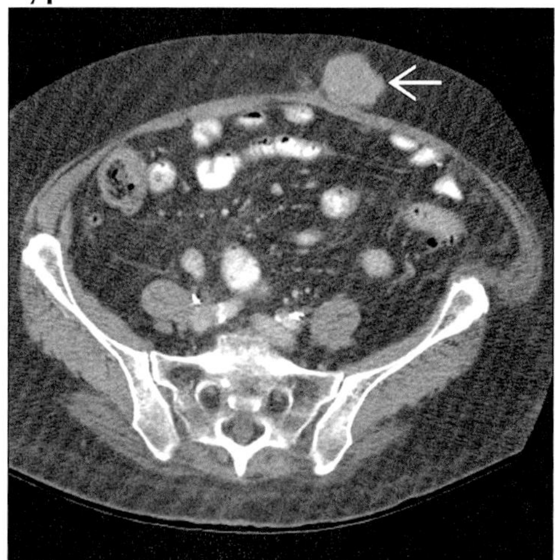

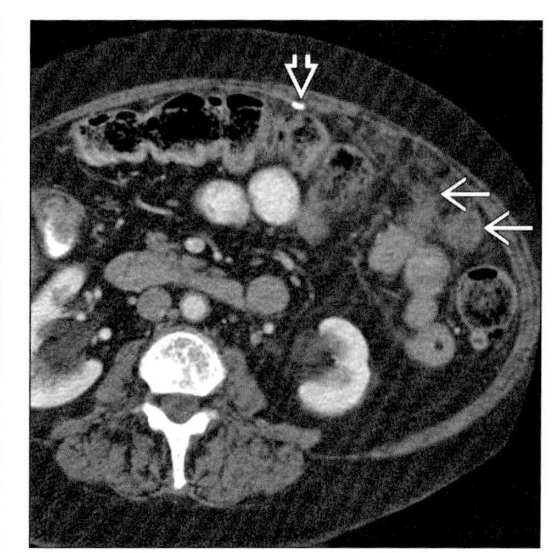

(Left) *Axial CECT shows desmoid in subcutaneous tissue (arrow) adjacent to scar from prior paramedian incision.* ***(Right)*** *Axial CECT shows multiple omental masses (arrows) near site of prior colon surgery. Note surgical clip (open arrow).*

Typical

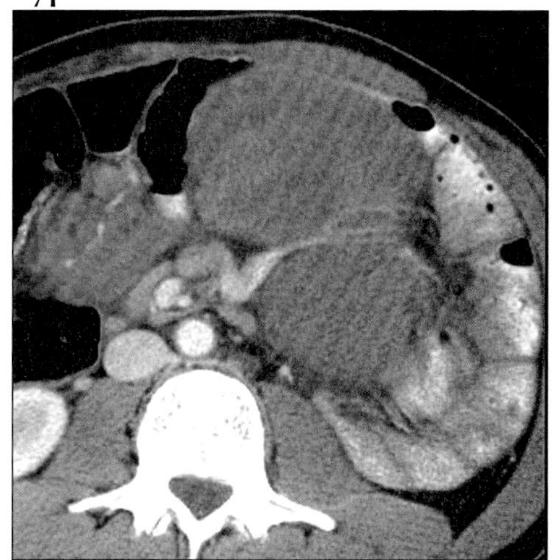

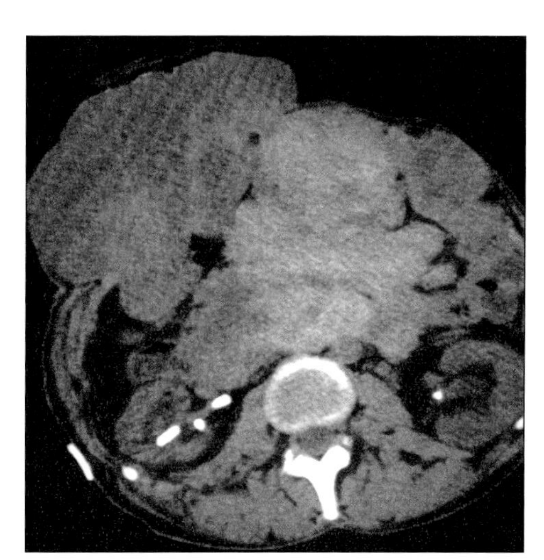

(Left) *Axial CECT in patient with Gardner syndrome shows bilobed large mesenteric desmoid tumor.* ***(Right)*** *Axial NECT in patient with Gardner syndrome shows desmoid tumors filling the abdomen, obstructing kidneys, and deforming the abdominal wall.*

Typical

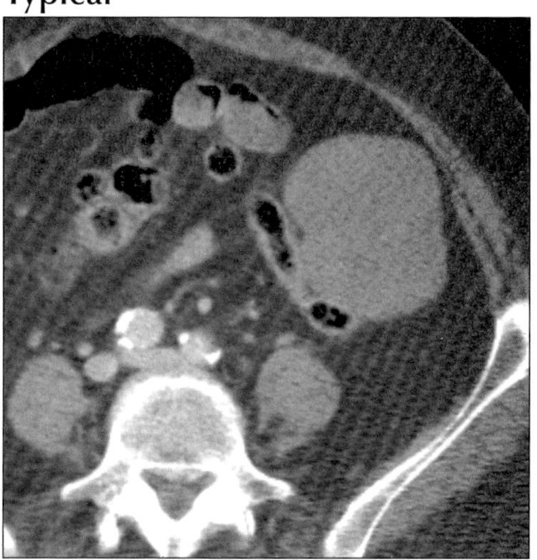

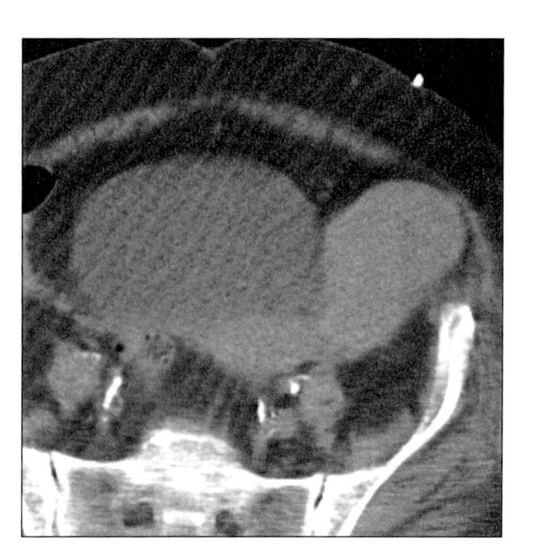

(Left) *Axial CECT in a 79 year old woman with a homogeneous omental mass; sporadic form of desmoid.* ***(Right)*** *Axial NECT in a 79 year old woman. CT-guided biopsy showed "rock hard" mass, but enough tissue to confirm desmoid tumor.*

MESOTHELIOMA

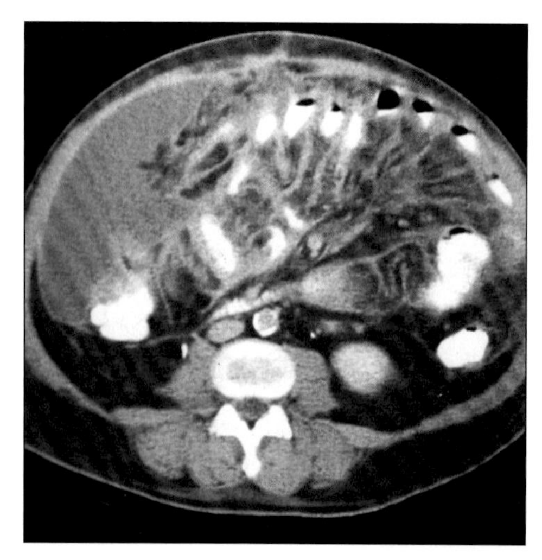

Axial CECT shows stellate thickened mesentery and diffusely thickened bowel loops with serosal implants.

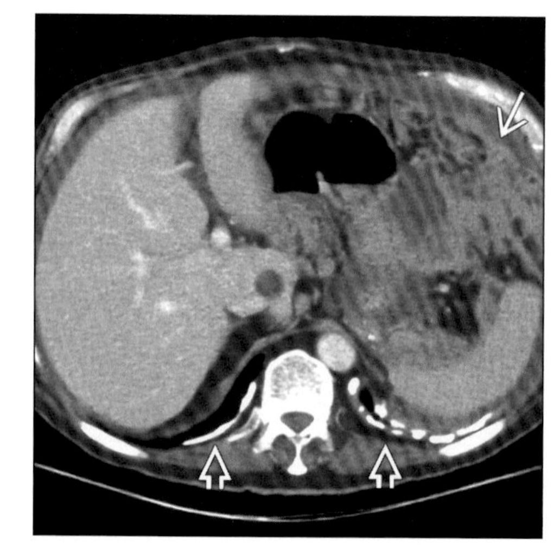

Axial CECT shows irregular, thickened peritoneum (arrow). Note bilateral calcified pleural plaques (open arrows).

TERMINOLOGY

Abbreviations and Synonyms

- Malignant mesothelioma (MM), peritoneal mesothelioma (PM)
- Variant: Benign cystic mesothelioma = multilocular peritoneal inclusion cyst = cystic mesothelioma of the peritoneum = multicystic peritoneal mesothelioma

Definitions

- Primary neoplasm arising from serosal lining of peritoneum

IMAGING FINDINGS

General Features

- Best diagnostic clue: Peritoneal masses or omental cake associated with calcified pleural plaques
- Location
  - 60% of malignant mesotheliomas arise in pleura
  - 20-30% of malignant mesotheliomas arise in peritoneum
  - May also arise in pericardium, tunica vaginalis
- Size
  - Involves peritoneal surfaces diffusely
  - Focal masses may be several mm to several cm
- Morphology
  - Two primary forms
    - Desmoplastic form: Diffuse disease thickening peritoneal surfaces and enveloping viscera
    - Focal form: Large tumor mass in upper abdomen with scattered peritoneal nodules

Radiographic Findings

- Radiography
  - Calcified pleural plaques in 50% of peritoneal mesotheliomas versus 20% of pleural mesotheliomas
    - Signifies heavier asbestos exposure in patients with peritoneal mesothelioma
- Fluoroscopy
  - Separation and fixation of bowel loops
  - Spiculation of loops when bowel wall invaded
  - Segmental stenoses with circumferential bowel invasion

CT Findings

- NECT
  - Calcified pleural plaques
  - Calcification in peritoneal masses very uncommon
- CECT
  - Stellate, thickened mesentery secondary to encasement and straightening of mesenteric vessels
  - Omental and peritoneal-based masses
  - Spreads along serosal surfaces and directly invades adjacent viscera, especially colon and liver

DDx: Peritoneal and Omental Masses

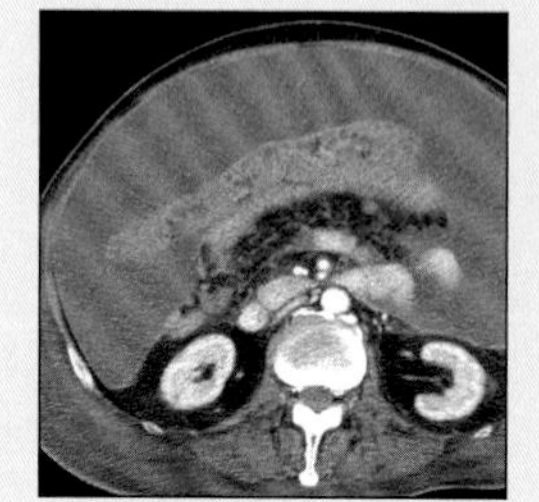

Carcinomatosis

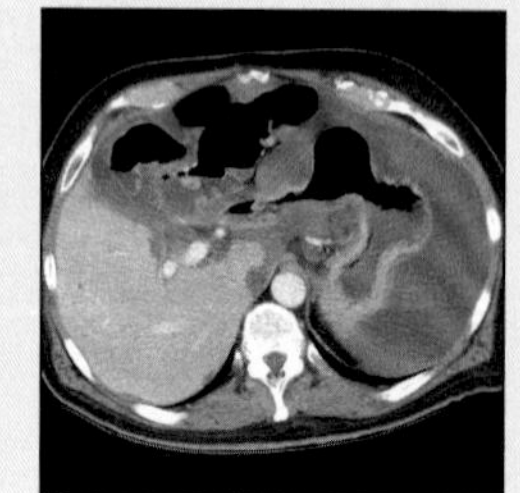

Pseudomyxoma

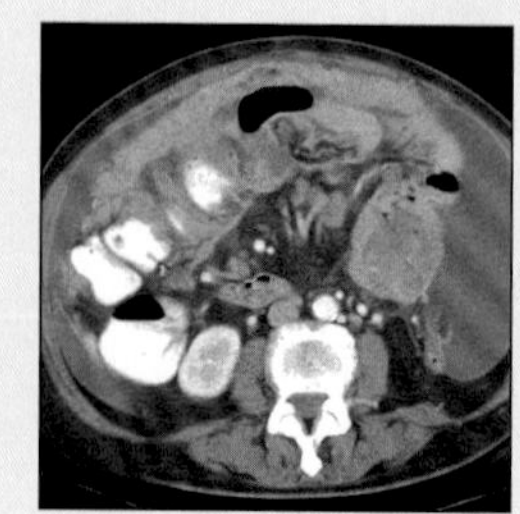

Lymphomatosis

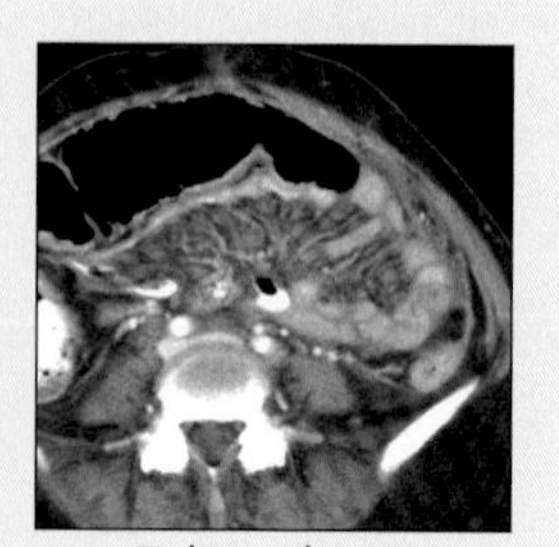

Tuberculosis

MESOTHELIOMA

Key Facts

Terminology
- Primary neoplasm arising from serosal lining of peritoneum

Imaging Findings
- Best diagnostic clue: Peritoneal masses or omental cake associated with calcified pleural plaques
- Stellate, thickened mesentery secondary to encasement and straightening of mesenteric vessels
- Omental and peritoneal-based masses
- Spreads along serosal surfaces and directly invades adjacent viscera, especially colon and liver
- Variable amount of ascites; massive ascites uncommon
- Benign cystic mesothelioma: Multiloculated thin-walled cystic pelvic mass with anechoic spaces

Pathology
- Malignant mesothelioma: Associated with asbestos exposure
- Rare: 1-2 cases/million
- Open biopsy rather than FNA often needed for diagnosis

Clinical Issues
- Extremely poor prognosis
- Remains confined to abdominal cavity and invades locally

Diagnostic Checklist
- Presence of distant metastases outside abdominal cavity makes PM unlikely

---

- Variable amount of ascites; massive ascites uncommon
- Benign cystic mesothelioma
  - Low density multiloculated cystic mass in pelvis
  - May have calcifications

MR Findings
- T1WI
  - Low-intermediate signal intensity omental and peritoneal masses with encasement of viscera
  - Benign cystic mesothelioma: Hypointense multicystic mass
- T2WI
  - High signal intensity peritoneal nodules
  - Fluid-fluid levels secondary to hemorrhage
  - Benign cystic mesothelioma: Intermediate to high signal intensity multiloculated cystic mass
- T1 C+
  - Enhancing peritoneal thickening and nodules
  - Improved conspicuity of enhancing masses with fat-saturation pulses

Ultrasonographic Findings
- Real Time
  - Hypoechoic, sheet-like peritoneal masses
  - Echogenic areas within hypoechoic masses representing entrapped mesenteric or omental fat
  - Thickened omentum
  - Benign cystic mesothelioma: Multiloculated thin-walled cystic pelvic mass with anechoic spaces

Nuclear Medicine Findings
- Gallium scan
  - Diffuse uptake

Imaging Recommendations
- Best imaging tool: Contrast-enhanced CT
- Protocol advice
  - Use water or oral contrast to distend small bowel loops
  - Coronal reformations useful for detecting implants near diaphragm

DIFFERENTIAL DIAGNOSIS

Neoplastic
- Peritoneal carcinomatosis
  - Most common cause of omental cake and peritoneal implants
  - Serosal spread of adenocarcinoma, especially ovarian, stomach, colon and pancreas
- Pseudomyxoma peritonei
  - Low attenuation peritoneal masses
  - Implants cause scalloping of visceral serosal surfaces
- Lymphoma
  - Concomitant lymphadenopathy
  - "Sandwich sign" = confluent mesenteric nodal masses surrounding mesenteric vessels
  - Ascites without loculation

Infectious
- Tuberculosis
  - Low attenuation lymphadenopathy in 40% of cases
  - High attenuation ascites (25-45 HU)
  - Lymphadenopathy most often mesenteric, omental and peripancreatic rather than retroperitoneal
  - Thickened cecum and terminal ileum
  - Smooth peritoneal thickening with pronounced enhancement
  - Transperitoneal permeation

Inflammatory
- Retractile mesenteritis

PATHOLOGY

General Features
- General path comments
  - Multifocal origin from mesothelial lining of abdomen and pelvis
  - Three histologic types
    - Epithelial 54%
    - Sarcomatoid 21%
    - Biphasic (mixed epithelial-sarcomatoid) 25%
- Genetics

MESOTHELIOMA

- Complex karyotypes
  - Deletions in 1p, 3p, 6q, 9p, 15q and 22q in various combinations
- Etiology
  - Malignant mesothelioma: Associated with asbestos exposure
    - 20-40 year latency between exposure and diagnosis
  - Simian virus-40 may be co-carcinogen
  - Benign cystic mesothelioma: Not asbestos related
- Epidemiology
  - Rare: 1-2 cases/million
  - 200-400 cases diagnosed annually in U.S.
  - Disease clusters around shipyards, docks, asbestos mines and factories
  - Non occupational exposure to asbestos and zeolites common in Turkey
- Associated abnormalities: Asbestos-related pleural and parenchymal lung disease

Gross Pathologic & Surgical Features

- Solid tumor masses growing along serosal surfaces
- Encasement and invasion of adjacent viscera
- Can recur along surgical and laparoscopy tracts

Microscopic Features

- Variable histologic appearance of tumor cells
  - Open biopsy rather than FNA often needed for diagnosis
- Positive immunostaining for calretinin, keratin, vimentin and thrombomodulin

CLINICAL ISSUES

Presentation

- Most common signs/symptoms
  - "Pain-predominant" type: Dominant tumor mass with little ascites
  - "Ascites-predominant" type: Abdominal distention
  - Other signs/symptoms
    - Weight loss, malaise, cramping, new-onset hernia
- Clinical profile: History of asbestos exposure for malignant mesothelioma

Demographics

- Age
  - Malignant mesothelioma: Usually 6th-7th decade, but can occur at any age
  - Benign cystic mesothelioma: 3rd-4th decade
- Gender
  - Malignant mesothelioma: M > > F
  - Benign cystic mesothelioma: M < < F

Natural History & Prognosis

- Extremely poor prognosis
  - Median survival 6 months; death usually within 1 year
- Solitary tumors have better prognosis than diffuse intra-abdominal disease
- Remains confined to abdominal cavity and invades locally
  - Does not disseminate hematogenously to brain, bone, or lung
- Benign cystic mesothelioma: Non-lethal but locally recurrent

Treatment

- Options, risks, complications
  - Cytoreductive surgery and peritonectomy combined with intra-peritoneal chemotherapy
    - Heated intra-operative intraperitoneal chemotherapy often used

DIAGNOSTIC CHECKLIST

Consider

- Consider PM in patients with diffuse peritoneal tumor on CT and stigmata of asbestos exposure
- Peritoneal carcinomatosis is much more common than mesothelioma

Image Interpretation Pearls

- Presence of distant metastases outside abdominal cavity makes PM unlikely

SELECTED REFERENCES

1. Wong WL et al: Best Cases from the AFIP: Multicystic Mesothelioma. Radiographics. 24(1): 247-250, 2004
2. Sheth S et al: Mesenteric neoplasms: CT appearances of primary and secondary tumors and differential diagnosis. Radiographics. 23(2):457-73; quiz 535-6, 2003
3. Hanbidge AE et al: US of the peritoneum. Radiographics. 23(3):663-84; discussion 684-5, 2003
4. Sugarbaker PH et al: A review of peritoneal mesothelioma at the Washington Cancer Institute. Surg Oncol Clin N Am. 12(3):605-21, xi, 2003
5. Puvaneswary M et al: Peritoneal mesothelioma: CT and MRI findings. Australas Radiol. 46(1): 91-6, 2002
6. Mohamed F et al: Peritoneal mesothelioma. Curr Treat Options Oncol. 3(5):375-86, 2002
7. DeVita V et al: Cancer: Principles and practice of Oncology. 6th ed. Baltimore, Lippincott Williams and Wilkins. 1937-1969, 2001
8. Haliloglu M et al: Malignant peritoneal mesothelioma in two pediatric patients: MR imaging findings. Pediatr Radiol. 30(4):251-5, 2000
9. Gore R et al: Textbook of Gastrointestinal Radiology. 2nd ed. Philadelphia, W.B. Saunders, 1980-1992, 2000
10. Kim Y et al: Peritoneal lymphomatosis: CT findings. Abdom Imaging. 23(1):87-90, 1998
11. Ozgen A et al: Giant benign cystic peritoneal mesothelioma: US, CT, and MRI findings. Abdom Imaging. 23(5): 502-4, 1998
12. Jadvar H et al: Still the great mimicker: abdominal tuberculosis. AJR Am J Roentgenol. 168(6):1455-60, 1997
13. Smith TR: Malignant peritoneal mesothelioma: marked variability of CT findings. Abdom Imaging. 19(1):27-9, 1994
14. Stoupis C et al: Bubbles in the belly: imaging of cystic mesenteric or omental masses. Radiographics. 14(4):729-37, 1994
15. Akhan O et al: Peritoneal mesothelioma: sonographic findings in nine cases. Abdom Imaging. 18(3): 280-2, 1993
16. Ros PR et al: Peritoneal mesothelioma. Radiologic appearances correlated with histology. Acta Radiol. 32(5):355-8, 1991
17. Cozzi G et al: Double contrast barium enema combined with non-invasive imaging in peritoneal mesothelioma. Acta Radiol. 30(1): 21-4, 1989

MESOTHELIOMA

IMAGE GALLERY

Typical

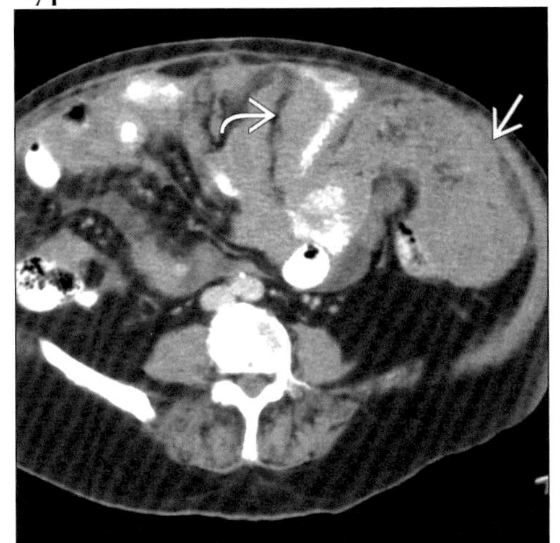

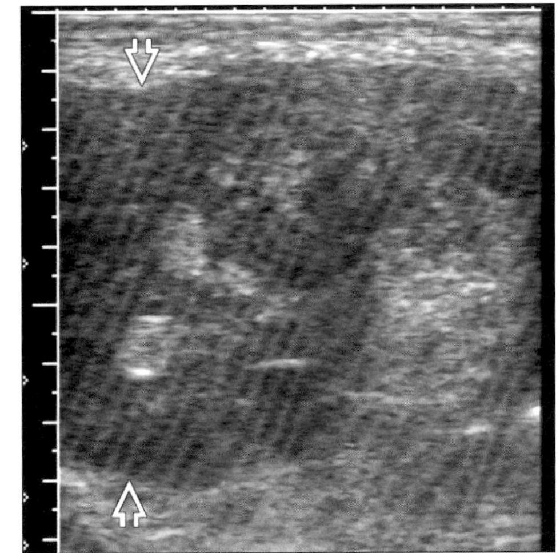

(Left) *Axial CECT shows marked omental thickening (arrow) and encasement of bowel loops (curved arrow).* ***(Right)*** *Axial US of left lower quadrant shows hypoechoic mass (open arrows) encasing rounded, echogenic loops of bowel and mesenteric leaves.*

Variant

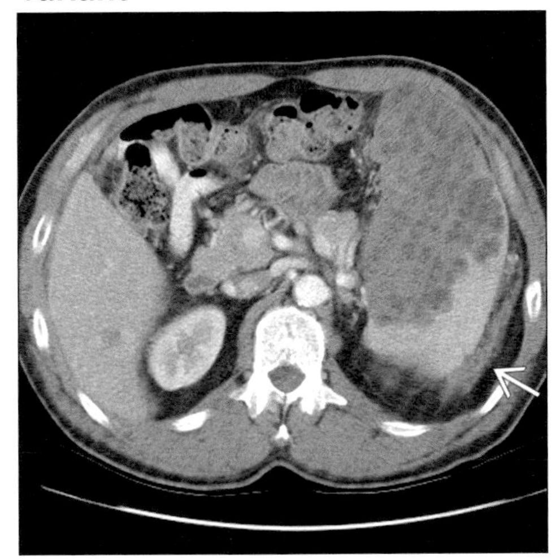

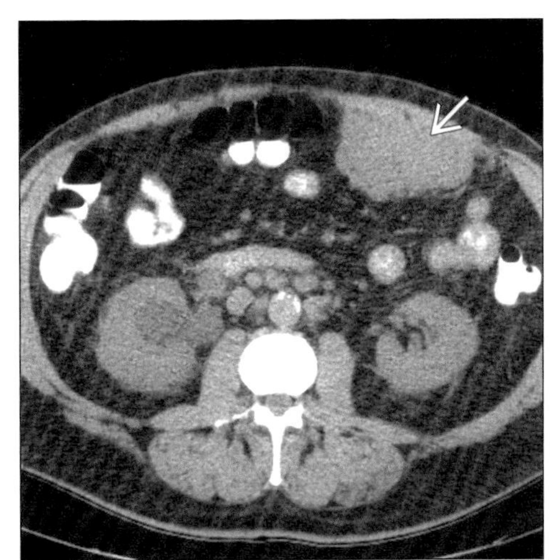

(Left) *Axial CECT shows large, complex mass invading the spleen. Peritoneal tumor (arrow) is also present.* ***(Right)*** *Axial NECT in patient with renal insufficiency shows lobulated mass (arrow) along peritoneal surface. Surgical biopsy confirmed malignant mesothelioma.*

Other

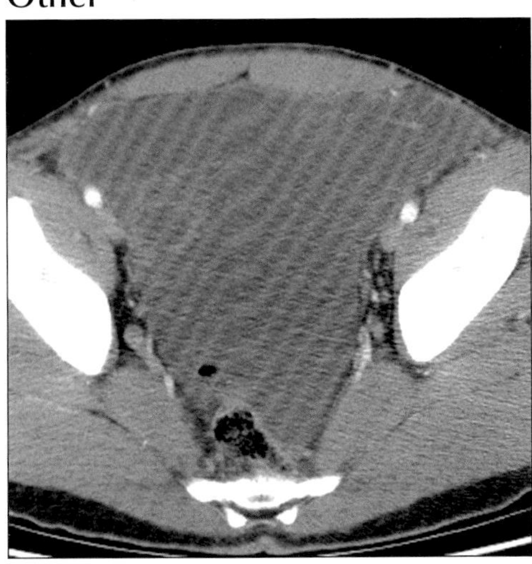

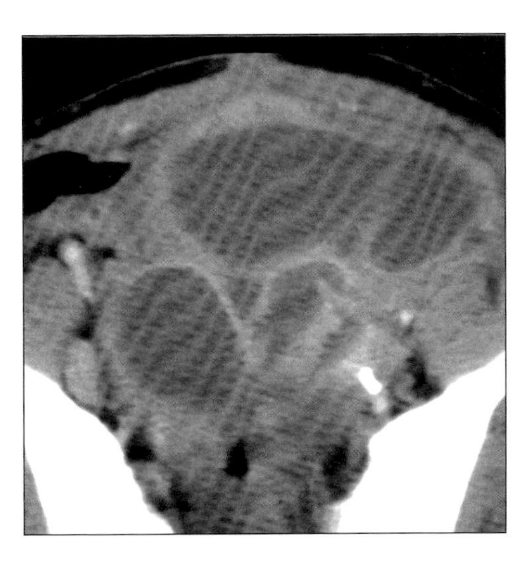

(Left) *Axial CECT shows multiloculated cystic mass filling the pelvis. Surgical excision proved the mass to be a benign cystic mesothelioma.* ***(Right)*** *Axial CECT shows multiloculated cystic mass, which proved to be a benign cystic mesothelioma.*

PERITONEAL METASTASES

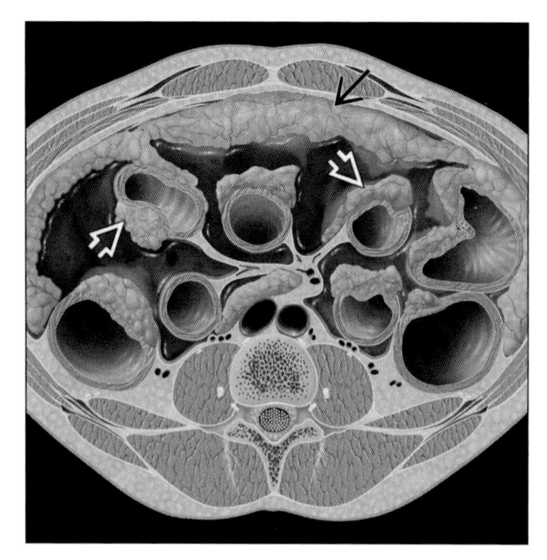

Axial anatomic rendering of peritoneal metastases. Note anterior omental cake (arrow) and serosal implants (open arrows).

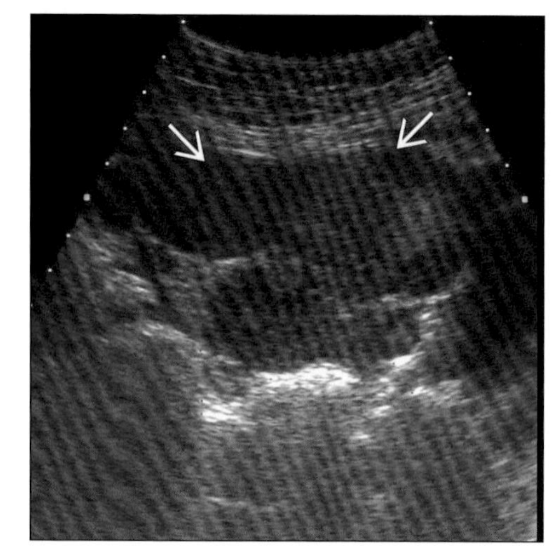

Axial US demonstrates large omental cake (arrows) from ovarian carcinoma.

TERMINOLOGY

Abbreviations and Synonyms

- Peritoneal carcinomatosis, peritoneal implants, omental caking

Definitions

- Metastatic disease to the omentum, peritoneal surface, peritoneal ligaments and/or mesentery

IMAGING FINDINGS

General Features

- Best diagnostic clue
  - Omental caking, soft tissue implants on peritoneal surface are best diagnostic clues
  - "Scalloped" contour of liver/spleen from pseudomyxoma peritonei
  - Cystic peritoneal masses with ovarian carcinoma on peritoneal surfaces
  - Ascites, mesenteric stranding, bowel obstruction
- Location: Peritoneum, mesentery, peritoneal ligaments
- Size: Variable; 5 mm nodules to large confluent omental masses ("caking")
- Morphology: Nodular, plaque-like or large omental mass

Radiographic Findings

- Radiography
  - Plain film findings of ascites
    - Medial displacement of cecum in 90% of patients with significant ascites
    - Pelvic "dog's ear" in 90% of patients with significant ascites
    - Medial displacement of lateral liver edge (Hellmer sign) in 80% of patients with significant ascites
    - Bulging of flanks, central displacement of bowel loops, indistinct psoas margin
  - Plain film findings of small bowel obstruction
    - Dilated small bowel > 3 cm
    - Fluid-fluid levels in small bowel on upright film
    - "String of pearls" sign
    - Collapsed gasless colon
- Fluoroscopy
  - Barium studies
    - Small bowel follow through (SBFT): Dilated bowel with transition zone; partial small bowel obstruction
    - Mural extrinsic filling defects due to serosal implants in small bowel
    - Spiculated extrinsic impression due to tethering of rectosigmoid from intraperitoneal mets to pouch of Douglas
    - Scalloping of cecum from peritoneal implants

DDx: Peritoneal Abnormalities

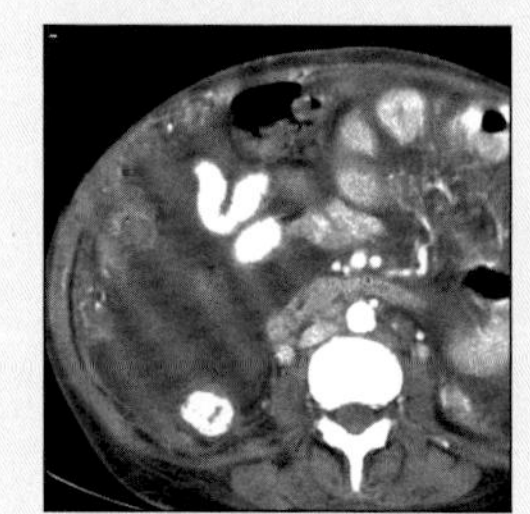

TB Peritonitis

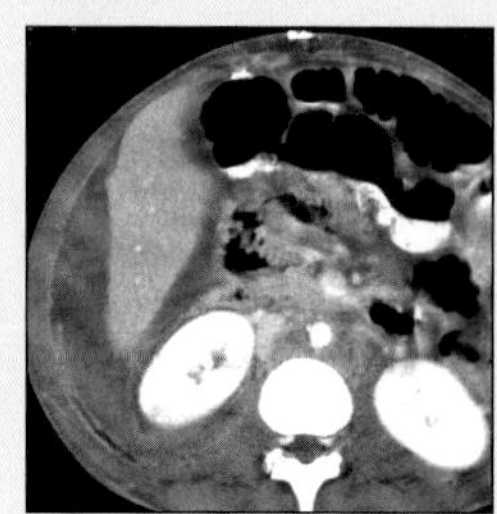

Papillary Serous CA

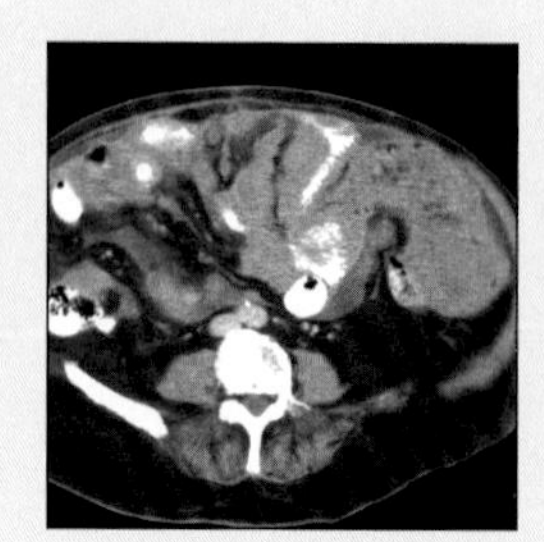

Mesothelioma

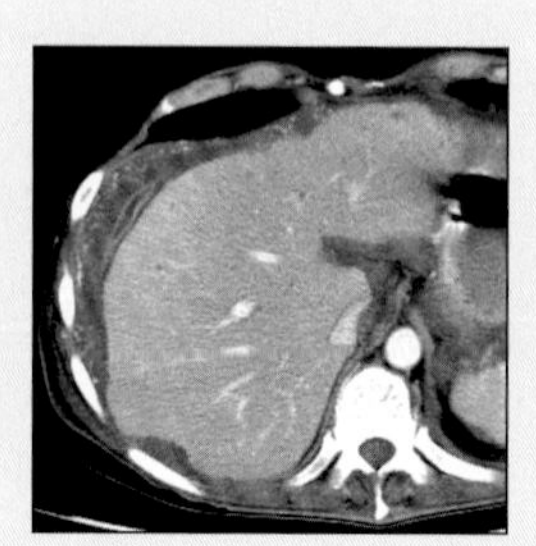

Pseudomyxoma

PERITONEAL METASTASES

Key Facts

Terminology

- Peritoneal carcinomatosis, peritoneal implants, omental caking
- Metastatic disease to the omentum, peritoneal surface, peritoneal ligaments and/or mesentery

Imaging Findings

- Omental caking, soft tissue implants on peritoneal surface are best diagnostic clues
- Location: Peritoneum, mesentery, peritoneal ligaments
- Ascites, nodular thickening/enhancement of peritoneum, hypovascular omental masses on CECT
- Best imaging tool: CECT or MRI
- Oral and I.V. contrast with CECT; gadolinium-enhanced GRE or T1 sequence with MRI

Top Differential Diagnoses

- TB peritonitis
- Papillary serous carcinoma of peritoneum
- Peritoneal mesothelioma

Pathology

- Peritoneal metastases indicate Stage IV disease

Clinical Issues

- Most common signs/symptoms: Abdominal distension and pain, weight loss; ascites may or may not be present
- Gender: More common in females than males, due to ovarian CA

Diagnostic Checklist

- TB peritonitis

    - "Omental caking" may cause invasion of transverse mesocolon with nodularity & spiculation of superior contour

CT Findings

- NECT: Ascites
- CECT
  - Ascites, nodular thickening/enhancement of peritoneum, hypovascular omental masses on CECT
  - Spiculated mesentery
  - Evidence of bowel obstruction with delineation of transition zone from dilated to non-dilated bowel

MR Findings

- T1WI: Low signal ascites, medium signal omental caking
- T2WI: Intermediate signal peritoneal mass and high signal ascites
- T1 C+: Abnormal enhancement of peritoneum with gadolinium, hypointense nodules and masses

Ultrasonographic Findings

- Real Time
  - Complex ascites with septations, hypoechoic omental masses (caking), peritoneal implants
  - Not sensitive for peritoneal implants in absence of ascites

Imaging Recommendations

- Best imaging tool: CECT or MRI
- Protocol advice
  - Oral and I.V. contrast with CECT; gadolinium-enhanced GRE or T1 sequence with MRI
    - 150 ml I.V. contrast at 2.5 ml/sec with 5 mm collimation and reconstruction at 5 mm intervals

DIFFERENTIAL DIAGNOSIS

TB peritonitis

- Abnormal enhancement of peritoneum and mesentery
- Nodular or symmetric thickening of peritoneum and mesentery
- Ascites, low attenuation mesenteric nodes
- Calcification in 14% of cases
- Ileo-cecal mural thickening
- Splenomegaly
- TB mimics CT appearance of peritoneal metastases

Papillary serous carcinoma of peritoneum

- Peritoneal metastases (implants, ascites, omental caking) without other source
- No ovarian or GI tract primary tumor
- Identical CT, US, MR findings to peritoneal metastases from ovarian CA

Peritoneal mesothelioma

- 1/5 of all mesotheliomas are peritoneal
- Large solid omental and mesenteric masses often infiltrating bowel and mesentery

Pseudomyxoma peritonei

- Diffuse accumulation of gelatinous masses within peritoneum
- CECT: Scalloping of lateral contour of liver and spleen
- Etiology related to perforation of mucinous neoplasm of appendix
- Treatment involves cytoreduction of peritoneal mass and intraperitoneal chemotherapy

PATHOLOGY

General Features

- General path comments
  - Ascites
  - Omental masses
  - Mesenteric masses
  - Nodular implants on peritoneal surface
- Genetics
  - Colorectal and ovarian CA related to Lynch II syndrome of hereditary nonpolypous colorectal cancer
  - GI and ovarian CA related to Lynch syndrome
  - GI cancers related to polyposis syndrome
- Etiology
  - Metastatic disease to peritoneal surfaces, omentum and mesentery

PERITONEAL METASTASES

- Peritoneal cavity spread of surface epithelium tumors such as ovarian CA
- Ovarian and GI tract adenocarcinomas most common etiologies
- Less common causes
  - Metastatic lung, breast and renal CA
  - Sarcoma, lymphoma less common causes
- Epidemiology: Varies according to primary tumor

Gross Pathologic & Surgical Features

- Infiltrating masses of peritoneal surfaces, omentum and mesentery
- Omental caking
- Ascites

Microscopic Features

- Varies according to primary tumor
  - Most commonly adenocarcinoma

Staging, Grading or Classification Criteria

- Peritoneal metastases indicate Stage IV disease

CLINICAL ISSUES

Presentation

- Most common signs/symptoms: Abdominal distension and pain, weight loss; ascites may or may not be present
- Clinical profile
  - No reliable lab data
  - Positive cytology on paracentesis
  - Positive FNA of omental mass

Demographics

- Age
  - Adults generally > 40 yrs
  - Younger patients with hereditary syndromes (i.e., Lynch II)
- Gender: More common in females than males, due to ovarian CA

Natural History & Prognosis

- Progressive if untreated
  - Bowel obstruction
- Pattern of peritoneal spread
  - Direct seeding along mesentery and ligaments
  - Intraperitoneal seeding along distribution of ascites
  - Lymphatic
  - Hematogenous
- Variable depending on primary tumor
- Poor prognosis in general

Treatment

- Cytoreductive surgery for ovarian metastases
- All others combination of systemic and intraperitoneal chemotherapy

DIAGNOSTIC CHECKLIST

Consider

- TB peritonitis
  - Causes symmetric thickening of peritoneum, ileo-cecal thickening, ascites, necrotic low attenuation nodes

Image Interpretation Pearls

- Omental caking
- Peritoneal and mesenteric implants
- Ascites

SELECTED REFERENCES

1. Jayne DG: The molecular biology of peritoneal carcinomatosis from gastrointestinal cancer. Ann Acad Med Singapore. 32(2):219-25, 2003
2. Park CM et al: Recurrent ovarian malignancy: patterns and spectrum of imaging findings. Abdom Imaging. 28(3):404-15, 2003
3. Pavlidis N et al: Diagnostic and therapeutic management of cancer of an unknown primary. Eur J Cancer. 39(14):1990-2005, 2003
4. Canis M et al: Risk of spread of ovarian cancer after laparoscopic surgery. Curr Opin Obstet Gynecol. 13(1):9-14, 2001
5. Tamsma JT et al: Pathogenesis of malignant ascites: Starling's law of capillary hemodynamics revisited. Ann Oncol. 12(10):1353-7, 2001
6. Sugarbaker PH: Review of a personal experience in the management of carcinomatosis and sarcomatosis. Jpn J Clin Oncol. 31(12):573-83, 2001
7. Chorost MI et al: The management of the unknown primary. J Am Coll Surg. 193(6):666-77, 2001
8. Raptopoulos V et al: Peritoneal carcinomatosis. Eur Radiol 11(11):2195-2206, 2001
9. Sugarbaker PH et al: Clinical pathway for the management of resectable gastric cancer with peritoneal seeding: best palliation with a ray of hope for cure. Oncology. 58(2):96-107, 2000
10. Seidman JD et al: Ovarian serous borderline tumors: a critical review of the literature with emphasis on prognostic indicators. Hum Pathol. 31(5):539-57, 2000
11. Leblanc E et al: Surgical staging of early invasive epithelial ovarian tumors. Semin Surg Oncol. 19(1):36-41, 2000
12. Canis M et al: Cancer and laparoscopy, experimental studies: a review. Eur J Obstet Gynecol Reprod Biol. 91(1):1-9, 2000
13. Patel SV et al: Supradiaphragmatic manifestations of papillary serous adenocarcinoma of the ovary. Clin Radiol. 54(11):748-54, 1999
14. Sugarbaker PH: Management of peritoneal-surface malignancy: the surgeon's role. Langenbecks Arch Surg. 384(6):576-87, 1999
15. Ohtani T et al: Early intraperitoneal dissemination after radical resection of unsuspected gallbladder carcinoma following laparoscopic cholecystectomy. Surg Laparosc Endosc. 8(1):58-62, 1998
16. Sugarbaker PH: Intraperitoneal chemotherapy and cytoreductive surgery for the prevention and treatment of peritoneal carcinomatosis and sarcomatosis. Semin Surg Oncol 14(3):254-61, 1998
17. Fecteau AH et al: Peritoneal metastasis of intracranial glioblastoma via a ventriculoperitoneal shunt preventing organ retrieval: case report and review of the literature. Clin Transplant. 12(4):348-50, 1998
18. Conlon KC et al: Laparoscopy and laparoscopic ultrasound in the staging of gastric cancer. Semin Oncol. 23(3):347-51, 1996
19. Averbach AM et al: Strategies to decrease the incidence of intra-abdominal recurrence in resectable gastric cancer. Br J Surg. 83(6):726-33, 1996

PERITONEAL METASTASES

IMAGE GALLERY

Typical

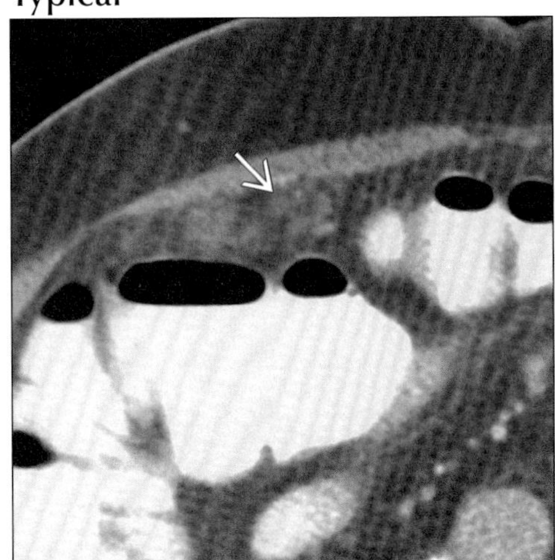

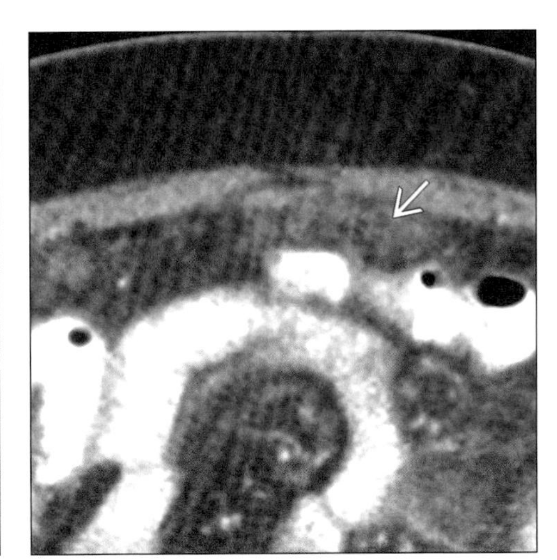

(Left) *Axial CECT demonstrates metastatic omental nodules from pancreatic carcinoma. Note soft tissue nodule anterior to hepatic flexure (arrow).* ***(Right)*** *Axial CECT shows metastatic omental nodules from pancreatic carcinoma. Note omental nodule anterior to transverse colon (arrow).*

Typical

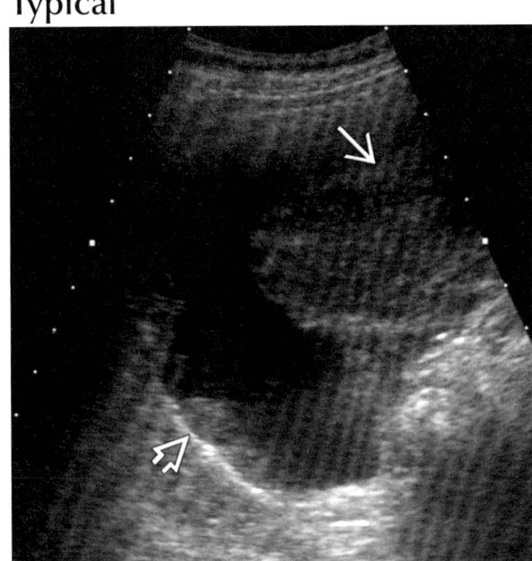

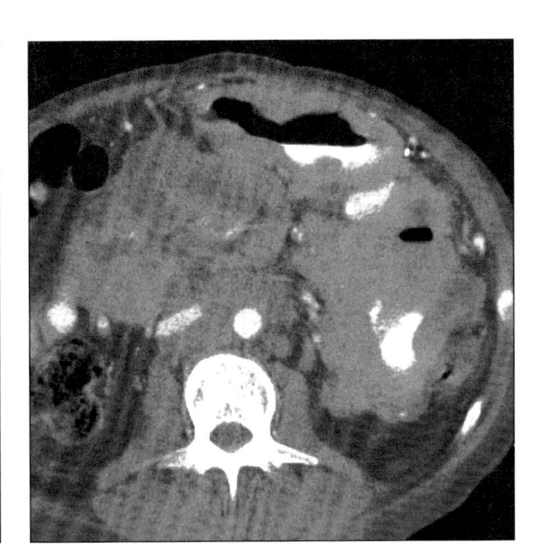

(Left) *Axial US demonstrates omental caking from ovarian carcinoma. Note echogenic mass within omentum (arrow), ascites and peritoneal implants (open arrow).* ***(Right)*** *Axial CECT shows peritoneal spread of lymphoma. Note diffuse infiltration of mesentery and colon by non-Hodgkin lymphoma.*

Typical

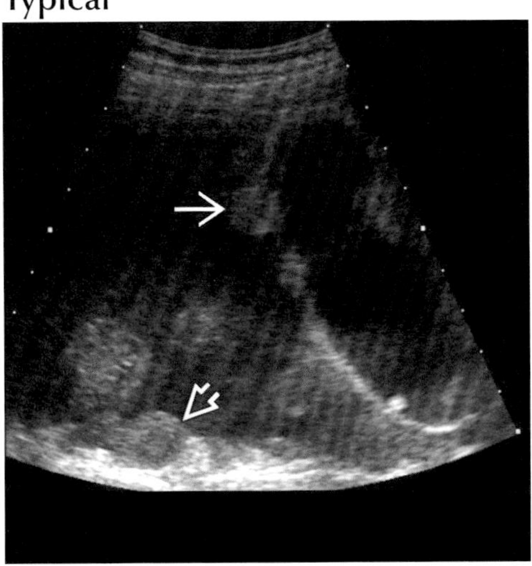

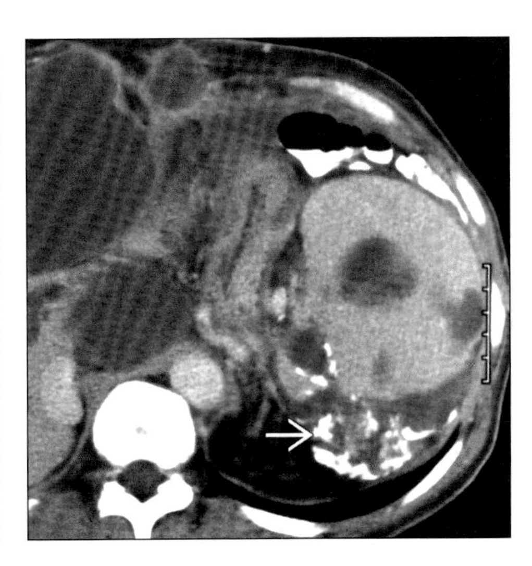

(Left) *Axial CECT demonstrates metastatic melanoma to peritoneum. Note echogenic implants on gallbladder (arrow) and peritoneum and massive ascites (open arrow).* ***(Right)*** *Axial CECT shows calcified peritoneal metastases from ovarian carcinoma. This image demonstrates calcified metastases (arrow) from serous carcinoma of the ovary adjacent to spleen.*

PSEUDOMYXOMA PERITONEI

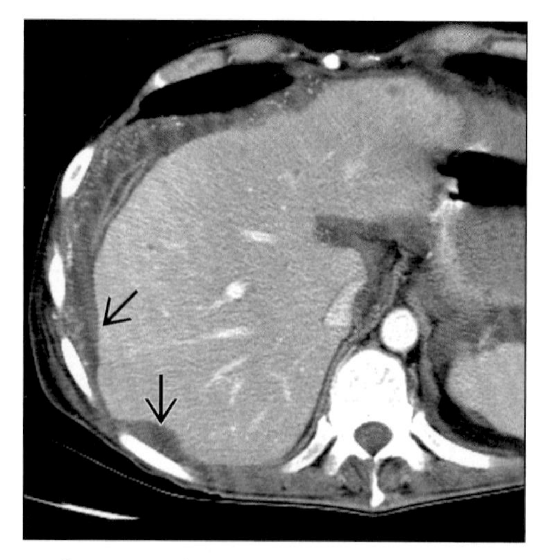

Axial CECT of pseudomyxoma peritonei. Note scalloping of liver contour (arrows).

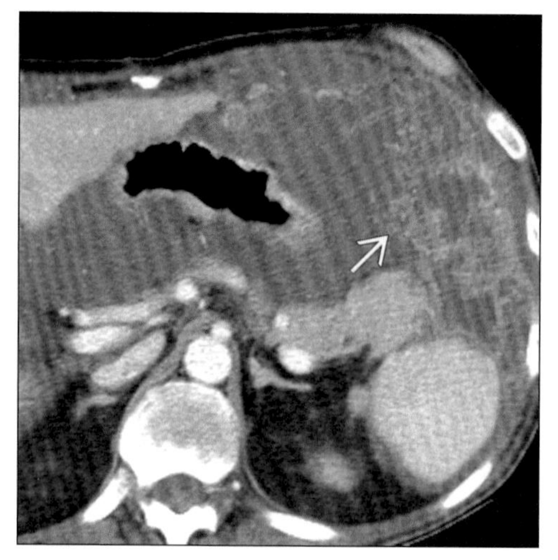

Axial CECT of pseudomyxoma peritonei. Note extensive low attenuation gelatinous masses involving lesser sac surrounding stomach with more nodular soft tissue infiltration laterally (arrow).

TERMINOLOGY

Abbreviations and Synonyms

- Pseudomyxoma peritonei (PMP)

Definitions

- Diffuse intraperitoneal accumulation of gelatinous ascites 2° to rupture of well-differentiated mucinous adeno CA of appendix
  - Rarely due to rupture of other mucinous tumors of colon, stomach, pancreas, gallbladder, fallopian tube
  - Previously, ovary was thought to be primary site, but ovarian lesions now felt to be metastatic from appendiceal primary

IMAGING FINDINGS

General Features

- Best diagnostic clue: Scalloping of contour of liver and spleen by low attenuation masses
- Location
  - Often diffuse throughout peritoneal cavity
  - Along mesenteries and ligaments
  - Extensive peritoneal involvement common
  - Subphrenic spaces
    - Perihepatic and perisplenic locations most common
- Size: Cystic implants vary in size
- Morphology: Gelatinous low attenuation masses

Radiographic Findings

- Radiography
  - Evidence for ascites
  - Lateral displacement of liver margin
  - Lateral displacement of cecum
  - Pelvic "dog's ear"
  - Displacement of bowel loops centrally within abdomen

CT Findings

- CECT
  - Low attenuation (< 20 HU) masses, mass effect on liver & spleen ("scalloping"), centrally displaced bowel loops on CECT
  - Appendiceal primary tumor, calcified mets, synchronous ovarian tumors in 44% of cases

MR Findings

- T1WI
  - Low signal intraperitoneal fluid collections
  - Cystic implants on ligaments
- T2WI
  - High signal intraperitoneal fluid collections
  - Cystic implants on ligaments
- T1 C+: Parenchymal scalloping of liver and spleen

DDx: Spectrum of Low Attenuation Peritoneal Diseases

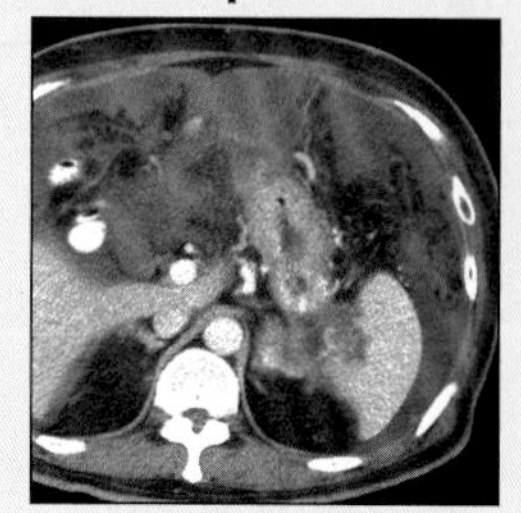

Carcinomatosis

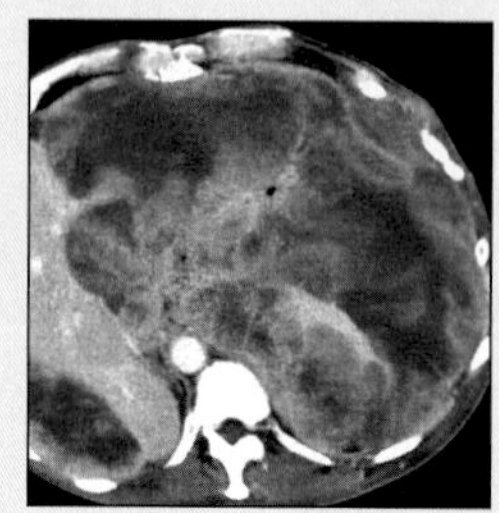

Leiomyosarcoma

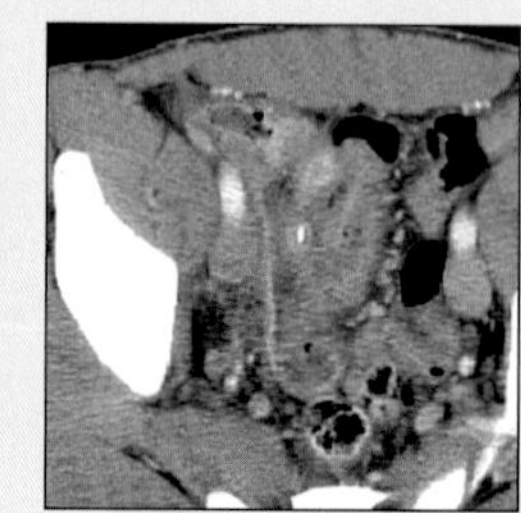

Bact. Peritonitis

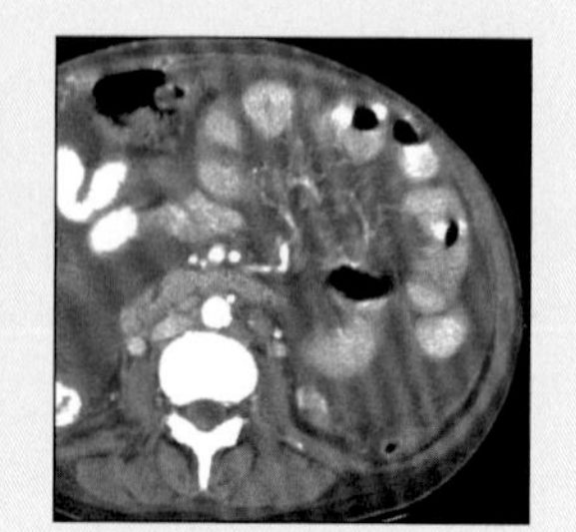

TB Peritonitis

PSEUDOMYXOMA PERITONEI

Key Facts

Terminology

- Pseudomyxoma peritonei (PMP)
- Diffuse intraperitoneal accumulation of gelatinous ascites 2° to rupture of well-differentiated mucinous adeno CA of appendix

Imaging Findings

- Best diagnostic clue: Scalloping of contour of liver and spleen by low attenuation masses
- Low attenuation (< 20 HU) masses, mass effect on liver & spleen ("scalloping"), centrally displaced bowel loops on CECT

Top Differential Diagnoses

- Carcinomatosis without mucinous ascites
- Peritoneal sarcomatosis
- Bacterial peritonitis
- TB peritonitis

Pathology

- Controversial etiology: Prior theory held that PMP 2° to rupture of mucinous tumors of appendix or ovary
- More recent theory holds that PMP is always appendiceal in origin, ovarian lesions are metastatic

Clinical Issues

- Slowly progressive, multiple bowel obstructions, 20% 5 year survival rate for very-well-differentiated adeno CA vs. 80% for well-differentiated adeno CA
- Cytoreductive surgery with extensive debulking of all intraperitoneal involvement (Sugarbaker procedure)
- Surgical treatment followed by infusion of heated intraperitoneal chemotherapy

Ultrasonographic Findings

- Real Time: Echogenic mucinous intraperitoneal masses, multiseptated peritoneal fluid on US
- Color Doppler: Cystic masses are avascular

Imaging Recommendations

- Best imaging tool: CECT
- Protocol advice
  - Oral and IV contrast CECT
    - 150 ml IV contrast injected at 2.5 ml/sec
    - 5 mm collimation with reconstruction at 5 mm intervals

DIFFERENTIAL DIAGNOSIS

Carcinomatosis without mucinous ascites

- Peritoneal mets from ovarian and GI tract primary tumors
- May occasionally cause scalloping of liver and spleen
- Omental caking
- Nodular soft tissue implants on peritoneum
- Variable amounts of ascites

Peritoneal sarcomatosis

- Solid and cystic peritoneal masses
- Hemorrhagic masses within peritoneum
- Most commonly leiomyosarcoma or GIST

Bacterial peritonitis

- Ascites and symmetrically smooth enhancing peritoneum
- Septations within ascites
- Bowel wall thickening and enhancing serosa
- "Spontaneous" peritonitis occurs in cirrhotic patients

TB peritonitis

- Nodular thickening of peritoneum
- Mesenteric and omental nodules
- Low attenuation mesenteric nodes
- Cecal and terminal ileum (TI) mural thickening

PATHOLOGY

General Features

- General path comments: Diffuse intraperitoneal mucinous ascites associated with mucinous tumors of appendix or GI tract
- Genetics: No known genetic association
- Etiology
  - Controversial etiology: Prior theory held that PMP 2° to rupture of mucinous tumors of appendix or ovary
  - More recent theory holds that PMP is always appendiceal in origin, ovarian lesions are metastatic

Gross Pathologic & Surgical Features

- Gelatinous intraperitoneal masses
- 44% have synchronous tumors of ovary (likely metastatic)
- Diffuse peritoneal involvement common
  - Most common sites of involvement
    - Right subphrenic space
    - Liver surface
    - Left subphrenic space
    - Spleen surface
    - Morison pouch
    - Left paracolic gutter
    - Pouch of Douglas or retrovesicle space
- Primary appendiceal tumor often impossible to identify

Microscopic Features

- Spectrum
- Cytologically bland adenomucinosis
- Adenocarcinoma with mucinous back grouped mixture of above findings

Staging, Grading or Classification Criteria

- Type I: Adenomucinous
  - Cytologically bland adenomatous cells and mucin
  - No frank adenocarcinoma
    - Some pathologists feel that it is not truly benign, but represents well-differentiated adenocarcinoma
  - Believed to be 2° to rupture of adenoma or mucocele

PSEUDOMYXOMA PERITONEI

  - Better prognosis
- Type II: Mucinous adenocarcinoma
  - Frank adenocarcinoma and mucin
  - Worse prognosis
- Type III: Intermediate
  - Mixture of types I and II
  - Combination of adenoma and adenocarcinoma cells with mucin

CLINICAL ISSUES

Presentation

- Most common signs/symptoms
  - Abdominal pain
  - Abdominal distension
  - Weight loss
  - New onset of hernia
- Clinical profile: Normal lab values common

Demographics

- Age: Adults, mean age 53
- Gender: M < F
- Ethnicity: No known association

Natural History & Prognosis

- Slowly progressive, multiple bowel obstructions, 20% 5 year survival rate for very-well-differentiated adeno CA vs. 80% for well-differentiated adeno CA
- Ultimately all patients die from this disease

Treatment

- Options, risks, complications
  - Cytoreductive surgery with extensive debulking of all intraperitoneal involvement (Sugarbaker procedure)
  - Surgical treatment followed by infusion of heated intraperitoneal chemotherapy

DIAGNOSTIC CHECKLIST

Consider

- Ovarian carcinomatosis

Image Interpretation Pearls

- Scalloped contour of liver and spleen by low attenuation masses
- Cystic implants on ligaments such as falciform and gastrohepatic ligament

SELECTED REFERENCES

1. van Ruth S et al: Pseudomyxoma peritonei: a review of 62 cases. Eur J Surg Oncol. 29(8):682-8, 2003
2. Moran BJ et al: The etiology, clinical presentation, and management of pseudomyxoma peritonei. Surg Oncol Clin N Am. 12(3):585-603, 2003
3. Park CM et al: Recurrent ovarian malignancy: patterns and spectrum of imaging findings. Abdom Imaging. 28(3):404-15, 2003
4. Pickhardt PJ et al: Primary neoplasms of the appendix: radiologic spectrum of disease with pathologic correlation. Radiographics. 23(3):645-62, 2003
5. Hanbidge AE et al: US of the peritoneum. Radiographics. 23(3):663-84; discussion 684-5, 2003
6. Harshen R et al: Pseudomyxoma peritonei. Clin Oncol (R Coll Radiol). 15(2):73-7, 2003
7. Lo NS et al: Mucinous cystadenocarcinoma of the appendix. The controversy persists: a review. Hepatogastroenterology. 50(50):432-7, 2003
8. van Ruth S et al: Prognostic value of baseline and serial carcinoembryonic antigen and carbohydrate antigen 19.9 measurements in patients with pseudomyxoma peritonei treated with cytoreduction and hyperthermic intraperitoneal chemotherapy. Ann Surg Oncol. 9(10):961-7, 2002
9. Rose MG et al: Typical clinical and radiographic findings in a patient with longstanding malignant pseudomyxoma peritonei secondary to a mucinous adenocarcinoma. Clin Colorectal Cancer. 2(1):59-60, 2002
10. Georgescu S et al: Mucinous digestive tumors. Case reports and review of the literature. Rom J Gastroenterol. 11(3):213-8, 2002
11. Nakao A et al: Appendiceal mucocele of mucinous cystadenocarcinoma with a cutaneous fistula. J Int Med Res. 30(4):452-6, 2002
12. Butterworth SA et al: Morbidity and mortality associated with intraperitoneal chemotherapy for Pseudomyxoma peritonei. Am J Surg. 183(5):529-32, 2002
13. Shappell HW et al: Diagnostic criteria and behavior of ovarian seromucinous (endocervical-type mucinous and mixed cell-type) tumors: atypical proliferative (borderline) tumors, intraepithelial, microinvasive, and invasive carcinomas. Am J Surg Pathol. 26(12):1529-41, 2002
14. Sulkin TV et al: CT in pseudomyxoma peritonei: a review of 17 cases. Clin Radiol. 57(7):608-13, 2002
15. Butterworth SA et al: Morbidity and mortality associated with intraperitoneal chemotherapy for Pseudomyxoma peritonei. Am J Surg. 183(5):529-32, 2002
16. Sherer DM et al: Pseudomyxoma peritonei: a review of current literature. Gynecol Obstet Invest. 51(2):73-80, 2001
17. Witkamp AJ et al: Rationale and techniques of intra-operative hyperthermic intraperitoneal chemotherapy. Cancer Treat Rev. 27(6):365-74, 2001
18. Elias DM et al: Intraperitoneal chemohyperthermia: rationale, technique, indications, and results. Surg Oncol Clin N Am. 10(4):915-33, xi, 2001
19. Yan H et al: Histopathologic analysis in 46 patients with pseudomyxoma peritonei syndrome: failure versus success with a second-look operation. Mod Pathol. 14(3):164-71, 2001
20. Esquivel J et al: Clinical presentation of the Pseudomyxoma peritonei syndrome. Br J Surg. 87(10):1414-8, 2000
21. Nawaz A et al: Pseudomyxoma peritonei manifesting as intestinal obstruction. South Med J. 93(9):891-3, 2000
22. Pestieau SR et al: Pleural extension of mucinous tumor in patients with pseudomyxoma peritonei syndrome. Ann Surg Oncol. 7(3):199-203, 2000
23. Wirtzfeld DA et al: Disseminated peritoneal adenomucinosis: a critical review. Ann Surg Oncol. 6(8):797-801, 1999
24. Hosch WP et al: Therapy of pseudomyxoma peritonei of appendiceal origin--surgical resection and intraperitoneal chemotherapy. Z Gastroenterol. 37(7):615-22, 1999
25. Hinson FL et al: Pseudomyxoma peritonei. Br J Surg. 85(10):1332-9, 1998

PSEUDOMYXOMA PERITONEI

IMAGE GALLERY

Typical

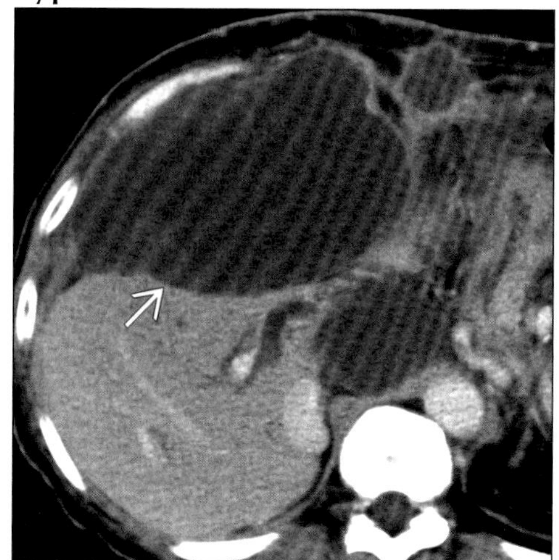

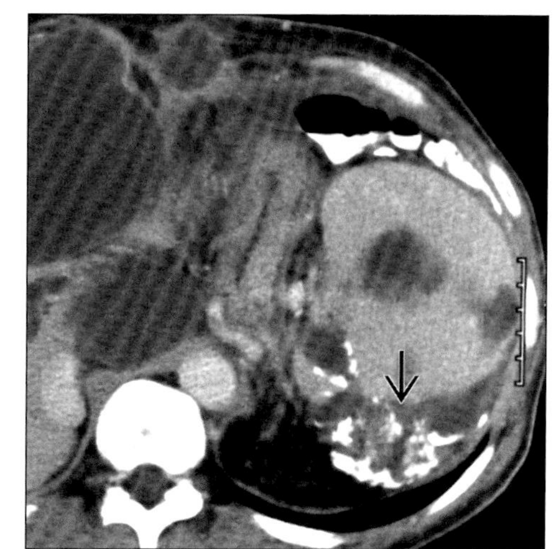

(Left) *Axial CECT shows cystic peritoneal implants from pseudomyxoma peritonei. Note large cystic mass indenting anterior aspect of liver (arrow).* ***(Right)*** *Axial CECT of cystic peritoneal implants from pseudomyxoma peritonei. Note calcified perisplenic implants (arrow).*

Typical

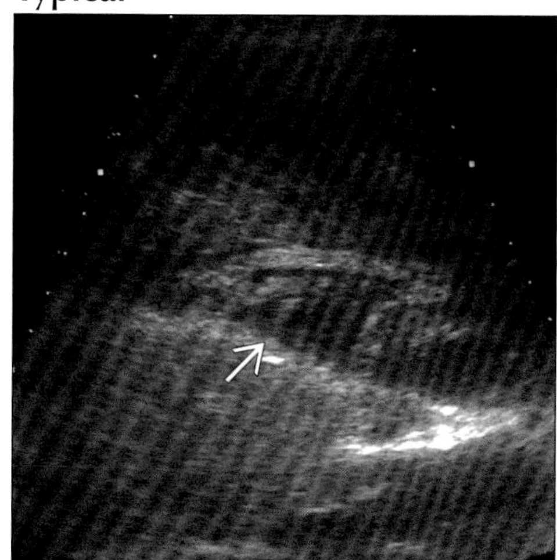

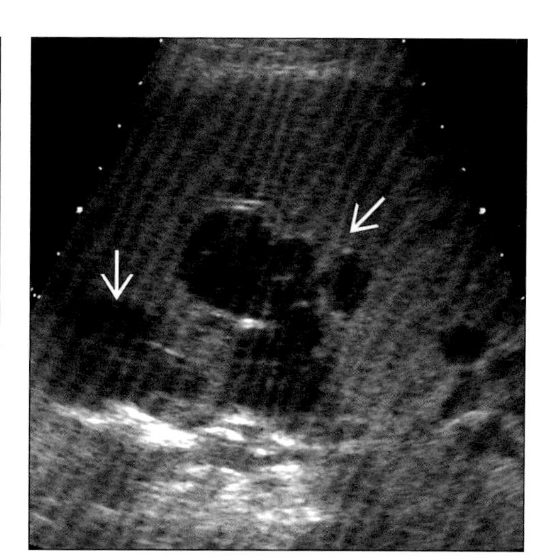

(Left) *Sagittal US findings in pseudomyxoma peritonei demonstrate echogenic perihepatic implant (arrow).* ***(Right)*** *Axial US of pseudomyxoma peritonei demonstrates cystic splenic and perisplenic implants (arrows).*

Typical

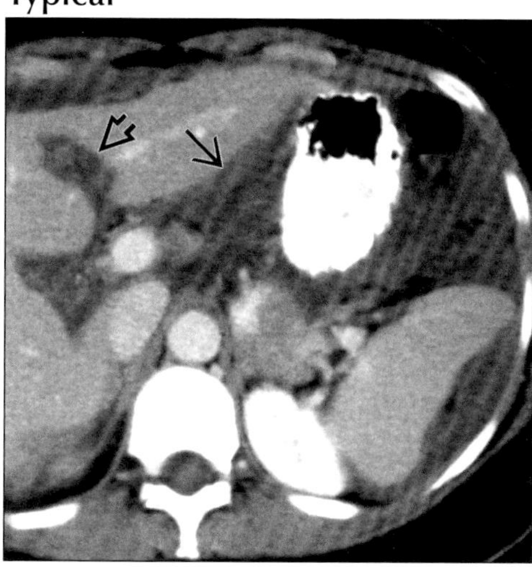

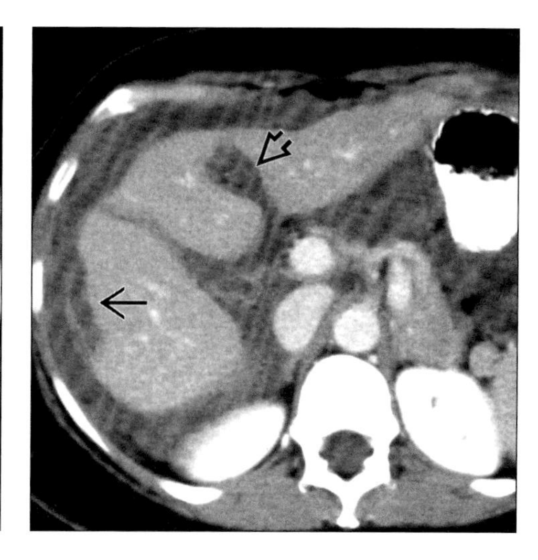

(Left) *Axial CECT demonstrates ligamentous involvement in pseudomyxoma peritonei. Note low density implants involving gastrohepatic (arrow) and falciform ligaments (open arrow).* ***(Right)*** *Axial CECT demonstrates scalloping of liver contour (arrow) and falciform ligament implant (open arrow) in pseudomyxoma peritonei.*

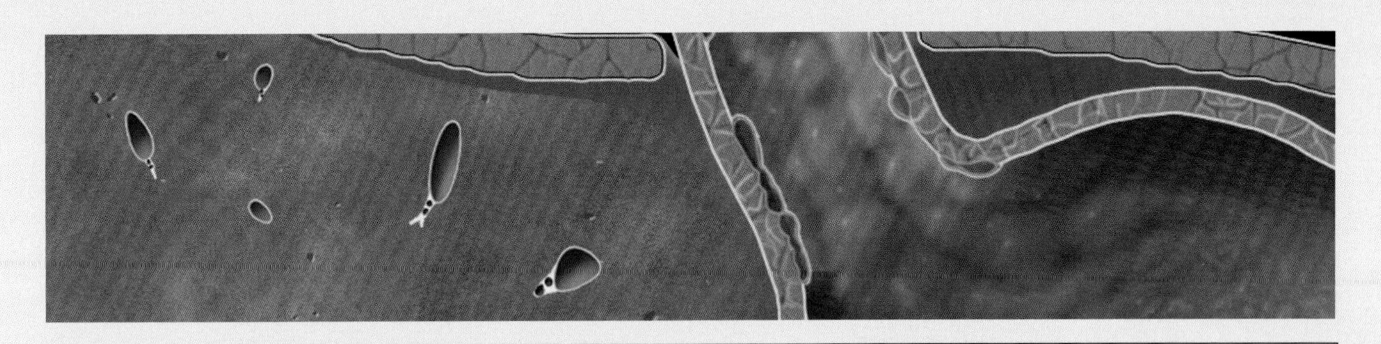

SECTION 2: Esophagus

Introduction and Overview

Congenital

Infection

Inflammation

Degenerative

Esophageal Diverticula

Trauma

Neoplasm, Benign

Neoplasm, Malignant

ESOPHAGUS ANATOMY AND IMAGING ISSUES

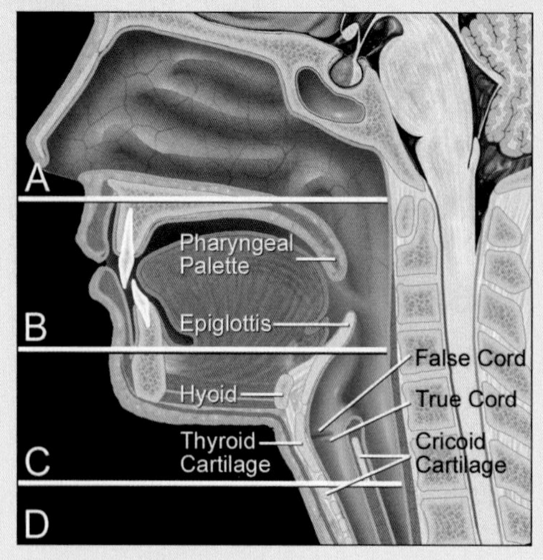

Graphic shows nasopharynx (base of skull to palate); oropharynx (palate to base of epiglottis); hypopharynx (epiglottis to cricopharyngeus muscle); esophagus (below cricopharyngeus).

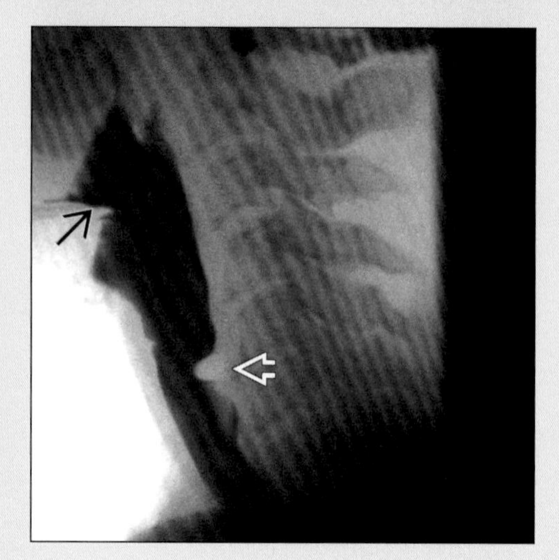

Lateral film from pharyngo-esophagram shows vallecula (arrow) and cricopharyngeal indentation (open arrow), usually located at the C5-C6 disk space.

TERMINOLOGY

Abbreviations and Synonyms

- Upper esophageal sphincter (UES)
  - Formed primarily by cricopharyngeal muscle
- Lower esophageal sphincter (LES)
  - Phrenic ampulla, esophageal vestibule
- "A" ring
  - Sporadically imaged indentation of esophageal lumen at the cephalic end of the lower esophageal sphincter (the "tubulo-vestibular junction")
- "B" ring
  - Transverse mucosal fold marking the esophagogastric junction and often corresponding to the mucosal junction between squamous + columnar epithelium
- GE
  - Gastroesophageal

IMAGING ANATOMY

- Pharynx
  - Nasopharynx
    - From base of skull to tip of soft palate
  - Oro (mesopharynx)
    - From soft palate to hyoid bone
  - Hypo (laryngopharynx)
    - From hyoid to bottom of cricopharyngeus muscle
- Upper esophageal sphincter
  - At pharyngoesophageal junction
  - Formed primarily by cricopharyngeus muscle
- Esophagus
  - Muscular tube 20-24 cm in length
  - Lined by stratified squamous epithelium
  - Outer longitudinal and inner circular muscle fibers
    - Striated muscle in upper third
    - Smooth muscle in distal two thirds
- Lower esophageal sphincter
  - Defined by manometric evidence of high resting tone or pressure
  - Essentially synonymous with "esophageal vestibule" or "phrenic ampulla"
  - Occasionally recognized radiographically as a 2-4 cm long luminal dilation between the esophageal "A" and "B" rings

ANATOMY-BASED IMAGING ISSUES

Normal Measurements

- Swallowing
  - Mastication and tongue motion propels food bolus into oropharynx
  - Sequential contraction of constrictor muscles of pharynx
  - Upper esophageal sphincter (including cricopharyngeus) opens to allow free passage of bolus
    - Failure of UES to open completely ("cricopharyngeal achalasia" or "dyskinesia") is a common cause of dysphagia and "food sticking in throat"
  - Cricopharyngeal dyskinesia often accompanies other malfunction of the pharynx and esophagus
- Primary peristalsis
  - Initiated by swallowing
  - Normal is a continuous aboral esophageal contraction wave, lasting 6-8 seconds, that propels the bolus to the stomach
  - Best evaluated by individual swallows of barium with patient in the prone oblique position
- Secondary peristalsis
  - Similar aboral contraction wave, but initiated by esophageal distention or gastric reflux, rather than by swallowing
- Tertiary peristalsis
  - Nonperistaltic, disorganized contractions
  - May be intermittent, weak, asymptomatic
  - May be persistent, repetitive, strong and produce dysphagia or pain

ESOPHAGUS ANATOMY AND IMAGING ISSUES

DIFFERENTIAL DIAGNOSIS

Benign masses
- Leiomyoma
- Squamous papilloma
- Glycogenic acanthosis
- Adenoma
- Fibrovascular polyps
- Granular cell tumor
- (Duplication cyst)

Malignant tumors
- Squamous cell carcinoma
- Adenocarcinoma
- Spindle cell carcinoma
- Small cell carcinoma
- Leiomyosarcoma
- Kaposi sarcoma
- Malignant melanoma (primary)
- Lymphoma
- Metastases

Mass lesions (common)
- Carcinoma
- Leiomyoma (+ other stromal)

Mass lesions (uncommon)
- Papilloma
- Adenoma
- Metastases/lymphoma
- Fibrovascular polyp
- Duplication cyst
- Inflammatory GE polyp

  - Increase in frequency with aging ("presbyesophagus")

Key Concepts or Questions
- Symptoms of "food sticking in back of throat", choking or concern for aspiration pneumonitis
  - Modified barium swallow is technique of choice
  - Videotaped monitoring while patient swallows barium suspensions varying in consistency from water to solid
- Symptoms of dysphagia, odynophagia,atypical chest pain, "heartburn", concern for GE reflux
  - Barium esophagram is techniques of choice
  - Double contrast views in upright position following ingestion of an effervescent agent and thick barium
  - Single contrast views in upright and prone position
  - Individual swallows to assess peristalsis
  - Full distention views with continuos swallowing to assess morphology (tumor, stricture, ulcer, etc.)
  - Videotaped swallowing of food or radiopaque tablet to assess stricture ("food sticking" symptoms)
- Symptoms of dyspepsia, early satiety, abdominal pain
  - Upper gastrointestinal (UGI) series is technique of choice
  - Double and single contrast views of stomach and duodenum before and after esophageal evaluation

Imaging Approaches
- Role of fluoroscopic imaging
  - Complementary to manometry for most causes of dysphagia
  - Complementary to pH monitoring for gastroesophageal reflux
  - Complementary to endoscopy for esophagitis and tumors
  - Complementary to CT (+ MR) for primary esophageal and mediastinal tumors
- CT (or MR)
  - Complementary to endoscopic ultrasound in staging esophageal cancer
- PET-CT
  - Best study for staging esophageal cancer
- Endoscopic sonography
  - Best study for depth of invasion by esophageal carcinoma

Imaging Pitfalls
- Accuracy of fluoroscopic evaluation of the pharynx and esophagus varies widely
- Depends mostly on the interest, skill, and thoroughness of the radiologist performing the exam

CLINICAL IMPLICATIONS

Clinical Importance
- Aspiration pneumonitis is an important cause of morbidity + mortality in hospitalized and debilitated patients
- "Modified barium swallow" is the most accurate means of evaluating patient's ability to safely ingest foods of varying consistency
- Opportunistic infections of the esophagus occur in immune-suppressed patients, though less commonly with newer pharmaceutical treatment protocols
- Gastroesophageal reflux is extremely common and has contributed to a dramatic increase in the prevalence of esophageal cancer

CUSTOM DIFFERENTIAL DIAGNOSIS

Submucosal mass lesions
- Leiomyoma
  - Smooth intramural mass
- Fibrovascular polyp
  - Large pedunculated mass
  - Arises in wall, but grows into lumen
  - Arises in cervical esophagus
- Lipoma
  - Fat density on CT
- Duplication or retention cyst
  - Near water density, no enhancement

Extrinsic impressions on esophagus
- Normal
  - Aortic arch

ESOPHAGUS ANATOMY AND IMAGING ISSUES

"A" ring
mucosal junction (z-line)
vestibule (lower esoph. sphincter)
"B" ring

Graphic shows normal esophageal landmarks and anatomy.

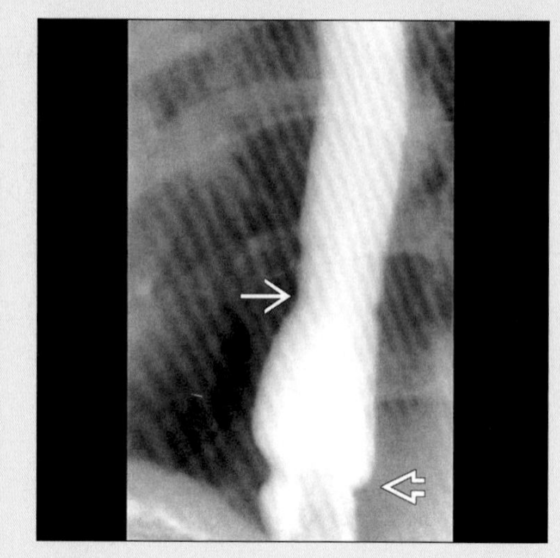

Barium esophagram shows "A" ring (arrow), "B" ring (open arrow), esophageal vestibule (between "A & B" rings) and a small hiatal hernia.

- Left main bronchus
- Heart
- Abnormal
  - Enlarged left atrium or ventricle
  - Tortuous aorta (distal esophagus)
  - Congenital vascular (e.g., aberrant subclavian artery)
  - Thyroid masses
  - Mediastinal masses
  - (Retraction to opposite side)

Esophageal ulceration

- Common
  - Reflux esophagitis
  - Candida esophagitis
  - Herpes esophagitis
  - Drug-induced esophagitis
- Uncommon
  - Radiation esophagitis
  - Caustic esophagitis
  - Tuberculous esophagitis
  - Cytomegalovirus esophagitis
  - HIV esophagitis
  - Crohn disease
  - Nasogastric intubation
  - Alkaline reflux esophagitis
  - Behcet disease
  - Epidermolysis bullosa dystrophica
  - Benign mucous membrane pemphigoid

Mucosal nodularity

- Common
  - Reflux esophagitis
  - Candida esophagitis
  - Glycogenic acanthosis
- Uncommon
  - Barrett esophagus
  - Radiation esophagus
  - Superficial spreading carcinoma
  - Esophageal papillomatosis
  - Acanthosis nigricans
  - Cowden disease
  - Leukoplakia

Esophageal strictures

- Distal esophagus
  - Peptic strictures
  - Barrett esophagus
  - Carcinoma (usually adenocarcinoma)
  - Nasogastric intubation
  - Crohn disease
  - Alkaline reflux strictures
- Midesophagus
  - Barrett esophagus
  - Carcinoma (usually squamous cell carcinoma)
  - Radiation
  - Caustic ingestion
  - Oral medications
  - Opportunistic infection (usually candidiasis)
  - Epidermolysis bullosa dystrophica
  - Benign mucous membrane pemphigoid
  - Eosinophilic esophagitis
  - Chronic graft-versus-host disease
  - Metastatic tumor

SELECTED REFERENCES

1. Dibble C et al: Detection of reflux esophagitis on double-contrast esophagrams and endoscopy using the histologic findings as the gold standard. Abdom Imaging. 29(4):421-5, 2004
2. Gupta S et al: Usefulness of barium studies for differentiating benign and malignant strictures of the esophagus. AJR Am J Roentgenol. 180(3):737-44, 2003
3. Dähnert W: Radiology Review Manual (4th ed). Philadelphia, Lippencott Williams + Wilkins. p615-721, 2000
4. Levine MS: Gastroesophageal reflux disease. In Gore RM, Levine MS (eds) Textbook of Gastrointestinal Radiology (2nd ed.) Philadelphia, WB Saunders. 329-49, 2000
5. Levine MS et al: Update on esophageal radiology. AJR Am J Roentgenol. 155(5):933-41, 1990
6. Ott DJ et al: Esophagogastric region and its rings. AJR Am J Roentgenol. 142(2):281-7, 1984

IMAGE GALLERY

Typical

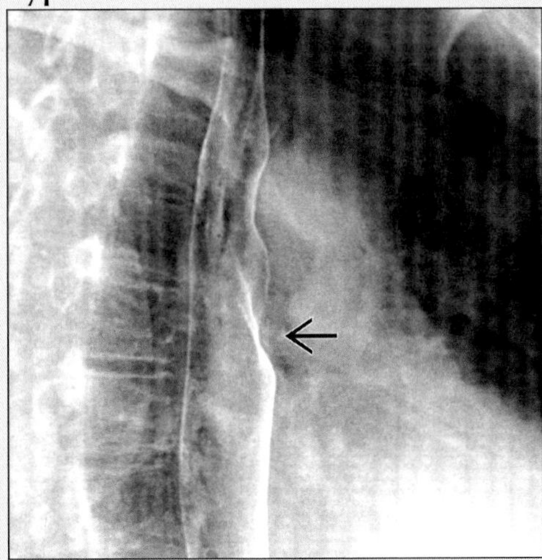

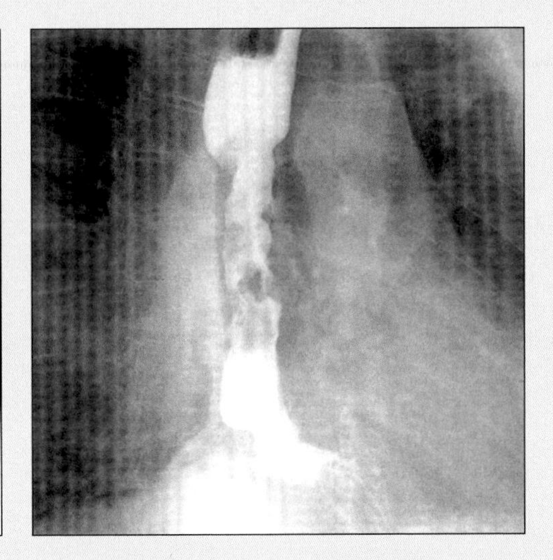

(Left) *Barium esophagram shows normal extrinsic indentations from aortic arch and left main bronchus (arrow).* ***(Right)*** *Esophagram shows typical malignant stricture (adenocarcinoma) with irregular mucosal surface, abrupt "shoulder" at proximal end of stricture.*

Typical

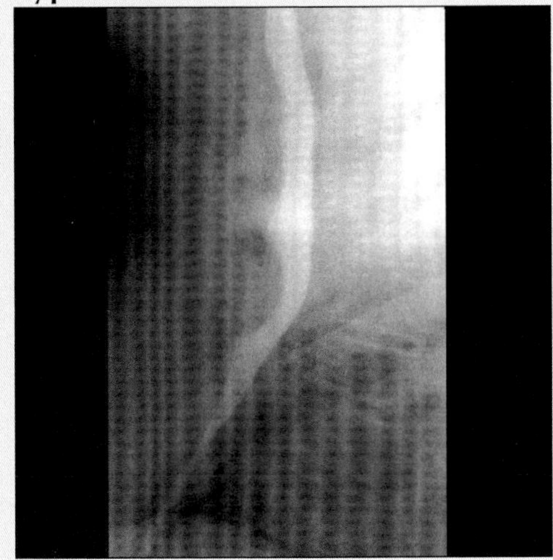

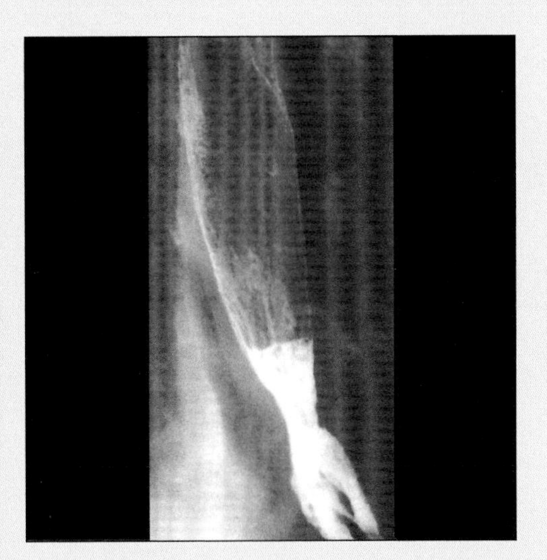

(Left) *Esophagram shows typical intramural or extrinsic mass effect; leiomyoma.* ***(Right)*** *Esophagram shows typical benign distal stricture with smooth taper and nodular mucosal surface, both due to reflux esophagitis.*

Typical

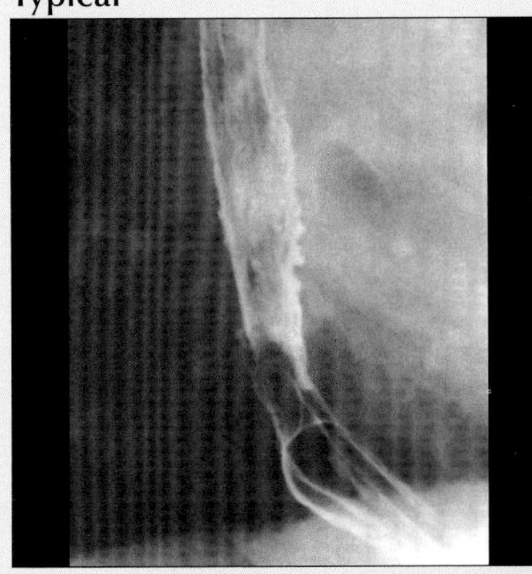

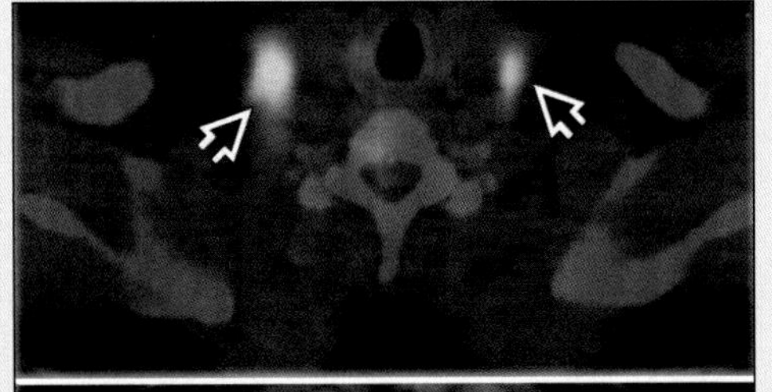

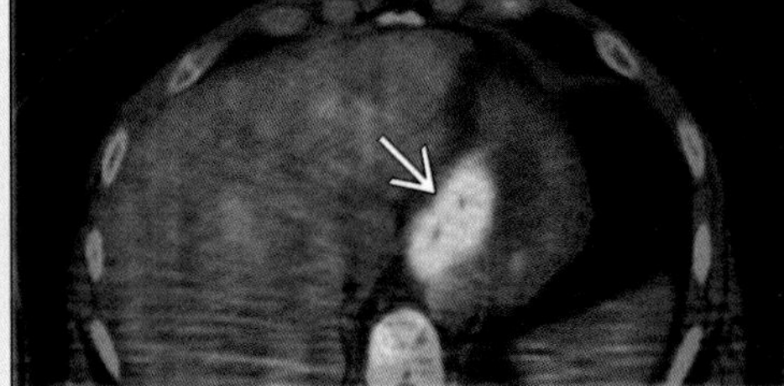

(Left) *Esophagram shows extensive ulcerations throughout distal esophagus due to reflux esophagitis.* ***(Right)*** *Fused PET-CT (axial) images show FDG-avid lesion at gastroesophageal junction (arrow) and multiple lesions in the neck (open arrows), representing metastatic esophageal adenocarcinoma.*

ESOPHAGEAL WEBS

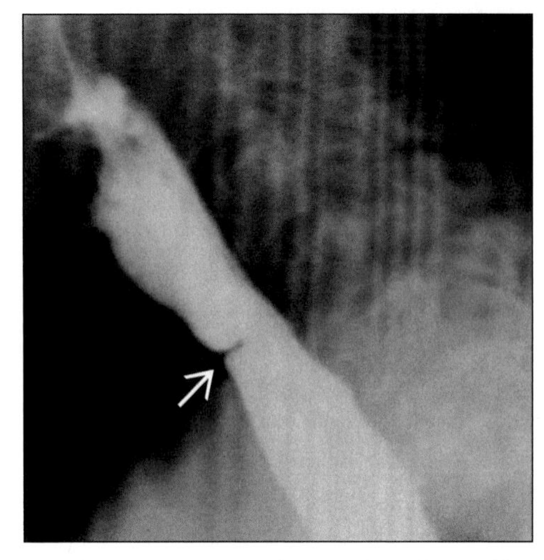

Lateral view of barium esophagram shows thin shelf-like indentation of the esophagus (arrow).

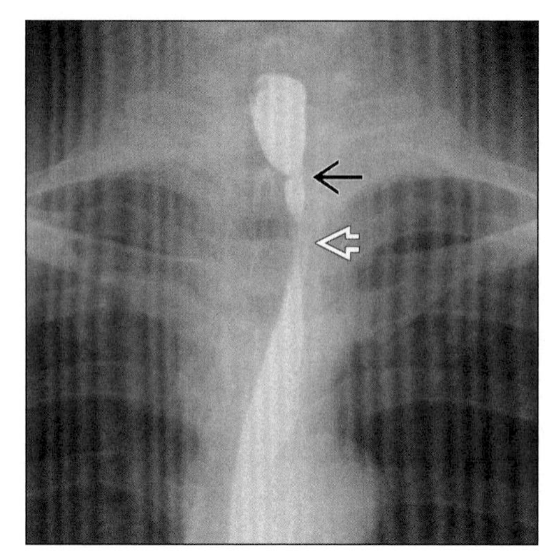

Frontal view of esophagram in a patient with Plummer-Vinson syndrome who had glossitis, iron deficiency anemia and dysphagia. Esophageal web (arrow) and stricture (open arrow) are noted.

TERMINOLOGY

Definitions

- Thin ring-like mucosal constriction projecting into lumen most frequently from anterior wall of proximal cervical esophagus

IMAGING FINDINGS

General Features

- Best diagnostic clue: A circumferential, radiolucent ring in proximal cervical esophagus on dynamic esophagogram
- Location
  - Cervical esophagus near cricopharyngeus (common)
  - Distal esophagus (uncommon)
- Morphology
  - Thin band of mucosa with or without submucosa
  - Usually composed of normal squamous epithelium & lamina propria
- Other general features
  - Congenital/normal variant/sequela of inflammation
  - Vary from small shelf-like lesion to hemispheric bar or circumferential ring
  - Seen as isolated findings in 3-8% of patients undergoing upper gastrointestinal examinations

Radiographic Findings

- Videofluoroscopic rapid sequence esophagram
  - Frontal, lateral & oblique views
    - 1-2 mm wide, shelf-like filling defect along anterior wall of cervical esophagus
    - Circumferential, radiolucent ring
    - Mild/moderate/severe luminal narrowing
  - Partial obstruction: Suggested by
    - Jet phenomenon: Barium spurting through ring
    - Dilatation of esophagus proximal to web

Imaging Recommendations

- Barium studies (including both frontal & lateral views)
- Dynamic examination reveals increase in percentage of webs than does a spot film alone
- Better demonstration achieved by large barium bolus
- Better visualized during maximal distention

DIFFERENTIAL DIAGNOSIS

Esophageal stricture

- Circumferential (> 2 mm) or vertical area of complete/incomplete narrowing with length of > 2 cm

Schatzki ring

- Lower (GE junction) esophageal mucosal or B ring
- Web-like constriction or shelf-like luminal projection

DDx: Short Segment Narrowing of Esophageal Lumen

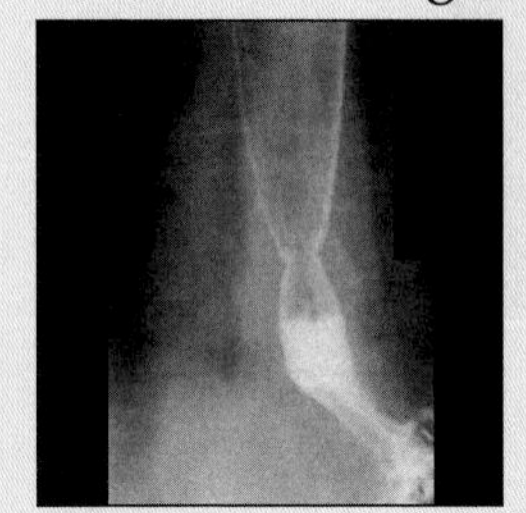

Peptic Stricture

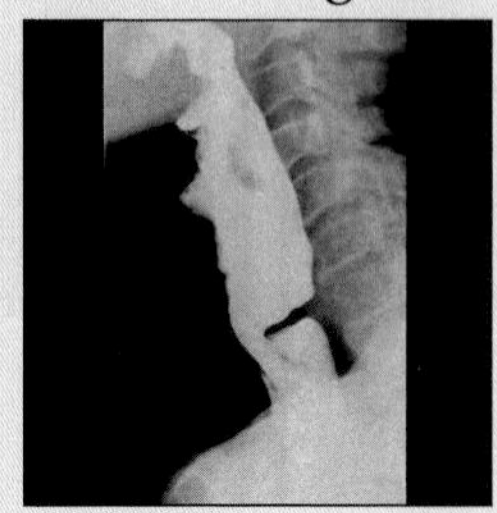

Crico. Achalasia

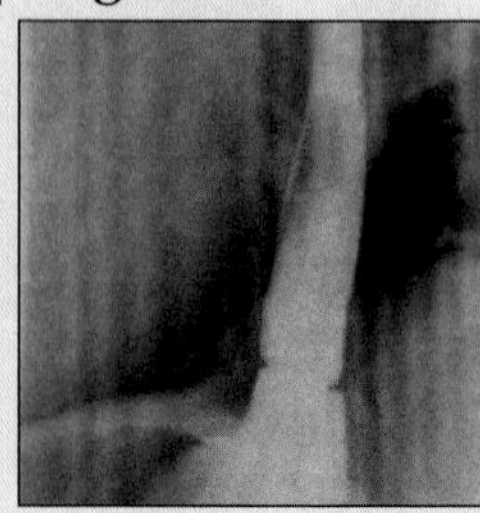

Schatzki Ring

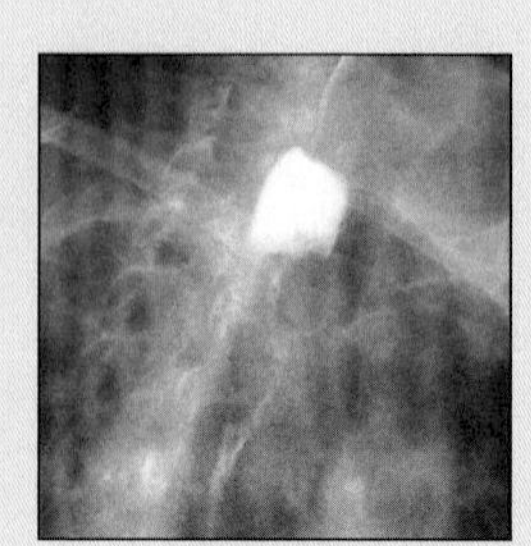

Eso. Cancer

ESOPHAGEAL WEBS

Key Facts

Terminology
- Thin ring-like mucosal constriction projecting into lumen most frequently from anterior wall of proximal cervical esophagus

Imaging Findings
- Best diagnostic clue: A circumferential, radiolucent ring in proximal cervical esophagus on dynamic esophagogram
- Jet phenomenon: Barium spurting through ring

Top Differential Diagnoses
- Esophageal stricture
- Schatzki ring
- Cricopharyngeal achalasia

Diagnostic Checklist
- Rapid sequence imaging during swallowing of a large bolus of barium

- Fixed, anatomic, nondistensible, transverse ring
- Symptomatic & acquired due to reflux esophagitis
- May be indistinguishable from distal esophageal web

Cricopharyngeal achalasia
- Appears as a round, broad-based protrusion from posterior pharyngeal wall at the level of pharyngoesophageal segment (C5-6 level)

PATHOLOGY

General Features
- Etiology
  - Cervical esophageal webs may be
    - Congenital
    - Idiopathic (normal variant)
    - Due to sequela of inflammation & scarring such as epidermolysis bullosa (postcricoid) or benign mucous membrane pemphigoid
  - Distal esophageal webs
    - Sequela of inflammation & scarring: GE reflux
- Epidemiology: One autopsy series showed 16% of patients had incidental cervical esophageal webs
- Associated abnormalities
  - Cervical esophageal webs may be associated with
    - Plummer-Vinson (Paterson-Kelly) syndrome
    - Glossitis, iron deficiency anemia, dysphagia (webs +/- strictures)
    - Rarely associated with GE reflux
  - Distal esophageal webs are associated with
    - Chronic gastroesophageal reflux

Gross Pathologic & Surgical Features
- Thin mucosal band projecting into esophageal lumen

Microscopic Features
- Normal variant: Squamous epithelium, lamina propria
- Inflammatory web: Inflammatory cells; scar tissue

CLINICAL ISSUES

Presentation
- Most common signs/symptoms
  - Mostly asymptomatic
  - Dysphagia: With impaction of meat above web
  - Odynophagia: Very rarely painful swallowing
- Plummer-Vinson syndrome findings
  - Esophageal webs, dysphagia
  - Iron deficiency anemia, stomatitis & glossitis
  - Spoon-shaped nails (koilonychia); ± thyroid disorder

Demographics
- Age: Usually middle-age group people
- Gender: More common in females than males

Natural History & Prognosis
- Complications
  - Cervical esophageal web
    - Cervical esophageal & hypopharyngeal carcinoma
- Prognosis
  - Usually good

Treatment
- Balloon dilatation; bougienage during esophagoscopy

DIAGNOSTIC CHECKLIST

Consider
- Rapid sequence imaging during swallowing of a large bolus of barium

SELECTED REFERENCES

1. Taylor AJ et al: The esophageal jet phenomenon revisited. AJR 155: 289-90, 1990
2. Nosher JL et al: The clinical significance of cervical esophageal and hypopharyngeal webs. Radiology 117: 45-7, 1975
3. Clements JL et al: Cervical esophageal webs: A roentgen-anatomic correlation. AJR 121: 221-31, 1974

IMAGE GALLERY

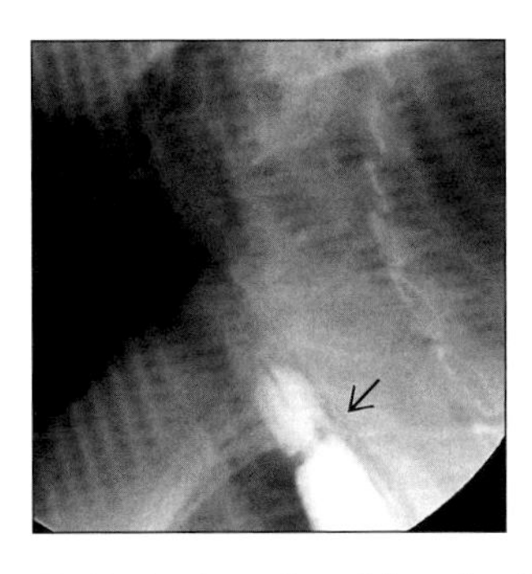

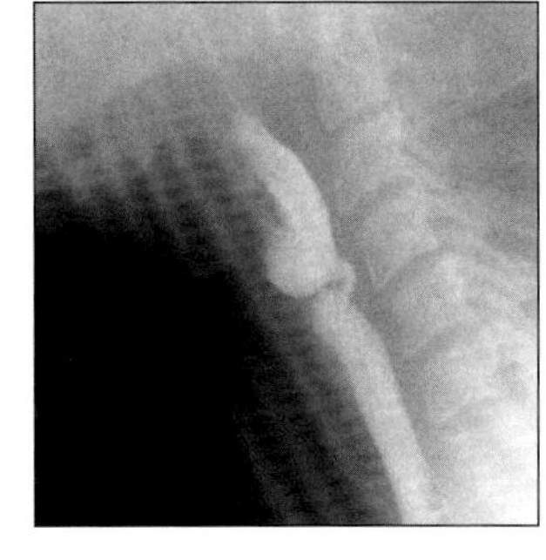

(Left) *A circumferential and relatively thick web (arrow) causes significant luminal narrowing.* ***(Right)*** *Lateral esophagram shows tight web-like narrowing at pharyngo-esophageal junction due to epidermolysis bullosa.*

CANDIDA ESOPHAGITIS

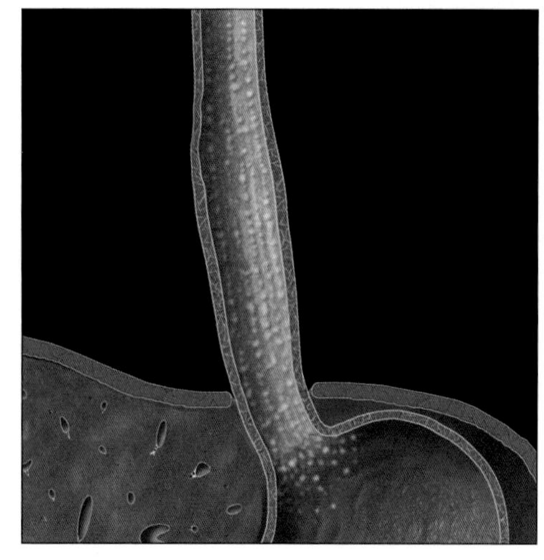

Illustration shows longitudinally-oriented mucosal plaques characteristic of Candida esophagitis.

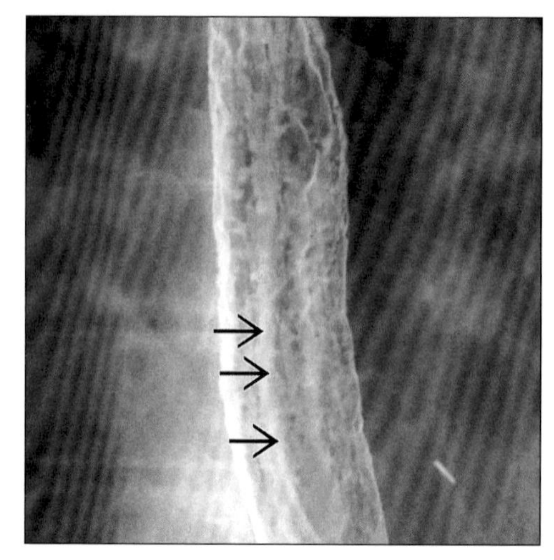

Double-contrast barium esophagram shows longitudinally-oriented filling defects representing Candida plaques (arrows).

TERMINOLOGY

Abbreviations and Synonyms

- Esophageal Candidiasis, moniliasis

Definitions

- Infectious esophagitis caused by fungi of Candida species, usually Candida albicans

IMAGING FINDINGS

General Features

- Best diagnostic clue: Mucosal plaques in immune-compromised patients
- Location
  - Predominantly mid or upper esophagus
  - Often spares distal esophagus
- Size: Plaques: Several mm in size; usually < 1 cm
- Morphology: Longitudinally oriented plaques with intervening normal mucosa

Radiographic Findings

- Fluoroscopy
  - Double-contrast esophagram: Several patterns
    - Discrete plaques in mid to upper esophagus with longitudinal orientation and intervening normal mucosa
    - Nodular or granular mucosa
    - Cobblestone or snakeskin appearance with confluent plaques
    - Severe: "Shaggy" esophagus; usually AIDS patients
    - Severe cases: Deep ulcers
    - "Foamy" esophagus
    - Discrete aphthoid ulcers that can mimic viral esophagitis
    - Nonspecific: Earliest finding is abnormal motility (tertiary contractions, dilatation or atony)
    - Nonspecific: Thickened longitudinal folds due to submucosal edema
    - Chronic: May lead to strictures (rare)
    - Rare: "Double-barreled esophagus" = intramural tracts
  - Single-contrast esophagram: Unreliable

CT Findings

- CECT
  - Uniform circumferential wall thickening (> 5 mm)
  - Usually involves relatively long segment
  - "Target" sign
    - Edematous wall shows hypodense submucosa
    - Inflamed mucosal segments show avid enhancement

Imaging Recommendations

- Best imaging tool: Double-contrast esophagram

DDx: Mimics of Candidiasis

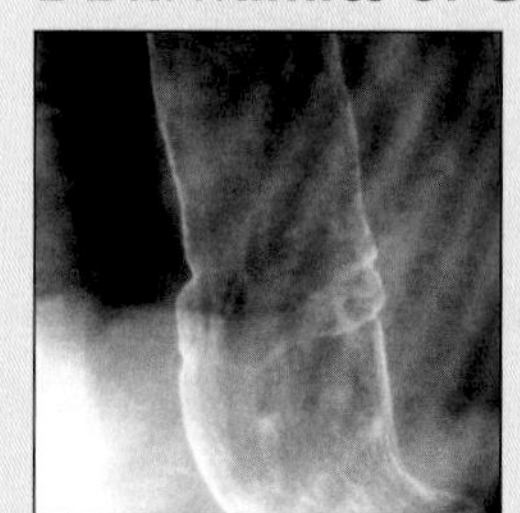

Reflux Ulcer

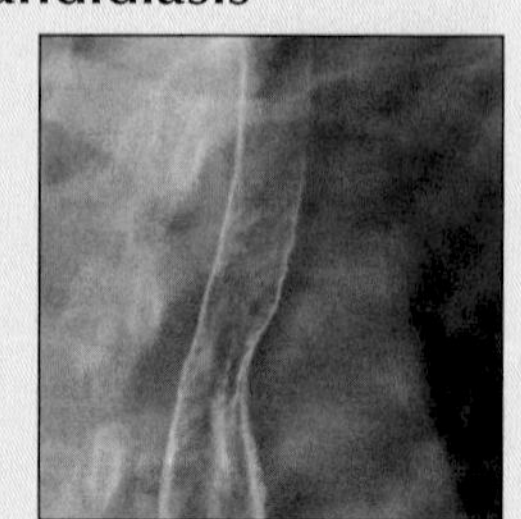

Herpes Ulcers

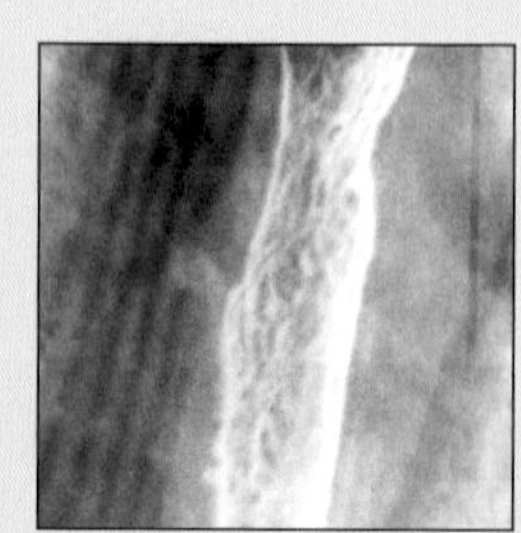

Superficial Ca.

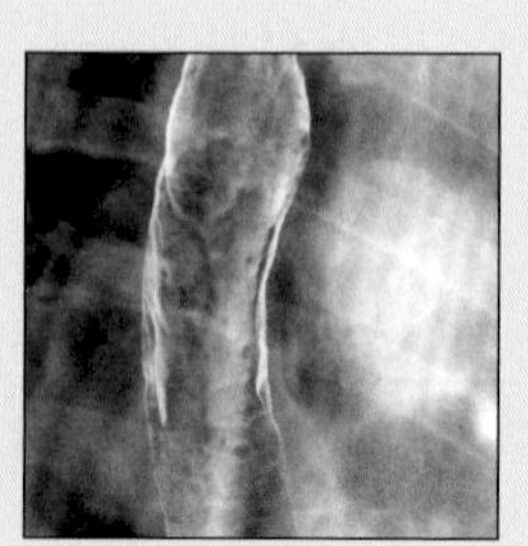

Glyc. Acanthosis

CANDIDA ESOPHAGITIS

Key Facts

Terminology
- Infectious esophagitis caused by fungi of Candida species, usually Candida albicans

Imaging Findings
- Best diagnostic clue: Mucosal plaques in immune-compromised patients
- Predominantly mid or upper esophagus
- Often spares distal esophagus
- Discrete plaques in mid to upper esophagus with longitudinal orientation and intervening normal mucosa
- Cobblestone or snakeskin appearance with confluent plaques

Top Differential Diagnoses
- Reflux esophagitis
- Viral esophagitis
- Superficial spreading carcinoma
- Drug-induced esophagitis

Pathology
- Most common cause of infectious esophagitis

Clinical Issues
- Immunocompromised patients
- Patients with physiologic or mechanical obstruction of esophagus and local stasis

Diagnostic Checklist
- Odynophagia with plaques in immunocompromised patient should suggest Candidiasis

- Protocol advice: Use rapid swallows instead of effervescent granules to achieve gaseous distention without artifacts

DIFFERENTIAL DIAGNOSIS

Reflux esophagitis
- Usually extends proximally from gastroesophageal junction
- Nodular or granular mucosa but with poorly defined borders
- Ulcers, strictures in distal esophagus
- Often associated with hiatal hernia

Viral esophagitis
- Herpes esophagitis
  - Usually multiple small discrete ulcers
  - Advanced disease may have larger plaque-like lesions indistinguishable from Candida
- CMV esophagitis
  - May have multiple small superficial ulcers
  - Giant (> 1 cm), flat ulcers in mid-distal esophagus more characteristic
  - AIDS patients

Superficial spreading carcinoma
- Confluent nodules without intervening areas of normal mucosa

Drug-induced esophagitis
- Tetracycline, KCl, quinidine, NSAIDs
- Focal contact esophagitis
  - Single or multiple superficial ulcers
  - Usually in mid-esophagus near aortic arch or left mainstem bronchus

Glycogenic acanthosis
- Asymptomatic; probably age-related benign degenerative condition
- Mucosal plaques or nodules more rounded and less well defined than Candidiasis

Technical artifacts
- Undissolved granules, air bubbles, debris
- Transient

Varices
- Serpiginous, longitudinal filling defects
- Best seen on mucosal relief views
- Appearance varies depending on respiration, intraluminal pressure

PATHOLOGY

General Features
- General path comments
  - Most common cause of infectious esophagitis
  - Candida species are normal oropharyngeal commensals
    - Can have esophageal colonization without infection
    - Mucosal biopsy, not brushings, necessary for definitive diagnosis
- Etiology
  - Probably caused by downward spread of Candida albicans to esophagus
  - Candida surface binding molecules → adhere to mucosa
    - Protease secretion → tissue breakdown
    - Adenosine secretion → blockage of neutrophil oxygen radical production → resistance to phagocytosis
- Epidemiology: Occurs in 15-20% of patients with AIDS
- Associated abnormalities
  - May coexist with herpes or CMV esophagitis
  - Oral thrush
    - Oral Candidiasis + esophageal symptoms ⇒ 71-100% positive predictive value for Candidal esophagitis
    - Absence of oral thrush does not rule out esophageal Candidiasis
    - Oral thrush may be present with esophagitis of viral etiology

CANDIDA ESOPHAGITIS

- Esophageal intramural pseudodiverticulosis
  - Tiny outpouchings representing dilated excretory ducts of esophageal submucosal glands
  - Usually seen with strictures
  - Candidiasis likely represents superinfection secondary to stasis in pseudodiverticula
- Chronic mucocutaneous candidiasis
  - Defect in cell-mediated immunity
  - Chronic fungal infection of skin, mucous membranes and nails
  - Esophageal strictures in this setting should suggest Candidal involvement

Gross Pathologic & Surgical Features

- Patchy, creamy white small (< 1 cm) plaques on friable erythematous mucosa
  - Plaques represent necrotic epithelial debris and/or Candida colonies
- Advanced cases
  - Ulcerated, necrotic mucosa with pseudomembranes
  - Shaggy or cobblestone mucosa
  - Narrowing

Microscopic Features

- Tissue invasion by fungal mycelia seen on endoscopic mucosal biopsy
- Budding yeast cells, hyphae and pseudohyphae on silver stain, PAS stain or Gram stain

CLINICAL ISSUES

Presentation

- Most common signs/symptoms
  - Odynophagia
  - Other signs/symptoms
    - Dysphagia
    - GI bleeding
  - Oral thrush in 50-75%, but may be absent
- Clinical profile
  - Immunocompromised patients
    - AIDS, transplant patients, diabetes, steroid use
  - Patients with physiologic or mechanical obstruction of esophagus and local stasis
    - Scleroderma
    - Achalasia
    - Esophageal strictures
  - Can occasionally occur in immunocompetent patients

Demographics

- Age: All
- Gender: Both

Natural History & Prognosis

- Usually self-limited with rapid response to oral therapy
- Radiographic findings may lag behind clinical response
- Chronic cases may develop strictures
- Severe cases: Hematogenously disseminated fungal infection
- Unusual complications
  - Esophageal perforation
  - Aortoesophageal fistula
  - Tracheoesophageal fistula
  - Lung abscess

Treatment

- Options, risks, complications
  - Oral antifungals: Ketoconazole or fluconazole
  - Amphotericin-B for treatment failures or recurrences

DIAGNOSTIC CHECKLIST

Consider

- Odynophagia with plaques in immunocompromised patient should suggest Candidiasis

Image Interpretation Pearls

- Plaques are raised mucosal lesions; if a central contrast collection is present ⇒ lesion is an ulcer

SELECTED REFERENCES

1. Luedtke P et al: Radiologic diagnosis of benign esophageal strictures: a pattern approach. Radiographics. 23(4): 897-909, 2003
2. Berkovich GY et al: CT findings in patients with esophagitis. AJR Am J Roentgenol. 175(5):1431-4, 2000
3. Gore R et al: Textbook of Gastrointestinal Radiology. 2nd ed. Philadelphia, W.B. Saunders, 1980-1992, 2000
4. Wilcox CM et al: Esophageal infections: etiology, diagnosis, and management. Gastroenterologist. 2(3): 188-206, 1994
5. Laine L et al: Esophageal disease in human immunodeficiency virus infection. Arch Intern Med. 154(14): 1577-82, 1994
6. Levine MS: Radiology of esophagitis: a pattern approach. Radiology. 179(1): 1-7, 1991
7. Levine MS et al: Update on esophageal radiology. AJR Am J Roentgenol. 155(5): 933-41, 1990
8. Levine MS: Radiology of the esophagus. Philadelphia, W.B. Saunders, 49-60, 1989
9. Raufman JP: Odynophagia/dysphagia in AIDS. Gastroenterol Clin North Am. 17(3): 599-614, 1988
10. Mathieson R et al: Candida esophagitis. Dig Dis Sci. 28(4): 365-70, 1983

CANDIDA ESOPHAGITIS

IMAGE GALLERY

Typical

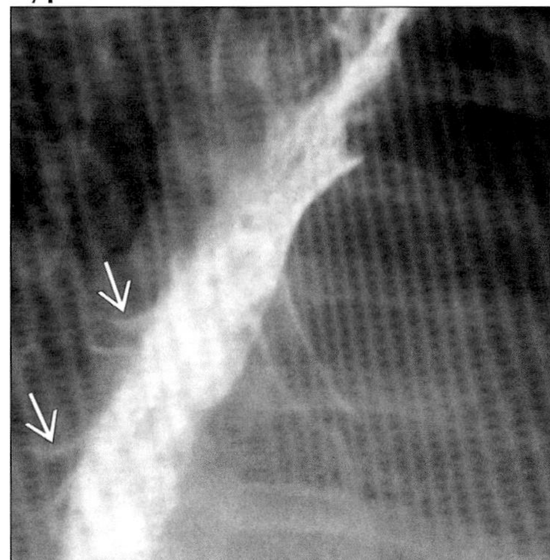

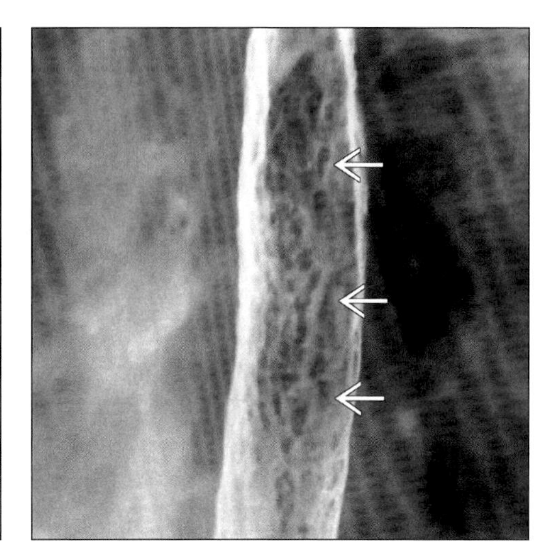

(Left) *Barium esophagram in a baby with AIDS shows a "shaggy" esophagus with areas of deep ulceration (arrows).* ***(Right)*** *Double-contrast esophagram shows longitudinally oriented plaques in mid-esophagus (arrows) (Courtesy of M. Nino-Murcia, MD).*

Variant

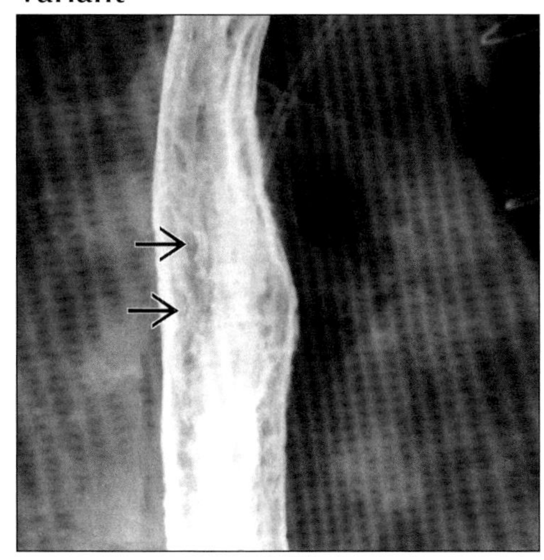

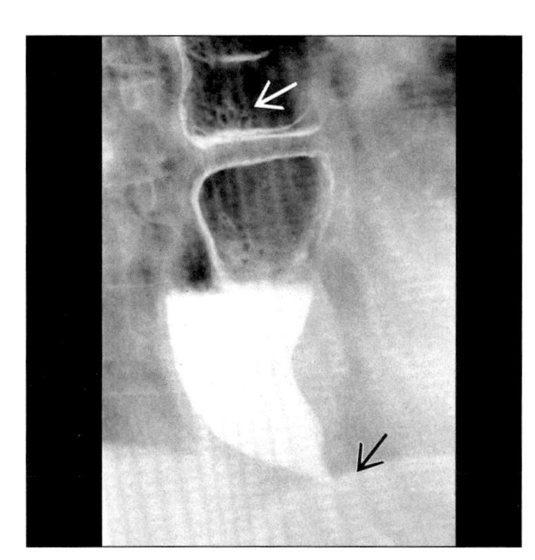

(Left) *Double-contrast esophagram in a patient with esophageal Candidiasis shows small shallow ulcerations indistinguishable from those of herpes esophagitis (arrows).* ***(Right)*** *Sagittal fluoroscopy shows Candida plaques (white arrow) in a patient with achalasia. Note dilated esophagus tapering to a beak (black arrow).*

Typical

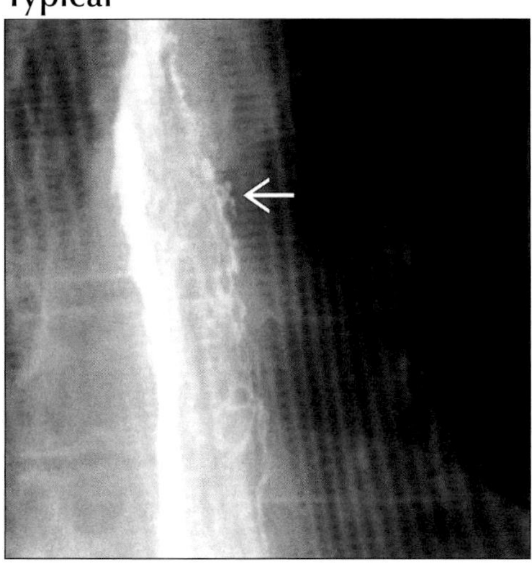

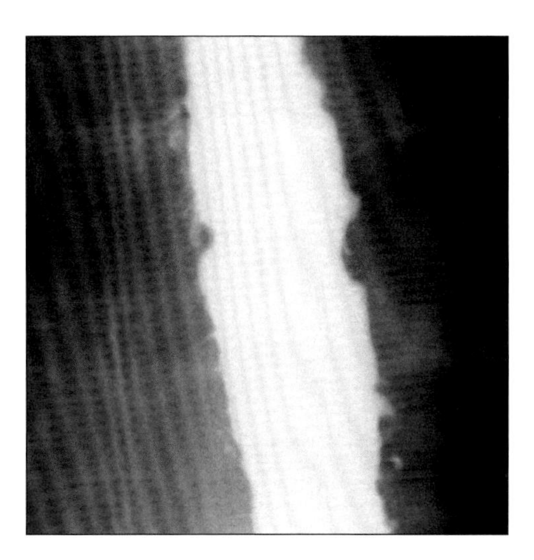

(Left) *Anteroposterior fluoroscopy shows Candida plaques and associated intramural pseudodiverticulosis (arrow) (Courtesy M. Federle, MD).* ***(Right)*** *Anteroposterior fluoroscopy shows extensive intramural pseudodiverticulosis in a patient with Candida esophagitis.*

VIRAL ESOPHAGITIS

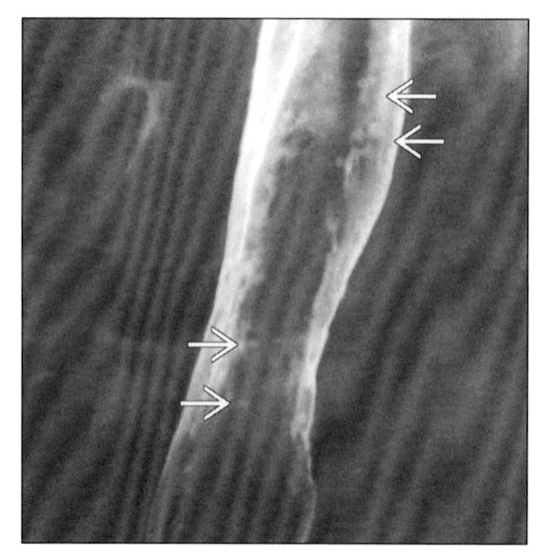

Anteroposterior fluoroscopy shows punctate ulcers of Herpes esophagitis en face (arrows).

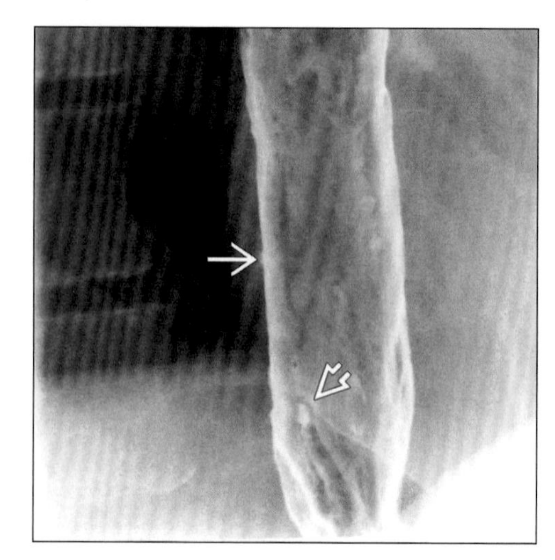

Lateral fluoroscopy shows tiny ulcers surrounded by radiolucent halo in patient with herpes esophagitis. Ulcers are seen en face (open arrow) and in profile (arrow).

TERMINOLOGY

Abbreviations and Synonyms

- Herpes simplex virus (HSV) esophagitis, herpetic esophagitis
- Cytomegalovirus (CMV) esophagitis
- Human immunodeficiency virus (HIV) esophagitis
- Varicella-zoster virus (VZV) esophagitis
- Ebstein-Barr virus (EBV) esophagitis
- Human papilloma virus (HPV) esophagitis, squamous papillomatosis

Definitions

- Inflammation of the esophagus of viral etiology

IMAGING FINDINGS

General Features

- Best diagnostic clue
  - Herpes: Multiple small, discrete punched-out ulcers on background of normal mucosa
  - CMV and HIV: One or more large, flat ulcers
    - Giant ulcers in HIV-positive patients most often caused by HIV rather than CMV
  - EBV: Esophagitis in patients with infectious mononucleosis
  - VZV: Esophagitis in immunocompromised children with chicken pox or adults with herpes zoster
- Location: Upper to mid-esophagus
- Size
  - Herpes: Small (< 1 cm) ulcers
  - CMV/HIV: Large ulcers
- Morphology
  - Herpes: Punctate or linear superficial ulcers with surrounding radiolucent halo
  - CMV: Large flat ulcer with surrounding lucent edema
    - Alternatively, small superficial ulcers resembling Herpetic lesions
  - HIV: Large, flat ulcers
  - EBV: Deep, linear ulcers
  - HPV: Multiple papillary excrescences

Radiographic Findings

- Double-contrast esophagram
  - Herpes
    - Multiple small discrete punctate, linear, or stellate ulcers
    - Surrounding radiolucent halo of edema
    - Background of normal mucosa
    - Plaque formation not typical
    - Advanced cases may have plaques, cobblestoning and "shaggy" appearance identical to Candida esophagitis
  - CMV

DDx: Mimics of Viral Esophageal Ulcers

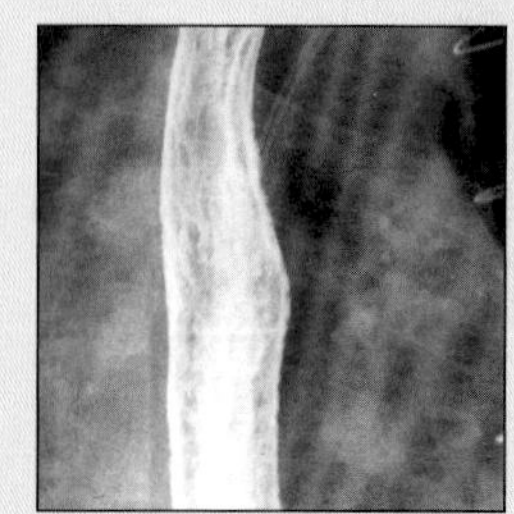

Candidiasis

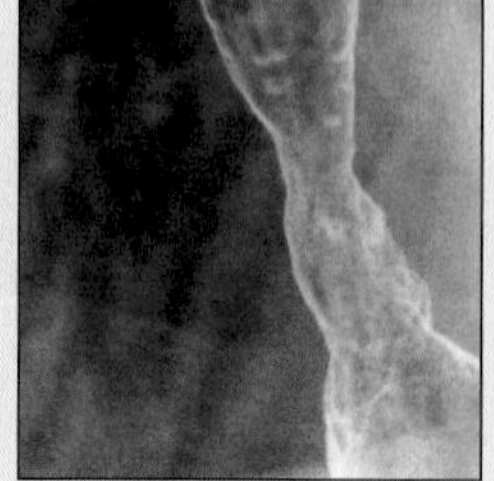

Reflux

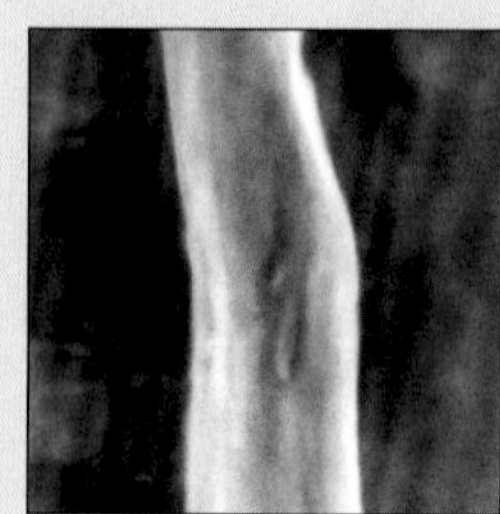

Drug Induced

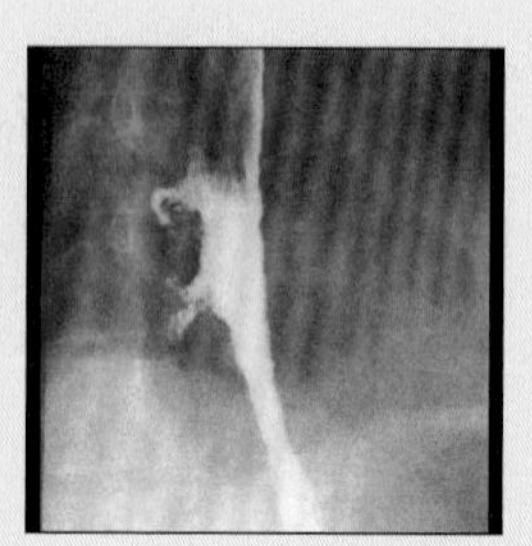

Lye Ingestion

VIRAL ESOPHAGITIS

Key Facts

Imaging Findings

- Herpes: Multiple small, discrete punched-out ulcers on background of normal mucosa
- CMV and HIV: One or more large, flat ulcers
- Best imaging tool: Double-contrast esophagram

Top Differential Diagnoses

- Reflux esophagitis
- Drug-induced esophagitis
- Crohn disease
- Radiation esophagitis, caustic esophagitis
- Technical artifacts

Pathology

- Viral esophagitis usually represents reactivation of latent virus rather than exogenous reinfection
- CMV may be transmitted via transplantation or transfusion
- Up to 40% of AIDS patients may develop symptoms of esophageal disease

Clinical Issues

- Odynophagia is most common presenting symptom
- Herpes: Usually in immunocompromised patients but can occur in otherwise healthy patients
- CMV, HIV esophagitis: Only in immunocompromised patients

Diagnostic Checklist

- Small discrete shallow ulcers should suggest viral esophagitis in immunocompromised patients with odynophagia

    - One or more giant (> 1 cm) flat ulcers in mid-distal esophagus
    - Discrete superficial ulcers indistinguishable from herpes
  - HIV
    - Giant, flat ulcers identical to CMV
    - May have small satellite ulcers
    - May cause fistulae
  - Squamous papillomatosis
    - Clusters of sessile polypoid lesions < 1 cm in size

CT Findings

- NECT: Thickened esophagus (> 5 mm)
- CECT
  - Thickened esophageal wall (> 5 mm)
  - "Target" sign
    - Edematous submucosa
    - Avidly enhancing mucosa

Imaging Recommendations

- Best imaging tool: Double-contrast esophagram
- Protocol advice
  - Prone oblique double-contrast views may be helpful
  - Use rapid swallows instead of effervescent granules to achieve gaseous distention without artifacts

DIFFERENTIAL DIAGNOSIS

Infectious

- Candida esophagitis may co-exist with viral esophagitis

Reflux esophagitis

- Usually in distal esophagus
- Often have distal stricture
- May see hiatal hernia
- Linear ulcers, fixed transverse folds ("stepladder esophagus")

Drug-induced esophagitis

- Focal contact esophagitis
- Usually in mid-esophagus secondary to compression by aortic arch or left mainstem bronchus
- Solitary or multiple shallow ulcers

Crohn disease

- Discrete aphthous ulcers
- Should have Crohn elsewhere in GI tract

Radiation esophagitis, caustic esophagitis

- Clinical history is diagnostic
- Shallow or deep ulcers
- Usually have stricture

Technical artifacts

- Gas bubbles, undissolved barium precipitates
- Should be transient

Differential diagnosis of giant ulcers

- Nasogastric intubation
- Sclerotherapy
- Caustic ingestion
- Mediastinal radiation
- Drugs

PATHOLOGY

General Features

- General path comments
  - Viral esophagitis usually represents reactivation of latent virus rather than exogenous reinfection
    - HSV: Latency in nerve ganglion cells
    - CMV: Latency in leukocytes
  - HPV infection found in many cases of esophageal papillomatosis
- Etiology
  - Impaired immune surveillance, radiation and chemotherapy render the esophageal mucosa vulnerable to infection
  - CMV may be transmitted via transplantation or transfusion
- Epidemiology
  - Up to 40% of AIDS patients may develop symptoms of esophageal disease
    - Candida, CMV and HSV most common pathogens

VIRAL ESOPHAGITIS

- Associated abnormalities: Oral lesions in 27-37% of HSV or HIV esophagitis

Gross Pathologic & Surgical Features

- Herpes esophagitis
  - Early disease: Esophageal blisters or vesicles rupture → discrete punched out ulcers
  - Advanced disease: Fibrinous exudate covers ulcers to form pseudomembranes
    - Mucosa in advanced disease indistinguishable from Candida esophagitis
- CMV: One or more shallow or deep ulcers
- HIV: Acute ulceration
  - Advanced disease: Deep penetrating ulcers

Microscopic Features

- Herpes
  - Cowdry type A intranuclear inclusions in epithelial cells
  - Ballooning of surface epithelial cells
  - Multinucleated giant cells
  - No intracytoplasmic inclusions
- CMV
  - "Owl's eye" inclusion bodies
  - Eosinophilic bodies within endothelial cells and fibroblasts
  - Ground glass nuclei with viral particles
  - Multinucleated giant cells
  - Cytoplasmic inclusions
- HIV
  - Viral particles morphologically similar to HIV on electron microscopy

CLINICAL ISSUES

Presentation

- Most common signs/symptoms
  - Odynophagia is most common presenting symptom
  - Other signs/symptoms
    - Dysphagia, chest pain, upper GI bleeding
  - Immunocompetent patients may have an influenza-type prodrome followed by acute onset odynophagia
  - HIV esophagitis may develop during acute HIV infection as well as after AIDS is established
    - May be associated with maculopapular rash
  - CMV may present with nausea and vomiting secondary to concomitant infection in stomach or intestine
- Clinical profile
  - Risk factors for viral esophagitis
    - Immunodeficiency
    - Chemotherapy
    - Organ transplantation
    - Diabetes
    - Alcoholism
    - Advanced age
  - Herpes: Usually in immunocompromised patients but can occur in otherwise healthy patients
    - Immunocompetent patients usually have history of exposure to sexual partners with herpetic ulcers
  - CMV, HIV esophagitis: Only in immunocompromised patients

Demographics

- Age: All
- Gender: Both

Natural History & Prognosis

- Herpes esophagitis
  - Usually self-limited in otherwise normal patients
  - Can be severe and prolonged in immunocompromised patients
- HIV esophagitis
  - Usually self-limited in immunocompetent patients
  - Advanced disease with deep ulcers may require treatment
- Varicella-zoster esophagitis
  - Usually resolves spontaneously
  - May cause necrotizing esophagitis in severely immunocompromised patients

Treatment

- Options, risks, complications
  - Treatment
    - Analgesics for odynophagia
    - Antiviral therapy for CMV, VZV and persistent herpes
    - Deep ulcers of HIV may require steroid or thalidomide treatment
  - Immunocompromised patients with HSV esophagitis can get herpes pneumonitis or disseminated infection
    - Severe esophagitis may cause strictures or hemorrhage

DIAGNOSTIC CHECKLIST

Consider

- Small discrete shallow ulcers should suggest viral esophagitis in immunocompromised patients with odynophagia

Image Interpretation Pearls

- Careful analysis of double-contrast patterns is necessary to distinguish plaques from ulcers

SELECTED REFERENCES

1. Mosca S et al: Squamous papilloma of the esophagus: long-term follow up. J Gastroenterol Hepatol. 16(8):857-61, 2001
2. Gore R et al: Textbook of Gastrointestinal Radiology. 2nd ed. Philadelphia, W.B. Saunders, 1980-1992, 2000
3. Berkovich GY et al: CT findings in patients with esophagitis. AJR Am J Roentgenol. 175(5):1431-4, 2000
4. Fenoglio-Preiser CM. Gastrointestinal pathology: An atlas and text. 2nd ed. Philadelphia, Lippincott-Raven. 62-65, 1999
5. Laine L et al: Esophageal disease in human immunodeficiency virus infection. Arch Intern Med. 154(14): 1577-82, 1994
6. Wilcox CM et al: Esophageal infections: etiology, diagnosis, and management. Gastroenterologist. 2(3): 188-206, 1994
7. Levine MS: Radiology of esophagitis: a pattern approach. Radiology. 179(1): 1-7, 1991

IMAGE GALLERY

Typical

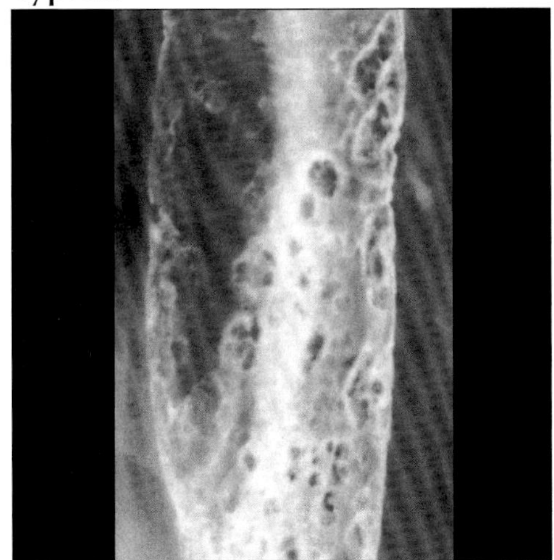

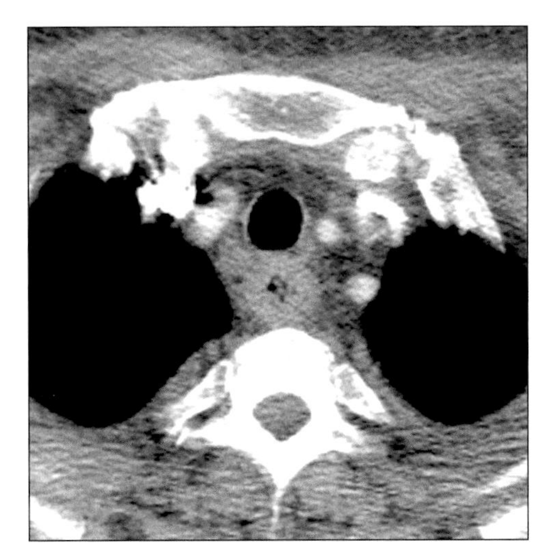

(Left) *Lateral fluoroscopy shows clusters of nodules typical of squamous papillomatosis.* ***(Right)*** *Axial CECT shows thickened irregular esophageal wall in patient with endoscopically proven CMV esophagitis.*

Variant

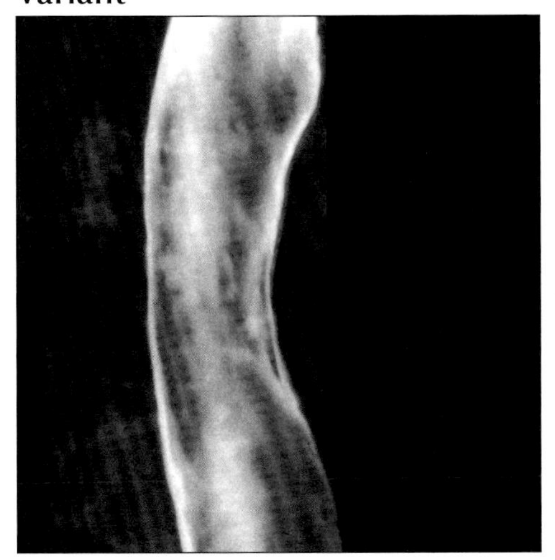

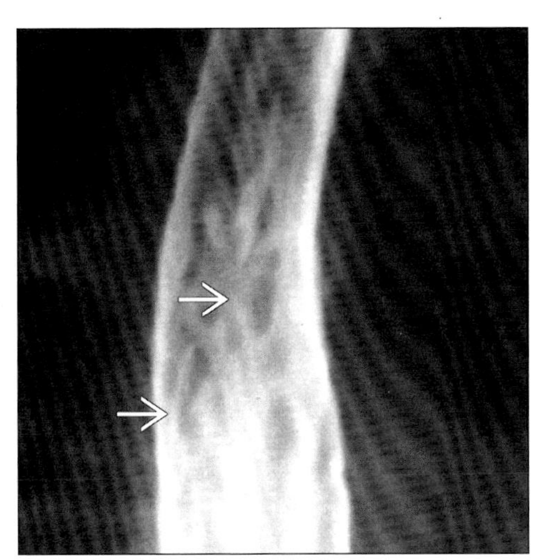

(Left) *Lateral fluoroscopy shows both ulcers and plaques in patient with herpes esophagitis.* ***(Right)*** *Lateral fluoroscopy shows elongated plaques (arrows) in patient with herpes esophagitis. Findings are indistinguishable from Candida esophagitis.*

Typical

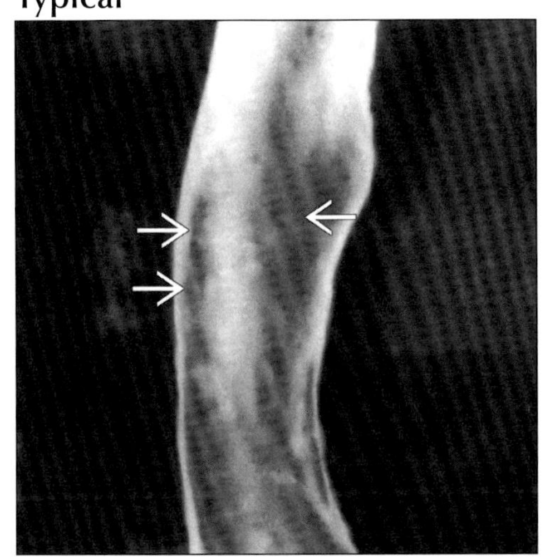

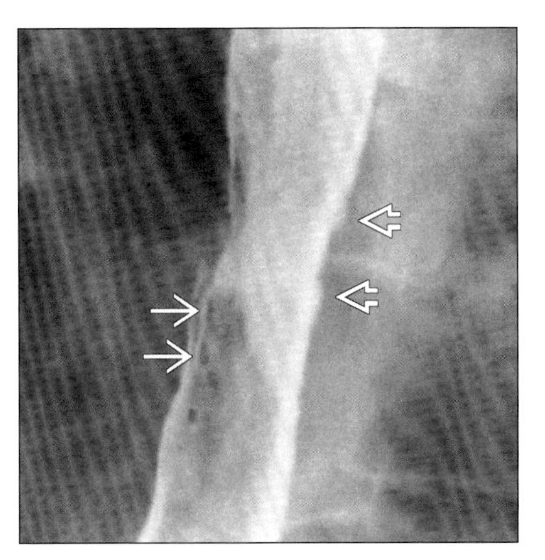

(Left) *Lateral fluoroscopy shows tiny ulcers (arrows) on background of normal mucosa, typical of herpes esophagitis.* ***(Right)*** *Lateral fluoroscopy shows small ulcers of herpes esophagitis in profile (open arrows) and en face (arrows).*

REFLUX ESOPHAGITIS

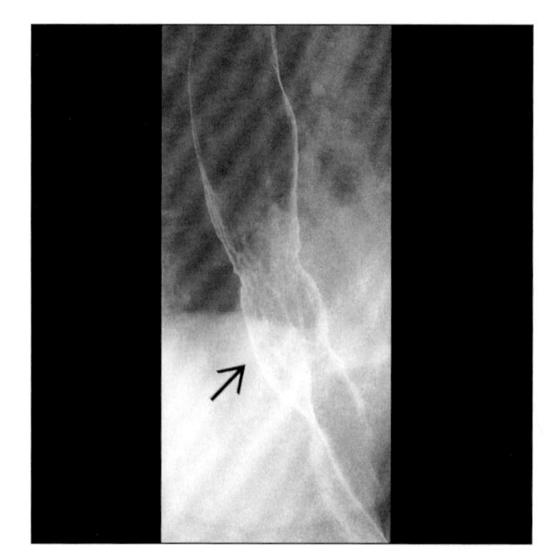

Double-contrast esophagram shows small hiatal hernia (arrow) with shallow round + linear esophageal ulcers.

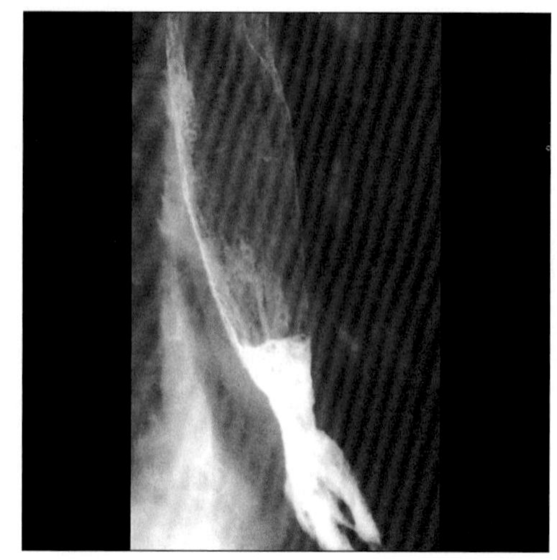

Double-contrast esophagram shows small hiatal hernia, distal stricture, and nodular, granular esophageal mucosa that simulates Barrett esophagus.

TERMINOLOGY

Definitions

- Inflammation of esophageal mucosa due to gastroesophageal reflux

IMAGING FINDINGS

General Features

- Best diagnostic clue: Irregular ulcerated mucosa of distal esophagus on barium esophagram
- Location: Usually involves distal third or half of thoracic esophagus
- Morphology
  - Varies from mild to severe
  - Various morphological changes of mucosa
    - Nodularity, ulceration, thickened folds
    - Inflammatory polyps, scarring & strictures
- Other general features
  - Usually a complication of gastroesophageal reflux disease (GERD)
  - Based on onset, classified clinically & radiologically
    - Acute reflux esophagitis
    - Chronic reflux esophagitis
  - Severity of reflux esophagitis
    - Depends on intrinsic resistance of mucosa
  - Sliding hiatal hernia
    - May represent an effect (complication) rather than a cause of reflux esophagitis

Radiographic Findings

- Double-contrast esophagraphy (en face/profile views)
  - Acute reflux esophagitis
    - May show increased frequency of nonperistaltic waves
    - Decreased or absent primary peristalsis
    - Mucosal nodularity: Fine nodular, granular or discrete plaque-like defects (pseudomembranes)
    - Ulcers: Multiple tiny collections of barium with surrounding mucosal edematous mounds, radiating folds & puckering
    - Thickened vertical or transverse folds (more than 3 mm): Best seen in mucosal relief views of collapsed esophagus
    - Inflammatory esophagogastric polyps: Smooth, ovoid elevations
  - Chronic or advanced reflux esophagitis
    - Decreased distal esophageal distensibility with irregular, serrated contour (due to ulceration/edema/spasm)
    - Sacculations & pseudodiverticula may be seen
    - Peptic stricture (1-4 cm length/0.2-2 cm width): Concentric smooth tapered narrowing of distal esophagus with proximal dilatation
    - "Stepladder" appearance: Transverse folds due to vertical scarring

DDx: Feline Esophagus

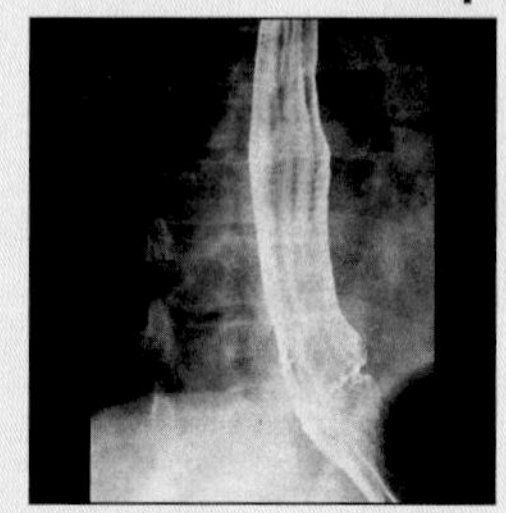

Feline Esoph.

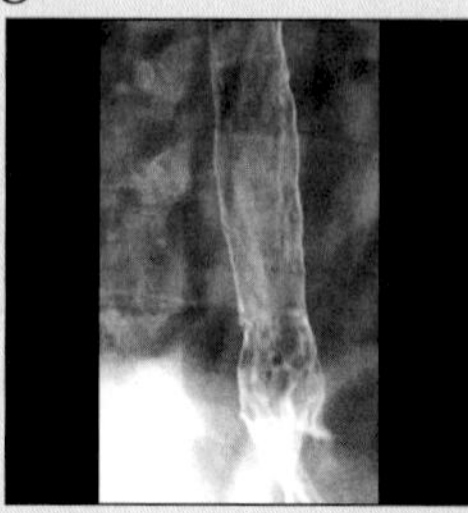

Candida

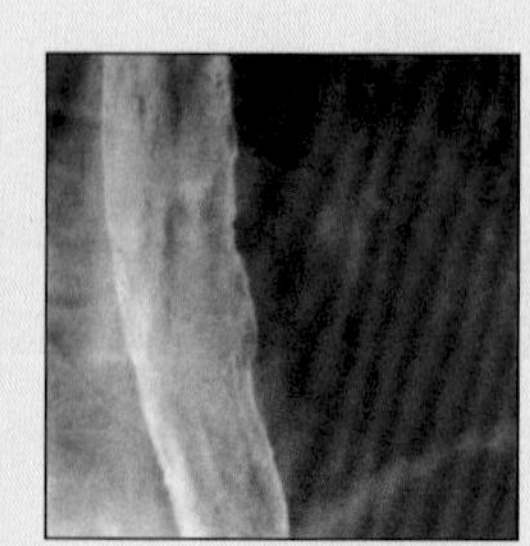

Herpes Esoph.

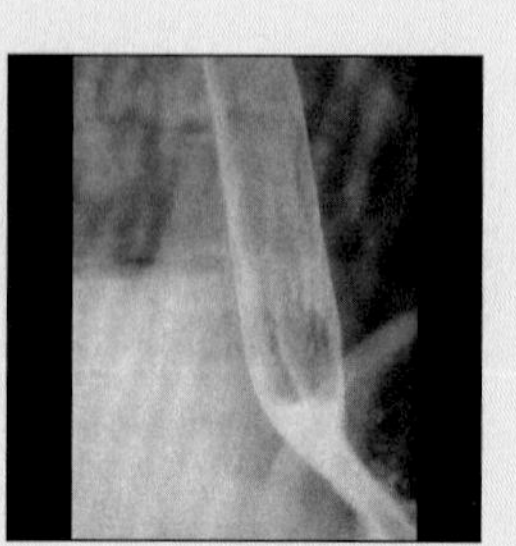

Crohn Esoph.

REFLUX ESOPHAGITIS

Key Facts

Imaging Findings
- Best diagnostic clue: Irregular ulcerated mucosa of distal esophagus on barium esophagram
- Mucosal nodularity: Fine nodular, granular or discrete plaque-like defects (pseudomembranes)
- Ulcers: Multiple tiny collections of barium with surrounding mucosal edematous mounds, radiating folds & puckering
- Thickened vertical or transverse folds (more than 3 mm): Best seen in mucosal relief views of collapsed esophagus
- Peptic stricture (1-4 cm length/0.2-2 cm width): Concentric smooth tapered narrowing of distal esophagus with proximal dilatation
- Hiatal hernia: Seen in more than 95% of peptic stricture patients

Top Differential Diagnoses
- Fungal & viral esophagitis
- "Feline esophagus"
- Caustic esophagitis
- Scleroderma
- Crohn disease

Pathology
- Gastroesophageal reflux disease (GERD)
- Lower esophageal sphincter (LES): Incompetent plus decreased tone leads to increased reflux

Diagnostic Checklist
- Differentiate from other types of esophagitis
- Smooth, tapered, concentric narrowing in distal esophagus above a sliding hiatal hernia is diagnostic of a peptic stricture due to reflux esophagitis

    - Hiatal hernia: Seen in more than 95% of peptic stricture patients

CT Findings
- NECT: Diffuse or focal circumferential esophageal wall thickening (≥ 5 mm)
- CECT: "Target" sign: Combination of mucosal enhancement & surrounding hypodense submucosa

Imaging Recommendations
- Videofluoroscopic double-contrast esophagram
  - En face & profile views
- Biphasic examination with upright double-contrast & prone single-contrast views of esophagus
  - Best radiologic technique for evaluation of patients with suspected reflux disease

DIFFERENTIAL DIAGNOSIS

Fungal & viral esophagitis
- Examples
  - Fungal: Candida albicans
  - Viral: Herpes, CMV & HIV
- Acute or early Candida esophagitis
  - "Foamy" esophagus: Multiple tiny, round lucencies on top of barium
  - Discrete linear, irregular plaque-like filling defects
  - Multiple fine nodular or granular appearance
  - "Cobblestone" or "snakeskin" appearance
- Severe or advanced Candidiasis
  - Grossly irregular or "shaggy" esophagus
  - Long tapered distal stricture due to scarring
    - Indistinguishable from reflux esophagitis stricture
  - Diagnosis: Endoscopy & biopsy
    - Budding yeast cells, hyphae, pseudohyphae
- Herpes esophagitis
  - Multiple small discrete superficial punctate, linear ulcers (often diamond shaped)
  - Extensive ulceration or plaque formation
- Cytomegalovirus (CMV) & HIV esophagitis
  - Location: Usually in mid-esophagus
  - CMV
    - Multiple discrete, superficial ulcers
    - One or more giant, flat ovoid/elongated ulcers more than 1 cm in size
  - HIV: One or more giant, flat ovoid/diamond-shaped ulcers more than 1 cm in size
  - Impossible to differentiate CMV ulcers from HIV
  - Diagnosis: Endoscopic biopsy

"Feline esophagus"
- Fine, barium-etched, thin transverse striations
  - Striations: Due to muscularis mucosae contractions
- Seen crossing entire esophageal luminal diameter
- This pattern is termed as "feline esophagus" or esophageal shiver; is transient
- Often associated with GE reflux

Caustic esophagitis
- Caustic agents: Liquid lye & strong acids
- Location: Usually involves entire thoracic esophagus
- CT imaging findings
  - Diffuse circumferential esophageal wall thickening (≥ 5 mm)
  - Acute phase: May show "target sign"
    - Mucosal enhancement & hypodense submucosa
  - Chronic phase
    - Luminal irregularity & narrowing
    - Esophageal perforation: Pneumomediastinum, pleural effusion
- Esophagography (water soluble contrast study)
  - Acute phase
    - Atonic dilated esophagus
    - Multiple shallow, irregular ulcers
  - Chronic phase
    - Segmental strictures, sacculations, pseudodiverticula
    - Thread-like/filiform appearance of mid-esophagus
- Indistinguishable from reflux esophagitis especially when lower esophageal segment is only involved

Scleroderma
- Collagen vascular disease, more common in women
- Pathogenesis
  - Smooth muscle atrophy/fibrosis; incompetent LES

- Increased risk of stasis, reflux esophagitis, Candida esophagitis & Barrett esophagus
- Barium study findings
  - Aperistalsis, dilated esophagus, patulous GE junction
  - GE reflux, mucosal findings of reflux esophagitis
  - Later leads to stricture + proximal dilatation
- Diagnosis: Anti-Scl 70 & anticentromere antibodies & endoscopic biopsy

Crohn disease

- Least common site of involvement of GI tract
- Always have associated disease in small bowel or colon
- Radiographic findings
  - Discrete, widely separated aphthous ulcers
  - Thickened folds, pseudomembranes
  - Longitudinal & transverse intramural tracks
    - Due to transmural involvement
    - Appearance: "Double-barreled" esophagus
  - Distal esophageal involvement may be indistinguishable from reflux esophagitis

PATHOLOGY

General Features

- General path comments
  - Sequelae of reflux esophagitis
    - Hyperemia, epithelial thinning, WBC infiltrate
    - Necrosis & finally ulceration
- Etiology
  - Gastroesophageal reflux disease (GERD)
    - Irritants: Drugs, alcohol, smoking, corrosive chemicals, radiation
  - Pathogenesis of reflux esophagitis
    - Lower esophageal sphincter (LES): Incompetent plus decreased tone leads to increased reflux
    - Hydrochloric acid (HCL) & pepsin: Increased synergistic effect producing more injury than HCL
- Epidemiology
  - GERD: Most common inflammatory disease
  - 15-20% of Americans have heartburn due to reflux

Gross Pathologic & Surgical Features

- Hyperemia, inflammation
- Superficial necrosis, white plaques, strictures

Microscopic Features

- Basal cell hyperplasia, edema
- Thinning of stratified squamous epithelium
- Submucosal polymorphonuclear leukocyte infiltrate
- Superficial necrosis & ulceration

CLINICAL ISSUES

Presentation

- Most common signs/symptoms
  - Heartburn, regurgitation, angina-like pain
  - Dysphagia, odynophagia
- Lab-data
  - Manometric/ambulatory pH-monitoring techniques
    - Reveal increase acid production
    - Assess gastroesophageal reflux
- Diagnosis
  - Endoscopy, biopsy & histological studies

Demographics

- Age: Usually middle & adult age group
- Gender: M = F

Natural History & Prognosis

- Complications
  - Ulceration, bleeding, stenosis
  - Sliding hiatal hernia: Due to
    - Severe inflammation & scarring
    - Longitudinal esophageal shortening
    - Disruption of ligaments surrounding GE junction
    - Pulls gastric fundus into thorax
  - Squamous or adeno carcinoma (Barrett mucosa)
- Prognosis
  - Acute reflux esophagitis: Good
  - Chronic reflux esophagitis: Poor

Treatment

- H2 receptor antagonists, proton-pump inhibitors
- Antacids, cessation of irritants
- Metaclopramide: Increases LES tone
- Surgery (fundoplication)

DIAGNOSTIC CHECKLIST

Consider

- Differentiate from other types of esophagitis

Image Interpretation Pearls

- Smooth, tapered, concentric narrowing in distal esophagus above a sliding hiatal hernia is diagnostic of a peptic stricture due to reflux esophagitis

SELECTED REFERENCES

1. Dibble C et al: Detection of reflux esophagitis on double-contrast esophagrams and endoscopy using the histologic findings as the gold standard. Abdom Imaging. 29:421-5, 2004
2. Hu C et al: Solitary ulcers in reflux esophagitis: radiographic findings. Abdom Imaging. 22(1):5-7, 1997
3. Levine MS: Reflux esophagitis and Barrett's esophagus. Semin Roentgenol. 29(4):332-40, 1994
4. Thompson JK et al: Detection of gastroesophageal reflux: Value of barium studies compared with 24-hr pH monitoring. AJR. 162: 621-6, 1994
5. Levine MS: Radiology of esophagitis: a pattern approach. Radiology. 179(1):1-7, 1991
6. Levine MS et al: Update on esophageal radiology. AJR Am J Roentgenol. 155(5):933-41, 1990
7. Mann NS et al: Barrett's esophagus in patients with symptomatic reflux esophagitis. Am J Gastroenterol. 84(12):1494-6, 1989
8. Levine MS et al: Pseudomembranes in reflux esophagitis. Radiology. 159(1):43-5, 1986
9. Levine MS et al: Fixed transverse folds in the esophagus: a sign of reflux esophagitis. AJR Am J Roentgenol. 143(2):275-8, 1984
10. Graziani L et al: Reflux esophagitis: radiologic-endoscopic correlation in 39 symptomatic cases. Gastrointest Radiol. 8(1):1-6, 1983
11. Creteur V et al: The role of single and double-contrast radiography in the diagnosis of reflux esophagitis. Radiology. 147(1):71-5, 1983

IMAGE GALLERY

Typical

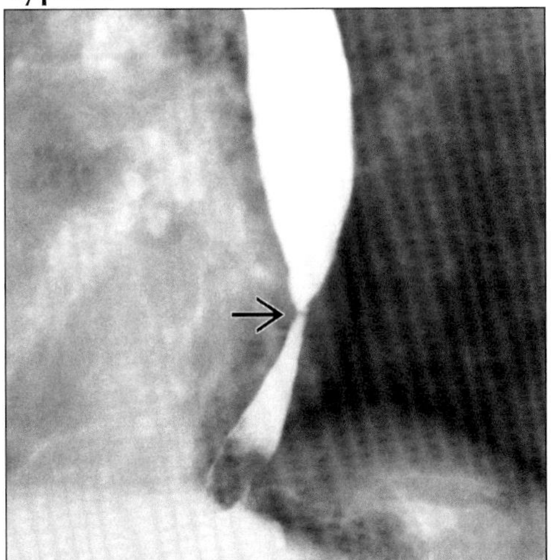

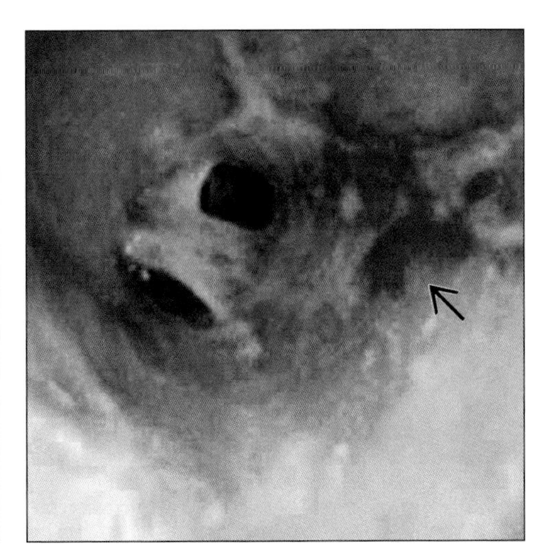

*(**Left**) Prone esophagram shows tight stricture (arrow) just above the GE junction, proximal dilatation. (**Right**) Endoscopic image of distal esophagus shows pseudo-membranes, mucosal ulceration (arrow), nodularity, and stricture.*

Typical

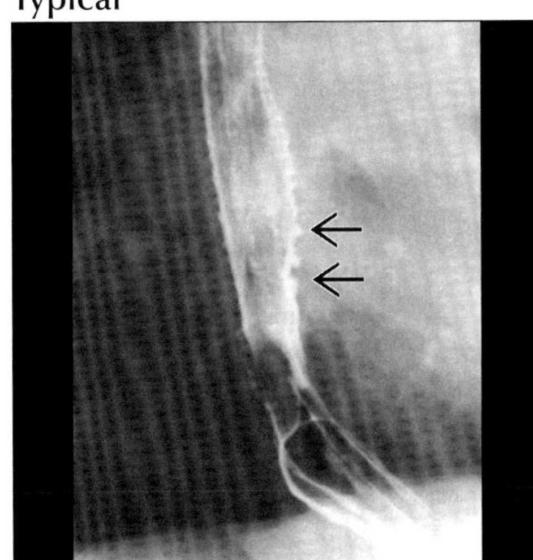

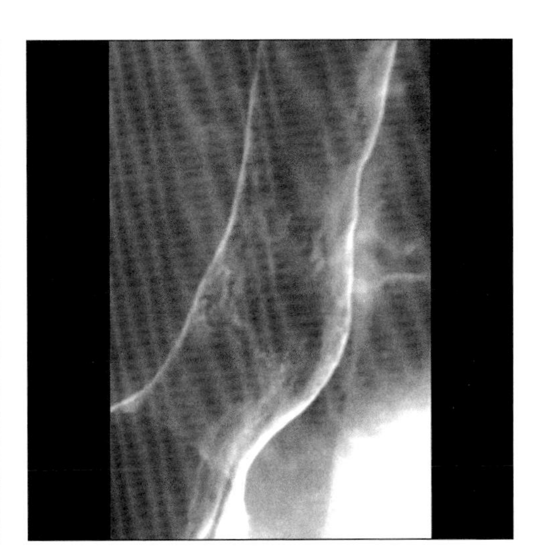

*(**Left**) Esophagram shows sliding hiatal hernia and multiple esophageal ulcers (arrows). (**Right**) Esophagram shows nodular distal esophageal mucosa.*

Typical

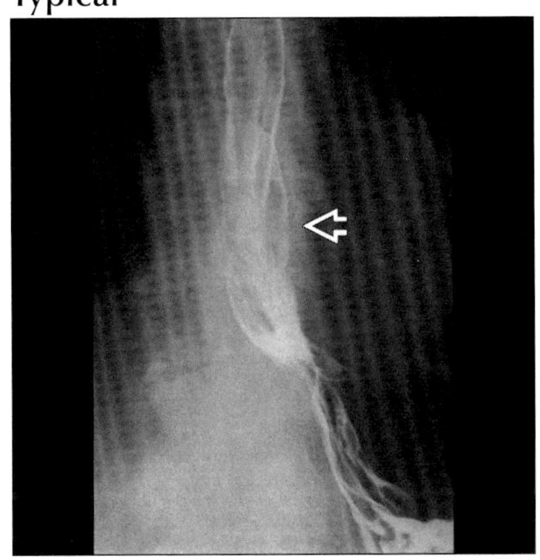

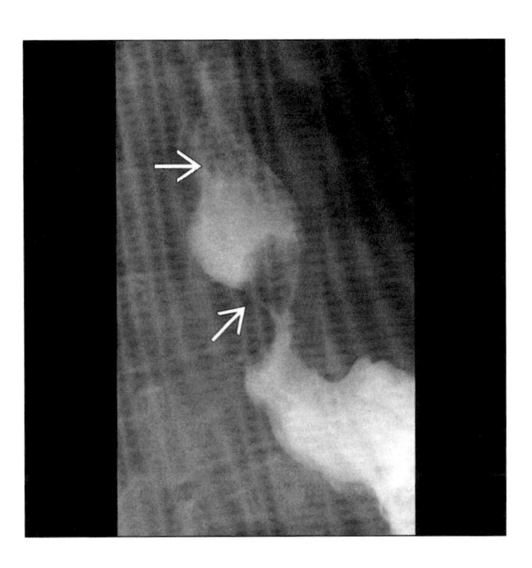

*(**Left**) Esophagram shows thickened, irregular longitudinal folds and a large ulcer (open arrow). (**Right**) Esophagram shows hiatal hernia, stricture at GE junction, and obstructed passage of food (walnuts) within lumen (arrows).*

BARRETT ESOPHAGUS

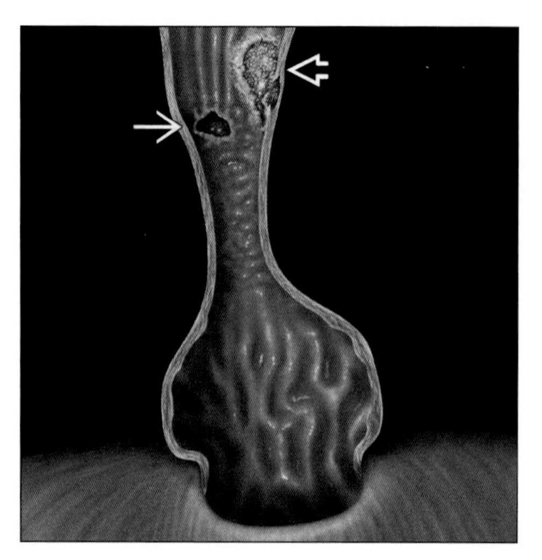

Graphic shows sliding hiatal hernia, distal esophageal stricture and nodular mucosal surface. Note discrete ulcer (arrow) and an adenocarcinoma (open arrow).

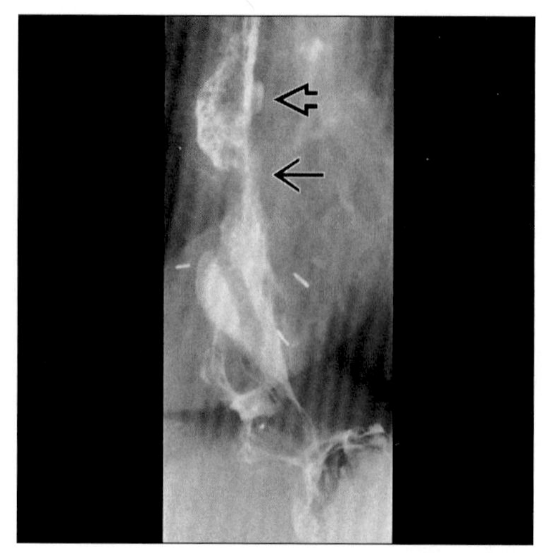

Esophagram shows hiatal hernia, strictures at GE junction (arrow) and higher. Also note discrete ulcer (open arrow).

TERMINOLOGY

Definitions

- Metaplasia of distal esophageal squamous epithelium to a columnar epithelium

IMAGING FINDINGS

General Features

- Best diagnostic clue: Mid-esophageal stricture with hiatal hernia & reflux is pathognomonic
- Location: Middle & distal third of esophagus
- Size: Length of esophageal stricture varies
- Morphology
  - Velvety, pinkish-red columnar mucosa which extends 3 cm or more above lower esophageal sphincter (LES)
    - Often seen as islands or tongues
  - Usually associated with hiatal hernia & reflux
- Other general features
  - Barrett esophagus is an acquired condition
  - Most common etiology: Chronic reflux esophagitis
  - Most common in adults but occasionally seen in children & infants
  - Premalignant condition associated with increased risk of esophageal adenocarcinoma
    - Risk of adenocarcinoma is 30-40 times more than in general population
    - 90-100% adenocarcinomas found to arise from Barrett mucosa
- Barrett esophagus is classified into two types based on endoscopy & histopathologic findings
  - Long-segment: Columnar epithelium more than 3 cm above gastroesophageal (GE) junction
    - Due to more severe reflux disease
    - Usually seen in mid-esophagus
    - Frequency less than short-segment
    - Hiatal hernia greater than 2 cm in 96% of patients
    - Increase risk of cancer than short-segment
    - Prevalence: Approximately 1% at endoscopy & 10% in symptomatic
  - Short-segment: Columnar epithelium 3 cm or less, above GE junction
    - Due to less severe reflux disease
    - Usually seen in distal esophagus
    - Frequency greater than long-segment
    - Hiatal hernia greater than 2 cm in 72% of patients
    - Less risk of cancer than long-segment
    - Prevalence: 2-12% at routine endoscopy

Radiographic Findings

- Double-contrast esophagography
  - Long-segment Barrett esophagus (mid-esophagus)
    - Early: Focal indentation/small concavity of wall
    - Severe: Ring-like constriction/tapered narrowing

DDx: Esophageal Stricture +/- Ulceration

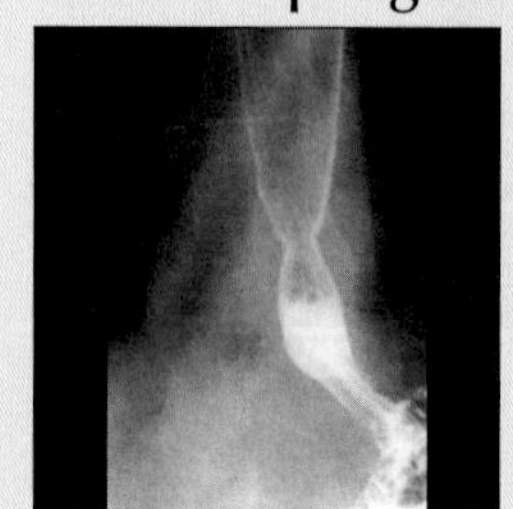

Reflux Stricture

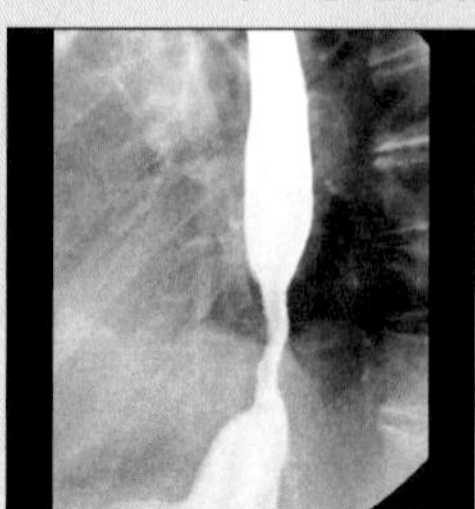

Esoph. Cancer

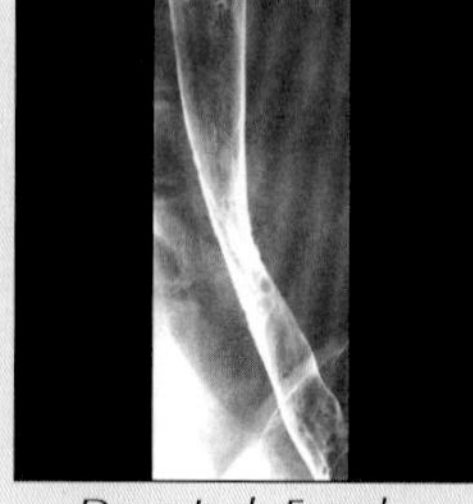

Drug-Ind. Esoph.

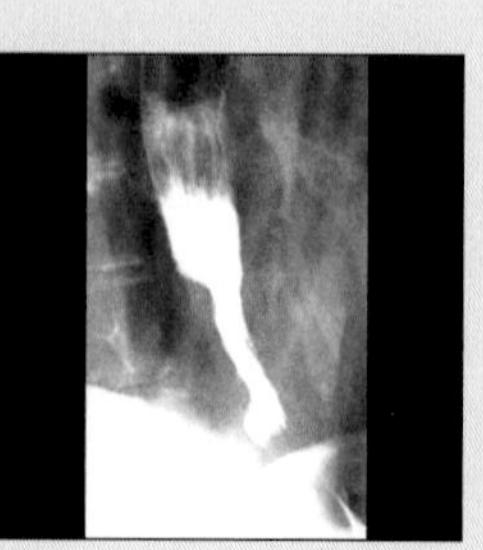

Scleroderma

BARRETT ESOPHAGUS

Key Facts

Terminology
- Metaplasia of distal esophageal squamous epithelium to a columnar epithelium

Imaging Findings
- Best diagnostic clue: Mid-esophageal stricture with hiatal hernia & reflux is pathognomonic
- Long-segment Barrett esophagus (mid-esophagus)
- Mid-esophageal Barrett ulcer (deep ulcer crater)
- Short-segment Barrett esophagus (distal esophagus)
- Granular or nodular mucosa, tiny ulcers
- Short tapered stricture: Symmetrical/asymmetrical

Top Differential Diagnoses
- Esophagitis
- Esophageal carcinoma
- Drug-induced injury
- Scleroderma

Pathology
- Recurrent GE reflux, inflammation, ulceration
- Re-epithelialization by pluripotent stem cells
- Differentiate into gastric or intestinal epithelium
- Increased risk: Adenocarcinoma in Barrett mucosa
- Velvety, pinkish-red mucosa (3 cm/more above LES)
- Long to mid-esophageal/short to distal esophageal stricture

Diagnostic Checklist
- Rule out other causes of esophageal stricture ± ulceration
- Long to mid-esophageal or short distal-esophageal stricture/ulcer associated with hiatal hernia/GE reflux

    - Mid-esophageal Barrett ulcer (deep ulcer crater)
    - Mid-esophageal stricture: More common & early
    - Sliding hiatal hernia
    - Gastroesophageal (GE) reflux
  - A mid-esophageal stricture/ulcer, hiatal hernia & GE reflux: Strongly suggests a Barrett esophagus
  - Short-segment Barrett esophagus (distal esophagus)
    - Granular or nodular mucosa, tiny ulcers
    - Vertical or transverse thickened, irregular folds (peptic scarring)
    - Short tapered stricture: Symmetrical/asymmetrical
    - Sliding hiatal hernia
    - Gastroesophageal reflux
    - May show intramural pseudodiverticula
  - Reticular mucosal pattern: Specific sign (distal to stricture)
    - Innumerable, tiny, barium-filled mucosal grooves/crevices
    - Usually seen adjacent to distal aspect of esophageal stricture
    - This finding is seen in only 5-30% of patients
  - Based on double-contrast esophagography & endoscopy
    - High risk patients: Mid-esophageal stricture, ulcer, reticular mucosa
    - Moderate risk: Distal peptic stricture & reflux esophagitis
    - Low risk: If none of above findings are present

Imaging Recommendations
- Videofluoroscopic double-contrast esophagography
- En face, profile, oblique & prone views

DIFFERENTIAL DIAGNOSIS

Esophagitis
- Reflux esophagitis
  - Mucosa: Fine nodular/granular or plaque-like
  - Ulcers: Multiple tiny, radiating folds & puckering
    - Barrett ulcers: Usually deep ulcer craters within columnar mucosa at a greater distance from GE junction
  - Folds: Thickened vertical or transverse (> 3 mm)
  - Peptic stricture: Smooth tapered narrowing of short distal segment
    - Barrett stricture: Classic mid-esophageal; when distally located indistinguishable from peptic type
  - Diagnosis: Endoscopic biopsy & histology
- Infectious esophagitis
  - Fungal esophagitis: Candida albicans
    - "Foamy" esophagus: Multiple tiny round lucencies
    - "Cobblestone" or "snakeskin" appearance
    - Grossly irregular or "shaggy" esophagus
  - Viral esophagitis: Herpes, CMV, HIV
    - Multiple small discrete superficial punctate ulcers
    - One or more giant, flat ovoid or elongated ulcers
  - Diagnosis: Endoscopic biopsy & histology
- Radiation esophagitis
  - Granular mucosa, decreased distensibility of irradiated segment
  - Mid-esophageal stricture indistinguishable from Barrett stricture
  - History of lung cancer with mediastinal irradiation
- Caustic esophagitis
  - Location: Any esophageal segment is involved
    - Usually middle & lower thirds (more common)
  - Atonic dilated esophagus; multiple shallow irregular ulcers
  - Acute: Diffuse narrowing, grossly irregular contour, ulceration
  - Chronic: Diffuse thread-like or filiform appearance
  - One/more segmental strictures in cervical/thoracic esophagus highly suggestive of caustic ingestion
  - Diagnosis: History & endoscopic biopsy

Esophageal carcinoma
- Asymmetric contour with abrupt proximal borders of narrowed distal segment (rat-tail appearance)
- Mid-esophageal stricture may mimic Barrett stricture
- Diagnosis: Endoscopic biopsy & history

Drug-induced injury
- Examples: Antibiotics (doxycycline & tetracycline)
  - Ulcers: Solitary, several discrete or multiple tiny
  - Clustered circumferentially in mid-esophagus

- Examples: KCL, quinidine, NSAIDs
  - Produce more severe esophagitis than antibiotics
  - KCL, quinidine: Large areas of ulceration & edema
    - Ulcer simulates ulcerated carcinoma
  - NSAIDs: Giant, flat, diamond-shaped ulcers (4-6 cm)
- Diagnosis: History of drug ingestion, biopsy

Scleroderma

- Collagen vascular disease; more common in women
- Pathogenesis
  - Smooth muscle atrophy/fibrosis; incompetent LES
  - Reflux esophagitis & high risk of Barrett esophagus
- Fluoroscopic barium study findings
  - Aperistalsis, dilated esophagus, patulous GE junction
  - GE reflux, peptic stricture in distal one third
- Diagnosis: Endoscopic biopsy

PATHOLOGY

General Features

- Genetics: Genomic instability in patients of Barrett esophagus may increase risk of adenocarcinoma
- Etiology
  - Chronic GE reflux & reflux esophagitis
    - Due to acid & pepsin reflux
  - Patients with total or partial gastrectomy
    - Due to bile reflux esophagitis
  - Contributing factors
    - Decreased LES pressure, transient LES relaxation
    - Hiatal hernia, decreased acid sensitivity
    - Alcohol, tobacco, chemotherapy, scleroderma
  - Pathogenesis
    - Recurrent GE reflux, inflammation, ulceration
    - Re-epithelialization by pluripotent stem cells
    - Differentiate into gastric or intestinal epithelium
- Epidemiology
  - Incidence: 0.3-4% in general population
  - Prevalence: 8-20% (overall 10%) in patients with reflux esophagitis
  - Increased risk: Adenocarcinoma in Barrett mucosa
    - Prevalence: 2.4-46% (overall about 15%)
    - Incidence: 0.5-1.5% per year (in long-segment)
- Associated abnormalities: GE reflux & hiatal hernia

Gross Pathologic & Surgical Features

- Velvety, pinkish-red mucosa (3 cm/more above LES)
- Long to mid-esophageal/short to distal esophageal stricture
- Hiatal hernia

Microscopic Features

- Proximally: Specialized columnar epithelium
  - Villous architecture with goblet cells
- Distal to above: Junctional-type epithelium
  - Cardiac mucous glands
- More distally: Fundic-type epithelium
  - Parietal & chief cells

CLINICAL ISSUES

Presentation

- Most common signs/symptoms
  - Reflux symptoms
    - Heartburn/regurgitation/angina-like pain/dysphagia
  - Long-segment Barrett esophagus patients
    - More severe reflux disease
  - Short-segment Barrett esophagus patients
    - Less severe reflux disease
  - 20-40% of patients are asymptomatic
- Lab-data
  - Manometric or ambulatory pH-monitoring
    - Assess gastroesophageal reflux
- Diagnosis
  - Endoscopy, biopsy & histopathology

Demographics

- Age: Mean: 55-65 years; prevalence increases with age
- Gender: M:F = 10:1
- Ethnicity: Caucasians more than African-Americans (10:1)

Natural History & Prognosis

- Complications
  - Ulceration, stricture, perforation, adenocarcinoma
- Prognosis
  - Usually good after early detection & treatment
  - Poor due to perforation or malignant transformation

Treatment

- Cessation of irritants (e.g., smoking & alcohol)
- H2 receptor antagonists; proton pump inhibitors
- Increase LES pressure: Metoclopramide
- Surgical resection in high grade dysplasia

DIAGNOSTIC CHECKLIST

Consider

- Rule out other causes of esophageal stricture ± ulceration

Image Interpretation Pearls

- Long to mid-esophageal or short distal-esophageal stricture/ulcer associated with hiatal hernia/GE reflux

SELECTED REFERENCES

1. Luedtke P et al: Radiologic diagnosis of benign esophageal strictures: a pattern approach. Radiographics. 23(4):897-909, 2003
2. Yamamoto AJ et al: Short-segment Barrett's esophagus: findings on double-contrast esophagography in 20 patients. AJR Am J Roentgenol. 176(5):1173-8, 2001
3. Rosch T: Gastroesophageal reflux disease and Barrett's esophagus. Endoscopy. 32(11):826-35, 2000
4. Chen MY et al: Barrett esophagus and adenocarcinoma. Radiol Clin North Am. 32(6):1167-81, 1994
5. Levine MS: Reflux esophagitis and Barrett's esophagus. Semin Roentgenol. 29(4):332-40, 1994
6. Glick SN: Barium studies in patients with Barrett's esophagus: importance of focal areas of esophageal deformity. AJR Am J Roentgenol. 163(1):65-7, 1994
7. Gilchrist AM et al: Barrett's esophagus: Diagnosis by double-contrast esophagography. AJR 150 (1): 97-102, 1988
8. Levine MS et al: Re: Reticular pattern as a radiologic sign of the Barrett esophagus. Radiology. 156(3):843-4, 1985

IMAGE GALLERY

Typical

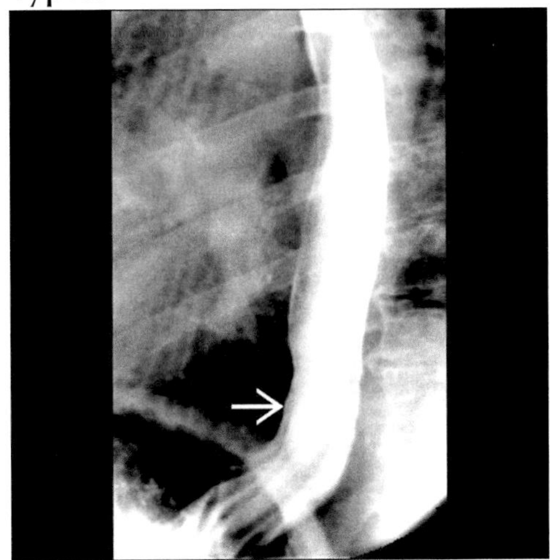

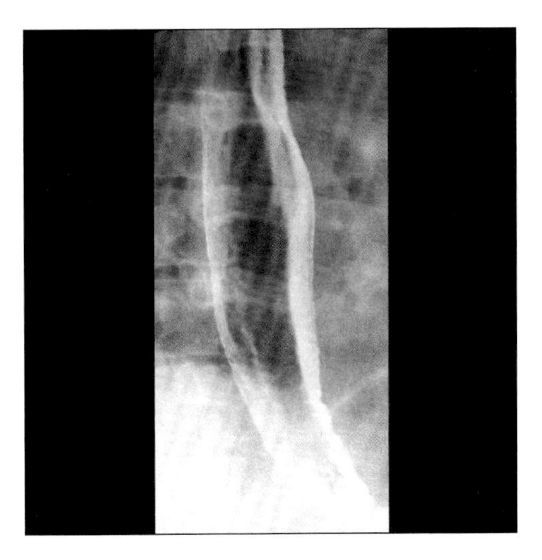

(Left) *Esophagram shows subtle raised plaque (arrow) in distal esophagus. Adenocarcinoma on Barrett mucosa.* ***(Right)*** *Esophagram shows diffuse nodularity of distal esophagus.*

Typical

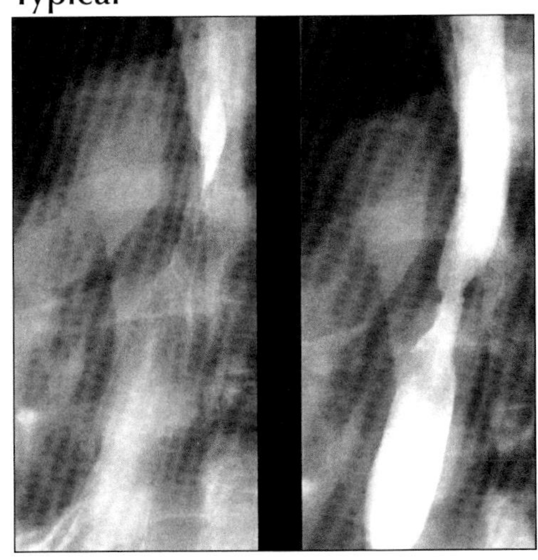

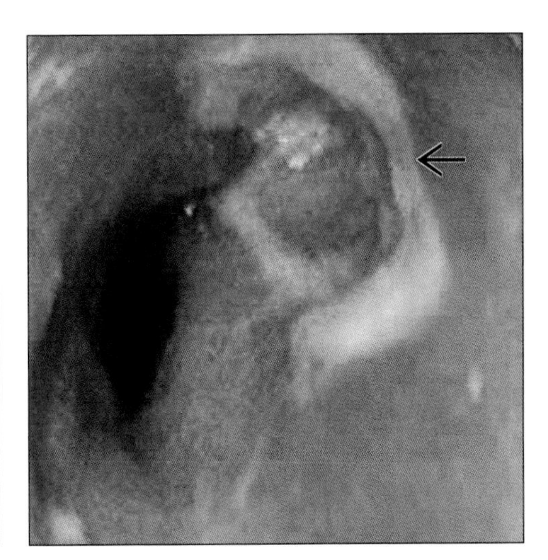

(Left) *Two views from esophagram show mid-esophageal stricture and ulcer in patient with small hernia and reflux.* ***(Right)*** *Endoscopic image shows large ulcer (arrow), velvet texture of Barrett mucosa and stricture.*

Typical

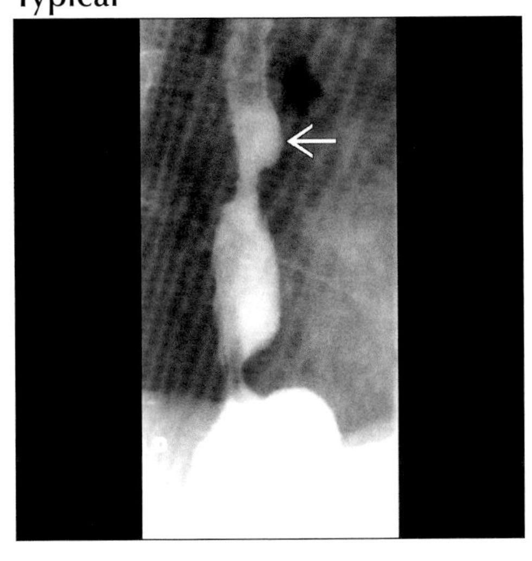

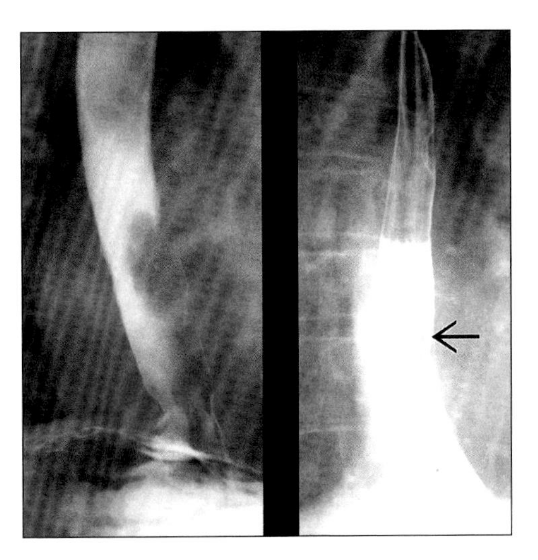

(Left) *Esophagram shows two strictures on an ulcer (arrow) and hiatal hernia.* ***(Right)*** *Two views from an esophagram show a polypoid mass (arrow); adenocarcinoma arising in Barrett mucosa.*

CAUSTIC ESOPHAGITIS

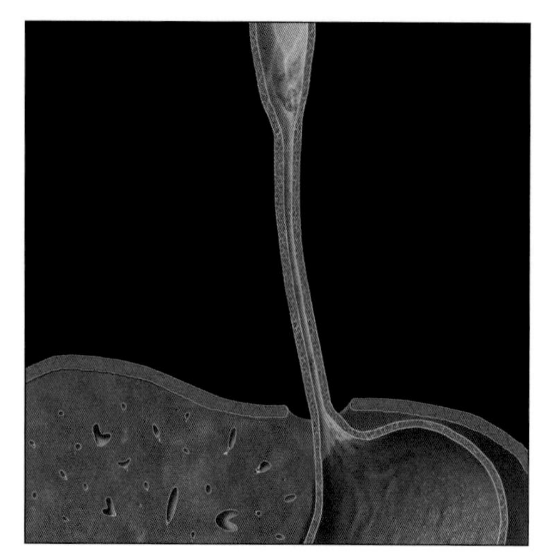

Graphic shows long stricture of esophagus and mucosal necrosis.

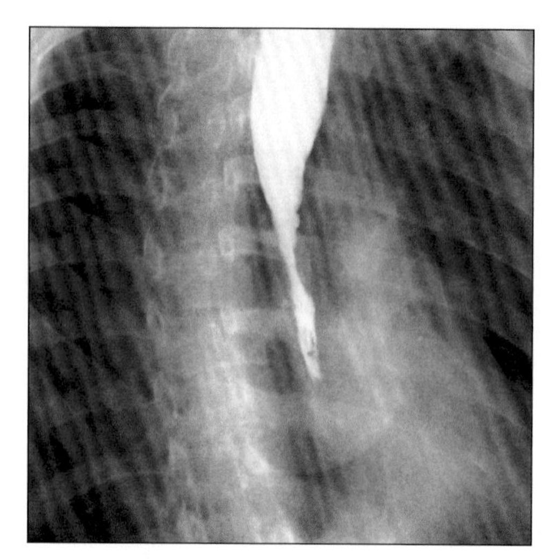

Esophagram shows long tapered high grade stricture of distal two-thirds of esophagus, several months after ingestion of lye in a suicide attempt.

TERMINOLOGY

Abbreviations and Synonyms

- Corrosive esophagitis

Definitions

- Esophageal inflammation/injury due to alkali or acid

IMAGING FINDINGS

General Features

- Best diagnostic clue: Nondistensible, rigid segment (stricture) of esophagus with ulcerated mucosa
- Location
  - Any esophageal segment is involved
  - Usually middle & lower thirds
- Other general features
  - Accidental or intentional ingestion of caustic agents cause
    - Mild or severe injury to upper GI tract: Esophagus/stomach/duodenum
  - Esophagus is most commonly injured in caustic ingestion
  - After esophagus, gastroduodenal injury is most common
  - Most commonly used caustic agent in US
    - Strong alkali: Liquid lye (concentrated sodium hydroxide) usually used as drain cleaner
  - Classification based on clinical & radiological findings
    - Acute & chronic phases
    - Mild & severe injury patterns

Radiographic Findings

- Chest PA & lateral views (acute)
  - Dilated, gas-filled esophagus
  - Esophageal perforation
    - Mediastinal widening, pneumomediastinum, pleural effusion
- Fluoroscopic guided water soluble or barium contrast studies
  - Acute mild phase
    - Atonic dilated esophagus
    - Multiple shallow, irregular ulcers
  - Acute severe phase
    - Extensive ulceration
    - Diffusely narrowed esophagus with irregular contour
    - Double-barreled appearance: Linear or streaky collections of barium in esophageal wall
  - Chronic phase
    - Sacculations, pseudodiverticula
    - Long or short segmental strictures: Smooth, concentric & symmetric or irregular, eccentric & asymmetric

DDx: Long Segmental Stricture

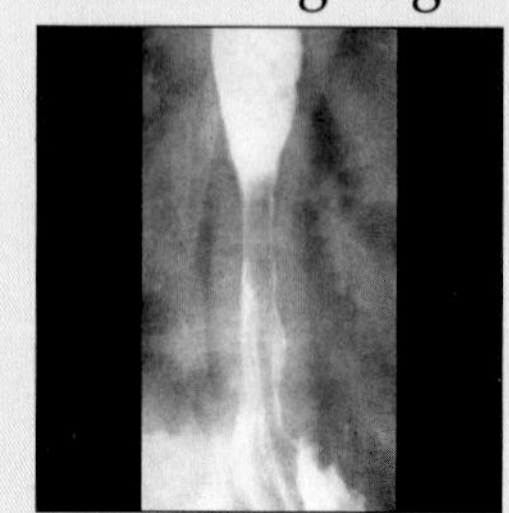

Peptic Stricture

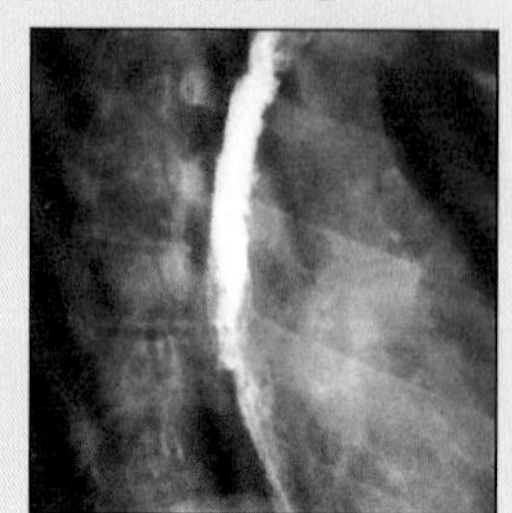

Candida

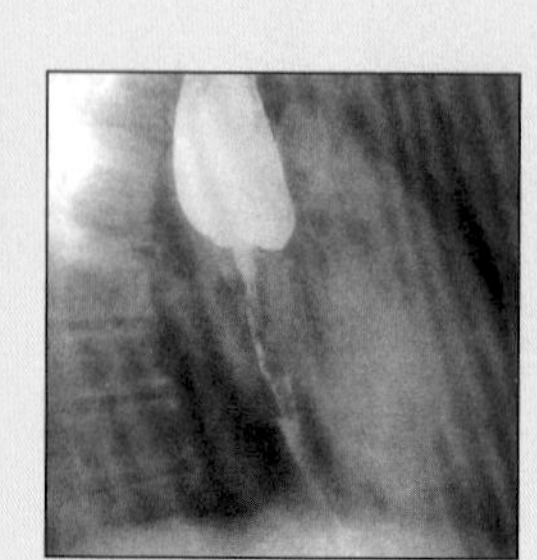

Carcinoma

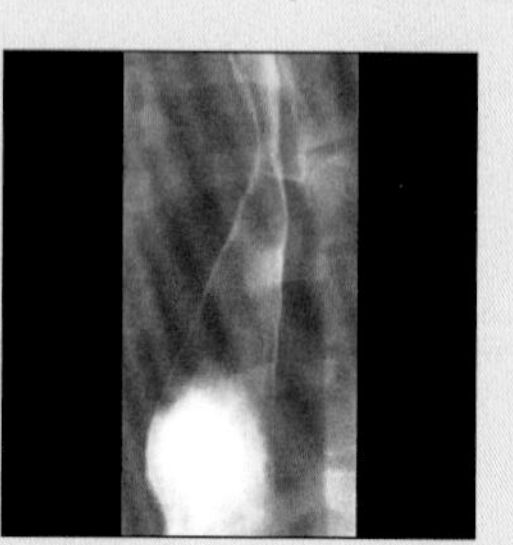

NG Intubation

CAUSTIC ESOPHAGITIS

Key Facts

Terminology

- Corrosive esophagitis
- Esophageal inflammation/injury due to alkali or acid

Imaging Findings

- Best diagnostic clue: Nondistensible, rigid segment (stricture) of esophagus with ulcerated mucosa
- Diffusely narrowed esophagus with irregular contour
- Double-barreled appearance: Linear or streaky collections of barium in esophageal wall
- Long or short segmental strictures: Smooth, concentric & symmetric or irregular, eccentric & asymmetric
- Diffuse long stricture: Thread-like or filiform appearance of entire thoracic esophagus (due to extensive scarring & fibrosis)

Top Differential Diagnoses

- Reflux esophagitis
- Infectious esophagitis
- Esophageal carcinoma
- Radiation esophagitis
- Nasogastric intubation

Diagnostic Checklist

- Rule out other inflammatory & noninflammatory pathology which can cause long segmental strictures
- Usually history of strong alkali or acid ingestion & endoscopy with biopsy yields definitive diagnosis
- Diffuse long segmental narrowing of thoracic esophagus with irregular contour & extensive ulceration giving rise to thread-like or filiform appearance highly suggestive of caustic ingestion

    - Diffuse long stricture: Thread-like or filiform appearance of entire thoracic esophagus (due to extensive scarring & fibrosis)

CT Findings

- Diffuse circumferential esophageal wall thickening (≥ 5 mm)
- Acute phase
  - "Target sign": Mucosal enhancement & hypodense submucosa
  - Esophageal perforation: Pneumomediastinum, pleural effusion
- Chronic phase: Luminal irregularity & narrowing

Imaging Recommendations

- Videofluoroscopic esophagram (en face/profile views)
- Acute phase: Water soluble, non-ionic contrast agent

DIFFERENTIAL DIAGNOSIS

Reflux esophagitis

- Acute reflux esophagitis
  - Mucosa
    - Fine nodular or granular
    - Pseudomembrane (plaque-like)
  - Ulcers
    - Multiple tiny ulcers
    - Mucosal edematous mounds, radiating folds, puckering
  - Folds
    - Thickened vertical or transverse (more than 3 mm)
  - Inflammatory polyps: Smooth, ovoid elevations
- Chronic reflux esophagitis
  - Sacculations or pseudodiverticula may be seen
  - Peptic stricture
    - Concentric, smooth tapered narrowing of short distal segment
    - May be indistinguishable from caustic stricture
    - History & associated sliding hernia differentiates reflux from caustic esophagitis

Infectious esophagitis

- Examples: Fungal & viral
  - Fungal: Candida albicans
  - Viral: Herpes, CMV & HIV
- Acute or early Candida esophagitis
  - "Foamy" esophagus
    - Multiple tiny, round lucencies on top of barium
  - "Cobblestone" or "snakeskin" appearance
- Severe or advanced Candidiasis
  - Grossly irregular or "shaggy" esophagus
  - Long tapered distal stricture due to scarring
    - May be indistinguishable from caustic stricture
    - Thread-like or filiform esophagus differentiates caustic from infectious stricture
- Viral esophagitis (herpes, CMV, HIV)
  - Multiple small discrete superficial punctate ulcers
  - One or more giant, flat ovoid or elongated ulcers more than 1 cm in size
  - Diagnosis: Endoscopic biopsy & history

Esophageal carcinoma

- Asymmetric contour with abrupt proximal borders of narrowed distal segment
  - Rat-tail appearance
- Lye stricture with irregular margins or abrupt borders
  - Indistinguishable from infiltrating carcinoma
- Diagnosis: Endoscopic biopsy & history

Radiation esophagitis

- Granular mucosa, decreased distensibility of irradiated segment
- Stricture: Usually smooth, tapered narrowing within area of radiation
- History of lung cancer with mediastinal irradiation

Nasogastric intubation

- Usually seen in patients with long standing intubation
- More commonly seen in institutionalized people
- Smooth narrowing of esophagus seen
  - Location: Esophageal lumen that comes in contact with distal part of nasogastric tube

CAUSTIC ESOPHAGITIS

PATHOLOGY

General Features

- Etiology
  - Strong alkaline agents: Liquid lye
    - Liquid lye is concentrated sodium hydroxide
    - Pathogenesis: Injury by liquefaction necrosis
  - Strong acids: Hydrochloric, sulfuric, acetic, oxalic, carbolic/nitric
    - Pathogenesis: Injury by coagulative necrosis
  - Ammonium chloride, phenols, silver nitrate
  - Degree of injury depends on
    - Nature/concentration/volume of agent & duration
  - Adults: Intentional (to commit suicide)
  - Children: Accidental
- Epidemiology: Increased incidence of caustic esophagitis by liquid lye in US
- Associated abnormalities: Associated gastroduodenal injuries seen in 5-10% cases

Gross Pathologic & Surgical Features

- Hyperemia/inflammation/necrosis/ulceration/strictures

Microscopic Features

- Thinning of stratified epithelium, inflammatory cells, basal cell hyperplasia

Staging, Grading or Classification Criteria

- Classification based on pathology
  - Stage I: Acute necrotic phase (1-4 days)
  - Stage II: Ulceration-granulation phase (5-28 days)
  - Stage III: Cicatrization & scarring (3-4 weeks)

CLINICAL ISSUES

Presentation

- Most common signs/symptoms
  - Pain, drooling, vomiting, hematemesis
  - Odynophagia, fever, shock
- Clinical profile: Patient with history of caustic ingestion & painful swallowing
- Diagnosis
  - History of alkali or acid ingestion
  - Endoscopy & water soluble contrast esophagraphy
    - Detects extent & severity of esophageal injury

Demographics

- Age: Any age group

Natural History & Prognosis

- Complications
  - Perforation, mediastinitis, peritonitis, fistulas, shock
  - Increased risk of cancer after 20-40 yrs
- Prognosis
  - Acute mild phase with early treatment: Good
  - Acute severe & chronic phases: Poor

Treatment

- Medical
  - Steroids, antibiotics, parenteral feedings
  - Esophageal bougienage
- Surgical
  - Esophageal bypass

DIAGNOSTIC CHECKLIST

Consider

- Rule out other inflammatory & noninflammatory pathology which can cause long segmental strictures
- Usually history of strong alkali or acid ingestion & endoscopy with biopsy yields definitive diagnosis

Image Interpretation Pearls

- Diffuse long segmental narrowing of thoracic esophagus with irregular contour & extensive ulceration giving rise to thread-like or filiform appearance highly suggestive of caustic ingestion

SELECTED REFERENCES

1. Dibble C et al: Detection of reflux esophagitis on double-contrast esophagrams and endoscopy using the histologic findings as the gold standard. Abdom Imaging. 29:421-5, 2004
2. Sam JW et al: The "foamy" esophagus: a radiographic sign of Candida esophagitis. AJR Am J Roentgenol. 174(4):999-1002, 2000
3. Berkovich GY et al: CT findings in patients with esophagitis. AJR Am J Roentgenol. 175(5):1431-4, 2000
4. Catalano O et al: Radiologic findings in chronic esophagitis dissecans. AJR Am J Roentgenol. 170(6):1671-2, 1998
5. Collazzo LA et al: Acute radiation esophagitis: radiographic findings. AJR Am J Roentgenol. 169(4):1067-70, 1997
6. Glick SN: Barium studies in patients with Candida esophagitis: pseudoulcerations simulating viral esophagitis. AJR Am J Roentgenol. 163(2):349-52, 1994
7. Thompson JK et al: Detection of gastroesophageal reflux: Value of barium studies compared with 24-hr pH monitoring. AJR. 162: 621-6, 1994
8. Levine MS: Radiology of esophagitis: a pattern approach. Radiology. 179(1):1-7, 1991
9. Levine MS et al: Update on esophageal radiology. AJR Am J Roentgenol. 155(5):933-41, 1990
10. Mann NS et al: Barrett's esophagus in patients with symptomatic reflux esophagitis. Am J Gastroenterol. 84(12):1494-6, 1989
11. Levine MS et al: Herpes esophagitis: sensitivity of double-contrast esophagography. AJR Am J Roentgenol. 151(1):57-62, 1988
12. Williams JM: Medication-induced esophagitis: diagnosis by double-contrast esophagography. AJR Am J Roentgenol. 149(3):646, 1987
13. Goldman LP et al: Corrosive substance ingestion: a review. American Journal of Gastroenterol. 79: 85-90, 1984
14. Creteur V et al: The role of single and double-contrast radiography in the diagnosis of reflux esophagitis. Radiology. 147(1):71-5, 1983
15. Graziani L et al: Reflux esophagitis: radiologic-endoscopic correlation in 39 symptomatic cases. Gastrointest Radiol. 8(1):1-6, 1983
16. Creteur V et al: Drug-induced esophagitis detected by double-contrast radiography. Radiology. 147(2):365-8, 1983
17. Muhletaler CA et al: Acid corrosive esophagitis: radiographic findings. AJR. 134: 1137-1140, 1980
18. Franken EA: Caustic damage of the gastrointestinal tract: Roentgen features. AJR. 118: 77-85, 1973
19. Martel W: Radiologic features of esophagogastritis secondary to extremely caustic agents. Radiology. 103: 31-36, 1972

CAUSTIC ESOPHAGITIS

IMAGE GALLERY

Typical

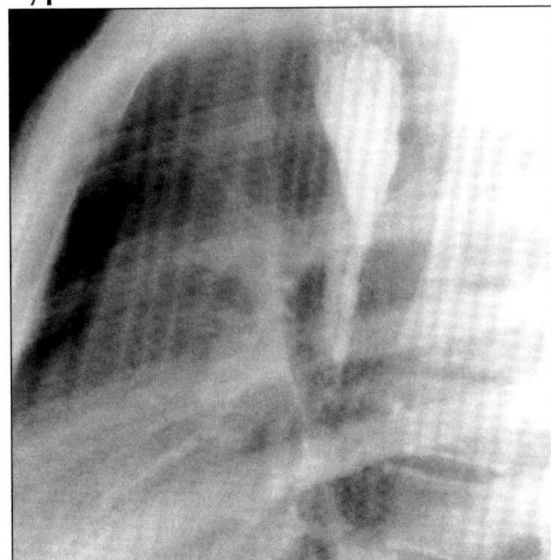

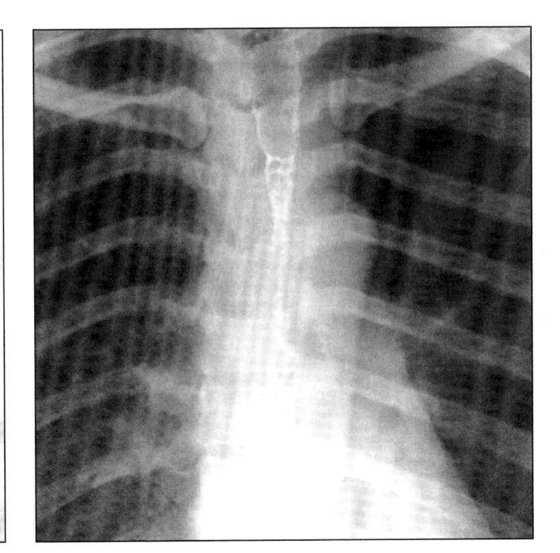

(Left) *Lateral view esophagram hours after caustic ingestion. Complete obstruction of mid-distal esophagus.* ***(Right)*** *Frontal chest radiograph following water soluble esophagram. Complete obstruction of esophagus with stasis. Acute injury.*

Typical

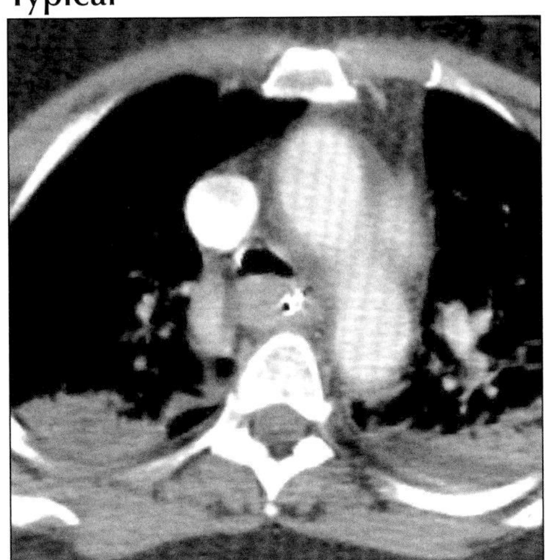

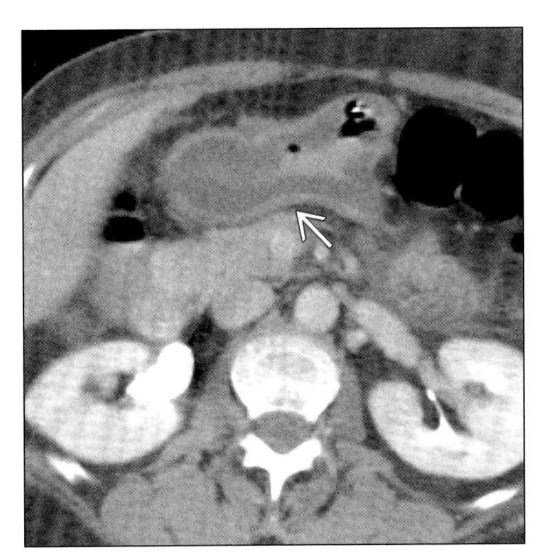

(Left) *Axial CECT of patient 2 hours after caustic ingestion. Marked thickening of esophageal wall.* ***(Right)*** *Axial CECT shows of patient following suicide attempt by caustic ingestion. Marked thickening of gastric wall with submucosal edema (arrow).*

Typical

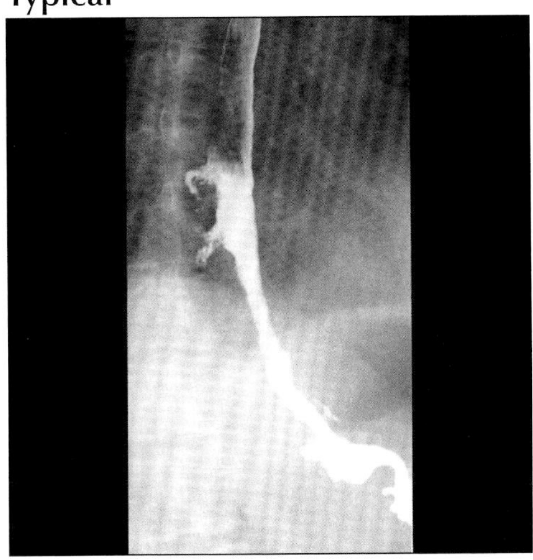

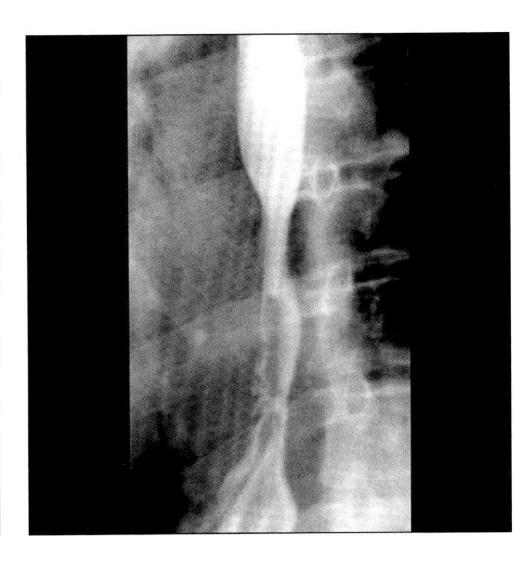

(Left) *Esophagram one week following a lye ingestion. Long stricture with deep ulcerations.* ***(Right)*** *Esophagram one month after caustic ingestion shows short esophagus, long stricture, and hiatal hernia. Reflux esophagitis may have similar appearance.*

DRUG-INDUCED ESOPHAGITIS

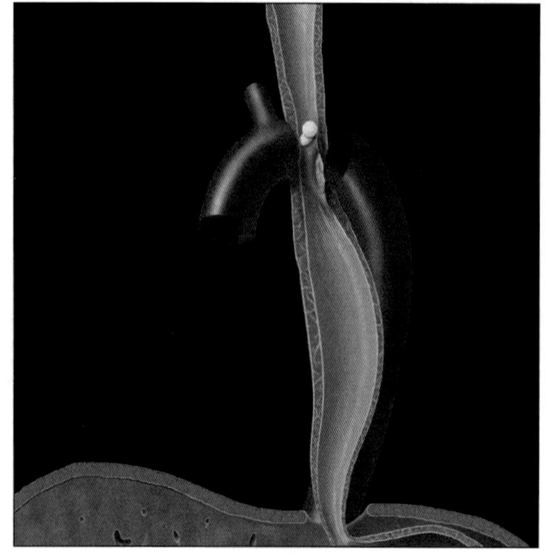

Graphic shows medication pills "stuck" at level of aortic arch with focal stricture and ulceration.

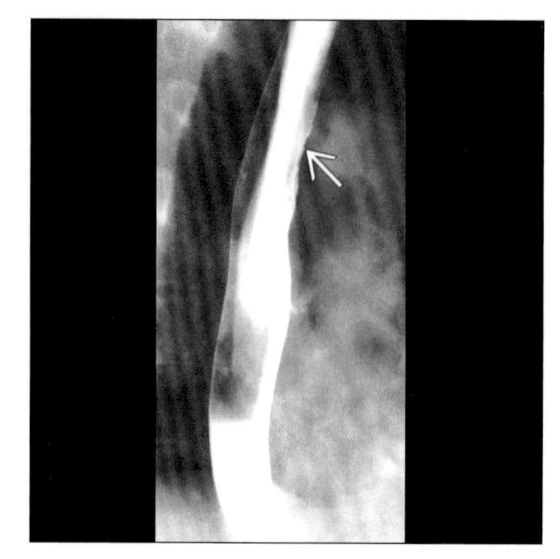

Esophagram shows broad shallow ulceration (arrow) at aortic arch level. Patient had odynophagia and recent tetracycline ingestion. Spontaneously resolved.

TERMINOLOGY

Abbreviations and Synonyms

- Drug-induced esophagitis (DIE)
- Pill-induced; medication-induced esophagitis

Definitions

- Iatrogenic esophageal injury induced by oral medication and direct contact

IMAGING FINDINGS

General Features

- Best diagnostic clue
  - Definite temporal relationship between ingestion of offending drug & onset of esophagitis
    - Healing of lesions after withdrawal of drug
- Location
  - Most common site; mid-esophagus
    - Near level of aortic arch or left main bronchus
- Morphology
  - Superficial ulcers
  - Solitary ring ulcer; several discrete ulcers; diamond-shaped ulcer
  - Deep ulcers & strictures

Radiographic Findings

- Double-contrast esophagography (en face & in profile)
- Superficial ulceration
  - Solitary or localized cluster of tiny ulcers distributed circumferentially on normal background mucosa
  - En face: Punctate, linear, stellate, serpiginous, or ovoid; collections of barium on esophageal mucosa
  - In profile; seen as shallow depressions
  - Thickening or distortion of adjacent esophageal folds; surrounding nodular mucosa
  - Usually with ingestion of doxycycline, tetracycline
- Giant, flat ulcers
  - Several centimeters or more in length
  - Larger areas of ulceration; with potassium chloride & quinidine; in patients with cardiomegaly
  - Mass effect surrounding ulcer; due to edema & inflammation; mimicking ulcerated carcinoma
  - Smooth, re-epithelialized depressions; seen with healing of these ulcers; mimicking active ulcer crater
- Strictures: Short segmental; concentric narrowing
  - Usually with potassium chloride & quinidine
  - Above level of enlarged left atrium, compressing distal esophagus; passage of pill impeded
    - Patients taking these pills often have cardiomegaly, mitral-valvular disease
- Severe esophagitis; may see stricture
- Repeat esophagram 7-10 days after withdrawal of drug
  - May show healing or disappearance of lesions

Imaging Recommendations

- Videofluoroscopic double-contrast esophagram

DDx: Superficial Esophageal Ulceration +/- Stricture

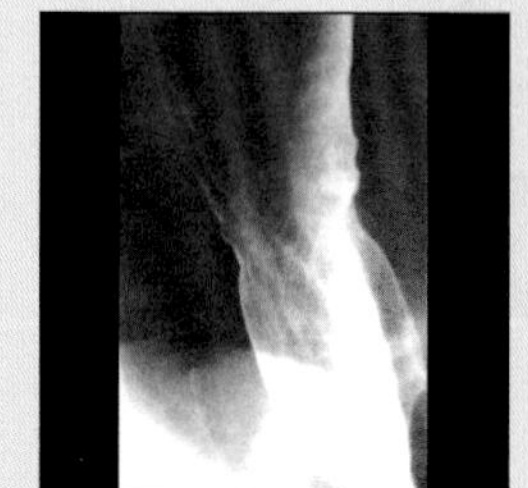

Reflux Esoph.

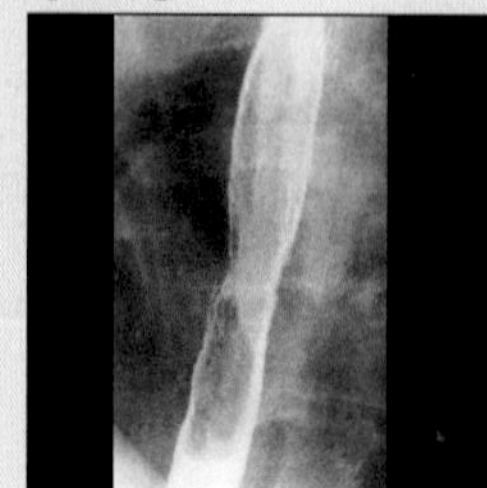

Herpes

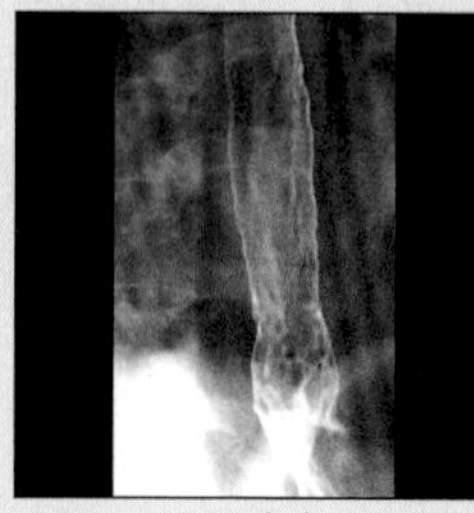

Candida

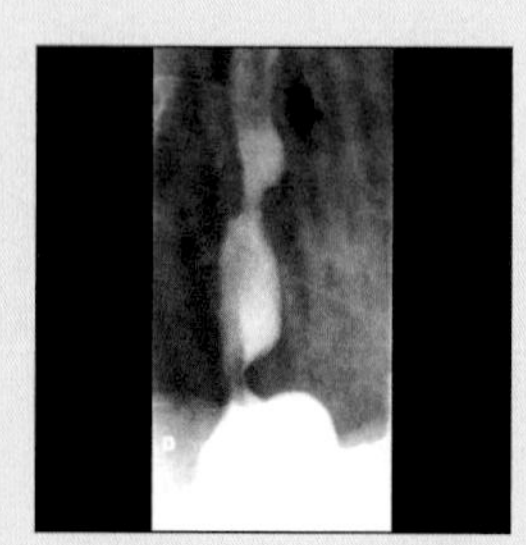

Barrett Esoph.

DRUG-INDUCED ESOPHAGITIS

Key Facts

Terminology
- Iatrogenic esophageal injury induced by oral medication and direct contact

Imaging Findings
- Most common site; mid-esophagus
- Superficial ulceration
- Giant, flat ulcers
- Strictures: Short segmental; concentric narrowing

Top Differential Diagnoses
- Reflux esophagitis
- Infectious esophagitis
- Esophageal carcinoma
- Barrett esophagus

Pathology
- Doxycycline, tetracycline, potassium chloride, quinidine, aspirin, nonsteroidal antiinflammatory

DIFFERENTIAL DIAGNOSIS

Reflux esophagitis
- Fine nodular; granular mucosa; superficial ulceration
- Peptic stricture: Concentric smooth; eccentric narrowing of distal esophagus
- In DIE, normal appearing mucosa below focal ulceration in mid-esophagus; unusual with reflux

Infectious esophagitis
- Herpes esophagitis; multiple, small, discrete, superficial ulcers; in upper or mid-esophagus
  - Immunocompromised patients; in healthy patients may be indistinguishable from DIE on imaging
- CMV + HIV; large superficial ulcers
- Clinical history & presentation help differentiate

Esophageal carcinoma
- Irregular narrowing; nodular, ulcerated mucosa

Barrett esophagus
- Mid-esophageal stricture; deep ulcer crater; hiatal hernia, reflux

PATHOLOGY

General Features
- General path comments
  - Prolonged contact of drug with esophageal mucosa
  - Secondary to drugs ability to alter local condition
  - Pill taken at bedtime, recumbent, without water
  - 22 times more frequent with capsules than with tablets; easier mucosal adhesion; slow disintegration
- Etiology
  - Doxycycline, tetracycline, potassium chloride, quinidine, aspirin, nonsteroidal antiinflammatory
  - Haloperidol, alendronate, ferrous sulfate, ascorbic acid, clindamycin, lincomycin, alprenolol chloride
- Epidemiology
  - DIE may be under-recognized
  - Doxycycline & tetracycline account for half the reported cases of DIE

Microscopic Features
- Chemical esophagitis; erosions or ulcerations, exudative inflammation

CLINICAL ISSUES

Presentation
- Odynophagia (94%), retrosternal burning pain (75%), dysphagia (56%), severe chest pain
- Elderly; those with pre-existing esophageal disease
- Symptoms resolve rapidly after withdrawal of offending medication; usually within 7-10 days
- Diagnosis: Endoscopy; presence & healing of lesions

Natural History & Prognosis
- Clinical course may be relatively uneventful, without any long term sequelae
- Complications: Persistent dysphagia due to stricture
  - Rarely, hemorrhage, perforation

Treatment
- Avoided by taking pills upright & with plenty of fluids
- Withdrawal of offending medication
- Proton-pump inhibitor & prokinetic; sucralfate

SELECTED REFERENCES

1. McCullough RW et al: Pill-induced esophagitis complicated by multiple esophageal septa. Gastrointest Endosc. 59(1):150-2, 2004
2. O'Neill JL et al: Drug-induced esophageal injuries and dysphagia. Ann Pharmacother. 37(11):1675-84, 2003
3. Levine MS: Drug-induced disorders of the esophagus. Abdom Imaging. 24(1):3-8, 1999

IMAGE GALLERY

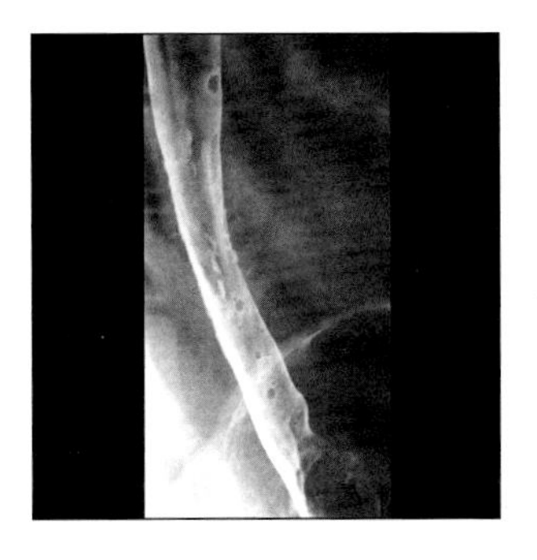

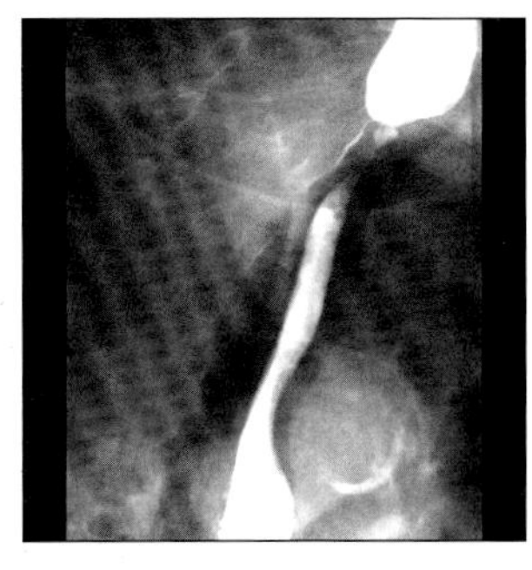

*(**Left**) Double-contrast esophagram shows cluster of ulcers in distal esophagus. (**Right**) Elderly patient with heart disease taking Quinidine and other medications. Stricture and ulceration at thoracic inlet, an unusual site for drug-induced esophagitis.*

RADIATION ESOPHAGITIS

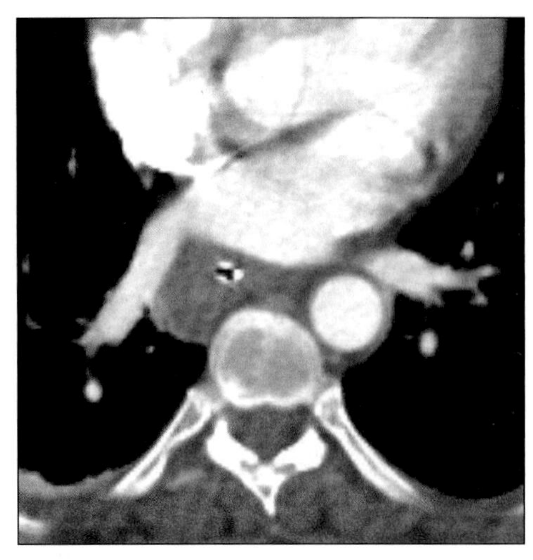

Axial CECT of a patient with dysphagia following radiation therapy for lung cancer shows marked thickening of esophageal wall surrounding the NG tube.

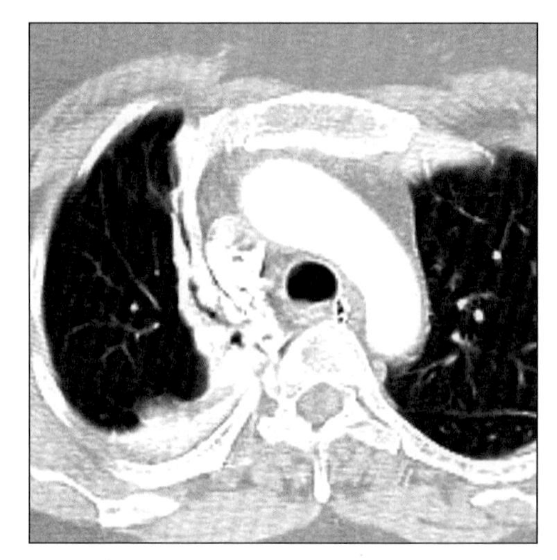

Axial CECT at lung windows shows radiation fibrosis of the lung in a patient with lung cancer and radiation esophagitis.

TERMINOLOGY

Abbreviations and Synonyms

- Radiation esophagitis (RE)

Definitions

- Inflammation of esophageal mucosa + wall induced by curative & palliative radiotherapy (RT)

IMAGING FINDINGS

General Features

- Location: Usually within prior radiation portal
- Morphology
  - Acute RE: Superficial ulcers, granular mucosa
  - Chronic RE: Deep ulcers, strictures, fistula

Radiographic Findings

- Fluoroscopy: Double-contrast esophagography
- Acute RE
  - Superficial ulcers; shallow, irregular collections of barium on esophageal mucosa
    - May be seen within 7-10 days of RT
    - Location: Usually conforms to radiation portal; sharp demarcation at inferior border of portal
    - Severe RE; grossly irregular, serrated contour; larger areas of ulcerations, mucosal sloughing
  - Granular appearance of mucosa
    - Due to punctate ulcers, edema, spasm
    - With ↓ distensibility of irradiated segment
  - Thickened mucosal folds due to edema
  - Slight luminal narrowing within radiation portal
  - Disordered motility; interruption of primary peristalsis at superior border of radiation portal
    - Numerous nonperistaltic contractions distal to point of disruption of primary wave
    - Usually 4-8 weeks after completion of RT; may present even decades after RT
- Chronic RE
  - Strictures: Concentric, smooth, tapered narrowing
    - Upper or mid-esophagus; within radiation portal
    - Usually 4-8 months after completion of RT
  - Ulcers at site of extrinsic compression by mediastinal lymphadenopathy, tumor
  - Late developing deep ulcers; ominous; may progress to fistula formation
  - Fistula: Left main bronchus; most frequent site
    - Radiation necrosis; usually within first year of RT
    - Avoid ionic; use nonionic water soluble contrast if suspected
  - Retraction of esophagus; related to lung resection & regional fibrosis

Imaging Recommendations

- Videofluoroscopic esophagram; en face + profile views
  - Start with nonionic water soluble
    - Follow with barium if no leak or fistula

DDx: Stricture and Mucosal Irregularity

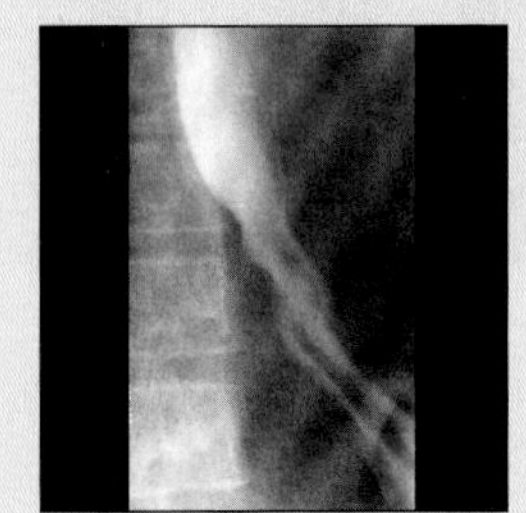

Peptic Stricture

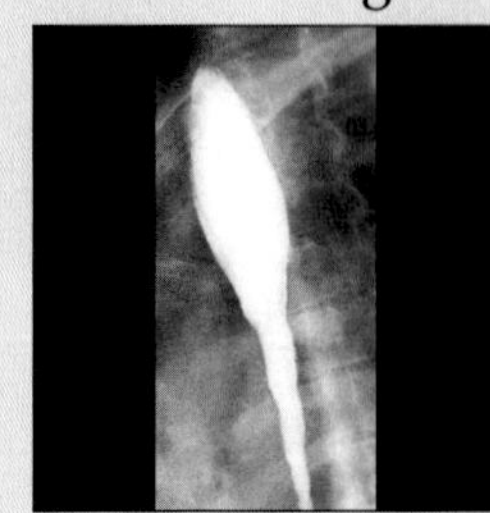

Lye Stricture

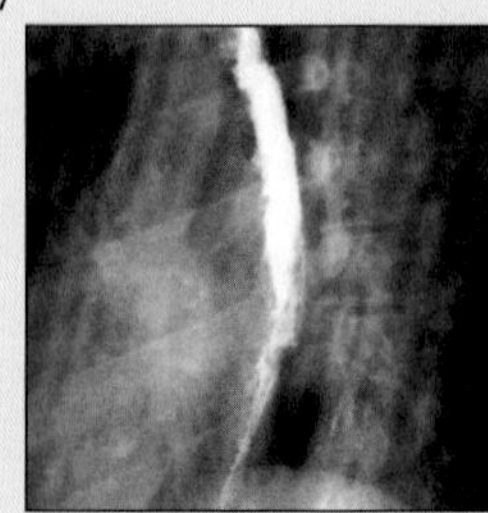

Candida

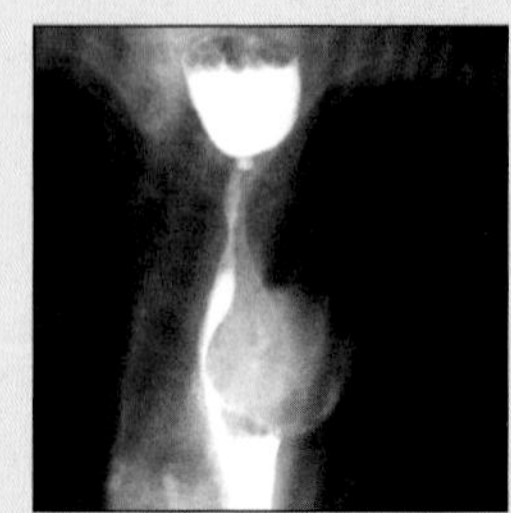

Drug-Ind. Strict.

RADIATION ESOPHAGITIS

Key Facts

Terminology
- Inflammation of esophageal mucosa + wall induced by curative & palliative radiotherapy (RT)

Imaging Findings
- Location: Usually within prior radiation portal
- Acute RE: Superficial ulcers, granular mucosa
- Chronic RE: Deep ulcers, strictures, fistula

Top Differential Diagnoses
- Reflux esophagitis
- Caustic esophagitis
- Infectious esophagitis
- Nasogastric intubation esophagitis
- Tumor recurrence

Clinical Issues
- Substantial morbidity

DIFFERENTIAL DIAGNOSIS

Reflux esophagitis
- Peptic stricture; concentric or eccentric smooth narrowing of distal esophagus + proximal dilatation

Caustic esophagitis
- Acute: Atonic dilated esophagus; multiple shallow, irregular ulcers
- Chronic: Long stricture, irregular contour

Infectious esophagitis
- Rarely causes stricture
- Candida; plaques + ulcers; shaggy surface

Nasogastric intubation esophagitis
- Unusually long, rapidly progressive stricture; long segment extensive ulceration; mid & distal esophagus

Tumor recurrence
- Irregular, eccentric narrowing; extrinsic mass effect
- Nodularity, "shaggy" ulcerated core; abrupt junction

PATHOLOGY

General Features
- Etiology
  - RT for adjacent thoracic & cervical neoplasms; Hodgkin disease, lung, mediastinum, spine
  - RT for primary esophageal carcinoma
- External beam; intraluminal brachytherapy
- Smaller doses; 2,000-4,500 rads → self-limited RE without permanent damage
- Doses of 4,500-6,000 rads → severe RE; irreversible
- Chemotherapy; Adriamycin; potentiates effect of RT
  - RE occurs earlier; even at low doses of RT

Microscopic Features
- Acute: Ulceration; necrosis; sloughing of mucosa
- Chronic: Marked thickening of submucosa; edema & fibrosis; cicatrization process; strictures

CLINICAL ISSUES

Presentation
- Acute RE: Substernal burning, dysphagia, odynophagia
  - Within 2-3 weeks of onset of RT
  - Usually abates within few weeks of cessation of RT
- Chronic RE: Motor dysfunction; even decades after RT
- Diagnosis: Acute RE on clinical grounds; temporal relationship between onset of RT & onset of symptoms
- Imaging & endoscopy + biopsy; not usually performed to confirm acute RE
  - In chronic RE; to detect strictures, tumor recurrence

Natural History & Prognosis
- Late reactions rare; might be severe, life-threatening
- Substantial morbidity

Treatment
- Early complications: Viscous lidocaine, indomethacin
- Strictures: Endoscopic dilatation, stent-implantation, endoscopic percutaneous gastrostomy
- Radio-protective: Pretreatment with amifostine

DIAGNOSTIC CHECKLIST

Consider
- Tumor infiltration of esophagus should be excluded
- With esophageal carcinoma, severe dysphagia before RT; clinical evidence of RE often masked

SELECTED REFERENCES

1. Collazzo LA et al: Acute radiation esophagitis: radiographic findings. AJR Am J Roentgenol. 169(4):1067-70, 1997
2. Seeman H et al: Esophageal motor dysfunction years after radiation therapy. Dig Dis Sci. 37(2):303-6, 1992
3. Chowhan NM: Injurious effects of radiation on the esophagus. Am J Gastroenterol. 85(2):115-20, 1990

IMAGE GALLERY

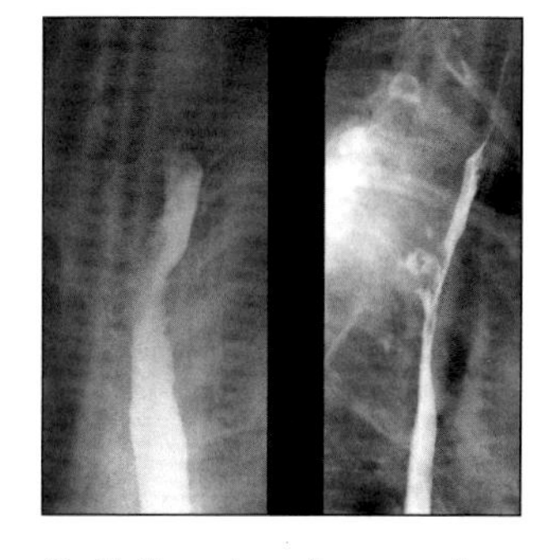

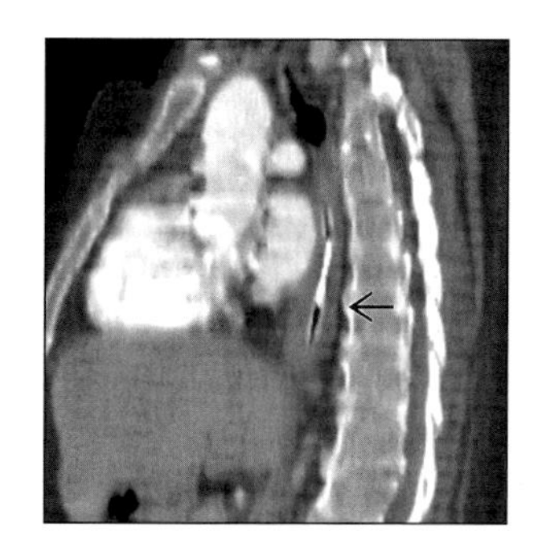

__(Left)__ Two views from esophagram show a mid-esophageal stricture following radiation therapy for lung cancer. __(Right)__ Sagittal reformation of CECT shows a long stricture and mural thickening (arrow) of esophagus following radiation for breast cancer.

ACHALASIA, CRICOPHARYNGEAL

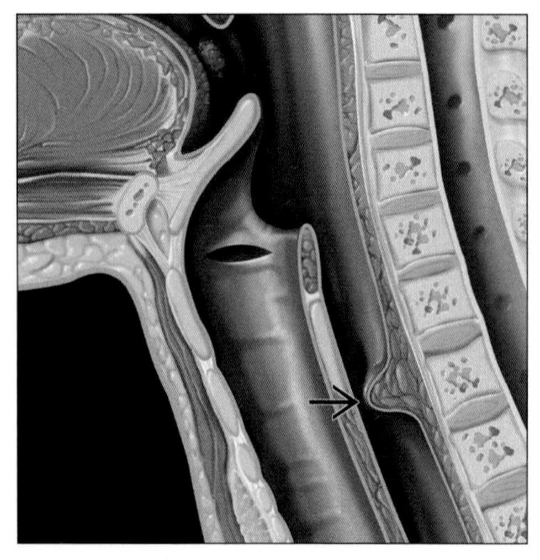

Graphic shows hypertrophied contracted cricopharyngeus muscle (arrow) at the pharyngo-esophageal junction (usually near the C5-6 cervical level).

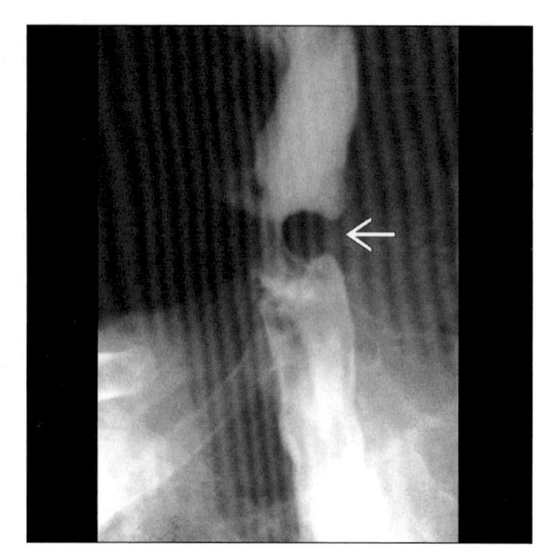

Lateral view from esophagram shows rounded, hypertrophied cricopharyngeus muscle impinging on the posterior lumen at the pharyngo-esophageal junction (arrow).

TERMINOLOGY

Definitions

- Failure of cricopharyngeal muscle (upper esophageal sphincter) relaxation due to hypertrophy or spasm

IMAGING FINDINGS

General Features

- Best diagnostic clue: Prominent cricopharyngeus muscle at pharyngoesophageal junction with retention of barium in pharynx on lateral view
- Location: Pharyngoesophageal junction: C5-6 level
- Morphology: Prominent or hypertrophied cricopharyngeus muscle
- Other general features
  - Activity of constrictor muscles must be coordinated with cricopharyngeal relaxation & luminal opening
  - Cricopharyngeus must relax & open completely to allow unimpeded passage of bolus
  - Complete opening of lumen at pharyngoesophageal junction is accomplished by several factors
    - Cricopharyngeal relaxation
    - Superior & anterior movement of larynx
    - Pharyngeal constriction, producing thrust
    - Intrabolus pressure

Radiographic Findings

- Videofluoroscopic pharyngoesophagraphy
  - Lateral view
    - Smoothly outlined shelf or lip-like projection posteriorly at level of cricoid (C5 6 level)
  - Prominent cricopharyngeus muscle
  - Prolonged closure of cricopharyngeus muscle
  - Jet effect: Noted below narrowing simulating a stenotic lesion
  - Pharyngeal retention of barium; ± aspiration
  - Distention of proximal cervical esophagus & distal pharynx

Imaging Recommendations

- Videofluoroscopic recording: Frontal/lateral/oblique
  - Rapid sequence required for demonstration

DIFFERENTIAL DIAGNOSIS

Cervical osteophytes (indentation)

- Large anterior cervical osteophytes can impinge on pharyngoesophageal junction simulating cricopharyngeal achalasia

Esophageal tumor

- Tumor at pharyngoesophageal junction may constrict the lumen concentrically or eccentrically

DDx: "Mass" at Pharyngo-Esophageal Junction

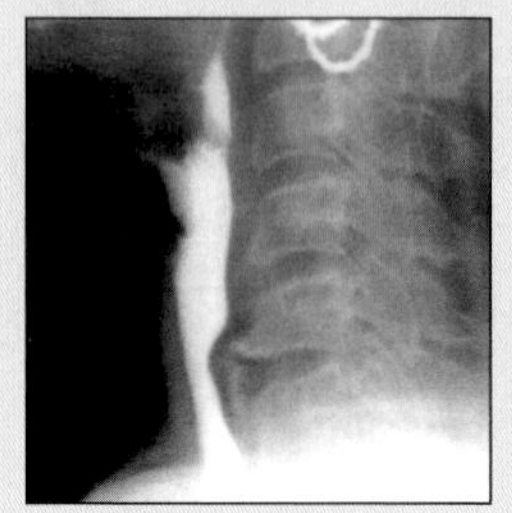

Osteophytes

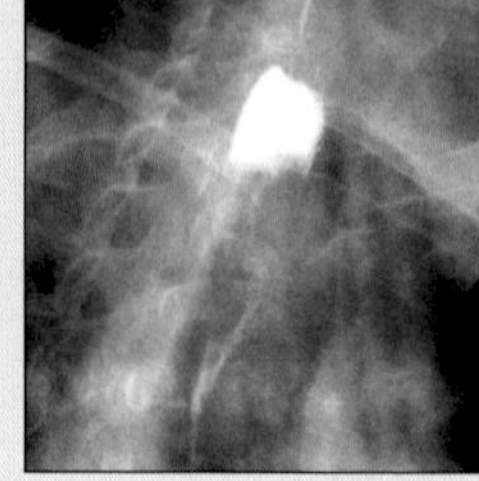

Carcinoma

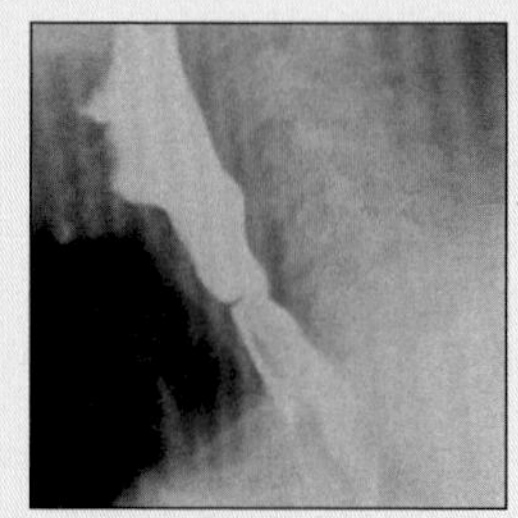

Esophageal Web

ACHALASIA, CRICOPHARYNGEAL

Key Facts

Terminology
- Failure of cricopharyngeal muscle (upper esophageal sphincter) relaxation due to hypertrophy or spasm

Imaging Findings
- Best diagnostic clue: Prominent cricopharyngeus muscle at pharyngoesophageal junction with retention of barium in pharynx on lateral view
- Smoothly outlined shelf or lip-like projection posteriorly at level of cricoid (C5-6 level)
- Jet effect: Noted below narrowing simulating a stenotic lesion
- Distention of proximal cervical esophagus & distal pharynx

Top Differential Diagnoses
- Cervical osteophytes (indentation)
- Esophageal tumor
    - Benign tumor: Smooth & well-defined margins
    - Malignant: Mucosal irregularity/ill-defined margins

PATHOLOGY

General Features
- General path comments
    - Embryology-anatomy
        - May be due to incomplete maturation of neurologic reflexes governing swallowing
- Etiology
    - Usually just "poor timing" of CP contraction
    - Due to "presbyesophagus" or other cause of dysmotility
    - Compensatory mechanism to GE reflux
    - Neuromuscular dysfunction of deglutition
        - Primary neural disorders: Brainstem disorder (bulbar poliomyelitis, amyotrophic lateral sclerosis, multiple sclerosis), cerebrovascular occlusive disease, Huntington chorea
        - Primary muscle disorder: Myotonic dystrophy, polymyositis, dermatomyositis, sarcoidosis, myopathies secondary to steroids, thyroid dysfunction, oculopharyngeal myopathy
        - Myoneural junction disorder: Myasthenia gravis, diphtheria, tetanus
- Epidemiology: Unusual in severe form
- Associated abnormalities
    - Pharyngeal paresis, gastroesophageal reflux
    - Esophageal motility disturbance: Spasm/achalasia
    - Aging, Zenker diverticulum

Gross Pathologic & Surgical Features
- Hypertrophy of cricopharyngeus muscle

CLINICAL ISSUES

Presentation
- Most common signs/symptoms: Intermittent symptoms: Dysphagia, food "sticking" in throat

Demographics
- Age: Any age group, more (10%) in elderly
- Gender: M = F

Natural History & Prognosis
- Natural history in infants
    - Prominent cricopharyngeal impression on barium-filled esophagus
    - Does not impede passage of fluids or food
    - Usually resolves within a few weeks of birth
- Complications
    - Rare: Aspiration, pneumonia, lung abscess
- Prognosis
    - Usually good

Treatment
- Cricopharyngeal myotomy
- Treat underlying problem
    - Such as reflux esophagitis with spasm
    - Neuromuscular disorder

DIAGNOSTIC CHECKLIST

Consider
- Persistent narrowing or just intermittent indentation

Image Interpretation Pearls
- Smoothly outlined lip-like projection posteriorly at C5-6 level with jet effect seen via narrowed lumen

SELECTED REFERENCES

1. Curtis DJ et al: The cricopharyngeal muscle: A videorecording review. AJR. 142: 497-500, 1984
2. Ekberg O et al: Dysfunction of the cricopharyngeal muscle. A cineradiographic study of patients with dysphagia. Radiology. 143: 481-6, 1982
3. Bergman AB et al: Complete esophageal obstruction from cricopharyngeal achalasia. Radiology. 123: 289-90, 1977

IMAGE GALLERY

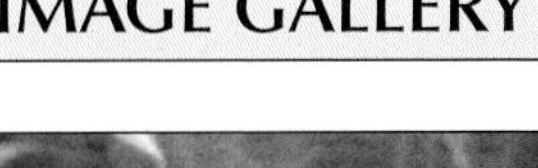

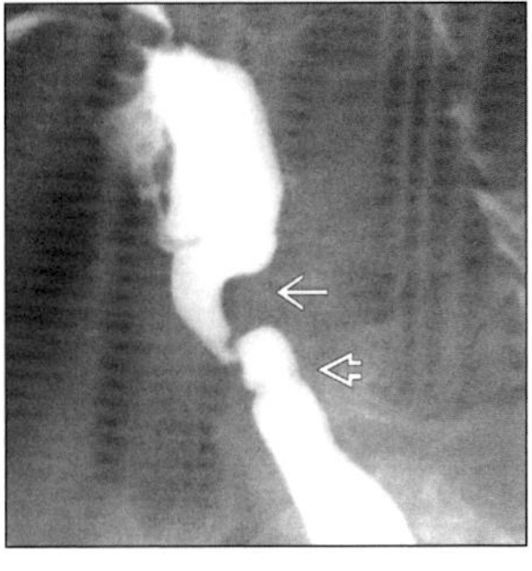

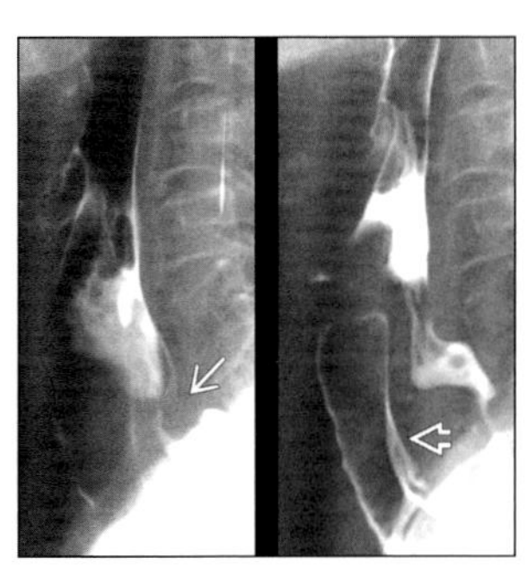

*(**Left**) Lateral esophagram shows prominent cricopharyngeus (arrow), pharyngeal distention, and an esophageal web (open arrow). (**Right**) Two views from rapid sequence esophagram show cricopharyngeal achalasia (arrow) and tracheal aspiration (open arrow).*

ACHALASIA, ESOPHAGUS

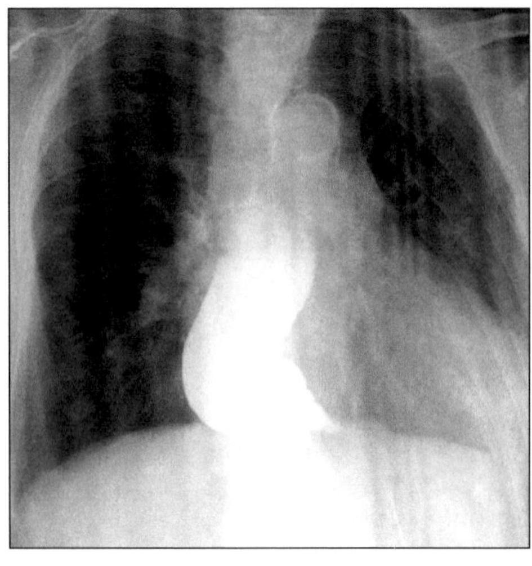

Upright frontal esophagram shows dilated esophagus with an abrupt taper ("bird beak") just above the GE junction. Note absent gastric air bubble.

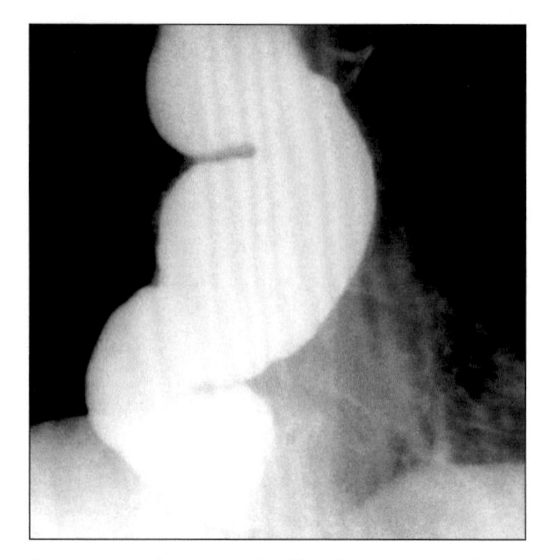

Esophagram shows grossly dilated, tortuous esophagus with "sigmoid" appearance.

TERMINOLOGY

Definitions

- Primary motility disorder, esophageal smooth muscle

IMAGING FINDINGS

General Features

- Best diagnostic clue: "Bird-beak" deformity-dilated esophagus with smooth, symmetric, tapered narrowing at esophagogastric region
- Morphology: Grossly dilated esophagus with smooth tapering at lower end of esophagus
- Other general features
  - Classified based on etiology
    - Primary (idiopathic)
    - Secondary (pseudoachalasia)
  - Manometric characteristics of achalasia
    - Absence of primary peristalsis
    - Increased or normal resting lower esophageal sphincter (LES) pressures
    - Incomplete or absent LES relaxation on swallowing
  - Variants of achalasia: Atypical manometric findings
    - Early: Characterized by aperistalsis with normal LES pressure
    - Vigorous: Simultaneous high-amplitude & repetitive contractions
    - Both variants are transitional & finally evolve into classic achalasia
  - Classic achalasia (primary): Simultaneous low-amplitude contractions
  - Motor function of pharynx & upper esophageal sphincter are normal

Radiographic Findings

- Radiography
  - Chest x-ray AP & lateral views: Advanced achalasia
    - Classic mediastinal widening; double contour of mediastinal borders
    - Outer borders represent dilated esophagus projecting beyond shadows of aorta & heart
    - Anterior tracheal bowing
    - Air-fluid level in mediastinum; small or absent gastric air bubble
    - Lower lobes: Decreased lung volume, linear opacities & tubular radiolucencies
- Videofluoroscopic barium study findings
  - Primary achalasia
    - Markedly dilated esophagus
    - Absent primary peristalsis
    - "Bird-beak" deformity: V-shaped conical & smooth, symmetric, tapered narrowing of distal esophagus extending to GE junction

DDx: Dilated Nonperistaltic Esophagus

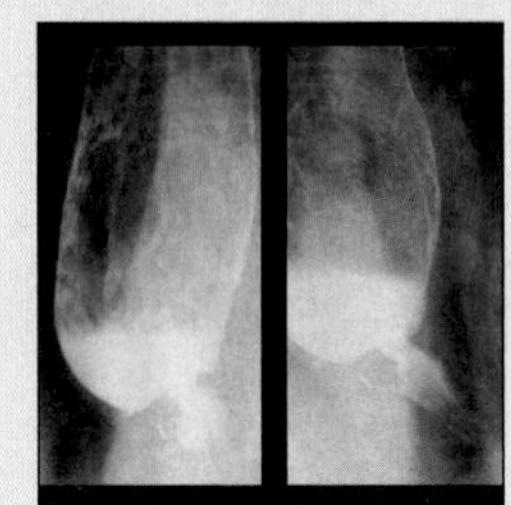

Scleroderma

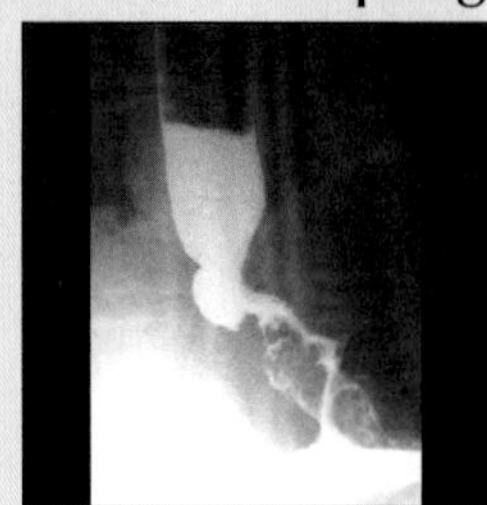

Gastric Ca.

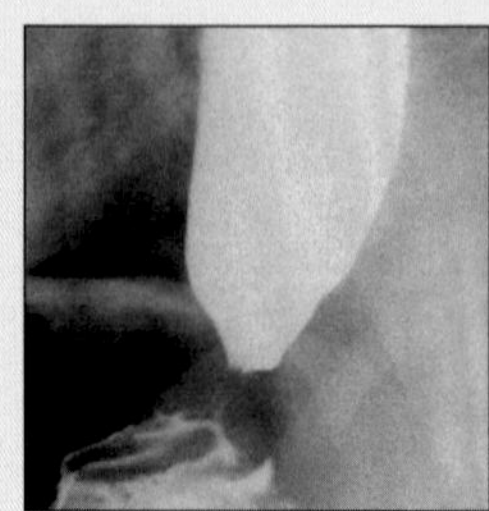

Peptic Stricture

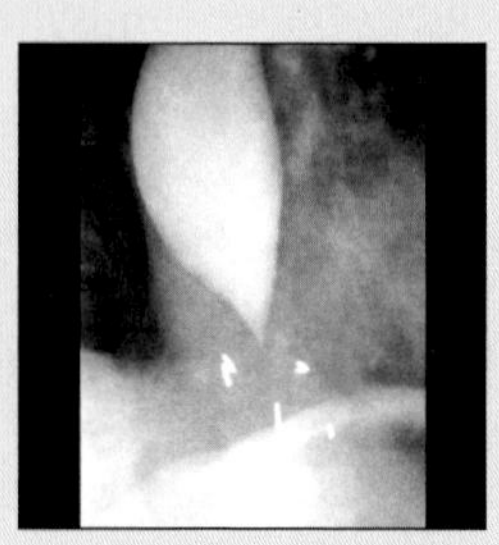

Post Vagotomy

ACHALASIA, ESOPHAGUS

Key Facts

Terminology
- Primary motility disorder, esophageal smooth muscle

Imaging Findings
- Best diagnostic clue: "Bird-beak" deformity-dilated esophagus with smooth, symmetric, tapered narrowing at esophagogastric region
- Absent primary peristalsis
- Timed barium swallow studies

Top Differential Diagnoses
- Scleroderma
- Esophageal carcinoma
- Gastric carcinoma
- Esophagitis with stricture
- Neuromuscular disorders
- Post surgical change (post vagotomy)

Pathology
- Primary achalasia: Idiopathic or neurogenic disorder
- Abnormality of myenteric ganglia (decrease in number) in Auerbach plexus
- Degenerative vagal nerve changes
- Secondary achalasia: Intrinsic/extrinsic neoplasm, peptic/fungal stricture, scleroderma, Chagas disease, postvagotomy effect

Clinical Issues
- Longstanding dysphagia, weight loss
- Regurgitation, foul breath
- Esophageal carcinoma (in 2-7% of cases)

Diagnostic Checklist
- Seek evidence of cancer, prior surgery, severe GERD

- Hurst phenomenon: Temporary transit via cardia when hydrostatic pressure of barium column is above tonic LES pressure
- Length of narrowed segment: Less than 3.5 cm; widest diameter is more than 4 cm
- Secondary achalasia (pseudoachalasia)
  - Mildly dilated esophagus (less than 4 cm at its widest point)
  - Decreased or absent peristalsis
  - Eccentricity, nodularity, shouldering of narrowed distal segment or
  - Smooth, symmetric, tapered narrowing of distal esophagus
  - Length of distal narrowed esophageal segment: More than 3.5 cm (approximately)

CT Findings
- Moderate to marked dilatation of esophagus with diameter more than 4 cm
- Decreased or normal wall thickness
- Air-fluid level within dilated esophagus
- Abrupt, smooth narrowing distal esophageal segment near gastroesophageal (GE) junction
- Squamous cell carcinoma of esophagus in longstanding achalasia
  - Mild dilatation of esophagus; irregular wall thickening
  - Eccentric narrowing of distal esophagus extending into GE junction
  - Involvement of periesophageal soft tissues & blood vessels
  - Enlarged mediastinal lymph nodes

Imaging Recommendations
- Videofluoroscopic barium studies
- Helical CT including sagittal reconstructions
- Timed barium swallow studies
  - Aid in diagnosis & management of achalasia
- Transit & emptying studies
  - Quantitate esophageal retention before & after therapy
  - Standing films taken after ingestion of 200 ml barium: At 1, 2, and 5 minutes

DIFFERENTIAL DIAGNOSIS

Scleroderma
- Collagen vascular disease; more common in women
- Pathogenesis
  - Smooth muscle atrophy/fibrosis; incompetent LES
  - Reflux esophagitis & high risk of peptic stricture & Barrett esophagus
- Uncomplicated cases: Dilated esophagus with a patulous esophagogastric region
- Fluoroscopic barium study findings
  - Aperistalsis, dilated esophagus
  - Peptic stricture in distal one third
    - May simulate primary achalasia
  - Diagnosis
    - Anti-Scl 70 & anticentromere antibodies
    - Endoscopic biopsy

Esophageal carcinoma
- Asymmetric contour with abrupt proximal borders of narrowed distal segment (rat-tail appearance)
- Mucosal irregularity, shouldering, mass effect
- Periesophageal & distal spread may be seen
- Occasionally, smooth, tapered narrowing of lower esophagus with aperistalsis, simulating achalasia
- Diagnosis: Endoscopic biopsy & history

Gastric carcinoma
- Malignancy involving gastric cardia with extension into distal esophagus shows
  - Smooth, tapered narrowing of lower esophagus with aperistalsis, simulating achalasia
- Growth may show mucosal irregularity & mass effect
- Diagnosis: Endoscopic biopsy & history

Esophagitis with stricture
- Peptic stricture: Smooth tapered narrowing of short distal segment
  - Almost always associated with hiatal hernia & gastroesophageal (GE) reflux
  - Rarely associated with aperistalsis or gross dilatation
- Mucosa: Fine nodular/granular or plaque-like
- Ulcers: Multiple tiny, radiating folds & puckering

ACHALASIA, ESOPHAGUS

- Folds: Thickened vertical or transverse
- Diagnosis: Endoscopic biopsy & history

Neuromuscular disorders

- Diffuse esophageal spasm (DES)
  - characterized by chest pain, radiation to shoulder simulating angina & dysphagia
  - Cause is unknown; may be neurogenic damage
  - Intermittent disruption of primary peristalsis associated with focally obliterative simultaneous contractions
  - Sometimes contractions are repetitive & esophageal lumen may show typical "corkscrew" or "rosary bead" appearance
  - Mostly normal LES function with complete sphincter relaxation during swallowing
  - Diagnosis: Clinical, radiographic & manometric

Post surgical change (post vagotomy)

- Stricture and neural damage may simulate achalasia

PATHOLOGY

General Features

- General path comments
  - Progressive dilatation of esophagus above LES
  - Marked esophageal wall thinning + risk of rupture
  - Following mucosal changes may be seen
    - Ulceroinflammatory lesions with white thickened patches (leukoplakia)
    - Superimposed infection (e.g., Candida esophagitis)
    - Sites of dysplasia or neoplasia
- Etiology
  - Primary achalasia: Idiopathic or neurogenic disorder
  - Pathogenesis of neurogenic disorder
    - Abnormality of myenteric ganglia (decrease in number) in Auerbach plexus
    - Degenerative vagal nerve changes
    - Decreased cell bodies in dorsal motor vagal nucleus: Primary defect is in extraesophageal parasympathetic nerve supply & esophageal changes are secondary
  - Secondary achalasia: Intrinsic/extrinsic neoplasm, peptic/fungal stricture, scleroderma, Chagas disease, postvagotomy effect
- Epidemiology: Uncommon primary motility disorder

Gross Pathologic & Surgical Features

- Massively dilated esophagus with smooth narrowed distal segment

Microscopic Features

- Decreased number of ganglion cells in myenteric plexus of esophagus

CLINICAL ISSUES

Presentation

- Most common signs/symptoms
  - Primary achalasia
    - Longstanding dysphagia, weight loss
    - Regurgitation, foul breath
  - Secondary achalasia
    - Short duration of dysphagia
    - Chest pain or odynophagia

Demographics

- Age
  - Primary achalasia: Younger patients (30-50 years)
  - Secondary achalasia: Older patients
- Gender: Equal in both males & females

Natural History & Prognosis

- Complications
  - Coughing, aspiration, pneumonia, lung abscess
  - Esophageal carcinoma (in 2-7% of cases)
- Prognosis
  - Treatment cannot correct abnormal esophageal motility & LES dysfunction
  - Aimed at improving esophageal emptying by disrupting increased LES pressure

Treatment

- Calcium channel blockers; botulinum toxin injection
- Pneumatic dilatation
- Heller myotomy (partial thickness incision of LES)
- Risks of treatment
  - Pneumatic dilatation: Perforation
  - Myotomy: Outpouching & reflux

DIAGNOSTIC CHECKLIST

Consider

- Seek evidence of cancer, prior surgery, severe GERD

Image Interpretation Pearls

- Markedly dilated esophagus, absence of primary peristalsis & "bird-beak" deformity of distal esophagus

SELECTED REFERENCES

1. Vaezi MF et al: Timed barium oesophagram: better predictor of long term success after pneumatic dilation in achalasia than symptom assessment. Gut. 50(6):765-70, 2002
2. Sabharwal T et al: Balloon dilation for achalasia of the cardia: experience in 76 patients. Radiology. 224(3):719-24, 2002
3. Adler DG et al: Primary esophageal motility disorders. Mayo Clin Proc. 76(2):195-200, 2001
4. Woodfield CA et al: Diagnosis of primary versus secondary achalasia: reassessment of clinical and radiographic criteria. AJR Am J Roentgenol. 175(3):727-31, 2000
5. de Oliveira JM et al: Timed barium swallow: a simple technique for evaluating esophageal emptying in patients with achalasia. AJR Am J Roentgenol. 169(2):473-9, 1997
6. Noh HM et al: CT of the esophagus: Spectrum of disease with emphasis on esophageal carcinoma. Radiographics 15: 1113-34, 1995
7. Schima W et al: Esophageal motor disorders: Videofluoroscopic and manometric evaluation-prospective study in 88 symptomatic patients. Radiology 185: 487-91, 1992

ACHALASIA, ESOPHAGUS

IMAGE GALLERY

Typical

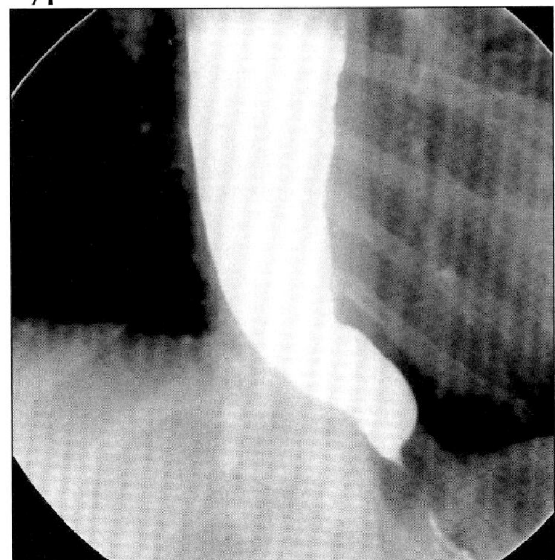

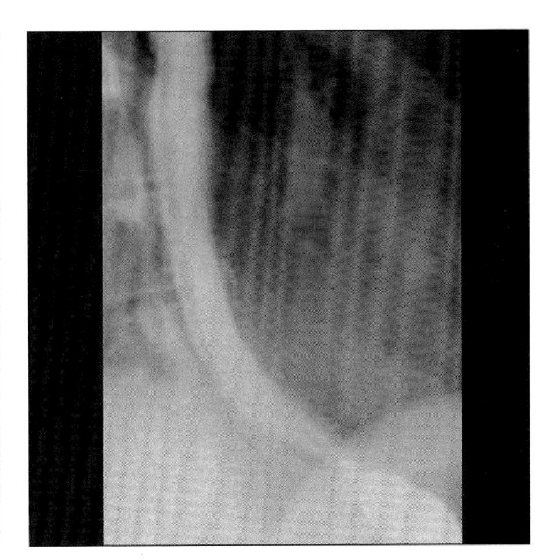

(Left) *Typical esophagram appearance of achalasia, pre Heller myotomy.* ***(Right)*** *Esophagram shows marked reduction of esophageal dilatation and retention following Heller myotomy.*

Typical

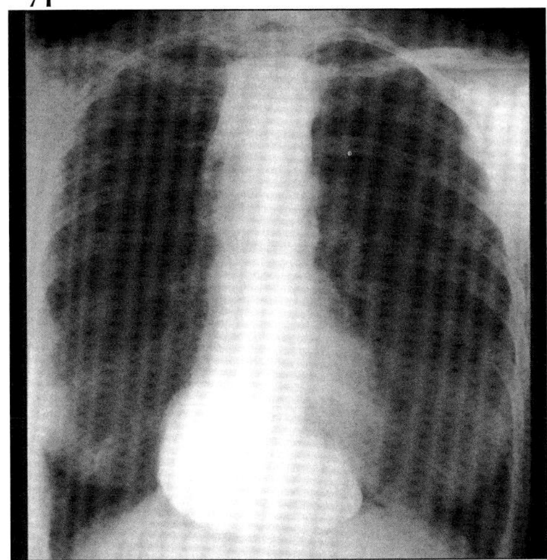

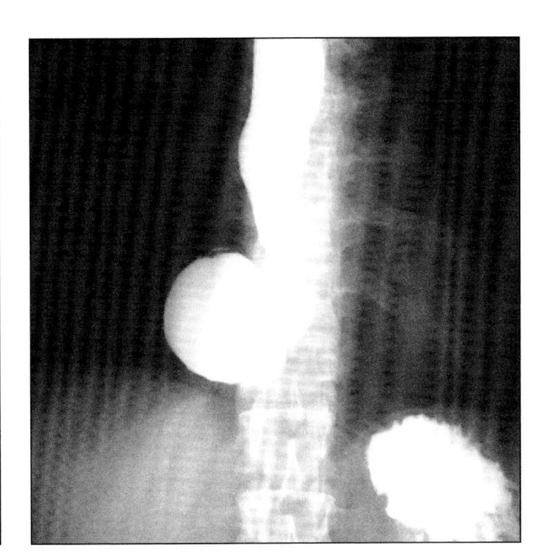

(Left) *Esophagram in upright position shows static column of barium at thoracic inlet; patient at a great risk for aspiration pneumonitis.* ***(Right)*** *Esophagram shows large pulsion epiphrenic diverticulum in a patient with achalasia.*

Typical

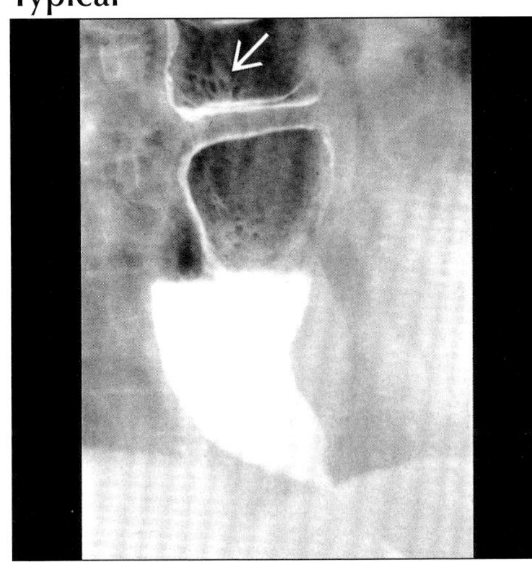

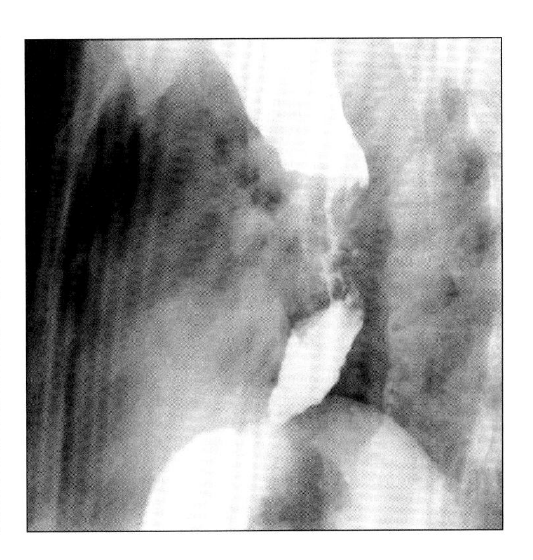

(Left) *Esophagram shows typical findings of achalasia plus numerous irregular plaques (arrow) due to Candida esophagitis.* ***(Right)*** *Esophagram in a patient with long-standing achalasia shows an "apple core" irregular constricting mass; squamous cell carcinoma.*

ESOPHAGEAL MOTILITY DISTURBANCES

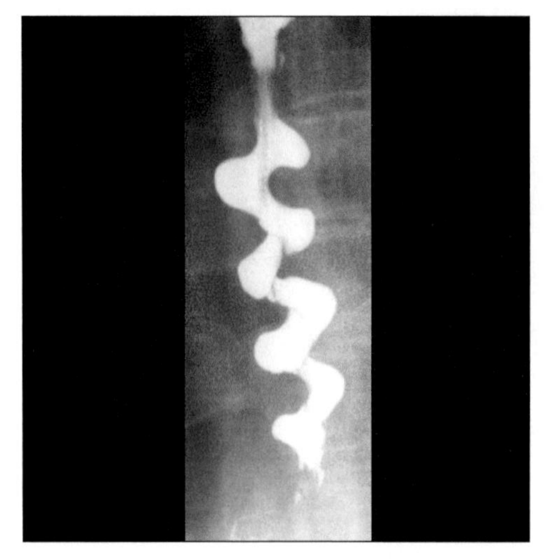

Esophagram shows "corkscrew" esophagus due to diffuse esophageal spasm.

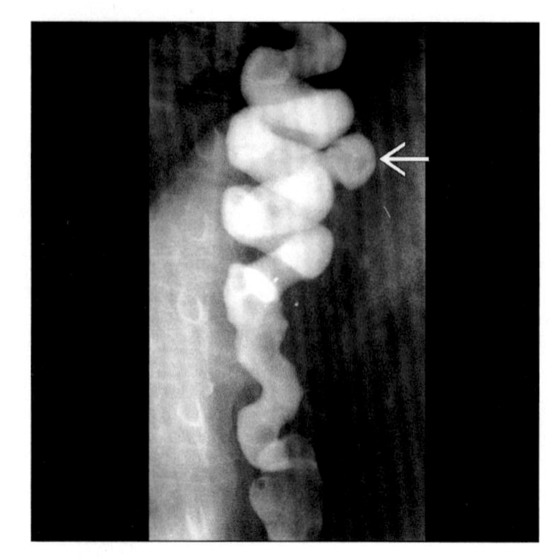

Esophagram shows diffuse esophageal spasm with corkscrew appearance. Also note pulsion diverticulum (arrow).

TERMINOLOGY

Abbreviations and Synonyms

- Diffuse esophageal spasm (DES)
- Presbyesophagus or nonspecific esophageal motility disorder (NEMD)

Definitions

- Primary & secondary motility disorders of esophageal smooth muscle

IMAGING FINDINGS

General Features

- Best diagnostic clue
  - Achalasia: "Bird-beak" deformity-dilated esophagus with smooth, tapered narrowing at GE junction
  - Scleroderma: Dilated atonic esophagus, distal stricture
- Other general features
  - Classification of esophageal motility disorders
    - Primary: Achalasia; diffuse esophageal spasm; presbyesophagus
    - Secondary: Scleroderma
  - Achalasia: Idiopathic or neurogenic disorder
    - Absence of primary peristalsis
    - Simultaneous low-amplitude contractions
    - Increased or normal resting lower esophageal sphincter (LES) pressures
    - Incomplete or absent LES relaxation on swallowing
    - Normal upper esophageal sphincter
  - Diffuse esophageal spasm: Related to varying degrees of neurogenic damage
    - Simultaneous contractions & intermittent primary peristalsis
    - Repetitive or prolonged-duration contractions
    - High amplitude & frequent spontaneous contractions
    - Normal LES function with complete sphincter relaxation during swallowing
    - Intermittent disruption of primary peristalsis associated with focally obliterative simultaneous contractions
  - Presbyesophagus: Esophageal motility dysfunction associated with aging
    - Also called nonspecific esophageal motility disorder (NEMD)
    - Decreased frequency of normal peristalsis
    - ↑ Frequency of nonperistaltic contractions
    - Less commonly, incomplete LES relaxation
  - Scleroderma: Multisystemic disorder of small vessels & connective tissue
    - Decreased or absent resting LES pressure
    - Absent peristalsis in lower two thirds of esophagus

DDx: Abnormal Contour or Contractions of Esophagus

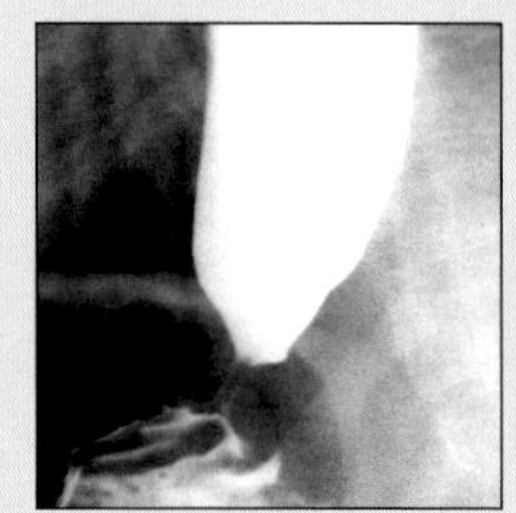

Peptic Stricture

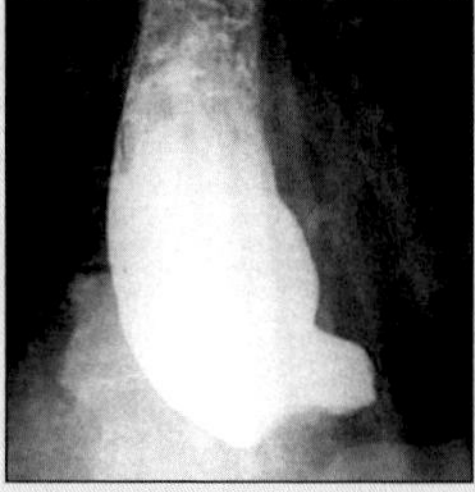

Post Fundoplication

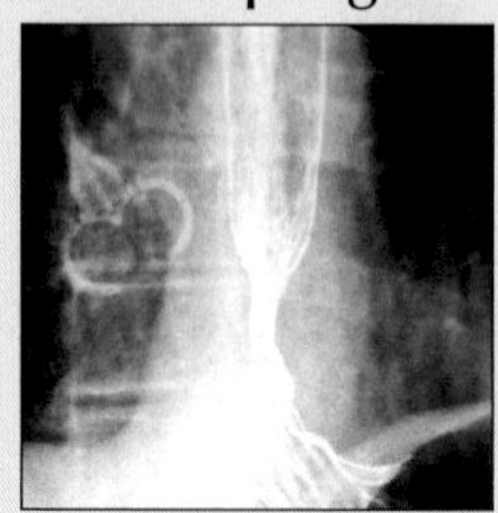

Esophageal Cancer

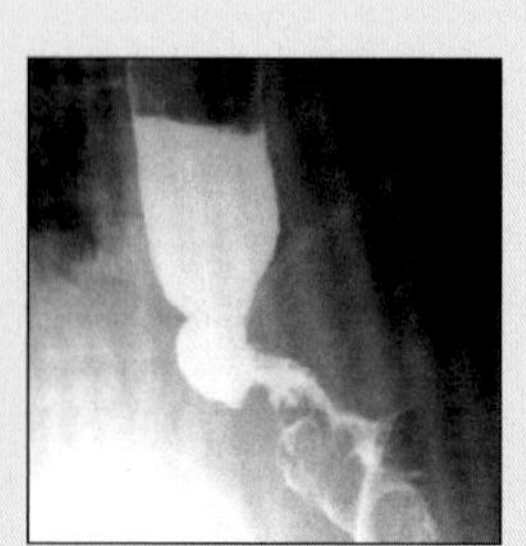

Gastric Cancer

ESOPHAGEAL MOTILITY DISTURBANCES

Key Facts

Terminology

- Diffuse esophageal spasm (DES)
- Presbyesophagus or nonspecific esophageal motility disorder (NEMD)
- Primary & secondary motility disorders of esophageal smooth muscle

Imaging Findings

- Achalasia: "Bird-beak" deformity-dilated esophagus with smooth, tapered narrowing at GE junction
- Scleroderma: Dilated atonic esophagus, distal stricture
- Intermittent absence of primary peristalsis in thoracic esophagus + focally obliterative contractions (DES)
- Presbyesophagus or NEMD: Multiple nonperistaltic contractions & disrupted primary peristalsis
- Scleroderma: 70% GE reflux → 37% Barrett esophagus

Top Differential Diagnoses

- Esophagitis with stricture
- Post-fundoplication
- Esophageal carcinoma

Diagnostic Checklist

- Differentiate from conditions which can simulate primary/secondary esophageal motility disorders
- Correlate: Clinical/radiographic/manometric findings
- Check family history of collagen vascular disorders
- Achalasia: Dilated esophagus, absence of primary peristalsis & "bird-beak" deformity of distal esophagus
- Scleroderma: Mild-moderate dilatation of esophagus with distal fusiform stricture + ↓ or absent peristalsis
- DES: Disrupted primary peristalsis; "corkscrew" or "rosary bead" pattern of esophagus

Radiographic Findings

- Videofluoroscopic barium studies
  - Achalasia
    - Markedly dilated esophagus
    - Absent primary peristalsis
    - "Bird-beak" deformity: V-shaped conical & smooth, tapered narrowing at esophagogastric region
  - Diffuse esophageal spasm
    - Primary peristalsis present in cervical esophagus
    - Intermittent absence of primary peristalsis in thoracic esophagus + focally obliterative contractions (DES)
    - Contractions are repetitive, esophageal lumen may show "corkscrew" or "rosary bead" pattern
  - Presbyesophagus or NEMD: Multiple nonperistaltic contractions & disrupted primary peristalsis
  - Scleroderma
    - Mild-moderate dilatation of proximal esophagus
    - Absence of peristalsis in lower 2/3rd of esophagus
    - Patulous GE region + reflux → fusiform distal peptic stricture; ± hiatal hernia
    - Erosions & superficial ulcers in distal esophagus
    - ± Wide-mouthed sacculations of esophagus
    - Scleroderma: 70% GE reflux → 37% Barrett esophagus

Imaging Recommendations

- Videofluoroscopic barium studies
- Timed barium swallow (transit & emptying) studies
  - Aid in diagnosis & management

DIFFERENTIAL DIAGNOSIS

Esophagitis with stricture

- Mucosa: Fine nodular/granular or plaque-like
- Ulcers: Multiple tiny ulcers + thickened folds
- Peptic stricture: Smooth, narrowing of distal segment
  - Usually associated with hiatal hernia & GE reflux
  - Occasionally may simulate achalasia in absence of hiatal hernia & GE reflux
  - Distinguished from motility disorders by normal peristalsis

Post-fundoplication

- Dilated esophagus, narrowed GE junction + delayed esophageal emptying
- Fairly common result or complication of fundoplication, especially in elderly

Esophageal carcinoma

- Asymmetric contour with abrupt proximal borders of narrowed distal segment (rat-tail appearance)
- Mucosal irregularity, shouldering, mass effect
- Periesophageal & distal spread may be seen
- Smooth, tapered narrowing of lower esophagus with aperistalsis simulates achalasia or scleroderma
- Gastric cancer invading esophagus submucosa may closely mimic achalasia
- Diagnosis: Endoscopic biopsy & history

PATHOLOGY

General Features

- General path comments
  - Achalasia
    - Progressive dilatation of esophagus above LES
    - Thin esophageal wall (↑ risk of rupture)
    - May have superimposed infection (e.g., Candida)
    - May show sites of dysplasia or neoplasia
  - Diffuse esophageal spasm
    - Esophageal muscle: Normal (< 4 mm) or thickened
  - Scleroderma
    - Initially smooth muscle atrophy & fragmentation
    - Followed by collagen deposition & fibrosis
- Genetics
  - Scleroderma
    - Localized: Associated with HLA-DR1, 4 & 5
    - Diffuse: Associated with HLA-DR5
- Etiology
  - Achalasia: Idiopathic or neurogenic disorder
    - Abnormality of myenteric ganglia in Auerbach plexus; degenerative vagal nerve changes

ESOPHAGEAL MOTILITY DISTURBANCES

- Diffuse esophageal spasm
  - Unknown
  - Related to varying degrees of neurogenic damage
- Presbyesophagus: Due to aging
- Scleroderma
  - Unknown; autoimmune + genetic predisposition
  - Environmental antigens: Silica, L-tryptophan
- Epidemiology
  - Primary motility disorders: Uncommon
  - Secondary motility disorder (scleroderma)
    - Incidence: 14.1 cases/million
    - Prevalence: 19-75/100,000 people
- Associated abnormalities: Scleroderma may be associated with SLE, polymyositis or dermatomyositis

Gross Pathologic & Surgical Features

- Achalasia: Massively dilated esophagus with smooth, narrowed distal segment
- DES: Esophageal muscle, normal or hypertrophied
- Scleroderma
  - Rubber-hose inflexibility: Lower 2/3rd esophagus
  - Thin & ulcerated mucosa + distal stricture

Microscopic Features

- Achalasia: ↓ Number of ganglia in myenteric plexus
- DES: Cellular hypertrophy of esophageal muscle
- Scleroderma: Atrophy + fragmentation of smooth muscle → collagen deposition + fibrosis

CLINICAL ISSUES

Presentation

- Most common signs/symptoms
  - Longstanding dysphagia for solids or liquids
  - Epigastric fullness, regurgitation, foul breath
  - Sensation of food sticking in chest; weight loss
  - DES
    - Chest pain ± radiation to shoulder or back simulating angina
    - Acute odynophagia due to food impaction
- Lab-data: Manometric findings
- Diagnosis
  - Clinical, radiographic & manometric findings
  - Endoscopic biopsy & histology

Demographics

- Age
  - Achalasia & scleroderma (young patients)
  - Presbyesophagus: Elderly people
- Gender
  - Primary motility disorders: M = F
  - Scleroderma
    - Young African-American females (> common); M:F = 1:3
- Ethnicity: Scleroderma: African-Americans more than Caucasians

Natural History & Prognosis

- Complications
  - Coughing, aspiration, pneumonia, lung abscess
  - Scleroderma: Barrett esophagus, adenocarcinoma
- Prognosis
  - Achalasia: Fair by improving esophageal emptying with balloon dilatation or myotomy
  - Scleroderma: Limited disease (good); diffuse (poor)

Treatment

- Achalasia: Ca++ channel blockers; botulinum toxin injection; pneumatic dilatation; Heller myotomy
- Scleroderma: Elevation of head end of bed; H-2 receptor antagonists; metoclopramide

DIAGNOSTIC CHECKLIST

Consider

- Differentiate from conditions which can simulate primary/secondary esophageal motility disorders
- Correlate: Clinical/radiographic/manometric findings
- Check family history of collagen vascular disorders

Image Interpretation Pearls

- Achalasia: Dilated esophagus, absence of primary peristalsis & "bird-beak" deformity of distal esophagus
- Scleroderma: Mild-moderate dilatation of esophagus with distal fusiform stricture + ↓ or absent peristalsis
- DES: Disrupted primary peristalsis; "corkscrew" or "rosary bead" pattern of esophagus
- Presbyesophagus: Nonperistaltic contractions in an elderly person

SELECTED REFERENCES

1. Adler DG et al: Primary esophageal motility disorders. Mayo Clin Proc. 76(2):195-200, 2001
2. Coggins CA et al: Wide-mouthed sacculations in the esophagus: a radiographic finding in scleroderma. AJR Am J Roentgenol. 176(4):953-4, 2001
3. Woodfield CA et al: Diagnosis of primary versus secondary achalasia: reassessment of clinical and radiographic criteria. AJR Am J Roentgenol. 175(3):727-31, 2000
4. Schima W et al: Radiographic detection of achalasia: diagnostic accuracy of videofluoroscopy. Clin Radiol. 53(5):372-5, 1998
5. Ott DJ: Esophageal motility disorders. Semin Roentgenol. 29(4):321-31, 1994
6. Schima W et al: Esophageal motor disorders: videofluoroscopic and manometric evaluation--prospective study in 88 symptomatic patients. Radiology. 185(2):487-91, 1992
7. Levine MS et al: Update on esophageal radiology. AJR 155: 933-41, 1990
8. Ott DJ et al: Radiologic evaluation of esophageal motility: results in 170 patients with chest pain. AJR Am J Roentgenol. 155(5):983-5, 1990
9. Ott DJ et al: Esophageal motility: assessment with synchronous video tape fluoroscopy and manometry. Radiology. 173(2):419-22, 1989
10. Chen YM et al: Diffuse esophageal spasm: radiographic and manometric correlation. Radiology. 170(3 Pt 1):807-10, 1989
11. Ott DJ et al: Esophageal radiography and manometry: correlation in 172 patients with dysphagia. AJR Am J Roentgenol. 149(2):307-11, 1987

IMAGE GALLERY

Typical

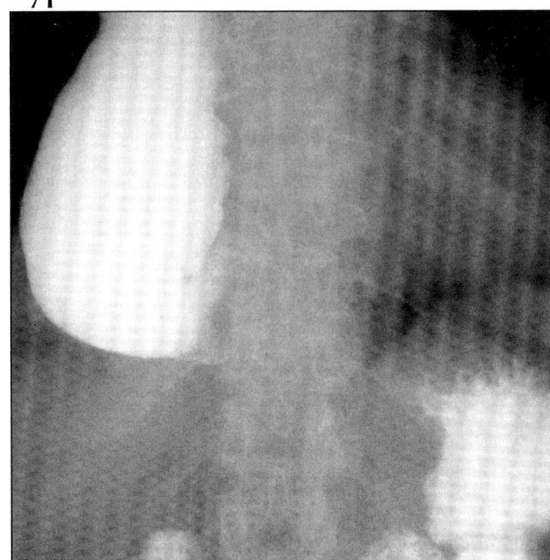

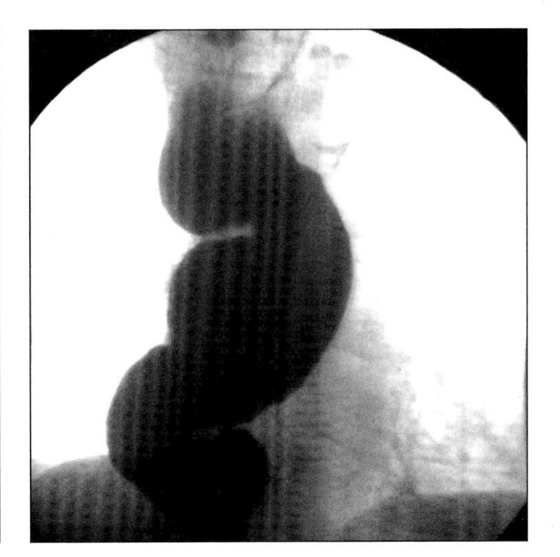

(Left) *Esophagram shows dilated atonic esophagus and complete spasm of LES; achalasia.* ***(Right)*** *Esophagram shows markedly dilated, elongated ("sigmoid") esophagus due to achalasia.*

Typical

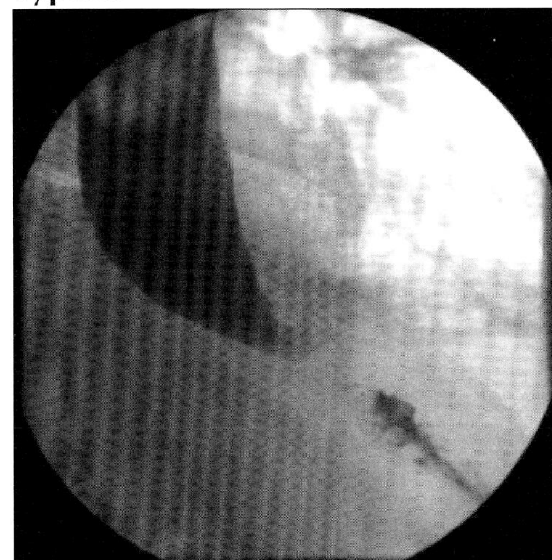

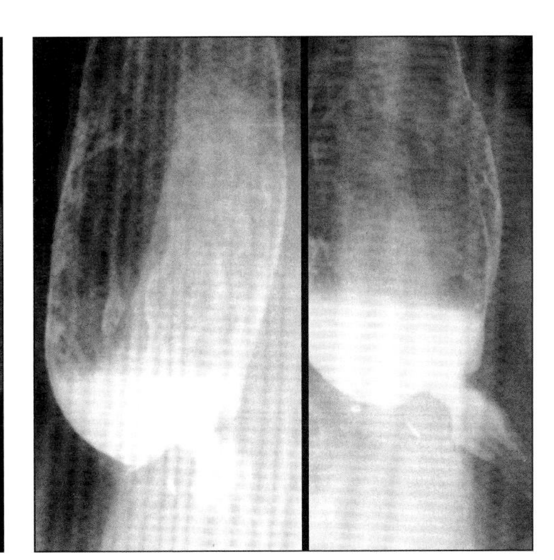

(Left) *Esophagram shows dilated, atonic esophagus with distal stricture due to scleroderma.* ***(Right)*** *Esophagram shows markedly dilated esophagus and short distal stricture due to scleroderma.*

Typical

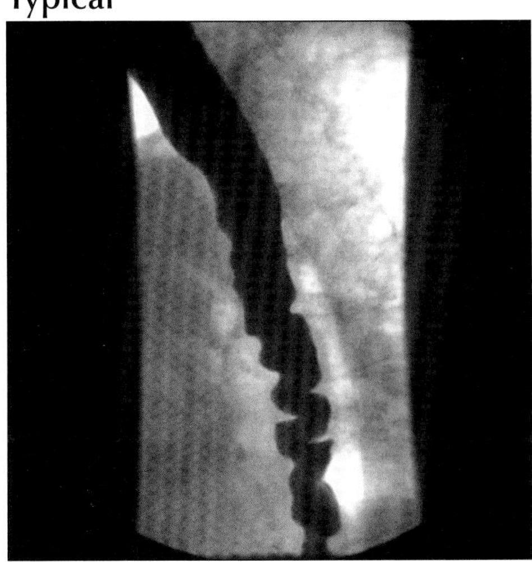

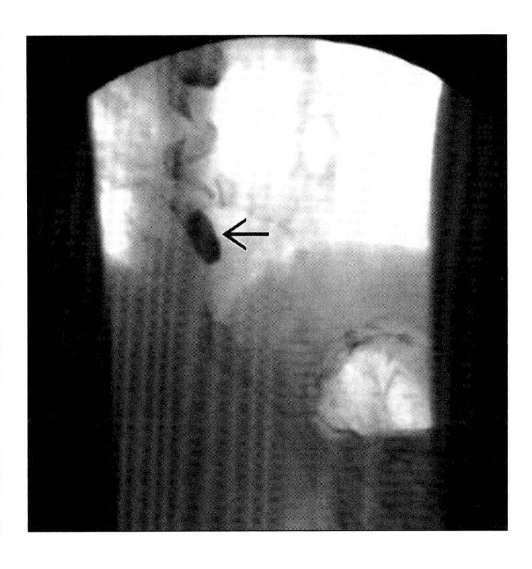

(Left) *Esophagram shows prominent tertiary, nonpropulsive contractions in an elderly patient ("presbyesophagus").* ***(Right)*** *Barium pill (arrow) is retained in LES after esophagram and swallow of water. Note underlying tertiary contractions/spasms.*

SCLERODERMA, ESOPHAGUS

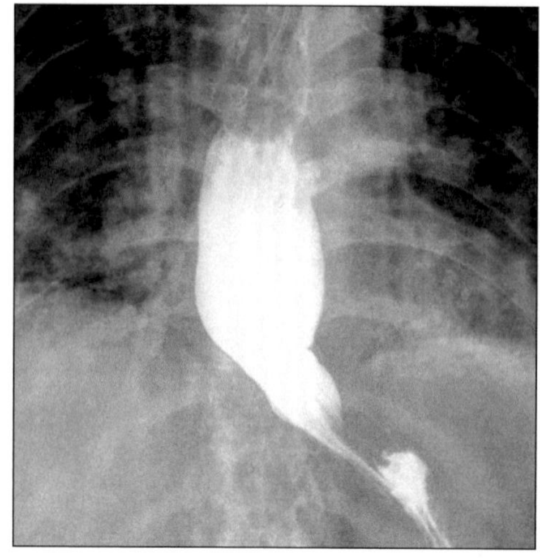
Upper GI shows dilated, atonic esophagus with distal stricture. Note underlying lung disease.

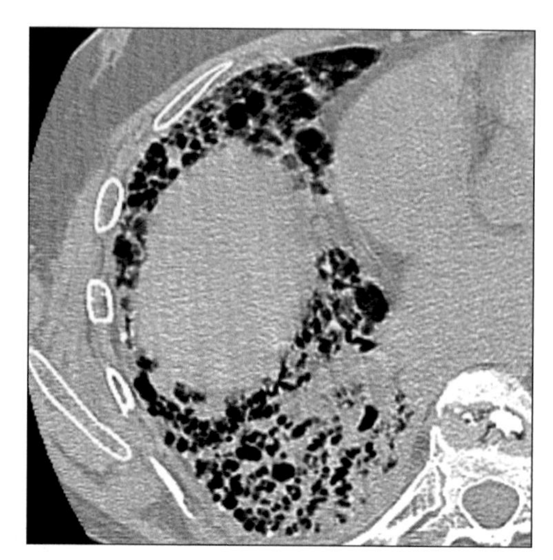
Axial NECT shows "honeycomb" lung. Interstitial fibrosis due to scleroderma.

TERMINOLOGY

Abbreviations and Synonyms

- Progressive systemic sclerosis (PSS)

Definitions

- Multisystemic disorder of small vessels & connective tissue (collagen vascular disease) of unknown etiology

IMAGING FINDINGS

General Features

- Best diagnostic clue: Dilated atonic esophagus with distal stricture
- Other general features
  - Multisystemic disorder with immunologic & inflammatory changes
  - Characterized by: Atrophy/fibrosis/sclerosis of skin, vessels & organs
  - Involves skin, synovium, & parenchyma of multiple organs like
    - Gastrointestinal tract (GIT), lungs, heart, kidneys & nervous system
  - Gastrointestinal tract scleroderma
    - 3rd most common manifestation after skin changes & Raynaud phenomenon
    - Seen in up to 90% of patients
    - Most common sites: Esophagus > anorectal > small bowel > colon
    - Most frequent cause of chronic intestinal pseudo-obstruction
  - Scleroderma is subclassified into two types
    - Diffuse scleroderma
    - CREST syndrome (benign course)
  - Diffuse scleroderma: Diffuse cutaneous + early visceral involvement
    - Severe interstitial pulmonary fibrosis
    - Tends to involve women/organ failure more likely
    - Associated with antitopoisomerase 1 antibody (anti-Scl 70)
  - CREST syndrome: Minimal cutaneous & late visceral involvement
    - C: Calcinosis of skin
    - R: Raynaud phenomenon
    - E: Esophageal dysmotility
    - S: Sclerodactyly (involvement of fingers)
    - T: Telangiectasia
    - Associated with anticentromere antibodies

Radiographic Findings

- Fluoroscopic guided (esophagography)
  - Normal peristalsis above aortic arch (striated muscle in proximal 1/3)
  - Hypotonia or atony or aperistalsis: Lower 2/3 esophagus (smooth muscle)
  - Mild-moderate dilatation of esophagus

DDx: Esophageal Stricture +/- Dysmotility

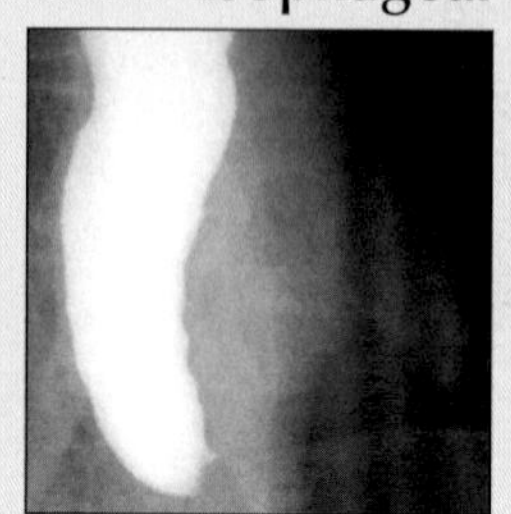
Achalasia

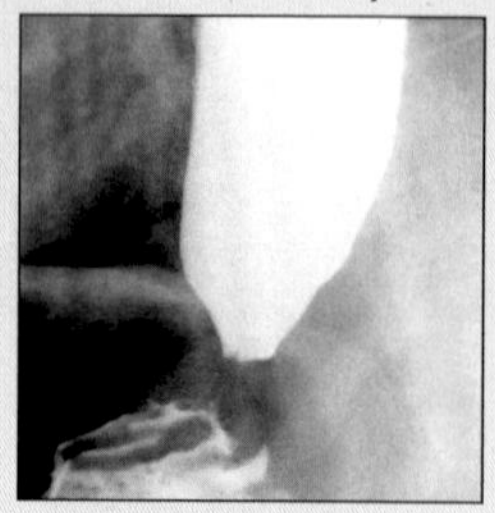
Peptic Stricture

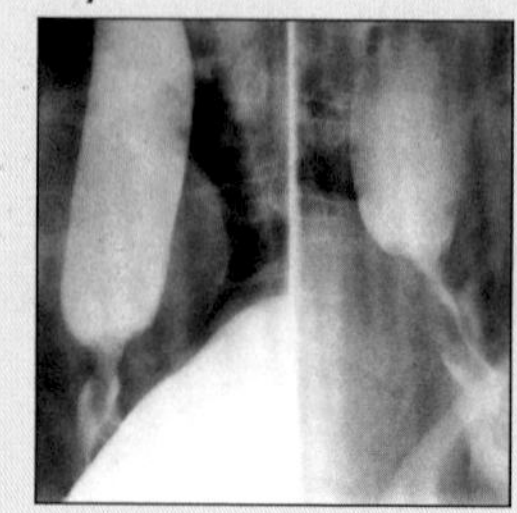
Esophageal Cancer

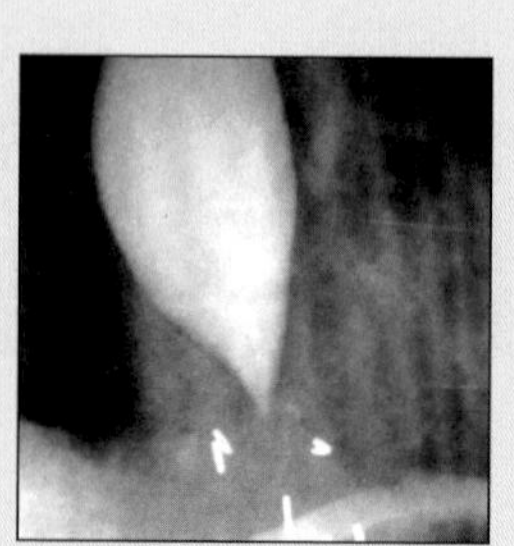
Post Vagotomy

SCLERODERMA, ESOPHAGUS

Key Facts

Terminology
- Multisystemic disorder of small vessels & connective tissue (collagen vascular disease) of unknown etiology

Imaging Findings
- Best diagnostic clue: Dilated atonic esophagus with distal stricture
- CREST syndrome (benign course)
- C: Calcinosis of skin
- R: Raynaud phenomenon
- E: Esophageal dysmotility
- S: Sclerodactyly (involvement of fingers)
- T: Telangiectasia
- Erosions, superficial ulcers, fusiform peptic stricture
- Gastroesophageal reflux (70% cases) → 37% develop Barrett esophagus

Top Differential Diagnoses
- Achalasia
- Peptic stricture
- Esophageal carcinoma
- Iatrogenic

Pathology
- Unknown; autoimmune condition with genetic predisposition
- Incidence: 14.1 cases/million

Diagnostic Checklist
- Rule out other pathologies that cause distal esophageal stricture ± dysmotility
- Check for family history of collagen vascular diseases
- Mild-moderate dilatation of esophagus with distal fusiform stricture + decreased or absent peristalsis

- Patulous lower esophageal sphincter (LES)
  - Early finding of scleroderma
- Erosions, superficial ulcers, fusiform peptic stricture
  - Due to reflux esophagitis
- Gastroesophageal reflux (70% cases) → 37% develop Barrett esophagus
- Hiatal hernia

- Fluoroscopic guided (barium meal)
  - Stomach: gastric dilatation + delayed emptying
- Fluoroscopic guided (small bowel study)
  - Marked dilatation of small-bowel (particularly 2nd, 3rd parts of duodenum & jejunum)
  - Pathognomonic: "Hidebound" sign
    - Dilated small-bowel (jejunum) with crowded normal circular folds
    - Seen in > 60% cases of scleroderma-related pseudo-obstruction
    - Due to muscle atrophy & its uneven replacement by collagen in longitudinal fibers + intense fibrosis of submucosa
  - Wide-mouthed sacculations (true diverticula) on antimesenteric border
  - Prolonged transit time with barium retention in duodenum up to 24 hrs
  - ± Pneumatosis intestinalis + pneumoperitoneum
  - ± Transient, nonobstructive intussusceptions
- Fluoroscopic guided (barium enema)
  - Sacculations on antimesenteric border (transverse & descending colon)
  - Marked dilatation (simulate Hirschsprung disease)
  - Chronic phase: Complete loss of haustrations
    - Simulates cathartic colon or chronic ulcerative colitis
  - Stercoral ulceration (from retained fecal material)
  - ± Benign pneumoperitoneum

Imaging Recommendations
- Fluoroscopic guided esophagography

DIFFERENTIAL DIAGNOSIS

Achalasia
- Best diagnostic clue
  - "Bird-beak" deformity: Grossly dilated esophagus with smooth beak-like tapering at lower end
  - Scleroderma shows mild-moderate dilatation of esophagus with fusiform stricture
- Absent primary peristalsis

Peptic stricture
- Concentric, smooth tapering of short distal esophageal segment with proximal dilatation
- Distinguished from scleroderma by normal peristalsis

Esophageal carcinoma
- Asymmetric contour with abrupt proximal borders of narrowed distal segment (rat-tail appearance)
- Mucosal irregularity, shouldering, mass effect
- Periesophageal & distal spread may be seen
- Absence of peristalsis in malignant stricture may simulate scleroderma
- Diagnosis: Endoscopic biopsy & history

Iatrogenic
- Example: Fundoplication & vagotomy
- Tight wrap narrows esophageal lumen
- Vagotomy, scarring decrease peristalsis

PATHOLOGY

General Features
- General path comments
  - Initial smooth muscle atrophy & fragmentation
  - Followed by collagen deposition & fibrosis
- Genetics
  - Localized: Anticentromere antibodies associated with HLA-DR1, 4 & 5
  - Diffuse: Antitopoisomerase 1 antibodies associated with HLA-DR5
- Etiology

- Unknown; autoimmune condition with genetic predisposition
- May be initiated by environmental antigens like silica & L-tryptophan
- Immunologic mechanism (delayed hypersensitivity reaction)
  - ↑ Production of cytokines (TNF-α or IL-1) → ↑ collagen production
  - Vascular damage & activation of fibroblasts
- Epidemiology
  - Incidence: 14.1 cases/million
  - Prevalence: 19-75/100,000 persons
- Associated abnormalities: Maybe SLE, polymyositis or dermatomyositis

Gross Pathologic & Surgical Features

- Rubber-hose inflexibility: Lower 2/3 esophagus
- Thin & ulcerated mucosa
- Dilated gas & fluid containing small bowel loops with sacculations

Microscopic Features

- Perivascular lymphocytic infiltrates
- Early capillary & arteriolar injury
- Atrophy + fragmentation of smooth muscle → collagen deposition + fibrosis

CLINICAL ISSUES

Presentation

- Most common signs/symptoms
  - Esophagus
    - Dysphagia, regurgitation
    - Epigastric fullness & burning pain
  - Small bowel
    - Bloating, abdominal pain
    - Weight loss, diarrhea, anemia
  - Colon
    - Chronic constipation & episodes of bowel obstruction
- Lab-data
  - Increased erythrocyte sedimentation rate (ESR)
  - Iron, B12 & folic acid deficiency anemias
  - Increased antinuclear antibodies (ANA)
  - CREST syndrome: Anticentromere antibodies
  - Diffuse scleroderma
    - Antitopoisomerase 1 antibody (anti-Scl 70)

Demographics

- Age: 30-50 years
- Gender: Young African-American females (> common); M:F = 1:3
- Ethnicity: African-Americans more than Caucasians

Natural History & Prognosis

- Complications
  - Barrett esophagus → adenocarcinoma
  - Bowel pseudoobstruction
- Prognosis
  - Limited disease with ANA bodies: Good prognosis
  - Diffuse disease: Poor with involvement of kidneys, heart & lungs rather than GI tract

Treatment

- Small frequent meals; elevation of head of the bed
- Avoid tea & coffee
- Cimetidine, ranitidine, omeprazole
- Metoclopramide, laxatives
- Patients with severe malabsorption
  - Parenteral hyperalimentation

DIAGNOSTIC CHECKLIST

Consider

- Rule out other pathologies that cause distal esophageal stricture ± dysmotility
- Check for family history of collagen vascular diseases

Image Interpretation Pearls

- Mild-moderate dilatation of esophagus with distal fusiform stricture + decreased or absent peristalsis

SELECTED REFERENCES

1. Mayes MD: Scleroderma epidemiology. Rheum Dis Clin North Am. 29(2):239-54, 2003
2. Goldblatt F et al: Antibody-mediated gastrointestinal dysmotility in scleroderma. Gastroenterology. 123(4):1144-50, 2002
3. Coggins CA et al: Wide-mouthed sacculations in the esophagus: a radiographic finding in scleroderma. AJR Am J Roentgenol. 176(4):953-4, 2001
4. Duchini A et al: Gastrointestinal hemorrhage in patients with systemic sclerosis and CREST syndrome. Am J Gastroenterol. 93(9):1453-6, 1998
5. Weston S et al: Clinical and upper gastrointestinal motility features in systemic sclerosis and related disorders. Am J Gastroenterol. 93(7):1085-9, 1998
6. Lock G et al: Gastrointestinal manifestations of progressive systemic sclerosis. Am J Gastroenterol. 92(5):763-71, 1997
7. Young MA et al: Gastrointestinal manifestations of scleroderma. Rheum Dis Clin North Am. 22(4):797-823, 1996
8. Ott DJ: Esophageal motility disorders. Semin Roentgenol. 29(4):321-31, 1994
9. Sjogren RW: Gastrointestinal motility disorders in scleroderma. Arthritis Rheum. 37(9):1265-82, 1994
10. Kahan A et al: Gastrointestinal involvement in systemic sclerosis. Clin Dermatol. 12(2):259-65, 1994
11. Levine MS et al: Update on esophageal radiology. AJR 155: 933-41, 1990
12. Marshall JB et al: Gastrointestinal manifestations of mixed connective tissue disease. Gastroenterology. 98(5 Pt 1):1232-8, 1990
13. Rohrmann CA Jr et al: Radiologic and histologic differentiation of neuromuscular disorders of the gastrointestinal tract: visceral myopathies, visceral neuropathies, and progressive systemic sclerosis. AJR Am J Roentgenol. 143(5):933-41, 1984
14. Berk RN: The radiology corner. Scleroderma of the gastrointestinal tract. Am J Gastroenterol. 61(3):226-31, 1974
15. Horowitz AL et al: The "hide-bound" small bowel of scleroderma: Characteristic mucosal fold pattern. AJR 119: 332-34, 1973

SCLERODERMA, ESOPHAGUS

IMAGE GALLERY

Typical

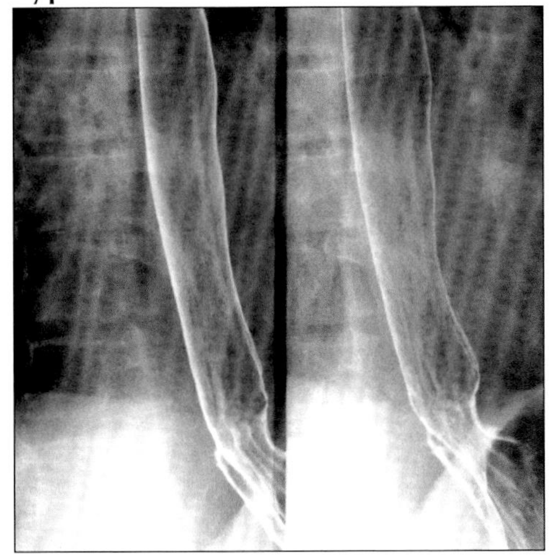

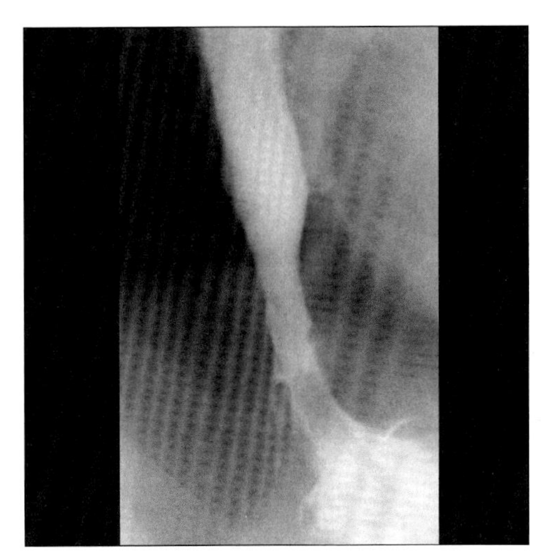

(Left) *Two views from air-contrast esophagram show atonic, but not dilated esophagus, patulous GE junction.* ***(Right)*** *Single-contrast esophagram shows hiatal hernia, shortening of the esophagus and a long smooth distal stricture.*

Typical

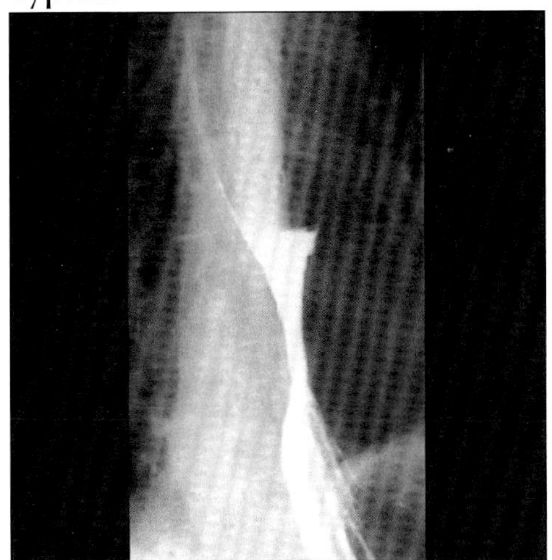

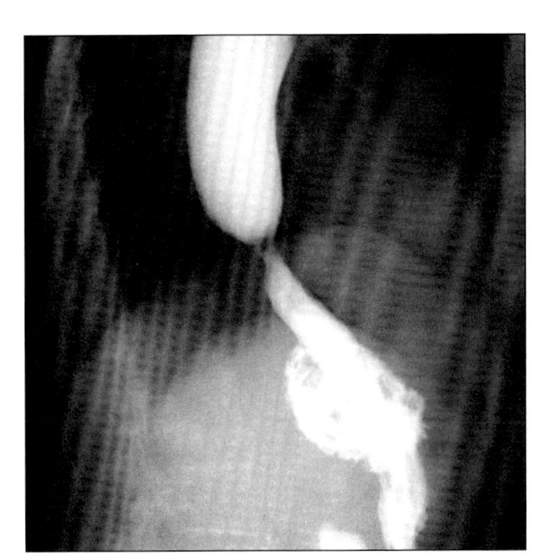

(Left) *Air-contrast esophagram shows dilated atonic esophagus with an air-contrast level and a long smooth distal stricture. Small hiatal hernia + "short esophagus".* ***(Right)*** *Single-contrast esophagram shows short smooth stricture + dilated atonic esophagus.*

Typical

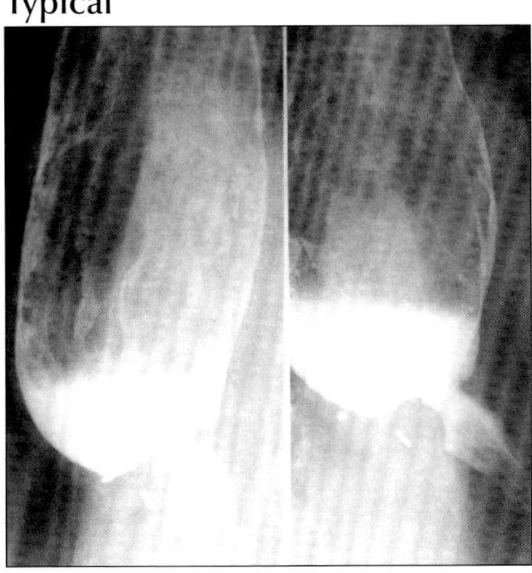

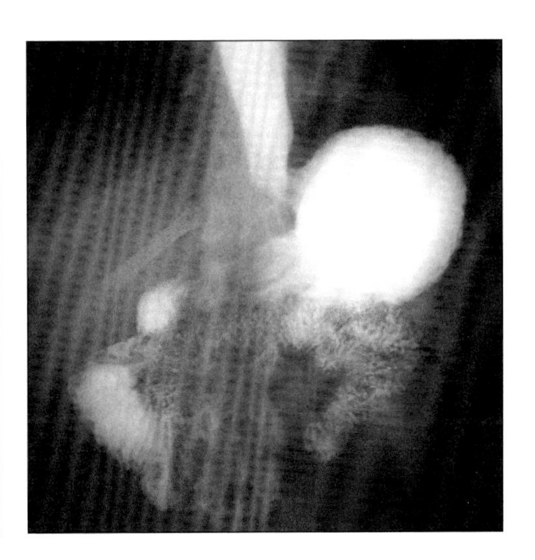

(Left) *Air-contrast esophagram shows markedly dilated esophagus and distal stricture due to scleroderma.* ***(Right)*** *Single-contrast upper GI shows dilated, atonic esophagus with patulous LES. Also dilated duodenum abruptly narrowed as it crosses spine. All due to scleroderma.*

ESOPHAGEAL VARICES

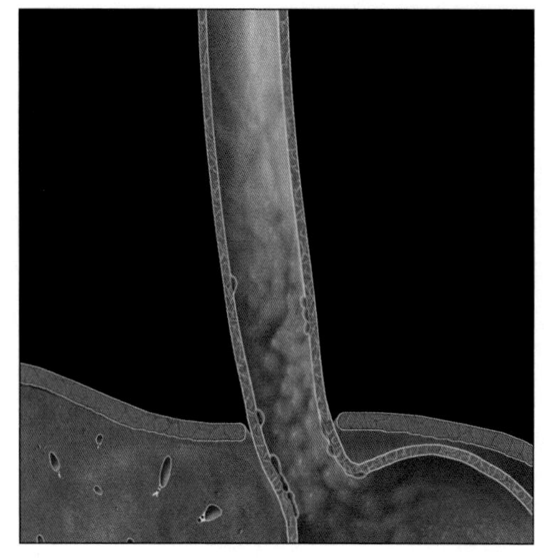

Graphic shows dilated, tortuous, submucosal collateral veins (varices).

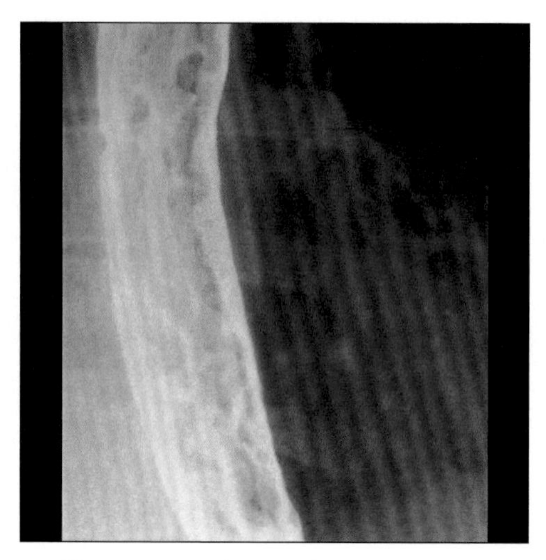

Double-contrast esophagram shows tortuous, nodular longitudinal "folds" typical of varices.

TERMINOLOGY

Definitions

- Dilated tortuous submucosal venous plexus of esophagus

IMAGING FINDINGS

General Features

- Best diagnostic clue: Tortuous or serpiginous longitudinal filling defects on esophagraphy
- Location
  - Uphill varices: Distal third or half of thoracic esophagus (more common)
  - Downhill varices: Upper or middle third of thoracic esophagus (less common)
- Morphology: Tortuous dilated veins in long axis of esophagus, protruding directly beneath mucosa or in periesophageal tissue
- Other general features
  - Usually due to portal HTN with cirrhosis or other liver diseases
  - Idiopathic varices: Very rare in patients with no portal HTN or SVC block
  - Classification of esophageal varices based on pathophysiology
    - Uphill varices: ↑ Portal venous pressure → upward venous flow via dilated esophageal collaterals to superior vena cava (SVC)
    - Downhill varices: Obstruction of SVC → downward venous flow via esophageal collaterals to portal vein & inferior vena cava (IVC)

Radiographic Findings

- Radiography
  - Chest x-ray
    - Retrocardiac posterior mediastinal lobulated mass
    - ± Mediastinal widening, abnormal azygoesophageal recess
- Fluoroscopic guided esophagography
  - Mucosal relief views
    - Tortuous, serpiginous, longitudinal radiolucent filling defects in collapsed or partially collapsed esophagus
  - Double-contrast study
    - Multiple radiolucent filling defects etched in white
  - Distended views of esophagus
    - Varices may be obscured
  - After sclerotherapy varices may appear as fixed, rigid filling defects

CT Findings

- NECT
  - Thickened esophageal wall, lobulated outer contour

DDx: Thickened Esophageal Folds

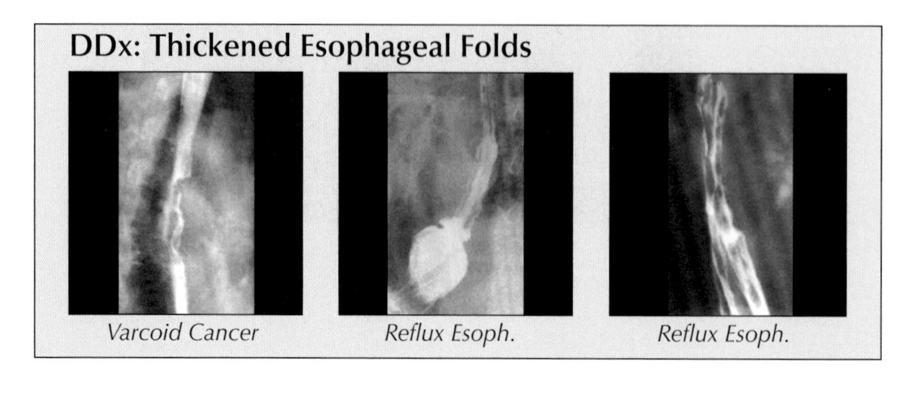

Varcoid Cancer | *Reflux Esoph.* | *Reflux Esoph.*

ESOPHAGEAL VARICES

Key Facts

Terminology
- Dilated tortuous submucosal venous plexus of esophagus

Imaging Findings
- Best diagnostic clue: Tortuous or serpiginous longitudinal filling defects on esophagraphy
- Uphill varices: Distal third or half of thoracic esophagus (more common)
- Downhill varices: Upper or middle third of thoracic esophagus (less common)
- Multiple radiolucent filling defects etched in white
- Varices may be obscured
- Scalloped esophageal mural masses
- Homogeneous HU & enhance to same degree as adjacent veins

Top Differential Diagnoses
- Varicoid esophageal carcinoma
- Esophagitis
- Lymphoma

Clinical Issues
- Asymptomatic until rupture or significant block
- Uphill varices: Hematemesis/mild bleeding (melena)
- Downhill varices: SVC syndrome
- Clinical profile: Patient with history of cirrhosis, portal HTN, hematemesis/melena, facial/arm swelling
- Alcoholic cirrhosis: Most prevalent cause in USA

Diagnostic Checklist
- Lack of change for thick "folds" should suggest esophagitis or cancer rather than varices

    - Scalloped esophageal mural masses
    - Uni-/bilateral soft tissue masses (paraesophageal varices)
- CECT
    - Well-defined round, tubular, or smooth serpentine structures
    - Homogeneous HU & enhance to same degree as adjacent veins
    - Location
        - Esophageal, coronary ± paraumbilical
        - Abdominal wall, perisplenic, perigastric, paraesophageal, retroperitoneal, omental, mesenteric

MR Findings
- T1 & T2WI
    - Multiple areas of flow void
- T1 C+
    - Portal venous phase (PVP)
        - Enhancement of varices seen

Ultrasonographic Findings
- Real Time: Increased esophageal wall thickness at least 5 mm with irregular wall surface
- Color Doppler: Hepatofugal venous flow within esophageal wall

Angiographic Findings
- Conventional
    - Portal venogram
        - Uphill varices: May show cavernous transformation of portal vein + reversal of blood flow via splenic vein → coronary vein → esophageal varices

Imaging Recommendations
- Helical NE + CECT
- Fluoroscopic guided esophagography
    - High density barium
    - Position: Prone right anterior oblique (RAO)
    - Mucosal relief views
    - Anticholinergic agents; avoid swallowing

DIFFERENTIAL DIAGNOSIS

Varicoid esophageal carcinoma
- May simulate varices, especially "downhill"
- Varicoid carcinoma
    - Rigid, fixed appearance; abrupt demarcation; well-defined borders
    - Produce thickened, tortuous folds in esophagus due to submucosal spread of tumor
    - Mostly squamous cell or adenocarcinomas
- Varices tend to change in size & shape with peristalsis, respiration & Valsalva maneuvers
- Clinically varicoid may cause dysphagia, whereas this symptom rarely occurs in patients with varices
- Diagnosis: Imaging & endoscopic biopsy

Esophagitis
- Occasionally esophagitis may mimic varices
- Due to associated submucosal edema & inflammation which can manifest as thickened, tortuous longitudinal folds simulating esophageal varices
- Diagnosis: Endoscopy & history

Lymphoma
- Esophagus: Least common site within GI tract
- Accounts for only about 1% of all cases
- Usually non-Hodgkin & less commonly Hodgkin
- Patients almost always have generalized lymphoma
- Primary esophageal lymphoma seen in AIDS cases
- Imaging
    - Usually contiguous spread from gastric cardia/fundus to distal esophagus
    - Polypoid or ulcerated mass or infiltrating stricture
    - Submucosal infiltration (less common)
        - Enlarged, tortuous longitudinal folds mimicking varices
- Diagnosis: Endoscopy with deep esophageal biopsy

PATHOLOGY

General Features
- General path comments

ESOPHAGEAL VARICES

- Normal esophageal venous drainage
  - Upper third of esophagus: Via intercostal, bronchial & inferior thyroid veins
  - Middle third: Via azygous & hemiazygous venous systems
  - Distal third: Via periesophageal plexus of veins → coronary vein → splenic vein
- Etiology
  - Uphill varices
    - Cirrhosis + portal HTN
    - Obstruction of hepatic veins/IVC; CHF; marked splenomegaly
    - Pathogenesis: Collateral blood flow from portal vein → azygos vein → SVC
  - Downhill varices
    - Obstruction of SVC distal to entry of azygos vein
    - Usually due to lung cancer, lymphoma, fibrosing mediastinitis, retrosternal goiter, thymoma
    - Pathogenesis: Collateral blood flow from SVC → azygos vein → IVC or portal system
  - Idiopathic varices: Exact mechanism is unknown
    - Postulated as a result of congenital weakness in venous channels of esophagus
- Epidemiology
  - Incidence: 30-70% cases of cirrhosis + portal HTN
  - Mortality: 20-50% of bleeding esophageal varices
- Associated abnormalities: Cirrhosis with portal HTN

Gross Pathologic & Surgical Features

- Tortuous dilated veins in long axis of esophagus

Microscopic Features

- Tortuous, serpiginous dilated veins protruding beneath mucosa
- ± Superficial ulceration, inflammation, blood clot

CLINICAL ISSUES

Presentation

- Most common signs/symptoms
  - Asymptomatic until rupture or significant block
  - Uphill varices: Hematemesis/mild bleeding (melena)
  - Downhill varices: SVC syndrome
    - Facial, periorbital, neck, bilateral arm swelling
    - Dilated superficial veins over chest
  - Lab-data: Guaiac positive stool or iron deficiency anemia
- Clinical profile: Patient with history of cirrhosis, portal HTN, hematemesis/melena, facial/arm swelling

Demographics

- Age: Middle & elderly age group
- Gender: M = F

Natural History & Prognosis

- Complications
  - Inflammation, ulceration, hemorrhage, hematemesis
  - Esophageal variceal hemorrhage
    - Common cause of acute upper GI bleeding
    - Alcoholic cirrhosis: Most prevalent cause in USA
    - Accounts for 20-50% of all deaths from cirrhosis
- Prognosis
  - Varices without bleeding
    - Usually good after treatment
  - Varices with massive bleeding
    - Poor with & without treatment

Treatment

- Bleeding varices
  - Vasopressin infusion
  - Balloon tamponade (Sengstaken-Blakemore tube)
  - Endoscopic sclerotherapy or variceal ligation
  - Portal shunt surgery
  - TIPS: Transjugular intrahepatic porto-systemic shunt

DIAGNOSTIC CHECKLIST

Consider

- Lack of change for thick "folds" should suggest esophagitis or cancer rather than varices

Image Interpretation Pearls

- Mucosal relief views: Tortuous, serpiginous, longitudinal radiolucent filling defects in collapsed esophagus
- Double-contrast study: Multiple radiolucent filling defects etched in white

SELECTED REFERENCES

1. Kang HK et al: Three-dimensional multi-detector row CT portal venography in the evaluation of portosystemic collateral vessels in liver cirrhosis. Radiographics. 22(5):1053-61, 2002
2. Matsumoto A et al: Three-dimensional portography using multislice helical CT is clinically useful for management of gastric fundic varices. AJR Am J Roentgenol. 176(4):899-905, 2001
3. Kishimoto R et al: Esophageal varices: evaluation with transabdominal US. Radiology. 206(3):647-50, 1998
4. Lee SJ et al: Computed radiography of the chest in patients with paraesophageal varices: diagnostic accuracy and characteristic findings. AJR Am J Roentgenol. 170(6):1527-31, 1998
5. Horton KM et al: Paraumbilical vein in the cirrhotic patient: imaging with 3D CT angiography. Abdom Imaging. 23(4):404-8, 1998
6. Cho KC et al: Varices in portal hypertension: evaluation with CT. Radiographics. 15(3):609-22, 1995
7. Itai Y et al: CT and MRI in detection of intrahepatic portosystemic shunts in patients with liver cirrhosis. J Comput Assist Tomogr. 18(5):768-73, 1994
8. Nelson RC et al: Splenic venous flow exceeding portal venous flow at Doppler sonography: relationship to portosystemic varices. AJR Am J Roentgenol. 161(3):563-7, 1993
9. Balthazar EJ et al: CT evaluation of esophageal varices. AJR Am J Roentgenol. 148(1):131-5, 1987
10. McCain AH et al: Varices from portal hypertension: correlation of CT and angiography. Radiology. 154(1):63-9, 1985
11. Ishikawa T et al: Detection of paraesophageal varices by plain films. AJR Am J Roentgenol. 144(4):701-4, 1985

ESOPHAGEAL VARICES

IMAGE GALLERY

Typical

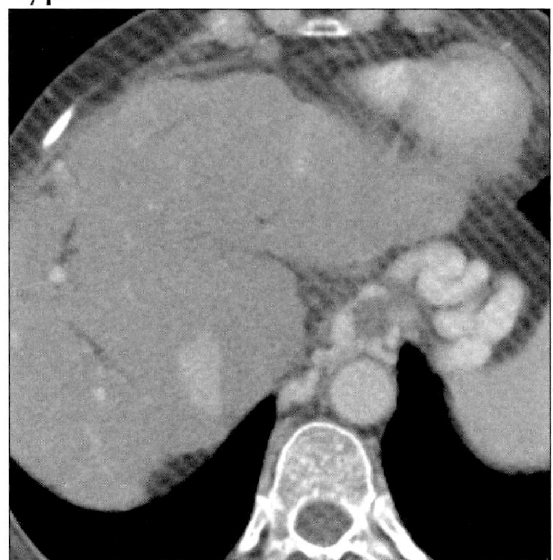

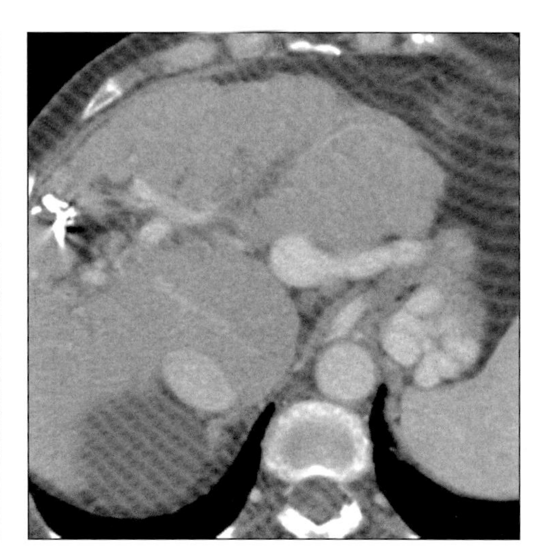

(Left) *Axial CECT in cirrhotic patient shows massive varices as tortuous enhanced vessels in the periesophageal region.* ***(Right)*** *Axial CECT in cirrhotic patient shows massive varices in the gastric fundus wall.*

Typical

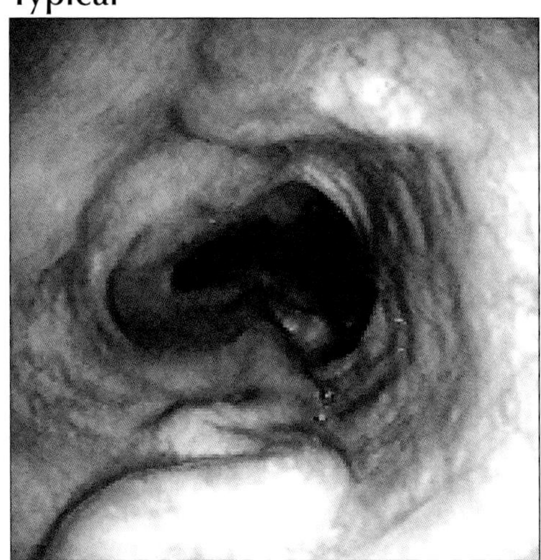

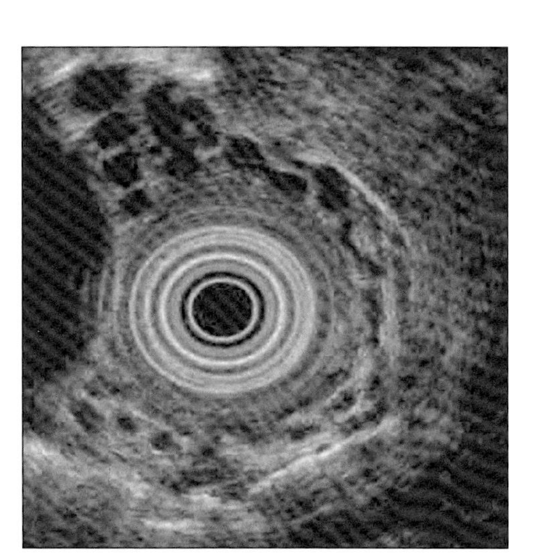

(Left) *Endoscopic image shows tortuous blue varices in the submucosal region of the distal esophagus.* ***(Right)*** *Endoscopic ultrasound shows multiple anechoic spaces with thin walls.*

Typical

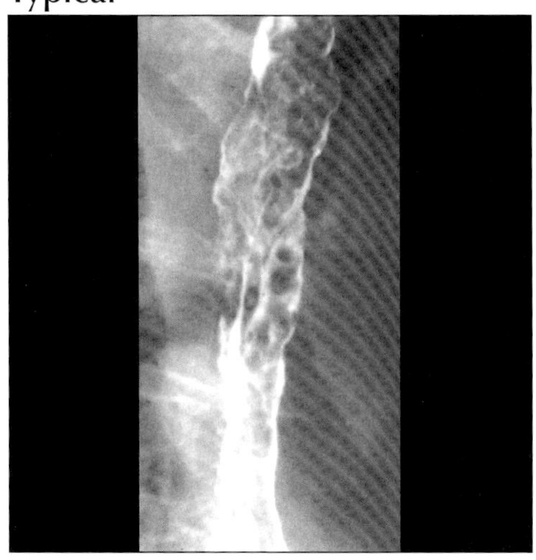

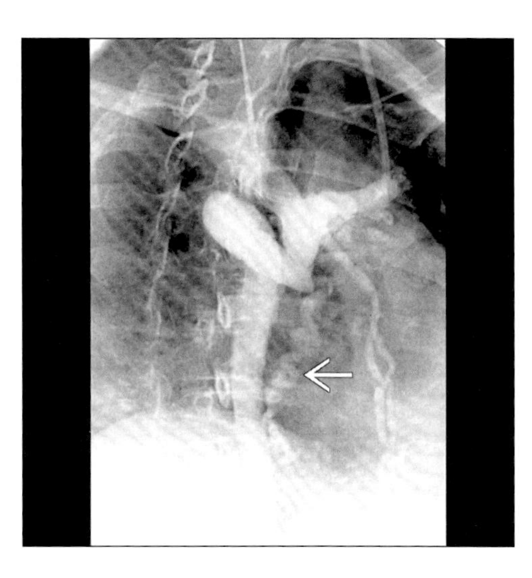

(Left) *Esophagram shows serpiginous submucosal filling defects in the proximal half of the esophagus; downhill varices.* ***(Right)*** *Catheter venacavogram shows obstruction of superior vena cava. The SVC is occluded above azygous arch. Downhill varices (arrow) in the periesophageal region are part of collateral venous drainage.*

SCHATZKI RING

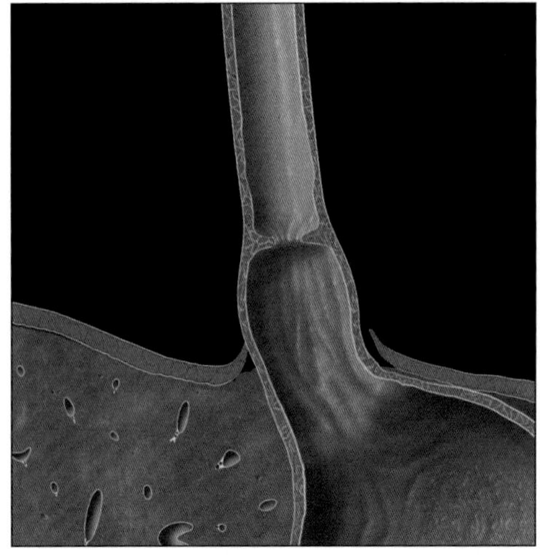

Graphic shows small hiatal hernia and an annular ring-like narrowing at the gastroesophageal junction

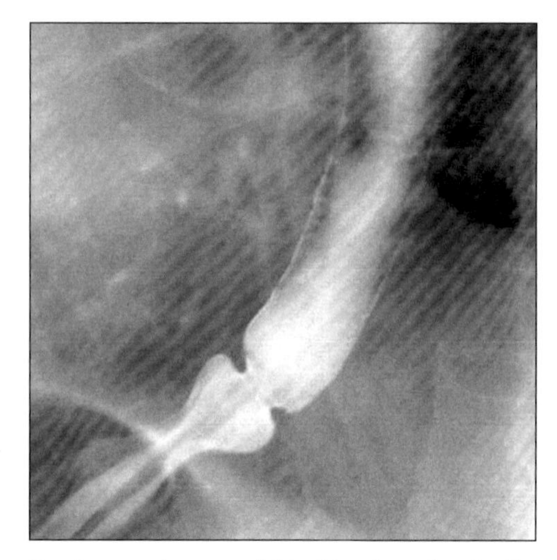

Esophagram shows small hiatal hernia and a ring-like narrowing at the gastroesophageal junction.

TERMINOLOGY

Definitions

- Annular, inflammatory, symptomatic narrowing of normal lower esophageal mucosal or "B" ring

IMAGING FINDINGS

General Features

- Best diagnostic clue: Thin, web-like or annular constriction at GE junction above hiatal hernia
- Location: Gastroesophageal (GE) junction
- Size: Less than 13 mm in diameter up to 20 mm
- Other general features
  - Schatzki ring: Symptomatic narrowing of normal lower esophageal mucosal or "B" ring
    - Most likely results from reflux esophagitis
  - Normal lower esophageal mucosal or "B" ring
    - Most common ring-like narrowing in distal esophagus that marks gastroesophageal junction
  - Classification based on ring diameter VS symptoms
    - Ring < 13 mm in diameter: Symptomatic
    - Ring 13-20 mm: Occasionally symptomatic
    - Ring > 20 mm in diameter: Asymptomatic

Radiographic Findings

- Fluoroscopic guided single-contrast barium study
  - Thin (2-4 mm in height), web-like constriction at GE junction
  - Margins: Smooth & symmetric
  - Fixed, anatomic, nondistensible, transverse ring with constant shape
  - Sliding hiatal hernia: Seen below the ring
- Schatzki ring best visualized
  - Lumen above & below ring is distended beyond caliber of ring
  - With adequate distention of GE junction
  - In prone: During arrested deep inspiration with Valsalva maneuver while barium column passes through GE junction
  - 50% of rings seen on prone single-contrast are not visualized on double-contrast

Imaging Recommendations

- Fluoroscopic guided single-contrast esophagram
- Position: Prone right anterior oblique (RAO)
  - Best position to demonstrate Schatzki ring
  - Best technique for distention of distal esophagus

DIFFERENTIAL DIAGNOSIS

Annular peptic stricture

- Irregular/asymmetric/greater height than Schatzki ring
- Accounts for 15% of peptic strictures in distal esophagus

DDx: Short Segment Narrowing at GE Junction

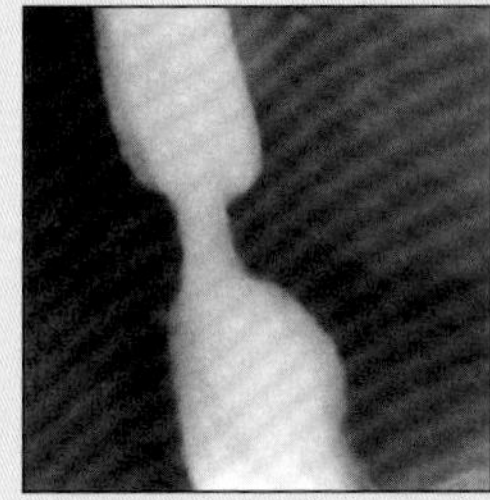

Peptic Stricture

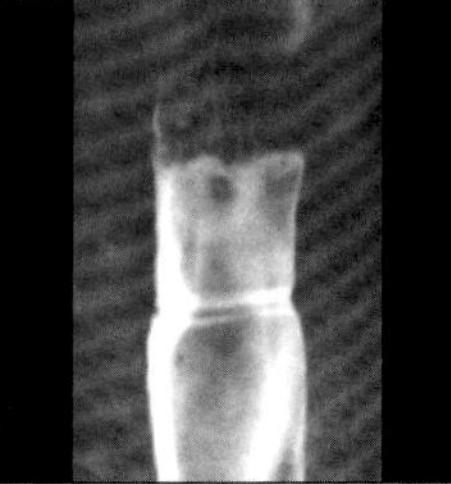

Esophageal Web

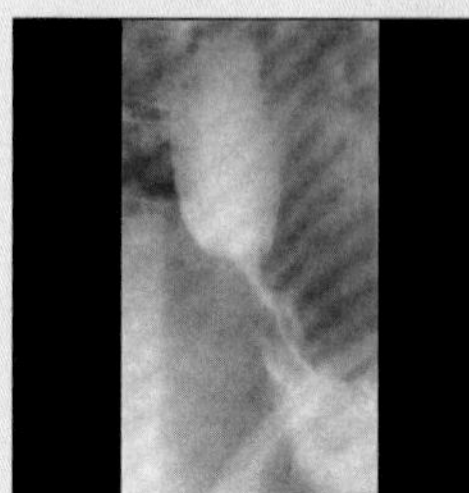

Cancer

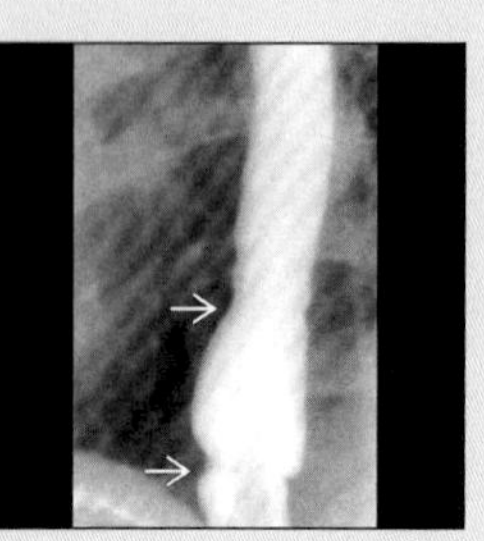

A + B Rings

SCHATZKI RING

Key Facts

Terminology
- Annular, inflammatory, symptomatic narrowing of normal lower esophageal mucosal or "B" ring

Imaging Findings
- Best diagnostic clue: Thin, web-like or annular constriction at GE junction above hiatal hernia
- Size: Less than 13 mm in diameter up to 20 mm
- Margins: Smooth & symmetric

Top Differential Diagnoses
- Annular peptic stricture
- Esophageal web
- Localized esophageal cancer
- Muscular or contractile or "A" ring

Diagnostic Checklist
- Need deep inspiration, Valsalva, distention to demonstrate Schatzki ring

Esophageal web
- Occasionally seen in distal esophagus
- Located several centimeters above GE junction
- More proximal location differentiates from Schatzki

Localized esophageal cancer
- Focal constriction may resemble a Schatzki ring
- Asymmetric, irregular borders within narrowed part

Muscular or contractile or "A" ring
- Caused by active muscular contraction
- Found much less frequently in distal esophagus
- Located at superior border of esophageal vestibule
- On esophagography
  - Broad, smooth area of narrowing
  - May vanish completely with esophageal distention
  - Varies in size/position due to esophageal contraction

PATHOLOGY

General Features
- Etiology
  - Gastroesophageal reflux disease, reflux esophagitis
  - Exaggeration of lower esophageal mucosal ring → ↑ mucosal fold thickening + ↑ scarring: Schatzki ring
- Epidemiology: Incidence: Seen in 6-14% of population
- Associated abnormalities: GERD, hiatal hernia

Gross Pathologic & Surgical Features
- Annular, ring-like mucosal structure at GE junction

Microscopic Features
- Squamous epithelium superiorly & columnar inferiorly

CLINICAL ISSUES

Presentation
- Most common signs/symptoms
  - Episodic dysphagia for solids
  - Minimal dysphagia or asymptomatic
  - "Steakhouse syndrome": Due to inadequately chewed piece of meat impacted above ring
  - Severe chest pain, sticking sensation

Demographics
- Age: More common in old age than young age
- Gender: M > F

Natural History & Prognosis
- Complications
  - Peptic stricture: With longstanding reflex
  - Perforation: Due to bolus obstruction & dilatation
- Prognosis
  - Decrease in caliber over 5 years (in 25-33%)

Treatment
- Advise to eat more slowly & chew food more carefully
- Recurrent dysphagia: Mechanical disruption of ring by
  - Endoscopic rupture, bougienage, pneumatic dilation
  - Electrocautery incision, rarely surgery

DIAGNOSTIC CHECKLIST

Consider
- Need deep inspiration, Valsalva, distention to demonstrate Schatzki ring

SELECTED REFERENCES

1. Ott DJ et al: Radiographic & endoscopic sensitivity in detecting lower esophageal mucosal ring. AJR 147: 261-5, 1986
2. Ott DJ et al: Esophagogastric region and its rings. AJR 142: 281-7, 1984
3. Schatzki R: The lower esophageal ring: Long term followup of symptomatic/asymptomatic rings. AJR 90: 805-10, 1963

IMAGE GALLERY

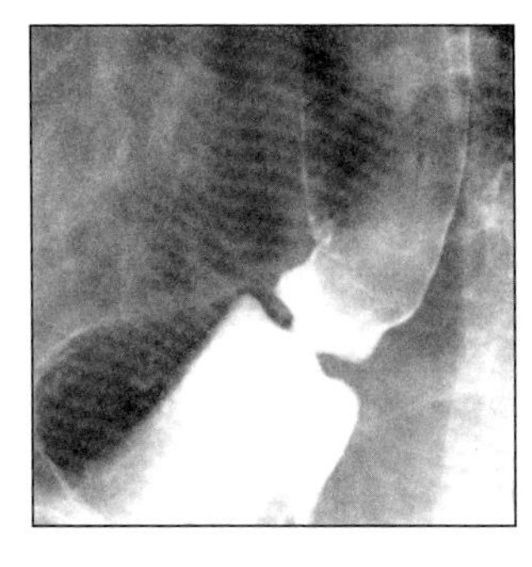

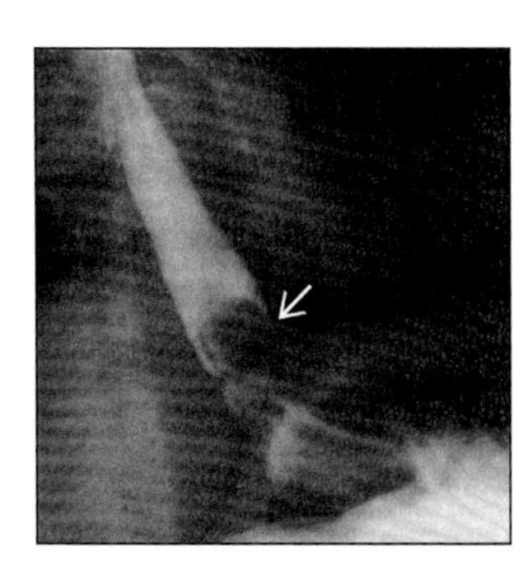

(Left) *Esophagram shows small hiatal hernia and a tight (7 mm diameter) Schatzki ring that was symptomatic (frequent dysphagia, food "sticking").* ***(Right)*** *Esophagram shows Schatzki ring with a piece of meat impacted (arrow) above it.*

HIATAL HERNIA

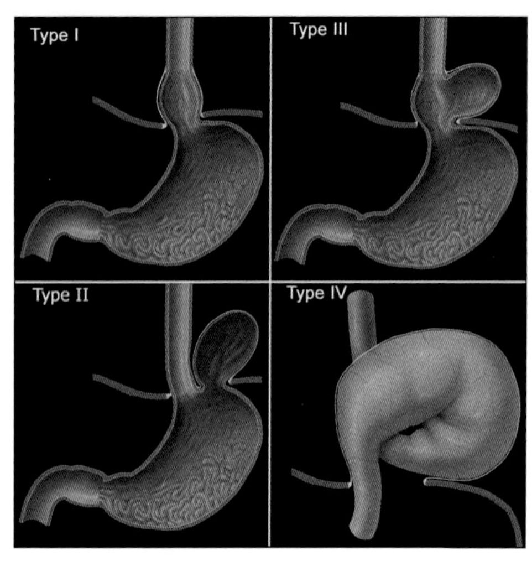

Surgical classification of hiatal hernias. (See text for details.) Type I = sliding hiatal hernia. Types II - IV = paraesophageal hernias.

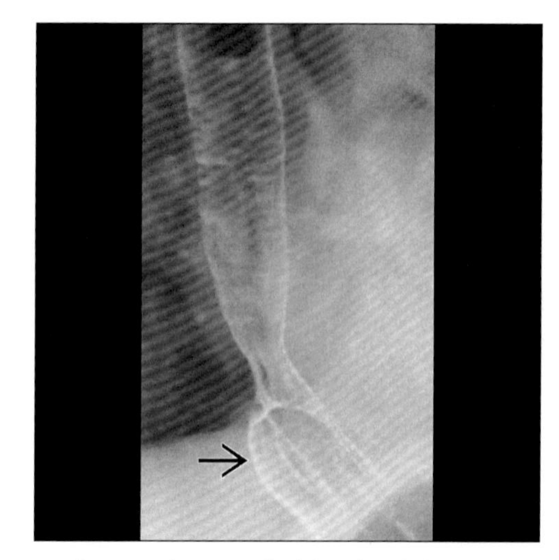

Esophagram shows small sliding hiatal hernia. Note thickened gastric folds (arrow) in hernia continuing into abdomen. "Feline" appearance of esophageal mucosa due to reflux.

TERMINOLOGY

Abbreviations and Synonyms

- Hiatal hernia (HH), hiatus hernia

Definitions

- Protrusion of a part of the stomach through the esophageal hiatus of the diaphragm

IMAGING FINDINGS

General Features

- Best diagnostic clue: Fluoroscopy after barium meal showing portion of stomach in thorax
- 2 types
  - Sliding (axial) hiatal hernia: Gastroesophageal (GE) junction and gastric cardia pass through esophageal hiatus of diaphragm into thorax
  - Paraesophageal (rolling) hernia: Gastric fundus ± other parts of stomach herniate into the chest

Radiographic Findings

- Fluoroscopic barium esophagram & upper GI
  - Sliding hiatal hernia
    - Lower esophageal mucosal ("B") ring observed 2 cm or more above diaphragmatic hiatus
    - Prominent diagonal notch may be seen on left lateral and superior aspect of hiatal hernia; due to crossing gastric sling fibers at cardiac incisura
    - ± Kink or narrowing of hiatal hernia at esophageal hiatus; extrinsic compression by diaphragm
    - Esophageal peristaltic wave stops at GE junction
    - Tortuous esophagus having an eccentric junction with hernia
    - Often reducible in erect position
    - Numerous (> 6) longitudinal gastric folds within hiatal hernia continue through hiatus into abdominal part of stomach
    - Areae gastricae pattern demonstrated within herniated portion of fundus
    - Gastric folds converging superiorly toward a point several centimeters above diaphragm
    - "Riding ulcers" at hiatal orifice: Repeated trauma of gastric mucosa on ridge riding over hiatus
    - Intra-hernial tumor: Well demarcated sessile polypoid mass with eccentric wall thickening
  - Paraesophageal hernia
    - Portion of stomach anterior or lateral to esophagus in chest
    - Frequently nonreducible
    - ± Gastric ulcer of lesser curvature at level of diaphragmatic hiatus
    - ± Entire stomach herniated into chest and inverted or upside-down configuration

DDx: Intrathoracic Position of Gastric Fundus

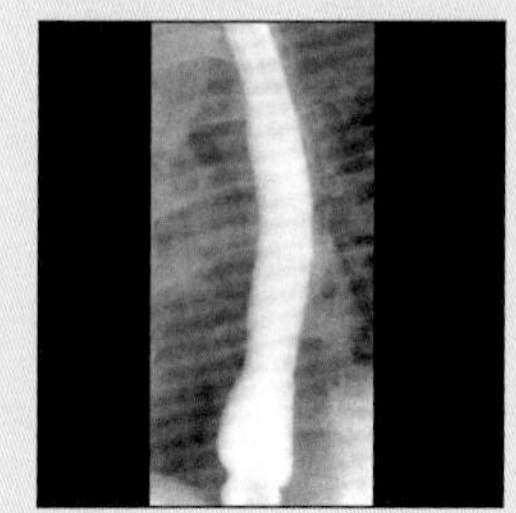

Phrenic Ampulla

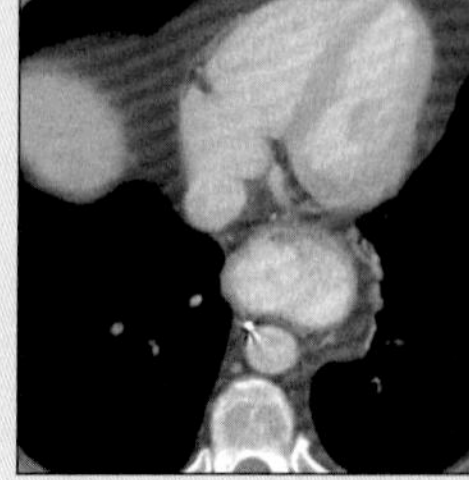

Post Esophagectomy

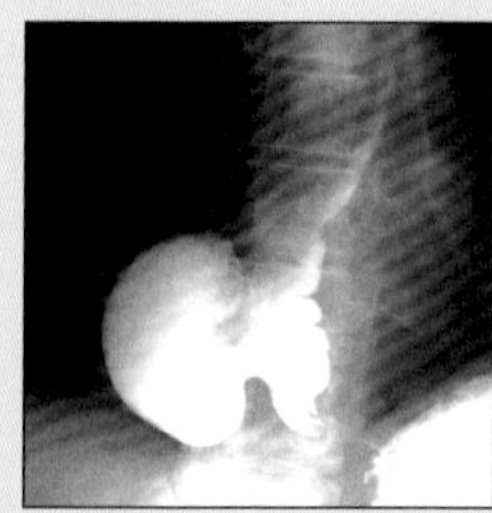

Epiphrenic 'Tic

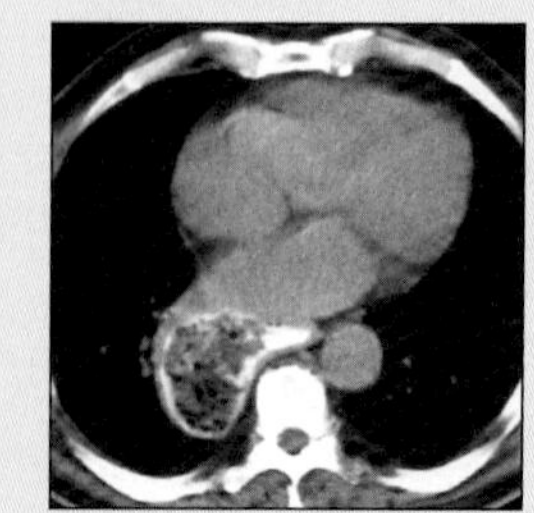

Epiphrenic 'Tic

HIATAL HERNIA

Key Facts

Imaging Findings
- Best diagnostic clue: Fluoroscopy after barium meal showing portion of stomach in thorax
- Sliding (axial) hiatal hernia: Gastroesophageal (GE) junction and gastric cardia pass through esophageal hiatus of diaphragm into thorax
- Paraesophageal (rolling) hernia: Gastric fundus ± other parts of stomach herniate into the chest
- Lower esophageal mucosal ("B") ring observed 2 cm or more above diaphragmatic hiatus
- Prominent diagonal notch may be seen on left lateral and superior aspect of hiatal hernia; due to crossing gastric sling fibers at cardiac incisura

Top Differential Diagnoses
- Phrenic ampulla
- Post-surgical change
- Epiphrenic diverticulum ('tic)

Pathology
- 99% of all hiatal hernias are sliding hiatal hernia
- 1% paraesophageal hernia

Clinical Issues
- Position dependent (supine worse) substernal or epigastric burning pain, regurgitation, dysphagia, hoarse voice
- Paraesophageal hernia may present with acute life-threatening complications

Diagnostic Checklist
- Esophagram and upper GI series remain best test to demonstrate and classify hiatal hernias

CT Findings
- Widening of esophageal hiatus
  - Dehiscence of diaphragmatic crura (> 15 mm); increased distance between crura and esophageal wall
- Pseudomass
  - Sliding hiatal hernia causing soft tissue mass involving gastric cardia
  - Usually filled with contrast material
  - Lying within or above esophageal hiatus
- Focal fat collection in middle compartment of lower mediastinum
  - Omentum herniates through phrenicoesophageal ligament
  - May see ↑ in fat surrounding distal esophagus
- CT clearly demonstrates paraesophageal hernia through widened esophageal hiatus; often as an incidental finding
  - Visualize size, contents, orientation of herniated stomach within lower thoracic cavity

Imaging Recommendations
- Best imaging tool: Fluoroscopic-guided barium esophagram and upper GI studies
- Protocol advice
  - Fluoroscopic-guided single contrast barium studies (patient prone right anterior oblique)
    - Obtain fully distended views in several positions, including upright
    - Film with full inspiration and Valsalva
  - Sensitivity; full-column technique 100%; mucosal relief 52%; double-contrast techniques 34%

DIFFERENTIAL DIAGNOSIS

Phrenic ampulla
- Saccular, slightly more distensible distal segment of esophagus that communicates with stomach
- Phrenic ampulla or vestibule corresponds to location of lower esophageal sphincter (LES)
  - 2-4 cm high pressure zone extends up from GE junction just into thorax
  - At upper end of sphincter; muscle coalescence called A ring; B ring at lower end of sphincter at GE junction
  - Ampulla when fully distended has bulbous configuration

Post-surgical change
- Gastric pull-up procedure: Stomach is mobilized and brought into chest to replace resected esophagus

Epiphrenic diverticulum ('tic)
- Usually large sac-like protrusion in epiphrenic region; tends to remain filled after most barium emptied
- From lateral esophageal wall of distal 10 cm
- Lateral chest film: May appear as soft tissue mass often containing air-fluid level; mimics hiatal hernia

PATHOLOGY

General Features
- Etiology
  - Complex multifactorial etiology
  - Acquired
    - Common cause for both types of hiatal hernia
    - Rupture or stretching of phrenoesophageal membrane; progressive wear and tear caused by constant swallowing; abdominal pressure > thoracic (pregnancy or obesity)
    - Nerve disease
    - Alteration in viscoelastic properties of distal esophagus
    - Increased strength of longitudinal muscle layers
    - Scarring from reflux esophagitis may cause longitudinal esophageal shortening; sliding hiatal hernia may be an effect rather than cause of reflux esophagitis; pulls gastric fundus into thorax
  - Family history
    - Familial occurrence of sliding hiatal hernia has been reported in more than 20 cases
- Epidemiology

- 99% of all hiatal hernias are sliding hiatal hernia
- 1% paraesophageal hernia
- Paraesophageal hernia: Sporadic pattern of incidence in most cases
- Associated abnormalities: Diverticulosis (25%), reflux esophagitis (25%), duodenal ulcer (20%) or gallstones (18%)

Staging, Grading or Classification Criteria

- Surgical classification of hiatal hernia
  - Type I: GE junction and gastric cardia are intrathoracic (sliding hiatal hernia)
  - Type II: GE junction intraabdominal gastric fundus intrathoracic (paraesophageal hernia)
  - Type III: Both GE junction and fundus are in chest (paraesophageal hernia)
  - Type IV: GE junction and all of stomach in chest (paraesophageal hernia)

CLINICAL ISSUES

Presentation

- Most common signs/symptoms
  - Sliding hiatal hernia
    - Symptoms of GE reflux disease & reflux esophagitis
    - Position dependent (supine worse) substernal or epigastric burning pain, regurgitation, dysphagia, hoarse voice
  - Paraesophageal hernia
    - Often asymptomatic; incidental finding on imaging
    - Severe anemia, abdominal pain, vomiting
    - Commonly have multiple other medical problems
    - GE reflux disease or reflux esophagitis are rare versus sliding hiatal hernia
- Diagnosis: Fluoroscopic-guided barium esophagram and upper GI studies
- Endoscopy
  - Cameron ulcers and erosions
    - In 5.2% of patients with hiatal hernia
    - Present as acute upper GI bleeding (6.3%)
    - Persistent, recurrent iron deficiency anemia (8.3%)
    - Pathogenesis: Mechanical trauma, ischemia, acid mucosal injury may play role

Demographics

- Age: Prevalence increases with age
- Gender: M < F

Natural History & Prognosis

- Complications of paraesophageal hernia
  - Paraesophageal hernia may present with acute life-threatening complications
  - May gradually enlarge
  - Gastric volvulus: Intrathoracic upside-down stomach
  - Traction or torsion of stomach at or near level of hiatus
    - May lead to obstruction, strangulation, infarction or perforation of intrathoracic stomach
    - Rapidly deteriorating clinical condition
    - Patients often undergo emergency surgery without pre-operative barium studies
  - Complications of nonsurgical treatment may be sudden and severe
- Associated transmural esophagitis and fibrosis of esophageal wall may cause hiatal hernia recurrence
- Prognosis
  - Laparoscopic repair: Low morbidity and mortality rates, shorter hospital stay, faster recovery, excellent results
  - Conventional open repair: Low mortality, but significant morbidity

Treatment

- Sliding hiatal hernia
  - Medical treatment and lifestyle modification; treatment same as GE reflux disease
  - Increasing use of laparoscopic fundoplication to treat GE reflux disease
- Paraesophageal hernia
  - Surgery, even without symptoms, is warranted
    - Hernial sac excision, crural closure & antireflux procedure (fundoplication or gastropexy)
  - Large or symptomatic patients or patients with giant paraesophageal hernia need surgical treatments

DIAGNOSTIC CHECKLIST

Consider

- Esophagram and upper GI series remain best test to demonstrate and classify hiatal hernias

SELECTED REFERENCES

1. Insko EK et al: Benign and malignant lesions of the stomach: evaluation of CT criteria for differentiation. Radiology. 228(1):166-71, 2003
2. Pierre AF et al: Results of laparoscopic repair of giant paraesophageal hernias: 200 consecutive patients. Ann Thorac Surg. 74(6):1909-15; discussion 1915-6, 2002
3. Luketich JD et al: Laparoscopic repair of giant paraesophageal hernia: 100 consecutive cases. Ann Surg. 232(4):608-18, 2000
4. Schauer PR et al: Comparison of laparoscopic versus open repair of paraesophageal hernia. Am J Surg. 176(6):659-65, 1998
5. Weston AP: Hiatal hernia with cameron ulcers and erosions. Gastrointest Endosc Clin N Am. 6(4):671-9, 1996
6. Vas W et al: Computed tomographic evaluation of paraesophageal hernia. Gastrointest Radiol. 14(4):291-4, 1989
7. Chen YM et al: Multiphasic examination of the esophagogastric region for strictures, rings, and hiatal hernia: evaluation of the individual techniques. Gastrointest Radiol. 10(4):311-6, 1985
8. Pupols A et al: Hiatal hernia causing a cardia pseudomass on computed tomography. J Comput Assist Tomogr. 8(4):699-700, 1984
9. Cassel DM et al: Hiatus hernia causing CT gastric mass lesions. Comput Radiol. 7(3):177-9, 1983

IMAGE GALLERY

Typical

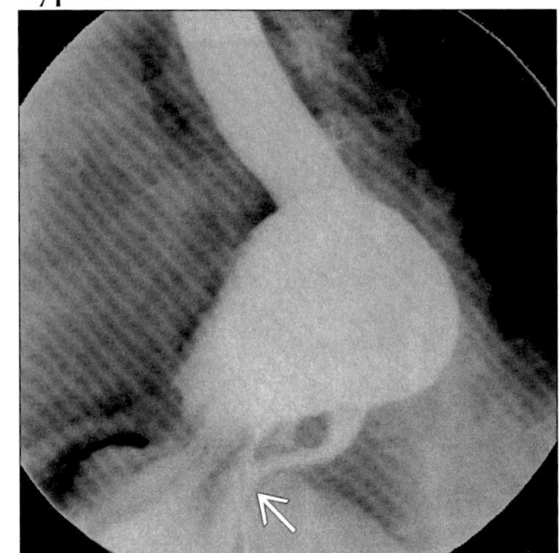

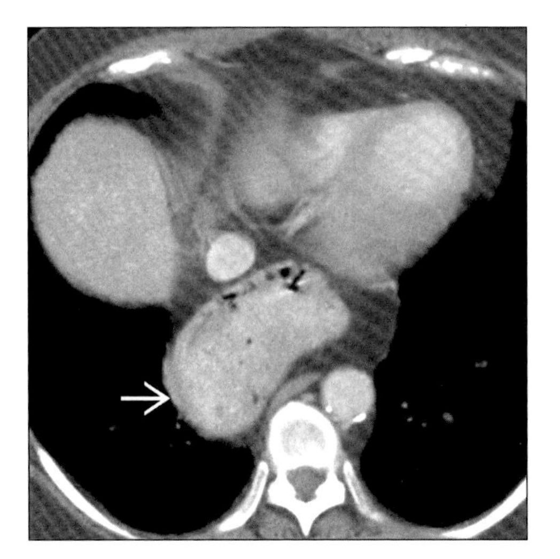

(Left) *Type II paraesophageal hernia. The gastric fundus has herniated into the chest, but the GE junction (arrow) remains below the diaphragm.* ***(Right)*** *Axial CECT shows a large type II paraesophageal hernia (arrow) with the fundus lying along side the esophagus within the thorax.*

Typical

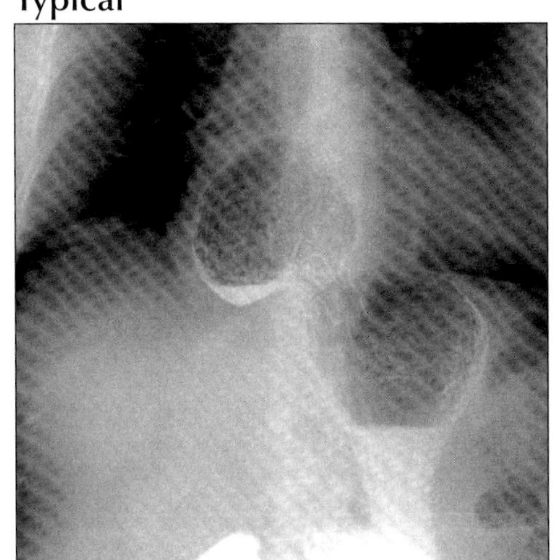

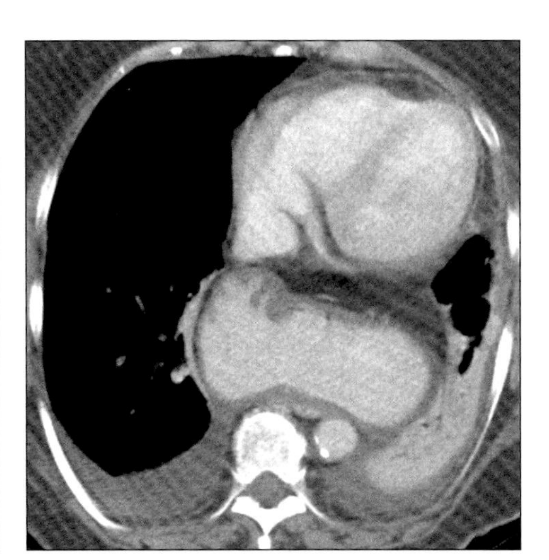

(Left) *Upright film from barium esophagram shows nonreducible type III paraesophageal, with fundus and GE junction above diaphragm.* ***(Right)*** *Axial CECT shows large type III paraesophageal hernia.*

Typical

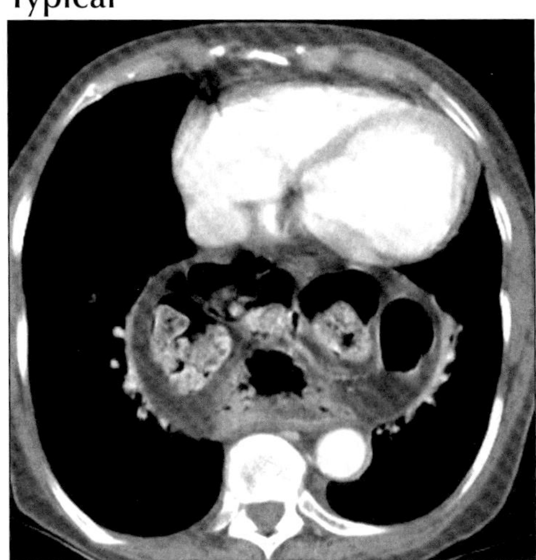

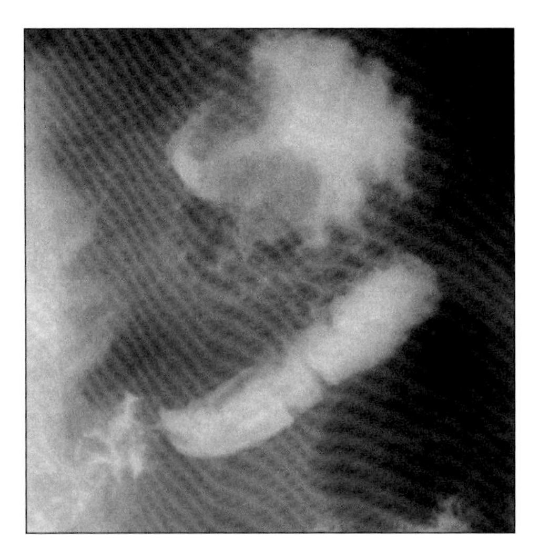

(Left) *Axial CECT shows herniation of stomach and colon through a grossly enlarged esophageal hiatus.* ***(Right)*** *Fluoroscopic upper GI series shows a type III paraesophageal hernia with a large fungation mass within the herniated stomach. Adenocarcinoma.*

ZENKER DIVERTICULUM

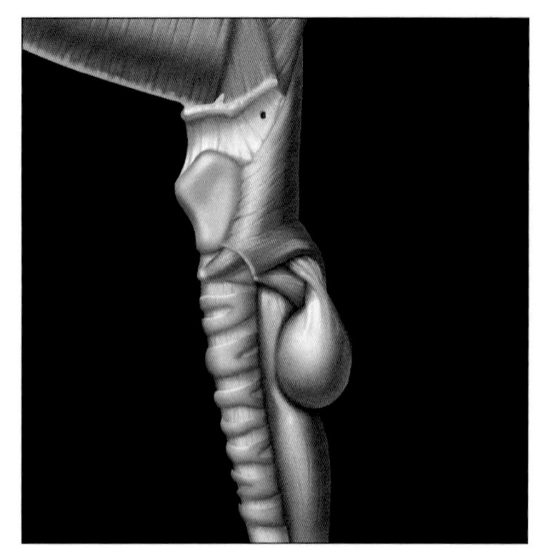

Graphic shows a pouch-like herniation through Killian dehiscence in the cricopharyngeal muscle.

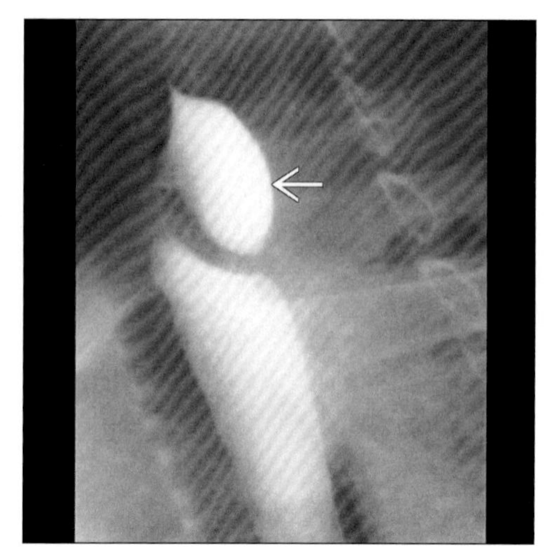

Lateral view of barium swallow shows large pouch (arrow) arising from the posterior pharyngoesophageal junction.

TERMINOLOGY

Abbreviations and Synonyms

- Pharyngoesophageal diverticulum or posterior hypopharyngeal diverticulum/outpouching

Definitions

- Mucosal herniation through an area of anatomic weakness in the region of cricopharyngeal muscle (Killian dehiscence)

IMAGING FINDINGS

General Features

- Best diagnostic clue: Lateral view: Barium-filled sac posterior to cervical esophagus
- Location
  - Location: Killian dehiscence (triangular anatomical area of weakness)
    - Midline posterior wall of pharyngoesophageal segment just above cricopharyngeus muscle (C5-C6 level)
    - Between oblique & horizontal fibers of cricopharyngeal muscle (most common site) or between thyro-/cricopharyngeal muscles
    - This area of weakness occurs in 1/3 of patients
- Size
  - Average maximal dimension: 2.5 cm
  - Range: 0.5-8 cm
- Morphology: Posterior hypopharyngeal saccular outpouching with neck opening above cricopharyngeus muscle
- General features
  - Posterior hypopharyngeal pulsion diverticulum
  - Usually acquired rather than congenital
  - Zenker diverticulum is a false diverticulum
    - Presence of mucosa & submucosa
    - Lack of muscle
  - Almost all patients have associated hiatal hernia & many of them have GE reflux & reflux esophagitis
  - On barium studies: Any irregularity of contour of Zenker diverticulum suggests either an inflammatory or a neoplastic complication

Radiographic Findings

- Radiography
  - Chest x-ray
    - May show air-fluid level in superior mediastinum
- Fluoroscopic guided barium study
  - Frontal view
    - Barium-filled sac below the level of hypopharynx
    - Large diverticulum: Extends inferiorly into mediastinum
  - Lateral or oblique view
    - Barium-filled sac posterior to cervical esophagus

DDx: Outpouching from Proximal Esophagus

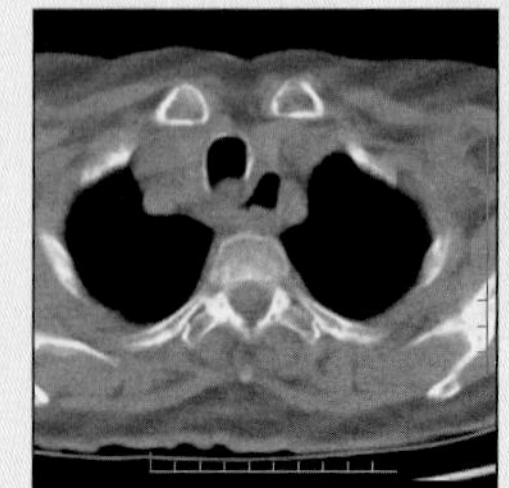

K-J Diverticulum

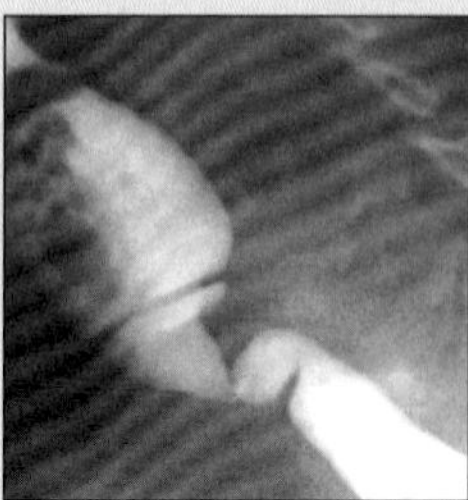

Esophageal Webs

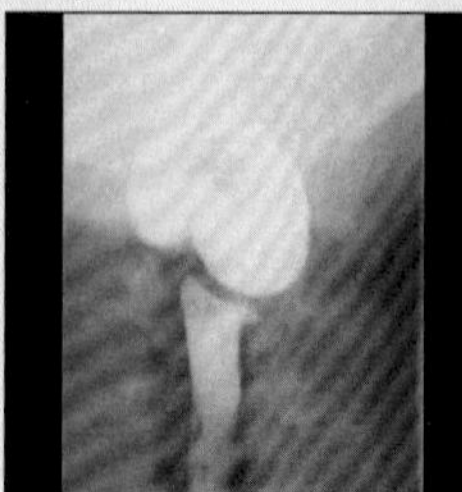

Web, Epi. Bull.

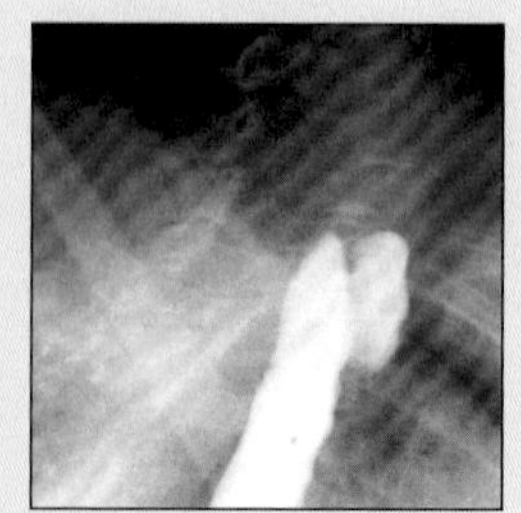

Pulsion 'Tic

ZENKER DIVERTICULUM

Key Facts

Terminology
- Pharyngoesophageal diverticulum or posterior hypopharyngeal diverticulum/outpouching
- Mucosal herniation through an area of anatomic weakness in the region of cricopharyngeal muscle (Killian dehiscence)

Imaging Findings
- Best diagnostic clue: Lateral view: Barium-filled sac posterior to cervical esophagus
- Average maximal dimension: 2.5 cm
- Barium-filled sac posterior to cervical esophagus
- Neck of diverticulum with its opening into posterior hypopharyngeal wall above cricopharyngeus muscle
- Luminal narrowing at pharyngoesophageal junction
- ± Nasopharyngeal regurgitation

Top Differential Diagnoses
- Killian-Jamieson diverticulum (K-J)
- Esophageal web
- Epidermolysis bullosa dystrophica

Pathology
- False diverticulum
- Cricopharyngeal dysfunction (achalasia, spasm, premature closure); ↑ intraluminal pressure

Diagnostic Checklist
- Differentiate from Killian-Jamieson diverticula which is also an outpouch at pharyngo-esophageal junction
- Lateral view: Barium-filled outpouch posterior to cervical esophagus with neck opening above cricopharyngeus muscle at C5-6 level

    - Neck of diverticulum with its opening into posterior hypopharyngeal wall above cricopharyngeus muscle
    - Prominent or thickened cricopharyngeal muscle
    - Luminal narrowing at pharyngoesophageal junction
    - ± Nasopharyngeal regurgitation
  - During swallowing
    - Diverticulum appears as a posterior bulging of distal pharyngeal wall above an anteriorly protruding pharyngoesophageal segment (cricopharyngeal muscle)
  - At rest
    - Barium-filled diverticulum extends below level of cricopharyngeal muscle & is posterior to cervical esophagus
    - Large diverticulum may protrude laterally to left or compress cervical esophagus
  - After swallowing
    - Regurgitation or emptying of barium into hypopharynx
  - Pseudo-Zenker diverticulum
    - Barium trapped between peristaltic wave & prominent cricopharyngeal muscle or early closure of upper cervical esophagus
    - Does not extend posteriorly beyond contour of cervical esophagus
    - After peristaltic wave has passed, during suspended respiration, trapped barium is cleared & pseudo-Zenker diverticulum is not evident

Imaging Recommendations
- Fluoroscopic guided pharyngo-esophagram
- AP, lateral & oblique views
- Lateral view during suspended respiration
  - To rule out pseudo-Zenkers diverticulum

DIFFERENTIAL DIAGNOSIS

Killian-Jamieson diverticulum (K-J)
- Transient or persistent protrusions of anterolateral cervical esophagus into Killian-Jamieson space
- Also known as lateral cervical esophageal pouches or diverticula
  - Killian-Jamieson pouches
- Morphology
  - Round to oval, smooth-surfaced outpouching
- Location
  - Seen in anterolateral wall of cervical esophagus (triangular area of weakness)
    - Just below cricopharyngeus muscle
- Size: 3-20 mm in diameter
- Less common & smaller than Zenker diverticulum
- Less likely to cause symptoms
- Less likely to be associated with overflow aspiration or gastroesophageal reflux than is Zenker diverticulum
- Fluoroscopic guided barium study
  - Frontal views
    - Appear as shallow & broad-based protrusions
    - Location: Lateral upper esophageal wall
    - Fill late during swallowing & empty after swallowing
  - Lateral views
    - Barium-filled saccular protrusions
    - Neck of diverticulum opening below level of cricopharyngeus muscle
    - Zenker diverticulum: Neck opens into posterior hypopharyngeal wall above cricopharyngeus muscle (C5-6 level) & sac extends inferiorly behind cervical esophagus

Esophageal web
- Radiolucent ring in proximal cervical esophagus near cricopharyngeus
- On imaging
  - 1-2 mm wide, shelf-like filling defect along anterior wall of cervical esophagus
  - Mild, moderate or severe luminal narrowing

Epidermolysis bullosa dystrophica
- Stricture + proximal dilatation of cervical esophagus
- Cervical esophageal webs near cricopharyngeus
- High esophageal strictures or webs in children or young adults with clinical history suggests diagnosis

ZENKER DIVERTICULUM

PATHOLOGY

General Features

- General path comments
  - False diverticulum
    - Only mucosa & submucosa are seen
    - No muscle tissue
- Etiology
  - Cricopharyngeal dysfunction (achalasia, spasm, premature closure); ↑ intraluminal pressure
  - Spasm, incoordination or abnormal relaxation of upper esophageal sphincter (achalasia)
  - Other contributing factors to development of Zenker diverticulum
    - Gastroesophageal reflux
    - Reflux esophagitis
    - Hiatal hernia
    - Esophageal spasm & achalasia
- Epidemiology: Prevalence: 0.01-0.11%
- Associated abnormalities
  - Gastroesophageal (GE) reflux
  - Reflux esophagitis, hiatal hernia

Gross Pathologic & Surgical Features

- Posterior hypopharyngeal saccular outpouching with broad or narrow neck

Microscopic Features

- Mucosal & submucosal layers of hypopharynx
- Lack of muscle

CLINICAL ISSUES

Presentation

- Most common signs/symptoms
  - Upper esophageal dysphagia
  - Regurgitation & aspiration of undigested food
  - Halitosis, choking, hoarseness, neck mass
  - Some patients are asymptomatic

Demographics

- Age
  - Higher in elderly people
  - 50% of cases seen in 7th-8th decade
- Gender: M > F

Natural History & Prognosis

- Complications
  - Aspiration pneumonia (in 30% of cases)
  - Bronchitis, bronchiectasis, lung abscess
  - Diverticulitis, ulceration, fistula formation
  - Risk of perforation during endoscopy or placement of nasogastric tube
  - Risk of carcinoma (seen in 0.3% of cases)
- Prognosis
  - Usually good after surgery
  - Poor prognosis: Neoplastic complication

Treatment

- Small asymptomatic diverticula: No treatment
- Asymptomatic & symptomatic large diverticula
  - Surgical diverticulectomy or endoscopic repair
- Associated motor disorder: Myotomy

DIAGNOSTIC CHECKLIST

Consider

- Differentiate from Killian-Jamieson diverticula which is also an outpouch at pharyngo-esophageal junction

Image Interpretation Pearls

- Lateral view: Barium-filled outpouch posterior to cervical esophagus with neck opening above cricopharyngeus muscle at C5-6 level

SELECTED REFERENCES

1. Postma GN: RE: endoscopic diverticulotomy of Zenker's diverticulum: management and complications (Dysphagia 17:34-39). Dysphagia. 18(3):227; author reply 227-8, 2003
2. Ibrahim IM: Zenker diverticulum. Arch Surg. 138(1):111, 2003
3. Veenker EA et al: Cricopharyngeal spasm and Zenker's diverticulum. Head Neck. 25(8):681-94, 2003
4. Sasaki CT et al: Association between Zenker diverticulum and gastroesophageal reflux disease: development of a working hypothesis. Am J Med. 115 Suppl 3A:169S-171S, 2003
5. Richtsmeier WJ: Endoscopic management of Zenker diverticulum: the staple-assisted approach. Am J Med. 115 Suppl 3A:175S-178S, 2003
6. Rubesin SE et al: Killian-Jamieson diverticula: radiographic findings in 16 patients. AJR Am J Roentgenol. 177(1):85-9, 2001
7. Siddiq MA et al: Pharyngeal pouch (Zenker's diverticulum). Postgrad Med J. 77(910):506-11, 2001
8. Sydow BD et al: Radiographic findings and complications after surgical or endoscopic repair of Zenker's diverticulum in 16 patients. AJR Am J Roentgenol. 177(5):1067-71, 2001
9. DeFriend DE et al: Sonographic demonstration of a pharyngoesophageal diverticulum. J Clin Ultrasound. 28(9):485-7, 2000
10. Achkar E: Zenker's diverticulum. Dig Dis. 16(3):144-51, 1998
11. Walters DN et al: Zenker's diverticulum in the elderly: a neurologic etiology? Am Surg. 64(9):909-11, 1998
12. Bremner CG: Zenker diverticulum. Arch Surg. 133(10):1131-3, 1998
13. Chin E et al: Zenker's diverticulum. Am J Gastroenterol. 92(4):720, 1997
14. Woodruff WW et al: Non-nodal neck masses. Semin Ultrasound CT MR. 18(3):182-204, 1997
15. Ponette E et al: Radiological aspects of Zenker's diverticulum. Hepatogastroenterology. 39(2):115-22, 1992
16. Fulp SR et al: Manometric aspects of Zenker's diverticulum. Hepatogastroenterology. 39(2):123-6, 1992
17. Cook IJ et al: Pharyngeal (Zenker's) diverticulum is a disorder of upper esophageal sphincter opening. Gastroenterology 103: 1229-35, 1992
18. Frieling T et al: Upper esophageal sphincter function in patients with Zenker's diverticulum. Dysphagia. 3(2):90-2, 1988
19. Bowdler DA et al: Carcinoma arising in posterior pharyngeal pulsion diverticulum (Zenker's diverticulum). Br J Surg. 74(7):561-3, 1987
20. Zitsch RP et al: Pharyngoesophageal diverticulum complicated by squamous cell carcinoma. Head Neck Surg. 9(5):290-4, 1987
21. Semenkovich JW et al: Barium pharyngography: Comparison of single and double contrast. AJR 144: 715-20, 1985

ZENKER DIVERTICULUM

IMAGE GALLERY

Typical

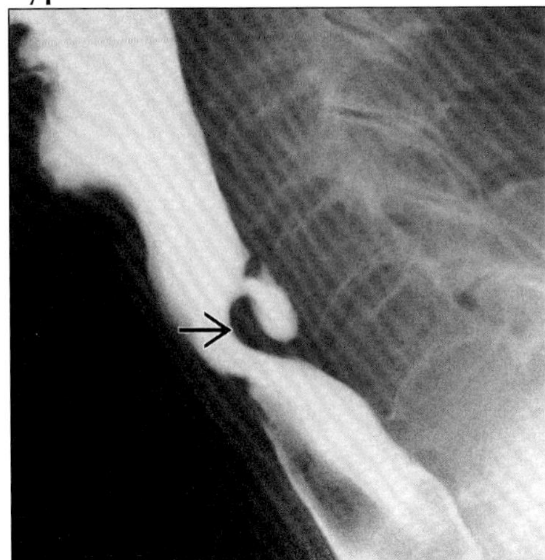

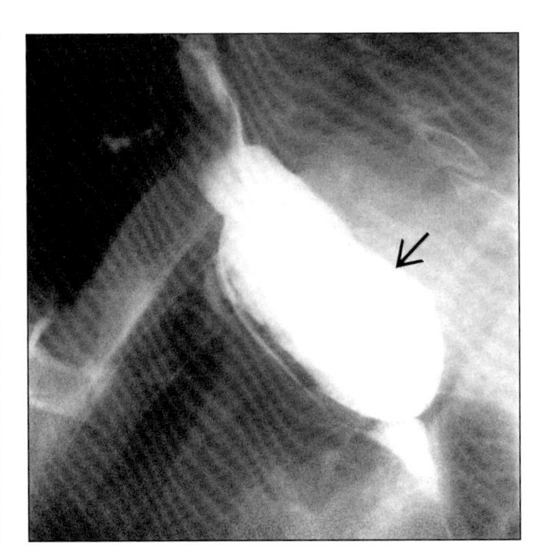

(Left) *Lateral esophagram shows small Zenker diverticulum at the C5-C6 level.* ***(Right)*** *Lateral view esophagram shows large diverticulum (arrow) displacing and compressing the posterior wall of the proximal esophagus.*

Typical

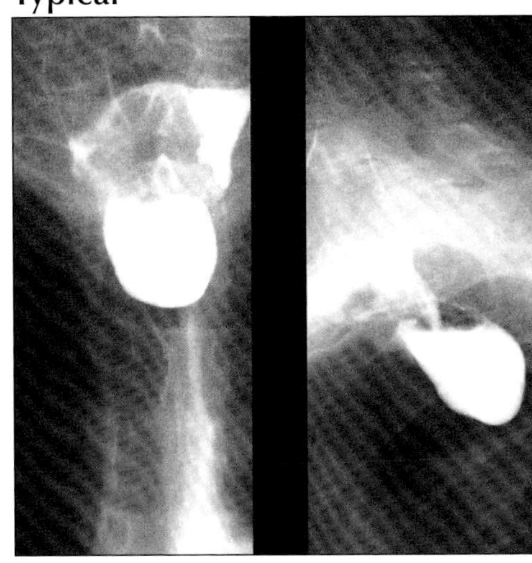

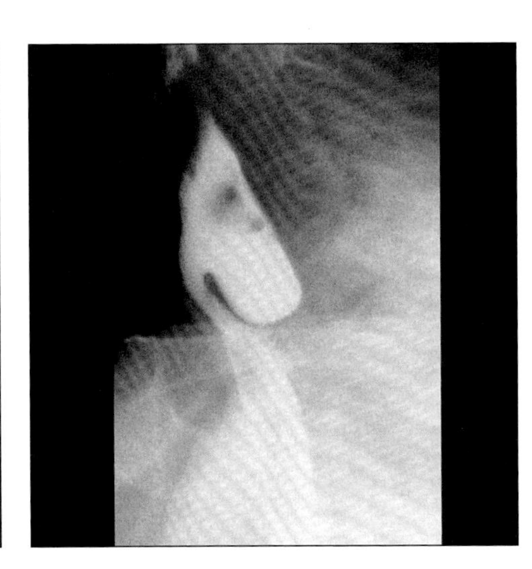

(Left) *Frontal and oblique views of barium esophagram show large Zenker diverticulum extending into thorax with retained barium after the bolus has passed.* ***(Right)*** *Lateral view esophagram shows a moderate size diverticulum displacing and compressing the posterior wall of the esophagus, just above the thoracic inlet.*

Variant

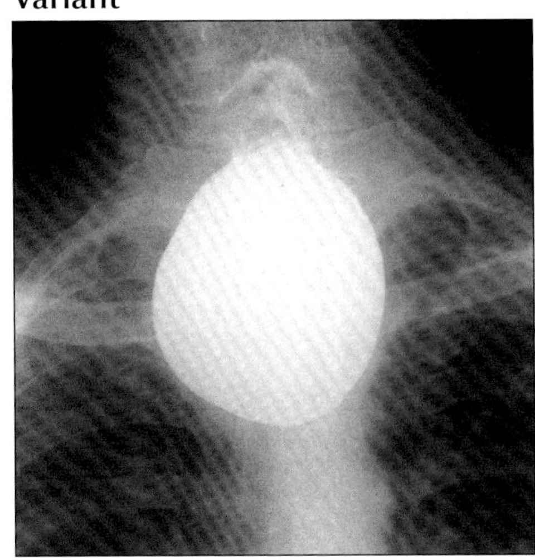

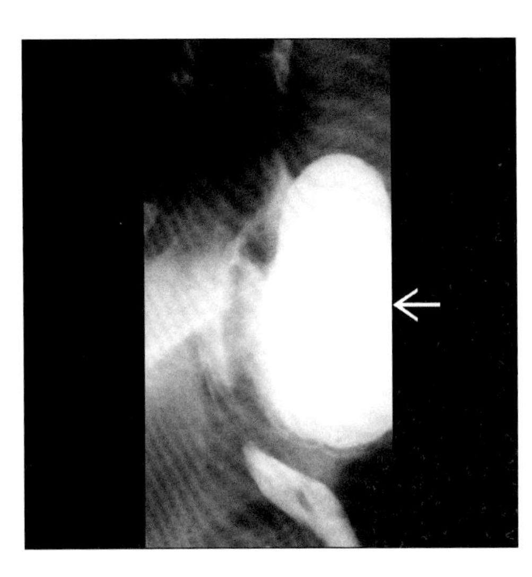

(Left) *Frontal view from esophagram shows large diverticulum with retained barium, extending into upper thorax.* ***(Right)*** *Lateral view esophagram shows unusually large Zenker diverticulum (arrow) that was symptomatic with obstructed swallowing, halitosis, and regurgitation of undigested food.*

TRACTION DIVERTICULUM

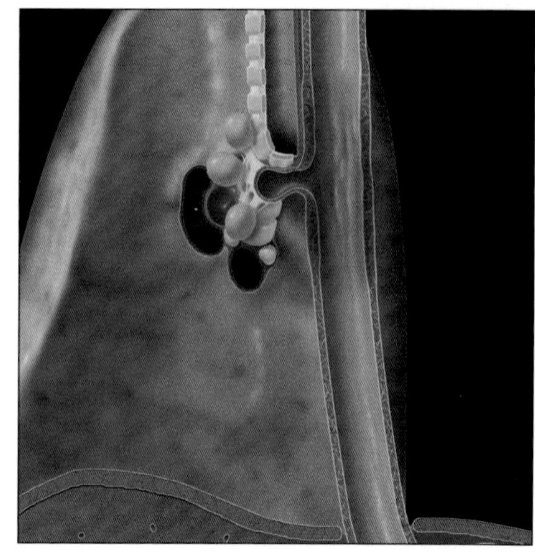

Graphic shows subcarinal lymph nodes that are adherent to the esophageal wall resulting in a diverticulum.

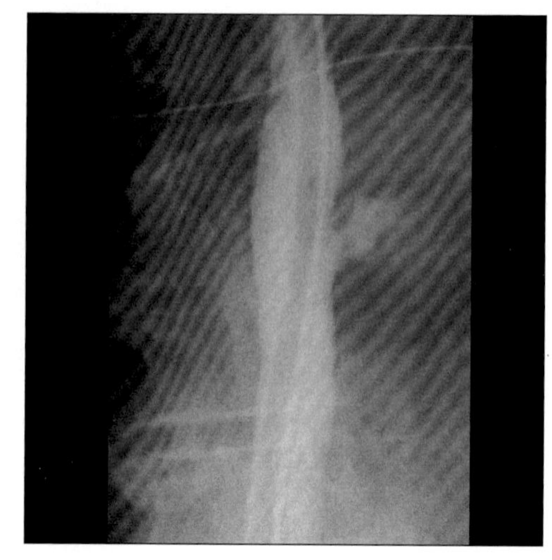

Esophagram shows barium-filled tented outpouching from midesophagus. Calcified subcarinal lymph nodes were evident on chest radiograph.

TERMINOLOGY

Abbreviations and Synonyms

- Esophageal saccular protrusion or outpouching

Definitions

- Outpouching of all layers of esophageal wall due to perihilar pathology

IMAGING FINDINGS

General Features

- Best diagnostic clue: Barium-filled tented or triangular shaped outpouching from mid-esophagus
- Location: Usually mid-esophagus
- Size: Varies from few millimeters to few centimeters
- Morphology: Tented or triangular configuration
- Other general features
  - Classification of esophageal diverticula based on mechanism of formation
    - Traction type (true diverticula): Herniation of all layers (mucosa, submucosa & muscularis propria)
    - Pulsion type (pseudodiverticula): Mucosal, submucosal herniation & lack of muscle
  - Traction diverticulum: Acquired condition due to subcarinal or perihilar granulomatous lymph node pathology

Radiographic Findings

- Radiography
  - Chest x-ray PA view
    - Calcification of perihilar nodes may be seen
  - Lateral view
    - Thickening of posterior tracheal stripe
- Videofluoroscopic esophagogram (barium studies)
  - Mid-esophagus: Traction diverticulum
    - Tented or triangular in shape with pointed tip
    - Diverticulum tends to empty when esophagus is collapsed (because it contains all layers)
  - Whereas pulsion diverticula tend to remain filled after most of barium is emptied (lack of muscle)

Imaging Recommendations

- Videofluoroscopic esophagram
  - Frontal, lateral & oblique views
- Better demonstrated by large bolus of barium
- Better visualized during maximal distention

DIFFERENTIAL DIAGNOSIS

Pulsion diverticulum

- Zenker diverticulum (posterior hypopharyngeal diverticulum)
  - Lateral view: Barium-filled sac opening into posterior hypopharynx above cricopharyngeus

DDx: Outpouching from Mid-Esophagus

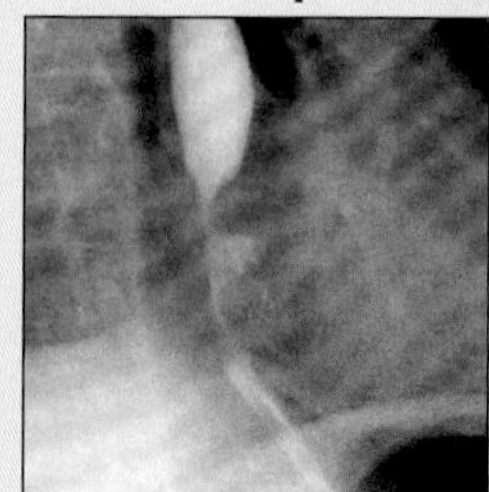

Eso. Perforation

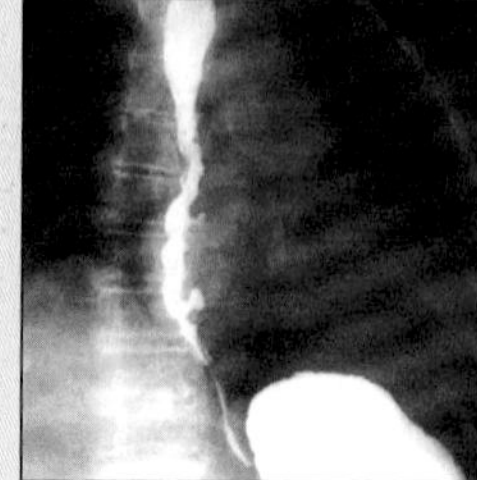

Pulsion 'Tics

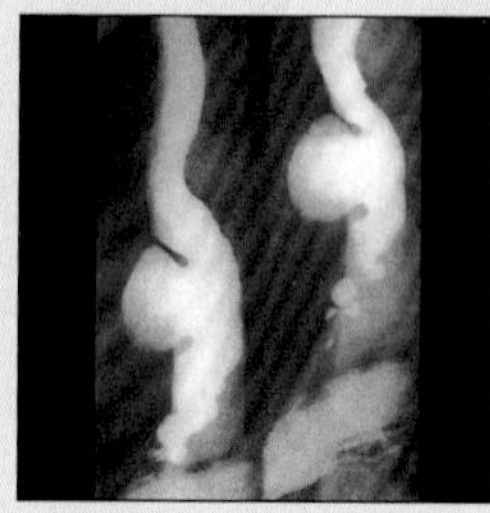

Pulsion 'Tics

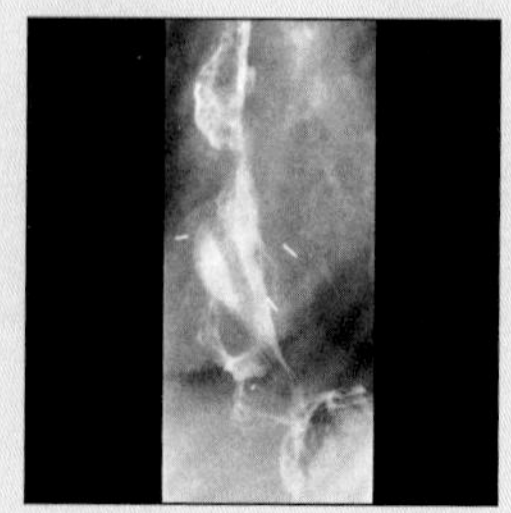

Stricture + Ulcer

TRACTION DIVERTICULUM

Key Facts

Imaging Findings
- Best diagnostic clue: Barium-filled tented or triangular shaped outpouching from mid-esophagus
- Diverticulum tends to empty when esophagus is collapsed (because it contains all layers)

Top Differential Diagnoses
- Pulsion diverticulum
- Esophageal perforation (mid-esophageal)
- Esophageal ulcer

Pathology
- Pulmonary tuberculosis, histoplasmosis, sarcoidosis

Diagnostic Checklist
- Differentiate from other outpouchings from mid-esophagus (in cases of mediastinitis or chest pain)
- Check for prior history of granulomatous diseases

- Mid & distal esophageal pulsion diverticula tend to remain filled after most of barium is emptied (lack of muscle)

Esophageal perforation (mid-esophageal)
- A sealed-off leak seen as self-contained extraluminal collection of contrast medium that communicates with adjacent esophagus
- May be indistinguishable from traction diverticulum without history

Esophageal ulcer
- Solitary ring-like or stellate shaped large ulcer
- Halo of edema and associated findings (e.g., stricture) differentiate from diverticulum

PATHOLOGY

General Features
- Etiology
  - Pulmonary tuberculosis, histoplasmosis, sarcoidosis
  - Pathogenesis: Acutely inflamed, enlarged subcarinal nodes indent and adhere to esophageal walls
  - As inflammation subsides, nodes shrink and retract the adherent esophagus
- Epidemiology
  - Less common compared to pulsion type
  - Common in areas of endemic histoplasmosis & TB

Gross Pathologic & Surgical Features
- Saccular outpouchings mostly in mid-esophagus
- Adherent to subcarinal lymph nodes

Microscopic Features
- All layers are seen (true diverticulum)

CLINICAL ISSUES

Presentation
- Most common signs/symptoms
  - Small diverticula: Usually asymptomatic
  - Large diverticula: ± Dysphagia or regurgitation

Demographics
- Age: Usually seen in elderly age group
- Gender: M = F

Natural History & Prognosis
- Complications
  - Erosion; inflammation; perforation; fistula
- Prognosis: Symptomatic-good after surgery

Treatment
- Large symptomatic diverticula: Diverticulectomy

DIAGNOSTIC CHECKLIST

Consider
- Differentiate from other outpouchings from mid-esophagus (in cases of mediastinitis or chest pain)
- Check for prior history of granulomatous diseases

Image Interpretation Pearls
- Traction diverticulum: Tends to empty when esophagus is collapsed (contain all esophageal layers)

SELECTED REFERENCES

1. Lopez A et al: Esophagobronchial fistula caused by traction esophageal diverticulum. Eur J Cardiothorac Surg. 23(1):128-30, 2003
2. Raziel A et al: Sarcoidosis and giant midesophageal diverticulum. Dis Esophagus. 13(4):317-9, 2000
3. Schima W et al: Association of midoesophageal diverticula with oesophageal motor disorders. Videofluoroscopy and manometry. Acta Radiol. 38(1):108-14, 1997
4. Savides TJ et al: Dysphagia due to mediastinal granulomas: diagnosis with endoscopic ultrasonography. Gastroenterology. 109(2):366-73, 1995
5. Duda M et al: Etiopathogenesis and classification of esophageal diverticula. Int Surg. 70(4):291-5, 1985

IMAGE GALLERY

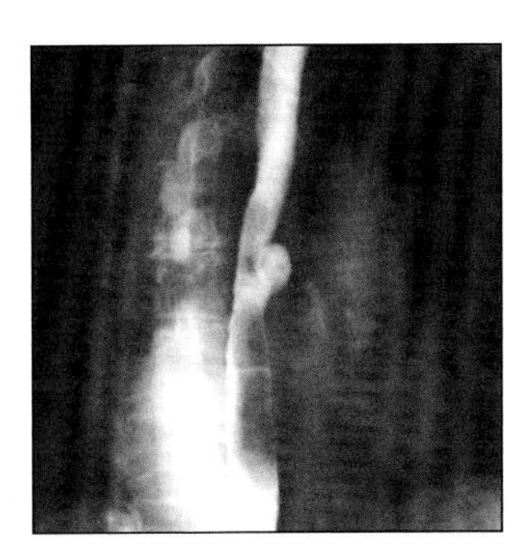

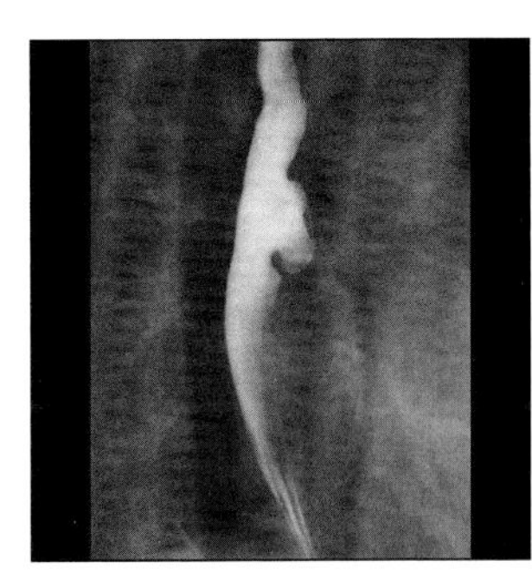

*(**Left**) Esophagram shows outpouching at the subcarinal level. (**Right**) Esophagram shows mid-esophageal diverticulum.*

PULSION DIVERTICULUM

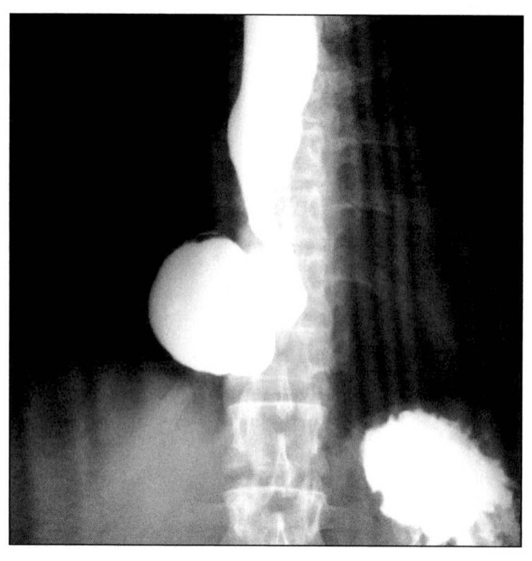

Esophagram shows large outpouching from esophagus just above diaphragm (epiphrenic). Esophagus was aperistaltic with narrowed GE junction (achalasia).

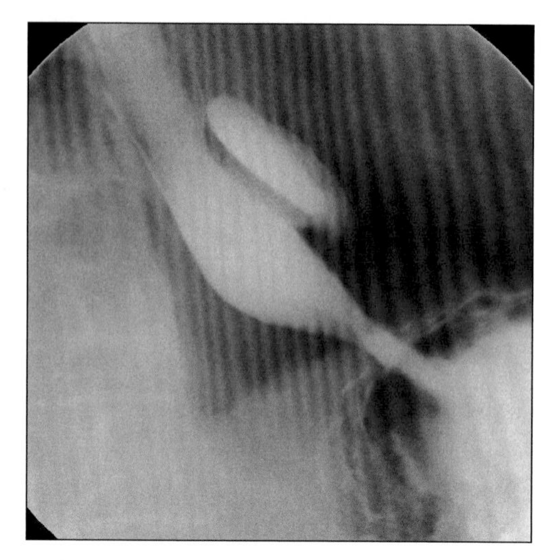

Esophagram shows sac-like protrusion from distal esophagus.

TERMINOLOGY

Abbreviations and Synonyms

- Esophageal saccular protrusion or outpouching or pseudodiverticulum

Definitions

- Outpouching or sac-like protrusion of one or more layers of esophageal wall

IMAGING FINDINGS

General Features

- Best diagnostic clue: Barium-filled outpouching from esophagus on esophagram
- Location
  - Posterior pharyngoesophageal junction (C5-6 level): Killian dehiscence (triangular anatomical area of weakness)
  - Mid-esophagus
  - Distal esophageal or epiphrenic
- Size: Varies from few millimeters to few centimeters
- Morphology
  - Saccular outpouching with broad or narrow neck
  - Pulsion type (pseudodiverticula): Mucosal & submucosal herniation through muscularis propria (common)
  - Traction type (true diverticula): Herniation of all layers
- Other general features
  - Classification of esophageal diverticula based on location & mechanism of formation
    - Pharyngoesophageal junction: Pulsion type (Zenker diverticulum)
    - Cervical esophagus (Killian-Jamieson space): Triangular area of weakness in anterolateral wall (Killian-Jamieson diverticula)
    - Mid-esophagus: Pulsion & traction type
    - Distal esophagus (epiphrenic): Pulsion type
    - Intramural pseudodiverticulosis: Small outpouchings of mucosal glands
  - Pulsion diverticula
    - Usually acquired rather than congenital
    - More common & are often associated with other radiographic evidence of motor dysfunction

Radiographic Findings

- Large epiphrenic pulsion diverticulum
  - Chest x-ray PA view: Prominent bulge along right or left border of heart
  - Lateral view
    - Large soft tissue mass, mimicking hiatus hernia
    - ± Air-fluid level
    - ± Retained foreign body or calcified enterolith
- Videofluoroscopic esophagram (barium studies)
  - Pharyngoesophageal junction: Zenker diverticulum

DDx: Outpouching from Distal Esophagus

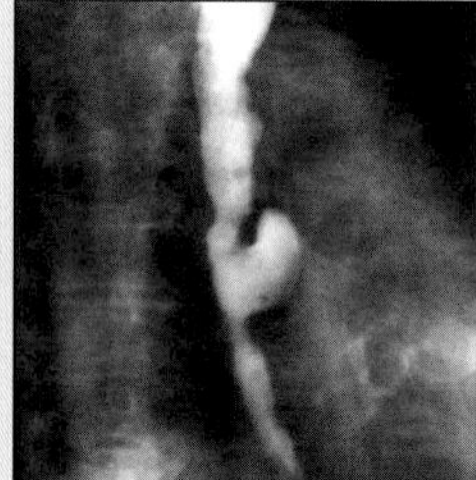

Traction Diverticulum

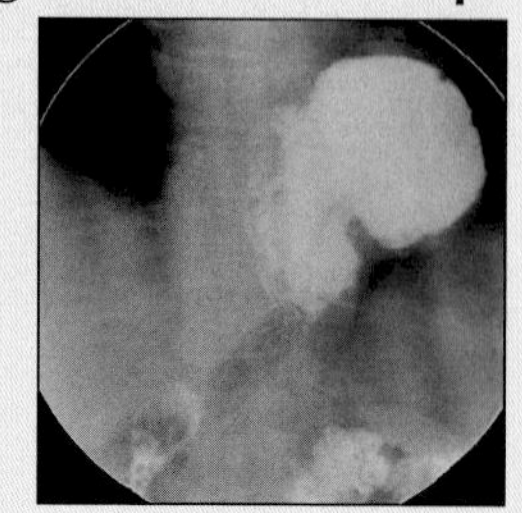

Hiatal Hernia

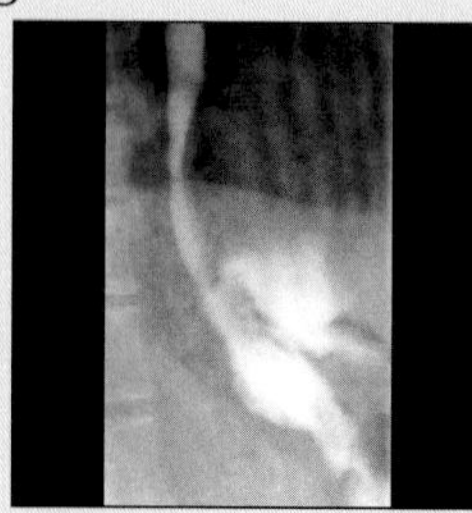

Boerhaave Syndrome

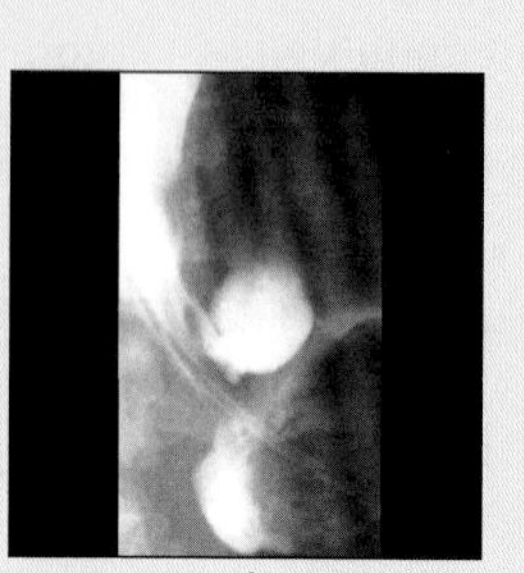

Eso. Perforation

PULSION DIVERTICULUM

Key Facts

Terminology
- Esophageal saccular protrusion or outpouching or pseudodiverticulum

Imaging Findings
- Best diagnostic clue: Barium-filled outpouching from esophagus on esophagram
- Lateral view: Barium-filled sac opening into posterior hypopharyngeal wall above cricopharyngeus-Zenker
- Diverticula tend to remain filled after most of barium is emptied (due to lack of muscle in wall)
- Midesophageal: Frequently multiple of varied sizes
- Usually large barium-filled sac in epiphrenic area
- Better visualized during maximal distention

Top Differential Diagnoses
- Traction diverticula (for mid-esophageal pulsion type)
- Hiatal hernia (for a large epiphrenic pulsion type)
- Esophageal perforation

Pathology
- Defect in muscular wall of esophagus
- Esophageal motility disorders
- Pathogenesis: Pulsion due to ↑ intraluminal pressure

Diagnostic Checklist
- Differentiate mid-esophageal pulsion pseudodiverticulum from traction (true) diverticulum
- Distal esophageal (epiphrenic) diverticulum may simulate hiatal hernia & Boerhaave syndrome
- Pulsion diverticula tend to remain filled after most of barium is emptied (lack of muscle/contraction)

        - Lateral view: Barium-filled sac opening into posterior hypopharyngeal wall above cricopharyngeus-Zenker
        - Size: 10-80 mm
    - Cervical esophagus: Anterolateral wall (Killian-Jamieson diverticula)
        - Frontal view: Round to oval, smooth-surfaced diverticula; 3-20 mm in size
        - Lateral view: Barium-filled transient or persistent pouches anterior to cervical esophagus opening below level of cricopharyngeus (Killian-Jamieson)
    - Mid-esophagus: Pulsion diverticula
        - Best seen in profile & recognized en face as ring shadows on double contrast studies
        - Barium-filled outpouchings from esophagus
        - Usually smooth, rounded contour & a wide neck
        - Diverticula tend to remain filled after most of barium is emptied (due to lack of muscle in wall)
        - Small diverticula (0.5-2 cm): Seen as transient outpouchings that develop only during peristalsis
        - Midesophageal: Frequently multiple of varied sizes
        - Evidence of diffuse esophageal spasm/motor dysfunction
    - Distal esophagus (epiphrenic): Pulsion diverticula
        - Usually large barium-filled sac in epiphrenic area
        - Usually from lateral esophageal wall (distal 10 cm)
        - Right side more common than left side
        - Often associated with achalasia or hiatal hernia

Imaging Recommendations
- Videofluoroscopic esophagram
    - Frontal, lateral & oblique views
- Better demonstration achieved by large boluses of barium
- Better visualized during maximal distention

DIFFERENTIAL DIAGNOSIS

Traction diverticula (for mid-esophageal pulsion type)
- Location: Mid-esophagus
- Etiology
    - Tuberculosis, histoplasmosis & sarcoidosis of perihilar/subcarinal nodes
- Pathogenesis
    - Due to fibrosis & scarring in periesophageal tissues
- Pathology: All layers are seen (true diverticula)
- Esophagography
    - Tented or triangular configuration with pointed tip
    - Usually have no neck
    - Diverticula tend to empty when esophagus is collapsed (because traction diverticula contain all layers of esophageal wall)
    - Whereas in pulsion type, diverticulum remains filled after most of barium is emptied from esophagus by peristalsis
        - Due to lack of muscle/contraction in diverticulum

Hiatal hernia (for a large epiphrenic pulsion type)
- About 99% of all hiatal hernias are sliding & remaining 1% are paraesophageal
- Phrenoesophageal membrane: Firm, elastic structure surrounding GE junction
    - Tethers distal esophagus to diaphragm & prevents herniation of proximal part of stomach via esophageal hiatus of diaphragm into chest
- Etiology: Gastroesophageal reflux disease (GERD), reflux esophagitis
- Pathogenesis
    - Progressive wear & tear of phrenoesophageal membrane
    - Stretching or rupture of membrane & axial herniation of stomach into chest
- Chest x-ray
    - Posteroanterior (PA) view
        - Prominent bulge along right border of heart
    - Lateral view: Large retrocardiac soft tissue mass mimicking the appearance of a epiphrenic diverticulum
- Esophagography
    - GE junction above esophageal hiatus of diaphragm
    - Lower esophageal mucosal ring or Schatzki ring demarcates anatomic location of GE junction

PULSION DIVERTICULUM

- Sliding hiatal hernia: Diagnosed on barium study when a mucosal ring is seen 2 cm or more above diaphragmatic hiatus
- Gastric rugae may be seen in herniated part
- May simulate a large epiphrenic pulsion diverticulum
- Best position for optimal demonstration
  - Prone right anterior oblique position (RAO)
- Avoid erect position: Hernia is reduced into abdomen & inadequately distended
- Barium: Thin, low density suspension

Esophageal perforation

- Iatrogenic: Post instrumentation (e.g., endoscopy)
- Boerhaave syndrome: Spontaneous perforation
  - Location: Lateral wall of distal esophagus just above GE junction
  - Morphology: Vertically oriented full-thickness linear tear measuring 1-4 cm in length
  - Etiology: Violent vomiting/retching (alcoholic binge, pregnancy, etc.)
- Small esophageal perforation
  - Extravasation of contrast medium from lateral wall of esophagus into adjacent mediastinum
- A sealed-off leak seen as self-contained extraluminal collection of contrast medium that communicates with adjacent esophagus
  - Extraluminal contrast is persistent
  - Wall is less distinct than epiphrenic diverticulum

PATHOLOGY

General Features

- General path comments
  - Embryology-anatomy
    - Cervical esophagus: Killian-Jamieson space is an anatomical triangular area of weakness just below cricopharyngeal muscle
- Etiology
  - Defect in muscular wall of esophagus
  - Pulsion diverticula
    - Chronic wear & tear forces (for most diverticula)
    - Esophageal motility disorders
    - Mechanical obstruction
    - Risk factors: GERD/reflux esophagitis/hiatal hernia
  - Pathogenesis: Pulsion due to ↑ intraluminal pressure
- Epidemiology
  - Pulsion (more common)
  - Traction (uncommon) except in areas with endemic histoplasmosis or tuberculosis
- Associated abnormalities: Esophageal motility disorders may be seen

Gross Pathologic & Surgical Features

- Saccular outpouchings with broad or narrow neck

Microscopic Features

- Pulsion diverticula (pseudodiverticula)
- Mucosa & submucosa are seen; lack of muscle

CLINICAL ISSUES

Presentation

- Most common signs/symptoms
  - Small diverticula: Usually asymptomatic
  - Large diverticula: ± Dysphagia or regurgitation

Demographics

- Age: Usually seen in elderly age group
- Gender
  - Zenker diverticulum: (M > F)
  - Mid-/distal esophageal diverticula (M = F)

Natural History & Prognosis

- Complications
  - Erosion with bleeding; inflammation with abscess
  - Perforation; fistula formation; retained foreign body
  - Neoplasm (carcinoma of epiphrenic diverticula)
- Prognosis: Usually good after surgery

Treatment

- Very large symptomatic diverticula
  - Surgical diverticulectomy
- Associated motor disorders: Distal myotomy

DIAGNOSTIC CHECKLIST

Consider

- Differentiate mid-esophageal pulsion pseudodiverticulum from traction (true) diverticulum
- Distal esophageal (epiphrenic) diverticulum may simulate hiatal hernia & Boerhaave syndrome

Image Interpretation Pearls

- Zenker diverticulum: Barium-filled outpouch posterior to cervical esophagus with neck opening above cricopharyngeus muscle at C5-6 level
- Distal esophageal epiphrenic diverticulum: Often associated with achalasia or hiatal hernia
- Pulsion diverticula tend to remain filled after most of barium is emptied (lack of muscle/contraction)

SELECTED REFERENCES

1. Fasano NC et al: Epiphrenic diverticulum: clinical and radiographic findings in 27 patients. Dysphagia. 18(1):9-15, 2003
2. Sasaki CT et al: Association between Zenker diverticulum and gastroesophageal reflux disease: development of a working hypothesis. Am J Med. 115 Suppl 3A:169S-171S, 2003
3. Sydow BD et al: Radiographic findings and complications after surgical or endoscopic repair of Zenker's diverticulum in 16 patients. AJR Am J Roentgenol. 177(5):1067-71, 2001
4. Rubesin SE et al: Killian-Jamieson diverticula: Radiographic findings in 16 patients. AJR 177: 85-9, 2001
5. Schima W et al: Association of midesophageal diverticula with esophageal motor disorders. Videofluoroscopy and manometry. Acta Radiol. 38(1):108-14, 1997
6. Ponette E et al: Radiological aspects of Zenker's diverticulum. Hepatogastroenterology. 39(2):115-22, 1992
7. Bruggeman LL et al: Epiphrenic diverticula. An analysis of 80 cases. Am J Roentgenol Radium Ther Nucl Med. 119(2):266-76, 1973

PULSION DIVERTICULUM

IMAGE GALLERY

Typical

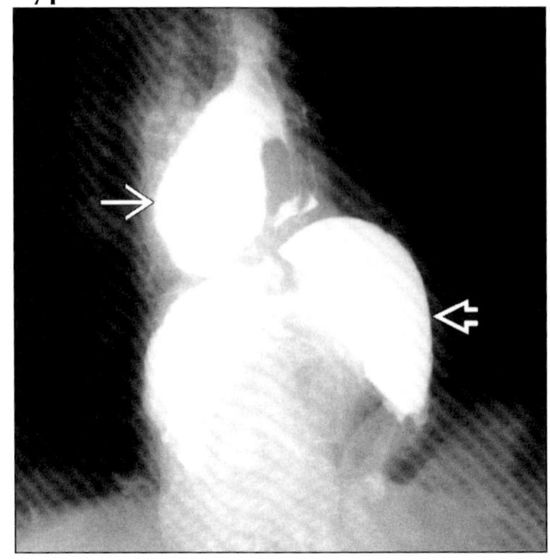

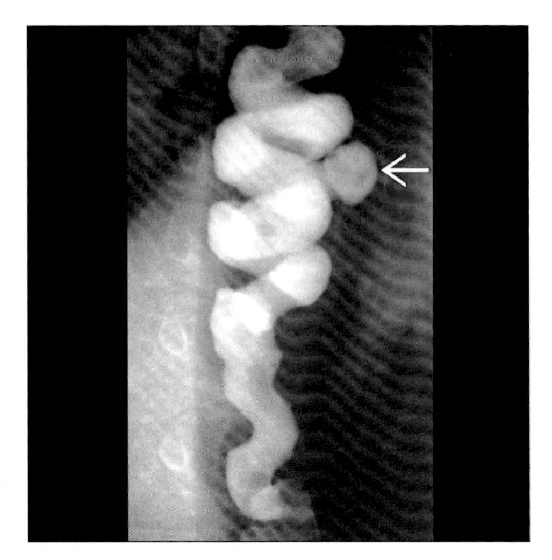

(Left) *Esophagram shows pulsion diverticulum (arrow) above a large paraesophageal hernia (open arrow).* ***(Right)*** *Esophagram shows severe distortion and dysmotility of esophagus ("corkscrew esophagus") and a pulsion diverticulum (arrow).*

Typical

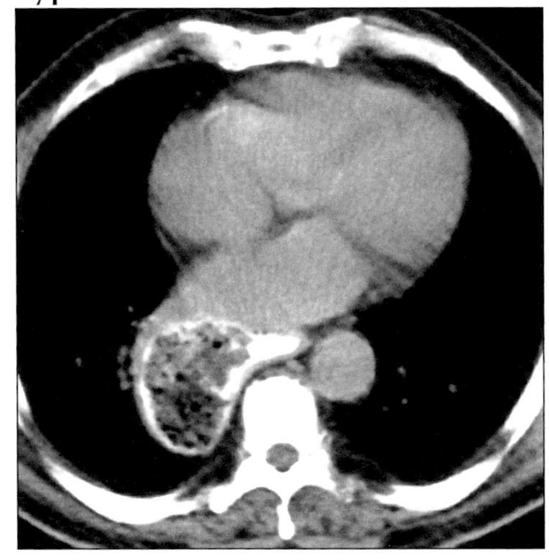

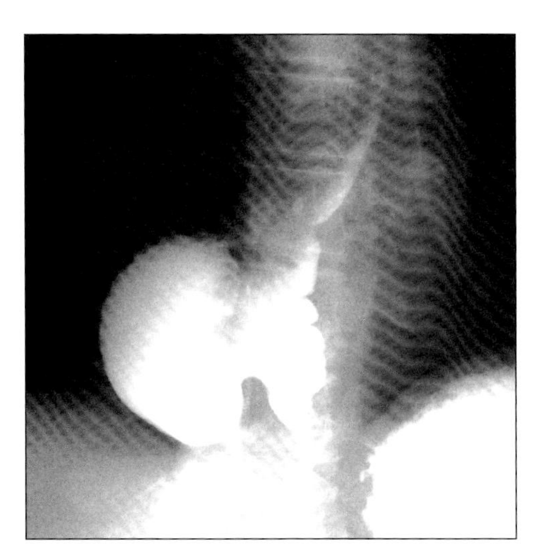

(Left) *Axial NECT shows a large epiphrenic diverticulum containing contrast and retained food.* ***(Right)*** *Esophagram shows large epiphrenic diverticulum and esophageal dysmotility with tertiary contractions.*

Typical

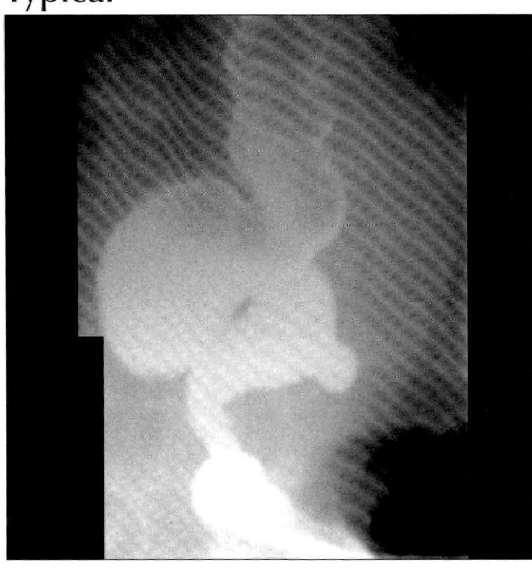

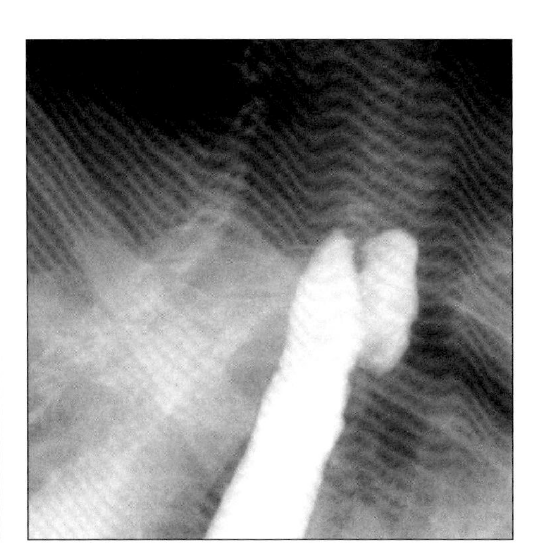

(Left) *Esophagram shows large right and small left epiphrenic diverticula and dilated esophagus with dysmotility.* ***(Right)*** *Oblique view of esophagram shows pulsion diverticulum near thoracic inlet.*

INTRAMURAL PSEUDODIVERTICULOSIS

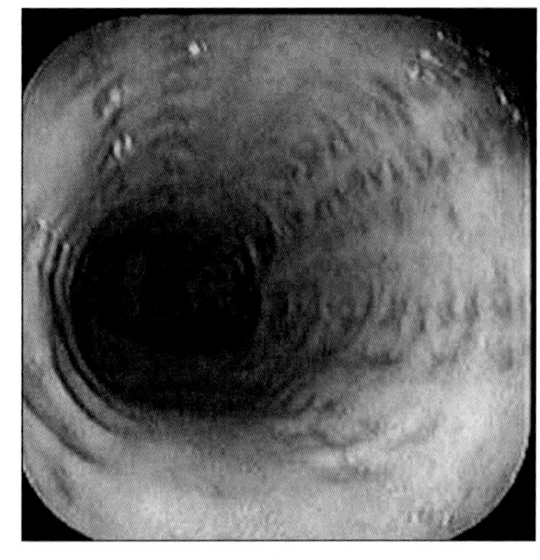

Endoscopic image shows the opening of innumerable pseudodiverticula, many arranged in orderly, longitudinal rows.

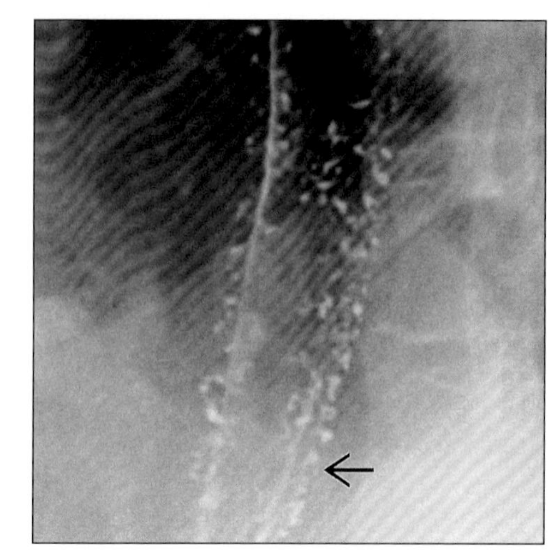

Double contrast esophagram shows innumerable tiny outpouchings from the distal two thirds of the esophagus. Bridging between adjacent pseudodivertula creates intramural tracks (arrow).

TERMINOLOGY

Abbreviations and Synonyms

- Intramural pseudodiverticulosis (IPD)

Definitions

- Rare benign disorder characterized by esophageal diverticulosis-like outpouchings
  - Caused by dilation of excretory ducts of deep mucous glands

IMAGING FINDINGS

General Features

- Best diagnostic clue: Radiologic examination; more sensitive than endoscopy
- Location
  - Pseudodiverticula in wall of thoracic esophagus
  - Diffuse (50%) or segmental involvement
- Size
  - Outpouchings: Tiny, usually 1-4 mm
  - Intramural tracks: Average length of 1.2 cm (length range, 0.3-7 cm)
- Morphology: Multiple; flask-shaped outpouchings

Radiographic Findings

- Outpouchings: Barium-filled; innumerable; tiny; in longitudinal rows parallel to long axis of esophagus
  - Neck; 1 mm or less in diameter
    - Incomplete filling may erroneously suggest lack of communication with esophageal lumen
  - Ductal obstruction by inflammatory material or debris may prevent barium from entering ducts resulting in failure to visualize pseudodiverticula
- Intramural tracks: Bridging between adjacent pseudodiverticula
  - More likely to occur with diffuse form of IPD
  - May have little or no known clinical significance
  - Mistaken for large flat ulcer or extramural collection
- Large, irregular extraluminal collection of barium
  - Due to massive ductal dilatation or sealed-off perforation of duct
- Stricture: Of varying length; pseudodiverticula often extend above & below level of stricture
  - Are not always benign; should be evaluated for radiologic signs of malignancy

CT Findings

- Thick esophageal wall, narrow lumen
- Intramural gas collections
- Peridiverticulitis & abscess formation
  - Due to perforation (rare)

Imaging Recommendations

- Esophagram: Frontal; lateral; oblique views

DDx: Numerous Small Outpouchings

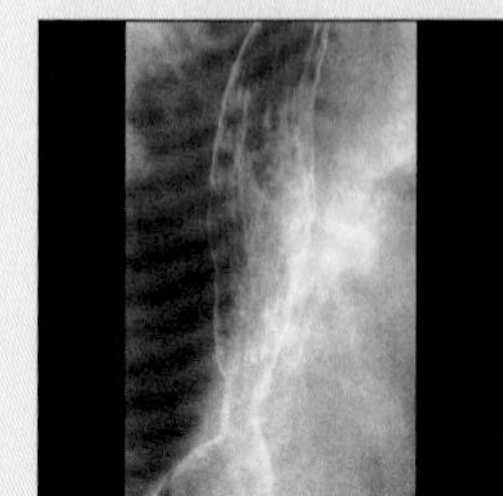

Reflux Stricture

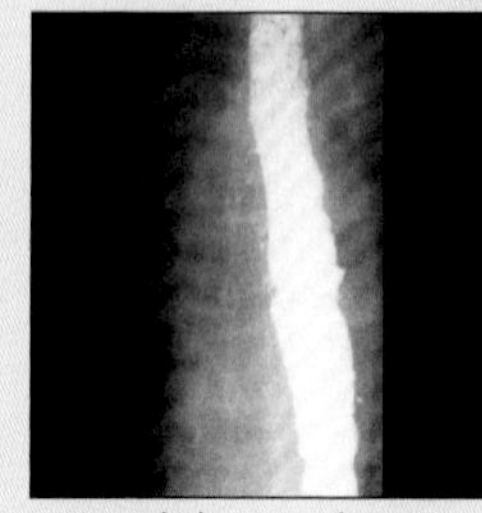

Candida Esophagitis

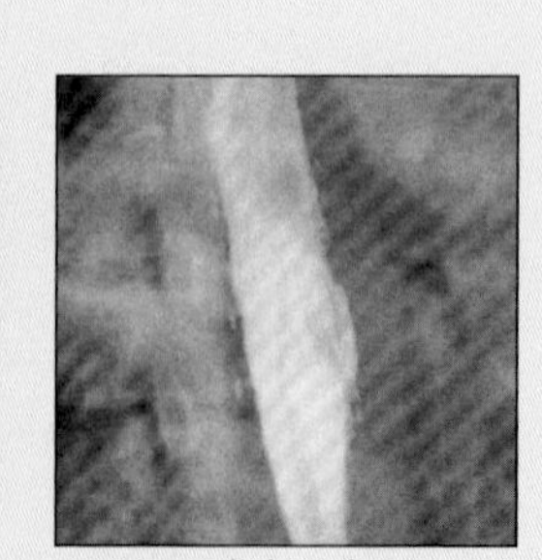

Candida Ulcers

INTRAMURAL PSEUDODIVERTICULOSIS

Key Facts

Terminology
- Rare benign disorder characterized by esophageal diverticulosis-like outpouchings
- Caused by dilation of excretory ducts of deep mucous glands

Imaging Findings
- Pseudodiverticula in wall of thoracic esophagus
- Morphology: Multiple; flask-shaped outpouchings

Top Differential Diagnoses
- Esophagitis

Diagnostic Checklist
- Periodic surveillance of patients with IPD for esophageal carcinoma may be worthwhile

DIFFERENTIAL DIAGNOSIS

Esophagitis
- Multiple discrete ulcers associated with various types of esophagitis
- Viewed en face; pseudodiverticula mistaken for ulcers
- True ulcers communicate directly with lumen
  - Pseudodiverticula seem to be floating outside; without apparent communication with lumen
- Ulcers, strictures and pseudodiverticula can occur together

PATHOLOGY

General Features
- Etiology
  - Its etiology & pathogenesis are largely unknown
  - Inflammation, resulting in periductal fibrosis & compression of duct orifices, may be causative factor
- Epidemiology: Rare condition; diagnosed in less than 1% of patients undergoing esophagography
- Associated abnormalities
  - Chronic esophagitis; gastroesophageal reflux
  - Candidiasis, motility disorders
  - Esophageal web formation; may be under-reported
  - Chronic alcoholism (15%), diabetes mellitus (20%)

Gross Pathologic & Surgical Features
- Numerous intramural pouches; stricture

CLINICAL ISSUES

Presentation
- Intermittent or slowly progressive dysphagia
- Associated diseases are almost always present
- Diagnosis: Esophagography; findings pathognomonic
  - Endoscopy; orifices of ducts difficult to visualize

Demographics
- Age: Seen in adults; most commonly 45-65 years
- Gender: M:F = 3:2

Natural History & Prognosis
- Pseudodiverticula themselves rarely cause problems
- Often noted to disappear after esophageal dilation, but may persist asymptomatically in some patients
- Complication: Perforation; peridiverticulitis; mediastinitis secondary to ruptured IPD
- Prevalence of IPD significantly higher in patients with esophageal carcinoma
  - Association implies increased risk of esophageal carcinoma in patients with IPD

Treatment
- Strictures: Balloon or bougie dilatation

DIAGNOSTIC CHECKLIST

Consider
- Periodic surveillance of patients with IPD for esophageal carcinoma may be worthwhile
  - Endoscopic; endosonographic; histological examination may be required to exclude malignancy

SELECTED REFERENCES

1. Bhattacharya S et al: Intramural pseudodiverticulosis of the esophagus. Surg Endosc. 16(4):714-5, 2002
2. Canon CL et al: Intramural tracking: a feature of esophageal intramural pseudodiverticulosis. AJR Am J Roentgenol. 175(2):371-4, 2000
3. Plavsic BM et al: Intramural pseudodiverticulosis of the esophagus detected on barium esophagograms: increased prevalence in patients with esophageal carcinoma. AJR Am J Roentgenol. 165(6):1381-5, 1995

IMAGE GALLERY

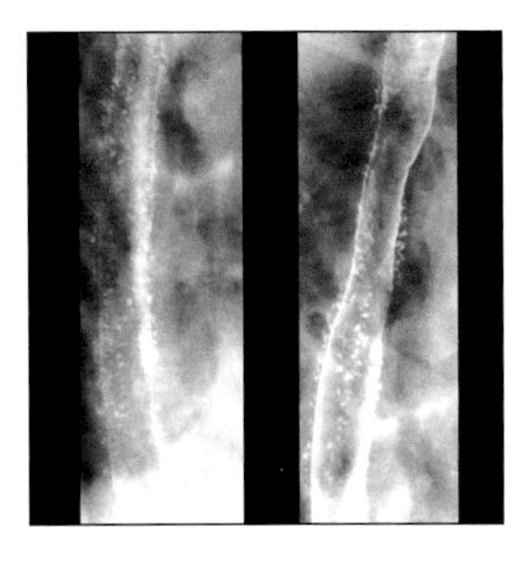

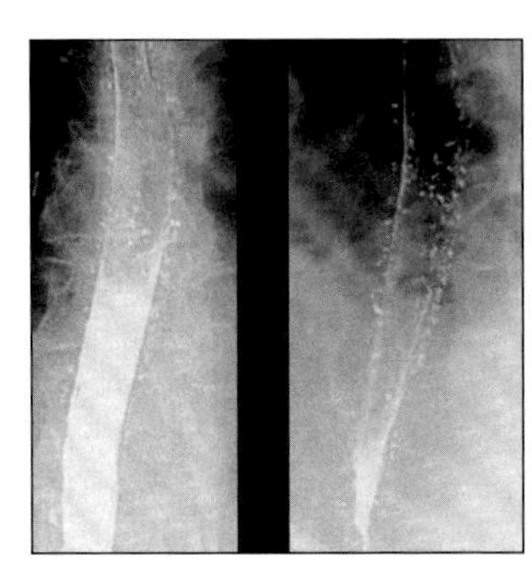

*(**Left**) Two views from esophagram show innumerable tiny flask-shaped outpouchings within the esophageal wall. (**Right**) Two views from esophagram show diffuse intramural pseudodiverticulosis. Note distal esophageal stricture.*

ESOPHAGEAL FOREIGN BODY

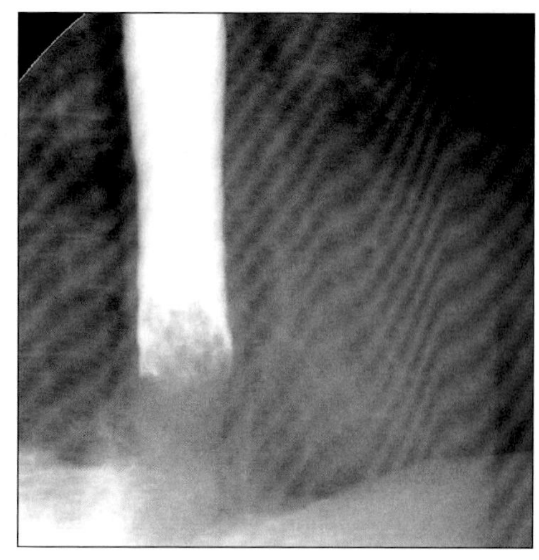

Esophagram shows complete obstruction of esophagus with distal filling defect (ingested meat). After endoscopic removal of foreign body, a Schatzki ring was found.

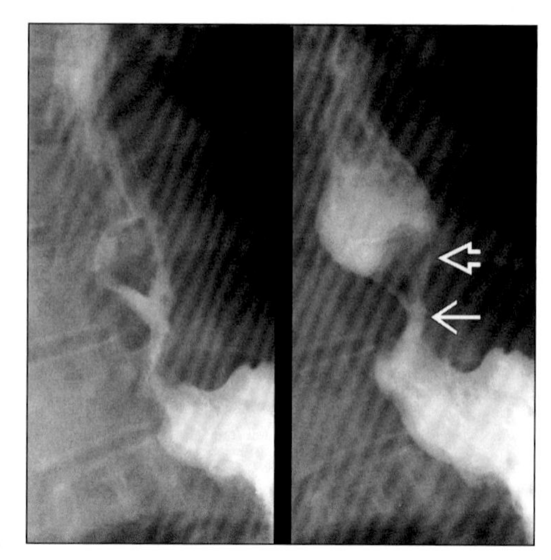

Two views from esophagram show hiatal hernia with a peptic stricture (arrow). Foreign body (walnut) is impacted above the stricture (open arrow).

TERMINOLOGY

Abbreviations and Synonyms

- Esophageal foreign body (FB)

Definitions

- Ingested foreign body impacted within esophagus

IMAGING FINDINGS

General Features

- Best diagnostic clue: History of ingestion followed by dysphagia or odynophagia
- Location
  - Gastroesophageal junction; area of indentation by aortic arch or left main bronchus
  - Above pre-existing stricture; Schatzki ring; tumor
  - Bones tend to lodge in cervical esophagus, just below level of cricopharyngeus muscle (C6 level)
- Size: Smooth objects measuring less than 1-2 cm in diameter may pass uneventfully
- Morphology
  - Radiolucent: Food, plastic, wood, medication etc.
  - Radiopaque: Coin, battery, pin, nail, needle etc.
  - Sharp or dull; pointed or blunt; toxic or nontoxic

Radiographic Findings

- Radiography
  - Lateral neck radiograph: Radiopaque FB
    - Ingested bone fragments; linear or slightly curved densities with well-defined margins
    - Radiolucent FB; indirect evidence of mucosal trauma; localized soft tissue emphysema, lump
    - Diffuse widening of retropharyngeal soft tissue
  - Chest radiograph: Coins; flat objects orient in coronal plane if within esophagus
- Fluoroscopy
  - Barium swallow; performed early; to determine presence of FB & obstruction
  - Animal, fish bone: Easily obscured by barium
    - May be seen as linear filling defect
  - Large food bolus; unchewed meat
    - May cause complete esophageal obstruction
    - Polypoid filling defect with irregular meniscus
    - Barium outlining superior border of food bolus
    - With incomplete obstruction; small amount of barium may trickle into distal esophagus, stomach
    - May erroneously suggest stricture; esophagus incompletely distended below level of impaction
  - Barium-soaked cotton ball, marshmallow; get entangled in FB; helps identify nonopaque, small FB
  - Follow up esophagram; after removal of FB
    - Underlying disease; motor function; lumen patency; mucosal injury induced by FB

CT Findings

- May detect faintly opaque bone chip

DDx: Filling Defect in Esophageal Lumen

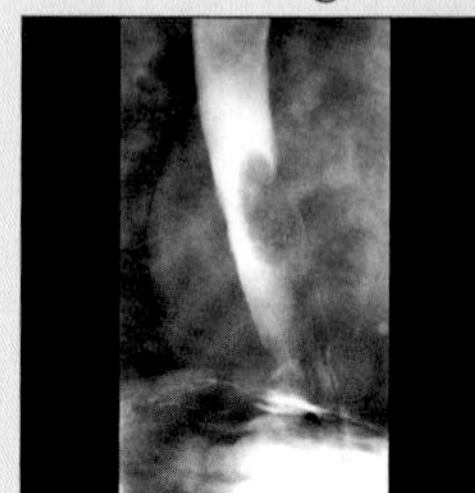

Eso. Cancer

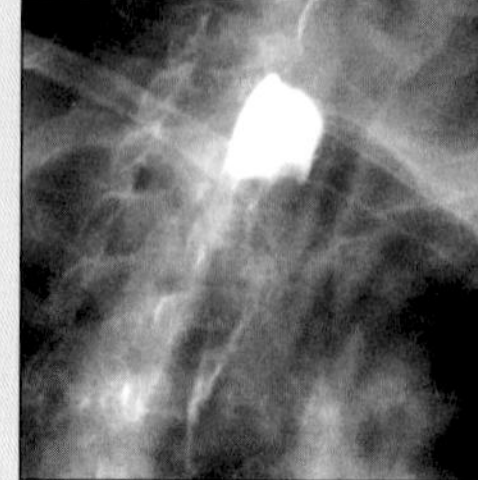

Eso. Cancer

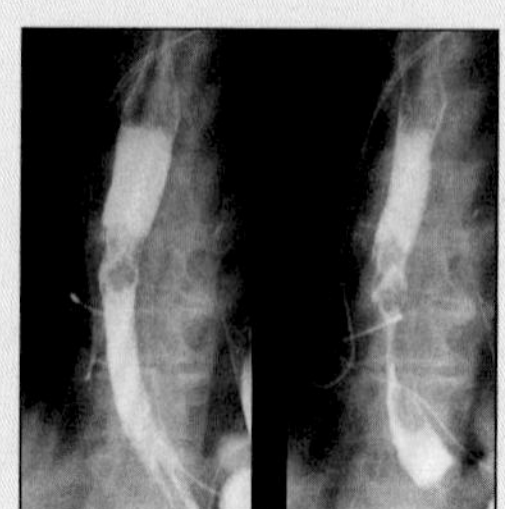

Candida Esoph.

ESOPHAGEAL FOREIGN BODY

Key Facts

Terminology
- Ingested foreign body impacted within esophagus

Imaging Findings
- Radiolucent: Food, plastic, wood, medication etc.
- Radiopaque: Coin, battery, pin, nail, needle etc.
- May be seen as linear filling defect
- May cause complete esophageal obstruction
- Polypoid filling defect with irregular meniscus
- May erroneously suggest stricture; esophagus incompletely distended below level of impaction

Top Differential Diagnoses
- Esophageal tumor
- Esophagitis

Clinical Issues
- High risk adults; underlying esophageal disease, prisoners, mentally retarded, psychiatric illness

- Localized soft tissue emphysema, edema, hematoma, or abscess; may see FB penetrating esophageal wall

Imaging Recommendations
- Protocol advice
  - Lateral film of neck; upright; neck well extended; phonation; more informative than anteroposterior
  - Water soluble iodinated contrast if perforation suspected; if no leakage seen, films with barium

DIFFERENTIAL DIAGNOSIS

Esophageal tumor
- May appear as filling defect in esophageal lumen
- Benign; submucosal; leiomyoma, lipoma, fibroma
- Malignant; esophageal cancer, metastases

Esophagitis
- Candida or viral; may cause fibrinous pseudotumor

PATHOLOGY

General Features
- Etiology
  - In adults; usually animal or fish bones or unchewed boluses of meat
  - In children; coins, toy parts, jewels, batteries, needles, pins, "large" amounts of food
- Epidemiology: 80% of FB impactions occur in children
- Associated abnormalities: FB impaction often caused by underlying webs, rings, or strictures (adults)

CLINICAL ISSUES

Presentation
- High risk adults; underlying esophageal disease, prisoners, mentally retarded, psychiatric illness
  - Carpenters, dressmakers; habit of holding nails, needles with their lips
- Transient symptoms at moment of ingestion; retrosternal pain, cyanosis, dysphasia
  - Acute dysphagia, odynophagia after choking or gagging episode; respiratory difficulty, wheezing
- Diagnosis: History; by patient or an observer
  - Children, impaired adults; unable to give accurate history; high index of suspicion must be maintained

Natural History & Prognosis
- Most FB pass spontaneously through gastrointestinal tract without difficulty
- Sharp, pointed, elongated FB; associated with greater risk of perforation, vascular penetration, fistula
- Recurrent FB ingestion; usually seen in mentally retarded & psychiatric patients
- Successful removal rates as high as 92-98%, with minimal or no complications

Treatment
- Management depends on type, size, location of FB
- Remove within 24 hours; risk of perforation later
- Conservative; natural elimination; smooth, small FB
- Endoscopic extraction; flexible fiberoptic (87%) or rigid esophagoscope
- Fluoroscopically guided; balloon-tipped Foley or Fogarty catheter; Dormia-type wire basket
- Surgical removal; rarely indicated; for complications

SELECTED REFERENCES

1. Arana A et al: Management of ingested foreign bodies in childhood and review of the literature. Eur J Pediatr. 160(8):468-72, 2001
2. Harned RK 2nd et al: Esophageal foreign bodies: safety and efficacy of Foley catheter extraction of coins. AJR Am J Roentgenol. 168(2):443-6, 1997
3. Brady PG: Esophageal foreign bodies. Gastroenterol Clin North Am. 20(4):691-701, 1991

IMAGE GALLERY

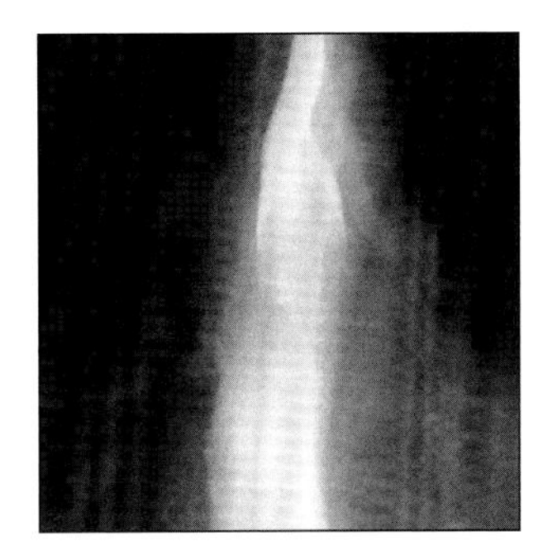

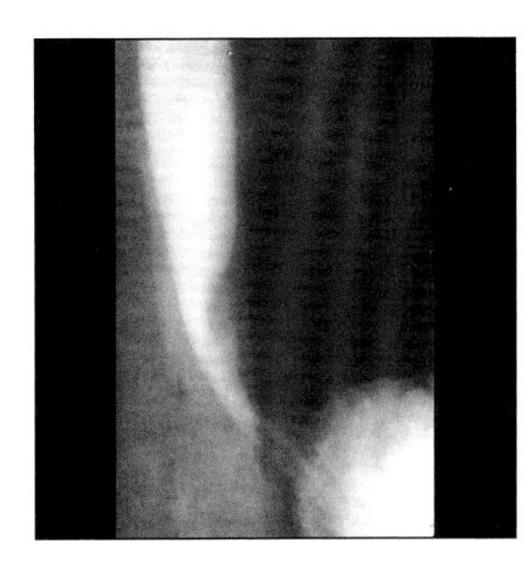

(Left) *Esophagram shows complete obstruction of mid-esophagus with a meniscus of barium outlining piece of ingested meat, impacted above a stricture in Barrett esophagus.* ***(Right)*** *Esophagram shows an intramural mass effect in distal esophagus due to an ingested chicken bone that was imbedded in wall, causing an inflammatory reaction.*

ESOPHAGEAL PERFORATION

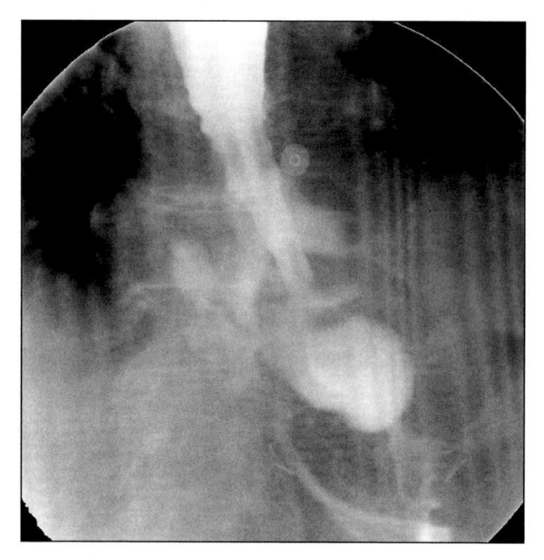

Water soluble esophagram shows extravasation into mediastinum due to esophageal perforation during laparoscopic hiatal hernia repair (fundoplication).

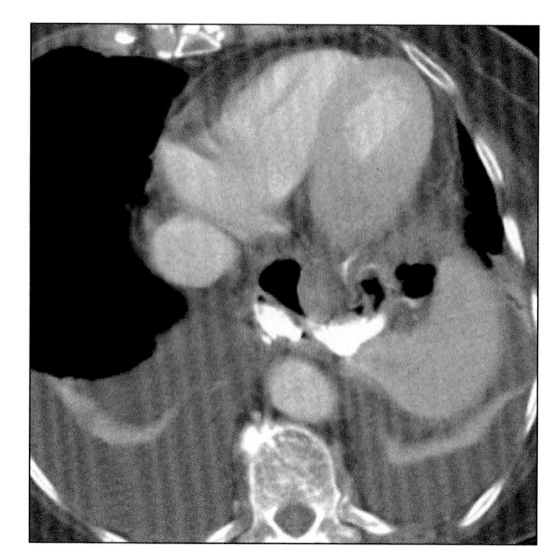

Axial CECT shows mediastinal collections of gas and oral contrast medium due to esophageal perforation during laparoscopic fundoplication. Also note pleural effusions and atelectasis.

TERMINOLOGY

Abbreviations and Synonyms

- Esophageal perforation (EP)
- Esophageal rupture or transection

Definitions

- Transmural esophageal tear

IMAGING FINDINGS

General Features

- Best diagnostic clue
  - Diagnosis depends on high degree of suspicion & recognition of clinical features
    - Confirmed by contrast esophagography or CT
- Location
  - Most endoscopic EP; occur on posterior wall of cervical esophagus; level of cricopharyngeus muscle
  - Thoracic EP; at or near gastroesophageal junction
    - Areas of anatomic narrowing; site of extrinsic compression by aortic arch or left main bronchus
    - At or above benign or malignant strictures; with biopsies or dilatation procedures
    - Site of ruptured anastomosis; after esophageal surgery

Radiographic Findings

- Radiography
  - Cervical EP: Anteroposterior, lateral films of neck
    - Subcutaneous emphysema
    - Lateral film: Widening of prevertebral space
    - Retropharyngeal abscess; mottled gas, air-fluid level
    - Air may dissect along fascial planes from neck into chest; pneumomediastinum
  - Thoracic EP: Chest radiograph
    - May be normal in 9-12%
    - Pneumomediastinum; radiolucent streaks of gas along left lateral border of aortic arch; descending aorta; right lateral border of ascending aorta; heart
    - V-shaped radiolucency seen through heart
    - Sympathetic left pleural effusion; atelectasis in basilar segment of left lung; with distal EP
    - Pleural effusion, hydropneumothorax, localized pneumonitis; due to esophageal-pleural fistula
    - Hydropneumothorax; 75% on left; 5% on right; 20% bilateral
    - Hydrothorax; usually unilateral; right sided with upper or mid EP; left-sided with distal EP
    - EP of intra-abdominal segment of distal esophagus; rare; below diaphragmatic hiatus
    - Abdominal plain film: Pneumoperitoneum
    - Gas in lesser sac or retroperitoneum
- Fluoroscopy

DDx: "Extraluminal" Contrast Material

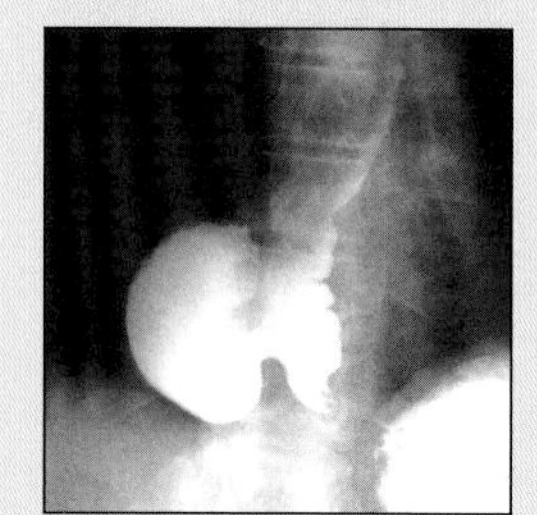

Pulsion Divertic.

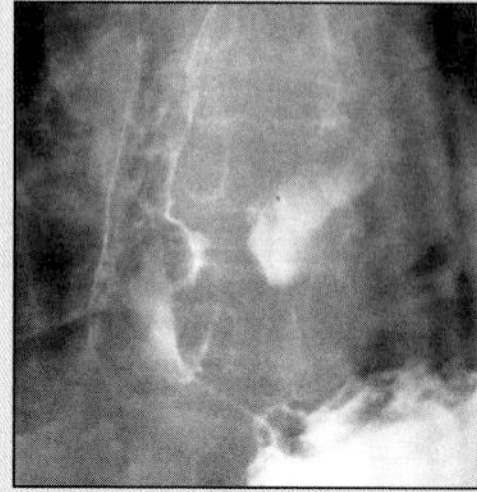

Boerhaave Syndrome

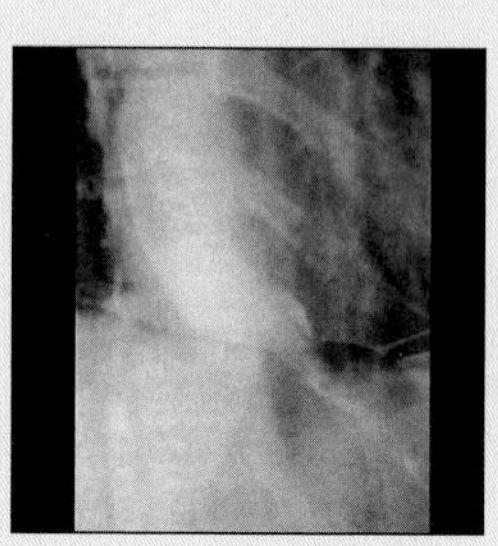

Heller Myotomy

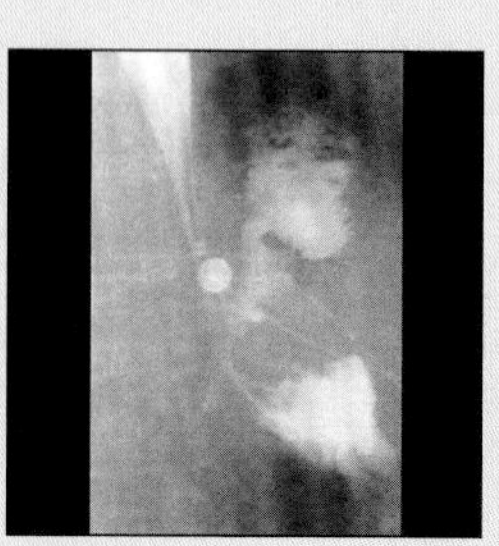

Boerhaave Syndome

ESOPHAGEAL PERFORATION

Key Facts

Terminology
- Transmural esophageal tear

Imaging Findings
- Diagnosis depends on high degree of suspicion & recognition of clinical features
- Confirmed by contrast esophagography or CT
- Subcutaneous emphysema
- Pneumomediastinum; radiolucent streaks of gas along left lateral border of aortic arch; descending aorta; right lateral border of ascending aorta; heart
- Pleural effusion, hydropneumothorax, localized pneumonitis; due to esophageal-pleural fistula
- Small EP: Localized extravasation of contrast medium from esophagus into neck or mediastinum
- Start with non-ionic water soluble contrast media
- If no leak or fistula; follow with barium

Top Differential Diagnoses
- Esophageal diverticulum
- Esophageal ulceration
- Boerhaave syndrome
- Post-operative
- Tracheobronchial aspiration

Pathology
- Instrumentation: Most common cause of EP
- Endoscopic procedures; responsible for 75-80% of all EP; rigid & fiberoptic endoscopy

Clinical Issues
- Mortality rates; cervical EP: 15%; thoracic EP: 25%

- Esophagography: To determine site & extent of EP
- Small EP: Localized extravasation of contrast medium from esophagus into neck or mediastinum
- EP near gastroesophageal junction
  - Contrast seen extravasating from left lateral aspect of distal esophagus into adjacent mediastinum
- Sealed-off EP: Self-contained extraluminal collection of contrast medium
  - Communicating with adjacent esophagus
- Larger EP: Free extravasation of contrast medium into mediastinum
  - Extension along fascial planes superiorly or inferiorly from site of EP
- Extension into pleural space is common

CT Findings
- CT scans optimally define extraluminal manifestations
- Extraesophageal air; most useful finding (92%)
- Extraluminal oral contrast medium
- Esophageal thickening, pleural effusion, single or multiple abscesses; acute mediastinitis
- Mediastinal fluid collections (92%); periesophageal, pleural, pericardial
- Esophagopleural fistula; site of communication between pleural space & esophagus may be seen
- Limitation: Inability to locate exact site of EP

Imaging Recommendations
- Protocol advice
  - Videofluoroscopic esophagography
    - Start with non-ionic water soluble contrast media
    - If no leak or fistula; follow with barium
    - Water soluble contrast agent; may fail to detect 15-25% of thoracic EP & 50% of cervical EP
    - Barium may detect small leak not visible initially

DIFFERENTIAL DIAGNOSIS

Esophageal diverticulum
- Mucosa lined pouch
- No free mediastinal gas or inflammation
- Extraluminal contrast, fluid + gas in mediastinal pocket in EP; may mimic epiphrenic (pulsion) diverticulum

Esophageal ulceration
- Mucosal inflammatory changes
- Deep; Barrett; large, flat ulcers; ulcerative carcinoma; ulcers with nasogastric intubation or caustic ingestion
- Extraluminal contrast; with communication with esophageal lumen; may mimic sealed-off perforation

Boerhaave syndrome
- Spontaneous distal EP; violent retching, vomiting, usually after an alcoholic binge
- Extraluminal gas + contrast material in lower mediastinum surrounding esophagus
- 1-4 cm, vertically oriented, linear tears on left lateral wall just above gastroesophageal junction

Post-operative
- Post esophagectomy anatomy can be misinterpreted
  - Near site of anastomosis; irregular contour
  - Post Heller myotomy-intramural linear collection

Tracheobronchial aspiration
- Contrast material in trachea or bronchi
- Differentiate esophageal-airway fistula from aspiration
  - Initial swallow in lateral projection with video recording of hypopharynx

PATHOLOGY

General Features
- Etiology
  - Iatrogenic
    - Instrumentation: Most common cause of EP
    - Endoscopic procedures; responsible for 75-80% of all EP; rigid & fiberoptic endoscopy
    - Biopsy, esophageal surgery, bouginage, breakdown of surgical anastomoses
    - Pneumatic balloon dilation (2-10%)
    - Nasogastric or endotracheal tubes; Celestin tubes
    - Sengstaken-Blakemore tubes (35%)

ESOPHAGEAL PERFORATION

- Esophageal obturator airways (2%)
- Flexible endoluminal prosthesis (10%)
- In infants & children; during placement of feeding tube or suctioning of oropharyngeal secretions
  - Trauma
    - Penetrating injuries; knife or bullet wounds
    - Blunt trauma; to chest or abdomen
  - Foreign bodies
    - Impacted animal or fish bones; sharp, pointed objects; caustic agents
  - Spontaneous
    - Boerhaave syndrome
    - ↑ Intrathoracic pressure; coughing, weightlifting, childbirth, status asthmaticus, seizures
  - Neoplastic
    - Esophageal carcinoma
  - Idiopathic
- Epidemiology: Incidence of EP has increased as use of endoscopic procedures has become more frequent

Gross Pathologic & Surgical Features

- Necrosis of mucosa; submucosal hemorrhage; transmural tear

CLINICAL ISSUES

Presentation

- Cervical EP: Acute onset dysphagia, neck pain, fever
  - Physical examination: Subcutaneous emphysema; crepitus in neck
- Thoracic EP: Sudden onset excruciating substernal or lower thoracic chest pain
  - Crepitus in soft tissues of anterior chest wall or neck
- Rapid onset of overwhelming sepsis; fever, tachycardia, hypotension, shock
  - Severe mediastinitis; as swallowed food, saliva, refluxed peptic acid may enter mediastinum
- Dysphagia, increased oral secretions, respiratory distress soon after endoscopy
  - Endoscopist may be unaware that EP has occurred at time of examination
  - Patient may subsequently present with neck pain, fever, dysphagia
- Atypical chest pain; referred to shoulder or back; epigastric pain; crepitus may not always be present
  - Clinical diagnosis may be mistaken
    - Perforated peptic ulcer, myocardial infarction, aortic dissection
- Presence of "signal" hemorrhage from gastrointestinal tract; vascular trauma caused by perforating object
- Diagnosis: Contrast esophagography
  - 90% of contrast esophagrams are positive

Demographics

- Age: Any age; EP in infants & children occurs more frequently than reported

Natural History & Prognosis

- Most serious & rapidly fatal type of perforation in gastrointestinal tract
- Life-threatening; associated with high morbidity & without intervention, high mortality
- ↑ Risk of EP; underlying esophageal disease, diverticulum, cervical lordosis, osteophytes
- With dilatation procedures, EP may be immediate or delayed for several days
- Complications: Cervical EP; retropharyngeal abscess associated with sepsis, shock
  - Thoracic EP: Mediastinitis, mediastinal abscess, pericarditis, pneumothorax, fistula
- Outcome after EP; dependent on cause; location of injury; presence of underlying esophageal disease
- Prognosis; directly related to interval between perforation & initiation of treatment
  - After 24 hours; 70% mortality rate for thoracic EP
  - Untreated thoracic EP; mortality rate nearly 100%
    - Fulminant mediastinitis
- Cervical EP has better prognosis than thoracic EP
  - Mortality rates; cervical EP: 15%; thoracic EP: 25%

Treatment

- Conservative: Parenteral fluids & antibiotics
  - Successful for limited esophageal injuries meeting proper selection criteria
  - Small cervical EP
  - Rarely; thoracic EP heal spontaneously without surgical intervention
- Surgical: Large cervical EP; cervical mediastinotomy & open drainage
  - Thoracic EP: Immediate thoracotomy
    - Primary closure of EP
    - Mediastinal drainage
  - Covered metallic stents for fistulas
- Nonsurgical interventional drainage techniques; transesophageal drainage of mediastinal abscesses

DIAGNOSTIC CHECKLIST

Consider

- Clinical & radiographic signs of EP may be subtle
  - Active investigation is needed to establish diagnosis

SELECTED REFERENCES

1. Nerot C et al: Esophageal perforation after fracture of the cervical spine: case report and review of the literature. J Spinal Disord Tech. 15(6):513-8, 2002
2. Buecker A et al: Esophageal perforation: comparison of use of aqueous and barium-containing contrast media. Radiology. 202(3):683-6, 1997
3. Kim IO et al: Perforation complicating balloon dilation of esophageal strictures in infants and children. Radiology. 189(3):741-4, 1993
4. Hoover EL: The diagnosis and management of esophageal perforations. J Natl Med Assoc. 83(3):246-8, 1991
5. Niezgoda JA et al: Pharyngoesophageal perforation after blunt neck trauma. Ann Thorac Surg. 50(4):615-7, 1990
6. Scholl DG et al: Esophageal perforation following the use of esophageal obturator airway. Radiology. 122(2):315-6, 1977

ESOPHAGEAL PERFORATION

IMAGE GALLERY

Typical

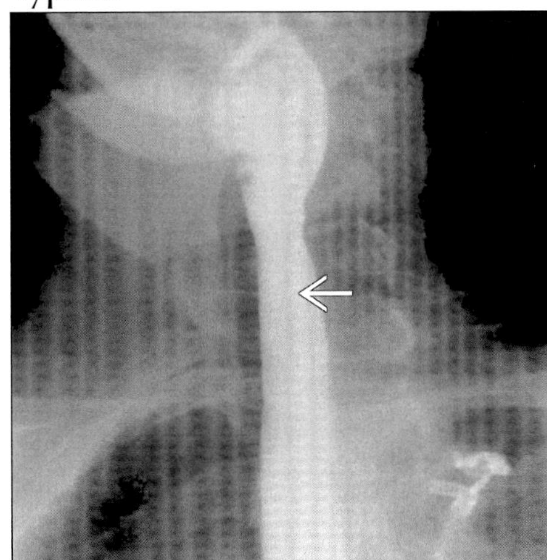
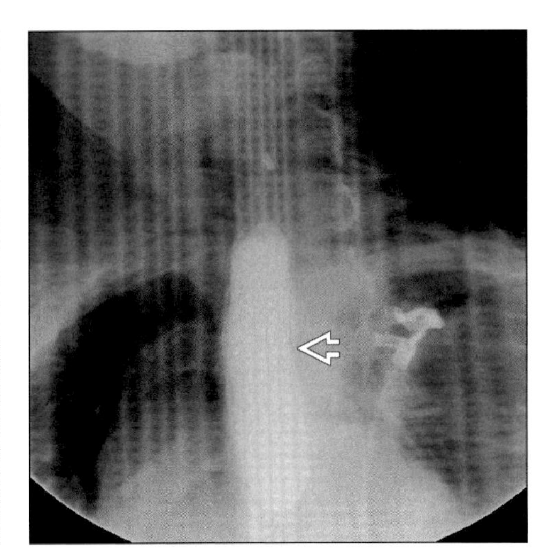

(Left) *Esophagram following upper endoscopy shows esophageal "dissection" with a mucosal flap (arrow) separating the true and false lumens. Perforation was at the pharyngo-esophageal junction.* ***(Right)*** *Esophagram following endoscopy shows esophageal "dissection". Contrast is retained within the false lumen (open arrow) long after the true lumen has cleared the contrast bolus.*

Typical

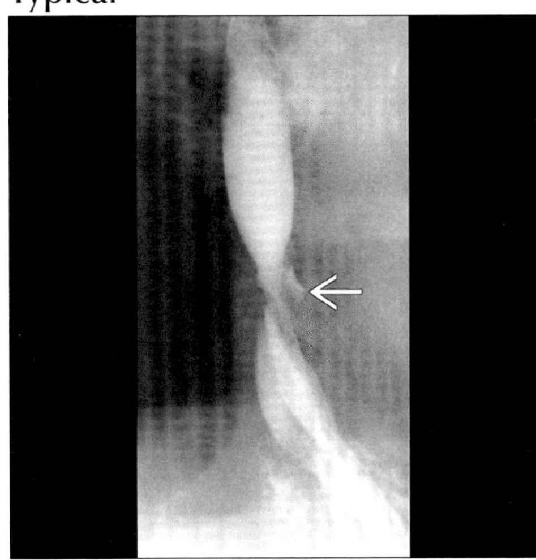
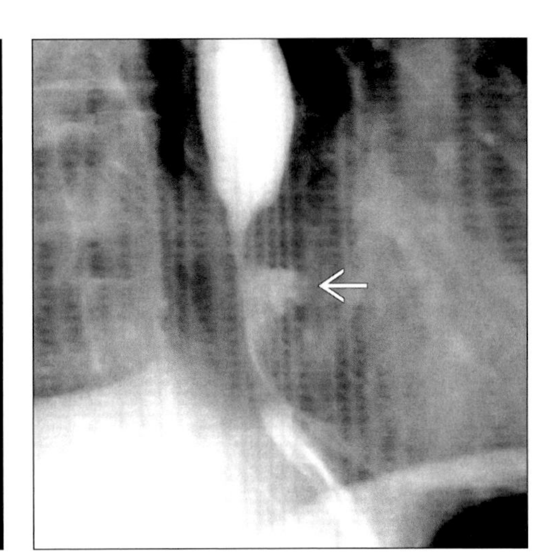

(Left) *Esophagram shows contained leak (arrow) following endoscopic biopsy of an esophageal stricture.* ***(Right)*** *Esophagram shows contained leak (arrow) following balloon dilatation of esophageal stricture.*

Typical

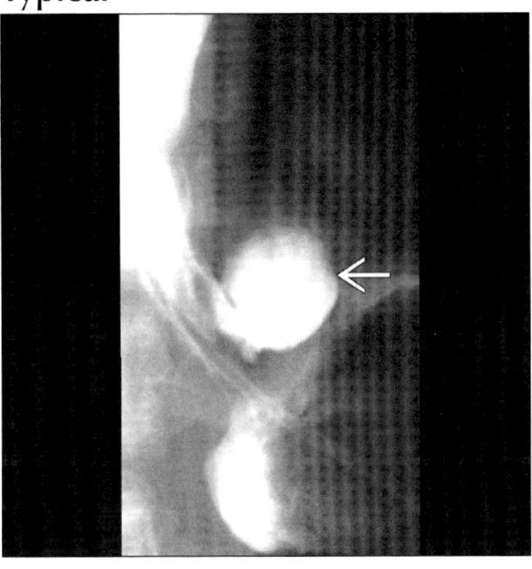
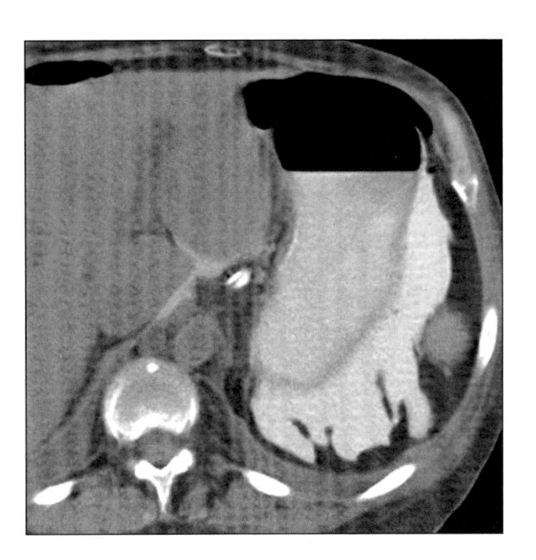

(Left) *Esophagram shows left mediastinal collection of contrast medium (arrow) following balloon dilatation for achalasia.* ***(Right)*** *Axial NECT shows extraluminal oral contrast medium in the upper abdomen folling placement of a nasogastric tube that perforated the esophago-gastric junction.*

BOERHAAVE SYNDROME

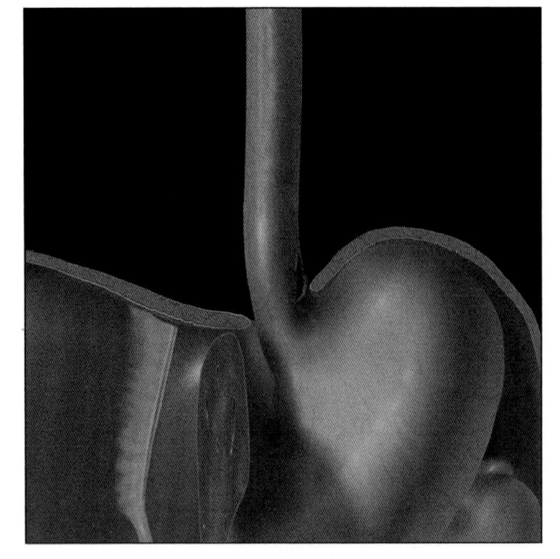

Graphic shows a vertically oriented laceration of the distal esophagus, just above the hiatus and GE junction.

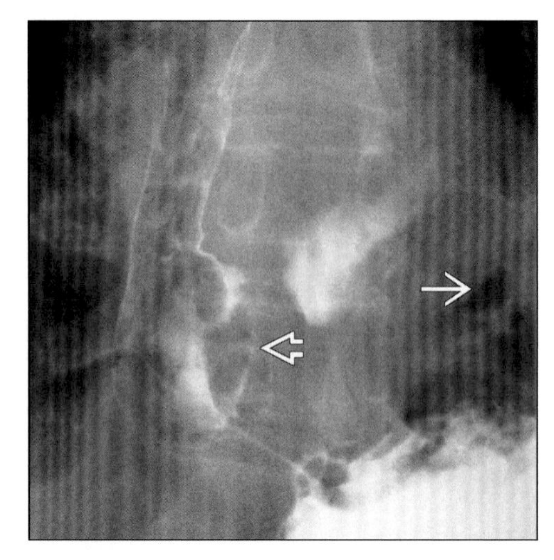

Esophagram shows irregular extraluminal contrast and gas (arrow) dissecting through mediastinum. Perforation of distal esophagus (open arrow). Immediate drainage was required.

TERMINOLOGY

Definitions

- Spontaneous distal esophageal perforation following vomiting or other violent straining

IMAGING FINDINGS

General Features

- Best diagnostic clue: Extraluminal gas & contrast material in lower mediastinum surrounding esophagus
- Location: Left lateral wall of distal esophagus just above gastroesophageal (GE) junction
- Morphology: Vertically oriented full-thickness linear tear of distal esophagus, 1-4 cm in length
- Other general features
  - Most serious & rapidly fatal type of perforation in gastrointestinal tract
  - Sudden increase in intraluminal pressure leads to full thickness esophageal perforation
  - Most cases: Violent retching/vomiting, usually after an alcoholic binge
  - Left side of distal thoracic esophagus
    - Most vulnerable (due to lack of supporting mediastinal structures)
    - Mortality rate: 70% due to fulminant mediastinitis if untreated
  - Right side is protected by descending thoracic aorta
  - Rarely, cervical/upper thoracic esophagus is involved
    - Cervical: Less devastating injury
    - Mortality rate: Less than 15%
    - Better prognosis than thoracic

Radiographic Findings

- Radiography
  - Chest x-ray
    - Mediastinal widening, pneumomediastinum
    - Left side pleural effusion or hydropneumothorax
    - Atelectasis in basilar segments of left lung
    - Radiolucent streaks of gas along left lateral border of aortic arch & descending thoracic aorta or along right lateral border of ascending aorta & heart
    - ± Gas streaks (supraclavicular area)
  - Abdomen x-ray
    - Rarely when perforation is below diaphragmatic hiatus, show collection of gas in lesser sac or retroperitoneum
- Fluoroscopic guided esophagography
  - Distal thoracic small esophageal perforation
    - Extravasation of contrast medium from left lateral wall of distal esophagus (laterally & superiorly) into adjacent mediastinum
  - Distal thoracic large esophageal perforation

DDx: "Extraluminal" Contrast at GE Junction

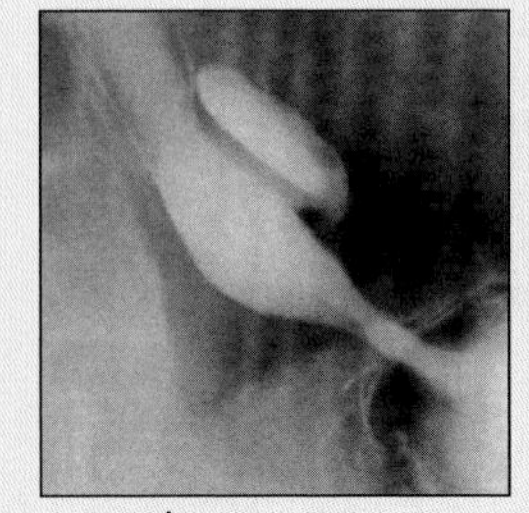

Epiphrenic Divertic.

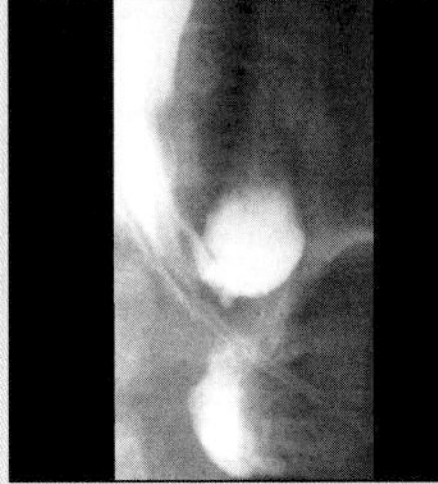

Post Dilation Leak

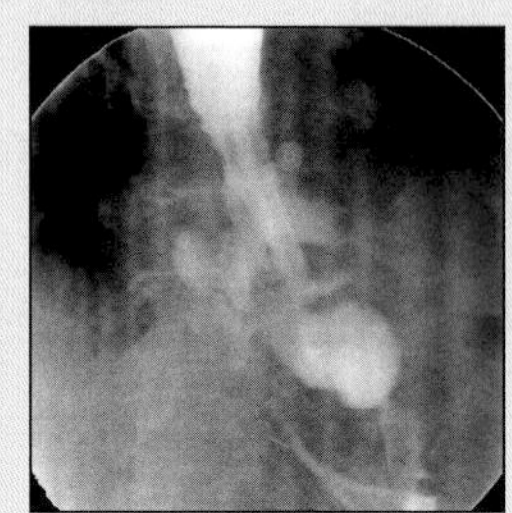

Post Surgical Leak

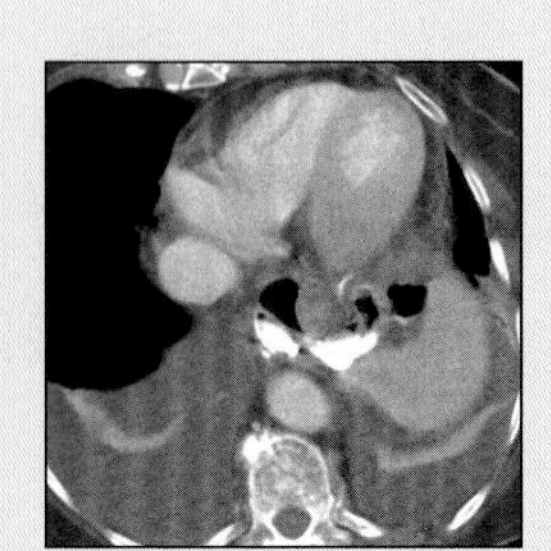

Post Surgical Leak

BOERHAAVE SYNDROME

Key Facts

Terminology

- Spontaneous distal esophageal perforation following vomiting or other violent straining

Imaging Findings

- Best diagnostic clue: Extraluminal gas & contrast material in lower mediastinum surrounding esophagus
- Location: Left lateral wall of distal esophagus just above gastroesophageal (GE) junction
- Extravasation of contrast medium from left lateral wall of distal esophagus (laterally & superiorly) into adjacent mediastinum
- Extraluminal gas (highly suggestive of esophageal perforation)
- Periesophageal, pleural, pericardial fluid collections

Top Differential Diagnoses

- Mallory-Weiss syndrome
- Epiphrenic (pulsion) diverticulum
- Iatrogenic injury (post instrumentation)

Pathology

- Rapid ↑ intraluminal pressure → spontaneous rupture of normal esophagus

Diagnostic Checklist

- Differentiate from other conditions in which "extraluminal" contrast material is seen near gastroesophageal junction
- Check for history of violent retching or vomiting
- Extraluminal air & contrast material around distal esophagus particularly on left side

    - Free extravasation of contrast medium into mediastinum & extension along fascial planes superiorly & inferiorly
    - Extravasation of contrast medium into pleural space (left more than right)
  - Upper cervical esophageal perforation
    - Extravasation of contrast medium into neck
  - Regardless of site of perforation
    - A sealed-off leak seen as self-contained extraluminal collection of contrast medium that communicates with adjacent esophagus

CT Findings

- Extraluminal gas (highly suggestive of esophageal perforation)
- Periesophageal, pleural, pericardial fluid collections
- Extravasation of oral contrast material, esophageal thickening
- Major limitation: Inability to locate exact site of perforation

Imaging Recommendations

- Helical CT, plain chest x-ray
- Esophagography with non-ionic water-soluble contrast agent
- Initial study with water-soluble contrast medium fails to show a leak; examination must be repeated immediately with barium to detect subtle leaks (more likely to be visualized with higher density contrast)

DIFFERENTIAL DIAGNOSIS

Mallory-Weiss syndrome

- Irregular linear mucosal tear or laceration in long axis of esophagus
- Location
  - Distal esophagus near GE junction or gastric cardia
- Size: 1-4 cm in length
- Pathology
  - Involves mucosa
  - Does not penetrate the wall
- Etiology
  - Violent retching or vomiting after an alcoholic binge or by protracted vomiting for any other reason
  - Rarely: Hiccuping, coughing, seizures, childbirth
- Pathogenesis
  - Sudden, rapid increase in intraesophageal pressure
- Imaging findings
  - Occasionally seen on double-contrast esophagrams
    - Shallow, longitudinally oriented, linear collection of barium in distal esophagus or near GE junction
    - Indistinguishable from a linear ulcer of reflux esophagitis in distal esophagus
    - Perforated Mallory-Weiss tear is Boerhaave

Epiphrenic (pulsion) diverticulum

- Synonym: Distal esophageal pulsion diverticulum
- Pulsion diverticulum is a pseudodiverticulum
  - Muscular layer is not seen
- Mucosa lined pouch from distal esophagus
- No linear mucosal tear or laceration
- No free mediastinal gas or inflammation
- Etiology: Chronic wear & tear forces, motility disorders
- Pathogenesis: Due to pulsion caused by increased intraluminal pressure
- Imaging findings
  - Chest x-ray PA view
    - Large epiphrenic diverticulum seen as prominent bulge along right or left border of heart
  - Chest x-ray lateral view
    - Large soft tissue mass mimicking hiatal hernia
    - ± Air-fluid level
  - Videofluoroscopic esophagography
    - Usually large barium-filled sac (epiphrenic area)
    - Location: Lateral esophageal wall of distal 10 cm
    - Side: Right side more common than left
    - Wide neck; fills & empties freely into esophagus
    - No linear tear or laceration
    - Often associated with achalasia or hiatal hernia

Iatrogenic injury (post instrumentation)

- Endoscopic procedures account for 75% of cases
- Location: Cervical (common), thoracic esophagus

BOERHAAVE SYNDROME

- Indistinguishable from Boerhaave syndrome especially when iatrogenic perforation occurs in left lateral wall of distal esophagus just above GE junction

PATHOLOGY

General Features

- Etiology
  - Commonly
    - Seen in bulimic patients
  - Rare causes
    - Coughing, weightlifting, childbirth, defecation
    - Seizures, status asthmaticus, blunt trauma to chest or upper abdomen
  - Rapid ↑ intraluminal pressure → spontaneous rupture of normal esophagus
  - Emetogenic injury of esophagus from sudden ↑ in intra-abdominal pressure + relaxation of distal esophageal sphincter in presence of a moderate to large amount of gastric contents
- Epidemiology: Accounts for 15% of total esophageal perforation cases

Gross Pathologic & Surgical Features

- Full-thickness linear tear on left lateral wall of distal esophagus just above GE junction

Microscopic Features

- Normal mucosa, submucosa & muscularis externa
- No inflammatory pathology

CLINICAL ISSUES

Presentation

- Most common signs/symptoms
  - Classic triad
    - Vomiting, severe substernal chest pain
    - Subcutaneous emphysema of chest wall & neck
  - Dysphagia or odynophagia
  - Rapid onset of overwhelming sepsis
    - Fever, tachycardia, ↓ in blood pressure, shock
- Clinical profile: Patient with history of chronic alcoholism, severe vomiting, sudden severe substernal chest pain & fall in blood pressure

Demographics

- Age: Usually adults
- Gender: M = F

Natural History & Prognosis

- Complications
  - Mediastinitis, sepsis & shock
- Prognosis
  - Large perforation
    - Without treatment after 24 hrs mortality rate 70%
    - After immediate surgical drainage: Good
  - Small perforation: Good
  - Small, self-contained perforation: Good

Treatment

- Large perforation
  - Immediate thoracotomy with surgical closure of perforation & mediastinal drainage
- Small perforation
  - May heal spontaneously without surgical intervention
- Small, self-contained perforation
  - Managed nonoperatively with broad spectrum antibiotics & parenteral alimentation

DIAGNOSTIC CHECKLIST

Consider

- Differentiate from other conditions in which "extraluminal" contrast material is seen near gastroesophageal junction
- Check for history of violent retching or vomiting

Image Interpretation Pearls

- Extraluminal air & contrast material around distal esophagus particularly on left side
- Left side pleural effusion or hydropneumothorax or pneumomediastinum

SELECTED REFERENCES

1. Rubesin SE et al: Radiologic diagnosis of gastrointestinal perforation. Radiol Clin North Am. v, 41(6):1095-115, 2003
2. Gimenez A et al: Thoracic complications of esophageal disorders. Radiographics. 22 Spec No:S247-58, 2002
3. Nehoda H et al: Boerhaave's Syndrome. New England Journal of Medicine 344: 138-9, 2001
4. Chang YC et al: Right-sided pleural effusion in spontaneous esophageal perforation. Ann Thorac Cardiovasc Surg. 6(1):73-6, 2000
5. Isserow JA et al: Spontaneous perforation of the cervical esophagus after an alcoholic binge: case report. Can Assoc Radiol J. 49(4):241-3, 1998
6. Buecker A et al: Esophageal perforation: comparison of use of aqueous and barium-containing contrast media. Radiology. 202(3):683-6, 1997
7. Lee S et al: The leaking esophagus: CT patterns of esophageal rupture, perforation, and fistulization. Crit Rev Diagn Imaging. 37(6):461-90, 1996
8. Gupta NM et al: Spontaneous esophageal perforation: atypical presentation. Indian J Gastroenterol. 14(1):29-30, 1995
9. Ooms HW et al: Esophageal perforation: role of esophagography and CT. AJR Am J Roentgenol. 162(4):1001-2, 1994
10. White CS et al: Esophageal perforation: CT findings. AJR Am J Roentgenol. 160(4):767-70, 1993
11. Ghahremani GG: Radiologic evaluation of suspected gastrointestinal perforations. Radiol Clin North Am. 31(6):1219-34, 1993
12. Backer CL et al: Computed tomography in patients with esophageal perforation. Chest. 98(5):1078-80, 1990
13. Henderson JA et al: Boerhaave revisited: spontaneous esophageal perforation as a diagnostic masquerader. Am J Med. 86(5):559-67, 1989
14. Allen KS et al: Perforation of distal esophagus with lesser sac extension: CT demonstration. J Comput Assist Tomogr. 10(4):612-4, 1986
15. Jeffrey RB et al: Value of computed tomography in detecting occult gastrointestinal perforation. J Comput Assist Tomogr. 7(5):825-7, 1983

IMAGE GALLERY

Typical

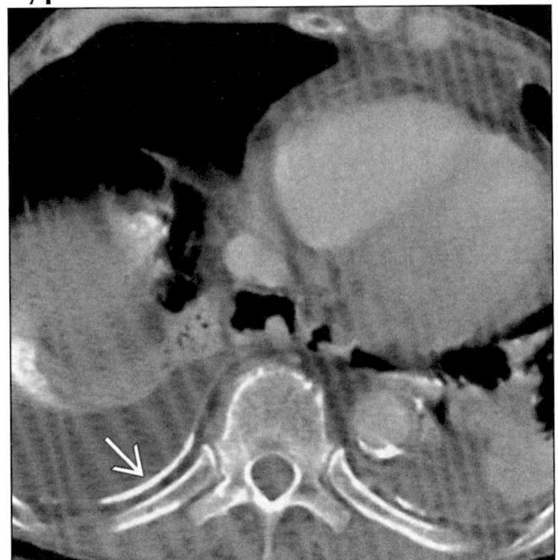

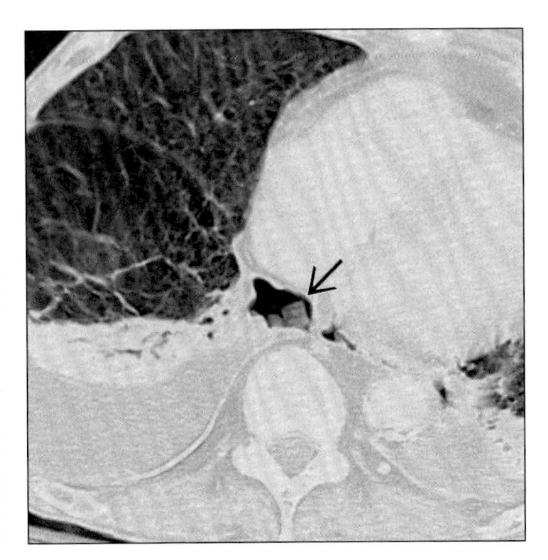

(Left) *Axial CECT shows bilateral pleural effusions containing high density that may be extravasated oral contrast medium (arrow) or pleural calcification.* ***(Right)*** *Axial CECT (lung window) shows food particles and gas (arrow) in mediastinum.*

Typical

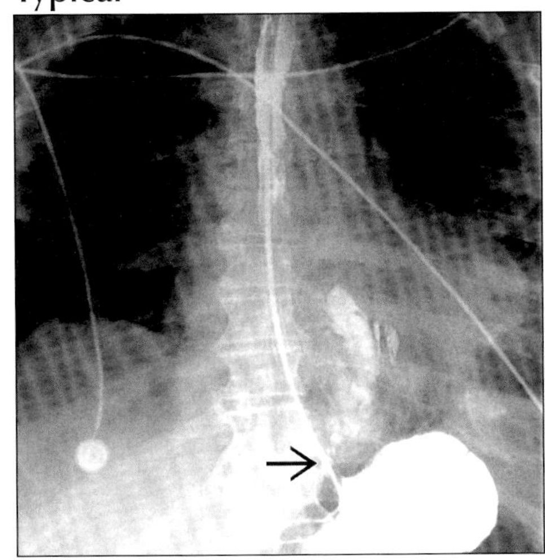

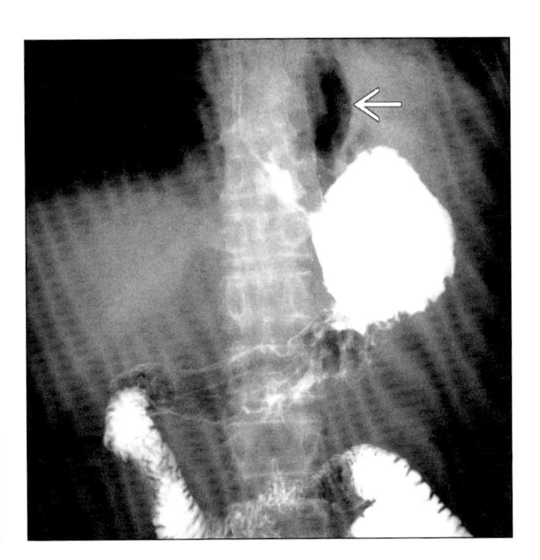

(Left) *Esophagram shows irregular collection of contrast material in mediastinum and source of perforation (arrow) in distal left side of esophagus.* ***(Right)*** *Esophagram shows mediastinal collection of gas (arrow) but no apparent leak of contrast material.*

Typical

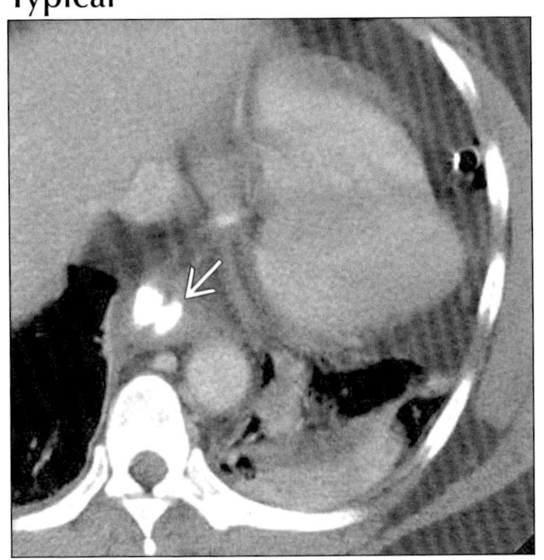

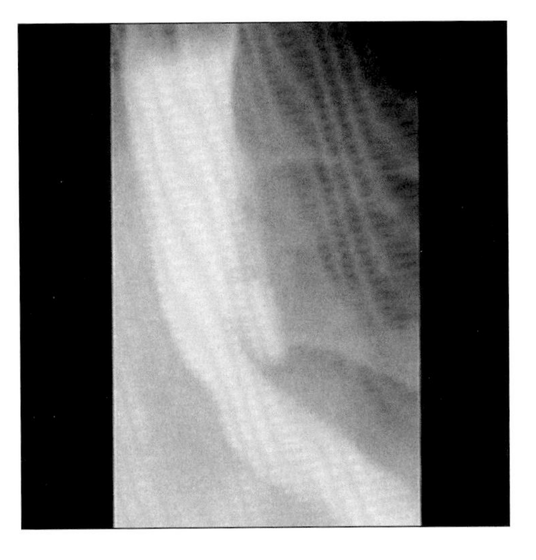

(Left) *Axial CECT shows localized perforation of distal left side of esophagus (arrow) with relatively mild periesophageal inflammatory changes.* ***(Right)*** *Esophagram shows localized perforation of distal left side of esophagus.*

INTRAMURAL BENIGN ESOPHAGEAL TUMORS

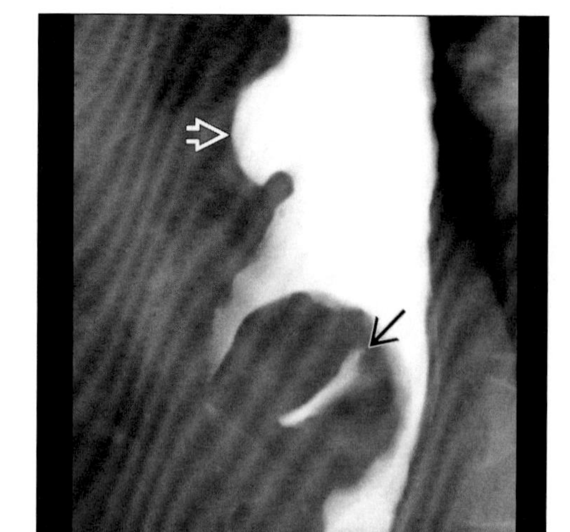

Single-contrast esophagram shows intramural mass in distal esophagus with central ulceration (arrow) due to leiomyoma. Incidental note of traction diverticulum (open arrow).

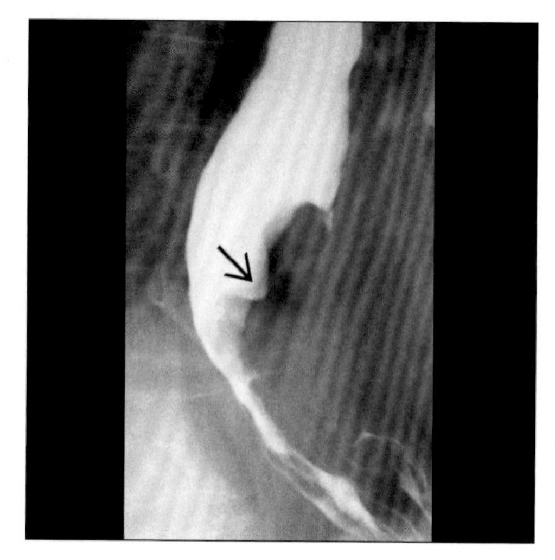

Single-contrast esophagram shows intramural mass (leiomyoma) with smooth surface, central ulcer (arrow) + right angle or obtuse angle with wall.

TERMINOLOGY

Definitions

- Benign mass composed of one or more tissue elements of the esophageal wall

IMAGING FINDINGS

General Features

- Best diagnostic clue: Intramural mass with smooth surface & slightly obtuse borders on barium studies
- Other general features
  - Types include leiomyoma (50%); granular cell; lipoma; hemangioma (2-3%); hamartoma

Radiographic Findings

- Fluoroscopic-guided barium studies
  - Discrete mass; solitary (most common) or multiple
  - Round or ovoid filling defects sharply outlined by barium on each side (en face view)
  - Overlying mucosa may ulcerate
  - Smooth surface lesion, with upper & lower borders of lesion forming right or slightly obtuse angles with adjacent esophageal wall (profile view)
  - Narrowed (tangential view) or stretched and widened (en face view) esophageal lumen
  - ± Varying degree of obstruction
  - Leiomyoma: ± Amorphous or punctate calcifications
  - Granular cell: Distal third > middle third; 0.5-2.0 cm
  - Hamartoma
    - Large & pedunculated; mimics fibrovascular polyp
    - Diffuse, nodular mucosa comprised of tiny, innumerable, hamartomatous polyps (hereditary)

CT Findings

- Discrete mass in wall; no signs of invasion/metastases
- Helps distinguish lipoma (fat density) and other mediastinal masses (e.g., mediastinal cyst)

Ultrasonographic Findings

- Endoscopic ultrasound
  - Leiomyoma: Hypoechoic & homogeneous mass with sharply demarcated margins in muscular layer
  - Granular cell: Hypo- to isoechoic mass in submucosa
  - Lipoma: Homogeneous hyperechoic mass with smooth outer margins in submucosa

Imaging Recommendations

- Best imaging tool: Barium studies followed by CT

DIFFERENTIAL DIAGNOSIS

Mediastinal tumor

- Extrinsically compress or indent the esophagus
- More obtuse, gently sloping borders (in profile view)

DDx: Mass Effect with Normal Mucosa

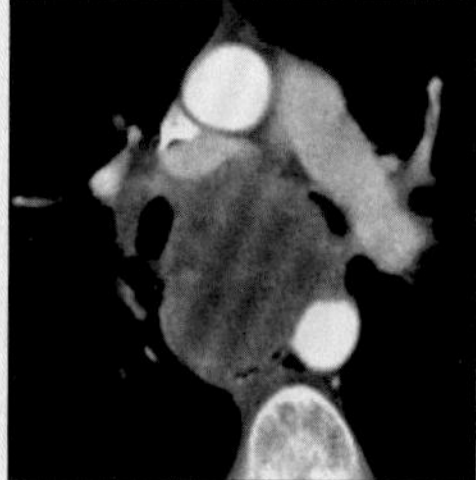

Mediastinal Tumor

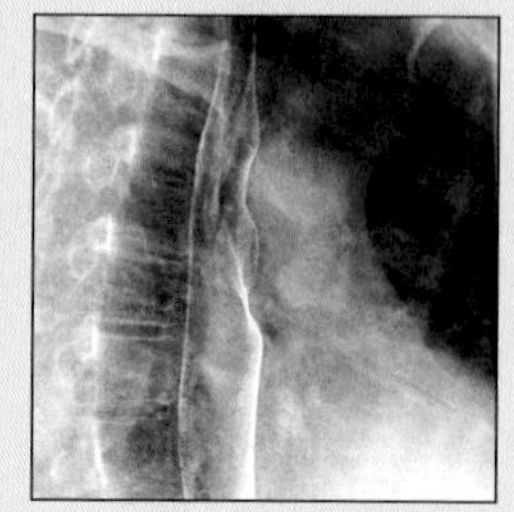

Aorta + Bronchus

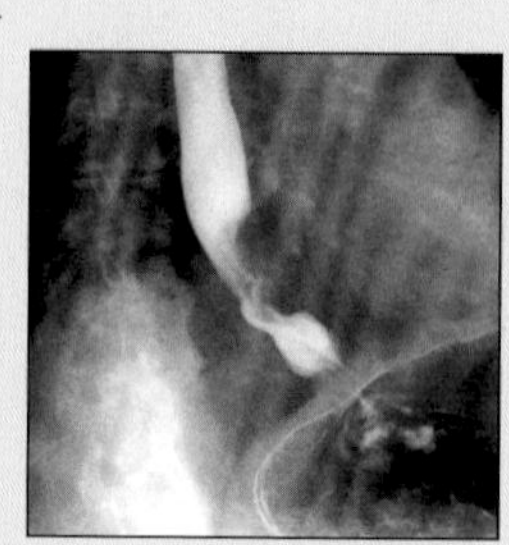

Esophageal Carcinoma

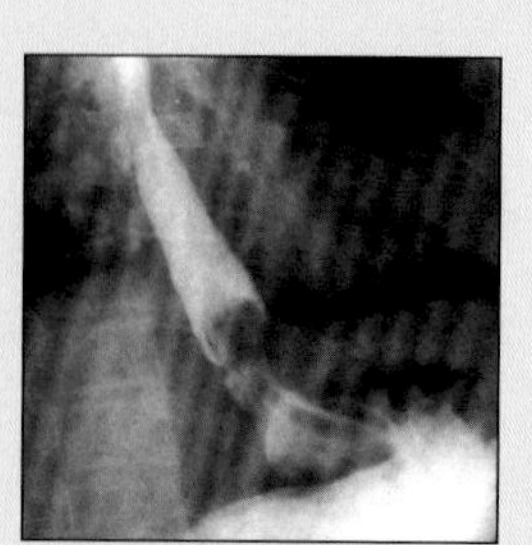

Impacted Food

INTRAMURAL BENIGN ESOPHAGEAL TUMORS

Key Facts

Imaging Findings

- Best diagnostic clue: Intramural mass with smooth surface & slightly obtuse borders on barium studies
- Best imaging tool: Barium studies followed by CT

Top Differential Diagnoses

- Mediastinal tumor
- Normal mediastinal structures
- Esophageal carcinoma
- Impacted food

Diagnostic Checklist

- Barium esophagram and CT are complementary
- Most intramural masses are benign (unlike gastric tumors)
- Calcifications suggest leiomyoma, almost never occurs in other benign/malignant esophageal tumors

- Use CT if barium studies are equivocal

Normal mediastinal structures

- Indentation by aorta, left main bronchus, aberrant or dilated vessels

Esophageal carcinoma

- Apple core or eccentric mucosal mass

Impacted food

- Intraluminal mass, often above stricture

PATHOLOGY

General Features

- Associated abnormalities
    - Leiomyoma: Uterine or vulva leiomyomas
    - Hemangioma: Osler-Weber-Rendu disease
    - Hamartoma: Cowden disease

Gross Pathologic & Surgical Features

- Leiomyoma: Firm, round, gray/yellow, unencapsulated
- Granular cell: Broad based, pinkish tan mass with normal overlying mucosa & rubbery consistency
- Lipoma: Smooth, yellow, encapsulated tumor composed of well-differentiated adipose tissue
- Hemangioma: Blue to red, nodular mass
- Hamartoma: Various elements including cartilage, bone, adipose and fibrous tissue and muscle
- Mediastinal foregut cyst
    - Thin-walled nonenhancing contents
    - Contents are water density (50%) to calcific

CLINICAL ISSUES

Presentation

- Most common signs/symptoms
    - Asymptomatic (most common)
    - Dysphagia, retrosternal pain, pyrosis, cough, odynophagia, weight loss and bleeding

Demographics

- Age: Leiomyoma: > 40 years of age
- Gender: Leiomyoma: M:F = 2:1

Natural History & Prognosis

- Complications: Hemorrhage, obstruction, ulceration
- Prognosis: Very good

Treatment

- No treatment; if symptomatic, do surgical enucleation

DIAGNOSTIC CHECKLIST

Consider

- Barium esophagram and CT are complementary
- Most intramural masses are benign (unlike gastric tumors)

Image Interpretation Pearls

- Calcifications suggest leiomyoma, almost never occurs in other benign/malignant esophageal tumors

SELECTED REFERENCES

1. Levine MS: Benign tumors of the esophagus: radiologic evaluation. Semin Thorac Cardiovasc Surg. 15(1):9-19, 2003
2. Rice TW: Benign esophageal tumors: esophagoscopy and endoscopic esophageal ultrasound. Semin Thorac Cardiovasc Surg. 15(1):20-6, 2003
3. Noh HM et al: CT of the esophagus: spectrum of disease with emphasis on esophageal carcinoma. Radiographics. 15(5):1113-34, 1995

IMAGE GALLERY

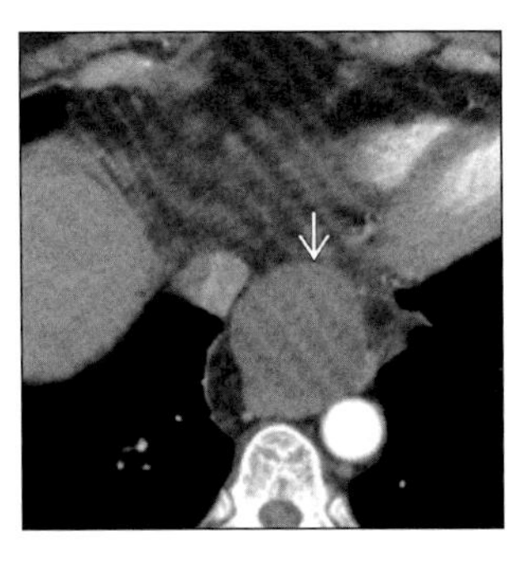

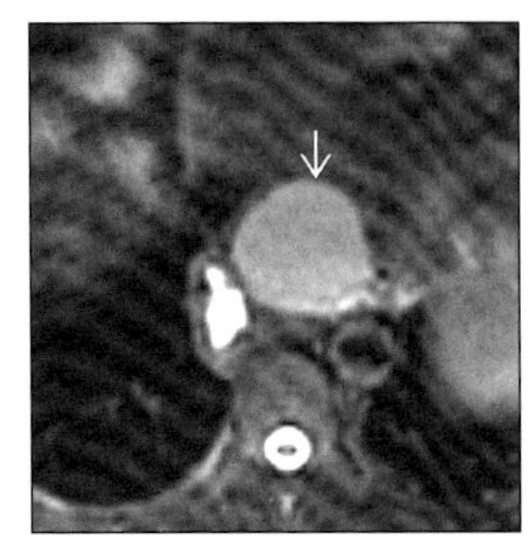

*(**Left**) Axial CECT shows water density mass (arrow) displacing distal esophagus (duplication or foregut cyst). (**Right**) Axial T2W MR shows duplication ("foregut") cyst (arrow) as complex fluid intensity.*

FIBROVASCULAR POLYP

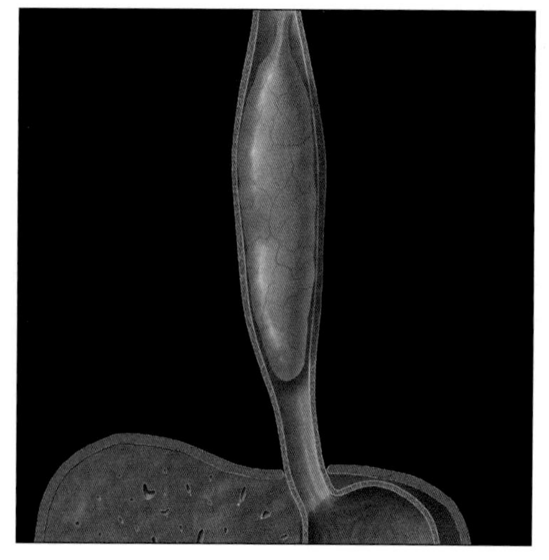

Graphic shows long, smooth, sausage-like mass arising from proximal esophageal wall.

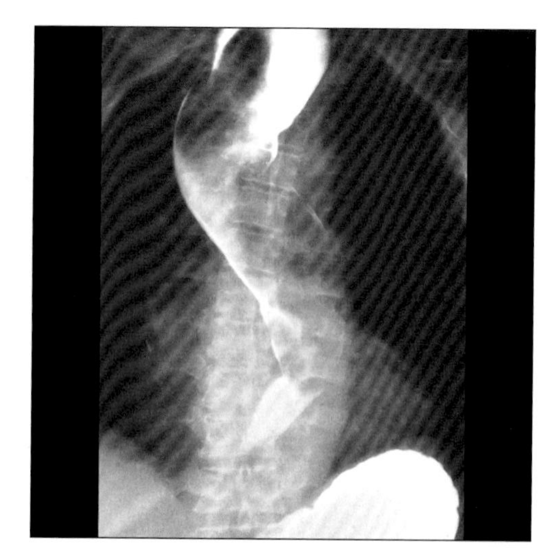

Barium esophagram shows expansile, sausage-shaped mass extending from cervical to distal esophagus (Courtesy M. Levine, MD).

TERMINOLOGY

Definitions

- Rare, benign tumor-like lesion of esophagus

IMAGING FINDINGS

General Features

- Best diagnostic clue: Giant, smooth, sausage-shaped, intraluminal, expansile mass
- Location: Cervical esophagus near cricopharyngeus
- Size: Variable in size (average length of 15 cm)
- Morphology
  - Usually originate in loose submucosal tissue
  - Pedunculated, intraluminal, gigantic mass
  - Extend inferiorly into middle or distal esophagus
  - Attached to cervical esophagus by a pedicle
- Other general features
  - Histologic classification of fibrovascular polyps based on predominant mesenchymal components
    - Hamartomas, fibromas, lipomas, fibrolipomas
    - Fibromyxomas, fibroepithelial polyps

Radiographic Findings

- Radiography: Superior mediastinal mass ± anterior tracheal bowing
- Fluoroscopic guided esophagography
  - Smooth, expansile, sausage-shaped & intraluminal
  - Cervical esophageal mass ± extension distally with prolapse into gastric fundus
  - Mass with lobulated contour mimics malignancy

CT Findings

- Intraluminal mass + surrounding thin rim of contrast
- Varied HU based on degree of fat, fibrovascular tissue
  - Fat density: Abundance of adipose tissue
  - Heterogeneous HU: Equal amount of fat, soft tissue
  - Soft tissue density: Abundance of soft tissue + no fat

MR Findings

- Fibrovascular polyp with abundant adipose tissue
  - Hyperintense (T1WI); hypointense (T2WI)

Ultrasonographic Findings

- Endoscopic sonography: Fibrovascular polyp with high fat content seen as increased echogenic lesion

DIFFERENTIAL DIAGNOSIS

Esophageal carcinoma (polypoid tumors)

- Spindle cell carcinoma
  - Usually large, polypoid intraluminal mass simulating fibrovascular polyp
  - Margins irregular & more lobulated compared to smooth contour & less lobulated fibrovascular polyp

DDx: Large Intramural or Mucosal Mass

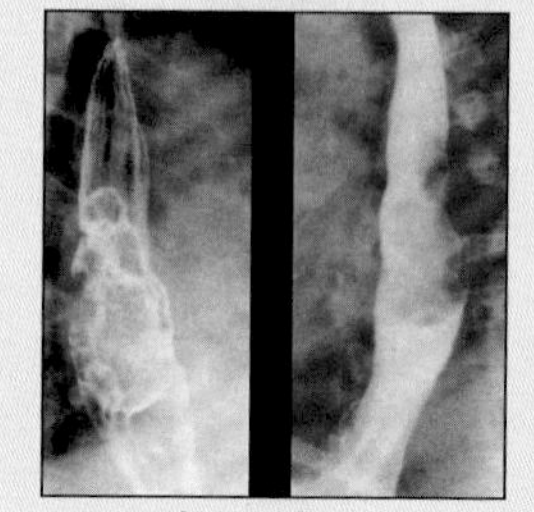

Esophageal Cancer

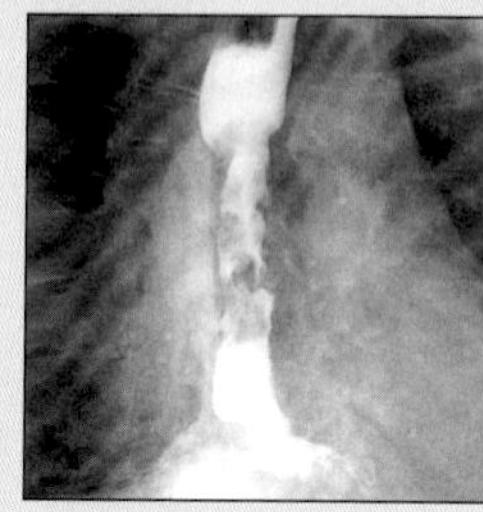

Esophageal Cancer

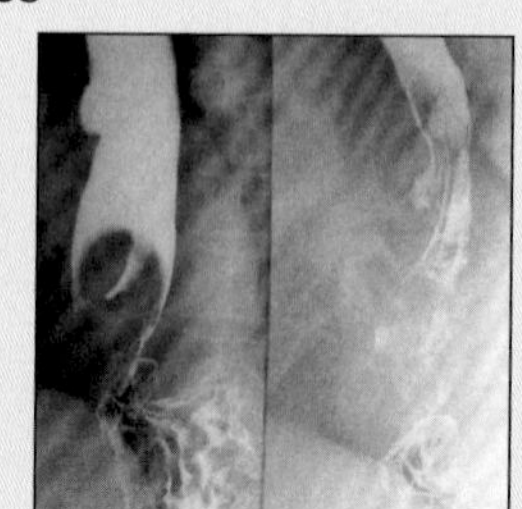

Leiomyoma

FIBROVASCULAR POLYP

Key Facts

Terminology
- Rare, benign tumor-like lesion of esophagus

Imaging Findings
- Best diagnostic clue: Giant, smooth, sausage-shaped, intraluminal, expansile mass
- Location: Cervical esophagus near cricopharyngeus
- Size: Variable in size (average length of 15 cm)
- Mass with lobulated contour mimics malignancy
- Varied HU based on degree of fat, fibrovascular tissue

Top Differential Diagnoses
- Esophageal carcinoma (polypoid tumors)
- Intramural (submucosal) benign tumors

Diagnostic Checklist
- Rule out other esophageal polypoidal masses
- Giant, smooth, sausage-shaped, expansile mass in cervical esophagus ± distal extension

  - Expand or dilate esophagus without obstruction
  - Rarely infiltrating or annular lesions
  - Location: Middle or distal third of esophagus
- Primary malignant melanoma
  - Bulky, polypoid intraluminal mass
  - Expand esophageal lumen without obstruction
  - Distinguished from fibrovascular polyp by its location (lower half) due to ↑ # of melanocytes
- Squamous cell & adenocarcinomas
  - May be polypoid mimicking fibrovascular polyp
  - Tend to narrow lumen rather than expand it
- On imaging: Abundant fat & fibrovascular tissue distinguishes fibrovascular polyp from other lesions
- Diagnosis: Biopsy & histology

Intramural (submucosal) benign tumors
- Leiomyoma & lipoma
  - En face: Round/ovoid filling defect + sharp margins
  - In profile: Smooth surface with obtuse borders
  - Pedunculated lipomas in cervical esophagus mimic fibrovascular polyp with abundant adipose tissue
  - Smooth extrinsic esophageal compression distinguishes from expansile fibrovascular polyp

PATHOLOGY

General Features
- Etiology: Unknown
- Epidemiology: Very rare

Gross Pathologic & Surgical Features
- Giant, smooth or lobulated expansile polyp with a discrete pedicle attached to cervical esophagus

Microscopic Features
- Varying amounts of fibrovascular & adipose tissue covered by normal squamous epithelium

CLINICAL ISSUES

Presentation
- Most common signs/symptoms
  - Long-standing dysphagia
  - Wheezing/inspiratory stridor (tracheal compression)
  - Rarely regurgitation of mass into pharynx or mouth

Demographics
- Age: Usually seen in elderly people
- Gender: Males more than females (M > F)

Natural History & Prognosis
- Complications
  - Laryngeal occlusion, asphyxia & sudden death
  - Fibrovascular polyps may bleed
  - Malignant degeneration: Extremely rare
- Prognosis: Usually good after resection

Treatment
- Small fibrovascular polyps: Endoscopic resection
- Gigantic fibrovascular polyps: Surgical resection

DIAGNOSTIC CHECKLIST

Consider
- Rule out other esophageal polypoidal masses

Image Interpretation Pearls
- Giant, smooth, sausage-shaped, expansile mass in cervical esophagus ± distal extension

SELECTED REFERENCES

1. Levine MS: Benign tumors of the esophagus: radiologic evaluation. Semin Thorac Cardiovasc Surg. 15(1):9-19, 2003
2. Ascenti G et al: Giant fibrovascular polyp of the esophagus: CT and MR findings. Abdom Imaging. 24(2):109-10, 1999
3. Levine MS et al: Fibrovascular polyps of the esophagus: clinical, radiographic, and pathologic findings in 16 patients. AJR Am J Roentgenol. 166(4):781-7, 1996

IMAGE GALLERY

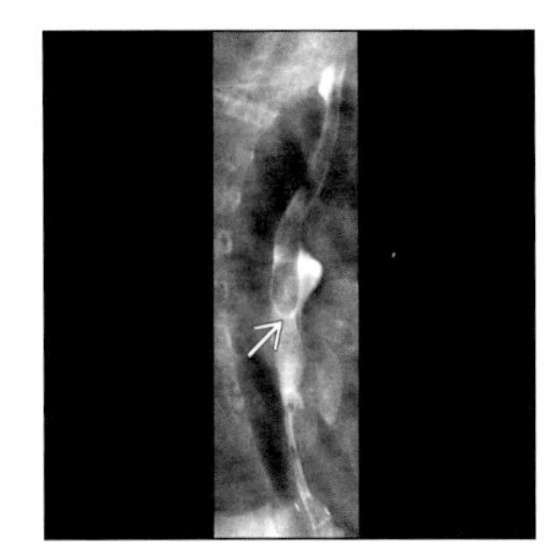
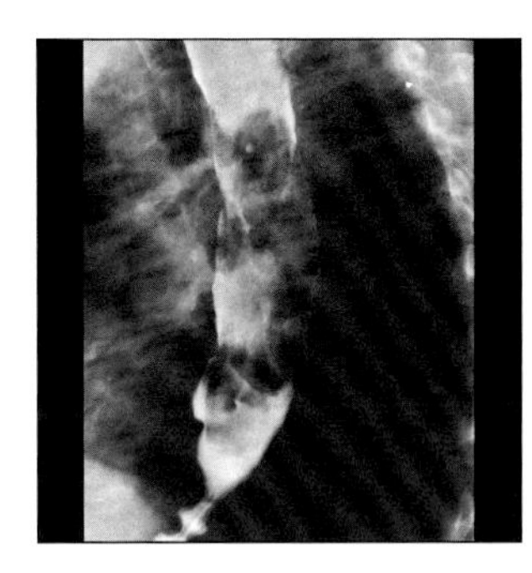

*(**Left**) Esophagram shows long, smooth polyp (arrow) extending from cervical esophagus into middle third; 67 year old woman (Courtesy M. Levine, MD). (**Right**) Esophagram shows large expansile mass filling proximal two thirds of esophagus (Courtesy M. Levine, MD).*

ESOPHAGEAL CARCINOMA

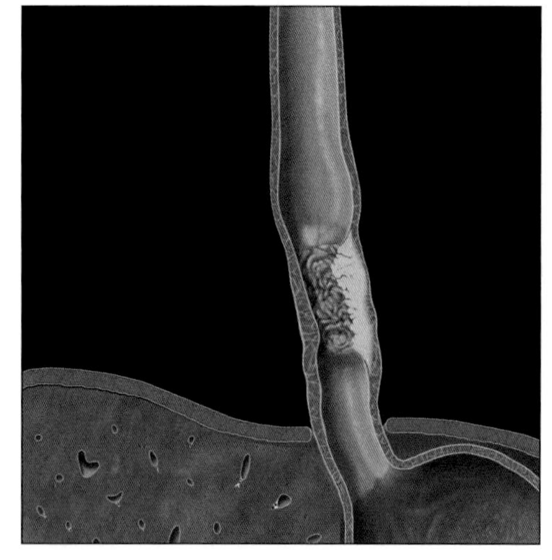

Graphic shows sessile polypoid mass with irregular surface that infiltrates the esophageal wall and narrows the lumen.

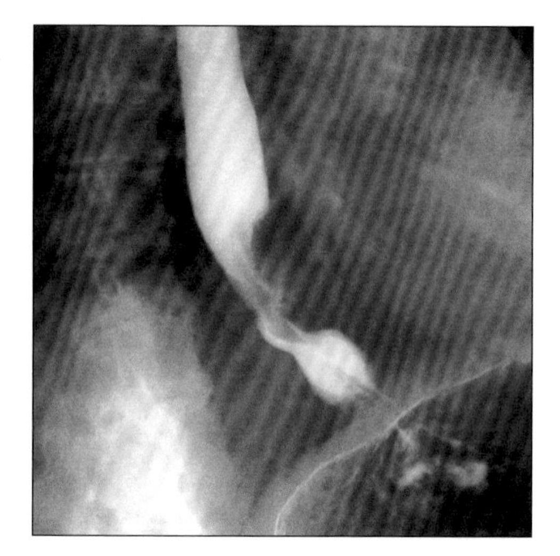

Esophagram shows polypoid mass of distal esophagus with irregular surface and luminal narrowing. Squamous cell carcinoma.

TERMINOLOGY

Definitions

- Squamous cell carcinoma: Malignant transformation of squamous epithelium
- Adenocarcinoma: Malignant dysplasia in columnar metaplasia (Barrett mucosa)

IMAGING FINDINGS

General Features

- Best diagnostic clue: Fixed irregular narrowing of esophageal lumen
- Location: Middle 3rd (50%), lower 3rd (30%), upper 3rd (20%)
- Size
  - Early esophageal cancer: Less than 3.5 cm
  - Advanced esophageal cancer: More than 3.5 cm
- Morphology
  - Histologically: Two main types
    - Squamous cell carcinoma
    - Adenocarcinoma (dysplasia in Barrett mucosa)
  - Classification of advanced esophageal cancer based on gross pathology & radiographic findings
    - Infiltrating, polypoid, ulcerative, varicoid lesions
- Other general features
  - Carcinoma is the most common tumor of esophagus
  - Squamous cell carcinoma
    - Accounts for 50-70% of all esophageal cancers
    - 1% of all cancers & 7% of all gastrointestinal cancers
    - Two major risk factors in US: Tobacco & alcohol abuse
    - Human papillomavirus: Synergistic increase risk factor in China & South Africa
  - Adenocarcinoma
    - Accounts for 30-50% of all esophageal cancers
    - 90-100% of cases arise from Barrett mucosa
    - Increasing in prevalence relative to squamous cell

Radiographic Findings

- Radiography
  - Chest x-ray (PA & lateral view): Advanced carcinoma
    - Hilar, retrohilar or retrocardiac mass
    - Anterior bowing of posterior tracheal wall
    - Retrotracheal stripe thickening more than 3 mm
- Double-contrast esophagography: En face/profile views
  - Early esophageal squamous cell cancer
    - Plaque-like lesions; small, sessile polyps or depressed lesions
  - Early adenocarcinoma in Barrett esophagus
    - Plaque-like lesions; flat, sessile polyps
    - Localized area of flattening/stiffening in wall of peptic stricture (common in distal 1/3)
  - Advanced esophageal squamous cell cancer

DDx: Irregular Luminal Narrowing

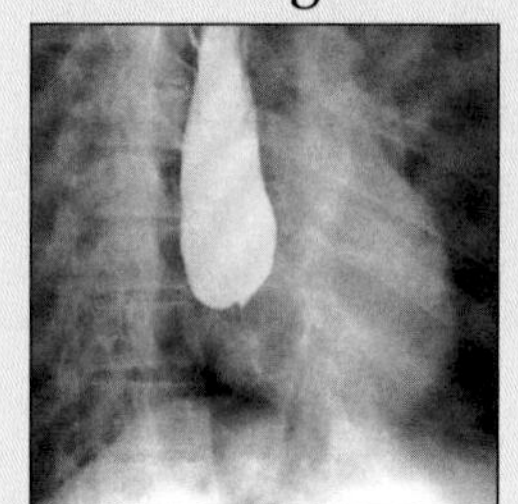

Peptic Stricture

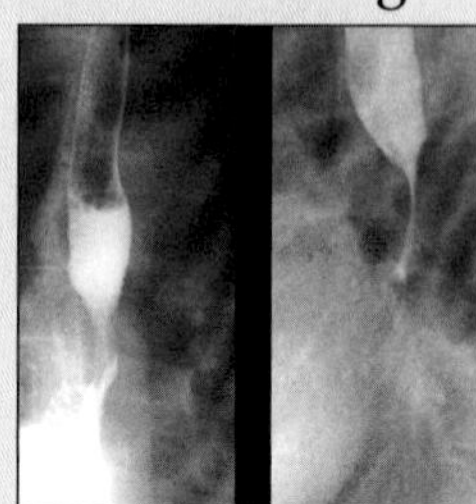

Lung Cancer

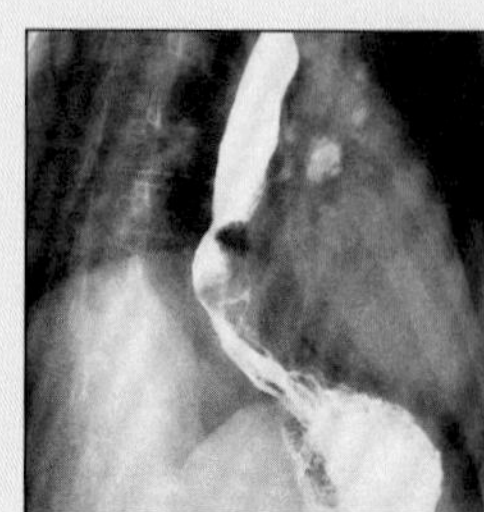

Eso. Leiomyoma

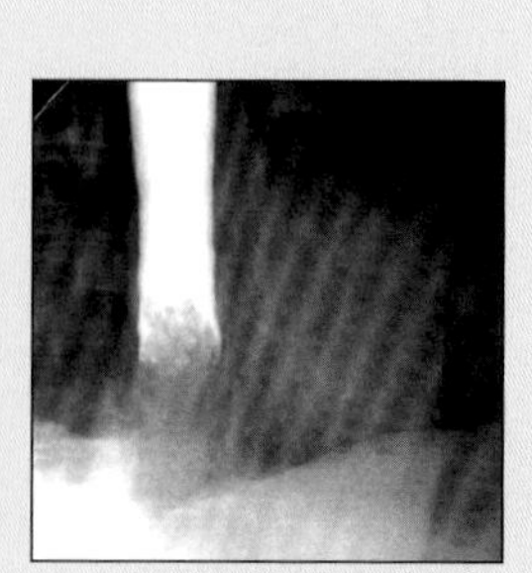

Foreign Body

ESOPHAGEAL CARCINOMA

Key Facts

Terminology
- Squamous cell carcinoma: Malignant transformation of squamous epithelium
- Adenocarcinoma: Malignant dysplasia in columnar metaplasia (Barrett mucosa)

Imaging Findings
- Best diagnostic clue: Fixed irregular narrowing of esophageal lumen
- Infiltrating, polypoid, ulcerative, varicoid lesions
- Plaque-like lesions; flat, sessile polyps
- Asymmetric contour with abrupt proximal borders of narrowed distal segment (rat-tail appearance)

Top Differential Diagnoses
- Inflammatory stricture
- Intramural primary esophageal tumor
- Extrinsic encasement of esophagus
- Esophageal foreign body

Pathology
- Smoking, alcohol, achalasia, lye strictures
- Adenocarcinoma: Barrett esophagus (increased risk factors-GERD, reflux esophagitis, motility disorders)
- Spread: Local, lymphatic, hematogenous

Clinical Issues
- Clinical profile: Elderly patient with history of difficulty in swallowing for solids & weight loss

Diagnostic Checklist
- Overlap of imaging findings with inflammatory causes of strictures and mucosal irregularity
- Endoscopic biopsy often required

    - Infiltrating lesion (most common): Irregular narrowing/luminal constriction (stricture) with nodular/ulcerated mucosa
    - Polypoid lesion: Lobulated/fungating intraluminal mass
    - Ulcerative lesion: Well-defined meniscoid ulcers with a radiolucent rim of tumor surrounding ulcer in profile view
    - Varicoid lesion: Thickened, tortuous, serpiginous longitudinal folds due to submucosal spread of tumor, mimicking varices (nonpliable)
  - Advanced adenocarcinoma in Barrett esophagus
    - Radiologically indistinguishable from squamous
    - Long infiltrating lesion in distal esophagus
  - Stricture in advanced carcinoma
    - Asymmetric contour with abrupt proximal borders of narrowed distal segment (rat-tail appearance)

CT Findings
- CT: Staging of esophageal carcinoma
  - Stage I: Localized wall thickening of 3-5 mm or intraluminal tumor
    - Not as accurate as endoscopic ultrasonography
  - Stage II: Localized wall thickening more than 5 mm & no mediastinal extension
  - Stage III: Tumor extends beyond esophagus into mediastinal tissues
    - Tracheobronchial invasion: Posterior wall indentation/bowing & tracheobronchial displacement/compression; ± collapse of lobes
    - Aortic invasion: Uncommon finding (2% of cases)
    - Pericardial invasion: Based on obliteration of fat plane/mass effect
    - Mediastinal adenopathy: Discrete/inseparable from primary tumor
  - Stage IV: Extends into mediastinum & distant areas
    - Liver, lungs, pleura, adrenals, kidneys & nodes
    - Subdiaphragmatic adenopathy: Seen in more than 2/3 of distal cancers

MR Findings
- MR imaging is limited by motion artifacts due to long acquisition times
- Axial cardiac-gated & sagittal scans
  - Esophageal mass indenting or displacing posterior tracheal wall

Ultrasonographic Findings
- Real Time
  - Endoscopic ultrasonography (EUS)
    - Useful technique for determining extent of esophageal wall invasion
    - Malignant nodes: Hypoechoic & well-defined
    - Benign nodes: Hyperechoic; indistinct borders

Nuclear Medicine Findings
- PET: 18F-Fluorodeoxyglucose Positron Emission Tomography (FDG PET): More sensitive & superior to CT in detecting regional & distant metastases

Imaging Recommendations
- Double-contrast esophagraphy (en face/profile views)
- Helical CT & PET for metastases
- EUS for local invasion

DIFFERENTIAL DIAGNOSIS

Inflammatory stricture
- Reflux esophagitis
  - Concentric, smooth tapering of short distal segment
    - Distinguished by normal peristalsis in benign type
    - Lack of peristalsis in malignant stricture
- Infectious esophagitis
  - Severe or advanced Candidiasis
    - Grossly irregular or "shaggy" esophagus
    - Long tapered distal stricture due to scarring
  - Viral: Herpes, CMV, HIV
    - Multiple small punctate or giant, flat ovoid ulcers
    - Usually strictures are not seen in viral esophagitis
- Caustic esophagitis
  - Chronic phase
    - Segmental strictures, sacculations, pseudodiverticula
    - Thread-like or filiform esophagus/stricture, differentiates from malignant stricture
    - History of caustic ingestion (strong alkali/acid)

ESOPHAGEAL CARCINOMA

- Diagnosis: Endoscopic biopsy & history

Intramural primary esophageal tumor

- Leiomyoma
  - Round or ovoid filling defect, outlined by barium
  - Borders form right angles/obtuse angles with wall
  - Lobulation/ulceration of mass suggests malignancy
- Fibrovascular polyp
  - Smooth, expansile, sausage-shaped mass

Extrinsic encasement of esophagus

- Example: Invasion by lung cancer
- History & imaging evidence of lung cancer

Esophageal foreign body

- Impacted meat bolus appears as polypoid filling defect
- Incompletely distended esophagus below impaction may be mistaken for a pathologic narrowing
- Esophagram after foreign body removal may show underlying normal esophagus, Schatzki ring, stricture

PATHOLOGY

General Features

- Genetics: Genomic instability in patients with Barrett esophagus may increase the risk of adenocarcinoma
- Etiology
  - Squamous cell carcinoma
    - Smoking, alcohol, achalasia, lye strictures
    - Celiac disease, head & neck tumor
    - Plummer-Vinson syndrome, radiation, tylosis
  - Adenocarcinoma: Barrett esophagus (increased risk factors-GERD, reflux esophagitis, motility disorders)
  - Spread: Local, lymphatic, hematogenous
- Epidemiology: Increased incidence in Turkey, Iran, India, China, S. Africa, France, Saudi Arabia

Gross Pathologic & Surgical Features

- Infiltrating, polypoid, ulcerative or varicoid lesions

Microscopic Features

- Squamous cell atypia; columnar glands
- Adeno & squamous components

Staging, Grading or Classification Criteria

- TNM Staging
  - Stage 0: Carcinoma in situ
  - Stage I: Lamina propria or submucosa
  - Stage IIA: Muscularis propria & adventitia
  - Stage IIB: Lamina propria, submucosa, muscularis propria & regional lymph nodes
  - Stage III: Adventitia, adjacent structures, regional lymph nodes + any other nodes
  - Stage IV: All layers, adjacent structures, regional lymph nodes + any other nodes & distant metastases

CLINICAL ISSUES

Presentation

- Most common signs/symptoms: Dysphagia (solids), odynophagia (painful swallowing), anorexia, weight loss, retrosternal pain
- Clinical profile: Elderly patient with history of difficulty in swallowing for solids & weight loss
- Lab-data
  - ± Hypochromic, microcytic anemia
  - ± Hemoccult positive stool or decreased albumin
- Diagnosis: Endoscopic biopsy & histology

Demographics

- Age: Usually above 50 years
- Gender: M:F = 4:1
- Ethnicity: African-Americans more than Caucasians (2:1)

Natural History & Prognosis

- Complications
  - Fistula to trachea (5-10%); bronchi/pericardium
- Prognosis
  - Early cancer: 5 year survival is 90%
  - Advanced cancer: 5 year survival is less than 10%

Treatment

- Curative treatment
  - Surgery, radiation (pre & post-op-radiation)
- Palliative treatment
  - Surgery, radiation, chemotherapy
  - Laser treatment, indwelling prosthesis

DIAGNOSTIC CHECKLIST

Consider

- Overlap of imaging findings with inflammatory causes of strictures and mucosal irregularity
- Endoscopic biopsy often required

Image Interpretation Pearls

- Irregular narrowing with nodular/ulcerated mucosa
- Asymmetric contour with abrupt proximal borders of narrowed distal segment (rat-tail appearance)

SELECTED REFERENCES

1. Gupta S et al: Usefulness of barium studies for differentiating benign and malignant strictures of the esophagus. AJR Am J Roentgenol. 180(3):737-44, 2003
2. Iyer RB et al: Diagnosis, staging, and follow-up of esophageal cancer. AJR Am. J. Roentgenol. 181: 785 - 793, 2003
3. Levine MS: Esophageal cancer. Radiologic diagnosis. Radiol Clin North Am. 35(2):265-79, 1997
4. Levine MS et al: Carcinoma of the esophagus and esophagogastric junction: sensitivity of radiographic diagnosis. AJR Am J Roentgenol. 168(6):1423-6, 1997
5. Levine MS et al: Fibrovascular polyps of the esophagus: clinical, radiographic, and pathologic findings in 16 patients. AJR Am J Roentgenol. 166(4):781-7, 1996
6. Glick SN: Barium studies in patients with Barrett's esophagus: importance of focal areas of esophageal deformity. AJR Am J Roentgenol. 163(1):65-7, 1994
7. Vilgrain V et al: Staging of esophageal carcinoma: Comparison of results with endoscopic sonography and CT. AJR 155: 277-81, 1990

IMAGE GALLERY

Typical

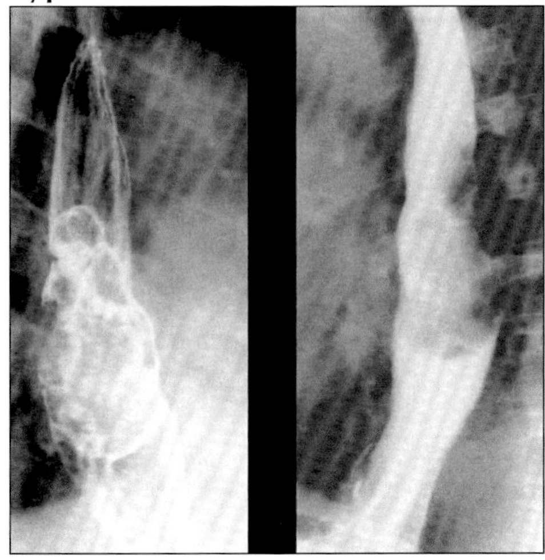

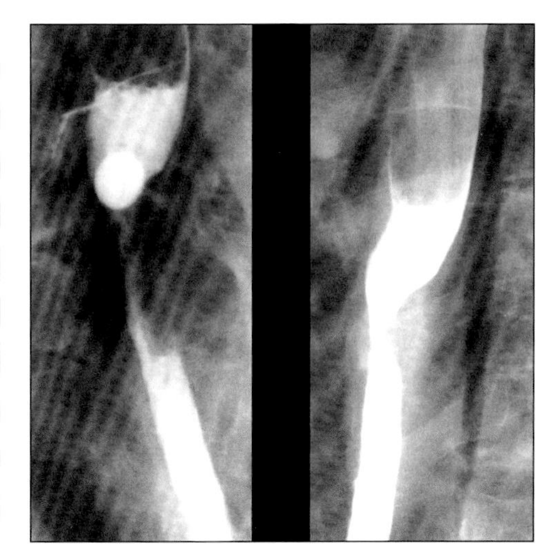

*(**Left**) Two views from esophagram show a large fungating polypoid mass with acute angle interface with esophageal wall. (**Right**) Two views from esophagram demonstrate delayed passage of barium pill, drawing attention to a stricture with subtle mucosal irregularity; squamous cell carcinoma.*

Typical

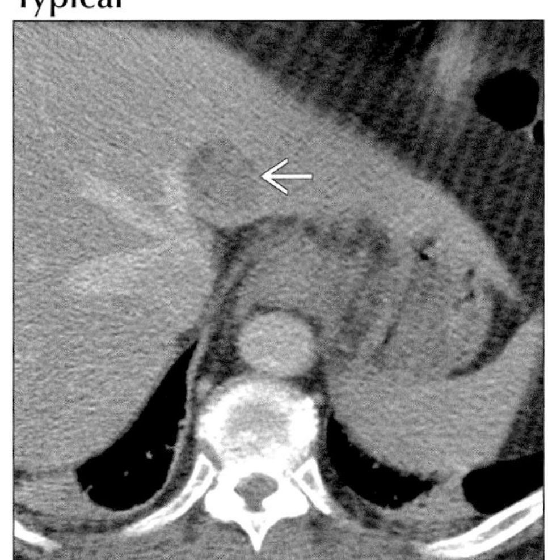

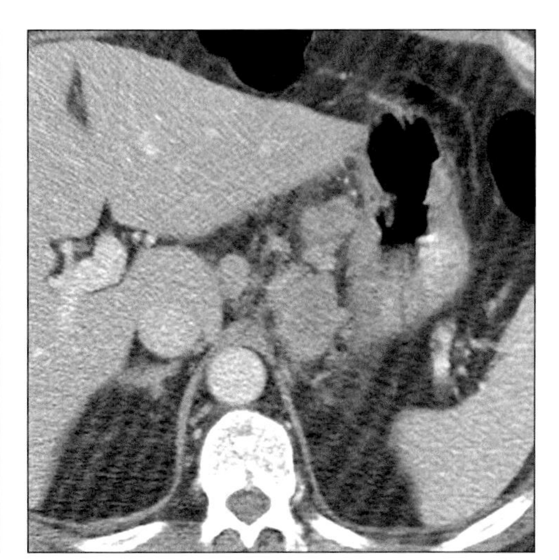

*(**Left**) Axial CECT shows mural thickening near GE junction due to carcinoma. Liver metastasis (arrow). (**Right**) Axial CECT shows shows extensive lymphadenopathy in gastrohepatic region due to metastases from distal esophageal carcinoma.*

Typical

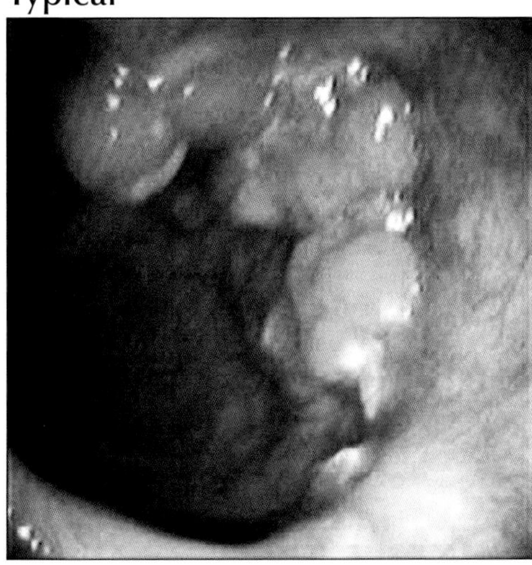

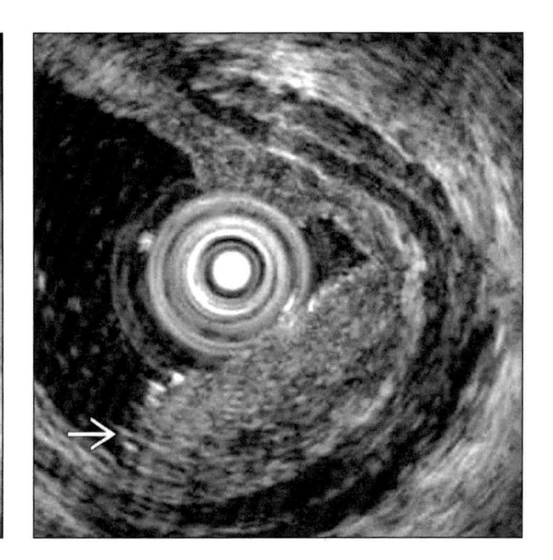

*(**Left**) Endoscopic image shows an irregular polypoid mass in distal esophagus; adenocarcinoma. (**Right**) Endoscopic sonography demonstrates an intraluminal mass (arrow) that does not penetrate the muscularis propria (T1A adenocarcinoma).*

ESOPHAGEAL METASTASES AND LYMPHOMA

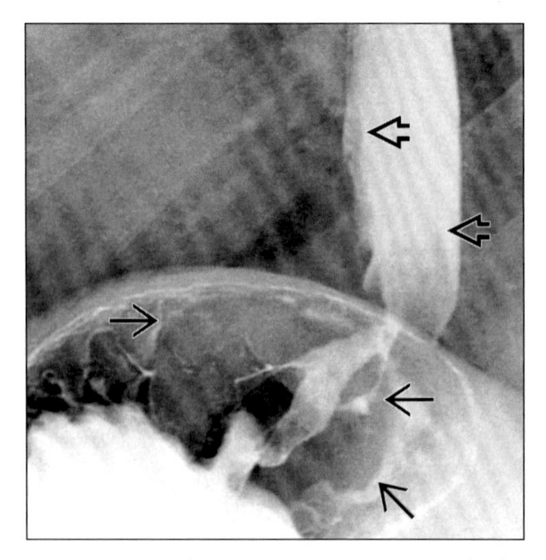

Upper GI series shows mass (arrows) in gastric fundus extending cephalad into the esophagus (open arrows). Gastric carcinoma.

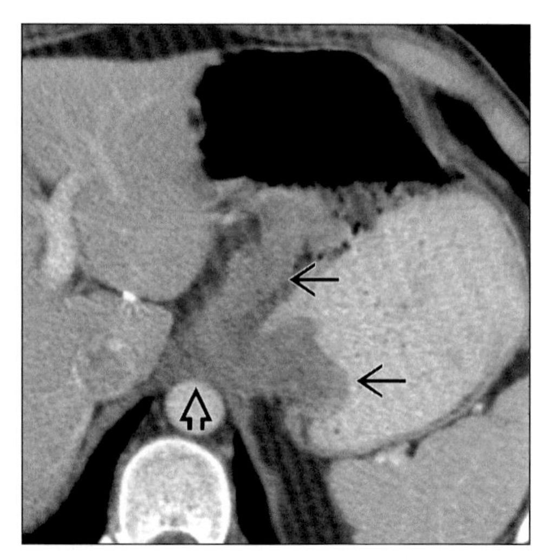

Axial CECT shows gastric carcinoma in fundus (arrows) extending cephalad into the esophagus (open arrow).

TERMINOLOGY

Definitions

- Metastases from primary cancer of other sites
- Lymphoma: Malignant tumor of lymphocytes

IMAGING FINDINGS

General Features

- Best diagnostic clue: Ulcerated or polypoid mass of gastric cardia extending into distal esophagus
- Other general features
  - Esophageal metastases
    - Types of spread: Direct, lymphatic, hematogenous
    - Direct invasion: Most common (carcinoma of stomach accounts for 50% of cases)
  - Esophageal lymphoma
    - Least common site of GIT (accounts for 1% cases)
    - Usually non-Hodgkin & less commonly Hodgkin
    - Secondary lymphoma (90%); primary (10%)
    - Primary esophageal lymphoma seen in AIDS

Radiographic Findings

- Fluoroscopic guided double contrast barium study
  - Direct invasion, gastric carcinoma: Distal esophagus
    - Ulcerated or polypoid mass of cardia or fundus extending into gastroesophageal junction
    - Irregular or smooth, tapered narrowing of distal esophagus ± discrete mass
    - Cardia obliteration + subtle esophageal extension
  - Direct invasion of larynx, pharynx, thyroid, lung cancer: Cervical or thoracic esophagus
    - Smooth or slightly irregular esophageal wall + soft tissue mass in adjacent neck or mediastinum
    - Serrated, scalloped, or nodular esophageal wall → narrowing or obstruction
    - Thyroid cancer: Expansile intraluminal mass
  - Contiguous involvement by mediastinal nodes (breast & lung cancer): Mid-esophagus
    - Smooth, lobulated esophageal indentation or ulceration at level of carina
  - Hematogenous spread: Mid-esophagus > common
    - Breast cancer > common: Short/eccentric strictures
    - Malignant melanoma (rare): Submucosal masses or centrally ulcerated "bull's eye" lesions
  - Secondary lymphoma: Spread from stomach
    - Polypoid or ulcerated lesion in cardia + irregular narrowing of distal esophagus
  - Primary intrinsic esophageal lymphoma
    - Polypoid mass/stricture mimicking esophageal cancer
    - Nodular "folds" mimicking varices
    - Submucosal mass simulating leiomyoma

CT Findings

- Shows primary tumor + esophageal extension

DDx: Intramural Esophageal Mass

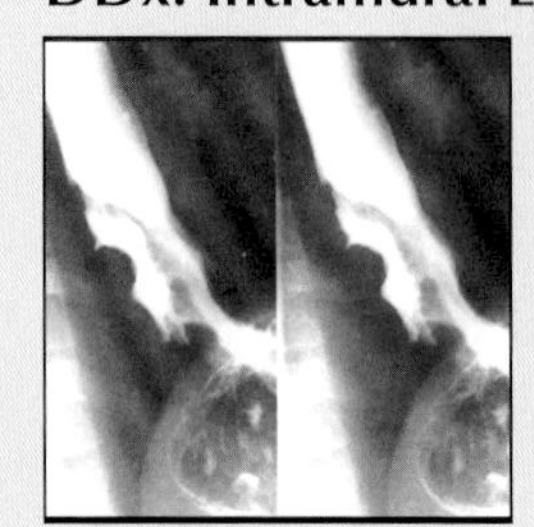

Esophageal Cancer

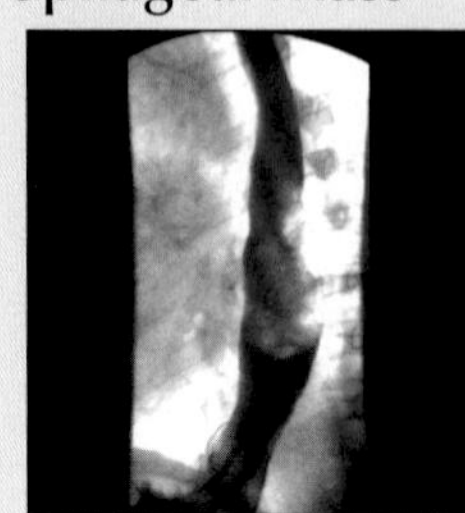

Esophageal Cancer

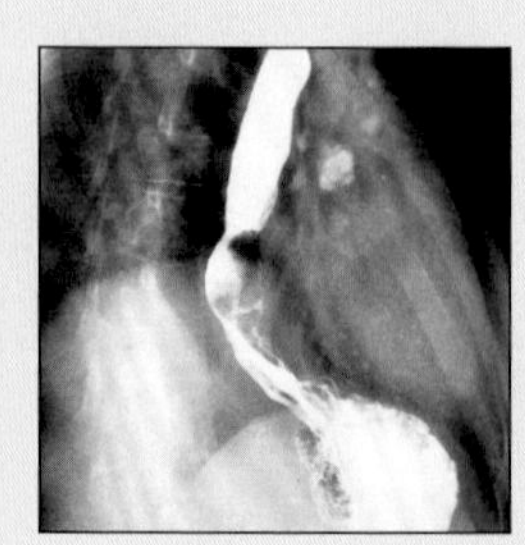

Leiomyoma

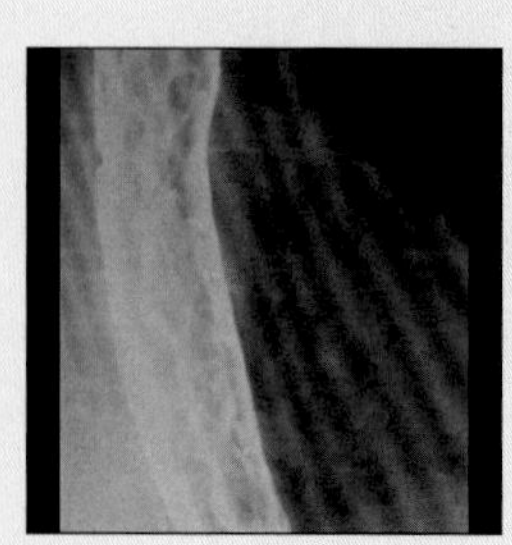

Varices

ESOPHAGEAL METASTASES AND LYMPHOMA

Key Facts

Imaging Findings
- Best diagnostic clue: Ulcerated or polypoid mass of gastric cardia extending into distal esophagus
- Direct invasion: Most common (carcinoma of stomach accounts for 50% of cases)
- Breast cancer > common: Short/eccentric strictures
- Polypoid mass/stricture mimicking esophageal cancer
- Submucosal mass simulating leiomyoma
- CT detects mediastinal lymphadenopathy + extent

Top Differential Diagnoses
- Intramural benign esophageal tumors
- Esophageal carcinoma
- Esophageal varices

Diagnostic Checklist
- Check for history of primary cancer, biopsy required
- Overlapping radiographic features of esophageal metastases, lymphoma & primary carcinoma

- CT detects mediastinal lymphadenopathy + extent

Ultrasonographic Findings
- Real Time
  - Endoscopic ultrasonography (EUS)
    - Hypoechoic mass disrupting normal wall layers
    - Selective/diffusely thickened echogenic wall layers

DIFFERENTIAL DIAGNOSIS

Intramural benign esophageal tumors
- Submucosal lesions arising within esophageal wall
- Leiomyoma
  - Round or ovoid filling defect, outlined by barium
  - Borders form right or obtuse angles with wall
  - Extrinsic compression & no invasion of wall
  - Lobulation/ulceration of mass suggests malignancy

Esophageal carcinoma
- Polypoid, ulcerated, infiltrative types simulate esophageal metastases or lymphoma
- Narrow distal segment + abrupt borders (rat-tail)
- Periesophageal & distal spread may be seen
- Diagnosis: Endoscopy & biopsy

Esophageal varices
- Serpiginous, longitudinal radiolucent filling defects mimic submucosal infiltration of lymphoma

PATHOLOGY

General Features
- Associated abnormalities: Primary carcinoma; generalized adenopathy; AIDS

Gross Pathologic & Surgical Features
- Solitary/multiple; polypoid/ulcerated masses; stricture

Microscopic Features
- Metastases: Varies based on primary cancer
- Lymphoma: Lymphoepithelial lesions

CLINICAL ISSUES

Presentation
- Most common signs/symptoms: Asymptomatic, dysphagia, weight loss, hematemesis

Natural History & Prognosis
- Complications: GI bleeding; perforation; obstruction
- Prognosis: Usually poor

Treatment
- Chemotherapy & surgical resection of lesions causing complications like obstruction & upper GI bleeding

DIAGNOSTIC CHECKLIST

Consider
- Check for history of primary cancer, biopsy required

Image Interpretation Pearls
- Overlapping radiographic features of esophageal metastases, lymphoma & primary carcinoma
- Imaging important to suggest & stage malignancy

SELECTED REFERENCES

1. Moreto M: Diagnosis of esophagogastric tumors. Endoscopy. 35(1):36-42, 2003
2. Coppens E et al: Primary Hodgkin's lymphoma of the esophagus. AJR Am J Roentgenol. 180(5):1335-7, 2003
3. Holyoke ED et al: Esophageal metastases and dysphagia in patients with carcinoma of the breast. J Surg Oncol 1:97-107, 1969

IMAGE GALLERY

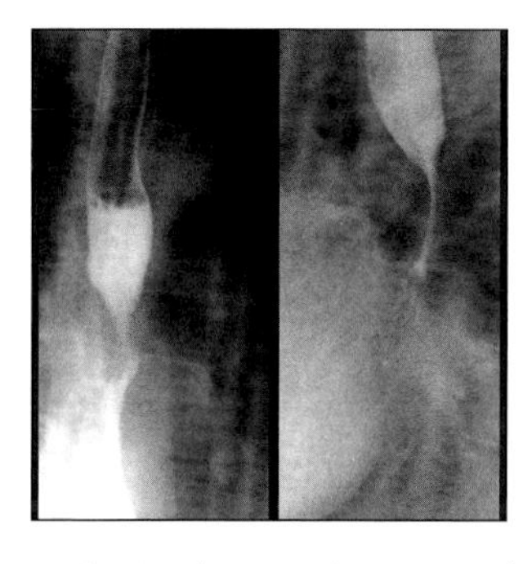

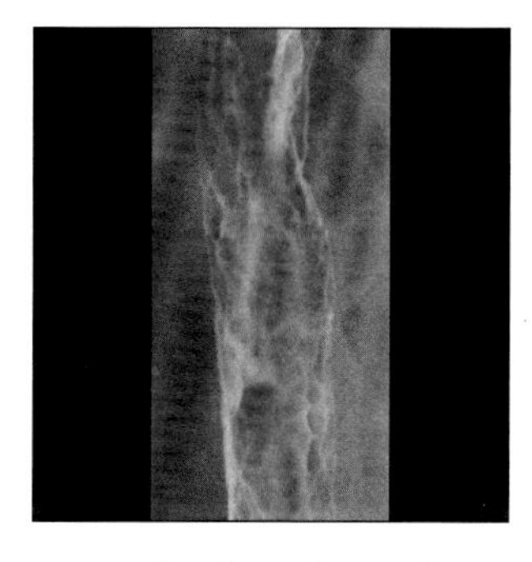

*(**Left**) Esophagram shows smooth stricture of mid esophagus due to direct invasion by lung cancer. (**Right**) Esophagram shows innumerable submucosal nodules, 3-10 mm in size due to lymphoma (Courtesy M. Levine, MD).*

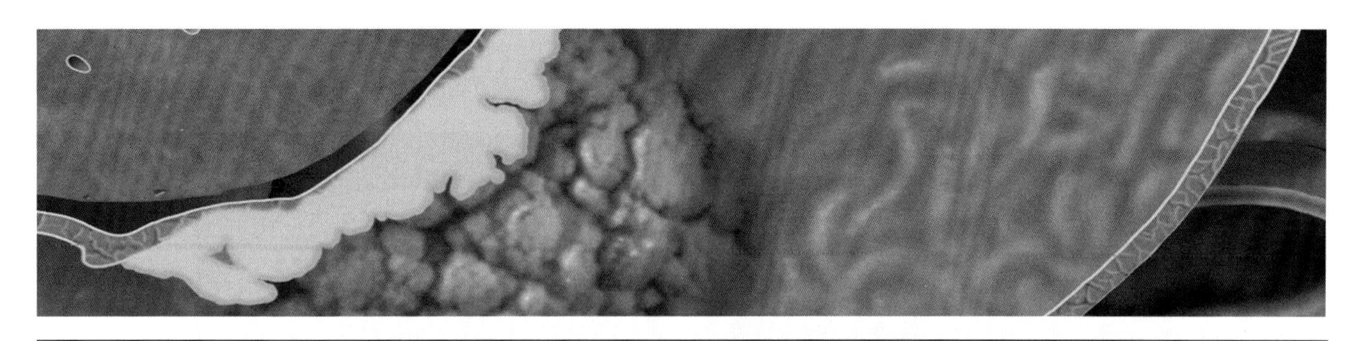

SECTION 3: Gastroduodenal

GASTRODUODENAL ANATOMY AND IMAGING ISSUES

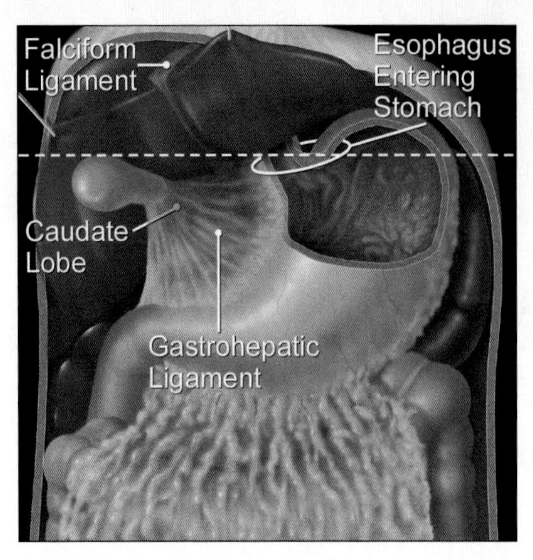

Graphic shows liver reflected up to reveal stomach and its ligamentous and omental attachments. Dotted line indicates plane of section through gastroesophageal junction.

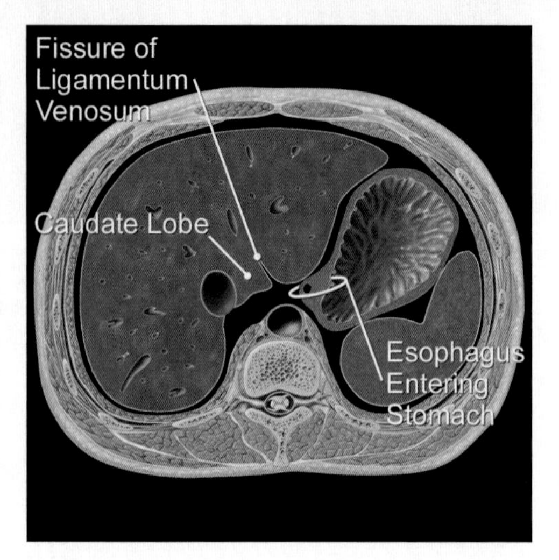

Graphic shows axial section through gastroesophageal junction, which usually lies at the level of the porta hepatis, or fissure of the ligamentum venosum.

TERMINOLOGY

Definitions

- Gastric cardia
  - Junction of the greater and lesser curvatures
  - Site of esophageal entry into stomach
  - Site of mucous secreting glands
- Gastric fundus
  - Uppermost section of stomach
  - Main site of pepsinogen secreting glands
- Hepatoduodenal ligament
  - Double layer of peritoneum; attached to pylorus and duodenum
  - Contains portal vein, hepatic artery, common bile duct
  - Forms margin of epiploic foramen (of Winslow); entry into lesser sac (behind stomach)
  - "Suspends" duodenum from underside of liver
- Greater omentum
  - Four layers of peritoneum
  - Passes from greater curvature as an "apron" covering bowel
  - Connects stomach to transverse colon (gastrocolic ligament)

ANATOMY-BASED IMAGING ISSUES

Imaging Approaches

- Role of fluoroscopic barium studies
  - Complementary to endoscopy for most cases of dyspepsia and abdominal pain (e.g., peptic ulcer, tumor)
  - Superior to endoscopy in evaluation of functional abnormalities (e.g., reflux, delayed emptying, submucosal masses/infiltrative processes)
- Role of computed tomography (CT)
  - Primary role in staging primary and metastatic tumors involving stomach
  - Complementary role in diagnosing gastritis and ulcers, especially complicated (e.g., perforation, abscess)
  - Primary role in diagnosing inflammatory processes that affect the stomach secondarily (e.g., pancreatitis)
- Role of endoscopy
  - Most accurate means of diagnosing gastric carcinoma and primary inflammatory conditions
  - May fail to detect submucosal gastric masses (normal overlying mucosa)

CLINICAL IMPLICATIONS

Clinical Importance

- Gastric mucosa and submucosa normally contain some lymphoid tissue
  - Chronic antigenic stimulation by H. pylori infection can result in proliferation of mucosa-associated lymphoid tissue (MALT)
  - If detected and eradicated (by antibiotics), treatment is curative
  - Otherwise may progress to gastric B cell lymphoma
- Duodenal wall contains mucus secreting (Brunner) glands
  - These may enlarge to simulate multiple polyps (Brunner gland hypertrophy)
  - May develop into a benign neoplastic mass (Brunner gland adenoma)
- Blood supply
  - Arterial blood supply to stomach and duodenum is quite variable; numerous collateral pathways arising from branches of the celiac and superior mesenteric arteries make these organs resistant to ischemic injury (and difficult to control by catheter embolotherapy in the setting of acute hemorrhage)
  - Gastric veins become enlarged collaterals (varices) commonly due to portal hypertension or splenic vein occlusion
- Post-operative stomach

GASTRODUODENAL ANATOMY AND IMAGING ISSUES

DIFFERENTIAL DIAGNOSIS

Gastric malignant tumors
- Adenocarcinoma
- Lymphoma
- GI stromal tumors (GIST)
- Carcinoid
- Kaposi sarcoma
- Metastases

Duodenal malignant tumors
- Carcinoma
- Ampullary carcinoma
- GI stromal tumors (GIST)
- Carcinoid
- Lymphoma/metastases
- Kaposi sarcoma

Gastric benign tumors
- Hyperplastic polyps
- Adenoma
- Hamartoma
- Stromal tumors
- Lipoma
- Villous adenoma

Duodenal benign tumors
- Villous adenoma
- Stromal tumors
- ⇒Lipoma, leiomyoma, etc.

  - In many radiology practices, most fluoroscopic exams of the stomach are performed following surgical procedures that alter gastric anatomy
  - Familiarity with surgical techniques, "normal" post-operative appearance, and complications is essential
- Most common procedures
  - Fundoplication
    - Indication: Gastroesophageal reflux
    - Technique: Various types of gastric fundus wraps around distal esophagus
    - Appearance: Extrinsic mass effect in fundus compressing distal esophagus
    - Complications: Perforation, esophageal obstruction, dehiscence or slip of wrap
  - Bariatric surgery
    - Indication: Morbid obesity
    - Technique: Reduction of gastric size ± bypass of proximal small bowel
    - Appearance: Vertical banding to create small pouch; surgical separation of proximal pouch with anastomosis to Roux limb (Roux-en-Y gastric bypass); extrinsic prosthetic adjustable band around fundus
    - Complications: Obstruction (esophagus, anastomosis, bowel); perforation (leak, abscess); internal hernia; ulceration (at anastomosis)
  - Partial gastrectomy
    - Indications: Intractable peptic ulcer disease; gastric tumor
    - Technique: Bilroth I and II (+ variations)
    - Appearance: Distal gastrectomy with duodenal anastomosis (Bil I); distal gastrectomy with anastomosis to jejunum or Roux loop (Bil II)
    - Complications: Recurrent ulcer or tumor; obstruction; perforation; gastric stasis; bezoar; dumping syndrome; intussusception

CUSTOM DIFFERENTIAL DIAGNOSIS

Gastric ulcers
- Erosions
  - Idiopathic
    - Varioliform erosions; top of antral rugal folds
  - Aspirin; nonsteroidal anti-inflammatory drugs (NSAIDs)
    - Linear, multiple, greater curvature body and antrum
  - Crohn disease
    - Antrum and body; contraction and stricture of stomach
- Ulcers
  - H. pylori
    - Lesser curvature or posterior wall of antrum
  - Aspirin (NSAIDs)
    - Distal body and antrum, greater curvature
  - Gastritis
    - Variable
  - Zollinger-Ellison
    - Gastric and duodenal
  - Gastric cancer
    - Nodular folds, mass surrounds ulcer
- Gastric bull's eye lesions
  - Metastases (especially melanoma)
  - Lymphoma
  - Kaposi sarcoma
  - Carcinoid
  - Adenocarcinoma
  - Ectopic pancreas (greater curve, antrum)

Thick gastric folds
- Common
  - (Normal, nondistended stomach)
  - Gastritis
  - Pancreatitis
  - Portal hypertension
  - Neoplastic
  - Varices
- Uncommon
  - Menetrier disease
  - Zollinger-Ellison syndrome
  - Caustic ingestion
  - Radiation gastritis
  - Eosinophilic gastritis
  - Amyloidosis

GASTRODUODENAL ANATOMY AND IMAGING ISSUES

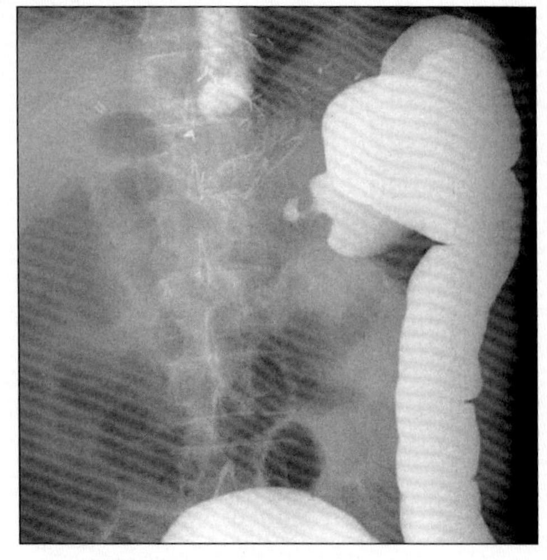

Barium enema shows obstruction to retrograde flow due to gastric carcinoma that invaded transverse colon via the gastrocolic ligament.

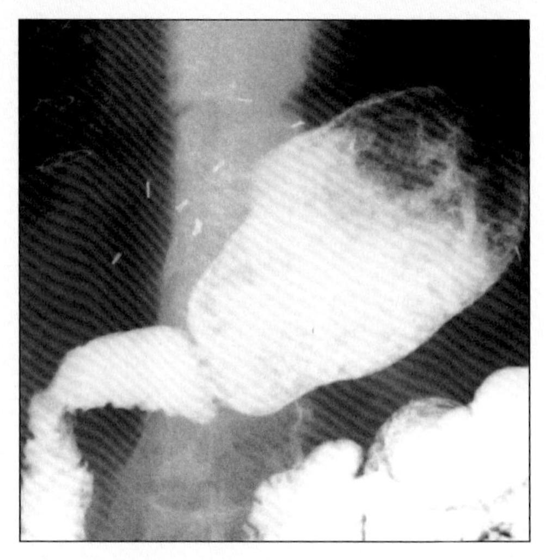

Upper GI series shows result of a Bilroth I type distal gastrectomy. Note bezoar within stomach.

Gastric antral narrowing

- Common
  - Gastritis
  - Carcinoma
  - Lymphoma/metastases
- Uncommon
  - Crohn disease
  - Tuberculous/fungal
  - Syphilis
  - Caustic ingestion
  - Radiation
  - Sarcoid and amyloid

Gastric dilation

- Gastric atony
  - Vagotomy, medications, post-op
  - Diabetes
  - Uremia
  - Scleroderma
- Outlet obstruction
  - Peptic ulcer
  - Antral stricture (Crohn, TB, etc.)
  - Pancreatitis
  - Tumor
  - Bezoar
  - Prolapsed antral polyp
  - Volvulus

Dilated (mega-) duodenum

- Mechanical obstruction
  - Pancreatitis
  - Tumor
  - SMA syndrome
  - Crohn disease
  - Peptic ulcer
- Scleroderma
- Acute ileus (post-op, metabolic, drugs)
- Hereditary visceral myopathy
- Hereditary visceral neuropathy

Duodenal filling defects

- Non-neoplastic
  - Prolapsed antral mucosa
  - Flexural pseudotumor
  - Heterotopic gastric mucosa
  - Brunner gland hyperplasia
  - Benign lymphoid hyperplasia
  - Choledochocele
  - Duplication cyst
  - Intramural hematoma
  - Intramural pseudocyst
- Neoplastic
  - Polyps (hyperplastic, hamartomatous, and adenomatous)
  - Isolated or polyposis syndromes

SELECTED REFERENCES

1. Horton KM et al: Current role of CT in imaging of the stomach. Radiographics. 23(1):75-87, 2003
2. Insko EK et al: Benign and malignant lesions of the stomach: evaluation of CT criteria for differentiation. Radiology. 228(1):166-71, 2003
3. Levine MS: Textbook of gastrointestinal radiology: Stomach and duodenum: differential diagnosis. 2nd ed. Philadelphia, WB Saunders. pp 698-702, 2000
4. Pattison CP et al: Helicobacter pylori and peptic ulcer disease: evolution to revolution to resolution. AJR Am J Roentgenol. 168(6):1415-20, 1997
5. Eisenberg RL: Gastrointestinal radiology: a pattern approach. 3rd ed. Philadelphia, JB Lippincott, 1996
6. Fishman EK et al: CT of the stomach: spectrum of disease. Radiographics. 16(5):1035-54, 1996
7. Levine MS et al: The Helicobacter pylori revolution: radiologic perspective. Radiology. 195(3):593-6, 1995
8. Dähnert W: Radiology review manual. 4th ed. Philadelphia, Lippincott, Williams, and Wilkins. p615-721, 2000
9. Reeder MM: Reeder and Felson's gamut's in radiology. 3rd ed. New York, Springer Verlag. 1993

IMAGE GALLERY

Typical

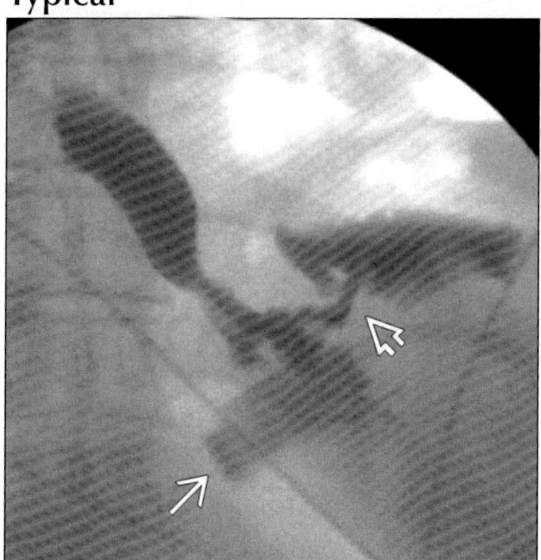

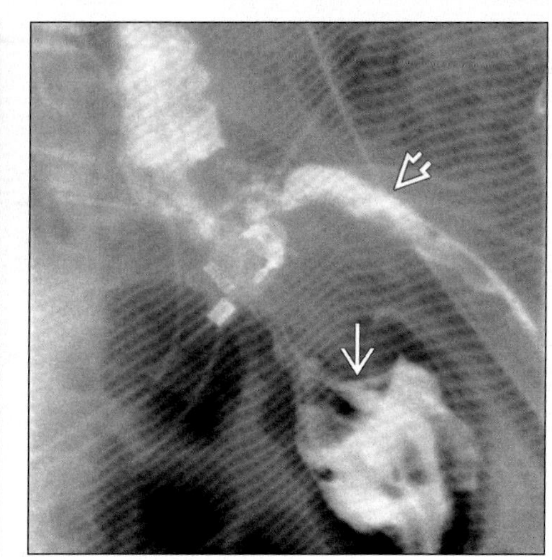

(Left) Esophagram shows laparoscopically placed band (arrow) around gastric fundus; an anti-obesity procedure. Note leak of contrast (open arrow). *(Right)* Upper GI following Nissen fundoplication shows intact wrap (arrow) but a leak from the fundus or esophagus (open arrow).

Typical

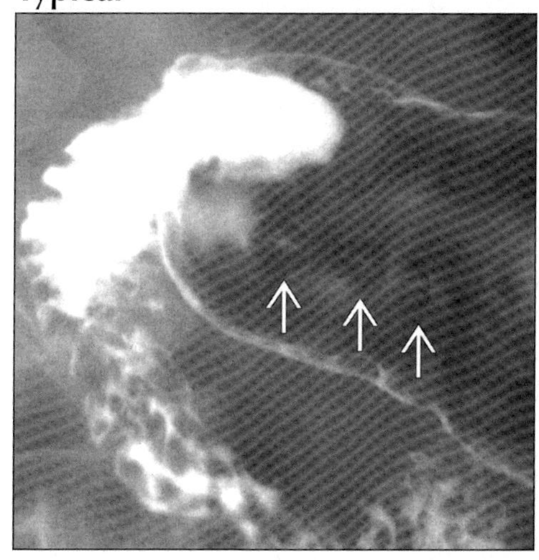

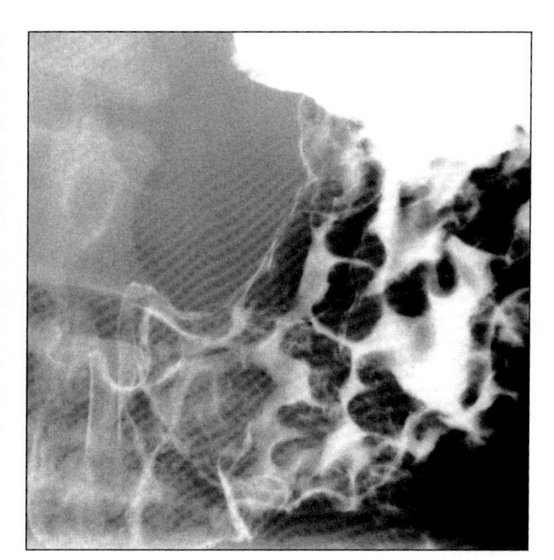

(Left) Upper GI series shows numerous aphthoid (varioliform) erosions (arrows) along antral folds; gastritis. *(Right)* Upper GI series shows massive nodular thickenings of gastric folds; lymphoma.

Typical

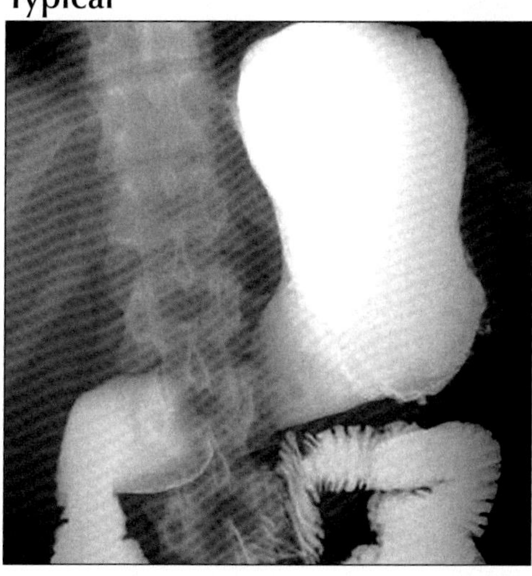

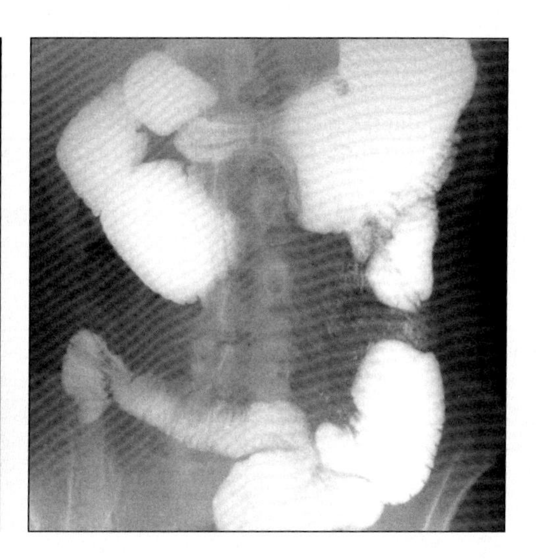

(Left) Upper GI series shows a featureless, contracted stomach with conical, narrowed antrum; Crohn disease. *(Right)* Upper GI series shows a "mega-duodenum"; scleroderma.

GASTRIC DIVERTICULUM

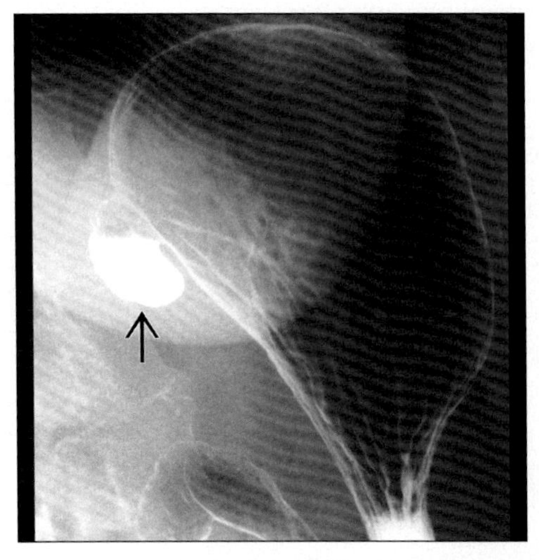

Upper GI series shows air-contrast level within a gastric diverticulum, arising near the gastroesophageal junction.

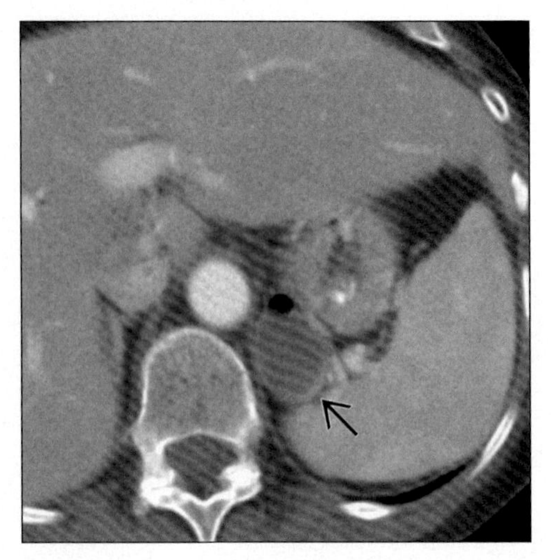

Axial CECT shows air-fluid level within gastric diverticulum (arrow) which lies medial + posterior to gastric fundus.

TERMINOLOGY

Definitions

- A pouch or sac opening from the stomach

IMAGING FINDINGS

General Features

- Best diagnostic clue: Barium-filled diverticulum from fundus, near gastroesophageal junction
- Other general features
  - 2 types of gastric diverticula
    - True gastric diverticula
    - Intramural or partial gastric diverticula (false)

Radiographic Findings

- Fluoroscopic-guided barium studies
  - True diverticula
    - Most (75%) are juxtacardiac diverticula: Diverticula near gastroesophageal junction, on posterior aspect of lesser curvature of the stomach
    - Usually 1-3 cm, up to 10 cm in diameter
    - Barium-filled diverticulum with air-fluid level
    - Pooling of barium; mimics ulceration
    - Large gastric diverticulum fails to fill with gas or barium; mimics smooth submucosal mass
    - In antrum (rare); mimics ulcer craters
  - Intramural or partial gastric diverticula
    - Most are prepyloric diverticula: Diverticula at greater curvature of the distal antrum
    - Tiny collection of barium extending outside the contour of the adjacent gastric wall; mimics ulcers
    - Movement of barium with peristalsis
    - Heaped-up area overlying diverticulum; mimics ectopic pancreatic rest on greater curvature

CT Findings

- Abnormal rounded soft tissue shadow
  - Often in suprarenal location; mimics adrenal mass
- Air-filled, fluid-filled or contrast-filled mass
- No enhancement of contents

Imaging Recommendations

- Best imaging tool: Fluoroscopic-guided barium studies
- Protocol advice
  - Juxtacardiac diverticula are best seen in lateral views on barium studies
  - Obtain CT in supine and prone position: Air will usually fill the diverticulum

DIFFERENTIAL DIAGNOSIS

Adrenal mass

- CT: Diverticular contents do not enhance; adrenal masses (except cyst) do

DDx: Mass or Pseudomass in LUQ

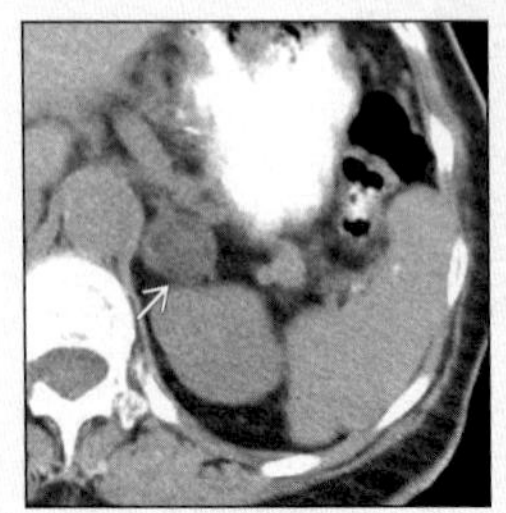

Adrenal Mass

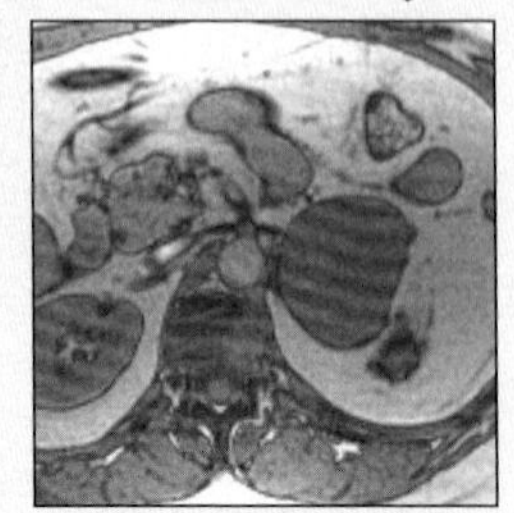

Adrenal Mass

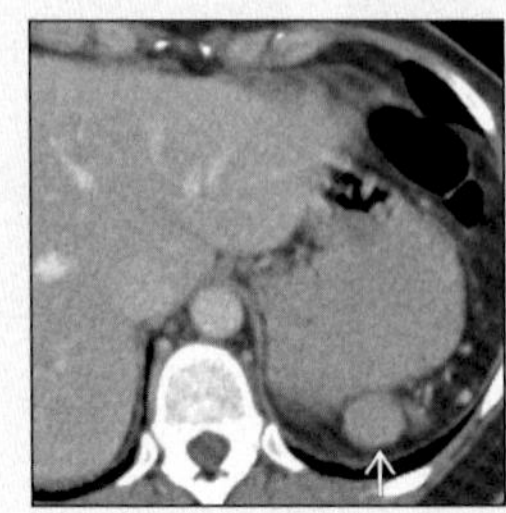

Splenosis

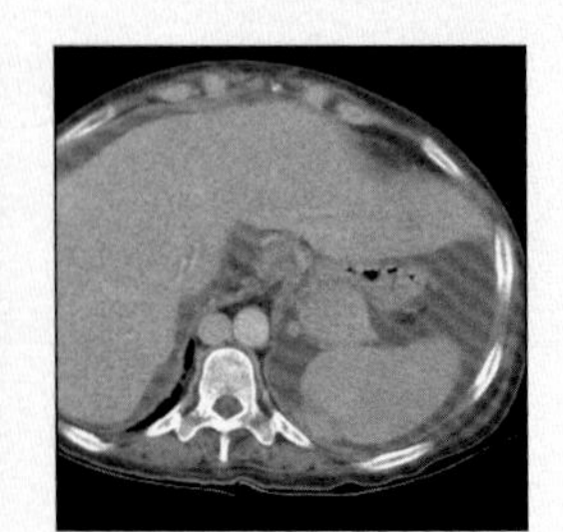

Polysplenia

GASTRIC DIVERTICULUM

Key Facts

Imaging Findings
- Best diagnostic clue: Barium-filled diverticulum from fundus, near gastroesophageal junction
- Best imaging tool: Fluoroscopic-guided barium studies

Top Differential Diagnoses
- Adrenal mass
- Abscess
- Pancreatic mass
- Splenic mass

Clinical Issues
- Asymptomatic (most common)

Diagnostic Checklist
- Often mistaken for adrenal mass on CT
- Barium studies or CT in supine/prone position

- Distinguished by barium studies

Abscess
- Air- or fluid-filled mass with a thick wall
- Distinguished by clinical history (i.e., fever)

Pancreatic mass
- Barium studies: Heaped-up area overlying prepyloric diverticulum

Splenic mass
- Splenosis, accessory spleen, polysplenia
- Distinguished by barium studies; isodensity to splenic tissue on all phases of CT contrast-enhancement

PATHOLOGY

General Features
- General path comments
  - 0.02% of autopsy specimens
  - 0.04% of upper gastrointestinal series
  - True gastric diverticula: 75% of gastric diverticula are juxtacardiac diverticula
  - Intramural or partial gastric diverticula: Rare
- Etiology
  - True gastric diverticula: Congenital
  - Intramural or partial gastric diverticula: Acquired
    - Associated with peptic ulcer disease, pancreatitis, cholecystitis, malignancy or outlet obstruction

Gross Pathologic & Surgical Features
- True gastric diverticula
  - Pouch/sac that includes 3 normal layers of bowel wall (i.e., mucosa, submucosa/muscularis propria)
- Intramural or partial gastric diverticula
  - Focal invagination of mucosa & submucosa into muscular layer of gastric wall; no muscular elements

CLINICAL ISSUES

Presentation
- Most common signs/symptoms
  - True gastric diverticula
    - Asymptomatic (most common)
    - Vague upper abdominal pain
  - Intramural or partial gastric diverticula
    - Asymptomatic
- Diagnosis
  - Fluoroscopic-guided barium studies

Demographics
- Age: Any age
- Gender: M:F = 1:1

Natural History & Prognosis
- Complications (rare): Bleeding, ulceration, carcinoma
- Prognosis: Very good

Treatment
- No treatment needed without complications
- If with complications, diverticulectomy or partial gastrectomy can be used to resect diverticulum

DIAGNOSTIC CHECKLIST

Consider
- Often mistaken for adrenal mass on CT

Image Interpretation Pearls
- Barium studies or CT in supine/prone position

SELECTED REFERENCES

1. Chasse E et al: Gastric diverticulum simulating a left adrenal tumor. Surgery. 133(4):447-8, 2003
2. Schwartz AN et al: Gastric diverticulum simulating an adrenal mass: CT appearance and embryogenesis. AJR Am J Roentgenol. 146(3):553-4, 1986
3. Dickinson RJ et al: Partial gastric diverticula: radiological and endoscopic features in six patients. Gut. 27(8):954-7, 1986

IMAGE GALLERY

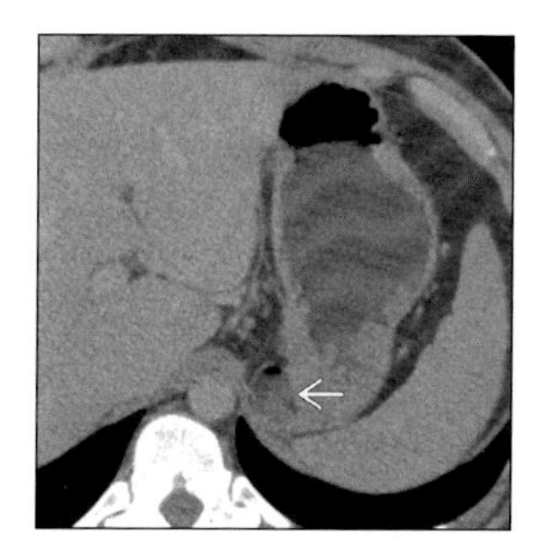

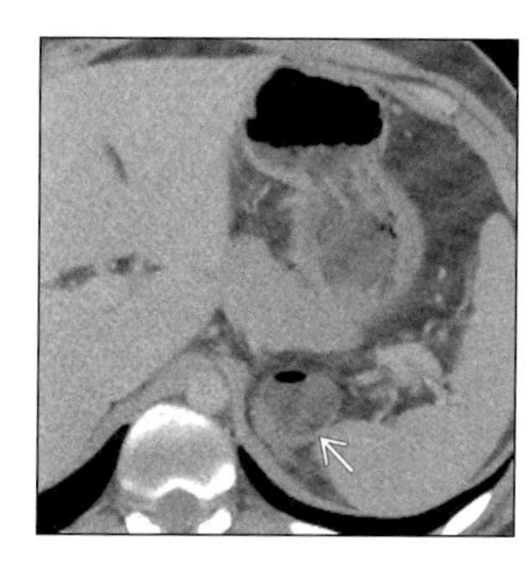

*(**Left**) Axial NECT shows gastric diverticulum (arrow) adjacent to fundus of stomach. (**Right**) Axial NECT shows gastric diverticulum (arrow) seemingly separated from stomach. Only air-fluid level allows recognition.*

DUODENAL DIVERTICULUM

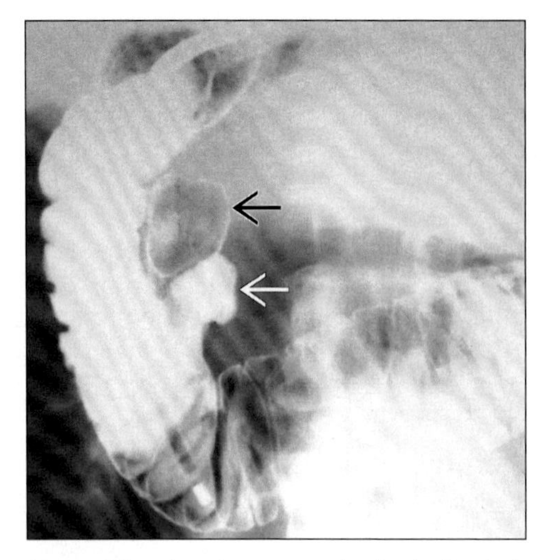

Upper GI series shows two duodenal diverticula (arrows) as rounded outpouchings from medial side of descending duodenum, one filled with air, one filled with barium.

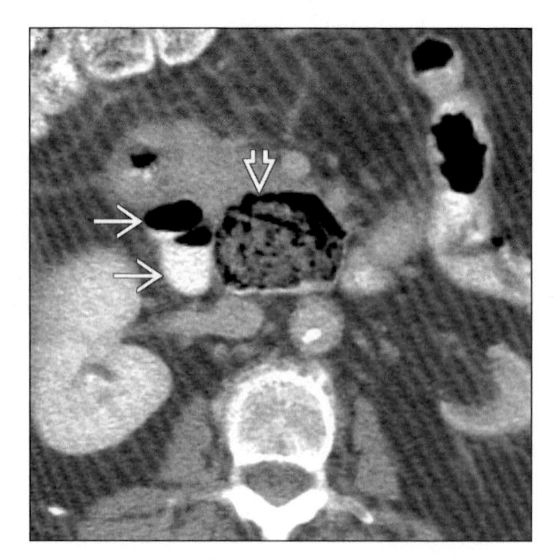

Axial CECT shows three duodenal diverticula, two with air-contrast levels (arrows) and one with food/particulate debris (open arrow).

TERMINOLOGY

Definitions

- A pouch or sac opening from the duodenum

IMAGING FINDINGS

General Features

- Best diagnostic clue: Smooth, rounded outpouching from medial descending duodenum
- Other general features
  - 3 types of duodenal diverticula
    - True diverticula
    - Pseudodiverticula
    - Intraluminal diverticula

Radiographic Findings

- Fluoroscopic-guided barium studies
  - True diverticula
    - Location: Medial (70%) descending duodenum in periampullary region, third or fourth portion (26%), lateral (4%) descending duodenum
    - Most (75%) are juxtapapillary diverticula: Diverticula within 2 cm of ampulla
    - Intradiverticular papilla: Papilla arises within
    - Multiple, smooth, rounded outpouchings
    - Diverticula change configuration during study
    - Air within the diverticulum
    - ± Filling defects (by food, blood clots or gas)
    - ± Bizarre, multiloculated or giant diverticula
  - Pseudodiverticula: Outpouching at base of bulb
  - Intraluminal diverticula
    - "Wind sock" appearance: Barium-filled, globular structure of variable length, originating in second portion of duodenum, fundus extending into third portion and filling defect in fourth portion; outlined by a thin, radiolucent line
    - "Halo" sign: Finger-like sac separated from contrast in adjacent duodenal lumen by a radiolucent band
    - Emptied of barium; mimics pedunculated polyps

CT Findings

- True diverticula: Air-fluid level within diverticulum
- Intraluminal diverticula
  - Contrast-opacified 2nd portion of duodenum
  - Air-filled fourth portion of duodenum
  - ± Debris within diverticulum; "halo" sign

Imaging Recommendations

- Best imaging tool: Fluoroscopic-guided barium studies

DIFFERENTIAL DIAGNOSIS

Pseudocyst in head of pancreas

- CT or US: Simulates fluid-filled duodenal diverticulum

DDx: Cystic Mass or Pseudomass Near Pancreatic Head

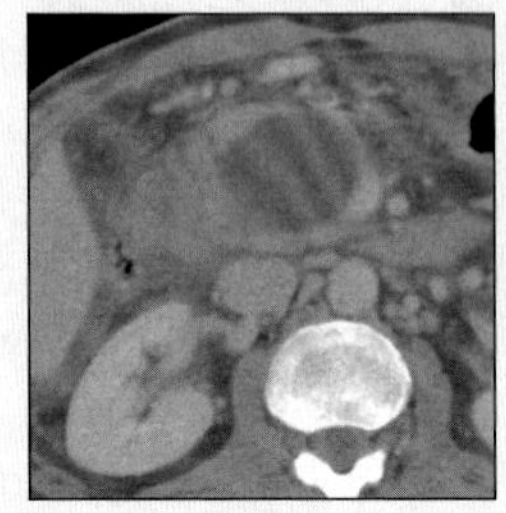

Pseudocyst

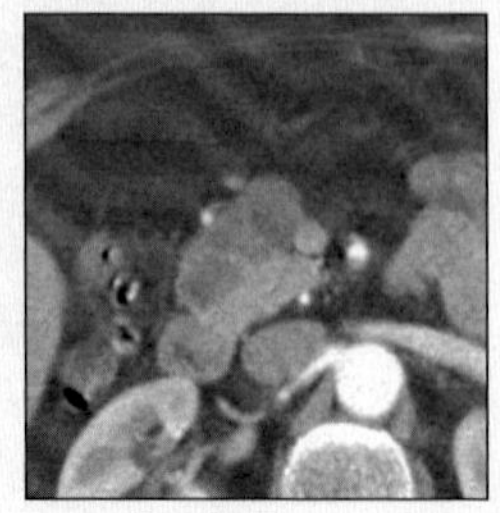

Cystic Tumor

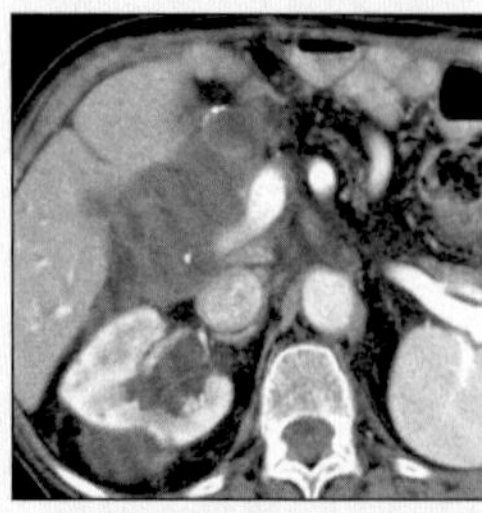

Mucinous Cyst

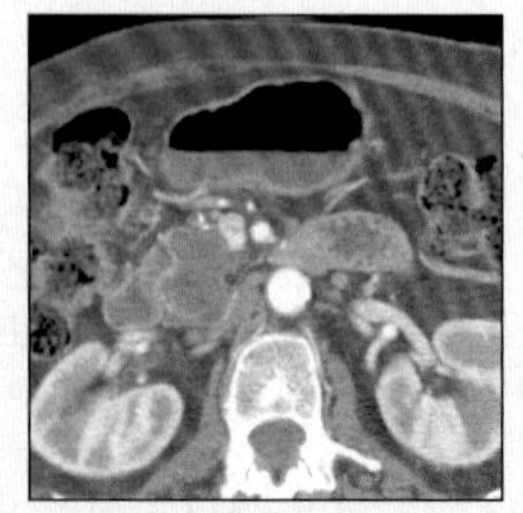

IPMT

DUODENAL DIVERTICULUM

Key Facts

Imaging Findings
- Best diagnostic clue: Smooth, rounded outpouching from medial descending duodenum
- Best imaging tool: Fluoroscopic-guided barium studies

Top Differential Diagnoses
- Pseudocyst in head of pancreas
- Pancreatic cystic tumor
- Perforated duodenal ulcer

Clinical Issues
- Asymptomatic (90%)

Diagnostic Checklist
- Periampullary diverticulum makes endoscopic sphincterotomy difficult or dangerous
- Use oral contrast and/or position changes to help identify

- Differentiate by history, signs of pancreatitis

Pancreatic cystic tumor
- Diverticulum may contain air or enteric contrast

Perforated duodenal ulcer
- Differentiate by inflammation of surrounding fat (gas and/or edema)

PATHOLOGY

General Features
- General path comments: 1-5% of upper GI series
- Etiology
  - True diverticula: Acquired
  - Pseudodiverticula: Acute or chronic duodenal ulcer
  - Intraluminal diverticula: Congenital duodenal diaphragm or web
  - Pathogenesis
    - True diverticula: Area of weakness where vessel penetrates the duodenal wall or where dorsal and ventral pancreas fuse in embryologic development
    - Intraluminal diverticula: Mechanical factors (i.e., forward pressure by food & peristalsis) → gradual elongation of duodenal web or diaphragm

Gross Pathologic & Surgical Features
- True diverticula: Sac of mucosal and submucosal layers herniated through a muscular defect
- Pseudodiverticula: Exaggerated outpouching of inferior & superior recesses of duodenal bulb
- Intraluminal diverticula
  - Sac of duodenal mucosa originating in the second portion of duodenum near papilla of Vater
  - Connected to part of or entire circumference of wall
  - Projecting distally as far as fourth part of duodenum; Often a second opening located eccentrically in sac

CLINICAL ISSUES

Presentation
- Most common signs/symptoms
  - True diverticula
    - Asymptomatic (90%)
  - Intraluminal diverticula: Nausea and vomiting
- Diagnosis: Fluoroscopic-guided barium studies

Demographics
- Age: True: 40-60 years of age; intraluminal: Any age

Natural History & Prognosis
- Complications: Diverticulitis, hemorrhage, outlet obstruction, perforation and pancreaticobiliary disease
- Prognosis: Very good

Treatment
- True diverticula: No treatment except in complications
- Intraluminal diverticula: Lateral duodenotomy and excision with preservation of ampulla of Vater

DIAGNOSTIC CHECKLIST

Consider
- Periampullary diverticulum makes endoscopic sphincterotomy difficult or dangerous

Image Interpretation Pearls
- Use oral contrast and/or position changes to help identify

SELECTED REFERENCES

1. Macari M et al: Duodenal diverticula mimicking cystic neoplasms of the pancreas: CT and MR imaging findings in seven patients. AJR Am J Roentgenol. 180(1):195-9, 2003
2. Lawler LP et al: Multidetector row computed tomography and volume rendering of an adult duodenal intraluminal "wind sock" diverticulum. J Comput Assist Tomogr. 27(4):619-21, 2003
3. Afridi SA et al: Review of duodenal diverticula. Am J Gastroenterol. 86(8):935-8, 1991

IMAGE GALLERY

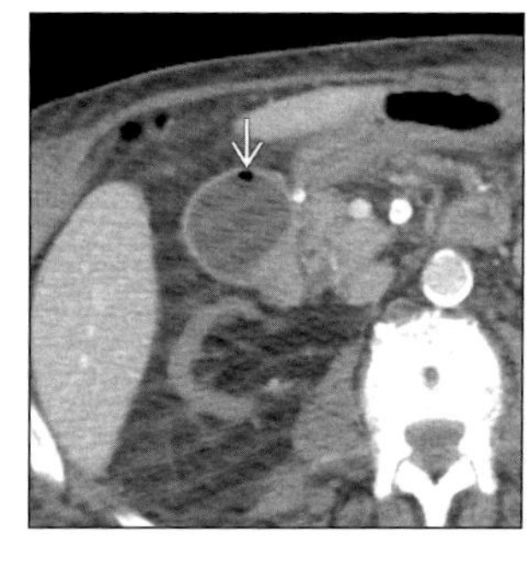

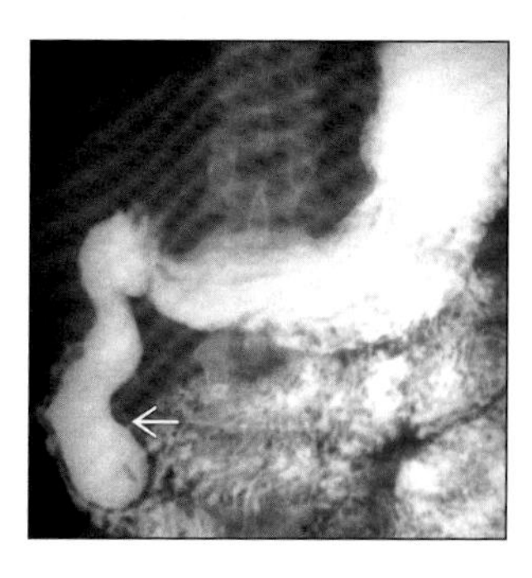

*(**Left**) Axial CECT shows unusual location of diverticulum, extending off lateral surface of duodenum (arrow). (**Right**) UGI series shows intraluminal diverticulum (arrow) having a "windsock" appearance, within the lumen of the duodenum.*

GASTRITIS

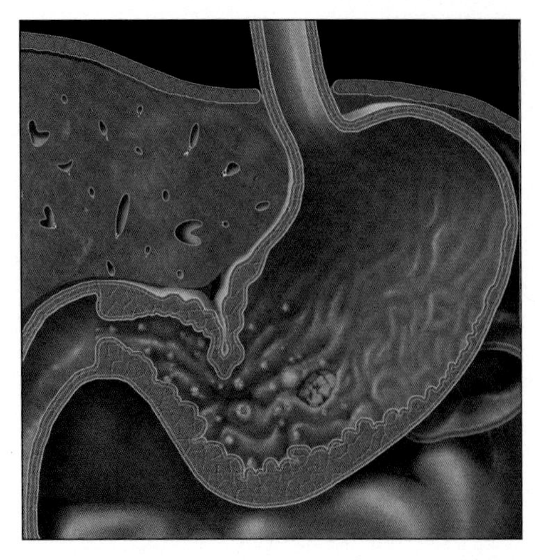

Graphic shows ulcer crater and numerous mucosal erosions, mostly in antrum along the "ridges" of hypertrophied folds.

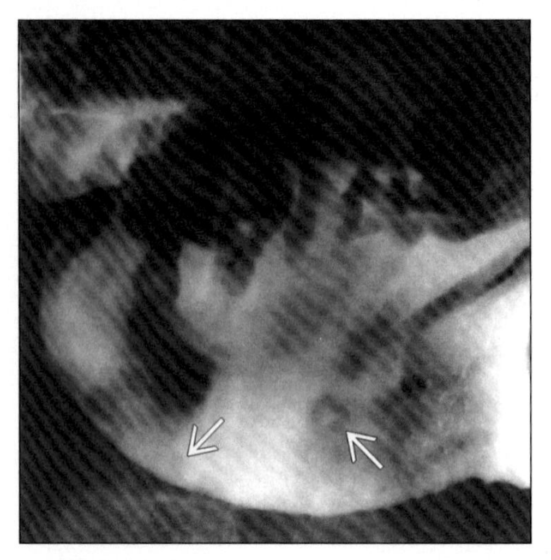

Upper GI series shows contracted antrum with thickened nodular folds and numerous varioliform erosions (arrows).

TERMINOLOGY

Definitions

- Inflammation of gastric mucosa induced by a group of disorders that differs in their etiological, clinical, histological and radiological findings

IMAGING FINDINGS

General Features

- Best diagnostic clue: Ulcers and thickened folds
- Other general features
  - Classification of gastritis
    - Erosive or hemorrhagic gastritis (2 types: Complete or varioliform and incomplete or "flat")
    - Antral gastritis
    - H. pylori gastritis
    - Hypertrophic gastritis
    - Atrophic gastritis (2 types: A and B)
    - Granulomatous gastritis (Crohn disease and tuberculosis)
    - Eosinophilic gastritis
    - Emphysematous gastritis
    - Caustic ingestion gastritis
    - Radiation gastritis
    - AIDS-related gastritis: Viral, fungal, protozoal and parasitic infections

Radiographic Findings

- Fluoroscopic-guided double contrast barium studies
  - Erosive gastritis, complete or varioliform erosions (most common type)
    - Location: Gastric antrum on crests of rugal folds
    - Multiple punctate or slit-like collections of barium
    - Erosions surrounded by radiolucent halos of edematous, elevated mucosa
    - Scalloped or nodular antral folds
    - Epithelial nodules or polyps (chronic)
  - Erosive gastritis, incomplete or "flat" erosions
    - Location: Antrum or body
    - Multiple linear streaks or dots of barium
    - No surrounding edematous mucosa
  - Erosive gastritis, NSAIDs induced
    - Linear or serpiginous erosions clustered in the body, on or near greater curvature
    - Varioliform or linear erosions in antrum
    - NSAIDs-related gastropathy: Subtle flattening and deformity of greater curvature of antrum
  - Antral gastritis
    - Thickened folds, spasm or decreased distensibility
    - Scalloped or lobulated folds oriented longitudinally or transverse folds
    - Crenulation or irregularity of lesser curvature
    - Hypertrophied antral-pyloric fold: Single lobulated fold on lesser curvature of distal antrum extends via pylorus to base of duodenal bulb (chronic)

DDx: Thickened Gastric Folds +/- Ulceration

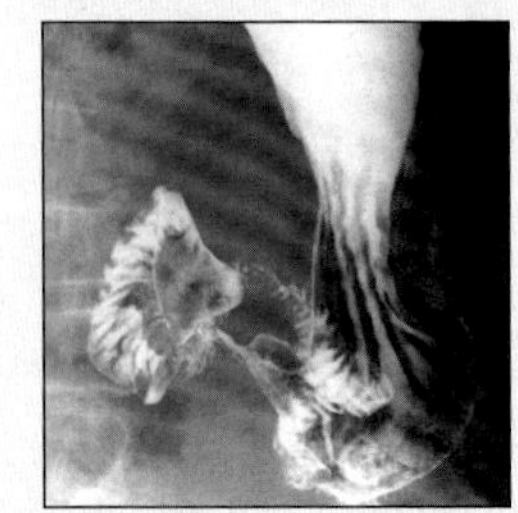

Antral Carcinoma

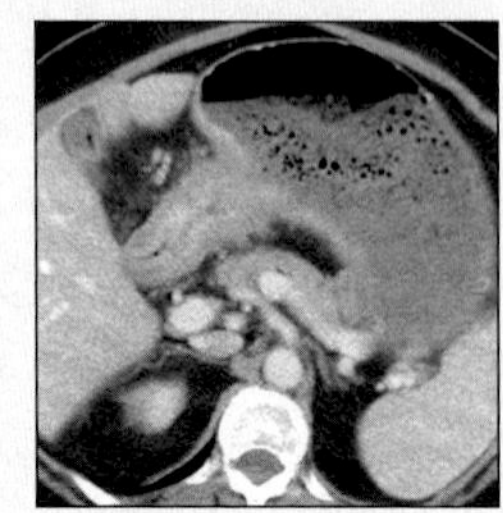

Antral Carcinoma

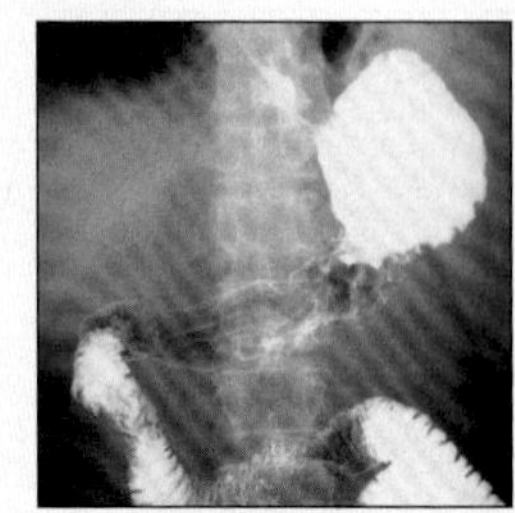

Pancreatitis

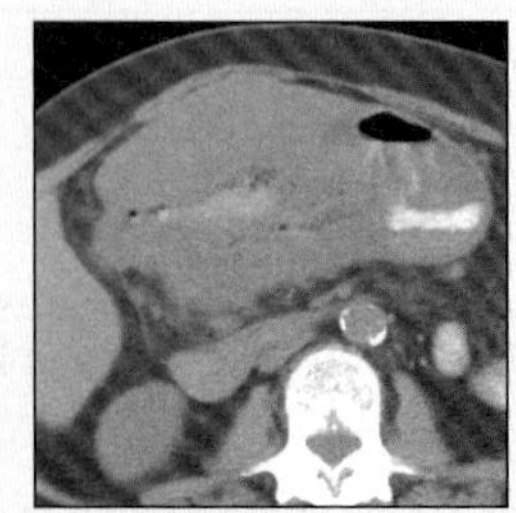

Lymphoma

GASTRITIS

Key Facts

Terminology

- Inflammation of gastric mucosa induced by a group of disorders that differs in their etiological, clinical, histological and radiological findings

Imaging Findings

- Best diagnostic clue: Ulcers and thickened folds
- Erosive gastritis, complete or varioliform erosions (most common type)
- Erosive gastritis, incomplete or "flat" erosions

Top Differential Diagnoses

- Gastric carcinoma
- Zollinger-Ellison syndrome
- Pancreatitis
- Gastric metastases and lymphoma

Pathology

- Erosive gastritis: Superficial acute inflammation or focal necrosis of mucosa
- H. pylori gastritis: Lymphoid nodules or increased neutrophils or plasma cells

Clinical Issues

- Asymptomatic
- Epigastric pain, nausea, vomiting or hematemesis

Diagnostic Checklist

- History and H. pylori infection
- H. pylori gastritis: Thickened, lobulated gastric folds with enlarged areae gastricae
- Erosive gastritis: Multiple collections of barium surrounded by radiolucent halos of edematous, elevated mucosa

- H. pylori gastritis
  - Location: Antrum, body or occasionally fundus; diffuse or localized
  - Thickened, lobulated gastric folds (polypoid gastritis)
  - Enlarged areae gastricae (≥ 3 mm in diameter)
- Hypertrophic gastritis
  - Location: Fundus and body
  - Markedly thickened, lobulated gastric folds
- Atrophic gastritis
  - Location of type A: Fundus and body
  - Location of type B: Antrum
  - Narrowed or tubular stomach
  - Smooth, featureless mucosa
  - Decreased distensibility
  - Decreased or absent mucosal folds
  - Small (1-2 mm in diameter) or absent areae gastricae
- Granulomatous gastritis, Crohn disease
  - Location: Antrum and body
  - Multiple aphthous ulcers
  - Indistinguishable from erosive gastritis, varioliform type
  - Advanced disease → large ulcers, thickened folds, nodular or cobblestone mucosa
  - "Ram's horn" sign: tubular, narrowed, funnel-shaped antrum
  - Single, continuous tubular structure involve antrum and duodenum; obliteration of normal anatomy
  - Severe disease → filiform polyps
- Granulomatous gastritis, tuberculosis
  - Location: Lesser curvature of antrum or pylorus
  - Antral narrowing → obstruction
- Eosinophilic gastritis
  - Location: Antrum and body
  - Mucosal nodularity, thickened folds, antral narrowing and rigidity
- Emphysematous gastritis (use water-soluble contrast)
  - Multiple streaks, bubbles, or mottled collections of gas in the wall of stomach, silhouetting the gastric shadow; do not alter with positional changes
- Caustic ingestion (use water-soluble contrast)
  - Location: Lesser curvature and distal antrum
  - Acute: Ulceration, thickened folds, gastric atony or mural defects
  - Chronic: Antral narrowing & deformity (scarring)
- Radiation gastritis
  - Acute: Ulceration, thickened folds, gastroparesis or spasm
  - Chronic: Antral narrowing & deformity (scarring)
- AIDS-related gastritis
  - Mucosal nodularity, erosions, ulcers, thickened folds or antral narrowing

CT Findings

- Decreased wall attenuation (edema or inflammation)
- Thickened gastric folds or wall
- Target or "halo": Mucosal enhancement and decreased HU of submucosa (edema)
- H. pylori gastritis: Circumferential antral wall thickening or focal thickening of posterior gastric wall along greater curvature
- Emphysematous gastritis: Thickened wall and collections of gas within the wall with or without gas in intrahepatic portal veins

Imaging Recommendations

- Best imaging tool: Fluoroscopic-guided double contrast barium or water-soluble contrast studies

DIFFERENTIAL DIAGNOSIS

Gastric carcinoma

- Differentiate from gastritis by loss of distensibility and decreased or absent peristalsis in involved portion
- Scirrhous carcinoma (linitis plastica)
  - Nodular, distorted mucosa
  - Thickened, irregular folds
  - Most important differential for atrophic gastritis

Zollinger-Ellison syndrome

- Thickened gastric folds in fundus and body (edema, inflammation, and hyperplasia)
- ↑ Fluids in lumen and ≥ 1 ulcers at unusual locations

GASTRITIS

Pancreatitis
- Common cause of gastric wall thickening; mimics thickened folds

Gastric metastases and lymphoma
- CT: Submucosal tumor is soft tissue density
- Metastases (e.g., malignant melanoma, breast cancer)
  - Simulate gastritis by thickened folds
  - Differentiate by loss of distensibility
- Gastric lymphoma
  - Variably sized, rounded, often confluent nodules
  - Mucosal nodularity is difficult to differentiate from enlarged areae gastricae
  - Markedly thickened and lobulated folds mimics antral, H. pylori and hypertrophic gastritis

PATHOLOGY

General Features
- Etiology
  - Erosive: NSAIDs, alcohol, steroids, stress, trauma, burns or infections
  - Atrophic: Fundus and body: Autoimmune; antral: H. pylori, bile or alcohol
  - Antral: Alcohol, tobacco, coffee or H. pylori
  - Granulomatous: Crohn disease, sarcoidosis, tuberculosis, syphilis or candidiasis
  - Emphysematous: E. coli, S. aureus, Clostridium perfringens or Proteus vulgaris
  - Caustic ingestion: Strong acids (hydrochloric, sulfuric, acetic, oxalic, carbolic or nitric acid)
  - Radiation: > 5,000 rads
  - AIDS-related: Cytomegalovirus, cryptosporidiosis, toxoplasmosis or strongyloidiasis
- Associated abnormalities
  - Atrophic gastritis: Underlying 90% of pernicious anemia patients
  - Hypertrophic gastritis: 66% of patients have duodenal ulcers

Gross Pathologic & Surgical Features
- Erosive gastritis: Areas of congested, edematous or ulcerated mucosa
- Atrophic gastritis: Thin smooth mucosa, flattened rugae or tubular stomach

Microscopic Features
- Erosive gastritis: Superficial acute inflammation or focal necrosis of mucosa
- H. pylori gastritis: Lymphoid nodules or increased neutrophils or plasma cells
- Atrophic gastritis: Thin mucosa, atrophy of mucosal glands, loss of parietal and chief cells or intestinal metaplasia

CLINICAL ISSUES

Presentation
- Most common signs/symptoms
  - Asymptomatic
  - Epigastric pain, nausea, vomiting or hematemesis
  - Atrophic gastritis: Neurologic symptoms from vitamin B12 deficiency
- Lab-Data
  - ↑ Leukocytes; Positive fecal occult blood test
  - Atrophic gastritis: ↓ Vitamin B12
  - Positive H. pylori (endoscopy, histology, cultures, urea breath and serologic tests)

Natural History & Prognosis
- Caustic ingestion gastritis: Acute necrotic phase (1-4 days); ulceration-granulation phase (5-28 days); cicatrization and scarring (3-4 weeks)
- Radiation gastritis: Inflammation (1-6 months); scarring and fibrosis (6 months)
- Complications
  - Gastric or duodenal ulcer, pernicious anemia, low grade MALT lymphoma or gastric carcinoma
  - Eosinophilic gastritis: Gastric outlet obstruction
  - Caustic ingestion gastritis: Gastric necrosis
- Prognosis
  - Erosive, antral, H. pylori and atrophic gastritis: Good after treatment
  - Eosinophilic gastritis: Chronic, relapsing disease with intermittent exacerbation and asymptomatic intervals
  - Emphysematous gastritis: 60-80% mortality

Treatment
- Stop offending agents: Alcohol, tobacco, NSAIDs, steroids and coffee
- H. pylori treatment: Metronidazole, bismuth and clarithromycin, amoxicillin or tetracycline
- Hypertrophic gastritis: Antisecretory agents (H2-receptor antagonists or proton-pump inhibitors)
- Atrophic gastritis: Replace vitamin B12
- Eosinophilic gastritis: Steroids
- Emphysematous gastritis: IV fluids, antibiotics, but no nasogastric tube
- Caustic ingestion gastritis: Steroids, antibiotics, parenteral feedings and ± surgery

DIAGNOSTIC CHECKLIST

Consider
- History and H. pylori infection

Image Interpretation Pearls
- H. pylori gastritis: Thickened, lobulated gastric folds with enlarged areae gastricae
- Erosive gastritis: Multiple collections of barium surrounded by radiolucent halos of edematous, elevated mucosa

SELECTED REFERENCES

1. Horton KM et al: Current role of CT in imaging of the stomach. Radiographics. 23(1):75-87, 2003
2. Bender GN et al: Double-contrast barium examination of the upper gastrointestinal tract with nonendoscopic biopsy: findings in 100 patients. Radiology. 202(2):355-9, 1997
3. Sohn J et al: Helicobacter pylori gastritis: radiographic findings. Radiology. 195(3):763-7, 1995

GASTRITIS

IMAGE GALLERY

Typical

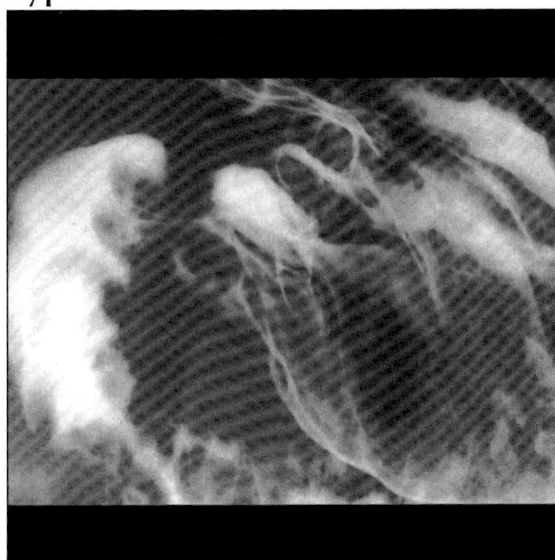

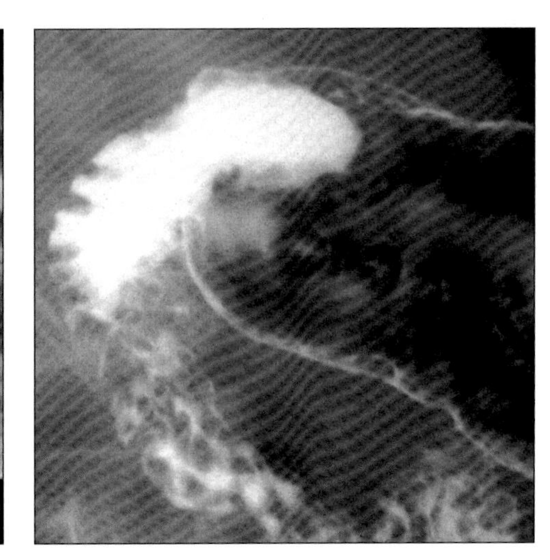

(Left) *Upper GI series shows contracted antrum with nodular thickened folds, some of which have prolapsed into the duodenum.* ***(Right)*** *Upper GI series shows rows of varioliform erosions along the top of hypertrophied gastric antral folds.*

Typical

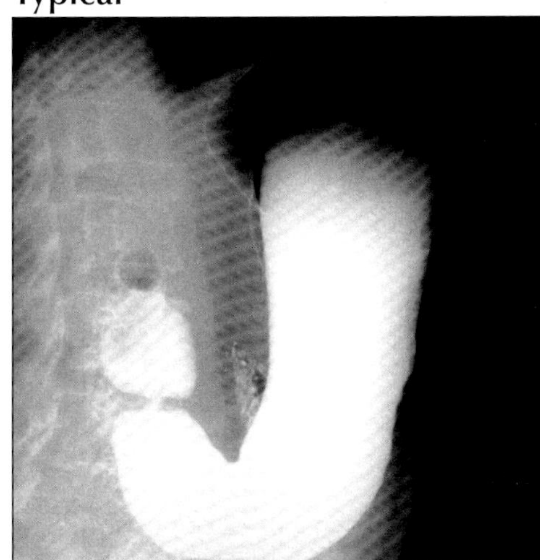

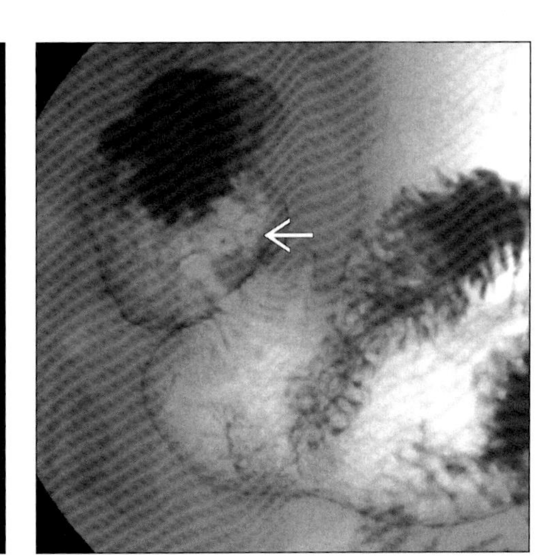

(Left) *Upper GI series shows almost complete absence of gastric folds in atrophic gastritis.* ***(Right)*** *Upper GI series shows numerous varioliform (aphthous) erosions in gastric antrum (arrow). (The duodenal bulb is collapsed and filled with barium).*

Typical

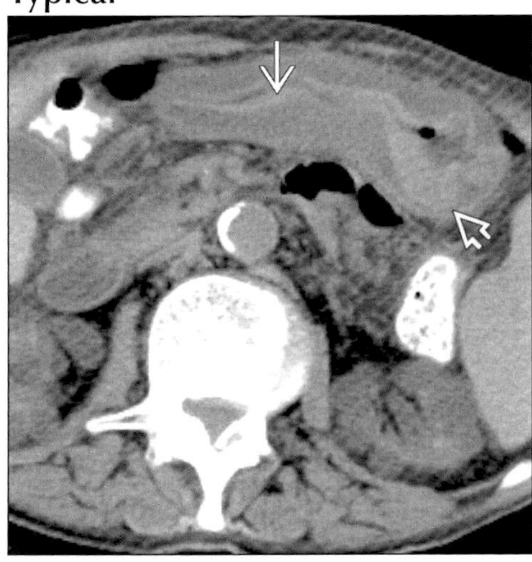

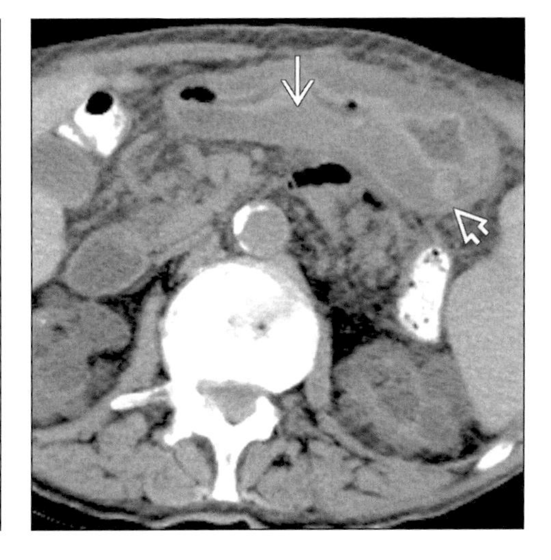

(Left) *Axial CECT in renal transplant recipient shows massive gastric wall thickening. The low density process (arrow) represents gastritis, while the soft tissue density (open arrow) is PTLD.* ***(Right)*** *Axial CECT in a renal transplant recipient shows gastritis (arrow) and PTLD (open arrow) (post-transplant lymphoproliferative disorder).*

GASTRIC ULCER

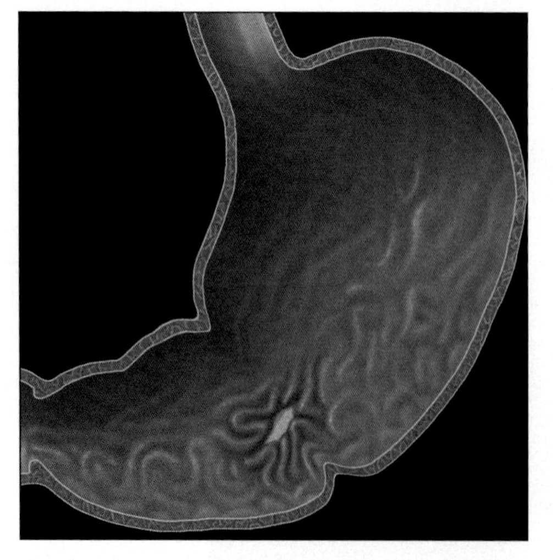

Graphic shows gastric ulcer with smooth gastric folds radiating to the edge of the ulcer crater. Also note infolding of the gastric wall "pointing" toward the ulcer.

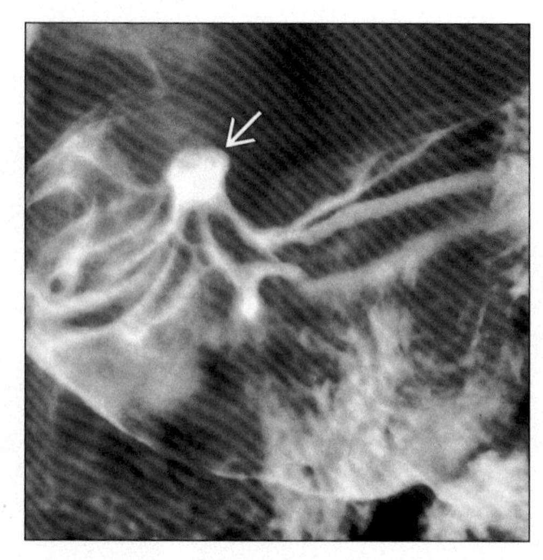

Upper GI series shows barium pool in ulcer crater (arrow), with smooth folds radiating to the edge of the ulcer.

TERMINOLOGY

Abbreviations and Synonyms

- Peptic ulcer disease

Definitions

- Mucosal lesion of stomach

IMAGING FINDINGS

General Features

- Best diagnostic clue: Sharply marginated barium collection and folds radiating to edge of ulcer crater on fluoroscopic-guided double contrast barium studies
- Location
  - Benign gastric ulcer
    - Usually lesser curvature or posterior wall of antrum or body
    - 3-11% on greater curvature; 1-7% on anterior wall
  - o Malignant gastric ulcer
    - Usually greater curvature
- Size
  - > 0.5 cm to be visualized
  - Most diagnosed ulcers are < 1 cm
  - Larger ulcers tend to be more proximal in stomach
  - Giant (> 3 cm) ulcers are mostly benign, but increased risk of complications
- Morphology
  - Lesser or greater curvature
    - Profile view: Can see size, shape, depth, Hampton line, ulcer collar, ulcer mound or radiating folds
  - Anterior or posterior wall
    - En face view: Best for radiating folds

Radiographic Findings

- Fluoroscopic-guided double contrast barium studies
  - Benign gastric ulcer - profile view
    - Ulcer crater: Round or ovoid collections of barium
    - Hampton line: Thin radiolucent line separating barium in gastric lumen from barium in crater
    - Ulcer mound: Smooth, bilobed hemispheric mass projecting into lumen on both sides of ulcer; outer borders form obtuse, gently-sloping angles with adjacent gastric wall (edema or inflammation)
    - Ulcer collar: Radiolucent rim of edematous mucosa around ulcer
    - Smooth, round ulcer projecting beyond lesser curvature
    - Smooth, symmetric radiating folds to edge of ulcer crater
    - Incisura defect: Smooth or narrow indentation on greater curvature opposite an ulcer on lesser curvature (muscle contraction)
    - Enlarged areae gastricae in adjacent mucosa (edema or inflammation)

DDx: Persistant Collection of Contrast with Mucosal Ulceration

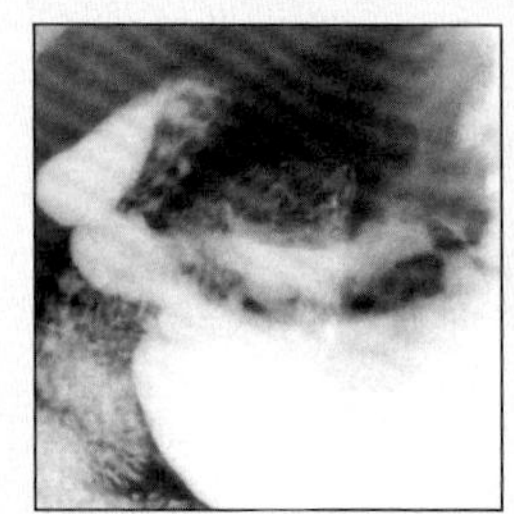

Gastric Cancer

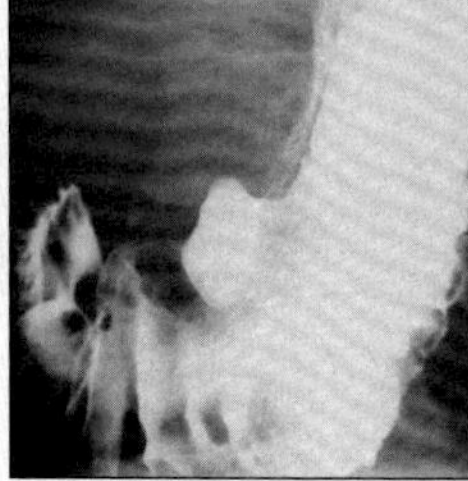

Gastric Lymphoma

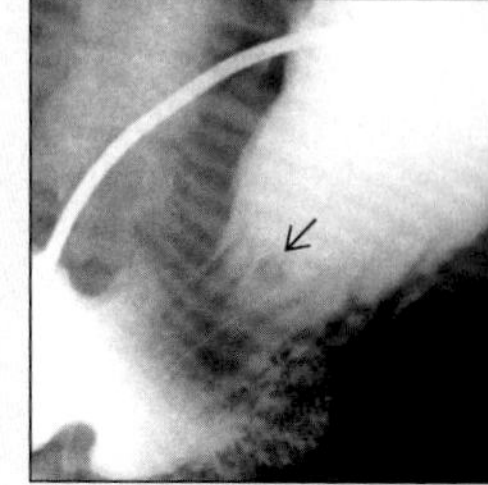

Melanoma Met.

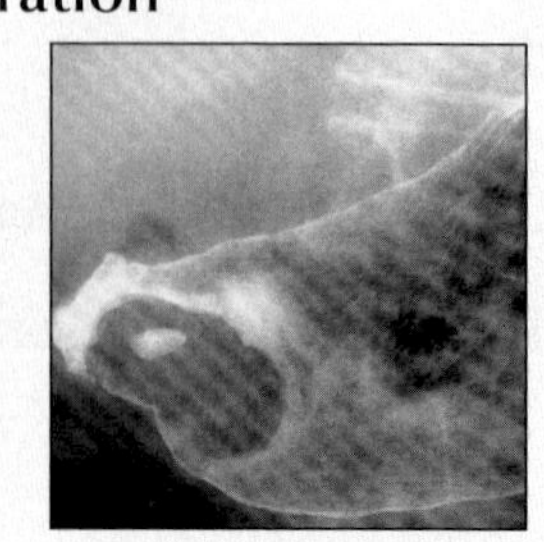

Leiomyoma

GASTRIC ULCER

Key Facts

Terminology

- Mucosal lesion of stomach

Imaging Findings

- Best diagnostic clue: Sharply marginated barium collection and folds radiating to edge of ulcer crater on fluoroscopic-guided double contrast barium studies
- Ulcer crater: Round or ovoid collections of barium
- Hampton line: Thin radiolucent line separating barium in gastric lumen from barium in crater
- Ulcer mound: Smooth, bilobed hemispheric mass projecting into lumen on both sides of ulcer; outer borders form obtuse, gently-sloping angles with adjacent gastric wall (edema or inflammation)
- Ulcer collar: Radiolucent rim of edematous mucosa around ulcer
- Smooth, symmetric radiating folds to edge of ulcer crater

Top Differential Diagnoses

- Gastritis
- Gastric metastases and lymphoma

Clinical Issues

- Burning, gnawing, or aching pain at the epigastrium
- < 2 hrs after meals; not relieved by food or antacids

Diagnostic Checklist

- Rule out malignant gastric ulcers
- Malignant gastric ulcers: "Carman meniscus" sign; nodular, blunted folds

    - Sump ulcers: Distal one half of greater curvature (NSAIDs)
    - Ulcer on greater curvature → area of mass effect and thickened irregular folds
    - Linear barium-coated ulcer: ↓ In depth (healing)
    - Splitting of 1 ulcer to 2 smaller ulcers (healing)
    - Central pit or depression, radiating folds, or retraction of adjacent gastric wall (scarring)
    - "Hourglass" stomach: Marked narrowing of body (scarring)
  - Benign gastric ulcer - en face view
    - Ring shadow: Shallow ulcer on anterior or posterior wall (barium coated rim and unfilled crater)
  - Malignant gastric ulcer - profile
    - "Carman meniscus" sign: Ulcer crater and radiolucent elevated border
    - Does not project beyond expected gastric contour
    - Discrete tumor mass forms acute angles
  - Malignant gastric ulcer - en face
    - Irregular crater eccentrically located within a tumor mass
    - Focal nodularity, distortion or obliteration of adjacent areae gastricae (tumor infiltration)
    - Nodular, clubbed, fused, or amputated folds (tumor infiltration)

CT Findings

- CECT (use water or water-soluble oral contrast)
  - Signs of complications
    - Wall thickening or luminal narrowing of stomach
    - Infiltration of surrounding fat or organs (pancreas)
    - Free air in abdomen or lesser sac

Imaging Recommendations

- Best imaging tool: Fluoroscopic-guided double-contrast barium studies (en face and profile views)
- Protocol advice: In addition to standard air-contrast views, prone compression views of gastric antrum and body or prone Trendelenburg position should be used to demonstrate anterior wall ulcers

DIFFERENTIAL DIAGNOSIS

Gastritis

- Markedly thickened gastric folds
- Helicobacter pylori (H. pylori) gastritis
  - Thickened gastric folds in antrum or body
  - Enlarged areae gastricae (≥ 3 mm)
- Hypertrophic gastritis
  - Glandular hyperplasia and ↑ secretion of acid
  - Thickened gastric folds in fundus and body

Ulcerated intramural primary tumor

- Leiomyoma or gastrointestinal stromal tumor (GIST)
  - Millimeters to enormous masses
  - In profile view, discrete submucosal mass with smooth surface that is etched in white and borders form right angles or slightly obtuse angles with adjacent gastric wall
  - Larger than 2 cm → ulcerated, central barium-filled crater within a smooth or slightly lobulated submucosal mass ("bull's eye" or "target" lesions)

Gastric metastases and lymphoma

- Malignant melanoma
  - Most common hematogenous metastasis to stomach
  - Necrotic; development of giant, cavitated lesion
  - Amorphous collection of barium (5-15 cm in size) that communicates with lumen
  - "Bull's eye" or "target" lesions also occur
- Metastatic Kaposi sarcoma
  - Increased incidence in homosexual AIDS patients
  - GI involvement in 50%
  - Almost always associated with cutaneous lesions
  - Rarely symptomatic
  - Elevated lesions; submucosal defects (0.5-3.0 cm)
  - As nodules enlarge, often ulcerate, producing 1 or more "bull's eye" or "target" lesions
- Gastric Lymphoma
  - More frequent than other parts of GI tract
  - 50% confined to stomach
  - Majority are non-Hodgkin lymphoma (B-cell origin)

GASTRIC ULCER

- Evidence suggests chronic H. pylori gastritis may lead to low grade mucosa-associated lymphoid tissue (MALT) lymphomas
- Low grade MALT lymphoma appearance is variably sized, rounded, often confluent nodules
- Mucosal nodularity is difficult to differentiate from enlarged areae gastricae

Artifactual

- Barium precipitates
  - Resemble tiny ulcers; differentiated by lack of projection beyond wall
  - Absence of mucosal edema or radiating fold
- "Stalactites"
  - Hanging droplets of barium (on anterior gastric wall)
  - Differentiated by transient nature on double-contrast barium studies

PATHOLOGY

General Features

- General path comments
  - One of two forms of peptic ulcer disease
  - Unequivocal benign gastric ulcers on double contrast studies: No further testing
  - Equivocal gastric ulcers (mixed features of benign and malignant)
    - Endoscopy and biopsy to exclude malignancy
    - If endoscopy and biopsy are negative, follow-up with double-contrast studies until complete healing or repeat endoscopy and biopsy
  - Multiplicity
    - 80% benign; most likely cause is NSAIDs
- Genetics
  - Genetic syndromes
    - Multiple endocrine neoplasia type 1 (MEN I)
    - Systemic mastocytosis
  - Greater concordance in monozygotic twins
  - Increased incidence with blood type O
- Etiology
  - 2 major risk factors: H. pylori (60-80%) and NSAIDs
  - Other risk factors: Steroids, tobacco, alcohol, coffee, stress, reflux of bile, delayed gastric emptying
  - Less common etiologies
    - Zollinger-Ellison syndrome
    - Hyperparathyroidism
    - Cushing ulcer: Head injuries (stress)
    - Curling ulcer: Burns
    - Gastritis
  - Pathogenesis
    - Normal or decreased levels of gastric acid
    - Breakdown in mucosal defense by H. pylori
- Epidemiology
  - 95% benign, 5% malignant
  - Multiplicity: 20-30% prevalence

Gross Pathologic & Surgical Features

- Round or oval; sharply punched-out and regular walls; flat adjacent mucosa

Microscopic Features

- Necrotic debris; zone of active inflammation; granulation and scar tissue

CLINICAL ISSUES

Presentation

- Most common signs/symptoms
  - Asymptomatic
  - Burning, gnawing, or aching pain at the epigastrium
  - < 2 hrs after meals; not relieved by food or antacids
  - Pain that awakens patients from sleep (33%)
  - Anorexia and weight loss (50%)
- Diagnosis: Endoscopy with biopsy

Demographics

- Age: > 40 years of age
- Gender: Equal in both males and females

Natural History & Prognosis

- Complications
  - Hemorrhage, perforation, obstruction and fistula
- Prognosis
  - Good with medical treatment and surgery

Treatment

- Ulcer without H. pylori: H2-receptor antagonists (cimetidine, ranitidine, or famotidine) or proton-pump inhibitors (omeprazole or lansoprazole)
- H. pylori treatment: Metronidazole, bismuth and clarithromycin, amoxicillin or tetracycline
- Ulcer with H. pylori: H. pylori treatment and H2-receptor antagonists or proton-pump inhibitors
- NSAID-induced: Misoprostol and stop NSAIDs
- Other agent: Sucralfate
- Surgery required for
  - Recurrent or intractable ulcers
  - Ulcer complications
  - Equivocal or suspicious findings on radiologic or endoscopic examinations
- Follow-up: 6-8 weeks after medical treatment
  - If not healed, suggests malignant gastric ulcer

DIAGNOSTIC CHECKLIST

Consider

- Rule out malignant gastric ulcers

Image Interpretation Pearls

- Benign gastric ulcers: Ulcer crater, Hampton line, ulcer mound and collar; smooth, radiating folds
- Malignant gastric ulcers: "Carman meniscus" sign; nodular, blunted folds

SELECTED REFERENCES

1. Horton KM et al: Current role of CT in imaging of the stomach. Radiographics. 23(1):75-87, 2003
2. Pattison CP et al: Helicobacter pylori and peptic ulcer disease: Evolution to revolution to resolution. AJR 168: 1415-20, 1997
3. Fishman EK et al: CT of the stomach: spectrum of disease. Radiographics. 16(5):1035-54, 1996
4. Levine MS et al: The Helicobacter pylori revolution: Radiologic perspective. Radiology 195: 593-6, 1995
5. Jacobs JM: Peptic ulcer disease: CT evaluation. Radiology 178: 745-8, 1991

GASTRIC ULCER

IMAGE GALLERY

Typical

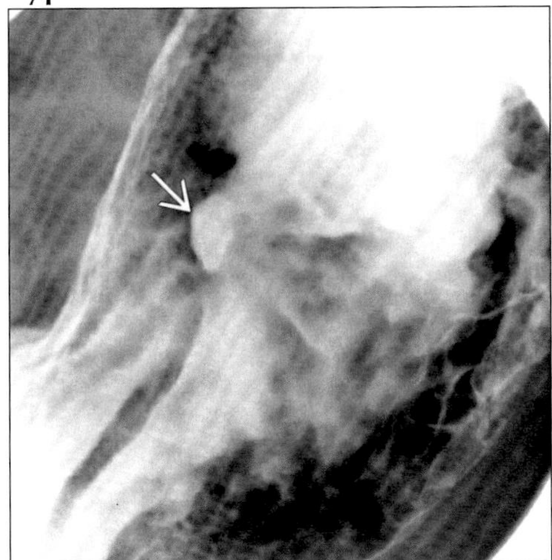

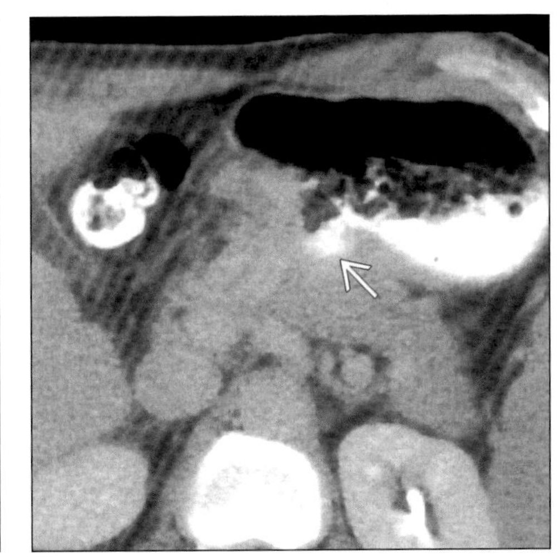

(Left) *Upper GI series shows ulcer crater (arrow) with radiating folds to the edge of the crater.* ***(Right)*** *Axial CECT shows large posterior wall gastric ulcer (arrow). Gastric folds are thickened.*

Typical

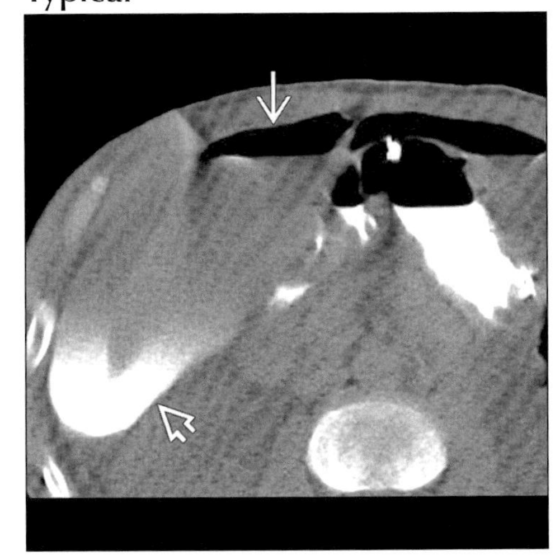

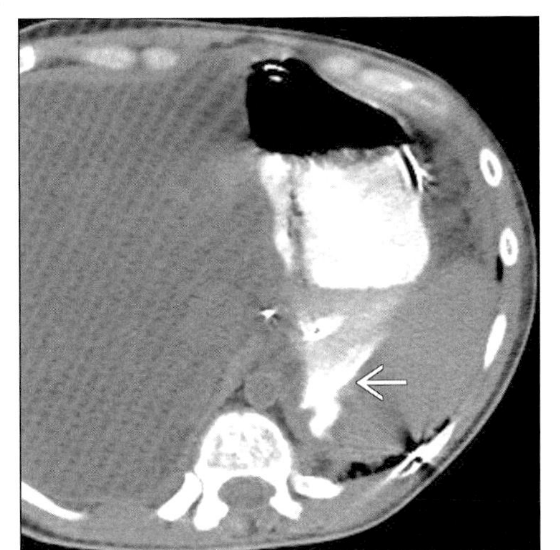

(Left) *Axial CECT shows perforated gastric antral ulcer resulting in intraperitoneal air (arrow), fluid and enteric contrast medium (open arrow).* ***(Right)*** *Axial CECT shows perforated posterior gastric wall ulcer with enteric (oral) contrast medium in lesser sac (arrow).*

Typical

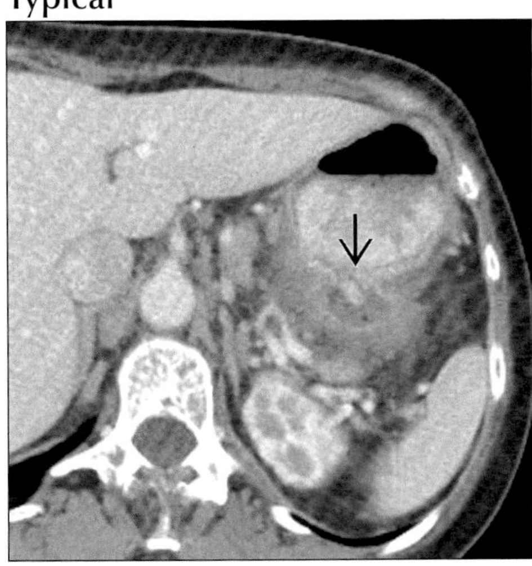

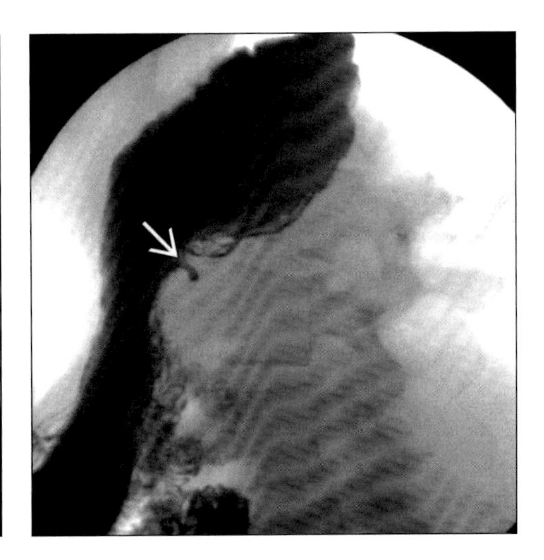

(Left) *Axial CECT shows posterior gastric wall ulcer with loculated fluid and gas in lesser sac (arrow).* ***(Right)*** *Lateral view of upper GI series shows deep posterior wall gastric ulcer (arrow) and a "mound" of edematous folds.*

DUODENAL ULCER

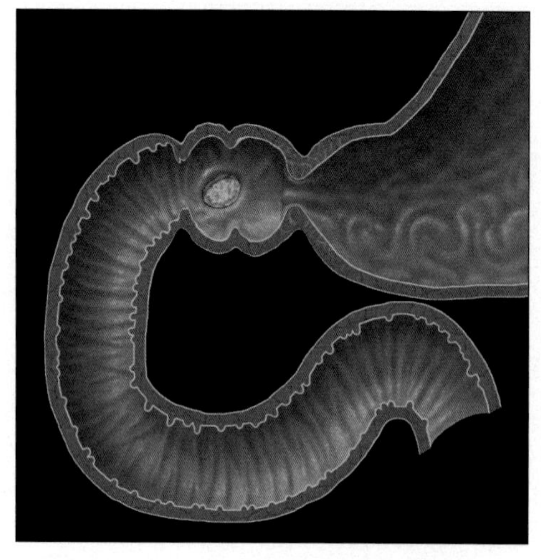

Graphic shows duodenal ulcer with deformed bulb due to converging folds and spasm.

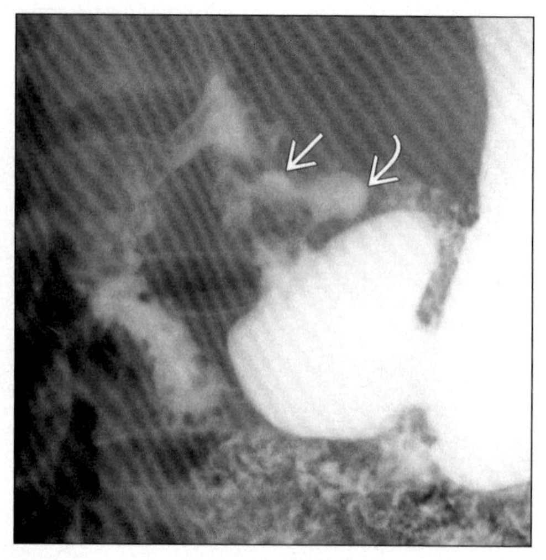

Upper GI series shows ulcer crater (arrow) and deformed bulb, including a pseudodiverticulum (curved arrow).

TERMINOLOGY

Abbreviations and Synonyms

- Peptic ulcer disease

Definitions

- Mucosal erosion of duodenum

IMAGING FINDINGS

General Features

- Best diagnostic clue: Sharply marginated barium collection with folds radiating to edge of ulcer crater on fluoroscopic-guided double-contrast barium study
- Location
  - 95% duodenal bulbar ulcers; 5% postbulbar ulcers
  - Bulbar ulcers are located at the apex, central portion, or base of the bulb
  - Postbulbar ulcers are located on the medial wall of the proximal descending duodenum above papilla of Vater
  - 50% of duodenal ulcers are located on anterior wall
- Size: Most ulcers are < 1 cm at time of diagnosis
- Morphology
  - Round or ovoid collections of barium
  - 5% of duodenal ulcers have linear configuration

Radiographic Findings

- Fluoroscopic-guided double-contrast barium studies
  - Bulbar ulcers
    - Persistent small round, ovoid or linear ulcer niche (collection of barium)
    - Ulcer mound: Smooth, radiolucent mound of edematous mucosa
    - Radiating folds converge centrally at the edge of the ulcer crater
    - Ring shadow: Barium coating rim of unfilled anterior wall ulcer crater (air contrast view)
    - Deformity of bulb (edema and spasm or scarring)
    - Residual depression of central portion of scar mimics active ulcer crater
    - Pseudodiverticula: Ballooning out between areas of fibrosis and spasm
    - "Cloverleaf" deformity: Multiple pseudodiverticula
  - Postbulbar ulcers
    - Smooth or rounded indentation on lateral wall opposite of ulcer crater (edema and spasm)
    - "Ring stricture": Eccentric narrowing (scarring)
  - Giant duodenal ulcers (> 2 cm)
    - Always located in duodenal bulb
    - Replace virtually entire bulb; mistaken for a scarred or normal bulb
    - Fixed or unchanging configuration is key clue
    - Focal narrowing → outlet obstruction (edema and spasm)

DDx: Duodenal Fixed Deformity +/- Contrast Collection

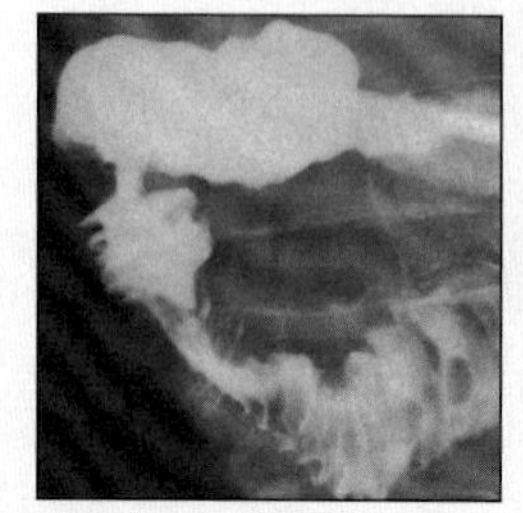

Radiation Duodenitis

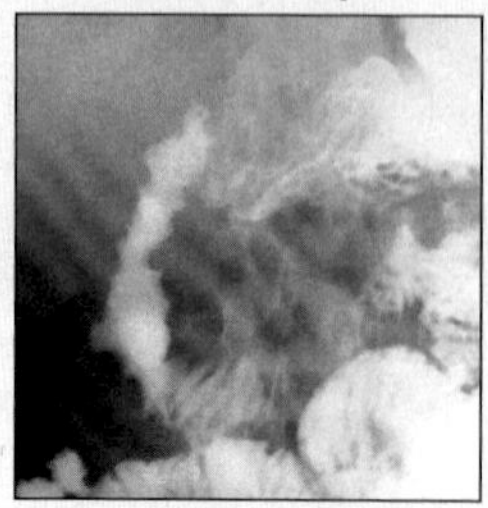

Crohn Duodenitis

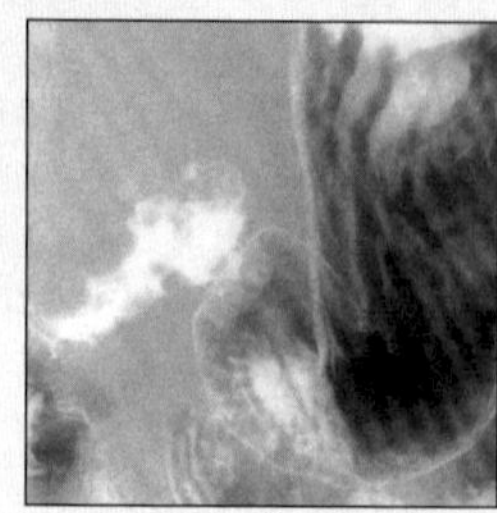

Duodenal Cancer

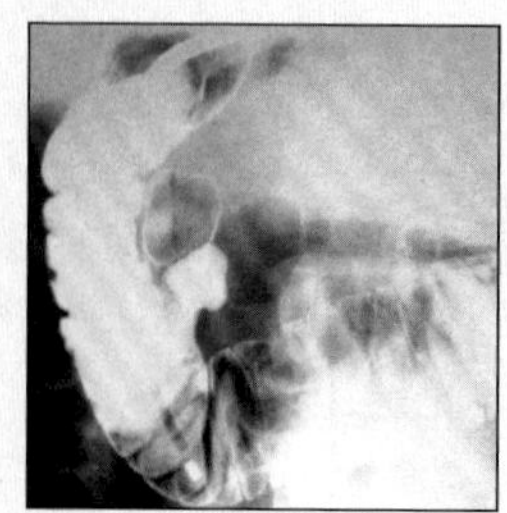

Duodenal Diverticula

DUODENAL ULCER

Key Facts

Terminology
- Mucosal erosion of duodenum

Imaging Findings
- Best diagnostic clue: Sharply marginated barium collection with folds radiating to edge of ulcer crater on fluoroscopic-guided double contrast barium study
- 95% duodenal bulbar ulcers; 5% postbulbar ulcers
- Persistent small round, ovoid or linear ulcer niche (collection of barium)
- Ulcer mound: Smooth, radiolucent mound of edematous mucosa
- Radiating folds converge centrally at the edge of the ulcer crater
- Ring shadow: Barium coating rim of unfilled anterior wall ulcer crater (air contrast view)

Top Differential Diagnoses
- Duodenal inflammation
- Duodenal stricture
- Duodenal carcinoma

Pathology
- 2-3 times more frequent than gastric ulcers

Clinical Issues
- Burning, gnawing, or aching pain at the epigastrium
- 2-4 hrs after meals; relieved by antacids or food

Diagnostic Checklist
- Eradication of H. pylori is the first step of treatment
- Check for deformity of the duodenal bulb
- Prone compression views are necessary to evaluate anterior wall duodenal ulcers

CT Findings
- CECT (use water or water-soluble oral contrast)
  - Signs of ulcer penetration and perforation
    - Wall thickening or luminal narrowing of duodenum
    - Infiltration of surrounding fat/organs (pancreas)
    - Extraluminal gas intra- or retro-peritoneal

Imaging Recommendations
- Best imaging tool
  - Fluoroscopic guided double contrast barium studies
  - High density barium for views of duodenal bulb
  - Low density barium for upright or prone compression views
- Protocol advice: Routinely obtain prone compression views of duodenum to observe anterior wall ulcers

DIFFERENTIAL DIAGNOSIS

Duodenal inflammation
- Duodenitis
  - Inflammation without frank ulceration
- Crohn disease
  - Usually with antral involvement
  - Aphthous ulcers are the earliest abnormality observed
  - Thickened, nodular folds; cobblestone appearance
  - Asymmetric duodenal narrowing with outward ballooning of duodenal wall between area of fibrosis
  - Smooth, tapered areas of narrowing; extend from apical portion of the bulb to descending duodenum
  - One or more strictures in second or third portions of duodenum → marked obstruction and proximal dilatation (megaduodenum)
- Tuberculosis
  - Usually with antral involvement
  - Ulcers, thickened folds, narrowing or fistula
  - Enlarged lymph nodes adjacent to duodenum → narrowing or obstruction of lumen

Duodenal stricture
- Pancreatitis
  - "Inverted 3" sign of Frostberg
    - Central limb of the 3: Point of fixation where pancreatic and common bile ducts insert into the papilla
    - Above and below the point reflect edema of major and minor papilla or smooth muscle spasm and edema in the duodenal wall
  - Thickened folds associated with medial compression or widening of duodenal sweep
  - Spiculation of mucosal folds (edema or inflammation)
- Gallstone erosion
  - Radiolucent filling defect in duodenum
  - Cause mucosal inflammation, ulceration, hemorrhage, perforation and obstruction
  - Barium reflux into gallbladder and bile ducts

Duodenal carcinoma
- < 1% of all gastrointestinal cancers
- Located in the postbulbar portion at or distal to papilla of Vater
- Polypoid, ulcerated, or annular lesions
- Narrowed lumen with thickened wall

Duodenal diverticulum
- 1-5% as incidental findings in barium studies
- Mostly located on medial border of descending duodenum in periampullary region
- Smooth, rounded outpouching from the medial border of the descending duodenum
- Multiple diverticula observed; configuration may change during course of study
- Differentiate from postbulbar ulcers by lack of inflammatory reaction and change in shape

Extrinsic invasion
- Pancreatic carcinoma
  - Widening of duodenal sweep
  - Mass effect → double contour effect on medial border of duodenum: Differential filling with interfold spaces along the inner aspect containing less barium than corresponding spaces along the outer aspect

- Displacement or frank splaying of the spikes: Tumor infiltrating the duodenal wall with traction and fixation of the folds
- Gallbladder carcinoma
  - Compression of bulb or proximal duodenum
- Metastases
  - Widening of duodenal sweep
  - Multiple submucosal masses or "bull's eye" lesion

Duodenal hematoma

- Radiolucent filling defects from blood clots
- Well-circumscribed intramural masses with discrete margins → stenosis and obstruction
- Diffuse hemorrhage → thickened, spiculated folds or thumbprinting

PATHOLOGY

General Features

- General path comments
  - Multiplicity
    - 15% of patients with duodenal ulcers
    - Ulcers located in duodenal bulb and beyond
    - Suspicious of Zollinger-Ellison syndrome
- Genetics
  - Genetic syndromes
    - Multiple endocrine neoplasia type 1 (MEN 1)
    - Systemic mastocytosis
  - Greater concordance in monozygotic twins
  - Increased incidence with blood type O
- Etiology
  - Two major risk factors: Helicobacter pylori (H. pylori) (95-100%) & NSAIDs
  - Other risk factors: Steroids, tobacco, alcohol, coffee, stress, reflux of bile, delayed gastric emptying
  - Less common etiologies
    - Zollinger-Ellison syndrome
    - Hyperparathyroidism
    - Chronic renal failure
    - Chronic obstructive pulmonary disease
  - Pathogenesis
    - H. pylori mediates or facilitates damage to gastric and duodenal mucosa
    - ↑ Gastric acid and ↑ gastric emptying → ↑ acidic exposure in the duodenum
- Epidemiology
  - Incidence: 200,000 cases per year
  - 2-3 times more frequent than gastric ulcers

Gross Pathologic & Surgical Features

- Round or oval; sharply punched-out and regular walls; flat adjacent mucosa

Microscopic Features

- Necrotic debris; zone of active inflammation; granulation and scar tissue

CLINICAL ISSUES

Presentation

- Most common signs/symptoms
  - Asymptomatic
  - Burning, gnawing, or aching pain at the epigastrium
  - 2-4 hrs after meals; relieved by antacids or food
  - Pain that awakens patients from sleep (66%)
  - Other signs/symptoms
    - Pain episodes that occur in clusters of days to weeks followed by longer pain-free intervals
    - Rarely anorexia and weight loss; hyperphagia and weight gain because eating relieves pain
- Lab-Data
  - Diagnostic tests (serology or urease breath test) for H. pylori
- Diagnosis: Endoscopy

Demographics

- Age: Adults of all ages
- Gender: Equal in both males and females

Natural History & Prognosis

- Complications:
  - Hemorrhage, perforation, obstruction and fistula
  - Giant duodenal ulcers have ↑ risks of complications
- Prognosis
  - Good with medical treatment and surgery

Treatment

- Ulcer without H. pylori: H2-receptor antagonists (cimetidine, ranitidine, or famotidine) or proton-pump inhibitors (omeprazole or lansoprazole)
- H. pylori treatment: Metronidazole, bismuth and clarithromycin, amoxicillin or tetracycline
- Ulcer with H. pylori: H.pylori treatment and H2-receptor antagonists or proton-pump inhibitors
- Other agent: Sucralfate
- Follow-up: Intractable ulcers and complications

DIAGNOSTIC CHECKLIST

Consider

- Eradication of H. pylori is the first step of treatment

Image Interpretation Pearls

- Check for deformity of the duodenal bulb
- Prone compression views are necessary to evaluate anterior wall duodenal ulcers

SELECTED REFERENCES

1. Jayaraman MV et al: CT of the duodenum: an overlooked segment gets its due. Radiographics. 21 Spec No:S147-60, 2001
2. Pattison CP et al: Helicobacter pylori and peptic ulcer disease: Evolution to revolution to resolution. AJR 168: 1415-20, 1997
3. Levine MS et al: The Helicobacter pylori revolution: Radiologic perspective. Radiology 195: 593-6, 1995
4. Jacobs JM: Peptic ulcer disease: CT evaluation. Radiology 178: 745-8, 1991

DUODENAL ULCER

IMAGE GALLERY

Typical

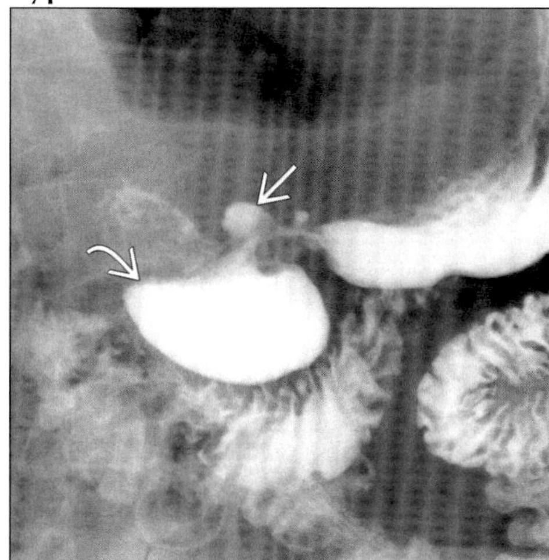

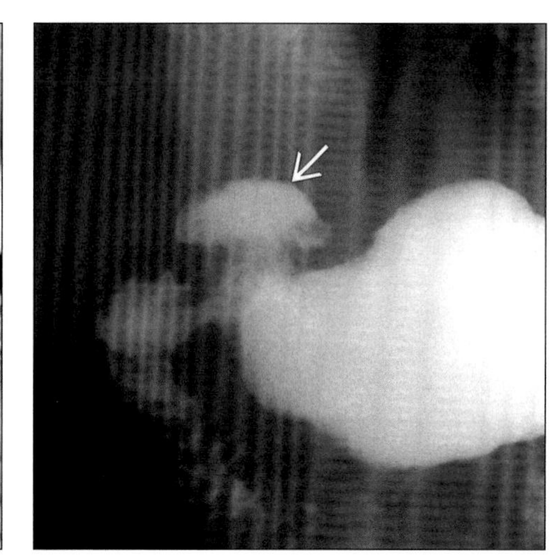

(Left) *Upper GI series shows a duodenal ulcer (arrow) and a large pseudodiverticulum (curved arrow), which changed shape during the exam.* ***(Right)*** *Upper GI series shows a "giant" duodenal ulcer (arrow) which did not change shape during the exam.*

Typical

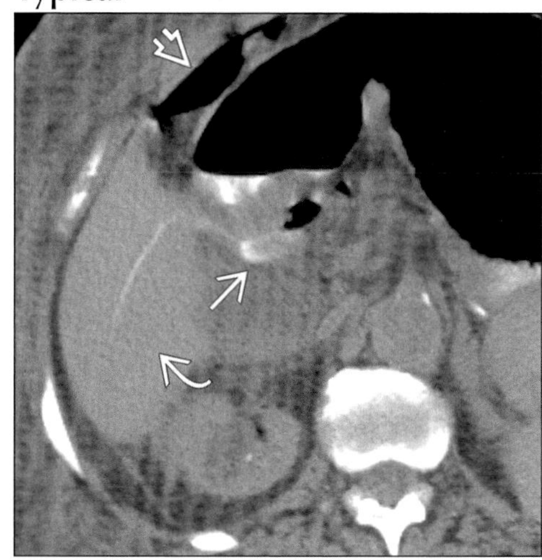

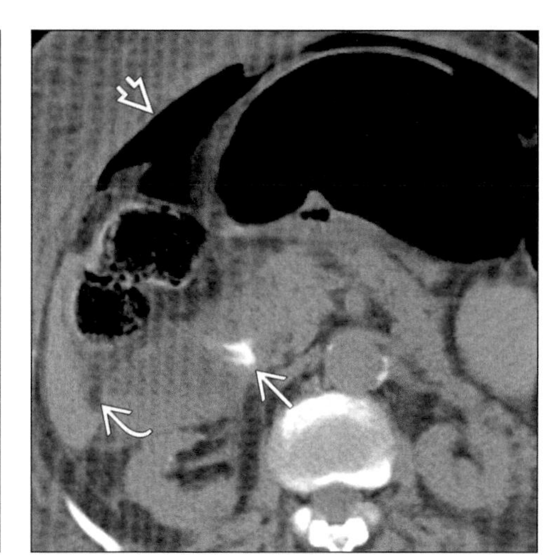

(Left) *Axial NECT shows free air (open arrow) and active extravasation of oral contrast medium (arrow) from a perforated duodenal ulcer. High density fluid accumulated in Morison pouch (curved arrow).* ***(Right)*** *Axial NECT shows free air (open arrow) and high density fluid in the right paracolic gutter (curved arrow) and retroperitoneum (arrow).*

Typical

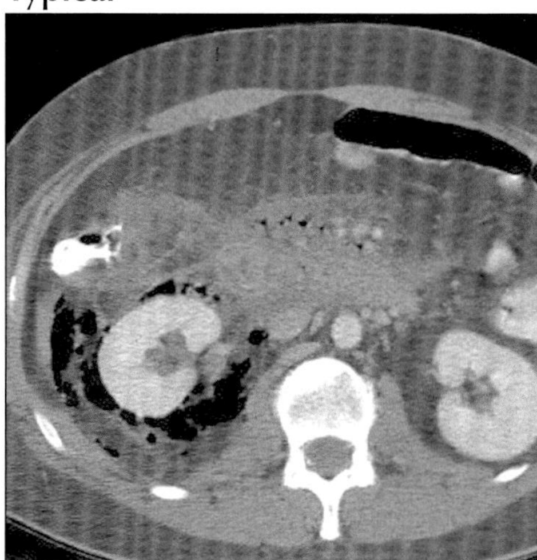

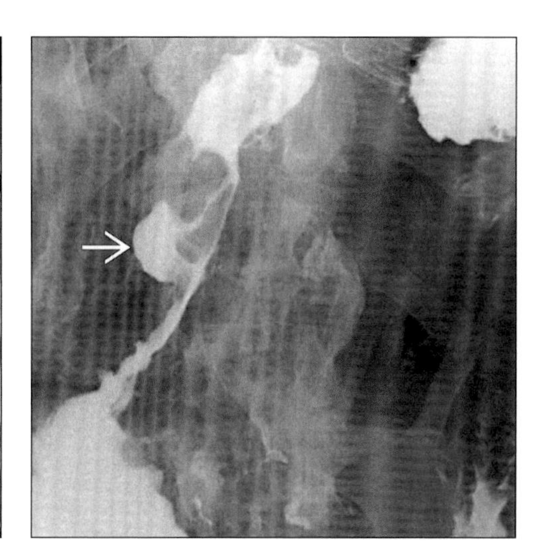

(Left) *Axial CECT shows retroperitoneal gas surrounding kidney and third portion of duodenum due to perforated ulcer.* ***(Right)*** *Upper GI series shows large post-bulbar ulcer (arrow) with marked narrowing of lumen.*

ZOLLINGER-ELLISON SYNDROME

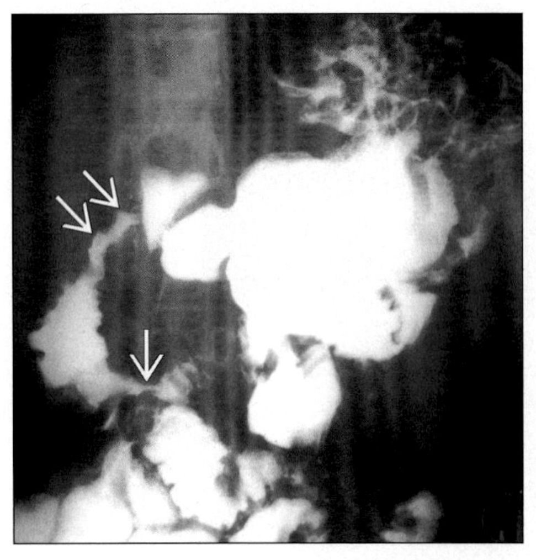

Upper GI series shows thickened gastric and duodenal folds and excess fluid in stomach. Several duodenal ulcers are present (arrows).

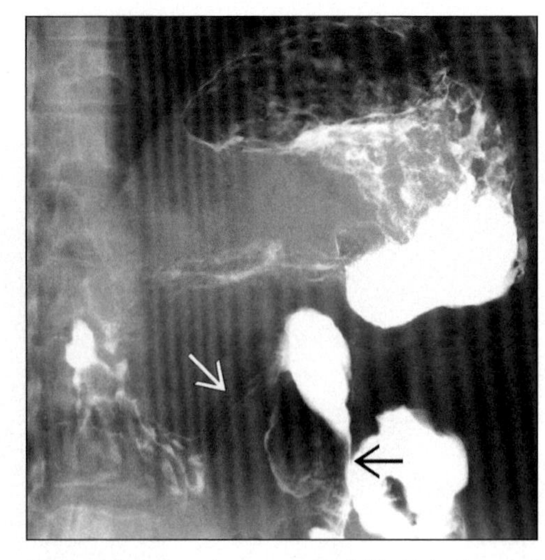

Upper GI series shows gastric fold thickening and excess fluid. Duodenal and jejunal strictures are present (arrows), probably from prior ulceration.

TERMINOLOGY

Abbreviations and Synonyms

- Zollinger-Ellison syndrome (ZES)

Definitions

- Severe peptic ulcer disease associated with marked increase in gastric acid due to gastrin producing islet cell tumor (gastrinoma) of pancreas

IMAGING FINDINGS

General Features

- Best diagnostic clue: Hypervascular pancreatic mass with multiple peptic ulcers & thickened folds
- Location
  - Gastrinoma: Pancreas (75%); duodenum (15%); liver & ovaries (10%)
  - Common site (gastrinoma): Gastrinoma triangle
    - Superiorly: Cystic & common bile ducts
    - Inferiorly: 2nd & 3rd parts of duodenum
    - Medially: Junction of neck & body of pancreas
  - Ulcers: Stomach & duodenal bulb (75%); postbulbar & jejunum (25%)
- Other general features
  - Usually due to non-β islet cell tumor (gastrinoma) of pancreas
  - Islet cell tumors are neuroendocrine tumors
  - Rare in comparison with tumors of exocrine pancreas
  - Gastrinomas are 2nd most common functioning islet cell tumors after insulinomas
  - Gastrinomas are multiple (60%); malignant (60%); metastases (30-50%)

Radiographic Findings

- Barium studies: Gastric, duodenal & proximal jejunum
  - Large volume of fluid dilutes barium & compromises mucosal coating
    - Hypersecretion of gastric acid → ↑ fluid collection
  - Markedly thickened folds
  - Peptic ulcers: Round or ovoid collections of barium surrounded by a thin or thick radiolucent rim (edematous mucosa) & radiating folds

CT Findings

- Gastrinomas
  - Heterogeneous density lesion; small or large
  - ± Cystic & necrotic areas; ± calcification
  - Liver metastases are common
  - Arterial & portal venous phase scans
    - Hypervascular (primary & secondary) lesions
    - ± Local or vascular invasion
  - Inflammatory changes in stomach, duodenum & proximal small-bowel
    - Thickened gastric, duodenal & jejunal folds

DDx: Gastric Wall Thickening +/- Ulceration

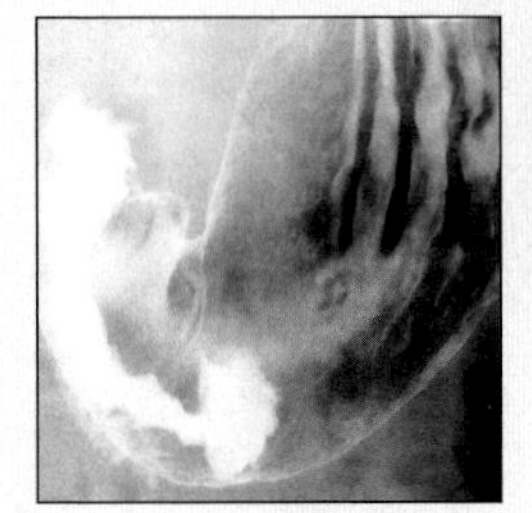

Gastritis

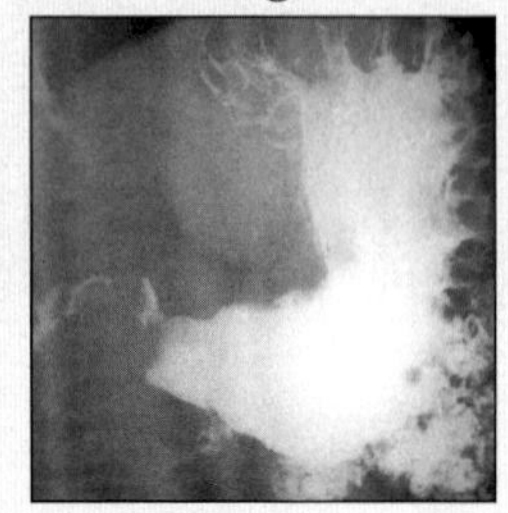

Gastritis

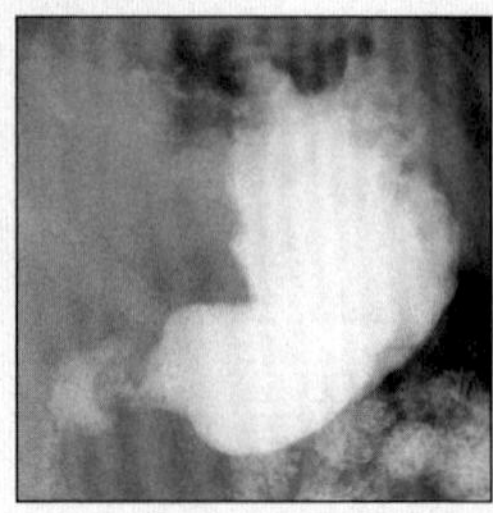

Carcinoma

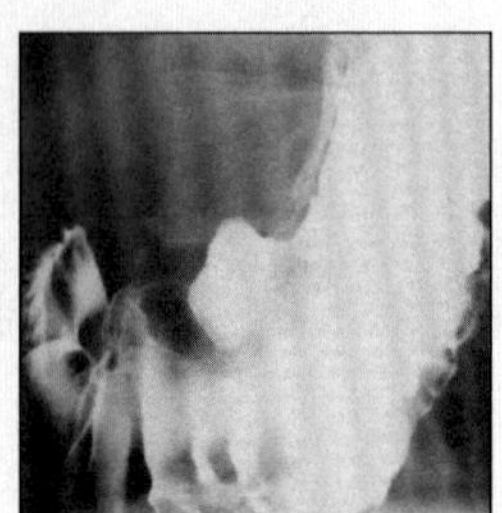

Lymphoma

ZOLLINGER-ELLISON SYNDROME

Key Facts

Terminology
- Zollinger-Ellison syndrome (ZES)
- Severe peptic ulcer disease associated with marked increase in gastric acid due to gastrin producing islet cell tumor (gastrinoma) of pancreas

Imaging Findings
- Best diagnostic clue: Hypervascular pancreatic mass with multiple peptic ulcers & thickened folds
- Common site (gastrinoma): Gastrinoma triangle
- Peptic ulcers: Round or ovoid collections of barium surrounded by a thin or thick radiolucent rim (edematous mucosa) & radiating folds
- Hypervascular (primary & secondary) lesions
- Thickened gastric, duodenal & jejunal folds
- Luminal narrowing of stomach & duodenum

Top Differential Diagnoses
- H. pylori gastritis
- Gastric carcinoma
- Gastric metastases & lymphoma
- Extrinsic inflammation
- Other gastritides

Pathology
- Gastrinoma: ↑ Gastrin levels → ↑ gastric acid secretions → peptic ulcers

Diagnostic Checklist
- Rule out other causes of gastric wall thickening & ulceration
- Hypervascular pancreatic tumor, liver metastases with multiple ulcers & thickened folds of stomach, duodenum & jejunum

  - Shows signs of ulcer penetration
    - Wall thickening
    - Luminal narrowing of stomach & duodenum
  - Shows signs of ulcer perforation
    - Free air in abdomen (duodenal/antral ulcer)
    - Lesser sac (gastric ulcer)

MR Findings
- T1WI: Fat-saturated sequence: Hypointense
- T2WI
  - Spin-echo sequence: Hyperintense
    - Both primary & secondaries
- T1 C+
  - Fat-saturated delayed spin-echo sequence
    - Hyperintense, hypervascular

Ultrasonographic Findings
- Real Time
  - Endoscopic ultrasonography (EUS)
    - Detects small gastrinomas, better than CT or MR
    - Homogeneously hypoechoic mass
  - Intraoperative ultrasonography
    - Detects very small tumors; sensitivity (75-100%)

Angiographic Findings
- Conventional
  - Gastrinomas & metastases: Hypervascular
  - Portal venous sampling: After intra-arterial secretin stimulation abnormal increase in gastrin levels

Imaging Recommendations
- Helical CT; MR & T1 C+; EUS; barium studies

DIFFERENTIAL DIAGNOSIS

H. pylori gastritis
- Helicobacter pylori: Gram-negative bacillus
- Most common cause of chronic active gastritis
- Location: Gastric antrum (most common site)
  - Proximal half or entire stomach may be involved
- Double contrast barium findings
  - Thickened gastric folds
  - Enlarged areae gastricae (≥ 3 mm in diameter)
  - Polypoid gastritis: Markedly thickened, lobulated gastric folds
- CT findings
  - Circumferential antral wall thickening
  - Focal thickening of posterior gastric wall along greater curvature
- Both barium & CT findings may mimic peptic ulcer disease of ZES
- Diagnosis: Endoscopic biopsy, culture, urea breath test

Gastric carcinoma
- Most common primary gastric tumor
- H. pylori (3-6 fold), pernicious anemia (2-3 fold) ↑ risk
- Double-contrast barium findings
  - Early gastric cancer
    - Superficial lesion: Mucosal nodularity, ulceration, plaque-like or localized thickened gastric folds
    - Indistinguishable from focal peptic ulcers of ZES
- CT findings
  - Early gastric cancer
    - Focal wall thickening with mucosal irregularity
  - May simulate focal peptic ulcer disease of ZES
- Diagnosis: Endoscopic biopsy & histology

Gastric metastases & lymphoma
- Gastric metastases: Most common organs of origin
  - Malignant melanoma, breast, lung, colon, pancreas
  - Breast cancer: Most common metastases to stomach
- Gastric lymphoma
  - Stomach most frequently involved organ in GI tract
    - Accounts 50% of all GI tract lymphomas
  - Majority are non-Hodgkin lymphomas (B-cell)
  - Arise from mucosa associated lymphoid tissue (MALT) in patients with chronic H. pylori gastritis
- Barium findings
  - Malignant melanoma: "Bull's eye" or "target" lesions
    - Centrally ulcerated submucosal masses
  - Lobular breast cancer metastases
    - Linitis plastica or "leather bottle" appearance: Loss of distensibility of antrum & body with thickened irregular folds
    - Mucosal nodularity, ulceration & spiculation (simulating peptic ulcers of ZES)

- Gastric lymphoma
  - Diffusely thickened irregular folds, discrete ulcers, ulcerated submucosal masses
  - Low grade MALT lymphoma: Confluent varying-sized nodules
- CT findings
  - Markedly thickened gastric wall & mucosal folds
  - "Bull's eye" or "target" or giant cavitated lesions
- Thickened gastric folds & ulcers may simulate ZES

Extrinsic inflammation
- Example: Pancreatitis
- Mimic ZES due to thickened gastric wall

Other gastritides
- Examples: Crohn & eosinophilic
- Early gastric Crohn: Multiple aphthous ulcers
- Eosinophilic: Mucosal nodularity, thickened folds

PATHOLOGY

General Features
- General path comments
  - Embryology-anatomy
    - Islet cell tumor: Originate from embryonic neuroectoderm
- Etiology
  - Gastrinomas
    - Arise from amine precursor uptake & decarboxylation (APUD) cells of islet of Langerhans
  - Pathogenesis
    - Gastrinoma: ↑ Gastrin levels → ↑ gastric acid secretions → peptic ulcers
- Epidemiology: Accounts 0.1-1% of pancreatic tumors
- Associated abnormalities
  - 20-60% are associated with multiple endocrine neoplasia (MEN I)
    - MEN I: Tumors of pituitary, parathyroid, adrenal cortex & pancreas

Gross Pathologic & Surgical Features
- Tumors: Encapsulated & firm; cystic, necrotic, Ca++
- Ulcers: Round or oval; sharply punched out walls

Microscopic Features
- Gastrinoma: Sheets of small round cells with uniform nuclei & cytoplasm
- Ulcers: Necrotic debris, zone of granulation tissue

CLINICAL ISSUES

Presentation
- Most common signs/symptoms: Pain, increased acidity, severe reflux, diarrhea, upper GI tract ulcers
- Lab-data: Secretin injection test
  - Paradoxical increase in serum gastrin to > 200 pg/ml above base levels in 90% of cases
- Diagnosis
  - Gastrinoma & peptic ulcers on imaging
  - Hypergastrinemia is hallmark of ZES
    - Serum gastrin levels of more than 1,000 pg/ml (virtually diagnostic of ZES)

Demographics
- Age: Any age group (more commonly 4th-5th decade)
- Gender: M > F

Natural History & Prognosis
- Complications
  - Gastrinoma: ↑ Risk of malignancy, metastases
  - Peptic ulcer: Perforation
- Prognosis
  - Good: After surgical resection of primary gastrinoma & stomach
  - Poor: Gastrinoma + liver metastases; post-operative recurrent ulcers

Treatment
- Medical: Cimetidine, ranitidine, famotidine, omeprazole
- Surgical: Gastrinoma resection; total gastrectomy
- Liver metastases: Chemotherapy & hepatic artery embolization

DIAGNOSTIC CHECKLIST

Consider
- Rule out other causes of gastric wall thickening & ulceration

Image Interpretation Pearls
- Hypervascular pancreatic tumor, liver metastases with multiple ulcers & thickened folds of stomach, duodenum & jejunum

SELECTED REFERENCES

1. Sheth S et al: Imaging of uncommon tumors of the pancreas. Radiol Clin North Am. 40(6):1273-87, vi, 2002
2. Sheth S et al: Helical CT of islet cell tumors of the pancreas: typical and atypical manifestations. AJR Am J Roentgenol. 179(3):725-30, 2002
3. Nino-Murcia M et al: Multidetector-row CT and volumetric imaging of pancreatic neoplasms. Gastroenterol Clin North Am. 31(3):881-96, 2002
4. Rodallec M et al: Helical CT of pancreatic endocrine tumors. J Comput Assist Tomogr. 26(5):728-33, 2002
5. Oshikawa O et al: Dynamic sonography of pancreatic tumors: comparison with dynamic CT. AJR Am J Roentgenol. 178(5):1133-7, 2002
6. Rodallec M et al: Helical CT of pancreatic endocrine tumors. J Comput Assist Tomogr. 26(5):728-33, 2002
7. Fidler JL et al: Imaging of neuroendocrine tumors of the pancreas. Int J Gastrointest Cancer. 30(1-2):73-85, 2001
8. Ichikawa T et al: Islet cell tumor of the pancreas: biphasic CT versus MR imaging in tumor detection. Radiology. 216(1):163-71, 2000
9. Buetow PC et al: Islet cell tumors of the pancreas: clinical, radiologic, and pathologic correlation in diagnosis and localization. Radiographics. 17(2):453-72; quiz 472A-472B, 1997
10. Van Hoe L et al: Helical CT for the preoperative localization of islet cell tumors of the pancreas: value of arterial and parenchymal phase images. AJR Am J Roentgenol. 165(6):1437-9, 1995
11. Eelkema EA et al: CT features of nonfunctioning islet cell carcinoma. AJR Am J Roentgenol. 143(5):943-8, 1984

IMAGE GALLERY

Typical

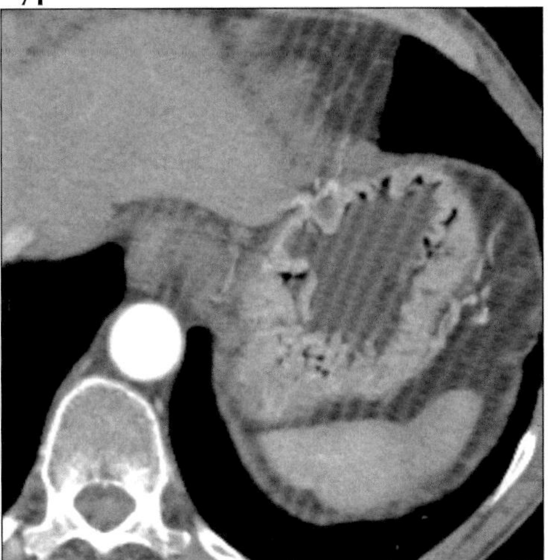

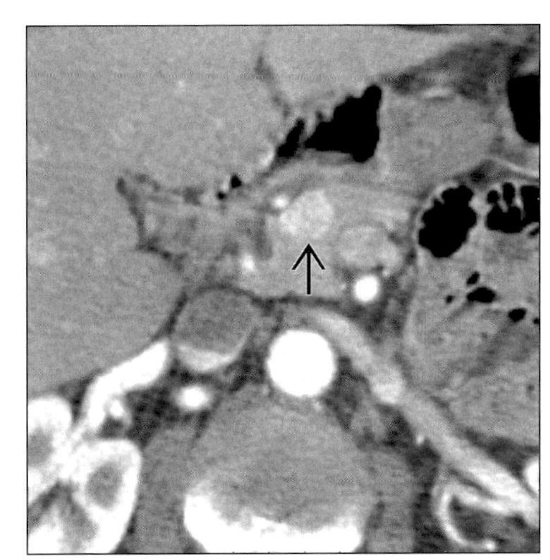

(Left) *Axial CECT shows thickened hypervascular gastric folds from Z-E syndrome.* ***(Right)*** *Axial CECT (arterial phase) shows small hypervascular mass (arrow) in pancreatic head, a gastrinoma.*

Typical

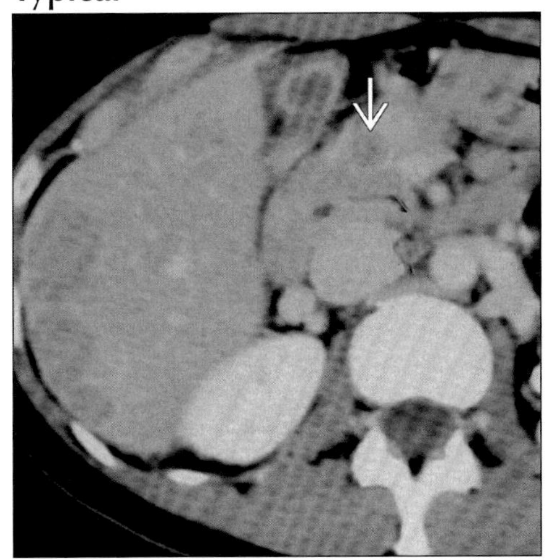

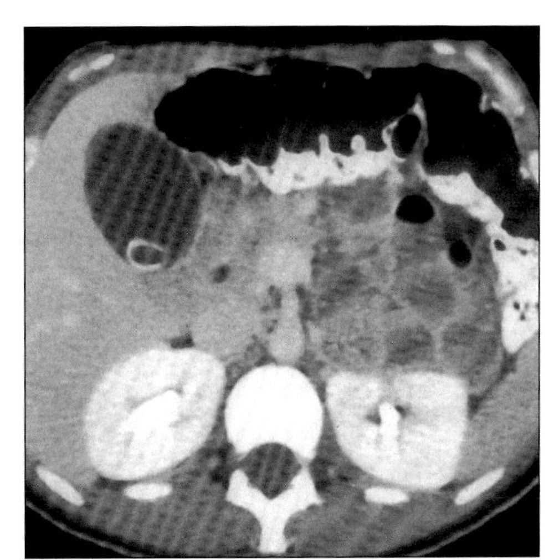

(Left) *Axial CECT (portal venous phase) shows small hypodense mass (arrow) in pancreatic head and multiple subtle liver metastases.* ***(Right)*** *Axial CECT shows fluid-distended intestine due to Zollinger-Ellison syndrome.*

Typical

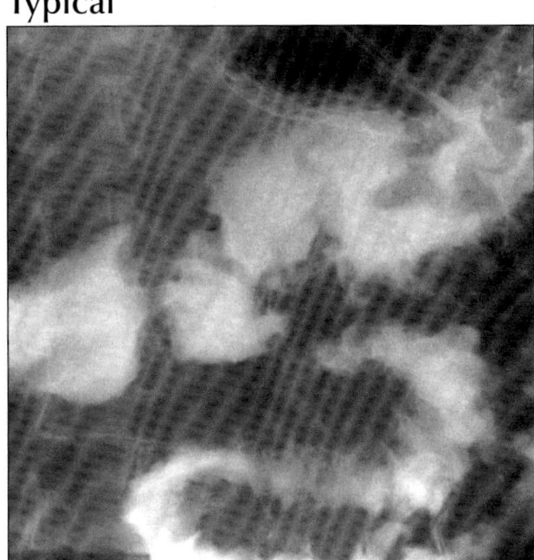

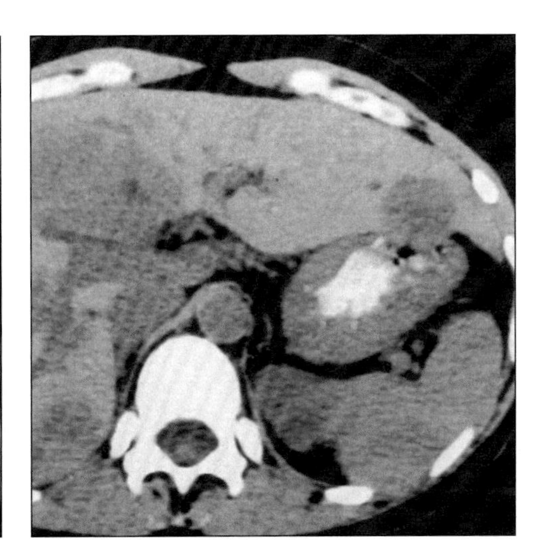

(Left) *Upper GI series shows markedly thickened folds in stomach, duodenum, and jejunum from Zollinger-Ellison syndrome.* ***(Right)*** *Axial NECT shows numerous hepatic metastases and thick gastric wall due to Zollinger-Ellison syndrome.*

MENETRIER DISEASE

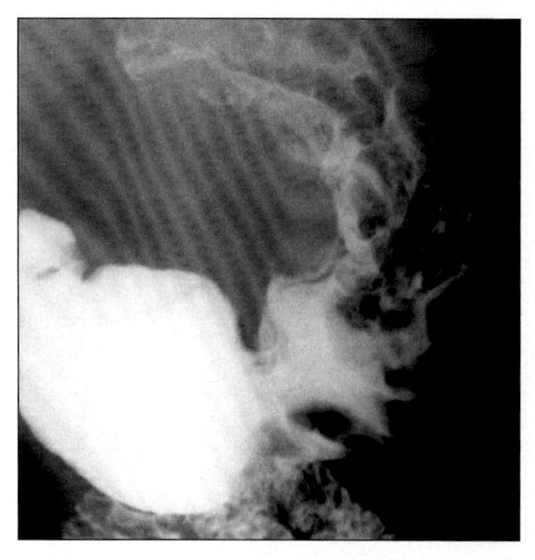

Upper GI series shows massive fold thickening in gastric fundus and body due to Menetrier disease.

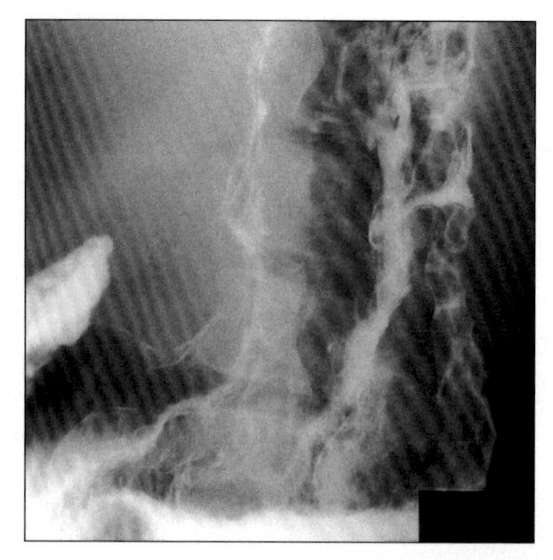

Upper GI series shows gross, tortuous gastric fold thickening in fundus and body with poor gastric coating by the barium.

TERMINOLOGY

Abbreviations and Synonyms

- Hyperplastic gastropathy, giant hypertrophic or cystic gastritis, giant mucosal hypertrophy

Definitions

- Characterized by large, tortuous gastric mucosal folds, which may be localized or may involve whole stomach

IMAGING FINDINGS

General Features

- Best diagnostic clue: Grossly thickened, lobulated folds in gastric fundus & body, with poor barium coating
- Location
  - Stomach
    - Throughout gastric fundus (most common)
    - Body (particularly along greater curvature)
    - Antrum (usually spared)
- Morphology: Large, thickened, tortuous gastric folds
- Other general features
  - Rare condition of unknown cause
  - Characterized by
    - Marked foveolar hyperplasia in stomach
    - Enlarged gastric rugae
    - Hypochlorhydria (HCl output ↓ in 75% of cases)
    - Hypoproteinemia: Protein loss (gastric mucosa)
  - Variations do occur: Mucosal hypertrophy may be associated with
    - Hyperproteinemia, hyperchlorhydria or normal protein & HCl levels
  - Menetrier disease: E.g., of hypertrophic gastropathy
  - Irreversible in most of adult patients, whereas in children it usually resolves spontaneously

Radiographic Findings

- Fluoroscopic guided double-contrast study
  - Grossly thickened, lobulated folds in gastric fundus & body with relative sparing of antrum
  - May show thickened gastric folds even in antrum
  - Focal area of rugal hypertrophy on greater curvature
  - Giant, mass-like elevation of folds on greater curvature of gastric body mimicking polypoid cancer
  - Stomach remains pliable & distensible
  - Excessive mucus may dilute barium & compromise mucosal coating
  - Rare variant of Menetrier disease: Thickened, nodular folds in proximal duodenum

CT Findings

- Markedly thickened gastric wall with mass-like elevations (giant, heaped-up folds)

Imaging Recommendations

- Barium double-contrast studies; helical CT

DDx: Diffuse or Focal Thickening of Gastric Wall

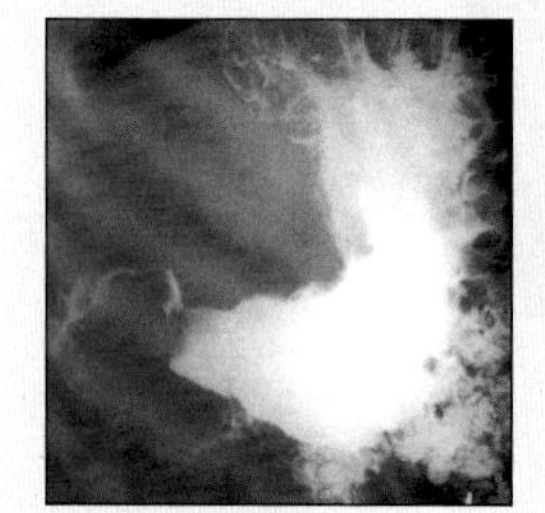

Gastritis

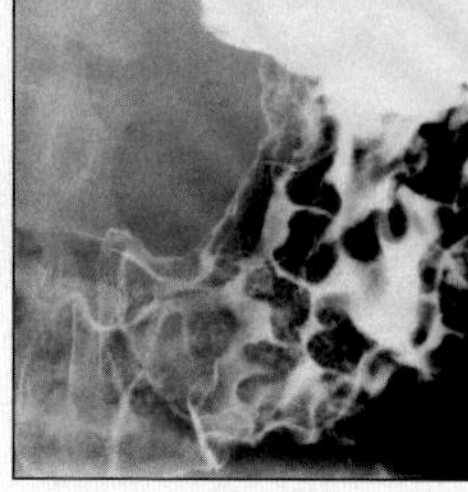

Lymphoma

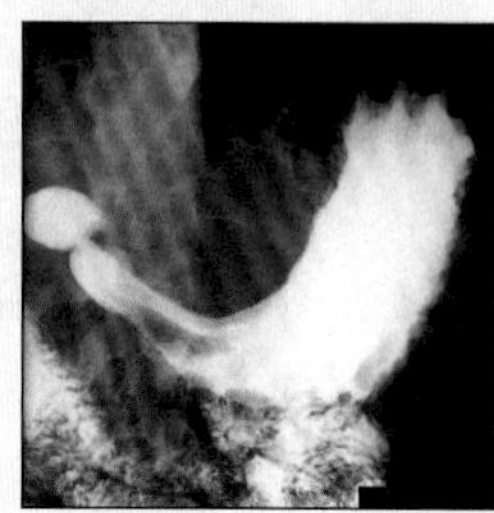

Gastric Carcinoma

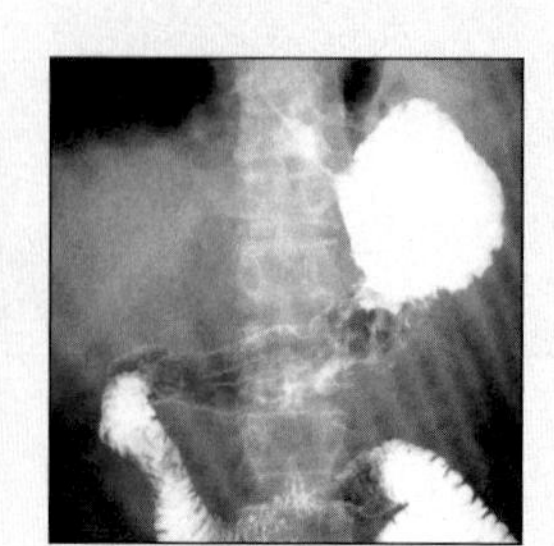

Pancreatitis

MENETRIER DISEASE

Key Facts

Terminology
- Hyperplastic gastropathy, giant hypertrophic or cystic gastritis, giant mucosal hypertrophy

Imaging Findings
- Grossly thickened, lobulated folds in gastric fundus & body with relative sparing of antrum
- Giant, mass-like elevation of folds on greater curvature of gastric body mimicking polypoid cancer

Top Differential Diagnoses
- H. pylori gastritis
- Gastric metastases & lymphoma
- Gastric carcinoma
- Extrinsic inflammation (e.g.,: Pancreatitis)
- Other gastritides

Diagnostic Checklist
- Check for hypoproteinemia & ↓ HCl with biopsy

DIFFERENTIAL DIAGNOSIS

H. pylori gastritis
- Thickened, lobulated folds favors antrum
- Diagnosis: Endoscopic biopsy; culture; urease test

Gastric metastases & lymphoma
- Lobular breast cancer metastases
- Gastric lymphoma
  - Thickened gastric folds with mucosal nodularity

Gastric carcinoma
- May cause mass-like elevated thickened folds on greater curvature; gastric peristalsis often absent

Extrinsic inflammation (e.g.,: Pancreatitis)
- Mimic Menetrier due to thickened gastric wall
- CT will show peripancreatic inflammation

Other gastritides
- Examples: Advanced Crohn & eosinophilic
- Thickened folds, large ulcers, mucosal nodularity
- Usually involves antrum or antrum & body

PATHOLOGY

General Features
- Etiology
  - Unknown
  - Mucosal thickening (massive foveolar hyperplasia)

Gross Pathologic & Surgical Features
- Large, thickened, tortuous gastric mucosal folds

Microscopic Features
- Cystic dilatation, elongated gastric mucous glands
- Atrophy of chief/parietal cells; deepening foveolar pits

CLINICAL ISSUES

Presentation
- Most common signs/symptoms
  - Epigastric pain, vomiting, diarrhea, weight loss
  - Occasionally peripheral edema (hypoproteinemia)
- Lab: ↓ Albumin; ↓ or absent HCl; ± fecal occult blood
- Diagnosis: Endoscopic full-thickness biopsy

Demographics
- Age: Usually occur in older people (range 20-70 years)
- Gender: M > F

Natural History & Prognosis
- Complications
  - Gastric carcinoma develops in about 10% of patients
  - Increased risk of deep venous thrombosis (DVT)
  - Risk of atrophic gastritis, gastric ulcer, GIT bleeding
- Prognosis
  - Prolonged illness with intractable symptoms
  - Spontaneous remission & few respond to treatment

Treatment
- Medical therapy: Anticholinergic agents, antibiotics
- Total gastrectomy & vagotomy (unresponsive cases)

DIAGNOSTIC CHECKLIST

Consider
- Check for hypoproteinemia & ↓ HCl with biopsy

SELECTED REFERENCES

1. Fishman EK et al: CT of the stomach: Spectrum of disease. RadioGraphics 16: 1035-54, 1996
2. Wolfsen HC et al: Menetrier's disease: A form of hypertrophic gastropathy (or) gastritis. Gastroenterology 104: 1310-9, 1993
3. Reese DF et al: Giant hypertrophy of the gastric mucosa (Menetrier's disease): A correlation of the roentgenographic, pathologic and clinical findings. AJR 88: 619-26, 1962

IMAGE GALLERY

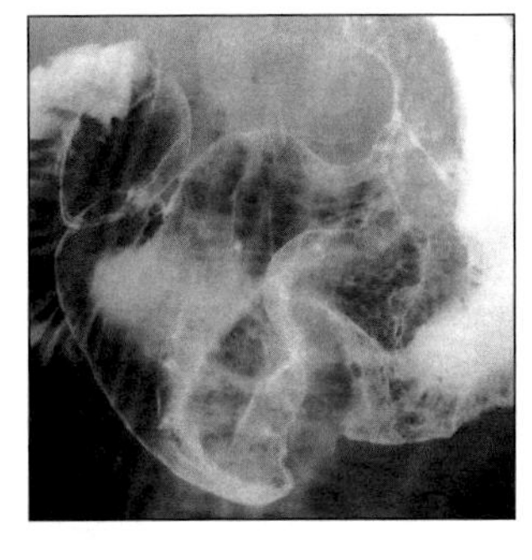

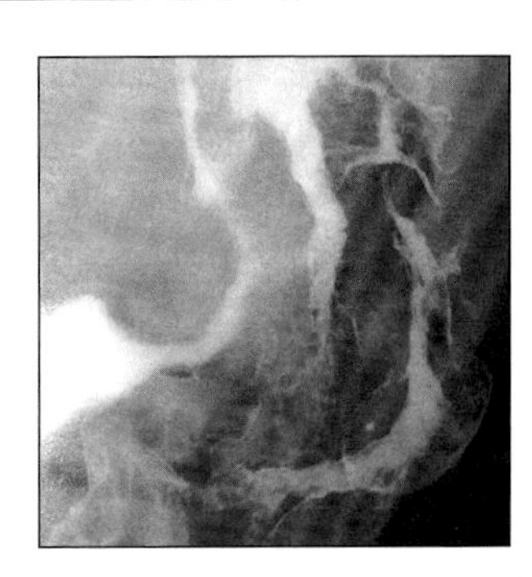

*(**Left**) Upper GI series shows massive fold thickening, sparing only the antrum. (**Right**) Upper GI series shows massive gastric fold thickening and poor coating by barium.*

CAUSTIC GASTRODUODENAL INJURY

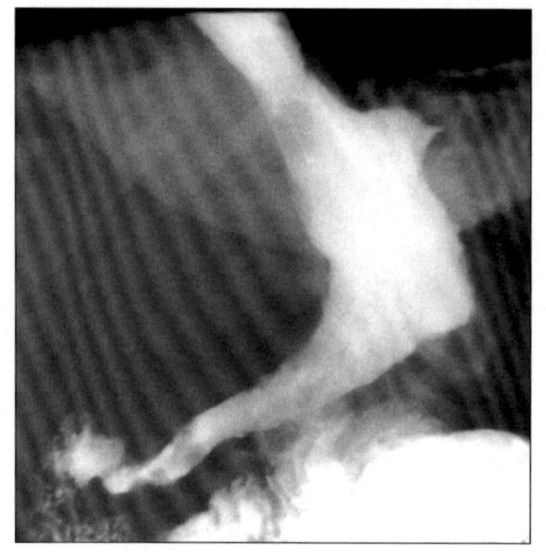

Upper GI series shows nondistensible, small nonperistaltic, featureless stomach due to prior ingestion of hydrochloric acid.

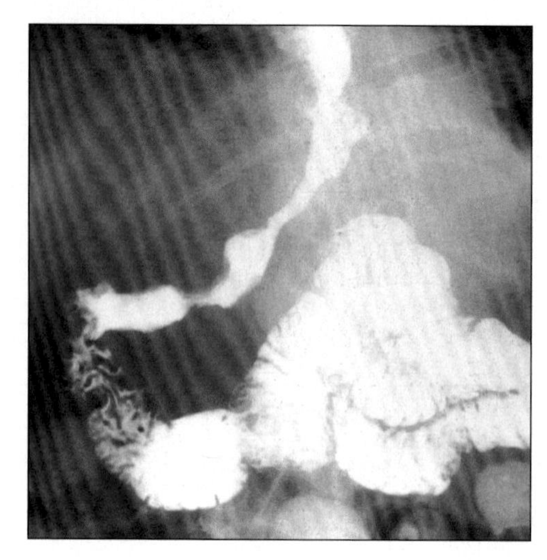

Upper GI series shows extremely small, non-distensible, distorted stomach due to ingestion of hydrochloric acid.

TERMINOLOGY

Abbreviations and Synonyms

- Corrosive gastroduodenitis

Definitions

- Gastroduodenal inflammation/injury due to acid or alkali

IMAGING FINDINGS

General Features

- Best diagnostic clue: Grossly abnormal stomach with intramural dissection of contrast & mural defects
- Location: Lesser curvature & distal antrum of stomach
- Other general features
  - Esophagus is most often injured within GI tract
    - Classically damaged by strong alkaline agents
    - Most commonly used alkali in US: Liquid lye
  - After esophagus, gastroduodenal injury > common
    - Most likely to be damaged by strong acids
    - Commonly used acids: Hydrochloric & sulfuric
    - Caustic agents cause intense pylorospasm, so duodenal injury is less common
  - Classification based on clinical/radiological findings
    - Acute & chronic phases
    - Mild & severe injury patterns

Radiographic Findings

- Radiography
  - Dilated & gas-filled stomach
  - Fulminating cases
    - Streaky, bubbly or mottled intramural gas
- Fluoroscopic guided water-soluble contrast studies
  - Acute mild phase
    - Atonic dilated stomach ± proximal duodenum
    - Multiple shallow, irregular ulcers
  - Acute severe phase
    - Thickened folds, extensive deep ulceration
    - Severe pylorospasm + delayed emptying
    - Mural defects (due to edema & hemorrhage)
    - Intramural dissection of contrast or loculated perigastric collections
    - ± Reveal free perforation into peritoneal cavity
  - Chronic phase
    - Narrowing/deformity of stomach ± duodenal bulb
    - Antrum may be smooth + tubular configuration
    - Gastric outlet obstruction (antral scarring/fibrosis)
    - Antral scarring mimics scirrhous carcinoma
    - Duodenal bulb & sweep may appear normal due to severe pylorospasm or marked antral scarring
    - Rarely strictures between bulb-ligament of Treitz

CT Findings

- Acute severe phase: Pneumoperitoneum (perforation)
- Chronic phase: Luminal irregularity & narrowing

DDx: Small, Non-Distensible Stomach

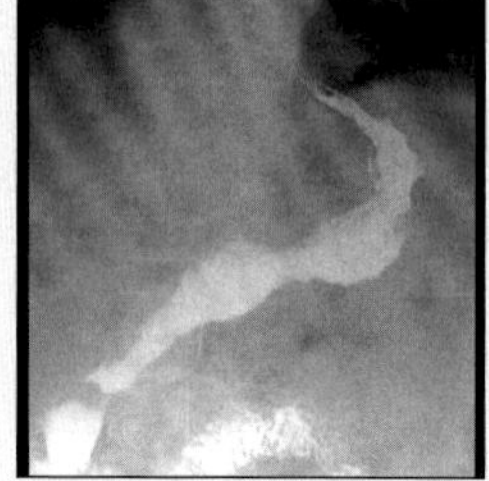

Gastric Carcinoma

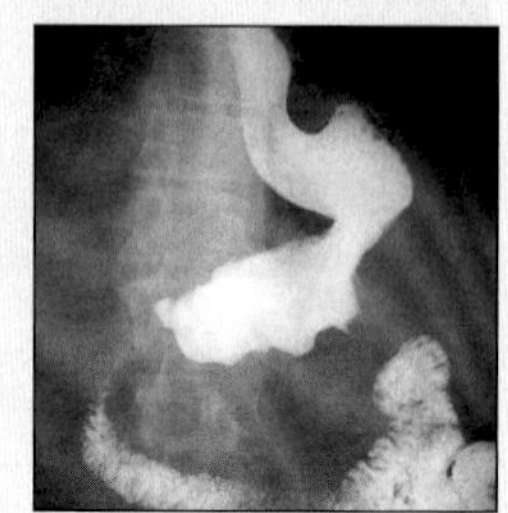

Gastric Carcinoma

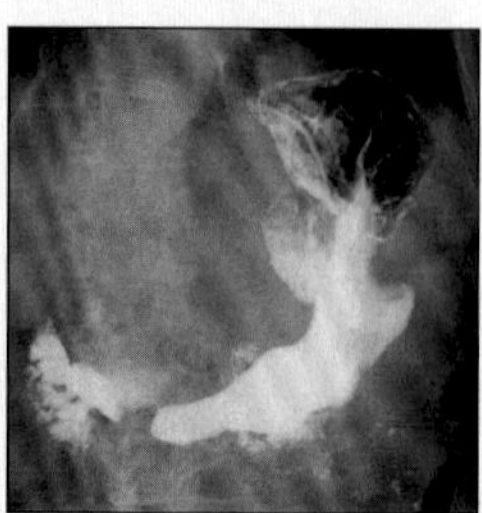

Gastric Lymphoma

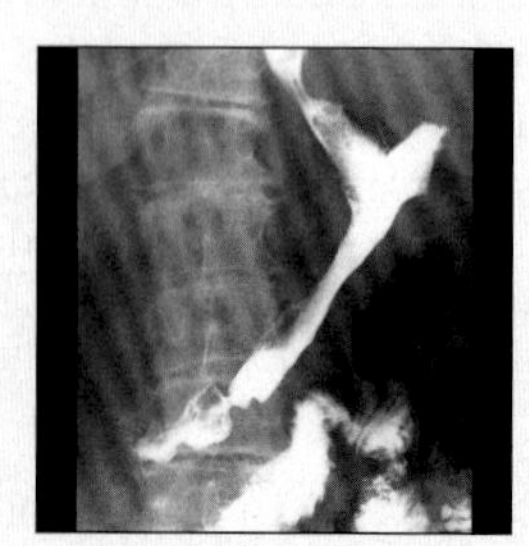

Post Freezing

CAUSTIC GASTRODUODENAL INJURY

Key Facts

Imaging Findings

- Best diagnostic clue: Grossly abnormal stomach with intramural dissection of contrast & mural defects
- Atonic dilated stomach ± proximal duodenum
- Thickened folds, extensive deep ulceration
- Severe pylorospasm + delayed emptying
- Narrowing/deformity of stomach ± duodenal bulb
- Antrum may be smooth + tubular configuration
- Antral scarring mimics scirrhous carcinoma

Top Differential Diagnoses

- Gastric carcinoma (scirrhous type)
- Gastric metastases & lymphoma
- Gastric thermal injury

Diagnostic Checklist

- Check for history of strong acid or alkali ingestion
- Thickened folds, ulceration, atony, spasm & stricture

Imaging Recommendations

- Videofluoroscopic water-soluble contrast studies

DIFFERENTIAL DIAGNOSIS

Gastric carcinoma (scirrhous type)

- Usually arise near pylorus & extend up
- Diffuse linitis plastica mimics caustic gastric injury
  - Nodularity, spiculation, ulceration, thickened folds
  - Esophageal injury & history favors caustic ingestion

Gastric metastases & lymphoma

- Example: Lobular breast carcinoma & non-Hodgkin
- Linitis plastica pattern simulate caustic injury
- Differentiated by breast primary, biopsy & history

Gastric thermal injury

- Example: Post freezing, when iced saline infusions used for bleeding varices

PATHOLOGY

General Features

- General path comments
  - Pathologically caustic injury occurs in three phases
    - Acute necrotic phase (after 1-4 days)
    - Ulceration-granulation phase (after 5-28 days)
    - Cicatrization & scarring (after 3-4 weeks)
- Etiology
  - Alkali: Liquid lye (concentrated sodium hydroxide)
    - Pathogenesis: Injury by liquefaction necrosis
  - Acids: HCl, sulfuric, acetic, oxalic, nitric, carbolic
    - Pathogenesis: Injury by coagulative necrosis
- Associated abnormalities: Esophageal injury

Gross Pathologic & Surgical Features

- Hyperemia/inflammation/necrosis/ulceration/strictures

Microscopic Features

- Thinning of epithelium, inflammatory cells, cellular hyperplasia & areas of necrosis

CLINICAL ISSUES

Presentation

- Most common signs/symptoms
  - Severe abdominal pain, nausea, vomiting
  - Hematemesis, fever & shock

Natural History & Prognosis

- Complications
  - Outlet obstruction, perforation, peritonitis, shock
  - Increased risk of cancer after 20-40 years
- Prognosis
  - Acute mild phase with early treatment: Good
  - Acute severe & chronic phases: Poor

Treatment

- Conservative treatment for stable patients
  - Antibiotics, steroids, parenteral feedings
- Gastric outlet obstruction
  - Gastroenterostomy or partial gastrectomy

DIAGNOSTIC CHECKLIST

Consider

- Check for history of strong acid or alkali ingestion

Image Interpretation Pearls

- Thickened folds, ulceration, atony, spasm & stricture

SELECTED REFERENCES

1. Muhletaler CA et al: Acid corrosive esophagitis: radiographic findings. AJR 134: 1137-1140, 1980
2. Franken EA: Caustic damage of the gastrointestinal tract: Roentgen features. AJR 118: 77-85, 1973
3. Martel W: Radiologic features of esophagogastritis secondary to extremely caustic agents. Radiology 103: 31-36, 1972

IMAGE GALLERY

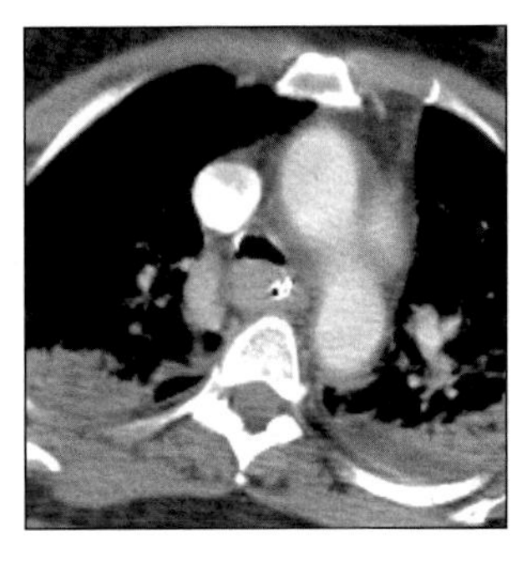

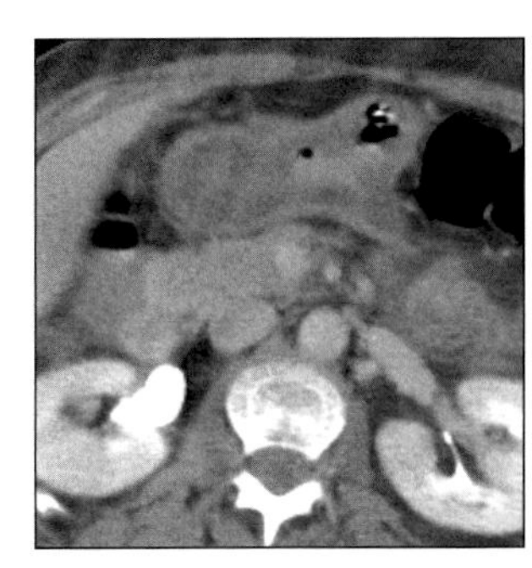

(Left) *Axial CECT shows dilated esophagus with thickened wall and aspiration pneumonitis. Lye ingestion.* ***(Right)*** *Axial CECT shows gastric wall thickening + submucosal edema due to lye ingestion.*

DUODENAL HEMATOMA AND LACERATION

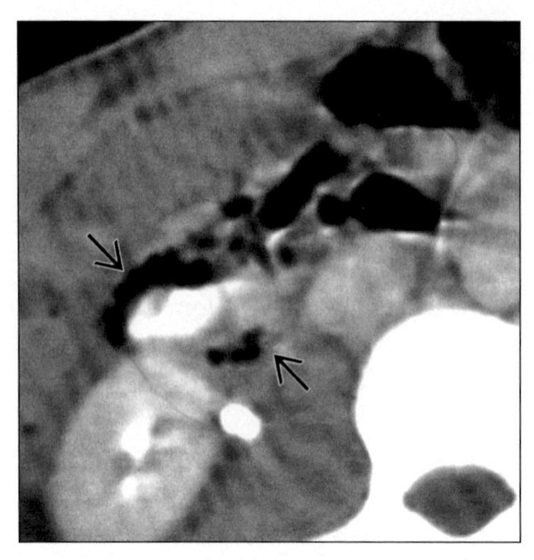

Axial CECT of duodenal perforation from blunt trauma. Note ectopic gas and fluid in right anterior pararenal space (arrows).

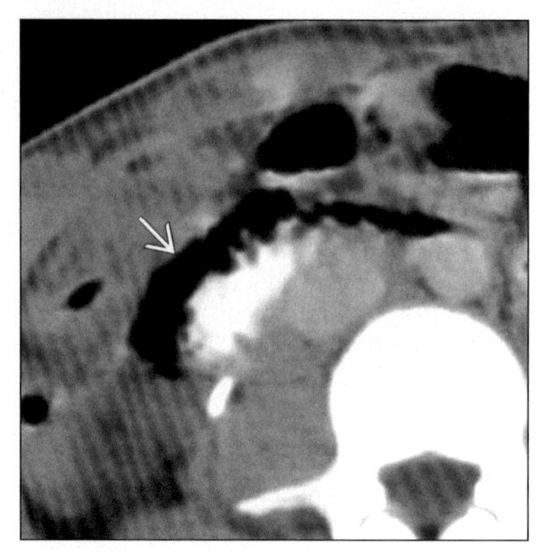

Axial CECT of duodenal traumatic perforation. Note large ectopic gas collection adjacent to duodenum (arrow).

TERMINOLOGY

Definitions

- Trauma to duodenum resulting in intramural hematoma or laceration

IMAGING FINDINGS

General Features

- Best diagnostic clue: High attenuation intramural hematoma, ectopic gas, fluid in peritoneal cavity or anterior pararenal space
- Morphology: "Dumbbell-shaped" intramural hematoma

Radiographic Findings

- Radiography
  - Free air (pneumoperitoneum or ectopic retroperitoneal gas)
  - Free fluid
- Fluoroscopy: GI series: Narrowing of duodenal lumen by intramural hematoma; extravasation of oral contrast into peritoneal cavity or retroperitoneum

CT Findings

- NECT
  - High attenuation intramural duodenal hematoma
  - Free air and fluid in peritoneal cavity or anterior pararenal space
- CECT: Nonenhancing intramural hematoma; active extravasation from gastroduodenal artery; interruption of duodenal wall; ectopic gas/fluid; periduodenal stranding

MR Findings

- T1WI: High signal intramural hematoma
- T2WI
  - High signal free fluid
  - High signal hematoma
- T1 C+
  - Wall thickening of duodenum
  - Nonenhancing hematoma

Ultrasonographic Findings

- Echogenic intramural mass representing hematoma

Angiographic Findings

- Conventional: Selective if active bleeding extravasation from gastroduodenal artery

Imaging Recommendations

- Best imaging tool: CECT, UGI

DDx: Lesions Mimicking Duodenal Hematoma

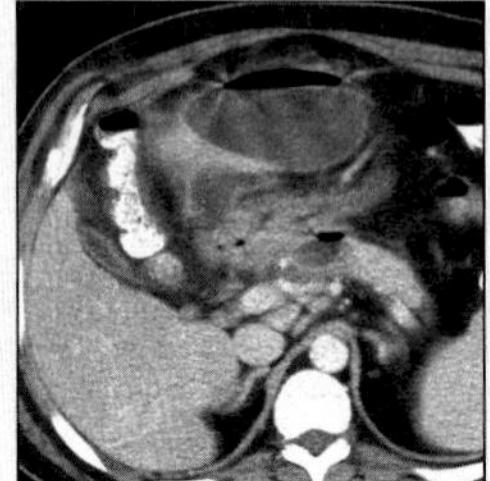

Perf. Duodenal Ulcer

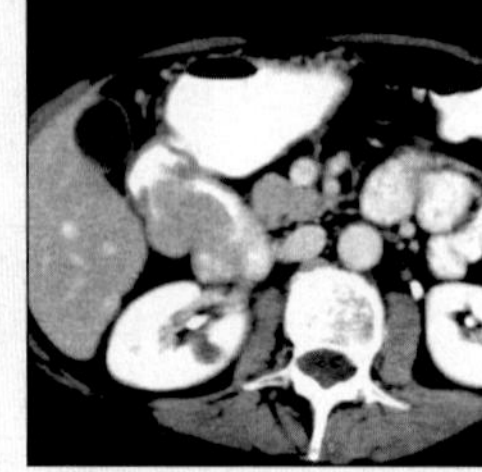

Villous Adenoma

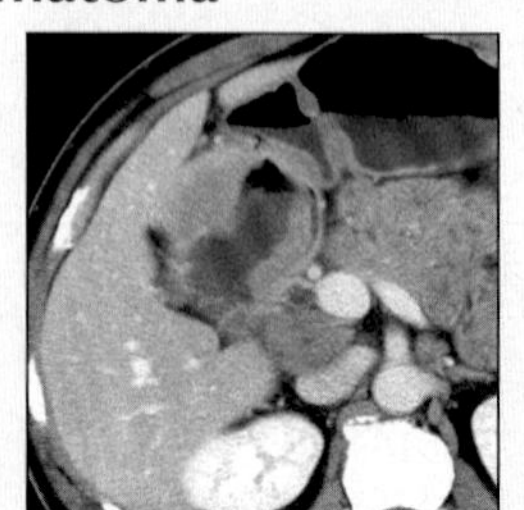

Duodenal Lymphoma

DUODENAL HEMATOMA AND LACERATION

Key Facts

Imaging Findings

- Best diagnostic clue: High attenuation intramural hematoma, ectopic gas, fluid in peritoneal cavity or anterior pararenal space
- Best imaging tool: CECT, UGI

Pathology

- 4th most common organ injury in children

Clinical Issues

- Clinical profile: Child with midepigastric blunt trauma, adult with high speed motor vehicle accident injuries
- Non-operative management for isolated duodenal hematoma with perforation
- Surgery for duodenal perforation and associated head of pancreas injury

DIFFERENTIAL DIAGNOSIS

Perforated duodenal ulcer

- Ectopic gas or fluid in peritoneal cavity or anterior pararenal space
- Periduodenal inflammatory changes
- Mural thickening of duodenum

Villous adenoma

- Polypoid mucosal mass 3-9 cm; rarely causes obstruction

Duodenal lymphoma

- Most often extension of gastric lymphoma
- Bulky submucosal mass

PATHOLOGY

General Features

- General path comments: Intramural duodenal hematoma
- Epidemiology
  - 4th most common organ injury in children
  - 2-10% of all blunt injuries
- Associated abnormalities
  - Pancreatic laceration (47%) or fracture
  - Liver or splenic laceration (16-32%)

Staging, Grading or Classification Criteria

- Isolated intramural hematoma
- Perforated duodenum
- Combined head of pancreas and duodenal injury

CLINICAL ISSUES

Presentation

- Most common signs/symptoms: Nausea, vomiting, abdominal pain/tenderness
- Clinical profile: Child with midepigastric blunt trauma, adult with high speed motor vehicle accident injuries

Natural History & Prognosis

- Isolated intramural hematoma has excellent prognosis with non-operative management
- Combined duodenal perforation with head of pancreas laceration has morbidity of 26%

Treatment

- Options, risks, complications
  - Non-operative management for isolated duodenal hematoma with perforation
  - Surgery for duodenal perforation and associated head of pancreas injury

DIAGNOSTIC CHECKLIST

Consider

- Perforated duodenal ulcer

Image Interpretation Pearls

- Ectopic gas/fluid in pararenal space

SELECTED REFERENCES

1. Desai KM et al: Blunt duodenal injuries in children. J Trauma. 54(4):640-5; discussion 645-6, 2003
2. Zissin R et al: Pictorial review. CT of duodenal pathology. Br J Radiol. 75(889):78-84, 2002
3. Degiannis E et al: Duodenal injuries. Br J Surg. 87(11):1473-9, 2000
4. Lorente-Ramos RM et al: Sonographic diagnosis of intramural duodenal hematomas. J Clin Ultrasound. 27(4):213-6, 1999
5. Weigelt JA: Duodenal injuries. Surg Clin North Am. 70(3):529-39, 1990

IMAGE GALLERY

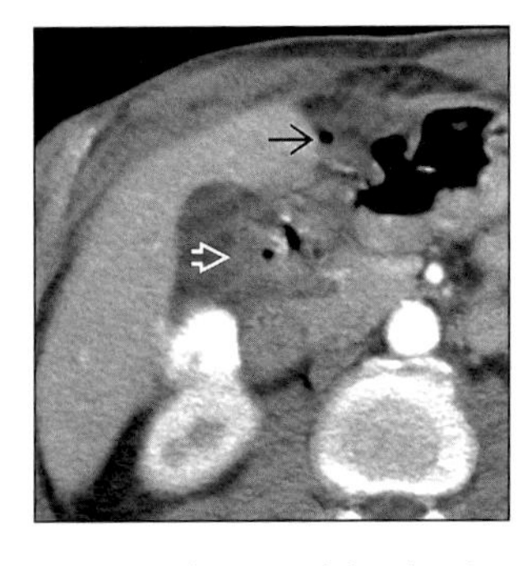

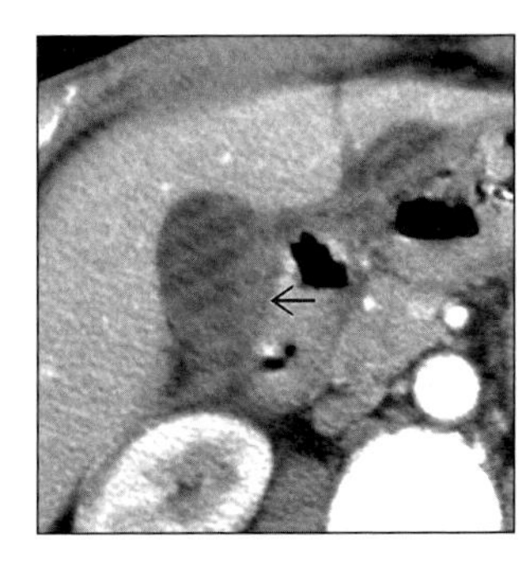

*(**Left**) Axial CECT of duodenal perforation secondary to trauma. Note mural thickening of duodenum (open arrow) and ectopic gas bubble (arrow). (**Right**) Axial CECT of duodenal perforation secondary to trauma. Note paraduodenal hematoma (arrow).*

GASTRIC POLYPS

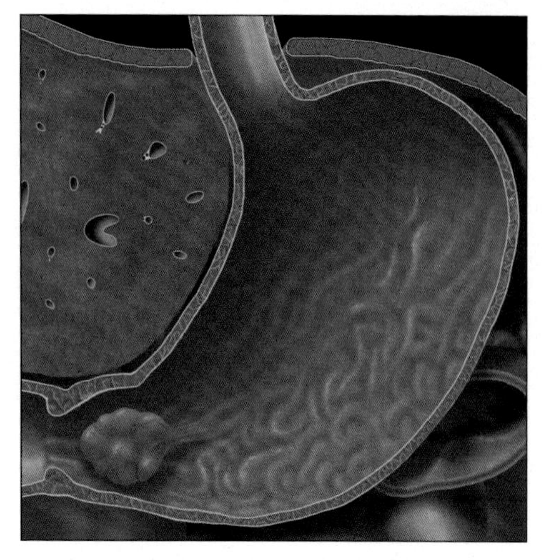

Graphic shows pedunculated polyp in gastric antrum, prone to prolapse through pylorus with peristalsis.

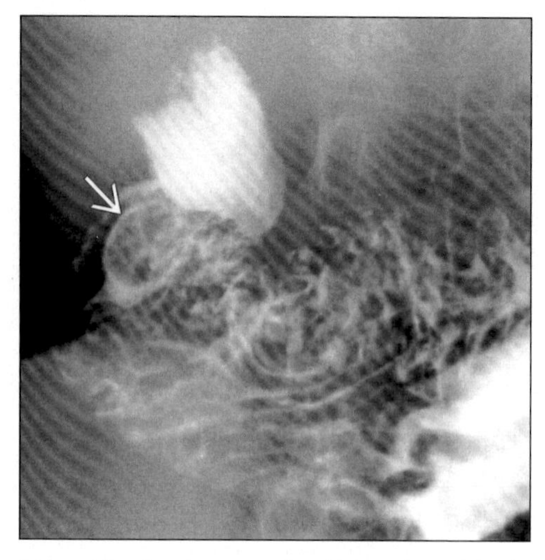

Upper GI series shows a polypoid mass (arrow) in the antrum that periodically prolapsed through the pylorus. Leiomyoma.

TERMINOLOGY

Definitions

- A protruding, space-occupying, epithelial lesion within stomach

IMAGING FINDINGS

General Features

- Best diagnostic clue: Radiolucent filling defects, ring shadows or contour defect on barium study
- Morphology
  - Hyperplastic polyps: Smooth, sessile, pedunculated
  - Adenomatous polyps: Usually single with lobulated or cauliflower-like surface
  - Hamartomas: Cluster of broad based polyps
- Other general features
  - 85-90% of gastric neoplasms are benign
    - 50% Mucosal & 50% submucosal
  - Gastric polyps are mucosal lesions
  - More common in hereditary polyposis syndromes
  - Polyps classified into three types based on pathology
    - Hyperplastic, adenomatous & hamartomatous
  - Hyperplastic polyps
    - Most common benign epithelial neoplasms of stomach (80-90%)
    - Typical & atypical (large & giant); virtually no malignant potential
    - Typical: Small, multiple, sessile (< 1 cm); location (fundus & body)
    - Atypical Large: Solitary, pedunculated (2-6 cm); location (body & antrum)
    - Atypical giant: Polyp (6-10 cm) multilobulated mass; location (antrum & body)
    - 8-28% associated with atrophic gastritis, pernicious anemia & cancer
    - Fundic gland polyps: Variant of hyperplastic polyps (< 1 cm)
  - Adenomatous polyps
    - Less common (< 20%); dysplastic lesions
    - ↑ Risk of malignant change via adenoma-carcinoma sequence
    - Usually solitary, occasionally multiple, > 1 cm; location (antrum)
    - Histologically: Tubular (75%); tubulovillous (15%); villous (10%)
    - Gastric adenomatous polyps 30 times < common than gastric cancer
    - Carcinoma in situ & invasive carcinoma: Seen in 50% of polyps > 2 cm
    - 30-40% associated with: Atrophic gastritis, pernicious anemia & cancer
    - ↑ Risk of coexisting gastric cancer more than risk of malignant change in polyp

DDx: Discrete Filling Defect(s) in Stomach

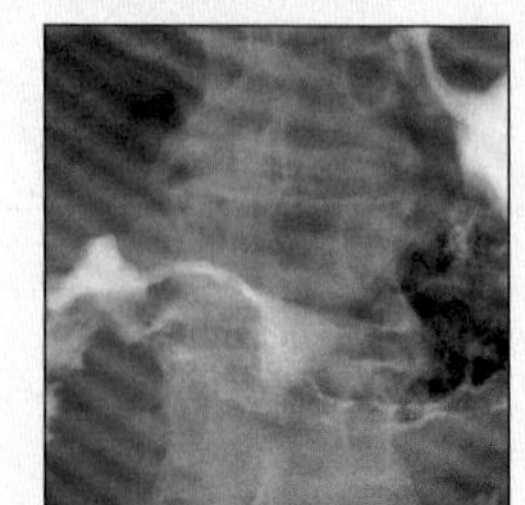

Gastric Carcinoma

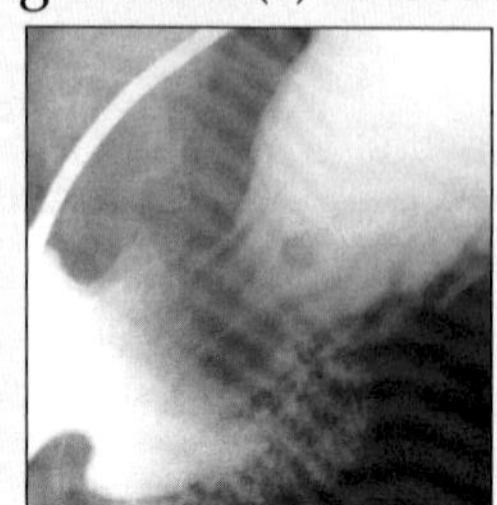

Met. Melanoma

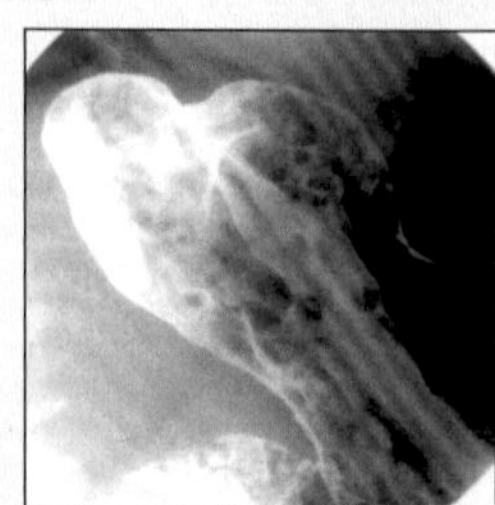

Carcinoid Tumors

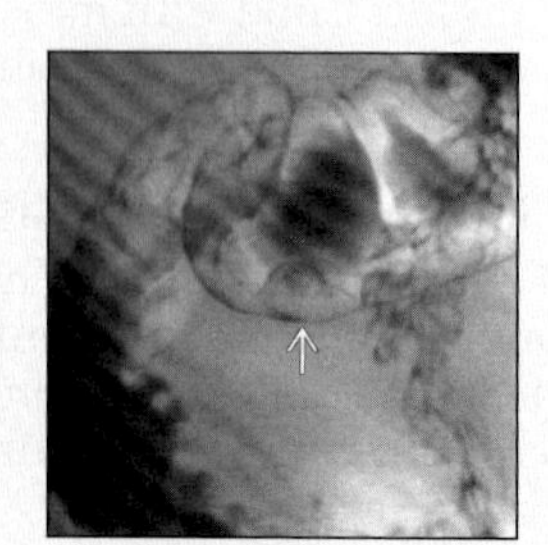

Ectopic Pancreas

GASTRIC POLYPS

Key Facts

Terminology

- A protruding, space-occupying, epithelial lesion within stomach

Imaging Findings

- Best diagnostic clue: Radiolucent filling defects, ring shadows or contour defect on barium study
- Dependent (posterior wall): Radiolucent filling defects
- Nondependent (anterior wall): Ring shadows + white rim (barium)
- "Mexican hat" sign: Characterized by a pair of concentric rings

Top Differential Diagnoses

- Retained food & pills
- Gastric carcinoma (polypoid type)
- Gastric metastases & lymphoma
- Gastric stromal tumor
- Ectopic pancreas

Pathology

- General path comments: Proliferation of mucosa
- Chronic atrophic & H. pylori gastritis
- Hereditary: Autosomal dominant (FAPS & PJS)

Diagnostic Checklist

- Check for family history of GI tract polyps
- Hyperplastic polyps (typical): Multiple, smooth, sessile, round or ovoid lesions, < 1 cm in size
- Adenomatous polyps: Solitary, sessile or pedunculated, more lobulated & > than 1 cm in size
- Large, solitary, sessile polyp with lobulated surface & basal indentation, highly suggests adenocarcinoma

- Polyposis syndromes involving stomach
    - Familial adenomatous polyposis syndrome (FAPS): Seen in > 50% cases of gastric polyps
    - Hamartomatous polyposis: Example: (PJS) Peutz-Jeghers syndrome (10-15% of gastric polyp cases)

Radiographic Findings

- Fluoroscopic guided double-contrast barium study
- Hyperplastic polyps
    - Typical: Multiple, smooth, sessile, round or ovoid lesions, < 1 cm in size
    - Based on location: Dependent & nondependent wall
        - Dependent (posterior wall): Radiolucent filling defects
        - Nondependent (anterior wall): Ring shadows + white rim (barium)
    - Variant: Fundic gland polyps (multiple up to 50 in fundus, < 1 cm in size)
        - Small rounded nodules, indistinguishable from hyperplastic polyps
    - Atypical: Large & giant
        - Large: Solitary, conglomerated, pedunculated, lobulated, 2-6 cm in size
        - Giant polyps: Multilobulated conglomerate mass + trapping of barium in interstices between lobules; 6-10 cm in size
        - Atypical antral large & giant pedunculated polyps may prolapse
        - Polyp prolapse → pylorus → duodenum leads to gastric outlet obstruction
- Adenomatous polyps
    - Usually solitary or rarely multiple; sessile or pedunculated; more lobulated; > 1 cm in size
    - Pedunculated polyp en face: Hanging from nondependent anterior wall
        - "Mexican hat" sign: Characterized by a pair of concentric rings
        - Outer ring: Represents head of polyp
        - Inner ring: Represents stalk of polyp
    - Lobulated polyp with basal indentation: ↑ Risk of adenocarcinoma
- Polyposis syndromes involving stomach
    - Familial adenomatous polyposis syndrome (FAPS)
        - Fundic gland polyps & adenomas (> 50% cases)
        - Seen as multiple small filling defects
    - Peutz-Jeghers syndrome (hamartomatous polyposis)
        - Cluster of polyps (10-15% of gastric polyp cases)

Imaging Recommendations

- Best imaging tool
    - Upper gastrointestinal double-contrast barium study
    - En face, profile & oblique views

DIFFERENTIAL DIAGNOSIS

Retained food & pills

- Filling defects in barium pool simulating polyps

Gastric carcinoma (polypoid type)

- Lobulated or fungating mass
- Barium study findings
    - Dependent or posterior wall: Filling defect
    - Nondependent or anterior wall
        - Etched in white by a thin layer of barium
- Indistinguishable from giant lobulated hyperplastic or adenomatous polyp
- Diagnosis: Endoscopic biopsy & histology

Gastric metastases & lymphoma

- Gastric metastases: Example: Malignant melanoma & squamous cell carcinoma
- Gastric lymphoma: Example: Low grade MALT lymphoma
    - MALT: Mucosa-associated lymphoid tissue
- Barium study findings
    - Malignant melanoma metastases
        - Initially: Submucosal masses seen as filling defects may mimic polyps
        - Ulcerated lesions: "Bull's eye" or "target" pattern
    - Low grade MALT lymphoma
        - Confluent varying-sized nodules (filling defects)
        - May be indistinguishable from gastric FAPS

Gastric stromal tumor

- Submucosal lesions

GASTRIC POLYPS

- Example: Leiomyoma
- Non-ulcerated leiomyoma
  - In profile
    - Smooth surface etched in white
    - Borders: Right or obtuse angles with adjacent wall
  - En face
    - Seen as a filling defect simulating polyp
    - Intraluminal surface: Abrupt well-defined borders
- Diagnosis: Endoscopic biopsy & histology

Ectopic pancreas

- Submucosal lesion
- Seen as smooth, broad-based submucosal mass
  - Indistinguishable from a gastric polyp
- Location: Greater curvature of distal antrum
- Often contain a central umbilication or dimple
  - Represents orifice of a primitive ductal system
- May present as a "bull's eye" appearance
- Rarely, due to barium reflux into rudimentary ducts may produce club shaped pouches (pathognomonic)

PATHOLOGY

General Features

- General path comments: Proliferation of mucosa
- Genetics
  - FAPS: Abnormal or deletion of APC gene located on chromosome 5q
  - Hamartomatous polyposis: Peutz-Jeghers syndrome
    - Spontaneous gene mutation on chromosome 19
- Etiology
  - Chronic atrophic & H. pylori gastritis
  - Hereditary: Autosomal dominant (FAPS & PJS)
    - Familial adenomatous polyposis syndrome
    - Hamartomatous polyposis syndromes
- Epidemiology
  - Incidence
    - Gastric polyps: 1-2% of all GI tract polyps
    - Giant hyperplastic polyps (2% of all hyperplastic)
    - FAPS & PJS: 1 in 10,000 people
- Associated abnormalities: Polyposis syndromes

Gross Pathologic & Surgical Features

- Hyperplastic polyps: Small, sessile nodules; smooth, dome-shaped contour
- Adenomatous polyps: Tubular (thin stalk + tufted head); villous (broad base)
- FAPS: Innumerable small-medium sized polyps
- PJS: Carpet, cluster-like or scattered polyps

Microscopic Features

- Hyperplastic polyps: Elongated, cystically dilated glandular structures
- Adenomatous polyps: Tubular, tubulovillous, villous pattern; dysplastic cells
- PJS: Muscularis mucosa core extends → lamina propria

CLINICAL ISSUES

Presentation

- Most common signs/symptoms
  - Usually asymptomatic
  - Ulcerated polyps: Low grade upper GI bleeding
  - FAPS: Rectal bleeding & diarrhea
  - PJS: Cramping pain, rectal bleeding or melena
  - Mostly incidental findings on imaging & endoscopy
  - Pedunculated polyps in antrum: Nausea & vomiting
    - Due to outlet obstruction
- Diagnosis: Endoscopic biopsy & histology

Demographics

- Age
  - Hyperplastic polyps: Middle & elderly age group
  - FAPS & PJS: 10-30 years
- Gender: Equal in both males & females

Natural History & Prognosis

- Complications
  - Risk of cancer in adenomatous polyp, FAPS & PJS
  - Gastric outlet obstruction
- Prognosis
  - Good: After removal of benign + cancer in situ polyp
  - Poor: Invasive carcinoma

Treatment

- Small < 1 cm & asymptomatic: Periodic surveillance
- Large > 1 cm; sessile or pedunculated; lobulated & symptomatic: Polypectomy

DIAGNOSTIC CHECKLIST

Consider

- Differentiate from other gastric discrete filling defects
- Check for family history of GI tract polyps
- Screen rest of GI tract to rule out associated hereditary polyposis syndromes

Image Interpretation Pearls

- Hyperplastic polyps (typical): Multiple, smooth, sessile, round or ovoid lesions, < 1 cm in size
- Adenomatous polyps: Solitary, sessile or pedunculated, more lobulated & > than 1 cm in size
- Large, solitary, sessile polyp with lobulated surface & basal indentation, highly suggests adenocarcinoma

SELECTED REFERENCES

1. Insko EK et al: Benign and malignant lesions of the stomach: evaluation of CT criteria for differentiation. Radiology. 228(1):166-71, 2003
2. Ba-Ssalamah A et al: Dedicated multidetector CT of the stomach: spectrum of diseases. Radiographics. 23(3):625-44, 2003
3. Cherukuri R et al: Giant hyperplastic polyps in the stomach: radiographic findings in seven patients. AJR Am J Roentgenol. 175(5):1445-8, 2000
4. Cho GJ et al: Peutz-Jeghers syndrome and the hamartomatous polyposis syndromes: radiologic-pathologic correlation. Radiographics. 17(3):785-91, 1997
5. Harned RK et al: Extracolonic manifestations of the familial adenomatous polyposis syndromes. AJR Am J Roentgenol. 156(3):481-5, 1991
6. Feczko PJ et al: Gastric polyps: radiological evaluation and clinical significance. Radiology. 155(3):581-4, 1985
7. Gordon R et al: Gastric polyps on routine double-contrast examination of the stomach. Radiology. 134(1):27-9, 1980

IMAGE GALLERY

Typical

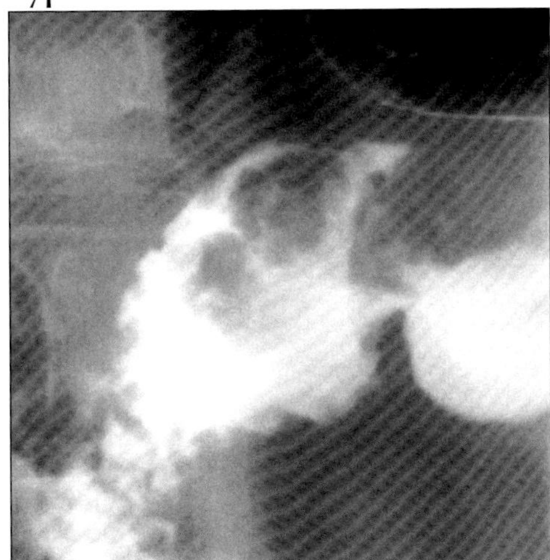

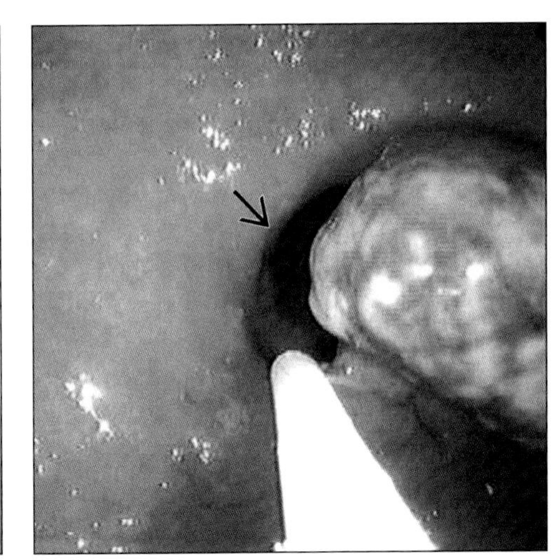

(Left) *Upper GI series shows polypoid mass in duodenal bulb that is a prolapsed gastric antral polyp (adenoma).* ***(Right)*** *Endoscopic photo shows antral polyp (adenoma) that intermittently prolapsed through the pylorus (arrow).*

Typical

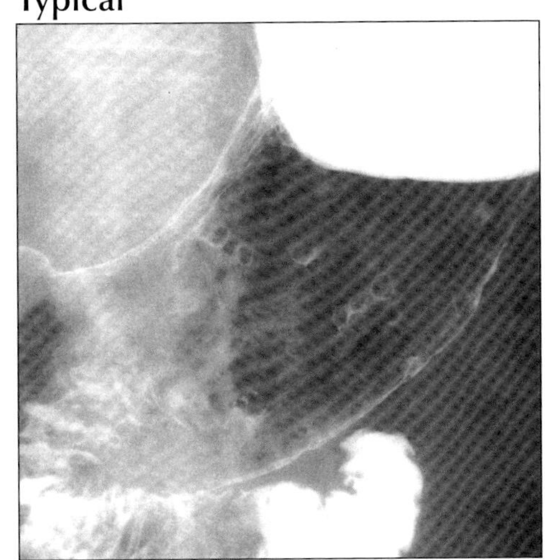

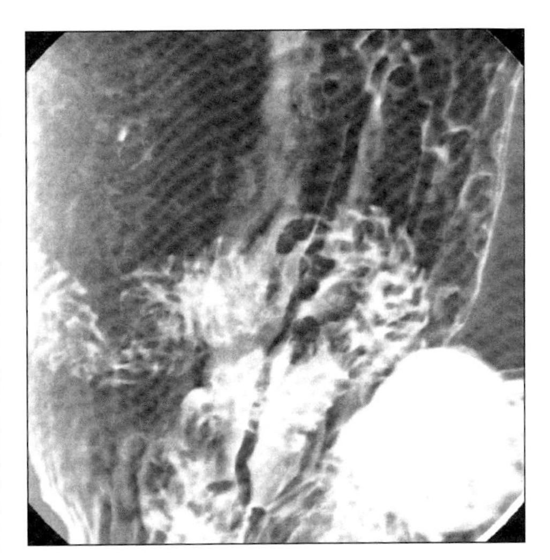

(Left) *Upper GI series shows dozens of small hyperplastic gastric polyps.* ***(Right)*** *Upper GI series shows multiple hyperplastic gastric polyps.*

Typical

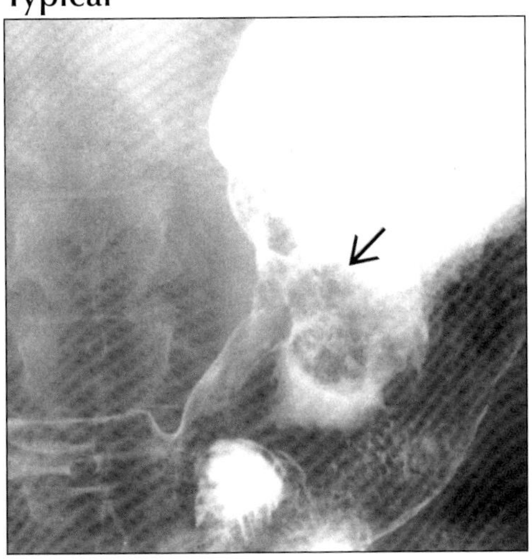

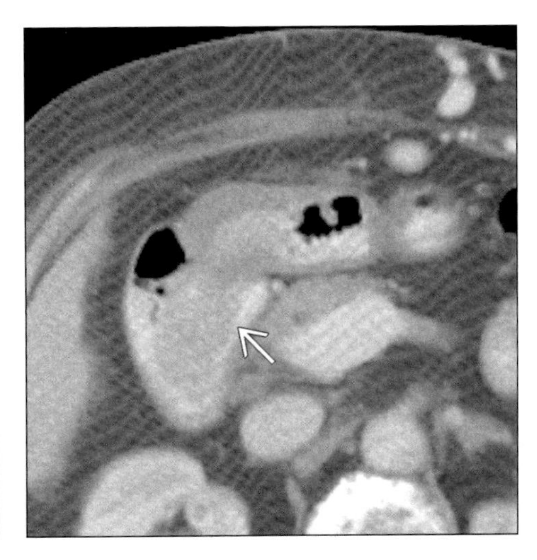

(Left) *Upper GI series shows large adenomatous gastric polyp (arrow).* ***(Right)*** *Axial CECT shows a large gastric adenomatous polyp (arrow) prolapsed into the duodenum.*

DUODENAL POLYPS

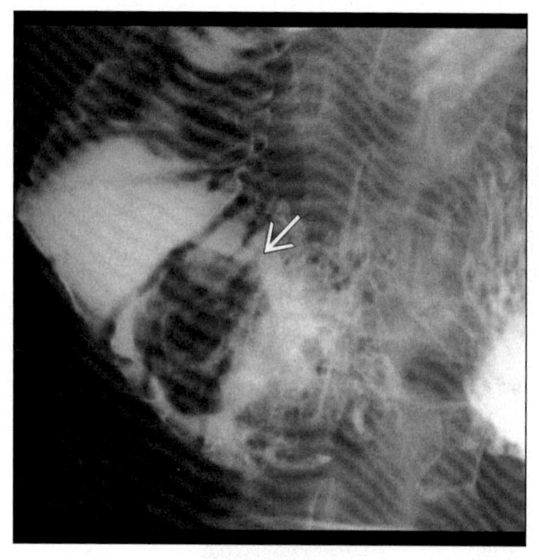

Upper GI series shows large adenomatous polyp (arrow) as a radiolucent filling defect.

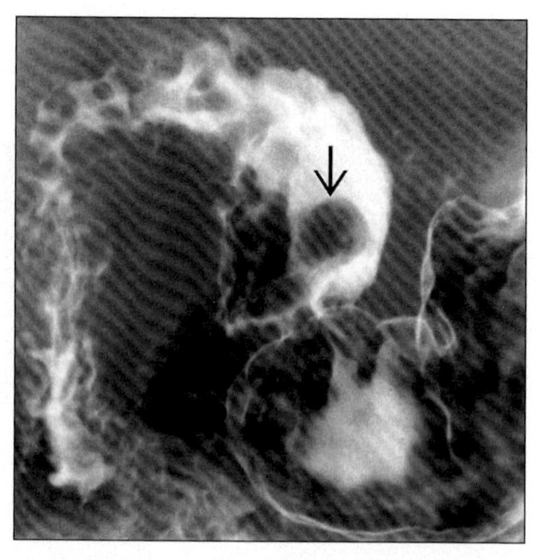

Upper GI series shows a polyp (arrow) in duodenal bulb, endoscopically resected and found to represent a carcinoid tumor.

TERMINOLOGY

Definitions

- Protruding, space-occupying, epithelial lesions

IMAGING FINDINGS

General Features

- Best diagnostic clue: Radiolucent filling defects, ring shadows or contour defect on barium study
- Location: Usually first & second parts of duodenum
- Size: Adenomatous polyps: Few mm to 2 cm
- Morphology
  - Adenomatous polyps (most common)
    - Usually single with lobulated or cauliflower-like
  - Hyperplastic polyps: Smooth, sessile, pedunculated
  - Hamartomas: Cluster of broad based polyps
- Other general features
  - Duodenal polyps are < common than gastric polyps
  - Polyps classified into three types based on pathology
    - Adenomatous, hyperplastic & hamartomatous
  - Adenomatous polyps
    - Most common polyps of duodenum
    - Typically arise from medial wall of duodenum
    - Occur in 47-72% of familial polyposis cases
    - Increased risk of malignant change via adenoma-carcinoma sequence
    - Duodenum 2nd most common site of familial adenomatous polyposis (FAPS) after colon
    - FAPS cases: Clustered around periampullary region
    - Non-FAPS cases: Bulbar distribution
    - Tubular (75%); tubulovillous (15%); villous (10%)
    - Duodenal carcinoma usually seen with adenoma
    - 4% of patients develop periampullary carcinoma in less than 5 years after colectomy
  - Hyperplastic polyps
    - Rare, benign epithelial neoplasms of duodenum
    - Virtually no malignant potential
  - Hamartomatous polyps
    - Usually seen in Peutz-Jegher syndrome (PJS)
    - Duodenum most common after jejunum/ileum

Radiographic Findings

- Fluoroscopic guided double contrast barium study
- Adenomatous polyps
  - Solitary or multiple tiny tubular adenomas
  - Sessile or pedunculated, > lobulated, 5 mm or less
  - Pedunculated polyp (en face)
    - "Mexican hat" sign: Pair of concentric rings
  - Lobulated polyp with basal indentation
    - Increased risk of adenocarcinoma
  - Fungating mass highly suggestive of carcinoma
    - May obstruct distal common bile duct
- Hyperplastic polyps
  - Typical: Small multiple, smooth, sessile, round or ovoid lesions, less than 2 cm in size

DDx: Duodenal Polyps

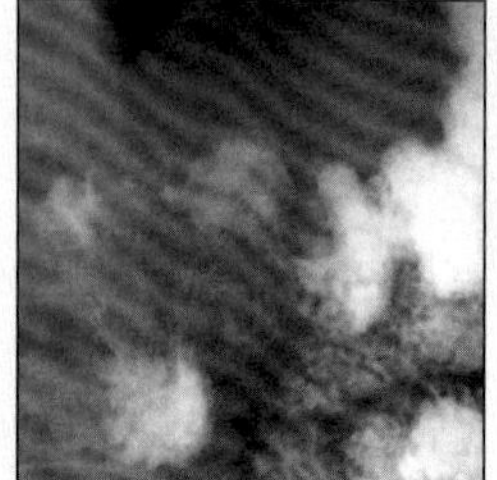

Brunner Glands

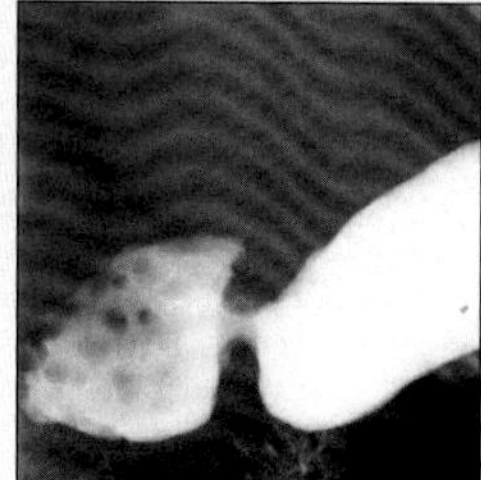

Brunner Glands

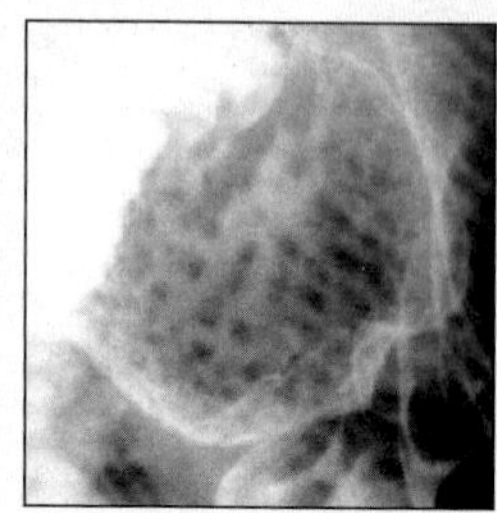

Brunner Glands

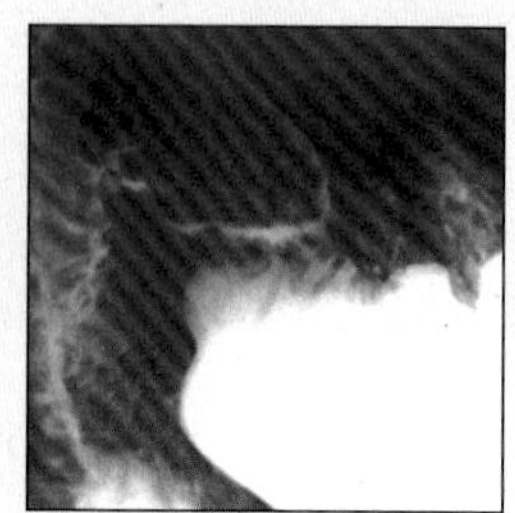

Ectop. Gastric Mucosa

DUODENAL POLYPS

Key Facts

Imaging Findings

- Best diagnostic clue: Radiolucent filling defects, ring shadows or contour defect on barium study
- Adenomatous polyps (most common)
- FAPS cases: Clustered around periampullary region
- Non-FAPS cases: Bulbar distribution
- Sessile or pedunculated, > lobulated, 5 mm or less
- "Mexican hat" sign: Pair of concentric rings
- Fungating mass highly suggestive of carcinoma

Top Differential Diagnoses

- Brunner gland hyperplasia
- Pseudopolyp
- Ectopic gastric mucosa

Diagnostic Checklist

- Check for family history of GI tract polyps
- Screen rest of GI tract to rule out polyposis syndromes

- Based on location: Dependent & nondependent wall
  - Dependent (posterior wall): Filling defect
  - Nondependent (anterior wall): Ring shadows + white rim (barium)
- Atypical: Large or giant pedunculated or lobulated

Imaging Recommendations

- Fluoroscopic guided double contrast barium study
  - En face, profile & oblique views

DIFFERENTIAL DIAGNOSIS

Brunner gland hyperplasia

- Multiple small, rounded nodules in duodenal bulb ("cobblestone" or "Swiss cheese" appearance)
- Brunner gland hamartomas
  - Submucosal or sessile lesions mimicking polyps
  - May also show large polypoid defects

Pseudopolyp

- Seen at apex of duodenal bulb due to acute bend
- Also seen in inflammatory & post inflammatory states

Ectopic gastric mucosa

- Discrete, angulated or polygonal 1-5 mm nodules (filling defects) near base of duodenal bulb

PATHOLOGY

General Features

- Genetics
  - FAPS: Abnormal APC gene on chromosome 5q
  - PJS: Spontaneous gene mutation on chromosome 19
- Etiology: Chronic duodenitis; hereditary (FAPS, PJS)
- Epidemiology: Less than 1% of all GI tract polyps
- Associated abnormalities: Polyposis syndromes

Gross Pathologic & Surgical Features

- Adenomatous polyps: Sessile or pedunculated

Microscopic Features

- Adenomatous: Tubular, tubulovillous, villous pattern

CLINICAL ISSUES

Presentation

- Most common signs/symptoms: Asymptomatic, low grade upper GI bleeding, obstructive jaundice

Demographics

- Age: FAPS & PJS: 10-30 years

Natural History & Prognosis

- Complications: Risk of cancer in adenomatous polyps
- Prognosis: Benign (good); invasive carcinoma (poor)

Treatment

- Small < 1 cm & asymptomatic: Periodic surveillance
- Large > 1 cm, lobulated, symptomatic: Polypectomy

DIAGNOSTIC CHECKLIST

Consider

- Check for family history of GI tract polyps
- Screen rest of GI tract to rule out polyposis syndromes

Image Interpretation Pearls

- Lobulated polyp + basal indentation (adenocarcinoma)

SELECTED REFERENCES

1. Waye JD et al: Approach to benign duodenal polyps. Gastrointest Endosc. 55(7):962-3, 2002
2. Harned RK et al: Extracolonic manifestations of the familial adenomatous polyposis syndromes. AJR Am J Roentgenol. 156(3):481-5, 1991

IMAGE GALLERY

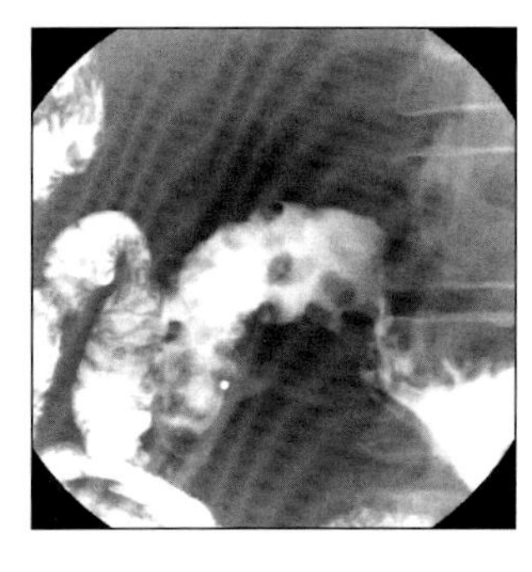
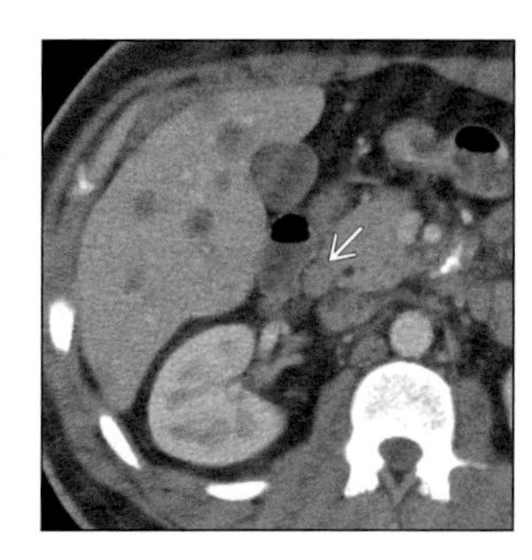

*(**Left**) Upper GI series in a patient with Gardner syndrome shows multiple adenomatous polyps in duodenal bulb. (**Right**) Axial CECT in a patient with Gardner syndrome shows a discrete ampullary tumor (arrow). Liver metastases are from concurrent colon cancer.*

INTRAMURAL BENIGN GASTRIC TUMORS

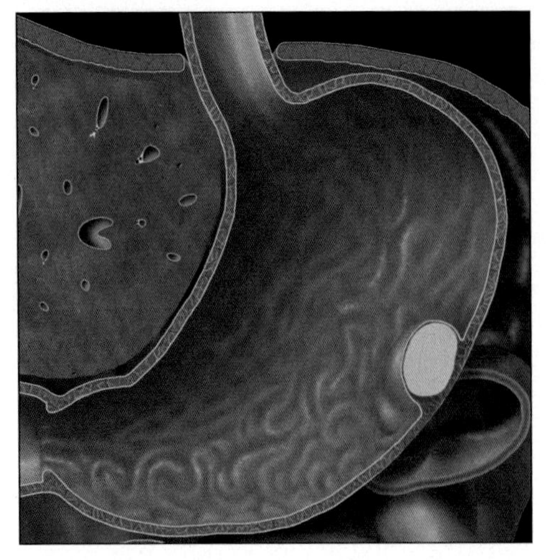

Graphic shows a "generic" intramural gastric mass with intact mucosa and acute to slightly obtuse angles at the interface.

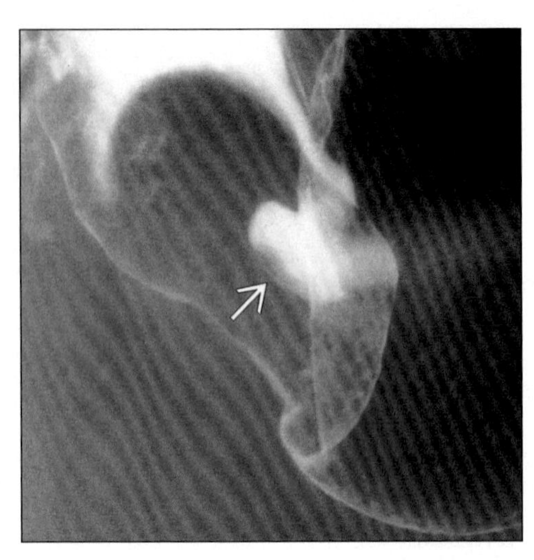

Upper GI series shows a sharply defined submucosal mass with intact mucosa, except for a central ulceration (arrow). Benign stromal tumor.

TERMINOLOGY

Definitions

- Benign mass composed of one or more tissue elements of the gastric wall

IMAGING FINDINGS

General Features

- Best diagnostic clue: Intramural mass with smooth surface & slightly obtuse borders on barium studies
- Other general features
  - Types of intramural benign gastric tumors
    - Gastrointestinal stromal tumor (GIST)
    - Leiomyoma, leiomyoblastoma, schwannoma, neurofibroma, lipoma, hemangioma, lymphangioma

Radiographic Findings

- Radiography
  - Mass indenting gastric air shadow; ± calcifications
  - Lipoma: Radiolucent shadow
  - Hemangioma: Phleboliths (pathognomonic)
- Fluoroscopic-guided barium studies
  - Discrete mass; solitary (very common) or multiple
  - Smooth surface lesion etched in white (double contrast) (profile view)
  - Borders form right angle or slightly obtuse angles with adjacent gastric wall (profile view)
  - Intraluminal surface of tumor has abrupt, well-defined borders (en face view)
  - Usually intact overlying mucosa; normal areae gastricae pattern
  - Focal areas of ulceration (60% of cases)
  - "Bull's eye" or "target" lesions: Central barium-filled crater within mass (ulceration)
  - Central dimple or spicule at apex of mass (exogastric); differentiate from extrinsic mass
  - Pedunculated; may prolapse into duodenum
  - ± Giant, cavitated lesions; may simulate gastric lymphoma and gastric metastases from melanoma
  - GIST
    - Most common; may occur anywhere in GI tract
    - Several mm to 30 cm
    - ± Extragastric extensions (86%): Gastrohepatic ligament, gastrosplenic ligament, lesser sac
  - Lipoma, lymphangioma: Tendency to change in size & shape by peristalsis or palpation
  - Schwannoma and neurofibroma: Multiple lesions with associated abnormalities

CT Findings

- GIST
  - Often large with central necrosis and ulceration of overlying mucosa

DDx: Intramural Gastric Mass

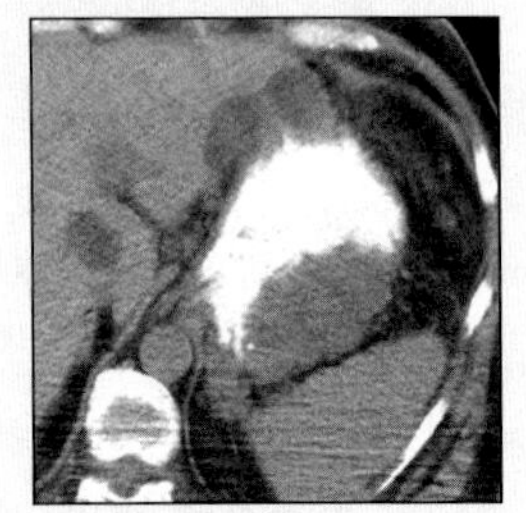

Pseudocyst

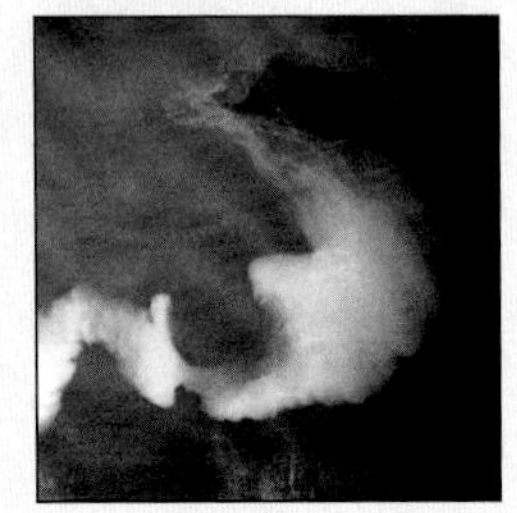

Pseudocyst

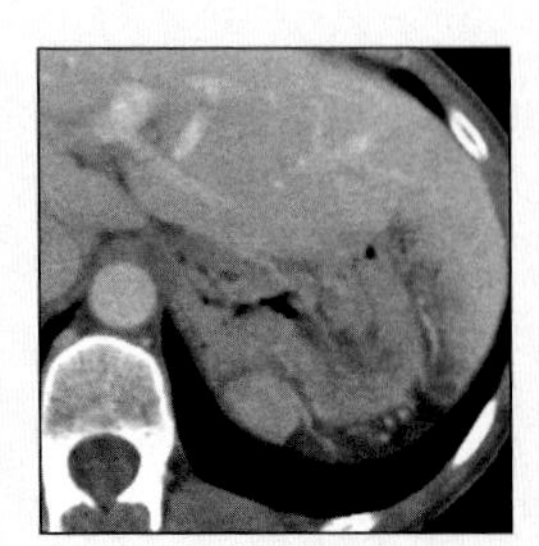

Splenosis

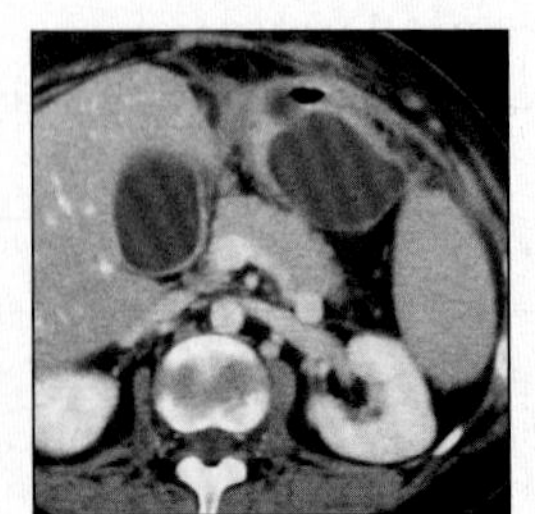

Seroma

INTRAMURAL BENIGN GASTRIC TUMORS

Key Facts

Terminology
- Benign mass composed of one or more tissue elements of the gastric wall

Imaging Findings
- Best diagnostic clue: Intramural mass with smooth surface & slightly obtuse borders on barium studies
- Discrete mass; solitary (very common) or multiple
- Smooth surface lesion etched in white (double contrast) (profile view)
- Borders form right angle or slightly obtuse angles with adjacent gastric wall (profile view)
- "Bull's eye" or "target" lesions: Central barium-filled crater within mass (ulceration)
- Best imaging tool: Barium studies followed by CT

Top Differential Diagnoses
- Gastric carcinoma
- Gastric metastases and lymphoma
- Ectopic pancreatic tissues
- Gastric or duodenal ulcer

Pathology
- Most diagnosed incidentally by imaging or autopsy

Clinical Issues
- Asymptomatic (most common)

Diagnostic Checklist
- GIST is most common; imaging criteria to separate from other intramural tumors are not well established, except for lipoma
- Smooth surface, right/slight obtuse angle with wall

  - Hypo- or hypervascular well-circumscribed submucosal mass (arterial phase)
  - Peripheral enhancement (92%)
  - Central area of low attenuation (hemorrhage, necrosis or cystic formation)
  - ± Cavitation that communicates with gastric lumen and contain air, air-fluid levels or oral contrast
  - ± Homogeneous enhancement (8%)
  - ± Calcification
- Lipoma
  - Located commonly in gastric antrum
  - Well-circumscribed areas of uniform fatty density (-80 to -120 HU); definitive diagnosis

Imaging Recommendations
- Best imaging tool: Barium studies followed by CT

DIFFERENTIAL DIAGNOSIS

Gastric carcinoma
- Usually appear as polypoid or circumferential mass with irregularity of luminal surface
- Large, lobulated, mucosal, hypovascular mass, ± ulceration
- Associated with perigastric or hepatoduodenal ligament and celiac lymphadenopathy
- Mucinous adenocarcinomas: Punctate, granular or finely stippled calcification

Gastric metastases and lymphoma
- Gastric metastases
  - Examples: Breast (most common metastases to stomach), colon, melanoma, lung, pancreas
  - Discrete nodules to linitis plastica
- Gastric lymphoma (e.g., non-Hodgkin B-cell)
  - In GI tract, stomach is the most common location
  - Associated with bulky adenopathy or adenopathy extends into lower abdomen and pelvis
- Multiple "bull's eye" lesions; unlike leiomyoma

Ectopic pancreatic tissues
- Located on greater curvature of distal antrum or proximal duodenum
- Broad-based smooth, extramucosal/intramural lesion
- Central barium collection present in orifice of primitive ductal system; may simulate ulceration
- Usually small (5-10 mm)

Pancreatic pseudocyst
- Gastric compression; may simulate leiomyoma
- No central dimple or spicule at apex of mass suggests mass is extrinsic, not intramural
- Use CT and US to help with diagnosis

Gastric or duodenal ulcer
- Gradual transition with adjacent mucosa
- Radiolucent mound of edema; may simulate ulcerated leiomyoma

Hematoma/seroma
- May follow gastrostomy tube

PATHOLOGY

General Features
- General path comments
  - 50% of all benign tumors in stomach and duodenum are intramural
  - Most diagnosed incidentally by imaging or autopsy
  - GIST
    - Distinguish by immunoreactivity for c-KIT (CD117), a tyrosine kinase growth factor receptor
    - Most common intramural primary masses
    - 70% of all GIST occur in stomach; 2-3% of all gastric tumors
    - Benign are 3x more common than malignant
  - Leiomyoblastoma
    - Predominantly in stomach, may affect small bowel, retroperitoneum, uterus
    - Also known as epithelioid leiomyomas
    - Most are benign, 10% malignant (usually > 6 cm)
  - Lipoma
    - 2-3% benign tumors in stomach, < in duodenum
    - No malignant degeneration
    - 5% of all GI lipomas are in stomach or duodenum
  - Hemangioma

INTRAMURAL BENIGN GASTRIC TUMORS

    - < 2% benign tumors in stomach, < in duodenum
    - Multiple hemangiomas in GI tract and/or skin
    - Classified as capillary or cavernous
    - Sarcomatous changes rarely occur
  - Lymphangioma: Affect anywhere, rare in GI tract
  - Schwannoma and neurofibroma
    - 5-10% of benign tumors in stomach
    - Schwannoma (most common): Neurilemoma, schwannoma or neuroma
    - Neurofibroma (less common): 10% undergo malignant degeneration
    - Neurofibroma: Arise from sympathetic nerves of Auerbach myenteric plexus (more common) or Meissner plexus
    - Sarcomatous changes rarely occur
- Etiology
  - GIST: KIT germ line mutations (52-85%)
  - Hemangioma: Possible congenital malformation
- Associated abnormalities
  - GIST: Carney triad and von Recklinghausen disease
  - Hemangioma: Telangiectasias of skin
  - Neurofibroma: von Recklinghausen disease

Gross Pathologic & Surgical Features

- May have central necrosis and ulceration
- GIST
  - Involves muscularis propria
  - Propensity for exogastric growth; mass arising from gastric wall and project into abdominal cavity
  - Mucosal ulceration on luminal surface (≤ 50%)
  - Well circumscribed mass that compresses adjacent tissue and lacks a true capsule
  - Pink, tan, or gray surface
  - ± Focal areas of hemorrhage, cystic degeneration, necrosis and cavitation
- Leiomyoma: Endogastric (80%), exogastric (15%) or "dumbbell-shaped" (5%)
- Leiomyoblastoma: Smooth muscle tumors
- Lipoma: Endogastric (95%) or exogastric (5%) lesions with superficial ulceration due to pressure necrosis
- Hemangioma: Numerous tiny vascular structures (capillary) or large blood spaces or sinusoids lined by endothelial tissue (cavernous)
- Lymphangioma: Cystic appearance with progressive accumulation of fluid

Microscopic Features

- GIST
  - Spindle cell (70-80%): Cigar-shaped cells, elongated nuclei, eosinophilic to basophilic cytoplasm
  - Epithelioid (20-30%): Round polygonal cells, centrally placed nuclei, cytoplasmic vacuolization
  - Variety of architectural patterns: Bundles of interlacing fascicles, nuclear palisading pattern, nesting organoid pattern and/or vascularity
  - Stromal portions of tumor may show extensive perivascular or stromal hyalinization, myxoid change or hemorrhage
  - < 5 cm in largest dimension with ≤ 5 mitoses per 50 consecutive high power fields (HPF): Benign
  - > 5 cm with ≥ 5 mitoses per 50 HPF: Malignant
- Lipoma: Mature fat cells surrounded by fibrous capsule
- Schwannoma: Bundled spindle-shaped cells with distinctive lymphoid cuff that may contain germinal centers; stain for S-100 protein

CLINICAL ISSUES

Presentation

- Most common signs/symptoms
  - Asymptomatic (most common)
  - Upper GI bleeding, nausea, vomiting, abdominal or epigastric pain, weight loss, abdominal distention

Demographics

- Age: GIST: > 45 years of age
- Gender: GIST: M:F = 1:1

Natural History & Prognosis

- Complications: Obstruction, intussusception, hemorrhage, catastrophic intraperitoneal bleeding
- Prognosis: Good, unless patients with recurrence or size > 5 cm

Treatment

- GIST
  - Surgical resection ± chemotherapy (Gleevec) for metastatic disease
  - Follow-up: Monitor indefinitely for recurrence
- Other types of tumors
  - No treatment if small and asymptomatic
  - Surgery if symptomatic/malignant; usually curative

DIAGNOSTIC CHECKLIST

Consider

- GIST is most common; imaging criteria to separate from other intramural tumors are not well established, except for lipoma

Image Interpretation Pearls

- Smooth surface, right/slight obtuse angle with wall

SELECTED REFERENCES

1. Levy AD et al: Gastrointestinal stromal tumors: radiologic features with pathologic correlation. Radiographics. 23(2):283-304, 456; quiz 532, 2003
2. Pidhorecky I et al: Gastrointestinal stromal tumors: current diagnosis, biologic behavior, and management. Ann Surg Oncol. 7(9):705-12, 2000
3. Suster S: Gastrointestinal stromal tumors. Semin Diagn Pathol. 13(4):297-313, 1996
4. Taylor AJ et al: Gastrointestinal lipomas: a radiologic and pathologic review. AJR Am J Roentgenol. 155(6):1205-10, 1990
5. Heiken JP et al: Computed tomography as a definitive method for diagnosing gastrointestinal lipomas. Radiology. 142(2):409-14, 1982
6. Appleman HD et al: Gastric epithelioid leiomyoma and leiomyosarcoma (leiomyoblastoma). Cancer. 38(2):708-28, 1976
7. Faegenburg D et al: Leiomyoblastoma of the stomach. Report of 9 cases. Radiology. 117(2):297-300, 1975
8. Kerekes ES: Gastric Hemangioma: A case report. Radiology 82:468-9, 1964

INTRAMURAL BENIGN GASTRIC TUMORS

IMAGE GALLERY

Typical

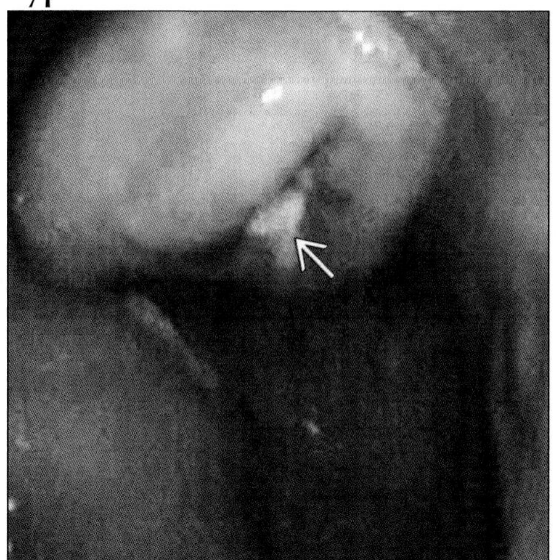

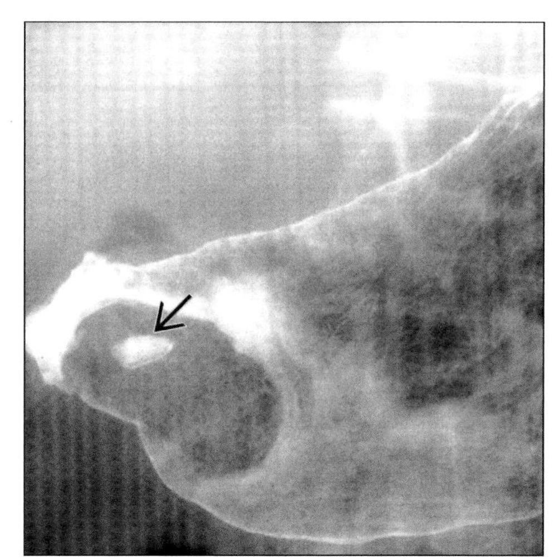

(Left) *Endoscopic photograph shows a submucosal benign gastric stromal tumor with central ulceration (arrow).* ***(Right)*** *Upper GI series shows gastric antral mass (stromal tumor) with intact mucosa, except for central ulcer (arrow).*

Typical

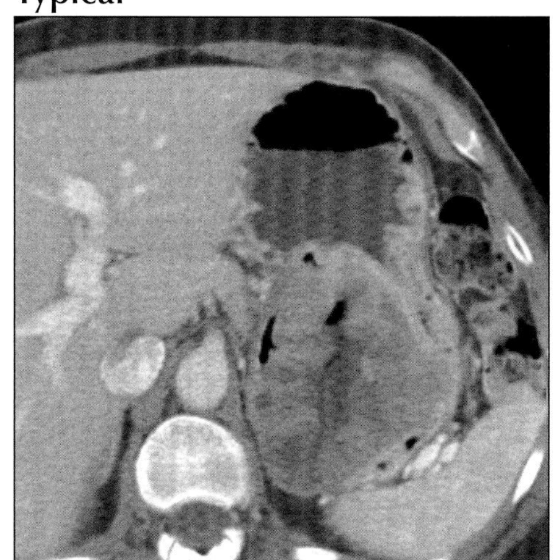

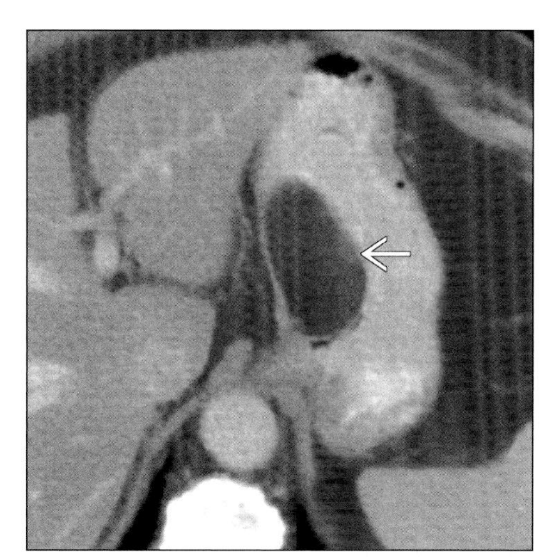

(Left) *Axial CECT shows benign gastric stromal tumor (GIST) as an intramural/exophytic mass that deforms the greater curvature/posterior wall.* ***(Right)*** *Axial CECT shows a discrete fat-density mass (arrow) within the gastric wall with intact, stretched mucosa (lipoma).*

Typical

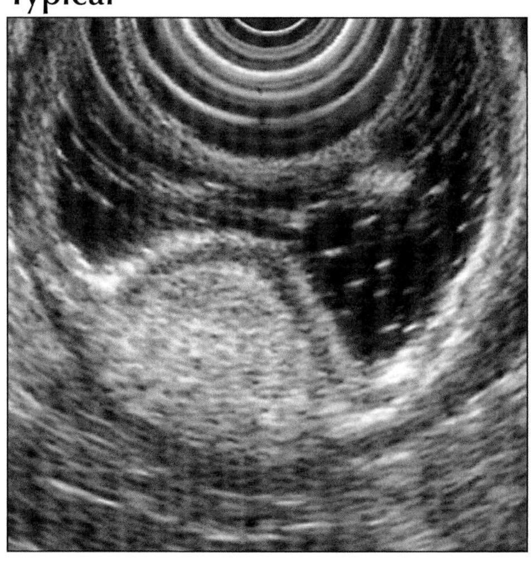

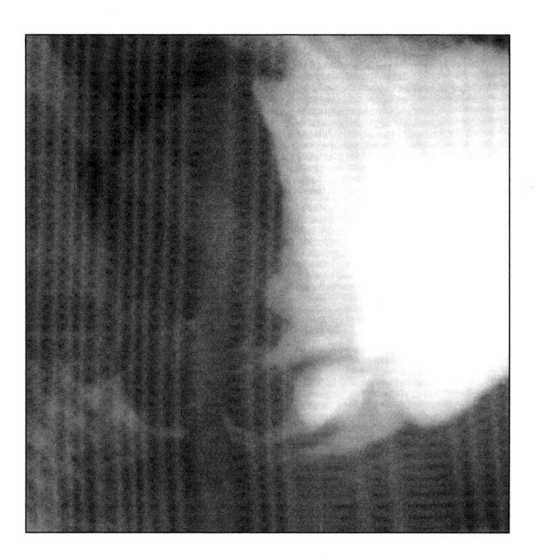

(Left) *Endoscopic sonography shows echogenic submucosal mass (lipoma).* ***(Right)*** *Upper GI series shows an antral submucosal mass prolapsing into the duodenum (lipoma).*

GASTRIC STROMAL TUMOR

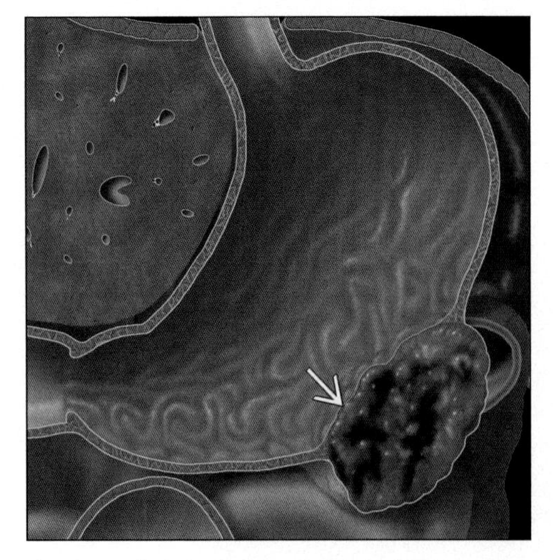

Anatomic depiction of gastric stromal tumor. Note exophytic submucosal mass (arrow) with internal necrosis.

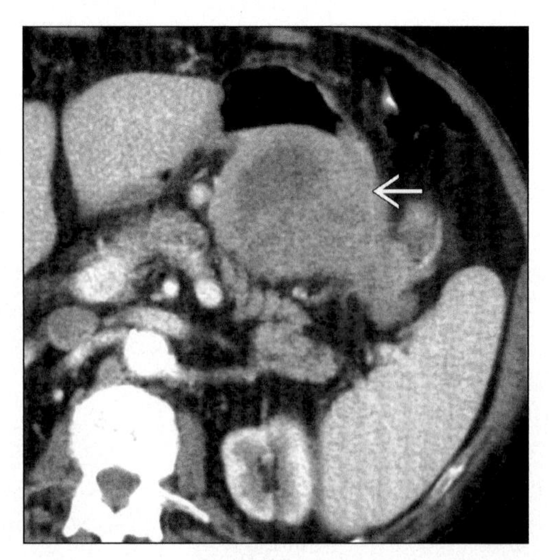

Axial CECT shows exophytic gastric GIST. Note heterogeneous mural mass with smooth interface with stomach (arrow).

TERMINOLOGY

Abbreviations and Synonyms

- Gastrointestinal stromal tumor (GIST)

Definitions

- Submucosal tumor of gastrointestinal (GI) tract derived from interstitial cells of Cajal

IMAGING FINDINGS

General Features

- Best diagnostic clue: Well-circumscribed submucosal mass extending exophytically from GI tract
- Location
  - Stomach most common site (2/3 of cases)
  - Small bowel (especially duodenum) next most common site
  - May occur anywhere in GI tract
  - Rarely occurring in esophagus (leiomyoma more common)
- Size
  - Variable
  - Large mass may be > 5 cm
- Morphology
  - Bulky, well-circumscribed and lobulated
  - Often exophytic, may have cystic element

Radiographic Findings

- Fluoroscopy
  - UGI
    - Rounded, exophytic, submucosal gastric mass
    - Ulcerations common in larger masses

CT Findings

- NECT: Calcifications in 25% of cases
- CECT
  - Hypo- or hypervascular well-circumscribed submucosal mass on arterial phase images; ulceration & necrosis common on CECT
  - Sensitivity 93%, specificity 100%

MR Findings

- T1WI: Isointense mass
- T2WI
  - Hypo- to isointense submucosal mass
  - Hyperintense areas of necrosis
- T2* GRE: Hyper- or hypointense with IV gadolinium on GRE sequences
- T1 C+
  - Variable vascularity; may be hyper- or hypovascular
  - Enhancement of solid areas
  - Nonenhancing necrotic or hemorrhagic areas

Ultrasonographic Findings

- Real Time: Hypoechoic mass

DDx: Spectrum of Gastric Lesions Mimicking GIST

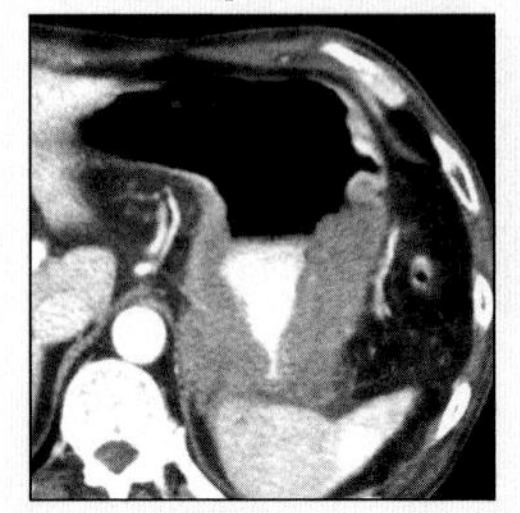

Lymphoma

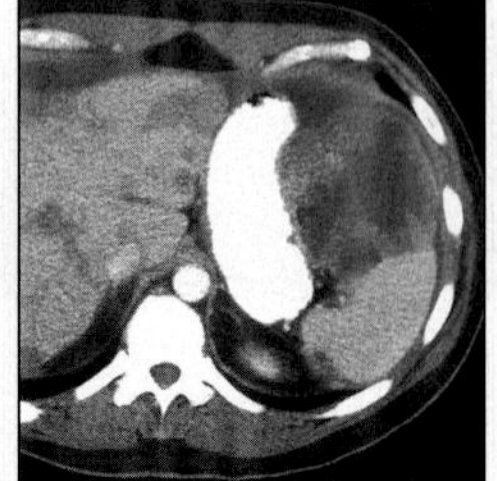

Sarcoma

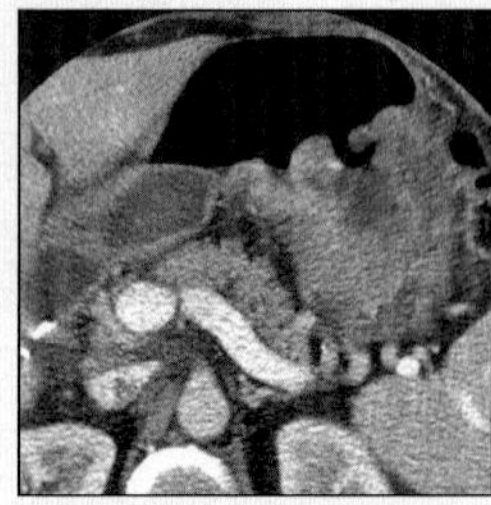

Carcinoma

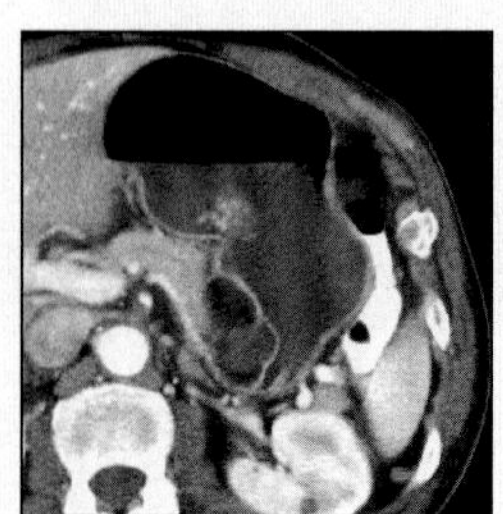

Lipoma

GASTRIC STROMAL TUMOR

Key Facts

Imaging Findings

- Hypo- or hypervascular well-circumscribed submucosal mass on arterial phase images; ulceration & necrosis common on CECT
- PET is superior to CT on predicting early response to Gleevec; hypermetabolic foci for both primary tumor & mets
- Best imaging tool: CECT, PET

Top Differential Diagnoses

- Gastric lymphoma
- Sarcoma invading stomach
- Exophytic gastric carcinoma
- Submucosal GI lipoma

Pathology

- GIST are distinct, not synonymous with leiomyoma/sarcoma, but may not be diagnosed by light microscopy alone

Clinical Issues

- Most common signs/symptoms: Mass effect from bulky tumor, GI bleed when ulcerated, nausea, vomiting, weight loss
- Excellent prognosis for completely resected benign lesions
- Good response to chemotherapy (Gleevec) in patients with metastatic disease and c-KIT mutation
- Prognosis often depends on tumor size; poor if > 5 cm

- Color Doppler: Variable vascularity on color Doppler

Nuclear Medicine Findings

- PET
  - PET is superior to CT on predicting early response to Gleevec; hypermetabolic foci for both primary tumor & mets
  - Sensitivity 86%, specificity 98%

Imaging Recommendations

- Best imaging tool: CECT, PET
- Protocol advice
  - Prior to scanning, distend stomach with 16-32 oz water
  - Use biphasic technique to cover entire liver
    - 150 ml IV contrast injected at 4-5 ml/sec
    - Arterial phase acquisition at 40 seconds, venous phase at 70 seconds
    - 2.5 mm collimation and 2.5-5 mm reconstruction interval

DIFFERENTIAL DIAGNOSIS

Gastric lymphoma

- Early stage polypoid type
- Nodular fold thickening on barium studies
- Exophytic mass without bowel obstruction
- Associated mesenteric and retroperitoneal adenopathy
- Bulky submucosal mass
- May ulcerate
- May be indistinguishable from GIST

Sarcoma invading stomach

- Bulky mass
- Heterogeneous on CECT
- Liposarcomas contain fat
- Secondary invasion of bowel mimics GIST
- Primary location in mesentery aids in differentiation
- Bowel obstruction common unlike GIST

Exophytic gastric carcinoma

- Hypodense mass less vascular than GIST
- May be bulky and exophytic on CT/MR
- Focal thickening of adjacent gastric wall and gastric outlet obstruction help differentiate from GIST
- Often causes obstruction when circumferential

Submucosal GI lipoma

- Fatty attenuation diagnostic

PATHOLOGY

General Features

- General path comments
  - Bulky submucosal mass
  - Central ulceration common
- Genetics
  - Express growth factor receptor with tyrosine kinase activity (c-KIT CD117)
  - Embryology-anatomy
    - Of mesenchymal origin, not related to leiomyomas or leiomyosarcomas
    - Derived from interstitial cells of Cajal that help regulate peristaltic activity (pacemaker function)
- Etiology: Unknown
- Epidemiology: Most common mesenchymal tumor of GI tract
- Associated abnormalities
  - Carney triad
    - Malignant epithelial gastric GIST
    - Pulmonary chondroma
    - Extra-adrenal paraganglioma
  - von Recklinghausen disease
    - Neurofibromatosis type 1

Gross Pathologic & Surgical Features

- Bulky submucosal mass
- Benign lesions typically small (< 3 cm)
- Malignant features include invasion, size > 5 cm, and evidence of metastases

Microscopic Features

- GIST are distinct, not synonymous with leiomyoma/sarcoma, but may not be diagnosed by light microscopy alone

GASTRIC STROMAL TUMOR

- Benign or malignant mesenchymal spindle cell or epithelioid neoplasm without muscle differentiation
- Malignant features include high mitotic rate (> 10 mitoses per 50 high power fields), high nuclear grade, and high cellularity

Staging, Grading or Classification Criteria

- Four tumor subtypes
  - Benign spindle cell GIST
  - Malignant spindle cell GIST
  - Benign epithelial GIST
  - Malignant epithelial GIST

CLINICAL ISSUES

Presentation

- Most common signs/symptoms: Mass effect from bulky tumor, GI bleed when ulcerated, nausea, vomiting, weight loss
- Clinical profile: No specific lab abnormality

Demographics

- Age: > 45 y
- Gender: No gender predilection

Natural History & Prognosis

- Metastasizes to liver, lungs and peritoneal cavity
- Excellent prognosis for completely resected benign lesions
- Good response to chemotherapy (Gleevec) in patients with metastatic disease and c-KIT mutation
- 50-80% 5-year survival
- Prognosis often depends on tumor size; poor if > 5 cm

Treatment

- Surgery with en bloc resection
- Tyrosine kinase inhibitor chemotherapy (Gleevec) for metastatic disease

DIAGNOSTIC CHECKLIST

Consider

- Consider lymphoma

Image Interpretation Pearls

- Exophytic hypervascular GI mass arising from submucosa with central ulceration

SELECTED REFERENCES

1. Logrono R et al: Recent Advances in Cell Biology, Diagnosis, and Therapy of Gastrointestinal Stromal Tumor (GIST). Cancer Biol Ther. 2004
2. Antoch G et al: Comparison of PET, CT, and Dual-Modality PET/CT Imaging for Monitoring of Imatinib (STI571) Therapy in Patients with Gastrointestinal Stromal Tumors. J Nucl Med. 45(3):357-365, 2004
3. Haider N et al: Gastric stromal tumors in children. Pediatr Blood Cancer. 42(2):186-9, 2004
4. Gayed I et al: The role of 18F-FDG PET in staging and early prediction of response to therapy of recurrent gastrointestinal stromal tumors. J Nucl Med. 45(1):17-21, 2004
5. Hu X et al: Primary malignant gastrointestinal stromal tumor of the liver. Arch Pathol Lab Med. 127(12):1606-8, 2003
6. Lin SC et al: Clinical manifestations and prognostic factors in patients with gastrointestinal stromal tumors. World J Gastroenterol. 9(12):2809-12, 2003
7. Bechtold RE et al: Cystic changes in hepatic and peritoneal metastases from gastrointestinal stromal tumors treated with Gleevec. Abdom Imaging. 28(6):808-14, 2003
8. Kinoshita K et al: Endoscopic ultrasonography-guided fine needle aspiration biopsy in follow-up patients with gastrointestinal stromal tumours. Eur J Gastroenterol Hepatol. 15(11):1189-93, 2003
9. Rossi CR et al: Gastrointestinal stromal tumors: from a surgical to a molecular approach. Int J Cancer. 107(2):171-6, 2003
10. Connolly EM et al: Gastrointestinal stromal tumours. Br J Surg. 90(10):1178-86, 2003
11. Wu PC et al: Surgical treatment of gastrointestinal stromal tumors in the imatinib (STI-571) era. Surgery. 134(4):656-65; discussion 665-6, 2003
12. Tateishi U et al: Gastrointestinal stromal tumor. Correlation of computed tomography findings with tumor grade and mortality. J Comput Assist Tomogr. 27(5):792-8, 2003
13. Reddy MP et al: F-18 FDG PET imaging in gastrointestinal stromal tumor. Clin Nucl Med. 28(8):677-9, 2003
14. Dong Q et al: Epithelioid variant of gastrointestinal stromal tumor: Diagnosis by fine-needle aspiration. Diagn Cytopathol. 29(2):55-60, 2003
15. Wong NA et al: Prognostic indicators for gastrointestinal stromal tumours: a clinicopathological and immunohistochemical study of 108 resected cases of the stomach. Histopathology. 43(2):118-26, 2003
16. Frolov A et al: Response markers and the molecular mechanisms of action of Gleevec in gastrointestinal stromal tumors. Mol Cancer Ther. 2(8):699-709, 2003
17. Besana-Ciani I et al: Outcome and long term results of surgical resection for gastrointestinal stromal tumors (GIST). Scand J Surg. 92(3):195-9, 2003
18. Ghanem N et al: Computed tomography in gastrointestinal stromal tumors. Eur Radiol. 13(7):1669-78, 2003
19. Duffaud F et al: Gastrointestinal stromal tumors: biology and treatment. Oncology. 65(3):187-97, 2003
20. Rosai J: GIST: an update. Int J Surg Pathol. 11(3):177-86, 2003
21. Burkill GJ et al: Malignant gastrointestinal stromal tumor: distribution, imaging features, and pattern of metastatic spread. Radiology. 226(2):527-32, 2003
22. Belloni M et al: Endoscopic ultrasound and Computed Tomography in gastric stromal tumours. Radiol Med (Torino). 103(1-2):65-73, 2002
23. Miettinen M et al: Evaluation of malignancy and prognosis of gastrointestinal stromal tumors: A review. Hum Pathol 33(5): 478-83, 2002
24. Kim CJ et al: Gastrointestinal stromal tumors: Analysis of clinical and pathologic factors. Am Surg 67(2): 135-7, 2001
25. Shojaku H et al: Malignant gastrointestinal stromal tumor of the small intestine: Radiologic-pathologic correlation. Radiat Med 5(3): 189-92, 1997

IMAGE GALLERY

Typical

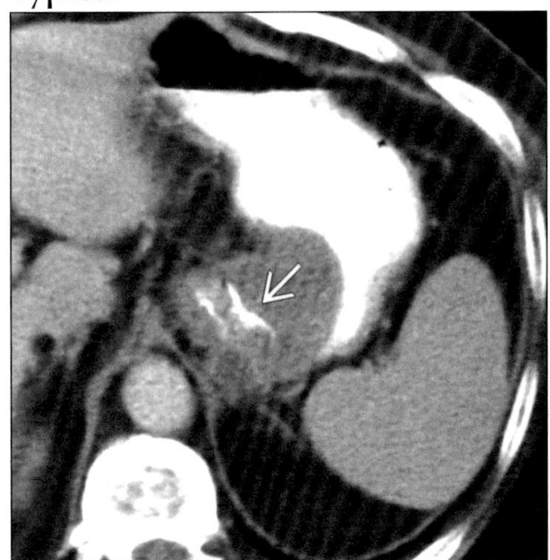

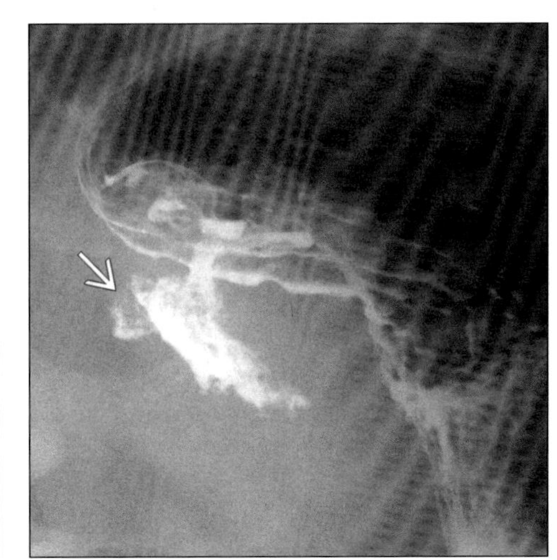

(Left) *Axial CECT shows ulcerated GIST. Note rounded mural mass with oral contrast extending into area of ulceration (arrow).* ***(Right)*** *Ulcerated GIST on lateral view of UGI. Note large accumulation of barium within ulceration (arrow).*

Typical

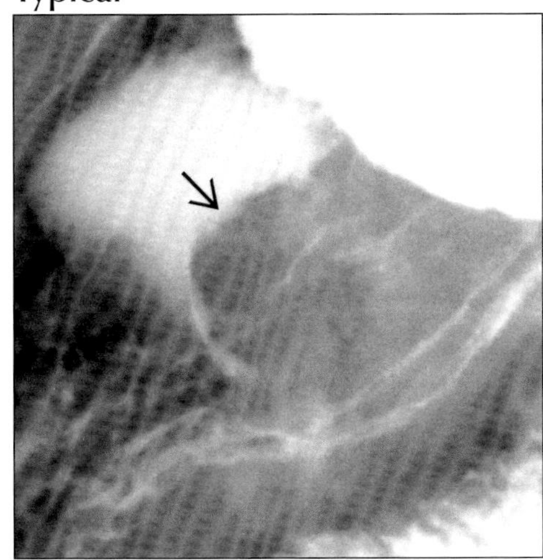

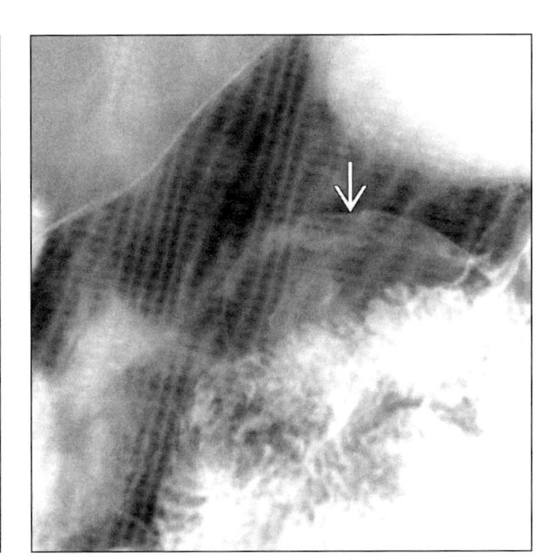

(Left) *GIST on AP view of UGI. Note smooth interface of mass with barium pool (arrow).* ***(Right)*** *AP air contrast view of GIST on UGI demonstrates slight lobulated contour of submucosal mass (arrow).*

Typical

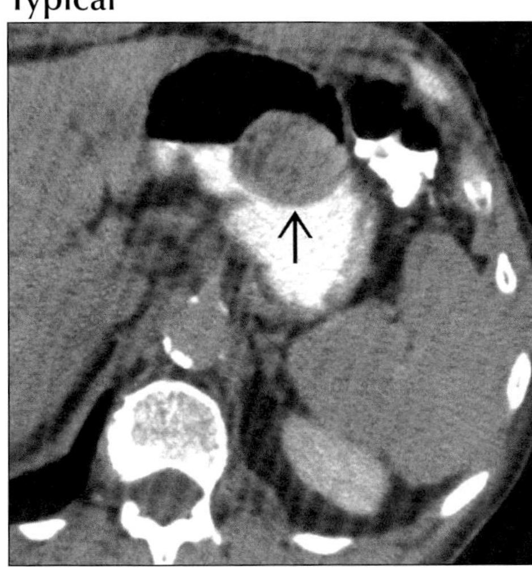

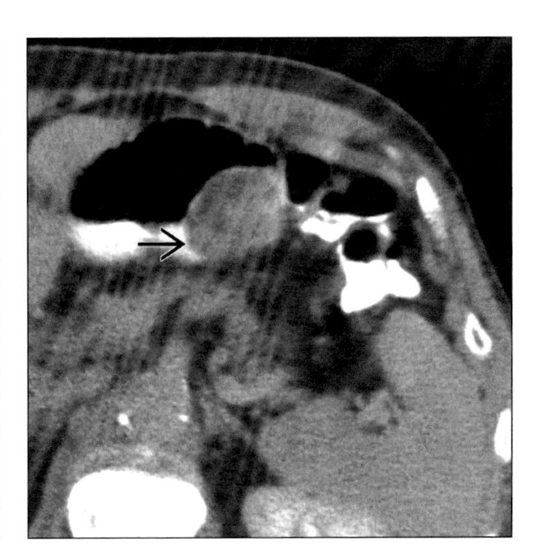

(Left) *Axial CECT of contrast-filled stomach demonstrates rounded mural mass (arrow).* ***(Right)*** *Axial CECT of GIST demonstrates homogeneously enhancing mural mass (arrow). The intraluminal polypoid component is less common than an exophytic extension.*

GASTRIC CARCINOMA

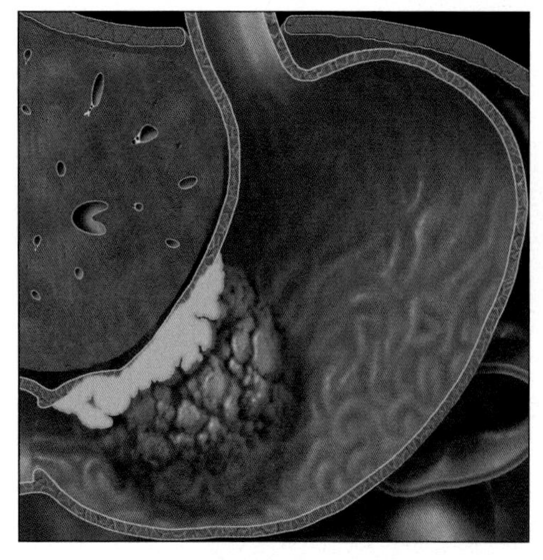

Graphic shows large mass with broad base and irregular surface.

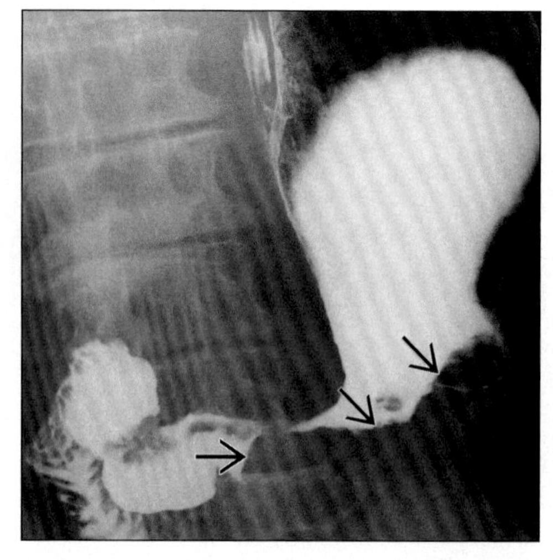

Upper GI series shows large mass (arrows) with a broad base and an irregular nodular surface.

TERMINOLOGY

Definitions

- Malignancy arising from gastric mucosa

IMAGING FINDINGS

General Features

- Best diagnostic clue: Polypoid or circumferential mass with no peristalsis through lesion
- Morphology: Polypoid, ulcerated, infiltrative lesions
- Other general features
  - 3rd most common GI malignancy after colorectal & pancreatic carcinoma
  - Adenocarcinoma (95%) is most common primary gastric tumor
  - Environmental factors have a major role in development of gastric cancer
  - Helicobacter pylori (3-6 fold ↑ risk); pernicious anemia (2-3 fold ↑ risk)
  - Spread of gastric carcinoma
    - Direct spread
    - Lymphatic (left supraclavicular Virchow node)
    - Hematogenous or transperitoneal: Krukenberg tumor (ovary); Blumer shelf (rectal wall)

Radiographic Findings

- Fluoroscopy
  - Early gastric cancer (elevated,superficial, shallow)
    - Type I: Elevated lesion-protrudes > 5 mm into lumen (polypoid)
    - Type II: Superficial lesion (plaque-like, mucosal nodularity, ulceration)
    - Type III: Shallow, irregular ulcer crater with adjacent nodular mucosa & clubbing/fusion/amputation of radiating folds
  - Advanced gastric cancer
    - Polypoid cancer can be lobulated or fungating
    - Lesion on dependent or posterior wall: Filling defect in barium pool
    - Lesion on nondependent or anterior wall: Etched in white by a thin layer of barium trapped between edge of mass & adjacent mucosa
    - Prolapsed polypoid antral carcinoma into duodenum: Seen as filling defect in barium pool
    - Ulcerated carcinoma (penetrating cancer): Accounts for 70% of all gastric cancers
  - Malignant ulcer (in profile)
    - Malignant ulcer has an intraluminal location within a tumor
    - Tumor surrounding ulcer forms acute angle with gastric wall
    - Clubbed/nodular folds seen radiating to edge of ulcer crater

DDx: Diffuse or Focal Thickening of Gastric Wall

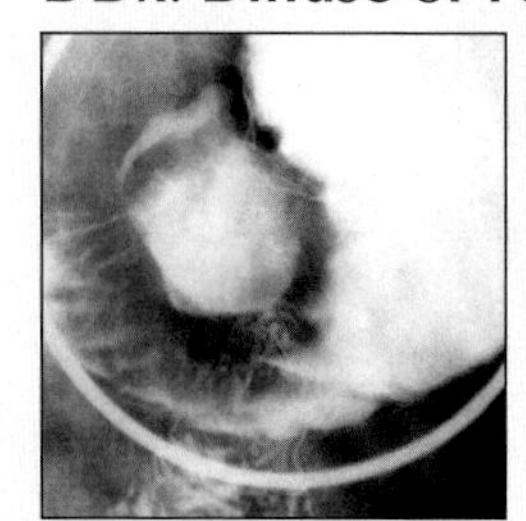

Benign Ulcer

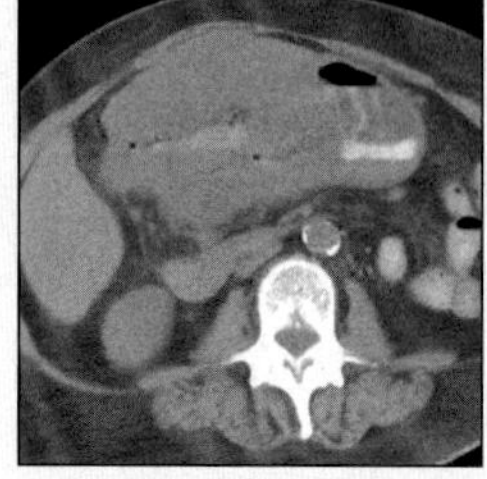

Lymphoma

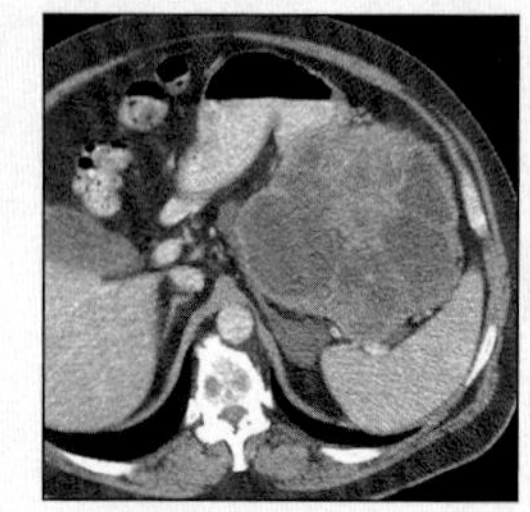

Gastric GIST

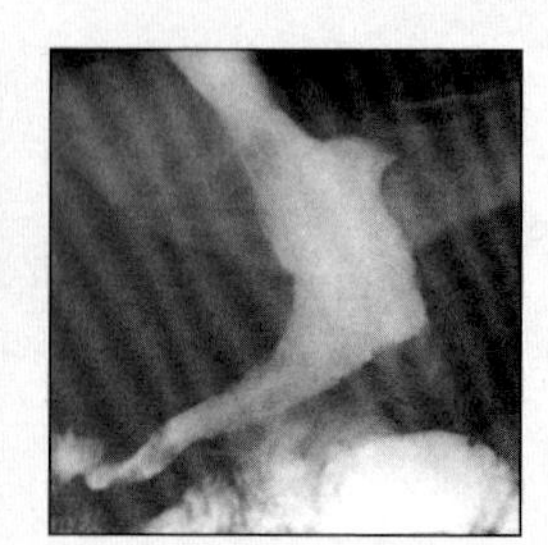

Caustic Gastritis

GASTRIC CARCINOMA

Key Facts

Terminology
- Malignancy arising from gastric mucosa

Imaging Findings
- Best diagnostic clue: Polypoid or circumferential mass with no peristalsis through lesion
- Early gastric cancer (elevated,superficial, shallow)
- Polypoid cancer can be lobulated or fungating
- Ulcer: Irregular, scalloped, angular, stellate borders
- Carman-Kirkland meniscus complex (lesser curvature antrum or body)
- Irregular narrowing of stomach + nodularity + mucosal spiculation
- Advanced cases: May cause gastric outlet obstruction
- Linitis plastica or "leather bottle": Irregular narrowing & rigidity (scirrhous carcinoma)

Top Differential Diagnoses
- Benign gastric (peptic) ulcer
- Gastritis
- Gastric metastases & lymphoma
- Gastric stromal tumor
- Pancreatitis (extrinsic inflammation)
- Menetrier disease

Pathology
- Risk factors: H. pylori, atrophic gastritis, pernicious anemia, adenomatous polyps, Menetrier, partial gastrectomy (Billroth II), blood type-A, smoking

Diagnostic Checklist
- Differentiate from other pathologies that can mimic gastric cancer on imaging; usually require deep biopsy

- Malignant ulcer (en face)
  - Ulcer: Irregular, scalloped, angular, stellate borders
  - Converging folds to ulcer: Blunted, nodular, clubbed, fused
  - Ulcer on nondependent or anterior wall: Double-ring shadow (outer ring represents edge of tumor & inner ring represents edge of ulcer)
  - Prone compression view: Demonstrate filling of ulcer crater within discrete tumor on anterior wall
- Carman-Kirkland meniscus complex (lesser curvature antrum or body)
  - Broad, flat lesion with central ulceration & elevated margins
  - Prone compression view (mass on anterior wall): Radiolucent halo (filling defect) due to elevated edges; meniscoid ulcer-convex inner border + concave outer border
- Infiltrating Ca: 5-15% of all gastric cancers
  - Irregular narrowing of stomach + nodularity + mucosal spiculation
  - Advanced cases: May cause gastric outlet obstruction
- Scirrhous carcinoma (5-15%): Usually arise near pylorus & extend up
  - Linitis plastica or "leather bottle": Irregular narrowing & rigidity (scirrhous carcinoma)
  - Localized scirrhous tumor: Short, annular lesion/shelf-like proximal borders in prepyloric region of antrum (fundus/body-40% cases)
  - Diffuse linitis plastica: Diffusely infiltrated by a scirrhous tumor (nodularity, spiculation, ulceration, or thickened irregular folds)

CT Findings
- Demonstration of lesions facilitated by: Negative contrast agents (water or gas)
  - A polypoid mass with or without ulceration
  - Focal wall thickening with mucosal irregularity or focal infiltration of wall
  - Ulceration: Gas-filled ulcer crater within mass
  - Infiltrating carcinoma: Wall thickening + loss of normal rugal fold pattern
    - Wisp-like perigastric soft tissue stranding: Perigastric fat extension
  - Scirrhous carcinoma: Markedly enhancing thickened wall on a dynamic CT scan
  - Mucinous carcinoma: Decreased attenuation of thickened wall (↑ mucin); calcification seen
  - Carcinoma of cardia: Irregular soft tissue thickening; lobulated mass

Ultrasonographic Findings
- Endoscopic ultrasonography (EUS)
  - Used to stage carcinoma; to assess depth of wall invasion & perigastric lymph nodes

Imaging Recommendations
- Double-contrast barium study, NE + CECT, EUS

DIFFERENTIAL DIAGNOSIS

Benign gastric (peptic) ulcer
- Round ulcer, smooth mound of edema, smooth radiating folds to ulcer edge
- Classic features of benign ulcer: Hampton line, ulcer collar, ulcer mound (diagnosis-endoscopic biopsy)

Gastritis
- Erosive gastritis
  - Varioliform erosions: Multiple punctate or slit-like collections of barium surrounded by radiolucent halos of edematous mucosa
  - Location: Typically in gastric antrum on crests of rugal folds
  - NSAID-related gastropathy: Subtle flattening & deformity of greater curvature of antrum
- Nonerosive or atrophic gastritis
  - Narrowed, tubular stomach; ↓/absent mucosal folds
- Antral, H. pylori & hypertrophic gastritis
  - Markedly thickened, lobulated gastric folds
- Granulomatous gastritis (Crohn disease)
  - Thickened nodular folds in antrum
  - "Ram's horn" sign: Smooth funnel-shaped narrowing
- Eosinophilic, radiation & AIDS-related gastritis

- Mucosal nodularity, erosions, ulcers, thickened folds & antral narrowing
- Chronic radiation gastritis
  - Antral narrowing may mimic scirrhous cancer

Gastric metastases & lymphoma

- Gastric metastases: Most common organs of origin
  - Colon, malignant melanoma, breast, lung, pancreas
  - Breast cancer: Most common metastases to stomach
- Gastric lymphoma (e.g., non-Hodgkin B-cell)
  - Stomach most commonly involved organ in GIT
- Barium findings
  - Colon cancer with gastric invasion: Mass effect, nodularity, spiculated & tethered mucosal folds
    - Transverse colon cancer invade stomach via gastrocolic ligament (vice versa)
  - Lobular breast cancer metastases
    - Linitis plastica or "leather bottle" appearance, thickened irregular folds (simulating scirrhous carcinoma)
  - Gastric lymphoma: Diffuse, thickened irregular folds
- CT findings
  - Markedly thickened gastric wall & mucosal folds
- Diagnosis: Endoscopic biopsy

Gastric stromal tumor

- Example: Leiomyosarcoma or GIST
  - Large, lobulated submucosal mass; ± cavitation
  - Intramural-50%, exogastric-35%, endogastric-15%

Caustic gastritis

- Subacute or chronic phase resembles linitis plastica

Pancreatitis (extrinsic inflammation)

- Mimic gastric cancer due to thickened gastric wall
- CT will show peripancreatic inflammation

Menetrier disease

- Markedly thickened, lobulated folds in gastric fundus & body usually sparing antrum may simulate cancer
- Diagnosis: Endoscopic full-thickness biopsy

PATHOLOGY

General Features

- Etiology
  - Risk factors: H. pylori, atrophic gastritis, pernicious anemia, adenomatous polyps, Menetrier, partial gastrectomy (Billroth II), blood type-A, smoking
  - Diet: Rich in nitrites or nitrates; salted, smoked, poorly-preserved food
- Epidemiology
  - Incidence
    - Uncommon & decreasing in US
    - Common in Japan, Chile, Finland, Poland, Iceland

Gross Pathologic & Surgical Features

- Polypoid, ulcerated, local or diffuse infiltrative & rarely multiple lesions

Microscopic Features

- Well-differentiated adenocarcinoma
- Signet ring cell, papillary, tubular, mucinous
- Early Ca: Limited to mucosa & submucosa
- Advanced: Mucosa, submucosa & muscularis propria

Staging, Grading or Classification Criteria

- CT staging of gastric cancer
  - I: Intraluminal mass
  - II: Intraluminal mass + gastric wall thickness > 1 cm
  - III: Adjacent structures + lymph nodes
  - IV: Distant metastases

CLINICAL ISSUES

Presentation

- Most common signs/symptoms
  - Asymptomatic, anorexia, weight loss, anemia, pain
  - Melena, enlarged left supraclavicular Virchow node
- Lab-data: Hypochromic, microcytic anemia; stool positive for occult blood
- Diagnosis: Endoscopic biopsy & histology

Demographics

- Age: Middle & elderly age group
- Gender: Males more than females (M:F = 2:1)

Natural History & Prognosis

- Complications
  - Gastric outlet obstruction in antral carcinoma
- Prognosis
  - 5 year survival rate
    - Early gastric cancer (85-100%)
    - Advanced cancer (3-21%)

Treatment

- Radiotherapy; chemotherapy
- Surgery: Subtotal or total gastrectomy

DIAGNOSTIC CHECKLIST

Consider

- Differentiate from other pathologies that can mimic gastric cancer on imaging; usually require deep biopsy

Image Interpretation Pearls

- Gastric carcinoma can be ulcerative, polypoid or infiltrative (scirrhous type) + local & distant metastases

SELECTED REFERENCES

1. Habermann CR et al: Preoperative staging of gastric adenocarcinoma: comparison of helical CT and endoscopic US. Radiology. 230(2):465-71, 2004
2. Insko EK et al: Benign and malignant lesions of the stomach: evaluation of CT criteria for differentiation. Radiology. 228(1):166-71, 2003
3. Ba-Ssalamah A et al: Dedicated multidetector CT of the stomach: spectrum of diseases. Radiographics. 23(3):625-44, 2003
4. Horton KM et al: Current role of CT in imaging of the stomach. Radiographics. 23(1):75-87, 2003
5. Fishman EK et al: CT of the stomach: spectrum of disease. Radiographics. 16(5):1035-54, 1996
6. Levine MS et al: The Helicobacter pylori revolution: Radiologic perspective. Radiology 195: 593-6, 1995
7. Urban BA et al: Helicobacter pylori gastritis mimicking gastric carcinoma at CT evaluation. Radiology. 179(3):689-91, 1991

GASTRIC CARCINOMA

IMAGE GALLERY

Typical

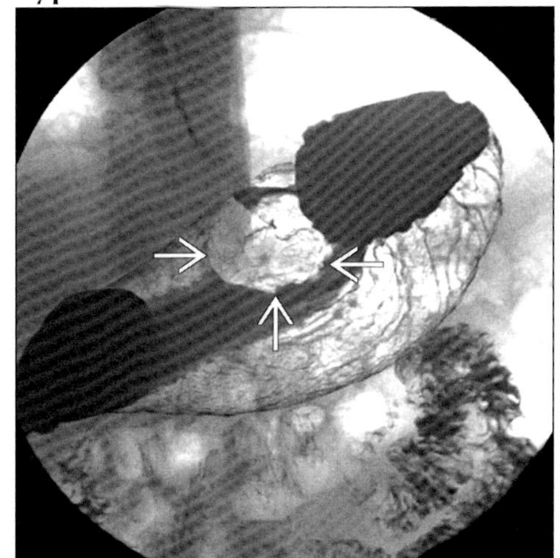

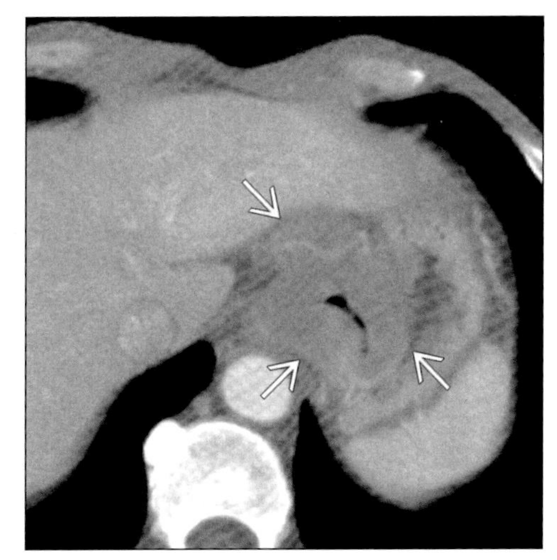

(Left) *Upper GI series shows mass (arrows) as a filling defect in the barium pool on this supine film.* ***(Right)*** *Axial CECT shows large mass (arrows) in gastric fundus.*

Typical

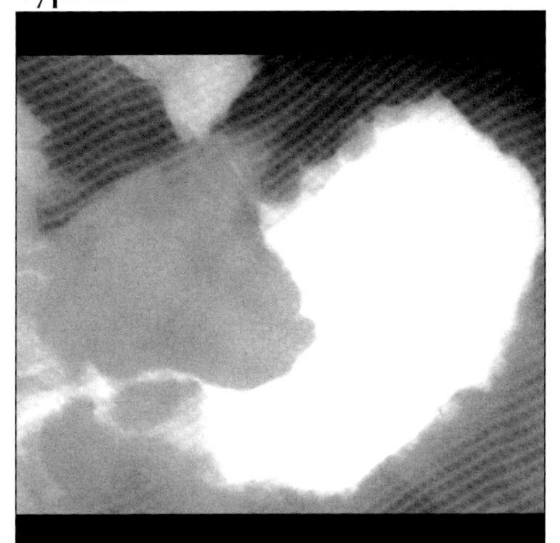

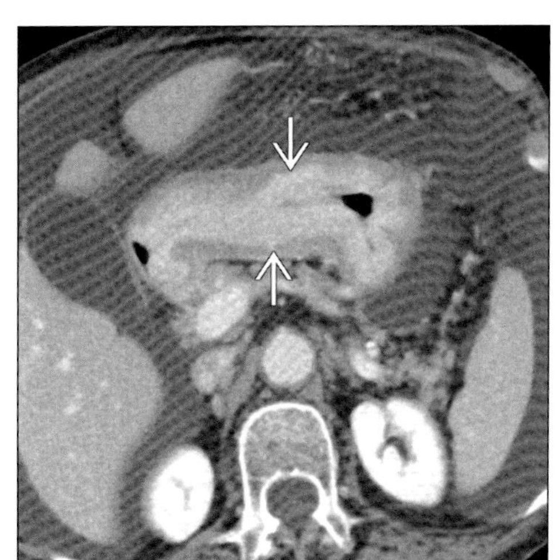

(Left) *Upper GI series shows advanced infiltrating carcinoma causing nodular thickened folds and limiting distensibility.* ***(Right)*** *Axial CECT shows scirrhous carcinoma with enhancing thickened wall (arrows) and malignant ascites.*

Typical

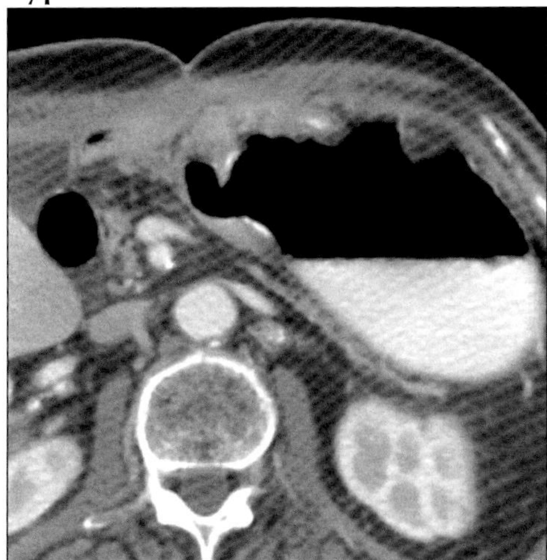

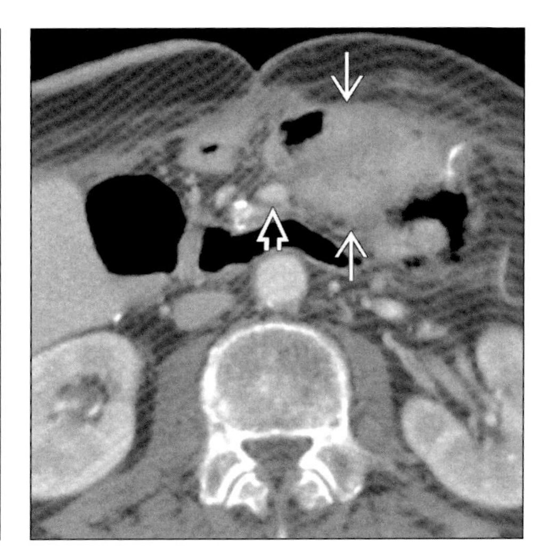

(Left) *Axial CECT shows nodular thickening of the ventral wall of the stomach and circumferential tumor of the antrum causing partial outlet obstruction.* ***(Right)*** *Axial CECT shows circumferential tumor encasing antrum. The tumor invades the anterior abdominal wall (arrows) and local lymph nodes (open arrow).*

GASTRIC LYMPHOMA AND METASTASES

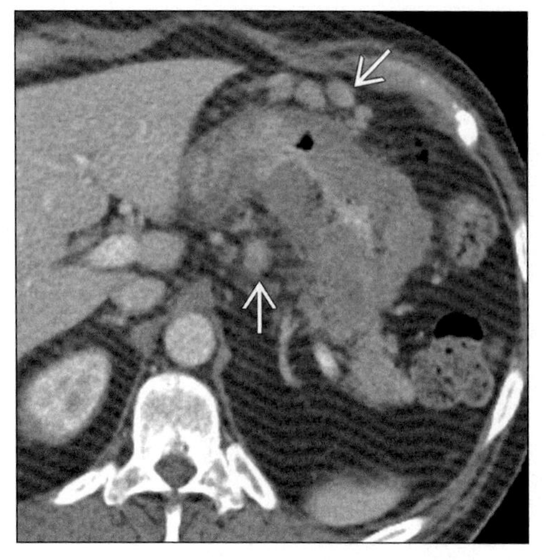

Axial CECT shows diffuse homogeneous thickening of the gastric wall and extensive perigastric lymphadenopathy (arrows). Gastric lymphoma

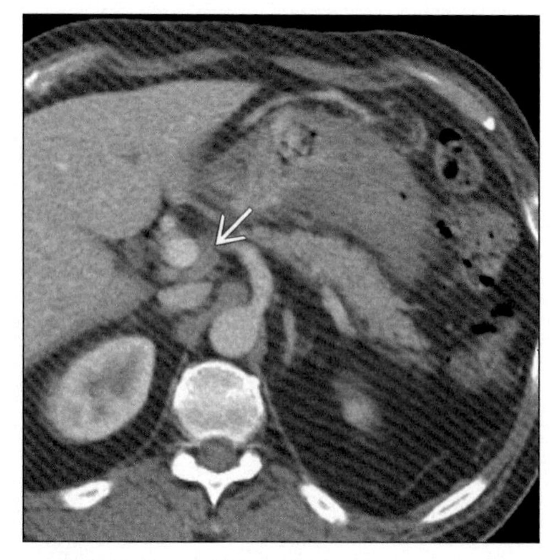

Axial CECT shows diffuse thickening of the gastric wall and porto-caval adenopathy (arrow). Gastric lymphoma.

TERMINOLOGY

Definitions

- Gastric metastases from primary cancer of other sites
- Lymphoma: Malignant tumor of B-lymphocytes

IMAGING FINDINGS

General Features

- Best diagnostic clue: "Bull's eye" lesions on imaging
- Gastric metastases
  - Seen at autopsy in < 2% patients who die of cancer
  - Various forms of metastatic spread to stomach
    - Hematogenous, lymphatic & direct invasion
  - Hematogenous (most common): Malignant melanoma, carcinoma (Ca) breast & lung
    - Malignant melanoma: Highest % of hematogenous metastases
    - Breast cancer (most common disease): Most common cause of metastases to stomach
    - Thyroid & testes: Rarely hematogenous spread to stomach
  - Lymphatic spread: Esophageal or colon cancer
  - Direct invasion or extension: Pancreatic & colon cancer, hepatocellular carcinoma
  - Direct gastric invasion via mesenteric reflections: Transverse colon carcinoma
    - Gastrocolic ligament, transverse mesocolon, greater omentum
  - Most patients with gastric metastases have a known underlying cancer
  - Occasionally may occur as initial manifestation of an occult primary tumor
  - Carcinoma of breast & kidney can metastasize to stomach many years after treatment of primary
- Gastric lymphoma
  - Stomach is most frequently involved part of GI tract
  - Accounts for 50% of all GI tract lymphomas, 25% of all extranodal lymphomas & 3-5% of all malignant tumors in stomach
  - More than 50% cases are primary, rest are secondary
  - Majority, non-Hodgkin lymphoma (B-cell) origin
  - Primary classified into two types based on pathology
    - Low grade MALT (mucosa-associated lymphoid tissue) lymphoma
    - High grade or advanced lymphoma

Radiographic Findings

- Fluoroscopic guided barium study
  - Malignant melanoma metastases
    - Solitary/multiple discrete submucosal masses
    - "Bull's eye" or "target" lesions: Centrally ulcerated submucosal masses
    - "Spoke-wheel" pattern: Radiating superficial fissures from central ulcer

DDx: Diffuse or Focal Thickening of Gastric Wall

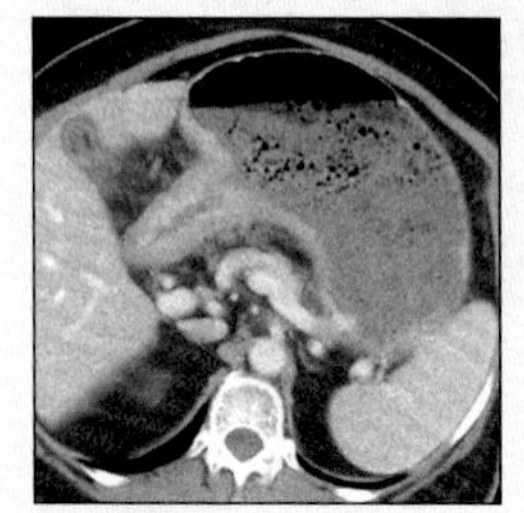

Gastric Carcinoma

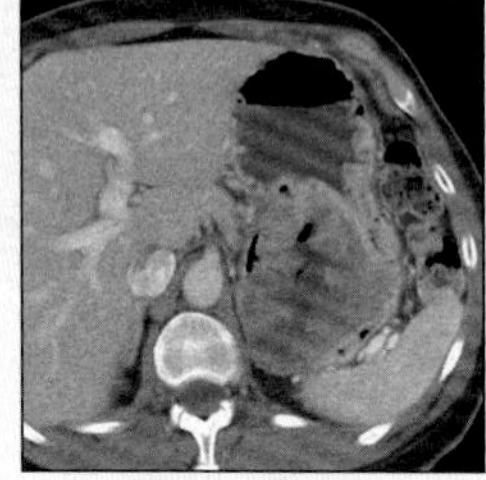

GIST Tumor

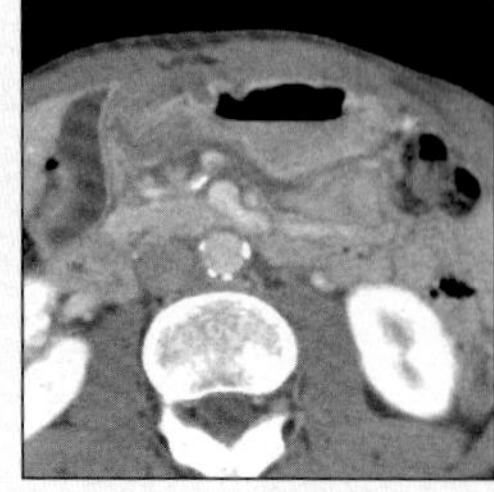

Gastritis

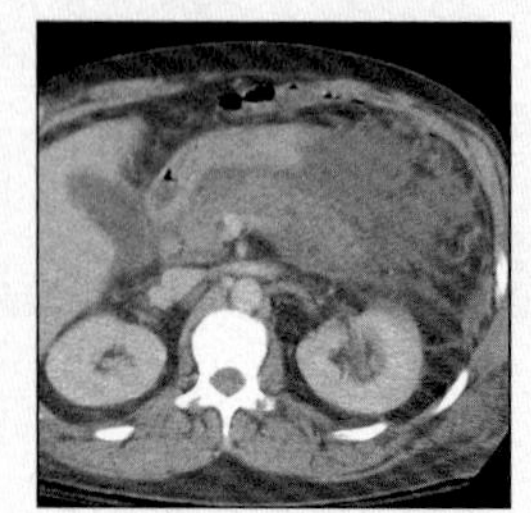

Pancreatitis

GASTRIC LYMPHOMA AND METASTASES

Key Facts

Imaging Findings

- Best diagnostic clue: "Bull's eye" lesions on imaging
- Solitary/multiple discrete submucosal masses
- Giant cavitated lesion: Large collection of barium (5-15 cm) communicating with lumen
- Lobular breast cancer: Linitis plastica or "leather bottle" appearance (loss of distensibility of antrum & body + thickened irregular folds)
- Direct invasion: Spiculated mucosal folds, nodular mass effect, ulceration, obstruction, rarely fistula
- Rounded, confluent nodules of low grade lymphoma (mimic enlarged areae gastricae of H. pylori gastritis)
- Polypoid lymphoma: Lobulated intraluminal mass
- Lacy reticular pattern to bulky masses (omental cake) displacing & causing gastric wall indentation
- Regional or widespread adenopathy

Top Differential Diagnoses

- Gastric carcinoma
- Gastric stromal tumor (leiomyosarcoma)
- Gastritis (erosive type)
- Pancreatitis (extrinsic inflammation)

Pathology

- Spread: Hematogenous, lymphatic, direct spread
- Example: Malignant melanoma; carcinoma of breast, lung, pancreas, colon, esophagus
- Primary: Non-Hodgkin B-cell type (> common)

Diagnostic Checklist

- Overlapping radiographic features of gastric metastases, lymphoma & primary carcinoma
- Imaging important to suggest & stage malignancy, but biopsy often required

    - Giant cavitated lesion: Large collection of barium (5-15 cm) communicating with lumen
    - Small or large lobulated masses
  - Breast carcinoma metastases
    - Lobular breast cancer: Linitis plastica or "leather bottle" appearance (loss of distensibility of antrum & body + thickened irregular folds)
    - Mucosal nodularity, spiculation, ulceration
  - Esophageal squamous cell metastasis
    - Large submucosal masses + central ulceration
  - Esophageal adenocarcinoma (from Barrett mucosa)
    - Large polypoid or ulcerated mass in gastric fundus
    - Subtle findings of cardia: Small ulcers & nodules
  - Pancreatic carcinoma
    - Carcinoma of head: Extrinsic compression of medial border of gastric antrum
    - Carcinoma of body or tail: Extrinsic compression of posterior wall of fundus or body
    - Direct invasion: Spiculated mucosal folds, nodular mass effect, ulceration, obstruction, rarely fistula
  - Omental metastases & transverse colon carcinoma: Gastric invasion via gastrocolic ligament more common by omental metastases than colon Ca
    - Greater curvature of antrum & body: Mass effect, nodularity, spiculation, mucosal fold tethering, gastrocolic fistula
  - Gastric low grade MALT lymphoma
    - Rounded, confluent nodules of low grade lymphoma (mimic enlarged areae gastricae of H. pylori gastritis)
    - Shallow, irregular ulcers with nodular surrounding mucosa
  - Gastric high grade or advanced lymphoma
    - Infiltrative lesions: Massively enlarged folds with distorted & nodular contour (stomach remains pliable/distensible)
    - Ulcerative lesions: Ulcers with surrounding nodular mucosa & thickened, irregular folds; some may appear as giant, cavitated lesions
    - Polypoid lymphoma: Lobulated intraluminal mass
    - Nodular lesions: Submucosal nodules or masses often ulcerate, resulting in "bull's eye" or target lesions

CT Findings

- Demonstration of lesions facilitated by negative contrast agents (water or gas)
- Hematogenous spread of metastases to stomach
  - Malignant melanoma
    - "Bull's eye" or "target" lesions (also seen in lymphoma, Kaposi sarcoma, carcinoid tumor)
    - Giant cavitated lesions
    - Location: Proximal stomach; antrum spared
  - Breast cancer: Linitis plastica or "leather bottle"
    - Markedly thickened gastric wall; folds preserved
    - Enhancement of thickened gastric wall
    - Mimics primary scirrhous carcinoma of stomach
    - Location: Proximal stomach; antrum spared
- Lymphatic spread of metastases to stomach
  - Esophageal Ca: Growth in gastric cardia or fundus
  - Multiple, well-defined, ↓ HU enlarged nodes
    - Characteristic of squamous cell metastases
    - Location: Paracardiac, lesser sac & celiac
- Direct invasion of stomach
  - Distal esophageal adenocarcinoma: Barrett mucosa
    - Polypoid, lobulated mass in gastric fundus
    - Indistinguishable from primary gastric carcinoma
  - Pancreatic carcinoma
    - Abnormal, irregular extrinsic gastric compression
    - Carcinoma of head: Greater curvature of antrum
    - Body & tail: Posterior wall (gastric fundus/body)
  - Transverse colon cancer → gastrocolic ligament → stomach greater curvature
    - Greater curvature: Thickened wall or mass
    - ± Gastrocolic fistulous tract
  - Omental metastases: Ovary, uterus, pancreas, breast
    - Omental masses as small as 1 cm can be seen
    - Lacy reticular pattern to bulky masses (omental cake) displacing & causing gastric wall indentation
- Gastric lymphoma
  - Markedly thickened gastric wall
  - Thickened rugal folds, but contour is preserved
  - Regional or widespread adenopathy
  - Transpyloric spread into duodenum may be seen

GASTRIC LYMPHOMA AND METASTASES

Ultrasonographic Findings

- Real Time
  - Endoscopic ultrasonography (EUS)
    - Hypoechoic mass disrupting normal wall layers
    - Selective/diffusely thickened echogenic wall layers

Imaging Recommendations

- Helical CT; barium (single/double) contrast studies

DIFFERENTIAL DIAGNOSIS

Gastric carcinoma

- Polypoid, ulcerated, infiltrative types indistinguishable from gastric metastases & lymphoma
- Linitis plastica appearance of primary scirrhous type mimics lobular metastatic breast cancer
- Loss of distensibility in scirrhous type differentiates from gastric non-Hodgkin lymphoma
- However gastric Hodgkin lymphoma indistinguishable from scirrhous due to similar desmoplastic response

Gastric stromal tumor (leiomyosarcoma)

- Usually occur as solitary lesions; mostly exophytic
- Also produce giant, cavitated lesions simulating gastric metastases of malignant melanoma & lymphoma

Gastritis (erosive type)

- Multiple punctate barium collections surrounded by thin radiolucent halos of edematous mucosa
- Occasionally surrounded by prominent mounds of edema resulting in "bull's eye" lesions simulating gastric metastases & lymphoma

Pancreatitis (extrinsic inflammation)

- Changes in greater curvature or posterior wall of stomach mimic omental metastatic invasion
- CT will show peripancreatic inflammation

PATHOLOGY

General Features

- Etiology
  - Gastric metastases
    - Spread: Hematogenous, lymphatic, direct spread
    - Example: Malignant melanoma; carcinoma of breast, lung, pancreas, colon, esophagus
  - Gastric lymphoma
    - Primary: Non-Hodgkin B-cell type (> common)
    - Arise from mucosa associated lymphoid tissue (MALT) in patients with chronic H. pylori gastritis containing cytotoxin-associated antigen (CagA)
    - Secondary lymphoma (generalized lymphoma)
- Epidemiology
  - Gastric metastases: Seen in < 2% who die of cancer
  - Gastric lymphoma: 3-5% of all gastric malignancies
- Associated abnormalities
  - Primary carcinoma in gastric metastases
  - Generalized adenopathy in secondary lymphoma

Gross Pathologic & Surgical Features

- Solitary/multiple; polypoid, ulcerated, cavitated masses or leather bottle appearance of stomach

Microscopic Features

- Metastases: Varies based on primary cancer
- Lymphoma: Lymphoepithelial lesions

Staging, Grading or Classification Criteria

- Ann Arbor staging of primary lymphoma
  - Stage I: Lesions involve gastric wall
  - Stage II: Involve regional lymph nodes in abdomen
  - Stage III: Nodes above & below diaphragm
  - Stage IV: Widely disseminated lymphoma

CLINICAL ISSUES

Presentation

- Most common signs/symptoms
  - Asymptomatic, pain, weight loss, palpable mass
  - Hematemesis, melena, acute abdomen (perforation)

Demographics

- Age: Usually middle & elderly age group
- Gender: Metastases (M = F); lymphoma (M > F)

Natural History & Prognosis

- Complications
  - Upper GI bleeding & perforation in ulcerated lesions
  - Antral lesion + pyloric extension: Outlet obstruction
- Prognosis: Poor

Treatment

- Chemotherapy & surgical resection of lesions causing complications like obstruction & upper GI bleeding

DIAGNOSTIC CHECKLIST

Consider

- Check for history of primary cancer/H. pylori gastritis

Image Interpretation Pearls

- Overlapping radiographic features of gastric metastases, lymphoma & primary carcinoma
- Imaging important to suggest & stage malignancy, but biopsy often required

SELECTED REFERENCES

1. Horton KM et al: Current role of CT in imaging of the stomach. Radiographics. 23(1):75-87, 2003
2. Ba-Ssalamah A et al: Dedicated multidetector CT of the stomach: spectrum of diseases. Radiographics. 23(3):625-44, 2003
3. Park MS et al: Radiographic findings of primary B-cell lymphoma of the stomach: low-grade versus high-grade malignancy in relation to the mucosa-associated lymphoid tissue concept. AJR Am J Roentgenol. 179(5):1297-304, 2002
4. McDermott VG et al: Malignant melanoma metastatic to the gastrointestinal tract. AJR Am J Roentgenol. 166(4):809-13, 1996
5. Fishman EK et al: CT of the stomach: spectrum of disease. Radiographics. 16(5):1035-54, 1996
6. Feczko PJ et al: Metastatic disease involving the gastrointestinal tract. Radiol Clin North Am. 31(6):1359-73, 1993

GASTRIC LYMPHOMA AND METASTASES

IMAGE GALLERY

Typical

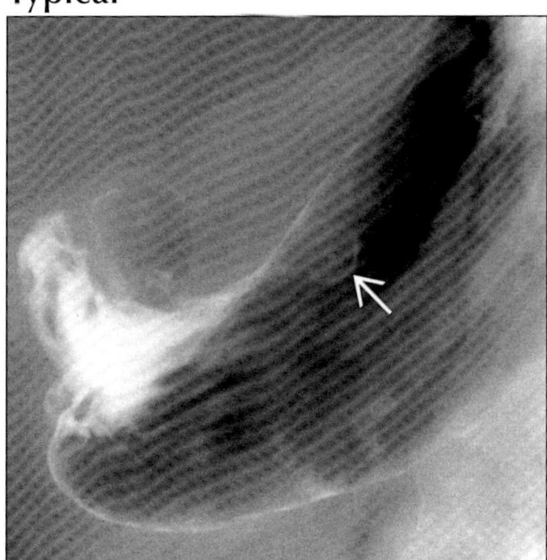

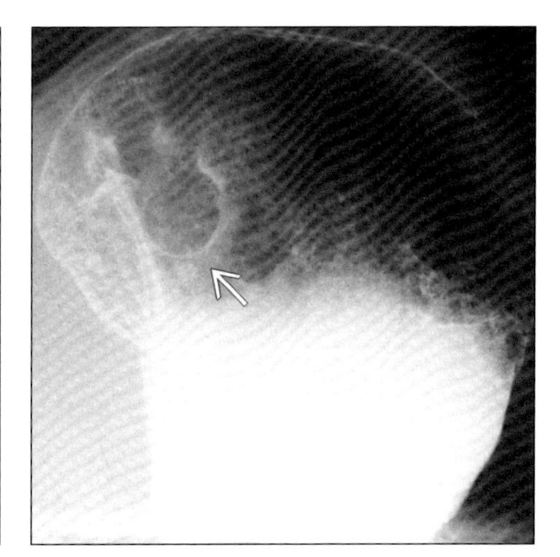

(Left) *Upper GI series shows "bull's eye" lesion (arrow), a discrete intramural polyp with central ulceration. Metastatic melanoma.* ***(Right)*** *Upper GI series shows a submucosal polypoid mass (arrow) from metastatic melanoma.*

Typical

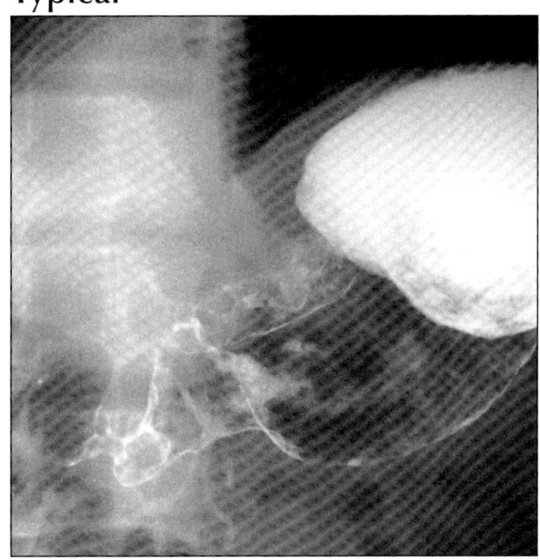

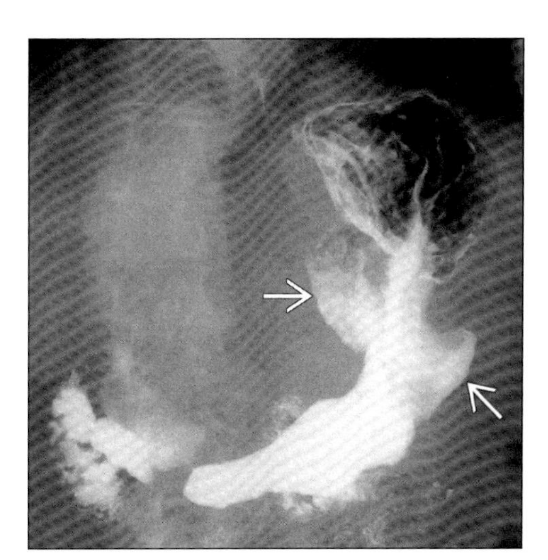

(Left) *Upper GI series shows circumferential massive thickening of gastric folds but no outlet obstruction. Lymphoma.* ***(Right)*** *Upper GI series shows gastric lymphoma. The stomach is encased by tumor with two large ulcerations (arrows), but no obstruction.*

Typical

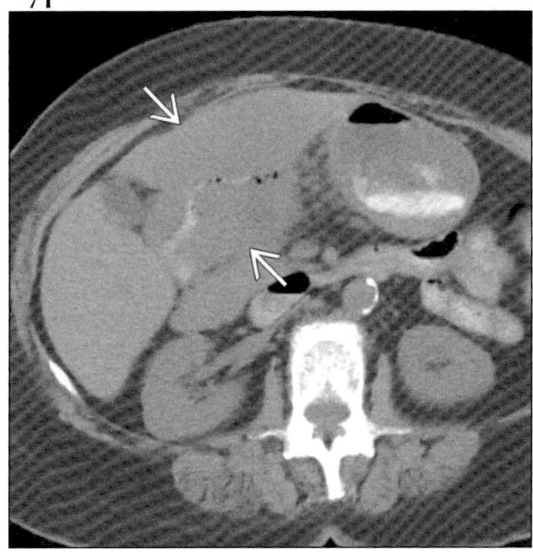

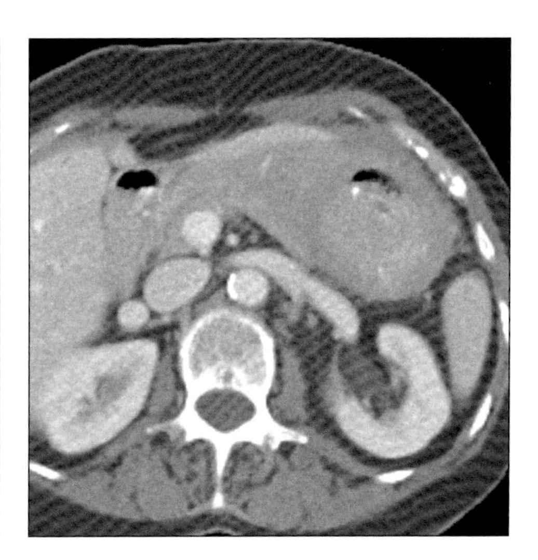

(Left) *Axial NECT shows massive circumferential thickening of gastric antral wall (arrows), but no obstruction. Lymphoma.* ***(Right)*** *Axial CECT shows gastric lymphoma. The entire stomach is involved with massive mural thickening.*

DUODENAL CARCINOMA

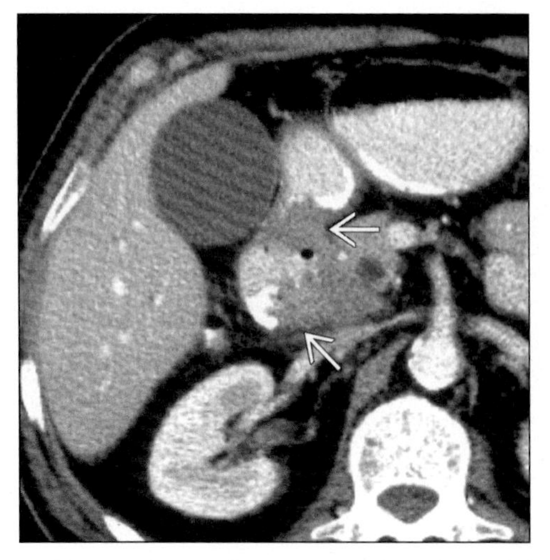

Axial CECT shows soft tissue mass along the medial border of the second portion of the duodenum, which proved to be duodenal carcinoma (Courtesy M. Nino-Murcia, MD).

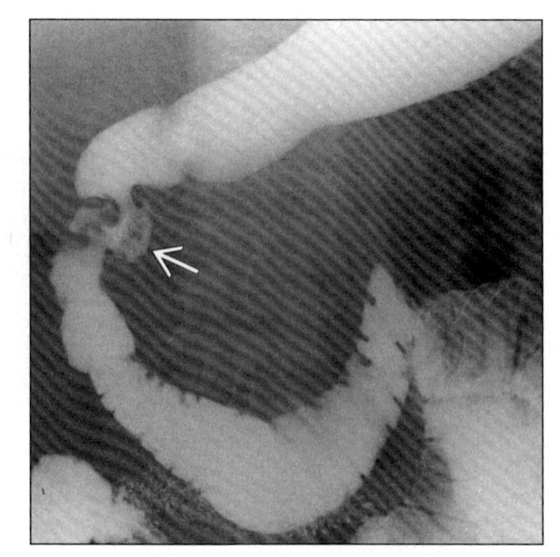

Single-contrast upper GI series shows ulcerated annular constricting mass in the descending duodenum (arrow), which proved to be duodenal carcinoma (Courtesy M. Nino-Murcia, MD).

TERMINOLOGY

Abbreviations and Synonyms

- Duodenal carcinoma (CA), duodenal adenocarcinoma

Definitions

- Primary neoplasm arising in duodenal mucosa

IMAGING FINDINGS

General Features

- Best diagnostic clue: Irregular intraluminal mass or apple-core lesion at or distal to ampulla of Vater
- Location
  - 15% in first portion of duodenum
  - 40% in 2nd portion of duodenum
  - 45% in distal duodenum
- Size: Usually < 8 cm
- Morphology
  - Polypoid, ulcerated, or annular constricting mass
  - Intraluminal mass with numerous frond-like projections for carcinomas arising in villous tumors

Radiographic Findings

- Radiography: Proximal obstruction pattern if lumen severely narrowed
- Fluoroscopy
  - May have various appearances
    - Ulcerated mass
    - Polypoid mass
    - Annular constricting "apple-core" lesion
    - "Soap-bubble" reticulated pattern for villous tumors

CT Findings

- CECT
  - Discrete mass or irregular thickening of duodenal wall
  - Concentric narrowing of duodenum
  - Polypoid intraluminal mass
  - Local lymphadenopathy
  - Infiltration of adjacent fat
  - Biliary or pancreatic duct dilatation with periampullary tumors

MR Findings

- MRCP
  - May see pancreatic or biliary ductal dilatation with periampullary duodenal carcinomas

Ultrasonographic Findings

- Real Time: Hypoechoic mass in duodenum with echogenic center: Pseudokidney sign
- Color Doppler: May see invasion of adjacent vascular structures

DDx: Duodenal Narrowing

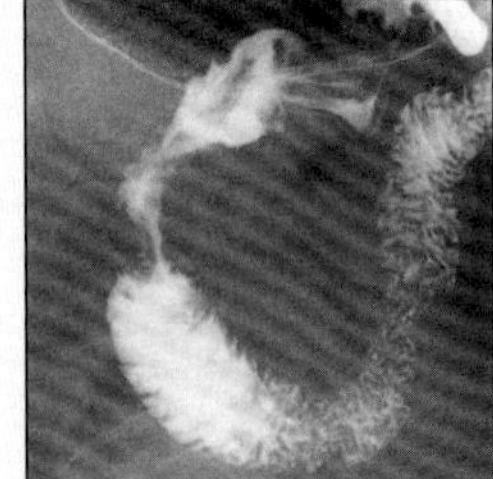

Colon Carcinoma

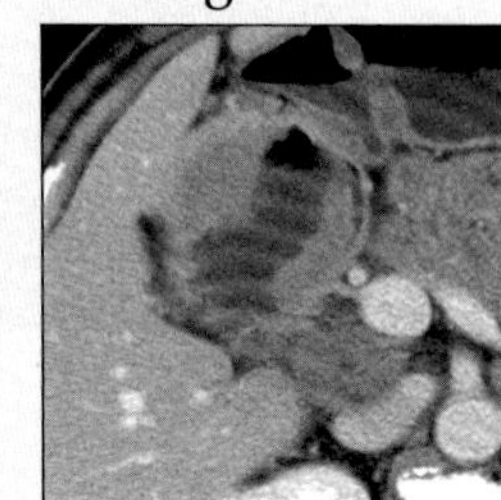

Lymphoma

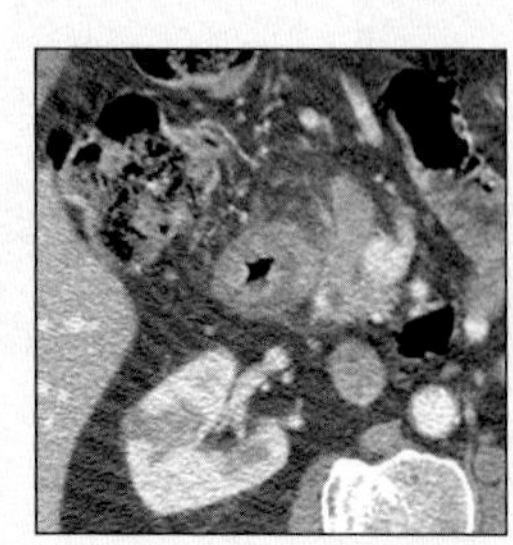

Duodenal Ulcer

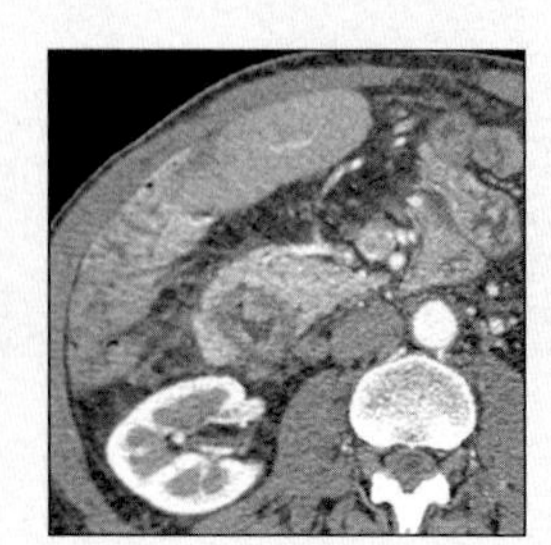

Annular Pancreas

DUODENAL CARCINOMA

Key Facts

Terminology
- Primary neoplasm arising in duodenal mucosa

Imaging Findings
- Best diagnostic clue: Irregular intraluminal mass or apple-core lesion at or distal to ampulla of Vater
- Biliary or pancreatic duct dilatation with periampullary tumors
- Best imaging tool: Thin-section CECT with water for luminal distention and dual-phase arterial and venous imaging
- Protocol advice: Multidetector CT with thin collimation generates best data set for multiplanar reformation

Pathology
- Adenocarcinomas represent 73-90% of malignant duodenal tumors
- 45% of small bowel adenocarcinomas arise in duodenum
- Rare: Represents < 1% of all gastrointestinal neoplasms
- Secondary cancers far more common than primary cancers in proximal small bowel
- Often difficult to distinguish primary duodenal CA from secondary GI adenocarcinoma even with special stains
- Proximal small bowel adenocarcinoma may be a marker for familial or multicentric cancer syndrome

Imaging Recommendations
- Best imaging tool: Thin-section CECT with water for luminal distention and dual-phase arterial and venous imaging
- Protocol advice: Multidetector CT with thin collimation generates best data set for multiplanar reformation

DIFFERENTIAL DIAGNOSIS

Neoplasms
- Ampullary and periampullary adenocarcinomas
  - Pancreatic adenocarcinoma
  - Ampullary carcinoma
  - Primary bile duct carcinoma
- Metastases
  - Contiguous spread from pancreatic, colon, kidney or gallbladder carcinoma
  - Hematogenous metastases from melanoma, Kaposi sarcoma
  - Periduodenal lymph node metastases from other malignancies
- Other duodenal primary neoplasms
  - Duodenal lymphoma
  - Malignant GI stromal tumor
  - Duodenal carcinoid

Inflammatory
- Benign post-bulbar peptic ulcers
- Zollinger-Ellison syndrome
  - Multiple post-bulbar ulcers, thickened folds, hypersecretion
- Crohn disease

Infectious
- Tuberculosis

Congenital
- Annular pancreas
- Duodenal duplication cyst

Trauma
- Duodenal hematoma

PATHOLOGY

General Features
- General path comments
  - Adenocarcinomas represent 73-90% of malignant duodenal tumors
  - Small bowel adenocarcinomas are rare, especially in relation to length of the small bowel
    - 45% of small bowel adenocarcinomas arise in duodenum
    - 25% of all malignant small bowel tumors occur in duodenum
- Genetics: Alterations in oncogenes erbB2, K-ras, cyclin D1 and p53
- Etiology
  - Adenoma-carcinoma sequence
    - Adenomatous polyps are most important risk factor
  - Risk factors
    - Familial polyposis syndromes
    - Crohn disease
    - Cigarette smoking
    - Alcoholism
- Epidemiology
  - Rare: Represents < 1% of all gastrointestinal neoplasms
  - Incidence rises with age

Gross Pathologic & Surgical Features
- Duodenal mass may be flat, stenosing, ulcerative, infiltrating or polypoid in growth pattern
- Secondary cancers far more common than primary cancers in proximal small bowel
  - Often difficult to distinguish primary duodenal CA from secondary GI adenocarcinoma even with special stains
- Proximal small bowel adenocarcinoma may be a marker for familial or multicentric cancer syndrome

DUODENAL CARCINOMA

Microscopic Features

- Similar histology to other GI adenocarcinomas
  - Cellular and nuclear pleomorphism
  - Dysplasia
  - Gland-in-gland appearance
  - Invasion into adjacent normal tissues
- Most duodenal carcinomas are moderately differentiated with variable mucin production
- 20% of duodenal carcinomas are poorly differentiated

Staging, Grading or Classification Criteria

- American Joint Committee on Cancer (AJCC) TNM staging system
  - Primary tumor (T)
    - T1: Tumor invades lamina propria or submucosa
    - T2: Tumor invades muscularis propria
    - T3: Tumor invades through muscularis propria and ≤ 2 cm into adjacent tissues
    - T4: Tumor perforates visceral peritoneum, directly invades other organs, or extends > 2 cm into adjacent tissues
  - Regional lymph nodes (N)
    - N0: No regional nodes involved
    - N1: Regional lymph node metastasis
  - Distant metastasis (M)
    - M0: No distant metastases
    - M1: Distant metastasis
  - Staging
    - Stage I: T1 or T2, N0, M0
    - Stage II: T3 or T4, N0, M0
    - Stage III: Any T, N1, M0
    - Stage IV: Any T, any N, M1

CLINICAL ISSUES

Presentation

- Most common signs/symptoms
  - Upper abdominal pain secondary to obstruction
  - Other signs/symptoms
    - Nausea and vomiting, weight loss, anemia, upper GI bleed
    - Periampullary tumors may present with jaundice
- Clinical profile
  - Increased incidence of duodenal CA in familial polyposis syndromes
    - Peutz-Jegher syndrome, Gardner syndrome

Demographics

- Age
  - 7th decade: Median age = 60 years
  - Low incidence in patients younger than 30
- Gender: Slight male predominance

Natural History & Prognosis

- Spreads by direct extension to adjacent organs and through serosa to peritoneal cavity
- Metastasizes hematogenously to liver, lungs, and bone
- Metastasizes via lymphatics to regional nodes
- 22-71% of patients have positive nodes at presentation
- Prognosis depends on resectability, lymph node involvement, and somewhat on histologic grade
- Vascular invasion makes lesion unresectable

Treatment

- Options, risks, complications
  - Surgery for resectable lesions
    - Pancreaticoduodenectomy for 1st and 2nd portion of duodenum lesions
    - Segmental duodenectomy and primary reanastomosis for 3rd and 4th portion of duodenum lesions
  - Unresectable tumors: Palliation with radiation, chemotherapy, stenting

DIAGNOSTIC CHECKLIST

Consider

- Check for vascular invasion, especially for lesions of 2nd and 3rd duodenum
- Look for regional lymph nodes and liver metastases

Image Interpretation Pearls

- Most duodenal carcinomas cause focal stenoses or obstruction; large mass with cavitation often is lymphoma
- Scrutinize duodenum when periduodenal lymphadenopathy is present on CT without obvious source

SELECTED REFERENCES

1. Lawler LP et al: Peripancreatic masses that simulate pancreatic disease: spectrum of disease and role of CT. Radiographics. 23(5):1117-31, 2003
2. Kim JH et al: Differential diagnosis of periampullary carcinomas at MR imaging. Radiographics. 22(6):1335-52, 2002
3. Korman MU: Radiologic evaluation and staging of small intestine neoplasms. Eur J Radiol. 42(3):193-205, 2002
4. Nagi B et al: Primary small bowel tumors: a radiologic-pathologic correlation. Abdom Imaging. 26(5):474-80, 2001
5. Ishida H et al: Duodenal carcinoma: sonographic findings. Abdom Imaging. 26(5):469-73, 2001
6. Iki K et al: Primary adenocarcinoma of the duodenum demonstrated by ultrasonography. J Gastroenterol. 36(3):195-9, 2001
7. Gore R et al: Textbook of Gastrointestinal Radiology. 2nd ed. Philadelphia, W.B. Saunders, 1980-1992, 2000
8. Buckley JA et al: CT evaluation of small bowel neoplasms: spectrum of disease. Radiographics. 18(2):379-92, 1998
9. Neugut AI et al: The epidemiology of cancer of the small bowel. Cancer Epidemiol Biomarkers Prev. 7(3):243-51, 1998
10. Maglinte DT et al: Small bowel cancer. Radiologic diagnosis. Radiol Clin North Am. 35(2):361-80, 1997
11. Buckley JA et al: The accuracy of CT staging of small bowel adenocarcinoma: CT/pathologic correlation. J Comput Assist Tomogr. 21(6):986-91, 1997
12. Buckley JA et al: Small bowel cancer. Imaging features and staging. Radiol Clin North Am. 35(2):381-402, 1997
13. Gore RM: Small bowel cancer. Clinical and pathologic features. Radiol Clin North Am. 35(2):351-60, 1997
14. Arber N et al: Molecular genetics of small bowel cancer. Cancer Epidemiol Biomarkers Prev. 6(9):745-8, 1997
15. Laurent F et al: CT of small-bowel neoplasms. Semin Ultrasound CT MR. 16(2):102-11, 1995

DUODENAL CARCINOMA

IMAGE GALLERY

Typical

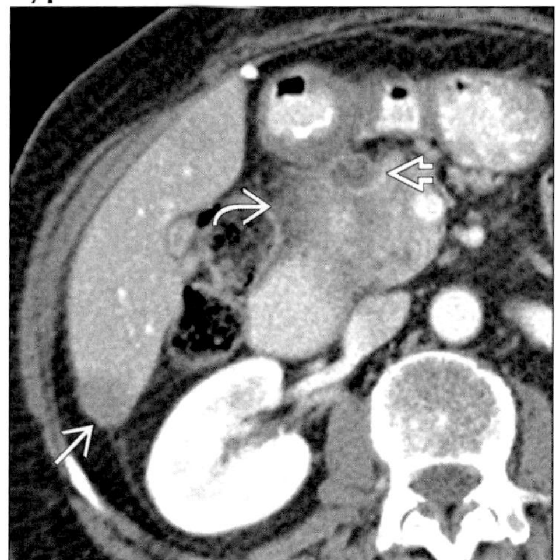

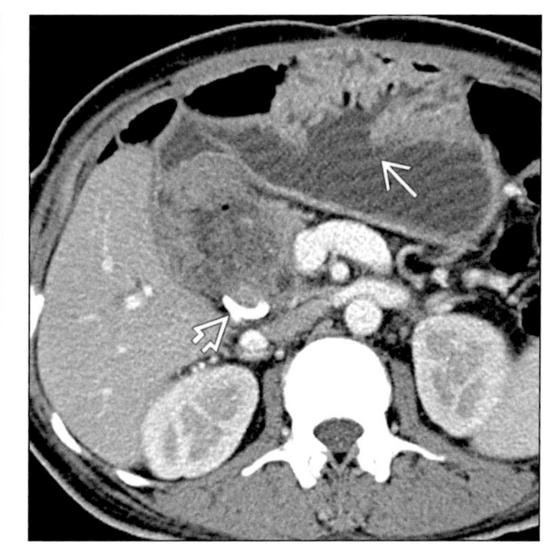

(Left) *Axial CECT shows duodenal wall thickening (curved arrow), low density lymph node (open arrow), and liver metastasis (arrow) in patient with duodenal carcinoma.* ***(Right)*** *Axial CECT shows bulky duodenal carcinoma in second portion of duodenum. Patient has a gastrojejunostomy (arrow) and biliary stent (open arrow) for palliation of obstruction.*

Typical

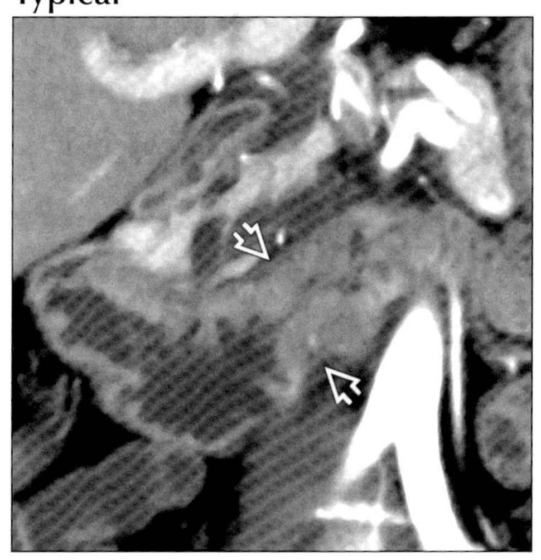

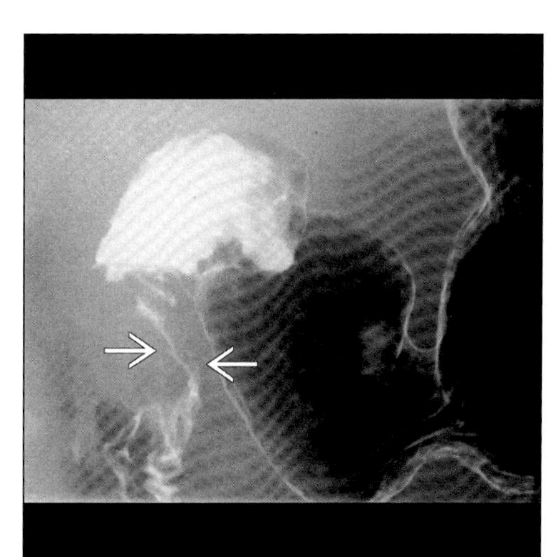

(Left) *Coronal CECT thin-slab-average image shows low attenuation annular constricting mass in transverse duodenum (arrows).* ***(Right)*** *Double-contrast upper GI series shows "apple-core" lesion of second portion of duodenum (arrows) representing duodenal carcinoma (Courtesy H. Harvin, MD).*

Typical

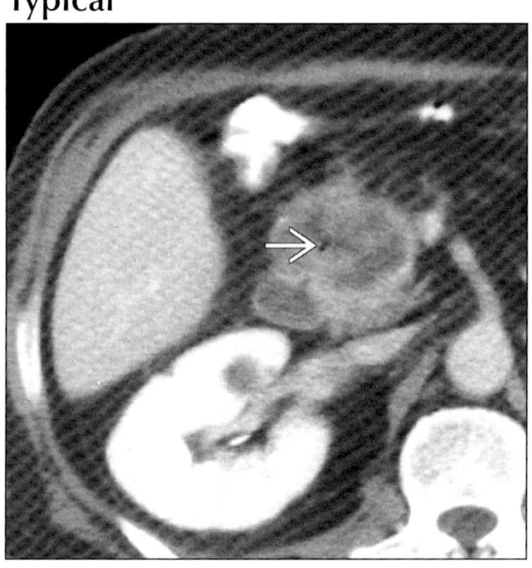

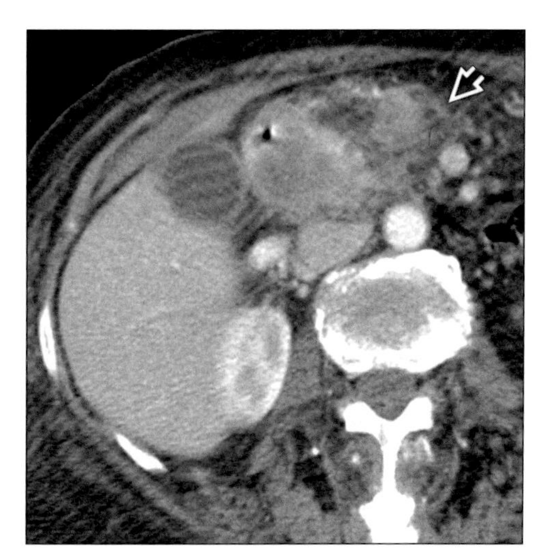

(Left) *Axial CECT shows irregular low attenuation mass in second portion of duodenum. Note central low density lumen (arrow) which shows the mass to be arising within duodenum rather than pancreas.* ***(Right)*** *Axial CECT shows irregular mass distorting duodenal lumen and extending into adjacent fat medially (open arrow).*

DUODENAL METASTASES AND LYMPHOMA

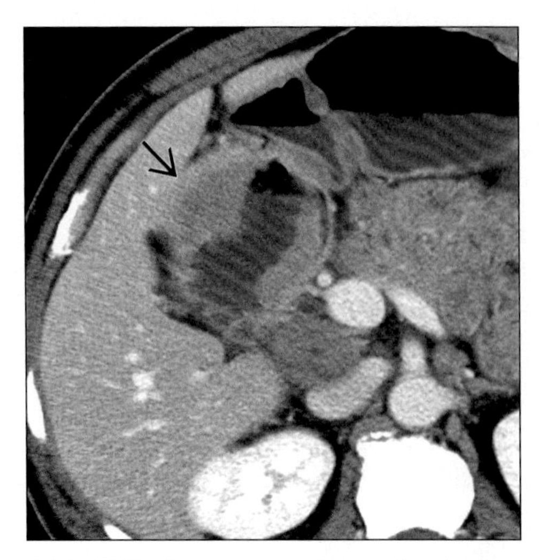

Axial CECT demonstrates submucosal soft tissue infiltrating mass (arrow) due to lymphoma.

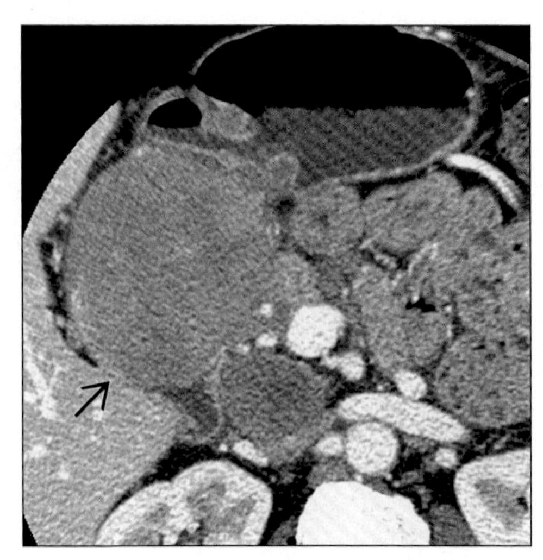

Axial CECT demonstrates bulky soft tissue mass involving duodenum (arrow). Biopsy revealed lymphoma.

TERMINOLOGY

Definitions

- Involvement of duodenum with malignant lymphoma or metastatic disease

IMAGING FINDINGS

General Features

- Best diagnostic clue
  - Lymphoma: Bulky submucosal mass extending through pylorus to secondarily invade duodenum
  - Mets: "Bull's eye" or "target lesion"; submucosal or polypoid mass
- Location: Pylorus and duodenum, submucosal lesion
- Size: 1-5 cm
- Morphology: Lymphoma: Smooth submucosal, often bulky, mass

Radiographic Findings

- Fluoroscopy
  - Lymphoma: Smooth or lobulated submucosal mass involving distal stomach and duodenum on UGI
  - Mets: "Target" or "bull's eye" lesion with rounded submucosal mass; ulceration common on UGI

CT Findings

- CECT
  - Lymphoma: Bulky hypovascular soft tissue mass infiltrating submucosa of stomach and duodenum on CECT
  - Hematogenous mets images as rounded submucosal mass; direct invasion mets shows involvement from primary tumor of pancreas, colon, kidney, gallbladder or retroperitoneal node on CECT

MR Findings

- T1WI: Low signal duodenal mass
- T2WI: Intermediate signal mass
- T1 C+: Variable enhancement: Adeno CA typically hypovascular, melanoma may be hypervascular

Imaging Recommendations

- Best imaging tool: UGI, CECT

DIFFERENTIAL DIAGNOSIS

Villous adenoma

- Bulky mucosal polypoid mass, 3-9 cm; rarely causes obstruction
- Risk of CA increases with size; 30-60% of tumors have malignant changes

DDx: Lesions Mimicking Lymphoma or Metastases

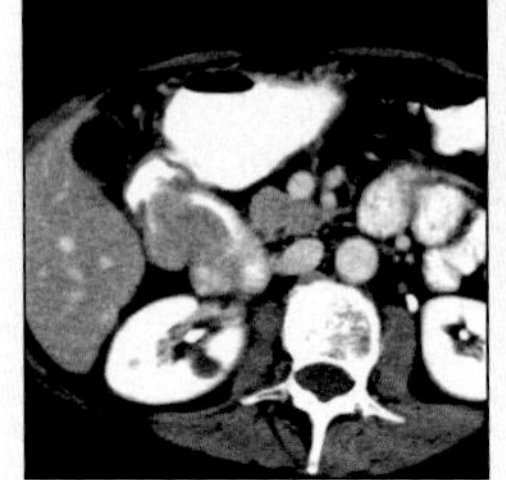

Villous Adenoma

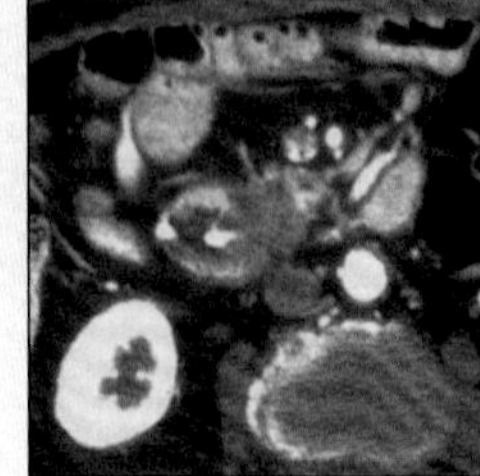

Duodenal Carcinoma

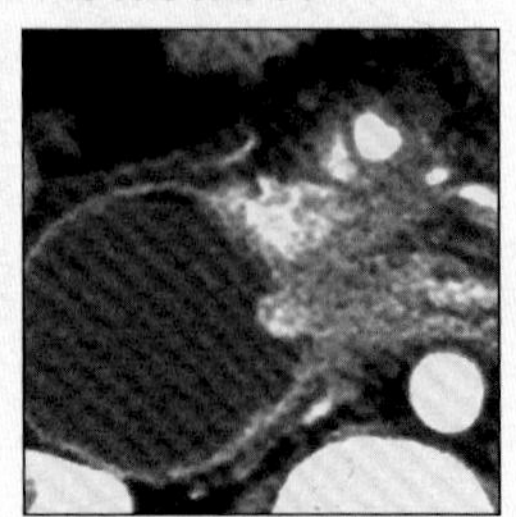

Secondary Invasion

DUODENAL METASTASES AND LYMPHOMA

Key Facts

Imaging Findings

- Lymphoma: Smooth or lobulated submucosal mass involving distal stomach and duodenum on UGI
- Mets: "Target" or "bull's eye" lesion with rounded submucosal mass; ulceration common on UGI
- Lymphoma: Bulky hypovascular soft tissue mass infiltrating submucosa of stomach and duodenum on CECT
- Hematogenous mets images as rounded submucosal mass; direct invasion mets shows involvement from primary tumor of pancreas, colon, kidney, gallbladder or retroperitoneal node on CECT

Clinical Issues

- Options, risks, complications: Best option for localized lymphoma is surgery; chemotherapy best for mets

Duodenal carcinoma

- Infiltrating mural mass; "apple core" annular lesion; may be polypoid; often ulcerated; more likely to obstruct lumen
- 1% of all GI neoplasms; increased incidence in Gardner syndrome, celiac disease, Crohn disease, neurofibromatosis
- Regional lymphadenopathy & pancreatic invasion common

Secondary duodenal invasion

- Most commonly due to pancreatic CA, colon CA or renal cell CA
- Large extramural mass; often asymptomatic but may lead to outlet obstruction

PATHOLOGY

General Features

- General path comments
  - Lymphoma: Non-Hodgkin lymphoma of B-cell origin or mucosa-associated lymphoid tissue (MALT)
  - Mets: Melanoma, CA of breast, lung, colon, pancreas, kidney
- Etiology: MALT lymphomas associated with H. pylori infection
- Associated abnormalities: Regional lymphadenopathy; mets may cause outlet obstruction

Gross Pathologic & Surgical Features

- Lymphoma: Most often associated with gastric lymphoma extending through pylorus into duodenum
- Mets: Polypoid mucosal/submucosal masses (melanoma or secondary extrinsic mass invading duodenum)

Staging, Grading or Classification Criteria

- GI lymphoma staging
  - I: Tumor confined to bowel wall
  - II: Limited nodal spread to local nodes
  - III: Widespread nodal mets
  - IV: Spread to bone marrow, solid viscera, i.e., liver

CLINICAL ISSUES

Presentation

- Most common signs/symptoms: Abd pain, nausea, vomiting, weight loss, palpable mass, UGI bleeding

Demographics

- Age: 55-60 years
- Gender: M < F

Treatment

- Options, risks, complications: Best option for localized lymphoma is surgery; chemotherapy best for mets

DIAGNOSTIC CHECKLIST

Consider

- Duodenal carcinoma

Image Interpretation Pearls

- Bulky submucosal mass without obstruction

SELECTED REFERENCES

1. Elliott LA et al: Metastatic breast carcinoma involving the gastric antrum and duodenum: computed tomography appearances. Br J Radiol. 68(813):970-2, 1995
2. Cirillo M et al: Primary gastrointestinal lymphoma: a clinicopathological study of 58 cases. Haematologica. 77(2):156-61, 1992
3. Najem AZ et al: Primary non-Hodgkin's lymphoma of the duodenum. Case report and literature review. Cancer. 54(5):895-8, 1984
4. Balthazar EJ: Duodenal Hodgkin's disease. Am J Gastroenterol. 68(3):306-11, 1977

IMAGE GALLERY

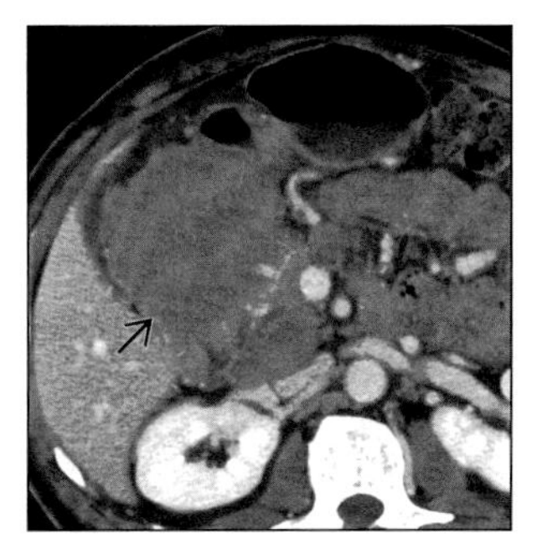

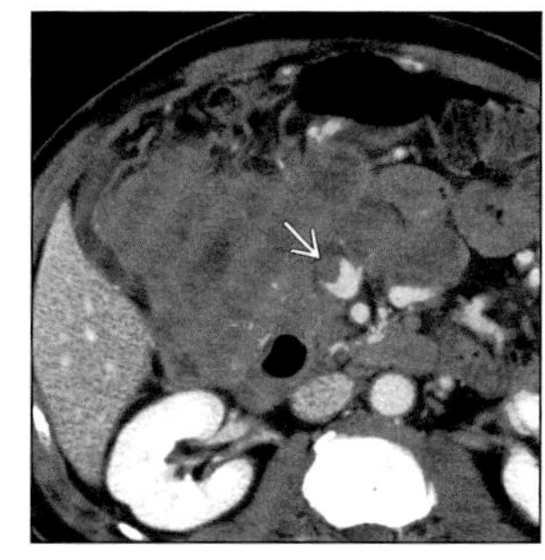

(Left) *Axial CECT of duodenal lymphoma. Note extensive infiltration of duodenum by soft tissue mass (arrow).* ***(Right)*** *Axial CECT of duodenal lymphoma. Bulky mass infiltrates duodenum, invades mesentery and extends into superior mesenteric vein (arrow).*

FUNDOPLICATION COMPLICATIONS

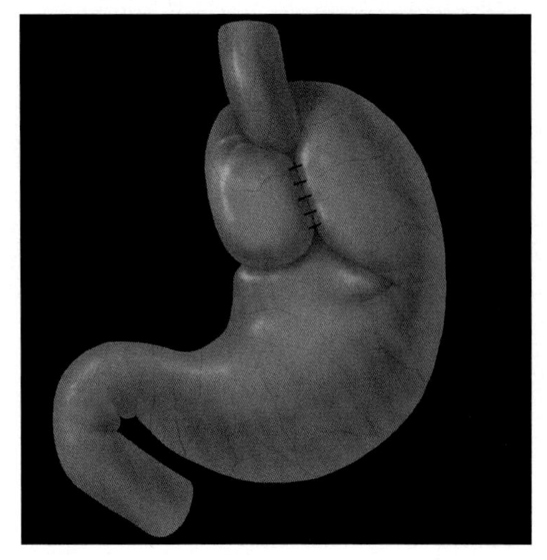

Graphic shows Nissen fundoplication with gastric fundus wrapped around the gastroesophageal (GE) junction.

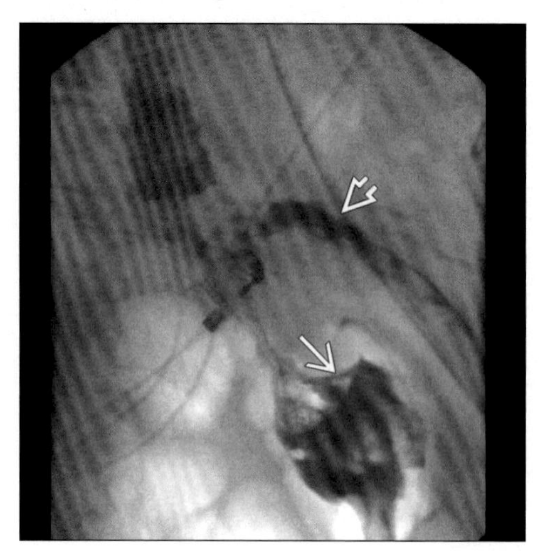

Upper GI series following surgery shows the expected fundoplication defect in the stomach (arrow), and extravasation of contrast material (open arrow) from the GE junction.

TERMINOLOGY

Abbreviations and Synonyms

- Fundoplication (FDP) complications

Definitions

- Complications of anti-reflux surgery for management of gastroesophageal reflux disease (GERD)
- Nissen FDP: Complete FDP
  - Approach: Laparoscopic or open FDP
  - Gastric fundus wrapped 360 degrees around intra-abdominal esophagus to create antireflux valve
  - Concomitant diaphragmatic hernia reduced; diaphragmatic esophageal hiatus sutured
- Toupet FDP: Partial FDP
  - 270 degree wrap; posterior hemivalve created
- Belsey Mark IV repair: Open surgical; 240-degree FDP wrap around left lateral aspect of distal esophagus
  - Fundus sutured to intra-abdominal esophagus; acute esophagogastric junction angle (angle of His)
  - Can also be done by minimally invasive techniques

IMAGING FINDINGS

General Features

- "Wrap" complications
  - Slipped or misplaced FDP
  - FDP disruption or breakdown
  - FDP herniation with intra thoracic migration
  - Too tight or too loose, or too long FDP
  - Herniation of stomach through re-opened diaphragmatic esophageal hiatus
- "Non-wrap" complications
  - Injury to intra-abdominal, intra-thoracic organs
  - Leaks; intra-abdominal, thoracic fluid collections
  - Fistulas; gastropericardial, gastrobronchial etc.
  - Pneumothorax, pneumonia, pancreatitis, incisional hernia, mesenteric & portal venous thrombosis
- Late complications
  - Recurrent paraesophageal herniation
  - Distal esophageal stricture

Radiographic Findings

- Fluoroscopy
- Normal post-operative appearance
  - Nissen FDP wrap: Well-defined "mass" in gastric fundus; smooth contour & surface
    - Distal esophagus tapers smoothly through center of symmetric compression by wrap
  - Pseudotumoral defect of gastric fundus; part of fundus wrapped around distal esophagus
    - Defect more pronounced for complete wrap of Nissen than partial wrap of Toupet; Belsey
  - Belsey Mark IV repair
    - Wrap produces smaller defect than Nissen FDP
    - 2 distinct angles form as esophagus passes FDP

DDx: Leak, Obstruction or Mass After Surgery

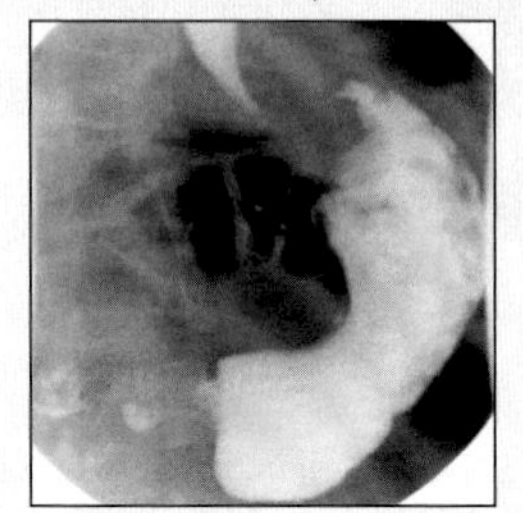

Normal Appearance

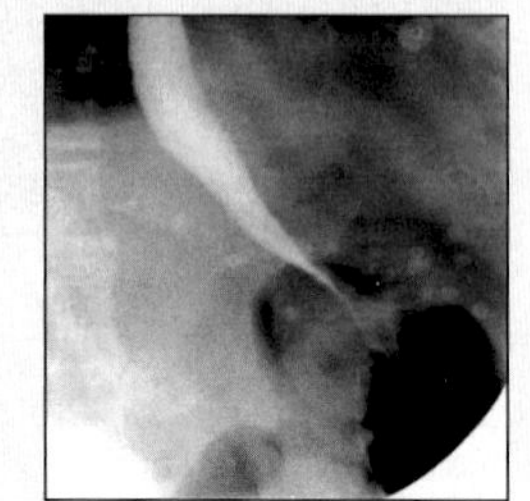

Tight; Edema

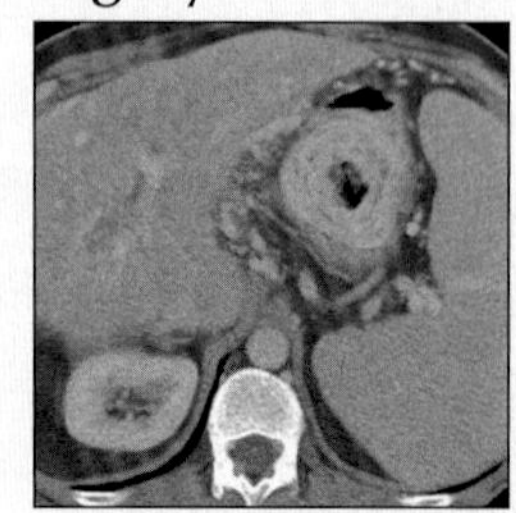

Bezoar

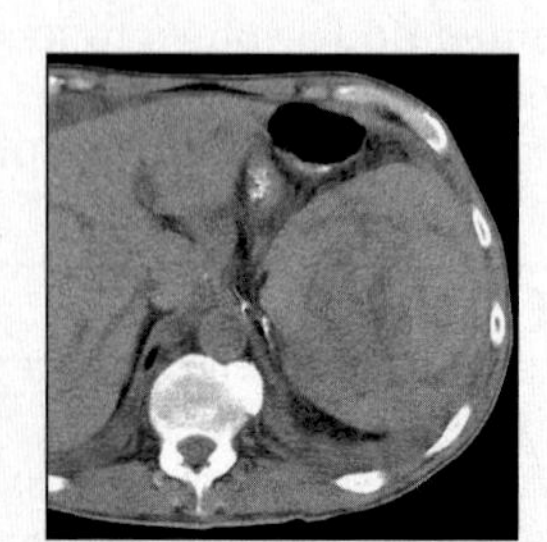

Splenic Laceration

FUNDOPLICATION COMPLICATIONS

Key Facts

Terminology
- Nissen FDP: Complete FDP
- Toupet FDP: Partial FDP

Imaging Findings
- "Wrap" complications
- Slipped or misplaced FDP
- FDP disruption or breakdown
- FDP herniation with intra thoracic migration
- Too tight or too loose, or too long FDP
- Herniation of stomach through re-opened diaphragmatic esophageal hiatus
- "Non-wrap" complications
- Injury to intra-abdominal, intra-thoracic organs
- Leaks; intra-abdominal, thoracic fluid collections
- Pneumothorax, pneumonia, pancreatitis, incisional hernia, mesenteric & portal venous thrombosis
- Recurrent paraesophageal herniation

Top Differential Diagnoses
- Post-op edema
- Bezoar
- Plication defect
- Extra-gastric complications

Diagnostic Checklist
- Post-operative fluoroscopic evaluation should be used liberally or even routinely
- CT for suspected leak or bleeding

    - Shallow upper angle; where esophagus, fundus, & diaphragm sutured together
    - Steep lower angle; where stomach pulled upward toward esophagus
- "Wrap" complications
  - Tight FDP wrap
    - Fixed narrowing of distal esophagus
    - Delayed emptying of barium into stomach
    - May also be caused by excessive closure of esophageal hiatus of diaphragm
  - Complete disruption of FDP sutures
    - Recurrent hiatal hernia & gastroesophageal reflux
    - Gastric outpouching above diaphragm
    - Expected mass of FDP wrap not present in fundus
  - Partial disruption of FDP sutures
    - Partially intact wrap; does not encircle esophagus
    - One or more small outpouchings from fundus
    - Hourglass stomach; as fundus slips through FDP
  - Slipped Nissen
    - Complete wrap may slide downward over stomach; hourglass configuration of stomach
  - Intrathoracic migration of wrap
    - Intact FDP wrap herniates partially or entirely through esophageal hiatus of diaphragm
    - Type I: Paraesophageal herniation of portion of wrap through esophageal hiatus (70%)
    - Type II: Herniation of entire FDP through hiatus
    - Gastroesophageal (GE) junction: In type 1, below diaphragm. In type 11, at or above diaphragm
    - In both types, wrap intact, without disruption
  - Inappropriate placement of FDP around gastric body
    - Hourglass appearance of stomach
- "Non-wrap" complications
  - Presence of leaks, fistula
  - Persistence of gastroesophageal reflux, gastric ulcer

CT Findings
- "Wrap": Soft tissue density area surrounding intra-abdominal esophagus at GE junction
  - Extending caudally about 4cm
  - Normal post surgical esophagus collapsed without gaseous distention of its distal part; no reflux
- Immediate post-operative period, gastric cardia wall may be thickened at area of operation; due to edema
- Wrap breakdown: Gastric circumferential thickening surrounding GE junction (due to wrap) is lacking
  - Distal esophagus may be distended
  - May see recurrent diaphragmatic hernia; reflux of contrast material into esophagus
- Herniation of an intact FDP through diaphragmatic hiatus; may be seen with coronal reformatted images
- Retraction injury to adjacent organs
  - During laparoscopic procedure; retraction of left hepatic lobe may result in liver or splenic laceration
  - Right ventricular laceration; cardiac tamponade
    - Trauma by liver retractor during laparoscopic FDP
  - Bleeding & hematoma in gastric wall or in peritoneal spaces adjacent to stomach & duodenum
- Fluid collections in abdomen or mediastinum
  - Herniated abdominal fluid; disrupted lymphatic drainage; hematoma; infection ± leak; abscess
  - Drainage under CT guidance; obviating surgical
- Visceral perforation: Extraluminal contrast; free air
  - Reported with open & laparoscopic FDP; correlates with surgeon experience
- Superior mesenteric vein & portal vein thrombosis
  - Rare; approximately 2 weeks after laparoscopic FDP

Imaging Recommendations
- Best imaging tool
  - Videofluoroscopic contrast-enhanced esophagram
    - Structural information; anatomical abnormalities
    - "Wrap" complications; leaks; persistence of reflux
  - CT; severe abdominal or chest pain; suspected visceral injury; abscess

DIFFERENTIAL DIAGNOSIS

Post-op edema
- Early post-operative period; edema of FDP wrap
- Large, smooth fundal mass; with smooth, tapered narrowing of intra-abdominal esophagus
- Delayed emptying of contrast material

FUNDOPLICATION COMPLICATIONS

- Edema usually subside, less compression of esophagus within 1-2 weeks
  - Repeat esophagram shows much smaller defect

Bezoar

- Intraluminal mass; mottled or streaked appearance
- May cause partial or complete obstruction

Plication defect

- Disruption of diaphragmatic sutures (not FDP sutures)
  - Recurrent hiatal hernia; above an intact FDP wrap
- Plication of diaphragm for eventration diaphragm may be complicated by traumatic diaphragmatic hernia
  - May see bowel herniating through diaphragmatic defect at site of previous diaphragm plication

Extra-gastric complications

- Abscess, retractor injury to spleen, liver, etc.

PATHOLOGY

General Features

- General path comments
  - Indications for anti-reflux surgery
    - Medical treatment ineffective
    - Side effects of long term medications
    - Complications of GERD; esophagitis, stricture, recurrent aspiration pneumonia, asthma etc.
  - Surgery also employs repair of large paraesophageal hernias associated with GERD
- Etiology: Surgeon inexperience; operative technique
- Epidemiology
  - Incidence of complications increasing; as many laparoscopic FDPs performed indiscriminately
    - Intrathoracic migration of wrap; seen in 30% after laparoscopic Nissen FDP; 9% after open procedure
    - Paraesophageal hernia; incidence higher after laparoscopic than open FDP

CLINICAL ISSUES

Presentation

- Dysphagia; transient in early post-operative period
- "Gas bloat" syndrome; upper abdominal fullness, inability to belch, early satiety, flatulence
- Nausea, vomiting, epigastric pain, diarrhea
- Intrathoracic wrap migrations; small, asymptomatic
  - 64% of radiologically visualized intrathoracic migrations without clinical manifestations
- Intrathoracic gastric herniation after FDP; uncommon; potentially life-threatening
  - May lead to gastric volvulus; intrathoracic incarceration of stomach; acute gastric perforation
- Too loose; disrupted FDP: Recurrent reflux symptoms
- Leaks: Pain, fever, leukocytosis
- Visceral injury: Pain, falling hematocrit

Natural History & Prognosis

- Advantages of laparoscopic FDP: Safe; effective; reduced length of hospital stay & recovery time
  - Effective even at long term follow-up; as effective as open procedures with lower morbidity rate
- Laparoscopic Toupet vs. Nissen FDP
  - Similar short term results
  - In longer follow-up; no difference in incidence of post FDP symptoms related to gas-bloat syndrome
  - Recurrence of GERD: Nissen; 8% symptomatic reflux; 4% by objective testing
    - Toupet: 20% symptomatic; 51% objective
  - Toupet FDP; higher incidence of proton pump inhibitor resumption, overall dissatisfaction
  - Superiority of total FDP over partial; even in setting of moderate decreases in esophageal motility
- Laparoscopic FDP: 3.5-5% rate of early post-operative complications
  - Surgical failure rate requiring re-operation: 2-17%
- Outcome: Good; as long as FDP remains intact
  - Keeping GE junction at hiatus, hiatus closed, preventing recurrence of hernia
  - Overall mortality rate: 0·3%
- Antireflux surgery undertaken primarily to improve quality of life by relieving symptoms of GERD
  - Small possibility of reflux symptoms becoming worse after FDP operation; 1% to 2% of patients
  - Creation of new symptoms due to side effects of surgery; may adversely impact quality of life

Treatment

- Minimize complications: Surgeon experience; training
  - Appropriate operative techniques
  - Low threshold for early laparoscopic reexploration, early radiological contrast studies
  - 5-10% of time; may need to change to open procedure while laparoscopic surgery in process
- Dilation of esophagus; reoperation to loosen wrap around esophagus; if dysphagia persists
- Redo laparoscopic Nissen can be performed safely after initial laparoscopic approach; low failure rate
- Prevent recurrent hernia after laparoscopic Nissen FDP
  - Appropriate closure of crura & anchoring suture between stomach & diaphragm are helpful
  - Reinforcement of hiatal crura using prosthetic mesh

DIAGNOSTIC CHECKLIST

Consider

- Post-operative fluoroscopic evaluation should be used liberally or even routinely
  - CT for suspected leak or bleeding

SELECTED REFERENCES

1. Graziano K et al: Recurrence after laparoscopic and open Nissen fundoplication: a comparison of the mechanisms of failure. Surg Endosc. 17(5):704-7, 2003
2. Hainaux B et al: Intrathoracic migration of the wrap after laparoscopic Nissen fundoplication: radiologic evaluation. AJR Am J Roentgenol. 178(4):859-62, 2002
3. Fernando HC et al: Outcomes of laparoscopic Toupet compared to laparoscopic Nissen fundoplication. Surg Endosc. 16(6):905-8, 2002
4. Pavlidis TE: Laparoscopic Nissen fundoplication. Minerva Chir. 56(4):421-6, 2001
5. Waring JP: Postfundoplication complications. Prevention and management. Gastroenterol Clin North Am. 28(4):1007-19, viii-ix, 1999

IMAGE GALLERY

Typical

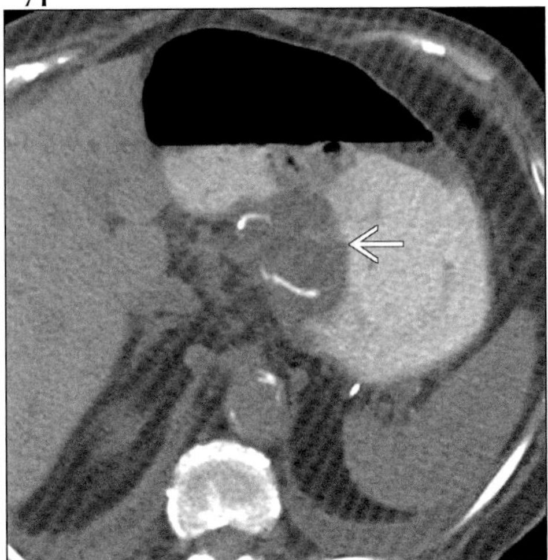

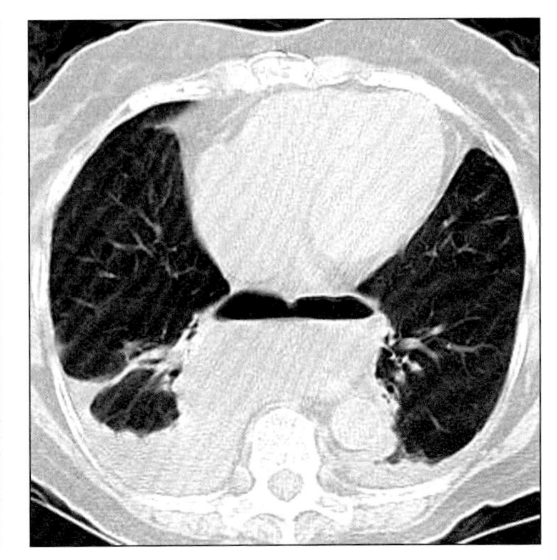

(Left) *Axial NECT shows intact fundoplication as a soft tissue density "mass" (arrow) in gastric fundus.* ***(Right)*** *Axial CECT shows large air-filled collection in mediastinum (sterile) following surgery.*

Typical

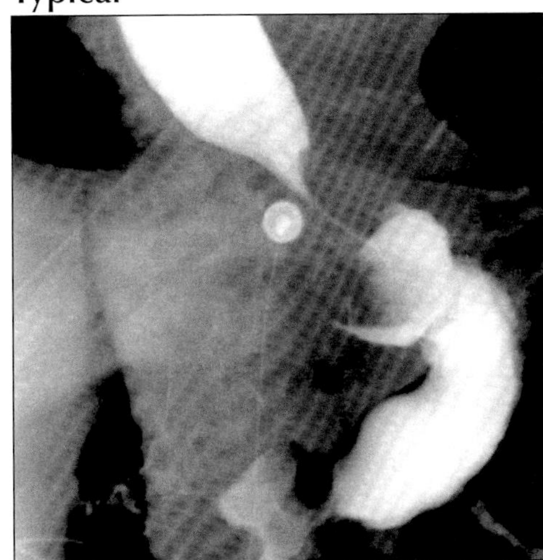

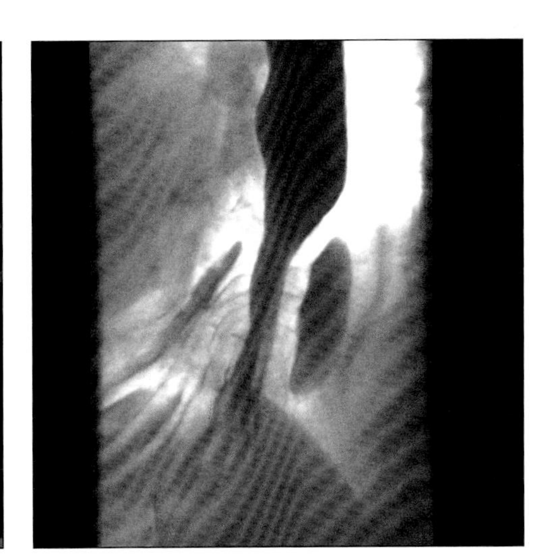

(Left) *Upper GI series shows an intact "tight" fundoplication with persistent dilation of the esophagus.* ***(Right)*** *Upper GI series shows intrathoracic migration of the intact fundoplication.*

Typical

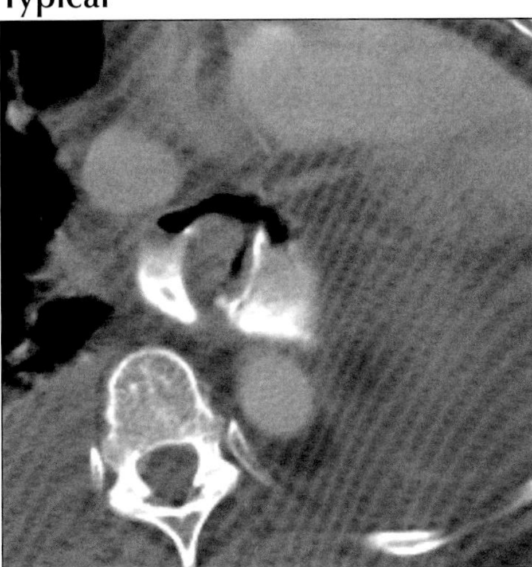

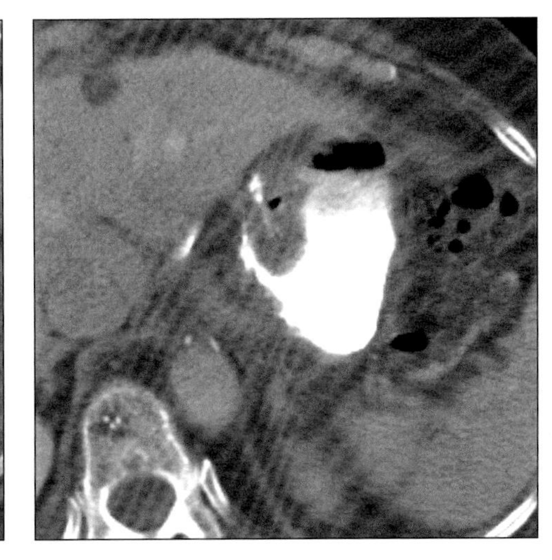

(Left) *Axial CECT shows distal esophagus surrounded by extravasated contrast medium within the mediastinum due to perforation of the esophageal wall.* ***(Right)*** *Axial CECT shows intact gastric fundus with oral contrast medium surrounding the fundoplication wrap.*

GASTRIC BYPASS COMPLICATIONS

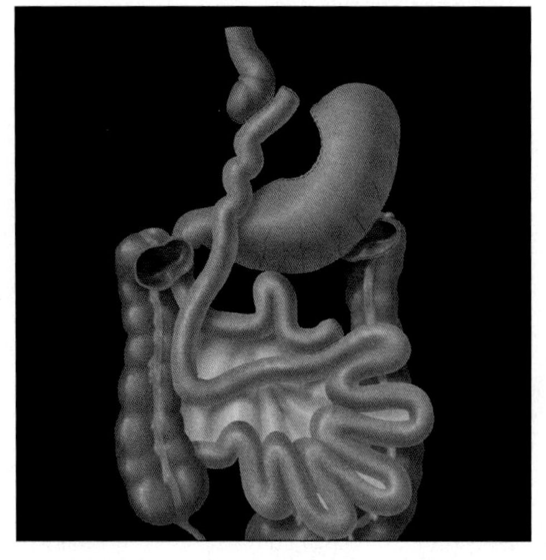

Graphic shows typical procedure for a Roux-en-y gastric bypass procedure, with a small gastric pouch anastomosed to a Roux limb (75 to 150 cm long).

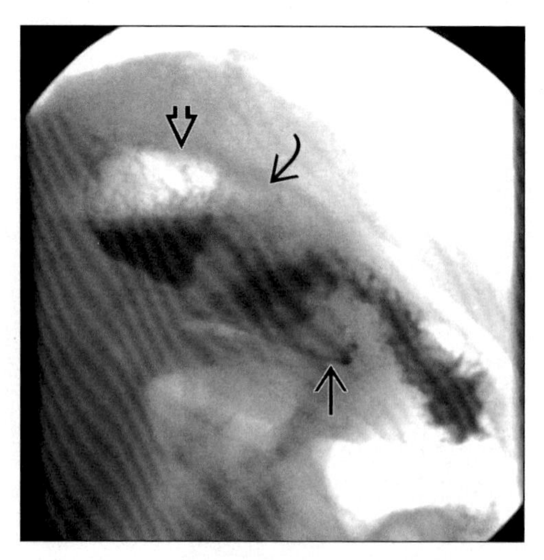

Upper GI series shows minor anastomotic leak, evident only as opacification of the surgical drain (arrow) placed near the gastric pouch (open arrow) - Roux anastomosis (curved arrow).

TERMINOLOGY

Definitions

- Complications of gastric bypass surgery (GBS) for morbid obesity

IMAGING FINDINGS

General Features

- Laparoscopic Roux-en-Y gastric bypass (RYGB); bariatric procedure of choice in North America
- RYGB procedure
  - Gastric pouch: 15-30 ml; along lesser curvature of proximal stomach; excluded from distal stomach
    - Anastomosed end to side to Roux-en-Y limb
    - Distal gastric remnant left in its normal anatomic position
  - Roux-en-Y limb; created by transection of jejunum at 35-45 cm distal to ligament of Treitz
    - 75 to 150 cm long
    - Anastomosed side to side with proximal jejunum
  - Roux limb may be brought through transverse mesocolon to be placed in retrocolic position
    - Or antecolic position; anterior to transverse colon
  - Mesenteric defects of jejunum, transverse colon sutured; closed with nonabsorbable sutures
- Gastrointestinal complications (> 10%)
  - Major complications; (require intervention; potentially life threatening) (9.5%)
    - Large anastomotic leak
    - Small bowel obstruction
    - Anastomotic stricture
    - Gastro-gastric; gastroenteric fistula
    - Gastrointestinal bleeding; abscess
  - Minor complications (6.7%)
    - Small leaks, marginal ulcers, pancreatitis, esophagitis, cholelithiasis
- Extra-enteric complications: Pulmonary embolism, pneumonia, lung atelectasis, wound infection

Radiographic Findings

- Fluoroscopy: Upper gastrointestinal (UGI) series
- Anastomotic leaks
  - Most commonly at gastrojejunal anastomosis
  - Less commonly at distal Roux anastomosis, bypassed stomach, esophagus, hypopharynx
  - Contrast material spills into peritoneal cavity
  - Opacification of surgical drain placed adjacent to anastomosis at surgery with contrast material
    - May be only clue to presence of leak; especially small leaks on first post operative day study
- Small bowel obstruction
  - Most common etiology: Internal hernias & adhesions
  - Other causes: Incarcerated ventral hernia
    - Gastric pouch bezoar formation

DDx: Leak, Obstruction or Mass After Surgery

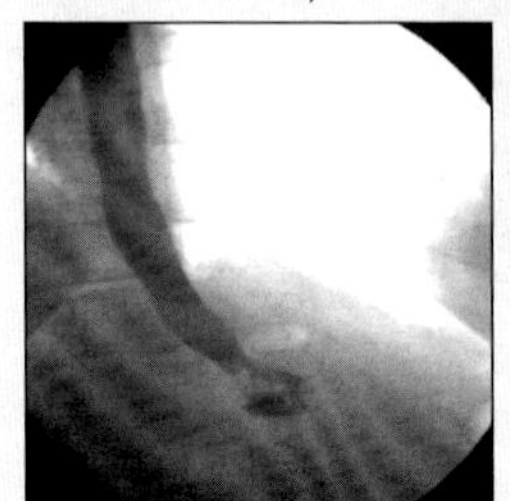

Anastomotic Edema

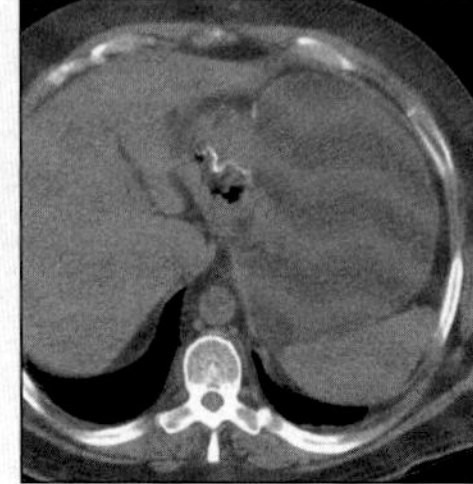

Distended Stomach

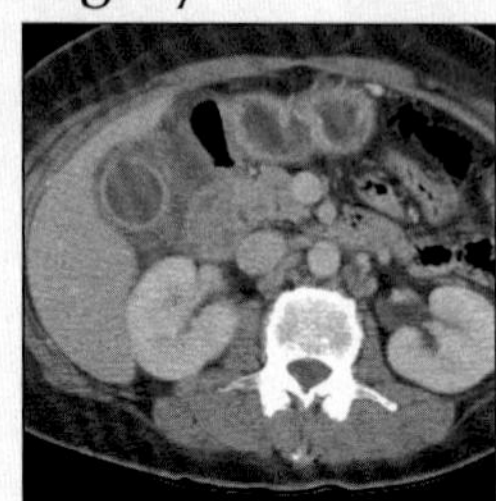

Cholecystitis

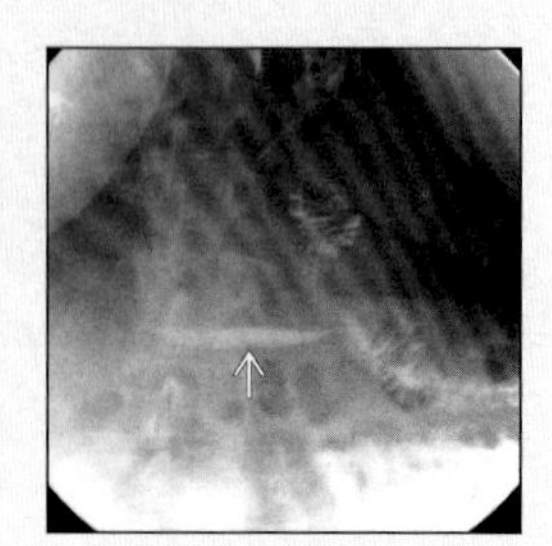

Reflux into Stomach

GASTRIC BYPASS COMPLICATIONS

Key Facts

Terminology

- Complications of gastric bypass surgery (GBS) for morbid obesity

Imaging Findings

- Laparoscopic Roux-en-Y gastric bypass (RYGB); bariatric procedure of choice in North America
- Major complications; (require intervention; potentially life threatening) (9.5%)
- Large anastomotic leak
- Small bowel obstruction
- Anastomotic stricture
- Internal hernia (IH)
- Relatively fixed cluster of small bowel loops; often seen in left upper quadrant or mid abdomen
- Cluster remaining high on erect radiographs

Top Differential Diagnoses

- Post-op anastomotic edema
- Post-op ileus
- Extra gastric complications
- Reflux into bypassed stomach

Clinical Issues

- Major complications: Require surgical intervention
- Minor complications; usually resolve spontaneously

Diagnostic Checklist

- CT & UGI series are important & complimentary in evaluation of these complications

    - Intussusception at entero-enterotomy site
  - Internal hernia (IH)
  - Overlap in UGI findings with adhesions & IH
    - Small-bowel segments: Clustered; distended (> 2.5 cm); abrupt angulation
    - Transition between dilated & nondilated segments
    - Stasis & delayed passage of contrast material
  - Findings that favor diagnosis of IH
    - Relatively fixed cluster of small bowel loops; often seen in left upper quadrant or mid abdomen
    - Cluster remaining high on erect radiographs
- Anastomotic stricture
  - Stomal stenosis of gastrojejunostomy (common)
    - Dilatation of gastric pouch; spherical shape; air-fluid-contrast material levels
    - Delayed passage of contrast material through anastomosis
  - Stenosis at jejuno-jejunal anastomosis is rare (0.9%)
- Less common complications
  - Gastro-gastric; gastro-cutaneous fistulas may develop rarely in cases with enteric content leak
  - Marginal ulcers; rate of 0.5-1.4% after RYGB
    - Result of exposure of gastrojejunal anastomosis to gastric acid, or ischemia
    - More common in "re-do" procedures

CT Findings

- Leaks: CT may demonstrate major & minor leaks; fluid collections not evident on UGI series
  - Fluid collections: Most commonly near anastomosis; in left upper abdomen, especially perisplenic area
    - May evolve into abscesses
    - Infected collections: Loculation; enhancing rim; air-fluid levels; gas bubbles
    - CT guided placement of drainage catheter into fluid collection obviates surgery in many cases
- Internal hernia: CT appearance depends on location
  - Abnormal clustering of small bowel loops; congestion & crowding of mesenteric vessels
    - Seen in all IH cases
  - Transmesenteric IH: Herniation through mesocolon, small bowel mesentery; most common
    - Small bowel cluster located posterior to remnant stomach exerting mass effect on its posterior wall
  - IH through small bowel mesentery
    - Cluster of small bowel pressed against anterior abdominal wall with no overlying omental fat
    - Causing central displacement of colon
  - Peterson type hernias; very difficult to diagnose
    - Herniation behind Roux-en-Y loop before passing through defect in transverse mesocolon
    - May not be apparent on CT; there is neither confining border nor characteristic location
    - Engorgement, crowding of mesenteric vessels & evidence of obstruction may be only clues
- Other less common complications
  - Obstruction & perforation of distal stomach
    - Markedly dilated distal stomach; free intraperitoneal air if perforation
    - Rarely seen with laparoscopic approach; fatal
  - Incisional & ventral hernias; infection of abdominal wall wound; seen with open procedure
    - Uncommon with laparoscopic RYGB

Imaging Recommendations

- Best imaging tool
  - CT & UGI radiography; complementary roles
    - Imaging of post-operative anatomy; complications
    - Allowing early diagnosis & treatment
- Protocol advice
  - UGI series with water-soluble contrast material; performed routinely; within 24 hours after surgery
    - May repeat study later if leak depicted or suspected clinically
    - Barium given subsequently after gastrointestinal extravasation excluded; delineate anatomy better
    - Evaluate pouch emptying; reflux into duodenum
  - CT used if small bowel obstruction or intra abdominal abscess suspected
    - In all patients with unexplained fever, pain, abdominal distension following RYGB

GASTRIC BYPASS COMPLICATIONS

DIFFERENTIAL DIAGNOSIS

Post-op anastomotic edema

- Delay in passage of contrast material at anastomotic site; early post-operative period obstruction; resolves

Post-op ileus

- Small + large bowel ± gastric distension
- Delayed but free passage of contrast material
- Usually resolves by 4th post-operative day

Extra gastric complications

- Cholecystitis, pulmonary embolism, etc.

Reflux into bypassed stomach

- Via retrograde passage of oral contrast through duodenojejunal segment
- Can simulate a leak

PATHOLOGY

General Features

- General path comments
  - Indications for GBS: Morbid obesity
    - Body mass index: 35 kg/m^2 with comorbidity
    - Or 40 kg/m^2 without comorbidity
  - Bariatric procedures: Restrictive & combination
    - Restrictive: Gastric capacity reduced; early sense of fullness after ingestion of small quantities of food
    - Combination: Part of digestive tract bypassed; causing ↓ absorption of nutrients & calories
- Etiology
  - Surgical technique
  - Leaks: Noncompliance of patients; premature ingestion of food or fluids early postoperative period
  - IH: Rapid massive weight reduction, results in ↓ intraperitoneal fat; enlarges mesenteric defect
- Epidemiology
  - During last 3 decades, incidence of overweight American adults nearly tripled to 35%
  - RYGB, combination procedure; most common bariatric procedure in North America

CLINICAL ISSUES

Presentation

- Leaks: Incidence of 1-6% after laparoscopic RYGB
  - Most dreaded complication of GBS surgery; may result in sepsis & even death
  - Leaks usually occur within first 10 days of surgery
  - May present with only tachycardia, abdominal discomfort, with no signs of peritonitis or fever
  - High index of suspicion; especially if respiratory distress & tachycardia > 120 beats per minute
- Small bowel obstruction
  - Reported in 4-5% of patients after laparoscopic GBS
  - Laparoscopic approach, associated with less trauma; fewer adhesions; higher prevalence of IH (2.8%)
  - Retrocolic placement of Roux limb, more frequently associated with IH
    - Regardless of suture closure of mesenteric defects
    - Antecolic approach has become more popular
  - Early obstructions, within 3 days to 3 months of surgery; more commonly due to adhesions
    - IHs develop later (in 93% > 1 month after surgery)
  - Clinical symptoms of IH: nonspecific, intermittent; nausea, distension, abdominal pain
    - High index of suspicion; in patients presenting with abdominal pain after surgery
    - Transmesenteric IH prone to volvulus & strangulation of small bowel
  - May result in closed loop obstruction; can be lethal
- Stenosis at gastrojejunostomy; due to relative ischemia
  - Incidence; up to 27% after RYGB
  - Dysphagia, vomiting, dehydration, excessive weight loss; diagnosis usually made with endoscopy
  - Late strictures; may present months after surgery

Natural History & Prognosis

- Advantages of bariatric surgery: Reliable, significant weight loss
  - Extended weight maintenance; control or reversal of some obesity-related health problems
- RYGB: Greater weight loss than other procedures
  - Good long term weight loss & patient tolerance
  - Acceptable short & long term complication rates
- Laparoscopic approach to RYGB: ↓ post-operative pain & complications; shorter hospital stay; faster recovery
  - Less invasive; especially benefits high-risk morbidly obese patients with multiple comorbidities
- Mortality: 0.4% after laparoscopic RYGB

Treatment

- Major complications: Require surgical intervention
  - Laparoscopy; excellent technique to treat these complications
- Anastomotic strictures: Endoscopic balloon dilatation
- Minor complications; usually resolve spontaneously

DIAGNOSTIC CHECKLIST

Consider

- Nonspecific clinical presentation of some of gastrointestinal complications of GBS
  - CT & UGI series are important & complimentary in evaluation of these complications

SELECTED REFERENCES

1. Champion JK et al: Small bowel obstruction and internal hernias after laparoscopic Roux-en-Y gastric bypass. Obes Surg. 13(4):596-600, 2003
2. Hamilton EC et al: Clinical predictors of leak after laparoscopic Roux-en-Y gastric bypass for morbid obesity. Surg Endosc. 17(5):679-84, 2003
3. Papasavas PK et al: Laparoscopic management of complications following laparoscopic Roux-en-Y gastric bypass for morbid obesity. Surg Endosc. 17(4):610-4, 2003
4. Blachar A et al: Gastrointestinal complications of laparoscopic roux-en-Y gastric bypass surgery in patients who are morbidly obese: findings on radiography and CT. AJR Am J Roentgenol. 179(6):1437-42, 2002
5. Blachar A et al: Gastrointestinal complications of laparoscopic Roux-en-Y gastric bypass surgery: clinical and imaging findings. Radiology. 223(3):625-32, 2002

IMAGE GALLERY

Typical

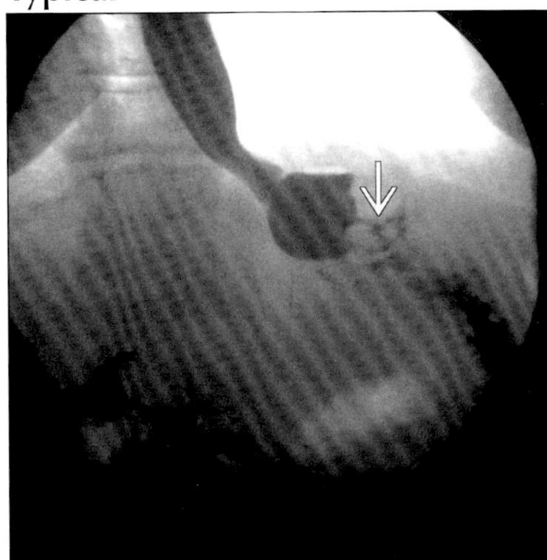

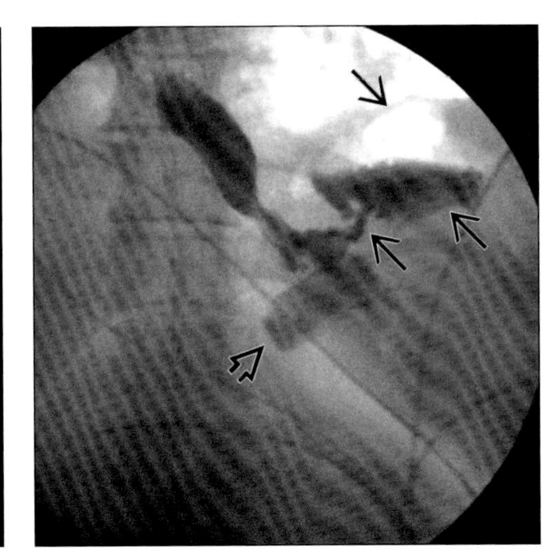

(Left) *Upper GI series shows anastomotic stricture (arrow) between the distended gastric pouch and the Roux limb.* ***(Right)*** *Upper GI series shows leak (arrows) following placement of a gastric band (open arrow) around the gastric fundus.*

Typical

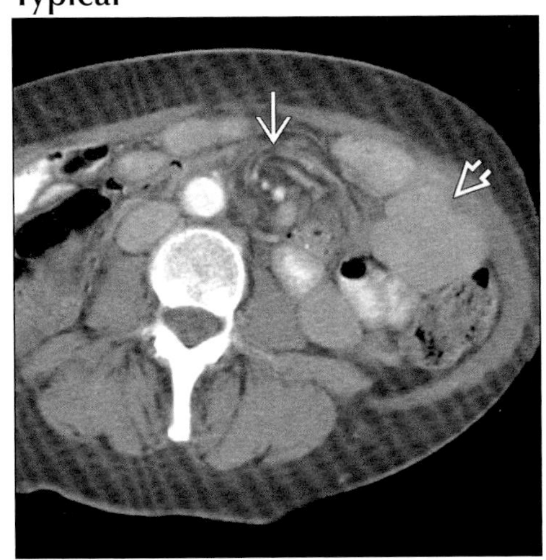

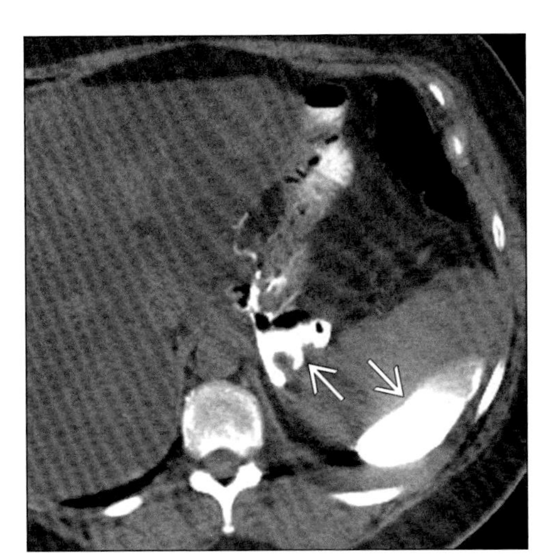

(Left) *Axial CECT shows internal hernia following Roux-en-Y gastric bypass (RYGB). The mesenteric vessels (arrow) to the herniated loops are crowded + swirled & the herniated bowel (open arrow) is dilated.* ***(Right)*** *Axial CECT shows major leak following RYGB with extravasated oral contrast seen (arrows).*

Typical

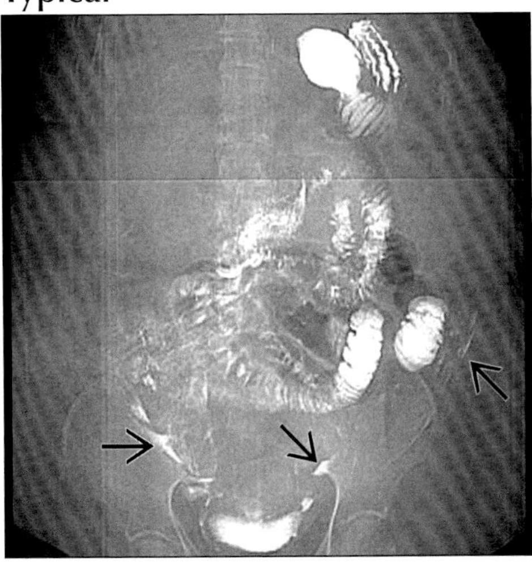

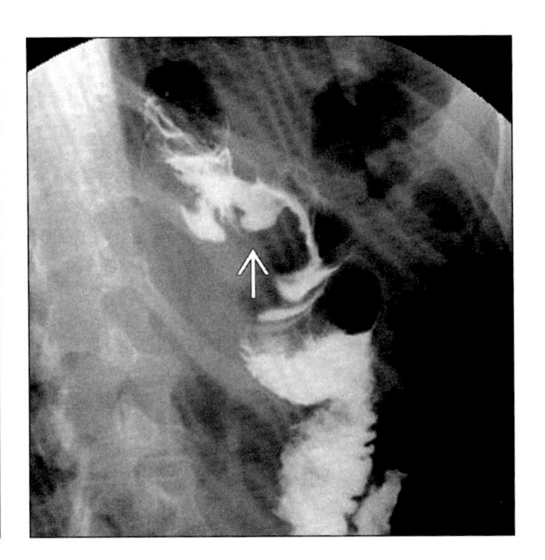

(Left) *Upper GI small bowel follow through (SBFT) shows contrast opacification of bowel following RYGB and intraperitoneal collections (arrows) due to leak at jejuno-jejunal anastomosis.* ***(Right)*** *Upper GI series shows marginal ulcer (arrow) within the Roux limb just beyond the anastomosis with the gastric pouch.*

GASTRIC BEZOAR

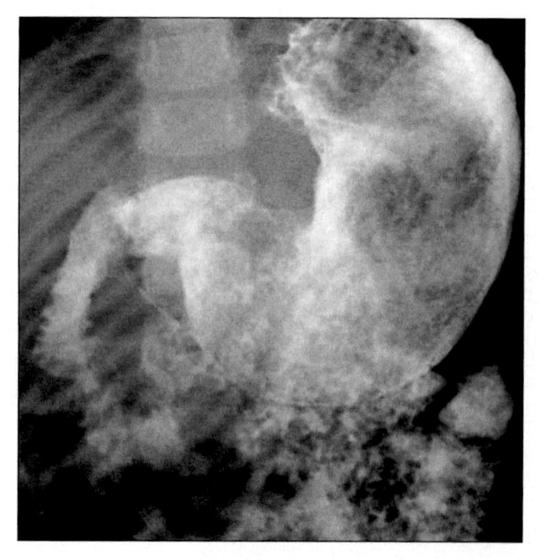

Upper GI series shows fixed filling defect in stomach with a swirled pattern-trichobezoar.

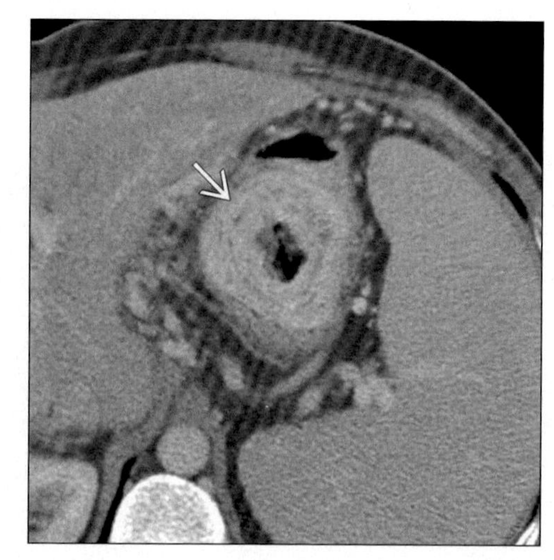

Axial CECT shows laminated mass (arrow) in stomach due to phytobezoar.

TERMINOLOGY

Definitions

- Intragastric mass composed of accumulated ingested (but not digested) material

IMAGING FINDINGS

General Features

- Best diagnostic clue: CT or fluoroscopy: Intraluminal mass containing mottled air pattern
- Location
  - Sites of impaction: Stomach, jejunum, ileum
    - Narrowest portion of small bowel 50-75 cm from ileocecal valve or valve itself
    - Any part can be affected; especially in patients with postoperative adhesions
- Morphology
  - Persistent concretions of foreign matter
    - Classified according to materials of which they are composed
  - Phytobezoar: Undigested vegetable matter
    - Poorly digested fibers; skin + seeds of fruits & vegetables
    - Diospyrobezoar: Persimmons
  - Trichobezoars: Accumulated, matted mass of hair
  - Trichophytobezoar: Both hair & vegetable matter
  - Lactobezoar: Undigested milk concretions
  - Pharmacobezoar: Bezoar comprised of medications

Radiographic Findings

- Radiography
  - Abdominal plain film: Soft-tissue mass floating in stomach at air-fluid interface
    - Mottled radiotransparencies in interstices of solid matter
  - ± Bowel obstruction
  - Insensitive test; bezoar identified in 10-18% of patients from radiographs alone
- Fluoroscopy
  - Intraluminal filling defect
    - With finely lobulated, villous-like surface
    - Without constant site of attachment to bowel wall
  - Barium or iodinated contrast outline bezoar; unattached intraluminal mass
  - Mottled or streaked appearance; contrast medium entering interstices of bezoar
  - Filling defect may occasionally appear completely smooth
    - Could be mistaken for an enormous gas bubble that is freely movable within stomach
  - Coiled-spring appearance (rare)
  - Partial or complete obstruction
    - Try to distinguish obstruction due to postoperative adhesions from bezoar-induced obstruction

DDx: Large Filling Defect in Stomach

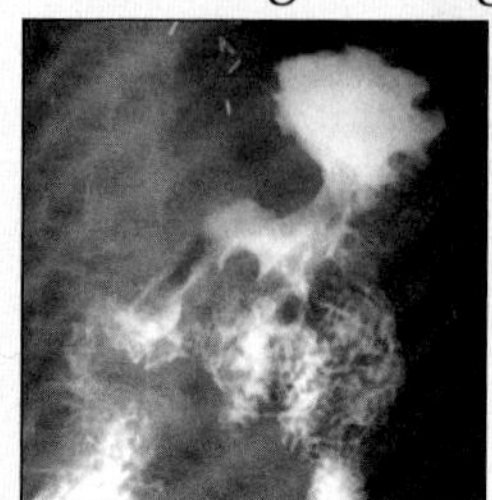

Gastric Cancer

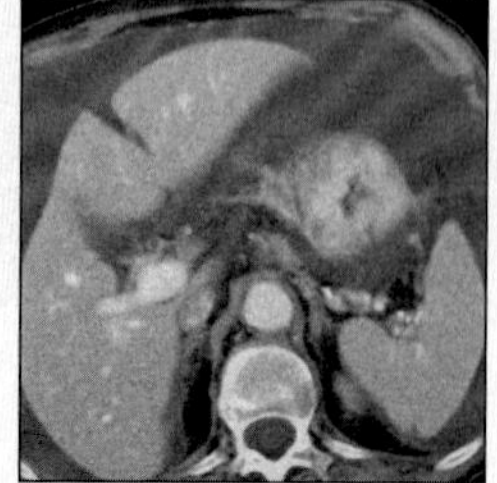

Gastric Cancer

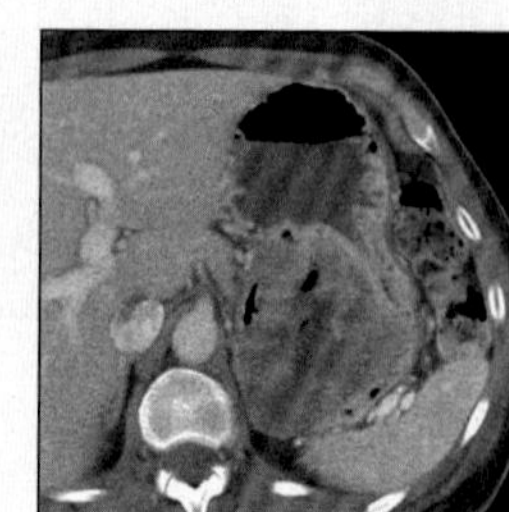

Gastric GIST

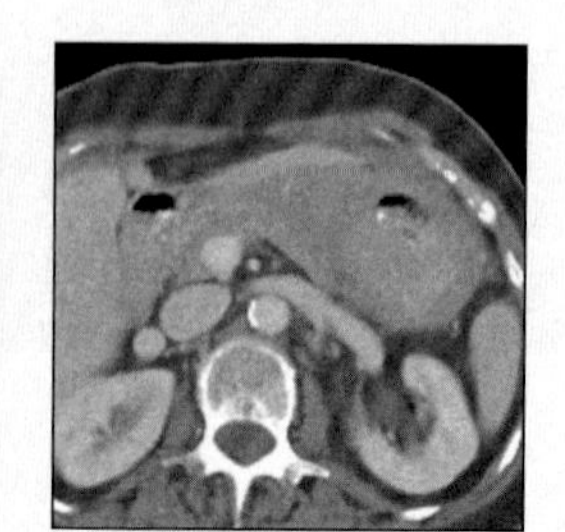

Gastric Lymphoma

GASTRIC BEZOAR

Key Facts

Terminology

- Intragastric mass composed of accumulated ingested (but not digested) material

Imaging Findings

- Phytobezoar: Undigested vegetable matter
- Trichobezoars: Accumulated, matted mass of hair
- Intraluminal filling defect
- With finely lobulated, villous-like surface
- Without constant site of attachment to bowel wall
- Mottled or streaked appearance; contrast medium entering interstices of bezoar
- Well-defined, oval, low-density, intraluminal mass
- Mottled air pattern
- Diagnose bezoar-induced obstruction
- Detect presence of additional gastric or intestinal bezoars

Top Differential Diagnoses

- Gastric carcinoma
- Post-prandial food
- Intramural mass

Clinical Issues

- In adults, bezoars are most frequently encountered after gastric operation
- Trichobezoars seen especially in those with schizophrenia or other mental instability
- Complications: Decubitus ulceration + pressure necrosis of bowel wall, perforation, peritonitis
- Symptomatic, large phytobezoars or trichobezoars require endoscopic fragmentation or surgical removal

CT Findings

- Well-defined, oval, low-density, intraluminal mass
  - Mottled air pattern
  - Mottled appearance is result of air bubbles retained in interstices of mass
  - Heterogeneous mass without post-contrast enhancement
    - Pockets of gas, debris, fluid scattered throughout; with no air-fluid level within lesion
- Small bezoars are rounded or ovoid; tend to float on water-air surface surrounded by gastric contents
  - Oral contrast material may be seen surrounding mass, establishing free intraluminal location
- Large bezoars tend to fill lumen

Ultrasonographic Findings

- Intraluminal mass with hyperechoic arc-like surface
  - With marked acoustic shadowing
- With US; identification of additional intestinal or gastric bezoars may be difficult

Imaging Recommendations

- Best imaging tool
  - CT; more accurate in confirming diagnosis of gastric bezoar suggested by other modalities
    - Diagnose bezoar-induced obstruction
    - Detect presence of additional gastric or intestinal bezoars
- Protocol advice
  - May go undetected if CT scan obtained at routine abdominal soft tissue window & level settings
    - Modifying window setting by reducing level to -100 HU makes it possible to better identify

DIFFERENTIAL DIAGNOSIS

Gastric carcinoma

- Filling defect in stomach; polypoid or fungating
- Lesion on dependent or posterior wall seen as filling defect in barium pool
- Wall thickening; ulceration; irregular narrowing & rigidity; amputation of folds; stenosis

Post-prandial food

- Fluoroscopy: Intraluminal filling defect
- Food usually less mass-like
- CT: Bezoar shows lower density than food particles
  - Occasionally difficult to differentiate bezoar from large amount of retained food

Intramural mass

- Stromal tumor (GIST); lymphoma; melanoma metastases
  - Lobulated or polypoid filling defects; arising from gastric wall
  - Infiltration of gastric wall; mucosal thinning or ulceration; submucosal mass

PATHOLOGY

General Features

- General path comments
  - Predisposing causes
    - Previous gastric surgery: Vagotomy, pyloroplasty, antrectomy, partial gastrectomy
    - Inadequate chewing, missing teeth, dentures
    - Overindulgence of foods with high fiber content
    - Altered gastric motility: Diabetes, mixed connective tissue disease, hypothyroidism
- Etiology
  - Material unable to exit stomach
    - Accumulated due to large size; indigestibility; gastric outlet obstruction; poor gastric motility
  - Phytobezoar: Unripe persimmon fruit, oranges
    - Persimmon contains tannin; coagulates on contact with gastric acid
    - Glue-like coagulum formed; traps seeds, skin, etc.
  - Medications reported to cause bezoars
    - Aluminum hydroxide gel, enteric-coated aspirin, sucralfate, guar gum, cholestyramine
    - Enteral feeding formulas, psyllium preparations, nifedipine XL, meprobamate
- Epidemiology
  - Incidence: 0.4% (large endoscopic series)

- Phytobezoar: 55% of all bezoars
- Phytobezoar responsible for 0.4-4% of all intestinal obstructions
- Associated abnormalities
  - Peptic ulcer; incidence high; especially with more abrasive phytobezoars
    - Trichobezoars associated with gastric ulcer in 24-70%
  - Concurrent gastric bezoar found in 17-53% of patients with small-bowel bezoar

Gross Pathologic & Surgical Features

- Conglomerates of food or fiber in alimentary tract
- Hairball

CLINICAL ISSUES

Presentation

- Most common signs/symptoms
  - Asymptomatic; incidentally found on imaging
  - Anorexia, bloating, early satiety
  - Crampy epigastric pain
  - Sense of dragging, heaviness in upper abdomen
  - With large bezoars; symptoms of pyloric obstruction
    - Can clinically simulate gastric carcinoma
  - May present with small bowel obstruction
  - Trichotillomania; impulse disorder to pull out hair from scalp, eyelashes, eyebrows, other parts of body
    - With trichotillomania; gastric trichobezoar may result in failure to gain weight
    - Iron deficiency anemia, painless epigastric mass
- Clinical profile
  - History of recent ingestion of pulpy foods
  - History of previous gastric surgery
  - Physical examination: Bald patches on patient's head or bald sibling as proof; with trichobezoar
- In adults, bezoars are most frequently encountered after gastric operation
  - In children, associated with pica, mental retardation, coexistent psychiatric disorders
- Trichobezoars seen especially in those with schizophrenia or other mental instability
  - Primarily girls who chew & swallow their own hair
- Lactobezoar, most often found in infants
  - Pre-term infants on caloric-dense formulas
  - Immature mechanism of gastric emptying

Demographics

- Age: Trichobezoar: 80% are in age less than 30 years
- Gender: Trichobezoars occur predominantly in females

Natural History & Prognosis

- Bezoars of any type most often occur in background of altered motility or anatomy of gastrointestinal tract
- Bezoars usually form in stomach
  - Fragment & enter small bowel where they absorb water, increase in size & become impacted
- Bezoars are an uncommon cause of acute gastric outlet obstruction
- Trichobezoar: Can enlarge to occupy entire lumen of stomach assuming shape of organ
  - Trichobezoars do not usually migrate toward small bowel
- Rapunzel syndrome, found characteristically in girls with varying gastrointestinal symptoms
  - Rare form of gastric trichobezoar extending throughout the bowel
  - Possessing "tail" which extends to or beyond ileo-cecal valve; causing intestinal obstruction
  - High comorbidity of serious pediatric psychiatric disorders
- Complications: Decubitus ulceration + pressure necrosis of bowel wall, perforation, peritonitis
  - Bleeding, obstructive jaundice, intussusception & appendicitis

Treatment

- Endoscopic lavage fragmentation + extraction presents safe method of bezoar resolution
- Symptomatic, large phytobezoars or trichobezoars require endoscopic fragmentation or surgical removal
- Diagnosis of bezoar as cause of obstruction important
  - Modifies approach to treatment; accelerating use of surgery
  - Bezoar-induced bowel obstruction rarely improves with conservative treatment
  - Early surgery required to secure definitive solution
- 9% of patients may require second operation
  - Recurrent bowel obstruction; caused by presence of residual bezoar
- Spontaneous expulsion of bezoar; uncommon

DIAGNOSTIC CHECKLIST

Consider

- Bezoar formation may be more common than previously thought
  - High index of suspicion could help avoid costly evaluations for obstructive symptoms
- When an intestinal bezoar is diagnosed, consider concomitant gastric bezoar
- Discrepancy between CT & surgical localization
  - May be caused by migration of bezoar during interval between imaging & surgery

SELECTED REFERENCES

1. Ripolles T et al: Gastrointestinal bezoars: sonographic and CT characteristics. AJR Am J Roentgenol. 177(1):65-9, 2001
2. DuBose TM 5th et al: Lactobezoars: a patient series and literature review. Clin Pediatr (Phila). 40(11):603-6, 2001
3. Morris B et al: An intragastric trichobezoar: computerised tomographic appearance. J Postgrad Med. 46(2):94-5, 2000
4. Gayer G et al: Bezoars in the stomach and small bowel--CT appearance. Clin Radiol. 54(4):228-32, 1999
5. West WM et al: CT appearances of the Rapunzel syndrome: an unusual form of bezoar and gastrointestinal obstruction. Pediatr Radiol. 28(5):315-6, 1998
6. Phillips MR et al: Gastric trichobezoar: case report and literature review. Mayo Clin Proc. 73(7):653-6, 1998
7. Newman B et al: Gastric trichobezoars--sonographic and computed tomographic appearance. Pediatr Radiol. 20(7):526-7, 1990

GASTRIC BEZOAR

IMAGE GALLERY

Typical

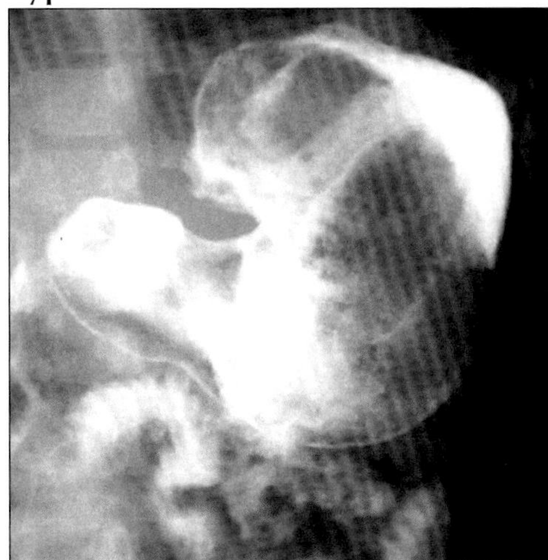

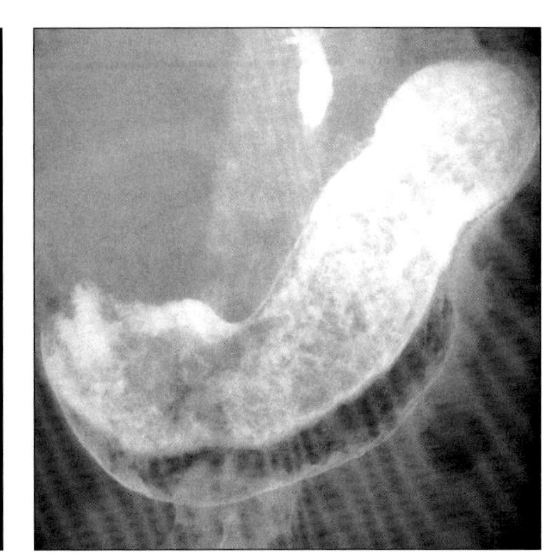

*(**Left**) Upper GI series in a 3 year old girl shows large mass in stomach: Trichobezoar. (**Right**) Upper GI series in edentulous adult shows mottled filling defects in stomach: Phytobezoars.*

Typical

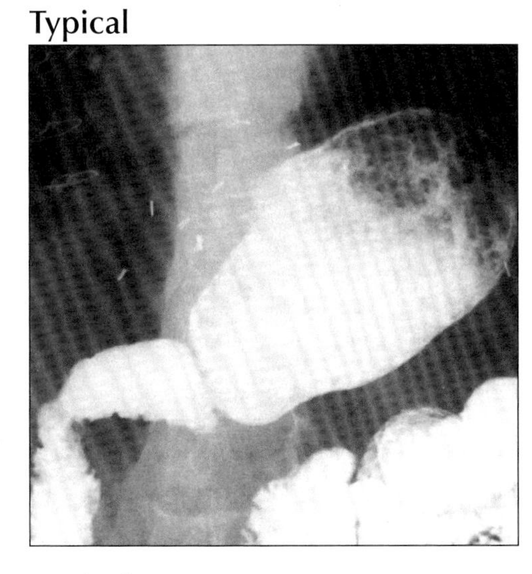

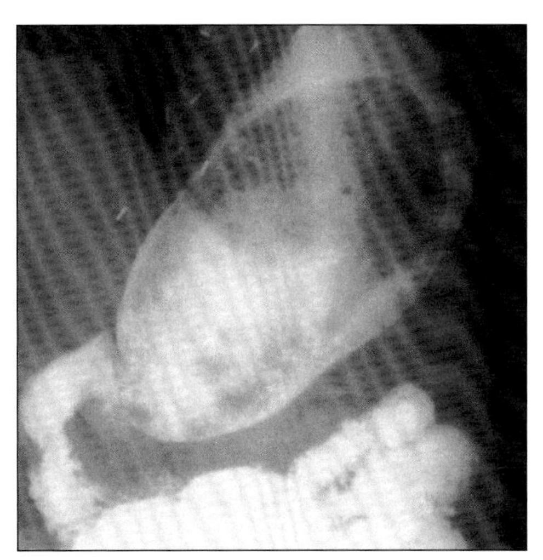

*(**Left**) Upper GI series in a patient with Bilroth 1 type partial gastrectomy shows bezoar in stomach. (**Right**) Upper GI series in a patient with Bilroth 1 type partial gastrectomy shows bezoar in stomach.*

Typical

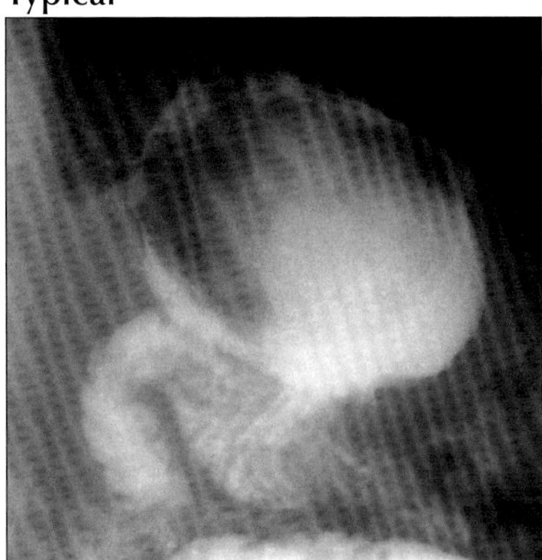

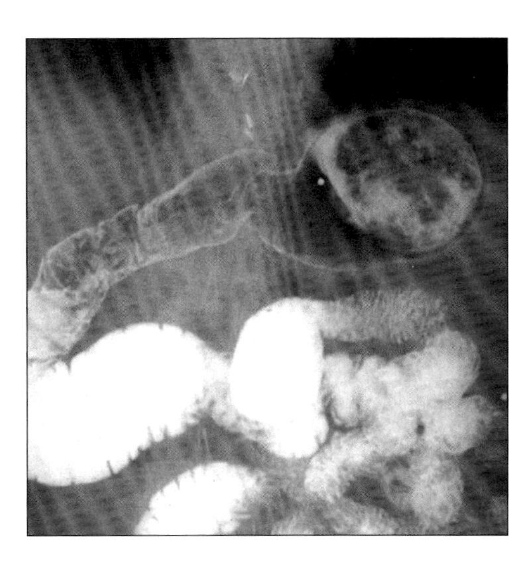

*(**Left**) Upper GI series in a patient with a Bilroth 2 type partial gastrectomy shows a large bezoar in stomach. (**Right**) Upper GI series in a patient with a Bilroth 1 type partial gastrectomy shows a bezoar in stomach.*

GASTRIC VOLVULUS

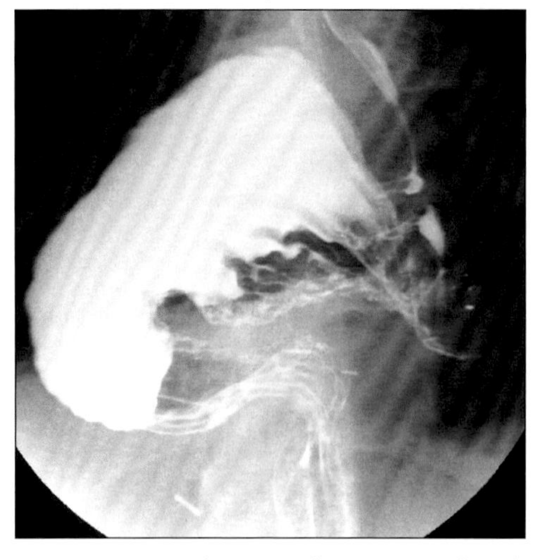

Upper GI series shows intrathoracic stomach with organoaxial volvulus and partial obstruction.

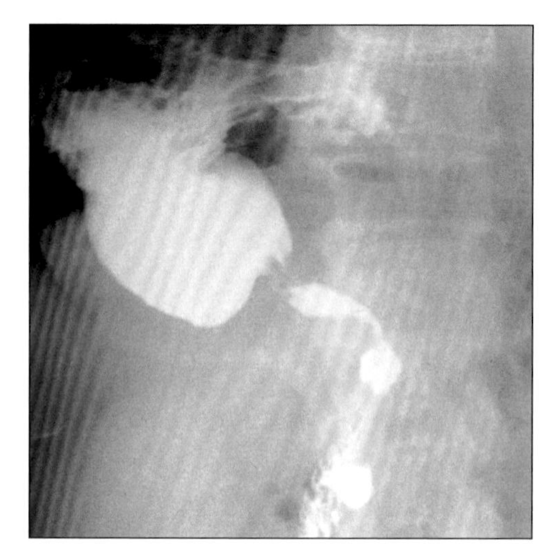

Upper GI series shows intrathoracic stomach with organoaxial volvulus, but no obstruction.

TERMINOLOGY

Abbreviations and Synonyms

- Gastric volvulus (GV)

Definitions

- Uncommon acquired twist of stomach on itself

IMAGING FINDINGS

General Features

- Morphology: Abnormal degree of rotation of one part of stomach around another part
- Types of GV: Organoaxial (most common); mesenteroaxial; mixed
- Organoaxial volvulus (OAV): Rotation of stomach around its longitudinal axis (most common form)
  - Around line extending from cardia to pylorus
  - Stomach twists either anteriorly or posteriorly
  - Antrum moves from an inferior to superior position
- Mesenteroaxial volvulus (MAV): Rotation of stomach about mesenteric axis
  - Axis running transversely across stomach at right angles to lesser & greater curvatures
  - Stomach rotates from right to left or left to right about long axis of gastrohepatic omentum
- Mixed volvulus: Combination of OAV & MAV

Radiographic Findings

- Radiography
  - Abdominal plain films; patient upright
    - Double air-fluid level
    - Large, distended stomach; seen as air & fluid-filled spheric viscus displaced upward & to left
    - Associated elevation of diaphragm
    - Usually small bowel collapsed; uncommon to see gas shadow beyond stomach
    - May see radiolucent line within gastric wall; caused by intramural emphysema
  - Chest film: Intrathoracic; up-side down stomach
    - Retrocardiac fluid level; two air-fluid interfaces at different heights; suggests intrathoracic GV
    - Simultaneous fluid levels above & below diaphragm are not required to make diagnosis
- Fluoroscopy
  - Massively distended stomach in left upper quadrant extending into chest
  - Inversion of stomach
    - Greater curvature above level of lesser curvature
    - Positioning of cardia & pylorus at same level
    - Downward pointing of pylorus & duodenum
  - OAV: 2 points of twist; luminal obstruction
  - Incomplete or absent entrance of contrast material into +/or out of stomach; acute obstructive GV
    - OAV: Failure of contrast to enter stomach; obstruction at esophagus or proximal stomach

DDx: Intrathoracic Stomach

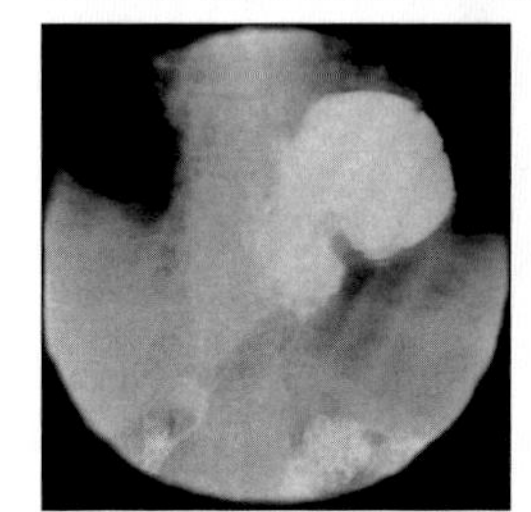

Hiatal Hernia

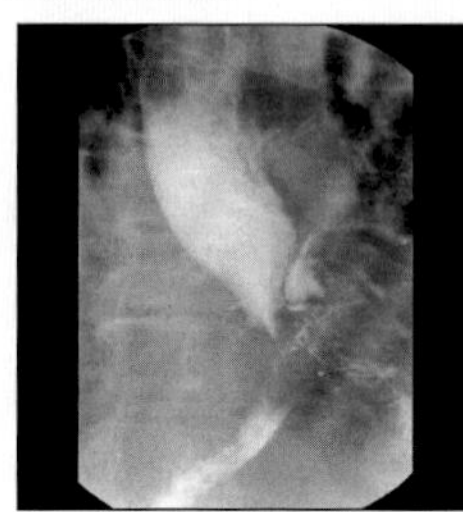

Post-Operative

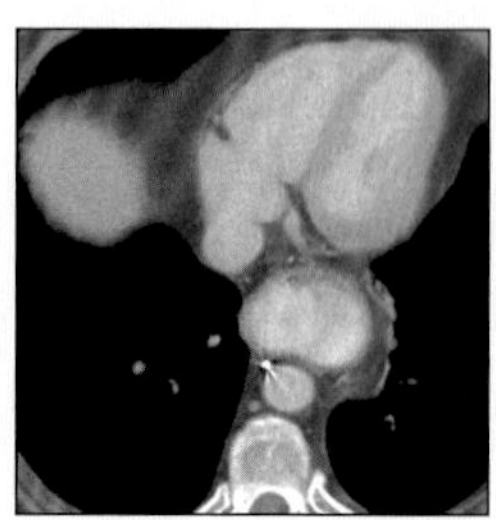

Post-Operative

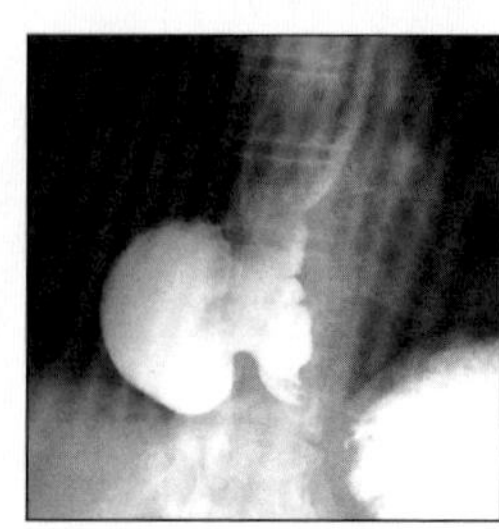

Epi. Diverticulum

GASTRIC VOLVULUS

Key Facts

Terminology

- Uncommon acquired twist of stomach on itself

Imaging Findings

- Organoaxial volvulus (OAV): Rotation of stomach around its longitudinal axis (most common form)
- Around line extending from cardia to pylorus
- Mesenteroaxial volvulus (MAV): Rotation of stomach about mesenteric axis
- Axis running transversely across stomach at right angles to lesser & greater curvatures
- Double air-fluid level
- Inversion of stomach
- Greater curvature above level of lesser curvature
- Positioning of cardia & pylorus at same level
- Downward pointing of pylorus & duodenum
- Incomplete or absent entrance of contrast material into +/or out of stomach; acute obstructive GV

Top Differential Diagnoses

- Hiatal hernia
- Post-operative

Pathology

- Large esophageal or paraesophageal hernia
- Diaphragmatic eventration or paralysis

Clinical Issues

- Complications: Intramural emphysema; perforation
- Mortality rate: 30%
- Detorse stomach
- Repair of associated defects
- Prevent recurrence

    - If contrast material does enter stomach, it may not pass beyond obstructed pylorus
  - May see "beaking" at point of twist
  - MAV: Antrum & pylorus lie above gastric fundus

CT Findings

- CT appearance may be variable
  - Depends upon extent of gastric herniation, points of torsion & final positioning of stomach
  - May see linear septum within gastric lumen; corresponding to area of torsion
- CT chest & abdomen; performed pre-operatively
  - To detect associated malformation or malposition & if possible, site, size, level of diaphragmatic gap
  - Presence of unattached herniated peritoneal sac
- Large hiatal hernia accompanied by partial GV; may mimic appearance of thrombus in inferior vena cava
  - "Pseudothrombosis" of inferior vena cava on CT

MR Findings

- Coronal images demonstrate 2 points of twisting
  - 2 different signal intensities reflect point of torsion

Angiographic Findings

- GV may present as acute upper gastrointestinal hemorrhage

Imaging Recommendations

- Best imaging tool
  - Fluoroscopic barium studies
    - Demonstrates area of twist; anatomic detail
  - Fluoroscopic guidance may help in advancing nasogastric tube into obstructed stomach
    - May allow decompression; stabilize patient
  - CT; complementary role

DIFFERENTIAL DIAGNOSIS

Hiatal hernia

- Stomach entering thorax through esophageal hiatus
- Gastroesophageal (GE) junction above diaphragmatic hiatus (type I, sliding)
- Herniation of fundus through hiatus; GE junction below diaphragm (type II; paraesophageal)
- Giant paraesophageal hernia: At least one third of stomach herniated into chest
  - Associated herniation of other abdominal viscera into chest ; including colon or small bowel
- Traction or torsion of stomach at or near level of hiatus (volvulus)

Post-operative

- Esophagectomy with gastric pull through procedure
- Complete mobilization of stomach, resection of lower esophagus, pyloroplasty, transhiatal dissection
  - Intrathoracic stomach

Epiphrenic diverticulum

- Epiphrenic diverticulum

PATHOLOGY

General Features

- General path comments
  - Point of anatomic fixation: Second portion of duodenum retains retroperitoneal position
    - Becomes fixed to posterior abdominal wall
  - Several ligaments normally anchor stomach within abdomen & limit free upward movement
    - 4 suspensory ligaments; gastrohepatic, gastrosplenic, gastrocolic, gastrophrenic
    - Gastrolienal ligaments also contribute to fixation of stomach
  - Due to sites of anatomic fixation, torsion of stomach may occur with significant degrees of herniation
  - Predisposing factors: Bands, adhesions
    - Rapid changes in intraabdominal pressure; degenerative changes; ↑ size of esophageal hiatus
    - Unusually long gastrohepatic + gastrocolic mesenteries
- Etiology
  - Primary GV: Stabilizing ligaments are too lax as a result of congenital or acquired causes
    - Absence of tethering gastric ligaments

    - One third of cases
  - Secondary GV: Paraesophageal hernia
    - Congenital or acquired diaphragmatic defects
    - In children; secondary to Morgagni hernia
  - Idiopathic; no apparent cause
- Epidemiology
  - In children, MAV most common type; associated anatomic defects are the rule
  - Five cases of combined organomesenteroaxial GV in children reported in world literature
- Associated abnormalities
  - Large esophageal or paraesophageal hernia
    - Permits part or all of stomach to assume intrathoracic position
  - Diaphragmatic eventration or paralysis
  - Wandering spleen: Absence of ligamentous connections between stomach, spleen
  - Hernia of colonic transverse loop with anterior OAV
    - In these cases; concomitant sliding hernia

Gross Pathologic & Surgical Features

- Partial or complete volvulus
- Term "gastric volvulus" used by some to identify abnormalities of gastric position without obstruction
  - Upside-down stomach"; gastric displacement through sliding & large paraesophageal hernias
- " True volvulus"; term used only when obstruction

CLINICAL ISSUES

Presentation

- GV can be asymptomatic if no outlet obstruction or vascular compromise; incidental finding on imaging
- Acute volvulus; associated interference of blood supply
  - Surgical emergency
  - Classic clinical triad (Borchardt triad)
    - Violent retching with production of little vomitus
    - Constant severe epigastric pain
    - Great difficulty in advancing nasogastric tube beyond distal esophagus
- Chronic GV: May present in chronic or recurrent form
  - Frequently not recognized early in its presentation
  - Vague & nonspecific symptoms suggestive of other abdominal processes; causing delay in diagnosis
  - May be discovered unexpectantly during clinical work-up for an unrelated condition
    - CT & MR often requested as first radiographic study during evaluation
- Symptomatic GV in infancy & childhood may not be as rare as is commonly assumed

Demographics

- Age
  - Seen in both pediatric & adult patients
  - Primarily after fourth decade of life

Natural History & Prognosis

- In small herniations, proximal portion of stomach enters hernia sac first
  - Obstruction or strangulation almost never occur at this stage
- As herniation progresses; body & variable portion of antrum come to lie above diaphragm
  - Stomach can become entirely intrathoracic organ; prone to volvulus
- Obstruction can occur at points of torsion or twisting
  - Or at points where stomach redescends through hiatus, fills & tightens in hernial ring
  - As much as 180 degrees of twisting may occur without obstruction or strangulation
  - Twisting beyond 180 degrees usually produces complete obstruction & clinically acute abdomen
  - OAV: Can obstruct; does not usually result in strangulation
    - MAV: Can occlude gastric vessels; strangulation
- " Upside-down stomach"
  - In typical case, sliding hernia & stomach (180 degrees OAV) pass through same diaphragmatic gap
    - An enlarged esophageal hiatus or Bochdalek defect
  - Presents with bleeding & anemia; does not usually induce obstruction or strangulation
- Vascular occlusion leads to necrosis, shock
- Complications: Intramural emphysema; perforation
  - Strangulation may lead to mucosal ischemia
    - Areas of focal necrosis; may permit gas to dissect into gastric wall
  - Perforation may result from full-thickness necrosis
- Prognosis: GV is potentially catastrophic condition
- Mortality rate: 30%

Treatment

- Goals: Early recognition & surgical repair
  - Detorse stomach
  - Repair of associated defects
    - Hiatal hernia repair
    - Gastropexy; may be prophylactic
  - Prevent recurrence
- Laparoscopic detorsion & percutaneous endoscopic gastropexy
- Gastric resection; for strangulation & necrosis
- Upside-down stomach: Balloon repositioning; fixation by percutaneous endoscopic gastrostomy

DIAGNOSTIC CHECKLIST

Consider

- Anatomical detail of stomach often better delineated on upper gastrointestinal studies
  - Identification of GV as incidental finding on CT
    - Should be excluded whenever stomach is noted not to be in normal anatomic position

SELECTED REFERENCES

1. Shivanand G et al: Gastric volvulus: acute and chronic presentation. Clin Imaging. 27(4):265-8, 2003
2. Tabo T et al: Balloon repositioning of intrathoracic upside-down stomach and fixation by percutaneous endoscopic gastrostomy. J Am Coll Surg. 197(5):868-71, 2003
3. Godshall D et al: Gastric volvulus: case report and review of the literature. J Emerg Med. 17(5):837-40, 1999
4. Schaefer DC et al: Gastric volvulus: an old disease process with some new twists. Gastroenterologist. 5(1):41-5, 1997
5. Chiechi MV et al: Gastric herniation and volvulus: CT and MR appearance. Gastrointest Radiol. 17(2):99-101, 1992

IMAGE GALLERY

Typical

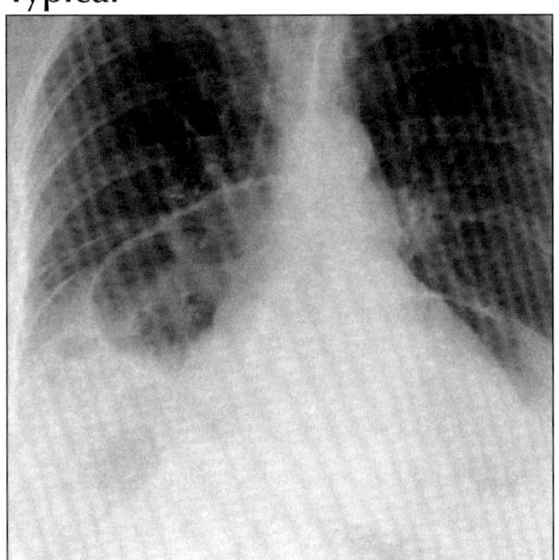

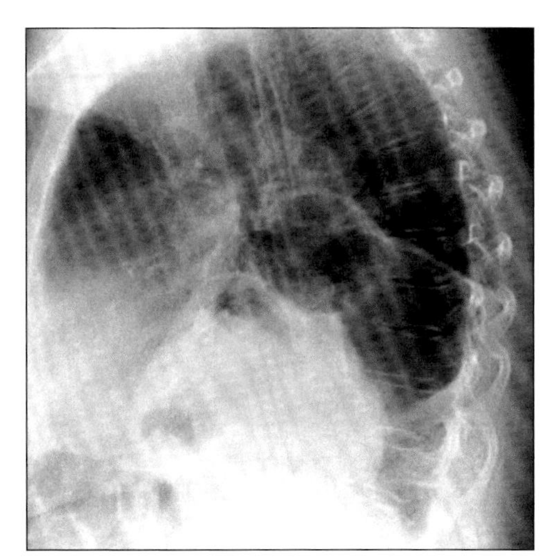

(Left) *PA chest radiograph shows distended intrathoracic stomach due to volvulus with acute obstruction.* ***(Right)*** *Lateral chest radiograph shows distended intrathoracic stomach due to volvulus with acute obstruction.*

Typical

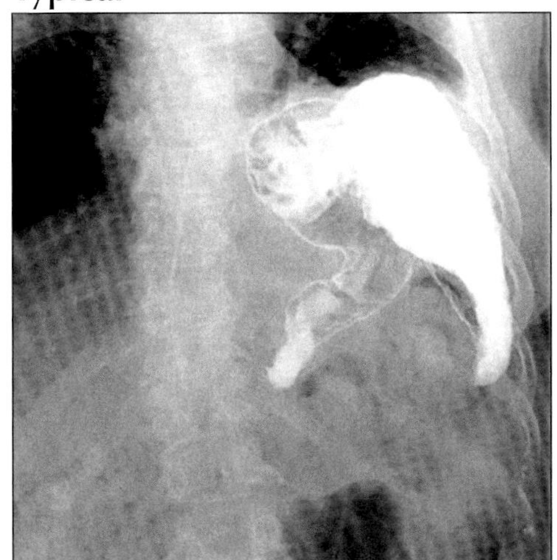

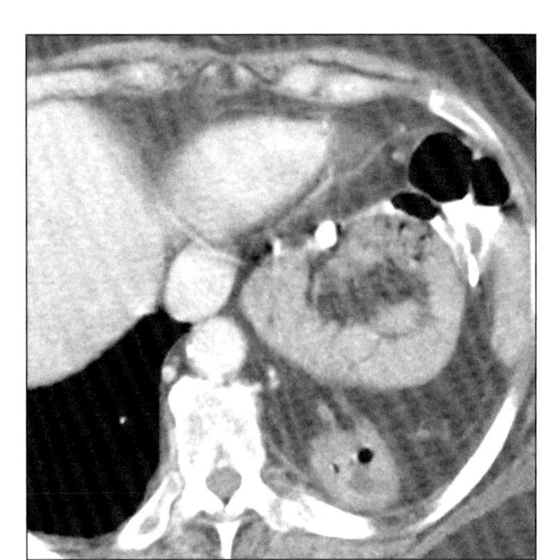

(Left) *Upper GI series shows eventration or paralysis of left diaphragm and mesenteroaxial gastric volvulus with obstruction.* ***(Right)*** *Axial CECT shows mesenteroaxial gastric volvulus with the stomach rotated left-to-right.*

Typical

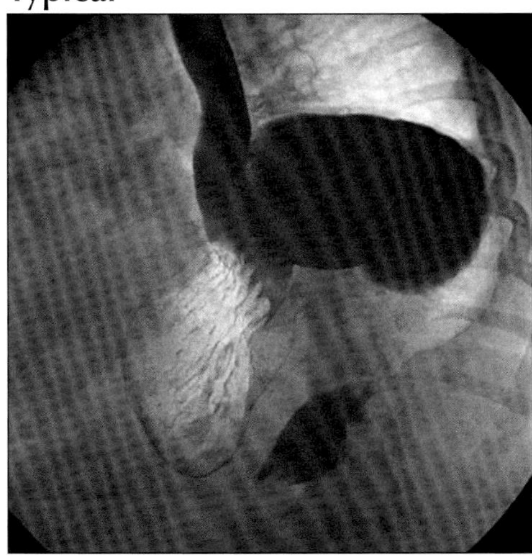

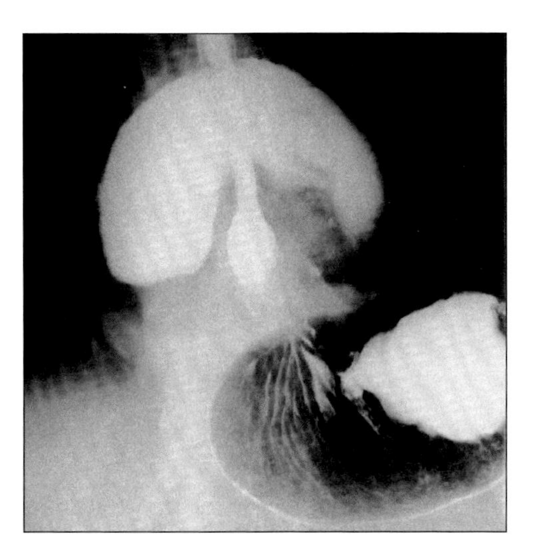

(Left) *Upper GI series shows intrathoracic stomach with organoaxial volvulus and obstruction.* ***(Right)*** *Upper GI series shows mixed organo- and mesenteroaxial gastric volvulus with obstruction.*

AORTO-ENTERIC FISTULA

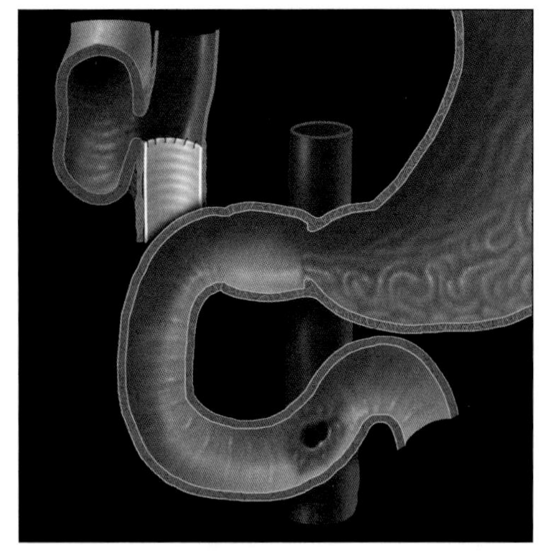

Graphic shows fistula between transverse duodenum and the aorta at the site of graft-aortic suture line.

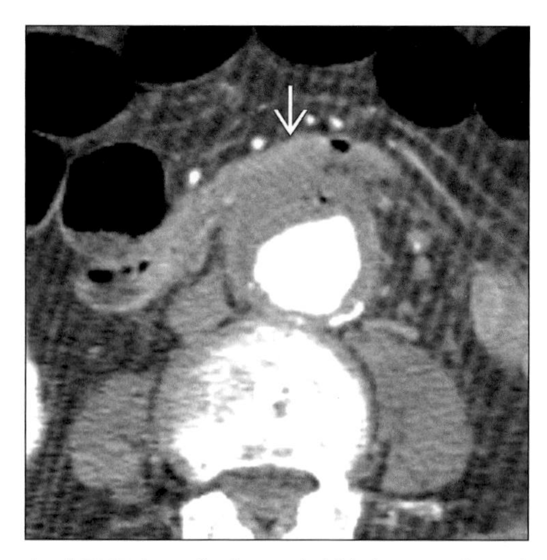

Axial CECT shows fluid & gas bubble between the graft lumen & aortic wall, which is wrapped around the graft. At surgery the graft was infected & a fistula was found to the duodenum (arrow).

TERMINOLOGY

Definitions

- Abnormal communication between aorta & gastrointestinal (GI) tract

IMAGING FINDINGS

General Features

- Best diagnostic clue: Inflammatory stranding and gas between abdominal aorta and third part of duodenum following aneurysm repair
- Location: Duodenum (80%) > jejunum and ileum (10-15%) > stomach and colon (5%)

Radiographic Findings

- Fluoroscopic-guided barium studies
  - Compression or displacement of third portion of duodenum by an extrinsic mass
  - Contrast extravasation: Wall of abdominal aorta outlined by extraluminal contrast medium tracking along the graft into periaortic space (rare)

CT Findings

- Ectopic gas: Microbubble of gas adjacent to aortic graft; may suggest perigraft infection
- Focal bowel wall thickening > 5 mm
- Perigraft soft tissue thickening > 5 mm (> 20 HU)
- Pseudoaneurysm formation
- Disruption of aneurysmal wrap
- ↑ Soft tissue between graft and aneurysmal wrap
- Contrast in pseudoaneurysm (arterial phase)
- Increased attenuation of intestinal lumen contents (arterial phase); decreased attenuation (delayed phase)

Nuclear Medicine Findings

- Tagged RBC within abdominal aorta & enters bowel

Imaging Recommendations

- Best imaging tool: CT: 94% sensitive & 85% specific

DIFFERENTIAL DIAGNOSIS

Periaortitis

- Also known as inflammatory perianeurysmal fibrosis
- Soft tissue attenuation encasing aorta, inferior vena cava and other structures

Retroperitoneal fibrosis

- Mantle of soft tissue enveloping aorta, IVC, ureters

Post-operation

- "Normal" scarring with fluid between graft & aorta

Post-endovascular stent

- Endoleak: Blood flow outside the stent, but within an aneurysm sac or adjacent vascular segment

DDx: Periaortic Inflammation

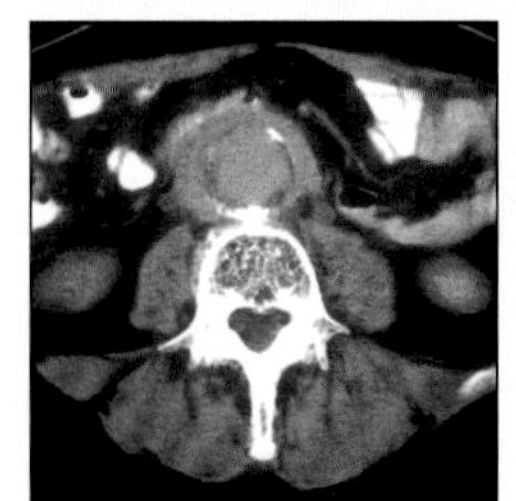

Periaortitis

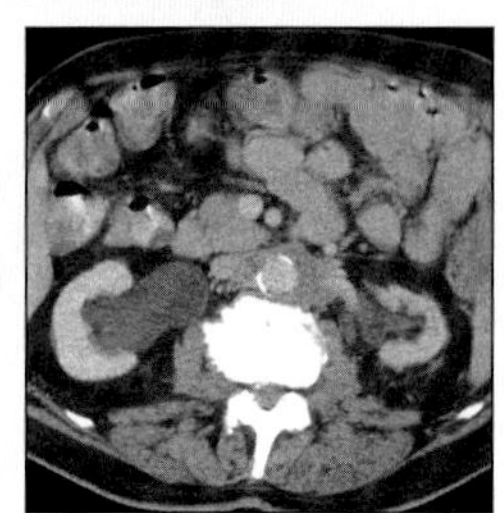

Retrop. Fibrosis

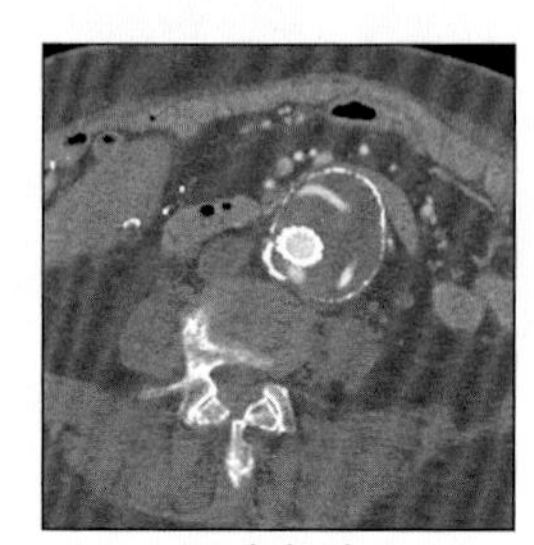

Endoleak

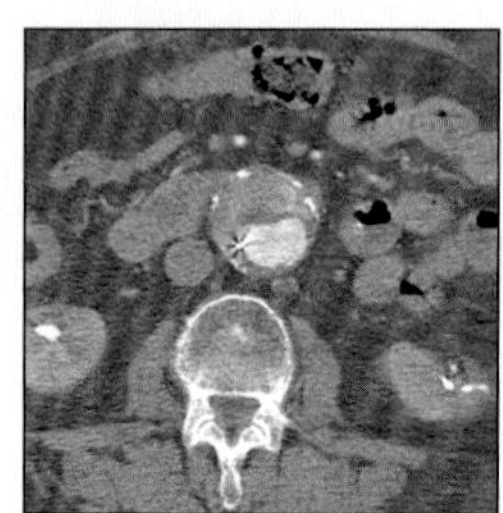

Endoleak

AORTO-ENTERIC FISTULA

Key Facts

Imaging Findings
- Best diagnostic clue: Inflammatory stranding and gas between abdominal aorta and third part of duodenum following aneurysm repair
- Best imaging tool: CT: 94% sensitive & 85% specific

Top Differential Diagnoses
- Periaortitis
- Post-operation
- Post-endovascular stent

Clinical Issues
- "Herald" GI bleeding, followed by hours, days or weeks by catastrophic hemorrhage (most common)

Diagnostic Checklist
- Clinical and past surgical history; diagnosis requires emergent surgery
- Perigraft infection ↑ suspicion of fistula

- May have couple of gas bubbles between stent-graft + aortic wall soon after placement

PATHOLOGY

General Features
- Etiology
  - Primary
    - Abdominal aortic aneurysms
    - Infectious aortitis
    - Penetrating peptic ulcer
    - Tumor invasion
    - Radiation therapy
  - Secondary
    - Aortic reconstructive surgery (most common)
  - Pathogenesis
    - Third portion of duodenum is fixed & apposed to anterior wall of aortic aneurysm → pressure necrosis
    - Surgery → blood supply compromised
    - Pseudoaneurysm formation with erosion
    - Graft & suture line infection → anastomotic breakdown
    - Intraoperative injury to adjacent bowel
- Epidemiology
  - Incidence: 0.6-1.5% after aortic surgery
  - Onset after surgery: 3 years; 21 days up to 14 years
- Associated abnormalities: Perigraft infection

CLINICAL ISSUES

Presentation
- Most common signs/symptoms
  - "Herald" GI bleeding, followed by hours, days or weeks by catastrophic hemorrhage (most common)
  - Abdominal or back pain, palpable and pulsatile mass
  - Intermittent rectal bleeding and recurrent anemia
  - Low-grade fever, fatigue, weight loss, leukocytosis (infection of graft and perigraft area)
- Diagnosis
  - Esophagogastroduodenoscopy: Exclude obvious causes of bleeding
  - Helical CT: Definitive diagnosis

Demographics
- Age: > 55 years of age
- Gender: M:F = 4-5:1

Natural History & Prognosis
- Prognosis: Very poor, up to 85% mortality

Treatment
- Extensive reconstructive surgery

DIAGNOSTIC CHECKLIST

Consider
- Clinical and past surgical history; diagnosis requires emergent surgery

Image Interpretation Pearls
- Perigraft infection ↑ suspicion of fistula

SELECTED REFERENCES

1. Perks FJ et al: Multidetector computed tomography imaging of aortoenteric fistula. J Comput Assist Tomogr. 28(3):343-7, 2004
2. Puvaneswary M et al: Detection of aortoenteric fistula with helical CT. Australas Radiol. 47(1):67-9, 2003
3. Lenzo NP et al: Aortoenteric fistula on (99m)Tc erythrocyte scintigraphy. AJR Am J Roentgenol. 177(2):477-8, 2001
4. Orton DF et al: Aortic prosthetic graft infections: radiologic manifestations and implications for management. Radiographics. 20(4):977-93, 2000
5. Low RN et al: Aortoenteric fistula and perigraft infection: evaluation with CT. Radiology. 175(1):157-62, 1990

IMAGE GALLERY

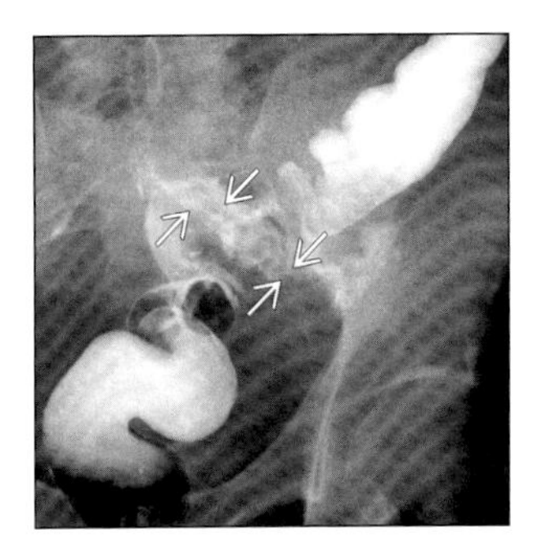
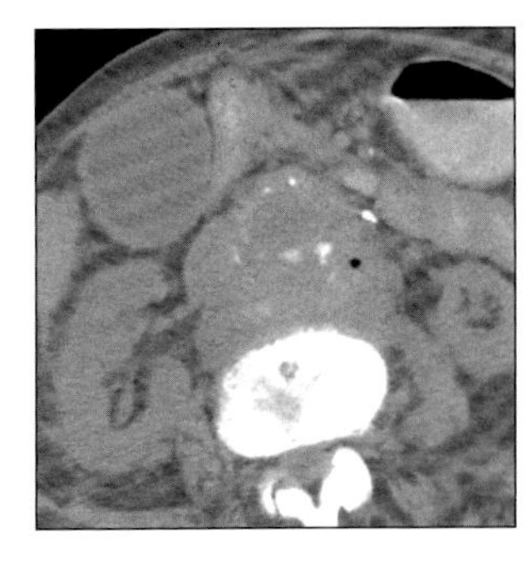

(Left) *Barium enema shows extravasation of contrast which outlines a left common iliac artery graft (arrows).* ***(Right)*** *Axial NECT shows mantle of soft tissue and gas surrounding abdominal aorta and graft. Fatal aorto-duodenal fistula and hemorrhage.*

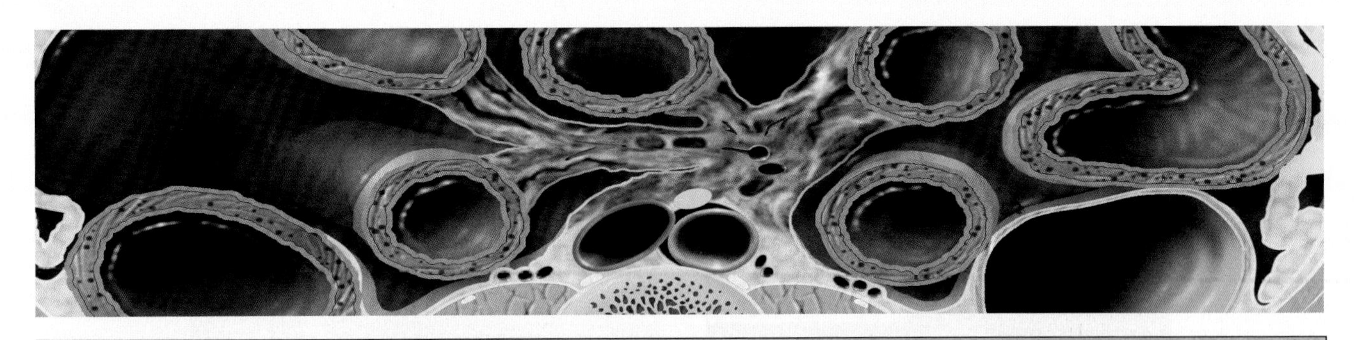

SECTION 4: Small Intestine

Introduction and Overview

Congenital

Infection

Inflammation

Vascular Disease

Trauma

Neoplasm, Benign

Neoplasm, Malignant

Miscellaneous

SMALL INTESTINE ANATOMY AND IMAGING ISSUES

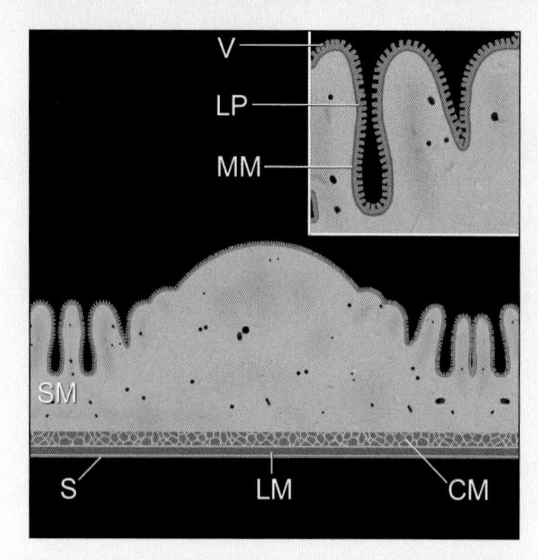

Graphic shows schematic cross section of small intestine wall and the effect of submucosal (SM) accumulation of fluid or cells to broaden the valvulae (V).

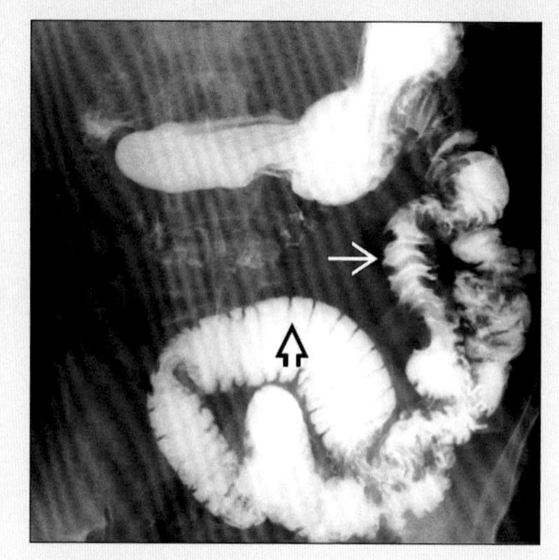

SBFT shows irregular segmental thickening of jejunal folds (arrow) due to hemorrhage in a patient with Henoch-Schönlein purpura. Compare with normal folds (open arrow).

TERMINOLOGY

Abbreviations and Synonyms

- Small bowel (SB) or intestine
- Serosa (S)
- Longitudinal muscle (LM)
- Circular muscle (CM)
- Lamina propria (LP)
- Muscularis mucosa (MM)

Definitions

- Small bowel follow through (SBFT)
  - Examination of the SB following barium ingestion
- Enteroclysis
  - Examination of SB following placement of a tube through the nose into the distal duodenum followed by sustained infusion of barium suspension to distend bowel

IMAGING ANATOMY

- Mesenteric SB begins at the duodenojejunal junction and ends at ileocecal valve
- Usually 20 to 22 feet long
- Suspended from posterior abdominal wall by fan-shaped mesentery, a double-layered fold of peritoneum that carries the blood vessels, lymphatics, and nerve supply to the SB
- Jejunum
  - Proximal 60% of the SB
  - Characterized by prominent crescentic folds of mucosa and submucosa, the valvulae conniventes (also called the plicae circulares, folds of Kerckring, or simply, small bowel folds)
  - Usually number 4 to 7 folds per inch in jejunum and 3 to 5 per inch in ileum
  - Prominence of folds and wall thickness depends on age (more in younger) and degree of bowel distention, and varies in numerous systemic and SB disease states
  - Distended SB has a wall thickness of only 1 or 2 mm
  - Luminal diameter
    - At rest, usually ≤ 2.5 cm
    - Distended after meal or upper GI series/SBFT, 3.5 cm for jejunum, 3 cm for ileum
    - Distended during enteroclysis, 4.5 cm upper jejunum, 4 cm mid jejunum, 3 cm ileum
- Ileum
  - Distal 40% of SB
  - SB fold are fewer and flatter, easily effaced by distension, as in enteroclysis
  - Prominent submucosal lymphoid follicles, especially in terminal ileum in younger individuals

ANATOMY-BASED IMAGING ISSUES

Key Concepts or Questions

- What are the advantages and disadvantages of the conventional barium meal, SBFT and enteroclysis?
  - SBFT advantages
    - Patient preference
    - Ease of performance
    - Ability to judge transit time
  - SBFT disadvantages
    - Length of exam (can be hours if slow gastric emptying or SB transit)
    - Dilution of barium by gastric contents
    - Lack of complete distention of SB
  - Advantages of enteroclysis
    - Shorter exam time
    - Better distention of SB
    - Greater positive and negative predictive value for wide range of SB pathology, including strictures, adhesions and intrinsic SB disease (e.g., sprue)
  - Disadvantages of enteroclysis
    - More radiologist time (intubation, etc.)
    - Patient discomfort
- What are the indications for fluoroscopic/barium examinations versus computed tomography (CT)?

SMALL INTESTINE ANATOMY AND IMAGING ISSUES

DIFFERENTIAL DIAGNOSIS

Small intestine benign tumors

- Stromal tumor ("GIST", leiomyoma, etc.)
- Lipoma
- Adenoma
- Hemangioma
- Neural tumor
- Hamartomas (polyposis syndromes)

Small intestine malignant tumors

- Stromal tumor ("GIST")
- Adenocarcinoma (duodenum > distal)
- Carcinoid (ileum > proximal)
- Lymphoma
- Metastases
- Kaposi sarcoma

Polyposis syndromes affecting SB

- Peutz-Jeghers (hereditary; stomach, SB, colon; hamartomas; mucocutaneous pigmentation)
- Cowden (hereditary; colon + SB; facial papillomas)
- Cronkhite-Canada (non-hereditary, inflammatory; ectodermal lesions)
- Gardner/familial polyposis (hereditary; adenomas; colon > SB)

Multiple target, "bull's eye" lesions

- Metastases (especially melanoma)
- Lymphoma
- Kaposi sarcoma

  - Enteroclysis is more accurate than CT in detecting SB ulceration, small luminal or mural masses, and subtle strictures or adhesions
  - CT is preferable in the setting of "acute abdomen" (abdominal pain +/or distention) and more accurately depicts extraluminal disease (e.g., abscess, fistulas, large masses, mesenteric and vascular disease)
- What are the advantages/disadvantages of capsule endoscopy versus barium studies of the SB?
  - Capsule endoscopy utilizes a tablet-size camera that is swallowed by the patient, recording images of the luminal GI tract at intervals until it is expelled by the patient
  - Advantages
    - Allows endoluminal photographic view of the SB beyond the reach of fiberoptic endoscopes
  - Disadvantages
    - Requires many hours for patient preparation time, SB transit and subsequent analysis of the images
    - Often difficult to localize a pathologic process that is visualized

Imaging Approaches

- Patient preparation for SBFT or enteroclysis
  - Preferable to clear right side of colon; to reduce SB transit time and filling defects in distal SB
  - Oral laxative (e.g., magnesium citrate) at noon on day prior to the exam
  - Plus 4 bisacodyl tablets (Dulcolax, Boehringer Ingelheim) the day prior to the exam
  - Low residue diet evening before; nothing by mouth after midnight
  - Metoclopramide (Reglan, AH Robbins) optional
    - 10 mg I.V. or 20 mg PO prior to intubation for enteroclysis (facilitates transpyloric passage of tube and SB peristalsis
- Upper GI and SBFT
  - Up to 500 ml of 42% weight/volume barium mixture is ingested
  - Fluoroscopic ± overhead radiographs at 15 to 30 minute intervals
  - Manually palpate each segment of SB
  - Continue until ileocecal valve is reached
  - Maintain a continuous column of barium in SB by having patient continue to drink
- Double contrast SBFT
  - When the barium has reached the colon, the patient may be given effervescent gas producing agent by mouth and positioned in a Trendelenburg, left side down position to distend the SB with gas: Additional fluoroscopy and films are obtained
- Peroral pneumocolon
  - After SBFT, a rectal tube is inserted and air is insufflated to distend the barium-filled distal SB loops and right colon
  - Helpful to administer 1 mg glucagon I.V. prior to insufflation to facilitate reflux through ileocecal valve and produce SB hypotonia
  - Good technique for suspected inflammatory or infectious process in ileocecal region (e.g., Crohn disease)
- Enteroclysis
  - Single contrast: Infusion of 30 to 40% weight/volume barium suspension at rate of 60 to 90 ml/min, suffices for most indications
  - "Double contrast" enteroclysis: Infusion of 60 to 95% weight/volume barium followed by infusion of air or methylcellulose to distend lumen, leaving only a thin coating of viscous barium
  - Double contrast technique theoretically superior for detection of subtle lesions, but prone to technical difficulties and patient discomfort
  - Reflux small bowel examination: Essentially a single contrast barium enema with addition of glucagon (1 mg I.V.) to facilitate ileocecal reflux
    - Can be fast and effective means to evaluate possible distal SB obstruction or inflammation

Imaging Pitfalls

- The length and redundancy of the SB and its normal variability in fold patterns and bowel position make radiographic visualization and characterization of SB pathology very challenging

SMALL INTESTINE ANATOMY AND IMAGING ISSUES

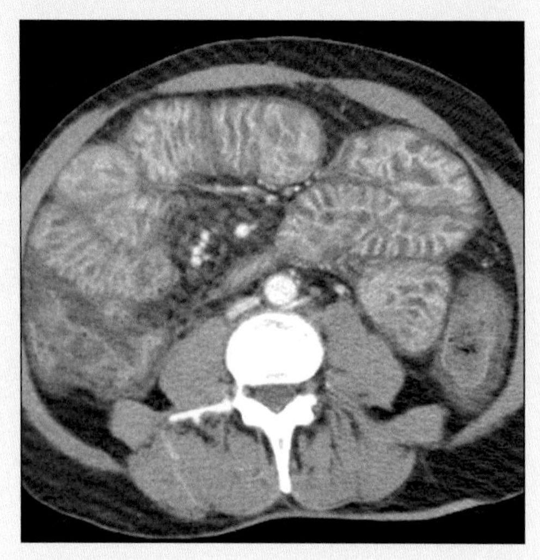

Axial CECT shows near-water density submucosal edema and intense mucosal enhancement in a patient with "shock bowel" following abdominal trauma.

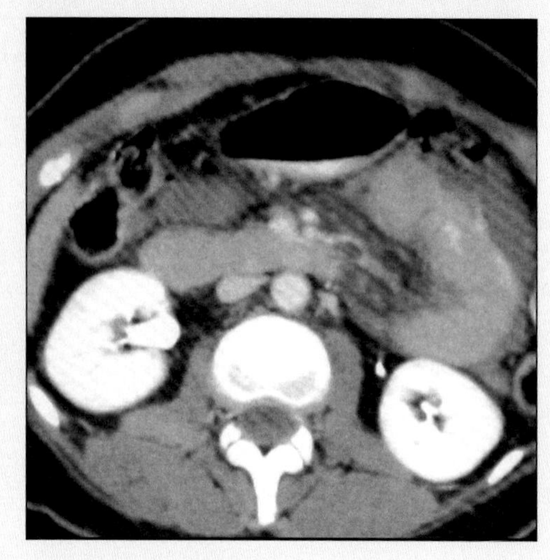

Axial CECT shows high density submucosal thickening (plus luminal narrowing and mesenteric hemorrhage) in a patient with jejunal hematoma due to blunt trauma.

CUSTOM DIFFERENTIAL DIAGNOSIS

SB folds evenly thickened

- Extensive
  - Edema (e.g., portal hypertension)
  - Infection (e.g., Giardiasis)
  - Eosinophilic gastroenteritis
- Focal
  - Hemorrhage (e.g., trauma, anticoagulation)
  - Ischemia
  - Vasculitis
  - Radiation enteritis

Micronodular SB fold pattern

- Whipple disease
- Mycobacterial disease
- Abetalipoproteinemia
- Histoplasmosis
- Lymphangiectasia
- Macroglobulinemia

Irregular diffuse SB fold thickening

- Lymphangiectasia
- Lymphatic obstruction
- Amyloidosis
- Mastocytosis
- Lymphoma
- Celiac-sprue

Terminal Ileum (TI) - fold thickening, luminal narrowing

- Crohn disease
- Acute bacterial (Yersinia, Campylobacter)
- Tuberculosis
- Cecal carcinoma (retrograde spread to TI)
- Lymphoma

"Aneurysmal" SB dilation

- Lymphoma
- Metastases (especially melanoma)
- Iatrogenic (side-to-side anastomosis)

Non-neoplastic obstructing lesion

- Congenital
  - Atresia: Stenosis
  - Malrotation
  - Bands
  - Hernias (internal and external)
- Inflammatory and acquired
  - Adhesions
  - Strictures
  - Hernias (internal and external)
  - Intussusception (may have neoplastic cause)
  - Inflammatory fibroid polyp
- Intraluminal
  - Gallstone
  - Meconium plug (or equivalent in adult cystic fibrosis)
  - Bezoar
  - Foreign body ingested
  - Worms, parasites
  - Inverted Meckel diverticulum
- Trauma/vascular
  - Hematoma
  - Vasculitis
  - Ischemia

SELECTED REFERENCES

1. Rubesin SE et al: Textbook of gastrointestinal radiology: Small bowel: differential diagnosis. 2nd ed. Philadelphia, WB Saunders. pp 884-890, 2000
2. Eisenberg RL: Gastrointestinal radiology: a pattern approach. 3rd ed. Philadelphia, JB Lippincott, 1996
3. Dähnert W: Radiology review manual. 4th ed. Philadelphia, Lippincott, Williams, and Wilkins. 2000
4. Herlinger H et al: Clinical radiology of the small intestine. New York, Springer Verlag, 1999
5. Reeder MM: Reeder and Felson's gamut's in radiology. 3rd ed. New York, Springer Verlag, 1993

IMAGE GALLERY

Typical

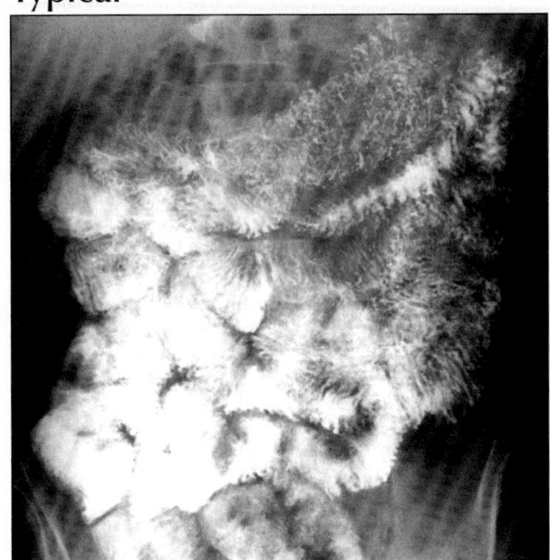

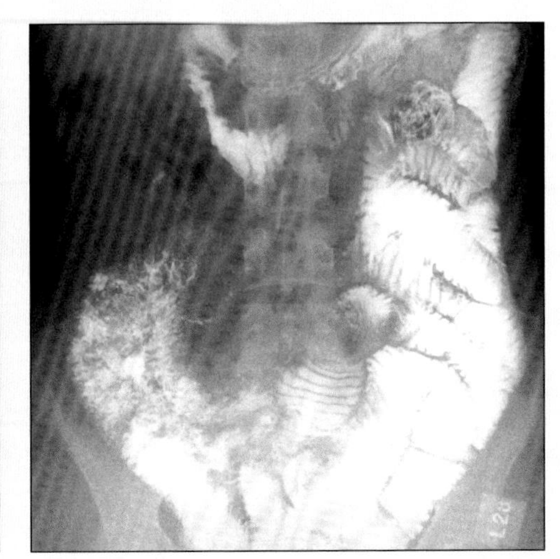

*(**Left**) SBFT shows diffuse uniform fold thickening in a patient with Giardiasis. (**Right**) SBFT shows luminal distention and reversal of fold pattern (atrophied jejunal, prominent ileal) in a patient with sprue.*

Typical

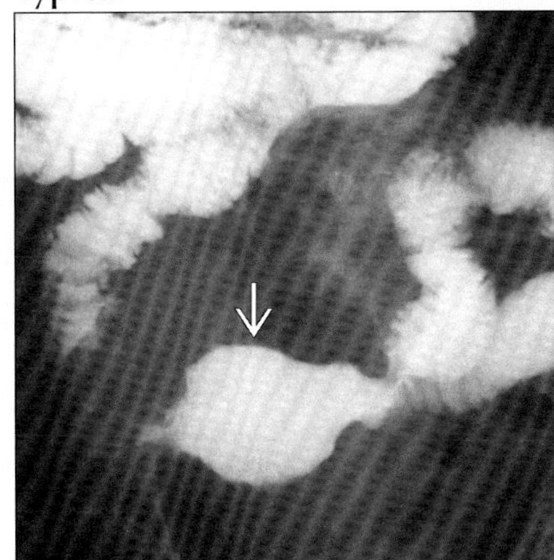

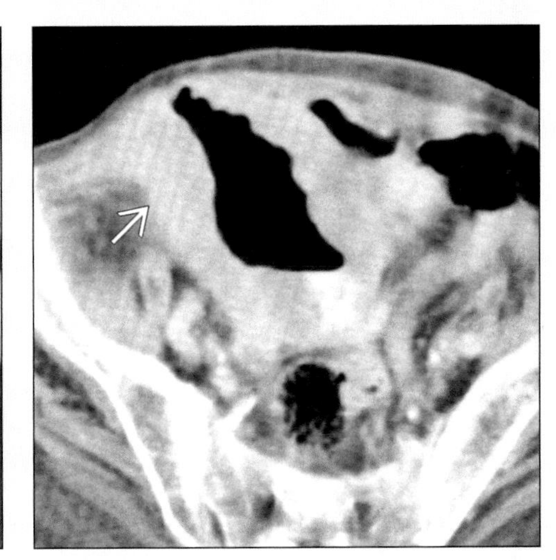

*(**Left**) SBFT shows aneurysmal dilation of SB and destruction of folds (arrow), along with mesenteric mass effect; lymphoma. (**Right**) Axial CECT shows aneurysmal dilation of the SB lumen and surrounding mass (arrow); lymphoma.*

Typical

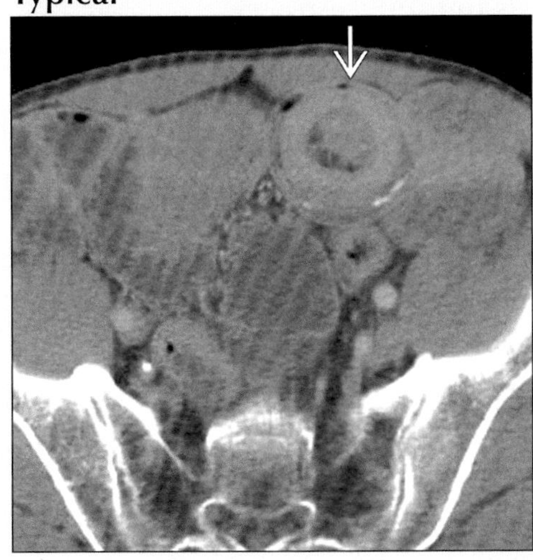

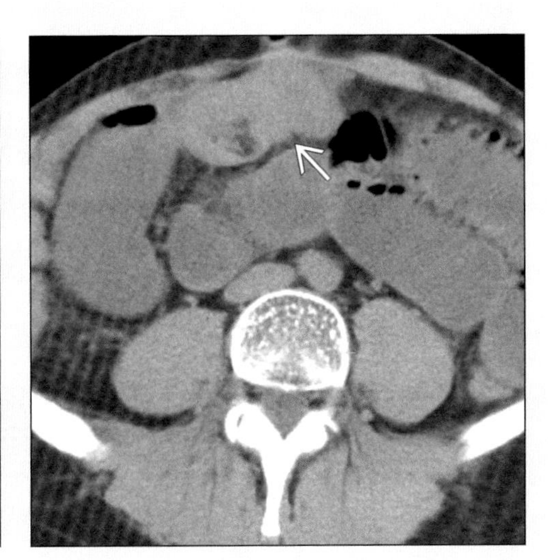

*(**Left**) Axial CECT shows small bowel distention, terminating at an intussusception (arrow); metastatic melanoma. (**Right**) Axial CECT shows a soft tissue mass (arrow) as the lead point of an intussusception; melanoma.*

MALROTATION

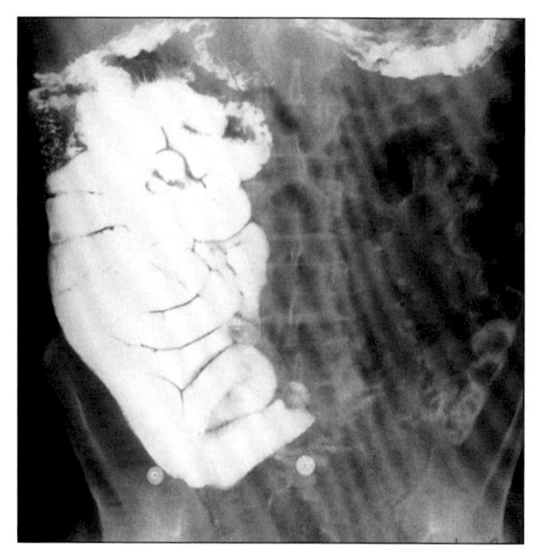
SBFT shows all of small intestine on right side of abdomen, colon on left, indicating nonrotation of gut. Duodenojejunal junction low + midline.

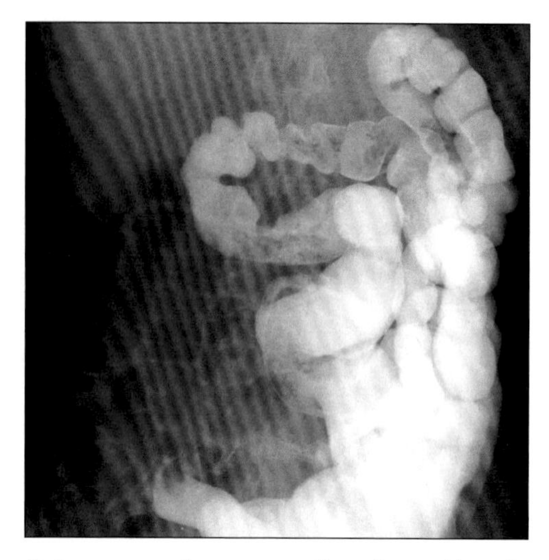
Barium enema shows nonrotation of gut with entire colon on left side of abdomen.

TERMINOLOGY

Definitions

- Rotational abnormality of the gut secondary to arrest of gut rotation & fixation in embryologic development

IMAGING FINDINGS

General Features

- Best diagnostic clue: Small bowel on right abdomen and large bowel on left abdomen
- Other general features
  - Classification of malrotation
    - Nonrotation: Midgut returns to peritoneal cavity without rotation
    - Incomplete rotation: Duodenojejunal loop fails to complete final 90° rotation
    - Reversed rotation: Cecocolic loop rotates first, thus unwinding counterclockwise rotation of duodenojejunal loop with 90° clockwise rotation

Radiographic Findings

- Radiography
  - "Double-bubble" sign: Enlarged stomach and duodenal bulb, little gas in remainder of small bowel
  - Gasless abdomen; distended & thickened bowel wall
- Fluoroscopic-guided barium studies
  - Nonrotation: Small bowel on right, large bowel on left
  - Incomplete rotation: Cecum inferior to pylorus
  - Reversed rotation: Duodenum anterior to superior mesenteric artery (SMA), SMA anterior to transverse colon
  - Duodenum and jejunum on right of spine
  - Redundancy of duodenum to right of spine
  - Z-shape configuration of duodenum and jejunum
  - Duodenojejunal junction (DDJ) low and in midline or medial to left pedicle; or over right pedicle
  - Contrast ends abruptly or in corkscrew pattern

CT Findings

- Superior mesenteric vein (SMV) on left of SMA
- Pancreatic aplasia or hypoplasia of uncinate process

Ultrasonographic Findings

- Superior mesenteric vein (SMV) on left of SMA

Imaging Recommendations

- Best imaging tool: Fluoroscopic-guided barium studies

DIFFERENTIAL DIAGNOSIS

Malposition

- E.g., extrinsic masses, indwelling nasojejunal tubes
- Distorted DDJ without malrotation

DDx: Abnormal Position of Gut

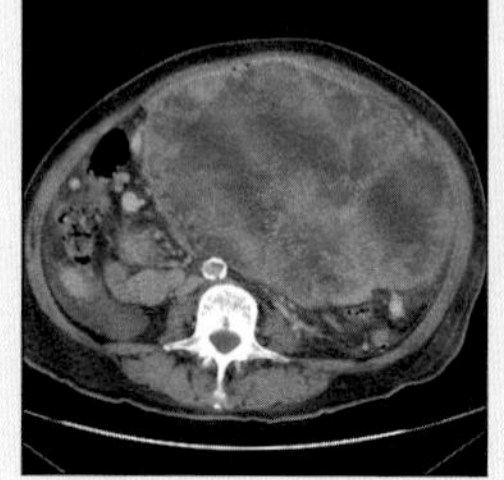
Displacement by Mass

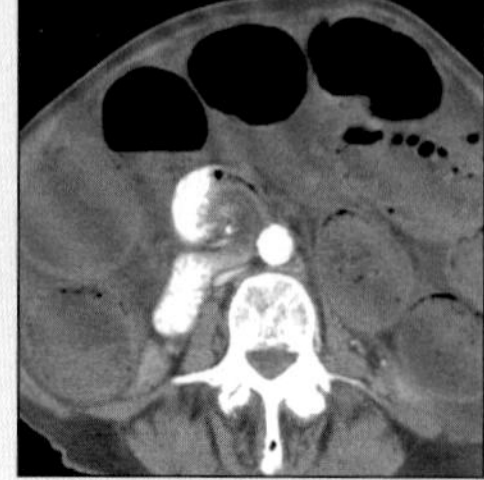
Volvulus

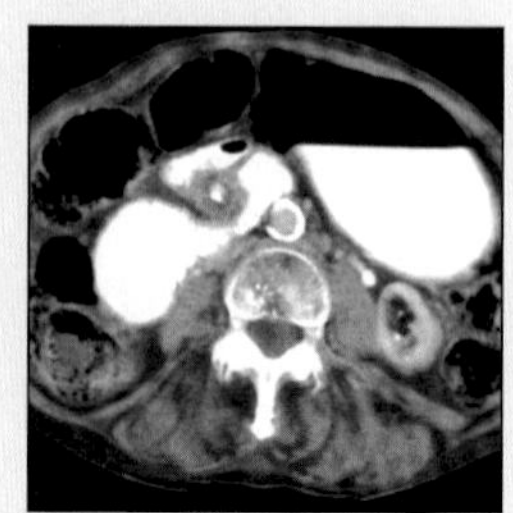
Volvulus

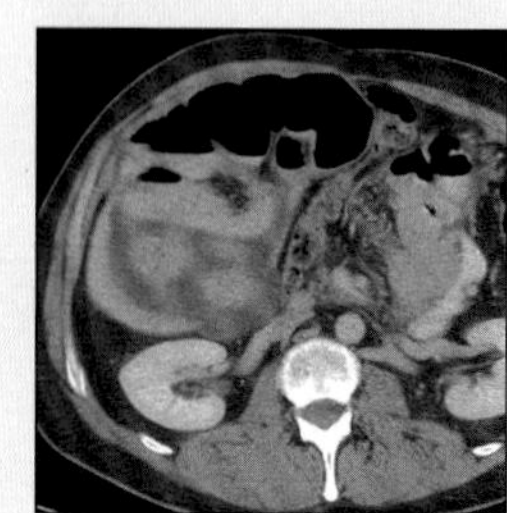
Internal Hernia

MALROTATION

Key Facts

Imaging Findings
- Best diagnostic clue: Small bowel on right abdomen and large bowel on left abdomen
- Superior mesenteric vein (SMV) on left of SMA
- Best imaging tool: Fluoroscopic-guided barium studies

Top Differential Diagnoses
- Malposition
- Volvulus
- Internal hernia

Clinical Issues
- Acute abdomen, abdominal distension

Diagnostic Checklist
- Any patient with small bowel obstruction
- Inversion of SMV and SMA

Volvulus
- Whirl-like pattern: Small bowel loops and adjacent mesenteric fat converging to point of torsion
- "Whirl" sign: SMV winds around SMA

Internal hernia
- E.g., paraduodenal hernia

PATHOLOGY

General Features
- General path comments
  - Embryology
    - 6-10th gestational week: Duodenojejunal loop rotates 180° counterclockwise, ending inferior to SMA; cecocolic loop rotates 90° counterclockwise, ending at anatomical left of SMA
    - 10th gestational week: Midgut returns to peritoneal cavity; duodenojejunal loop rotates additional 90° counterclockwise → positions DDJ to left of spine at level of stomach, jejunum in left upper quadrant (LUQ) and ileum in right lower quadrant (RLQ); cecocolic loop rotates additional 180° counterclockwise → cecum in RLQ
    - Broad mesenteric base, from LUQ to RLQ
    - Fixation of gut by attachment at cecal base and ligament of Treitz at DDJ
- Epidemiology: 1 in 500 live births in U.S.
- Associated abnormalities
  - Omphalocele, gastroschisis, diaphragmatic hernia
  - Asplenia & polysplenia syndrome, duodenal stenosis or atresia, annular pancreas, Hirschsprung disease
  - Preduodenal portal vein, inferior vena cava anomalies

CLINICAL ISSUES

Presentation
- Most common signs/symptoms
  - Acute abdomen, abdominal distension
  - Chronic vomiting, failure to thrive

Demographics
- Age: Up to 40% diagnosed by 1 week of age, 50% by 1 month, 75% by 1 year, 25% after 1 year
- Gender: Overall M = F; < 1 year of age: 2:1

Natural History & Prognosis
- Complications: Obstruction, volvulus, Ladd bands
- Prognosis: Good, if treated promptly

Treatment
- Ladd procedure: Reduction of volvulus, division of mesenteric bands, placement of small bowel on right and large bowel on left and appendectomy

DIAGNOSTIC CHECKLIST

Consider
- Any patient with small bowel obstruction

Image Interpretation Pearls
- Inversion of SMV and SMA

SELECTED REFERENCES

1. Boudiaf M et al: Ct evaluation of small bowel obstruction. Radiographics. 21(3):613-24, 2001
2. Berrocal T et al: Congenital anomalies of the small intestine, colon, and rectum. Radiographics. 19(5):1219-36, 1999
3. Long FR et al: Intestinal malrotation in children: tutorial on radiographic diagnosis in difficult cases. Radiology. 198(3):775-80, 1996
4. Long FR et al: Radiographic patterns of intestinal malrotation in children. Radiographics. 16(3):547-56; discussion 556-60, 1996

IMAGE GALLERY

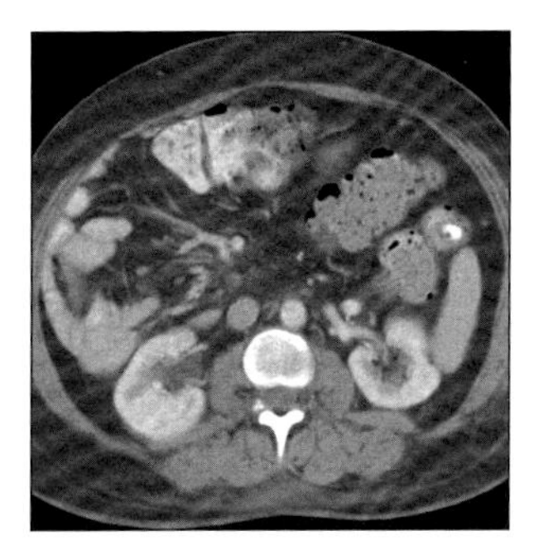

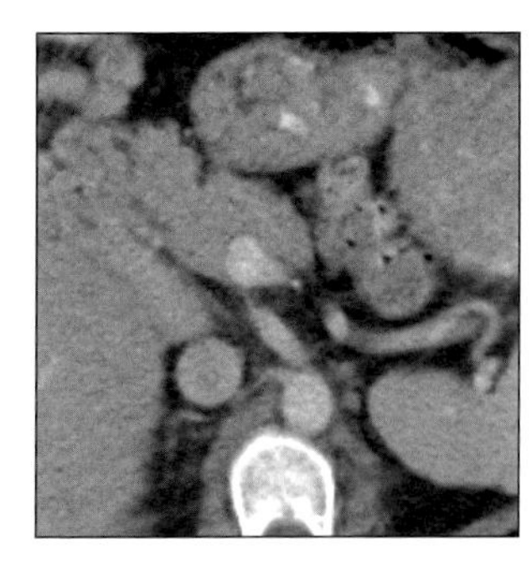

(Left) *Axial CECT shows malrotation with intestine on right, colon on left.* ***(Right)*** *Axial CECT in patient with bowel malrotation shows agenesis of dorsal pancreas + uncinate process.*

MECKEL DIVERTICULUM

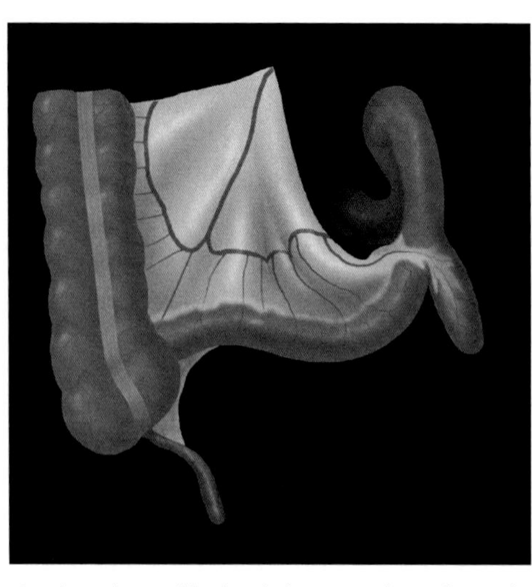

Graphic shows blind-ended outpouching from the antimesenteric border of the distal ileum.

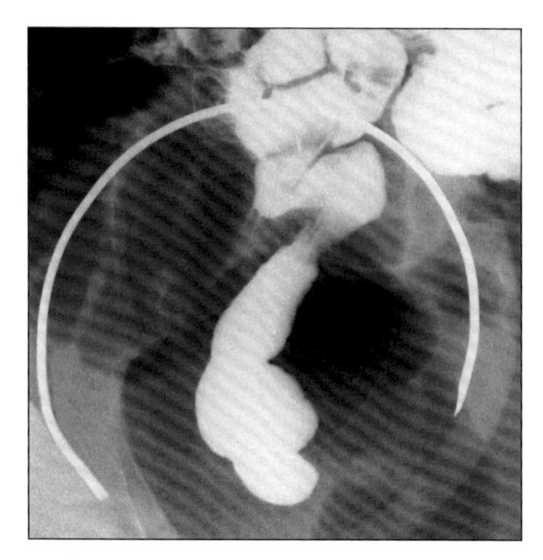

Small bowel follow through (SBFT) shows blind-ended outpouching from distal ileum.

TERMINOLOGY

Abbreviations and Synonyms

- Meckel diverticulum (MD)

Definitions

- An ileal outpouching, due to persistence of the congenital omphalomesenteric or vitelline duct

IMAGING FINDINGS

General Features

- Best diagnostic clue: Blind-ended sac or outpouching on antimesenteric border of ileum
- Location
  - Within 50-60 cm of ileocecal valve
    - May vary from RLQ to mid abdomen
- Size: 4-10 cm in length
- Morphology: Tubular outpouching of ileum
- Other general features
  - Most common congenital anomaly of GI tract
  - True diverticulum (contains all layers of bowel wall)
  - Arises from antimesenteric border of distal ileum & has a separate blood supply
  - Formed by incomplete obliteration of ileal end of vitelline duct
  - Usually located within 50-60 cm of ileocecal valve
  - 50% of Meckel diverticula, contain ectopic gastric mucosa
    - Pancreatic, duodenal & colonic mucosa can also be found
  - 90% of cases that present with bleeding contain gastric mucosa
  - A fibrous band, representing obliterated part of vitelline duct may connect apex of diverticulum to umbilicus
  - Rule of 2s
    - Seen in 2% of population (approximately)
    - Located within 2 feet of ileocecal valve
    - Length of 2 inches (on average)
    - Symptomatic usually before age 2
  - Two main complications in adults: Diverticulitis (20%); intestinal obstruction (40%)

Radiographic Findings

- Radiography
  - X-ray A-P abdomen
    - Round collection of gas & ± solitary or multiple calcified densities (enteroliths) within it in right lower quadrant (RLQ)
- Fluoroscopic guided enteroclysis
  - Superior due to maximum luminal distention
  - Blind-ended sac on antimesenteric border of ileum with either a broad base or a narrow neck
  - Broad based diverticulum

DDx: Right Lower Quadrant Inflammation

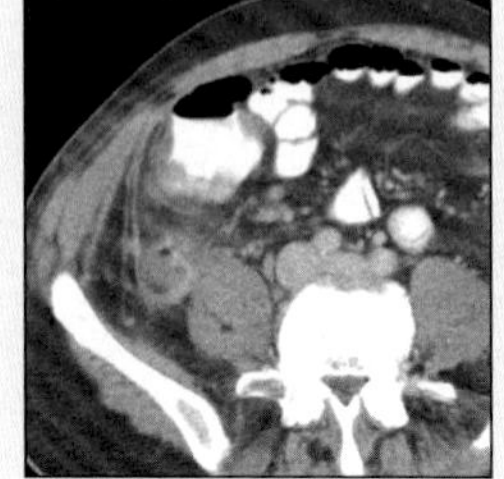

Appendicitis

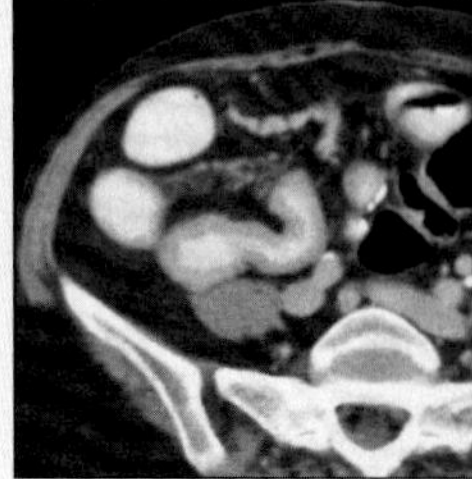

Crohn Disease

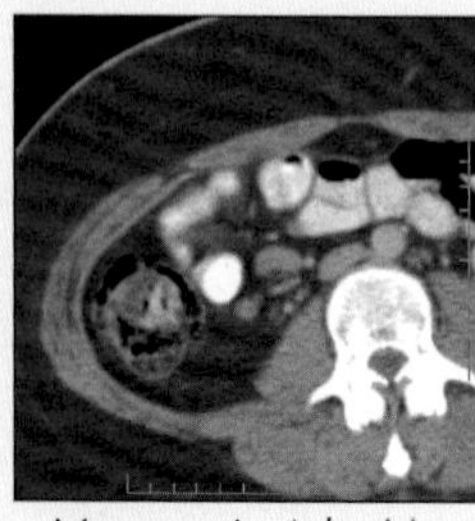

Mesenteric Adenitis

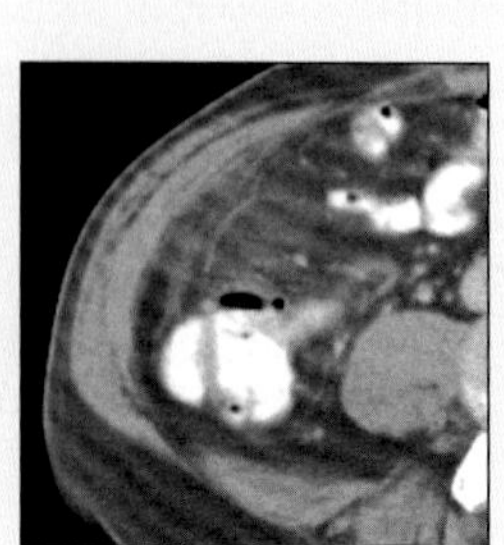

Cecal Diverticulitis

MECKEL DIVERTICULUM

Key Facts

Terminology

- An ileal outpouching, due to persistence of the congenital omphalomesenteric or vitelline duct

Imaging Findings

- True diverticulum (contains all layers of bowel wall)
- Seen in 2% of population (approximately)
- Located within 2 feet of ileocecal valve
- Length of 2 inches (on average)
- Symptomatic usually before age 2
- Two main complications in adults: Diverticulitis (20%); intestinal obstruction (40%)
- CT for symptomatic (pain, tenderness) adults
- Tc99m scintigraphy in bleeding cases (children)

Top Differential Diagnoses

- Appendicitis
- Crohn disease
- Mesenteric adenitis
- Cecal diverticulitis

Pathology

- Composed of all layers of GI tract (true diverticulum)
- Ectopic gastric or pancreatic mucosa may be seen

Clinical Issues

- Children: Present with GI bleeding before age 2
- Adults: Present with diverticulitis or obstruction

Diagnostic Checklist

- Enteroclysis: Blind-ended sac on antimesenteric border of ileum with a broad base or a narrow neck
- Meckel diverticulitis: CT shows mural thickening, perimesenteric fat infiltration & fluid ± enteroliths

    - Enteroclysis shows distinctive triangular junctional fold pattern at site of origin
  - Narrow neck diverticulum
    - Diagnosis depends on demonstration of blind end of diverticulum & its antimesenteric origin
  - Initially appears small but fills more completely with increased distention of lumen
  - Inverted Meckel diverticulum
    - Seen in 20% of cases
    - Solitary, elongated, smoothly marginated, often club-shaped intraluminal mass parallel to long axis of distal ileum; may lead to intussusception
- Double contrast barium enema
  - Occasionally demonstrates MD by reflux into ileum
- Conventional small bowel follow-through
  - Rarely demonstrates diverticulum

CT Findings

- Blind-ending pouch of bowel attenuation within RLQ
- Rarely, calcified enteroliths within diverticular lumen
- Communication with small-bowel present
- Meckel diverticulitis
  - Blind-ending pouch + fluid, air or particulate matter
  - Inflamed: No oral contrast within diverticular lumen
  - Mural thickening: Diverticulum + adjacent bowel
  - Mesenteric fat infiltration & fluid ± nodes
  - ± Partial or complete small bowel obstruction
  - Shows mural enhancement on contrast study

Ultrasonographic Findings

- Real Time
  - Hypoechoic mass with ± echogenic calculi in right lower quadrant (nonspecific)
  - Inverted Meckel diverticulum: Target-like mass with a central area of increased echogenicity

Angiographic Findings

- Conventional
  - Superselective catheterization of distal ileal arteries
  - Active bleeding: Blush of contrast medium in RLQ
  - Pathognomonic: Vitelline artery (anomalous end branch of SMA)

Nuclear Medicine Findings

- Tc99m sulfur colloid (5-10 min serial images for 1 hr)
  - Most widely used method (diagnosing bleeding MD)
  - Sensitivity > 85%; specificity > 95%; accuracy > 85-88%
  - Accumulation of isotope in RLQ on positive scans
  - Pentagastrin used to stimulate MD gastric mucosal uptake (initial scintigraphy is equivocal or normal)
  - False-positive results: Gastric ectopia, appendicitis, inflammatory bowel disease
  - False-negative results: Absent or minimal ectopic gastric mucosa

Imaging Recommendations

- Fluoroscopic compression during enteroclysis
- CT for symptomatic (pain, tenderness) adults
- Tc99m scintigraphy in bleeding cases (children)
- Angiography in active bleeding cases

DIFFERENTIAL DIAGNOSIS

Appendicitis

- Best imaging clue on CT
  - Appendicolith (usually calcified) within distended tubular appendix
- Distended enhancing appendix with surrounding inflammation (fat stranding)
- Wall thickening of cecum or terminal ileum
- Right lower quadrant (RLQ) lymphadenopathy
- In perforated cases
  - Fluid collection most commonly in RLQ or dependent pelvis (Cul-de-sac)
  - Abscess, small-bowel obstruction
- Ultrasound findings
  - Echogenic appendicolith with posterior shadowing
  - Noncompressible blind-ending tubular structure over 6 mm in diameter
  - Fluid or abscess collection in RLQ
- Clinically & radiographically may mimic MD
- Differentiation often made only at surgery

Crohn disease

- Synonym(s)
  - Terminal ileitis, regional enteritis, ileocolitis
- Chronic, recurrent, segmental, granulomatous inflammatory bowel disease
- Location: Anywhere along gut from mouth to anus
  - Most common site: Terminal ileum
- Characterized by
  - Skip lesions: Segmental or discontinuous
  - Transmural inflammation, noncaseating granulomas
  - Cobblestone mucosa, fissures, fistulas, sinus tracts
- Clinically may be indistinguishable from Meckel diverticulitis when disease is localized in terminal ileum with history of RLQ pain & melena
- Diagnosis: Endoscopic mucosal biopsy

Mesenteric adenitis

- Other common cause of RLQ pain
  - Predominantly in children & adolescents
- Pathology: Benign inflammation of mesenteric nodes
- Imaging findings
  - Enlarged, clustered adenopathy in mesentery & RLQ
  - May also have ileal wall thickening
- Usually resolves spontaneously within 2 days

Cecal diverticulitis

- Perforation + localized pericolic inflammation/abscess
  - Most common complication of diverticulosis
- Location: Right lower quadrant (RLQ)
- Imaging findings
  - Bowel wall thickening, fat stranding, free fluid & air
  - Bowel outpouching filled with air, contrast, feces
  - Acute: Enhancement of thickened colonic wall
  - Pericecal changes: Abscess, sinus or fistula
  - ± Gas or thrombus in mesenteric or portal veins
  - ± Liver abscesses
- Clinical presentation
  - RLQ pain, fever, tenderness
  - More common in middle & elderly age group
- May mimic complicated MD clinically & radiologically

PATHOLOGY

General Features

- General path comments
  - Embryology-anatomy
    - Persistent omphalomesenteric or vitelline duct, which usually obliterates by 5th embryonic week
- Etiology
  - Early fetal life: Primitive midgut communicates with yolk sac through vitelline duct & failure of usual complete regression leads to anomalies
  - MD (> common); bands, fistula, cyst (< common)
- Epidemiology: Incidence (0.3-3%)

Gross Pathologic & Surgical Features

- Solitary diverticulum located approximately 50-60 cm proximal to ileocecal valve on antimesenteric border

Microscopic Features

- Composed of all layers of GI tract (true diverticulum)
- Ectopic gastric or pancreatic mucosa may be seen

CLINICAL ISSUES

Presentation

- Most common signs/symptoms
  - Mostly asymptomatic
  - 4.2% likelihood of symptomatic during lifetime
    - Pain in right lower quadrant (RLQ)
  - Children: Present with GI bleeding before age 2
  - Adults: Present with diverticulitis or obstruction

Demographics

- Age: Any age group
- Gender: M:F = 3:1

Natural History & Prognosis

- Complications: Life time complication rate (4%)
  - GI bleeding, diverticulitis, perforation, abscess
  - Enteroliths, rarely malignancy
  - Obstruction due to intussusception or volvulus
  - Extrusion of diverticulum into an inguinal hernia
    - Hernia of Littre
  - Inverted MD: Intussusception → bowel obstruction
  - Giant MD: More likely obstruction than bleeding
- Prognosis: Good after surgery

Treatment

- Asymptomatic: No treatment
- Symptomatic: Surgical resection

DIAGNOSTIC CHECKLIST

Consider

- Rule out other inflammatory pathologies of RLQ

Image Interpretation Pearls

- Enteroclysis: Blind-ended sac on antimesenteric border of ileum with a broad base or a narrow neck
- Meckel diverticulitis: CT shows mural thickening, perimesenteric fat infiltration & fluid ± enteroliths

SELECTED REFERENCES

1. Bennett GL et al: CT of Meckel's diverticulitis in 11 patients. AJR Am J Roentgenol. 182(3):625-9, 2004
2. Murakami R et al: Strangulation of small bowel due to Meckel diverticulum: CT findings. Clin Imaging. 23(3):181-3, 1999
3. Mitchell AW et al: Meckel's diverticulum: angiographic findings in 16 patients. AJR Am J Roentgenol. 170(5):1329-33, 1998
4. Pantongrag-Brown L et al: Meckel's enteroliths: clinical, radiologic, and pathologic findings. AJR Am J Roentgenol. 167(6):1447-50, 1996
5. Pantongrag-Brown L et al: Inverted Meckel diverticulum: clinical, radiologic, and pathologic findings. Radiology. 199(3):693-6, 1996
6. Rossi P et al: Meckel's diverticulum: imaging diagnosis. AJR Am J Roentgenol. 166(3):567-73, 1996
7. Maglinte DD et al: Meckel diverticulum: radiologic demonstration by enteroclysis. AJR Am J Roentgenol. 134(5):925-32, 1980

MECKEL DIVERTICULUM

IMAGE GALLERY

Typical

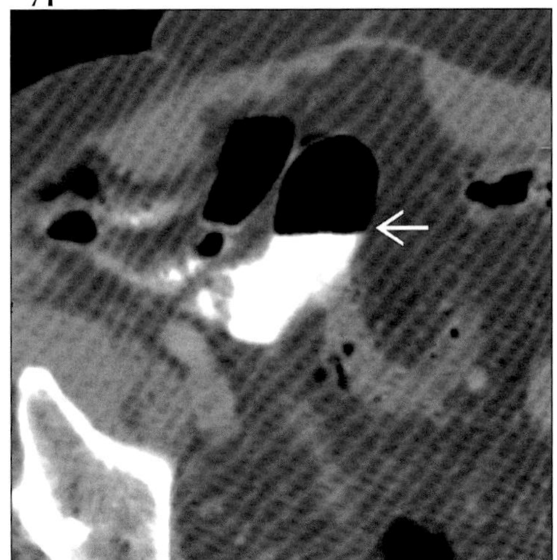

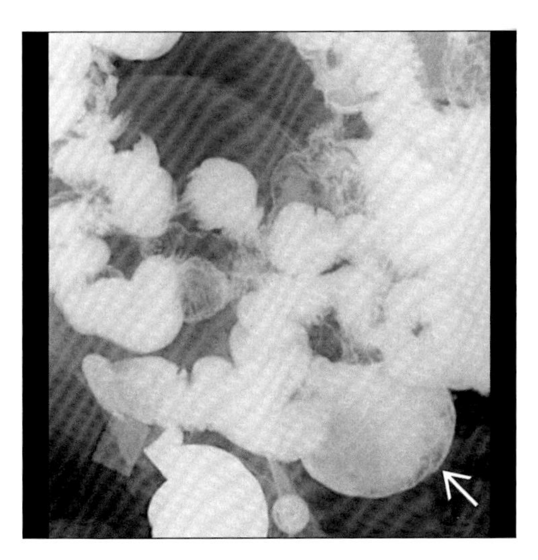

(Left) *Axial CECT shows air-contrast level within a blind-ended "sac" (arrow) in the right lower quadrant.* ***(Right)*** *SBFT shows blind-ended sac (arrow) arising from distal ileum.*

Typical

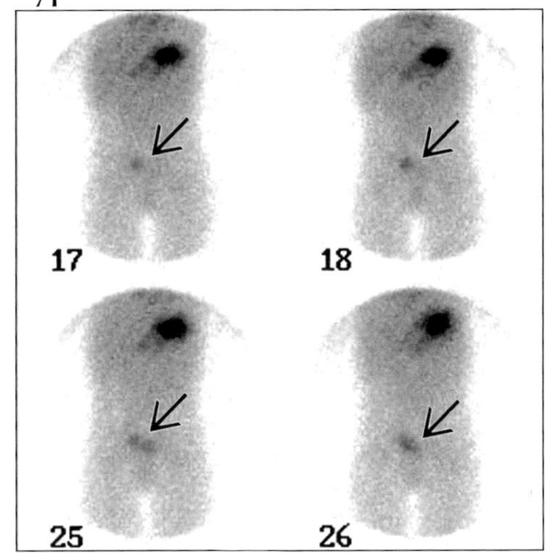

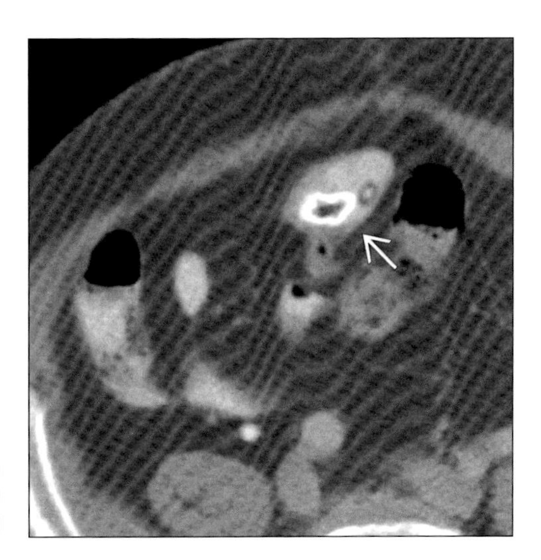

(Left) *Tc99m sulfur colloid scan in a young child with GI bleeding shows accumulation of isotope in RLQ (arrows) due to ectopic gastric mucosa within Meckel diverticulum.* ***(Right)*** *Axial CECT shows enteroliths within blind-ended sac (arrow) in RLQ.*

Typical

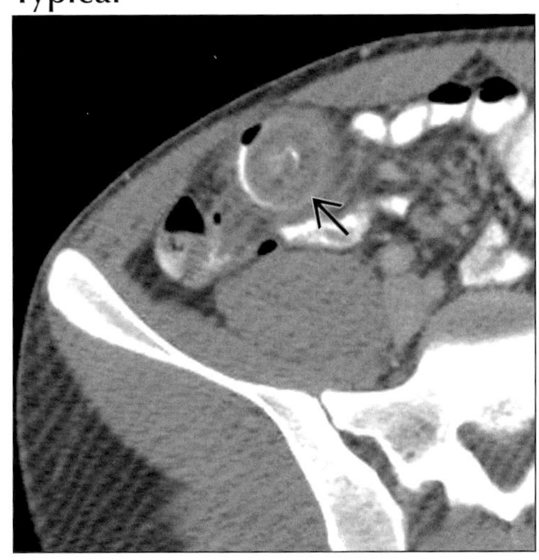

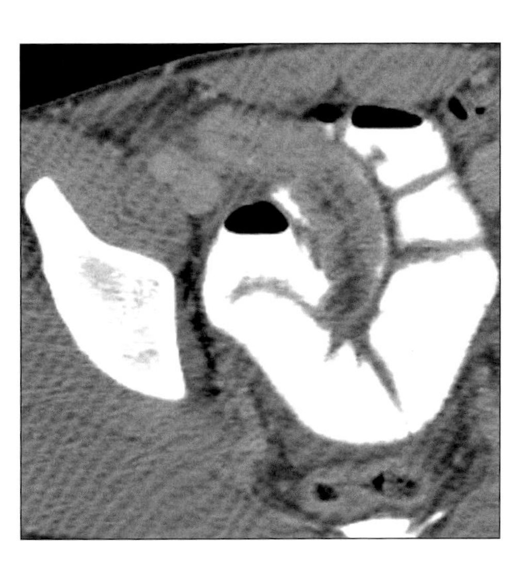

(Left) *Axial CECT in 17 year old girl with crampy abdominal pain + RLQ tenderness shows distal ileal intussusception (arrow) due to an inverted Meckel diverticulum.* ***(Right)*** *Axial CECT in 17 year old girl shows distal SB intussusception with an inverted Meckel diverticulum as the lead mass.*

INTESTINAL PARASITIC DISEASE

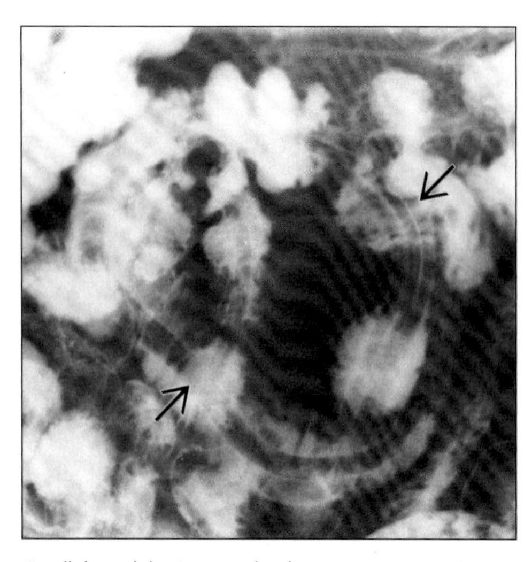

Small bowel barium study demonstrates ascaris as a longitudinal filling defect in the distal small bowel (arrows).

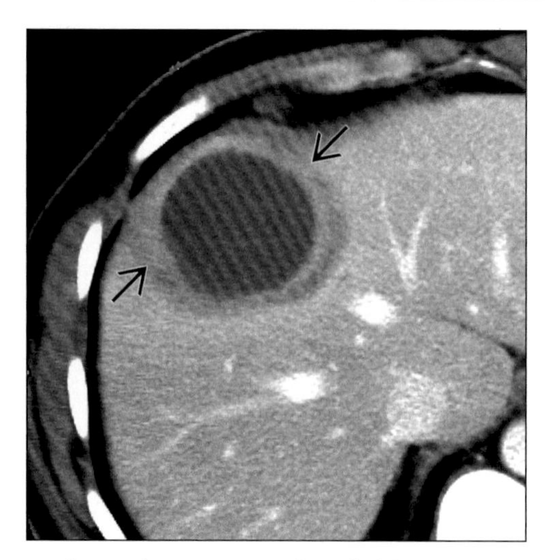

Axial CECT demonstrates amebic right lobe liver abscess with peripheral zone of edema (arrows).

TERMINOLOGY

Definitions

- Ascariasis: Enteric infection with roundworm ascaris lumbricordes
- Giardiasis: Enteric protozoal infection with Giardia lamblia; amebiasis: Enteric protozoal infection with entamoeba histolytica

IMAGING FINDINGS

General Features

- Best diagnostic clue
  - Ascariasis: Linear filling defect on small bowel follow through (SBFT)
  - Giardiasis: Thickened duodenal and jejunal folds on SBFT
  - Amebiasis: Diffuse ulcerating colitis, right lobe liver abscess on CECT
- Location: Ascariasis: Small bowel (SB), colon, common bile duct (CBD) or pancreatic duct; giardiasis: Duodenum and jejunum; amebiasis: Colon, liver

Radiographic Findings

- Radiography: Ascariasis: Soft tissue mass from coiled worms at ileocecal valve, may progress to small bowel obstruction (SBO)
- Fluoroscopy
  - Barium studies
    - Ascariasis: Linear filling defects up to 35 cm in SB; giardiasis: Thickened duodenal & jejunal folds; amebiasis: Colon ulcerations

CT Findings

- CECT: Ascariasis: SBO; amebiasis: Thickened colonic wall, rounded right lobe liver abscess with peripheral zone of edema; giardiasis: Thickened SB +/- excess fluid, distension

Ultrasonographic Findings

- Real Time: Amebiasis: Rounded hypoechoic right lobe liver abscess with low-level internal echoes, little distal acoustic enhancement

Imaging Recommendations

- Best imaging tool: SBFT for ascariasis and giardiasis; barium enema and CECT for amebiasis

DIFFERENTIAL DIAGNOSIS

Crohn disease

- Apthous ulcers; skip areas of SB and colon; sinus tracts into mesentery; fibrofatty extramural masses

DDx: Intestinal Disorders Mimicking Parasitic Disease

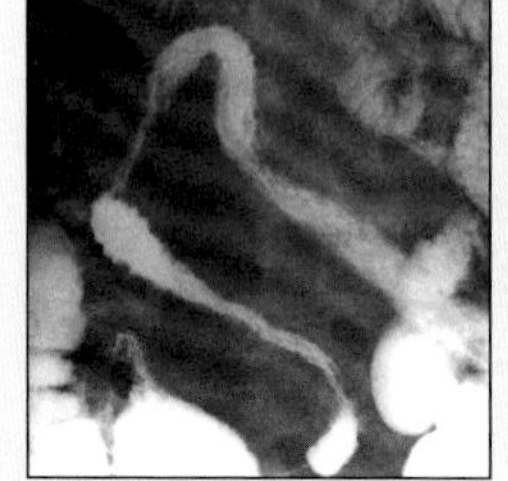

Crohn Disease

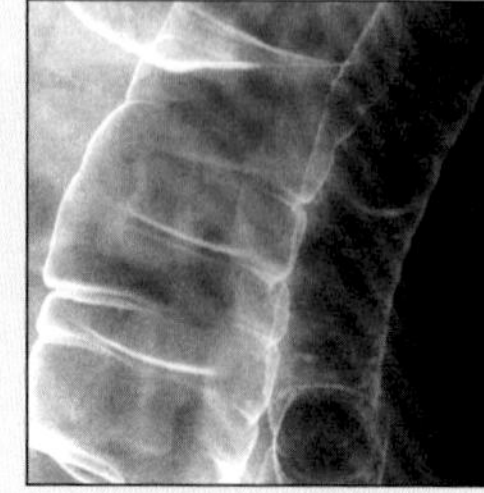

Ulcerative Colitis

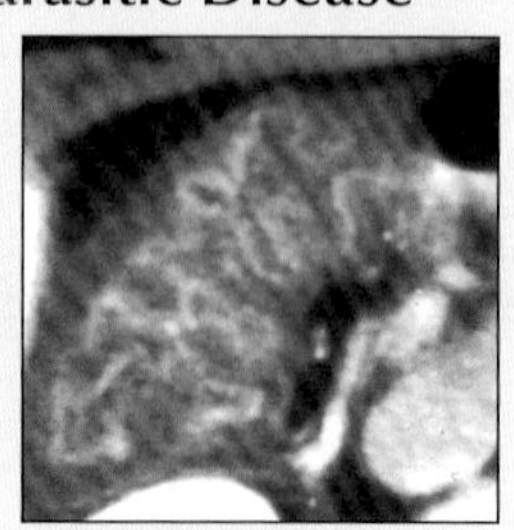

Pseudomem. Colitis

INTESTINAL PARASITIC DISEASE

Key Facts

Imaging Findings
- Ascariasis: Linear filling defect on small bowel follow through (SBFT)
- Giardiasis: Thickened duodenal and jejunal folds on SBFT
- Amebiasis: Diffuse ulcerating colitis, right lobe liver abscess on CECT
- Best imaging tool: SBFT for ascariasis and giardiasis; barium enema and CECT for amebiasis

Clinical Issues
- Ethnicity: African patients

Diagnostic Checklist
- Consider Crohn disease, ulcerative colitis
- Linear filling defects, thickened proximal small bowel folds, ileocecal ulcerations

Ulcerative colitis
- Superficial ulcers; granular mucosa; long segment strictures; rectal involvement

Pseudomembranous colitis
- "Accordion sign" of haustral edema

PATHOLOGY

General Features
- General path comments
  - Ascariasis: SBO, pancreatobiliary obstruction
  - Giardiasis: Medium to moderate blunting of SB villi
  - Amebiasis: Colitis with flask-shaped ulcers
- Epidemiology
  - Ascariasis infests 25% of world population
  - Giardiasis is most common protozoal disease in US
    - 2-7% of US population; higher in immunocompromised patients
  - Amebiasis infests 10% of world population

Gross Pathologic & Surgical Features
- Ascariasis: SBO due to mass of worms; appendicitis, pancreatitis, cholangitis
- Giardiasis: Often grossly normal small bowel
- Amebiasis: Ulcerating acute colitis; toxic megacolon; large inflammatory masses (ameboma) mimic colon CA

Microscopic Features
- Ascariasis: Mucosal destruction at worm attachment sites; inflammatory changes in appendix, pancreas, biliary tree
- Giardiasis: Villous blunting, inflammatory cells in lamina propria
- Amebiasis: Colitis with neutrophilic infiltrate and deep ulcers into submucosa

CLINICAL ISSUES

Presentation
- Most common signs/symptoms
  - Ascariasis: Abdominal pain, diarrhea; SBO; appendicitis, pancreatitis; growth retardation
  - Giardiasis: Abdominal pain, diarrhea; nausea, vomiting, distension; weight loss; malabsorption
  - Amebiasis: Diarrhea, fever, GI bleeding

Demographics
- Ethnicity: African patients

Natural History & Prognosis
- Ascariasis: SBO, appendicitis, pancreatitis
- Giardiasis & amebiasis: Continued diarrhea until effective treatment

Treatment
- Options, risks, complications
  - Ascariasis: Antihelminthic chemotherapy with mebendazole, albendazole or pyrantel pamoate
  - Giardiasis: Nitazoxanide or metronidazole
  - Amebiasis: Metronidazole

DIAGNOSTIC CHECKLIST

Consider
- Consider Crohn disease, ulcerative colitis

Image Interpretation Pearls
- Linear filling defects, thickened proximal small bowel folds, ileocecal ulcerations

SELECTED REFERENCES

1. Ali SA et al: Giardia intestinalis. Curr Opin Infect Dis. 16(5):453-60, 2003
2. McDonald V: Parasites in the gastrointestinal tract. Parasite Immunol. 25(5):231-4, 2003
3. Jong E: Intestinal parasites. Prim Care. 29(4):857-77, 2002

IMAGE GALLERY

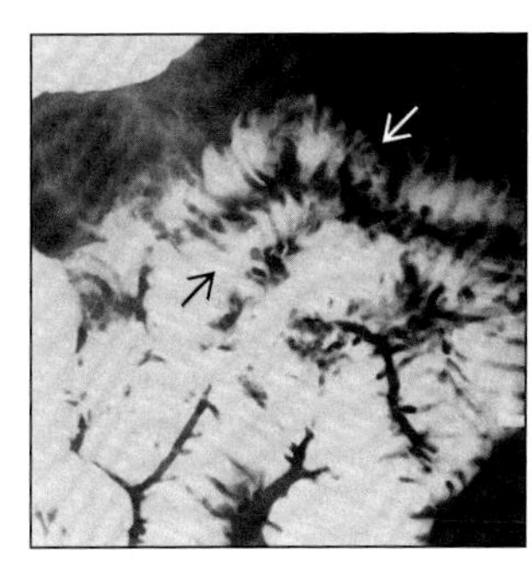

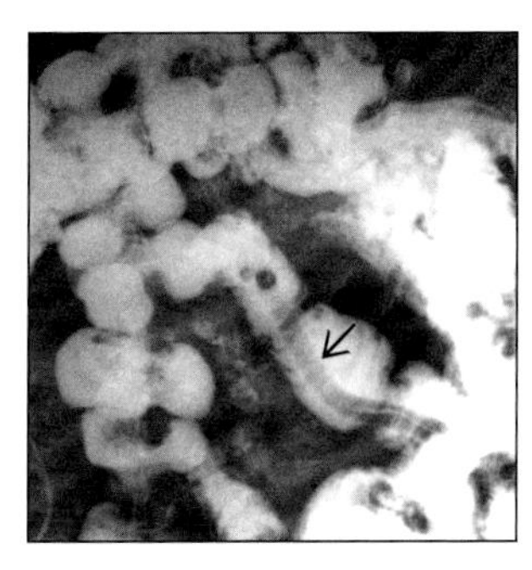

(Left) *Anteroposterior view of SBFT demonstrates thickened folds in proximal jejunum (arrows) from giardiasis.* ***(Right)*** *Spot film of SBFT demonstrates ascaris in distal small bowel (arrow).*

OPPORTUNISTIC INTESTINAL INFECTIONS

AP compression spot film from barium enema demonstrates deep cecal ulcer (arrow) from CMV colitis.

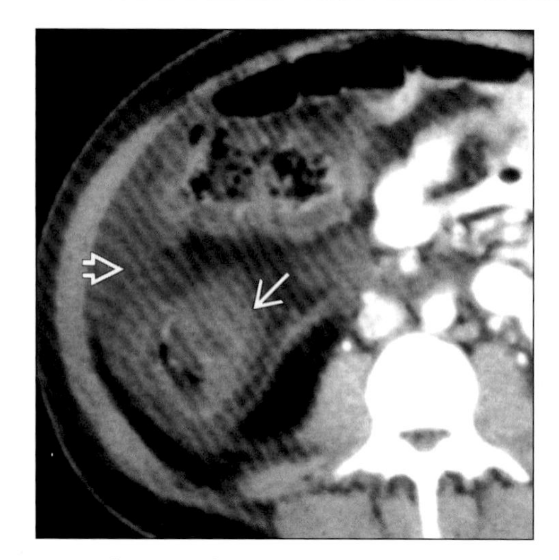

CMV colitis on axial CECT. Note edematous right colon (arrow) with pericolonic soft tissue stranding (open arrow).

TERMINOLOGY

Abbreviations and Synonyms

- Cytomegalovirus (CMV), mycobacterium avium-intracellulare (MAI), atypical mycobacterial infection, cryptosporidiosis, cryptosporidium parvum

Definitions

- Gastrointestinal infection of immune-compromised host by virus (CMV), protozoa (cryptosporidium), or mycobacterium (MAI)

IMAGING FINDINGS

General Features

- Best diagnostic clue
  - CMV: Thickened folds with deep ulcerations on barium studies of small bowel or colon
  - MAI: Thickened small bowel folds; mesenteric or peri-portal adenopathy with low attenuation nodes
  - Cryptosporidiosis: Secretory enteritis with thickened small bowel folds and increased fluid in bowel
- Location
  - CMV: Small bowel or colon, stomach
  - MAI: Small bowel, nodes
  - Cryptosporidiosis: Small bowel

Radiographic Findings

- Fluoroscopy
  - CMV: Barium enema
    - Diffuse colitis
    - Early stage resembles ulcerative colitis
    - Aphthous ulcers
    - Terminal ileum characteristically involved with thickened folds and/or ulceration
    - Later stages demonstrate deep ulceration
  - CMV: Small bowel follow through (SBFT)
    - Thickened edematous folds
    - Discrete ulcerations, deep ulcers in sinus tracts
  - MAI: SBFT
    - Diffuse enteritis with thickened folds
    - Micronodular mucosal pattern
  - Cryptosporidiosis: SBFT
    - Secretory enteritis
    - Thickened folds, increased fluid in bowel

CT Findings

- CECT
  - CMV
    - Mural thickening of colon, stomach, small bowel, especially terminal ileum
    - Mesenteric stranding due to deep ulcers
  - MAI
    - Low attenuation mesenteric adenopathy due to necrosis or caseation

DDx: Gastrointestinal Lesions Mimicking Opportunistic Infections

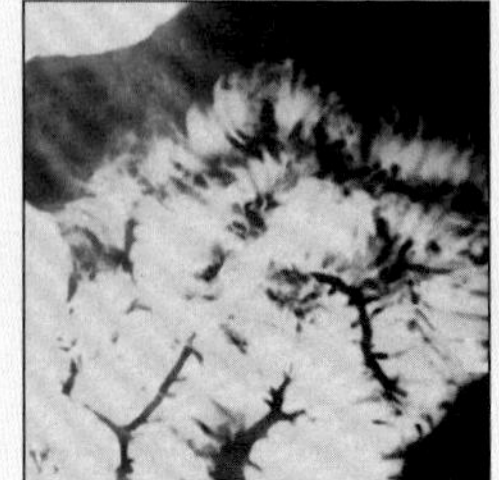

Giardiasis

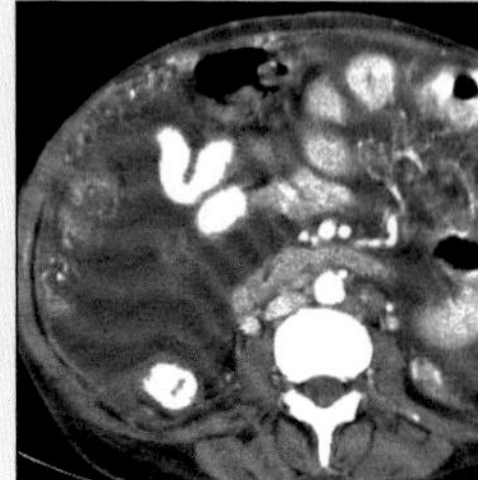

TB Peritonitis

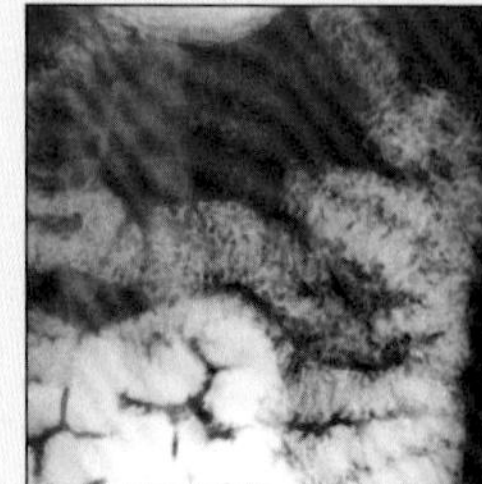

Lymphoma

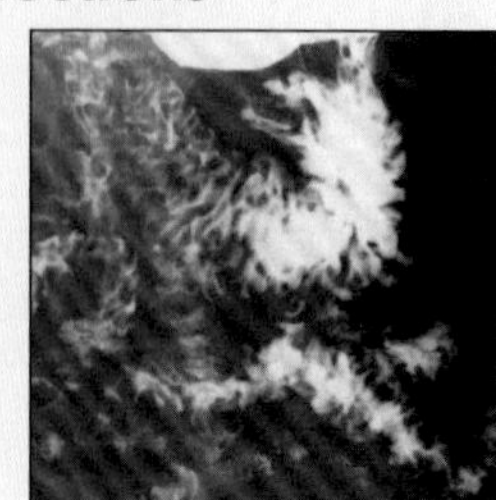

Whipple Disease

OPPORTUNISTIC INTESTINAL INFECTIONS

Key Facts

Terminology
- Cytomegalovirus (CMV), mycobacterium avium-intracellulare (MAI), atypical mycobacterial infection, cryptosporidiosis, cryptosporidium parvum
- Gastrointestinal infection of immune-compromised host by virus (CMV), protozoa (cryptosporidium), or mycobacterium (MAI)

Imaging Findings
- CMV: Thickened folds with deep ulcerations on barium studies of small bowel or colon
- MAI: Thickened small bowel folds; mesenteric or peri-portal adenopathy with low attenuation nodes
- Cryptosporidiosis: Secretory enteritis with thickened small bowel folds and increased fluid in bowel
- Best imaging tool: SBFT, CECT
- Protocol advice: CECT: 150 ml I.V. contrast (2.5 ml/sec) with 5 mm collimation

Top Differential Diagnoses
- Giardiasis
- Tuberculosis
- Gastrointestinal lymphoma
- Whipple disease

Clinical Issues
- Most common signs/symptoms: Abdominal pain; nausea, vomiting, diarrhea; fever; GI bleeding
- Clinical profile: HIV or immunocompromised patient
- CMV and cryptosporidiosis respond to nitazoxanide in early stage
- MAI in AIDS patients often difficult to treat

    - Thickened small bowel folds
  - Cryptosporidiosis
    - Thickened folds
    - Spotty lymphadenopathy without necrosis

Imaging Recommendations
- Best imaging tool: SBFT, CECT
- Protocol advice: CECT: 150 ml I.V. contrast (2.5 ml/sec) with 5 mm collimation

DIFFERENTIAL DIAGNOSIS

Giardiasis
- Small bowel infection by protozoal flagellate Giardia lamblia; water-borne infection resulting in abdominal pain, diarrhea
- Diagnosis made by stool culture, duodenal mucosal biopsy
- Small bowel barium studies: Thickened folds, increased fluid in duodenum & jejunum, and sparing of ileum

Tuberculosis
- May be 2° to pulmonary source or ingestion of bovine bacillus
- Most common symptoms: Chronic abdominal pain, weight loss, fever, diarrhea, palpable masses
- Ileocecal involvement mimics Crohn disease
- Peritonitis results in ascites, nodular thickening of omentum and peritoneum
- Low attenuation mesenteric and peri-portal nodes
- Marked narrowing of ileocecal area = Stierlin sign; hypertrophy of ileocecal valve = Fleishner sign
- Skip areas of colonic involvement with ulceration of structures, inflammatory polyps
- Diagnosis with endoscopic biopsy

Gastrointestinal lymphoma
- Most commonly non-Hodgkin lymphomas that are high grade large cell or immunoblastic cell types; Burkitt more common in pediatric patients
- Most common symptoms: Abdominal pain, weight loss, fever, anemia
- Different patterns on barium studies: Nodular (due to submucosal infiltration); infiltrating, polypoid, and aneurysmal
- CECT: Focal bowel wall thickening without obstruction; "sausage-shaped" mass

Whipple disease
- Multisystem disease caused by Thropheryma whippelii bacillus
  - Results in periodic-acid-Schiff positive glycoprotein-laden macrophages infiltrating tissues
- Small bowel, central nervous system, heart valves, joint capsules frequently involved
- CECT: SB fold thickening, mesenteric and retroperitoneal adenopathy
- Enteroclysis: Submucosal fold thickening with micronodular pattern

PATHOLOGY

General Features
- General path comments
  - CMV
    - DNA herpes virus
    - Causes vasculitis leading to gastrointestinal ulceration, ischemia, bleeding, perforation
  - MAI
    - Atypical mycobacterium
    - Macrophages infiltrate lamina propria of small bowel distending villi
  - Cryptosporidiosis
    - Protozoan organisms attach between microvilli of small intestine
    - Leads to mucosal damage and secretory enteritis
- Epidemiology: HIV or other immunocompromised patients

Microscopic Features
- CMV
  - Cytoplasmic inclusion bodies in enterocytes, macrophages, fibroblasts, and endothelial cells
- MAI

- Coarsely granular mucosa
- Sheets of foamy macrophages infiltrate lamina propria
- Positive acid-fast or Fite stain for organisms

- Cryptosporidiosis
  - Organisms proliferate from apex of enterocyte
  - Villous atrophy
  - Crypt hyperplasia
  - Inflammatory infiltrate

CLINICAL ISSUES

Presentation

- Most common signs/symptoms: Abdominal pain; nausea, vomiting, diarrhea; fever; GI bleeding
- Clinical profile: HIV or immunocompromised patient

Demographics

- Age: Any age
- Gender: M = F

Natural History & Prognosis

- CMV and cryptosporidiosis respond to nitazoxanide in early stage
- MAI in AIDS patients often difficult to treat

Treatment

- Options, risks, complications
  - CMV: Antiviral therapy with acyclovir or gancyclovir
  - MAI: Antituberculous chemotherapy
  - Cryptosporidiosis: Chemotherapy with nitazoxanide

DIAGNOSTIC CHECKLIST

Consider

- Tuberculosis, giardiasis or lymphoma

Image Interpretation Pearls

- CMV: Deep ulcerations & focal enteritis or colitis
- MAI: Enteritis & low attenuation nodes
- Cryptorsporidiosis: Thickened bowel wall and edematous folds

SELECTED REFERENCES

1. Ishida T et al: The management of gastrointestinal infections caused by cytomegalovirus. J Gastroenterol. 38(7):712-3, 2003
2. Streetz KL et al: Acute CMV-colitis in a patient with a history of ulcerative colitis. Scand J Gastroenterol. 38(1):119-22, 2003
3. Hunter PR et al: Epidemiology and clinical features of Cryptosporidium infection in immunocompromised patients. Clin Microbiol Rev. 15(1):145-54, 2002
4. Masur H et al: Guidelines for preventing opportunistic infections among HIV-infected persons--2002. Recommendations of the U.S. Public Health Service and the Infectious Diseases Society of America. Ann Intern Med. 137(5 Pt 2):435-78, 2002
5. Ukarapol N et al: Cytomegalovirus-associated manifestations involving the digestive tract in children with human immunodeficiency virus infection. J Pediatr Gastroenterol Nutr. 35(5):669-73, 2002
6. Oldfield EC 3rd: Evaluation of chronic diarrhea in patients with human immunodeficiency virus infection. Rev Gastroenterol Disord. 2(4):176-88, 2002
7. Brantsaeter AB et al: CMV disease in AIDS patients: incidence of CMV disease and relation to survival in a population-based study from Oslo. Scand J Infect Dis. 34(1):50-5, 2002
8. von Reyn CF et al: Sources of disseminated Mycobacterium avium infection in AIDS. J Infect. 44(3):166-70, 2002
9. Pollok RC: Viruses causing diarrhoea in AIDS. Novartis Found Symp. 238:276-83; discussion 283-8, 2001
10. Huh JJ et al: Mycobacterium avium complex peritonitis in an AIDS patient. Scand J Infect Dis. 33(12):936-8, 2001
11. Clemente CM et al: Gastric cryptosporidiosis as a clue for the diagnosis of the acquired immunodeficiency syndrome. Arq Gastroenterol. 37(3):180-2, 2000
12. Chamberlain RS et al: Ileal perforation caused by cytomegalovirus infection in a critically ill adult. J Clin Gastroenterol. 30(4):432-5, 2000
13. Wallace MR et al: Gastrointestinal manifestations of HIV infection. Curr Gastroenterol Rep. 2(4):283-93, 2000
14. Sharpstone D et al: Small intestinal transit, absorption, and permeability in patients with AIDS with and without diarrhoea. Gut. 45(1):70-6, 1999
15. Drew WL et al: Cytomegalovirus: disease syndromes and treatment. Curr Clin Top Infect Dis. 19:16-29, 1999
16. Sprinz E et al: AIDS-related cryptosporidial diarrhoea: an open study with roxithromycin. J Antimicrob Chemother. 41 Suppl B:85-91, 1998
17. Manabe YC et al: Cryptosporidiosis in patients with AIDS: correlates of disease and survival. Clin Infect Dis. 27(3):536-42, 1998
18. Rossi P et al: Gastric involvement in AIDS associated cryptosporidiosis. Gut. 43(4):476-7, 1998
19. Lumadue JA et al: A clinicopathologic analysis of AIDS-related cryptosporidiosis. AIDS. 12(18):2459-66, 1998
20. Sesin GP et al: New trends in the drug therapy of localized and disseminated Mycobacterium avium complex infection. Am J Health Syst Pharm. 53(21):2585-90, 1996
21. Julander I: Clinical manifestations and treatment of mycobacterium avium-intracellulare complex infection in HIV-infected patients. Scand J Infect Dis Suppl. 98:19-20, 1995
22. Benson CA: Disease due to the Mycobacterium avium complex in patients with AIDS: epidemiology and clinical syndrome. Clin Infect Dis. 18 Suppl 3:S218-22, 1994
23. Torriani FJ et al: Autopsy findings in AIDS patients with Mycobacterium avium complex bacteremia. J Infect Dis. 170(6):1601-5, 1994
24. Hellyer TJ et al: Gastro-intestinal involvement in Mycobacterium avium-intracellulare infection of patients with HIV. J Infect. 26(1):55-66, 1993
25. Cappell MS et al: Gastrointestinal obstruction due to Mycobacterium avium intracellulare associated with the acquired immunodeficiency syndrome. Am J Gastroenterol. 87(12):1823-7, 1992

IMAGE GALLERY

Typical

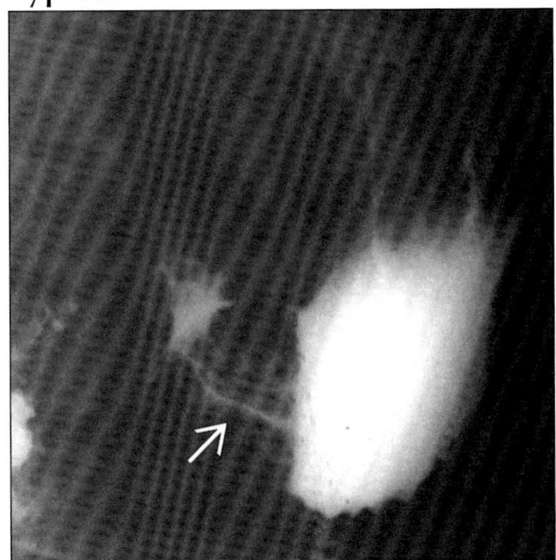

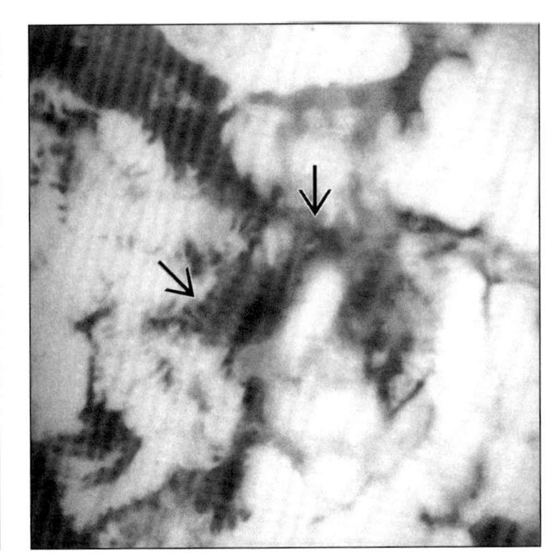

(Left) *Gastrointestinal CMV. Lateral view of stomach from upper GI series shows marked antral narrowing and thickened folds (arrow) from antral CMV gastritis.* ***(Right)*** *Gastrointestinal CMV. Small bowel follow through demonstrates enteritis with thickened folds (arrows).*

Typical

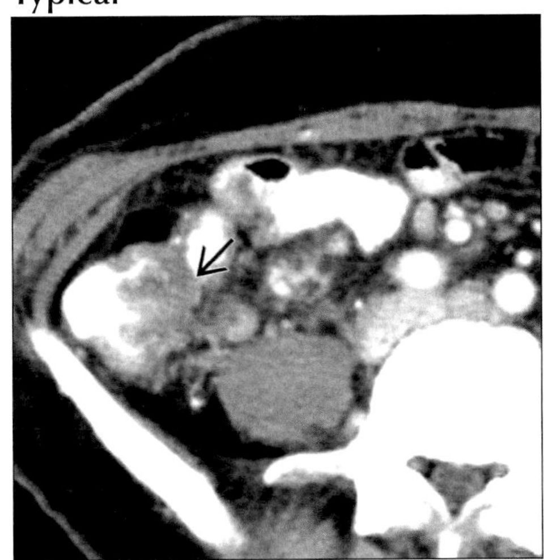

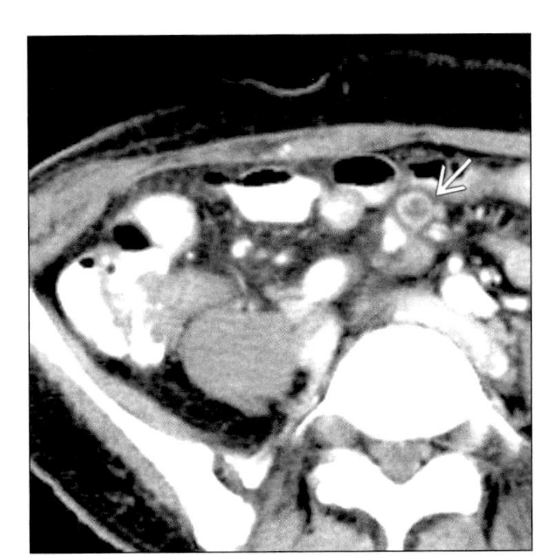

(Left) *MAI on CECT demonstrates mural thickening of cecum and terminal ileum (arrow).* ***(Right)*** *Axial CECT of MAI. Note low attenuation mesenteric lymph node (arrow).*

Typical

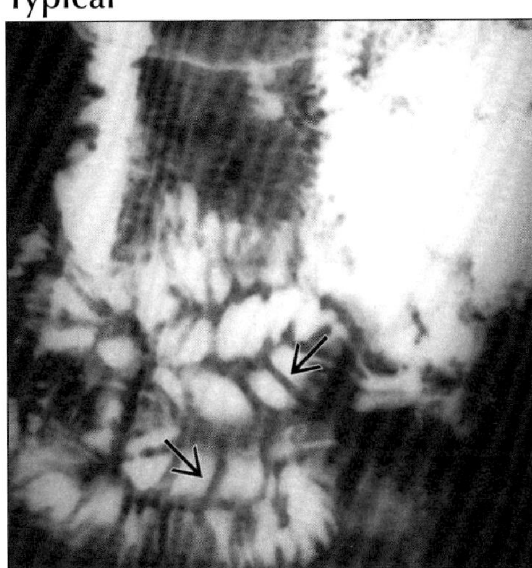

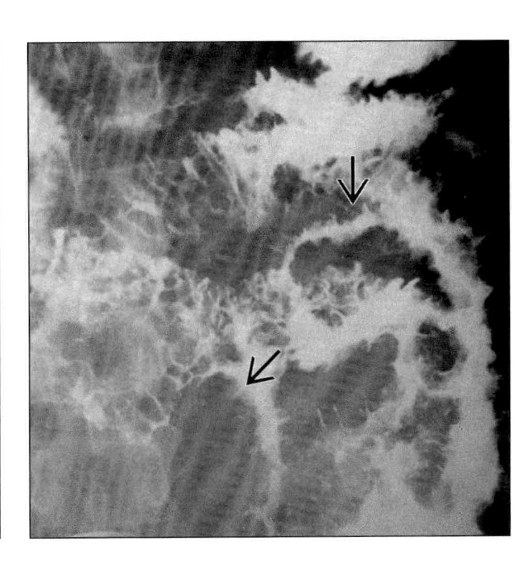

(Left) *Early cryptosporidiosis of the small bowel. Small bowel follow through demonstrates mild fold thickening (arrows).* ***(Right)*** *Advanced cryptosporidiosis of the small bowel. Note marked mural narrowing and extensive fold thickening (arrows) on small bowel follow through.*

SPRUE-CELIAC DISEASE

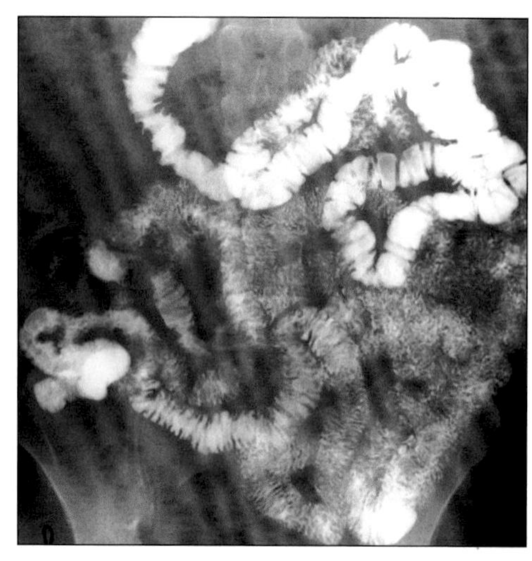

Small bowel follow through (SBFT) shows decreased size + number of jejunal folds and increased number and size of ileal folds (reversal pattern).

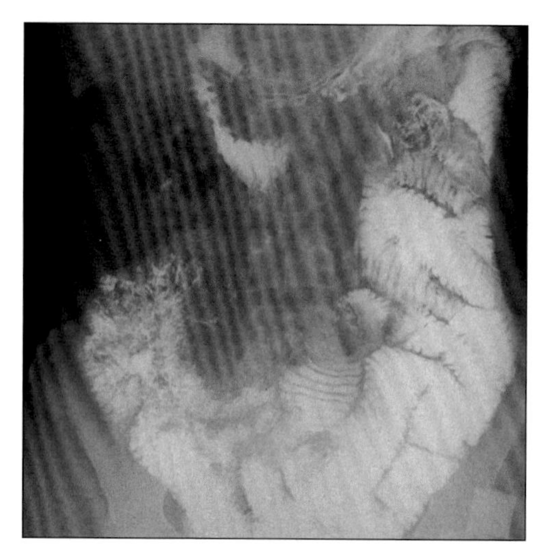

SBFT shows dilated lumen + dilution of barium within the small intestine and a reversal of the normal jejunal + ileal fold patterns.

TERMINOLOGY

Abbreviations and Synonyms

- Nontropical sprue; celiac sprue or disease; gluten sensitive enteropathy

Definitions

- Nontropical sprue (Celiac disease): Malabsorption due to intolerance to gluten protein
- Tropical sprue: Malabsorption seen in inhabitants of tropical countries

IMAGING FINDINGS

General Features

- Best diagnostic clue: ↓ # of jejunal folds (< 3 inch) & ↑ # of ileal folds (4-6 inch) on enteroclysis
- Location
  - Celiac disease: More proximal small bowel
  - Tropical sprue: Entire small bowel
- Morphology: Celiac disease-dilated small bowel, thickened wall + valvulae, reversal of jejunoileal folds
- Other general features
  - Most common small bowel disease producing malabsorption syndrome
  - Due to sensitivity of small bowel to α-gliadin, a component of gluten
  - Has a familial susceptibility with genetic basis
  - Evidence suggesting existence of immune reaction in bowel mucosa
  - Classic presentation: Steatorrhea & weight loss or failure to thrive
  - Variants of celiac sprue
    - Refractory; collagenous; unmasked celiac sprue

Radiographic Findings

- Fluoroscopic guided small bowel follow through
  - Dilatation of small bowel (jejunum): > 3 cm
  - Valvulae conniventes: May exhibit 5 patterns
    - Normal: In most patients valvulae look normal
    - Squared ends: Ends at margin are squared off rather than rounded
    - Reversed jejunoileal fold pattern: Decreased jejunal folds & increased ileal folds
    - Absence of valvulae: "Moulage sign" (cast) characteristic of sprue
    - Thickening: In severe disease & hypoproteinemia
  - "Colonization of jejunum": Loss of jejunal folds → colon-like haustrations
  - Hypersecretion-related artifacts: Due to excess fluid
    - Flocculation: Coarse granular appearance of small clumps of disintegrated barium due to excess fluid; mainly with steatorrhea
    - Segmentation: Breakup of normal continual column of barium creating large masses of barium separated by string-like strands

DDx: Fold Thickening, Increased Fluid

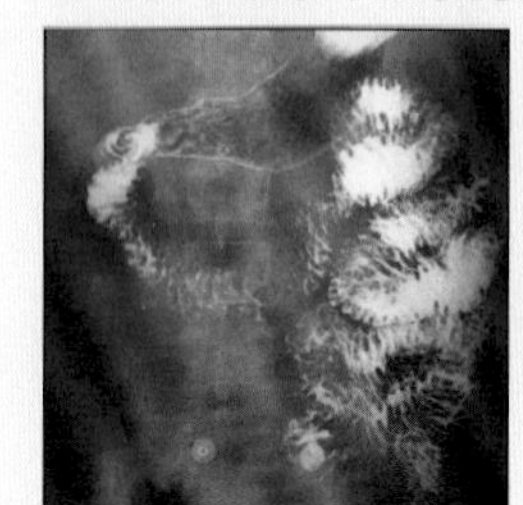

Whipple Disease

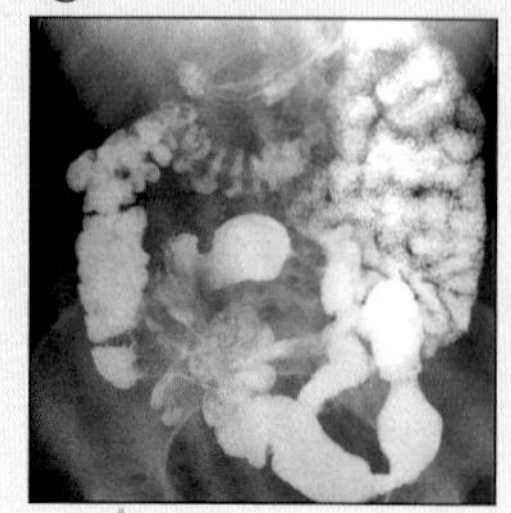

Crohn Disease

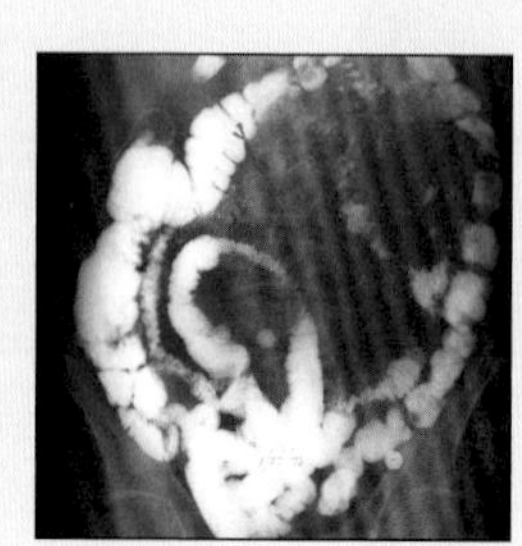

Ischemia

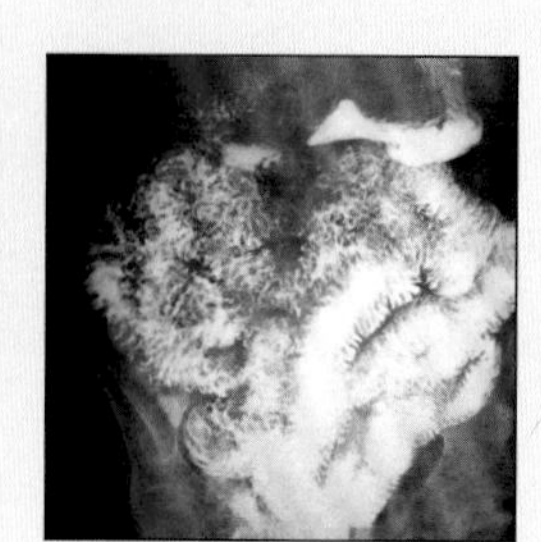

Waldenstrom Macro.

SPRUE-CELIAC DISEASE

Key Facts

Terminology
- Nontropical sprue; celiac sprue or disease; gluten sensitive enteropathy
- Nontropical sprue (Celiac disease): Malabsorption due to intolerance to gluten protein

Imaging Findings
- Best diagnostic clue: ↓ # of jejunal folds (< 3 inch) & ↑ # of ileal folds (4-6 inch) on enteroclysis
- Absence of valvulae: "Moulage sign" (cast) characteristic of sprue
- "Colonization of jejunum": Loss of jejunal folds → colon-like haustrations
- Mosaic pattern: Due to total villous atrophy
- "Bubbly" duodenum: Nodular pattern in mucosa
- ± Lymphadenopathy

Top Differential Diagnoses
- Whipple disease
- Crohn disease
- Opportunistic infection
- Ischemia
- Cystic fibrosis
- Immunologic disorders

Clinical Issues
- ↑ Risk of T-cell lymphoma & carcinoma of jejunum

Diagnostic Checklist
- Check for history of gluten diet & differentiate from other simulating pathologies
- Reversal of jejunoileal fold pattern; jejunal colonization; ↑ separation or absence of jejunal folds

- Transit time: Long, short or normal time
- Nonpropulsive peristalsis (flaccid & poorly contracting bowel loops)
- Normal or thickened or effaced mucosal folds: Based on degree of hypoproteinemia
- Painless, transient intussusceptions often seen on fluoroscopic studies

- Fluoroscopic guided enteroclysis
  - Facilitates diagnosis or exclusion of celiac sprue in higher percentage of cases
  - Jejunal folds
    - Decreased number of proximal jejunal folds (< 3 inch); normal (5 or > folds/inch)
    - Increased separation & absence of folds
    - Ileal appearance of jejunum
  - Ileal folds
    - Increased number of folds in distal ileum (4-6 inch); normal (2-4 folds/inch)
    - Increased fold thickness ≥ 1 mm: "Jejunization" of ileum seen in 78% cases (ileum' ability to gain functions normally performed by jejunum)
  - Jejunoileal fold pattern reversal: Sensitivity in diagnosing disease increased to 83%
  - Mosaic pattern: Due to total villous atrophy
    - 1-2 mm islands of mucosa surrounded by barium-filled grooves
  - Duodenal changes
    - Decreased number & irregular folds, especially in distal duodenum
    - "Bubbly" duodenum: Nodular pattern in mucosa

CT Findings
- Dilated, fluid-filled small bowel loops
- ± Ileal mucosal thickening
- ± Small bowel intussusception (classic "target" lesion)
- ± Lymphadenopathy

Ultrasonographic Findings
- Real Time
  - Increased echo free intraluminal fluid; flaccid & mildly dilated bowel loops
  - Moderately thickened small-bowel wall & valvulae conniventes
  - ± Increase in caliber of superior mesenteric artery
  - ± Enlarged mesenteric lymph nodes; ± ascites

Imaging Recommendations
- Enteroclysis; small bowel follow through; CT

DIFFERENTIAL DIAGNOSIS

Whipple disease
- Thickened proximal small-bowel folds mimic sprue
- Micronodules in jejunum on enteroclysis
- Thickening of mesentery & lymphadenopathy
- Periodic acid-Schiff (PA S) stain positive
- Electron microscopy: Tropheryma whippleii

Crohn disease
- Predominantly involves distal ileum & colon
- Classic sign on barium exam: "String sign"
- Skip lesions/fistulae/fissures/transmural inflammation
- Diagnosis: Biopsy & histology

Opportunistic infection
- AIDS (e.g., cryptosporidiosis; tuberculosis; CMV)
- Cryptosporidiosis
  - Most common cause of enteritis in AIDS patients
  - Pathology: Mucosal damage & secretory diarrhea
  - Thickening of folds & bowel wall; ↑ fluid in lumen
  - Diagnosis: Oocysts in stool & mucosal biopsy
- Mycobacterial tuberculosis
  - Example: (Atypical) Mycobacterium avium-intracellulare
  - Small bowel shows thickened folds, fine nodularity
  - CT: Shows low density (caseated) lymph nodes
  - Diagnosis: Mucosal biopsy
- Cytomegalovirus (CMV)
  - Causes terminal ileitis in AIDS patients
  - Narrow lumen, thickened folds, spiculation, ulcers
  - Diagnosis: Round intranuclear inclusion bodies

Ischemia
- Due to vascular insufficiency
  - Superior mesenteric artery (SMA) clot or narrowing
    - Segmental thickening of bowel wall > 3 mm

- ± Gas in portal vein, bowel wall

Cystic fibrosis
- Duodenum (2nd part): Thickened or flattened folds, nodular filling defects, sacculation of lateral border
- Thickened proximal jejunal folds & reticular mucosal pattern may simulate celiac disease

Immunologic disorders
- Waldenstrom macroglobulinemia
  - Cancer of B lymphocytes
  - Overproduction of IgM
- IgA deficiency
  - Often accompanied by opportunistic infections (e.g., Giardia)
  - Both can cause nodular SB folds + malabsorption pattern

PATHOLOGY

General Features
- General path comments
  - Nontropical sprue (Celiac disease)
    - Affects proximal small bowel
    - Atrophy of jejunal villi, crypt hyperplasia, lengthening & flattening of mucosa, chronic inflammation of lamina propria
    - ↑ Production of IgA & IgM antigliadin antibodies
    - Increased number of intraepithelial lymphocytes & activation of T-cells in lamina propria
  - Tropical sprue
    - Similar to celiac disease; affects entire small bowel
- Genetics: Celiac disease- class II human leukocyte antigens (HLA) - HLA-DR3 & HLA-DQw2
- Etiology
  - Celiac disease: Allergic, immunologic, or toxic reaction to gliadin component of gluten
  - Tropical sprue: Unknown etiology; may be due to enterotoxigenic E. coli
- Epidemiology
  - Celiac disease: Increased incidence in Ireland & Northern Europe; unknown in Africa, China, Japan
  - Tropical sprue: Increased incidence in tropics, especially in Vietnam & Puerto Rico
- Associated abnormalities
  - Dermatitis herpetiformis; IgA deficiency
  - Hyposplenism; benign adenopathy
  - Cavitary mesenteric lymph node syndrome (rare)
    - Hyposplenism, nodal cavitary masses

Gross Pathologic & Surgical Features
- Dilated small bowel, thickened wall + valvulae, reversal of jejunoileal folds

Microscopic Features
- Villous atrophy, thickened lamina propria, increased number of crypts + cellular infiltrate
- Immunoperoxidase shows immunocytes with IgA & IgM antigliadin antibodies

CLINICAL ISSUES

Presentation
- Most common signs/symptoms
  - Malabsorption, steatorrhea, abdominal distension
  - Diarrhea, weight loss, glossitis, anemia
- Clinical profile: Young patient with history of gluten diet, steatorrhea, abdominal distension & diarrhea
- Sensitive predictor in asymptomatic: ↑ Levels of antireticulin antibody (ARA)
- Lab-data
  - Specific screening tests: IgG antigliadin & IgA antiendomysial antibodies
  - Positive Sudan stain for fecal fat & ↓ D-xylose absorption
  - ↓ Iron, folate, Ca++, K+, albumin, cholesterol levels
- Diagnosis
  - Duodenojejunal mucosal biopsy
  - Clinical & imaging response to gluten free diet

Demographics
- Age
  - Childhood by age of 2 years
  - 2nd peak in 3rd & 4th decades
- Gender: M = F

Natural History & Prognosis
- Natural history
  - Adult disease: Extension childhood form; new onset
- Complications
  - Ulcerative jejunoileitis
  - ↑ Risk of T-cell lymphoma & carcinoma of jejunum
- Prognosis
  - Celiac disease: Improvement within 48 hours; full remission (weeks to months)
  - Tropical sprue: Improvement in 4-7 days; complete recovery (6-8 weeks)

Treatment
- Nontropical sprue or Celiac disease
  - Lifelong gluten-free diet
- Tropical sprue
  - Broad spectrum antibiotics (tetracycline) & folates

DIAGNOSTIC CHECKLIST

Consider
- Check for history of gluten diet & differentiate from other simulating pathologies

Image Interpretation Pearls
- Reversal of jejunoileal fold pattern; jejunal colonization; ↑ separation or absence of jejunal folds

SELECTED REFERENCES

1. Bosch HCM et al: Celiac disease: Small-bowel enteroclysis findings in adult patients treated with a gluten-free diet. Radiology. 201: 803-8, 1996
2. Strobl PW et al: CT diagnosis of celiac disease. Journal of Computer Assisted Tomography. 19 (2): 319-20, 1995
3. Rubesin SE et al: Adult celiac disease and its complications. RadioGraphics. 9: 1045-66, 1989

IMAGE GALLERY

Typical

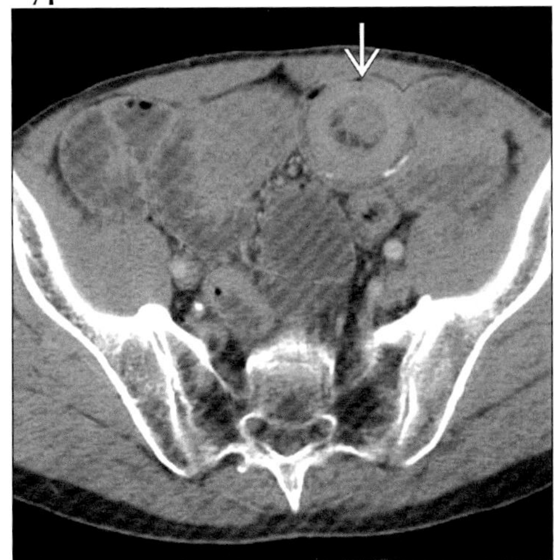

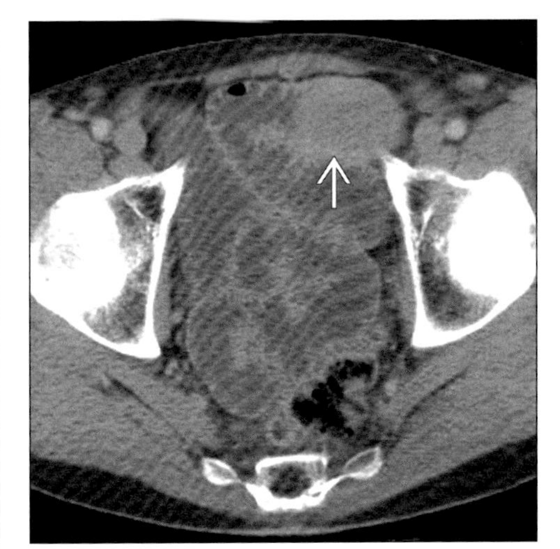

*(**Left**) Axial CECT shows fluid-distended small bowel, prominent folds in ileum, and a short segment intussusception (arrow). (**Right**) Axial CECT shows shows dilated fluid-distended bowel and the bottom of an intussusception (arrow). Note prominent folds in ileum.*

Typical

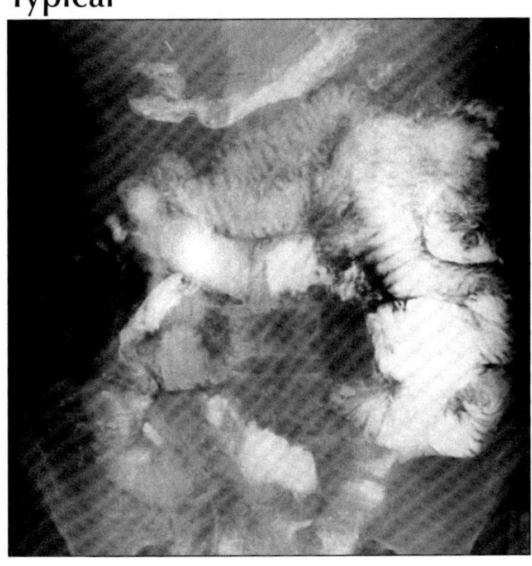

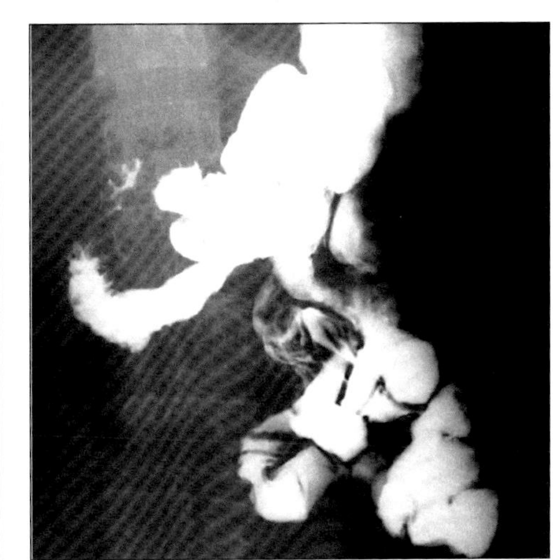

*(**Left**) SBFT shows dilated bowel, dilution of barium, reduced number + size of jejunal folds. (**Right**) SBFT shows severe loss of folds in duodenum + jejunum; the "moulage" pattern.*

Typical

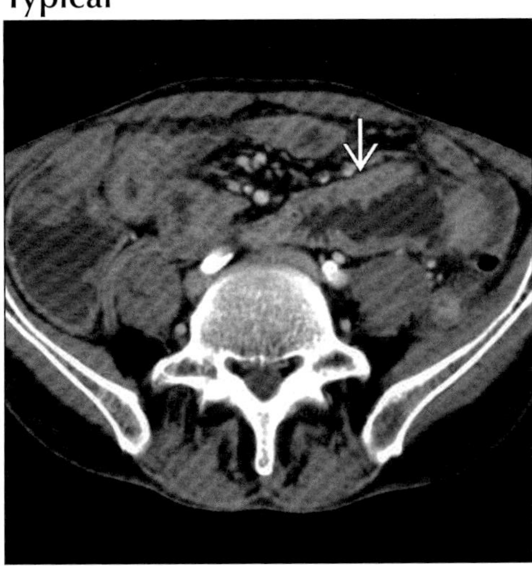

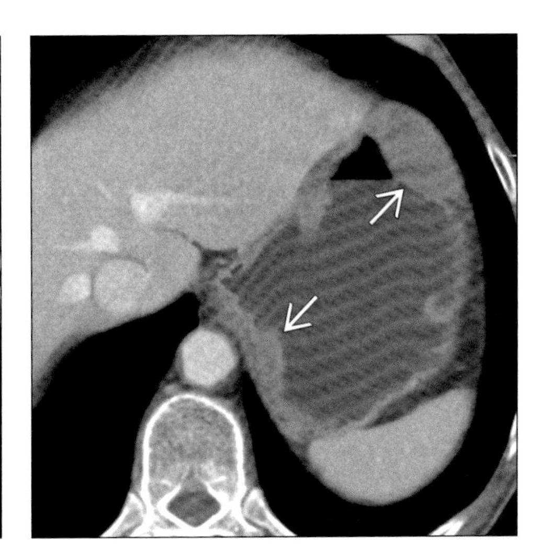

*(**Left**) Axial CECT in patient with sprue. CT shows fluid-distended bowel. One segment of jejunum has focal thickening (arrow) of the wall found to be lymphoma. (**Right**) Axial CECT of patient with sprue and jejunal lymphoma. Gastric wall thickening (arrows) was also due to lymphoma.*

WHIPPLE DISEASE

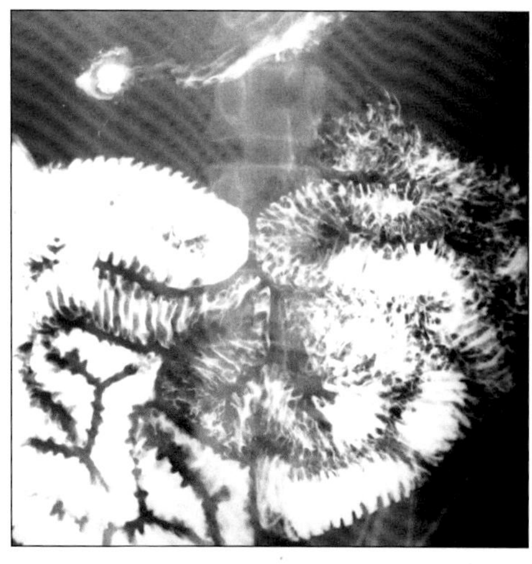

Small bowel follow through (SBFT) shows micronodular pattern of jejunum in a 40 year old man with Whipple disease.

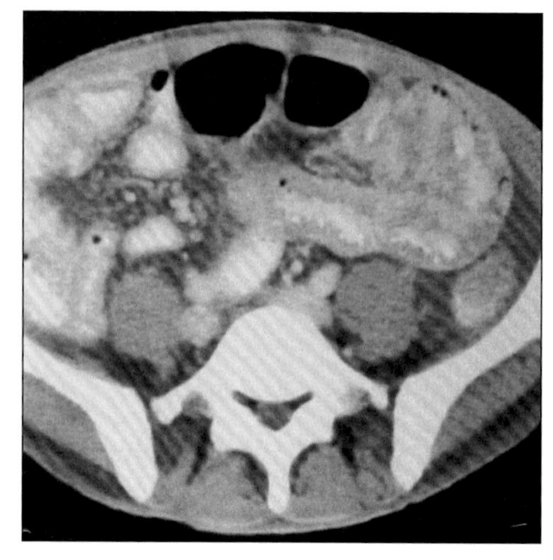

Axial CECT shows mural thickening of long segment of jejunum.

TERMINOLOGY

Abbreviations and Synonyms

- Synonym: Intestinal lipodystrophy

Definitions

- Rare, chronic, bacterial infectious & systemic disease leading to chronic diarrhea & malabsorption

IMAGING FINDINGS

General Features

- Best diagnostic clue: Thickened proximal small bowel folds with micronodularity (1-2 mm) & low density mesenteric adenopathy (near fat HU)
- Location: Small bowel: Jejunum & duodenojejunal junction
- Morphology: Small intestinal mucosa laden with distended macrophages in lamina propria, which contain rod-shaped bacilli

Radiographic Findings

- Fluoroscopic guided enteroclysis
  - Micronodules (1-2 mm) in proximal small bowel
  - Thickened folds (especially jejunum)
  - ± Thickened mesentery & separation of bowel loops
  - Small bowel lumen may be normal or mildly dilated
  - Distal small bowel may be involved in severe cases

CT Findings

- Low density (near fat HU), large, mesenteric & retroperitoneal lymphadenopathy
  - Fatty material derived from digested Whipple bacilli
- Thickened proximal small bowel folds
  - Submucosal edema due to hypoalbuminemia
- ± Ascites, splenomegaly, pneumatosis intestinalis

Imaging Recommendations

- Enteroclysis & CECT

DIFFERENTIAL DIAGNOSIS

Opportunistic infection in AIDS

- Giardiasis
  - Duodenum & jejunum
    - Thickened irregular folds with hypermotility
    - Luminal narrowing & increased secretions
  - Ileum: Usually appears normal
- Cryptosporidiosis
  - Duodenum & proximal jejunum: Thickened folds
  - Distal small bowel shows areas of flocculation
  - CT may show small lymph nodes
- Mycobacterium avium-intracellulare (MAI) infection
  - May have low density nodes similar to Whipple
  - Wet bowel pattern may be similar, but no nodules

DDx: Abnormal Small Bowel Fold Pattern

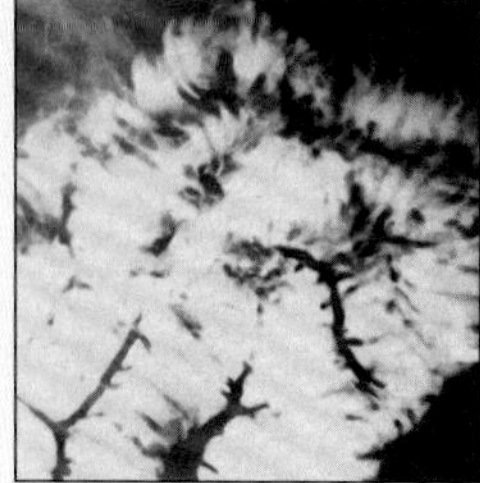

Giardiasis

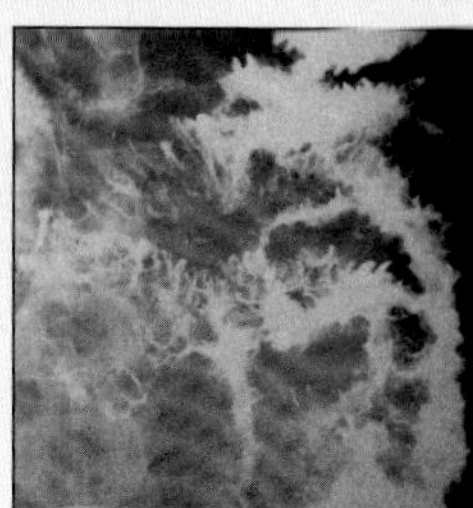

Cryptosporidiosis

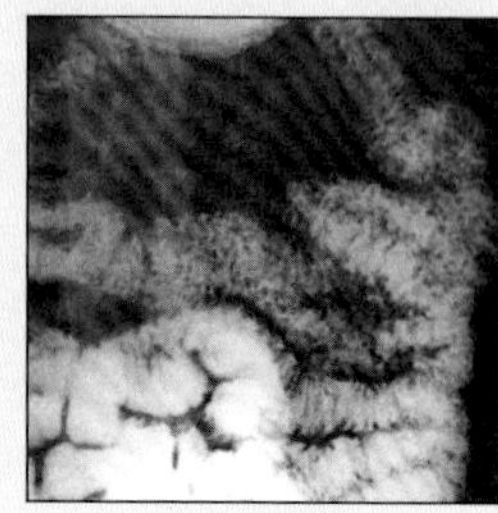

Lymphoma

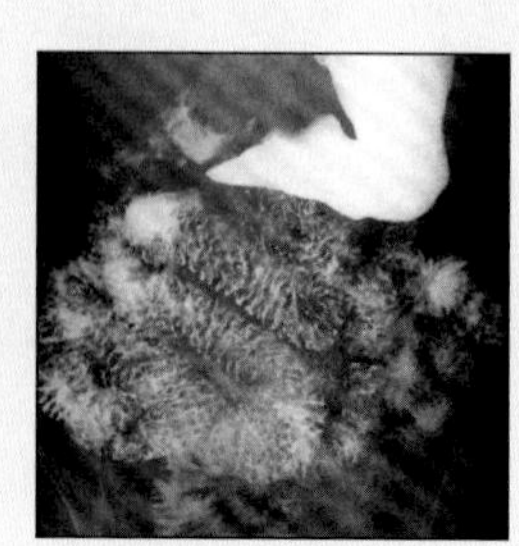

Waldenstrom

WHIPPLE DISEASE

Key Facts

Terminology
- Synonym: Intestinal lipodystrophy
- Rare, chronic, bacterial infectious & systemic disease leading to chronic diarrhea & malabsorption

Imaging Findings
- Best diagnostic clue: Thickened proximal small bowel folds with micronodularity (1-2 mm) & low density mesenteric adenopathy (near fat HU)

Top Differential Diagnoses
- Opportunistic infection in AIDS
- Dysgammaglobulinemia
- Lymphoma (non-Hodgkin)

Pathology
- Genetics: May be associated with HLA-B27
- Etiology: Tropheryma whippelii gram positive bacilli
- Periodic acid-Schiff (PAS) stain positive

Dysgammaglobulinemia
- E.g., Waldenstrom macroglobulinemia
- B-cell neoplasm of older adults with ↑ serum IgM spike
- ↑ Monoclonal IgM protein → hyperviscosity syndrome
- Lymphadenopathy, hepatosplenomegaly, anemia
- Barium studies: Micronodular mucosal pattern mimicking Whipple (massive deposition of IgM protein in lamina propria → lymphatic obstruction)

Lymphoma (non-Hodgkin)
- Nodular, polypoid, infiltrating, mesenteric invasive
- Focally infiltrating form of terminal ileum

PATHOLOGY

General Features
- Genetics: May be associated with HLA-B27
- Etiology: Tropheryma whippelii gram positive bacilli
- Epidemiology: One of the causes of malabsorption in US and Europe
- Associated abnormalities: Pleuropericarditis, sacroiliitis, CNS abnormalities

Gross Pathologic & Surgical Features
- Duodenum & proximal jejunum
  - Marked thickening of intestinal villi
  - Thickened, irregular mucosal folds

Microscopic Features
- Periodic acid-Schiff (PAS) stain positive
  - Villi distended with macrophages full of bacilli in lamina propria & mesenteric lymph nodes

CLINICAL ISSUES

Presentation
- Most common signs/symptoms
  - Diarrhea, fever, steatorrhea, adenopathy
  - Usually present with malabsorption syndrome
  - Other symptoms & signs
    - Arthralgia, increased skin pigmentation, anemia
    - Pleuritis, pericarditis or CNS symptoms
- Diagnosis: PAS positive from small bowel biopsy

Demographics
- Age: 40-49 years old (usually middle-aged Caucasian men)
- Gender: M:F = 10:1

Natural History & Prognosis
- Whipple disease is fatal without therapy
- Relapses may occur, even after long term therapy
- CNS relapse is resistant to antibiotic therapy

Treatment
- Antibiotic therapy for 1 or more years (often curative)
  - Parental penicillin + streptomycin initially
  - Followed by 1 year oral treatment with Bactrim
    - Trimethoprim/sulfamethoxazole

DIAGNOSTIC CHECKLIST

Consider
- Rule out other causes of malabsorption syndrome

Image Interpretation Pearls
- Enteroclysis: Diffuse or patchy micronodules (1-2 mm in diameter), predominantly in proximal small bowel
- CT: Mesenteric & retroperitoneal nodes of near fat HU

SELECTED REFERENCES

1. Herlinger H et al: Whipple disease malabsorption states. In Clinical imaging of the small intestine. Eds Herlinger H et al. 2nd ed. p 357-9, 1999
2. Horton KM et al: Uncommon inflammatory diseases of the small bowel: CT findings. AJR Am J Roentgenol. 170(2): 385-8, 1998
3. Rubesin SE et al: Small bowel malabsorption: clinical and radiologic perspectives. How we see it. Radiology. 184(2):297-305, 1992

IMAGE GALLERY

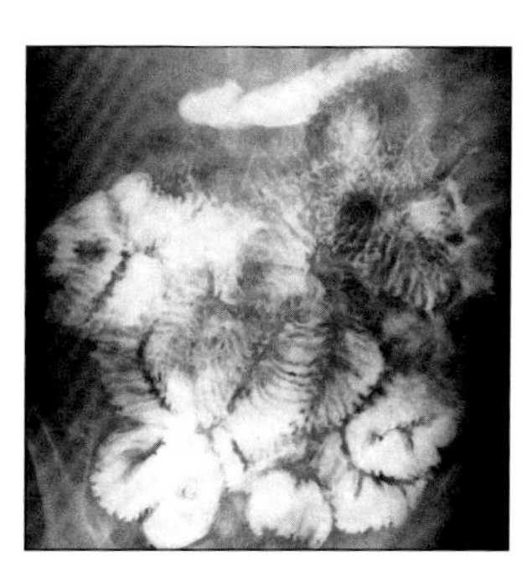

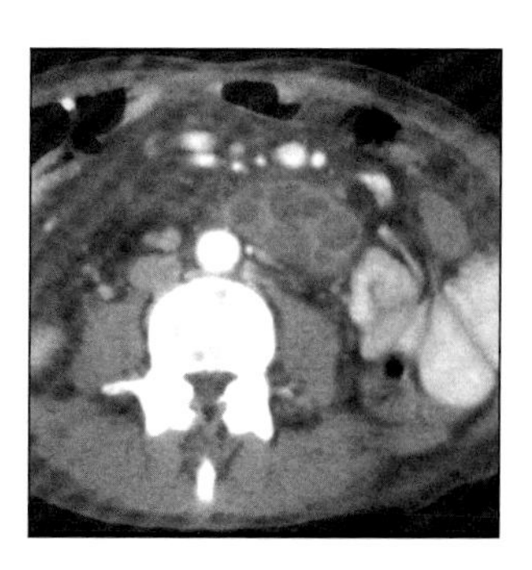

*(**Left**) SBFT shows micronodular fold thickening of most of the SB, but no excess fluid. (**Right**) Axial CECT shows very low density nodes in the retroperitoneum and mesentery.*

MASTOCYTOSIS

Axial CECT shows hepatomegaly, biliary ductal dilation, thickened bowel wall and fluid distention of lumen.

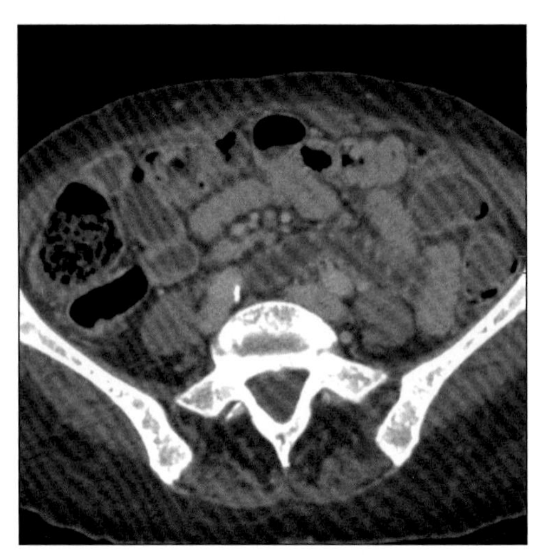

Axial CECT shows sclerotic lesions in pelvic bones due to mastocytosis.

TERMINOLOGY

Abbreviations and Synonyms

- Systemic mast cell disease (SMCD)

Definitions

- Rare disorder characterized by mast cell proliferation in skin, bones, lymph nodes & parenchymal organs

IMAGING FINDINGS

General Features

- Best diagnostic clue: Multiple peptic ulcers, thickened bowel, mucosal nodularity & "bull's-eye" lesions
- Location: GIT: Stomach, duodenum & small bowel
- Other general features
  - Usually more transient & self-limited in children compared to adults
  - Systemic disease with mast cell proliferation
    - Most commonly involves skin
    - Urticaria pigmentosa (UP-most common form)
  - Gastrointestinal tract is involved in 16% of cases
    - Peptic ulcer disease & malabsorption
  - SMCD can present as isolated hematologic disease or accompanied by other hematologic malignancies
    - Myelodysplastic syndrome or acute leukemia

Radiographic Findings

- Fluoroscopic guided barium studies
  - Stomach
    - Multiple ulcers: Round or ovoid collections of barium & folds radiating to edge of ulcer crater
  - Duodenum & small bowel
    - Multiple duodenal ulcers; ↓ barium transit time
    - Mucosal nodularity (2-3 mm sandlike nodules)
    - Bowel wall: Diffuse thickening
    - Folds: Thickened, irregular, distorted
    - "Bull's-eye" lesions; dilated bowel loops
  - Colon: Diffuse inflammatory changes
- Central skeleton: Mixed sclerotic & lytic lesions

CT Findings

- Thickened bowel wall & mucosal folds
- Dilated fluid-filled small bowel loops
- Lymphadenopathy; thickened omentum & mesentery
- Hepatomegaly (seen in 40% of adult cases)
- Splenomegaly with nodular deposits (50% of cases)
- ± Esophageal varices, ascites (due to portal HTN)
- ± Budd-Chiari hepatic veno-occlusive disease

Nuclear Medicine Findings

- Bone scan: Lytic lesions, osteoporosis or osteosclerosis

Imaging Recommendations

- Best imaging tool: Double contrast barium studies

DDx: SB Fold Thickening + Fluid Distention

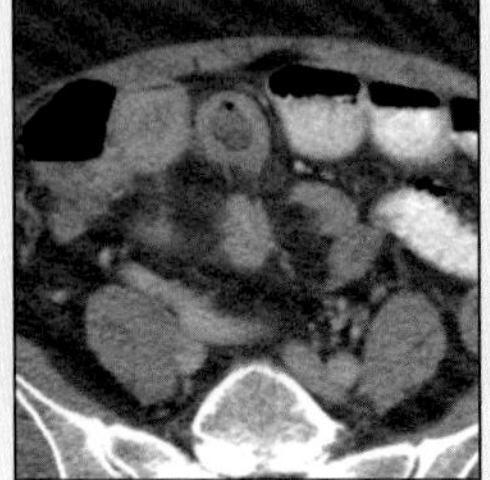

Carcinoid

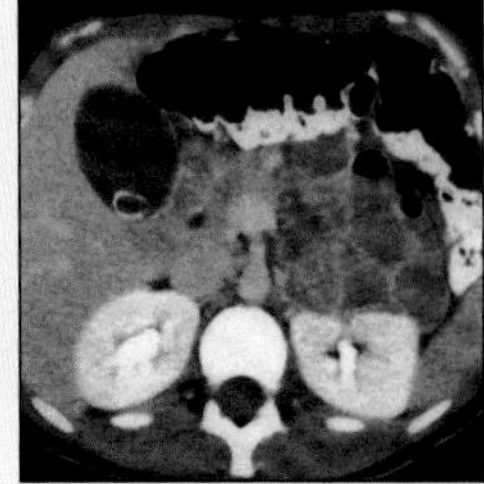

Zollinger-Ellison

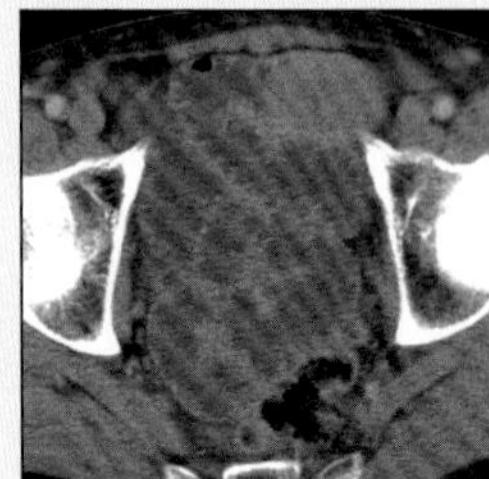

Sprue

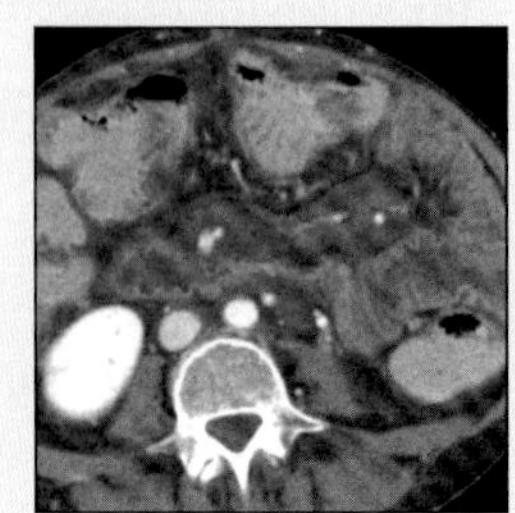

Lymphangiectasia

MASTOCYTOSIS

Key Facts

Terminology
- Systemic mast cell disease (SMCD)

Imaging Findings
- Best diagnostic clue: Multiple peptic ulcers, thickened bowel, mucosal nodularity & "bull's-eye" lesions
- Central skeleton: Mixed sclerotic & lytic lesions
- Lymphadenopathy; thickened omentum & mesentery
- Hepatomegaly (seen in 40% of adult cases)
- Splenomegaly with nodular deposits (50% of cases)

Top Differential Diagnoses
- Carcinoid syndrome
- Zollinger-Ellison syndrome (ZES)
- Celiac sprue (nontropical)

Diagnostic Checklist
- Check for skin & bony lesions, hematologic disorders
- Correlate: Clinical, biochemical & imaging findings

DIFFERENTIAL DIAGNOSIS

Carcinoid syndrome
- Flushing, diarrhea, etc., indicated hepatic metastases
- Urticaria pigmentosa seen in SMCD, but not carcinoid
- Lab-data: Serotonin levels are increased in carcinoid

Zollinger-Ellison syndrome (ZES)
- Gastrinoma (pancreatic tumor) + peptic ulcers
- Lab-data: Elevated gastrin levels in ZES

Celiac sprue (nontropical)
- Malabsorption due to intolerance to gluten protein
- Reversed jejunoileal folds (↓ jejunal & ↑ ileal folds)
- Lab-data: Increased IgA & IgM antigliadin antibodies

Malabsorption conditions
- E.g., Whipple disease, lymphangiectasia

PATHOLOGY

General Features
- Genetics: Mutations in c-kit proto-oncogene (Codon-816 c-kit)
- Etiology
  - Hyperplastic response to an unknown stimulus rather than a neoplastic condition in some cases
  - Pathogenesis: Histamine, prostaglandins, proteases
- Epidemiology: Extremely rare disorder
- Associated abnormalities: Urticaria pigmentosa, bony lesions, hematopoietic disorders

Gross Pathologic & Surgical Features
- Ulcers, thickened wall & folds, mucosal lesions

Microscopic Features
- Cells: Spindle-shaped nucleus, eosinophilic granules

CLINICAL ISSUES

Presentation
- Most common signs/symptoms
  - Malabsorption: Pain, diarrhea, nausea, vomiting
  - Skin: Macules, papules, nodules & plaques
  - Bone pain, hepatomegaly & splenomegaly
  - Lab-data
    - CBC: Anemia, leukopenia, thrombocytopenia
    - Tryptase > 20 ng/mL; histamine twice of normal

Demographics
- Age: Up to childhood (75%); adults 30-49 years (25%)
- Gender: Equal in both males & females
- Ethnicity: Caucasians > African-Americans

Natural History & Prognosis
- Complications
  - GE reflux; hemorrhage & perforation; chloroma
  - Hematologic malignant transformation rate is 30%

Treatment
- Epinephrine, steroids, H1 + H2 antagonists, proton pump inhibitors, anticholinergics, chemotherapy
- Symptomatic therapy; no curative therapy exists

DIAGNOSTIC CHECKLIST

Consider
- Check for skin & bony lesions, hematologic disorders
- Correlate: Clinical, biochemical & imaging findings

SELECTED REFERENCES

1. Avila NA et al: Systemic mastocytosis: CT and US features of abdominal manifestations. Radiology. 202(2):367-72, 1997
2. Metcalfe DD: The liver, spleen, and lymph nodes in mastocytosis. J Invest Dermatol. 96(3):45S-46S, 1991
3. Johnson AC et al: Systemic mastocytosis and mastocytosis-like syndrome: radiologic features of gastrointestinal manifestations. South Med J. 81(6):729-33, 750, 1988

IMAGE GALLERY

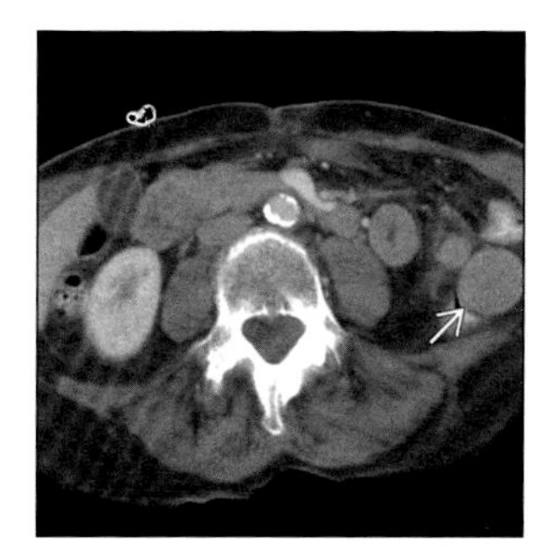

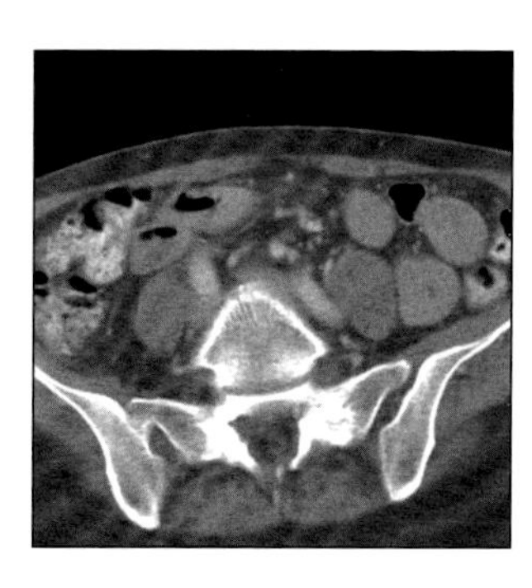

*(**Left**) Axial CECT shows thickened bowel wall and intussusception (arrow) due to mastocytosis. (**Right**) Axial CECT shows thickened bowel wall and sclerotic bone lesions.*

CROHN DISEASE

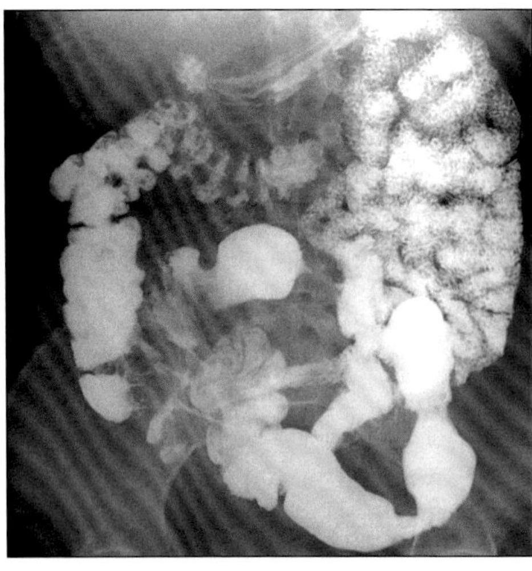

Small bowel follow through (SBFT) shows severe SB strictures, fistulas and ulceration with skip areas.

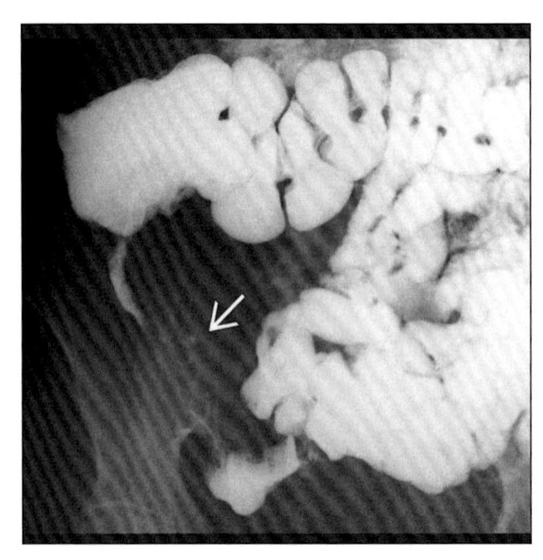

SBFT shows "string sign", severe luminal narrowing of terminal ileum and ascending colon. Also sinus tract (arrow) + indirect evidence of mesenteric fibrofatty proliferation.

TERMINOLOGY

Abbreviations and Synonyms

- Terminal ileitis, regional enteritis, ileocolitis

Definitions

- Chronic, recurrent, segmental, granulomatous inflammatory bowel disease

IMAGING FINDINGS

General Features

- Best diagnostic clue: Segmental areas of ileo-colonic ulceration + wall thickening on barium study
- Location
  - Anywhere along gut from mouth to anus
    - Most common: Terminal ileum & proximal colon
  - Distribution
    - Terminal ileum (95%); colon (22-55%)
    - Rectum (14-50%)
- Morphology
  - Characterized by
    - Skip lesions (segmental or discontinuous)
    - Transmural, granulomas (noncaseating type)
    - Cobblestone mucosa, fissures & fistulas
    - "String sign" on barium enema
- Other general features
  - Idiopathic inflammatory bowel disease
  - There is a familial disposition
  - Disease with prolonged & unpredictable course
  - Risk factors
    - Caucasian race, Jewish (8-fold ↑)
    - Urban, family history & smoking (4-fold ↑)
  - Crohn disease is less common than ulcerative colitis

Radiographic Findings

- Fluoroscopic guided barium study
  - Early changes
    - Lymphoid hyperplasia: 1-3 mm mucosal elevations; no ring shadow
    - Aphthoid ulcerations: "Target" or "bull's eye" appearance (punctate shallow central barium collections surrounded by a halo of edema)
    - Cobblestone pattern: Combination of longitudinal & transverse ulcers
    - Deep ulcerations (fissuring ulcers)
    - Mural thickening: Transmural inflammation, fibrosis (Crohn more than ulcerative colitis)
  - Late changes
    - Skip lesions: Segmental/normal intervening areas
    - Sacculations: Seen on antimesenteric border (↑ luminal pressure)
    - Postinflammatory pseudopolyps, loss of haustra, intramural abscess
    - "String sign": Luminal narrowing + ileal stricture
    - Sinus tracts, fissures, fistulas: Hallmark of disease

DDx: Thickened Wall, Narrowed Lumen at Ileocecal Junction

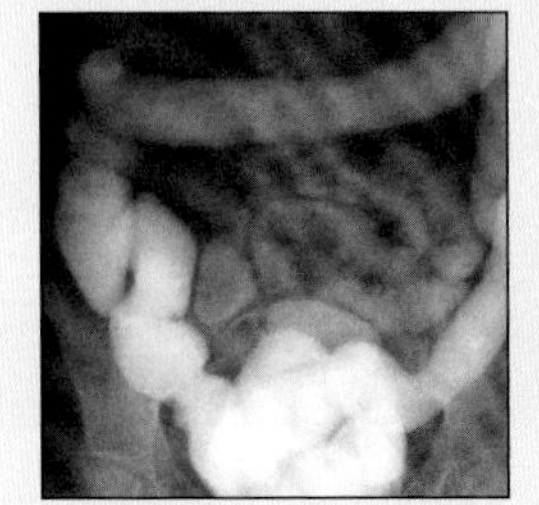

Ulcerative Colitis

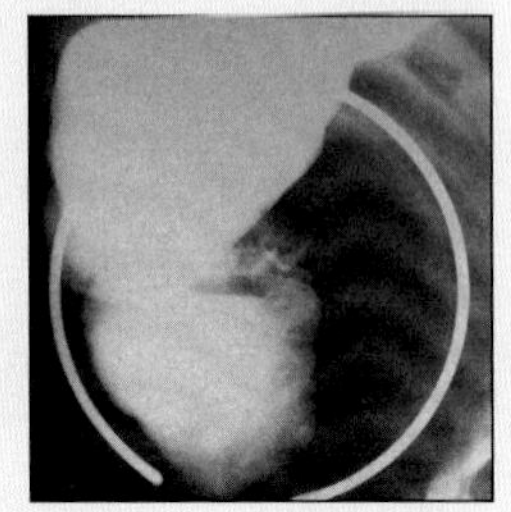

Yersinia

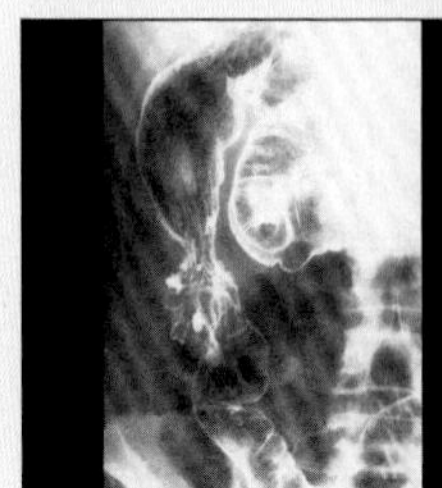

Tuberculosis

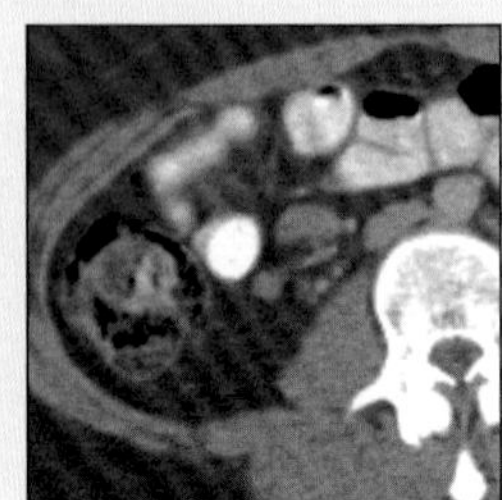

Mesenteric Adenitis

CROHN DISEASE

Key Facts

Terminology
- Terminal ileitis, regional enteritis, ileocolitis
- Chronic, recurrent, segmental, granulomatous inflammatory bowel disease

Imaging Findings
- Best diagnostic clue: Segmental areas of ileo-colonic ulceration + wall thickening on barium study
- Cobblestone pattern: Combination of longitudinal & transverse ulcers
- Deep ulcerations (fissuring ulcers)
- Mural thickening: Transmural inflammation, fibrosis (Crohn more than ulcerative colitis)
- Skip lesions: Segmental/normal intervening areas
- "String sign": Luminal narrowing + ileal stricture
- Sinus tracts, fissures, fistulas: Hallmark of disease

Top Differential Diagnoses
- Ulcerative colitis ("backwash" ileitis)
- Infection
- Ischemia
- Radiation enteritis
- Metastases & lymphoma
- Mesenteric adenitis

Pathology
- Exact etiology unknown
- Genetic, environmental, infectious & psychologic
- Immunologic: Antibody & cell-mediated types

Diagnostic Checklist
- Check for associated findings (cholangitis, arthritis)
- CT: Small-bowel wall thickening, mesenteric fat proliferation & hyperemia very suggestive of Crohn

    - Anorectal lesions: Ulcers, fissures, abscess, hemorrhoids, stenosis

CT Findings
- Discontinuous & asymmetric bowel wall thickening (more than 1 cm)
- Acute or noncicatrizing phase: Minimal narrowing
    - Mural stratification: Intact
        - Distinct mucosa, submucosa, muscularis propria
    - Inner ring: Soft tissue density (mucosa)
    - Middle ring: Low density (submucosal edema/fat)
    - Outer ring: Soft tissue density (muscularis propria-serosa)
    - Proliferation of mesenteric fat ± lymphadenopathy
    - "Target" or "double halo" sign on CECT
        - Intense enhancement: Inner mucosa + outer muscularis propria
        - ↓ Attenuation: Edematous thickened submucosa
- Chronic or cicatrizing phase: ↑ Luminal narrowing + no "target" sign
    - Mural stratification lost (indistinct mucosa, submucosa, muscularis propria)
    - Homogeneous attenuation of thickened bowel wall on CECT (indicating irreversible transmural fibrosis)
    - Abscesses, fistulas, sinus tracts
    - Mesenteric changes: Abscess, fibrofatty areas, nodes
    - Perianal disease, enlarged mesenteric lymph nodes
    - "Comb" sign: Mesenteric hypervascularity (dilatation, tortuosity & wide spacing)
        - Indicates active disease

MR Findings
- Breath-holding (FLASH), fat suppression & Gd-DTPA
    - Show extent, mural thickening & severity of disease
    - Perianal Crohn disease
        - MR sensitive in detecting fistulas/sinuses/abscesses

Ultrasonographic Findings
- Real Time
    - Transrectal sonography
        - Mural thickening, abscesses, fistulas
        - Anal sphincter heterogeneity

Imaging Recommendations
- Barium enema, enteroclysis
- Helical NE + CECT
- MR for perianal & rectal Crohn disease

DIFFERENTIAL DIAGNOSIS

Ulcerative colitis ("backwash" ileitis)
- 25% of UC cases have terminal ileal pathology
- Widely patent ileocecal valve, slightly thickened folds
- Mucosa may have a nodular or granular pattern
- No strictures/ulcerations; adjacent colonic lesions seen
- Usually lesions of UC: Continuous, non-transmural, pseudopolyps, ↑ risk of colon cancer

Infection
- Yersiniosis: Yersinia enterocolitica (gram negative rod)
    - Common location: Terminal ileum
    - Thickened mucosal folds, nodules, aphthous ulcers
    - Lumen narrowing: Rare; Crohn disease (common)
    - Ulceration: Superficial; Crohn disease (deep)
    - Resolution: Over 6-8 weeks; Crohn (almost never)
- Particularly seen in AIDS patients
    - Example: Mycobacterial, cryptosporidiosis, CMV
- Typical: Mycobacterium tuberculosis
    - Ileocecal (> common): Transmural, stenosis, fistulas
    - Horizontal ulcers, nodular mucosal thickening
    - Cecal contraction with widely patent ileocecal valve
    - CT shows pericecal lymphadenopathy
- Atypical: Mycobacterium avium-intracellulare (MAI)
    - Small-bowel: Most common site of infection
    - Diffusely thickened folds, micronodular mucosa
    - CT: Mesenteric adenopathy & abscess
- Cryptosporidiosis
    - Most common cause of enteritis in AIDS patients
    - Thickening of folds & bowel wall; ↑ fluid in lumen
    - CT: May show small lymph nodes
    - Diagnosis: Oocysts in stool & mucosal biopsy
- Cytomegalovirus (CMV)
    - Terminal ileitis indistinguishable from Crohn
    - Diagnosis: Round intranuclear inclusion bodies

Ischemia

- Due to vascular insufficiency
- Superior mesenteric artery (SMA) clot or narrowing
  - Pneumatosis: "Bubble" or "band-like" air in affected small-bowel wall + segmental thickening > 3 mm
  - ± Gas in mesenteric or portal vein

Radiation enteritis

- Cause: Therapeutic or excessive abdominal irradiation
- Location: Terminal ileum + adjacent colon & rectum
- Thickened bowel wall, narrow pelvic bowel loops
- ± Strictures, sinuses, fistulas simulating Crohn disease

Metastases & lymphoma

- Metastases (small-bowel)
  - E.g., from malignant melanoma, lung/breast cancer
  - Location: Antimesenteric border (due to rich vascular submucosal plexus)
  - Malignant melanoma
    - Smoothly polypoid lesions of different sizes
    - Polypoid lesion with ulcers & radiating folds form a typical "spoke-wheel" pattern
  - Bronchogenic carcinoma
    - Single/multiple intramural lesions (flat/polypoid)
    - Frequently ulcerated, narrowing & obstruction
  - Breast carcinoma
    - Highly cellular submucosal masses
- Non-Hodgkin lymphoma: More common
  - Distribution: Stomach (51%), small-bowel (33%)
  - Nodular, polypoid, infiltrating, mesenteric invasive
  - Focally infiltrating form of terminal ileum
    - Widened segment devoid of folds
    - Sausage-shaped thickening of affected bowel wall
    - May be indistinguishable from Crohn disease

Mesenteric adenitis

- Common cause of RLQ pain in children & adolescents
- Enlarged mesenteric nodes; ileal wall thickening
- Usually resolves spontaneously within 2-4 days

PATHOLOGY

General Features

- General path comments
  - Classified into three stages based on pathology
    - Early stage: Hyperplasia of lymphoid tissue & obstructive lymphedema in submucosa → shallow mucosal erosions (aphthoid ulcers)
    - Intermediate stage: Transmural extension in mucosa & submucosa → marked fold thickening
    - Advanced stage: Transmural extension to serosa & beyond → deep linear clefts of ulceration/fissures
- Genetics
  - Common in monozygotic twins & siblings
  - Polygenic inheritance pattern
- Etiology
  - Exact etiology unknown
  - Possible factors considered
    - Genetic, environmental, infectious & psychologic
    - Immunologic: Antibody & cell-mediated types
    - Nutritional, hormonal, vascular & traumatic
- Epidemiology: Incidence, 0.6-6.3 cases/100,000 people
- Associated abnormalities
  - Arthritis, gallstones, sclerosing cholangitis, uveitis
  - Ankylosing spondylitis

Gross Pathologic & Surgical Features

- Skip lesions more common in distal ileum
- Edema, inflammation, fibrosis, luminal narrowing
- Adhesions, fistulae, fissures, strictures

Microscopic Features

- Transmural inflammation, lymphoid aggregates, noncaseating granulomas

CLINICAL ISSUES

Presentation

- Most common signs/symptoms
  - Diarrhea, pain, melena, weight loss, fever
  - Malabsorption; fissures & fistulas (perianal area)
- Diagnosis: Mucosal biopsy

Demographics

- Age: Age: 15-25 years (small peak at 50-80 y)
- Gender: M = F
- Ethnicity: More in Caucasians & Jews

Natural History & Prognosis

- Complications
  - Fistula, sinus, toxic megacolon
  - Obstruction, perforation, malignancy
- Prognosis
  - 10-20% lead symptom free lives
  - Recurrence: 30-53% after surgical resection
    - Usually proximal side of anastomosis

Treatment

- Medical
  - Steroids, azathioprine, mesalamine
  - Metronidazole, antibody treatment
- Surgical
  - Resection of diseased bowel
  - Strictureplasty, primary fistulotomy

DIAGNOSTIC CHECKLIST

Consider

- Check for associated findings (cholangitis, arthritis)

Image Interpretation Pearls

- CT: Small-bowel wall thickening, mesenteric fat proliferation & hyperemia very suggestive of Crohn

SELECTED REFERENCES

1. Wold PB et al: Assessment of small bowel Crohn disease: noninvasive peroral CT enterography compared with other imaging methods and endoscopy--feasibility study. Radiology. 229(1):275-81, 2003
2. Antes G: Inflammatory disease of the small intestine and colon: Contrast enema and CT. Radiology. 38: 41-5, 1998
3. Gore RM et al: CT features of ulcerative colitis and Crohn's disease. AJR. 167: 3-15, 1996
4. Hizawa K et al: Crohn disease: early recognition and progress of aphthous lesions. Radiology. 190: 451-4, 1994

CROHN DISEASE

IMAGE GALLERY

Typical

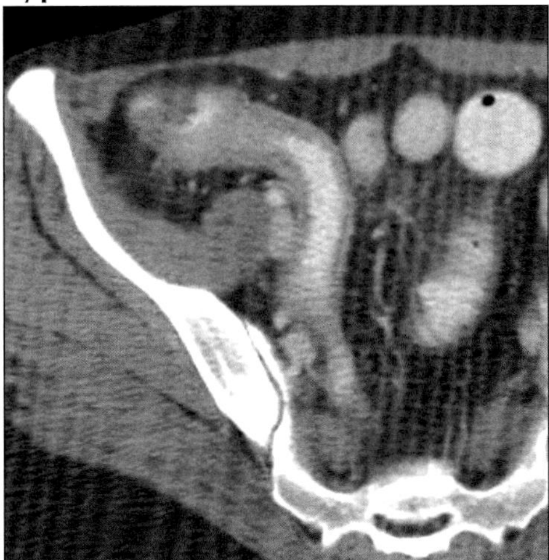

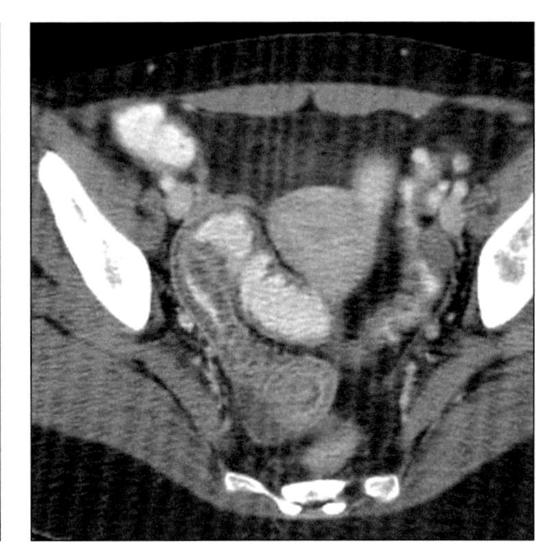

(Left) *Axial CECT shows mural thickening of ileum with submucosal edema + mesenteric hypervascularity ("comb sign") indicating active disease.* ***(Right)*** *Axial CECT shows pelvic SB segment with submucosal edema + "comb sign" of active inflammation.*

Typical

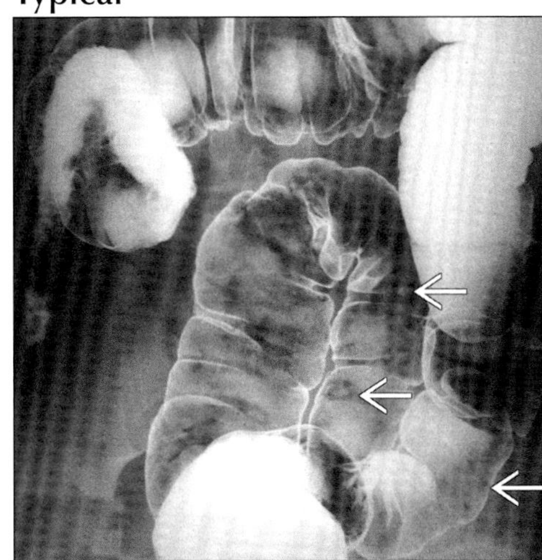

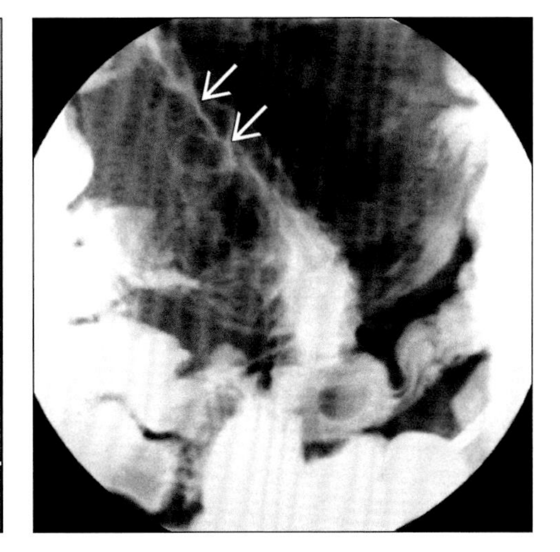

(Left) *Air-contrast BE shows Crohn (granulomatous) colitis with multiple aphthous ulcerations (arrows) throughout the colon.* ***(Right)*** *SBFT shows "cobblestone" appearance of terminal ileum, due to longitudinal, transverse ulcerations (arrows).*

Typical

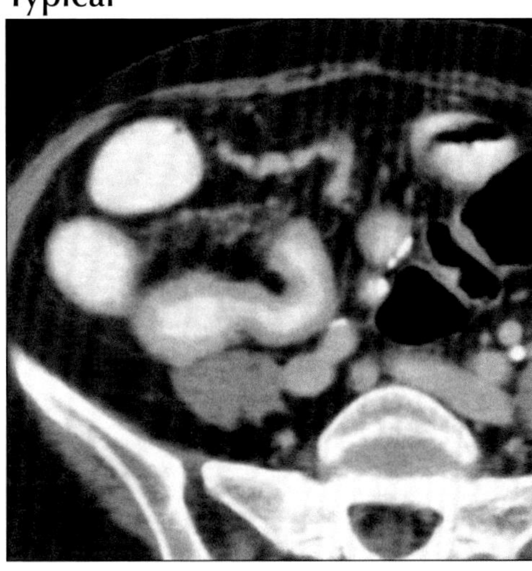

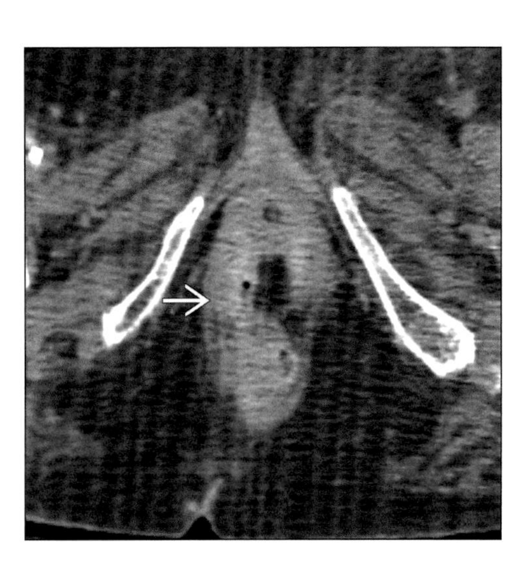

(Left) *Axial CECT shows recurrent Crohn disease several years following resection. Mural thickening and mesenteric fibrofatty proliferation.* ***(Right)*** *Axial CECT shows recto-vaginal fistula (arrow) due to Crohn disease.*

SCLERODERMA, INTESTINAL

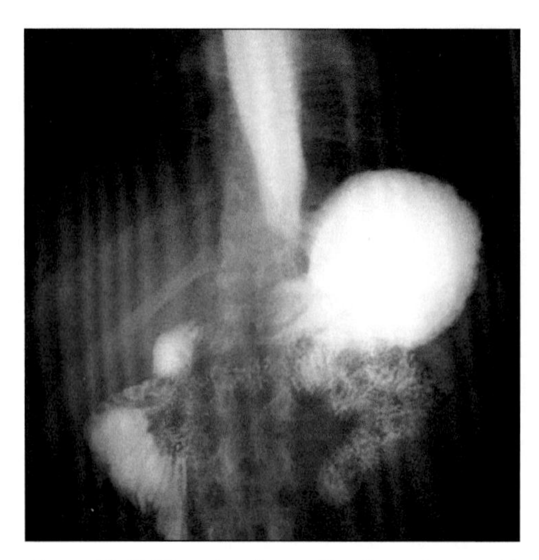

Upper GI (UGI) series shows dilated esophagus with patulous GE junction and dilated duodenum with abrupt "cut off" as it crosses the midline.

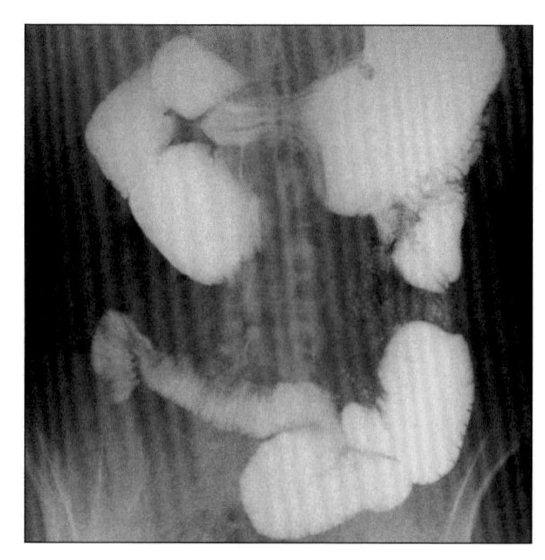

UGI series + small bowel follow through (SBFT) shows dilated duodenum and jejunum with thin, closely spaced valvulae; the "hidebound" pattern.

TERMINOLOGY

Abbreviations and Synonyms

- Progressive systemic sclerosis (PSS)

Definitions

- Multisystemic disorder of small vessels & connective tissue (collagen vascular disease) of unknown etiology

IMAGING FINDINGS

General Features

- Best diagnostic clue: Dilated atonic small bowel with crowded folds & wide-mouthed sacculations
- Other general features
  - Multisystemic disorder with immunologic & inflammatory changes
  - Characterized by atrophy, fibrosis, sclerosis of skin, vessels & organs
  - Involves skin, synovium, & parenchyma of multiple organs
    - GI tract, lungs, heart, kidneys & nervous system
  - Gastrointestinal tract (GIT) scleroderma
    - 3rd most common manifestation after skin changes & Raynaud phenomenon
    - Seen in up to 90% of patients
    - Most common sites: Esophagus > anorectal > small bowel > colon
    - Most frequent cause of chronic intestinal pseudo-obstruction
  - Scleroderma is subclassified into two types
    - Diffuse scleroderma
    - CREST syndrome (benign course)
  - Diffuse scleroderma: Diffuse cutaneous + early visceral involvement
    - Severe interstitial pulmonary fibrosis
    - Tends to involve women; organ failure more likely
    - Associated with antitopoisomerase 1 antibody (anti-Scl 70)
  - CREST syndrome: Minimal cutaneous & late visceral involvement
    - Calcinosis of skin
    - Raynaud phenomenon
    - Esophageal dysmotility
    - Sclerodactyly (involvement of fingers)
    - Telangiectasia
    - Associated with anticentromere antibodies

Radiographic Findings

- Fluoroscopic guided (esophagography)
  - Normal peristalsis above aortic arch (due to striated muscle in proximal one-third)
  - Hypotonia, atony, aperistalsis: Lower two-thirds esophagus (smooth muscle)
  - Mild-moderate dilatation of esophagus

DDx: Dilated Small Intestine +/- Abnormal Fold Pattern

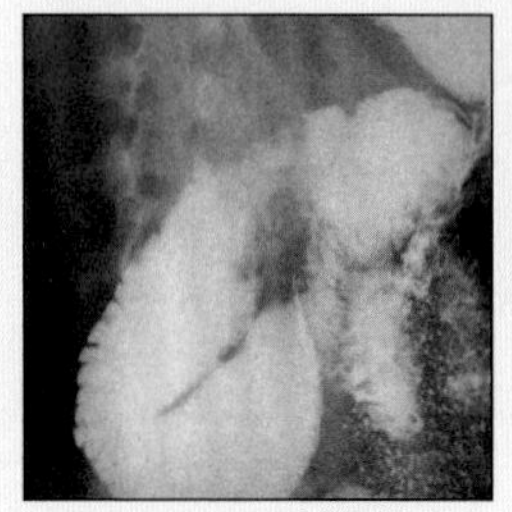

SMA Syndrome

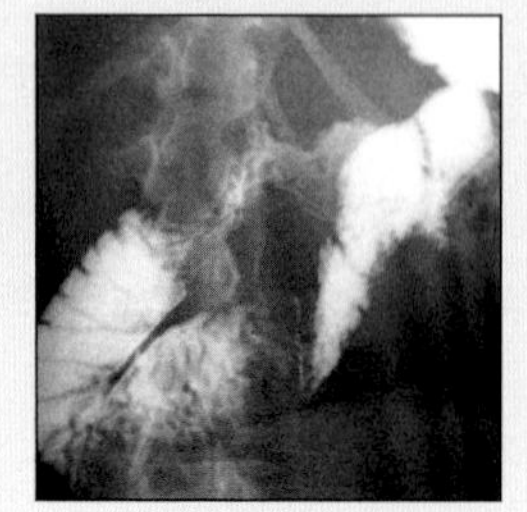

SMA Syndrome

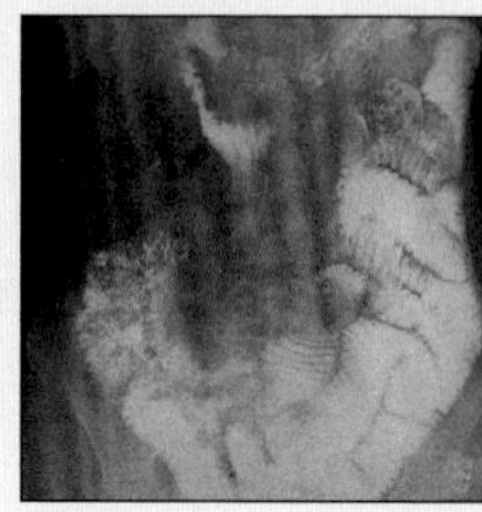

Celiac Sprue Disease

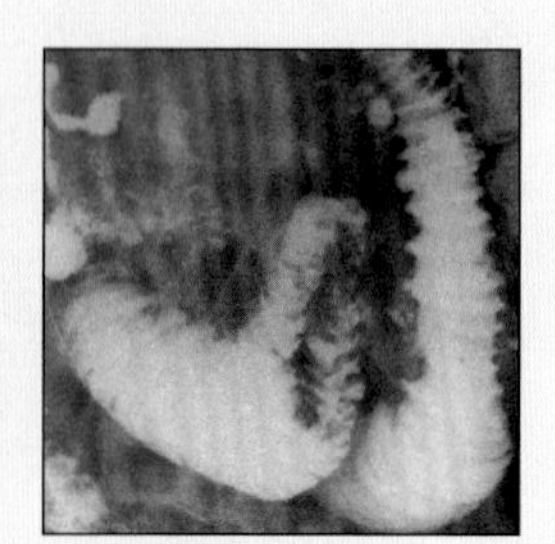

Hemorrhage

SCLERODERMA, INTESTINAL

Key Facts

Terminology

- Progressive systemic sclerosis (PSS)
- Multisystemic disorder of small vessels & connective tissue (collagen vascular disease) of unknown etiology

Imaging Findings

- Best diagnostic clue: Dilated atonic small bowel with crowded folds & wide-mouthed sacculations
- CREST syndrome (benign course)
- Calcinosis of skin
- Raynaud phenomenon
- Esophageal dysmotility
- Sclerodactyly (involvement of fingers)
- Telangiectasia
- Marked dilatation of small bowel (particularly 2nd, 3rd parts of duodenum & jejunum)
- Dilated small bowel (jejunum) with crowded thin circular folds ("hidebound" sign)

Top Differential Diagnoses

- Superior mesenteric root syndrome or SMA
- Celiac sprue
- Small bowel ileus

Pathology

- Initially smooth muscle atrophy & fragmentation
- Followed by collagen deposition & fibrosis
- Unknown; autoimmune with genetic predisposition

Diagnostic Checklist

- Rule out other pathologies that cause dilatation of small bowel ± abnormal fold pattern
- Check for family history of collagen vascular diseases

    - Patulous lower esophageal sphincter (LES): Early finding
    - Erosions, superficial ulcers, fusiform peptic stricture (reflux esophagitis)
    - Gastroesophageal reflux (70%) → (37%) develop Barrett esophagus or peptic stricture
        - Distal stricture + dilated nonperistaltic esophagus can simulate achalasia
    - Hiatal hernia
- Fluoroscopic guided (barium meal)
    - Stomach: Gastric dilatation + delayed emptying
- Fluoroscopic guided (small bowel study)
    - Marked dilatation of small bowel (particularly 2nd, 3rd parts of duodenum & jejunum)
    - Pathognomonic: "Hidebound" sign
        - Dilated small bowel (jejunum) with crowded thin circular folds ("hidebound" sign)
        - Seen in > 60% cases of scleroderma-related pseudo-obstruction
        - Due to muscle atrophy & its uneven replacement by collagen in longitudinal fibers + intense fibrosis of submucosa
    - Wide-mouthed sacculations (true diverticula) on antimesenteric border
    - Prolonged transit time with barium retention in duodenum up to 24 hours
    - ± Pneumatosis intestinalis + pneumoperitoneum
    - ± Transient, nonobstructive intussusceptions
- Fluoroscopic guided (barium enema)
    - Sacculations on antimesenteric border (transverse & descending colon)
    - Marked dilatation (may simulate Hirschsprung disease)
    - Chronic phase: Complete loss of haustrations
        - Simulates cathartic colon or chronic ulcerative colitis
    - Stercoral ulceration (from retained fecal material)
    - ± Benign pneumoperitoneum

Imaging Recommendations

- Fluoroscopic guided single & double contrast barium studies
    - Upper GI & barium enema

DIFFERENTIAL DIAGNOSIS

Superior mesenteric root syndrome or SMA

- Superior mesenteric artery (SMA)
- Narrowing of angle between SMA & aorta may compress 3rd part of duodenum causing proximal dilatation
    - Usually seen in asthenic persons who rapidly lose weight & retroperitoneal fat
    - Due to prolonged bed rest or immobilization
        - Example: Patients with whole body burns; body casts; spinal injury or surgery
- Barium study findings
    - Marked dilatation of 1st & 2nd parts of duodenum
    - Vertical, linear, extrinsic, bandlike defect in transverse part of duodenum overlying spine
    - Occasionally, obstruction is partially relieved in prone position
- CT may demonstrate beak-like compression of 3rd part of duodenum between SMA & aorta
- Transient delay of barium in 3rd part of duodenum seen in normal person may also mimic SMA syndrome

Celiac sprue

- Segmental dilatation of small intestine
- Excess fluid in lumen (flocculation of barium)
- Atrophic duodenal + jejunal folds, relative hypertrophy of ileal folds
- Transient intussusceptions

Small bowel ileus

- Markedly dilated proximal duodenum (atony) due to acute upper abdominal inflammatory process
    - Example: Acute pancreatitis, cholecystitis, peptic ulcer
    - Small bowel folds in ileus usually normal to thick, rather than thin + crowded in scleroderma

Intramural hemorrhage

- Bleeding into the submucosa
- Etiology: Trauma, anticoagulation
- Thick valvulae with ↓ space between folds
- "Stack of coins" pattern

SCLERODERMA, INTESTINAL

PATHOLOGY

General Features

- General path comments
  - Initially smooth muscle atrophy & fragmentation
  - Followed by collagen deposition & fibrosis
- Genetics
  - Diffuse: Antitopoisomerase 1 antibodies associated with HLA-DR5
  - Localized: Anticentromere antibodies associated with HLA-DR 1, 4 & 5
- Etiology
  - Unknown; autoimmune with genetic predisposition
  - May be initiated by environmental antigens like silica & L-tryptophan
  - Immunologic mechanism (delayed hypersensitivity reaction)
    - ↑ Production of cytokines (TNF-α or IL-1) → ↑ collagen production
    - Vascular damage & activation of fibroblasts
- Epidemiology
  - Incidence: 14.1 cases/million
  - Prevalence: 19-75/100,000 persons
- Associated abnormalities: May be associated with systemic lupus erythematosus (SLE); polymyositis; dermatomyositis

Gross Pathologic & Surgical Features

- Rubber-hose inflexibility: Lower 2/3 esophagus
- Thin & ulcerated mucosa
- Dilated gas & fluid containing small-bowel loops with sacculations

Microscopic Features

- Perivascular lymphocytic infiltrates
- Early capillary & arteriolar injury
- Atrophy + fragmentation of smooth muscle → collagen deposition + fibrosis

CLINICAL ISSUES

Presentation

- Most common signs/symptoms
  - Esophagus
    - Dysphagia, regurgitation
    - Epigastric fullness & burning pain
  - Small-bowel
    - Bloating, abdominal pain
    - Weight loss, diarrhea, anemia
  - Colon
    - Chronic constipation & episodes of bowel obstruction
- Lab-data
  - Increased erythrocyte sedimentation rate (ESR)
  - Iron, B12 & folic acid deficiency anemias
  - Increased antinuclear antibodies (ANA)

Demographics

- Age: 30-50 years
- Gender: Young African-American females; M:F = 1:3
- Ethnicity: African-Americans more than Caucasians

Natural History & Prognosis

- Complications
  - Barrett esophagus → adenocarcinoma
  - Bowel pseudoobstruction
- Prognosis
  - Limited disease with ANA bodies: Good prognosis
  - Diffuse disease: Poor with involvement of kidneys, heart & lungs rather than GI tract

Treatment

- Small frequent meals; elevation of head of the bed
- Avoid tea & coffee
- Cimetidine, ranitidine, omeprazole
- Metoclopramide, laxatives
- Patients with severe malabsorption
  - Parenteral hyperalimentation

DIAGNOSTIC CHECKLIST

Consider

- Rule out other pathologies that cause dilatation of small bowel ± abnormal fold pattern
- Check for family history of collagen vascular diseases

Image Interpretation Pearls

- Markedly dilated atonic small bowel with thin, crowded circular folds & delayed barium transit time

SELECTED REFERENCES

1. Neef B et al: Image of the month: "hide-bound" bowel sign in scleroderma. Gastroenterology. 124(5):1179, 1567, 2003
2. Mayes MD: Scleroderma epidemiology. Rheum Dis Clin North Am. 29(2):239-54, 2003
3. Goldblatt F et al: Antibody-mediated gastrointestinal dysmotility in scleroderma. Gastroenterology. 123(4):1144-50, 2002
4. Coggins CA et al: Wide-mouthed sacculations in the esophagus: a radiographic finding in scleroderma. AJR Am J Roentgenol. 176(4):953-4, 2001
5. Weston S et al: Clinical and upper gastrointestinal motility features in systemic sclerosis and related disorders. Am J Gastroenterol. 93(7):1085-9, 1998
6. Lock G et al: Gastrointestinal manifestations of progressive systemic sclerosis. Am J Gastroenterol. 92(5):763-71, 1997
7. Ott DJ: Esophageal motility disorders. Semin Roentgenol. 29(4):321-31, 1994
8. Levine MS et al: Update on esophageal radiology. AJR. 155: 933-41, 1990
9. Rohrmann CA Jr et al: Radiologic and histologic differentiation of neuromuscular disorders of the gastrointestinal tract: visceral myopathies, visceral neuropathies, and progressive systemic sclerosis. AJR Am J Roentgenol. 143(5):933-41, 1984
10. Silver TM et al: Radiological features of mixed connective tissue disease and scleroderma-systemic lupus erythematous overlap. Radiology. 120: 269-275, 1976
11. Berk RN: The radiology corner. Scleroderma of the gastrointestinal tract. Am J Gastroenterol. 61(3):226-31, 1974
12. Horowitz AL et al: The "hide-bound" small bowel of scleroderma: Characteristic mucosal fold pattern. AJR. 119: 332-34, 1973
13. Miercort RD et al: Pneumatosis and pseudo-obstruction in scleroderma. Radiology. 92(2):359-62, 1969

IMAGE GALLERY

Typical

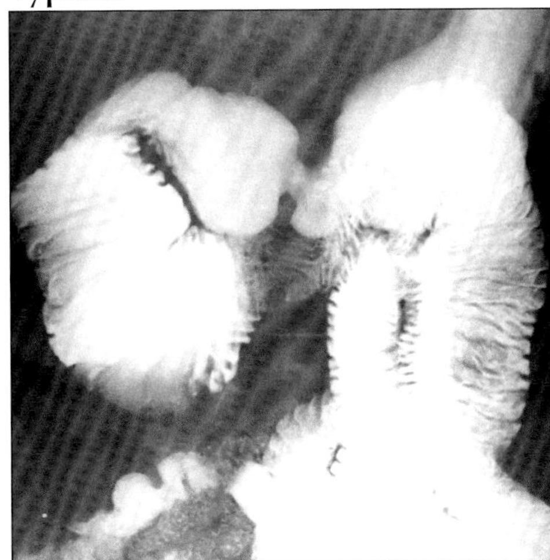

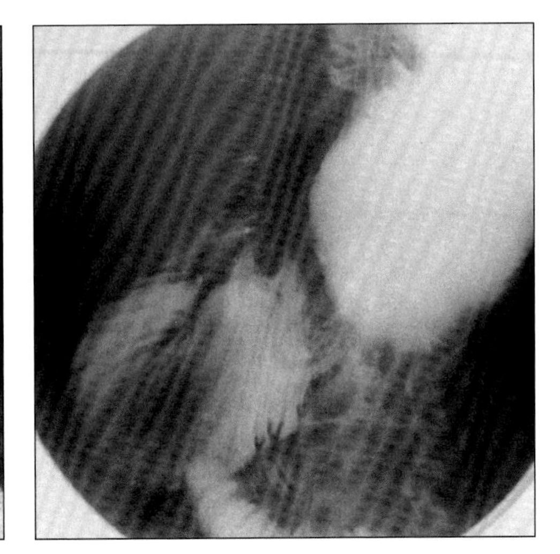

(Left) *SBFT shows typical dilated duodenum, hidebound fold pattern of jejunum.* ***(Right)*** *SBFT shows dilated + hidebound appearance of duodenum.*

Typical

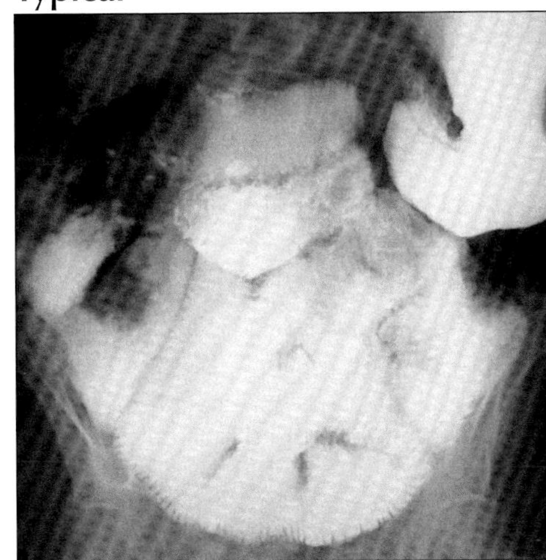

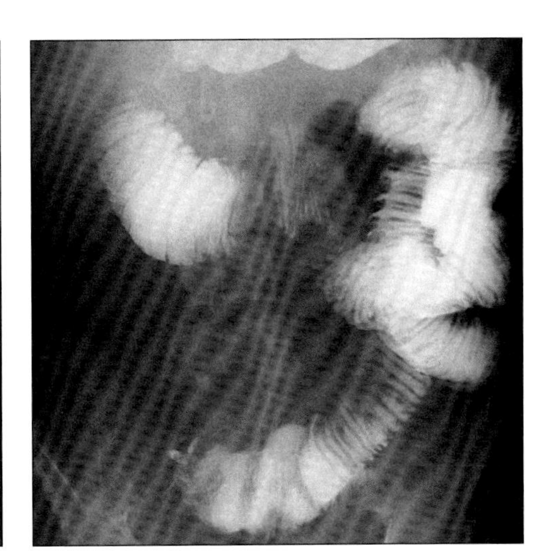

(Left) *SBFT shows dilated, atonic bowel with hidebound fold pattern.* ***(Right)*** *SBFT shows dilated duodenum, "pseudo SMA syndrome", + hidebound jejunal folds.*

Typical

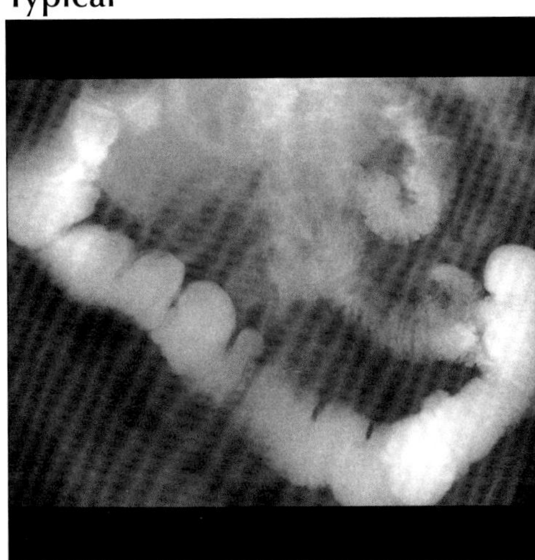

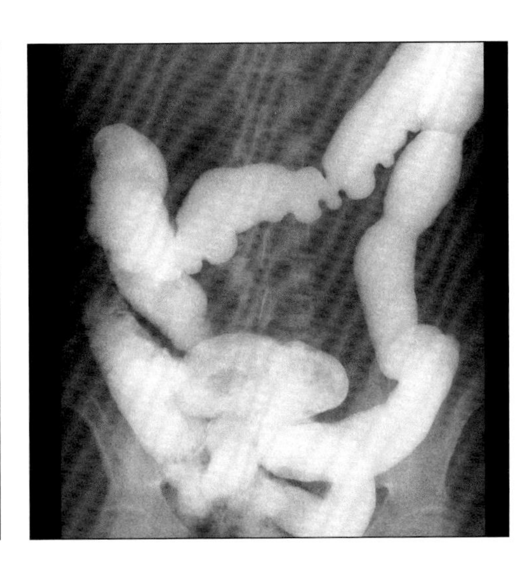

(Left) *SBFT shows pseudosacculation + abnormal folds in the small bowel.* ***(Right)*** *Barium enema shows pseudosacculation of the transverse colon due to scleroderma.*

PNEUMATOSIS OF THE INTESTINE

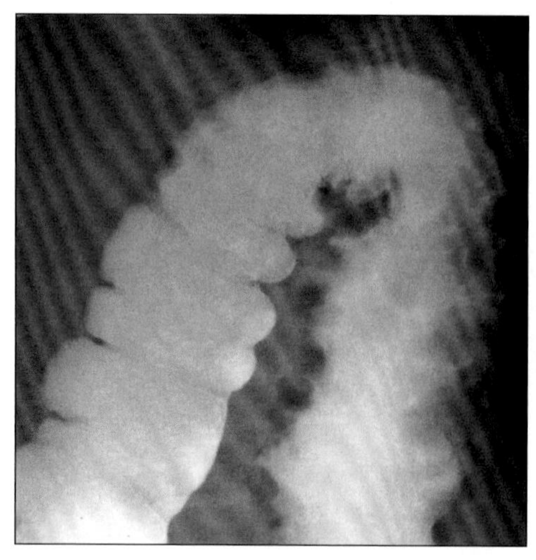

Single contrast BE shows gas "cysts" in wall of colon; primary colonic pneumatosis in asymptomatic patient.

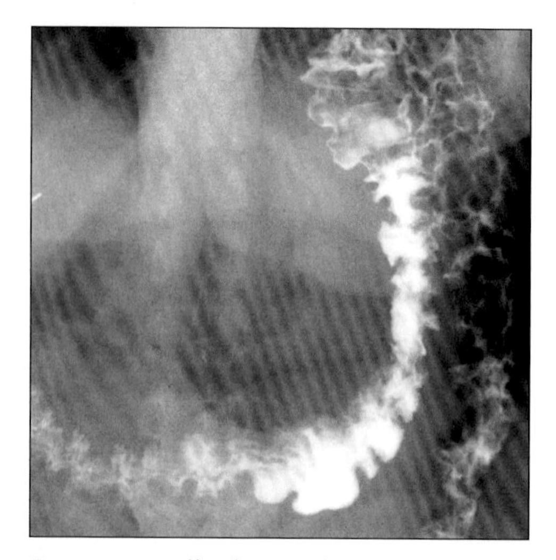

Post-evacuation film from single contrast BE shows extensive gas within the wall of the descending colon due to primary colonic pneumatosis.

TERMINOLOGY

Abbreviations and Synonyms

- Pneumatosis cystoides intestinalis; pneumatosis intestinalis; intestinal gas cysts

Definitions

- Cystic or linear collections of gas in subserosal or submucosal layers of gastrointestinal tract wall

IMAGING FINDINGS

General Features

- Best diagnostic clue: Cystic or linear distribution of gas along bowel wall on helical CT
- Other general features
  - Classification of pneumatosis intestinalis
    - Primary
    - Secondary

Radiographic Findings

- Radiography
  - Primary
    - Location: Colon
    - Cystic gas collections along bowel wall
  - Secondary
    - Location: Small bowel or colon
    - Linear distribution of gas; dilated bowel loops
- Fluoroscopic-guided barium studies
  - Primary
    - Radiolucent clusters of cysts along contours of colon; resemble polyps
    - Multiple, large gas-filed cysts with scalloped defects in bowel wall; resemble inflammatory pseudopolyps or thumbprinting seen with intramural hemorrhage
    - Cysts can concentrically compress the lumen
    - Differentiate from other conditions by the striking lucency of the gas-filled cysts versus the soft tissue density of an intramural or intraluminal lesion or the compressibility of the cyst on palpation
    - Usually insignificant, asymptomatic, idiopathic
  - Secondary
    - Mottled, bubbly or linear collection of gas in bowel wall (feces-like appearance)
    - Dilated bowel loops ± thumbprinting
    - Usually due to ischemia, medication, or other known etiology

CT Findings

- CECT
  - Primary
    - "Bubble-like": Isolated bubbles of air or clusters of cysts in the left colonic wall
  - Secondary
    - "Band-like": Bands or linear distribution of air in affected bowel wall

DDx: Pneumatosis of the Small Bowel

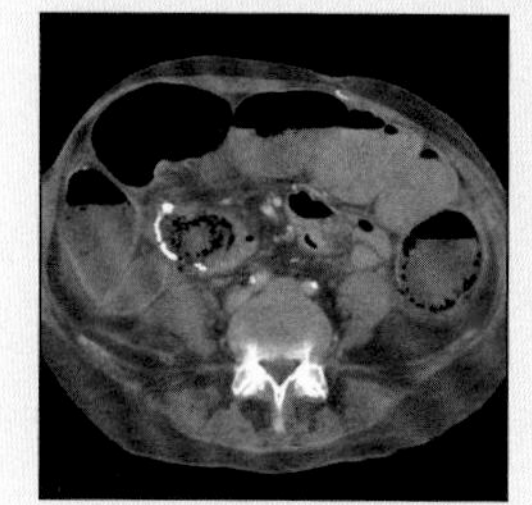

Bowel Necrosis

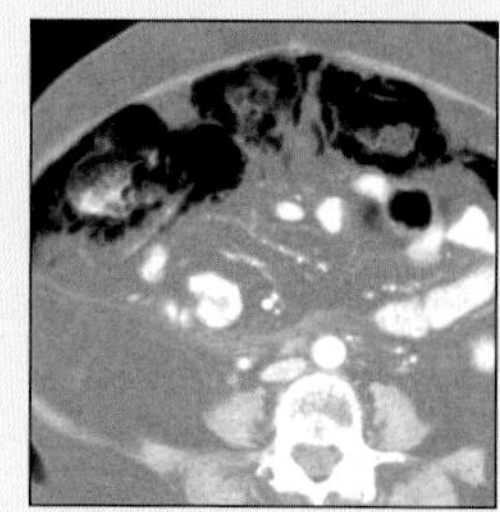

Medication Induced

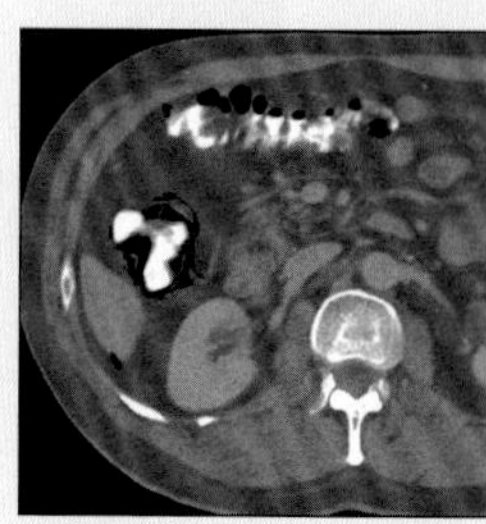

Autoimmune Disease

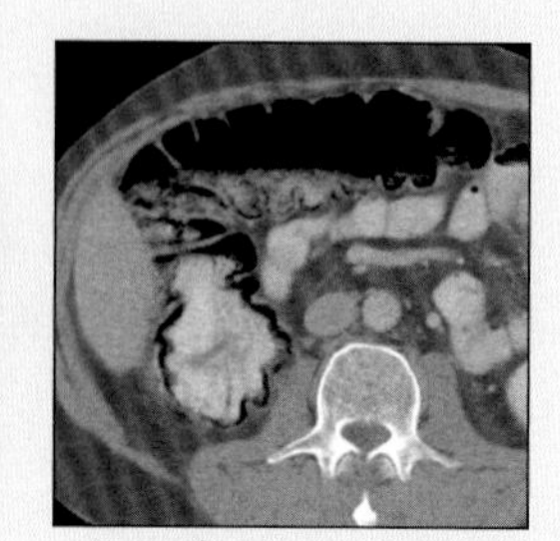

Pseudopneumatosis

PNEUMATOSIS OF THE INTESTINE

Key Facts

Terminology
- Pneumatosis cystoides intestinalis; pneumatosis intestinalis; intestinal gas cysts

Imaging Findings
- Best diagnostic clue: Cystic or linear distribution of gas along bowel wall on helical CT
- Lung window: Detect intramural & venous gas

Top Differential Diagnoses
- Bowel necrosis
- Post-endoscopy
- Post-operation
- Medication-induced
- Autoimmune disease
- Pulmonary disease

Pathology
- Pneumatosis intestinalis is a sign, not a disease

Clinical Issues
- Abdominal pain, distension, melena, fever, vomiting and cough
- Dependent on underlying etiology

Diagnostic Checklist
- Bowel necrosis is a surgical emergency
- Prognosis depends on underlying cause, not imaging findings
- Other causes of pneumatosis intestinalis usually are asymptomatic; little clinical significance
- Important to recognize pneumatosis intestinalis, but significance depends on etiology and clinical setting

    - Linear or curvilinear shape
    - In ischemic etiology, bowel lumen will be dilated (ileus), wall thickened & abnormally enhancing
    - ± Pneumoperitoneum or pneumoretroperitoneum
    - ± Mesenteric or portal venous gas (portal venous gas: Collects in periphery of liver; biliary gas: Collects in central ducts near porta hepatis)
    - ± Mesenteric arterial or venous thrombosis

Imaging Recommendations
- Best imaging tool
  - Helical CT
    - Lung window: Detect intramural & venous gas
- Protocol advice
  - Water for "oral contrast"
    - Facilitates CT angiography
  - Helical CT at 1.5 to 3 mm collimation
    - Image delay approximately 35 seconds
    - Repeat venous phase after 80 seconds
    - Intravenous contrast at 3-4 ml/second
    - Multiplanar reformation (CT angiography)

DIFFERENTIAL DIAGNOSIS

Bowel necrosis
- Example: ischemic enteritis, volvulus or necrotizing enterocolitis
- Mucosal damage → entry of bacteria (mainly enteric organisms) into bowel wall → gas in wall
- Necrotizing enterocolitis
  - Location: Ileum and right colon
  - Feces-like appearance in right bowel
  - Age: Premature or debilitated infants
  - Prognosis: Very low survival rate
  - Gas in intrahepatic branches of portal vein: Catastrophic sign
- Ischemic colitis or enteritis
  - Colonic ischemia
    - Often due to hypoperfusion in elderly, debilitated
  - Small bowel infarction: Often due to embolus or thrombus of large vessels (superior mesenteric artery or vein)
    - Associated mesenteric infiltration
  - Late phase of ischemia → diffuse or localized pneumatosis intestinalis ± mesenteric or portal venous gas
    - Pneumatosis intestinalis can occur with either transmural or partial mural ischemia

Post-endoscopy
- Mucosal disruption and ↑ pressure → ↑ distension → air dissection into wall

Post-operation
- Example: Intestinal bypass
- Mucosal disruption and ↑ pressure → ↑ distension → air dissection into wall

Medication-induced
- Example: Steroids or immunosuppressives
- ↑ Mucosal permeability and ↓ immune system → bacterial gas in wall

Autoimmune disease
- Example: Systemic lupus erythematosus
- ↑ Mucosal permeability and ↓ immune system → bacterial gas in wall

Pseudopneumatosis
- Gas trapped against mucosal surface of bowel by semisolid feces
- Most common in ascending colon

Pulmonary disease
- Example: Chronic obstructive pulmonary disease and ventilator (barotrauma)
- Partial bronchial obstruction & coughing → alveolar rupture → air dissection into peribronchial & perivascular tissue planes of the mediastinum → hiatus of esophagus & aorta → retroperitoneum → mesentery → subserosa and submucosa of the bowel wall

PATHOLOGY

General Features
- General path comments

PNEUMATOSIS OF THE INTESTINE

- Pneumatosis intestinalis is a sign, not a disease
- 15% of pneumatosis intestinalis is primary
- 85% of pneumatosis intestinalis is secondary
- Portal venous gas
  - Often accompanies pneumatosis intestinalis
  - Amount of gas does not correlate with etiology, prognosis or need for treatment
  - Not always due to bowel infarction
  - Increased likelihood of transmural infarction from ischemic enteritis versus ischemic colitis

- Etiology
  - Secondary pneumatosis intestinalis (at least 50 reported causes)
    - Bowel necrosis (most common): Necrotizing enterocolitis, bowel infarction or caustic ingestion
    - Mucosal disruption: Endoscopy, ulcers, obstruction, inflammatory bowel disease (IBD) or bowel anastomoses
    - Increased mucosal permeability: Steroids, chemotherapy, immunosuppressive therapy, immunodeficiency states
    - Auto-immune: Systemic lupus erythematosus, scleroderma or other collagen vascular diseases
    - Pulmonary: Asthma, chronic obstructive pulmonary disease, positive pressure ventilation, pneumothorax or trauma
  - Portal venous gas
    - Intestinal wall lesions: Ischemia or IBD
    - Bowel distention: Endoscopy, obstruction or trauma
    - Sepsis: Diverticulitis, cholecystitis, appendicitis, colitis including Clostridium difficile infection
  - Pathogenesis
    - Intraluminal gastrointestinal gas: Intramural pressure or mucosal injury → gas entering wall
    - Bacterial gas production: Bacterial invasion → has high tension → gas diffusion
    - Pulmonary gas: Air dissection

Gross Pathologic & Surgical Features

- Primary: Gas cysts in otherwise normal colon
- Secondary: Bowel is often abnormal or ischemic

Microscopic Features

- Primary
  - Multiple thin-walled, noncommunicating, gas-filled cysts in subserosal or submucosal layer of the bowel
  - Normal muscularis and mucosa
- Secondary
  - Linear streaks of gas parallel to bowel wall
  - Necrotic, inflammatory, ulcerative or ischemic features

CLINICAL ISSUES

Presentation

- Most common signs/symptoms
  - Primary: Asymptomatic; insignificant
  - Secondary
    - Abdominal pain, distension, melena, fever, vomiting and cough
    - Depends on etiology

Demographics

- Age
  - Primary: Adults
  - Secondary: Any age
- Gender: M = F

Natural History & Prognosis

- Complications
  - Spontaneous rupture of pneumatosis → pneumoperitoneum
- Prognosis
  - Benign or catastrophic prognosis cannot be distinguished by imaging
  - Primary: Good
  - Secondary: Depends on etiology of gas; not extent
  - Detection of pneumatosis and portomesenteric venous gas
    - Positive on radiography → 75% mortality
    - Positive on CT → 25% mortality (more sensitive)
  - Pneumatosis associated with pneumoperitoneum or portal venous gas can be "benign" and transient

Treatment

- Primary
  - No treatment; resolves spontaneously
- Secondary
  - Dependent on underlying etiology
  - Oxygen may be beneficial: ↓ Gas tension in tissues

DIAGNOSTIC CHECKLIST

Consider

- Bowel necrosis is a surgical emergency
- Prognosis depends on underlying cause, not imaging findings
- Other causes of pneumatosis intestinalis usually are asymptomatic; little clinical significance

Image Interpretation Pearls

- Important to recognize pneumatosis intestinalis, but significance depends on etiology and clinical setting

SELECTED REFERENCES

1. See C et al: Images in clinical medicine. Pneumatosis intestinalis and portal venous gas. N Engl J Med. 350(4):e3, 2004
2. Sherman SC et al: Pneumatosis intestinalis and portomesenteric venous gas. J Emerg Med. 26(2):213-5, 2004
3. St Peter SD et al: The spectrum of pneumatosis intestinalis. Arch Surg. 138(1):68-75, 2003
4. Kernagis LY et al: Pneumatosis intestinalis in patients with ischemia: correlation of CT findings with viability of the bowel. AJR Am J Roentgenol. 180(3):733-6, 2003
5. Sebastia C et al: Portomesenteric vein gas: Pathologic mechanisms, CT findings and prognosis. Radiographics. 20: 1213-24, 2000
6. Pear BL: Pneumatosis intestinalis: A review. Radiology. 207: 13-19, 1998
7. Faberman RS et al: Outcome of 17 patients with portal venous gas detected by CT. AJR. 169: 1535-8, 1997

IMAGE GALLERY

Typical

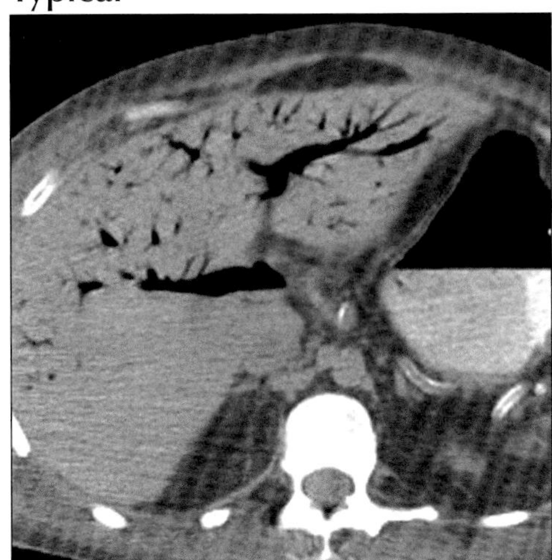

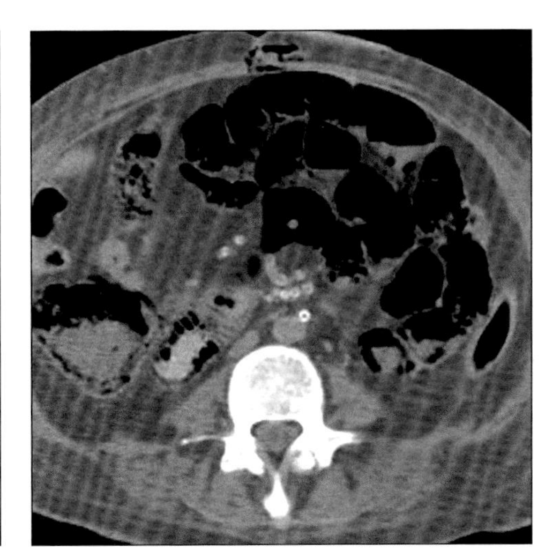

(Left) *Axial NECT shows extensive portal venous gas in a patient with fatal bowel infarction.* ***(Right)*** *Axial NECT in a patient with fatal bowel infarction shows pneumatosis throughout the small intestine and right colon.*

Typical

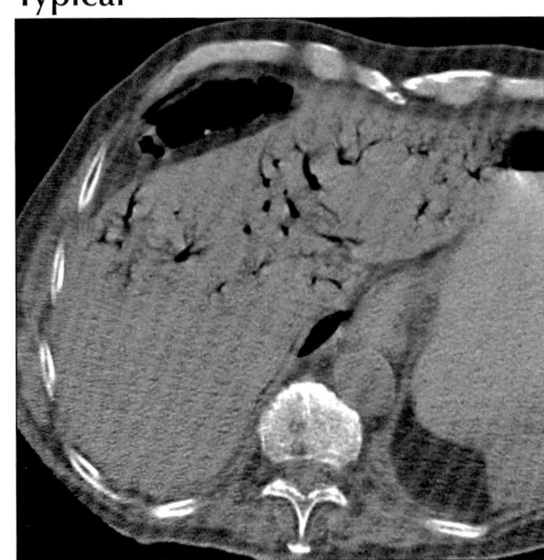

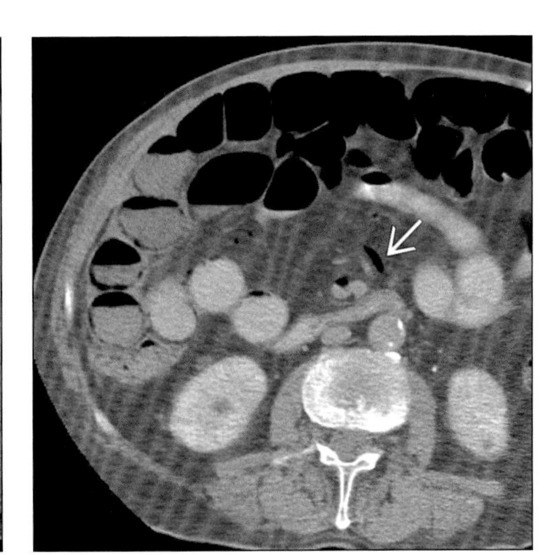

(Left) *Axial NECT shows extensive portal venous gas in a patient with bowel infarction who recovered completely following bowel resection.* ***(Right)*** *Axial CECT shows pneumatosis and mesenteric venous gas (arrow). Surgery revealed infarcted bowel + patient recovered completely.*

Typical

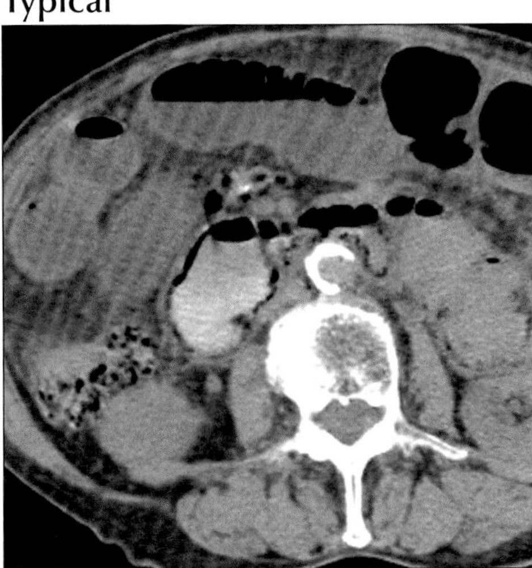

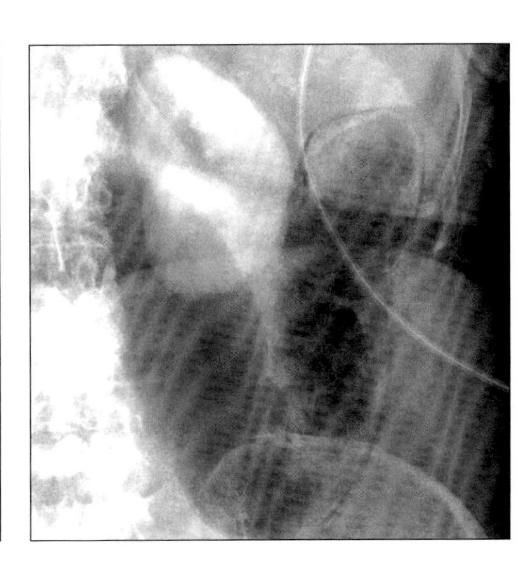

(Left) *Axial CECT shows extensive pneumatosis and ileus or obstruction. Patient had resection for infarcted bowel + recovered completely.* ***(Right)*** *Supine radiograph shows linear and bubbly appearance of gas in wall of dilated bowel. Also note persistent nephrogram following prior hypotensive episode.*

ISCHEMIC ENTERITIS

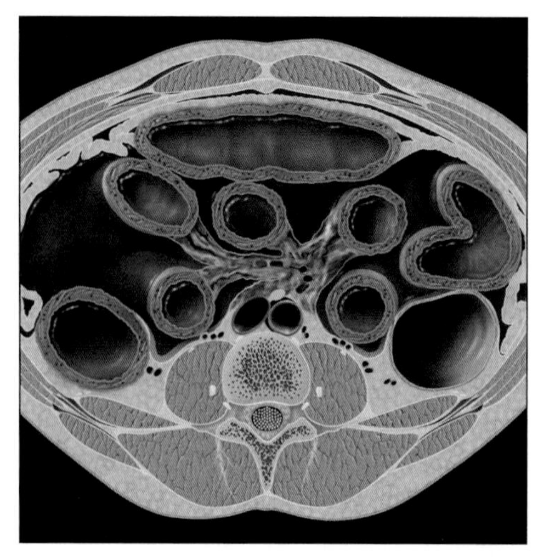

Graphic shows dilated small intestine with thickened wall, ascites, + edematous mesentery; findings seen typically with occlusion of the superior mesenteric vein.

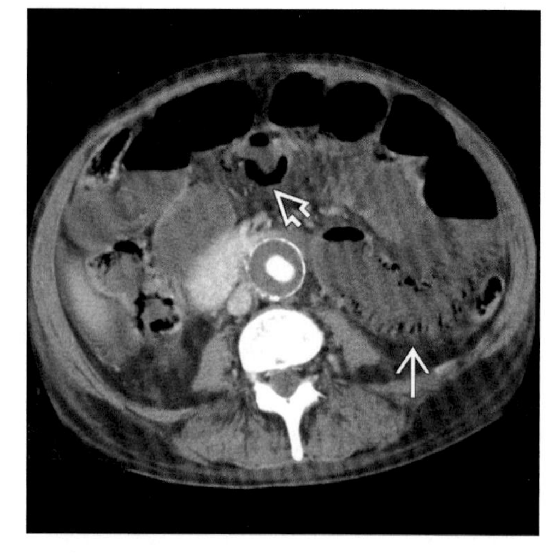

Axial CECT shows severely atherosclerotic aorta. Small bowel is dilated with extensive gas in bowel wall (arrow) + mesenteric veins (open arrow).

TERMINOLOGY

Abbreviations and Synonyms

- Small bowel ischemia; mesenteric ischemia

Definitions

- Mesenteric arterial or venous narrowing or occlusion leading to inadequate supply of nutrients and oxygen to the small intestine

IMAGING FINDINGS

General Features

- Best diagnostic clue: Clot or narrowing of superior mesenteric artery (SMA) or superior mesenteric vein (SMV) with bowel wall thickening
- Other general features
  - Imaging findings vary: Acute vs. chronic; arterial vs. venous thrombosis
  - Classification of ischemic enteritis
    - Acute occlusive ischemia (arterial or venous)
    - Acute nonocclusive ischemia
    - Chronic ischemia: Older "vasculopaths"
  - Late phase of ischemia can lead to diffuse or localized "pneumatosis intestinalis"

Radiographic Findings

- Radiography
  - Multiple air-fluid levels; ileus pattern
  - Thickening of valvulae conniventes
  - Linear distribution of gas (pneumatosis intestinalis)
- Fluoroscopic-guided barium studies
  - Thickening of valvulae conniventes
  - "Thumbprinting": Intramural accumulation of blood distending the submucosa → focally rounded mesenteric folds, especially along the mesenteric border of bowel
  - "Stack of coins" appearance: Enlarged, smooth, straight, parallel folds perpendicular to longitudinal axis of the small bowel (submucosal edema)
  - Strictures can be seen with proximal bowel dilation
  - Mottled, frothy, bubbly or linear collections of gas in bowel wall (pneumatosis intestinalis)

CT Findings

- CECT
  - Clot or reduced lumen in SMA, SMV or other mesenteric vessels
  - Segmental thickening of bowel wall (> 3 mm); average 8 mm, up to 20 mm
  - Emboli usually observed in origin of SMA or within 3-10 cm of SMA distal to middle colic artery
  - Compromised arterial blood flow → lack of mucosal enhancement
  - "Misty mesentery": Mesenteric fat infiltrated by edema; more with venous thrombosis

DDx: Thick Bowel Wall + Infiltrated Mesentery

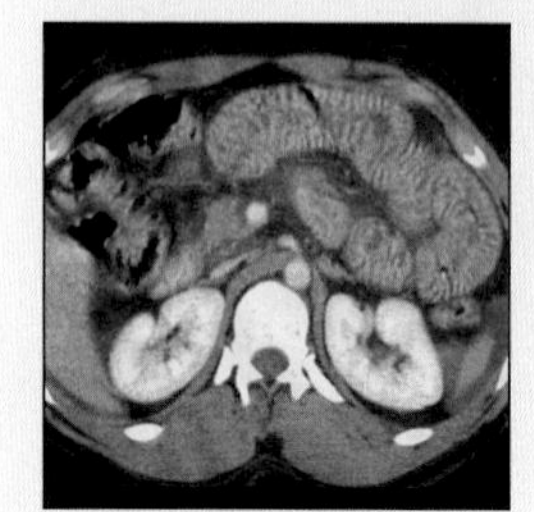

Shock Bowel

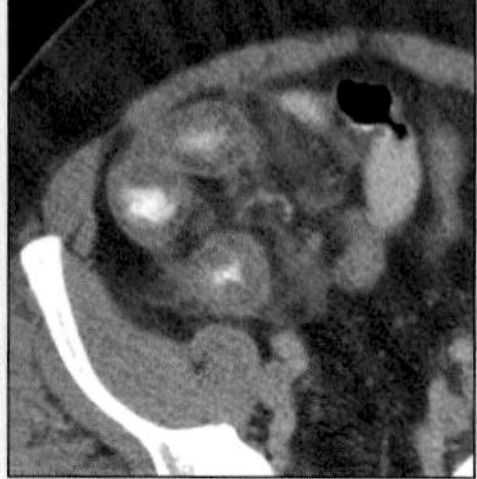

Crohn Disease

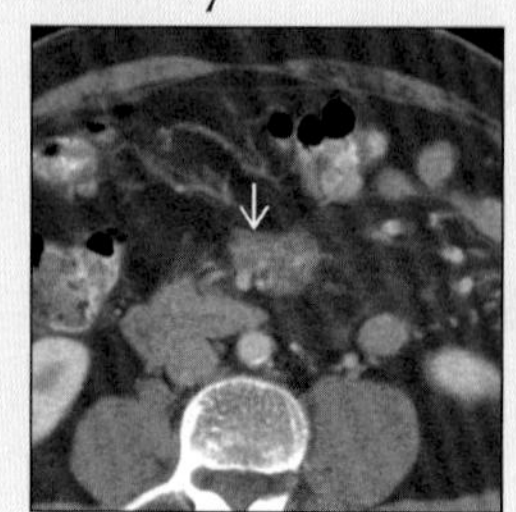

Fibrosing Mesenteritis

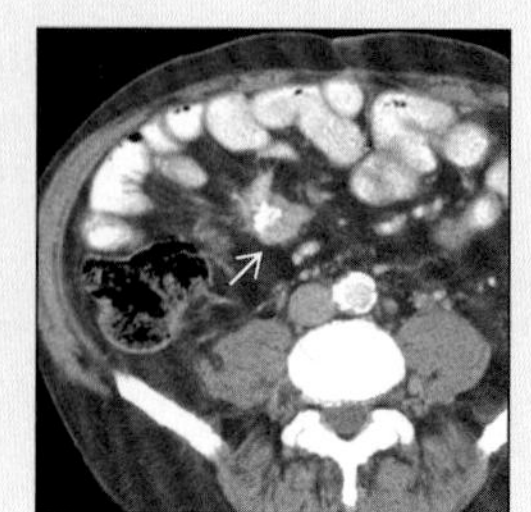

Carcinoid Tumor

ISCHEMIC ENTERITIS

Key Facts

Terminology
- Small bowel ischemia; mesenteric ischemia

Imaging Findings
- Best diagnostic clue: Clot or narrowing of superior mesenteric artery (SMA) or superior mesenteric vein (SMV) with bowel wall thickening
- Late phase of ischemia can lead to diffuse or localized "pneumatosis intestinalis"
- Segmental thickening of bowel wall (> 3 mm); average 8 mm, up to 20 mm

Top Differential Diagnoses
- "Shock bowel"
- Crohn disease
- Fibrosing mesenteritis
- Carcinoid tumor

Pathology
- Vascular occlusion: Embolic events (atrial fibrillation or endocarditis), thrombotic events (atherosclerosis) or mechanical obstruction (strangulation or tumor)

Clinical Issues
- Unremitting abdominal pain out of proportion to physical exam findings
- Intestinal angina: Postprandial abdominal pain that subsides 1-2 hours after meal

Diagnostic Checklist
- Small bowel ischemia is a clinico-radiological diagnosis
- Prognosis depends on underlying cause, not imaging
- Gas-filled dilated intestinal loops with multiple air-fluid levels; bowel wall thickening

- Submucosal hemorrhage or hyperemia → increased attenuation (venous > arterial thrombosis)
- Pneumatosis intestinalis (venous > arterial thrombus)
  - "Band-like" (linear) or "bubble-like" (cystic) appearance in the bowel wall
  - Linear, curvilinear or cystic, gas-filled spaces
  - ± Gas in mesenteric or portal vein
  - Bowel loops are partly fluid-filled

Ultrasonographic Findings
- Duplex Doppler
  - Mainly used to assess degree of narrowing or occlusion in chronic ischemia
  - Narrowed or occluded vessels → ↓ blood flow

Angiographic Findings
- Conventional
  - Acute arterial ischemia: Clot or stenosis of SMA or its branches
  - Acute venous ischemia: SMV occlusion with collaterals
  - Nonocclusive ischemia: Slow flow in SMA
  - Chronic ischemia: Narrowing or occlusion of celiac artery, SMA or inferior mesenteric artery
    - Increased collateral arteries

Imaging Recommendations
- Best imaging tool
  - Multidetector CT with CT angiography
    - More sensitive in assessing strangulation of bowel
    - Lung window setting: For pneumatosis intestinalis
  - Angiography
    - Diagnostic confirmation and treatment

DIFFERENTIAL DIAGNOSIS

"Shock bowel"
- Ischemia ± reperfusion of small bowel, usually following trauma or other cause of hypotension
- Intense mucosal enhancement
- Submucosal and mesenteric edema
- Reversible with resuscitation

Crohn disease
- Location: Usually distal small bowel
- Differentiate from ischemia
  - Asymmetric, discontinuous, thickened bowel wall with proliferation of mesenteric fat
  - Abnormalities of the folds persist; changes occur within days to weeks in ischemic enteritis
  - Focal, inflammatory, mucosal ulceration with patchy submucosal fibrosis → distortion and interruption of folds

Fibrosing mesenteritis
- Idiopathic process of inflammation and fibrosis involving fatty tissue of mesentery
- Bowel wall thickening uncommon, due to fibrotic constriction of mesenteric veins and lymphatics
- "Misty mesentery" appearance with halo of fat surrounding mesenteric vessels without displacing these vessels
- Mesenteric soft tissue mass (advanced)
- Calcifications seen histologically; rare radiologically

Carcinoid tumor
- Focal bowel wall thickening with focal desmoplastic reaction
- Encasement of mesenteric vessels → edema, ischemia
- Multiple hepatic metastases can usually be found

PATHOLOGY

General Features
- General path comments
  - Accounts for 1% of acute abdomen
  - Arterial-to-venous occlusive ischemia: 9:1
  - 60-70% of acute ischemia is from arterial occlusion
  - 5-10% of acute ischemia is from venous occlusion
  - 20-30% of acute ischemia is nonocclusive ischemia
- Etiology
  - Vascular occlusion: Embolic events (atrial fibrillation or endocarditis), thrombotic events (atherosclerosis) or mechanical obstruction (strangulation or tumor)
  - Closed loop obstruction is especially dangerous

- Hypoperfusion (more common in ischemic colitis): Low flow states, hypotension, sepsis or heart failure
- Hypercoagulable states: Oral contraceptives, protein C deficiency, factor V Leiden deficiency
- Inflammatory: Pancreatitis, peritonitis or vasculitis
  - Common cause of ischemia in younger patients
  - Systemic lupus erythematosus, polyarteritis nodosa or other collagen vascular diseases
  - Vasculitis may affect kidneys and other organs
  - Angiography may show microaneurysms and occluded vessels
- Iatrogenic causes: Chemotherapy, radiation therapy, legal drugs (digitalis, dopamine or vasopressin) and illegal drugs (heroin or cocaine)
- Risk factors of chronic ischemia: Hypertension, coronary artery disease or cerebrovascular disease
- Pathogenesis of pneumatosis intestinalis
  - Ischemia → mucosa damage → necrosis and entry of bacteria (mainly enteric organisms) into bowel wall → gas in bowel wall

Gross Pathologic & Surgical Features

- Discolored (purple), infarcted small bowel

Microscopic Features

- Necrotic, inflammatory or ischemic features in the small bowel wall

CLINICAL ISSUES

Presentation

- Most common signs/symptoms
  - Acute ischemia
    - Clinical triad: Sudden onset of abdominal pain, diarrhea and vomiting
    - Unremitting abdominal pain out of proportion to physical exam findings
    - Abdominal distention, tenesmus and passage of bloody stool
    - Guarding and rebound (infarction or perforation)
    - Venous ischemia has a more gradual onset
  - Chronic ischemia
    - Intestinal angina: Postprandial abdominal pain that subsides 1-2 hours after meal
    - Nausea, vomiting, diarrhea and weight loss
    - Intense pain → fear of eating (sitophobia)
  - Lab-Data
    - ↑ WBC: 75%; Acidosis: 50%; ↑ amylase: 25%
  - Diagnosis
    - High clinical suspicion is key to early diagnosis

Demographics

- Age: Majority > 50 years of age
- Gender: M = F

Natural History & Prognosis

- Complications
  - Stricture, infarction, necrosis and perforation
- Prognosis
  - Acute ischemia
    - Depends on promptness of diagnosis and the amount of small bowel that can be saved
    - After surgical resection, results of patients with venous ischemia are generally better
    - 50-90% mortality
  - Chronic ischemia
    - Key to survival is dependent on degree of collateral circulation
    - Diagnosis requires at least two major mesenteric arteries to be occluded and a third artery to be narrowed
  - Infarction: 69% mortality (in recent series)
  - Pneumatosis intestinalis: Poor, dependent on etiology of gas
  - Pneumatosis intestinalis with postmesenteric venous gas
    - Positive on radiography: 75% mortality
    - Positive on CT: 25% mortality (more sensitive)

Treatment

- Surgical treatment
  - Exploratory laparotomy, bowel resection and mesenteric bypass (re-establish blood flow)
  - Main treatment modality for acute ischemia, chronic ischemia and complications
- Endovascular intervention
  - Intra-arterial thrombolysis, percutaneous transluminal angioplasty ± stent placement
  - Thrombolytics (streptokinase or urokinase)
  - Vasodilators (papaverine): Reduce vasospasm
- Systemic anticoagulation (warfarin or heparin) for venous occlusion

DIAGNOSTIC CHECKLIST

Consider

- Small bowel ischemia is a clinico-radiological diagnosis
- Talk to referring physician for history, symptoms and key lab values

Image Interpretation Pearls

- Prognosis depends on underlying cause, not imaging
- Gas-filled dilated intestinal loops with multiple air-fluid levels; bowel wall thickening

SELECTED REFERENCES

1. Chou CK et al: CT of small bowel ischemia. Abdom Imaging, 24-30, 2003
2. Segatto E et al: Acute small bowel ischemia: CT imaging findings. Semin Ultrasound CT MR. 24(5):364-76, 2003
3. Tendler DA: Acute intestinal ischemia and infarction. Semin Gastrointest Dis. 14(2):66-76, 2003
4. Burns BJ et al: Intestinal ischemia. Gastroenterol Clin North Am. 32(4):1127-43, 2003
5. Wiesner W et al: CT of acute bowel ischemia. Radiology. 226(3):635-50, 2003
6. Horton KM et al: Computed tomography evaluation of intestinal ischemia. Semin Roentgenol. 36(2):118-25, 2001
7. Horton KM et al: Multi-detector row CT of mesenteric ischemia: can it be done? Radiographics. 21(6):1463-73, 2001
8. Singer A et al: Acute small bowel ischemia: Spectrum of computed tomographic findings. Emer Radiol. 7: 302-307, 2000

IMAGE GALLERY

Typical

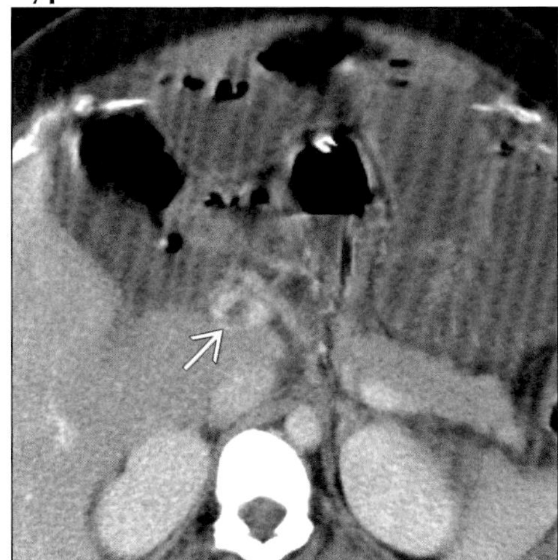

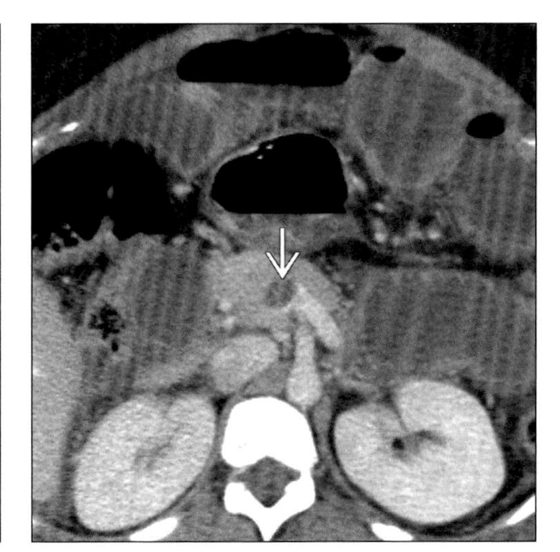

(Left) *Axial CECT shows dilated fluid-filled small intestine + clot in portal vein (arrow).* ***(Right)*** *Axial CECT in 23 year old woman with hypercoagulable state + bowel ischemia. Dilated fluid-filled small bowel + thrombosis of the SMV (arrow).*

Typical

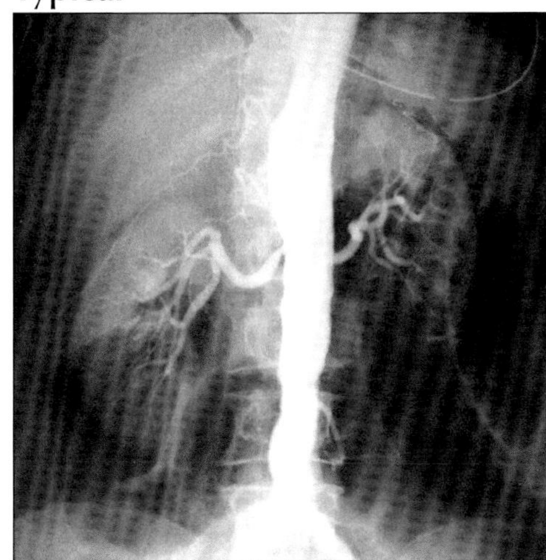

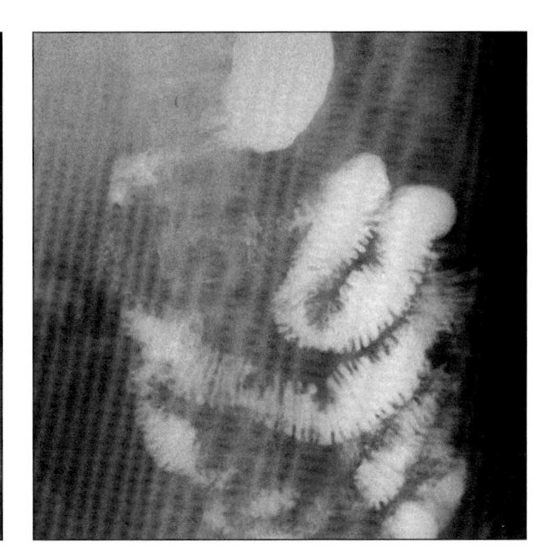

(Left) *Aortography shows irregular narrowing of aorta and occluded celiac + superior mesenteric arteries. Dilated gas-filled ischemic small bowel.* ***(Right)*** *SBFT shows "stack of coins" small bowel fold pattern due to ischemia, intramural hemorrhage.*

Typical

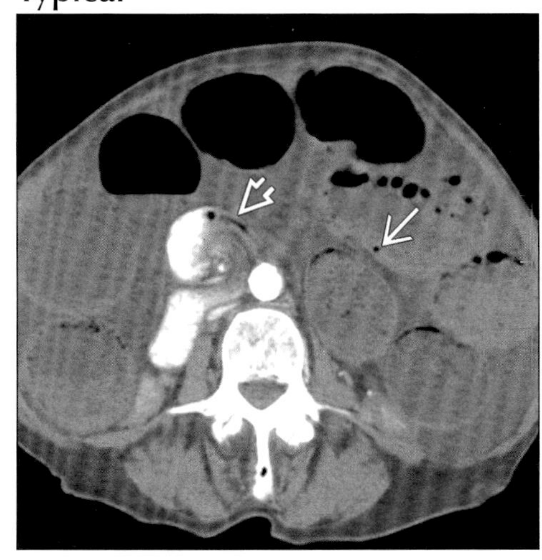

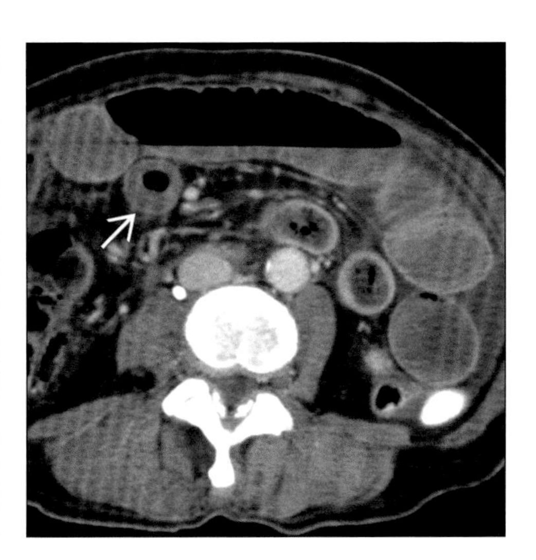

(Left) *Axial CECT shows dilated fluid-filled small bowel, pneumatosis (arrow) + ascites. Duodenum + jejunum are twisted with a "whirl sign" (open arrow) due to mid-gut volvulus.* ***(Right)*** *Axial CECT shows dilated small bowel with areas of wall thickening (arrow). Patient had severe acute abdominal pain + acidosis. Bowel infarction from atrial fibrillation.*

VASCULITIS, SMALL INTESTINE

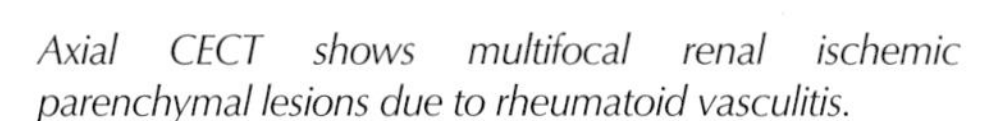

Axial CECT shows multifocal renal ischemic parenchymal lesions due to rheumatoid vasculitis.

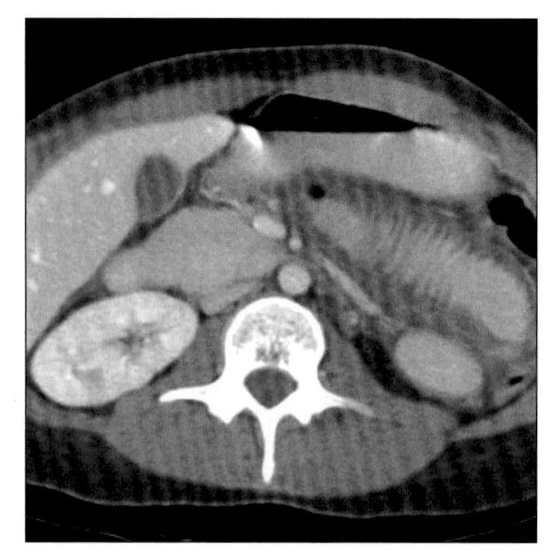

Axial CECT shows marked SB wall thickening and renal ischemic parenchymal lesions due to rheumatoid vasculitis.

TERMINOLOGY

Definitions

- Inflammation of the blood vessels of small intestine caused by a large group of rare, systemic conditions

IMAGING FINDINGS

General Features

- Best diagnostic clue: Straight thickened folds with luminal dilatation of the small bowel
- Other general features
  - Categories of vasculitis
    - Large-vessel vasculitis: Aorta, main visceral arteries (e.g., superior mesenteric artery)
    - Medium-vessel vasculitis: Main visceral arteries and their branches
    - Small-vessel vasculitis: Arterioles, venules, capillaries
  - Different sizes of blood vessels are affected by various systemic conditions; however, imaging findings overlap
  - Most common systemic conditions include
    - Polyarteritis nodosa: Small- to medium-vessel
    - Henoch-Schönlein purpura (HSP): Small-vessel
    - Systemic lupus erythematosus (SLE): Small-vessel
    - Behçet syndrome: Small-vessel

Radiographic Findings

- Fluoroscopic-guided barium studies
  - Segmental or extensive intestinal involvement
  - Aphthous ulcers
  - Straight, thickened folds ± dilatation of bowel lumen
  - Concentric filling defects (submucosal hemorrhage)
  - Ulceration and stricture (small-vessel)
  - ± Thumbprinting; pneumatosis intestinalis
  - SLE
    - Nodularity of small bowel folds
    - Motility disorder of lower esophagus
    - Esophagitis and gastritis
  - Behçet syndrome
    - Involves ileocecal area, especially terminal ileum; may simulate Crohn disease
    - Narrowing and irregularity of terminal ileum with slight dilatation proximally
    - Large ovoid or irregular ulcers with marked mucosal thickening of surrounding intestinal wall
    - Small, multiple, discrete, "punched-out" ulcers

CT Findings

- Thickened bowel wall ± target sign
- Contrast enhancement of bowel wall
- Submucosal hemorrhage and edema
- Polyarteritis nodosa

DDx: Thick Bowel Wall and Ileus

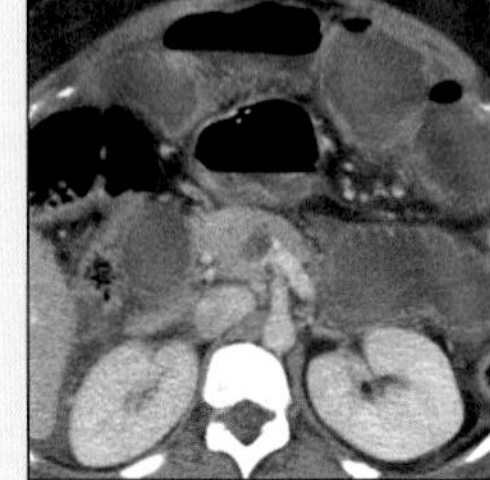

Bowel Ischemia

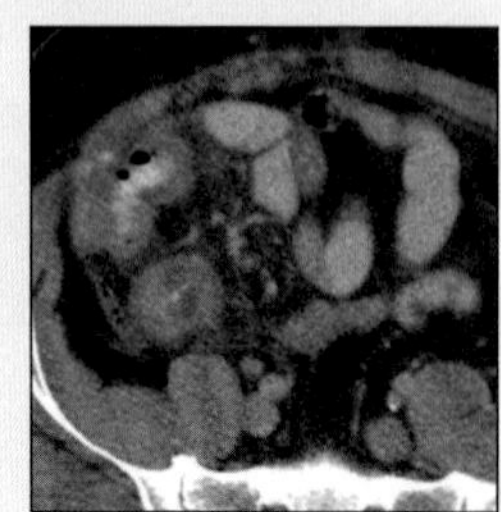

Crohn Disease

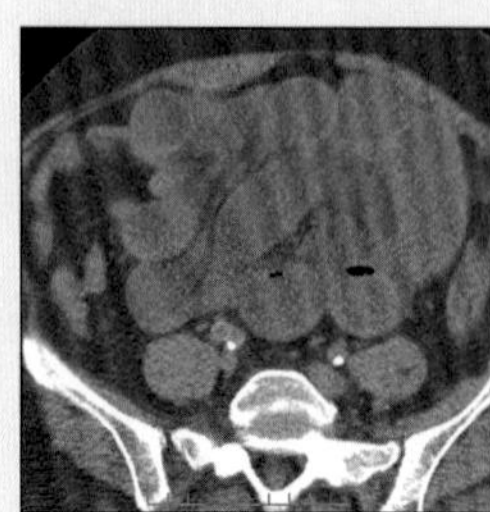

Closed Loop SBO

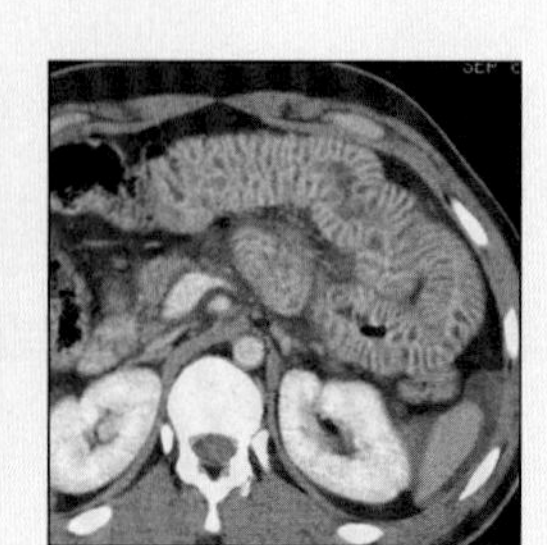

Shock Bowel

VASCULITIS, SMALL INTESTINE

Key Facts

Terminology

- Inflammation of the blood vessels of small intestine caused by a large group of rare, systemic conditions

Imaging Findings

- Best diagnostic clue: Straight thickened folds with luminal dilatation of the small bowel
- Different sizes of blood vessels are affected by various systemic conditions; however, imaging findings overlap
- Segmental or extensive intestinal involvement
- Aphthous ulcers
- Ulceration and stricture (small-vessel)
- Submucosal hemorrhage and edema
- Aneurysm formation

Top Differential Diagnoses

- Ischemic enteritis
- Crohn disease
- Small bowel obstruction (SBO)
- "Shock bowel"

Clinical Issues

- Abdominal pain, fever, nausea, vomiting, weight loss, diarrhea or constipation

Diagnostic Checklist

- Differentiate by associated extraintestinal pattern of involvement and by size of vessels affected
- Biopsy and clinical findings essential for diagnosis
- Imaging can suggest vasculitis as a cause of small bowel disease, but angiography and other tests are essential to diagnosis

  - Lobulated renal contour and irregular thinning (cortical infarcts)
  - Multiple hypoattenuating bands of kidney (arterial occlusion), or "striated nephrogram"
- SLE
  - Abnormal bowel wall enhancement; dilated bowel
  - Comb sign: Engorged mesenteric vessels in a comb-like arrangement
  - Ascites, lymphadenopathy
  - Multiple linear hypoattenuating bands of kidney
  - Hepatomegaly, splenomegaly
- Behçet syndrome
  - Concentric bowel wall thickening or polypoid mass
  - ± Perienteric or pericolonic infiltration

Angiographic Findings

- Conventional
  - Aneurysm formation
    - May be seen in patients with polyarteritis nodosa, SLE, Wegener granulomatosis,, rheumatoid vasculitis, Churg-Strauss syndrome, drug abuse
  - Polyarteritis nodosa visceral involvement
    - Renal (80-90%), GI tract (50-70%), heart (65%), liver (50-60%), spleen (45%), pancreas (25-35%), central nervous system (rare)
    - Small intestine is most commonly affected in GI tract; followed by mesentery and colon
    - Multiple aneurysms (50-60% of cases), typically at branching points
    - 1-5 mm saccular aneurysms (more common) or fusiform aneurysms
    - Stenoses or occlusions of arteries

Imaging Recommendations

- Best imaging tool: Helical CECT and angiography

DIFFERENTIAL DIAGNOSIS

Ischemic enteritis

- Multiple causes including embolus, thrombosis, volvulus
- Distinguish by observing clot or narrowing of superior mesenteric artery, superior mesenteric vein or other mesenteric vessels
- Imaging features usually indistinguishable from vasculitis, especially large-vessel vasculitis

Crohn disease

- Skip lesions, transmural inflammation, non-caseating granulomas, cobblestone mucosa and fistulas
- Differentiate by irregular, prominent mural thickening, fused and distorted folds

Small bowel obstruction (SBO)

- Closed loop → segmental obstruction and ischemia
- Air-fluid levels; smooth beaking
- Bowel wall thickening
- ± Portal venous gas, pneumatosis intestinalis

"Shock bowel"

- Ischemia ± reperfusion of small bowel, usually following trauma or other cause of hypotension
- Intense mucosal enhancement
- Submucosal and mesenteric edema
- Reversible with resuscitation

PATHOLOGY

General Features

- Etiology
  - Large-vessel
    - Giant cell arteritis
    - Takayasu disease
  - Medium-vessel
    - Polyarteritis nodosa: > 50% involves GI
    - Kawasaki disease
    - Primary granulomatous central nervous system vasculitis
  - Antineutrophil cytoplasmic autoantibody (ANCA)-associated small-vessel vasculitis
    - Microscopic polyangiitis
    - Wegener granulomatosis
    - Churg-Strauss syndrome
  - Immune-complex small-vessel vasculitis

- HSP: > 50% involves GI
- SLE: 10-60% involves GI
- Behçet syndrome: 10-40% involves GI
- Cryoglobulinemic vasculitis
- Rheumatoid vasculitis
- Sjögren syndrome
- Hypocomplementemic urticarial vasculitis
- Goodpasture syndrome
- Serum sickness
- Drug-induced
- Infection-induced

- Paraneoplastic small-vessel vasculitis
  - Lymphoproliferative neoplasm–induced
  - Myeloproliferative neoplasm–induced
  - Carcinoma-induced vasculitis
- Inflammatory bowel disease small-vessel vasculitis
- Risk factors: HSP: Bacterial or viral infection, allergies, insect sting, drugs, certain foods

- Associated abnormalities
  - Polyarteritis nodosa: Hepatitis B infection
  - SLE: Hematologic, immunologic and neurologic involvement, photosensitivity, oral ulceration
  - HSP: Renal involvement
  - Behçet syndrome: Neurologic involvement

Gross Pathologic & Surgical Features

- Segmental fibrinoid necrotizing vasculitis
- Nonspecific ulceration or inflammation
- Polyarteritis nodosa: Panmural necrotizing arterial vasculitis; mucoid degeneration
- Behçet's syndrome: Discrete, "punched-out" ulcers and irregular perforations

Microscopic Features

- Polyarteritis nodosa
  - Acute: Polymorphonuclear cell infiltrate in all layers of arterial wall and perivascular tissue
  - Chronic: Mononuclear cell infiltrate with intimal proliferation, thrombosis and perivascular inflammation
- SLE: Local deposition of antigen-antibody complexes
- HSP: Immunoglobulin A deposited in vessel wall (direct immunofluorescence)
- Behçet syndrome: Immune complexes deposited in vessel wall

CLINICAL ISSUES

Presentation

- Most common signs/symptoms
  - Abdominal pain, fever, nausea, vomiting, weight loss, diarrhea or constipation
  - Polyarteritis nodosa: Peripheral neuropathies
  - HSP: Palpable purpura, arthritis, GI bleeding
  - SLE: Cough (serositis), oral ulcers, polyarthritis, malar rash, discoid rash
  - Behçet's syndrome: Oral and genital ulcers, arthritis, uveitis, erythema nodosum
- Lab-Data
  - Polyarteritis nodosa: Cryoglobulin, positive for hepatitis B surface antigen
  - HSP: Hematuria, proteinuria
  - SLE: Antinuclear antibody, anti-Smith antibody
- Diagnosis: Biopsy of involved tissue may help establish diagnosis

Demographics

- Age
  - Polyarteritis nodosa: 18-81 years of age
  - HSP: 3-10 years of age (most common), > 20 years of age (up to 30% of cases)
  - SLE: 16-41 years of age
  - Behçet syndrome: 11-30 years of age
- Gender
  - Polyarteritis nodosa: M:F = 2:1
  - HSP: M:F = 2:1
  - SLE: M:F = 1:10
  - Behçet syndrome: M:F = 2:1

Natural History & Prognosis

- Complications
  - Paralytic ileus, ischemia, hemorrhage, perforation, stricture, fistula, peritonitis, sepsis
  - Polyarteritis nodosa: Renal failure, congestive heart failure, myocardiac infarction, cirrhosis, hepatic carcinoma
  - HSP: Intussusception in children, renal failure
  - SLE: Renal failure
  - Prognosis: Good, unless left untreated with complications

Treatment

- Polyarteritis nodosa: Corticosteroid ± cyclophosphamide
- HSP: Spontaneous resolution
- SLE: Corticosteroid, non-steroid anti-inflammatory drugs, hydroxychloroquine
- Behçet syndrome: Corticosteroid, sulfasalazine

DIAGNOSTIC CHECKLIST

Consider

- Differentiate by associated extraintestinal pattern of involvement and by size of vessels affected
- Biopsy and clinical findings essential for diagnosis

Image Interpretation Pearls

- Imaging can suggest vasculitis as a cause of small bowel disease, but angiography and other tests are essential to diagnosis

SELECTED REFERENCES

1. Ha HK et al: Radiologic features of vasculitis involving the gastrointestinal tract. Radiographics. 20(3):779-94, 2000
2. Rha SE et al: CT and MR imaging findings of bowel ischemia from various primary causes. Radiographics. 20(1):29-42, 2000
3. Byun JY et al: CT features of systemic lupus erythematosus in patients with acute abdominal pain: emphasis on ischemic bowel disease. Radiology. 211(1):203-9, 1999
4. Ha HK et al: Intestinal Behcet syndrome: CT features of patients with and patients without complications. Radiology. 209(2):449-54, 1998
5. Jeong YK et al: Gastrointestinal involvement in Henoch-Schonlein syndrome: CT findings. AJR Am J Roentgenol. 168(4):965-8, 1997

IMAGE GALLERY

Typical

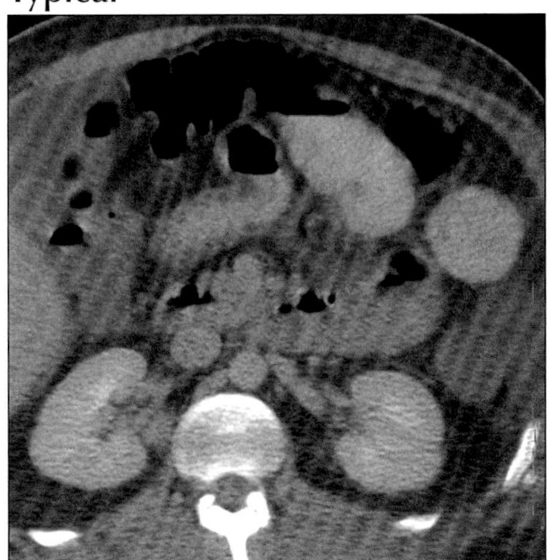

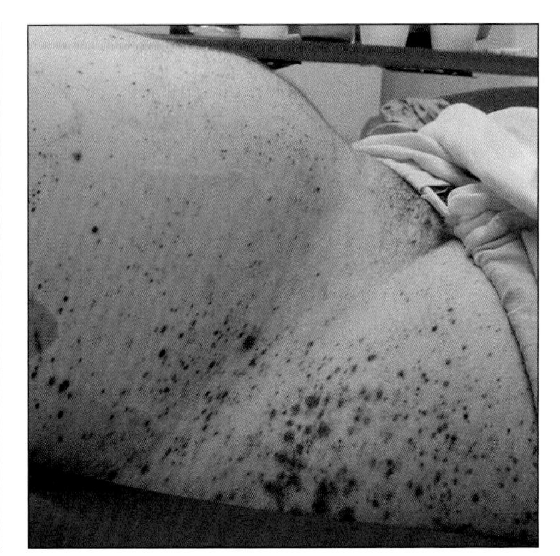

(Left) *Axial CECT shows decreased renal enhancement, multifocal SB mural hemorrhage, and hemorrhagic ascites in a 24 year old man with Henoch-Schönlein purpura (HSP).* ***(Right)*** *Clinical photograph of purpuric skin rash in a 24 year old man with HSP.*

Typical

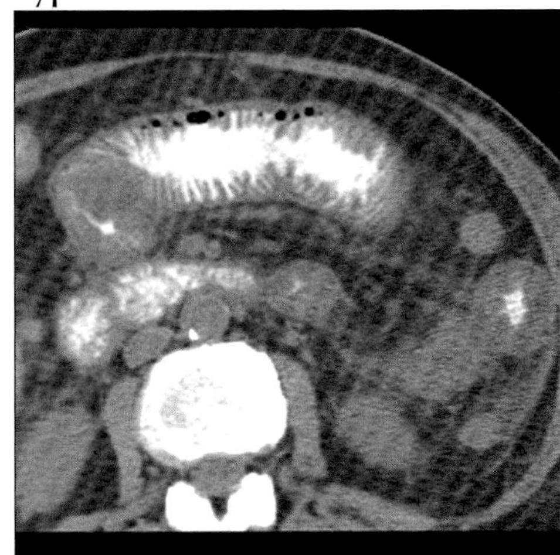

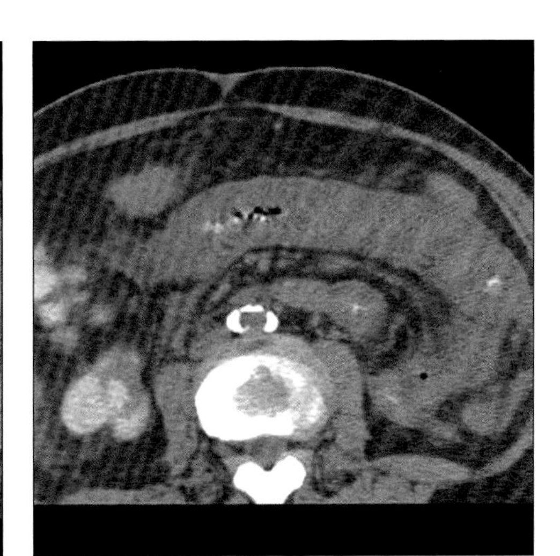

(Left) *Axial NECT shows multifocal SB mural hemorrhage due to Sjögren syndrome.* ***(Right)*** *Axial NECT shows long segmental mural thickening in a patient with Sjögren vasculitis.*

Typical

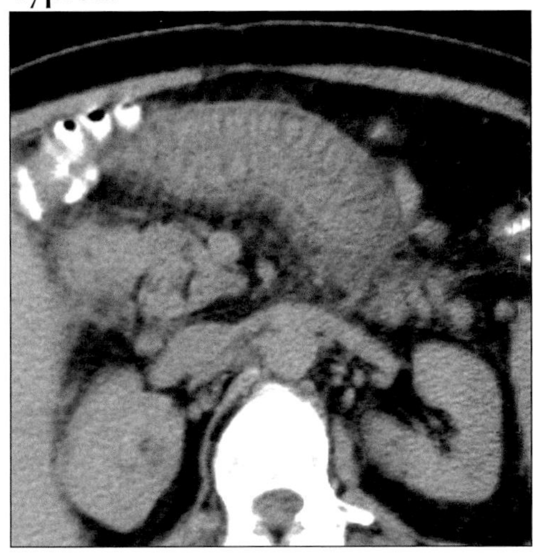

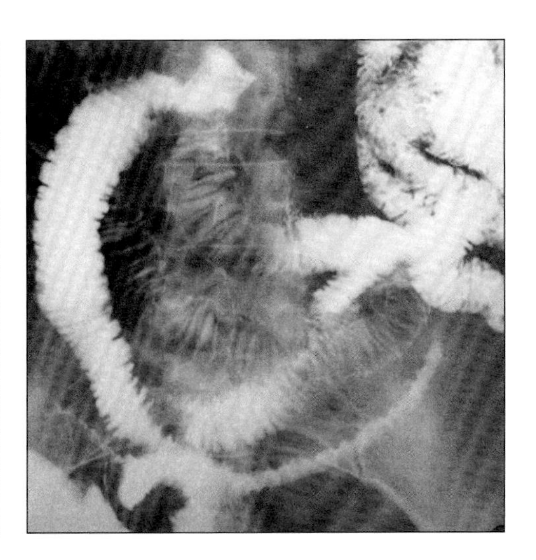

(Left) *Axial NECT shows long segmental mural thickening attributed to vasculitis in this I.V. drug abuser.* ***(Right)*** *SBFT shows multifocal segmental SB wall thickening and luminal narrowing attributed to vasculitis in a patient with systemic lupus.*

INTESTINAL TRAUMA

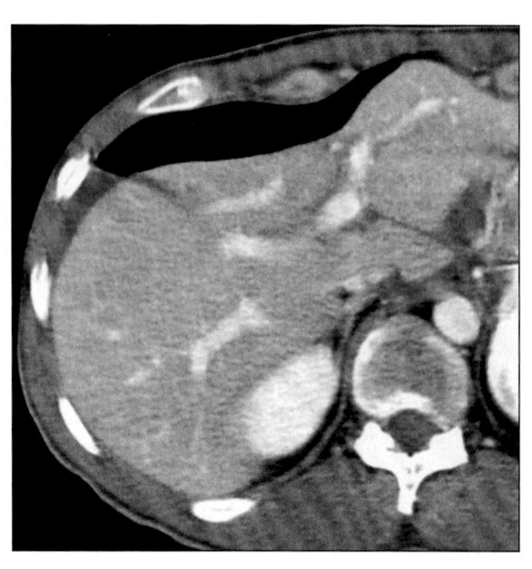

Axial CECT shows free air + blood in perihepatic location. Distal small bowel and sigmoid transections.

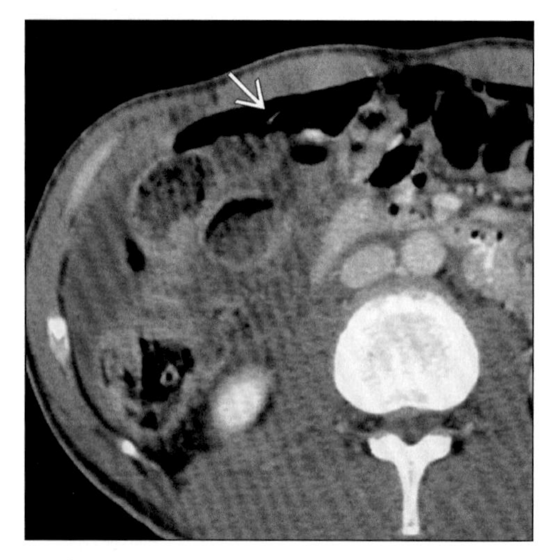

Axial CECT shows fluid (blood + bowel contents) in paracolic gutter + mesentery, along with free air (arrow); transected small bowel.

TERMINOLOGY

Definitions

- Injury to bowel (duodenum, small bowel, colon)

IMAGING FINDINGS

General Features

- Best diagnostic clue: Bowel wall thickening, mesenteric infiltration ± extravasation of enteric or vascular contrast medium
- Location: Duodenum + proximal jejunum (most common)
- Other general features
  - Abdominal trauma: Leading cause of death in United States (< 40 yrs old)
  - Children: ↑ Incidence of intramural hematoma
  - Adults: ↑ Incidence of bowel wall transection
  - Bowel & mesenteric injuries: Seen in 5% of patients with blunt trauma at laparotomy
  - Most common causes
    - Motor vehicle accidents (MVA), falls & assault
  - Impact injuries
    - Crushing of bowel against spine
    - Location: Small bowel of limited mobility (duodenum, near ligament of Treitz & near ileocecal valve)
    - Transverse tears of mesentery → hematoma → bowel infarction
  - Rapid deceleration injuries
    - Caused by abrupt forward movement of proximal jejunum from its fixation by ligament of Treitz
    - Shearing force between restricted & mobile bowel: Cause transection at duodenojejunal flexure
  - Gastric injury
    - More common in children than in adults
    - ↑ Risk: Distended stomach after eating
    - Associated injuries: Rupture of spleen & left-sided thoracic injury
  - Duodenal injury
    - Common findings: Hematoma; ectopic air or contrast (perforation)
    - Location: Descending 2nd & horizontal 3rd part
    - 3rd part compressed against spine by a direct blow
    - Associated injury: Pancreatic head, left lobe of liver
  - Jejunal & ileal injury
    - Common findings: Hematoma, bowel wall discontinuity, thickening
    - At or near ligament of Treitz & ileocecal valve
    - Clinically: Symptoms & signs develop slowly (due to neutral pH & relative absence of bacteria)
  - Colon injury
    - Cause: Compression of upper abdomen (steering wheel & seat belts)

DDx: Hemorrhage or Edema in the Bowel Wall

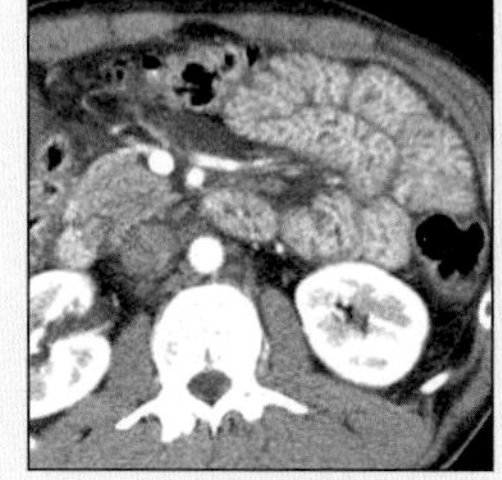

Shock Bowel

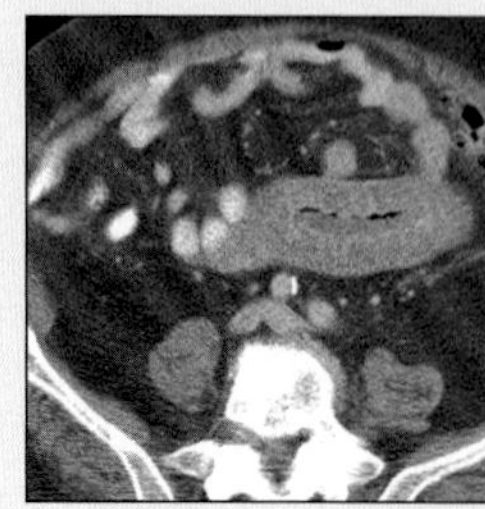

Jejunal Hematoma

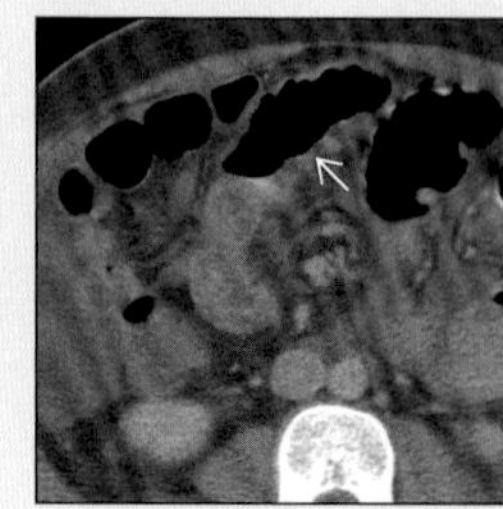

Henoch-Schonlein

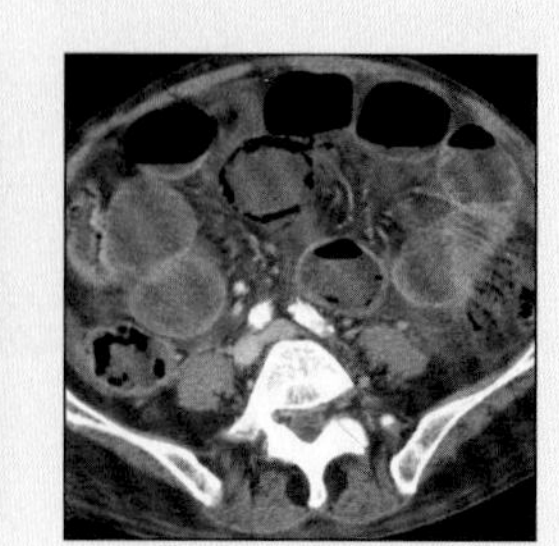

Ischemic Bowel

INTESTINAL TRAUMA

Key Facts

Terminology
- Injury to bowel (duodenum, small bowel, colon)

Imaging Findings
- Best diagnostic clue: Bowel wall thickening, mesenteric infiltration ± extravasation of enteric or vascular contrast medium
- Bowel discontinuity: Injury primary finding, unusual
- Extraluminal oral contrast material: 100% specific for bowel perforation, but uncommon
- Intramural air, extraluminal air, interloop free fluid
- Bowel wall thickening: More than 3 mm (seen in 75% of transmural injuries)
- Mesenteric infiltration or "stranding"
- Hematoma (> 60 HU); liquefied blood (35-50 HU)
- Active bleeding: Isodense with enhanced vessels

Top Differential Diagnoses
- Shock bowel
- Coagulopathy
- Vasculitis
- Ischemic enteritis

Clinical Issues
- Clinical profile: Patient with history of MVA, abdominal pain, distension, tenderness & guarding
- Diagnostic peritoneal lavage (DPL): Positive, severe injury

Diagnostic Checklist
- Check for MVA history or other abdominal injury
- CT evidence of extraluminal air/contrast, bowel wall thickening, free fluids & mesenteric "stranding" highly suggestive of bowel injury

        - Location: Transverse colon, sigmoid colon, cecum
        - Transverse: Intramural hematoma or serosal tear
        - Ascending or descending: Mesenteric avulsion, full-thickness laceration, transection, ischemia
        - Complications: Ischemic stricture or perforation
    - Mesenteric injury
        - Hematoma: Most common GIT injury seen on CT
        - Complications: Disruption of mesenteric vasculature, hemorrhage & GIT perforation
        - Active mesenteric bleeding requires surgery

Radiographic Findings
- Radiography
    - "Flank-stripe" sign: ↑ Density zone (> 800 ml abdominal fluid) separates vertical colon segments from properitoneal fat & peritoneal reflection
    - "Dog's-ear" sign: Pelvic fluid collections displace bowel from urinary bladder
- Fluoroscopy (water soluble contrast) study
    - Fold thickening, luminal narrowing, extravasation
    - Mainly for duodenal hematoma & laceration

CT Findings
- Must view at "abdominal" & "lung" windows
- Bowel discontinuity: Injury primary finding, unusual
- Extraluminal oral contrast material: 100% specific for bowel perforation, but uncommon
- Extraluminal air: Intra-or retroperitoneal air
- Extraluminal gas not diagnostic of bowel perforation (also seen in barotrauma & mechanical ventilation)
- Location
    - Perihepatic, perisplenic regions
    - Trapped in leaves of mesentery; omental interstices
    - Trapped by adhesions or ligaments (e.g., falciform)
- Intramural air, extraluminal air, interloop free fluid
    - Indicates full-thickness tear
- Bowel wall thickening: More than 3 mm (seen in 75% of transmural injuries)
    - Circumferential or eccentric thickening: Due to intramural hematoma, mesenteric trauma (arterial or venous injury)
- Bowel-wall enhancement: More than HU of psoas muscle or equal to blood vessels
    - Enhancement + thickening + free fluid: Strongly suggests perforation
- Mesenteric infiltration or "stranding"
    - Small hemorrhages: Streaky soft tissue infiltration of mesenteric fat
    - "Sentinel clot" sign: Localized > 60 HU mesenteric hematoma at site of bleeding
- Intra-/retroperitoneal free fluid: Hemoperitoneum or bowel contents
    - Polygonal fluid collections between folds of mesentery & bowel loops
    - Hematoma (> 60 HU); liquefied blood (35-50 HU)
    - Bowel content; extravasated enteric contrast
    - Free intraperitoneal fluids: Common in bowel & mesenteric injuries
    - Hemoperitoneum: Common in intraperitoneal bowel or mesenteric injury
    - Active bleeding: Isodense with enhanced vessels
    - Bowel rupture: At sites of oral contrast extravasation

Ultrasonographic Findings
- Real Time: Free fluid in abdomen & pelvis

Angiographic Findings
- Conventional: Vascular transection, laceration, pseudoaneurysm, arteriovenous fistula

Imaging Recommendations
- Helical CECT ± oral contrast: Modality of choice
    - I.V. contrast ≥ 3 ml/sec

DIFFERENTIAL DIAGNOSIS

Shock bowel
- Intense mucosal enhancement, submucosal edema (not blood)
- Often diffuse mesenteric edema + hypovolemia signs
    - Signs of hypovolemia: Collapsed IVC & renal veins
- Is a reversible sign of recent hypotension
- Resolves quickly with fluid resuscitation

Coagulopathy
- Spontaneous, or anticoagulant treatment

INTESTINAL TRAUMA

- Spontaneous bleeding: Example: Idiopathic thrombocytopenic purpura, leukemia, hemophilia
- Abdominal pain, melena, intestinal obstruction
- Barium studies or CT of small bowel
  - Segmental, extensive, or localized changes
  - Uniform, regular thickening of valvulae conniventes with a symmetric, spike-like configuration & reduced luminal diameter simulating a stack of coins
  - Localized bleeding may be seen as intramural mass

Vasculitis

- Polyarteritis nodosa
  - Bowel: Diffuse or segmental ischemia/hemorrhage
  - Renal & liver involvement are frequent
  - Angiography: Small aneurysms of branches of SMA
- Systemic lupus erythematosus
  - Causes small vessel arteritis in 10 to 60% of cases
  - Segmental bowel lesions → necrosis & perforation
- Henoch-Schonlein purpura
  - Children, young & middle-aged; GI tract (> 50%)
  - Present with clinical triad
    - Palpable purpura, arthritis & abdominal pain
- Barium studies or CT of small bowel in vasculitis
  - Extensive fold thickening + luminal narrowing
  - May show thumbprinting on mesenteric border
  - Intussusceptions may be seen in childhood purpura

Ischemic enteritis

- Cause: Superior mesenteric vessel clot or narrowing
- Barium studies of small bowel
  - Markedly thickened valvulae conniventes
  - Thumbprinting: Thick, rounded folds along mesenteric border (intramural blood collection)
- CT findings
  - Shows clot or reduced lumen in SMA or SMV
  - Segmental bowel wall thickening (> 3 mm)
  - Later phase: Pneumatosis (focal or diffuse)
    - Gas within small bowel wall & venous radicles

PATHOLOGY

General Features

- Etiology: Blunt or penetrating trauma; falls; assault
- Epidemiology: 5% of blunt trauma at laparotomy
- Associated abnormalities: Hepatic, splenic, renal & pancreatic injuries

Gross Pathologic & Surgical Features

- Contusion, laceration, bowel discontinuity
- Wall thickening, blood clot, rupture

CLINICAL ISSUES

Presentation

- Most common signs/symptoms
  - Abdominal pain, distension, tenderness, guarding
  - Hypotension, tachycardia
  - Loss of consciousness, shock: Due to ↑ loss of blood
- Clinical profile: Patient with history of MVA, abdominal pain, distension, tenderness & guarding
- Lab-data
  - Altered CBC, electrolytes, BUN, creatinine, amylase, PT, PTT & hematocrit
- Diagnostic peritoneal lavage (DPL): Positive, severe injury
  - RBC > 150,000/mm3; WBC > 500/mm3
  - Food, bile or bacteria on Gram stain from aspirate

Demographics

- Age: Any age group
- Gender: M = F

Natural History & Prognosis

- Complications
  - Perforation → sepsis → abdominal abscess → peritonitis → shock → death
- Prognosis
  - Good: In early diagnosis & treatment
  - Poor: In delayed diagnosis & treatment
    - Increased morbidity & mortality up to 65%

Treatment

- Minor: Airway, I.V. fluids, monitor vital signs, blood transfusion, antibiotics
- Major: Surgery (perforation or active bleeding)

DIAGNOSTIC CHECKLIST

Consider

- Check for MVA history or other abdominal injury

Image Interpretation Pearls

- CT evidence of extraluminal air/contrast, bowel wall thickening, free fluids & mesenteric "stranding" highly suggestive of bowel injury

SELECTED REFERENCES

1. Hanks PW et al: Blunt injury to mesentery and small bowel: CT evaluation. Radiol Clin North Am. 41(6):1171-82, 2003
2. Hawkins AE et al: Evaluation of bowel and mesenteric injury: role of multidetector CT. Abdom Imaging. 28(4):505-14, 2003
3. Butela ST et al: Performance of CT in detection of bowel injury. AJR Am J Roentgenol. 176(1):129-35, 2001
4. Brody JM et al: CT of blunt trauma bowel and mesenteric injury: typical findings and pitfalls in diagnosis. Radiographics. 20(6):1525-36; discussion 1536-7, 2000
5. Federle MP: Diagnosis of intestinal injuries by computed tomography and the use of oral contrast medium. Ann Emerg Med. 31(6):769-71, 1998
6. Levine CD et al: CT findings of bowel and mesenteric injury. J Comput Assist Tomogr. 21(6):974-9, 1997
7. Nghiem HV et al: CT of blunt trauma to the bowel and mesentery. AJR Am J Roentgenol. 160(1):53-8, 1993
8. Rizzo MJ et al: Bowel and mesenteric injury following blunt abdominal trauma: evaluation with CT. Radiology. 173(1):143-8, 1989
9. Wing VW et al: The clinical impact of CT for blunt abdominal trauma. AJR Am J Roentgenol. 145(6):1191-4, 1985

IMAGE GALLERY

Typical

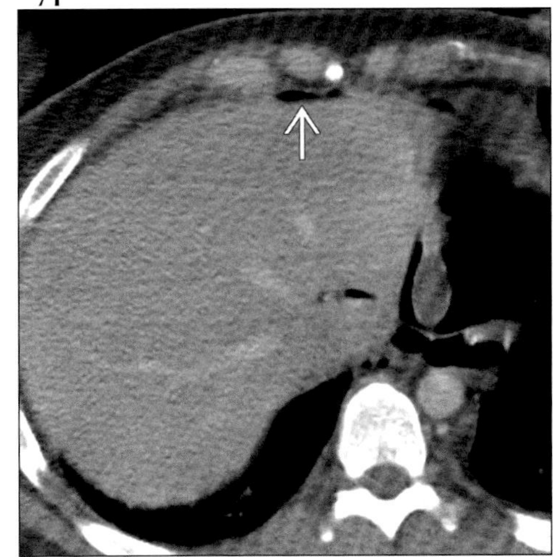

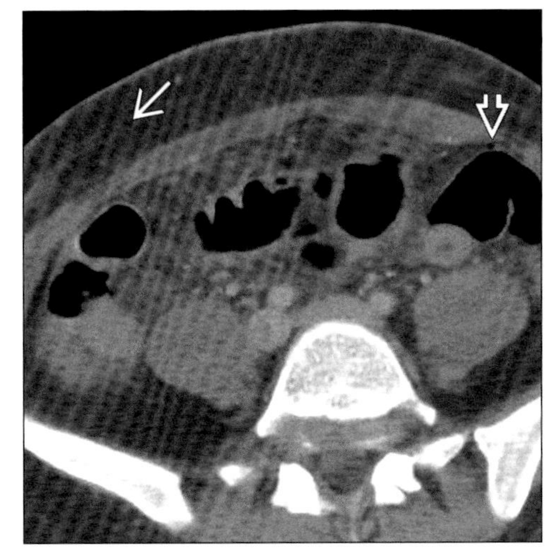

(Left) *Axial CECT shows free air (arrow) from jejunal transection.* ***(Right)*** *Axial CECT shows seat belt contusion (arrow), mesenteric infiltration, bowel wall thickening, and free air (open arrow). Jejunal transection.*

Typical

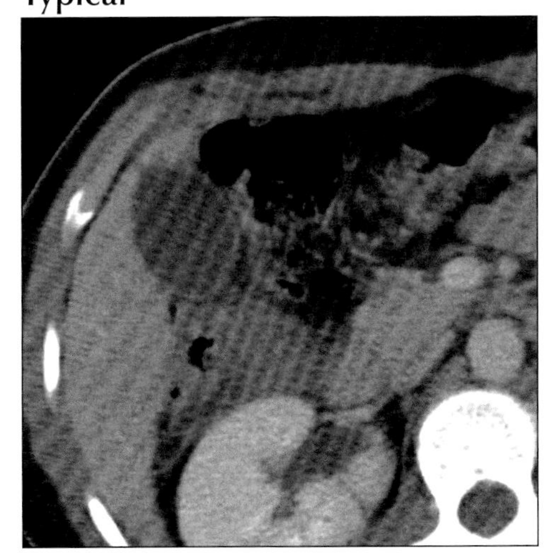

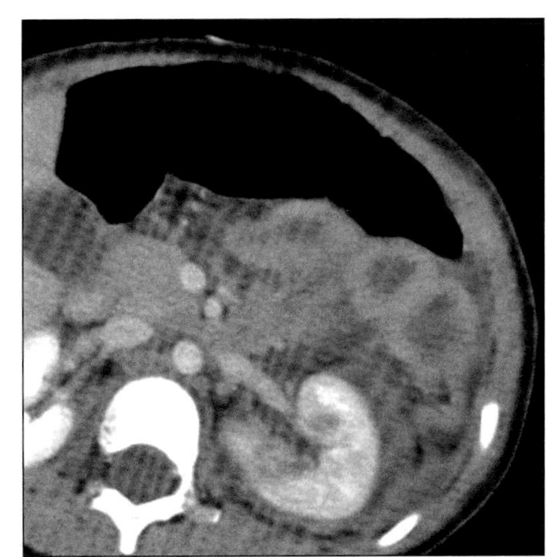

(Left) *Axial CECT shows fluid in anterior pararenal space. Duodenal laceration.* ***(Right)*** *Axial CECT shows thick-walled jejunum + mesenteric blood. Surgery: Jejunal transection, splenic + renal lacerations.*

Typical

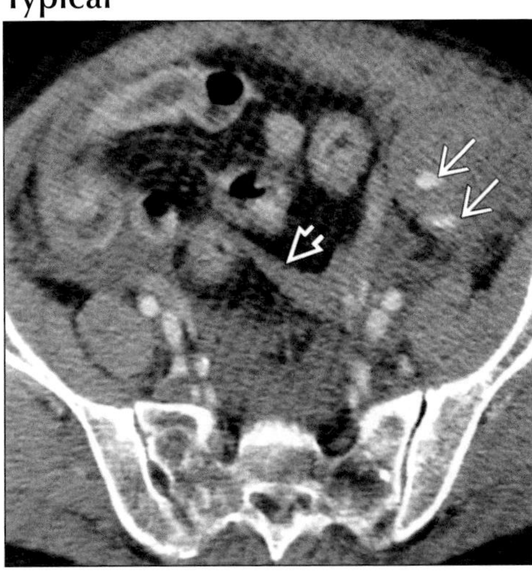

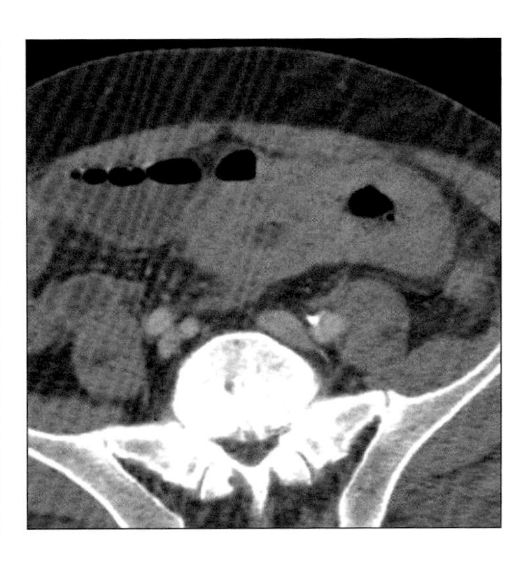

(Left) *Axial CECT shows thick-walled distal SB, mesenteric blood (open arrow) and two sites of active mesenteric bleeding (arrows).* ***(Right)*** *Axial CECT shows intramural hematoma of jejunum.*

GASTROINTESTINAL BLEEDING

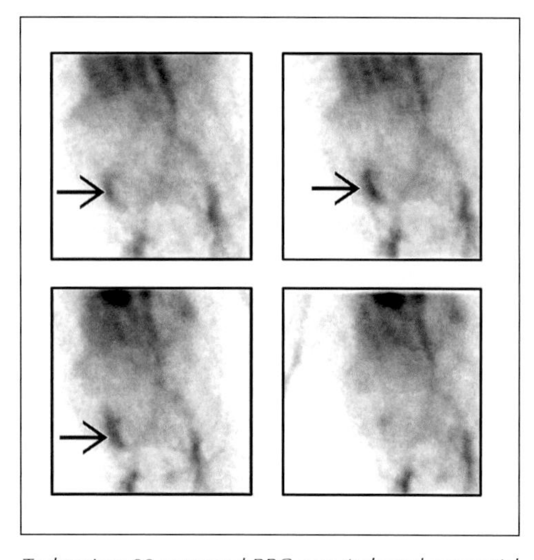

Technetium 99 m tagged RBC scan (selected sequential images) shows accumulation of radiotracer within the cecum and ascending colon (arrows) indicating active bleeding.

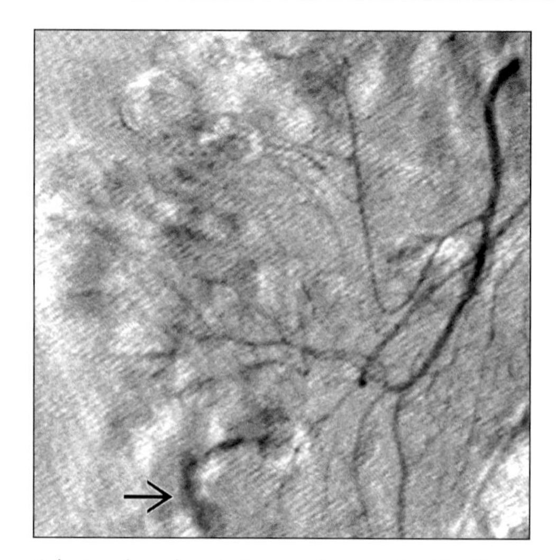

Selective ileocolic arteriogram shows active bleeding in the cecum (arrow). Bleeding diverticulum.

TERMINOLOGY

Definitions

- Acute or chronic bleeding from gastrointestinal tract

IMAGING FINDINGS

General Features

- Location
  - Upper GI bleed: Proximal to ligament of Treitz
  - Lower GI bleed: Distal to ligament of Treitz
- Key concepts
  - GI bleeding may originate anywhere from mouth to anus & may be overt or occult
  - Classification based on location & presentation
    - Upper GI bleed: Hematemesis; melena (may also seen in small bowel or right colon bleeding)
    - Lower GI bleed: Hematochezia (may also result from vigorous upper GI bleeding, duodenal ulcer)
    - Bleed anywhere in GI tract: Occult bleeding detected by chemical testing of stool
  - Classification based on onset & presentation
    - Acute: Hematemesis, hematochezia, melena
    - Chronic: Iron deficiency anemia
  - Upper GI bleed accounts for 76% of GI hemorrhage
    - Most common cause: Peptic ulcers (> 50% cases)
    - Usually present with hematemesis, melena
  - Lower GI bleed accounts for 24% of GI hemorrhage
    - Most common cause: Diverticulosis (50% cases)
    - Usually present with hematochezia
  - Endoscopy
    - First line of procedure in upper GI bleeding
    - Elective procedure; detect any rate of bleeding
    - 90-95% accurate diagnosis
  - Capsule endoscopy: Pitfalls
    - May fail in bowel with strictures or prior surgery
    - Long study time: 8 hrs prior preparation & 8 hrs to compile & interpret data
    - Hard to localize lesions & can also miss lesions
  - Uncommon site of hemorrhage: Small bowel between 2nd part of duodenum & ileocecal valve
    - Accounts for 3-5% of all GI tract bleeding
    - Difficult location to diagnose bleeding source
    - Present with prolonged undiagnosed iron deficiency anemia or episodes of melena with normal upper endoscopy & colonoscopy
  - Choice of imaging varies based on availability, expertise, severity of hemorrhage, patient condition & clinically suspected origin of bleeding

CT Findings

- NECT: Hyperattenuating hematoma within bowel
- CECT & CTA
  - Active bleeding
    - Linear, pooled or swirled focal collection of hyperdense intraluminal contrast extravasation

DDx: Endoscopic Manifestations of GI Bleeding

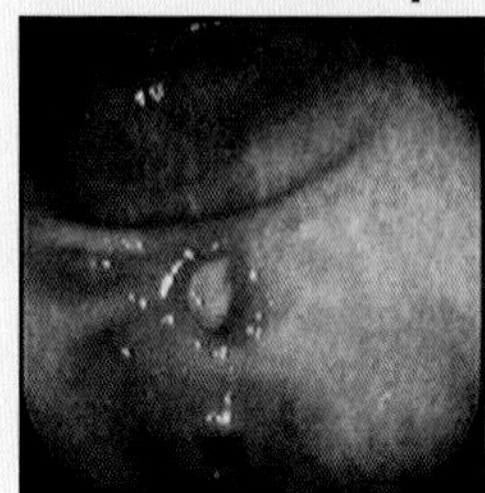

Gastric Ulcer

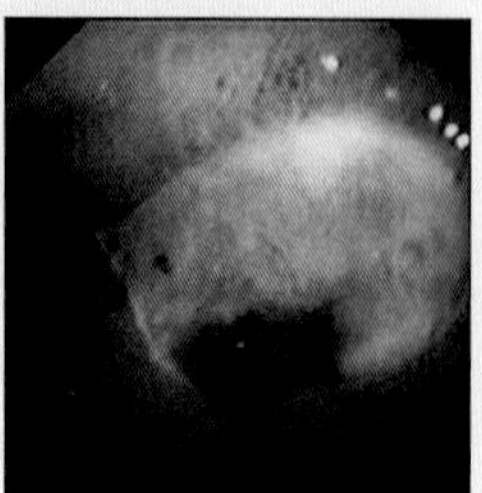

Duodenal Ulcer

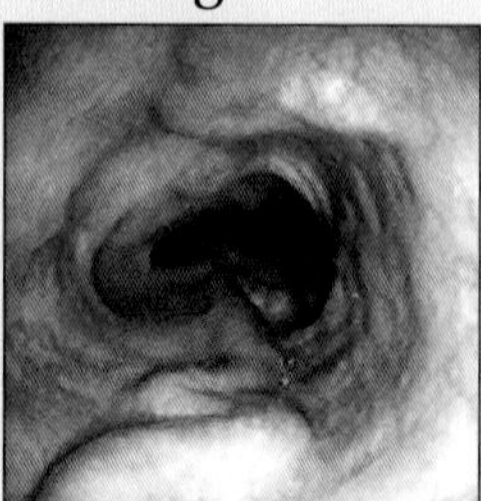

Esophageal Varices

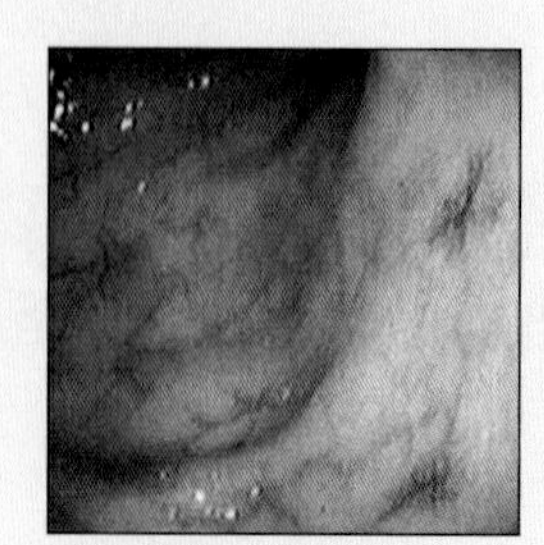

Colonic AVMs

GASTROINTESTINAL BLEEDING

Key Facts

Imaging Findings
- Upper GI bleed: Proximal to ligament of Treitz
- Lower GI bleed: Distal to ligament of Treitz
- Linear, pooled or swirled focal collection of hyperdense intraluminal contrast extravasation
- Mesenteric angiography: > 0.5 mL/min
- Non-selective aortic angiography: 6 mL/min
- Criteria for a positive study: Radiolabeled RBCs outside normal areas of blood pool
- Extravasation of isotope at active bleeding site

Top Differential Diagnoses
- Peptic ulceration
- Erosive gastritis
- Esophageal varices
- Diverticulosis
- Colonic angiodysplasia (AVM)

Pathology
- Duodenal ulcer (24%), gastric erosions (23%)
- Gastric ulcer (21%), varices (10%)
- Diverticulosis (43%), vascular ectasia (20%)

Clinical Issues
- Hematemesis: Bloody vomitus; red, coffee grounds
- Melena: Black, tarry stools (100-200 mL of blood in upper GI tract is required to produce melena)
- Hematochezia: Red blood per rectum
- Occult blood detected by stool chemical testing

Diagnostic Checklist
- Scintigraphy for hemodynamically stable cases
- Angiography for hemodynamically unstable cases

Angiographic Findings
- Conventional
  - Usually preceded by Tc99m-labeled RBC scan
  - Bleeding rate (required to be detected)
    - Mesenteric angiography: > 0.5 mL/min
    - Non-selective aortic angiography: 6 mL/min
  - Diagnostic accuracy: 70-95%

Nuclear Medicine Findings
- Tc99m-labeled red blood cell (RBC) scans
  - Bleeding rate: Requires 0.2 mL/min to be detected
    - Less sensitive than Tc99m sulfer colloid (SC) scan
  - Minimal amount of 5-10 mL of extravasated blood must be present to be identified
  - Sensitivity (85-95%); specificity (70-85%)
    - Recent new technique: ↑ Efficiency & sensitivity
  - Continuous dynamic imaging (large FOV camera)
    - 15 min dynamic image sequence of 60 images
    - Repeated until bleeding identified/study stopped
  - Criteria for a positive study: Radiolabeled RBCs outside normal areas of blood pool
  - Advantages of RBC scan
    - Ability to monitor over a prolonged period of time
    - ↑ Likelihood of detecting intermittent bleeding
  - Disadvantages of RBC scan
    - Origin of bleed unclear on delayed scans
    - Vascular organs may interfere with detection
    - Loss of tag can produce false +/-
- Technetium Tc99m SC scans
  - Bleeding rate: 0.05-0.1mL/min is detected
    - More sensitive than Tc99m labeled RBC scan
  - Extravasation of isotope at active bleeding site
  - Highly sensitive & specific test in detecting active bleeding
  - Intravascular half-life: 2.5 minutes
    - By 12-15 min, injected tracer is cleared from vascular system, producing significant contrast between bleeding site & surrounding tissue
  - Usually valid, only for localizing lower GI bleeding

Imaging Recommendations
- Angiography
  - Procedure of choice
    - Hemodynamically unstable patients
    - In active or massive GI bleeding
  - Upper GI bleeding
    - When endoscopy inconclusive
    - Anticipation of transcatheter intervention
  - Lower GI bleeding
    - Angiography may be the procedure of choice
- Nuclear scintigraphy
  - Procedure of choice
    - Hemodynamically stable patients
    - In active gastrointestinal hemorrhage
    - Delineate obscure sources: Small bowel, intermittent bleeding
    - Enhance the efficacy of angiography

DIFFERENTIAL DIAGNOSIS

Peptic ulceration
- Round or ovoid collections of barium
- Giant ulcers, usually located in duodenal bulb (↑ risk of hemorrhage)
- Wall thickening or luminal narrowing

Erosive gastritis
- Complete or varioliform
  - Multiple punctate or slit-like collections of barium
  - Scalloped or nodular antral folds
  - Location: Gastric antrum
- Incomplete or flat erosions
  - Location: Antrum or body
  - Multiple linear streaks or dots of barium
- Linear or serpiginous erosions clustered in body or near greater curvature (NSAIDs induced)

Esophageal varices
- Mucosal relief views
  - Tortuous, serpiginous, longitudinal radiolucent filling defects in collapsed esophagus
- Double-contrast study: Multiple radiolucent filling defects etched in white
- CT: Scalloped esophageal mural masses

GASTROINTESTINAL BLEEDING

Diverticulosis

- 75% of tics-left colon; 70% bleeding tics-right colon
- 80% resolve spontaneously
- Up to 20% of cases bleed, 5% massively
- Flask-like protrusion, long or large neck
- Diverticulum with long & narrow neck: Mimic pedunculated polyp on air-contrast enema
- Diverticulum with large neck: Mimics sessile polyp
- En face: Ring shadow or round barium collection
- "Bowler hat" sign: Dome of hat points away from bowel wall (diverticulum); toward lumen (polyp)
- CT imaging
  - Mural thickening of colon (4-15 mm)
  - Multiple air or contrast or stool-containing outpouchings (diverticula)

Colonic angiodysplasia (AVM)

- Arteriovenous malformations-cecum, ascending colon
- Angiography
  - Cluster (tangle) of small arteries in arterial phase
  - Early filling & delayed emptying of dilated veins

PATHOLOGY

General Features

- Etiology
  - Upper GI bleeding
    - Duodenal ulcer (24%), gastric erosions (23%)
    - Gastric ulcer (21%), varices (10%)
    - Mallory-Weiss tear (7%), esophagitis (6%)
    - Neoplasm (3%), other causes (11%)
  - Lower GI bleeding
    - Diverticulosis (43%), vascular ectasia (20%)
    - Idiopathic (12%), neoplasia (9%)
    - Colitis: Radiation (6%), ischemic (2%), ulcerative (1%)
  - Risk factors of upper GI bleeding
    - Alcohol, tobacco, anticoagulants
    - Aspirin, non-steroidal anti-inflammatory drugs
- Epidemiology
  - More than 400,000 hospitalizations annually in US
  - Lower GI bleeding
    - Accounts for < 1% of all hospital admissions in US
    - Annual incidence: 20.5/100,000 (M > F)

Gross Pathologic & Surgical Features

- Varies based on underlying pathology

Microscopic Features

- Varies depending on underlying cause

CLINICAL ISSUES

Presentation

- Most common signs/symptoms
  - Upper GI bleeding
    - Hematemesis: Bloody vomitus; red, coffee grounds
    - Melena: Black, tarry stools (100-200 mL of blood in upper GI tract is required to produce melena)
  - Lower GI bleeding
    - Hematochezia: Red blood per rectum
    - May also result from upper GI bleed (> 1000 cc)
  - Bleed anywhere in GI tract
    - Occult blood detected by stool chemical testing
  - Symptoms & signs of blood loss
    - Dizziness, tachycardia, hypotension, shock
  - Symptoms & signs of underlying pathology
- Lab data
  - Fresh blood in vomitus or stool
  - Occult blood in stool, iron deficiency anemia
  - ↓ CBC count, hematocrit, serum electrolytes
  - Abnormal coagulation profile (aPTT, PT, platelet count, bleeding time)
  - Serum blood urea nitrogen to creatine ratio > 25
    - Suggests upper GI hemorrhage

Demographics

- Age: More common in older age group
- Gender: Males more than females (M > F)

Natural History & Prognosis

- Complications
  - Acute massive GI bleeding: Shock & death
  - Complications of underlying cause
- Prognosis
  - Early detection, resuscitation & treatment: Good
  - Delayed detection, resuscitation & treatment: Poor
  - Mortality rate
    - Upper GI bleeding: Esophageal varices (30-50%) & varies based on Rockall risk score
    - Lower GI bleeding: Ranges from 0-21%

Treatment

- Medical: Resuscitation (fluids, electrolytes, blood)
- Endoscopic therapy
  - Topical: Tissue adhesives, collagen, clotting factors
  - Injection: Sclerosant agents & vasoconstrictors
  - Mechanical: Clips, balloons, sutures
  - Thermal: Laser photo & electrocoagulation
- Transjugular intrahepatic portosystemic shunt (TIPS)
- Interventional: Embolotherapy (Gelfoam & coils)
- Surgical treatment

DIAGNOSTIC CHECKLIST

Consider

- Scintigraphy for hemodynamically stable cases
- Angiography for hemodynamically unstable cases

SELECTED REFERENCES

1. Tew K et al: MDCT of acute lower gastrointestinal bleeding. AJR Am J Roentgenol. 182(2):427-30, 2004
2. Hastings GS: Angiographic localization and transcatheter treatment of gastrointestinal bleeding. Radiographics. 20(4):1160-8, 2000
3. Maurer AH et al: Effects of in vitro versus in vivo red cell labeling on image quality in gastrointestinal bleeding studies. J Nucl Med Technol. 26:87-90, 1998
4. Whitaker SC et al: The role of angiography in the investigation of acute or chronic gastrointestinal hemorrhage. Clin Radiol. 47:382-8, 1993
5. Bunker SR: Cine scintigraphy of gastrointestinal bleeding. Radiology. 187:877-8, 1993

IMAGE GALLERY

Typical

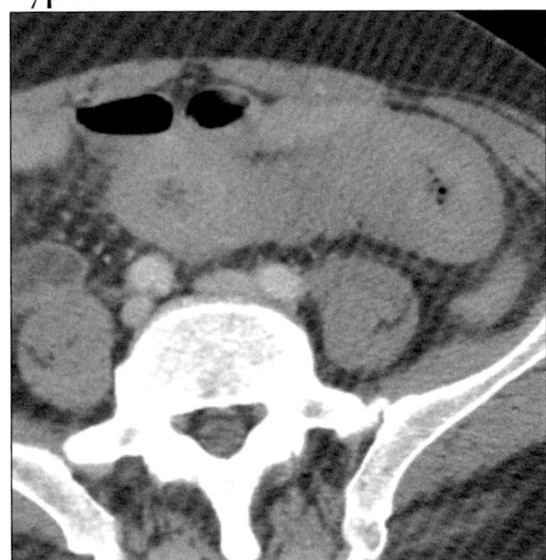

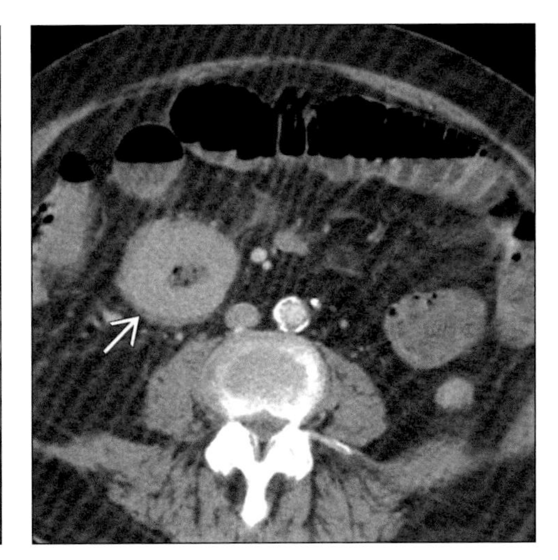

(Left) *Axial CECT shows massive thickening of the jejunal wall due to spontaneous bleeding in an anticoagulated patient.* ***(Right)*** *Axial CECT in a patient with abdominal pain and GI bleeding following heart transplantation shows small bowel hematoma (arrow) resulting in partial SB obstruction.*

Typical

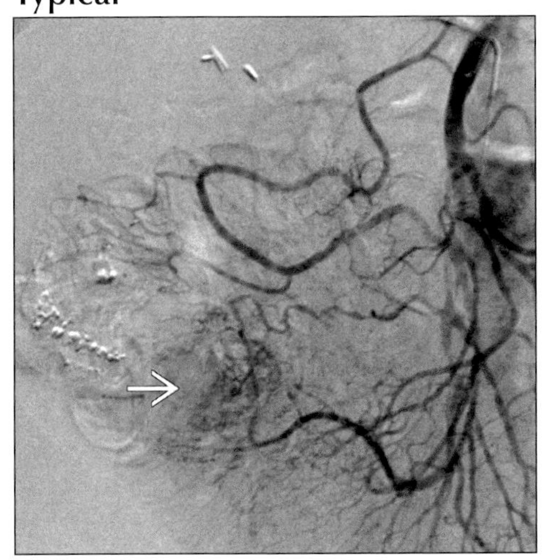

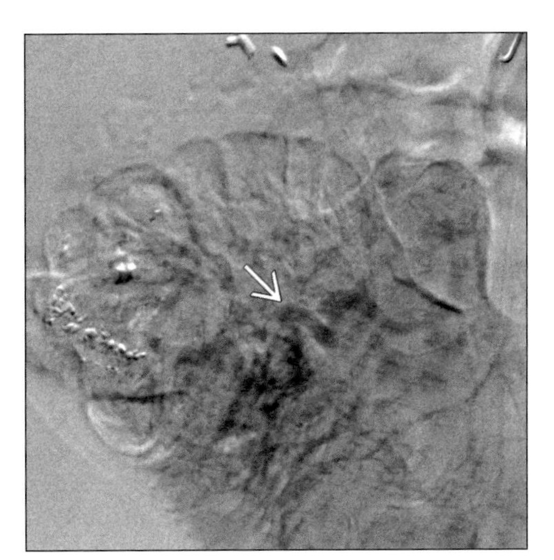

(Left) *Superior mesenteric artery (SMA) arteriogram shows a tangled cluster of small vessels (arrow) in cecum; angiodysplasia.* ***(Right)*** *SMA arteriogram shows early draining vein (arrow) from angiodysplasia.*

Typical

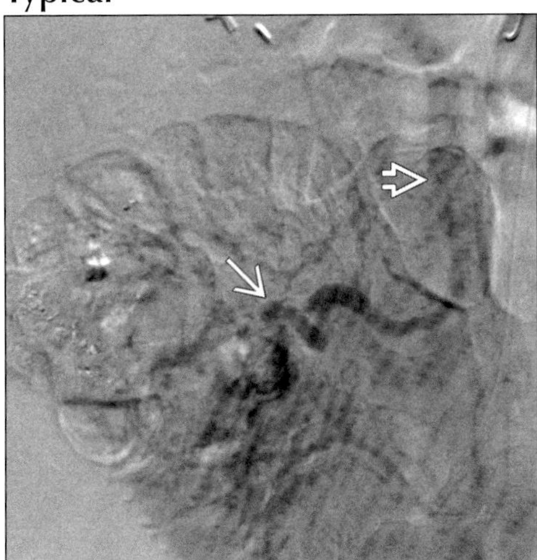

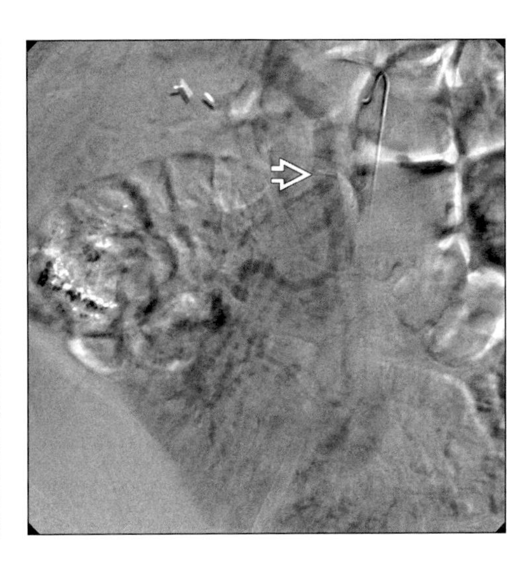

(Left) *SMA arteriogram shows enlarged early draining vein (arrow) from angiodysplasia and premature filling of SM vein (open arrow).* ***(Right)*** *SMA arteriogram shows early draining vein from angiodysplasia and premature filling of SM vein (open arrow).*

INTRAMURAL BENIGN INTESTINAL TUMORS

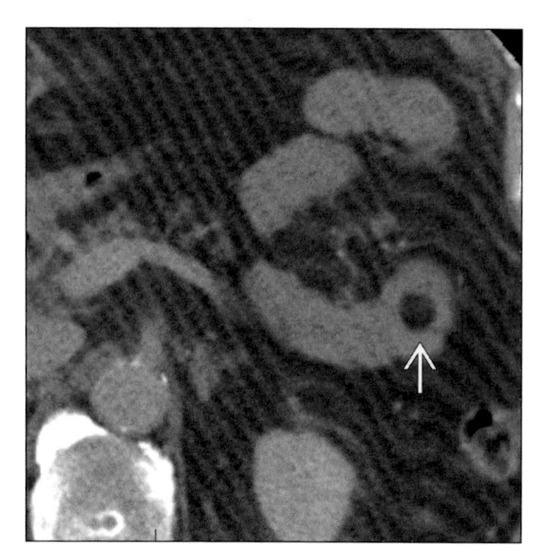
Axial CECT shows fat density intramural/luminal mass (arrow); jejunal lipoma.

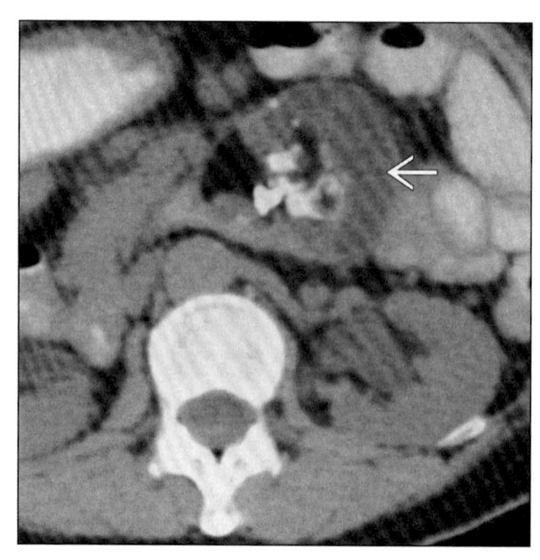
Axial CECT shows duodenal teratoma (arrow) that is comprised of fat, calcification, and soft tissue.

TERMINOLOGY

Definitions

- Benign mass composed of one or more tissue elements of the small bowel wall

IMAGING FINDINGS

General Features

- Best diagnostic clue: Intramural mass with smooth, oval or round luminal defects on barium studies
- Other general features
  - Types of intramural benign intestinal tumors
    - Gastrointestinal stromal tumor (GIST)
    - Leiomyoma, lipoma, hemangioma

Radiographic Findings

- Fluoroscopic-guided barium studies
  - GIST
    - Most common; intestinal dilatation
    - Circumscribed, lobulated mass; few mm to 30 cm
    - Sharply defined margins; mucosal surface may show luminal irregularity/focal ulceration (≤ 50%)
    - Extraserosal component: Mass effect often large
    - ± Cavity with fistula to intestinal lumen; irregular gas collection
  - Leiomyoma
    - Luminal protrusion & displaces adjacent bowel
    - Smooth, oval/round luminal defects (endoenteric)
  - Lipoma
    - Most common in ileum; sharply demarcated, pedunculated mass conforms to bowel lumen
    - Configuration changes in peristalsis/compression
  - Hemangioma
    - Millimeters in size; ± calcified phleboliths
    - Multiple, intraluminal/intramural nodular defects

CT Findings

- GIST
  - Hypo-/hypervascular mass; ± calcification
  - Enhancing mass/polyp with areas of low attenuation from hemorrhage, necrosis or cyst formation
  - ± Extends into mesentery & encases other structures
- Leiomyoma: 1-10 cm sharply defined spherical mass with homogeneous enhancement; ± focal calcification
- Lipoma: 1-6 cm solitary mass with attenuation (-80 to -120 HU) similar to fat; ± soft tissue stranding

Imaging Recommendations

- Best imaging tool: Barium studies followed by CT

DIFFERENTIAL DIAGNOSIS

Intestinal metastases and lymphoma

- E.g., malignant melanoma, lung, breast cancer

DDx: Focal Wall Thickening

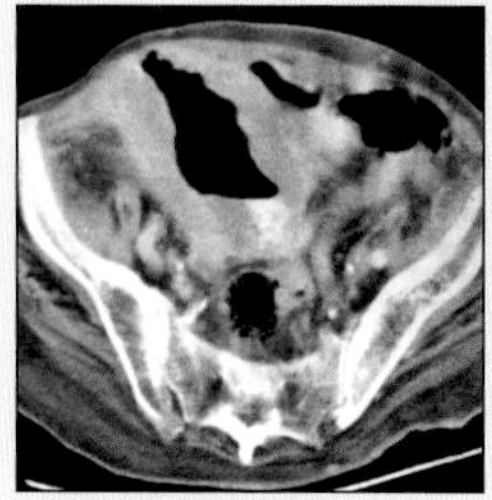
Lymphoma

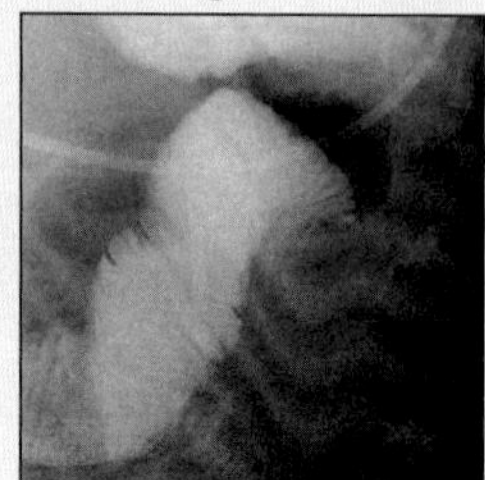
Melanoma

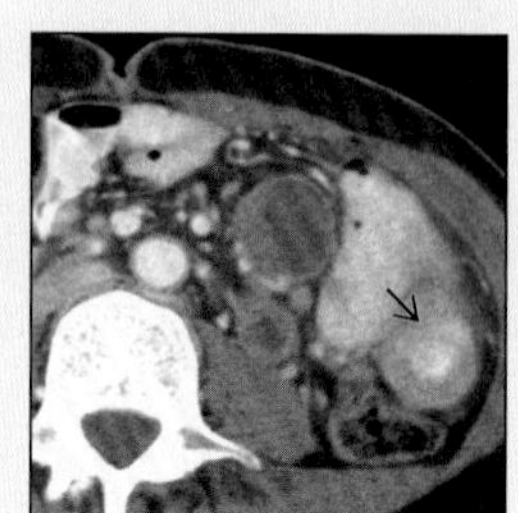
Carcinoma

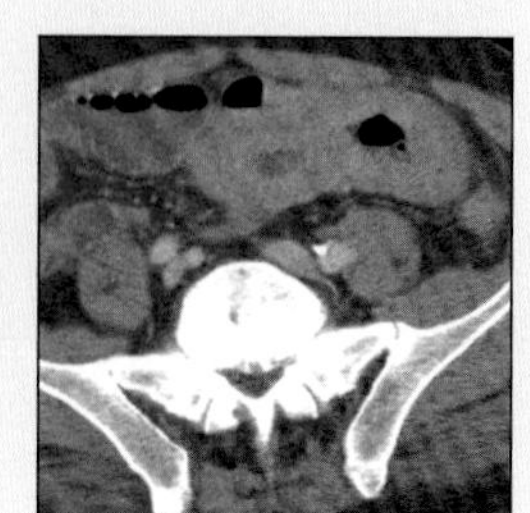
Hematoma

INTRAMURAL BENIGN INTESTINAL TUMORS

Key Facts

Imaging Findings
- Best diagnostic clue: Intramural mass with smooth, oval or round luminal defects on barium studies
- Sharply defined margins; mucosal surface may show luminal irregularity/focal ulceration (≤ 50%)
- Best imaging tool: Barium studies followed by CT

Top Differential Diagnoses
- Intestinal metastases and lymphoma
- Intestinal carcinoma
- Intramural hematoma

Diagnostic Checklist
- GIST is most common; imaging criteria to separate from other intramural tumors have not been established, except lipoma
- Smooth intramural mass, may be "pulled" into lumen by peristalsis

- Lymphoma (non-Hodgkin): Large masses that ulcerate, cavitate, extend into mesentery; ± lymphadenopathy

Intestinal carcinoma
- Adenocarcinoma: "Apple core" lesion; rigid stricture
- Carcinoid tumor: Crowded folds, bowel wall kinking, luminal narrowing (mesenteric extension)

Intramural hematoma
- Jejunum (most common) > ileum
- Large, concentric submucosal filling defects
- Anticoagulation (10-35%), bleeding disorders, trauma

PATHOLOGY

General Features
- General path comments
  - Most diagnosed incidentally by imaging or autopsy
  - GIST
    - Immunoreactivity to c-KIT (CD117)
    - Most common intramural primary masses
    - 20-30% of all GISTs occur in small intestine
    - More aggressive than gastric GISTs with same size
- Etiology: GIST: KIT germ line mutations (52-85%)
- Associated abnormalities
  - GIST: von-Recklinghausen disease
  - Hemangioma: Cutaneous lesions, tuberous sclerosis, Turner syndrome, Osler-Weber-Rendu disease

Gross Pathologic & Surgical Features
- GIST: Well-circumscribed mass compressing adjacent tissue and lacks capsule; pink, tan or gray surface
- Leiomyoma: Endoenteric, exoenteric or bidirectional
- Lipoma: Well-circumscribed proliferation of fat

CLINICAL ISSUES

Presentation
- Most common signs/symptoms
  - Asymptomatic (most common)
  - GI bleeding, intestinal obstruction, intussusception
  - Nausea, vomiting, weight loss, palpable mass
  - Hemangioma: 80% symptomatic; acute, severe & intermittent GI bleeding (most common)

Demographics
- Age: > 45 years of age
- Gender: M = F

Natural History & Prognosis
- Complications: Intussusception, obstruction, bleeding
- Prognosis: Good, unless with recurrence or size > 5 cm

Treatment
- GIST
  - Surgical resection ± chemotherapy (Gleevec) for metastatic disease
- Other types of tumors: Usually no treatment

DIAGNOSTIC CHECKLIST

Consider
- GIST is most common; imaging criteria to separate from other intramural tumors have not been established, except lipoma

Image Interpretation Pearls
- Smooth intramural mass, may be "pulled" into lumen by peristalsis

SELECTED REFERENCES

1. Levy AD et al: Gastrointestinal stromal tumors: radiologic features with pathologic correlation. Radiographics. 23(2):283-304, 456; quiz 532, 2003
2. Laurent F et al: CT of small-bowel neoplasms. Semin Ultrasound CT MR. 16(2):102-11, 1995
3. Solomon A et al: Computed tomographic investigation of serosal and intramural gastrointestinal pathology. Gastrointest Radiol. 12(1):13-7, 1987

IMAGE GALLERY

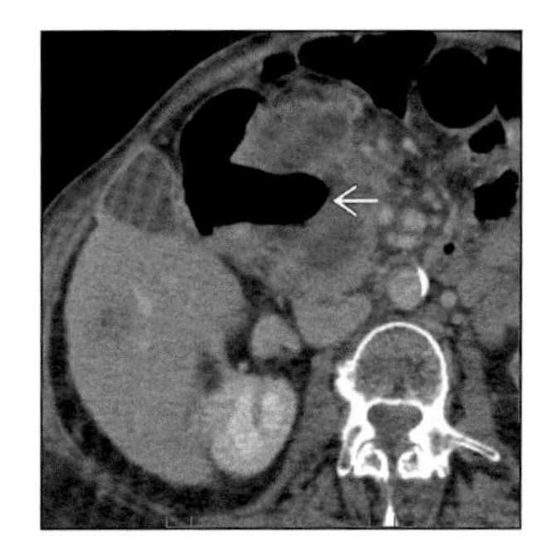

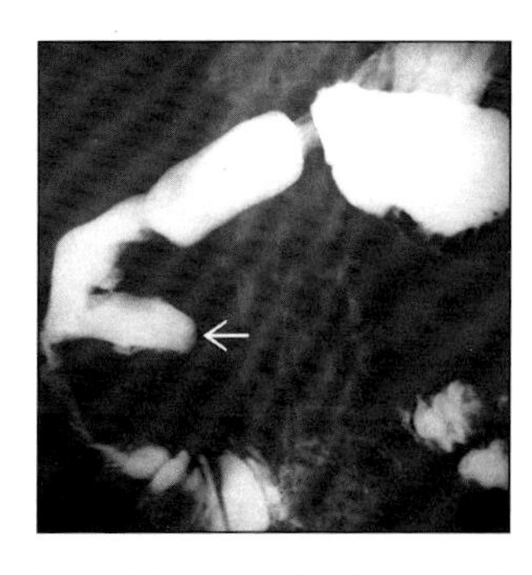

(Left) *Axial CECT shows large mass in medial wall of duodenum with large central ulceration (arrow) (GIST).* ***(Right)*** *Upper GI series shows large duodenal mass with central ulceration (arrow) (GIST).*

HAMARTOMATOUS POLYPOSIS (P-J)

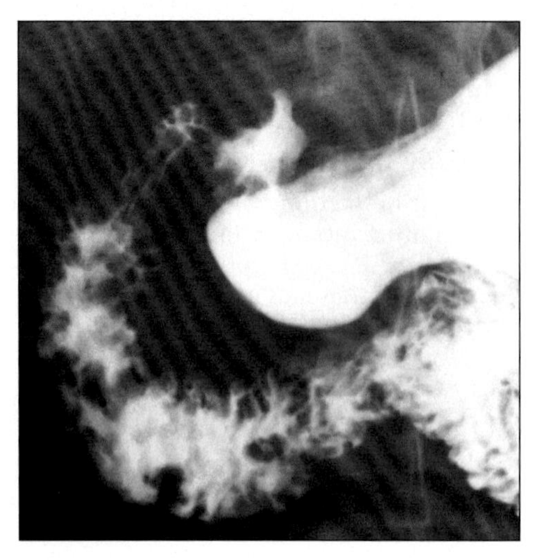

Upper GI (UGI) series shows multiple polyps in duodenum in a patient with Peutz-Jeghers syndrome.

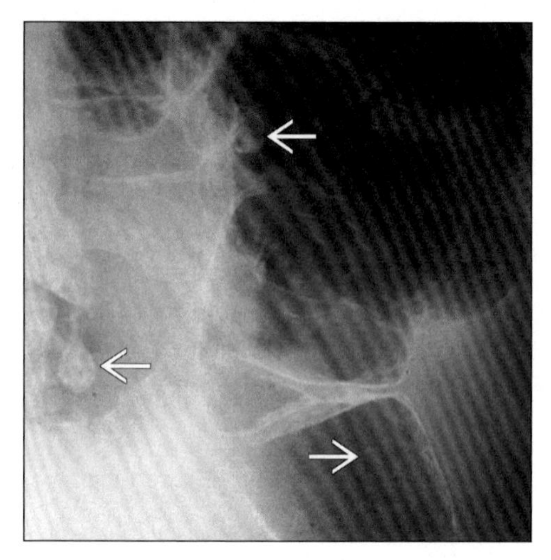

Air-contrast BE shows numerous polyps (arrows) in a patient with Peutz-Jeghers syndrome.

TERMINOLOGY

Abbreviations and Synonyms

- Peutz-Jeghers syndrome (PJS)
- Multiple hamartoma syndrome/Cowden disease (MHS)
- Juvenile polyposis (JP)
- Cronkhite-Canada syndrome (CCS)

Definitions

- Spectrum of both hereditary & nonhereditary polyposis syndromes characterized by gastrointestinal tract (GIT) polyps & other associated lesions

IMAGING FINDINGS

General Features

- Best diagnostic clue: Cluster of small radiolucent filling defects in small bowel (PJS)
- Location
  - Peutz-Jeghers syndrome
    - Jejunum + ileum > duodenum > colon > stomach
  - MHS + JP: Most polyps in rectosigmoid colon
  - CCS: Stomach-100%, colon-100%, small bowel-50%
- Size: Varied size (small, medium, large)
- Morphology
  - Sessile or pedunculated polypoid lesions
  - Pattern: Carpet-like, cluster-like or scattered
- Other general features
  - Classification: Hamartomatous polyposis syndromes
    - Peutz-Jeghers syndrome; MHS; JP; CCS
    - Bannayan-Riley-Ruvalcaba syndrome
  - PJS: Autosomal dominant, characterized by
    - Hamartomatous GI tract polyps; mucocutaneous pigmentation of lips, oral mucosa, palms & soles
    - Risk for cancer (10%): Stomach, duodenum, colon
    - Extra-GIT cancers: Pancreas, breast, reproductive
  - MHS: Autosomal dominant (AD) genodermatosis
    - Mucocutaneous: Facial papules, oral papillomas, keratosis
    - Breast: Fibrocystic (50%); ductal type cancer (30%)
    - Thyroid (65%): Adenomas, goiter, follicular cancer
    - Clinically: Bird-like face, high arched palate
  - Juvenile polyposis (JP): Sub classified into 2 types
    - Isolated juvenile polyps of childhood
    - Juvenile polyposis of colon or entire GIT
  - CCS: Inflammatory polyps + ectodermal defects

Radiographic Findings

- Fluoroscopic guided double-contrast studies
  - Multiple varied-size radiolucent filling defects
- Polyps in Peutz-Jeghers syndrome (PJS)
  - Stomach to rectum (mouth & esophagus spared)
  - Small bowel (> 95%)
    - Usually multiple & broad-based polyps
    - PJS polyps occur in clusters > carpeting bowel
    - Large polyps: Characteristic lobulated surface

DDx: Multiple Bowel Wall Lesions

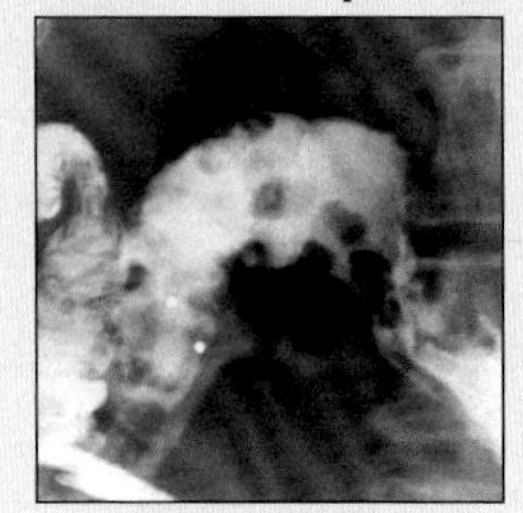

Gardner Syndrome

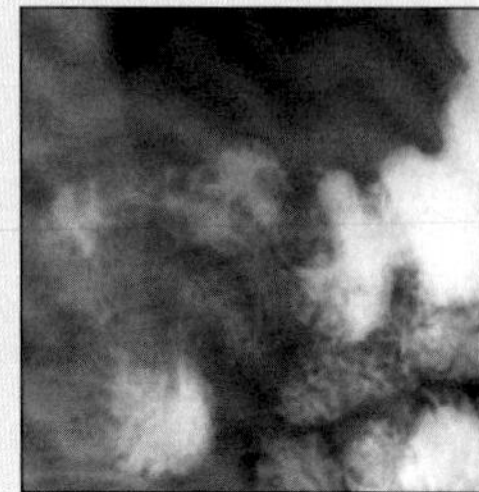

Brunner Glands

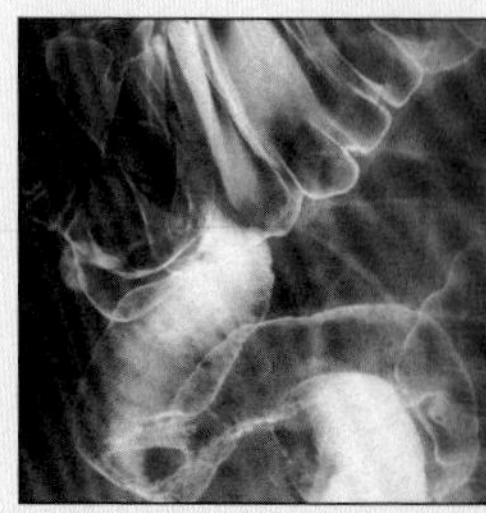

Lymphoid Follicles

Lymphoma

HAMARTOMATOUS POLYPOSIS (P-J)

Key Facts

Terminology

- Spectrum of both hereditary & nonhereditary polyposis syndromes characterized by gastrointestinal tract (GIT) polyps & other associated lesions

Imaging Findings

- Best diagnostic clue: Cluster of small radiolucent filling defects in small bowel (PJS)
- Jejunum + ileum > duodenum > colon > stomach
- PJS polyps occur in clusters > carpeting bowel

Top Differential Diagnoses

- Familial adenomatous polyposis
- Brunner gland hyperplasia (hamartoma)
- Lymphoid follicles (hyperplasia)
- Metastases & lymphoma (GI tract)

Pathology

- Hereditary (AD): PJS, MHS, 25% of JP

  - Colorectal: (30%): Multiple scattered; no carpeting
  - Stomach + duodenum (25%): Diffuse involvement

Imaging Recommendations

- Double-contrast barium studies (multiple views)

DIFFERENTIAL DIAGNOSIS

Familial adenomatous polyposis

- 500-2,500 polyps carpeting colonic mucosa
- Tubular or tubulovillous; colorectal cancer risk 100%

Brunner gland hyperplasia (hamartoma)

- Location: Duodenal bulb & descending part
- Hyperplasia: Multiple nodules (Swiss cheese pattern)
- Hamartomas: Simulate hamartomatous polyps
- Associated thickened, irregular folds differentiates

Lymphoid follicles (hyperplasia)

- Innumerable small or tiny radiolucent nodules
- Usually generalized (duodenum, small bowel, colon)
- Distinguished by clinical history & generalized pattern

Metastases & lymphoma (GI tract)

- Metastases: May be polypoid mimicking polyps
- Lymphoma: Small/bulky polypoid, mimicking polyps
  - Thickened bowel wall & folds; adenopathy seen

PATHOLOGY

General Features

- Genetics: Spontaneous mutation of gene on chromosome 19 (PJS) & 10 (MHS)
- Etiology
  - Hereditary (AD): PJS, MHS, 25% of JP
  - Nonhereditary: CCS & 75% of JP
- Epidemiology: PJS incidence: 1 in 10,000 people

Gross Pathologic & Surgical Features

- Sessile/pedunculated; carpet or cluster-like or scattered

Microscopic Features

- Extensive smooth muscle arborization of polyps

CLINICAL ISSUES

Presentation

- Most common signs/symptoms: PJS: Pain, mucocutaneous pigmentation, melena
- Lab-data: Hypochromic anemia; positive stool guaiac test

Demographics

- Age: PJS (10-30); MHS (30-40); CCS (above 60) years
- Gender: PJS (M = F); MHS (M < F); JP (M > F); CCS (M < F)

Natural History & Prognosis

- Complications: Intussusception, SBO, cancer risk (PJS)
- Prognosis: Risk of cancer, 40% by 40 years of age

Treatment

- Follow-up & surveillance; surgery in malignant cases

DIAGNOSTIC CHECKLIST

Image Interpretation Pearls

- PJS: Small bowel polyps, mucocutaneous pigmentation

SELECTED REFERENCES

1. Cho GJ et al: Peutz-Jeghers syndrome and hamartomatous polyposis syndromes: Radiologic-pathologic correlation. RadioGraphics. 17: 785-91, 1997
2. Harned RK et al: The hamartomatous polyposis syndromes: Clinical and radiologic features. AJR. 164: 565-71, 1995
3. Buck JL et al: From the archives of the AFIP: Peutz-Jeghers syndrome. RadioGraphics. 12: 365-78, 1992

IMAGE GALLERY

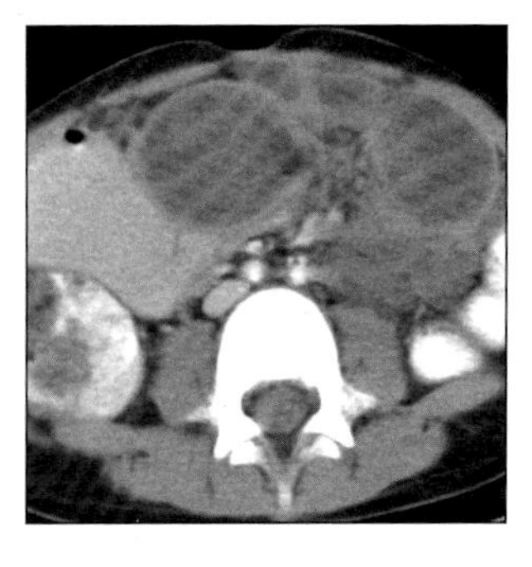

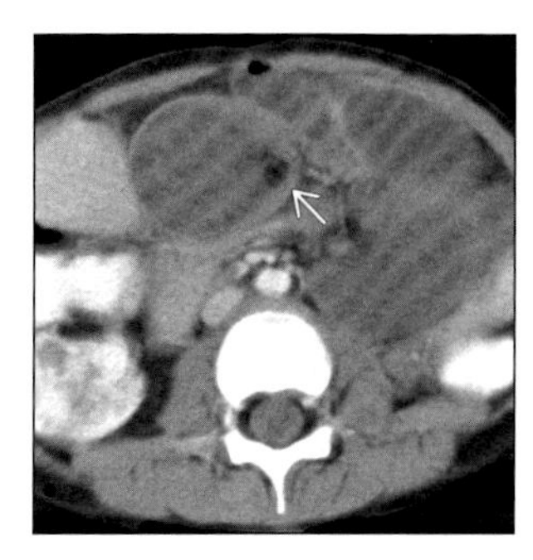

(Left) *Axial CECT shows markedly dilated small bowel (SB) due to intussusception.* ***(Right)*** *Axial CECT in a 4 year old boy shows SB intussusception (arrow) proved at surgery to be due to hamartomatous polyps.*

SMALL BOWEL CARCINOMA

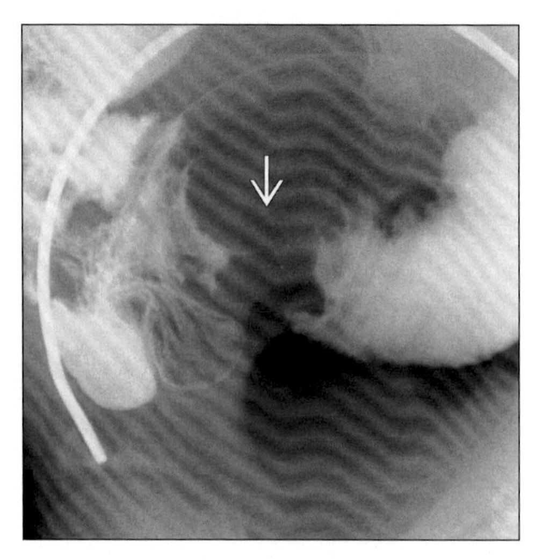

Spot film from small bowel follow through (SBFT) shows "apple core" stricture of terminal ileum (arrow) and mass effect on cecum.

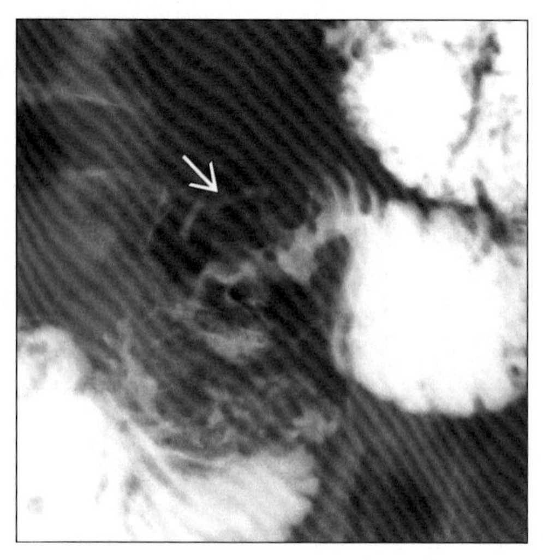

Spot film from SBFT shows jejunal mass (arrow) with mucosal destruction and luminal narrowing.

TERMINOLOGY

Abbreviations and Synonyms

- Small bowel carcinoma (SBC)

Definitions

- Primary adenocarcinoma of small intestine

IMAGING FINDINGS

General Features

- Best diagnostic clue: Annular or plaque-like polypoid mass on enteroclysis
- Location
  - Proximal jejunum: Most common within first 30 cm beyond ligament of Treitz
  - Ileum: Less than 15%
  - Duodenum: Most common (> 50%) when included in small bowel
- Size: Tumor size varies from 3-8 cm in diameter
- Morphology
  - Infiltrating tumor
  - Small plaque-like polypoid adenocarcinoma
  - Pedunculated polyp

Radiographic Findings

- Fluoroscopic guided enteroclysis
  - Infiltrating tumor: "Apple-core" or annular lesion
    - Short, well-demarcated, circumferential narrowing
    - Irregular lumen, overhanging edges, ± ulceration
    - Narrow, rigid stricture with prestenotic dilatation
  - Polypoid sessile tumor: Small plaque-like growth
  - Pedunculated polypoid adenocarcinoma (rare)

CT Findings

- Annular, ulcerative lesion or discrete nodular mass
- Circumferential thickened wall ± mesenteric invasion
- Soft tissue mass ± luminal narrowing & obstruction
- Heterogeneous density; ± enlarged mesenteric nodes
- Growth shows moderate enhancement on CECT
- ± Metastases: Liver, peritoneal surfaces, ovaries

Ultrasonographic Findings

- Real Time: Thickened small bowel wall

Angiographic Findings

- Tumor that displaces feeding arteries

Imaging Recommendations

- Best imaging tool: Enteroclysis; upper GI series & small bowel follow-through

DIFFERENTIAL DIAGNOSIS

Lymphoma & leiomyosarcoma

- Lymphoma
  - Non-Hodgkin lymphoma most common

DDx: Small Intestinal Mass

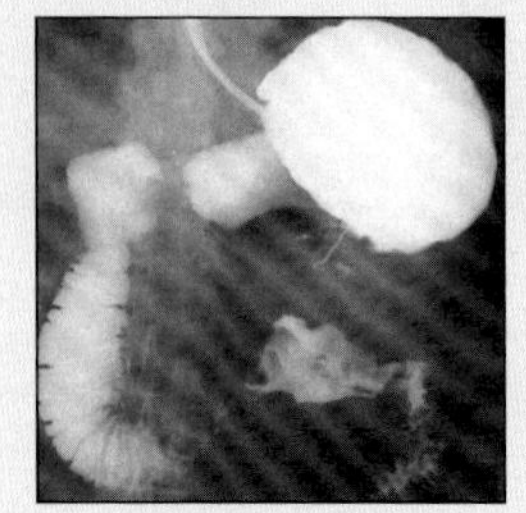

Jejunal Lymphoma

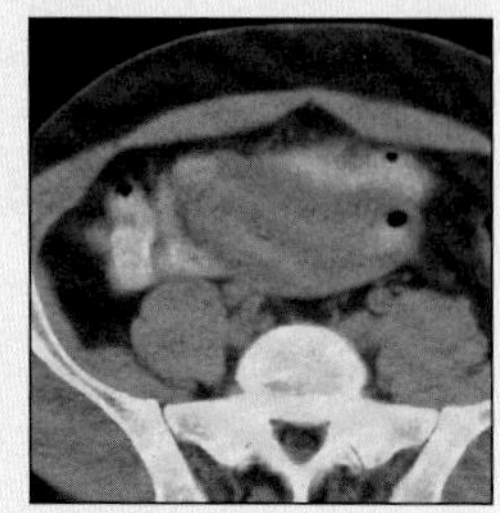

Ileal Lymphoma

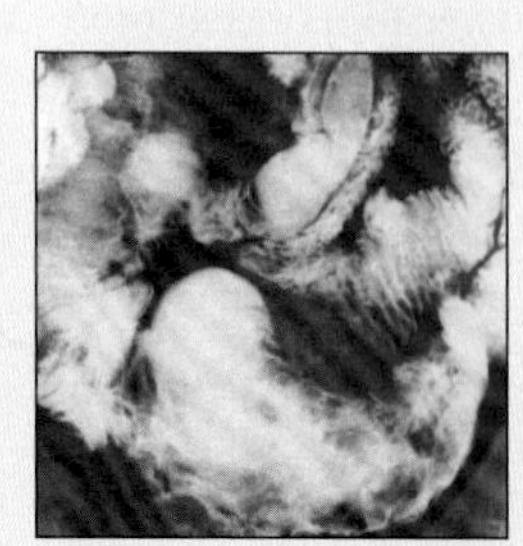

Metastatic Melanoma

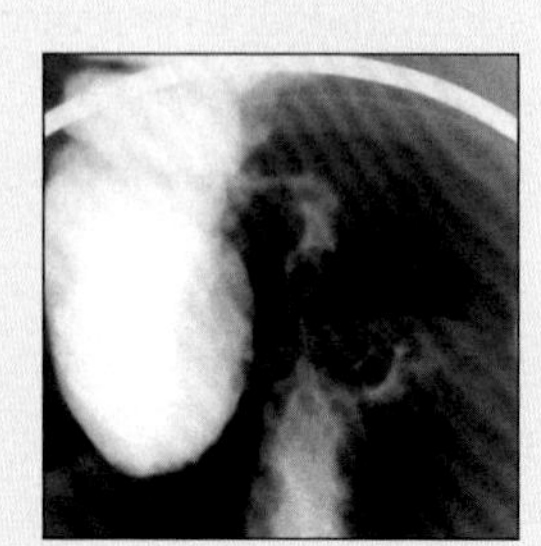

Crohn Disease

SMALL BOWEL CARCINOMA

Key Facts

Imaging Findings

- Best diagnostic clue: Annular or plaque-like polypoid mass on enteroclysis
- Proximal jejunum: Most common within first 30 cm beyond ligament of Treitz
- Circumferential thickened wall ± mesenteric invasion
- Soft tissue mass ± luminal narrowing & obstruction
- Heterogeneous density; ± enlarged mesenteric nodes
- ± Metastases: Liver, peritoneal surfaces, ovaries

Top Differential Diagnoses

- Lymphoma & leiomyosarcoma
- Metastases
- Carcinoid tumor
- Crohn disease

Diagnostic Checklist

- MDCT, 3D imaging in diagnosing, surgical planning
- Annular, discrete nodular or ulcerative lesion on CT

  - Marked luminal dilatation is characteristic
  - Nodular, polypoid, infiltrating, mesenteric invasive
- GI stromal tumor: Sharply defined, exophytic & bulky

Metastases

- Usually from carcinoma of colon or melanoma
- Lesions often longer, more pronounced narrowing & obstruction
- Ulceration tends to produce irregular cavity

Carcinoid tumor

- Malignant neuroendocrine tumor of small bowel
- Hypervascular submucosal mass; mesenteric invasion
- Mesenteric mass: Ca++ & desmoplastic reaction
- More common in ileum & displaces bowel loops
- Ill-defined homogeneous mass, rarely annular lesion

Crohn disease

- Usually distal ileum with long area of wall thickening
- Aphthoid ulceration, cobblestone appearance
- Skip lesions, transmural, fistulae, sinuses, fissures

PATHOLOGY

General Features

- Etiology: Most probably arise from adenoma (adenoma-carcinoma sequence)
- Epidemiology: Prevalence, 0.5-3/100,000 population
- Associated abnormalities: Adult celiac disease, Crohn disease, Peutz-Jeghers syndrome (↑ incidence of SBC)

Gross Pathologic & Surgical Features

- Infiltrating or plaque-like polypoid tumor ± stricture

Microscopic Features

- Moderate to well-differentiated neoplastic cells

CLINICAL ISSUES

Presentation

- Most common signs/symptoms
  - Symptomatic at the time of diagnosis
    - Abdominal pain, obstruction or both (90%)
    - Bleeding or anemia (50%)
  - Small palpable abdominal mass (30%)

Demographics

- Age: Usually seen in elder age group
- Gender: M > F

Natural History & Prognosis

- Complications
  - SBO; intussusception; GI bleed; perforation (rarely)
- Prognosis
  - 5 year survival rate: Jejunum carcinoma - 46%; ileal - 20%

Treatment

- Surgical resection (localized); chemotherapy (spread)

DIAGNOSTIC CHECKLIST

Consider

- MDCT, 3D imaging in diagnosing, surgical planning

Image Interpretation Pearls

- Annular, discrete nodular or ulcerative lesion on CT
- Infiltrating annular or polypoid lesion on enteroclysis

SELECTED REFERENCES

1. Horton KM et al: Multidetector-row computed tomography and 3-dimensional computed tomography imaging of small bowel neoplasms: current concept in diagnosis. J Comput Assist Tomogr. 28(1):106-16, 2004
2. Buckley JA et al: CT evaluation of small bowel neoplasms: spectrum of disease. Radiographics. 18(2):379-92, 1998
3. Laurent F et al: CT of small-bowel neoplasms. Semin Ultrasound CT MR. 16(2):102-11, 1995

IMAGE GALLERY

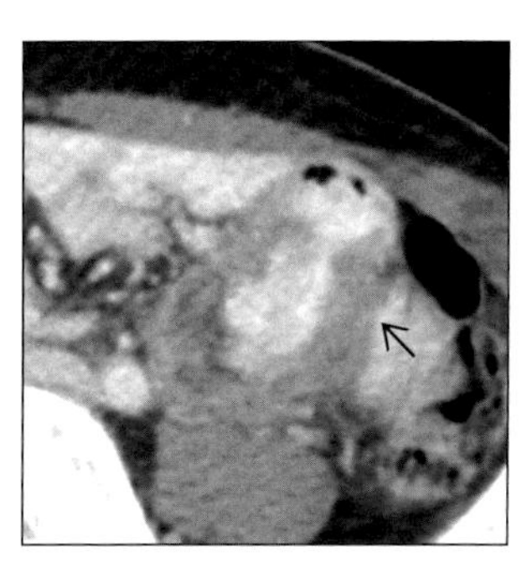
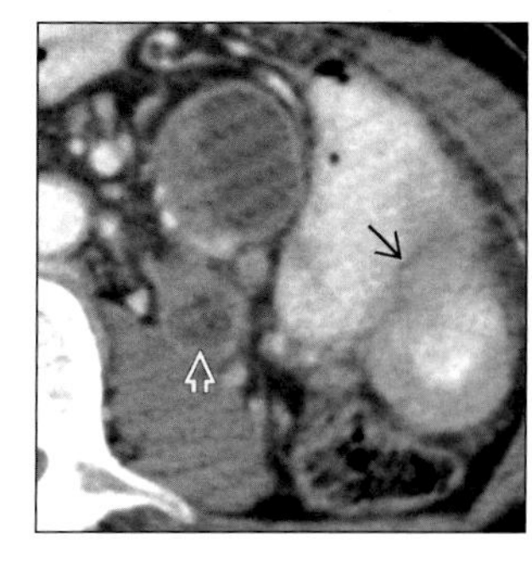

(Left) *Axial CECT shows jejunal mass (arrow) with thickened wall.* ***(Right)*** *Axial CECT shows jejunal mass (arrow) and regional lymphadenopathy (open arrow).*

CARCINOID TUMOR

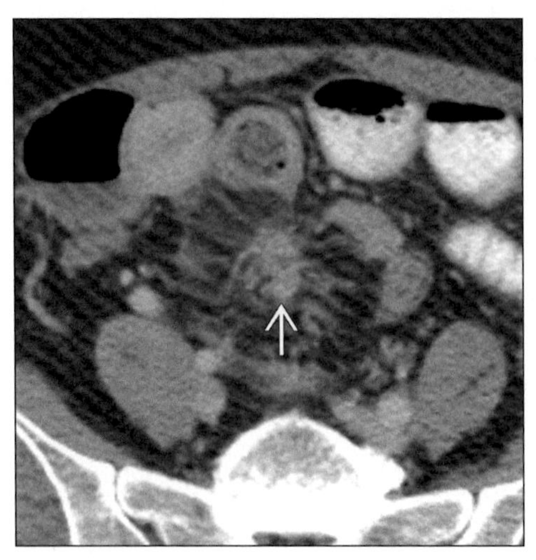

Axial CECT shows stellate mesenteric mass (arrow) and retraction of small bowel in right lower quadrant (RLQ).

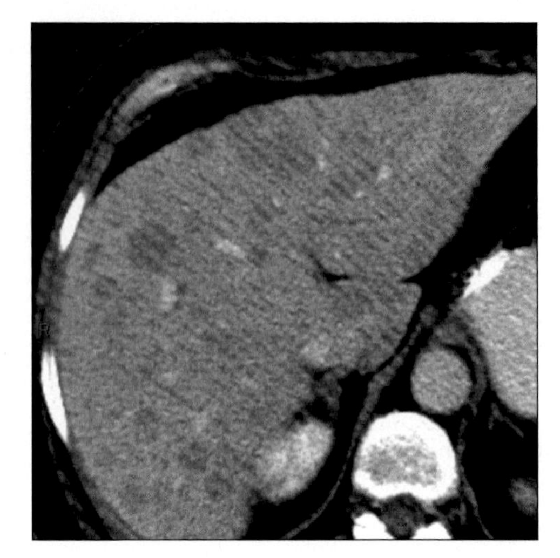

Axial CECT shows innumerable liver metastases in a patient with "carcinoid syndrome".

TERMINOLOGY

Abbreviations and Synonyms

- Gastrointestinal carcinoid (neuroendocrine tumor)

Definitions

- Primary malignant neoplasm of small bowel that arises from enterochromaffin cells of Kulchitsky

IMAGING FINDINGS

General Features

- Best diagnostic clue: Solitary, well or ill-defined, enhancing distal ileal mass + mesenteric infiltration
- Size: Varies from less than 1 cm to a few cm
- Other general features
  - Most common primary small bowel tumor beyond ligament of Treitz
  - Slow-growing tumors, but are potentially malignant
  - 2nd most common small bowel malignancy after adenocarcinoma
  - 85% of all carcinoid tumors arise within GI tract
    - Appendix (50%), incidental at appendectomy
    - Small bowel (33%); gastric, colon & rectum (2%)
    - 90% of small bowel carcinoids arise in distal ileum
  - 15% of all carcinoids arise from pancreas, lungs, biliary tree, liver, genitourinary tract & thymus
  - 30% of small bowel carcinoids are multiple
    - Associated with other malignant neoplasm, usually within GI tract in 29-53% of patients
  - 40-80% of GIT carcinoids spread to mesentery
- Key concepts
  - Carcinoid syndrome (metastatic spread to liver)
    - Spectrum of symptoms (flushing, diarrhea, asthma, pain, right heart failure)
    - Often misdiagnosed for years
    - Indicates hepatic metastases, usually from small bowel tumor
  - Symptoms require systemic circulation of secretory factors produced by carcinoid
    - Serotonin, histamine, dopamine, somatostatin
    - Vasoactive intestinal polypeptide, substance P

Radiographic Findings

- Fluoroscopic guided small bowel series or enteroclysis
  - Submucosal: Solitary/multiple, smooth filling defect
  - Ulcerated submucosal tumor: "Target lesion"
  - Thickening of wall & mucosal folds (extension)
  - Mesenteric infiltration: Small bowel loops show angulation, tethering, fixation & retraction
  - Dilated & thickened bowel loops due to ischemia

CT Findings

- Submucosal tumors
  - Solitary or multiple, well-defined enhancing lesion

DDx: Mesenteric Mass +/- Small Bowel Abnormality

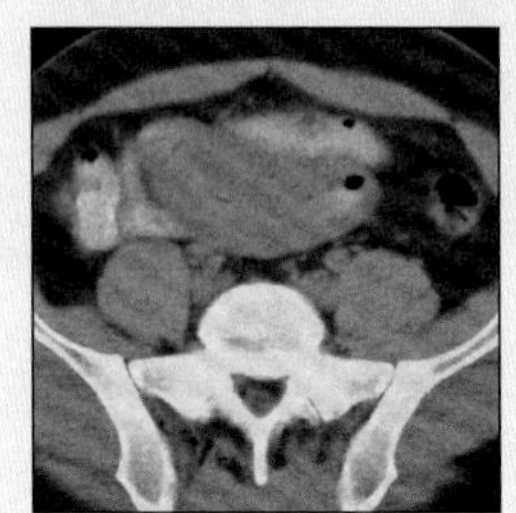

Lymphoma

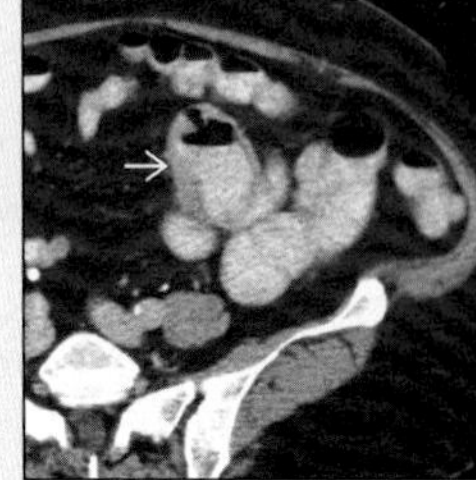

Metastasis

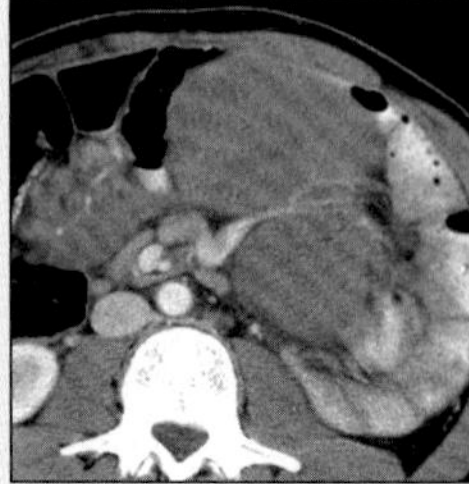

Desmoid (Gardner)

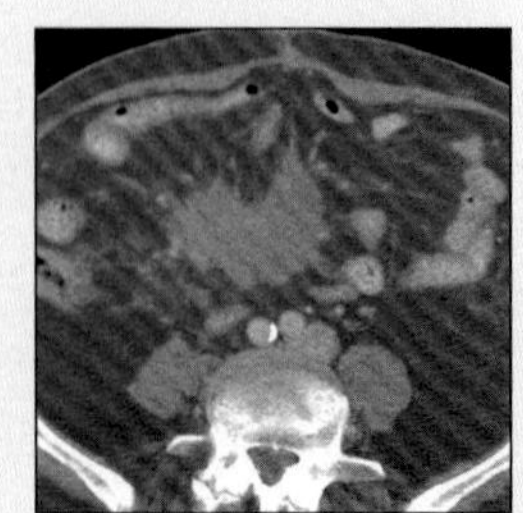

Fibros. Mesenteritis

CARCINOID TUMOR

Key Facts

Terminology
- Primary malignant neoplasm of small bowel that arises from enterochromaffin cells of Kulchitsky

Imaging Findings
- Best diagnostic clue: Solitary, well or ill-defined, enhancing distal ileal mass + mesenteric infiltration
- 90% of small bowel carcinoids arise in distal ileum
- Calcification within mesenteric mass (up to 70% of cases)

Top Differential Diagnoses
- Small bowel metastases & lymphoma
- Desmoid tumor
- Hematoma
- Fibrosing mesenteritis
- Small bowel carcinoma

Pathology
- Epidemiology: Rare (accounts 2% of GI tract tumors)
- Cardiac abnormalities (incidence 60-70%)

Clinical Issues
- Mostly asymptomatic
- Some patients symptomatic for 2-7 years before diagnosis made
- Carcinoid syndrome: Episodic cutaneous flushing, wheezing & diarrhea
- ↑ Blood levels of serotonin or 5 hydroxytryptophan
- 24-hour urine: Increased 5-HIAA levels (5x normal)

Diagnostic Checklist
- Enhancing submucosal mass in distal ileum
- Mesenteric, discrete soft-tissue mass with calcification, desmoplastic reaction ± liver metastases

- Visualization of enhancing mural mass is better with enteric water as contrast agent
- Mesenteric extension of tumor
  - Ill-defined, heterogeneous mesenteric mass
  - Calcification within mesenteric mass (up to 70% of cases)
  - Occasionally tumor may be of cystic density
  - Tumor may show spiculation with a stellate pattern
  - ± Tethering, fixation, retraction of small bowel loops
    - Due to mesenteric fibrosis & desmoplastic reaction
  - Desmoplastic reaction: "Finger-like" projections of mass into adjacent mesentery
  - ± Encasement & narrowing of mesenteric vessels
- Liver metastases
  - Arterial phase: Intense enhancement (↑ vascularity)
  - Delayed imaging: Lesions may be isodense with liver
- Three-dimensional CT angiography
  - Detects mesenteric mass & its relationship to vessels
  - Shows encasement/occlusion of mesenteric vessels
  - ± Bowel wall thickening & submucosal edema
    - Due to ischemia of involved small bowel loops
  - Small bowel mesenteric mass with calcification & desmoplastic reaction favors carcinoid tumor
    - Must be differentiated from treated lymphoma & retractile mesenteritis due to similar CT findings

MR Findings
- Submucosal tumor
  - T1WI: Isointense to muscle
  - T2WI: Hyperintense or isointense to muscle
  - T1 C+: Homogeneous enhancement
- Bowel wall thickening
  - T1WI & T2WI: Isointense to muscle
  - T1 C+: Shows enhancement
- Mesenteric extension of tumor
  - T1WI & T2WI
    - Mass & spiculation: Isointense to muscle
    - Desmoplastic strands: Hypointense
    - Calcification: Cannot be detected
  - T1 C+: Shows intense enhancement
- Liver metastases
  - T1WI: Hypointense
  - T2WI: Mild-moderately hyperintense
  - Arterial phase: Homogeneous enhancement
  - Portal venous phase: Isointense to liver
  - Larger metastases: Heterogeneous enhancement
    - Due to areas of necrosis
  - Occasionally enhancement may be peripheral, with progressive fill-in or delayed

Angiographic Findings
- Conventional
  - Primary bowel mass: Focal blush of enhancement
  - Mesenteric vessels, due to extension of tumor: Retracted, beading, tortuous or occluded
  - Liver metastases: Hypervascular
  - Hormonal assays can be done by selective portal & systemic venous blood sampling

Nuclear Medicine Findings
- In-111octreotide or somatostatin receptor scintigraphy
  - Positive for GI tract carcinoids & liver metastases
- Whole-body fluorine-18 dopa PET
  - Detects primary tumor, nodal & distant organ metastases by increased uptake
- 131I-labeled MIBG
  - ↑ Uptake by GI tract, nodal & liver metastases

Imaging Recommendations
- Helical CECT with enteric water is best imaging
  - 125 mL IV contrast at 4 mL/sec: Arterial (35 sec) & venous (70 sec) delay scans through liver
- In-111octreotide or somatostatin receptor scintigraphy
  - Sensitivity (75%) & specificity (100%)

DIFFERENTIAL DIAGNOSIS

Small bowel metastases & lymphoma
- Metastases: Recurrent colon cancer, mesothelioma, ovarian cancer mimic mesenteric carcinoid tumor
- Lymphoma
  - Most common tumor to involve mesentery is NHL
  - Bulky mass that encases bowel in "sandwich sign"
  - Associated retroperitoneal adenopathy confirms

Desmoid tumor
- Well or ill-defined soft tissue mesenteric mass
- When at root of mesentery, mimics carcinoid

Hematoma
- Cause: Blunt trauma, excessive anticoagulation, thrombocytopenia
- Acute hematoma
  - Typically quite dense (50-60 HU)
  - Focal or dispersed between leaves of mesentery
  - Density ↓ attaining of water HU by two weeks
  - Old mesenteric hematoma may calcify (rare)

Fibrosing mesenteritis
- Mesentery becomes thickened & inflamed
- "Misty" mesentery appearance seen with halo of fat surrounding mesenteric vessels

Small bowel carcinoma
- More common in jejunum than in ileum
- Spread to mesenteric nodes/liver mimicking carcinoid

PATHOLOGY

General Features
- General path comments: Belong to tumors called apudomas (amine precursor uptake & decarboxylation tumors)
- Etiology: Malignant tumor of small bowel that arises from enterochromaffin cells of Kulchitsky in crypts of Lieberkuhn
- Epidemiology: Rare (accounts 2% of GI tract tumors)
- Associated abnormalities
  - Cardiac abnormalities (incidence 60-70%)
    - Pulmonary & tricuspid stenosis or insufficiency
    - Enlargement of right heart & septal irregularities
  - May be associated with other malignant neoplasms

Gross Pathologic & Surgical Features
- Firm, yellow, submucosal nodules

Microscopic Features
- Small round cells, round nucleus, clear cytoplasm
- Tumor infiltration along neurovascular bundles
- 70% of mesenteric infiltrated tumors show Ca++
- Desmoplastic reaction

CLINICAL ISSUES

Presentation
- Most common signs/symptoms
  - Mostly asymptomatic
  - Some patients symptomatic for 2-7 years before diagnosis made
  - Carcinoid syndrome: Episodic cutaneous flushing, wheezing & diarrhea
    - Implies liver metastases with subsequent systemic venous drainage of carcinoid secretory factors
  - Other signs/symptoms
    - Abdominal pain: Secondary to intestinal ischemia
    - Right heart failure & murmurs (valvular defects)
- Lab-data
  - ↑ Blood levels of serotonin or 5 hydroxytryptophan
  - 24-hour urine: Increased 5-HIAA levels (5x normal)

Demographics
- Age: Most occur in 5th or 6th decade of life
- Gender: M:F = 2:1

Natural History & Prognosis
- GI carcinoid, no lymph node or liver metastases
  - Excellent prognosis with surgical resection
- 5 year survival rate for small bowel carcinoids is 90%
- 5 year survival rate with hepatic metastases is 50%

Treatment
- Distal small bowel tumors: Surgical resection of bowel & mesentery often with right hemicolectomy
- Proximal small bowel tumors
  - Pancreaticoduodenectomy
- Liver metastases
  - Palliative surgery of primary tumor often completed
  - Localized to single segment/lobe: Surgical resection
  - Chemoembolization; radiofrequency ablation
- Somatostatin analogue: Octreotide relieve symptoms
- Chemotherapy: No role, except in bone metastases

DIAGNOSTIC CHECKLIST

Consider
- Carcinoid syndrome indicates liver metastases

Image Interpretation Pearls
- Enhancing submucosal mass in distal ileum
- Mesenteric, discrete soft tissue mass with calcification, desmoplastic reaction ± liver metastases

SELECTED REFERENCES

1. Horton KM et al: Multidetector-row computed tomography and 3-dimensional computed tomography imaging of small bowel neoplasms: current concept in diagnosis. J Comput Assist Tomogr. 28(1):106-16, 2004
2. Horton KM et al: Carcinoid tumors of the small bowel: a multitechnique imaging approach. AJR Am J Roentgenol. 182(3):559-67, 2004
3. Maccioni F et al: Magnetic resonance imaging of an ileal carcinoid tumor. Correlation with CT and US. Clin Imaging. 27(6):403-7, 2003
4. Sheth S et al: Mesenteric neoplasms: CT appearances of primary and secondary tumors and differential diagnosis. Radiographics. 23(2):457-73; quiz 535-6, 2003
5. Buckley JA et al: CT evaluation of small bowel neoplasms: spectrum of disease. Radiographics. 18(2):379-92, 1998
6. Whitfill CH et al: Primary carcinoid of the duodenum: detection and characterization by magnetic resonance imaging. J Magn Reson Imaging. 8(5):1175-6, 1998
7. Mindelzun RE et al: The misty mesentery on CT: Differential diagnosis. AJR 167:61-5, 1996
8. Laurent F et al: CT of small-bowel neoplasms. Semin Ultrasound CT MR. 16(2):102-11, 1995
9. Pantongrag-Brown L et al: Calcification and fibrosis in mesenteric carcinoid tumor: CT findings and pathologic correlation. AJR Am J Roentgenol. 164(2):387-91, 1995

IMAGE GALLERY

Typical

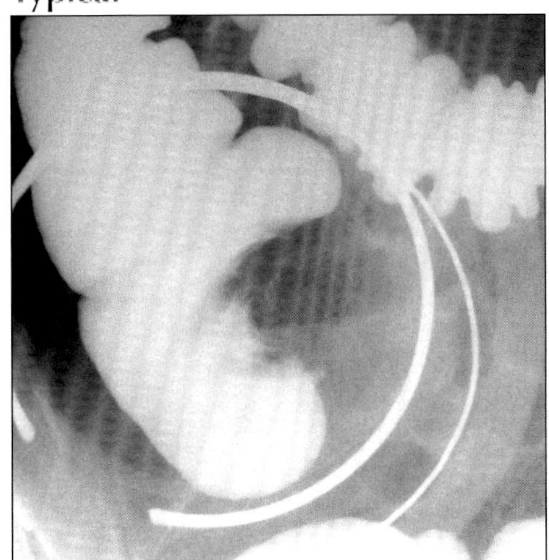

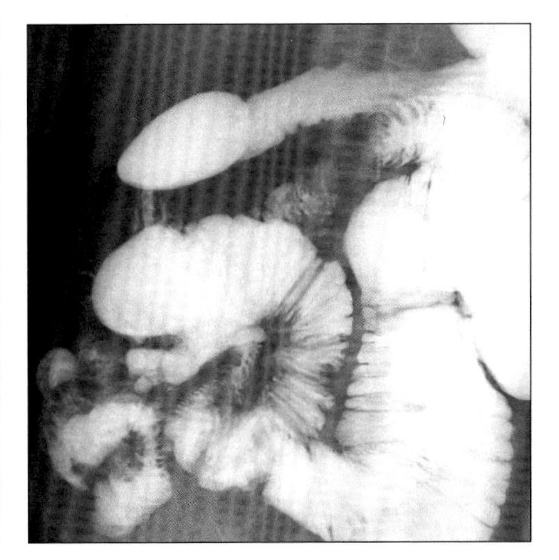

(Left) Barium enema shows mass effect on medial cecum due to carcinoid of ileum. *(Right)* Small bowel follow through (SBFT) shows partial SB obstruction, plus angulation, spiculation and narrowing of ileal loops.

Typical

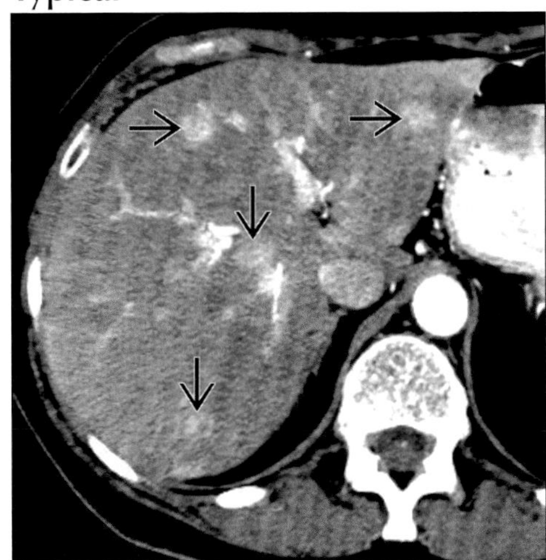

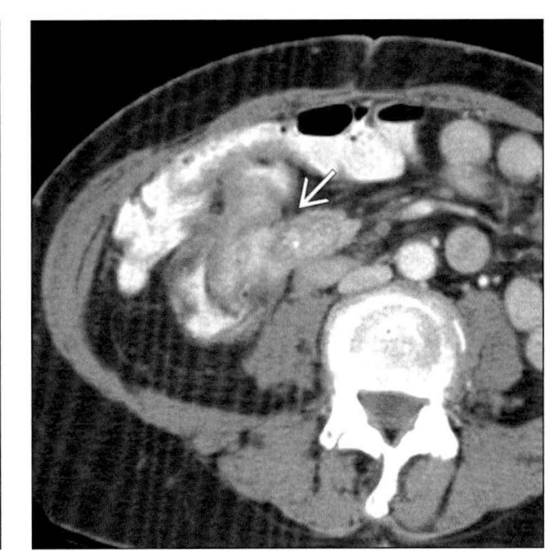

(Left) Axial CECT (arterial phase) shows multiple hypervascular liver metastases (arrows) that could not be detected on parenchymal phase images. *(Right)* Axial CECT shows mass in ileum and mesentery (arrow).

Typical

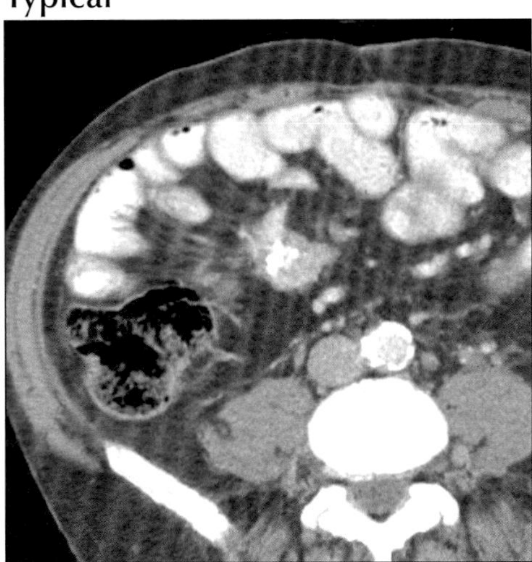

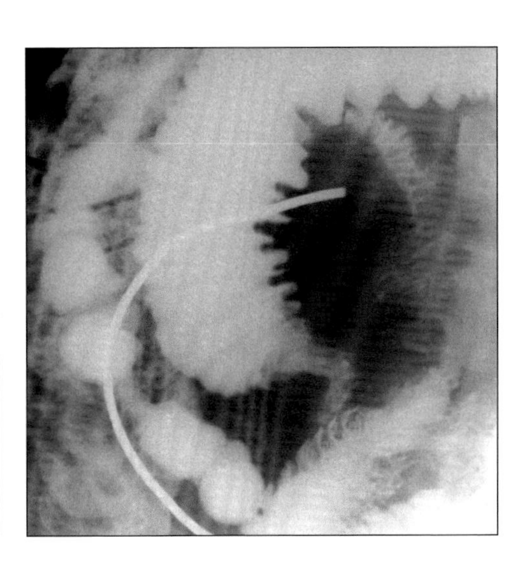

(Left) Axial CECT shows stellate mesenteric mass with central calcification. *(Right)* SBFT shows partial SB obstruction with luminal narrowing, mesenteric mass, and thick angled SB folds.

INTESTINAL METASTASES AND LYMPHOMA

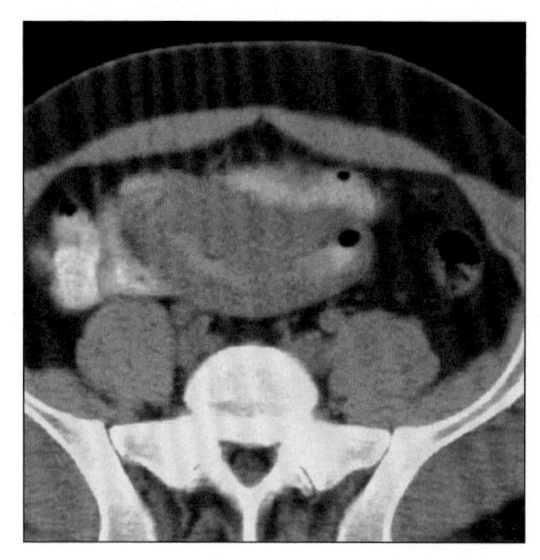

Axial CECT shows massive SB wall thickening of one ileal segment with soft tissue density, due to lymphoma.

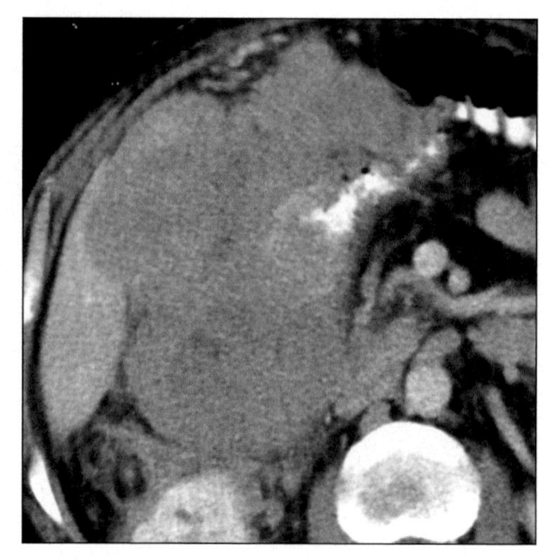

Axial CECT shows soft tissue mass in mesentery and wall of colon due to lymphoma (post-transplant lymphoproliferative disorder).

TERMINOLOGY

Definitions

- Intestinal metastases from other primary cancer site
- Lymphoma: Malignant tumor of B-lymphocytes

IMAGING FINDINGS

General Features

- Best diagnostic clue: "Bull's-eye" or "target" lesions
- Sprue: Causes 1° intestinal lymphoma & carcinoma
- Intestinal metastases
  - Usually incidental finding with known carcinoma
  - Various forms of metastatic spread to intestine
    - Intraperitoneal spread or seeding
    - Hematogenous & lymphatic spread
    - Direct extension from contiguous neoplasms
  - Intraperitoneal spread
    - E.g., primary mucinous tumors of ovary, appendix, colon & breast cancer
    - Due to natural flow & accumulation of ascitic fluid within peritoneal recesses; influences serosal implantation of cancer cells
    - Common sites: Ileocecal region, small bowel mesentery & posterior pelvic cul-de-sac
  - Hematogenous spread
    - E.g., melanoma (> common), lung, breast cancer
    - Reaches antimesenteric border of small bowel via small mesenteric arterial branches
    - Small bowel & mesentery most common sites of GIT metastases from melanoma after lung & liver
    - Malignant melanoma: At autopsy 35-58% of cases
    - Lung cancer: At autopsy 11% (39% are large cell)
    - Breast cancer: Stomach, duodenum & colon are more often involved than mesenteric small bowel
    - Breast, melanoma metastases may come to clinical attention many years after primary tumor removal
  - Lymphatic spread: E.g., colon, ovarian, breast, lung cancer, carcinoid & melanoma
  - Direct invasion
    - Pancreatic cancer: 2nd & 3rd parts of duodenum
    - Cecal & gynecologic malignancy: Distal ileum
- Intestinal lymphoma
  - Most common malignant small bowel tumor
  - Lymphoma accounts for one-half of all primary malignant small bowel tumors
  - Small bowel is 2nd most frequent site of GI tract
    - Stomach (51%); small bowel (33%); colon (16%) & esophagus (< 1%)
    - Ileum (51%); jejunum (47%) & duodenum (2%)
  - More than 50% cases are primary, rest are secondary
  - Majority, non-Hodgkin lymphoma (B-cell) origin
  - Most small bowel non-Hodgkin lymphomas: High grade of large cell or immunoblastic cell types
    - 30-50% harbor disease in mesenteric lymph nodes

DDx: Multifocal Bowel Wall Thickening

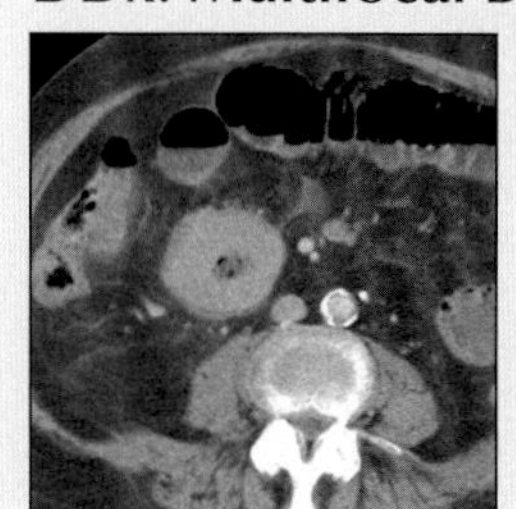

Hematoma

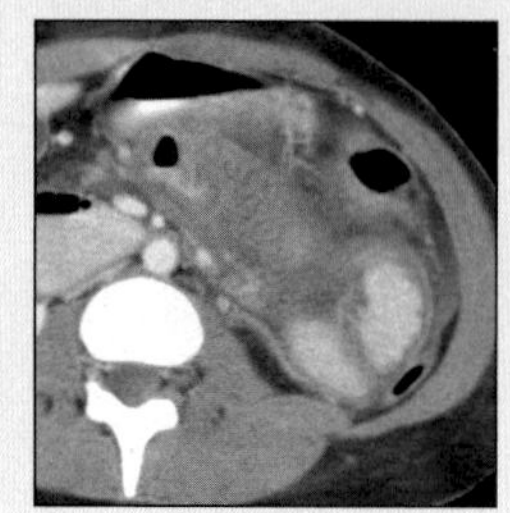

Vasculitis

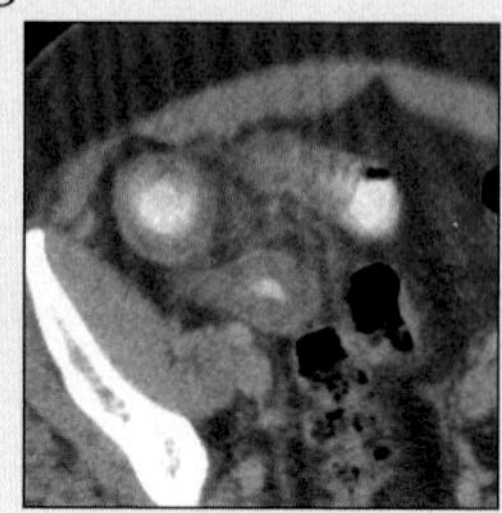

Crohn Disease

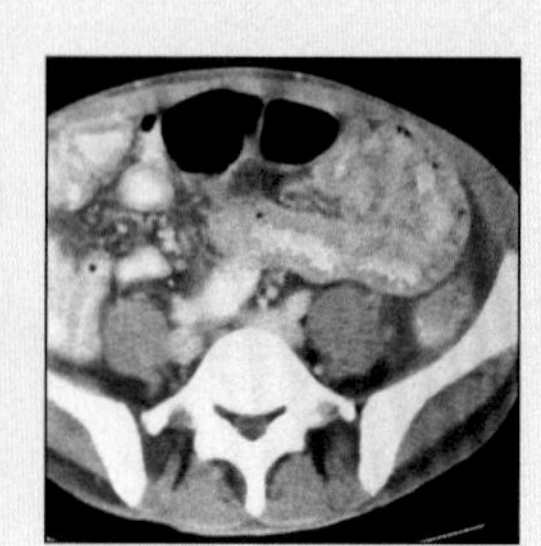

Whipple Disease

INTESTINAL METASTASES AND LYMPHOMA

Key Facts

Imaging Findings

- Best diagnostic clue: "Bull's-eye" or "target" lesions
- Solitary or multiple discrete submucosal masses
- Giant cavitated mass: Large barium collection contiguous with lumen (melanoma, lymphoma)
- Direct invasion: Spiculated mucosal folds, nodular mass effect, ulceration, obstruction, rarely fistula
- Lymphoma types: Infiltrative, polypoid, nodular, endoexoenteric, mesenteric
- Infiltrating lymphoma (most frequent type)
- Circumferential thickening & effacement of folds
- Aneurysmal dilatation on antimesenteric border (replacement of muscularis propria by lymphoma)
- "Sandwich sign": Mildly enhancing, multiple, rounded masses encasing mesenteric vessels (mesenteric lymphoma)

Top Differential Diagnoses

- Hemorrhage
- Vasculitis (small intestine)
- Crohn disease
- Other inflammatory (Whipple disease)
- Opportunistic infection

Pathology

- Malignant melanoma; breast, lung, ovarian cancer
- Primary: Non-Hodgkin (most common)
- Secondary: Generalized lymphoma

Diagnostic Checklist

- Check for history of primary cancer or enteropathy
- Overlapping radiographic features of intestinal metastases, lymphoma & primary carcinoma
- Imaging important to suggest & stage malignancy

Radiographic Findings

- Fluoroscopic guided enteroclysis
  - Intraperitoneal metastatic spread
    - Pelvic ileal mass with fixation, deformity & tethering along mesenteric border
    - Multiple fixed segments of pelvic ileum + shallow nodular indentations on mesenteric border
    - Dilated jejunum & proximal ileum (obstruction)
  - Malignant melanoma metastases
    - Solitary or multiple discrete submucosal masses
    - "Bull's-eye or "target" lesions: Centrally ulcerated submucosal masses
    - "Spoke-wheel" pattern: Radiating superficial fissures from central ulcer
    - Giant cavitated mass: Large barium collection contiguous with lumen (melanoma, lymphoma)
    - Small or large lobulated masses
    - Nonobstructive, large intraluminal mass favors melanoma metastasis
  - Bronchogenic carcinoma metastases
    - Solitary/multiple, flat/polypoid intramural masses
    - Ulceration; narrowing, obstruction (desmoplastic)
    - ± Localized extravasation or free perforation (due to marked tendency to penetrate bowel wall)
  - Breast carcinoma metastases
    - Submucosal masses ± ulceration
    - May be seen as multiple strictures
  - Direct invasion: Spiculated mucosal folds, nodular mass effect, ulceration, obstruction, rarely fistula
  - Intestinal lymphoma (types)
    - Lymphoma types: Infiltrative, polypoid, nodular, endoexoenteric, mesenteric
  - Infiltrating lymphoma (most frequent type)
    - Circumferential thickening & effacement of folds
    - Luminal dilatation, narrowing, stricture
    - Aneurysmal dilatation on antimesenteric border (replacement of muscularis propria by lymphoma)
  - Polypoid lymphoma
    - Single/multiple, mucosal/submucosal masses
    - "Bull's-eye" lesion: Polypoid mass + ulceration
    - Rare form: Lymphomatous polyposis (follicular mantle cell origin)
  - Nodular lymphoma
    - Multiple small submucosal nodular defects
  - Endoexoenteric (cavitary form): Localized perforation into a sealed-off mesenteric space
    - Barium extravasates into an exoenteric space
    - Barium, air & debris-filled cavity along mesenteric border of small bowel
    - ± Ulcer, fistulae, aneurysmal dilatation
  - Mesenteric lymphoma
    - Displace, compress & obstruct small bowel loops

CT Findings

- Demonstration of lesions facilitated by negative contrast agents (water or gas)
- Intestinal metastases
  - Intraperitoneal metastatic spread
    - Mesenteric tethering of terminal ileum in RLQ
    - Enhancing focal masses within mesenteric leaves
    - "Stellate" appearance: Mesenteric fat infiltration
    - Ovarian carcinoma: Calcified mesenteric/omental masses + large calcified primary pelvic mass
  - Malignant melanoma
    - "Bull's-eye" or "target" lesions (also seen in lymphoma, Kaposi sarcoma, carcinoid tumor)
    - Enhancing mural nodules protruding into lumen or focal thickening of intestinal wall
    - Enhancing masses within small bowel mesentery
    - Lobulated submucosal or giant cavitated lesions
    - Location: Distal small bowel (usually ileum)
  - Bronchogenic carcinoma metastases
    - Flat or polypoid intramural masses
    - ± Ulceration; narrowing & obstruction
  - Breast carcinoma metastases
    - Submucosal masses or multiple strictures
  - Direct invasion of duodenum, jejunum & ileum
    - Pancreatic head carcinoma: Medial wall changes in 2nd or 3rd parts of duodenum
    - Cecal, gynecologic cancer: Distal ileal changes
- Intestinal lymphoma
  - Infiltrating form (most common)
    - Circumferential type: Sausage-shaped mass of homogeneous density + minimal enhancement
    - Aneurysmal dilatation on antimesenteric border

INTESTINAL METASTASES AND LYMPHOMA

- Polypoid form
  - "Bull's-eye" or "target" lesion: Mass + ulceration
- Mesenteric form
  - "Sandwich sign": Mildly enhancing, multiple, rounded masses encasing mesenteric vessels (mesenteric lymphoma)
  - Large, lobulated, "cake-like" heterogeneous mass + areas of necrosis displacing small bowel loops
  - Ill-defined mesenteric fat infiltration
  - Retroperitoneal adenopathy favors lymphoma

Imaging Recommendations

- Helical CECT with negative enteric contrast agents
- Fluoroscopic guided enteroclysis

DIFFERENTIAL DIAGNOSIS

Hemorrhage

- E.g., Coumadin, trauma
- Localized bleeding may be seen as intramural mass

Vasculitis (small intestine)

- Henoch-Schönlein purpura
  - Triad: Palpable purpura, arthritis, abdominal pain
- Systemic lupus erythematosus (SLE)
  - Segmental bowel lesions → necrosis & perforation
- Barium studies or CT of small bowel in vasculitis
  - Extensive fold thickening + luminal narrowing
  - May show thumbprinting on mesenteric border
  - Intussusceptions may be seen in childhood purpura

Crohn disease

- Predominantly involves distal ileum
- Skip lesions, transmural, fistulae, fissures
- Aphthoid ulceration, cobblestoning, "string sign"

Other inflammatory (Whipple disease)

- Thickened proximal small bowel folds
- Micronodules in jejunum on enteroclysis
- Thickened mesentery & lymphadenopathy

Opportunistic infection

- Giardiasis
  - Duodenum & jejunum
    - Thickened irregular folds with hypermotility
    - Luminal narrowing & increased secretions
  - Ileum: Usually appears normal
- Mycobacterium avium-intracellulare (MAI)
  - Small bowel shows thickened folds, fine nodularity
  - CT: Low density (caseated) lymph nodes
- Cytomegalovirus (CMV)
  - Often causes terminal ileitis in AIDS patients
  - Thickened folds, spiculation, ulcers, narrow lumen

PATHOLOGY

General Features

- Etiology
  - Intestinal metastases
    - Malignant melanoma; breast, lung, ovarian cancer
    - Appendix, colon, pancreatic cancer; carcinoid
  - Intestinal lymphoma
    - Primary: Non-Hodgkin (most common)
    - Secondary: Generalized lymphoma
    - Enteropathy-associated lymphoma: Celiac disease
    - Mediterranean type: Arabs, Middle Eastern Jews
    - Burkitt lymphoma in children involve ileocecal areas
- Epidemiology
  - Metastases: Most common in melanoma & carcinoid
  - Lymphoma: Most common small bowel neoplasm
- Associated abnormalities
  - Primary carcinoma in intestinal metastases
  - Generalized adenopathy in secondary lymphoma

Gross Pathologic & Surgical Features

- Solitary/multiple; polypoid, ulcerated, cavitated

Microscopic Features

- Metastases: Varies based on primary cancer
- Lymphoma: Lymphoepithelial lesions

CLINICAL ISSUES

Presentation

- Most common signs/symptoms
  - Asymptomatic, pain, weight loss, palpable mass
  - Malabsorption, diarrhea
  - Acute abdomen: Obstruction, perforation

Natural History & Prognosis

- Complications: Bleeding, perforation, obstruction
- Prognosis: Poor

Treatment

- Chemotherapy & surgical resection of lesions causing complications like obstruction & GI bleeding

DIAGNOSTIC CHECKLIST

Consider

- Check for history of primary cancer or enteropathy

Image Interpretation Pearls

- Overlapping radiographic features of intestinal metastases, lymphoma & primary carcinoma
- Imaging important to suggest & stage malignancy

SELECTED REFERENCES

1. Horton KM et al: Multidetector-row computed tomography and 3-dimensional computed tomography imaging of small bowel neoplasms: current concept in diagnosis. J Comput Assist Tomogr. 28(1):106-16, 2004
2. Sheth S et al: Mesenteric neoplasms: CT appearances of primary and secondary tumors and differential diagnosis. Radiographics. 23(2):457-73; quiz 535-6, 2003
3. Buckley JA et al: CT evaluation of small bowel neoplasms: spectrum of disease. Radiographics. 18(2):379-92, 1998
4. Balthazar EJ et al: CT of small-bowel lymphoma in immunocompetent patients and patients with AIDS: comparison of findings. AJR Am J Roentgenol. 168(3):675-80, 1997
5. Rubesin SE et al: Non-Hodgkin lymphoma of the small intestine. Radiographics. 10(6):985-98, 1990

INTESTINAL METASTASES AND LYMPHOMA

IMAGE GALLERY

Typical

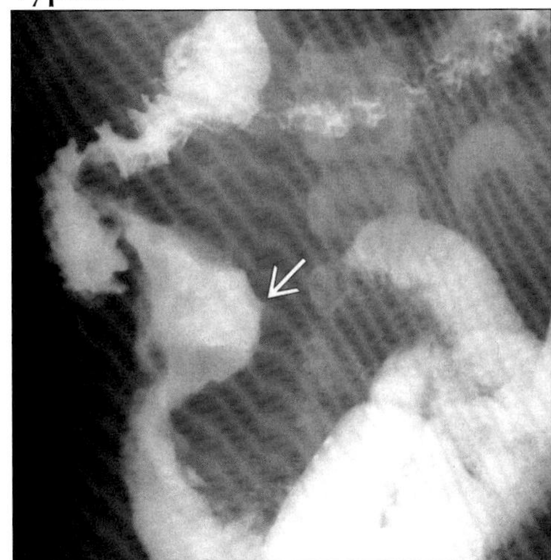

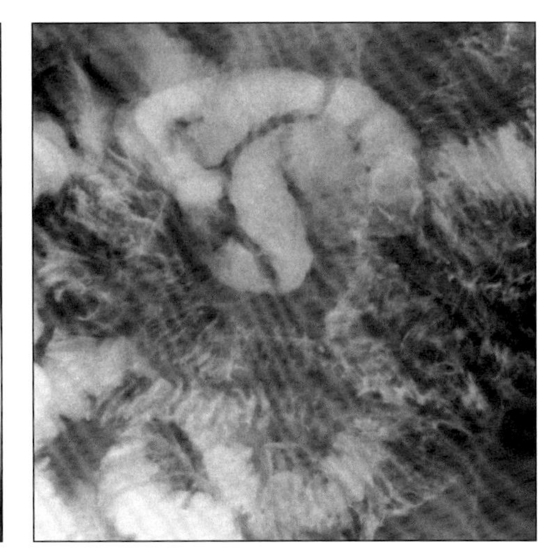

(Left) *SBFT shows aneurysmal dilation (arrow) of the lumen of terminal ileum along with mesenteric mass effect (lymphoma).* ***(Right)*** *SBFT shows diffuse nodular fold thickening of most of the SB (lymphoma).*

Typical

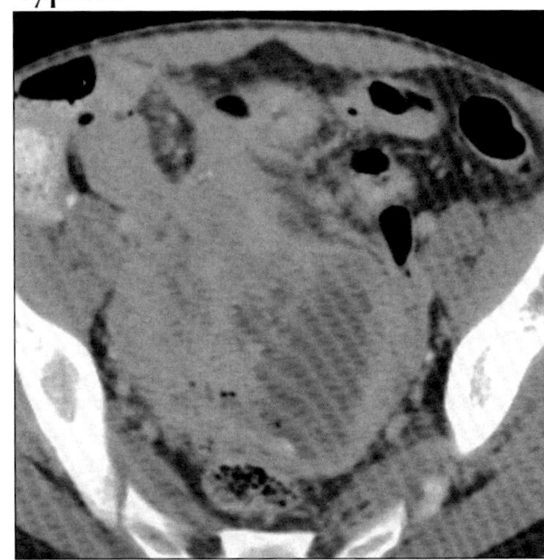

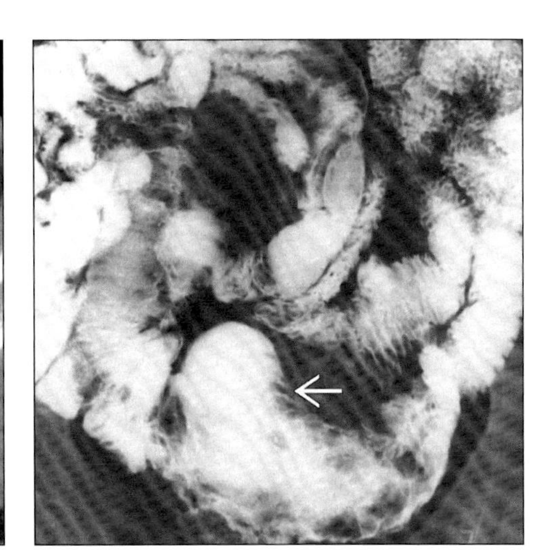

(Left) *Axial CECT shows partially necrotic mass that envelopes, but does not obstruct, multiple SB segments (melanoma).* ***(Right)*** *SBFT shows aneurysmal dilation (arrow) and mucosal destruction of distal SB segment (melanoma).*

Typical

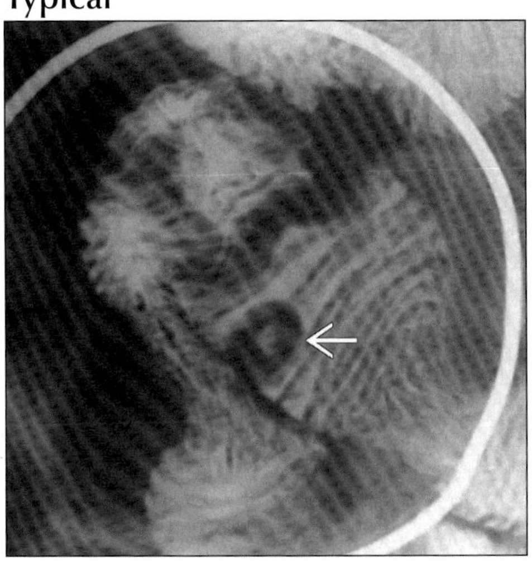

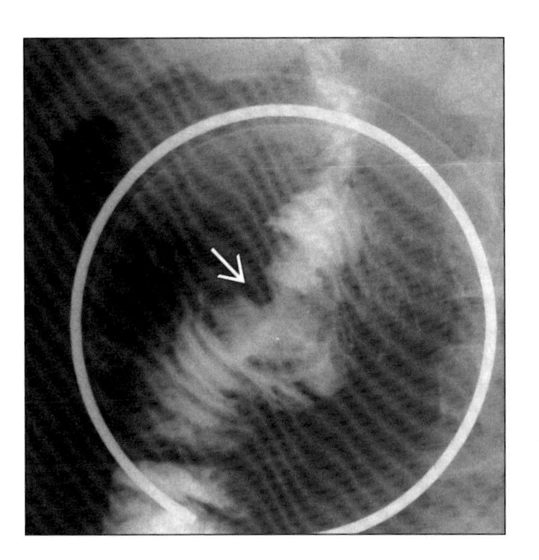

(Left) *Fluoroscopic spot film from SBFT shows bull's-eye lesion (arrow) due to metastatic melanoma.* ***(Right)*** *SBFT spot film shows intramural mass (arrow) with distorted mucosa, due to melanoma.*

SMALL BOWEL OBSTRUCTION

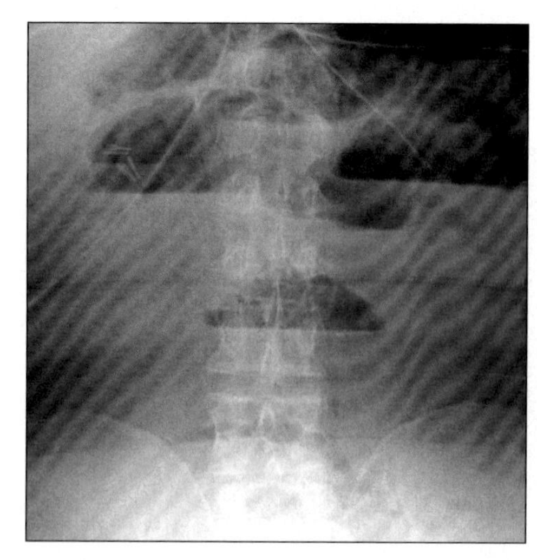

Upright radiograph shows dilated small bowel with air-fluid levels; no colonic gas.

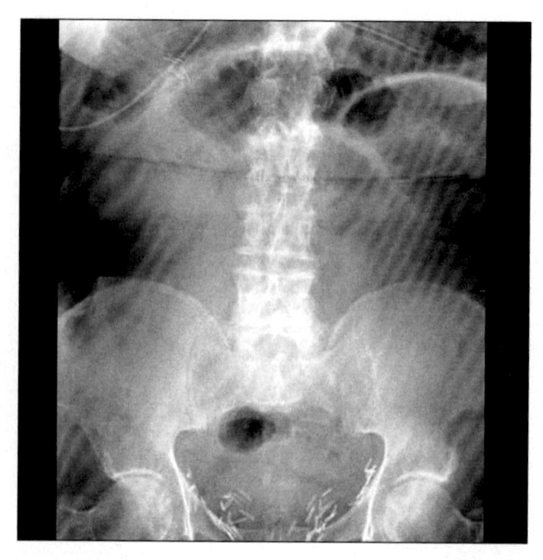

Supine radiograph shows dilated small bowel, no colonic gas. Note surgical clips in pelvis as clue to prior surgery and probable adhesions.

TERMINOLOGY

Abbreviations and Synonyms

- Small bowel obstruction (SBO)

Definitions

- Obstruction or blockage of small bowel loops

IMAGING FINDINGS

General Features

- Best diagnostic clue: Dilated small bowel loops with air-fluid levels on upright film
- Size: Small bowel, proximal to obstruction: > 2.5 cm
- Key concepts
  - Most common causes
    - Adhesion & hernias (> 80% of all cases)
  - Accounts for 20% of surgical admissions of patients with acute abdomen
  - Classification based on mechanism of obstruction
    - Mechanical: Extrinsic, intrinsic or intraluminal lesions
    - Non-mechanical: Adynamic ileus; dynamic or spastic ileus (due to neuromuscular disturbances)
  - Mechanical SBO is 4-5 times more common than large bowel obstruction
  - SBO classified into two types: Simple & complicated
  - Simple: Sub-classified based on degree of obstruction
    - Intermittent, incomplete or partial or low grade obstruction (more common)
    - Prolonged, complete or high grade obstruction
  - Complicated: Sub-classified into two types
    - Closed-loop or incarcerated obstruction: Adhesive bands > internal or external hernia
    - Strangulation: Most common cause of closed-loop obstruction; indicates vascular compromise
  - Major predisposing causes of intestinal obstruction
    - SBO: Adhesions (75%); external hernia (10%); neoplasm (5%)
    - Adhesions: Post surgery (> 80%), inflammation (15%) & congenital (5%)
    - Large bowel: Carcinoma (55%); volvulus (11%); diverticulitis (9%)
  - Clinical onset of bowel obstruction
    - SBO: Acute in onset
    - Large bowel obstruction: Subacute or chronic
  - Transition zone between normal and abnormal bowel critical to define site and cause of obstruction

Radiographic Findings

- Radiography
  - Need supine + upright or decubitus views
  - Dilated proximal small bowel loops with multiple air-fluid levels & collapsed distal bowel
    - Pneumoperitoneum is a sign of bowel perforation

DDx: Small Bowel Distension

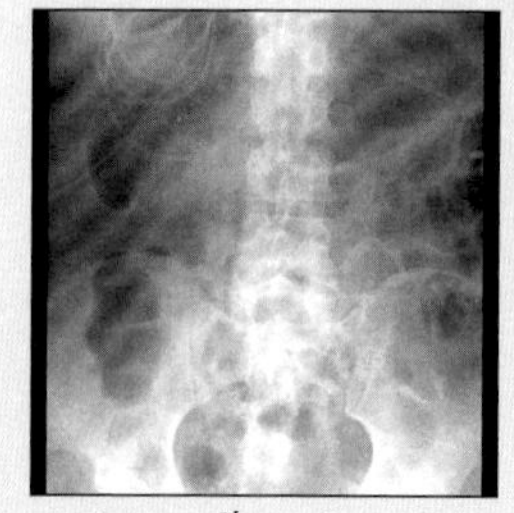

Ileus

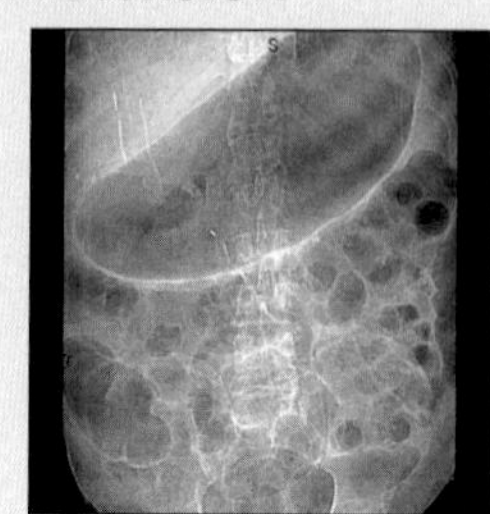

Aerophagia

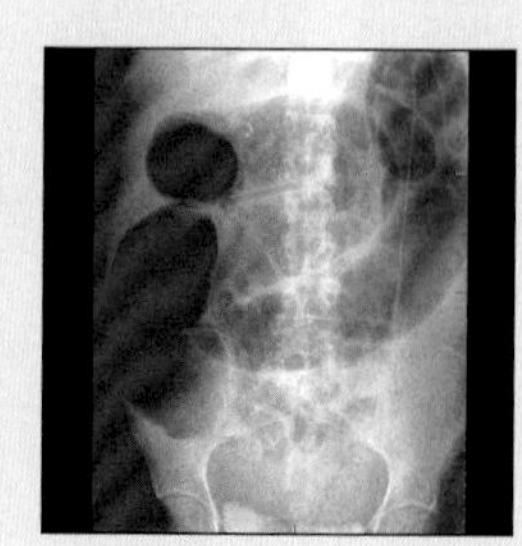

Ascites + Ileus

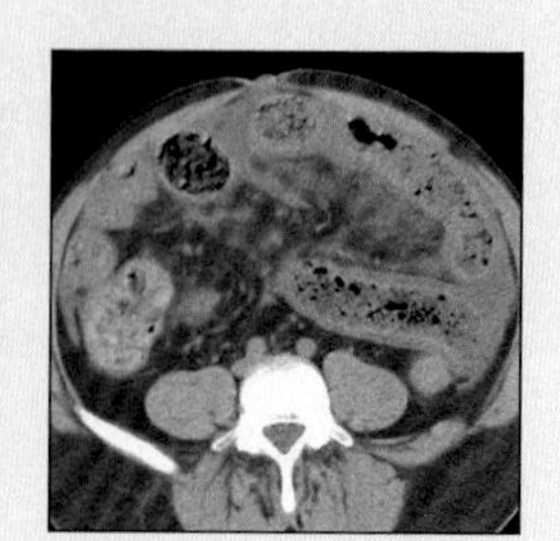

Cystic Fibrosis

SMALL BOWEL OBSTRUCTION

Key Facts

Imaging Findings

- "Small-bowel feces" sign: Gas bubbles mixed with particulate matter in dilated loops proximal to SBO
- Dilated fluid-filled small-bowel loops (> 2.5 cm) proximal to collapsed loops
- Gas filled bowel loops, mesenteric fat, vessels in inguinal canal or other external hernia
- Thickened enhancing wall & luminal narrowing at transition zone
- Intussusception: Target sign; sausage-shaped or reniform mass
- ± Pneumatosis intestinalis; ± portomesenteric venous gas (strangulation)
- Mesenteric vessels: Haziness, obliteration, congestion or hemorrhage; ascites (strangulation)

Top Differential Diagnoses

- Adynamic or paralytic ileus
- Aerophagia
- Ascites
- Cystic fibrosis (CF)

Pathology

- Most common: Adhesions (~ 60%), hernias (15%), + tumors (~ 15%; metastases > primary tumor)
- Pathogenesis: Obstruction of small-bowel leads to proximal dilatation due to accumulation of GI secretions & swallowed air

Diagnostic Checklist

- CT best to determine presence, site, & cause of SBO + any complications

- Radiography can "miss" SBO (fluid, distended bowel not evident)
- "String of pearls"
    - On supine radiographs
    - Small air bubbles within fluid, distended bowel
- Fluoroscopic guided enteroclysis or small-bowel series
    - Incomplete or partial or low grade obstruction
        - Sufficient flow of contrast through point of obstruction
    - Complete or high grade obstruction
        - Stasis or delay in flow of contrast beyond point of obstruction
    - Transitions in contrast column can define location & degree of obstruction

CT Findings

- Dilated small bowel loops (> 2.5 cm) ± air-fluid levels
- "Small-bowel feces" sign: Gas bubbles mixed with particulate matter in dilated loops proximal to SBO
    - Less common but reliable indicator of SBO
- Extrinsic lesions
    - Adhesions
        - Dilated fluid-filled small-bowel loops (> 2.5 cm) proximal to collapsed loops
        - ± Transition zone, minimal mural thickening & enhancement
        - Uncomplicated adhesive bands: Typically unidentified on CT (diagnosis of exclusion)
    - Hernia
        - Gas filled bowel loops, mesenteric fat, vessels in inguinal canal or other external hernia
        - Strangulated hernia: Thickened + ↑ attenuation of bowel wall
        - Internal hernia: Cluster of dilated loops + crowding or twisting of mesenteric vessels
    - Peritoneal carcinomatosis: Omental masses; dilated bowel loops; multiple transition zones
    - Appendicitis: RLQ inflammatory mass, dilated loops, fluid collection, abscess
    - Diverticulitis
        - Complicated: Abscess, peritonitis, obstruction, dilated bowel loops
    - Intrinsic lesions: Adenocarcinoma, Crohn, TB, radiation enteropathy
        - Thickened enhancing wall & luminal narrowing at transition zone
        - Fluid & gas filled dilated bowel loops proximal to collapsed loops
        - Intussusception: Target sign; sausage-shaped or reniform mass
    - Intraluminal lesions: Gallstones, foreign bodies, bezoars, ascaris worms
        - Classic triad: Ectopic calcified stone, gas in GB or biliary tree, obstruction (gallstone ileus)
        - Bezoar: Intraluminal mass + air in interstices; dilated fluid-filled loops
    - Closed-loop obstruction: Obstruction at two points + involves mesentery
    - Relatively little dilatation of bowel proximal to closed loop obstruction
        - Fluid distended bowel, minimal gas (closed-loop)
        - Volvulus: C-shaped, U-shaped or "coffee bean" configuration of bowel loop
        - Stretched mesenteric vessels converging toward site of torsion
        - "Beak sign": Fusiform tapering at point of torsion or obstruction (closed-loop)
        - "Whirl sign": Due to tightly twisted mesentery with volvulus
    - Strangulating obstruction: Blood flow to obstructed bowel is blocked
        - "Target" or "halo" sign: Circumferentially thickened bowel wall + ↑ wall attenuation
        - "Serrated beak sign": Twisting of bowel, mesenteric edema, bowel wall thickening
        - ± Pneumatosis intestinalis; ± portomesenteric venous gas (strangulation)
        - Absence or ↓ or delayed bowel wall enhancement in affected loops
        - Mesenteric vessels: Haziness, obliteration, congestion or hemorrhage; ascites (strangulation)

Imaging Recommendations

- Helical CECT: Acutely ill; suspected ischemia; history of cancer or inflammatory bowel disease

- Accuracy (95%), specificity (96%) in high grade SBO
- Enteroclysis: Intermittent, chronic or low grade SBO
- Suspected perforation: Water soluble contrast agent

DIFFERENTIAL DIAGNOSIS

Adynamic or paralytic ileus

- Etiology: Post-op, medications, post injury, ischemia
- Dilated small and large bowel loops with no transition point, fluid levels seen but aperistaltic
- CT shows absence of obstruction

Aerophagia

- Excessive air swallowing associated with prominent belching, flatulence & abdominal distention
- Air swallowing: Independent or with eating/drinking
- Etiology: Unknown; may be functional, behavioral, neurological or psychiatric
- Small-bowel loops are dilated simulating SBO
  - Associated gastric & colonic distension without air-fluid levels differentiates from SBO

Ascites

- Pathologic accumulation of fluid in peritoneal cavity
- Plain abdominal film
  - Medial displacement + collapse of ascending & descending colon
  - Separation & centralization of gas-filled small-bowel loops may simulate small-bowel obstruction

Cystic fibrosis (CF)

- Small-bowel may be functionally obstructed due to thick, viscous bowel contents
- CT findings
  - Shows fatty replacement of pancreas often with small-bowel feces sign (chronic low grade SBO)

PATHOLOGY

General Features

- Etiology
  - Most common: Adhesions (~ 60%), hernias (15%), + tumors (~ 15%; metastases > primary tumor)
  - Extrinsic: Adhesions; external & internal hernias, tumor, abscess, aneurysm
  - Intrinsic: Tumors, inflammatory, vascular (ischemic), metabolic, radiation enteropathy
  - Intraluminal: Gallstones, bezoars, foreign bodies, ascaris worms
  - Pathogenesis: Obstruction of small-bowel leads to proximal dilatation due to accumulation of GI secretions & swallowed air
    - Bowel dilatation stimulates secretory activity resulting in more fluid accumulation
- Epidemiology
  - Incidence: 20% of "acute abdomen" presentations
  - Mortality: Simple SBO 5-8%; strangulation 20-37%

Gross Pathologic & Surgical Features

- Dilated proximal & distal collapsed loop + transition point; (dilated small-bowl more than 2.5 cm)

CLINICAL ISSUES

Presentation

- Most common signs/symptoms
  - Variable from mild abdominal pain to vomiting, constipation, fever & signs of acute abdomen
    - Abdominal distention, tenderness, guarding
    - Bowel sounds high pitched or absent (late sign)

Natural History & Prognosis

- Complications
  - Bowel strangulation, infarction, gangrene, perforation, peritonitis & sepsis
- Prognosis
  - Simple obstruction (good); complicated (poor)
  - Mortality 25%: Surgery postponed beyond 36 hrs
  - Mortality ↓ to 8%: Surgery performed within 36 hrs
  - Mortality 100%: Untreated strangulated obstructions

Treatment

- Nasogastric suction, decompression, I.V. fluids, NPO
- Incomplete or low grade SBO: Conservative treatment
- Complete or high grade SBO: Immediate surgery

DIAGNOSTIC CHECKLIST

Consider

- CT best to determine presence, site, & cause of SBO + any complications
- Difficult to distinguish partial + complete obstruction by imaging alone

Image Interpretation Pearls

- Dilated small bowel loops with "small bowel feces" sign on CT & "string of pearls" sign on supine film

SELECTED REFERENCES

1. Khurana B et al: Bowel obstruction revealed by multidetector CT. AJR Am J Roentgenol. 178(5):1139-44, 2002
2. Maglinte DD et al: Small bowel obstruction: Optimizing radiologic investigation and nonsurgical management. Radiology. 218: 39-46, 2001
3. Furukawa A et al: Helical CT in the diagnosis of small bowel obstruction. Radiographics. 21(2):341-55, 2001
4. Caoili EM et al: CT of small bowel obstruction: Another perspective using multiplanar reformations. AJR. 174: 993-8, 2000
5. Nevitt PC: The string of pearls sign. Radiology. 214(1):157-8, 2000
6. Maglinte DD et al: The role of radiology in the diagnosis of small bowel obstruction. AJR. 168: 1171-80, 1997
7. Mayo-Smith WW et al: The CT small bowel feces sign: description and clinical significance. Clin Radiol. 50(11):765-7, 1995

IMAGE GALLERY

Typical

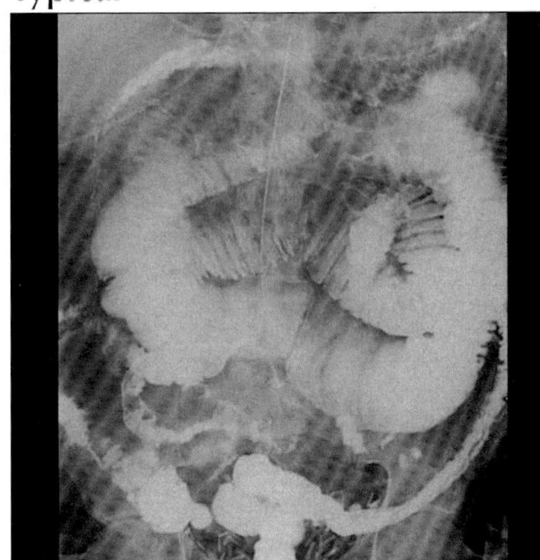

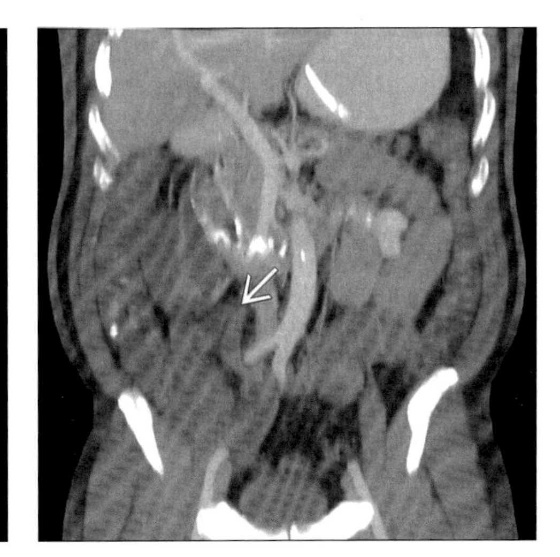

(Left) SBFT shows barium distended proximal small bowel, collapsed distal bowel and colon. **(Right)** Coronal reconstruction of CECT shows diluted fluid-filled small-bowel (not evident on supine radiographs). Note acutely angulated distal SB (arrow) due to adhesions, which caused SBO.

Typical

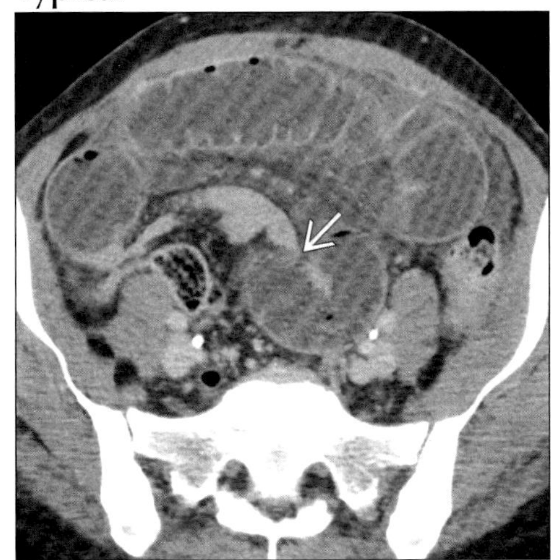

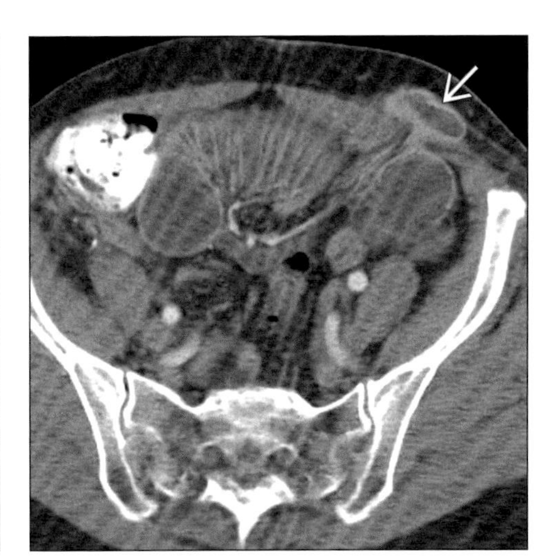

(Left) Axial CECT shows abrupt transition from dilated to non-dilated bowel (arrow). SBO due to adhesions. **(Right)** Axial CECT shows Spigellian hernia (arrow) with dilated bowel leading into the hernia, collapsed bowel leaving

Typical

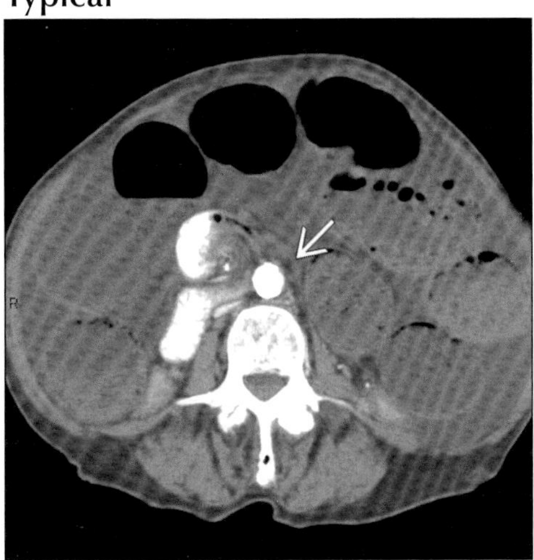

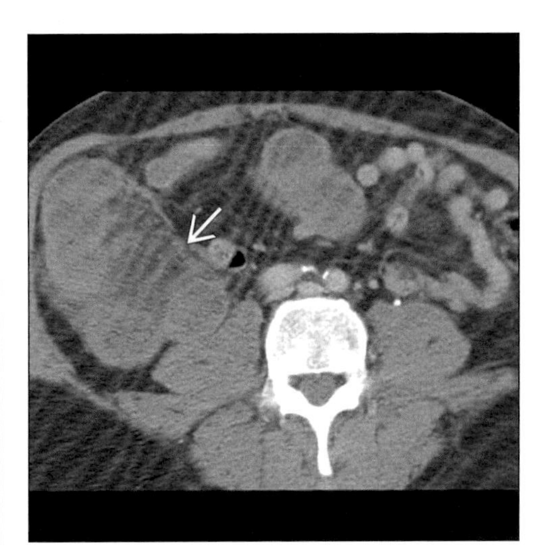

(Left) Closed loop obstruction due to midgut volvulus. CECT shows fluid-distended small-bowel, ascites, and twisting of the root of the mesentery (arrow). **(Right)** Closed loop obstruction CECT shows cluster of dilated fluid-distended loops of bowel (arrow), with mesenteric infiltration, and blurred engorged blood vessels. Plain radiographs were "normal".

GALLSTONE ILEUS

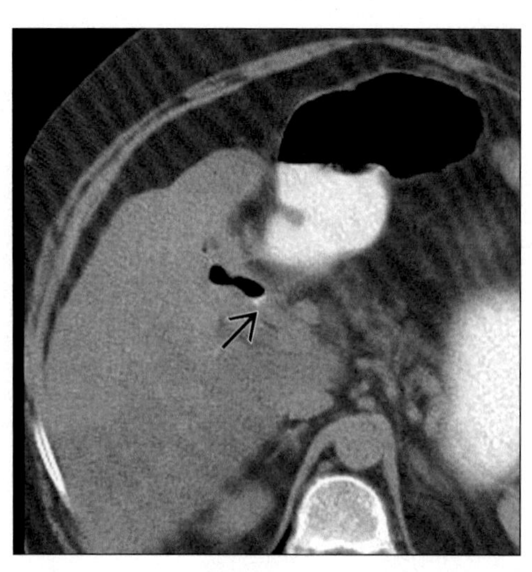

Axial CECT shows gas in collapsed gallbladder, fistula to duodenum (arrow).

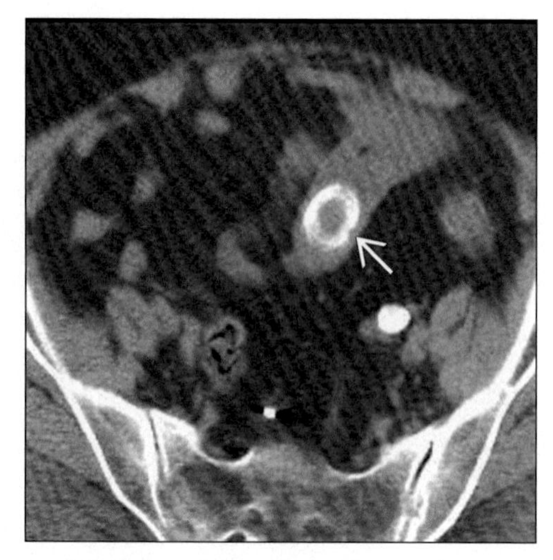

Axial CECT shows large gallstone (arrow) within dilated small bowel; bowel distal to gallstone is collapsed.

TERMINOLOGY

Abbreviations and Synonyms

- Gallstone Ileus (GSI)

Definitions

- Mechanical intestinal obstruction caused by impaction of one or more gallstones in intestine

IMAGING FINDINGS

General Features

- Best diagnostic clue: Small bowel obstruction + gas in biliary tree + ectopic gallstone; (Rigler triad)
- Location
  - Gallstone may be "hung up" at narrow portions
    - Duodenum; ligament of Treitz; ileocecal valve; sigmoid colon; any area of stricture
- Size: Large gallstone; ≥ 2.5 cm

Radiographic Findings

- Radiography
  - Plain abdominal film
  - Dilated proximal bowel
  - Gas in shrunken gallbladder; bile ducts or both
    - Gas in biliary tree: One of three cases; branching pattern; gas more prominent centrally
  - Ectopic calcified gallstone (15-25%)
    - Radiopaque gallstone surrounded by intestinal gas in obstructed bowel loop
    - Difficult visualization: Cholesterol; located over shadow of sacrum; obscured by dilated bowel
  - Rigler triad present in only 38% of cases
  - Change in position of previously identified gallstone
- Fluoroscopy
  - Upper gastrointestinal series or barium enema
  - Well-contained localized barium collection lateral to first portion of duodenum
    - Barium filled collapsed gallbladder, biliary ducts
  - Fistulous communication: Cholecystoduodenal (60%); choledochoduodenal; cholecystocolic
    - Choledochocolic; cholecystogastric

CT Findings

- CT better reveals gallstone as cause of obstruction
  - May see stone surrounded by intestinal gas in obstructed bowel loop
  - Cholesterol stones are usually low density (near water), but often calcified rim
- Will show collapsed gallbladder, pneumobilia

DIFFERENTIAL DIAGNOSIS

Intussusception

- "Coiled spring appearance"; sausage-shaped mass

DDx: Small Bowel Obstruction Plus "Mass"

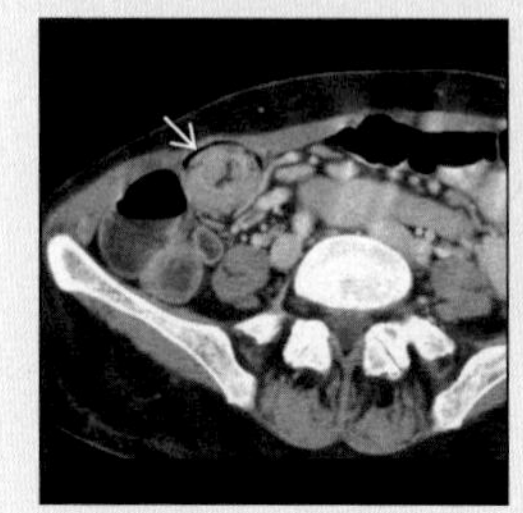

Intussusception

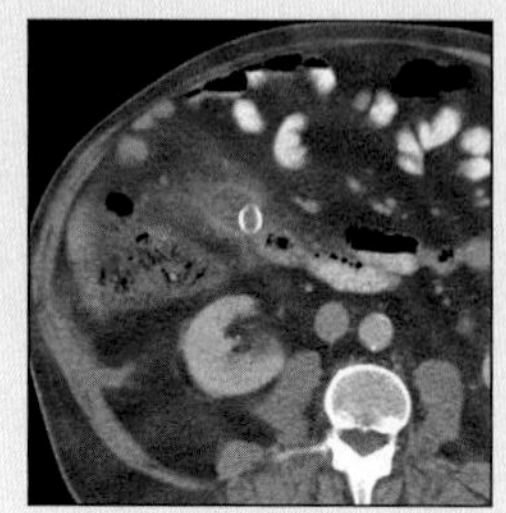

Dropped Gallstone

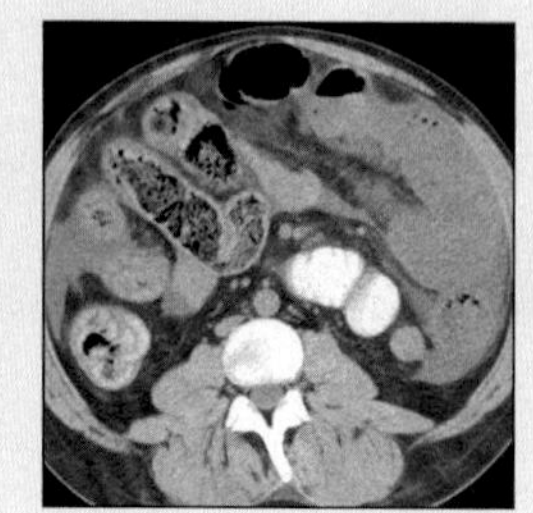

Ischemia

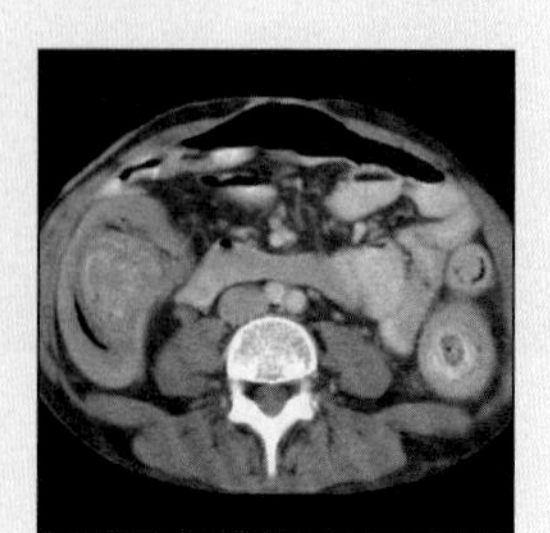

Pseudoobstruction

GALLSTONE ILEUS

Key Facts

Terminology
- Mechanical intestinal obstruction caused by impaction of one or more gallstones in intestine

Imaging Findings
- Best diagnostic clue: Small bowel obstruction + gas in biliary tree + ectopic gallstone; (Rigler triad)

Top Differential Diagnoses
- Intussusception
- Dropped gallstone
- Small bowel tumor
- Pseudoobstruction

Clinical Issues
- Age: Risk ↑ with age; average 65-75 years

Dropped gallstone
- In peritoneal cavity, dropped during laparoscopic cholecystectomy

Small bowel tumor
- Leiomyoma: 1-4 cm range; lipoma; adenoma

Pseudoobstruction
- Intestinal neurological dysfunction
- May have laminated enteroliths simulating gallstones

Bowel ischemia
- Thickened wall, proximal bowel dilated

PATHOLOGY

General Features
- Etiology
  - GSI occurs in setting of chronic cholecystitis
  - Following multiple ERCPs for biliary calculi
  - Complication of endoscopic sphincterotomy for large common bile duct stones
- Epidemiology
  - 0.4-5% of all intestinal obstructions
  - In < 1% of patients with cholelithiasis
- Although called ileus; actually mechanical obstruction

CLINICAL ISSUES

Presentation
- Intermittent, acute colicky abdominal pain (20-30%); nausea, vomiting, fever, distension, obstipation
- In elderly females; frequently an underlying pathological condition at site of obstruction in colon
- Delayed complication of ERCP; up to 2 months
- Small bowel obstruction following endoscopic sphincterotomy for very large bile duct calculi
- Diagnosis: Frequently delayed or missed

Demographics
- Age: Risk ↑ with age; average 65-75 years
- Gender: M:F = 1:4-7

Natural History & Prognosis
- Gallstone erodes inflamed gallbladder wall; passes into gastrointestinal tract; causes bowel obstruction
  - Usually erodes directly into duodenum
- Very large stones can pass into duodenum; apparently after "unsuccessful" sphincterotomy
- Recurrence: 5-10% (additional silent proximal calculi)
- Prognosis: High mortality; operative mortality 19%

Treatment
- Surgical therapy to relieve bowel obstruction
- Cholecystectomy & biliary fistula excision; to prevent recurrence
- Staged laparoscopic management of GSI & associated cholecystoduodenal fistula; feasible & safe

DIAGNOSTIC CHECKLIST

Consider
- Middle-aged or elderly female; recurrent episodes of RUQ pain with most recent episode being more severe & associated with prolonged vomiting
  - With most recent episode being more severe & associated with prolonged vomiting
- If barium studies are performed; attempt should be made to identify biliary-enteric fistula

SELECTED REFERENCES

1. Vaidya JS et al: Gallstone ileus. Lancet. 362(9390):1105, 2003
2. Lyburn ID et al: Gall-stone ileus: imaging features. Hosp Med. 63(7):434-5, 2002
3. Gandhi A et al: Gallstone ileus following endoscopic sphincterotomy. Br J Hosp Med. 54(5):229-30, 1995

IMAGE GALLERY

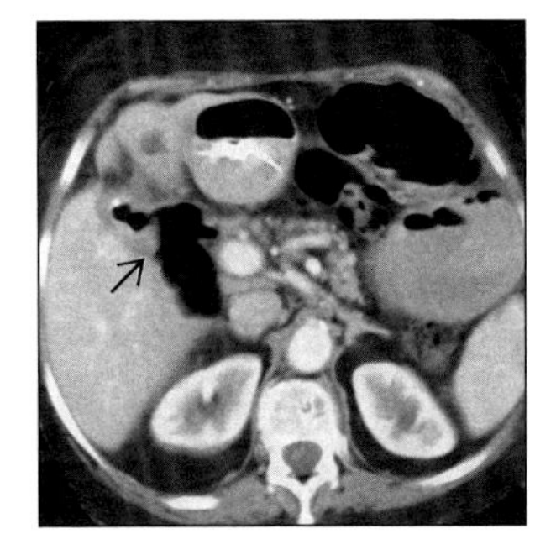

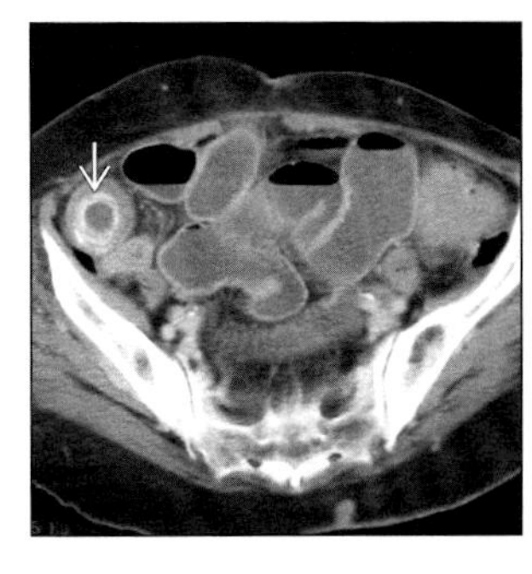

*(**Left**) Axial CECT shows gas in thick-walled collapsed gallbladder and fistula (arrow) to duodenum. (**Right**) Axial CECT shows small bowel obstruction with large laminated gallstone (arrow) "stuck" near ileocecal valve.*

MALABSORPTION CONDITIONS

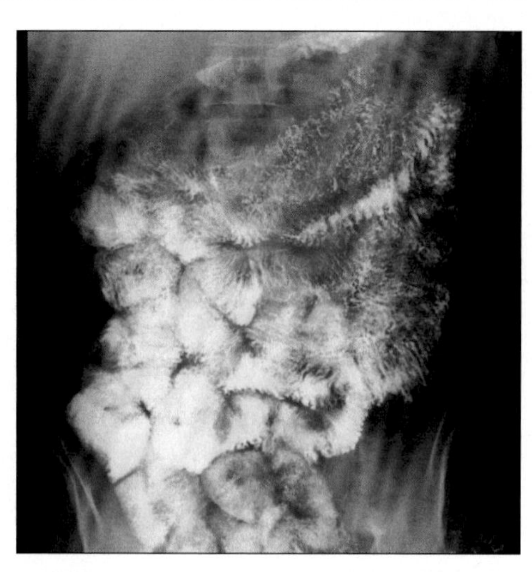

Small bowel follow through (SBFT) shows mild dilation of SB lumen, dilution of barium, and nodular fold thickening due to congenital IgA deficiency and giardiasis.

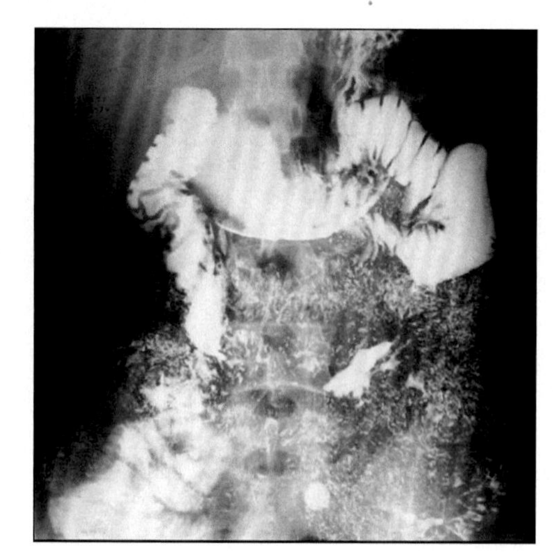

SBFT shows dilation of SB lumen, marked dilution and flocculation of the barium. Malabsorption related to IgA deficiency.

TERMINOLOGY

Definitions

- Impaired intestinal absorption of dietary constituents

IMAGING FINDINGS

General Features

- Best diagnostic clue: Segmental dilatation & spasm of small-bowel with excess fluid + abnormal fold pattern
- Location
  - Celiac disease: Proximal small-bowel
  - Tropical sprue: Entire small-bowel
  - Crohn: Usually terminal ileum
- Key concepts
  - Hallmark: Steatorrhea (↑ fecal fat excretion)
    - Normal: Less than 6 g/24 hrs
  - Abnormal fecal excretion of fat, fat-soluble vitamins (ADEK), proteins, carbohydrates, minerals & water
  - Celiac sprue: Most common small-bowel disease producing malabsorption
  - Classification of malabsorption
    - Maldigestion
    - Malabsorption at mucosal level
    - Malassimilation
    - Malabsorption caused by bacterial overgrowth
  - Maldigestion
    - Chronic pancreatitis, cholestasis, ileal resection
    - Disaccharidase deficiency, Zollinger-Ellison syndrome (ZES)
  - Malabsorption at mucosal level
    - Celiac disease, Crohn, tropical sprue
    - Short bowel syndrome, cystic fibrosis
    - Eosinophilic gastroenteritis, Whipple disease
    - Amyloidosis, hypogammaglobulinemia, mastocytosis
  - Malassimilation
    - Primary & secondary lymphangiectasia
    - Abeta & hypobetalipoproteinemia
  - Malabsorption caused by bacterial overgrowth
    - Idiopathic pseudo-obstruction, systemic sclerosis
    - Multiple, large small-bowel diverticula

Radiographic Findings

- Enteroclysis or small bowel follow through
  - Luminal dilatation
    - Usually more than 3 cm
    - May be segmental or uniform
  - Changes in fold pattern
    - Valvulae conniventes: Thickened (more than 2 mm); uniform, irregular, distorted or nodular
    - Increased, decreased or absent (varies by etiology)
  - Bowel wall thickening
    - Focal or diffuse thickening (> 1 cm)
  - Mucosal nodulation & ulceration
    - Nodulation: Diffuse, punctate or sand-like

DDx: Dilated Bowel, Excess Fluid

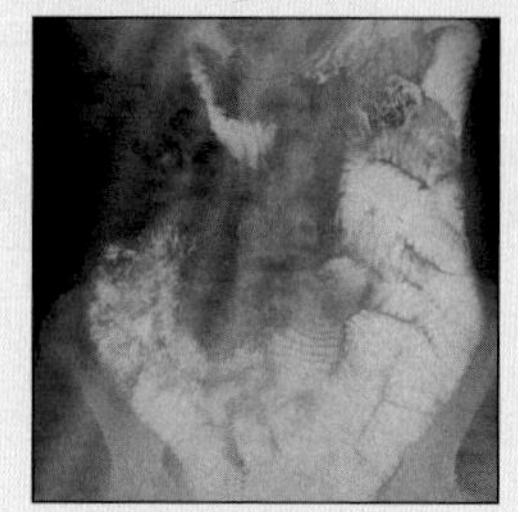

Celiac Sprue

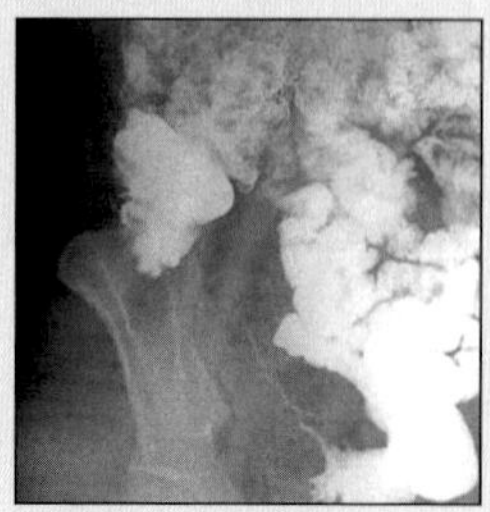

Crohn Disease

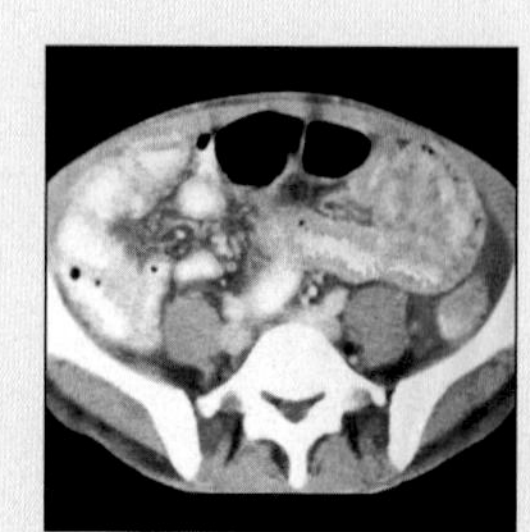

Whipple Disease

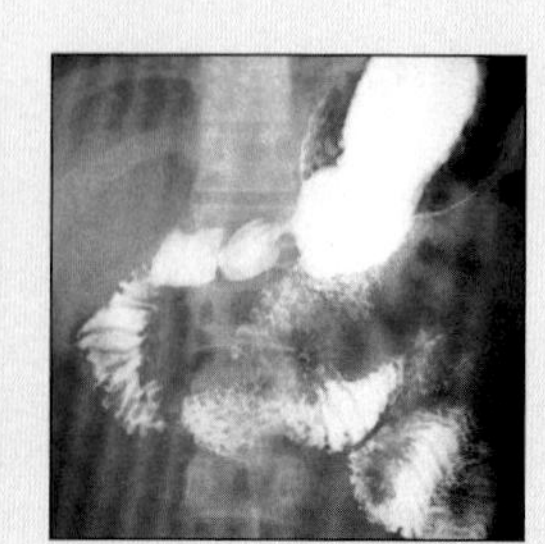

Giardiasis

MALABSORPTION CONDITIONS

Key Facts

Terminology

- Impaired intestinal absorption of dietary constituents

Imaging Findings

- Best diagnostic clue: Segmental dilatation & spasm of small-bowel with excess fluid + abnormal fold pattern
- Luminal dilatation
- Changes in fold pattern
- Bowel wall thickening
- Mucosal nodulation & ulceration
- Motility changes
- Luminal narrowing
- Increased amount of intestinal fluid
- Small-bowel intussusception (target lesion on CT)
- Mesenteric or retroperitoneal lymph nodes
- Mesenteric hypervascularity on contrast study

Top Differential Diagnoses

- Celiac sprue
- Crohn disease
- Whipple disease (intestinal lipodystrophy)
- Opportunistic infection
- Pancreatic disease

Pathology

- Celiac disease: Allergic, immunologic or toxic reaction to gluten protein

Diagnostic Checklist

- Check history of small-bowel diseases & food allergies
- Dilated small-bowel with fold thickening, mucosal nodularity or ulceration, motility changes & ↑ fluid
- Imaging findings often suggest malabsorption; specific diagnosis is difficult

    - Ulceration: Aphthoid or cobblestoning
  - Motility changes
    - Normal, short or long transit time in small-bowel
    - Normal time: For barium to travel through small-bowel to cecum can range from 1-2 hours
  - Luminal narrowing
    - Focal or generalized
  - Increased amount of intestinal fluid
    - Due to diminished absorption
    - Due to increased secretions
  - Multiple or large SB diverticula

CT Findings

- Dilated, fluid-filled small-bowel loops
- Thickened bowel wall & mesentery
- Mucosal fold thickening
- Small-bowel intussusception (target lesion on CT)
- Discontinuous, asymmetric thickened bowel wall
- Mesenteric or retroperitoneal lymph nodes
- Mesenteric hypervascularity on contrast study

Imaging Recommendations

- Enteroclysis or small-bowel follow through; helical CT

DIFFERENTIAL DIAGNOSIS

Celiac sprue

- Small bowel follow through
  - Small bowel dilatation > 3 cm: Mid & distal jejunum
  - Valvulae conniventes: May exhibit various patterns
    - Normal: In most patients
    - Reversed jejunoileal pattern: Decreased jejunal folds & increased ileal folds
    - Absence of valvulae: "Moulage sign" (cast)-characteristic of sprue
    - Thickening: Severe disease & hypoproteinemia
  - "Colonization" of jejunum
    - Loss of jejunal folds lead to colon-like haustrations
  - Transit time: Normal, short or long
- Enteroclysis: Facilitates diagnosis/exclusion of disease
  - Jejunal folds
    - Decreased number of proximal jejunal folds (< 3 inch); normally 5 or > folds/inch
    - Increased separation or absence of folds
    - Ileal appearance of jejunum
  - Ileal folds
    - Increased number of folds in distal ileum (4-6 inch); normally 2-4 folds/inch
    - Increased fold thickness (≥ 1 mm): "Jejunization" of ileum (seen in 78% cases)
- Diagnosis
  - Duodenojejunal mucosal biopsy
  - ↑ IgA & IgM antigliadin antibodies

Crohn disease

- Predominantly involves distal ileum
  - Can affect from mouth to anal canal
- Aphthoid ulceration, cobblestoning
- Skip lesions, transmural, string sign
- Fistulae, fissures, sinuses
- Diagnosis: Mucosal biopsy

Whipple disease (intestinal lipodystrophy)

- Caused by Tropheryma whippelii (PAS positive bacilli)
- Presents with chronic diarrhea & malabsorption
- Duodenum & proximal jejunum
  - Dilated with thickened mucosal folds
  - Micronodules: 1-2 mm; separation of loops
  - Mesenteric lymphadenopathy
- Diagnosis: Mucosal biopsy

Opportunistic infection

- AIDS with giardiasis & cryptosporidiosis
- Giardiasis
  - Giardia lamblia: A protozoan intestinal flagellate
  - Cause of travelers diarrhea & enteritis in AIDS
  - Clinical: Acute self-limited diarrhea, or may develop chronic diarrhea, malabsorption & weight loss
  - Duodenum & jejunum
    - Thickened irregular folds with hypermotility
    - Luminal narrowing & increased secretions
  - Ileum: Usually appears normal
  - Diagnosis: Mucosal biopsy; stool (ova/trophozoites)
- Cryptosporidiosis
  - Most common cause of enteritis in AIDS patients

- Severe cholera-like illness & malabsorption
  - Diarrhea, crampy abdominal pain, vomiting
- Pathology: Mucosal damage
- Radiographic findings
  - Duodenum & proximal jejunum: Thickened folds
  - Distal loops show incomplete barium coating with areas of flocculation
- CT may reveal small (< 0.5 cm) lymph nodes

Pancreatic disease

- Example: Zollinger-Ellison syndrome (ZES); cystic fibrosis
- Zollinger-Ellison syndrome
  - Stomach
    - Thickened folds, enlarged areae gastricae, erosions
  - Duodenum & proximal jejunum
    - Dilated, thickened nodular folds, erosions & ulcers
  - Large volume of fluid dilutes barium & compromises mucosal coating
  - CT: Shows other associated lesions
    - Islet cell tumor: Gastrinoma (75% in pancreas)
    - Hypervascular on contrast scan
    - Hypervascular metastases (common in liver)
  - Diagnosis: Serum gastrin levels > 1000pg/ml
- Cystic fibrosis
  - Autosomal recessive genetic disorder
  - Defect in CFTR gene on chromosome 7q
  - Small-bowel barium study
    - Duodenum: Thickened or flattened folds, nodular filling defects, sacculation along lateral border of descending duodenum
    - Ileum: Thickened folds + reticular mucosal pattern
  - CT will show fatty replacement of pancreas

PATHOLOGY

General Features

- Genetics: Celiac sprue- class II human leukocyte antigens (HLA-DR3/HLA-DQw2)
- Etiology
  - Celiac disease: Allergic, immunologic or toxic reaction to gluten protein
  - Tropical sprue: Unknown etiology; may be due to enterotoxigenic E. coli
  - Crohn disease: Genetic, environmental, infectious, immunologic, psychologic
  - Whipple disease: Tropheryma whippelii (bacilli)
  - ZES: Gastrinoma → ↑ gastric acid secretion → severe peptic ulcer disease
  - Eosinophilic gastroenteritis (EGE): Self-limited disease; affect patients with allergic disorders
- Epidemiology: Celiac- ↑ incidence Ireland & N. Europe

Gross Pathologic & Surgical Features

- Varies by etiology
- Celiac: Dilated, thickened bowel; reversal fold pattern

Microscopic Features

- Varies by etiology

CLINICAL ISSUES

Presentation

- Most common signs/symptoms
  - Diarrhea, steatorrhea, flatulence
  - Abdominal distension, weight loss, anemia
- Lab-data
  - Decreased levels of vitamins (ADEK, B12)
  - Decreased levels of iron, folates, albumin, bile salts
  - Increased fecal fat (15-30 g/24 hrs)
  - Abnormal D-Xylose breath test: Bacterial overgrowth
  - Hydrogen test for lactase deficiency
  - Abnormal CBC & liver function tests
- Diagnosis: Mucosal biopsy & histology

Demographics

- Age: Any age group

Natural History & Prognosis

- Complications
  - Micro or macrocytic anemia, bony changes
  - Edema, ascites, failure to thrive
- Prognosis
  - Malabsorption after treatment
    - Celiac disease: Improvement within 48 hrs
    - Tropical sprue: 4-7 days
    - Crohn: 10-20% symptom free; 30-53% post surgical recurrence

Treatment

- Celiac: Gluten-free diet
- Tropical sprue: Antibiotics
- Crohn disease
  - Medical (steroids, azathioprine, mesalamine)
  - Surgical resection

DIAGNOSTIC CHECKLIST

Consider

- Check history of small-bowel diseases & food allergies

Image Interpretation Pearls

- Dilated small-bowel with fold thickening, mucosal nodularity or ulceration, motility changes & ↑ fluid
- Imaging findings often suggest malabsorption; specific diagnosis is difficult

SELECTED REFERENCES

1. Farrell RJ et al: Celiac sprue. N Engl J Med. 346(3):180-8, 2002
2. Koch J et al: Small intestine pathogens in AIDS: conventional and opportunistic. Gastrointest Endosc Clin N Am. 8(4):869-88, 1998
3. Antes G: Inflammatory disease of the small intestine and colon: contrast enema and CT. Radiology. 38: 41-45, 1998
4. Horton KM et al: Uncommon inflammatory diseases of the small bowel: CT findings. AJR Am J Roentgenol. 170(2):385-8, 1998
5. Maglinte DD et al: Current status of small bowel radiography. Abdom Imaging. 21(3):247-57, 1996
6. Rubesin SE et al: Small bowel malabsorption: clinical and radiologic perspectives. How we see it. Radiology. 184(2):297-305, 1992

IMAGE GALLERY

Typical

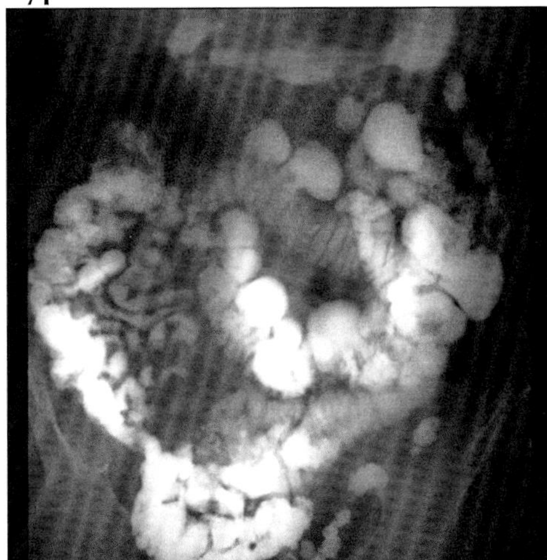

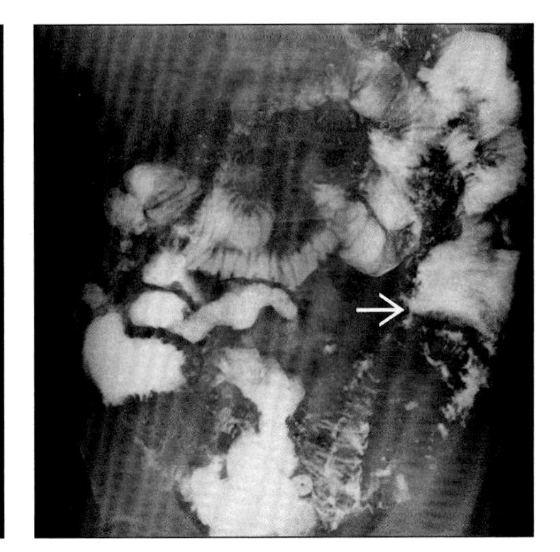

(Left) *SBFT shows numerous large diverticula throughout the jejunum which can lead to stasis, bacterial overgrowth and malabsorption.* ***(Right)*** *SBFT in patient with sprue shows segmental dilation + spasm of bowel + dilution + flocculation of barium. Also note transient intussusception (arrow) of jejunum with "coiled spring" appearance.*

Typical

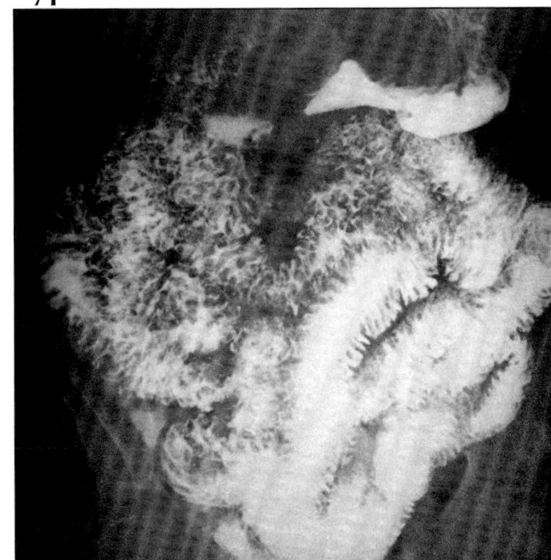

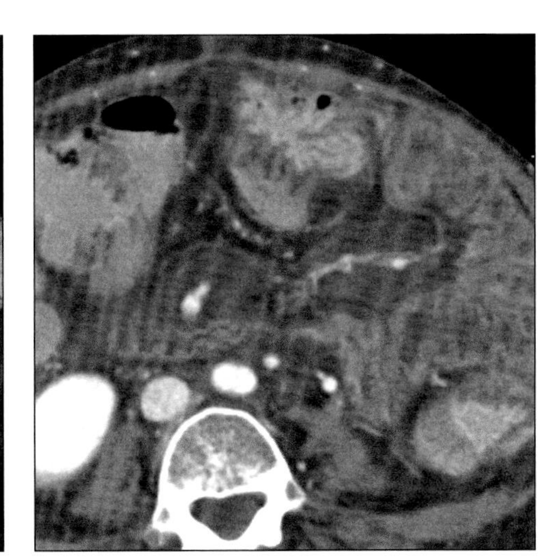

(Left) *SBFT shows nodular SB fold pattern + dilated lumen. Waldenstrom macroglobulinemia.* ***(Right)*** *Axial CECT shows SB fold thickening, excess fluid in lumen, mesenteric engorgement due to primary lymphangiectasia.*

Typical

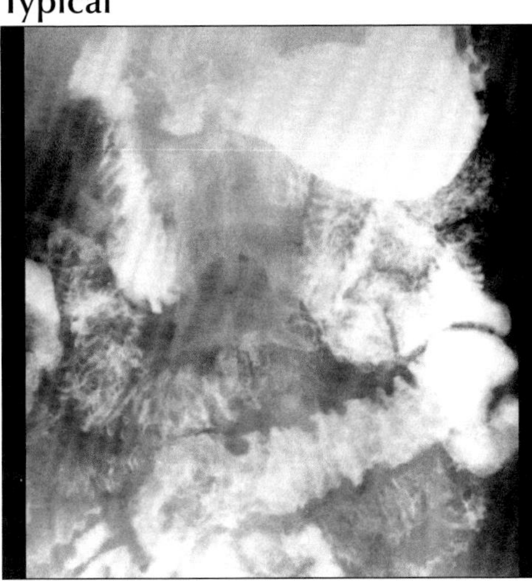

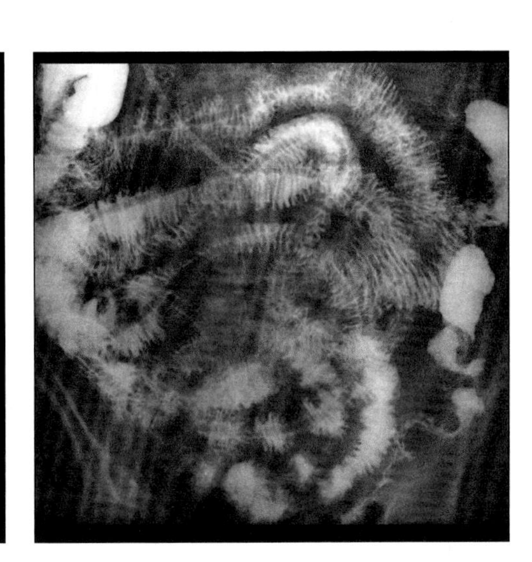

(Left) *SBFT shows distorted nodular SB folds, diluted barium. Dysgammaglobulinemia.* ***(Right)*** *Enteroclysis shows symmetrical fold thickening due to hypoproteinemia.*

INTUSSUSCEPTION

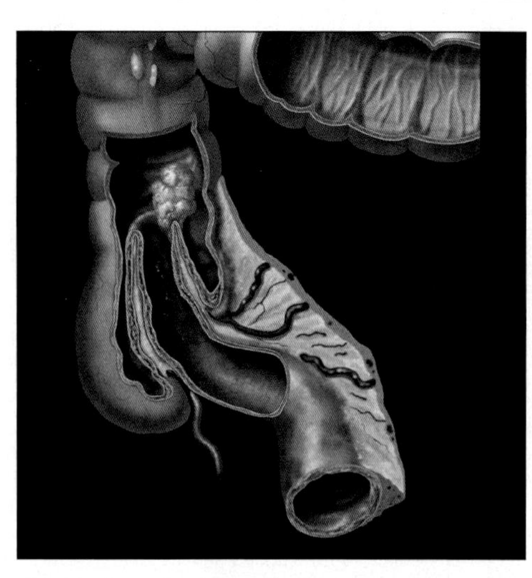

Graphic shows ileocolic intussusception with a tumor in the bowel wall as the "lead mass". Note vascular compromise and ischemia.

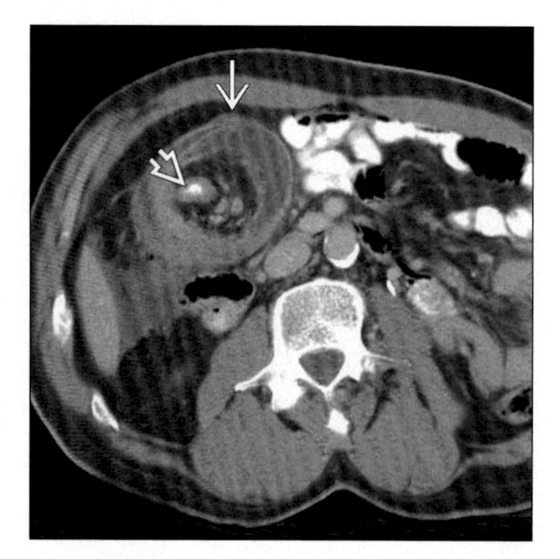

Axial CECT shows ileocolic intussusception. The outer ring (arrow) is the wall of the intussuscipiens (colon, here) while the intussusceptum is the small intestinal segment (open arrow).

TERMINOLOGY

Definitions

- Invagination or telescoping of a proximal segment of bowel (intussusceptum) into lumen of a distal segment (intussuscipiens)

IMAGING FINDINGS

General Features

- Best diagnostic clue: Bowel within bowel, "coiled spring" appearance
- Location: Ileoileal > ileocolic > colocolic
- Key concepts
  - Rarely symptomatic in adults: 0.003-0.02% of all hospital admissions
  - Adults: A different entity than in children
  - Accounts for 5% of all intussusceptions & 1% of all bowel obstructions
  - May be transient or persistent
  - Classified into two types in adults
    - Short-segment, non-obstructing intussusception: Usually self-limited without a lead mass (idiopathic, adhesions, bowel wall thickening)
    - Long-segment, obstructing intussusception: Mass
  - Small-bowel: Benign tumors more common than malignant
  - Colon: Malignant tumors more common than benign
  - Infants & children
    - Accounts for 95% of all intussusceptions
    - 90% of cases, cause is idiopathic (lymphoid hyperplasia)
    - Ranks 2nd to appendicitis in children as a cause of acute abdomen
    - Location: Usually small bowel in adults; ileocolic in children
  - Imaging findings (barium studies & CT) are pathognomonic for intussusception

Radiographic Findings

- Radiography
  - Findings of bowel obstruction may be seen
    - Air-fluid levels; proximal bowel dilatation
    - Absence of gas in distal collapsed bowel
- Fluoroscopic guided barium study
  - Classic "coiled spring" appearance
    - Due to trapping of contrast between folds of intussusceptum & intussuscipiens
  - Bowel obstruction, proximal dilatation & distal collapsed loops

CT Findings

- Seen as three different patterns on axial CT scans
  - "Target" sign: Earliest stage of intussusception
    - Outer layer represents intussuscipiens

DDx: Bowel Obstruction with Mass

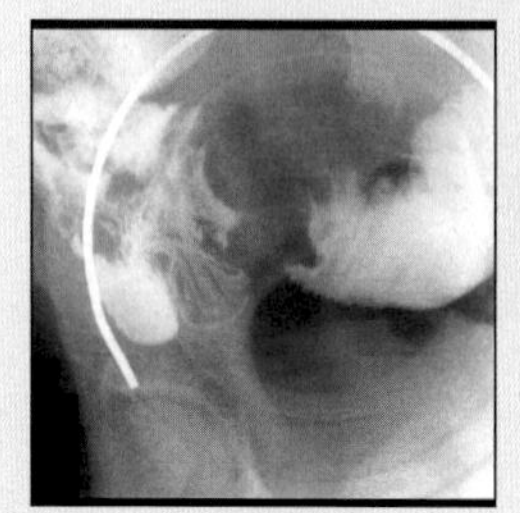

Carcinoma

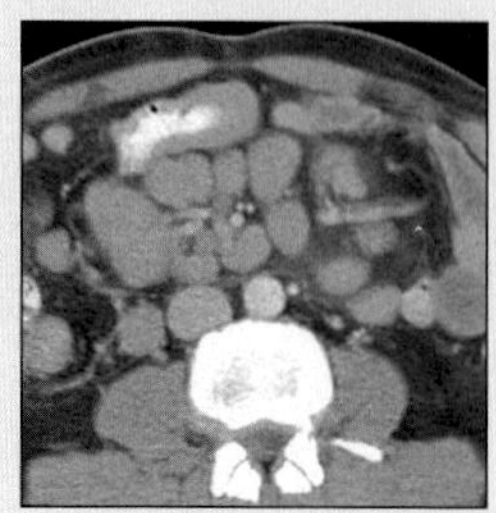

Lymphoma

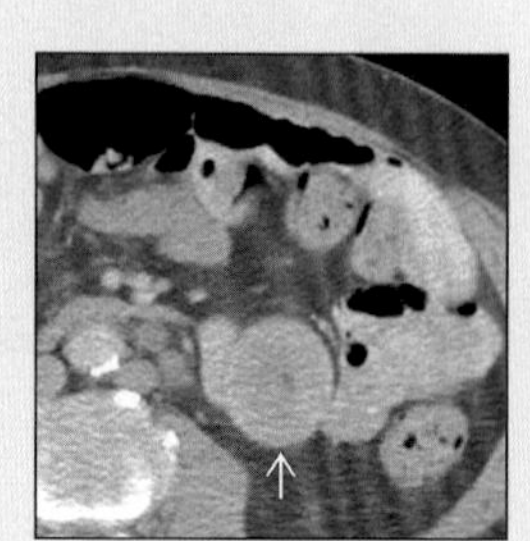

Melanoma

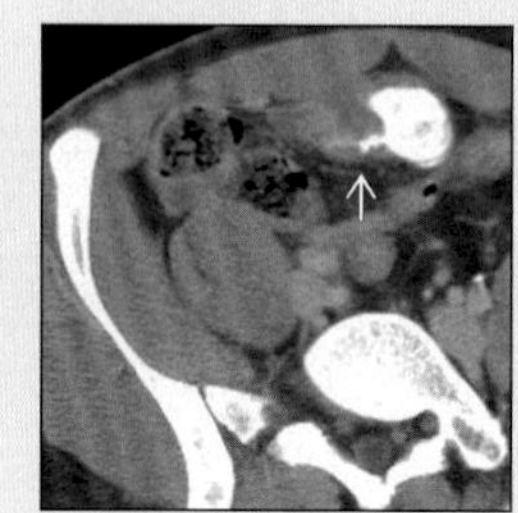

Meckel Diverticulum

INTUSSUSCEPTION

Key Facts

Terminology
- Invagination or telescoping of a proximal segment of bowel (intussusceptum) into lumen of a distal segment (intussuscipiens)

Imaging Findings
- Best diagnostic clue: Bowel within bowel, "coiled spring" appearance
- Location: Ileoileal > ileocolic > colocolic
- Short-segment, non-obstructing intussusception: Usually self-limited without a lead mass (idiopathic, adhesions, bowel wall thickening)
- Long-segment, obstructing intussusception: Mass
- "Target" sign: Earliest stage of intussusception
- Sausage-shaped mass: A layering pattern (later phase)
- Reniform mass: Due to edema or mural thickening (vascular compromise)

Top Differential Diagnoses
- Primary bowel tumor
- Metastases & lymphoma
- Endometrial implant
- Meckel diverticulum

Pathology
- Tumor related lead point: Benign & malignant

Diagnostic Checklist
- Short segment, non-obstructing intussusceptions are common in adults + require no therapy
- "Coiled spring" appearance due to trapped barium
- Lead point: Lobulated mass etched in white

    - Inner layer represents intussusceptum
  - Sausage-shaped mass: A layering pattern (later phase)
    - Alternating layers of low-attenuation (mesenteric fat) & high-attenuation areas (bowel wall)
    - Enhancing mesenteric vessels
  - Reniform mass: Due to edema or mural thickening (vascular compromise)
    - Vascular compromise: Seen in returning wall of intussusceptum as hypodense layer in middle of inner part of thick bowel wall & crescent-shaped fluid or gas collections
  - Features of intestinal obstruction
    - Air-fluid levels; proximal bowel distension

MR Findings
- Bowel-within-bowel or coiled-spring appearance
- Best seen on turbo spin-echo T2WI

Ultrasonographic Findings
- Real Time
  - Transverse US: Target, doughnut or "bull's eye" sign
    - Peripheral hypoechoic halo: Edematous wall of intussuscipiens
    - Intermediate hyperechoic area: Space between intussuscipiens & intussusceptum
    - Internal hypoechoic ring: Due to intussusceptum
  - Longitudinal US: "Pseudokidney" or hay fork sign
    - Multiple, thin, parallel, hypoechoic & echogenic stripes
- Color Doppler: Shows mesenteric vessels dragged between entering and returning wall of intussusceptum

Imaging Recommendations
- Depends on patient age & presentation
- Helical CT; barium studies; US

DIFFERENTIAL DIAGNOSIS

Primary bowel tumor
- Example: Carcinoid tumor, adenocarcinoma, stromal tumor, lipoma & adenoma
- Enteroclysis
  - Best technique for detecting mass

Metastases & lymphoma
- Metastases (small bowel)
  - Example: From malignant melanoma, lung & breast cancer
  - Location: Antimesenteric border
  - Malignant melanoma
    - Smoothly polypoid lesions of different sizes
    - "Spoke-wheel" pattern: Polypoid lesion with ulcers & radiating folds
  - Bronchogenic carcinoma
    - Single/multiple intramural lesions (flat/polypoid)
    - Frequently ulcerated, narrowing & obstruction
  - Breast carcinoma
    - Highly cellular submucosal masses
    - Multiple strictures + intervening bowel dilatation
    - Intussusception of ulcerated mural lesions
- Non-Hodgkin lymphoma (more common)
  - Distribution: Stomach (51%), small-bowel (33%)
  - Nodular, polypoid, infiltrating, mesenteric invasive
  - Focal infiltration: Sausage-shaped thickening of affected bowel wall simulating intussusception
  - May cause intussusception

Endometrial implant
- Endometrial tissue outside the myometrium
- Common location: Pelvic organs
  - Bowel is involved in 37% cases: Rectosigmoid (95%); small-bowel (7%) predominantly terminal ileum
- Crenulation of folds or plaque-like deformities
- High grade or low grade small-bowel obstruction
  - Usually due to fibrosis & rarely intussusception
- Diagnosis: Laparoscopy

Meckel diverticulum
- Most frequent congenital anomaly of GI tract
- Ileal outpouching (2 feet from ileocecal valve)
- Causes of small-bowel obstruction
  - Torsion associated with a persistent vitelline band
  - Extrusion of diverticulum into an inguinal hernia (hernia of Littre)
  - Intussusception of an inverted diverticulum

INTUSSUSCEPTION

PATHOLOGY

General Features

- Etiology
  - Most adult intussusceptions are short segment, transient, non-obstructing + not associated with a lead tumor mass
  - Tumor related lead point: Benign & malignant
  - Benign
    - Polyp, leiomyoma, lipoma, adenoma of appendix
    - Appendiceal stump granuloma (> common in small bowel)
  - Malignant
    - Primary (more common in colon)
    - Metastases & lymphoma (> common-small bowel)
  - Postoperative: Risk factors (> common-small bowel)
    - Suture lines, ostomy closure sites
    - Adhesions, long intestinal tubes
    - Bypassed intestinal segments, submucosal edema
    - Abnormal bowel motility, electrolyte imbalance
    - Chronic dilated loop
  - Miscellaneous
    - Meckel diverticulum; celiac & Whipple disease
    - Colitis (eosinophilic & pseudomembranous)
    - Epiploic appendagitis
  - Idiopathic
- Epidemiology
  - Incidence
    - Adults (uncommon); children (more common)

Gross Pathologic & Surgical Features

- Three layers are seen
  - Intussusceptum: Entering or inner tube + returning or middle tube
  - Intussuscipiens: Sheath or outer tube

Microscopic Features

- Early: Inflammatory changes; late-ischemic necrosis + mucosal sloughing

CLINICAL ISSUES

Presentation

- Most common signs/symptoms
  - Adults
    - Intermittent pain, vomiting, red blood in stool
  - Children
    - Acute pain, palpable oblong mass in abdomen
    - "Red currant jelly" stools

Demographics

- Age: Any age group
- Gender: M = F

Natural History & Prognosis

- Complications
  - Obstruction, infarction & necrosis
  - Hemorrhage, perforation & peritonitis
- Prognosis
  - Early: Good
    - After reduction, surgical resection
    - Recurrence very rare
  - Late: Poor
    - Severe vascular compromise
    - Gangrene & perforation

Treatment

- None for transient , non-obstructing
- Ileocolic, ileocecocolic & colocolic: Resection
- Children
  - Hydrostatic or pneumatic reduction
  - Surgical reduction or resection

DIAGNOSTIC CHECKLIST

Consider

- Short segment, non-obstructing intussusceptions are common in adults + require no therapy

Image Interpretation Pearls

- Lumen of intussusceptum as a narrow, tubular structure lined by twisted mucosal folds
- "Coiled spring" appearance due to trapped barium
- Lead point: Lobulated mass etched in white

SELECTED REFERENCES

1. Huang BY et al: Adult intussusception: diagnosis and clinical relevance. Radiol Clin North Am. 41(6):1137-51, 2003
2. Lvoff N et al: Distinguishing features of self-limiting adult small-bowel intussusception identified at CT. Radiology. 227(1):68-72, 2003
3. Saenz De Ormijana J et al: Idiopathic enteroenteric intussusceptions in adults. Abdom Imaging. 28(1):8-11, 2003
4. Gayer G et al: Pictorial review: adult intussusception--a CT diagnosis. Br J Radiol. 75(890):185-90, 2002
5. Fujimoto T et al: Unenhanced CT findings of vascular compromise in association with intussusceptions in adults. AJR Am J Roentgenol. 176(5):1167-71, 2001
6. Warshauer DM et al: Adult intussusception detected at CT or MR imaging: clinical-imaging correlation. Radiology. 212(3):853-60, 1999
7. Catalano O: Transient small bowel intussusception: CT findings in adults. Br J Radiol. 70(836):805-8, 1997
8. Lorigan JG et al: The computed tomographic appearances and clinical significance of intussusception in adults with malignant neoplasms. Br J Radiol. 63(748):257-62, 1990
9. Merine D et al: Enteroenteric intussusception: CT findings in nine patients. AJR Am J Roentgenol. 148(6):1129-32, 1987

INTUSSUSCEPTION

IMAGE GALLERY

Typical

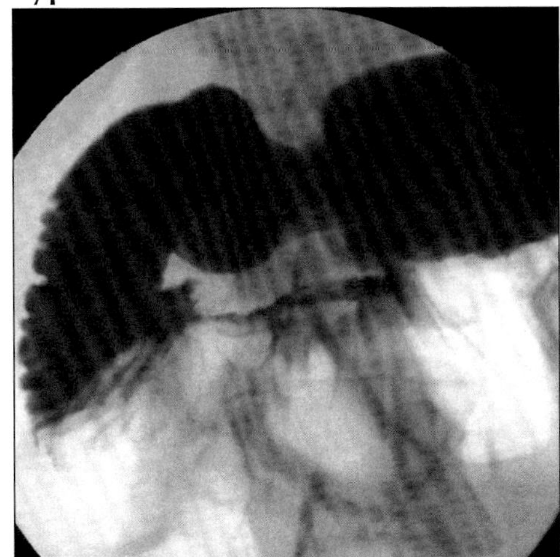

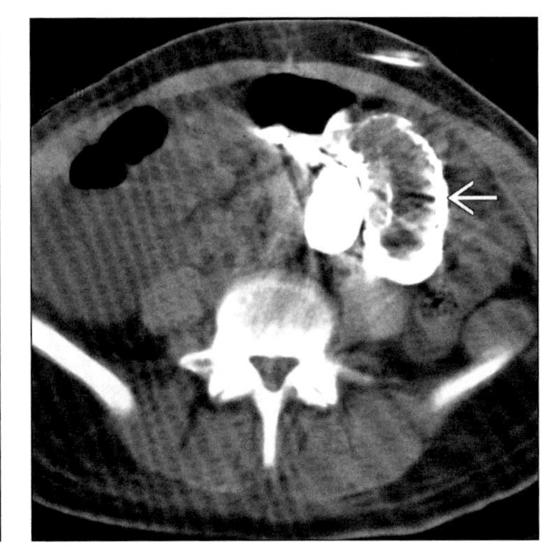

*(**Left**) SBFT shows "coiled spring" + "bowel-in-bowel" appearance of SB intussusception. Lead mass was melanoma metastatic to bowel wall. (**Right**) Axial NECT following SBFT in patient with metastatic melanoma shows bowel-in-bowel appearance (arrow) of intussusception.*

Typical

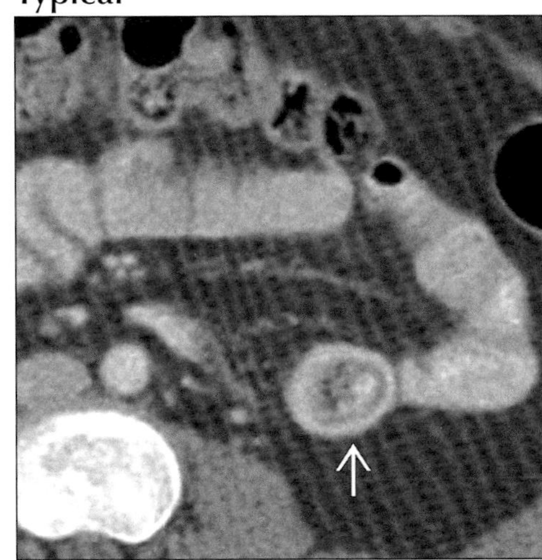

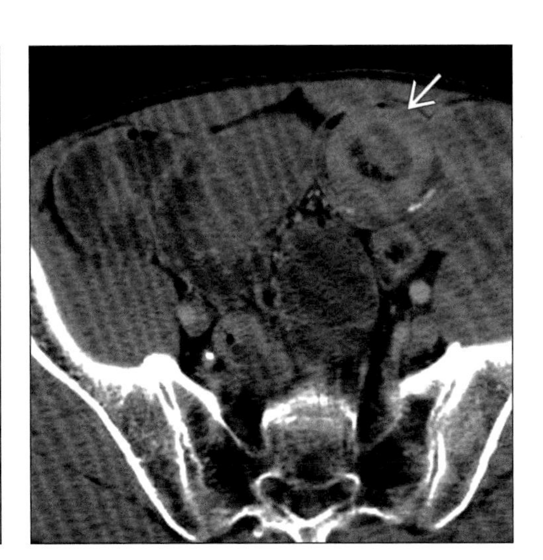

*(**Left**) Axial CECT in a patient with cystic fibrosis shows a short segment, non-obstructing SB intussusception (arrow). (**Right**) Axial CECT shows SB intussusception (arrow) with crescent of mesenteric fat accompanying the intussusceptum. Bowel lumen is dilated due to sprue, not obstruction.*

Typical

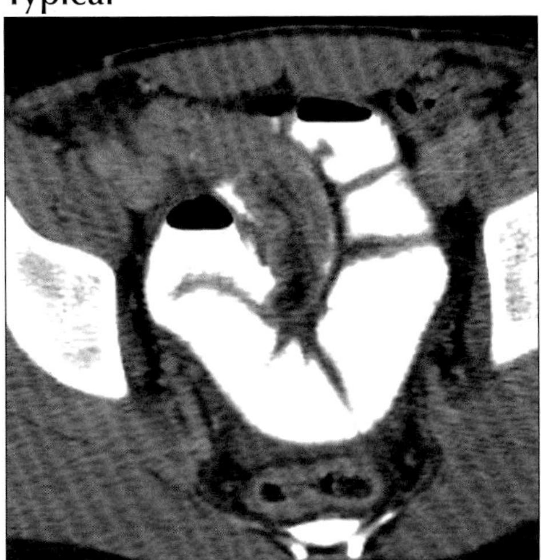

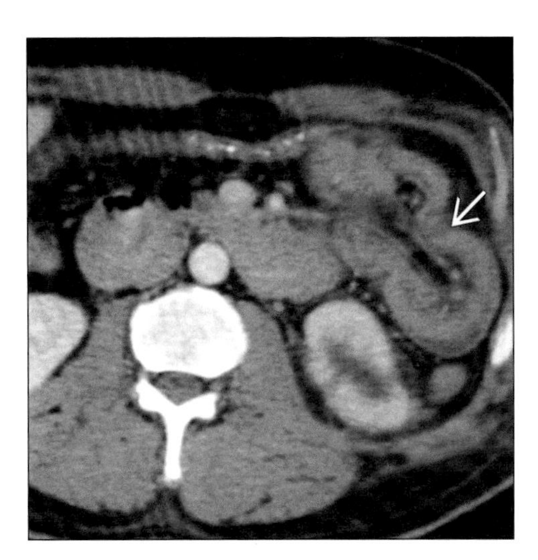

*(**Left**) Axial CECT shows sausage-like mass within the lumen of the terminal ileum due to inverted, intussuscepting Meckel diverticulum. (**Right**) Axial CECT shows "reniform" (kidney-shaped) small bowel (arrow) due to jejunal intussusception.*

RADIATION ENTERITIS

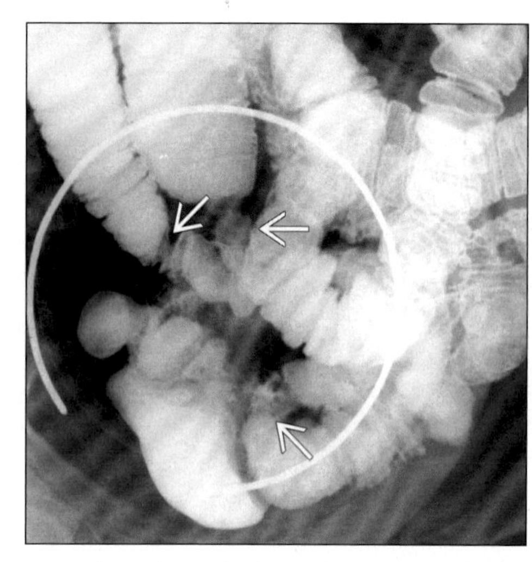

Enteroclysis shows fixed segments of SB luminal narrowing + fold distortion (arrows) due to radiation treatment for cervical cancer.

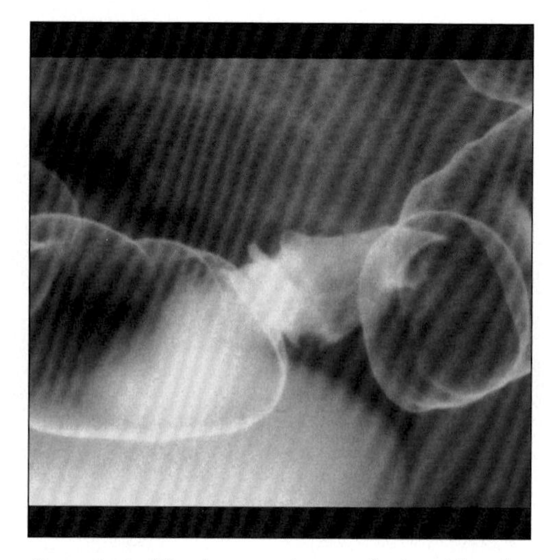

Air-contrast BE shows stricture of sigmoid colon following radiation therapy for prostate cancer. Mucosa is intact.

TERMINOLOGY

Definitions

- Damage of small bowel mucosa + wall due to therapeutic or excessive abdominal irradiation

IMAGING FINDINGS

General Features

- Best diagnostic clue: Mural thickening and luminal narrowing of pelvic bowel loops
- Location
  - Small bowel (ileum more common than jejunum)
  - Adjacent colon (radiation colitis) and rectum (radiation proctitis)
- Key concepts
  - Permanent radiographic findings: 1 month to 2 years after radiation therapy
  - Classification of radiation enteritis
    - Acute stage: Concurrent with or < 2 months after treatment
    - Subacute stage: 2-12 months after treatment
    - Chronic stage: > 12 months after treatment

Radiographic Findings

- Fluoroscopic-guided enteroclysis
  - Acute radiation enteritis
    - Bowel loops appear spastic (↓ lumen diameter) with thickened folds (edema)
  - Chronic radiation enteritis
    - Thickened valvulae conniventes and intestinal wall (edema or fibrosis)
    - Thickened folds appear straight and parallel
    - "Stack of coins" appearance: Enlarged smooth, straight, parallel folds perpendicular to the longitudinal axis of small bowel (submucosal hemorrhage)
    - Spiky appearance: Barium trapped between thickened folds (in profile view); folds are thickened and held closely to one another, where thickness exceeds the distance between folds
    - Narrowed or stenotic lumen → small bowel obstruction with dilation of proximal bowel loops
    - Single or multiple stenoses (stricture) of varying length (up to several cm)
    - Adhesions → angulation between adjacent loops, fixation of loops or "mucosal tacking"
    - "Mucosal tacking": Angulation, spiking and distortion of the mucosal folds on antimesenteric border; usually seen in terminal ileum and adjacent bowel loops
    - Peristaltic activity is ↓ or absent
    - ± Large, deep ulcers; difficult to detect shallow ulcers

DDx: Long Segment Luminal Narrowing

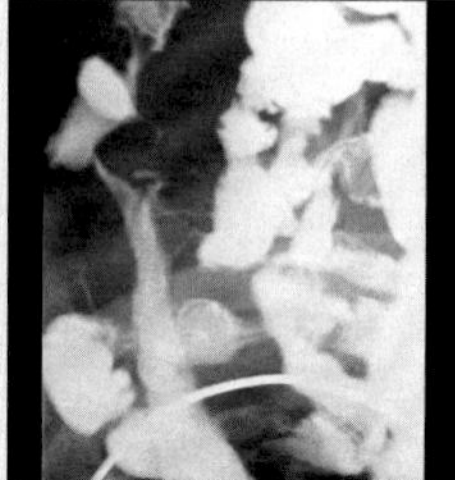

Crohn Disease

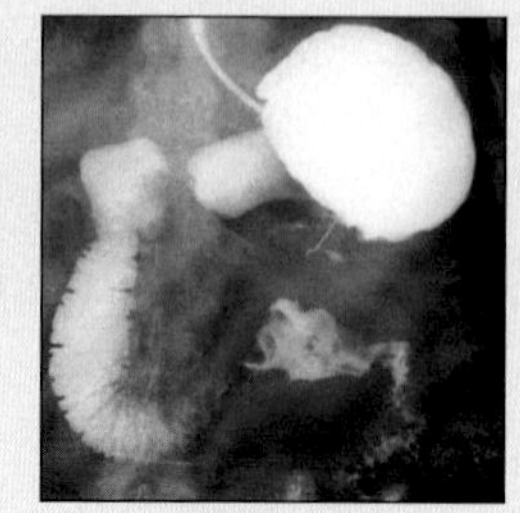

Lymphoma

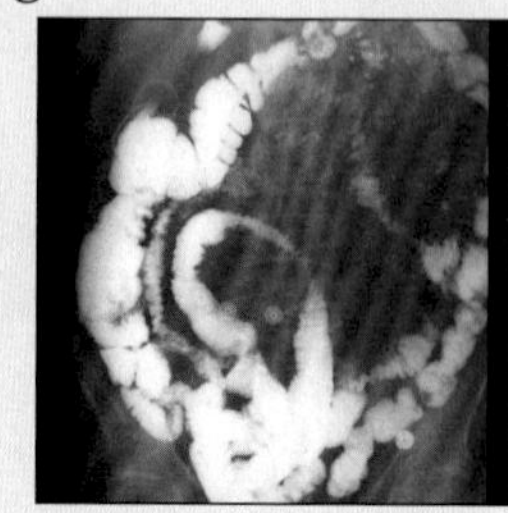

Ischemia

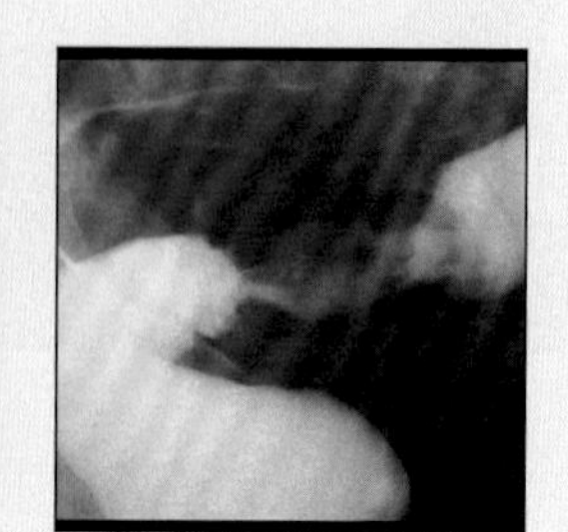

Colon Carcinoma

RADIATION ENTERITIS

Key Facts

Terminology

- Damage of small bowel mucosa + wall due to therapeutic or excessive abdominal irradiation

Imaging Findings

- Best diagnostic clue: Mural thickening and luminal narrowing of pelvic bowel loops
- Permanent radiographic findings: 1 month to 2 years after radiation therapy
- Thickened folds appear straight and parallel
- Narrowed or stenotic lumen → small bowel obstruction with dilation of proximal bowel loops
- Adhesions → angulation between adjacent loops, fixation of loops or "mucosal tacking"
- Peristaltic activity is ↓ or absent
- Best imaging tool: Fluoroscopic-guided enteroclysis; fluoroscopic-guided barium enema; helical CT

Top Differential Diagnoses

- Crohn disease
- Metastases and lymphoma
- Ischemic enteritis

Clinical Issues

- Colicky abdominal pain, nausea, vomiting, tenesmus, bloody diarrhea, steatorrhea (10-40 g per day) and weight loss

Diagnostic Checklist

- History of radiation therapy
- Small bowel obstruction; wall thickening with fixed and angulated loops of bowel; reduced peristalsis

        - Sinuses and fistulas (especially at damaged, surgical anastomotic site caused by radiation)
        - Effacement of valvulae conniventes (late, atrophic feature)
- Fluoroscopic-guided barium enema
    - Acute radiation colitis and proctitis
        - ± Disrupted or distorted mucosal pattern (edema or hemorrhage)
    - Chronic radiation colitis and proctitis
        - Diffuse or focal narrowing with tapered margins
        - Rectal stricture or rectovaginal fistula can be seen
        - Presacral space widened (in lateral view)

CT Findings

- Bowel wall thickening, luminal narrowing
- Fibrosis surrounding the bowel ± strictures
- ± Small bowel obstruction with multiple air-fluid levels and dilation of proximal bowel loops
- ± Sinuses or fistulas

MR Findings

- T2WI
    - Thick, high signal intensity layer suggests submucosal edema; not tumor invasion
    - "Bull's eye" pattern: Thickened high signal intensity submucosa surrounded by the low signal intensity muscularis propria and muscularis mucosae
    - ± Bowel fistula: Fluid in tract appears as high signal, contrasted by soft tissue and fat

Imaging Recommendations

- Best imaging tool: Fluoroscopic-guided enteroclysis; fluoroscopic-guided barium enema; helical CT

DIFFERENTIAL DIAGNOSIS

Crohn disease

- Skip lesions, transmural inflammation, granulomas, cobblestone mucosa and fistulas
- Differentiate by irregular thickening, fused and distorted folds; more prominent mural thickening

Metastases and lymphoma

- Metastases (e.g., malignant melanoma, breast cancer)
    - Cause small bowel obstruction and narrowing
    - Irradiation of abdomen is a treatment modality
        - Difficult to differentiate radiologically because both can coexist
        - Interval growth or biopsy diagnostic
- Non-Hodgkin lymphoma
    - Distribution: Small bowel (33%) and colon (16%)
    - Infiltrative form appears similar to radiation enteritis
        - Circumferential infiltration involves variable length of small intestine → thickening → effacement of folds
    - However, non-Hodgkin lymphoma rarely has same radiologic features
        - Widened lumen of bowel and rare stricture formation
        - Other classic features include multiple nodular defects, polypoid and mesenteric invasive form
        - CT: Demonstration of bulky mesenteric node involvement

Ischemic enteritis

- Ischemic changes of small bowel → small bowel obstruction mimicking radiation enteritis
- Ischemia and radiation can both cause submucosal hemorrhage → "stack of coins" appearance
- Distinguish by observing clot or narrowing of superior mesenteric artery, superior mesenteric vein or other mesenteric vessels

Primary bowel tumor

- Can cause irregular stricture
- Usually more mass effect than with radiation enteritis

PATHOLOGY

General Features

- General path comments
    - Acute stage
        - Mucosa: Thinning (reduction of crypt cell mitoses) and edema

- Submucosa: Edema
- Hyperemia and ulceration
- Subacute stage
  - Mucosa: Process of healing
  - Submucosa: Obliterative changes in arterioles and fibrotic thickening
- Chronic stage
  - Muscularis propria: Fibrosis
  - Serosa: Diffuse hyaline change → adhesions between bowel loops
  - Ulceration and fibrosis → strictures
- Radiation tolerance
  - Duodenum, jejunum, ileum, transverse colon, sigmoid colon, esophagus and rectum (from highest to lowest tolerance)
  - Tolerance dose (TD 5/5): Total dose that produce radiation damage in 5% of patients within 5 years
  - TD 5/5: 4,500 cGy (1 centigray = 1 rad) in small bowel and colon; 5,000 cGy in rectum
- Etiology
  - Radiation therapy for cancer of cervix, uterus, ovary, cecum, colon, rectum and bladder
  - Risk factors of chronic radiation enteritis
    - Prior abdominal surgery → adhesions
    - Peritonitis prior to radiation treatment
    - High radiation dose over a short period of time
    - Hypertension, atherosclerosis or diabetes mellitus
    - Chemotherapy given with radiation → increase radiation damage
  - Pathogenesis (2 methods)
    - Direct cytotoxic effect: Radiation → free radicals interact with DNA → elimination of replication, transcription and protein synthesis → cell disruption and death
    - Ischemic changes: Medial wall thickening and subendothelial proliferation → endarteritis obliterans → fibrosis
- Epidemiology
  - 5-15% treated with radiation will develop chronic radiation enteritis
  - Radiation enteritis: Usually within 12 years after radiation therapy (can occur during radiation treatment, but up to 25 years latency)
  - Radiation colitis: Within 2 years after radiation

Gross Pathologic & Surgical Features

- Thickened folds, ulceration, stricture or adhesions

Microscopic Features

- Shortening of villi, inflammatory cells and megalocytosis of epithelial cells can be seen

CLINICAL ISSUES

Presentation

- Most common signs/symptoms
  - Acute
    - Abdominal cramping, nausea, vomiting, tenesmus and watery diarrhea
    - Dehydration and malabsorption (change in small bowel flora)
    - Severity of symptoms proportional to dose and volume of irradiation
    - Radiation proctitis: Mucoid rectal discharge, rectal pain and rectal bleeding
  - Chronic
    - Colicky abdominal pain, nausea, vomiting, tenesmus, bloody diarrhea, steatorrhea (10-40 g per day) and weight loss
- Lab-data
  - Hypocalcemia; ↓ iron and vitamin B12 levels
- Diagnosis
  - Radiation enteritis: Radiologic appearances on fluoroscopic-guided enteroclysis and helical CT
  - Radiation colitis: Radiologic appearances on fluoroscopic-guided barium enema and helical CT

Demographics

- Gender: M < F

Natural History & Prognosis

- Complications
  - Fistula, stricture, small bowel obstruction, hemorrhage, abscess and perforation
- Prognosis
  - Good, after medical treatment with reduction or cessation of radiation
  - Poor, if patients have chronic radiation injury with complications

Treatment

- Reduction or cessation of radiation; medical treatment and low-residue diet
- Surgery indicated when medical treatment fails

DIAGNOSTIC CHECKLIST

Consider

- Rule out recurrent tumors
- History of radiation therapy
  - Dose rate, fractionation of therapeutic doses, total dose, radiation portal, type of radiation and length of time after radiation therapy

Image Interpretation Pearls

- Small bowel obstruction; wall thickening with fixed and angulated loops of bowel; reduced peristalsis

SELECTED REFERENCES

1. Low RN et al: Distinguishing benign from malignant bowel obstruction in patients with malignancy: findings at MR imaging. Radiology. 228(1):157-65, 2003
2. Bismar MM et al: Radiation enteritis. Curr Gastroenterol Rep. 4(5):361-5, 2002
3. Horton KM et al: CT of nonneoplastic diseases of the small bowel: spectrum of disease. J Comput Assist Tomogr. 23(3):417-28, 1999
4. Capps GW et al: Imaging features of radiation-induced changes in the abdomen. Radiographics. 17(6):1455-73, 1997
5. Bluemke DA et al: Complications of radiation therapy: CT evaluation. Radiographics. 11(4):581-600, 1991
6. Mendelson RM et al: The radiological features of chronic radiation enteritis. Clin Radiol. 36(2):141-8, 1985

RADIATION ENTERITIS

IMAGE GALLERY

Typical

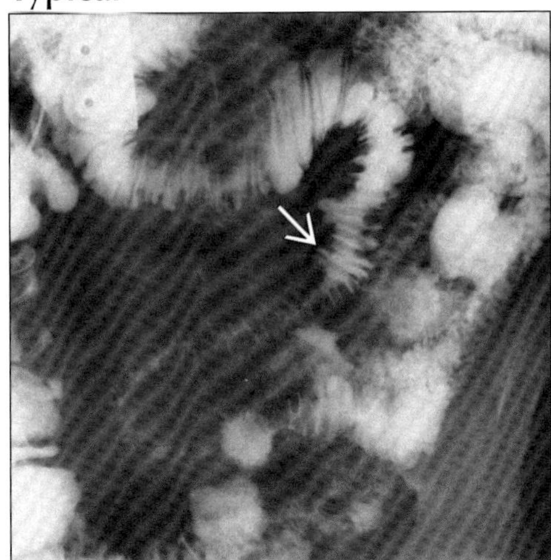

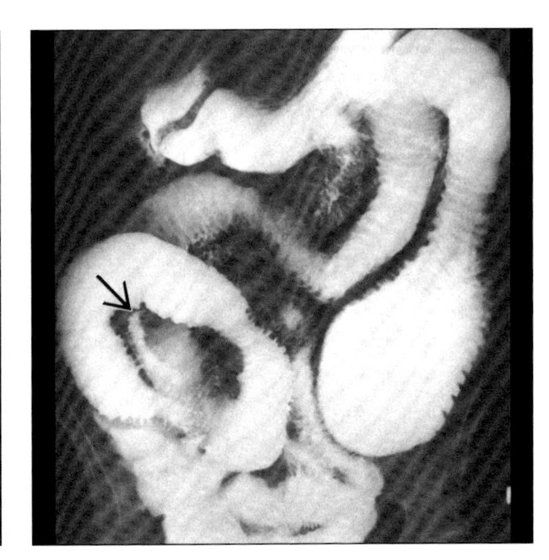

(Left) *Small bowel follow through (SBFT) shows distortion, angulation + thickening of folds with a "spike" appearance (arrow).* ***(Right)*** *SBFT shows luminal dilation proximal to stricture (arrow) + indirect evidence of bowel wall thickening, with a "stack of coins" appearance of the valvulae.*

Typical

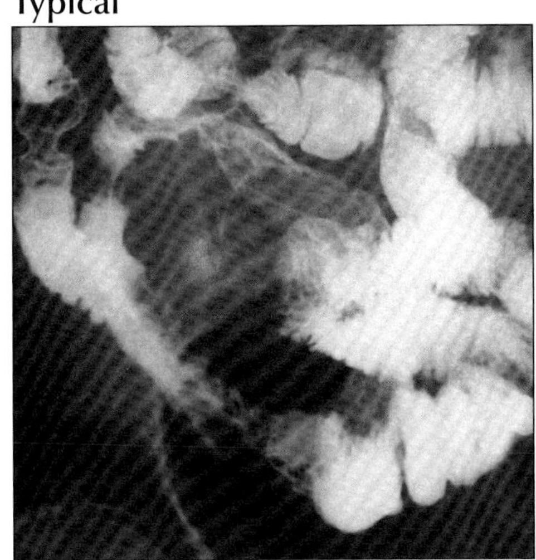

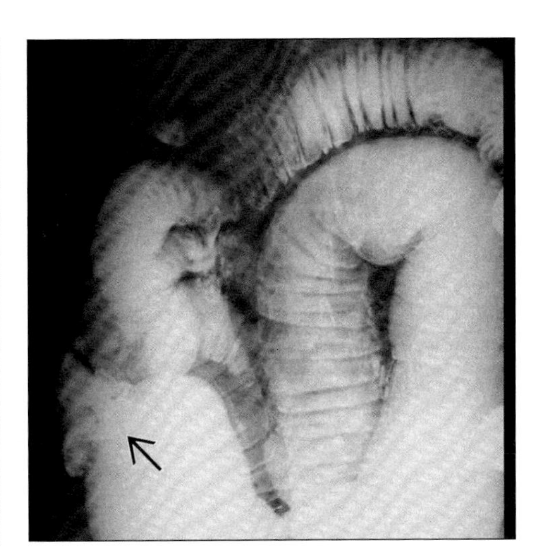

(Left) *SBFT shows luminal narrowing, fold thickening and nodularity.* ***(Right)*** *SBFT shows distal SB stricture (arrow) + proximal dilation of SB following surgery and radiation therapy for cecal carcinoma.*

Typical

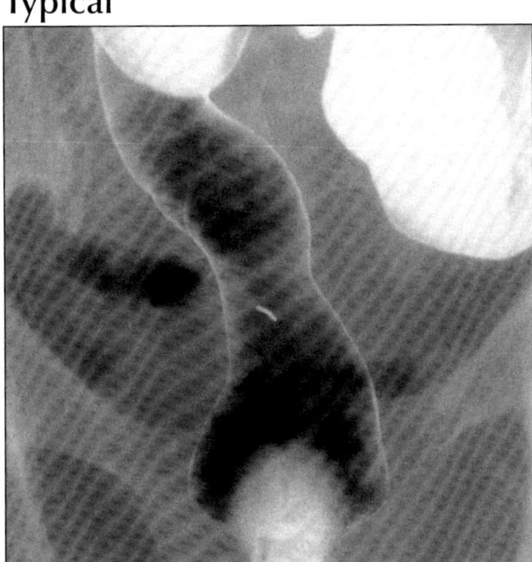

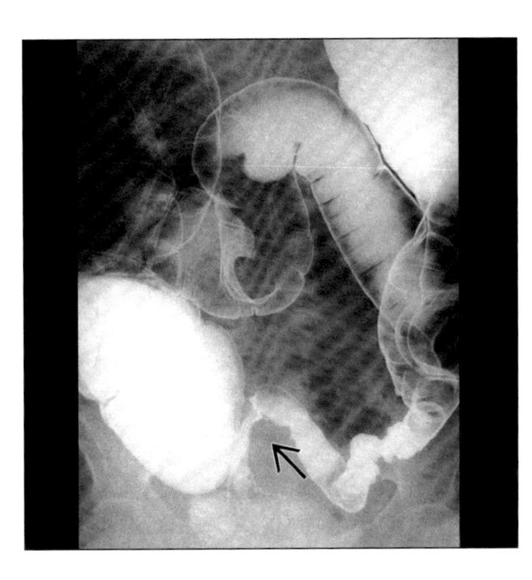

(Left) *Air-contrast BE shows radiation proctitis with granular mucosal pattern and luminal narrowing.* ***(Right)*** *Air-contrast BE shows long segment stricture of rectosigmoid colon (arrow) months after radiation therapy for endometrial carcinoma.*

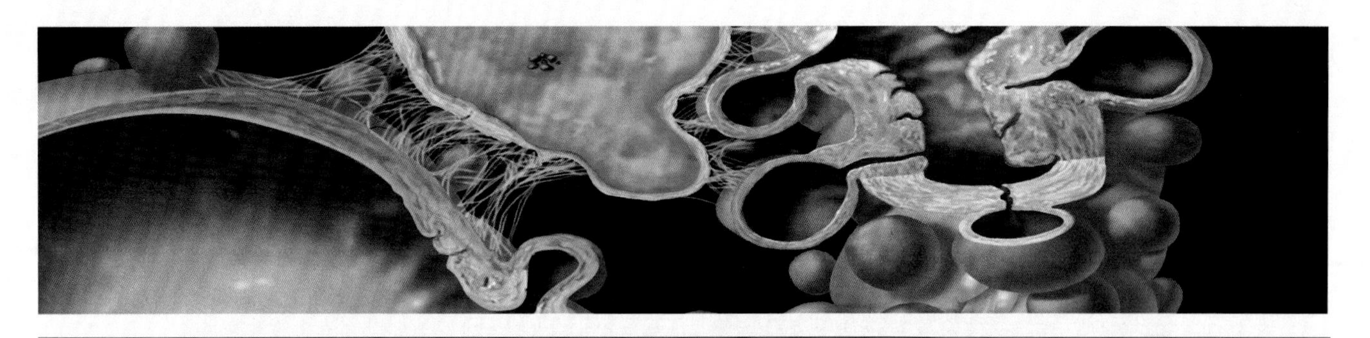

SECTION 5: Colon

Introduction and Overview

Infection

Inflammation and Ischemia

Neoplasm

Miscellaneous

COLON ANATOMY AND IMAGING ISSUES

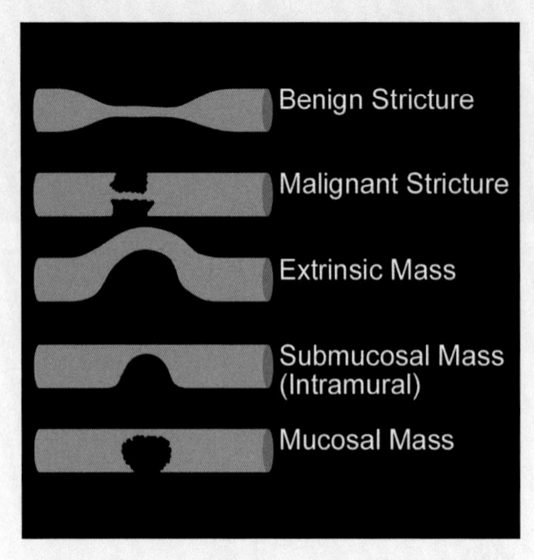

Graphic shows schematic representation of various processes that may narrow the lumen of the colon (or any other part of the gut).

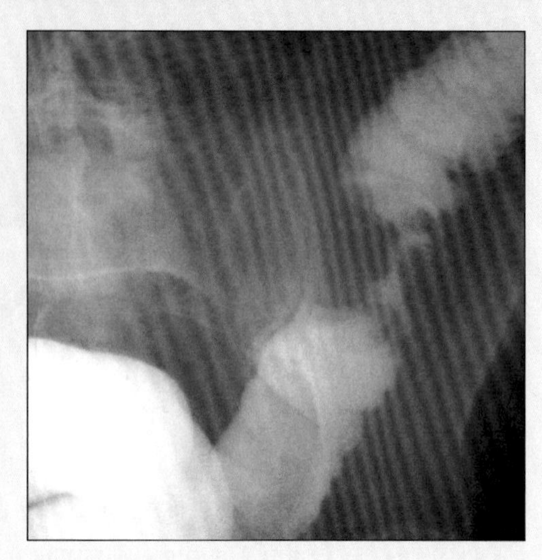

Barium enema shows malignant stricture ("apple core") of the colon; primary carcinoma.

TERMINOLOGY

Abbreviations and Synonyms

- Barium Enema (BE)
- Small Bowel (SB)

IMAGING ANATOMY

- Anatomic splenic flexure is the point at which the descending colon becomes retroperitoneal (distal to radiologic splenic flexure)
- Sigmoid colon
  - Intraperitoneal colonic segment bridging the retroperitoneal descending colon and the rectum

ANATOMY-BASED IMAGING ISSUES

Key Concepts or Questions

- Advantages of double-contrast BE over single-contrast BE
  - Detection of small polypoid lesions
  - Detection of superficial ulcerations
  - Subtle changes from endometriosis and metastases
- Advantages of single-contrast BE
  - Patient comfort
  - Elderly or arthritic patients
  - Detection of strictures
  - Known or suspected diverticular disease
  - Detection of large masses
  - Evaluation for obstruction
  - Evaluation for ischemia or other submucosal pathology
- Indication for water soluble contrast enema
  - Possible perforation
  - Possible fistula
  - Pre-operative emergent study
  - "Therapeutic" (obstipation)
- What criteria are useful to distinguish ulcerative colitis and Crohn disease?
  - Ulcerative colitis
    - Presenting symptoms: Diarrhea, rectal bleeding, pain
    - Pathology: Mucosa and submucosa; crypt abscesses; punctate and collar button ulcers
    - Radiology: Continuous circumferential involvement starting distally; shortened, ahaustral colon; colonic strictures (late); SB involvement only by backwash; no fistulas, sinus tracts, or abscesses; colon cancer and toxic megacolon are serious risks
  - Crohn disease
    - Presenting symptoms: Diarrhea, pain, weight loss, palpable mass
    - Pathology: Transmural; granulomas and enlarged lymphoid follicles; aphthous ulcers; linear and transverse ulcers; perianal fistulas
    - Radiology: Discontinuous eccentric colonic and SB involvement; fibrofatty proliferation in mesentery; fistulas, sinus tracts, abscesses; colon cancer and toxic megacolon rare
- How do you distinguish among the various causes of colonic luminal narrowing?
  - Benign stricture: Smooth taper, both ends
  - Malignant stricture: Irregular, abrupt narrowing, apple core, shoulders at one or both ends
  - Extrinsic: Intact mucosa, whole lumen is displaced, oblique angles for mass effect
  - Submucosal: Intact mucosa, almost right angle interface with luminal surface
  - Mucosal: Irregular mucosal surface, acute angle interface with luminal surface
- How common are colonic polyps (detection rate is good measure of adequacy of examination technique)?
  - Varies from 3% (age 20 to 30) , to 25% (age 80 to 90)
  - More than half are present in rectum and sigmoid
- How do you distinguish a barium-coated polyp from a barium-lined diverticulum on an air-contrast barium enema?

COLON ANATOMY AND IMAGING ISSUES

DIFFERENTIAL DIAGNOSIS

Benign tumors
- Hyperplastic polyp
- Adenomatous polyp
- Villous adenoma
- Hamartoma
- Spindle cell tumor
- ⇒(Lipoma, leiomyoma, etc.)
- Carcinoid tumor

Malignant tumors
- Carcinoma
- Lymphoma
- Metastases
- Kaposi sarcoma
- Squamous cell carcinoma
- ⇒(Anal)

POLYPOSIS SYNDROMES

Adenomatous polyps
- Familial polyposis coli
- Gardner syndrome
- Turcot syndrome
- Attenuated adenomatous polyposis coli

Hamartomatous polyps
- Peutz-Jeghers syndrome
- Juvenile polyposis
- Cronkhite-Canada syndrome
- Cowden syndrome
- Bannayan-Riley-Ruvalcaba syndrome

  - Varies with location of polyp (dependent or non-dependent wall) and whether seen in profile or "en face"
  - Easiest when polyp appears as filling defect in barium pool; diverticulum fills with barium and projects off surface of colon
  - Look for "bowler hat" (sessile polyp) or "Mexican hat" (pedunculated polyp) signs
  - Polyp has sharp inner margins and fuzzy (indistinct) outer margins
  - Diverticulum has sharp outer margins, fuzzy inner
- How do you distinguish colon carcinoma from diverticular disease on imaging?
  - Carcinoma: Luminal narrowing is short (< 10 cm), abrupt, irregular and eccentric, may resemble apple core; CT may show lymphadenopathy, metastases
  - Diverticulosis: Luminal narrowing is long (> 10 cm), transverse folds are thick, irregular, resemble "cog wheel" (circular muscle hypertrophy): No pericolonic disease
  - Diverticulitis: Luminal narrowing is long (> 10 cm), asymmetric with combination of circular muscle hypertrophy, spasm, pericolonic inflammation and mass (abscess); CT shows pericolonic inflammation ± pericolonic extraluminal gas, abscess, fistula
- What is the current role of CT colonography?
  - Competitive with barium enema and endoscopy as a screening procedure for colonic polyps
  - Must be performed and interpreted with expertise to achieve comparable results
  - Main rationale is to provide screening for patients who are resistant to, or poor candidates for barium enema or colonoscopy

CLINICAL IMPLICATIONS

Clinical Importance
- Normal stratified squamous epithelium of the anal canal can be infected by human papilloma virus (sexually transmitted)
  - May develop benign condyloma (locally invasive)
  - May develop malignant anal tumors (squamous, basaloid, etc.)

CUSTOM DIFFERENTIAL DIAGNOSIS

Heredity nonpolyposis colon cancer syndrome (HNPCC)
- Five times more common than familial polyposis
- Lynch I
  - Early onset (< 50), right-sided, often multiple colon cancers
- Lynch II
  - Lynch I + extracolonic tumors
- Muir-Torre
  - Similar to Lynch II + skin lesions

Length of colon involvement
- Cancer
  - Short (< 10 cm)
- Diverticulitis
  - Segmental (> 10 cm), usually sigmoid, spares rectum
- Ulcerative colitis
  - Long segmental, usually distal, includes rectum
- Crohn (granulomatous) colitis
  - Segmental, usually proximal, perirectal involvement
- Ischemia
  - Segmental (90%), usually splenic flexure or sigmoid
- Infectious colitis (e.g., C. difficile)
  - Long segmental or pancolitis, involves rectum
- Neutropenic colitis (typhlitis)
  - Segmental, ascending colon + cecum

Aphthoid ulcers
- Amebic colitis
- Crohn disease
- CMV + herpes colitis
- Salmonella + Shigella colitis
- Myotonic dystrophy
- Behçet disease
- Lymphoma

COLON ANATOMY AND IMAGING ISSUES

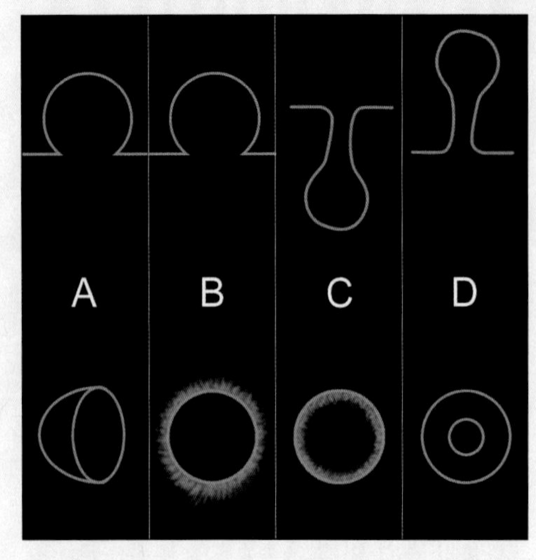

Graphic shows the profile and en face appearance of various polyps (A,B,D) and a diverticulum (C) on an air contrast barium enema (lower row of pictures).

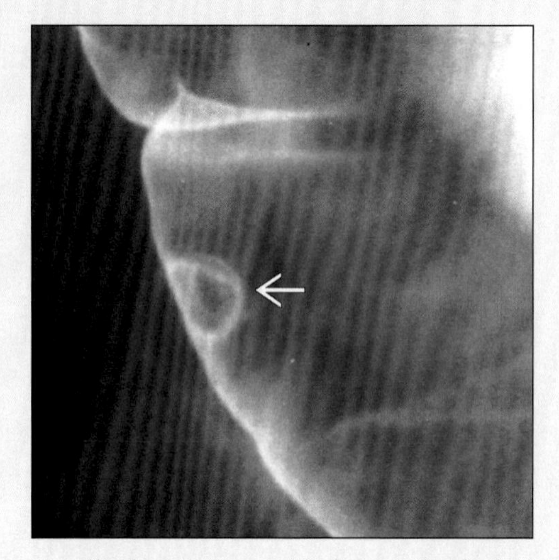

Spot film from air-contrast BE shows a "bowler hat" appearance of a small sessile polyp (arrow).

Colonic (or small bowel) submucosal thickening

- Air density = pneumatosis
  - E.g., bowel infarct, "benign" pneumatosis
- Fat density
  - E.g., chronic inflammatory bowel disease (IBD), cytoreductive therapy, obesity
- Near-water density
  - E.g., acute inflammation, ischemia, "shock bowel"
- Soft tissue density
  - E.g., tumor, inflammation, ischemia
- Higher density
  - E.g., hemorrhage in bowel wall

Colonic Ileus (pseudo-obstruction)

- Systemic acute inflammatory/traumatic
  - Pneumonia, myocardial infarction
  - Pancreatitis, appendicitis, peritonitis
  - Trauma; spinal injury
  - Post-operative
- Drug effect
  - Narcotics
  - Antidepressants
  - Antipsychotics
  - Antiparkinsonian
  - Anticholinergics
- Endocrine disorder
  - Hypothyroidism
  - Hypoparathyroidism
  - Diabetes
- Neuromuscular disorders
  - Parkinson disease

Colonic thumbprinting

- Vascular lesions
  - Ischemic colitis
  - Intramural hemorrhage (anticoagulation, trauma)
  - Vasculitis (e.g., Henoch-Schönlein purpura)
  - Hereditary angioneurotic edema
- Inflammatory
  - Ulcerative colitis
  - Crohn disease
- Infectious
  - Pseudomembranous colitis
  - Neutropenic colitis (typhlitis)
  - CMV colitis
  - Other rare
- Neoplastic
  - Lymphoma
  - Metastases
- Miscellaneous
  - Pneumatosis cystoides coli
  - Endometriosis
  - Cirrhosis (portal hypertension)

Ahaustral (smooth) colon

- Normal (descending colon, elderly)
- Ulcerative colitis (> Crohn)
- Cathartic abuse
- Radiation colitis (late)
- Ischemic colitis (late)

SELECTED REFERENCES

1. Koeller KK, et al (eds) Radiologic Pathology (2nd ed) Washington, DC, Armed Forces Institute of Pathology, 2003
2. Gore RM: Colon: Differential Diagnosis. In Gore RM, Levine MS (eds) Textbook of Gastrointestinal Radiology. 2nd ed. Philadelphia, WB Saunders. 1159-65, 2000
3. Dähnert W: Radiology Review Manual (4th ed), Philadelphia: Lippincott, 2000
4. Eisenberg RL: Gastrointestinal Radiology: A Pattern Approach (3rd ed). Philadelphia: JB Lippincott, 1996
5. Reeder MM: Reeder and Felson's Gamuts in Radiology (3rd ed) New York: Springer Verlag, 1993

IMAGE GALLERY

Typical

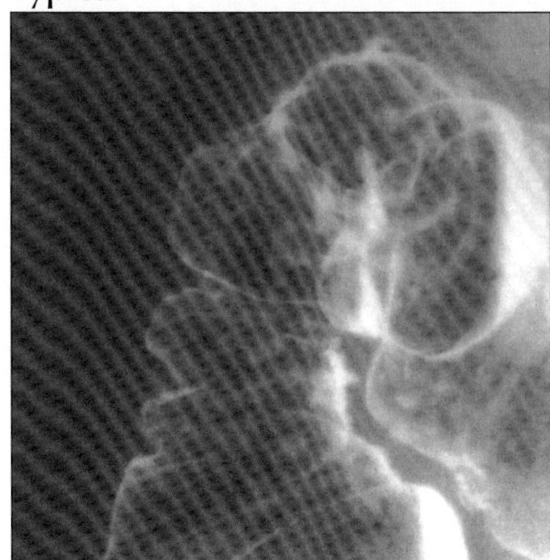

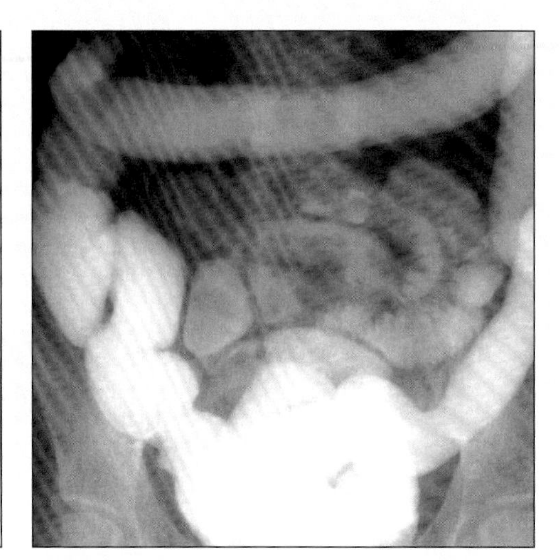

(Left) *Air-contrast BE shows many shallow aphthoid ulcerations of the hepatic flexure; Crohn (granulomatous) colitis.* ***(Right)*** *BE shows ahaustral colon due to chronic ulcerative colitis.*

Typical

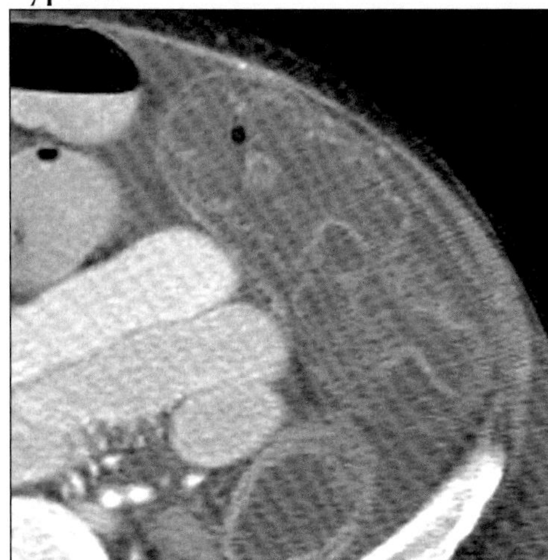

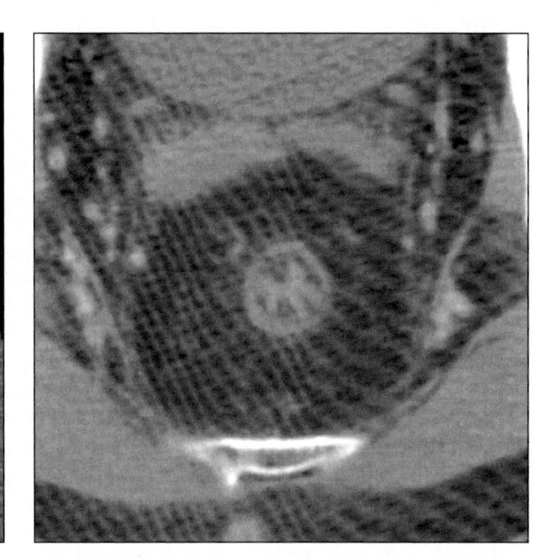

(Left) *Axial CECT shows marked water density submucosal colonic wall thickening; pseudomembranous (C. difficile) colitis.* ***(Right)*** *Axial CECT shows fat density submucosal thickening of the rectal wall; chronic ulcerative colitis.*

Typical

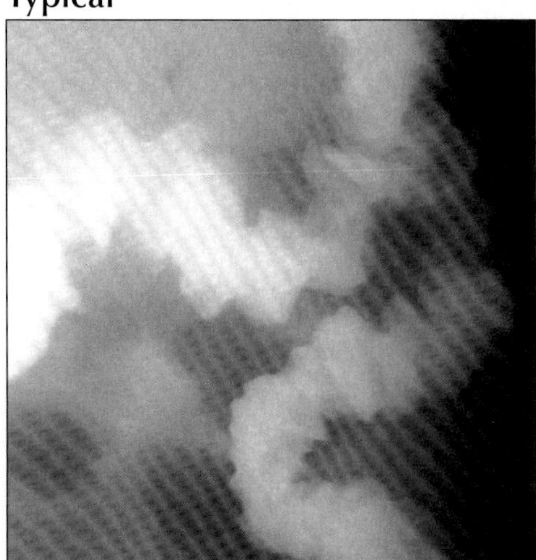

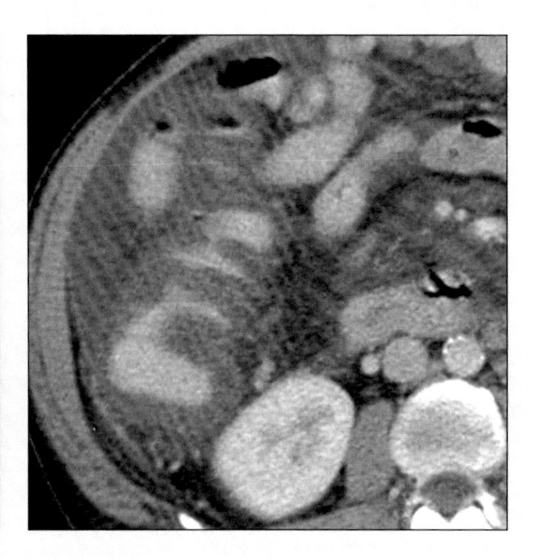

(Left) *BE shows "thumbprinting" of the colonic wall near the splenic flexure; ischemic colitis.* ***(Right)*** *Axial CECT shows "thumbprinting" of ascending colon; cirrhosis and portal hypertension with colonic edema.*

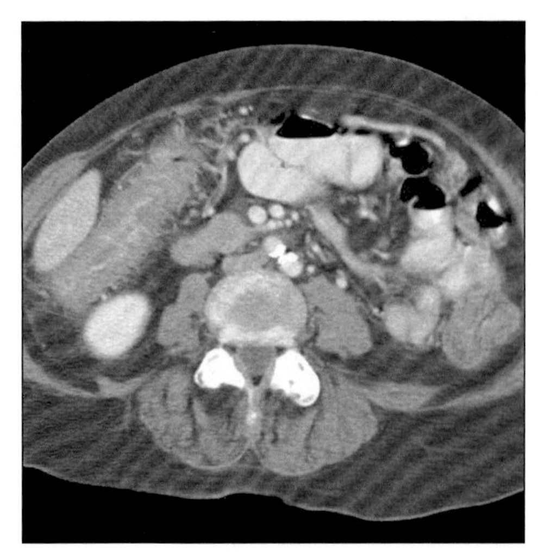

Axial CECT shows pancolitis with colonic wall thickening and mesenteric hyperemia. Campylobacter colitis in 78 year old woman.

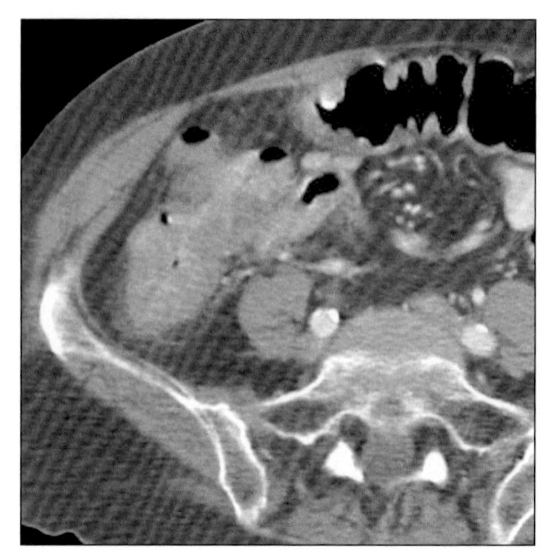

Axial CECT shows mural thickening of ascending + transverse colon plus dilated mesenteric vessels. Campylobacter colitis.

TERMINOLOGY

Definitions

- Inflammation of the colon caused by bacterial, viral, fungal, or parasitic infections

IMAGING FINDINGS

General Features

- Best diagnostic clue: Focal or diffuse colonic wall thickening with mucosal ulcerations
- Location: Dependent on etiology

Radiographic Findings

- Fluoroscopic-guided barium enema
  - Focal or diffuse; segmental colitis or pancolitis
  - Lumen narrowing & loss of haustra (edema/spasm)
  - Thickened folds & colonic wall (edema)
  - Ulceration → mucosal irregularity
  - Superficial or deep "collar button" ulcers
  - Discrete punctate, aphthous or large oval ulcers; may simulate Crohn disease
  - ± Small nodules or inflammatory polyps
  - ± Diffuse, mucosal granularity; may simulate ulcerative colitis
  - ± Extrinsic mass with inflammatory changes → distortion, short strictures; may simulate carcinoma
  - ± Thumbprinting; may simulate ischemic colitis
  - ± Fistulas or sinus tracts
  - Typhoid fever (Salmonellosis)
    - Cecum or right colon; invariably in ileum
    - Ileal fold thickening and ulceration
  - Shigellosis: Predominantly in left colon; mucosal granularity of rectum
  - Campylobacteriosis: Small bowel and colon
  - Yersinia enterocolitis: Predominantly in right colon, occasionally in left; invariably in terminal ileum
  - E. coli colitis: Transverse colon; extends to right, left or both sides of colon
  - Tuberculosis
    - Right & proximal transverse colon, involves ileum
    - Oval/circumferential, transverse ulcers, loss of ileum & right colon anatomic demarcation
    - Fleischner sign: Right-angle intersection between ileum and cecum with marked hypertrophy of ileocecal valve
    - Exuberant mural thickening; > than Crohn disease
    - Can cause "apple core" colonic stricture; indistinguishable from carcinoma
  - Actinomycosis: Rectosigmoid colon (intrauterine devices) or ileocecal region (appendectomy)
  - Gonorrheal, Chlamydia, Herpesvirus colitis: Rectosigmoid colon
  - Cytomegalovirus (CMV) colitis: Cecum & proximal colon; extends to distal ileum

DDx: Long Segment Wall Thickening

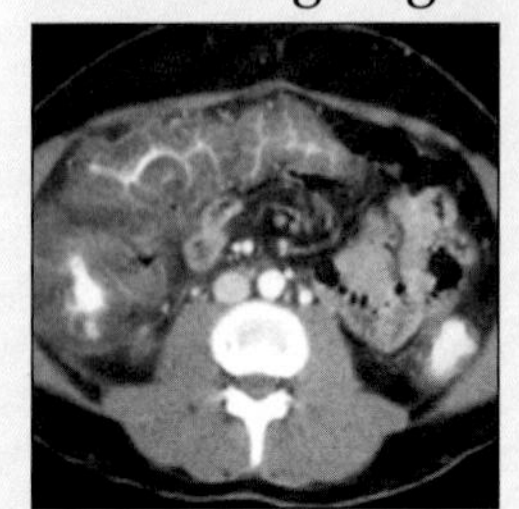

Pseudomem. Colitis

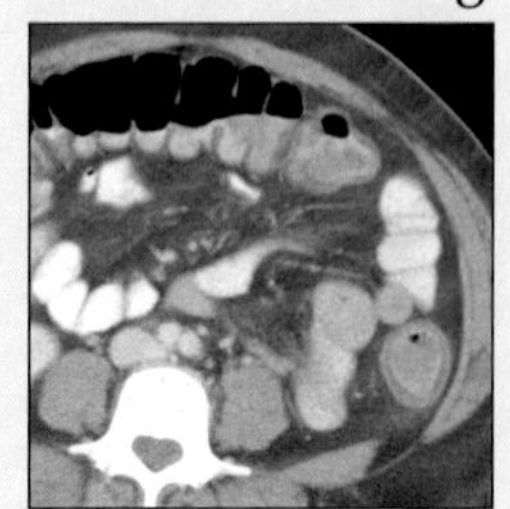

Granulom. Colitis

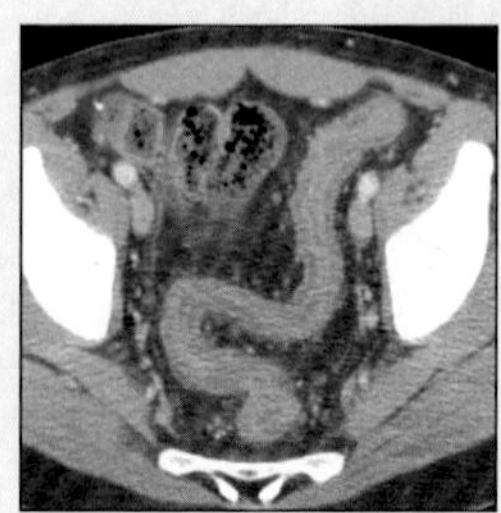

Ulcerative Colitis

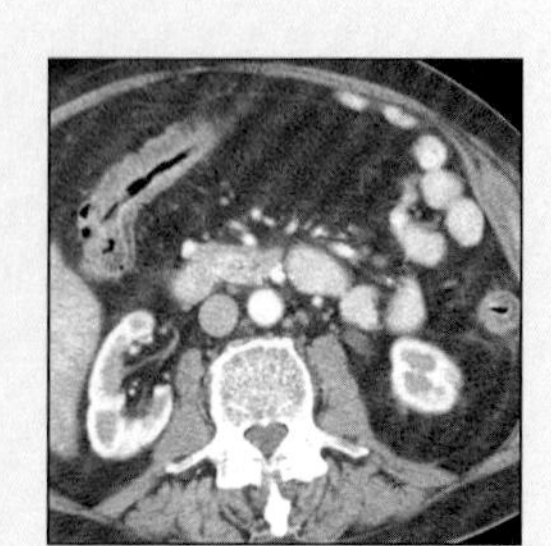

Ischemic Colitis

INFECTIOUS COLITIS

Key Facts

Terminology
- Inflammation of the colon caused by bacterial, viral, fungal, or parasitic infections

Imaging Findings
- Best diagnostic clue: Focal or diffuse colonic wall thickening with mucosal ulcerations
- Lumen narrowing & loss of haustra (edema/spasm)
- Discrete punctate, aphthous or large oval ulcers; may simulate Crohn disease
- ± Diffuse, mucosal granularity; may simulate ulcerative colitis
- ± Thumbprinting; may simulate ischemic colitis
- ± Fistulas or sinus tracts
- Best imaging tool: Fluoroscopic-guided barium enema

Top Differential Diagnoses
- Pseudomembranous colitis
- Granulomatous colitis (Crohn disease)
- Ulcerative colitis
- Ischemic colitis

Clinical Issues
- Usually acute in onset, except tuberculosis (chronic)
- Watery or bloody diarrhea
- Crampy abdominal pain and tenderness

Diagnostic Checklist
- Diagnosis by clinical presentation; lab tests
- Barium enema or CT detects colitis; need clinical confirmation of specific type

---

- Histoplasmosis
  - Ileocecal region; polyps in rectum
  - Pericecal masses; may simulate appendicitis
- Mucormycosis: Right colon; polypoid mass
- Anisakiasis: Occasionally in right colon, rarely in transverse colon
- Amebiasis
  - Right colon; terminal ileum spared
  - Skip lesions (in 95%); may simulate granulomatous colitis
  - Ameboma (in 10%): Marked granulation in short segments of bowel, located in right colon
  - Can produce "apple core" type colonic stricture
  - Discrete ulcers appearing as marginal defects or granularity with barium flecks
  - Residual deformity and strictures after treatment
- Schistosomiasis
  - Left or sigmoid colon
  - Hallmark is inflammatory polyps (granulation response to eggs deposited in bowel wall)
- Trichuriasis
  - Clumping & granularity of barium (excessive mucus)
  - Wavy, linear 3-5 cm lucencies, occasionally terminates in ring shape with central barium collection (worm)

CT Findings
- Wall thickening and low attenuation (edema)
- Mucosal and serosal enhancement; ascites
- Multiple air-fluid levels; inflammatory pericolic fat
- Salmonellosis: ± Small bowel thickening & effacement
- Tuberculosis
  - Changes in lungs, but usually from ingestion
  - Low density, marked enlargement of lymph nodes
- Actinomycosis: Large inflammatory masses
- CMV colitis
  - Deep ulcers & marked wall thickening (advanced)
  - Enhancement of mucosa and serosa with hypodense thickening of intervening bowel wall (edema)
  - Hemorrhage causes ↑ attenuation in wall
- Histoplasmosis
  - Changes in lungs and skin
  - Mesenteric adenopathy, hepatosplenomegaly with or without calcifications
- Mucormycosis: Changes in sinuses, lungs & central nervous system
- Schistosomiasis
  - Changes in mesenteric or hemorrhoidal vein, urinary tract, terminal ileum
  - ± Calcification of bowel wall or liver

Imaging Recommendations
- Best imaging tool: Fluoroscopic-guided barium enema

DIFFERENTIAL DIAGNOSIS

Pseudomembranous colitis
- Radiography: Colonic wall thickening, nodularity; may simulate infectious colitis
- CT: "Accordion sign": Trapped oral contrast between thickened colonic haustral folds
- Usually results in more colonic wall thickening than other colitides

Granulomatous colitis (Crohn disease)
- Concurrent small bowel (distal ileum) disease
- Barium enema
  - Cobblestoning: Longitudinal & transverse ulcerations produce a paving stone appearance
  - Transmural, skip lesions, sinuses, fistulas; may simulate infectious colitis
- Disease is chronic, compared with infectious colitis

Ulcerative colitis
- Barium enema
  - Pancolitis with decreased haustration & multiple ulcerations; may simulate infectious colitis
  - "Mucosal islands" or "inflammatory pseudopolyps"
  - Diffuse & symmetric wall thickening of colon
  - Chronic phase → "lead-pipe" colon

Ischemic colitis
- Usually located in watershed areas; focal or diffuse
- Barium enema: Thumbprinting, ulcerations (1-3 weeks after onset of disease); strictures (later)

- CT
  - ± Pneumatosis, portomesenteric venous gas
  - ± Thrombus within splanchnic vessels
- Differentiate by clinical presentation

PATHOLOGY

General Features

- Etiology
  - Bacterial organisms (most common in Western countries): Salmonella, Shigella, Campylobacter, Yersinia, Staphylococcus, Escherichia coli (O157:H7), M. tuberculosis, Actinomyces, Chlamydia trachomatis, C. gonorrhea
    - Chlamydia is the causative agent for "lymphogranuloma venereum"
  - Viral organisms: Herpes virus, CMV, Norwalk virus, Rotavirus
  - Fungal organisms: Histoplasma, Mucor
  - Parasitic organisms (most common in underdeveloped countries): Anisakis, Amoeba, Schistosoma, Strongyloides, Trichuriasis
  - Risk factors
    - Salmonella, Shigella: Outbreaks, warm weather
    - E. coli: Travel, nursing homes (O157:H7)
    - Tuberculosis, CMV: AIDS
    - Actinomycosis: Intrauterine devices, appendectomy
    - Histoplasma, Mucor: Chronic debilitation or immunosuppression
    - Strongyloides: Severe debilitation
  - Pathogenesis
    - Ingestion of pathogenic organisms (often fecal-oral route)
    - Chlamydia, Gonorrhea, Herpes virus: Direct inoculation of rectum (anal intercourse)

Gross Pathologic & Surgical Features

- Varies based on etiology

Microscopic Features

- Varies based on etiology

CLINICAL ISSUES

Presentation

- Most common signs/symptoms
  - Usually acute in onset, except tuberculosis (chronic)
  - Watery or bloody diarrhea
  - Crampy abdominal pain and tenderness
  - Fever, headache, nausea, vomiting, weight loss
  - Palpable abdominal mass, anemia, malaise, rash
  - Arthritis, pneumonitis, seizures, peripheral neuropathy, microangiopathy
  - Varied incubation period
  - E. colli colitis: Traveler's diarrhea, hemolytic-uremic syndrome (0157:H7)
  - Schistosomiasis: Hepatosplenomegaly → portal hypertension
- Lab-Data
  - Bacterial organisms: Increased neutrophilic count
  - Viral organisms: ↑ Lymphocytes (↓ in AIDS)
  - Fungal, parasitic organisms: Eosinophilia
- Diagnosis
  - Stool cultures, blood cultures, endoscopic biopsy, serology studies

Demographics

- Age: Any age, but incidence ↑ with age
- Gender: M:F = 1:1

Natural History & Prognosis

- Complications
  - Toxic megacolon, bacteremia, sepsis, death
  - Hemorrhage, perforation, obstruction
  - Yersinia enterocolitis: Hepatic abscess
  - E. coli colitis: Hemolytic-uremic syndrome
  - Amebiasis: Liver and lung abscesses
- Prognosis
  - Usually very good, after treatment
  - Campylobacteriosis: 25% recurrence if untreated
  - E. coli O157:H7 colitis: ↑ Morbidity, 33% mortality
  - CMV colitis: Hemorrhage and ischemia can be fatal
  - Mucormycosis, Strongyloidiasis: Fatal

Treatment

- Bacterial organisms: Mostly self-limiting, last 1-2 weeks; up to 1 month
  - Salmonellosis: Parenteral cephalosporins if severe
  - Shigellosis: Ampicillin in severe cases
  - Yersinia enterocolitis: Lasts several months; no treatment available
  - E. coli O157:H7 colitis: Supportive treatment, isolation procedures
  - Tuberculosis: Antituberculosis drugs, no steroids
- Viral organisms: Mostly self-limiting
  - CMV: Treat underlying AIDS
- Parasitic organisms: Antihelminthic drugs
  - Anisakiasis: Mostly self-limiting, last 7-10 days
- Fungal organisms: Antifungal drugs

DIAGNOSTIC CHECKLIST

Consider

- Diagnosis by clinical presentation; lab tests

Image Interpretation Pearls

- Barium enema or CT detects colitis; need clinical confirmation of specific type

SELECTED REFERENCES

1. Thielman NM et al: Clinical practice. Acute infectious diarrhea. N Engl J Med. 350(1):38-47, 2004
2. Horton KM et al: CT evaluation of the colon: inflammatory disease. Radiographics. 20(2):399-418, 2000
3. Philpotts LE et al: Colitis: use of CT findings in differential diagnosis. Radiology. 190(2):445-9, 1994
4. Schmitt SL et al: Bacterial, fungal, parasitic, and viral colitis. Surg Clin North Am. 73(5):1055-62, 1993
5. Wall SD et al: Gastrointestinal tract in the immunocompromised host: opportunistic infections and other complications. Radiology. 185(2):327-35, 1992

IMAGE GALLERY

Typical

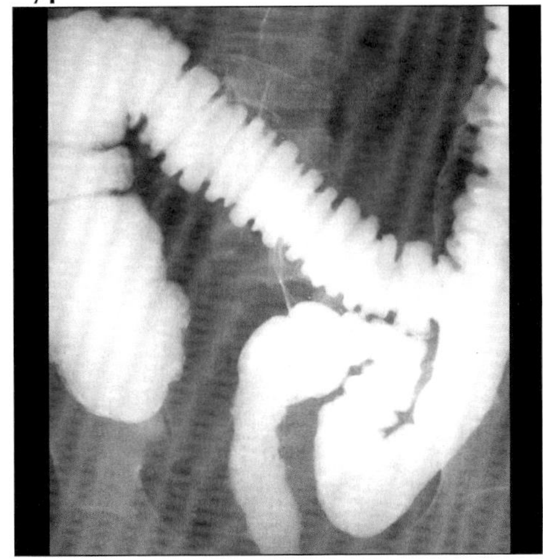

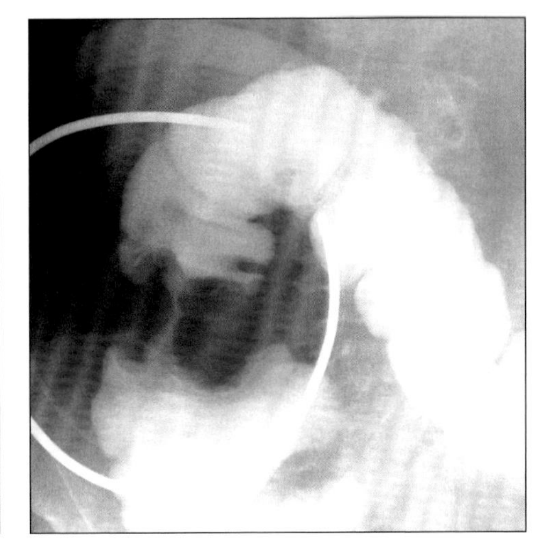

(Left) *Single-contrast BE shows rectal stricture with mucosal irregularity due to "lymphogranuloma venereum" (Chlamydia trachomatis).* ***(Right)*** *Single-contrast BE shows "apple core" lesion of ascending colon due to Mycobacterium tuberculosis.*

Typical

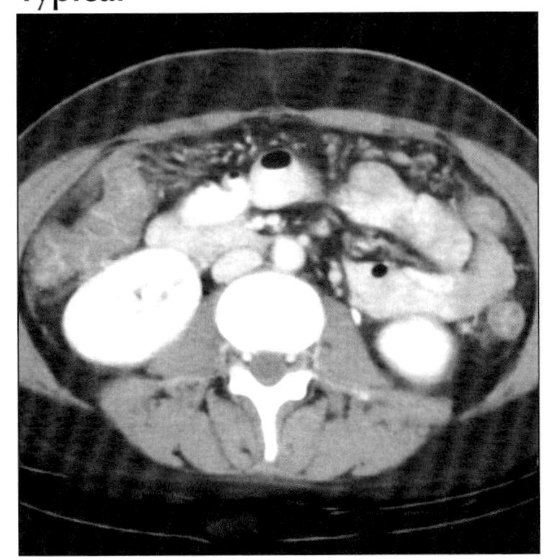

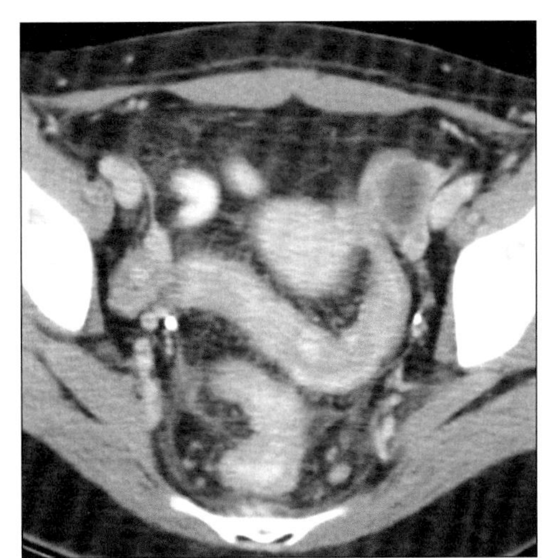

(Left) *Axial CECT shows pancolitis due to Cytomegalovirus (CMV) in a patient with AIDS.* ***(Right)*** *Axial CECT shows proctocolitis with mural thickening and mesenteric hyperemia in a 32 year old woman due to CMV colitis.*

Typical

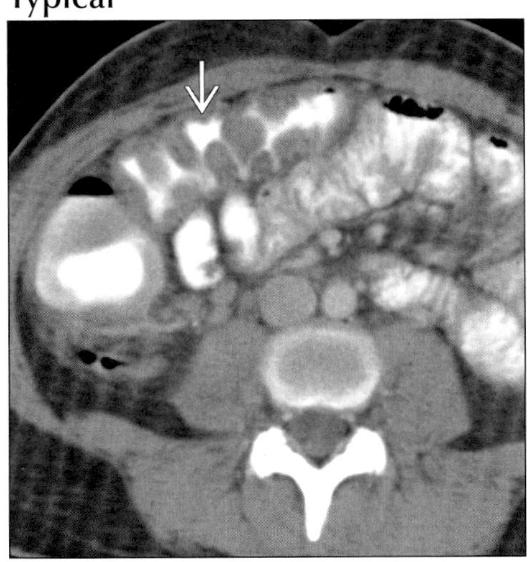

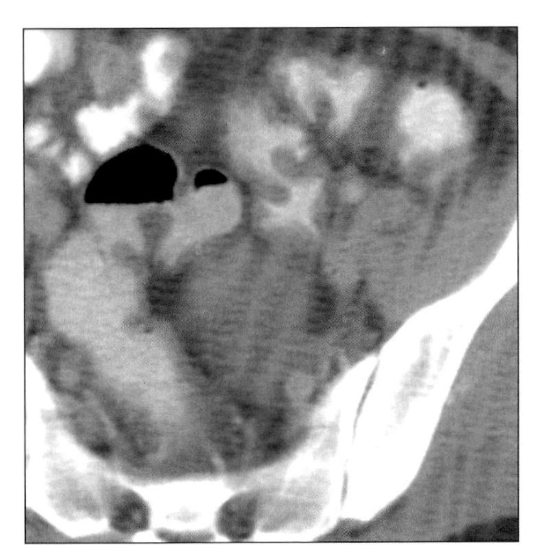

(Left) *Axial CECT shows Campylobacter pancolitis in a previously healthy 26 year old woman. Note "thumbprinting" of colonic wall (arrow).* ***(Right)*** *Axial CECT of 26 year old woman with Campylobacter colitis shows marked mural thickening of sigmoid colon.*

PSEUDOMEMBRANOUS COLITIS

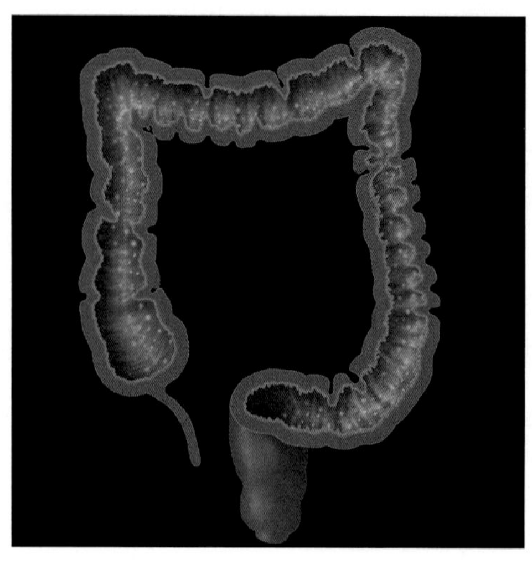

Graphic shows pancolitis with marked mural thickening with multiple elevated yellow-white plaques (pseudomembranes).

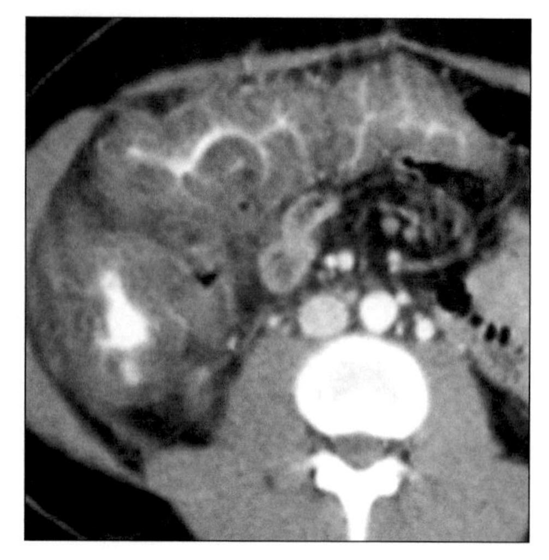

Axial CECT shows pancolitis with marked mural thickening and enteric contrast trapped between haustra ("accordion sign").

TERMINOLOGY

Abbreviations and Synonyms

- Pseudomembranous colitis (PMC)
- Antibiotic colitis, Clostridium difficile colitis

Definitions

- Acute inflammation of colon caused by toxins produced by Clostridium difficile bacteria

IMAGING FINDINGS

General Features

- Best diagnostic clue: Marked submucosal edema over a long segment of colon
- Location
  - Usually entire colon (pancolitis)
  - Rectum & sigmoid colon (typically involved in 80-90% of cases)
  - Confined to more proximal colon (10% of cases)
- Morphology: Plaque-like adhesions of fibrinopurulent necrotic debris & mucus on damaged colonic mucosa with submucosal edema
- Other general features
  - PMC is usually associated with antibiotic use, especially clindamycin
  - Clindamycin causes diarrhea in 20% & pseudomembranous colitis in 10% of patients
  - C. difficile infection of colon follows insult to gut by antibiotic or chemotherapy
  - C. difficile infection is responsible for virtually all cases of PMC
  - C. difficile toxins: Cytotoxic + enterotoxic effects
  - Complications range from watery diarrhea to colonic perforation & death

Radiographic Findings

- Radiography
  - Colonic ± small bowel ileus
  - Gaseous distension of colon + nodular haustral thickening
  - Thumbprinting: Unusual, wide transverse bands due to thickening of haustral folds
    - Most prominent in transverse colon
  - Severe cases: Polypoid mucosal thickening
    - Represent pseudomembranous plaques protruding into air-containing lumen
  - Fulminant cases
    - Toxic megacolon
    - Pneumoperitoneum
- Fluoroscopic guided contrast enema studies
  - Enema is contraindicated in severe PMC (due to increased risk of perforation)
  - Limited role in diagnosis of PMC

DDx: Other Causes of Colitis

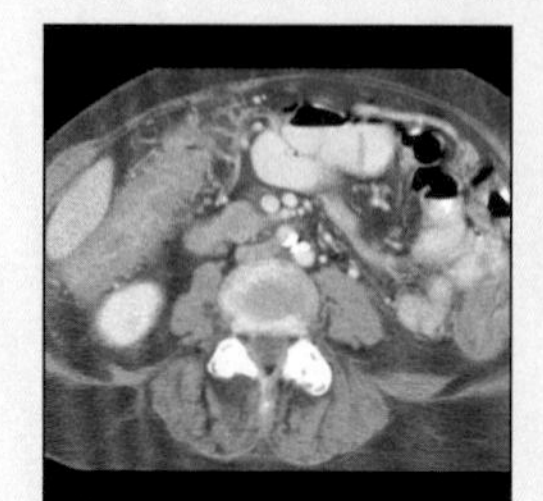

Campylobacter

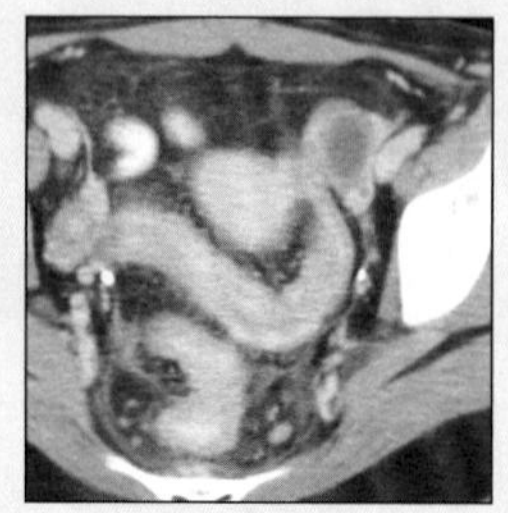

CMV Colitis

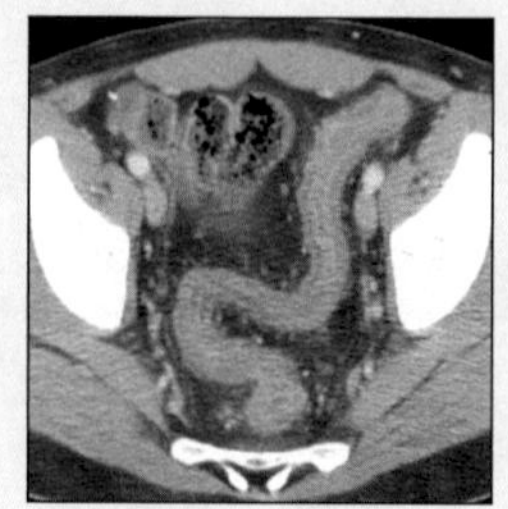

Ulcerative Colitis

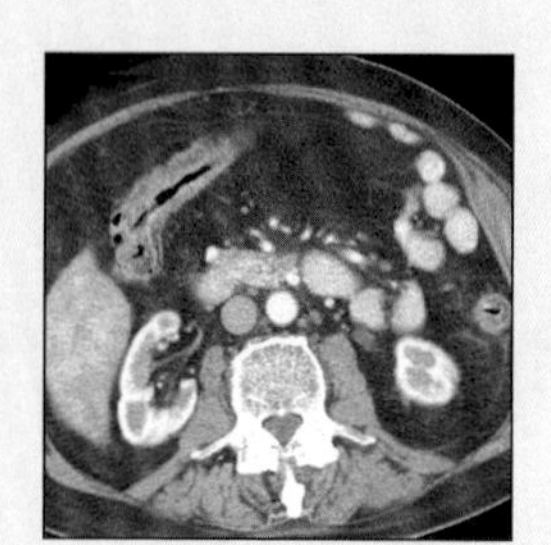

Ischemic Colitis

PSEUDOMEMBRANOUS COLITIS

Key Facts

Terminology
- Pseudomembranous colitis (PMC)
- Antibiotic colitis, Clostridium difficile colitis
- Acute inflammation of colon caused by toxins produced by Clostridium difficile bacteria

Imaging Findings
- Best diagnostic clue: Marked submucosal edema over a long segment of colon
- Usually entire colon (pancolitis)
- Rectum & sigmoid colon (typically involved in 80-90% of cases)
- "Accordion sign": Represents trapped enteric contrast between thickened colonic haustral folds
- Pericolonic stranding
- Ascites common in severe PMC

Top Differential Diagnoses
- Granulomatous colitis (Crohn disease)
- Ulcerative colitis
- Ischemic colitis
- Neutropenic enterocolitis

Pathology
- Antibiotic therapy (clindamycin most common) due to overgrowth of resistant enteric C. difficile

Clinical Issues
- Clinical profile: Patient with history of watery diarrhea after antibiotic use or hospitalization

Diagnostic Checklist
- Check history of antibiotic use or debilitating diseases
- Suspect in any hospitalized patient with acute colitis

  - Findings vary depending on severity & extent of disease
  - Marked mural thickening & wide haustral folds due to intramural edema

CT Findings
- CECT with oral contrast
  - Colonic wall thickening & nodularity
    - Thickening is more irregular & shaggy compared to symmetric & homogeneous in Crohn
  - "Accordion sign": Represents trapped enteric contrast between thickened colonic haustral folds
    - Alternating bands of high attenuation (contrast) + low attenuation (edematous haustra)
    - Usually seen in advanced cases
    - Highly suggestive of PMC
  - "Target sign"
    - Mucosa: Intense enhancement (hyperemia)
    - Submucosa: Thickened, non-enhancing, ↓ HU (edema)
  - Extent of disease
    - Pancolitis is most common
    - Rectum & sigmoid colon (80-90% of cases)
    - Right colon involved exclusively in some cases
  - Pericolonic stranding
    - Usually mild (due to primary mucosal & submucosal nature of PMC)
    - Relative paucity of pericolonic inflammation + marked colonic wall thickening differentiates PMC from other colitides
  - Ascites common in severe PMC
    - Uncommon in other inflammatory bowel diseases
  - ± Pneumatosis coli or air in intrahepatic portal vein
  - Small pleural effusions & subcutaneous edema
    - May be due to primary disease or debilitated state

Imaging Recommendations
- Best imaging approach is CECT with oral contrast
  - 125 ml I.V contrast at ≥ 2.5 ml/sec

DIFFERENTIAL DIAGNOSIS

Other infectious colitis
- Campylobacter, cytomegalovirus, etc.
- May be indistinguishable from pseudomembranous colitis
- Often have less severe colonic wall thickening

Granulomatous colitis (Crohn disease)
- Concurrent small bowel (distal ileum) disease usually
- Cobblestoning: Longitudinal & transverse ulcerations produce a paving stone appearance
- Segmental distribution
- Transmural, skip lesions, sinuses, fissures, fistulas
- CT shows fibrofatty proliferation of mesentery & enlarged mesenteric lymph nodes

Ulcerative colitis
- Classic imaging appearance
  - Pancolitis with decreased haustration & multiple ulcerations on barium enema
- Colorectal narrowing; ↑ presacral space > 1.5 cm
- "Mucosal islands" or "inflammatory pseudopolyps"
- Diffuse & symmetric wall thickening of colon
- Backwash ileitis: Distal ileum involvement (10-40%)
- Chronic phase
  - "Lead-pipe" colon: Rigid colon with loss of haustra

Ischemic colitis
- Usually seen in watershed areas; focal or diffuse
  - Left side colon: Typical in elderly (hypoperfusion)
    - Splenic flexure: Junction of SMA & IMA
  - Right-side colon: Younger patients
    - Due to decreased collateral blood supply
- Barium findings
  - Thumbprinting: Submucosal edema or hemorrhage
  - Ulceration: 1-3 weeks after onset of disease
  - Stricture: Seen in late phase
- CT findings
  - Bowel wall thickening ± luminal dilatation
  - ± Pneumatosis, portomesenteric venous gas
  - ± Thrombus within splanchnic vessels
- Has less wall thickening than PMC

PSEUDOMEMBRANOUS COLITIS

Neutropenic enterocolitis

- Clinical history of neutropenia & immunosuppression
- Usually focal disease in right colon & cecum
- Mural thickening limited to right colon & distal ileum
- Thumbprinting, luminal narrowing, ulceration
- Immunocompromised with PMC mimics neutropenic colitis when localized to cecum & right colon

PATHOLOGY

General Features

- Etiology
  - Antibiotic therapy (clindamycin most common) due to overgrowth of resistant enteric C. difficile
    - Antibiotic therapy usually within 2 days to 2 weeks & rarely up to 6 months
    - Ampicillin, tetracycline, erythromycin, penicillin (less common)
  - Other causes
    - Abdominal surgery, colonic obstruction, uremia, prolonged hypotension or hypoperfusion of bowel
    - Severe debilitating diseases (e.g., lymphoma, leukemia, AIDS)
  - Pathogenesis
    - Antibiotic therapy (clindamycin)
    - Inhibits & alters normal intestinal microflora
    - Overgrowth of resistant enteric C. difficile
    - Enterotoxin (toxin A) & cytotoxin (toxin B) → mucosal damage
- Epidemiology
  - 1-10 cases per 1,000 patient discharges from hospital
  - 1 case per 10,000 antibiotic prescriptions written outside hospital

Gross Pathologic & Surgical Features

- Inflamed colon with multiple elevated, yellow-white plaques (pseudomembranes)
- When removed during endoscopy, reveals erythematous & inflamed mucosa

Microscopic Features

- Colonization of colon by clostridium difficile
- Mild-early: Focal necrosis of surface epithelial cells in glandular crypts with neutrophilic infiltration & fibrin plugging of capillaries in lamina propria and mucus hypersecretion in adjacent crypts
- Moderate: Crypt abscesses
- Severe-late: Necrosis & denudation of mucosa with thrombosis of submucosal venules

CLINICAL ISSUES

Presentation

- Most common signs/symptoms
  - Mild cases: Watery diarrhea
  - Severe cases: Acute abdomen
    - Fever, abdominal pain & tenderness, tachycardia
    - Dehydration, leukocytosis & sepsis
- Clinical profile: Patient with history of watery diarrhea after antibiotic use or hospitalization
- Diagnosis
  - Demonstration of C. difficile toxins in stool
    - Typically takes 48 hours to confirm
  - Proctosigmoidoscopy or colonoscopy
    - Adherent yellow plaques 2-10 mm in diameter

Demographics

- Age
  - Elderly age group > young age group
    - Elderly are at higher risk for developing PMC & recurrent PMC than young
- Gender: Equal in both males & females (M = F)

Natural History & Prognosis

- Complications
  - Toxic megacolon, sepsis, perforation & death
- Prognosis
  - If treated early, full recovery expected
  - Recurrence rate higher in women & elderly
  - Severe cases may need colectomy
  - Untreated cases can lead to perforation, acute abdomen & death (mortality rate 1.1-3.5%)

Treatment

- Mild cases: Discontinue offending antibiotic therapy
- Severe cases
  - Metronidazole (drug of choice) or oral vancomycin
  - Fulminant & toxic megacolon: Colectomy

DIAGNOSTIC CHECKLIST

Consider

- Check history of antibiotic use or debilitating diseases
- Suspect in any hospitalized patient with acute colitis

Image Interpretation Pearls

- Marked submucosal edema over long segment of colon
- "Accordion sign": Trapped oral contrast between thickened colonic haustral folds
- Usually pancolitis; rectum & sigmoid colon involved in 80-90% of cases

SELECTED REFERENCES

1. Gore RM et al: Inflammatory conditions of the colon. Semin Roentgenol. 36(2):126-37, 2001
2. Kirkpatrick ID et al: Evaluating the CT diagnosis of Clostridium difficile colitis: should CT guide therapy? AJR Am J Roentgenol. 176(3):635-9, 2001
3. Horton KM et al: CT evaluation of the colon: inflammatory disease. Radiographics. 20(2):399-418, 2000
4. Kawamoto S et al: Pseudomembranous colitis: spectrum of imaging findings with clinical and pathologic correlation. Radiographics. 19(4):887-97, 1999
5. Macari M et al: The accordion sign at CT: a nonspecific finding in patients with colonic edema. Radiology. 211(3):743-6, 1999
6. Ros PR et al: Pseudomembranous colitis. Radiology. 198(1):1-9, 1996
7. Gore RM et al: Radiologic investigation of acute inflammatory and infectious bowel disease. Gastroenterol Clin North Am. 24(2):353-84, 1995
8. Fishman EK et al: Pseudomembranous colitis: CT evaluation of 26 cases. Radiology. 180(1):57-60, 1991
9. Rubesin SE et al: Pseudomembranous colitis with rectosigmoid sparing on barium studies. Radiology. 170(3 Pt 1):811-3, 1989

IMAGE GALLERY

Typical

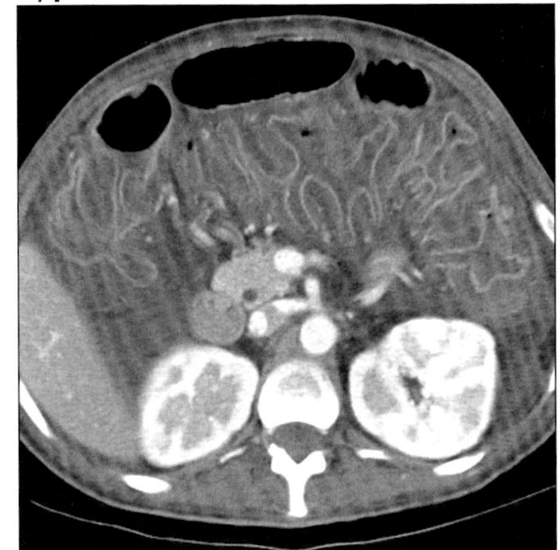

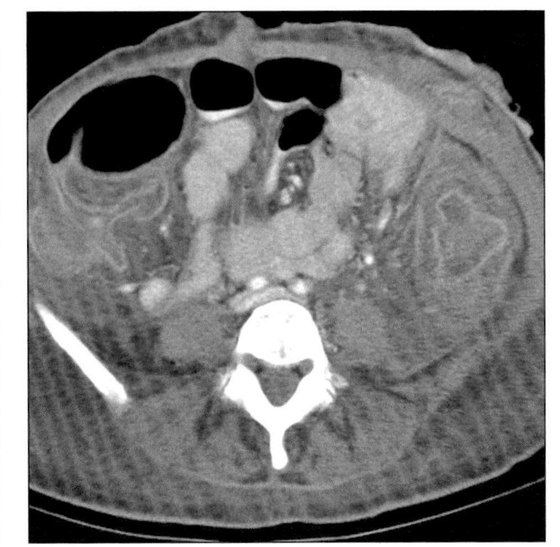

*(**Left**) Axial CECT shows massive submucosal edema of the transverse colon with luminal narrowing + striking mucosal enhancement. (**Right**) Axial CECT shows transmural involvement of entire colon, plus ascites, raising the concern for perforation.*

Typical

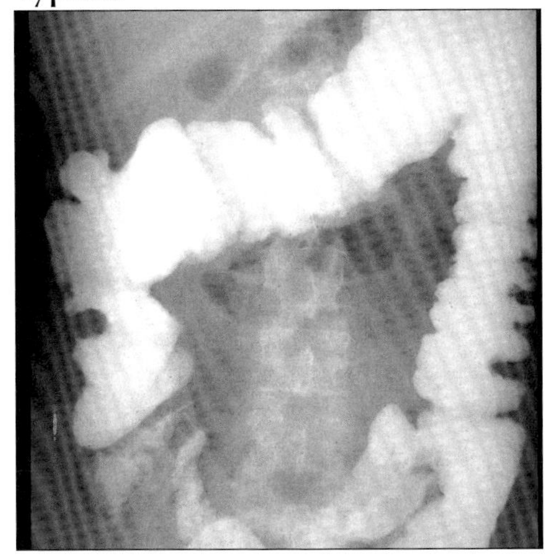

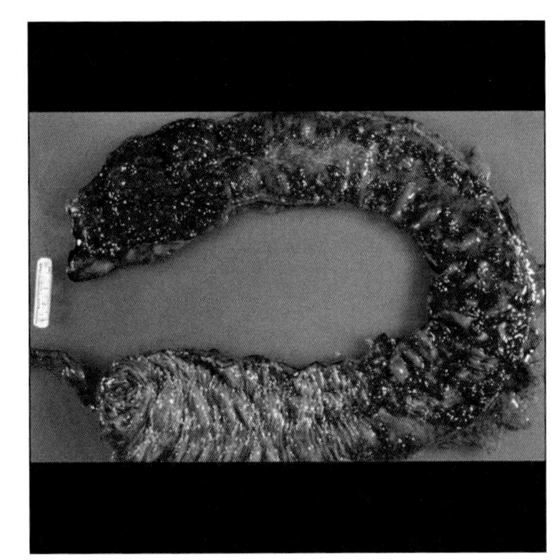

*(**Left**) Single-contrast BE shows pancolitis, with "thumbprinting" indicating submucosal edema or hemorrhage. (**Right**) Photo of opened resected colon shows sloughed, necrotic mucosa and raised yellow plaques or pseudomembranes.*

Typical

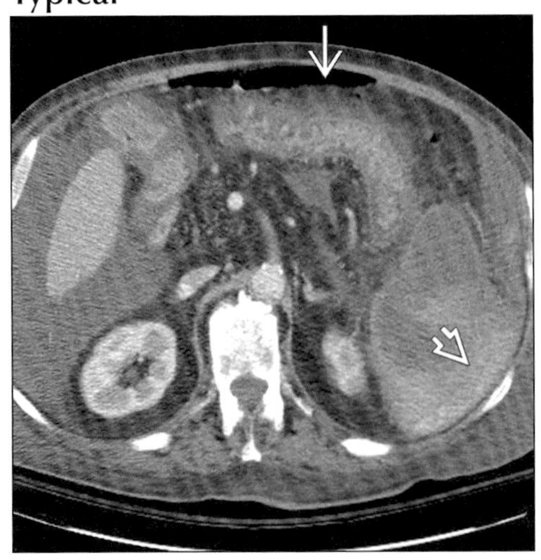

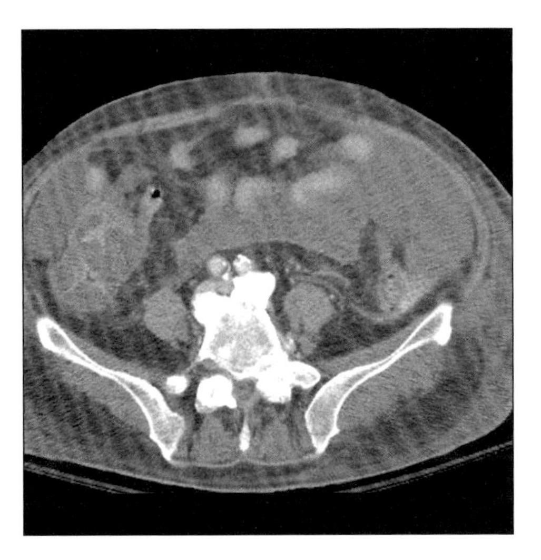

*(**Left**) Axial CECT shows marked mural thickening of transverse colon, plus extraluminal gas (arrow) and enteric contrast media (open arrow). Fatal colonic perforation. (**Right**) Axial CECT shows marked wall thickening of ascending colon with high density ascites. C. difficile colitis with perforation.*

TYPHLITIS

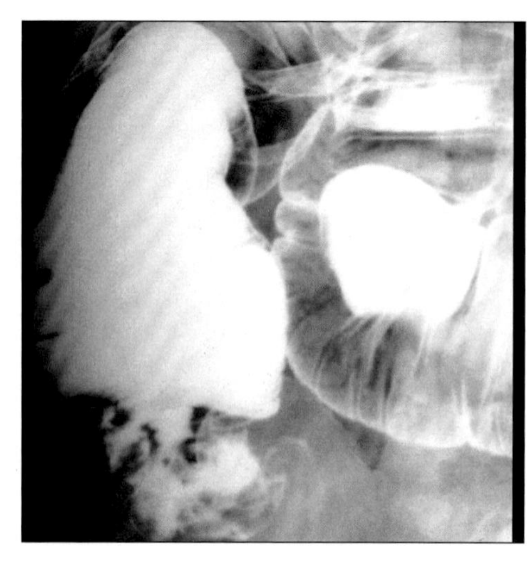

Single contrast BE shows marked irregular narrowing of the lumen of the cecum. Small bowel is dilated.

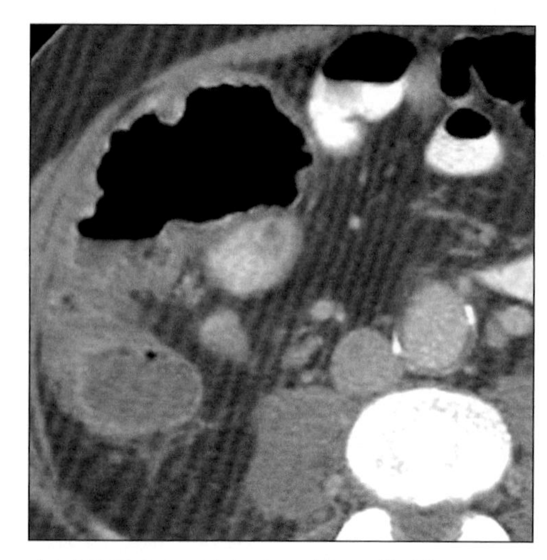

Axial CECT shows thickening of the wall of the cecum + ascending colon. The lumen of the cecum is narrowed; ascending colon dilated.

TERMINOLOGY

Abbreviations and Synonyms

- Synonym(s): Neutropenic colitis, ileocecal syndrome, cecitis, necrotizing enteropathy

Definitions

- Inflammatory or necrotizing process involving cecum, ascending colon & occasionally distal ileum/appendix

IMAGING FINDINGS

General Features

- Best diagnostic clue: Massive mural thickening of cecal ± ascending colon wall
- Location: Cecum + ascending colon (more common)
- Morphology: Dilated or narrow lumen, thickened wall
- Other general features
  - Usually seen in severely neutropenic patients
    - Post chemotherapy or bone marrow transplant
  - More common in children than adults
  - Clinical syndrome of fever & right lower quadrant tenderness in immunosuppressed host

Radiographic Findings

- Radiography
  - Ileocecal dilatation with air-fluid levels
  - RLQ soft tissue mass
  - Thumbprinting due to bowel edema
  - ± Pneumatosis: Speckled or linear pattern
- Fluoroscopic guided water-soluble contrast enema
  - Usually not recommended (possible perforation)
  - Mural thickening & mucosal thumbprinting
  - Luminal narrowing or dilatation of cecum
  - ± Dilated adjacent bowel loops (paralytic ileus)
  - Shallow or deep ulcerations

CT Findings

- NECT
  - Cecal luminal distention or narrowing
  - Circumferential wall thickening of cecum ± ascending colon & distal ileum
  - ↓ Bowel-wall attenuation (due to edema)
  - Pericecal fat stranding + thickened fascial planes
  - Pericolonic fluid collection
  - ± Pneumatosis, pneumoperitoneum
  - ± Dilated adjacent bowel loops (paralytic ileus)
- CECT: Heterogeneous enhancement of bowel wall

Ultrasonographic Findings

- Real Time
  - Hypoechoic or hyperechoic thickened bowel wall
  - Anechoic free fluid; ± mixed echoic abscess

Imaging Recommendations

- Helical CT: Study of choice for diagnosis of typhlitis
- Water-soluble contrast; colonoscopy (contraindicated)

DDx: Fold Thickening, Contraction of Cecum

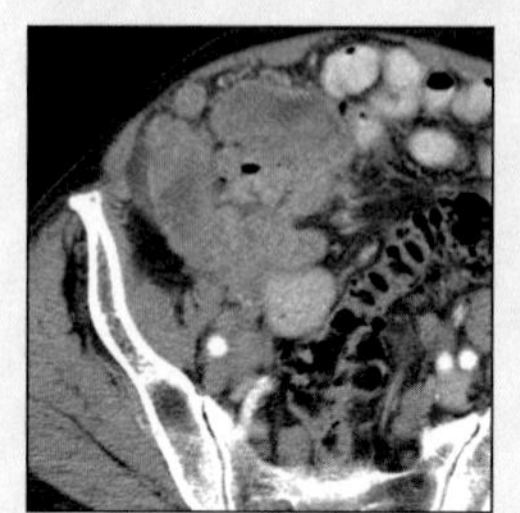

Cecal Carcinoma

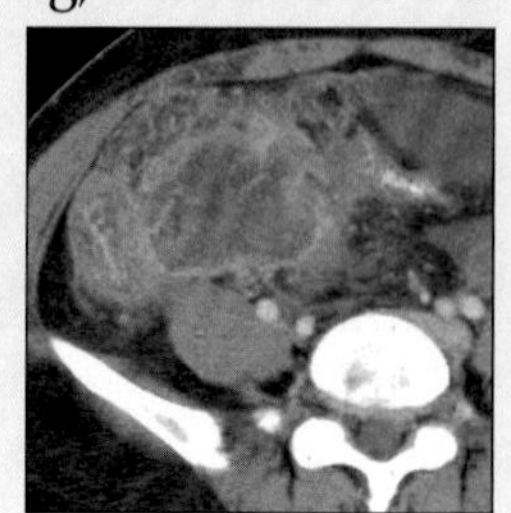

App. Abscess

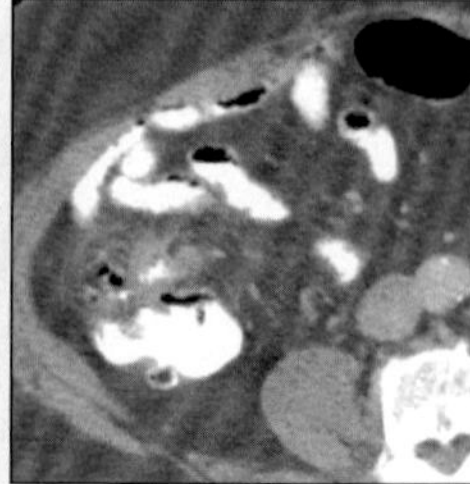

Diverticulitis

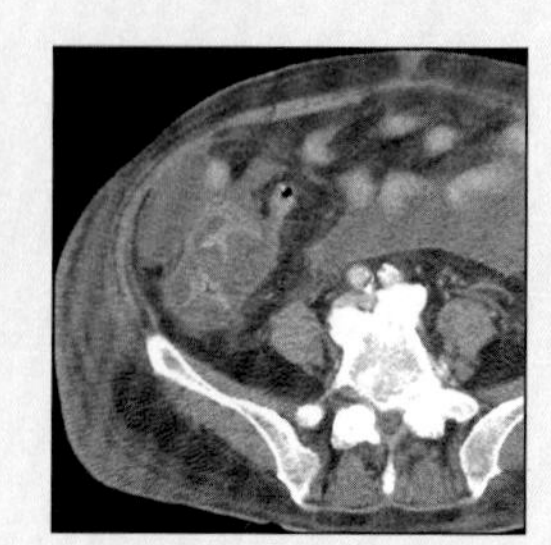

Pseudomem. Colitis

TYPHLITIS

Key Facts

Imaging Findings
- Best diagnostic clue: Massive mural thickening of cecal ± ascending colon wall
- Pericecal fat stranding + thickened fascial planes
- CECT: Heterogeneous enhancement of bowel wall

Top Differential Diagnoses
- Cecal carcinoma
- Appendicitis
- Cecal diverticulitis
- Crohn disease
- Pseudomembranous colitis

Pathology
- Hemorrhagic, thick, boggy cecum & adjacent colon

Diagnostic Checklist
- Check for history of chemotherapy for leukemia or bone marrow transplantation

DIFFERENTIAL DIAGNOSIS

Cecal carcinoma
- "Apple core" lesion: Narrow lumen; irregular mucosa

Appendicitis
- Thickened cecal wall adjacent to inflamed appendix

Cecal diverticulitis
- Bowel wall thickening, fat stranding, free fluid or air
- Cecal outpouching differentiates from typhlitis

Crohn disease
- Segmental, transmural, cobblestone mucosa
- Luminal narrowing, typically seen in terminal ileum

Pseudomembranous colitis
- Due to Clostridium difficile bacteria
- Usually affects all or most of colon

PATHOLOGY

General Features
- Etiology
  - Neutropenic conditions
    - Leukemia, lymphoma, any malignancy, organ or bone marrow transplant cases on chemotherapy
    - AIDS; viral, bacterial & fungal infections
    - Idiopathic, aplastic anemia, ischemia, antibiotics
  - Mechanism
    - Chemotherapy/antibiotics → immunosuppression → neutropenia → infection → typhlitis
- Epidemiology: Incidence: Children > adults
- Associated abnormalities: Neutropenic diseases

Gross Pathologic & Surgical Features
- Hemorrhagic, thick, boggy cecum & adjacent colon

Microscopic Features
- Inflammatory, ischemic, necrotic, ulcerative changes

CLINICAL ISSUES

Presentation
- Most common signs/symptoms
  - Fever, RLQ pain, watery diarrhea, ± hematochezia
  - Fullness; palpable mass; RLQ tenderness (± rebound)
- Lab data: Neutropenia, leukopenia; ± blood in stool
- Diagnosis: Imaging, clinical & lab correlation

Demographics
- Age: Children & young adults > older adults
- Gender: Equal in both males & females (M=F)

Natural History & Prognosis
- Complications: Abscess, necrosis, perforation, sepsis
- Prognosis: Early stage (good); late stage (poor)

Treatment
- Medical: High doses of antibiotics & IV fluids
- Complicated case
  - CT signs of perforation: Surgical resection
  - Granulocyte transfusions

DIAGNOSTIC CHECKLIST

Consider
- Check for history of chemotherapy for leukemia or bone marrow transplantation

Image Interpretation Pearls
- Cecal wall thickening & pericolonic inflammation in severely neutropenic patients

SELECTED REFERENCES

1. Horton KM et al: CT evaluation of the colon: Inflammatory disease. RadioGraphics. 20: 399-418, 2000
2. Adams GW et al: CT detection of typhlitis. Journal of Computed Assisted Tomography. 9: 363-5, 1985
3. Frick MP et al: Computed tomography of neutropenic colitis. AJR. 143: 763-5, 1984

IMAGE GALLERY

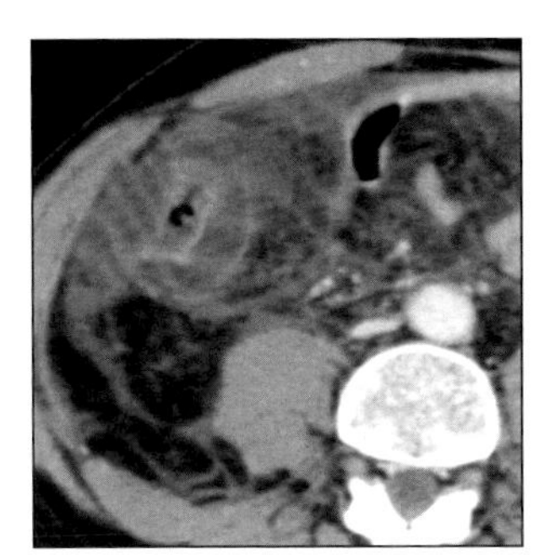
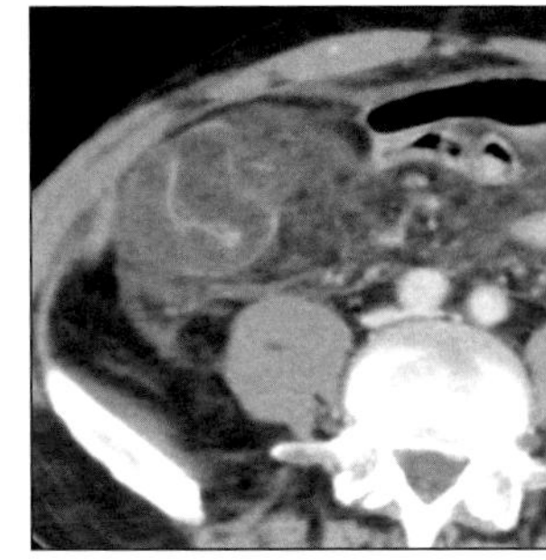

(Left) *Axial CECT in leukemic patient. Cecal wall is massively thickened, lumen narrowed, with pericolonic infiltration.* ***(Right)*** *Axial CECT shows cecal wall thickening, obliteration of lumen.*

ULCERATIVE COLITIS

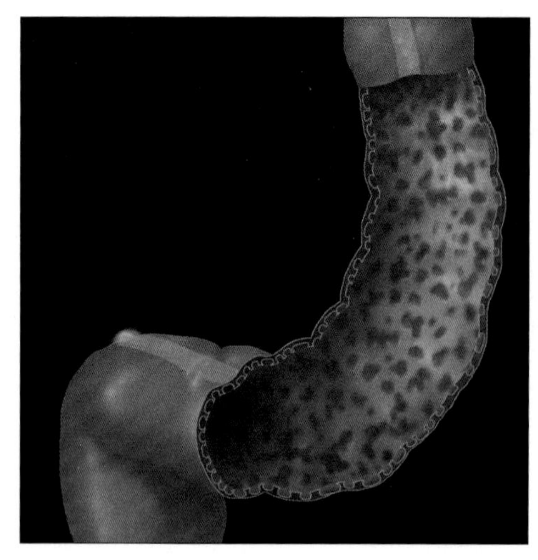

Graphic shows innumerable "collar button" ulcers and loss of haustra throughout descending and sigmoid colon.

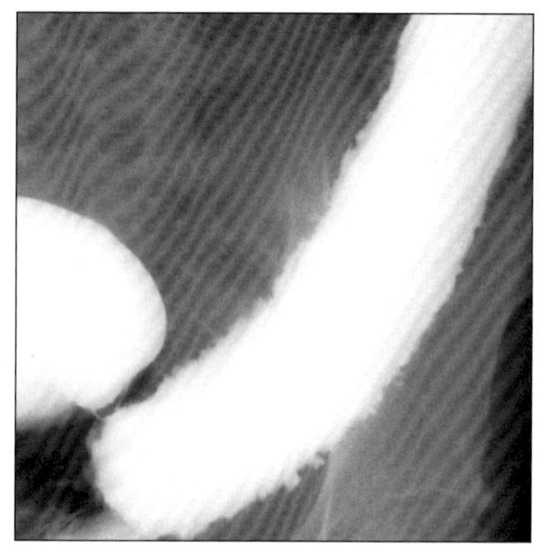

Single-contrast barium enema (BE) shows innumerable "collar button" ulcers and loss of haustra throughout descending colon.

TERMINOLOGY

Abbreviations and Synonyms

- Ulcerative colitis (UC)

Definitions

- Chronic, idiopathic diffuse inflammatory disease that primarily involves colorectal mucosa & submucosa

IMAGING FINDINGS

General Features

- Best diagnostic clue: Pancolitis with ↓ haustration + multiple ulcerations on barium enema
- Location: Rectum (30%); rectum + colon (40%); pancolitis (30%)
- Morphology
  - Narrow lumen, superficial ulcers, pseudopolyps
  - "Lead-pipe" colon & lack of haustra in chronic phase
- Other general features
  - Chronic relapsing inflammatory bowel disease with acute features
  - Disease with continuous concentric + symmetric colonic involvement
  - Inflammation limited to mucosa & submucosa
  - Characterized by pseudopolyps/crypt microabscesses
  - Begins in rectum & extends proximally to involve part or all of the colon
  - Backwash ileitis: 10-40% of chronic UC patients, distal ileum is inflamed
  - Ulcerative colitis > common than Crohn disease
  - Incidence: First-degree relatives 30-100 times > general population
  - ↑ Risk of colorectal cancer in UC than Crohn colitis
    - Annual incidence: 10% after first decade of UC
    - 75-80% who develop colon cancer have pancolitis
    - 25% UC cases have multiple carcinomas (often flat & scirrhous, difficult to image)

Radiographic Findings

- Fluoroscopic guided barium contrast enema
  - Acute changes
    - Colorectal narrowing + incomplete filling (spasm + irritability)
    - Fine mucosal granular pattern (edema/hyperemia)
    - Mucosal stippling: Punctate barium collections (crypt abscesses erode → ulcers & barium collection)
    - "Collar button" ulcers (flask-like): Due to undermining of ulcers (ulcers enlarge → configuration lost → mucosal islands + polyps)
    - Haustra: Edematous & thickened
    - Polyps: Inflammatory & postinflammatory pseudopolyps (remnants of mucosa & submucosa)
  - Chronic changes

DDx: Ulceration, Wall Thickening of Colon

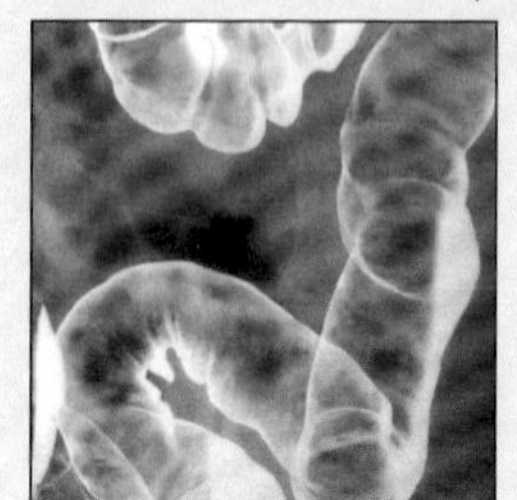

Granulomatous Colitis

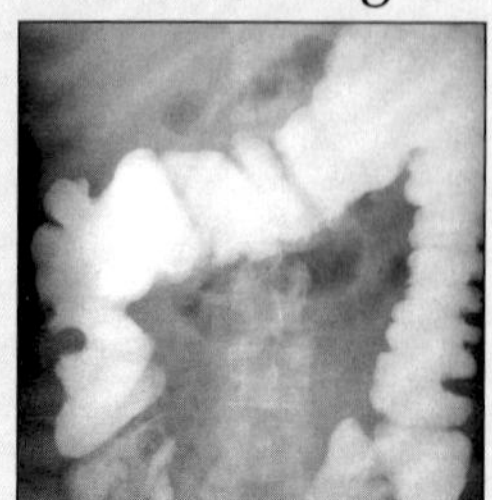

Pseudomem. Colitis

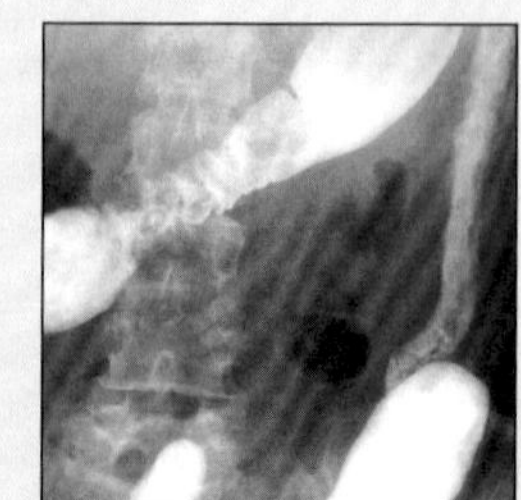

Ischemic Colitis

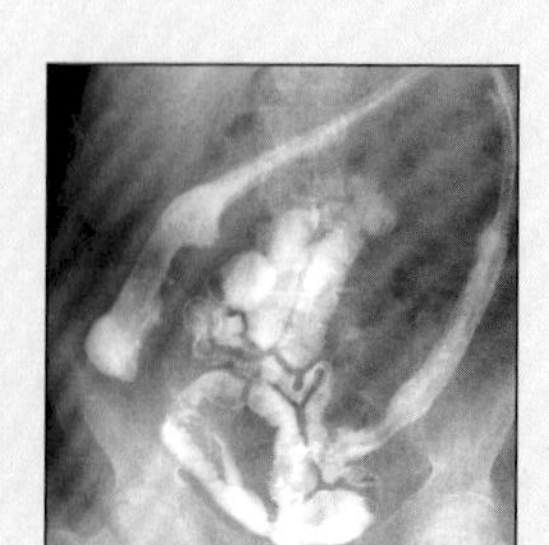

Cathartic Colon

ULCERATIVE COLITIS

Key Facts

Terminology
- Chronic, idiopathic diffuse inflammatory disease that primarily involves colorectal mucosa & submucosa

Imaging Findings
- Best diagnostic clue: Pancolitis with ↓ haustration + multiple ulcerations on barium enema
- Location: Rectum (30%); rectum + colon (40%); pancolitis (30%)
- Fine mucosal granular pattern (edema/hyperemia)
- Mucosal stippling: Punctate barium collections (crypt abscesses erode → ulcers & barium collection)
- "Collar button" ulcers (flask-like): Due to undermining of ulcers (ulcers enlarge → configuration lost → mucosal islands + polyps)
- Haustra: Edematous & thickened
- Polyps: Inflammatory & postinflammatory pseudopolyps (remnants of mucosa & submucosa)
- "Lead-pipe" colon: Rigidity + luminal narrowing
- Widening of presacral space: > 1.5 cm
- Diffuse + symmetric wall thickening of colon

Top Differential Diagnoses
- Granulomatous colitis (Crohn disease)
- Pseudomembranous colitis (PMC)
- Ischemic colitis
- Neutropenic enterocolitis
- Diverticulitis

Diagnostic Checklist
- Continuous concentric & symmetric involvement
- Consider UC in any patient with sclerosing cholangitis

    - Shortening of colon with depression of flexures (reversible)
    - "Lead-pipe" colon: Rigidity + luminal narrowing
    - Haustrations: Blunted or complete loss
    - Backwash ileitis: Distal 5-25 cm of ileum is inflamed (seen in 10-40% cases)
    - Luminal narrowing & widened presacral space (more than 1.5 cm)
    - Benign strictures: Local sequelae of UC (seen in 10% of patients)
  - Rectal valve abnormalities (double contrast study)
    - Lateral rectal view: Normally at least one rectal valve should be visible
    - Fold is usually seen at the level of S3 & S4 (less than 5 mm thick)
    - Proctitis: Valve thickness > 6.5 mm or absent

CT Findings
- NECT
  - Colorectal narrowing
  - Widening of presacral space: > 1.5 cm
    - Due to perirectal fibrofatty proliferation
  - Diffuse + symmetric wall thickening of colon
    - Less than 10 mm (average 7.8 mm)
  - Mural thickening & luminal narrowing
    - Seen in subacute & chronic ulcerative colitis
- CECT
  - "Target" or "halo" sign
    - Enhancing inner ring of bowel wall (mucosa)
    - Nonenhancing middle ring of bowel wall (submucosa): Due to edema in acute or halo of fat in chronic phase
    - Enhancing outer ring of bowel wall (muscularis propria)
  - Enhancement of
    - "Mucosal islands" or inflammatory "pseudopolyps"
    - Inflammatory pericolonic stranding

Imaging Recommendations
- Barium enema (single & double contrast studies); helical NE + CECT

DIFFERENTIAL DIAGNOSIS

Granulomatous colitis (Crohn disease)
- Barium enema findings
  - Aphthae: Punctate central collections of barium
  - Cobblestoning: Longitudinal & transverse ulcerations produce a paving stone
  - Segmental distribution
    - Involve both colon & small bowel (60% cases)
    - Isolated to colon (20% cases)
  - Transmural, skip lesions, sinuses, fissures, fistulas
  - In late stage indistinguishable from ulcerative colitis due to haustral loss & pseudopolyps
- CT shows
  - Bowel wall thickening (1-2 cm)
  - "Creeping fat" or fibrofatty proliferation of mesentery
  - Enlarged mesenteric lymph nodes
  - "Comb" sign: Mesenteric hypervascularity
    - Indicates active disease

Pseudomembranous colitis (PMC)
- Synonym(s): Antibiotic colitis or C. difficile colitis
- Usually involves entire colon (pancolitis)
- CT findings
  - Colonic wall thickening & nodularity
  - "Accordion" sign: Represents trapped enteric contrast between thickened colonic folds
  - Ascites common in PMC & unusual in other IBD

Ischemic colitis
- Usually seen in watershed areas; focal or diffuse
  - Left side colon: Typical in elderly (hypoperfusion)
    - Splenic flexure: Junction of SMA & IMA
  - Right-side colon: Young patients
    - Due to decreased collateral blood supply
- Barium findings
  - Thumbprinting: Submucosal edema or hemorrhage
  - Ulceration: 1-3 Weeks after onset of disease
  - Stricture: Seen in late phase
- CT findings
  - Bowel wall thickening, ± luminal dilatation
  - ± Pneumatosis, portomesenteric venous gas

- ± Thrombus within splanchnic vessels

Cathartic colon

- Due to long term use/abuse of laxatives + cathartics
- Appearance of ahaustral "rigid" colon simulates ulcerative colitis

Neutropenic enterocolitis

- Clinical history: Neutropenia & immunosuppression
- Usually focal disease in right colon & cecum
- Imaging findings
  - Mural thickening limited to right colon ± distal ileum
  - Thumbprinting: Due to bowel edema
  - Luminal narrowing
  - Shallow or deep ulcerations ± pneumatosis

Diverticulitis

- CT findings
  - Location: Most common in sigmoid colon
  - Bowel wall & fascial thickening; fat stranding; free fluid & air
  - Pericolic inflammatory changes
    - Abscess, sinuses, fistulas
  - "Arrowhead" sign: Due to diverticular orifice edema
  - Focal area of eccentric luminal narrowing
  - Diverticulosis uncommon in patients with ulcerative colitis

PATHOLOGY

General Features

- Genetics
  - ↑ Frequency in monozygotic twins
  - HLA B5, BW52 & DR2 linked to UC
- Etiology
  - Genetic, familial, environmental, neural, hormonal
  - Infectious, nutritional, immunological, vascular
  - Traumatic, psychological & stress factors
  - Smoking decreases risk factor
- Epidemiology: Incidence 2-10 cases/100,000 people
- Associated abnormalities
  - Primary sclerosing cholangitis (PSC), uveitis
  - Ankylosing spondylitis, rheumatoid arthritis
  - Pyoderma gangrenosum, sacroiliitis

Gross Pathologic & Surgical Features

- Rectum + colon involved
- Continuous involvement
- Superficial ulcers, pseudopolyps

Microscopic Features

- Inflammatory infiltrate, crypt microabscesses
- Limited to mucosa & submucosa

CLINICAL ISSUES

Presentation

- Most common signs/symptoms
  - Relapsing bloody mucus diarrhea
  - Fever, weight loss, abdominal pain & cramps
  - Systemic manifestations
- Lab-data: Blood & mucus in stool
- Diagnosis: Mucosal biopsy & histology

Demographics

- Age: 15-25 years (small peak at 55-65 years)
- Gender: Males less than females (M < F)
- Ethnicity: More common in Caucasians & Jews

Natural History & Prognosis

- Complications
  - Toxic megacolon, colorectal cancer, strictures
  - Increased incidence of colon carcinoma up to 50% after 25 years of disease
- Prognosis
  - Improves with diagnosis & management
  - Mortality: First 2 years of UC in > 40 years old
    - Males (2.1%); females (1.5%)

Treatment

- Medical
  - Sulfasalazine, steroids, azathioprine
  - Methotrexate, LTB4 inhibitors
- Surgical: Total or proctocolectomy + Brooke or continent ileostomy (Kock pouch)

DIAGNOSTIC CHECKLIST

Consider

- Rule out other inflammatory diseases of colon

Image Interpretation Pearls

- Colorectal narrowing + punctate & collar button ulcers
- Continuous concentric & symmetric involvement
- "Lead-pipe" (rigid) colon & haustral loss (late phase)
- Consider UC in any patient with sclerosing cholangitis

SELECTED REFERENCES

1. Carucci LR et al: Radiographic imaging of inflammatory bowel disease. Gastroenterol Clin North Am. 31(1):93-117, ix, 2002
2. Horton KM et al: CT evaluation of the colon: inflammatory disease. Radiographics. 20(2):399-418, 2000
3. Kawamoto S et al: Pseudomembranous colitis: spectrum of imaging findings with clinical and pathologic correlation. Radiographics. 19(4):887-97, 1999
4. Balthazar EJ et al: Ischemic colitis: CT evaluation of 54 cases. Radiology. 211(2):381-8, 1999
5. Antes G: Inflammatory disease of the small intestine and colon: Contrast enema and CT. Radiology. 38: 41-5, 1998
6. Gore RM et al: CT features of ulcerative colitis and Crohn's disease. AJR. 167: 3-15, 1996
7. Jacobs JE et al: CT of inflammatory disease of the colon. Semin Ultrasound CT MR. 16(2):91-101, 1995
8. Gore RM et al: CT findings in ulcerative, granulomatous, and indeterminate colitis. AJR Am J Roentgenol. 143(2):279-84, 1984
9. Kelvin FM et al: Double contrast barium enema in Crohn's disease and ulcerative colitis. AJR Am J Roentgenol. 131(2):207-13, 1978
10. Laufer I et al: The radiological differentiation between ulcerative and granulomatous colitis by double contrast radiology. Am J Gastroenterol. 66(3):259-69, 1976

IMAGE GALLERY

Typical

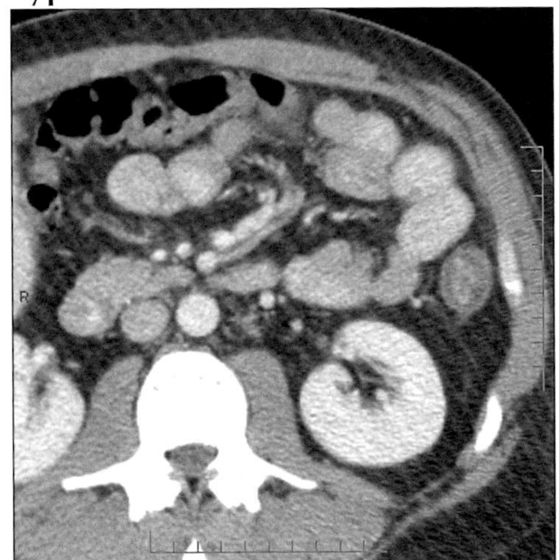

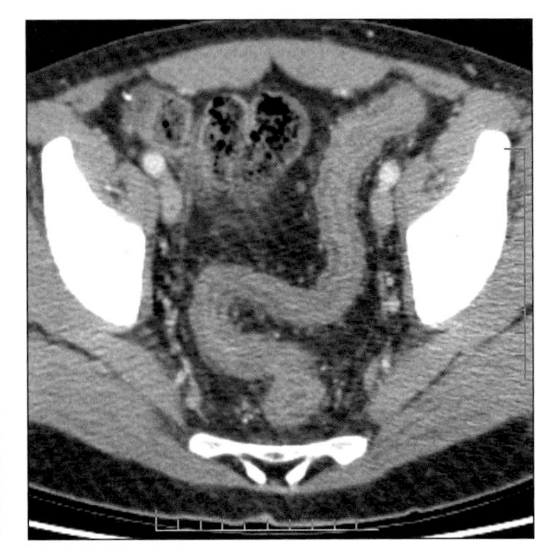

(Left) *Axial CECT shows narrowed lumen and thickened wall of descending colon. Submucosal halo of low density (edema) and engorged blood vessels indicate active disease.* ***(Right)*** *Axial CECT shows narrowed lumen and thickened wall of sigmoid colon with submucosal edema and engorged vessels.*

Typical

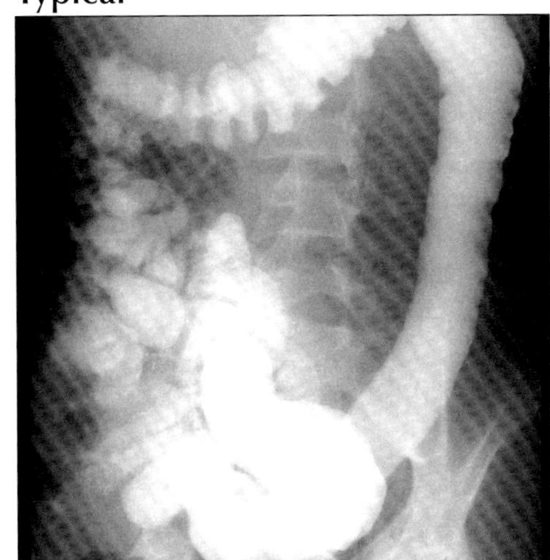

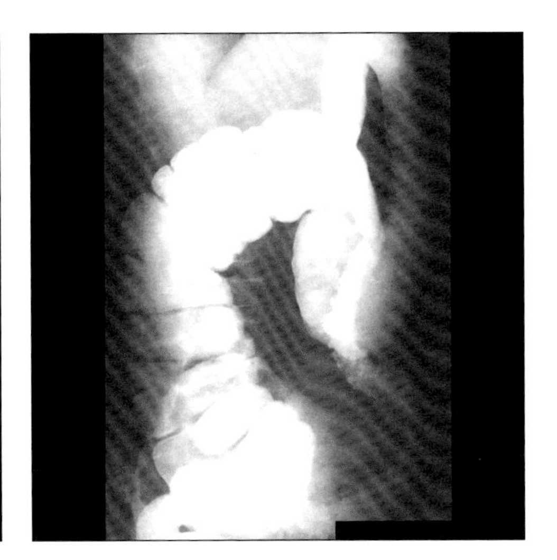

(Left) *Single-contrast BE shows prominent, thickened haustra in right colon, but diminished haustra in left colon.* ***(Right)*** *Oblique new single-contrast BE shows narrowed lumen, ahaustral left colon with diffuse ulceration (collar button + flask-shaped).*

Typical

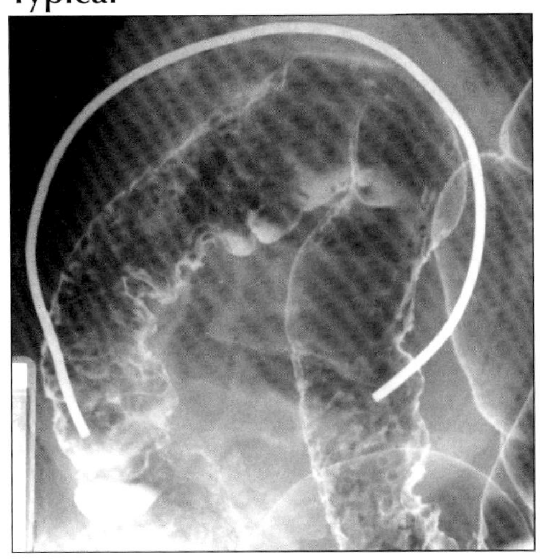

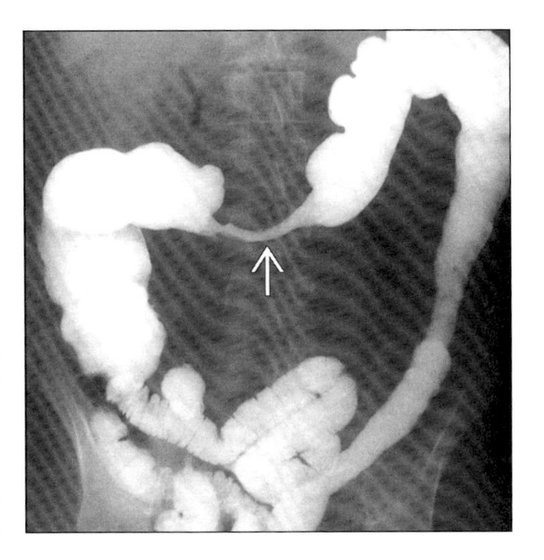

(Left) *Double-contrast BE shows filiform polyps in a patient with chronic UC, now in remission.* ***(Right)*** *Single-contrast BE shows ahaustral colon due to chronic UC. Apple core stricture of transverse colon (arrow) due to adenocarcinoma.*

TOXIC MEGACOLON

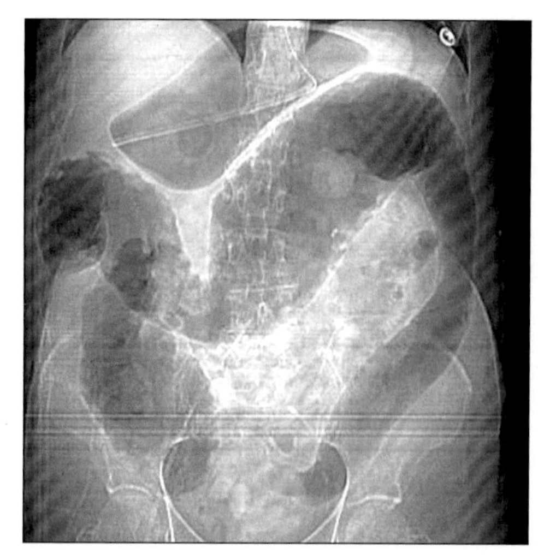

Supine radiograph shows ahaustral colon in an acutely ill patient with chronic UC. The transverse colon is dilated and "shaggy" in appearance due to sloughed mucosa and pseudopolyps.

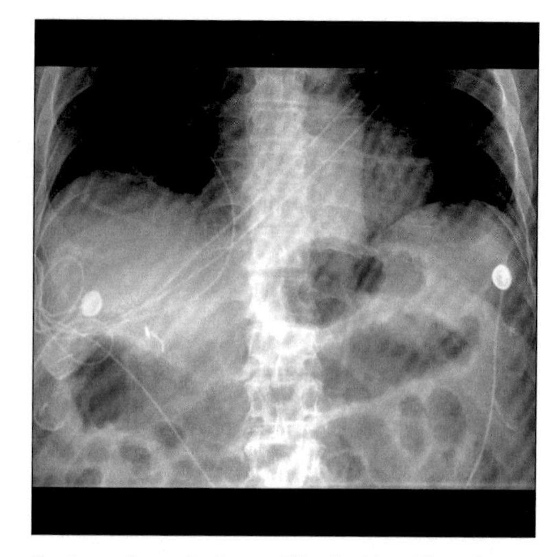

Supine radiograph shows diffusely dilated bowel in an acutely ill patient with ulcerative colitis. The transverse colon is dilated, ahaustral with an irregular mucosal surface.

TERMINOLOGY

Definitions

- Acute transmural fulminant colitis with neuromuscular degeneration & colonic dilatation

IMAGING FINDINGS

General Features

- Best diagnostic clue: Dilated ahaustral colon with pseudopolyps & air-fluid levels
- Location: Transverse colon (least dependent part in supine position)
- Other general features
  - Most severe, life-threatening complication of inflammatory bowel disease
  - More common in ulcerative colitis (1.6-13% cases)
  - May be the initial manifestation of ulcerative colitis
  - Most common cause of death directly related to ulcerative colitis
  - Diagnosis: Based on clinical status of patient + radiographic evidence
  - Precipitating factors of toxic megacolon
    - Endoscopy; use of opiates & anticholinergic drugs
    - Progressive metabolic alkalosis; aerophagia

Radiographic Findings

- Radiography
  - Marked colonic dilatation is hallmark
    - Transverse colon dilatation most common (because least dependent on supine view)
    - ↑ Colon caliber on serial radiographs
    - Mean diameters of dilated segments (8.2-9.2 cm)
    - Absolute colonic diameter: Not a diagnostic criterion
  - "Mucosal islands" or "pseudopolyps": Common finding (indicate severe disruption of mucosa)
  - Radiologically thick bowel wall (due to subserosal + omental edema), pathologically, wall is thin
  - Radiolucent stripe parallel to colon: Pericolic fat line
  - Loss of haustral pattern
    - Due to profound inflammation + ulceration
    - Presence of normal haustra excludes diagnosis
  - ± Air-fluid levels in colon; ± Small bowel distention
  - Pneumatosis coli ± pneumoperitoneum

CT Findings

- Distended colon filled with air, fluid, blood
- Distorted or absent haustral pattern
- Irregular nodular contour of colonic wall
- Intramural air ± blood
- ± Mesenteric abscess or pneumoperitoneum

DDx: Dilated Transverse Colon

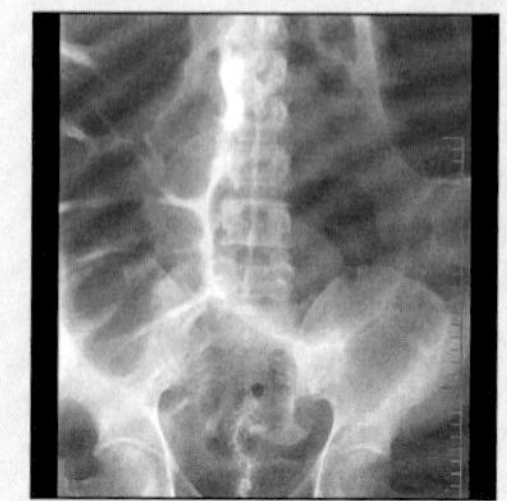

Sigmoid Volvulus

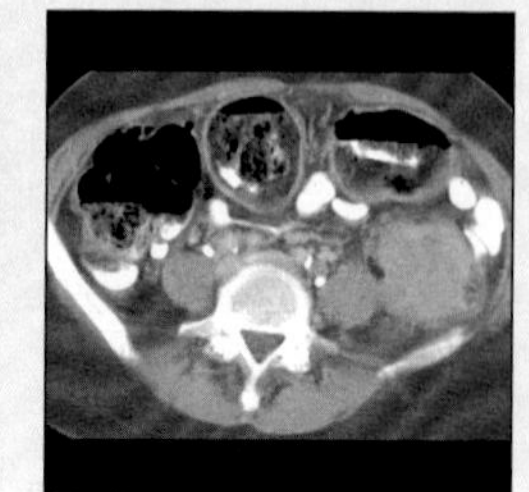

Colon Cancer

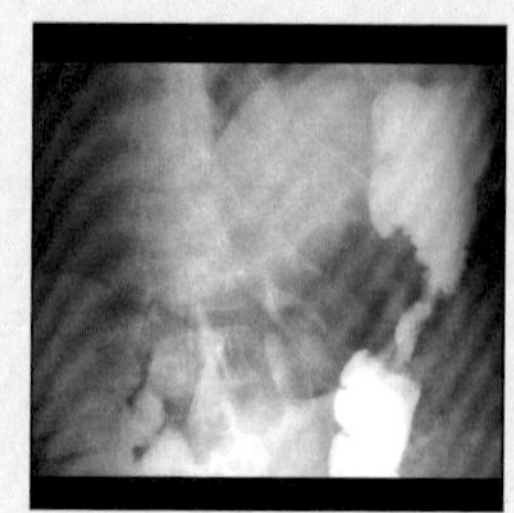

Colon Cancer

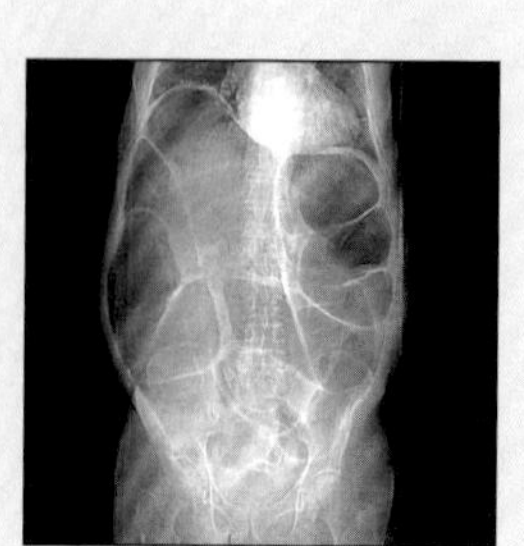

Ileus

TOXIC MEGACOLON

Key Facts

Terminology
- Acute transmural fulminant colitis with neuromuscular degeneration & colonic dilatation

Imaging Findings
- Best diagnostic clue: Dilated ahaustral colon with pseudopolyps & air-fluid levels
- Presence of normal haustra excludes diagnosis
- (Barium enema: Contraindicated, ↑ risk perforation)

Top Differential Diagnoses
- Colonic obstruction
- Adynamic or paralytic ileus

Pathology
- Ulcerative colitis (most common), other colitides

Diagnostic Checklist
- Check prior history of underlying colonic pathology

Imaging Recommendations
- Plain x-ray abdomen: Supine & lateral decubitus views
- Helical NECT
- (Barium enema: Contraindicated, ↑ risk perforation)

DIFFERENTIAL DIAGNOSIS

Colonic obstruction
- Etiology: Carcinoma (55%); volvulus (11%)
- Usually subacute or chronic in onset
- Gas + stool-filled colon to point of obstruction
- Retained haustral pattern excludes toxic megacolon

Adynamic or paralytic ileus
- Etiology: Post-op, medications, post injury, ischemia
- Dilated small & large bowel loops up to rectum
- Normal haustral pattern excludes toxic megacolon

PATHOLOGY

General Features
- Etiology
  - Ulcerative colitis (most common), other colitides
  - Pseudomembranous & ischemic colitis
  - Amebiasis, strongyloidiasis, bacillary dysentery
  - Typhoid fever, cholera, Behcet's syndrome
- Epidemiology
  - Incidence: Seen in 1.6-13% of ulcerative colitis cases
  - Medical & surgical mortality: 21.5%

Gross Pathologic & Surgical Features
- Grossly dilated colon + air & fluid; mucosal ulceration
- Absence of haustral pattern (thin bowel wall 2-3 mm)

Microscopic Features
- Transmural inflammation
- Large areas of denuded mucosa + edema
- Fissuring ulcers with extension to serosa

CLINICAL ISSUES

Presentation
- Most common signs/symptoms: Fever, pain, tenderness, abdominal distension, bloody diarrhea
- Lab-data: ↑ WBC; ↑ ESR; + ve fecal occult blood test

Demographics
- Age: 20-35 years
- Gender: Males less than females (M < F)

Natural History & Prognosis
- Complications: Perforation, abscess, peritonitis
- Prognosis
  - Good: After colectomy without complications
  - Poor: With perforation & complications

Treatment
- Surgery (colectomy); treat complications

DIAGNOSTIC CHECKLIST

Consider
- Check prior history of underlying colonic pathology

Image Interpretation Pearls
- Extensive ahaustral colonic dilatation with air-fluid levels & "mucosal islands" or "pseudopolyps"

SELECTED REFERENCES

1. Halpert RD: Toxic dilatation of the colon. Radiologic Clinics of North America 25: 147-155, 1987
2. Truelove SC et al: Toxic megacolon: Part I. Pathogenesis, diagnosis & treatment. Clinical Gastroenterology 10: 107: 114, 1981
3. Fazio VW: Toxic megacolon in ulcerative colitis and Crohn disease. Clinical Gastroenterology 9: 389-407, 1980

IMAGE GALLERY

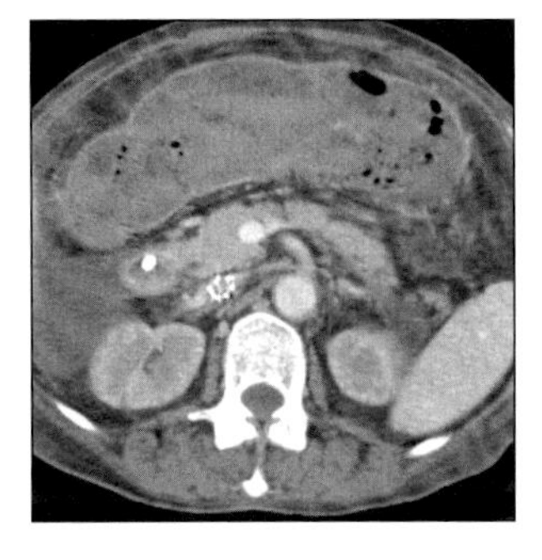

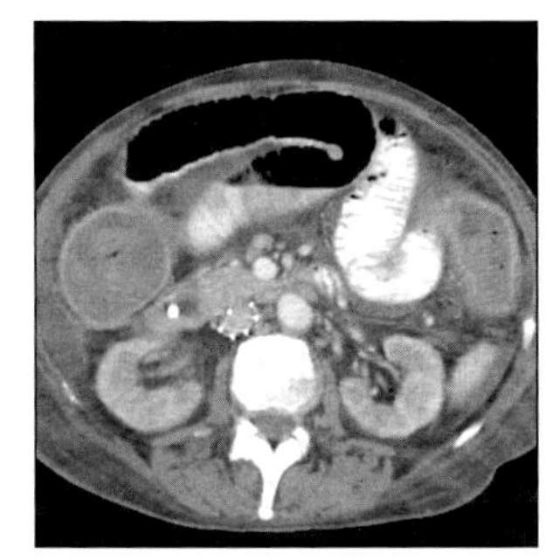

(Left) *Axial CECT shows dilated transverse colon with pneumatosis, intraluminal bleeding, sloughed mucosa. Toxic megacolon due to C. difficile colitis.* ***(Right)*** *Axial CECT shows generalized ileus. Ascending + descending colon are distended with blood + sloughed mucosa. C. difficile colitis.*

APPENDICITIS

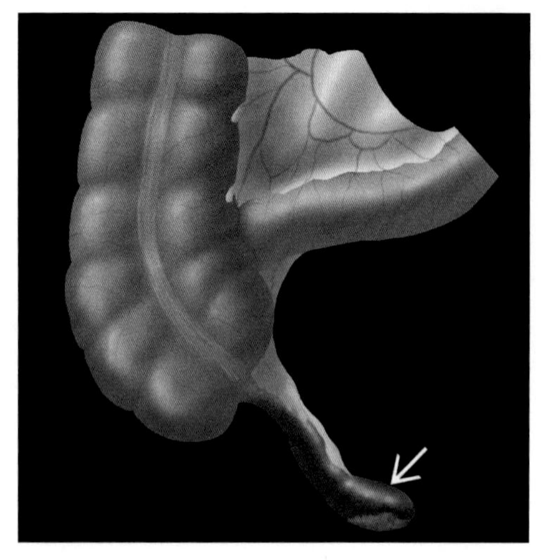

Anatomic drawing of acute appendicitis. Note enlarged, inflamed appendix (arrow).

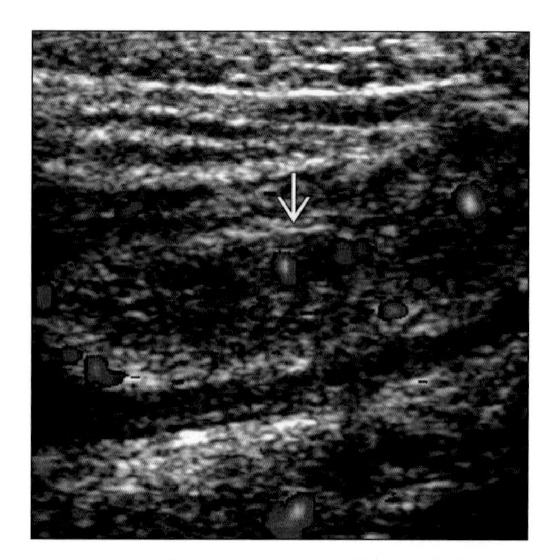

Acute appendicitis on sonography. Sagittal color Doppler sonogram of enlarged non-compressible appendix demonstrates abnormal mural flow (arrow) consistent with appendicitis.

TERMINOLOGY

Definitions

- Acute appendiceal inflammation due to luminal obstruction and superimposed infection

IMAGING FINDINGS

General Features

- Best diagnostic clue
  - Distended non-compressible appendix (≥ 7 mm) on US or CECT
  - Abnormal mural enhancement of appendix on CECT
  - Periappendiceal fat stranding on CECT
- Location: Cecal tip
- Size
  - Noncompressible appendix > 6 mm has sensitivity of 100%, but specificity of only 64%
  - Noncompressible appendix > 7 mm has sensitivity of 94% and specificity of 88%
  - Noncompressible appendix 6-7 mm equivocal size; increased flow on color Doppler in appendix indicates positive study
- Morphology: Tip of appendix is often first site of inflammation and appendiceal perforation

Radiographic Findings

- Radiography
  - Appendicolith in 5-10% of patients
  - Air-fluid levels within bowel in RLQ
  - Splinting
  - Loss of right psoas margin
  - Free peritoneal air very uncommon
  - With perforation
    - Small bowel obstruction
    - RLQ extraluminal gas
    - Displacement of bowel loops from RLQ
- Fluoroscopy
  - Barium Enema
    - Non-filling of appendix (normal in 1/3 of patients)
    - Focal mural thickening of medial wall of cecum ("arrowhead deformity")

CT Findings

- NECT
  - Dilated appendix ≥ 7 mm
  - Periappendiceal fat stranding
  - Appendicolith
    - May be incidental finding
    - Seen much more frequently on CT than on radiography
  - With perforation
    - Small bowel obstruction

DDx: Mimics of Appendicitis

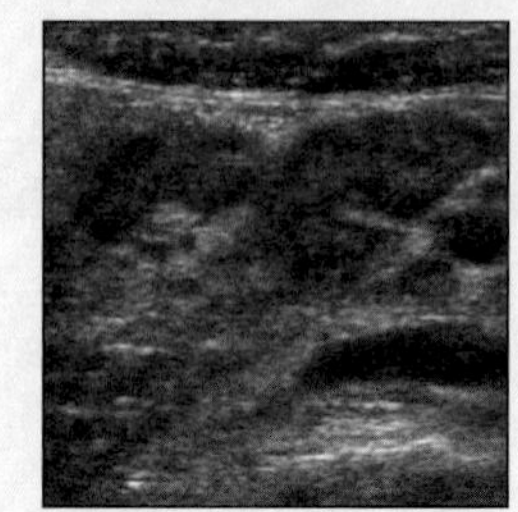

Mesenteric Adenitis

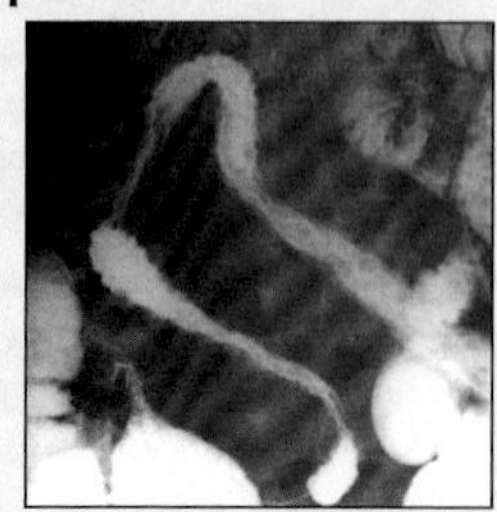

Ileocolitis

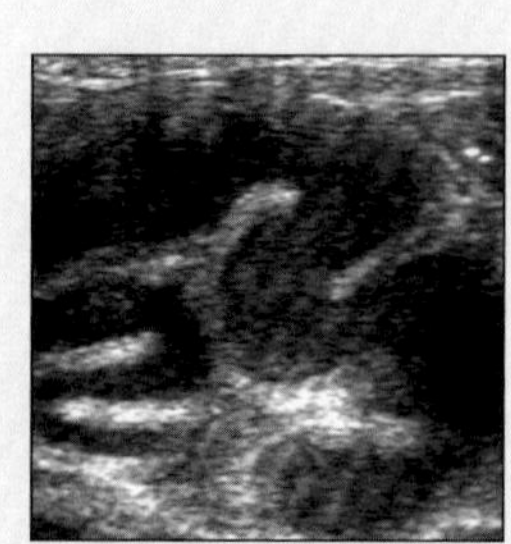

PID

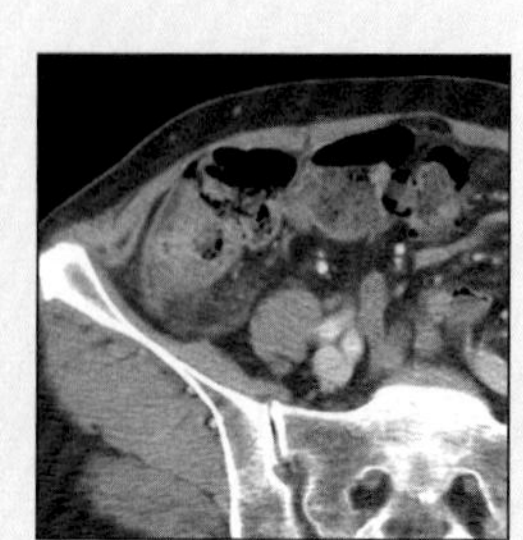

Diverticulitis

APPENDICITIS

Key Facts

Terminology
- Acute appendiceal inflammation due to luminal obstruction and superimposed infection

Imaging Findings
- Dilated appendix ≥ 7 mm; abnormal enhancement of appendiceal wall on CECT; appendicolith may or may not be present; focal bowel wall thickening of cecal tip
- In pediatric patients, thin young adults & pregnant patients: US first imaging method, to avoid excessive radiation
- CT performed for patients with inconclusive US, if perforation suspected or if obese

Top Differential Diagnoses
- Mesenteric adenitis
- Ileocolitis
- Pelvic inflammatory disease
- Cecal diverticulitis

Pathology
- General path comments: Obstructed appendiceal lumen: Appendicolith or hypertrophied Peyer patches; pus-filled lumen; thickened appendiceal wall with infiltration by inflammatory cells

Clinical Issues
- Periumbilical pain migrating to RLQ; peritoneal irritation @ McBurney point; atypical signs in 1/3 of patients
- Nonspecific presentation more common in young children

    - Inflammatory fluid collections demonstrating mass effect, most commonly in RLQ or dependent pelvis (cul-de-sac)
- CECT
  - Dilated appendix ≥ 7 mm; abnormal enhancement of appendiceal wall on CECT; appendicolith may or may not be present; focal bowel wall thickening of cecal tip
    - Sensitivity 95%, specificity 95%

Ultrasonographic Findings
- Real Time
  - Non-compressible appendix ≥ 7mm
  - Sonographic "McBurney sign" with focal pain over appendix
  - Shadowing, echogenic appendicolith
  - RLQ fluid, phlegmon, abscess
- Color Doppler
  - Flow within wall of appendix is abnormal, indicating inflammation
    - Sensitivity 85%, specificity 90%

Imaging Recommendations
- Best imaging tool
  - In pediatric patients, thin young adults & pregnant patients: US first imaging method, to avoid excessive radiation
  - CT performed for patients with inconclusive US, if perforation suspected or if obese
  - CT procedure of choice for
    - Elderly: Consider cecal or appendiceal tumor
    - Subacute symptoms or palpable mass
    - Helps differentiate inflammation, abscess, tumor
- Protocol advice
  - Oral contrast alone or rectal contrast alone may be given
  - NECT may be performed in patients with ample intraperitoneal fat
  - CECT
    - Visualize early appendicitis (abnormal mural enhancement)
    - Diagnose perforation with non-enhancement of appendix & surrounding inflammation or abscess

DIFFERENTIAL DIAGNOSIS

Mesenteric adenitis
- Enlarged and clustered lymphadenopathy in mesentery and RLQ
- Normal appendix
- May have ileal wall thickening due to GI involvement
- Pain when pressure applied with US transducer over nodes
- Diagnosis of exclusion as appendicitis (especially perforated appendicitis) may have enlarged mesenteric nodes

Ileocolitis
- Crohn disease or infectious (e.g., Yersinia)
- US: Mural thickening of cecum and terminal ileum; increased mural flow on color Doppler
- CECT: Submucosal edema of cecum and terminal ileum; surrounding cecal inflammation

Pelvic inflammatory disease
- Complex adnexal mass
- Dilated fallopian tube with fluid-fluid level (pyosalpinx)
- "Indefinite uterus" sign with obscuration of posterior wall of myometrium

Cecal diverticulitis
- Cecal diverticulum with mural thickening
- Pericecal inflammatory changes
- Thickening of lateral conal fascia
- Abscess in anterior pararenal space

Appendiceal tumor
- Soft tissue density mass infiltrating and/or obstructing appendix
- Usually little surrounding infiltration
- Carcinoma; lymphoma; carcinoid

Cecal carcinoma
- May obstruct appendiceal orifice
- Appendix is dilated but no periappendiceal inflammation

- Circumferential cecal mass and lymphadenopathy suggest tumor rather than appendicitis

PATHOLOGY

General Features

- General path comments: Obstructed appendiceal lumen: Appendicolith or hypertrophied Peyer patches; pus-filled lumen; thickened appendiceal wall with infiltration by inflammatory cells
- Etiology: Obstruction of appendiceal lumen by appendicolith or Peyer patches
- Epidemiology: 7% of all individuals in western world develop appendicitis during their lifetime

Gross Pathologic & Surgical Features

- Distended appendix with or without appendicolith
- Surrounding adhesions

Microscopic Features

- Pus in lumen
- Leukocyte infiltration of appendiceal wall
- Mucosal ulceration
- Necrosis if gangrenous

Staging, Grading or Classification Criteria

- Nonperforated
  - No evidence for necrosis and/or perforation
- Perforated
  - May have surrounding periappendiceal abscess or soft tissue inflammation of mesentery and omentum

CLINICAL ISSUES

Presentation

- Most common signs/symptoms
  - Periumbilical pain migrating to RLQ; peritoneal irritation @ McBurney point; atypical signs in 1/3 of patients
  - Other signs/symptoms
    - Anorexia, nausea, vomiting, diarrhea, possible fever
    - Nonspecific presentation more common in young children
- Clinical profile
  - Highly variable and not reliable
  - WBC may or may not be elevated

Demographics

- Age: All ages affected
- Gender: M = F

Natural History & Prognosis

- Treatment
  - Surgery if non-perforated or if minimal perforation
  - Percutaneous drainage if well-localized abscess > 3 cm
  - Antibiotic therapy if periappendiceal soft tissue inflammation and no abscess
- Complications
  - Gangrene and perforation; abscess formation
  - Peritonitis; septicemia; liver abscess
  - Pyelophlebitis
- Prognosis
  - Excellent with early surgery

DIAGNOSTIC CHECKLIST

Consider

- Mesenteric adenitis if appendix normal and nodes enlarged

Image Interpretation Pearls

- Distended non-compressible appendix ≥ 7 mm
- May or may not have appendicolith
- Periappendiceal fat stranding on contrast enhancement

SELECTED REFERENCES

1. Andersson RE: Meta-analysis of the clinical and laboratory diagnosis of appendicitis. Br J Surg. 91(1):28-37, 2004
2. O'Malley ME et al: US of gastrointestinal tract abnormalities with CT correlation. Radiographics. 23(1):59-72, 2003
3. Paulson EK et al: Clinical practice. Suspected appendicitis. N Engl J Med. 348(3):236-42, 2003
4. Wijetunga R et al: The CT diagnosis of acute appendicitis. Semin Ultrasound CT MR. 24(2):101-6, 2003
5. Jacobs JE et al: CT imaging in acute appendicitis: techniques and controversies. Semin Ultrasound CT MR. 24(2):96-100, 2003
6. Lee JH: Sonography of acute appendicitis. Semin Ultrasound CT MR. 24(2):83-90, 2003
7. Horrow MM et al: Differentiation of perforated from nonperforated appendicitis at CT. Radiology. 227(1):46-51, 2003
8. Morgan AC: Unveiling appendicitis. Contemp Nurse. 15(1-2):114-7, 2003
9. Puylaert JB: Ultrasonography of the acute abdomen: gastrointestinal conditions. Radiol Clin North Am. 41(6):1227-42, vii, 2003
10. Macari M et al: The acute right lower quadrant: CT evaluation. Radiol Clin North Am. 41(6):1117-36, 2003
11. Neumayer L et al: Imaging in appendicitis: a review with special emphasis on the treatment of women. Obstet Gynecol. 102(6):1404-9, 2003
12. Dixon MR et al: An assessment of the severity of recurrent appendicitis. Am J Surg. 186(6):718-22; discussion 722, 2003
13. Morris KT et al: The rational use of computed tomography scans in the diagnosis of appendicitis. Am J Surg. 183(5):547-50, 2002
14. Raman SS et al: Accuracy of nonfocused helical CT for the diagnosis of acute appendicitis: a 5-year review. AJR Am J Roentgenol. 178(6):1319-25, 2002
15. Albiston E: The role of radiological imaging in the diagnosis of acute appendicitis. Can J Gastroenterol. 16(7):451-63, 2002
16. See TC et al: Appendicitis: spectrum of appearances on helical CT. Br J Radiol. 75(897):775-81, 2002
17. Bendeck SE et al: Imaging for suspected appendicitis: negative appendectomy and perforation rates. Radiology. 225(1):131-6, 2002
18. van Breda Vriesman AC et al: Epiploic appendagitis and omental infarction. Eur J Surg. 167(10):723-7, 2001
19. Jones PF: Suspected acute appendicitis: trends in management over 30 years. Br J Surg. 88(12):1570-7, 2001
20. Rosendahl K et al: Imaging strategies in children with suspected appendicitis. Eur Radiol. 2004

IMAGE GALLERY

Typical

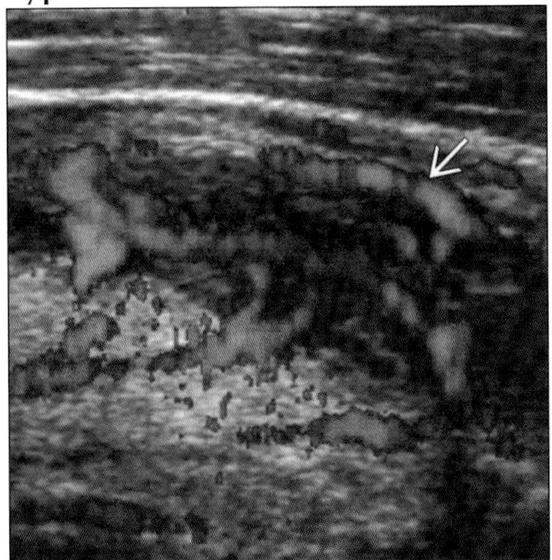

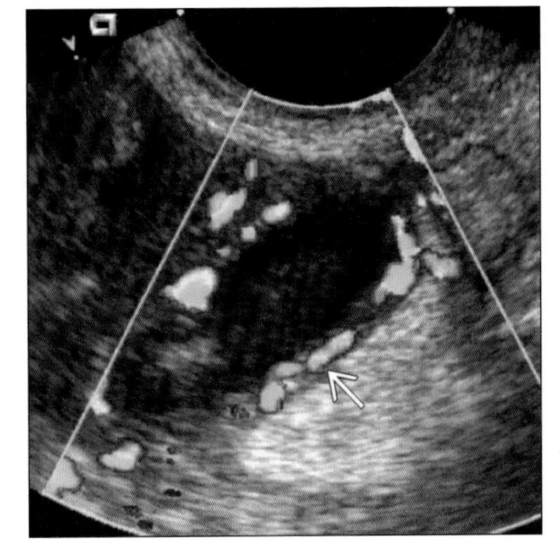

(Left) *Color Doppler sonography in acute appendicitis demonstrates marked hyperemia in wall of appendix (arrow) consistent with acute appendicitis.* ***(Right)*** *Endovaginal coronal view of right adnexa demonstrates hyperemia of appendix (arrow), consistent with pelvic appendicitis.*

Typical

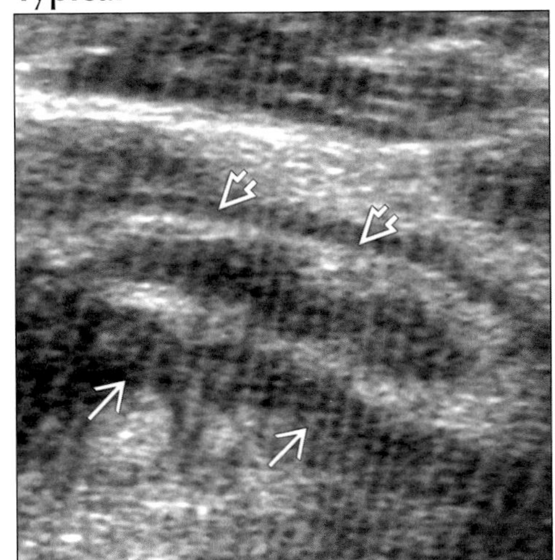

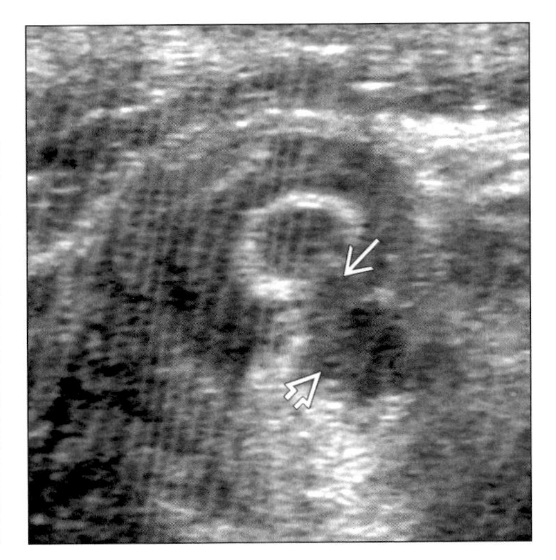

(Left) *Longitudinal sonogram demonstrates enlarged (10 mm) appendix (open arrows) with adjacent hypoechoic inflammation (arrows).* ***(Right)*** *Transverse sonogram of appendix demonstrates focal necrosis of appendiceal wall (arrow) and small adjacent abscess (open arrow).*

Typical

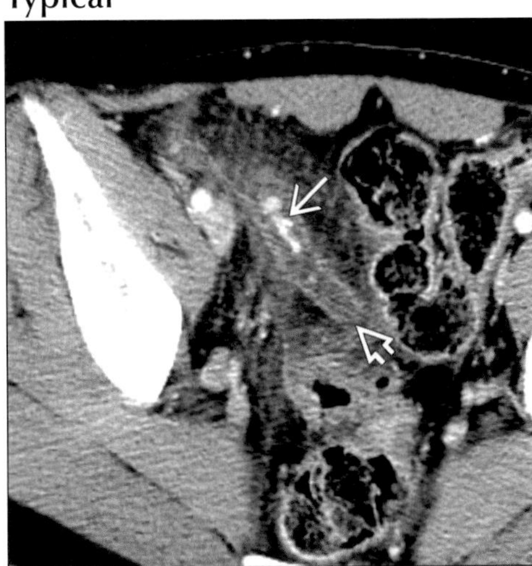

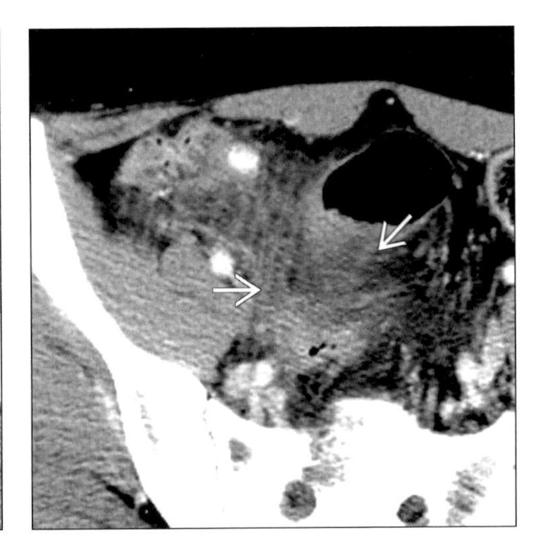

(Left) *Axial CECT of perforated appendicitis. Note multiple calcified appendicoliths (arrow) and lack of enhancement of appendiceal tip (open arrow).* ***(Right)*** *Axial CECT of perforated appendicitis. Note marked surrounding periappendiceal inflammation (arrows).*

MUCOCELE OF THE APPENDIX

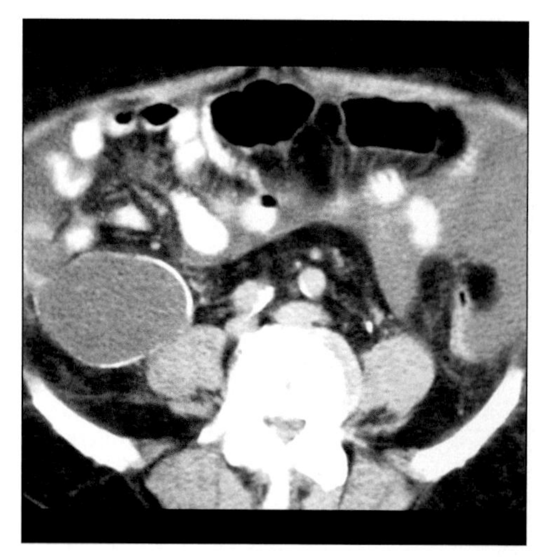

Axial CECT shows oval, thin-walled, calcified mass at the tip of the cecum.

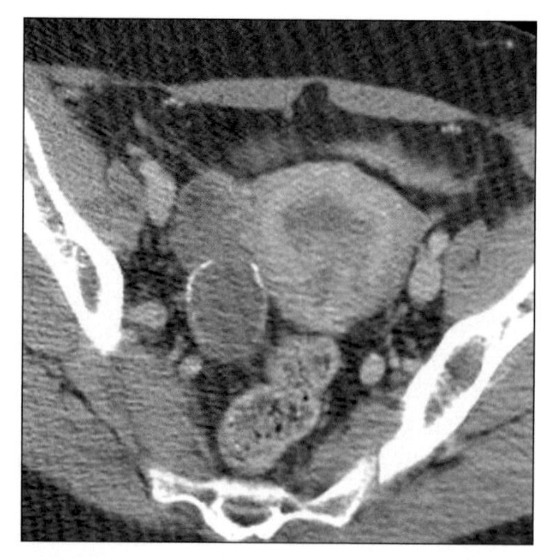

Axial CECT shows oval, partially calcified thin-walled "cyst" near tip of cecum.

TERMINOLOGY

Definitions

- Chronic cystic dilatation of appendiceal lumen by mucin accumulation

IMAGING FINDINGS

General Features

- Best diagnostic clue: Round or oval, thin-walled, cystic mass near tip of cecum
- Size: 3-6 cm in diameter
- Other general features
  - Mucocele of appendix is a rare entity
  - Classified into three groups based on histology
    - Focal or diffuse mucosal hyperplasia
    - Mucinous cystadenoma
    - Mucinous cystadenocarcinoma
  - Focal or diffuse mucosal hyperplasia
    - Resembles hyperplastic polyp of colon
    - Does not perforate
  - Mucinous cystadenoma
    - A benign neoplasm
    - Most common type of mucocele
    - 20% of cases perforate with mucus seeding
  - Mucinous cystadenocarcinoma
    - One fifth as common as cystadenomas
    - ↑ Risk of perforation, forming peritoneal implants
  - Pseudomyxoma peritonei
    - Due to rupture: Malignant > benign mucocele
    - Peritoneal cavity filled with mucus seedlings
  - Myxoglobulosis
    - Rare variant with multiple small globules
    - Calcify & produce 1-10 mm mobile calcifications
    - Differentiate from phleboliths & calcified nodes

Radiographic Findings

- Fluoroscopic guided barium enema
  - Appendix: Fails to fill on barium enema
  - Cecum: Indented on its medial aspect by smooth-walled globular mass
  - Ileum: Terminal part is displaced

CT Findings

- NECT
  - Mucocele
    - Well-defined cystic mass RLQ (near water HU)
    - Calcification (curvilinear) within wall or lumen
  - Mucinous cystadenoma
    - Encapsulated low attenuation cyst
    - Indistinguishable from retention mucocele
  - Mucinous cystadenocarcinoma
    - Large irregular mass with thickened nodular wall
    - Components: Solid & cystic; Ca++ in solid area
  - Pseudomyxoma peritonei
    - Massive ascites + septations (heterogeneous HU)

DDx: Dilation or Mass of Appendix; RLQ Cystic Mass

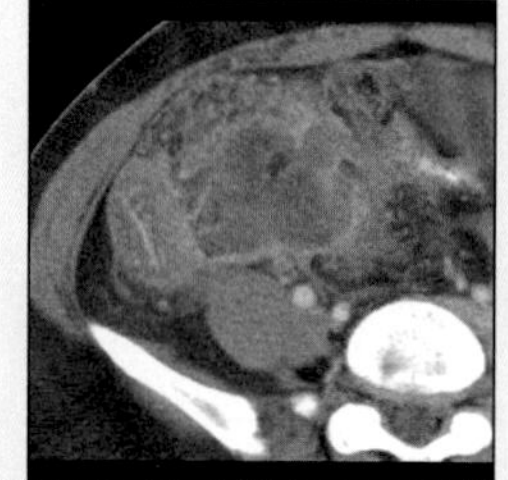

Appendiceal Abscess

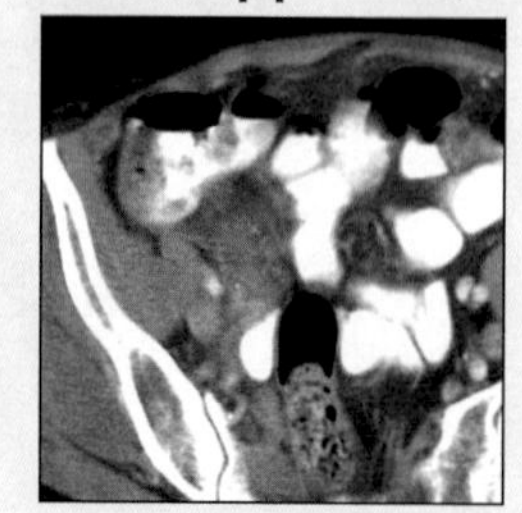

Appendiceal Ca

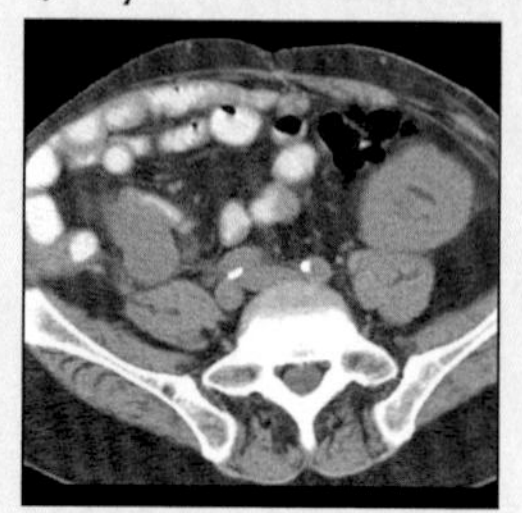

Appy. Lymphoma

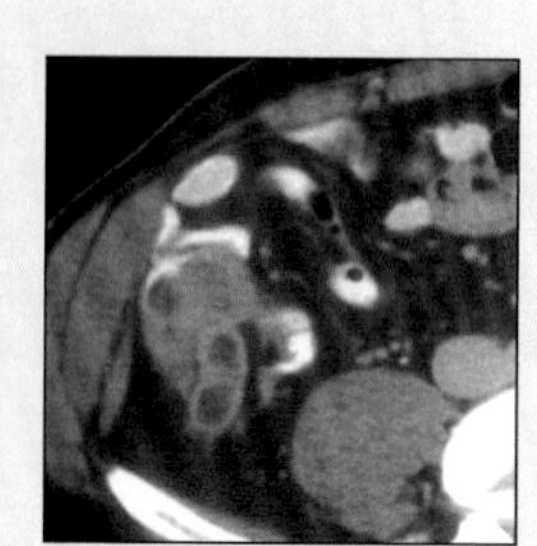

Cecal Carcinoma

MUCOCELE OF THE APPENDIX

Key Facts

Terminology
- Chronic cystic dilatation of appendiceal lumen by mucin accumulation

Imaging Findings
- Best diagnostic clue: Round or oval, thin-walled, cystic mass near tip of cecum
- Calcification (curvilinear) within wall or lumen

Top Differential Diagnoses
- Acute appendicitis (abscess)
- Appendiceal carcinoma
- Ovarian cystic mass

Pathology
- Obstructing lesions can cause mucocele formation
- Associated abnormalities: Colonic adenocarcinoma (6-fold risk)

- CECT: Loculated ascites; scalloped surface of liver + spleen

MR Findings
- Mucocele with ↑ fluid content: Long T1 & T2
  - T1WI: Hypointense
  - T2WI: Hyperintense
- Mucocele with ↑ mucin content: Short T1 & long T2
  - T1WI & T2WI: Mucocele appears hyperintense

Ultrasonographic Findings
- Real Time
  - Anechoic or cystic + internal echoes (septations)
  - Increased through transmission is characteristic
  - Complex cystic mass; ± calcification
  - Gravity-dependent echoes (inspissated mucus)

Imaging Recommendations
- NE + CECT, MR, US

DIFFERENTIAL DIAGNOSIS

Acute appendicitis (abscess)
- More inflammatory changes
- Thick irregular abscess wall

Appendiceal carcinoma
- Irregular mixed density mass (solid & cystic)

Appendiceal lymphoma
- Soft tissue mass near tip of cecum

Cecal carcinoma
- May cause dilated appendix

Ovarian cystic mass
- Distinguish by relation to broad ligament vs. cecal tip

PATHOLOGY

General Features
- Etiology
  - Obstructing lesions can cause mucocele formation
    - Postappendicitis scarring (most common)
    - Fecalith, appendiceal carcinoma, endometrioma
    - Carcinoid, polyp, volvulus, Ca of cecum & colon
- Epidemiology: Seen in 0.3% appendectomy specimens
- Associated abnormalities: Colonic adenocarcinoma (6-fold risk)

Gross Pathologic & Surgical Features
- Mucocele: Thin-walled, mucin-filled cystic structure

Microscopic Features
- Mucoid material; malignant cells-cystadenocarcinoma

CLINICAL ISSUES

Presentation
- Most common signs/symptoms: Asymptomatic; pain & tenderness RLQ; palpable mass
- Complications: Rupture, torsion, bowel obstruction

Demographics
- Age: Mean age: 55 years
- Gender: M:F = 1:4

Natural History & Prognosis
- Mucocele & cystadenoma (good); carcinoma (poor)

Treatment
- Surgical resection

SELECTED REFERENCES

1. Lim HK et al: Primary mucinous cystadenocarcinoma of the appendix: CT findings. AJR. 173: 1071-4, 1999
2. Kim SH et al: Mucocele of the appendix: Ultrasonographic and CT findings. Abdominal Imaging. 23: 292-6, 1998
3. Madwell D et al: Mucocele of the appendix: Imaging findings. AJR. 159: 69-72, 1992

IMAGE GALLERY

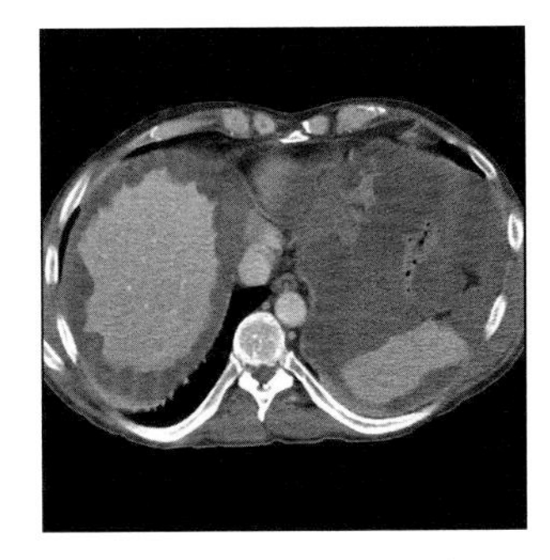

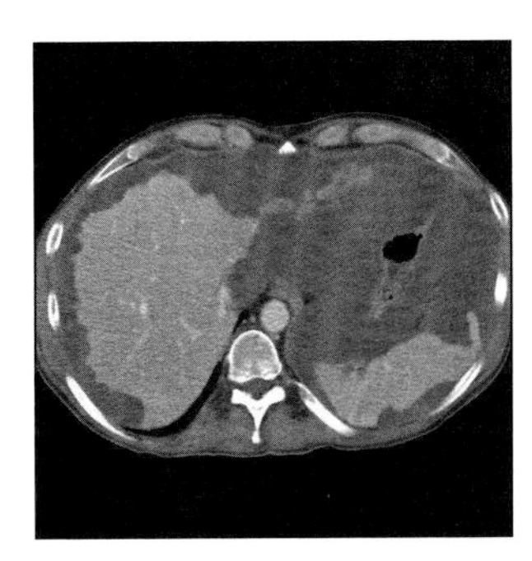

*(**Left**) Axial CECT shows complex ascites with scalloped surface of liver + spleen. Pseudomyxoma peritonei due to ruptured mucinous cystadenocarcinoma of appendix. (**Right**) Axial CECT shows pseudomyxoma peritonei.*

DIVERTICULITIS

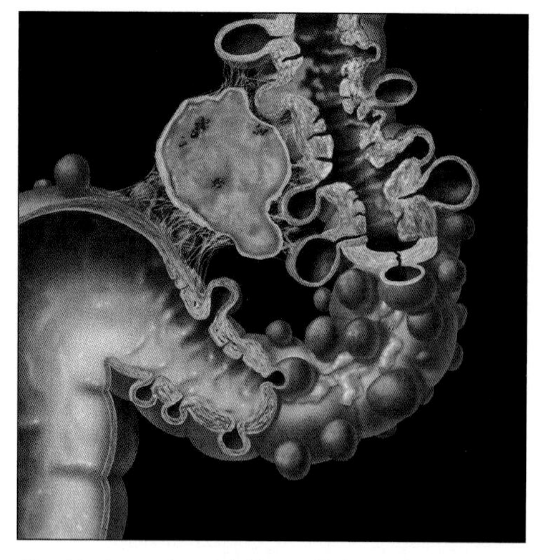

Graphic shows sigmoid diverticula, luminal narrowing + wall thickening (circular muscle hypertrophy). Pericolic abscess due to perforated diverticulum. Rectum spared.

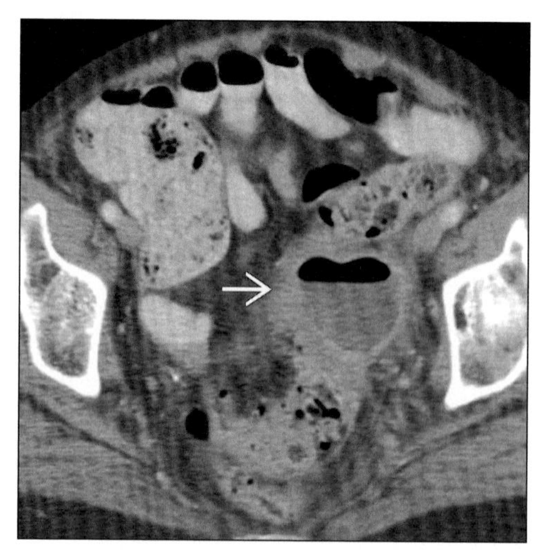

Axial CECT shows a pericolonic abscess (arrow) adjacent to the sigmoid colon, with luminal narrowing, gas-filled diverticula, and pericolonic fat infiltration.

TERMINOLOGY

Definitions

- Inflammation or perforation of colonic diverticula, which are acquired herniations of mucosa and submucosa through muscular layers of bowel wall

IMAGING FINDINGS

General Features

- Best diagnostic clue: Small colonic outpouchings with irregular wall thickening & pericolic fat stranding
- Location: Most common in sigmoid colon
- Size: Diverticula: Usually about 0.5-1.0 cm
- Morphology: Saccular outpouchings of colon with perforation, inflammation & abscess formation
- Other general features
  - Most common colonic disease in Western world
  - Diverticula occur mainly where vasa recta vessels pierce muscularis propria, between mesenteric & antimesenteric taeniae
  - Colonic diverticula are pseudodiverticula
    - Mucosa + submucosa; no muscularis propria
  - Diverticular disease of colon represents a collection or sequence of events
    - Prediverticular phase: Circular muscular thickening of colonic wall (myochosis)
    - Diverticulosis: Frank outpouchings (diverticula)
    - Diverticulitis: Perforation + localized pericolic inflammation or abscess

Radiographic Findings

- Diverticulosis: Fluoroscopic guided barium enema (single contrast preferred)
  - Immature diverticula
    - En face: Resemble punctate ulcer
    - In profile: Conical or triangular (1-2 mm high)
  - Mature: Shape varies based on angle & degree of barium filling
    - In profile: Flask-like protrusion, long or large neck
    - Diverticulum with long & narrow neck: Mimic pedunculated polyp on air-contrast enema
    - Diverticulum with large neck: Mimics sessile polyp
    - En face: Ring shadow or round barium collection
    - "Bowler hat" sign: Dome of hat points away from bowel wall (diverticulum); toward lumen (polyp)
  - In progressive disease: Due to muscular hypertrophy, diverticula are irregular, lumen narrowed with serrated or "cog-wheel" appearance
- Diverticulitis: Fluoroscopic guided water soluble contrast enema (not recommended)
  - Focal area of eccentric luminal narrowing caused by pericolic or intramural inflammatory mass (abscess) + mucosal tethering
  - Marked thickening + distortion of haustral folds
  - Extraluminal contrast (due to peridiverticulitis)

DDx: Wall Thickening, Pericolonic Infiltration

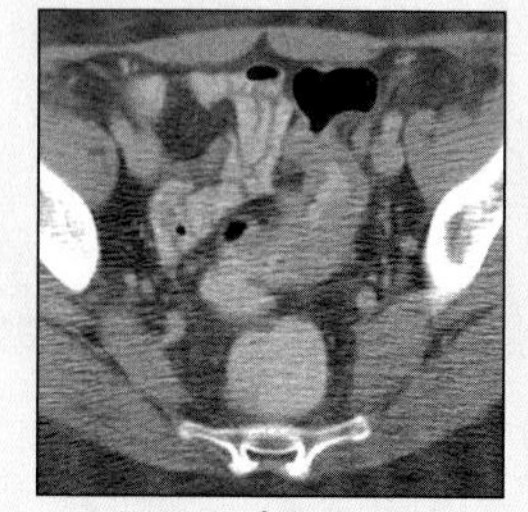

Sigmoid Cancer

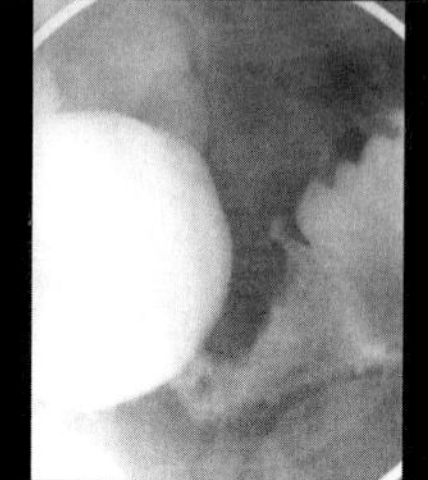

Radiation Colitis

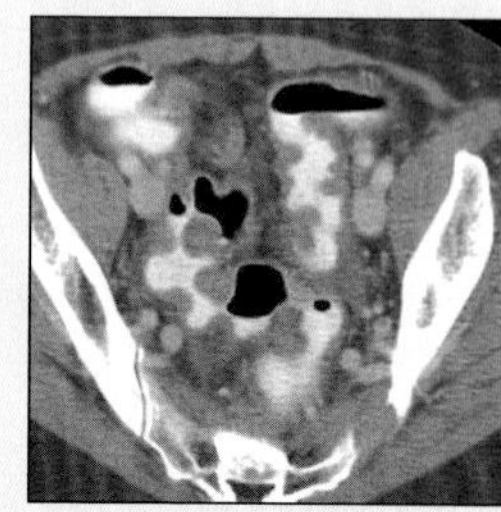

Pseudomem. Colitis

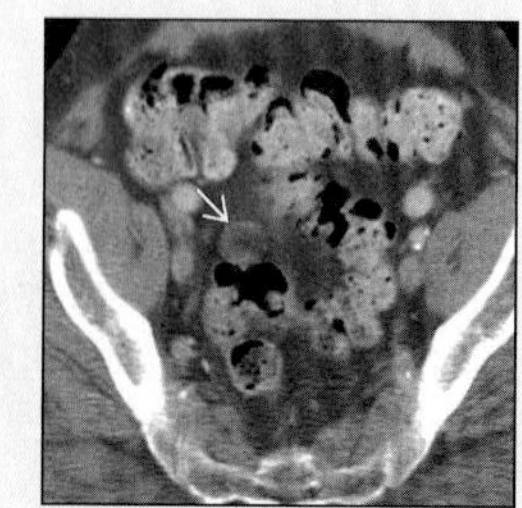

Epiploic Appendagitis

DIVERTICULITIS

Key Facts

Terminology
- Inflammation or perforation of colonic diverticula, which are acquired herniations of mucosa and submucosa through muscular layers of bowel wall

Imaging Findings
- Best diagnostic clue: Small colonic outpouchings with irregular wall thickening & pericolic fat stranding
- Location: Most common in sigmoid colon
- CT is very accurate in diagnosis (> 95%)
- Helical CT: Oral & IV ± rectal contrast for acutely ill

Top Differential Diagnoses
- Colon carcinoma
- Radiation colitis
- Ischemic colitis
- Pseudomembranous colitis (PMC)

Pathology
- Most common complication of diverticulosis, in 30% of patients with moderate diverticulosis
- Very common in Western society, rare in less developed countries due to more processed food & less fiber in diet

Clinical Issues
- Percutaneous abscess drainage can eliminate surgery or allow elective one-step procedure in most cases

Diagnostic Checklist
- Check whether patient has signs & symptoms of diverticulitis
- Long segment colonic involvement, extensive inflammatory changes & absence of nodes or metastases favors diverticulitis over colon cancer

    - "Double-tracking": Longitudinal intramural fistulous tract (connecting ruptured diverticula) is parallel to sigmoid lumen
    - Pericolonic fistulous tracts: Bladder, bowel, vagina
    - Pericolonic collection: Abscess compresses colon
  - Sigmoid colon obstruction with zone of transition can mimic cancer
    - Tethered or saw-toothed luminal configuration suggests diverticular disease

CT Findings
- Diverticulosis
  - Mural thickening of colon (4 to 15 mm)
  - Multiple air or contrast or stool-containing outpouchings (diverticula)
- Diverticulitis
  - CT is very accurate in diagnosis (> 95%)
  - Bowel wall thickening, fat stranding, thickened base of sigmoid mesocolon, free fluid
  - Long segment (> 10 cm) of colonic involvement
  - Pericolic abscess, sinus tracts, fistulas
  - Intramural or abdomino-pelvic abscess
  - "Arrowhead" sign: Due to edema at orifice of inflamed diverticulum
  - Inflammation usually localized to pericolonic area
  - Free air + peritonitis (less common)
    - Omentum acts as "band-aid" to limit spread
    - Immunocompromised at ↑ risk: Peritonitis/sepsis
  - ± Gas or thrombus in mesenteric & portal veins
    - Follow course of inferior mesenteric vein
  - ± Liver abscesses

Ultrasonographic Findings
- Real Time
  - Diverticulosis
    - Thickened bowel wall (> 4 mm)
    - Diverticula: Round or oval hypo-/hyperechoic foci protruding from colonic wall with focal disruption of normal layer ± acoustic shadows
  - Diverticulitis
    - Pericolic inflammation: ↑ Echogenicity ± ill-defined hypoechoic areas
    - Pericolic abscess: Hypoechoic ± internal echoes

Imaging Recommendations
- Helical CT: Oral & IV ± rectal contrast for acutely ill
- Diverticulosis: Single contrast barium enema
- Double contrast barium enema: Hard to distinguish polyps from diverticula

DIFFERENTIAL DIAGNOSIS

Colon carcinoma
- Asymmetric bowel wall thickening ± irregular surface
- Wall thickening, fat stranding & pericolonic infiltration mimics diverticulitis
- CT findings favoring cancer
  - Short segment involvement (< 10 cm)
  - Wall thickness: More than 2 cm
  - Mesenteric lymphadenopathy
  - Metastases

Radiation colitis
- Barium enema findings
  - Acute radiation colitis & proctitis
    - Disrupted or distorted mucosal pattern (due to edema or hemorrhage)
  - Chronic radiation colitis & proctitis
    - Diffuse or focal narrowing with tapered margins
    - Colonic stricture or fistula may be seen
    - Widened presacral space (in profile view)
- CT findings
  - More uniform wall thickening + luminal narrowing, less pericolonic inflammation than diverticulitis
  - Colonic luminal narrowing or stricture
  - ± Sinuses or fistulas
- Diagnosis: History of radiation therapy

Ischemic colitis
- Usual sites
  - Splenic flexure > recto-sigmoid junction
- Barium enema findings
  - Thumbprinting (usually within 24 hrs after insult)
    - Due to submucosal edema or hemorrhage
  - Ulceration: Sloughing of mucosa (46-60% cases)
    - Usually develops 1-3 weeks after onset of disease

DIVERTICULITIS

- CT findings
  - More uniform, extensive wall thickening & less pericolonic infiltration than diverticulitis
  - Bowel wall attenuation
    - Hypoattenuation: Submucosal or diffuse edema
    - Hyperattenuation: Submucosal or diffuse bleeding
  - ± Pneumatosis & portomesenteric venous gas
- Diagnosis
  - History of nonocclusive vascular disease
    - Hypoperfusion in elderly people
    - Example: CHF, arrhythmia, shock & drugs

Pseudomembranous colitis (PMC)

- Synonym(s): Antibiotic colitis or C. difficile colitis
- CT findings
  - Massive wall thickening, usually pancolonic
  - Often transmural with pericolonic infiltration
  - "Accordion" sign: Represents trapped enteric contrast between thickened colonic folds
  - Full recovery with early diagnosis, discontinuation of offending antibiotic & metronidazole treatment

PATHOLOGY

General Features

- General path comments
  - Diverticulitis
    - Most common complication of diverticulosis, in 30% of patients with moderate diverticulosis
- Etiology
  - Diverticulitis: Due to fecal impaction at mouth of diverticulum with subsequent perforation
  - Contributing factors to development of diverticula
    - Pressure gradient between lumen & serosa (sigmoid): Narrowest of colon + ↑ pressure + dehydrated stool
    - Bowel wall weakness: Between mesenteric & antimesenteric taeniae
- Epidemiology
  - Incidence
    - 33-50% cases, over 50 years old have diverticulosis
    - More than 50% have diverticulosis after 80 years
    - Can occur in young adults (< 30 years old)
  - Very common in Western society, rare in less developed countries due to more processed food & less fiber in diet
- Associated abnormalities: Liver abscesses

Gross Pathologic & Surgical Features

- Outpouchings from sigmoid colon between taenia coli

Microscopic Features

- Diverticula: Mucosal herniation through a defect in circular layer of muscle
- Diverticulitis: Perforation with inflammation & micro-/macroabscess

CLINICAL ISSUES

Presentation

- Most common signs/symptoms
  - Diverticulosis
    - Asymptomatic; pain & rectal bleeding (30% cases)
    - Alternating constipation + diarrhea due to luminal narrowing (circular muscle hypertrophy)
  - Diverticulitis
    - LLQ colicky pain, tenderness & palpable mass
    - Fever, altered bowel habits
  - Lab-data
    - ↑ WBC count; anemia; ± blood in stool

Demographics

- Age: 5th to 8th decade (peak); not rare in younger
- Gender: Equal in both males & females (M = F)

Natural History & Prognosis

- Complications
  - Perforation, pericolonic abscess, fistula, sinus
  - Obstruction, hemorrhage
  - Peritonitis & sepsis (uncommon)
  - Pylephlebitis (portal vein thrombus); liver abscesses
    - Via mesenteric + portal vein → liver abscesses
- Prognosis: Early stages & after surgery (good)

Treatment

- High-fiber diet (preventive)
- Antibiotics, IV fluids, bowel rest
- Percutaneous abscess drainage can eliminate surgery or allow elective one-step procedure in most cases

DIAGNOSTIC CHECKLIST

Consider

- Check whether patient has signs & symptoms of diverticulitis

Image Interpretation Pearls

- Bowel wall thickening, pericolic infiltration & fat stranding affecting sigmoid colon
- Long segment colonic involvement, extensive inflammatory changes & absence of nodes or metastases favors diverticulitis over colon cancer

SELECTED REFERENCES

1. Jang HJ et al: Acute diverticulitis of the cecum and ascending colon: the value of thin-section helical CT findings in excluding colonic carcinoma. AJR Am J Roentgenol. 174(5):1397-402, 2000
2. Horton KM et al: CT evaluation of the colon: inflammatory disease. Radiographics. 20(2):399-418, 2000
3. Gore RM et al: Helical CT in the evaluation of the acute abdomen. AJR 174: 901-13, 2000
4. Chintapalli KN et al: Diverticulitis versus colon cancer: differentiation with helical CT findings. Radiology. 210(2):429-35, 1999
5. Rao PM et al: Colonic diverticulitis: evaluation of the arrowhead sign and the inflamed diverticulum for CT diagnosis. Radiology. 209(3):775-9, 1998
6. Padidar AM et al: Differentiating sigmoid diverticulitis from carcinoma on CT scans: mesenteric inflammation suggests diverticulitis. AJR Am J Roentgenol. 163(1):81-3, 1994
7. Balthazar EJ et al: Limitations in the CT diagnosis of acute diverticulitis: comparison of CT, contrast enema, and pathologic findings in 16 patients. AJR Am J Roentgenol. 154(2):281-5, 1990

IMAGE GALLERY

Typical

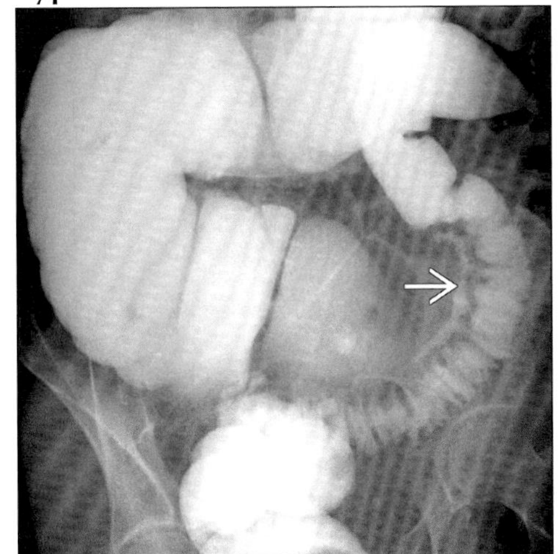

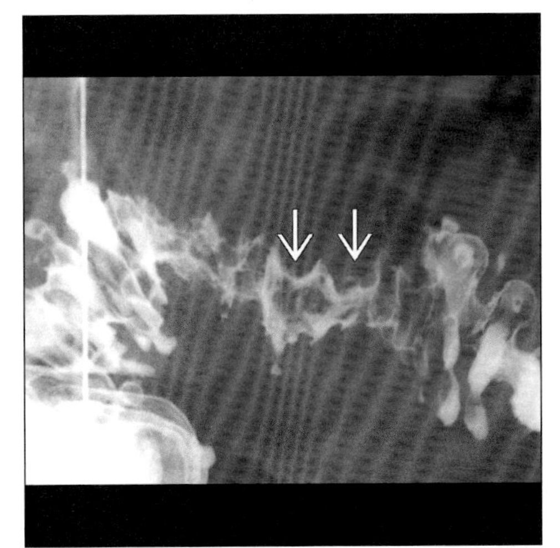

(Left) *Single contrast BE shows an intramural track of barium (arrow) paralleling the sigmoid lumen, due to submucosal spread of infection from perforated diverticulum.* ***(Right)*** *Single contrast water soluble enema shows marked distortion of sigmoid lumen mostly due to circular muscle hypertrophy. Broad-based intramural + extrinsic mass effect (arrows) due to diverticulitis.*

Typical

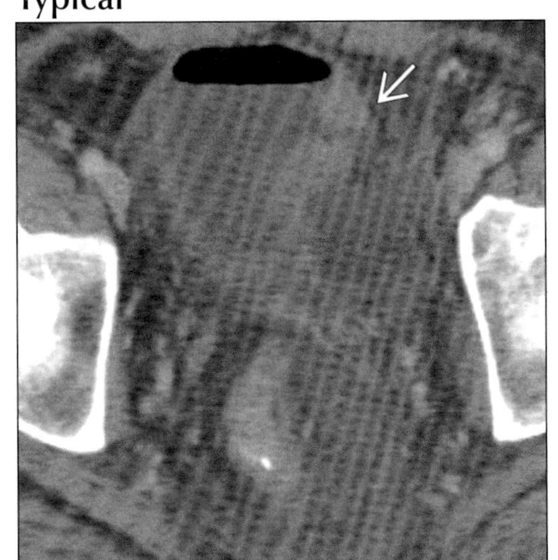

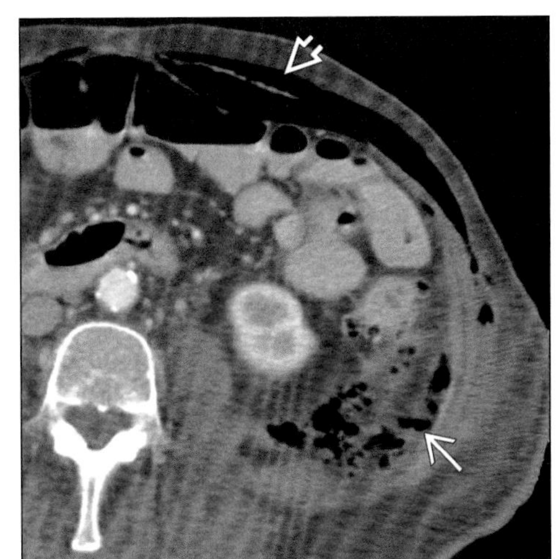

(Left) *Axial CECT shows shows extensive infiltration of pelvic/pericolic fat. Bladder has gas-fluid level and a fistula (arrow) to the sigmoid colon.* ***(Right)*** *Axial CECT shows diverticulosis of descending colon. Perforated diverticulitis resulted in extensive abscess in retroperitoneum (arrow), with dissection throughout the abdominal wall (open arrow).*

Typical

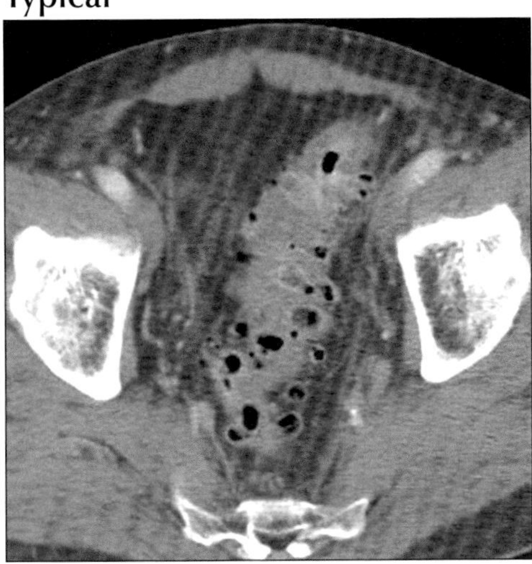

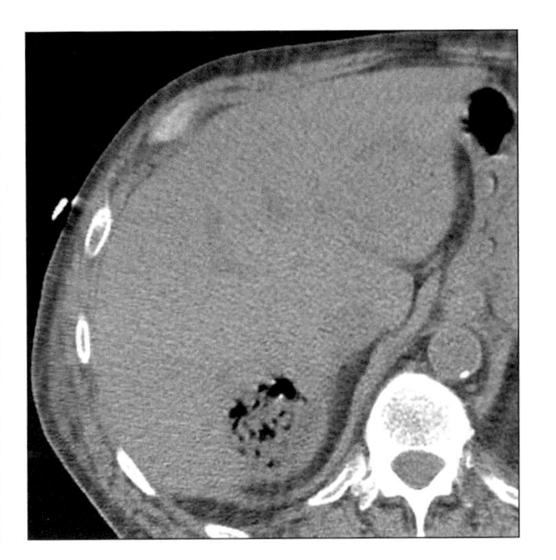

(Left) *Axial CECT shows sigmoid diverticulitis with irregular luminal narrowing + mild pericolic infiltration; subacute diverticulitis.* ***(Right)*** *Axial CECT in patient with subacute diverticulitis shows a pyogenic liver abscess due to bacterial seeding from inferior mesenteric-portal vein.*

EPIPLOIC APPENDAGITIS

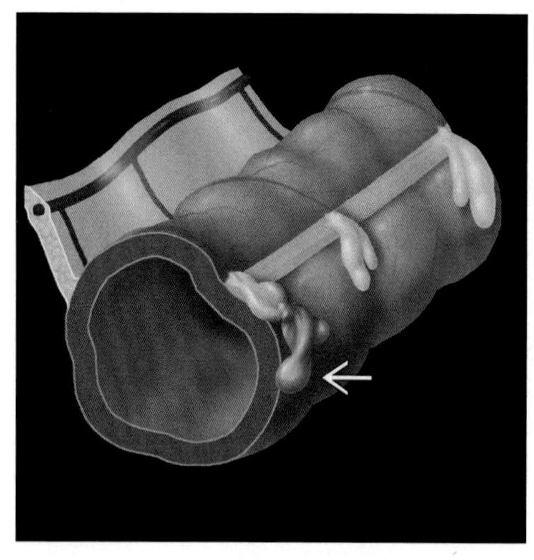

Graphic shows two normal epiploic appendages and one that is twisted and infarcted (arrow).

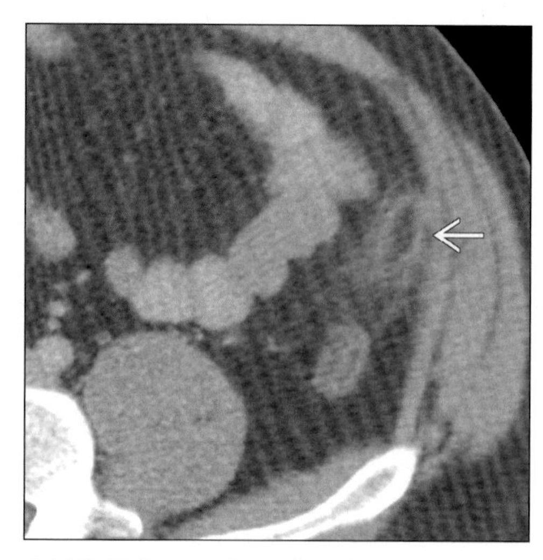

Axial CECT shows oval pericolonic fatty nodule (arrow) with hyperdense ring and surrounding inflammation.

TERMINOLOGY

Abbreviations and Synonyms

- Epiploic appendagitis (EA)

Definitions

- Acute inflammation or infarction of epiploic appendages

IMAGING FINDINGS

General Features

- Best diagnostic clue: Small oval pericolonic fatty nodule with hyperdense ring + surrounding inflammation
- Location
  - Left lower quadrant > right lower quadrant
    - Rectosigmoid junction (57%); ileocecal (26%); ascending colon (9%)
    - Transverse colon (6%); descending colon (2%); occasionally appendix
- Morphology: Epiploic appendages: Small adipose structures protruding from serosal surface of colon
- Other general features
  - Uncommon inflammatory & ischemic condition
  - Uncommon cause of acute abdomen
  - Typically seen in obese people in 2nd-5th decades of life, can occur in children
  - Benign self-limiting disease
  - Torsion of epiploic appendage
  - Spontaneous venous thrombosis of draining appendageal vein
  - Rarely diagnosed clinically but has highly characteristic CT features
  - Radiological (CT) potential misdiagnosis of EA as: Diverticulitis or appendicitis

CT Findings

- Normal epiploic appendages
  - Small lobulated masses of pericolonic fat
    - Rectosigmoid most evident
  - Seen on CT scans only when outlined by ascites
- 1-4 cm, oval-shaped, fat density paracolic lesion with adjacent fat stranding
- Thickened & compressed bowel wall
- Thickened visceral & parietal peritoneum
- ± Central increased attenuation "dot" within inflamed appendage (indicates thrombosed vein)
- Hyperattenuating ring sign: Characteristic finding of EA on postcontrast
  - Pericolonic round fat-containing mass + thin hyperattenuating ring
  - Ring: Thickened visceral peritoneum of inflamed epiploic appendage
  - May calcify when infarcted

DDx: Pericolonic Infiltration

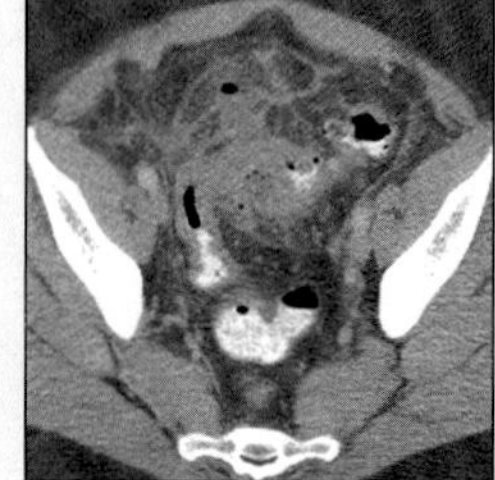

Diverticulitis

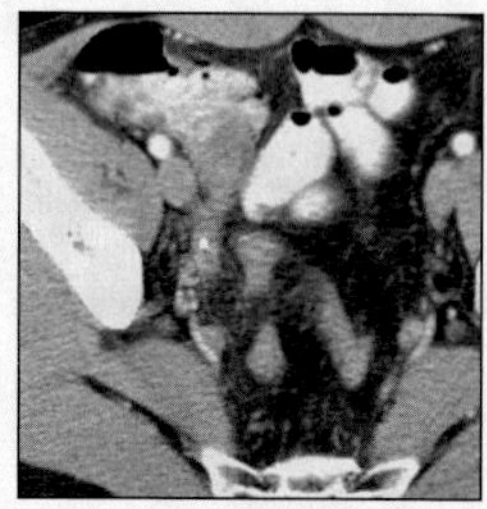

Appendicitis

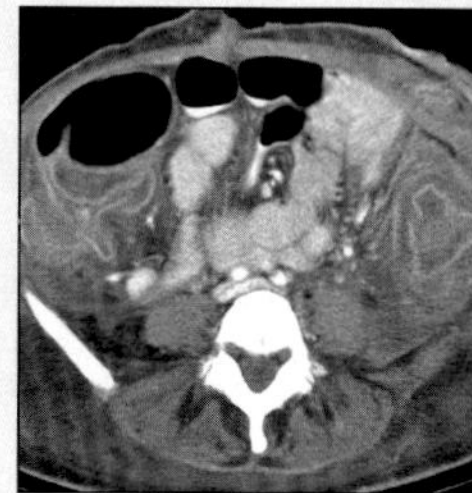

Pseudomem. Colitis

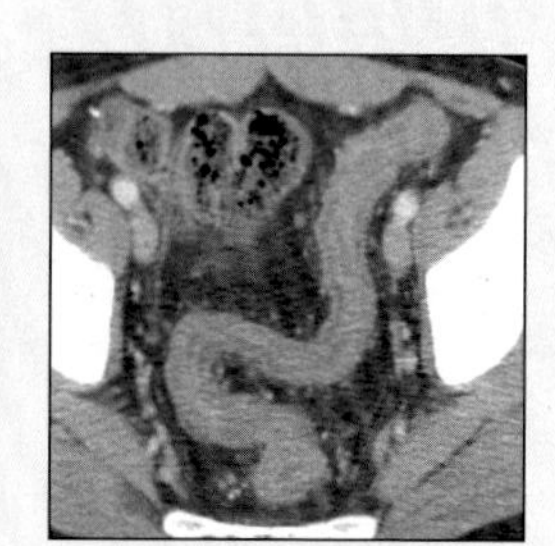

Ulcerative Colitis

EPIPLOIC APPENDAGITIS

Key Facts

Terminology
- Acute inflammation or infarction of epiploic appendages

Imaging Findings
- Best diagnostic clue: Small oval pericolonic fatty nodule with hyperdense ring + surrounding inflammation
- Left lower quadrant > right lower quadrant
- ± Central increased attenuation "dot" within inflamed appendage (indicates thrombosed vein)
- Pericolonic round fat-containing mass + thin hyperattenuating ring
- Infarcted EA: Probably accounts for otherwise unexplained smooth calcified "stones" occasionally found in dependent peritoneal recesses

Top Differential Diagnoses
- Diverticulitis
- Appendicitis
- Pseudomembranous colitis (PMC)
- Ulcerative colitis

Pathology
- Appendages: Small pouches of peritoneum protruding from serosal surface of colon filled with fat + small vessels

Diagnostic Checklist
- Differentiate epiploic appendagitis especially from diverticulitis (LLQ) & appendicitis (RLQ)
- Pericolonic round fatty mass (1-4 cm) with hyperdense rim (most common in rectosigmoid area)
- Not limited to left colon or elderly

- Infarcted EA: Probably accounts for otherwise unexplained smooth calcified "stones" occasionally found in dependent peritoneal recesses

MR Findings
- T1 & T2WI breath-hold spoiled gradient echo (SGE) images
  - Increased signal lesion + hypointense central dot + thin hypointense ring
- T1 C+ fat suppressed gradient echo image
  - Increased enhancement of ring

Ultrasonographic Findings
- Real Time
  - Solid hyperechoic noncompressible ovoid mass adherent to colonic wall
  - Surrounded by a hypoechoic ring (corresponds to $\uparrow$ HU ring on CT scan)

Imaging Recommendations
- Helical CECT

DIFFERENTIAL DIAGNOSIS

Diverticulitis
- Most common complication of diverticulosis
- Barium enema findings
  - Focal eccentric luminal narrowing
  - Marked thickening & distortion of haustral folds
  - Colonic obstruction with zone of transition
  - "Double-track": Intramural fistulous tract
- CT findings
  - Location: Most common in sigmoid colon
  - Bowel wall & fascial thickening, luminal narrowing
  - Pericolonic fat stranding, free fluid & air
  - Pericolic inflammatory changes
    - Abscess, sinuses, fistulas
  - "Arrowhead" sign: Due to diverticular orifice edema
- Clinically simulates epiploic appendagitis

Appendicitis
- Best imaging clue on CT
  - Appendicolith (usually calcified) within distended tubular appendix
- Distended enhancing appendix with surrounding inflammation (fat stranding)
- Wall thickening of cecum or terminal ileum
- Right lower quadrant (RLQ) lymphadenopathy
- In perforated cases
  - Fluid collection most commonly in RLQ or in dependent pelvis (Cul-de-sac)
  - Abscess, small-bowel obstruction
- Ultrasound findings
  - Echogenic appendicolith with posterior shadowing
  - Noncompressible blind-ending tubular structure over 7 mm in diameter
  - Fluid or abscess collection in RLQ
- Right colonic EA clinically may simulate appendicitis

Pseudomembranous colitis (PMC)
- Synonym(s): Antibiotic colitis or C. difficile colitis
- Usually involves entire colon (pancolitis)
- CT findings
  - Colonic wall thickening, nodularity, thumbprinting
  - "Accordion" sign: Represents trapped enteric contrast between thickened colonic folds
  - Ascites common in PMC
  - Full recovery with early diagnosis, discontinuation of offending antibiotic & treatment with metronidazole

Ulcerative colitis
- Pathology: Continuous, not transmural, pseudopolyps, crypt microabscesses
- Classic imaging appearance
  - Pancolitis with decreased haustration & multiple ulcerations on barium enema
- Colorectal narrowing; $\uparrow$ presacral space > 1.5 cm
- "Mucosal islands" or "inflammatory pseudopolyps"
- Diffuse & symmetric wall thickening of colon
- Backwash ileitis: Distal ileum involvement (10-40%)
- Chronic phase
  - "Lead-pipe" colon: Rigid colon with loss of haustra

EPIPLOIC APPENDAGITIS

PATHOLOGY

General Features

- General path comments
  - Appendages: Small pouches of peritoneum protruding from serosal surface of colon filled with fat + small vessels
    - Seen along free tenia & tenia omentalis between cecum & sigmoid colon
- Etiology
  - Torsion & venous thrombosis of appendages
  - Predisposing factors for torsion & infarction of epiploic appendages
    - Precarious blood supply from colic arterial branches
    - Pedunculated morphology; ↑ mobility & obesity
- Epidemiology
  - Though uncommon, not as rare as assumed
  - Seen in 2.3-7.1% of clinically suspected colonic diverticulitis
  - Reported in 1.0% of suspected appendicitis cases

Gross Pathologic & Surgical Features

- Round fat containing paracolic lesion, fat stranding, thickened wall

Microscopic Features

- Visceral peritoneal lining of inflamed epiploic appendage covered with a fibrinoleukocytic exudates
- Fat necrosis within appendage

CLINICAL ISSUES

Presentation

- Most common signs/symptoms
  - Sudden onset of focal abdominal pain
    - Usually left or right lower quadrant
  - Pain worsening with: Coughing, deep breathing, abdominal stretching
  - Symptoms usually subside within one week of onset
  - Physical exam
    - Localized tenderness, some guarding, no rigidity
  - Lab-data
    - WBC count (normal or slightly ↑ in most cases)

Demographics

- Age: 2nd-5th decades (obese people)
- Gender: Equal in both males & females (M = F)

Natural History & Prognosis

- Complications of epiploic appendages
  - Recurrent episodes of inflammation (unusual)
  - Intraperitoneal loose bodies
  - Infarction
- Prognosis
  - Benign self-limiting process with spontaneous resolution within 1 week
  - Good: After medical or surgical treatment

Treatment

- Medical: Conservative treatment with analgesics
- Surgical: Simple ligation & excision of infarcted epiploic appendage
  - Rarely required if accurately diagnosed

DIAGNOSTIC CHECKLIST

Consider

- Differentiate epiploic appendagitis especially from diverticulitis (LLQ) & appendicitis (RLQ)

Image Interpretation Pearls

- Pericolonic round fatty mass (1-4 cm) with hyperdense rim (most common in rectosigmoid area)
- Not limited to left colon or elderly

SELECTED REFERENCES

1. van Breda Vriesman AC: The hyperattenuating ring sign. Radiology. 226(2):556-7, 2003
2. Ghosh BC et al: Primary epiploic appendagitis: diagnosis, management, and natural course of the disease. Mil Med. 168(4):346-7, 2003
3. Chowbey PK et al: Torsion of appendices epiploicae presenting as acute abdomen: laparoscopic diagnosis and therapy. Indian J Gastroenterol. 22(2):68-9, 2003
4. Hollerweger A et al: Primary epiploic appendagitis: sonographic findings with CT correlation. J Clin Ultrasound. 30(8):481-95, 2002
5. Son HJ et al: Clinical diagnosis of primary epiploic appendagitis: differentiation from acute diverticulitis. J Clin Gastroenterol. 34(4):435-8, 2002
6. van Breda Vriesman AC et al: Epiploic appendagitis and omental infarction: pitfalls and look-alikes. Abdom Imaging. 27(1):20-8, 2002
7. Chung SP et al: Primary epiploic appendagitis. Am J Emerg Med. 20(1):62, 2002
8. Sirvanci M et al: Primary epiploic appendagitis: MRI findings. Magn Reson Imaging. 20(1):137-9, 2002
9. Legome EL et al: Epiploic appendagitis: the emergency department presentation. J Emerg Med. 22(1):9-13, 2002
10. Horton KM et al: CT evaluation of the colon: inflammatory disease. Radiographics. 20(2):399-418, 2000
11. Rao PM et al: Case 6: primary epiploic appendagitis. Radiology. 210(1):145-8, 1999
12. Habib FA et al: Laparoscopic approach to the management of incarcerated hernia of appendices epiploicae: report of two cases and review of the literature. Surg Laparosc Endosc. 8(6):425-8, 1998
13. Rao PM et al: Misdiagnosis of primary epiploic appendagitis. Am J Surg. 176(1):81-5, 1998
14. Rao PM et al: Primary epiploic appendagitis: evolutionary changes in CT appearance. Radiology. 204(3):713-7, 1997
15. Rioux M et al: Primary epiploic appendagitis: clinical, US, and CT findings in 14 cases. Radiology. 191(2):523-6, 1994
16. Ghahremani GG et al: Appendices epiploicae of the colon: radiologic and pathologic features. Radiographics. 12(1):59-77, 1992
17. Derchi LE et al: Appendices epiploicae of the large bowel. Sonographic appearance and differentiation from peritoneal seeding. J Ultrasound Med. 7(1):11-4, 1988

EPIPLOIC APPENDAGITIS

IMAGE GALLERY

Typical

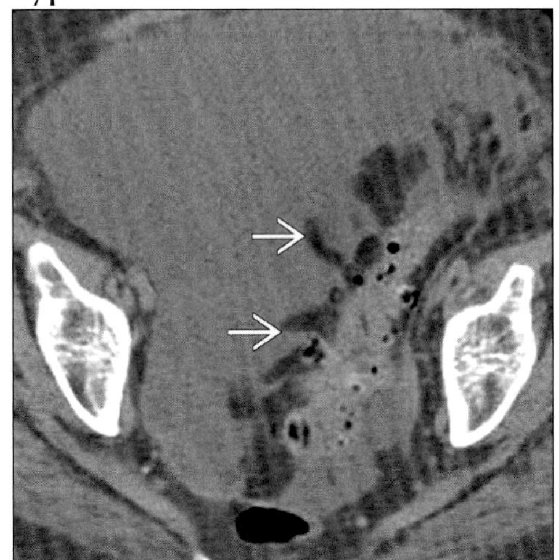

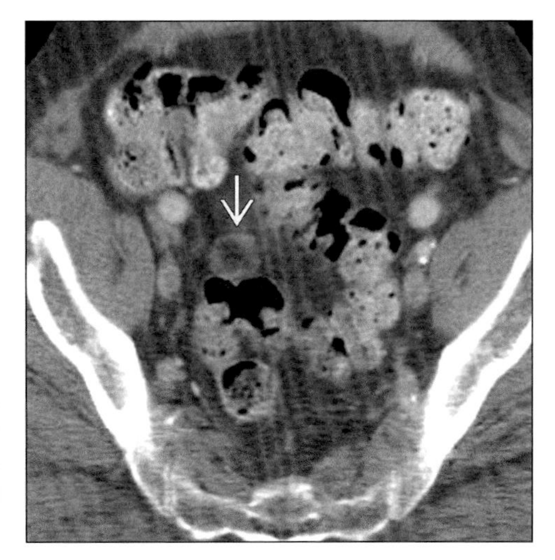

*(**Left**) Axial CECT shows ascites outlining the fat density of normal epiploic appendages (arrows) of the sigmoid colon. (**Right**) Axial CECT in a patient with suspected diverticulitis shows an oval pericolonic fat density nodule (arrow) with a hyperdense ring; epiploic appendagitis.*

Typical

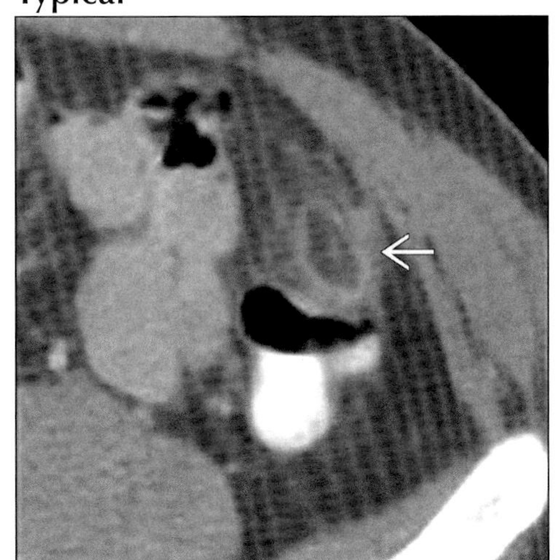

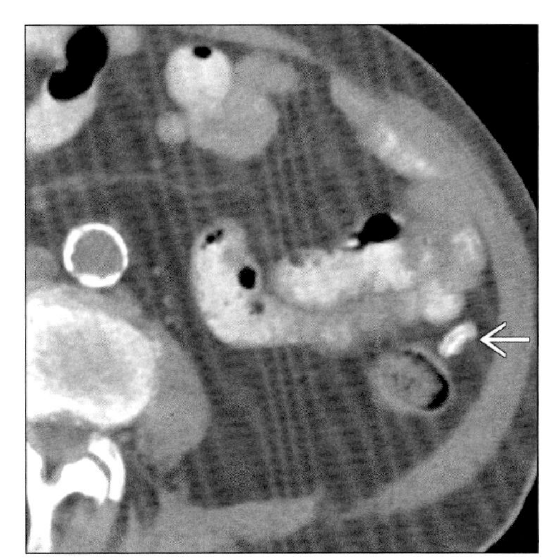

*(**Left**) Axial CECT shows a fat density nodule (arrow) with a hyperdense ring and surrounding inflammation. (**Right**) Axial NECT shows calcified epiploic appendage (arrow) of the descending colon. Such infarcted appendages may detach from the colon and result in loose bodies in the peritoneal cavity.*

Typical

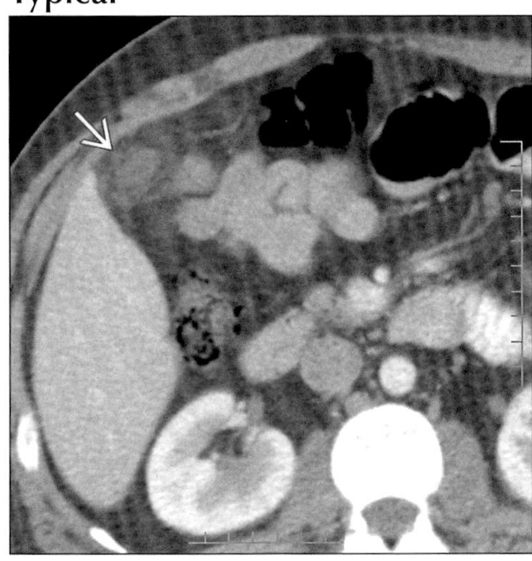

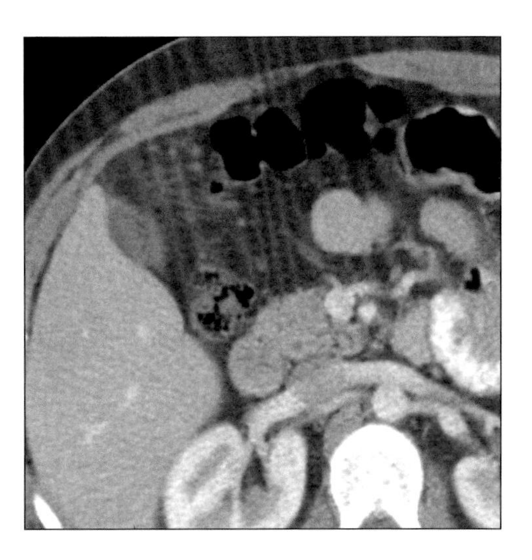

*(**Left**) Axial CECT shows epiploic appendagitis of the hepatic flexure (arrow) with typical findings. (**Right**) Axial CECT shows infiltrated fat near the hepatic flexure just cephalan to the inflamed or infarcted epiploic appendage.*

ISCHEMIC COLITIS

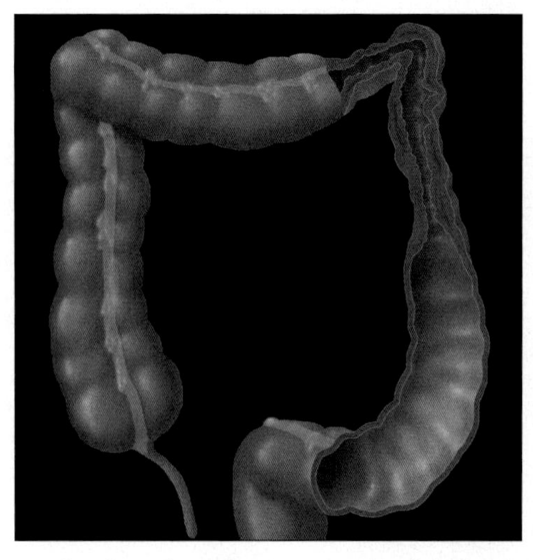

Graphic shows luminal narrowing and wall thickening near the splenic flexure, the "watershed" area between the vascular distribution of the SMA and IMA.

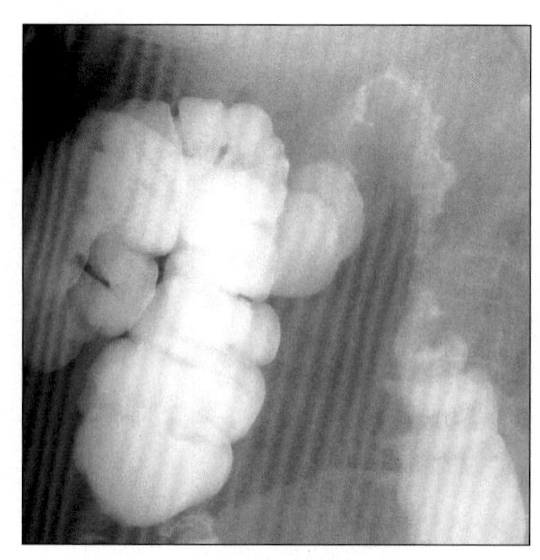

Single-contrast BE shows narrowed lumen of the splenic flexure with "thumbprinting" (thickened haustral folds) due to submucosal edema or hemorrhage. Elderly patient with heart disease.

TERMINOLOGY

Definitions

- Compromise of mesenteric blood supply leading to colonic injury

IMAGING FINDINGS

General Features

- Best diagnostic clue: Evidence of pneumatosis, mesenteric venous gas, symmetric bowel wall thickening or thumbprinting on CT
- Location
  - Commonly watershed segments of colon
    - Splenic flexure: Junction of SMA & IMA (Griffith point)
    - Rectosigmoid: Junction of IMA & hypogastric artery (Sudeck point)
    - Left-side colon: Typical in elderly with ↓ perfusion
    - Right-side: Young patients (↓ collateral blood supply); chronic renal failure
- Other general features
  - Most common vascular disorder of GI tract
  - Most common cause of colitis in elderly & is often self limiting
  - Major predisposing cause in elderly: Nonocclusive vascular disease (hypoperfusion)
    - Hemorrhagic, septic or hypovolemic shock
    - Congestive heart failure (CHF); drugs like digitalis
  - 20% Colonic ischemia are proximal to obstruction
    - Colon cancer, volvulus, closed loop obstruction
  - Common cause of abdominal pain in elderly with history of heart disease
  - Most common form is segmental (90%) or pancolitis
  - Usually a partial mural (nontransmural) & superficial mucosal ischemia
  - Spectrum of diseases caused by colonic ischemia
    - Reversible or transient ischemic colitis (> frequent)
    - Colonic stricture, gangrene of colon & perforation

Radiographic Findings

- Radiography
  - Plain x-ray abdomen (supine view)
    - Normal or nonspecific ileus
    - Thumbprinting (submucosal edema or bleeding)
    - Luminal narrowing or transverse ridging (spasm)
    - Ahaustral loops (rare)
- Fluoroscopic guided barium enema
  - Hallmark: Serial change on studies performed over days, weeks or months
  - Thumbprinting (usually within 24 hrs after insult)
    - Smooth, round, polypoid scalloped filling defects along lumen (submucosal edema or hemorrhage)
    - Most consistent & characteristic finding-75% cases
    - Occurs within first 24 hrs, resorbs in less than a week or may persist for weeks

DDx: Wall Thickening; Pericolonic Infiltration

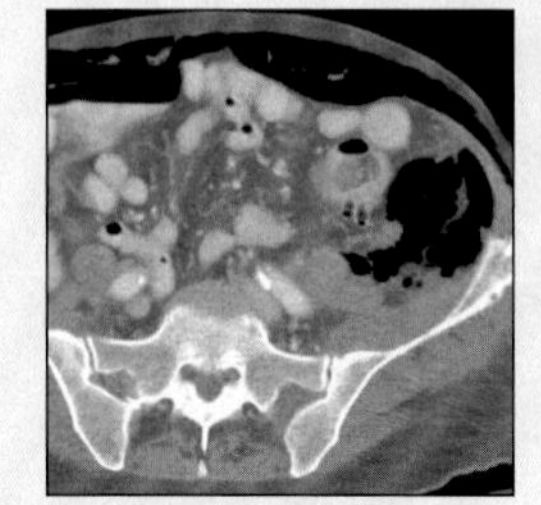

Diverticulitis

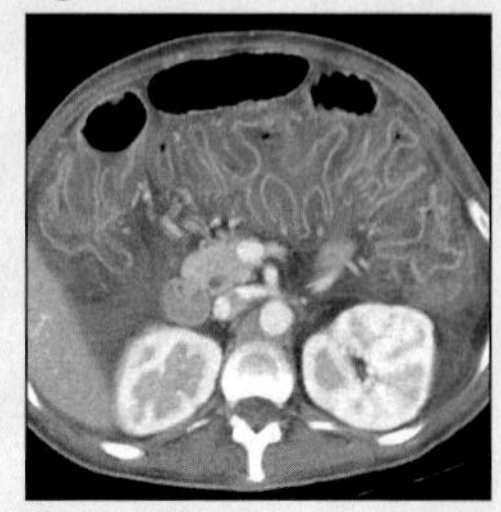

Pseudomem. Colitis

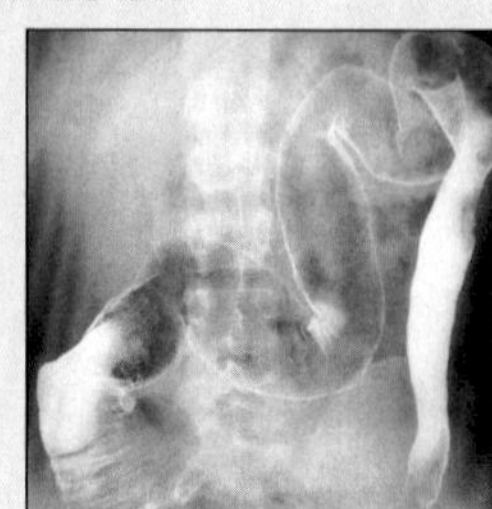

Ulcerative Colitis

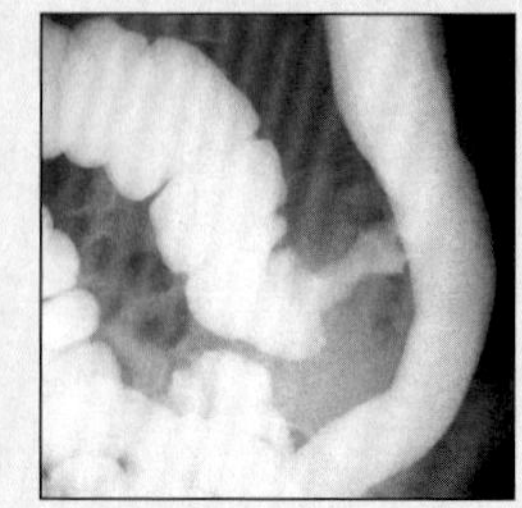

Colon Cancer

ISCHEMIC COLITIS

Key Facts

Terminology
- Compromise of mesenteric blood supply leading to colonic injury

Imaging Findings
- Best diagnostic clue: Evidence of pneumatosis, mesenteric venous gas, symmetric bowel wall thickening or thumbprinting on CT
- Commonly watershed segments of colon
- Thumbprinting (usually within 24 hrs after insult)
- Ulceration: Sloughing of mucosa (46-60% cases)
- Bowel wall thickening (normal range 3-5 mm)
- Hypoattenuation: Submucosal or diffuse edema
- Hyperattenuation: Submucosal or diffuse bleeding
- ± Pneumatosis
- ± Portomesenteric venous gas

Top Differential Diagnoses
- Diverticulitis
- Pseudomembranous colitis (PMC)
- Ulcerative colitis (UC)
- Granulomatous colitis (Crohn disease)
- Colon carcinoma

Pathology
- Nonocclusive vascular disease (in elderly people)
- Hypoperfusion: Predisposing factors
- Hypotensive episodes: Hemorrhagic, cardiogenic or septic shock
- CHF, arrhythmia, drugs, trauma

Diagnostic Checklist
- Check for history of cardiac, bowel, renal problems & hypotensive medication use in elderly people

    - May also seen in other inflammatory bowel diseases or infectious colitides
  - Transverse ridging: Less common finding-13% cases
    - Parallel, symmetric thickened folds running perpendicular to bowel lumen
    - Caused by edema or spasm; early finding & usually resolves rapidly
  - Ulceration: Sloughing of mucosa (46-60% cases)
    - Longitudinal/discrete; superficial/deep; small/large
    - Usually develop 1-3 weeks after onset of disease
  - Intramural barium: Unusual (sloughing of necrotic portion of wall → tracking of barium intramurally)
  - Stricture: 12% cases heal with stricture formation

CT Findings
- NECT
  - Bowel wall thickening (normal range 3-5 mm)
    - Circumferential, symmetric wall thickening ± thumbprinting
    - Due to submucosal edema or hemorrhage
  - Bowel wall attenuation
    - Hypoattenuation: Submucosal or diffuse edema
    - Hyperattenuation: Submucosal or diffuse bleeding
    - Heterogeneous: Outer serosa & muscular layers
  - ± Luminal narrowing or dilatation & air-fluid levels
  - Loss of haustral pattern (rare); pericolic streakiness; paracolic fluid collections
  - ± Pneumatosis
    - Small gas bubbles within ischemic bowel wall
    - Circumferential or band like pneumatosis
  - ± Portomesenteric venous gas
    - Portal venous gas collects in periphery of liver
- CECT
  - Double halo or target sign: Concentric layers of low & high attenuation
    - Enhancement of mucosa & serosa (hyperemia or hyperperfusion during recovery)
    - Nonenhancement of submucosa (due to submucosal edema or hemorrhage)
  - ± Thrombus within splanchnic vessels

Ultrasonographic Findings
- Color Doppler
  - Hypoechoic thickening of bowel wall
  - Absence of arterial flow in wall of ischemic colon

Angiographic Findings
- Usually not helpful in diagnosis
  - Ischemic colitis: Usually nonocclusive ischemia

Imaging Recommendations
- Helical NE + CECT; plain x-ray abdomen
- Single contrast barium enema (for chronic disease)

DIFFERENTIAL DIAGNOSIS

Diverticulitis
- Most common complication of diverticulosis
- Barium enema findings
  - Focal eccentric luminal narrowing
  - Marked thickening & distortion of haustral folds
  - Colonic obstruction with zone of transition
  - "Double-tracking": Longitudinal intramural fistulous tract
- CT findings
  - Location: Most common in sigmoid colon
  - Bowel wall & fascial thickening; fat stranding; free fluid & air
  - Pericolic inflammatory changes
    - Abscess, sinuses, fistulas
  - "Arrowhead" sign: Due to diverticular orifice edema
  - Focal area of eccentric luminal narrowing

Pseudomembranous colitis (PMC)
- Synonym(s): Antibiotic colitis or C. difficile colitis
- Usually involves entire colon (pancolitis)
- Barium enema (contraindicated in acutely ill)
  - Small, irregular plaques on mucosal surface
    - Represent pseudomembranes
  - Small, subtle elevated, round nodules
- Single contrast study: Shows thumbprinting indistinguishable from ischemic colitis
- CT findings
  - Colic wall thickening & nodularity
  - "Accordion" sign: Represents trapped enteric contrast between thickened colonic folds

- Ascites common in PMC
- Full recovery with early diagnosis, discontinuation of offending antibiotic & treatment with metronidazole

Ulcerative colitis (UC)

- Pathology: Continuous, not transmural, pseudopolyps, crypt microabscesses
- Classic imaging appearance
  - Pancolitis with decreased haustration & multiple ulcerations on barium enema
- Colorectal narrowing; ↑ presacral space > 1.5 cm
- "Mucosal islands" or "inflammatory pseudopolyps"
- Diffuse & symmetric wall thickening of colon
  - Ischemic colitis usually shows segmental (watershed areas) bowel wall thickening & thumbprinting
- Backwash ileitis: Distal ileum involvement (10-40%)
- Chronic phase
  - "Lead-pipe" colon: Rigid colon with loss of haustra

Granulomatous colitis (Crohn disease)

- Barium enema findings
  - Cobblestoning: Longitudinal & transverse ulceration produce a paving stone appearance
  - Segmental in distribution
    - Involve both colon & small-bowel (60% cases)
    - Isolated to colon (20% cases)
  - Transmural, skip lesions, sinuses, fissures, fistulas
- CT findings
  - Bowel wall thickening (1-2 cm)
  - "Creeping fat" or mesenteric fibrofatty proliferation
  - Enlarged mesenteric lymph nodes
  - "Comb" sign: Hypervascularity (active disease)

Colon carcinoma

- Asymmetric mural thickening with irregular surface
- Classic annular "apple core" lesion
  - Circumferential bowel narrowing + mucosal destruction with shelf-like, overhanging borders
  - High grade obstruction + ischemia shows proximal bowel dilatation with thumbprinting
- Extracolonic tumor extension
  - Strands of soft tissue: Serosal surface → pericolic fat
  - Loss of fat planes between colon & adjacent muscles

PATHOLOGY

General Features

- General path comments
  - Normal mesenteric vascular anatomy
    - Superior mesenteric artery (SMA): Vascular supply from 3rd part of duodenum to splenic flexure
    - Inferior mesenteric artery (IMA): Splenic flexure to rectum
- Etiology
  - Nonocclusive vascular disease (in elderly people)
  - Hypoperfusion: Predisposing factors
    - Hypotensive episodes: Hemorrhagic, cardiogenic or septic shock
    - CHF, arrhythmia, drugs, trauma
    - Arteriosclerotic disease, chronic renal failure
    - Vasculitis, colonic obstruction
- Epidemiology: Mortality rate: 7% of cases

Gross Pathologic & Surgical Features

- Segmental or focal; localized or diffuse
- Thick bowel wall; dark red or purple
  - Edematous, hemorrhagic, ulcerated

Microscopic Features

- Mucosal erosions, ulceration, necrosis
- Submucosal edema, hemorrhage

CLINICAL ISSUES

Presentation

- Most common signs/symptoms
  - Mild or severe abdominal pain
  - Rectal bleeding, bloody diarrhea, hypotension
- Lab-data
  - ↑ Leukocytosis; positive guaiac stool test
  - Negative blood cultures; EKG changes may be seen

Demographics

- Age: Usually elderly age group (> 50 years)
- Gender: Equal in both males & females (M = F)

Natural History & Prognosis

- Complications
  - Transmural bowel infarction → perforation → death
- Prognosis
  - Partial mural ischemia: Good prognosis
  - Transmural infarction: Poor prognosis

Treatment

- Partial mural ischemia (nonocclusive type)
  - Conservative medical treatment
- Transmural infarction: Surgical resection

DIAGNOSTIC CHECKLIST

Consider

- Check for history of cardiac, bowel, renal problems & hypotensive medication use in elderly people

Image Interpretation Pearls

- Segmental bowel wall thickening in watershed areas, thumbprinting, pneumatosis, portal venous gas

SELECTED REFERENCES

1. Wiesner W et al: CT of acute bowel ischemia. Radiology. 226(3):635-50, 2003
2. Horton KM et al: Volume-rendered 3D CT of the mesenteric vasculature: normal anatomy, anatomic variants, and pathologic conditions. Radiographics. 22(1):161-72, 2002
3. Horton KM et al: Multi-detector row CT of mesenteric ischemia: can it be done? Radiographics. 21(6):1463-73, 2001
4. Horton KM et al: CT evaluation of the colon: inflammatory disease. Radiographics. 20(2):399-418, 2000
5. Balthazar EJ et al: Ischemic colitis: CT evaluation of 54 cases. Radiology. 211(2):381-8, 1999
6. Iida M et al: Ischemic colitis: serial changes in double-contrast barium enema examination. Radiology. 159(2):337-41, 1986

IMAGE GALLERY

Typical

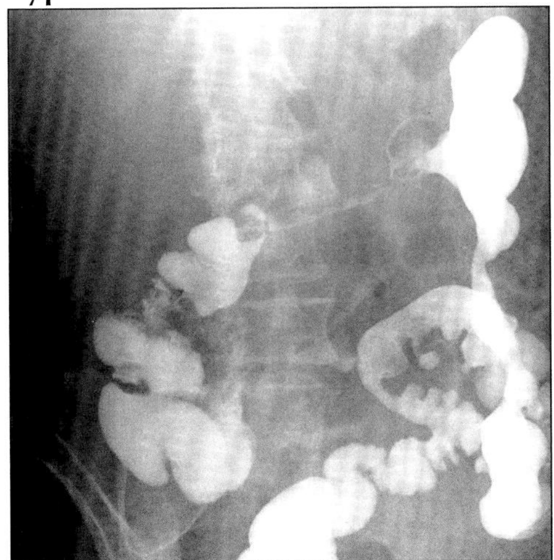

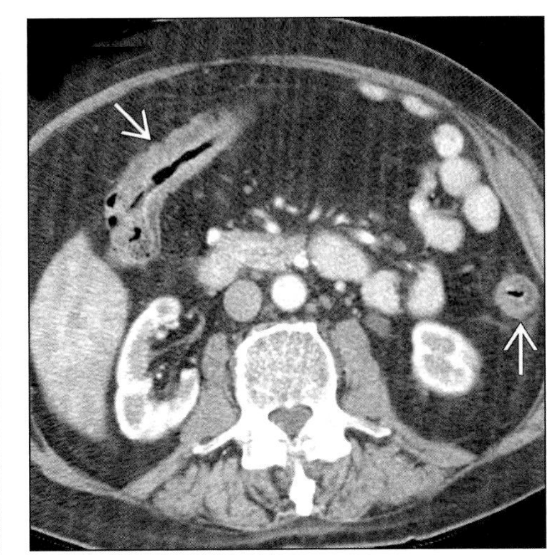

*(**Left**) Single-contrast BE in a 60 year old man with chronic heart disease, shows strictures of distal transverse + proximal descending colon due to subacute colonic ischemia. (**Right**) Axial CECT in a 60 year old patient with subacute colonic ischemia shows wall thickening (arrows), submucosal edema, and luminal narrowing of the colon.*

Variant

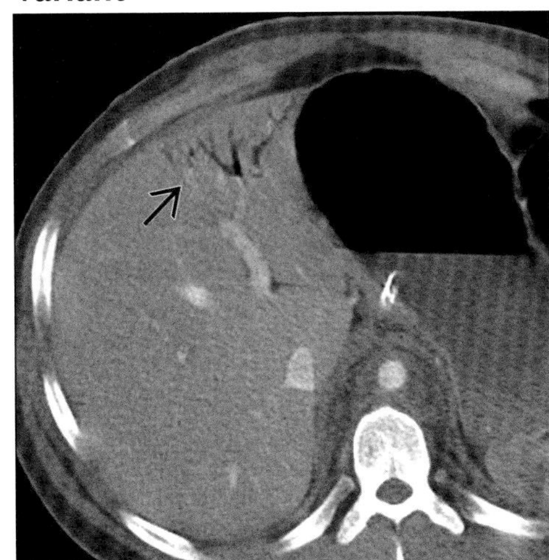

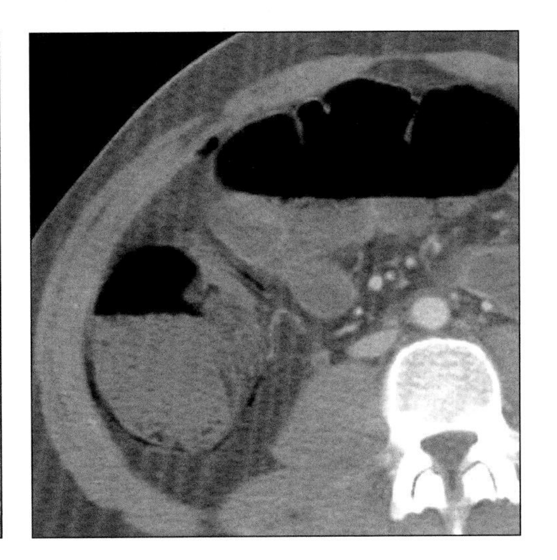

*(**Left**) Axial CECT of a patient 24 hours post abdominal trauma (motor vehicle crash) shows portal venous gas (arrow). (**Right**) Axial CECT shows intramural and mesenteric venous gas. At surgery, patient had "degloving" injury (serosal tear + devascularization) with cecal infarction.*

Variant

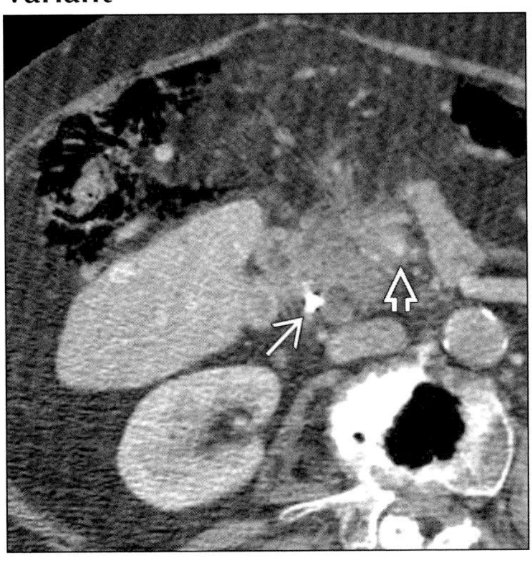

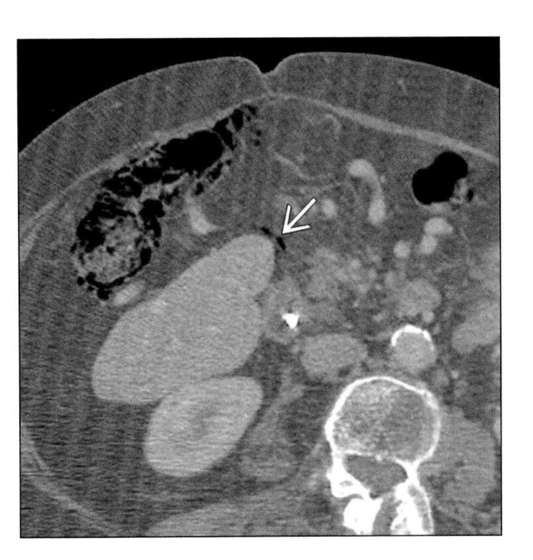

*(**Left**) Axial CECT shows a mass in the pancreatic head with a biliary stent (arrow). The superior mesenteric artery + vein (open arrow) are encased and narrowed. Gas is present in the colon wall. (**Right**) Axial CECT in patient with pancreatic cancer. Intramural + mesenteric venous (arrow) gas are present due to colon infarction.*

COLONIC POLYPS

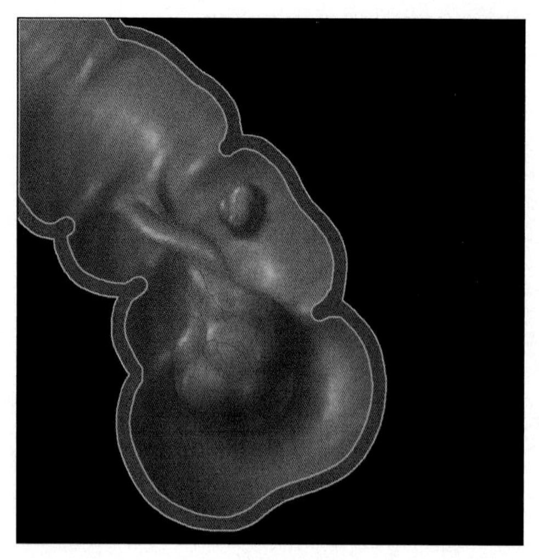

Graphic shows tubulovillous adenoma on a long stalk and a small sessile polyp.

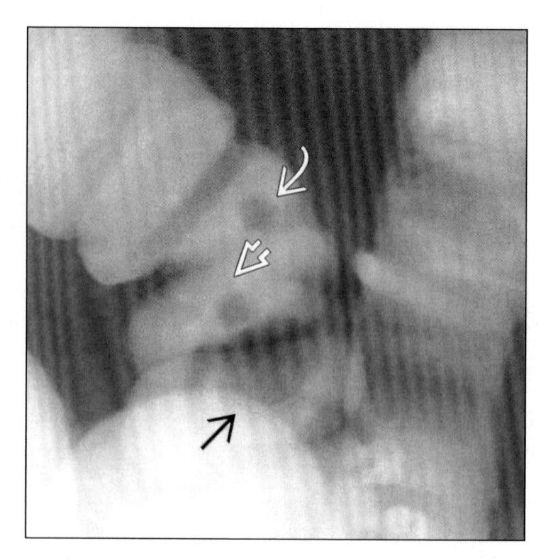

Single contrast BE shows tubulovillous adenoma with a large "head" (arrow) and a long stalk (open arrow). Small sessile polyp (curved arrow) also noted.

TERMINOLOGY

Definitions

- A protruding, space-occupying lesion within the colonic lumen

IMAGING FINDINGS

General Features

- Best diagnostic clue: Radiolucent filling defect, contour defect or ring shadow
- Location: Cecum (4%); ascending colon (6%); hepatic flexure (4%); transverse (2%); splenic flexure (8%); descending (20%); sigmoid (41%); rectum (23%)
- Morphology
  - Sessile polyps: Broad base with little or no stalk
  - Pedunculated polyps: Arise from narrow stalk
- Other general features
  - 2 Types of colon polyps
    - Neoplastic: Adenomatous (tubular, tubulovillous & villous)
    - Non-neoplastic: Hyperplastic, hamartomatous and inflammatory

Radiographic Findings

- Fluoroscopic-guided double contrast barium enema
  - Sessile polyps
    - Dependent wall: Radiolucent filling defect
    - Nondependent wall: Ring shadow with barium-coated white rim
    - "Bowler hat" sign: Dome of hat points toward lumen of bowel (en face view); brim and dome of hat represents base and head of polyp
  - Pedunculated polyps
    - "Mexican hat" sign: Characterized by a pair of concentric rings; outer and inner ring represents head and stalk of polyp
  - Tubular adenomatous polyps
    - Small size; pedunculated polyps
    - Minor degree of villous changes
  - Tubulovillous adenomatous polyps
    - Medium size; sessile polyps
    - Fine nodular or reticular surface pattern
    - Filling of barium within interstices of adenoma
  - Villous adenomatous polyps
    - Larger size; sessile polyps
    - Barium trapped between frond-like projections → polypoid lesion with granular or reticular pattern
    - "Carpet" lesion: Flat, lobulated; localized or diffuse
  - "Carpet" lesion
    - Location: Rectum, cecum & ascending colon
    - Subtle alteration in surface texture of colon with little or no protrusion into lumen
    - Irregular contour in contrast to smooth, fine contour of adjacent normal bowel (profile view)

DDx: Filling Defect in Colon

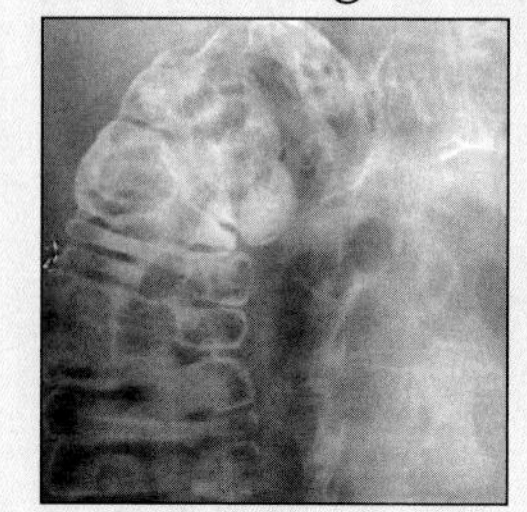

Feces

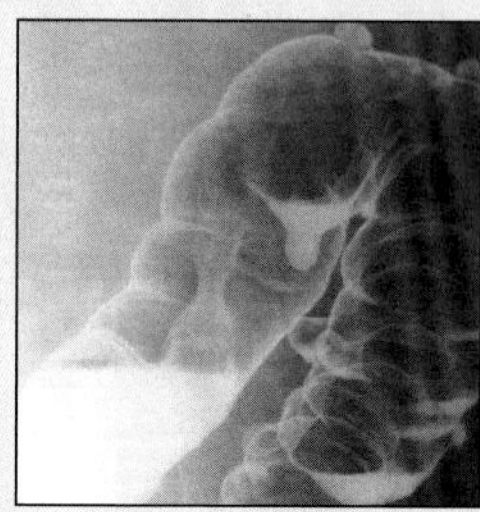

Divertics. + Polyp

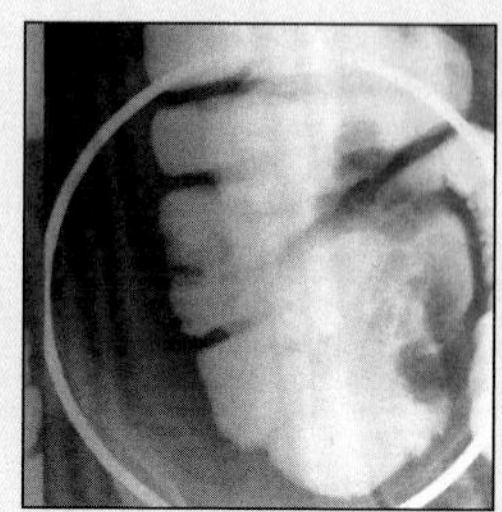

Cancer + Lipoma

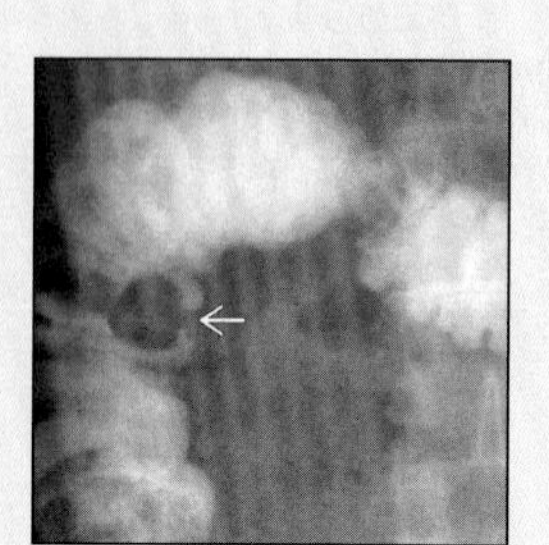

Spasm + Feces

COLONIC POLYPS

Key Facts

Terminology
- A protruding, space-occupying lesion within the colonic lumen

Imaging Findings
- Best diagnostic clue: Radiolucent filling defect, contour defect or ring shadow
- Sessile polyps: Broad base with little or no stalk
- Pedunculated polyps: Arise from narrow stalk
- "Carpet" lesion: Flat, lobulated; localized or diffuse
- Best imaging tool: Air-contrast barium enema

Top Differential Diagnoses
- Retained fecal debris
- Colonic diverticula
- Colon carcinoma
- Intramural mass

Pathology
- Spectrum of adenoma: Tubular ↔ tubulovillous ↔ villous
- Adenoma-carcinoma sequence (7-10 years): Benign adenoma → malignant transformation

Clinical Issues
- Asymptomatic (75%)
- Colonoscopic polypectomy if polyps > 1 cm
- Colonoscopy or fluoroscopic-guided double contrast barium enema for periodic surveillance

Diagnostic Checklist
- Family history of colonic polyps & colon carcinoma
- Polypectomy if changes noted on follow-up imaging
- If patient has known diverticulosis, single contrast barium enema is easier for polyp detection

---

        - Tiny, coalescent nodules and plaques → finely nodular or reticular pattern with sharply demarcated border (en face view)
    - Hyperplastic polyps
        - Location: Rectosigmoid colon
        - Smooth round sessile nodules; < 5 mm (common)
        - Lobulated or pedunculated; > 1 cm (occasional)
    - Hamartomatous polyps
        - Multiple, scattered radiolucent filling defects
        - Vary in size; no "carpet" lesion
    - Inflammatory polyps
        - Islands of elevated, inflamed, edematous mucosa surrounded by ulceration (inflammatory)
        - Small & round, long & filiform or bush-like; simulate villous adenoma (postinflammatory)

CT Findings
- CT "virtual colonoscopy"
    - Small or large, sessile or pedunculated lesions extending from colonic wall
    - Polyps ≥ 10 mm: Sensitivity 90%
    - Adenoma ≥ 10 mm: Sensitivity 94%
    - Advantages: Shorter procedural time, ↓ risk to patient and no IV sedation

Ultrasonographic Findings
- Real Time Transrectal Ultrasonography
    - Determine depth of invasion by a sessile polyp

Imaging Recommendations
- Best imaging tool: Air-contrast barium enema
- Protocol advice
    - Patient rotated 180° or in upright position
        - Confirm presence of a pedunculated polyp
        - Visualize stalk in profile view

DIFFERENTIAL DIAGNOSIS

Retained fecal debris
- Mobile & on dependent surface in barium pool
- Inconsistent location; irregular configuration; impregnated with barium
- Adherent stool can be difficult to differentiate; repeat fluoroscopic-guided barium enema
- Proper cleansing of bowel can reduce confusion

Colonic diverticula
- "Bowler hat" sign: Dome of hat points away from lumen of bowel
- Nondependent wall: Ring shadow with barium-coated white rim (en face view); simulates polyps
    - Rotate patient 90° to see outpouchings from wall versus protrusion into lumen (profile view)
- Inverted diverticula can be difficult to differentiate

Colon carcinoma
- Sessile or pedunculated polyps seen in early cancer
- Biopsy is necessary to differentiate

Intramural mass
- Example: Leiomyoma, lipoma
- Leiomyoma
    - Filling defect mimics villous adenoma (en face view)
    - Abrupt well-defined borders of bowel wall
- Lipoma
    - Commonly arises near ileocecal valve
    - Soft + deformable with compression
    - CT diagnostic with fat density
- Usually single; polyps often multiple
- Biopsy is necessary for diagnosis

PATHOLOGY

General Features
- General path comments
    - Neoplastic colonic polyps
        - From proliferative dysplasia → adenoma
        - Slow growing (doubling every 10 years)
        - Single or multiple (more common)
        - Spectrum of adenoma: Tubular ↔ tubulovillous ↔ villous
    - Tubular adenoma
        - 80-86% of neoplastic polyps (most common)
        - > 80% of glands are branching, tubule type
    - Tubulovillous adenoma

COLONIC POLYPS

- 8-16% of neoplastic polyps
- Villous adenoma
  - 3-16% of neoplastic polyps
  - > 80% of glands are villiform (shaggy surface)
- Non-neoplastic colonic polyps
  - From abnormal mucosal maturation, architecture or inflammation
  - 90% of all epithelial polyps
  - Small; occur at distal colon
- Hyperplastic polyps
  - Almost never undergo malignant degeneration
- Hamartomatous polyps
  - Varied polyp appearances & wide-range of ages, depends on etiology
- Inflammatory polyps
  - Also known as "pseudopolyps"
  - 2 Types: Inflammatory and postinflammatory
  - Postinflammatory: Mucosal healing → overgrowth

- Etiology
  - Family history
    - Adenomatous polyps (e.g., hereditary nonpolyposis colorectal cancer syndrome, familial polyposis, Gardner syndrome & Turcot syndrome)
    - Hamartomatous polyps (e.g., Peutz-Jeghers syndrome and juvenile polyposis)
  - Acquired
    - Adenomatous polyps (e.g., sporadic adenoma)
    - Hyperplastic polyps
    - Hamartomatous polyps (e.g., Cronkhite-Canada syndrome)
    - Inflammatory polyps (e.g., ulcerative colitis)
  - Risk factors: Diet, alcohol, smoking and obesity
  - Pathogenesis
    - Adenoma: Precursor to colon carcinoma
    - Adenoma-carcinoma sequence (7-10 years): Benign adenoma → malignant transformation
- Epidemiology
  - Incidence of colon polyps: 3% in third decade; 5% in fourth; 7% in fifth; 11% in sixth; 10% in seventh; 18% in eighth; 26% in ninth
  - ↑ Age → incidence of polyps shifts to right colon
  - Hyperplastic polyps increases with age
- Associated abnormalities
  - Colon carcinoma (adenocarcinoma)
    - Polyps < 1 cm: 1% adenocarcinoma
    - Polyps 1-2 cm: 10-20% adenocarcinoma
    - Polyps > 2 cm: 40-50% adenocarcinoma
    - ↑ Villous changes or "carpet" lesion → ↑ risk
    - Lobulated contour or a basal indentation → ↑ risk
    - Tubular adenoma: < 1 cm: 1% with cancer; 1-2 cm: 10%; > 2 cm: 35%
    - Tubulovillous adenoma: < 1 cm: 4% with cancer; 1-2 cm: 7%; > 2 cm: 46%
    - Villous adenoma: < 1 cm: 10% with cancer; 1-2 cm: 10%; > 2 cm: 53%

Gross Pathologic & Surgical Features

- Tubular adenoma: Thin stalk and tufted head
- Villous adenoma: "Cauliflower-like" with broad base

Microscopic Features

- Adenomatous polyps
  - Tubular, tubulovillous or villous structure lined by columnar epithelium
  - Tubular: Tubular glands with smooth surface
  - Tubulovillous: Mixture of tubular & villous
  - Villous: Surface consists of frondlike structures
  - ± Cellular atypia, mitosis or loss of normal polarity
- Hyperplastic polyps
  - Colonic crypts are elongated and epithelial cells assume papillary configuration
  - No cytologic atypia; epithelium is well-differentiated

CLINICAL ISSUES

Presentation

- Most common signs/symptoms
  - Asymptomatic (75%)
  - Lower abdominal pain, rectal bleeding and diarrhea

Demographics

- Age
  - Adenomatous polyps: 24-47% > 50 years of age
  - Hyperplastic polyps: 50% > 60 years of age
- Gender
  - Adenomatous polyps
    - M:F = 2:1

Natural History & Prognosis

- Good, after resect benign or carcinoma in situ polyps
- Poor, with invasive colon carcinoma

Treatment

- Colonoscopic polypectomy if polyps > 1 cm
  - Completely resect villous adenoma or "carpet" lesion
  - In patients with neoplastic polyps caused by genetic mutations, prophylactic colectomy is required prior to malignant transformation
- Follow-up (20% recur at 5 years; 50% recur at 15 years)
  - Colonoscopy or fluoroscopic-guided double contrast barium enema for periodic surveillance

DIAGNOSTIC CHECKLIST

Consider

- Family history of colonic polyps & colon carcinoma
- Polypectomy if changes noted on follow-up imaging

Image Interpretation Pearls

- If patient has known diverticulosis, single contrast barium enema is easier for polyp detection

SELECTED REFERENCES

1. Yee J et al: Colorectal neoplasia: Performance characteristics of CT colonography for detection in 300 patients. Radiology 219: 685-92, 2001
2. Macari M et al: Comparison of time-efficient CT colonography with two and three-dimensional colonic evaluation for detecting colorectal polyps. AJR 174: 1543-9, 2000
3. Levine MS et al: Diagnosis of colorectal neoplasms at double-contrast barium enema examination. Radiology 216: 11-8, 2000

IMAGE GALLERY

Typical

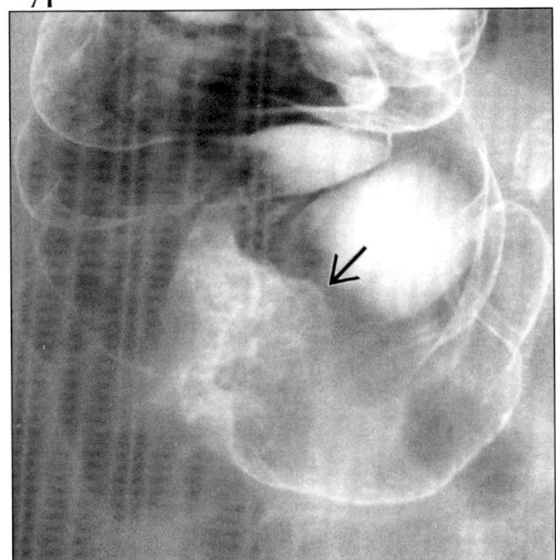

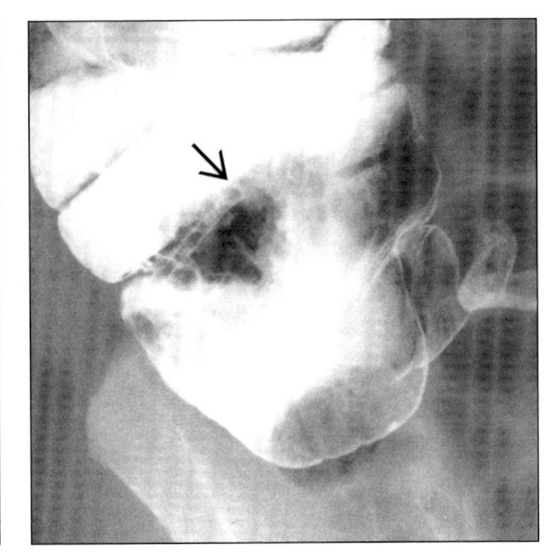

(Left) *Air contrast BE shows a large sessile polyp (arrow) in the cecum; villous adenoma.* ***(Right)*** *Single contrast BE shows a cauliflower-like polypoid cecal mass (arrow); villous adenoma.*

Typical

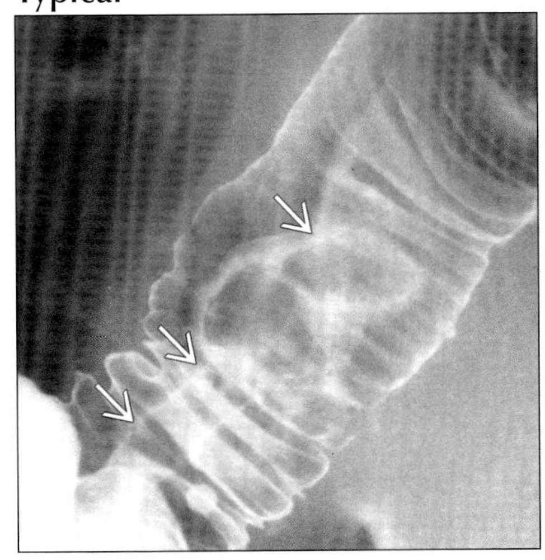

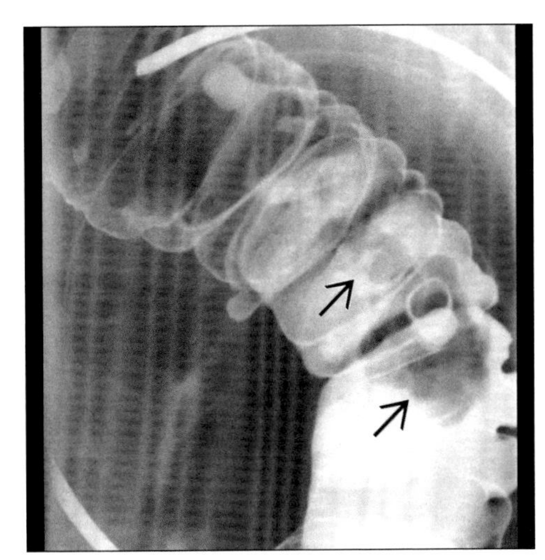

(Left) *Air contrast BE shows a large pedunculated polyp (arrows) in sigmoid colon; tubulovillous adenoma.* ***(Right)*** *Air contrast BE shows numerous diverticula; also a large pedunculated polyp on a stalk (arrows); tubulovillous adenoma.*

Typical

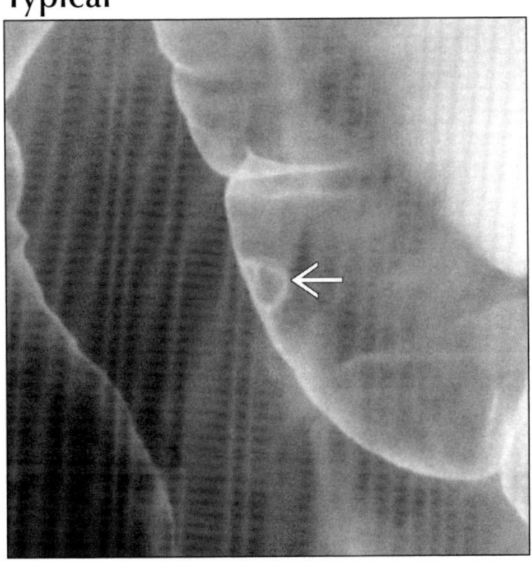

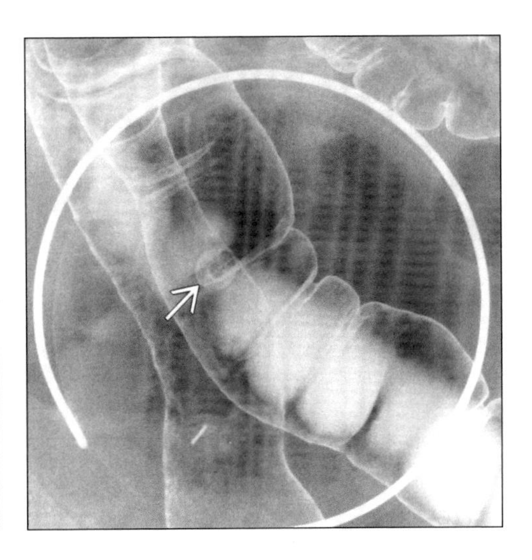

(Left) *Air contrast BE shows small sessile tubular adenoma (arrow). The dome of the "bowler hat" points toward the colonic lumen.* ***(Right)*** *Air contrast BE shows a small polyp on a short stalk (arrow). The outer rim of the "Mexican hat" is the head of the polyp; the inner ring is the stalk.*

COLON CARCINOMA

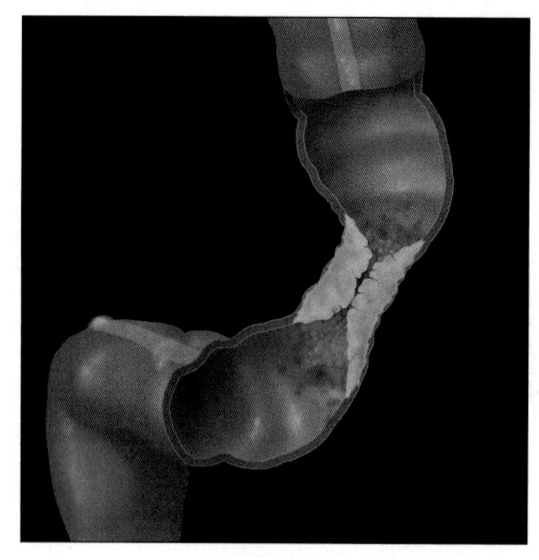

Graphic shows "apple core" constricting tumor of sigmoid colon with circumferential narrowing of the lumen and a nodular tumor surface.

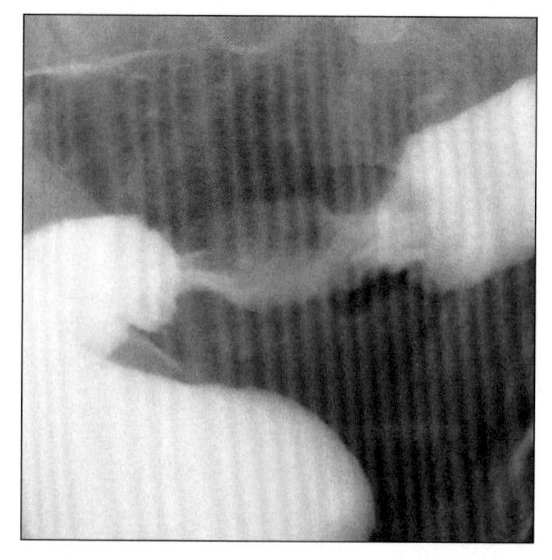

Single contrast BE shows classic "apple core" lesion of colon. There is a short segment irregular, circumferential narrowing of the lumen with destroyed mucosa and nodular "shoulders".

TERMINOLOGY

Definitions

- Malignant transformation of colon mucosa

IMAGING FINDINGS

General Features

- Best diagnostic clue: Short segment luminal wall thickening
- Location: Cecum (10%); ascending colon (15%); transverse colon (15%); descending colon (5%); sigmoid colon (25%); rectosigmoid colon (10%); rectum (20%)
- Morphology
  - Early cancer: Sessile or pedunculated tumors
  - Advanced cancer: Annular, semiannular, polypoid or carpet tumors
- Other general features
  - Radiology is critical for screening, diagnosis, treatment and follow-up of colon cancer
  - Screening: Fluoroscopic-guided double contrast barium enema or CT "virtual colonoscopy" are comparable to colonoscopy for cancer detection

Radiographic Findings

- Fluoroscopic-guided double contrast barium enema
  - Early cancer: Sessile (plaquelike) lesion
    - Typical early colon cancer
    - Flat, protruding lesion with a broad base and little elevation of mucosa (in profile view)
    - Discrete borders and shallow central ulcers (in profile view)
    - Curvilinear or undulating lines (in en face view)
  - Early cancer: Pedunculated lesion
    - Short and thick polyp stalk
    - Irregular or lobulated head of polyp
  - Advanced cancer: Polypoid lesion (large)
    - Dependent wall: Filling defect in barium pool
    - Nondependent wall: Etched in white
  - Advanced cancer: Semi-annular (saddle) lesion
    - Transition to annular carcinoma: Polypoid → semi-annular → annular
    - Convex barium-etched margins (in profile view)
  - Advanced cancer: Annular (apple-core) lesion
    - Circumferential narrowing of bowel; Shelf-like, overhanging borders (mucosal destruction)
    - High-grade obstruction and ischemia: Thumbprinting of dilated proximal colon
  - Advanced cancer: Carpet lesion
    - Malignant villous tumor may appear as carpet lesion with minimal protrusion into lumen
    - Radiolucent nodules surrounded by barium-filled grooves; finely nodular or reticular pattern

DDx: Luminal Narrowing, Wall Thickening of Colon

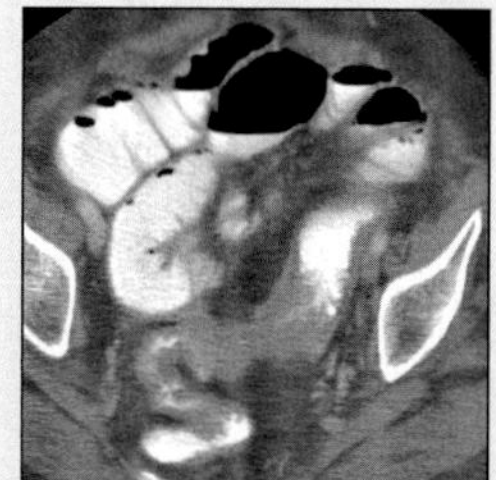

Diverticulitis

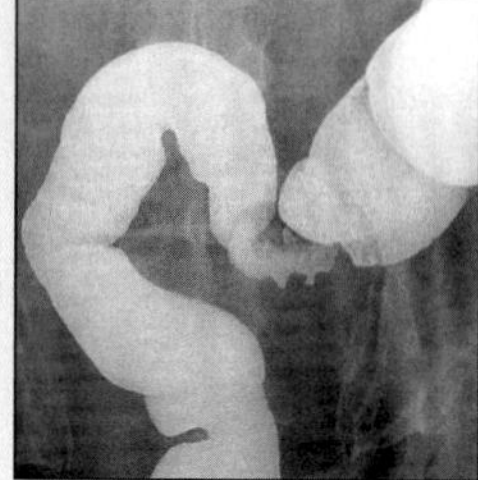

Ischemic Stricture

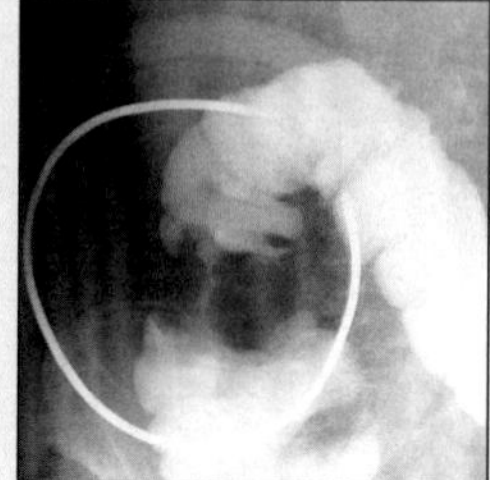

Tuberculosis

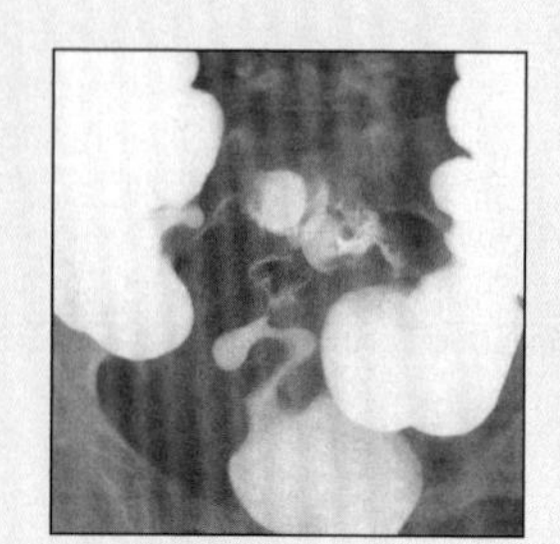

Endometriosis

COLON CARCINOMA

Key Facts

Imaging Findings
- Best diagnostic clue: Short segment luminal wall thickening
- Early cancer: Sessile or pedunculated tumors
- Advanced cancer: Annular, semiannular, polypoid or carpet tumors
- Asymmetric mural thickening ± irregular surface
- Hepatic metastases most common
- Detection: Fluoroscopic-guided double contrast barium enema
- Staging: Helical CT
- Tumor recurrence and surveillance: PET-CT

Top Differential Diagnoses
- Diverticulitis
- Ischemic colitis
- Infectious colitis

Pathology
- ↓ Fiber + ↑ fat and animal protein diet

Clinical Issues
- Melena, hematochezia, iron deficiency
- Overall 5 year survival is 50%
- CT: Follow-up 3-4 months after surgery, then every 6 months for 2-3 years, then annually for 5 years
- PET-CT is best for recurrence and surveillance

Diagnostic Checklist
- Evaluate entire colon for synchronous lesions
- Tumor mass with irregular margins; apple-core lesion; pericolonic extension and distant metastases

CT Findings
- Asymmetric mural thickening ± irregular surface
- Wall thickness: < 3 mm: Normal; 3-6 mm: Indeterminate; > 6 mm: Abnormal
- Tumor within lumen
  - Smooth outer bowel margins
- Extracolonic tumor extension
  - Mass with irregular border
  - Strands of soft tissue extending from serosal surface into perirectal or pericolic fat
  - Loss of tissue fat planes between colon and surrounding muscles
- Metastases to mesenteric nodes, peritoneum
- Hepatic metastases most common

Nuclear Medicine Findings
- PET: Fluorine 18-labeled deoxyglucose uptake is 2 fold higher in tumors than normal or nonmalignant lesions
- PET-CT
  - Combines morphologic information (of CT) with metabolic information (of PET)

Imaging Recommendations
- Best imaging tool
  - Detection: Fluoroscopic-guided double contrast barium enema
  - Staging: Helical CT
  - Tumor recurrence and surveillance: PET-CT

DIFFERENTIAL DIAGNOSIS

Diverticulitis
- CT findings
  - Location: Most common in sigmoid colon
  - Bowel wall and fascial thickening; fat stranding; free air and fluid
  - Pericolic inflammatory changes: Abscess, sinuses, fistulas or strictures

Ischemic colitis
- Usually seen in watershed areas; focal or diffuse
- Fluoroscopic-guided double contrast barium enema
  - "Thumbprinting" (submucosal edema or bleeding)
  - Stricture: Smooth, tapered margins, but no mass effect (chronic)
- CT: Bowel wall thickening; ± pneumatosis, portomesenteric venous gas

Infectious colitis
- Example: Tuberculosis and Amebiasis
- Rare in the U.S.; usually in proximal colon
- Stricture formation may simulate carcinoma (chronic)

Ulcerative colitis
- Significant etiology of colon carcinoma
- Bowel wall thickening, luminal narrowing
- Fluoroscopic-guided double contrast barium enema
  - Punctate and collar button ulcers
  - Continuous, concentric and symmetric colonic involvement
  - "Lead pipe" colon and absent haustra (chronic)

Extrinsic lesions
- Endometriosis, ovarian cancer, "drop" metastases
- Smooth, eccentric, obtuse angles with colonic wall

PATHOLOGY

General Features
- General path comments
  - Most common cancer of gastrointestinal tract
  - Second most common cancer mortality
  - 5% of colon cancers have synchronous colonic tumors
- Genetics
  - Mutations of genes
    - Proto-oncogene (K-ras): 50% of colon cancers
    - Tumor suppressor genes (APC, DCC, SMAD4, p53, TGF-β1 RII): 70-75% of colon cancers
    - DNA mismatch repair genes: 15% of colon cancers
- Etiology
  - Risk factors
    - ↓ Fiber + ↑ fat and animal protein diet
    - History of colon adenoma or carcinoma

    - Benign polyps > 1 cm
    - Inflammatory bowel disease
    - Family history
  - Family history
    - Colon cancer in first-degree relatives (↑ 2-3 fold)
    - Familial Adenomatous Polyposis (< 1%), Gardner, familial juvenile polyposis and hereditary nonpolyposis colorectal cancer
  - Pathogenesis
    - Adenoma-carcinoma sequence (7-10 years): Benign adenoma → malignant transformation
    - Inflammatory bowel disease: Inflammation → dysplasia → carcinoma
    - De novo colon carcinoma: Normal mucosa → small, aggressive, ulcerated tumors
  - Hereditary nonpolyposis colorectal cancer (HNPCC)
    - 6% of colon cancer
    - Discrete adenoma, not polyposis
    - Autosomal dominant with high penetration
- Epidemiology
  - > Common in N. America, Europe & New Zealand
    - Incidence in U.S. is 135,000 per year
    - Mortality in U.S. is 57,000 per year
  - ↑ Age → incidence of cancers shifts to right colon
- Associated abnormalities
  - 5% of colon cancers: Metachronous carcinomas
  - 33% of colon cancers: Adenomatous polyps

Gross Pathologic & Surgical Features

- Cecum and proximal colon: Bulky and polypoid, outgrowing their blood supply → necrosis
- Distal colon and rectum: Annular constriction or "napkin-ring" appearance → obstruction & ulceration

Microscopic Features

- Adenocarcinoma (> 95% of colon cancers)
  - Mucin-producing glands
  - Mucinous: "Signet-ring" cells
  - Colloid (15%): Large lakes of mucin contain scattered collections of tumor cells
- Squamous cell carcinoma (< 5%)

Staging, Grading or Classification Criteria

- Surgical-pathologic (modified Dukes) staging of colon cancer with TNM correlation
  - Stage A (T1N0M0): Limited to mucosa ± submucosa
  - Stage B (T2 or 3 & N0M0): Limited to serosa or into adjacent tissues
  - Stage C (T2 or 3 & N1M0): Lymph node metastases
  - Stage D (any T and N, M1): Distant metastases

CLINICAL ISSUES

Presentation

- Most common signs/symptoms
  - Melena, hematochezia, iron deficiency
  - Abdominal pain & changes in bowel habit
  - Colonic obstruction (50%, most common cause)
  - Weight loss, fever and weakness
- Lab-Data
  - Positive or negative fecal occult blood test
  - ± Micro- to normocytic anemia
  - Carcinoembryonic antigen (CEA) > 2.5 μg/L
- Diagnosis: Colonoscopy with mucosal biopsy

Demographics

- Age: > 50 years of age; peak at 70 years of age
- Gender: M:F = 3:2

Natural History & Prognosis

- Complications
  - Hemorrhage, obstruction, perforation and fistula
- Prognosis
  - Overall 5 year survival is 50%
    - Duke's stage A: 81-85%
    - Duke's stage B: 64-78%
    - Duke's stage C: 27-33%
    - Duke's stage D: 5-14%

Treatment

- Complete surgical resection (≥ 5 cm on each side of tumor) with removal of lymphatic drainage vessels ± adjuvant chemotherapy
- Pre- & post-operative radiation therapy (selected cases)
- Follow-up
  - CT: Follow-up 3-4 months after surgery, then every 6 months for 2-3 years, then annually for 5 years
  - CEA titer: If elevated, CT is indicated
  - PET-CT is best for recurrence and surveillance

DIAGNOSTIC CHECKLIST

Consider

- Evaluate entire colon for synchronous lesions

Image Interpretation Pearls

- Tumor mass with irregular margins; apple-core lesion; pericolonic extension and distant metastases

SELECTED REFERENCES

1. Bar-Shalom R et al: Clinical performance of PET/CT in evaluation of cancer: additional value for diagnostic imaging and patient management. J Nucl Med. 44(8):1200-9, 2003
2. Cohade C et al: Direct comparison of (18)F-FDG PET and PET/CT in patients with colorectal carcinoma. J Nucl Med. 44(11):1797-803, 2003
3. Pickhardt PJ et al: Computed tomographic virtual colonoscopy to screen for colorectal neoplasia in asymptomatic adults. N Engl J Med. 349(23):2191-200, 2003
4. Levine MS et al: Diagnosis of colorectal neoplasms at double-contrast barium enema examination. Radiology 216: 11-8, 2000
5. Gazelle GS et al: Screening for colorectal cancer. Radiology. 215(2):327-35, 2000
6. Horton KM et al: Spiral CT of colon cancer: imaging features and role in management. Radiographics. 20(2):419-30, 2000
7. Thoeni RF: Colorectal cancer: cross-sectional imaging for staging of primary tumor and detection of local recurrence. AJR Am J Roentgenol. 156(5):909-15, 1991
8. Balthazar EJ et al: Carcinoma of the colon: detection and preoperative staging by CT. AJR Am J Roentgenol. 150(2):301-6, 1988
9. Kelvin FM et al: Colorectal carcinoma detected initially with barium enema examination: site distribution and implications. Radiology. 169(3):649-51, 1988

IMAGE GALLERY

Typical

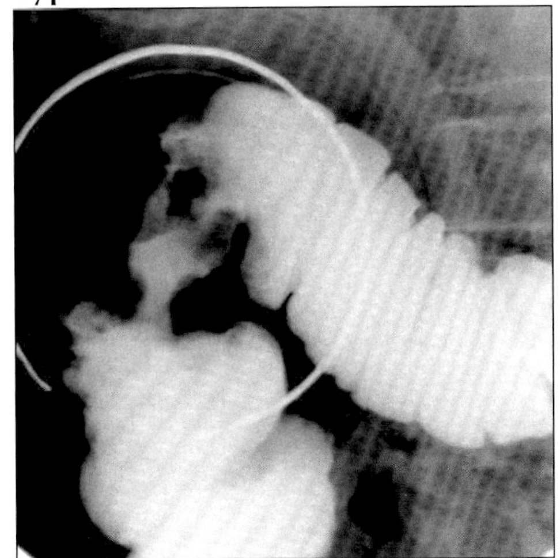

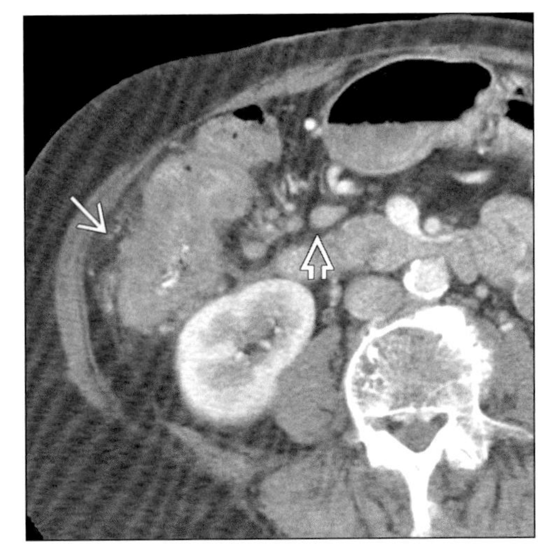

(Left) *Single contrast BE shows bulky mass at hepatic flexure with abrupt transition to normal colon. Little or no obstruction in spite of large mass + luminal narrowing.* ***(Right)*** *Axial CECT shows mass at hepatic flexure with circumferential wall thickening, narrowed lumen. Infiltrated fat (arrow) + mesenteric adenopathy (open arrow) indicate local spread of disease.*

Typical

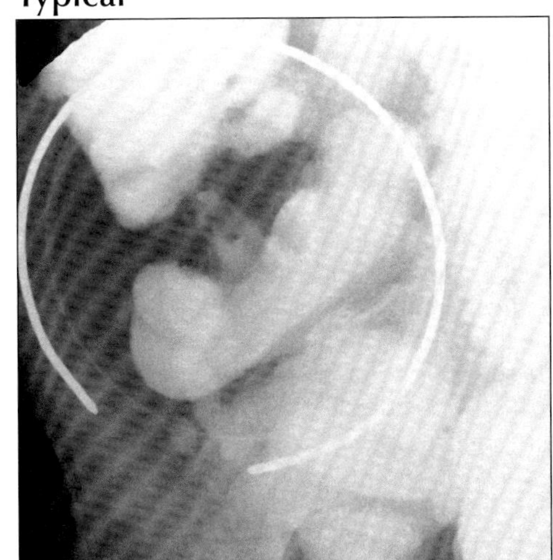

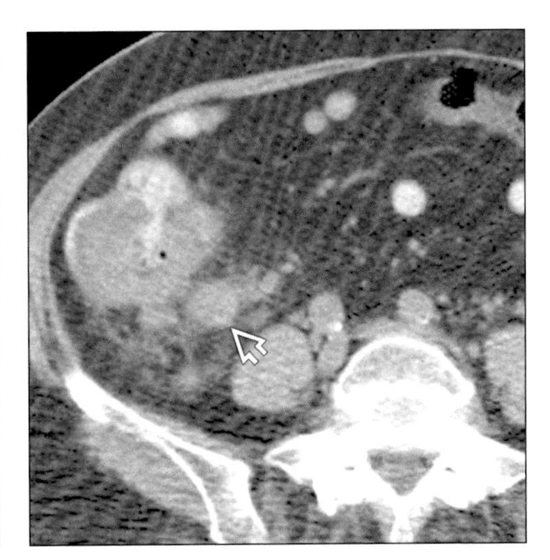

(Left) *Single contrast BE shows short segment luminal narrowing of ascending colon with "apple core" appearance, but no obstruction.* ***(Right)*** *Axial CECT shows large eccentric mass in ascending colon with extensive infiltration of pericolonic fat + lymphadenopathy (open arrow).*

Variant

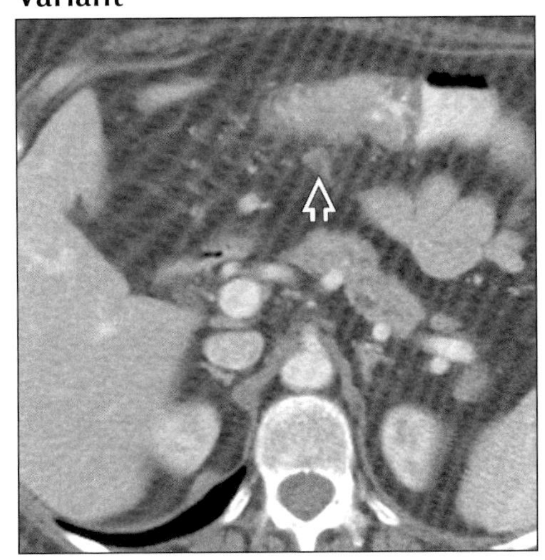

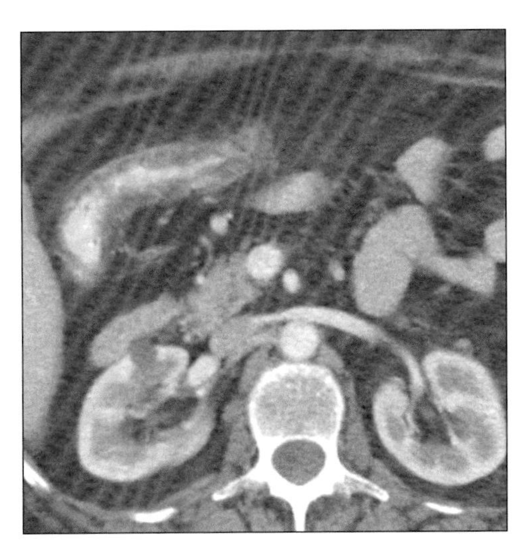

(Left) *Axial CECT shows wall thickening + luminal narrowing of transverse colon near splenic flexure with abrupt transition to normal colon. Mesenteric adenopathy (open arrow).* ***(Right)*** *Axial CECT of patient with colon cancer near splenic flexure. Long segment of wall thickening + luminal narrowing with submucosal low density. At resection found to represent ischemic colitis.*

RECTAL CARCINOMA

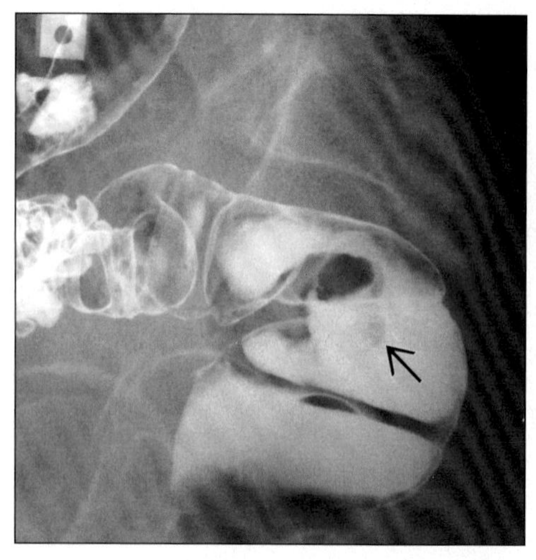

Single contrast BE shows mass (arrow) arising from anterior rectal wall as a filling defect in the barium pool.

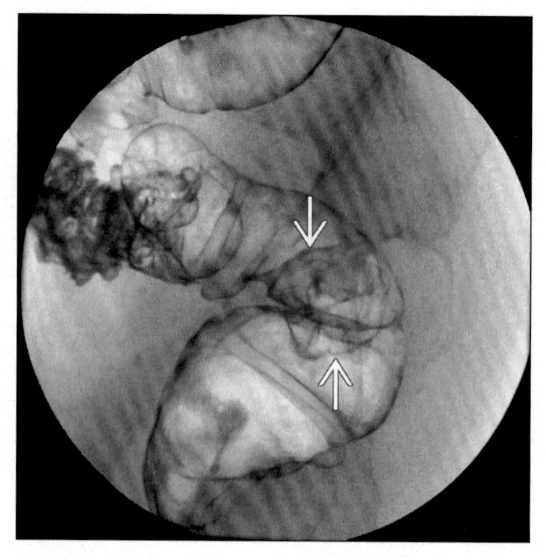

Double contrast BE shows rectal mass (arrows) outlined by a coating of barium.

TERMINOLOGY

Definitions

- Malignant transformation of rectal mucosa

IMAGING FINDINGS

General Features

- Best diagnostic clue: Polypoid mass with irregular surface
- Morphology
  - Early cancer: Sessile or pedunculated tumors
  - Advanced cancer: Annular, semiannular, polypoid or carpet tumors
  - Most common in rectum: Sessile and polypoid
- Other general features
  - Radiologic features are similar to colon carcinoma
  - Transrectal ultrasonography for tumor staging
  - Types of rectal cancer: Adenocarcinoma (80%) and squamous cell carcinoma (20%)

Radiographic Findings

- Fluoroscopic-guided barium enema
  - Early cancer: Sessile (plaque-like) lesion
    - Most typical early colorectal cancer
    - Flat, protruding lesion with a broad base and little elevation of mucosa (profile view)
    - Discrete borders and shallow central ulcers (profile view)
    - Curvilinear or undulating lines (en face view)
  - Early cancer: Pedunculated lesion
    - Short and thick polyp stalk
    - Irregular or lobulated head of polyp
  - Advanced cancer: Polypoid lesion
    - Dependent wall: Filling defect in barium pool
    - Nondependent wall: Etched in white
  - Advanced cancer: Semi-annular (saddle) lesion
    - Transition to annular carcinoma
    - Convex barium-etched margins (profile view)
  - Advanced cancer: Annular (apple-core) lesion
    - Circumferential narrowing of bowel; shelf-like, overhanging borders (mucosal destruction)
    - High grade obstruction and ischemia: Thumbprinting of dilated proximal colon
  - Advanced cancer: Carpet lesion
    - Malignant villous tumor may appear as carpet lesion with minimal protrusion into lumen
    - Radiolucent nodules surrounded by barium-filled grooves; finely nodular or reticular pattern

CT Findings

- Mass & focal or circumferential wall thickening
- Asymmetric mural thickening ± irregular surface
- Wall thickness: < 3 mm: Normal; 3-6 mm: Indeterminate; > 6 mm: Abnormal
- Tumor within lumen

DDx: Rectal Mass or Luminal Narrowing

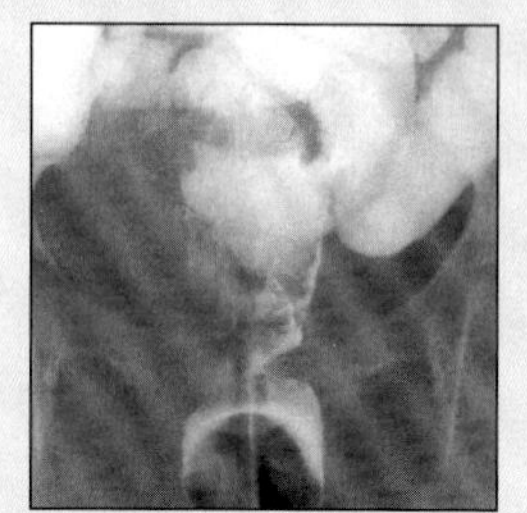

Cervical Cancer

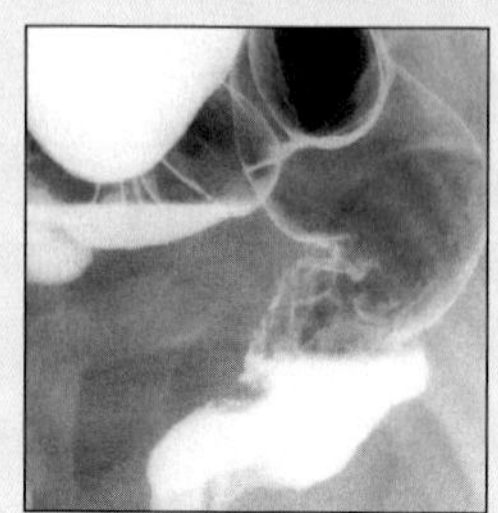

Rectal Ulcer

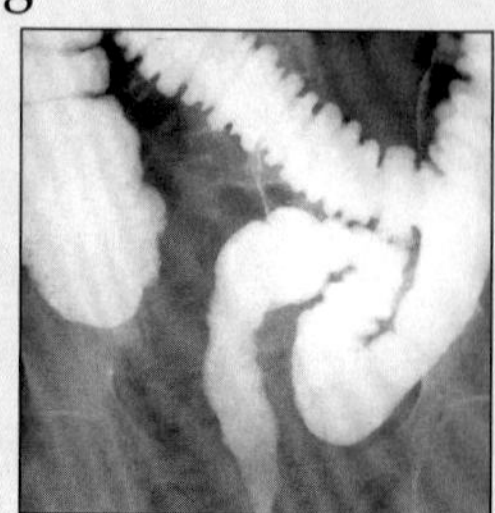

Lymphogranuloma

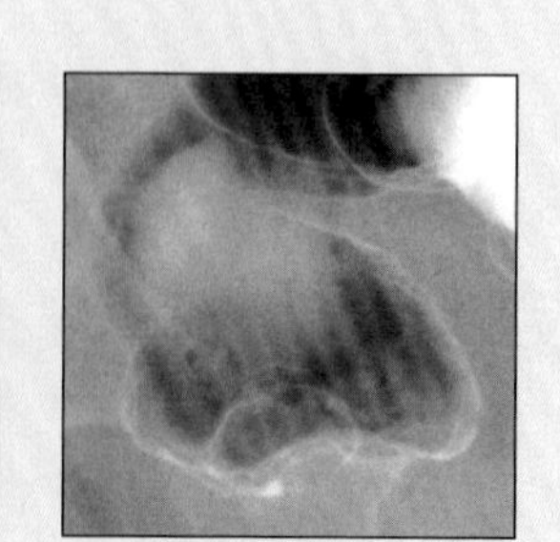

Hemorrhoids

RECTAL CARCINOMA

Key Facts

Imaging Findings
- Best diagnostic clue: Polypoid mass with irregular surface
- Radiologic features are similar to colon carcinoma
- Types of rectal cancer: Adenocarcinoma (80%) and squamous cell carcinoma (20%)
- May have lung and bone metastases before liver metastases
- Metastases to lymph nodes: Spherical, hypoechoic & distinct margins
- Transrectal ultrasonography: Visualize layers of rectal wall & depth of tumor penetration

Top Differential Diagnoses
- Local invasion
- Villous adenoma
- Trauma
- Infection

Pathology
- ↓ Fiber + ↑ fat and animal protein diet
- HIV positive homosexual males
- Human papilloma virus (HPV): Type 16, 18, 45, 46

Clinical Issues
- Hematochezia, rectal pain, change in bowel habits
- Anal pain, anal discharge & tenesmus
- Overall 5 year survival is 50%

Diagnostic Checklist
- Evaluate entire colon for synchronous lesions
- Image detection of perirectal tumor spread is vital; requires pre-operative radiation ± chemotherapy

---

  - Smooth outer bowel margins
- Extracolonic tumor extension
  - Mass with irregular border
  - Strands of soft tissue extending from serosal surface into perirectal fat
  - Loss of tissue fat planes between rectum and surrounding muscles
- Metastasis to lymph nodes at external iliac and para-aortic chain, inguinal, retroperitoneum or porta hepatis
- May have lung and bone metastases before liver metastases

MR Findings
- Mass; pericolonic infiltration, lymphadenopathy shown slightly better than by CT
- Endorectal coil - improves resolution, but may not be worth the effort

Ultrasonographic Findings
- Real time transrectal ultrasonography
  - Hypoechoic mass with disruption of wall segments
  - Focal or circumferential wall thickening
  - Metastases to lymph nodes: Spherical, hypoechoic & distinct margins
  - Rings of different echogenicities (center → outer)
    - Innermost ring: Hyperechoic; interface between balloon and mucosa
    - Second ring: Hypoechoic; muscularis mucosae
    - Third ring: Hyperechoic; submucosa
    - Fourth ring: Hypoechoic; muscularis propria
    - Fifth ring: Hyperechoic; perirectal fat or serosa
  - Sonographic staging based on TNM classification
    - T1: Confined to mucosa/submucosa; middle echogenic layer intact
    - T2: Confined to rectal wall; outermost echogenic layer is intact
    - T3: Penetrates into perirectal fat; disrupting outer hyperechoic ring

Nuclear Medicine Findings
- PET: Fluorine 18-labeled deoxyglucose uptake is 2 fold higher in tumors than normal or nonmalignant lesions
- PET-CT
  - Combines morphologic information (of CT) with metabolic information (of PET)

Imaging Recommendations
- Best imaging tool
  - Detection: Fluoroscopic-guided barium enema
  - Staging: Helical CT and transrectal ultrasonography
  - Transrectal ultrasonography: Visualize layers of rectal wall & depth of tumor penetration
  - Tumor recurrence, surveillance: PET-CT
- Protocol advice: Transrectal ultrasonography: Pass transducer proximal to tumor into the colon for complete assessment of mural and nodal pathology

DIFFERENTIAL DIAGNOSIS

Local invasion
- Example: Carcinoma of the cervix, prostate, bladder
- Direct extension to pelvic sidewall and adjacent structures including rectum
- Circumferential narrowing ± lymphadenopathy
- Depends on size of tumor; can be hard to differentiate

Villous adenoma
- Polypoid lesion with a granular or reticular appearance
- High risk of malignant degeneration
- Similar to "carpet" lesion in advanced rectal cancer

Trauma
- Penetrating injuries: Anal intercourse & insertion of foreign bodies
- Fibrosis and stricture (chronic) can simulate cancer
- Perianal and rectal mucosa ulceration

Infection
- Mucosal ulceration or granular mucosal pattern
- Mechanism: Anal sex, spread from vaginal discharge or lymphatic extension from inguinal lymph nodes
- Most common: C. trachomatis → lymphatic tissue infection → Lymphogranuloma venereum (LGV)
- Other STDs include N. gonorrhoeae, HSV and syphilis

RECTAL CARCINOMA

- Progress to fistula, perirectal abscess or stricture (chronic); similar in complications of rectal cancer

PATHOLOGY

General Features

- General path comments
  - Colon cancer: Rectum (20%) & rectosigmoid (15%)
  - Rectal cancer tends to invade locally (lack serosa)
  - Metastases: Upper 2/3 of rectum
    - Portal system → liver
    - Batson vertebral venous plexus → lumbar & thoracic vertebra
  - Metastases: Lower 1/3 of rectum
    - Superior hemorrhoidal vein → portal → liver
    - Middle hemorrhoidal vein → IVC → lung
- Genetics: Adenocarcinoma: Mutation in proto-oncogene, tumor suppressor genes or DNA mismatch repair genes
- Etiology
  - Adenocarcinoma
    - ↓ Fiber + ↑ fat and animal protein diet
    - History of colorectal adenoma or carcinoma
    - Benign polyps > 1 cm
    - Family history & Inflammatory bowel disease
  - Squamous cell carcinoma
    - HIV positive homosexual males
    - Human papilloma virus (HPV): Type 16, 18, 45, 46
    - Lubricants, cleansers & mechanical irritation
  - Pathogenesis
    - Adenocarcinoma: Adenoma-carcinoma sequence
    - Squamous cell carcinoma: Squamous metaplasia → dysplasia → carcinoma
- Epidemiology: Adenocarcinoma: More common in N. America, Europe & New Zealand

Gross Pathologic & Surgical Features

- Flat, infiltrative, annular or ulcerative & rolled borders
- Annular constriction or "napkin-ring" appearance → obstruction, ulceration and intramural spread
- Squamous cell carcinoma: Mass from epithelium of anorectal junction (dentate line)

Microscopic Features

- Adenocarcinoma: Mucin-producing glands
- Squamous cell (cloacogenic) carcinoma
  - Mixture of basaloid cell, transitional cell with squamous differentiation, adenoid cyst and mucoepithelial cell

Staging, Grading or Classification Criteria

- Surgical-pathologic (modified Dukes) staging of colon cancer with TNM correlation
  - Stage A (T1N0M0): Limited to mucosa ± submucosa
  - Stage B (T2 or 3 & N0M0): Limited to or invades adjacent tissues
  - Stage C (T2 or 3 & N1M0): Lymph node metastases
  - Stage D (any T and N, M1): Distant metastases

CLINICAL ISSUES

Presentation

- Most common signs/symptoms
  - Hematochezia, rectal pain, change in bowel habits
  - Perineal or sacral pain (chronic)
  - Squamous cell carcinoma:
    - Anal pain, anal discharge & tenesmus
- Lab-Data
  - ± HIV (PCR) test
  - Carcinoembryonic antigen (CEA) > 2.5 μg/L
- Diagnosis
  - Sigmoidoscopy with mucosal biopsy

Demographics

- Age: Adenocarcinoma: Age: > 50 years of age; peak at 70 years of age
- Gender: M:F = 3:2

Natural History & Prognosis

- Complications
  - Hemorrhage, obstruction, perforation and fistula
- Prognosis
  - Overall 5 year survival is 50%
    - Duke's stage A: 81-85%
    - Duke's stage B: 64-78%
    - Duke's stage C: 27-33%
    - Duke's stage D: 5-14%

Treatment

- Surgical resection (depends on location) & removal of lymphatic drainage vessels ± adjuvant chemotherapy
- Pre- & post-operative radiation ± chemotherapy therapy (selected cases)
- Follow-up
  - CT: Follow-up 3-4 months after surgery, then every 6 months for 2-3 years, then annually for 5 years
  - CEA titer: If elevated, CT is indicated (PET-CT best)

DIAGNOSTIC CHECKLIST

Consider

- Evaluate entire colon for synchronous lesions

Image Interpretation Pearls

- Image detection of perirectal tumor spread is vital; requires pre-operative radiation ± chemotherapy

SELECTED REFERENCES

1. Fuchsjager MH et al: Comparison of transrectal sonography and double-contrast MR imaging when staging rectal cancer. AJR Am J Roentgenol. 181(2):421-7, 2003
2. Winawer SJ et al: A comparison of colonoscopy and double-contrast barium enema for surveillance after polypectomy. New Eng J Med 342: 1766-72, 2000
3. Maier AG et al: Transrectal sonography of anal sphincter infiltration in lower rectal carcinoma. AJR Am J Roentgenol. 175(3):735-9, 2000
4. Levine MS et al: Diagnosis of colorectal neoplasms at double-contrast barium enema examination. Radiology 216: 11-8, 2000
5. Thompson WM et al: Computed tomography of the rectum. Radiographics. 7(4):773-807, 1987
6. Cohan RH et al: Computed tomography of epithelial neoplasms of the anal canal. AJR Am J Roentgenol. 145(3):569-73, 1985

IMAGE GALLERY

Typical

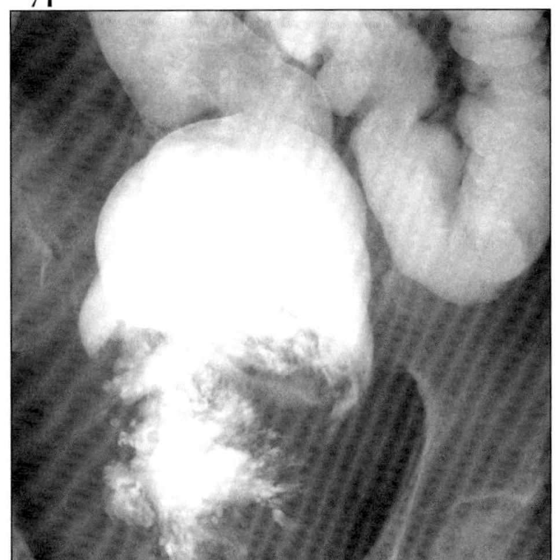

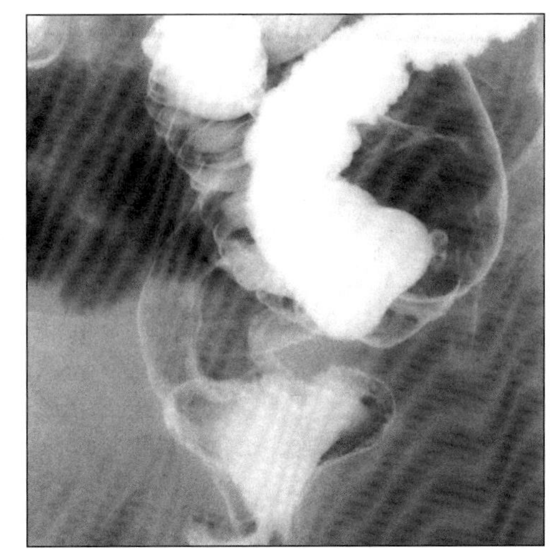

(Left) *Single contrast BE shows large rectal mass with markedly irregular surface; carcinoma arising from villous adenoma.* ***(Right)*** *Air contrast BE shows large mass arising from lateral wall of the rectum.*

Typical

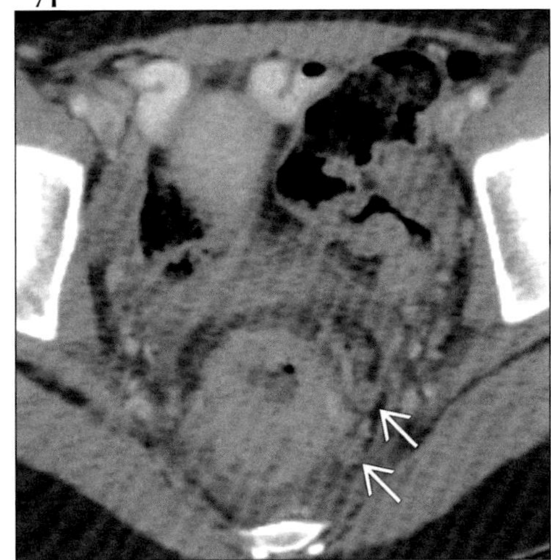

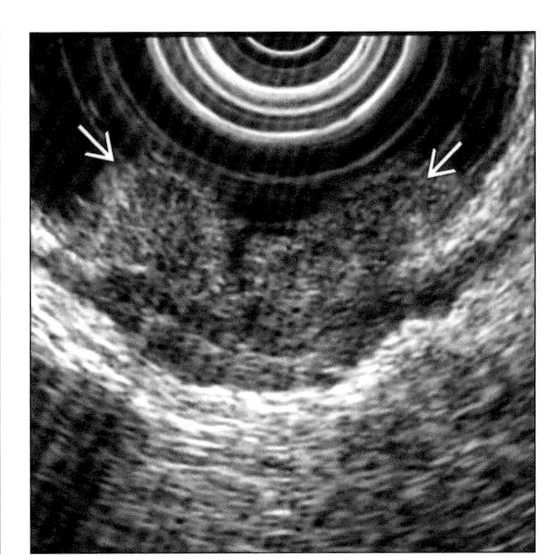

(Left) *Axial CECT shows large mass that fill the rectal lumen and infiltrates the perirectal fat. Extensive lymphadenopathy (arrows).* ***(Right)*** *Transrectal ultrasonography shows a bulky rectal mass (arrows) with invasion through submucosa; T3 stage.*

Typical

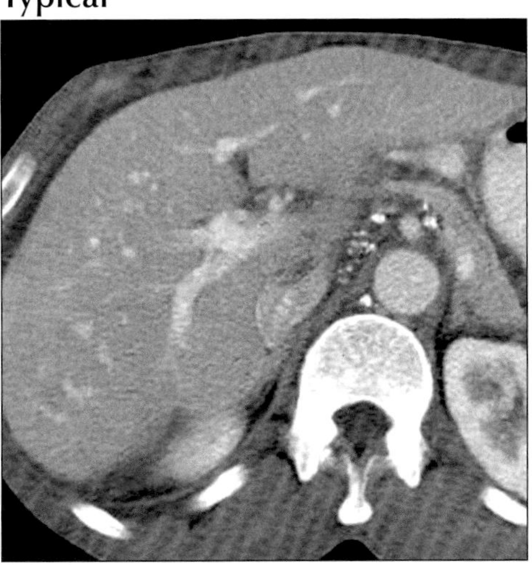

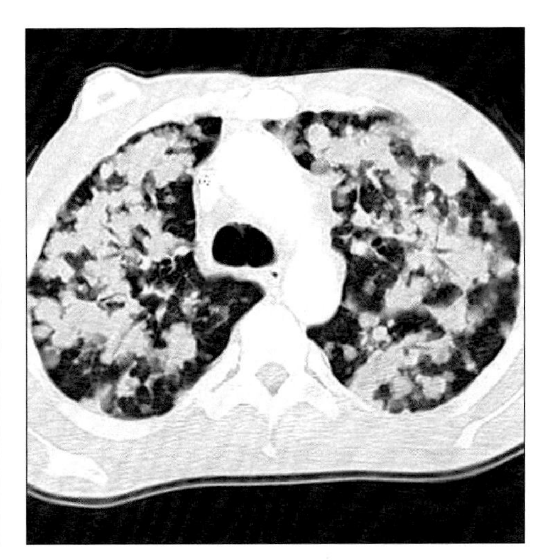

(Left) *Axial CECT shows calcified periaortic and retrocrural nodes, no liver metastases. Mucinous rectal adenocarcinoma.* ***(Right)*** *Axial CECT shows extensive pulmonary metastases from rectal cancer in a patient with no liver metastases.*

VILLOUS ADENOMA

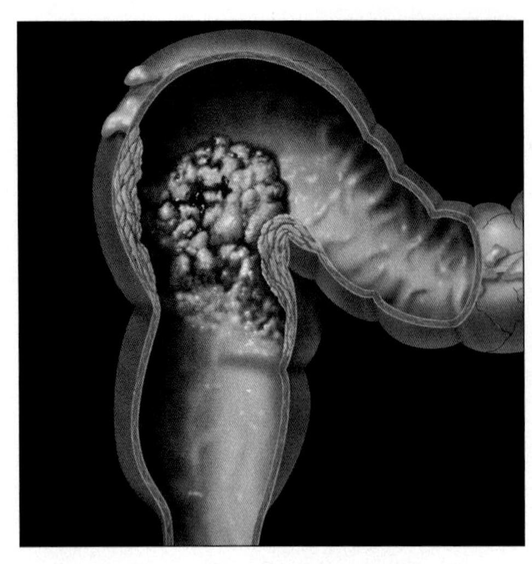

Graphic shows polypoid mass in rectosigmoid colon having a shaggy, nodular surface.

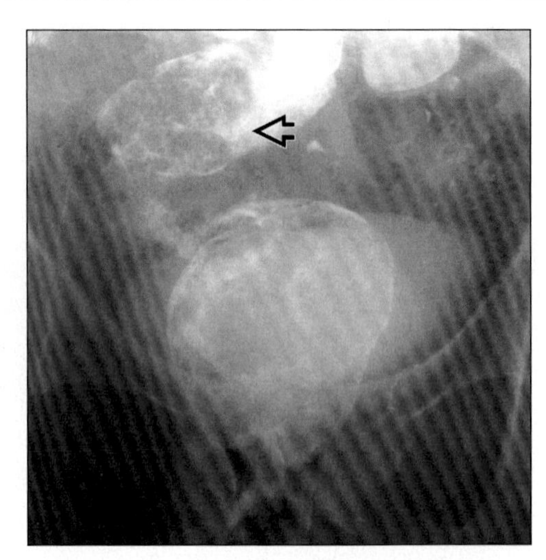

Single contrast BE shows a polypoid mass in the rectosigmoid colon (arrow) with a very nodular surface. Barium within the rectum is diluted by mucous secreted by the tumor.

TERMINOLOGY

Abbreviations and Synonyms

- Villous tumor

Definitions

- Adenomatous polyp that contains predominantly villous elements ("villous" means "shaggy surface")

IMAGING FINDINGS

General Features

- Best diagnostic clue: Polypoid lesion with a nodular or frond-like surface on barium enema
- Location: Rectosigmoid > cecum > ascending colon > stomach > duodenum
- Size
  - Range from < 1-10 cm in diameter
  - Giant villous tumor: 10-15 cm
- Morphology: "Cauliflower-like" sessile growth with a broad base or flat "carpet" lesion
- Other general features
  - Villous adenoma is one of the histological types of adenomatous polyps (true neoplasms)
    - Tubular adenoma: 75% of neoplastic polyps; villous change less than 25%
    - Villous adenoma: 10% of neoplastic polyps; villous change more than 75%
    - Tubulovillous adenoma: 15%; villous change between 25-50%
  - As adenoma increases in size, degree of villous change usually increases
  - Villous adenoma has highest risk of malignant degeneration
  - Risk of cancer is related to proportion of villous change in adenoma
  - Greater risk of carcinoma in villous tumors of stomach & duodenum than colon
    - Stomach: Carcinoma in 50% of lesions 2-4 cm & 80% in more than 4 cm
    - Duodenum: Carcinoma in 30-60% of villous tumors more than 4 cm
    - Colon: Carcinoma in situ in 10% & invasive carcinoma in up to 45% of cases

Radiographic Findings

- Fluoroscopic guided double contrast barium enema
  - Two types of villous adenomas
    - Polypoid mass
    - "Carpet" lesion
  - Polypoid mass
    - May look like a cauliflower within colon
    - Nodular, lace or soap bubble pattern
    - Due to trapping of barium between frond-like projections (interstices)

DDx: Irregular Solitary Filling Defect

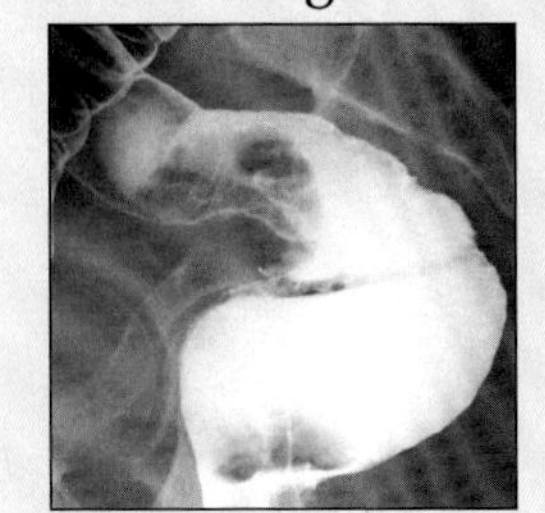

Rectal Cancer

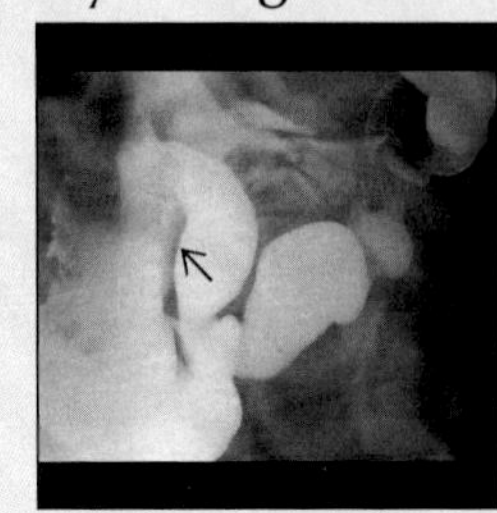

Colon Cancer

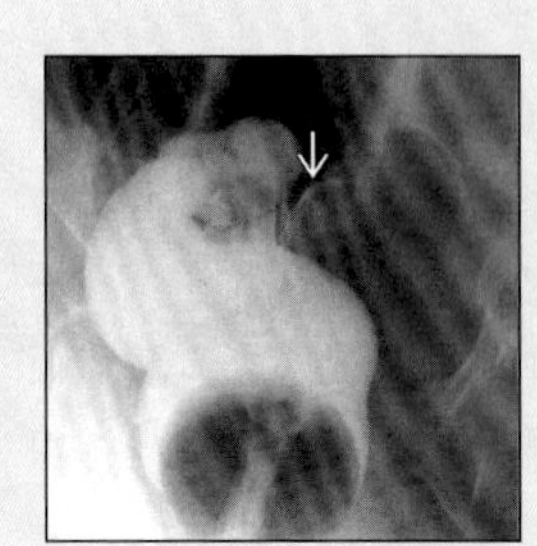

Fecal Mass

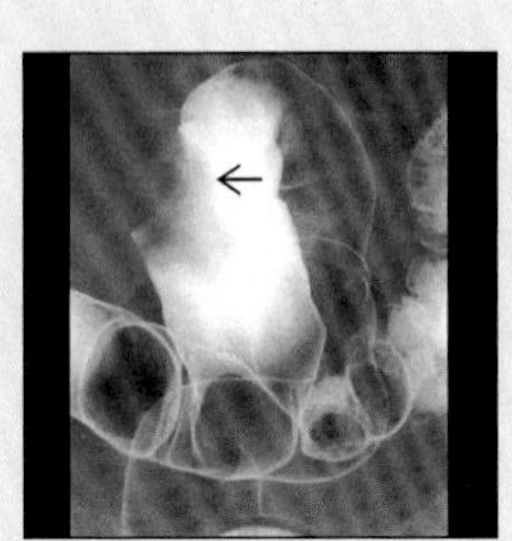

Colon Lipoma

VILLOUS ADENOMA

Key Facts

Terminology
- Adenomatous polyp that contains predominantly villous elements ("villous" means "shaggy surface")

Imaging Findings
- Best diagnostic clue: Polypoid lesion with a nodular or frond-like surface on barium enema
- Location: Rectosigmoid > cecum > ascending colon > stomach > duodenum
- Range from < 1-10 cm in diameter
- Malignant transformation in a bulky adenoma: Annular lesion with shelf-like, overhanging borders
- Localized "carpet" lesion: Subtle alteration in surface texture
- Extensive "carpet" lesion: Involves a large area of colon, encircling lumen

Top Differential Diagnoses
- Colon carcinoma
- Fecal mass
- Intramural mass

Pathology
- Family history, idiopathic inflammatory disease
- Malignant potential: 5% in lesions < 1 cm; 10% in lesions 1-2 cm; 53% in > 2 cm lesions
- Gray-tan lesion

Diagnostic Checklist
- Check for family history of colonic polyps & evaluate entire colon for synchronous lesions
- Cauliflower-like sessile mass with a broad base or carpet lesion with reticular or soap-bubble surface pattern

---

    - Malignant transformation in a bulky adenoma: Annular lesion with shelf-like, overhanging borders
  - "Carpet" lesion
    - Flat, lobulated lesion
    - Localized or extensive
  - Localized "carpet" lesion: Subtle alteration in surface texture
  - Extensive "carpet" lesion: Involves a large area of colon, encircling lumen
    - En face: Fine nodular, reticular pattern with sharply demarcated border
    - Profile: Irregular contour in contrast to smooth, fine contour of adjacent normal bowel
  - Malignant transformation in "carpet" lesion (↑ risk)
    - Radiolucent nodules surrounded by barium-filled grooves (produce fine nodular or reticular pattern)
    - Polypoid carcinoma with surrounding mucosal change represents underlying adenoma
    - Seen in rectum, cecum, ascending colon, stomach & duodenum

CT Findings
- Large villous adenoma
  - Low-attenuation irregular polypoid mass
  - Convolutional gyral enhancement pattern
  - Corrugated, feathery appearance due to trapping of oral contrast in interstices of villous adenoma

MR Findings
- Large villous adenoma
  - T1WI: Low signal intensity mass with multiple frond-like projections & central cord-like structure
  - T2WI: Frond-like projections will be more prominent
- Villous adenoma with ↑ mucin producing cells
  - Short T1 & long T2 times
  - T1WI & T2WI: Adenoma appears hyperintense

Ultrasonographic Findings
- Transrectal sonography
  - Determine depth of invasion into colonic wall by adenoma

Imaging Recommendations
- Double-contrast barium enema
  - En face, profile & oblique views
- Transrectal US; NE + CECT

DIFFERENTIAL DIAGNOSIS

Colon carcinoma
- Barium enema findings
  - Early cancer: Sessile (plaquelike) lesion
    - Typical early colon cancer
    - Flat, protruding lesion with a broad base & little elevation of mucosa (in profile view)
  - Early cancer: Pedunculated lesion
    - Short & thick polyp stalk
    - Irregular or lobulated head of polyp
  - Advanced cancer: Polypoid lesion (large)
    - Dependent wall: Filling defect in barium pool
    - Nondependent wall: Etched white
  - Sessile & pedunculated polypoid cancers may be indistinguishable from villous adenoma
  - Advanced cancer: Semi-annular (saddle) lesion
  - Advanced cancer: Annular (apple-core) lesion
    - Circumferential narrowing of bowel
    - Shelf-like, overhanging borders (mucosal destruction)
- CT findings
  - Asymmetric mural thickening ± irregular surface
  - Extracolonic tumor extension
    - Mass with irregular borders
    - Extension from serosa to pericolic fat
    - Loss of fat planes: Colon & adjacent muscles
  - Metastases to mesenteric nodes
  - Metastases to liver more common
- Diagnosis: Biopsy & histology

Fecal mass
- Large, irregular colonic fecal impaction
  - Most common location: Rectum
- Mimic large cauliflower-like sessile polyp
- May cause bowel obstruction + proximal dilatation

VILLOUS ADENOMA

- Usually seen in elderly, sedentary patients
- Diagnosis: Clinical history & colonoscopy

Intramural mass

- Example: Stromal tumors (leiomyoma, sarcoma or GIST)
- Leiomyoma
  - In profile
    - Smooth surface etched in white
    - Borders: Right or obtuse angles with adjacent wall
  - En face
    - Seen as a filling defect simulating polypoid type of villous adenoma
    - Intraluminal surface: Abrupt well-defined borders
- Leiomyosarcoma
  - Bulky stromal tumors most frequently seen in rectum
  - Broad based mass simulating large villous adenoma
  - Large tumors show ulceration or cavitation
  - CT shows pericolonic extension (large extraluminal mass), liver & peritoneal metastases
- Hypervascular on angiography
- Diagnosis: Biopsy

PATHOLOGY

General Features

- Etiology
  - Villous adenoma or tumor
    - Family history, idiopathic inflammatory disease
    - Malignant potential: 5% in lesions < 1 cm; 10% in lesions 1-2 cm; 53% in > 2 cm lesions
- Epidemiology: Incidence: Least common (10%) of all neoplastic adenomatous polyps

Gross Pathologic & Surgical Features

- Usually sessile
  - May be polypoid, broad, flat or carpet-like lesion
  - Gray-tan lesion
- May have a short, broad stalk & focal areas of hemorrhage or ulceration

Microscopic Features

- Frond-like papillary projections of adenomatous epithelium
- ± Well-differentiated areas
- Carcinoma in situ; invasive cancer

CLINICAL ISSUES

Presentation

- Most common signs/symptoms
  - Asymptomatic, diarrhea, pain, rectal bleeding or melena
  - Lesion closer to rectum: More likely to have diarrhea, electrolyte loss
- Lab-data
  - Guaiac positive stool
  - Iron deficiency anemia
  - Decreased protein, K+, Na+
  - ± Increased direct bilirubin levels (due to obstruction of ampulla of Vater (duodenum) by adenoma)
- Diagnosis: Endoscopy, biopsy & histology

Demographics

- Age: 60-70 years of age or older
- Gender: Equal in both males & females (M = F)

Natural History & Prognosis

- Complications
  - Malignant transformation or invasion; hemorrhage
- Prognosis
  - Good: After removal of benign & carcinoma in situ adenoma
  - Poor: Invasive carcinoma

Treatment

- Colonoscopic, endoscopic or surgical resection

DIAGNOSTIC CHECKLIST

Consider

- Check for family history of colonic polyps & evaluate entire colon for synchronous lesions

Image Interpretation Pearls

- Cauliflower-like sessile mass with a broad base or carpet lesion with reticular or soap-bubble surface pattern

SELECTED REFERENCES

1. Smith TR et al: CT appearance of some colonic villous tumors. AJR Am J Roentgenol. 177(1):91-3, 2001
2. Levine MS et al: Diagnosis of colorectal neoplasms at double-contrast barium enema examination. Radiology 216: 11-8, 2000
3. Cunnane ME et al: Small flat umbilicated tumors of the colon: radiographic and pathologic findings. AJR Am J Roentgenol. 175(3):747-9, 2000
4. Chung JJ et al: Large villous adenoma in rectum mimicking cerebral hemispheres. AJR Am J Roentgenol. 175(5):1465-6, 2000
5. Iida M et al: Endoscopic features of villous tumors of the colon: correlation with histological findings. Hepatogastroenterology. 37(3):342-4, 1990
6. Iida M et al: Villous tumor of the colon: correlation of histologic, macroscopic, and radiographic features. Radiology. 167(3):673-7, 1988
7. Galandiuk S et al: Villous and tubulovillous adenomas of the colon and rectum. A retrospective review, 1964-1985. Am J Surg. 153(1):41-7, 1987
8. Galandiuk S et al: Villous and tubulovillous adenomas of the colon and rectum. A retrospective review, 1964-1985. Am J Surg. 153(1):41-7, 1987
9. Ott DJ et al: Single-contrast vs double-contrast barium enema in the detection of colonic polyps. AJR Am J Roentgenol. 146(5):993-6, 1986
10. de Roos A et al: Colon polyps and carcinomas: prospective comparison of the single- and double-contrast examination in the same patients. Radiology. 154(1):11-3, 1985
11. Delamarre J et al: Villous tumors of the colon and rectum: double-contrast study of 47 cases. Gastrointest Radiol. 5(1):69-73, 1980

VILLOUS ADENOMA

IMAGE GALLERY

Typical

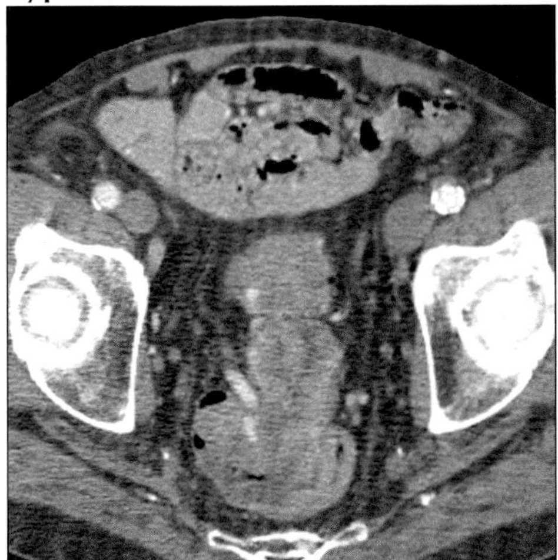

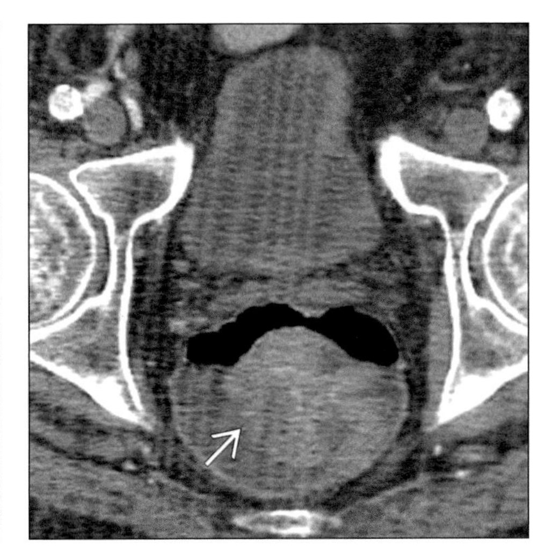

(Left) *Axial CECT shows a large mass that fills the rectosigmoid colon with dilated stool-filled colon, noted more proximally.* ***(Right)*** *Axial CECT shows a large polypoid mass (arrow) within the rectum.*

Typical

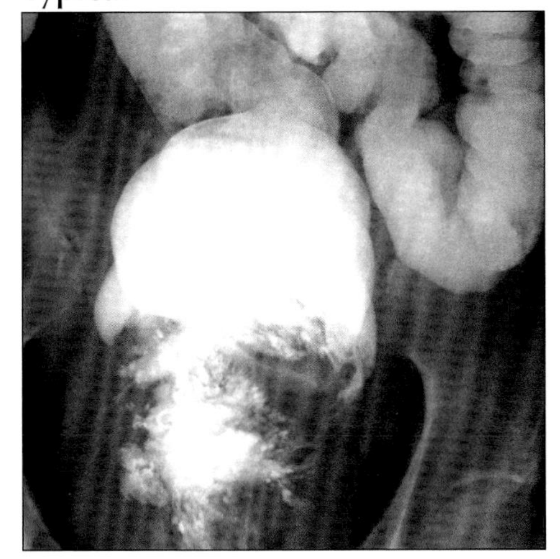

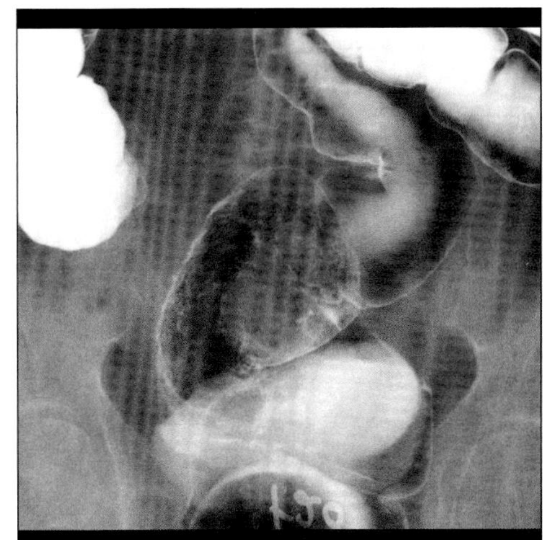

(Left) *Single contrast BE shows large rectal mass with frond-like surface. Note absence of colonic obstruction.* ***(Right)*** *Air contrast BE shows rectal mass with nodular surface that fills, but does not obstruct, the rectal lumen.*

Typical

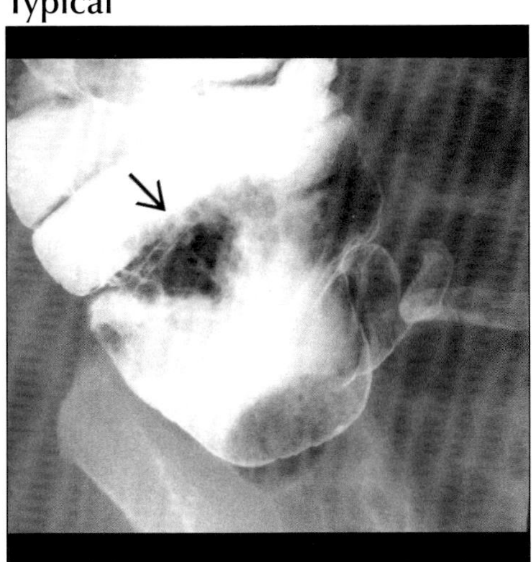

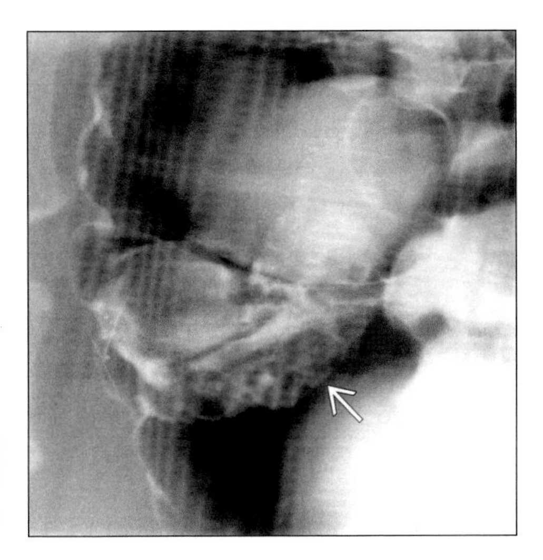

(Left) *Single contrast BE shows a cauliflower-like mass (arrow) in the cecum.* ***(Right)*** *Air contrast BE shows cauliflower-like cecal mass; (arrow) villous adenoma.*

FAMILIAL POLYPOSIS

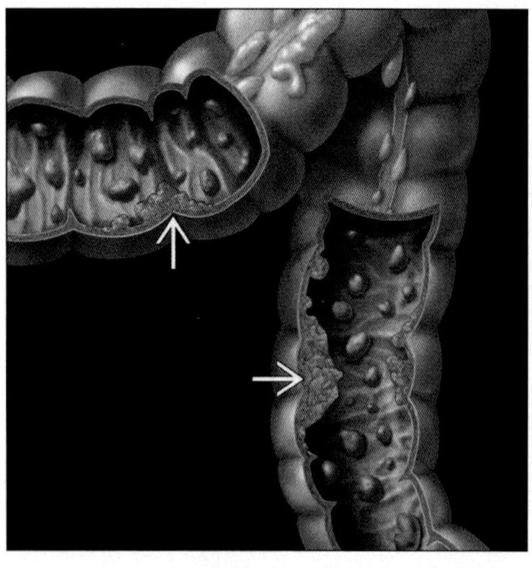

Graphic shows innumerable small polyps and multifocal carcinomas (arrows).

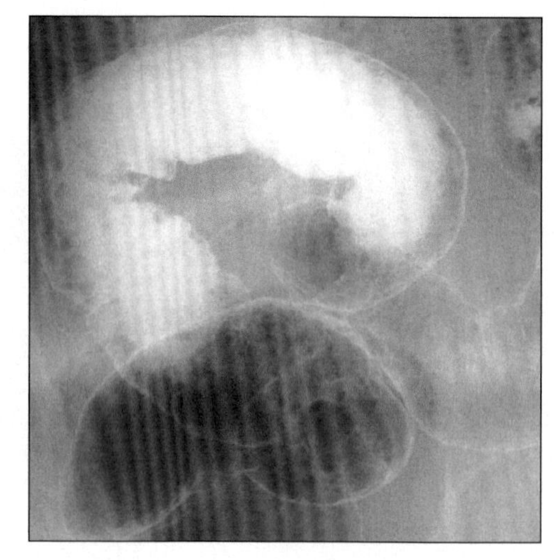

Air contrast BE shows innumerable small polyps in sigmoid colon.

TERMINOLOGY

Abbreviations and Synonyms

- Familial adenomatous polyposis syndrome (FAPS)

Definitions

- Spectrum of autosomal dominant disease characterized by innumerable adenomatous colonic polyps & other associated lesions

IMAGING FINDINGS

General Features

- Best diagnostic clue: Innumerable colonic filling defects or ring shadows ± extraintestinal lesions
- Location
  - Most common in colon (↑ predilection-left colon)
    - Colon > stomach > duodenum > small bowel
- Size: Varies from pin point to > 1 cm
- Morphology: Sessile or pedunculated polypoid lesions
- Other general features
  - FAPS is a rare condition, but is most common of polyposis syndromes
  - Two varied expressions of FAPS
    - Familial polyposis coli: Multiple colonic adenomatous polyps
    - Gardner syndrome
  - Familial polyposis coli: Multiple colonic adenomatous polyps
    - Entire colonic mucosa is carpeted with polyps
  - Gardner syndrome: Combination of
    - Familial polyposis coli, osteomas, epidermoid (sebaceous) cyst
    - Soft tissue tumors: Desmoid, mesenteric fibromatosis, lipoma
    - Dental abnormalities; periampullary, duodenal & thyroid carcinomas
  - 500-2500 polyps present carpeting colonic mucosa
  - Polyps appear around puberty & onset of symptoms in 3rd or 4th decade
  - Most polyps are tubular & tubulovillous, occasionally villous adenomas
  - FAPS adenomas small (80% < 5 mm) & sessile
  - Colorectal cancer develops in almost 100% of untreated patients
  - 2/3 of afflicted cases are inherited & 1/3 are sporadic
  - Abnormal gene has high penetrance (80-100%)
  - Extracolonic GI tract manifestations of FAPS
    - Stomach, duodenum, jejunum & ileum
  - Stomach
    - Fundic gland polyps & adenomas in > 50% cases
  - Duodenum: Adenomas of 2nd part & periampullary in > 47% cases
    - Periampullary cancer: 2nd most frequent site of cancer outside colon seen in 12% of FAPS patients

DDx: Multiple Colonic Filing Defects

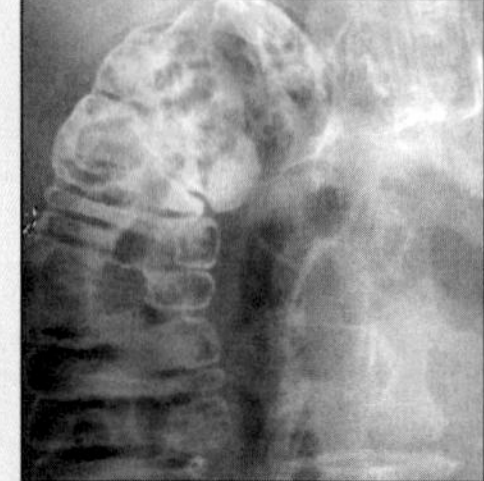

Retained Feces

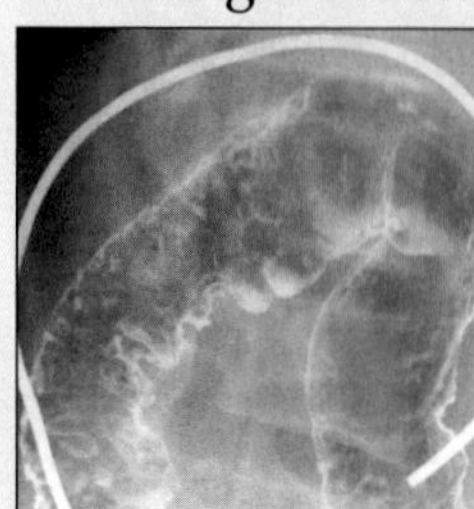

Ulcerative Colitis

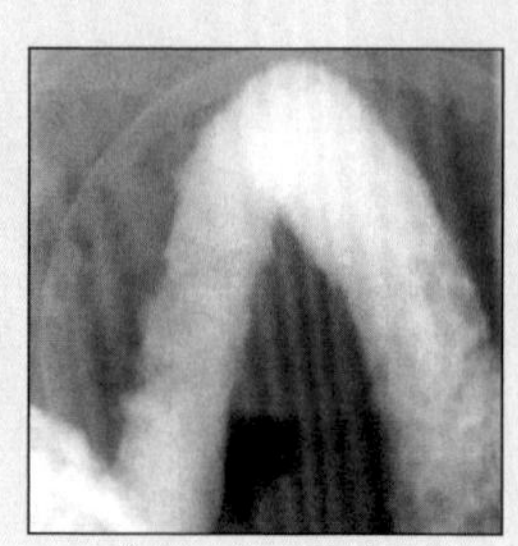

Ulcerative Colitis

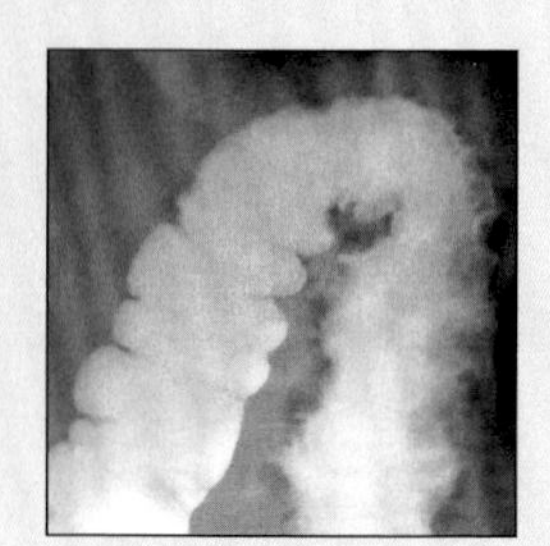

Primary Pneumatosis

Key Facts

Terminology

- Familial adenomatous polyposis syndrome (FAPS)
- Spectrum of autosomal dominant disease characterized by innumerable adenomatous colonic polyps & other associated lesions

Imaging Findings

- Best diagnostic clue: Innumerable colonic filling defects or ring shadows ± extraintestinal lesions
- Familial polyposis coli, osteomas, epidermoid (sebaceous) cyst
- Soft tissue tumors: Desmoid, mesenteric fibromatosis, lipoma
- Dental abnormalities; periampullary, duodenal & thyroid carcinomas
- FAPS adenomas small (80% < 5 mm) & sessile

Top Differential Diagnoses

- Retained feces & food
- Lymphoid hyperplasia
- Metastases & lymphoma
- Pseudopolyps
- Primary colonic pneumatosis

Pathology

- Virtually all untreated patients develop colon cancer
- FAPS is inherited as an autosomal dominant trait

Diagnostic Checklist

- Check for family history: Colonic polyps, abdominal soft tissue tumors & malignancies at a young age
- 500-2500 polyps carpeting entire colon-rectosigmoid
- Gardner syndrome: Soft tissue tumors, bony osteomas, dental defects & periampullary cancer

- Jejunum & ileum
  - Adenomas, lymphoid hyperplasia in > 20% cases
- Associated with ↑ incidence: Malignant CNS tumors

Radiographic Findings

- Fluoroscopic guided double contrast barium enema
  - Innumerable varied sized radiolucent filling defects
  - Carpet entire colon particularly rectosigmoid region
  - May be widely scattered radiolucent filling defects
- Fluoroscopic guided double contrast UGI, small bowel
  - Multiple small filling defects in stomach, duodenum, jejunum & ileum

CT Findings

- NECT
  - Imaging appearance varies due to relative amounts of fibroblast proliferation/fibrosis/fat/collagen content & vascularity of tumor
  - Desmoid tumor & mesenteric fibromatosis
    - Well or ill-defined; homo-/heterogeneous density
    - ± Displacement or compression of bowel loops
    - ± Areas of necrosis
    - Desmoid location: Mesentery & abdominal wall
- CECT: Both desmoid & mesenteric fibromatosis: Higher attenuation than muscle
- CT colonography after colonic air insufflation: Endoluminal images show
  - Small or large, sessile or pedunculated polyps extending from colonic wall
  - Polyps 10 mm & above: 90% Sensitivity

MR Findings

- Desmoid tumor & mesenteric fibromatosis
  - T1WI: ↓ Signal intensity relative to muscle
  - T2WI: Variable signal intensity (low, medium or high) relative to muscle
  - T1C+: Marked homo-/heterogeneous enhancement

Imaging Recommendations

- Double contrast barium studies (for polyps)
  - En face, profile & oblique views
- NE + CECT; MR & T1C+ (for abdominal tumors)
- CT colonography (for polyps)

DIFFERENTIAL DIAGNOSIS

Retained feces & food

- Filling defects in barium pool mimicking polyps

Lymphoid hyperplasia

- Lymphoid follicles
  - Aggregates of lymphocytes in muscularis mucosae
  - Seen in 50% of barium studies (kids); 13% (adults)
  - Enlarged or hyperplastic: Infectious, neoplastic, immunologic & inflammatory diseases
- Barium studies
  - Innumerable small or tiny radiolucent nodules
  - Usually generalized (duodenum, small-bowel, colon)
  - Simulate small size adenomatous polyposis
- Distinguished by clinical history & generalized pattern

Metastases & lymphoma

- Metastases (e.g., malignant melanoma, breast, lung)
  - Malignant melanoma
    - Smooth polypoid submucosal lesions of different sizes seen as filling defects may mimic polyps
    - Polypoid lesion with ulcers & radiating folds form a typical "spoke-wheel" pattern
  - Breast carcinoma: Mural nodules simulating polyps
  - Bronchogenic carcinoma
    - Single or multiple intramural lesions (flat or polypoid) indistinguishable from polyps
    - Frequently ulcerated, narrowing & obstruction
- Lymphoma
  - Distribution: Stomach (51%), small-bowel (33%), colon (16%), esophagus (<1%)
  - Low grade MALT lymphoma
    - Seen only in stomach due to H. pylori gastritis
    - Confluent varying-sized nodules (filling defects)
    - May be indistinguishable from gastric polyps
  - Non-Hodgkin lymphoma
    - Small or bulky polypoid masses may mimic polyps
    - Smooth surface, broad base, sessile lesions ± central depressions or ulcerations
    - Bull's-eye" sign: Polypoid mass with ulceration
    - Markedly thickened bowel wall & folds
    - Regional or widespread adenopathy seen

FAMILIAL POLYPOSIS

Pseudopolyps
- Example: Ulcerative colitis (common); granulomatous colitis
- Two types of pseudopolyps
  - Inflammatory pseudopolyps
  - Postinflammatory pseudopolyps
- Inflammatory pseudopolyps
  - Islands of elevated, inflamed, edematous mucosa surrounded by ulceration appear as pseudopolyps
    - Represent remnants of pre-existing mucosa & submucosa rather than new growths
  - Natural progression of collar button ulcers, which extend, interconnect & mimics pseudopolyps
- Postinflammatory pseudopolyps (mucosal overgrowth)
  - Regenerated mucosa results in polypoid lesions
    - May be small & rounded; long & filiform or bushlike structure simulating a villous adenoma
  - Seen during mucosal healing, so they are termed postinflammatory pseudopolyps
  - Also seen after ischemia or after any severe infection

Primary colonic pneumatosis
- Cystic intramural collections of gas in colon
- Asymptomatic
- Not due to ischemia

PATHOLOGY

General Features
- General path comments
  - Proliferation of mucosa
  - Polyps begin in rectosigmoid & spread entire colon
  - Polyps are indistinguishable from sporadic adenomatous polyps
  - Virtually all untreated patients develop colon cancer
- Genetics: Abnormal or deletion of APC gene located on chromosome 5q
- Etiology
  - FAPS is inherited as an autosomal dominant trait
  - Occasionally due to spontaneous mutations
- Epidemiology: FAPS affects 1 in 10,000 people in US

Gross Pathologic & Surgical Features
- Innumerable polyps carpeting colonic mucosa
- Desmoid tumor
  - Confined to muscle, fascia or deeply infiltrate
  - Size: 5-20 cm; firm & gritty texture; lack capsule
  - Cut surface: Glistening white + trabeculated

Microscopic Features
- Adenomas
  - Tubular, tubulovillous & villous; ± atypia or mitosis
- Desmoid tumor
  - Spindle shaped cells & dense bands of collagen

CLINICAL ISSUES

Presentation
- Most common signs/symptoms
  - Rectal bleeding & diarrhea (75% cases)
  - Asymptomatic, pain, mucus discharge
  - Family history of colonic polyps (66% cases)
- Extraintestinal manifestations: Gardner syndrome
  - Epidermoid cyst, lipoma, fibroma, desmoid tumors (3-29%), mesenteric fibromatosis, peritoneal adhesions, retroperitoneal fibrosis
  - Osteomas: Membranous bone-50%; mandible-80%
  - Teeth: Odontoma, unerupted supernumerary teeth
  - Thyroid cancer: Papillary type more common in girls & young women

Demographics
- Age: Mean age 16 years; by 35 years 95% have polyps
- Gender: Equal in both males & females

Natural History & Prognosis
- Complications
  - Polyps: Malignant transformation
    - Colon > periampullary > stomach > jejunum
  - Colon carcinoma by age 34-43 years
- Prognosis
  - Bad if abdominal desmoids, colonic carcinoma or ampullary carcinoma develop

Treatment
- Prophylactic total colectomy at about 20 years of age
- Permanent ileostomy, Kock pouch
- Continent endorectal pull-through pouch

DIAGNOSTIC CHECKLIST

Consider
- Check for family history: Colonic polyps, abdominal soft tissue tumors & malignancies at a young age

Image Interpretation Pearls
- 500-2500 polyps carpeting entire colon-rectosigmoid
- Gardner syndrome: Soft tissue tumors, bony osteomas, dental defects & periampullary cancer

SELECTED REFERENCES

1. Spigelman AD: Extracolonic polyposis in familial adenomatous polyposis: so near and yet so far. Gut. 53(3):322, 2004
2. Macari M et al: Diagnosis of familial adenomatous polyposis using two-dimensional and three-dimensional CT colonography. AJR Am. J. Roentgenol. 173: 249-250, 1999
3. Hizawa K et al: Desmoid tumors in familial adenomatous polyposis/Gardner's syndrome. J Clin Gastroenterol. 25(1):334-7, 1997
4. Harned RK et al: The hamartomatous polyposis syndromes: clinical and radiologic features. AJR Am J Roentgenol. 164(3):565-71, 1995
5. Rustgi AK: Hereditary gastrointestinal polyposis and nonpolyposis syndromes. N Engl J Med. 331(25):1694-702, 1994
6. Casillas J et al: Imaging of intra- and extraabdominal desmoid tumors. RadioGraphics 11: 959-968, 1991
7. Harned RK et al: Extracolonic manifestations of the familial adenomatous polyposis syndromes. AJR Am J Roentgenol. 156(3):481-5, 1991
8. Bartram CI et al: Colonic polyp patterns in familial polyposis. AJR 142: 305-308, 1984

FAMILIAL POLYPOSIS

IMAGE GALLERY

Typical

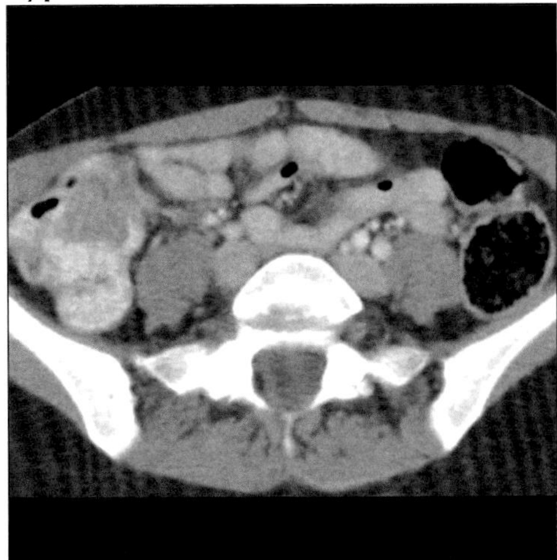

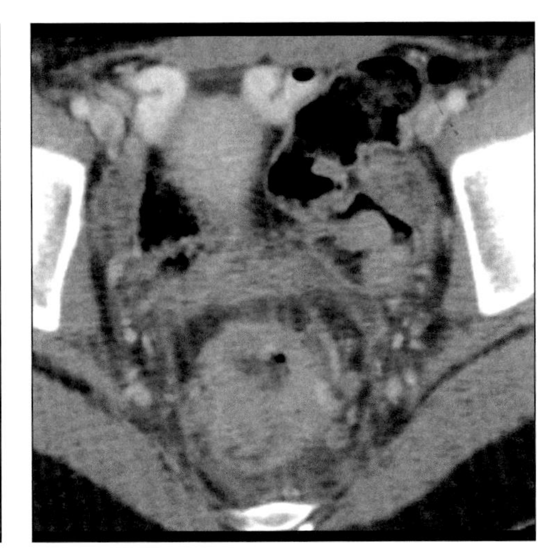

(Left) *Axial CECT shows subtle polypoid thickening of descending colon, large mass in cecum.* ***(Right)*** *Axial CECT, 25 year old woman with familial polyposis. Extensive rectal cancer with local invasion and lymphadenopathy.*

Typical

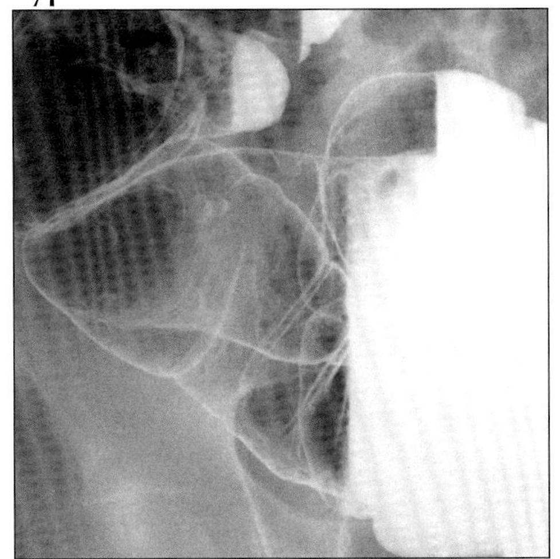

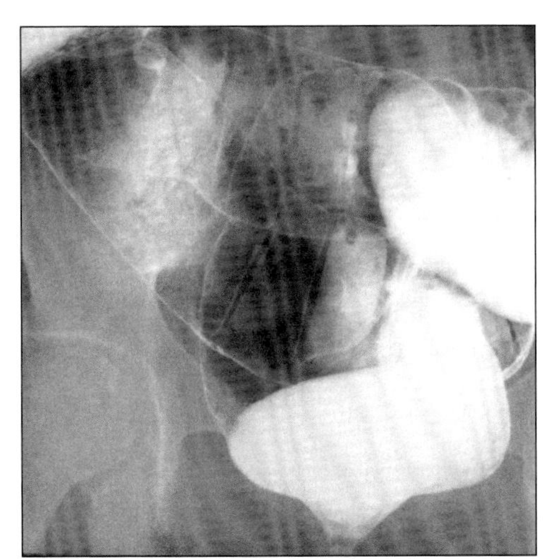

(Left) *Air contrast BE shows innumerable polyps in rectosigmoid colon.* ***(Right)*** *Air contrast BE shows innumerable polyps in rectosigmoid colon.*

Typical

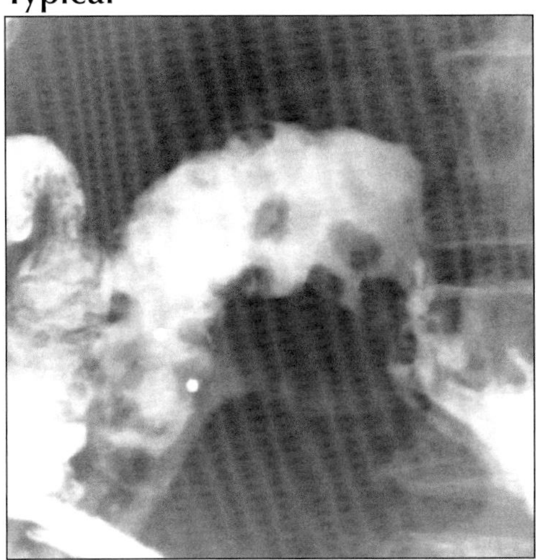

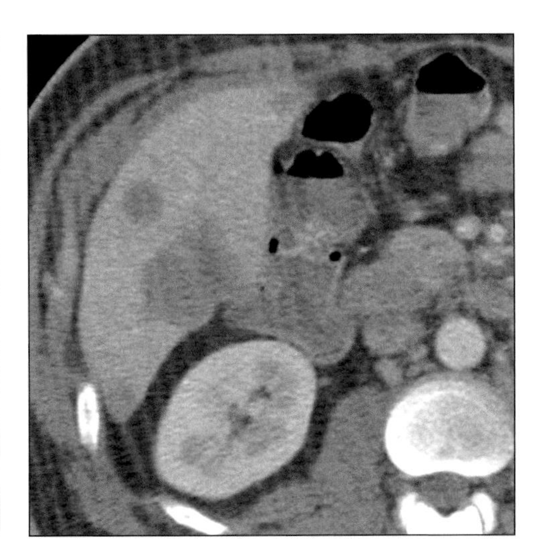

(Left) *Upper GI series shows polyps in duodenum. Familial polyposis-Gardner syndrome.* ***(Right)*** *Axial CECT shows extensive liver metastases in 23 year old man with familial polyposis.*

GARDNER SYNDROME

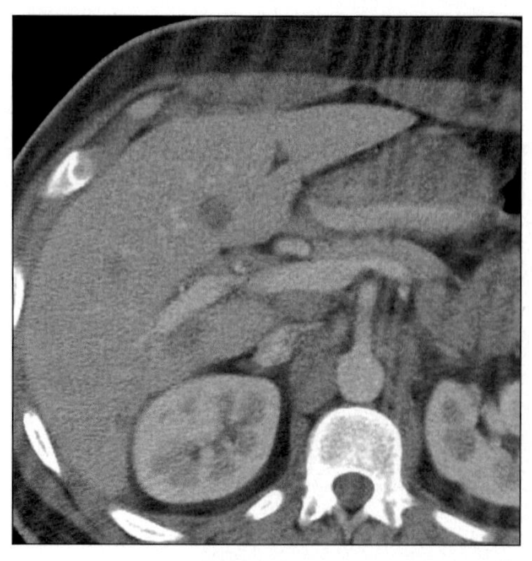

Axial CECT in a 30 year old man with Gardner syndrome shows multiple hepatic metastases.

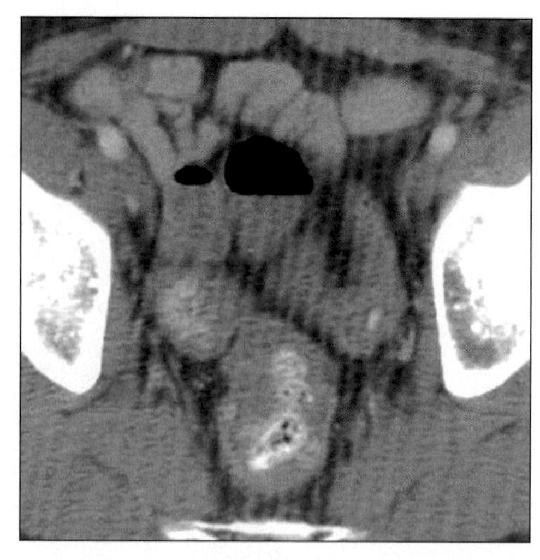

Axial CECT shows thickening of the rectosigmoid wall due to innumerable polyps + invasive rectal carcinoma.

TERMINOLOGY

Definitions

- Familial adenomatous polyposis + extracolonic lesions

IMAGING FINDINGS

General Features

- Best diagnostic clue: Innumerable, colonic, radiolucent filling defects with extraintestinal lesions
- Other general features
  - Combination of familial polyposis coli and
    - Osteomas; dental abnormalities
    - Desmoid tumor & mesenteric fibromatosis
    - Epidermoid cysts & fibromas of skin
    - Adrenal, thyroid & liver carcinomas
    - Congenital pigmented lesions of retina
  - Not all extracolonic lesions occur in same patient

Radiographic Findings

- Radiography
  - Osteomas
    - Cortical thickening of angle of mandible, sinuses, outer table of skull, flat bones & long bones
    - Size: Indiscernible to several cm
    - Single to dozens; localized or diffuse
- Fluoroscopic-guided barium enema & UGI
  - Familial polyposis: Innumerable varied sized radiolucent filling defects

CT Findings

- NECT
  - Desmoid tumors & mesenteric fibromatosis: Well or ill-defined; homo- or heterogeneous masses
  - Other carcinomas: Isodense or heterogenous attenuation of the mass
- CECT: Desmoid tumors & mesenteric fibromatosis: Increased attenuation; greater than muscles
- CT colonography
  - Familial polyposis: Small or large, sessile or pedunculated polyps extending from wall
  - May see colon carcinoma, liver mets

Imaging Recommendations

- Best imaging tool
  - Familial polyposis: Fluoroscopic-guided barium enema or CT colonography
  - Osteoma: Radiography
  - Desmoid tumors & mesenteric fibromatosis: CT

DIFFERENTIAL DIAGNOSIS

Retained fecal debris

- Filling defects on dependent surface in barium pool mimicking polyps

DDx: Multiple Colonic Filing Defects

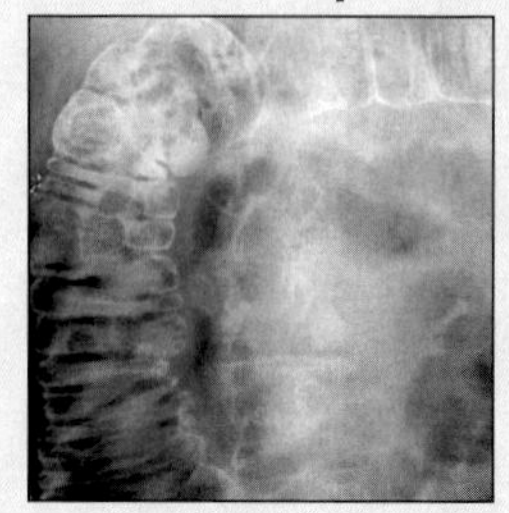

Retained Feces

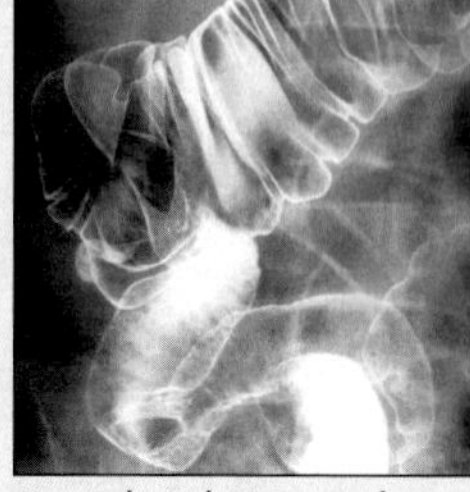

Lymphoid Hyperplasia

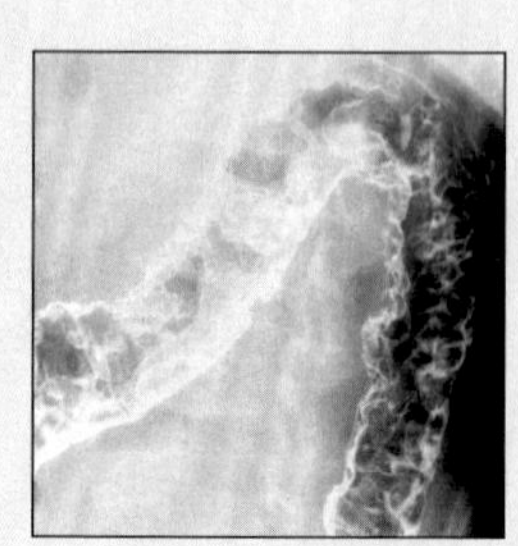

Granulom. Colitis

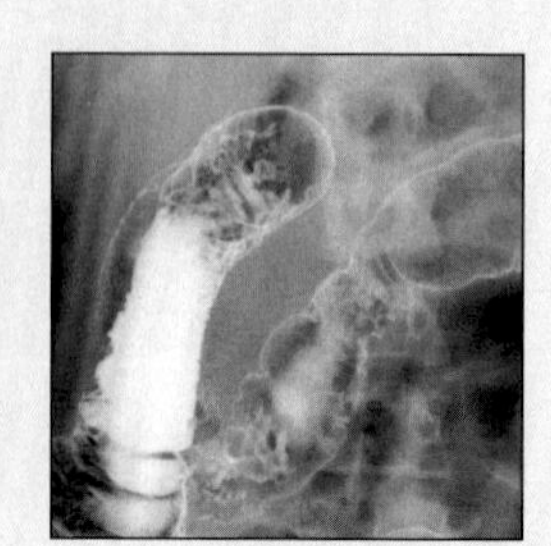

Ulcerative Colitis

GARDNER SYNDROME

Key Facts

Imaging Findings
- Best diagnostic clue: Innumerable, colonic, radiolucent filling defects with extraintestinal lesions

Top Differential Diagnoses
- Retained fecal debris
- Lymphoid hyperplasia
- Pseudopolyps
- Metastases & lymphoma

Pathology
- A variant of familial polyposis (very rare)
- Etiology: Autosomal dominant inheritance

Diagnostic Checklist
- Family history; colectomy to prevent colon carcinoma
- Innumerable polyps carpeting entire colon with extraintestinal manifestations

Lymphoid hyperplasia
- Enlarged or hyperplastic lymphoid follicles
- Innumerable small or tiny radiolucent nodules
- Usually generalized (duodenum, small bowel or colon)

Pseudopolyps
- Inflammatory pseudopolyps
  - Examples: Ulcerative or granulomatous colitis
  - Islands of elevated, inflamed, edematous mucosa surrounded by ulceration
- Postinflammatory pseudopolyps
  - Mucosal healing; mimicking villous adenoma
  - Small & round; long & filiform; bushlike

Metastases & lymphoma
- Metastases (e.g., Malignant melanoma, breast or lung)
- Lymphoma (e.g., low grade MALT or non-Hodgkin)

PATHOLOGY

General Features
- General path comments
  - A variant of familial polyposis (very rare)
  - Dental abnormalities: Unerupted or supernumerary teeth, dentigerous cysts & odontomas
  - Epidermoid (sebaceous) cysts & fibromas of skin: Common on legs, face, scalp & arms; mm to cm
  - Congenital pigmented lesions of retina
    - Single, multiple, bilateral; 0.1 to 1.0 disc diameter
    - Darkly pigmented; round, oval or kidney-shaped
- Genetics: Mutation in APC gene at 5q22
- Etiology: Autosomal dominant inheritance
- Associated abnormalities: Colonic adenomatous polyps → colon carcinoma in 100% if not treated

CLINICAL ISSUES

Presentation
- Most common signs/symptoms
  - rectal bleeding, diarrhea
  - Skin, dental or retinal abnormalities

Demographics
- Age: Mean age of diagnosis is 22 years of age
- Gender: M:F = 1:1

Natural History & Prognosis
- Congenital pigmented lesions of retina may be earliest clinically detectable lesion
- Osteomas, dental & retinal abnormalities & epidermal cysts occur prior to puberty & familial polyposis
- Desmoid tumors & mesenteric fibromatosis usually occur post-operation
  - Histologically benign but aggressive growth; ↑ morbidity and mortality

Treatment
- Prophylactic colectomy to prevent colon carcinoma

DIAGNOSTIC CHECKLIST

Consider
- Family history; colectomy to prevent colon carcinoma

Image Interpretation Pearls
- Innumerable polyps carpeting entire colon with extraintestinal manifestations

SELECTED REFERENCES

1. Van Epps KJ et al: Epidermoid inclusion cysts seen on CT of a patient with Gardner's syndrome. AJR Am J Roentgenol. 173(3):858-9, 1999
2. Kawashima A et al: CT of intraabdominal desmoid tumors: is the tumor different in patients with Gardner's disease? AJR Am J Roentgenol. 162(2):339-42, 1994
3. Nannery WM et al: Familial polyposis coli & Gardner's syndrome. N J Med. 87(9):731-3, 1990

IMAGE GALLERY

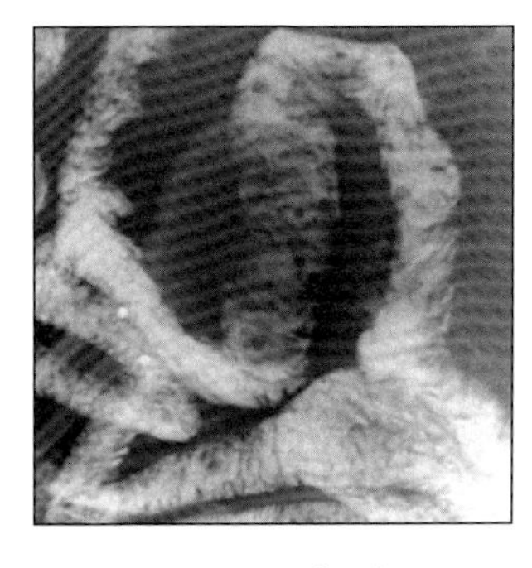

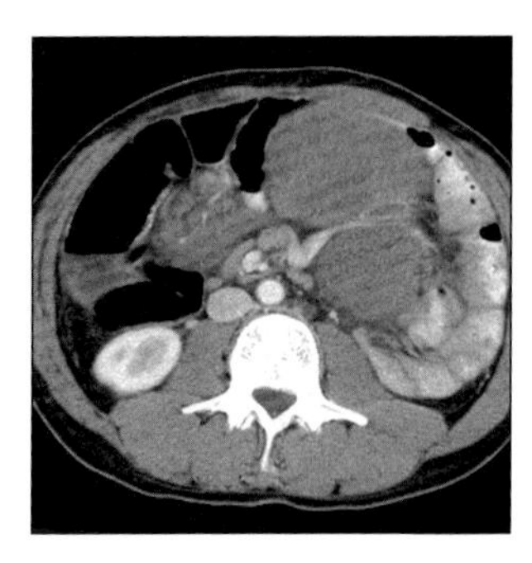

(Left) *Barium study shows numerous jejunal polyps (adenomas).* ***(Right)*** *Axial CECT in a patient who had colectomy for Gardner polyposis. Large rapidly-growing mesenteric masses are desmoid tumors.*

SIGMOID VOLVULUS

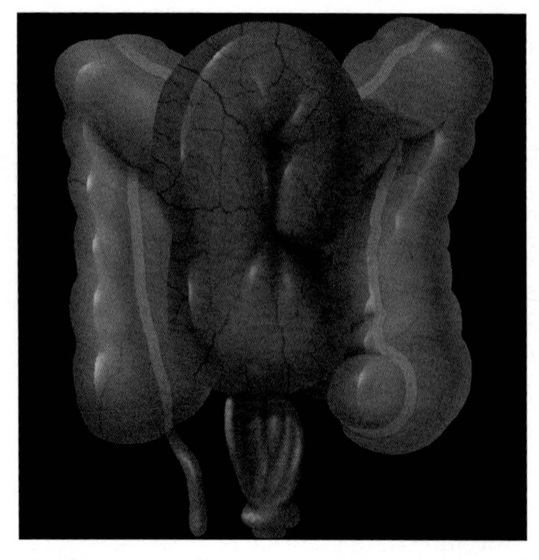

Graphic shows dilated, twisted, elongated sigmoid colon with venous engorgement + colonic obstruction.

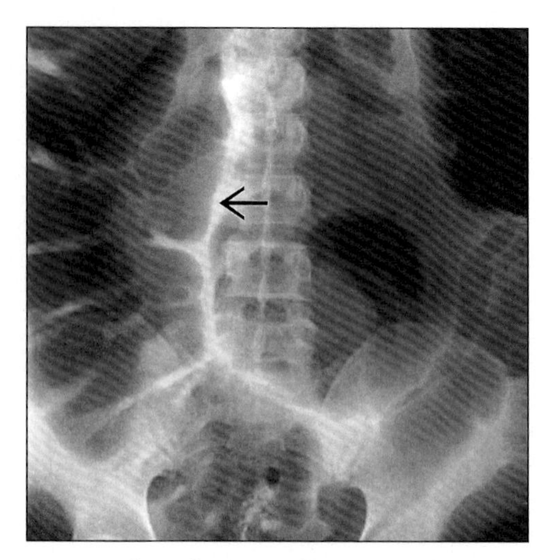

Supine radiograph shows dilation of entire colon. Vertical white line (arrow) represents the apposed walls of the dilated, inverted sigmoid colon and points toward the mesenteric volvulus.

TERMINOLOGY

Abbreviations and Synonyms

- Volvulus of sigmoid colon

Definitions

- Torsion or twisting of sigmoid colon around its mesenteric axis

IMAGING FINDINGS

General Features

- Best diagnostic clue: Dilated sigmoid colon with inverted U configuration and absent haustra
- Location: At midline; directed toward RUQ or LUQ → elevation of hemidiaphragm
- Other general features
  - Types of colonic volvulus
    - Sigmoid volvulus: 60-75%
    - Cecal volvulus: 22-33%
    - Transverse colon volvulus: 2-4%
    - Splenic flexure volvulus: < 1%
    - Compound volvulus: Very rare
  - Colonic volvulus is rare in children
    - Radiography may be interpreted as normal; used to exclude other causes of abdominal pain and free air
  - Twist > 360° do not resolve spontaneously

Radiographic Findings

- Radiography
  - Sigmoid volvulus
    - Vertical dense white line: Apposed inner walls of sigmoid colon pointing toward the pelvis
    - Closed loop obstruction: Segment of bowel obstructed at two points
    - Gas in proximal small intestine and colon; absence of gas in rectum
    - Absent rectal gas in spite of prone or decubitus views
    - "Northern exposure" sign: Dilated, twisted sigmoid colon projects above transverse colon
    - Apex above T10 vertebra and under left hemidiaphragm; directed toward right shoulder
  - Cecal volvulus
    - Dilated air-filled cecum in an ectopic location
    - Cecal apex in LUQ
    - Kidney or coffee bean-shaped gas-filled cecum
    - One or two haustral markings usually seen
    - Markedly distended gas or fluid-filled small bowel; little gas in distal colon
  - Splenic flexure volvulus
    - Dilated, featureless, air-filled bowel loop in LUQ; separate from stomach
  - Compound volvulus

DDx: Dilated Colon

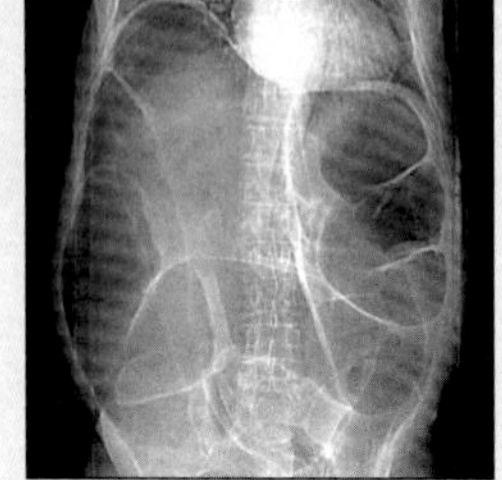

Acute Ileus

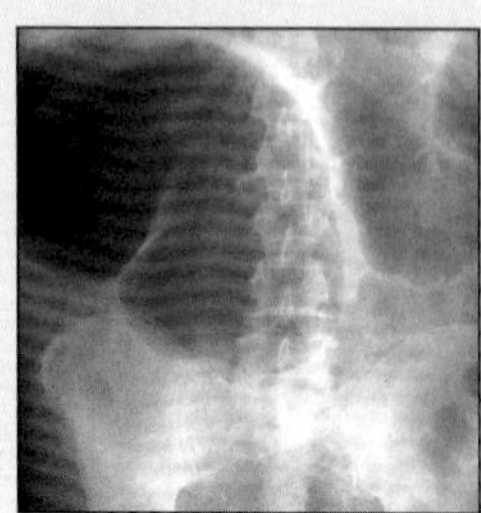

Ogilvie Syndrome

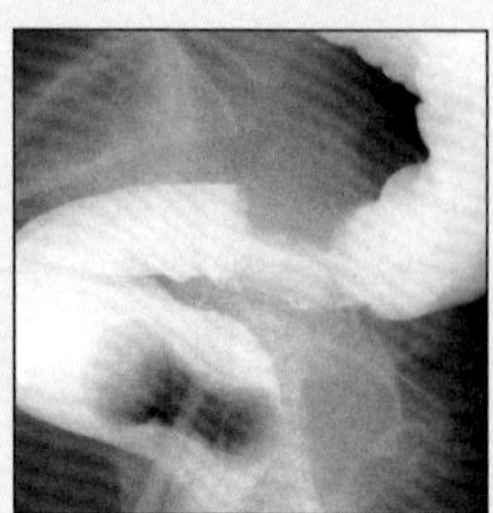

Colon Cancer

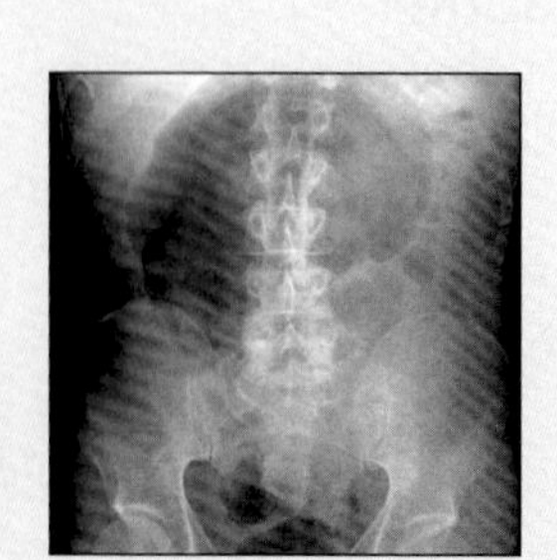

Cecal Volvulus

SIGMOID VOLVULUS

Key Facts

Terminology
- Torsion or twisting of sigmoid colon around its mesenteric axis

Imaging Findings
- Best diagnostic clue: Dilated sigmoid colon with inverted U configuration and absent haustra
- Location: At midline; directed toward RUQ or LUQ → elevation of hemidiaphragm
- Gas in proximal small intestine and colon; absence of gas in rectum
- Absent rectal gas in spite of prone or decubitus views
- "Northern exposure" sign: Dilated, twisted sigmoid colon projects above transverse colon
- "Beaking": Smooth, tapered narrowing or point of torsion at rectosigmoid junction
- "Whirl" sign: Tightly twisted mesentery and bowel
- Radiography (supine, upright, prone and decubitus views)

Top Differential Diagnoses
- Acute ileus
- Functional megacolon
- Distal colon obstruction

Clinical Issues
- Acute or insidious in onset
- Abdominal pain (< 33%), vomiting and distension

Diagnostic Checklist
- Acute abdomen; rule out other causes of obstruction
- Dilated sigmoid colon in inverted U configuration; absent haustra; "beaking"; "whirl" sign

---

    - Dilated sigmoid loop in mid-abdomen extending to RLQ with distended small bowel
    - Medially deviated distal left colon
- Fluoroscopic-guided water-soluble contrast enema
  - Sigmoid volvulus (can use low pressure barium enema without balloon inflation)
    - "Beaking": Smooth, tapered narrowing or point of torsion at rectosigmoid junction
    - Mucosal folds often show a corkscrew pattern at point of torsion
    - Shouldering: Localized wall thickening at site of twist (chronic)
  - Cecal volvulus
    - "Beaking" at mid-ascending colon
  - Transverse colon volvulus
    - "Beaking" at level of transverse colon
    - Two air-fluid levels in dilated transverse colon (helpful in distinguishing from cecal volvulus)
  - Splenic flexure volvulus
    - "Beaking" at LUQ

CT Findings
- CECT
  - Progressive tapering of afferent and efferent limbs leading into the twist or "beaking"
  - "Whirl" sign: Tightly twisted mesentery and bowel
  - Compound volvulus
    - Medial deviation of distal left colon with pointed appearance of its medial border

Imaging Recommendations
- Radiography (supine, upright, prone and decubitus views)
- Fluoroscopic-guided water-soluble contrast enema; helical CT

DIFFERENTIAL DIAGNOSIS

Acute ileus
- Post-op, medication, post-traumatic injury and ischemia
- Dilated large bowel with no transition point
- Air-fluid levels observed, but no peristalsis
- No colonic obstruction

Functional megacolon
- Gross constipation without organic cause
- Markedly dilated, ahaustral, air or stool-filled colon
- Ogilvie Syndrome - non-obstructive dilation of cecum

Distal colon obstruction
- Change in stool caliber over several months
- Gas-filled intestinal loops proximal to obstruction; no gas seen distally
- Abrupt transition at site of obstruction
- Malignancy
  - Most common (55%) cause of colonic obstruction
  - Insidious in onset
  - Weakness, weight loss and anorexia
  - "Apple-core" configuration with destruction of mucosa
  - Positive fecal occult blood test is highly suggestive of colon cancer
- Stricture secondary to diverticulitis
  - Second most common (12%) cause of colonic obstruction
  - History of recurrent attacks of diverticulitis
  - Other diverticula are present

PATHOLOGY

General Features
- Etiology
  - Major predisposing factors for colonic volvulus
    - Redundant segment of bowel that is freely moveable within the peritoneal cavity
    - Close approximation of points of bowel fixation
  - Sigmoid volvulus
    - Diet: ↑ Fiber → ↑ bulk of stool and elongates colon
    - Chronic constipation and obtundation from medications → gaseous distension
    - Degree of rotation relative to chance of nonsurgical decompression: 180°:35%; 360°:50%; 540°:10%

SIGMOID VOLVULUS

- Cecal volvulus
  - Congenital defect in attachment
  - Postpartum ligamentous laxity and mobile cecum
  - Colon distension (pseudo-obstruction, distal tumor, endoscopy, enema or postoperative ileus)
- Transverse colon volvulus
  - Failure of normal fixation of mesentery → ↑ mobility of right colon and hepatic flexure
- Splenic flexure volvulus
  - Postoperative adhesions
  - Congenital or surgical removal of normal attachments to abdominal wall
- Compound volvulus
  - Also known as ileosigmoid knot
  - Hyperactive ileum winding around narrow pedicle of a passive sigmoid colon
- Etiology in children
  - Malrotation and other mesenteric attachment abnormalities
  - Constipation (mental retardation, Hirschsprung's disease, cystic fibrosis or aerophagia)

- Epidemiology
  - Third most common (10%) cause of colonic obstruction
  - Incidence of colonic volvulus
    - U.S. and other western countries: 1-4% of intestinal obstructions
    - Africa and Asia: 20-25% of intestinal obstructions
  - Incidence of sigmoid volvulus
    - U.S.: 1-2% of intestinal obstructions
    - Increased incidence in elderly men and residents of nursing homes and mental hospitals (constipation and obtundation)
    - Increased significantly in South America and Africa (↑ fiber in diet)
  - Colonic volvulus in children
    - Mean is 7 years of age
    - Boys to girls: 2-3:1
- Associated abnormalities
  - Comorbid disease in sigmoid volvulus
    - 30% with psychiatric disease; 13% are institutionalized at time of diagnosis

Gross Pathologic & Surgical Features

- Twisted narrow segment with marked proximally dilated bowel loop

Microscopic Features

- Localized thickening of mucosal folds; ischemic and necrotic changes

CLINICAL ISSUES

Presentation

- Most common signs/symptoms
  - Acute or insidious in onset
  - Abdominal pain (< 33%), vomiting and distension
  - Transverse colon volvulus
    - Severe vomiting (compression of duodenojejunal junction at root of mesentery)
  - Compound volvulus
    - Rapid deterioration (greater than other colonic volvulus)
    - Pain out of proportion to physical findings and absolute constipation
  - Diagnosis
    - Sigmoid volvulus and cecal volvulus: Diagnosed by radiography (75%)
    - Transverse colon and splenic flexure volvulus: Diagnosed by fluoroscopic-guided water-soluble contrast enema

Demographics

- Age
  - Sigmoid volvulus: 60-70 years of age
  - Cecal volvulus: Younger age than sigmoid

Natural History & Prognosis

- Complications
  - Closed loop obstruction → strangulation
  - Ischemia, necrosis (15-20%) and perforation
  - Ileosigmoid knot → strangulation and gangrene of small bowel within hours
- Prognosis
  - Uncomplicated: Good; complicated: Poor
  - Colonic volvulus
    - 8% mortality (from gangrenous bowel)
  - Sigmoid volvulus
    - 40-50% recurrence after nonoperative reduction
    - 3% recurrence after nonoperative and operative reduction
  - Transverse colon volvulus: Up to 33% mortality

Treatment

- Sigmoid volvulus
  - Nonoperative: Proctoscopic or colonoscopic decompression of obstruction ± stabilization by inserting rectal tube (successful 70-80% of attempts)
  - Nonoperative and operative: Decompression, mechanical cleansing and elective sigmoid resection
- Complicated cases: Surgical emergency
- Follow-up
  - Sigmoid volvulus: Fluoroscopic-guided water-soluble contrast enema to rule out underlying colon cancer

DIAGNOSTIC CHECKLIST

Consider

- Acute abdomen; rule out other causes of obstruction

Image Interpretation Pearls

- Dilated sigmoid colon in inverted U configuration; absent haustra; "beaking"; "whirl" sign

SELECTED REFERENCES

1. Moore CJ et al: CT of cecal volvulus: unraveling the image. AJR Am J Roentgenol. 177(1):95-8, 2001
2. Lee SH et al: The ileosigmoid knot: CT findings. AJR Am J Roentgenol. 174(3):685-7, 2000
3. Dulger M et al: Management of sigmoid colon volvulus. Hepatogastroenterology. 47(35):1280-3, 2000
4. Javors BR et al: The northern exposure sign: A newly described finding in sigmoid volvulus. AJR 173:571-574, 1999
5. Catalano O: Computed tomographic appearance of sigmoid volvulus. Abdominal Imaging 21:314-317, 1996

SIGMOID VOLVULUS

IMAGE GALLERY

Typical

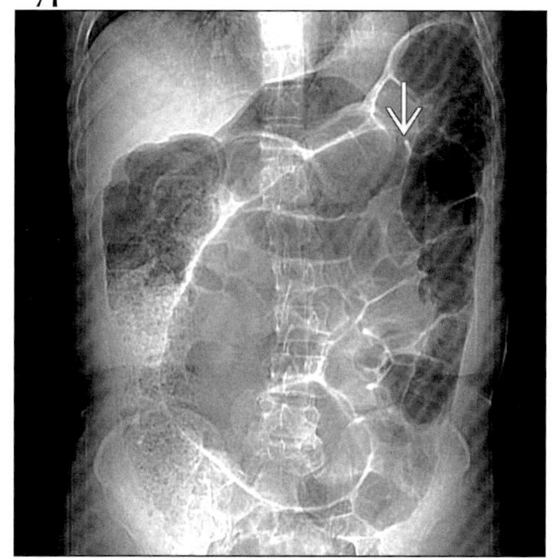

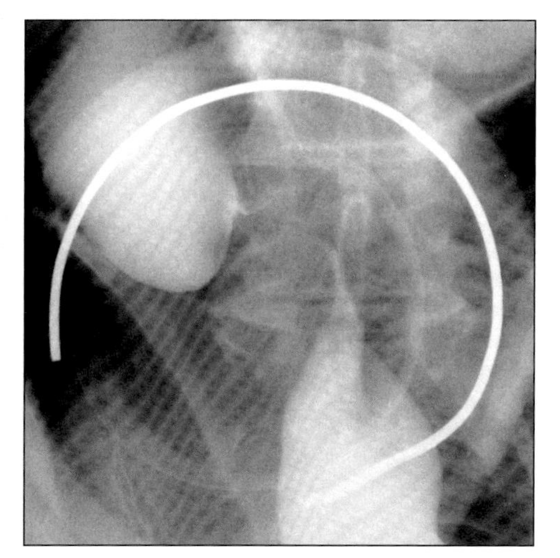

(Left) *Supine radiograph shows dilated colon. The apex of the sigmoid colon (arrow) is above the transverse colon, the "northern exposure" sign of sigmoid volvulus.* ***(Right)*** *Single contrast BE shows a smooth tapered beak obstructing the lumen of the sigmoid colon.*

Typical

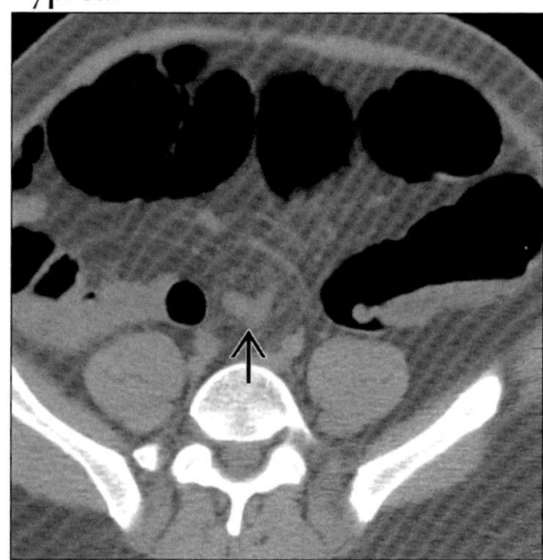

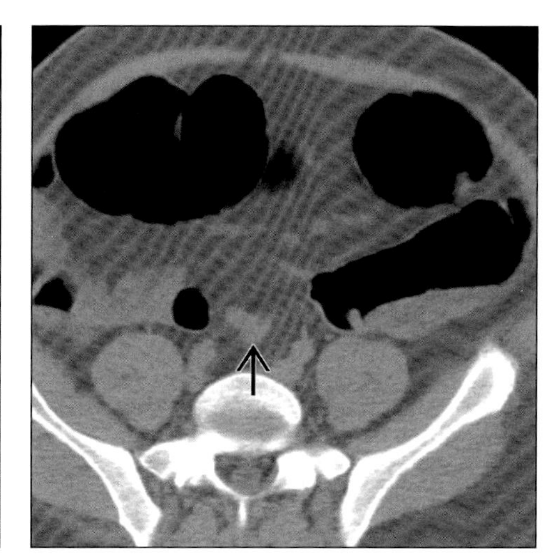

(Left) *Axial CECT of sigmoid volvulus shows a diffuse dilation of colon. There is a swirl of sigmoid mesocolic blood vessels that converge at the site of volvulus (arrow).* ***(Right)*** *Axial CECT shows swirl of mesocolic vessels at the base of the volvulus (arrow).*

Typical

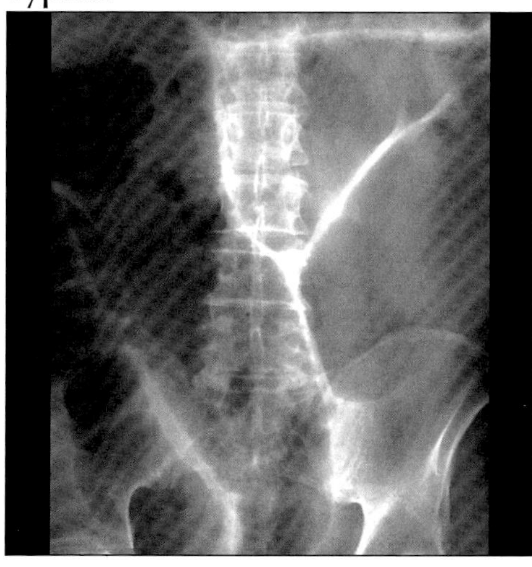

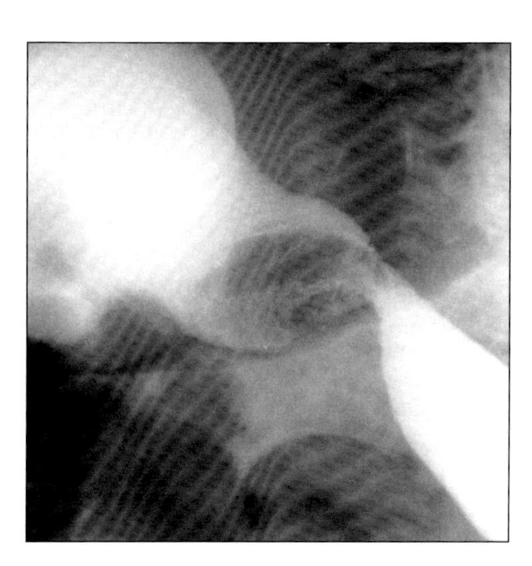

(Left) *Supine radiograph shows dilated, inverted "U" shaped sigmoid colon.* ***(Right)*** *Single contrast BE shows twist + beak at point of volvulus with dilated colon beyond the twist.*

CECAL VOLVULUS

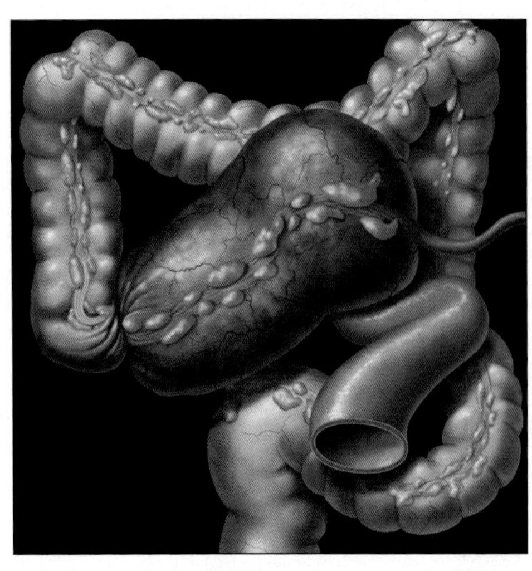

Graphic shows twist (volvulus) of ascending colon, obstructing lumen and blood supply. Cecum, on a mesentery, dilated + displaced toward left upper quadrant.

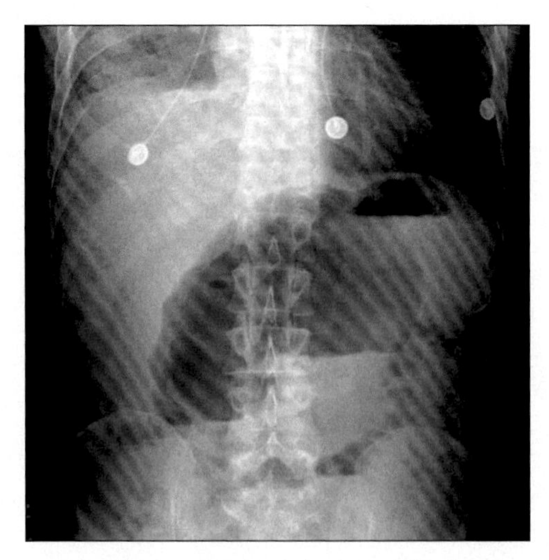

Upright abdominal radiograph shows dilated cecum with air-fluid level pointing toward left upper quadrant. Remainder of colon collapsed.

TERMINOLOGY

Abbreviations and Synonyms

- Volvulus of cecum, ascending colon

Definitions

- Rotational twist of the right colon on its axis; associated with folding of the right colon

IMAGING FINDINGS

General Features

- Best diagnostic clue: Dilated, twisted cecum with tip pointing to left upper quadrant
- Location
  - Twist is distal to ileocecal valve (term is a misnomer)
  - Cecum is located in mid-abdomen or LUQ
  - Terminal ileum swings around the dilated bowels

Radiographic Findings

- Radiography
  - Dilated air-filled cecum in an ectopic location
  - Single, long air-fluid level
  - Cecal apex at LUQ
  - Medially placed ileocecal valve produces soft tissue indentation → kidney or coffee bean-shaped gas-filled cecum
  - One or two haustral markings usually seen
  - Markedly distended gas or fluid-filled small bowel; little gas in distal colon
- Fluoroscopic-guided water-soluble contrast enema
  - "Bird's beak" sign: Point of torsion at mid-ascending colon

CT Findings

- CECT
  - Progressive tapering of afferent and efferent limbs leading into the twist or "beaking"
  - "Whirl" sign: Tightly twisted mesentery and bowel at right mid-abdomen or RUQ

Imaging Recommendations

- Best imaging tool: Fluoroscopic-guided water-soluble contrast enema

DIFFERENTIAL DIAGNOSIS

Sigmoid volvulus

- Dilated, ahaustral sigmoid loop with inverted U configuration
- Dilated proximal colon; Gas-less distally

Acute ileus

- Dilated colon to rectum with haustra pattern

Distal colon obstruction

- Gas and stool-filled colon

DDx: Marked Colonic Distention

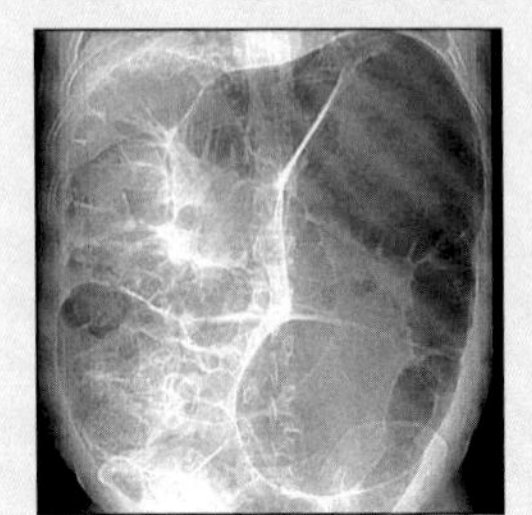

Sigmoid Volvulus

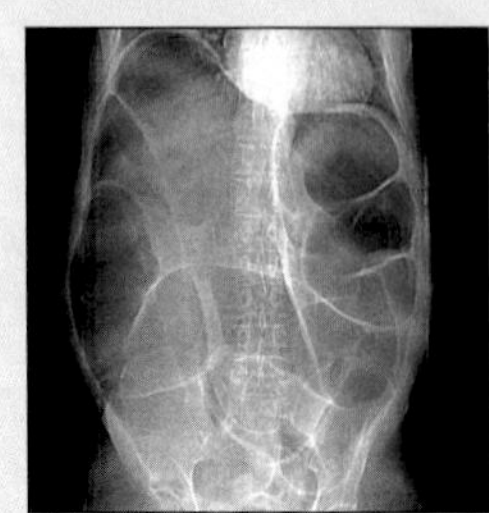

Acute Ileus

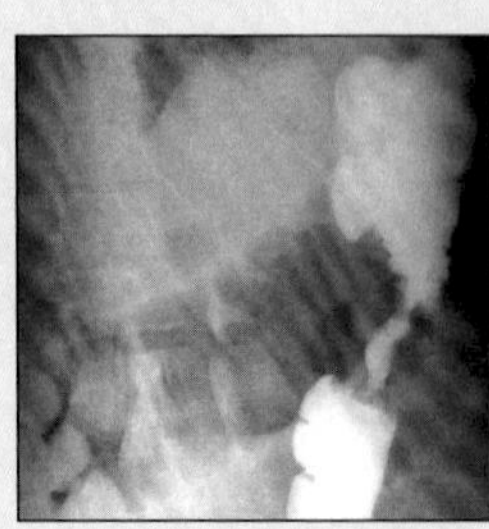

Colon Cancer

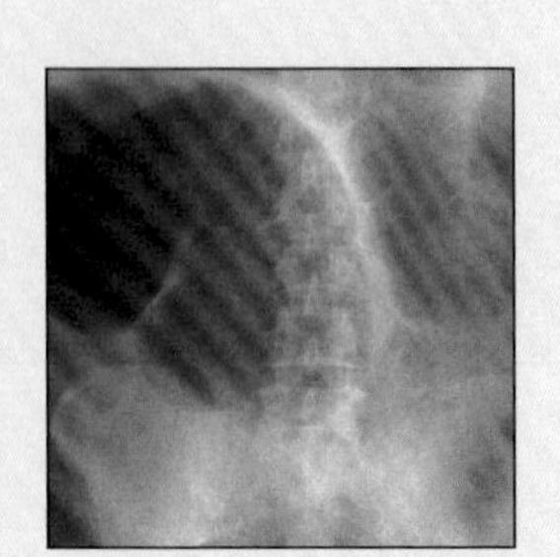

Olgilvie Syndrome

CECAL VOLVULUS

Key Facts

Imaging Findings
- Best diagnostic clue: Dilated, twisted cecum with tip pointing to left upper quadrant
- Single, long air-fluid level
- Medially placed ileocecal valve produces soft tissue indentation → kidney or coffee bean-shaped gas-filled cecum
- Markedly distended gas or fluid-filled small bowel; little gas in distal colon

Top Differential Diagnoses
- Sigmoid volvulus

Clinical Issues
- Acute or insidious onset
- Abdominal pain, distension and vomiting

Diagnostic Checklist
- Rule out ileus and Ogilvie syndrome

Functional megacolon
- Gross constipation without organic cause
- Markedly dilated, ahaustral, air or stool-filled colon

Ogilvie syndrome
- Colonic pseudo-obstruction without mechanical cause

PATHOLOGY

General Features
- General path comments
  - Cecal bascule
    - Anterior folding (not twisting) of cecum positioned at mid-abdomen
    - Possibly due to adhesive band from previous abdominal surgery
  - Embryology-Anatomy
    - Right colon is incompletely fused to posterior parietal peritoneum (10-37% adults)
- Etiology
  - Congenital defect in attachment of right colon
  - Postpartum ligamentous laxity and a mobile cecum
  - Colon distension (pseudo-obstruction, distal tumor, endoscopy, enema, or postoperative ileus)
  - Chronic constipation and laxative use
- Epidemiology
  - One-third of all cases of colonic volvulus
  - 2-3% of colonic obstructions
  - 22-33% of colonic volvulus; second to sigmoid
- Associated abnormalities
  - One third of patients have concomitant partially obstructing lesion located more distally in the colon
  - Malrotation and long mesentery

Gross Pathologic & Surgical Features
- Twisted, markedly dilated segment with moderate dilation of small intestine

Microscopic Features
- Localized thickening of mucosal folds; ischemic and necrotic changes

CLINICAL ISSUES

Presentation
- Most common signs/symptoms
  - Acute or insidious onset
  - Abdominal pain, distension and vomiting
- Diagnosis: 75% by radiography

Demographics
- Age: Younger patients than sigmoid volvulus

Natural History & Prognosis
- Complications
  - Ischemia, necrosis (15-20%) and perforation
- Prognosis
  - Uncomplicated: Good; Complicated: Poor

Treatment
- Colonoscopy to reduce volvulus (higher risk of perforation than sigmoid volvulus)
- Complicated cases: Surgical emergency
- Surgical options: Cecopexy, cecostomy and resection

DIAGNOSTIC CHECKLIST

Consider
- Rule out ileus and Ogilvie syndrome

Image Interpretation Pearls
- Massively dilated cecum at mid-abdomen, distended loops of small bowel, "bird's beak" sign

SELECTED REFERENCES

1. Moore CJ et al: CT of cecal volvulus. AJR. 177:95-98, 2001
2. Hemingway AP: Cecal volvulus: A new twist to the barium enema. British Journal Radiol. 53:806-807, 1980

IMAGE GALLERY

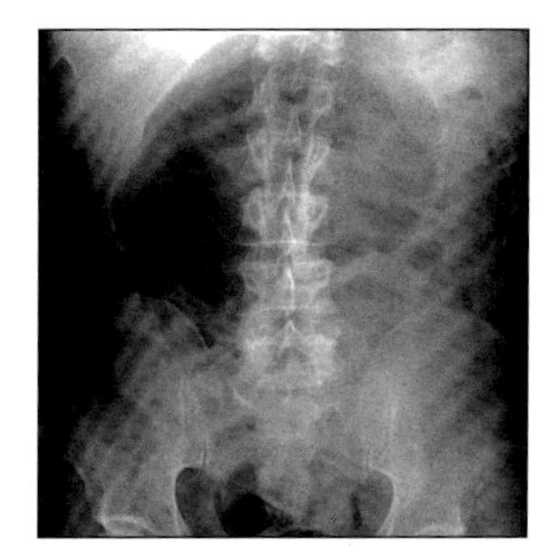

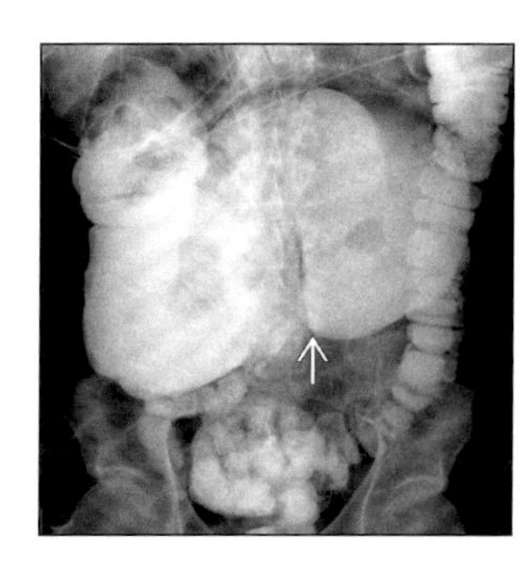

*(**Left**) Supine radiograph shows dilated cecum in mid-abdomen, pointed toward left upper quadrant. (**Right**) Cecal bascule. Enema fills markedly dilated cecum (ascending colon) which is folded acutely + displaced. Note ileo-cecal valve (arrow).*

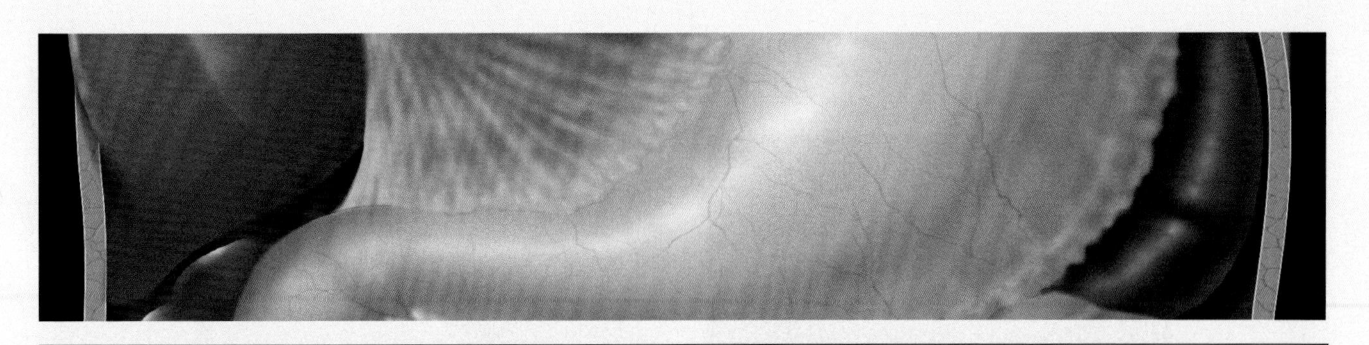

SECTION 6: Spleen

SPLEEN ANATOMY AND IMAGING ISSUES

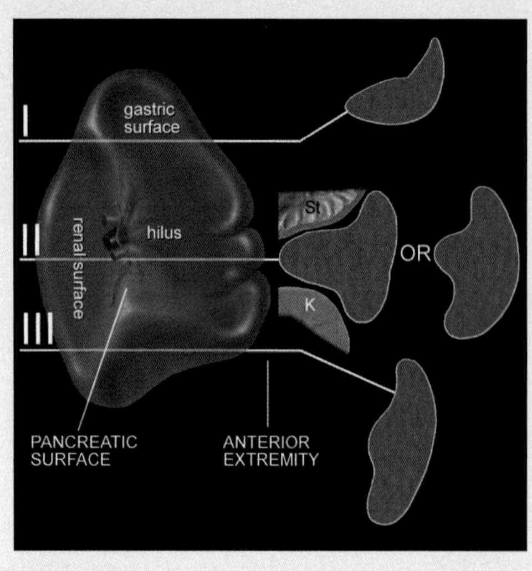

Graphic shows some of the variations in splenic shape on axial sections.

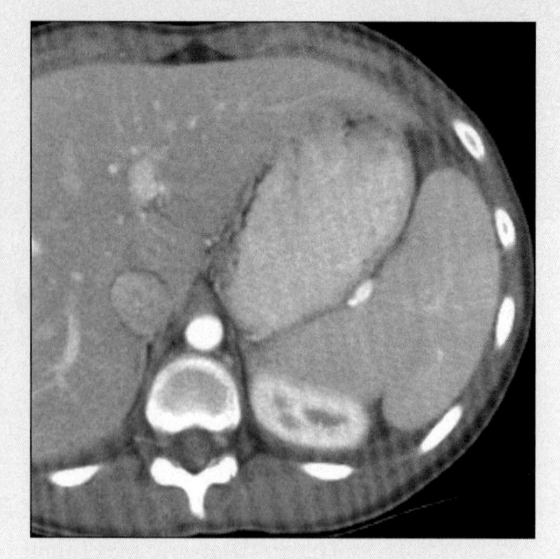

Axial CECT shows prominent medial lobulation of the spleen, a common variant.

IMAGING ANATOMY

Location

- Usually left upper quadrant
- Intraperitoneal
- Supported by gastrosplenic and splenorenal ligaments

Anatomic Relationships

- Diaphragm
  - Supero-laterally
- Left kidney
  - Posterio-medially
- Pancreas
  - Medially
- Stomach
  - Anterio-medially
- Tail of pancreas inserts into spleno-renal ligament
  - Pancreatic tail becomes intraperitoneal
  - Pancreatitis involving tail can spread directly to the spleen, lead to intrasplenic pseudocyst

ANATOMY-BASED IMAGING ISSUES

Normal Measurements

- Size can vary widely among individuals
  - Range 100 to 250 cm^3 , mean 150 cm^3 in adults
- Size varies even in one person
  - Age, state of nutrition, body habitus, blood volume
  - Average about 12 cm length, 7 cm breadth by 4 cm width
  - Length x width x breath should not exceed 470 cm^3
  - Can measure volume accurately by CT computation and summation of splenic area on sequential scans (limited clinical value)
  - Splenomegaly often results in convexity of (usually concave) visceral surface
- Structure: Branching trabeculae subdivide the spleen into communicating compartments
  - Branches of arteries, veins, nerves, lymphatics travel through trabeculae
- Splenic red pulp
  - Comprises the vascular tissue of the spleen
    - Splenic cords (plates of cells) lie between sinusoids; red pulp veins drain sinuses
  - Most common source of primary nonhematolymphoid tumors
  - Mottled enhancement, typical of early phase enhanced CT or MR is due to variable flow rates through cords and sinuses of red pulp
    - Can simulate or hide splenic pathology on arterial phase imaging
- Splenic white pulp
  - Comprises the lymphatic tissues, organized similar to lymph nodes
  - Gives rise to lymphatic tumors
    - Most common splenic tumor, lymphoma
- Frequently has notches and indentations on surface
  - On axial CT/MR sections may simulate laceration
  - Key differential feature is absence of perisplenic hemorrhage

Key Concepts or Questions

- MR appearance of spleen
  - Spleen has relatively long T1 and T2 relaxation times
  - Appears dark relative to liver on T1 WI
    - Similar to renal cortex
    - Becomes abnormally dark with iron deposition (transfusion hemochromatosis)
  - Liver metastases often similar in signal to normal spleen
- Spleen texture
  - Soft and pliable, relatively mobile
  - Easily indented and displaced by masses and even loculated fluid collections
  - Changes position in response to resection of adjacent organs
    - (E.g., post nephrectomy)
- Splenic lymphatic tumors
  - Most common
  - Lymphoma, leukemia

SPLEEN ANATOMY AND IMAGING ISSUES

DIFFERENTIAL DIAGNOSIS

Primary benign splenic tumors
- Hemangioma
- Lymphangioma
- Hamartoma
- Lipoma
- (Inflammatory pseudotumor)

Splenic tumors: Malignant - primary
- Lymphoma
- Hemangiopericytoma
- Angiosarcoma
- Hemangioendothelioma

Splenic tumors: Malignant - metastatic
- Melanoma (50% of cases)
- Breast, lung, ovary, etc.

Splenic cyst - solitary or multiple
- Congential cyst
- Post-traumatic cyst
- Parasitic cyst (hydatid)
- Abscess
- Metastasis
- Intrasplenic pseudocyst

Multiple complex or solid masses
- Lymphoma, leukemia
- Abscesses (immunocompromised)
- Candida, mycobacterial, Pneumocystis, etc.
- Angiosarcoma
- Sarcoidosis
- Metastases
- Multiple hemangiomas or lymphangioma

 - Imaging appearance: Splenomegaly (often massive in leukemia, NHL)
- Splenic metastases
  - Usually multiple and part of widespread disease
  - Variety of sources, especially melanoma
  - Contiguous spread (stomach, pancreas)
  - Retrograde spread through splenic vein
- Primary vascular tumors
  - Hemangioma
    - Peripheral or solid enhancement
  - Hamartoma
    - Homogeneous hypervascular
  - Lymphangioma
    - Multicystic, subcapsular
  - Littoral cell angioma
    - Splenomegaly, multiple nodules
  - Peliosis
    - Multifocal, heterogeneous masses
  - Hemangiopericytoma and hemangioendothelioma
    - Solid mass with necrosis
  - Angiosarcoma
    - Multiple, heterogeneous, hypervascular; also in liver
- Incidental splenic mass
  - Patient with known malignancy
    - Very aggressive tumor (e.g., melanoma), or tumor affecting portal venous system (e.g., pancreatic cancer); suspect metastasis
  - No known primary tumor
    - At high risk for lymphoma (e.g., AIDS, transplant recipient, associated lymphadenopathy); suspect lymphoma
    - Immunosuppressed, chronically ill; suspect opportunistic infection, peliosis, lymphoma
  - Asymptomatic, healthy adult
    - Echogenic or peripherally or uniform enhancing mass; probable hemangioma
    - Subcapsular, multicystic mass; probably lymphangioma
  - Symptomatic splenic mass not meeting these criteria
    - Probably primary vascular tumor
    - Probably requires splenectomy for diagnosis and management
- Splenic infection
  - Histoplasmosis and tuberculosis (TB) commonly affect spleen
  - Otherwise, uncommon, except in immunocompromised patients
    - AIDS, transplant recipients, leukemic, alcoholic
  - Multiple small abscesses
    - Candida (and other fungal), TB, Pneumocystis
  - Single large abscess
    - Usually bacterial
  - Calcification
    - Seen in treated abscesses (TB, fungal, Pneumocystis)
- Splenic Infarction
  - Relatively common cause of acute left upper quadrant pain
  - Appears as sharply marginated, wedge-shaped, poorly-enhancing lesions abutting splenic capsule
  - Etiologies
    - Sickle cell and other hemoglobinopathies
    - "Spontaneous" in any cause of splenomegaly
    - Embolic (e.g., I.V. drug abuse, endocarditis, atrial fibrillation)

Imaging Pitfalls
- Heterogeneous enhancement of the spleen may simulate or hide pathology
- Any mass or splenic parenchymal lesion suspected on the basis of an arterial phase CT/MR image should be confirmed on venous/parenchymal phase scans

EMBRYOLOGY

Embryologic Events
- from dorsal mesogastrium during fifth fetal week
- Normally rotates to left
- Usually fixed into left subphrenic location by peritoneal reflections linking it to the diaphragm, abdominal wall, kidney, stomach
- Usually develops as one main "fused" mass of tissue

SPLEEN ANATOMY AND IMAGING ISSUES

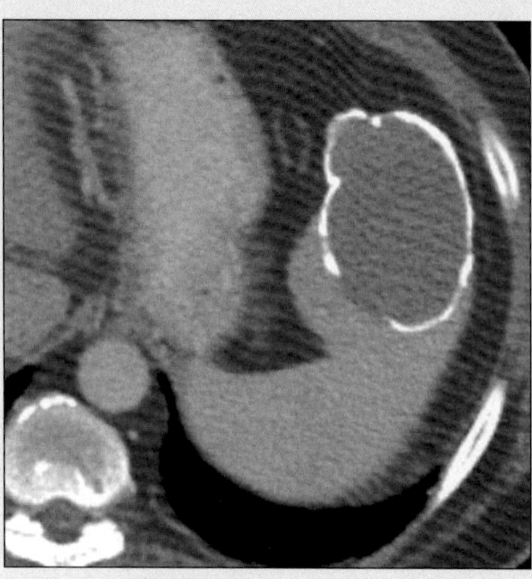
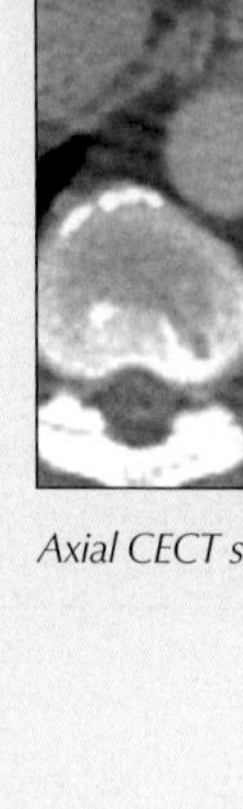

Axial CECT shows a splenic cyst with a calcified wall.

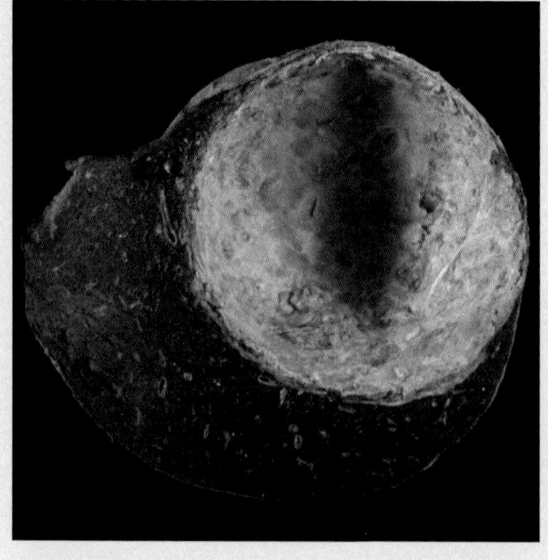

Surgical photograph of splenectomy specimen shows a splenic cyst with a calcified wall.

Practical Implications

- Failure of fusion
  - Accessory spleen found in 10 to 30% of population
  - Usually small (pea size) and near splenic hilum
  - Accessory spleen may be aberrant in location, may increase in size, especially after splenectomy
    - May enlarge after splenectomy performed for hypersplenism or tumor, resulting in relapse
    - Enlarged or ectopic accessory spleen can simulate tumor
    - May present as mass in pancreatic tail
    - Diagnosis by radionuclide sulfur colloid or tagged RBC scan (more sensitive)
- Spleen may be on long mesentery
  - "Wandering spleen" may result
  - Spleen may be found in any intraperitoneal location of the abdomen or pelvis
  - May simulate a mass
  - May torse and cause acute splenic infarction
- Failure to develop
  - Asplenia often associated with other congenital anomalies including situs inversus and cardiac anomalies
    - High mortality, especially early death from sepsis
- Polysplenia
  - Also associated with cardiac and other anomalies, azygous continuation of IVC
  - Also associated with early mortality
  - Can be simulated by splenosis (heterotopic implantation and subsequent growth of splenic tissue following traumatic rupture of spleen)

CUSTOM DIFFERENTIAL DIAGNOSIS

Multiple splenic lesions: Calcified

- Common
  - Granulomas (TB, histoplasmosis)
  - Opportunistic infection
    - Pneumocystis, mycobacterial
  - Splenic artery ± aneurysm
- Uncommon
  - Infarcts
  - Hydatid disease
  - Cysts
  - Amyloid
  - Healed abscesses
  - Hamartomas
  - Sickle cell anemia

Splenomegaly

- Congestive
  - E.g., cirrhosis, heart failure
- Neoplastic
  - E.g., lymphoma, leukemia, metastases
- Infection
  - E.g., hepatitis, HIV, malaria
- Hemolytic anemias
  - E.g., thalassemia, heterozygous sickle cell
- Extramedullary hematopoiesis
  - E.g., polycythemia vera, myelofibrosis
- Collagen vascular disease
  - E.g., Felty syndrome, rheumatoid arthritis
- Storage diseases
  - E.g., Gaucher, amyloid

SELECTED REFERENCES

1. Abbott RM et al: From the archives of the AFIP: primary vascular neoplasms of the spleen: radiologic-pathologic correlation. Radiographics. 24(4):1137-63, 2004
2. Gore RM et al: Textbook of gastrointestinal radiology: Spleen: differential diagnosis. 2nd ed. Philadelphia, WB Saunders. pp 1925-1928, 2000
3. Dähnert W: Radiology review manual. 4th ed, Philadelphia, Lippincott, Williams and Wilkins, 2000
4. Reeder MM: Reeder and Felson's gamuts in radiology .3rd ed. New York, Springer Verlag, 1993
5. Warnke R et al: Tumors of the lymph nodes and spleen. Washington, DC, Armed Forces Institute of Pathology. 1995

IMAGE GALLERY

Typical

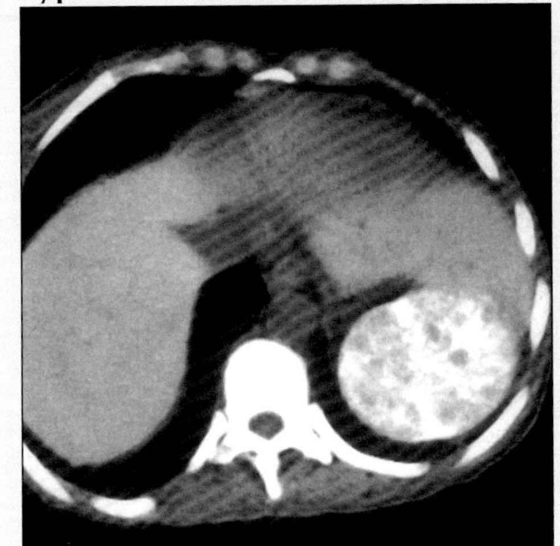

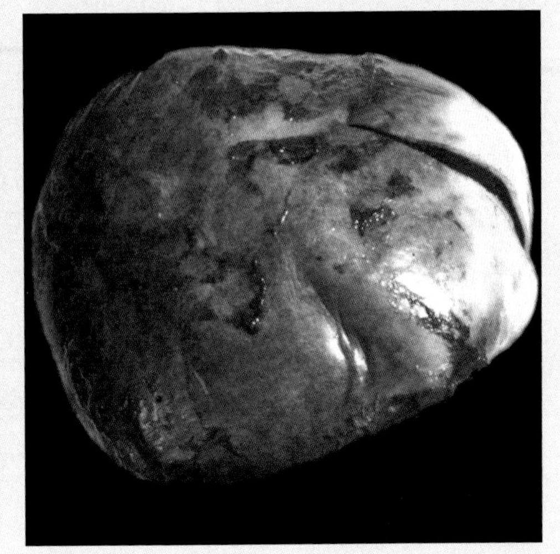

(Left) *Axial NECT shows heterogeneously calcified spleen in a patient with sickle cell anemia.* ***(Right)*** *Photograph of surgical specimen shows discoloration of the splenic parenchyma and capsule due to chronic infarction and calcification.*

Typical

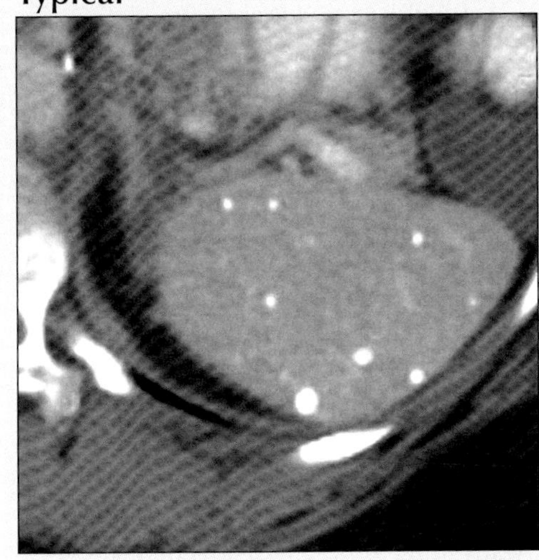

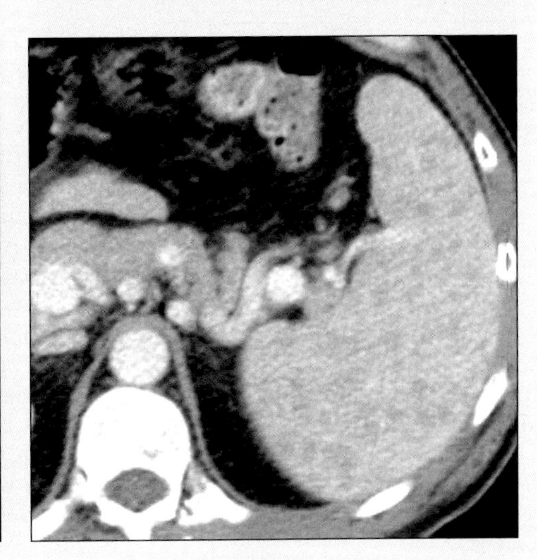

(Left) *Axial CECT shows multiple calcified granulomas from healed histoplasmosis.* ***(Right)*** *Axial CECT shows innumerable focal splenic lesions and splenomegaly; sarcoidosis.*

Typical

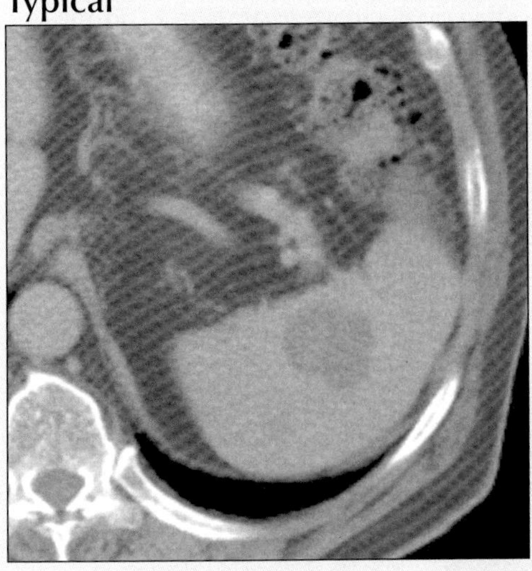

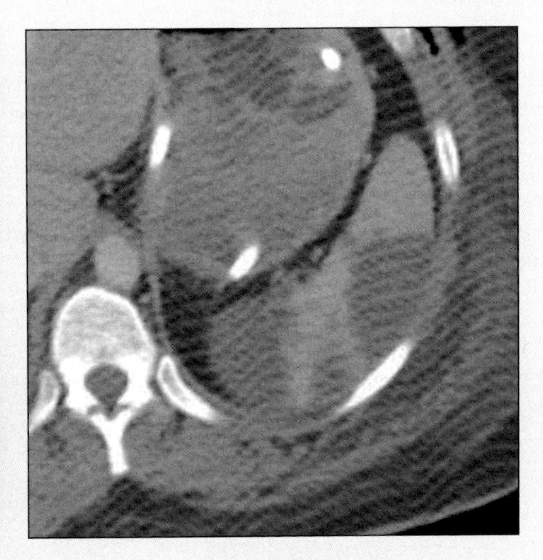

(Left) *Axial CECT shows one of several focal splenic lesions; metastatic melanoma.* ***(Right)*** *Axial CECT shows multiple wedge-shaped, capsular-based nonenhancing lesions; infarcts due to cardiac assist pump.*

ASPLENIA AND POLYSPLENIA

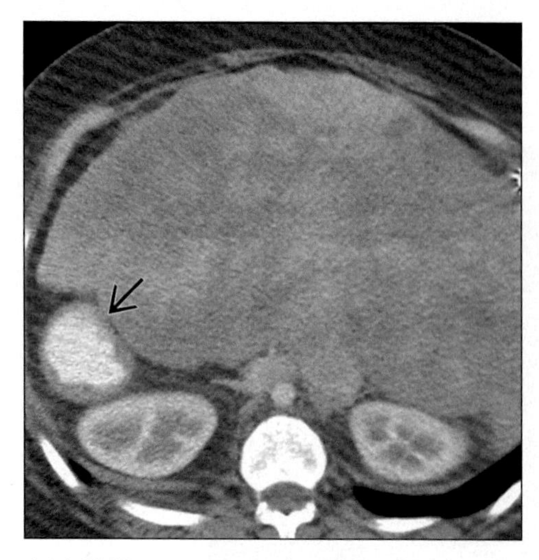

Axial CECT in a patient with asplenia shows midline enlarged (congested) liver + right-sided stomach (arrow). Extensive cardiac anomalies.

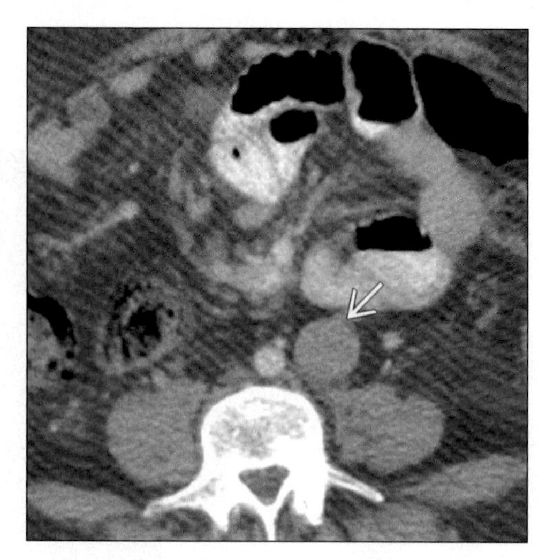

Axial CECT in patient with asplenia shows left-sided IVC (arrow).

TERMINOLOGY

Abbreviations and Synonyms

- Abbreviations: Asplenia (ASP); polysplenia (PSP)
- Synonyms
  - ASP: Asplenia syndrome; Ivemark syndrome; bilateral right-sidedness
  - PSP: Polysplenia syndrome; bilateral left-sidedness

Definitions

- ASP: Congenital absence of splenic tissue, situs ambiguous & associated anomalies
- PSP: Congenital abnormality characterized by multiple small splenic masses, situs ambiguous & associated anomalies

IMAGING FINDINGS

General Features

- Best diagnostic clue
  - ASP: Absence of spleen, abdominal aorta & IVC on same side (usually right) & bilateral distribution of right-sided viscera
  - PSP: Multiple small splenic masses, intrahepatic interruption of IVC with continuation of azygos vein, bilateral distribution of left-sided viscera
- Location: PSP, right & left upper abdominal quadrants
- Size: PSP: Varied size of splenic masses
- Morphology
  - Polysplenia
    - PSP: Number of spleens varies from 2-16
- Key concepts
  - Asplenia syndrome: Right isomerism
    - Situs ambiguous + bilateral right-sidedness
    - Cardiovascular malformations (50%): Total anomalous pulmonary venous return (almost 100%); endocardial cusion defect (85%); single ventricle (51%), transposition of great vessels (58%), pulmonary stenosis or atresia (70%), dextrocardia(42%), mesocardia, ventricular septal defect, single atrioventricular valve, bilateral superior vena cavae; absent coronary sinus
    - Pulmonary: Abnormal distribution of lobes, bilateral trilobed lungs
    - ASP-gastrointestinal anomalies: Situs inversus, imperforate anus, ectopic liver, annular pancreas, esophageal varices, GB agenesis, Hirschsprung disease, duplication/hypoplasia of stomach
    - Genitourinary (15%): Horseshoe kidney, bilobed urinary bladder, hydroureter, double collecting system, cystic kidney
    - Miscellaneous: Cleft palate, cleft lip, fused or horseshoe adrenal, absent left adrenal, scoliosis, bicornuate uterus, single umbilical artery, lumbar myelomeningocele

DDx: More or Less Than One Spleen

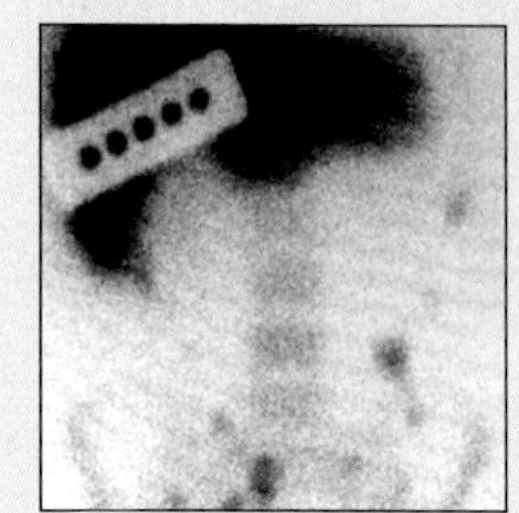

Splenosis

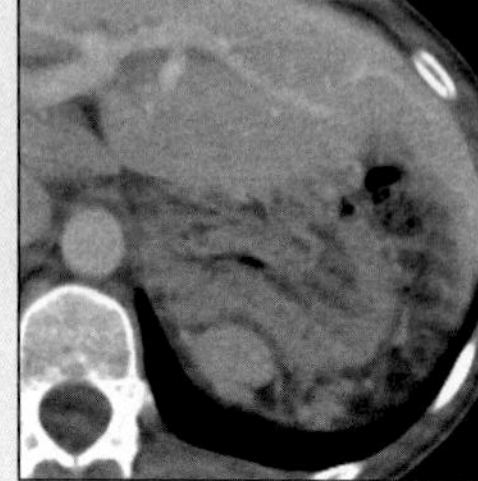

Splenosis

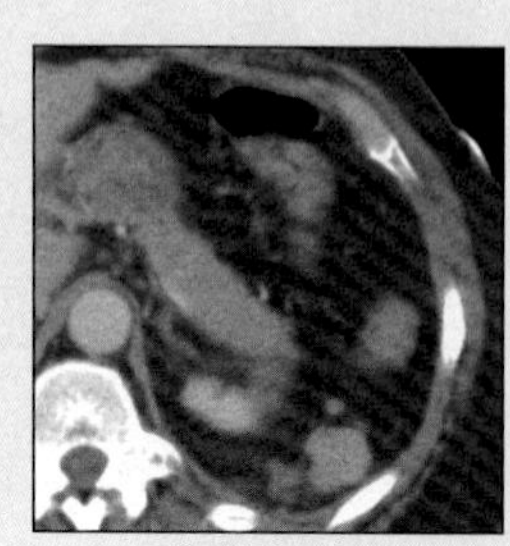

Splenosis

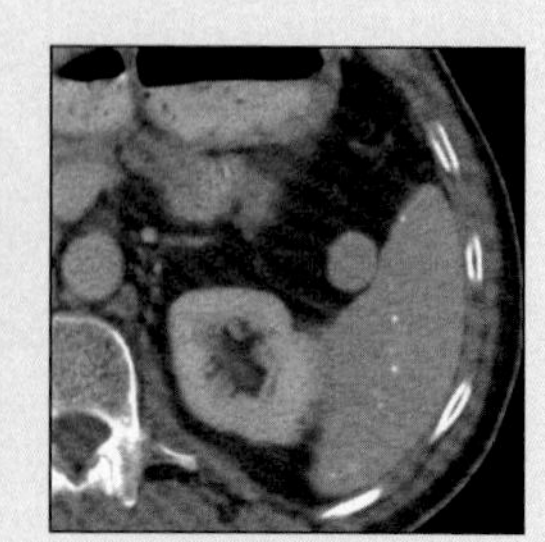

Accessory Spleen

ASPLENIA AND POLYSPLENIA

Key Facts

Terminology

- Abbreviations: Asplenia (ASP); polysplenia (PSP)
- ASP: Asplenia syndrome; Ivemark syndrome; bilateral right-sidedness
- PSP: Polysplenia syndrome; bilateral left-sidedness

Imaging Findings

- ASP: Absence of spleen, abdominal aorta & IVC on same side (usually right) & bilateral distribution of right-sided viscera
- PSP: Multiple small splenic masses, intrahepatic interruption of IVC with continuation of azygos vein, bilateral distribution of left-sided viscera
- Location: PSP, right & left upper abdominal quadrants
- PSP: Number of spleens varies from 2-16
- ASP-gastrointestinal anomalies: Situs inversus, imperforate anus, ectopic liver, annular pancreas, esophageal varices, GB agenesis, Hirschsprung disease, duplication/hypoplasia of stomach
- PSP-gastrointestinal: Esophageal/duodenal atresia, TEF, semiannular pancreas, gastric duplication, short bowel, absent GB, biliary atresia, malrotation

Top Differential Diagnoses

- Splenectomy
- Splenosis
- Accessory spleens

Diagnostic Checklist

- Centrally located left lobe of liver can simulate as spleen on US & diagnosis of ASP may be missed
- Differentiate PSP from accessory spleens & splenosis

- Polysplenia syndrome: Left isomerism
  - Situs ambiguous + bilateral left-sidedness
  - Cardiac: Continuation of azygos vein (65%); transposition of great vessels (13%), double outlet right ventricle (13%), pulmonary valvular stenosis (23%), subaortic stenosis or atresia
  - PSP-gastrointestinal: Esophageal/duodenal atresia, TEF, semiannular pancreas, gastric duplication, short bowel, absent GB, biliary atresia, malrotation
  - GU anomalies: Renal agenesis, renal/ovarian cysts

Radiographic Findings

- Radiography
  - Asplenia: Show situs ambiguous, situs solitus or situs inversus
    - Spleen: No distinct visible splenic contour
    - Liver: Symmetric & midline in its position
    - Stomach: Right, left or central in position
    - Malrotation of bowel
    - Cardia: Mesocardia or dextrocardia
    - Right-sided bronchial pattern & a minor fissure may be seen bilaterally
    - Superior mediastinal widening (due to bilateral superior vena cava)
    - Both pulmonary arteries anterior to trachea (on lateral chest film)
    - Bronchography: Bilateral eparterial bronchi on frontal view (pulmonary arteries inferior to bronchi)
  - Polysplenia
    - Frontal view: Paratracheal soft tissue prominence (dilated azygos or hemiazygos vein) mimicking a mediastinal mass
    - Chest lateral view: Both pulmonary arteries posterior to trachea & absence of IVC

CT Findings

- Asplenia
  - NECT
    - Absence of spleen
    - Situs abnormalities: Liver, gallbladder, stomach, bowel, heart, trilobed lungs
  - CECT
    - Vascular anatomy well depicted
    - IVC & abdominal aorta lie on same side of spine (usually right side with aorta lying posteriorly)
- Polysplenia
  - NECT
    - 2-16 splenic masses in right & left upper quadrants
    - ± Asymmetric liver & midgut malrotation
    - ± Abdominal situs solitus or situs inversus
  - CECT
    - Absence of IVC between renal & hepatic veins with independent drainage of hepatic veins into right atrium
    - Prominent azygos or hemiazygos vein
    - Crossing of IVC in front of aorta to enter a common atrium on right ("crossover", not unique to ASP)

Angiographic Findings

- Conventional
  - Asplenia
    - Splenic artery absent; entire celiac axis may arise from superior mesenteric artery
    - IVC & abdominal aorta lie on same side (usually right): Virtually pathognomonic feature of ASP
  - Polysplenia
    - Multiple spleens + common splenic/celiac artery
    - Variations in course of IVC: Absence of intrahepatic segment of IVC; crossover of IVC in front of aorta

Nuclear Medicine Findings

- Asplenia
  - Tc sulfur colloid or tagged RBC scan
    - Absence of spleen
    - Hepatic symmetry, prominent left lobe of liver
- Polysplenia: Multiple splenic tissues
  - Tc99m labeled heat-damaged RBC scan
    - More sensitive in detecting splenic tissue
  - Hepatobiliary imaging
    - Differentiate hepatic from splenic tissue
    - Presence & position of gallbladder

ASPLENIA AND POLYSPLENIA

Imaging Recommendations

- Best imaging tool
  - Asplenia: Helical CT & Tc99m RBC scan
  - Polysplenia
    - CT; US; MR can demonstrate size, position, number of spleens & relationship to liver & bowel

DIFFERENTIAL DIAGNOSIS

Splenectomy

- No splenic visualization after surgical splenectomy

Splenosis

- Traumatized splenic tissue scattered in abdominal cavity, where it attaches to adjacent peritoneal surface
- Multiple small encapsulated sessile implants (few mm-3 cm in size)

Accessory spleens

- Range from one to six in number
- Usually found near splenic hilum along course of splenic vessels or within layers of omentum
- Imaging findings identical to normal splenic tissue

PATHOLOGY

General Features

- Etiology
  - Uncertain
  - Asplenia: Delayed embryonic body curvature
  - Polysplenia: Accelerated embryonic body curvature
  - Altered timing in development of embryonic body curvature leads to visceroatrial situs abnormalities
  - Pressure of adjacent structures may interfere with splenic blood supply
- Epidemiology: Asplenia: 1 in 40,000 live births
- Associated abnormalities: Pulmonary, cardiovascular, gastrointestinal, genitourinary & miscellaneous anomalies

Gross Pathologic & Surgical Features

- ASP: Congenital absence of spleen
- PSP: Multiple splenic tissues in upper quadrants

Microscopic Features

- ASP: Heinz or Howell-Jolly bodies (absent in PSP)
  - RBC inclusions

CLINICAL ISSUES

Presentation

- Most common signs/symptoms
  - ASP: Cardiopulmonary disease (83%); cyanosis in neonatal period or infancy; bowel obstruction (17%)
  - PSP
    - Cardiac disease; acyanotic left-to-right shunts such as septal defects
    - Heart murmur, congestive heart failure, occasional cyanosis, heart block
    - Jaundice (extrahepatic biliary obstruction)
    - Bowel malrotation, obstruction, pain (infarction)
    - 10-15% may not present clinically until adulthood
    - Acute abdominal pain: Due to splenic infarction

Demographics

- Age
  - ASP: Newborn or infant
  - PSP: Infant or adult age group
- Gender
  - Asplenia: M > F
  - Polysplenia: M < F

Natural History & Prognosis

- Asplenia
  - Patients are prone to overwhelming septicemia post-operatively
  - Prognosis: Usually poor
  - Mortality rate: 80% die by end of 1st year of life due to cardiac failure & post-operative complications
- Polysplenia
  - Prognosis: Usually good
  - Mortality rate
    - 50-60% mortality in first year of life
    - 25% of patients live up to 5 years of age
    - 10% survive to midadolescence

Treatment

- Asplenia: Prophylactic antibiotics (not needed in PSP)
  - Associated cardiac disease: Surgical correction

DIAGNOSTIC CHECKLIST

Image Interpretation Pearls

- Centrally located left lobe of liver can simulate as spleen on US & diagnosis of ASP may be missed
- Differentiate PSP from accessory spleens & splenosis

SELECTED REFERENCES

1. Combs LS et al: Evaluation of spleen in children with heterotaxia and congenital heart disease. Tenn Med. 97(4):161-3, 2004
2. Fulcher AS et al: Abdominal manifestations of situs anomalies in adults. Radiographics. 22(6):1439-56, 2002
3. Paterson A et al: A pattern-oriented approach to splenic imaging in infants and children. Radiographics. 19(6):1465-85, 1999
4. Applegate KE et al: Situs revisited: imaging of the heterotaxy syndrome. Radiographics. 19(4):837-52; discussion 853-4, 1999
5. Ruscazio M et al: Interrupted inferior vena cava in asplenia syndrome and a review of the hereditary patterns of visceral situs abnormalities. Am J Cardiol. 81(1):111-6, 1998
6. Nakada K et al: Digestive tract disorders associated with asplenia/polysplenia syndrome. J Pediatr Surg. 32(1):91-4, 1997
7. Freeman JL et al: CT of congenital and acquired abnormalities of the spleen. Radiographics. 13(3):597-610, 1993
8. Chitayat D et al: Prenatal diagnosis of asplenia/polysplenia syndrome. Am J Obstet Gynecol. 158(5):1085-7, 1988

IMAGE GALLERY

Typical

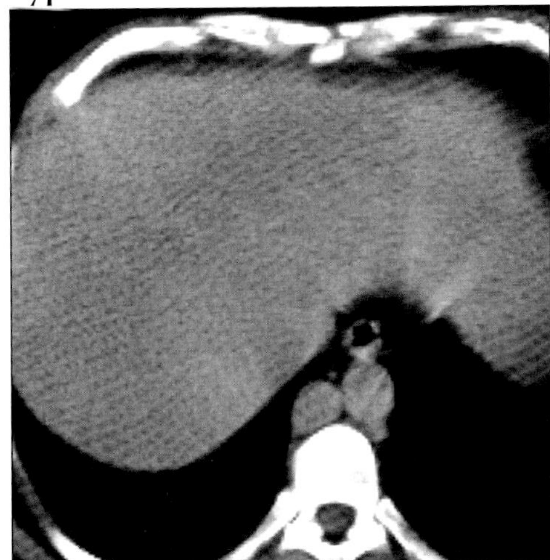

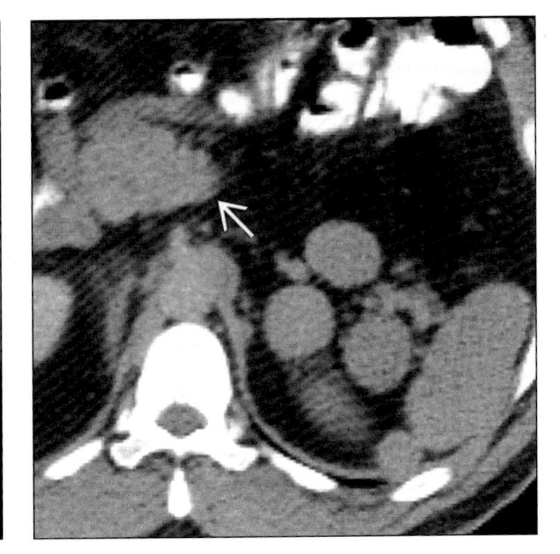

*(**Left**) Axial NECT in patient with polysplenia shows absent IVC, azygous continuation. (**Right**) Axial CECT shows polysplenia and congenital absence of all but the head of the pancreas (arrow).*

Typical

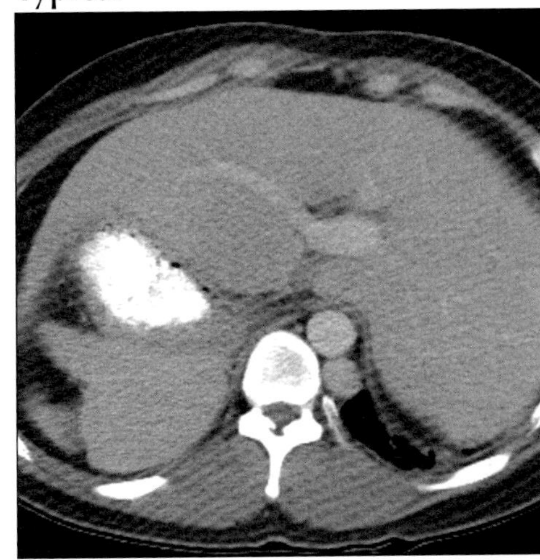

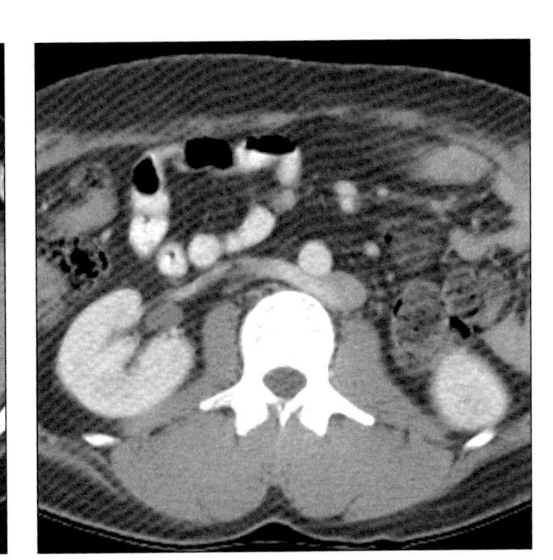

*(**Left**) Axial CECT in patient with polysplenia shows situs ambiguous, absent IVC, azygous continuation. (**Right**) Axial CECT in a patient with polysplenia shows retroaortic right renal vein leading to azygous continuation.*

Typical

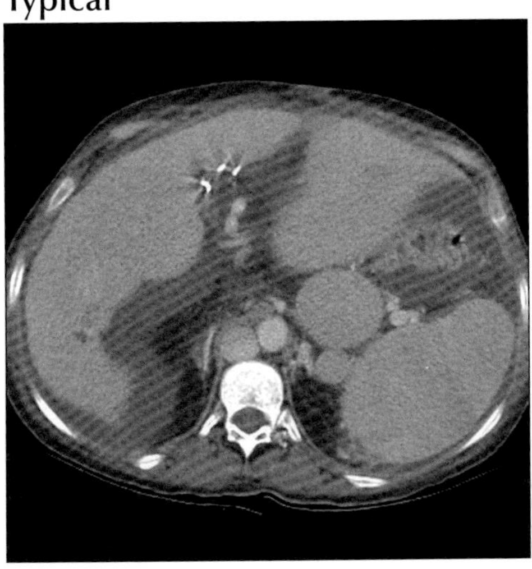

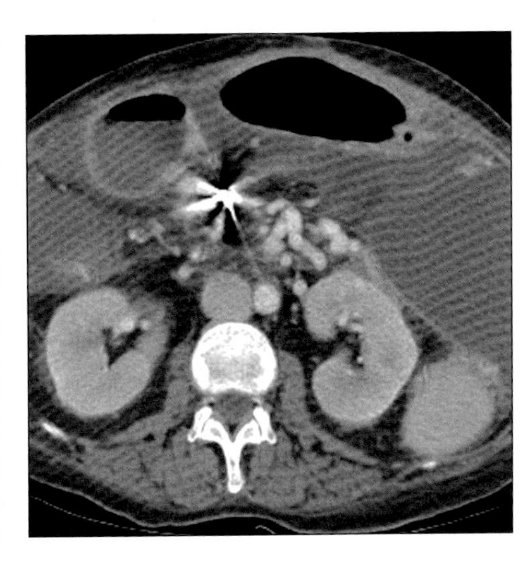

*(**Left**) Axial CECT in patient with polysplenia shows liver abnormalities and azygous continuation. (**Right**) Axial CECT in patient with polysplenia shows prior resection of small intestine due to volvulus and infarction.*

ACCESSORY SPLEEN

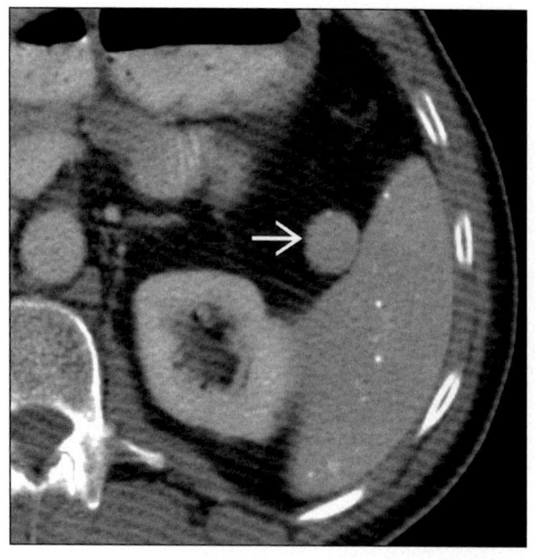

Axial CECT shows small spherical accessory spleen (arrow) near splenic hilum.

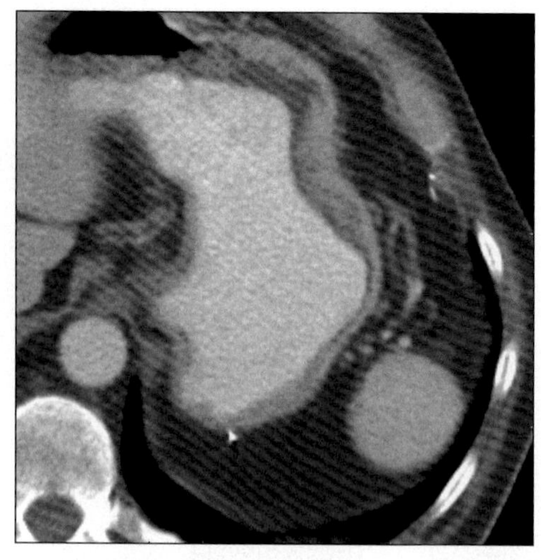
Axial CECT shows hypertrophied accessory spleen, following splenectomy.

TERMINOLOGY

Definitions

- Ectopic splenic tissue of congenital origin

IMAGING FINDINGS

General Features

- Best diagnostic clue: Small nodule with same texture and enhancement of normal spleen
- Location
  - In or near splenic hilum or ligaments (most cases)
  - Anywhere in abdomen or retroperitoneum (20% of cases), especially around tail of pancreas
  - Embedded within pancreatic tail (rare)
  - Usually left upper quadrant, above renal pedicle; may also be in paratesticular, diaphragmatic, pararenal and gastric sites
  - 1 focus (88%), 2 foci (9%), > 2 foci (3%); multiple foci usually clustered in 1 location
- Size: Varies from mm to several cm, usually < 2.5 cm
- Morphology: Same texture as main spleen

CT Findings

- NECT: Single/multiple, round/ovoid, uniform soft tissue
- CECT
  - Same enhancement as normal main spleen
  - ± Supplying branch of splenic artery

MR Findings

- T1WI: Hypointense; T2WI: Hyperintense

Ultrasonographic Findings

- Real Time
  - Splenic artery and vein (90% of cases)
  - Parenchymal bridges between main spleen and accessory spleen

Angiographic Findings

- Conventional: Blood supplied by a branch of splenic artery and drained into splenic veins

Nuclear Medicine Findings

- Technetium sulfur colloid
  - Functional splenic tissue, differentiate from bleeding by consistent size, shape & location

Imaging Recommendations

- Best imaging tool: CT followed by nuclear scintigraphy

DIFFERENTIAL DIAGNOSIS

Splenosis

- Usually result of trauma, portions of disrupted spleen implant anywhere including abdomen, pelvis, chest

DDx: Left Upper Quadrant "Mass"

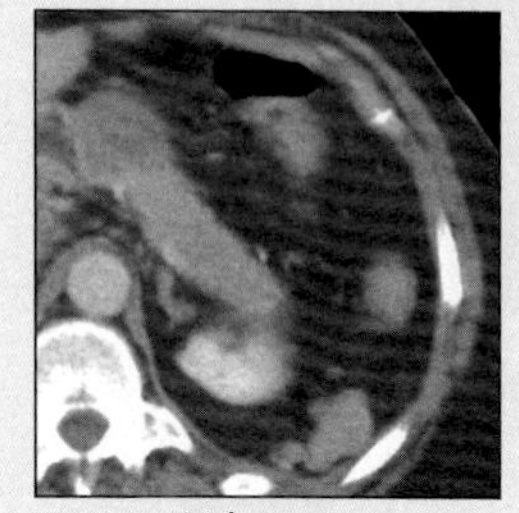
Splenosis

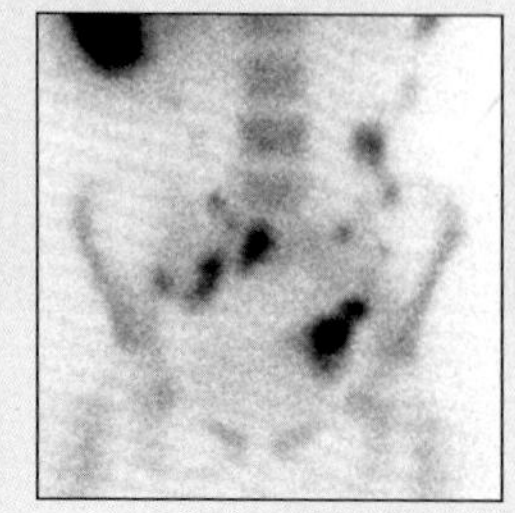
Splenosis

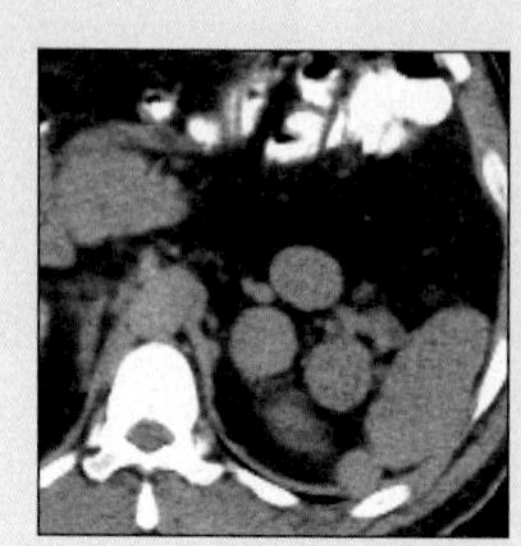
Polysplenia

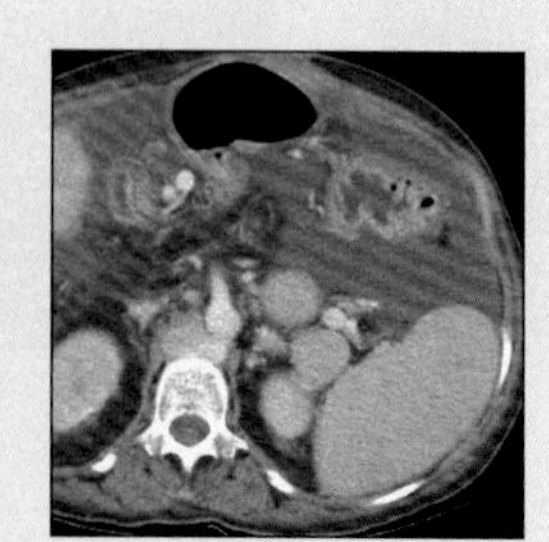
Polysplenia

ACCESSORY SPLEEN

Key Facts

Terminology
- Ectopic splenic tissue of congenital origin

Imaging Findings
- Best diagnostic clue: Small nodule with same texture and enhancement of normal spleen
- Best imaging tool: CT followed by nuclear scintigraphy

Top Differential Diagnoses
- Metastases
- Visceral mass

Diagnostic Checklist
- Accessory spleen is common, can be mistaken for tumor
- Hypertrophic accessory spleen may occur after splenectomy

- Implanted splenic tissue continues to function; enlarged "splenules"

Polysplenia
- Congenital disorder with multiple small spleens, bilateral "left-sidedness" of abdominal viscera, cardiovascular anomalies

Metastases
- E.g., omental, peritoneal metastases

Visceral mass
- E.g., intramural benign gastric tumors, renal, adrenal or primary retroperitoneal mass, mass in tail of pancreas

PATHOLOGY

General Features
- Etiology
  - Congenital
  - Pathogenesis
    - Failure of embryonic splenic buds to unite within dorsal mesogastrium
    - Extreme lobulation of spleen with pinching off of splenic tissue
- Epidemiology: Incidence: 10-30% patients at autopsy
- Associated abnormalities
  - Hypertrophied accessory spleen: Enlarged due to disease and/or splenectomy
  - Splenic neoplasms (e.g., recurrent lymphoma)
  - Hematologic diseases (e.g., idiopathic thrombocytopenic purpura, hereditary spherocytosis, acquired autoimmune hemolytic anemia): Recurs after splenectomy

Gross Pathologic & Surgical Features
- Structurally normal splenic tissues

CLINICAL ISSUES

Presentation
- Most common signs/symptoms: Asymptomatic (most cases)
- Diagnosis
  - Incidentally at surgery, autopsy or imaging
  - May be mistaken for tumor
  - Hypertrophic accessory spleen may be site of residual lymphoma or hypersplenism

Natural History & Prognosis
- Complications: Spontaneous rupture, infarction, torsion
- Prognosis: Good

Treatment
- Surgical resection: Complications, recurrence of lymphoma or hypersplenism

DIAGNOSTIC CHECKLIST

Consider
- Accessory spleen is common, can be mistaken for tumor

Image Interpretation Pearls
- Hypertrophic accessory spleen may occur after splenectomy

SELECTED REFERENCES

1. Harris GN et al: Accessory spleen causing a mass in the tail of the pancreas: MR imaging findings. AJR Am J Roentgenol. 163(5):1120-1, 1994
2. Freeman JL et al: CT of congenital and acquired abnormalities of the spleen. Radiographics. 13(3):597-610, 1993
3. Ambriz P et al: Accessory spleen compromising response to splenectomy for idiopathic thrombocytopenic purpura. Radiology. 155(3):793-6, 1985

IMAGE GALLERY

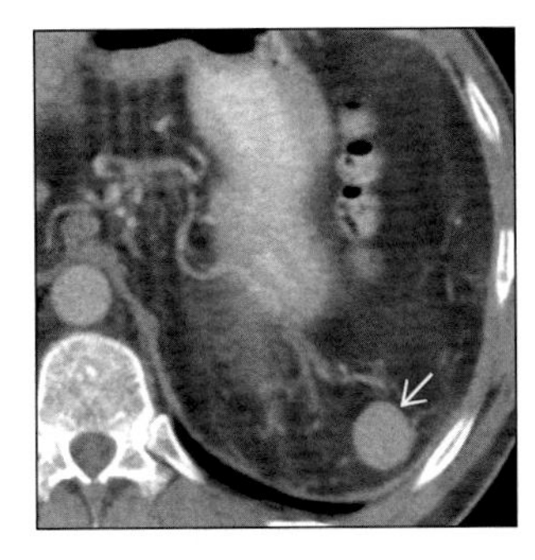

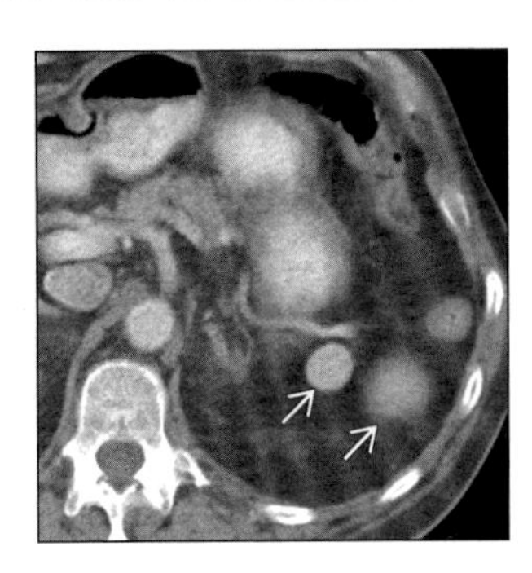

(Left) *Axial CECT shows one (arrow) of several accessory spleens, hypertrophied following splenectomy.* ***(Right)*** *Axial CECT shows 2 of several accessory spleens (arrows), hypertrophied following splenectomy.*

SPLENIC INFECTION AND ABSCESS

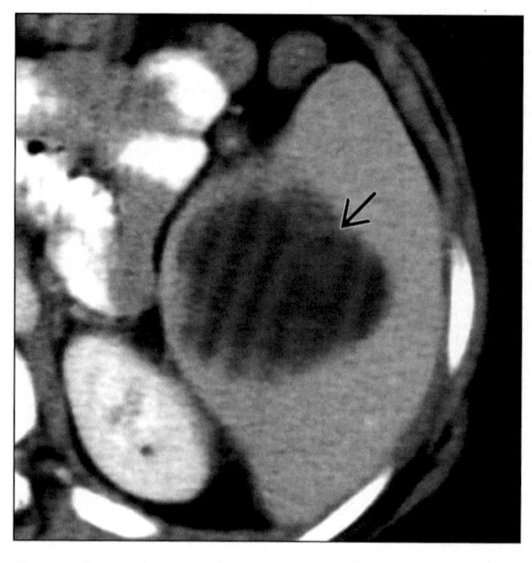

Pyogenic splenic abscess on CECT. Note low attenuation abscess bulging splenic parenchyma (arrow).

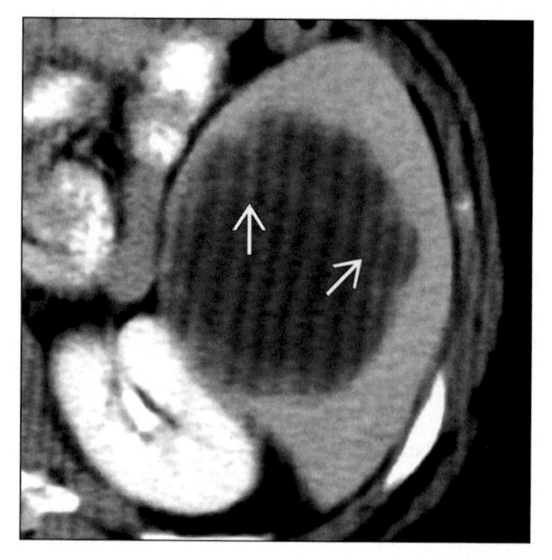

Pyogenic splenic abscess on axial CECT. Note thin septations within abscess (arrows).

TERMINOLOGY

Definitions

- Focal collection of liquified pus within splenic parenchyma

IMAGING FINDINGS

General Features

- Best diagnostic clue: Rounded low attenuation complex fluid collection with mass effect
- Location: Variable; may be located anywhere within splenic parenchyma
- Size: Variable; typically 3-5 cm for pyogenic abscesses; microabscesses (often fungal) < 1.5 cm
- Morphology
  - Rounded or with irregular borders
  - May have multiple locules similar to hepatic "cluster sign" of pyogenic abscess
  - Mass effect on splenic capsule
  - Internal septations common

Radiographic Findings

- Radiography
  - Rarely gas bubbles within abscess
  - Associated with left lower lobe atelectasis and left pleural effusion on chest x-ray

CT Findings

- NECT
  - Low attenuation ill-defined lesion within splenic parenchyma
  - May rarely contain gas bubbles or air-fluid levels
- CECT: Low attenuation, nonenhancing complex fluid collection; may extend to subcapsular location, rarely causes splenic rupture with generalized peritonitis

MR Findings

- T1WI: Low or intermediate signal lesion
- T2WI: High signal lesion
- T1 C+: Low signal lesion with peripheral enhancement

Ultrasonographic Findings

- Real Time
  - Typical pyogenic abscess
    - Hypoechoic with internal septations, low-level echoes representing pus or debris
    - May have little distal acoustic enhancement
  - Atypical pyogenic abscess
    - Reverberation artifacts from gas
    - Echogenic
  - Microabscesses
    - Target or "bull's eye" appearance similar to hepatic microabscesses
- Color Doppler
  - Typical pyogenic abscess shows no internal flow

DDx: Splenic Lesions Mimicking Infection

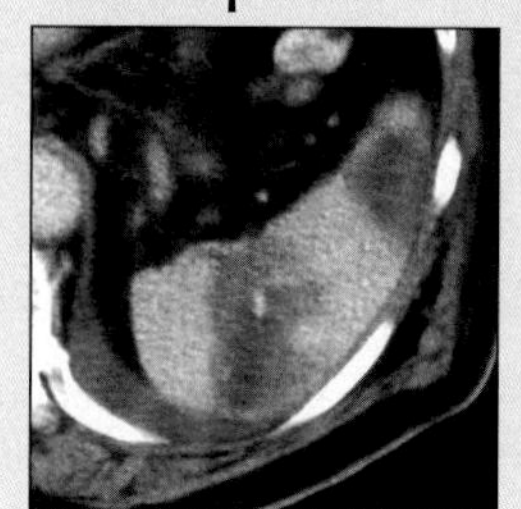

Splenic Infarct

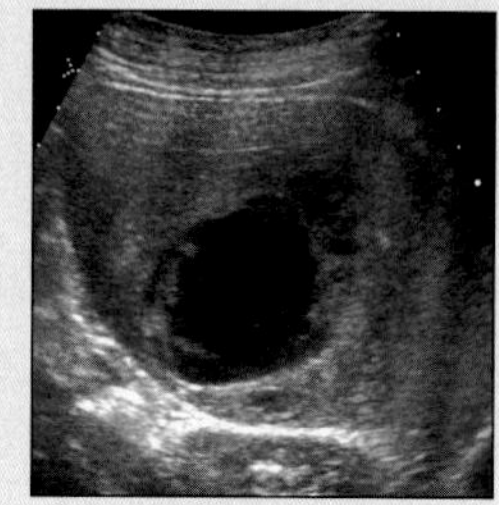

Splenic Tumor

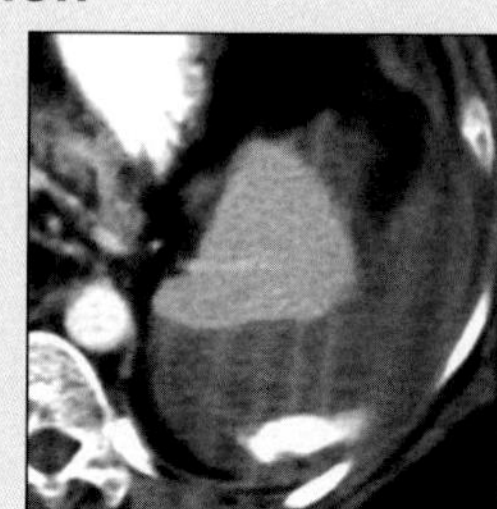

Splenic Trauma

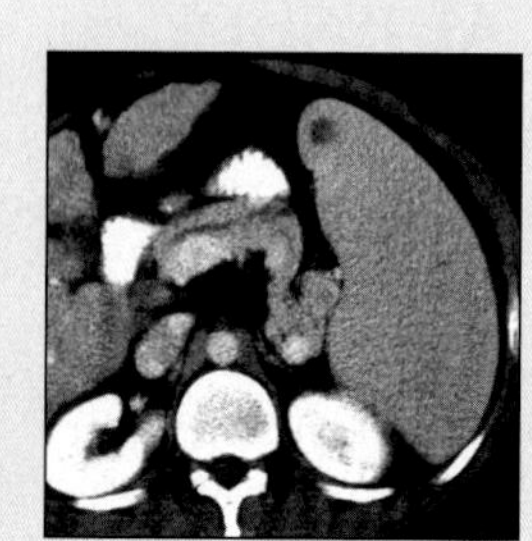

Sarcoidosis

SPLENIC INFECTION AND ABSCESS

Key Facts

Terminology
- Focal collection of liquified pus within splenic parenchyma

Imaging Findings
- Best diagnostic clue: Rounded low attenuation complex fluid collection with mass effect
- CECT: Low attenuation, nonenhancing complex fluid collection; may extend to subcapsular location, rarely causes splenic rupture with generalized peritonitis
- Protocol advice: 150 ml of I.V. contrast injected at 2.5 ml/sec; 5 mm slice thickness with 5 mm reconstruction interval

Top Differential Diagnoses
- Splenic infarct
- Splenic tumor
- Splenic trauma
- Infiltrating disorders

Pathology
- General path comments: Liquified pus, splenomegaly
- Genetics: Hemoglobinopathies (sickle cell) predispose

Clinical Issues
- Percutaneous drainage for unilocular unruptured abscesses
- Splenectomy for multiple pyogenic abscesses and/or abscess rupture

Diagnostic Checklist
- Infarct; necrotic or cystic mets; lymphoma

  - Hypoechoic nodular avascular microabscesses

Nuclear Medicine Findings
- PET: Increased isotope uptake from hypermetabolic focus
- WBC Scan
  - Increased isotope uptake

Imaging Recommendations
- Best imaging tool: CECT, US
- Protocol advice: 150 ml of I.V. contrast injected at 2.5 ml/sec; 5 mm slice thickness with 5 mm reconstruction interval

DIFFERENTIAL DIAGNOSIS

Splenic infarct
- Wedge-shaped, but may occasionally be rounded in configuration
- Low attenuation
- Peripheral location
- Nonenhancement with contrast

Splenic tumor
- Single or multiple lesions
- Solid: Lymphoma, melanoma
- Cystic: Ovarian carcinoma, germ cell tumors or sarcoma
- Benign: Lymphangioma, hemangioma
- Rounded with variable enhancement

Splenic trauma
- History of blunt injury
- Associated with perisplenic hematoma and hemoperitoneum
- May have active arterial extravasation
- Arterial clot adjacent to spleen in small lacerations

Infiltrating disorders
- Sarcoid
  - Multiple low attenuation lesions
- Gaucher disease
  - Multiple low attenuation lesions

PATHOLOGY

General Features
- General path comments: Liquified pus, splenomegaly
- Genetics: Hemoglobinopathies (sickle cell) predispose
- Etiology
  - Generalized septicemia
  - Septic emboli
    - Endocarditis
  - Immunosuppression
    - Fungal microabscesses
  - Secondary infection of traumatic splenic hematoma or infarct
  - Hematologic disorders
- Epidemiology
  - Rare: 0.2% of reported autopsies
  - 25% are immunocompromised patients
- Associated abnormalities
  - Post-operative state
  - Endocarditis
  - Immunocompromised state
  - Pancreatitis, colon cancer

Gross Pathologic & Surgical Features
- Necrotic areas of liquified pus

Microscopic Features
- Liquefactive necrosis
- Pus with leukocyte debris
- Gram stain for pyogenic abscess
  - Fungal stain for mycotic organism
  - TB stains for tuberculous abscesses
  - Pyogenic: 57% aerobic
    - Staphylococcus
    - Strep E. coli
    - Salmonella
  - Fungal
    - Candida most common
    - Aspergillus and cryptococcus
    - Tuberculosis (TB) and mycobacterium avium intracellulari (MAI) in AIDS patient

SPLENIC INFECTION AND ABSCESS

Staging, Grading or Classification Criteria

- Pyogenic
  - Unilocular (65%)
  - Multilocular or multiple (20%)
- Fungal
  - Microabscesses < 1.5 cm (25%)
- Parasitic
  - Echinococcus granulosa

CLINICAL ISSUES

Presentation

- Most common signs/symptoms
  - Fever
  - Chills
  - LUQ pain
  - Splenomegaly
- Clinical profile
  - Lab data
    - Leukocytosis
    - Positive blood cultures

Demographics

- Age: Adult patients with predisposing factors
- Gender: M = F
- Ethnicity: No known predilection

Natural History & Prognosis

- Variable
- Excellent prognosis for pyogenic abscesses in immunocompetent patient
- Guarded prognosis in immunocompromised patients with fungal microabscesses

Treatment

- Options, risks, complications
  - Percutaneous drainage for unilocular unruptured abscesses
    - Reported success rate of 67-100%
  - Splenectomy for multiple pyogenic abscesses and/or abscess rupture
    - Mortality post-splenectomy 6%

DIAGNOSTIC CHECKLIST

Consider

- Infarct; necrotic or cystic mets; lymphoma

Image Interpretation Pearls

- Single or multiple low attenuation lesions in febrile patient
- Morphology variable

SELECTED REFERENCES

1. Tasar M et al: Computed tomography-guided percutaneous drainage of splenic abscesses. Clin Imaging. 28(1):44-8, 2004
2. Chiang IS et al: Splenic abscesses: review of 29 cases. Kaohsiung J Med Sci. 19(10):510-5, 2003
3. Kaushik R et al: Splenic abscess. Trop Doct. 32(4):246-7, 2002
4. Thanos L et al: Percutaneous CT-guided drainage of splenic abscess. AJR Am J Roentgenol. 179(3):629-32, 2002
5. Ng KK et al: Splenic abscess: diagnosis and management. Hepatogastroenterology. 49(44):567-71, 2002
6. Loualidi A et al: Splenic abscess caused by Peptostreptococcus species, diagnosed with the aid of abdominal computerized tomography and treated with percutaneous drainage and antibiotics: a case report. Neth J Med. 59(6):280-5, 2001
7. Green BT: Splenic abscess: report of six cases and review of the literature. Am Surg. 67(1):80-5, 2001
8. Smyrniotis V et al: Splenic abscess. An old disease with new interest. Dig Surg. 17(4):354-7, 2000
9. Poggi SH et al: Puerperal splenic abscess. Obstet Gynecol. 96(5 Pt 2):842, 2000
10. Mehanna D et al: Cat scratch disease presenting as splenic abscess. Aust N Z J Surg. 70(8):622-4, 2000
11. Drevelengas A: The spleen in infectious disorders. JBR-BTR. 83(4):208-10, 2000
12. Murray AW et al: A case of multiple splenic abscesses managed non-operatively. J R Coll Surg Edinb. 45(3):189-91, 2000
13. Frumiento C et al: Complications of splenic injuries: expansion of the nonoperative theorem. J Pediatr Surg. 35(5):788-91, 2000
14. Nakao A et al: Portal venous gas associated with splenic abscess secondary to colon cancer. Anticancer Res. 19(6C):5641-4, 1999
15. Bernabeu-Wittel M et al: Etiology, clinical features and outcome of splenic microabscesses in HIV-infected patients with prolonged fever. Eur J Clin Microbiol Infect Dis. 18(5):324-9, 1999
16. Duggal RK et al: Splenic abscess as a complication of acute pancreatitis. J Assoc Physicians India. 47(3):338-9, 1999
17. Alterman P et al: Splenic abscess in geriatric care. J Am Geriatr Soc. 46(11):1481-3, 1998
18. de Bree E et al: Splenic abscess: a diagnostic and therapeutic challenge. Acta Chir Belg. 98(5):199-202, 1998
19. Al-Salem AH et al: Splenic abscess and sickle cell disease. Am J Hematol. 58(2):100-4, 1998
20. Wang Y et al: CT findings in splenic tuberculosis. J Belge Radiol. 81(2):90-1, 1998
21. Kumar N et al: Splenic abscess caused by Clostridium difficile. Eur J Clin Microbiol Infect Dis. 16(12):938-9, 1997
22. Vleminckx WG et al: Splenic abscess with Clostridium novyi bacteraemia and sepsis. Eur J Gastroenterol Hepatol. 9(3):303-5, 1997
23. Rypens F et al: Splenic parenchymal complications of pancreatitis: CT findings and natural history. J Comput Assist Tomogr. 21(1):89-93, 1997
24. Yelon JA et al: Splenic abscess associated with osteomyelitis. Eur J Surg. 162(11):913-4, 1996
25. Liang JT et al: Splenic abscess: a diagnostic pitfall in the ED. Am J Emerg Med. 13(3):337-43, 1995

SPLENIC INFECTION AND ABSCESS

IMAGE GALLERY

Typical

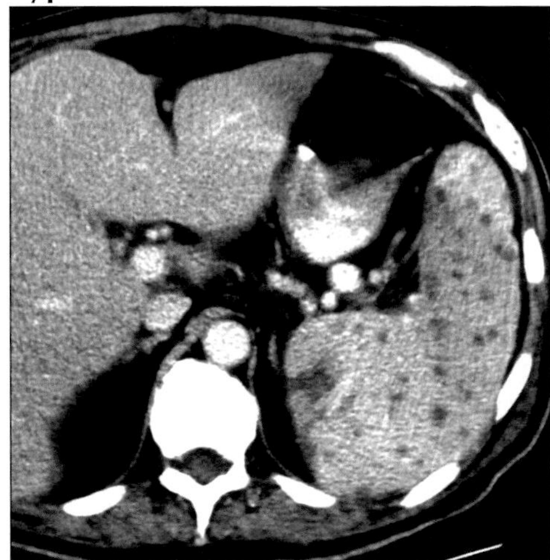

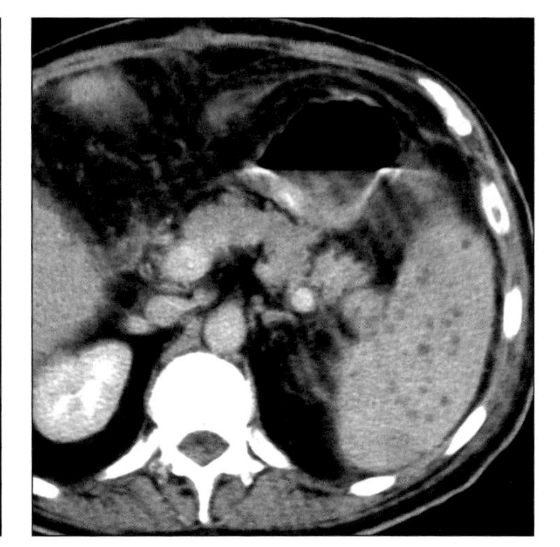

*(**Left**) Axial CECT of fungal microabscesses. Note numerous hypodense lesions; cultures grew Candida. (**Right**) Axial CECT demonstrates splenic microabscesses. Note small < 1 cm lesions diffusely throughout the spleen.*

Typical

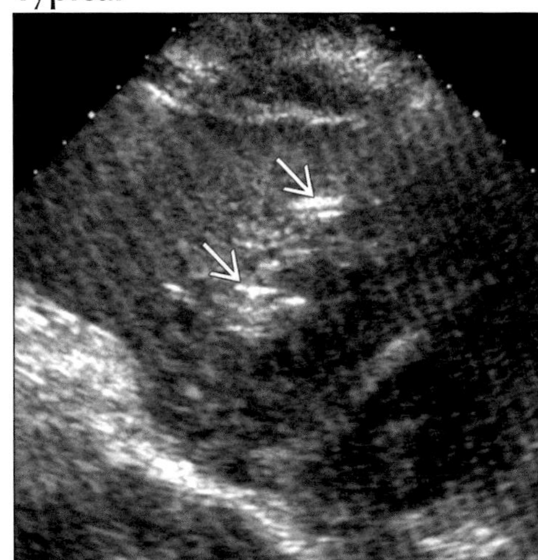

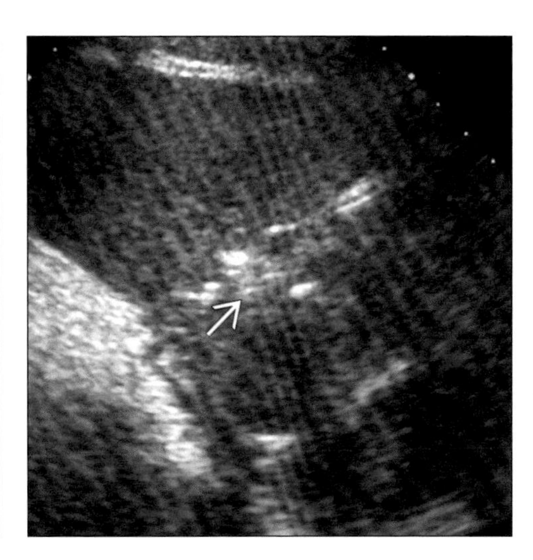

*(**Left**) Gas-forming splenic pyogenic abscess on transverse sonogram. Note linear high amplitude echoes representing gas (arrows). (**Right**) Gas-forming pyogenic splenic abscess on transverse sonogram. Note ring down artifacts from gas bubbles (arrow).*

Typical

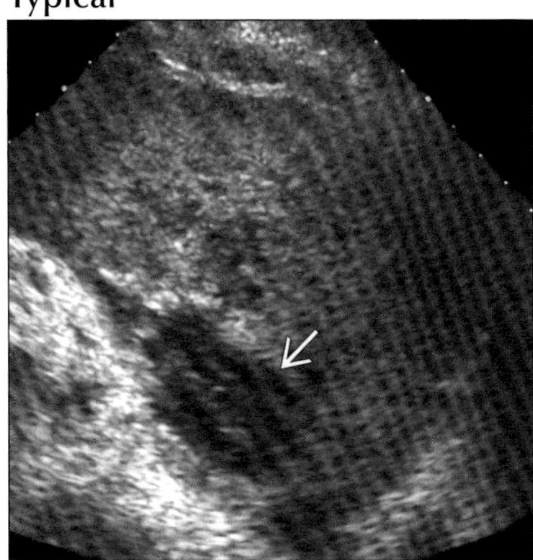

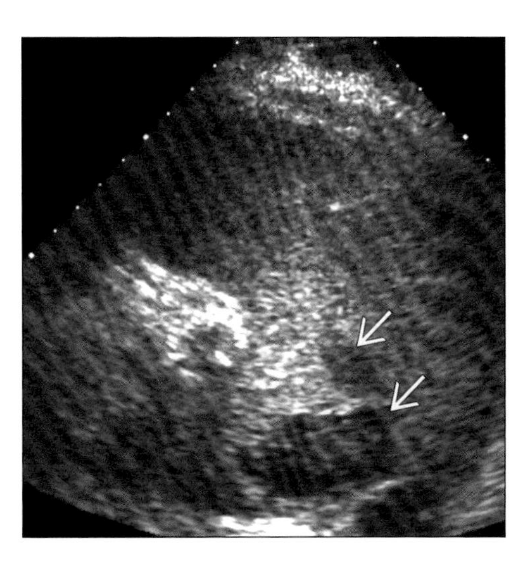

*(**Left**) Pyogenic splenic abscess on transverse sonogram. Note hypoechoic abscess (arrow) with little distal acoustic enhancement. (**Right**) Multiple pyogenic splenic abscesses on transverse sonogram. Note multiple hypoechoic round lesions (arrows).*

SPLENIC INFARCTION

Axial CECT of embolic splenic infarction. Note apical thrombus in left ventricle (arrow).

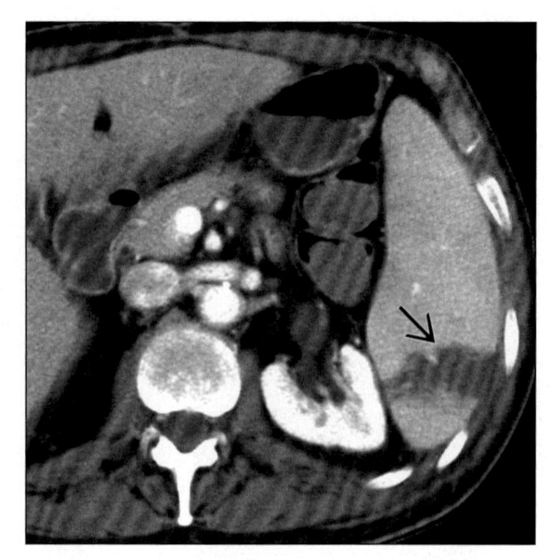

Axial CECT of embolic splenic infarct. Note low attenuation defect with linear margins (arrow).

TERMINOLOGY

Definitions

- Global or segmental parenchymal splenic ischemia & necrosis caused by vascular occlusion

IMAGING FINDINGS

General Features

- Best diagnostic clue: Peripheral wedge-shaped nonenhancing areas on CECT within splenic parenchyma in patient with LUQ pain
- Location
  - Variable
  - Entire spleen may be infarcted or more commonly segmental areas
- Size
  - Variable; global or segmental
  - Spleen may or may not demonstrate splenomegaly
- Morphology
  - Most commonly wedge-shaped when segmental
  - Straight margins indicating vascular lesion
  - May be rounded (atypical)

Radiographic Findings

- Radiography: May be associated with left pleural effusion on chest x-ray

CT Findings

- NECT: Poorly visualized without contrast
- CECT
  - Segmental: Wedge-shaped or rounded low attenuation area on CECT
  - Global: Complete nonenhancement of spleen with or without "cortical rim sign" on CECT

MR Findings

- T1WI
  - High signal areas of hemorrhagic infarction if recent
  - Low signal if chronic
- T2WI: High signal within area of infarct
- T1 C+: Wedge-shaped area of low signal

Ultrasonographic Findings

- Real Time: Hypoechoic or anechoic wedge-shaped or rounded parenchymal defect
- Color Doppler: Absent flow in areas of infarction with color Doppler

Angiographic Findings

- Conventional: Main splenic artery occlusion or segmental emboli

Imaging Recommendations

- Best imaging tool: CECT
- Protocol advice: 150 ml of I.V. contrast at 2.5 ml/sec; 5 mm collimation with 5 mm reconstruction interval

DDx: Splenic Lesions Mimicking Infarction

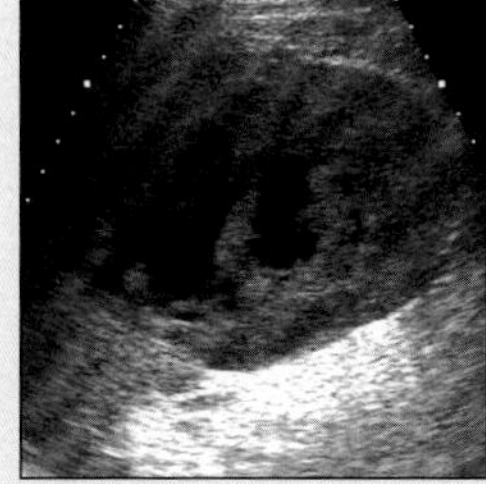

Splenic Abscess

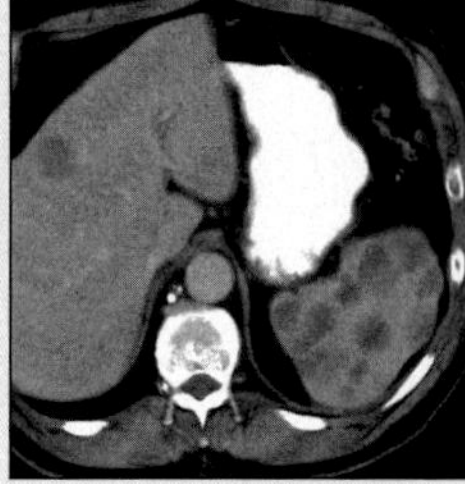

Splenic Lymphoma

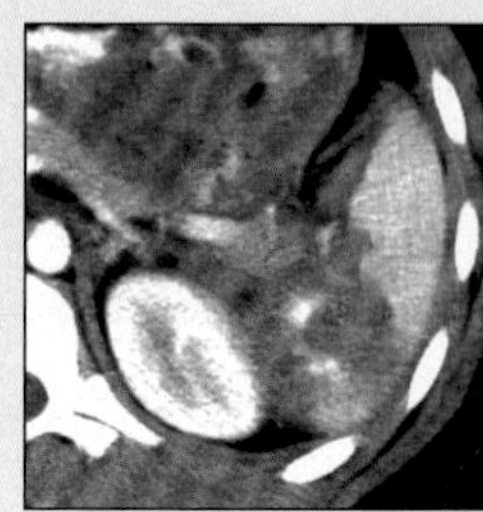

Splenic Laceration

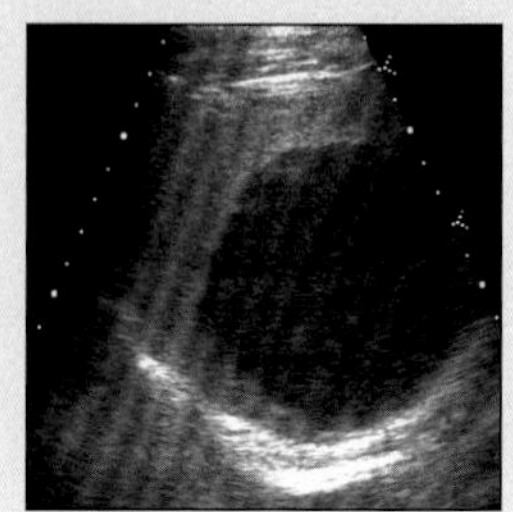

Splenic Cyst

SPLENIC INFARCTION

Key Facts

Terminology
- Global or segmental parenchymal splenic ischemia & necrosis caused by vascular occlusion

Imaging Findings
- Best diagnostic clue: Peripheral wedge-shaped nonenhancing areas on CECT within splenic parenchyma in patient with LUQ pain
- Segmental: Wedge-shaped or rounded low attenuation area on CECT
- Global: Complete nonenhancement of spleen with or without "cortical rim sign" on CECT
- Protocol advice: 150 ml of I.V. contrast at 2.5 ml/sec; 5 mm collimation with 5 mm reconstruction interval

Top Differential Diagnoses
- Splenic abscess
- Splenic tumor
- Splenic laceration

Pathology
- Genetics: Predisposition among some hematologic causes such as sickle cell disease, sickle cell trait

Clinical Issues
- LUQ pain and chills
- Asymptomatic: No treatment
- Symptomatic: Splenectomy for increasing pain or splenic rupture

Diagnostic Checklist
- Consider splenic abscess or tumor

DIFFERENTIAL DIAGNOSIS

Splenic abscess
- Complex fluid collection
- Internal septations and debris
- Low level echoes on sonography
- Multiple gas bubbles
- Rounded, mass effect
- Multiple small lesions (microabscesses) in fungal infections in immunocompromised patients

Splenic tumor
- Primary malignant
  - Angiosarcoma
  - Hypervascular lesions with prominent areas of necrosis
- Primary benign
  - Hemangioma
  - May have contrast-enhancement pattern similar to hepatic hemangiomas
- Secondary
  - Lymphoma
  - Melanoma
  - Ovarian carcinoma
  - Often complex cystic masses
  - May have perisplenic cystic implants

Splenic laceration
- History of trauma
- Associated hemoperitoneum
- May have high attenuation active arterial extravasation on CECT
- High attenuation perisplenic hematoma
- Intra-parenchymal low attenuation hematoma on CECT

Splenic cyst
- Non-neoplastic cysts divided into two categories
  - True epithelial cysts ("primary")
  - "Pseudocysts" or "secondary cysts" lacking an epithelial lining
- Epidermoid cysts are 10-25% of all splenic cysts
- Secondary cysts most often due to infection, infarction or trauma
- Primary cysts may be due to parasitic infection (echinococcal) or epidermoid cysts
- Calcification of cyst wall in 14% of primary cysts, 50% of secondary cysts
- Low-level echoes and thin septations on US in both primary and secondary cysts

PATHOLOGY

General Features
- General path comments: Liquefactive necrosis
- Genetics: Predisposition among some hematologic causes such as sickle cell disease, sickle cell trait
- Etiology
  - Embolic
    - Atrial fibrillation
    - Aortic atherosclerotic disease
    - Aortic valve emboli from subacute bacterial endocarditis
  - Hematologic
    - Sickle hemoglobinopathies
    - Myelofibrosis
    - Any cause of hypersplenism
    - Hypercoagulable states
    - Leukemia and lymphoma
  - Splenic vascular disease
    - Aneurysm
    - Aortic dissection
    - Splenic venous thrombosis
  - Anatomic causes
    - Splenic torsion
    - Torsion secondary to wandering spleen
  - Miscellaneous
    - Pancreatic disease, pseudocysts
    - Collagen vascular disease
    - Gastric tumors invading gastro-splenic ligament
- Epidemiology
  - Embolic
    - Elderly cardiac patients with atrial fibrillation
  - Hematologic

SPLENIC INFARCTION

- Younger patients with sickle hemoglobinopathy or myeloproliferative disease

Gross Pathologic & Surgical Features

- Acute infarction
  - Hemorrhagic or bland necrosis
- Chronic infarction
  - Fibrous scar
  - Rarely calcifies

Microscopic Features

- Coagulative necrosis
- Hemorrhage

Staging, Grading or Classification Criteria

- Segmental
  - Wedge-shaped or round segmental lesion
  - Straight margins typical
- Global
  - Entire spleen is avascular
  - May demonstrate "cortical rim" sign

CLINICAL ISSUES

Presentation

- Most common signs/symptoms
  - LUQ pain and chills
  - 69% of patients have fever in embolic infarction
- Clinical profile
  - Anemia in 53% of patients
  - Leukocytosis in 41% of patients
  - Elevated platelet count in 7% of patients

Demographics

- Age: 2-87 yrs, mean age 54
- Gender: Occurs with equal frequency in males and females

Natural History & Prognosis

- Highly variable
  - May require no treatment
  - Surgery for increased pain or rupture

Treatment

- Options, risks, complications
  - Asymptomatic: No treatment
  - Symptomatic: Splenectomy for increasing pain or splenic rupture

DIAGNOSTIC CHECKLIST

Consider

- Consider splenic abscess or tumor

Image Interpretation Pearls

- Wedge-shaped peripheral area of nonenhancement

SELECTED REFERENCES

1. Wilkinson NW et al: Splenic infarction following laparoscopic Nissen fundoplication: management strategies. JSLS. 7(4):359-65, 2003
2. Sodhi KS et al: Torsion of a wandering spleen: acute abdominal presentation. J Emerg Med. 25(2):133-7, 2003
3. Romero JR et al: Wandering spleen: a rare cause of abdominal pain. Pediatr Emerg Care. 19(6):412-4, 2003
4. Gorg C et al: Chronic recurring infarction of the spleen: sonographic patterns and complications. Ultraschall Med. 24(4):245-9, 2003
5. Sodhi KS et al: Torsion of a wandering spleen: acute abdominal presentation. J Emerg Med. 25(2):133-7, 2003
6. Hatipoglu AR et al: A rare cause of acute abdomen: splenic infarction. Hepatogastroenterology. 48(41):1333-6, 2001
7. Toth PP et al: Spontaneous splenic infarction secondary to diabetes-induced microvascular disease. Arch Fam Med. 9(2):195-7, 2000
8. Barzilai M et al: Noninfectious gas accumulation in an infarcted spleen. Dig Surg. 17(4):402-4, 2000
9. Andrews MW: Ultrasound of the spleen. World J Surg. 24(2):183-7, 2000
10. Nores M et al: The clinical spectrum of splenic infarction. Am Surg. 64(2):182-8, 1998
11. Argiris A: Splenic and renal infarctions complicating atrial fibrillation. Mt Sinai J Med. 64(4-5):342-9, 1997
12. Rypens F et al: Splenic parenchymal complications of pancreatitis: CT findings and natural history. J Comput Assist Tomogr. 21(1):89-93, 1997
13. Beeson MS: Splenic infarct presenting as acute abdominal pain in an older patient. J Emerg Med. 14(3):319-22, 1996
14. Frippiat F et al: Splenic infarction: report of three cases of atherosclerotic embolization originating in the aorta and retrospective study of 64 cases. Acta Clin Belg. 51(6):395-402, 1996
15. Collie DA et al: Case report: computed tomography features of complete splenic infarction, cavitation and spontaneous decompression complicating pancreatitis. Br J Radiol. 68(810):662-4, 1995
16. Chin JK et al: Liver/spleen scintigraphy for diagnosis of splenic infarction in cirrhotic patients. Postgrad Med J. 69(815):715-7, 1993
17. Valentine RJ et al: Splenic infarction after splenorenal arterial bypass. J Vasc Surg. 17(3):602-6, 1993
18. Orringer EP et al: Case report: splenic infarction and acute splenic sequestration in adults with hemoglobin SC disease. Am J Med Sci. 302(6):374-9, 1991
19. Ting W et al: Splenic septic emboli in endocarditis. Circulation. 82(5 Suppl):IV105-9, 1990
20. Goerg C et al: Splenic infarction: sonographic patterns, diagnosis, follow-up, and complications. Radiology. 174(3 Pt 1):803-7, 1990
21. Haft JI et al: Computed tomography of the abdomen in the diagnosis of splenic emboli. Arch Intern Med. 148(1):193-7, 1988
22. O'Keefe JH Jr et al: Thromboembolic splenic infarction. Mayo Clin Proc. 61(12):967-72, 1986
23. Jaroch MT et al: The natural history of splenic infarction. Surgery. 100(4):743-50, 1986
24. Shirkhoda A et al: Computed tomography and ultrasonography in splenic infarction. J Can Assoc Radiol. 36(1):29-33, 1985
25. Balthazar EJ et al: CT of splenic and perisplenic abnormalities in septic patients. AJR Am J Roentgenol. 144(1):53-6, 1985

SPLENIC INFARCTION

IMAGE GALLERY

Typical

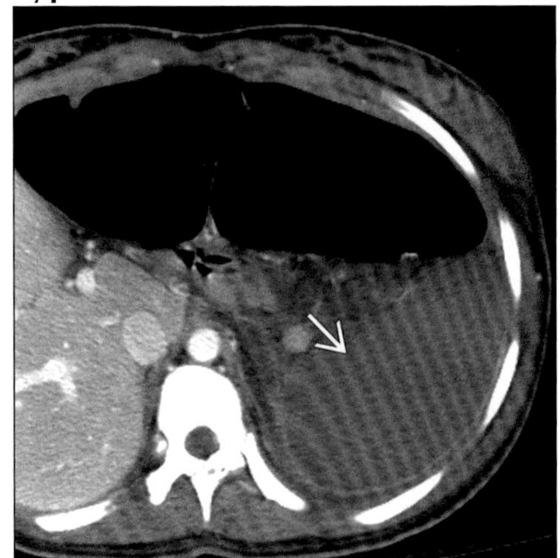

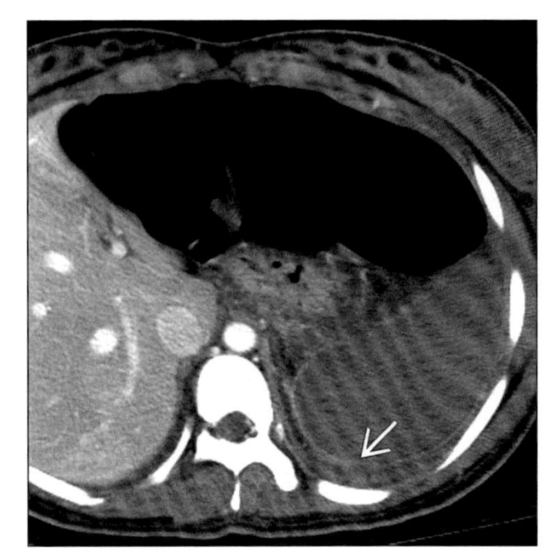

(Left) *Global splenic infarction on axial CECT. Note complete lack of enhancement of splenic parenchyma (arrow).* **(Right)** *Axial CECT of global splenic infarction demonstrates peripheral "cortical rim sign".*

Typical

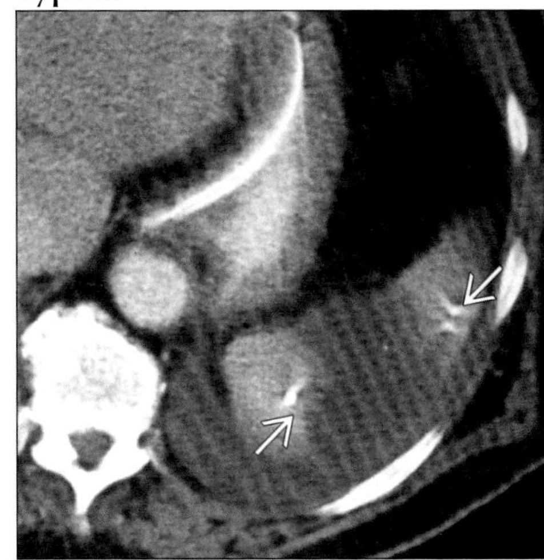

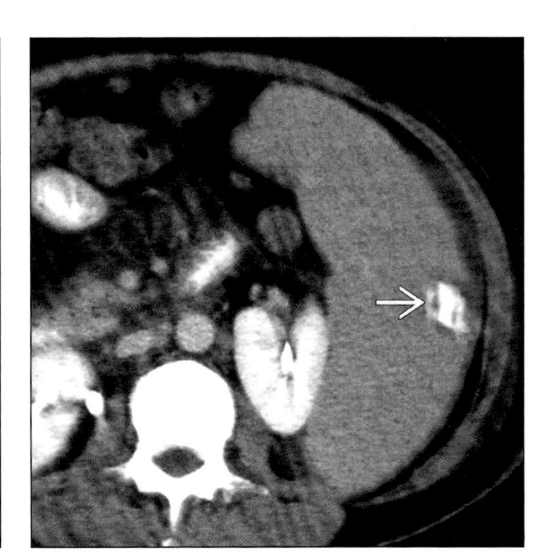

(Left) *Axial CECT of acute splenic infarction demonstrates peripheral wedge-shaped infarct (arrow).* **(Right)** *Axial CECT of chronic splenic infarction demonstrates calcification (arrow).*

Typical

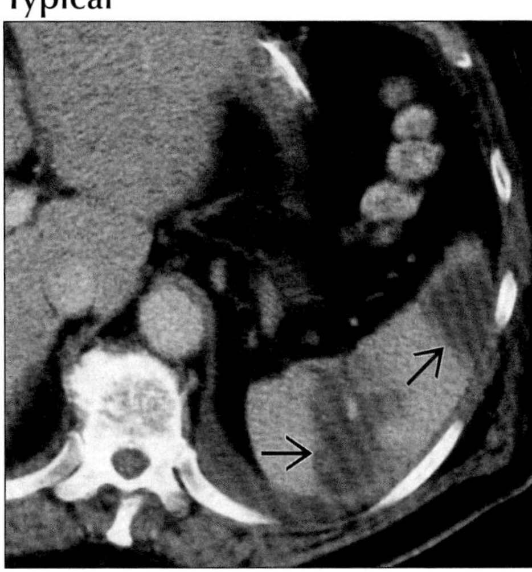

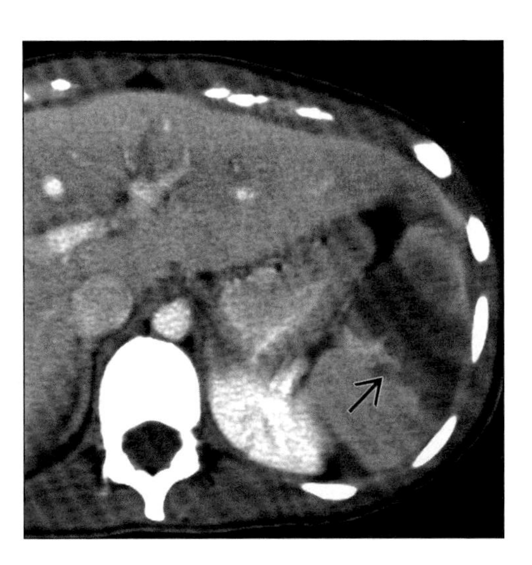

(Left) *Axial CECT of splenic infarcts demonstrates multiple peripheral wedge-shaped emboli (arrows).* **(Right)** *Axial CECT demonstrates lack of enhancement of upper pole of spleen (arrow).*

SPLENIC TRAUMA

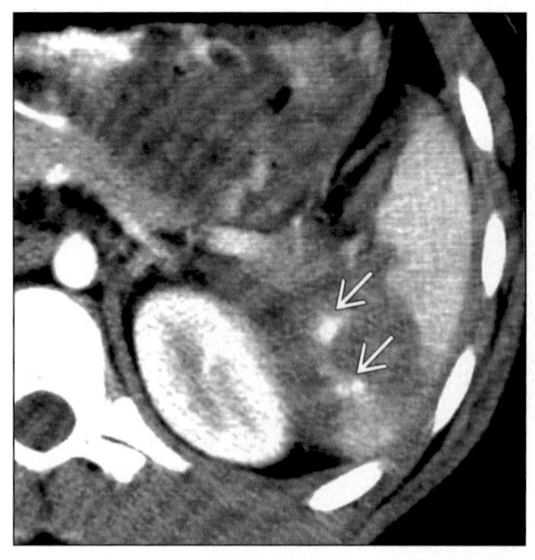

Axial CECT of splenic fracture with active bleeding. Note area of high attenuation arterial extravasation (arrows).

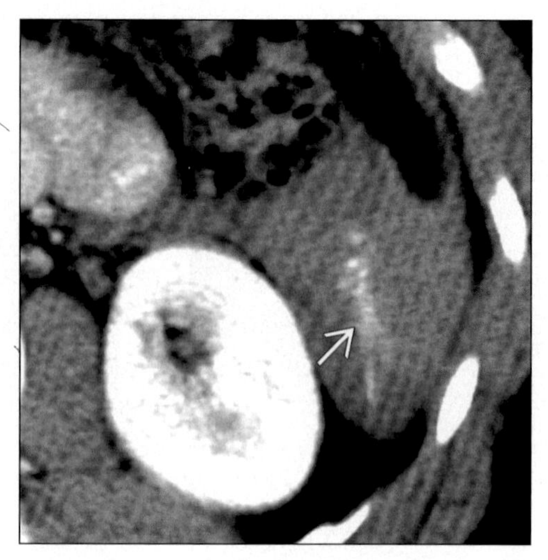

Axial CECT of splenic fracture shows jet of active hemorrhage in left paracolic gutter (arrow).

TERMINOLOGY

Abbreviations and Synonyms

- Splenic laceration, splenic fracture, subcapsular hematoma of spleen

Definitions

- Parenchymal injury to spleen with or without capsular disruption

IMAGING FINDINGS

General Features

- Best diagnostic clue: Low attenuation splenic laceration with high density active bleeding
- Morphology: Lacerations: Linear or jagged edges; subcapsular hematoma: Flattened contour of splenic parenchyma; fracture: Laceration extending from outer cortex to hilum

CT Findings

- NECT: High attenuation (> 30 HU) hemoperitoneum or perisplenic clot (> 45 HU)
- CECT
  - Subcapsular hematoma: Compresses lateral margin of parenchyma
  - Parenchymal laceration: Jagged linear area of nonenhancement due to hematoma
  - Splenic fracture: Deep laceration extending from outer capsule through splenic hilum
  - Active arterial extravasation: High attenuation focus isodense with aorta; surrounded by lower attenuation clot or hematoma

Ultrasonographic Findings

- Real Time: Hypo- or isoechoic hematoma or laceration

Angiographic Findings

- Avascular parenchymal laceration; flattened lateral contour 2° subcapsular hematoma; rounded contrast collections (pseudoaneurysms); amorphous parenchymal extravasation

Imaging Recommendations

- Best imaging tool: CECT
- Protocol advice: 150 ml I.V. contrast at 2.5 ml/sec with 5 mm collimation

DIFFERENTIAL DIAGNOSIS

Splenic abscess

- Rounded, irregular, low attenuation lesion; clinical signs of infection

Splenic infarct

- Wedge-shaped area of low attenuation; associated with splenomegaly; systemic embolization

DDx: Splenic Lesions Mimicking Trauma

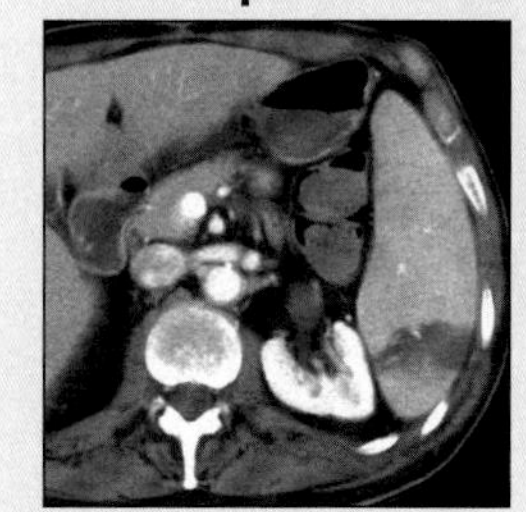

Infarct

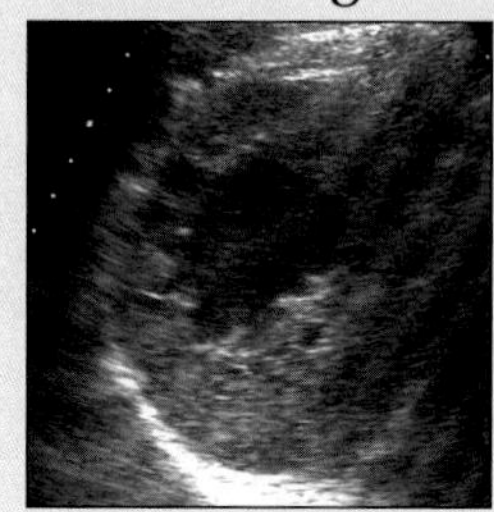

Abscess

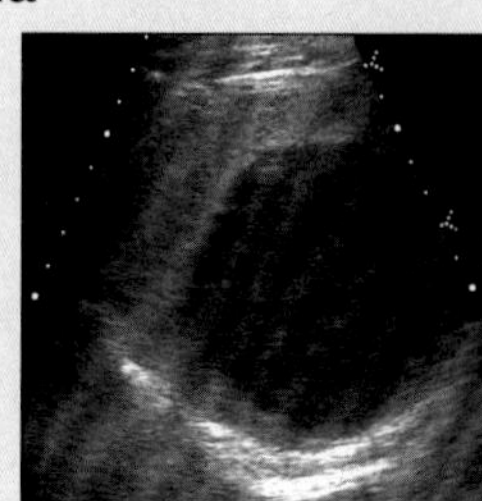

Cyst

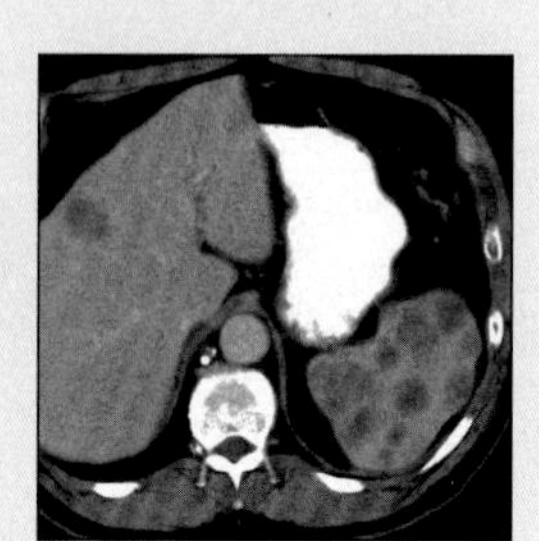

Lymphoma

SPLENIC TRAUMA

Key Facts

Terminology
- Parenchymal injury to spleen with or without capsular disruption

Imaging Findings
- Best imaging tool: CECT
- Protocol advice: 150 ml IV contrast @ 2.5 ml/sec with 5 mm collimation

Pathology
- Associated abnormalities: Injuries to left thorax, tail of pancreas, left liver lobe and/or mesentery

Diagnostic Checklist
- Congenital cleft if no hemoperitoneum
- Innocuous injury may lead to life-threatening delayed hemorrhage, especially with anticoagulation

Splenic cyst
- Rounded hypoechoic lesion on US; definable cyst wall; no internal enhancement on CECT

Lymphoma
- Single or multiple hypodense lesions; splenomegaly

PATHOLOGY

General Features
- General path comments: Laceration, fractures or subcapsular hematoma
- Etiology: Blunt trauma with blow to LUQ
- Epidemiology: Most common abdominal organ injury requiring surgery
- Associated abnormalities: Injuries to left thorax, tail of pancreas, left liver lobe and/or mesentery

Gross Pathologic & Surgical Features
- Varies according to extent of injury

Microscopic Features
- Necrotic injured tissue with surrounding hematoma

Staging, Grading or Classification Criteria
- Grading may be misleading; minor injuries may go on to devastating delayed bleed
  - Grade 1: Subcapsular hematoma or laceration < 1 cm
  - Grade 2: Subcapsular hematoma or laceration 1-3 cm
  - Grade 3: Capsular disruption; hematoma > 3 cm; parenchymal hematoma > 3 cm
  - Grade 4A: Active parenchymal or subcapsular bleeding, pseudoaneurysm or arteriovenous fistula; shattered spleen
  - Grade 4B: Active intraperitoneal bleed

CLINICAL ISSUES

Presentation
- Most common signs/symptoms: Blunt abdominal trauma; LUQ pain; hypotension

Natural History & Prognosis
- Prone to develop delayed hemorrhage; excellent prognosis with early diagnosis & intervention (surgery or embolization)

Treatment
- Non-operative management for minor injuries; angiographic embolization if active arterial extravasation on CT; splenectomy or splenorrhaphy when surgery required

DIAGNOSTIC CHECKLIST

Consider
- Congenital cleft if no hemoperitoneum

Image Interpretation Pearls
- Innocuous injury may lead to life-threatening delayed hemorrhage, especially with anticoagulation

SELECTED REFERENCES

1. Jeffrey RB Jr et al: Detection of active intraabdominal arterial hemorrhage: value of dynamic contrast-enhanced CT. AJR Am J Roentgenol. 156(4):725-9, 1991
2. Jeffrey RB Jr: CT diagnosis of blunt hepatic and splenic injuries: a look to the future. Radiology. 171(1):17-8, 1989
3. Federle MP et al: Splenic trauma: evaluation with CT. Radiology. 162(1 Pt 1):69-71, 1987

IMAGE GALLERY

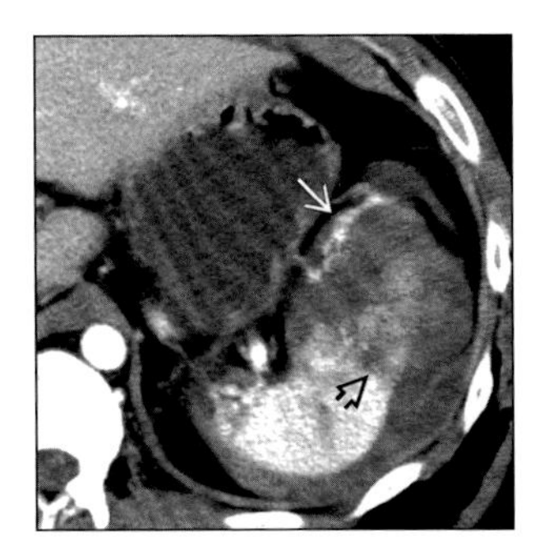

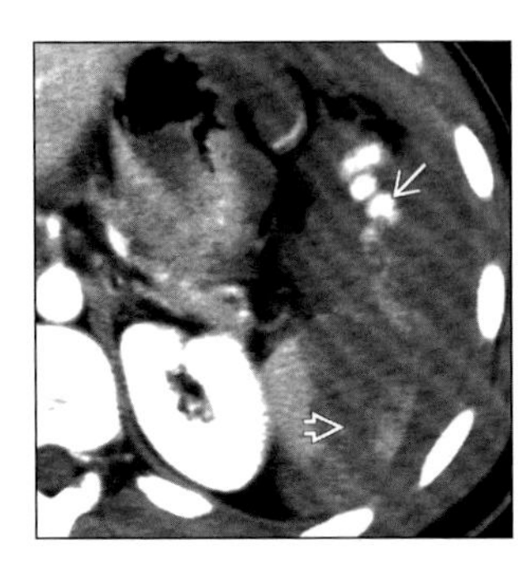

*(**Left**) Axial CECT of splenic laceration with active bleeding. Note low attenuation area of parenchymal laceration (open arrow) with adjacent active bleeding (arrow). (**Right**) Axial CECT of splenic fracture with active bleeding. Note non-enhancing area of splenic fracture (open arrow) and high-density active bleeding (arrow).*

SPLENIC CYST

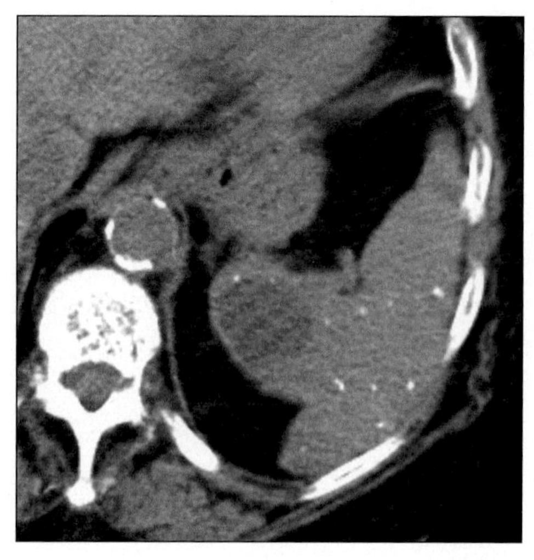

Axial CECT shows a water density cyst, plus calcified granulomas (old histoplasmosis).

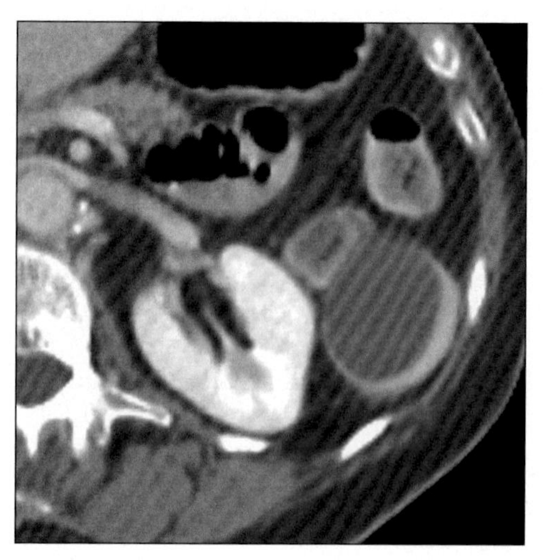

Axial CECT shows water density splenic cyst.

TERMINOLOGY

Definitions

- Cystic masses of spleen

IMAGING FINDINGS

General Features

- Best diagnostic clue: Sharply defined spherical lesion of water density
- Location: Usually in lower pole; subcapsular (65%)
- Size: Varies (congenital usually large, acquired small)
- Key concepts
  - Cystic masses do not commonly occur in spleen
  - Classification of splenic cysts based on etiology
    - Congenital (primary or true) cyst: Epidermoid
    - Acquired (false or pseudo) cyst: Post-traumatic
  - Congenital (primary or true) cyst: Epidermoid
    - Inner cellular lining (endothelial lining) present
    - Account for 10-25% of all splenic cysts
  - Acquired (secondary or false or pseudo) cyst: Post-traumatic (end stage of hematoma)
    - Inner cellular lining absent, but has a fibrous wall
    - Accounts for 80% of splenic cysts
    - Wall calcification seen in 38-50% of cases
    - Cystic nature: Due to liquefactive necrosis

Radiographic Findings

- Radiography
  - Large acquired splenic cysts
    - Curvilinear or plaque-like wall calcification

CT Findings

- Congenital (primary or true) cyst: Epidermoid
  - Solitary, well-defined, spherical, unilocular, cystic lesion (water HU)
  - Thin wall + sharp interface to normal splenic tissue
  - Hemorrhagic, infected, ↑ protein: ↑ Attenuation
  - No rim or intracystic enhancement
  - May rarely have calcified wall
- Acquired (false or pseudo) cyst: Post-traumatic
  - False or pseudocyst (end stage of splenic hematoma)
    - Usually small, solitary, sharply defined, water HU
    - ± Wall calcification (may resemble eggshell)
  - Hematoma (evolving)
    - ↓ HU, clear cut margins, nonspecific cystic lesion
    - CECT: No enhancement of contents

MR Findings

- Congenital (primary or true) cyst: Epidermoid
  - T1WI: Hypointense
    - Variable intensity: Infected or hemorrhagic
  - T2WI: Hyperintense
- Acquired (false or pseudo) cyst: Post-traumatic
  - T1WI: Hypointense; variable intensity (blood)

DDx: Low Density Splenic Mass

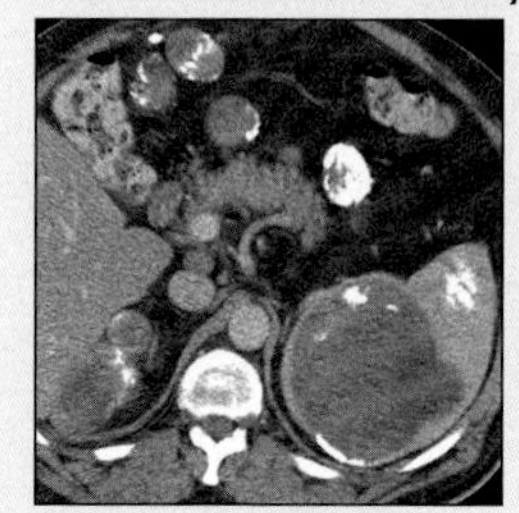

Hydatid Cyst

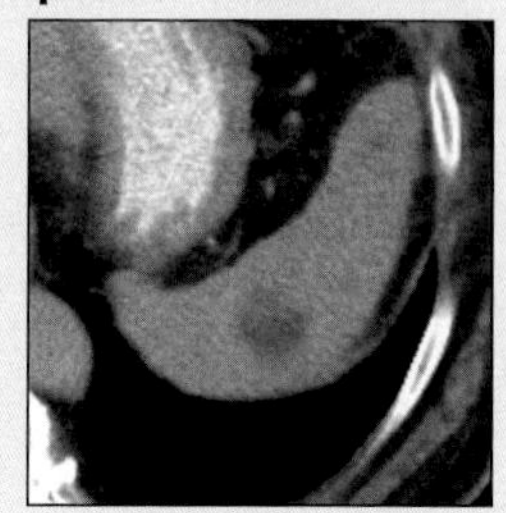

Metastatic Melanoma

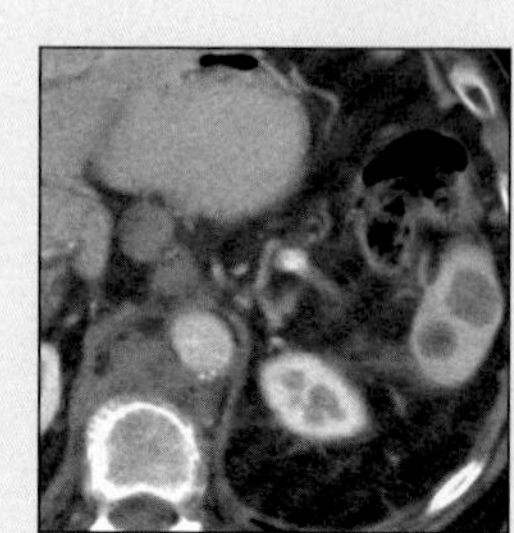

Lymphoma

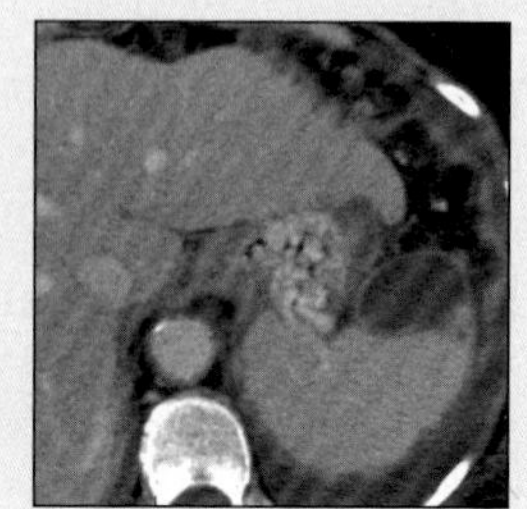

Pseudocyst

SPLENIC CYST

Key Facts

Imaging Findings
- Best diagnostic clue: Sharply defined spherical lesion of water density
- Location: Usually in lower pole; subcapsular (65%)
- Congenital (primary or true) cyst: Epidermoid
- Acquired (false or pseudo) cyst: Post-traumatic
- Anechoic, smooth borders, non-detectable walls ± trabeculation (36%) congenital
- Small, anechoic or mixed with internal echoes, echogenic wall (calcification) acquired

Top Differential Diagnoses
- Inflammatory or infection
- Vascular
- Neoplastic
- Intrasplenic pseudocyst

Pathology
- Congenital (true) epidermoid: Genetic defect of mesothelial migration
- Post-traumatic: End stage of splenic hematoma
- Pathogenesis: Liquefactive necrosis, cystic change
- Epidemiology: 7.6 per 10,000 people
- Congenital (true) cyst: Endothelial lining present
- Post-traumatic (false) cyst: Endothelial lining absent

Diagnostic Checklist
- Rule out infective, vascular & neoplastic cystic lesions
- Congenital: Large, well-defined, water density with thin wall & no rim or intracystic enhancement
- Acquired (post-traumatic): Usually small, sharply defined, near water HU with thick wall ± calcification
- Differentiation by imaging alone is often impossible

- T2WI: Hyperintense
- Calcification or hemosiderin deposited in wall
  - Hypointense (both T1 & T2WI)
- Hematoma: Varied intensity based on age & evolution of blood products
  - After 3 weeks appears as a cystic mass: T1WI hypointense; T2WI hyperintense

Ultrasonographic Findings
- Real Time
  - Congenital (primary or true) cyst: Epidermoid
    - Anechoic, smooth borders, non-detectable walls ± trabeculation (36%) congenital
    - No septations or nodules
    - Complicated: Septations, internal echoes (debris), thickened wall ± calcification
  - Acquired (false or pseudo) cyst: Post-traumatic
    - Small, anechoic or mixed with internal echoes, echogenic wall (calcification) acquired
    - ± Trabeculation of cyst wall (15%)

Angiographic Findings
- Conventional: Avascular mass; stretched or normal appearing capsular & intrasplenic vessels

Imaging Recommendations
- Helical NE + CECT; US; MR

DIFFERENTIAL DIAGNOSIS

Inflammatory or infection
- Pyogenic abscess
  - Solitary, multiple, well-defined, irregular borders
  - 20-40 HU (due to proteinaceous material)
  - CECT: ± Rim & no central enhancement
  - T1WI: Hypointense; variable intensity (protein or hemorrhagic content)
  - T2WI: Hyperintense
- Fungal abscess
  - E.g., Candida, Aspergillus, Cryptococcal
  - Usually microabscesses: Multiple, small, well-defined, low attenuation, no rim-enhancement
  - T1WI hypointense; T2WI hyperintense
- Granulomatous abscesses
  - E.g., Mycobacterium & atypical TB; cat-scratch
  - Multiple, small, well-defined, low attenuation, no rim-enhancement
  - T1WI hypointense; T2WI hyperintense
- Parasitic: Echinococcal or hydatid cyst
  - Pathology: Inner germinal layer + outer pericyst
  - Large unilocular or multilocular; well-defined near water density cyst
  - ± Area of ↑ density within cyst (hydatid sand)
  - Daughter cysts: ↓ Density than mother cyst
  - Curvilinear ring-like calcification (common)
  - CECT: Enhancement of cyst wall + septations
  - T1WI
    - Matrix: Hypointense, rarely hyperintense (↓ water)
    - Rim: Hypointense (fibrosis); daughter cyst ↓ signal
  - T2WI
    - Mother & daughter cysts: Hyperintense
    - Floating membrane: Low-intermediate signal
  - Ultrasound findings
    - Well-defined anechoic cyst ± hydatid sand (internal echoes)
    - Multiseptate cyst + daughter cysts
    - "Water lily" sign: Cyst + floating membrane + detached endocyst
  - Look for "cysts" in liver & peritoneal cavity

Vascular
- Infarction (arterial or venous)
  - Acute phase: Well-defined areas of ↓ attenuation
  - Subacute & chronic phases: Near water HU
    - Due to liquefactive necrosis
  - CECT: No enhancement
  - MR findings
    - Acute & subacute phases: Vary with age & evolution of blood products
    - Chronic phase: Cystic intensity (due to necrosis)
    - T1WI hypointense; T2WI hyperintense
- Peliosis
  - Multiple, low HU, rounded lesions of varied size
  - CECT: Early peripheral nodular & delayed centripetal enhancement
  - T1WI hypointense; T2WI hyperintense

- T1 C+: Same as CECT

Neoplastic

- Benign: E.g., hemangioma & lymphangioma
  - Hemangioma
    - Have overlapping & variable CT features preventing accurate diagnosis
    - Usually not confused with simple cyst or metastases
  - Lymphangioma
    - NECT: Solitary or multiple well-defined, ↓ HU
    - CECT: No enhancement
    - T1WI hypointense; T2WI hyperintense
    - ↑ Signal on T1WI: Subacute hemorrhage or proteinaceous fluid
- Malignant: E.g., Lymphoma & metastases
  - Lymphoma
    - NECT: Miliary multifocal or solitary, ↓ attenuation
    - CECT: Mild enhancement
    - Necrosis within lesion: Near water density (rare)
    - T1WI hypointense; T2WI hyperintense
    - ↑ Signal on T2WI: Areas of necrosis
  - Metastases
    - Relatively common
    - E.g., Malignant melanoma, pancreatic & ovarian cancer may cause "cystic" splenic metastases
    - NECT: Ill-/well-defined, unilocular ↓ or water HU
    - CECT: Peripheral & septal enhancement
    - T1WI hypointense; T2WI hyperintense
    - ↑ Signal on T1WI: Subacute hemorrhage, melanin
    - ↑ Signal on T2WI: Water, necrosis, old bleed

Intrasplenic pseudocyst

- Pancreatitis → intrasplenic pseudocyst or abscess
- Seen in 1.1-5% of patients with pancreatitis
- Pathogenesis
  - Direct extension of pancreatic pseudocyst
  - Secondary to digestive effects of enzymes on splenic vessels or parenchyma along splenorenal ligament
- Imaging
  - Well-defined rounded cystic mass + enlarged spleen
  - Density varies: Amount of debris + hemorrhage
  - Associated inflammatory changes of pancreas seen
    - Peripancreatic fluid collection (especially near tail)

PATHOLOGY

General Features

- Etiology
  - Congenital (true) epidermoid: Genetic defect of mesothelial migration
  - Post-traumatic: End stage of splenic hematoma
    - Pathogenesis: Liquefactive necrosis, cystic change
- Epidemiology: 7.6 per 10,000 people

Gross Pathologic & Surgical Features

- Congenital (true) epidermoid cyst
  - Usually large, glistening smooth walls
- Post-traumatic (false or pseudocyst)
  - Smaller than true cysts, debris, wall calcification

Microscopic Features

- Congenital (true) cyst: Endothelial lining present
- Post-traumatic (false) cyst: Endothelial lining absent

CLINICAL ISSUES

Presentation

- Most common signs/symptoms
  - Asymptomatic; mild pain, palpable mass in LUQ
  - Tenderness in LUQ; splenomegaly

Demographics

- Age: 2/3rd below 40 years old
- Gender: M:F = 2:3

Natural History & Prognosis

- Complications: Hemorrhage, rupture, infection
- Prognosis
  - Good: Noncomplicated cases; after surgical removal
  - Poor: Complicated cases

Treatment

- Small & asymptomatic: No treatment
- Small & symptomatic: Surgery
- Large (> 6 cm): Surgical removal (debatable)

DIAGNOSTIC CHECKLIST

Consider

- Rule out infective, vascular & neoplastic cystic lesions

Image Interpretation Pearls

- Congenital: Large, well-defined, water density with thin wall & no rim or intracystic enhancement
- Acquired (post-traumatic): Usually small, sharply defined, near water HU with thick wall ± calcification
- Differentiation by imaging alone is often impossible

SELECTED REFERENCES

1. Ito K et al: MR imaging of acquired abnormalities of the spleen. AJR Am J Roentgenol. 168(3):697-702, 1997
2. Urrutia M et al: Cystic masses of the spleen: radiologic-pathologic correlation. Radiographics. 16(1):107-29, 1996
3. Shirkhoda A et al: Imaging features of splenic epidermoid cyst with pathologic correlation. Abdom Imaging. 20(5):449-51, 1995
4. Freeman JL et al: CT of congenital and acquired abnormalities of the spleen. Radiographics. 13(3):597-610, 1993
5. von Sinner WN et al: Hydatid disease of the spleen. Ultrasonography, CT and MR imaging. Acta Radiol. 33(5):459-61, 1992
6. Maves CK et al: Splenic and hepatic peliosis: MR findings. AJR Am J Roentgenol. 158(1):75-6, 1992
7. Dachman AH et al: Nonparasitic splenic cysts: a report of 52 cases with radiologic-pathologic correlation. AJR Am J Roentgenol. 147(3):537-42, 1986
8. Chintapalli K et al: Differential diagnosis of low-attenuation splenic lesions on computed tomography. J Comput Tomogr. 9(4):311-9, 1985
9. Faer MJ et al: Traumatic splenic cyst: radiologic-pathologic correlation from the Armed Forces Institute of Pathology. Radiology. 134(2):371-6, 1980

SPLENIC CYST

IMAGE GALLERY

Typical

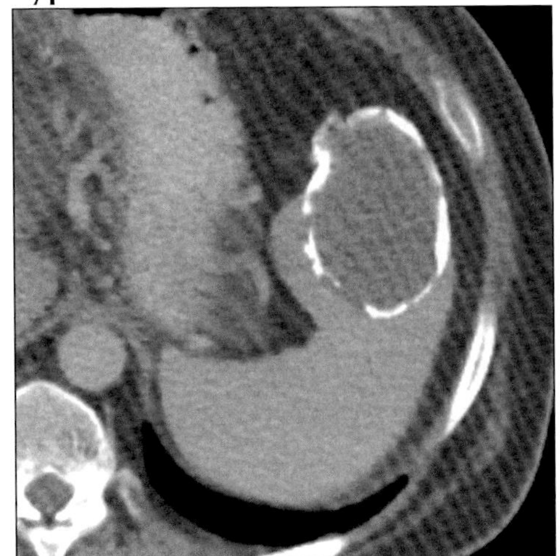

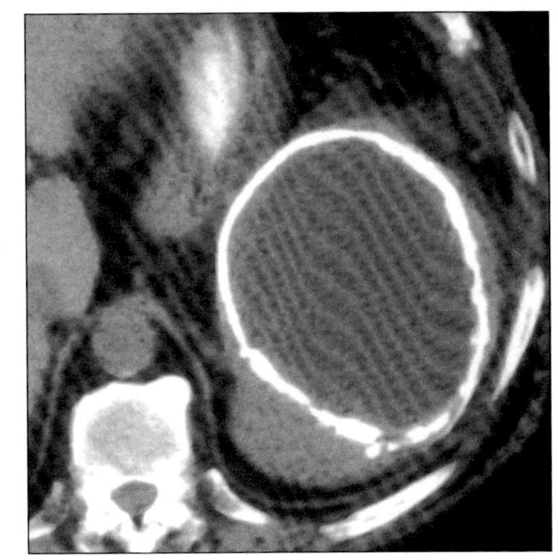

(Left) *Axial CECT shows a splenic cyst with a calcified capsule.* ***(Right)*** *Axial NECT shows a rim-calcified splenic cyst.*

Typical

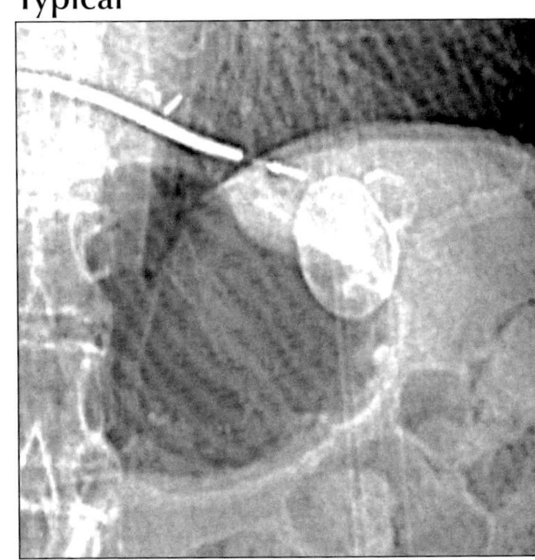

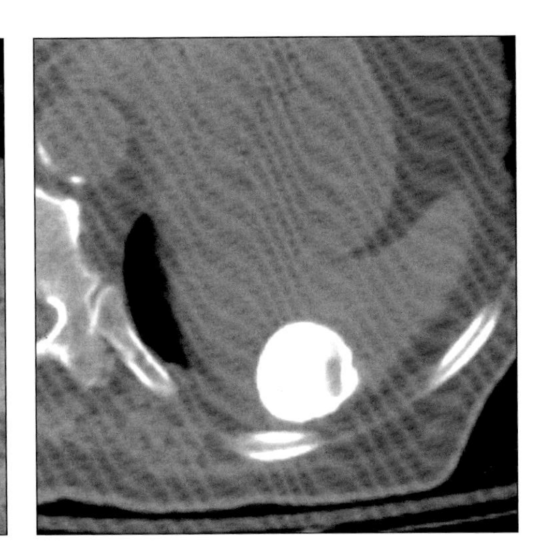

(Left) *Radiograph shows "eggshell" calcification of a splenic cyst.* ***(Right)*** *Axial NECT shows "eggshell" calcification of a splenic cyst.*

Typical

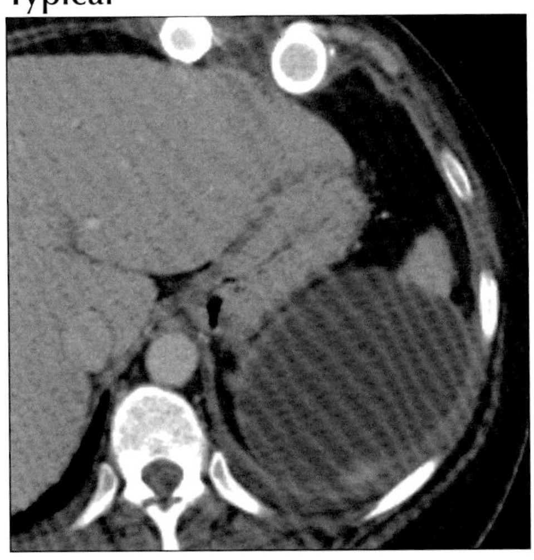

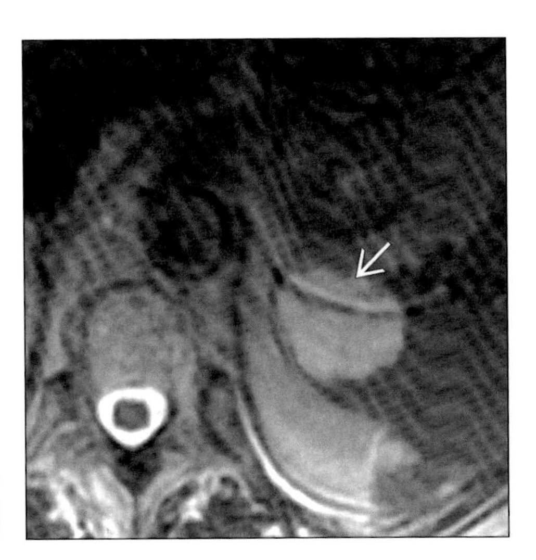

(Left) *Axial CECT in a woman with cardiomyopathy shows acquired splenic cyst that developed after prior splenic infarctions.* ***(Right)*** *Axial T2WI MR shows hyperintense splenic cyst (arrow).*

SPLENIC TUMORS

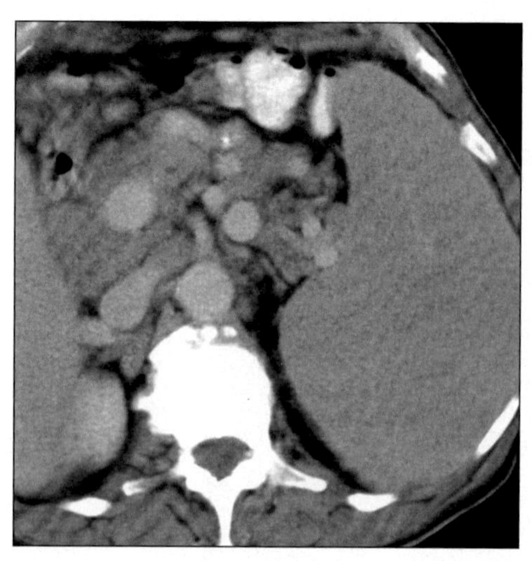

Axial CECT shows splenomegaly due to non-Hodgkin lymphoma.

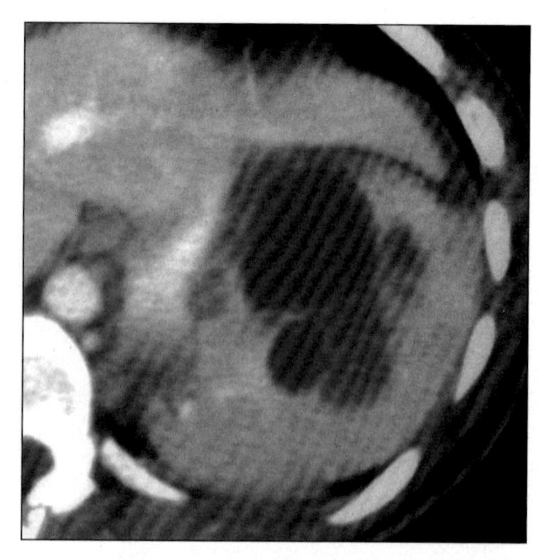

Axial CECT shows a multi-cystic mass; lymphangioma.

TERMINOLOGY

Abbreviations and Synonyms

- Splenic mass or lesion

Definitions

- Space occupying benign or malignant tumor of spleen

IMAGING FINDINGS

General Features

- Best diagnostic clue: Solid or cystic, solitary or multiple splenic masses
- Key concepts
  - Classification based on pathology & histology
    - Benign & malignant tumors
  - Benign tumors
    - Hemangioma, hamartoma, lymphangioma
  - Hemangioma
    - Most common primary benign neoplasm of spleen
    - Autopsy incidence: 0.03-14%
    - Usually incidental on radiologic studies
    - Primarily affect adults; peak age 35-55 y
    - Multiple as part of a generalized angiomatosis (Klippel-Trenaunay-Weber syndrome)
    - Hemangiomatosis: Diffuse splenic hemangiomas
  - Hamartoma
    - Rare benign tumor of spleen; no sex predilection
    - Autopsy incidence: 0.13%; solitary or multiple
    - Incidental at autopsy & exploratory laparotomy
    - Contain anomalous mixtures of normal elements of splenic tissue
  - Lymphangioma
    - Rare benign splenic neoplasm (common in neck)
    - Solitary/multiple; usually subcapsular in location
    - Lymphangiomatosis: Diffuse lymphangiomas
    - Most lymphangiomas occur in childhood
  - Malignant tumors
    - Lymphoma, AIDS-related lymphoma
    - Angiosarcoma, metastases
    - Rare malignant splenic tumors: Malignant fibrous histiocytoma, leiomyosarcoma & fibrosarcoma
  - Lymphoma
    - Most common malignant tumor of spleen
    - Hodgkin (HD) & non-Hodgkin lymphoma (NHL)
    - Presence or absence of splenic involvement may determine type of therapy
    - Spleen: Considered as "nodal organ" in Hodgkin & "extranodal organ" in non-Hodgkin lymphoma
    - Manifest: Focal lesions (> 1 cm) or diffuse (typical)
    - Staging laparotomy or splenectomy: Uncommonly necessary to determine splenic involvement
    - Initial involvement: 23-39% (Hodgkin); 30-40% (non-Hodgkin)
    - Primary lymphoma; secondary (more common)

DDx: Splenic Mass

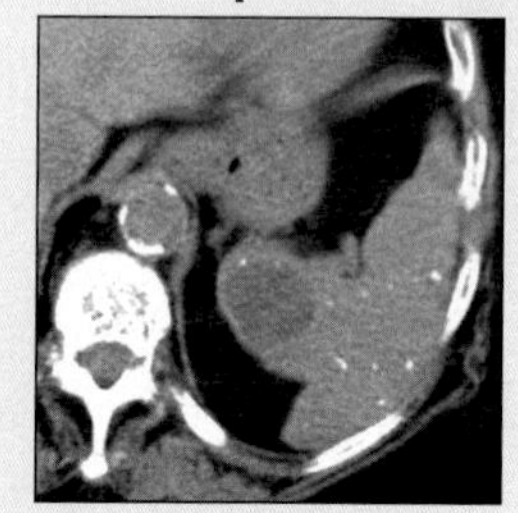

Cyst (+ Granulomas)

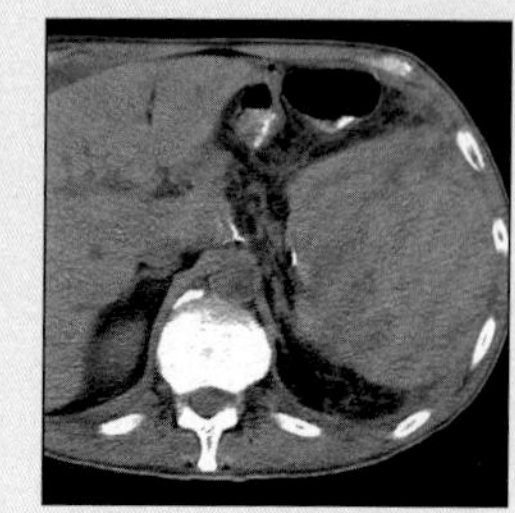

Hematoma

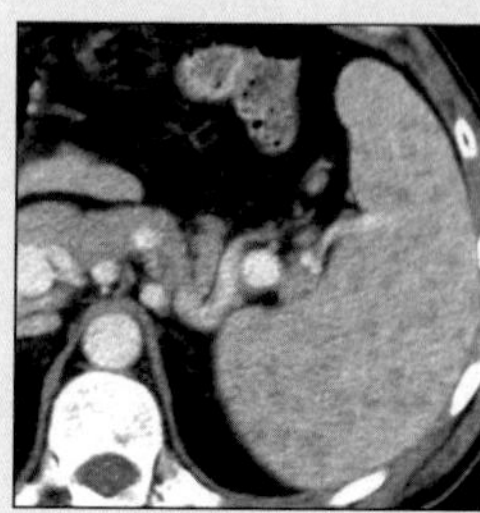

Sarcoid

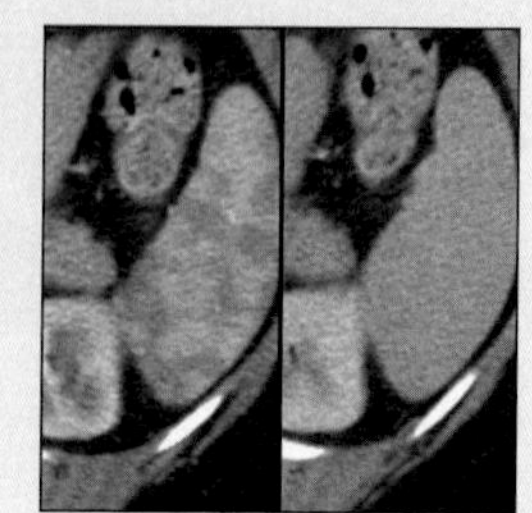

Artifact (Perfusion)

SPLENIC TUMORS

Key Facts

Imaging Findings

- Best diagnostic clue: Solid or cystic, solitary or multiple splenic masses
- "Cystic" splenic metastases: Melanoma; adenocarcinoma of breast, ovary & endometrium
- Early peripheral & late central enhancement (not as reliable as for liver hemangioma)
- Small: Isodense or hypodense (hamartoma)
- Solitary mass; multifocal or diffuse infiltration (lymphoma)
- Solitary or multiple, nodular, irregular margins (angiosarcoma)
- Multiple, solid (common) or cystic, ↓ attenuation (metastases)
- Metastases: Iso-/hypo-/hyperechoic, "target"/"halo"

Top Differential Diagnoses

- Splenic cyst
- Splenic hematoma
- Splenic infection
- Splenic sarcoidosis

Clinical Issues

- Most common signs/symptoms: LUQ pain, palpable mass, splenomegaly, fever, weight loss
- Complications: Hemorrhage, rupture
- Prognosis: Good (benign tumors); poor (malignant)

Diagnostic Checklist

- Must rely on clinical setting & biopsy, if necessary
- Imaging findings alone are unreliable in distinguishing solid from pseudocystic splenic masses

    - Primary splenic lymphoma: Typically represents NHL (B-cell origin)
    - Age: Primary (young age); secondary (older age)
  - AIDS-related lymphoma
    - Intra-abdominal involvement: Seen in 2/3rd cases
    - 86% extranodal disease, half involve GI tract
    - Histologic subtypes: Small noncleaved (Burkitt & non-Burkitt type)
  - Angiosarcoma
    - Very rare malignant tumor of spleen; seen in patients with previous exposure to Thorotrast
    - Most common primary nonlymphoid vascular malignant tumor of spleen
    - Poor prognosis with early, widespread metastases
    - Metastases: Liver (70%); lung, pleura, nodes, bone, brain (30%)
    - Usually affect older individuals (mean age at diagnosis is 50s; M = F)
    - Most patients die within 1 year of diagnosis
  - Metastases
    - Relatively uncommon (autopsy incidence, 7.1% with malignancy)
    - May be multiple (60%), solitary (31.5%), nodular & diffuse (8.5%)
    - Common route: Hematogenous spread (splenic arterial blood flow)
    - Retrograde (< common): Via splenic vein (portal HTN) & lymphatics
    - Common primary sites & metastases to spleen: Breast (21%), lung (18%), ovary (8%), stomach (7%), melanoma (6%), prostate (6%)
    - Frequency of splenic metastases by primary tumor: Melanoma (34%), breast (12%), lung (9%)
    - "Cystic" splenic metastases: Melanoma; adenocarcinoma of breast, ovary & endometrium
  - Peritoneal implants to surface of spleen: Carcinoma of ovary, GIT, pancreas
  - Direct invasion of spleen: Uncommon
    - E.g., gastric, colonic, pancreatic tail, left renal cancer, retroperitoneal sarcoma

CT Findings

- Benign tumors
  - Hemangioma
    - Homogeneous, hypodense, solid or cystic masses
    - Central punctate or peripheral curvilinear Ca++
    - Early peripheral & late central enhancement (not as reliable as for liver hemangioma)
    - Heterogeneous: Cystic + avascular components
  - Hamartoma
    - Small: Isodense or hypodense (hamartoma)
    - Large: Area of ↓ HU (scar or necrosis); calcification
    - Enhancement: Variable; uniform on delayed scans
  - Lymphangioma
    - Thin-walled low density lesions; sharp margins
    - Wall enhances; usually subcapsular in location
- Malignant tumors
  - Lymphoma
    - Homogeneously enlarged spleen, no discrete mass
    - Solitary mass; multifocal or diffuse infiltration (lymphoma)
    - Focal lesions: AIDS-related lymphoma > common
    - ↓ HU lesions + minimal enhancement; ± ascites
    - Rarely necrosis ± air or cystic mimicking abscess
    - Lymphadenopathy: Abdominal or retroperitoneal
  - Angiosarcoma
    - Solitary or multiple, nodular, irregular margins (angiosarcoma)
    - Heterogeneous density; variable enhancement
    - Enlarged spleen; ± hemorrhage & calcification
    - ± Hematoma: Intrasplenic/subcapsular/perisplenic
    - ± Multiple liver or distant metastases
  - Metastases
    - Multiple, solid (common) or cystic, ↓ attenuation (metastases)
    - Malignant melanoma: Solid or cystic
    - Ovary, breast & endometrium: Hypodense, solid
    - Central or peripheral enhancement
    - Calcification (rare) except mucinous colon cancer

MR Findings

- Benign tumors
  - Hemangioma
    - T1WI: Hypointense; T2WI: Hyperintense
    - T1 C+: Uniform or heterogeneous enhancement
  - Hamartoma

- T1WI: Isointense; T2WI: Hypo-to hyperintense
- T1 C+: Variable; uniform on delayed scans
  - Lymphangioma
    - T1WI: Hypointense; T2WI: Markedly hyperintense
    - Septa: T2WI; hypointense; enhances on T1 C+
- Malignant tumors
  - Lymphoma
    - MR not reliable due to similar T1, T2 relaxation times & proton densities of spleen/lymphoma
    - SPIO-enhanced MR: Altered signal intensity of spleen + overall less uptake of SPIO
  - Angiosarcoma
    - T1 & T2WI: Variable signal due to hemorrhage, necrosis & calcification
    - T1 C+: Variable or ring-like enhancement
  - Metastases
    - T1WI: Isointense to hypointense
    - T2WI: Hyperintense
    - T1 C+: Enhancement depends on type of primary

Ultrasonographic Findings

- Real Time
  - Hemangioma
    - Echogenic masses with areas of complex echoes
    - Complex masses: Solid & cystic areas
  - Hamartoma
    - Well-defined, homogeneous echogenic mass
  - Lymphangiomas
    - Grossly enlarged spleen; multicystic appearance
    - Intracystic internal echoes: Proteinaceous material
  - Lymphoma
    - Typically diffuse or focal hypoechoic lesions
    - Anechoic/mixed echoic; small or large nodules
  - Angiosarcoma: Solid, mixed echogenic mass
  - Metastases: Iso-/hypo-/hyperechoic, "target"/"halo"

Angiographic Findings

- Conventional
  - Hemangioma
    - "Cotton wool" appearance: Pooling of contrast
    - No neovascularity or arteriovenous shunting
    - Typically retain contrast beyond venous phase
  - Hamartoma & angiosarcoma: Hypervascular

Nuclear Medicine Findings

- Hemangioma
  - Tc-99m labeled RBC scan with SPECT
    - Early dynamic scan: Focal defect or less uptake
    - Delayed scans (over 30-50 min): Persistent filling

Imaging Recommendations

- Helical NE + CECT; MR with T1 C+

DIFFERENTIAL DIAGNOSIS

Splenic cyst

- Low density, sharp margins, no enhancement
- Rim of dense calcification; anechoic on ultrasound

Splenic hematoma

- Subcapsular: Low attenuation, crescentic fluid collection along lateral surface of spleen
- Intrasplenic: Irregular low density fluid collection

Splenic infection

- Tuberculosis, fungal (microabscesses) & pyogenic
- Solitary or multiple, small or large, low density lesions
- May show minimal peripheral enhancement

Splenic sarcoidosis

- Low density splenic nodules ranging from 0.3-2.0 cm
- Periaortic & retrocrural adenopathy (small & discrete)
- Associated mediastinal, hilar nodes & lung lesions

Artifact

- Heterogeneous enhancement during arterial phase of imaging

PATHOLOGY

General Features

- Etiology: Primary splenic tumor (unknown); metastases (underlying malignancy)
- Epidemiology: Incidence varies based on type of tumor

Gross Pathologic & Surgical Features

- Varies depending on type of tumor

Microscopic Features

- Varies based on histology of tumor

CLINICAL ISSUES

Presentation

- Most common signs/symptoms: LUQ pain, palpable mass, splenomegaly, fever, weight loss

Natural History & Prognosis

- Complications: Hemorrhage, rupture
- Prognosis: Good (benign tumors); poor (malignant)

Treatment

- Surgical resection: Primary benign, malignant tumors

DIAGNOSTIC CHECKLIST

Consider

- Must rely on clinical setting & biopsy, if necessary

Image Interpretation Pearls

- Imaging findings alone are unreliable in distinguishing solid from pseudocystic splenic masses

SELECTED REFERENCES

1. Abott RM et al: Primary vascular neoplasms of the spleen: radiologic-pathologic correlation. Radiographics. 24: 1137-63, 2004
2. Urban BA et al: Helical CT of the spleen. Am. J. Roentgenol. 170: 997-1003, 1998
3. Rabushka LS et al: Imaging of the spleen: CT with supplemental MR examination. Radiographics. 14(2):307-32, 1994

IMAGE GALLERY

Typical

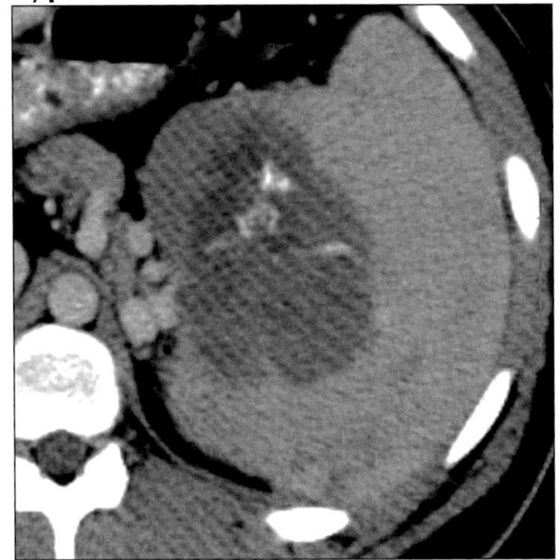

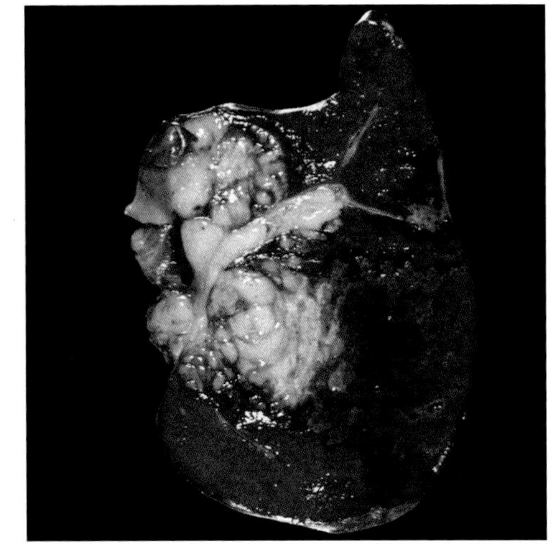

(Left) *Axial CECT in a 20 year old man shows partially calcified splenic mass; non-Hodgkin lymphoma (untreated).* ***(Right)*** *Surgical photograph of specimen of spleen from a 20 year old man with non-Hodgkin lymphoma.*

Typical

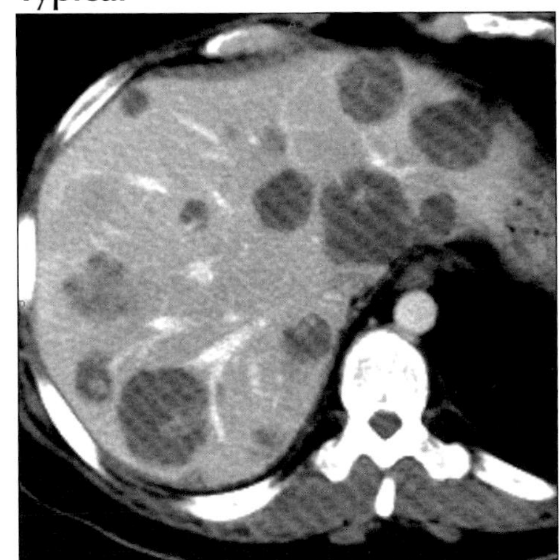

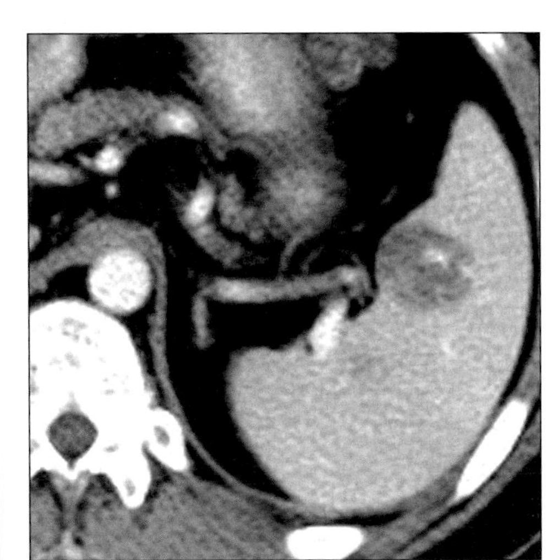

(Left) *Axial CECT shows innumerable hepatic masses; angiosarcoma in a young man.* ***(Right)*** *Axial CECT of a young man with hepatic angiosarcoma shows one of several splenic tumors with similar appearance.*

Typical

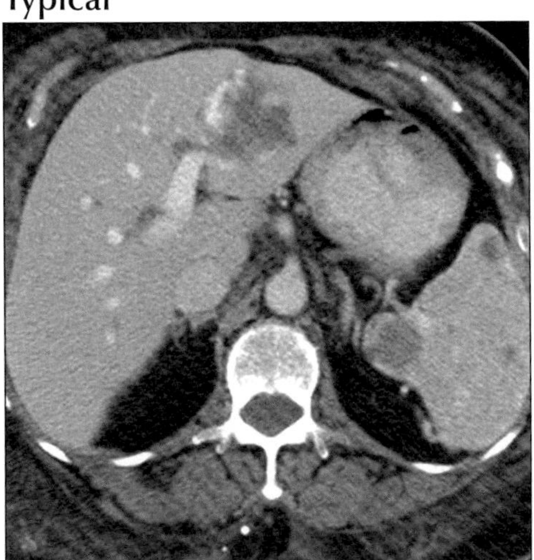

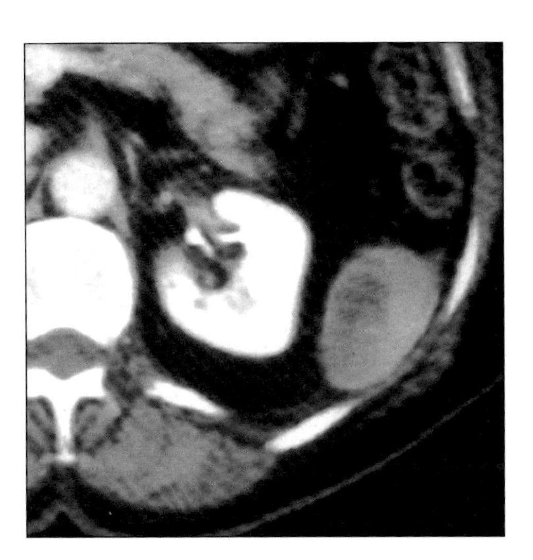

(Left) *Axial CECT in a patient with cavernous hemangiomas in the spleen, liver, and body wall.* ***(Right)*** *Axial CECT shows malignant fibrous histiocytoma in the spleen.*

SPLENIC METASTASES AND LYMPHOMA

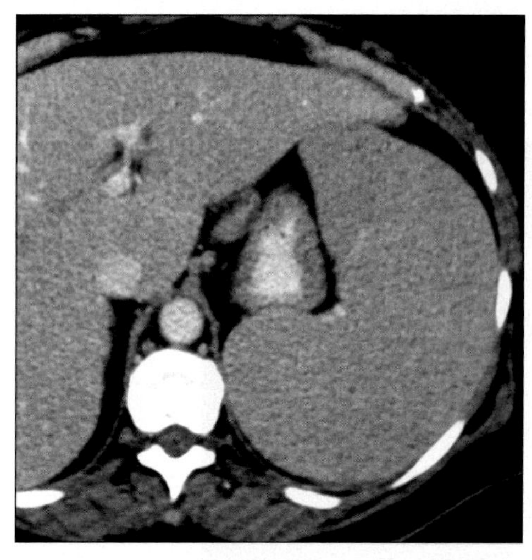

Axial CECT in June, 2003 shows splenomegaly due to non-Hodgkin lymphoma (NHL).

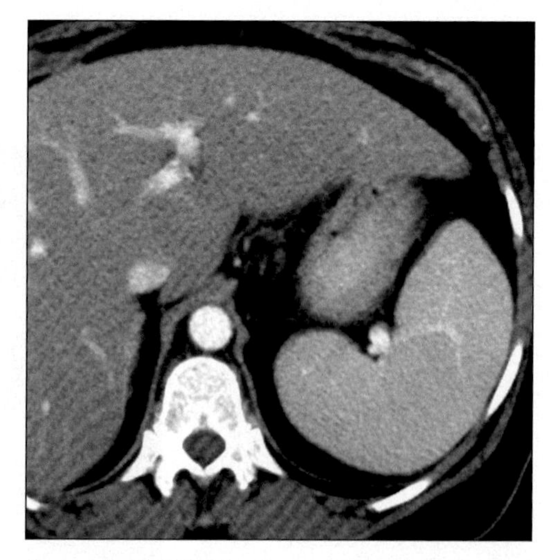

Repeat Axial CECT in August, 2003 shows marked reduction of splenic size following treatment for NHL.

TERMINOLOGY

Definitions

- Splenic metastases from other primary cancer site
- Splenic lymphoma: Malignant tumor of lymphocytes

IMAGING FINDINGS

General Features

- Best diagnostic clue: Solitary or multiple solid lesions (metastases); diffuse infiltrative lesions (lymphoma)
- Size
  - Splenic metastases: One-third are microscopic nodules, two-thirds are grossly visible at autopsy
  - Splenic lymphoma: Solitary mass or multifocal lesions 1 cm or greater are detectable on CT
- Key concepts
  - Splenic metastases
    - Relatively uncommon (autopsy incidence, 7.1%)
    - May be multiple (60%), solitary (31.5%), nodular or diffuse infiltrative lesions (8.5%)
    - Common route: Hematogenous (splenic artery)
    - Retrograde (< common): Via splenic vein (portal HTN) & lymphatics
    - Common primary sites & metastases to spleen: Breast (21%), lung (18%), ovary (8%), stomach (7%), melanoma (6%), prostate (6%)
    - Frequency of splenic metastases by primary tumor: Melanoma (34%), breast (12%), lung (9%)
    - "Cystic" metastases: Melanoma, adenocarcinoma of breast, ovary & endometrium
    - Serosal implants: Peritoneal carcinomatosis secondary to ovarian, GIT, pancreatic cancers
    - Direct invasion: Uncommon; may occur in large gastric, colon, pancreatic tail cancer, left renal cell cancer, neuroblastoma or retroperitoneal sarcoma
  - Splenic lymphoma
    - Most common malignant tumor of spleen
    - Hodgkin (HD) & non-Hodgkin lymphoma (NHL)
    - Primary lymphoma; secondary (more common)
    - Primary splenic lymphoma: NHL (B-cell origin)
    - Initial splenic involvement: Hodgkin (23-34%); non-Hodgkin (30-40%)
    - Spleen: Considered as "nodal organ" in Hodgkin & "extranodal organ" in non-Hodgkin lymphoma
    - Manifest: Focal lesions (> 1 cm) or diffuse (typical)
    - Age: Primary (young age); secondary (older age)
    - Presence or absence of splenic involvement may determine type of therapy
    - Staging laparotomy or splenectomy: Uncommonly necessary to determine splenic involvement
  - AIDS-related lymphoma
    - Histologic subtypes: Small noncleaved (Burkitt & non-Burkitt type)
    - Splenic involvement: NHL > common than HD

DDx: Multiple Low Density Splenic Lesions

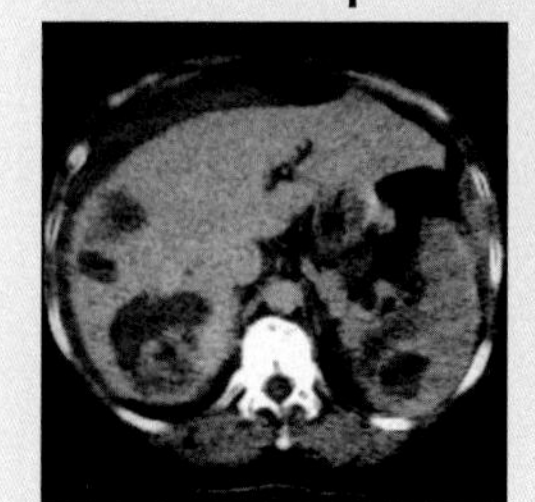

Angiosarcoma

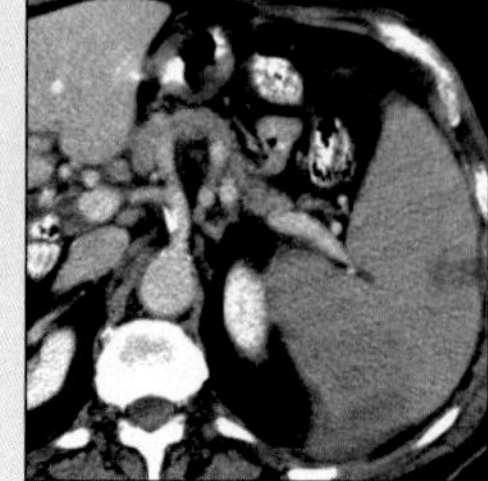

Infarcts

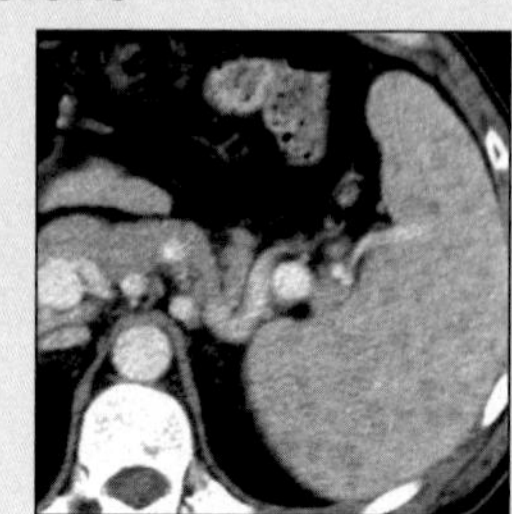

Sarcoid

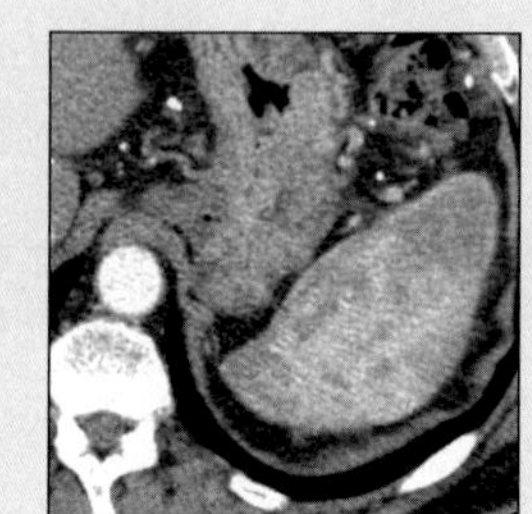

Artifact (Perfusion)

SPLENIC METASTASES AND LYMPHOMA

Key Facts

Imaging Findings
- Best diagnostic clue: Solitary or multiple solid lesions (metastases); diffuse infiltrative lesions (lymphoma)
- Central or peripheral enhancement (metastases)
- Calcification (rare) except mucinous colon cancer
- Solitary mass; multifocal or diffuse infiltration (lymphoma)
- Focal lesions: AIDS-related lymphoma > common
- Splenic hilum nodes: NHL (59%); HD (uncommon)
- Lymphadenopathy: Abdominal or retroperitoneal
- Splenic lymphoma (MR nonspecific)

Top Differential Diagnoses
- Splenic primary tumor
- Splenic infection
- Granulomatous

Pathology
- Splenic metastases: Breast (21%), lung (18%), ovary (8%), stomach (7%), melanoma (6%), prostate (6%)
- Hodgkin (HD) & non-Hodgkin lymphoma (NHL)
- Primary lymphoma; secondary (more common)
- Primary: Usually non-Hodgkin lymphoma (NHL)
- Secondary: Extension from generalized lymphoma

Diagnostic Checklist
- To look for primary cancer, generalized adenopathy
- HD (45-66%); NHL (70%) show diffuse infiltration making difficult to detect by US, CT & MR
- Splenic size is not a reliable clinical & imaging indicator of presence or absence of disease
- Imaging important to suggest & stage malignancy, evaluate tumor volume/monitor response to therapy

CT Findings
- Splenic metastases
  - Decreased attenuation in relation to normal spleen
  - Usually appear as well-defined masses
  - Multiple, solid (more common) or "cystic", decreased attenuation (metastases)
  - Malignant melanoma: Solid or "cystic"
  - Ovary, breast & endometrium: "Cystic" (homogeneous & hypodense)
  - Central or peripheral enhancement (metastases)
  - Calcification (rare) except mucinous colon cancer
- Splenic lymphoma
  - Homogeneously enlarged spleen, no discrete mass
  - Solitary mass; multifocal or diffuse infiltration (lymphoma)
  - Focal lesions: AIDS-related lymphoma > common
  - ↓ HU lesions + minimal enhancement; ± ascites
  - Organ specific contrast agent: Ethiodol-Oil-Emulsion (EOE-13) potentially increase detection rate
  - Rarely necrosis ± air or cystic mimicking abscess
  - Splenic hilum nodes: NHL (59%); HD (uncommon)
  - Lymphadenopathy: Abdominal or retroperitoneal

MR Findings
- Splenic metastases
  - T1WI: Isointense to hypointense
  - T2WI: Hyperintense
  - T1 C+: Enhancement depends on type of primary
    - Conspicuity ↑ with SPIO-enhanced imaging
- Splenic lymphoma (MR nonspecific)
  - MR not reliable due to similar T1 & T2WI relaxation times, proton densities of normal spleen/lymphoma

Ultrasonographic Findings
- Real Time
  - Splenic metastases
    - Iso-/hypoechoic, hyperechoic, "target" or "halo"
    - Larger lesions more complex than smaller ones
    - Echogenic lesions (rare): E.g., plasmacytoma, hepatoma, melanoma, prostate & ovarian cancer
  - Splenic lymphoma
    - Typically diffuse or focal hypoechoic lesions
    - Anechoic/mixed echoic; small or large nodules
    - Large & necrotic: Heterogeneous appearance
    - AIDS-related lymphoma: Uniform decreased echogenicity or focal hypoechoic lesions

Angiographic Findings
- Conventional
  - Splenic metastases
    - Typically hypovascular or avascular
    - Less frequently hypervascular
  - Splenic lymphoma
    - Nonspecific findings: May be normal; single or multiple parenchymal defects

Nuclear Medicine Findings
- Tc99m sulfur colloid scintigraphy (liver-spleen scan)
  - Splenic metastases: Areas of focal defects
  - Splenic lymphoma: Nonspecific findings
    - Areas of decreased tracer uptake within spleen
    - Total lack of uptake ("functional asplenia")
    - Altered splenic contour (encasement by tumor)

Imaging Recommendations
- Best imaging tool: Helical CT

DIFFERENTIAL DIAGNOSIS

Splenic primary tumor
- Benign tumor
  - Hemangioma
    - Homogeneously solid or multiple cystic masses
    - Slightly hypodense on NECT & isodense on CECT
    - Central punctate or peripheral curvilinear Ca++
- Malignant tumor
  - Angiosarcoma
    - Heterogeneous density; variable enhancement
    - Solitary or multiple, nodular, irregular margins
    - ± Multiple liver or distant metastases

Splenic infarction
- Wedge-shaped peripheral defect

Splenic infection
- Pyogenic (bacterial): Abscess

SPLENIC METASTASES AND LYMPHOMA

  - Low density lesion with thick, irregular dense rim
  - ± Gas within fluid collection, left pleural effusion
- Fungal (e.g., Candida, Aspergillus, Cryptococcus)
  - Solitary/multiple; small/large; low density lesions
  - May show minimal peripheral enhancement
  - Candida microabscesses
    - Well-defined nonenhancing ↓ attenuation lesions
    - Typically 5-10 mm in size (smaller than 2 cm)
    - Wheel within wheel pattern: ↑ HU central focus
- AIDS: Pneumocystis carinii infection
  - Focal low-attenuation splenic lesions
  - Large lesions: Calcification, rim-like or punctate type

Granulomatous

- Tuberculosis
  - Splenomegaly (SMG)
  - Micro & macronodular lesions of low attenuation
  - ± Abdominal adenopathy, high density ascites
- Sarcoidosis
  - SMG, low density nodules ranging from 0.3-2.0 cm
  - Periaortic & retrocrural adenopathy (small, discrete)
  - Associated mediastinal, hilar nodes & lung lesions

Artifact

- Early phase of of I.V. bolus contrast injection
  - Heterogeneous enhancement, mimic splenic tumor

PATHOLOGY

General Features

- General path comments
  - Splenic metastases
    - Usually seen in widespread tumor dissemination
    - Isolated splenic metastases are rarely seen
- Etiology
  - Splenic metastases: Breast (21%), lung (18%), ovary (8%), stomach (7%), melanoma (6%), prostate (6%)
  - Splenic lymphoma
    - Hodgkin (HD) & non-Hodgkin lymphoma (NHL)
    - Primary lymphoma; secondary (more common)
    - Primary: Usually non-Hodgkin lymphoma (NHL)
    - Secondary: Extension from generalized lymphoma
- Epidemiology
  - Splenic metastases
    - 7.1% of patients with malignancy at autopsy
  - Splenic lymphoma
    - Primary splenic lymphoma: 1% of all cases of NHL
    - Hodgkin (23-34%); non-Hodgkin (30-40%)
- Associated abnormalities
  - Primary carcinoma in splenic metastases
  - Generalized adenopathy in secondary lymphoma

Gross Pathologic & Surgical Features

- Splenic metastases: Solitary or multiple
- Splenic lymphoma
  - Solitary, multifocal, diffuse infiltration (> common)
  - Homogeneously enlarged spleen, no discrete mass

Microscopic Features

- Splenic metastases: Varies based on primary carcinoma
- Splenic lymphoma: Hodgkin type
  - Nodular sclerosing; mixed cellular
  - Lymphocyte predominance; lymphocyte depletion

Staging, Grading or Classification Criteria

- Ann Arbor staging: Anatomic extent of HD & NHL

CLINICAL ISSUES

Presentation

- Most common signs/symptoms
  - Splenic metastases
    - Asymptomatic; LUQ pain, mass, splenomegaly
    - Acute pain: Splenic infarct due to tumor emboli
  - Splenic lymphoma
    - Fever, weight loss, night sweats, malaise
    - LUQ pain, palpable mass, splenomegaly
- Diagnosis: Imaging + clinical; guided splenic biopsy

Demographics

- Age
  - Primary splenic lymphoma: Older age group
  - Splenic metastases: Any age group
- Gender
  - HD: Young age, M:F = 4:1; older age, M:F = 2:1
  - NHL: M:F = 1.4:1

Natural History & Prognosis

- Complications
  - Splenic metastases: Splenic vein thrombosis, rupture
- Prognosis
  - Splenic metastases: Poor prognosis
  - Splenic lymphoma: Early stage (good); late (poor)

Treatment

- Chemotherapy, radiation or surgery in isolated lesion

DIAGNOSTIC CHECKLIST

Consider

- To look for primary cancer, generalized adenopathy

Image Interpretation Pearls

- HD (45-66%); NHL (70%) show diffuse infiltration making difficult to detect by US, CT & MR
- Splenic size is not a reliable clinical & imaging indicator of presence or absence of disease
- Imaging important to suggest & stage malignancy, evaluate tumor volume/monitor response to therapy

SELECTED REFERENCES

1. Urban BA et al: Helical CT of the spleen. AJR Am J Roentgenol. 170(4):997-1003, 1998
2. Rabushka LS et al: Imaging of the spleen: CT with supplemental MR examination. Radiographics. 14(2):307-32, 1994
3. Freeman JL et al: CT of congenital and acquired abnormalities of the spleen. Radiographics. 13(3):597-610, 1993
4. Taylor AJ et al: CT of acquired abnormalities of the spleen. AJR Am J Roentgenol. 157(6):1213-9, 1991
5. Hahn PF et al: MR imaging of focal splenic tumors. AJR Am J Roentgenol. 150(4):823-7, 1988

IMAGE GALLERY

Typical

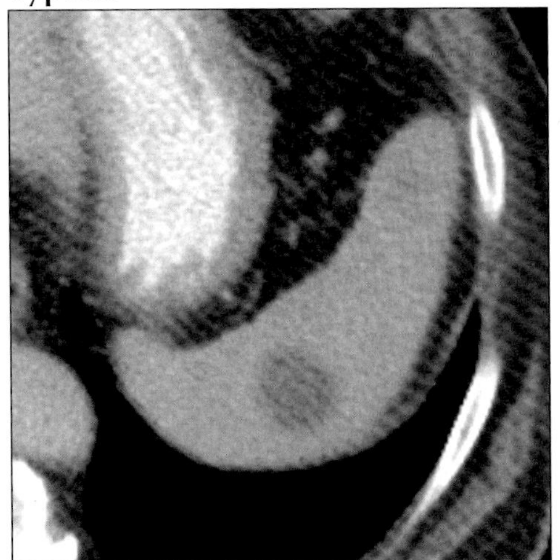

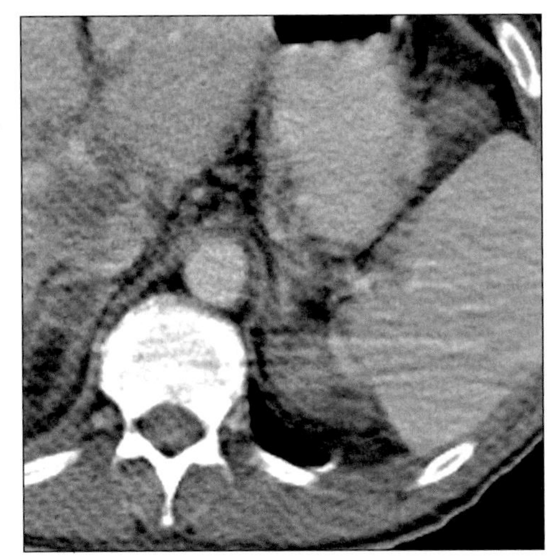

(Left) *Axial CECT shows splenic metastases from melanoma.* ***(Right)*** *Axial CECT shows splenic, liver and adrenal metastases from primary sinus carcinoma.*

Typical

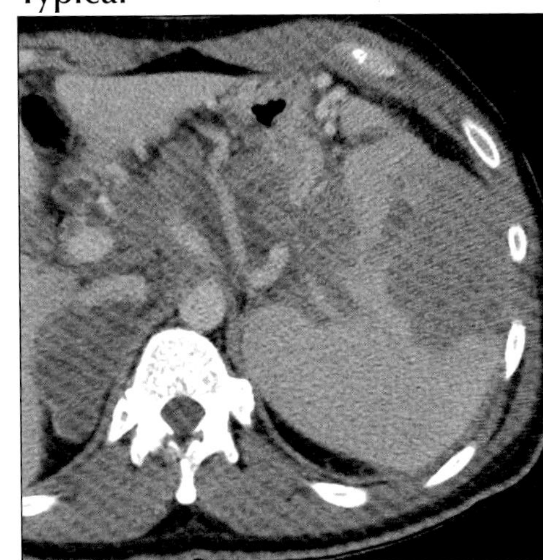

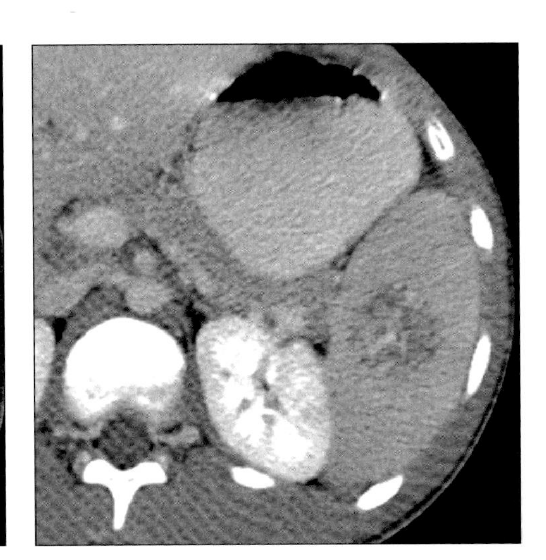

(Left) *Axial CECT shows splenic, nodal and adrenal masses due to NHL.* ***(Right)*** *Axial CECT shows splenic metastasis from choriocarcinoma.*

Typical

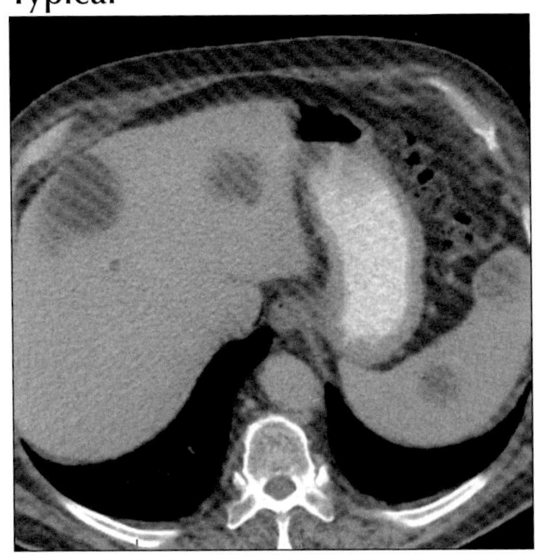

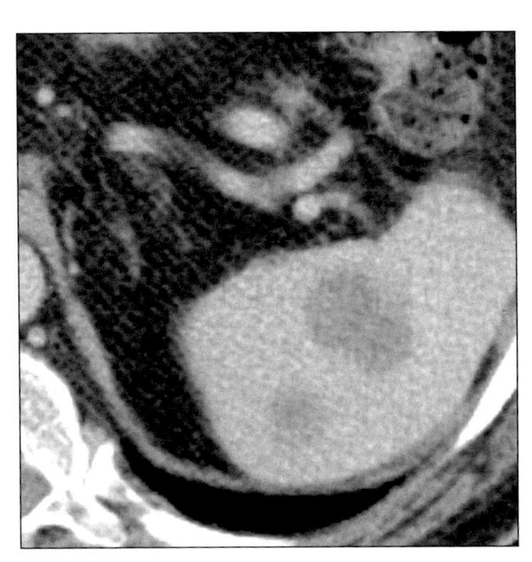

(Left) *Axial CECT shows splenic and liver metastases from melanoma.* ***(Right)*** *Axial CECT shows splenic metastases from melanoma.*

SPLENOMEGALY AND HYPERSPLENISM

Axial CECT shows heterogeneous enlargement of liver and spleen due to amyloidosis.

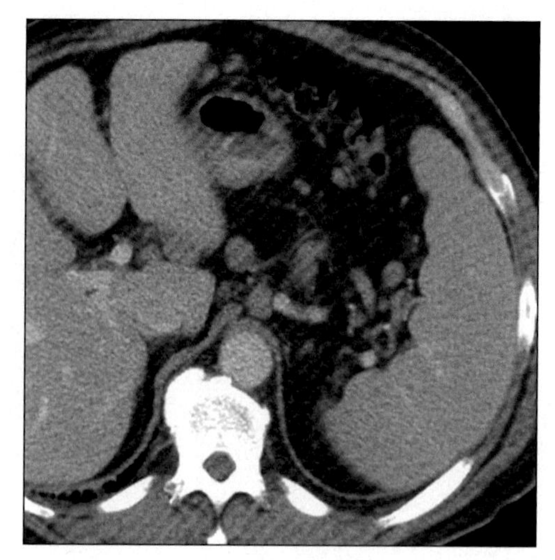

Axial CECT shows splenomegaly due to chronic lymphocytic leukemia.

TERMINOLOGY

Abbreviations and Synonyms

- Splenomegaly (SMG); hypersplenism (HS)

Definitions

- Splenomegaly: Enlarged spleen; volume > 500 cm^3
- Hypersplenism: Syndrome consisting of splenomegaly & pancytopenia in which bone marrow is either normal or hyperreactive

IMAGING FINDINGS

General Features

- Best diagnostic clue: Increased volume of spleen with convex medial border
- Location: Spleen occupies LUQ with tip extending inferiorly below 12th rib
- Size
    - Normal spleen in adult measures up to 13 cm; enlarged if it is 14 cm or longer
    - Splenic index: Normally 120-480 cm^3 (product of length, breadth & width of spleen)
    - Splenic weight: Splenic index x 0.55
        - Normal weight: 100-250g
    - SMG: Anteroposterior (AP) diameter > two-thirds distance of AP diameter of abdominal cavity
- Morphology
    - Enlarged spleen tends to become directed anteriorly
    - Splenic tip extends below tip of right lobe of liver
    - Mild, moderate or marked splenomegaly

Radiographic Findings

- Radiography
    - Splenic tip below 12th rib
    - Marked SMG may displace stomach medially
    - Displacement of splenic flexure of colon (splenic flexure usually anterior to spleen)
    - Calcification within or adjacent to spleen

CT Findings

- SMG: Medial margin of spleen is convex on CT
- Congestive SMG
    - Portal hypertension: SMG with varices, nodular shrunken liver, ascites
    - Splenic vein occlusion or thrombosis (often secondary to pancreatitis or pancreatic tumors)
    - Sickle-cell disease: Splenic sequestration
        - Peripheral ↓ HU areas + areas ↑ attenuation
        - Represent areas of infarct & hemorrhage
- Storage disorders
    - Gaucher disease
        - Spleen may have abnormal low attenuation
        - Marked SMG, often extending into pelvis
    - Amyloidosis
        - NECT & CECT: Generalized or focal ↓ density

DDx: Splenic or Left Upper Quadrant Mass

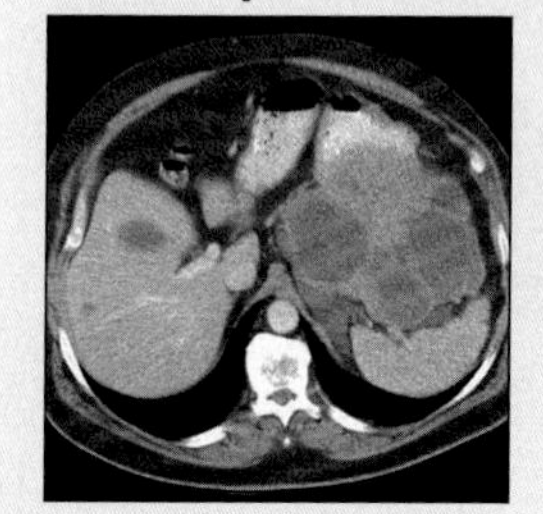

Gastric Tumor

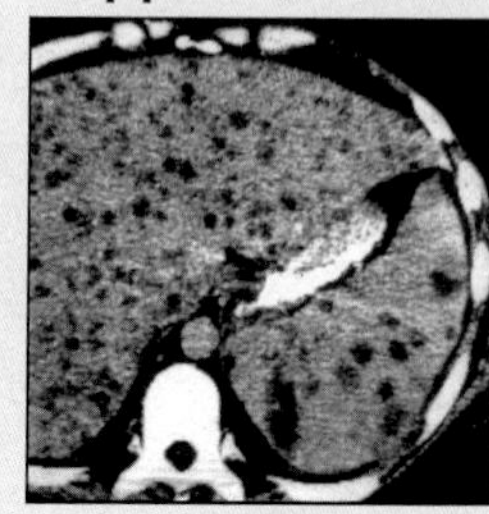

Candida Abscesses

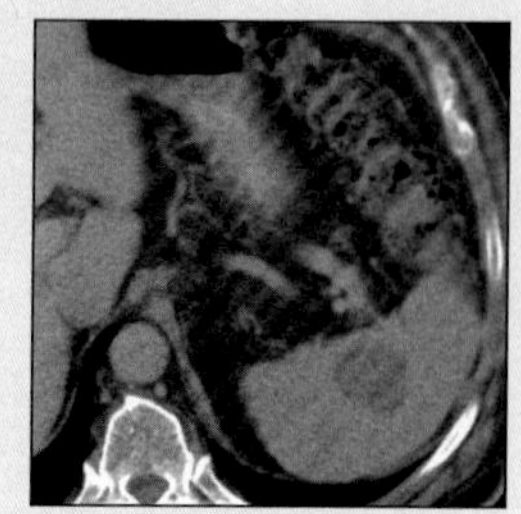

Melanoma Metastases

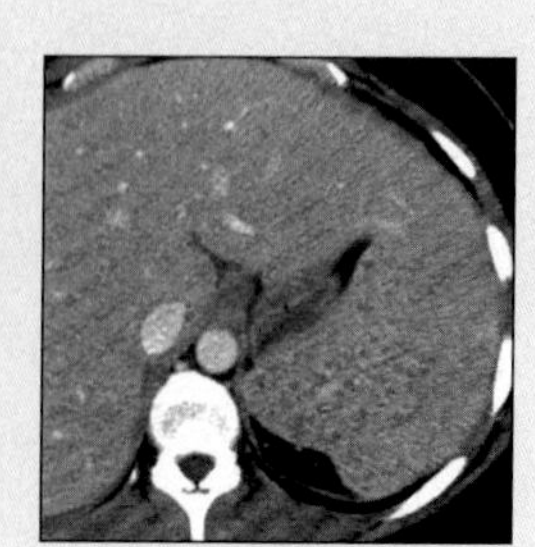

Splenic MAI, AIDS

SPLENOMEGALY AND HYPERSPLENISM

Key Facts

Terminology
- Splenomegaly (SMG); hypersplenism (HS)
- Splenomegaly: Enlarged spleen; volume > 500 cm^3
- Hypersplenism: Syndrome consisting of splenomegaly & pancytopenia in which bone marrow is either normal or hyperreactive

Imaging Findings
- Best diagnostic clue: Increased volume of spleen with convex medial border
- Normal spleen in adult measures up to 13 cm; enlarged if it is 14 cm or longer
- Splenic tip below 12th rib
- Marked SMG may displace stomach medially
- Displacement of splenic flexure of colon (splenic flexure usually anterior to spleen)

Top Differential Diagnoses
- Other LUQ masses
- Lymphoma & metastases
- Primary splenic tumor

Diagnostic Checklist
- SMG, most common cause of left upper quadrant mass
- SMG, usually a systemic cause rather than primary
- US can confirm presence of enlarged spleen or space occupying lesions
- CT & MR can further clarify abnormalities in size, shape & define parenchymal pathology
- Radioisotope scanning can diagnose HS & provide functional status of spleen

---

 - Primary hemochromatosis
   - Density of spleen is normal (unlike that of liver)
 - Secondary hemochromatosis
   - Increased attenuation values in liver & spleen
- Space occupying lesions: Cysts, abscess, tumor
- Splenic infarction (veno-occlusion caused by sickling)
  - SMG with focal infarcts; peripheral areas of low attenuation & hemorrhage associated with SMG
- Hemosiderosis
  - ↑ Attenuation of spleen (hemosiderin deposition)
  - Consequence of multiple blood transfusions (thalassemia, hemophilia)
- Splenic trauma
  - Splenic laceration or subcapsular hematoma
  - Surrounding perisplenic hematoma (> 30 HU)
- Extramedullary hematopoiesis
  - Spleen may be diffusely enlarged
  - CECT: Focal masses of hematopoietic tissue that are isoattenuating relative to normal splenic tissue

MR Findings
- Congestive SMG
  - Portal hypertension
    - Multiple tiny (3-8 mm) foci of decreased signal
    - Hemosiderin deposits; organized hemorrhage (Gamna-Gandy bodies or siderotic nodules)
  - Sickle cell disease (splenic sequestration)
    - Areas of abnormal signal intensity
    - Hyperintense with dark rim on T1WI (subacute hemorrhage)
  - Hemochromatosis
    - Primary: Normal signal & size of spleen
    - Secondary: Marked signal loss; enlarged spleen
  - Gaucher disease: Increased signal intensity on T1WI
  - Infarction
    - Peripheral, wedge-shaped areas of abnormal signal
    - Low signal resulting from iron deposition
  - Hemosiderosis
    - ↓ Signal intensity of spleen on both T1 & T2WI
  - Extramedullary hematopoiesis
    - Focal hypointense nodules

Ultrasonographic Findings
- Real Time
  - SMG with normal echogenicity
    - Infection, congestion (portal HT), early sickle cell
    - H. spherocytosis, hemolysis, Felty syndrome
    - Wilson disease, polycythemia, myelofibrosis, leukemia
  - SMG with hyperechoic pattern
    - Leukemia, post chemo & radiation therapy
    - Malaria, TB, sarcoidosis, polycythemia
    - Hereditary spherocytosis, portal vein thrombosis, hematoma, metastases
  - SMG with hypoechoic pattern
    - Lymphoma, multiple myeloma, chronic lymphocytic leukemia
    - Congestion from portal HT, noncaseating granulomatous infection
  - Sickle cell disease: Immediately after sequestration, peripheral hypoechoic areas seen
  - Gaucher disease: Multiple, well-defined, discrete hypoechoic lesions; fibrosis or infarction

Nuclear Medicine Findings
- Chromium 51-labeled RBCs or platelets
  - HS is diagnosed if injected RBCs exhibit shortened half-life (average half-life of 25-35 days)
- Tc99m sulfur colloid scan: Detect splenic function

Imaging Recommendations
- Best imaging tool: Helical CT

DIFFERENTIAL DIAGNOSIS

Other LUQ masses
- E.g., gastric, renal, adrenal tumor

Lymphoma & metastases
- Lymphoma: Various patterns of splenic involvement
  - Homogeneous enlargement without discrete mass
  - Solitary or multifocal lesions: Discrete ↓ HU lesions
  - Diffuse infiltration with SMG
- Metastases

- ↓ Attenuation; "cystic" or solid masses; SMG
- "Cystic" metastases: "Homogeneous & hypodense"
  - E.g., melanoma, ovary, breast & endometrium

Primary splenic tumor

- Benign tumor
  - Hemangioma
    - Homogeneously solid or multiple cystic masses
    - Central punctate or peripheral curvilinear Ca++
    - Slight hypodense on NECT & isodense on CECT
  - Hamartoma
    - Iso-hypodense on NECT; variable enhancement
    - Central star-like scar or necrosis, focal calcification
  - Lymphangioma
    - Thin-walled low density lesions; sharp margins
    - CECT: Enhancement of walls but not contents
    - Usually subcapsular in location; ± calcification
- Malignant tumor
  - Angiosarcoma
    - Solitary or multiple, nodular, irregular margins
    - Heterogeneous density; variable enhancement
    - Enlarged spleen; ± hemorrhage & calcification
    - ± Multiple liver or distant metastases
    - ± Hematoma: Intrasplenic/subcapsular/perisplenic

Splenic infection

- Pyogenic (bacterial)
  - Abscess
    - Low density lesion with thick, irregular dense rim
    - ± Gas within fluid collection, left pleural effusion
- Fungal (e.g., Candida, Aspergillus, Cryptococcus)
  - Solitary/multiple; small/large; low density lesions
  - May show minimal peripheral enhancement
  - Candida: Microabscesses (nonenhancing lesions)
- AIDS: Pneumocystis, mycobacterial (e.g., MAI)
  - Focal low attenuation splenic lesions
  - Large lesions: Calcification, rim-like or punctate type

Granulomatous

- TB: Micro & macronodular lesions of low attenuation
- Sarcoidosis
  - SMG with low density intrasplenic lesions
  - Abdominal & pelvic lymphadenopathy
- Mycobacterium avium intracellulare (in AIDS patients)
  - Low density splenic lesions; marked SMG

PATHOLOGY

General Features

- Etiology
  - Congestive SMG
    - CHF, portal HT, cirrhosis, cystic fibrosis, splenic vein thrombosis, sickle cell (SC) sequestration
  - Neoplasm: Leukemia, lymphoma, metastases, primary neoplasm, Kaposi sarcoma
  - Storage disease: Gaucher, Niemann-Pick, gargoylism, amyloidosis, DM, hemochromatosis, histiocytosis
  - Infection: Hepatitis, malaria, mononucleosis, TB, typhoid, kala-azar, schistosomiasis, brucellosis
  - Hemolytic anemia: Hemoglobinopathy, hereditary spherocytosis, primary neutropenia, thrombocytopenic purpura
  - Extramedullary hematopoiesis: Osteopetrosis, myelofibrosis
  - Collagen disease: SLE, RA, Felty syndrome
  - Splenic trauma; sarcoidosis; hemodialysis
- Associated abnormalities: HS seen in association with hemoglobinopathies & autoimmune diseases

Microscopic Features

- Varies depending on underlying etiology

CLINICAL ISSUES

Presentation

- Most common signs/symptoms
  - Asymptomatic, splenomegaly, abdominal pain
  - Signs & symptoms related to underlying cause
- Lab data: Abnormal CBC, LFT, antibody titers, cultures or bone marrow exam

Natural History & Prognosis

- Complications
  - Splenic rupture, shock & death
- HS: Usually develops as a result of SMG
  - Hyperfunctioning spleen removes normal RBC, WBC & platelets from circulation
- Prognosis
  - Splenic rupture, sequestration in SC disease: Poor

Treatment

- Treatment varies based on underlying condition
- Splenectomy in symptomatic & complicated cases

DIAGNOSTIC CHECKLIST

Consider

- SMG, most common cause of left upper quadrant mass
- SMG, usually a systemic cause rather than primary

Image Interpretation Pearls

- US can confirm presence of enlarged spleen or space occupying lesions
- CT & MR can further clarify abnormalities in size, shape & define parenchymal pathology
- Radioisotope scanning can diagnose HS & provide functional status of spleen

SELECTED REFERENCES

1. Peck-Radosavljevic M: Hypersplenism. Eur J Gastroenterol Hepatol. 13(4):317-23, 2001
2. McCormick PA et al: Splenomegaly, hypersplenism and coagulation abnormalities in liver disease. Clin Gastroenterol. 14(6):1009-31, 2000
3. Paterson A et al: A pattern-oriented approach to splenic imaging in infants and children. Radiographics. 19(6):1465-85, 1999
4. Bowdler AJ: Splenomegaly and hypersplenism. Clin Haematol. 12(2):467-88, 1983
5. Mittelstaedt CA et al: Ultrasonic-pathologic classification of splenic abnormalities: gray-scale patterns. Radiology. 134(3):697-705, 1980

SPLENOMEGALY AND HYPERSPLENISM

IMAGE GALLERY

Typical

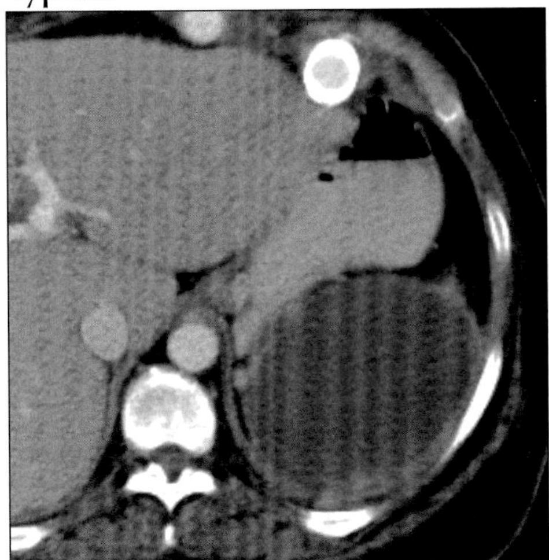

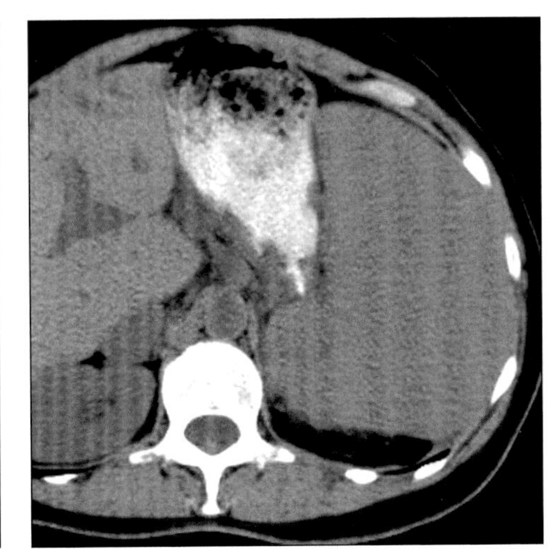

(Left) *Axial CECT shows acquired splenic cyst following infarct (cardiac assist device).* ***(Right)*** *Axial NECT shows splenomegaly due to myeloproliferative disorder.*

Typical

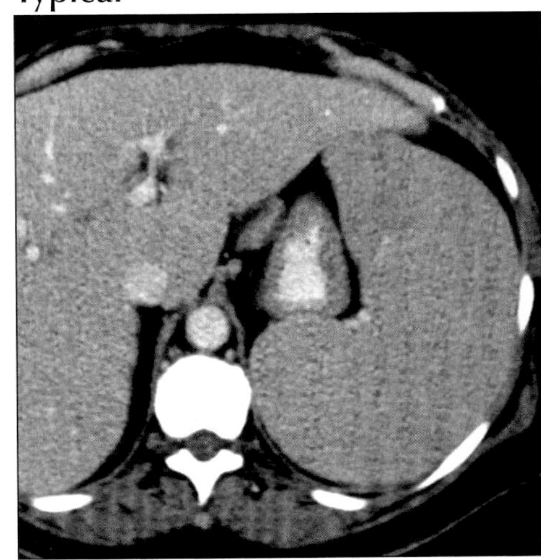

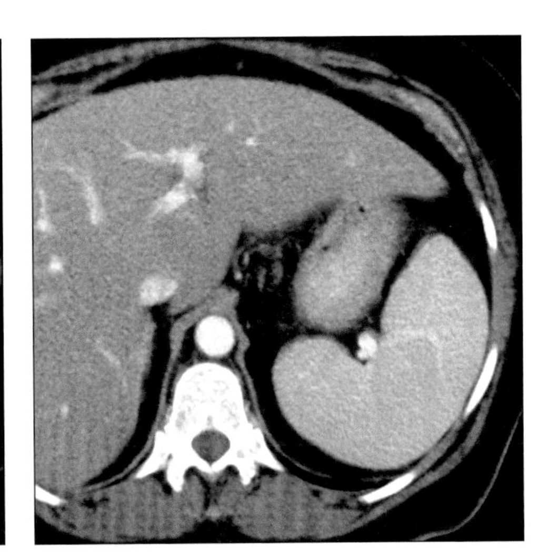

(Left) *Axial CECT in June, 2003 shows splenomegaly due to non-Hodgkin lymphoma.* ***(Right)*** *Axial CECT in August, 2003 shows marked reduction in splenic size following treatment for lymphoma.*

Typical

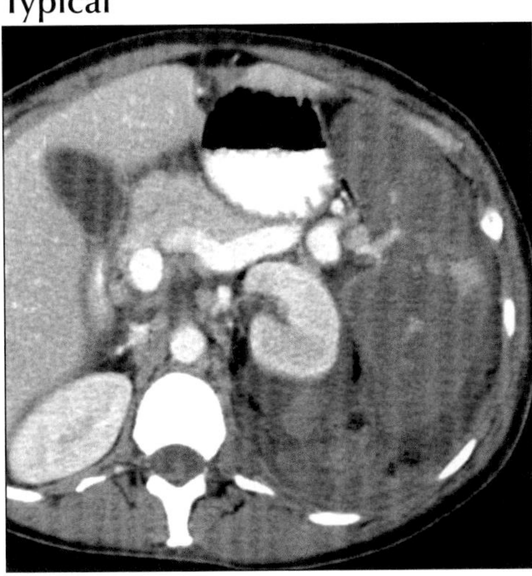

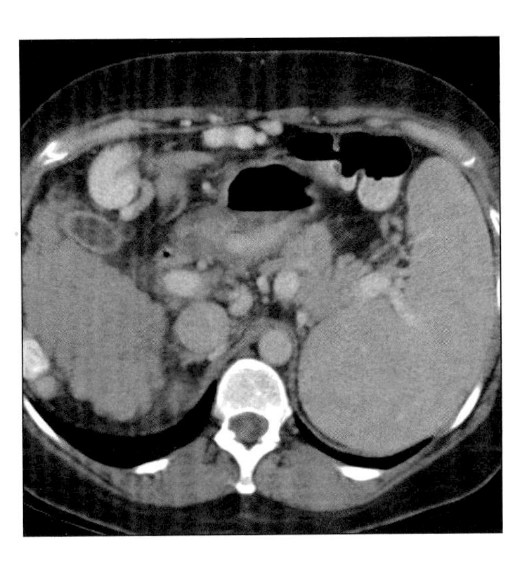

(Left) *Axial CECT shows spontaneous rupture of spleen due to leukemia.* ***(Right)*** *Axial CECT shows splenomegaly due to cirrhosis and portal hypertension.*

PART II

Hepatobiliary and Pancreas

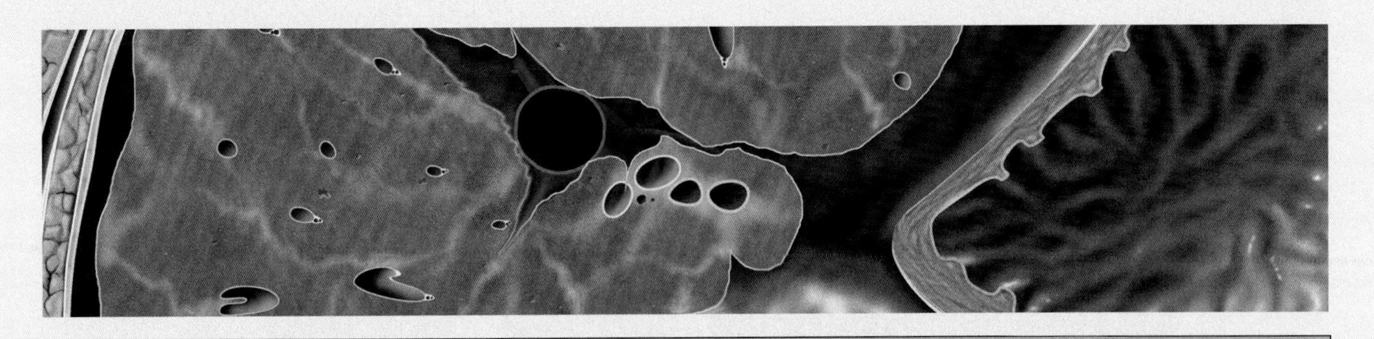

SECTION 1: Liver

LIVER ANATOMY AND IMAGING ISSUES

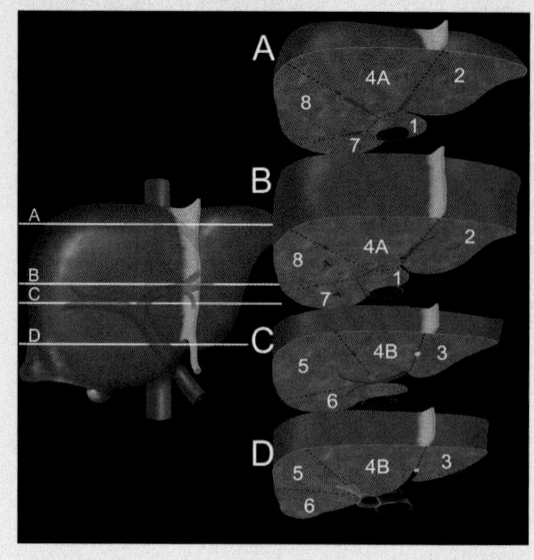

Graphic shows 4 sections through the liver depicting the 8 segments of the liver which are separated by vertical planes through the hepatic veins & horizontal plane through the portal vein.

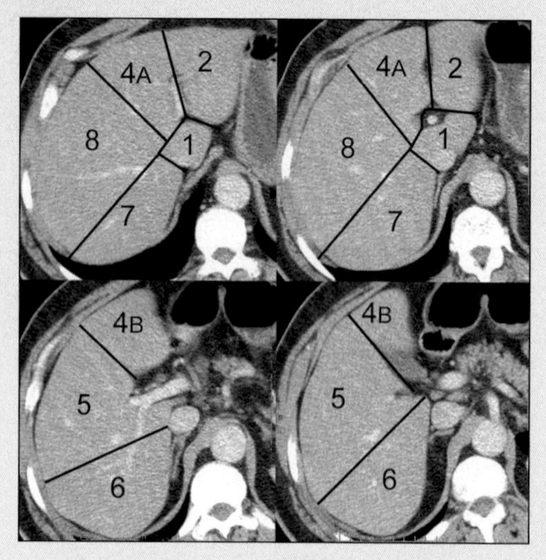

Axial CECT sections correspond to levels in graphic. Liver segments are numbered. Falciform ligament plane separates medial (seg. 4) from lateral (segs. 2 &3) left lobe. Seg. 3 not shown.

TERMINOLOGY

Definitions

- Phases of dynamic CT/MR imaging following a rapid bolus infusion of contrast material
  - Early arterial (~ 18-25 sec)
    - Indications: Best for hepatic arterial angiography
  - Late arterial (portal venous inflow) (~ 35-40 sec)
    - Indications: Best for detecting + characterizing hypervascular tumors
  - Portal venous (parenchymal) (~ 70 sec)
    - Indications: Should be obtained in almost all abdominal CT/MR scans
  - Equilibrium (delayed) (2-10 minutes)
    - Indications: As an added sequence to characterize a known or suspected hepatic mass
    - Especially for hemangioma or cholangiocarcinoma

IMAGING ANATOMY

Location

- Liver is divided into 8 segments (Bismuth system)
  - Vertical planes through the course of hepatic veins
  - Horizontal plane through the right and left portal vein

ANATOMY-BASED IMAGING ISSUES

Key Concepts or Questions

- What is the overall density (attenuation) of the liver?
  - Should be about 8-10 HU greater than that of spleen or muscle
  - Abnormally decreased density usually due to fatty infiltration (steatosis), but may occur with diffuse tumor (e.g., lymphoma), infection (e.g., acute severe viral hepatitis, opportunistic infections) or acute drug toxicity
  - Abnormally increased density usually due to iron deposition (primary or secondary hemochromatosis) or amiodarone Rx (antiarrhythmic drug)
- Is there evidence of cirrhosis?
  - Look for widened fissures, relative increase of caudate to right lobe width ratio (> 0.6), nodular surface contour, signs of portal hypertension (splenomegaly, ascites, varices)
- What is the status of the major hepatic vessels?
  - Hepatic veins dilated?
    - Consider passive congestion (cardiac dysfunction), arteriovenous shunts (e.g., Osler-Weber-Rendu)
  - Hepatic vein(s) occluded?
    - Consider Budd-Chiari, hypercoagulable state, tumor encasement
  - Portal vein dilated?
    - Consider "early" portal hypertension, tumor invasion (especially hepatocellular carcinoma)
  - Portal vein occluded?
    - Consider "late" portal hypertension with shunts, hypercoagulable state, septic thrombophlebitis (e.g., diverticulitis), tumor encasement
  - Hepatic artery dilated?
    - Consider cirrhosis (portal hyperention), hypervascular liver tumor, congenital hepatic fibrosis, portal vein occlusion, arteriovenous shunts
- What is the morphology, vascularity, and density of a tumor relative to liver, preferably defined on nonenhanced and multiphasic enhanced scans (CT or MR)?
  - Most tumors are spherical, hypovascular, and hypodense to normal liver on all phases of imaging
  - Simple cysts are near water density, do not enhance, have no mural nodularity
  - Abscesses are above water density, do not enhance (contents), have mural irregularities +/- septations
  - All tumors (viable) enhance and almost always show some solid component

LIVER ANATOMY AND IMAGING ISSUES

DIFFERENTIAL DIAGNOSIS

Benign Liver Masses

Hepatocellular origin
- Hepatic adenoma
- Focal nodular hyperplasia
- Hyperplastic + dysplastic nodules

Cholangiocellular origin
- Simple hepatic cyst
- Congenital hepatic fibrosis/polycystic

Mesenchymal origin
- Cavernous hemangioma
- Lipoma, angiomyolipoma
- Mesenchymal hamartoma
- Lymphangioma
- Leiomyoma, fibroma; heterotopic tissue

Malignant Liver Tumors

Hepatocellular origin
- Hepatocellular carcinoma (HCC)
- Fibrolamellar HCC
- Hepatoblastoma
- Clear cell + giant cell carcinoma
- Carcinosarcoma

Cholangiocellular origin
- Cholangiocarcinoma
- Biliary cystadenocarcinoma

Mesenchymal origin
- Epithelioid hemangioendothelioma
- Angio-,leiomyo-, fibro-, osteosarcoma
- Primary lymphoma + MFH

  - Hypervascular tumors have diffuse or partial hyperdensity (or intensity, MR) compared with liver on arterial phase images
  - Tumors with fibrous stroma (e.g., cholangiocarcinoma) or large vascular spaces (e.g., cavernous hemangioma) are hyperdense to normal liver on delayed imaging
- Are nonenhanced CT (or MR) images needed in addition to enhanced images?
  - Not for "acute abdomen" indications (e.g., trauma, abscess, etc.)
  - Not for most follow-up scans in most oncology patients
  - Yes, for patients with liver dysfunction or new diagnosis of cancer
  - Nonenhanced (+ enhanced) images help to detect and distinguish fatty infiltration and benign hepatic lesions with greater confidence (e.g., small lesion less than blood density on nonenhanced scan that shows no change in size or apparent enhancement is a cyst, not a "lesion too small to characterize")
- Which patients require evaluation by multiphasic CT (or MR)?
  - Patients with known or suspected cirrhosis, primary hepatic mass, known hypervascular primary tumor (e.g., endocrine cancer)
- What are the technical requirements and timing necessary to obtain multiphase CT (or MR) scans of the liver?
  - Adequate volume + rate of IV contrast medium
  - For CT, ≥ 125 ml or 40 gm of iodine, at ≥ 3 ml/sec
  - Arterial phase (= portal venous inflow phase) at about 35 sec delay (assuming normal cardiac output)
  - Portal venous phase (= hepatic parenchymal phase) at about 70 sec delay (longer for elderly patients, shorter for injection rates of ≥ 4 ml/sec)
- How can you determine the effective phase and adequacy of an hepatic CT scan?
  - Arterial phase: Arteries densely opacified, portal veins moderately enhanced, liver parenchyma + hepatic veins not, or minimally enhanced
  - Parenchymal phase (portal venous): Portal and hepatic veins densely opacified and hyperdense to liver parenchyma which is maximally enhanced
  - Equilibrium (or delayed phase) should never be obtained alone, only in addition to other phases: Liver, vessels, nodes, most masses, etc. all isodense

EMBRYOLOGY

Embryologic Events
- Anomalies may occur during embryologic development of the ductal plate that surrounds the portal vein
  - May result in variety of "fibropolycystic" defects (e.g., polycystic disease, Caroli, congenital hepatic fibrosis)

CUSTOM DIFFERENTIAL DIAGNOSIS

Focal hyperdense (noncalcified) lesion
- Noncontrast CT
- "Any" mass in a fatty liver
  - Even focal sparing
- Mucinous metastases
- Acute hemorrhage
  - Trauma
  - Bleeding tumor
  - Anticoagulated state
- Budd-Chiari + primary sclerosing cholangitis
  - "Spared" hypertrophied segments

Focal hypervascular liver lesion
- Benign
  - Focal nodular hyperplasia
  - Arterioportal shunt or THAD
  - Hepatic adenoma
  - Dysplastic or regenerative nodule*
- Malignant
  - Hepatocellular carcinoma (HCC)
  - Fibrolamellar HCC
  - Cholangiocarcinoma*

LIVER ANATOMY AND IMAGING ISSUES

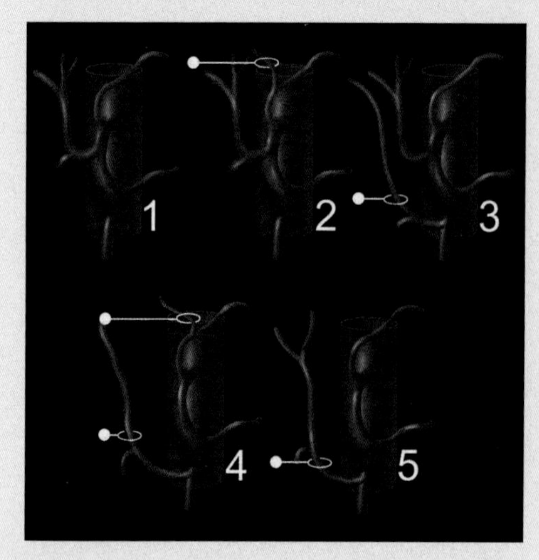

Graphic shows variation of hepatic arterial (HA) anatomy. 1 = conventional; 2 = accessory LHA; 3 = accessory RHA; 4 = replaced RHA from SMA; 5 = totally replaced hepatic artery from SMA

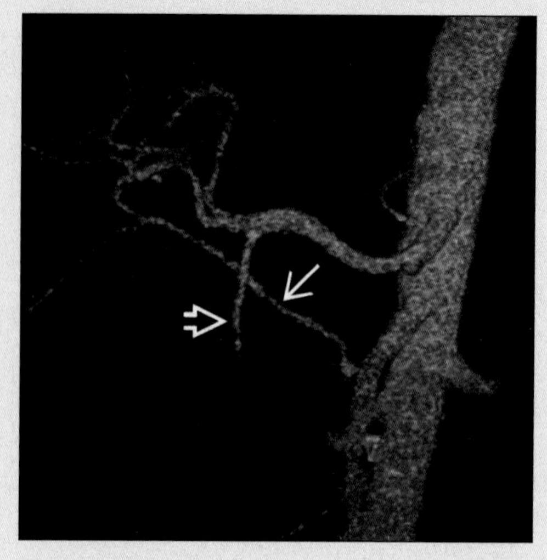

CT Angiogram shows accessory RHA (arrow) arising from SMA. Other hepatic arteries supplied by the celiac trunk. Gastroduodenal artery (open arrow).

- Metastases
    - Endocrine (islet cell, thyroid, carcinoid)
    - Renal cell carcinoma
    - Sarcoma
    - Breast* + Melanoma*
    - * Uncommonly
- Hyperperfusion abnormalities (transient hepatic attenuation difference = THAD)
  - Small, subcapsular
    - Idiopathic, probably small portal vein (PV) thrombus
    - Usually insignificant
  - Subsegmental
    - Arterioportal (AP) fistula following biopsy
    - Cirrhosis with spontaneous AP shunt
    - Adjacent hypervascular mass or abscess
    - PV obstruction or thrombus (often malignant)
  - Segmental/lobar
    - PV obstruction or thrombus (often malignant)
    - Mass effect on PV
    - HCC with AP shunt

Fat containing liver masses

- Hepatocellular carcinoma
- Hepatic adenoma
- Metastasis (liposarcoma, teratoma)
- Angiomyolipoma, lipoma
- (Focal fatty infiltration)

Complex cystic mass

- Abscess
  - Pyogenic, amebic, hydatid
- Cystic or necrotic metastases
- Necrotic or ablated HCC
- Biliary cystadenoma/carcinoma
- Hemorrhagic cyst
- Biloma or old hematoma
- Intrahepatic pseudocyst

Focal lesion with capsular retraction

- Metastasis (usually post-treatment)
- Cholangiocarcinoma
- Focal confluent fibrosis
- Epithelioid hemangioendothelioma
- Primary sclerosing cholangitis
- Hemangioma (rarely)

Liver mass with scar

- FNH (focal nodular hyperplasia)
- HCC - fibrolamellar
- Cavernous hemangioma (large)
- Cholangiocarcinoma
- Adenoma (rare)
- HCC - conventional*
- (Metastasis)*
- * Usually central necrosis, not scar

Periportal lucency or edema

- (Biliary dilation)
- Overhydration
- Congestive heart failure
- Acute hepatitis
- Obstructed lymphatics
  - E.g., porta hepatis tumor
- Liver transplantation
- Liver tumor

SELECTED REFERENCES

1. Dähnert W: Radiology Review Manual (4th ed). Baltimore: Williams & Wilkins, 1999
2. Venbrux AC, Friedman AC: Diffuse hepatocellular diseases, portal hypertension, and vascular diseases. In Friedman AC, Dachman AH (eds): radiology of the Liver, Biliary Tract, and Pancreas. St. Louis: CV Mosby, 49-168, 1994
3. Friedman AC, Frazier S, Hendrix TM, et al: Focal disease. In Friedman AC, Dachman AH (eds): radiology of the LIver, Biliary Tract, and Pancreas. St Louis: CV Mosby, 169-328, 1994
4. Reeder, MM: Reeder and Felson's Gamuts in Radiology (3rd ed). New York: Springer-Verlag, 1993

IMAGE GALLERY

Typical

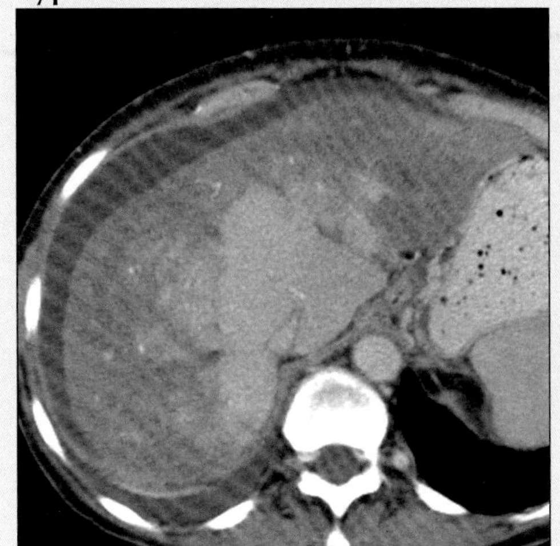

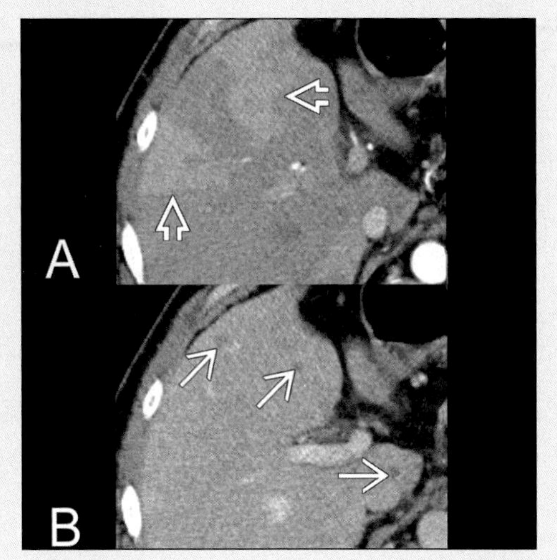

(Left) *Axial CECT in parenchymal phase, shows "hyperdense" enlarged caudate lobe which is normal hypertrophied liver in this patient with Budd-Chiari syndrome.* ***(Right)*** *Axial CECT arterial phase (A) + parenchymal phase (B) show multiple THAD (open arrows) due to portal venous branch occlusions from metastases (arrows).*

Typical

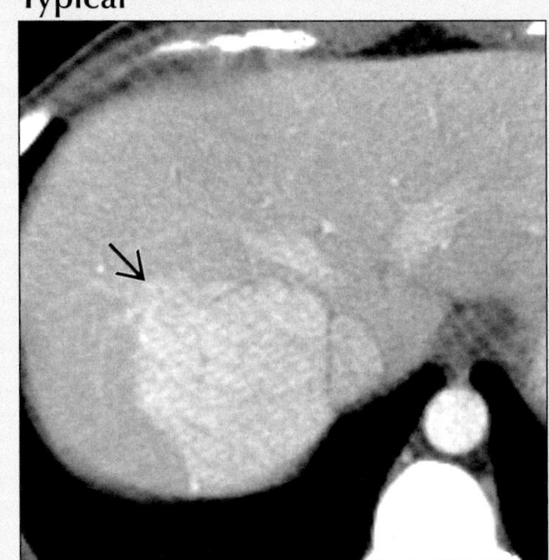

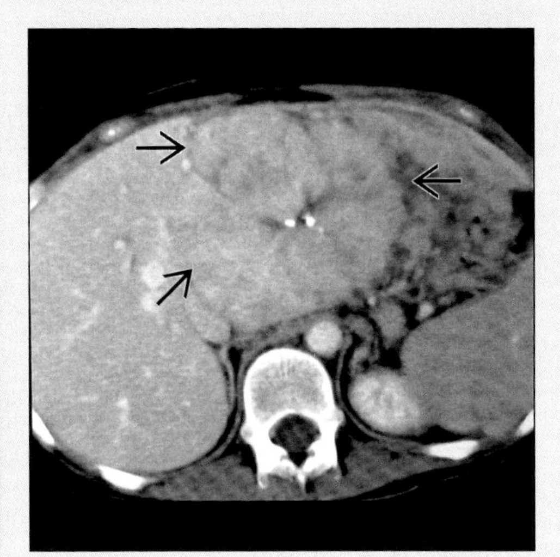

(Left) *Axial CECT in arterial phase, shows prototype of hypervascular liver mass, a focal nodular hyperplasia (FNH) (arrow).* ***(Right)*** *Axial CECT in parenchymal phase, shows prototype tumor with scar, a fibrolamellar HCC (arrows).*

Typical

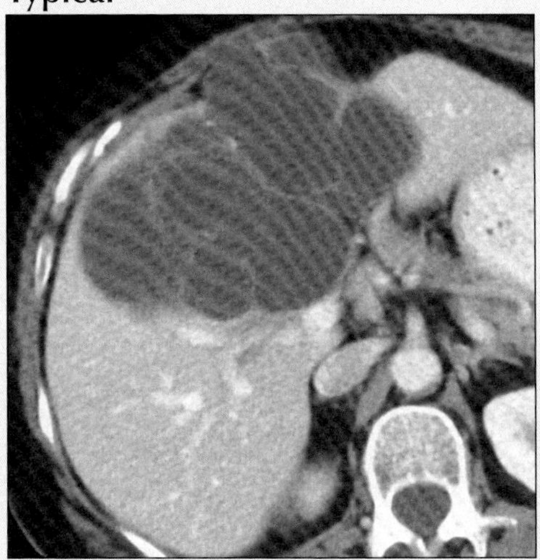

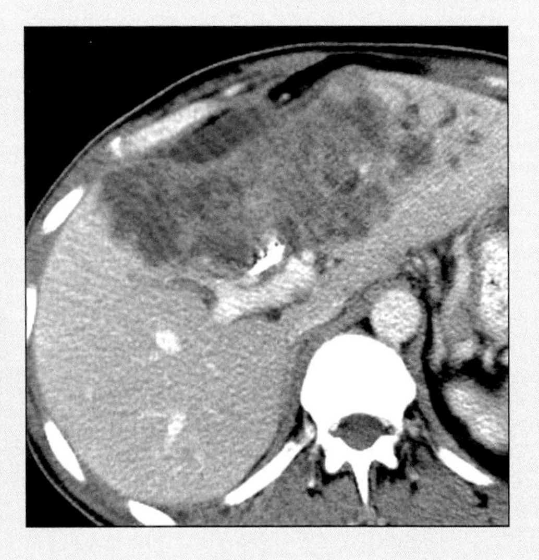

(Left) *Axial CECT in parenchymal phase, shows prototype complex cystic mass, biliary cystadenoma.* ***(Right)*** *Axial CECT in parenchymal phase shows prototype mass with capsular retraction, cholangiocarcinoma.*

CONGENITAL ABSENCE OF HEPATIC SEGMENTS

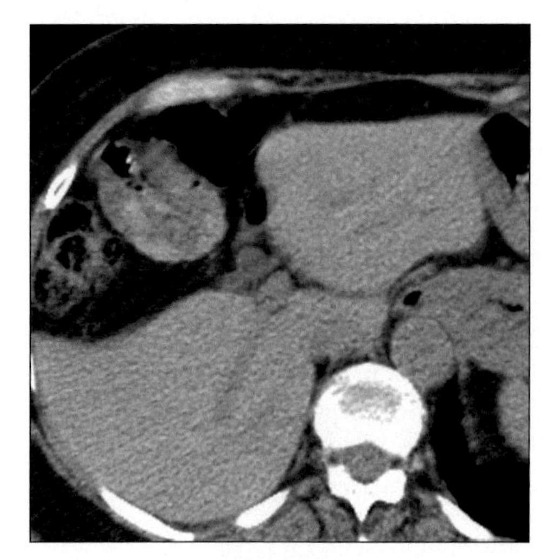

Axial NECT shows absence of anterior and medial segments of the liver.

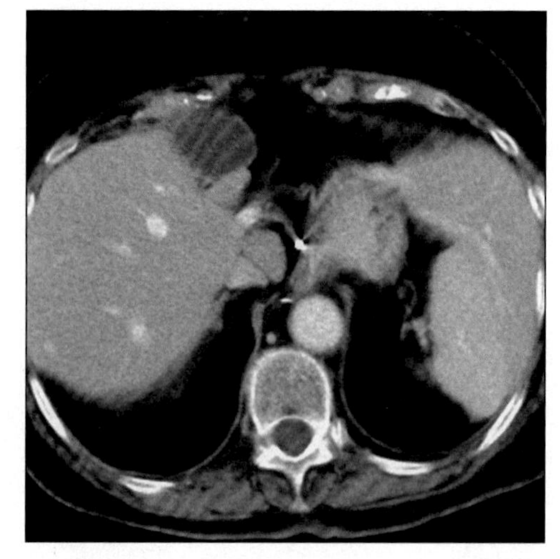

Axial CECT shows congenital absence of the medial segment with the lateral segment of the liver contiguous to the spleen.

TERMINOLOGY

Abbreviations and Synonyms

- Agenesis or hypoplasia of hepatic lobes/segments

Definitions

- Lobar agenesis is a rare developmental anomaly with absence of liver tissue to the right or left of gallbladder fossa without prior surgery or disease

IMAGING FINDINGS

General Features

- Best diagnostic clue
  - Absence of right or left hepatic vein, portal vein & its branches, & intrahepatic ducts
    - If none of these structures are visible, agenesis is substantiated
- Location
  - Right lobe: Segments V & VIII (anterior segments)
    - Segments VI & VII (posterior segments)
  - Left lobe: Segments II & III (lateral segment)
    - Segment IVa & IVb: Medial segment
  - Segment I is caudate lobe
- Size: Compensatory hypertrophy of remaining segments
- Key concepts
  - Severe distortion of hepatic morphology
    - Ectopy of gall bladder

CT Findings

- Altered normal topography of upper abdomen
- Right lobar agenesis: Absence of liver tissue, right of main interlobar plane
  - Absence of right hepatic vein, portal vein & its branches, & right intrahepatic ducts
  - Suprahepatic/subdiaphragmatic/infrahepatic location of gallbladder
  - Colonic interposition/ high position of right kidney/U- or hammock-shaped stomach
  - Direct contact of inferior vena cava to posterior surface of medial segment of left lobe
- Agenesis of left lobe: Absence of liver tissue to the left of main interlobar plane
  - Failure to visualize falciform ligament or ligamentum teres
  - Stomach & splenic flexure of colon migrate superiorly & medially, low-lying hepatic flexure
  - High position of duodenal bulb/U-shaped stomach

Angiographic Findings

- Conventional: Absence of right/left hepatic artery & portal vein & its branches

Other Modality Findings

- Cholangiography: Absence of right/left hepatic duct

DDx: Absence of Hepatic Segment(s)

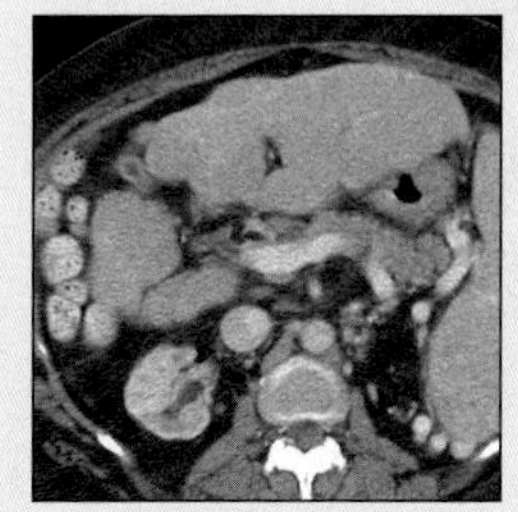

Cirrhotic Atrophy

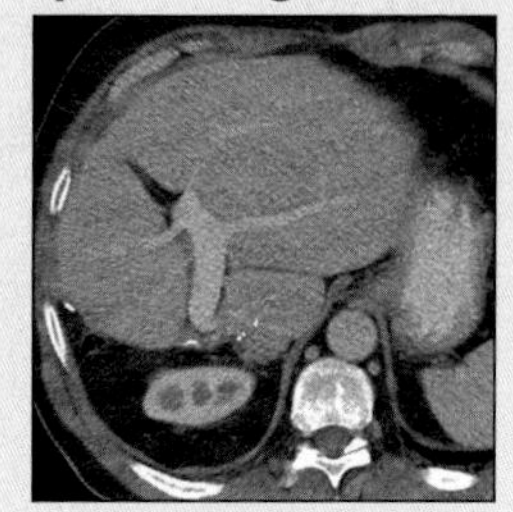

Right Hepatectomy

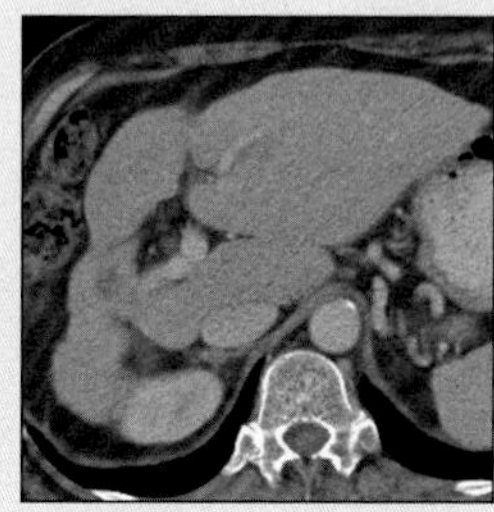

Chemoembolization

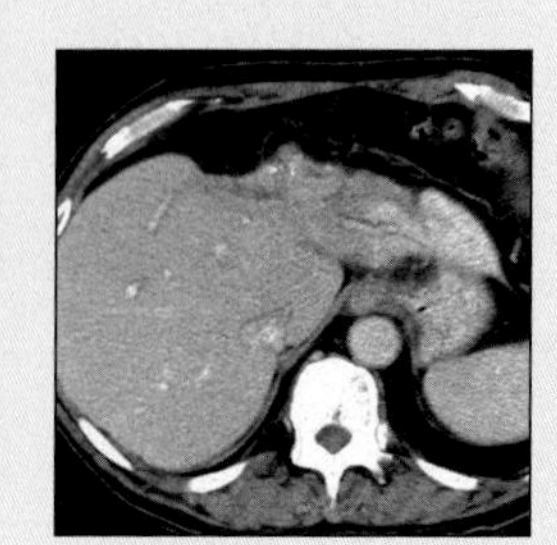

Chemoembolization

CONGENITAL ABSENCE OF HEPATIC SEGMENTS

Key Facts

Terminology
- Lobar agenesis is a rare developmental anomaly with absence of liver tissue to the right or left of gallbladder fossa without prior surgery or disease

Imaging Findings
- Absence of right or left hepatic vein, portal vein & its branches, & intrahepatic ducts

Clinical Issues
- Symptoms of associated: Biliary tract disease, portal hypertension & volvulus of stomach

Diagnostic Checklist
- Conditions including cirrhosis, atrophy secondary to biliary obstruction, hepatic surgery & trauma can mimic agenesis, & should first be ruled out

Imaging Recommendations
- Best imaging tool: CT is most commonly used
- Protocol advice: NECT + CECT

DIFFERENTIAL DIAGNOSIS

Acquired atrophy after infarction, fibrosis
- At least one of these structures (hepatic vein, portal vein & dilated intrahepatic ducts) is recognizable
- Atrophy of anterior + medial segments commonly follows development of focal confluent fibrosis in cirrhosis

Post surgical resection
- For hepatic resection, incisions can be made along longitudinal or transverse scissurae or both combined

Post chemoembolization
- Hyperattenuation of atrophic liver parenchyma

PATHOLOGY

General Features
- Etiology
  - Agenesis of right lobe is thought to result from: Either failure of right portal vein to develop or an error in mutual induction between septum transversum (primitive diaphragm) & endodermal diverticulum (primitive liver)
  - Left lobe agenesis results from extension of obliterative process that closes ductus venosus to lt. branch of portal vein
- Epidemiology
  - Incidence of lobar agenesis: 0.005% of 19,000 autopsy cases; about 42 cases of agenesis of right lobe are reported in literature
  - Left lobe agenesis slightly more common than right-sided anomalies
- Associated abnormalities
  - With agenesis of right lobe: Partial or complete absence of right hemidiaphragm
  - Intestinal malrotation/choledochal cysts/agenesis of gall bladder/intrahepatic venovenous shunt
  - With agenesis of left lobe: Partial or complete absence of left hemidiaphragm & gastric volvulus

Gross Pathologic & Surgical Features
- Unusual enlargement of remaining segments, resultant displacement of gallbladder & change in axis of fissure of ligamentum venosum

CLINICAL ISSUES

Presentation
- Discovered incidentally on imaging studies
- Symptoms of associated: Biliary tract disease, portal hypertension & volvulus of stomach

Natural History & Prognosis
- Calculus formation & biliary malignancy are very rare

DIAGNOSTIC CHECKLIST

Consider
- Conditions including cirrhosis, atrophy secondary to biliary obstruction, hepatic surgery & trauma can mimic agenesis, & should first be ruled out

SELECTED REFERENCES

1. Gathwala G et al: Agenesis of the right lobe of liver. Indian J Pediatr. 70(2):183-4, 2003
2. Sato N et al: Agenesis of the right lobe of the liver: report of a case. Surg Today. 28(6):643-6, 1998
3. Chou CK et al: CT of agenesis and atrophy of the right hepatic lobe. Abdom Imaging. 23(6):603-7, 1998

IMAGE GALLERY

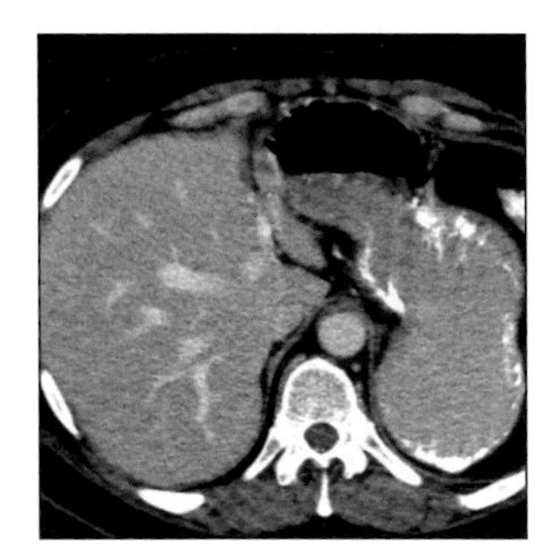

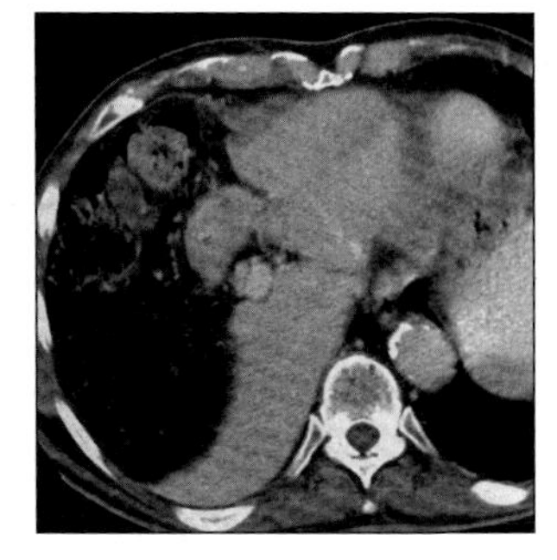

(Left) *Axial CECT shows congenital hypoplasia of left hepatic lobe.* ***(Right)*** *Axial CECT shows congenital absence of anterior and medial segments of the liver.*

CONGENITAL HEPATIC FIBROSIS

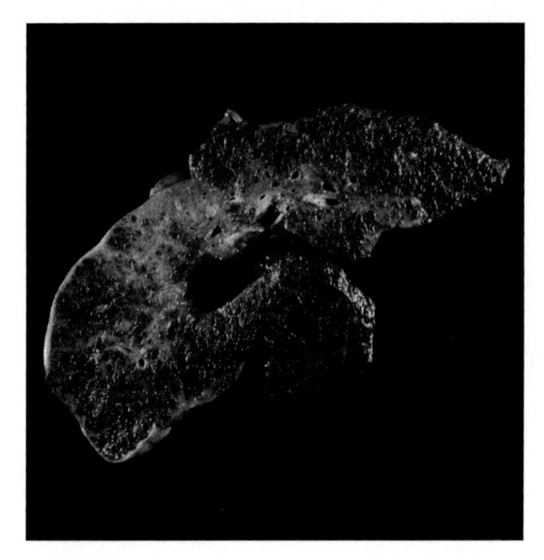

Cross section of explanted liver shows parenchymal distortion with extensive fibrosis, especially in anterior and medial segments.

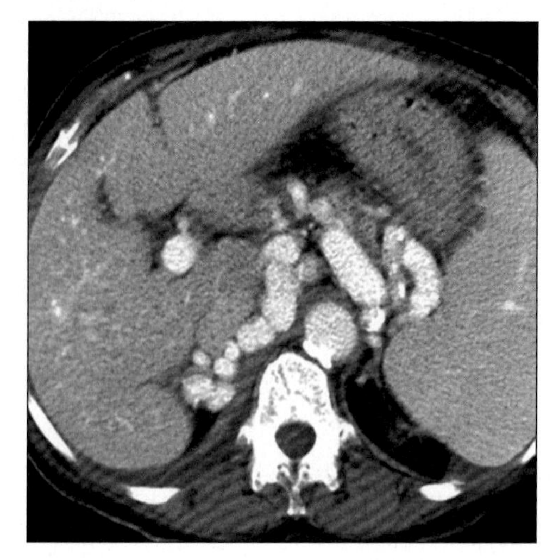

Axial CECT shows dysmorphic liver with varices and splenomegaly, large left lobe, atrophic right.

TERMINOLOGY

Abbreviations and Synonyms

- Congenital hepatic fibrosis (CHF), also referred as fibropolycystic liver disease

Definitions

- Part of a spectrum of congenital abnormalities resulting in variable degrees of fibrosis & cystic anomalies of liver & kidneys

IMAGING FINDINGS

General Features

- Best diagnostic clue: Combination of cystic dilatation of bile ducts + renal collecting duct ectasia
- Location: Both lobes of liver
- Size: Hepatic cystic size varies based on severity of CHF
- Other General Features
  - Always present in autosomal recessive polycystic kidney disease (ARPKD)
    - Sometimes with autosomal dominant PKD
  - Two constant features of ARPKD involve
    - Kidney: Tubular ectasia (cysts) & fibrosis
    - Liver: CHF (dilated bile ducts; portal tracts enlarged/fibrotic) & multiple cysts
  - Relative severity of organ involvement varies
    - Severe renal disease & mild CHF/liver cysts
    - Severe liver (CHF/liver cysts) & mild renal disease
  - All patients with ARPKD have findings of CHF at liver biopsy
    - Not all patients with CHF have ARPKD
  - Variants of congenital hepatic fibrosis (CHF)
    - Biliary hamartomas; polycystic liver disease & Caroli disease
  - CHF: Exists by itself (very rare)
    - Coexists more commonly other with polycystic liver diseases

Radiographic Findings

- ERCP
  - ± Dilatation of intrahepatic bile ducts

CT Findings

- Mild CHF
  - Liver may appear normal
- Moderate to severe CHF
  - Bile ducts: Normal to irregularly dilated
  - Enlarged dysmorphic liver; left lobe hypertrophy, right lobe atrophy
  - Splenomegaly
  - Varices; increased size and number of hepatic arteries
  - May develop hypervascular benign large regenerative nodules
    - Similar to those seen in Budd-Chiari syndrome

DDx: Dysmorphic Liver with Abnormal Bile Ducts

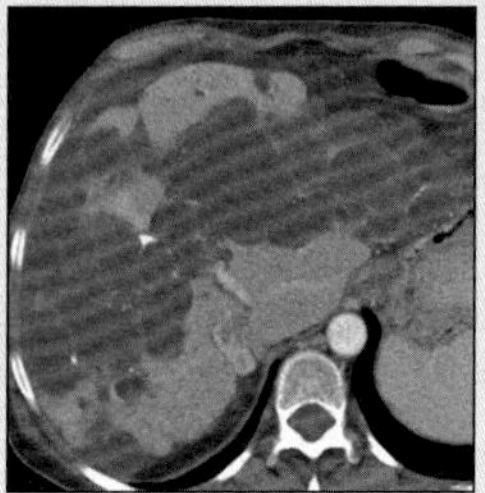

ADPLD

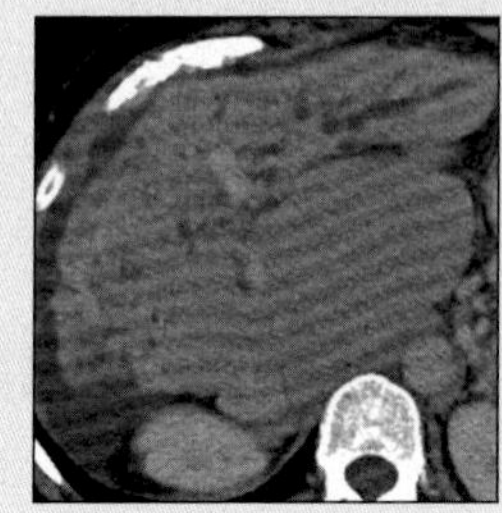

Sclerosing Cholangitis

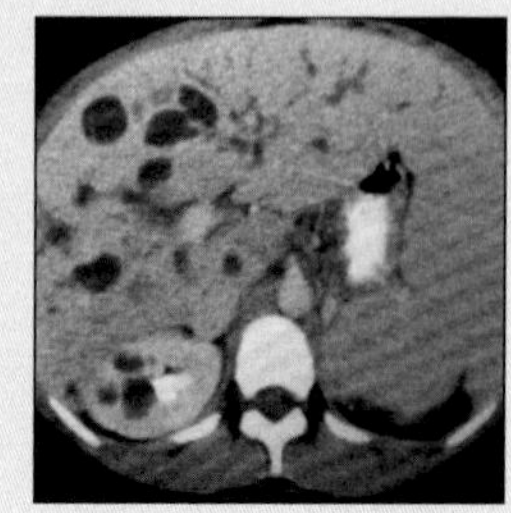

Caroli Disease

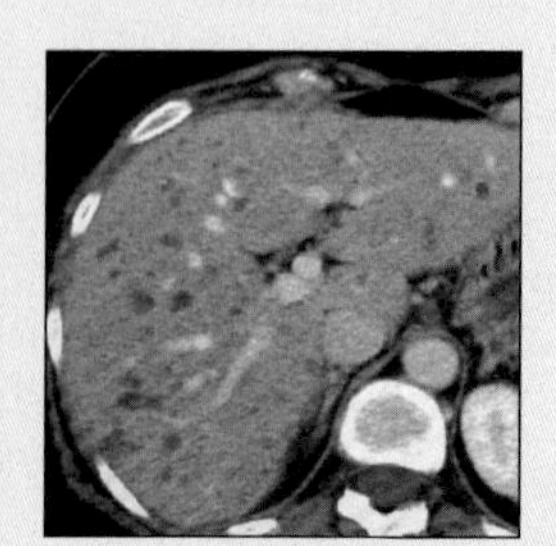

Biliary Hamartomas

CONGENITAL HEPATIC FIBROSIS

Key Facts

Terminology
- Part of a spectrum of congenital abnormalities resulting in variable degrees of fibrosis & cystic anomalies of liver & kidneys

Imaging Findings
- Best diagnostic clue: Combination of cystic dilatation of bile ducts + renal collecting duct ectasia
- Always present in autosomal recessive polycystic kidney disease (ARPKD)
- Severe renal disease & mild CHF/liver cysts
- Severe liver (CHF/liver cysts) & mild renal disease

Top Differential Diagnoses
- Isolated polycystic liver disease
- Primary sclerosing cholangitis
- Caroli disease
- Biliary hamartomas

Pathology
- Periportal fibrosis → portal HTN → hepatosplenomegaly → esophageal varices
- Sporadic or autosomal recessive inheritance pattern
- Embryological: Ductal plate malformation
- ARPKD: 100%; sometimes autosomal dominant PKD

Clinical Issues
- CHF is variable in its severity/age at presentation/clinical manifestations

Diagnostic Checklist
- No further evaluation needed in a child with: Hepatic ductal dilatation + enlarged portal tracts + hepatic/renal cysts

- Associated polycystic disease of liver & kidney
  - Multiple hypodense (water density) hepatic/renal cysts of varied size
  - CECT: No enhancement (simple/complicated cysts)

MR Findings
- MR Cholangiography (MRC)
  - Dilated intrahepatic bile ducts & biliary cysts
- Associated polycystic disease liver/kidney
  - T1WI: Hypointense
  - T2WI: Hyperintense
  - T1 C+: No enhancement
  - Heavily T2WI: ↑ Signal intensity due to pure fluid content

Ultrasonographic Findings
- Real Time
  - Moderate to severe CHF
    - Bile ducts: Moderate to severe dilatation
    - Liver & spleen: Enlarged
    - Splenic & portal veins: Dilated
    - Portal tracts: Distinct ↑ echogenicity
  - Associated polycystic disease liver & kidney
    - Uncomplicated: Anechoic lesions/smooth borders/non-detectable walls
    - Complicated: Septations/internal echoes/wall thickening
- Color Doppler
  - Depicts collaterals/direction & velocity of blood flow in splenic/portal veins
    - Direction: Hepatofugal or hepatopetal

Imaging Recommendations
- Best imaging tool: High-resolution CT or MRC
- Protocol advice
  - Helical CT: 5 mm collimated scans reconstructed every 2.5 mm
  - MR: Heavily T2WI/MRC

DIFFERENTIAL DIAGNOSIS

Isolated polycystic liver disease
- May have autosomal dominant polycystic liver disease (ADPLD) without congenital hepatic fibrosis

Primary sclerosing cholangitis
- Irregular strictures/dilation of intra-/extra-hepatic bile ducts
- Often leads to cirrhotic, dysmorphic liver
- Often associated with inflammatory bowel disease

Caroli disease
- Simple type: Cystic dilatation of bile ducts without periportal fibrosis
- Periportal type
  - Ductal dilatation + cysts + periportal fibrosis
  - Indistinguishable from congential hepatic fibrosis
- Best imaging clue on CECT
  - Enhancing tiny dot (portal radicle) within dilated cystic intrahepatic ducts

Biliary hamartomas
- Rare benign/congenital malformation of bile ducts
- Location: Subcapsular or intraparenchymal
- Innumerable subcentimeter nodules in both lobes of liver
- Diagnosis: Biopsy & histologic exam

PATHOLOGY

General Features
- General path comments
  - Pathophysiology
    - Periportal fibrosis → portal HTN → hepatosplenomegaly → esophageal varices
  - Periportal fibrosis: Fetal type (more common); adult type (rare)
- Genetics
  - Sporadic or autosomal recessive inheritance pattern
  - Probably linked to gene on chromosome 6p
- Etiology

CONGENITAL HEPATIC FIBROSIS

- Embryological: Ductal plate malformation
  - Abnormal bile duct formation & resorption
- Epidemiology: Incidence varies due to various degrees of expression
- Associated abnormalities
  - ARPKD: 100%; sometimes autosomal dominant PKD
  - Medullary sponge kidney: 80%
  - Caroli disease; vaginal atresia; tuberous sclerosis
  - Juvenile nephronophthisis
  - Meckel-Gruber syndrome

Gross Pathologic & Surgical Features

- Liver: Normal/enlarged/lobulated; dilated bile ducts/cysts

Microscopic Features

- Periportal fibrosis; malformed, dilated, nonobstructive bile ducts

CLINICAL ISSUES

Presentation

- Most common signs/symptoms
  - Mild CHF: Asymptomatic
  - Moderate to severe CHF
    - Childhood type (more common); adult type (rare)
    - Portal HTN/hepatosplenomegaly/bleeding varices
- Clinical profile
  - CHF is variable in its severity/age at presentation/clinical manifestations
  - Usually by adolescence have portal hypertension
- Lab
  - Normal liver function tests
  - ± Leukopenia/thrombocytopenia/anemia
    - Due to hypersplenism
- Diagnosis
  - Hepatic/renal cysts (ARPKD)
  - ± Hepatosplenomegaly
  - Normal liver function tests; liver biopsy

Demographics

- Age: Usually 5-13 years & rarely adult age group
- Gender: M = F

Natural History & Prognosis

- Rare developmental malformation of ductal plate
- CHF predominant features: ± intra-hepatic bile duct (IHBD) + periportal fibrosis
  - ± Ductal dilatation + periportal fibrosis
- Variable in its age of onset, severity & clinical presentation
- Complications
  - Biliary: Cholangitis/obstruction/sepsis
  - Portal: Hypertension & bleeding varices
- Prognosis
  - Early intervention
    - Slow progress or arrest of disease
    - Increased life expectancy & quality of life
  - Those who survive childhood
    - Relatively good prognosis & live into midlife

Treatment

- Options
  - Mild CHF (asymptomatic): No treatment
  - Moderate to severe CHF
    - Variceal sclerotherapy
    - Portosystemic shunting
    - Liver transplantation
  - Hypersplenism: Splenectomy

DIAGNOSTIC CHECKLIST

Consider

- High-resolution helical CT or MRC
  - To detect early stage of CHF
  - To differentiate communicating/noncommunicating biliary abnormalities
- Hepatic fibrosis, polycystic disease (liver and kidney), Caroli and biliary hamartomas can occur in isolation or in any combination

Image Interpretation Pearls

- No further evaluation needed in a child with: Hepatic ductal dilatation + enlarged portal tracts + hepatic/renal cysts

SELECTED REFERENCES

1. Brancatelli G et al: Fibropolycystic liver disease: CT and MR imaging findings. Scientific exhibit. Radiol. Soc. North America, Chicago, Illinois, 2003
2. Khan K et al: Morbidity from congenital hepatic fibrosis after renal transplantation for autosomal recessive polycystic kidney disease. Am J Transplant. 2(4):360-5, 2002
3. Lonergan GJ et al: Autosomal recessive polycystic kidney disease: radiologic-pathologic correlation. Radiographics. 20(3):837-55, 2000
4. Ernst O et al: Congenital hepatic fibrosis: findings at MR cholangiopancreatography. AJR Am J Roentgenol. 170(2):409-12, 1998
5. Desmet VJ: Ludwig symposium on biliary disorders--part I. Pathogenesis of ductal plate abnormalities. Mayo Clin Proc. 73(1):80-9, 1998
6. Kaplan BS et al: Variable expression of autosomal recessive polycystic kidney disease and congenital hepatic fibrosis within a family. Am J Med Genet. 29(3):639-47, 1988
7. Premkumar A et al: The emergence of hepatic fibrosis and portal hypertension in infants and children with autosomal recessive polycystic kidney disease. Initial and follow-up sonographic and radiographic findings. Pediatr Radiol. 18(2):123-9, 1988
8. Proesmans W et al: Association of bilateral renal dysplasia and congenital hepatic fibrosis. Int J Pediatr Nephrol. 7(2):113-6, 1986

CONGENITAL HEPATIC FIBROSIS

IMAGE GALLERY

Typical

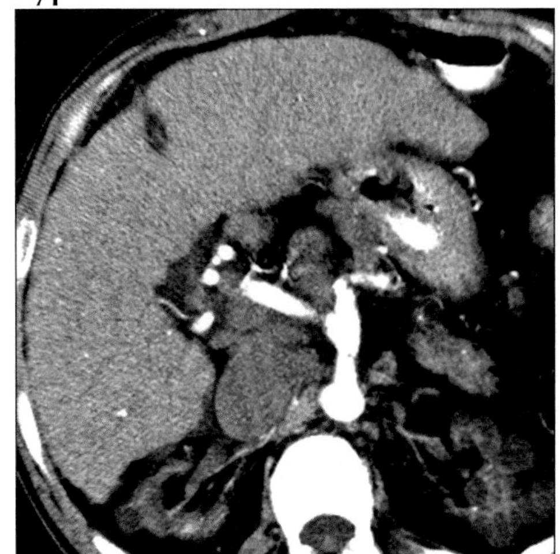

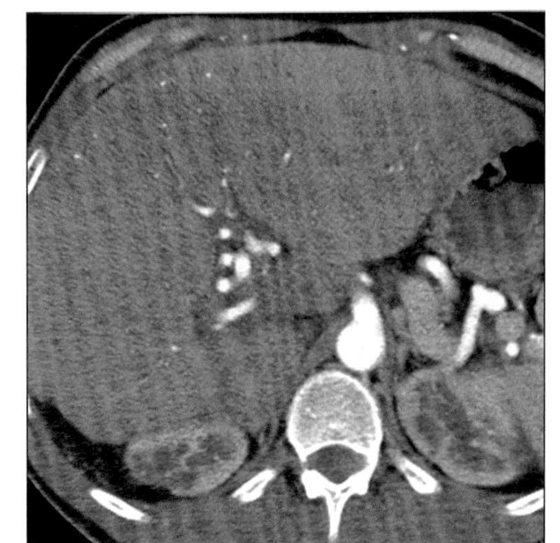

(Left) *Axial CECT shows dysmorphic liver with enlarged multiple hepatic arteries and fibrotic multicystic kidneys.* ***(Right)*** *Axial CECT shows large dysmorphic liver with enlarged multiple hepatic arteries. Kidneys were scarred and had multiple cysts.*

Typical

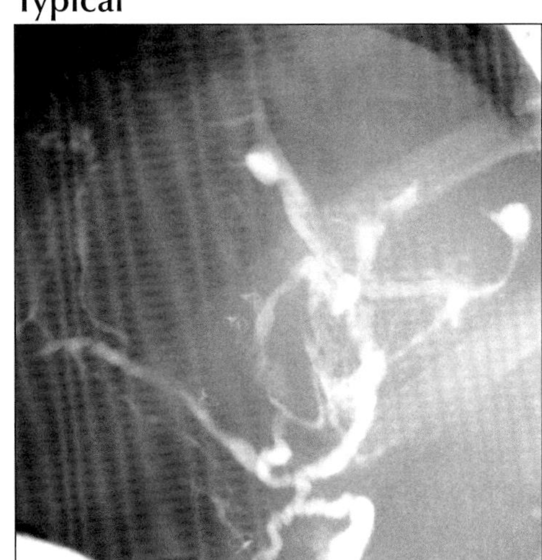

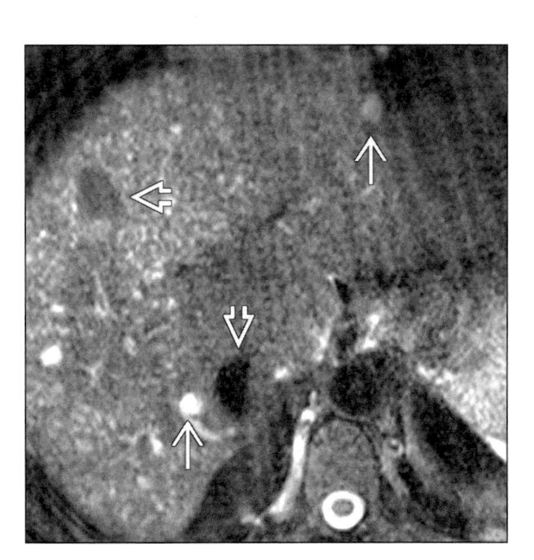

(Left) *Cholangiogram shows fusiform and cystic dilatation in intrahepatic bile ducts (Caroli disease) along with distortion and deviation due to hepatic fibrosis.* ***(Right)*** *Axial T2WI MR shows innumerable biliary hamartomas (arrows) and large regenerative nodules (open arrows). The liver is large and dysmorphic.*

Typical

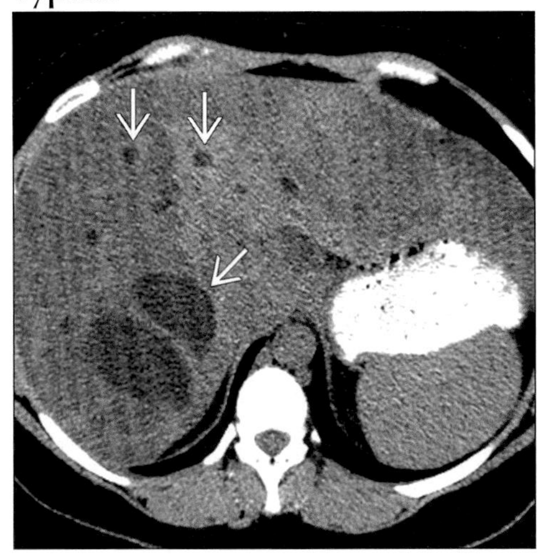

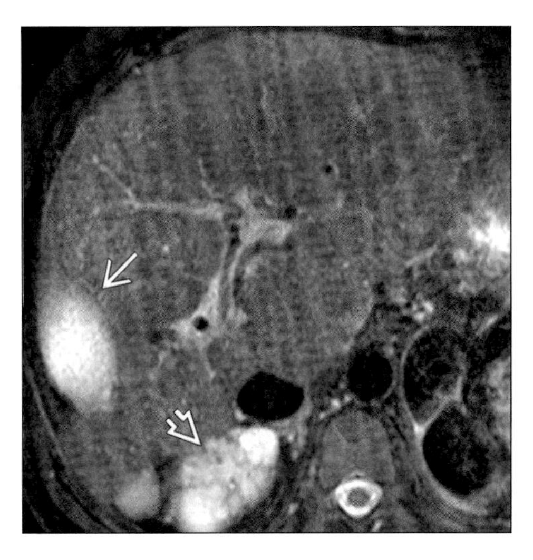

(Left) *Axial NECT shows large dysmorphic liver in patient with congenital hepatic fibrosis with intrahepatic "cysts" (arrows) due to associated Caroli disease.* ***(Right)*** *Axial T2WI MR shows large left hepatic lobe, tiny right lobe. Note position of gallbladder (arrow) and multiple renal cysts (open arrow).*

AD POLYCYSTIC DISEASE, LIVER

Gross pathology photograph of hepatectomy specimen shows numerous cysts replacing liver parenchyma. Cysts range in size from microscopic to 5.0 cm in greatest dimension, contain clear fluid.

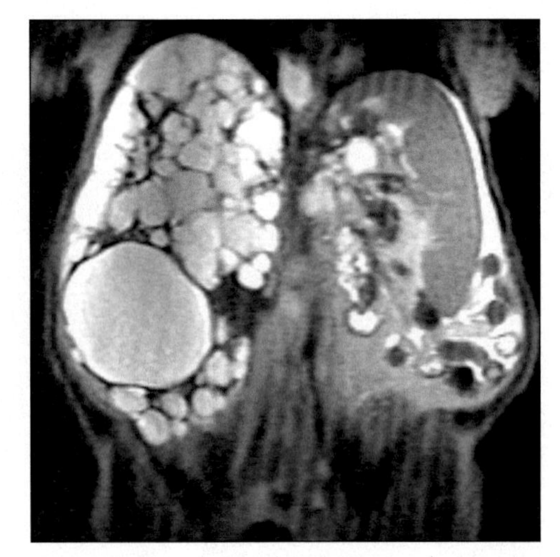

Coronal T2WI MR shows innumerable cysts of varying size, of high signal intensity, nearly completely replacing hepatic parenchyma. Multiple cysts within left kidney are also visualized.

TERMINOLOGY

Abbreviations and Synonyms

- Autosomal dominant polycystic liver disease (ADPLD) or adult PLD

Definitions

- Polycystic liver disease is a rare inherited disorder
- Part of spectrum of fibropolycystic liver disease; constitutes group of related lesions of liver & biliary tract caused by abnormal development of embryological ductal plate

IMAGING FINDINGS

General Features

- Best diagnostic clue: Multiple cysts of varying size
- Location
  - Extent of hepatic involvement ranges from limited sporadic area of cystic disease to diffuse involvement of all lobes of liver
    - ± Cysts in kidneys & in other organs
- Size: Range from < 1 mm to > 12 cm
- Key concepts
  - Numerous large & small cysts coexist with fibrosis
    - Round-or oval shape, smooth thin wall, absence of internal structures

Radiographic Findings

- ERCP
  - No communication with biliary tree & cysts do not opacify

CT Findings

- NECT
  - Multiple to innumerable, homogeneous & hypoattenuating cystic lesions
    - Cyst contents often greater than water density due to hemorrhage (less commonly, infection)
    - Calcification in cyst wall often seen: Due to old hemorrhage
- CECT
  - No wall or content enhancement
  - Cysts complicated by infection or hemorrhage may have septations &/or internal debris, as well as enhancement of wall
  - Cysts may contain fluid levels

MR Findings

- T1WI: Cysts have very low signal intensity
- T2WI
  - Owing to their pure fluid content, homogeneous high signal intensity is demonstrated on T2WI & heavily T2WI
    - Intracystic hemorrhage

DDx: Multiple Hepatic Cysts

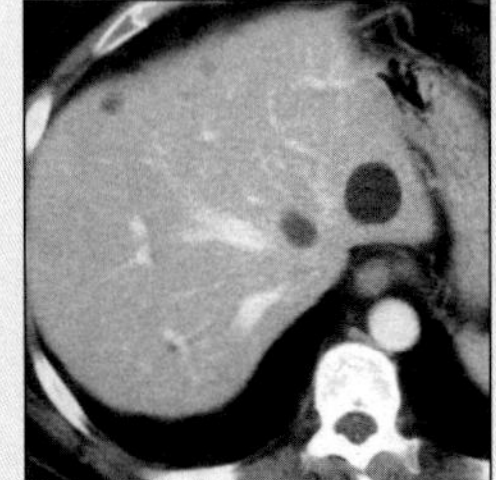

Hepatic Cysts

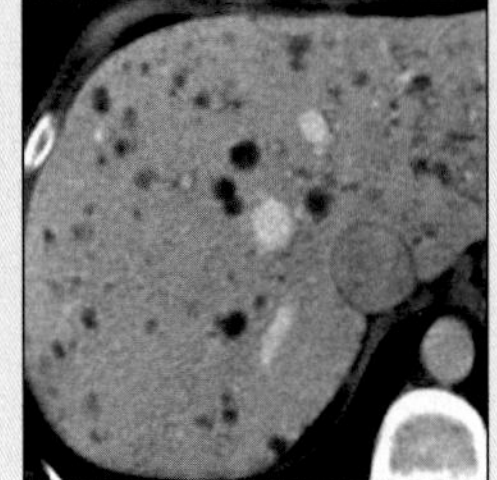

Biliary Hamartomas

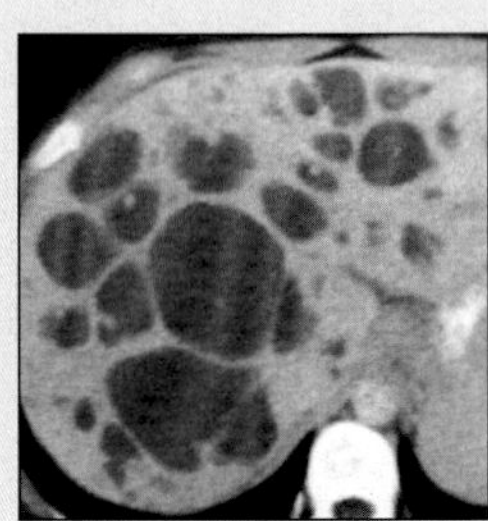

Caroli Disease

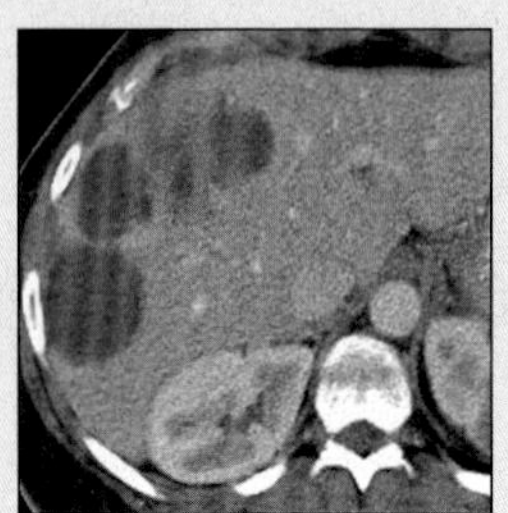

Cystic Metastases

AD POLYCYSTIC DISEASE, LIVER

Key Facts

Terminology
- Polycystic liver disease is a rare inherited disorder

Imaging Findings
- Size: Range from < 1 mm to > 12 cm
- Morphology:
- Multiple to innumerable, homogeneous & hypoattenuating cystic lesions
- Cyst contents often greater than water density due to hemorrhage (less commonly, infection)
- Calcification in cyst wall often seen: Due to old hemorrhage
- MR Cholangiography (MRC)
- No communication with each other or biliary tree

Pathology
- Hepatobiliary & renal anomalies frequently coexist in various combinations suggesting an expression of a common underlying genetic abnormality
- Isolated PLD is distinct genetic disease, unlinked to polycystic kidney disease (PKD) 1 & PKD 2
- Due to ductal plate malformation of small intrahepatic bile ducts
- Ducts lose communication with biliary tree

Clinical Issues
- Often causes massive hepatomegaly
- Complications: Spontaneous intracystic hemorrhage, rupture, infection

- T1 C+: Nonenhancing after administration of gadolinium contrast material
- MR Cholangiography (MRC)
  - No communication with each other or biliary tree

Ultrasonographic Findings
- Real Time
  - Anechoic masses, with smooth borders, thin-walled, no septations or mural nodularity
  - Acoustic enhancement beyond each cyst may produce impression of an abnormal liver pattern in addition to cysts

Nuclear Medicine Findings
- Tc-99m DISIDA scintigraphy: Permits differential diagnosis between Caroli disease & PLD
  - In Caroli disease: Areas of focally ↑ radiotracer accumulation that persist more than 120 minutes
  - In PLD: Areas of focally ↓ radiotracer accumulation with normal liver washout & biliary excretion

Imaging Recommendations
- Best imaging tool: Although diagnosis is easily made with both CT & MR imaging, MR is more sensitive for detection of complicated cysts
- Protocol advice: Heavily weighted T2WI, MRC, T1 gadolinium-enhanced sequences

DIFFERENTIAL DIAGNOSIS

Hepatic (bile duct) cysts
- Unilocular cyst lined by cuboidal, bile duct epithelium, containing serous fluid
  - Wall is 1 mm or less in thickness (nearly imperceptible)
    - Adjacent liver is normal
- Multiple, round or ovoid, well-defined, nonenhancing, water density lesions
- Homogeneous very low signal intensity on T1WI & homogeneous very high signal intensity on T2WI
  - Owing to their fluid content, an ↑ in signal intensity seen on heavily T2WI
- Usually solitary, but can number fewer than 10
  - When more than 10 cysts are seen, one of fibropolycystic disease should be considered

Biliary hamartomas
- Solitary or multiple (more common), well-defined nodules of varied density/subcapsular or intraparenchymal, scattered in both lobes of liver
  - Typically measuring less than 1.5 cm in diameter
- Varied enhancement based on cystic/solid components of lesions
  - Predominantly cystic (water density) lesions: No enhancement
  - Predominantly solid (fibrous stroma) lesions: Enhance & become isodense with liver parenchyma
- MRC: Markedly hyperintense nodules, no communication with biliary tree

Caroli disease
- Congenital communicating cavernous ectasia of biliary tract, autosomal recessive
- Multiple small rounded hypodense/hypointense saccular dilatation of intrahepatic bile ducts, multiple intrahepatic calculi
- "Central dot" on CECT: Enhancing tiny dots (portal radicles) within dilated cystic intrahepatic ducts
- MRC: Communicating bile duct abnormality

Cystic metastases
- Hypervascular metastases from neuroendocrine tumors, sarcoma, melanoma, subtypes of lung & breast carcinoma with necrosis & cystic degeneration
  - CECT & MR show multiple lesions with strong contrast-enhancement of peripheral viable & irregularly defined tissue
- Cystic metastases with mucinous adenocarcinoma (pancreatic or ovarian)
  - Cystic serosal implants on visceral peritoneal surface of liver & parietal peritoneum of diaphragm

AD POLYCYSTIC DISEASE, LIVER

PATHOLOGY

General Features

- General path comments: Hepatic cysts are pathologically identical to simple or bile duct cysts
- Genetics
  - Autosomal dominant
  - Hepatobiliary & renal anomalies frequently coexist in various combinations suggesting an expression of a common underlying genetic abnormality
    - Isolated PLD is distinct genetic disease, unlinked to polycystic kidney disease (PKD) 1 & PKD 2
- Etiology
  - Due to ductal plate malformation of small intrahepatic bile ducts
    - Ducts lose communication with biliary tree
- Epidemiology
  - Incidence is difficult to determine because of various degrees of expression; variable degrees of fibrosis & cystic anomalies
  - In patients with autosomal dominant PKD, there is hepatic involvement in approximately 30-40%
    - Approximately 70% of patients with PLD also have PKD
- Associated abnormalities
  - Biliary hamartomas
  - Congenital hepatic fibrosis is part of spectrum of hepatic cystic diseases
  - Often coexists with PKD

Gross Pathologic & Surgical Features

- Presence of multiple cysts in liver may distort normal liver architecture considerably
  - Liver surrounding cysts frequently contains biliary hamartomas & ↑ fibrous tissue

Microscopic Features

- Cuboidal & flat monolayer epithelium with no dysplasia in wall of cysts

CLINICAL ISSUES

Presentation

- Most common signs/symptoms
  - Asymptomatic; dull abdominal pain; abdominal distention; dyspnea; cachexia
  - Other signs/symptoms
    - Often causes massive hepatomegaly
- Clinical profile
  - Extrinsic compression of intrahepatic bile ducts
  - Hepatic venous outflow obstruction: Mechanical compression by cysts & associated formation of thrombi in small hepatic vein tributaries
  - Transudative ascites, portal hypertension due to distortion of portal venules by cysts & fibrosis
  - Lab data: ADPLD rarely affects liver function

Demographics

- Age: Adult manifestation
- Gender: Females have a significantly higher mean cyst score than male patients

Natural History & Prognosis

- Liver gradually enlarges as it is replaced by cysts
- In advanced disease: Liver failure, or Budd-Chiari syndrome
- Complications: Spontaneous intracystic hemorrhage, rupture, infection
- Prognosis: Surgical intervention has significant morbidity & inconsistent long term palliation
  - Orthotopic liver transplantation has excellent long term results, but substantial morbidity & mortality

Treatment

- Options, risks, complications
  - Simple unroofing; cyst fenestration alone; fenestration combined with resection
  - Total hepatectomy & orthotopic liver transplantation for patients with severe ADPLD
  - Ultrasound-guided percutaneous aspiration; multiple cyst punctures & alcohol sclerotherapy in patients with high surgical risk
  - Combined liver & kidney transplantation because of renal cystic involvement with renal insufficiency
    - One year survival rate: 89% with excellent symptomatic relief & improved quality of life

DIAGNOSTIC CHECKLIST

Consider

- Isolated ADPLD is underdiagnosed & genetically distinct from PLD associated with ADPKD but with similar pathogenesis, manifestations & management
- Clinical implications of & therapeutic strategies for cystic focal liver lesions vary according to their causes
  - Understanding of classic CT & MR appearances of cystic focal liver lesions will allow more definitive diagnosis & shorten diagnostic work-up

Image Interpretation Pearls

- MR findings that are important to recognize to differentiate cystic lesions of liver are: Size of lesion; presence & thickness of wall; presence of septa, calcifications, or internal nodules; enhancement pattern; MRC appearance; & signal intensity spectrum

SELECTED REFERENCES

1. Qian Q et al: Clinical profile of autosomal dominant polycystic liver disease. Hepatology. 37(1):164-71, 2003
2. Mortele KJ et al: Cystic focal liver lesions in the adult: differential CT and MR imaging features. Radiographics. 21(4):895-910, 2001
3. Steinberg ML et al: MRI and CT features of polycystic liver disease. N J Med. 90(5):398-400, 1993
4. Wan SK et al: Sonographic and computed tomographic features of polycystic disease of the liver. Gastrointest Radiol. 15(4):310-2, 1990
5. Wilcox DM et al: MR imaging of a hemorrhagic hepatic cyst in a patient with polycystic liver disease. J Comput Assist Tomogr. 9(1):183-5, 1985
6. Segal AJ et al: Computed tomography of adult polycystic disease. J Comput Assist Tomogr. 6(4):777-80, 1982

AD POLYCYSTIC DISEASE, LIVER

IMAGE GALLERY

Typical

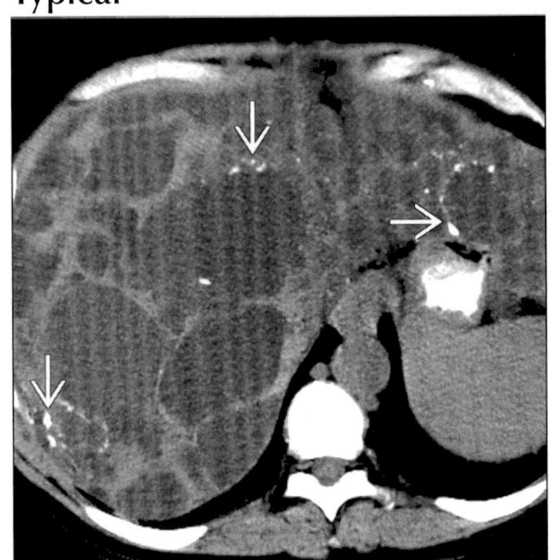

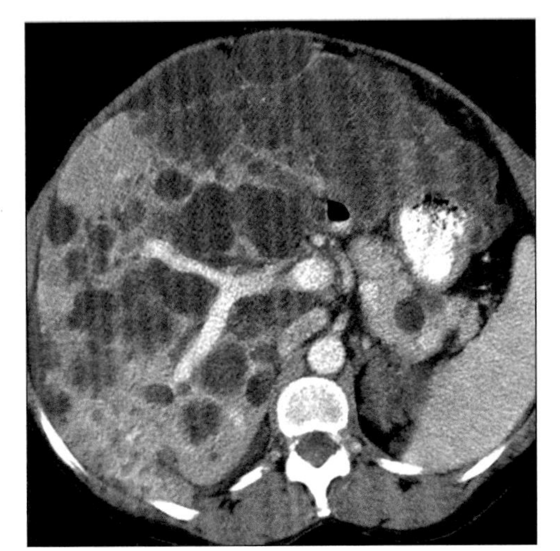

(Left) *Axial NECT shows innumerable, homogeneous & hypoattenuating cystic lesions with smooth thin-walls & absence of internal structures. Note peripheral calcification in cyst wall (arrows).* ***(Right)*** *Axial CECT shows innumerable cysts on portal venous phase. These are uncomplicated cysts with no wall or content enhancement & absence of internal debris/septations. Pancreatic cysts are also seen.*

Typical

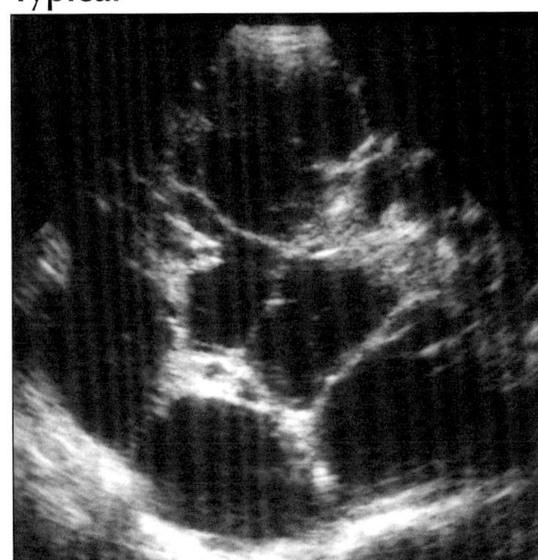

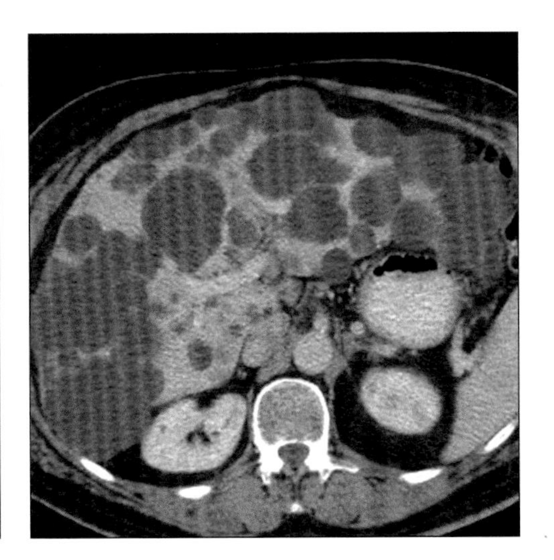

(Left) *Ultrasound of RUQ demonstrating anechoic, multiple hepatic cysts, with smooth borders, thin-walled, no septations or mural nodularity, in patient with ADPLD.* ***(Right)*** *Axial CECT through liver & kidneys. Note lack of involvement of kidneys in a patient with isolated ADPLD.*

Typical

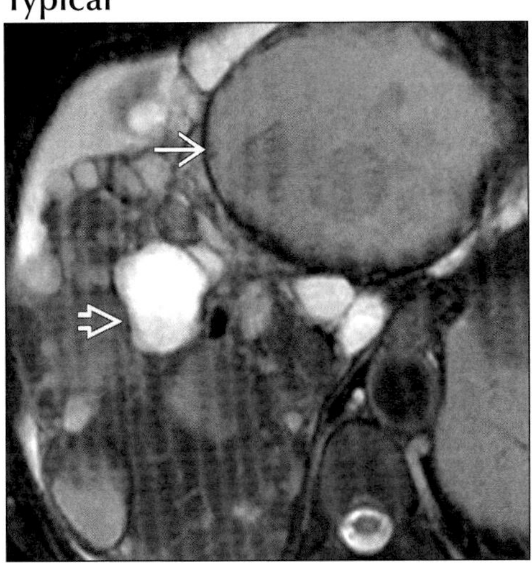

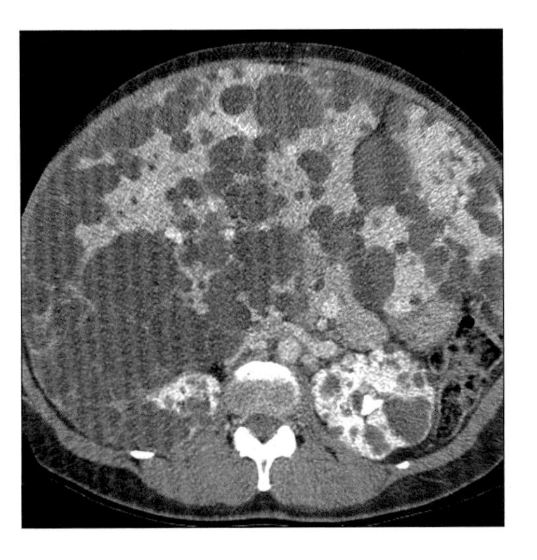

(Left) *Axial T2WI MR shows hemorrhagic cyst (arrow) in patient with ADPLD being less hyperintense than noncomplicated cyst (open arrow).* ***(Right)*** *Axial CECT shows that liver is markedly enlarged as it is replaced with innumerable cysts. Note also involvement of both kidneys; PLD coexisting with PKD.*

HEPATITIS

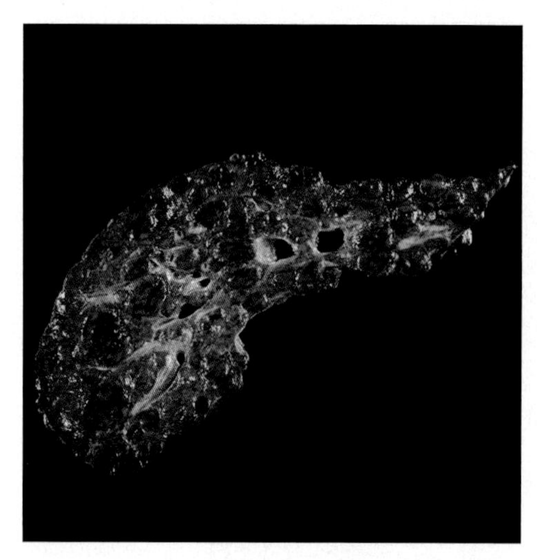

Cross section of explanted specimen shows shrunken, nodular liver with bridging fibrosis due to chronic active hepatitis B.

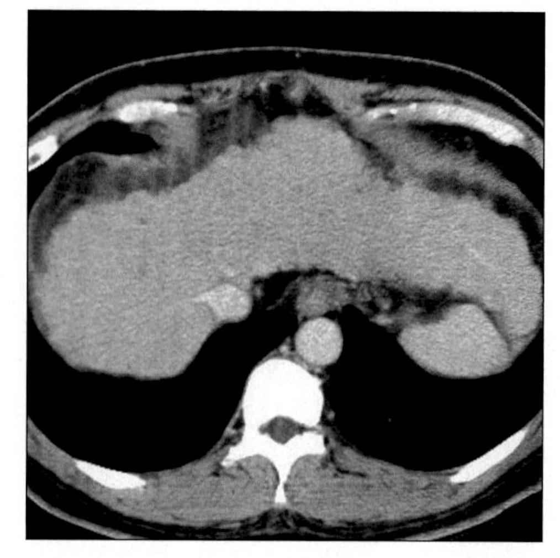

Axial CECT in a patient with cirrhosis due to chronic viral hepatitis shows nodular contour of liver with right lobe atrophy and lateral segment hypertrophy.

TERMINOLOGY

Definitions

- Nonspecific inflammatory response of liver to various agents

IMAGING FINDINGS

General Features

- Best diagnostic clue
  - Acute viral hepatitis on US
    - "Starry-sky" appearance: ↑ Echogenicity of portal venous walls
    - Hepatomegaly & periportal lucency (edema)
- Location: Diffusely; involving both lobes
- Size
  - Acute: Enlarged liver
  - Chronic: Decrease in size of liver
- Other general features
  - Leading cause of hepatitis is viral infection
  - In medical practice, hepatitis refers to viral infection
  - Viral hepatitis
    - Infection of liver by small group of hepatotropic viruses
    - Stages: Acute, chronic active hepatitis (CAH) & chronic persistent hepatitis
    - Responsible for 60% of cases of fulminant hepatic failure in U.S.
  - Alcoholic hepatitis: Acute & chronic
  - Nonalcoholic steatohepatitis (NASH)
    - Significant cause of acute & progressive liver disease
    - May be an underlying cause of cryptogenic cirrhosis
  - Imaging of viral/alcoholic hepatitis done to exclude
    - Obstructive biliary disease/neoplasm
    - To evaluate parenchymal damage noninvasively

CT Findings

- NECT
  - Acute viral hepatitis
    - Hepatomegaly, gallbladder wall thickening
    - Periportal hypodensity (fluid/lymphedema)
  - Chronic active hepatitis
    - Lymphadenopathy in porta hepatis/gastrohepatic ligament & retroperitoneum (in 65% of cases)
    - Hyperdense regenerating nodules
  - Acute alcoholic hepatitis
    - Hepatomegaly
    - Diffuse hypodense liver (due to fatty infiltration)
    - Fatty infiltration may be focal/lobar/segmental
  - Chronic alcoholic hepatitis
    - Mixture of steatosis & early cirrhotic changes
    - Steatosis: Liver-spleen attenuation difference will be less than 10 HU

DDx: Diffuse Hepatomegaly

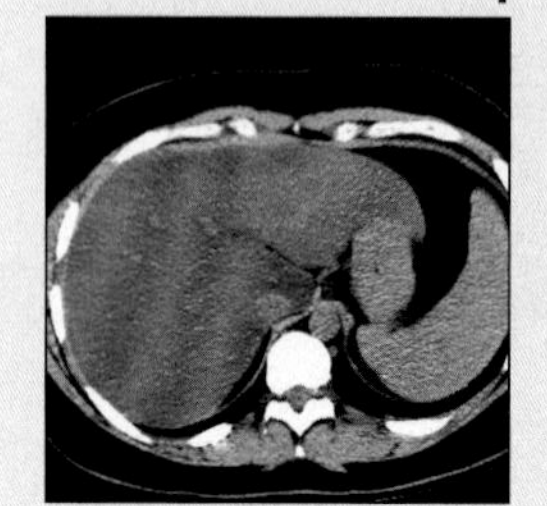

Steatosis

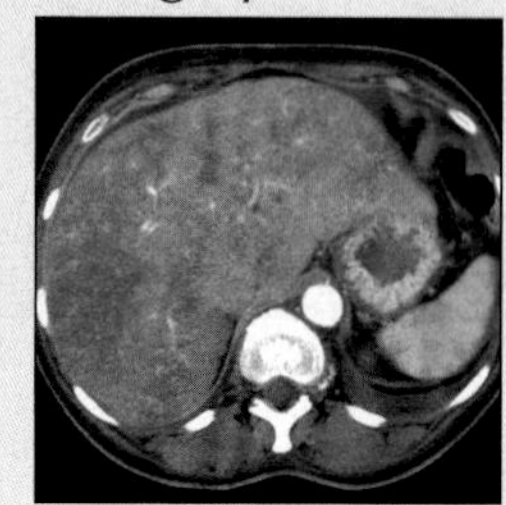

Passive Congestion

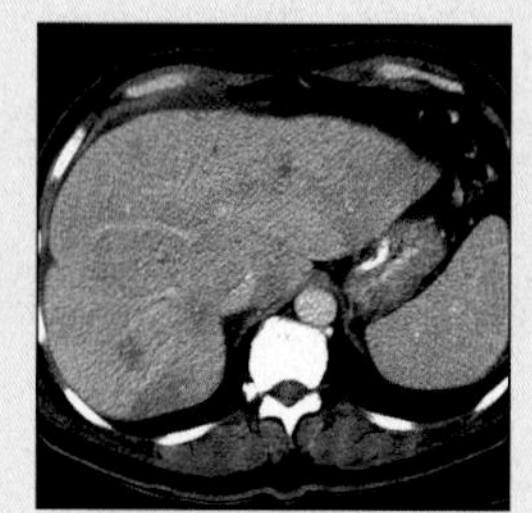

Diffuse Lymphoma

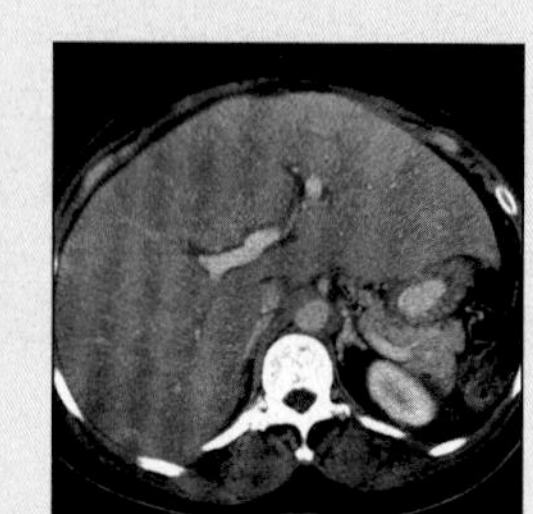

Amyloidosis

HEPATITIS

Key Facts

Imaging Findings
- "Starry-sky" appearance: ↑ Echogenicity of portal venous walls
- Hepatomegaly, gallbladder wall thickening
- Periportal hypodensity (fluid/lymphedema)
- Hyperdense regenerating nodules
- Increase in T1 & T2 relaxation times of liver
- Alcoholic steatohepatitis (diffuse fatty infiltration)

Top Differential Diagnoses
- Steatosis (fatty liver)
- Passive hepatic congestion
- Diffuse lymphoma
- Myeloproliferative & infiltrative disorders

Pathology
- Cellular dysfunction, necrosis, fibrosis, cirrhosis
- HBV: Sensitized cytotoxic → T cells hepatocyte necrosis → tissue damage
- Alcoholic hepatitis: Inflammatory reaction leads to acute liver cell necrosis

Clinical Issues
- Malaise/anorexia/fever/pain/hepatomegaly/jaundice
- Lab data: ↑ Serologic markers; ↑ liver function tests
- Age: Any age group (particularly teen-/middle-age)
- Self-limited; more progressive & chronic in nature
- Cirrhosis: 10% of HBV & 20-50% of HCV
- HCC: Particularly among carriers of HBsAg

Diagnostic Checklist
- Ruling out other causes of "diffuse hepatomegaly"
- Two most consistent findings in acute hepatitis: Hepatomegaly & periportal edema

- Normal liver has slightly ↑ attenuation than spleen
  - Nonalcoholic steatohepatitis (NASH)
    - Indistinguishable from alcoholic hepatitis
- CECT
  - Acute & chronic viral hepatitis
    - ± Heterogeneous parenchymal enhancement
  - Chronic hepatitis: Regenerating nodules may be isodense with liver

MR Findings
- Viral hepatitis
  - Increase in T1 & T2 relaxation times of liver
  - T2WI: High signal intensity bands paralleling portal vessels (periportal edema)
- Alcoholic steatohepatitis (diffuse fatty infiltration)
  - T1WI in-phase GRE image: Increased signal intensity of liver than spleen or muscle
  - T1WI out-of-phase GRE image: Decreased signal intensity of liver (due to lipid in liver)

Ultrasonographic Findings
- Real Time
  - Acute viral hepatitis
    - ↑ In liver & spleen size; ↓ echogenicity of liver
    - "Starry-sky" appearance: Increased echogenicity of portal venous walls
    - Periportal hypo-/anechoic area (hydropic swelling of hepatocytes)
    - Thickening of GB wall; hypertonic GB
  - Chronic viral hepatitis
    - Increased echogenicity of liver & coarsening of parenchymal texture
    - "Silhouetting" of portal vein walls (loss of definition of portal veins)
    - Adenopathy in hepatoduodenal ligament
  - Acute alcoholic hepatitis
    - ↑ Echogenicity (fatty infiltration), ↑ size of liver
  - Alcoholic steatohepatitis
    - Liver parenchyma: Increased echogenicity & sound attenuation
    - Indistinguishable from liver fibrosis
  - Late stage of alcoholic hepatitis
    - Liver is atrophic & micronodular pattern

Imaging Recommendations
- Best imaging tool
  - Helical NECT
  - MR (in & out-of-phase GRE images)
    - Alcoholic steatohepatitis
- Protocol advice: NECT and CECT, or MR with in and out of phase GRE

DIFFERENTIAL DIAGNOSIS

Steatosis (fatty liver)
- Diffuse decreased attenuation of enlarged liver
- T1WI out-of-phase GREI: Decreased signal of liver
- Normal vessels course through "lesion" (fatty infiltration)

Passive hepatic congestion
- Diffuse hepatomegaly
- Early enhancement of dilated IVC & hepatic veins
- Doppler: Loss of triphasic pattern in IVC/hepatic veins

Diffuse lymphoma
- Hepatomegaly due to diffuse infiltration
- Large, lobulated low density discrete masses
- More common in immune-suppressed patients
- Examples: AIDS & organ transplant recipients

Myeloproliferative & infiltrative disorders
- Sickle cell: Diffuse hepatomegaly (due to congestion)
- Amyloidosis: Hepatomegaly & low attenuation areas

PATHOLOGY

General Features
- General path comments
  - Different stages of hepatitis
    - Cellular dysfunction, necrosis, fibrosis, cirrhosis
  - HBV: Sensitized cytotoxic → T cells hepatocyte necrosis → tissue damage

HEPATITIS

- Alcoholic hepatitis: Inflammatory reaction leads to acute liver cell necrosis
- Etiology
  - Viral hepatitis: Caused by one of 5 viral agents
    - Hepatitis A (HAV), B (HBV), C (HCV) viruses
    - Hepatitis D (HDV), E (HEV) viruses
  - Other causes of hepatitis
    - Alcohol abuse
    - Bacterial or fungal
    - Autoimmune reactions; metabolic disturbances
    - Drug induced injury; exposure to environmental agents; radiation therapy
- Epidemiology
  - HBV (serum)
    - In U.S. incidence is 13.2 cases/100,000 population
    - In U.S. & Europe, carrier rate is < 1%
    - In Africa & Asia, carrier rate is 10%
    - Endemic areas: HCC accounts 40% of all cancers

Gross Pathologic & Surgical Features

- Acute viral hepatitis: Enlarged liver + tense capsule
- Chronic fulminant hepatitis: Atrophic liver
- Alcoholic steatohepatitis: Enlarged, yellow, greasy liver

Microscopic Features

- Acute viral: Coagulative necrosis with ↑ eosinophilia
- Chronic viral: Lymphocytes/macrophages/plasma cells/piecemeal necrosis
- Alcoholic hepatitis: Neutrophils/necrosis/Mallory bodies (alcoholic hyaline)

Staging, Grading or Classification Criteria

- Hepatitis A (HAV)
  - Virus: ssRNA
  - Transmission: Fecal-oral
  - Incubation period: 2-6 weeks
  - No carrier & chronic phase
- Hepatitis B (HBV)
  - Virus: DNA
  - Transmission: Parenteral + sexual
  - Incubation period: 1-6 months
  - Carrier & chronic phase present
- Hepatitis C (HCV)
  - Virus: RNA
  - Transmission: Blood transfusion
  - Incubation period: 2-26 weeks
  - Carrier & chronic phase present
- Hepatitis D (HDV)
  - Virus: RNA
  - Transmission: Parenteral + sexual
  - Incubation period: 1-several months
  - Carrier with HBV; chronic phase present
- Hepatitis E (HEV)
  - Virus: ssRNA
  - Transmission: Water-borne
  - Incubation period: 6 weeks
  - No carrier & chronic phase

CLINICAL ISSUES

Presentation

- Most common signs/symptoms
  - Acute & chronic hepatitis
    - Malaise/anorexia/fever/pain/hepatomegaly/jaundice
  - Acute HBV: May present with serum sickness-like syndrome
- Clinical profile: Teenage or middle-aged patient with history of fever, RUQ pain, hepatomegaly & jaundice
- Lab data: ↑ Serologic markers; ↑ liver function tests
- Diagnosis: Based on
  - Serologic markers; virological; clinical findings
  - Liver function tests & liver biopsy

Demographics

- Age: Any age group (particularly teen-/middle-age)
- Gender: M = F

Natural History & Prognosis

- Hepatitis can be
  - Self-limited; more progressive & chronic in nature
- Complications
  - Relapsing & fulminant hepatitis
  - Of chronic viral (HBV, HCV) & alcoholic hepatitis
    - Cirrhosis: 10% of HBV & 20-50% of HCV
    - HCC: Particularly among carriers of HBsAg
- Prognosis
  - Acute viral & alcoholic: Good
  - Chronic persistent hepatitis: Good
  - Chronic active hepatitis (CAH): Not predictable
  - Fulminant hepatitis: Poor

Treatment

- Acute viral hepatitis: No specific treatment; prophylaxis-IG, HBIG, vaccine
- Chronic viral hepatitis: Interferon for HBV & HCV
- Alcoholic hepatitis: Alcohol cessation & good diet

DIAGNOSTIC CHECKLIST

Consider

- Ruling out other causes of "diffuse hepatomegaly"
- Liver biopsy for diagnosis and staging

Image Interpretation Pearls

- Two most consistent findings in acute hepatitis: Hepatomegaly & periportal edema

SELECTED REFERENCES

1. Mortele KJ et al: Imaging of diffuse liver disease. Seminars In Liver Disease 21, number 2: 195-212, 2001
2. Okada Y et al: Lymph nodes in the hepatoduodenal ligament: US appearance with CT and MR correlation. Clin Radiol. 51 (3): 160-6, 1996
3. Murakami T et al: Liver necrosis and regeneration after fulminant hepatitis: pathologic correlation with CT and MR findings. Radiology. 198(1):239-42, 1996
4. Kurtz AB et al: Ultrasound findings in hepatitis. Radiology. 136: 717-23, 1980

HEPATITIS

IMAGE GALLERY

Typical

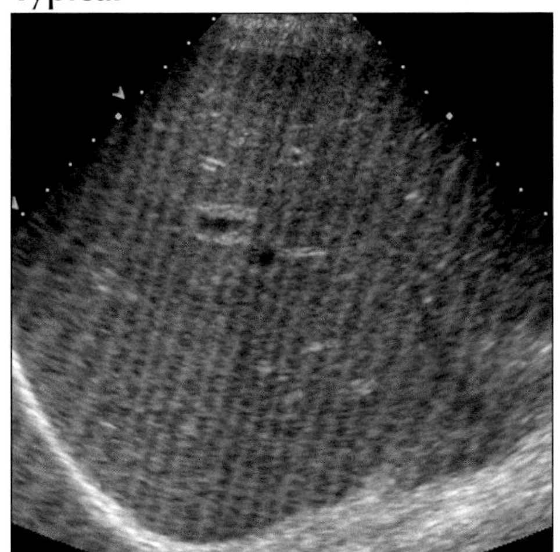

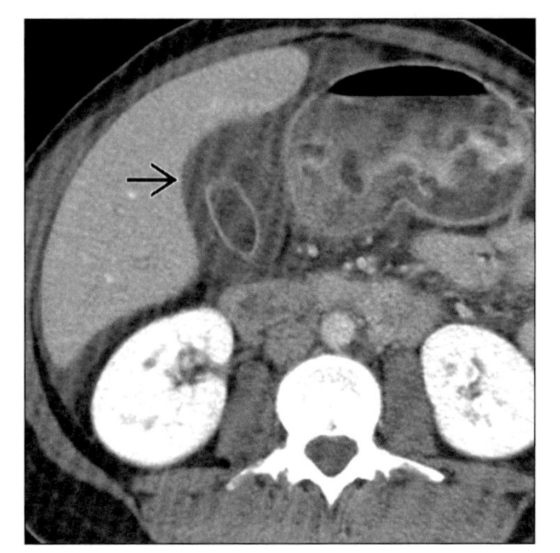

*(**Left**) Sagittal sonogram in a patient with acute hepatitis shows enlarged hypoechoic liver with increased echogenicity of portal venous walls ("starry sky"). (**Right**) Axial CECT shows marked gallbladder wall edema (arrow) and ascites in Morison pouch.*

Typical

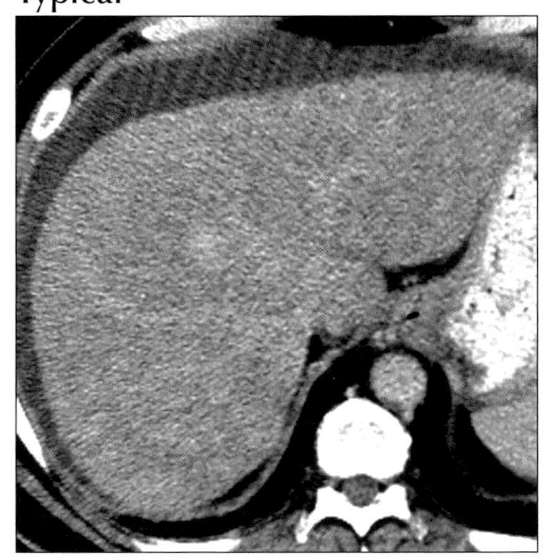

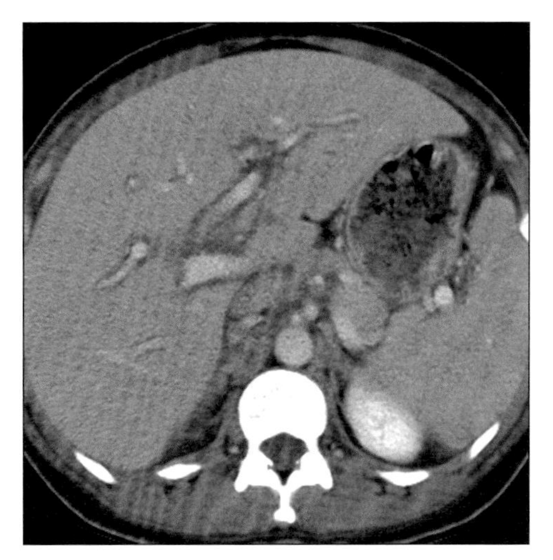

*(**Left**) Axial CECT shows heterogeneous enhancement of the liver and ascites in a patient with acute hepatitis. (**Right**) Axial CECT shows hepatosplenomegaly, periportal lucency (lymphedema) and lymphadenopathy.*

Typical

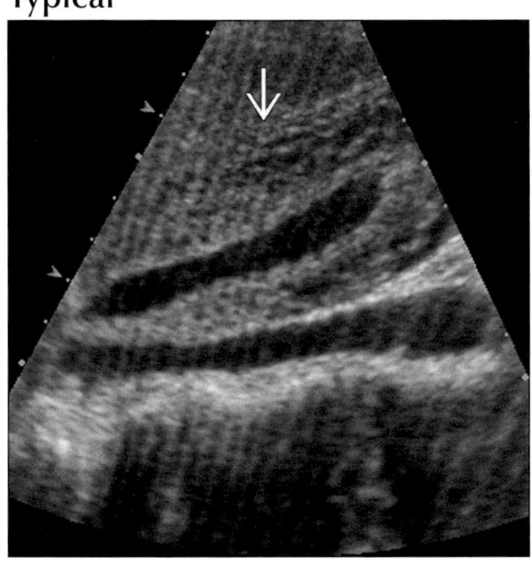

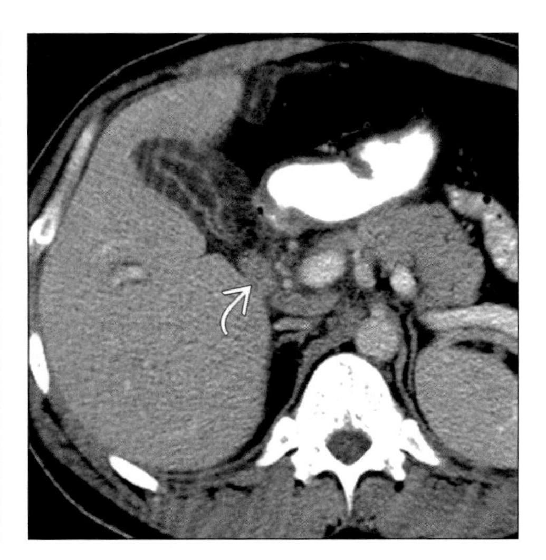

*(**Left**) Sagittal sonogram in patient with acute viral hepatitis shows marked thickening of gallbladder wall (arrow). (**Right**) Axial CECT in a patient with chronic active viral hepatitis shows marked gallbladder wall edema and portacaval enlarged lymph nodes (curved arrow).*

HEPATIC CANDIDIASIS

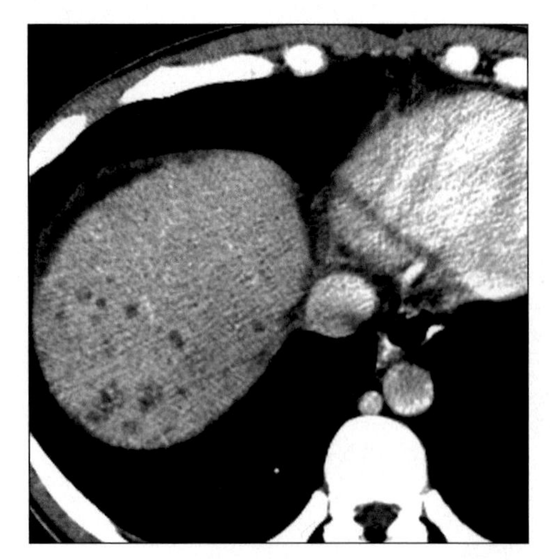

Axial CECT in an immunocompromised patient shows numerous hypodense "microabscesses" proven to represent hepatic candidiasis.

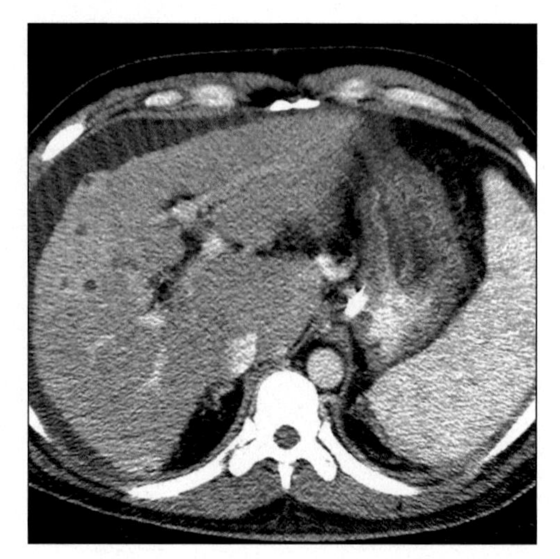

Axial CECT shows numerous "microabscesses" with peripheral enhancement. Also ascites.

TERMINOLOGY

Definitions

- A systemic fungal infection (Candida albicans) that often affects abdominal viscera

IMAGING FINDINGS

General Features

- Best diagnostic clue: Multiple well-defined, rounded microabscesses in liver
- Location: Both lobes of liver
- Size: Less than 1 cm (microabscesses)
- Other general features
  - Most common fungal infection in immunocompromised patients
  - More common in patients with
    - Acquired immunodeficiency syndrome (AIDS)
    - Intensive chemotherapy
    - Acute leukemia (50-70%) recovering from profound neutropenia
    - Lymphoma (50%) at the time of autopsy
    - Chronic granulomatous disease of childhood
    - Renal transplant
  - Chronic disseminated candidiasis
    - Involvement of several organs

CT Findings

- NECT
  - Multiple small hypodense lesions
  - ± Periportal areas of increased attenuation (fibrosis)
  - ± Scattered areas of calcific density (healing phase)
- CECT
  - Nonenhancing hypodense areas
  - ± Peripheral enhancement
  - Central or eccentric "dot" felt to represent hyphae

MR Findings

- T1WI: Hypointense
- T2WI: Hyperintense
- STIR: Short T1 inversion recovery (STIR): Hyperintense
- T1 C+: Nonenhancing hypointense lesions
- Contrast-enhanced FLASH (fast low-angle shot) images
  - Detect more lesions

Ultrasonographic Findings

- Real Time
  - Four major patterns of hepatic Candidiasis are seen
    - "Wheel within a wheel": Peripheral zone surrounds inner echogenic wheel, in turn surrounds a central hypoechoic nidus (early stage)
    - "Bull's eye": 1-4 mm lesion with a hyperechoic center surrounded by a hypoechoic rim (seen when neutrophil count returns to normal)

DDx: Innumerable Hypodense Liver Lesions

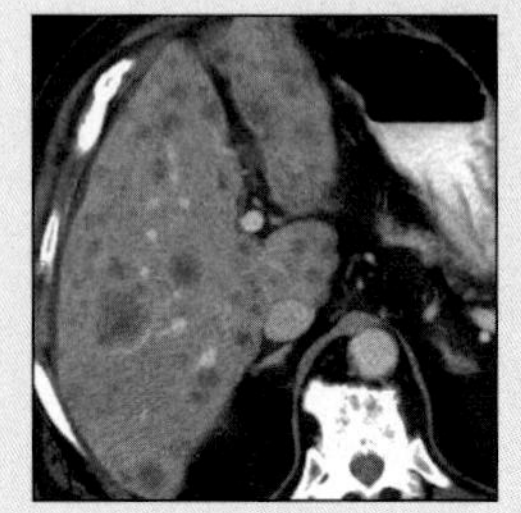

Metastases

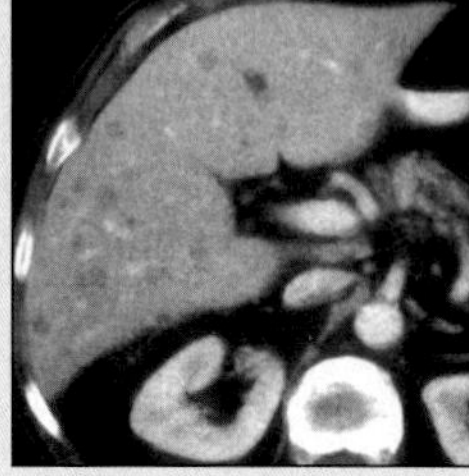

Lymphoma/Leukemia

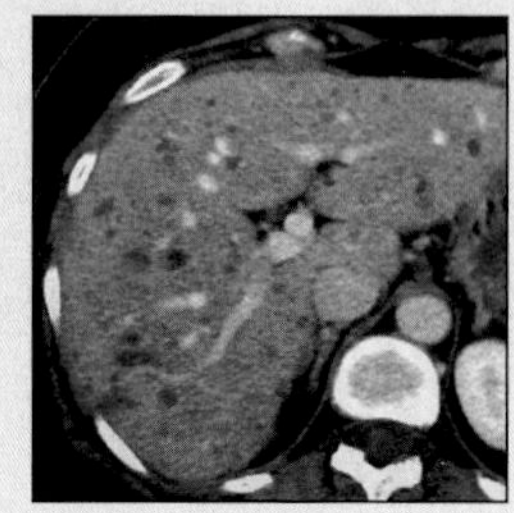

Biliary Hamartomas

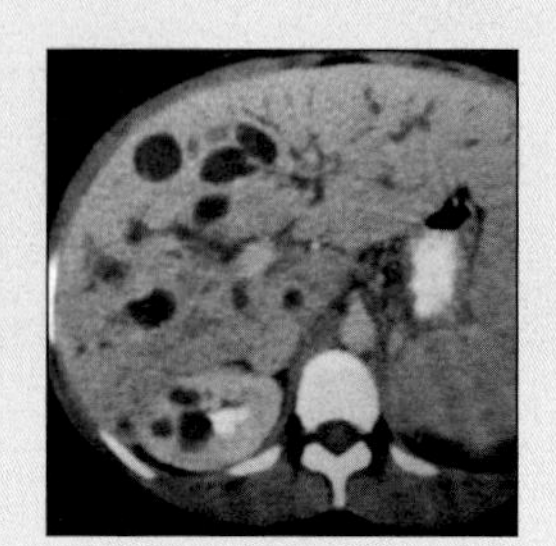

Caroli Disease

HEPATIC CANDIDIASIS

Key Facts

Terminology
- A systemic fungal infection (Candida albicans) that often affects abdominal viscera

Imaging Findings
- Best diagnostic clue: Multiple well-defined, rounded microabscesses in liver
- Location: Both lobes of liver
- Size: Less than 1 cm (microabscesses)
- Most common fungal infection in immunocompromised patients

Top Differential Diagnoses
- Metastases
- Lymphomatous/leukemic foci in liver
- Biliary hamartomas
- Caroli disease

Pathology
- Candida albicans
- Originates from intestinal seeding of portal & venous circulation

Clinical Issues
- Asymptomatic or abdominal pain
- Erythematous papules on skin
- Clinical profile: Immunocompromised patients recovering from neutropenia (examples: Acute leukemia, lymphoma, AIDS, chemotherapy & organ transplant recipient)
- Antifungal therapy (amphotericin B & fluconazole)

Diagnostic Checklist
- Rule out other "innumerable hypodense liver lesions"
- Biopsy & send specimen for histology/microbiology

        - "Uniformly hypoechoic": Most common appearance (due to fibrosis & debris)
        - "Echogenic": Caused by scar formation
    - After antifungal therapy: Lesions
        - Increase in echogenicity
        - Decrease in size or often disappear altogether

Nuclear Medicine Findings
- Candida microabscesses
    - Technetium sulfur colloid
        - Cold lesions (due to decreased uptake)
    - Gallium scan
        - Cold lesions (due to diminished uptake)

Imaging Recommendations
- Best imaging tool: Helical CT or MR
- Protocol advice
    - CT: Thin sections ($\leq$ 5 mm)
    - MR: FLASH sequences
    - Both pre-contrast & post-contrast studies

DIFFERENTIAL DIAGNOSIS

Metastases
- Less numerous, larger & usually do not affect spleen
- Epithelial metastases: Rim-enhancement
- Can be cystic or calcified

Lymphomatous/leukemic foci in liver
- Less well-defined; less numerous; larger
- Usually foci can also be seen in spleen

Biliary hamartomas
- Rare benign congenital malformation of bile ducts
- Location: Intraparenchymal or subcapsular
- Innumerable subcentimeter nodules in both lobes
- Diagnosis: Biopsy & histologic exam

Caroli disease
- Best imaging clue on CECT
    - "Central dot" sign: Enhancing tiny dot (portal radicle) within dilated cystic intrahepatic ducts
- Two types: Simple & periportal
    - Simple: Cystic dilatation of bile ducts without periportal fibrosis
    - Periportal: Ductal dilatation, cysts & periportal fibrosis

PATHOLOGY

General Features
- Etiology
    - Candida albicans
        - Most common cause of Candidiasis
    - Candida tropicalis
        - Accounts for 1/3 of deep candidiasis cases
        - Usually in tropical countries
    - Originates from intestinal seeding of portal & venous circulation
- Epidemiology: More commonly seen in areas with endemic AIDS
- Associated abnormalities
    - Acquired immunodeficiency syndrome
    - Underlying malignancy
    - Leukemia
    - Lymphoma
    - Neutropenia due to other causes
        - Chemotherapy
        - Post radiation therapy
        - Organ transplant

Gross Pathologic & Surgical Features
- Multiple microabscesses of liver

Microscopic Features
- Simple media: Oval, budding cells
- Special culture
    - Hyphae
    - Elongated branching called pseudohyphae
- In serum
    - Germ tubes
    - Thick-walled spores called chlamydospores

HEPATIC CANDIDIASIS

CLINICAL ISSUES

Presentation

- Most common signs/symptoms
  - Asymptomatic or abdominal pain
  - Fever
  - Erythematous papules on skin
    - Acute candidemia (neutropenic patients)
  - Rarely hepatomegaly
- Clinical profile: Immunocompromised patients recovering from neutropenia (examples: Acute leukemia, lymphoma, AIDS, chemotherapy & organ transplant recipient)
- Fever in neutropenic patients whose WBC count is returning to normal
- Lab data: ↑ Alkaline phosphatase
- Diagnosis: By histologic section of biopsy specimens
  - Pseudohyphae in central necrotic portion of lesion
  - Difficult on clinical grounds because blood cultures are positive in only 50% of affected patients

Demographics

- Age: Any age group
- Gender: M = F

Natural History & Prognosis

- Systemic fungal infection
- Origin: Intestinal seeding
- Liver lesions via portal & venous circulation
- Affected individuals; particularly immunocompromised patients
- Complications (rare)
  - Rupture of microabscesses
  - Cholangitis due to candidiasis of biliary tract
- Prognosis
  - Usually good with prompt diagnosis & treatment

Treatment

- Liver microabscesses
  - Antifungal therapy (amphotericin B & fluconazole)
  - Very rarely surgical or percutaneous drainage

DIAGNOSTIC CHECKLIST

Consider

- Rule out other "innumerable hypodense liver lesions"
- Biopsy & send specimen for histology/microbiology

Image Interpretation Pearls

- Both pre- & post-contrast studies
  - CT & MR (FLASH) sequences show
    - Multiple small, rounded lesions

SELECTED REFERENCES

1. Wig JD et al: Cholangitis due to candidiasis of the extra-hepatic biliary tract. HPB Surg. 11(1):51-4, 1998
2. Semelka RC et al: Hepatosplenic fungal disease: Diagnostic accuracy and spectrum of appearances on MR imaging. AJR 169:1311-6, 1997
3. Giamarellou H et al: Epidemiology, diagnosis, and therapy of fungal infections in surgery. Infect Control Hosp Epidemiol. 17(8):558-64, 1996
4. Lamminen AE et al: Infectious liver foci in leukemia: Comparison of short-inversion-time inversion-recovery, T1-weighted spin-echo, and dynamic gadolinium-enhanced MR imaging. Radiology. 191:539-43, 1994
5. Meunier F: Candidiasis. Eur J Clin Microbiol Infect Dis. 8(5):438-47, 1989
6. Pastakia B et al: Hepatosplenic candidiasis: Wheels within wheels. Radiology. 166:417-21, 1988
7. Cunha BA: Systemic infections affecting the liver. Some cause jaundice, some do not. Postgrad Med. 84(5):148-58, 161-3, 166-8, 1988
8. Maxwell AJ et al: Fungal liver abscesses in acute leukaemia--a report of two cases. Clin Radiol. 39(2):197-201, 1988
9. Thaler M et al: Hepatic candidiasis in cancer patients: the evolving picture of the syndrome. Ann Intern Med. 108(1):88-100, 1988
10. Tashjian LS et al: Focal hepatic candidiasis: a distinct clinical variant of candidiasis in immunocompromised patients. Rev Infect Dis. 6(5):689-703, 1984

IMAGE GALLERY

Typical

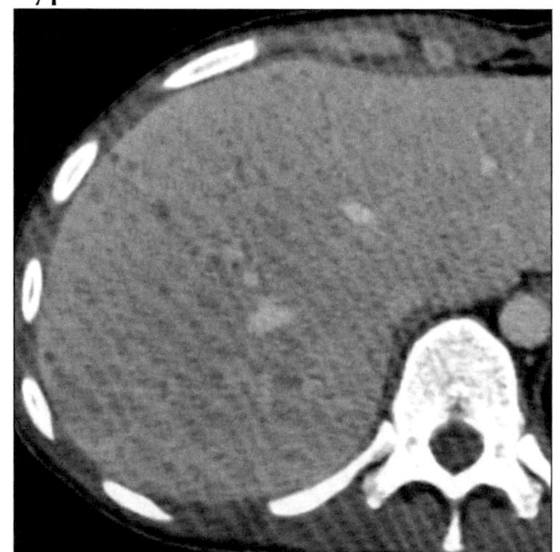

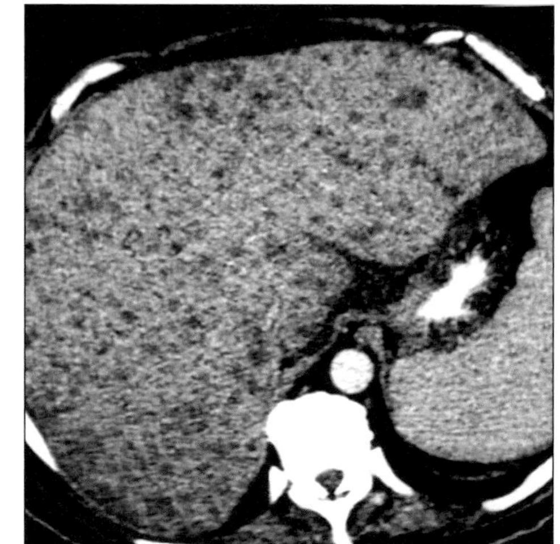

*(**Left**) Axial CECT in patient with AIDS shows innumerable hypodense "microabscesses" from hepatic Candidiasis. (**Right**) Axial CECT shows innumerable hypodense "microabscesses" scattered throughout the liver.*

Typical

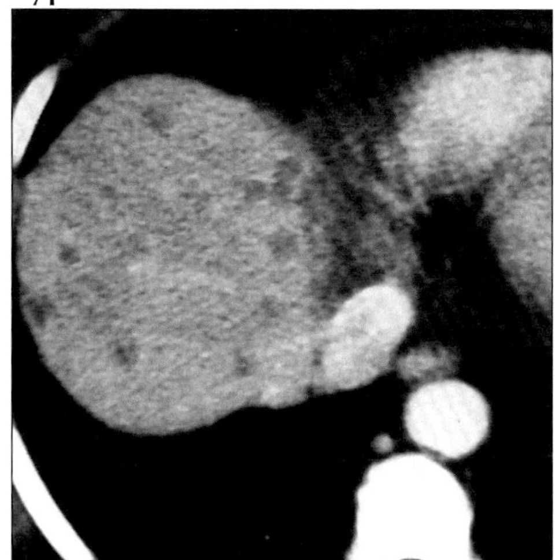

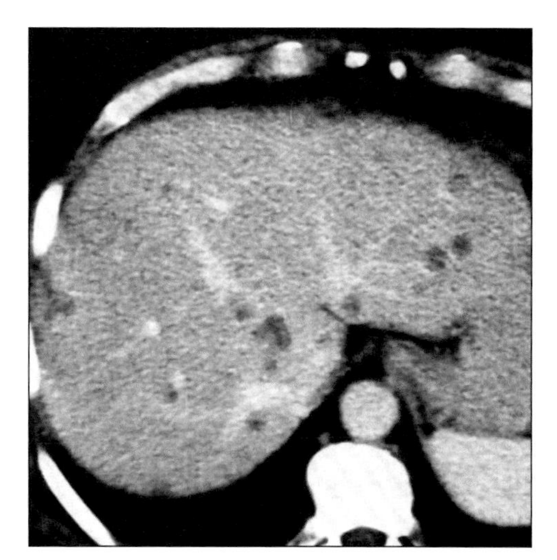

*(**Left**) Axial CECT shows small hypodense lesions within the hepatic dome, some which demonstrate peripheral enhancement. (**Right**) Axial CECT demonstrating some of the small hypodense "microabscesses" having an eccentric "dot" representing hyphae.*

Typical

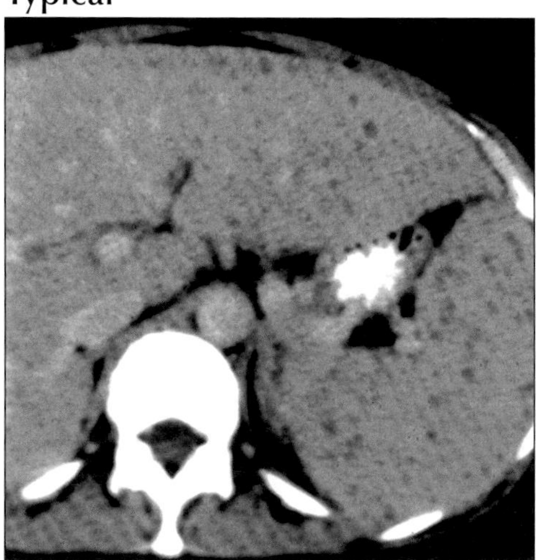

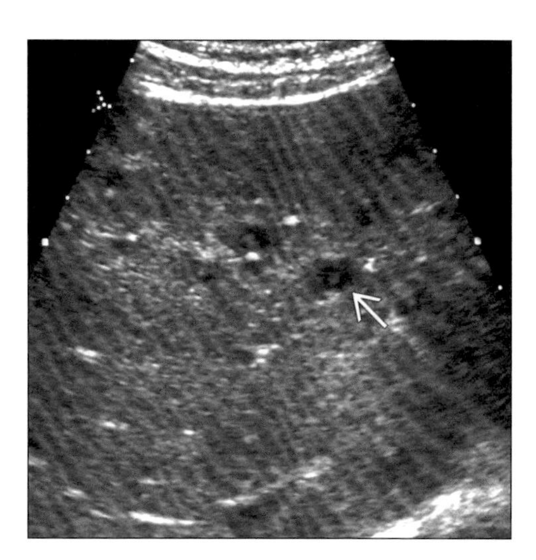

*(**Left**) Axial CECT in patient with AIDS shows diffuse microabscesses in liver and spleen. (**Right**) Axial US in immunocompromised patient shows multiple, small hypoechoic masses (arrow) from hepatic Candida lesions some with bull's eye appearance.*

HEPATIC PYOGENIC ABSCESS

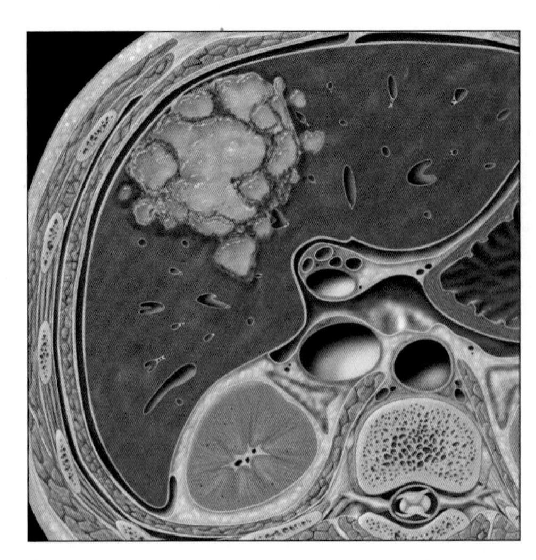

Graphic shows peripheral multiloculated collections of pus with surrounding inflamed liver.

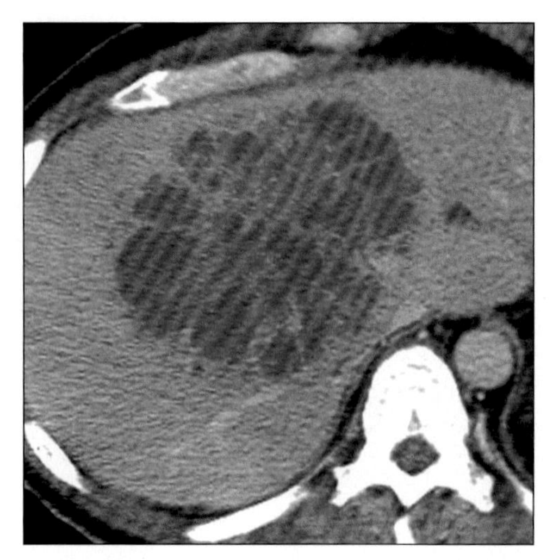

Axial CECT shows multiple coalescing cystic lesions with enhancing septa. Pyogenic abscess resulted from diverticulitis

TERMINOLOGY

Abbreviations and Synonyms

- Liver pyogenic abscess

Definitions

- Localized collection of pus in liver due to bacterial infectious process with destruction of hepatic parenchyma & stroma

IMAGING FINDINGS

General Features

- Best diagnostic clue: "Cluster" sign - cluster of small pyogenic abscesses coalesce into a single large cavity
- Location
  - Varies based on origin
    - Portal origin: Right lobe (65%); left lobe (12%); both lobes (23%)
    - Biliary tract origin: 90% involve both lobes
- Size: Varies from few millimeters to 10 centimeters
- Other General Features
  - Western countries: Liver abscess
    - Usually pyogenic (bacterial in origin)
    - Typically due to complication of infection elsewhere
  - Among all liver abscesses
    - Pyogenic accounts: 88% (bacterial)
    - Amebic: 10% (Entamoeba histolytica)
    - Fungal: 2% (Candida albicans)
  - Most common causes of pyogenic abscess
    - Diverticulitis
    - Ascending cholangitis
    - Infection of infarcted tissue (e.g., post liver transplantation, necrotic tumor)
  - Pyogenic abscesses may be single or multiple
    - Biliary tract origin: Multiple small abscesses
    - Portal origin: Usually solitary larger abscess
    - Direct extension & trauma: Solitary large abscess
  - Developing countries: Liver abscesses
    - Mostly due to parasitic infections
    - Amebic, echinococcal or other protozoal/helminthic

Radiographic Findings

- Radiography
  - Chest x-ray
    - Elevation of right hemidiaphragm
    - Right lower lobe atelectasis
    - Infiltrative lesions, right pleural effusion
  - Plain x-ray abdomen
    - Hepatomegaly, intrahepatic gas, air-fluid level
  - Contrast studies of gut & urinary tract: May show cause of abscess
    - Diverticulitis, perforated ulcer & renal abscess
- ERCP

DDx: Cystic Liver Lesion with/without Gas

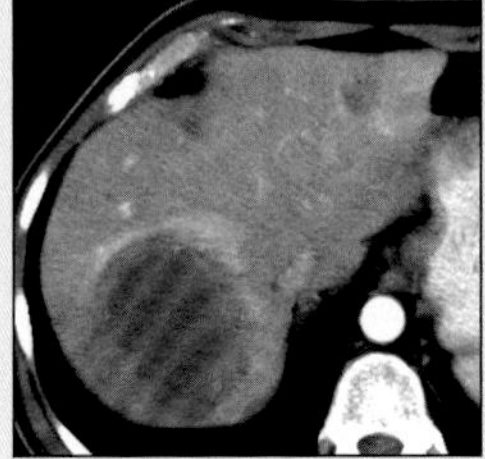

Cystic Metastases

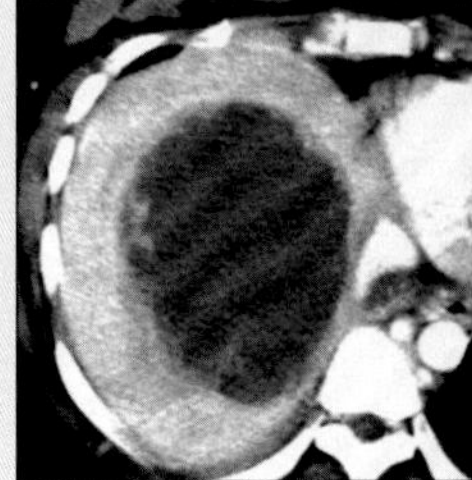

Amebic Abscess

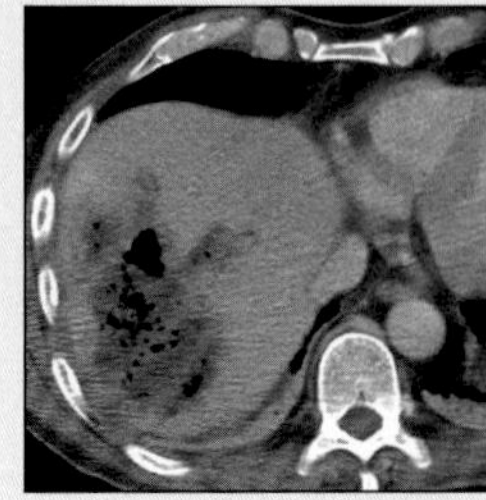

Liver Infarction (OLT)

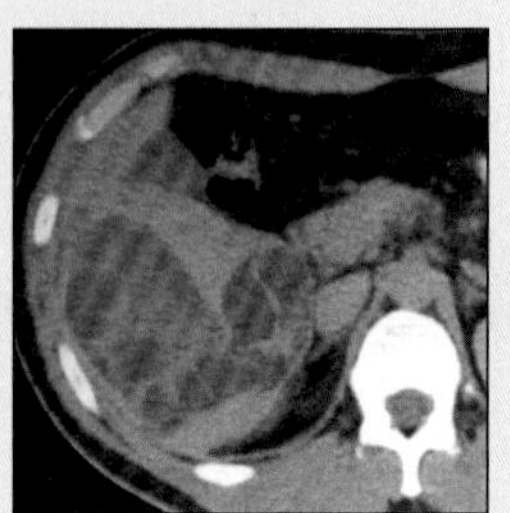

Hydatid Cyst

HEPATIC PYOGENIC ABSCESS

Key Facts

Terminology
- Localized collection of pus in liver due to bacterial infectious process with destruction of hepatic parenchyma & stroma

Imaging Findings
- Best diagnostic clue: "Cluster" sign - cluster of small pyogenic abscesses coalesce into a single large cavity
- Portal origin: Right lobe (65%); left lobe (12%); both lobes (23%)
- Biliary tract origin: 90% involve both lobes
- Pyogenic abscesses may be single or multiple
- Right lower lobe atelectasis

Top Differential Diagnoses
- Metastases (especially after treatment)
- Hepatic amebic abscess
- Infarction in liver transplant (OLT)
- Hepatic hydatid cyst
- Biliary cystadenocarcinoma

Pathology
- Pyogenic: Accounts 88% of all liver abscesses
- E. coli (adults) & S. aureus (children)
- Diverticulitis, appendicitis

Clinical Issues
- Fever, RUQ & usually left lower quadrant pain

Diagnostic Checklist
- Rule out: Amebic/fungal liver abscesses; cystic tumors
- Check for history of transplantation or ablation/chemotherapy for liver tumor

    - Accurately define level & cause of biliary obstruction

CT Findings
- NECT
    - Simple pyogenic abscess
        - Well-defined, round, hypodense mass (0-45 HU)
    - "Cluster" sign
        - Small abscesses aggregate to coalesce into a single big cavity, usually septated
    - Complex pyogenic abscess: "Target" lesion
        - Hypodense rim
        - Isodense periphery
        - Decreased HU in center
    - Specific sign: Abscess with central gas
        - Seen as air bubbles or an air-fluid level
        - Present in less than 20% of cases
    - Large air-fluid or fluid-debris level
        - Often associated with gut communication or necrotic tissue
- CECT
    - Sharply-defined, round, hypodense mass
    - Rim- or capsule- and septal-enhancement
    - Right lower lobe atelectasis & pleural effusion
    - Non-liquified infection may simulate hypervascular tumor

MR Findings
- T1WI: Hypointense
- T2WI
    - Hyperintense mass
    - High signal intensity perilesional edema
- T1 C+
    - Hypointense mass
    - Rim or capsule enhancement
    - Small abscesses less than 1 cm
        - May show homogeneous enhancement
        - Mimicking hemangiomas
- MRCP
    - Highly specific in detecting
        - Obstructive biliary pathology
        - Leading cause of cholangitis → pyogenic abscess

Ultrasonographic Findings
- Real Time
    - Variable in shape & echogenicity
    - Usually spherical or ovoid in shape
    - Wall: Irregular hypoechoic/mildly echogenic
    - Echogenicity of abscesses
        - Anechoic (50%), hyperechoic (25%), hypoechoic (25%)
    - ± Septa or fluid level within abscess
    - ± Debris & posterior enhancement
    - Early lesions tend to be echogenic & poorly demarcated
    - May evolve into well-demarcated, nearly anechoic lesions
    - Gas in an abscess seen as brightly echogenic foci with posterior artefacts

Nuclear Medicine Findings
- Hepato biliary & sulfur colloid scans
    - Rounded, cold areas
    - Occasionally, communication between abscess cavity & biliary system can be seen
- Gallium scan (Gallium citrate Ga 67)
    - Hot lesions
    - Mixed lesion: Cold center & hot rim
- WBC Scan
    - Hot lesions (due to WBC accumulation)
    - Highly specific for pyogenic abscesses compared to any nuclear or cross-sectional imaging

Imaging Recommendations
- Best imaging tool: CECT
- Image guided aspiration

DIFFERENTIAL DIAGNOSIS

Metastases (especially after treatment)
- Usually do not appear as a cluster or septated cystic mass
- Usually no elevation of diaphragm or atelectasis
- No fever or ↑ WBC with metastases
- Treated necrotic metastases may be indistinguishable from abscess

HEPATIC PYOGENIC ABSCESS

Hepatic amebic abscess
- Compared to pyogenic: Amebic abscesses are
  - Usually peripheral, round or oval shape
  - Sharply-defined hypoechoic or low attenuation
- Most often solitary (85%)
- Affects right lobe more often (72%) than left lobe (13%)
- Abuts liver capsule
  - US shows homogeneous echoes + distal enhancement
- More common in recent immigrants, institutionalized, homosexuals
- Dark, reddish-brown, consistency of anchovy paste

Infarction in liver transplant (OLT)
- Hepatic artery thrombosis (HAT) → hepatic and biliary necrosis
- Indistinguishable from pyogenic abscess

Hepatic hydatid cyst
- Large cystic liver mass + peripheral daughter cysts
- ± Curvilinear or ring-like pericyst calcification
- ± Dilated intrahepatic bile ducts: Due to mass effect and/or rupture into bile ducts

Biliary cystadenocarcinoma
- Rare, multiseptated, water density cystic mass
- No surrounding "inflammatory changes"

PATHOLOGY

General Features
- General path comments
  - Pyogenic abscess can develop via five major routes
  - Biliary: Ascending cholangitis from
    - Choledocholithiasis
    - Benign or malignant biliary obstruction
  - Portal vein: Pylephlebitis from
    - Appendicitis, diverticulitis
    - Proctitis, inflammatory bowel disease
    - Right colon infection spreads via: Superior mesenteric vein → portal vein → liver
    - Left colon infection via: Inferior mesenteric vein → splenic vein → portal vein → liver
  - Hepatic artery: Septicemia from bacterial endocarditis, pneumonitis, osteomyelitis
  - Direct extension
    - Perforated gastric or duodenal ulcer
    - Subphrenic abscess, pyelonephritis
  - Traumatic: Blunt or penetrating injuries
- Etiology
  - Pyogenic: Accounts 88% of all liver abscesses
  - Most common bacterial organisms
    - E. coli (adults) & S. aureus (children)
- Epidemiology: Incidence rate is increasing in Western countries due to ascending cholangitis & diverticulitis
- Associated abnormalities
  - Diverticulitis, appendicitis
  - Benign or malignant biliary obstruction
  - Perforated gastric or duodenal ulcer
  - Bacterial endocarditis, pneumonitis, osteomyelitis

Gross Pathologic & Surgical Features
- Pyogenic abscess: Multiple or solitary lesions

CLINICAL ISSUES

Presentation
- Most common signs/symptoms
  - Fever, RUQ pain, rigors, malaise
  - Nausea, vomiting, weight loss, tender hepatomegaly
  - If subphrenic then atelectasis and pleural effusion possible
- Clinical profile
  - Middle-aged/elderly patient with history of
    - Fever, RUQ & usually left lower quadrant pain
    - Tender hepatomegaly & increased WBC count
- Lab data
  - Increased leukocytes & serum alk phosphatase
- Diagnosis: Fine needle aspiration cytology (FNAC)

Natural History & Prognosis
- Complications
  - Spread of infection to subphrenic space
    - Causes atelectasis & pleural effusion
- Prognosis
  - Good after medical therapy & aspiration
  - Catheter drainage failure rate 8.4%
  - Recurrent abscess rate 8%

Treatment
- Antibiotics
- Percutaneous aspiration + parenteral antibiotics
- Percutaneous catheter drainage
- Surgical drainage

DIAGNOSTIC CHECKLIST

Consider
- Rule out: Amebic/fungal liver abscesses; cystic tumors
  - Amebic: Entamoeba histolytica
  - Fungal: Candida albicans
  - Hepatic hydatid or simple cyst, biliary cystadenoma
- Check for history of transplantation or ablation/chemotherapy for liver tumor

Image Interpretation Pearls
- "Cluster" sign: Small abscesses coalesce into big cavity
- Specific sign: Presence of central gas or fluid level
- Elevation of right hemidiaphragm
- Right lower lobe atelectasis & pleural effusion
- Non-liquified abscess may simulate solid tumor

SELECTED REFERENCES

1. Giorgio A et al: Pyogenic liver abscesses: 13 years of experience in percutaneous needle aspiration with US guidance. Radiology. 195: 122-4, 1995
2. Mendez RZ et al: Hepatic abscesses: MR imaging findings. Radiology. 190: 431-6, 1994
3. Jeffrey RB et al: CT small pyogenic hepatic abscesses: The cluster sign. AJR. 151(3): 487-9, 1988

IMAGE GALLERY

Typical

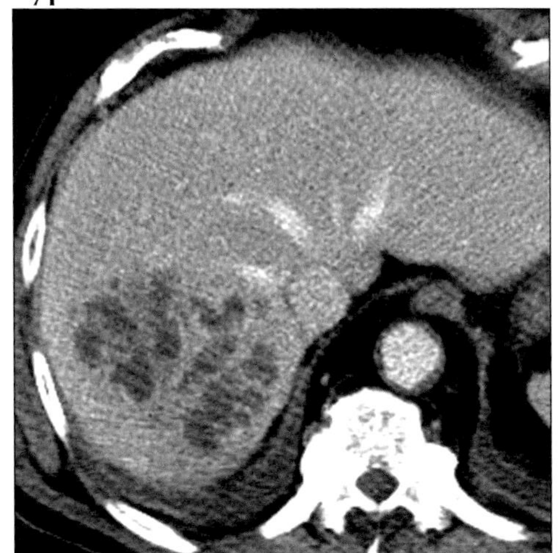

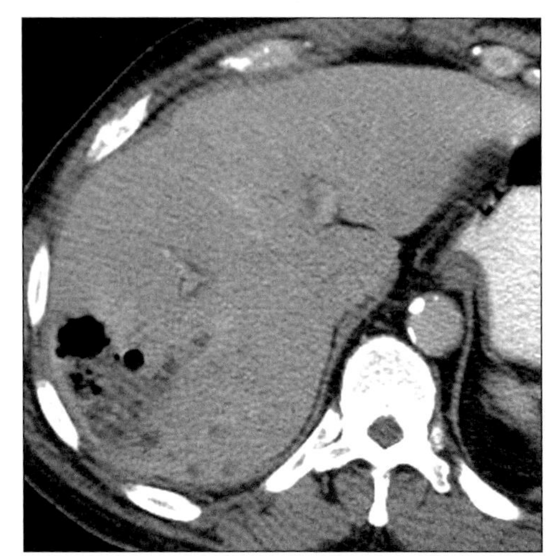

(Left) *Axial CECT shows cluster of small abscesses coalescing into large septated mass. Note pleural effusion. (Source, cholangitis).* ***(Right)*** *Axial CECT shows cluster of peripheral hypodense abscesses, some containing gas (source, diverticulitis).*

Typical

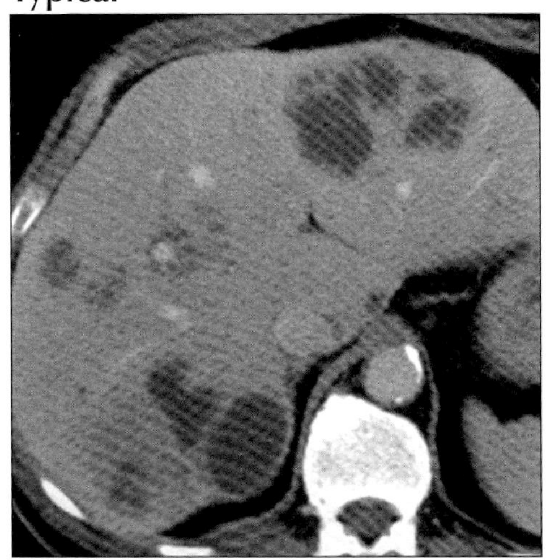

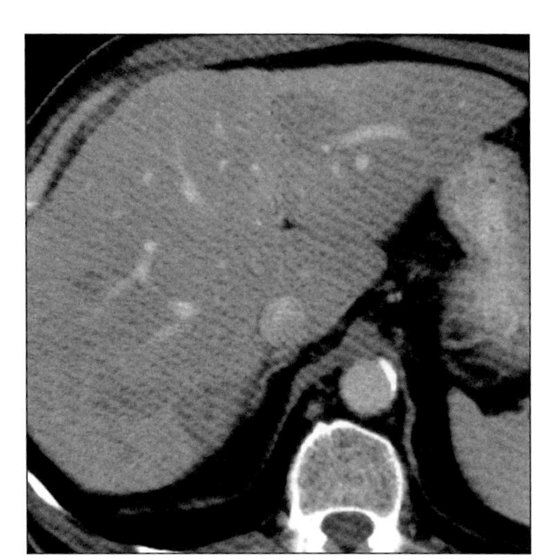

(Left) *Axial CECT shows cluster of abscesses in both lobes of the liver in a patient with prior history of diverticulitis.* ***(Right)*** *Axial CECT shows almost complete resolution of multiple bilobar pyogenic abscesses following antibacterial treatment .*

Variant

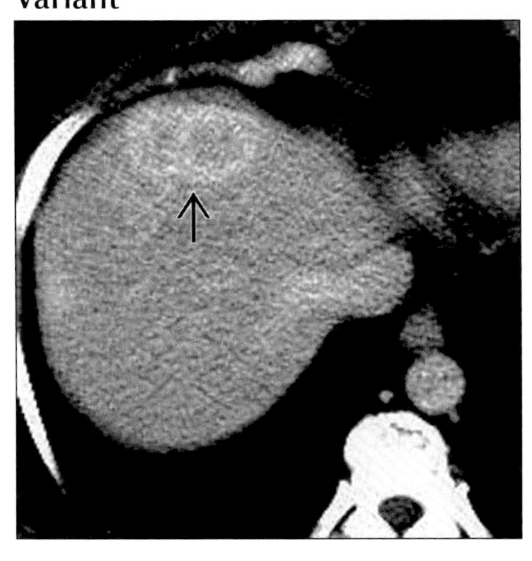

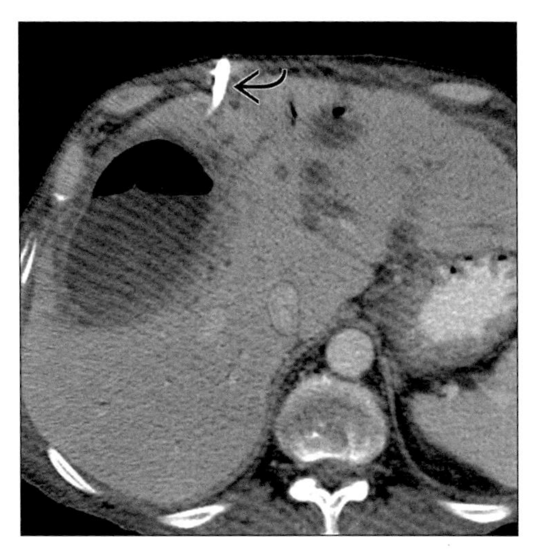

(Left) *Axial CECT shows shows early, non-liquified pyogenic abscess (arrow) in a patient with diverticulitis. Mass resembles a hypervascular tumor.* ***(Right)*** *Axial CECT shows multiple abscesses with gas-fluid levels following Whipple procedure for pancreatic carcinoma. Catheter drainage (curved arrow) was therapeutically effective.*

HEPATIC AMEBIC ABSCESS

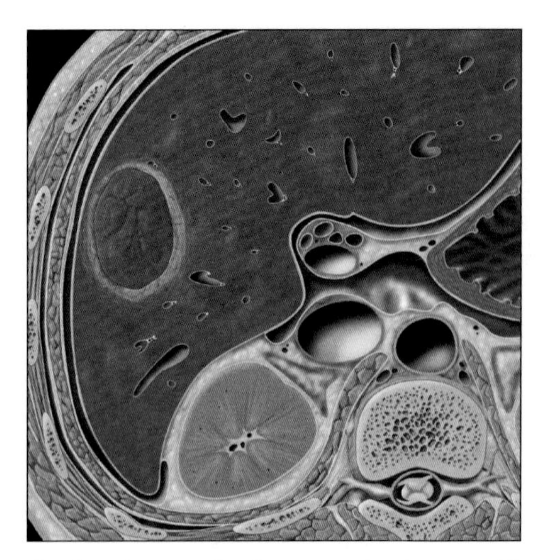

Graphic shows unilocular encapsulated mass with "anchovy paste" contents.

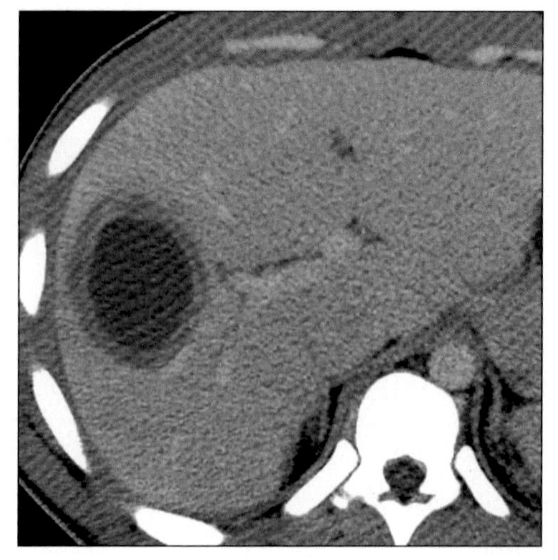

Axial CECT shows homogeneous hypodense nonenhancing mass with thick capsule or wall.

TERMINOLOGY

Definitions

- Localized collection of pus in liver due to entamoeba histolytica with destruction of hepatic parenchyma & stroma

IMAGING FINDINGS

General Features

- Best diagnostic clue: Peripherally located, sharply-defined, round, hypodense mass with enhancing capsule
- Location
  - Right lobe: 72%
  - Left lobe: 13%
  - Usually peripheral
- Size: Varies from few millimeters to several centimeters
- Other General Features
  - Most common extraintestinal manifestation of amebic infestation
  - Most common in developing countries
  - Western nations: High risk groups are
    - Recent immigrants, institutionalized & homosexuals
  - Most often solitary (85%)
  - Primary source of infection
    - Human carriers who pass amebic cysts into stool
  - May become secondarily infected with pyogenic bacteria

Radiographic Findings

- Radiography
  - Elevation of right hemidiaphragm
  - Right lower lobe atelectasis or infiltrate
  - Right pleural effusion
  - Ruptured amebic abscess into chest may show
    - Lung abscess, cavity, hydropneumothorax
    - Pericardial effusion
  - Barium enema often shows changes of amebic colitis

CT Findings

- NECT: Peripheral, round or oval hypodense mass (10-20 HU)
- CECT
  - Lesions may appear unilocular or multilocular
  - May demonstrate nodularity of margins
  - Show rim- or capsule-enhancement
  - Extrahepatic abnormalities
    - Right lower lobe atelectasis
    - Right pleural effusion
    - Usually colonic & rarely gastric changes

MR Findings

- T1WI: Hypointense abscess
- T2WI

DDx: Complex Cystic Mass

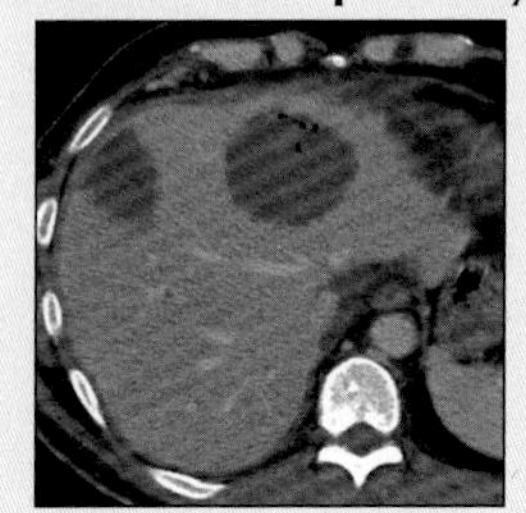

Treated Metastases

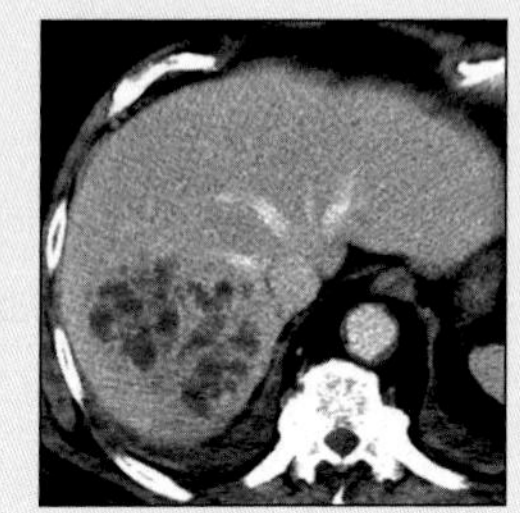

Pyogenic Abscess

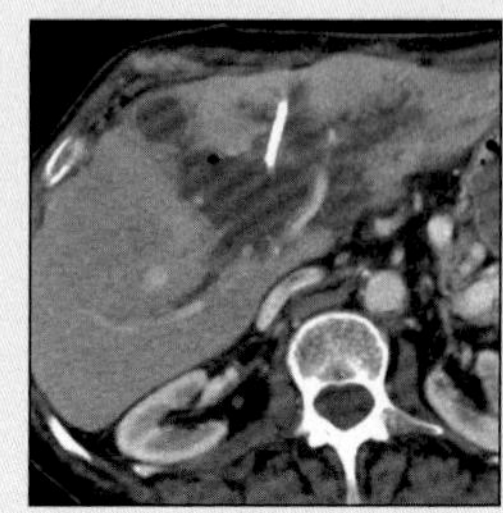

Post Transplant HAT

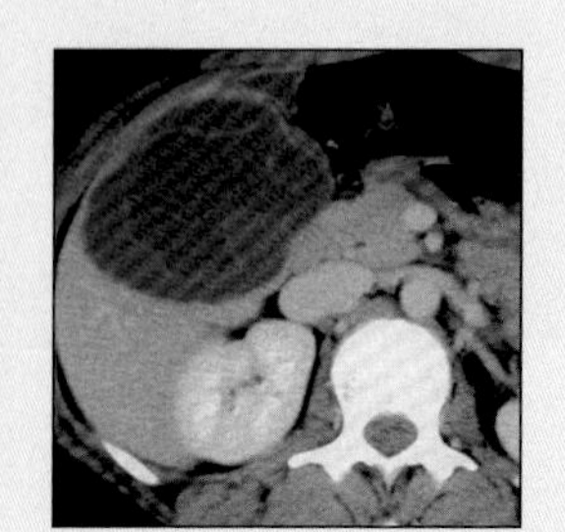

Hydatid Cyst

HEPATIC AMEBIC ABSCESS

Key Facts

Terminology
- Localized collection of pus in liver due to entamoeba histolytica with destruction of hepatic parenchyma & stroma

Imaging Findings
- Best diagnostic clue: Peripherally located, sharply-defined, round, hypodense mass with enhancing capsule
- Most often solitary (85%)
- Right lower lobe atelectasis or infiltrate
- Right pleural effusion

Top Differential Diagnoses
- Treated (cystic or necrotic) metastases
- Hepatic pyogenic abscess
- Infarcted liver after transplantation
- Hepatic hydatid cyst
- Biliary cystadenocarcinoma

Pathology
- Entamoeba histolytica

Clinical Issues
- Clinical profile: Patient with history of diarrhea (mucus), RUQ pain & tender hepatomegaly
- Indirect hemagglutination positive in 90% cases

Diagnostic Checklist
- Rule out other liver pathologies: Pyogenic or fungal abscess & cystic lesions, which may simulate amebic abscess on imaging
- Check for history of transplantation & ablation or chemotherapy for liver tumor or metastasis

---

- Hyperintense abscess
- Perilesional edema: High signal intensity
- T1 C+
  - Abscess contents: No enhancement
  - Rim or capsule: Shows enhancement

Ultrasonographic Findings
- Real Time
  - Usually round or oval, sharply-defined hypoechoic mass
  - Abuts liver capsule with homogeneous echoes & distal enhancement
  - Compared to pyogenic
    - Amebic is more likely to have a round or oval shape (82:60%)
    - Hypoechoic with fine internal echoes (58:36%)

Nuclear Medicine Findings
- Hepatobiliary scan (HIDA)
  - Cold lesion with a hot periphery
- Technetium sulfur colloid
  - Cold defects
- WBC Scan
  - Cold center & hot rim

Imaging Recommendations
- Best imaging tool: CECT
- Protocol advice: Scan to include lung bases through pelvis

DIFFERENTIAL DIAGNOSIS

Treated (cystic or necrotic) metastases
- May be indistinguishable from amebic abscess
- Usually no elevation of diaphragm or atelectasis
- No fever or increased WBC

Hepatic pyogenic abscess
- Simple pyogenic abscess
  - Well-defined round, hypodense mass (0-45 HU)
  - "Cluster" sign
    - Aggregation of small abscesses, sometimes coalesce into a single septated cavity
- Specific sign: Abscess with central gas
  - Air bubbles or an air-fluid level

Infarcted liver after transplantation
- Hepatic artery thrombosis (HAT) causes biliary & hepatic necrosis
- Can look exactly like an abscess with or without gas

Hepatic hydatid cyst
- Large well-defined cystic liver mass
- Numerous peripheral daughter cysts
- May show curvilinear or ring-like pericyst calcification
- Intrahepatic duct dilatation may be seen

Biliary cystadenocarcinoma
- Rare, multiseptated, water density cystic mass
- No surrounding inflammatory changes

PATHOLOGY

General Features
- General path comments
  - Cystic form of E. histolytica gains access to body via contaminated water
  - Mature cysts resistant to gastric acid, pass unchanged into intestine
  - Cyst wall is digested by trypsin & invasive trophozoites are released
  - Trophozoites enter mesenteric venules & lymphatics
  - Usually spread from colon to liver
    - Via portal vein (most common) & lymphatics
  - Rarely direct spread
    - Colonic wall to peritoneum
    - Peritoneum to liver capsule & finally liver
- Etiology
  - Entamoeba histolytica
  - May become secondarily infected with pyogenic bacteria
- Epidemiology: Approximately 10% of world's population is infected with E. histolytica
- Associated abnormalities: Amebic colitis

HEPATIC AMEBIC ABSCESS

Gross Pathologic & Surgical Features

- Usually solitary abscess
- Predominantly in right lobe
- Fluid-dark, reddish-brown
- Consistency of "anchovy paste"

Microscopic Features

- Blood, destroyed hepatocytes
- Necrotic tissue & rarely trophozoites

CLINICAL ISSUES

Presentation

- Most common signs/symptoms
  - RUQ pain, tender hepatomegaly
  - Diarrhea with mucus
- Clinical profile: Patient with history of diarrhea (mucus), RUQ pain & tender hepatomegaly
- Lab data
  - Stool exam: Usually nonspecific or negative
  - Indirect hemagglutination positive in 90% cases

Demographics

- Age
  - More common in 3rd-5th decade
  - Can occur in any age group
- Gender: M:F = 4:1

Natural History & Prognosis

- Complications
  - Pleuropulmonary amebiasis (20-35%)
    - Pulmonary consolidation or abscess
    - Effusion, empyema or hepatobronchial fistula
  - Peritoneal amebiasis (2-7.5%)
  - Pericardial or renal amebiasis
- Prognosis
  - Usually good after amebicidal therapy
  - Poor in individuals who develop complications
  - Mortality rate in US: < 3%
    - < 1% when confined to liver
    - 6% with extension into chest
    - 30% with extension into pericardium

Treatment

- 90% respond to antimicrobial therapy
  - Metronidazole or chloroquine
- 10% require aspiration & drainage

DIAGNOSTIC CHECKLIST

Consider

- Rule out other liver pathologies: Pyogenic or fungal abscess & cystic lesions, which may simulate amebic abscess on imaging
- Check for history of transplantation & ablation or chemotherapy for liver tumor or metastasis

Image Interpretation Pearls

- On CT: Peripheral, round or oval hypodense mass with rim or capsule enhancement
- On US: Abuts liver capsule with homogeneous echoes & distal enhancement

SELECTED REFERENCES

1. Ralls PW: Inflammatory disease of the liver. Clin Liver Dis. 6(1):203-25, 2002
2. Balci NC et al: MR imaging of infective liver lesions. Magn Reson Imaging Clin N Am. 10(1):121-35, vii, 2002
3. Sharma MP et al: Management of amebic and pyogenic liver abscess. Indian J Gastroenterol. 20 Suppl 1:C33-6, 2001
4. Hughes MA et al: Amebic liver abscess. Infect Dis Clin North Am. 14(3):565-82, viii, 2000
5. Natarajan A et al: Ruptured liver abscess with fulminant amoebic colitis: case report with review. Trop Gastroenterol. 21(4):201-3, 2000
6. Das P et al: Molecular mechanisms of pathogenesis in amebiasis. Indian J Gastroenterol. 18(4):161-6, 1999
7. Rajak CL et al: Percutaneous treatment of liver abscesses: needle aspiration versus catheter drainage. AJR Am J Roentgenol. 170(4):1035-9, 1998
8. Ralls PW: Focal inflammatory disease of the liver. Radiol Clin North Am. 36(2):377-89, 1998
9. Kimura K et al: Amebiasis: modern diagnostic imaging with pathological and clinical correlation. Semin Roentgenol. 32(4):250-75, 1997
10. Fujihara T et al: Amebic liver abscess. J Gastroenterol. 31(5):659-63, 1996
11. Takhtani D et al: Intrapericardial rupture of amebic liver abscess managed with percutaneous drainage of liver abscess alone. Am J Gastroenterol. 91(7):1460-2, 1996
12. Giorgio A et al: Pyogenic liver abscesses: 13 years of experience in percutaneous needle aspiration with US guidance. Radiology. 195: 122-124, 1995
13. Mendez RZ et al: Hepatic abscesses: MR imaging findings.Radiology. 190: 431-436, 1994
14. Van Allan RJ et al: Uncomplicated amebic liver abscess: prospective evaluation of percutaneous therapeutic aspiration. Radiology. 183(3):827-30, 1992
15. Gibney EJ: Amoebic liver abscess. Br J Surg. 77(8):843-4, 1990
16. Ken JG et al: Perforated amebic liver abscesses: successful percutaneous treatment Radiology. 170: 195-197, 1989
17. Sarda AK et al: Intraperitoneal rupture of amoebic liver abscess. Br J Surg. 76(2):202-3, 1989
18. Singh JP et al: A comparative evaluation of percutaneous catheter drainage for resistant amebic liver abscesses. Am J Surg. 158(1):58-62, 1989
19. Ken JG et al: Perforated amebic liver abscesses: successful percutaneous treatment. Radiology. 170(1 Pt 1):195-7, 1989
20. Rustgi AK et al: Pyogenic and amebic liver abscess. Med Clin North Am. 73(4):847-58, 1989
21. Frey CF et al: Liver abscesses. Surg Clin North Am. 69(2):259-71, 1989
22. Jeffrey RB et al: CT small pyogenic hepatic abscesses: The cluster sign. AJR. 151(3): 487-9, 1988
23. Greenstein AJ et al: Pyogenic and amebic abscesses of the liver. Semin Liver Dis. 8(3):210-7, 1988
24. G Elizondo et al: Amebic liver abscess: diagnosis and treatment evaluation with MR imaging Radiology. 165: 795-800, 1987
25. Ralls PW et al: Amebic liver abscess: MR imaging Radiology. 165: 801-804, 1987

HEPATIC AMEBIC ABSCESS

IMAGE GALLERY

Typical

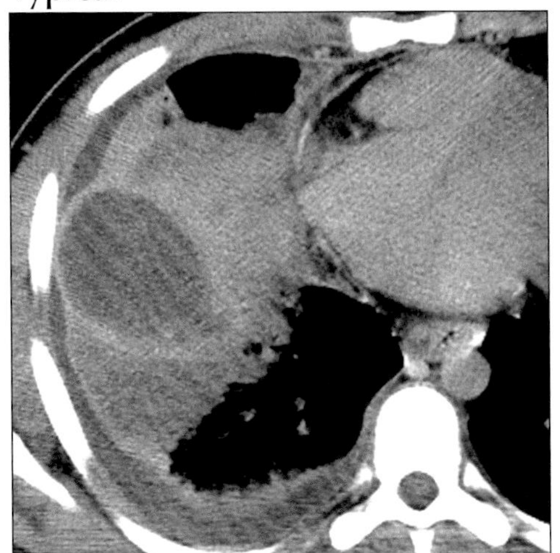

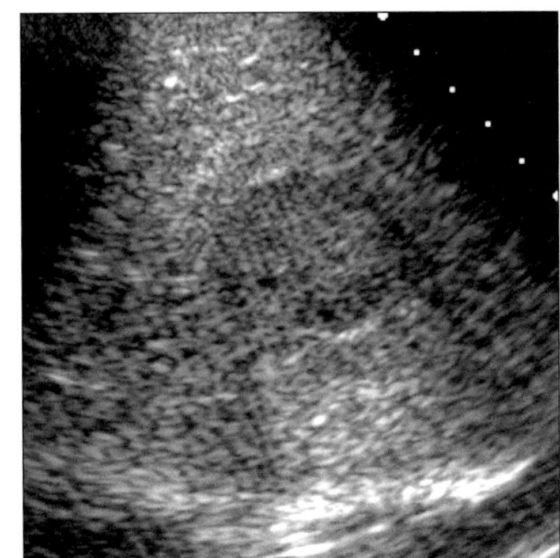

(Left) *Axial CECT shows typical peripheral hypodense mass abutting hepatic capsule. Elevation of hemidiaphragm, atelectasis and pleural effusion.* ***(Right)*** *Sagittal sonogram shows hypoechoic mass with fine internal echoes and posterior acoustic enhancement.*

Variant

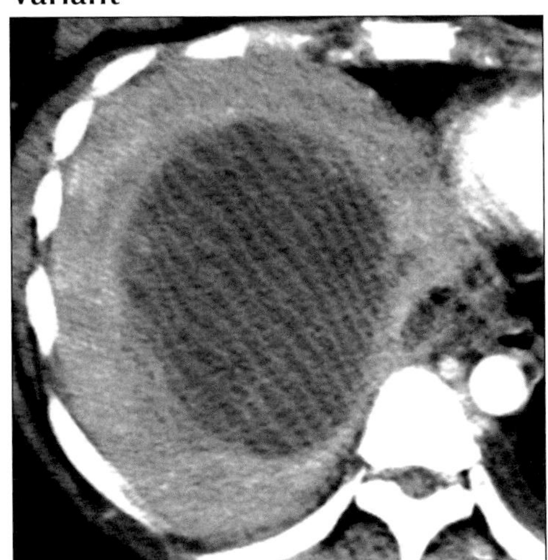

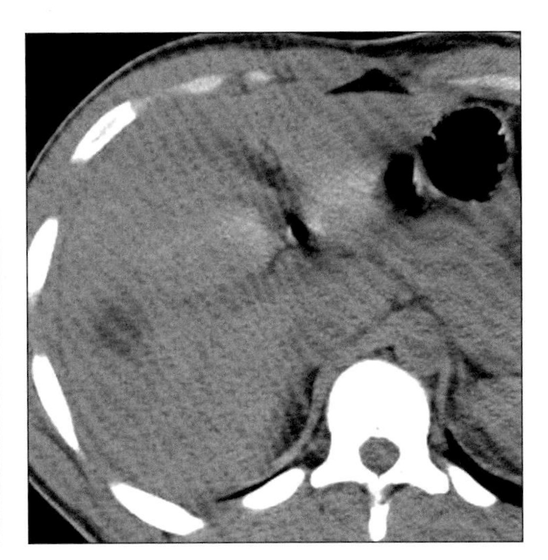

(Left) *Axial CECT in an Asian immigrant shows unusually large amebic abscess of the liver. Note shaggy wall and no prominent septations.* ***(Right)*** *Axial NECT shows unusually small isolated amebic abscess.*

Other

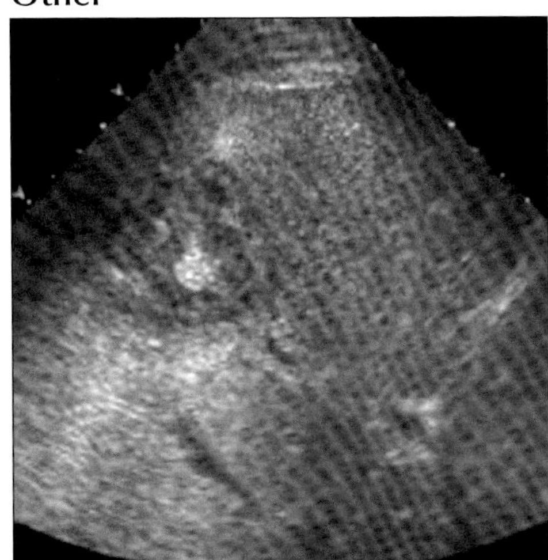

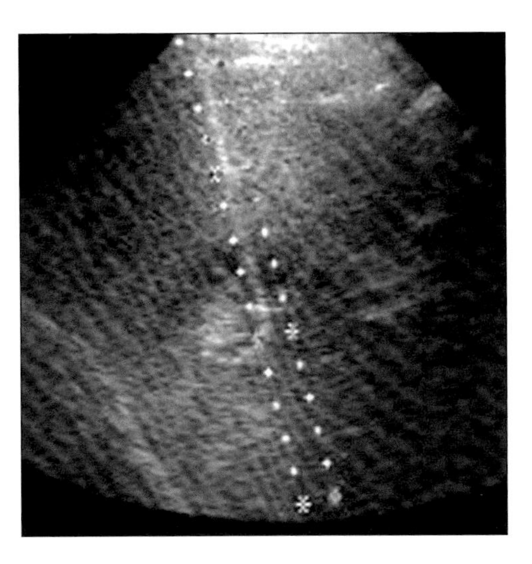

(Left) *Sagittal US shows complex mass with thick capsule and coarse internal echoes in a patient with amebiasis.* ***(Right)*** *Sagittal US guided fine needle aspiration yielded reddish-brown thick fluid. Microbiological results were positive for E. histolytica.*

HEPATIC HYDATID CYST

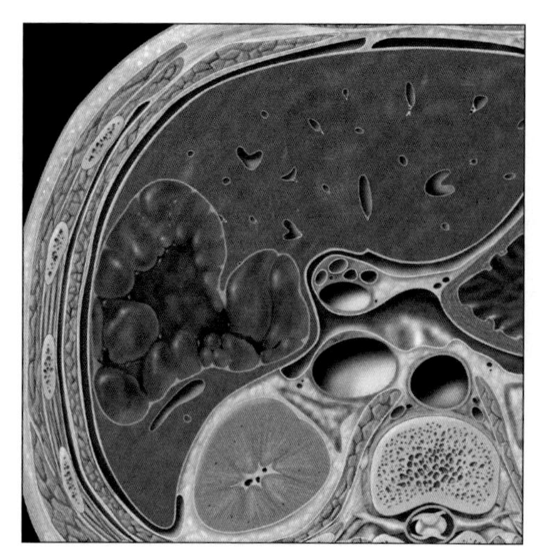

Graphic shows eccentric cystic mass with numerous peripheral daughter cysts.

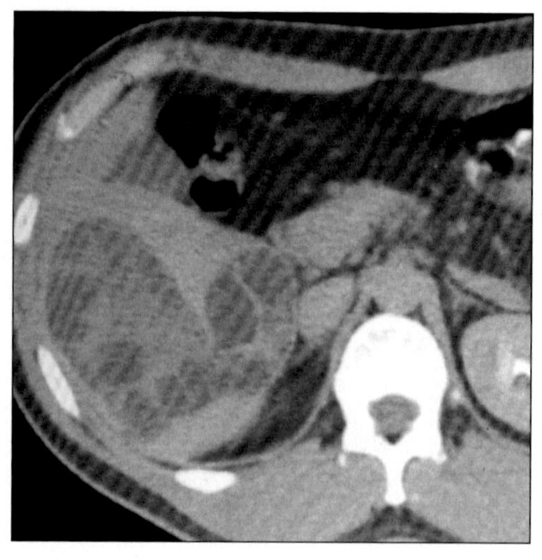

Axial CECT shows oblong hypodense cystic mass with peripheral "daughter" cysts.

TERMINOLOGY

Abbreviations and Synonyms

- Echinococcal or hydatid disease; echinococcosis

Definitions

- Infection of humans caused by larval stage of Echinococcus granulosus or multilocularis

IMAGING FINDINGS

General Features

- Best diagnostic clue: Large well-defined cystic liver mass with numerous peripheral daughter cysts
- Location: Right lobe more than left lobe of liver
- Size
  - Varies
    - Average size: 5 cm
    - Maximum size: Up to 50 cm
    - May contain up to 15 liters of fluid
- Key concepts
  - E. granulosus: Most common form of hydatid disease
    - Up to 60% of cysts are multiple
  - E. multilocularis (alveolaris): Less common but aggressive form
  - Most common sites for hydatid cyst
    - Liver & lungs

Radiographic Findings

- Radiography
  - E. granulosus
    - Curvilinear or ring-like pericyst calcification
    - Seen in 20-30% of abdominal plain films
  - E. multilocularis (alveolaris)
    - Microcalcifications in 50% of cases
- ERCP
  - Hydatid cyst may communicate with biliary tree
    - Right hepatic duct (55%); left hepatic duct (29%)
    - Common hepatic duct (9%)
    - Gallbladder (6%) & common bile duct (1%)

CT Findings

- NECT
  - E. granulosus
    - Large unilocular/multilocular well-defined hypodense cysts
    - Contains multiple peripheral daughter cysts of less density than mother cyst
    - Curvilinear ring-like calcification
    - Calcified wall: Usually indicates no active infection if completely circumferential
    - Dilated intrahepatic bile duct (IHBD): Due to compression/rupture of a cyst into bile ducts
  - Dilated ducts within vicinity of a cyst
  - E. multilocularis (alveolaris)

DDx: Complex or Septated Cystic Hepatic Mass

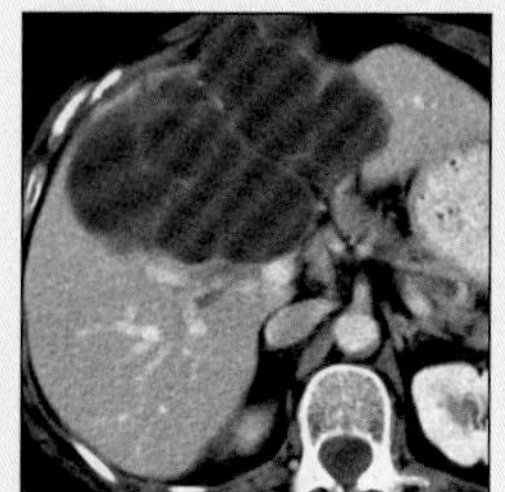

Cystadenocarcinoma

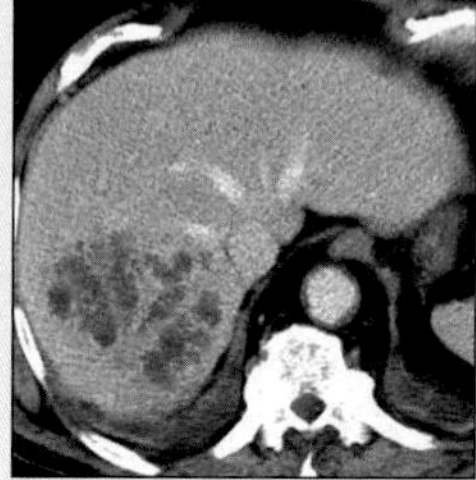

Pyogenic Abscess

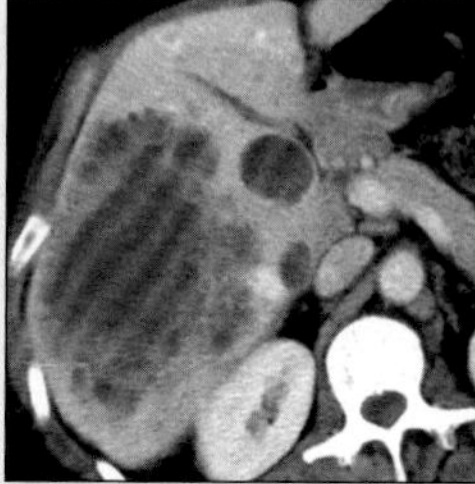

Cystic Metastases

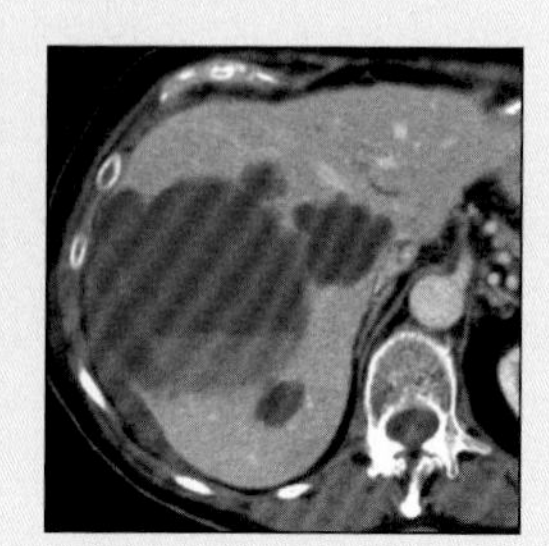

Hemorrhagic Cyst

HEPATIC HYDATID CYST

Key Facts

Terminology

- Echinococcal or hydatid disease; echinococcosis
- Infection of humans caused by larval stage of Echinococcus granulosus or multilocularis

Imaging Findings

- Best diagnostic clue: Large well defined cystic liver mass with numerous peripheral daughter cysts
- Location: Right lobe more than left lobe of liver
- Average size: 5 cm
- Curvilinear or ringlike pericyst calcification

Top Differential Diagnoses

- Biliary cystadenocarcinoma
- Complex pyogenic abscess
- "Cystic" metastases
- Hemorrhagic or infected cyst

Pathology

- Larvae → portal vein → liver (75%)
- Develop into hydatid stage (4-5 days) within liver
- Hydatid cysts grow to 1 cm during first 6 months
- 2-3 cm annually

Clinical Issues

- Cysts: Initially asymptomatic
- Symptomatic when size ↑/infected/ruptured
- Pain, fever, jaundice, hepatomegaly
- Serologic tests positive in more than 80% of cases

Diagnostic Checklist

- Daughter cysts can float freely within mother cyst
- Altering patient's position may change position of daughter cysts
- Confirms diagnosis of echinococcal disease

    - Extensive, infiltrative cystic and solid masses of low density (14-40 HU)
    - Margins are irregular/ill-defined
    - Amorphous type of calcification
    - Can simulate a primary or secondary tumor
- CECT
  - E. granulosus
    - Enhancement of cyst wall and septations
  - E. multilocularis
    - Minimal enhancement of noncalcified portion

MR Findings

- T1WI
  - Rim (pericyst): Hypointense (fibrous component)
  - Mother cyst (hydatid matrix)
    - Usually intermediate signal intensity
    - Rarely hyperintense: Due to reduction in water content
  - Daughter cysts: Less signal intensity than mother cyst (matrix)
  - Floating membrane: Low signal intensity
  - Calcifications: Difficult to identify on MR images
    - Display low signal on both T1 & T2WI
- T2WI
  - Rim (pericyst): Hypointense (fibrous component)
  - First echo T2WI: Increased signal intensity
    - Mother cysts more than daughter cysts
  - Strong T2WI: Hyperintense
    - Mother & daughter cysts have same intensity
  - Floating membrane
    - Low-intermediate signal intensity
- T1 C+
  - E. granulosus
    - Enhancement of cyst wall and septations
  - E. multilocularis
    - Minimal enhancement of noncalcified portion
    - ± Transdiaphragmatic spread to: Pleura, lung, pericardium & heart
- MRCP
  - ± Demonstrate communication with biliary tree

Ultrasonographic Findings

- Real Time
  - Hepatic hydatid cyst manifests in different ways
    - Based on stage of evolution & maturity
  - E. granulosus
    - A well-defined anechoic cyst
    - An anechoic cyst except for hydatid "sand"
    - A multiseptate cyst with daughter cysts & echogenic material between cysts (characteristic)
    - "Water lily" sign: A cyst with a floating, undulating membrane with a detached endocyst
    - A densely calcified mass
  - E. multilocularis
    - Single/multiple echogenic lesions
    - Usually right lobe of liver
    - Irregular necrotic regions & microcalcifications
    - ± Intrahepatic bile duct dilatation
  - US also used to monitor efficacy of
    - Medical antihydatid therapy
  - Positive response findings include
    - Reduction in cyst size
    - Membrane detachment
    - Progressive increase in cyst echogenicity
    - Mural calcification

Imaging Recommendations

- Best imaging tool: Helical NECT + CECT
- Protocol advice
  - Multiplanar imaging show
    - Extrahepatic extension of E. multilocularis

DIFFERENTIAL DIAGNOSIS

Biliary cystadenocarcinoma

- Rare, multiseptated water density cystic mass
- No surrounding inflammatory changes

Complex pyogenic abscess

- "Cluster of grapes": Confluent complex cystic lesions

"Cystic" metastases

- E.g., cystadenocarcinoma of pancreas or ovary
- May present with debris, mural nodularity, rim-enhancement

HEPATIC HYDATID CYST

Hemorrhagic or infected cyst
- Complex cystic heterogeneous mass
- Septations, fluid-levels & mural nodularity
- Calcification may or may not be seen

PATHOLOGY

General Features
- General path comments
  - Definitive host: Dog or fox
  - Intermediate host: Human, sheep or wild rodents
  - Germinal layer (endocyst) → scolices → larval stage
  - Hydatid sand: Free floating brood capsules & scolices form a white sediment
  - Larvae → portal vein → liver (75%)
    - Lungs (15%); other tissues (10%)
  - E. granulosus
    - Develop into hydatid stage (4-5 days) within liver
    - Hydatid cysts grow to 1 cm during first 6 months
    - 2-3 cm annually
  - E. multilocularis
    - Larvae proliferate & penetrate surrounding tissue
    - Cause a diffuse & infiltrative process
    - Simulates a malignancy
    - Induce a granulomatous reaction
    - Necrosis → cavitation → calcification
- Etiology
  - Caused by two types of parasites
    - E. granulosus & E. multilocularis
  - Hydatid disease
    - Caused by larval stage of Echinococcus tapeworm
- Epidemiology
  - E. granulosus: Mediterranean region, Africa, South America, Australia & New Zealand
  - E. multilocularis: France, Germany, Austria, USSR, Japan, Alaska & Canada

Gross Pathologic & Surgical Features
- E. granulosus
  - Large unilocular/multilocular cystic mass
- E. multilocularis or alveolaris
  - Multilocular or irregular solid mass

Microscopic Features
- E. granulosus: Pericyst; ectocyst; endocyst
- E. alveolaris: Lamellated wall/liver necrosis + giant cells + lymphocytes

CLINICAL ISSUES

Presentation
- Most common signs/symptoms
  - Cysts: Initially asymptomatic
    - Symptomatic when size ↑/infected/ruptured
  - Pain, fever, jaundice, hepatomegaly
  - Allergic reaction; portal hypertension
- Clinical profile
  - Middle-aged patient with
    - RUQ pain, palpable mass, jaundice
    - Eosinophilia, urticaria + anaphylaxis
- Lab data
  - Eosinophilia; ↑ serologic titers
  - ± ↑ Alkaline phosphatase
  - ± ↑ Gamma-glutamyl transpeptidase (GGTP)
- Diagnosis
  - Serologic tests positive in more than 80% of cases
  - Percutaneous aspiration of cyst fluid
    - Danger of peritoneal spill & anaphylactic reaction

Demographics
- Age
  - Hydatid disease usually acquired in childhood
  - Not diagnosed until 30-40 years of age
- Gender: M = F

Natural History & Prognosis
- Complications
  - Compression/infection or rupture into biliary tree
  - Rupture into peritoneal or pleural cavity
  - Spread of lesions to lungs, heart, brain & bone
- Prognosis
  - E. granulosus: Good
  - E. alveolaris: Fatal - left untreated within 10-15 years

Treatment
- E. granulosus
  - Medical: Albendazole/mebendazole
  - Direct injection of scolicidal agents
  - Percutaneous aspiration & drainage of cyst
  - Surgical: Segmental or lobar hepatectomy
- E. multilocularis
  - Partial hepatectomy/hepatectomy + liver transplant
- Surgical: For exophytic groth of hydatid cyst

DIAGNOSTIC CHECKLIST

Consider
- Rule out other complex or septated cystic liver masses
  - Biliary cystadenoma, pyogenic liver abscess, cystic metastases & hemorrhagic or infected cyst
  - E. multilocularis imaging and clinical behavior simulates solid malignant neoplasm

Image Interpretation Pearls
- Daughter cysts can float freely within mother cyst
  - Altering patient's position may change position of daughter cysts
  - Confirms diagnosis of echinococcal disease

SELECTED REFERENCES

1. Polat P et al: Hydatid disease from head to toe. Radiographics. 23(2):475-94; quiz 536-7, 2003
2. Mortele KJ et al: Cystic focal liver lesions in the adult: differential CT and MR imaging features. Radiographics. 21(4):895-910, 2001
3. Pedrosa I et al: Hydatid disease: Radiologic and pathologic features and complications. RadioGraphics. 20: 795-817, 2000
4. Taourel P et al: Hydatid cyst of the liver: Comparison of CT and MRI. Journal of Computer Assisted Tomography. 17(1): 80-5, 1993

HEPATIC HYDATID CYST

IMAGE GALLERY

Typical

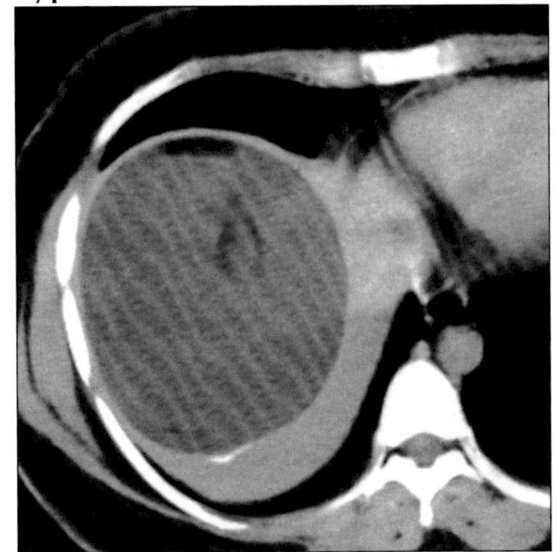

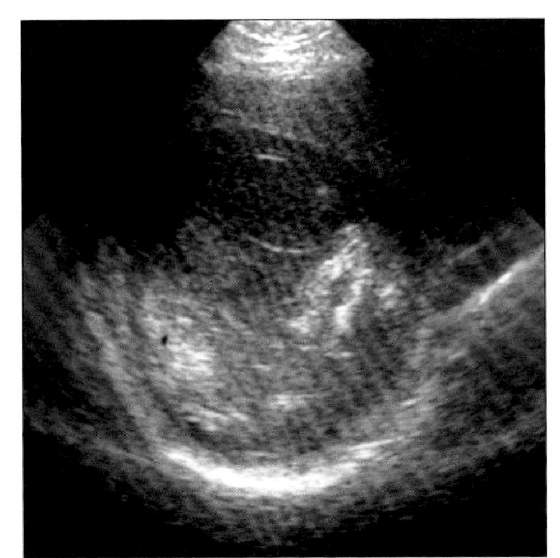

(Left) *Axial CECT shows large cystic mass with partially calcified wall. Note hypodense septa and floating debris (scolices).* ***(Right)*** *Sagittal sonogram shows complex echogenic mass with enhanced transmission.*

Typical

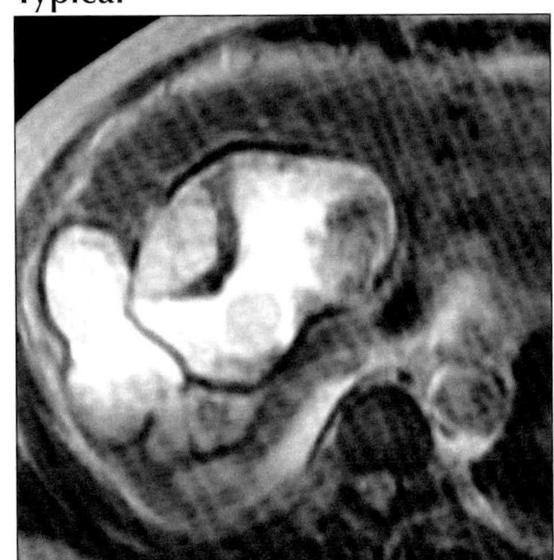

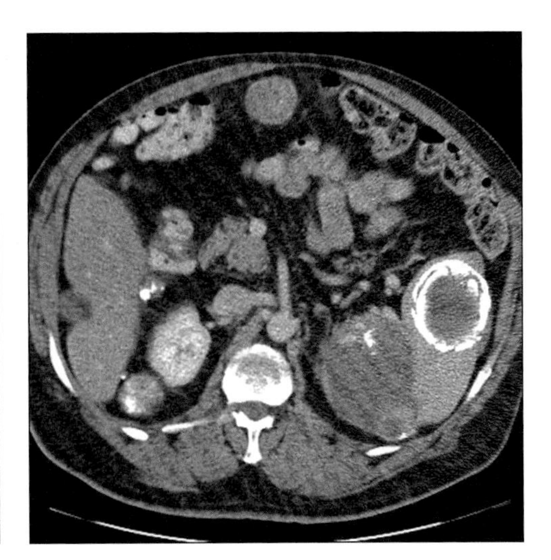

(Left) *Axial T2WI MR shows complex cystic mass with peripheral daughter cysts.* ***(Right)*** *Axial CECT shows disseminated hydatid disease with cystic masses in the spleen, liver and throughout the peritoneal cavity. Note the calcified wall especially in the splenic cystic masses.*

Typical

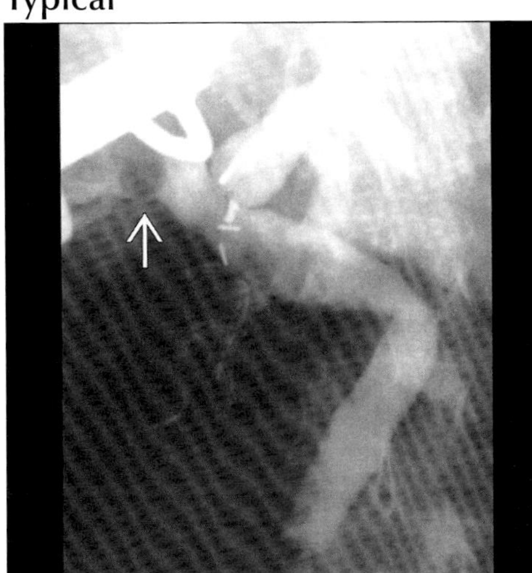

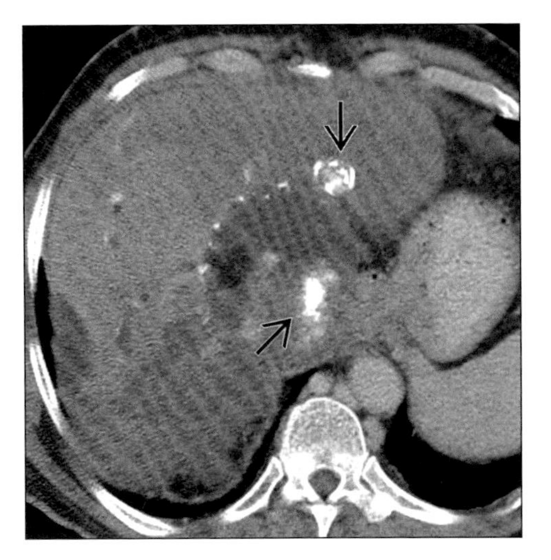

(Left) *Intra-operative cholangiogram shows dilated biliary tree with filling defects (arrow) due to rupture of hydatic cyst into intrahepatic bile ducts.* ***(Right)*** *Axial CECT in a Mediterranean immigrant with E. multilocularis, shows extensive cystic and solid infiltrative mass with ill-defined margins and foci of calcification (arrows).*

STEATOSIS (FATTY LIVER)

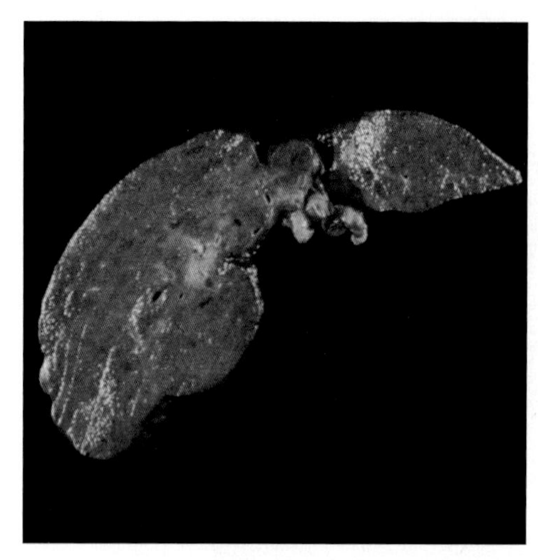

Cut section of explanted liver shows yellowish, greasy, pale appearance due to steatosis.

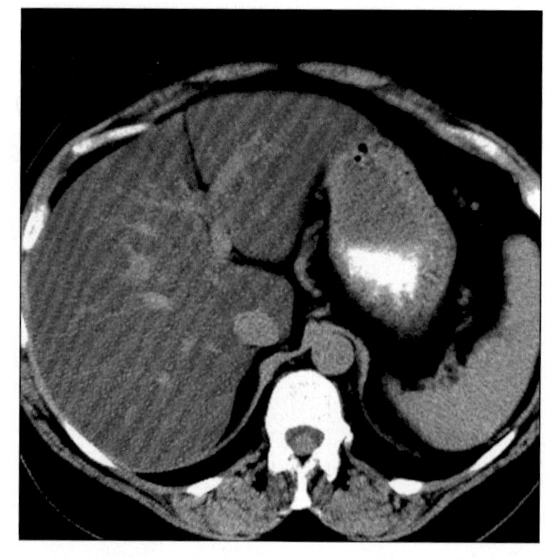

Axial NECT shows diffuse low attenuation of liver. Note relative hyperdensity of vessels and spleen.

TERMINOLOGY

Abbreviations and Synonyms

- Hepatic steatosis or hepatic fatty metamorphosis

Definitions

- Steatosis is a metabolic complication of a variety of toxic, ischemic & infectious insults to liver

IMAGING FINDINGS

General Features

- Best diagnostic clue: Decreased signal intensity of liver on T1W out-of-phase gradient echo images
- Location: Both lobes of liver
- Size: Diffuse fatty infiltration: Enlarged liver
- Key concepts
  - Diffuse (more common) or focal fatty infiltration
  - Often lobar, segmental or wedge shaped
  - Rarely, unifocal or multifocal spherical lesions; simulating metastases or primary tumor
  - Fatty replacement occurs where glycogen is depleted from liver
  - Key on all imaging modalities
    - Presence of normal vessels coursing through "lesion" (fatty infiltration)
  - Imaging features of fatty liver: Variable based on
    - Amount of fat deposited in liver
    - Fat distribution within liver
    - Presence of associated hepatic disease

CT Findings

- NECT
  - Diffuse or focal
    - Decreased attenuation of liver compared to spleen
    - Normal: Liver 8-10 HU more than spleen on NECT
  - Focal nodular fatty infiltration: Low attenuation
    - Common location: Adjacent to falciform ligament
    - Cause: Due to nutritional ischemia
    - Because it is a vascular watershed area
  - Lobar, segmental or wedge-shaped fatty infiltration
    - Decreased attenuation
    - May have a straight-line margin
    - Extending to liver capsule without mass effect
- CECT
  - Detect fatty infiltration due to variations in
    - Degrees of liver & splenic relative enhancement
  - Normal vessels course through "lesion" (fatty infiltration)
  - CECT has lower sensitivity in detecting fatty liver

MR Findings

- T1WI out-of-phase gradient echo image
  - Decrease or loss of signal intensity of fatty liver
- T1WI in-phase gradient echo image
  - Increased signal intensity of fatty liver than spleen

DDx: Diffuse or Geographic Hypodense Liver

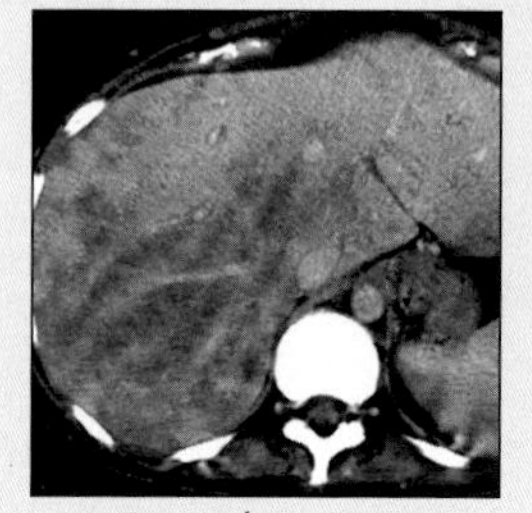

Steatohepatitis

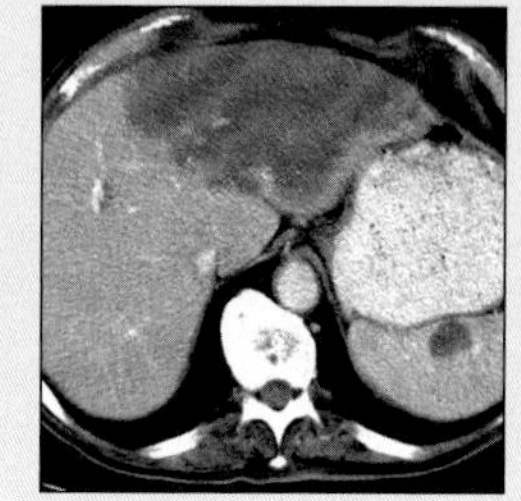

Diffuse Lymphoma

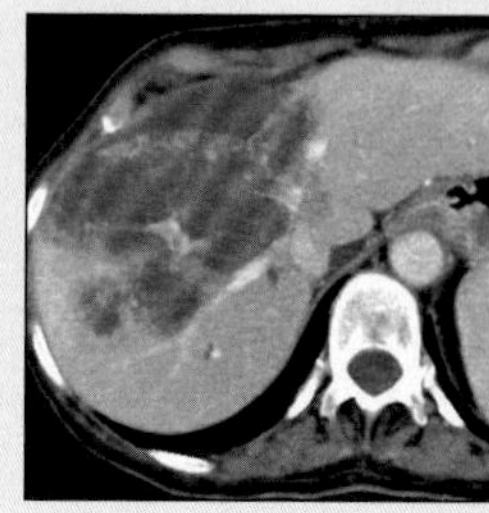

Cholangiocarcinoma

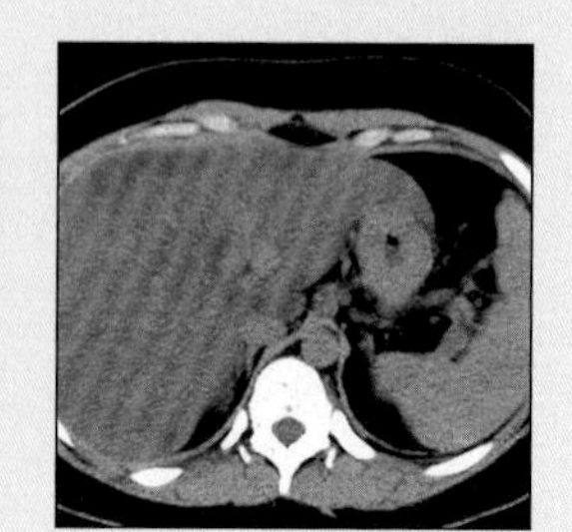

Acute Hepatitis

STEATOSIS (FATTY LIVER)

Key Facts

Terminology

- Steatosis is a metabolic complication of a variety of toxic, ischemic & infectious insults to liver

Imaging Findings

- Best diagnostic clue: Decreased signal intensity of liver on T1W out-of-phase gradient echo images
- Diffuse increased hepatic echogenicity
- Focal fatty sparing: Pseudotumor variations

Top Differential Diagnoses

- Alcoholic steatohepatitis
- Diffuse lymphoma or other tumor

Pathology

- Most frequently seen on liver biopsies of alcoholics
- Seen in up to 50% of patients with diabetes mellitus
- Quite prevalent in general population with obesity

Clinical Issues

- Asymptomatic, but often with abnormal LFTs
- 2/3 alcoholics: RUQ pain, tenderness, hepatomegaly
- Removal of alcohol or offending toxins
- Correction of metabolic disorders

Diagnostic Checklist

- Rule out other liver pathologies which may mimic focal or diffuse steatosis (fatty liver)
- Key on all imaging modalities is presence of normal vessels coursing through "lesion" (fatty infiltration)

- T1 C+ out-of-phase GRE image
  - Paradoxical decreased signal intensity of liver
- STIR (short T1 inversion recovery)
  - Shows fatty areas as low signal intensity
- MR spectroscopy (MRS)
  - Fatty liver demonstrates an increase in intensity of lipid resonance peak
  - Used for quantitative assessment of fatty infiltration of liver

Ultrasonographic Findings

- Real Time
  - Diffuse fatty infiltration
    - Diffuse increased hepatic echogenicity
    - Increased attenuation of ultrasound beam (feature of fat, not fibrosis)
  - Hepatic steatosis & fibrosis frequently coexist
    - Produce similar sonographic findings
  - Poor visualization of portal & hepatic veins
  - Focal fatty infiltration
    - Hyperechoic nodule
    - Multiple confluent hyperechogenic lesions
  - Focal fatty sparing: Pseudotumor variations
    - Target lesion: Hypoechoic area with a central hyperechoic core
    - Ovoid or spherical hypoechoic area in an otherwise echogenic liver
    - Usually seen in segment IV of fatty liver
    - Often borders of gallbladder fossa

Nuclear Medicine Findings

- Technetium Tc 99m sulfur colloid
  - Differentiates true space occupying lesion from focal fat
  - Fat does not displace reticuloendothelial cells
  - Diffuse fatty infiltration
    - Inhomogeneous radionuclide uptake
- Xenon 133
  - Highly fat soluble
  - Accumulation of isotope in fatty areas of liver
    - Specific sign of hepatic steatosis

Imaging Recommendations

- Best imaging tool: NECT or T1W in & out-of-phase gradient echo images

DIFFERENTIAL DIAGNOSIS

Alcoholic steatohepatitis

- Acute phase
  - Hepatomagaly
  - Diffuse hypodense liver (due to fatty infiltration)
  - Fatty infiltration: May be focal, lobar, segmental
- Chronic phase
  - Mixture of steatosis & early cirrhotic changes
  - Liver-spleen attenuation difference more than 10 HU
- Nonalcoholic steatohepatitis (NASH)
  - Indistinguishable from alcoholic hepatitis

Diffuse lymphoma or other tumor

- Diffuse lymphoma infiltration: Indistinguishable from normal liver or steatosis
- Confluent tumor distorts vessels and bile ducts
- Usually secondary deposits are multiple, well-defined, low density masses

Acute hepatitis

- Diffuse hypodensity of the liver
- Associated gallbladder wall and periportal edema
- Clinical presentation suggests diagosis

PATHOLOGY

General Features

- General path comments
  - Fat is deposited in liver due to
  - Ethanol; increased hepatic synthesis of fatty acids
  - Carbon tetrachloride & high dose tetracycline; decreased hepatic oxidation or utilization of fatty acids
  - Starvation, steroids & alcohol
    - Impaired release of hepatic lipoproteins

STEATOSIS (FATTY LIVER)

- Excessive mobilization of fatty acids from adipose tissue
- Segmental areas of fatty infiltration occurs where glycogen is depleted from liver
  - Due to traumatic & ischemic insults
  - Decreased nutrients & insulin → decreased glycogen
  - Causes: Secondary to a mass, Budd-Chiari syndrome or tumor thrombus
- Etiology
  - Metabolic derangement
    - Poorly controlled diabetes mellitus (50%)
    - Obesity & hyperlipidemia
    - Severe hepatitis & protein malnutrition
    - Parenteral hyperalimentation
    - Malabsorption (jejunoileal bypass)
    - Pregnancy, trauma
    - Inflammatory bowel disease
    - Cystic fibrosis, Reye syndrome
  - Hepatotoxins
    - Alcohol (> 50%)
    - Carbon tetrachlorides, phosphorus
  - Drugs
    - Tetracycline, amiodarone, corticosteroids
    - Salicylates, tamoxifen, calcium channel blockers
- Epidemiology
  - Most frequently seen on liver biopsies of alcoholics
  - Seen in up to 50% of patients with diabetes mellitus
  - Quite prevalent in general population with obesity
  - Seen in 25% of nonalcoholics
    - Healthy adult males meeting accidental deaths
- Associated abnormalities
  - Nonalcoholic steatohepatitis (NASH)
    - Seen in patients with hyperlipidemia & diabetes
    - May lead to "cryptogenic" cirrhosis

Gross Pathologic & Surgical Features

- Liver may weigh 4-6 kg
- Soft, yellow, greasy cut surface

Microscopic Features

- Macrovesicular fatty liver (most common type)
  - Hepatocytes with large cytoplasmic fat vacuoles displacing nucleus peripherally
    - Examples: Alcohol & diabetes mellitus
- Microvesicular
  - Fat is present in many small vacuoles
  - Example: Reye syndrome

CLINICAL ISSUES

Presentation

- Most common signs/symptoms
  - Asymptomatic, but often with abnormal LFTs
  - Enlarged liver in obese or diabetic patient
  - Alcoholic patients
    - 1/3 Asymptomatic
    - 2/3 alcoholics: RUQ pain, tenderness, hepatomegaly
- Clinical profile
  - Asymptomatic obese or diabetic patient with enlarged liver
- Lab data
  - Asymptomatic fatty liver
    - Normal to mildly elevated liver function tests
  - Alcoholic patients
    - Abnormal liver function tests
  - Steatohepatitis
    - May have markedly abnormal liver functions
- Diagnosis
  - Biopsy & histology

Natural History & Prognosis

- Complications
  - Acute fatty liver
    - Alcoholic binge, pregnancy, CCL4 exposure
    - Present with jaundice, acute hepatic failure & encephalopathy
- Prognosis
  - Alcoholics: Gradual disappearance of fat from liver after 4-8 weeks of adequate diet & abstinence from alcohol
  - Resolves in 2 weeks after discontinuation of parenteral hyperalimentation
  - Steatohepatitis
    - May progress to acute or chronic liver failure

Treatment

- Removal of alcohol or offending toxins
- Correction of metabolic disorders
- Lipotropic agents like choline when indicated
  - Patient must avoid alcohol & control diabetes

DIAGNOSTIC CHECKLIST

Consider

- Rule out other liver pathologies which may mimic focal or diffuse steatosis (fatty liver)

Image Interpretation Pearls

- Decreased attenuation of liver compared to spleen
- Key on all imaging modalities is presence of normal vessels coursing through "lesion" (fatty infiltration)

SELECTED REFERENCES

1. Rubaltelli L et al: Target appearance of pseudotumors in segment IV of the liver on sonography. AJR. 178: 75-7, 2002
2. Kemper J et al: CT and MRI findings of multifocal hepatic steatosis mimicking malignancy. Abdom Imaging. 27(6):708-10, 2002
3. Outwater EK et al: Detection of lipid in abdominal tissues with opposed-phase gradient-echo images at 1.5 T: Techniques and diagnostic importance. RadioGraphics. 18: 1465-80, 1998
4. Thu HD et al: Value of MR imaging in evaluating focal fatty infiltration of the liver: preliminary study. Radiographics. 11(6):1003-12, 1991

IMAGE GALLERY

Typical

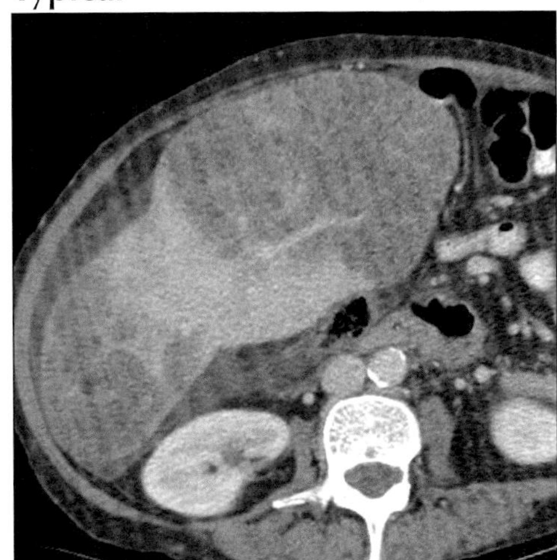

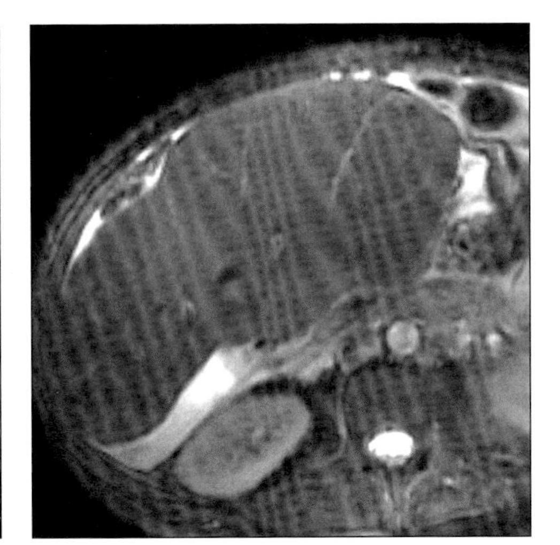

*(**Left**) Axial CECT with focal steatosis shows large hypodense "masses" within liver. Note vessels traversing "masses". (**Right**) Axial T2WI MR in patient with focal steatosis shows no apparent mass with normal branching of intrahepatic vessels.*

Typical

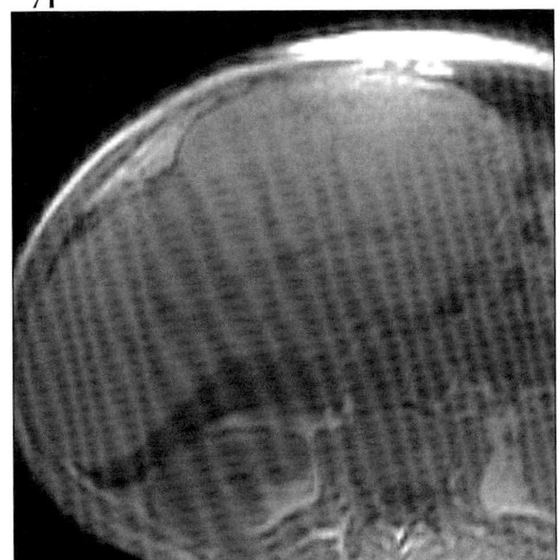

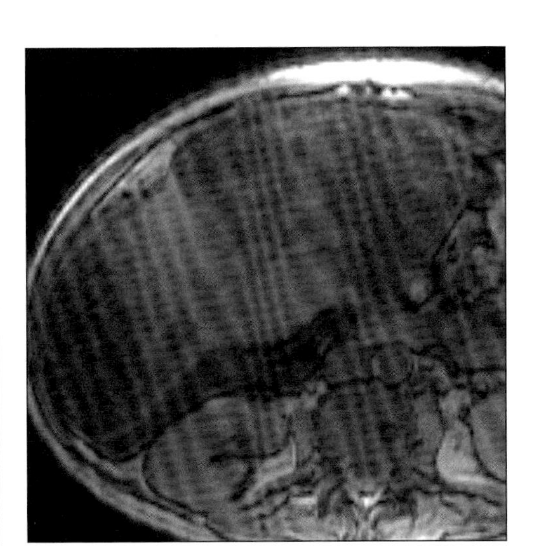

*(**Left**) Axial T1WI GRE MR in-phase shows no clear "mass" appearance in areas of steatosis. (**Right**) Axial T1WI GRE MR out-of-phase image shows striking signal loss from areas of hepatic fatty infiltration.*

Typical

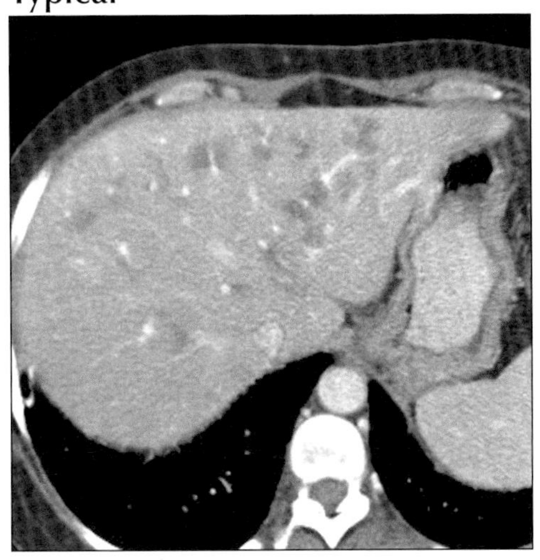

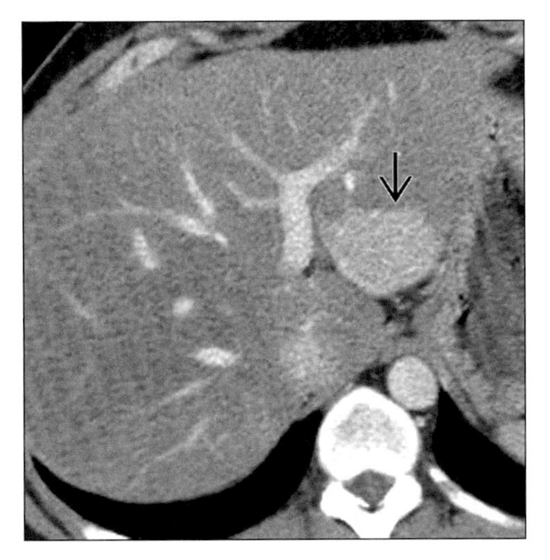

*(**Left**) Axial CECT shows multiple focal low density lesions mimicking metastases. Normal appearing blood vessels withing "lesions" are clue to multifocal fatty infiltration. (**Right**) Axial CECT shows diffuse fatty infiltration of the liver (decreased attenuation). "Hyperdense mass" along dorsal surface of left lobe (arrow) is normal liver (focal sparing).*

CIRRHOSIS

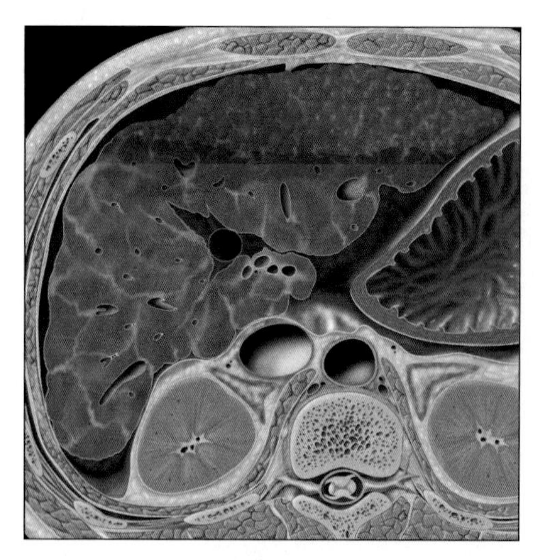

Graphic illustrates nodular surface of liver, fibrosis, relative enlargement of caudate lobe and lateral segment.

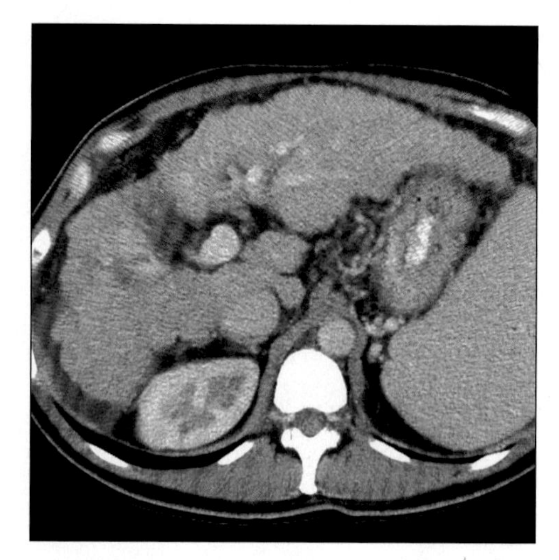

Axial CECT shows nodular surface of scarred liver with fibrotic, small right lobe and enlarged caudate lobe and lateral segment. Splenomegaly, varices and ascites also noted.

TERMINOLOGY

Definitions

- Chronic liver disease characterized by diffuse parenchymal necrosis with extensive fibrosis & regenerative nodule formation

IMAGING FINDINGS

General Features

- Best diagnostic clue: Nodular contour, widened fissures & hyperdense nodules on NECT that disappear on CECT (cirrhosis with siderotic nodules)
- Location: Diffuse liver involving both lobes
- Size: Liver usually reduced in size
- Key concepts
  - Common end response of liver to a variety of insults and injuries
  - Classification of cirrhosis based on morphology, histopathology & etiology
  - Classification
    - Micronodular (Laennec) cirrhosis: Alcoholism (60-70% cases in U.S.)
    - Macronodular (postnecrotic) cirrhosis: Viral hepatitis (10% in U.S.; majority of cases worldwide)
    - Mixed cirrhosis
  - Alcohol abuse is most common cause in West
  - One of 10 leading causes of death in Western world
  - 6th leading cause of death in U.S.

CT Findings

- Nodular liver contour
- Atrophy of right lobe & medial segment of left lobe
- Enlarged caudate lobe & lateral segment of left lobe
- Widened fissures between segments/lobes
- Regenerative nodules; fibrotic & fatty changes
- Varices, ascites, splenomegaly & peribiliary cysts
- Siderotic regenerative nodules
  - NECT: Increased attenuation due to iron content
  - CECT: Nodules disappear after contrast
    - Nodules & parenchyma enhance to same level
- Dysplastic regenerative nodules
  - NECT
    - Large nodules: Hyperdense (↑ iron + ↑ glycogen)
    - Small nodules: Isodense with liver (undetected)
  - CECT
    - Usually enhance as normal liver
    - Sometimes hypervascular
- Fibrotic & fatty changes
  - NECT
    - Fibrosis: Diffuse lacework, thick bands & mottled areas of decreased density
    - Fatty changes: Mottled areas of low attenuation
  - CECT

DDx: Nodular Dysmorphic Liver

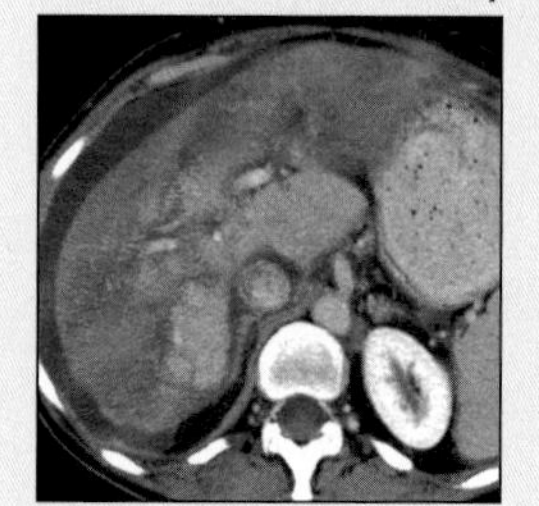

Budd-Chiari Syndrome

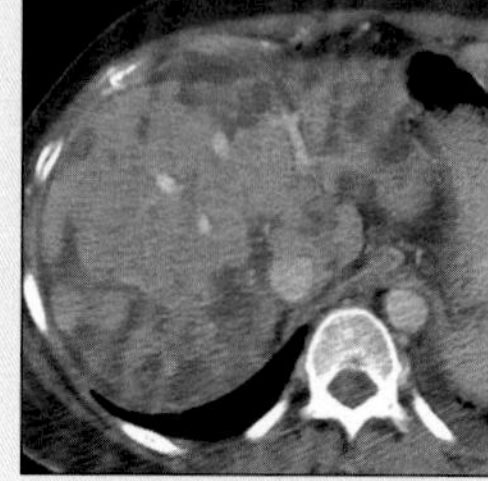

Treated Metastases

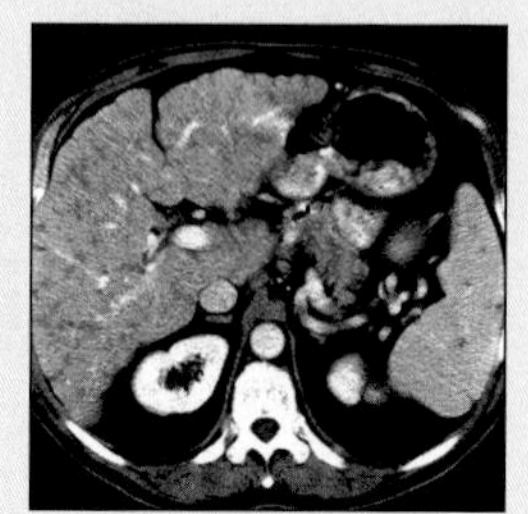

Sarcoidosis

CIRRHOSIS

Key Facts

Imaging Findings
- Best diagnostic clue: Nodular contour, widened fissures & hyperdense nodules on NECT that disappear on CECT (cirrhosis with siderotic nodules)
- Atrophy of right lobe & medial segment of left lobe
- Enlarged caudate lobe & lateral segment of left lobe
- Regenerative nodules; fibrotic & fatty changes
- Varices, ascites, splenomegaly & peribiliary cysts
- Gamna-Gandy bodies (siderotic nodules in spleen)

Top Differential Diagnoses
- Budd-Chiari syndrome
- Treated metastatic disease
- Hepatic sarcoidosis

Pathology
- Micronodular (Laennec) cirrhosis: Alcohol
- Macronodular (postnecrotic) cirrhosis: Viral hepatitis
- Mixed cirrhosis
- Steatosis → hepatitis → cirrhosis
- Alcohol (60-70%), chronic viral hepatitis B/C (10%)
- 3rd leading cause of death for men 34-54 years
- U.S: Hepatitis C (cirrhosis) causes 30-50% of HCC
- Japan: Hepatitis C (cirrhosis) 70% of HCC cases

Clinical Issues
- Splenomegaly, varices, caput medusae
- Fatigue, jaundice, ascites, encephalopathy
- Gynecomastia & testicular atrophy in males
- Virilization in females
- Advanced stage: Liver transplantation

Diagnostic Checklist
- Rule out other causes of "nodular dysmorphic liver"

    - Confluent fibrosis: May show delayed persistent enhancement
    - Fatty changes: Areas of low attenuation
- Cirrhosis-induced hepatocellular carcinoma (HCC)
  - NECT: Hypodense or heterogeneous; ± fat
  - CECT
    - Intense or heterogeneous enhancement on arterial phase; usually iso- to hypodense on venous and delayed phase scans
    - ± Capsule enhancement

MR Findings
- Siderotic regenerative nodules: Paramagnetic effect of iron within nodules
  - T1WI: Hypointense
  - T2WI: Increased conspicuity of low signal intensity
  - T2 Gradient-echo & fast low-angle shot (FLASH) images
    - Markedly hypointense
  - Gamna-Gandy bodies (siderotic nodules in spleen)
    - Seen in cirrhotic patients with portal hypertension
    - Caused by hemorrhage into splenic follicles
    - Composed of fibrous tissue encrusted with hemosiderin & calcium
    - T1 & T2WI: Hypointense
    - T2 GRE & FLASH images: Markedly hypointense
- Dysplastic regenerative nodules
  - T1WI: Hyperintense compared to liver parenchyma
  - T2WI: Hypointense relative to liver parenchyma
- HCC nodule
  - T1WI: Isointense or hypointense
  - T2WI: Hyperintense
  - T1C+: Increased enhancement
- Fibrotic & fatty changes
  - T1WI
    - Fibrosis: Hypointense
    - Fat: Hyperintense
  - T2WI
    - Fibrosis: Hyperintense
    - Fat: Hypointense
- MR angiography
  - Varices: Tortuous structures of high signal intensity
  - Major collateral channels in portal hypertension
    - Transhepatic, gastroesophageal, paraesophageal
    - Paraumbilical, intrahepatic, splenorenal

Ultrasonographic Findings
- Real Time
  - Increased liver echogenicity/loss of normal triphasic hepatic vein Doppler tracing/increased pulsatility of portal vein Doppler tracing and same as CT findings
  - Nodular liver contour
  - Increased liver echogenicity
  - Enlarged caudate lobe & lateral segment of left lobe
  - Atrophy of right lobe & medial segment of left lobe
  - Regenerating nodules
  - Features of portal hypertension (PHT)
    - Portal vein (> 13 mm), splenic vein (> 11 mm)
    - Superior mesenteric vein (> 12 mm)
    - Coronary veins (> 7 mm)
    - Increased pulsatility of portal vein Doppler tracing
    - Dilated hepatic & splenic arteries with increased flow
    - Portal cavernoma, ascites, splenomegaly & varices
    - Siderotic nodules
- Color Doppler
  - Used to determine portal vein patency & direction of flow
  - When portal venous flow pattern is hepatofugal
    - Patient is not a candidate for splenorenal shunt
    - Must undergo a total shunt (portacaval or mesocaval)
  - To guide shunt procedures & to assess blood flow
    - Transjugular intrahepatic portosystemic shunt

Imaging Recommendations
- Best imaging tool: Helical NECT & CECT

DIFFERENTIAL DIAGNOSIS

Budd-Chiari syndrome
- Liver damaged, but no bridging fibrosis
- Occluded or narrowed IVC ± hepatic veins
- Chronic phase: "Large regenerative nodules"
- Central hypertrophy, peripheral atrophy

CIRRHOSIS

- Chronic phase: "Large regenerative nodules"
- Central hypertrophy, peripheral atrophy

Treated metastatic disease

- Example: Breast cancer metastases to liver
  - May shrink and fibrose with treatment
  - Simulating nodular contour of cirrhotic liver

Hepatic sarcoidosis

- Systemic noncaseating granulomatous disorder
- Hypoattenuating nodules (size: Up to 2 cm)
- Hypointense nodules on T1 & T2WI MR

PATHOLOGY

General Features

- General path comments
  - Micronodular (Laennec) cirrhosis: Alcohol
  - Macronodular (postnecrotic) cirrhosis: Viral hepatitis
  - Mixed cirrhosis
  - Catalase oxidation of ethanol → damage cellular membranes & proteins
  - Cellular antigens → inflammatory cells → immune mediated cell damage
  - Steatosis → hepatitis → cirrhosis
  - Regenerative (especially siderotic) nodules → dysplastic nodules → HCC
    - Dysplastic nodules considered premalignant
- Etiology
  - Alcohol (60-70%), chronic viral hepatitis B/C (10%)
  - Primary biliary cirrhosis (5%)
  - Hemochromatosis (5%)
  - Primary sclerosing cholangitis, drugs, cardiac causes
  - Malnutrition, hereditary (Wilson), cryptogenic
  - In children: Biliary atresia, hepatitis, α-1 antitrypsin deficiency
- Epidemiology
  - 3rd leading cause of death for men 34-54 years
  - Risk of HCC
    - U.S.: Hepatitis C (cirrhosis) causes 30-50% of HCC
    - Japan: Hepatitis C (cirrhosis) 70% of HCC cases
    - 2.5 times higher in cirrhotic hepatitis B positive
    - Alcohol & primary biliary cirrhosis: 2-5 fold ↑ risk
  - Mortality due to complication
    - Ascites (50%)
    - Variceal bleeding (25%)
    - Renal failure (10%)
    - Bacterial peritonitis (5%)
    - Complications of ascites therapy (10%)

Gross Pathologic & Surgical Features

- Alcoholic cirrhosis
  - Early stage: Large, yellow, fatty, micronodular liver
  - Late stage: Shrunken, brown-yellow, hard organ with macronodules
- Postnecrotic cirrhosis
  - Macronodular (> 3 mm-1 cm); fibrous scars

Microscopic Features

- Portal-central, portal-portal fibrous bands
- Micro & macronodules; mononuclear cells
- Abnormal arteriovenous interconnections

CLINICAL ISSUES

Presentation

- Most common signs/symptoms
  - Alcoholic cirrhosis
    - May be clinically silent
    - 10-40% cases found at autopsy
  - Nodular liver, anorexia, malnutrition, weight loss
  - Signs of portal hypertension
    - Splenomegaly, varices, caput medusae
  - Fatigue, jaundice, ascites, encephalopathy
  - Gynecomastia & testicular atrophy in males
  - Virilization in females
- Clinical profile: Patient with history of alcoholism, nodular liver, jaundice, ascites & splenomegaly
- Lab data: Increase in liver function tests; anemia
  - Alcoholic cirrhosis: Severe increase in AST (SGOT)
  - Viral: Severe increase in ALT (SGPT)

Demographics

- Age: Middle & elderly age group
- Gender: Males more than females

Natural History & Prognosis

- Complications
  - Ascites, variceal hemorrhage, renal failure, coma
  - HCC: Due to hepatitis B, C & alcoholism
- Prognosis
  - Alcoholic cirrhosis: 5 year survival in less than 50%
  - Advanced disease: Poor prognosis
  - Liver transplantation: Increases survival period

Treatment

- Alcoholic cirrhosis
  - Abstinence; decrease protein diet; multivitamins
  - Prednisone; diuretics (for ascites)
- Management limited to
  - Treatment of complications & underlying cause
- Advanced stage: Liver transplantation

DIAGNOSTIC CHECKLIST

Consider

- Rule out other causes of "nodular dysmorphic liver"

Image Interpretation Pearls

- Nodular liver contour; lobar atrophy & hypertrophy
- Regenerative nodules, ascites, splenomegaly, varices

SELECTED REFERENCES

1. Krinsky GA et al: Hepatocellular carcinoma and dysplastic nodules in patients with cirrhosis: Prospective diagnosis with MR imaging & explantation correlation. Radiology. 219:445-54, 2001
2. Lim JH et al: Detection of hepatocellular carcinomas and dysplastic nodules in cirrhotic livers. AJR. 175:693-8, 2000
3. Dodd GD 3rd et al: End-stage primary sclerosing cholangitis: CT findings of hepatic morphology in 36 patients. Radiology. 211(2):357-62, 1999
4. Dodd GD et al: Spectrum of imaging findings of the liver in end-stage cirrhosis: Part I, gross morphology and diffuse abnormalities. AJR. 173:1031-1036, 1999

CIRRHOSIS

IMAGE GALLERY

Typical

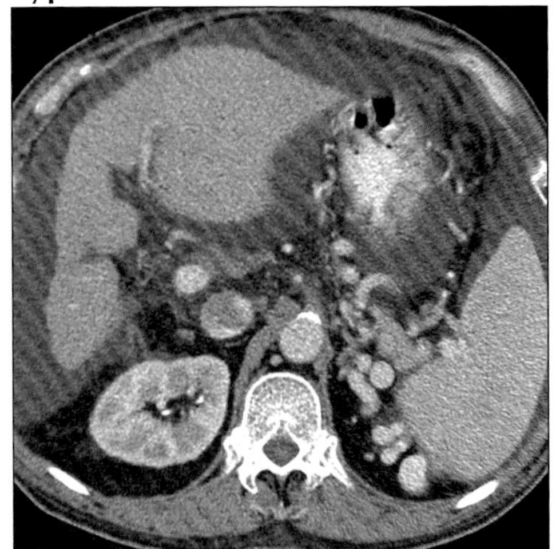

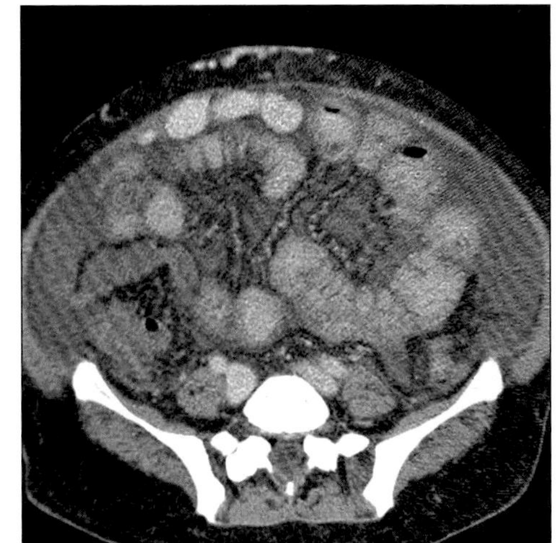

*(**Left**) Axial CECT shows shrunken dysmorphic liver, ascites, large varices. (**Right**) Axial CECT shows large periumbilical varices ("caput medusae"), ascites, mesenteric edema; all manifestations of portal venous hypertension.*

Typical

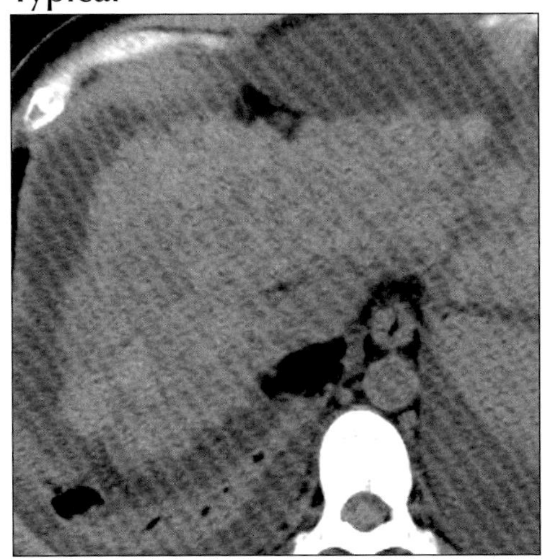

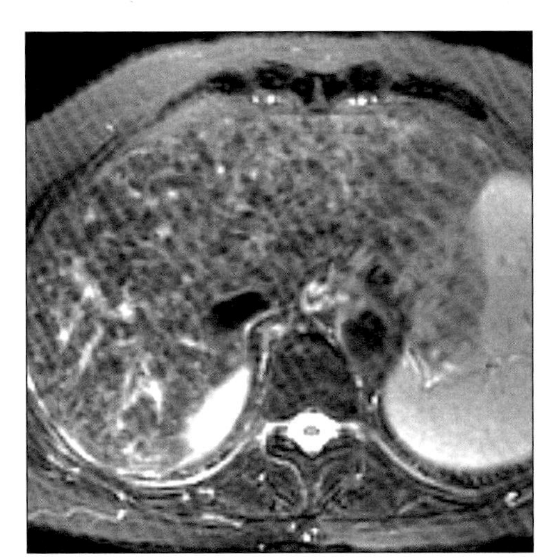

*(**Left**) Axial NECT shows dysmorphic liver and ascites, along with dozens of hyperdense nodules, 0.5-2 cm diameter, representing siderotic regenerative nodules. (**Right**) Axial T2WI MR shows innumerable subcentimeter hypointense lesions throughout a cirrhotic liver, representing siderotic nodules.*

Typical

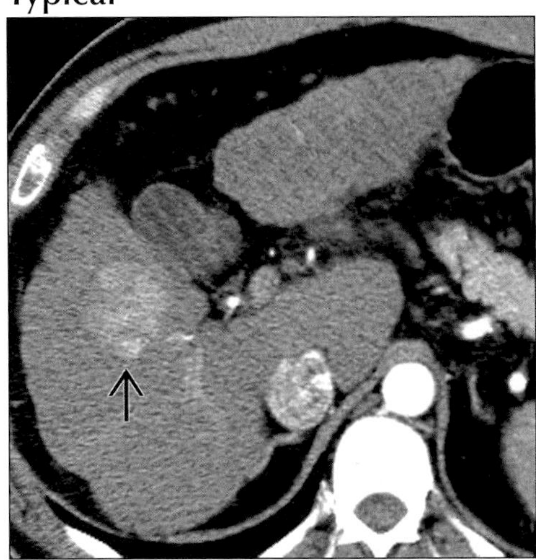

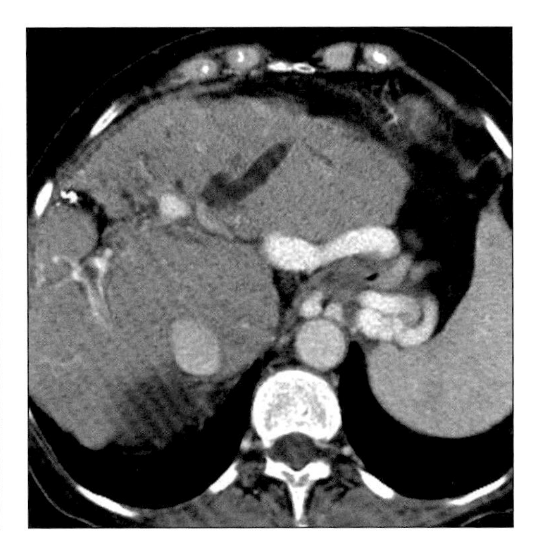

*(**Left**) Axial CECT in arterial phase shows dysmorphic liver with widened fissures. Heterogeneous hypervascular lesion (arrow) is hepatocellular carcinoma (HCC). (**Right**) CECT shows dysmorphic liver with right lobe and medial segment atrophy, hypertrophy of caudate, and irregular dilatation of intrahepatic bile ducts. Cirrhosis due to primary sclerosing cholangitis.*

FOCAL CONFLUENT FIBROSIS

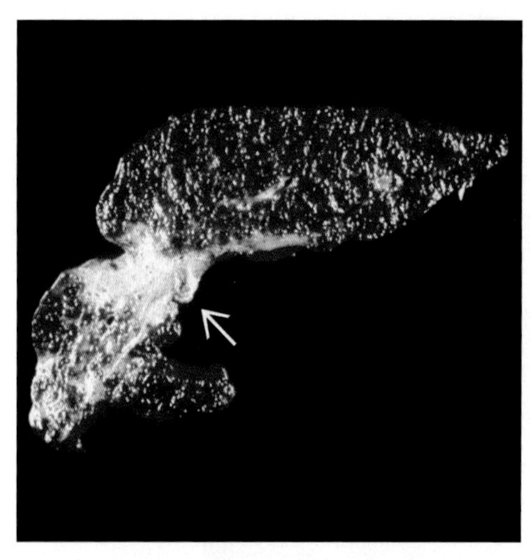

Cut section of liver shows nodular, cirrhotic morphology and area of confluent fibrosis, the pale yellow tissue (arrow) with overlying capsular retraction.

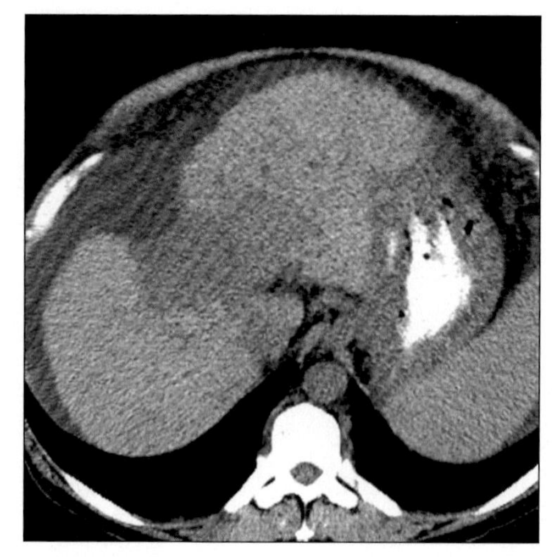

Axial NECT shows cirrhotic morphology, ascites, and hypodense lesion (confluent fibrosis) with overlying capsular retraction.

TERMINOLOGY

Abbreviations and Synonyms

- Confluent hepatic fibrosis (CHF)

Definitions

- Mass-like fibrosis in advanced cirrhosis

IMAGING FINDINGS

General Features

- Best diagnostic clue: Pre-contrast CT showing hypoattenuating lesion with volume loss that becomes isoattenuating or minimally hypoattenuating at post-contrast CT, especially if wedge-shaped, located in medial segment of left lobe &/or anterior segment of right lobe, in patients with advanced cirrhosis
- Location
  - Wedge-shaped lesions radiate from porta hepatis & extend to hepatic capsule
  - 90% of wedge-shaped fibrosis involve medial segment of left lobe &/or anterior segment of right lobe, with sparing of caudate lobe
  - Peripheral lesions are remote from porta hepatis
  - Lobar or segmental involvement, most commonly in lateral segment of left lobe
- Size: May range from 2 x 1.5 cm to 15 x 6 cm
- Key concepts
  - Wedge-shaped; peripheral are band-shaped
  - Total lobar or segmental fibrosis

CT Findings

- NECT
  - Wedge-shaped area of lesser attenuation than adjacent liver parenchyma
    - Retraction of overlying liver capsule (90%)
  - Peripheral band-like hypoattenuating lesion
  - Total lobar or segmental involvement
    - Seen as areas of low attenuation involving entire segment or lobe, with marked shrinkage
  - In advanced cirrhosis, there may be no apparent medial segment of left lobe or anterior segment of right lobe, producing bizarre contour of liver at CT
- CECT
  - Lesions are isoattenuating to adjacent liver parenchyma post-contrast (80%)
    - May appear minimally or substantially hypoattenuating post-contrast
    - Or may be of higher attenuation compared to surrounding parenchyma (delayed scans)
  - Mechanism of variability in contrast enhancement of confluent fibrosis relates to relative vascularity & extent of fibrosis
    - May show delayed persistent enhancement like other fibrotic liver lesions

DDx: Focal Hepatic Lesions with Capsular Retraction

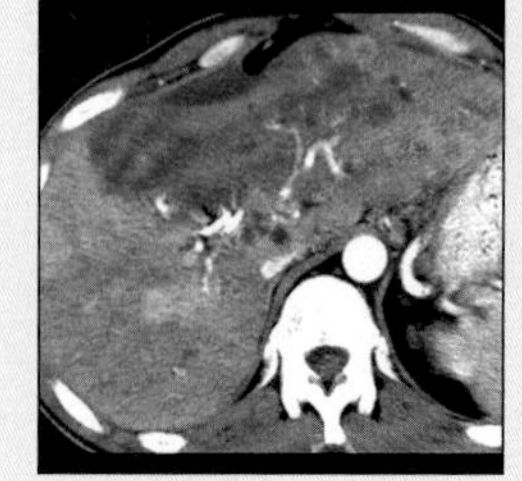

Cholangiocarcinoma

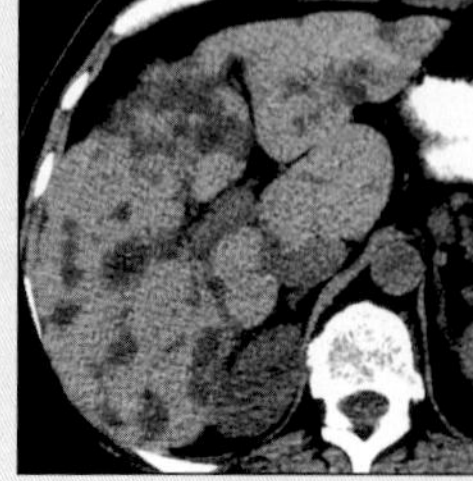

Treated Metastases

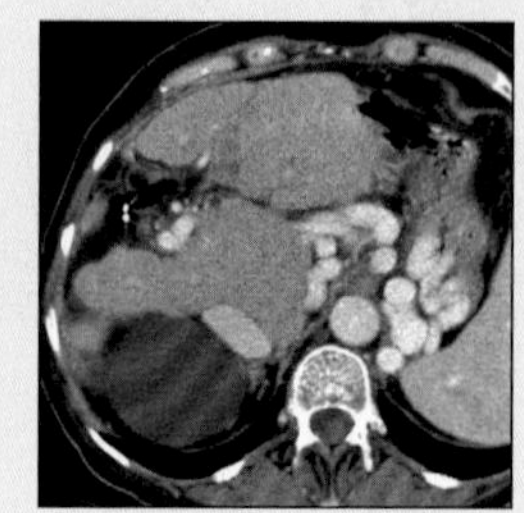

Sclerosing Cholangitis

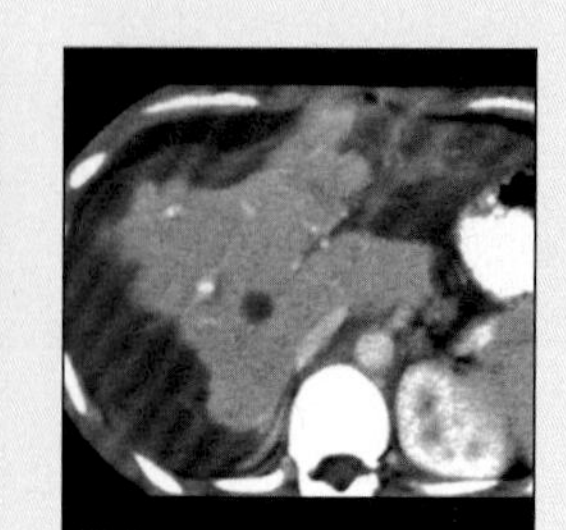

EHE of Liver

FOCAL CONFLUENT FIBROSIS

Key Facts

Terminology
- Confluent hepatic fibrosis (CHF)

Imaging Findings
- Best diagnostic clue: Pre-contrast CT showing hypoattenuating lesion with volume loss that becomes isoattenuating or minimally hypoattenuating at post-contrast CT, especially if wedge-shaped, located in medial segment of left lobe &/or anterior segment of right lobe, in patients with advanced cirrhosis
- Wedge-shaped lesions radiate from porta hepatis & extend to hepatic capsule
- Peripheral lesions are remote from porta hepatis
- Lobar or segmental involvement, most commonly in lateral segment of left lobe
- MR imaging does show morphological changes & characteristic locations that suggest diagnosis, but no more so than CT

Pathology
- Can be seen on imaging in approximately 14% of patients with advanced cirrhosis who are candidates for liver transplantation
- Associated volume loss seen as retraction of overlying hepatic capsule or total shrinkage of segment or lobe

Diagnostic Checklist
- Consider cholangiocarcinoma or treated malignancy in differential diagnosis

- Greater enhancement than adjacent liver parenchyma in arterial phase (13%), may relate to ↑ hepatic arterial flow
- Crowding of blood vessels, or bile ducts, within collapsed area of hepatic parenchyma

MR Findings
- T1WI
  - Lesions appear as regions of hypointense signal relative to adjacent liver parenchyma
    - May be isointense (less common)
- T2WI
  - Lesions are hyperintense
    - Due to prominent edema & approximation of remnant portal triads within fibrotic areas
- STIR: Hyperintense lesions
- T1 C+
  - Lesions are slightly hypointense to liver on immediate post-gadolinium sequences (80%)
    - During later dynamic phase, portions of fibrotic lesions may become isointense with liver
    - Delayed progressive increased enhancement on portal venous & equilibrium phase images
  - May be slightly hyperintense, due to pooling of contrast material within fibrotic stroma
- Ferumoxide-enhanced MR: Wedge-shaped area of high signal intensity (corresponds to distribution of fibrosis) with internal focal areas of low signal intensity (correspond to residual functioning liver parenchyma)

Imaging Recommendations
- Best imaging tool
  - CT or MR
    - MR imaging does show morphological changes & characteristic locations that suggest diagnosis, but no more so than CT
  - Lesion conspicuity better at pre-contrast than post-contrast CT
- Protocol advice: NECT & CECT or MR & CEMR

DIFFERENTIAL DIAGNOSIS

Cholangiocarcinoma (CC)
- May cause segmental volume loss, capsular retraction, delayed enhancement
  - Capsular retraction; because these tumors have prominent fibrous stroma & because they often cause intrahepatic bile duct obstruction
- Look for biliary obstruction
  - In confluent fibrosis, unlike in cholangiocarcinoma, bile ducts within affected segments are not dilated
- Peripheral form of CC, most often associated with capsular retraction, is usually spherical, hypodense mass on NECT
  - CECT: ↑ & prolonged enhancement (fibrosis)
- Clinical clues: History of primary sclerosing cholangitis or other chronic bile duct inflammation
- May require biopsy for diagnosis

Treated malignancies
- Treatment (chemotherapy, ablation, etc.) may result in volume loss + fibrosis of tumor & surrounding liver
- May be indistinguishable from confluent fibrosis
- Check for prior imaging or clinical evidence of tumor

Intrahepatic biliary obstruction
- May present as focal hepatic lesion with capsular retraction
- Malignant or benign biliary obstruction leads to hepatic atrophy of segments drained by obstructed bile ducts
  - Chronic primary sclerosing cholangitis results in disproportionate atrophy of periphery of liver with confluent fibrosis & capsular retraction
  - Compensatory hypertrophy of deep right lobe & caudate lobe may result in "pseudotumor" appearance
  - CT: Irregular strictures & focal dilatations of intrahepatic bile ducts

Cavernous hemangiomas
- Hepatic hemangioma, especially in cirrhotic livers may have retraction of liver capsule (24%)

FOCAL CONFLUENT FIBROSIS

- Cirrhotic liver hemangiomas often undergo progressive fibrosis & diminution in size, often resulting in hyalinized scar that no longer maintains typical radiologic & pathologic features of hemangioma
- Hemangiomas in cirrhotic livers: May be subcapsular/demonstrate exophytic growth/peripheral progressive nodular enhancement/& near isoattenuation with blood vessels

Epithelioid hemangioendothelioma (EHE)

- Predominantly involves peripheral portion of liver
- Capsular retraction is frequently seen
- Usually occurs in young patients without cirrhosis
- Almost always multiple & more nodular

PATHOLOGY

General Features

- Etiology: Cause of liver cirrhosis: Viral infection/alcohol abuse/biliary disease (primary biliary cirrhosis, sclerosing cholangitis, biliary atresia), autoimmune hepatitis, α1-antitrypsin deficiency, hemochromatosis, cryptogenic or uncertain
- Epidemiology
  - Can be seen on imaging in approximately 14% of patients with advanced cirrhosis who are candidates for liver transplantation
  - It occurs most commonly in cirrhosis secondary to primary sclerosing cholangitis (56%)
- Associated abnormalities
  - Wedge-shaped fibrosis seen frequently with alcoholic cirrhosis (19%), only 6% in cirrhosis due to viral infection; reason is unclear
  - Lobar or segmental atrophy of liver parenchyma associated with primary sclerosing cholangitis

Gross Pathologic & Surgical Features

- CHF appears as a regional mass; area of yellowish color with little intervening liver parenchyma
- Associated volume loss seen as retraction of overlying hepatic capsule or total shrinkage of segment or lobe

Microscopic Features

- Fibrosis with prominent edema & approximation of remnant portal triads with little intervening regenerating nodules
- Bile-duct proliferation & lymphocyte infiltration

CLINICAL ISSUES

Presentation

- Most common signs/symptoms
  - Signs/symptoms relate to cause + extent of cirrhosis
  - CHF seen incidentally in patients with advanced cirrhosis who undergo pretransplant imaging
- Clinical profile
  - Lab: ↑ LFT; alcoholic cirrhosis- ↑↑ AST; viral- ↑↑ ALT
  - Diagnosis: Liver biopsy to differentiate CHF from hepatic malignancy, as some overlap of findings are seen at imaging

Demographics

- Age
  - Adults, mean age: 51 years
  - Case report of CHF in children, in hepatic damage associated with anti-tuberculous drugs
- Gender: M > F (related to cirrhosis)

Natural History & Prognosis

- Severe fibrosis with capsular retraction can result in thinning of the involved segments to such a degree that they are no longer present
- Medial segment of left lobe & anterior segment of right lobe are most easily damaged, this might be related to impaired portal microcirculation
- Massive fibrosis takes long time to develop
- Complications: Cirrhotic patients are at high risk of developing hepatocellular carcinoma
- Advanced cirrhosis: Poor prognosis - ↑ survival period with liver transplantation

Treatment

- Management limited to treatment of complications & underlying cause of cirrhosis
- Advanced disease: Liver transplantation

DIAGNOSTIC CHECKLIST

Consider

- Fibrosis is present in all cirrhotic livers
  - Confluent is just one pattern that is evident on imaging + gross pathology
  - Notable because it may simulate tumor, especially cholangiocarcinoma
- MR is not complementary to CT; shows same features

Image Interpretation Pearls

- Characteristic location (medial segment of left lobe, anterior segment of right lobe or both) & shape (wedge-shape with capsular retraction & volume loss) - enables correct diagnosis & may prevent unnecessary biopsy
- Consider cholangiocarcinoma or treated malignancy in differential diagnosis

SELECTED REFERENCES

1. Blachar A et al: Hepatic capsular retraction: spectrum of benign and malignant etiologies. Abdom Imaging. 27(6):690-9, 2002
2. Matsuo M et al: Confluent hepatic fibrosis in cirrhosis: ferumoxides-enhanced MR imaging findings. Abdom Imaging. 26(2):146-8, 2001
3. Ooi CG et al: Confluent hepatic fibrosis in monozygotic twins. Pediatr Radiol. 29(1):53-5, 1999
4. Ahn IO et al: Early hyperenhancement of confluent hepatic fibrosis on dynamic MR imaging. AJR Am J Roentgenol. 171(3):901-2, 1998
5. Ohtomo K et al: Confluent hepatic fibrosis in advanced cirrhosis: appearance at CT. Radiology. 188(1):31-5, 1993
6. Ohtomo K et al: Confluent hepatic fibrosis in advanced cirrhosis: evaluation with MR imaging. Radiology. 189(3):871-4, 1993

FOCAL CONFLUENT FIBROSIS

IMAGE GALLERY

Other

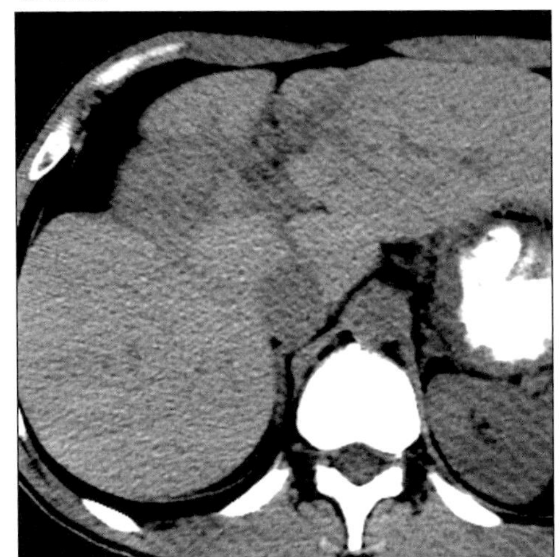

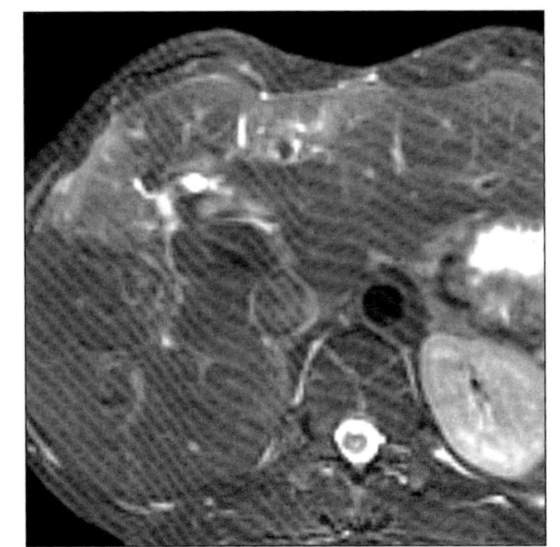

(Left) *Axial NECT shows hypodense lesion in anterior and medial segments with capsular retraction.* ***(Right)*** *Axial T2WI MR shows wedge-shaped hyperintense lesion in anterior and medial segments with capsular retraction, representative of focal confluent fibrosis.*

Other

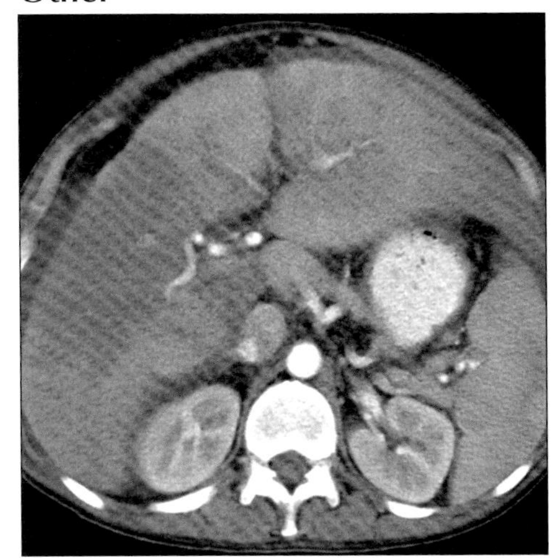

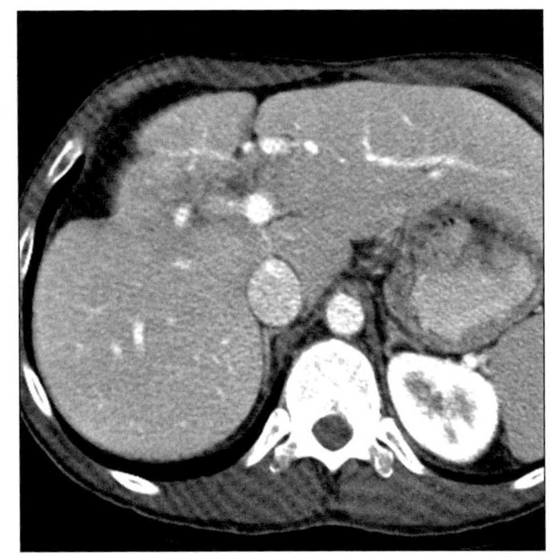

(Left) *Axial CECT shows cirrhotic morphology and subtle low attenuation throughout the anterior right lobe from early focal confluent fibrosis.* ***(Right)*** *Axial CECT 3 months following prior image shows marked volume loss of anterior segment and capsular retraction.*

Typical

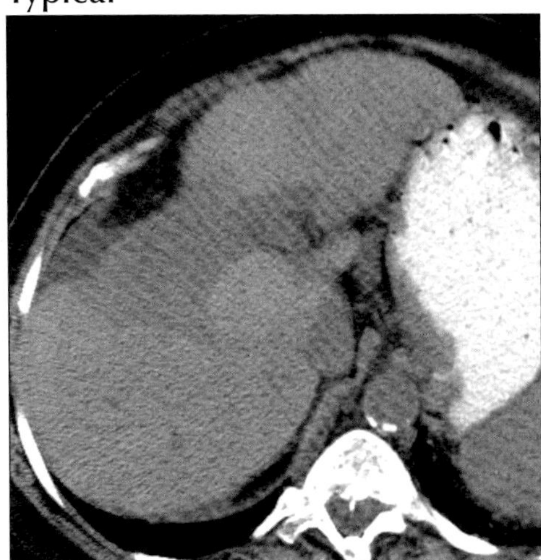

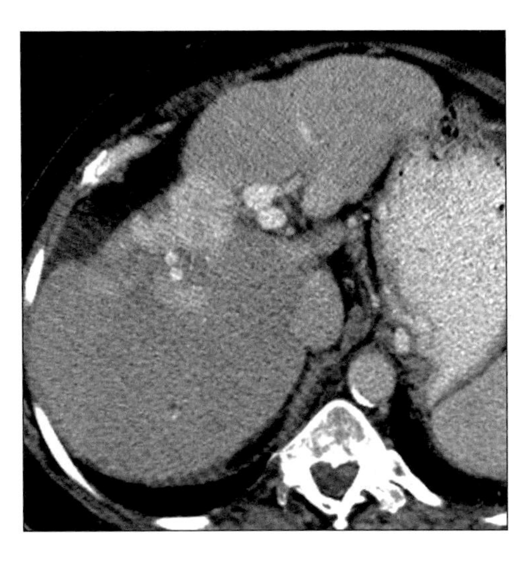

(Left) *Axial NECT shows wedge-shaped focal confluent fibrosis with capsular retraction in a patient with cirrhosis.* ***(Right)*** *Axial CECT in portal venous phase shows heterogeneous enhancement of fibrotic lesion.*

PRIMARY BILIARY CIRRHOSIS

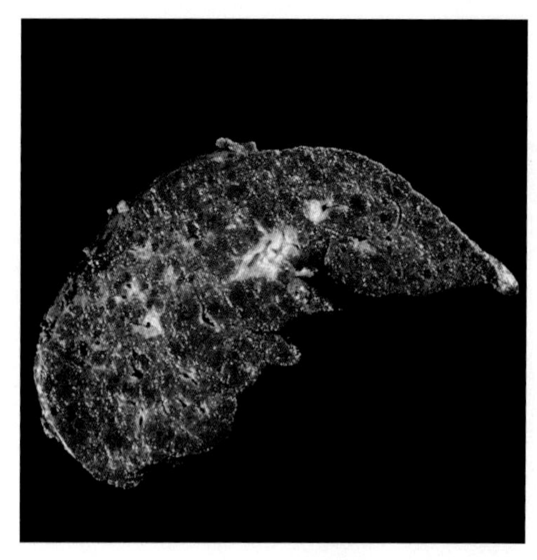

Transverse cut section of explanted liver shows regenerating nodules and lace-like fibrosis.

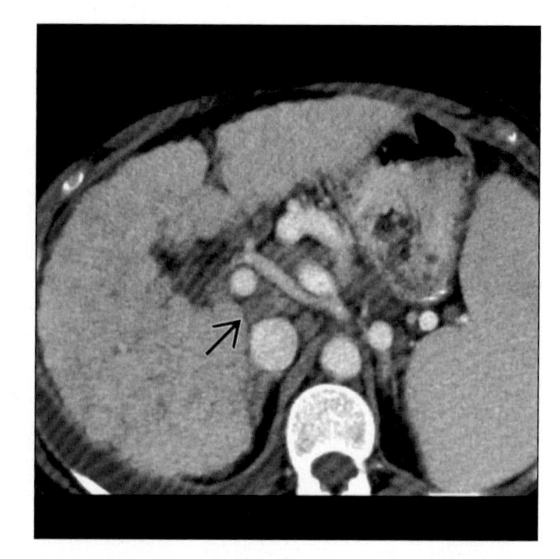

Axial CECT shows nodular heterogeneous cirrhotic liver with prominent porta hepatis lymphadenopathy (arrow). Note enlarged spleen secondary to portal hypertension.

TERMINOLOGY

Abbreviations and Synonyms

- Primary biliary cirrhosis (PBC)

Definitions

- PBC is a chronic progressive cholestatic liver disease characterized by non-suppurative destruction of interlobular bile ducts leading to advanced fibrosis, cirrhosis, & liver failure

IMAGING FINDINGS

General Features

- Best diagnostic clue: Liver biopsy; to identify cause of cirrhosis & histopathology stage
- Location: Diffuse involvement
- Size
  - Normal or ↑ hepatic volume at time of diagnosis
  - Trend toward decreasing hepatic volume with more advanced disease; some patients with advanced disease may have grossly enlarged liver
  - Mean nodule diameter in liver: Having nodular contour is 0.5 cm, with smooth contour it is 0.3 cm
  - Lymph nodes: Moderate enlargement: 1.5-2 cm
- Key concepts
  - In less advanced disease, liver is enlarged & smooth, as disease progresses, becomes more nodular & eventually grossly cirrhotic
  - Smooth liver contour in 71% of less advanced & 43% of advanced PBC
  - Nodular contour in 29% of less advanced & 57% of advanced PBC cases

CT Findings

- NECT
  - Heterogeneously attenuating liver parenchyma in both advanced & less advanced cases of PBC
    - In less advanced PBC, liver parenchyma may be homogeneously attenuating
    - ↑ Caudate lobe to right lobe ratio; 43-48% with advanced & less advanced PBC
    - Focal or diffuse atrophy; 65% with advanced PBC, 52% with less advanced PBC
    - Global or segmental hypertrophy; 51% with advanced, 67% in less advanced PBC
    - Segmental hypertrophy typically in lateral & caudate segments, segmental atrophy of right lobe & medial segment.
  - Fibrosis: Lace-like pattern of thin or thick bands of low attenuation that surround regenerating nodules (seen in one-third of patients, regardless of stage, & seems to be characteristic of PBC)

DDx: Heterogeneous Hepatomegaly with Lymphadenopathy

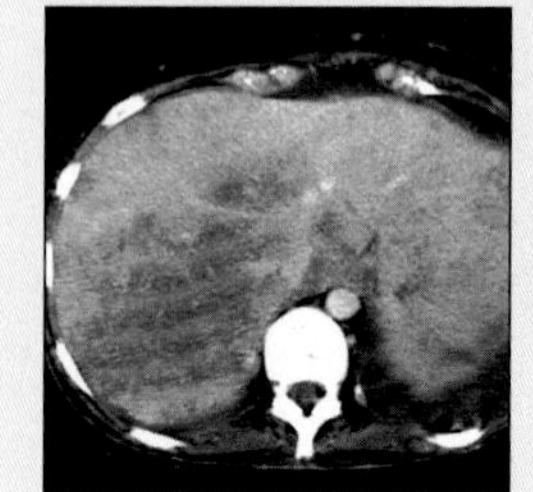

Early ETOH Cirrhosis

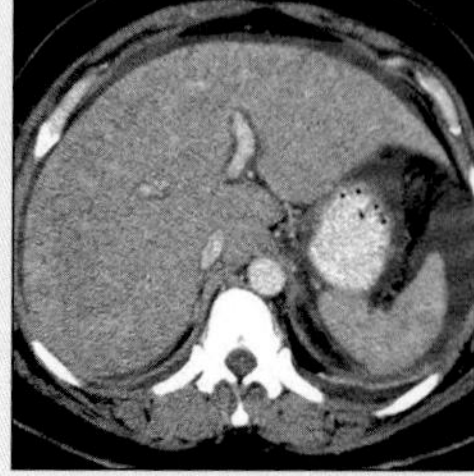

Diffuse Metastases

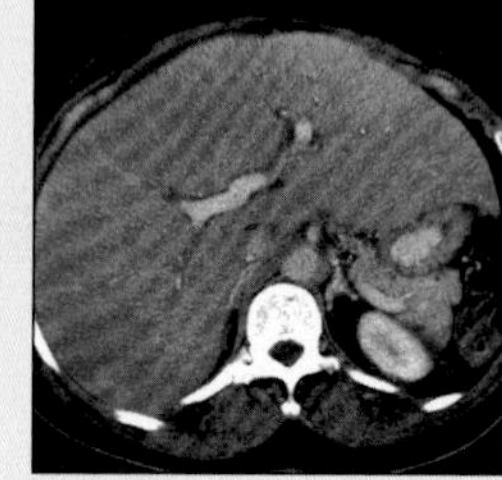

Amyloidosis

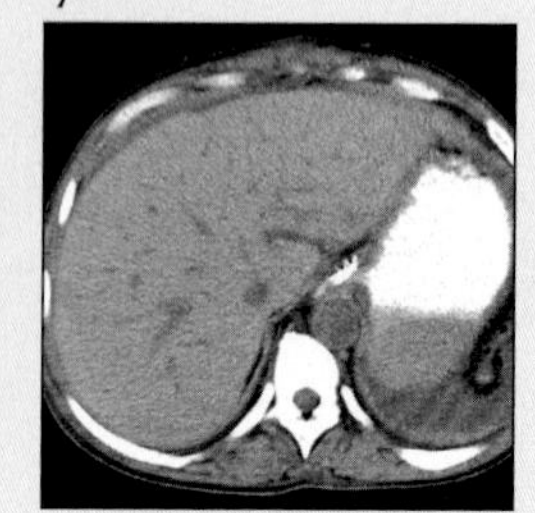

Sickle Cell Disease

PRIMARY BILIARY CIRRHOSIS

Key Facts

Terminology
- PBC is a chronic progressive cholestatic liver disease characterized by non-suppurative destruction of interlobular bile ducts leading to advanced fibrosis, cirrhosis, & liver failure

Imaging Findings
- Fibrosis: Lace-like pattern of thin or thick bands of low attenuation that surround regenerating nodules (seen in one-third of patients, regardless of stage, & seems to be characteristic of PBC)
- Lymphadenopathy (88%): Portacaval spaces (81%), porta hepatis (72%), paracardiac (24%)
- Regenerating nodules (43%): Small spherical hyperattenuating foci

Pathology
- General path comments: PBC is characterized by destruction of small intrahepatic bile ducts, portal inflammation, & progressive scarring
- Accounts for 0.6-2.0% of deaths from cirrhosis

Clinical Issues
- Fatigue (78%) & pruritus (60%)
- Portal hypertension: Varices, splenomegaly, ascites
- Serum antimitochondrial-antibody tests are highly sensitive & specific; positive in 95%
- Gender: Approximately 95% of patients are female
- Complication: Liver failure
- Liver transplantation improves survival & quality of life for patients with advanced PBC

    - Patchy poorly defined regions of low attenuation; pattern noted slightly more frequently in advanced PBC (30%)
    - Confluent hepatic fibrosis may be seen in advanced PBC (3%)
  - Lymphadenopathy (88%): Portacaval spaces (81%), porta hepatis (72%), paracardiac (24%)
  - Regenerating nodules (43%): Small spherical hyperattenuating foci
  - CT signs of portal hypertension: More common in advanced PBC, including splenomegaly (88%), varices(87%), ascites (44%)
    - Less advanced cases show splenomegaly in 71%, varices in 62% & ascites in 24% of patients
  - Cirrhosis-induced hepatocellular carcinoma (HCC), 0.5-7.5%: Less frequent than with other causes of cirrhosis
- CECT
  - Regenerative nodules are isoattenuating to liver on portal venous phase
  - Fibrosis is not as evident; isodense to liver
  - Lymph nodes show moderate & homogeneous contrast-enhancement
  - HCC: Intense/heterogeneous enhancement in arterial phase/ portal venous phase; ± capsule

MR Findings
- Regenerative nodules (siderotic): Hypointense on T1WI/↑ conspicuity of low signal intensity on T2WI/markedly hypointense on GRE images
- Fibrosis: Hypointense on T1WI/hyperintense on T2WI

Ultrasonographic Findings
- Real Time: Major bile ducts are patent/signs of portal hypertension/hepatomegaly (early)/↑ liver echogenicity/gall stones/lymphadenopathy

Nuclear Medicine Findings
- Technetium-99m iminodiacetic acid scan
  - Diffuse, uniform hepatic isotope retention & normal major bile ducts
  - Normal visualization of gallbladder, but gallbladder ejection fraction & ejection rate are reduced

Other Modality Findings
- Cholangiography: Tortuous intrahepatic (interlobar & septal) bile ducts with narrowing + caliber variation
  - ↓ Arborization = "tree-in-winter" appearance

Imaging Recommendations
- Best imaging tool: Triphasic helical CT
- Protocol advice: NECT & CECT; or MR & CEMR

DIFFERENTIAL DIAGNOSIS

Early viral or alcoholic (ETOH) cirrhosis
- Hepatomegaly, micro/macronodular/mixed cirrhosis
- Fatty changes: Mottled areas of ↓ attenuation, coexist with characteristic features of cirrhosis
- Alcoholic cirrhosis: Male > female, serum AST ↑ markedly
- Viral: Positive hepatitis antibodies or antigens, ↑↑ ALT

Hepatic lymphoma (+ metastases)
- Homogeneous/heterogeneous hepatomegaly + hypodense focal lesions
- Secondary lymphoma is either multinodular or diffusely infiltrative with metastatic lymphadenopathy

Sarcoidosis (+ amyloidosis)
- Hepatosplenomegaly/diffuse parenchymal heterogeneity/multiple low-attenuation nodules

Myeloproliferative disorders
- Hepatomegaly associated with splenomegaly & generalized lymphadenopathy (e.g., in chronic myelogenous leukemia/agnogenic myeloid metaplasia)

Opportunistic infection
- In immunocompromised hosts
- Infection by fungi, mycobacteria, viruses etc.
  - Heterogeneous hepatosplenomegaly ± abdominal lymphadenopathy may be indistinguishable by imaging alone

PRIMARY BILIARY CIRRHOSIS

PATHOLOGY

General Features

- General path comments: PBC is characterized by destruction of small intrahepatic bile ducts, portal inflammation, & progressive scarring
- Genetics
  - Genetic factors play a part in development but it is not inherited in any recessive/dominant pattern
  - Prevalence in families with one affected member is 1000 times higher than in general population
- Etiology: Cause of PBC is unknown, but it is probably due to an inherited abnormality of immunoregulation
- Epidemiology
  - Accounts for 0.6-2.0% of deaths from cirrhosis
  - Third most common indication for liver transplantation in adults
  - Prevalence: 19 to 151 cases per million population
  - Incidence: 3.9 to 15 per million population per year
- Associated abnormalities: May have at least one other autoimmune disease (84%); thyroiditis, scleroderma, rheumatoid arthritis, Sjögren syndrome

Gross Pathologic & Surgical Features

- Enlarged & smooth/finely nodular/grossly cirrhotic liver with nodular regeneration, fibrosis, shrinkage of hepatic parenchyma, prominent lymphadenopathy
- Gall bladder & bile ducts are grossly normal
- Hypersplenism & evidence of portal hypertension

Microscopic Features

- Stage 1: Damage to epithelial cells of bile duct, presumably mediated by lymphocytes that surround & often infiltrate the duct
  - Necrotic bile ducts are often located at center of large granuloma-like lesions that consist of histiocytes, lymphocytes, plasma cells, eosinophils, & occasionally true giant cells
  - Inflammation remains confined to portal triads
- Stage II: Many portal triads become scarred, inflammatory cells spill out of triads into surrounding periportal parenchyma, &, atypical, poorly formed, tortuous bile ducts with no obvious lumens are seen
- Stage III: Scarring progresses, fibrous septa link many adjoining portal triads
- Stage IV: Frank cirrhosis
- All stages may be seen in single biopsy specimen

CLINICAL ISSUES

Presentation

- Most common signs/symptoms
  - Fatigue (78%) & pruritus (60%)
  - Asymptomatic (48%)/diffuse hepatomegaly (11%)
  - Portal hypertension: Varices, splenomegaly, ascites
  - Xanthomas (25%)/cholelithiasis (39%)
  - Jaundice is a later manifestation of disease, but in some patients it may be seen at presentation
  - Osteopenia due to hepatic osteodystrophy, erosive arthritis, intraosseous lytic defects, osteoporosis
- Clinical profile
  - Lab data: Cholestatic pattern: Alkaline phosphatase & glutamyltransferase levels are disproportionately higher than aminotransferase, ↑ serum IgM levels
    - Serum antimitochondrial-antibody tests are highly sensitive & specific; positive in 95%
    - Elevated serum bilirubin is sign of poor prognosis
  - Diagnosis: Percutaneous liver biopsy; findings provide confirmatory information & assist in determination of histologic stage of disease

Demographics

- Age: Onset between ages of 30 & 65
- Gender: Approximately 95% of patients are female

Natural History & Prognosis

- Natural history of PBC is not fully understood; although disease progresses in almost all patients, rate of progression is variable
- Hepatic size & parenchymal heterogeneity may evolve over course of PBC
- Varices may develop & bleed relatively early in course of PBC, well before jaundice or true cirrhosis manifest
- HCC in PBC, tends to be well differentiated; not possible to determine whether any PBC-specific risk factors other than cirrhosis per se exist for development of HCC
- Complication: Liver failure
- Prognosis: Average length of survival 11.9 years
- Prognostic factors: Patient's age, serum bilirubin level & prothrombin time & histopathology

Treatment

- Liver transplantation improves survival & quality of life for patients with advanced PBC
- 1 year survival rate post transplant has increased to 90%
- PBC can recur after transplantation; recurrence rate: About 1-2% per year of survival after transplantation

DIAGNOSTIC CHECKLIST

Consider

- In middle-aged woman who reports unexplained itching, fatigue, hyperpigmentation, jaundice, or unexplained weight loss, with discomfort in right upper quadrant & an unaccountable elevation of serum alkaline phosphatase

Image Interpretation Pearls

- Presence of lace-like fibrosis, prominent lymphadenopathy & hepatomegaly (early) along with positive anti-mitochondrial antibody test

SELECTED REFERENCES

1. Kita H et al: Pathogenesis of primary biliary cirrhosis. Clin Liver Dis. 7(4):821-39, 2003
2. MacQuillan GC et al: Liver transplantation for primary biliary cirrhosis. Clin Liver Dis. 7(4):941-56, ix, 2003
3. Blachar A et al: Primary biliary cirrhosis: clinical, pathologic, and helical CT findings in 53 patients. Radiology. 220(2):329-36, 2001

PRIMARY BILIARY CIRRHOSIS

IMAGE GALLERY

Typical

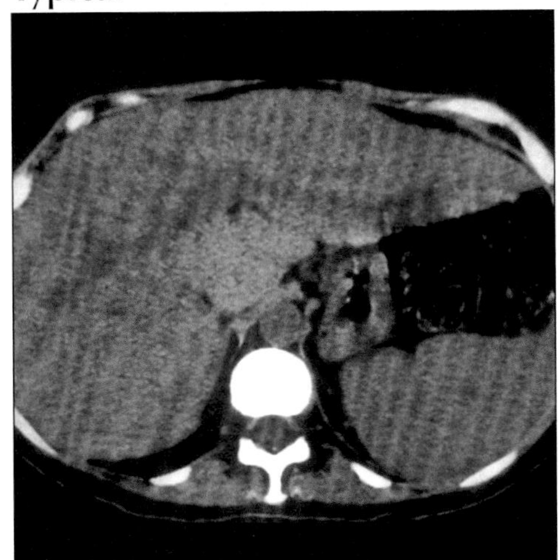

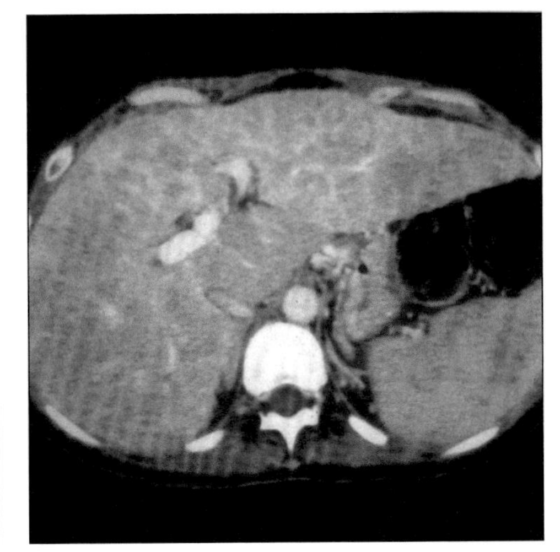

*(**Left**) Axial NECT shows hepatomegaly and thin interconnecting bands of low attenuation from lace-like from fibrosis. (**Right**) Axial CECT in portal venous phase shows heterogeneity of the liver, with enhancing bands of fibrosis. There is no discrete mass.*

Typical

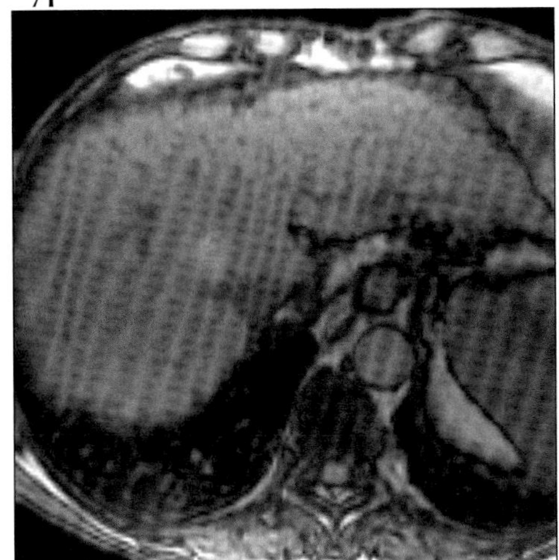

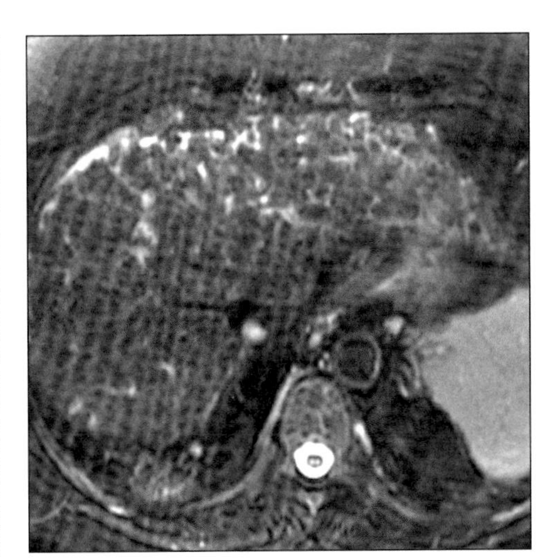

*(**Left**) Axial T1WI MR shows lace-like hypointense fibrosis of the liver. (**Right**) Axial T2WI MR shows innumerable subcentimeter hypointense regenerating nodules surrounded by thin bands of hyperintense fibrosis.*

Typical

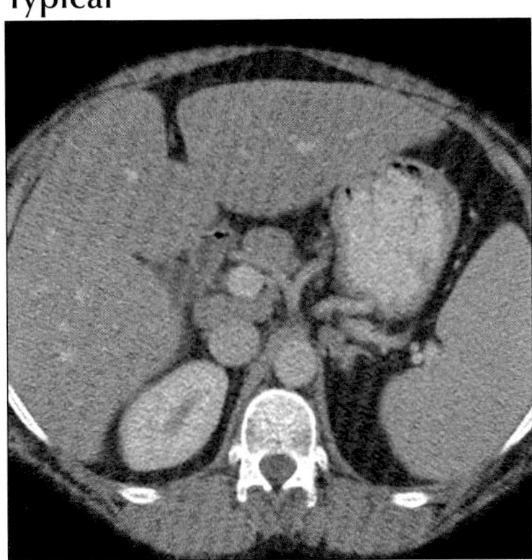

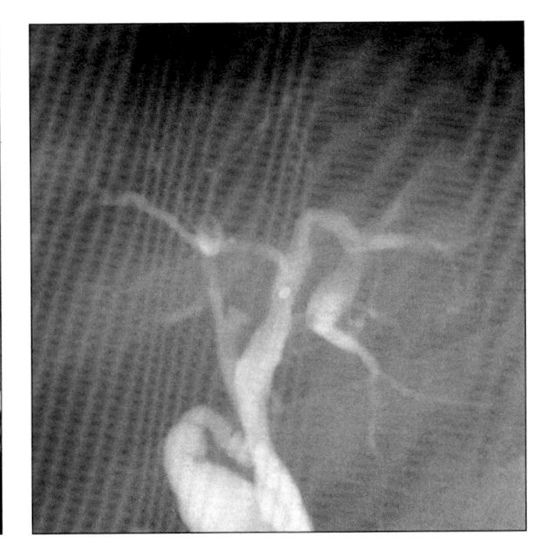

*(**Left**) Axial CECT shows cirrhotic morphology of the liver with wide fissures; prominent porta hepatis, lymphadenopathy and splenomegaly. Despite cirrhosis, there is a smooth liver contour. (**Right**) ERCP shows pruned, intrahepatic bile ducts with decreased arborization; "tree in winter" appearance.*

NODULAR REGENERATIVE HYPERPLASIA

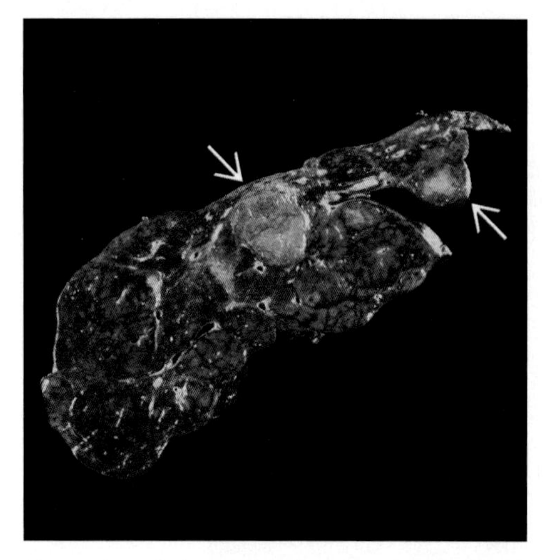

Cut section of dysmorphic liver from a patient with Budd-Chiari syndrome shows caudate hypertrophy, lateral segment atrophy, large and numerous orange regenerative nodules (arrows).

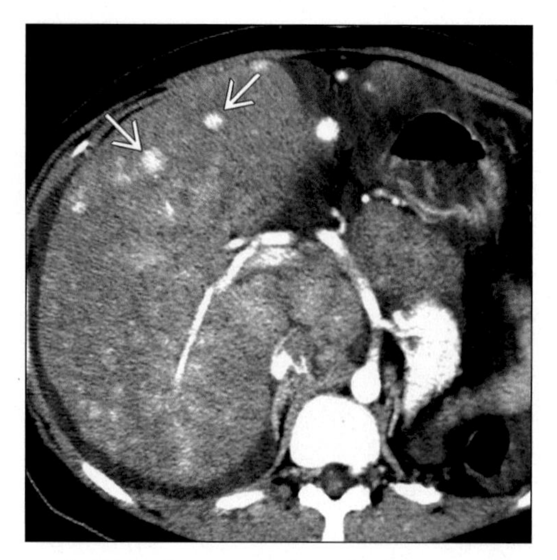

Axial CECT of patient with Budd-Chiari syndrome, shows dysmorphic liver with numerous hypervascular lesions (arrows) representing regenerative nodules.

TERMINOLOGY

Abbreviations and Synonyms

- Nodular regenerative hyperplasia (NRH); large (multiacinar) regenerative nodules

Definitions

- NRH of liver is a rare disorder characterized by diffuse micronodular transformation of hepatic parenchyma without fibrous septa between nodules
- Larger lesions are called multiacinar (large) regenerative nodules

IMAGING FINDINGS

General Features

- Best diagnostic clue: Liver biopsy; multiple mono- or multiacinar regenerative nodules of hyperplastic hepatocytes
- Location: Diffuse involvement; microscopic nodules predominantly distributed in periportal region
- Size
  - Monoacinar lesions in NRH are only about 1 mm in diameter, with clusters of lesions up to 10 mm
  - Large regenerative nodules: 0.5-4 cm
- Key concepts
  - Multiple nodules of hyperplastic hepatocytes with atrophy or compression of intervening parenchyma

CT Findings

- NECT
  - Due to small size of nodules & preserved framework, imaging examinations may show normal findings
  - Nodules are usually isoattenuating to normal liver
  - Diffuse low attenuation in Budd-Chiari syndrome or steatosis or aggregation of blood-filled spaces may result in hyperattenuation of nodules
- CECT
  - Helical CT may show solitary or multiple uniformly hypervascular, & hyperattenuating lesions on both arterial & portovenous phases
  - Large regenerative nodules have an ↑ arterial supply corresponding to their bright enhancement
    - ± Perinodular hypoattenuating rim
  - Large regenerative nodules can be recognized as discrete focal hypervascular lesions on CECT
- Pseudotumoral presentation: Peripheral rim of enhancement due to peliosis surrounding nodules
- Portal hypertension: Varices, ascites, splenomegaly due to underlying liver disease (e.g., Budd-Chiari)

MR Findings

- T1WI: Larger nodules: Hyperintense on T1WI (75%)
- T2WI
  - Isointense or hypointense nodules on T2WI

DDx: Multiple Hypervascular Lesions in Dysmorphic Liver

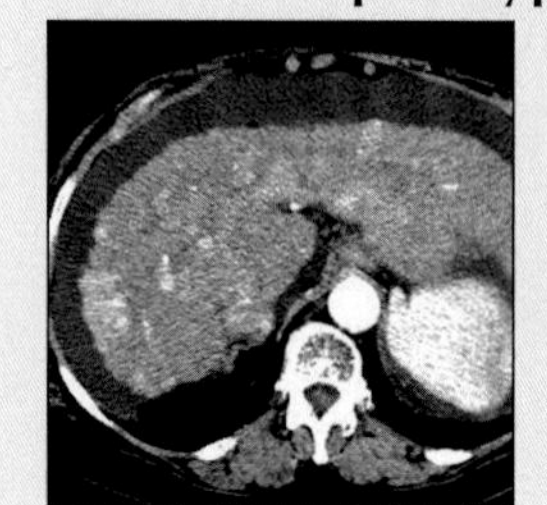

Multifocal HCC

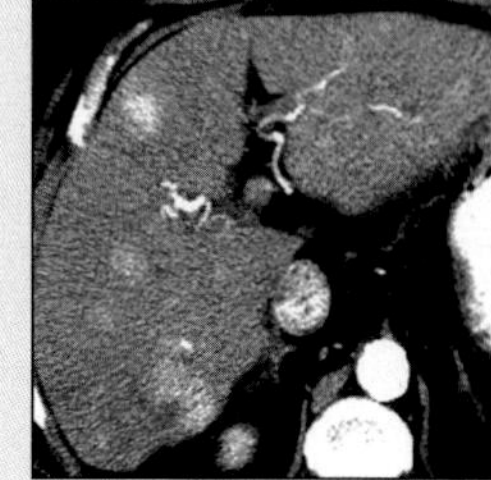

Multifocal HCC

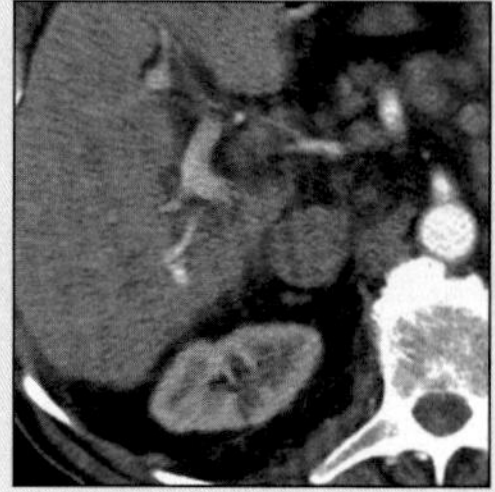

AP Shunts in Cirrhosis

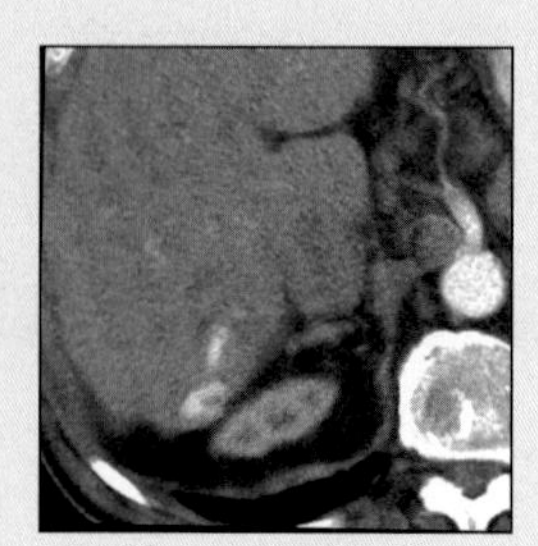

AP Shunts in Cirrhosis

NODULAR REGENERATIVE HYPERPLASIA

Key Facts

Terminology

- NRH of liver is a rare disorder characterized by diffuse micronodular transformation of hepatic parenchyma without fibrous septa between nodules

Imaging Findings

- Large regenerative nodules: 0.5-4 cm
- Helical CT may show solitary or multiple uniformly hypervascular, & hyperattenuating lesions on both arterial & portovenous phases
- Bright homogeneous enhancement
- ± Ring (halo) enhancement

Pathology

- Local hyperplastic response of hepatocytes, probably due to chronic ischemia
- Chronic Budd-Chiari syndrome (most common setting)
- Monoacinar lesions present in liver that is not fibrotic or cirrhotic

Clinical Issues

- Clinical profile: Diagnosis: Liver biopsy of large regenerative nodules shows hyperplastic liver tissue, ± inflammation, scar (similar to FNH)

  - May appear hyperintense (due to infarction)
  - "Halo sign": Nodule surrounded by peliosis
- Multiphasic enhanced MR
  - Bright homogeneous enhancement
  - ± Ring (halo) enhancement
- MR with gadobenate dimeglumine: Uptake & delayed clearance (prolonged enhancement) due to metaplastic bile ductules in some large regenerative nodules

Ultrasonographic Findings

- Real Time: Nodules may appear as hypoechoic (38%)/isoechoic (10%)/hyperechoic (53%) lesions
- Color Doppler: Occlusion or narrowing of IVC or hepatic veins (Budd-Chiari); obliteration or congenital absence or portal vein

Angiographic Findings

- Conventional: Nodules may fill from periphery on angiography, are vascular, & sometimes contain small hypovascular areas due to hemorrhage or scar

Nuclear Medicine Findings

- Technetium sulfur colloid
  - Nodules take up technetium sulfur colloid

Imaging Recommendations

- Best imaging tool: NECT + CECT; or MR
- Protocol advice: NE & multiphasic CECT or MR

DIFFERENTIAL DIAGNOSIS

Multifocal hepatocellular carcinoma (HCC)

- Hypoattenuating to liver on NECT & delayed CT
- Hyperattenuating on arterial phase, & may be hypo-, iso-, or hyperattenuating on portovenous phase
- Hypointense on T1WI/ hyperintense on T2WI
- Other characteristics of HCC: Heterogeneity, multiplicity, encapsulation, venous invasion

Multiple arterioportal (AP) shunts in cirrhosis

- Transsinusoidal arterioportal shunting in advanced cirrhosis
- Multiple hypervascular lesions in dysmorphic (cirrhotic) liver
  - Hypervascular & hyperattenuating on both arterial & portovenous phases (isodense to blood vessels)

PATHOLOGY

General Features

- General path comments
  - Large regenerative nodules are caused by vascular derangement of liver due to decreased portovenous or hepatovenous flow
    - Resulting multifocal hepatic arterial dilatation leads to focal hyperplasia & proliferation of hepatocytes
  - Existence of familial cases of NRH of liver
    - Occurring without underlying or associated systemic disease & is characterized by poor clinical course & often associated with renal failure
- Etiology
  - Unknown etiology of NRH; various theories
    - Local hyperplastic response of hepatocytes, probably due to chronic ischemia
    - Prolonged exposure to hepatopoietins
    - Antiphospholipid antibodies may play a role
    - Coexpression of interleukins, IL-6 & soluble IL-6R
    - Drugs: Azathioprine (AZA), steroids, & Thorotrast
- Epidemiology
  - Rare entity; incidence: 0.6-2.6% on autopsy series
  - NRH is seldom reported, most reports have been single cases, prevalence is not exactly known
  - Probably underdiagnosed owing to a lack of recognition of entity & limited sampling by biopsy
- Associated abnormalities
  - Chronic Budd-Chiari syndrome (most common setting)
  - Restrictive cardiomyopathy, congenital absence of portal vein, idiopathic superior mesenteric arteriovenous fistula
  - Systemic lupus erythematosus, systemic sclerosis, rheumatoid arthritis, polyarteritis nodosa
  - Hematologic diseases (myeloma or lymphoma)

NODULAR REGENERATIVE HYPERPLASIA

- Hepatic metastases from breast carcinoma
- Common variable immunodeficiency
- Secondary to organ transplantation

Gross Pathologic & Surgical Features

- Round, orange-brown, well-demarcated, soft to firm nodules, scattered in dysmorphic liver
- Obliteration of portal veins, portal hypertension, chronic hepatovenous outflow obstruction

Microscopic Features

- Multi-acinar nodules, consist of different-sized hepatocytes one or two plates wide & narrow sinusoids organized to form large regenerative nodules
- Interspersed between nodules are areas of centrilobular atrophy with curvilinear areas of sinusoidal dilation, marked congestion & paucity of fibrosis
- Monoacinar lesions present in liver that is not fibrotic or cirrhotic
- Larger (multi-acinar) nodules may have hepatic fibrosis between them
- May have central scar; indistinguishable from FNH
- ± Proliferation of bile ductules or metaplastic transformation of hepatocytes into bile ductules
- ± Mineral deposits & copper accumulation in nodules

CLINICAL ISSUES

Presentation

- Most common signs/symptoms: NRH generally asymptomatic; signs, symptoms & biochemical abnormalities relate to underlying liver disease
- Clinical profile: Diagnosis: Liver biopsy of large regenerative nodules shows hyperplastic liver tissue, ± inflammation, scar (similar to FNH)

Demographics

- Age: Adults; rare entity, especially in children
- Gender: No gender predilection known

Natural History & Prognosis

- Large regenerative nodules are prone to congestion & infarction if venous drainage is impaired, as in Budd-Chiari syndrome or cardiomyopathy
- NRH may bleed, may be associated with portal hypertension in one-half of cases, & often associated with systemic diseases (myelo- or lymphoproliferative)
- Diagnosis of NRH is often missed; present with secondary complications of underlying liver disease
- Variceal bleeding is main source of mortality
- Liver failure is uncommon due to satisfactory preservation of liver function
- NRH associated with azathioprine (AZA) represents a risk factor for HCC; AZA should be stopped in patients with NRH & patients should be screened for HCC
- Prognosis is related to consequences of portal hypertension & severity of associated diseases
    - Prognosis in absence of portal hypertension is good

Treatment

- Management directed to portal hypertension & variceal bleeding, with beta-blockers, sclerotherapy, mesenteric-caval shunt & transjugular intrahepatic portosystemic shunt (TIPS)
- Orthotopic liver transplantation for progressive hepatic failure & clinical end-stage liver disease

DIAGNOSTIC CHECKLIST

Consider

- NRH (with large multiacinar nodules) is easily confused with other masses such as HCC in cirrhosis
    - Important to recognize underlying liver disorder (e.g., Budd-Chiari) + characteristic appearance of NRH to avoid mistakes

SELECTED REFERENCES

1. Maetani Y et al: Benign hepatic nodules in Budd-Chiari syndrome: radiologic-pathologic correlation with emphasis on the central scar. AJR Am J Roentgenol. 178(4):869-75, 2002
2. Brancatelli G et al: Benign regenerative nodules in Budd-Chiari syndrome and other vascular disorders of the liver: radiologic-pathologic and clinical correlation. Radiographics. 22(4):847-62, 2002
3. Brancatelli G et al: Large regenerative nodules in Budd-Chiari syndrome and other vascular disorders of the liver: CT and MR imaging findings with clinicopathologic correlation. AJR Am J Roentgenol. 178(4):877-83, 2002
4. Horita T et al: Significance of magnetic resonance imaging in the diagnosis of nodular regenerative hyperplasia of the liver complicated with systemic lupus erythematosus: a case report and review of the literature. Lupus. 11(3):193-6, 2002
5. Trenschel GM et al: Nodular regenerative hyperplasia of the liver: case report of a 13-year-old girl and review of the literature. Pediatr Radiol. 30(1):64-8, 2000
6. Zhou H et al: Multiple macroregenerative nodules in liver cirrhosis due to Budd-Chiari syndrome. Case reports and review of the literature. Hepatogastroenterology. 47(32):522-7, 2000
7. Soler R et al: Benign regenerative nodules with copper accumulation in a case of chronic Budd-Chiari syndrome: CT and MR findings. Abdom Imaging. 25(5):486-9, 2000
8. Rha SE et al: Nodular regenerative hyperplasia of the liver in Budd-Chiari syndrome: CT and MR features. Abdom Imaging. 25(3):255-8, 2000
9. Grazioli L et al: Congenital absence of portal vein with nodular regenerative hyperplasia of the liver. Eur Radiol. 10(5):820-5, 2000
10. Vilgrain V et al: Hepatic nodules in Budd-Chiari syndrome: imaging features. Radiology. 210(2):443-50, 1999
11. Clouet M et al: Imaging features of nodular regenerative hyperplasia of the liver mimicking hepatic metastases. Abdom Imaging. 24(3):258-61, 1999
12. Morla RM et al: Nodular regenerative hyperplasia of the liver and antiphospholipid antibodies: report of two cases and review of the literature. Lupus. 8(2):160-3, 1999
13. Casillas C et al: Pseudotumoral presentation of nodular regenerative hyperplasia of the liver: imaging in five patients including MR imaging. Eur Radiol. 7(5):654-8, 1997
14. Wanless IR: Micronodular transformation (nodular regenerative hyperplasia) of the liver: a report of 64 cases among 2,500 autopsies and a new classification of benign hepatocellular nodules. Hepatology. 11(5):787-97, 1990

NODULAR REGENERATIVE HYPERPLASIA

IMAGE GALLERY

Typical

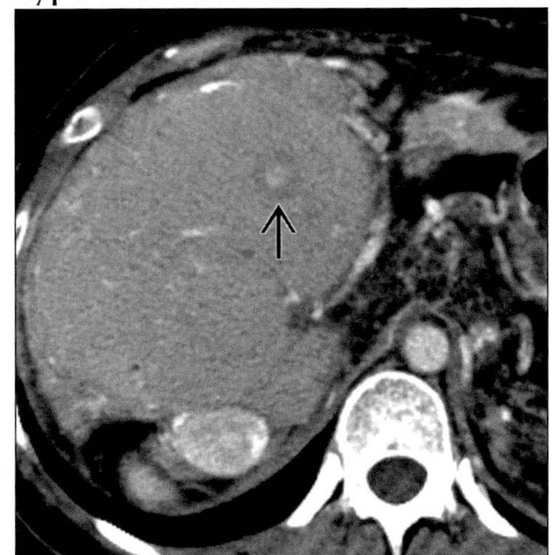

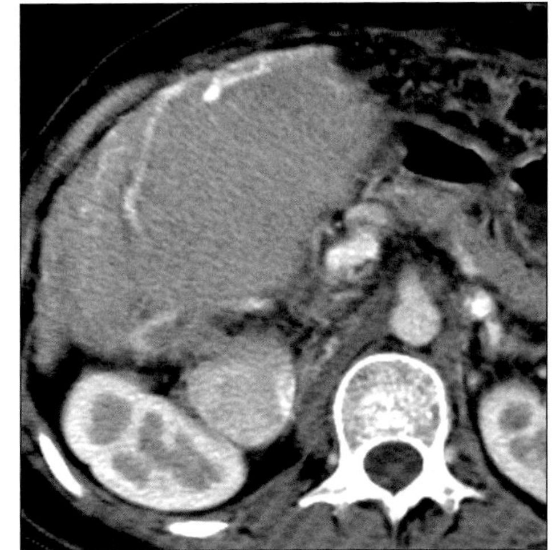

*(**Left**) Axial CECT shows dysmorphic liver with collateral blood vessels on the surface of the liver. Hypervascular lesion with hypodense ring (arrow) represents a focus of nodular regenerative hyperplasia (**Right**) Axial CECT shows dysmorphic liver, intra and extrahepatic collaterals bypassing occluded portal vein.*

Typical

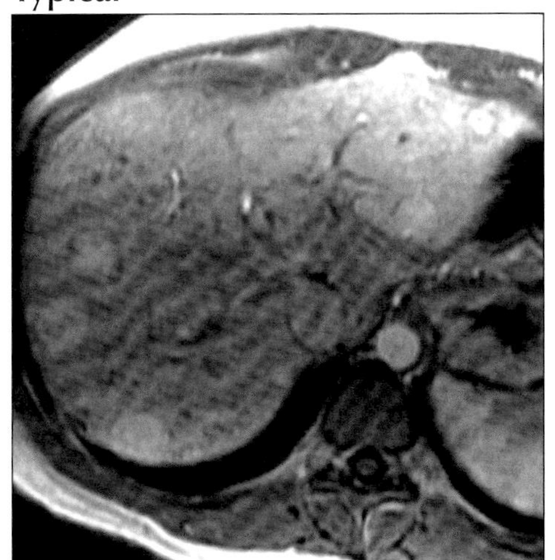

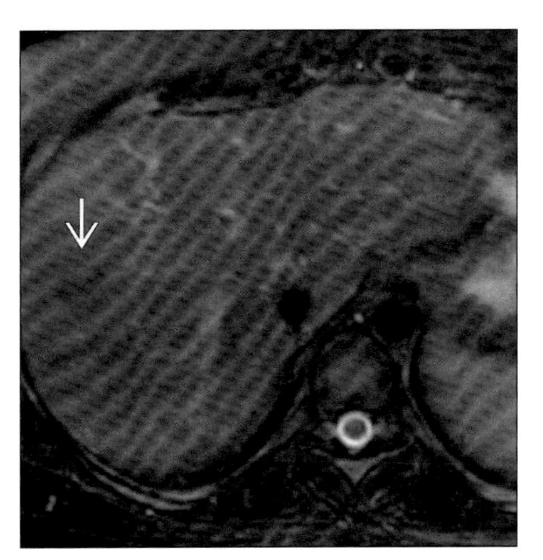

*(**Left**) Axial T1 C+ MR shows numerous 2 cm hyperintense lesions in a patient with Budd-Chiari syndrome. (**Right**) Axial T2WI MR demonstrates inconspicuous hypointense foci of nodular regenerative hyperplasia (arrow) in the right hepatic lobe.*

Typical

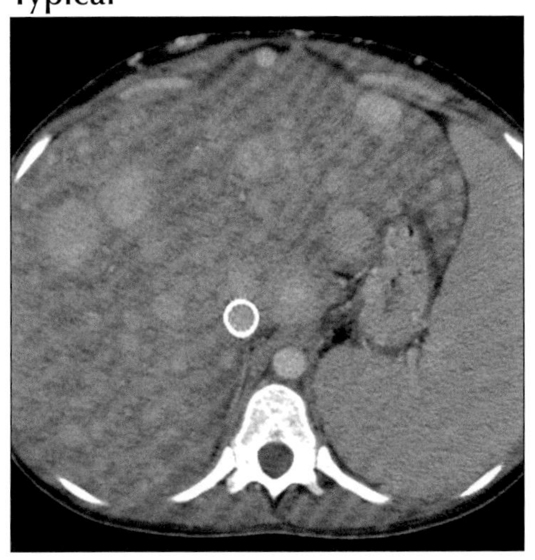

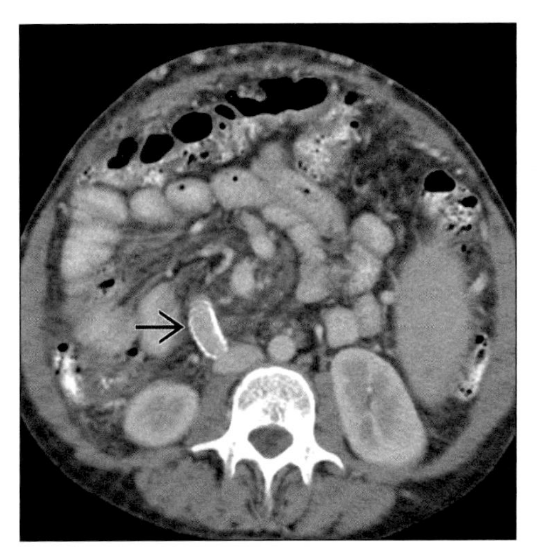

*(**Left**) Axial CECT shows in patient with Budd-Chiari syndrome (note IVC stent). Innumerable hypervascular foci in liver are large regenerative hyperplastic nodules. (**Right**) Axial CECT shows synthetic mesocaval shunt (arrow) and subcutaneous collateral veins. Hepatic imaging demonstrated dysmorphic morphology with multiple large regenerative hyperplastic nodules.*

HEPATIC SARCOIDOSIS

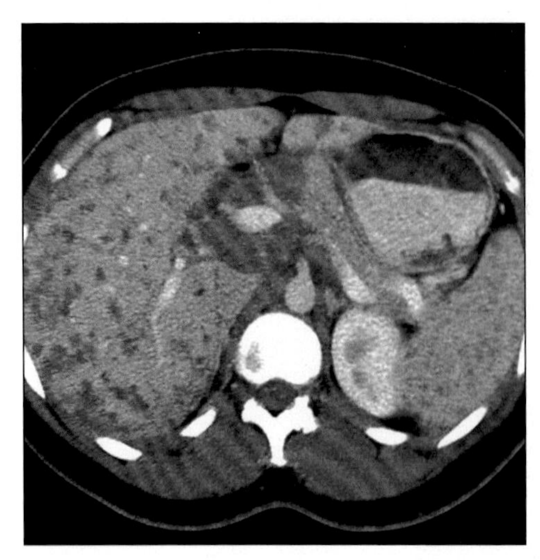

Axial CECT shows multiple small hypodense lesions in liver and spleen. Note porta hepatis lymphadenopathy.

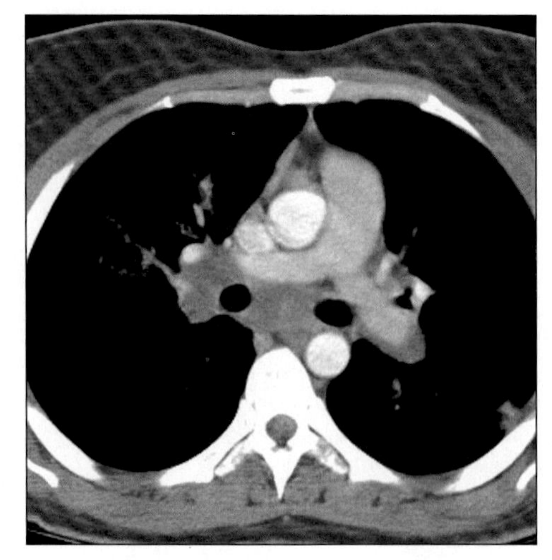

Axial CECT shows bilateral hilar and mediastinal lymphadenopathy.

TERMINOLOGY

Abbreviations and Synonyms

- Boeck sarcoid

Definitions

- Sarcoidosis is a relatively common, chronic, multisystem disease of unknown origin characterized by presence of noncaseating epithelioid granulomas

IMAGING FINDINGS

General Features

- Best diagnostic clue: Liver biopsy showing diffuse small noncaseating granulomas
- Location
  - Sarcoidosis can affect almost every organ
  - Most common site of involvement is lung
  - Lymph nodes, spleen, liver, eyes, skin, salivary glands, bones, nervous system, heart
- Size
  - Granulomas are generally 50 to 300 μm in size, & even though they may aggregate into larger clusters, they usually remain smaller than 2 mm in diameter
    - Large nodules may be up to 2 cm
- Key concepts
  - Granulomas are usually inconspicuous & featureless
  - It has been suggested that radiologically visible nodules do not simply represent increasing numbers of microscopic granulomas
    - May instead represent a more vigorous immunologic response to unknown causative agent of sarcoidosis
    - Larger coalescence of granulomas are coupled with exuberant deposition of reticulin, evolving to fibrosis around granulomatous aggregates

CT Findings

- NECT
  - Most common finding is nonspecific hepatosplenomegaly
  - Diffuse parenchymal heterogeneity or multinodular pattern in liver, spleen, or both
    - Multiple low-attenuation nodules
  - Upper abdominal lymphadenopathy is often present
- CECT
  - Low density nodules before contrast agent injection, usually become rapidly isodense with rest of liver parenchyma on enhanced scans
  - Advanced disease may cause or simulate cirrhosis

MR Findings

- T1WI: When visualized, nodules appear hypointense to adjacent liver parenchyma
- T2WI: Hypointense

DDx: Heterogeneous Hepatomegaly with Lymphadenopathy

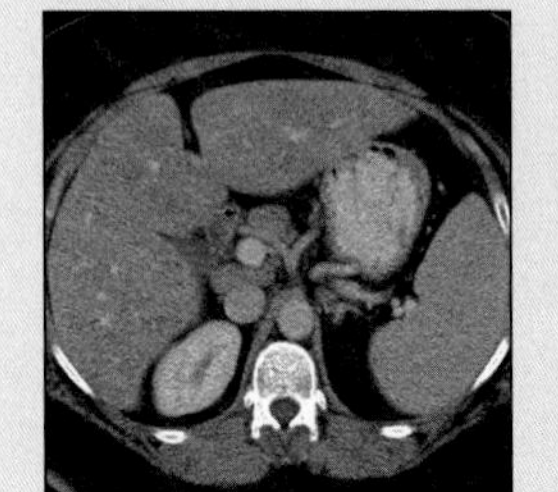

Biliary Cirrhosis

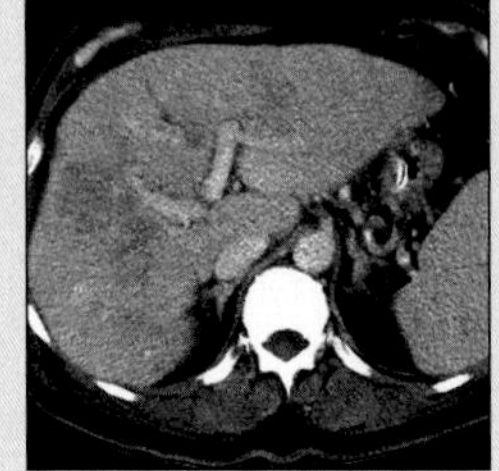

Lymphoma

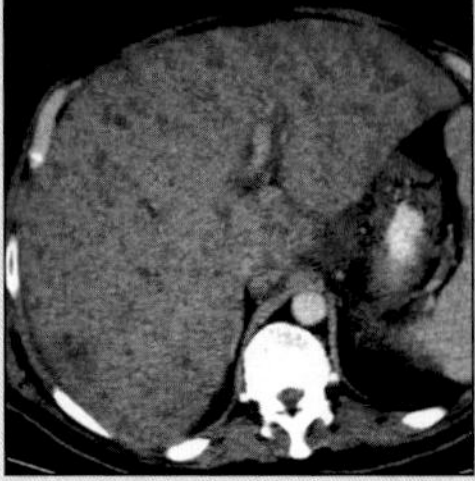

Opportunistic Infection

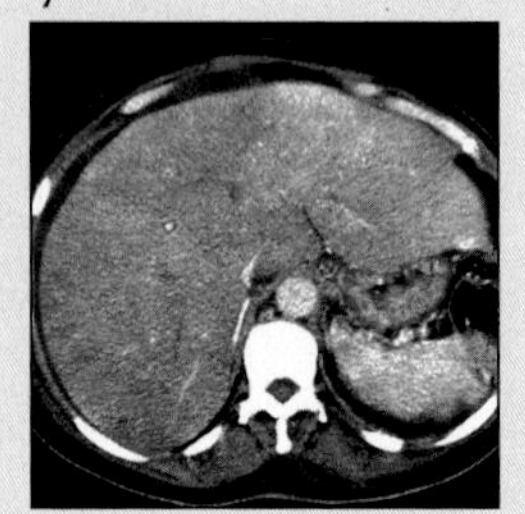

Amyloidosis

HEPATIC SARCOIDOSIS

Key Facts

Terminology

- Sarcoidosis is a relatively common, chronic, multisystem disease of unknown origin characterized by presence of noncaseating epithelioid granulomas

Imaging Findings

- Sarcoidosis can affect almost every organ
- Most common finding is nonspecific hepatosplenomegaly
- Diffuse parenchymal heterogeneity or multinodular pattern in liver, spleen, or both
- Upper abdominal lymphadenopathy is often present
- Low density nodules before contrast agent injection, usually become rapidly isodense with rest of liver parenchyma on enhanced scans
- Advanced disease may cause or simulate cirrhosis

Clinical Issues

- 3 well-recognized clinical syndromes: Chronic intrahepatic cholestasis, portal hypertension & Budd-Chiari syndrome are quite sporadic
- Complications: Hepatic failure is due to intrahepatic cholestasis & portal hypertension

Diagnostic Checklist

- Presence of hepatic nodules at imaging is not correlated with advanced pulmonary disease
- Sarcoidosis can appear in an atypical fashion, & it should be kept in mind in differential diagnosis of focal & diffuse liver disease
- Do not assume that heterogeneous hepatomegaly & abdominal lymphadenopathy are always malignant

Ultrasonographic Findings

- Real Time
  - Nonspecific, diffuse parenchymal heterogeneity
  - Granulomas: Hypoechoic nodules

Nuclear Medicine Findings

- Gallium scan
  - Gallium-67 localizes in areas of granulomatous infiltrates; however, it is nonspecific

Other Modality Findings

- Cholangiography may be useful in evaluation of patients with cholestatic sarcoid liver disease
  - Diffuse intrahepatic biliary strictures

Imaging Recommendations

- Best imaging tool
  - CT & MR are more sensitive than US
    - Able to show multinodular pattern
- Protocol advice: NECT & CECT followed by biopsy

DIFFERENTIAL DIAGNOSIS

Primary biliary cirrhosis (PBC)

- Disease of hepatic parenchyma with epithelioid cell granulomas affecting intra-hepatic biliary tree
- Idiopathic, progressive, nonsuppurative, destructive cholangitis of interlobar bile ducts/nodular regeneration/shrinkage of hepatic parenchyma
- PBC may also have heterogeneous hepatomegaly & upper abdominal lymphadenopathy
- "Lace-like" diffuse fibrosis & subtle high density nodules more characteristic of PBC
- Mitochondrial antibody test can differentiate among these; test is negative in sarcoid & usually positive in primary biliary cirrhosis

Lymphoma

- Secondary lymphoma is either multinodular or diffusely infiltrative
- Homogeneous hepatomegaly &/or hypoechoic focal nodules
- Lower frequency of retrocrural adenopathy in sarcoidosis than in non-Hodgkin lymphoma (NHL)
- Mean nodal size is greater in NHL & nodes tend to be more confluent in NHL, discrete in sarcoidosis

Opportunistic infection

- In immunocompromised hosts
- Infection by fungi, mycobacteria, viruses, etc.
- Heterogeneous hepatosplenomegaly ± abdominal lymphadenopathy may be indistinguishable by imaging alone

Gaucher disease

- Glucocerebroside accumulates in reticuloendothelium
- Well-defined hypoechoic areas on US/low-density lesions lesions on CT
- Early onset of significant hepatosplenomegaly

Amyloidosis

- Heterogeneous hepatomegaly, ± lymphadenopathy
- Generalized or focal ↓ liver density, both on pre & post-contrast-enhanced scans

PATHOLOGY

General Features

- General path comments
  - Sarcoidosis is a main cause of hepatic granulomas
  - Although involvement of abdominal organs is frequent in course of systemic sarcoidosis, its clinical manifestations are usually documented after diagnosis has been made on basis of thoracic manifestations
- Etiology: Mechanisms that initiate formation of sarcoid granulomas are unknown
- Epidemiology
  - Prevalence: 1-6:100,000
  - 24-79% of patients have liver involvement
- Associated abnormalities
  - Association between sarcoidosis & primary sclerosing cholangitis has been suggested
  - Coexistence of sarcoidosis & a wide range of autoimmune disorders

HEPATIC SARCOIDOSIS

Gross Pathologic & Surgical Features

- Hepatomegaly (18-29%)
- Scattered nodular lesions (5%)

Microscopic Features

- Noncaseating epithelioid granulomas with multinucleated giant cells of Langhans type are scattered throughout liver
- Characteristic inclusions in giant cells (for example, Schaumann bodies & asteroid bodies) are not seen in all cases & are not pathognomonic
- Confluent granulomas & fibrosis can be present in cases with severe hepatic involvement

CLINICAL ISSUES

Presentation

- Most common signs/symptoms
  - Overt clinical manifestations are uncommon
  - Asymptomatic
  - Hepatosplenomegaly in about 20% of cases
  - Abdominal &/or pelvic lymphadenopathy
  - Rarely it can lead to chronic inflammation, chronic hepatitis, & cirrhosis, or nodular hyperplasia
    - Cirrhosis & focal fibrosis may be caused by ischemia secondary to primary granulomatous phlebitis of portal & hepatic veins
- Clinical profile
  - Diverse clinical presentations
  - Lab: Mild elevation of liver enzymes (4%)
    - Hypercalcemia, hypercalciuria, hypergammaglobenemia, anemia, leukopenia
    - Angiotensin-converting enzyme; elevated in 60% of patients with sarcoidosis - nonspecific & generally not useful in following course of disease
  - Diagnosis: Liver biopsy; showing diffuse small noncaseating granulomas, usually < 2 mm in size
  - Fine-needle biopsy of palpable or radiologically visible lesions has been proposed recently as reliable, cost-effective method for diagnosis of sarcoidosis

Demographics

- Age: 20-40 years
- Gender: M:F = 1:3
- Ethnicity: African-Americans:Caucasians = 14:1

Natural History & Prognosis

- Variable natural history; small granulomas may heal without a trace, but confluent granulomas can result in extensive, irregular scarring
- 3 well-recognized clinical syndromes: Chronic intrahepatic cholestasis, portal hypertension & Budd-Chiari syndrome are quite sporadic
  - Occlusion of intrahepatic portal vein branches by granulomatous inflammation probably accounts for development of portal hypertension in some cases
  - Granulomatous cholangitis leading to ductopenia seems to be underlying pathogenetic mechanism of chronic cholestatic syndrome of sarcoidosis
- Life-threatening situations are extremely rare
  - May be due to failure of vital organs--lungs, heart, kidney, liver; & usually due to irreversible fibrosis
- Complications: Hepatic failure is due to intrahepatic cholestasis & portal hypertension
- Rare complications: Budd-Chiari syndrome & obstructive jaundice, attributable to hepatic hilar lymphadenopathy or strictures of bile ducts
- Sarcoidosis-lymphoma syndrome: Sarcoidosis complicated by non-Hodgkin lymphoma (infrequent but well-described event)
- Prognosis: 10% mortality (cor pulmonale/CNS/lung fibrosis/liver cirrhosis)

Treatment

- Spontaneous remission
- Corticosteroids, anti-inflammatory agents & cytotoxic drugs: Prednisone, chloroquine, methotrexate
- Follow-up US &/or CT show good correlation with improvement in liver enzyme levels after steroid therapy & normalization of liver pattern

DIAGNOSTIC CHECKLIST

Consider

- Most granulomas at pathology are small; imaging studies depict nodular changes in only approximately one-third of affected patients
- Marked abdominal CT findings are uncommon in sarcoidosis & correlate with disease activity but not chest radiographic stage
  - Presence of hepatic nodules at imaging is not correlated with advanced pulmonary disease
- Important to differentiate sarcoidosis from other causes of hepatic granulomas, such as infectious diseases, in which treatment with corticosteroids could be fatal
- Sarcoidosis can appear in an atypical fashion, & it should be kept in mind in differential diagnosis of focal & diffuse liver disease
- US can play useful role in reaching diagnosis & monitoring response to treatment, despite its negligible usefulness for most patients with sarcoidosis who have typical pulmonary & nodal manifestations

Image Interpretation Pearls

- Do not assume that heterogeneous hepatomegaly & abdominal lymphadenopathy are always malignant
- Biopsy is key to diagnosis

SELECTED REFERENCES

1. Amarapurkar DN et al: Hepatic sarcoidosis. Indian J Gastroenterol. 22(3):98-100, 2003
2. Sartori S et al: Sonographically guided biopsy and sonographic monitoring in the diagnosis and follow-up of 2 cases of sarcoidosis with hepatic nodules and inconclusive thoracic findings. J Ultrasound Med. 21(9):1035-9, 2002
3. Scott GC et al: CT patterns of nodular hepatic and splenic sarcoidosis: a review of the literature. J Comput Assist Tomogr. 21(3):369-72, 1997
4. Warshauer DM et al: Abdominal CT findings in sarcoidosis: radiologic and clinical correlation. Radiology. 192(1):93-8, 1994
5. Britt AR et al: Sarcoidosis: abdominal manifestations at CT. Radiology. 178(1):91-4, 1991

HEPATIC SARCOIDOSIS

IMAGE GALLERY

Typical

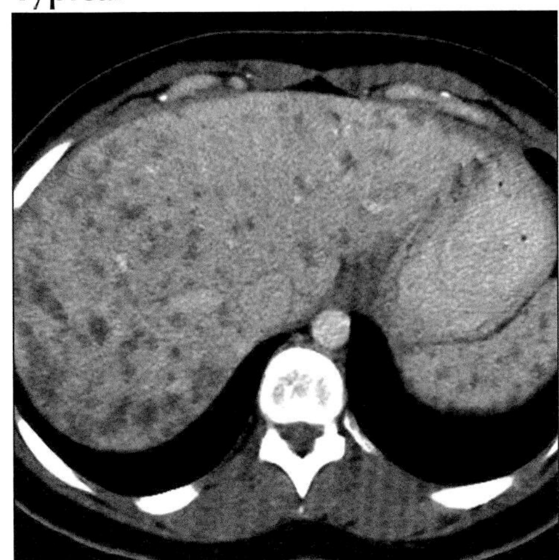

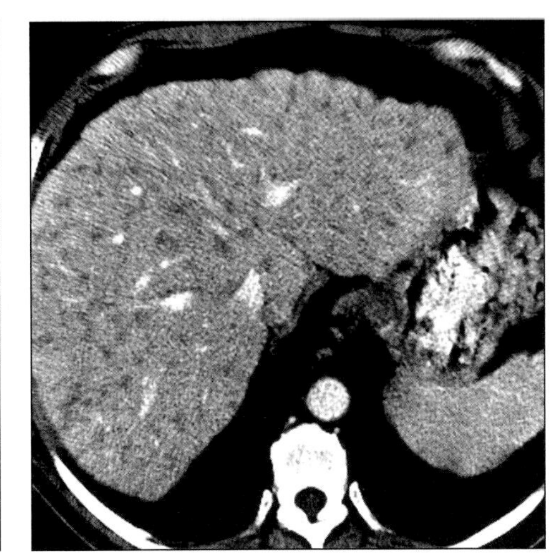

(Left) *Axial CECT shows innumerable hypodense nodules in liver and spleen.* ***(Right)*** *Axial CECT shows multiple hypodense nodules and some fibrosis, suggested by irregular contour of the liver.*

Typical

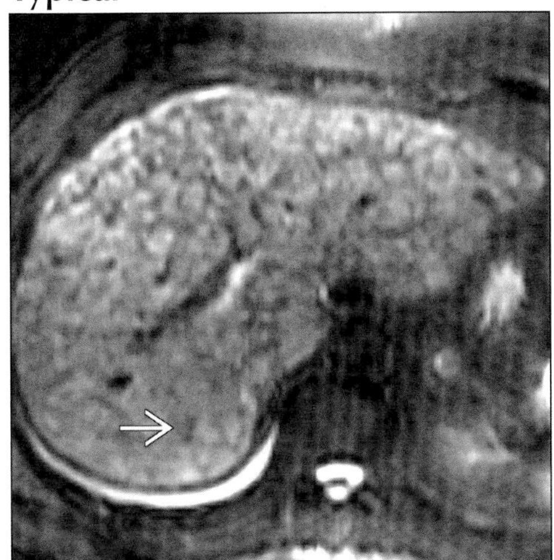

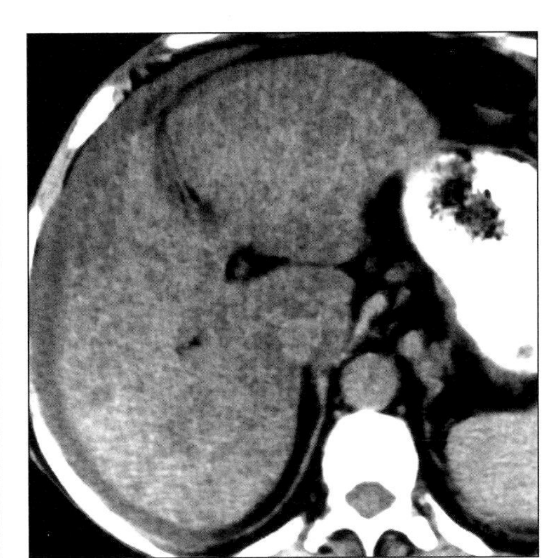

(Left) *Axial T2WI MR shows heterogeneous liver with nodular surface and parenchyma secondary to underlying sarcoidosis. Granulomas appear hypointense to surrounding liver parenchyma (arrow).* ***(Right)*** *Axial CECT shows cirrhotic morphology (wide fissures, enlarged caudate and lateral segment), ascites and multinodular liver, all due to sarcoidosis.*

Typical

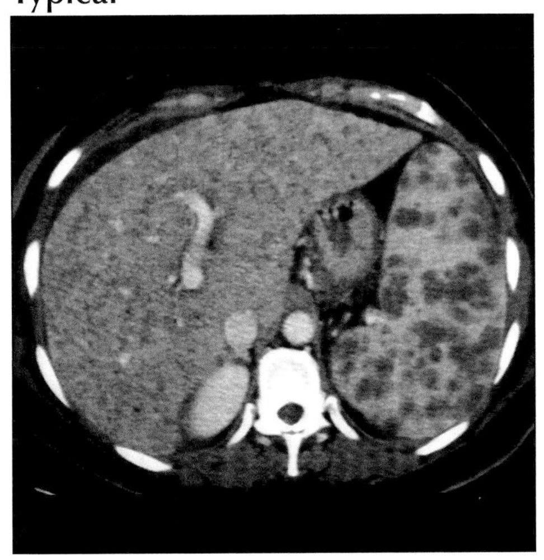

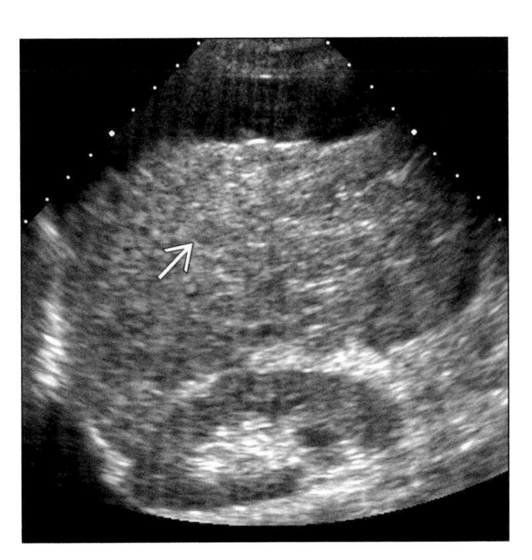

(Left) *Axial CECT shows multinodular liver and spleen with splenic lesions substantially larger and more evident.* ***(Right)*** *Sagittal sonogram shows heterogeneous liver with innumerable subcentimeter hypoechoic nodules (arrow), ascites.*

HEPATIC AV MALFORMATION (O-W-R)

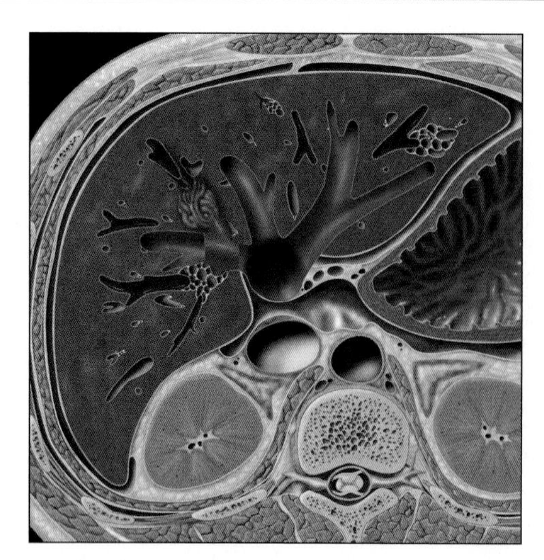

Graphic shows dilated hepatic veins and arteries with direct intraparenchymal communication through tortuous vascular channels.

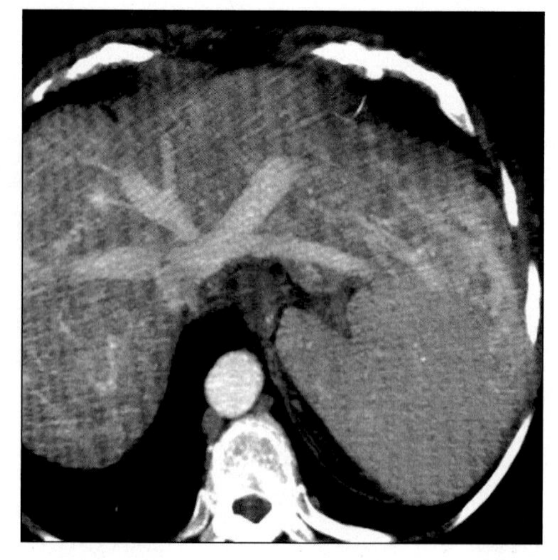

Axial CECT in late arterial phase shows early filling of dilated hepatic veins and innumerable irregular vascular channels connecting arteries and veins, more apparent in lateral segment.

TERMINOLOGY

Abbreviations and Synonyms

- Arterio-venous malformation (AVM)
- Hereditary hemorrhagic telangiectasia (HHT); Osler-Weber-Rendu disease (O-W-R)

Definitions

- Hereditary multiorgan disorder that results in fibrovascular dysplasia with development of telangiectasias & AVMs
- Direct connection between arteries & veins with absence of capillaries (telangiectasias are small AVMs)

IMAGING FINDINGS

General Features

- Best diagnostic clue: Dilated hepatic/portal veins and arteries with direct intraparenchymal communication through tortuous vascular channels
- Location: Skin, lungs, liver, mucus membranes, gastrointestinal tract (GIT) & brain
- Size: Few mm to several cm; composed of punctate spots with diameter of 1-4 mm diameter
- Key concepts
  - Focal sinusoidal ectasia, arteriovenous (AV) shunts through abnormal direct communications between arterioles & ectatic sinusoids, & portovenous shunts due to frequent & large communications between portal veins & ectatic sinusoids

CT Findings

- CECT
  - Prominent extra-hepatic or both extra- & intra-hepatic hepatic artery; dilated hepatic and/or portal veins
  - Early filling of portal venous or hepatic venous trunks on helical CT indicates AV shunt (52%)
  - Intrahepatic AV fistulas may also be identified
  - ± Focal bile duct dilatation (external compression)
  - Arterial phase: Tortuous, irregular & poorly defined hepatic arterial branches
  - Venous phase: Early opacification of ectatic veins
  - Parenchymal phase: Hepatogram is heterogeneous
  - Transient hepatic parenchymal enhancement (65%); indirect sign of presence of arterioportal shunt
  - Telangiectases (63%): Small vascular spots; more readily recognizable on reconstructed multiplanar reformatted & MIP images
  - Large confluent vascular masses (25%); appear as larger vascular pools with early & persistent enhancement during arterial phases
    - These are large areas of multiple telangiectases that coalesce or large shunts

DDx: Hepatic Arteriovenous (AV) Shunts

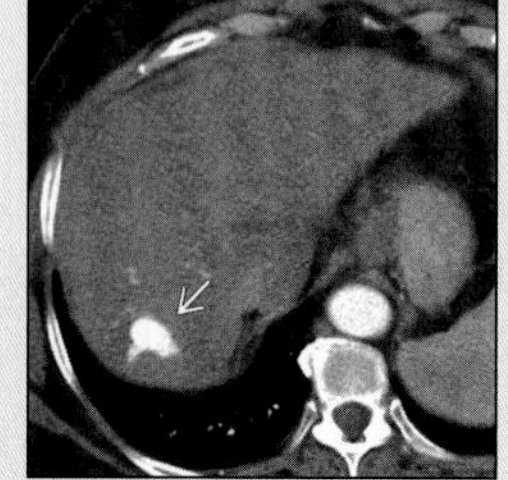

AV Shunt in Cirrhosis

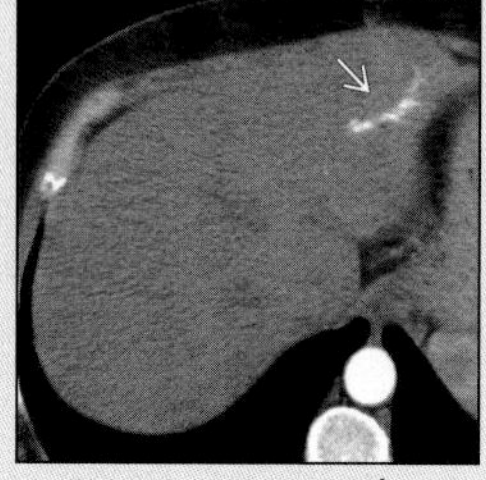

Post Bx AV Fistula

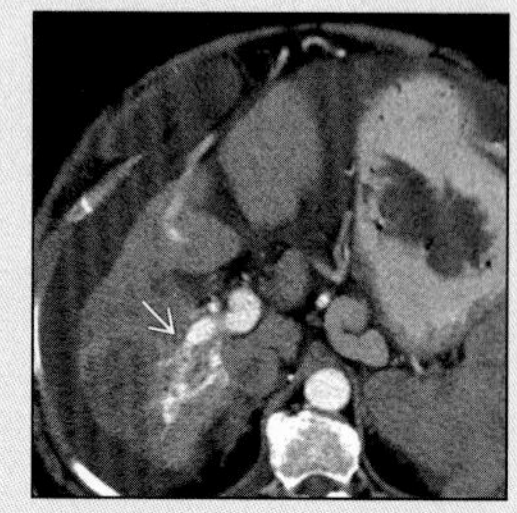

HCC with AV Shunting

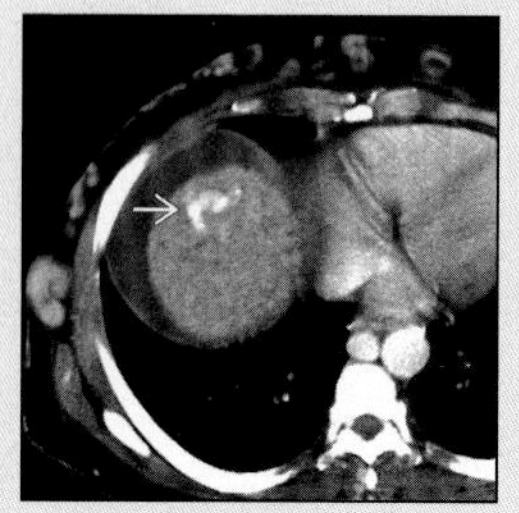

Budd-Chiari Syndrome

HEPATIC AV MALFORMATION (O-W-R)

Key Facts

Terminology
- Hereditary multiorgan disorder that results in fibrovascular dysplasia with development of telangiectasias & AVMs

Imaging Findings
- Best diagnostic clue: Dilated hepatic/portal veins and arteries with direct intraparenchymal communication through tortuous vascular channels
- Numerous irregular areas of dense contrast accumulation throughout liver parenchyma

Top Differential Diagnoses
- Sinusoidal & arterial changes in cirrhosis
- Traumatic intrahepatic arteriovenous fistulas
- Arteriovenous shunting with tumors
- Budd-Chiari syndrome

Pathology
- Pulmonary AVMs are more likely to cause symptoms & complications in patients with O-W-R
- Hepatic angiodysplastic vascular changes include telangiectasias, cavernous hemangiomas, aneurysm of intraparenchymal branches of hepatic artery & intraparenchymal hepato-portal & arterio-venous fistulas

Clinical Issues
- Clinical profile: Diagnostic criteria: Family history, epistaxis, mucocutaneous telangiectasias, AVMs
- Complications: High-output congestive heart failure, portal hypertension, hepatic portosystemic encephalopathy, biliary ischemia & liver failure

---

- CTA: Multiple ectatic vessels & AVMs

MR Findings
- T1WI
  - Network of vessels with flow voids on spin echo
  - Telangiectasias: Hypo- to isointense
- STIR
  - Best for extent of malformations
    - MRA alone may underestimate extent
- T1 C+
  - Telangiectasias: Small homogeneously enhancing
  - Early enhancement of peripheral portal veins & wedge-shaped transient parenchymal enhancement during hepatic arterial phase
  - Dynamic gradient echo after GD-DTPA for analysis of filling kinetics
- MRA
  - Number & size of feeding arteries & draining veins/depict map of anomalous vessels
  - Flow tagging makes it possible to define flow direction & to estimate flow velocity
  - Time between early arterial phase & enhancement of malformation used to distinguish high- & low-flow lesions
    - High-flow AVMs show early, intense enhancement
    - Venous malformations: Either not visible or show late enhancement of veins

Ultrasonographic Findings
- Real Time: Dilated hepatic arteries, multiple arteriovenous malformations & abnormal hepatic echogenicity
- Pulsed Doppler
  - High hepatic artery velocities = 153 +/- 65.2 cm/sec
  - Hepatic artery to portal vein shunts cause pulsatility of portal flow with phasic or continuous reversal
  - Hepatic artery to hepatic vein shunts show significant changes in Doppler waveform of hepatic vein; only in severe stages of disease
- Color Doppler: Tangled masses of enlarged tortuous arteries or multiple aneurysms of hepatic arteriole branches within liver

Angiographic Findings
- Conventional
  - Tortuous dilated hepatic arterial branches
  - Numerous irregular areas of dense contrast accumulation throughout liver parenchyma
  - Diffuse angiectases & diffuse mottled capillary blush
  - Early filling of hepatic or portal vein in shunts
  - Appearance depends on stage of development: All findings are present if shunting is severe
    - Isolated parenchymal modifications are found only in case of mild intrahepatic shunt

Imaging Recommendations
- Best imaging tool
  - Color Doppler as non-invasive screening modality
  - Angiography useful to delineate extent before surgical or angiographic interventions
  - Multi-detector row helical CT & multiplanar & angiographic reconstructions depict complex hepatic vascular alterations typical of HHT
- Protocol advice: Combine dynamic contrast-enhanced 3D gradient-echo MRI with STIR sequences

DIFFERENTIAL DIAGNOSIS

Sinusoidal & arterial changes in cirrhosis
- Dilation & ↑ number of hepatic arteries; 2 or 3 branches run parallel: "Duplication" &" trifurcation"
- "Corkscrew" appearance produced by combination of ↑ in arterial flow & ↓ in liver size
- Transsinusoidal arterioportal shunting in advanced cirrhosis; intrasegmental hepatofugal portal flow

Traumatic intrahepatic arteriovenous fistulas
- Causes: Biopsy (Bx), transhepatic biliary drainage, blunt or penetrating injury, rupture of hepatic aneurysm into portal vein
- Dilatation of feeding artery, early opacification of draining vein & poor visualization of artery distal to fistula due to steal of blood by fistula

HEPATIC AV MALFORMATION (O-W-R)

Arteriovenous shunting with tumors

- With hepatomas & metastatic tumors, shunting suggests venous invasion by tumors

Budd-Chiari syndrome

- Obstruction of hepatic venous outflow; collateral channels develop between hepatic, portal & systemic venous systems
- CT: Heterogeneous parenchymal density with periportal & peripheral enhancement, caudate lobe enlargement
- Arteriography: Stretching & attenuation of intrahepatic arteries & inhomogeneous parenchyma

PATHOLOGY

General Features

- General path comments
  - Pathogenesis is not known, involves several factors: Special formation of venules, capillaries & arterioles, abnormal perivascular connective tissue & endothelial cells
  - Small telangiectasis = focal dilatation of post capillary venules with prominent stress fibers in pericytes along luminal borders
  - Fully developed telangiectasis = markedly dilated & convoluted venules with excessive layers of smooth muscle without elastic fibers directly connecting to dilated arterioles
- Genetics
  - Autosomal dominant trait & exhibits high penetrance & great genetic heterogeneity
  - HHT phenotypes: HHT1, mutations at chromosome 9 alter protein endoglin & in HHT2, mutations at chromosome 12 alter protein activine or ALK-1
- Etiology: Gene encoding a protein that binds transforming growth factor
- Epidemiology: 10-20:100,000
- Associated abnormalities
  - 60% of pulmonary AVMs occur in patients with O-W-R; 15% of patients with O-W-R will have pulmonary AVMs
  - Pulmonary AVMs are more likely to cause symptoms & complications in patients with O-W-R

Gross Pathologic & Surgical Features

- Hepatic angiodysplastic vascular changes include telangiectasias, cavernous hemangiomas, aneurysm of intraparenchymal branches of hepatic artery & intraparenchymal hepato-portal & arterio-venous fistulas

Microscopic Features

- Clusters of dilated small blood vessels, lined by a single layer of endothelium

CLINICAL ISSUES

Presentation

- Most common signs/symptoms
  - Asymptomatic; anemia due to recurrent bleeds
  - Multiple mucocutaneous telangiectasias with multiorgan involvement
    - Nasal mucosa: Recurrent epistaxis
    - Skin: Lips, tongue, palate, face, conjunctiva & nail bed
    - CNS (cerebral or spinal AVM): Seizures, paraparesis, subarachnoid hemorrhage
    - Gastrointestinal: GI bleed & angiodysplasias
    - Pulmonary: Cyanosis, polycythemia, dyspnea on effort, clubbing, bruit
  - Liver: Often asymptomatic; rarely causes parenchymal fibrosis, biliary ischemia, liver failure
    - Hepatic parenchyma: Fibrosis, atypical cirrhosis, chronic active hepatitis (rare)
- Clinical profile: Diagnostic criteria: Family history, epistaxis, mucocutaneous telangiectasias, AVMs

Demographics

- Age: Onset: Adult life; hepatic involvement diagnosed 10-20 years after first appearance of telangiectasias
- Gender: M = F

Natural History & Prognosis

- Complications: High-output congestive heart failure, portal hypertension, hepatic portosystemic encephalopathy, biliary ischemia & liver failure
  - With extrahepatic involvement: Hemoptysis, hemothorax, cerebrovascular accident, cerebral abscess
- Prognosis: Usually good

Treatment

- Supportive: Iron/blood transfusion
- Hepatic arterial coil embolization, surgical ligation of hepatic artery, liver transplantation

DIAGNOSTIC CHECKLIST

Consider

- Due to high prevalence of pulmonary & cerebral AVMs, all patients with HHT should be screened for their presence
- Existence of arteriosystemic or arterioportal intrahepatic shunts that are not correlated with other pathologic conditions (i.e., neoplasia, cirrhosis, trauma) should raise suspicion of HHT
- Relatives of patients with HHT should be investigated for presence of disease

SELECTED REFERENCES

1. Hashimoto M et al: Angiography of hepatic vascular malformations associated with hereditary hemorrhagic telangiectasia. Cardiovasc Intervent Radiol. 26(2):177-80, 2003
2. Larson AM: Liver disease in hereditary hemorrhagic telangiectasia. J Clin Gastroenterol. 36(2):149-58, 2003
3. Hatzidakis AA et al: Hepatic involvement in hereditary hemorrhagic telangiectasia (Rendu-Osler-Weber disease). Eur Radiol. 12 Suppl 3:S51-5, 2002
4. Matsumoto S et al: Intrahepatic porto-hepatic venous shunts in Rendu-Osler-Weber disease: imaging demonstration. Eur Radiol. 2003

IMAGE GALLERY

Typical

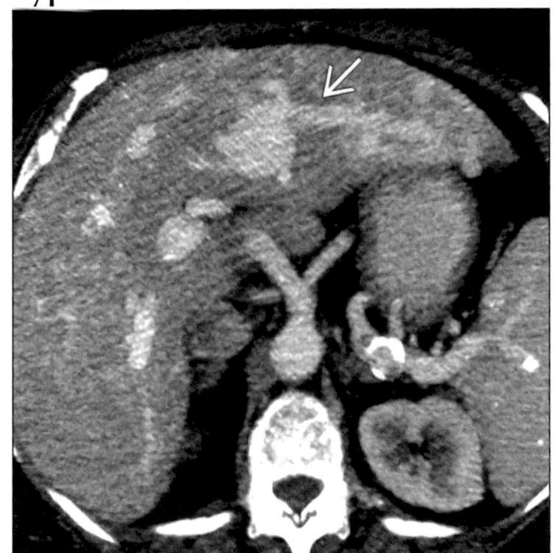

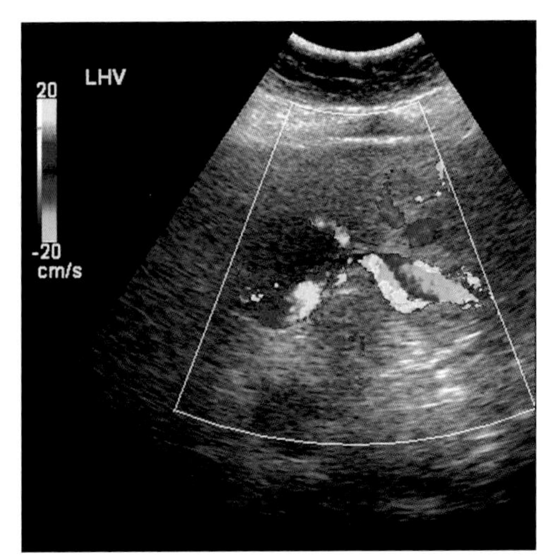

(Left) *Thick section axial CECT during arterial phase shows dilated hepatic artery, early opacification of large left portal vein and large arterio-portal fistula (arrow) in lateral segment.* ***(Right)*** *Color Doppler sonography shows tangled vascular mass along lateral segment of the liver, representative of intraparenchymal arterio-portal shunt.*

Typical

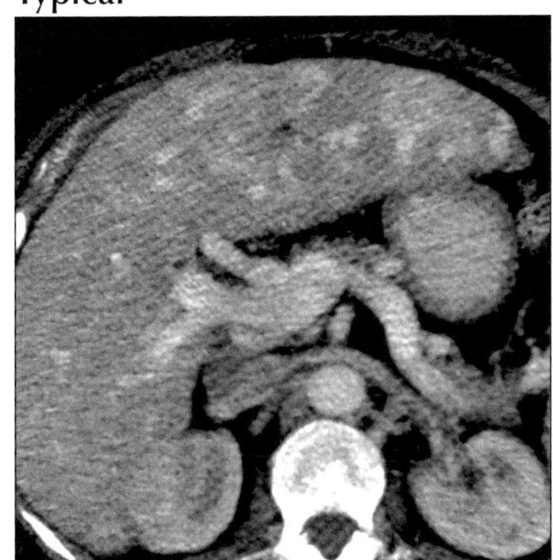

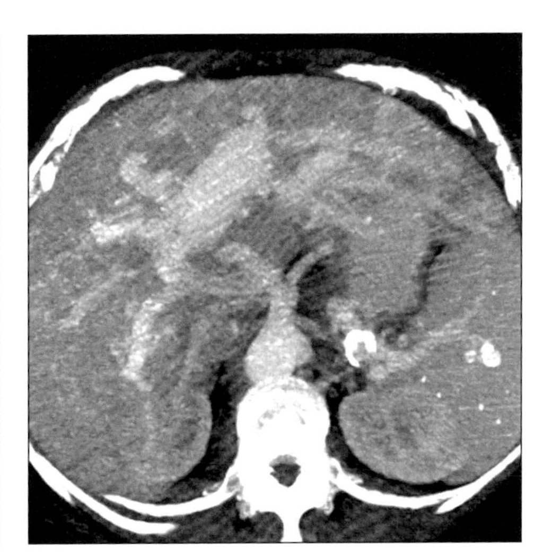

(Left) *Thick section axial CECT shows dilated hepatic artery and portal vein as well as early enhancement of diffusely dilated intrahepatic veins.* ***(Right)*** *Thick section axial CECT shows massive dilatation and early filling of hepatic and portal veins due to vascular malformations.*

Typical

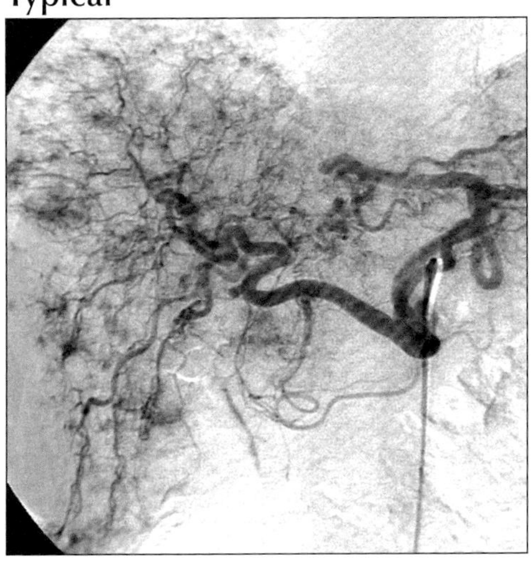

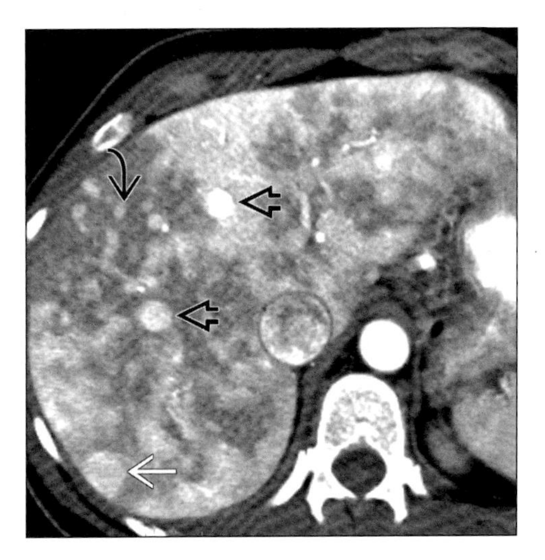

(Left) *Angiography shows dilated, tortuous hepatic arteries and early filling of innumerable telangiectasias.* ***(Right)*** *CECT (arterial phase) shows heterogeneous enhancement, early filling of dilated hepatic veins (open arrows), small telangiectasias (curved arrow), and larger confluent vascular masses (arrow).*

BUDD-CHIARI SYNDROME

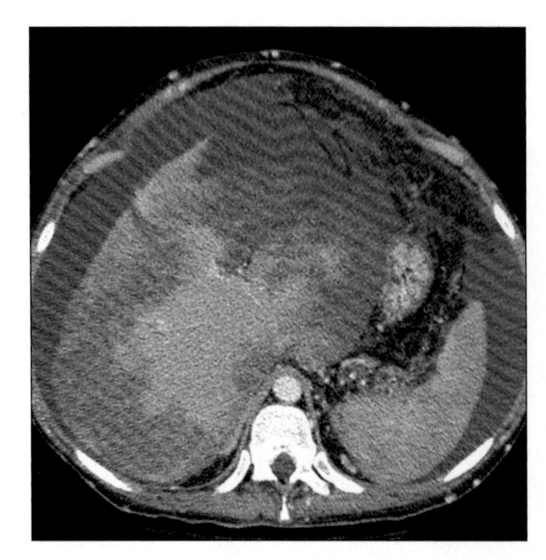

Axial CECT shows dysmorphic liver, ascites and subcutaneous venous collaterals. Central liver enhances normally and is hypertrophied while peripheral liver is hypodense and scarred.

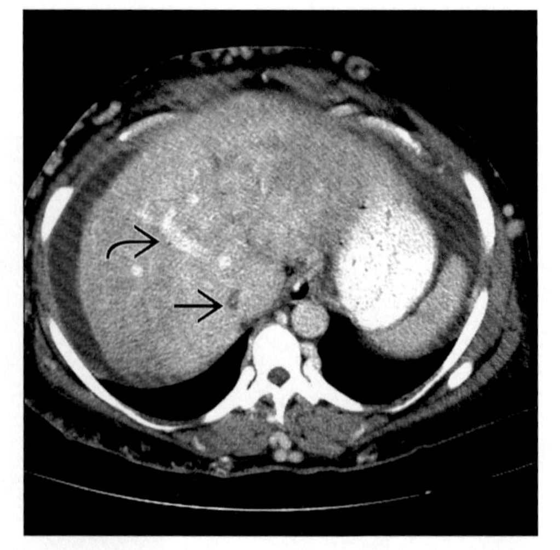

Axial CECT shows subcutaneous collaterals and ascites. Liver enhances heterogeneously. Note thrombosed IVC (arrow) and intrahepatic collateral (curved arrow), a veno-venous shunt.

TERMINOLOGY

Abbreviations and Synonyms

- Hepatic venous outflow obstruction

Definitions

- Global or segmental hepatic venous outflow obstruction (at level of large hepatic veins or suprahepatic segment of IVC)

IMAGING FINDINGS

General Features

- Best diagnostic clue: "Bicolored" hepatic veins (due to intrahepatic collateral pathways) pathognomonic of chronic Budd-Chiari on color Doppler sonography
- Location: Hepatic veins, IVC or centrilobar veins
- Size
  - Acute phase: Markedly enlarged liver
  - Chronic phase
    - Right & left lobes of liver: Atrophy
    - Caudate lobe: Hypertrophy
- Key concepts
  - Hepatic venous outflow obstruction
  - Characteristic finding of Budd-Chiari syndrome
    - Large regenerative nodules (nodular regenerative hyperplasia) in dysmorphic liver
- Other general features
  - Budd-Chiari is a rare syndrome
  - Classified based on cause & pathophysiology
  - Primary type: Congenital, injury, infection
    - Secondary type: Usually due to thrombosis

CT Findings

- NECT
  - Acute phase
    - Diffusely hypodense enlarged liver
    - Narrowed IVC + hepatic veins & ascites
    - Hyperdense IVC & hepatic veins (due to ↑ attenuation of thrombus)
  - Chronic phase
    - Diffusely hypodense liver
    - Non-visualization of IVC & hepatic veins
    - Hypertrophy of caudate lobe
    - Atrophy of peripheral segments
    - Ratio of caudate width to right lobe: ≥ 0.55:1
- CECT
  - Acute phase
    - Classic "flip-flop" pattern is seen
    - Early enhancement of caudate lobe & central portion around IVC, with decreased liver enhancement peripherally
    - Later decreased enhancement centrally with increased enhancement peripherally
    - Narrowed hypodense hepatic veins & IVC with hyperdense walls

DDx: Dysmorphic Liver with Lobular Contour

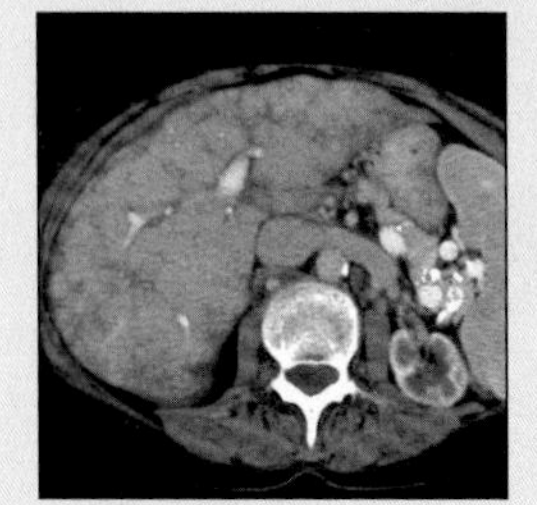

Cardiac Cirrhosis

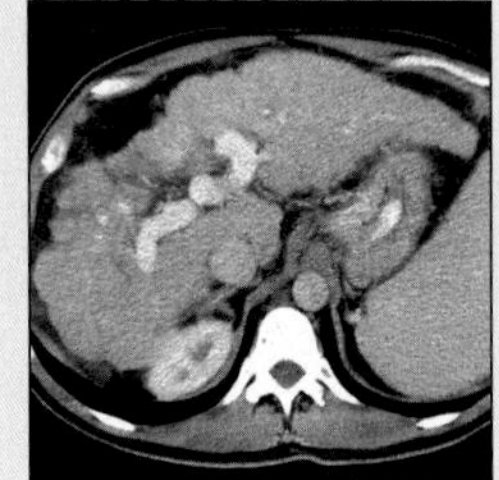

Cirrhosis

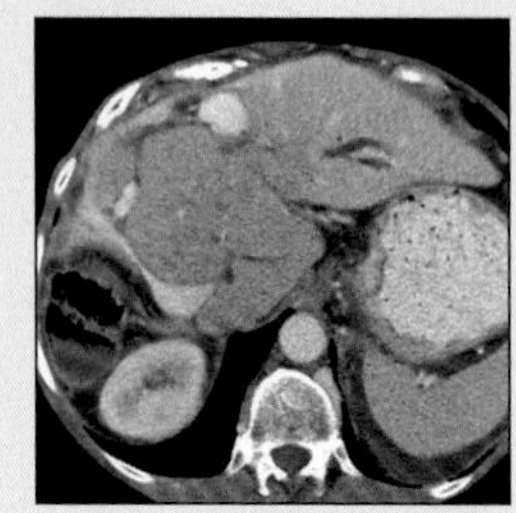

Sclerosing Cholangitis

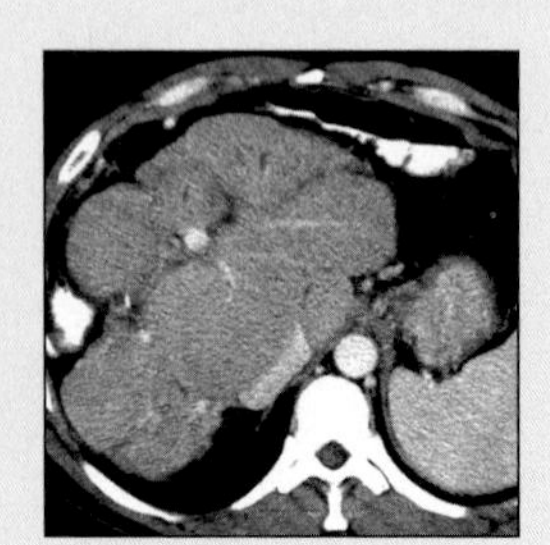

Sclerosing Cholangitis

BUDD-CHIARI SYNDROME

Key Facts

Terminology
- Hepatic venous outflow obstruction
- Global or segmental hepatic venous outflow obstruction (at level of large hepatic veins or suprahepatic segment of IVC)

Imaging Findings
- Best diagnostic clue: "Bicolored" hepatic veins (due to intrahepatic collateral pathways) pathognomonic of chronic Budd-Chiari on color Doppler sonography
- Non-visualization of IVC & hepatic veins
- Hypertrophied caudate vein
- "Spider web" pattern of hepatic venous collaterals

Top Differential Diagnoses
- Hepatic cirrhosis
- Primary sclerosing cholangitis (PSC)

Pathology
- Primary: Venous outflow membranous obstruction
- Secondary: Thrombotic; rarely nonthrombotic
- Type I: Occlusion of IVC ± hepatic veins
- Type II: Occlusion of major hepatic veins ± IVC
- Type III: Occlusion of small centrilobar veins

Diagnostic Checklist
- Rule out cirrhosis & primary sclerosing cholangitis
- Absent, reversed or flat flow in hepatic veins & reversed flow in IVC on color-Doppler sonography
- Check for hypercoagulable conditions, prior chemotherapy or marrow transplant

---

  - Chronic phase
    - Total obliteration of IVC & hepatic veins
    - "Large regenerative nodules": Nodular regenerative hyperplasia
    - Enhancing 1-4 cm hyperdense nodules ± hypodense ring
- CTA: Hepatic venous outflow obstruction

MR Findings
- T1WI
  - Increased intensity of liver centrally with peripheral heterogeneity
  - Narrowed or absent hepatic veins & IVC
  - Hyperintense nodules & enlarged caudate lobe
- T2WI
  - Fail to visualize hepatic veins & IVC
  - Isointense or hypointense regenerative nodules
- T2* GRE: Fails to show flow in hepatic veins or IVC
- T1 C+
  - Tumor thrombus (rare cause) may show contrast-enhancement
  - Acute phase
    - Involved parenchyma enhances less than surrounding liver
  - Congested liver with ↑ water content
    - Peripheral liver enhances less than central liver due to increased parenchymal pressure & decreased blood supply
  - Chronic phase
    - Enhancement is more variable & may be increased
    - Nodules: Intense homogeneous enhancement
- MRA: Depicts thrombus & level of venous obstruction

Ultrasonographic Findings
- Real Time
  - Hepatic veins
    - Narrowed, not visualized or filled with thrombus
    - Hypertrophied caudate vein
- Color Doppler
  - Hepatic veins & IVC
    - Absent or flat flow in hepatic veins
    - Reversed flow in hepatic veins or IVC
    - "Bicolored" hepatic veins: Due to intrahepatic collateral pathways
    - Sensitivity: 87.5%
  - Portal vein
    - Slow hepatofugal flow: < 11 cm/sec
    - Congestion index: > 0.1
  - Hepatic artery: Resistive index ≥ 0.75

Angiographic Findings
- Inferior venacavography or hepatic venacavography
  - "Spider web" pattern of hepatic venous collaterals
  - Thrombus in hepatic veins or IVC
  - Long segmental compression of IVC
    - Acute phase: Due to diffuse hepatomegaly
    - Chronic phase: Hypertrophy of caudate lobe
  - Hepatic arteries
    - Acute phase: Narrowing, stretching, bowing
    - Chronic phase: Dilated & arterioportal shunts

Imaging Recommendations
- Best imaging tool: Color Doppler sonography or angiography

DIFFERENTIAL DIAGNOSIS

Hepatic cirrhosis
- Hypertrophy: Caudate & lateral segment of left lobe
- Atrophy: Right lobe & medial segment of left lobe
- Varices, ascites, splenomegaly
- Patent hepatic veins & IVC
- Regenerative nodules
  - Usually small in size compared to Budd-Chiari
  - Cirrhotic nodules often have increased iron
    - Usually hypovascular; decreased signal on T2WI

Primary sclerosing cholangitis (PSC)
- Chronic cholestatic disease of unknown cause
- 70% of cases: Associated with ulcerative colitis
- Bile ducts on cholangiography: Segmental strictures, beading, pruning, nodular thickening, skip dilatations
- Atrophy/hypertrophy of lobes; ± cirrhotic changes

BUDD-CHIARI SYNDROME

PATHOLOGY

General Features

- General path comments
  - Embryology-anatomy
    - Primary type: Total or incomplete membranous obstruction of hepatic venous outflow
    - Deviations of complex embryologic process of IVC
- Etiology
  - Primary: Venous outflow membranous obstruction
    - Controversial etiology
    - Congenital, injury or infection
  - Secondary: Thrombotic; rarely nonthrombotic
    - Obstruction of central & sublobular veins: Chemotherapy & radiation
    - Obstruction of major hepatic veins: Hypercoagulable states (e.g., oral contraceptives, polycythemia, protein C deficiency)
    - Obstruction of small centrilobular veins (veno-occlusive disease): Bone marrow transplantation & antineoplastic drugs
    - Nonthrombotic: Hepatic & extrahepatic masses
- Epidemiology
  - Primary (congenital-membranous type)
    - Common in Japan, India, Israel & South Africa
  - Secondary (thrombotic)
    - Most common in Western countries
    - Usually due to hypercoagulable state
  - Secondary (nonthrombotic)
    - 2nd most common in Western countries

Gross Pathologic & Surgical Features

- Acute phase
  - Liver enlarged, congested
  - Occlusion of hepatic veins & IVC
- Chronic phase
  - Liver: Nodular, shrunken, may be cirrhotic
  - Hypertrophy of caudate lobe & atrophy other lobes

Microscopic Features

- Centrilobular congestion, dilated sinusoids
- Fibrosis, necrosis & cell atrophy

Staging, Grading or Classification Criteria

- Classified into three types based on location
  - Type I: Occlusion of IVC ± hepatic veins
  - Type II: Occlusion of major hepatic veins ± IVC
  - Type III: Occlusion of small centrilobar veins
    - Defined as "veno-occlusive disease"

CLINICAL ISSUES

Presentation

- Most common signs/symptoms
  - Acute phase
    - Rapid onset RUQ pain, tender liver, hypotension
  - Chronic phase
    - RUQ pain, hepatomegaly, splenomegaly
    - Jaundice, ascites, varices
- Lab data
  - Acute
    - Liver function tests: Mild to markedly increased
    - Clotting factors: Decreased
  - Chronic:
    - Transaminases: Normal or moderately increased
    - Albumin & clotting factors: Decreased

Demographics

- Age: Any age group
- Gender: Females more than males

Natural History & Prognosis

- Complications
  - Acute: Liver failure, emboli from IVC thrombus
  - Chronic: Variceal bleeding (cirrhosis), portal HTN
  - Membranous obstruction of IVC
    - Complicated by hepatocellular carcinoma in 20-40% cases in Japan & South Africa
- Prognosis
  - Based on rate/degree of hepatic outflow obstruction
    - Mild & moderate obstruction: Good
    - Severe obstruction: Poor
  - Acute early phase (good); acute late phase (poor)
  - Chronic phase: Poor (with or without treatment)
  - Veno-occlusive disease: Varies from fulminant failure & death to mild with complete recovery

Treatment

- Medical management
  - Steroids, nutritional therapy, anticoagulants
- Membranous occlusion of IVC & hepatic veins
  - Balloon angioplasty, lasers, stent insertion
- TIPS (transjugular intrahepatic portosystemic shunt)
- Surgical alternatives
  - Membranotomy, membranectomy
  - Cavoplasty, liver transplantation

DIAGNOSTIC CHECKLIST

Consider

- Rule out cirrhosis & primary sclerosing cholangitis

Image Interpretation Pearls

- Absent, reversed or flat flow in hepatic veins & reversed flow in IVC on color-Doppler sonography
- Characteristic large benign regenerative nodules
- Check for hypercoagulable conditions, prior chemotherapy or marrow transplant

SELECTED REFERENCES

1. Brancatelli G et al: Benign regenerative nodules in Budd-Chiari syndrome and other vascular disorders of the liver: Radiologic-pathologic and clinical correlation. RadioGraphics. 22: 847-62, 2002
2. Vilgrain V et al: Hepatic nodules in Budd-Chiari syndrome: Imaging features. Radiology. 210: 443-50, 1999
3. Kane R et al: Diagnosis of Budd-Chiari syndrome: comparison between sonography and MR angiography. Radiology. 195(1):117-21, 1995
4. Millener P et al: Color Doppler imaging findings in patients with Budd-Chiari syndrome: correlation with venographic findings. AJR Am J Roentgenol. 161(2):307-12, 1993
5. Ralls PW et al: Budd-Chiari syndrome: detection with color Doppler sonography. AJR Am J Roentgenol. 159(1):113-6, 1992

BUDD-CHIARI SYNDROME

IMAGE GALLERY

Typical

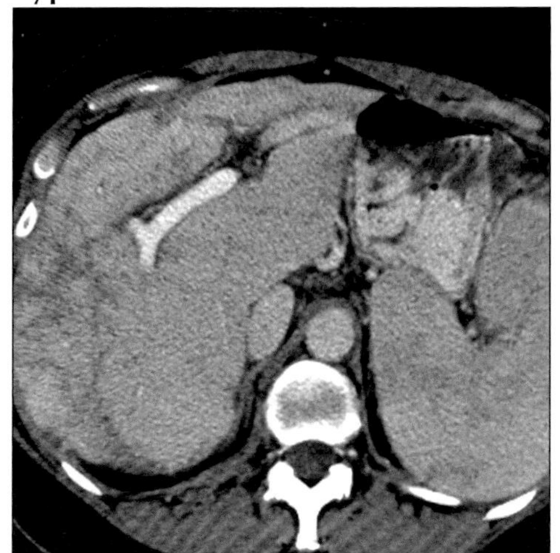

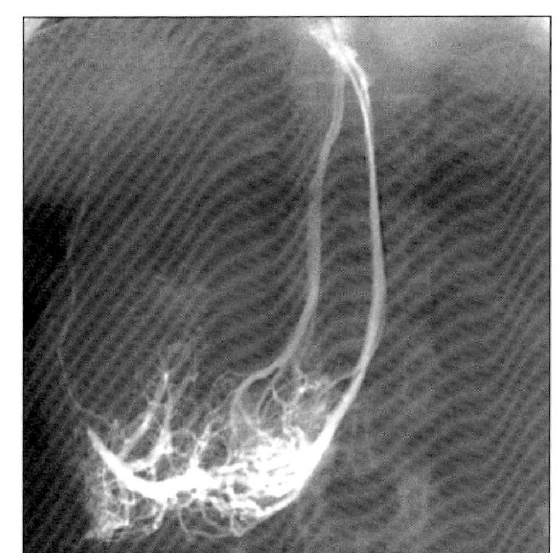

(Left) *Axial CECT shows caudate hypertrophy, peripheral scarring and heterogeneous enhancement.* ***(Right)*** *Hepatic venography shows no patency of hepatic veins; filling of collateral veins and "spider web" pattern of intrahepatic collateral vessels.*

Typical

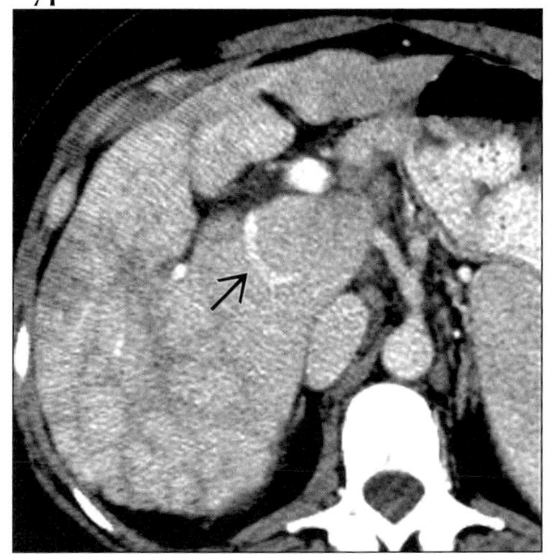

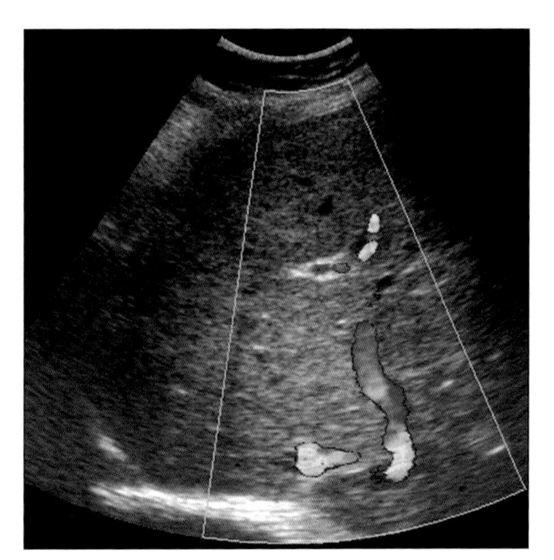

(Left) *Axial CECT shows caudate hypertrophy, large caudate collateral vein (arrow), and peripheral atrophy and heterogeneity.* ***(Right)*** *Color Doppler US shows large "bicolored" intrahepatic collateral vein.*

Typical

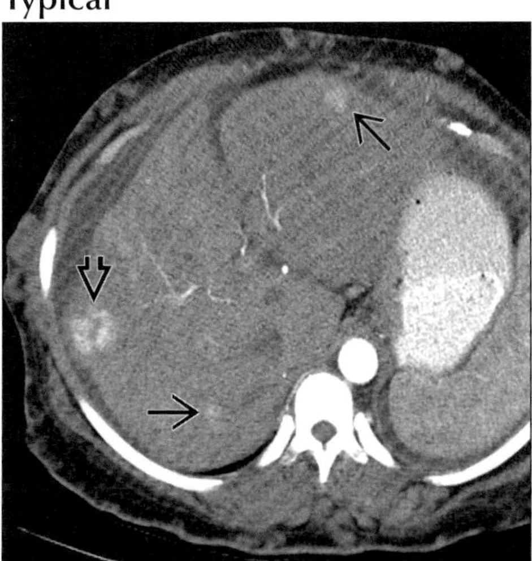

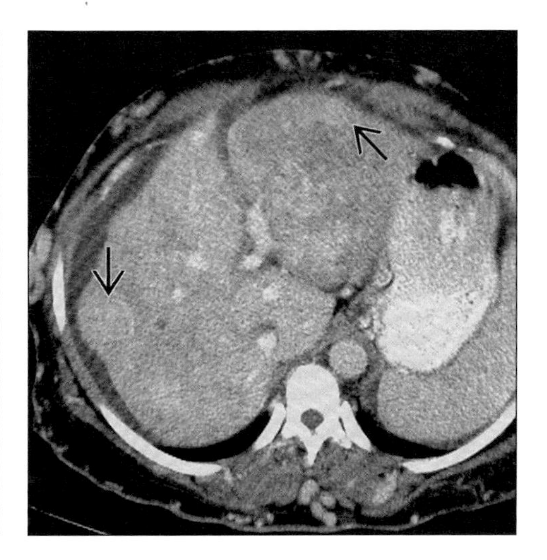

(Left) *Axial CECT in arterial phase shows dysmorphic liver, subcutaneous collaterals and ascites. Also note hypervascular nodules (arrows), the largest of which resembles FNH, with central scar (open arrow).* ***(Right)*** *Axial CECT in portal venous phase shows less apparent hypervascular nodules (arrows) nearly isodense to liver.*

PASSIVE HEPATIC CONGESTION

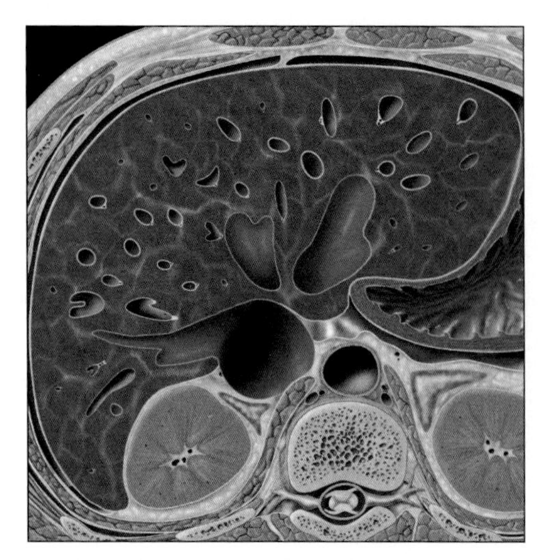

Graphic shows massive diffuse dilatation of hepatic veins and mildly heterogeneous liver parenchyma.

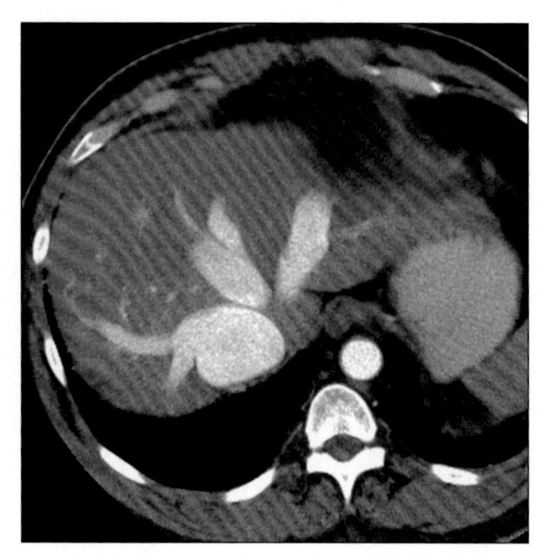

Axial CECT in arterial phase shows early filling, by reflux through heart, of dilated hepatic veins and IVC.

TERMINOLOGY

Abbreviations and Synonyms

- Congested liver in cardiac disease

Definitions

- Definition: Stasis of blood within liver parenchyma as a result of impaired hepatic venous drainage

IMAGING FINDINGS

General Features

- Best diagnostic clue: Dilated hepatic veins with to-and-fro blood flow on color Doppler
- Location: Liver, hepatic veins & IVC
- Size
  - Acute phase
    - Increase in liver size
  - Chronic phase
    - Decrease in liver size ("cardiac cirrhosis")
- Key concepts
  - Manifestations of liver in cardiac disease
    - Acute or early manifestation: Enlarged, heterogeneous liver
    - Chronic or late manifestation: Small cirrhotic liver (may resemble cirrhosis of other causes)
  - Passive hepatic congestion usually secondary to
    - Congestive heart failure (CHF)
    - Constrictive pericarditis
    - Tricuspid insufficiency
    - Right heart failure (e.g., pulmonary artery obstruction caused by lung cancer)
  - Characteristic sign on physical exam
    - Hepatojugular reflux

CT Findings

- Early enhancement of dilated IVC & hepatic veins
  - Due to contrast reflux from right atrium into IVC
- Heterogeneous, mottled, reticulated mosaic parenchymal pattern
- Linear & curvilinear areas of poor enhancement
  - Due to delayed enhancement of small & medium-sized hepatic veins
- Peripheral large patchy areas of poor/delayed enhancement
- Periportal low-attenuation (perivascular lymphedema)
- Decreased attenuation around intrahepatic IVC
- Hepatomegaly & ascites
- Chest findings
  - Cardiomegaly
  - ± Pericardial or pleural effusions

MR Findings

- T2WI: Periportal high signal intensity (periportal edema)

DDx: Hepatomegaly with Heterogeneous Enhancement

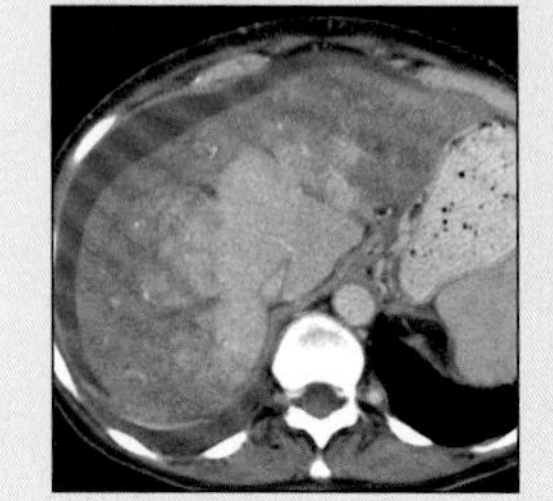

Budd-Chiari Syndrome

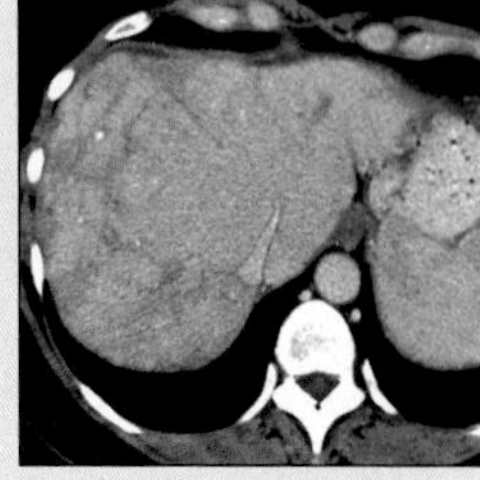

Budd-Chiari Syndrome

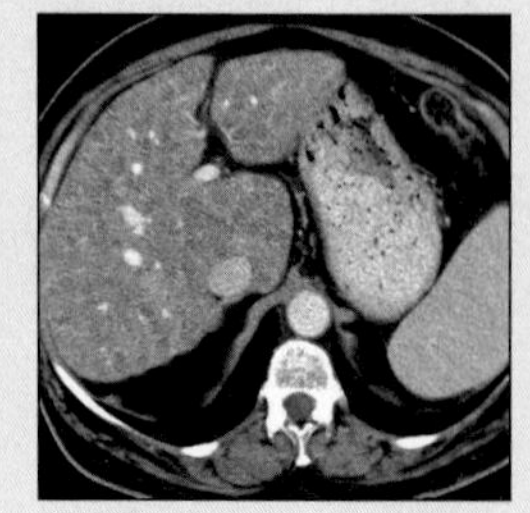

Cirrhosis with Steatosis

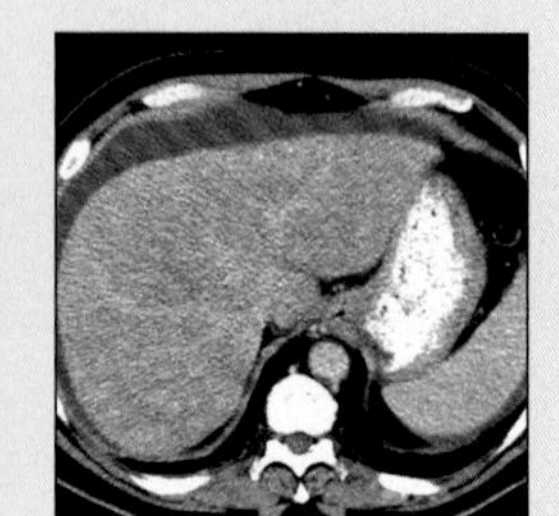

Acute Hepatitis

PASSIVE HEPATIC CONGESTION

Key Facts

Terminology
- Congested liver in cardiac disease
- Definition: Stasis of blood within liver parenchyma as a result of impaired hepatic venous drainage

Imaging Findings
- Best diagnostic clue: Dilated hepatic veins with to-and-fro blood flow on color Doppler
- Early enhancement of dilated IVC & hepatic veins
- Cardiomegaly
- ± Pericardial or pleural effusions
- Dilated IVC/hepatic veins; hepatomegaly; ± ascites

Top Differential Diagnoses
- Budd-Chiari syndrome
- Hepatic cirrhosis with steatosis
- Acute viral hepatitis

Pathology
- CHF, right heart failure, constrictive pericarditis
- Increased right atrial central venous pressure
- Pressure transmitted to IVC & hepatic veins
- Enlarged reddish-purple color liver
- "Nutmeg liver"

Clinical Issues
- Liver enlarged, tender
- Positive hepatojugular reflux
- Clinical profile: A cardiac disease patient with hepatomegaly & positive hepatojugular reflux

Diagnostic Checklist
- Differentiate acute Budd-Chiari syndrome, acute viral hepatitis from acute passive hepatic congestion, & viral or alcoholic cirrhosis from cardiac cirrhosis

- T2* GRE: Slow or even absent antegrade flow within IVC
- T1 C+
  - Liver enhancement pattern
    - Reticulated mosaic pattern of low signal intensity linear markings
    - Within 1-2 minutes liver becomes more homogeneous
  - Hepatic veins & suprahepatic IVC
    - Early enhancement due to reflux from atrium
  - Portal vein
    - Diminished, delayed or absent enhancement
  - Fast low-angle shot (FLASH) contrast-enhanced MR images
    - Early reflux of contrast into dilated hepatic veins & IVC
- MRA: Slow or absent antegrade flow within IVC

Ultrasonographic Findings
- Real Time
  - Dilated IVC/hepatic veins; hepatomegaly; ± ascites
  - Diameter of hepatic vein
    - Normal: 5.6 to 6.2 mm
    - Mean diameter: 8.8 mm (in passive congestion)
    - Increases up to 13 mm with pericardial effusion
- Color Doppler
  - Spectral velocity pattern (IVC & hepatic veins)
    - Loss of normal triphasic flow pattern
    - Spectral signal may have an "M" shape
    - Cardiac cirrhosis: Flattening of Doppler wave form in hepatic veins
  - Spectral velocity pattern (portal vein)
    - Increased pulsatility of portal venous Doppler signal
    - Normal continuous flow pattern
    - Mild respiratory variation
  - To-and-fro motion in hepatic veins & IVC
  - Tricuspid regurgitation: Normal triphasic hepatic vein shows
    - Decrease In size of antegrade systolic wave
    - Systolic/diastolic flow velocity ratio less than 0.6 (normal more than 4.0)

Imaging Recommendations
- Best imaging tool: Color Doppler sonography
- Protocol advice: Bi-phasic CT or MR to evaluate extent of liver damage

DIFFERENTIAL DIAGNOSIS

Budd-Chiari syndrome
- Global or segmental hepatic venous outflow obstruction at level of hepatic veins or IVC
- Acute phase
  - Diffuse hypodense enlarged liver
  - Narrowed IVC or hepatic veins & ascites
  - Classic "flip-flop" pattern is seen
    - Early enhancement of caudate lobe & central portion around IVC
    - Decreased enhancement peripherally
    - Later decreased enhancement centrally & increased enhancement peripherally
- Chronic phase
  - Nodular regenerative hyperplasia ("large regenerative nodules")
  - Enhancing 1 to 4 cm hyperdense nodules
- Color Doppler
  - Absent or flat or reversed flow in hepatic veins
  - Bicolored" hepatic veins
    - Due to intrahepatic collateral pathways
    - Pathognomonic of chronic phase of Budd-Chiari

Hepatic cirrhosis with steatosis
- Nodular liver contour
- Hepatic veins: Normal caliber & flow pattern
- Portal vein: May be large; possible hepatofugal flow
- Atrophy of right lobe & medial segment of left lobe
- Enlarged caudate lobe & lateral segment of left lobe
- Regenerative nodules are hypovascular
- Diagnosis: Biopsy & histology

Acute viral hepatitis
- Hepatomegaly
- Periportal hypodensity (fluid/lymphedema)
- Hepatic & portal veins: Normal caliber & flow pattern

PASSIVE HEPATIC CONGESTION

- Gallbladder wall thickening
- "Starry-sky" appearance on sonography
- Acute HBV
  - May present with serum sickness-like syndrome
    - Urticaria/arthritis/vasculitis/glomerulonephritis
- Lab data
  - Markedly elevated liver function tests
  - Elevated serologic markers

PATHOLOGY

General Features

- General path comments
  - CHF, right heart failure, constrictive pericarditis
  - Increased right atrial central venous pressure
  - Pressure transmitted to IVC & hepatic veins
  - Engorgement & dilatation of hepatic sinusoids with blood
  - Enlarged liver
- Etiology
  - Congestive heart failure (CHF)
  - Constrictive pericarditis
  - Pericardial effusion
  - Right-sided valvular diseases
    - Tricuspid & pulmonary
  - Cardiomyopathy

Gross Pathologic & Surgical Features

- Enlarged reddish-purple color liver
- "Nutmeg liver"
  - Congestion of central veins
  - Congestion of centrilobular hepatic sinusoids

Microscopic Features

- Acute or early phase
  - Centrilobular congestion
  - With or without sinusoidal dilatation
- Chronic or late phase
  - Atrophy
  - Centrilobular necrosis
  - Fibrosis
  - Finally sclerosis

CLINICAL ISSUES

Presentation

- Most common signs/symptoms
  - Liver enlarged, tender
  - Right upper quadrant pain due to stretching of Glisson capsule
  - Positive hepatojugular reflux
  - Pulsatile liver in acute phase
  - Splenomegaly in late phase
  - Rarely cardiomyopathy patient may present with hepatic failure before cardiac disease is diagnosed
  - Lab data
    - Acute: Mild abnormal liver function tests (LFT)
    - Chronic: Grossly abnormal LFT
  - Diagnosis
    - Based on radiological, pathological & clinical findings
    - Usually not made clinically because signs & symptoms of cardiac failure overshadow those of liver disease
- Clinical profile: A cardiac disease patient with hepatomegaly & positive hepatojugular reflux

Demographics

- Age: Any age group
- Gender: M = F

Natural History & Prognosis

- Complications
  - Hepatic failure
  - Cardiac cirrhosis
- Prognosis
  - Acute phase: Good
  - Chronic phase: Poor

Treatment

- Acute or early phase
  - Full recovery once patient's cardiac disease is corrected
- Chronic or late phase
  - Cardiac cirrhosis may be irreversible, even with correction of cardiac function

DIAGNOSTIC CHECKLIST

Consider

- Differentiate acute Budd-Chiari syndrome, acute viral hepatitis from acute passive hepatic congestion, & viral or alcoholic cirrhosis from cardiac cirrhosis

Image Interpretation Pearls

- Cardiomegaly
- Inferior venacava & hepatic veins
  - Dilated & early enhancement (due to reflux)
  - To-and-fro motion on color Doppler
  - Loss of normal triphasic velocity flow pattern

SELECTED REFERENCES

1. Gore RM et al: Passive hepatic congestion: cross-sectional imaging features. AJR Am J Roentgenol. 162(1):71-5, 1994
2. Abu-Yousef MM et al: Pulsatile portal vein flow: a sign of tricuspid regurgitation on duplex Doppler sonography. AJR Am J Roentgenol. 155: 785-788, 1990
3. Holley HC et al: Inhomogeneous enhancement of liver parenchyma secondary to passive congestion: contrast-enhanced CT. Radiology. 170(3 Pt 1):795-800, 1989
4. Moulton JS et al: Passive hepatic congestion in heart failure: CT abnormalities. AJR Am J Roentgenol. 151: 939-942, 1988
5. Tani I et al: MR imaging of diffuse liver disease. AJR Am J Roentgenol. 2000

PASSIVE HEPATIC CONGESTION

IMAGE GALLERY

Typical

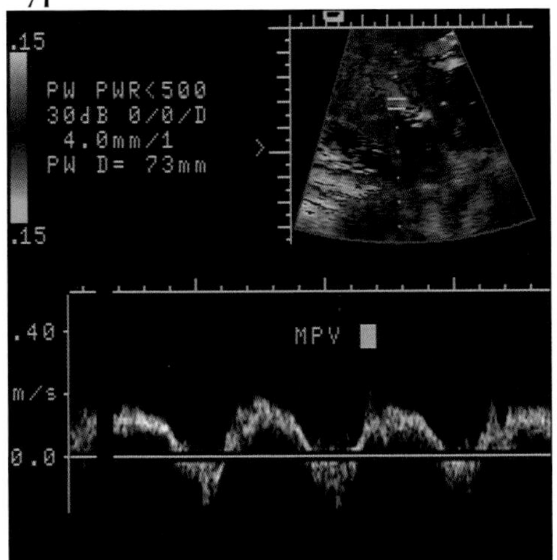

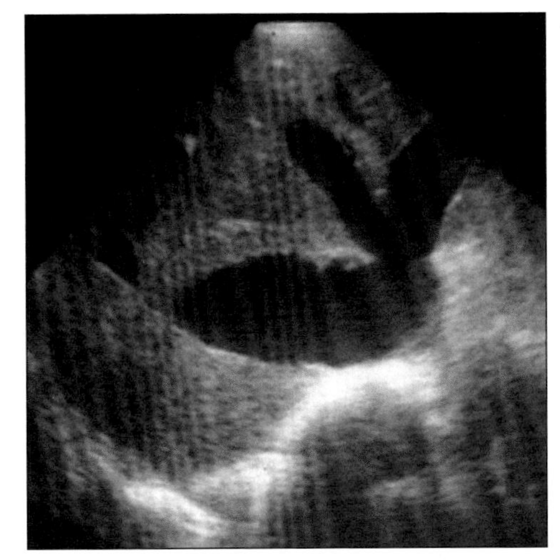

(Left) *Increased pulsatility of portal vein (MPV) Doppler signal is demonstrated in this patient with passive hepatic congestion secondary to tricuspid insufficiency.* ***(Right)*** *Axial grayscale US shows dilated hepatic veins and IVC in a patient with passive hepatic congestion.*

Typical

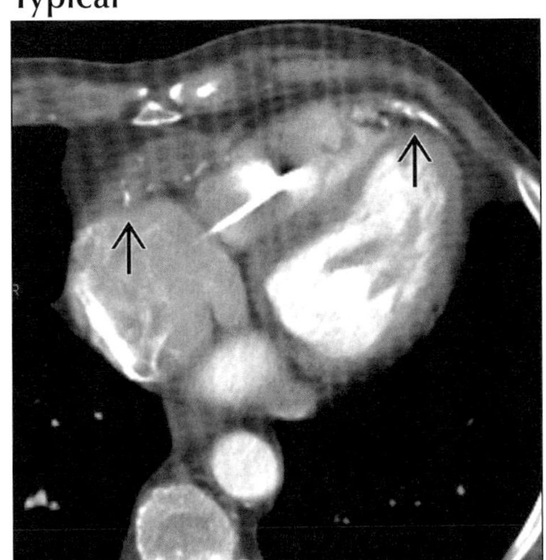

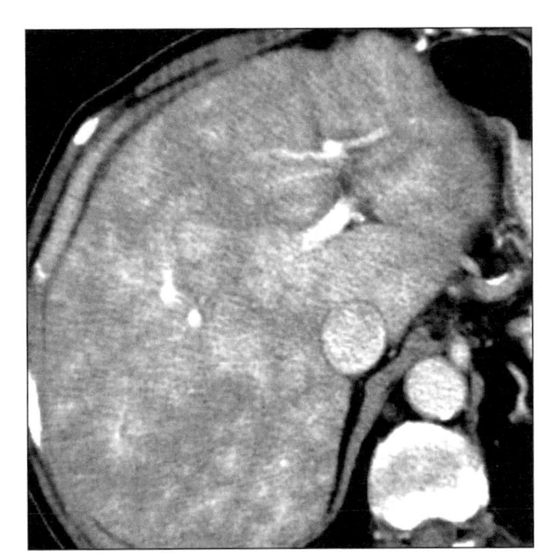

(Left) *Axial CECT shows chronic constrictive pericarditis with soft tissue and calcified thickening of pericardium (arrows), deviation of interventricular septum.* ***(Right)*** *Axial CECT in portal venous phase shows mottled enhancement of liver, halo of lymphedema around IVC. This patient presented with passive hepatic congestion secondary to constrictive pericarditis.*

Typical

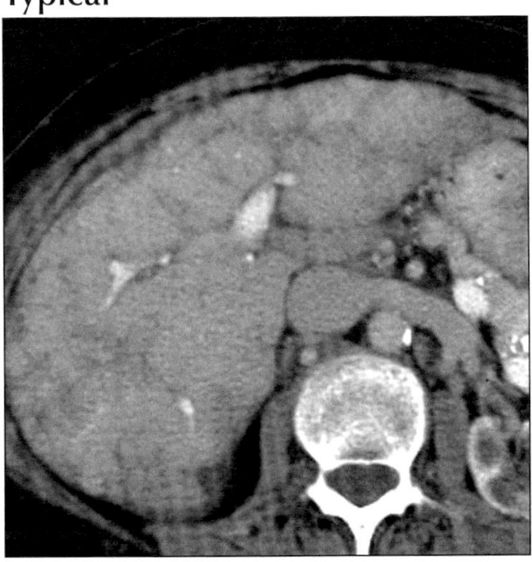

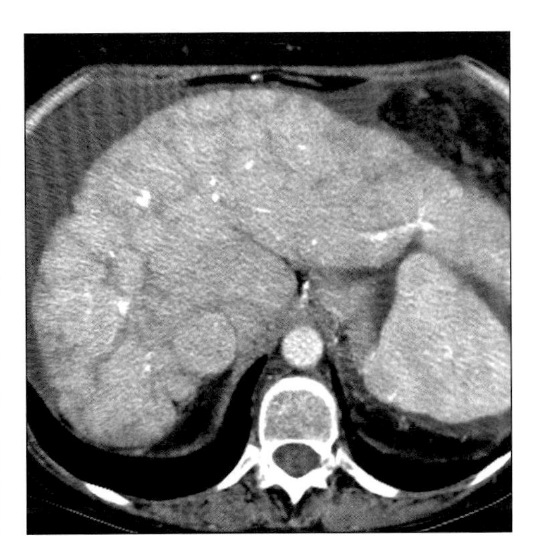

(Left) *Axial CECT shows typical changes from cardiac cirrhosis. Liver is small and dysmorphic with heterogeneous enhancement.* ***(Right)*** *Axial CECT shows dysmorphic liver with atrophic right lobe, hypertrophied lateral segment and heterogeneous enhancement. Ascites.*

HELLP SYNDROME

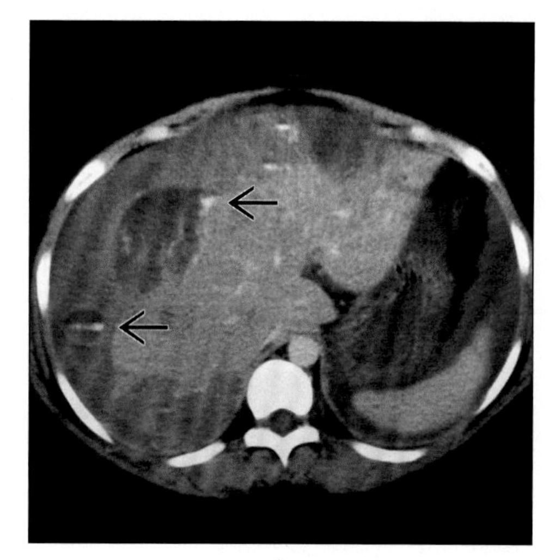

Axial CECT shows massive hemoperitoneum, liver parenchymal hemorrhage, and active extravasation (arrows) in a female patient during third trimester pregnancy.

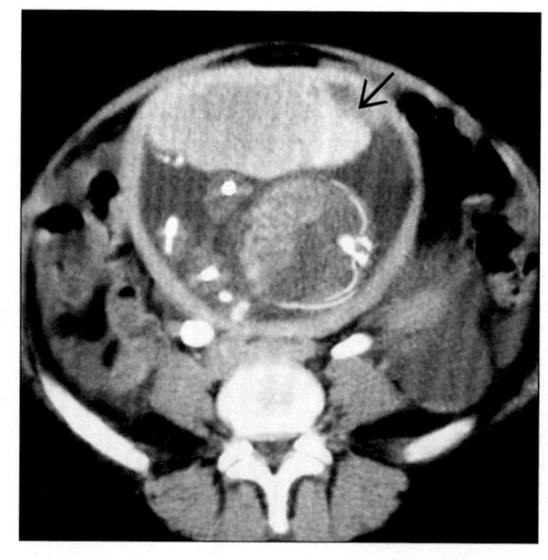

Axial CECT shows hemoperitoneum. Normal gravid uterus, placenta (arrow), and full term fetus.

TERMINOLOGY

Abbreviations and Synonyms

- Hemolysis, elevated liver enzymes, low platelets (HELLP)

Definitions

- HELLP syndrome: A severe variant of preeclampsia

IMAGING FINDINGS

General Features

- Best diagnostic clue: Intrahepatic or subcapsular fluid collection (hematoma) on US or CT
- Location
  - Liver
    - Subcapsular or intraparenchymal
- Key concepts
  - HELLP syndrome is a variant of toxemia in primigravidas
    - Usually preeclampsia & occasionally eclampsia
    - Usually before birth in 3rd trimester (antepartum)
    - Occasionally soon after birth (postpartum)
    - Rarely seen in multiparous patients
  - Preeclampsia: Leading cause of maternal death in USA & Europe
  - Based on classification of American college of Obstetricians/Gynecologists
    - Bilirubin: More than 1.2 mg/dL
    - Lactate dehydrogenase: More than 600U/L
    - Aspartate aminotransferase: More than 70U/L
    - Platelet count: Less than 100,000/mm3
  - Radiologically
    - US features may be seen before increase in biological markers (41% cases)
    - Helps in differentiating from other medical & surgical conditions

CT Findings

- Liver hematomas
  - Well-defined hyperdense or hypodense
  - Nonenhancing
  - Acute: Hyperattenuating (first 24-72 hours)
  - Chronic: Decreased attenuation (after 72 hours)
  - Location: Subcapsular or intraparenchymal
- Liver infarction
  - Small or large areas of low attenuation
  - Usually peripheral & wedge-shaped
- Occasionally active contrast extravasation sites or ascites

MR Findings

- T1WI & T2WI
  - Varied signal intensity depending on
    - Degree & age of hemorrhage or infarct

DDx: Diffuse or Focal Liver Lesion with Hemorrhage

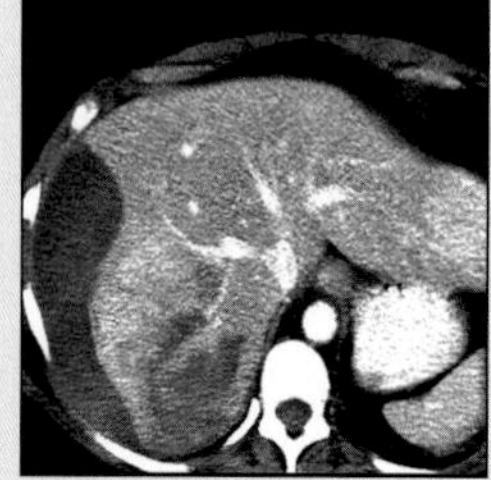

Hemorrhagic Adenoma

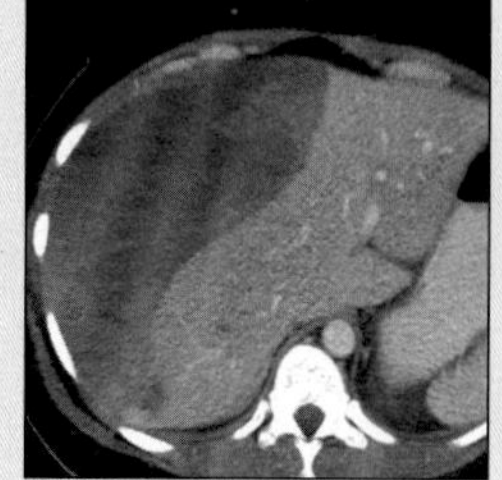

Coagulopathy

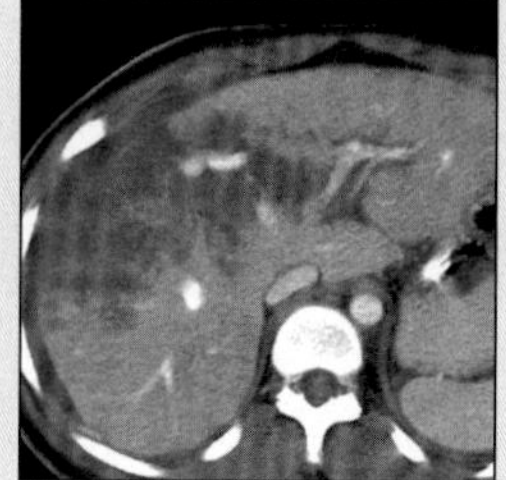

Hepatic Trauma

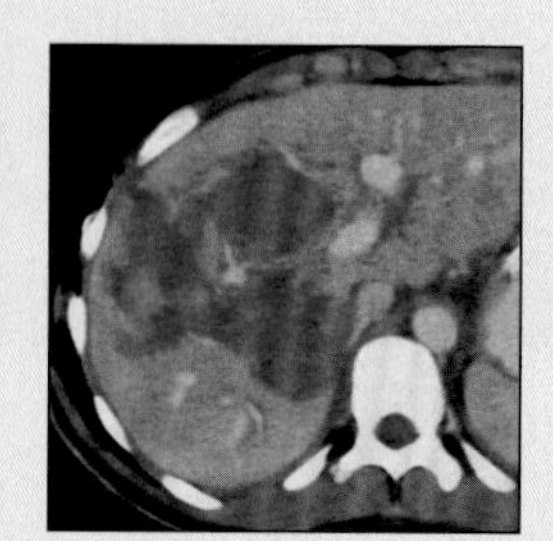

Hepatic Trauma

HELLP SYNDROME

Key Facts

Terminology

- Hemolysis, elevated liver enzymes, low platelets (HELLP)
- HELLP syndrome: A severe variant of preeclampsia

Imaging Findings

- Best diagnostic clue: Intrahepatic or subcapsular fluid collection (hematoma) on US or CT
- Acute: Hyperattenuating (first 24-72 hours)
- Chronic: Decreased attenuation (after 72 hours)

Top Differential Diagnoses

- Bleeding hepatic tumor (adenoma or HCC)
- Spontaneous bleed (coagulopathy)
- Hepatic trauma

Clinical Issues

- Acute epigastric & RUQ pain: Present in 90% of cases
- Clinical profile: Black female, primigravida with features of preeclampsia & lab data showing findings of hemolysis, elevated liver enzymes & low platelets
- Age: 2nd & 3rd decades

Diagnostic Checklist

- Rule out bleeding liver tumors like adenoma, HCC & other liver pathologies like acute viral hepatitis & acute fatty liver of pregnancy
- Clinically can mimic: Cholecystitis, biliary colic & hepatitis
- Very rarely can occur without classic preeclampsia triad: Hypertension, proteinuria & edema

        - Degree of necrosis & steatosis
- Greater degree of edema & cellular necrosis
    - T1WI: Low signal intensity
    - T2WI: High signal intensity

Ultrasonographic Findings

- Real Time
    - Liver hemorrhage or infarct
        - Irregular or wedge shaped
        - Increased echogenicity
        - Location: Usually peripheral
    - Periportal halo sign
        - Hyperechoic thickening of periportal area
    - Subcapsular hematoma
        - Complex echogenicity of fluid collection
    - Enlarged liver (predominantly right lobe)
    - Occasionally ascites

Imaging Recommendations

- Best imaging tool: Ultrasonography
- Try to avoid CT because of radiation to fetus

DIFFERENTIAL DIAGNOSIS

Bleeding hepatic tumor (adenoma or HCC)

- May bleed & present as fluid collection on US or CT
    - Intraparenchymal or subcapsular
    - Indistinguishable from HELLP syndrome
- Look for enhancing heterogeneous spherical hepatic mass

Spontaneous bleed (coagulopathy)

- History of bleeding disorder
- Lab data: Abnormal bleeding time, clotting time, prothrombin time & partial thromboplastin time
- On imaging
    - Subcapsular or intrahepatic blood collection
    - Occasionally active extravasation site may be seen
    - Indistinguishable from HELLP syndrome without history

Hepatic trauma

- History of injury to liver
- On imaging
    - Intraparenchymal or subcapsular hematomas
    - Lacerations, wedge-shaped areas of infarction
    - Areas of active hemorrhage (isodense with vessels)
    - Hemoperitoneum & pseudoaneurysm

Acute fatty liver of pregnancy

- Usually diffuse increase echogenicity of liver on US
- No intraparenchymal or subcapsular fluid collection
    - Indicates no hemorrhage

PATHOLOGY

General Features

- General path comments
    - Pathophysiology of HELLP syndrome: Begins in placental bed
        - Arteriolar vasospasm → endothelial damage → fibrin deposition
        - Platelet deposition on fibrin aggregates → decrease number of circulating platelets
    - RBC destruction by fibrin aggregates (hemolytic anemia)
        - Abnormal cells in peripheral smear (burr cells & schistocytes)
        - Elevated indirect bilirubin levels & anemia
    - Hepatocyte destruction: Due to hepatic microemboli (↑ LFT levels)
        - Distention of liver: Due to impeded blood flow (RUQ pain)
        - Severe cases: Liver rupture & subcapsular hematoma
    - Pathophysiology of preeclampsia
        - Primary site: Increased size of glomerular endothelial cells
        - Abnormal vasoconstriction + hyperactive vascular smooth muscle
        - Hypertension → proteinuria → edema
- Etiology
    - Variant of severe preeclampsia & occasionally eclampsia
    - Preeclampsia & eclampsia: May be due to

HELLP SYNDROME

- Coagulation abnormalities
- Hormonal factors
- Uteroplacental ischemia
- Immune mechanisms
- Epidemiology
  - Prevalence
    - 4-12% of patients with severe preeclampsia
    - 1 per 150 live births
    - Toxemia occurs in 6% of pregnancies
  - Maternal mortality rate (MMR) in severe preeclampsia due to HELLP syndrome is 3.5%

Gross Pathologic & Surgical Features

- Enlarged liver
- Parenchymal hemorrhage or infarct
- Subcapsular hematoma

Microscopic Features

- Periportal necrosis
- Microthrombi
- Fibrin deposits in sinusoids & portal veins

CLINICAL ISSUES

Presentation

- Most common signs/symptoms
  - Acute epigastric & RUQ pain: Present in 90% of cases
  - Other signs/symptoms
    - Malaise, nausea, vomiting, weight gain
    - Edema, headache, visual impairment, jaundice
  - Preeclampsia: Classic triad
    - Hypertension, proteinuria & edema
  - Eclampsia
    - Classic triad of preeclampsia
    - Associated with convulsions & coma
  - Clinical differential diagnosis
    - Viral hepatitis, gallstones, peptic ulcer
    - Pancreatitis, acute fatty liver
    - Hemolytic uremic syndrome
    - Idiopathic thrombocytopenic purpura (ITP)
- Clinical profile: African-American female, primigravida with features of preeclampsia & lab data showing findings of hemolysis, elevated liver enzymes & low platelets
- Lab data
  - Hemoglobin: Less than 11 g/dL
  - Bilirubin: More than 1.2 mg/dL
  - Lactate dehydrogenase: More than 600 U/L
  - Aspartate aminotransferase: More than 70 U/L
  - Platelet count: Less than 100,000/mm3

Demographics

- Age: 2nd & 3rd decades
- Gender: Females
- Ethnicity: More frequent in African-Americans

Natural History & Prognosis

- Usually seen in primigravidas with preeclampsia
- Occasionally seen in eclampsia patients
- Maternal risk factors
  - Nulliparity, young age (2nd & 3rd decades)
  - African-American females; familial
  - Underlying diseases
    - Hypertension, diabetes, renal disease
- Complications
  - Rupture of subcapsular hematoma
  - Hepatic necrosis
  - Disseminated intravascular coagulation (DIC)
  - Abruptio placenta & renal failure
  - Pulmonary edema & hypoglycemia
  - Maternal mortality rate: 3.5% (due to liver rupture)
    - In delayed diagnosis & treatment

Treatment

- Majority of cases: Supportive treatment
- Standard treatment: Expeditious delivery of fetus
- Hepatic rupture & intra-abdominal bleeding
  - Surgery & selective embolization

DIAGNOSTIC CHECKLIST

Consider

- Rule out bleeding liver tumors like adenoma, HCC & other liver pathologies like acute viral hepatitis & acute fatty liver of pregnancy
- Preeclampsia & HELLP syndrome
  - Must be routinely checked for in all pregnant women with acute abdominal (epigastric/RUQ) pain
- HELLP syndrome
  - Clinically can mimic: Cholecystitis, biliary colic & hepatitis
  - Very rarely can occur without classic preeclampsia triad: Hypertension, proteinuria & edema

Image Interpretation Pearls

- Subcapsular hematoma
  - Complex echogenicity of fluid collection
  - Indistinguishable from fluid collections of bleeding tumors like adenoma & HCC
    - Look for heterogeneous enhancing spherical liver tumors
- Liver hemorrhage or infarct
  - Usually peripheral, irregular or wedge shaped
  - Increased echogenicity

SELECTED REFERENCES

1. Di Salvo DN: Sonographic imaging of maternal complications of pregnancy. J Ultrasound Med. 22(1):69-89, 2003
2. Suarez B et al: Abdominal pain and preeclampsia: sonographic findings in the maternal liver. J Ultrasound Med. 21(10):1077-83; quiz 1085-6, 2002
3. Casillas VJ et al: Imaging of nontraumatic hemorrhagic hepatic lesions. Radiographics. 20(2):367-78, 2000
4. Barton JR et al: Hepatic imaging in HELLP syndrome (hemolysis, elevated liver enzymes, and low platelet count). Am J Obstet Gynecol. 174(6):1820-5; discussion 1825-7, 1996
5. Peitz U et al: Sonographic findings of liver and gallbladder in early hemolysis, elevated liver enzymes, and low platelet count syndrome. J Clin Ultrasound. 21(8):557-60, 1993
6. Kronthal AJ et al: Hepatic infarction in preeclampsia. Radiology. 177(3):726-8, 1990

HELLP SYNDROME

IMAGE GALLERY

Typical

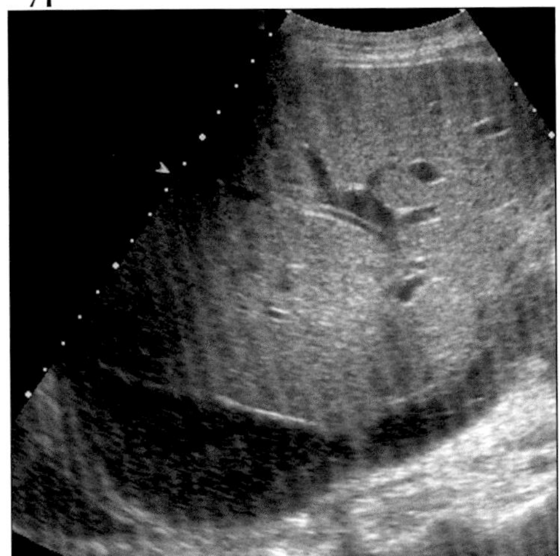

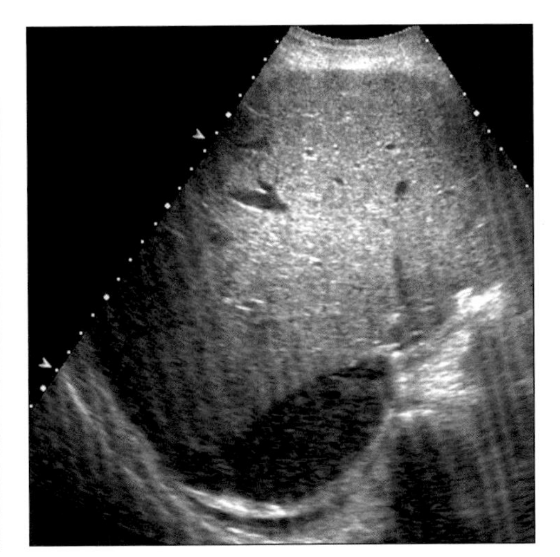

(Left) *Oblique sonogram shows peripheral lentiform subcapsular hematoma with low level echogenicity.* ***(Right)*** *Sagittal sonogram in patient with HELLP syndrome shows peripheral subcapsular hematoma.*

Typical

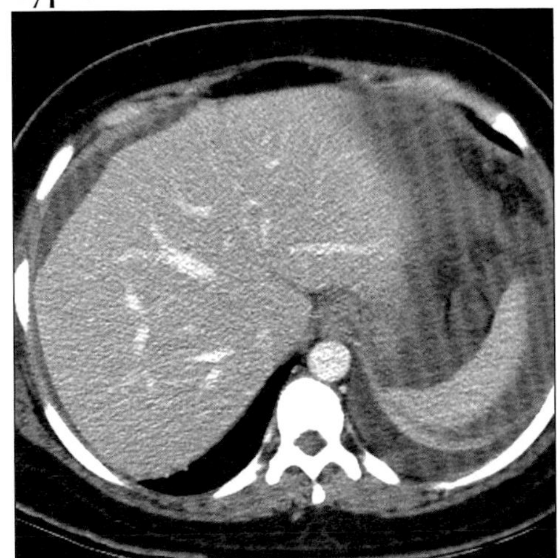

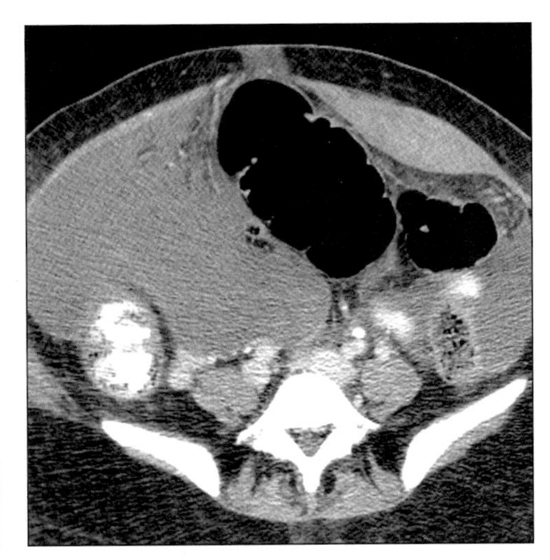

(Left) *Axial CECT shows lentiform subcapsular hematoma deforming lateral contour of the liver and hemoperitoneum.* ***(Right)*** *Axial CECT shows large hemoperitoneum and left rectus sheath hematoma in patient with HELLP syndrome.*

Typical

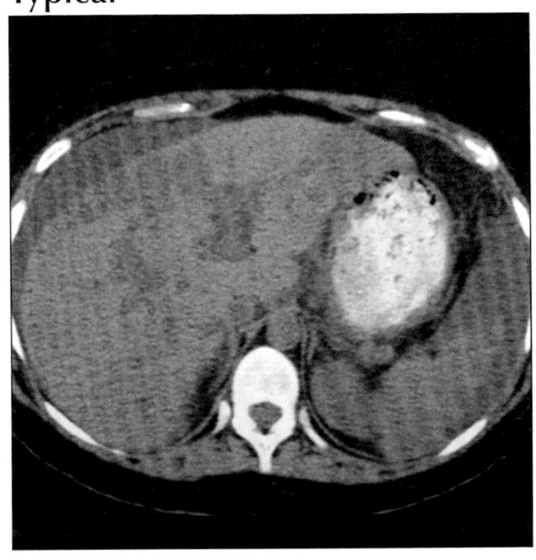

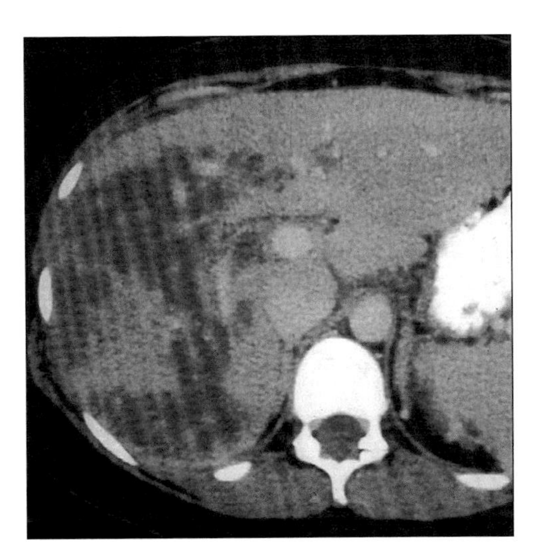

(Left) *Axial CECT shows heterogeneous liver parenchyma consistent with bleeding and/or infarction, and hemoperitoneum.* ***(Right)*** *Axial CECT shows large areas of nonenhancing liver, consistent with liver infarction or "old" hemorrhage.*

HEPATIC INFARCTION

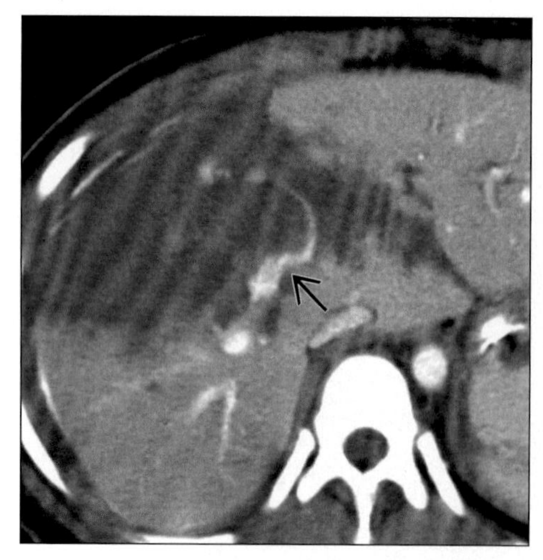

Axial CECT following blunt trauma shows no enhancement of anterior right lobe. Hepatic artery to this segment is transected with acute extravasation (arrow).

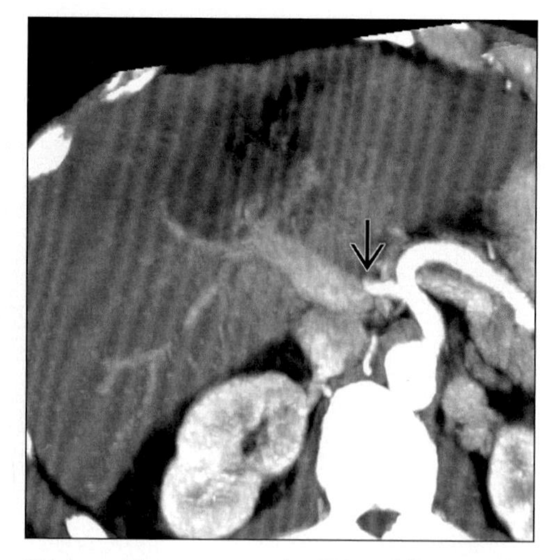

Thick axial reconstructed CECT following liver transplantation. The hepatic artery is thrombosed at the anastomosis (arrow) with a large liver infarction.

TERMINOLOGY

Abbreviations and Synonyms

- Liver infarction

Definitions

- Development of area of coagulation necrosis due to local ischemia resulting from obstruction of circulation to the area; most commonly by a thrombus or embolus

IMAGING FINDINGS

General Features

- Best diagnostic clue: Peripheral wedge shaped, rounded or ovoid low attenuation areas with absent or heterogeneous enhancement
- Location
  - Usually peripheral and wedge-shaped
  - Can be more central and rounded
- Size: Variable: Few mm to centimeters
- Key concepts
  - Usually single and focal
  - Can be multiple or diffuse

CT Findings

- NECT
  - Wedge-shaped, rounded or oval, or irregularly shaped low attenuation areas paralleling bile ducts
    - Acute: Poorly demarcated low density lesions
    - Subacute: Confluent with more distinct margins
  - ± Gas formation within sterile or infected infarcts
  - Bile lakes seen as late sequela: Cystic changes
- CECT
  - Lesions may have geographic segmental distribution with straight margins
  - Lesions on NECT are more conspicuous after enhancement (perfusion defects)
  - Heterogeneous patchy enhancement with zones of enhancement equal to liver parenchyma
  - Components of lesions remaining hypodense on arterial, portal venous & delayed phase
    - Represent regions of necrotic tissue, hemorrhage or fibrous tissue with no or minimum revascularization on histology
  - Lesions isoenhancing with surrounding liver parenchyma on portal venous phase
    - Histologically consistent with retained viable tissue or fibrotic tissue with revascularization

MR Findings

- T1WI
  - Small, relatively well-defined, hypointense
  - Edema of infarction: Lower signal intensity on T1
- T2WI
  - Heterogeneous appearance of liver parenchyma

DDx: Segmental/Lobar Hypodensity or Decreased Enhancement

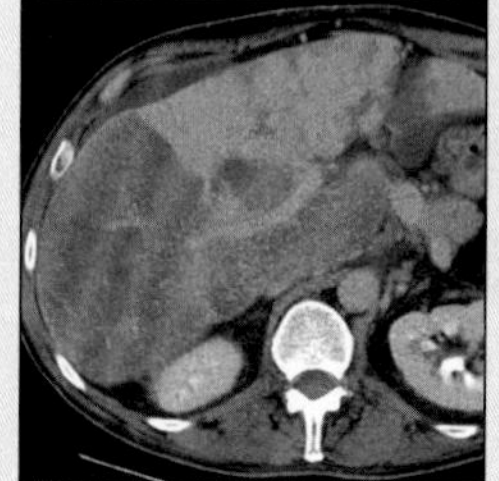

Focal Steatosis

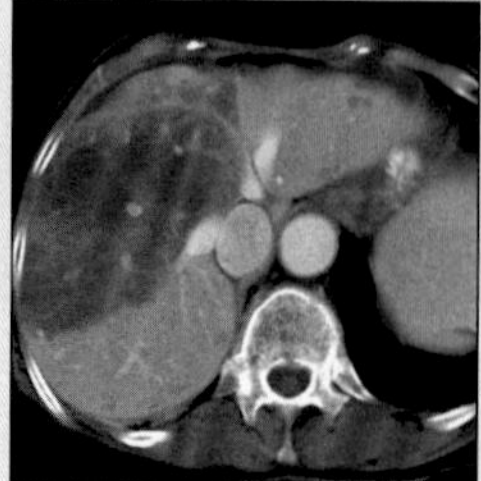

Focal Steatosis

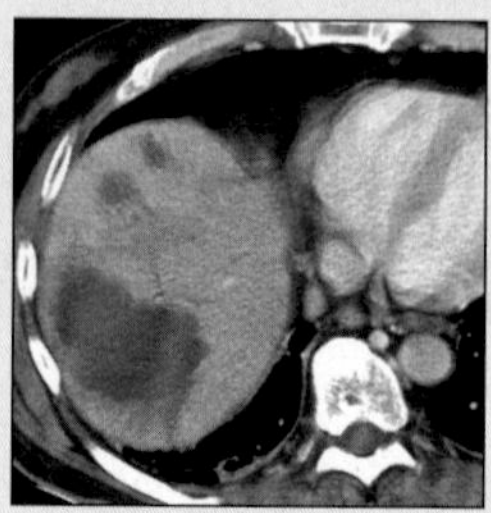

Hepatic Abscess

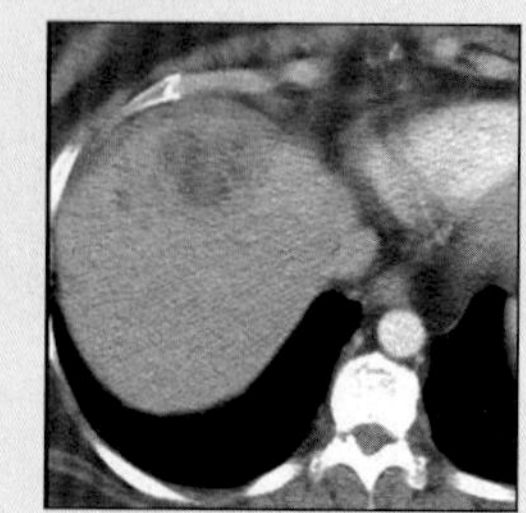

Hepatic Abscess

HEPATIC INFARCTION

Key Facts

Imaging Findings

- Best diagnostic clue: Peripheral wedge shaped, rounded or ovoid low attenuation areas with absent or heterogeneous enhancement
- Lesions may have geographic segmental distribution with straight margins
- Lesions on NECT are more conspicuous after enhancement (perfusion defects)
- CT or MR angiography can be diagnostic
- Catheter angiography may be necessary for diagnosis and treatment

Top Differential Diagnoses

- Focal steatosis
- Hepatic abscess

Pathology

- Rarity of hepatic infarction due to dual blood supply from hepatic artery & portal vein & extensive collateral pathways

Clinical Issues

- Infarction is serious complication of liver transplantation with significant morbidity & mortality & often requiring retransplantation

Diagnostic Checklist

- New focal liver lesion with branching pattern in transplant patient with deteriorating function suggests infarction (usually hepatic artery thrombosis)

  - Edema of infarction: Higher signal intensity on T2
- T1 C+
  - Heterogeneous parenchymal enhancement & areas of perfusion defect
  - Necrotic areas: Predominantly hypointense compared with enhancing parenchyma in arterial, portal venous & delayed phases

Ultrasonographic Findings

- Real Time
  - In native liver
    - Early: Hypoechoic lesion with indistinct margins (when sufficient edema & round cell infiltration)
    - Small bile duct cysts; large bile duct lakes (as necrotic tissue is resorbed)
  - In liver transplant recipients
    - Geographic areas hypoechoic with preservation of portal tracts (early sign of ischemia)
    - Development of transient small hyperechoic lesions (progression to true infarction)
- Color Doppler
  - Hepatic artery thrombosis: Absence of normal hepatic artery signal
    - Hepatic artery thrombosis much more common than portal vein thrombosis
  - Transplant vasculature or portal vein thrombosis
  - Porto-systemic shunting, collateral supply

Angiographic Findings

- Conventional: To confirm occlusion of hepatic artery suggested by US, CT, or MR

Nuclear Medicine Findings

- Hepato biliary scan
  - Peripheral wedge shaped sharply defined lesion
  - Communication with bile lakes for infarcts following transplantation
- Technetium sulfur colloid
  - Photopenic area
- Cholescintigraphy: Communication with bile lakes for infarcts following transplantation

Imaging Recommendations

- Best imaging tool
  - Real time B mode & Doppler: Often first modality to evaluate allograft dysfunction/post-operative complications
  - Triphasic helical CT with CT angiography
    - CT or MR angiography can be diagnostic
- Protocol advice
  - CECT + CTA or dynamic contrast-enhanced gradient-echo & contrast-enhanced TI weighted spin-echo images in axial plane with MRA
  - Catheter angiography may be necessary for diagnosis and treatment

DIFFERENTIAL DIAGNOSIS

Focal steatosis

- May be geographic, wedge-shaped
- Preserved patent vessels; preservation of enhancing vessels within "lesion"
- Characteristic suppression of signal on opposed-phased GRE MR

Hepatic abscess

- Usually spherical, often septated
- Central nonenhancing contents, enhancing rim

PATHOLOGY

General Features

- General path comments
  - Rarity of hepatic infarction due to dual blood supply from hepatic artery & portal vein & extensive collateral pathways
  - In most cases superimposition of portal thrombosis on hepatic arterial occlusion results in chronic insufficiency & infarction
  - Infarcted regenerative nodules in cirrhosis develop from hypoperfusion of liver followed by ischemic necrosis of nodules that are vulnerable to hypoxia
  - Hepatic artery thrombosis in liver allograft recipients more likely to lead to infarction as collateral supply is severed during transplant

HEPATIC INFARCTION

- Etiology
  - Iatrogenic
    - Cholecystectomy, hepatobiliary surgery, intrahepatic chemoembolization, transjugular intrahepatic portosystemic shunt (TIPS) procedure
  - Liver transplantation
    - Hepatic artery stenosis or thrombosis
  - Blunt trauma
    - Hepatic artery & portal vein laceration
  - Hypercoagulable states
    - Sickle cell/antiphospholipid antibody syndrome
  - Vasculitis
    - Polyarteritis, lupus, etc.
  - Infection
    - Rare "emphysematous hepatitis"
    - Following sepsis & shock
- Epidemiology
  - Hepatic infarction is uncommon
  - Hepatic artery thrombosis following transplant reported in 3% adults, 12% children or in 7-8% of mixed population

Gross Pathologic & Surgical Features

- Liver at autopsy: Atrophic, hard & irregularly surfaced
- Focal, multiple necrotic areas, peripheral collapse of parenchymal tissue with fibrosis

Microscopic Features

- Central congestion & centrilobular necrosis surrounded by hemorrhagic rims
- Infarcted nodules have central core of amorphous eosinophilic material representing remnants of necrotic hepatocytes
  - Cells with foamy cytoplasm representing macrophages surround necrotic core
  - Ultimate replacement by fibrovascular tissue

CLINICAL ISSUES

Presentation

- Most common signs/symptoms
  - Diagnosed at laparotomy, autopsy or imaging
  - Asymptomatic, nonspecific: Right upper quadrant or back pain, fever
  - Massive infarction: Coma, ascites, jaundice, renal failure
- Clinical profile
  - Lab: Leukocytosis,abnormal liver function tests
  - In pregnancy: Associated with hemolytic anemia with elevated liver enzymes & low platelets (HELLP), pre-eclampsia, eclampsia

Demographics

- Age: Any age group
- Gender: M = F

Natural History & Prognosis

- Parenchymal atrophy & scarring, progressive liquefaction, or both; affects center of hepatic lobule (venous) most prominently with relative sparing of portal (arterial) end
- Infarction is serious complication of liver transplantation with significant morbidity & mortality & often requiring retransplantation
- Complications:
  - Native liver: Liver failure, fibrosis
  - Transplanted liver: Biliary strictures, bilomas, abscess

Treatment

- Options: Revascularization, retransplantation, spontaneous resolution

DIAGNOSTIC CHECKLIST

Consider

- Pre-TIPS evaluation of arterial supply to liver by Doppler/ angiography; sufficient arterial perfusion crucial to avoid infarction
- Post TIPS: If pain develops in right upper quadrant, fever, shock & disseminated intravascular coagulation
- Recognize infarction as separate entity among spectrum of pregnancy-related liver disorders to avoid delay in diagnosis & treatment
- Ultrasound & CT suggest diagnosis, angiography often necessary for confirmation
- Ischemia alone may produce "typical sonographic features of infarction"; if recognized early enough, may be reversible

Image Interpretation Pearls

- Preservation of portal tracts: Feature worthy of emphasis as it helps differentiate infarction from other causes of focal hypoechoic areas in post transplant liver e.g., abscess, biloma or hematoma following biopsy
- New focal liver lesion with branching pattern in transplant patient with deteriorating function suggests infarction (usually hepatic artery thrombosis)

SELECTED REFERENCES

1. Blachar A et al: Acute fulminant hepatic infection causing fatal "emphysematous hepatitis": case report. Abdom Imaging. 27(2):188-90, 2002
2. Mayan H et al: Fatal liver infarction after transjugular intrahepatic portosystemic shunt procedure. Liver. 21(5):361-4, 2001
3. Quiroga S et al: Complications of orthotopic liver transplantation: spectrum of findings with helical CT. Radiographics. 21(5):1085-102, 2001
4. Kim T et al: Infarcted regenerative nodules in cirrhosis: CT and MR imaging findings with pathologic correlation. AJR Am J Roentgenol. 175(4):1121-5, 2000
5. Smith GS et al: Hepatic infarction secondary to arterial insufficiency in native livers: CT findings in 10 patients. Radiology. 208(1):223-9, 1998
6. Holbert BL et al: Hepatic infarction caused by arterial insufficiency: spectrum and evolution of CT findings. AJR Am J Roentgenol. 166(4):815-20, 1996

HEPATIC INFARCTION

IMAGE GALLERY

Typical

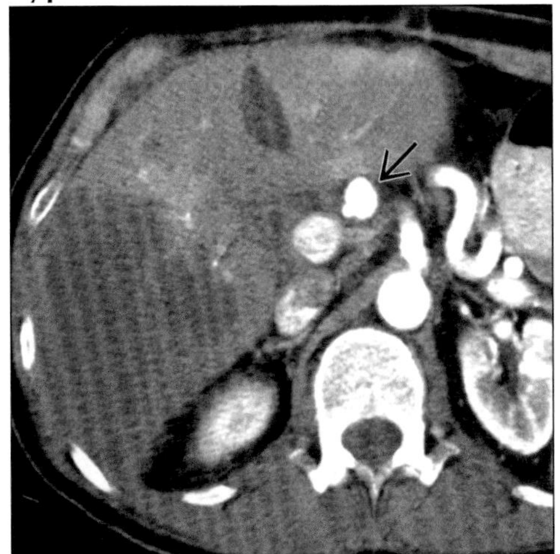

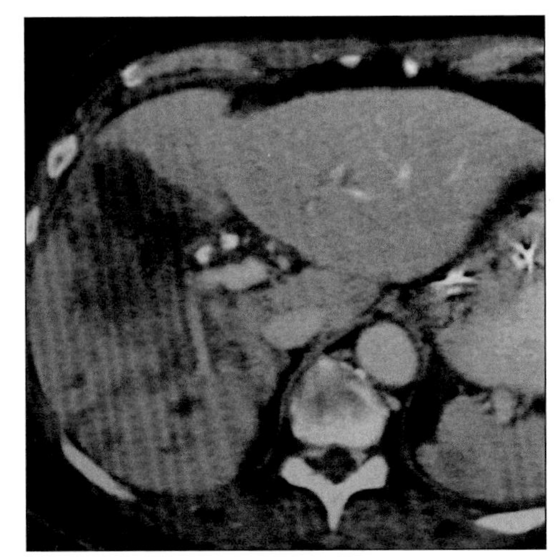

(Left) *Axial CECT shows minimal enhancement of infarcted posterior right lobe. Hepatic artery pseudoaneurysm (arrow) with embolic occlusion of right artery.* ***(Right)*** *Axial CECT in a patient with iatrogenic infarction following laparoscopic cholecystectomy (occluded right hepatic artery).*

Typical

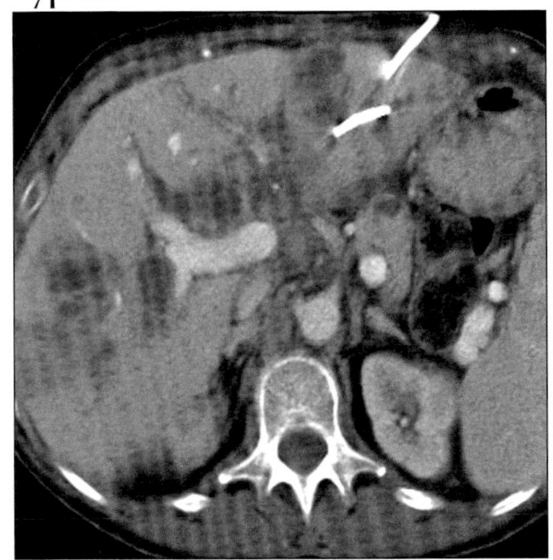

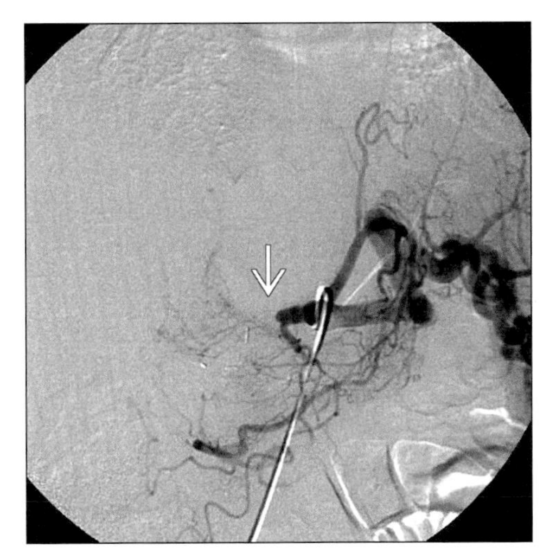

(Left) *Axial CECT in liver transplant recipient shows rounded and branching hypodense liver lesions due to hepatic infarction and biliary necrosis. Pigtail catheter placed to drain biloma.* ***(Right)*** *Celiac arteriogram shows lack of arterial blood supply to the liver with occlusion of hepatic artery (arrow) in patient who had undergone recent liver transplant.*

Variant

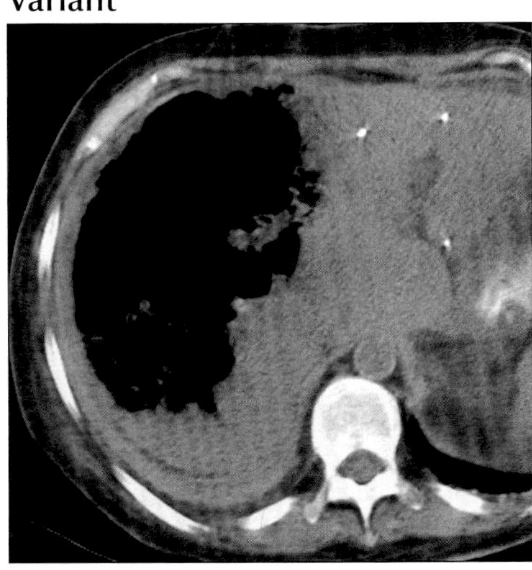

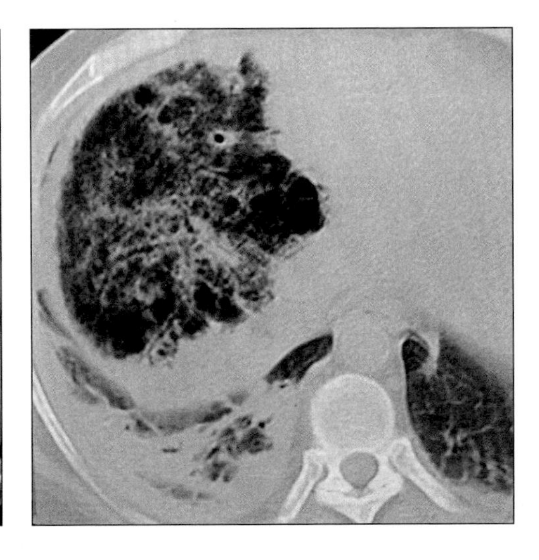

(Left) *Axial NECT shows gas replacing right lobe of liver due to spontaneous infarction and infection in a diabetic patient with sepsis.* ***(Right)*** *Axial NECT using lung window settings shows nearly complete replacement of liver parenchyma with gas ("emphysematous hepatitis") and no apparent purulent collections.*

PELIOSIS HEPATIS

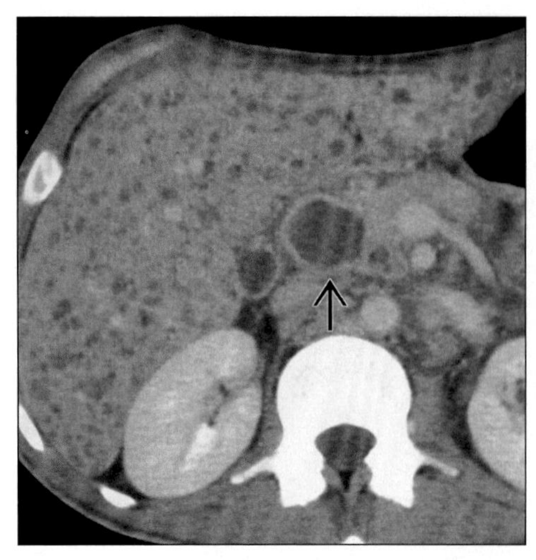

CECT in a patient with AIDS and biopsy proven peliosis due to Bartonella infection. Innumerable hypodense liver lesions with peripheral enhancement. Hypodense porta hepatis nodes (arrow).

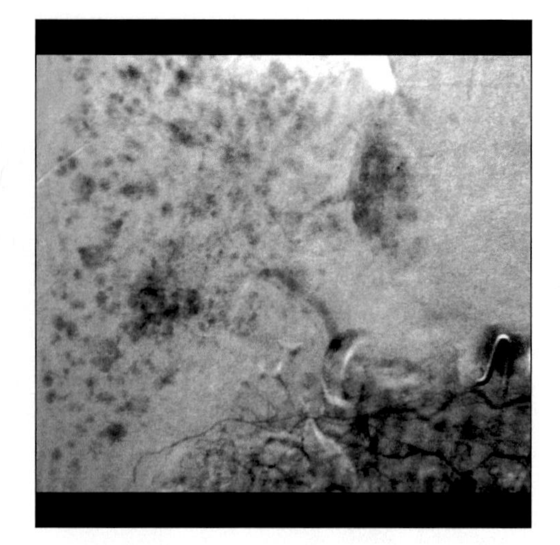

Delayed arterial phase of celiac arteriogram shows persistent "puddling" of contrast in innumerable vascular hepatic lesions. Lesions resolved completely with antibiotics

TERMINOLOGY

Abbreviations and Synonyms

- Hepatic peliosis

Definitions

- Rare benign disorder causing sinusoidal dilatation & presence of multiple blood filled lacunar spaces within liver

IMAGING FINDINGS

General Features

- Best diagnostic clue
  - Strong contrast-enhancement on delayed imaging with "branching" appearance caused by vascular component
  - Spherical lesion with centrifugal or centripetal enhancement
- Location
  - No preferential location within hepatic lobule
  - Spleen, bone marrow, lymph node, lungs, pleura, kidneys, adrenals, stomach, ileum
- Size: Varies from 1 mm to several centimeters
- Key concepts
  - Irregularly shaped blood-filled hepatic cavities

CT Findings

- NECT
  - Multiple hepatic areas of low attenuation
  - CT findings differ with size of lesions, presence or absence of thrombus within cavity & presence of hemorrhage
    - If peliotic cavities < 1 cm diameter, CT findings may appear normal
- CECT
  - Larger cavities communicating with sinusoids have same attenuation as blood vessels
  - Thrombosed cavities will have same appearance as nonenhancing nodules
  - Arterial phase: Early globular vessel-like enhancement
    - Multiple small accumulations of contrast, hyperdense in center or periphery of lesion
  - Portal phase: Centrifugal or centripetal enhancement without mass effect on hepatic vessels
  - Delayed phase: Late diffuse homogenous hyperattenuation characteristic of phlebectatic type

MR Findings

- T1WI
  - Hypointense
  - ↑ Signal due to presence of subacute blood suggestive of hemorrhagic necrosis
- T2WI

DDx: Heterogeneous Hypervascular Mass(es)

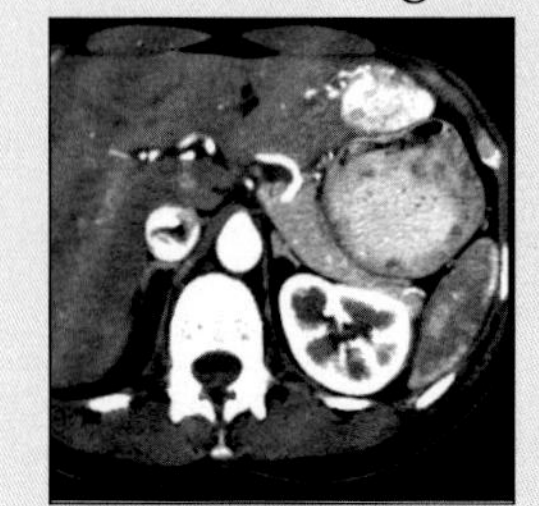

Hepatic Adenoma

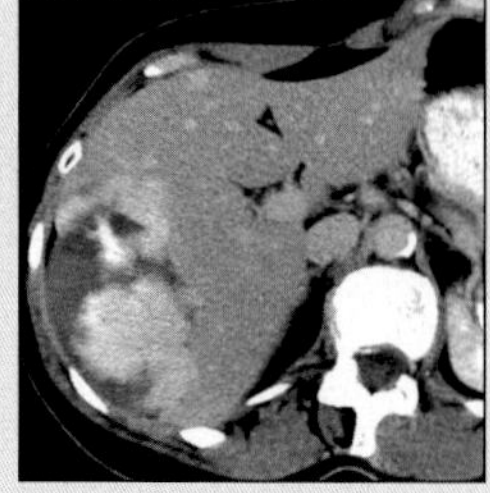

Hepatic Hemangioma

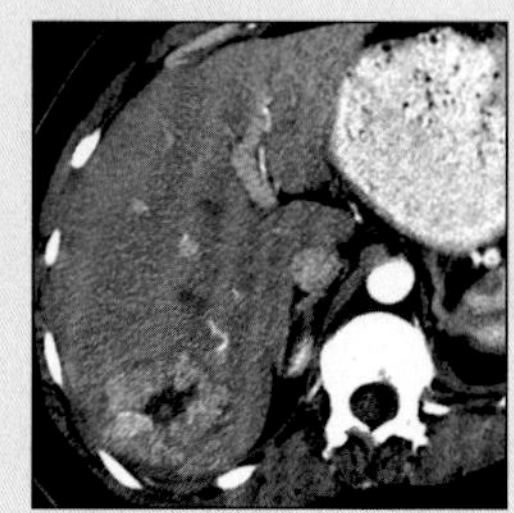

Atypical FNH

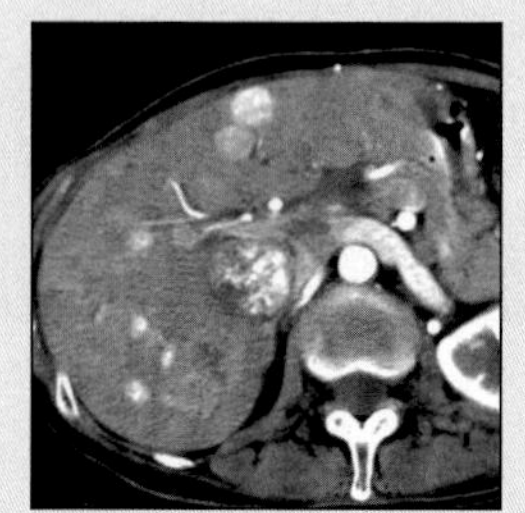

Metastases

PELIOSIS HEPATIS

Key Facts

Terminology
- Rare benign disorder causing sinusoidal dilatation & presence of multiple blood filled lacunar spaces within liver

Imaging Findings
- Spherical lesion with centrifugal or centripetal enhancement
- Size: Varies from 1 mm to several centimeters
- Best imaging tools: Multiphase helical CT and/or MRI

Top Differential Diagnoses
- Hepatic adenoma
- Hepatic cavernous hemangioma
- Focal nodular hyperplasia (FNH)
- Hypervascular metastases

Pathology
- Associated with chronic wasting diseases
- Associated with steroid medications, sprue, diabetes, vasculitis, hematological disorders
- Bacillary peliosis hepatis caused by Bartonella species in HIV-positive patients

Clinical Issues
- Complications: Liver failure/cholestasis/portal hypertension/liver rupture leading to shock

Diagnostic Checklist
- Multiphase enhanced CT or MR showing heterogeneous liver lesion with centrifugal or centripetal enhancement

---

  - Hyperintense
  - Multiple foci of ↑ signal due to presence of subacute blood
- T1 C+
  - Lesions usually show contrast-enhancement
  - Cystic cavity with enhancing rim representing hematoma
  - Strong contrast-enhancement with "branching" appearance caused by vascular component on fat-suppressed T1 in delayed imaging

Ultrasonographic Findings
- Real Time
  - Heterogeneous hepatic echopattern with hyper/hypoechoic regions
    - Homogenous hypoechoic lesions (in patients with steatosis)
    - But also heterogeneous hypoechoic (complicated with hemorrhage)
    - Or hyperechoic (in patients with normal liver) patterns

Angiographic Findings
- Conventional
  - Multiple nodular vascular lesions; accumulations of contrast material on late arterial phase
    - ± Simultaneous opacification of hepatic veins
    - More prominent during parenchymal phase & persist on venous phase
    - Angiographic evaluation may be diagnostic in difficult cases

Imaging Recommendations
- Protocol advice
  - Multiphase enhanced helical CT imaging
  - Dynamic T1 C+ MR with fat-suppression
- Best imaging tools: Multiphase helical CT and/or MRI

DIFFERENTIAL DIAGNOSIS

Hepatic adenoma
- Might also be associated with long term use of estrogen
- Differentiate from peliosis by contrast-enhancement pattern on triphasic CT & dynamic MRI
- Presence of fatty contents helpful in narrowing diagnosis
- Uncommonly, totally hyperdense in arterial phase, becoming isodense to liver in portal phase
- Biopsy often necessary

Hepatic cavernous hemangioma
- Typical enhanced pattern (peripheral enhancement with centripetal progression)
- Enhancement similar to peliosis; but discontinuous nodular or globular for hemangioma, continuous ring for peliosis
- Larger lesions produce mass effect on hepatic vessels

Focal nodular hyperplasia (FNH)
- Homogenous hypervascular mass on arterial phase, isodense to liver on portal & delayed phases
  - Central scar with ↓ attenuation on arterial & portal phases & enhancement on delayed images

Hypervascular metastases
- Usually totally hypodense or isodense in delayed phase because of rapid washout of contrast material

Hepatic abscess
- Important to differentiate from peliosis to avoid percutaneous drainage of peliotic lesions which can be dangerous & fatal
- Pyogenic abscess
  - Multiseptated mass; "cluster of grapes" appearance
  - Nonenhancing contents
  - Typical clinical presentation; sepsis

PATHOLOGY

General Features
- General path comments
  - Pathogenesis remains uncertain
  - Pathogenesis theories
    - Outflow obstruction at sinusoidal level
    - Hepatocellular necrosis leading to cyst formation

PELIOSIS HEPATIS

- Dilatation of portion of central vein of hepatic lobule
- Direct lesions of sinusoidal barrier

- Etiology
  - Anabolic steroids, corticosteroids, tamoxifen, oral contraceptive, diethylstilbestrol, Azathioprine
  - After renal/cardiac transplant
  - Polyvinyl chloride/arsenic/thorium oxide exposure
  - Congenital-angiomatous malformation
- Epidemiology
  - Peliosis is a rare entity
  - Increasing incidence of cases of bacillary peliosis & angiomatosis in immunocompromised patients
- Associated abnormalities
  - Associated with chronic wasting diseases
    - TB, leprosy, malignancy (HCC), AIDS
  - Associated with steroid medications, sprue, diabetes, vasculitis, hematological disorders
  - Bacillary peliosis hepatis caused by Bartonella species in HIV-positive patients

Gross Pathologic & Surgical Features

- Irregularly shaped blood-filled hepatic cavities

Microscopic Features

- Cystic dilated sinusoids filled with red blood cells & bound by cords of liver cells
- Phlebectatic type: Endothelial lined blood filled spaces & aneurysmal dilatation of central vein & sinusoids
- Parenchymal type: Not lined by endothelium & usually associated with hemorrhagic parenchymal necrosis

CLINICAL ISSUES

Presentation

- Most common signs/symptoms
  - Asymptomatic
  - Other signs/symptoms
    - ± Hepatomegaly/ascites/portal hypertension
    - Lymphadenopathy with Bartonella henselae & neurological symptoms with Bartonella quintana bacillary peliosis
- Clinical profile: Found incidentally at autopsy

Demographics

- Age: Fetal life (rare) to adult life
- Gender: M = F

Natural History & Prognosis

- Regression after drug withdrawal, cessation of steroid therapy, resolution of associated infectious disease
- Pseudotumoral & hemorrhagic evolution
- Complications: Liver failure/cholestasis/portal hypertension/liver rupture leading to shock
- If untreated may be rapidly fatal

Treatment

- Options, risks, complications
  - Withdrawal of inciting agents
  - Surgical resection of involved liver section
  - Resolves spontaneously (uncommon)
  - Clinical improvement with antibiotics (erythromycin) in HIV related peliosis hepatis caused by Bartonella henselae

DIAGNOSTIC CHECKLIST

Consider

- Clinical setting (e.g., AIDS, chronic illness, medications)

Image Interpretation Pearls

- Multiphase enhanced CT or MR showing heterogeneous liver lesion with centrifugal or centripetal enhancement

SELECTED REFERENCES

1. Resto-Ruiz S et al: The role of the host immune response in pathogenesis of Bartonella henselae. DNA Cell Biol. 22(6):431-40, 2003
2. Chomel BB et al: Clinical impact of persistent Bartonella bacteremia in humans and animals. Ann N Y Acad Sci. 990:267-78, 2003
3. Gouya H et al: Peliosis hepatis: triphasic helical CT and dynamic MRI findings. Abdom Imaging. 26(5):507-9, 2001
4. Wang SY et al: Hepatic rupture caused by peliosis hepatis. J Pediatr Surg. 36(9):1456-9, 2001
5. Ferrozzi F et al: Peliosis hepatis with pseudotumoral and hemorrhagic evolution: CT and MR findings. Abdom Imaging. 26(2):197-9, 2001
6. Dehio C: Bartonella interactions with endothelial cells and erythrocytes. Trends Microbiol. 9(6):279-85, 2001
7. Vignaux O et al: Hemorrhagic necrosis due to peliosis hepatis: imaging findings and pathological correlation. Eur Radiol. 9(3):454-6, 1999
8. Walter E et al: Images in clinical medicine. Peliosis hepatis. N Engl J Med. 337(22):1603, 1997
9. Muradali D et al: Peliosis hepatis with intrahepatic calcifications. J Ultrasound Med. 15(3):257-60, 1996
10. Saatci I et al: MR findings in peliosis hepatis. Pediatr Radiol. 25(1):31-3, 1995
11. Jamadar DA et al: Case report: radiological appearances in peliosis hepatis. Br J Radiol. 67(793):102-4, 1994
12. Toyoda S et al: Magnetic resonance imaging of peliosis hepatis: a case report. Eur J Radiol. 16(3):207-8, 1993
13. Maves CK et al: Splenic and hepatic peliosis: MR findings. AJR Am J Roentgenol. 158(1):75-6, 1992
14. Radin DR: Spontaneous resolution of peliosis of the liver and spleen in a patient with HIV infection. AJR Am J Roentgenol. 158(6):1409, 1992
15. Tsukamoto Y et al: CT and angiography of peliosis hepatis. AJR Am J Roentgenol. 142(3):539-40, 1984
16. Lyon J et al: Peliosis hepatis: diagnosis by magnification wedged hepatic venography. Radiology. 150(3):647-9, 1984
17. Smathers RL et al: Computed tomography of fatal hepatic rupture due to peliosis hepatis. J Comput Assist Tomogr. 8(4):768-9, 1984

PELIOSIS HEPATIS

IMAGE GALLERY

Typical

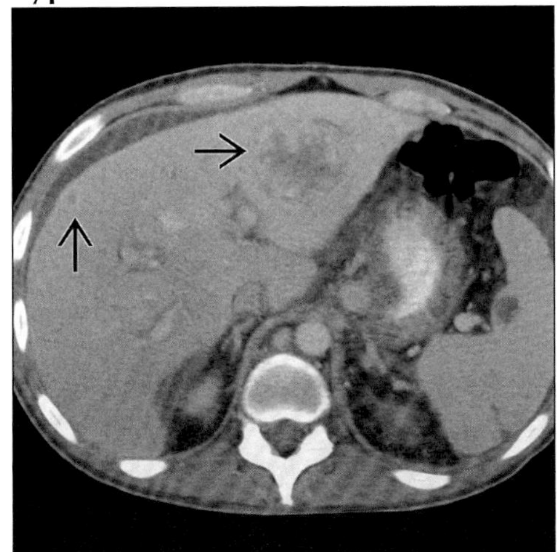

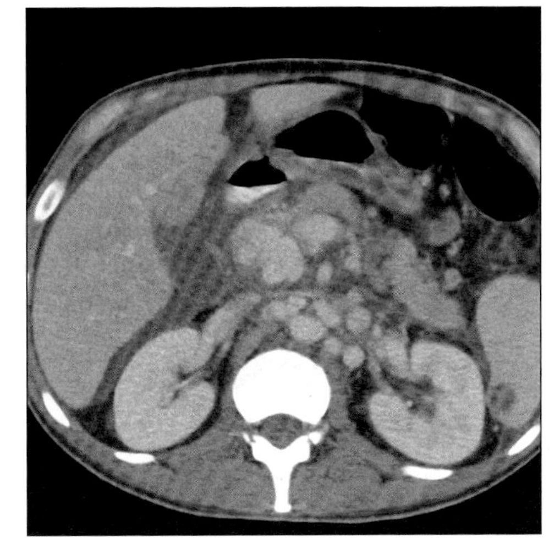

(Left) *Axial CECT shows multiple hypodense hepatic lesions (arrows) with peripheral enhancement; peliosis due to Bartonella infection which resolved after antibiotic treatment. Spleen is also involved.* ***(Right)*** *Axial CECT shows extensive brightly enhancing lymphadenopathy in this patient with peliosis hepatis due to Bartonella infection.*

Typical

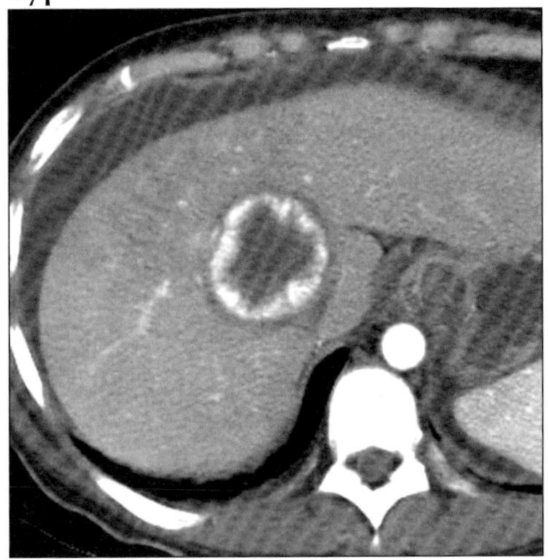

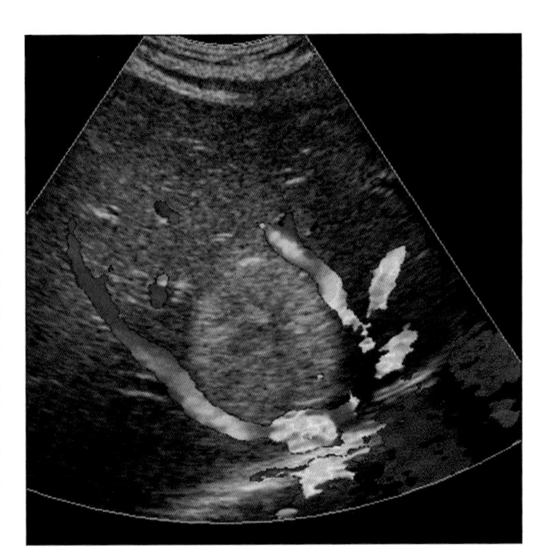

(Left) *Axial CECT during arterial phase shows hypodense lesion with bright continuous peripheral enhancement. (Centripetal progression of enhancement on venous phase). Biopsy proven peliosis hepatis.* ***(Right)*** *Axial color Doppler sonogram show hyperechoic liver mass without prominent vascularity.*

Variant

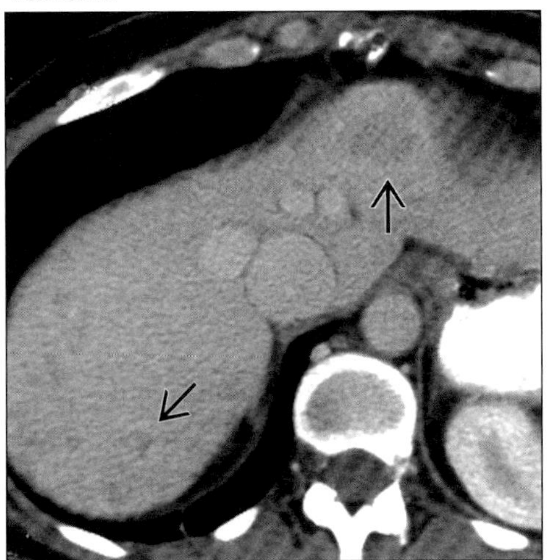

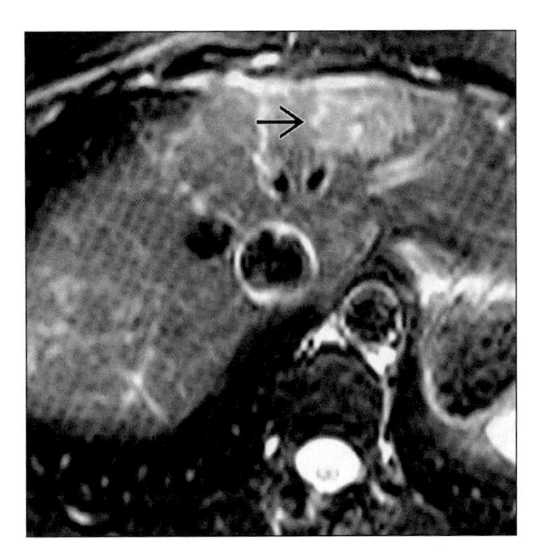

(Left) *Axial CECT in venous phase shows multiple hypodense lesions (arrows) with enhanced periphery. 42 year old woman with 25 year use of oral contraceptives.* ***(Right)*** *Axial T2WI MR shows hyperintense lesion in left lobe (arrow); biopsy proven peliosis. Other liver lesions had similar appearance. These partially resolved after discontinuation of contraceptives.*

HEMOCHROMATOSIS

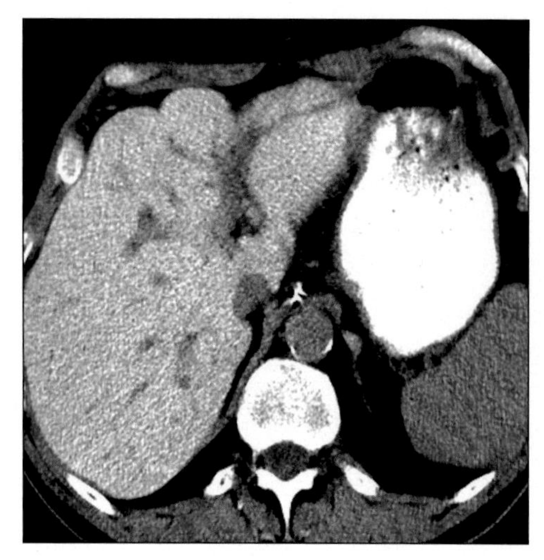

Axial NECT shows liver parenchyma of much higher attenuation than spleen (or muscle); primary hemochromatosis.

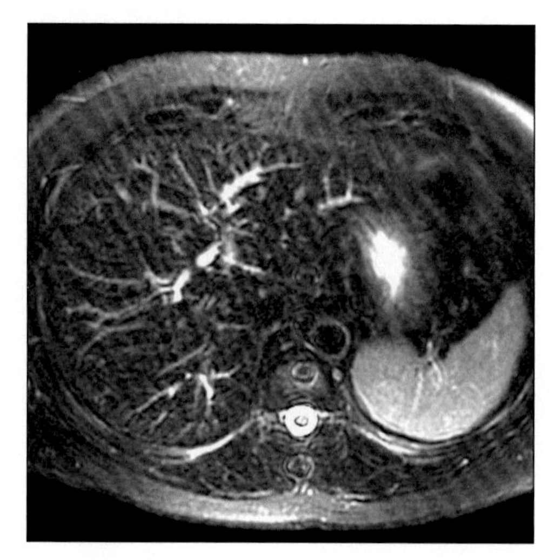

Axial T2WI MR shows marked hypointensity throughout liver; primary hemochromatosis.

TERMINOLOGY

Definitions

- Iron overload disorder in which there is structural & functional impairment of involved organs (total body iron may be 50-60 grams)

IMAGING FINDINGS

General Features

- Best diagnostic clue: Hyperdense liver on NECT & markedly hypointense on T2WI
- Location
  - Primary hemochromatosis
    - Parenchymal cells of liver, pancreas & heart
  - Secondary hemochromatosis
    - Initially reticuloendothelial system (RES)
    - After saturation of RES, parenchymal cells of liver, pancreas, myocardium, kidneys & endocrine glands
- Size
  - Precirrhotic stage: Increase in liver size
  - Cirrhotic stage: Decrease in liver size
- Key concepts
  - Hemochromatosis: Classified into two types
    - Primary (idiopathic): Inherited autosomal recessive disorder
    - Secondary: Due to increased iron intake, ineffective erythropoiesis, multiple blood transfusions, alcoholic cirrhosis & after portacaval shunts
    - Total body iron may be 50-60 grams
  - Hemosiderosis
    - Increased iron deposition without organ damage
    - Usually seen with body iron stores of 10-20 g
  - Normal body iron storage: 2 to 6 g of iron
    - 80% Functional iron: Hemoglobin, myoglobin & iron containing enzymes
    - 20% In storage form: Hemosiderin or ferritin
    - Liver contains up to one third of total body store of iron

CT Findings

- NECT
  - Homogeneously increased liver density
    - Up to 75-135 HU (normal 45-65 HU)
  - Prominent low attenuated hepatic & portal veins
  - Dual energy CT (at 80 & 120 kVp) technique used to
    - Establish diagnosis if attenuation is borderline
    - To quantitate amount of iron deposition in liver
    - To follow efficacy of therapy
  - Late stage
    - Liver shows cirrhotic features
- CECT
  - Decreases inherent contrast differences
    - Between liver, blood vessels & tumor

DDx: Diffusely Increased Liver Density

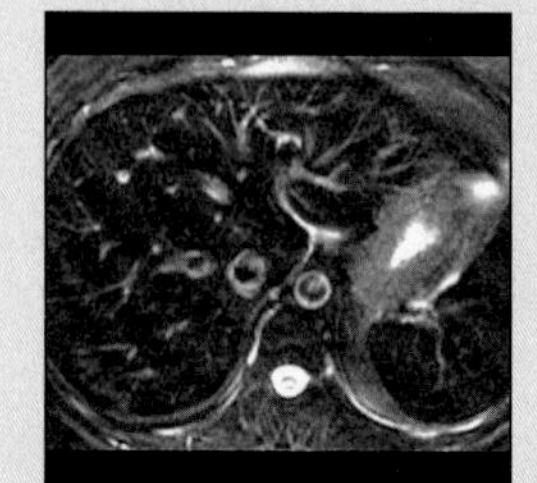

Hemosiderosis

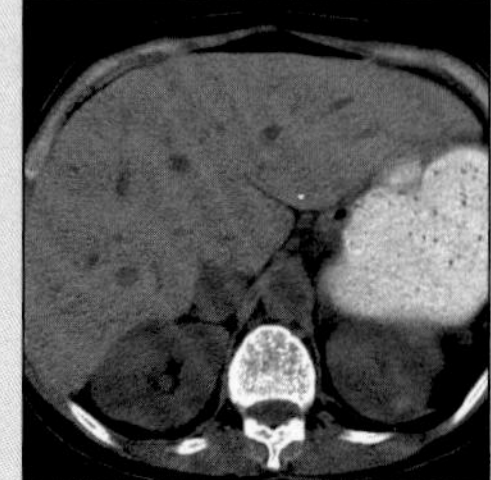

Hemosiderosis

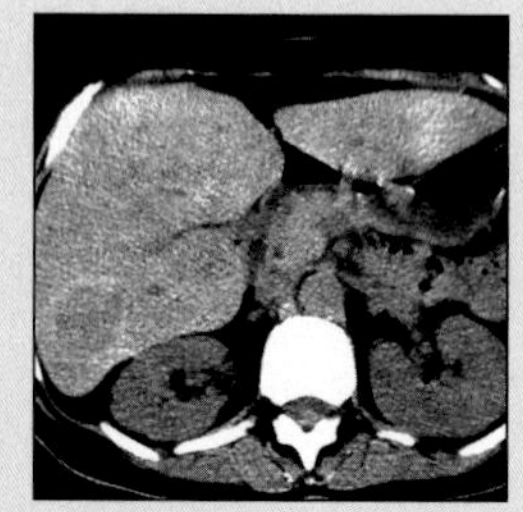

Glycogen Storage

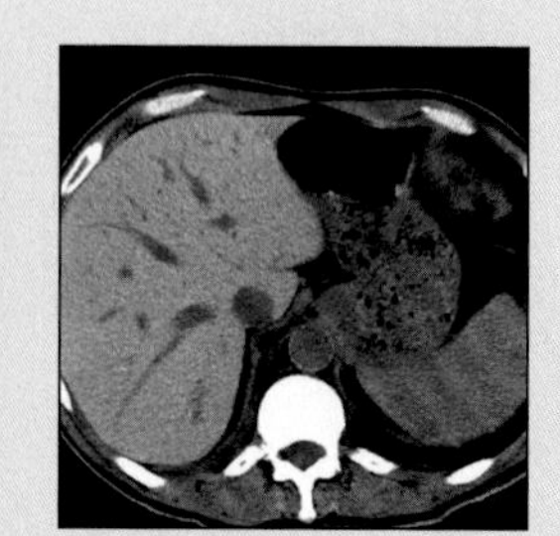

Amiodarone Therapy

HEMOCHROMATOSIS

Key Facts

Terminology
- Iron overload disorder in which there is structural & functional impairment of involved organs (total body iron may be 50-60 grams)

Imaging Findings
- Best diagnostic clue: Hyperdense liver on NECT & markedly hypointense on T2WI

Top Differential Diagnoses
- Hemosiderosis
- Glycogen storage disease
- Amiodarone therapy

Pathology
- Primary: Gene is human leukocyte antigen (HLA-A3 & B14) linked located on short arm of chromosome 6
- Secondary hemochromatosis

Clinical Issues
- Clinical profile: Patient with family history, hepatomegaly, diabetes mellitus, hyperpigmentation & elevated blood iron/ferritin levels

Diagnostic Checklist
- Rule out other conditions like hemosiderosis, glycogen storage disease, amiodarone & gold therapy which can cause diffusely hyperdense liver on NECT simulating hemochromatosis
- On T2WI: Marked signal loss of liver in primary type & marked signal loss of both liver/spleen in secondary type of hemochromatosis
- MR advantage: Other disorders do not simulate appearance of liver iron overload at MR like CT

MR Findings
- T1WI
  - Primary hemochromatosis
    - Decreased signal intensity in liver
- T2WI
  - Primary: Marked signal loss in liver
  - Secondary: Marked signal loss in both liver & spleen
- T2* GRE
  - Signal intensity ratios of liver/muscle or liver/fat
    - Establishes direct correlation with liver iron content better than T2 relaxation measurements
    - Accurate in quantifying liver iron content

Ultrasonographic Findings
- Real Time: Has no role in the diagnosis of hepatic iron overload

Imaging Recommendations
- Best imaging tool: MR T2* GRE for diagnosing hepatic hemochromatosis
- Protocol advice
  - For estimation of hepatic iron concentration
    - T2 GRE image (18/5, 10° flip angle)
    - Heavily T2W fast spin-echo sequence

DIFFERENTIAL DIAGNOSIS

Hemosiderosis
- Decrease signal intensity in both liver & spleen
- Early stage: Indistinguishable from hemochromatosis

Glycogen storage disease
- Increase or decrease attenuation of liver on NECT
- Associated with multiple hepatic adenomas (60%)

Amiodarone therapy
- Iodine containing anti-arrhythmic medication
- Diffuse homogeneous dense liver on NECT

PATHOLOGY

General Features
- Genetics
  - Primary: Gene is human leukocyte antigen (HLA-A3 & B14) linked located on short arm of chromosome 6
  - Mutations in HFE gene responsible for common form of HLA-linked hereditary hemochromatosis
- Etiology
  - Primary hemochromatosis
    - Autosomal recessive disorder
    - Relatively common & underdiagnosed cause of liver disease
    - Abnormal increase iron absorption by mucosa of duodenum & jejunum
    - Excess iron stored as cytoplasmic ferritin & lysosomal hemosiderin
    - Organs affected: Parenchymal cells (liver, pancreas, heart); joints, endocrine glands & skin
    - Does not affect Kupffer cells & reticuloendothelial cells of bone marrow, spleen
  - Secondary hemochromatosis
    - Patients with increased iron intake: Increased consumption of medicinal iron, iron laden wine, Kaffir beer & multiple blood transfusions
    - Anemic patients with infective erythropoiesis & multiple blood transfusions (e.g., thalassemia major, sideroblastic anemia)
    - Patients with alcoholic cirrhosis & after portacaval shunts
    - Initially iron deposition in RES, sparing parenchymal cells
    - After saturation of RES iron accumulates in parenchymal cells of liver, pancreas, myocardium
- Epidemiology
  - Primary or idiopathic
    - Increase prevalence in non-Jewish Caucasians of northern European origin (1:220)
    - Homozygote frequency: 0.25-0.50%
    - Heterozygote carriers: More than 10%

HEMOCHROMATOSIS

Gross Pathologic & Surgical Features

- Early stage
  - Liver is slightly larger & dense
  - Chocolate brown (ferritin)
  - Golden yellow granules (hemosiderin)
- Late stage
  - Decrease in liver size
  - Cirrhotic micronodules & fibrous septa
- Pancreas
  - Skin pigmentation ("bronze diabetes"), atrophy & fibrosis

Microscopic Features

- Prussian blue staining
  - Hemosiderin deposits in hepatocytes, Kupffer cells & lysosomes
- In late stages
  - Hepatocellular necrosis, scarring, fibrosis & cirrhosis

CLINICAL ISSUES

Presentation

- Most common signs/symptoms
  - Asymptomatic during 1st decade of disease
  - Hepatomegaly in 95% of cases
  - Splenomegaly in 50% of case
  - Classic triad of hemochromatosis
    - Micronodular cirrhosis
    - Diabetes mellitus
    - Hyperpigmentation of skin
  - Other signs/symptoms
    - Congestive heart failure, arrhythmias
    - Arthralgias
    - Loss of libido, impotence
    - Amenorrhea, testicular atrophy
- Clinical profile: Patient with family history, hepatomegaly, diabetes mellitus, hyperpigmentation & elevated blood iron/ferritin levels
- Lab data
  - Serum iron: Above 250 mg/DL (normal 50-150 mg/DL)
  - Serum ferritin: Above 500 ng/DL (normal below 150 ng/DL)
  - Transferrin saturation: Approaches 100% (normal 25-30%)
    - Earliest & most sensitive indicator of increased iron stores
  - Liver iron index: More than 2
  - Increased blood glucose
  - Urine analysis: Glycosuria
- Complications
  - Periportal fibrosis leads to cirrhosis in late stage
    - If iron concentration: Above 22,000 μg/g of tissue
  - Hepatocellular carcinoma (14-30%)
  - IDDM (30-60%)
  - Hepatic coma (15%); hematemesis (14%)
  - Cardiac failure (30%)

Demographics

- Age
  - Primary: Usually present in 4th or 5th decade
  - Secondary: Usually present at earlier age
- Gender
  - M:F = 10:1
  - Women are usually protected from this disorder
    - Due to iron loss during normal menstruation, pregnancy & lactation

Natural History & Prognosis

- Normal life expectancy with early diagnosis & treatment
- Life expectancy of untreated patients: 4.4 years

Treatment

- Deferoxamine (iron chelation therapy)
- Phlebotomies in precirrhotic stage

DIAGNOSTIC CHECKLIST

Consider

- Rule out other conditions like hemosiderosis, glycogen storage disease, amiodarone & gold therapy which can cause diffusely hyperdense liver on NECT simulating hemochromatosis

Image Interpretation Pearls

- On T2WI: Marked signal loss of liver in primary type & marked signal loss of both liver/spleen in secondary type of hemochromatosis
- MR advantage: Other disorders do not simulate appearance of liver iron overload at MR like CT

SELECTED REFERENCES

1. Kim MJ et al: Hepatic iron deposition on magnetic resonance imaging: correlation with inflammatory activity. J Comput Assist Tomogr. 26(6):988-93, 2002
2. Pomerantz S et al: MR imaging of iron depositional disease. Magn Reson Imaging Clin N Am. 10(1):105-20, vi, 2002
3. Bonkovsky HL et al: Hepatic iron concentration: Noninvasive estimation by means of MR imaging techniques. Radiology. 212: 227-34, 1999
4. Ito K et al: Hepatocellular carcinoma: Association with increased iron deposition in cirrhotic liver at MR imaging. Radiology. 212: 235-40, 1999
5. Press RD et al: Hepatic iron overload: direct HFE (HLA-H) mutation analysis vs quantitative iron assays for the diagnosis of hereditary hemochromatosis. Am J Clin Pathol. 109(5):577-84, 1998
6. Ernst O et al: Hepatic iron overload: diagnosis and quantification with MR imaging. AJR Am J Roentgenol. 168(5):1205-8, 1997
7. Siegelman ES et al: Abdominal iron deposition: metabolism, MR findings, and clinical importance. Radiology. 199(1):13-22, 1996
8. Gandon Y et al: Hemochromatosis: diagnosis and quantification of liver iron with gradient-echo MR imaging. Radiology. 193(2):533-8, 1994
9. Siegelman ES et al: Idiopathic hemochromatosis: MR imaging findings in cirrhotic and precirrhotic patients. Radiology. 188(3):637-41, 1993
10. Siegelman ES et al: Parenchymal versus reticuloendothelial iron overload in the liver: distinction with MR imaging. Radiology. 179(2):361-6, 1991
11. Guyader D et al: Evaluation of computed tomography in the assessment of hepatic iron overload. Gastroenterology. 97: 747-53, 1989

IMAGE GALLERY

Typical

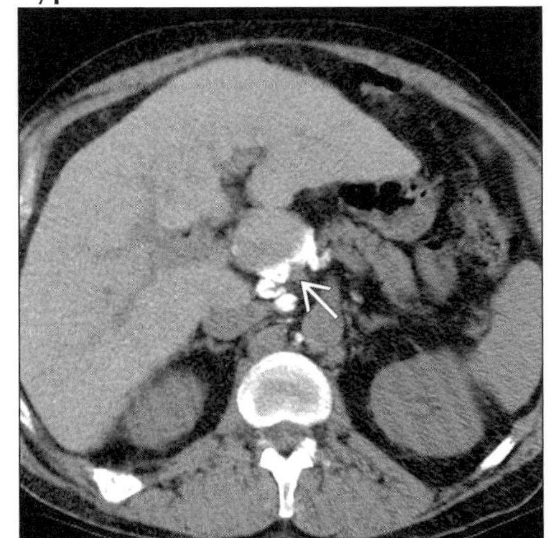

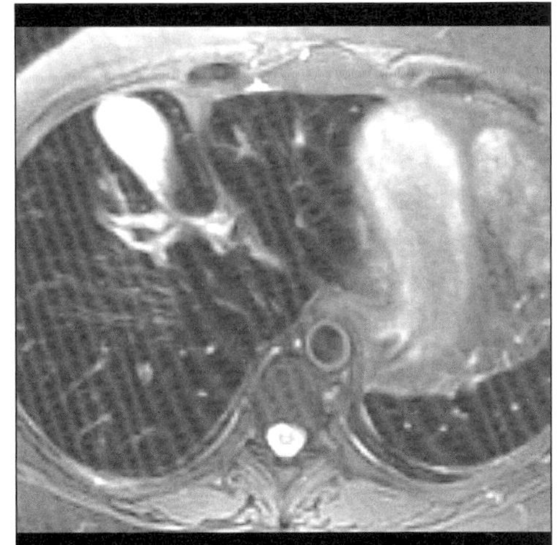

(Left) *Axial NECT shows hyperdense liver and very dense lymph nodes (arrow).* ***(Right)*** *Axial T2* GRE MR shows decreased signal intensity of liver and spleen when compared with that of paraspinal muscle; secondary hemochromatosis.*

Typical

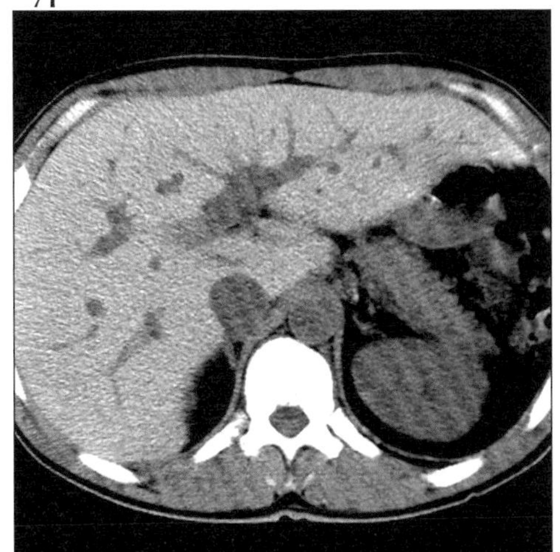

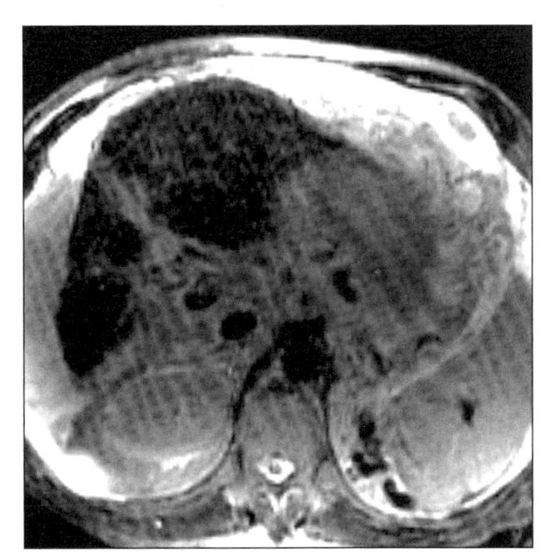

(Left) *Axial NECT shows marked diffuse increased density in liver; the spleen is surgically absent; secondary hemochromatosis from multiple transfusions.* ***(Right)*** *Axial T2WI MR shows shrunken cirrhotic, markedly hypointense liver, ascites, varices; primary hemochromatosis.*

Typical

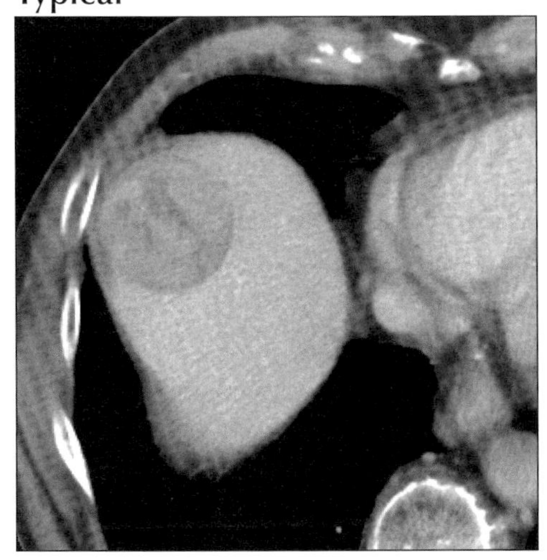

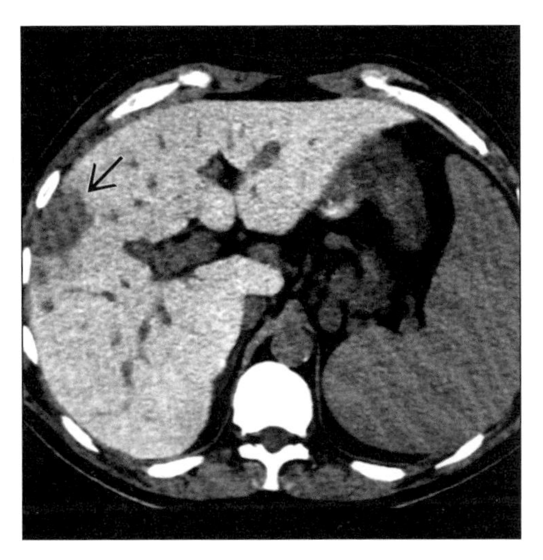

(Left) *Axial NECT in patient with primary hemochromatosis and cirrhosis shows dense liver with mass representing hepatocellular carcinoma (HCC).* ***(Right)*** *Axial NECT in patient with primary hemochromatosis. Hyperdense liver with focal HCC (arrow). Note the attenuation difference of the liver when compared to that of the enlarged spleen.*

WILSON DISEASE

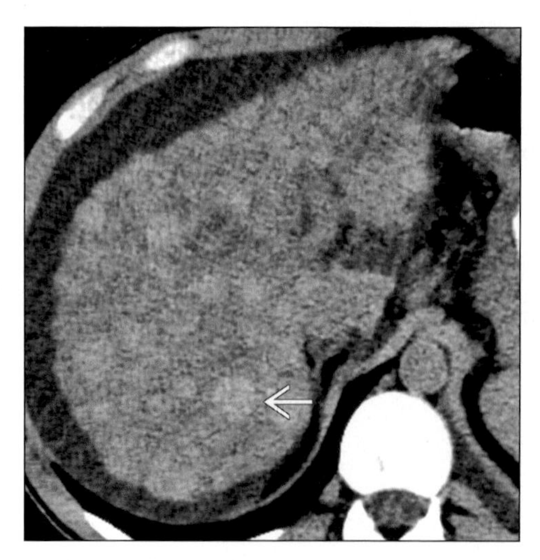

Axial NECT shows cirrhotic morphology and multiple discrete hyperdense regenerating nodules (arrow) which were more apparent than on CECT.

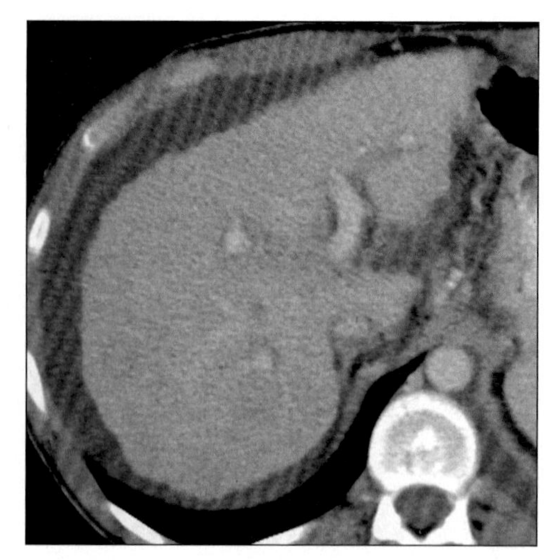

Axial CECT shows cirrhotic morphology and ascites. Regenerating nodules are isodense with liver and undedectable. Patient with Wilson disease being evaluated for liver transplantation.

TERMINOLOGY

Abbreviations and Synonyms

- Wilson disease (WD)
- Hepatolenticular degeneration

Definitions

- Autosomal recessive disorder in which copper (Cu) accumulates pathologically primarily within liver & subsequently in neurologic system & other tissues

IMAGING FINDINGS

General Features

- Best diagnostic clue: Liver biopsy for copper analysis
- Location: Early on, diffuse distribution of Cu in liver cytoplasm, later on within lysosomes & then throughout liver nodules
- Size: Diffuse involvement
- Key concepts
  - Fatty infiltration, acute or chronic active hepatitis, cirrhosis or massive liver necrosis

CT Findings

- Spectrum of hepatic injury is nonspecific; changes of fatty infiltration or cirrhosis frequently indistinguishable from those of other etiologies
- Although Cu has high atomic number & can cause elevation of liver density on CT, this is an unusual finding perhaps because coexisting fatty infiltration diminishes hepatic parenchymal attenuation
- Multiple, small, dysplastic nodules enhancing at arterial phase & thickened perihepatic fat layer; unusual finding
- Multiple hyperdense regenerating nodules (on NECT)

MR Findings

- Copper deposition has no ferromagnetic effect on MR imaging
- Iron in regenerative nodules cause hypointensity on T1 and T2WI

Ultrasonographic Findings

- Most commonly diffusely ↑ hepatic echogenicity (cirrhosis)

Imaging Recommendations

- Best imaging tool: CT & MR; however no major role in diagnosing WD
- Imaging alone cannot distinguish WD from other forms of hepatitis or cirrhosis

DDx: Diffusely Decreased Liver Density

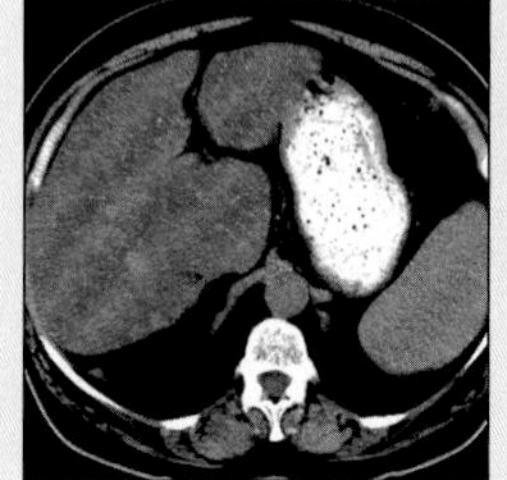

Steatosis

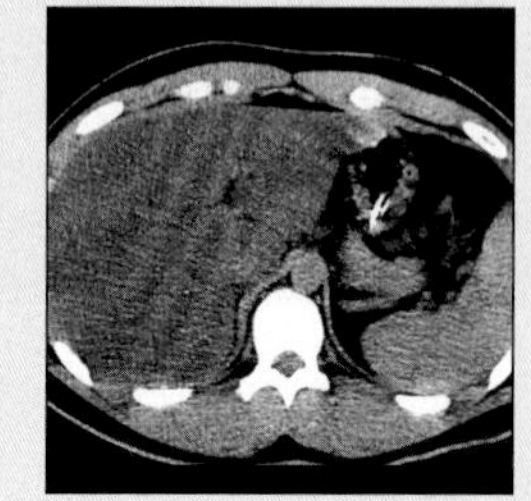

Acute Hepatitis

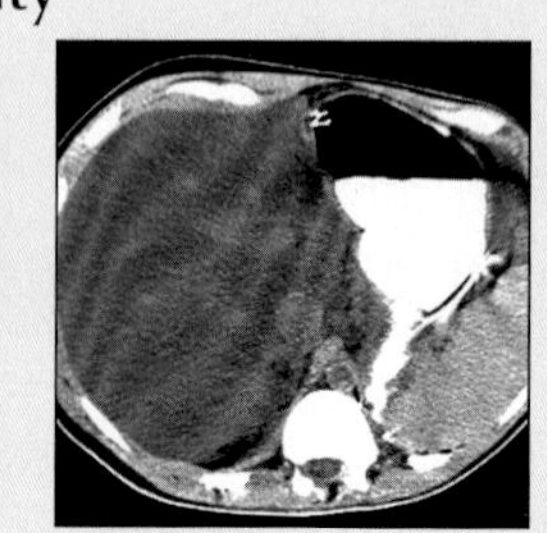

Acute Hepatitis

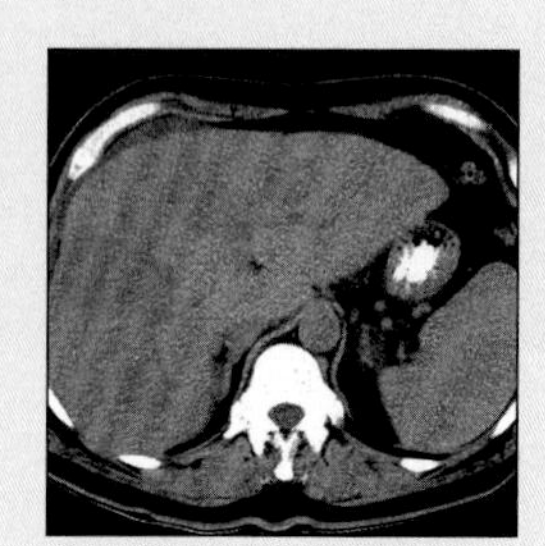

Diffuse Lymphoma

WILSON DISEASE

Key Facts

Terminology

- Autosomal recessive disorder in which copper (Cu) accumulates pathologically primarily within liver & subsequently in neurologic system & other tissues

Imaging Findings

- Spectrum of hepatic injury is nonspecific; changes of fatty infiltration or cirrhosis frequently indistinguishable from those of other etiologies

Pathology

- Hepatic sinusoidal and periportal deposition of Cu
- Cu deposition incites inflammatory reaction leading to cirrhosis

Clinical Issues

- Most common signs/symptoms:
- Lab data: Serum ceruloplasmin < 20 mg/dL
- Diagnosis: Liver biopsy & Cu quantitation

DIFFERENTIAL DIAGNOSIS

Steatosis

- NECT: Density of hepatic parenchyma less than spleen
- Presence of normal vessels coursing through

Hepatitis

- Marked hepatomegaly, ascites, or both in conjunction with decreased attenuation of liver parenchyma on CT

Lymphoma

- Diffuse or focal hepatic hypodensity

PATHOLOGY

General Features

- Genetics: Autosomal recessive disorder
- Etiology: ↓ Biliary excretion of Cu, ↑ intestinal absorption of Cu, abnormal urinary excretion of Cu
- Epidemiology: Prevalence: 1:30,000 individuals

Gross Pathologic & Surgical Features

- Steatosis, followed by fibrosis & ultimately cirrhosis

Microscopic Features

- Hepatic sinusoidal and periportal deposition of Cu
- Cu deposition incites inflammatory reaction leading to cirrhosis

CLINICAL ISSUES

Presentation

- Clinical profile
  - Chronic hepatitis, cirrhosis, acute liver failure
  - Acute fulminant hepatitis: Presents acutely with signs of jaundice, ascites that progresses to encephalopathy, & liver failure
  - Lab data: Serum ceruloplasmin < 20 mg/dL
  - Diagnosis: Liver biopsy & Cu quantitation
    - Hepatic Cu content > 250 ug/g dry weight
    - Presence of Kayser-Fleisher rings & low level of ceruloplasmin is sufficient to diagnose WD

Demographics

- Gender: Acute fulminant presentation of WD is most often seen in females (M:F = 1:2)

Natural History & Prognosis

- Deposition to toxic levels occurs in basal ganglia, renal tubules, cornea, bones, joints
- Patients with cirrhosis associated with WD are also predisposed to hepatocellular carcinoma
- Acute & early presentations like fulminant hepatic failures have poor outcome
- Timely & appropriate utilization of current modes of treatment offer patients excellent long term survival

Treatment

- Initial & maintenance therapy with Cu-chelator
- Liver transplantation

DIAGNOSTIC CHECKLIST

Consider

- Myriad manifestations of WD make its diagnosis dependent on a high index of suspicion

SELECTED REFERENCES

1. Akhan O et al: Unusual imaging findings in Wilson's disease. Eur Radiol. 12 Suppl 3:S66-9, 2002
2. Ko S et al: Unusual liver MR findings of Wilson's disease in an asymptomatic 2-year-old girl. Abdom Imaging. 23(1):56-9, 1998
3. Garg RK et al: Wilson's disease: unusual clinical and radiological features. J Assoc Physicians India. 42(3):253-4, 1994

IMAGE GALLERY

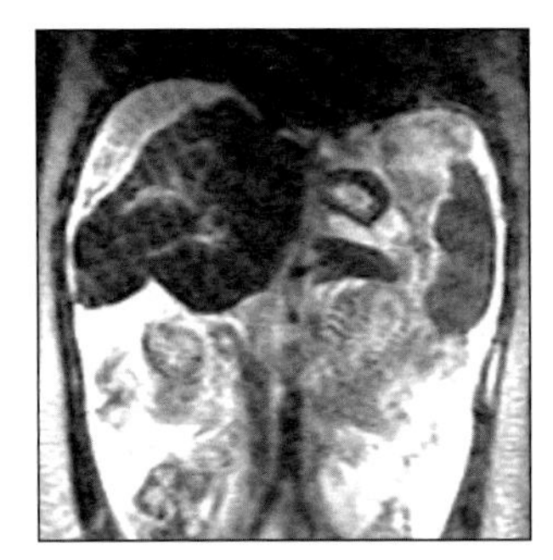

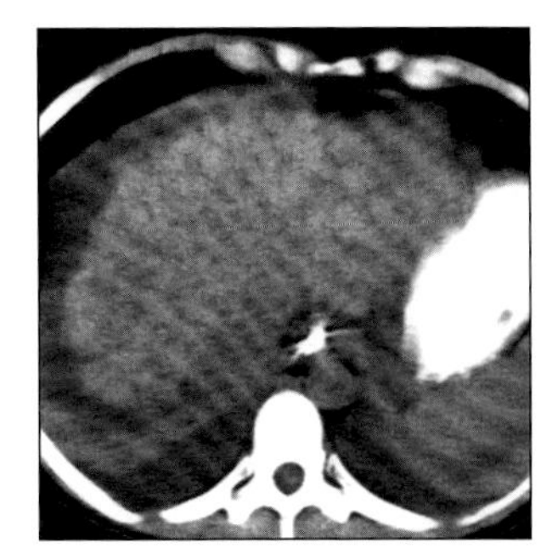

(Left) *Coronal T2WI MR shows diffusely hypointense and shrunken cirrhotic liver. Ascites.* ***(Right)*** *Axial NECT shows multiple hyperdense regenerating nodules within a cirrhotic liver. Patient with Wilson disease and acute fulminant hepatitis.*

HEPATIC TRAUMA

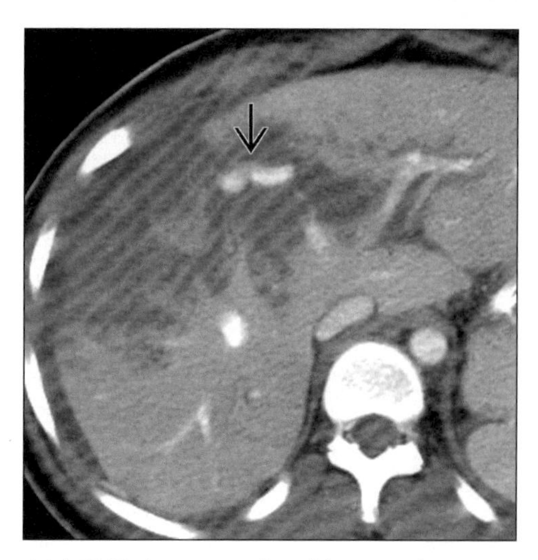

Axial CECT shows parenchymal laceration/hematoma, with active bleeding (arrow).

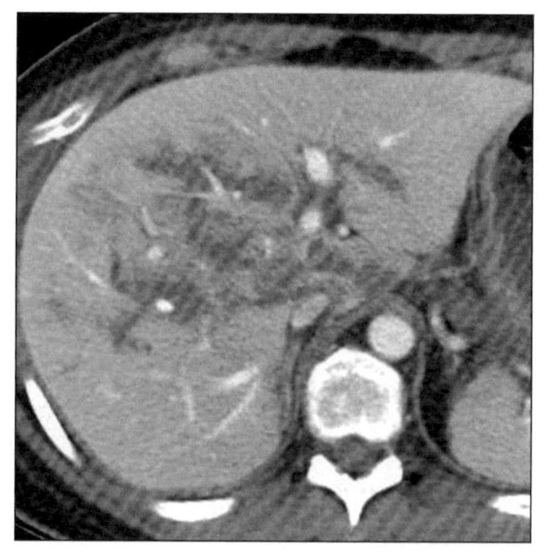

Axial CECT shows multiple linear and stellate planes of laceration but no active bleeding. There is minimal hemoperitoneum.

TERMINOLOGY

Abbreviations and Synonyms

- Liver or hepatic injury

IMAGING FINDINGS

General Features

- Best diagnostic clue: CT evidence of irregular parenchymal lesions with intra & perihepatic hemorrhage
- Location
  - Right lobe (75%); left lobe (25%)
    - Intraparenchymal or subcapsular
- Key concepts
  - Liver 2nd most frequently injured solid intra-abdominal organ after spleen
    - Due to its anterior & partially subcostal location
  - Most common causes of hepatic trauma
    - Blunt (more common), penetrating & iatrogenic injuries
  - Iatrogenic injury due to liver biopsy
    - Most common cause of subcapsular hematoma in US
  - Abdominal trauma
    - Leading cause of death in United States (< 40 yrs)

CT Findings

- Lacerations
  - Simple or stellate (parallel to portal/hepatic vein branches)
    - Simple: Hypodense solitary linear laceration
    - Stellate: Hypodense branching linear lacerations
- Parenchymal & subcapsular hematomas (lenticular configuration)
  - Unclotted blood (35-45 HU) soon after injury
    - NECT: May be hyperdense relative to normal liver
    - CECT: Hypodense compared to enhancing normal liver tissue
  - Clotted blood (60-90 HU)
    - More dense than unclotted blood & normal liver
    - May be more dense than unenhanced liver
- Active hemorrhage or pseudoaneurysm
  - CECT: Active hemorrhage
    - Isodense to enhanced vessels
    - Seen as contrast extravasation (85-350 HU)
    - Extravasated contrast material & surrounding decreased attenuation clot
- Hemoperitoneum: Perihepatic and peritoneal recess collections of blood
- Periportal tracking: Linear, focal or diffuse periportal zones of decreased HU
  - Due to dissecting blood, bile or dilated periportal lymphatics
  - DDx: Overhydration (check for distended IVC)

DDx: Focal Liver Lesion with Hemorrhage

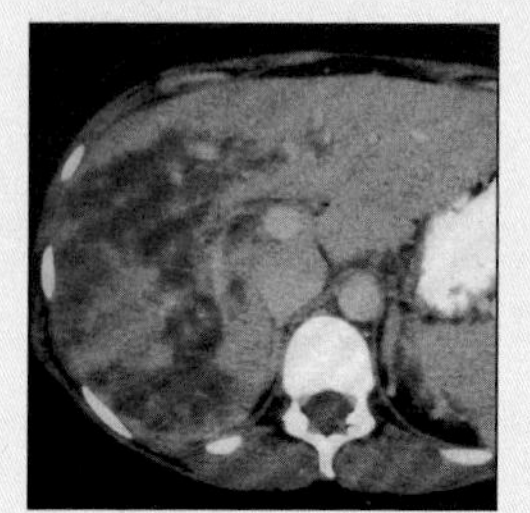

HELLP Syndrome

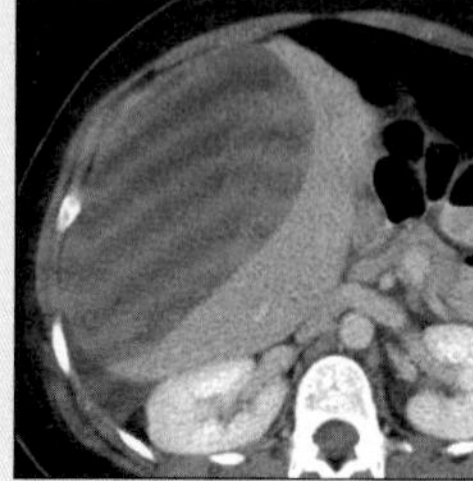

Coagulopathy

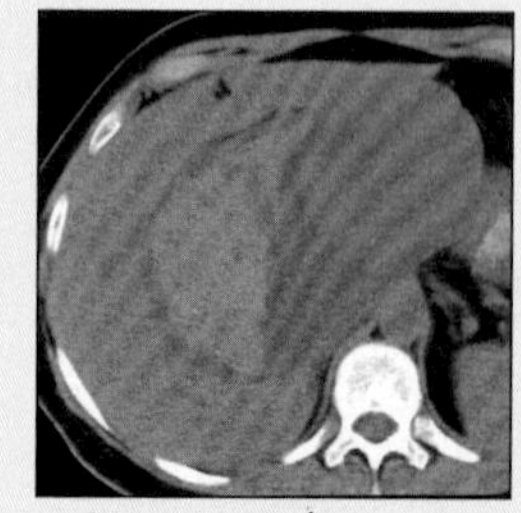

Hepatic Adenoma

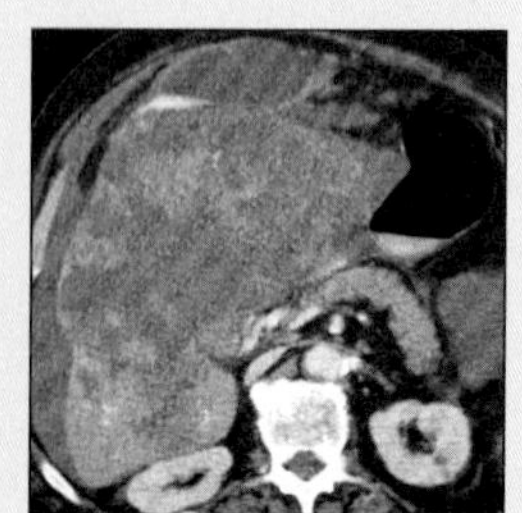

Metastases

HEPATIC TRAUMA

Key Facts

Imaging Findings
- Best diagnostic clue: CT evidence of irregular parenchymal lesions with intra & perihepatic hemorrhage
- Right lobe (75%); left lobe (25%)
- Intraparenchymal or subcapsular
- Morphology:

Top Differential Diagnoses
- HELLP syndrome
- Spontaneous hemorrhage (coagulopathy)
- Bleeding hepatic tumor (e.g.,: HCC or adenoma)

Pathology
- Blunt trauma (more common)

Clinical Issues
- Clinical profile: Patient with history of motor vehicle accident, RUQ tenderness, guarding & hypotension
- Mortality: 10-20%

Diagnostic Checklist
- Differentiate from HELLP syndrome; spontaneous hemorrhage (coagulopathy) & bleeding hepatic tumors like HCC or adenoma
- CT evidence of active extravasation (intra- or extra-hepatic collection, isodense with vessels) usually indicates need for embolization or surgery regardless of "grade" of injury
- Laceration of left hepatic lobe often associated with bowel and pancreatic injury

        - Elevated venous pressure & transudation
- Areas of infarction
    - Small or large areas of low attenuation
    - Usually wedge-shaped; segmental or lobar
    - Intrahepatic/subcapsular gas (due to hepatic necrosis)
- CT diagnosis of liver trauma
    - Accuracy: 96%
    - Sensitivity: 100%
    - Specificity: 94%

MR Findings
- T1WI & T2WI
    - Varied signal intensity depending on
        - Degree & age of hemorrhage or infarct

Ultrasonographic Findings
- Real Time
    - Subcapsular hematoma: Lentiform or curvilinear fluid collection
        - Initially: Anechoic
        - After 24 hrs: Echogenic
        - After 4-5 days: Hypoechoic
    - After 1-4 weeks: Internal echoes & septations develop within hematoma
    - Intraparenchymal hematoma
        - Rounded echogenic or hypoechoic foci
    - Bilomas
        - Rounded/ellipsoid, anechoic, loculated structures
        - Well-defined sharp margins, close to bile ducts
    - Parenchymal tears
        - Irregular defects
        - Abnormal echotexture relative to normal liver

Angiographic Findings
- Conventional
    - Demonstrate
        - Active extravasation, pseudoaneurysm
        - A-V, arteriobiliary or portobiliary fistulas

Imaging Recommendations
- Best imaging tool
    - Helical CECT: In hemodynamically stable cases
    - Angiography: To localize active hemorrhage & embolization
- Protocol advice: Helical CECT: Include lung bases and pelvis

DIFFERENTIAL DIAGNOSIS

HELLP syndrome
- Severe variant of preeclampsia
- HELLP: Hemolysis, elevated liver enzymes & low platelets
- On imaging
    - Intrahepatic or subcapsular fluid collection (hematoma)
    - Wedge-shaped areas of infarction
    - Occasionally active extravasation

Spontaneous hemorrhage (coagulopathy)
- History of bleeding disorder
- Lab data: Abnormal hematologic coagulation values
- On imaging
    - Subcapsular or intrahepatic blood collection
    - Indistinguishable from hepatic trauma without history

Bleeding hepatic tumor (e.g.,: HCC or adenoma)
- Spherical enhancing parenchymal masses
- Hepatocellular carcinoma
    - Vascular, nodal & visceral invasion (common)

PATHOLOGY

General Features
- Etiology
    - Blunt trauma (more common)
        - Motor vehicle accidents (more common)
        - Falls and assaults
    - Penetrating injuries
        - Gunshot and stab injuries
    - Iatrogenic

HEPATIC TRAUMA

- Liver biopsy, chest tubes, transhepatic cholangiography
- Epidemiology
  - 5-10% blunt abdominal trauma have liver injury
  - Mortality from hepatic trauma: 10-20%
- Associated abnormalities
  - Splenic injury (45%); bowel injury (5%); rib fractures
  - Left hepatic lobe laceration often associated with bowel or pancreatic injury

Gross Pathologic & Surgical Features

- Laceration or contusion
- Subcapsular or intraparenchymal hematoma

Staging, Grading or Classification Criteria

- Clinical classification based on American Association for Surgery of Trauma (AAST)
  - Grade I
    - Subcapsular hematoma: Less than 10% surface area
    - Laceration: Capsular tear, less than 1 cm parenchymal depth
  - Grade II
    - Subcapsular hematoma: 10-50% surface area
    - Intraparenchymal hematoma: Less than 10 cm diameter
    - Laceration: 1-3 cm parenchymal depth, less than 10 cm in length
  - Grade III
    - Subcapsular hematoma: More than 50% surface area; expanding/ruptured subcapsular or parenchymal hematoma
    - Intraparenchymal hematoma: More than 10 cm or expanding
    - Laceration: Parenchymal fracture more than 3 cm deep
  - Grade IV
    - Laceration: Parenchymal disruption involving 25-75% of hepatic lobe or 1-3 Couinaud segments within a single lobe
  - Grade V
    - Laceration: Parenchymal disruption involving > 75% of hepatic lobe or > 3 Couinaud segments within a single lobe
    - Vascular: Juxtahepatic venous injuries (retrohepatic venacava, major hepatic veins)
  - Grade VI
    - Vascular: Hepatic avulsion

CLINICAL ISSUES

Presentation

- Most common signs/symptoms
  - RUQ pain, tenderness, guarding, rebound tenderness
  - Hypotension, tachycardia, jaundice
  - Hematemesis or melena (due to hemobilia)
- Clinical profile: Patient with history of motor vehicle accident, RUQ tenderness, guarding & hypotension
- Lab data
  - Decreased hematocrit (not acutely)
  - Increased direct/indirect bilirubin
  - Increased alkaline phosphatase levels

Natural History & Prognosis

- Complications
  - Hemobilia, bilomas, A-V fistula, pseudoaneurysm
- Prognosis
  - Grade I, II & III: Good
  - Grade IV, V & VI: Poor
  - May not necessarily correlate with AAST grading
  - Mortality: 10-20%
    - 50% due to liver injury itself
    - Rest from associated injuries

Treatment

- Grade I, II, III
  - Conservative management for almost all injuries diagnosed on CT
    - Implies some degree of clinical stability
- Grade IV, V, VI
  - Surgical intervention for shock & peritonitis
    - Control hemorrhage, drainage & repair
  - Embolization for active extravasation

DIAGNOSTIC CHECKLIST

Consider

- Differentiate from HELLP syndrome; spontaneous hemorrhage (coagulopathy) & bleeding hepatic tumors like HCC or adenoma

Image Interpretation Pearls

- CT evidence of active extravasation (intra- or extra-hepatic collection, isodense with vessels) usually indicates need for embolization or surgery regardless of "grade" of injury
- Laceration of left hepatic lobe often associated with bowel and pancreatic injury

SELECTED REFERENCES

1. Yao DC et al: Using contrast-enhanced helical CT to visualize arterial extravasation after blunt abdominal trauma: incidence and organ distribution. AJR Am J Roentgenol. 178(1):17-20, 2002
2. Patten RM et al: CT detection of hepatic and splenic injuries: usefulness of liver window settings. AJR Am J Roentgenol. 175(4):1107-10, 2000
3. Poletti PA et al: CT criteria for management of blunt liver trauma: correlation with angiographic and surgical findings. Radiology. 216(2):418-27, 2000
4. Becker CD et al: Blunt hepatic trauma in adults: correlation of CT injury grading with outcome. Radiology. 201(1):215-20, 1996
5. Mirvis SE et al: Blunt hepatic trauma in adults: CT-based classification and correlation with prognosis and treatment. Radiology. 171(1):27-32, 1989

IMAGE GALLERY

Typical

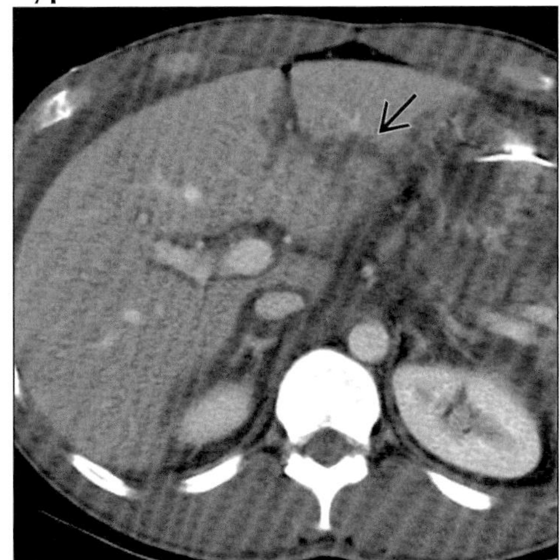

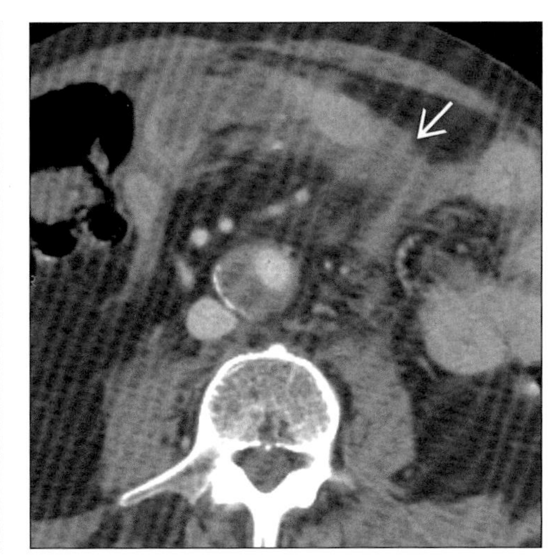

(Left) *Axial CECT shows laceration of lateral segment (arrow) and hemoperitoneum.* ***(Right)*** *Axial CECT shows jejunal injury with clotted blood ("sentinel clot") (arrow) in mesentery between bowel loops. Imaging through the liver demonstrated lateral segment laceration.*

Typical

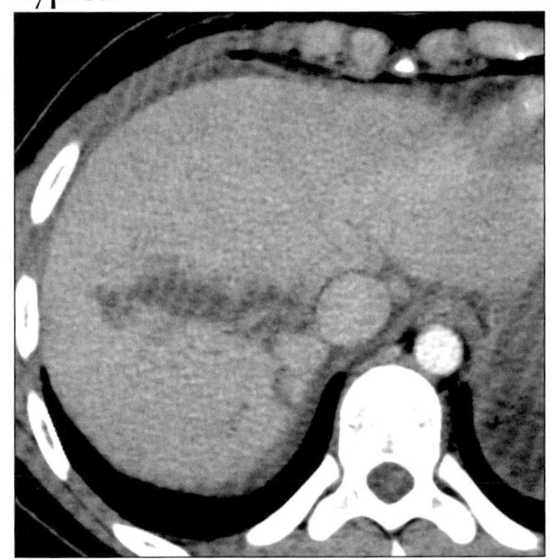

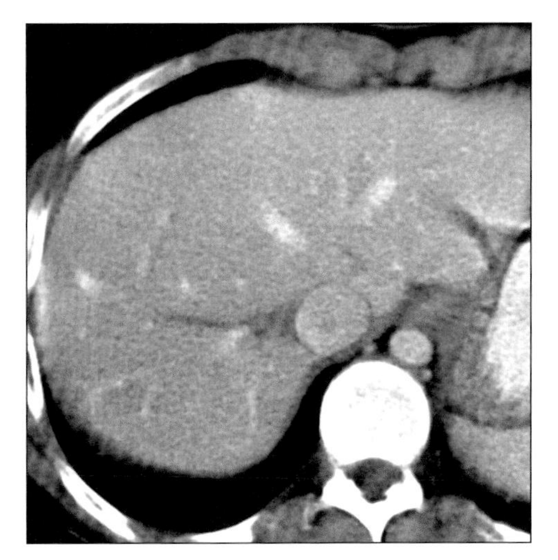

(Left) *Axial CECT shows deep linear laceration of right hepatic lobe in a patient who was managed conservatively.* ***(Right)*** *Axial CECT obtained two weeks after blunt trauma demonstrates considerable healing of deep hepatic laceration.*

Typical

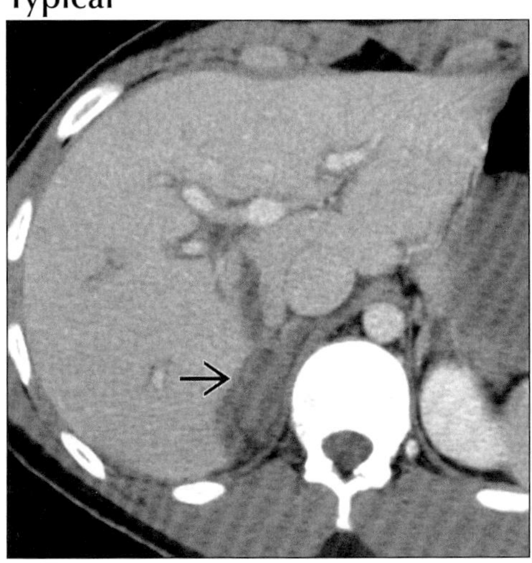

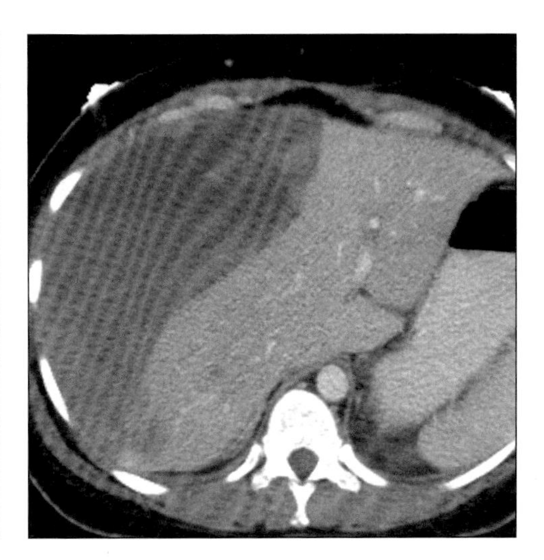

(Left) *Axial CECT shows linear laceration through base of caudate lobe extending to the bare area of the liver and resulting in retroperitoneal hematoma (arrow).* ***(Right)*** *Axial CECT shows large subcapsular hematoma. Patient had previous recent motor vehicle accident without medical evaluation, and self medicated with aspirin and ibuprofen.*

BILIARY TRAUMA

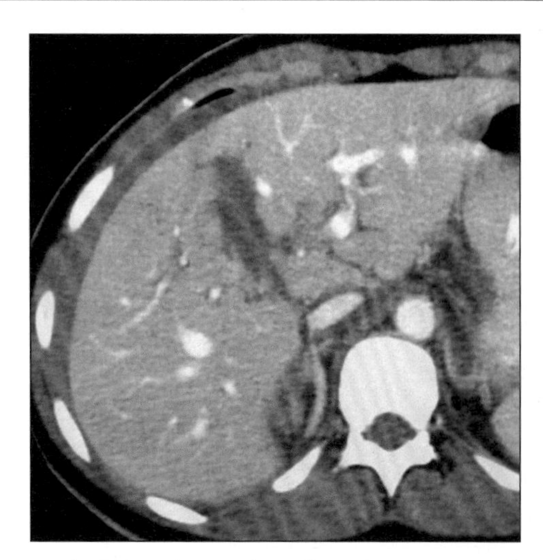

Axial CECT shows deep liver laceration, small hemoperitoneum. The severity of the injury transected intrahepatic bile ducts.

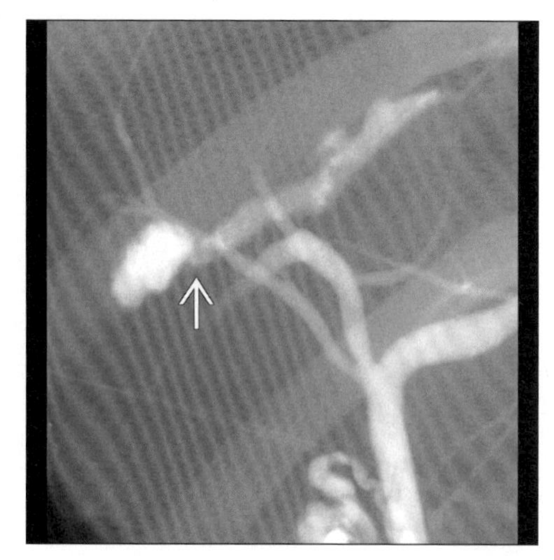

Cholangiogram performed in patient who had developed peritoneal symptoms shows extravasation of bile (arrow) from transected intrahepatic duct. Successfully treated with biliary stenting.

TERMINOLOGY

Abbreviations and Synonyms

- Bile duct injury

Definitions

- Hemobilia: Bleeding into biliary tract
- Bilhemia: Condition in which bile enters veins of liver, & is rare

IMAGING FINDINGS

General Features

- Best diagnostic clue
  - Clinical history can suggest diagnosis
  - Percutaneous transhepatic (PTC) or endoscopic retrograde cholangiography (ERCP): "Gold standard'" for diagnosis of bile duct injuries
- Location
  - In major bile duct injuries, common bile duct/common hepatic duct are most frequently injured (61.1%)
  - Post-operative strictures: Common hepatic duct (45-64%); hepatic hilus (20-33%)
  - Biliary fistulas: Internal (communication with duodenum, colon, bronchi, etc.) or external (skin)
    - Biliary-vascular fistulas: To portal vein, hepatic artery, hepatic veins
- Size: Focal or diffuse involvement
- Morphology: Bile leakage, strictures, biliary tree obstruction, various types of biliary fistulas, hemobilia

Radiographic Findings

- Radiography: Biliary-enteric fistula: Pneumobilia
- Fluoroscopy
  - Biliary-enteric fistula: Barium filling of biliary tree
  - Nonionic or oil-based contrast material is indicated when biliary-bronchial fistula is suspected
- ERCP
  - To evaluate: Level, length, contour of strictures
    - Posttraumatic strictures are typically focal, smooth areas of narrowing with proximal dilation
  - ERCP can facilitate definitive diagnosis & treatment of bile leaks & simple strictures
  - Visualization & cannulation of fistula orifice & permits good-quality cholangiographic evaluation
  - May see active bleeding coming from major papilla

CT Findings

- Helical CT cholangiography or helical CT after administration of biliary i.v. contrast material to verify & localize bile duct leakage; may help avoid ERCP
- Biliary-enteric fistula: Oral contrast media in both bowel & biliary tree

DDx: Biliary "Leak"

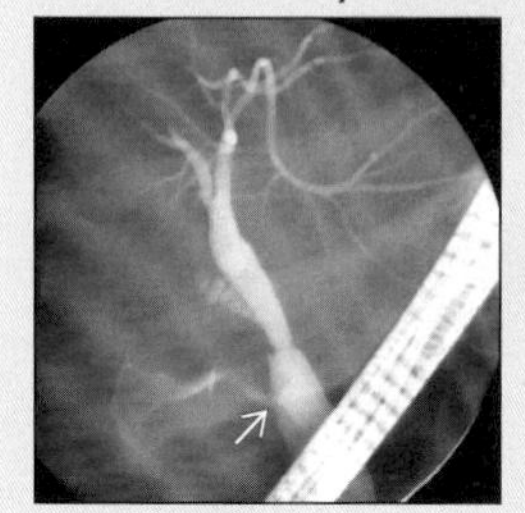

Biliary Leak (OLT)

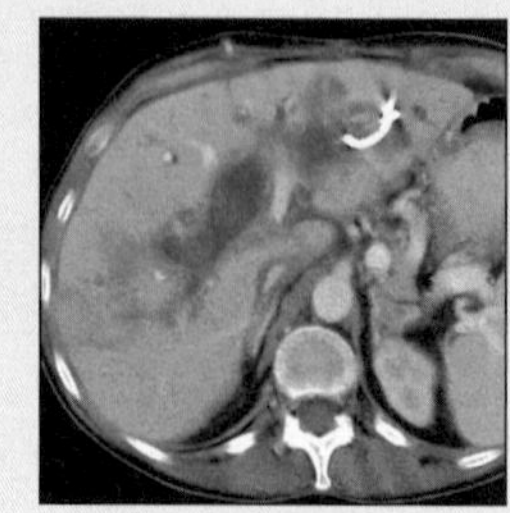

HAT/Biliary Necrosis

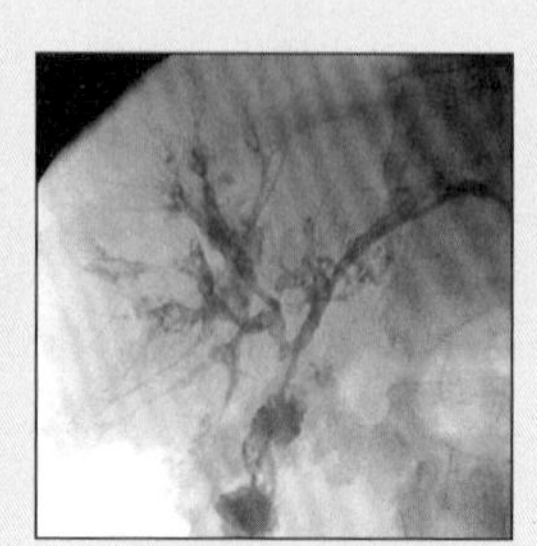

HAT/Biliary Necrosis

BILIARY TRAUMA

Key Facts

Terminology

- Bile duct injury
- Hemobilia: Bleeding into biliary tract

Imaging Findings

- Clinical history can suggest diagnosis
- Percutaneous transhepatic (PTC) or endoscopic retrograde cholangiography (ERCP): "Gold standard'" for diagnosis of bile duct injuries
- Morphology: Bile leakage, strictures, biliary tree obstruction, various types of biliary fistulas, hemobilia
- Posttraumatic strictures are typically focal, smooth areas of narrowing with proximal dilation
- ERCP can facilitate definitive diagnosis & treatment of bile leaks & simple strictures
- Hemobilia: Diagnosis is most commonly confirmed by selective hepatic arteriography, demonstrating extravasation of contrast material into biliary tree
- Best imaging tool: PTC is generally more valuable than ERCP; in that it defines anatomy of proximal biliary tree that is to be used in surgical reconstruction

Pathology

- Trauma: Blunt or penetrating
- Iatrogenic: Laparoscopic or conventional open cholecystectomy

Clinical Issues

- Biliary cirrhosis from long-standing obstruction

- Extent & localization of parenchymal destruction in bilio-vascular fistula
- Presence of biliary dilatation, configuration of injured bile duct, & ancillary abdominal findings
- Hemobilia: Blood may appear as high-attenuation material (> 50 HU) in ducts/gallbladder
  - Liver laceration, hematoma, other potential sources of blood may also be detected
- CT-guided drainage; nonoperative management of parenchymal & (retro)-peritoneal collections

MR Findings

- MRCP: Noninvasive, non-ionizing tool for diagnosis of wide variety of bile duct injuries
- Can use hepatobiliary contrast agent (Mangafodipir) to directly visualize leak or stricture

Ultrasonographic Findings

- Proximal biliary dilation with gradual tapering of duct diameter, without a surrounding mass present
- Blood: Echogenic material in bile-ducts
- Bilioma: Echo-free well-marginated fluid collection
- Ultrasound-guided drainage of collections

Angiographic Findings

- Hemobilia: Diagnosis is most commonly confirmed by selective hepatic arteriography, demonstrating extravasation of contrast material into biliary tree
  - Excludes hepatic artery aneurysm/pseudoaneurysm (10% of cases of biliary bleeding)
  - Facilitates embolic occlusion therapy
- If vessel injury, or hemobilia are suspected on CT scan, angiography should be carried out, even in active bleeding; can be therapeutic, thereby avoiding surgery

Nuclear Medicine Findings

- Technetium 99m dimethyliminodiacetic acid (HIDA) scan to detect occult ductal injuries, confirm bile leak
- Hepatobiliary scintigraphy is safe means of investigating a possible biliary-bronchial fistula

Other Modality Findings

- Cholangiography: May demonstrate clotted blood as cast-like filling defect in bile ducts & may reveal other potential causes of hemorrhage

Imaging Recommendations

- Best imaging tool: PTC is generally more valuable than ERCP; in that it defines anatomy of proximal biliary tree that is to be used in surgical reconstruction
- Protocol advice: ERCP/PTC/MRCP; angiography

DIFFERENTIAL DIAGNOSIS

Iatrogenic injury

- After liver transplantation (OLT); bile leak at T-tube site, in nonanastomotic location in donor biliary tree or at the anastomosis

Hepatic artery thrombosis (HAT)

- Ischemia of bile ducts, nonanastomotic leak in hilar region or intrahepatic ducts
- Ballooning of necrotic ducts into irregular cystic spaces from distended, debris-filled, necrotic bile ducts

PATHOLOGY

General Features

- General path comments
  - Iatrogenic injury occurs as a result of technical errors or misidentification of biliary anatomy
    - Inexperience, inflammation, aberrant anatomy
- Etiology
  - Trauma: Blunt or penetrating
  - Iatrogenic: Laparoscopic or conventional open cholecystectomy
    - Other: Percutaneous liver puncture; percutaneous biliary drainage; during biliary tract exploration; hepatic artery embolization; infusion of chemotherapeutic agents; gastrectomy; hepatic resection; cautery-induced; etc.
- Epidemiology

BILIARY TRAUMA

- 95% of strictures are secondary to surgical injury & > 80% from trauma during cholecystectomy
- ↑ Incidence over past decade with introduction of laparoscopic cholecystectomy; affecting ≈ 2,000 patients annually in United States
  - Higher incidence during laparoscopic than open cholecystectomy; at least double the rate
- 50% of cases of hemobilia are due to blunt trauma

Gross Pathologic & Surgical Features

- Post-laparoscopic cholecystectomy: Spectrum of injury ranges from cystic duct stump leakage to partial obstruction to complete occlusion of ducts & common hepatic or common bile duct ischemic strictures

Microscopic Features

- Disruption of duct epithelium, communication between ducts & other organs, narrowing of lumen

CLINICAL ISSUES

Presentation

- Most common signs/symptoms
  - Patients with biliary stricture after blunt abdominal trauma may exhibit a delayed onset of symptoms
  - Post-operative bile duct injuries: May present early with obstructive jaundice or evidence of a bile leak
    - In patients presenting months to years after surgery, cholangitis is most common symptom
    - In only 10-25% of patients with postcholecystectomy injury is the problem recognized within first week, but nearly 70% are recognized within first 6 months
  - Triad of GI blood loss, biliary colic, jaundice suggests presence of hemobilia, although both pain & jaundice may be absent
- Clinical profile
  - Lab data: An excessively high serum level of direct bilirubin & only moderately elevated liver enzymes indicate bilhemia in trauma patients
  - Thoracentesis: Presence of bile in pleural cavity is considered proof of pleural-biliary fistula

Demographics

- Age: Iatrogenic trauma more common in adults
- Gender
  - Males: More blunt trauma
  - Females: More iatrogenic injuries (more frequent cholecystectomies)

Natural History & Prognosis

- Major, profuse hemobilia is rare but may be life-threatening; minor hemobilia is more frequent & often clinically silent
  - Bleeding may often be delayed by 3-4 weeks & even by as much as 12 weeks after liver injury
- Causes of delayed complications are multiple & include: Abnormal or insufficient injury healing process; retention of necrotic tissue; secondary infection of initially sterile collections; underestimation of injury severity
- Biliary cirrhosis from long-standing obstruction
- Significant morbidity & mortality associated with non-surgical trauma to extrahepatic biliary tract

Treatment

- Principles: Definition of anatomy, relief of any impedance to biliary flow, & drainage of collections
- In minor leaks, endoscopic diversion by sphincterotomy or stenting
- More severe injury, ERCP (or PTC) to assess injury and plan operative repair
  - Surgical treatment for strictures or major leak: Roux-en-Y hepaticojejunostomy
- Hemobilia: Conservatively (if minor bleeding); transarterial embolization; surgery
- Lesions detected during cholecystectomy should be repaired immediately; biliary-enteric anastomosis (41.8%) or t-tube or stent (27.5%)
- Bile duct fistula: Spontaneous closure, suture of fistula & T-tube drainage, decompression

DIAGNOSTIC CHECKLIST

Consider

- Patients with major bile duct injuries should be evaluated for concomitant hepatic arterial injury; management & outcome may be influenced by absence of arterial blood flow to injured ducts & liver
- High index of suspicion is mandatory in patients complaining of discomfort several days after surgery

SELECTED REFERENCES

1. Federle MP et al: Complications of liver transplantation: imaging and intervention. Radiol Clin North Am. 41(6):1289-305, 2003
2. Familiari L et al: An endoscopic approach to the management of surgical bile duct injuries: nine years' experience. Dig Liver Dis. 35(7):493-7, 2003
3. Wong YC et al: Magnetic resonance imaging of extrahepatic bile duct disruption. Eur Radiol. 12(10):2488-90, 2002
4. Goffette PP et al: Traumatic injuries: imaging and intervention in post-traumatic complications (delayed intervention). Eur Radiol. 12(5):994-1021, 2002
5. Green MH et al: Haemobilia. Br J Surg. 88(6):773-86, 2001
6. Yoon KH et al: Biliary stricture caused by blunt abdominal trauma: clinical and radiologic features in five patients. Radiology. 207(3):737-41, 1998
7. Slanetz PJ et al: Imaging and interventional radiology in laparoscopic injuries to the gallbladder and biliary system. Radiology. 201(3):595-603, 1996

IMAGE GALLERY

Typical

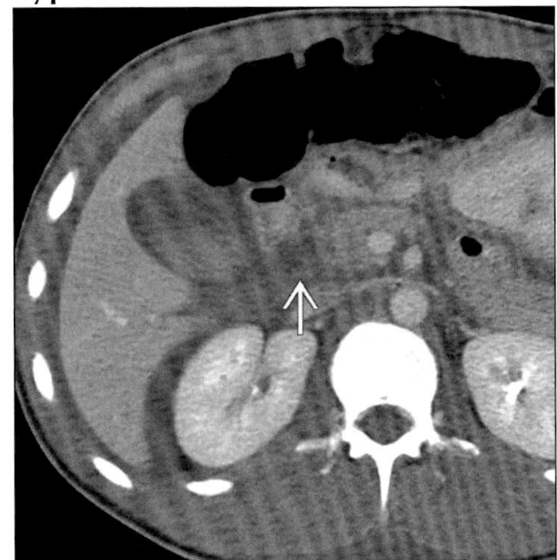

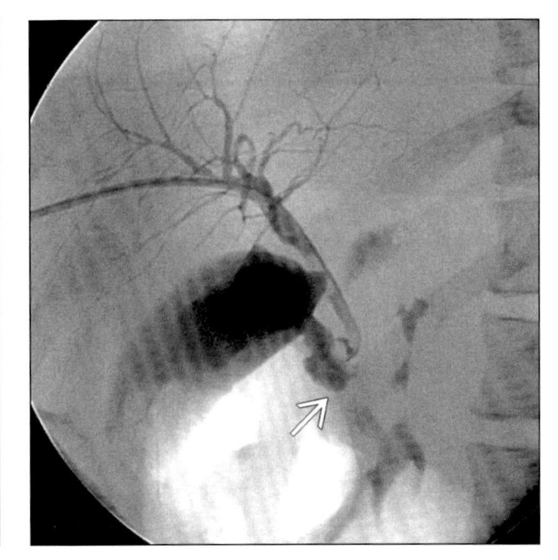

(Left) *Axial CECT following blunt trauma shows hemoperitoneum including hematoma (arrow) between duodenum and pancreatic head. Further evaluation demonstrated transection of the distal common bile duct.* ***(Right)*** *Percutaneous transhepatic cholangiogram shows active extravasation from transected common bile duct (arrow); no flow of contrast into duodenum.*

Typical

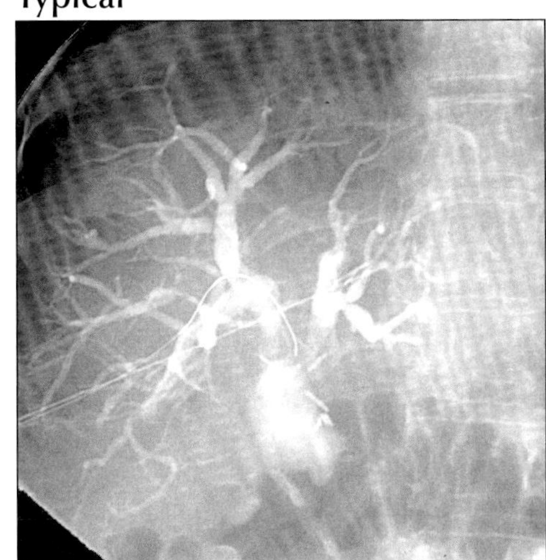

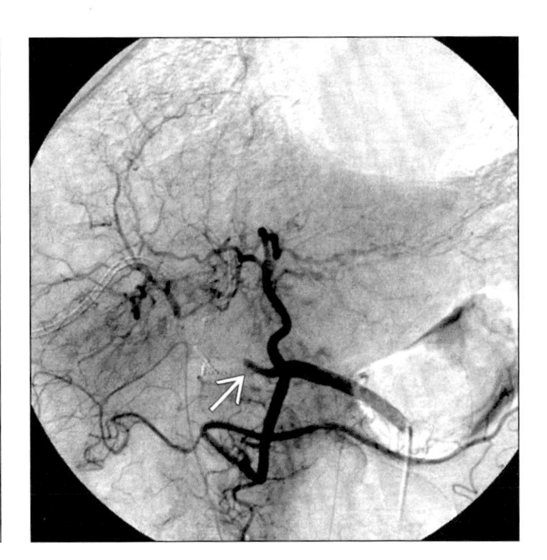

(Left) *Percutaneous transhepatic cholangiogram shows complete obstruction and contrast extravasation from common hepatic duct following laparoscopic cholecystectomy.* ***(Right)*** *Hepatic arteriogram shows complete occlusion beyond the origin of the right hepatic artery (arrow) in a patient who also had iatrogenic biliary injury following laparoscopic cholecystectomy.*

Typical

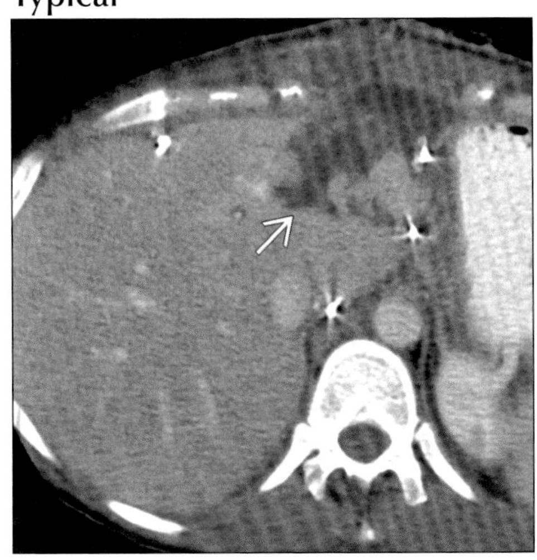

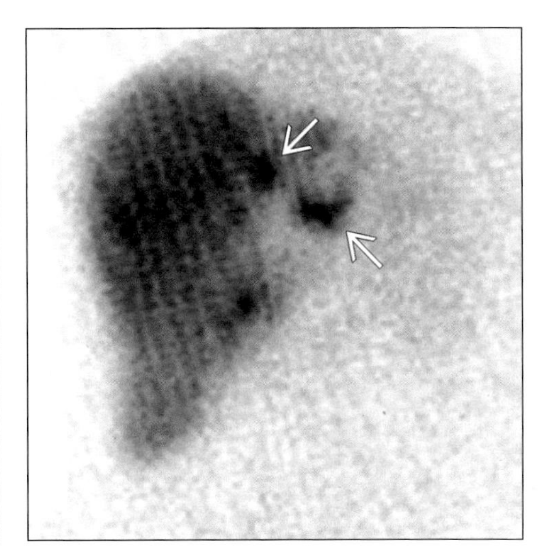

(Left) *Axial CECT following surgery for blunt abdominal trauma. Stellate fracture of lateral segment (arrow) is seen. Intrahepatic biliary injury was not appreciated during initial laparotomy.* ***(Right)*** *Tc HIDA scan shows foci of increased activity (arrows) due to bile extravasation from lateral segment in a patient who had prior blunt abdominal trauma.*

HEPATIC CYST

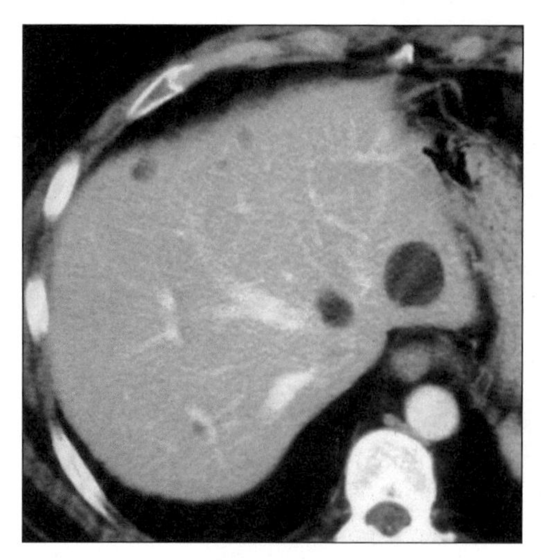

Axial CECT shows multiple hypodense lesions in liver. The largest lesion is sharply defined and had a ROI of 2 HU, while the smaller lesions are too small to characterize with confidence.

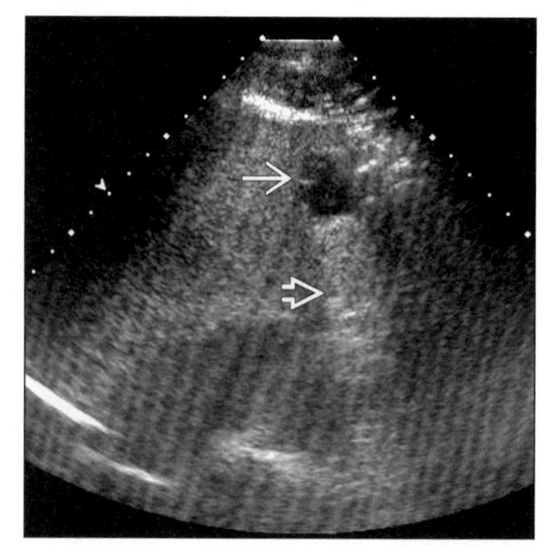

Sagittal ultrasound shows anechoic lesion (arrow) with no visible wall. Through transmission of sound (open arrow).

TERMINOLOGY

Abbreviations and Synonyms

- Simple hepatic or bile duct cyst

Definitions

- Simple hepatic cyst is a benign congenital developmental lesion derived from biliary endothelium

IMAGING FINDINGS

General Features

- Best diagnostic clue: Anechoic lesion with through transmission & no mural nodularity on US
- Location
  - Simple cyst
    - Typically occurs beneath the surface of liver
    - Some may occur deeper
- Size: Varies from few mm to 10 cm
- Key concepts
  - Current theory
    - True hepatic cysts arise from hamartomatous tissue
  - 2nd most common benign hepatic lesion after cavernous hemangioma
  - Classified based on etiology & pathogenesis
  - Congenital or developmental: Simple hepatic or bile duct cyst
    - Often solitary
    - Occasionally multiple: Less than 10
    - No communication with bile ducts
    - More prevalent in women
    - Usually asymptomatic
  - When more than 10 in number, one of fibropolycystic diseases must be considered
    - Example: Autosomal dominant polycystic liver disease (ADPLD) or biliary hamartomas
  - Acquired cyst-like hepatic lesions
    - Trauma (seroma or biloma)
    - Infection: Pyogenic or parasitic
    - Neoplasm: Primary or metastatic

CT Findings

- NECT
  - Simple liver or bile duct cyst
    - Sharply defined margins
    - Smooth, thin walls
    - Water density (-10 to +10 HU)
    - Usually no septations (rarely up to 2 thin septa)
    - No fluid-debris levels
    - No mural nodularity or wall calcification
  - Hemorrhage into cyst may be indistinguishable from tumor
    - Mural nodularity
    - With or without calcification & fluid level

DDx: Cystic Hepatic Lesion

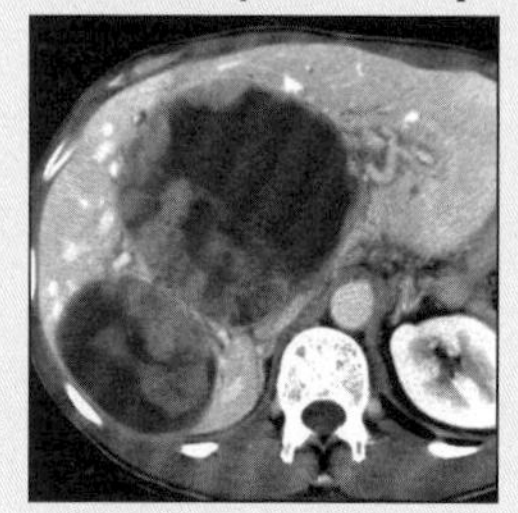

Cystic Metastases

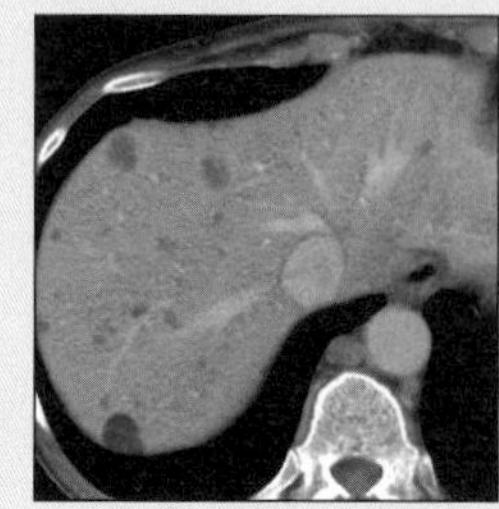

Cystic Metastases

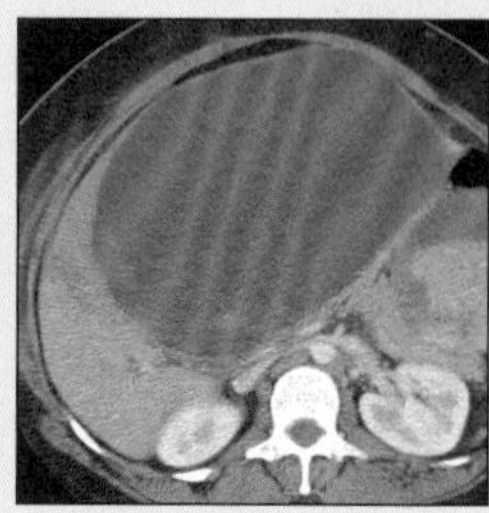

Cystadenocarcinoma

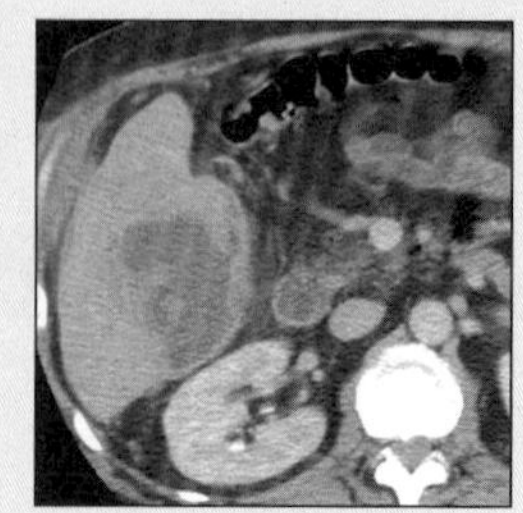

Hepatic Abscess

HEPATIC CYST

Key Facts

Terminology
- Simple hepatic cyst is a benign congenital developmental lesion derived from biliary endothelium

Imaging Findings
- Best diagnostic clue: Anechoic lesion with through transmission & no mural nodularity on US
- Morphology:

Top Differential Diagnoses
- Cystic or necrotic metastases
- Biliary cystadenocarcinoma
- Typical hepatic hemangioma
- Pyogenic abscess
- Hepatic hydatid cyst

Pathology
- Defective development of intrahepatic biliary duct
- No communication with bile ducts

Clinical Issues
- Usually asymptomatic
- Pain & fever (intracystic hemorrhage or infection)

Diagnostic Checklist
- Rule out cyst-like hepatic lesions (infection, tumor)
- CT: Nonenhancing, well-defined, round, homogeneous, water-density lesion
- Small lesion less than blood density on NECT is probably a cyst

---

- Autosomal dominant polycystic liver disease
    - Multiple cysts (more than 10)
    - Vary in size: 1-10 cm
    - Water or hemorrhagic density
    - Calcification of some cyst walls is seen
    - No septations or mural nodularity
    - Occasionally fluid levels seen
    - Liver often distorted by innumerable cysts
- CECT
  - Simple cyst
    - Uncomplicated: No enhancement
    - Complicated (infected): No enhancement
  - Autosomal dominant polycystic liver disease
    - Uncomplicated: No enhancement
    - Complicated (infected): No enhancement

MR Findings
- Simple hepatic cyst & ADPLD
  - T1WI: Hypointense
  - T2WI: Hyperintense
  - Heavily T2W images
    - Markedly increased signal intensity due to pure fluid content
    - Sometimes indistinguishable from a typical hemangioma
  - MRCP: No communication with bile duct
- Complicated (hemorrhagic) cyst
  - T1WI & T2WI
    - Varied signal intensity (due to mixed blood products)
    - With or without a fluid level
- T1C+
  - Uncomplicated cysts
    - No enhancement
  - Complicated cysts
    - No enhancement
- MRCP
  - Simple hepatic cyst & ADPLD
    - Shows no communication with bile ducts

Ultrasonographic Findings
- Real Time
  - Uncomplicated simple (bile duct) cyst
    - Anechoic mass
    - Smooth borders
    - Thin or non-detectable wall
    - No or few septations
    - No mural nodules
    - No wall calcification
  - Hemorrhagic or infected hepatic cyst
    - Septations
    - Internal debris
    - Thickened wall
    - With or without calcification

Imaging Recommendations
- Best imaging tool: Ultrasonography or NE + CECT

DIFFERENTIAL DIAGNOSIS

Cystic or necrotic metastases
- Demonstrate (e.g., ovarian cystadenocarcinoma & metastatic sarcoma)
  - Debris
  - Mural nodularity
  - Thick septa
  - Wall enhancement

Biliary cystadenocarcinoma
- Usually large in size
- Homogeneous, hypodense, water density mass
  - Rarely nonseptated
  - Indistinguishable from large simple cyst
- Almost always has septations
- May show fine mural or septal calcifications
- On contrast study
  - Unilocular or multilocular
  - Enhancement of capsule, septa & mural nodules

Typical hepatic hemangioma
- Well-defined margins
- NECT: Isodense to blood
- CECT: Early peripheral nodular & delayed centripetal enhancement isodense to enhanced vessels
- T1W: Hypointense
- T2WI: Markedly hyperintense

HEPATIC CYST

- Sometimes indistinguishable from simple hepatic cyst on MR images

Pyogenic abscess

- Complex cystic mass
- Heterogeneous density
- Thick or thin multiple septations
- May show mural nodularity
- With or without hemorrhage
- May show fluid-debris levels
- Wall enhancement may be seen
- Absent or heterogeneous enhancement of lesion

Hepatic hydatid cyst

- Large well-defined cystic liver mass with numerous peripheral daughter cysts
- With or without calcification & dilated bile ducts

PATHOLOGY

General Features

- Etiology
  - Congenital simple hepatic cyst
    - Defective development of intrahepatic biliary duct
  - Acquired hepatic cyst: Secondary to
    - Trauma, inflammation
    - Neoplasia, parasitic infestation
- Epidemiology
  - Reported to occur in 2.5% of population
  - Incidence: 1-14% in autopsy series
- Associated abnormalities
  - Autosomal dominant polycystic liver disease
    - 50% have polycystic kidney disease
    - M:F = 1:2
  - Polycystic kidney disease: 40% have hepatic cysts
  - Tuberous sclerosis

Gross Pathologic & Surgical Features

- Simple hepatic cyst
  - Cyst wall: ≤ 1 mm thick
  - Seen beneath the surface of liver
  - Some are deeper

Microscopic Features

- True simple hepatic cyst
  - Single unilocular cyst with serous fluid
  - Lined by
    - Cuboidal bile duct epithelium
    - A thin underlying rim of fibrous stroma
  - No communication with bile ducts

CLINICAL ISSUES

Presentation

- Most common signs/symptoms
  - Uncomplicated simple cysts & ADPLD
    - Usually asymptomatic
  - Complicated cyst
    - Pain & fever (intracystic hemorrhage or infection)
  - Large cysts present with symptoms of mass effect
    - Abdominal pain, jaundice, palpable mass
  - Advanced disease of ADPLD patients present with
    - Hepatomegaly, liver failure (rarely)
    - Budd-Chiari syndrome
- Clinical profile: Asymptomatic patient with incidental detection of simple hepatic cyst on imaging or at time of autopsy
- Lab data
  - Patients with large hepatic cyst & mass effect: ↑ Direct bilirubin levels
  - Patients with advanced disease of ADPLD: ↑ LFTs
- Diagnosis
  - Fine needle aspiration & cytology (rarely necessary)

Demographics

- Age
  - Seen in any age group
  - Usually discovered incidentally in 5th-7th decades
- Gender: M:F = 1:5

Natural History & Prognosis

- Complications
  - Infection
  - Hemorrhage
  - Large cyst: Compression of IHBD & jaundice
- Prognosis
  - Small & large hepatic cysts: Good prognosis
  - Advanced disease of ADPLD: Good prognosis

Treatment

- Asymptomatic simple hepatic cyst & ADPLD
  - No treatment
- Large, symptomatic, infected hepatic cyst
  - Percutaneous aspiration & sclerotherapy with alcohol
  - Surgical resection or marsupialization
- Advanced disease of ADPLD
  - Partial liver resection
  - Liver transplantation

DIAGNOSTIC CHECKLIST

Consider

- Rule out cyst-like hepatic lesions (infection, tumor)

Image Interpretation Pearls

- US: Anechoic, thin wall, through transmission
- CT: Nonenhancing, well-defined, round, homogeneous, water-density lesion
  - Small lesion less than blood density on NECT is probably a cyst

SELECTED REFERENCES

1. Mortele KJ et al: Cystic focal liver lesions in the adult: Differential CT and MR imaging features. RadioGraphics. 21: 895-910, 2001
2. Martin DR et al: Imaging of benign and malignant focal liver lesions. Magn Reson Imaging Clin N Am. 9(4):785-802, vi-vii, 2001
3. Casillas VJ et al: Imaging of nontraumatic hemorrhagic hepatic lesions. RadioGraphics. 20: 367-78, 2000
4. Horton KM et al: CT and MR imaging of benign hepatic and biliary tumors. Radiographics. 19(2):431-51, 1999
5. Murphy BJ et al: The CT appearance of cystic masses of the liver. RadioGraphics. 9: 307-22, 1989

IMAGE GALLERY

Typical

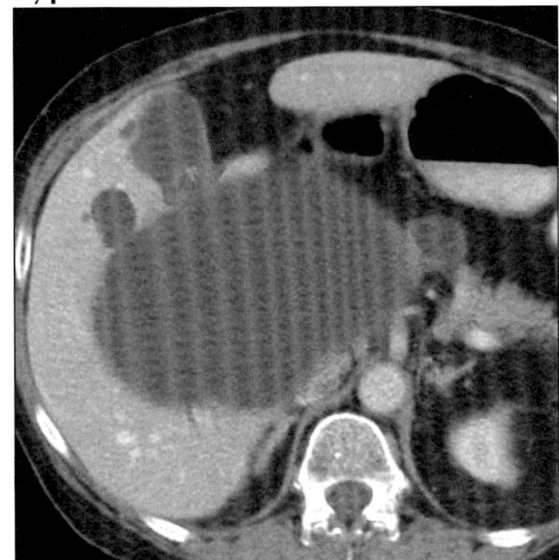

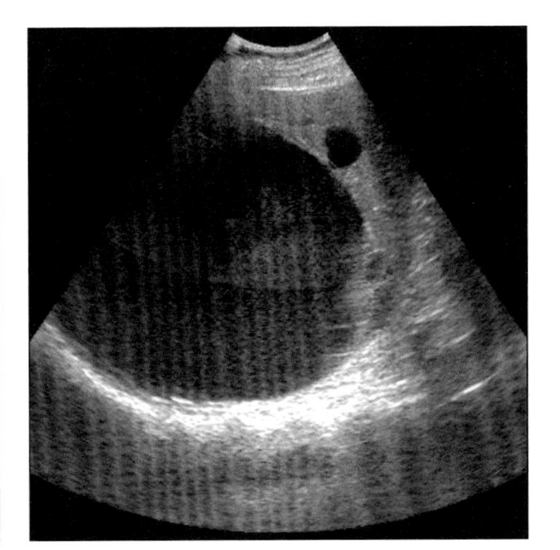

*(**Left**) Axial CECT shows multiple hepatic cysts of varying size. Water density, no enhancement. (**Right**) Sagittal sonogram shows anechoic lesions with thin walls, through transmission, no mural nodularity.*

Typical

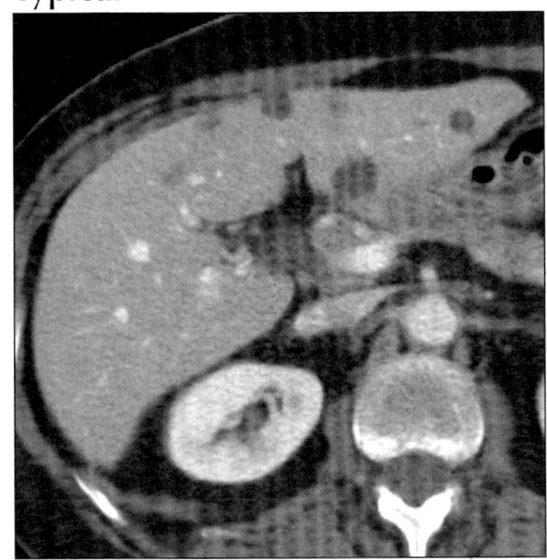

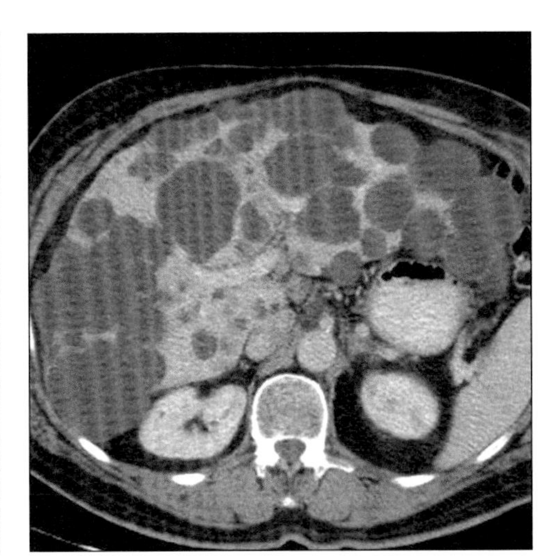

*(**Left**) Axial CECT shows multiple simple cysts. Smaller ones appear of higher than water density due to partial volume averaging. (**Right**) Axial CECT of patient with autosomal dominant polycystic liver disease. Innumerable hepatic cysts; no renal cysts.*

Variant

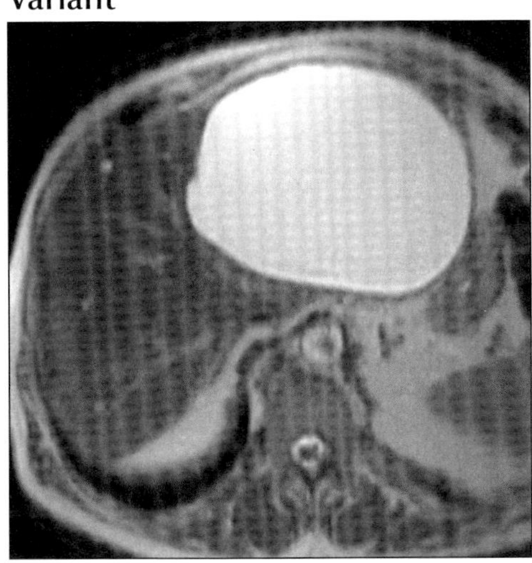

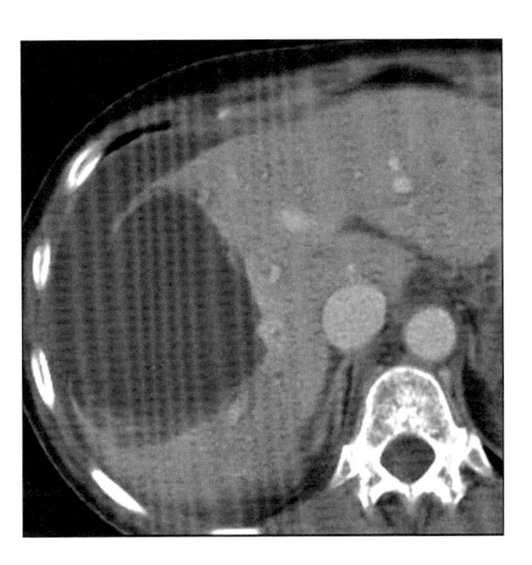

*(**Left**) Axial T2WI MR shows a large and very hyperintense hepatic cyst with no mural nodularity or septations. (**Right**) Axial CECT following blunt trauma. Peripheral water density contents have ruptured through the capsule of the liver. Higher density hemorrhage is present in dependent aspect of cyst.*

HEPATIC CAVERNOUS HEMANGIOMA

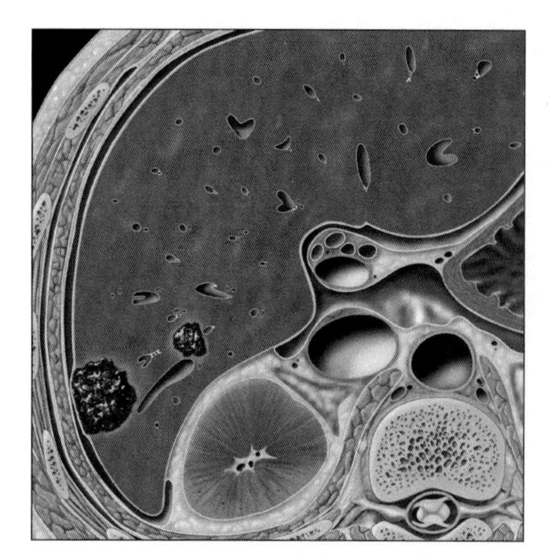
Graphic shows nonencapsulated collections of blood within enlarged sinusoidal spaces. Otherwise normal liver.

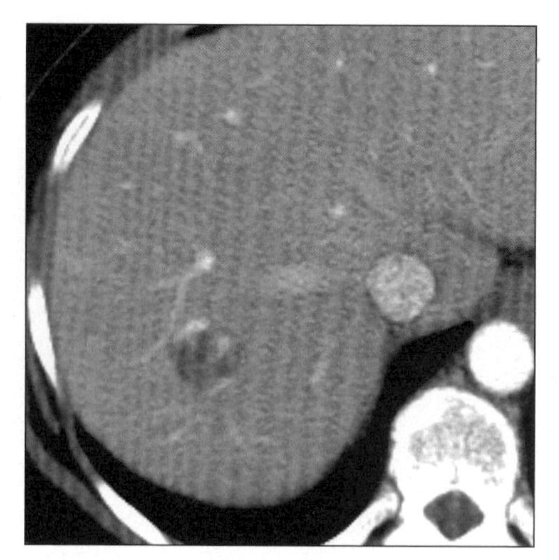
Axial CECT in venous-parenchymal phase shows spherical mass with nodular discontinuous peripheral enhancement that is nearly isodense to blood vessels.

TERMINOLOGY

Abbreviations and Synonyms

- Cavernous hemangioma of liver; capillary hemangioma (small lesion)

Definitions

- Benign tumor composed of multiple vascular channels lined by a single layer of endothelial cells supported by a thin fibrous stroma

IMAGING FINDINGS

General Features

- Best diagnostic clue: Peripheral nodular enhancement on arterial phase (AP) scan with slow progressive centripetal enhancement isodense to vessels
- Location: Common in subcapsular area in posterior right lobe of liver
- Size
  - Vary from few millimeters to more than 20 cm
  - Giant hemangiomas: Larger than 10 cm (arbitrary)
- Morphology
  - Most common benign tumor of liver
  - Second most common liver tumor after metastases
  - More commonly seen in postmenopausal women
  - Usually solitary & grow slowly
  - May be multiple in up to 50% of cases
  - Calcification is rare (less than 10%)
    - Usually in scar of giant hemangioma

CT Findings

- NECT
  - Small (1-2 cm) & typical hemangioma (2-10 cm)
    - Well-circumscribed, spherical to ovoid mass isodense to blood
  - Giant hemangioma (more than 10 cm)
    - Heterogeneous hypodense mass
    - Central decreased attenuation (scar)
- CECT
  - Small hemangiomas ("capillary"): Less than 2 cm
    - Arterial & venous phases: Show homogeneous enhancement ("flash filling")
  - Typical hemangiomas: 2-10 cm in diameter
    - Arterial phase: Early peripheral, nodular or globular, discontinuous enhancement
    - Venous phase: Progressive centripetal enhancement to uniform filling, still isodense to blood vessels
    - Delayed phase: Persistent complete filling
  - Giant hemangioma: More than 10 cm in diameter
    - Arterial phase: Typical peripheral nodular or globular enhancement
    - Venous & delayed phases: Incomplete centripetal filling of lesion (scar does not enhance)
  - Atypical hemangioma: Inside to outside pattern

DDx: Focal Lesion with Persistent or Delayed Enhancement

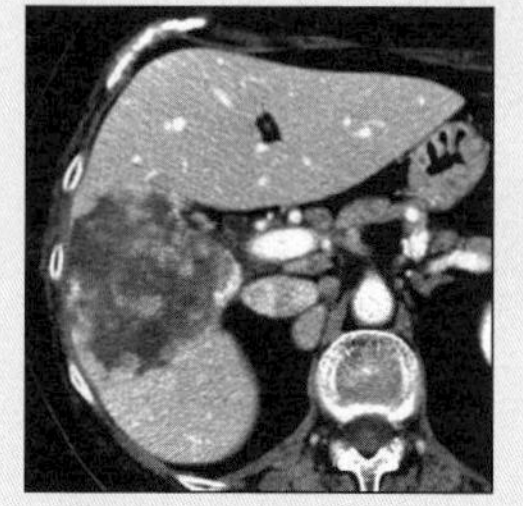
Cholangiocarcinoma

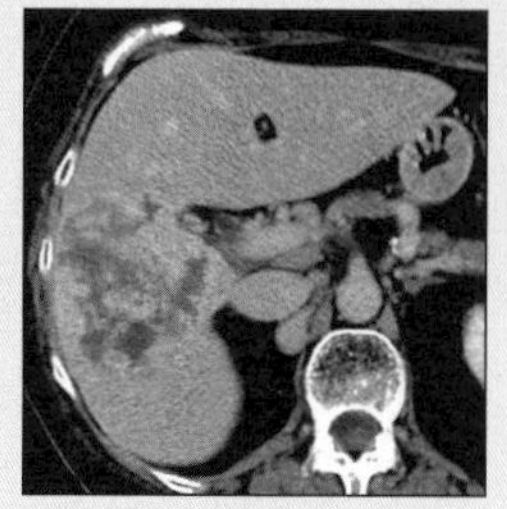
Cholangiocarcinoma

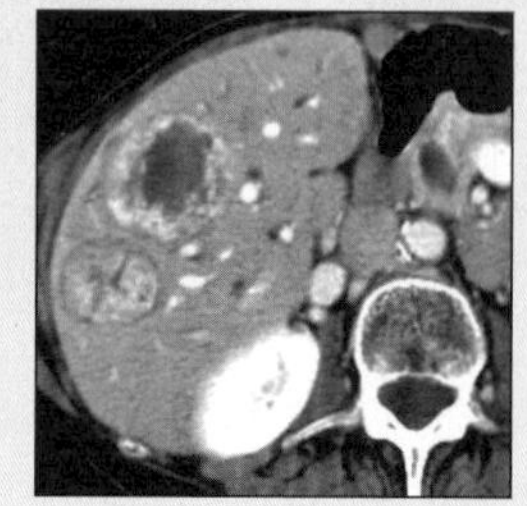
Metastases

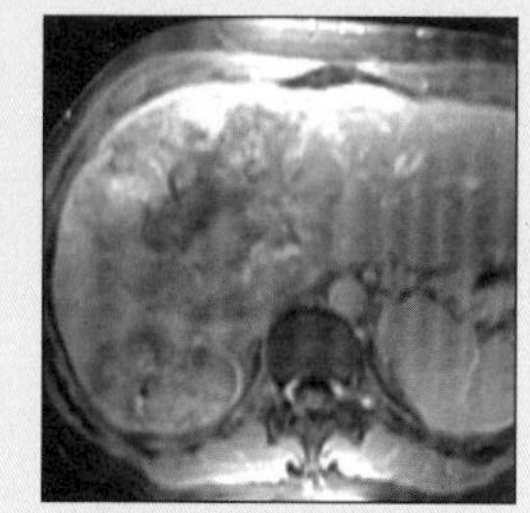
Metastases

HEPATIC CAVERNOUS HEMANGIOMA

Key Facts

Terminology
- Benign tumor composed of multiple vascular channels lined by a single layer of endothelial cells supported by a thin fibrous stroma

Imaging Findings
- Best diagnostic clue: Peripheral nodular enhancement on arterial phase (AP) scan with slow progressive centripetal enhancement isodense to vessels
- Most common benign tumor of liver
- Second most common liver tumor after metastases
- May be multiple in up to 50% of cases

Top Differential Diagnoses
- Peripheral (intrahepatic) cholangiocarcinoma
- Hypervascular metastases

Pathology
- Hemangiomas occur sporadically
- Associated with focal nodular hyperplasia (FNH)

Clinical Issues
- Usually asymptomatic
- More common in postmenopausal age group

Diagnostic Checklist
- Small hepatocellular carcinomas & hypervascular metastases can mimic small hemangiomas by their uniform homogeneous enhancement pattern
- Hemangiomas: Remain isodense to blood vessels on portal venous & delayed phases of enhancement
- Other benign & malignant liver masses: Usually become hypodense to blood vessels & liver (except cholangiocarcinoma)

    - Arterial phase: No significant enhancement
    - Venous & delayed phases: Gradual enhancement from center to periphery (centrifugal filling)
  - Hyalinized (sclerosed) hemangioma
    - Shows minimal enhancement
    - Cannot be diagnosed with confidence by imaging
  - Hemangioma in cirrhotic liver
    - Flash-filling of small lesions
    - May lose characteristic enhancement pattern
    - Capsular retraction
    - Decrease in size over time

MR Findings
- T1WI
  - Small & typical hemangiomas
    - Well marginated
    - Isointense to blood or hypointense
  - Giant hemangioma
    - Hypointense mass
    - Central cleft like area of marked decreased intensity (scar or fibrous tissue)
- T2WI
  - Small & typical hemangiomas
    - Hyperintense, similar to CSF
  - Giant hemangioma
    - Hyperintense mass
    - Marked hyperintense center (scar or fibrosis)
    - Hypointense internal septa
- T1 C+
  - Small hemangiomas (less than 2 cm)
    - Homogeneous enhancement in arterial & portal phases
  - Typical & giant hemangiomas
    - Arterial phase: Peripheral, nodular discontinuous enhancement
    - Venous phase: Progressive centripetal filling
    - In both phases: Isointense to blood
    - Central scar: No enhancement & remains hypointense

Ultrasonographic Findings
- Real Time
  - Small hemangioma
    - Well-defined, hyperechoic lesion
    - Size: Less than 2 cm
  - Typical hemangioma
    - Homogeneous hyperechoic mass with acoustic enhancement
    - Size: 2-10 cm
  - Giant hemangioma
    - Lobulated heterogeneous mass with echoic border
    - Size: More than 10 cm
  - Atypical hemangioma
    - Well-defined
    - Iso-/hypoechoic mass with echoic rim
- Color Doppler
  - Show filling vessels in periphery of tumor
  - No significant color Doppler flow in center of lesion
- Power Doppler
  - May detect flow within hemangiomas
  - Flow pattern is nonspecific
  - Similar flow pattern may be seen in hepatocellular carcinoma & metastases

Angiographic Findings
- Conventional
  - Dense opacification of lesion
  - "Cotton wool" appearance
    - Pooling of contrast medium within hemangioma
  - Normal-sized feeders
  - No neovascularity
  - No arteriovenous shunting
  - Typically retain contrast beyond venous phase

Nuclear Medicine Findings
- Tc-99m labeled RBC scan with SPECT (95% accuracy)
  - Early dynamic scan: Focal defect or less uptake
  - Delayed scans (over 30-50 min): Persistent filling
  - Vascular tumors (adenoma, HCC & FNH)
    - All exhibit early uptake rather than a defect
    - May have persistent uptake on delayed scan
  - Rarely angiosarcomas exhibit hemangioma pattern
    - Early defect & late uptake of isotope

Imaging Recommendations
- Best imaging tool: Helical NE + CECT or MR
- Protocol advice: Arterial, venous & delayed scans

HEPATIC CAVERNOUS HEMANGIOMA

DIFFERENTIAL DIAGNOSIS

Peripheral (intrahepatic) cholangiocarcinoma

- Delayed persistent enhancement, "fill in" may mimic hemangioma
- Often heterogeneous, not isodense with vessels on CT
- Not bright on T2WI
- Often invades/obstructs vessels & bile ducts

Hypervascular metastases

- Usually multiple
- Hyperdense in late arterial phase images
- Hypo-or isodense on NECT & portal venous phase
- Treated metastases may mimic hemangioma on imaging (e.g., breast)
- Not isodense to vessels on NECT or CECT
- Examples: Islet cell, carcinoid, thyroid, renal carcinomas, pheochromocytoma & some breast cancers

PATHOLOGY

General Features

- Etiology
  - Hemangiomas occur sporadically
  - No well-defined predisposing factors
- Epidemiology
  - Incidence
    - Ranging from 5-20% of population
    - Increases with multiparity
  - Prevalence: Uniform worldwide
- Associated abnormalities
  - Associated with focal nodular hyperplasia (FNH)
  - Kasabach-Merritt syndrome
    - Hemangioma with thrombocytopenia

Gross Pathologic & Surgical Features

- Solitary, well-defined, blood-filled, soft nodule
  - Size ranging from 2-20 cm
- Cut section: Giant hemangioma
  - Areas of fibrosis, necrosis & cystic spaces

Microscopic Features

- Large vascular channels lined by single layer of endothelial cells separated by thin fibrous septa
- No bile ducts
- Thrombosis of vascular channels resulting
  - Fibrosis & calcification

CLINICAL ISSUES

Presentation

- Most common signs/symptoms
  - Small & typical hemangioma
    - Usually asymptomatic
    - Commonly seen on routine examination & autopsy
  - Giant hemangioma
    - Asymptomatic
    - Liver enlargement, abdominal discomfort & pain
- Lab data: Normal liver function tests
- Diagnosis
  - Helical CECT, CEMR or RBC scan with SPECT imaging are highly diagnostic
  - Atypical hemangioma
    - Percutaneous or fine needle aspiration biopsy

Demographics

- Age
  - All age groups
  - More common in postmenopausal age group
  - Uncommonly diagnosed in children
- Gender: M:F = 1:5

Natural History & Prognosis

- Complications (extremely rare)
  - Spontaneous rupture
  - Abscess formation
- Prognosis: Usually good
  - Often show slow growth

Treatment

- Asymptomatic: Usually ignore
- Symptomatic large lesions: Surgical resection

DIAGNOSTIC CHECKLIST

Consider

- Small hepatocellular carcinomas & hypervascular metastases can mimic small hemangiomas by their uniform homogeneous enhancement pattern
- Hemangiomas: Remain isodense to blood vessels on portal venous & delayed phases of enhancement
- Other benign & malignant liver masses: Usually become hypodense to blood vessels & liver (except cholangiocarcinoma)

Image Interpretation Pearls

- Peripheral nodular or globular enhancement on arterial phase & centripetal enhancement on venous phase is a useful discriminating feature of hemangiomas from other lesions

SELECTED REFERENCES

1. Danet IM et al: Giant hemangioma of the liver: MR imaging characteristics in 24 patients. Magn Reson Imaging. 21(2):95-101, 2003
2. Brancatelli G et al: Hemangioma in the cirrhotic liver: Diagnosis and natural history. Radiology. 219: 69-74, 2001
3. Kim T et al: Discrimination of small hepatic hemangiomas from hypervascular malignant tumors smaller than 3 cm with three-phase helical CT. Radiology. 219: 699-706, 2001
4. Jeong MG et al: Hepatic cavernous hemangioma: temporal peritumoral enhancement during multiphase dynamic MR imaging. Radiology. 216(3):692-7, 2000
5. Vilgrain V et al: Imaging of atypical hemangiomas of the liver with pathologic correlation. RadioGraphics. 20: 379-97, 2000
6. Leslie DF et al: Distinction between cavernous hemangiomas of the liver and hepatic metastases on CT: value of contrast enhancement patterns. AJR Am J Roentgenol. 164(3):625-9, 1995

IMAGE GALLERY

Typical

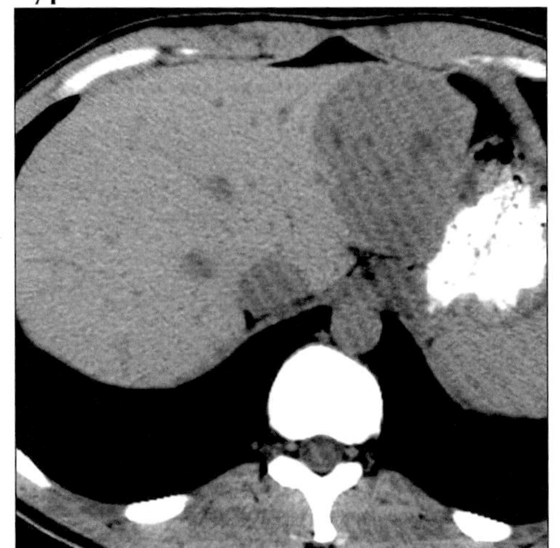

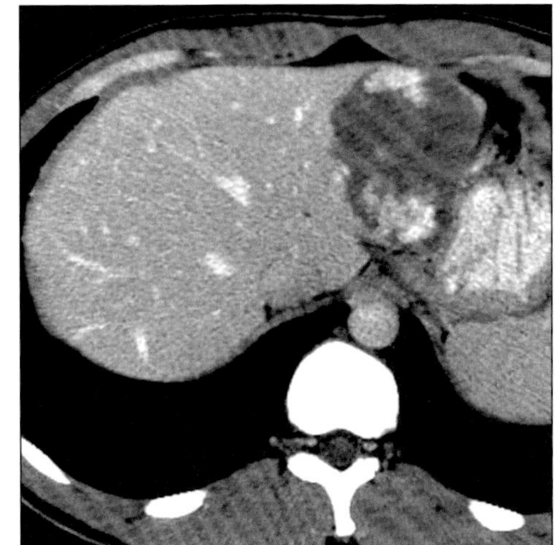

(Left) *Axial NECT shows large mass in lateral segment, most of which is isodense to blood except for hypodense foci of scar.* ***(Right)*** *Axial CECT in venous parenchymal phase shows cloud-like peripheral enhancement that is isodense to vessels.*

Typical

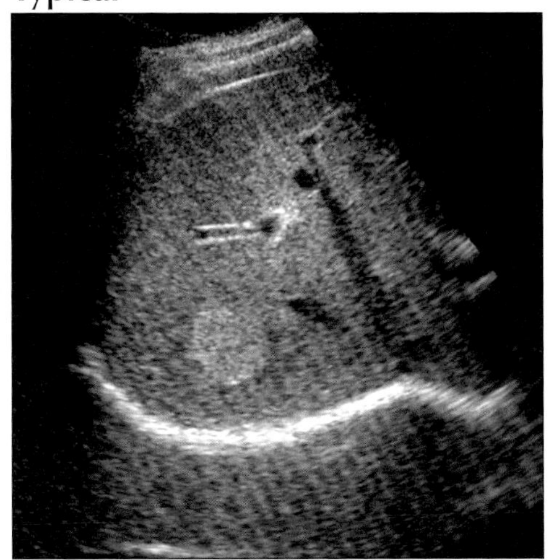

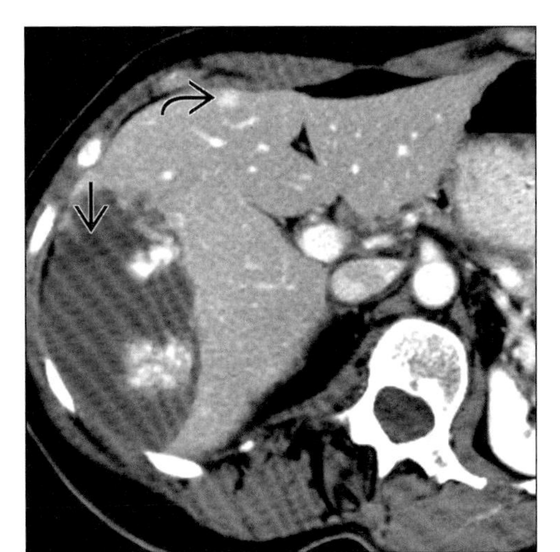

(Left) *Sagittal sonogram shows uniformly hyperechoic lesion in peripheral right lobe.* ***(Right)*** *Axial CECT in venous phase shows typical large hemangioma with nodular peripheral enhancement and nonenhancing scar (arrow). Capillary hemangioma (curved arrow) isodense to vessels in all phases.*

Variant

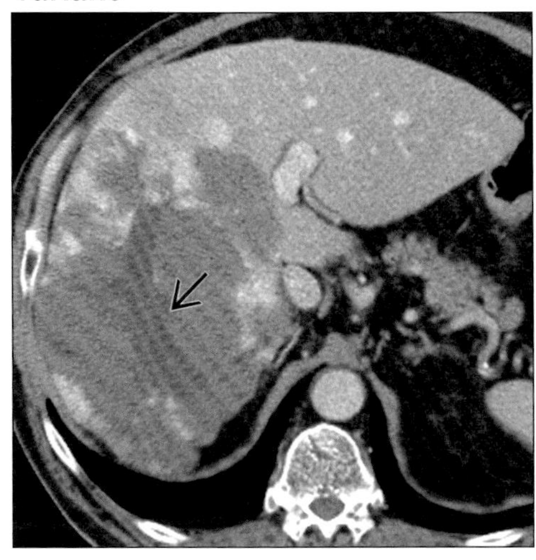

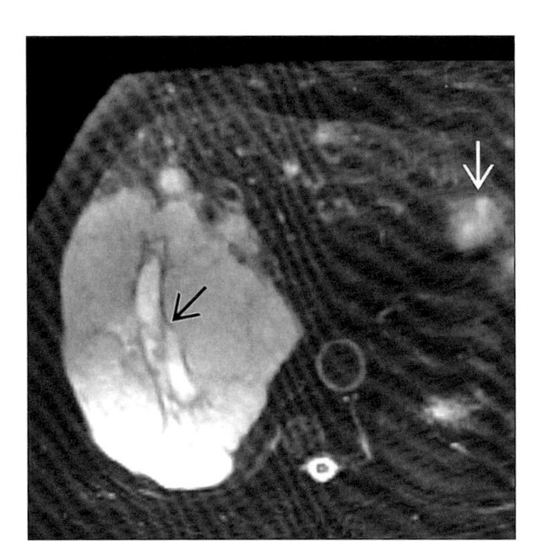

(Left) *Axial CECT in venous-parenchymal phase shows nodular peripheral enhancement and nonenhancing central scar (arrow) in a very large hepatic cavernous hemangioma.* ***(Right)*** *Axial T2WI MR shows large mass with central hyperintense scar (black arrow) that is even more intense. Several other hemangiomas were noted (white arrow).*

FOCAL NODULAR HYPERPLASIA

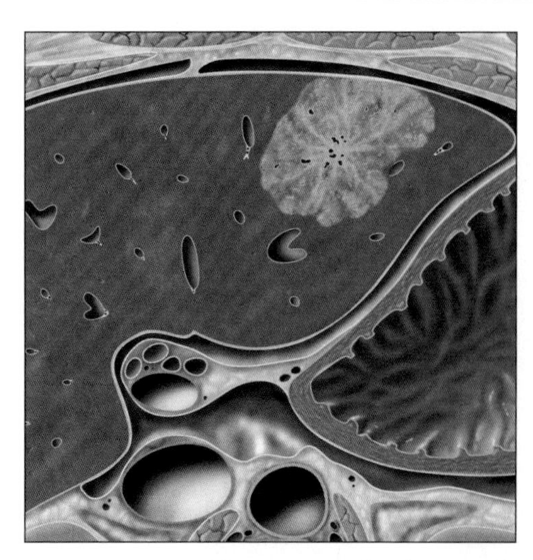

Homogeneous vascular, nonencapsulated mass with central scar and thin radiating septa dividing mass into hyperplastic nodules. Otherwise normal liver.

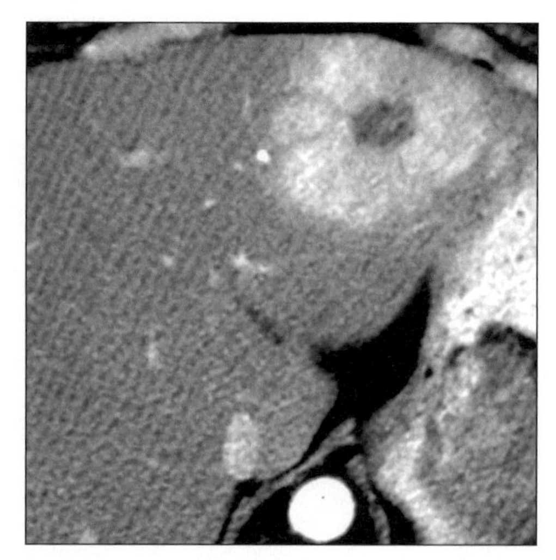

Axial CECT in arterial phase shows homogeneous hypervascular mass with central scar and thin radiating septa.

TERMINOLOGY

Abbreviations and Synonyms

- Focal nodular hyperplasia (FNH)

Definitions

- Benign tumor of liver caused by hyperplastic response to a localized vascular abnormality

IMAGING FINDINGS

General Features

- Best diagnostic clue: Brightly and homogeneously enhancing mass in arterial phase CT or MR with delayed enhancement of central scar
- Location
  - More common in right lobe
  - Right lobe to left lobe: 2:1
  - Usually subcapsular & rarely pedunculated
- Size
  - Majority are smaller than 5 cm (85%)
  - Mean diameter at time of diagnosis is 3 cm
- Key concepts
  - 2nd most common benign tumor of liver
  - Benign congenital hamartomatous malformation
  - Accounts for 8% of primary hepatic tumors in autopsy series
  - Usually a solitary lesion (80%); multiple in 20%
  - Multiple FNHs associated with multiorgan vascular malformations and with certain brain neoplasms

CT Findings

- NECT: Isodense or hypodense to normal liver
- CECT
  - Hepatic arterial phase (HAP) scan
    - Transient intense hyperdensity
  - Portal venous phase (PVP) scan
    - Hypodense or isodense to normal liver
  - Delayed scans
    - Mass: Isodense to liver
    - Central scar: Hyperdense
    - Scar visible in 2/3rd of large & 1/3rd of small FNH

MR Findings

- T1WI
  - Mass: Isointense to slightly hypointense
  - Central scar: Hypointense
- T2WI
  - Mass: Slightly hyperintense to isointense
  - Central scar: Hyperintense
- T1 C+
  - Arterial Phase: Hyperintense (homogeneous)
  - Portal Venous: Isointense
  - Delayed phase
    - Mass: Isointense
    - Scar: Hyperintense

DDx: Uniformly Hypervascular Liver Mass

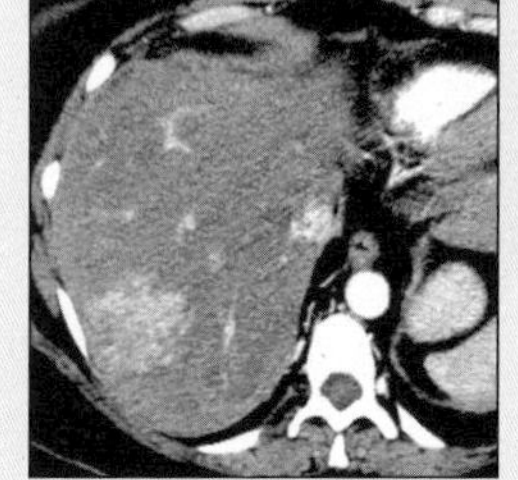

Hepatic Adenoma

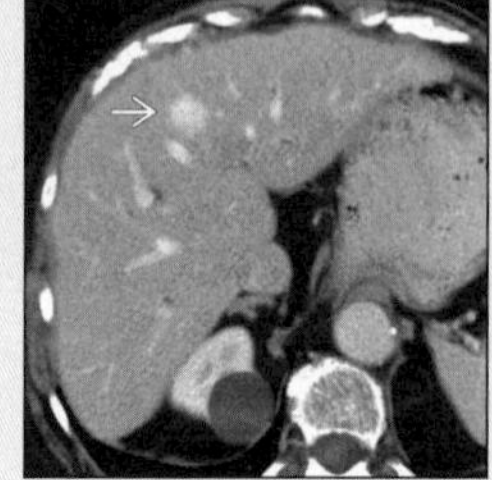

Small Hemangioma

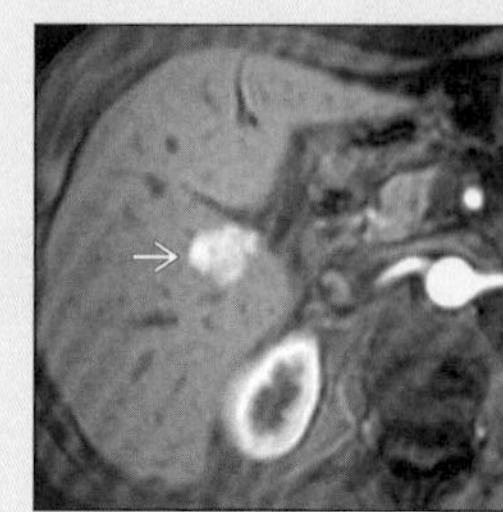

HCC

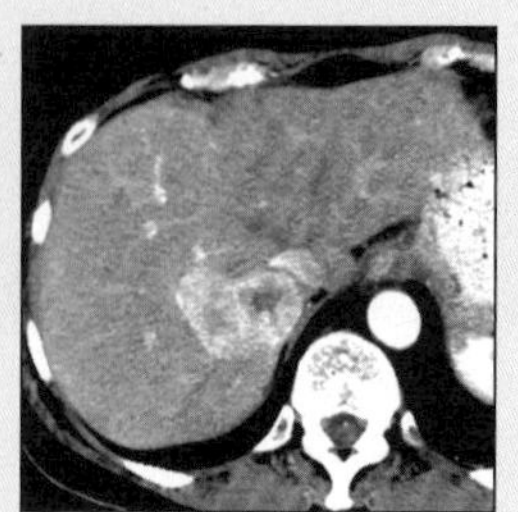

Metastases

FOCAL NODULAR HYPERPLASIA

Key Facts

Terminology
- Benign tumor of liver caused by hyperplastic response to a localized vascular abnormality

Imaging Findings
- Best diagnostic clue: Brightly and homogeneously enhancing mass in arterial phase CT or MR with delayed enhancement of central scar
- Usually subcapsular & rarely pedunculated

Top Differential Diagnoses
- Hepatic adenoma
- Cavernous hemangioma
- Fibrolamellar carcinoma
- Hepatocellular carcinoma
- Hypervascular metastasis

Pathology
- Oral contraceptives don't cause FNH, but have trophic effect on growth
- No intratumoral calcification, hemorrhage or necrosis
- Size: Less than 5 cm (in 85%)

Clinical Issues
- Often asymptomatic (in 50-90% incidental finding)
- 3rd-4th decades of life

Diagnostic Checklist
- Classic FNH looks like a cross-section of an orange (central "scar", radiating septa)
- Radiologically FNH may mimic fibrolamellar HCC, which is usually a large lesion (> 12 cm), has evidence of calcification (in 68%) & metastases in 70% cases

- Specific hepatobiliary MR contrast agents
  - T2WI with superparamagnetic iron oxide (SPIO)
    - FNH shows decreased signal due to uptake of iron oxide particles by Kupffer cells within lesion
    - Degree of signal loss in FNH is greater than other focal liver lesions (metastases, adenoma & HCC)
  - Gadobenate dimeglumine (Gd-BOPTA)
    - Bright homogeneous enhancement of FNH
    - Prolonged enhancement of FNH on delayed scan (due to malformed bile ductules)
    - Delayed scan: Significant enhancement of scar

Ultrasonographic Findings
- Real Time
  - Well-demarcated liver lesion
  - Mass: Mostly homogeneous & isoechoic to liver
    - Occasionally hypoechoic or hyperechoic
  - Central scar: Hypoechoic
  - Prominent draining veins or displacement of vessels
- Color Doppler
  - "Spoke-wheel" pattern
    - Large central feeding artery with multiple small vessels radiating peripherally
  - Large draining veins at tumor margins
  - High-velocity Doppler signals
    - Due to increased blood flow or arteriovenous shunts

Angiographic Findings
- Conventional
  - Arterial phase
    - Tumor: Hypervascular
    - Scar: Hypovascular
    - Enlargement of main feeding artery with a centrifugal blood supply
    - "Spoke-wheel" pattern" as on color-Doppler
  - Venous phase: Large draining veins
  - Capillary phase
    - Intense & nonhomogeneous stain
    - No avascular zones

Nuclear Medicine Findings
- Technetium Sulfur Colloid
  - Normal or increased uptake
  - Only FNH has both Kupffer cells & bile ductules
  - Almost PATHOGNOMONIC in 60% of cases
- Tc-HIDA scan (hepatic iminodiacetic acid)
  - Normal or increased uptake
  - Prolonged enhancement (80%)
- Tc 99m-Tagged RBC scan (not useful)
  - Early isotope uptake & late defect

Imaging Recommendations
- Helical CT or MR (multi-phase studies); Tc-99m-sulfur colloid scan

DIFFERENTIAL DIAGNOSIS

Hepatic adenoma
- Large tumor
- Symptomatic due to hemorrhage in 50%, scar atypical
- Usually heterogeneous due to hemorrhage, necrosis or fat

Cavernous hemangioma
- Only small ones with rapid enhancement simulate FNH
- NECT: Isodense with blood vessels
- CECT: Peripheral enhanced areas stay isodense with blood vessels

Fibrolamellar carcinoma
- Large (more than 12 cm) heterogeneous mass
- Biliary, vascular & nodal invasion
- Metastases (70% of cases)
- Fibrous scar
  - Large & central or eccentric with fibrous bands & calcification (68%)
  - Hypointense scar on T2WI

Hepatocellular carcinoma
- Heterogeneous mass within cirrhotic liver
- Necrosis & hemorrhage
- Vascular & nodal invasion

Hypervascular metastasis
- Multiple lesions; older patient

FOCAL NODULAR HYPERPLASIA

- Hypodense during portal venous phase

PATHOLOGY

General Features

- Genetics
  - In genetic hemochromatosis patients, FNH cells were homozygous for Cys282Tyr mutation
  - Ki-67 antigen positive in 4% of FNH hepatocytes
- Etiology
  - Ischemia caused by an occult occlusion of intrahepatic vessels
  - Localized arteriovenous shunting caused by anomalous arterial supply
  - Hyperplastic response to an abnormal vasculature
  - Oral contraceptives don't cause FNH, but have trophic effect on growth
- Epidemiology
  - 4% of all primary hepatic tumors in pediatric population
  - 3-8% in adult population
- Associated abnormalities
  - Hepatic hemangioma (in 23%)
  - Multiple lesions of FNH are associated with
    - Brain neoplasms: Meningioma, astrocytoma
    - Vascular malformations of various organs

Gross Pathologic & Surgical Features

- Localized, well-delineated, usually solitary (80%), subcapsular mass
- No true capsule, frequently central fibrous scar
- No intratumoral calcification, hemorrhage or necrosis
- Multiple masses (in 20%), rarely pedunculated
- Size: Less than 5 cm (in 85%)

Microscopic Features

- Normal hepatocytes with large amounts of fat, triglycerides & glycogen
- Thick-walled arteries in fibrous septa radiating from center to periphery
- Proliferation & malformation of bile ducts lead to slowing of bile excretion
- Absent portal triads & central veins
- Difficult differentiation from regenerative cirrhotic nodule & liver adenoma

CLINICAL ISSUES

Presentation

- Most common signs/symptoms
  - Often asymptomatic (in 50-90% incidental finding)
  - Vague abdominal pain (10-15%) due to mass effect
  - Other signs/symptoms
    - Hepatomegaly & abdominal mass (very rare)
  - Lab data: Usually normal liver function tests
  - Diagnosis
    - Characteristic imaging findings
    - Core needle biopsy (include central scar)

Demographics

- Age
  - Common in young to middle-aged women
    - 3rd-4th decades of life
  - Range: 7 months to 75 years
- Gender: M:F = 1:8

Natural History & Prognosis

- Excellent

Treatment

- Discontinuation of oral contraceptives
- FNH seldom requires surgery

DIAGNOSTIC CHECKLIST

Consider

- To rule out other benign & malignant liver lesions particularly fibrolamellar hepatocellular carcinoma

Image Interpretation Pearls

- Immediate, intense, homogeneously enhancing lesion on arterial phase followed rapidly by isodensity on venous phase with delayed enhancement of scar
- Classic FNH looks like a cross-section of an orange (central "scar", radiating septa)
- Radiologically FNH may mimic fibrolamellar HCC, which is usually a large lesion (> 12 cm), has evidence of calcification (in 68%) & metastases in 70% cases
- Atypical FNH (telangiectatic FNH): Lack of central scar, heterogeneous lesion, hyperintense on T1WI, markedly hyperintense on T2WI & has persistent contrast-enhancement on delayed CECT & T1 C+
  - Probably can not make this diagnosis by imaging

SELECTED REFERENCES

1. Attal P et al: Telangiectatic focal nodular hyperplasia: US, CT, and MR imaging findings with histopathologic correlation in 13 cases. Radiology. 228(2):465-72, 2003
2. Vilgrain V et al: Prevalence of hepatic hemangioma in patients with focal nodular hyperplasia: MR imaging analysis. Radiology. 229(1):75-9, 2003
3. Brancatelli G et al: Focal nodular hyperplasia: CT findings with emphasis on multiphasic helical CT in 78 patients. Radiology. 219: 61-8, 2001
4. Grazioli L et al: Focal nodular hyperplasia: Morphologic and functional information from MR imaging with gadobenate dimeglumine. Radiology. 221: 731-9, 2001
5. Casillas VJ et al: Imaging of nontraumatic hemorrhagic hepatic lesions. Radiographics. 20(2):367-78, 2000
6. Leconte I et al: Focal nodular hyperplasia: Natural course observed with CT and MRI. Journal of computer assisted tomography. 24(1): 61-6, 2000
7. Horton KM et al: CT and MR imaging of benign hepatic and biliary tumors. Radiographics. 19(2):431-51, 1999
8. Buetow PC et al: Focal nodular hyperplasia of the liver: radiologic-pathologic correlation. Radiographics. 16(2):369-88, 1996
9. Caseiro-Alves F et al: Calcification in focal nodular hyperplasia: a new problem for differentiation from fibrolamellar hepatocellular carcinoma. Radiology. 198(3):889-92, 1996

HEPATIC ADENOMA

Key Facts

Terminology
- Hepatocellular adenoma (HCA) or liver cell adenoma
- Benign tumor that arises from hepatocytes arranged in cords that occasionally form bile

Imaging Findings
- Best diagnostic clue: Heterogeneous, hypervascular mass with hemorrhage in a young woman
- Subcapsular region of right lobe of liver (75%)
- Average size: 8-10 cm

Top Differential Diagnoses
- Hepatocellular carcinoma (HCC)
- Fibrolamellar hepatocellular carcinoma
- Focal nodular hyperplasia (FNH)
- Hypervascular metastases

Pathology
- Hemorrhage, necrosis & fatty change
- No scar within tumor
- ↑ Risk in oral contraceptives & anabolic steroid users

Clinical Issues
- RUQ pain (40%): Due to hemorrhage
- May be mistaken clinically/pathologically for HCC
- Clinical profile: Woman on oral contraceptives

Diagnostic Checklist
- Rule out other benign & malignant liver tumors which have similar imaging features, particularly HCC or FNH
- Check for history of oral contraceptives & glycogen storage disease (in case of multiple adenomas)

        - Increased signal intensity (due to fat & recent hemorrhage), more evident on MR than CT
        - Decreased signal intensity (necrosis, calcification, old hemorrhage)
    - Rim (fibrous pseudocapsule): Hypointense
- T2WI
    - Mass: Heterogeneous signal intensity
        - Increased signal intensity (old hemorrhage/necrosis)
        - Decreased signal intensity (fat, recent hemorrhage)
    - Rim (fibrous pseudocapsule): Hypointense
- T1 C+
    - Gadolinium arterial phase
        - Mass: Heterogeneous enhancement
    - Delayed phase
        - Pseudocapsule: Hyperintense to liver & adenoma
- Superparamagnetic iron oxide (SPIO)
    - No uptake in adenoma
    - Few cases take up SPIO (due to active Kupffer cells)
        - Resulting in a decreased signal on T2WI
        - Indistinguishable from FNH for these cases
        - Uptake varies based on number of Kupffer cells
- Gd-BOPTA (gadobenate dimeglumine)
    - Hepatocellular-specific contrast agent
    - Adenoma
        - No substantial uptake
        - Hypointense even on delayed images

Ultrasonographic Findings
- Real Time
    - Well-defined, solid, echogenic mass
    - Complex hyper & hypoechoic heterogeneous mass with anechoic areas
        - Due to fat, hemorrhage, necrosis & calcification
- Color Doppler
    - Hypervascular tumor
    - Large peripheral arteries & veins
    - Intratumoral veins present
        - Absent in FNH
        - Useful discriminating feature for HCA

Angiographic Findings
- Conventional
    - Hypervascular mass with centripetal flow
    - Enlarged hepatic artery with feeders at tumor periphery (50%)
    - Hypovascular; avascular regions
        - Due to hemorrhage & necrosis

Nuclear Medicine Findings
- Technetium Sulfur Colloid
    - Usually "cold" (photopenic): In 80%
    - Uncommonly "warm": In 20%
        - Due to uptake in sparse Kupffer cells
- HIDA scan
    - Increased activity
- Gallium Scan
    - No uptake

Imaging Recommendations
- T2WI; T1WI with dynamic enhanced multiphasic; GRE in - and opposed-phase images

DIFFERENTIAL DIAGNOSIS

Hepatocellular carcinoma (HCC)
- May have identical imaging features as hepatic adenoma
- Histologically: May be difficult to distinguish well-differentiated HCC from adenoma
- Biliary, vascular, nodal invasion & metastases establish that lesion is malignant

Fibrolamellar hepatocellular carcinoma
- Large, lobulated mass with scar & septa
- Vascular, biliary, nodal invasion common
- Heterogeneous on all imaging

Focal nodular hyperplasia (FNH)
- No malignant degeneration or hemorrhage
- T2WI: Scar is typically hyperintense
- Arterial phase
    - FNH is homogeneously enhancing mass

HEPATIC ADENOMA

- Central scar: Hypoattenuating/hypointense
- Delayed phase
  - Central scar: Hyperdense/hyperintense
- Small (≤ 3 cm) FNH without scar indistinguishable from adenoma

Hypervascular metastases

- Usually multiple & look for primary tumors
  - Breast, thyroid, kidney and endocrine
- Arterial phase: Heterogeneous enhancement
- Portal & delayed phases: Iso-/hypodense
- T1WI: Hypointense
- T2WI: Markedly hyperintense

PATHOLOGY

General Features

- General path comments
  - HCA: Surrounded by a fibrous pseudocapsule
    - Due to compression of adjacent liver tissue
  - High incidence of
    - Hemorrhage, necrosis & fatty change
  - No scar within tumor
- Etiology
  - ↑ Risk in oral contraceptives & anabolic steroid users
  - Pregnancy
    - Increase tumor growth rate and tumor rupture
  - Diabetes mellitus
  - Von-Gierke type Ia glycogen storage disease
    - Multiple adenomas: 60%
- Epidemiology
  - Estimated incidence in oral contraceptive users
    - 4 Adenomas per 100,000 users

Gross Pathologic & Surgical Features

- Well-circumscribed mass on external surface of liver
- Soft, pale or yellow tan
- Frequently bile-stained nodules
- Large areas of hemorrhage or infarction
- "Pseudocapsule" & occasional "pseudopods"

Microscopic Features

- Sheets or cords of hepatocytes
- Absence of portal & central veins & bile ducts
- Increased amounts of glycogen & lipid
- Scattered, thin-walled, vascular channels

Staging, Grading or Classification Criteria

- Typical hepatocellular adenoma (HCA)
  - Type I: Estrogen associated HCA
  - Type II: Spontaneous HCA in women
  - Type III: Spontaneous HCA in men
  - Type IV: Spontaneous HCA in children
  - Type V: Metabolic disease associated HCA
- Anabolic steroid-associated HCA
- Multiple hepatocellular adenomas (adenomatosis)

CLINICAL ISSUES

Presentation

- Most common signs/symptoms
  - RUQ pain (40%): Due to hemorrhage
  - Asymptomatic (20%)
  - May be mistaken clinically/pathologically for HCC
- Clinical profile: Woman on oral contraceptives
- Lab data: Usually normal liver function tests
- Diagnosis: Biopsy & histology

Demographics

- Age
  - Young women of childbearing age group
  - Predominantly in 3rd & 4th decades
- Gender
  - 98% seen in females (M:F = 1:10)
  - Not seen in males unless on anabolic steroids or with glycogen storage disease

Natural History & Prognosis

- Complications
  - Hemorrhage: Intrahepatic or intraperitoneal (40%)
  - Rupture: Increased risk in pregnancy
  - Risk of malignant transformation
    - When size is more than 10 cm (in 10%)
- Prognosis
  - Usually good
    - After discontinuation of oral contraceptives
    - After surgical resection of large/symptomatic
  - Poor
    - Intraperitoneal rupture
    - Rupture during pregnancy
    - Adenomatosis (> 10 adenomas)
    - Malignant transformation

Treatment

- Adenoma less than 6 cm
  - Observation & discontinue oral contraceptives
- Adenoma more than 6 cm & near surface
  - Surgical resection
- Pregnancy should be avoided due to increased risk of rupture

DIAGNOSTIC CHECKLIST

Consider

- Rule out other benign & malignant liver tumors which have similar imaging features, particularly HCC or FNH
- Check for history of oral contraceptives & glycogen storage disease (in case of multiple adenomas)

Image Interpretation Pearls

- Spherical well-defined hypervascular & heterogeneous mass due to hemorrhage & fat, most evident on MR
- Adenomas with uptake of SPIO agent mimic FNH on T2WI as decreased signal

SELECTED REFERENCES

1. Grazioli L et al: Hepatic adenomas: Imaging and pathologic findings. RadioGraphics. 21: 877-94, 2001
2. Ichikawa T et al: Hepatocellular adenoma: Multiphasic CT and histopathologic findings in 25 patients. Radiology. 214: 861-8, 2000
3. Grazioli L et al: Liver adenomatosis: Clinical, pathologic and imaging findings in 15 patients. Radiology. 216: 395-402, 2000

HEPATIC ADENOMA

IMAGE GALLERY

Typical

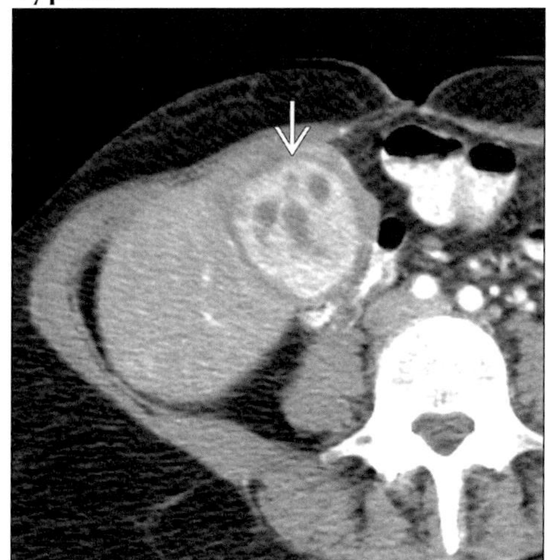

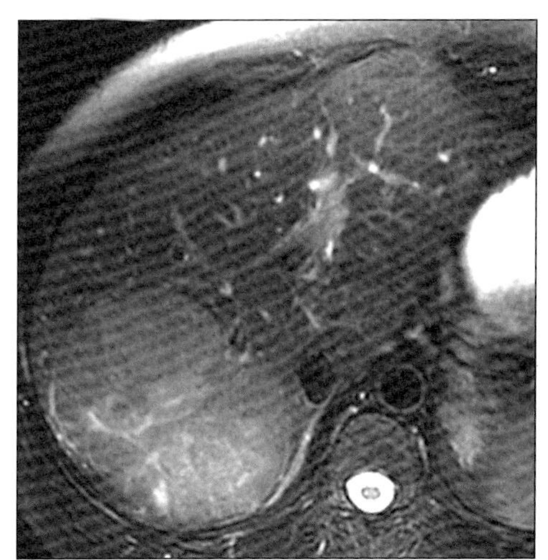

(Left) *Axial CECT shows hypervascular mass (arrow) with nonenhancing foci. Hepatic adenoma which required surgical resection.* ***(Right)*** *Axial T2WI MR shows large encapsulated hepatic mass with heterogeneously increased signal intensity.*

Typical

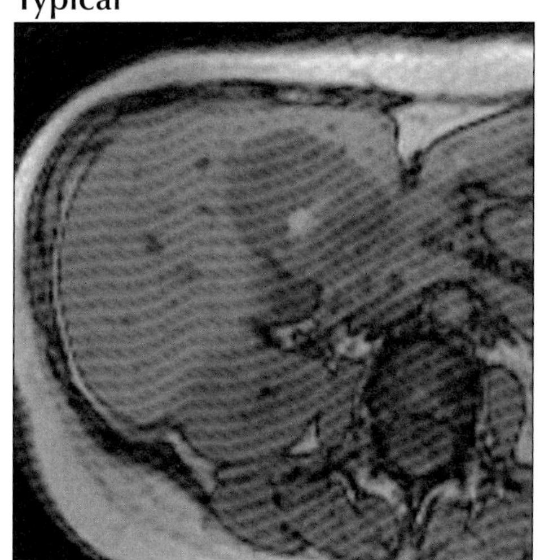

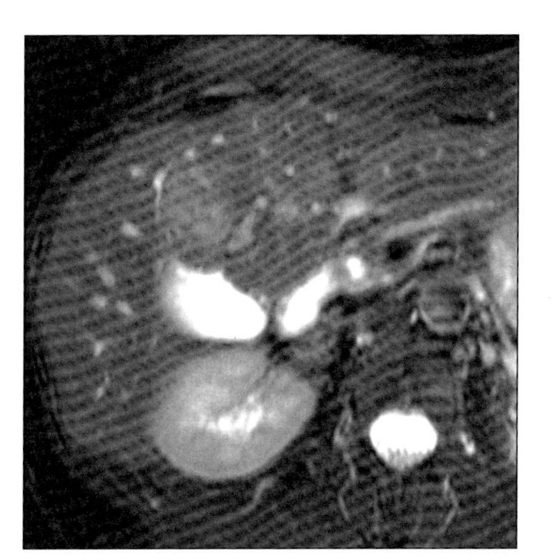

(Left) *Axial T1WI MR shows hypointense encapsulated mass with hyperintense foci (hemorrhage or fat).* ***(Right)*** *Axial T2WI MR shows mass nearly isointense to liver with central focus of hyperintensity (hemorrhage).*

Variant

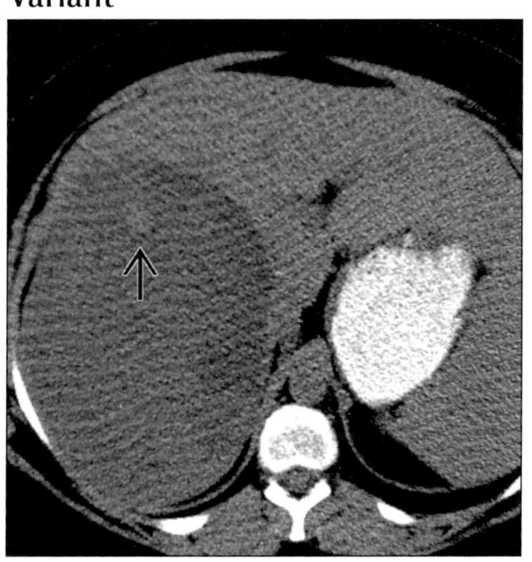

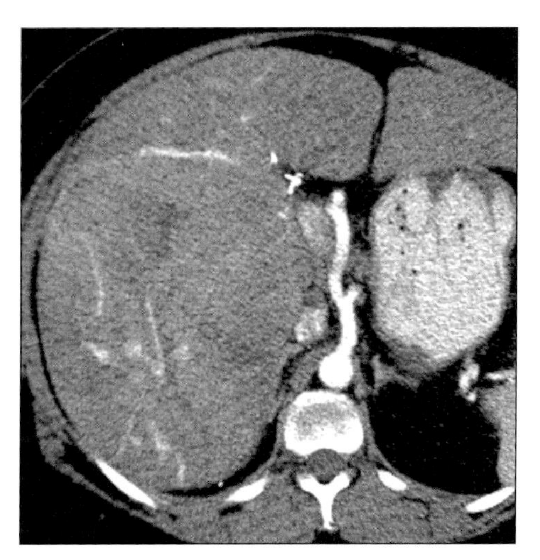

(Left) *Axial NECT shows very large, mostly homogeneous mass with small focus of hemorrhage (arrow).* ***(Right)*** *Axial CECT in arterial phase shows hypervascularity with enlarged vessels within and on surface of tumor.*

BILIARY HAMARTOMA

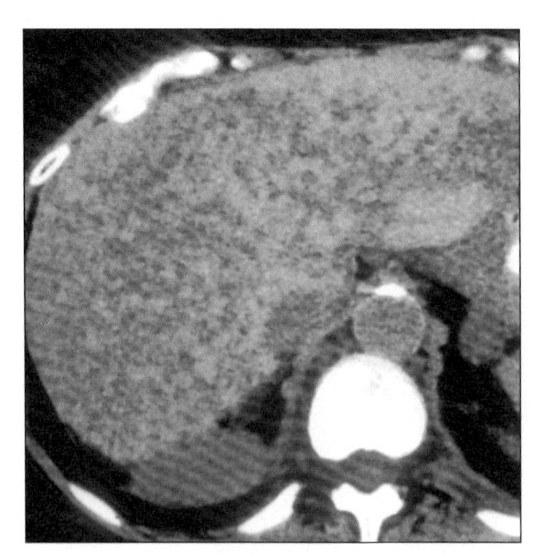

Axial NECT in an asymptomatic patient with biliary hamartomas shows innumerable subcentimeter hypodense foci in both lobes of the liver.

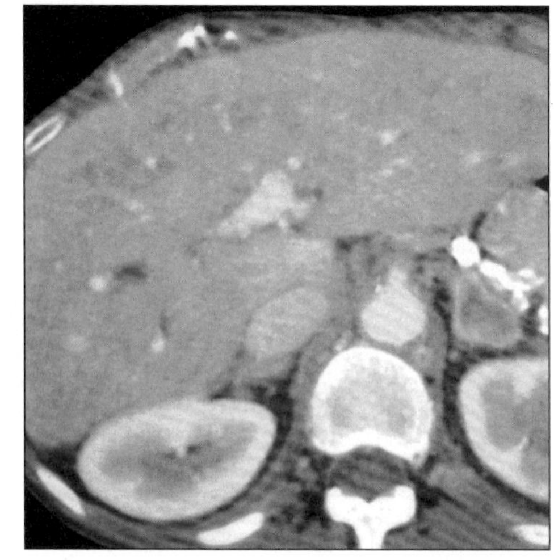

Axial CECT of same patient. Many of the hamartomas are now isodense with liver.

TERMINOLOGY

Abbreviations and Synonyms

- Von Meyenburg complexes
- Multiple biliary hamartomas
- Multiple bile duct hamartomas

Definitions

- Rare benign malformations of biliary tract

IMAGING FINDINGS

General Features

- Best diagnostic clue: Multiple near water density/intensity liver lesions < 1.5 cm diameter
- Location
  - Subcapsular or intraparenchymal in location
  - Scattered throughout both lobes of liver
- Size: Varies from less than 1.0 to 1.5 cm
- Key facts
  - Typically well-circumscribed but not encapsulated
- May occur as
  - Multiple lesions (more common) or
  - Isolated solitary lesion

CT Findings

- NECT
  - Solitary or multiple small, well-defined nodules of varied density
    - Multiple hamartomas: Seen as an aggregation of tiny lesions
  - Density of lesions: Depending on predominant cystic/solid component
    - Predominantly cystic: Water density
    - Predominantly solid (fibrous stroma): Decreased attenuation
  - Distribution & size
    - Both uniform & nonuniform in distribution
    - Relatively uniform compared to nonuniform metastases
    - This pattern is nonspecific
    - Biopsy is needed to diagnose
    - Varied size: 2-15 mm
- CECT
  - Varied enhancement based on
    - Cystic or solid components of lesions
  - Predominantly cystic (water density) lesions
    - No enhancement
  - Predominantly solid (fibrous stroma) lesions
    - Enhance & become isodense with liver tissue

MR Findings

- T1WI: Hypointense (both cystic & solid lesions)
- T2WI
  - Hyperintense (cystic lesions)
  - Intermediate intensity (solid lesions)

DDx: Multiple Hepatic "Cystic" Lesions

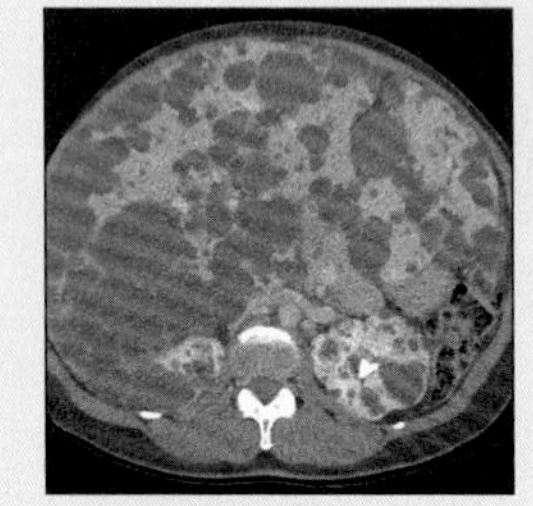

ADPLD

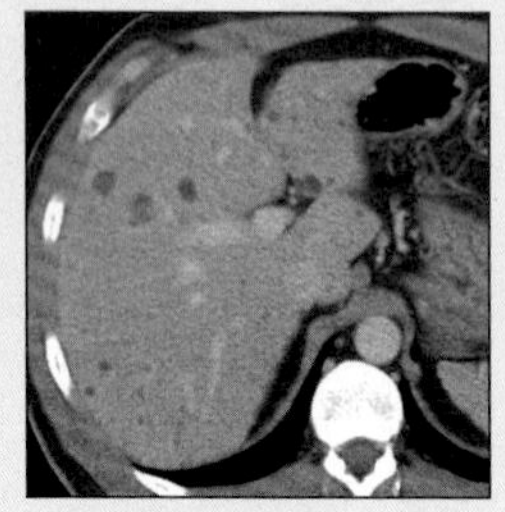

Hepatic Cysts

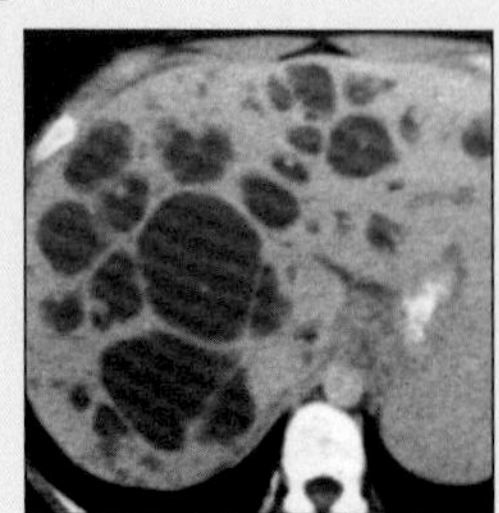

Caroli Disease

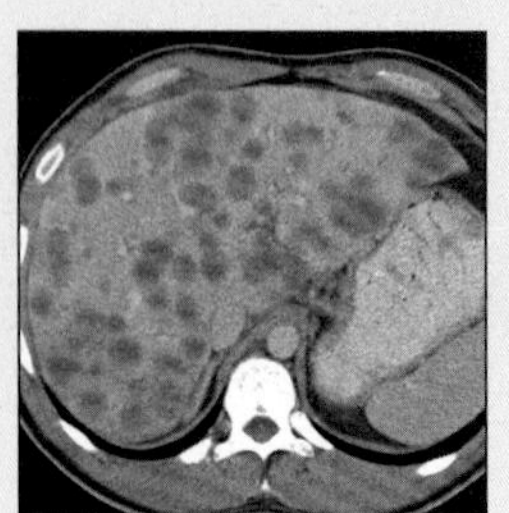

Liver Metastases

BILIARY HAMARTOMA

Key Facts

Terminology
- Von Meyenburg complexes
- Multiple biliary hamartomas
- Multiple bile duct hamartomas
- Rare benign malformations of biliary tract

Imaging Findings
- Best diagnostic clue: Multiple near water density/intensity liver lesions < 1.5 cm diameter
- Subcapsular or intraparenchymal in location

Top Differential Diagnoses
- Autosomal dominant polycystic disease (ADPLD)
- Multiple simple hepatic cysts
- Caroli's disease
- Multiple/solitary small metastatic lesions

Pathology
- Biliary hamartomas are one of the variants of congenital hepatic fibropolycystic disease
- Other variants include: Polycystic liver disease & Caroli disease
- Congenital (embryological/developmental) malformation

Clinical Issues
- Varies; asymptomatic → RUQ pain
- Malignant transformation to cholangiocarcinoma (very rare)

Diagnostic Checklist
- No further evaluation needed when seen as isolated finding in a healthy nononcologic patient

---

- Heavily T2WI
  - Signal intensity increases further (equal to fluid)
- T1 C+
  - Predominantly cystic lesions: No enhancement
    - ± Thin rim-enhancement on early & late post-gadolinium images
    - Compressed surrounding liver tissue & inflammatory cell infiltrate
  - Predominantly solid lesions: Enhancement seen
    - Due to fibrous stroma
- MR Cholangiography (MRC)
  - Markedly hyperintense nodules
  - Typically measuring less than 1.5 cm
  - No communication with biliary tree

Ultrasonographic Findings
- Real Time
  - Small & well-circumscribed lesions
  - Scattered throughout liver
  - Hypo-/hyperechoic or mixed-echoic
    - Based on solid, cystic or both contents

Angiographic Findings
- Conventional: Lesions show abnormal vascularity: Grape-like clusters of small rings

Imaging Recommendations
- Best imaging tool: High-resolution CT or MRC
- Protocol advice
  - Helical CT
    - 5 mm collimated scans reconstructed every 2.5 mm
  - MR: Heavily T2WI/MRC

DIFFERENTIAL DIAGNOSIS

Autosomal dominant polycystic disease (ADPLD)
- Usually large & numerous hepatic cysts
- Do not communicate with each other or biliary tract
- Do not opacify at
  - ERCP: Endoscopic retrograde cholangiopancreatography
  - PTC: Percutaneous transhepatic cholangiography
- Usually have cysts in kidneys & other organs
- ± Family history

Multiple simple hepatic cysts
- Multiple well-defined, round, nonenhancing, water-density lesions
- Cysts are usually seen in a normally appearing liver
- Usually solitary, can be multiple as in ADPLD
- US: Anechoic lesion with through transmission; no mural nodularity

Caroli's disease
- Inherited as an autosomal recessive pattern
- Multiple small rounded hypodense/hypointense saccular dilatation of IHBDs
- "Central dot" sign on CECT
  - Enhancing tiny dots (portal radicles) within dilated cystic intrahepatic ducts
- ERCP & microscopically: Communicating bile duct abnormality

Multiple/solitary small metastatic lesions
- Metastases: Usually nonuniform in size & distribution of lesions
- Biopsy & histology: Distinguishes hamartomas from metastases

PATHOLOGY

General Features
- General path comments
  - Rare benign malformations of bile ducts
  - Usually diagnosed
    - At autopsy or imaging as an incidental finding
  - Biliary hamartomas are one of the variants of congenital hepatic fibropolycystic disease
  - Other variants include: Polycystic liver disease & Caroli disease
  - Increase prevalence of bile duct hamartomas in polycystic liver disease cases

BILIARY HAMARTOMA

  - Malignant transformation: Hamartoma to cholangiocarcinoma (very rare)
- Genetics: No genetic predisposition
- Etiology
  - Congenital (embryological/developmental) malformation
    - Due to failure of involution of embryonic bile ducts
  - Same pathogenesis for polycystic liver disease
    - Prevalence of bile duct hamartomas is greater in patients with polycystic liver disease than in those without
    - These two entities may coexist
- Epidemiology: Very rare; autopsy incidence (0.69%)

Gross Pathologic & Surgical Features

- Multiple grayish white nodules of varying sizes ranging from solid to cystic
- Subcapsular or intraparenchymal
- Biliary dilatation within hamartomas: Narrow/mild/prominent

Microscopic Features

- Proliferation of dilated bile ducts (bile duct duplication)
- Surrounded by dense, hyalinized fibrous stroma
- No communication exists with biliary system
- Cystic hamartomas: Lined by cuboidal/flattened epithelium

Staging, Grading or Classification Criteria

- Classification based on degree of hamartoma consistency/biliary dilatation (not widely used)
  - Class 1: Predominantly solid lesions + narrow bile channels
  - Class 2: Intermediate lesions (both solid & cystic foci) + mild dilatation
  - Class 3: Predominantly cystic lesions + prominent dilated bile channels

CLINICAL ISSUES

Presentation

- Most common signs/symptoms
  - Asymptomatic; occasionally dull pain & fullness in right upper quadrant (RUQ)
  - Abdomen: Soft/nontender/no palpable masses or hepatosplenomegaly
- Clinical profile
  - Varies; asymptomatic → RUQ pain
  - Lab data
    - Normal liver function tests
    - Normal bilirubin assays
    - Normal carcinoembryonic antigen (CEA) levels
  - Diagnosis: Wedge or core-needle biopsy & histologic exam

Demographics

- Age: Any age group (usually incidental finding)
- Gender: M = F

Natural History & Prognosis

- Rare congenital, noncommunicating bile duct abnormality
- Usually occur as multiple lesions (more common)
- Scattered throughout both lobes of liver
- Follow-up imaging usually shows no change in appearance
- Complications
  - Infection: Microabscess formation (rare)
  - Malignant transformation to cholangiocarcinoma (very rare)
- Prognosis: Usually good

Treatment

- Options
  - Asymptomatic: No treatment
  - Symptomatic: Exploratory laparotomy & surgical resection

DIAGNOSTIC CHECKLIST

Consider

- MRC to differentiate biliary hamartomas abnormalities
  - From other biliary (communicating/noncommunicating)

Image Interpretation Pearls

- No further evaluation needed when seen as isolated finding in a healthy nononcologic patient
- Possibility of misdiagnosing as multiple small liver metastases

SELECTED REFERENCES

1. Orii T et al: Cholangiocarcinoma arising from preexisting biliary hamartoma of liver--report of a case. Hepatogastroenterology. 50(50):333-6, 2003
2. Mortele B et al: Hepatic bile duct hamartomas (von Meyenburg Complexes): MR and MR cholangiography findings. J Comput Assist Tomogr. 26(3):438-43, 2002
3. Allgaier HP et al: Ampullary hamartoma: A rare cause of biliary obstruction. Digestion. 60(5):497-500, 1999
4. von Schweinitz D et al: Mesenchymal hamartoma of the liver--new insight into histogenesis. J Pediatr Surg. 34(8):1269-71, 1999
5. Semelka RC et al: Biliary hamartomas: solitary and multiple lesions shown on current MR techniques including gadolinium enhancement. J Magn Reson Imaging. 10(2):196-201, 1999
6. Principe A et al: Bile duct hamartomas: diagnostic problems and treatment. Hepatogastroenterology. 44(16):994-7, 1997
7. Cheung YC et al: MRI of multiple biliary hamartomas. Br J Radiol. 70(833):527-9, 1997
8. Iha H et al: Biliary hamartomas simulating multiple hepatic metastasis on imaging findings. Kurume Med J. 43(3):231-5, 1996
9. Lev-Toaff AS et al: The radiologic and pathologic spectrum of biliary hamartomas. AJR Am J Roentgenol. 165(2):309-13, 1995
10. Powers C et al: Primary liver neoplasms: MR imaging with pathologic correlation. Radiographics. 14(3):459-82, 1994

IMAGE GALLERY

Typical

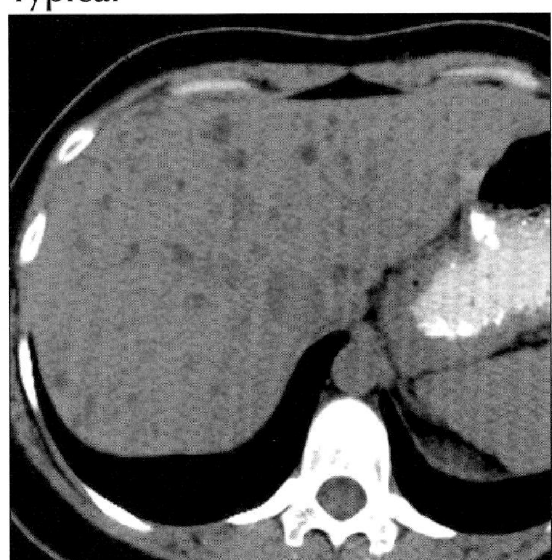

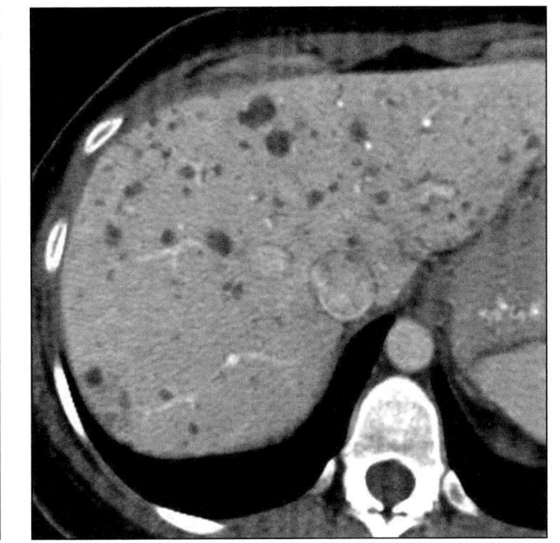

(Left) *Axial NECT shows multiple small hypodense lesions in liver.* ***(Right)*** *Axial CECT of same patient shows multiple small nonenhancing lesions of varied sizes distributed predominantly in the right lobe of liver, consistent with biliary hamartomas.*

Typical

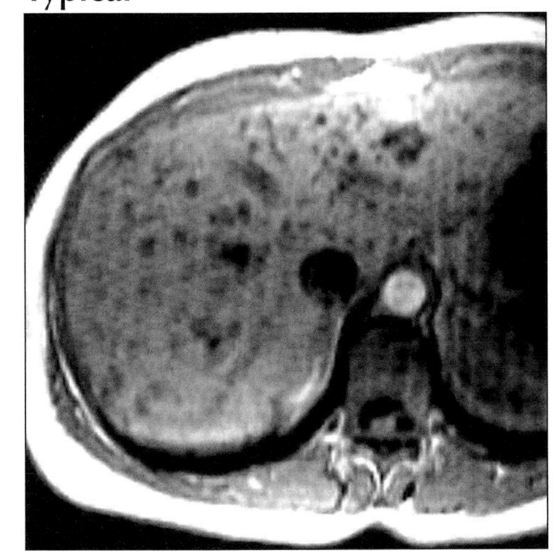

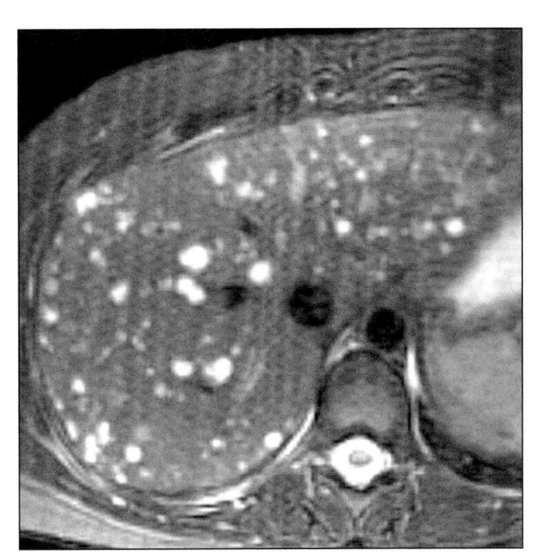

(Left) *Axial T1 C+ MR shows numerous small nonenhancing hypointense lesions diffusely involving the liver.* ***(Right)*** *Axial T2WI MR in the same patient demonstrates multiple small hyperintense foci of varied sizes throughout both lobes of liver (biliary hamartomas).*

Typical

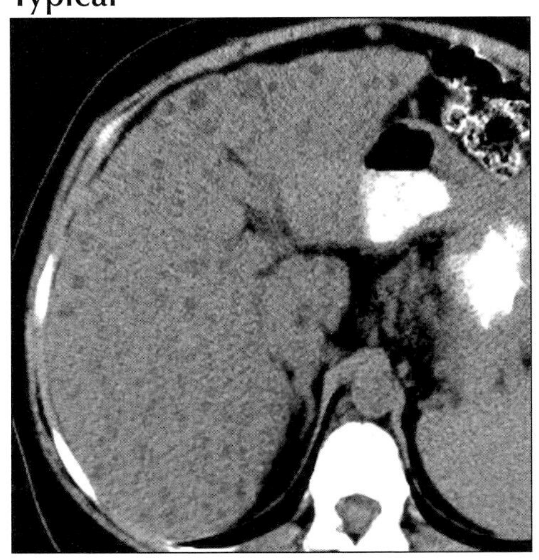

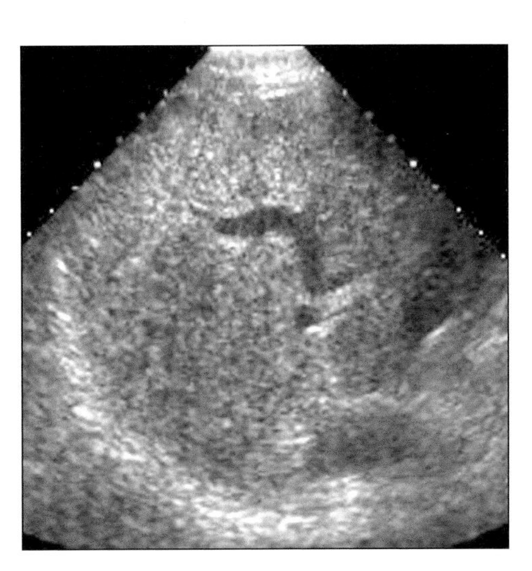

(Left) *Axial NECT in a cirrhotic liver shows multiple subcentimeter hypodense lesions (biopsy proven biliary hamartomas).* ***(Right)*** *Transverse US of the liver shows diffuse replacement of normal parenchyma with small hyperechoic foci consistent with biliary hamartomas.*

HEPATIC ANGIOMYOLIPOMA

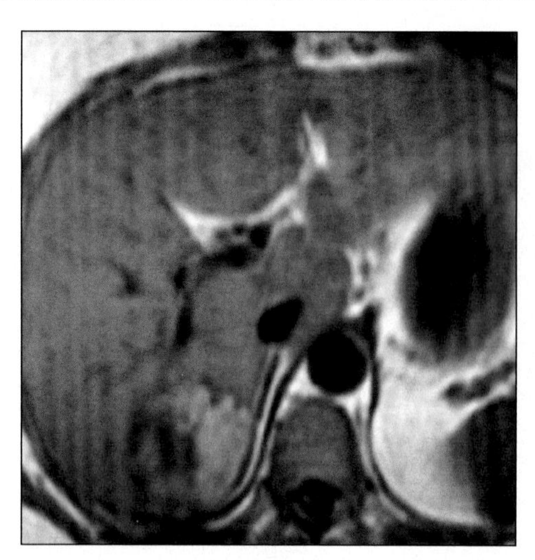

Axial T1WI MR shows heterogeneous hepatic mass with some hyperintense foci indicating fat component.

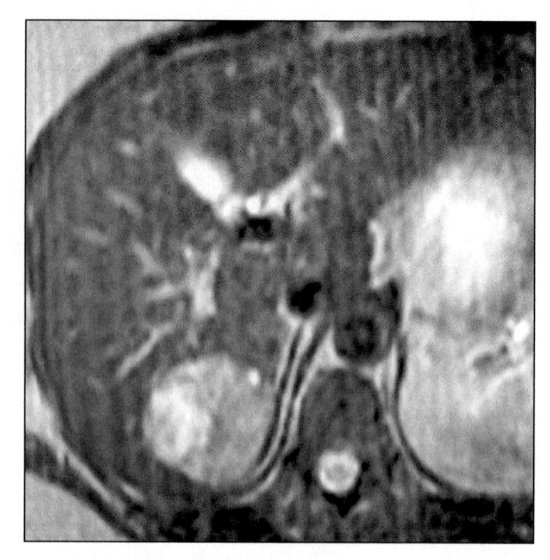

Axial T2WI MR shows heterogeneous mass with fatty component nearly isointense to subcutaneous fat; other tumor foci are typically hyperintense.

TERMINOLOGY

Abbreviations and Synonyms

- Hepatic angiomyolipoma (AML), benign liver hamartoma

Definitions

- Benign mesenchymal tumor composed of variable amounts of smooth muscle (myoid), fat (lipoid), & proliferating blood vessels (angioid) components

IMAGING FINDINGS

General Features

- Best diagnostic clue: Well-circumscribed fatty mass; successful diagnosis relies on identification of intratumoral fat at imaging
- Location
  - Liver: Second most common site (first: Kidney)
  - In liver: Right lobe is most common site
- Size: Variable; range from 0.3-36 cm in diameter
- Key concepts
  - Round or lobulated solitary mass or as multiple lesions with variable shape

CT Findings

- NECT
  - Well-defined mass with heterogeneous attenuation values due to presence of fat & soft tissue densities
  - May be predominantly low density mass
- CECT
  - Arterial phase: Prominent enhancement of lesion
  - Portal phase: Lesion shows hypoattenuation
  - On early phase of dynamic study, enhancement is higher than on late phase, but it has prolonged enhancement (due to proliferation of blood vessels)
- CTA: Central vessels within lesion

MR Findings

- T1WI
  - Hypointensity or hyperintensity on T1WI
  - Fatty component of tumor results in hyperintense (high signal) foci on T1WI
  - Relative loss of signal intensity on opposed-phase images compared with in-phase; qualitative assessment of relatively small amounts of lipid
  - Frequency-selective fat saturation techniques useful
- T2WI
  - High signal intensity of fatty components
  - Heterogeneous hyperintensity
- T1 C+: Soft tissue elements within lesion enhance

Ultrasonographic Findings

- Real Time
  - Homogeneous or heterogeneous echogenic mass due to fat

DDx: Fat Containing Liver Mass

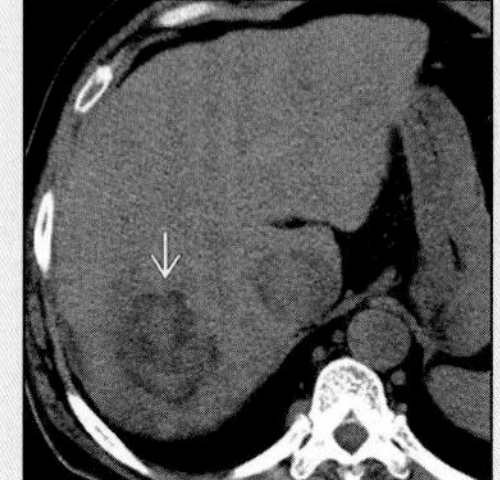

HCC with Fat

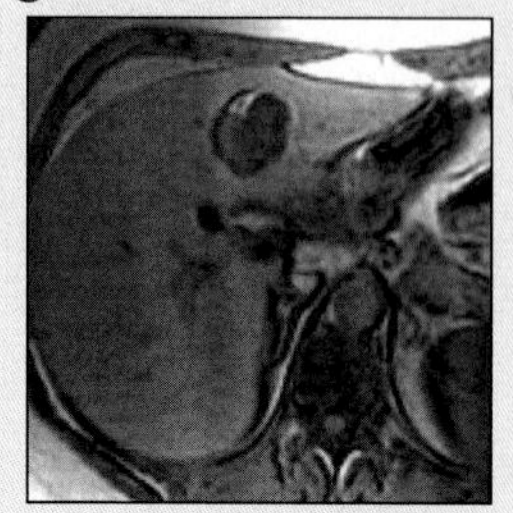

Focal Steatosis

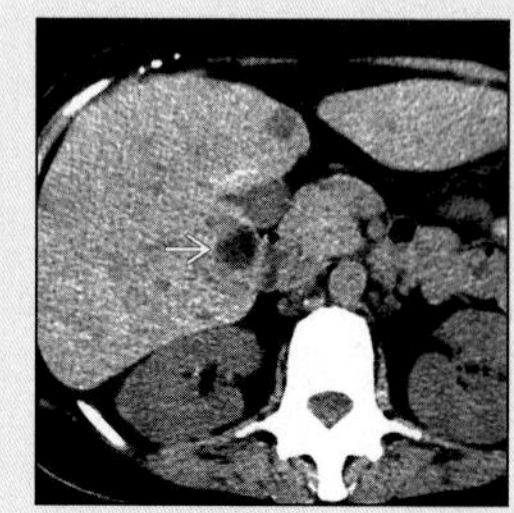

Hepatic Adenoma

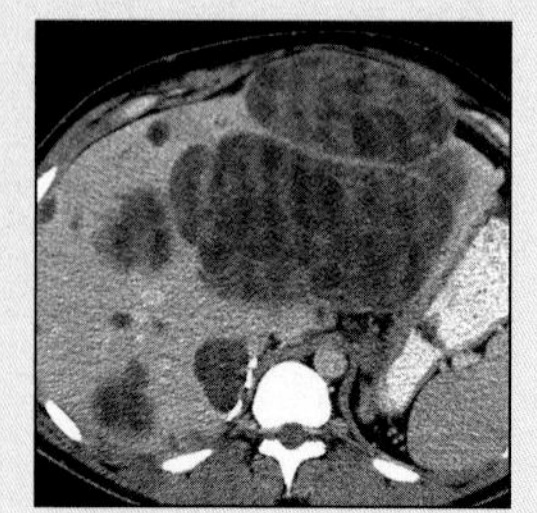

Metastatic Teratoma

HEPATIC ANGIOMYOLIPOMA

Key Facts

Terminology

- Benign mesenchymal tumor composed of variable amounts of smooth muscle (myoid), fat (lipoid), & proliferating blood vessels (angioid) components

Imaging Findings

- Best diagnostic clue: Well-circumscribed fatty mass; successful diagnosis relies on identification of intratumoral fat at imaging
- On early phase of dynamic study, enhancement is higher than on late phase, but it has prolonged enhancement (due to proliferation of blood vessels)

Pathology

- Associated with tuberous sclerosis

Clinical Issues

- Complication: Spontaneous hemorrhage or rupture

 - If muscle, vascular elements or hemorrhage predominate, lesion may be hypoechoic

Angiographic Findings

- Heterogeneously hypervascular tumor

Imaging Recommendations

- Best imaging tool: MR; fat suppression ± opposed phase GRE imaging

DIFFERENTIAL DIAGNOSIS

Hepatocellular carcinoma (HCC)

- Fat within tumor may be localized or show a diffusely scattered or a mosaic pattern
- Fat is usually a minor component of HCC tumor mass

Focal steatosis

- Usually poorly defined & not well circumscribed
- Shows blood vessels traversing lesion

Hepatic adenoma

- Well defined as surrounded by capsule
- Hypoattenuating mass with heterogeneous attenuation; areas of hemorrhage & infarction within

Hepatic lipoma

- No enhancement on incremental bolus dynamic CT

Metastases (teratoma or liposarcoma)

- Fat containing; ± fluid, calcification in teratoma
- Most liposarcomas are large, well-circumscribed; vascular structures with soft tissue attenuation

PATHOLOGY

General Features

- Associated with tuberous sclerosis

Gross Pathologic & Surgical Features

- Fat content varies; < 10% to > 90% of tumor volume
- Usually yellow-to-light tan, secondary to fat content

Microscopic Features

- Epithelioid smooth muscle cells, admixture of mature fat cells & proliferating blood vessels

CLINICAL ISSUES

Presentation

- Asymptomatic & discovered incidentally at imaging
- Pain results from intratumoral hemorrhage

Natural History & Prognosis

- Complication: Spontaneous hemorrhage or rupture
- No malignant potential

Treatment

- Conservative, embolization, surgical resection

DIAGNOSTIC CHECKLIST

Consider

- Small fat-density hepatic mass in tuberous sclerosis patient is benign

SELECTED REFERENCES

1. Takayama Y et al: Hepatic angiomyolipoma: radiologic and histopathologic correlation. Abdom Imaging. 27(2):180-3, 2002
2. Yoshimura H et al: Angiomyolipoma of the liver with least amount of fat component: imaging features of CT, MR, and angiography. Abdom Imaging. 27(2):184-7, 2002
3. Balci NC et al: Hepatic angiomyolipoma: demonstration by out of phase MRI. Clin Imaging. 26(6):418-20, 2002

IMAGE GALLERY

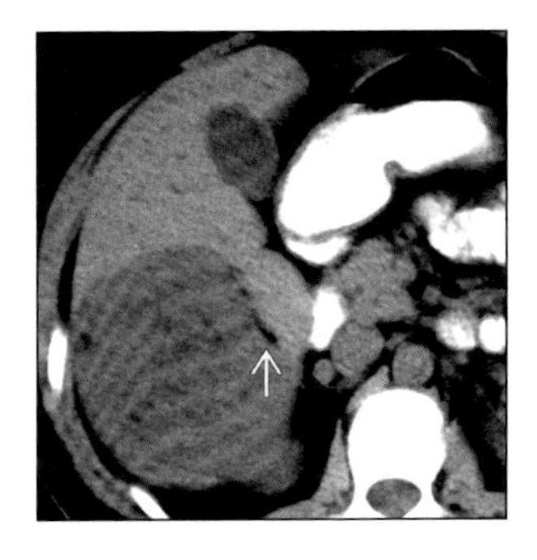

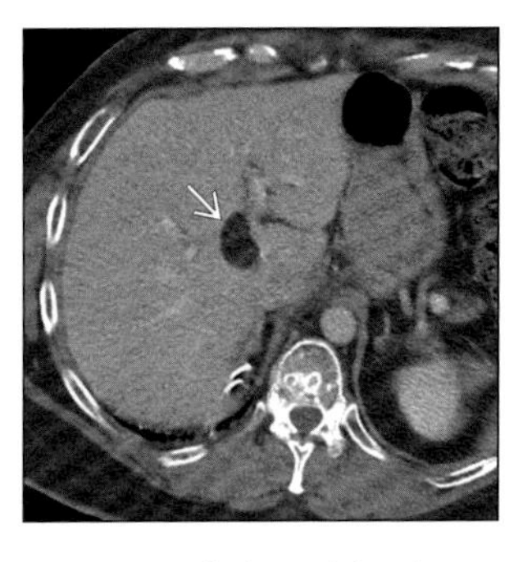

(Left) *Axial NECT shows large hepatic mass with foci of fat density (arrow); angiomyolipoma.* ***(Right)*** *Axial CECT in patient with tuberous sclerosis (prior right nephrectomy for angiomyolipoma). Almost pure fat density liver mass (arrow).*

HEPATOCELLULAR CARCINOMA

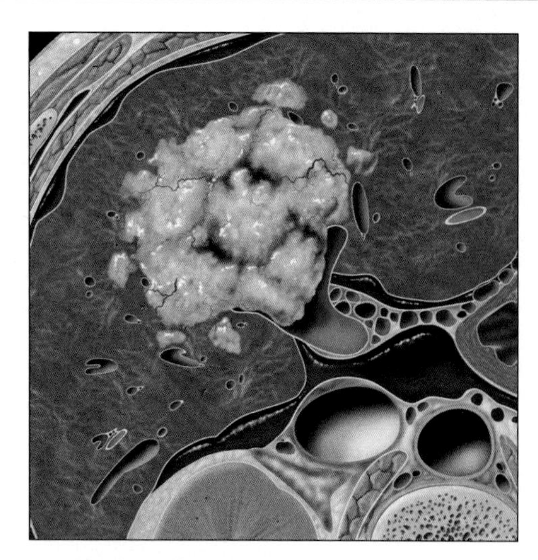

Graphic shows heterogeneous vascular mass invading portal vein. Surrounding liver is cirrhotic with fibrosis. Varices and ascites.

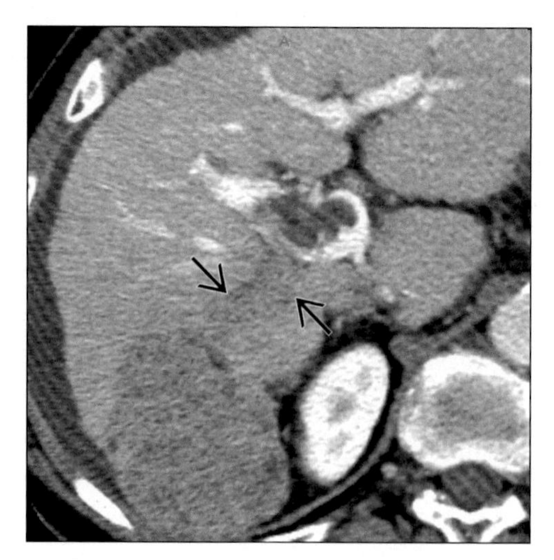

Axial CECT in venous phase shows hypodense mass in cirrhotic liver which invades, and occludes the posterior branch of right portal vein (arrows). Nonocclusive thrombus in main portal vein.

TERMINOLOGY

Abbreviations and Synonyms

- Hepatocellular carcinoma (HCC); hepatoma or primary liver cancer

Definitions

- Most common primary malignant liver tumor usually arising in cirrhotic liver due to chronic viral hepatitis (HBV/HCV) or alcoholism

IMAGING FINDINGS

General Features

- Best diagnostic clue: Large heterogeneous hypervascular mass with portal vein invasion
- Location
  - Most often right lobe of liver (solitary)
  - Both hepatic lobes (multicentric small nodular)
  - Throughout liver in a diffuse manner (diffuse small foci)
- Size
  - Small tumors: Less than 3 cm
  - Large tumors: More than 5 cm
  - Diffuse or cirrhotomimetic: Subcentimeter to few cms
- Key concepts
  - Most frequent primary visceral malignancy in world
    - Accounts 80-90% of all primary liver malignancies
  - 2nd most common malignant liver tumor in children after hepatoblastoma
  - Growth patterns of HCC: Three major types
    - Solitary, often large mass
    - Nodular or multifocal
    - Diffuse or cirrhotomimetic

CT Findings

- NECT
  - In noncirrhotic liver
    - Solitary HCC: Large hypodense mass; ± necrosis, fat, calcification
    - Multifocal HCC: Multiple hypodense lesions rarely with a central necrotic portion
    - Dominant hypodense mass with decreased attenuation satellite nodules
    - Encapsulated HCC: Well-defined, rounded, hypodense mass
  - In cirrhotic liver
    - Iso-/hypodense mass
    - Nodular cirrhotic liver
    - Ascites & varices
- CECT
  - Hepatic arterial phase (HAP) scan
    - Heterogeneous enhancement

DDx: Hypervascular Mass(es) in Dysmorphic Liver

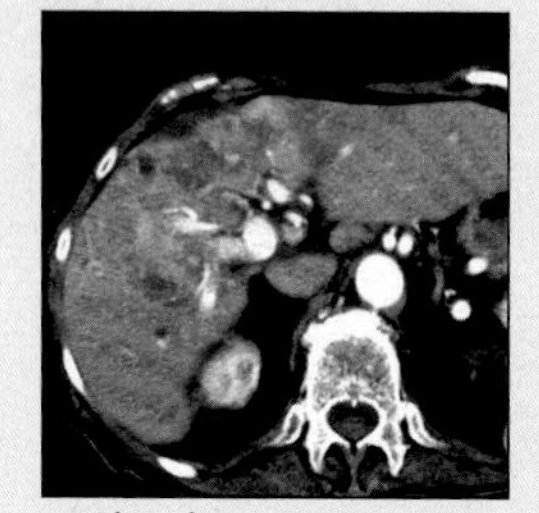

Choalgiocarcinoma

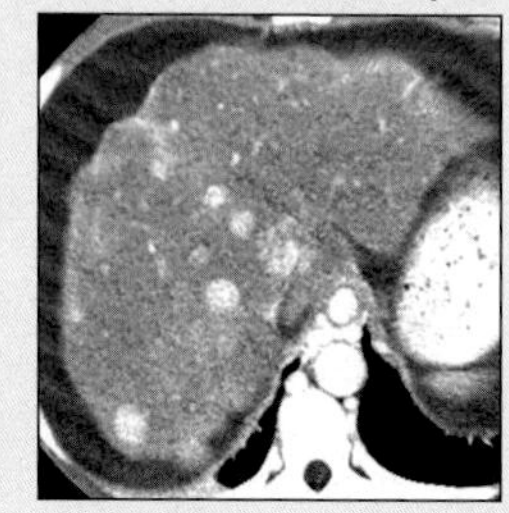

Regenerative Nodules

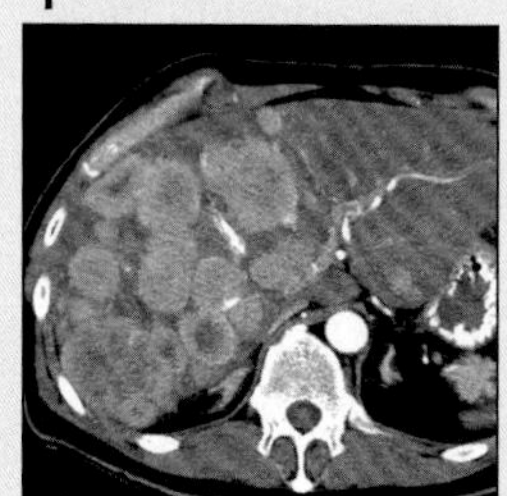

Metastases

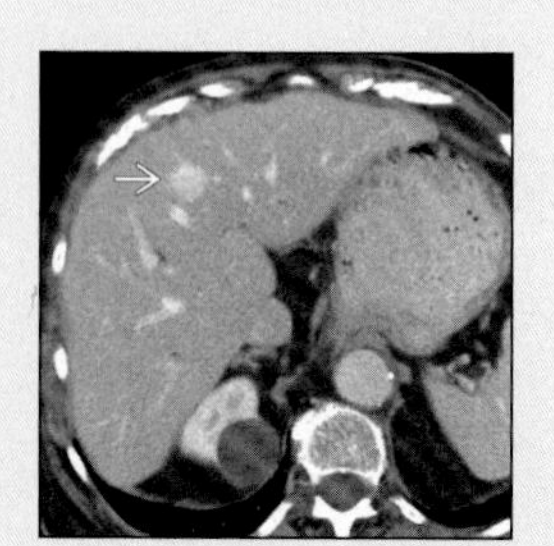

Small Hemangioma

HEPATOCELLULAR CARCINOMA

Key Facts

Terminology

- Most common primary malignant liver tumor usually arising in cirrhotic liver due to chronic viral hepatitis (HBV/HCV) or alcoholism

Imaging Findings

- Best diagnostic clue: Large heterogeneous hypervascular mass with portal vein invasion
- Most often right lobe of liver (solitary)

Top Differential Diagnoses

- Cholangiocarcinoma
- Hypervascular metastases
- Focal nodular hyperplasia (FNH)
- Small hepatic hemangioma

Pathology

- Invasion: Vascular (common) & biliary (uncommon)
- Clear cell carcinoma: HCC with large amounts of fat
- Cirrhosis (60-90%): Due to chronic viral hepatitis (HBV, HCV) or alcoholism
- Worldwide highest incidence is in Japan (4.8%)
- North America: 40% of HCC in non-cirrhotic livers

Clinical Issues

- Clinical profile: Elderly patient with history of cirrhosis, ascites, weight loss, RUQ pain & ↑ AFP

Diagnostic Checklist

- HCC: Hypervascular mass invading portal vein
- Small HCC may mimic hemangioma or metastasis in cirrhotic liver

        - Wedge-shaped areas of increased density on HAP: Perfusion abnormality due to portal vein occlusion by tumor thrombus & increased arterial flow
    - Portal venous phase (PVP) scan
        - Decreased attenuation with heterogeneous areas of contrast accumulation
    - Delayed scan: Hypodense to surrounding liver
    - Small hypervascular HCC
        - Early & late arterial phases: Hyperattenuating, more on late phase
        - CT hepatic arteriography: Lesions show intense enhancement
        - CT during arterial portography: No enhancement

MR Findings

- Variable intensity depending on degree of fatty change, fibrosis, necrosis
- T1WI
    - Noncirrhotic liver
        - Hypointense; iso-/hyperintense
    - Cirrhotic liver
        - HCC: Hypointense
        - Cirrhotic nodules: Increased signal intensity
- T2WI
    - Noncirrhotic liver: Slightly hyperintense
    - Cirrhotic liver
        - HCC: Hyperintense
        - Cirrhotic nodules: Iso to hypointense
    - HCC arising within a siderotic nodule
        - "Nodule within a nodule" pattern
        - HCC appears as a small focus of increased signal intensity within decreased signal intensity nodule
- T1 C+ (gadolinium)
    - Large HCC in noncirrhotic liver: Nonspecific
        - Central or peripheral enhancement
        - Homogeneous or rim-enhancement
    - HCC nodules (hypervascular)
        - Arterial phase: Hyperintense
- SPIO (superparamagnetic iron oxide)
    - FLASH & long TR sequences
    - HCC: Higher signal than surrounding liver
        - Liver takes up SPIO more than lesion
    - Increases sensitivity of MR in diagnosing
        - Small HCCs in cirrhotic livers
- Mangafodipir trisodium (Mn-DPDP)
    - T1WI: Increased signal in well-differentiated HCC
    - Differentiates HCC from nonhepatocellular tumors

Ultrasonographic Findings

- Real Time
    - Mixed echogenicity due to tumor necrosis & hypervascularity
    - Hypoechoic: Due to solid tumor
    - Hyperechoic: Due to fatty metamorphosis
        - Small hyperechoic HCC simulate hemangioma
    - Capsule in encapsulated HCC
        - Thin hypoechoic band
- Color Doppler
    - Shows hypervascularity & tumor shunting
    - Small HCC: Indistinguishable from small hemangiomas & metastases

Angiographic Findings

- Conventional
    - Hypervascular tumor
        - Marked neovascularity & AV shunting
        - Large hepatic artery & vascular invasion
    - "Threads & streaks" sign
        - Sign of tumor thrombus in portal vein

Nuclear Medicine Findings

- Hepato Biliary Scan
    - Uptake in 50% of lesions
- Technetium Sulfur Colloid
    - HCC in a cirrhotic liver: Seen as a defect
    - HCC in a noncirrhotic liver: Heterogeneous uptake
- Gallium Scan
    - HCC is gallium avid in 90% of cases

Imaging Recommendations

- Helical triphasic CT (NE, arterial & venous phases) or MR & CEMR; angiography

HEPATOCELLULAR CARCINOMA

DIFFERENTIAL DIAGNOSIS

Cholangiocarcinoma

- Peripheral tumor often obstructs bile ducts
- Capsular retraction; volume loss
- Delayed enhancement

Nodular regenerative hyperplasia (as in Budd-Chiari syndrome)

- Called "large regenerative nodules"
- Small nodules: Not detectable
- Large nodules: Homogeneously hypervascular
- Usually 1-4 cm in size

Hypervascular metastases

- Mimic small nodular or multifocal HCC
- Less likely to invade portal vein

Focal nodular hyperplasia (FNH)

- Homogeneous hypervascular mass with central scar
- On nonenhanced & delayed CECT & CEMR almost isodense/isointense to liver

Small hepatic hemangioma

- Well-defined, spherical mass isodense to blood
- CECT: "Flash filling" (still isodense with blood)
- On US: Usually hyperechoic nodule
- Angiography: Characteristic "cotton wool" appearance

PATHOLOGY

General Features

- General path comments
  - Soft tumor; may have necrosis & hemorrhage
  - Invasion: Vascular (common) & biliary (uncommon)
  - Clear cell carcinoma: HCC with large amounts of fat
- Genetics: HBV DNA integrated into host's genomic DNA in tumor cells
- Etiology
  - Cirrhosis (60-90%): Due to chronic viral hepatitis (HBV, HCV) or alcoholism
  - Carcinogens
    - Aflatoxins, siderosis, thorotrast, androgens
  - α-1-antitrypsin deficiency, hemochromatosis
  - Wilson disease, tyrosinosis
- Epidemiology
  - High incidence: Africa & Asia (HBV & aflatoxins)
  - Low incidence: Western hemisphere
  - Worldwide highest incidence is in Japan (4.8%)
  - HCC in cirrhosis due to hepatitis C virus
    - United States: 30-50% of cases of HCC
    - Japan: 70% of cases of HCC
  - North America: 40% of HCC in non-cirrhotic livers

Gross Pathologic & Surgical Features

- Solitary, nodular or multifocal, diffuse, encapsulated
- Soft tumor with or without necrosis, hemorrhage, calcification, fat, vascular invasion

Microscopic Features

- Histologic appearances: Solid (cellular) or acinar
- Increased fat & glycogen in cytoplasm

CLINICAL ISSUES

Presentation

- Clinical profile: Elderly patient with history of cirrhosis, ascites, weight loss, RUQ pain & ↑ AFP
- Lab data: Increased alpha-fetoprotein (AFP) & LFTs
- Diagnosis: Biopsy & histology

Demographics

- Age
  - Low incidence areas: 6th-7th decade
  - High incidence areas: 30-45 years
- Gender
  - Low incidence areas (M:F = 2.5:1)
  - High incidence areas (M:F = 8:1)

Natural History & Prognosis

- Complications
  - Spontaneous rupture & massive hemoperitoneum
- Prognosis
  - More than 90% mortality rate; 17% resectability rate
  - 6 Months average survival time; 30% 5 year survival

Treatment

- Radiofrequency & alcohol ablation
  - Small isolated tumors
- Intraarterial chemoembolization
  - Multifocal unresectable tumor
- Surgical resection
  - Limited by inadequate hepatic reserve

DIAGNOSTIC CHECKLIST

Image Interpretation Pearls

- HCC: Hypervascular mass invading portal vein
- Small HCC may mimic hemangioma or metastasis in cirrhotic liver

SELECTED REFERENCES

1. Laghi A et al: Hepatocellular carcinoma: detection with triple-phase multi-detector row helical CT in patients with chronic hepatitis. Radiology. 226(2):543-9, 2003
2. Brancatelli G et al: Hepatocellular carcinoma in noncirrhotic liver: CT, clinical and pathologic findings in 39 U.S residents. Radiology. 222: 89-94, 2002
3. Kim T et al: Discrimination of small hepatic hemangiomas from hypervascular malignant tumors smaller than 3 cm with three-phase helical CT. Radiology. 219(3):699-706, 2001
4. Murakami T et al: Hyper vascular hepatocellular carcinoma: Detection with double arterial phase multi-detector row helical CT. Radiology. 218: 763-7, 2001
5. Peterson MS et al: Pretransplantation surveillance for possible hepatocellular carcinoma in patients with cirrhosis: Epidemiology and CT-based tumor detection rate in 430 cases with surgical pathologic correlation. Radiology. 217: 743-9, 2000
6. Oliver JH 3rd et al: Detecting hepatocellular carcinoma: value of unenhanced or arterial phase CT imaging or both used in conjunction with conventional portal venous phase contrast-enhanced CT imaging. AJR. 167(1):71-7, 1996

IMAGE GALLERY

Typical

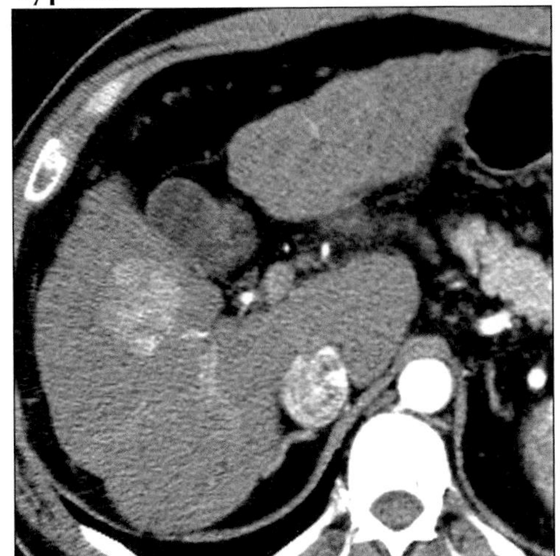

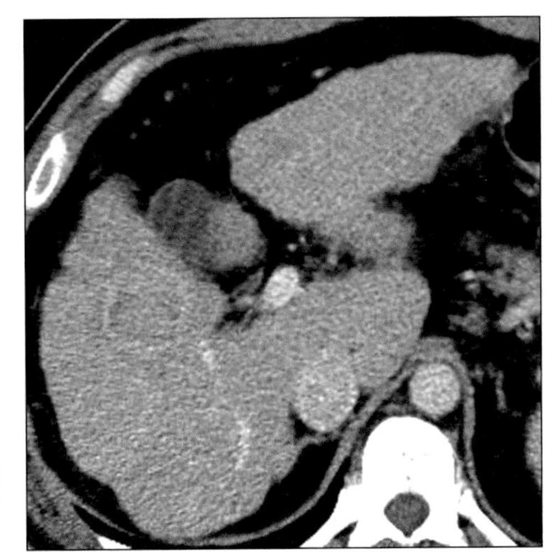

(Left) *Axial CECT in arterial phase shows hypervascular mass in right lobe, cirrhotic liver.* ***(Right)*** *Axial CECT in portal venous phase. Mass is isodense to liver; hyperdense capsule delineates tumor.*

Typical

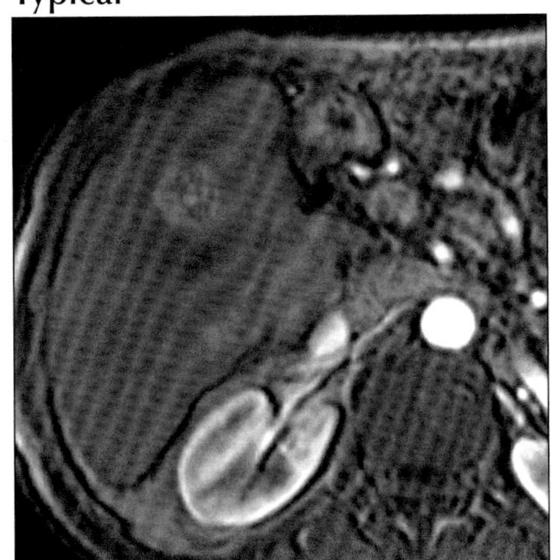

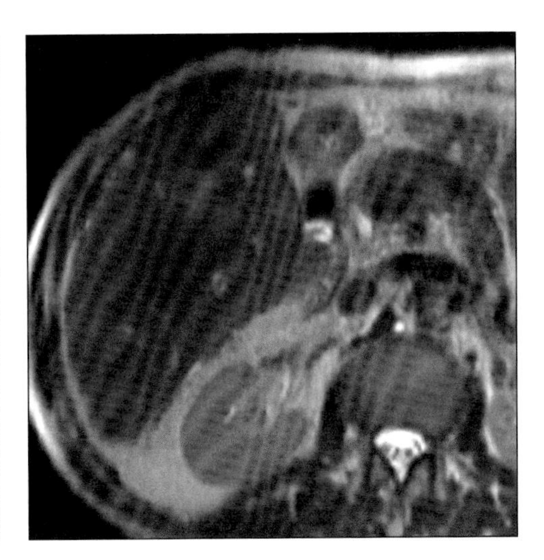

(Left) *Axial T1 C+ MR in arterial phase shows heterogeneous hypervascular mass.* ***(Right)*** *Axial T2WI MR barely detects mass as subtle hyperintense lesion.*

Variant

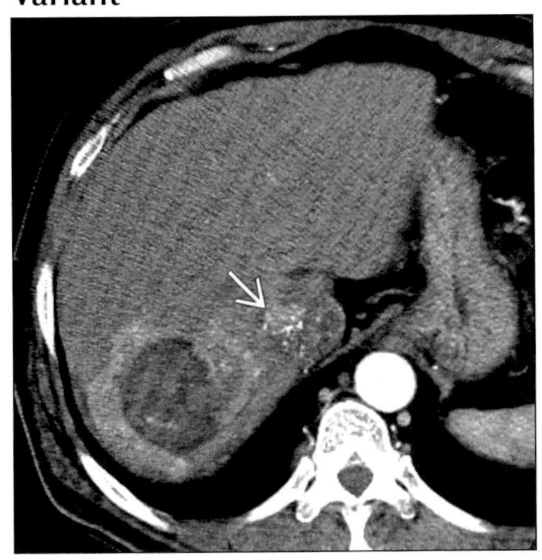

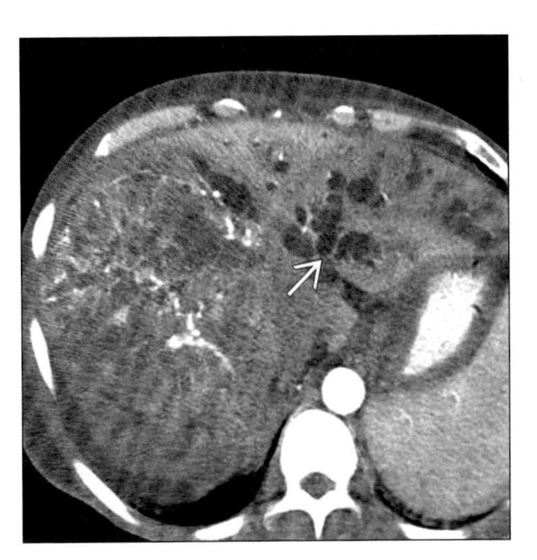

(Left) *Axial CECT in arterial phase shows heterogeneous mass in right lobe with enhancing tumor thrombus in IVC (arrow) and right hepatic vein. The liver mass had foci of fat most evident on NECT.* ***(Right)*** *Axial CECT in arterial phase shows large heterogeneous hypervascular mass that occupies the right lobe, and causes intrahepatic biliary obstruction (arrow).*

FIBROLAMELLAR HCC

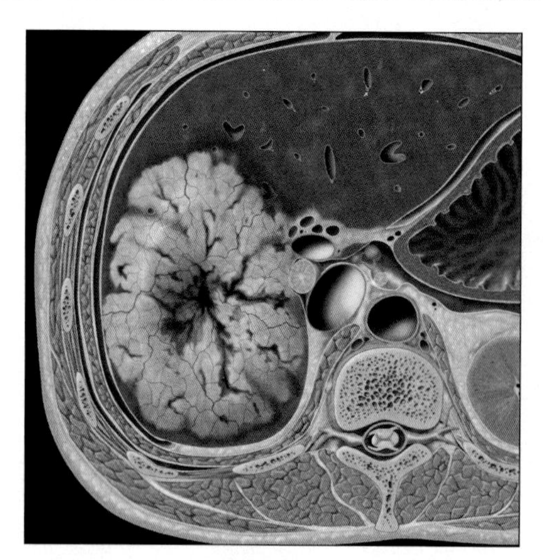

Graphic shows large heterogeneous hypervascular mass with central scar and porta hepatis lymphadenopathy.

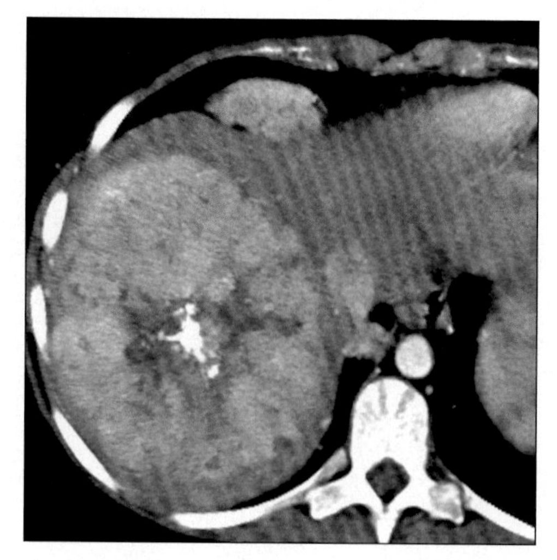

Axial CECT shows large heterogeneous hypervascular mass with large calcified central scar and cardiophrenic lymphadenopathy.

TERMINOLOGY

Abbreviations and Synonyms

- Fibrolamellar carcinoma (FLC) of liver

Definitions

- Uncommon malignant hepatocellular tumor with distinct clinical, histopathologic & imaging differences from conventional hepatocellular carcinoma (HCC)

IMAGING FINDINGS

General Features

- Best diagnostic clue: Heterogeneously-enhancing, large, lobulated mass with hypointense central scar and radial septa on T2WI
- Location
  - Intrahepatic (80%)
  - Pedunculated (20%)
- Size: Vary from 5-20 cm (mean 13 cm)
- Key concepts
  - Slow-growing tumor that usually arises in a normal liver
  - In less than 5% cases it may occur with underlying cirrhosis
  - Satellite nodules are often present
  - Characteristic microscopic pattern
    - Eosinophilic malignant hepatocytes containing prominent nuclei
  - Absence of pathologic markers like inclusions of alpha-fetoprotein bodies which are present in typical HCC
  - Better prognosis than conventional HCC, but still locally invasive and frequently metastatic

CT Findings

- NECT
  - Mass
    - Well-defined contour
    - Hypoattenuating and heterogeneous
  - Central scar & septa: Markedly hypodense
  - Calcification & necrosis are common
  - Hemorrhage is rarely seen
- CECT
  - Arterial phase
    - Mass: Heterogeneous & hyperattenuating
  - Portal phase
    - Mass: Iso-/hypoattenuating
  - Delayed phase (10 min)
    - Mass: Isodense
    - Scar/septa/capsule: Hyperdense
  - Malignant features
    - Biliary & vessel invasion
    - Nodal metastases (2/3rd cases)
    - Lung metastases

DDx: Hepatic Mass with Central Scar

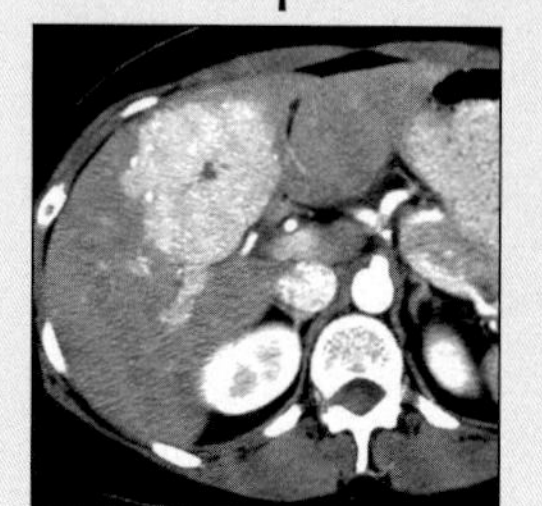

FNH

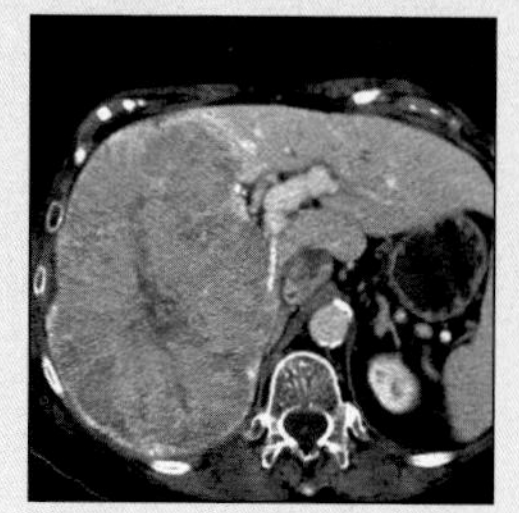

HCC

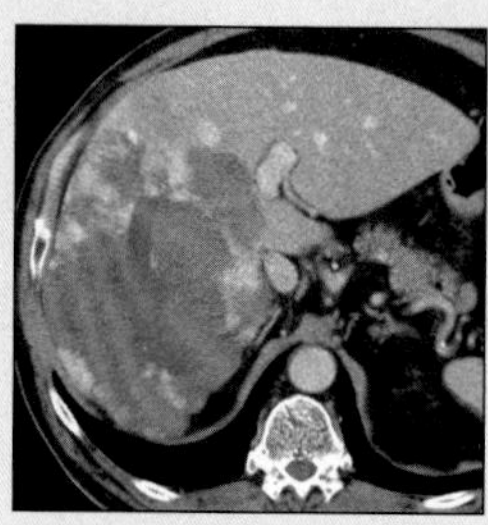

Giant Hemangioma

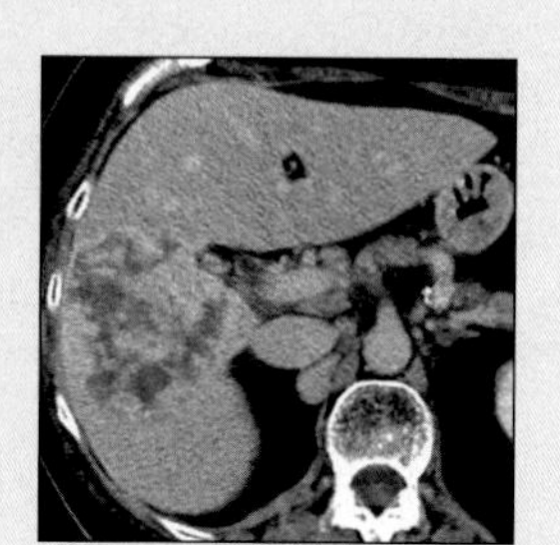

Cholangiocarcinoma

FIBROLAMELLAR HCC

Key Facts

Terminology

- Fibrolamellar carcinoma (FLC) of liver
- Uncommon malignant hepatocellular tumor with distinct clinical, histopathologic & imaging differences from conventional hepatocellular carcinoma (HCC)

Imaging Findings

- Best diagnostic clue: Heterogeneously-enhancing, large, lobulated mass with hypointense central scar and radial septa on T2WI
- Size: Vary from 5-20 cm (mean 13 cm)

Top Differential Diagnoses

- Focal nodular hyperplasia (FNH)
- Conventional (HCC)
- Cavernous hemangioma
- Intrahepatic cholangiocarcinoma

Pathology

- No specific risk factors
- FLC accounts for 1-9% of HCC overall

Clinical Issues

- Usually α-fetoprotein levels are normal

Diagnostic Checklist

- FLC: Bigger, more heterogeneous mass frequently with calcified central/eccentric scar & features of malignancy (vessel/biliary obstruction, nodal invasion & lung metastases)
- Conventional HCC can be differentiated from FLC by underlying cirrhosis (more common) & lack of scar

MR Findings

- T1WI
  - Mass: Homogeneous & slightly hypointense
  - Scar & septa: Hypointense
- T2WI
  - Mass: Heterogeneous & hyperintense
  - Scar & septa: Hypointense
- T1 C+
  - Arterial & portal phases
    - Mass: Intense heterogeneous enhancement
    - Scar: No enhancement
  - Delayed phase
    - Mass: More homogeneous enhancement
    - Scar & septa: Delayed partial enhancement

Ultrasonographic Findings

- Real Time
  - Mass
    - Large, solitary, well-defined & lobulated
    - Variable echotexture
  - Central scar: Hyperechoic

Angiographic Findings

- Conventional
  - Mass
    - Hypervascular (neovascularity)
    - Enlarged feeding arteries
    - Dense tumor blush
    - No A-V or A-P shunting
  - Septa
    - Multiple serpiginous hypovascular areas
  - Central scar: Avascular
  - Satellite nodule
    - May be seen in capillary phase

Nuclear Medicine Findings

- Tc-99m-labeled sulfur colloid
  - Solitary
    - Single photopenic defect
  - Multifocal
    - Multiple defects
    - Difficult to differentiate from multifocal HCC/metastases
- Tagged red blood cell scan (not useful)
  - FLC: Early uptake & late defect
  - Hemangioma: Early defect & late uptake

Imaging Recommendations

- Multiphasic helical CT (NECT plus arterial and venous) or multiphasic MR

DIFFERENTIAL DIAGNOSIS

Focal nodular hyperplasia (FNH)

- Marked homogeneous enhancement on arterial phase CT or MR
- Scar: Hyperintense on T2WI
- Nonencapsulated & no calcification
- Substantial uptake of SPIO
- Tc-99m-labeled colloid: Highly specific for FNH
- Usually asymptomatic
- Microscopic pattern
  - Normal hepatocytes; disorganized bile ductules

Conventional (HCC)

- Usually underlying cirrhosis is seen
- Vascular, nodal & visceral invasion: Common
- Intratumoral hemorrhage, necrosis, calcification & fat (more common)
- May be multifocal
- Conventional HCC in noncirrhotic liver mimics FLC
- Pathologic markers
  - Inclusions of alpha-fetoprotein bodies are present

Cavernous hemangioma

- Giant hemangioma
  - Heterogeneous hypodense mass
  - Central decreased attenuation (scar)
  - Size: Usually more than 10 cm
  - Arterial phase: Peripheral nodular or globular enhancement
  - Venous & delayed phases
    - Incomplete centripetal filling of lesion
    - No enhancement of scar
- Typical hemangioma: Well-circumscribed, spherical to ovoid mass isodense to blood on both NECT & CECT

FIBROLAMELLAR HCC

Intrahepatic cholangiocarcinoma

- May be central (hilar) or peripheral
- Peripheral: Hypodense solitary or satellite lesions
- Intrahepatic bile duct dilatation seen in both
- No central scar but extensive fibrosis
- Early rim enhancement with progressive, central and persistent patchy enhancement
- Often causes hepatic volume loss and capsular retraction

PATHOLOGY

General Features

- Etiology
  - No specific risk factors
  - Usually no underlying cirrhosis or liver disease
  - Occasionally hepatitis & cirrhosis may be present
    - In less than 5% of cases
- Epidemiology
  - FLC
    - Increased prevalence in US
    - Less common in Europe
    - Rare in Japan & China
  - FLC accounts for 1-9% of HCC overall
  - Represents 35% of HCC under 50 years of age

Gross Pathologic & Surgical Features

- Large, single, well-demarcated, lobulated, nonencapsulated mass
- Cut section: Tan, brown or brownish green with streaks of fibrous tissue
- Infiltrating fibrous septa; central scar (45-60%)
- Rarely encapsulated
- Solitary mass (80-90%)
- Peripheral satellite lesions (10-15%)
- Intrahepatic (80%); pedunculated (20%)
- Size: Average size is 13 cm, most vary from 5-20 cm

Microscopic Features

- Large eosinophilic, polygonal cells
  - Arranged in sheets/cords/trabeculae
  - Separated by parallel sheets of fibrous tissue (i.e., lamellae)
- Large nuclei with prominent nucleoli
- Granular-appearing cytoplasm

CLINICAL ISSUES

Presentation

- Most common signs/symptoms
  - Pain, hepatomegaly, palpable RUQ mass, cachexia
  - Occasionally jaundice when invades biliary tract
  - Symptoms
    - Usually present for 3-12 months before diagnosis
  - Rarely present with
    - Metastatic disease, fever, gynecomastia
    - Venous thrombosis (hepatic, portal & IVC)
- Clinical profile: Healthy young adult with large liver mass
- Laboratory data
  - Usually α-fetoprotein levels are normal
  - In 10% cases: Mild increase in levels (< 200 ng/μl)
  - Rarely marked increase in levels (10,000 ng/μl) similar to conventional HCC
- Diagnosis: Biopsy & histology

Demographics

- Age
  - Adolescents/young adults
    - Age range of 5-69 years (mean 23 years)
  - Most patients present in 2nd/3rd decade of life
- Gender: M:F = 1:1

Natural History & Prognosis

- Resectability rate: 48%
- FLC is frequently recurrent
- Average survival time: 32 months
- 5-year survival: 67%
- Better prognosis compared to conventional HCC

Treatment

- Localized tumor
  - Surgical resection of hepatic mass & regional nodes
  - May resect isolated lung metastases
- Inoperable cases: Chemotherapy

DIAGNOSTIC CHECKLIST

Consider

- Differentiate FLC from FNH & conventional HCC
- FLC simulates FNH due to presence of central scar in both tumors

Image Interpretation Pearls

- FLC: Bigger, more heterogeneous mass frequently with calcified central/eccentric scar & features of malignancy (vessel/biliary obstruction, nodal invasion & lung metastases)
- Scar on T2WI: Hypointense (FLC); hyperintense (FNH)
- Conventional HCC can be differentiated from FLC by underlying cirrhosis (more common) & lack of scar

SELECTED REFERENCES

1. Ichikawa T et al: Fibrolamellar hepatocellular carcinoma: Pre- and posttherapy evaluation with CT and MR imaging. Radiology. 217: 145-51, 2000
2. Mclarney J et al: Fibrolamellar carcinoma of the liver: Radiologic-pathologic correlation. RadioGraphics. 19: 453-71, 1999
3. Ichikawa T et al: Fibrolamellar hepatocellular carcinoma: Imaging and pathologic findings in 31 recent cases. Radiology. 213: 352-61, 1999
4. Schlitt HJ et al: Recurrence patterns of hepatocellular and fibrolamellar carcinoma after liver transplantation. J Clin Oncol. 17(1):324-31, 1999
5. Stevens WR et al: Fibrolamellar hepatocellular carcinoma: stage at presentation and results of aggressive surgical management. AJR Am J Roentgenol. 164(5):1153-8, 1995
6. Brandt DJ et al: Imaging of fibrolamellar hepatocellular carcinoma. AJR Am J Roentgenol. 151(2):295-9, 1988
7. Titelbaum DS et al: Fibrolamellar hepatocellular carcinoma: pitfalls in nonoperative diagnosis. Radiology. 167(1):25-30, 1988

IMAGE GALLERY

Typical

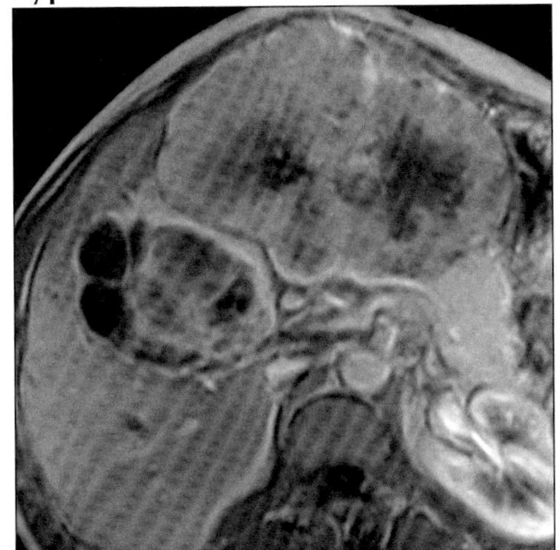

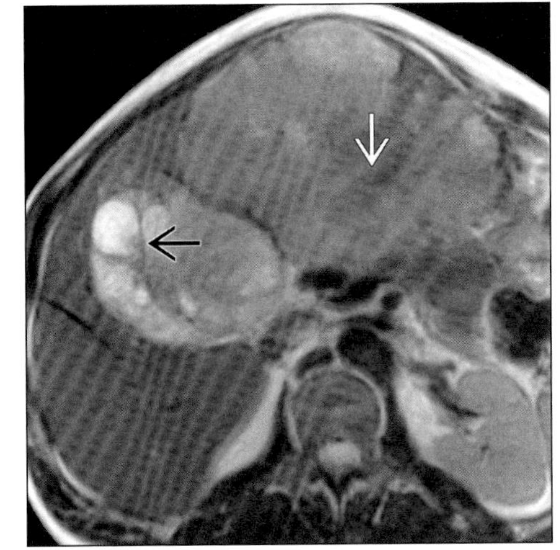

(Left) *Axial T1 C+ MR in arterial phase shows dominant and satellite masses that are well-demarcated, lobulated, heterogeneous and hypervascular.* ***(Right)*** *Axial T2WI MR shows hypointense eccentric scar (white arrow) and hyperintense foci of necrosis (black arrow).*

Typical

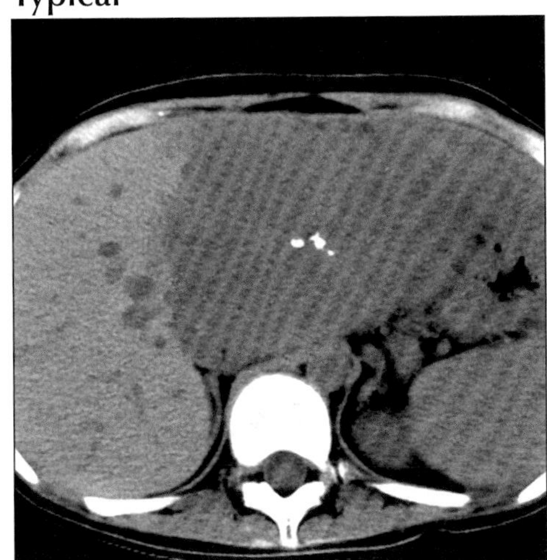

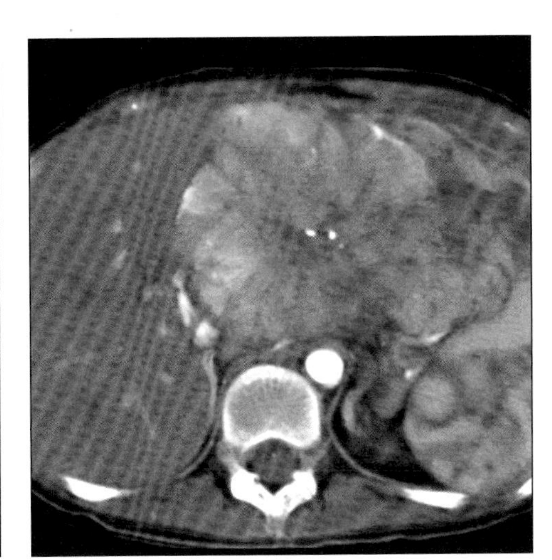

(Left) *Axial NECT shows in 17 year old male with palpable epigastric mass. Large hypodense mass with calcified central scar.* ***(Right)*** *Axial CECT in arterial phase shows heterogeneous, hypervascular enhancement. Note radiating septa converging at central calcified scar.*

Typical

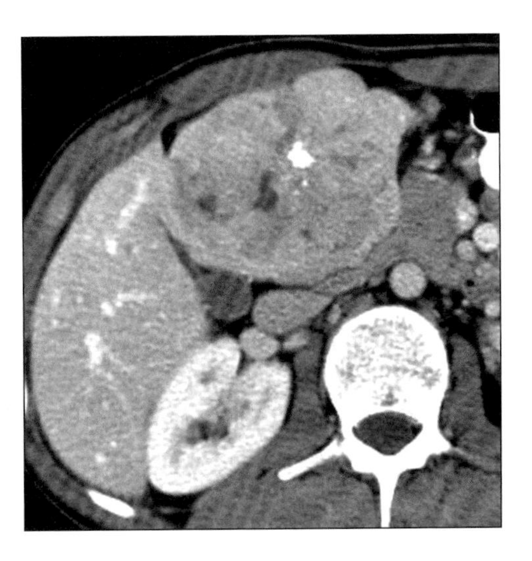

(Left) *Cut section of resected tumor shows well-demarcated, heterogeneous tumor with bile staining and central/eccentric fibrous scars (arrows).* ***(Right)*** *Axial CECT in venous phase shows large, heterogeneous, lobulated mass with calcified central scar.*

CHOLANGIOCARCINOMA (PERIPHERAL)

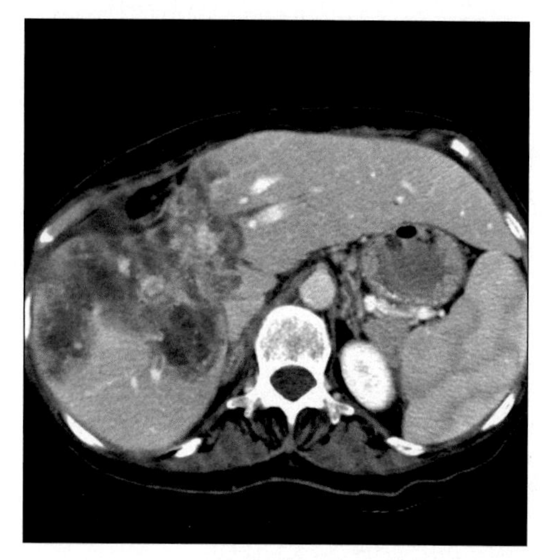

Axial CECT in portal venous phase shows heterogeneous infiltrative mass with intrahepatic biliary obstruction and volume loss with capsular retraction.

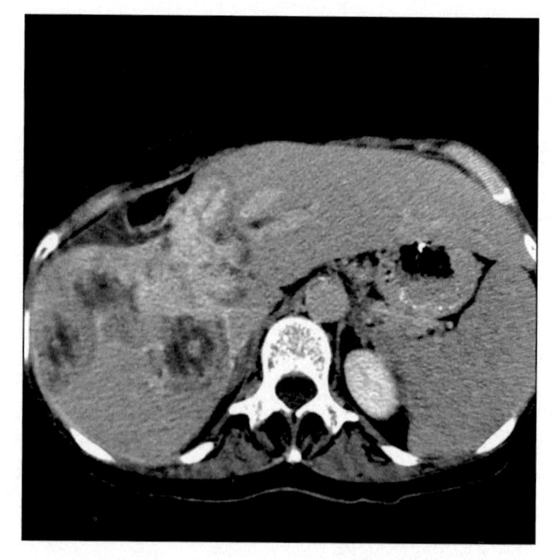

Axial CECT in delayed phase shows increased and persistent enhancement of the tumor due to its fibrous stroma.

TERMINOLOGY

Abbreviations and Synonyms

- Cholangiocellular carcinoma, intrahepatic cholangiocarcinoma, peripheral cholangiocarcinoma (PCC)

Definitions

- Cholangiocarcinoma (CC) is an adenocarcinoma that arises from bile duct epithelium
- Tumor that arises peripheral to secondary bifurcation of left or right hepatic duct is considered to be peripheral cholangiocarcinoma (PCC)

IMAGING FINDINGS

General Features

- Best diagnostic clue: Infiltrative hepatic mass with capsular retraction and delayed persistent enhancement (CECT and MR)
- Location: Originates from interlobular bile ducts (i.e., bile ducts distal to second-order branches)
- Size: Mass-forming PCC is usually large, 5-15 cm in diameter
- Key concepts
  - Mass-forming PCC: Well-circumscribed with lobulated margins
    - Multicentricity, especially around main tumor
  - Periductal-infiltrating CC: Grows along bile ducts & is elongated, spiculated, or branch-like
  - Intraductal-growing CC: Small, sessile, or polypoid, often spreading superficially along mucosal surface & resulting in multiple tumors (papillomatosis) along various segments of bile ducts

CT Findings

- NECT
  - Well-defined, single, predominantly homogeneous hypodense mass
    - With lobular margins
    - Hypodense satellite, daughter nodules (65%)
    - Punctate, stippled, chunky calcifications (18%)
  - Intrahepatic bile duct dilatation (IHBD) peripheral to tumor
    - May not be a constant finding even if tumor arises from one of the intrahepatic ducts
- CECT
  - Mass-forming PCC
    - Thin or thick, rim-like enhancement frequently seen around periphery of tumor on arterial phase images
    - Progressive, gradual & concentric filling (centripetal) on delayed phase images (usually not isodense to blood, unlike cavernous hemangioma)
    - Substantial delayed enhancement (i.e., greater than that of liver parenchyma) is common (74%)

DDx: Hepatic Mass with Biliary Obstruction

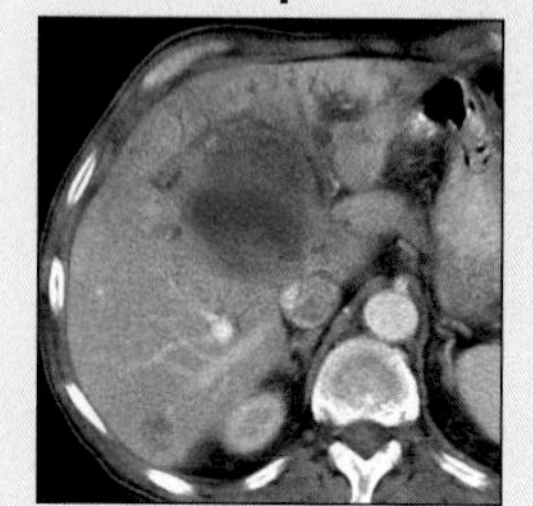

Metastases

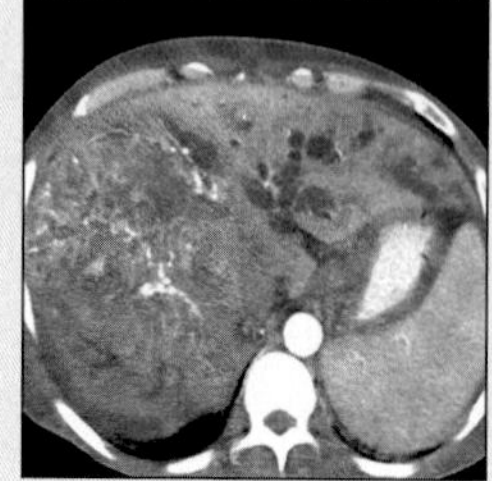

HCC

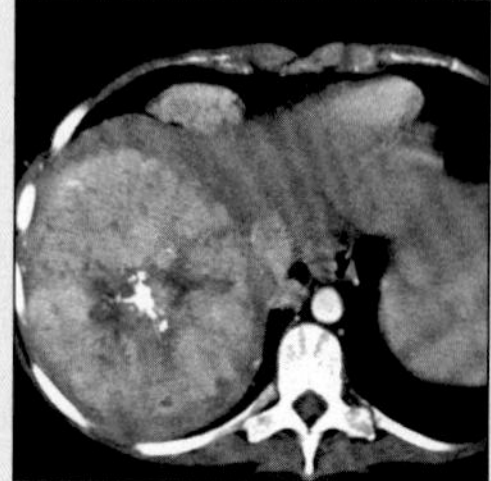

Fibrolamellar HCC

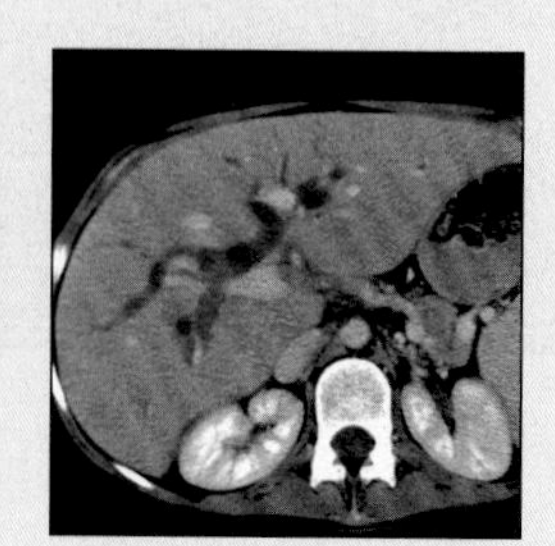

Fibrolamellar HCC

CHOLANGIOCARCINOMA (PERIPHERAL)

Key Facts

Terminology

- Cholangiocarcinoma (CC) is an adenocarcinoma that arises from bile duct epithelium
- Tumor that arises peripheral to secondary bifurcation of left or right hepatic duct is considered to be peripheral cholangiocarcinoma (PCC)

Imaging Findings

- Best diagnostic clue: Infiltrative hepatic mass with capsular retraction and delayed persistent enhancement (CECT and MR)
- Location: Originates from interlobular bile ducts (i.e., bile ducts distal to second-order branches)
- Size: Mass-forming PCC is usually large, 5-15 cm in diameter
- Best imaging tool: Helical CT/MR + MRC/ERCP

Top Differential Diagnoses

- Metastases
- Hepatocellular carcinoma (HCC)
- Fibrolamellar hepatocellular carcinoma

Pathology

- Large, firm, white tumor with dense fibrosis, irregular margins & capsular retraction

Clinical Issues

- PCC presents as large mass because tumor does not cause clinical symptoms in its early stages

Diagnostic Checklist

- Hepatocellular-cholangiocarcinoma should be considered when a hepatic tumor has features of both

---

        - Fibrotic component within contributes to delayed tumoral contrast-enhancement
        - Homogeneous hyperattenuating enhancement
        - Entire mass may be enhanced only on delayed phase images
        - Only evidence of tumor may be on delayed images; it may be missed without them
        - Delayed imaging demonstrates tumor margins more optimally
        - Tumors with delayed enhancement tend to be fibrous; however degree of contrast material retention does not always correlate with fibrous content of tumors at histopathology
        - ± Capsular retraction (frequent), with parenchymal atrophy of liver segments peripheral to tumor
    - Periductal-infiltrating CC: Bile ducts proximal to PCC are dilated & involved bile ducts are diffusely narrow or obliterated without an identifiable mass
        - Ill-defined, branch-like, low-attenuation
    - Intraductal-growing CC: Presents with focal or segmental bile duct dilatation
        - Tumor may not be depicted when it is small & isoattenuating to adjacent hepatic parenchyma or when complex orientation of dilated bile ducts obscures presence of mass

MR Findings

- T1WI: Large central heterogeneous hypointense mass
- T2WI
    - Hyperintense periphery (cellular tumor) + large central hypointensity (fibrosis)
    - Hyperintense foci in center may represent necrosis, mucin
- T1 C+
    - Central hypointense areas exhibiting homogeneous, heterogeneous, or no enhancement
        - Regions of fibrosis display enhancement, whereas those of coagulative necrosis, cell debris & mucin show no enhancement
    - Dynamic MR: Minimal or moderate rim enhancement with progressive & concentric filling with contrast material
        - Intratumoral fibrous stroma display marked or prolonged enhancement on delayed phase scans
        - Some cases of PCC exhibiting little fibrosis may show early enhancement on dynamic studies
- MRA: Displacement or encasement of adjacent vessels

Ultrasonographic Findings

- Real Time
    - Mass forming PCC: Homogeneous or heterogeneous mass with irregular borders & satellite nodules
        - Hyperechoic (75%); iso- and/or hypoechoic (14%) mass
    - IHBD of involved hepatic segment may contain calculi or intraductal mass (echogenic): Mucin is echo-free

Angiographic Findings

- Avascular, hypo- or hypervascular mass
- Stretched, encased arteries (frequent); neovascularity in 50%; venous invasion (rarely)

Nuclear Medicine Findings

- Cold lesion on sulfur colloid scans
- May show uptake on gallium scan

Other Modality Findings

- ERCP/percutaneous transhepatic cholangiography:
    - Periductal-infiltrating CC: Lumen of bile duct may be completely obstructed or stringlike, severely narrowed bile duct may be seen
    - Intraductal CC: Biliary tree is dilated (partial obstruction); diffusely, lobarly, or segmentally, or aneurysmally

Imaging Recommendations

- Best imaging tool: Helical CT/MR + MRC/ERCP
- Protocol advice: Delayed contrast-enhanced images, obtained 5-20 minutes after contrast injection; ideal timing of delayed images has not been established

CHOLANGIOCARCINOMA (PERIPHERAL)

DIFFERENTIAL DIAGNOSIS

Metastases

- Hepatic colorectal metastases: Metastatic adenocarcinoma histologically same as PCC, mimic mass-forming PCC on imaging
- Look for IHBD/bile duct disease: Clonorchiasis/sclerosing cholangitis; may indicate PCC

Hepatocellular carcinoma (HCC)

- Hypervascularity on arterial phase, hypodense on portal venous & delayed-phase CT
- Delayed enhancement within fibrous capsule
- Satellite lesions, venous invasion, IHBD, regional lymphadenopathy

Fibrolamellar hepatocellular carcinoma

- Large, lobulated, heterogeneous mass with central scar
- Delayed partial enhancement of fibrous scar & septa
- Calcification (scar), lymphadenopathy are common

PATHOLOGY

General Features

- Etiology: Associated with several etiological factors: Primary sclerosing cholangitis, bile stasis, repeated cholangitis, clonorchiasis, congenital cystic disease of liver, hepatolithiasis, Thorotrast deposition
- Epidemiology
  - 8-13% of all CC are peripheral/intrahepatic
  - PCC is relatively rare cancer; world wide it accounts for an estimated 15% of liver cancer

Gross Pathologic & Surgical Features

- Large, firm, white tumor with dense fibrosis, irregular margins & capsular retraction

Microscopic Features

- 90% are adenocarcinomas
- Tendency to spread between hepatocyte plates, along duct walls, & adjacent to nerves
- Mucin production is often abundant
- Most of mass-forming PCC are poorly-differentiated; most periductal-infiltrating are well-differentiated; most intraductal are papillary adenocarcinomas

Staging, Grading or Classification Criteria

- Based on growth characteristics: Mass-forming (exophytic/nodular); periductal-infiltrating (sclerosing); intraductal-growing (polypoid/papillary)

CLINICAL ISSUES

Presentation

- Most common signs/symptoms
  - Abdominal pain (84%), weight loss (77%), painless jaundice (28%), palpable mass (18%), fatigue
  - PCC presents as large mass because tumor does not cause clinical symptoms in its early stages
- Clinical profile
  - Lab data: Moderate anemia, leucocytosis, mild ↑ AST & ALT, ↑ carcinoembryonic antigen
  - Diagnosis: Biopsy; delayed enhancement can be helpful as target for CT-guided biopsy

Demographics

- Age: 50-60 years, rarely occurring in younger than 40

Natural History & Prognosis

- Tumoral spread
  - Local extension along duct
  - Local infiltration of liver substance
  - Metastases to regional lymph nodes
- Vascular or lymphatic invasion
- Perineural invasion
- Prognosis: Poor; < 20% resectable; 30% 5 year survival

Treatment

- Surgical resection remains primary treatment
- Palliative: Biliary catheter drainage, biliary stenting
- Adjuvant: Radiation & chemotherapy
- Liver transplantation (not considered appropriate in most cases, high recurrence)

DIAGNOSTIC CHECKLIST

Consider

- Hepatocellular-cholangiocarcinoma should be considered when a hepatic tumor has features of both
- Delayed tumoral contrast-enhancement is typical feature of PCC & may aid in detection & characterization, however, delayed images must be interpreted in conjunction with clinical information, as well as unenhanced & dynamic contrast-enhanced images, as specificity of these findings is uncertain

Image Interpretation Pearls

- In suspected cholangiocarcinoma (e.g., history of PSC, or mass with capsular retraction, biliary obstruction) obtain delayed enhanced scans

SELECTED REFERENCES

1. Ebied O et al: Hepatocellular-cholangiocarcinoma: helical computed tomography findings in 30 patients. J Comput Assist Tomogr. 27(2):117-24, 2003
2. Lim JH: Cholangiocarcinoma: morphologic classification according to growth pattern and imaging findings. AJR Am J Roentgenol. 181(3):819-27, 2003
3. Maetani Y et al: MR imaging of intrahepatic cholangiocarcinoma with pathologic correlation. AJR Am J Roentgenol. 176(6):1499-507, 2001
4. Lacomis JM et al: Cholangiocarcinoma: delayed CT contrast enhancement patterns. Radiology. 203(1):98-104, 1997
5. Kim TK et al: Peripheral cholangiocarcinoma of the liver: two-phase spiral CT findings. Radiology. 204(2):539-43, 1997
6. Soyer P et al: Imaging of intrahepatic cholangiocarcinoma: 1. Peripheral cholangiocarcinoma. AJR Am J Roentgenol. 165(6):1427-31, 1995

CHOLANGIOCARCINOMA (PERIPHERAL)

IMAGE GALLERY

Typical

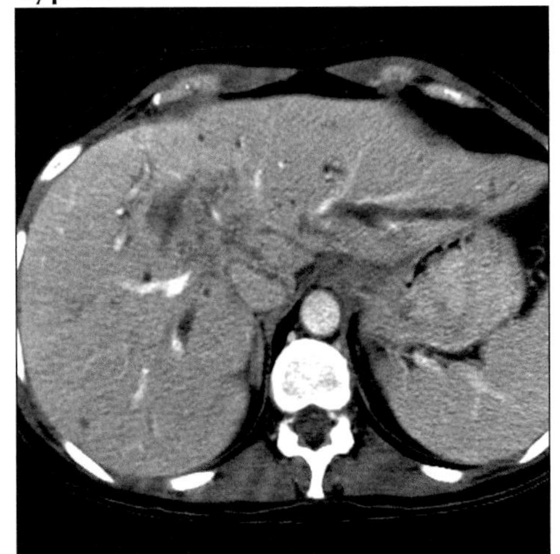

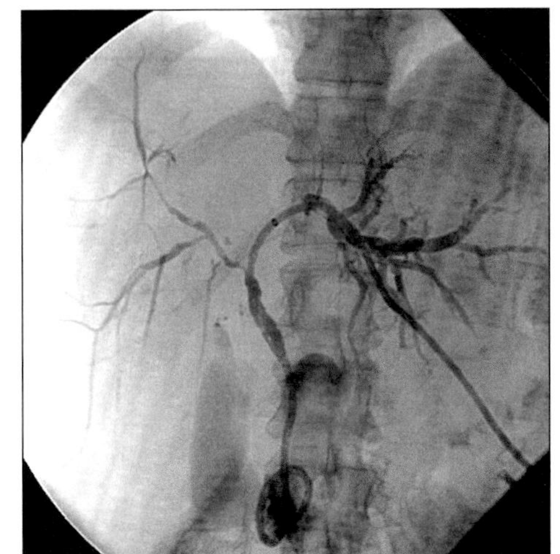

(Left) *Axial CECT shows a heterogeneous infiltrative mass causing intrahepatic biliary obstruction.* ***(Right)*** *Catheter cholangiogram shows long segmental stenosis of left main bile duct with occlusion of multiple side branches and dilatation of peripheral biliary ducts from intrahepatic cholangiocarcinoma.*

Typical

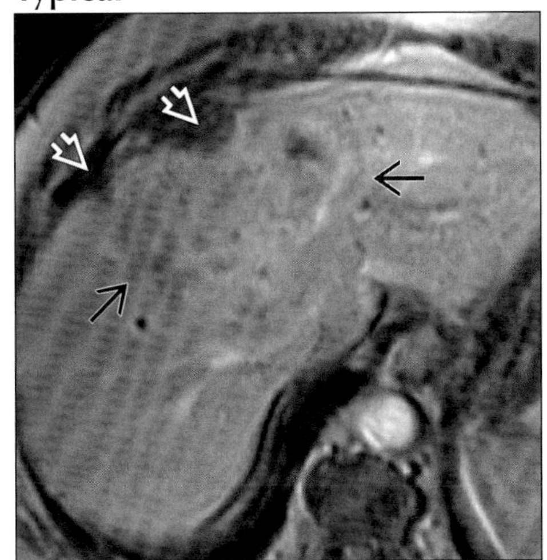

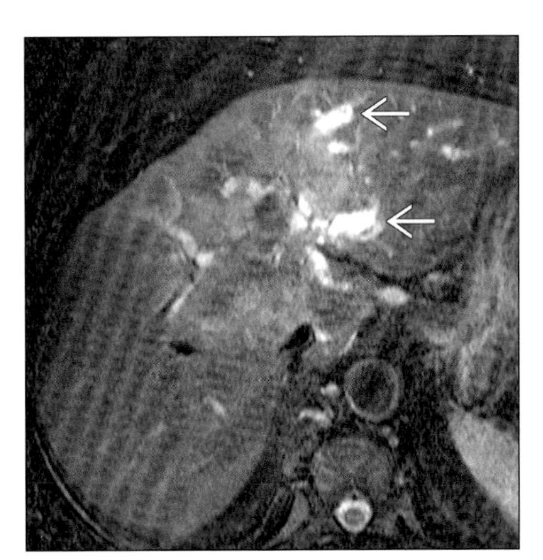

(Left) *Axial T1 C+ MR shows large but subtle heterogeneous mass (arrows). Capsular retraction (open arrows).* ***(Right)*** *Axial T2WI MR shows mild hyperintensity within hepatic mass. Dilated intrahepatic ducts (arrows).*

Typical

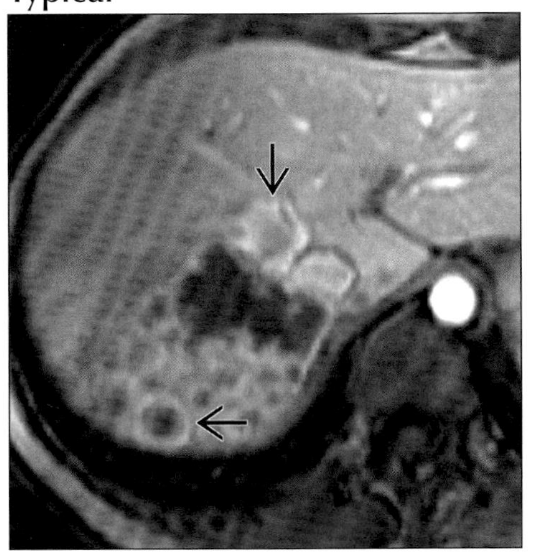

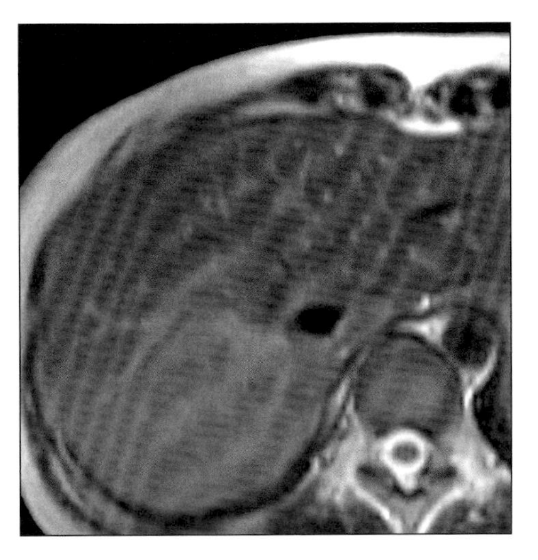

(Left) *Axial T1 C+ GRE MR shows large hepatic mass with ring enhancing components (arrows) along with low intensity, nonenhancing (fibrotic/necrotic) areas.* ***(Right)*** *Axial T2WI MR scarcely shows right hepatic lobe mass, although absence of normal ducts and vessels within mass is a clue.*

EPITHELIOID HEMANGIOENDOTHELIOMA

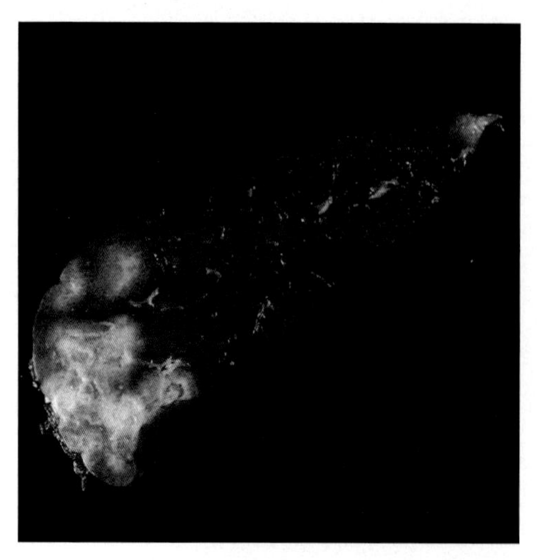

Cut section of explanted liver shows multifocal confluent masses with extensive fibrous stroma, causing volume loss and capsular retraction in right lobe.

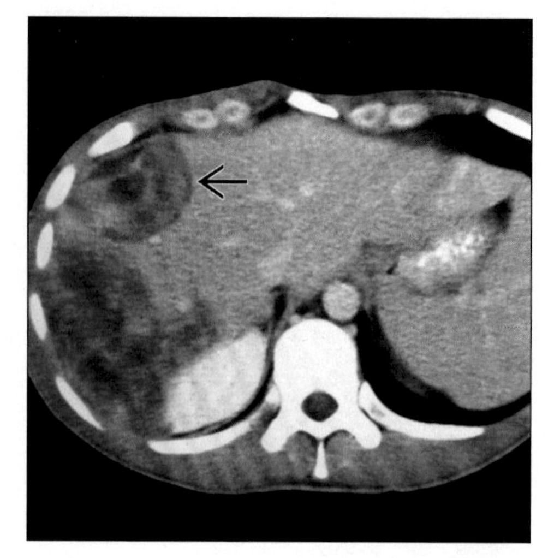

Axial CECT in venous phase shows multifocal confluent masses, "target" enhancement (arrow) and capsular retraction.

TERMINOLOGY

Abbreviations and Synonyms

- Epithelioid hemangioendothelioma (EHE); hepatic epithelioid hemangioendothelioma

Definitions

- Primary malignant tumor of liver arising from vascular elements of mesenchymal tissue

IMAGING FINDINGS

General Features

- Best diagnostic clue: Coalescent peripheral hepatic nodules with target appearance & capsular retraction
- Location
  - Liver: Periphery (more than 75% of lesions) with extension to capsule
  - Locations other than liver
    - Soft tissues, bone & lung
    - Lung: Diagnosed as "intravascular bronchioalveolar tumor"
- Size: Varies from small tumor nodules to large confluent masses
- Key concepts
  - Rare primary malignant (low grade) vascular tumor of liver in adults
  - Other primary malignant vascular tumors of liver
    - Angiosarcoma (2% of all primary malignant liver tumors)
    - Kaposi sarcoma: Metastatic vascular tumor in AIDS and transplant patients
  - All hepatic malignant vascular tumors share
    - Histologic characteristics
    - Grow around & into vessels
    - Multifocal
  - Clinical course
    - Between benign cavernous hemangiomas & malignant angiosarcomas
    - Variable & unpredictable
    - Identical to that of extrahepatic EHE
    - Metastatic in 40% cases (spleen, mesentery, lymph nodes, lung, bone)

CT Findings

- Spectrum of growth in lesions may be seen
  - Nodular form: Multiple nodules (more common)
    - Multiple liver nodules coalesce to form large confluent masses
  - Diffuse or extensive form (very rare)
- Usually located at periphery with extension to capsule
- Typical "capsular retraction" (due to tumor fibrosis & ischemia) or flattening
- Occasionally calcification within tumor
- Compensatory hypertrophy
  - Uninvolved liver (predominantly left lobe)

DDx: Hepatic Mass with Capsular Retraction

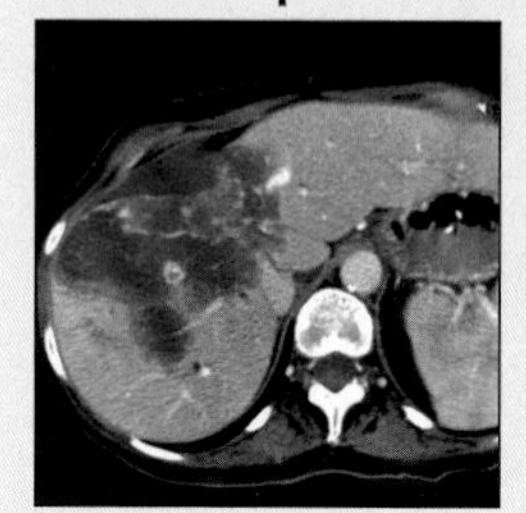

Cholangiocarcinoma

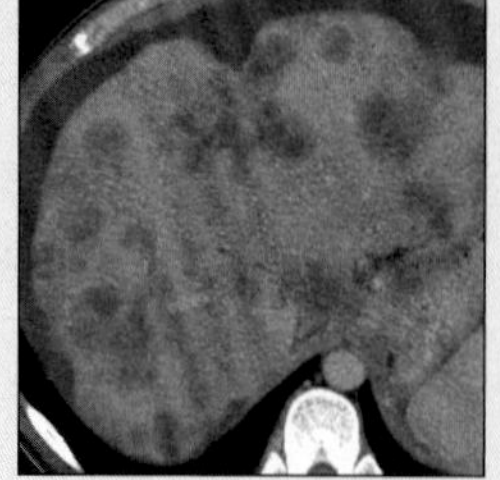

Treated Metastases

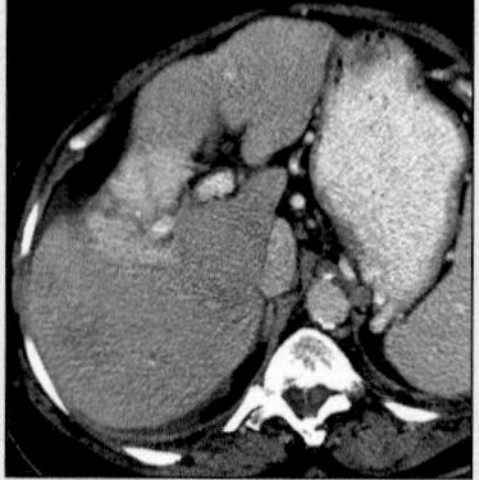

Confluent Fibrosis

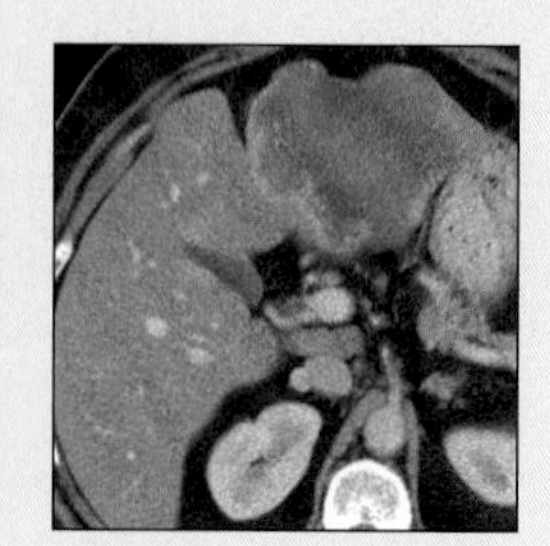

Hepatic Hemangioma

EPITHELIOID HEMANGIOENDOTHELIOMA

Key Facts

Terminology

- Epithelioid hemangioendothelioma (EHE); hepatic epithelioid hemangioendothelioma
- Primary malignant tumor of liver arising from vascular elements of mesenchymal tissue

Imaging Findings

- Best diagnostic clue: Coalescent peripheral hepatic nodules with target appearance & capsular retraction
- Size: Varies from small tumor nodules to large confluent masses

Top Differential Diagnoses

- Cholangiocarcinoma (peripheral)
- Treated malignancy (HCC or metastases)
- Focal confluent fibrosis
- Hemangioma (especially in cirrhotic liver)

Pathology

- Slowly progressing low-grade malignant vascular tumor of liver
- Must not be confused with infantile hemangioendothelioma
- Exact etiology: Unknown

Clinical Issues

- Abdominal pain, jaundice, hepatosplenomegaly
- Gender: Females more than males

Diagnostic Checklist

- Rule out other hepatic lesions that typically cause "capsular retraction"
- Differentiated from other lesions by tumor cells staining positive for factor VIII-related antigen

- May have metastatic lesions & ascites
- NECT
  - Tumor nodules
    - Foci of homogeneous decreased attenuation (due to myxoid stroma) compared to normal liver parenchyma
  - Conspicuity & extent of lesions
    - Superior on NECT than CECT
- CECT
  - "Target" like enhancement pattern of tumor
    - Nonenhancing central part of tumor (myxoid & hyalinized stroma)
    - Enhancing (hyperemic) peripheral inner rim (increased vascularity)
    - Nonenhancing peripheral outer rim or halo (avascular rim)

MR Findings

- T1WI
  - Hypointense centrally
  - Peripheral thin hypointense rim
- T2WI
  - Hyperintense centrally
  - Peripheral thin hypointense rim
- T1 C+
  - Target pattern: Three concentric layers of alternating signal intensity (analogous to CECT)
    - Central: Hypointense
    - Peripheral: Thick enhancing inner rim & thin nonenhancing outer rim

Ultrasonographic Findings

- Real Time
  - Tumor nodules show varied echogenicity pattern
    - Predominantly: Hypoechoic
    - Occasionally: Hyperechoic or iso-/hypoechoic lesions relative to liver
    - Hyper-/isoechoic lesions may have peripheral hypoechoic rims

Angiographic Findings

- Conventional
  - Hypervascular, hypovascular or avascular lesions
    - Based on degree of sclerosis & hyalinization
  - Invasion or occlusion of intrahepatic portal & hepatic veins

Imaging Recommendations

- Best imaging tool: Helical NE + CECT, or MR + CEMR
- Protocol advice: Multiphasic CT or MR

DIFFERENTIAL DIAGNOSIS

Cholangiocarcinoma (peripheral)

- Heterogeneous mass with capsular retraction
- Satellite lesions may be seen
- Often invades or obstructs vessels & bile ducts
- Intrahepatic bile duct dilatation

Treated malignancy (HCC or metastases)

- Capsular retraction
- Heterogeneous enhancement pattern
- History of ablation or chemotherapy for liver tumor
- Treated metastatic nodules may show
  - Cystic or necrotic changes
    - Debris, mural nodularity
    - Thick septa & wall enhancement

Focal confluent fibrosis

- Common in advanced cirrhosis
- NECT: Areas of lower attenuation than adjacent liver
- CECT: Isoattenuating or minimally hypo-/hyperattenuating
- MR: Hypointense on T1WI; hyperintense on T2WI
- T1C+: Isointense or delayed enhancement
- Associated volume loss seen
  - Capsular retraction adjacent to lesion
  - Segmental or lobar shrinkage
- Shape & location: Usually wedge shaped lesions radiating from porta affecting anterior & medial segments

Hemangioma (especially in cirrhotic liver)

- Capsular retraction seen in large lesion with scar; hyalinization
- Decrease in size over time as cirrhosis progresses
- Rest of liver shows cirrhotic changes

EPITHELIOID HEMANGIOENDOTHELIOMA

PATHOLOGY

General Features

- General path comments
  - Slowly progressing low-grade malignant vascular tumor of liver
    - Histologically: Composed of epithelioid-appearing endothelial cells
    - Abundant matrix of myxoid & fibrous stroma
    - Positive factor VIII-associated antigen staining
    - Malignant cells infiltrate into hepatic sinusoids
    - Compress surrounding hepatocytes
    - Capsular retraction
    - Invade hepatic & portal veins
    - Infarction of tumor & central fibrosis
  - Must not be confused with infantile hemangioendothelioma
    - Histologically: Benign primary vascular liver tumor
    - Seen in infants & young children
    - Resolves spontaneously in many cases
- Etiology
  - Exact etiology: Unknown
  - Possibly associated with oral contraceptives or exposure to vinyl chloride
- Epidemiology
  - Rare vascular tumor of liver
    - Exact incidence is not known

Gross Pathologic & Surgical Features

- Multiple solid nodules
  - Tan, white, firm, varied size
  - Coalesce more peripherally
- Tumor nodules with hyperemic rim
- Lesions close to capsule cause retraction

Microscopic Features

- Dendritic spindle-shaped or epithelioid cells
- Matrix: Myxoid or fibrous stroma
- Epithelioid cells
  - Stain positive for factor VIII-related antigen

CLINICAL ISSUES

Presentation

- Most common signs/symptoms
  - Abdominal pain, jaundice, hepatosplenomegaly
  - Occasionally asymptomatic
  - Rarely hemoperitoneum & Budd-Chiari syndrome
    - Due to hepatic vein invasion
- Clinical profile: Middle aged patient with history of RUQ pain, hepatomegaly & tumor cells stained positive for factor VIII-related antigen
- Lab data
  - Liver enzymes mildly increased
  - α-Fetoprotein & CEA levels: Normal
- Diagnosis
  - Tumor cells staining positive for factor VIII-related antigen

Demographics

- Age: 25-58 yrs (average age 45 yrs)
- Gender: Females more than males

Natural History & Prognosis

- Complications
  - Rupture & hemoperitoneum
  - Budd-Chiari syndrome
  - Liver failure
- Prognosis
  - Most patients survive 5-10 years after diagnosis
  - 20% die within first 2 years after diagnosis
  - 20% survive for 5-28 years
    - With or without treatment

Treatment

- Radical resection or liver transplantation

DIAGNOSTIC CHECKLIST

Consider

- Rule out other hepatic lesions that typically cause "capsular retraction"
- Differentiated from other lesions by tumor cells staining positive for factor VIII-related antigen

Image Interpretation Pearls

- Usually located at periphery with extension to capsule
- Typical "capsular retraction" of peripheral tumor (due to fibrosis & ischemia)
- Target appearance on CECT or MR

SELECTED REFERENCES

1. Uchimura K et al: Hepatic epithelioid hemangioendothelioma. J Clin Gastroenterol. 32(5):431-4, 2001
2. Kehagias DT et al: Hepatic epithelioid hemangioendothelioma: MR imaging findings. Hepatogastroenterology. 47(36):1711-3, 2000
3. Lauffer JM et al: Epithelioid hemangioendothelioma of the liver. A rare hepatic tumor. Cancer. 78(11):2318-27, 1996
4. Buetow PC et al: Malignant vascular tumors of the liver: radiologic-pathologic correlation. Radiographics. 14(1):153-66; quiz 167-8, 1994
5. Furuta K et al: Epithelioid hemangioendothelioma of the liver diagnosed by liver biopsy under laparoscopy. Am J Gastroenterol. 87(6):797-800, 1992
6. Miller WJ et al: Epithelioid hemangioendothelioma of the liver: imaging findings with pathologic correlation. AJR Am J Roentgenol. 159(1):53-7, 1992
7. Van Beers B et al: Epithelioid hemangioendothelioma of the liver: MR and CT findings. J Comput Assist Tomogr. 16(3):420-4, 1992
8. Furui S et al: Hepatic epithelioid hemangioendothelioma: report of five cases. Radiology. 171(1):63-8, 1989
9. Radin DR et al: Hepatic epithelioid hemangioendothelioma. Radiology. 169(1):145-8, 1988
10. Scoazec JY et al: Epithelioid hemangioendothelioma of the liver. Diagnostic features and role of liver transplantation. Gastroenterology. 94(6):1447-53, 1988
11. Marino IR et al: Treatment of hepatic epithelioid hemangioendothelioma with liver transplantation. Cancer. 62(10):2079-84, 1988

IMAGE GALLERY

Typical

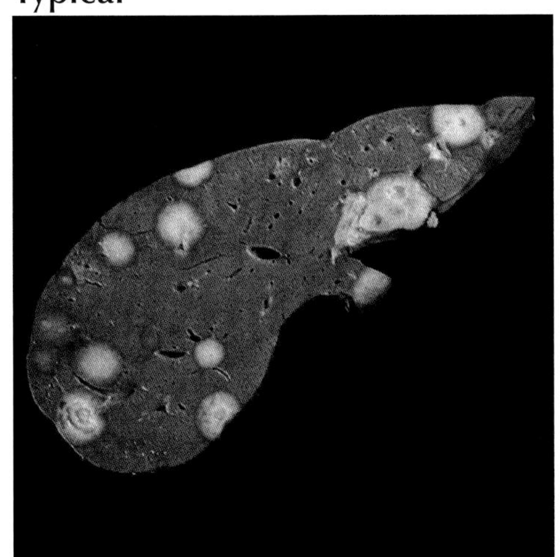

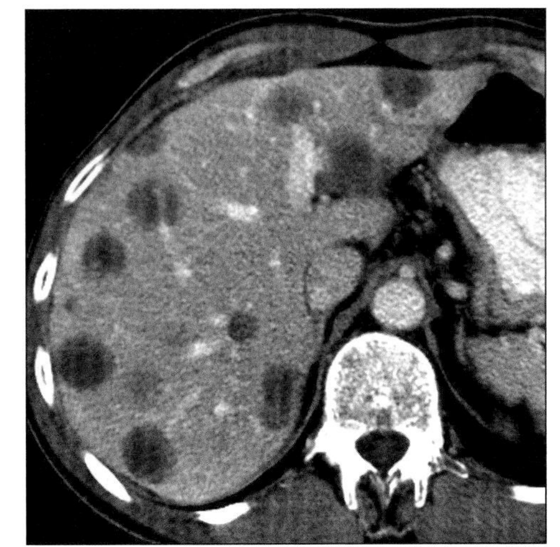

(Left) *Cut section of explanted liver shows multifocal tumor nodule with extensive fibrous stroma, target appearance and capsular retraction.* ***(Right)*** *Axial CECT in venous phase shows multifocal tumor nodules, some with target appearance and capsular retraction.*

Typical

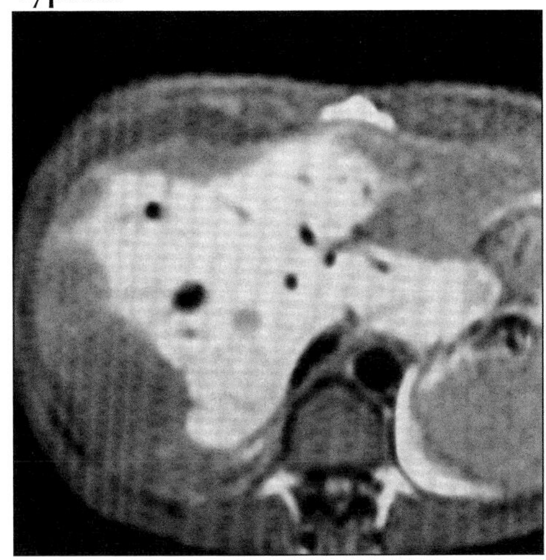

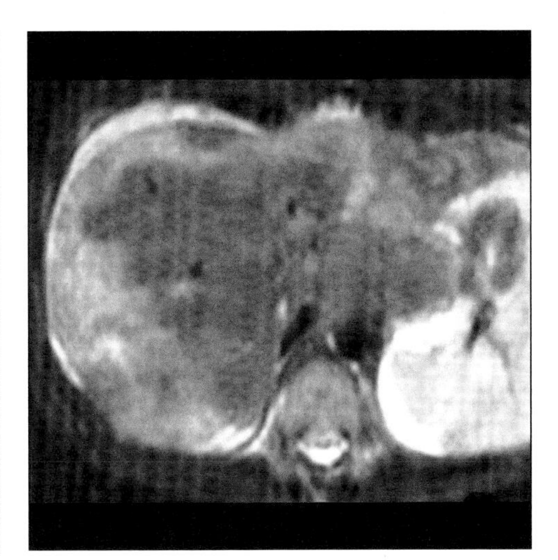

(Left) *Axial T1WI MR shows extensive confluent hypointense tumor in peripheral liver causing volume loss. Compensatory hypertrophy of uninvolved liver, including caudate lobe.* ***(Right)*** *Axial T2WI MR shows heterogeneous, hyperintense confluent liver tumor.*

Typical

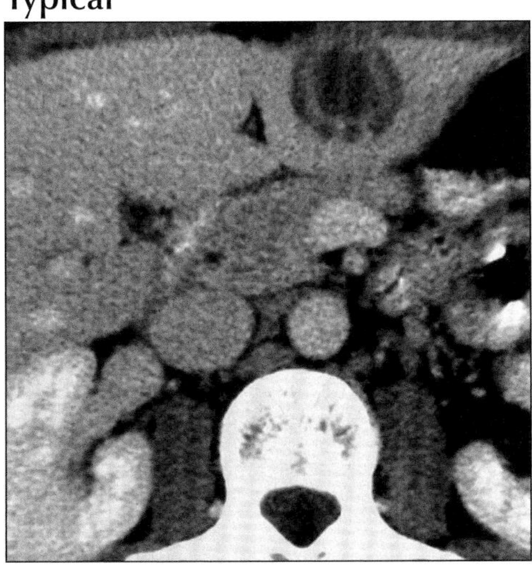

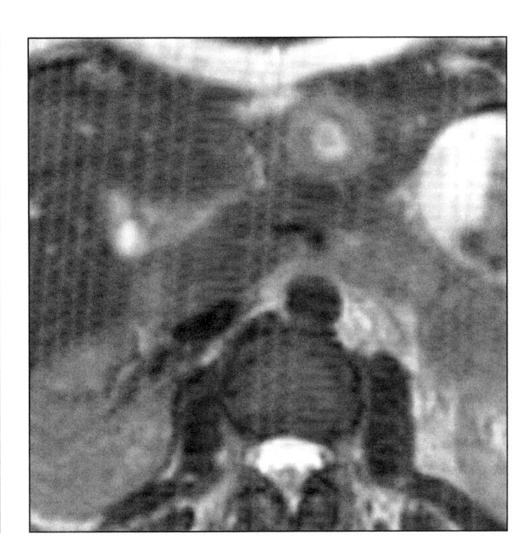

(Left) *Axial CECT in venous phase shows classic target appearance of EHE, with hypodense center, inner rim of hypervascular enhancement, and outer rim of avascular hypodensity.* ***(Right)*** *Axial T2WI MR shows target appearance of EHE with central hyperintensity, inner rim of hypointensity and outer rim of hyperintensity.*

BILIARY CYSTADENOCARCINOMA

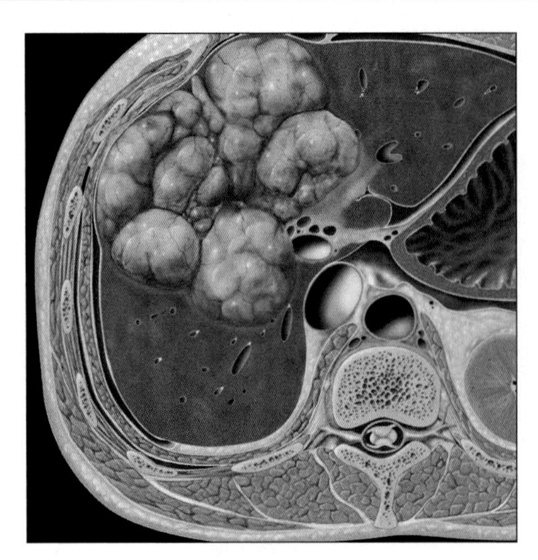

Graphic shows lobulated complex cystic mass with vascularized wall and septa.

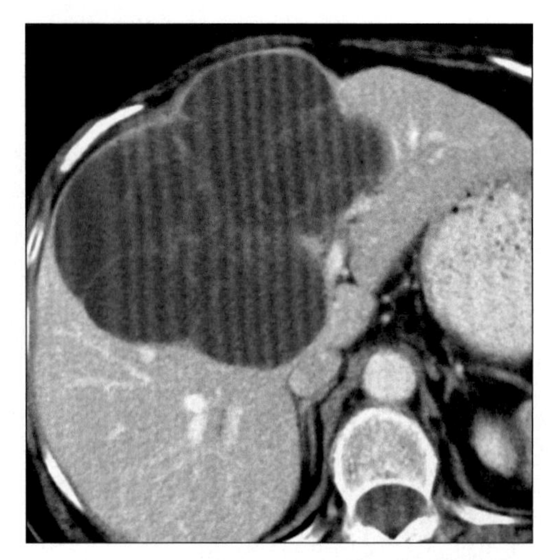

Axial CECT shows complex cystic mass with lobulated margins, enhancing wall and septa.

TERMINOLOGY

Abbreviations and Synonyms

- Bile duct cystadenocarcinoma, cystadenoma

Definitions

- Rare malignant or premalignant, unilocular or multilocular cystic tumor that may arise from IHBD within liver (common site) & very rarely from extrahepatic biliary tree or gallbladder

IMAGING FINDINGS

General Features

- Best diagnostic clue: Complex multiloculated cystic mass in liver with septations & mural calcifications
- Location
  - Right lobe (55%), left lobe (29%), both lobes (16%)
  - Intrahepatic biliary ducts (IHBD): 83%
  - Extrahepatic bile ducts: 13%
  - Gallbladder: 0.02%
- Size: Varies from 1.5-25 cm in diameter
- Key concepts
  - Biliary cystadenocarcinoma
    - Malignant transformation of benign biliary cystadenoma
    - Typically solitary tumor, usually multilocular, but sometimes unilocular
    - Tumor is well-encapsulated
    - Usually seen in middle-aged women
    - May recur after excision
  - Benign biliary cystadenoma
    - Probably congenital in origin due to presence of aberrant bile ducts
    - May recur after excision
    - Malignant potential to develop into cystadenocarcinoma even after years of stability
  - Benign & malignant lesions together account for only 5% of all intrahepatic lesions of bile duct origin
  - " Microcystic" cystadenoma variant
    - Composed of multiple small cysts
    - Glycogen rich cystadenoma
    - Typical papillary & mesenchymal stromal features are not seen
    - Lined by a single layer of cuboidal epithelial cells
    - Resembles serous microcystic adenoma of pancreas in pathology & on imaging

CT Findings

- NECT
  - Large, well-defined, homogeneous, hypodense, water density mass
  - Large, well-defined, heterogeneous mass (cystic & hemorrhagic areas)
  - Cystadenocarcinoma: Septations & nodularity

DDx: Complex Cystic Mass

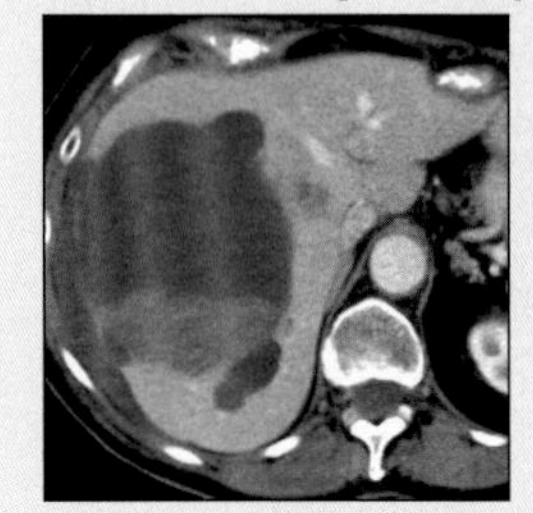

Hemorrhagic Cyst

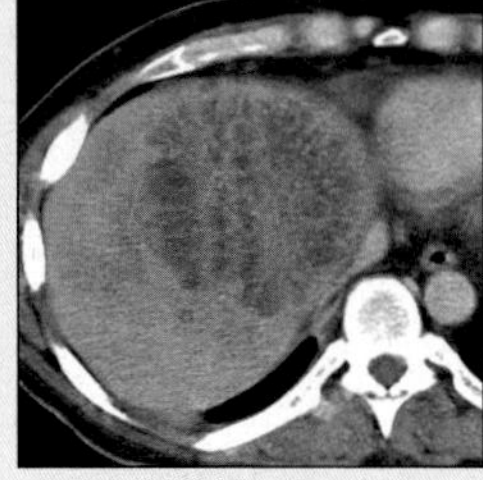

Pyogenic Abscess

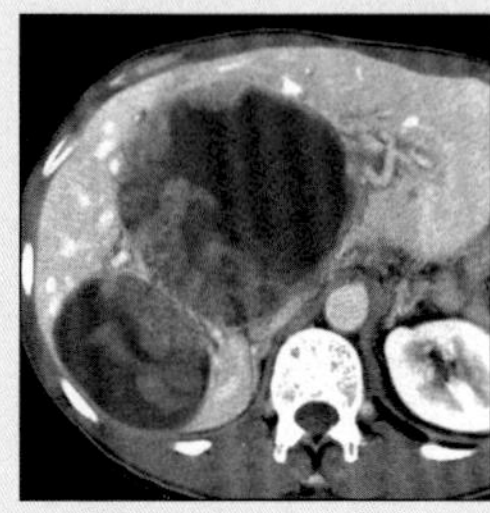

Cystic Metastases

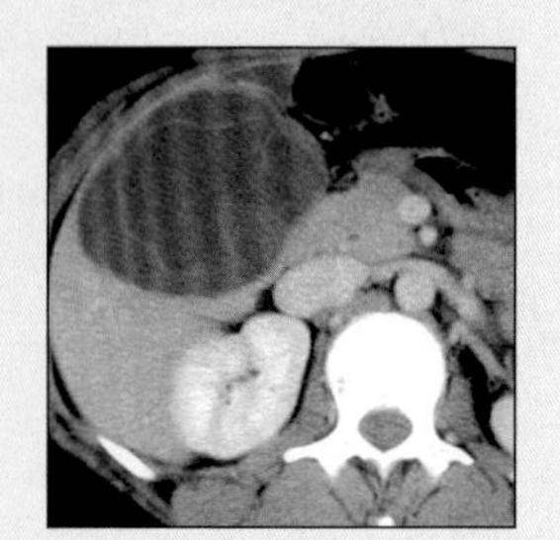

Hydatid Cyst

BILIARY CYSTADENOCARCINOMA

Key Facts

Terminology
- Bile duct cystadenocarcinoma, cystadenoma
- Rare malignant or premalignant, unilocular or multilocular cystic tumor that may arise from IHBD within liver (common site) & very rarely from extrahepatic biliary tree or gallbladder

Imaging Findings
- Best diagnostic clue: Complex multiloculated cystic mass in liver with septations & mural calcifications
- Size: Varies from 1.5-25 cm in diameter

Top Differential Diagnoses
- Hemorrhagic or infected hepatic cyst
- Hepatic pyogenic abscess
- Cystic metastases
- Hydatid cyst

Pathology
- Varying degrees of mural & septal nodularity or thickening seen
- Papillary excrescences project into cystic spaces
- Cystadenocarcinomas: Thick, coarse, mural & septal calcifications
- Cystadenomas: Fine septal calcifications
- Malignant transformation of benign biliary cystadenoma by invasion of capsule

Diagnostic Checklist
- Rule out other "complex cystic" masses of liver
- Large well-defined homogeneous or heterogeneous "complex cystic" mass with septations & nodularity
- May mimic hemorrhagic or infected hepatic cyst

  - Cystadenoma: Septations without nodularity
  - Fine mural or septal calcifications
  - Biliary dilatation (due to pressure effect)
- CECT
  - Multilocular tumor
    - Nonenhancing cystic spaces (decreased attenuation)
    - Enhancement of internal septa, capsule & nodules
    - Enhancement of mural & septal nodules
    - Enhancement of papillary excrescences
    - Fine mural or septal calcifications
    - Less commonly, honeycomb or sponge appearance (microcystic variant)
    - May or may not show metastases or adenopathy
  - Unilocular tumor
    - Large or small nonenhancing cystic space
    - Enhancement of outer capsule & papillary excrescences
    - Fine mural calcifications

MR Findings
- T1WI
  - Variable signal intensity locules depending on content of cystic fluid
  - Increased signal intensity (mucoid fluid)
  - Decreased signal intensity (serous fluid)
  - Septal or mural calcifications: Hypointense
- T2WI
  - Decreased signal intensity (mucoid fluid)
  - Increased signal intensity (serous fluid)
  - Septations are well-delineated
  - Septal or mural calcifications: Hypointense
- T1 C+: Enhancement of capsule & septa

Ultrasonographic Findings
- Real Time
  - Large, well-defined, multiloculated, anechoic mass
  - Highly echogenic septations
  - Tumor nodules or papillary growths
  - Mural or septal calcifications or fluid levels
  - Complex fluid: Areas of anechoic + internal echoes (cystic + hemorrhagic)

Angiographic Findings
- Conventional
  - Avascular mass with small clusters of peripheral abnormal vessels
  - Stretching & displacement of vessels

Imaging Recommendations
- NE + CECT or MR + CEMR

DIFFERENTIAL DIAGNOSIS

Hemorrhagic or infected hepatic cyst
- Complex heterogeneous cystic mass
- Multiple thick or thin septations
- May show mural nodularity & fluid-level
- Calcification may or may not be seen
- No enhancement on CECT

Hepatic pyogenic abscess
- Simple pyogenic abscess
  - Well-defined, round, hypodense mass (0-45 HU)
- "Cluster" sign: Small abscesses aggregate, sometimes coalesce into a single big septated cavity
- Complex pyogenic abscess: "Target" lesion
  - Hypodense rim
  - Isodense periphery & decreased HU in center
- Often associated with diaphragmatic elevation, atelectasis & right side pleural effusion

Cystic metastases
- Usually from ovarian cystadenocarcinoma & metastatic sarcoma
- Show debris & mural nodularity
- May have thick septa & wall enhancement

Hydatid cyst
- Large well-defined cystic liver mass
- Often has numerous peripheral daughter cysts or scolices of different density or intensity
- May show curvilinear or ring-like pericyst calcification
- Occasionally dilated intrahepatic bile ducts
  - Due to pressure effect or rupture into ducts

BILIARY CYSTADENOCARCINOMA

PATHOLOGY

General Features

- General path comments
  - Varying degrees of mural & septal nodularity or thickening seen
  - Papillary excrescences project into cystic spaces
  - Cystic cavities are filled with mucinous or serous or necrotic or blood content
  - Cystadenocarcinomas: Thick, coarse, mural & septal calcifications
  - Cystadenomas: Fine septal calcifications
- Etiology
  - Malignant transformation of benign biliary cystadenoma by invasion of capsule
  - Biliary cystadenoma
    - Probably derived from ectopic nests of primitive biliary tissue
- Epidemiology
  - Very rare malignant biliary tumor
  - Incidence: 5% of all intrahepatic cystic masses of biliary origin

Gross Pathologic & Surgical Features

- Multiloculated cystic tumor with well-defined thick capsule containing
  - Serous, mucinous, bilious, hemorrhagic or mixed fluid
- Surface is shiny, smooth or bosselated
- Polypoid excrescences & septations may be seen

Microscopic Features

- Single layer of cuboidal or tall columnar biliary type epithelium with papillary projections
- Malignant epithelial cells line the cysts
- Subepithelial stroma resembles that of ovary
- Usually mucinous, but serous type is also seen
- Goblet cells, Paneth cells & argyrophilic endocrine cells may be seen

CLINICAL ISSUES

Presentation

- Most common signs/symptoms
  - Abdominal pain, obstructive jaundice, nausea, vomiting
  - Abdominal swelling with palpable mass (90%)
- Diagnosis: Fine needle aspiration & cytology

Demographics

- Age: Peak incidence in 5th decade
- Gender
  - Usually occur in middle aged women
  - M:F = 1:4
- Ethnicity: Predominantly seen in Caucasians

Natural History & Prognosis

- Complications
  - Rupture into peritoneum or retroperitoneum
  - Recurrence common
- Prognosis
  - Tumors with ovarian stroma found in women have an indolent course & good prognosis
  - Tumors without ovarian stroma found in both sexes have an aggressive clinical course & poor prognosis

Treatment

- Surgical resection

DIAGNOSTIC CHECKLIST

Consider

- Rule out other "complex cystic" masses of liver

Image Interpretation Pearls

- Large well-defined homogeneous or heterogeneous "complex cystic" mass with septations & nodularity
- May mimic hemorrhagic or infected hepatic cyst

SELECTED REFERENCES

1. Levy AD et al: Benign tumors and tumorlike lesions of the gallbladder and extrahepatic bile ducts: Radiologic-pathologic correlation. RadioGraphics. 22: 387-413, 2002
2. Mortele KF et al: Cystic focal liver lesions in the adult: Differential CT and MR imaging features. Radiographics. 21: 895-910, 2001
3. Hwang IK et al: Huge biliary cystadenoma mimicking cholecystic lymphangioma in subhepatic space. J Comput Assist Tomogr. 24(4):652-4, 2000
4. Gabata T et al: Biliary cystadenoma with mesenchymal stroma of the liver: correlation between unusual MR appearance and pathologic findings. J Magn Reson Imaging. 8(2):503-4, 1998
5. Singh Y et al: Multiloculated cystic liver lesions: Radiologic-pathologic differential diagnosis. RadioGraphics. 17: 219-24, 1997
6. Buetow PC et al: Biliary cystadenoma and cystadenocarcinoma: clinical-imaging-pathologic correlations with emphasis on the importance of ovarian stroma. Radiology. 196(3):805-10, 1995
7. Devaney K et al: Hepatobiliary cystadenoma and cystadenocarcinoma. A light microscopic and immunohistochemical study of 70 patients. Am J Surg Pathol. 18(11):1078-91, 1994
8. Wang YJ et al: Primary biliary cystic tumors of the liver. Am J Gastroenterol. 88(4):599-603, 1993
9. Agildere AM et al: Biliary cystadenoma and cystadenocarcinoma. AJR Am J Roentgenol. 156(5):1113, 1991
10. Korobkin M et al: Biliary cystadenoma and cystadenocarcinoma: CT and sonographic findings. AJR Am J Roentgenol. 153(3):507-11, 1989
11. Choi BI et al: Biliary cystadenoma and cystadenocarcinoma: CT and sonographic findings. Radiology. 171(1):57-61, 1989
12. Genkins SM et al: Biliary cystadenoma with mesenchymal stroma: CT and angiographic appearance. J Comput Assist Tomogr. 12(3):527-9, 1988
13. Kokubo T et al: Mucin-hypersecreting intrahepatic biliary neoplasms. Radiology. 168(3):609-14, 1988
14. Forrest ME et al: Biliary cystadenomas: sonographic-angiographic-pathologic correlations. AJR Am J Roentgenol. 135(4):723-7, 1980

BILIARY CYSTADENOCARCINOMA

IMAGE GALLERY

Typical

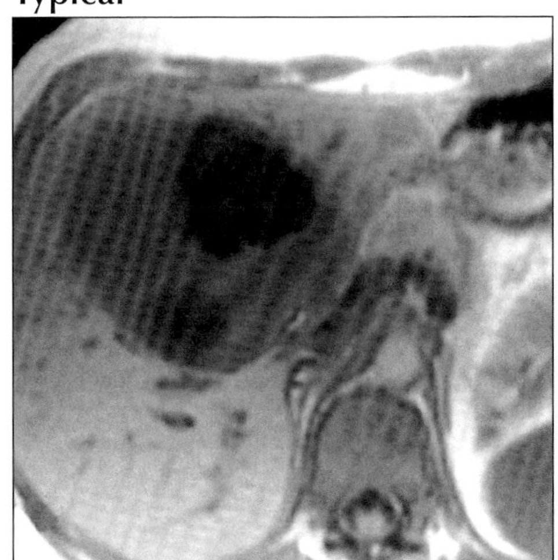

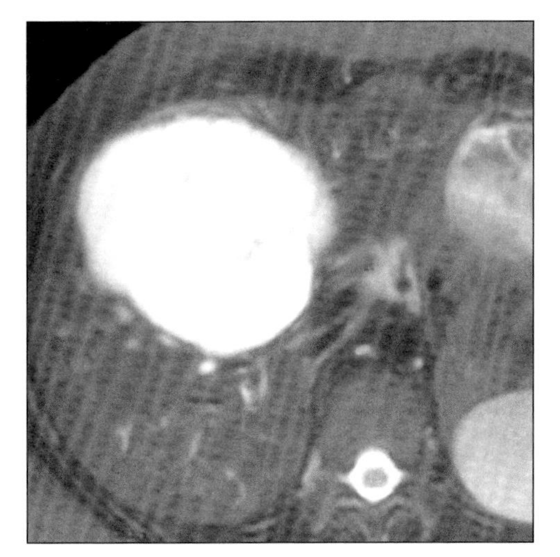

*(**Left**) Axial T1WI MR shows non-communicating hypointense cystic spaces within liver mass. Heterogeneous hypointensity is due to fluid and mucin content. (**Right**) Axial T2WI MR shows multiloculated cystic liver mass.*

Typical

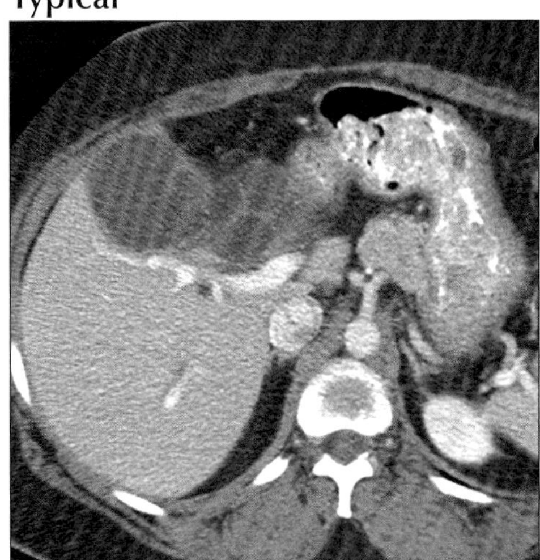

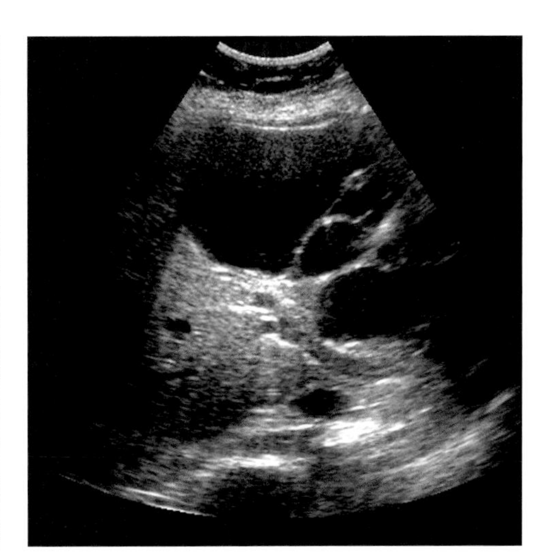

*(**Left**) Axial CECT in a young female who had prior left hepatectomy shows recurrent multiloculated mass with large cystic spaces, visible wall and septa. Biliary Cystadenocarcinoma. (**Right**) Transverse sonogram shows anechoic cystic spaces separated by thin and thick septa.*

Variant

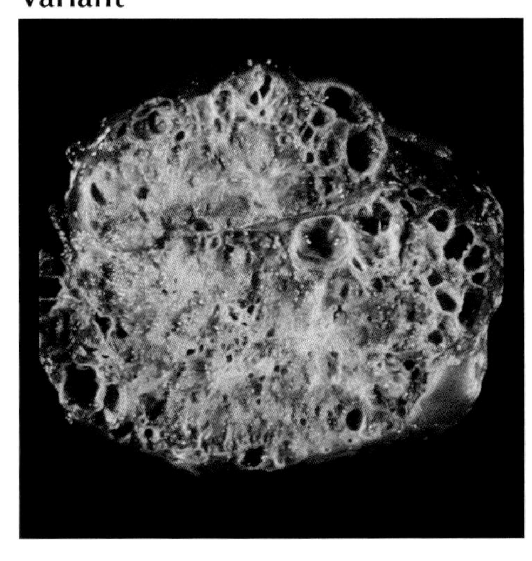

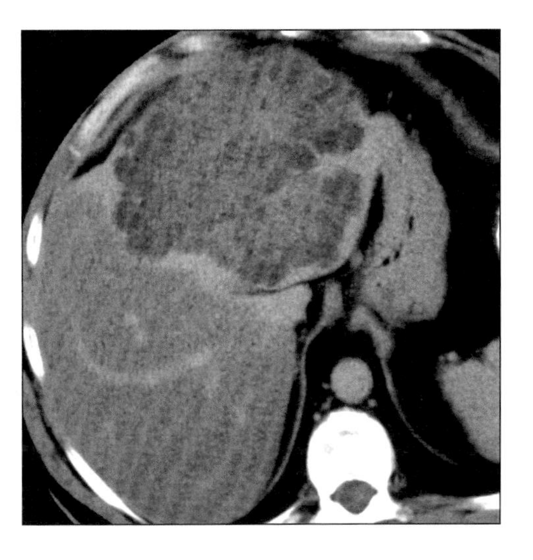

*(**Left**) Cut surface of resected mass shows innumerable small cystic spaces with a honeycomb or sponge appearance. "Microcystic" variant of biliary cystadenocarcinoma. (**Right**) Axial CECT of microcystic cystadenocarcinoma shows innumerable tiny cystic spaces in honeycomb or sponge appearance.*

ANGIOSARCOMA, LIVER

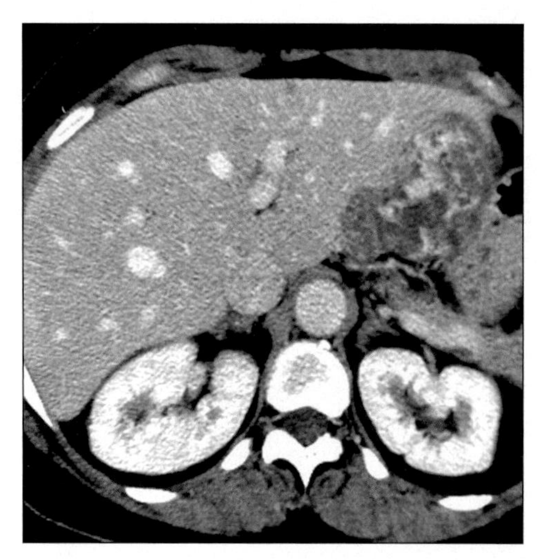

Axial CECT shows mass in lateral segment with central and peripheral, progressive enhancement, isodense with vessels; simulating hemangioma.

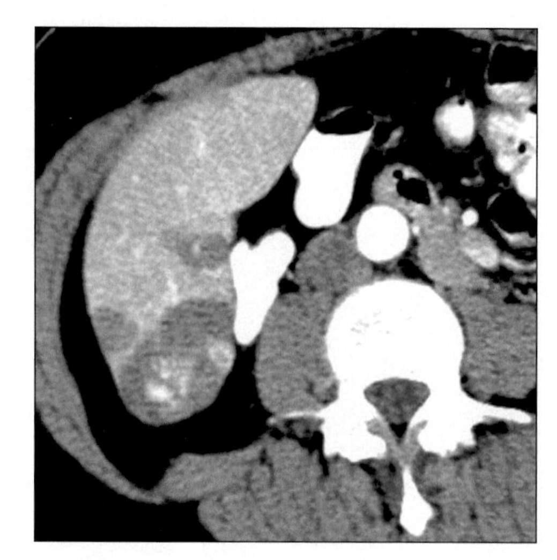

Axial CECT shows multifocal tumors, some with nodular central or peripheral enhancement.

TERMINOLOGY

Abbreviations and Synonyms

- Angiosarcoma (AGS)
- Hemangioendothelial sarcoma, hemangiosarcoma, Kupffer-cell sarcoma

Definitions

- Angiosarcoma (AGS) is a malignant spindle cell tumor of endothelial cell derivation that can form poorly organized vessels, grow along preformed vascular channels, be arranged in sinusoidal or large cavernous spaces or form solid nodules or masses

IMAGING FINDINGS

General Features

- Best diagnostic clue: MR imaging demonstrates hemorrhagic, heterogeneous, & hypervascular nature
- Location: Skin, soft tissue, breast, liver & spleen
- Size
  - Variable; as lesion can be micronodular (few mm) or massive (several centimeters in diameter) or diffuse
    - Vascular channels within: Variable size from capillary to cavernous
- Key concepts
  - Multifocal or multinodular (more common: 71%), or large solitary mass, or as diffusely infiltrating lesion

Radiographic Findings

- Radiography: If Thorotrast exposure: Localized areas of ↑ (metallic) density in patchy or circumferential pattern

CT Findings

- NECT
  - Single or multiple hypodense masses
    - Hyperdense areas of fresh hemorrhage
  - Reticular pattern of deposition of Thorotrast in liver, spleen, mesenteric, celiac lymph nodes
  - Circumferential displacement of Thorotrast in periphery of a nodule is characteristic
- CECT
  - Heterogeneous pattern of enhancement (typical) likely represents heterogeneity of microscopic vascular patterns within each tumor
  - Peripheral nodular enhancement with centripetal progression in a dominant mass (less typical)
  - Usually hypointense with nodular enhancement
    - Bizarre or ring-enhancement possible
  - Vascular channels show persistence of contrast
  - Portal & delayed phase demonstrate progressive enhancement over time

DDx: Heterogeneous Hypervascular Liver Mass(es)

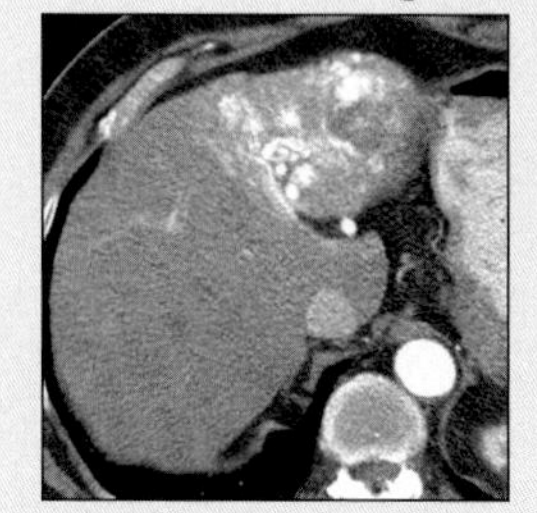

Hemangioma

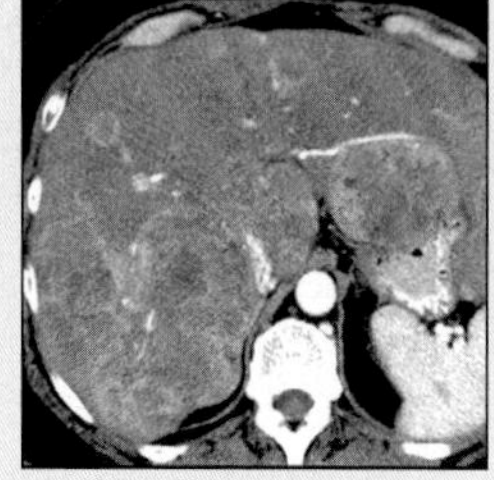

Metastases

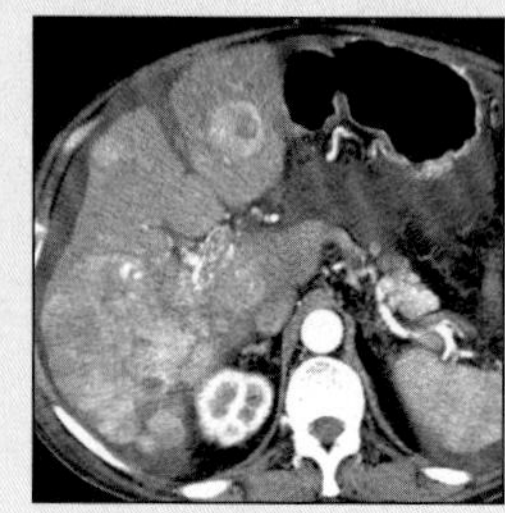

HCC

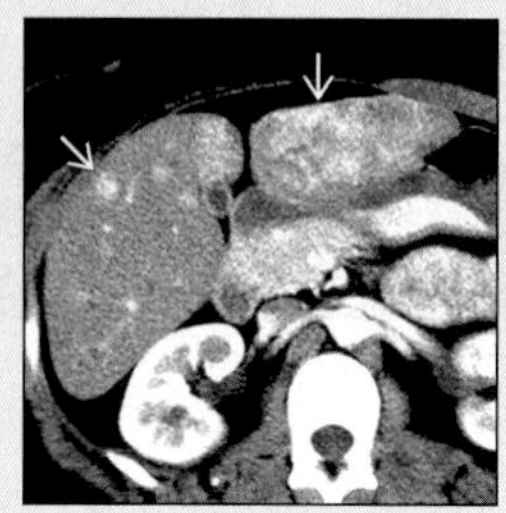

Hepatic Adenomas

ANGIOSARCOMA, LIVER

Key Facts

Terminology
- Angiosarcoma (AGS) is a malignant spindle cell tumor of endothelial cell derivation that can form poorly organized vessels, grow along preformed vascular channels, be arranged in sinusoidal or large cavernous spaces or form solid nodules or masses

Imaging Findings
- Heterogeneous pattern of enhancement (typical) likely represents heterogeneity of microscopic vascular patterns within each tumor
- Bizarre or ring-enhancement possible
- Portal & delayed phase demonstrate progressive enhancement over time

Top Differential Diagnoses
- Hemangioma
- Hepatic metastases
- Hepatocellular carcinoma (HCC)
- Hepatic adenoma(s)
- Focal nodular hyperplasia (atypical)

Clinical Issues
- Rapid & early metastatic spread: Spleen (16%), lung, bone marrow, porta hepatis nodes, peritoneum
- Tend to be multifocal, to recur, & to metastasize
- Complication: Rupture & acute hemoperitoneum

Diagnostic Checklist
- Multiphasic contrast-enhanced helical CT & dynamic MR showing progressive enhancement of multiple heterogeneous liver lesions

- CTA: ± Portal vein thrombus by malignant endothelial cells/intrahepatic arterial encasement

MR Findings
- T1WI
  - Large mass, multiple nodules of low signal intensity
  - Areas of hemorrhage: Irregular hyperintense regions
  - Thorotrast does not produce recognizable MR signal, may be easily missed
  - Micronodular diffusely infiltrative pattern (less common), seen as diffuse signal heterogeneity throughout liver
- T2WI
  - Heterogeneous or compartmentalized appearance: Predominantly high signal on T2 with central septum-like or rounded areas of low signal
    - Areas of low signal intensity may reflect hemosiderin, fibrosis ,or fresh hemorrhage
    - Areas of high intensity represent hemorrhage or necrosis
    - Fluid-fluid levels on T2WI is another finding which reflects hemorrhagic nature of AGS
- T1 C+
  - Dynamic enhancement of dominant mass: Heterogeneous enhancement on early phase images
  - Progressive enhancement on delayed images
  - May show peripheral nodular enhancement which progresses centripetally
  - Center of lesion remaining unenhanced may represent fibrous tissue or deoxyhemoglobin
  - Areas of abundant, freely anastomosing vascular channels enhance quickly & contrast persists while dilated cavernous vascular spaces may show slowly progressive enhancement

Ultrasonographic Findings
- Hyperechoic masses or nodules
- Heterogeneous echotexture; due to hemorrhage of various ages

Angiographic Findings
- Conventional
  - Moderately hypervascular tumor, diffuse puddling of contrast material that persists into venous phase
  - Fed by large peripheral vessels, & centripetal flow
  - If rupture: Demonstrate bleeding/hemoperitoneum

Nuclear Medicine Findings
- Tagged RBC: Early as well as late persistent uptake
- Gallium scan: Increased gallium uptake

Imaging Recommendations
- Best imaging tool: Triphasic helical CT & dynamic MR
- Protocol advice: TI & T2WI, using fast spin-echo with dynamic TI weighted three-dimensional fast spoiled-gradient echo technique

DIFFERENTIAL DIAGNOSIS

Hemangioma
- Centripetal nodular enhancement that approximates density of contrast-opacified blood in aorta or hepatic artery during all phases & unenhanced imaging
- Inhomogeneity on T2WI; seen in AGS, may not be seen in typical hemangiomas (homogeneously hyperintense)
- Hemangiomas are more often solitary, & when multiple, are rarely as numerous as seen with AGS

Hepatic metastases
- Multiple; scattered randomly throughout liver
- Hypervascular lesions: Hyperdense in late arterial phase; may have internal necrosis without uniform hyperdense enhancement

Hepatocellular carcinoma (HCC)
- Heterogeneous hypervascular mass(es); vascular + nodal invasion; necrosis + hemorrhage

Hepatic adenoma(s)
- Blush of homogeneous enhancement in arterial phase & nearly isointense in later phases of dynamic scans
- Heterogeneous due to hemorrhage, necrosis or fat

Focal nodular hyperplasia (atypical)
- Immediate, intense, hyperdensity on arterial phase, followed rapidly by isodensity on portal phase

ANGIOSARCOMA, LIVER

- May show heterogeneous enhancement, but not isodense to vessels on multiphasic CT or MR

PATHOLOGY

General Features

- General path comments
  - Growth pattern: Multi-nodular, massive (large dominant mass), mixed (multi-nodular & massive) & diffuse infiltrative micronodular
  - Compartmentalization within tumor
- Etiology
  - Environmental carcinogens:
    - Polyvinyl chloride (plastic resin), 45-fold ↑ risk of AGS, latent period: 4-28 years
    - Thorotrast (radiocontrast used from 1928 to 1950), 38.3% of Thorotrast-related hepatic malignancies are AGS, latent period: 15-37 years
    - Arsenicals (in some pesticides)
  - Drugs: Cyclophosphamide, anabolic steroids
  - Post radiation (median latency: 74 months)
  - Most of AGS occur either in absence of known risk factors or with cirrhosis, cause is not apparent
- Epidemiology
  - Prevalence varies from 0.14 to 0.25 per million
  - Most common mesenchymal tumor of liver
  - Up to 2% of all primary malignant liver tumors
  - 30 times less common than HCC
  - Approximately 10 to 20 new cases are diagnosed every year in United States
- Associated abnormalities
  - Hemochromatosis
  - Von Recklinghausen disease

Gross Pathologic & Surgical Features

- Begin as small, well-demarcated red nodules evolving into fleshy, grey-white, soft tissue masses
- Dominant large mass without capsule & containing large cystic areas filled with bloody debris & necrosis

Microscopic Features

- Malignant endothelial cells lining vascular channels
- Vascular channels with varied patterns
- Fibrosis & hemosiderin in solid portions of tumor
- Thorotrast particles can be found within malignant endothelial cells in case of Thorotrast induced AGS

CLINICAL ISSUES

Presentation

- Most common signs/symptoms
  - Weakness, weight loss, abdominal pain, hepatomegaly, ascites, jaundice
    - Anemia (62%),thrombocytopenia (54%), disseminated intravascular coagulation (31%), & microangiopathic hemolytic anemia (23%)
- Clinical profile
  - Rapid & early metastatic spread: Spleen (16%), lung, bone marrow, porta hepatis nodes, peritoneum
    - Portal vein invasion/ hemorrhagic ascites
  - Lab data: Elevation of serum neuron-specific enolase; no elevation of α-fetoprotein

Demographics

- Age: Commonly 60-70 years of age, can occur in younger patients & in childhood
- Gender: Strong male predominance, M:F = 4:1

Natural History & Prognosis

- Tend to be multifocal, to recur, & to metastasize
- Complication: Rupture & acute hemoperitoneum
- Prognosis is poor
- Most patients die within one year of diagnosis
- Median survival time is 6 months

Treatment

- Surgical resection: Primary modality when tumor confined to one lobe of liver without metastases
  - Outcome is poor, 5 year survival rate is 37%
- Systemic or hepatic arterial chemotherapy: Antiangiogenic therapy
- Liver transplantation abandoned as treatment because recurrence rate 64%, & low survival rate

DIAGNOSTIC CHECKLIST

Consider

- Pleomorphic histopathology of AGS correlates with various patterns of tumor enhancement
- Multiphasic helical CT, faster scanning techniques & temporal assessment of lesion enhancement relative to blood vessels (aorta), helps differentiate AGS from hemangioma
- Environmental exposure is now rare, detection of Thorotrast accumulation on CT will become increasingly rare, with few additional Thorotrast-induced tumors detected

Image Interpretation Pearls

- Multiphasic contrast-enhanced helical CT & dynamic MR showing progressive enhancement of multiple heterogeneous liver lesions

SELECTED REFERENCES

1. Kitami M et al: Diffuse hepatic angiosarcoma with a portal venous supply mimicking hemangiomatosis. J Comput Assist Tomogr. 27(4):626-9, 2003
2. Yu R et al: Hepatic angiosarcoma: CT findings. Chin Med J (Engl). 116(2):318-20, 2003
3. Koyama T et al: Primary hepatic angiosarcoma: findings at CT and MR imaging. Radiology. 222(3):667-73, 2002
4. Peterson MS et al: Hepatic angiosarcoma: findings on multiphasic contrast-enhanced helical CT do not mimic hepatic hemangioma. AJR Am J Roentgenol. 175(1):165-70, 2000
5. White PG et al: The computed tomographic appearances of angiosarcoma of the liver. Clin Radiol. 48(5):321-5, 1993

ANGIOSARCOMA, LIVER

IMAGE GALLERY

Typical

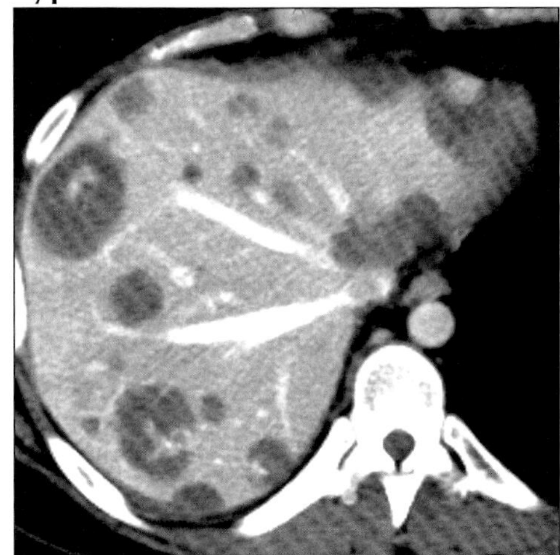

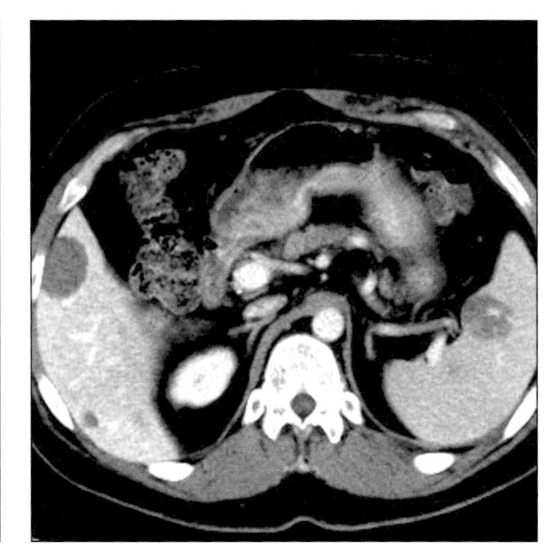

(Left) *Axial CECT in venous phase shows multifocal tumors, many with central, progressive enhancing channels nearly isodense to vessels.* ***(Right)*** *Axial CECT shows multifocal liver and splenic tumors, many of which had similar nodular enhancement patterns.*

Typical

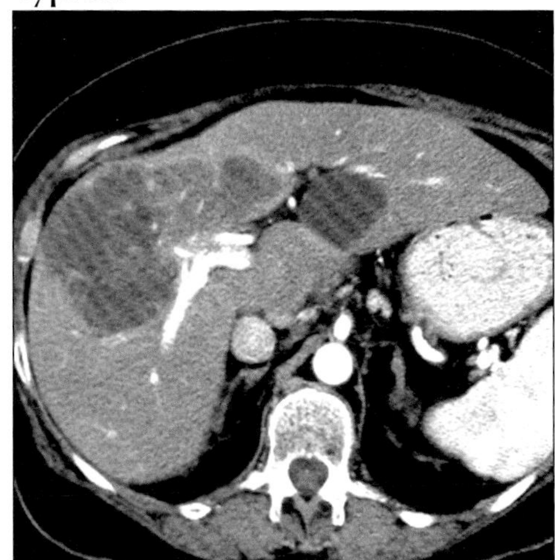

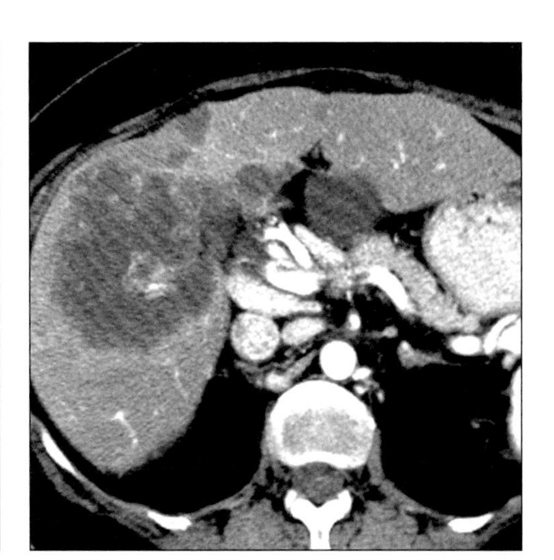

(Left) *Axial CECT shows heterogeneous mass that encases the right hepatic artery and the anterior branch of the right portal vein.* ***(Right)*** *Axial CECT shows heterogeneous masses with foci of hypervascularity within the tumors, and markedly hypervascular lymphadenopathy in the porta hepatis.*

Variant

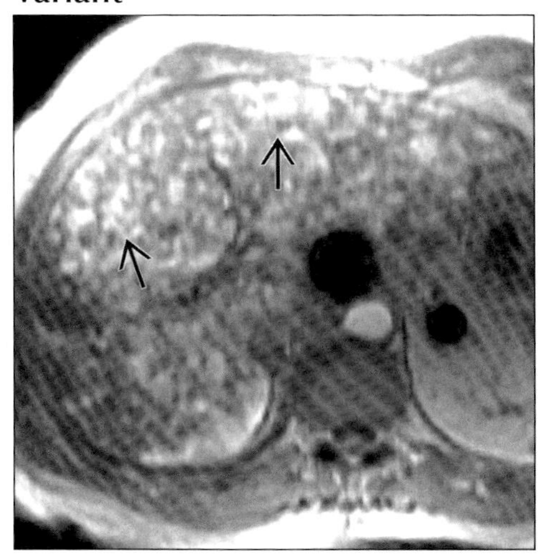

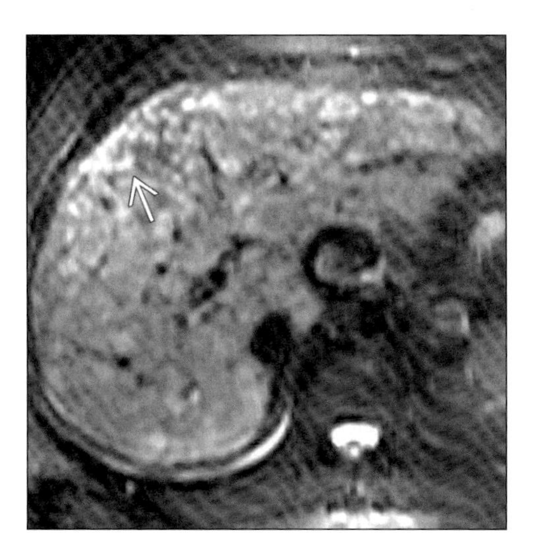

(Left) *Axial T1 C+ MR in arterial phase shows diffuse micronodular hypervascular liver tumor (arrows) in a patient with underlying cirrhosis.* ***(Right)*** *Axial T2WI MR shows diffuse micronodular hyperintense tumor (arrow) in underlying cirrhotic liver.*

HEPATIC METASTASES AND LYMPHOMA

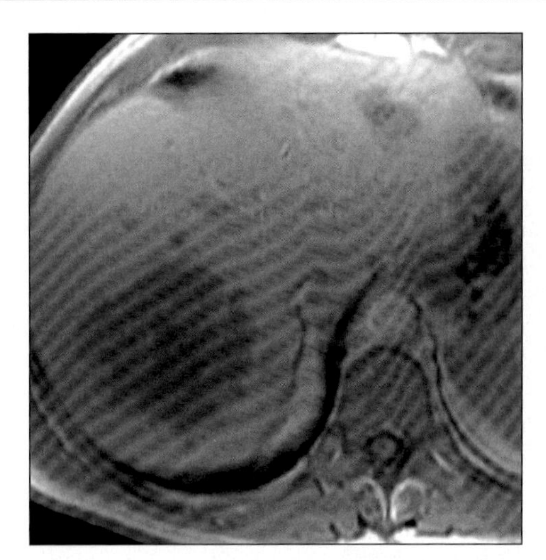

Axial T1 C+ MR shows large heterogeneous hypointense mass. Metastatic colon cancer.

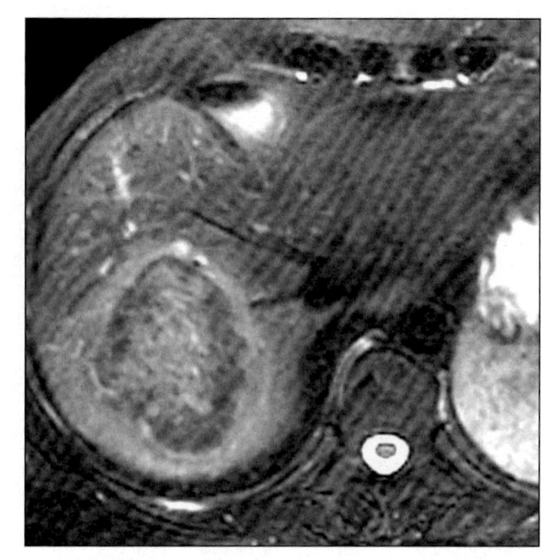

Axial T2WI MR shows heterogeneous intensity within the metastatic mass (colon primary) and hyperintensity of surrounding parenchyma, perhaps due to edema and/or compression of liver.

TERMINOLOGY

Abbreviations and Synonyms

- Hodgkin disease (HD), non-Hodgkin lymphoma (NHL)

Definitions

- Lymphoma: Neoplasm of lymphoid tissues
- Metastases: Malignant spread of neoplasm to hepatic parenchyma

IMAGING FINDINGS

General Features

- Best diagnostic clue
  - Lymphoma: Lobulated low density masses
  - Metastases: Multiple hypo- or hyperdense lesions scattered throughout liver in random distribution
- Location
  - Both lobes of liver
  - Lymphoma (HD & NHL) arises in periportal areas due to high content of lymphatic tissue
- Size: Variable; from few millimeters to centimeters
- Key concepts
  - Hepatic lymphoma
    - Primary (rare)
    - Secondary (more common): Seen in more than 50% of patients with HD or NHL
    - Transplant recipients & AIDS patients (high risk)
  - Types of lymphoma
    - Hodgkin disease (HD)
    - Non-Hodgkin lymphoma (NHL)
  - Liver metastases
    - Most common malignant tumor of liver
    - Compared to primary malignant tumors: 18:1
    - Liver is second only to regional lymph nodes as a site of metastatic disease
    - Autopsy studies reveal 55% of oncology patients have liver metastases
    - May be hypovascular or hypervascular

CT Findings

- NECT
  - May be normal
  - Primary lymphoma
    - Isodense or hypodense to liver
  - Secondary lymphoma
    - Multiple well-defined, large, homogeneous lobulated low density masses
    - Diffuse infiltration: Indistinguishable from normal liver or steatosis
  - Metastases
    - Isodense, hypodense or hyperdense
    - Calcified: Mucinous adenocarcinoma (colon), treated metastases (breast), malignant teratoma
  - Cystic metastases (less than 20 HU)
    - Fluid levels, debris, mural nodules

DDx: Multiple Heterogeneous Hypodense Focal Masses

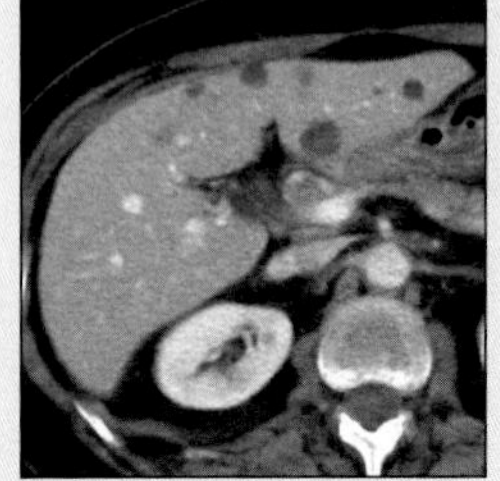

Hepatic Cysts

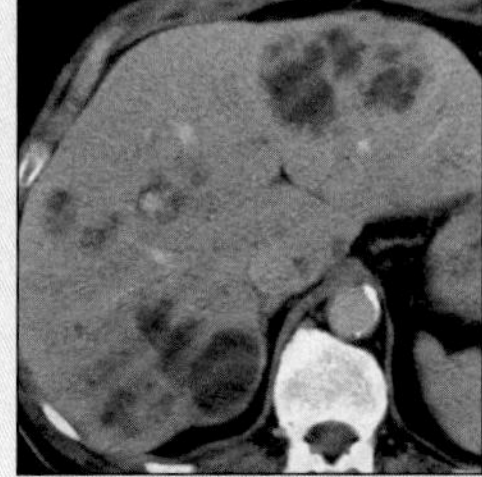

Pyogenic Abscesses

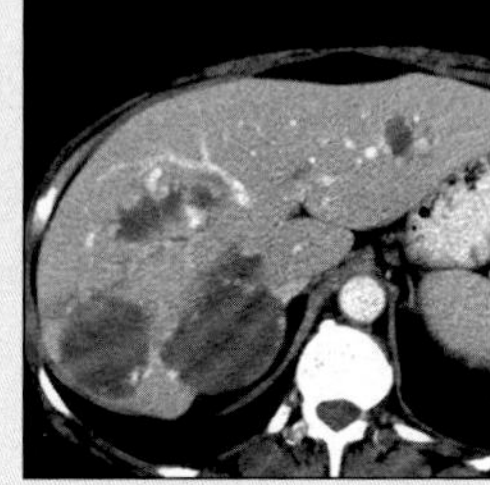

Hemangiomas

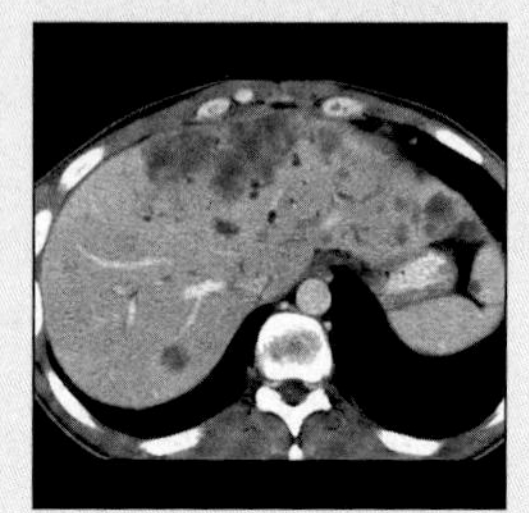

Cholangiocarcinoma

HEPATIC METASTASES AND LYMPHOMA

Key Facts

Imaging Findings

- Lymphoma: Lobulated low density masses
- Metastases: Multiple hypo- or hyperdense lesions scattered throughout liver in random distribution
- Calcified: Mucinous adenocarcinoma (colon), treated metastases (breast), malignant teratoma
- "Light bulb" sign: Very high signal intensity (e.g., cystic & neuroendocrine metastases)

Top Differential Diagnoses

- Multiple hepatic cysts
- Multiple liver abscesses
- Multiple hemangiomas
- Multifocal fatty infiltration (steatosis)
- Multifocal HCC or cholangiocarcinoma (CC)

Pathology

- Metastases: Depends on underlying primary tumor
- Epidemiology: Over 50,000 deaths per year in US alone due to metastatic spread of colorectal cancer
- Usually AIDS in lymphoma

Clinical Issues

- Asymptomatic, RUQ pain, tender hepatomegaly
- Weight loss, jaundice or ascites
- 20-40% have good 5 year survival rate if resectable

Diagnostic Checklist

- Rule out other multiple liver lesions like hepatic cysts, abscesses, hemangiomas which can mimic metastases
- Epithelial metastases: Vascular rim-enhancement

  - Cystic metastases (less than 20 HU)
    - Fluid levels, debris, mural nodules
    - Thickened walls or septations may be seen
    - Usually cystadenocarcinoma or sarcoma (pancreatic, GI or ovarian primaries)
- CECT
  - Lymphoma
    - Homogeneous low density discrete masses
    - More common in immunosuppressed patients
    - Examples: AIDS & organ transplant recipients
  - Hypovascular metastases
    - Low attenuation center with peripheral rim enhancement (e.g., epithelial metastases)
    - Indicates vascularized viable tumor in periphery & hypovascular or necrotic center
    - Rim enhancement may also be due to compressed normal parenchyma
  - Hypervascular metastases
    - Hyperdense in late arterial phase images
    - May have internal necrosis without uniform hyperdense enhancement
    - Hypo- or isodense on NECT & portal venous phase images
    - Examples: Islet cell, carcinoid, thyroid, renal carcinomas & pheochromocytoma

MR Findings

- T1WI
  - Lymphoma: Hypointense lesions
  - Metastases: Multiple low signal lesions
- T2WI
  - Lymphoma: Hyperintense
  - Metastases
    - Moderate to high signal
    - "Light bulb" sign: Very high signal intensity (e.g., cystic & neuroendocrine metastases)
    - Mimic cysts or hemangiomas due to high signal "light bulb" appearance
- T1 C+
  - Hypovascular metastases
    - Similar with gadolinium enhancement to CECT
    - Low signal in center and peripheral rim-enhancement
  - Hypervascular metastases
    - Hyperintense enhancement on arterial phase
- Superparamagnetic iron oxide (SPIO)
  - Metastases: Bright signal on T2WI
    - Free of reticuloendothelial system (RES)
  - Rest of normal liver: Decreased signal
    - Due to SPIO particles phagocytized by RES of liver

Ultrasonographic Findings

- Real Time
  - Hepatic lymphoma
    - Multiple well-defined hypoechoic lesions
    - Diffuse form: May detect innumerable subcentimeter hypoechoic foci or indistinguishable from normal liver
  - Hypoechoic metastases
    - Usually from hypovascular tumors
  - Hyperechoic metastases
    - GI tract malignancy
    - Vascular metastases from islet cell tumors, carcinoid, choriocarcinoma & renal cell carcinoma
  - "Bull's eye" or "target" metastatic lesions
    - Alternating layers of hyper- & hypoechoic tissue
    - Solid mass with hypoechoic rim or halo
    - Usually from aggressive primary tumors
    - Example: Bronchogenic carcinoma
  - Cystic metastases
    - Cystadenocarcinoma of pancreas & ovary
    - Treated metastases, sarcomas & squamous cell carcinoma
  - Calcified metastases
    - Markedly echogenic with acoustic shadowing
    - Example: Mucinous adenocarcinoma of colon, treated metastases (e.g., breast) malignant teratoma

Nuclear Medicine Findings

- PET
  - Metastases
    - Multiple increased metabolic foci
    - Fluorodeoxyglucose (18-FDG) avid
  - Hepatic lymphoma
    - Good concordance with CT & MR

HEPATIC METASTASES AND LYMPHOMA

Imaging Recommendations

- NE + CECT or MR + CEMR
- MR with liver specific contrast agents (e.g., SPIO or mangafodipir) if resection or ablation of metastases is considered

DIFFERENTIAL DIAGNOSIS

Multiple hepatic cysts

- No peripheral rim or central enhancement
- May have increased density or intensity due to prior bleed or infection (e.g., polycystic liver)
- No mural nodules or debris

Multiple liver abscesses

- "Cluster sign" on CT for pyogenic abscesses
- Often with atelectasis & right pleural effusion
- Typical systemic signs of infection

Multiple hemangiomas

- Typical peripheral nodular discontinuous enhancement on CECT or CEMR
- Isodense with blood vessels on NECT & CECT
- Markedly hyperintense on T2WI
- Uniformly hyperechoic on US

Multifocal fatty infiltration (steatosis)

- Focal signal dropout on opposed-phase T1 GRE MR
- Vessels course through "lesions" without disruption

Multifocal HCC or cholangiocarcinoma (CC)

- HCC: Cirrhotic liver, vascular invasion
- CC: Capsular retraction, delayed enhancement

PATHOLOGY

General Features

- General path comments
  - Lymphoma
    - Early disease: Miliary lesions
    - Late disease: Multiple nodules
  - Metastases: Depends on underlying primary tumor
- Etiology
  - Hypovascular liver metastases
    - Lung, GI tract, pancreas & most breast cancers
    - Lymphoma, bladder & uterine malignancy
  - Hypervascular liver metastases
    - Endocrine tumors, renal & thyroid cancers
    - Some breast cancers, sarcomas & melanoma
- Epidemiology: Over 50,000 deaths per year in US alone due to metastatic spread of colorectal cancer
- Associated abnormalities
  - Primary malignant tumor for metastases
  - Usually AIDS in lymphoma

Gross Pathologic & Surgical Features

- Lymphoma: Miliary, nodular or diffuse form
- Metastases
  - Vary in size, consistency & vascularity
  - Nodular, infiltrative, expansile or miliary

Microscopic Features

- Hodgkin disease (HD)
  - Typical Reed-Sternberg cells
- Non-Hodgkin lymphoma (NHL)
  - Follicular small cleaved-cells (most common)
  - Small noncleaved cells (Burkitt lymphoma - rare)

CLINICAL ISSUES

Presentation

- Most common signs/symptoms
  - Asymptomatic, RUQ pain, tender hepatomegaly
  - Weight loss, jaundice or ascites
- Lab data: Elevated LFTs; normal in 25-50% of patients
- Diagnosis: Imaging, core biopsy and FNA

Demographics

- Age: Usually middle & older age group
- Gender: Depends on underlying primary tumor

Natural History & Prognosis

- Depends on primary tumor site
- 20-40% have good 5 year survival rate if resectable
- In patients with metastatic colon cancer
  - 3 year survival rate
    - In 21% of patients with solitary lesions
    - In 6% with multiple lesions in one lobe
    - In 4% with widespread disease

Treatment

- Resection or ablation for colorectal liver metastases
- Chemoembolization: Carcinoid/endocrine metastases
- Chemotherapy for all others

DIAGNOSTIC CHECKLIST

Consider

- Rule out other multiple liver lesions like hepatic cysts, abscesses, hemangiomas which can mimic metastases

Image Interpretation Pearls

- Liver metastases: Hypovascular or hypervascular
- Epithelial metastases: Vascular rim-enhancement
- "Light bulb" sign on T2WI: Cystic & neuroendocrine metastases

SELECTED REFERENCES

1. Valls C et al: Hepatic metastases from colorectal cancer: preoperative detection and assessment of resectability with helical CT. Radiology. 218(1):55-60, 2001
2. Helmberger T et al: new contrast agents for imaging the liver. Magn Reson Imaging Clin N Am. 9(4):745-66, 2001
3. Blake SP et al: Liver metastases from melanoma: detection with multiphasic contrast-enhanced CT. Radiology. 213(1):92-6, 1999
4. Nazarian LN et al: Size of colorectal liver metastases at abdominal CT: comparison of precontrast and postcontrast studies. Radiology. 213(3):825-30, 1999
5. Paulson EK et al: Carcinoid metastases to the liver: role of triple-phase helical CT. Radiology. 206(1):143-50, 1998

HEPATIC METASTASES AND LYMPHOMA

IMAGE GALLERY

Typical

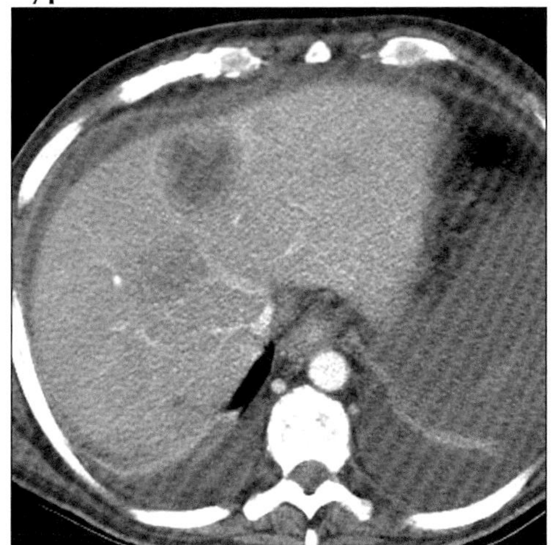

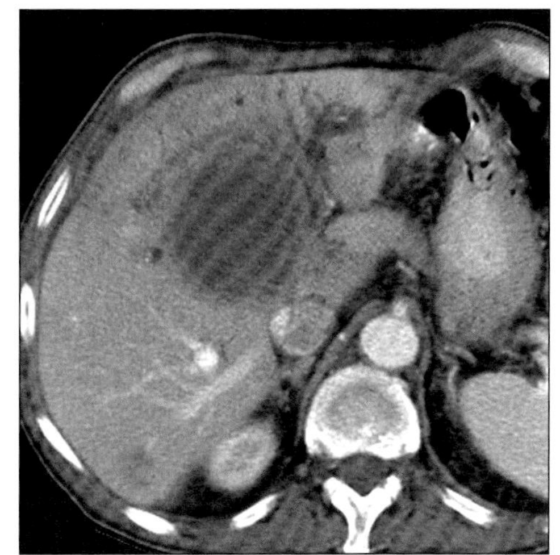

(Left) *Axial CECT shows multiple hypodense masses; some almost isodense to liver in a patient with AIDS and diffuse lymphoma.* ***(Right)*** *Axial CECT shows necrotic metastasis with shaggy enhancing wall causing extrinsic compression and obstruction of left lobe bile ducts. Metastatic colon carcinoma.*

Typical

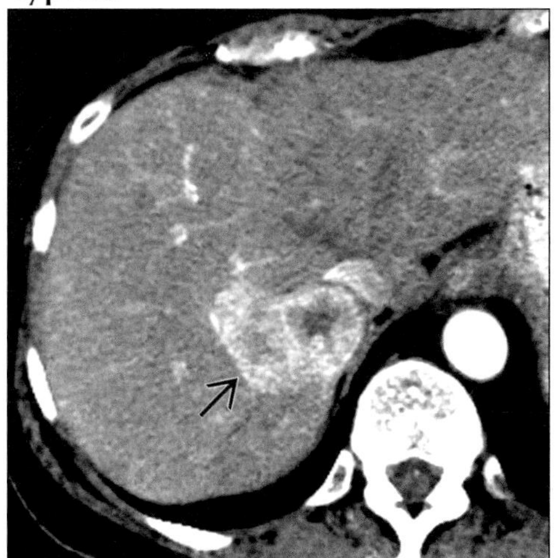

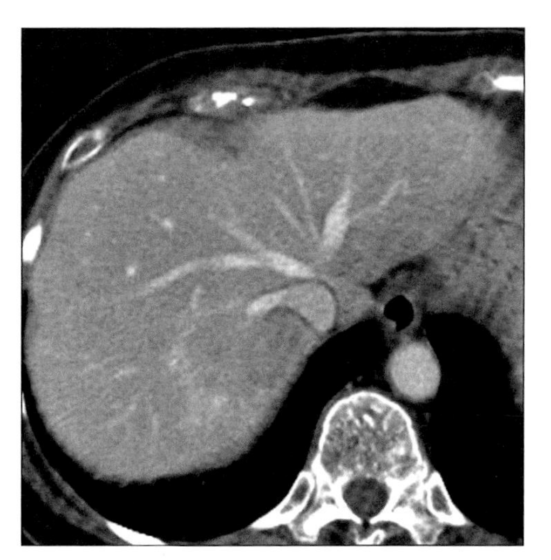

(Left) *Axial CECT in arterial phase shows hypervascular mass (arrow) adjacent to IVC. Metastatic carcinoid tumor.* ***(Right)*** *Axial CECT in venous parenchymal phase. Mass adjacent to IVC is nearly isodense to liver and difficult to recognize.*

Typical

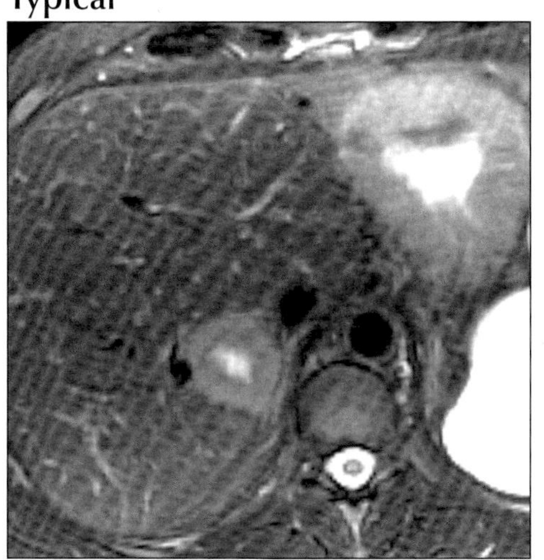

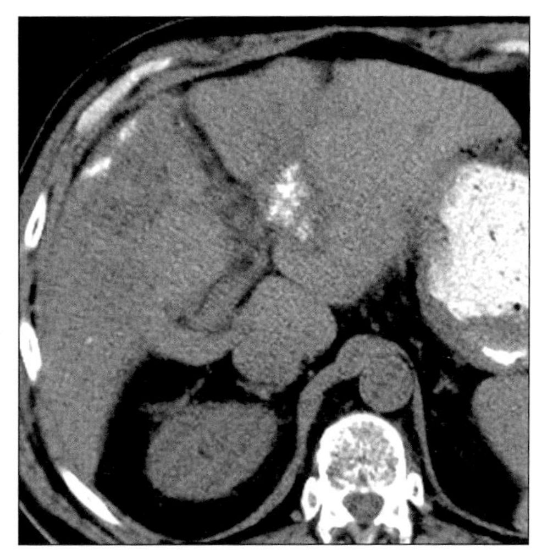

(Left) *Axial T2WI MR shows bright signal ("light bulb") in center of two cystic/necrotic metastases from sarcoma of gastrointestinal tract.* ***(Right)*** *Axial NECT shows several focal masses with amorphous calcification, characteristic of mucinous adenocarcinoma (colon primary).*

RADIATION HEPATITIS

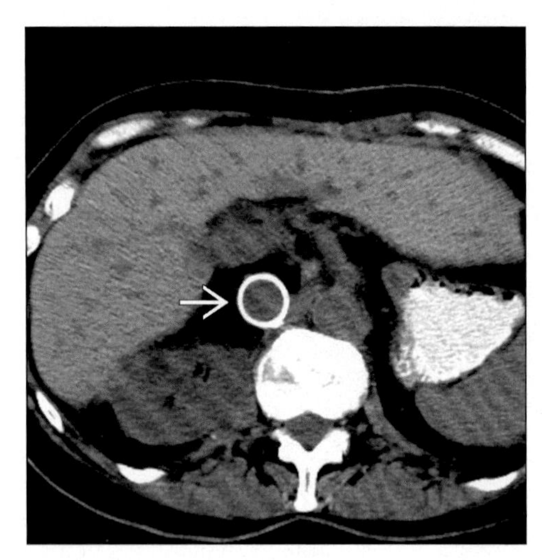

Axial NECT following resection of primary sarcoma of IVC and prior to radiation therapy. Note synthetic graft (arrow).

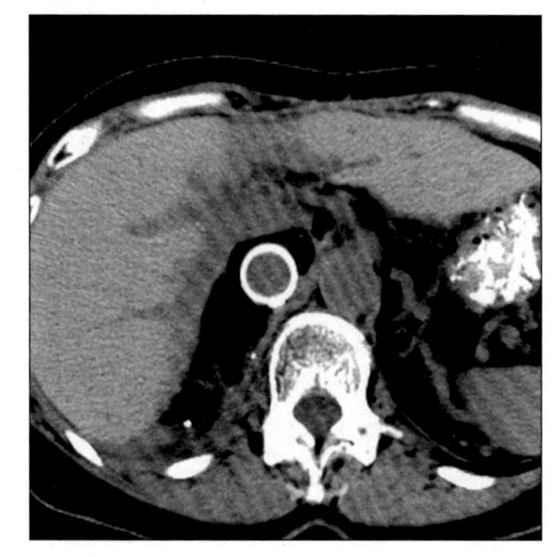

Axial NECT several months after surgical resection and radiation therapy for sarcoma of IVC. Band of low attenuation and volume loss in left lobe corresponds to radiation port.

TERMINOLOGY

Abbreviations and Synonyms

- Radiation-induced liver disease (RILD)

Definitions

- Radiation-induced liver disease (RILD), often called radiation hepatitis, is a syndrome characterized by development of anicteric ascites approximately 2 weeks to 4 months after hepatic irradiation
- RILD is a form of veno-occlusive disease due to fibrous obliteration of terminal hepatic venules leading to postsinusoidal obstruction

IMAGING FINDINGS

General Features

- Best diagnostic clue: Sharp line of demarcation between normal & abnormal parenchyma corresponds to radiation port or vascular distribution of yttrium - 90 microspheres

CT Findings

- NECT
  - CT performed several months after radiation therapy shows sharply defined band of low attenuation corresponding to treatment port
    - Due to localized edema or hepatic congestion
  - If hepatic congestion is severe, patchy congestion simulating tumor nodules may be seen
  - In patients with fatty infiltration of liver, irradiated area may appear as a region of increased attenuation
    - May be due to loss of fat in irradiated hepatocytes or regional edema, with water content demonstrating higher attenuation than fatty liver
  - Over a period of weeks, the initially sharp borders of irradiated zone become more irregular & indistinct (peripheral parenchyma regenerates)
  - Eventually, irradiated area may be atrophic
- CECT
  - Enhancement pattern of irradiated liver may vary depending on pre-existing hepatic pathology
  - Intense enhancement of irradiated parenchyma compared with normal; attributed to ↑ arterial flow secondary to reduced portal flow; seen in patients treated for hepatocellular carcinoma (HCC)
  - Region of radiation damage is hypodense on portal venous phase & becomes hyperdense with marked prolonged enhancement on delayed phase
    - Due to ↓ vascular perfusion & ↓ hepatic venous drainage & subsequent stasis of contrast medium
  - Narrowing & irregularity of hepatic vessels (sinusoidal congestion & perisinusoidal edema)

MR Findings

- T1WI: Geographic areas of low signal on T1WI

DDx: Segmental or Geographic Hypodense Liver

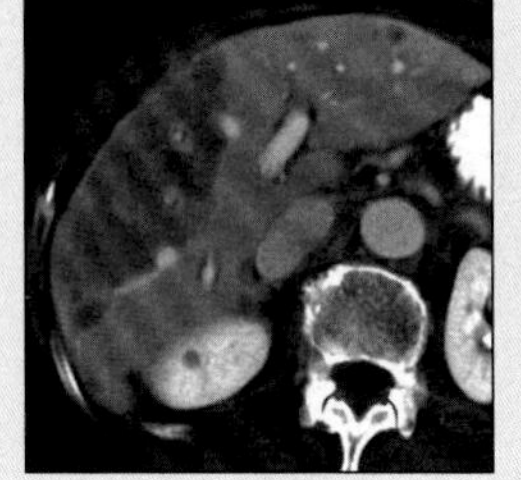

Focal Steatosis

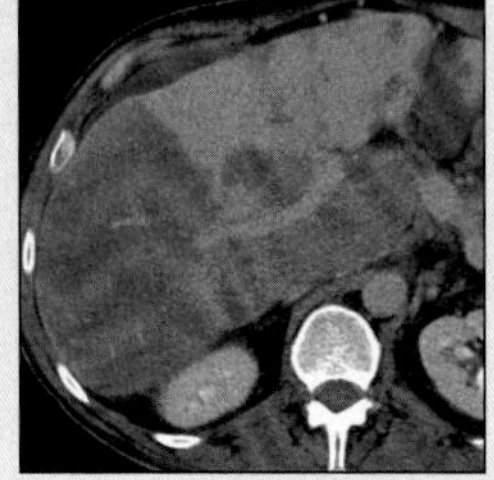

Focal Steatosis

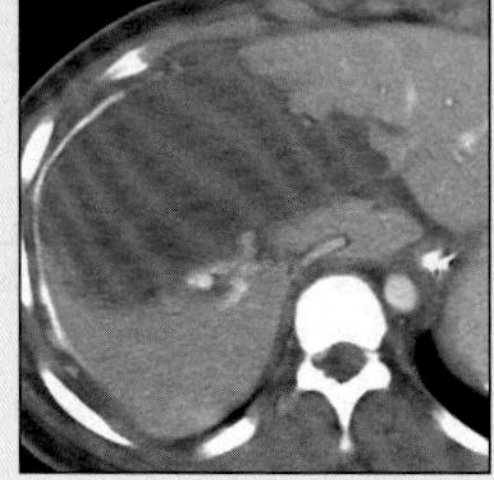

Hepatic Infarction

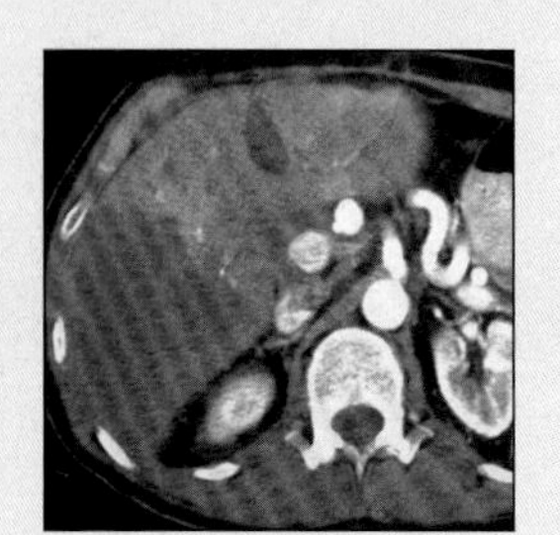

Hepatic Infarction

RADIATION HEPATITIS

Key Facts

Terminology
- Radiation-induced liver disease (RILD)

Imaging Findings
- CT performed several months after radiation therapy shows sharply defined band of low attenuation corresponding to treatment port
- Best imaging tool: NECT and CECT; or MR T1WI GRE with in- and out-of-phase

Top Differential Diagnoses
- Focal steatosis
- Hepatic infarction

Pathology
- Veno-occlusive disease

Clinical Issues
- Usually presents 2-16 weeks after treatment

Imaging Recommendations
- Best imaging tool: NECT and CECT; or MR T1WI GRE with in- and out-of-phase

DIFFERENTIAL DIAGNOSIS

Focal steatosis
- May be geographic, band or wedge-shaped
- Preservation of enhancing vessels within "lesion"
- Suppression of signal on opposed-phase GRE MR

Hepatic infarction
- Segmental or geographic hypodense area with straight margins with absent or heterogeneous enhancement

PATHOLOGY

General Features
- Etiology
  - Unintentional & occurs when liver is unavoidably included in the treatment portal designed to encompass tumors & adjacent organs
  - Patients who receive a single 1200-rad dose of external beam radiation or a 4000- to 5500-rad fractionated dose over 6 weeks can develop RILD
  - Investigational use of hepatic arterial administration of Yttrium-90 glass microspheres
    - Emit radiation to perfused hepatic area persisting for 64 hour half-life
- Epidemiology: Now more commonly seen with advent of three-dimensional treatment planning & bone marrow transplantation with total body radiation

Microscopic Features
- Veno-occlusive disease
- Massive panlobar congestion, hyperemia, hemorrhage, & mild proliferative change in sublobar central veins
  - Stasis secondary to injury of these veins

CLINICAL ISSUES

Presentation
- Hepatomegaly/ascites/fatigue/rapid weight gain
- Usually presents 2-16 weeks after treatment
- May present as late as 7 months

Natural History & Prognosis
- Complete clinical recovery is typically seen within 60 days, but there may be permanent hepatocyte loss, fat deposition, fibrosis & obliteration of central veins
- Chronic changes: Atrophy of involved segments & rarely cirrhosis

DIAGNOSTIC CHECKLIST

Consider
- Variability in liver damage influenced by factors: Irradiated liver volume, radiation fraction size, cytotoxic agents & nutritional status
- Modification of arterial enhancement pattern in irradiated region by concomitant HCC

SELECTED REFERENCES

1. Mori H et al: Radiation-induced liver injury showing low intensity on T2-weighted images noted in Budd-Chiari syndrome. Radiat Med. 20(2):69-76, 2002
2. Willemart S et al: Acute radiation-induced hepatic injury: evaluation by triphasic contrast enhanced helical CT. Br J Radiol. 73(869):544-6, 2000
3. Unger EC et al: CT and MR imaging of radiation hepatitis. J Comput Assist Tomogr. 11(2):264-8, 1987

IMAGE GALLERY

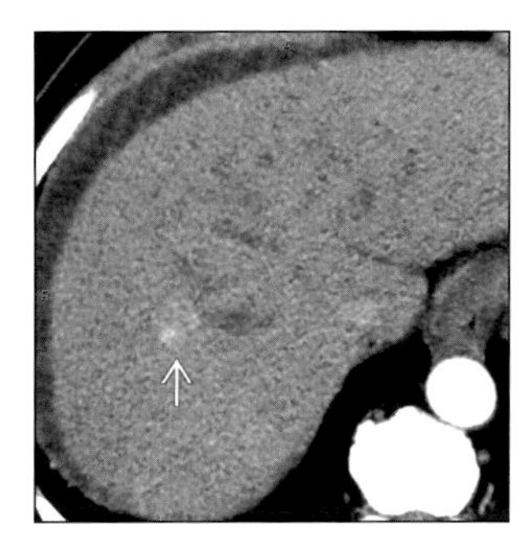

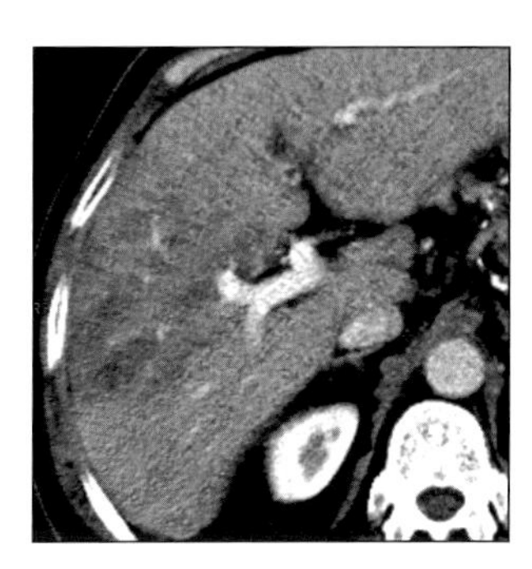

*(**Left**) Axial CECT shows heterogeneous mass, (arrow) enhancing during arterial phase; hepatocellular carcinoma (HCC). (**Right**) Axial CECT in patient with HCC, following hepatic arterial embolization of Yttrium-90 microspheres. Wedge of hypodensity in segments 5 + 8 represents radiation hepatitis.*

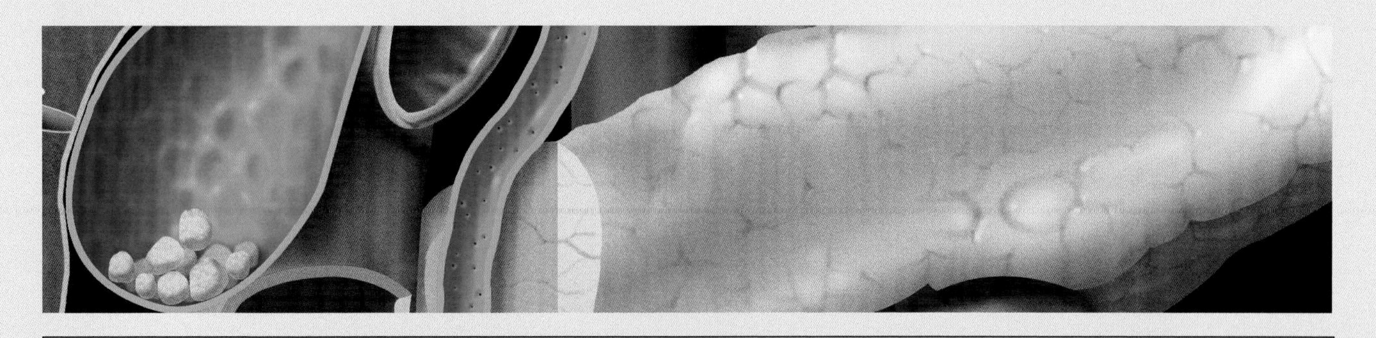

SECTION 2: Biliary System

Introduction and Overview

Congenital

Infection

Inflammation

Neoplasm, Primary

Treatment Related

BILIARY SYSTEM ANATOMY AND IMAGING ISSUES

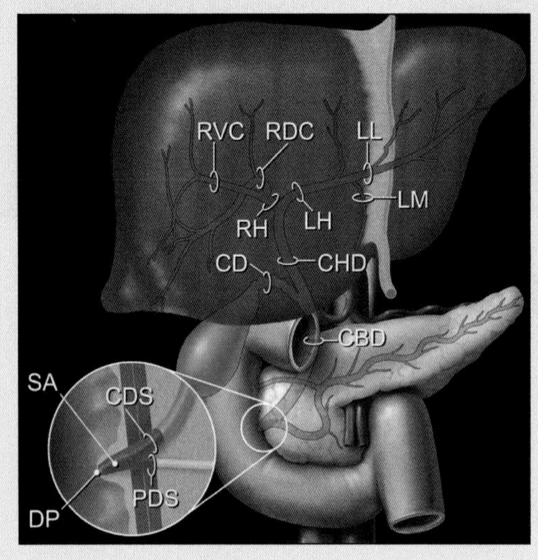

Graphic shows normal biliary tree, with detailed view of papilla of Vater. (See text for abbreviation keys).

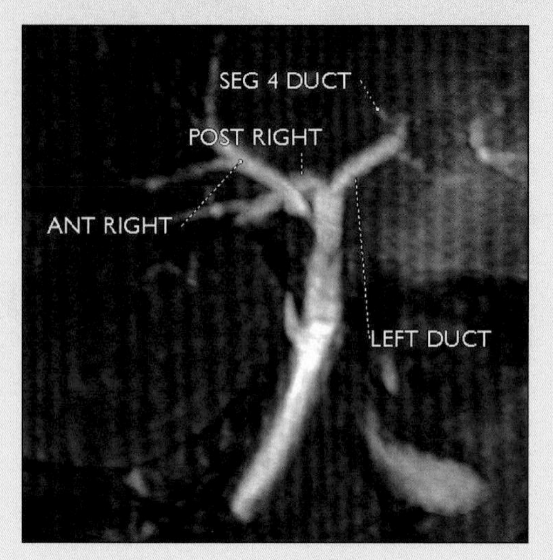

Coronal view of mangafodipir-enhanced MR showing normal biliary tree. Some branches are not in this plane of section.

TERMINOLOGY

Abbreviations and Synonyms

- Gallbladder (GB)
- Common hepatic duct (CHD)
- Right ventral - cephalic duct (RVC)
- Right dorsal - caudal duct (RDC)
- Left lateral duct (LL) and medial duct (LM)
- Duodenal papilla (DP)
- Choledochal sphincter (CDS)
- Pancreatic duct sphincter (PDS)
- Portal Vein (PV)
- Ultrasound (US)
- MR cholangiopancreatography (MRCP)

IMAGING ANATOMY

Location

- GB is located in GB fossa, indentation on undersurface of liver
  - "Intrahepatic" in < 10% of cases
  - Attached to liver by short veins and bile ducts (of Lushka) and covered by parietal peritoneum
- GB fundus
  - Rounded distal tip; projects below edge of liver
  - "Phrygian cap"; fundus partially septated and folded upon itself
- GB body
  - Midportion; often in contact with duodenum and hepatic flexure of colon
- GB neck
  - Lies between GB body and cystic duct
  - Bears a constant relationship to the main interlobar fissure (hepatic) and undivided right PV
- Cystic duct
  - 2 to 4 cm long
  - Contains tortuous spiral folds (valves of Heister)
  - Highly variable point of entry into CHD
- CBD usually joins pancreatic duct (PD) as a common channel within the duodenal wall
- Sphincter of Oddi (ampullary sphincter)
  - Smooth muscle sheath surrounding the common channel (CBD + PD)
- Sphincter of Boyden
  - Choledochal sphincter; surrounds distal CBD
  - Contraction of sphincter of Oddi or Boyden can simulate CBD stricture or stone, though usually transient

ANATOMY-BASED IMAGING ISSUES

Imaging Approaches

- Current role of oral and intravenous (IV) cholecystography
  - Largely replaced by newer modalities
  - IV contrast agents may be used to augment CT or MR cholangiography (e.g., mangafodipir; IV manganese-based contrast agent that has hepatobiliary excretion and can increase sensitivity and specificity of hepatic and biliary MR studies)
- Role of cholescintigraphy
  - IV administration of Tc-99m iminodiacetic acid compounds ("Tc-HIDA" scans)
  - Undergoes rapid uptake by liver and excretion into bile
  - Allows visualization of CBD and gallbladder
  - Main uses: To confirm Dx of acute cholecystitis (nonfilling GB) and bile leaks (e.g., after surgery or trauma)
- Role of ultrasonography
  - Primary imaging modality for most GB + CBD lesions
  - Detection of gallstones (near 100% accuracy)
  - Acute cholecystitis (~ 95% accuracy)
    - Based on finding stones and focal tenderness ("sonographic Murphy sign"), ± GB wall thickening

BILIARY SYSTEM ANATOMY AND IMAGING ISSUES

DIFFERENTIAL DIAGNOSIS

Tumor Classification

Epithelial tumors

- Adenoma
- Cystadenoma
- Papillomatosis
- Adenocarcinoma
- Cystadenocarcinoma

Nonepithelial tumors

- Stromal (leiomyoma, lipoma, hemangioma, lymphangioma)
- Granular cell tumor
- Neurofibroma, ganglioneuroma
- Lymphoma/metastasis
- Carcinoid

Tumor-like lesions

- Adenomatous hyperplasia
- Cholesterol polyp
- Inflammatory polyp
- Xanthogranulomatous cholangitis
- Primary sclerosing cholangitis

Diffuse GB wall thickening

- Common
- ⇒ Cholecystitis
- ⇒ Hepatitis
- ⇒ Cardiac, renal, liver failure

- Uncommon
- ⇒ Tumor (primary or metastatic)
- ⇒ Adenomatous hyperplasia
- ⇒ Xanthogranulomatous cholecystitis

  - Complications of cholecystitis (perforation, gangrene)
  - GB mass lesions (polyps, primary and metastatic tumor)
- Role of computed tomography (CT)
  - Less sensitive than US in detecting gallstones (stone attenuation varies from less than water to densely calcified, by chemical composition of stone)
  - Accurate in diagnosis of complicated cholecystitis (emphysematous, gangrenous, perforated, abscess)
  - Primary role in diagnosis and staging of GB carcinoma
- Role of magnetic resonance (MR)
  - MRCP now a primary tool in evaluation of biliary obstruction (calculi, intrinsic and extrinsic masses)
  - Utilizes heavily T2 weighted sequences to show bile as bright signal fluid
  - Can be combined with other sequences to yield comprehensive evaluation of liver, biliary tree, and pancreas
  - Produces images of intra- and extrahepatic bile ducts (+ pancreatic duct) that rival those of endoscopic or transhepatic cholangiograms
- Direct cholangiography
  - Percutaneous transhepatic cholangiography (PTC)
    - For known or suspected biliary obstruction, especially when endoscopic techniques are unsuccessful (e.g., following prior biliary diversion) or to diagnose and treat intrahepatic or proximal extrahepatic biliary obstruction (e.g., Klatskin tumor)
  - Endoscopic retrograde cholangiopancreatography (ERCP)
    - For known or suspected biliary obstruction, especially distal obstruction (e.g., pancreatic carcinoma) that may require endoscopic placement of biliary stent, endoscopic retrieval of stones, or biopsy/brushing of biliary tumors
    - Also modality of choice for diagnosis and treatment of traumatic/post-surgical bile leaks; will usually resolve following endoscopic placement of a biliary stent
  - Post-operative (T-tube) cholangiography
    - When CBD has been manipulated at surgery (e.g., choledocholithiasis, liver transplantation) a T-tube is usually left in place within the CBD, allowing safe access to the external limb
    - Allows convenient and safe diagnosis of retained stones, leaks, or strictures

CLINICAL IMPLICATIONS

Clinical Importance

- Imaging goals (jaundice, abnormal LFTs)
  - Determine presence, level, and cause of biliary obstruction
- Bile duct dilation criteria
  - Intrahepatic: Continuous arborization
    - Or > 40% diameter of adjacent vein
  - Extrahepatic
    - CHD at porta, > 6 mm
    - CBD, 8 to 10 mm
    - ↑ With aging, post-cholecystectomy
- Level of biliary obstruction
  - Intrahepatic
    - Primary sclerosing cholangitis (PSC)
    - Liver mass
  - Porta hepatic
    - Cholangiocarcinoma
    - PSC
    - GB carcinoma
    - Metastases
    - Iatrogenic (e.g., lap. cholecystectomy)
  - Intrapancreatic
    - Pancreatic carcinoma
    - Pancreatitis (usually chronic)
    - CBD stones
    - Ampullary stenosis
    - Ampullary or duodenal carcinoma
    - Cholangiocarcinoma
- Criteria for malignant obstruction
  - Abrupt transition (dilated to stricture)
  - Eccentric duct wall thickening
  - Mass in or around duct (e.g., pancreatic cancer)
  - Presence of nodes, metastases, vessel invasion

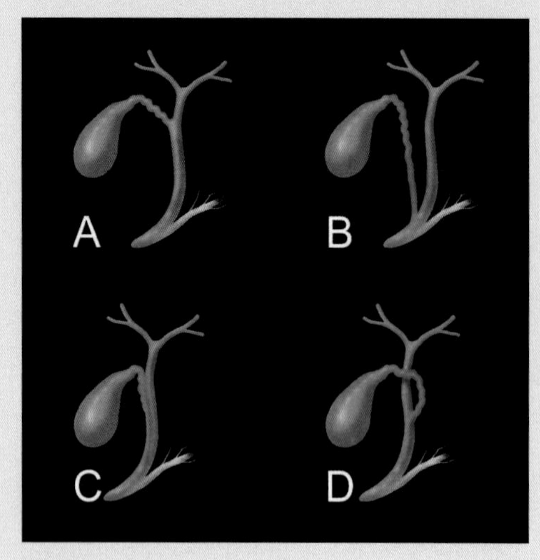

Graphic shows common variations of cystic duct entry into common duct.

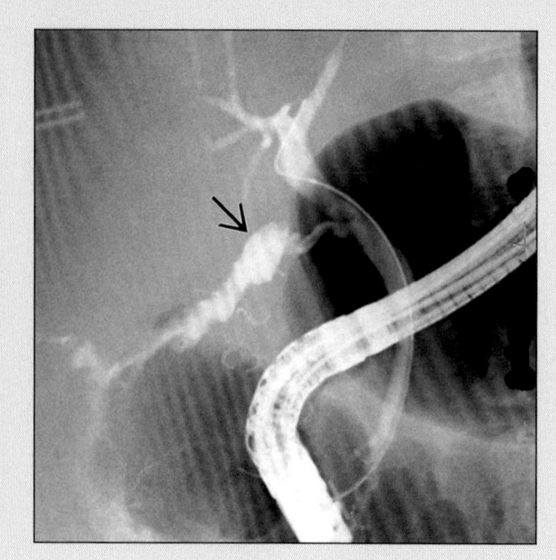

ERCP following cholecystectomy shows leak of bile (contrast) from cystic duct remnenat (arrow). Leak resolved following placement of biliary stent.

- Criteria for benign obstruction
  - Tapered transition
    - E.g., chronic pancreatitis, PSC
  - Concentric wall thickening
  - Presence of ductal calculi

EMBRYOLOGY

Embryologic Events

- Bile ducts develop from embryologic ductal plate which surrounds the PV
  - Errors that occur during involution/remodeling of ductal plate lead to variety of fibrocystic abnormalities of the liver
    - Polycystic liver (autosomal recessive and dominant)
    - Caroli disease
    - Biliary hamartomas
    - Choledochal cyst
    - Congenital hepatic fibrosis
  - Often associated with similar fibrocystic diseases of the kidney

CUSTOM DIFFERENTIAL DIAGNOSIS

Bile duct stricture

- Primary sclerosing cholangitis
- Iatrogenic (e.g., laparoscopic surgery)
- Cholangiocarcinoma
- AIDS cholangiopathy
- Chemotherapy cholangitis
- Recurrent pyogenic (Oriental) cholangitis

Gas in biliary tree

- Common
  - Sphincterotomy; post-surgical
  - Biliary stent
  - Patulous sphincter
    - Elderly, prior CBD stone
- Uncommon
  - Emphysematous cholecystitis
  - Parasitic biliary infestation
  - CBD entry into duodenal diverticulum
  - Fistula from gut to CBD or GB
    - Crohn disease
    - Tumor of GB, gut, pancreas
    - Gallstone erosion into duodenum (origin of "gallstone ileus")

Sonographic echogenic and shadowing lesion in GB wall

- Adherent or impacted gallstone
- Spiral valve folds
- Cholesterol polyp
- Stone or crystal in Rokitansky-Aschoff sinus
- Intramural gas
- Gas in duodenum
- Refraction from folds in GB neck

SELECTED REFERENCES

1. Krause D et al: MRI for evaluating congenital bile duct abnormalities. J Comput Assist Tomogr. 26(4):541-52, 2002
2. Gore RM: Gallbladder and Biliary Tract: Differential Diagnosis. In Gore RM, Levine MS (eds) Textbook of Gastrointestinal Radiology. 2nd ed. Philadelphia, WB Saunders. 1408-14, 2000
3. Dohke M et al: Anomalies and anatomic variants of the biliary tree revealed by MR cholangiopancreatography. AJR Am J Roentgenol. 173(5):1251-4, 1999
4. Koeller KK, et al (eds): Radiologic pathology: 2nd ed. Washington, DC, Armed Forces Institute of Pathology. 2003
5. Dähnert W: Radiology review manual 4th ed, Philadelphia, Lippincott, Williams and Wilkins. 2000
6. Eisenberg RL: Gastrointestinal radiology: A pattern approach: 3rd ed. Philadelphia, JB Lippincott. 1996
7. Schulte SJ: Embryology, Normal Variation and Congenital Anomalies of the Gallbladder and Biliary Tract. In: Freeny PC, Stevenson GW (eds): Alimentary Tract Radiology. 5th ed. St. Louis, Mosby 1994

IMAGE GALLERY

Typical

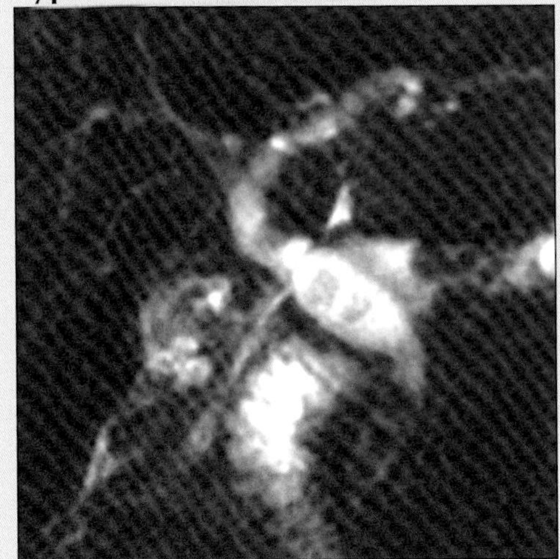

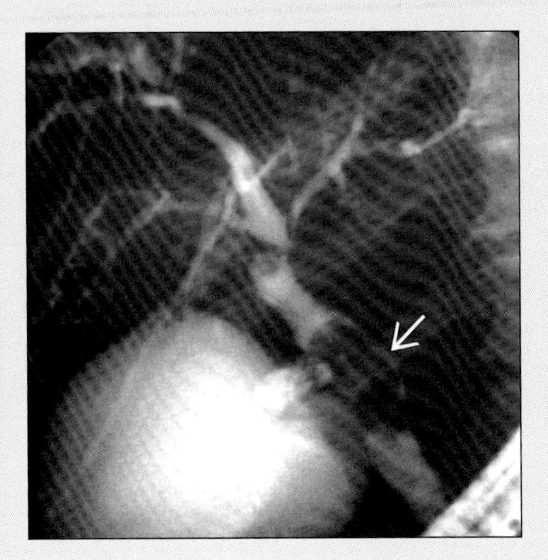

*(**Left**) MRCP in coronal plane shows multiple calculi in dilated CBD. (**Right**) ERCP shows irregular arborization of intrahepatic ducts with multiple strictures due to primary sclerosing cholangitis (PSC). Polypoid mass (arrow) in CBD is cholangiocarcinoma.*

Typical

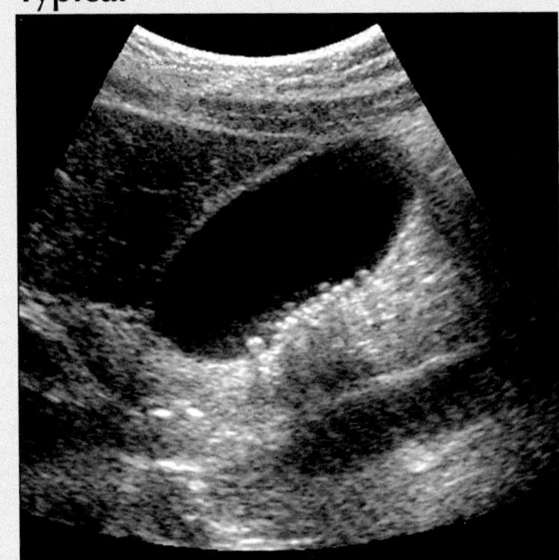

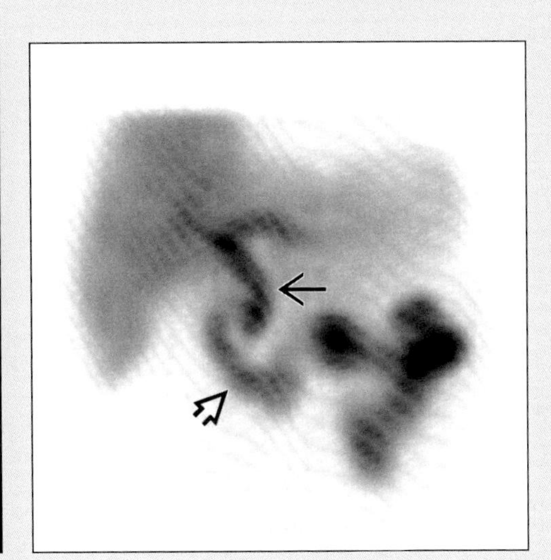

*(**Left**) Sagittal sonogram in 40 year old woman with pain shows distended GB and multiple small stones. (**Right**) Tc-HIDA scan shows flow of radiotracer from liver into CBD (arrow) and bowel (open arrow), but not the GB. Acute cholecystitis.*

Typical

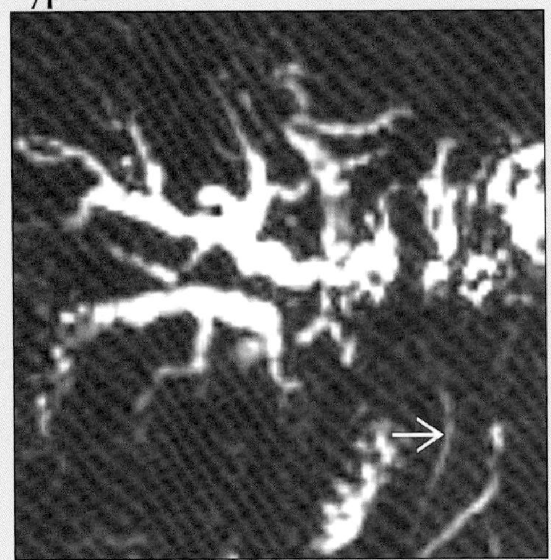

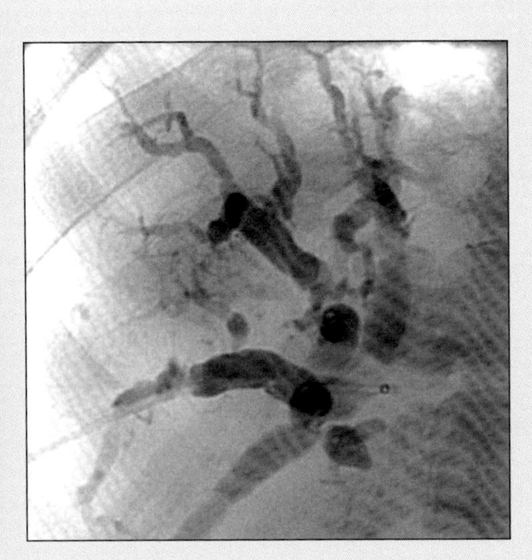

*(**Left**) MRCP shows dilated intrahepatic ducts, normal CBD (arrow); Klatskin tumor. (**Right**) Transhepatic cholangiogram shows dilated intrahepatic ducts, complete obstruction at confluence of main ducts; Klatskin tumor.*

CAROLI DISEASE

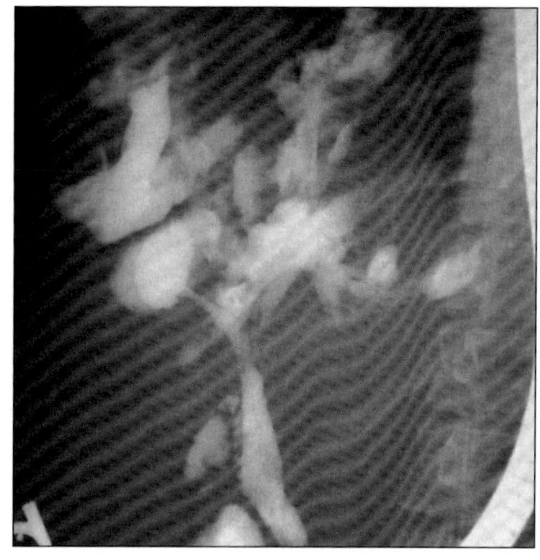

ERCP shows saccular dilatation of intrahepatic bile ducts.

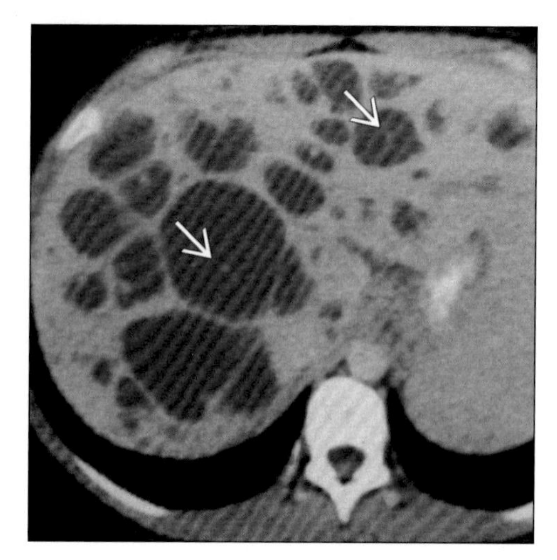

Axial CECT shows massive dilatation of intrahepatic bile ducts. Note the "central" or "eccentric" dot within many of the cystic structures, representing portal radicles (arrows).

TERMINOLOGY

Abbreviations and Synonyms

- Communicating cavernous biliary ectasia

Definitions

- Caroli disease: Congenital, multifocal, segmental, saccular dilatation of intrahepatic bile ducts (IHBD)
- Caroli syndrome: Cystic bile duct dilatation plus hepatic fibrosis

IMAGING FINDINGS

General Features

- Best diagnostic clue: "Central dot" sign: Strong, enhancing, tiny dots (portal radicles) within dilated intrahepatic bile ducts on CECT
- Location: Liver: Diffuse, lobar, or segmental
- Size: Varies from few millimeters to few centimeters
- Morphology
  - One of the variants of fibropolycystic disease
  - Other variants of fibropolycystic disease
    - Congenital hepatic fibrosis
    - Autosomal dominant polycystic liver disease
    - Biliary hamartomas
    - Choledochal cyst
  - Based on Todani classification
    - Type V: Represents Caroli disease
    - Cystic dilatation of intrahepatic bile ducts
  - Caroli disease is of two types
    - Simple type
    - Periportal fibrosis type
    - Frequently associated with renal tubular ectasia
    - Have an autosomal recessive inheritance pattern
  - Usually manifests in adolescence, also seen in newborns & infants

Radiographic Findings

- Endoscopic retrograde cholangiopancreatogram (ERCP) findings
  - Saccular dilatations of IHBDs, stones, strictures
  - May show communicating hepatic abscesses

CT Findings

- NECT: Multiple, rounded, hypodense areas inseparable from dilated IHBD
- CECT: Enhancing tiny dots (portal radicles) within dilated IHBD

MR Findings

- T1WI: Multiple, small, hypointense, saccular dilatations of IHBD
- T2WI: Hyperintense
- Coronal half-Fourier rapid acquisition with relaxation enhancement (RARE)

DDx: Hepatic Cysts with or without Dilated Ducts

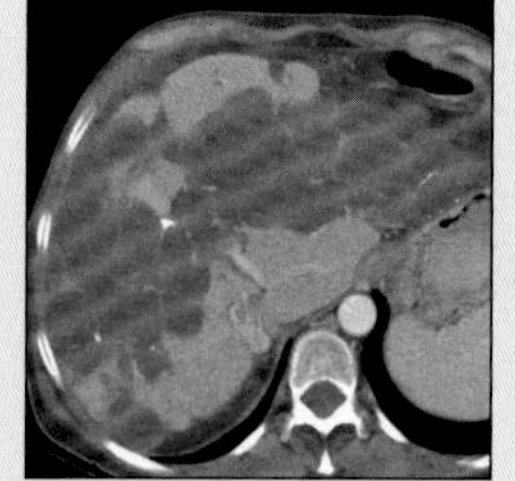

ADPLD

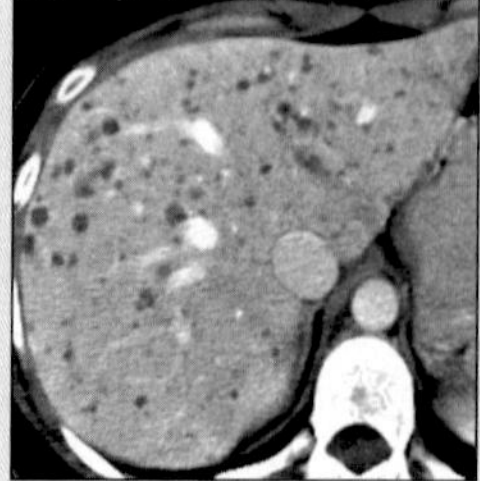

Biliary Hamartomas

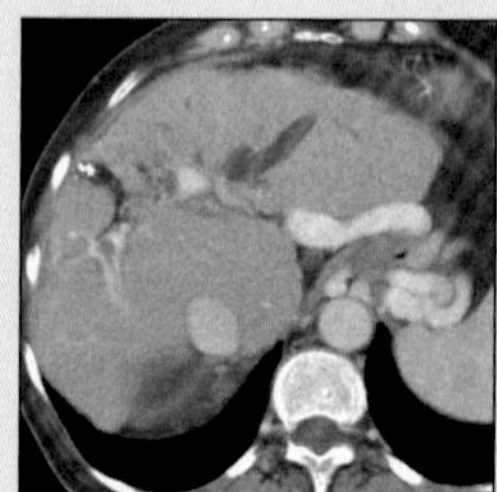

Sclerosing Cholangitis

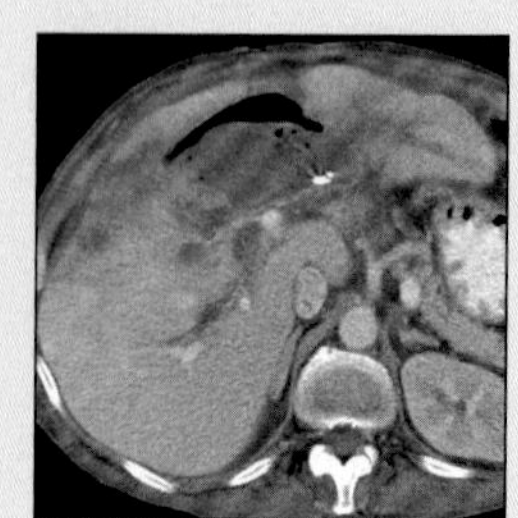

RPC

CAROLI DISEASE

Key Facts

Terminology

- Communicating cavernous biliary ectasia
- Caroli disease: Congenital, multifocal, segmental, saccular dilatation of intrahepatic bile ducts (IHBD)
- Caroli syndrome: Cystic bile duct dilatation plus hepatic fibrosis

Imaging Findings

- Best diagnostic clue: "Central dot" sign: Strong, enhancing, tiny dots (portal radicles) within dilated intrahepatic bile ducts on CECT
- Frequently associated with renal tubular ectasia

Top Differential Diagnoses

- Polycystic liver disease
- Biliary hamartomas
- Primary sclerosing cholangitis (PSC)
- Ascending cholangitis
- Recurrent pyogenic cholangitis (RPC)

Pathology

- Genetics: Inherited as an autosomal recessive pattern
- Simple type
- Periportal fibrosis type

Clinical Issues

- Recurrent attacks of cholangitis, fever & jaundice

Diagnostic Checklist

- Rule out other liver diseases which have hepatic cysts with or without dilated bile ducts
- ERCP: Saccular dilatations show communication with IHBD which differentiates Caroli from other variants of fibropolycystic disease

---

  - Kidney: Multiple fluid-containing foci in papillae (e.g., medullary sponge kidney or renal tubular ectasia)
- T1 C+
  - Enhancement of portal radicles within dilated IHBD
- MR Cholangiopancreatography (MRCP)
  - Multiple hyperintense oval-shaped structures
  - Shows continuity with biliary tree
  - Luminal contents of bile ducts appear hyperintense in contrast to portal vein, which appears as signal void

Ultrasonographic Findings

- Real Time
  - Dilated intrahepatic bile ducts
  - May show intraductal calculi
  - Echogenic septa completely or incompletely traversing dilated lumen of bile ducts (referred to as intraductal bridging)
  - Small portal venous branches partially or completely surrounded by dilated IHBD

Nuclear Medicine Findings

- Hepato Biliary Scan
  - Unusual pattern of retained activity throughout liver
- Technetium Sulfur Colloid
  - Multiple cold defects

Imaging Recommendations

- Best imaging tool: ERCP or 3D MRCP

DIFFERENTIAL DIAGNOSIS

Polycystic liver disease

- Hepatic cysts
  - Numerous (> 10; usually hundreds)
  - Do not communicate with each other or biliary tract
  - Not associated with biliary ductal dilatation
  - Do not opacify at ERCP or percutaneous transhepatic cholangiography (PTC)
- Patients with this disease often harbor renal cysts - not confined to medulla

Biliary hamartomas

- One of the variants of fibropolycystic disease
- Rare benign, congenital malformation of bile ducts
- Location: Intraparenchymal
- Innumerable subcentimeter nodules in liver
- Varied enhancement
  - Cystic: No enhancement
  - Solid: Enhance & become isodense with liver

Primary sclerosing cholangitis (PSC)

- Strictures of both intra- & extrahepatic bile ducts
- Ductal dilatation is not as great as Caroli disease & not saccular type
- PSC often shows isolated obstructions of IHBD; Caroli disease does not
- Usually more prominent findings
  - Strictures & ductal irregularity
- Often progresses to cirrhosis

Ascending cholangitis

- Intrahepatic abscesses communicate with bile ducts
  - Mimics Caroli disease
- Margins of abscesses are irregular
- Extrahepatic bile duct dilatation
  - Noted due to an obstructing stone or tumor

Recurrent pyogenic cholangitis (RPC)

- Dilatation of both intra- & extrahepatic bile ducts; usually of cylindrical and not saccular type
- Biliary calculi of RPC
  - Cast-like
  - Often fill ductal lumen

PATHOLOGY

General Features

- General path comments
  - Embryology-anatomy
    - Ductal plate malformation: Incomplete remodeling of ductal plate leads to persistence of embryonic biliary ductal structures
- Genetics: Inherited as an autosomal recessive pattern

CAROLI DISEASE

- Etiology
  - Simple type
    - Malformation of ductal plate of large central IHBD
    - More common in adults
  - Periportal fibrosis type
    - Malformation of ductal plates of central IHBD & smaller peripheral bile ducts, latter leading to development of fibrosis
    - More common in infants & children
- Epidemiology: Rare disease
- Associated abnormalities
  - Medullary sponge kidney (renal tubular ectasia)
  - Autosomal dominant polycystic kidney disease

Gross Pathologic & Surgical Features

- Saccular dilatations of intrahepatic bile ducts
- Diffuse, lobar or segmental

Microscopic Features

- Simple type
  - Dilatation of segmental IHBD
  - Normal hepatic parenchyma
- Periportal fibrosis type
  - Segmental dilatation of IHBD
  - Proliferation of bile ductules & fibrosis

CLINICAL ISSUES

Presentation

- Most common signs/symptoms
  - Simple type
    - RUQ pain
    - Recurrent attacks of cholangitis, fever & jaundice
  - Periportal fibrosis type
    - Pain, hepatosplenomegaly
    - Hematemesis (due to varices)
  - Can be asymptomatic at an early stage
- Lab data
  - May show elevated liver enzymes & bilirubin levels
- Diagnosis
  - ERCP
  - MRCP

Demographics

- Age
  - Childhood and 2nd-3rd decade
  - Occasionally in infancy
- Gender: M:F = 1:1

Natural History & Prognosis

- Complications
  - Simple type
    - Stone formation (95%): Calcium bilirubinate
    - Recurrent cholangitis
    - Hepatic abscesses
  - Periportal fibrosis type
    - Cirrhosis & portal hypertension
    - Varices & hemorrhage
  - Cholangiocarcinoma in 7% of patients
- Prognosis
  - Long-term prognosis for Caroli disease is poor

Treatment

- Localized to lobe or segment
  - Hepatic lobectomy or segmentectomy
- Diffuse disease
  - Conservative
  - Decompression of biliary tract: External drainage or biliary-enteric anastomoses are effective
  - Extracorporeal shock wave lithotripsy
  - Oral bile salts
  - Liver transplantation

DIAGNOSTIC CHECKLIST

Consider

- Rule out other liver diseases which have hepatic cysts with or without dilated bile ducts

Image Interpretation Pearls

- Cholangiography: Bulbous dilatations of peripheral intrahepatic bile ducts
- ERCP: Saccular dilatations show communication with IHBD which differentiates Caroli from other variants of fibropolycystic disease

SELECTED REFERENCES

1. Guy F et al: Caroli's disease: magnetic resonance imaging features. Eur Radiol. 12(11):2730-6, 2002
2. Levy AD et al: Caroli's disease: radiologic spectrum with pathologic correlation. AJR Am J Roentgenol. 179(4):1053-7, 2002
3. Krause D et al: MRI for evaluating congenital bile duct abnormalities. J Comput Assist Tomogr. 26(4):541-52, 2002
4. Fulcher AS et al: Case 38: Caroli disease and renal tubular ectasia. Radiology. 220(3):720-3, 2001
5. Mortele KJ et al: Cystic focal liver lesions in the adult: differential CT and MR imaging features. Radiographics. 21(4):895-910, 2001
6. Akin O et al: An unusual sonographic finding in Caroli's disease. AJR Am J Roentgenol. 171(4):1167, 1998
7. Gorka W et al: Value of Doppler sonography in the assessment of patients with Caroli's disease. J Clin Ultrasound. 26(6):283-7, 1998
8. Asselah T et al: Caroli's disease: a magnetic resonance cholangiopancreatography diagnosis. Am J Gastroenterol. 93(1):109-10, 1998
9. Pavone P et al: Caroli's disease: evaluation with MR cholangiography. AJR Am J Roentgenol. 166(1):216-7, 1996
10. Miller WJ et al: Imaging findings in Caroli's disease. AJR Am J Roentgenol. 165(2):333-7, 1995
11. Zangger P et al: MRI findings in Caroli's disease and intrahepatic pigmented calculi. Abdom Imaging. 20(4):361-4, 1995
12. Rizzo RJ et al: Congenital abnormalities of the pancreas and biliary tree in adults. Radiographics. 15(1):49-68; quiz 147-8, 1995
13. Choi BI et al: Caroli disease: central dot sign in CT. Radiology. 174(1):161-3, 1990
14. Murphy BJ et al: The CT appearance of cystic masses of the liver. Radiographics. 9(2):307-22, 1989
15. Marchal GJ et al: Caroli disease: high-frequency US and pathologic findings. Radiology. 158(2):507-11, 1986

IMAGE GALLERY

Typical

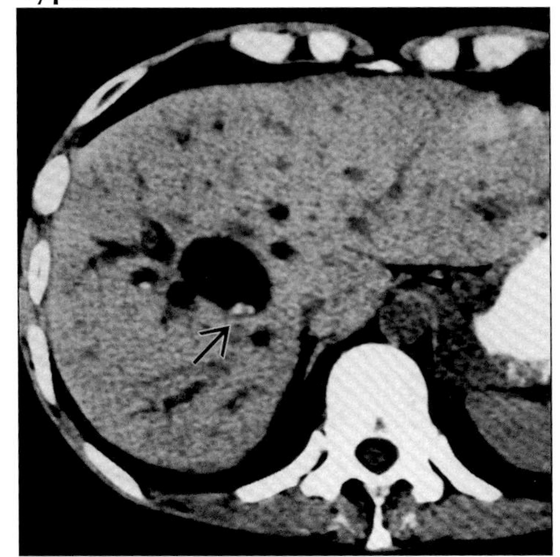
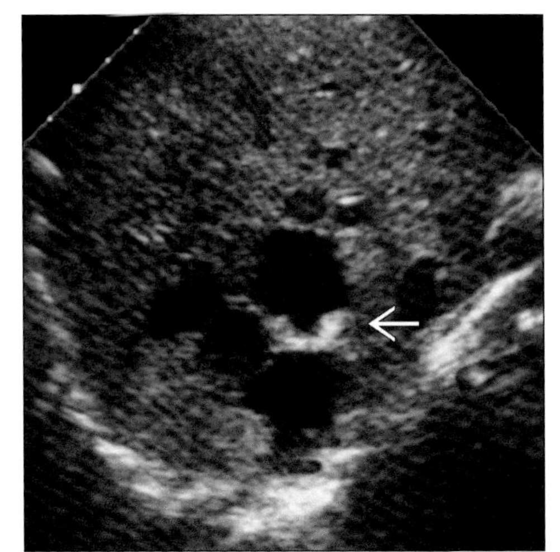

*(**Left**) Axial NECT shows cystic and fusiform dilatation of IHBDs, worse in right lobe. Note calcified stones in dependent position (arrow). (**Right**) Sagittal sonogram shows cystic lesions in liver; some with hyperechoic calculi (arrow).*

Typical

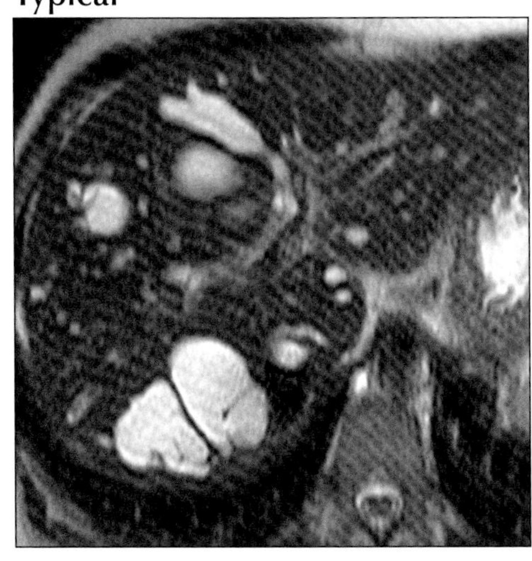
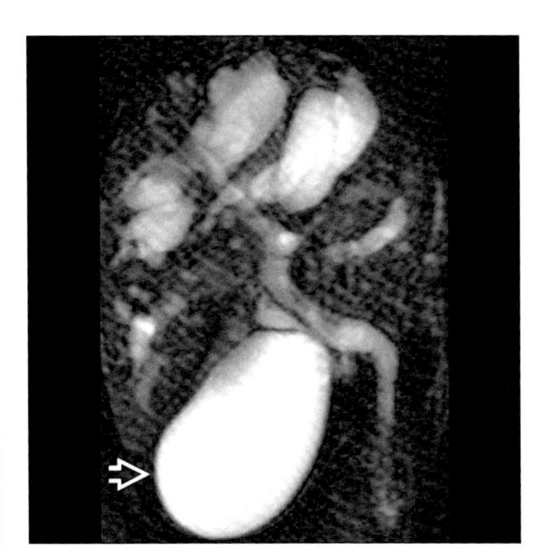

*(**Left**) Axial T2WI MR shows cystic and fusiform dilatation of IHBD. (**Right**) MRCP shows communication between the saccular hyperintense spaces and the biliary tree. Gallbladder (open arrow).*

Typical

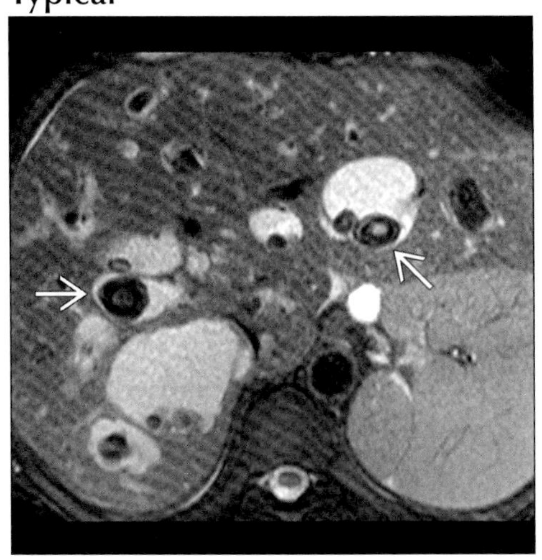
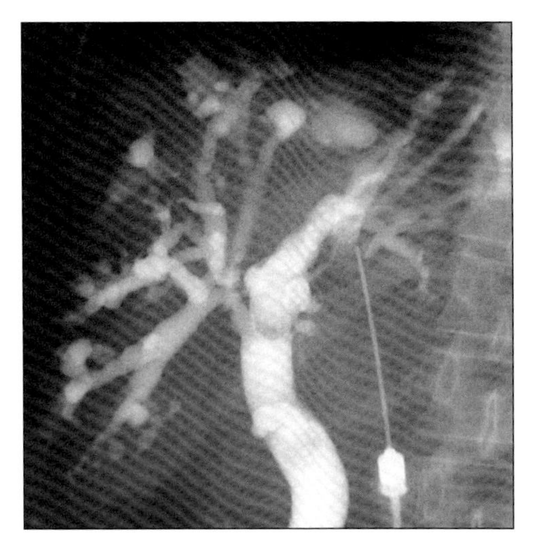

*(**Left**) Axial T2WI MR shows saccular dilatation of IHBD, many containing hypointense calculi (arrows). (**Right**) Percutaneous transhepatic cholangiogram shows generalized saccular dilatation of IHBD.*

CHOLEDOCHAL CYST

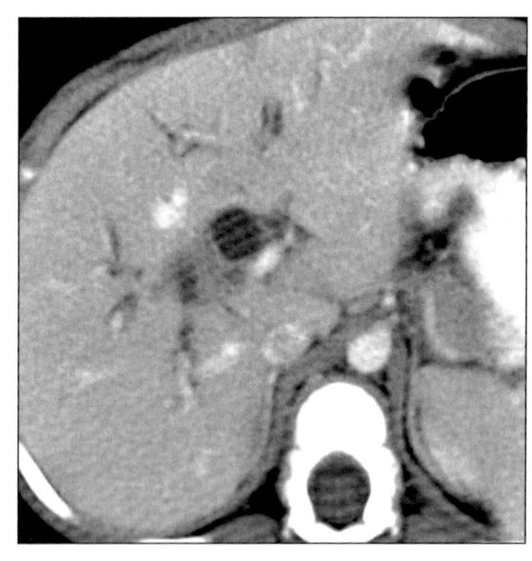

Type IV-a choledochal cyst. Axial CECT shows marked dilatation of common hepatic duct and moderate dilatation of IHBD.

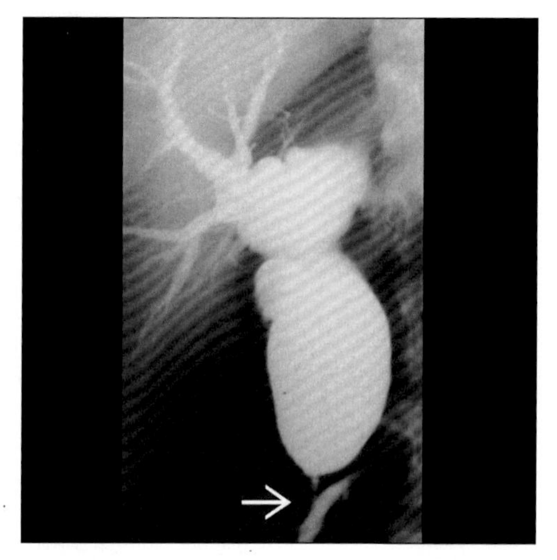

Type IV-a choledochal cyst. ERCP shows fusiform dilatation of common duct, dilated IHBD, and near perpendicular entrance of CBD into pancreatic duct (arrow).

TERMINOLOGY

Definitions

- Congenital segmental aneurysmal dilatation of any portion of bile ducts, most commonly main portion of common duct (CD)

IMAGING FINDINGS

General Features

- Best diagnostic clue: MRCP: Type I choledochal cyst - hyperintense fusiform dilatation of extrahepatic bile duct
- Location: Usually extrahepatic (80-90%); CD
- Size: Varies from 2-15 cm
- Morphology
  - Most common congenital lesion of large bile ducts
  - Maintains continuity with biliary tree
  - Often co-exist with other cystic & fibrotic disorders of liver
  - Rare & usually manifest in infancy and childhood
  - Increased female predominance
  - Todani classification of choledochal cysts
    - Type I: Solitary fusiform - extrahepatic (80-90%); CD
    - Type II: Extrahepatic supraduodenal diverticulum
    - Type III: Intraduodenal diverticulum; choledochocele
    - Type IVa: Fusiform & intrahepatic cysts
    - Type IVb: Multiple extrahepatic cysts
    - Type V: Multiple intrahepatic cysts; Caroli disease

Radiographic Findings

- Radiography
  - Upper gastrointestinal series may show
    - Anterior displacement of second part of duodenum & antrum
    - Inferior displacement of duodenum
    - Widening of duodenal sweep
- ERCP
  - Demonstrates all types of choledochal cysts
  - Cystic or fusiform dilatation of common duct (CD)
  - Shows CD mucosal diaphragm & aberrant insertion of CBD into pancreatic duct

CT Findings

- NECT
  - Well-defined water density cystic lesion along course of bile ducts
  - May show dilatation of intra-/extrahepatic bile ducts
- CECT
  - Nonenhancing hypodense cystic lesion
  - May show dilatation of intra-/extrahepatic bile ducts

DDx: Grossly Dilated Bile Ducts or Cysts Simulating Dilated Bile Ducts

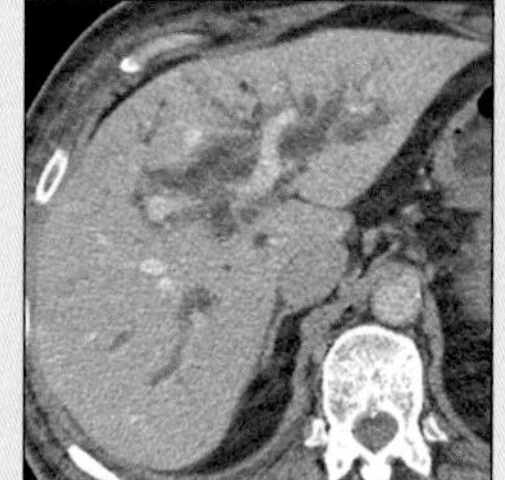

Pancreatic Carcinoma

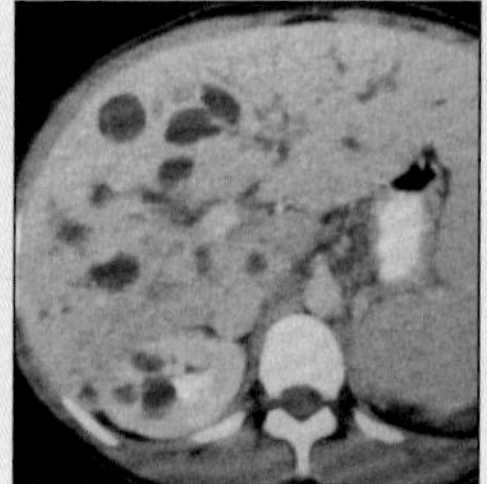

Caroli Disease

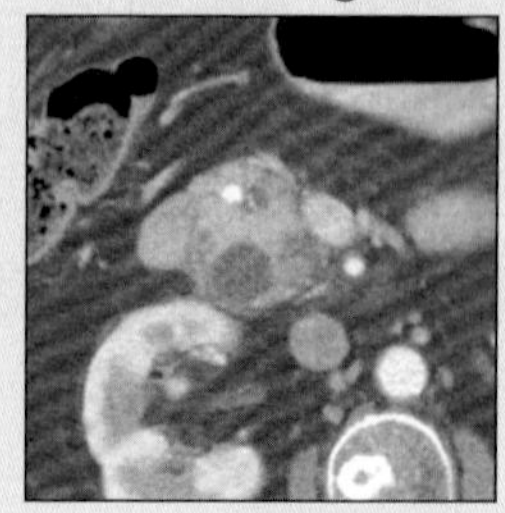

Pancreatic Pseudocyst

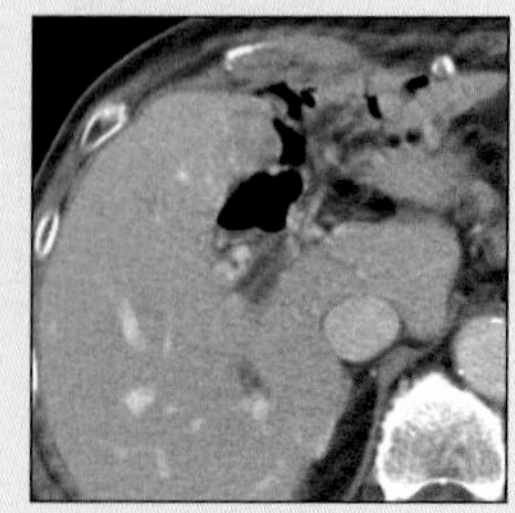

RPC

CHOLEDOCHAL CYST

Key Facts

Terminology

- Congenital segmental aneurysmal dilatation of any portion of bile ducts, most commonly main portion of common duct (CD)

Imaging Findings

- Best diagnostic clue: MRCP: Type I choledochal cyst - hyperintense fusiform dilatation of extrahepatic bile duct
- Size: Varies from 2-15 cm
- May show dilatation of intra-/extrahepatic bile ducts

Top Differential Diagnoses

- Malignant CBD obstruction
- Caroli disease
- Pancreatic pseudocyst
- Recurrent pyogenic cholangitis (RPC)

Pathology

- Due to ductal plate malformation of large bile ducts
- Anomalous junction of CBD & pancreatic duct proximal to duodenal papilla forming long common channel strongly associated with type I & IV choledochal cysts of Todani classification
- Incidence: Asians more than Western countries

Clinical Issues

- Triad: Recurrent RUQ pain, jaundice, palpable mass

Diagnostic Checklist

- Rule out other pathologies which can cause "marked biliary dilatation"
- MRCP or ERCP: Large, solitary cystic or fusiform dilatation of CBD with intra- & extrahepatic bile duct dilatation

MR Findings

- T1WI: Hypointense
- T2WI: Hyperintense
- T1 C+: Nonenhancing lesion
- MRCP (MR cholangiopancreatography)
  - Bile appears hyperintense in contrast to portal vein
    - Type I: Solitary cystic or fusiform dilatation of common duct (CD)
    - Type II: Extrahepatic supraduodenal diverticulum
    - Type III: Bulbous dilatation of intramural segment of distal CD
    - Type IVa: Marked dilatation of entire extrahepatic bile duct plus central intrahepatic bile ducts (IHBD) on coronal oblique images

Ultrasonographic Findings

- Real Time
  - Transverse scan
    - Large anechoic cyst in subhepatic area
  - Oblique scan: Large, marked cystic & fusiform dilatation of IHBD plus CBD (type IVa cyst)
  - Obstetric-ultrasound
    - Earliest diagnosis at 25 weeks of pregnancy
    - Right-sided large cyst in fetal abdomen plus adjacent dilated hepatic ducts
    - DDx: Duodenal atresia; cyst of liver, mesentery, omentum or ovary

Nuclear Medicine Findings

- Hepato Biliary Scan
  - Large photopenic area in liver
  - Shows late filling & prolonged stasis of isotope
  - Excludes all other differential diagnoses

Imaging Recommendations

- ERCP or MRCP (coronal & oblique views)
- NE + CECT (coronal reconstruction images)

DIFFERENTIAL DIAGNOSIS

Malignant CBD obstruction

- Extrahepatic: Short stricture or small polypoid mass
- Dilatation of both intra- & extrahepatic bile ducts proximal to obstruction
- NECT
  - Small growth: Poorly sensitive
  - Large growth: Hypodense mass & IHBD dilatation
- CECT
  - Persistent enhancing tumor (due to fibrous stroma)
- ERCP
  - Exophytic intraductal tumor mass (2-5 mm)
  - Infiltrating type
    - Frequently long & rarely short focal stricture

Caroli disease

- Inherited as an autosomal recessive pattern
- Due to ductal plate malformation of large intrahepatic bile ducts
- Multiple small rounded hypodense or hypointense saccular dilatations of intrahepatic bile ducts
- CECT: "Central dot" sign
  - Enhancing tiny dots (portal radicles) within dilated cystic intrahepatic bile ducts
- ERCP, MRCP & microscopic finding
  - Communicating bile duct abnormality
- According to Todani classification Caroli disease represents type V

Pancreatic pseudocyst

- No communication with bile ducts
- Cystic mass with infiltration of peripancreatic fat planes
- NECT
  - Round hypodense lesion of near water density
  - Lobulated, heterogeneous, mixed density lesion (due to hemorrhage or infection)
- CECT
  - Enhancement of thin rim of fibrous capsule
- MRCP: Hyperintense cyst contiguous with dilated pancreatic duct
- ERCP: Pseudocyst communicating with pancreatic duct seen in 70% cases

Recurrent pyogenic cholangitis (RPC)

- Dilatation of both intra- & extrahepatic bile ducts
- Stones, sludge, pneumobilia & abscess

CHOLEDOCHAL CYST

- Biliary calculi: Cast-like & often fill ductal lumen
- MRCP: Skip dilatations, strictures of intra- & extrahepatic bile ducts
- Etiology: Parasites, stones, gram negative bacteria
- Most commonly seen in Asians

PATHOLOGY

General Features

- General path comments
  - Embryology-anatomy
    - Due to ductal plate malformation of large bile ducts
    - Faulty budding of primitive pancreatic duct
- Etiology
  - Anomalous junction of CBD & pancreatic duct proximal to duodenal papilla forming long common channel strongly associated with type I & IV choledochal cysts of Todani classification
  - Higher pressure in pancreatic duct & absent ductal sphincter
  - Free reflux of enzymes into CBD causes weakening of CBD wall & dilatation
- Epidemiology
  - Prevalence: 1:13,000 admissions
  - Incidence: Asians more than Western countries
- Associated abnormalities
  - Gallbladder: Aplasia or double gallbladder
  - Biliary anomalies
    - Biliary atresia or stenosis
    - Congenital hepatic fibrosis
  - Annular pancreas

Gross Pathologic & Surgical Features

- Cystic/fusiform dilated sac with bile, stones or sludge
- Long ectatic common channel with pancreatic duct
  - Normal length: 0.2-1.0 cm; average: 0.5 cm

Microscopic Features

- Widespread ulceration & a denuded mucosa in dilated common bile duct (CBD)
- Thickened ductal wall consists of chronic inflammatory cells & fibrous tissue

Staging, Grading or Classification Criteria

- Classification of anomalous pancreaticobiliary ductal junction (APBD)
  - Type (P-B): Perpendicular insertion of pancreatic duct into CBD (fusiform)
  - Type (B-P): Perpendicular insertion of CBD into pancreatic duct (cystic)
  - Two major duct unions are associated with type I choledochal cyst
- Classification according to angle of ductal union
  - Right-angled union: Cystic dilatation of CBD
  - Acute angled union: Fusiform dilatation of CBD

CLINICAL ISSUES

Presentation

- Most common signs/symptoms
  - Triad: Recurrent RUQ pain, jaundice, palpable mass
    - Abdominal pain (fusiform); palpable mass or jaundice (cystic)
  - Infants: Intermittent jaundice & abdominal mass
  - Children & adults: Intermittent fever, vomiting, jaundice, pain & pruritus

Demographics

- Age
  - Usually seen in infancy & childhood
  - Can present from birth to old age
  - 25% detected before age 1
  - 80% diagnosed in childhood
    - 60% of patients present before age 10
  - 20% in adults
- Gender: M:F = 1:4

Natural History & Prognosis

- Complications
  - Calculi, cholangitis, pancreatitis
  - Rupture, bile peritonitis, abscess & hemorrhage
  - Rarely malignant degeneration
    - Cholangiocarcinoma
- Prognosis
  - Usually good after surgical repair
  - Poor: Rupture, peritonitis & malignant degeneration

Treatment

- Surgical excision & reconstruction by Roux-en-Y hepaticojejunostomy

DIAGNOSTIC CHECKLIST

Consider

- Rule out other pathologies which can cause "marked biliary dilatation"

Image Interpretation Pearls

- MRCP or ERCP: Large, solitary cystic or fusiform dilatation of CBD with intra- & extrahepatic bile duct dilatation

SELECTED REFERENCES

1. Dohke M et al: Anomalies and anatomic variants of the biliary tree revealed by MR cholangiopancreatography. AJR Am J Roentgenol. 173(5):1251-4, 1999
2. Govil S et al: Choledochal cysts: evaluation with MR cholangiography. Abdom Imaging. 23(6):616-9, 1998
3. Irie H et al: Value of MR cholangiopancreatography in evaluating choledochal cysts. AJR Am J Roentgenol. 171(5):1381-5, 1998
4. Matos C et al: Choledochal cysts: comparison of findings at MR cholangiopancreatography and endoscopic retrograde cholangiopancreatography in eight patients. Radiology. 209(2):443-8, 1998
5. Kim OH et al: Imaging of the choledochal cyst. Radiographics. 15(1):69-88, 1995
6. Rizzo RJ et al: Congenital abnormalities of the pancreas and biliary tree in adults. Radiographics. 15(1):49-68; quiz 147-8, 1995
7. Ebel KD et al: Choledochal cysts: classification and cholangiographic appearance. AJR Am J Roentgenol. 159(3):674-5, 1992

IMAGE GALLERY

Typical

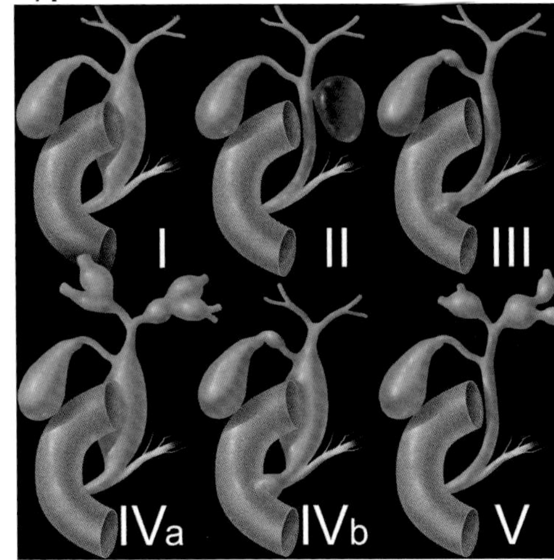

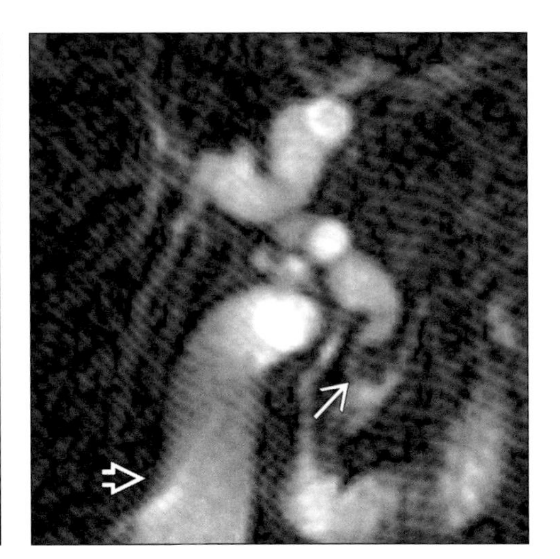

*(**Left**) Todani classification of choledochal cysts. (**Right**) Type IVa choledochal cyst. MRCP shows fusiform dilatation of intra- and extrahepatic bile ducts and abnormal entrance of CBD into pancreatic duct (arrow). Gallbladder (open arrow).*

Typical

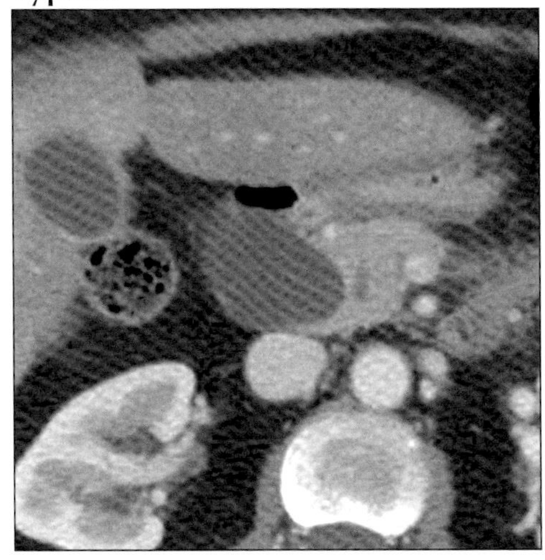

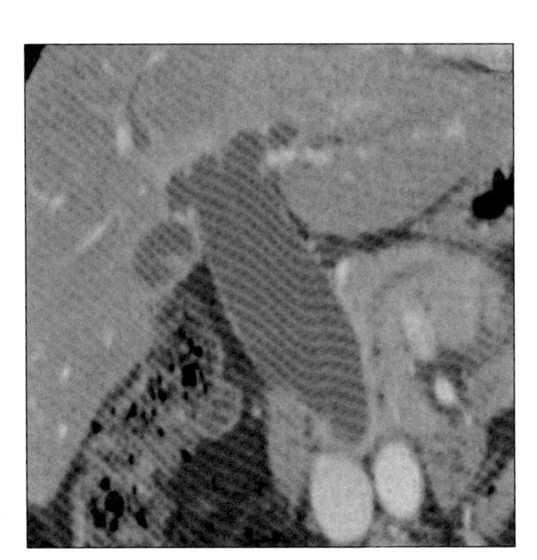

*(**Left**) Axial CECT shows markedly dilated extra- and intrapancreatic CBD. Type I choledochal cyst. (**Right**) Coronal reformation of CECT shows fusiform dilatation of common duct. Type I choledochal cyst.*

Variant

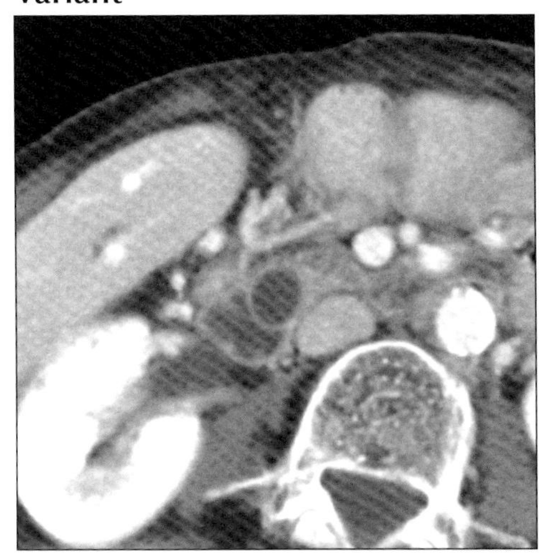

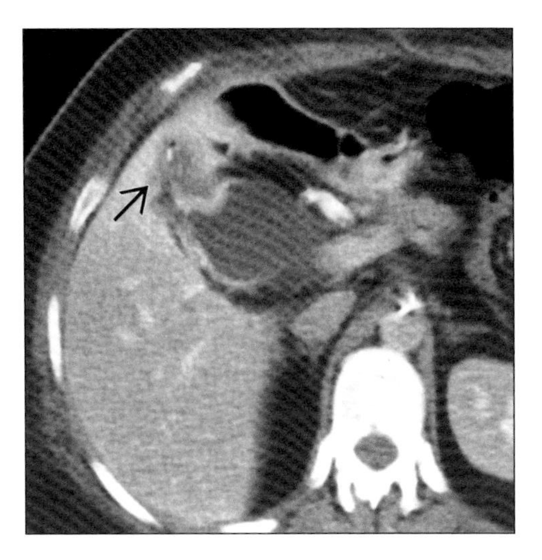

*(**Left**) Axial CECT shows a spherical cystic lesion within the medial wall of the duodenum, while the rest of the biliary tree was normal. Type III; choledochocele. (**Right**) Axial CECT shows Type I choledochal cyst with cholangiocarcinoma. Dilated common duct with irregular wall thickening and a mass (arrow) invading adjacent liver.*

ASCENDING CHOLANGITIS

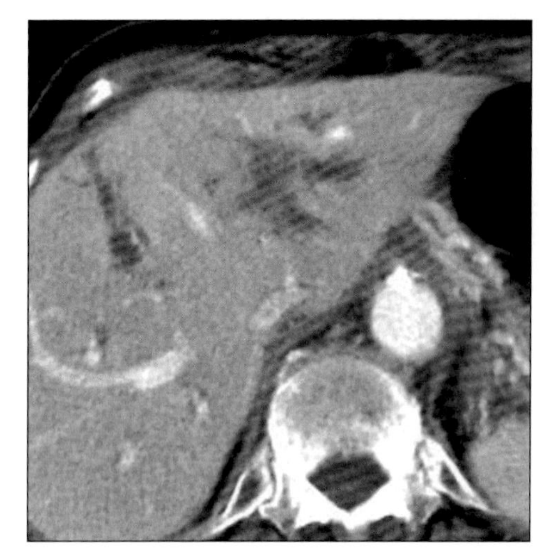

Axial CECT shows dilated IHBD with indistinct margins.

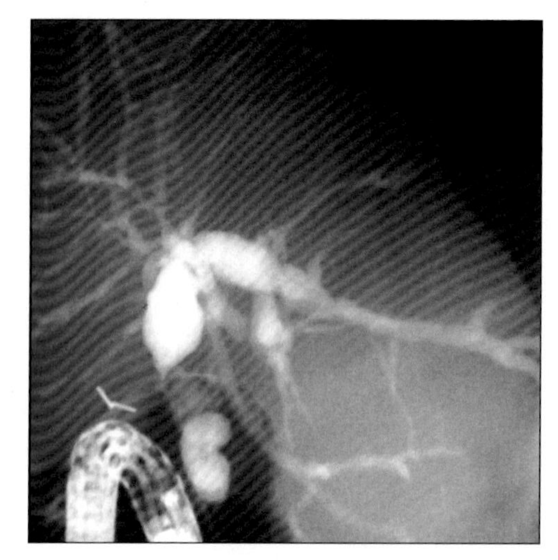

ERCP shows dilated left IHBD with abnormal arborization and tapering.

TERMINOLOGY

Abbreviations and Synonyms

- Ascending (bacterial) cholangitis

Definitions

- Inflammation of intra-/extrahepatic bile duct walls, usually due to ductal obstruction and infection

IMAGING FINDINGS

General Features

- Best diagnostic clue: Irregular contour, branching pattern & dilatation of bile ducts
- Location: Intra-/extrahepatic bile ducts
- Morphology
  - Usually secondary to gallstones & infection in industrialized countries
  - Often due to poor nutrition & parasitic infestation in developing countries
  - Classification of cholangitis (etiology/pathogenesis)
    - Primary sclerosing cholangitis (PSC)
    - Secondary sclerosing cholangitis
    - Secondary nonsclerosing cholangitis
  - Secondary sclerosing cholangitis
    - Ascending (bacterial) cholangitis
    - Recurrent pyogenic (parasitic) cholangitis (RPC)
    - AIDS-related cholangitis
    - Chemotherapy-induced cholangitis
    - Ischemic cholangitis
  - Secondary nonsclerosing cholangitis
    - Malignant or benign liver/biliary pathology
  - Based on onset, classified into acute & chronic

Radiographic Findings

- Cholangiography
  - Ascending (bacterial) cholangitis
    - Stone: Radiolucent filling defect
    - Irregular & thick bile duct lumen/wall
    - Ductal stricture, obstruction & proximal dilatation
    - IHBD may show communicating hepatic abscesses
  - Secondary sclerosing & nonsclerosing may mimic primary sclerosing cholangitis at cholangiography

CT Findings

- Obstructing stone: Calcific/soft tissue/water density
- "Bull's eye" sign: Rim of bile surrounding a stone
- Dilatation of intra-/extrahepatic bile ducts
- High density intraductal material (purulent bile)
- Communicating small hepatic abscesses may be seen

MR Findings

- T1WI
  - Stones (hypointense); bile (hypointense)
  - Dilatations, strictures, thickening of bile duct wall
- T2WI: Stones (hypointense); bile (hyperintense)

DDx: Irregularly Dilated Bile Ducts

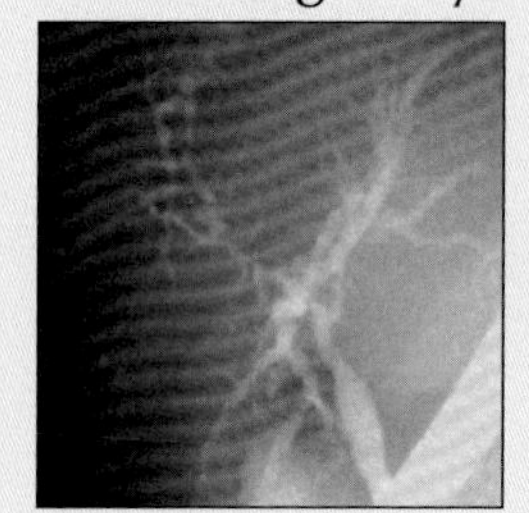

Sclerosing Cholangitis

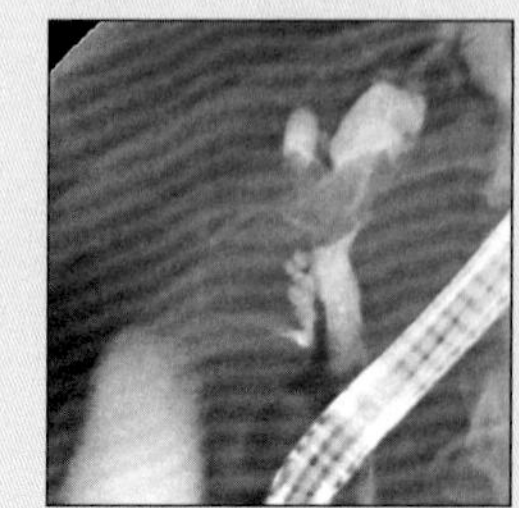

RPC

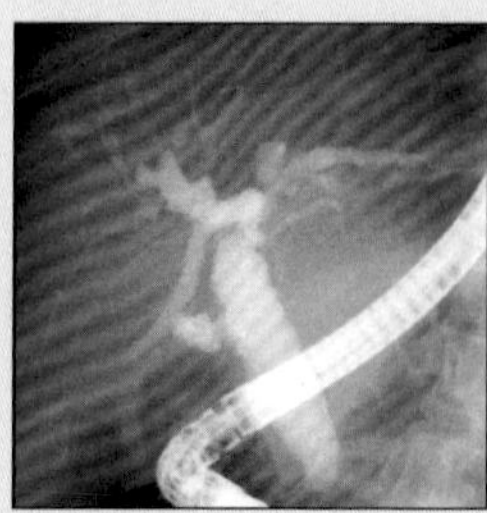

AIDS Cholangiopathy

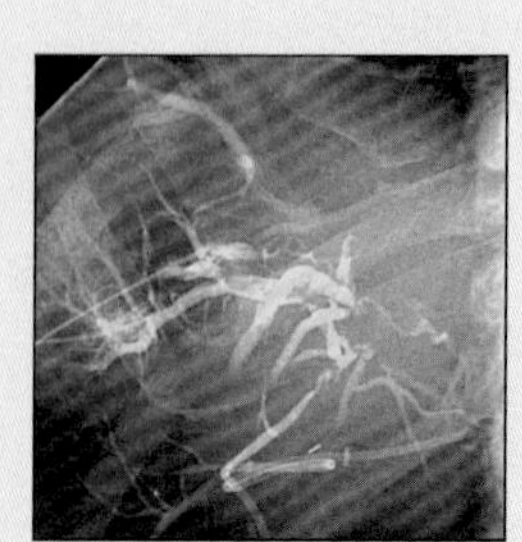

Chemo. Cholangitis

ASCENDING CHOLANGITIS

Key Facts

Terminology
- Ascending (bacterial) cholangitis

Imaging Findings
- Best diagnostic clue: Irregular contour, branching pattern & dilatation of bile ducts
- "Bull's eye" sign: Rim of bile surrounding a stone
- High density intraductal material (purulent bile)
- Communicating small hepatic abscesses may be seen

Top Differential Diagnoses
- Primary sclerosing cholangitis (PSC)
- Recurrent pyogenic (RPC), AIDS, chemotherapy cholangitis

Diagnostic Checklist
- Correlate with clinical & lab data to achieve an accurate cholangiographic interpretation

- MRCP
  - Low signal filling defects (stones) within increased signal bile
  - Irregular strictures, proximal dilatation of bile ducts

Ultrasonographic Findings
- Real Time
  - Dilatation, stenosis & thickening of bile duct walls
  - Purulent bile: Intraluminal echogenic material
  - Gallbladder: Thickened wall with or without calculi

Imaging Recommendations
- Ultrasonography & MRCP
- Cholangiography (T-tube, retrograde, PTC)

DIFFERENTIAL DIAGNOSIS

Primary sclerosing cholangitis (PSC)
- Segmental strictures, beaded and pruned ducts
- Involves both intrahepatic & extrahepatic ducts
- End-stage: Liver (lobular, hypertrophy & atrophy)

Recurrent pyogenic (RPC), AIDS, chemotherapy cholangitis
- Clinical setting supports diagnosis

PATHOLOGY

General Features
- Etiology
  - Due to bile duct calculi, stricture & papillary stenosis
  - Pathogenesis: Stone/obstruction/bile stasis/infection
- Epidemiology: Most common type of cholangitis in Western countries
- Associated abnormalities: Usually gallstones

CLINICAL ISSUES

Presentation
- Most common signs/symptoms: Charcot triad (pain, fever, jaundice)
- Lab data
  - Increased WBC count & bilirubin levels
  - Increased alkaline phosphatase
  - Positive blood cultures in toxic phase

Demographics
- Age: 20-50 years
- Gender: M:F = 1:1

Natural History & Prognosis
- Complications: Small liver abscesses & sepsis
- Prognosis: 100% mortality if not decompressed

Treatment
- Antibiotics to cover gram negative organisms
- Interventional management of stones/strictures

DIAGNOSTIC CHECKLIST

Consider
- Correlate with clinical & lab data to achieve an accurate cholangiographic interpretation

Image Interpretation Pearls
- Cholangiography: Strictures, dilatations, intraluminal filling defects due to stones

SELECTED REFERENCES

1. Arai K et al: Dynamic CT of acute cholangitis: early inhomogeneous enhancement of the liver. AJR Am J Roentgenol. 181(1):115-8, 2003
2. Song HH et al: Eosinophilic cholangitis: US, CT, and cholangiography findings. J Comput Assist Tomogr. 21(2):251-3, 1997
3. Balthazar EJ et al: Acute cholangitis: CT evaluation. J Comput Assist Tomogr. 17(2):283-9, 1993

IMAGE GALLERY

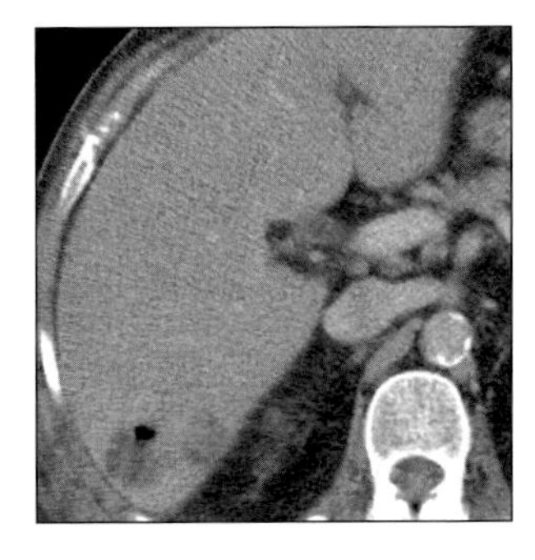

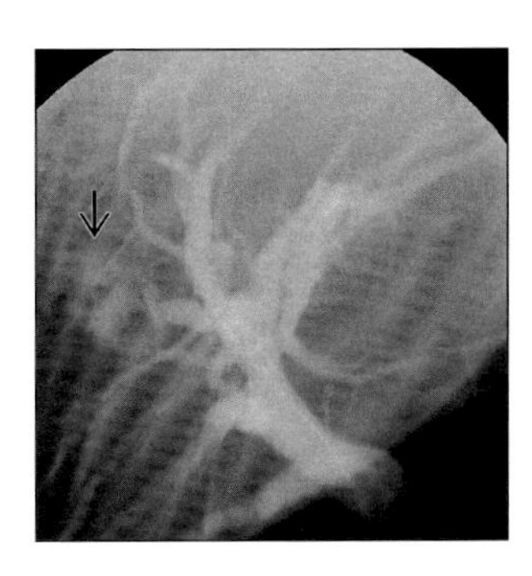

(Left) *Axial CECT shows an abscess in right hepatic lobe due to ascending cholangitis.* ***(Right)*** *ERCP shows moderate dilatation of bile ducts, extravasation into a liver abscess (arrow). CBD stone was extracted endoscopically earlier.*

RECURRENT PYOGENIC CHOLANGITIS

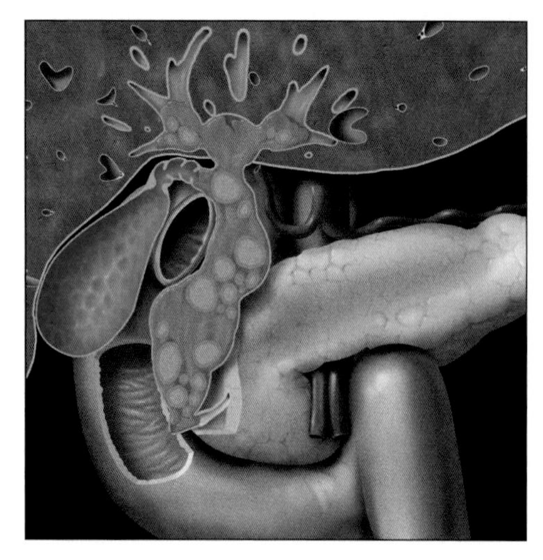

Graphic demonstrates marked dilation of intrahepatic bile ducts with multiple common bile duct and intrahepatic stones.

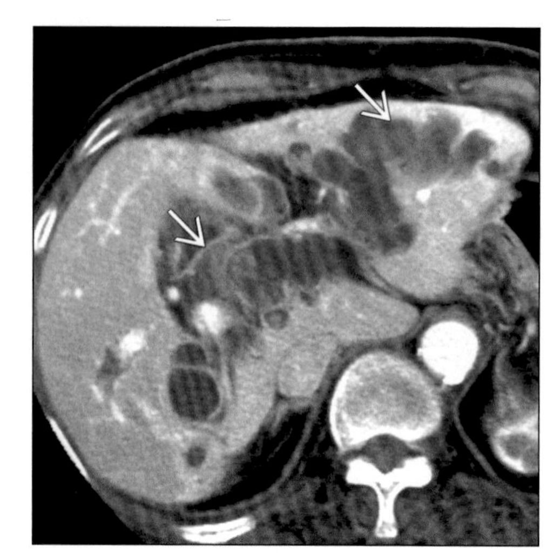

Axial CECT of recurrent pyogenic cholangitis demonstrates marked intrahepatic biliary dilatation with numerous intrahepatic stones (arrows).

TERMINOLOGY

Abbreviations and Synonyms

- Recurrent pyogenic cholangitis (RPC), oriental cholangitis, oriental cholangiohepatitis

Definitions

- Intra- and extrahepatic biliary pigment stones occurring in patients and immigrants from SE Asia

IMAGING FINDINGS

General Features

- Best diagnostic clue: Intra- and extrahepatic biliary stones without stones in gallbladder (GB)
- Location: Confined to left lobe (often lateral segment) or involving all biliary ductal segments & common bile duct (CBD)
- Size: Stones are typically 1-4 cm in size
- Morphology: Combination of pigment stones and biliary sludge

Radiographic Findings

- ERCP
  - Dilated intra- and extrahepatic bile ducts
  - Common duct stones and intrahepatic duct stones without stones in gallbladder
  - Rapid tapering of dilated intrahepatic ducts with "arrowhead" configuration
  - Non-filling of biliary ductal segments due to strictures of intrahepatic ducts

CT Findings

- NECT: Biliary stones may be high attenuation or isodense to liver
- CECT
  - Dilated intra- and extrahepatic biliary ducts within involved segments on CECT
  - CBD may be markedly enlarged
  - May be associated with low attenuation pyogenic liver abscesses, fatty liver atrophy of segments with chronic biliary obstruction

MR Findings

- T1WI: Hypointense dilated ducts and intermediate intensity biliary calculi; may have hyperintense rim
- T2WI: Hyperintense bile within obstructed ducts and low signal calculi
- T1 C+: Hypointense dilated bile ducts with low to intermediate signal biliary calculi
- MRCP
  - Dilated intra- and extrahepatic ducts with low signal filling defects representing stones; intrahepatic ducts taper rapidly ("arrowhead sign")

DDx: Spectrum of Biliary Lesions Mimicking RPC

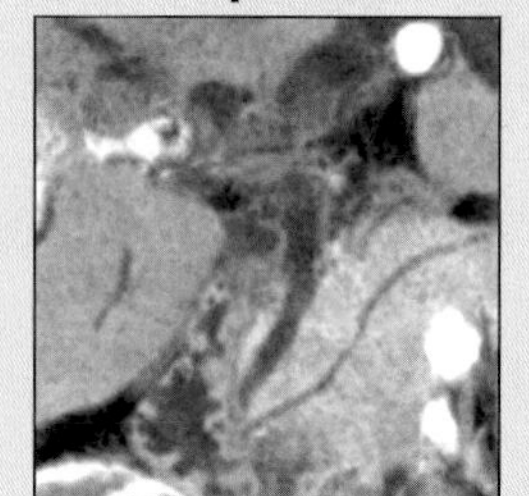

Hepatic Stones

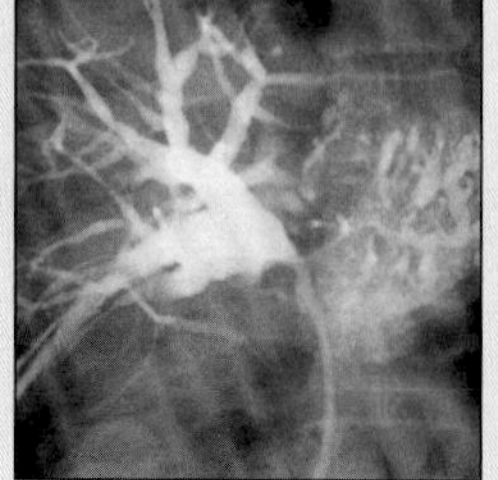

Cholangitis

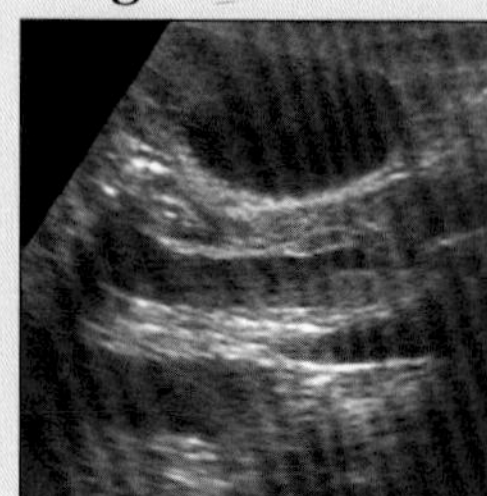

Bact Cholangitis

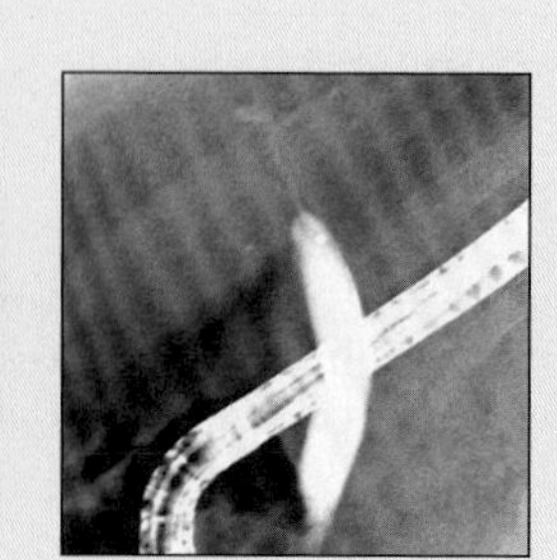

Cholangiocarcinoma

RECURRENT PYOGENIC CHOLANGITIS

Key Facts

Terminology
- Intra- and extrahepatic biliary pigment stones occurring in patients and immigrants from SE Asia

Imaging Findings
- Best diagnostic clue: Intra- and extrahepatic biliary stones without stones in gallbladder (GB)
- Location: Confined to left lobe (often lateral segment) or involving all biliary ductal segments & common bile duct (CBD)
- Dilated intra- and extrahepatic biliary ducts within involved segments on CECT

Top Differential Diagnoses
- Intrahepatic stones secondary to biliary stricture
- Sclerosing cholangitis
- Bacterial cholangitis
- Cholangiocarcinoma

Pathology
- Intraductal pigment calculi within intra- and extrahepatic ducts, proliferative fibrosis of CBD walls, periductal abscesses
- Associated with biliary parasitic infection with Clonorchis sinensis and/or ascaris lumbricoides

Clinical Issues
- RUQ pain, recurrent fevers, jaundice
- Leukocytosis, elevated alkaline phosphatase and bilirubin

Diagnostic Checklist
- Intra- & extrahepatic bile duct dilatation/stones in SE Asian patients

Ultrasonographic Findings
- Real Time: Dilated intrahepatic ducts with echogenic debris; calculi may/may not cause acoustic shadowing
- Color Doppler: No flow within dilated bile ducts

Nuclear Medicine Findings
- WBC Scan
  - Positive for cholangitic liver abscesses

Other Modality Findings
- Cholangiographic findings: Dilated intra- and extrahepatic ducts with filling defects (stones)
  - "Arrowhead" deformity of rapidly tapering intrahepatic ducts
  - Similar to ERCP

Imaging Recommendations
- Best imaging tool: CECT
- Protocol advice
  - 150 ml IV contrast at 2.5 ml/sec for CECT
    - 5 mm collimation and reconstruction at 5 mm intervals
  - Heavily T2WI/MRCP and gadolinium-enhanced breathheld GRE

DIFFERENTIAL DIAGNOSIS

Intrahepatic stones secondary to biliary stricture
- Stricture may be due to prior surgery, trauma or chemotherapy
- Non-Asian patient
- Similar clinical presentation with RUQ pain, fever and chills

Sclerosing cholangitis
- Diffuse thickening of CBD
- Multiple intrahepatic strictures
- Stones form distal to strictures
- Associated with inflammatory bowel disease

Bacterial cholangitis
- Dilated intra- and extrahepatic ducts
- Stones, sludge and pus in bile ducts
- Intra- or extrahepatic strictures

Cholangiocarcinoma
- Associated with sclerosing cholangitis, choledochal cyst, RPC, clonorchiasis
- Infiltrative type at confluence of right and left ducts most common
- Ductal dilatation of involved segments with or without parenchymal mass
- CECT demonstrates delayed contrast-enhancement within parenchymal mass components

PATHOLOGY

General Features
- General path comments
  - Intraductal pigment calculi within intra- and extrahepatic ducts, proliferative fibrosis of CBD walls, periductal abscesses
  - End stage biliary cirrhosis
- Genetics: No known genetic predisposition
- Etiology
  - Associated with biliary parasitic infection with Clonorchis sinensis and/or ascaris lumbricoides
  - Associated with E. coli infection of bile ducts
    - Bacterial production of beta-glucuronidase
    - Leads to hydrolysis of bilirubin, development of calcium bilirubinate stones within intra- & extrahepatic bile ducts
  - Associated with poor general nutrition
- Epidemiology
  - Primarily within SE Asia and immigrants from SE Asia
  - Endemic in SE Asian patients and immigrants from SE Asia (China, Vietnam, Philippines)
- Associated abnormalities: Poor nutrition

RECURRENT PYOGENIC CHOLANGITIS

Gross Pathologic & Surgical Features

- Dilated bile ducts with brown mud-like pigment stones, pus
- May have parasitic infection in biliary ducts with Clonorchis or ascaris

Microscopic Features

- Periductal inflammatory changes with infiltration of periportal spaces with inflammatory cells leading to periductal fibrosis and ultimately biliary cirrhosis
  - Inflammatory cells leading to periductal fibrosis and ultimately biliary cirrhosis
- Localized segmental hepatic atrophy
- Fatty changes in liver

Staging, Grading or Classification Criteria

- Classification based on distribution of affected biliary segment
  - May be isolated to left lobe, particularly lateral segment
  - May involve all biliary segments, as well as CBD

CLINICAL ISSUES

Presentation

- Most common signs/symptoms
  - RUQ pain, recurrent fevers, jaundice
  - Other signs/symptoms
    - Hypotension, shaking, chills
    - Related to gram-negative septicemia
- Clinical profile
  - Leukocytosis, elevated alkaline phosphatase and bilirubin
  - Diagnosis by CT, US or cholangiography

Demographics

- Age: Over 40
- Gender: Affects males and females equally
- Ethnicity: Chinese and SE Asian population

Natural History & Prognosis

- Repeated episodes of cholangitis
- Complications
  - Benign: Liver abscesses, biliary stricture and biliary stones
    - Long-term repeated episodes of cholangitis & stricture formation lead to biliary cirrhosis
  - Malignant: Cholangiocarcinoma

Treatment

- Options, risks, complications
  - Endoscopic sphincterotomy
  - Surgical
    - Biliary drainage with hepatico-jejunostomy
    - Subcutaneous jejunal ostomy for biliary access
    - Left hepatic lobe resection if isolated left lobe disease
  - Interventional radiology
    - Percutaneous biliary drainage of affected segments
    - Basket and removal of pigment stones
    - Balloon dilation of biliary strictures
    - Repeated percutaneous procedures to clear pigment stones & mud-like biliary debris
  - Medical therapy
    - Long-term suppressive antibiotic therapy

DIAGNOSTIC CHECKLIST

Consider

- CECT, MRCP

Image Interpretation Pearls

- Intra- & extrahepatic bile duct dilatation/stones in SE Asian patients
- Massive CBD dilatation
- Rapid tapering of intrahepatic ducts ("arrowhead" sign)

SELECTED REFERENCES

1. Lee WJ et al: Radiologic spectrum of cholangiocarcinoma: emphasis on unusual manifestations and differential diagnoses. Radiographics. 21 Spec No:S97-S116, 2001
2. Park MS et al: Recurrent pyogenic cholangitis: comparison between MR cholangiography and direct cholangiography. Radiology. 220(3):677-82, 2001
3. Kim MJ et al: MR imaging findings in recurrent pyogenic cholangitis. AJR Am J Roentgenol. 173(6):1545-9, 1999
4. Cosenza CA et al: Current management of recurrent pyogenic cholangitis. Am Surg. 65(10):939-43, 1999
5. Harris HW et al: Recurrent pyogenic cholangitis. Am J Surg. 176(1):34-7, 1998
6. Leow CK et al: Re: Biliary access procedure in the management of oriental cholangiohepatitis. Am Surg. 64(1):99, 1998
7. Lo CM et al: The changing epidemiology of recurrent pyogenic cholangitis. HKMJ. 3(3):302-4, 1997
8. Lee DW et al: Biliary infection. Baillieres Clin Gastroenterol. 11(4):707-24, 1997
9. Sperling RM et al: Recurrent pyogenic cholangitis in Asian immigrants to the United States: natural history and role of therapeutic ERCP. Dig Dis Sci. 42(4):865-71, 1997
10. Kirby CL et al: US case of the day. Oriental cholangiohepatitis. Radiographics. 15(6):1503-6, 1995
11. Mack E: Pathogenesis and Clinical Presentation of Bile Duct Calculi. Semin Laparosc Surg. 2(2):76-84, 1995
12. Reynolds WR et al: Oriental cholangiohepatitis. Mil Med. 159(2):158-60, 1994
13. Enriquez G et al: Intrahepatic biliary stones in children. Pediatr Radiol. 22(4):283-6, 1992
14. Kusano S et al: Oriental cholangiohepatitis: correlation between portal vein occlusion and hepatic atrophy. AJR Am J Roentgenol. 158(5):1011-4, 1992
15. Goldberg HI et al: Diagnostic and interventional procedures for the biliary tract. Curr Opin Radiol. 3(3):453-62, 1991
16. Lim JH: Oriental cholangiohepatitis: pathologic, clinical, and radiologic features. AJR Am J Roentgenol. 157(1):1-8, 1991
17. Schulte SJ et al: CT of the extrahepatic bile ducts: wall thickness and contrast enhancement in normal and abnormal ducts. AJR Am J Roentgenol. 154(1):79-85, 1990
18. Kashi H et al: Recurrent pyogenic cholangiohepatitis. Ann R Coll Surg Engl. 71(6):387-9, 1989
19. vanSonnenberg E et al: Oriental cholangiohepatitis: diagnostic imaging and interventional management. AJR Am J Roentgenol. 146(2):327-31, 1986
20. Federle MP et al: Recurrent pyogenic cholangitis in Asian immigrants. Use of ultrasonography, computed tomography, and cholangiography. Radiology. 143(1):151-6, 1982

RECURRENT PYOGENIC CHOLANGITIS

IMAGE GALLERY

Typical

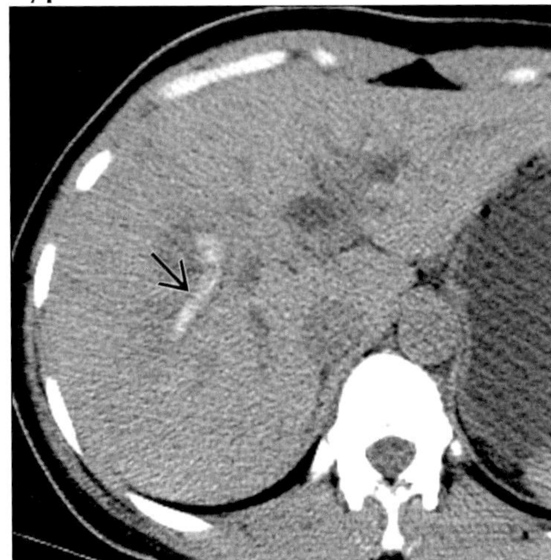

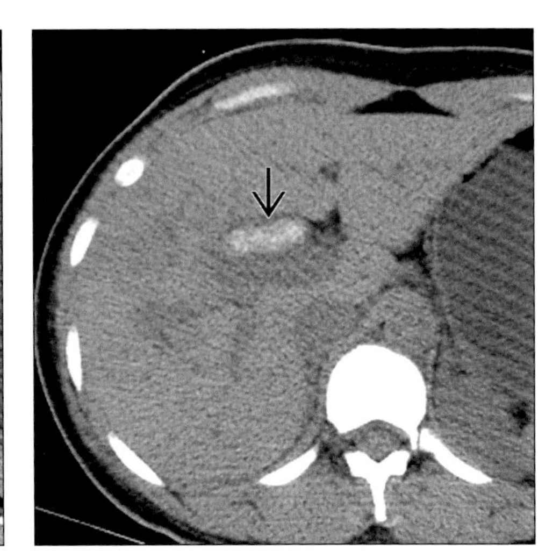

(Left) *Axial NECT of intrahepatic stones in recurrent pyogenic cholangitis. Note high attenuation stones in right lobe (arrow).* ***(Right)*** *Axial NECT of intrahepatic stones in recurrent pyogenic cholangitis. Note high attenuation stones in left lobe (arrow).*

Typical

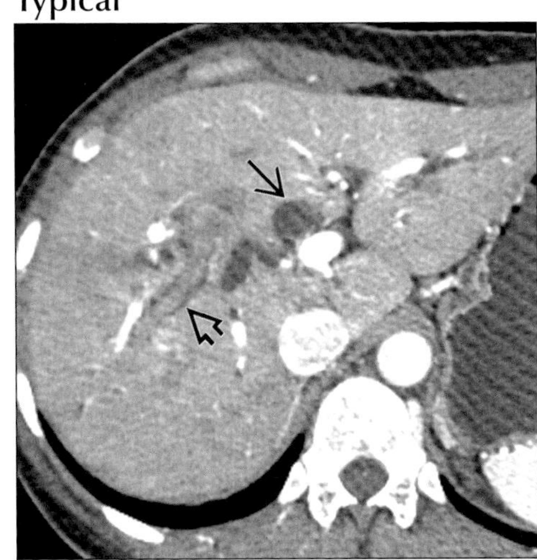

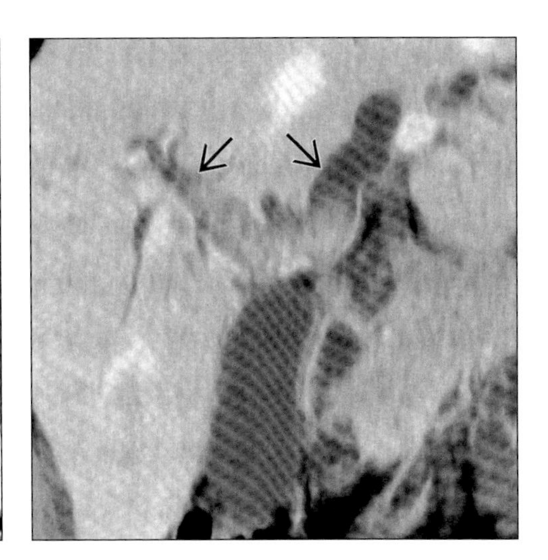

(Left) *Axial CECT of recurrent pyogenic cholangitis. Note dilated intrahepatic ducts (arrow) with numerous right lobe intrahepatic calculi (open arrow).* ***(Right)*** *Coronal CECT reformation of recurrent pyogenic cholangitis demonstrates lack of stones in gallbladder but numerous intrahepatic stones (arrows).*

Typical

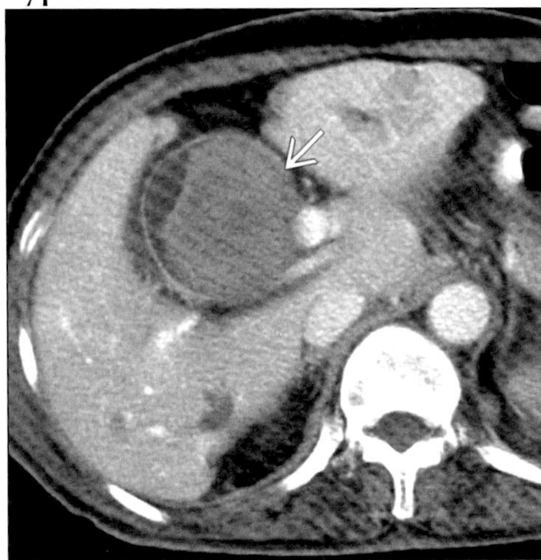

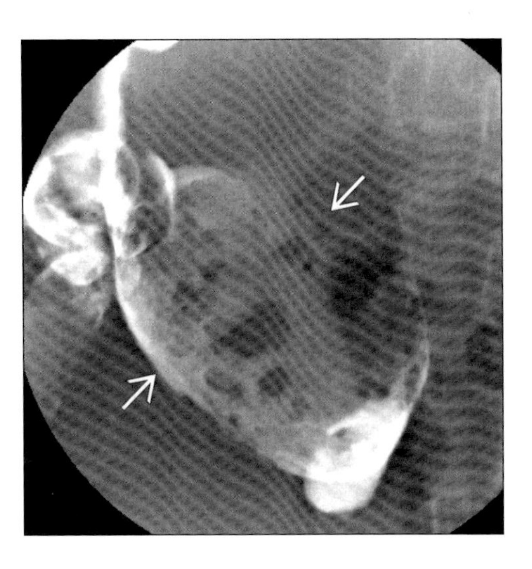

(Left) *Axial CECT demonstrates massive dilatation of common bile duct due to stones from recurrent pyogenic cholangitis. Note large pigment stone in common bile duct (arrow).* ***(Right)*** *ERCP of common bile duct in recurrent pyogenic cholangitis demonstrates massive filling defect in common bile duct due to stone (arrows).*

PANCREATO-BILIARY PARASITES

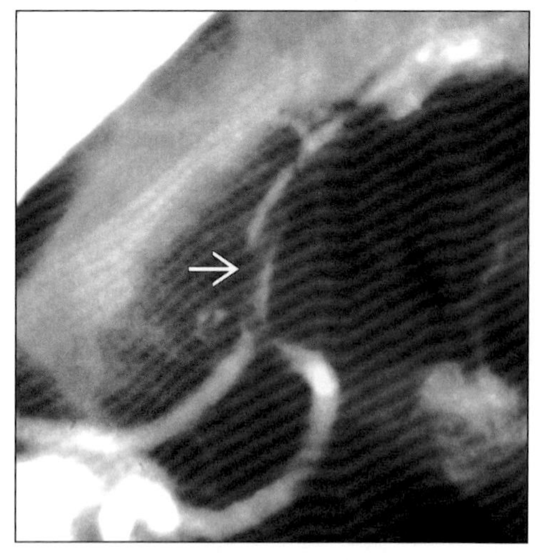

ERCP of ascaris in main pancreatic duct (arrow).

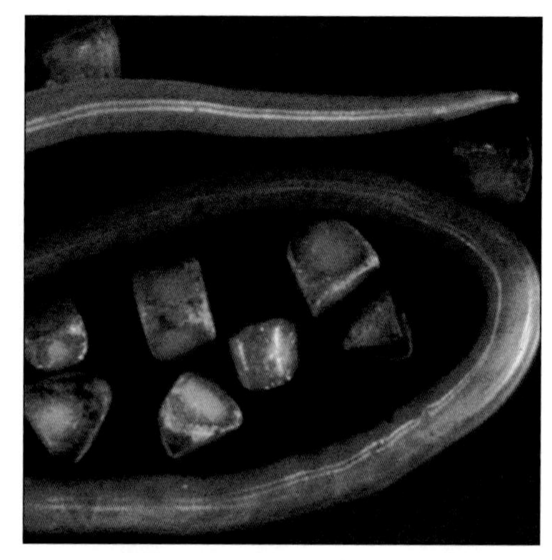

Ascaris retrieved from bile duct with multiple stones.

TERMINOLOGY

Abbreviations and Synonyms

- Biliary ascariasis, biliary clonorchiasis

Definitions

- Biliary & pancreatic duct involvement with parasitic infection from ascariasis, clonorchiasis

IMAGING FINDINGS

General Features

- Best diagnostic clue: Longitudinal filling in bile or pancreatic ducts on ERCP in ascariasis
- Location: Gallbladder (GB), common bile duct (CBD), intrahepatic or pancreatic ducts
- Size: 2-10 cm
- Morphology: Linear or rounded

Radiographic Findings

- ERCP
  - Linear, elliptical, rounded filling defects on ERCP
    - Ascariasis may involve entire biliary tract & pancreatic duct
    - Clonorchiasis typically involves peripheral intrahepatic ducts, not GB or CBD

CT Findings

- CECT: Intraductal increased density areas 2° to biliary worms, fluids; peripancreatic inflammation 2° to pancreatitis

MR Findings

- T1WI: Low signal branching dilated bile ducts
- T2WI: High signal fluid 2° to dilated ducts with low signal filling defects
- MRCP
  - Dilated bile ducts with low signal linear or rounded filling defects within high signal bile

Ultrasonographic Findings

- Real Time
  - US: "Bull's eye" appearance 2° to echogenic filling defect
    - Motility of worms may be evident; central anechoic area (digestive tract of worm)

Nuclear Medicine Findings

- Hepato Biliary Scan: Lack of GB filling in ascariasis-related cholecystitis

Imaging Recommendations

- Best imaging tool: US, ERCP

DDx: Biliary/Pancreatic Lesions Mimicking Parasites

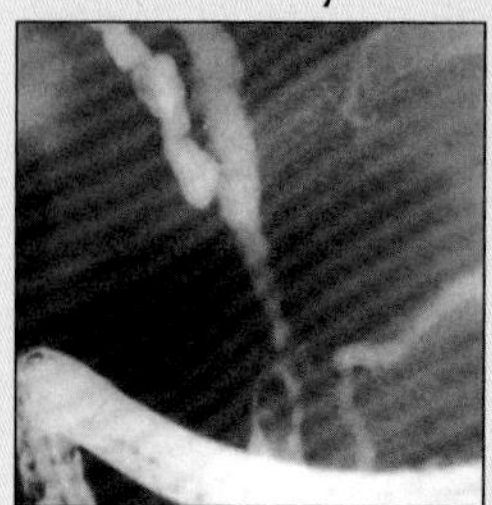

Biliary Lithiasis

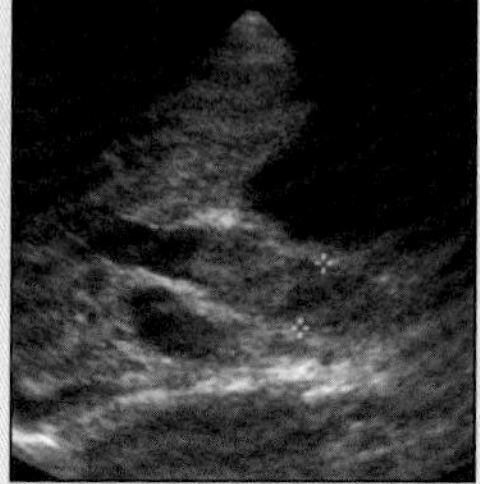

Cholangiocarcinoma

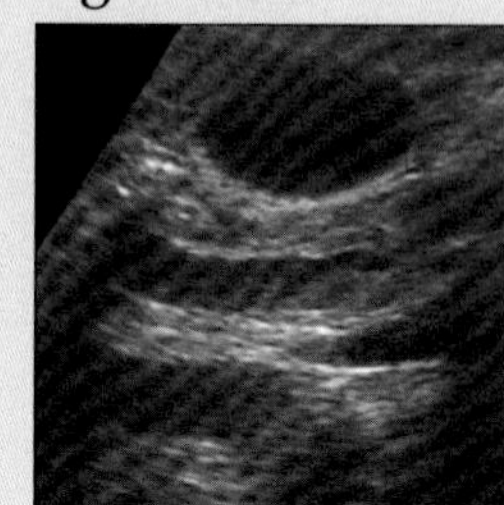

Bact Cholangitis

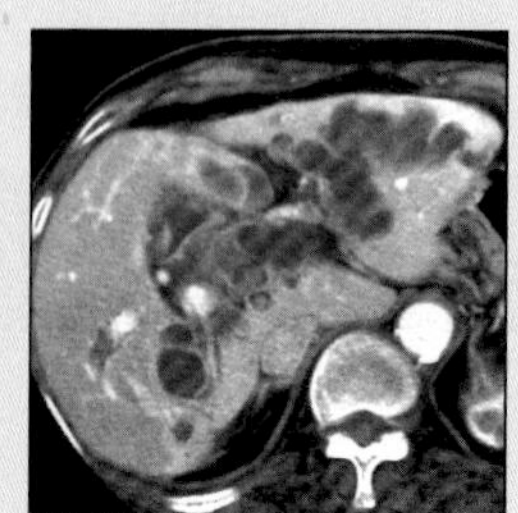

RPC

PANCREATO-BILIARY PARASITES

Key Facts

Terminology
- Biliary & pancreatic duct involvement with parasitic infection from ascariasis, clonorchiasis

Imaging Findings
- Linear, elliptical, rounded filling defects on ERCP
- US: "Bull's eye" appearance 2° to echogenic filling defect

Pathology
- General path comments: Cholangitis, periductal fibrosis, acute pancreatitis

Clinical Issues
- RUQ pain, jaundice, anorexia

DIFFERENTIAL DIAGNOSIS

Biliary lithiasis
- Echogenic ductal foci with acoustic shadowing on US

Cholangiocarcinoma
- Polypoid intraductal mass

Bacterial cholangitis
- Low-level echoes (pus) in bile ducts

Recurrent pyogenic cholangitis (RPC)
- Marked biliary dilatation with multiple intraductal pigment stones

PATHOLOGY

General Features
- General path comments: Cholangitis, periductal fibrosis, acute pancreatitis
- Etiology: Ascaris infestation: Ova ingested, pass through liver, lungs; worms develop in small bowel
- Epidemiology
  - Ascariasis most prevalent helminth infection worldwide
  - Clonorchiasis endemic in Asia, present in Western world 2° to travel, immigration
- Associated abnormalities: Cholangio CA w/Clonorchis

Gross Pathologic & Surgical Features
- Worms & flukes in biliary tree, cholangitis, acute cholecystitis, acute pancreatitis

Microscopic Features
- Periductal round cell infiltrate & fibrosis; pancreatic edema & infiltration

CLINICAL ISSUES

Presentation
- Most common signs/symptoms
  - RUQ pain, jaundice, anorexia
  - Weight loss, diarrhea, palpitations; small bowel obstruction common with ascariasis
- Clinical profile: Patient with poor nutrition

Demographics
- Age: Ascariasis common in children
- Ethnicity: Higher incidence of clonorchiasis in Asian patients

Natural History & Prognosis
- May be asymptomatic with early infection; medication therapy generally effective

Treatment
- Options, risks, complications
  - Pyrantel embonate 90% effective for ascariasis
  - Praziquantel moderately effective for clonorchiasis; 20% cure rate with single dose

DIAGNOSTIC CHECKLIST

Consider
- Cholangiocarcinoma

Image Interpretation Pearls
- Linear or rounded intraductal filling defects in pancreatic or common bile ducts

SELECTED REFERENCES

1. Haseeb AN et al: Evaluation of excretory/secretory Fasciola (Fhes) antigen in diagnosis of human fascioliasis. J Egypt Soc Parasitol. 33(1):123-38, 2003
2. Amjad N et al: An unusual presentation of acute cholecystitis: biliary ascariasis. Hosp Med. 62(6):370-1, 2001
3. Capallo DV et al: Biliary ascariasis. South Med J. 77(9):1201-2, 1984

IMAGE GALLERY

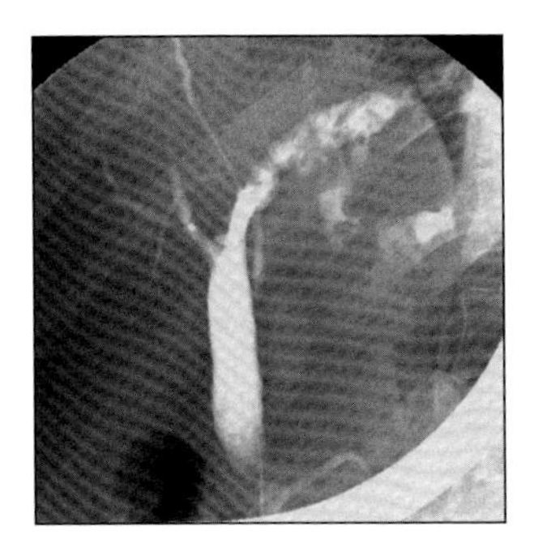

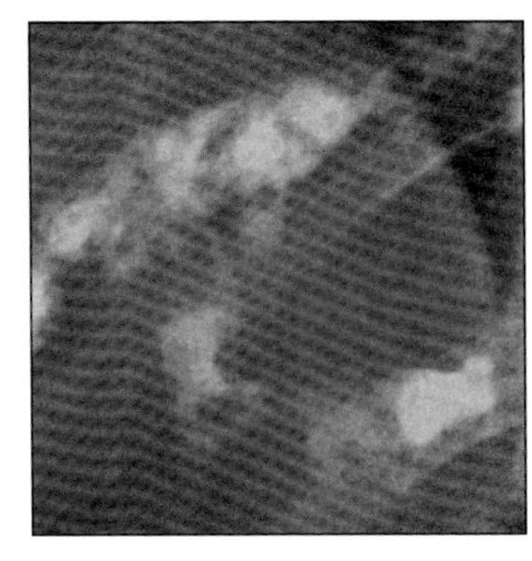

*(**Left**) ERCP demonstrates left duct dilatation and filling defects from intrahepatic flukes. (**Right**) Magnified view of ERCP demonstrates rounded filling defects in left ducts from Clonorchis.*

AIDS CHOLANGIOPATHY

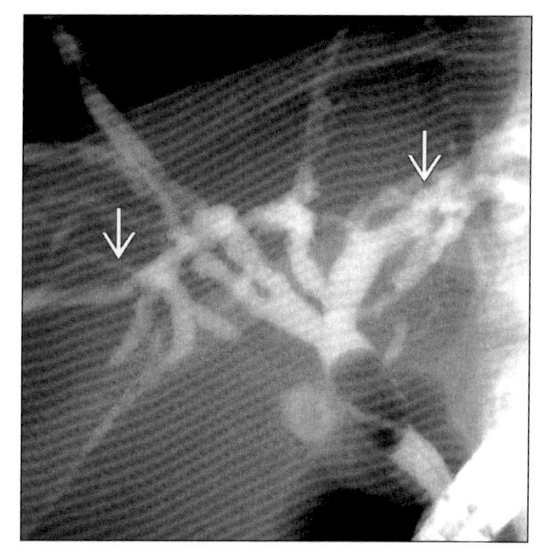

ERCP of AIDS cholangitis demonstrates multiple intrahepatic biliary strictures and irregular ductal contours (arrows).

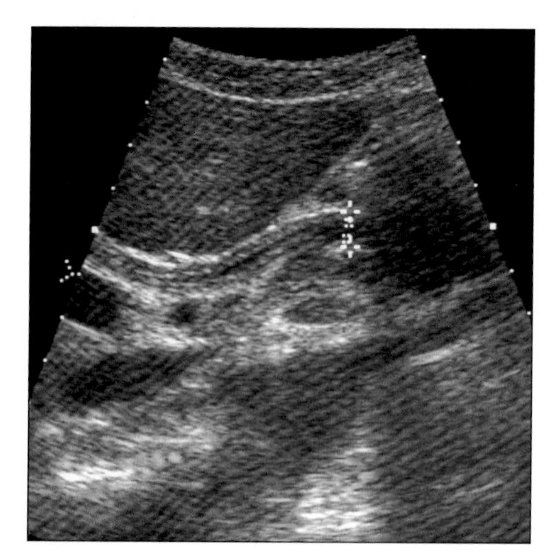

Sagittal US of common bile duct demonstrates diffuse thickening representing edema from AIDS cholangitis.

TERMINOLOGY

Abbreviations and Synonyms

- AIDS cholangitis, AIDS-related sclerosing cholangitis

Definitions

- Spectrum of biliary inflammatory lesions caused by AIDS-related opportunistic infections leading to biliary stricture/obstruction or cholecystitis

IMAGING FINDINGS

General Features

- Best diagnostic clue: AIDS patient with multiple intrahepatic strictures, ampullary stenosis, or gallbladder (GB) wall thickening from cholecystitis
- Location
  - GB, common bile duct (CBD)
  - Intrahepatic ducts may be involved
- Size: Varies from short focal biliary strictures to longer segment strictures
- Morphology: Irregular intrahepatic strictures mimic sclerosing cholangitis

Radiographic Findings

- ERCP
  - Ampullary stenosis with CBD dilation, CBD ulcerations, mult intrahepatic strictures on ERCP

CT Findings

- CECT
  - Ductal dilatation of CBD and intrahepatic ducts
  - Ductal dilatation may asymmetrically involve intrahepatic bile ducts

MR Findings

- T1WI: Low signal dilated bile ducts
- T2WI: High signal dilated bile ducts and CBD; GB wall thickening
- T1 C+
  - Dilated bile ducts and CBD; GB wall thickening; pericholecystic inflammatory changes
    - Patients with acute acalculous cholecystitis
- MRCP
  - High signal dilated ducts; intra- and extrabiliary strictures on MRCP

Ultrasonographic Findings

- Real Time
  - Diffuse GB thickening; periductal hyper- or hypoechoic areas; dilated intrahepatic ducts; diffuse CBD thickening on US
    - Due to edema and ulceration of CBD
- Color Doppler: No flow within dilated intrahepatic ductal segments

DDx: Spectrum of Biliary Lesions Mimicking AIDS Cholangitis

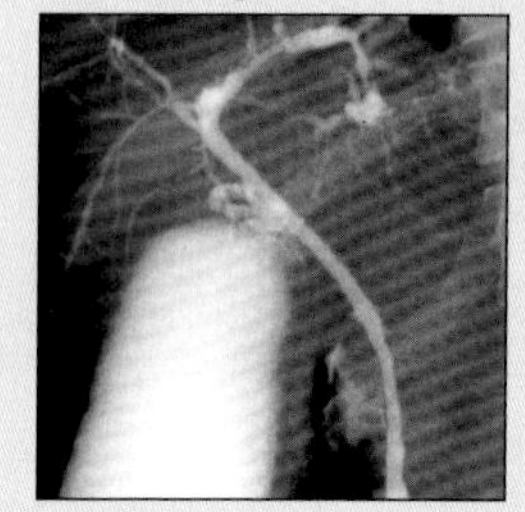

Sclerosing Chol

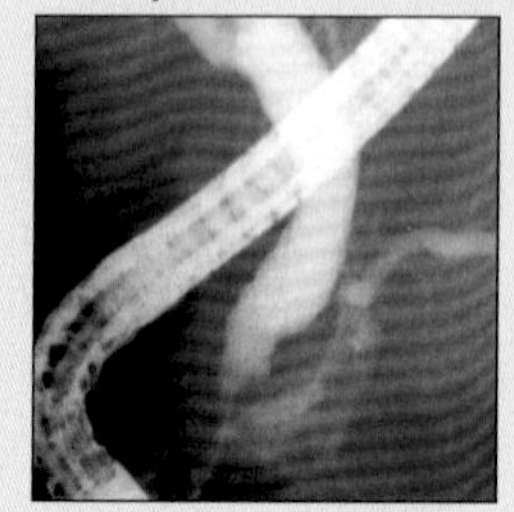

Ampullary Stenosis

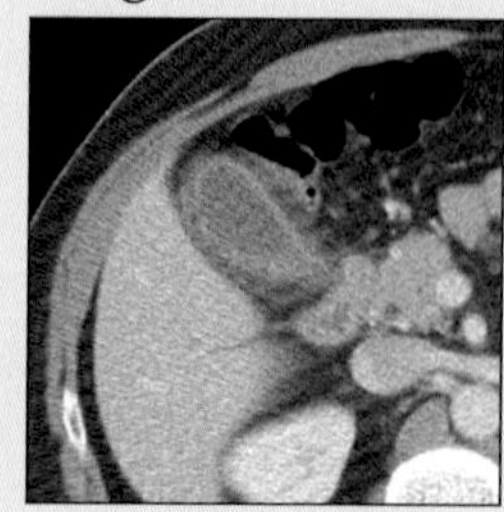

Acalculous Chole

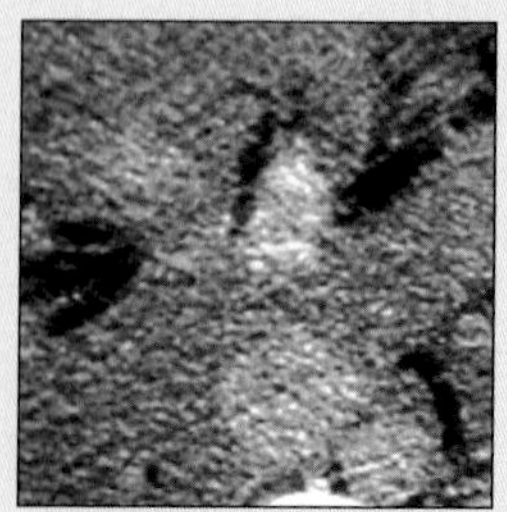

Cholangiocarcinoma

AIDS CHOLANGIOPATHY

Key Facts

Terminology

- Spectrum of biliary inflammatory lesions caused by AIDS-related opportunistic infections leading to biliary stricture/obstruction or cholecystitis

Imaging Findings

- Best diagnostic clue: AIDS patient with multiple intrahepatic strictures, ampullary stenosis, or gallbladder (GB) wall thickening from cholecystitis
- Ampullary stenosis with CBD dilation, CBD ulcerations, mult intrahepatic strictures on ERCP
- High signal dilated ducts; intra- and extrabiliary strictures on MRCP
- Diffuse GB thickening; periductal hyper- or hypoechoic areas; dilated intrahepatic ducts; diffuse CBD thickening on US
- US protocol: Parasagittal view of CBD to demonstrate mural thickening
- MRC protocol: Axial and coronal SSFSE; obliques, heavily weighted T2WI

Pathology

- General path comments: Opportunistic infection of GB, bile ducts from cryptosporidium & CMV; periductal inflammation, acalculous cholecystitis

Clinical Issues

- Poor prognosis due to advanced AIDS presentation

Diagnostic Checklist

- AIDS patient with distal ampullary stenosis, intrahepatic strictures or acalculous cholecystitis

Imaging Recommendations

- Best imaging tool
  - US
  - MRCP
  - ERCP
- Protocol advice
  - US protocol: Parasagittal view of CBD to demonstrate mural thickening
  - MRC protocol: Axial and coronal SSFSE; obliques, heavily weighted T2WI

DIFFERENTIAL DIAGNOSIS

Sclerosing cholangitis

- Beading of CBD
- Pseudodiverticula of CBD
- Multiple intrahepatic strictures with asymmetric involvement
- Long segment stricture of CBD
- Involvement of cystic duct characteristic

Ampullary stenosis

- Distal CBD fibrous stricture not associated with infection
- May result from passage of CBD stones or chronic pancreatitis
- Smooth stricture not associated with ulceration or irregularities

Acalculous cholecystitis

- GB wall thickened by sterile inflammation
- Not associated with AIDS or opportunistic infection
- Often secondary to ischemic injury from low flow state
  - Post-cardiac surgery
- Percutaneous cholecystomy for poor operative risk patients
- Late in clinical course
  - Secondary to bacterial invasion

Cholangiocarcinoma

- Progressive biliary obstruction
- Associated with primary sclerosing cholangitis (PSC), recurrent pyogenic cholangitis (RPC), choledochal cyst
- Infiltrates along ductal epithelium, invades hepatic parenchyma
- Delayed CECT useful to demonstrate intrahepatic component

PATHOLOGY

General Features

- General path comments: Opportunistic infection of GB, bile ducts from cryptosporidium & CMV; periductal inflammation, acalculous cholecystitis
- Genetics: No known predisposition
- Etiology
  - AIDS-related opportunistic infection of biliary tract
  - Cryptosporidium, cytomegalovirus (CMV) most common pathologies
  - CMV has been implicated in vasculitis of central nervous system (CNS), retinal or gastrointestinal (GI) tract
    - Biliary strictures may be secondary to vasculitis associated with CMV
- Epidemiology: Late stage AIDS patients (CDC stage IV AIDS based on T4 counts)
- Associated abnormalities
  - Other AIDS-related GI infections
    - Enteritis and colitis from CMV, cryptosporidium, other opportunistic infections
    - MAI causing ileo-cecal inflammation and necrotic low attenuation mesenteric nodes
    - Hepatic and splenic microabscesses from fungal infections (Candida, cryptococcus)
  - AIDS-related malignancies
    - Non-Hodgkin lymphoma of liver, stomach, spleen, mesentery or retroperitoneal nodes
    - Kaposi sarcoma of retroperitoneal nodes

Gross Pathologic & Surgical Features

- Acalculous cholecystitis, biliary strictures involving CBD, intrahepatic duct
- Biliary strictures involving CBD, intrahepatic duct

AIDS CHOLANGIOPATHY

Microscopic Features

- CMV inclusions, cryptosporidium organisms may be found on biopsy
- Fibrotic strictures of CBD, intrahepatic ducts

Staging, Grading or Classification Criteria

- Type I: Distal CBD stricture from ampullary stenosis
  - 15-20% of patients
- Type II: Diffuse intrahepatic biliary strictures
  - 20% of patients
- Type III: Combined types I, II
  - 50% of patients
- Type IV: Long segment stricture of CBD with possible ulceration
  - 15% of patients
- Type V: Acalculous cholecystitis, GB wall thickening, pericholecystic inflammation

CLINICAL ISSUES

Presentation

- Most common signs/symptoms
  - Fever
  - RUQ pain
  - Jaundice
- Clinical profile
  - Elevated alkaline phosphatase
  - Cryptosporidium in stool or duodenum

Demographics

- Age: > 20 yrs
- Gender: M < F
- Ethnicity
  - Parallel demographics for AIDS patients
    - Incidence has decreased substantially with newer anti-retroviral therapy

Natural History & Prognosis

- Sphincterotomy provides some pain relief
  - Doesn't alter intrahepatic disease
- Anti-CMV therapy not effective
- Poor prognosis due to advanced AIDS presentation

Treatment

- Options, risks, complications
  - Asymptomatic
    - Observation
  - Symptomatic
    - Sphincterotomy for pain relief

DIAGNOSTIC CHECKLIST

Consider

- Sclerosing cholangitis

Image Interpretation Pearls

- AIDS patient with distal ampullary stenosis, intrahepatic strictures or acalculous cholecystitis

SELECTED REFERENCES

1. Ko WF et al: Prognostic factors for the survival of patients with AIDS cholangiopathy. Am J Gastroenterol. 98(10):2176-81, 2003
2. Enns R: AIDS cholangiopathy: "an endangered disease". Am J Gastroenterol. 98(10):2111-2, 2003
3. Chen XM et al: Cryptosporidiosis and the pathogenesis of AIDS-cholangiopathy. Semin Liver Dis. 22(3):277-89, 2002
4. Kumar KS et al: Isolated intrahepatic biliary dilatation in a patient with acquired immune deficiency syndrome (AIDS): AIDS cholangiopathy versus incidental unilobar Caroli's disease. J Clin Gastroenterol. 32(1):79-81, 2001
5. Mukhopadhyay S et al: AIDS cholangiopathy as initial presentation of HIV infection. Trop Gastroenterol. 22(1):29-30, 2001
6. Mahajani RV et al: Cholangiopathy in HIV-infected patients. Clin Liver Dis. 3(3):669-84, x, 1999
7. Cello JP: AIDS-Related biliary tract disease. Gastrointest Endosc Clin N Am. 8(4):963, 1998
8. Kumar M et al: AIDS associated cholangiopathy. Trop Gastroenterol. 19(4):155-6, 1998
9. Boige N et al: Hydrops-like cholecystitis due to cryptosporidiosis in an HIV-infected child. J Pediatr Gastroenterol Nutr. 26(2):219-21, 1998
10. Wilcox CM et al: Hepatobiliary diseases in patients with AIDS: focus on AIDS cholangiopathy and gallbladder disease. Dig Dis. 16(4):205-13, 1998
11. Misra A et al: AIDS cholangiopathy. Indian J Gastroenterol. 17(3):104-5, 1998
12. Keaveny AP et al: Hepatobiliary and pancreatic infections in AIDS: Part II. AIDS Patient Care STDS. 12(6):451-6, 1998
13. Castiella A et al: AIDS-associated cholangiopathy in a series of ten patients. Rev Esp Enferm Dig. 90(6):419-30, 1998
14. Keaveny AP et al: Hepatobiliary and pancreatic infections in AIDS: Part one. AIDS Patient Care STDS. 12(5):347-57, 1998
15. Fulcher AS et al: Magnetic resonance cholangiopancreatography: a new technique for evaluating the biliary tract and pancreatic duct. Gastroenterologist. 6(1):82-7, 1998
16. Lefkowitch JH: The liver in AIDS. Semin Liver Dis. 17(4):335-44, 1997
17. Liberman E et al: Foamy macrophages in acquired immunodeficiency syndrome cholangiopathy with Encephalitozoon intestinalis. Arch Pathol Lab Med. 121(9):985-8, 1997
18. Castiella A et al: Ursodeoxycholic acid in the treatment of AIDS-associated cholangiopathy. Am J Med. 103(2):170-1, 1997
19. Daly CA et al: Sonographic prediction of a normal or abnormal ERCP in suspected AIDS related sclerosing cholangitis. Clin Radiol. 51(9):618-21, 1996
20. Cacciarelli AG et al: Biliary fistula in a patient with HIV cholangiopathy. Gastrointest Endosc. 44(3):345-8, 1996
21. Willson R et al: Human immunodeficiency virus 1-associated necrotizing cholangitis caused by infection with Septata intestinalis. Gastroenterology. 108(1):247-51, 1995
22. Cello JP et al: Long-term follow-up of endoscopic retrograde cholangiopancreatography sphincterotomy for patients with acquired immune deficiency syndrome papillary stenosis. Am J Med. 99(6):600-3, 1995
23. Carmody E et al: Cytomegalovirus cholangitis after renal transplantation. Can Assoc Radiol J. 45(6):473-5, 1994
24. Chung CJ et al: Hepatobiliary abnormalities on sonography in children with HIV infection. J Ultrasound Med. 13(3):205-10, 1994

IMAGE GALLERY

Typical

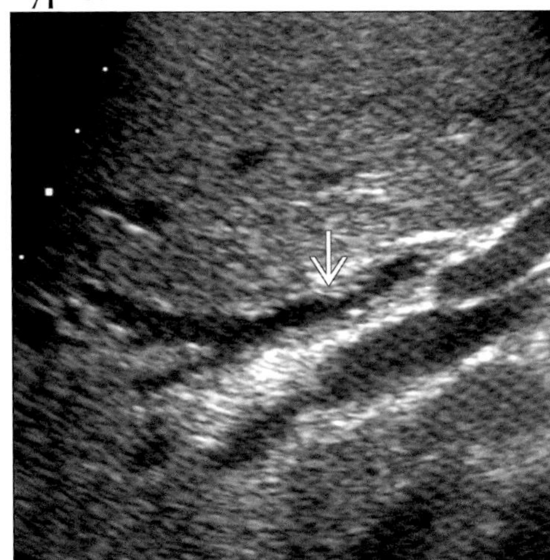

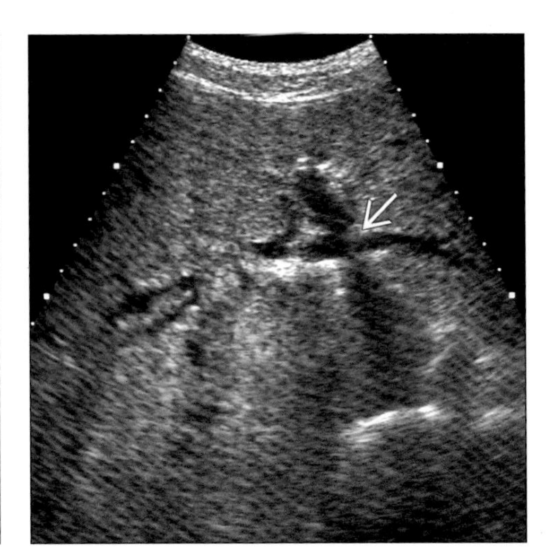

*(**Left**) Sagittal US of common bile duct demonstrates irregular thickening (arrow) from AIDS cholangitis. (**Right**) Axial US of common bile duct shows irregular thickening (arrow) from AIDS cholangitis.*

Typical

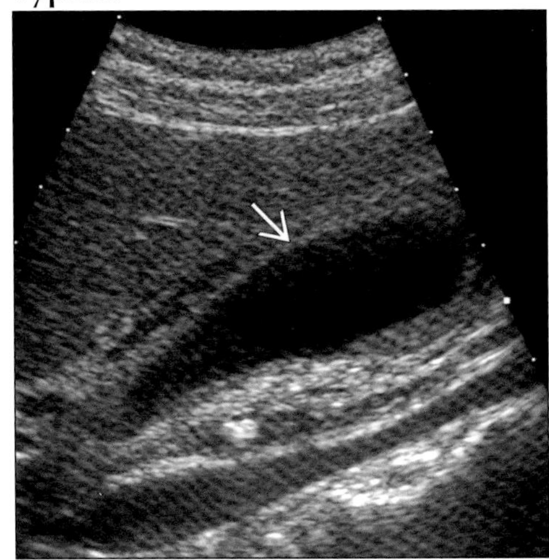

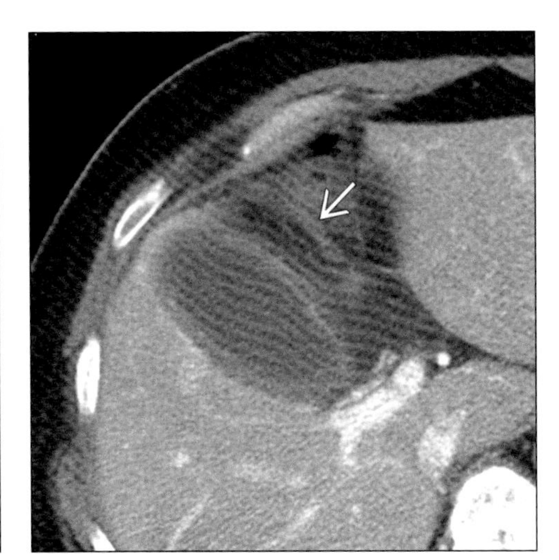

*(**Left**) Sagittal US demonstrates diffuse gallbladder wall thickening (arrow) in CMV acalculous cholecystitis. (**Right**) Axial CECT demonstrates gallbladder wall edema and pericholecystic inflammatory changes (arrow) in CMV acalculous cholecystitis.*

Typical

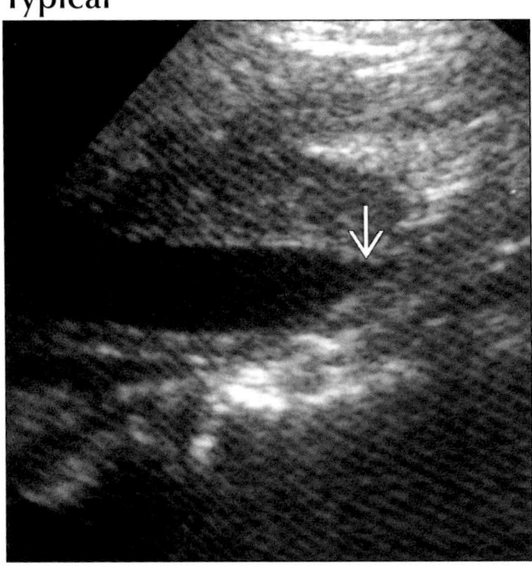

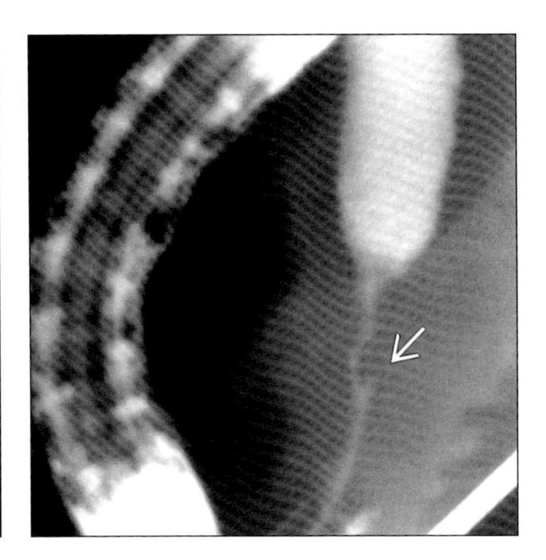

*(**Left**) Sagittal US of distal common bile duct demonstrating focal mural thickening (arrow) in ampullary stenosis from cryptosporidium in AIDS cholangitis. (**Right**) ERCP of patient with ampullary stenosis from cryptosporidium in AIDS cholangitis demonstrates distal common bile duct stricture (arrow).*

CHOLEDOCHOLITHIASIS

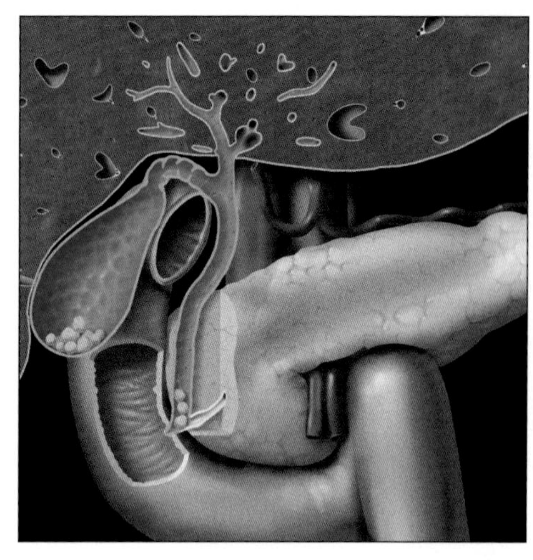

Coronal graphic shows multiple small nonobstructive stones in distal CBD and gallbladder.

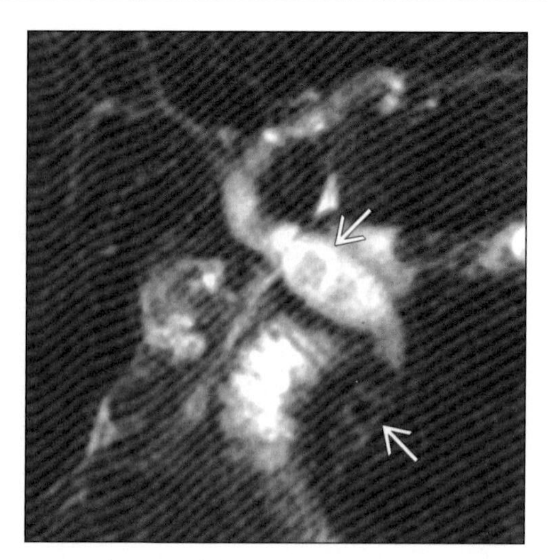

MRCP shows multiple hypointense calculi (arrows) within a dilated CBD.

TERMINOLOGY

Abbreviations and Synonyms

- Cholangiolithiasis or biliary calculi

Definitions

- Intra- &/or extrahepatic stones or calculi

IMAGING FINDINGS

General Features

- Best diagnostic clue: MRC: Discrete low signal filling defects within bile ducts
- Location: Intra- & extrahepatic bile ducts (more common in CBD)
- Size: Varies from 1-15 mm
- Morphology
  - Most frequent cause of biliary obstruction without ductal dilatation
  - Classified into two types based on etiology
    - Primary duct stones
    - Secondary duct stones
  - Primary duct stones: Form within bile ducts
    - Accounts only 5% of CBD stones in USA
    - Major composition: Pigment
    - Far less common than secondary stones in Western countries
  - Secondary duct stones due to passage of gallstones into CBD
    - Most common duct stones in Western countries
    - Accounts 95% of CBD stones in USA
    - Within CBD anywhere between porta hepatis & ampulla of Vater
    - Major composition: Cholesterol 70-80%; pigment 20-30% in Western countries & calcium bilirubinate in Eastern countries
    - 15% of gallstone patients also have CBD stones
    - 95% of patients with CBD stones have or have had gallstones
    - 15-25% acute calculous cholecystitis patients have CBD stones
    - 12-15% who undergo cholecystectomy have choledocholithiasis

Radiographic Findings

- ERCP (both diagnostic & therapeutic)
  - Opacification of both extra- & intrahepatic duct system + pancreatic duct
  - Stones: Seen as radiolucent filling defects
- Intraoperative & postoperative (T tube) cholangiography
  - Direct tests for detection of CBD stones
  - A meniscus of contrast material clearly outlines margins of stones

DDx: Obstrucion of Common Bile Duct

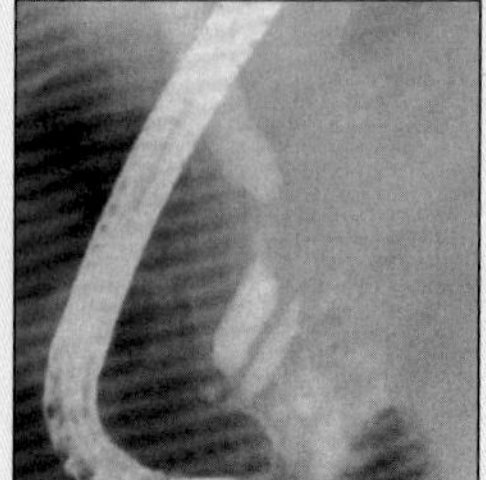

Pancreatic Carcinoma

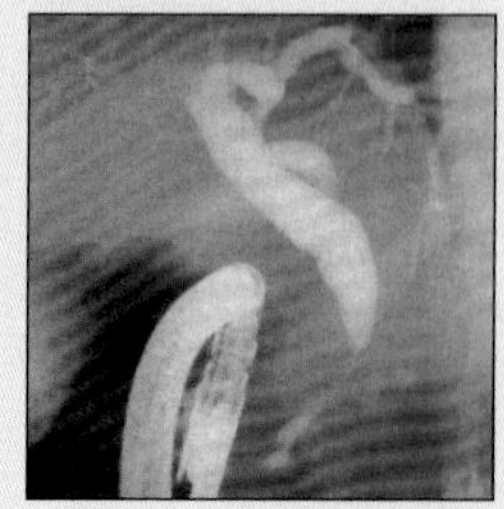

Chronic Pancreatitis

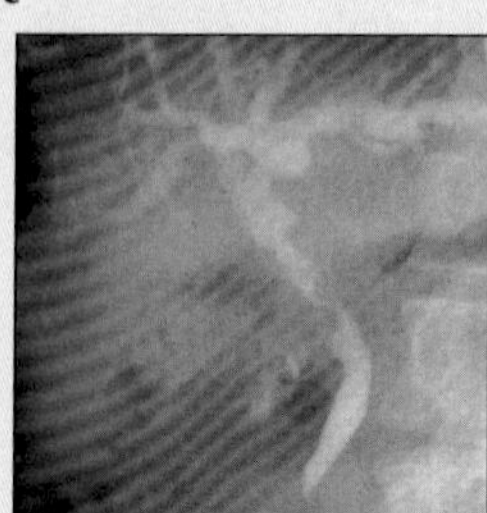

Cholangiocarcinoma

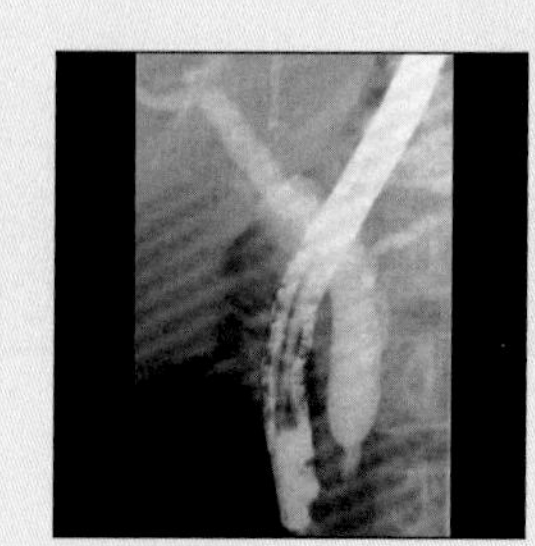

Papillary Stenosis

CHOLEDOCHOLITHIASIS

Key Facts

Terminology

- Intra- &/or extrahepatic stones or calculi

Imaging Findings

- Best diagnostic clue: MRC: Discrete low signal filling defects within bile ducts
- "Bull's eye" sign: Rim of bile surrounding a stone within duct
- Thin meniscus of water density bile around stone posteriorly

Top Differential Diagnoses

- Pancreatic or ampullary cancer
- Chronic pancreatitis
- Cholangiocarcinoma
- Papillary stenosis or dysfunction
- Primary sclerosing cholangitis (PSC)

Pathology

- Obstruction, dilatation, sclerosis, stricture
- Bile stasis/infection: Bilirubinate stone formation
- Primary duct stones (5%): Form within bile ducts
- Secondary duct stones (95%): Gallstones into CBD
- Approximately 25 million Americans have gallstones
- Associated abnormalities: Gall stones

Clinical Issues

- Acute RUQ pain, pruritus, jaundice, pancreatitis
- Increased alkaline phosphatase & direct bilirubin
- Gender: Females (middle age) more than males

Diagnostic Checklist

- Rule out other causes of "CBD obstruction"
- MRC & ERCP: Discrete filling defects or obstruction & prestenotic dilatation of CBD/intrahepatic bile ducts

CT Findings

- NECT
  - Attenuation of calculi varies from less than water density, through soft tissue, to dense calcification
  - Mixed stones: Predominantly cholesterol & calcium bilirubinate (usually calcified rim or central nidus)
    - Increased attenuation (75-85% stones due to sufficient Ca++ bilirubinate)
    - "Bull's eye" sign: Rim of bile surrounding a stone within duct
    - Most accurate sign (60-80% sensitivity) for bile duct stones on NECT
    - Thin meniscus of water density bile around stone posteriorly
    - Some stones are isodense to soft tissue
  - Pure cholesterol stones are rare
    - Isodense with bile (indistinguishable)
  - Abrupt termination of CBD (complete obstruction by a large stone)
    - Stone isodense to bile or pancreas (DDx: Malignant stricture & carcinoma of ampulla)
    - Less accurate than "bull's eye" sign
  - CBD &/or IHBD dilatation: Varies based on
    - Stone size, degree & duration of obstruction
    - Water density tubular branching structures

MR Findings

- MRC (MR cholangiography)
  - Bile: Very bright signal
  - Ductal stones: Decreased signal intensity foci
  - Low-signal filling defects within increased signal intensity bile

Ultrasonographic Findings

- Real Time
  - Echogenic focus with posterior acoustic shadowing
    - Stone within CBD
  - 10% stones: No acoustic shadow
    - Small size, soft & porous composition
    - DDx: Intraductal clot, infection, sludge ball, tumor, parasite
  - CBD/intrahepatic bile duct dilatation (IHBD) based on stone size, degree & duration of obstruction
    - CBD: 4-6 mm (normal size); 6-7 mm (equivocal); more than 8 mm (dilatation)
    - Common hepatic duct: 4-5 mm (normal size)
    - IHBD: 1-2 mm (usually not visible)

Nuclear Medicine Findings

- Hepato-biliary scan (HIDA)
  - Diagnose early, low grade obstruction
  - Stones with intermittent obstruction
    - Retention & delayed passage of isotope

Imaging Recommendations

- Best imaging tool
  - ERCP: Gold standard for detection of CBD stones in the absence of T tube
  - MRC: Sensitivity: 81-100%; specificity: 85-100%
- Protocol advice
  - MRC: Using two techniques
    - RARE: Single-slab rapid acquisition with relaxation enhancement
    - HASTE: Multislice half-Fourier acquisition single-shot turbo spin-echo
- Helical NECT: Using thin sections (≤ 5 mm)
  - Sensitivity: 75-85%
- Ultrasonography: High frequency transducer (5 MHz) sensitivity 50-60%
  - Proximal CBD: Parasagittal scan in supine/left posterior oblique position
  - Distal CBD
    - Semierect (approx. 60° to vertical)
    - Right posterior oblique (45°) position in transverse plane provides a good acoustic window
  - If gas obscures; have patient drink 6-12 oz of water
    - Keep patient in right decubitus position for 2-3 minutes & rescan in semierect position
  - Postcholecystectomy patients with persistent RUQ pain: CBD imaged
    - After a fast & 45 mins to 1 hr after a fatty meal
    - CBD dilates more than 2 mm above baseline in partial stone obstruction

CHOLEDOCHOLITHIASIS

DIFFERENTIAL DIAGNOSIS

Pancreatic or ampullary cancer

- Hypodense mass in head of pancreas or ampulla
- "Double duct" sign
  - Obstruction & dilatation of pancreatic duct/CBD
- Heterogeneous poorly enhancing mass
- Contiguous organ invasion may be seen
  - Duodenum, stomach & mesenteric root
- Duodenal distention with water: Helpful for visualization with CT

Chronic pancreatitis

- Focal or diffuse atrophy of gland; enlarged head
- Dilated main pancreatic duct with ductal calculi
- Distal CBD long stricture causes prestenotic dilatation
- Thickening of peripancreatic fascia & fat necrosis

Cholangiocarcinoma

- Extrahepatic: CBD growth (stricture or polypoid mass)
- Obstruction & dilatation of CBD/IHBD
- ERCP: Depicts stricture or intraductal tumor mass

Papillary stenosis or dysfunction

- Causes dilatation of CBD & intrahepatic bile ducts
- No mass or filling defect

Primary sclerosing cholangitis (PSC)

- Idiopathic or autoimmune reaction or genetic
- CBD always involved; IHBD & extrahepatic (68-89%)
- ERCP: Classic "beaded appearance"
  - Alternating segments of dilatation & focal strictures

PATHOLOGY

General Features

- General path comments
  - Mechanism of stones in CBD & IHBD
    - Obstruction, dilatation, sclerosis, stricture
    - Bile stasis/infection: Bilirubinate stone formation
    - Infection: E. coli, Klebsiella & other gram negative organisms with β-glucuronidase activity
- Etiology
  - Primary duct stones (5%): Form within bile ducts
    - Chronic hemolytic disease, recurrent cholangitis
    - Congenital anomalies of bile ducts (e.g., Caroli disease)
    - Motor disorder of sphincter of Oddi
    - Low fat & protein diet, foreign body (suture material)
    - Parasites: Clonorchis sinensis & ascaris (major causes in Asia)
  - Secondary duct stones (95%): Gallstones into CBD
    - Obesity, Crohn disease & ileal resection
    - Hemolytic anemias (sickle cell anemia & hereditary spherocytosis)
    - Increased triglycerides, hyperalimentation, Native American heritage
- Epidemiology
  - Approximately 25 million Americans have gallstones
  - Secondary duct stones
    - Accounts for 95% of choledocholithiasis in USA
- Associated abnormalities: Gall stones

CLINICAL ISSUES

Presentation

- Most common signs/symptoms
  - Acute RUQ pain, pruritus, jaundice, pancreatitis
  - Small stones spontaneously pass with/without pain
- Clinical profile: Fat, fertile, forty year old female with history of acute or intermittent RUQ pain & jaundice
- Lab data
  - Increased alkaline phosphatase & direct bilirubin
  - Late phase: Increased AST & ALT levels
- Diagnosis
  - Cholangiography: MRC, ERCP or intra-operative/post-operative T tube

Demographics

- Age: Usually adults, can be seen in any age group
- Gender: Females (middle age) more than males

Natural History & Prognosis

- Complications: Cholangitis, obstructive jaundice, pancreatitis, secondary biliary cirrhosis
- Prognosis
  - Cholecystectomy patients
    - Undetected duct stones left behind in 1-5% cases

Treatment

- Stones < 3 mm: Usually spontaneously pass
- Stones 3-10 mm: Endoscopic sphincterotomy
  - Stone retrieval balloon to sweep duct
  - Basket to snare stones
- Stones more than 10-15 mm
  - Require fragmentation by mechanical lithotripsy

DIAGNOSTIC CHECKLIST

Consider

- Rule out other causes of "CBD obstruction"

Image Interpretation Pearls

- MRC & ERCP: Discrete filling defects or obstruction & prestenotic dilatation of CBD/intrahepatic bile ducts

SELECTED REFERENCES

1. Kim TK et al: Diagnosis of intrahepatic stones: superiority of MR cholangiopancreatography over endoscopic retrograde cholangiopancreatography. AJR Am J Roentgenol. 179(2):429-34, 2002
2. Soto JA et al: Detection of choledocholithiasis with MR cholangiography: comparison of three-dimensional fast spin-echo and single- and multisection half-Fourier rapid acquisition with relaxation enhancement sequences. Radiology. 215(3):737-45, 2000
3. Vitellas KM et al: MR cholangiopancreatography of bile and pancreatic duct abnormalities with emphasis on the single-shot fast spin-echo technique. RadioGraphics. 20: 939-957, 2000
4. Fulcher AS et al: MR cholangiography: technical advances and clinical applications. Radiographics. 19(1):25-41; discussion 41-4, 1999

IMAGE GALLERY

Typical

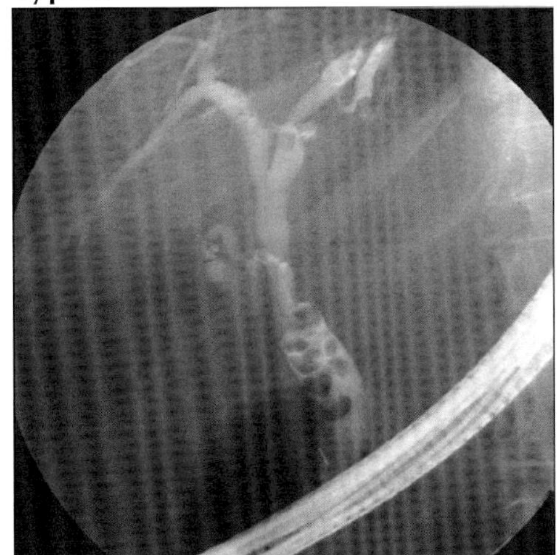

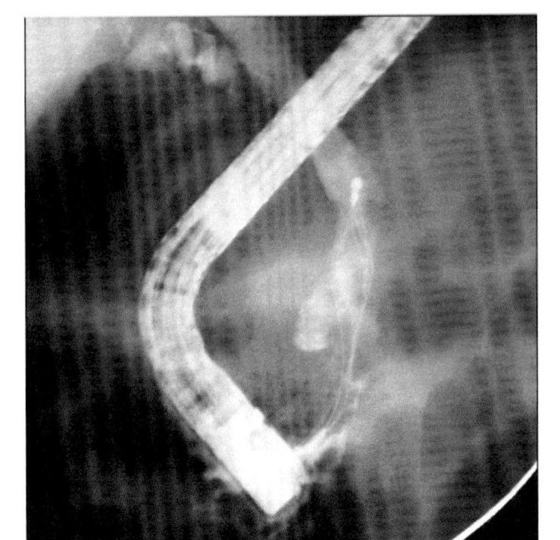

(Left) ERCP shows multiple calculi (filling defects) within cystic and common bile ducts. *(Right)* ERCP following endoscopic papillotomy shows a wire basket being used to fragment, snare and extract biliary calculi.

Typical

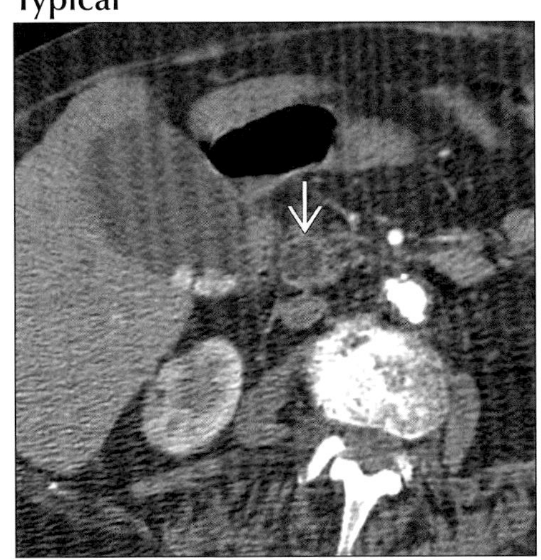

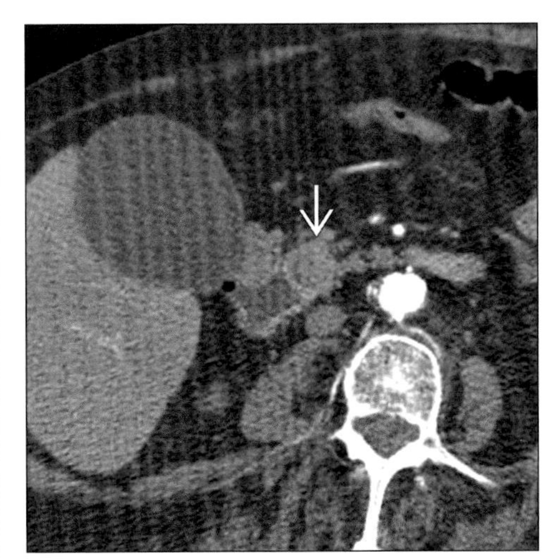

(Left) Axial CECT shows calcific density within the dependent portion of the gallbladder. Dilated CBD (arrow) from obstructing biliary calculus. *(Right)* Axial CECT shows soft tissue density stone (arrow) causing abrupt obstruction of CBD. Note thin meniscus of bile around stone, within CBD.

Typical

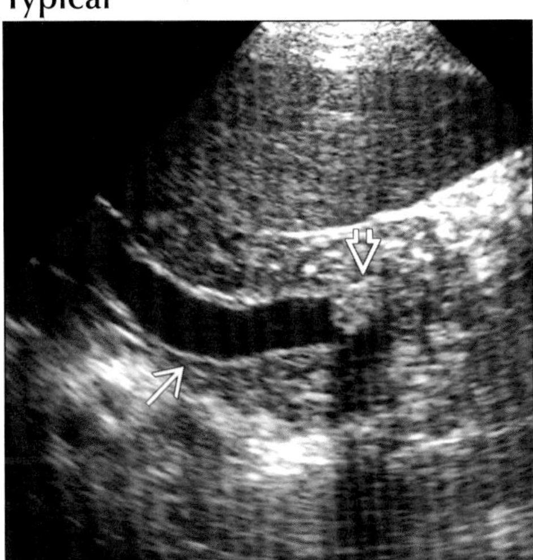

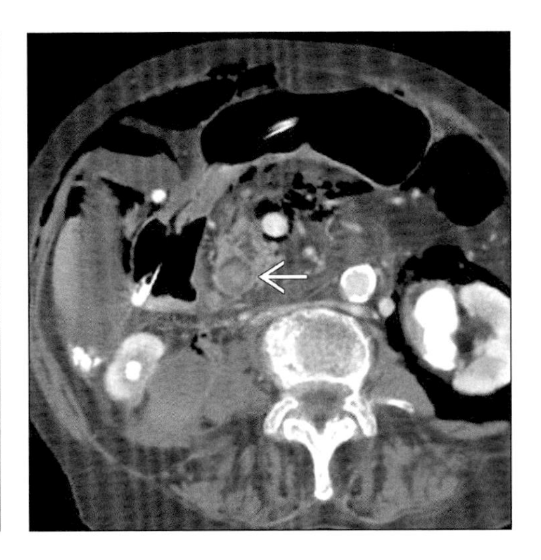

(Left) Oblique sagittal sonogram shows dilated CBD (arrow) and obstructing stone (open arrow). Note acoustic shadow behind stone. *(Right)* Axial CECT following attempted ERCP extraction of CBD stone. The stone (arrow) remains within the distal CBD. Extraluminal gas in retroperitoneum and peritoneal cavity due to duodenal perforation.

MIRIZZI SYNDROME

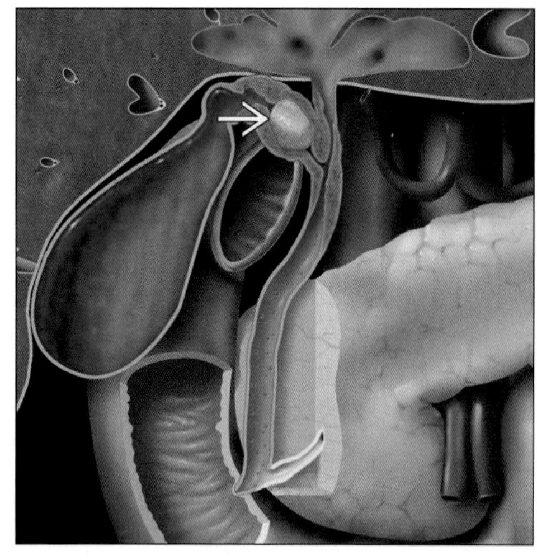

Mirizzi syndrome. Graphic depicts large cystic duct stone (arrow) causing dilatation of common hepatic duct.

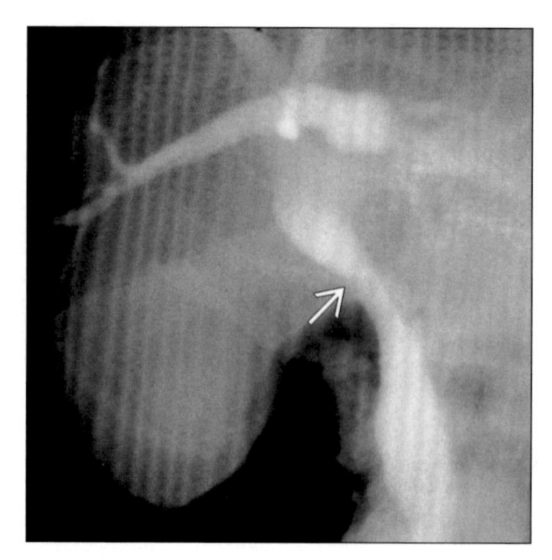

ERCP of Mirizzi syndrome. Note smooth extrinsic mass effect by a stone compressing common hepatic duct and non-filling of gallbladder (arrow).

TERMINOLOGY

Definitions

- Partial or complete obstruction of common hepatic duct (CHD) due to gallstone impacted in cystic duct or infundibulum of gallbladder (GB)

IMAGING FINDINGS

General Features

- Best diagnostic clue: Impacted cystic duct stone on US with proximal dilatation of intrahepatic duct
- Location: Gallstone in cystic duct
- Size: Gallstones typically 1-3 cm
- Morphology: Smooth extrinsic filling defect on ERCP at level of CHD

Radiographic Findings

- ERCP: Extrinsic narrowing of CHD; dilated intrahepatic ducts; lack of GB filling

CT Findings

- CECT: Dilated intrahepatic bile ducts; stone in neck of GB; normal distal CBD; acute/chronic cholecystitis with pericholecystic inflammatory changes

MR Findings

- T2WI
    - High signal dilated intrahepatic bile ducts
    - Low signal filling defects representing gallstones in GB, cystic duct
- T1 C+: Low signal dilated intrahepatic bile ducts, gallstones in cystic duct, normal distal CBD
- MRCP
    - Dilated intrahepatic ducts, filling defect in CHD

Ultrasonographic Findings

- Real Time: Gallstone impacted in cystic duct; dilated intrahepatic ducts on US

Nuclear Medicine Findings

- Hepato Biliary Scan
    - Non-filling GB, dilated intrahepatic ducts

Imaging Recommendations

- Best imaging tool: US, CECT, ERCP
- Protocol advice: Color Doppler to detect dilated intrahepatic bile ducts

DIFFERENTIAL DIAGNOSIS

Cystic duct stones

- No dilation of intrahepatic ducts
- Biliary colic or acute cholecystitis

Common hepatic obstruction from nodes

- Porta hepatis nodes obstructing dilated ducts

DDx: Lesions Mimicking Mirizzi Syndrome

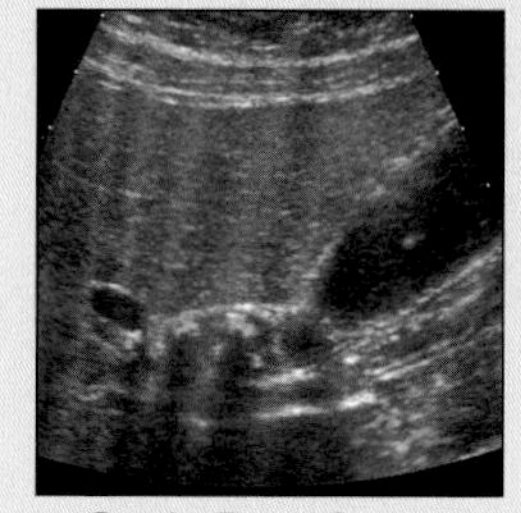

Cystic Duct Stones

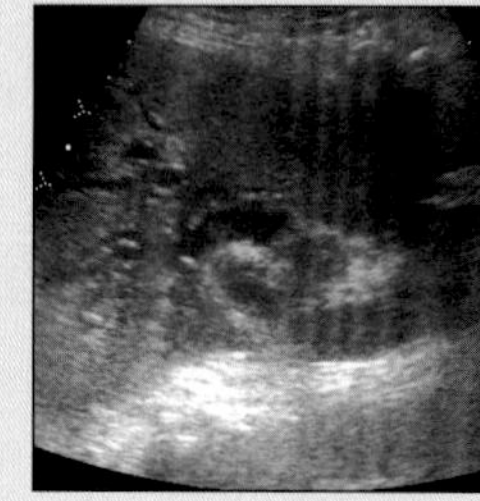

Porta Hepatis Node

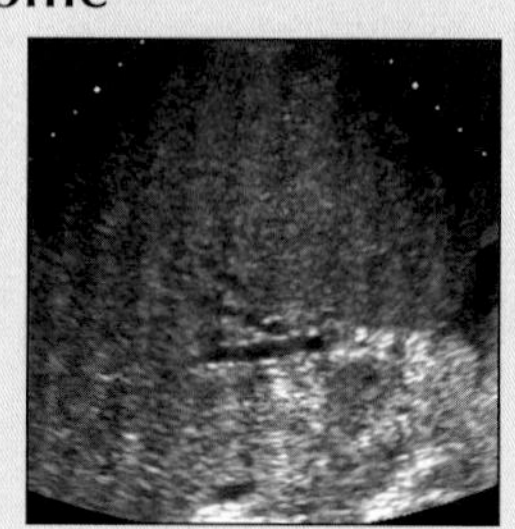

Cholangiocarcinoma

MIRIZZI SYNDROME

Key Facts

Terminology

- Partial or complete obstruction of common hepatic duct (CHD) due to gallstone impacted in cystic duct or infundibulum of gallbladder (GB)

Imaging Findings

- Best diagnostic clue: Impacted cystic duct stone on US with proximal dilatation of intrahepatic duct
- ERCP: Extrinsic narrowing of CHD; dilated intrahepatic ducts; lack of GB filling
- CECT: Dilated intrahepatic bile ducts; stone in neck of GB; normal distal CBD; acute/chronic cholecystitis with pericholecystic inflammatory changes
- Real Time: Gallstone impacted in cystic duct; dilated intrahepatic ducts on US

Diagnostic Checklist

- Porta hepatis obstruction from nodes

- Hypoechoic masses without shadowing

Cholangiocarcinoma

- Soft tissue mass porta hepatis obstructing bile ducts
- Dilated intrahepatic ducts

PATHOLOGY

General Features

- General path comments
  - Acute/chronic cholecystitis
  - May have cholecysto-choledochal fistula
  - Repeated episodes of cholangitis lead to biliary cirrhosis
- Etiology
  - Cystic duct anatomically oriented parallel to CHD
  - Impaction of stone in cystic duct or GB neck
  - Partial or complete obstruction of CHD
- Epidemiology: Tracks prevalence of gallstones

Gross Pathologic & Surgical Features

- Acute/chronic cholecystitis
- May have cholecysto-choledochal fistula

Staging, Grading or Classification Criteria

- Type I: Extrinsic compression of CHD due to cystic duct stone
- Type II: Cholecystobiliary fistula < 1/3 circumference of ductal wall
- Type III: Cholecystobiliary fistula with 2/3 of ductal wall involvement
- Type IV: Cholecystobiliary fistula with entire ductal wall circumference involvement

CLINICAL ISSUES

Presentation

- Most common signs/symptoms: Fever, jaundice, RUQ pain

Demographics

- Age: Adults > 40 yrs
- Gender: M:F = 1:2

Natural History & Prognosis

- Jaundice & cholangitis if not treated promptly

Treatment

- Cholecystectomy with careful dissection of cystic duct to avoid injury to CHD

DIAGNOSTIC CHECKLIST

Consider

- Porta hepatis obstruction from nodes

Image Interpretation Pearls

- Stone impacted in cystic duct with dilated ducts

SELECTED REFERENCES

1. Abou-Saif A et al: Complications of gallstone disease: Mirizzi syndrome, cholecystocholedochal fistula, and gallstone ileus. Am J Gastroenterol. 97(2):249-54, 2002
2. Haritopoulos KN et al: Mirizzi syndrome: a case report and review of the literature. Int Surg. 87(2):65-8, 2002
3. Gomez G: Mirizzi Syndrome. Curr Treat Options Gastroenterol. 5(2):95-99, 2002
4. Xiaodong H et al: Diagnosis and treatment of the Mirizzi syndrome. Chin Med Sci J. 14(4):246-8, 1999

IMAGE GALLERY

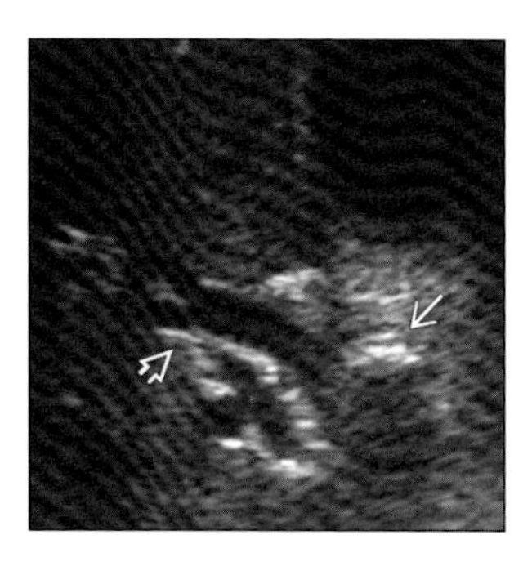

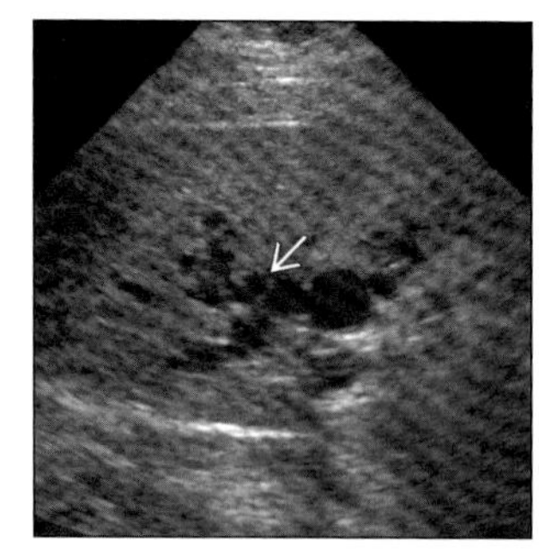

*(**Left**) Sagittal US of neck of gallbladder demonstrates impacted stone (arrow) with dilation of common hepatic duct (open arrow). (**Right**) Sagittal US of right lobe of liver. Note dilated intrahepatic duct (arrow).*

HYPERPLASTIC CHOLECYSTOSES

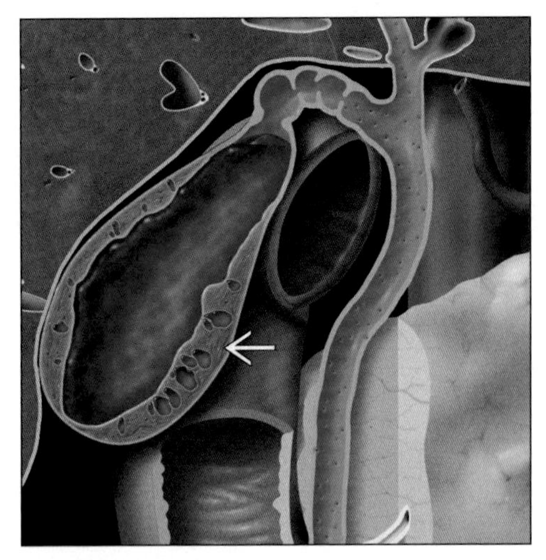

Schematic drawing of adenomyomatosis. Note thickened gallbladder wall with multiple intramural cystic spaces (arrow).

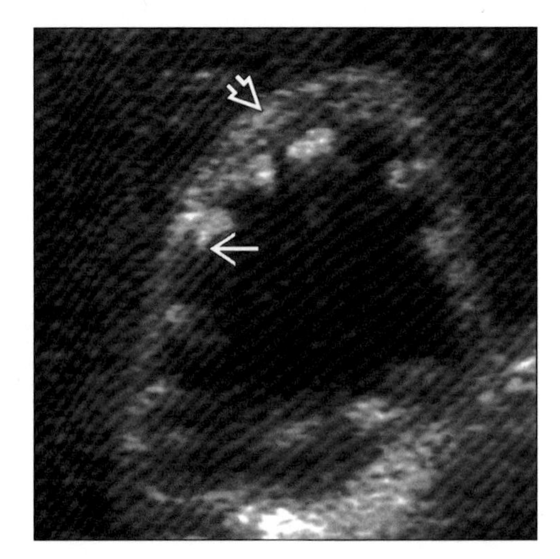

Transverse sonogram of adenomyomatosis. Note thickened gallbladder wall (open arrow) and multiple echogenic foci with "comet tail" reverberation artifacts (arrow).

TERMINOLOGY

Abbreviations and Synonyms

- Cholesterolosis; strawberry gallbladder (GB), cholesterol polyp; adenomyomatosis

Definitions

- General: Idiopathic non-neoplastic & non-inflammatory proliferative disorder resulting in GB wall thickening
- Adenomyomatosis: Mural GB wall thickening 2° to exaggeration of normal luminal epithelial folds (Rokitansky-Aschoff sinuses) in conjunction with smooth muscle proliferation
- Cholesterolosis: Deposition of foamy cholesterol-laden histiocytes in subepithelium of GB; numerous small accumulations (strawberry GB) or larger polypoid deposit (cholesterol polyp)

IMAGING FINDINGS

General Features

- Best diagnostic clue
  - Adenomyomatosis: Fundal, diffuse or mid-body GB wall thickening with intramural high amplitude echoes & "comet tail" reverberation artifacts
  - Cholesterolosis: Multiple GB polyps
- Location: GB wall
- Size: Polyps typically 5-10 mm

Radiographic Findings

- Radiography: OCG: "Pearl necklace" GB with multiple contrast-filled intramural diverticula

CT Findings

- CECT: Thickened GB wall (segmental or diffuse); often brisk enhancement of wall post-contrast; cystic nonenhancing spaces within GB wall corresponding to intramural diverticula

MR Findings

- T2WI: High signal cystic spaces within thickened GB wall
- T1 C+: Nonenhancing cystic spaces within thickened GB wall

Ultrasonographic Findings

- Real Time
  - Adenomyomatosis: Focal or diffuse GB wall thickening; intramural high amplitude echoes with "comet tail" artifacts
  - Cholesterolosis: Multiple small (< 10 mm) polyps
- Color Doppler: Avascular or hypervascular areas

Imaging Recommendations

- Best imaging tool: US, CECT

DDx: Spectrum of GB Disease Mimicking Cholecystoses

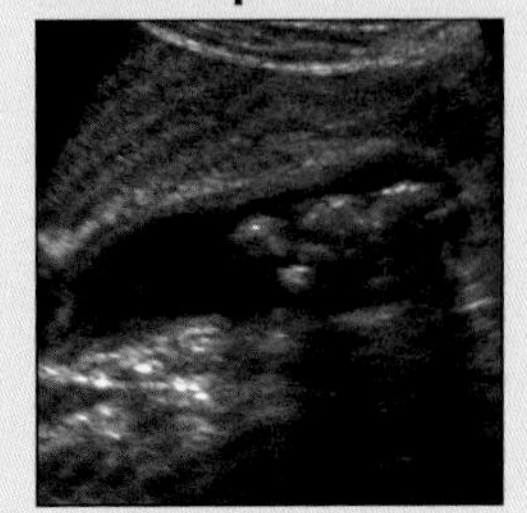

Cholecystitis

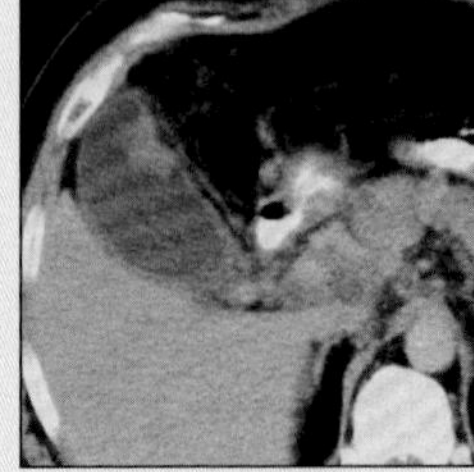

Gallbladder CA

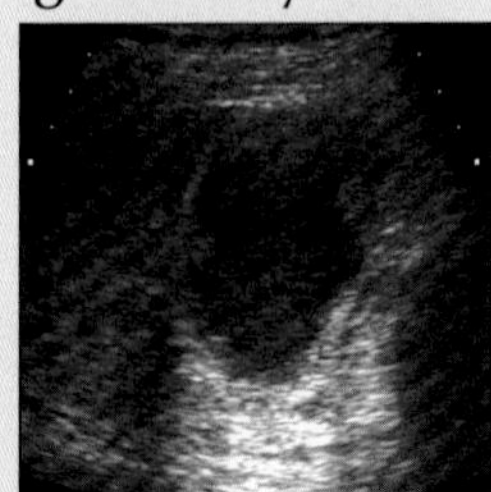

GB Polyps

HYPERPLASTIC CHOLECYSTOSES

Key Facts

Imaging Findings
- Adenomyomatosis: Fundal, diffuse or mid-body GB wall thickening with intramural high amplitude echoes & "comet tail" reverberation artifacts
- Cholesterolosis: Multiple GB polyps
- CECT: Thickened GB wall (segmental or diffuse); often brisk enhancement of wall post-contrast; cystic nonenhancing spaces within GB wall corresponding to intramural diverticula
- Best imaging tool: US, CECT

Clinical Issues
- Most common signs/symptoms: Most often asymptomatic, but may present with RUQ pain

Diagnostic Checklist
- Consider chronic cholecystitis
- "Comet tail" reverberation artifacts within thickened wall

DIFFERENTIAL DIAGNOSIS

Chronic cholecystitis
- Generalized wall thickening; gallstones; no mural "comet tail" artifact

GB carcinoma
- Polypoid mass > 2 cm; diffuse asymmetric GB wall thickening; hepatic invasion

Adenomatous polyp
- Polypoid mass 5-15 mm

PATHOLOGY

General Features
- General path comments: Adenomyomatosis: Diffuse or segmental GB wall thickening with multiple cystic spaces
- Etiology: Idiopathic
- Epidemiology: Occurs in 5-25% of resected GB patients
- Associated abnormalities: Gallstones in 25-75% of patients

Gross Pathologic & Surgical Features
- Focal or diffuse GB wall thickening without inflammatory changes

Microscopic Features
- Adenomyomatosis
  - Mural thickening 2° to smooth muscle proliferation & exaggerated folds of Rokitansky-Aschoff sinuses
- Cholesterolosis
  - Subepithelium deposition of cholesterol-laden histiocytes with villus-like mucosal protrusions; may coalesce into polyps

CLINICAL ISSUES

Presentation
- Most common signs/symptoms: Most often asymptomatic, but may present with RUQ pain

Demographics
- Age: > 35 yrs
- Gender: M:F = 1:3

Natural History & Prognosis
- Usually incidental finding; little clinical importance

Treatment
- Options, risks, complications: Cholecystectomy only if symptomatic

DIAGNOSTIC CHECKLIST

Consider
- Consider chronic cholecystitis

Image Interpretation Pearls
- "Comet tail" reverberation artifacts within thickened wall

SELECTED REFERENCES

1. Owen CC et al: Gallbladder polyps, cholesterolosis, adenomyomatosis, and acute acalculous cholecystitis. Semin Gastrointest Dis. 14(4):178-88, 2003
2. Ghersin E et al: Twinkling artifact in gallbladder adenomyomatosis. J Ultrasound Med. 22(2):229-31, 2003
3. Yoshimitsu K et al: Radiologic diagnosis of adenomyomatosis of the gallbladder: comparative study among MRI, helical CT, and transabdominal US. J Comput Assist Tomogr. 25(6):843-50, 2001

IMAGE GALLERY

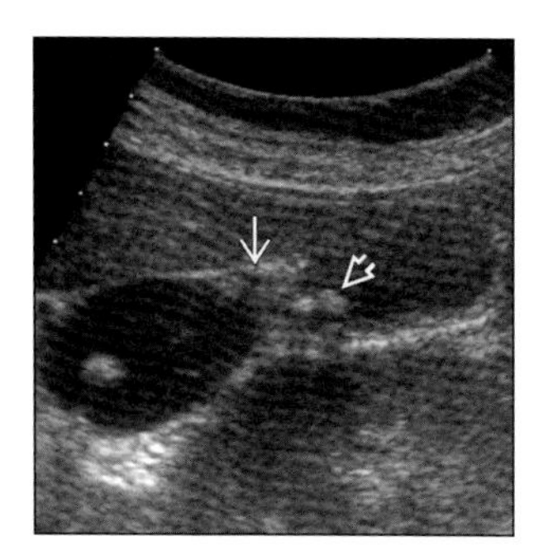

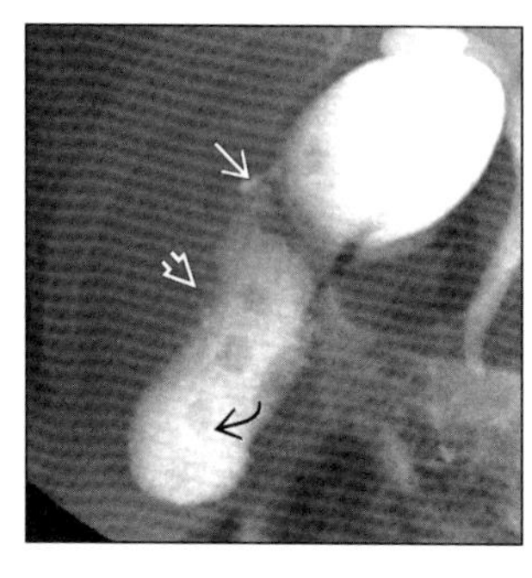

*(**Left**) Sagittal sonogram of segmental adenomyomatosis demonstrates focal gallbladder wall thickening in mid-body ("hourglass deformity" - arrow). Note gallstones (open arrow). (**Right**) ERCP demonstrates focal narrowing ("hourglass deformity" - arrow), multiple intramural diverticula (open arrow), and gallstones (curved arrow).*

CHOLECYSTITIS

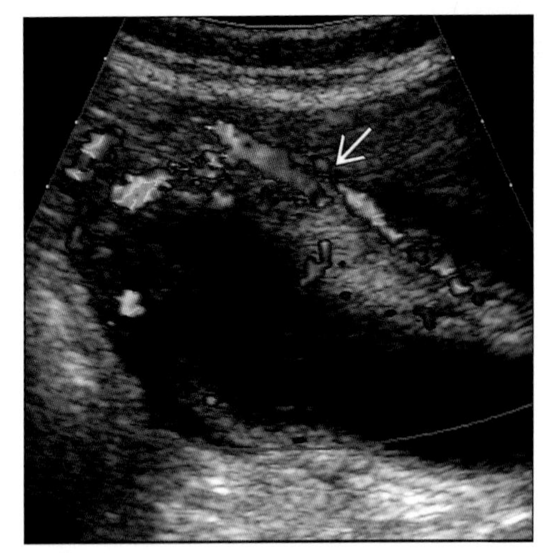

Transverse color Doppler sonogram demonstrates marked hyperemia (arrow) and gallbladder wall thickening from acute cholecystitis.

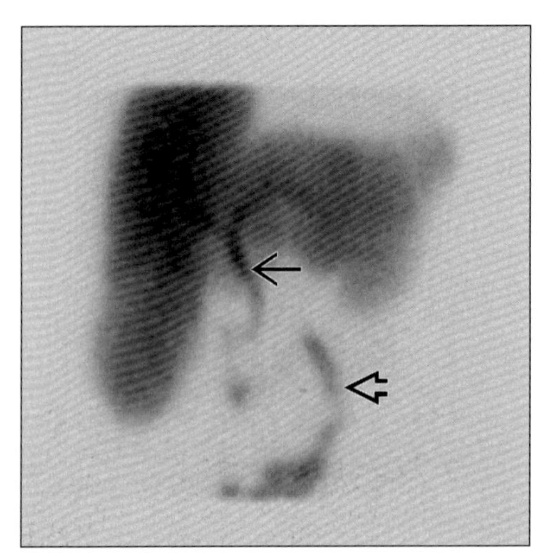

Acute cholecystitis on biliary scintigram. Note isotope filling common bile duct (arrow) and small bowel (open arrow) without filling of gallbladder.

TERMINOLOGY

Abbreviations and Synonyms

- Acute calculous cholecystitis, acute acalculous cholecystitis

Definitions

- Acute inflammation of gallbladder (GB), 95% 2° to calculus obstructing cystic duct; acalculous in 5% 2° to ischemia, secondary inflammation/infection

IMAGING FINDINGS

General Features

- Best diagnostic clue
  - Impacted gallstone in cystic duct
  - Positive sonographic Murphy sign
  - Gallbladder wall thickening
- Location: Stone impacted in GB neck or cystic duct
- Size: Distended GB (> 5 cm transverse diameter)
- Morphology: Distended GB more rounded in shape than normal "pear-shaped" configuration

Radiographic Findings

- Radiography: Calcified stones in only 15-20% of patients with cholecystitis
- ERCP
  - May document common bile duct (CBD) stones in patients with associated cholangitis
  - No filling of gallbladder

CT Findings

- NECT
  - Distended GB
  - Edematous pericholecystic fat with stranding
  - Calcified gallstones (15%)
- CECT
  - Uncomplicated cholecystitis
    - GB wall thickening
    - Increased mural enhancement
    - Pericholecystic fat stranding
    - Cholesterol stones typically not visible
  - Complicated cholecystitis
    - Intramural or pericholecystic abscesses leading to asymmetric GB wall thickening
    - Gas in lumen and/or wall of gallbladder
    - High attenuation gallbladder hemorrhage
    - Focal interruption of GB wall due to necrosis
    - Adherent omentum

MR Findings

- T2WI
  - Distended GB with stones
  - High signal pericholecystic fat
- T1 C+

DDx: Spectrum of Diseases Mimicking Cholecystitis

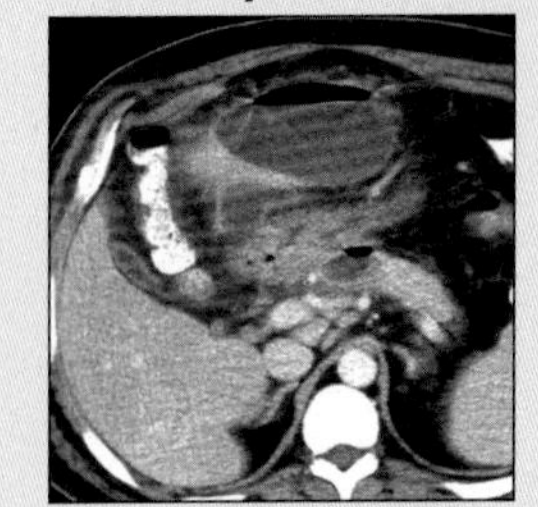

PUD

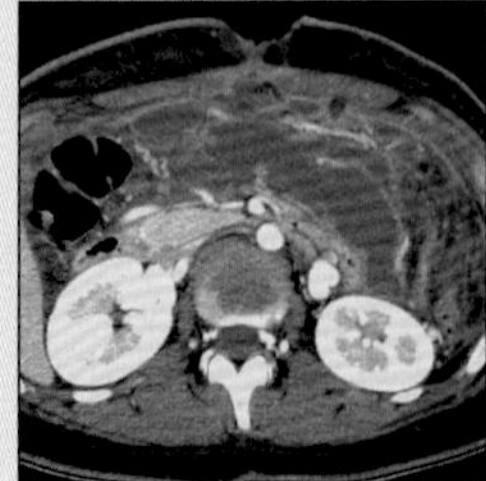

Pancreatitis

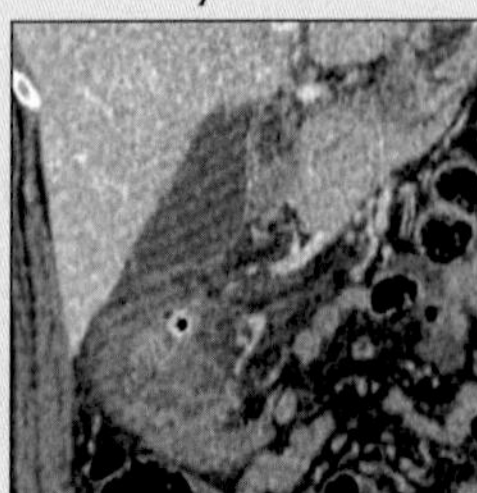

Diverticulitis

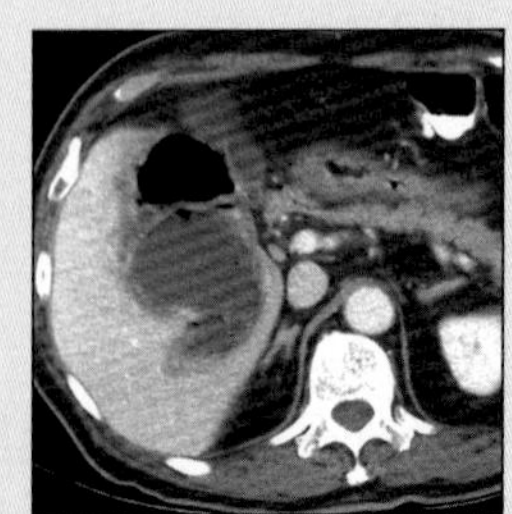

Liver Abscess

CHOLECYSTITIS

Key Facts

Terminology

- Acute inflammation of gallbladder (GB), 95% 2° to calculus obstructing cystic duct; acalculous in 5% 2° to ischemia, secondary inflammation/infection

Imaging Findings

- Uncomplicated: Gallstones; positive sonographic Murphy sign; gallstone impacted in neck of GB or cystic duct; thickened GB wall (> 4 cm) on US
- Complicated: Gallstones; pericholecystic fluid/abscess; intraluminal membranes; gas in GB wall/lumen; sonographic Murphy sign absent in 1/3 of patients; asymmetric wall thickening on US
- Best imaging tool: US or biliary scintigraphy
- Protocol advice: Longitudinal and transverse images of GB, parasagittal images of GB neck region & cystic duct in LPO position to detect impacted gallstones (i.e., immobile)

Clinical Issues

- Most common signs/symptoms: Acute RUQ pain, fever
- May progress to gangrenous cholecystitis and perforation if untreated
- Excellent prognosis in uncomplicated cases or with prompt surgery

Diagnostic Checklist

- Perforated ulcer or pancreatitis with secondary GB wall thickening

  - "Rim sign" of increased hepatic enhancement in patient with gangrenous cholecystitis
  - Focal interruption of enhancement

Ultrasonographic Findings

- Real Time
  - Uncomplicated: Gallstones; positive sonographic Murphy sign; gallstone impacted in neck of GB or cystic duct; thickened GB wall (> 4 cm) on US
  - Complicated: Gallstones; pericholecystic fluid/abscess; intraluminal membranes; gas in GB wall/lumen; sonographic Murphy sign absent in 1/3 of patients; asymmetric wall thickening on US

Nuclear Medicine Findings

- Hepato Biliary Scan
  - 99m Tc iminodiacetic acid derivatives
  - Non-visualization of GB at 4 hours has 99% specificity
  - Increased uptake in gallbladder fossa during arterial phase due to hyperemia in 80% of patients
  - "Rim sign" seen in 34% of patients is due to increased uptake in gallbladder fossa
    - Positive predictive value of 57% for gangrenous cholecystitis

Imaging Recommendations

- Best imaging tool: US or biliary scintigraphy
- Protocol advice: Longitudinal and transverse images of GB, parasagittal images of GB neck region & cystic duct in LPO position to detect impacted gallstones (i.e., immobile)

DIFFERENTIAL DIAGNOSIS

Peptic ulcer disease (PUD)

- Thickened duodenum
- Ectopic gas if perforated ulcer
- Periduodenal inflammatory changes in anterior pararenal space
- Secondary GB wall thickening

Acute pancreatitis

- Enlarged pancreas
- Peripancreatic fluid or inflammatory changes
- Nonenhancing areas of necrosis

Hepatic flexure diverticulitis

- Colonic diverticula
- Pericolonic inflammation
- Fecalith

Liver abscess

- CECT
  - "Cluster sign" of multiloculated pyogenic abscesses
  - Air-fluid level from gas-forming organism
- US: Irregular, hypoechoic mass with little enhancement through transmission

PATHOLOGY

General Features

- General path comments
  - Distended GB
  - Thickened, inflamed GB wall
  - Pericholecystic adhesions to omentum
- Genetics
  - Increased incidence of gallstones in selected population
    - Hispanics, Pima Indians
- Etiology
  - 95% calculous
    - Obstructing stone in cystic duct
  - 5% acalculous
    - Ischemia with secondary inflammation/infection
    - AIDS patients have opportunistic GB infection
- Epidemiology
  - Incidence parallels prevalence of gallstones
    - 25 million Americans have gallstones
    - M:F = 1:3

Gross Pathologic & Surgical Features

- Gallstones in gallbladder neck or cystic duct
- Thickened GB wall with hyperemia of wall

- Omental adhesions

Microscopic Features

- Lumen: Gallstones, sludge
- GB mucosa: Ulcerations
- GB wall: Acute polymorphonuclear (PMN) infiltration
- Bacterial cultures positive in 40-70% of patients

Staging, Grading or Classification Criteria

- Non-perforated
  - GB wall intact on CT and/or US
- Gangrenous
  - US: Pericholecystic fluid, intraluminal membranes, asymmetric GB wall thickening
- Perforated
  - CECT: Pericholecystic abscess, GB wall necrosis with lack of enhancement

CLINICAL ISSUES

Presentation

- Most common signs/symptoms: Acute RUQ pain, fever
- Clinical profile
  - Increased WBC
  - May have mild elevation in liver enzymes

Demographics

- Age: Typically > 25 y
- Gender: M:F = 1:3

Natural History & Prognosis

- May progress to gangrenous cholecystitis and perforation if untreated
- Excellent prognosis in uncomplicated cases or with prompt surgery
- Complications
  - Mirizzi syndrome and Bouveret syndrome (gallstone erodes into duodenum causing obstruction)

Treatment

- Prompt cholecystectomy
  - Laparoscopic surgery for uncomplicated cases
- Percutaneous cholecystectomy
  - Useful for poor operative risk patients with GB empyema
- Percutaneous drainage
  - Well-defined, well-localized pericholecystic abscesses

DIAGNOSTIC CHECKLIST

Consider

- Perforated ulcer or pancreatitis with secondary GB wall thickening

Image Interpretation Pearls

- Stone impacted in cystic duct
- Sonographic Murphy sign must be unequivocal to be considered positive

SELECTED REFERENCES

1. Yusoff IF et al: Diagnosis and management of cholecystitis and cholangitis. Gastroenterol Clin North Am. 32(4):1145-68, 2003
2. Bennett GL et al: Ultrasound and CT evaluation of emergent gallbladder pathology. Radiol Clin North Am. 41(6):1203-16, 2003
3. Browning JD et al: Gallstone disease and its complications. Semin Gastrointest Dis. 14(4):165-77, 2003
4. Barie PS et al: Acute acalculous cholecystitis. Curr Gastroenterol Rep. 5(4):302-9, 2003
5. Pazzi P et al: Biliary sludge: the sluggish gallbladder. Dig Liver Dis. 35 Suppl 3:S39-45, 2003
6. Roth T et al: Acute acalculous cholecystitis associated with aortic dissection: report of a case. Surg Today. 33(8):633-5, 2003
7. Cheema S et al: Timing of laparoscopic cholecystectomy in acute cholecystitis. Ir J Med Sci. 172(3):128-31, 2003
8. Ozaras R et al: Acute viral cholecystitis due to hepatitis A virus infection. J Clin Gastroenterol. 37(1):79-81, 2003
9. Ko CW et al: Gastrointestinal disorders of the critically ill. Biliary sludge and cholecystitis. Best Pract Res Clin Gastroenterol. 17(3):383-96, 2003
10. Gandolfi L et al: The role of ultrasound in biliary and pancreatic diseases. Eur J Ultrasound. 16(3):141-59, 2003
11. Trowbridge RL et al: Does this patient have acute cholecystitis? JAMA. 289(1):80-6, 2003
12. Fayad LM et al: Functional magnetic resonance cholangiography (fMRC) of the gallbladder and biliary tree with contrast-enhanced magnetic resonance cholangiography. J Magn Reson Imaging. 18(4):449-60, 2003
13. Pedrosa I et al: The interrupted rim sign in acute cholecystitis: a method to identify the gangrenous form with MRI. J Magn Reson Imaging. 18(3):360-3, 2003
14. Oh KY et al: Limited abdominal MRI in the evaluation of acute right upper quadrant pain. Abdom Imaging. 28(5):643-51, 2003
15. Gore RM et al: Imaging benign and malignant disease of the gallbladder. Radiol Clin North Am. 40(6):1307-23, vi, 2002
16. Bingener-Casey J et al: Reasons for conversion from laparoscopic to open cholecystectomy: a 10-year review. J Gastrointest Surg. 6(6):800-5, 2002
17. Kitano S et al: Laparoscopic cholecystectomy for acute cholecystitis. J Hepatobiliary Pancreat Surg. 9(5):534-7, 2002
18. Indar AA et al: Acute cholecystitis. BMJ. 325(7365):639-43, 2002
19. Merchant SS et al: Staphylococcus aureus cholecystitis: a report of three cases with review of the literature. Yale J Biol Med. 75(5-6):285-91, 2002
20. Tanaka M: Bile duct clearance, endoscopic or laparoscopic? J Hepatobiliary Pancreat Surg. 9(6):729-32, 2002
21. Abou-Saif A et al: Complications of gallstone disease: Mirizzi syndrome, cholecystocholedochal fistula, and gallstone ileus. Am J Gastroenterol. 97(2):249-54, 2002
22. Adusumilli S et al: MR imaging of the gallbladder. Magn Reson Imaging Clin N Am. 10(1):165-84, 2002
23. Kalimi R et al: Diagnosis of acute cholecystitis: sensitivity of sonography, cholescintigraphy, and combined sonography-cholescintigraphy. J Am Coll Surg. 193(6):609-13, 2001

IMAGE GALLERY

Typical

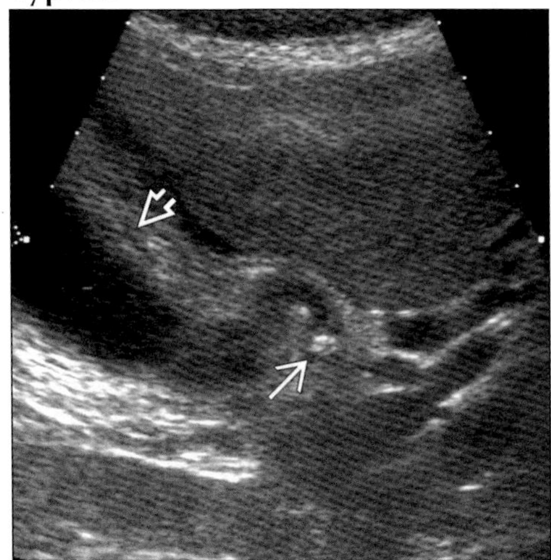

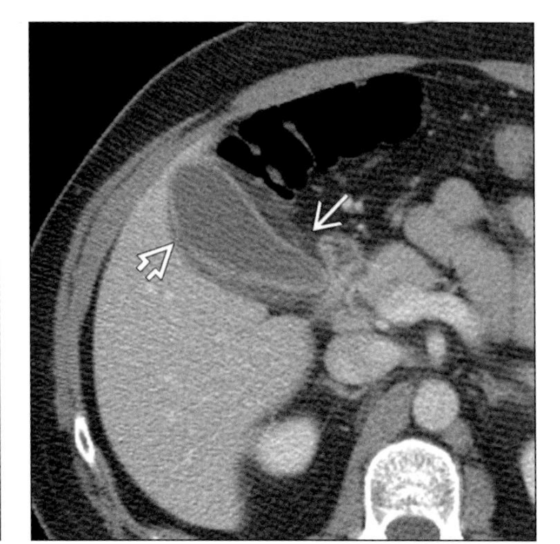

(Left) *Transverse sonogram of calculous cholecystitis. Note cystic duct stone (arrow) and marked gallbladder wall thickening (open arrow).* ***(Right)*** *Axial CECT of cholecystitis. Note pericholecystic stranding (arrow) and focal interruption of enhancement of gallbladder wall (open arrow).*

Typical

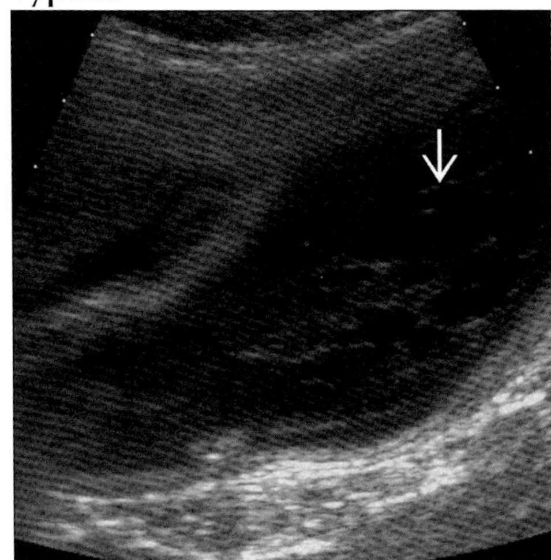

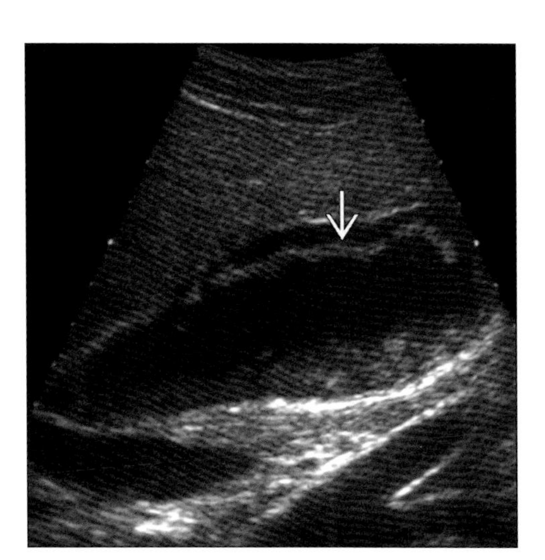

(Left) *Sagittal US of gangrenous cholecystitis demonstrates intraluminal membranes from fibrinous debris (arrow).* ***(Right)*** *Sagittal US of gangrenous cholecystitis demonstrating sloughed mucosa (arrow).*

Typical

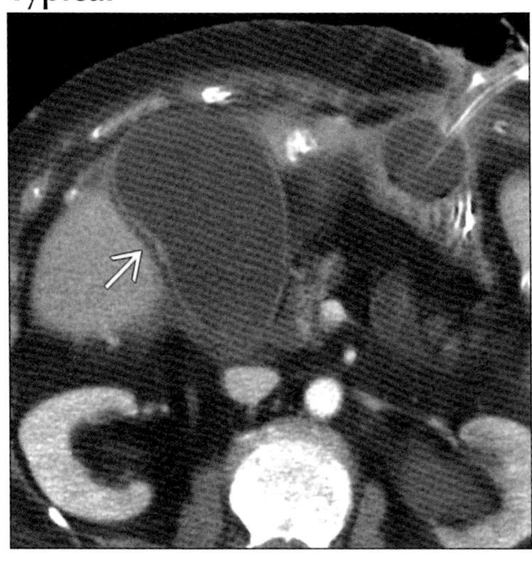

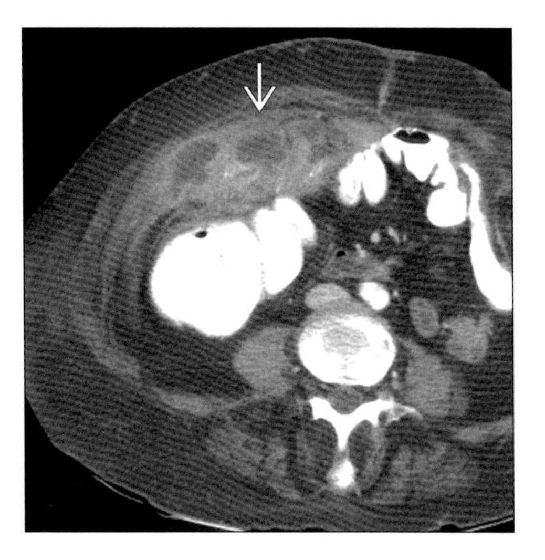

(Left) *Axial CECT of pericholecystic abscess demonstrates distended gallbladder with edematous wall (arrow).* ***(Right)*** *Axial CECT of pericholecystic abscess demonstrates walled-off abscess from fundal perforation (arrow).*

PORCELAIN GALLBLADDER

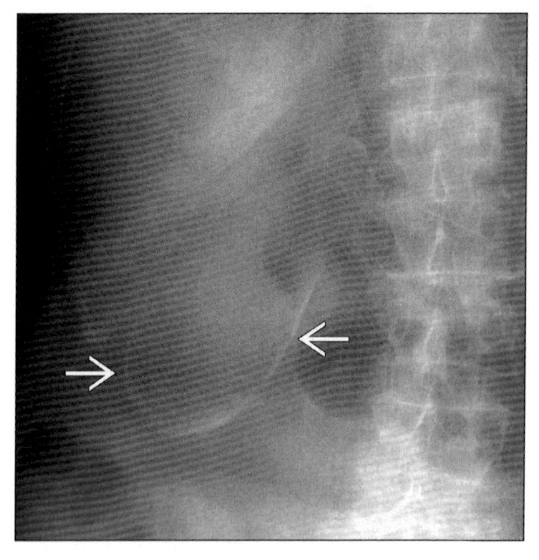

Anteroposterior radiography shows thin rim of calcification (arrows) in right upper quadrant conforming to the shape of the gallbladder.

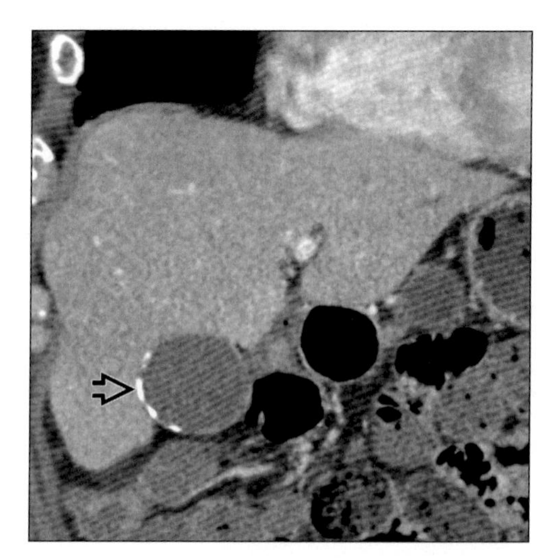

Coronal CECT shows rim of interrupted calcifications (arrow) in the lateral aspect of the gallbladder wall.

TERMINOLOGY

Abbreviations and Synonyms

- Calcified gallbladder (GB), calcifying cholecystitis, cholecystopathia chronica calcarea

Definitions

- Calcification of the gallbladder wall

IMAGING FINDINGS

General Features

- Best diagnostic clue: Rim of calcification in right upper quadrant conforming to shape of the gallbladder
- Location: Gallbladder wall
- Size: May involve all or part of the gallbladder wall
- Morphology: Two patterns: Selective mucosal calcification and diffuse intramural calcification

Radiographic Findings

- Radiography
  - Curvilinear or granular calcification in GB wall
  - May involve entire wall or just a segment
- Fluoroscopy: Usually non-functional GB on oral cholecystograms

CT Findings

- NECT: Calcification in GB wall
- CECT: Calcification in GB wall

Ultrasonographic Findings

- Real Time
  - Echogenic curvilinear structure in GB fossa with acoustic shadowing
  - Coarse foci of calcification with acoustic shadowing in GB wall

Imaging Recommendations

- Best imaging tool: CT
- Protocol advice: Thin-sections may facilitate detection and localization of calcifications

DIFFERENTIAL DIAGNOSIS

Large gallstone

- Mimics porcelain GB on all modalities

Emphysematous cholecystitis (ultrasound)

- Echogenic crescent in gallbladder with acoustic shadowing on ultrasound mimics GB wall calcification
- CT can best distinguish gas from calcium

Iatrogenic

- Iodized oil in GB wall following hepatic chemo-embolization

DDx: Mimics of Porcelain Gallbladder

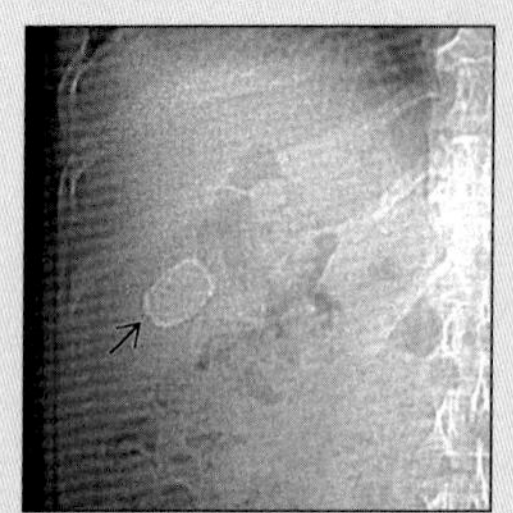

Large Gallstone

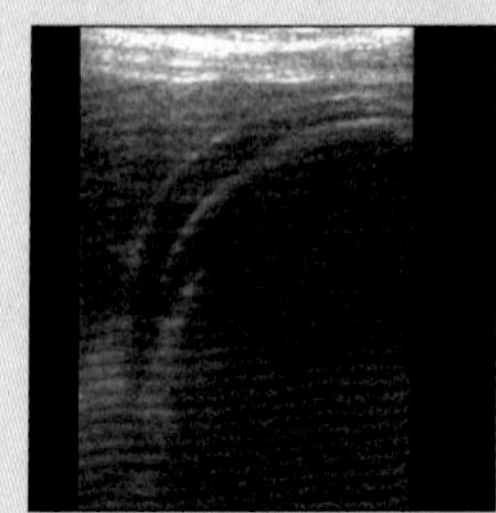

Large Gallstone

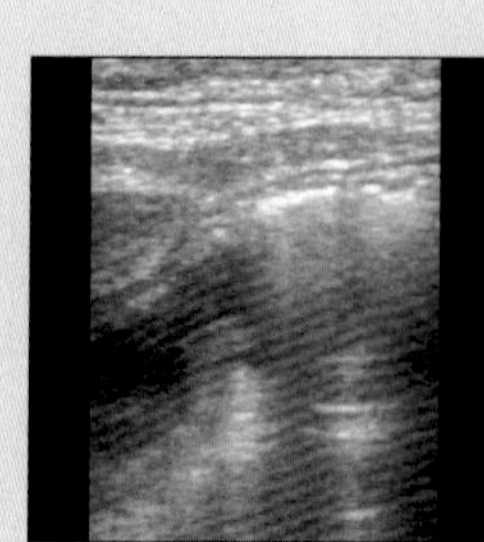

Gallbladder Gas

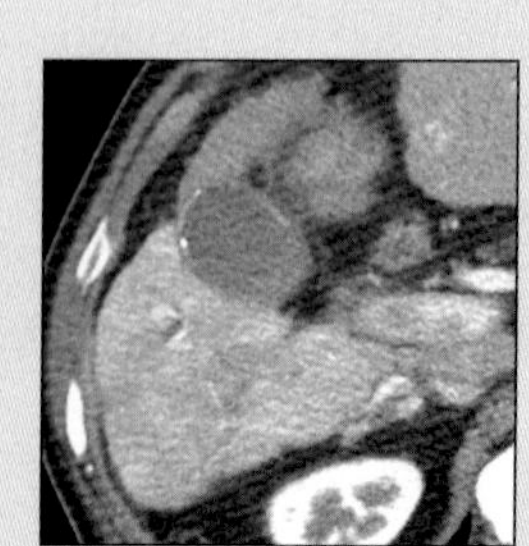

Chemoembolization

PORCELAIN GALLBLADDER

Key Facts

Imaging Findings
- Curvilinear or granular calcification in GB wall
- May involve entire wall or just a segment

Top Differential Diagnoses
- Large gallstone
- Emphysematous cholecystitis (ultrasound)

Pathology
- General path comments: Risk factor for gallbladder carcinoma
- Associated abnormalities: Gallstones in 90-95%

Clinical Issues
- Options, risks, complications: Prophylactic cholecystectomy is current consensus recommendation

PATHOLOGY

General Features
- General path comments: Risk factor for gallbladder carcinoma
- Etiology
  - Several theories of pathogenesis
    - Intermittently obstructed GB → supersaturated bile → calcium carbonate accumulates → precipitates in wall
    - Intramural hemorrhage from cholecystitis → mural calcification
    - Dystrophic mural calcification related to chronic inflammation of GB wall
- Epidemiology: Rare: 0.06-0.8% of cholecystectomy specimens
- Associated abnormalities: Gallstones in 90-95%

Gross Pathologic & Surgical Features
- Brittle gallbladder wall with bluish discoloration
- Fibrotic wall and brittle consistency makes laparoscopic cholecystectomy technically challenging

Microscopic Features
- Two histopathologic forms
  - Coarse plaques of calcium in muscularis of GB wall
  - Punctate foci of mucosal calcification

CLINICAL ISSUES

Presentation
- Most common signs/symptoms
  - Usually asymptomatic
  - Other signs/symptoms
    - Right upper quadrant pain
    - Palpable right upper quadrant mass

Demographics
- Age: Occurs in 6th decade; mean age = 54 years
- Gender: M:F = 1:5

Natural History & Prognosis
- Incidence of cancer in porcelain GB usually quoted as 12-62% based on retrospective data from 1950s-1960s
- More recent reviews suggest much weaker association: 0-5% incidence
- Risk of gallbladder cancer may depend on pattern of calcification
  - Diffuse intramural calcification → no risk of cancer
  - Flecks of calcium in gallbladder mucosa → significant (5%) risk of cancer

Treatment
- Options, risks, complications: Prophylactic cholecystectomy is current consensus recommendation

DIAGNOSTIC CHECKLIST

Consider
- Look for gallbladder mass on CT if porcelain GB identified

Image Interpretation Pearls
- Wall-echo-shadow (WES) sign on ultrasound can differentiate gallstones from porcelain GB

SELECTED REFERENCES

1. Opatrny L: Porcelain gallbladder. CMAJ. 166(7): 933, 2002
2. Gore RM et al: Imaging benign and malignant disease of the gallbladder. Radiol Clin North Am. 40(6): 1307-23, vi, 2002
3. Stephen AE et al: Carcinoma in the porcelain gallbladder: a relationship revisited. Surgery. 129(6): 699-703, 2001
4. Towfigh S et al: Porcelain gallbladder is not associated with gallbladder carcinoma. Am Surg. 67(1): 7-10, 2001
5. Rybicki FJ: The WES sign. Radiology. 214(3): 881-2, 2000

IMAGE GALLERY

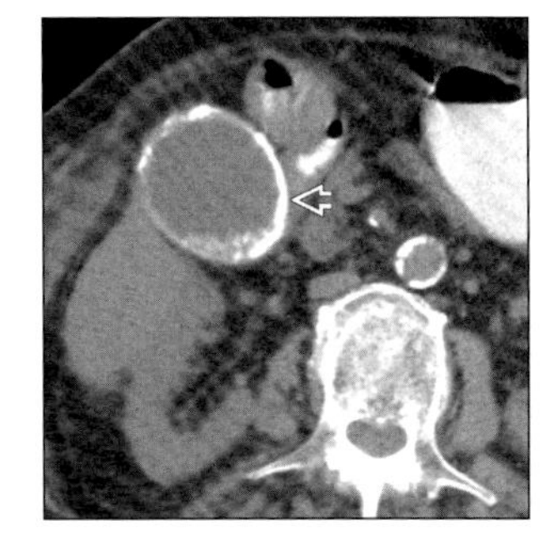

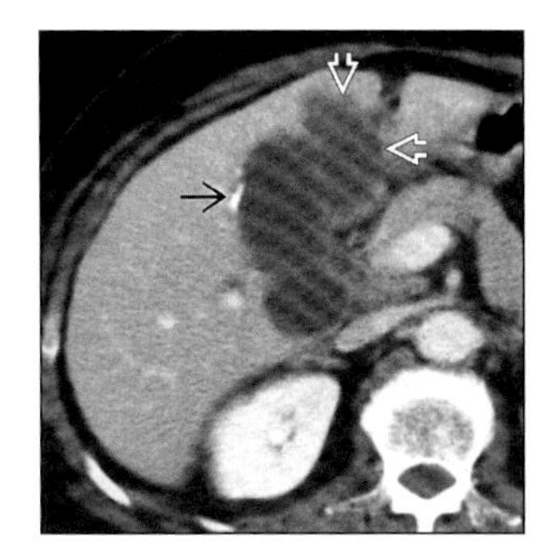

(Left) *Axial NECT shows thick rind of calcification in gallbladder wall (open arrow). Note layer of calcified stones in dependent aspect of gallbladder.* ***(Right)*** *Axial CECT shows curvilinear calcification in GB wall (black arrow) and low attenuation area in adjacent liver representing gallbladder carcinoma (white arrows) (Courtesy M. Nino-Murcia, MD).*

MILK OF CALCIUM BILE

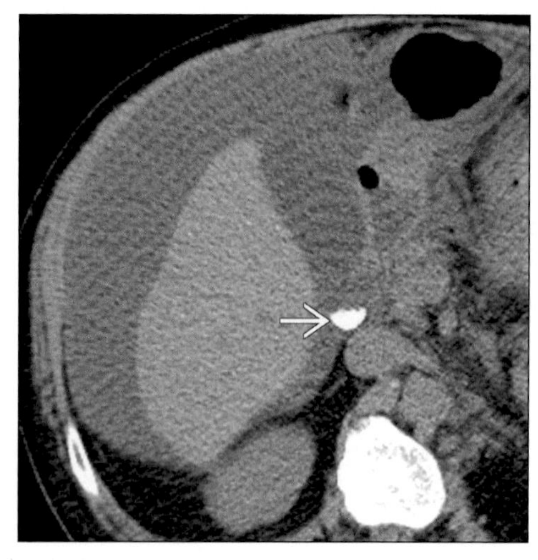

Axial NECT of milk of calcium bile in asymptomatic patient. Note high attenuation liquid layering posteriorly in gallbladder (arrow).

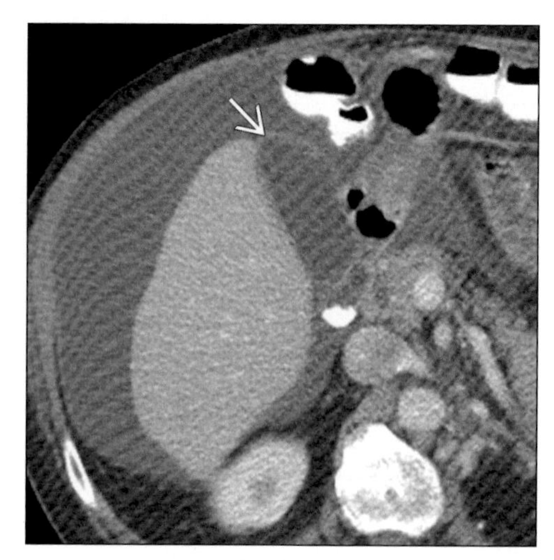

Axial CECT of milk of calcium bile in asymptomatic patient. Note normal gallbladder wall (arrow) & lack of inflammatory changes.

TERMINOLOGY

Abbreviations and Synonyms

- Limy bile syndrome, calcium carbonate bile

Definitions

- Calcium carbonate precipitate within gallbladder (GB) lumen

IMAGING FINDINGS

General Features

- Best diagnostic clue: Identification of calcified liquid within GB
- Location: Within GB

Radiographic Findings

- Radiography: Calcified liquid in GB, fluid-fluid level on upright film
- ERCP
  - Milk of calcium may obstruct cystic duct, prevent filling of GB

CT Findings

- NECT: High attenuation (> 150 HU) liquid in gallbladder; rarely occurring in common bile duct (CBD)
- CECT: GB wall may enhance normally or be thickened in patients with associated cholecystitis

MR Findings

- T2WI: Low signal calcium carbonate layers dependently in GB
- T1 C+: GB wall may enhance normally or be thickened in cholecystitis
- MRCP
  - Low signal fluid layering dependently in GB

Ultrasonographic Findings

- Real Time: Medium- to low-level echogenic fluid within GB with acoustic shadowing; may have associated gallstones, thickened GB wall on US
- Color Doppler: Hyperemia of GB wall in patients with acute cholecystitis

Nuclear Medicine Findings

- Hepato Biliary Scan
  - Non-filling of GB if cystic duct obstruction

Imaging Recommendations

- Best imaging tool: US, NECT
- Protocol advice: Tissue harmonic US may improve visualization of acoustic shadowing

DDx: Gallbladder Lesions Mimicking Milk of Calcium Bile

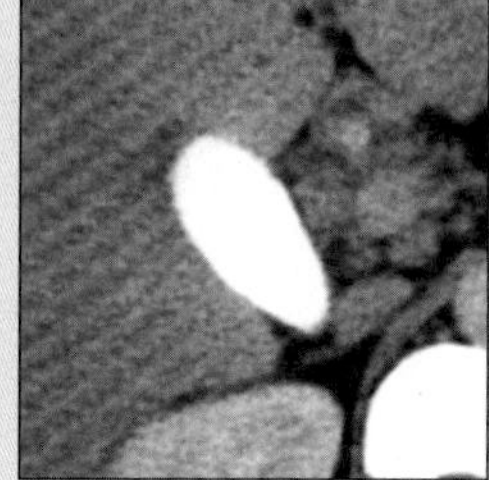

Vicarious Excretion

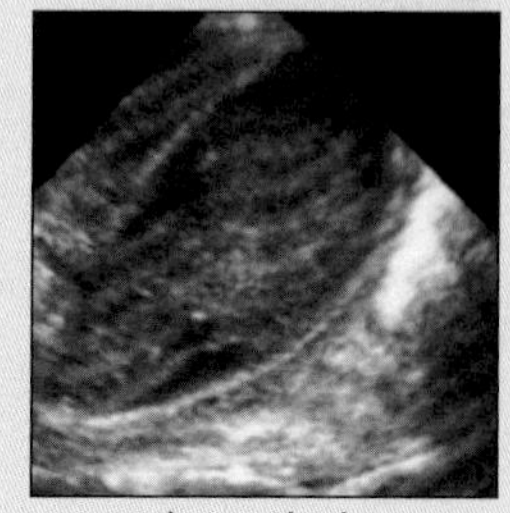

Biliary Sludge

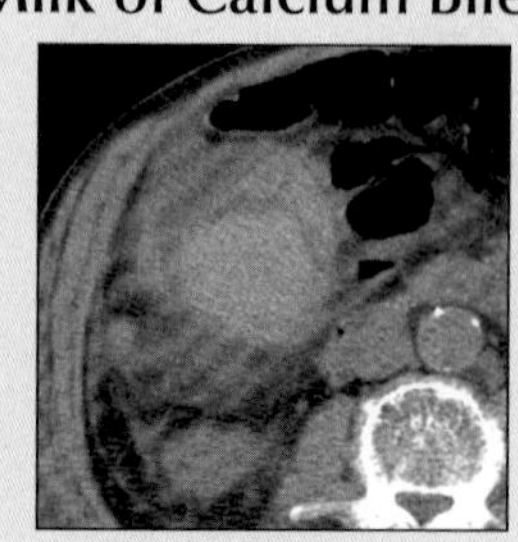

GB Heme

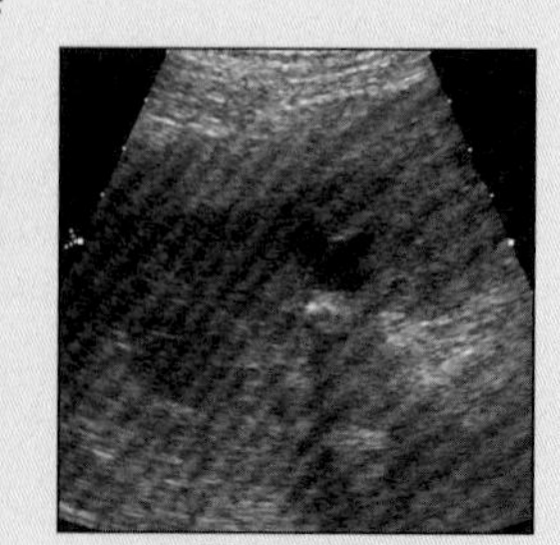

Gallbladder CA

MILK OF CALCIUM BILE

Key Facts

Terminology
- Calcium carbonate precipitate within gallbladder (GB) lumen

Imaging Findings
- NECT: High attenuation (> 150 HU) liquid in gallbladder; rarely occurring in common bile duct (CBD)
- Real Time: Medium- to low-level echogenic fluid within GB with acoustic shadowing; may have associated gallstones, thickened GB wall on US
- Best imaging tool: US, NECT
- Protocol advice: Tissue harmonic US may improve visualization of acoustic shadowing

Clinical Issues
- Incidental finding; may be asymptomatic

DIFFERENTIAL DIAGNOSIS

Vicarious excretion of contrast
- Hepatic excretion of iodinated contrast
- Normally seen 24 hours post-contrast administration

GB sludge
- Echogenic luminal debris, no distal acoustic enhancement
- May be "congested" as tumefactive sludge
- Avascular with color Doppler

GB hemorrhage
- High attenuation (> 30 HU) fluid in GB
- Echogenic on ultrasound, no distal acoustic shadowing

GB carcinoma
- Polypoid hypoechoic mass > 1 cm arising from GB mucosa with internal flow on color Doppler
- Large hypoechoic mass infiltrating GB fossa & invading liver
- Associated with gallstones, chronic cholecystitis, porcelain GB

PATHOLOGY

General Features
- Etiology: GB stasis
- Epidemiology: Incidence of 0.27% in patients undergoing cholecystectomy

Gross Pathologic & Surgical Features
- Calcium carbonate in bile, thickened GB wall

CLINICAL ISSUES

Presentation
- Most common signs/symptoms: RUQ pain, but may be asymptomatic

Demographics
- Age: 42-66 yrs
- Gender: M < F

Natural History & Prognosis
- Incidental finding; may be asymptomatic
- Calcium carbonate, associated gallstones with cystic duct obstruction & acute/chronic cholecystitis

Treatment
- Options, risks, complications
  - No treatment
  - Laparoscopic or open cholecystectomy
  - Bile leak, abscess, hematoma related to cholecystectomy

DIAGNOSTIC CHECKLIST

Consider
- GB sludge or hemorrhage

Image Interpretation Pearls
- Echogenic fluid similar to sludge but with acoustic shadowing

SELECTED REFERENCES

1. Itoh H: Management of limy bile syndrome: no therapy, laparotomy or endoscopic treatment? Intern Med. 42(1):1-2, 2003
2. Moreaux J et al: Limy bile. A surgical experience in 16 patients. Gastroenterol Clin Biol. 18(6-7):550-5, 1994
3. Fowler CL et al: Limy bile syndrome. J Pediatr Surg. 28(12):1568-9, 1993

IMAGE GALLERY

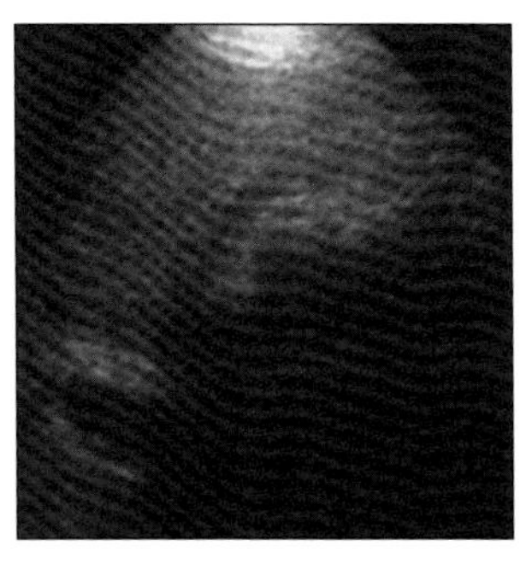
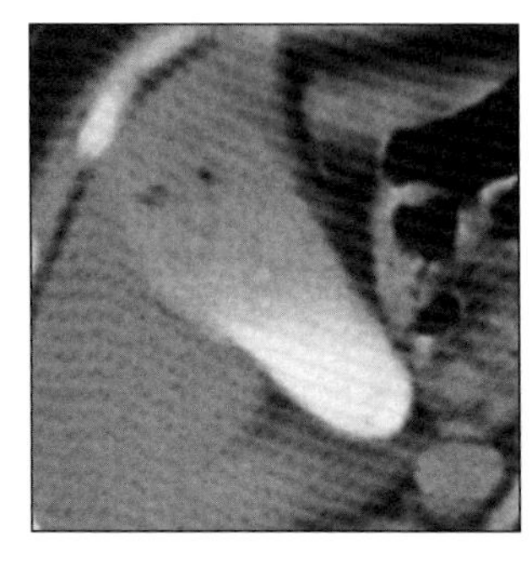

*(**Left**) Axial US of milk of calcium bile shows echogenic material with acoustic shadowing. (**Right**) Axial NECT reveals high attenuation milk of calcium bile in gallbladder.*

PRIMARY SCLEROSING CHOLANGITIS

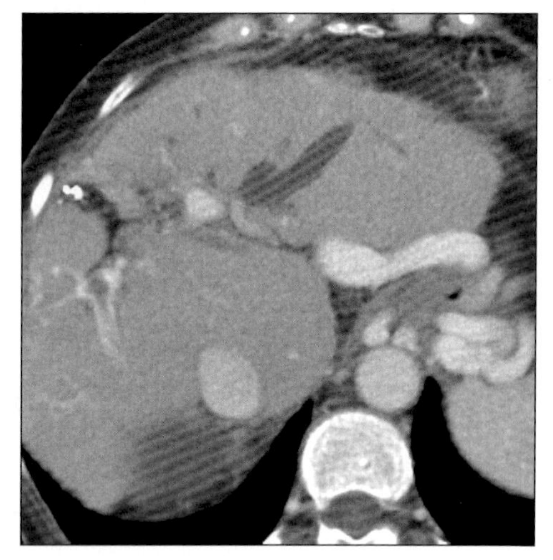

Advanced PSC induced liver disease. Axial CECT shows dysmorphic liver with caudate hypertrophy and peripheral atrophy. IHBD are irregularly dilated, especially in left lobe.

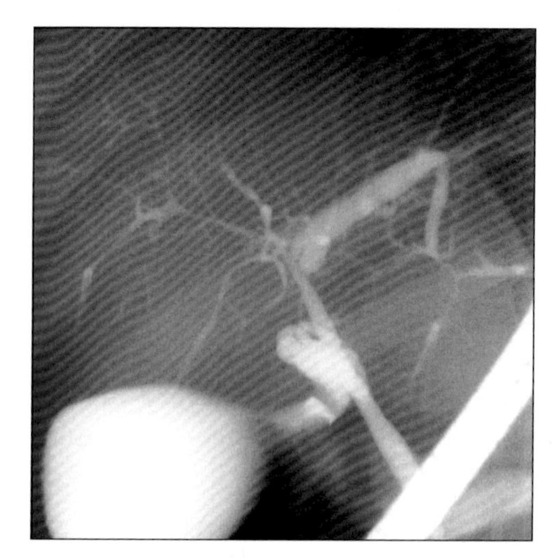

Typical cholangiographic findings of PSC. ERCP shows abnormal arborization of IHBD with multifocal strictures, marked dilatation of left hepatic ducts.

TERMINOLOGY

Abbreviations and Synonyms

- Primary sclerosing cholangitis (PSC)

Definitions

- Primary sclerosing cholangitis (PSC) is a chronic idiopathic inflammatory process of bile ducts

IMAGING FINDINGS

General Features

- Best diagnostic clue
  - Classic "beaded appearance"- alternating segments of dilatation & focal circumferential strictures
    - Combined findings of multifocal short strictures, beading, pruning, diverticula, & mural irregularity
- Location
  - CBD (almost always involved); IHBD + extrahepatic (68-89%); Intrahepatic ducts only (11-25%); extrahepatic ducts only (2-3%); cystic duct (15-18%)
  - Most severely affected segments of biliary tree in PSC are main right & left bile ducts
- Morphology
  - Liver in patients with PSC-induced end-stage cirrhosis is markedly deformed
    - Contour is grossly lobulated; caudate lobe hypertrophy (98%)
  - Accentuated lobulation of hepatic contour is result of asymmetric atrophy & marked focal hypertrophy

Radiographic Findings

- ERCP
  - Multifocal strictures, mural irregularity, diverticula
  - Biliary ductal dilatation is a typical finding (> 80%)

CT Findings

- NECT
  - Scattered, dilated intrahepatic ducts with no apparent connection to main bile ducts
  - Skip dilatations, stenosis, beading, pruning & thickening of duct wall
  - Periportal fibrosis: Low-attenuation soft tissue adjacent to major ducts & portal vein branches
  - Intrahepatic biliary calculi: Areas of calcific density adjacent to portal veins (8%)
  - End-stage PSC: Abnormal, rounded, lobular liver contour; atrophy of lateral & posterior segments
    - Low-attenuation rind-like appearance of right lobe in presence of hypertrophy of caudate lobe
    - Rind accentuates hypertrophied (& higher-attenuation) caudate lobe; creating effect of pseudotumor (seen best on NECT)

DDx: Irregularly Dilated Bile Ducts

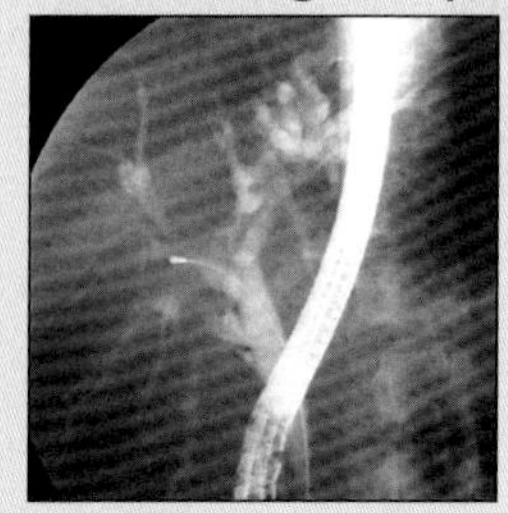

Ascending Cholangitis

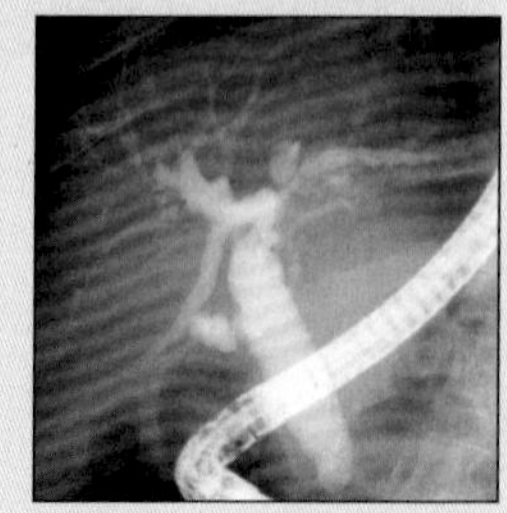

AIDS Cholangiopathy

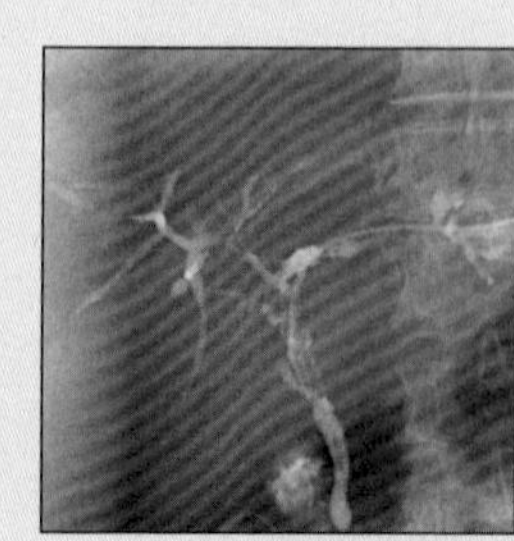

Cholangiocarcinoma

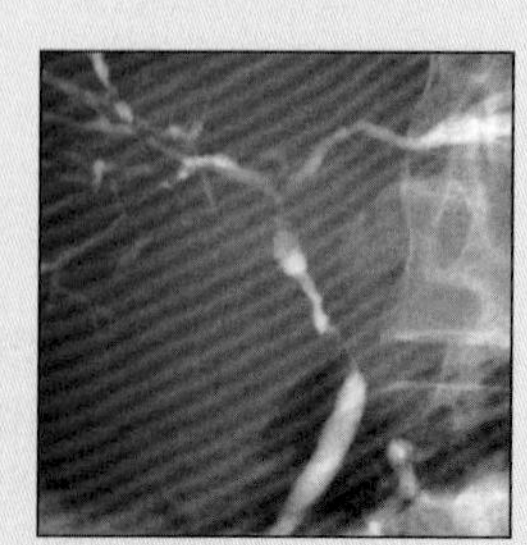

Liver Mets (Gastric CA)

PRIMARY SCLEROSING CHOLANGITIS

Key Facts

Terminology

- Primary sclerosing cholangitis (PSC) is a chronic idiopathic inflammatory process of bile ducts

Imaging Findings

- Classic "beaded appearance"- alternating segments of dilatation & focal circumferential strictures
- Combined findings of multifocal short strictures, beading, pruning, diverticula, & mural irregularity
- Periportal fibrosis: Low-attenuation soft tissue adjacent to major ducts & portal vein branches
- Intrahepatic biliary calculi: Areas of calcific density adjacent to portal veins (8%)
- End-stage PSC: Abnormal, rounded, lobular liver contour; atrophy of lateral & posterior segments
- Low-attenuation rind-like appearance of right lobe in presence of hypertrophy of caudate lobe

Clinical Issues

- Age: 70% are younger than 45 years
- 70% of patients are male
- Multicentric nature & chronic obstruction of bile ducts leads to cirrhosis in many patients
- Complications: Biliary cirrhosis, portal hypertension, cholangiocarcinoma (6-15%)

Diagnostic Checklist

- Pseudotumor of caudate lobe in PSC should be recognized as such & not misinterpreted as neoplasm

    - Result of different amounts of fibrosis & hepatic parenchyma in the two regions; atrophic right lobe contains more fibrosis & is thus of lower attenuation than hypertrophied caudate lobe
  - Cholangiocarcinoma (CC): Progressive biliary dilatation; periportal soft tissue hypoattenuating mass of 1.5 cm or greater, suspicious for malignancy
    - Mural thickening of extrahepatic ducts of 5 mm or greater is presumptive evidence of CC
- CECT
  - Mural enhancement of bile ducts (may be seen in normal ducts & in other diseases as well)
  - Endstage PSC: Cirrhotic liver; hypertrophy of caudate & deep right lobe; atrophy of peripheral liver
  - CC: Delayed contrast-enhancement

MR Findings

- T1WI
  - ↓ Signal intensity along bile ducts & portal veins (periportal fibrosis)
  - ↓ Signal intensity-periportal infiltrating soft tissue mass (CC)
- T2WI
  - ↑ Signal intensity (periportal fibrosis)
  - ↑ Signal intensity soft tissue mass (CC)
- Peripheral wedge-shaped areas of parenchymal atrophy consistent with confluent fibrosis
- Biliary ductal dilatation
- May see focal areas of parenchymal edema & hyperperfusion secondary to focal inflammation (also seen in other inflammatory bile duct diseases: Infectious/recurrent pyogenic cholangitis)
- MRCP: Irregular strictures with segmental dilatations of intra- & extrahepatic ducts

Ultrasonographic Findings

- Dilatation + stenosis + irregular fibrous thickening of walls of bile ducts (2-5 mm)
- Brightly echogenic portal triads
- Gallbladder abnormalities (40%): Gallstones (26%), adenoma & adenocarcinoma (4%)
  - Direct involvement of gallbladder wall (15%): Asymmetric, uniform, symmetric thickening of wall
- Echogenic biliary casts punctate coarse calcifications along portal vein

Other Modality Findings

- Cholangiography: Multiple segmental strictures involving both intra- & extrahepatic bile ducts
  - "Beaded appearance" - alternating segments of dilatation & focal circumferential strictures
  - Stricture length: Can vary from 1-2 mm (band strictures) to several cms (commonly 1-1.5 cm)
- "Pruned-tree" appearance: Opacification of central ducts + obliterated peripheral smaller radicles
- Diverticular outpouchings (25%): 1-2 mm to 1 cm; saccular; herniations adjacent to strictures, mucosal extensions of thickened duct wall
- Mural irregularity (50%): Fine, brush-border to coarse, shaggy or frankly nodular appearance
- Intraluminal filling defects (5-10%): Small, 2-5 mm diameter; 1 cm or larger occur in 50% of PSC complicated by cholangiocarcinoma
- Cholangiographic findings which suggest malignant degeneration include markedly dilated ducts or ductal segments (100% in PSC + CC vs. 24% in PSC alone)
  - Polypoid mass (46% in PSC + CC vs. 7% in PSC alone); 1 cm or greater in diameter with CC
  - Progressive stricture formation or ductal dilatation: New stricture + lengthening of strictures between 6 months & 6 years (< 20%)
  - Dominant stricture: Often near hilum, more severely strictured than remainder of ducts (25%)
- Gallbladder carcinoma: Discrete or infiltrative mass

Imaging Recommendations

- Best imaging tool
  - Helical NE + CECT, MR + MRCP, cholangiography (T-tube & retrograde)
  - Diagnosis is usually established with ERCP & less often with percutaneous transhepatic cholangiography (PTC)
- Protocol advice

PRIMARY SCLEROSING CHOLANGITIS

- Coronal-oblique half-Fourier RARE MR cholangiogram using of both thin- & thick-section
  - Thin-section (5 mm): Useful in depicting focal strictures & subtle irregularities of ductal segments on multiple images
  - Thick-section (40 mm): Provides comprehensive view of entire biliary tract on single image

DIFFERENTIAL DIAGNOSIS

Ascending cholangitis

- Presence of bile duct stones; prior sphincterotomy

AIDS cholangiopathy

- ERCP findings closely simulate PSC
- Strictures of distal CBD and IHBD

Sclerosing cholangiocarcinoma

- Intraductal mass > 1 cm; ductal dilatation proximal to dominant stricture
- Progressive cholangiographic changes within 0.5-1.5 years of initial diagnosis

Extrinsic compression

- Cirrhosis: Peripheral dilated bile ducts & irregular strictures; centrally obstructed by large nodules
- Liver masses: Hepatic metastases + lymph nodes in porta hepatis cause extrinsic compression of ducts

Chemotherapy cholangitis

- Segmental strictures of variable length
- Duct beading & intrahepatic duct involvement are relatively less common in chemotherapy cholangitis
- Gallbladder & cystic duct are usually more severely involved in chemotherapy cholangitis than in PSC

PATHOLOGY

General Features

- Etiology: Idiopathic; hypersensitivity reaction (genetic & immunologic)
- Associated abnormalities: Ulcerative colitis (70%), Crohn disease (13%), pancreatitis, sicca complex, Riedel struma, retroperitoneal/mediastinal fibrosis

Gross Pathologic & Surgical Features

- Dilated + stenotic bile ducts, periportal fibrosis, intraductal calculi
- Endstage PSC: Lobular liver, atrophy (periphery liver), caudate hypertrophy

Microscopic Features

- Nonsuppurative, non granulomatous destruction of bile ducts
- Portal hepatitis or cholangitis (stage 1); periportal hepatitis or fibrosis (stage 2); septal fibrosis or bridging necrosis or both (stage 3); cirrhosis (stage 4)

CLINICAL ISSUES

Presentation

- Progressive chronic & intermittent obstructive jaundice
- Fatigue, pruritus, RUQ pain, hepatosplenomegaly
- Lab: Elevation of serum alkaline phosphatase

Demographics

- Age: 70% are younger than 45 years
- Gender
  - 70% of patients are male
  - 70% of patients have ulcerative colitis

Natural History & Prognosis

- Variable natural history; usually progressive downhill
- Involvement of small intrahepatic bile ducts (so-called small-duct PSC) may be the only &/or earliest manifestation of PSC
  - Classic obliterative fibrosis of medium-sized & large ducts (so-called large-duct PSC) occurs at later stage
    - Results in cholestasis with progression to secondary biliary cirrhosis & hepatic failure
  - Multicentric nature & chronic obstruction of bile ducts leads to cirrhosis in many patients
- Biliary calculi develop in obstructed ducts as a consequence of biliary stasis & possible secondary infection
- Atrophy/hypertrophy complex may even occur in absence of cirrhosis
- Parenchymal atrophy: Due to chronic obstruction of segmental bile ducts, & is a sequela of parenchymal necrosis & fibrosis
- Regional hypertrophy: In areas with absent or less severe biliary obstruction; in response to ↓ hepatic function that accompanies parenchymal atrophy
- Complications: Biliary cirrhosis, portal hypertension, cholangiocarcinoma (6-15%)
- Prognosis: 5 year survival is 88%; median survival is 11.9 years from time of diagnosis

Treatment

- Liver transplantation

DIAGNOSTIC CHECKLIST

Consider

- History of ulcerative colitis; identification of morphologic findings seen more frequently in PSC induced end-stage cirrhosis & known ductal abnormalities of PSC may allow one to strongly suggest PSC as cause of cirrhosis
- Pseudotumor of caudate lobe in PSC should be recognized as such & not misinterpreted as neoplasm

SELECTED REFERENCES

1. Bader TR et al: MR imaging features of primary sclerosing cholangitis: patterns of cirrhosis in relationship to clinical severity of disease. Radiology. 226(3):675-85, 2003
2. Campbell WL et al: Using CT and cholangiography to diagnose biliary tract carcinoma complicating primary sclerosing cholangitis. AJR Am J Roentgenol. 177(5):1095-100, 2001
3. Fulcher AS et al: Primary sclerosing cholangitis: evaluation with MR cholangiography-a case-control study. Radiology. 215(1):71-80, 2000

IMAGE GALLERY

Typical

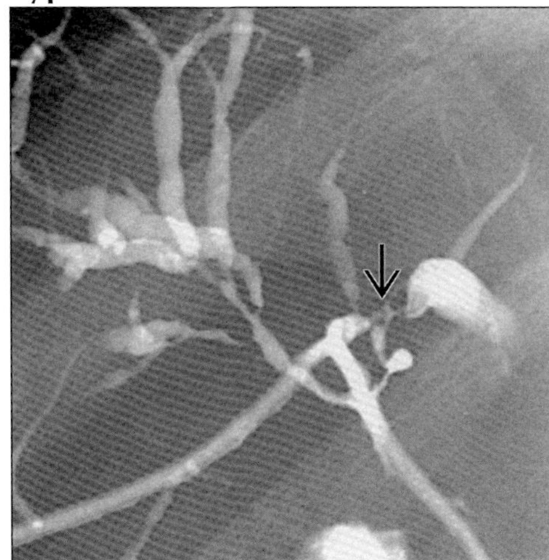

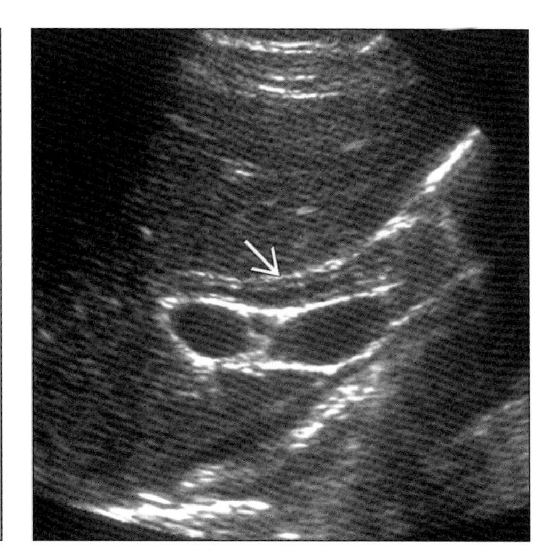

(Left) *Catheter cholangiogram shows multifocal strictures, moderate dilatation of right IHBD. Tight stricture of main left hepatic duct (arrow) with partial opacification of very dilated left IHBD.* ***(Right)*** *Oblique sagittal sonogram shows thickened, hyperechoic wall of common duct (arrow) in patient with PSC.*

Typical

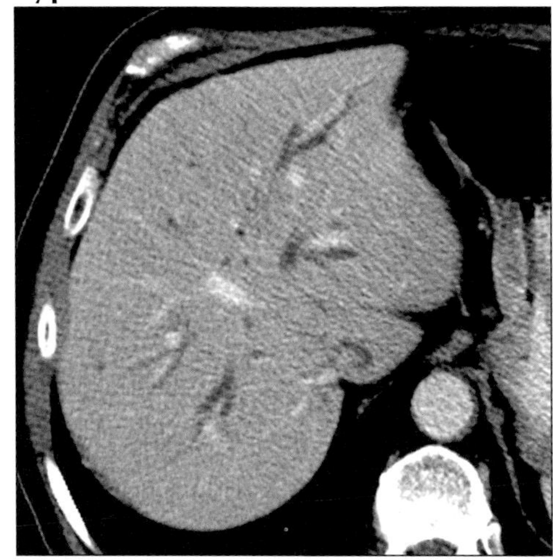

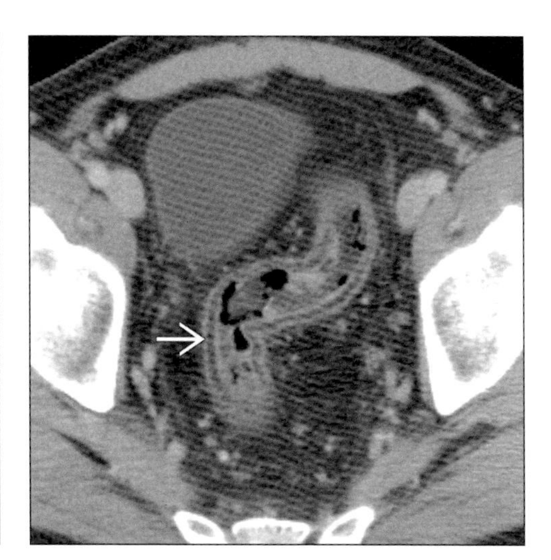

(Left) *Axial CECT shows irregular dilatation of IHBD and abnormal arborization with no apparent connection to the central biliary tree.* ***(Right)*** *Axial CECT in patient with PSC and ulcerative colitis shows thick-walled recto-sigmoid colon with fat density proliferation in submucosa (arrow).*

Other

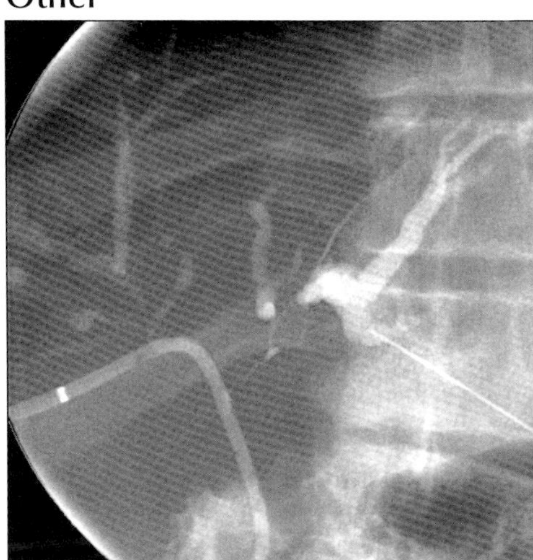

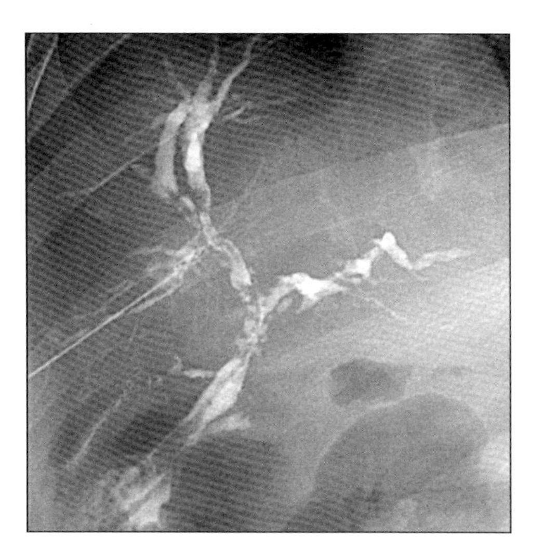

(Left) *PTC shows multifocal strictures and dilatations of right IHBD. There is a long dominant stricture of the main left duct and marked dilatation of the left IHBD raising concern for cholangiocarcinoma.* ***(Right)*** *PTC of patient with liver transplant for endstage PSC-induced liver disease shows markedly irregular mucosal surface with multifocal strictures and filling defects within ducts. Recurrent PSC.*

GALLBLADDER CARCINOMA

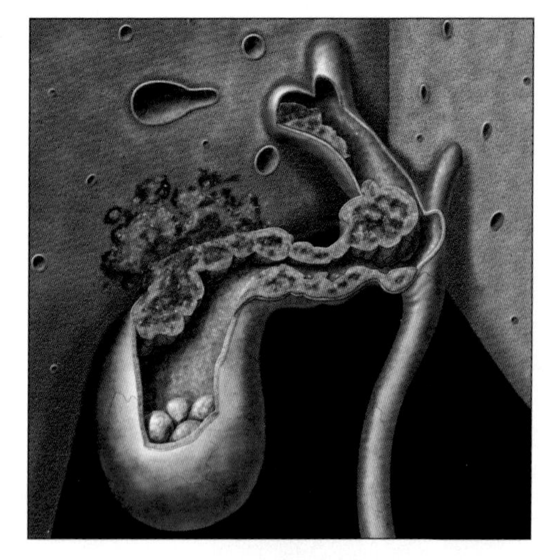

Schematic drawing of gallbladder carcinoma. Note gallstones and focal mural mass arising from the gallbladder wall that invades the adjacent liver and obctructs the common hepatic duct.

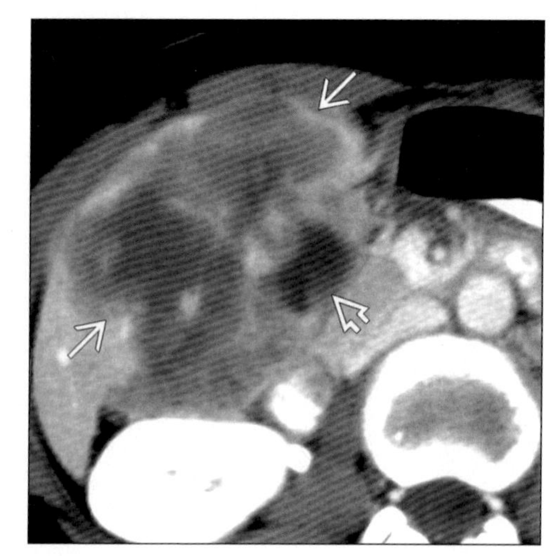

Axial CECT demonstrates large hypodense mass infiltrating gallbladder fossa and invading liver (arrows). Open arrow denotes gallbladder.

TERMINOLOGY

Definitions

- Malignant epithelial neoplasm arising from gallbladder (GB) mucosa

IMAGING FINDINGS

General Features

- Best diagnostic clue
  - Large GB mass infiltrating gallbladder fossa extending into liver
  - Polypoid intraluminal mass
  - Diffuse or focal irregular mural thickening mimics chronic cholecystitis
- Location: GB fundus and body; uncommon in cystic duct
- Size: Variable; smaller polypoid mass in early CA, large infiltrating lesions typical
- Morphology: Large soft tissue mass infiltrating GB fossa; polypoid mucosal mass in GB

Radiographic Findings

- Radiography
  - Plain abdominal radiographs
    - Calcified gallstones or porcelain GB
  - Oral cholecystogram (OCG)
    - Non-visualization of GB
    - Rarely pneumobilia 2° to GB enteric fistula
- ERCP
  - Non-visualization of GB
  - Common hepatic duct obstruction
  - Dilated intrahepatic ducts

CT Findings

- NECT: Calcification of GB wall (porcelain GB); calcified gallstones
- CECT
  - Hypovascular mass infiltrating GB fossa, invading liver along main lobar fissure; porta hepatis adenopathy on CECT
  - Nodal mets to peripancreatic area may simulate pancreatic carcinoma
  - Invasion of liver and porta hepatis
  - Calcified stones or porcelain GB

MR Findings

- T1WI
  - Iso- or hypointense GB fossa mass with increased signal compared to normal liver
  - Hypovascular after IV gadolinium
- T2WI: Mass slightly increased in signal intensity compared to liver
- T1 C+: Hypovascular GB fossa mass invading liver
- MRCP

DDx: Spectrum of GB Diseases Mimicking Carcinoma

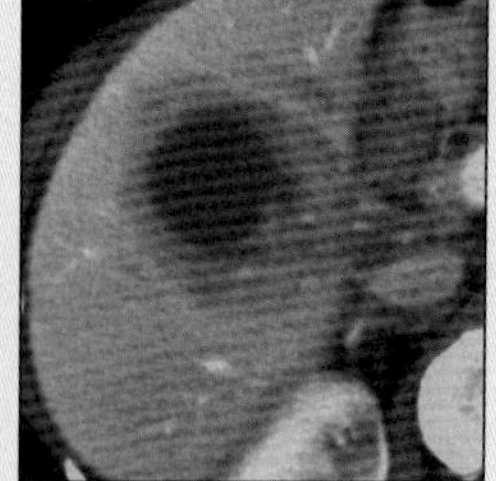

Cholecystitis

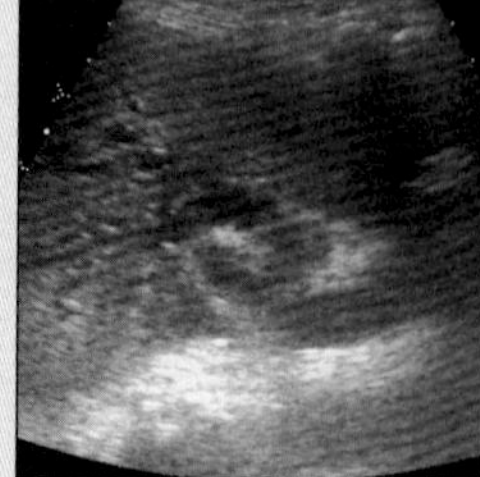

Porta Hepatis Nodes

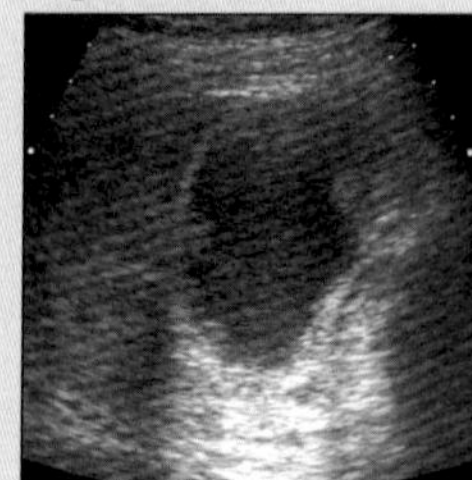

GB Polyps

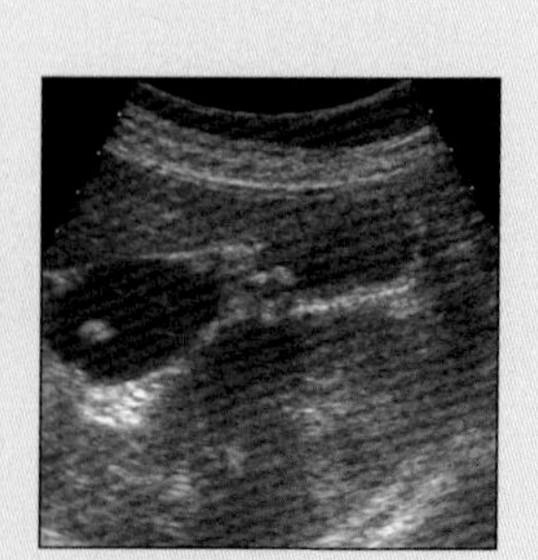

Adenomyomatosis

GALLBLADDER CARCINOMA

Key Facts

Imaging Findings

- Morphology: Large soft tissue mass infiltrating GB fossa; polypoid mucosal mass in GB
- NECT: Calcification of GB wall (porcelain GB); calcified gallstones
- Hypovascular mass infiltrating GB fossa, invading liver along main lobar fissure; porta hepatis adenopathy on CECT
- Real Time: Gallstones, calcified GB wall mass infiltrating GB fossa, porcelain GB, moderately echogenic polypoid mucosal mass (> 1 cm) on US
- Best imaging tool: US, CECT

Top Differential Diagnoses

- Complicated cholecystitis
- Metastatic disease to GB fossa
- GB polyp
- Adenomyomatosis

Pathology

- Associated with porcelain GB & chronic inflammation 2° to gallstones; malignant degeneration of adenomatous mucosal polyps
- Scirrhous infiltrating mass extending from GB wall to obliterate GB fossa & invade liver; porta hepatis adenopathy

Clinical Issues

- Most common signs/symptoms: RUQ pain, weight loss, jaundice, vomiting

  - Dilated bile ducts due to common hepatic duct obstruction

Ultrasonographic Findings

- Real Time: Gallstones, calcified GB wall mass infiltrating GB fossa, porcelain GB, moderately echogenic polypoid mucosal mass (> 1 cm) on US
- Color Doppler: Areas of increased vascularity

Nuclear Medicine Findings

- Hepato Biliary Scan
  - Non-filling of GB

Imaging Recommendations

- Best imaging tool: US, CECT
- Protocol advice: Longitudinal & transverse images of GB fossa with grayscale and color Doppler

DIFFERENTIAL DIAGNOSIS

Complicated cholecystitis

- Gallstones
- Thick-walled GB, pericholecystic abscess
- GB may be contracted
- May be indistinguishable from carcinoma

Metastatic disease to GB fossa

- Most often nodal distribution around portal vein
- Melanoma may directly metastasize to GB mucosa
- Hepatoma and other hepatic tumors may secondarily spread to GB via duct invasion
- Porta hepatis lymphadenopathy
  - Lymphoma and GI tract carcinoma most common

GB polyp

- Non-shadowing, mucosal mass
  - Moderately echogenic without shadowing
- Non-mobile, attached to wall
- Typically < 1 cm for cholesterol polyp
- No flow on color Doppler

Adenomyomatosis

- Localized fundal GB wall thickening
- Gallstones
- Focal thickening of midportion of GB ("hourglass GB")
- May demonstrate diffuse wall thickening
- Intramural cholesterol crystals as bright echoes with "comet tail" reverberation echoes

PATHOLOGY

General Features

- General path comments
  - 90% adenocarcinoma
    - Early stage: Polypoid mucosal mass
    - Late stage: Mass infiltrating GB fossa
  - 10% squamous or anaplastic
- Genetics: No known association
- Etiology
  - Associated with porcelain GB & chronic inflammation 2° to gallstones; malignant degeneration of adenomatous mucosal polyps
  - 75% have gallstones
  - Porcelain GB predisposes to GB carcinoma
- Epidemiology
  - Most common type of biliary cancer
    - 6500 deaths per year in US
  - 75% are women
  - Average age of presentation is 70 yrs
  - Fifth most common GI cancer, 9 times more common than extrahepatic cholangiocarcinoma
- Associated abnormalities
  - Gallstones in > 65%
  - Chronic cholecystitis
  - Porcelain GB (4-60%)
  - Ulcerative colitis; rarely Crohn disease
  - Primary sclerosing cholangitis
  - Familial polyposis coli

Gross Pathologic & Surgical Features

- Scirrhous infiltrating mass extending from GB wall to obliterate GB fossa & invade liver; porta hepatis adenopathy
- Direct invasion of liver, duodenum, stomach, bile duct, pancreas, R kidney

- Lymphatic spread to porta hepatis, peripancreatic & retroperitoneal nodes
- Intraperitoneal spread common with ascites, omental nodules & peritoneal implants
- Hematogenous spread (late in clinical course) to lungs, liver & bones
- Perineural invasion common

Microscopic Features

- Adenocarcinoma (90%)
- Squamous or anaplastic carcinoma (10%)

Staging, Grading or Classification Criteria

- Stage I: Carcinoma confined to mucosa
- Stage II: Carcinoma involves mucosa & muscularis
- Stage III: Carcinoma extends to serosa
- Stage IV: Transmural involvement with positive nodes
- Stage V: Liver or distant metastases

CLINICAL ISSUES

Presentation

- Most common signs/symptoms: RUQ pain, weight loss, jaundice, vomiting
- Clinical profile: Elevated bilirubin, elevated alkaline phosphatase with biliary obstruction

Demographics

- Age: Mean 70 years
- Gender: M:F = 1:3

Natural History & Prognosis

- Spreads by local invasion to liver, nodal spread to porta hepatis and para-aortic nodes, hematogeneous spread to liver
- Very poor prognosis; 4% 5 yr survival rate, 75% of patients have mets at time of diagnosis

Treatment

- Cholecystectomy for lesions confined to GB wall without liver invasion
- Radical cholecystectomy and/or partial hepatectomy with regional node dissection for lesions infiltrating porta hepatis

DIAGNOSTIC CHECKLIST

Consider

- Adenomyomatosis with GB wall thickening
  - Benign adenomatous polyp < 2 cm

Image Interpretation Pearls

- Porcelain GB
- Mass infiltrating GB fossa
- Large polypoid GB mucosal mass with flow

SELECTED REFERENCES

1. Yun EJ et al: Gallbladder carcinoma and chronic cholecystitis: differentiation with two-phase spiral CT. Abdom Imaging. 29(1):102-8, 2003
2. Enomoto T et al: Xanthogranulomatous cholecystitis mimicking stage IV gallbladder cancer. Hepatogastroenterology. 50(53):1255-8, 2003
3. Pandey M: Risk factors for gallbladder cancer: a reappraisal. Eur J Cancer Prev. 12(1):15-24, 2003
4. Misra S et al: Carcinoma of the gallbladder. Lancet Oncol. 4(3):167-76, 2003
5. Goindi G et al: Risk factors in the aetiopathogenesis of carcinoma of the gallbladder. Trop Gastroenterol. 24(2):63-5, 2003
6. Kokudo N et al: Strategies for surgical treatment of gallbladder carcinoma based on information available before resection. Arch Surg. 138(7):741-50; dis 750, 2003
7. Yamamoto T et al: Early gallbladder carcinoma associated with primary sclerosing cholangitis and ulcerative colitis. J Gastroenterol. 38(7):704-6, 2003
8. Varshney S et al: Incidental carcinoma of the gallbladder. Eur J Surg Oncol. 28(1):4-10, 2002
9. Cunningham CC et al: Primary carcinoma of the gall bladder: a review of our experience. J La State Med Soc. 154(4):196-9, 2002
10. Doty JR et al: Cholecystectomy, liver resection, and pylorus-preserving pancreaticoduodenectomy for gallbladder cancer: report of five cases. J Gastrointest Surg. 6(5):776-80, 2002
11. Corvera CU et al: Role of laparoscopy in the evaluation of biliary tract cancer. Surg Oncol Clin N Am. 11(4):877-91, 2002
12. Rashid A: Cellular and molecular biology of biliary tract cancers. Surg Oncol Clin N Am. 11(4):995-1009, 2002
13. Gore RM et al: Imaging benign and malignant disease of the gallbladder. Radiol Clin North Am. 40(6):1307-23, vi, 2002
14. Xu AM et al: Multi-slice three-dimensional spiral CT cholangiography: a new technique for diagnosis of biliary diseases. Hepatobiliary Pancreat Dis Int. 1(4):595-603, 2002
15. Towfigh S et al: Porcelain gallbladder is not associated with gallbladder carcinoma. Am Surg. 67(1):7-10, 2001
16. Levy AD et al: Gallbladder carcinoma: radiologic-pathologic correlation. Radiographics. 21(2):295-314; questionnaire, 549-55, 2001
17. Dixit VK et al: Aetiopathogenesis of carcinoma gallbladder. Trop Gastroenterol. 22(2):103-6, 2001
18. Tazuma S et al: Carcinogenesis of malignant lesions of the gall bladder. The impact of chronic inflammation and gallstones. Langenbecks Arch Surg. 386(3):224-9, 2001
19. Stewart CJ et al: Brush cytology in the assessment of pancreatico-biliary strictures: a review of 406 cases. J Clin Pathol. 54(6):449-55, 2001
20. Pandey M et al: Carcinoma of the gallbladder: a retrospective review of 99 cases. Dig Dis Sci. 46(6):1145-51, 2001
21. Kaushik SP: Current perspectives in gallbladder carcinoma. J Gastroenterol Hepatol. 16(8):848-54, 2001
22. Fujii H et al: Small cell carcinoma of the gallbladder: a case report and review of 53 cases in the literature. Hepatogastroenterology. 48(42):1588-93, 2001
23. Eriguchi N et al: Xanthogranulomatous cholecystitis. Kurume Med J. 48(3):219-21, 2001
24. Donohue JH: Present status of the diagnosis and treatment of gallbladder carcinoma. J Hepatobiliary Pancreat Surg. 8(6):530-4, 2001
25. Narula IM: Historical review of carcinoma of the gallbladder. Indian J Hist Med. 16:6-11, 1971

GALLBLADDER CARCINOMA

IMAGE GALLERY

Typical

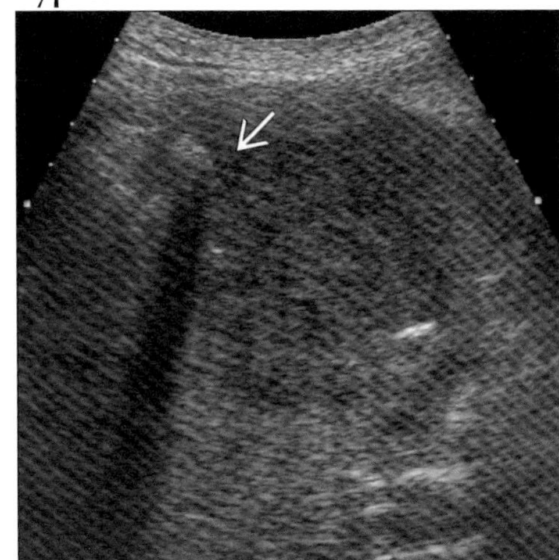

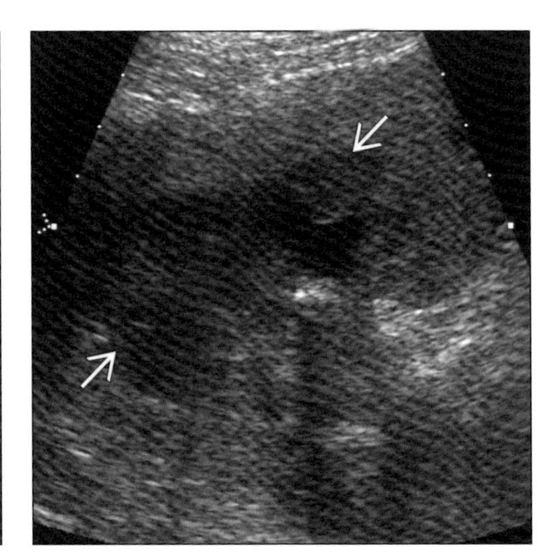

(Left) *Axial US of gallbladder carcinoma. Note large hypoechoic mass obliterating gallbladder. Note gallstone (arrow).* ***(Right)*** *Axial US of gallbladder carcinoma invading liver. Note hypoechoic mass infiltrating gallbladder fossa and invading right and left lobes of liver (arrows).*

Typical

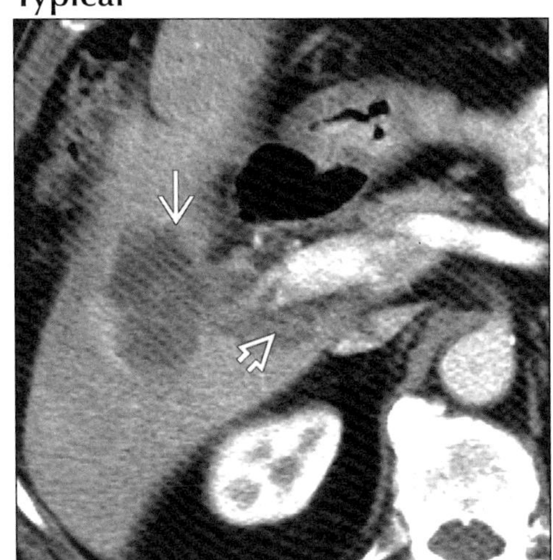

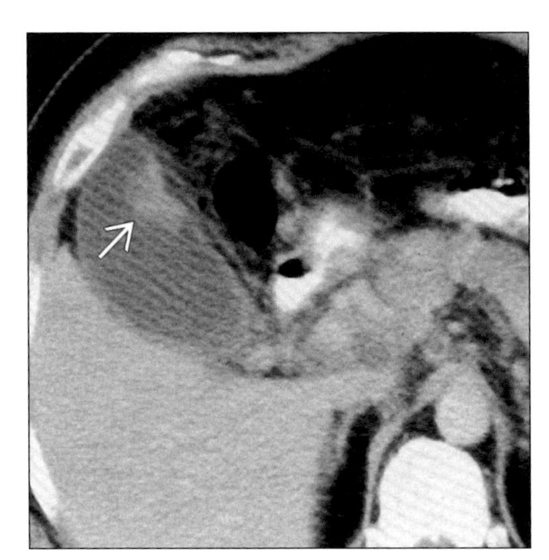

(Left) *Axial CECT of invasive gallbladder carcinoma. Note hypodense mass invading liver (arrow) infiltrating along portal vein (open arrow).* ***(Right)*** *Axial CECT demonstrates gallbladder carcinoma as polypoid mass (arrow).*

Typical

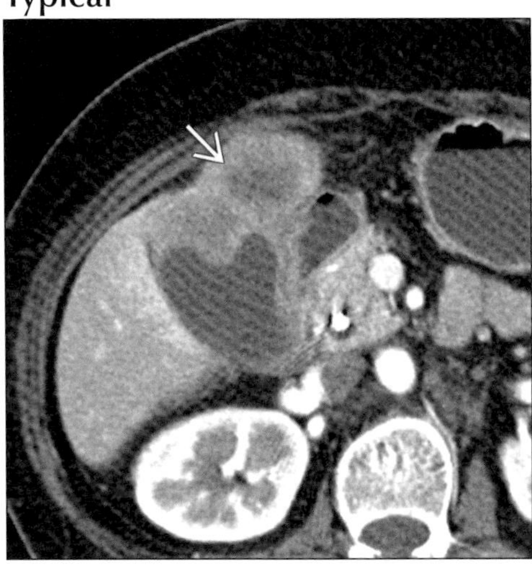

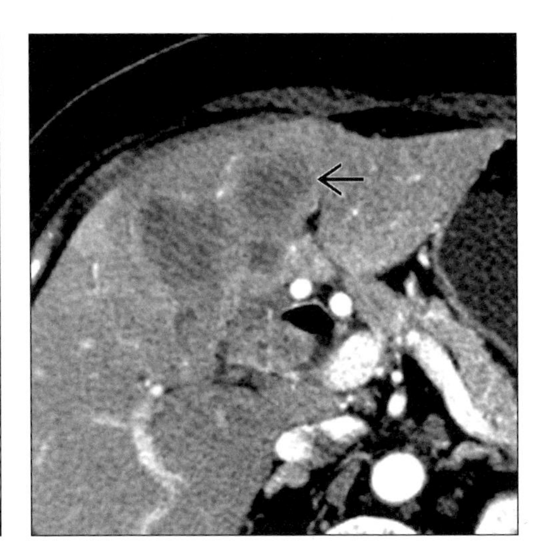

(Left) *Axial CECT demonstrates hypodense fundal mass (arrow).* ***(Right)*** *Axial CECT demonstrates invasion of liver (arrow).*

CHOLANGIOCARCINOMA

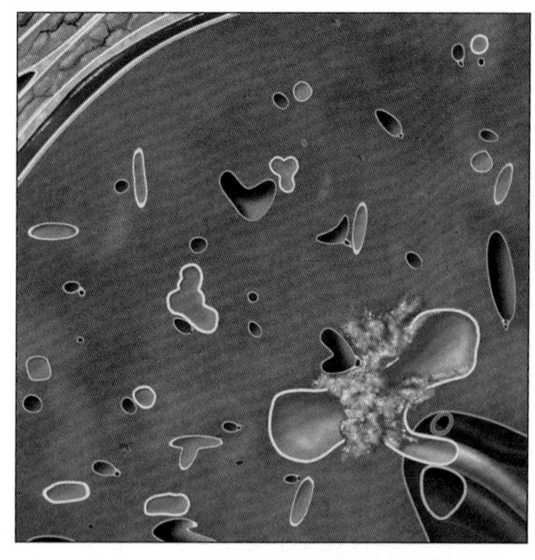

Axial graphic shows Klatskin tumor, a small mass at the confluence of the main right and left bile ducts, that invades adjacent liver and hepatic vein.

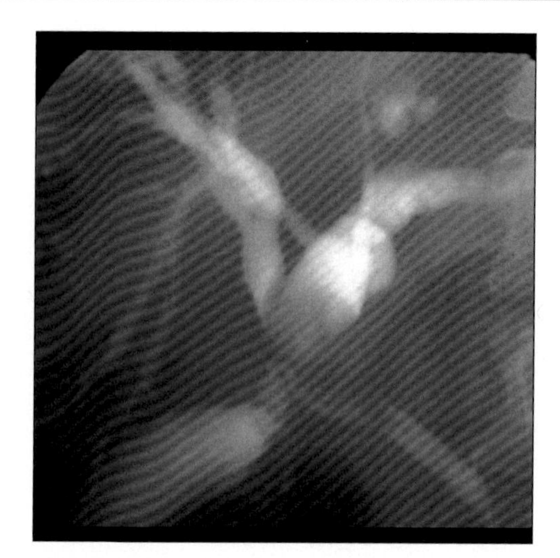

Cholangiogram shows mass at confluence of main right and left ducts with marked dilatation of IHBD. Common hepatic duct is involved but cystic and common bile ducts are not.

TERMINOLOGY

Abbreviations and Synonyms

- Cholangiocellular or bile duct adenocarcinoma

Definitions

- Malignancy that arises from intrahepatic bile duct (IHBD) or extrahepatic bile duct epithelium

IMAGING FINDINGS

General Features

- Best diagnostic clue: Klatskin tumor - small hilar mass obstructing bile ducts on CT or ERCP
- Location
  - Distal CBD (30-50%); common hepatic duct (14-37%)
  - Proximal CBD (15-30%); confluence of hepatic ducts (10-26%)
  - Left & right hepatic duct (8-13%); cystic duct (6%)
- Size: Intrahepatic mass (5-20 cm); extrahepatic - smaller
- Morphology
  - 2nd most common primary hepatic tumor after hepatoma
  - Manifests with various histologic types and growth patterns
  - Two types, based on anatomy & radiography
    - Intrahepatic (peripheral/central); extrahepatic
  - Intrahepatic
    - Peripheral (intrahepatic bile ducts): May be exophytic, polypoid or infiltrative
    - Central or hilar (confluence of right & left hepatic ducts and proximal common hepatic duct): Klatskin tumor (small mass in liver hilus)
  - Extrahepatic
    - Common duct: Distal common hepatic duct (CBD)
    - May arise as short stricture or small polypoid mass

Radiographic Findings

- Cholangiography (PTC/ERCP)
  - Exophytic intraductal tumor mass (2-5 mm in diameter)
  - Infiltrating type: Frequently long, rarely short concentric focal stricture
  - Ductal wall irregularities; prestenotic diffuse/focal biliary dilatation
  - Hilar strictures (due to Klatskin tumor): Proximal bile duct dilatation

CT Findings

- NECT
  - Intrahepatic
    - Peripheral-hypodense solitary or satellite lesions & IHBD dilatation

DDx: Obstruction of Extrahepatic Bile Duct

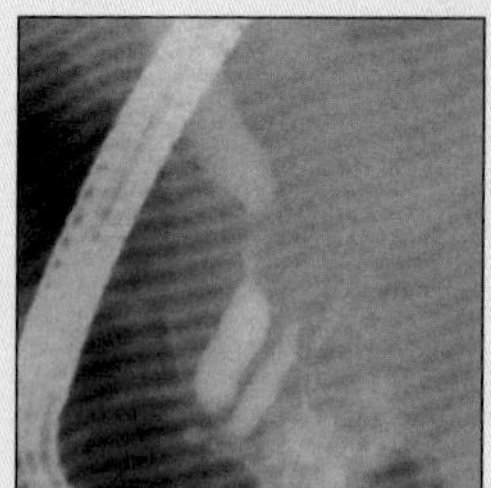

Pancreatic Carcinoma

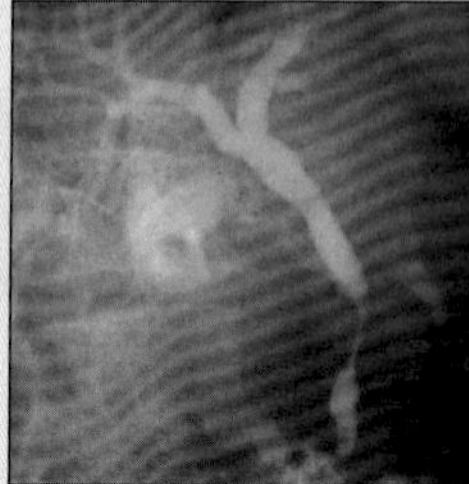

Chronic Pancreatitis

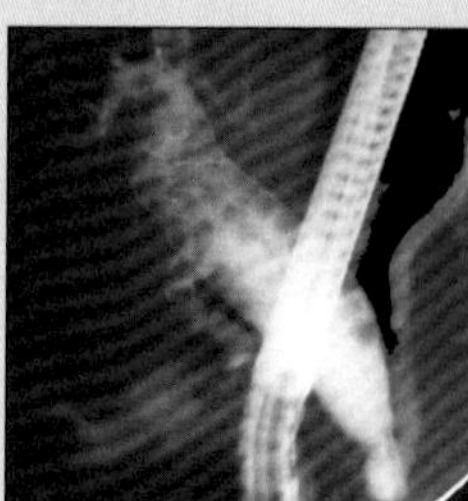

Choledocholithiasis

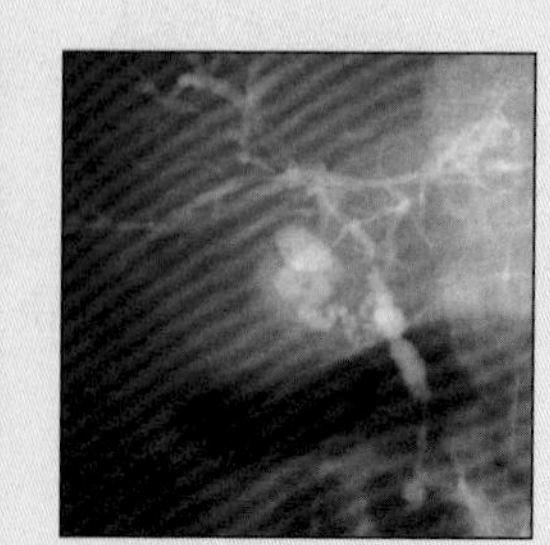

Sclerosing Cholangitis

CHOLANGIOCARCINOMA

Key Facts

Terminology
- Cholangiocellular or bile duct adenocarcinoma
- Malignancy that arises from intrahepatic bile duct (IHBD) or extrahepatic bile duct epithelium

Imaging Findings
- Best diagnostic clue: Klatskin tumor - small hilar mass obstructing bile ducts on CT or ERCP
- Intrahepatic (peripheral/central); extrahepatic
- Capsular retraction may be seen

Top Differential Diagnoses
- Pancreatic carcinoma
- Chronic pancreatitis
- Choledocholithiasis
- Primary sclerosing cholangitis (PSC)
- Porta hepatis tumor

Pathology
- Pre-existing bile duct diseases
- Inflammatory bowel disease (10x increased risk)
- One third of all malignancies originating in the liver
- Obstructive type: U-/V-shaped obstruction (70-85%)
- Ductal sclerosing adenocarcinoma (2/3 cases)

Diagnostic Checklist
- Rule out other biliary & pancreatic pathologies that can mimic cholangiocarcinoma by obstructing extrahepatic bile duct
- Cholangiography & MRCP: Usually long & rarely short focal stricture, irregular ductal wall, stenosis & prestenotic biliary ductal dilatation
- Klatskin tumor: Small tumor at confluence of right/left hepatic & proximal common hepatic ducts

        - Central (hilar)-hypodense mass at confluence & IHBD dilatation
        - Capsular retraction may be seen
    - Extrahepatic: Common duct
        - Small growth (poorly sensitive)
        - Large growth (seen as hypodense mass) & IHBD dilatation
- CECT
    - Arterial phase
        - Early rim-enhancement with progressive, central patchy enhancement & IHBD dilatation
    - Portal phase
        - Minimal enhancement of irregular thickened bile duct wall
        - Invasion of portal vein seen with intrahepatic type
        - Enlarged portal lymph nodes may be seen
    - Delayed phase
        - Persistent enhancing tumor (due to fibrous stroma)

MR Findings
- T1WI: Iso-/hypointense
- T2WI: Hyperintense periphery (viable) & central hypointensity (fibrosis)
- T1 C+: Superior to CT in detecting small hilar tumors, intrahepatic and periductal tumor infiltration
- T1WI fat suppressed image
    - Shows tumor of intrapancreatic portion of CBD as low signal intensity against high signal intensity head of pancreas
- MRCP
    - Reveals site & extension of tumor growth
    - Shows location of obstruction & IHBD dilatation

Ultrasonographic Findings
- Real Time
    - Mixed echoic, homo-/heterogeneous mass & dilated IHBD
    - Dilated intra & extrahepatic bile ducts if lesion is in CBD

Angiographic Findings
- Conventional
    - Avascular, hypo-/hypervascular
    - Poor or absent tumor stain
    - Hepatic artery & portal vein
        - Displacement, encasement or occlusion

Nuclear Medicine Findings
- Hepato Biliary Scan
    - Cold lesion
- Technetium Sulfur Colloid
    - Cold lesion
- Gallium Scan
    - May show uptake

Imaging Recommendations
- Best imaging tool: MRCP or ERCP
- Protocol advice: MRCP three-dimensional maximum intensity projection (MIP) reconstruction images

DIFFERENTIAL DIAGNOSIS

Pancreatic carcinoma
- Arises from ductal epithelium of exocrine pancreas
- Irregular, heterogeneous, poorly enhancing mass
- Abrupt obstruction of pancreatic and/or distal CBD
    - Distal CBD block mimics cholangiocarcinoma
- Dilated pancreatic duct & obliteration of retropancreatic fat
- Location: Head (60%), body (20%), tail (15%)
- 65% of patients present with advanced local disease & distant metastases

Chronic pancreatitis
- NECT
    - Focal or diffuse atrophy of gland; calcification
    - Dilated main pancreatic duct (MPD) & intraductal calculi
    - Intra & peripancreatic cysts
    - Thickening of peripancreatic fascia
    - Small hypodense focal masses (fat & fibrosis)
    - Splenic vein thrombosis, splenomegaly, varices
- CECT: Heterogeneous enhancement
- MRCP: Dilated MPD plus radicles; may show long tapered stricture of distal CBD & dilated bile ducts

CHOLANGIOCARCINOMA

Choledocholithiasis

- Intra-/extrahepatic bile duct stones
- 60-70% stones, increased attenuation (Ca++); 20-30%, less than water or soft tissue density
- "Bull's-eye" sign: Rim of bile surrounding a calcified stone within duct
  - Most accurate sign on NECT
- CBD obstruction & intrahepatic duct dilatation

Primary sclerosing cholangitis (PSC)

- Dilatation of both intra- & extrahepatic bile ducts
- PSC often shows isolated obstructions of IHBDs
- ERCP: Skip dilatations, strictures, beading, pruning & thickening of ductal wall
- PSC strictures indistinguishable from scirrhous infiltrating cholangiocarcinoma

Porta hepatis tumor

- Bulky primary (HCC) & secondary liver tumors
- HCC & metastases may invade or obstruct IHBD

PATHOLOGY

General Features

- General path comments
  - Almost all cholangiocarcinomas are adenocarcinomas arising from bile duct epithelium
  - Tumor types
    - Exophytic intrahepatic masses
    - Scirrhous infiltrating neoplasms: Cause stricture
    - Polypoid neoplasms of ductal wall: Bulge into bile duct lumen
  - Patterns of dissemination
    - Local extension along duct
    - Local infiltration of liver
    - Spread to lymph nodes
- Etiology
  - Pre-existing bile duct diseases
    - E.g., biliary lithiasis, clonorchiasis, recurrent pyogenic cholangitis & PSC
  - Inflammatory bowel disease (10x increased risk)
  - Caroli disease, Thorotrast exposure
  - Familial polyposis & choledochal cyst
- Epidemiology
  - More frequent in Asia
  - One third of all malignancies originating in the liver

Gross Pathologic & Surgical Features

- Intrahepatic: Mass (5-20 cm)
- Satellite nodules in 65%, biliary calculi
- Extrahepatic (from common duct): Growth pattern
  - Obstructive type: U-/V-shaped obstruction (70-85%)
  - Stenotic type: Strictured rigid lumen with irregular margins (10-25%)
  - Polypoid or papillary type
    - Intraluminal filling defect (5-6%)

Microscopic Features

- Desmoplastic reaction (fibrosis)
- Mucin; glandular & tubular structures
- Most common histologic type
  - Ductal sclerosing adenocarcinoma (2/3 cases)

CLINICAL ISSUES

Presentation

- Most common signs/symptoms
  - Varies with location
    - Intrahepatic: Pain, palpable mass, weight loss, painless jaundice
    - Extrahepatic: Pain, enlarged tender liver, obstructive jaundice, anorexia
- Lab data
  - Increased bilirubin & alkaline phosphatase

Demographics

- Age: Peak age: 6-7th decade
- Gender: M:F = 3:2

Natural History & Prognosis

- Intrahepatic
  - 5 year survival (30%)
- Extrahepatic
  - Median survival of 5 months
  - 5 year survival (1.6%)
- Recurrence after transplantation: Quite common

Treatment

- Surgical resection (less than 20% resectable)
- Radiation; laser therapy & biliary stenting
- Liver transplantation (controversial)

DIAGNOSTIC CHECKLIST

Consider

- Rule out other biliary & pancreatic pathologies that can mimic cholangiocarcinoma by obstructing extrahepatic bile duct

Image Interpretation Pearls

- Cholangiography & MRCP: Usually long & rarely short focal stricture, irregular ductal wall, stenosis & prestenotic biliary ductal dilatation
- Klatskin tumor: Small tumor at confluence of right/left hepatic & proximal common hepatic ducts

SELECTED REFERENCES

1. Kim YH: Extrahepatic cholangiocarcinoma associated with clonorchiasis: CT evaluation. Abdom Imaging. 28(1):68-71, 2003
2. Han JK et al: Cholangiocarcinoma: pictorial essay of CT and cholangiographic findings. Radiographics. 22(1):173-87, 2002
3. Campbell WL et al: Using CT and cholangiography to diagnose biliary tract carcinoma complicating primary sclerosing cholangitis. AJR Am J Roentgenol. 177(5):1095-100, 2001
4. Lee WJ et al: Radiologic spectrum of cholangiocarcinoma: emphasis on unusual manifestations and differential diagnoses. Radiographics. 21 Spec No:S97-S116, 2001
5. Maetani Y et al: MR imaging of intrahepatic cholangiocarcinoma with pathologic correlation. AJR Am J Roentgenol. 176(6):1499-507, 2001
6. Han JK et al: Hilar cholangiocarcinoma: thin-section spiral CT findings with cholangiographic correlation. Radiographics. 17(6):1475-85, 1997

CHOLANGIOCARCINOMA

IMAGE GALLERY

Typical

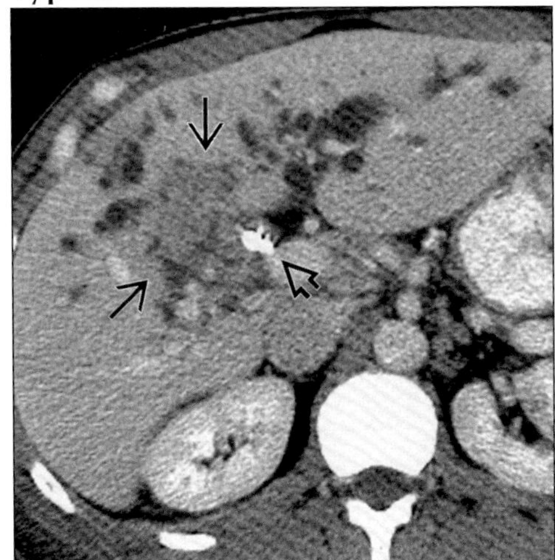

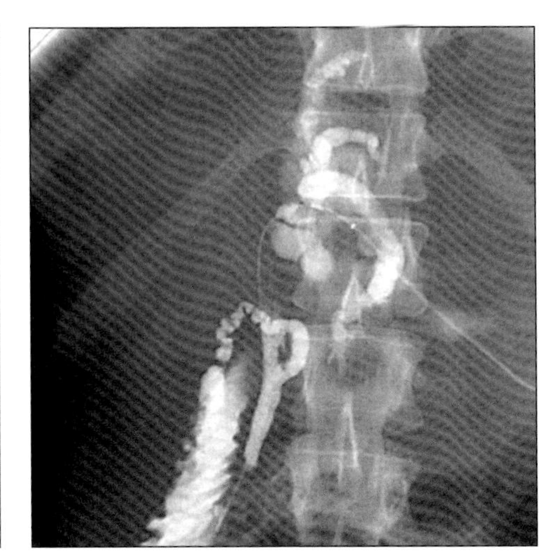

(Left) *Axial CECT shows a mass (arrows) arising near the confluence of the main right and left bile ducts. A biliary stent (open arrow) is in place, but the IHBD are still dilated.* ***(Right)*** *Klatskin tumor. Cholangiogram through left hepatic internal-external stent shows marked dilatation of the left biliary ductal system, no filling of the right and normal cystic and common bile ducts.*

Typical

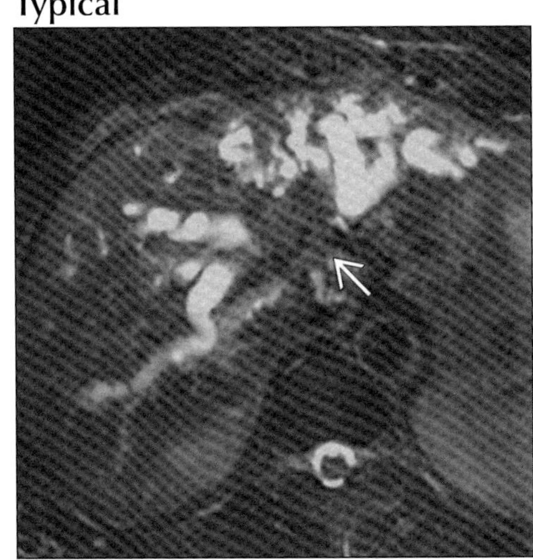

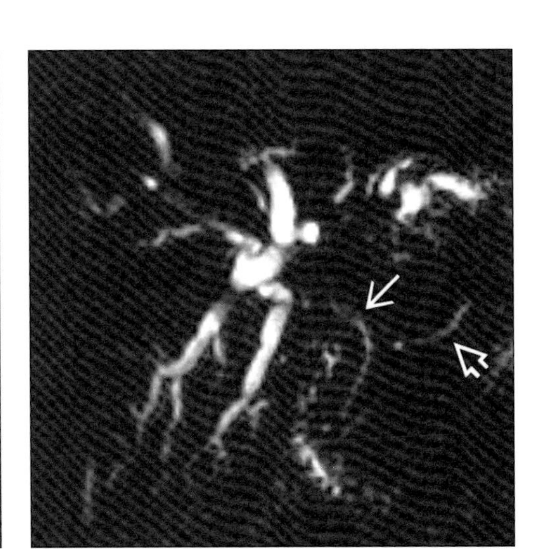

(Left) *Klatskin tumor. Axial T2WI MR shows massively dilated ducts ending abruptly at a hypointense mass (arrow) at the confluence of the main hepatic bile ducts.* ***(Right)*** *Klatskin tumor. MRCP in coronal plane shows dilated IHBD, normal CBD (arrow) and pancreatic duct (open arrow).*

Typical

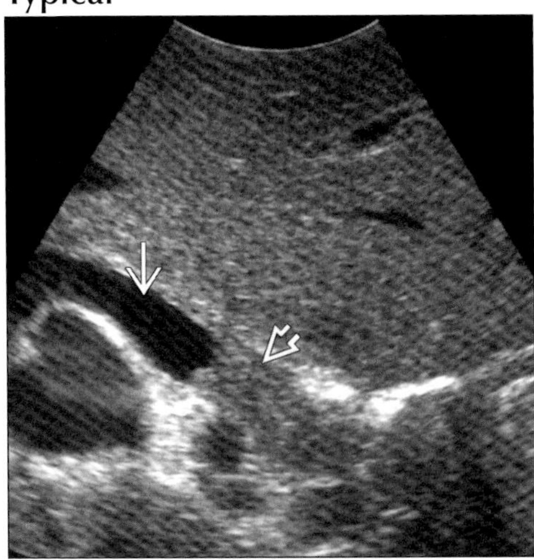

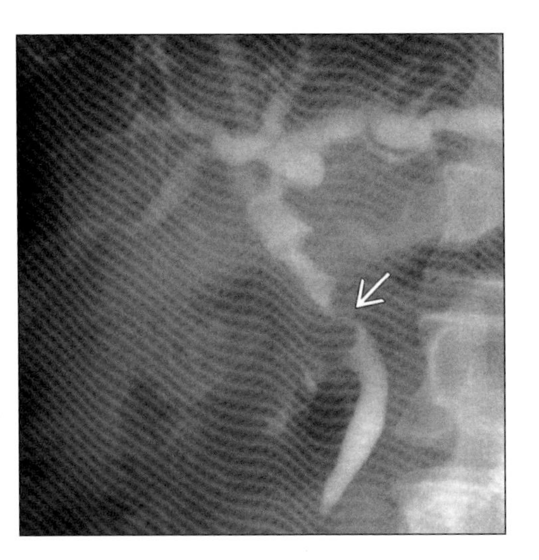

(Left) *Sagittal oblique sonogram shows markedly dilated CBD (arrow) obstructed by a mass (open arrow) with homogeneous echogenicity and no acoustic shadowing.* ***(Right)*** *ERCP shows an "apple core" stricture (arrow) of the common duct due to cholangiocarcinoma.*

AMPULLARY CARCINOMA

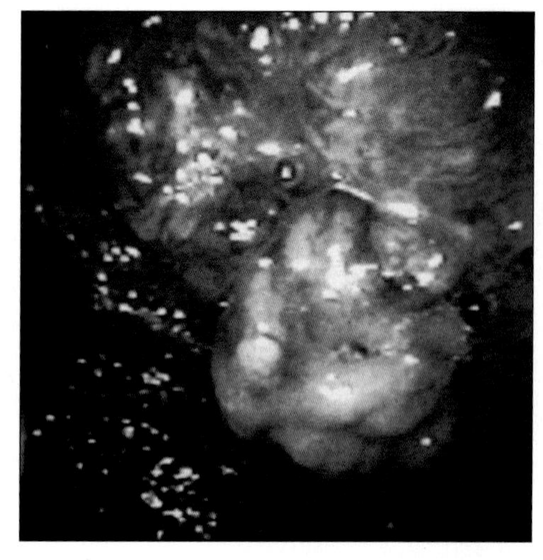

Endoscopic image of ampulla demonstrates soft tissue ampullary carcinoma.

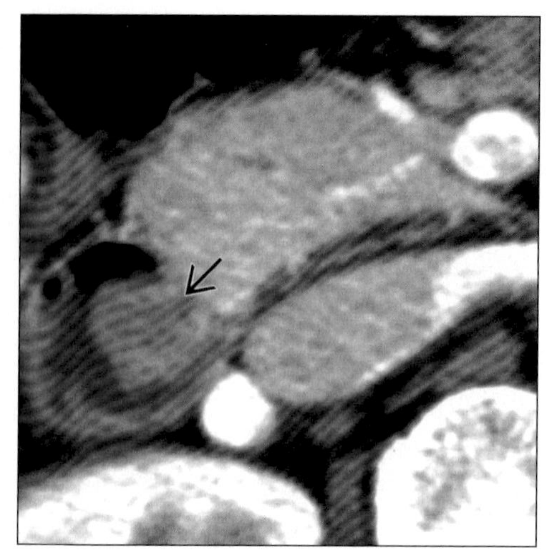

Axial CECT with water distension of duodenum demonstrates lobulated, round soft tissue mass (arrow) arising from ampulla.

TERMINOLOGY

Definitions

- Malignant epithelial neoplasm (adenocarcinoma) arising from ampulla of Vater

IMAGING FINDINGS

General Features

- Best diagnostic clue
  - Soft tissue mass involving ampulla
  - "Double duct" sign with obstruction of both common bile duct (CBD) and pancreatic duct (PD)
  - Lesion best visualized on CECT when duodenum distended with water
- Location: Ampulla of Vater; medial wall of duodenum
- Size: 1-4 cm in diameter; mean 2.7 cm
- Morphology: Often lobulated mass

Radiographic Findings

- Fluoroscopy: UGI: Filling defect in second part of duodenum in region of ampulla of Vater
- ERCP
  - Visible ampullary mass
  - Obstruction of CBD and PD
  - Useful for biopsy

CT Findings

- CECT: Ampullary mass w/variable attenuation (most often hypodense) distinct from pancreas w/dilated CBD & PD; nodal or liver mets in advanced cases

MR Findings

- T1WI: Isointense with pancreas; low signal on fat-saturated T1 compared to normal pancreas
- T2WI
  - Intermediate signal ampullary mass
  - Dilated main PD and CBD
- T1 C+: Enhancing soft tissue mass of lower signal than pancreas
- MRCP
  - Dilated PD and CBD
- Fat suppressed T1WIs
  - Adenocarcinomas low signal-intensity
  - Low signal-intensity compared to normal-enhancing pancreas following gadolinium administration and breathheld gradient echo imaging

Ultrasonographic Findings

- Real Time
  - Dilated CBD and PD
  - Ampullary mass usually not visible
  - Liver mets in advanced cases
- Color Doppler: No flow in dilated hypoechoic CBD and PD

DDx: Spectrum of Ampullary Lesions Mimicking Carcinoma

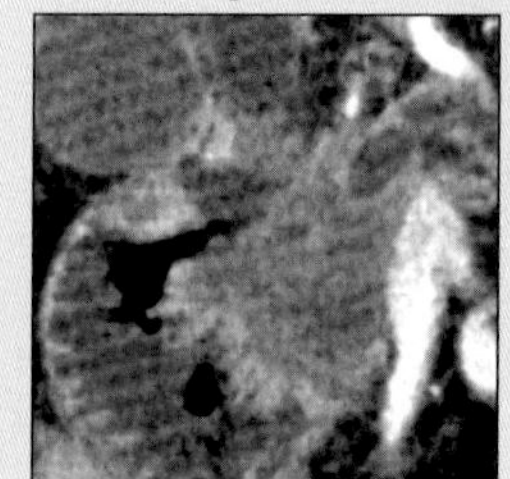

Pancreatic CA

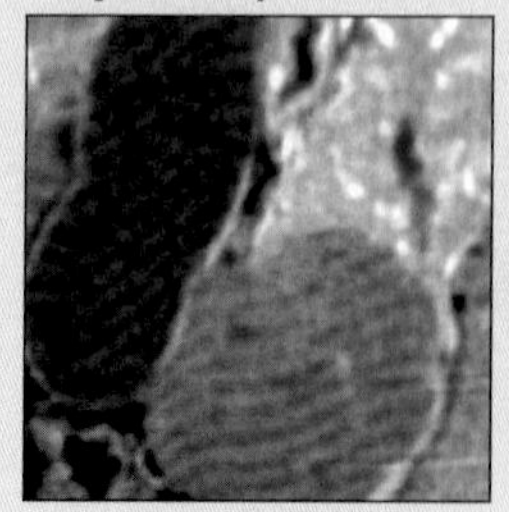

Duodenal Adenoma

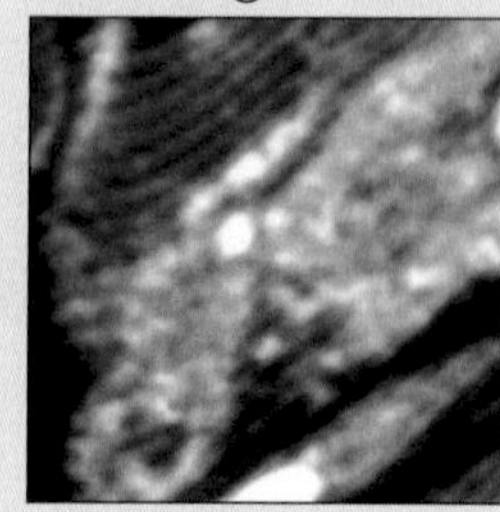

Amp Schwannoma

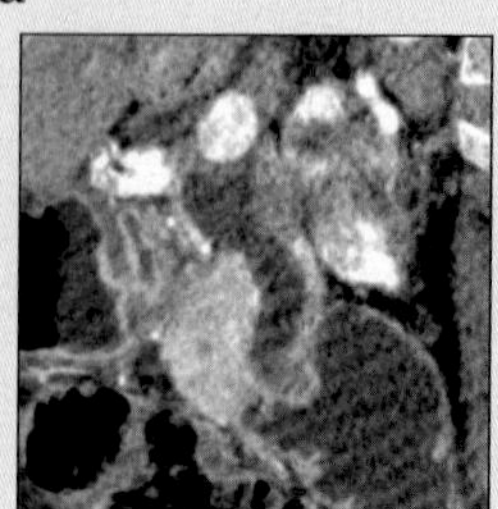

CBD Stone

AMPULLARY CARCINOMA

Key Facts

Terminology
- Malignant epithelial neoplasm (adenocarcinoma) arising from ampulla of Vater

Imaging Findings
- Soft tissue mass involving ampulla
- "Double duct" sign with obstruction of both common bile duct (CBD) and pancreatic duct (PD)
- CECT: Ampullary mass w/variable attenuation (most often hypodense) distinct from pancreas w/dilated CBD & PD; nodal or liver mets in advanced cases

Top Differential Diagnoses
- Pancreatic head carcinoma invading ampulla
- Adenoma of ampulla
- Mesenchymal tumor of ampulla
- Duodenal carcinoma (adenocarcinoma)

Pathology
- Associated abnormalities: Familial adenomatosis coli, Gardner syndrome, colon cancer

Clinical Issues
- Jaundice (71%), weight loss (61%), back pain (46%) are most common symptoms
- Prognosis: 5-year survival rate 38% in resected patients

Diagnostic Checklist
- Duodenal distension with water on CECT key to identifying lesion; ampullary mass and double duct sign also key indicators
- Perform dedicated pancreatic protocol when ampullary lesion suspected

- Endoscopic US
  - Useful for staging and biopsy
  - Detection of nodal mets

Angiographic Findings
- Conventional: Superselective injection of gastroduodenal artery demonstrates hypovascular mass

Nuclear Medicine Findings
- PET: May demonstrate liver mets
- Hepato Biliary Scan
  - Dilated bile ducts

Imaging Recommendations
- Best imaging tool: CECT with dedicated biphasic pancreatic protocol
- Protocol advice
  - Patient to drink 16 oz water immediately prior to CT
  - Arterial phase acquisition: Rapid bolus injection of 150 ml IV contrast (4-5 ml/sec); 1.25 mm collimation after 10 sec delay
  - Venous phase acquisition at 70 seconds with 5 mm collimation
  - Reconstruct pancreas images with 20 cm field of view
  - Additional reformations including curved planar reformat of PD and CBD useful

DIFFERENTIAL DIAGNOSIS

Pancreatic head carcinoma invading ampulla
- Hypoattenuating mass on late arterial phase CECT
- Obstructed CBD and PD
- Mass on fat-saturated T1 and after IV gadolinium

Adenoma of ampulla
- Indistinguishable from carcinoma on CT
- "Double duct" sign with dilated CBD and PD
- Variable in size, from 1-5 cm

Mesenchymal tumor of ampulla
- May be hypervascular ampullary mass on CT if neurogenic, i.e., schwannoma or carcinoid
- High T2 signal on MR if neurogenic origin

Duodenal carcinoma (adenocarcinoma)
- Soft tissue mass arising from duodenal mucosa secondarily invading ampulla and/or pancreas
- May not have "double duct" sign
- May present with gastrointestinal bleeding

PATHOLOGY

General Features
- General path comments: Lobulated or infiltrating mass (adenocarcinoma) arising from ampulla of Vater
- Genetics: No known association
- Etiology: Unknown adenocarcinoma arising from adenomatous epithelium of ampulla
- Epidemiology
  - Associated with history of smoking (30%) and diabetes (17%)
  - Rare tumor representing 0.2% of GI tract malignancies
- Associated abnormalities: Familial adenomatosis coli, Gardner syndrome, colon cancer

Gross Pathologic & Surgical Features
- Lobulated soft tissue mass arising from ampulla of Vater

Microscopic Features
- Malignant ductal epithelial cells; varying degrees of differentiation and necrosis
- Intestinal type
  - Simple and cribriform glands present with pseudostratified oval nuclei with varying degrees of nuclear atypia
- Pancreatobiliary type
  - Similar histology to ductal adenocarcinoma with single layer of round markedly atypical nuclei, micropapillary areas

AMPULLARY CARCINOMA

- Spectrum of histology: Dysplasia, CA in situ, frank adeno CA

Staging, Grading or Classification Criteria

- TNM staging system related to nodal and distant metastases
  - Nodal metastases outside of peripancreatic region considered M1 lesion
- T1: Lesion confined to ampulla
- T2: Tumor invading duodenal wall
- T3: Pancreatic invasion < 2 cm deep
- T4: Pancreatic invasion > 2 cm deep

CLINICAL ISSUES

Presentation

- Most common signs/symptoms
  - Jaundice (71%), weight loss (61%), back pain (46%) are most common symptoms
  - Other signs/symptoms
    - Gastrointestinal bleeding with heme positive stool
    - Clay-colored stool
    - Fever and chills from cholangitis
- Clinical profile
  - Elevated bilirubin and alkaline phosphatase
  - May have both elevated CEA or CA19-9 tumor markers
    - Elevated pre-operative tumor markers associated with poor outcome

Demographics

- Age: Mean age 65 years
- Gender: M:F = 2:1
- Ethnicity: No known ethnic predilection

Natural History & Prognosis

- Depends on nodal and distal metastases at time of presentation
- Prognosis: 5 year survival rate 38% in resected patients
  - Best survival in patients with negative surgical margins, negative nodes and well-differentiated tumors

Treatment

- Pancreatoduodenal resection (Whipple procedure) in good operative risk patients
- Local resection prone to recurrence of tumor

DIAGNOSTIC CHECKLIST

Consider

- Primary pancreatic cancer invading ampulla

Image Interpretation Pearls

- Duodenal distension with water on CECT key to identifying lesion; ampullary mass and double duct sign also key indicators
- Perform dedicated pancreatic protocol when ampullary lesion suspected

SELECTED REFERENCES

1. Martin JA et al: Ampullary adenoma: clinical manifestations, diagnosis, and treatment. Gastrointest Endosc Clin N Am. 13(4):649-69, 2003
2. Duffy JP et al: Improved survival for adenocarcinoma of the ampulla of Vater: fifty-five consecutive resections. Arch Surg. 138(9):941-8; discussion 948-50, 2003
3. Clements WM et al: Ampullary carcinoid tumors: rationale for an aggressive surgical approach. J Gastrointest Surg. 7(6):773-6, 2003
4. Smith TR et al: Prolapse of the common bile duct with small ampullary villous adenocarcinoma into third part of the duodenum. AJR Am J Roentgenol. 181(2):599-600, 2003
5. Lindell G et al: Management of cancer of the ampulla of Vater: does local resection play a role? Dig Surg. 20(6):511-5, 2003
6. Nakano K et al: Combination therapy of resection and intraoperative radiation for patients with carcinomas of extrahepatic bile duct and ampulla of Vater: prognostic advantage over resection alone? Hepatogastroenterology. 50(52):928-33, 2003
7. Trimbath JD et al: Attenuated familial adenomatous polyposis presenting as ampullary adenocarcinoma. Gut. 52(6):903-4, 2003
8. Handra-Luca A et al: Adenomyoma and adenomyomatous hyperplasia of the Vaterian system: clinical, pathological, and new immunohistochemical features of 13 cases. Mod Pathol. 16(6):530-6, 2003
9. Kim JH et al: Differential diagnosis of periampullary carcinomas at MR imaging. Radiographics. 22(6):1335-52, 2002
10. Kaiser A et al: The adenoma-carcinoma sequence applies to epithelial tumours of the papilla of Vater. Z Gastroenterol. 40(11):913-20, 2002
11. Rodriguez C et al: How accurate is preoperative diagnosis by endoscopic biopsies in ampullary tumours? Rev Esp Enferm Dig. 94(10):585-92, 2002
12. Irie H et al: MR imaging of ampullary carcinomas. J Comput Assist Tomogr. 26(5):711-7, 2002
13. Jordan PH Jr et al: Treatment of ampullary villous adenomas that may harbor carcinoma. J Gastrointest Surg. 6(5):770-5, 2002
14. Skordilis P et al: Is endosonography an effective method for detection and local staging of the ampullary carcinoma? A prospective study. BMC Surg. 2(1):1, 2002
15. Yoshida T et al: Hepatectomy for liver metastasis from ampullary cancer after pancreatoduodenectomy. Hepatogastroenterology. 49(43):247-8, 2002
16. Eriguchi N et al: Carcinoma of the ampulla of Vater associated with other organ malignancies. Kurume Med J. 48(4):255-9, 2001
17. Nikfarjam M et al: Local resection of ampullary adenocarcinomas of the duodenum. ANZ J Surg. 71(9):529-33, 2001
18. Hirata S et al: Periampullary choledochoduodenal fistula in ampullary carcinoma. J Hepatobiliary Pancreat Surg. 8(2):179-81, 2001
19. Wittekind C et al: Adenoma of the papilla and ampulla--premalignant lesions? Langenbecks Arch Surg. 386(3):172-5, 2001
20. Wagle PK et al: Pancreaticoduodenectomy for periampullary carcinoma. Indian J Gastroenterol. 20(2):53-5, 2001
21. Yeo CJ et al: Periampullary adenocarcinoma. Ann Surg 227(6): 821-31, 1998
22. Talamini MA et al: Adenocarcinoma of the ampulla of Vater. Ann Surg 225(5): 590-600, 1997

IMAGE GALLERY

Typical

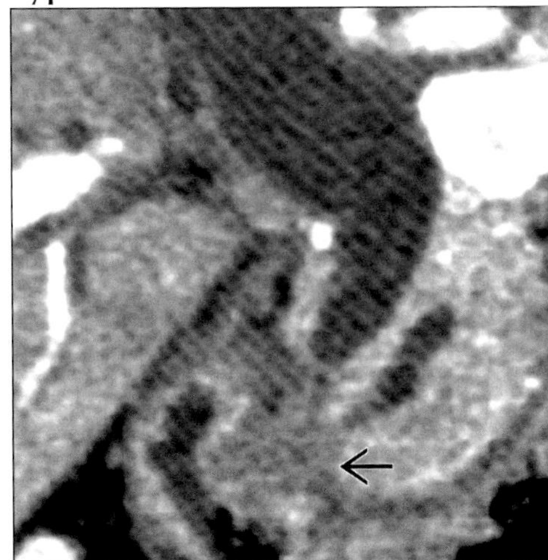

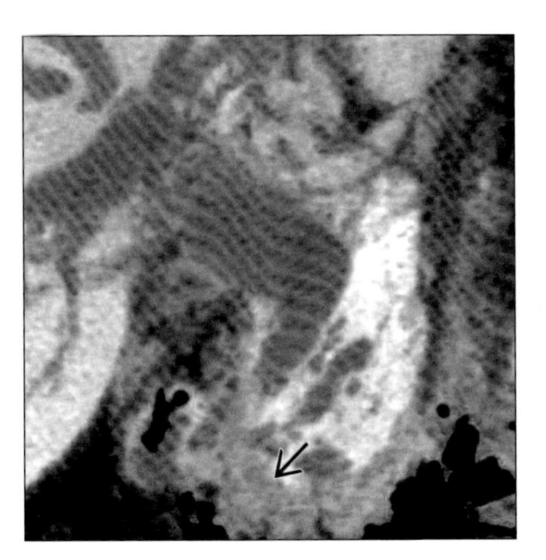

(Left) *Coronal CECT curved planar reformation of common bile duct shows ampullary CA invading head of pancreas. Note hypodense mass obstructing pancreatic and common bile ducts (arrow).* ***(Right)*** *Coronal CECT minimum intensity image of ampullary CA invading head of pancreas demonstrates hypodense mass (arrow).*

Typical

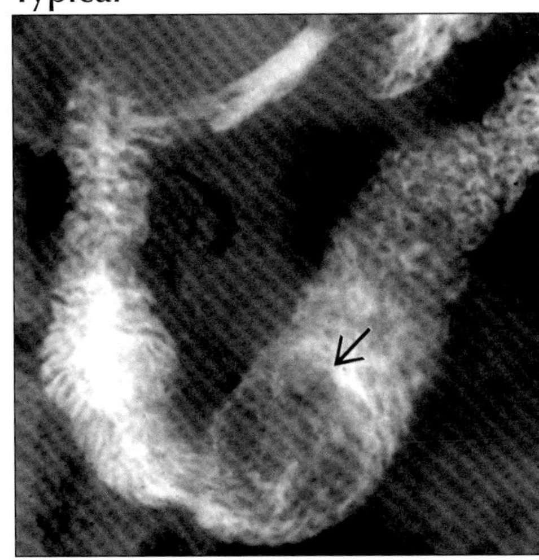

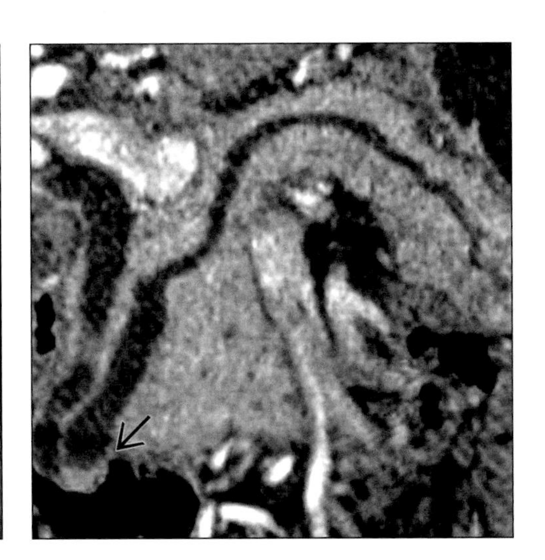

(Left) *Anteroposterior spot film from upper GI series demonstrates rounded filling defect from ampullary CA (arrow).* ***(Right)*** *Coronal CECT curved planar reformation of pancreatic duct demonstrates very small ampullary tumor (arrow) obstructing pancreatic & common bile ducts.*

Typical

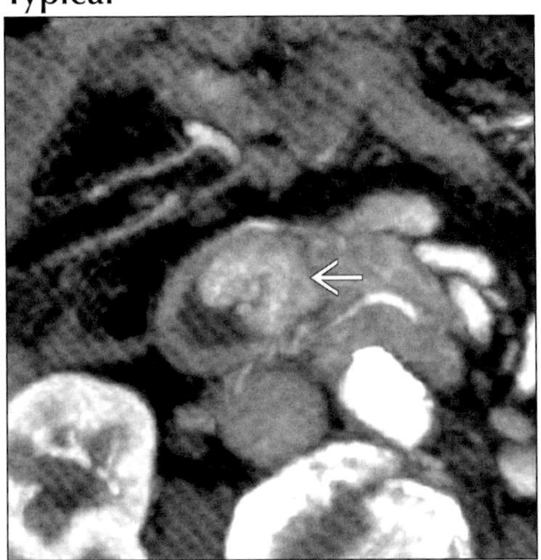

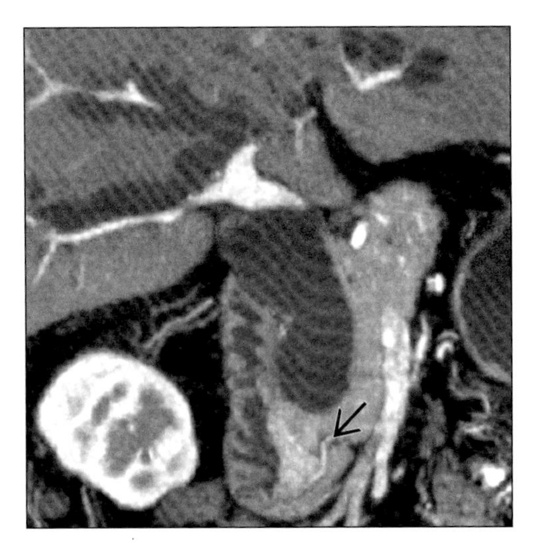

(Left) *Axial CECT thick slab image demonstrates ampullary CA with an unusual degree of increased vascularity (arrow).* ***(Right)*** *Coronal CECT of ampullary CA demonstrates marked common bile duct obstruction secondary to mass (arrow).*

IPMT, BILIARY

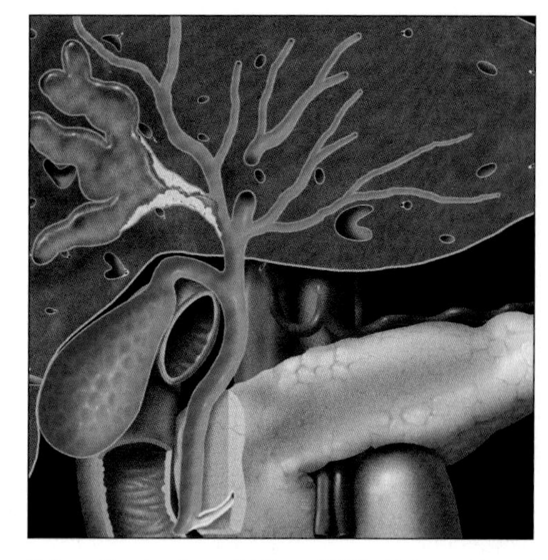

Schematic drawing of biliary IPMT. Note segmental dostension of right lobe intra-hepatic ducts filled with mucin and containing mucosal mass arising from ductal epithelium.

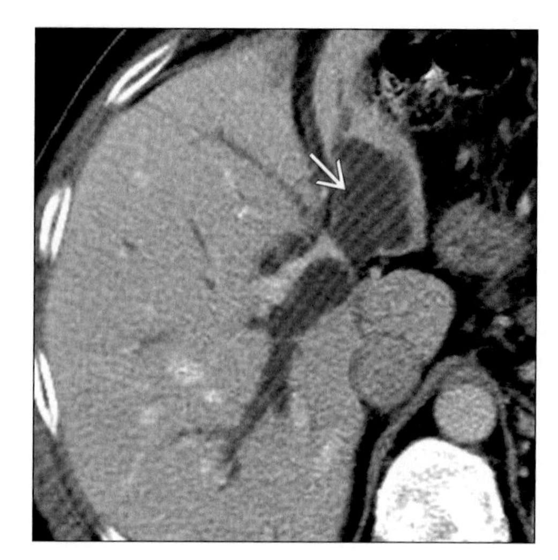

Axial CECT of biliary IPMT. Note marked distended intrahepatic ducts with "aneurysmal" dilatation of left intrahepatic duct (arrow) distended with mucin.

TERMINOLOGY

Abbreviations and Synonyms

- Intraductal papillary mucinous tumor of biliary ducts (IPMT)

Definitions

- Mucin-producing papillary neoplasm of biliary mucosa

IMAGING FINDINGS

General Features

- Best diagnostic clue: Diffuse segmental "aneurysmal" dilation of bile ducts with polypoid or nodular intraductal mass
- Location: Intra- or extrahepatic bile ducts
- Size: Marked dilation of bile ducts
- Morphology: "Aneurysmal" dilatation of mucin-distended ducts

Radiographic Findings

- ERCP
  - Dilated ducts with intraluminal filling defects representing mucin plugs or tumor
  - Segmental ductal obstruction

CT Findings

- CECT: Markedly dilated intra- or extrahepatic bile ducts with intraluminal enhancing fungating mass

MR Findings

- T2WI: Markedly dilated high signal bile duct with low signal intraluminal filling defects
- MRCP
  - Diffuse or segmental biliary ductal dilatation
  - Intraductal filling defects

Ultrasonographic Findings

- Real Time
  - Complex "mass" of aneurysmally dilated bile ducts
  - Echogenic intraductal masses; anechoic mucin
- Color Doppler: No flow in dilated bile ducts

Imaging Recommendations

- Best imaging tool: CECT, MRCP, US
- Protocol advice: 150 ml contrast injected at 2.5 ml/sec; 5 mm collimation; 5 mm reconstruction; use 1.25-2.5 mm reconstructions for improved reformations

DIFFERENTIAL DIAGNOSIS

Cholangiocarcinoma

- Polypoid intraductal mass; biliary stricture; not mucin producing

DDx: Hepato-Biliary Lesions Mimicking IPMT

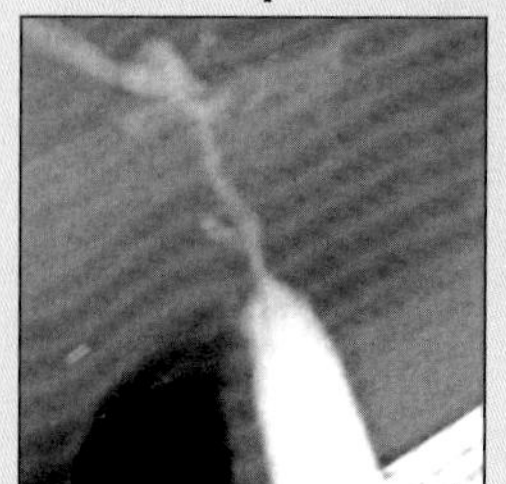

Cholangiocarcinoma

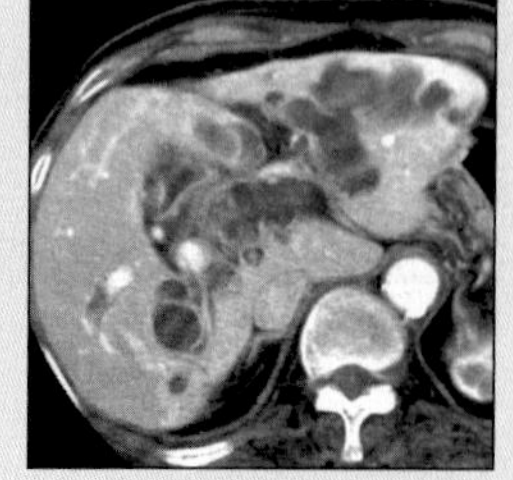

RPC

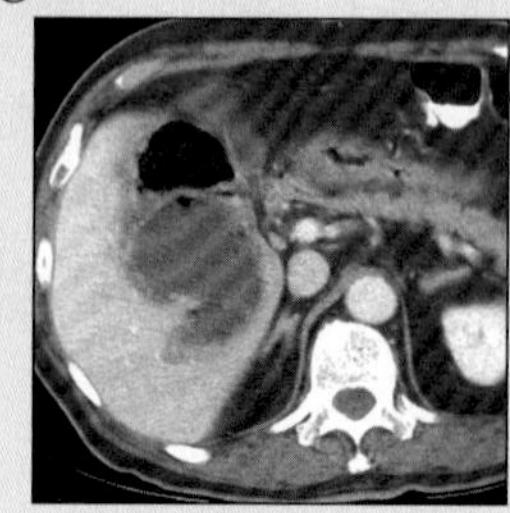

Liver Abscess

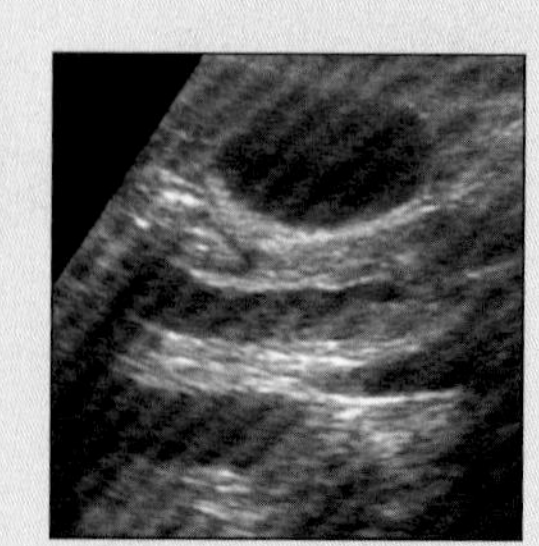

Bact Cholangitis

IPMT, BILIARY

Key Facts

Terminology

- Mucin-producing papillary neoplasm of biliary mucosa

Imaging Findings

- Best diagnostic clue: Diffuse segmental "aneurysmal" dilation of bile ducts with polypoid or nodular intraductal mass
- CECT: Markedly dilated intra- or extrahepatic bile ducts with intraluminal enhancing fungating mass
- Protocol advice: 150 ml contrast injected at 2.5 ml/sec; 5 mm collimation; 5 mm reconstruction; use 1.25-2.5 mm reconstructions for improved reformations

Clinical Issues

- Most common signs/symptoms: Intermittent abdominal pain, fever, chills, jaundice

Recurrent pyogenic cholangitis (RPC)

- Intrahepatic & common bile duct (CBD) stones; markedly dilated intra- or extrahepatic ducts

Pyogenic liver abscess

- "Cluster sign" of low attenuation locules of abscess; no dilated bile ducts, mucin production or intraductal mass

Bacterial cholangitis

- Pus within dilated bile ducts; low-level echoes in CBD; thickened CBD walls

PATHOLOGY

General Features

- General path comments
  - Extensive mucin formation in dilated ducts
  - Nodular intraductal tumors
- Genetics: No known genetic disposition
- Etiology: Unknown
- Epidemiology: Adult East Asians
- Associated abnormalities: Rupture with intraperitoneal mucinous mass

Gross Pathologic & Surgical Features

- Markedly distended mucin-filled bile ducts with frond-like papilloma

Microscopic Features

- Mucinous papillary tumor of bile duct mucosa
  - Columnar epithelial cells arranged in innumerable papillary fronds distending bile ducts with mucin
- Spectrum: Adenomatous dysplasia to frank invasive adenocarcinoma
- Multiple tumors: Papillomatosis

CLINICAL ISSUES

Presentation

- Most common signs/symptoms: Intermittent abdominal pain, fever, chills, jaundice
- Clinical profile: Asian patient with recurrent episodes of abdominal pain and fever

Demographics

- Age: Fifth through seventh decade
- Ethnicity: Eastern Asia

Natural History & Prognosis

- Repeated episodes of cholangitis
- Good prognosis for adenomas or dysplasia; invasive CA prognosis depends on nodal status, presentation

Treatment

- Surgical resection of involved lobe or segment curative for adenoma of dysplasia

DIAGNOSTIC CHECKLIST

Consider

- Cholangiocarcinoma

Image Interpretation Pearls

- Aneurysmal dilation of segmental bile ducts with nodular enhancing intraductal tumor

SELECTED REFERENCES

1. Lim JH et al: Intraductal papillary mucinous tumor of the bile ducts. Radiographics. 24(1):53-66; discussion 66-7, 2004
2. Sugiyama M et al: Magnetic resonance cholangiopancreatography for postoperative follow-up of intraductal papillary-mucinous tumors of the pancreas. Am J Surg. 185(3):251-5, 2003
3. Oshikiri T et al: Mucin-secreting bile duct adenoma--clinicopathological resemblance to intraductal papillary mucinous tumor of the pancreas. Dig Surg. 19(4):324-7, 2002

IMAGE GALLERY

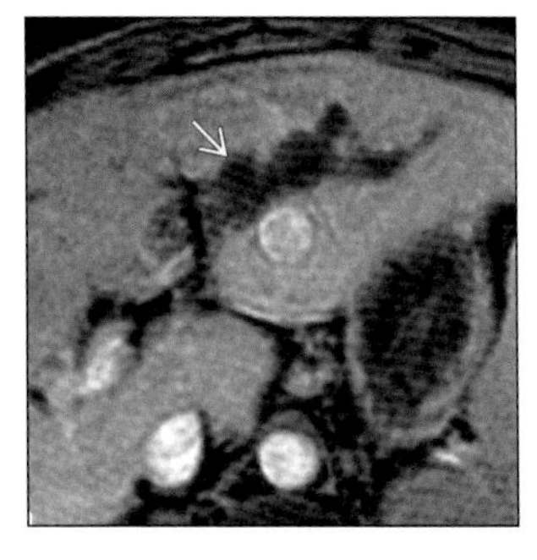

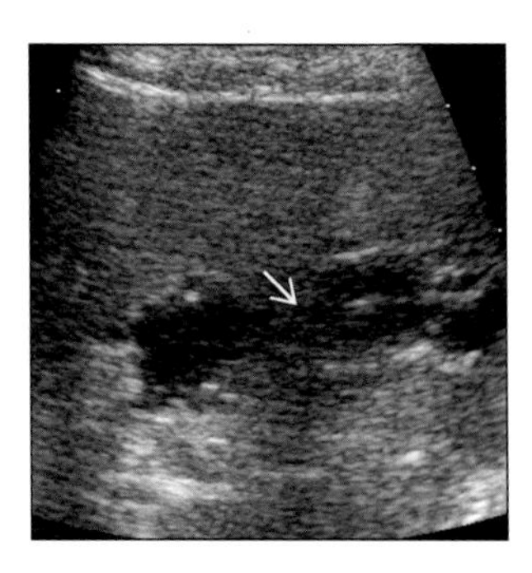

*(**Left**) Left bile duct IPMT on gadolinium-enhanced GRE axial image. Note marked dilatation of left bile ducts. (**Right**) Left bile duct IPMT on transverse sonogram. Note distended bile duct with echogenic mucin (arrow).*

CHEMOTHERAPY CHOLANGITIS

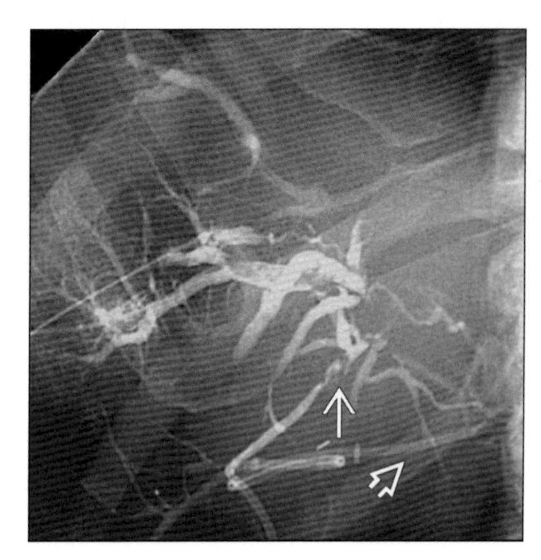

Transhepatic cholangiogram in patient who had intra-arterial chemotherapy through infusion pump catheter (open arrow); dilated right IHBD with multiple central biliary strictures (arrow).

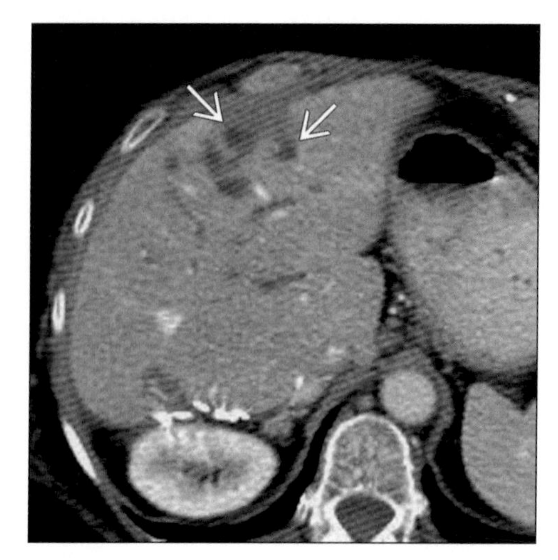

Axial CECT in patient who had right hepatectomy for metastatic colon cancer. Mildly dilated IHBD are noted while some focally dilated ducts (arrows) could be mistaken for metastases.

TERMINOLOGY

Definitions

- Iatrogenic cholangitis following intra-arterial chemotherapy for treatment of primary hepatic tumors and metastases

IMAGING FINDINGS

General Features

- Best diagnostic clue: Clinical setting; if tumor progression excluded by CT, MR, US or other means
- Location
  - Strictures of common hepatic duct are characteristic
  - Strictures frequently involve biliary bifurcation
  - Intrahepatic duct strictures are less common
  - Sparing of distal common bile duct; hallmark
  - Gallbladder & cystic duct may be involved
  - Distribution of strictures in biliary tree reflects hepatic arterial supply to bile ducts
- Morphology: Extrahepatic biliary sclerosis; common hepatic duct stenosis or complete obstruction; intrahepatic bile duct (IHBD) dilatation without a well-recognized obstruction

Radiographic Findings

- ERCP
  - Abnormalities range from minimal duct wall irregularity to marked duct wall thickening with near obliteration of lumen

CT Findings

- CT may be necessary to differentiate cholangitis from extrinsic duct compression by lymph nodes or tumor
- Shows mildly dilated IHBD

Other Modality Findings

- Cholangiography: Segmental strictures of variable length (similar to those seen in primary sclerosing cholangitis)

Imaging Recommendations

- Best imaging tool: CT & ERCP; may be used in concert to differentiate this entity from other causes of jaundice (hepatic replacement by tumor, porta hepatis adenopathy & chemotherapy hepatotoxicity)

DIFFERENTIAL DIAGNOSIS

Primary sclerosing cholangitis (PSC)

- Skip dilatations, stenosis, beading, pruning, irregular thickening of wall of intra- & extrahepatic bile ducts
  - Duct beading & intrahepatic duct involvement are relatively less common in chemotherapy cholangitis

DDx: Irregularly Dilated Bile Ducts

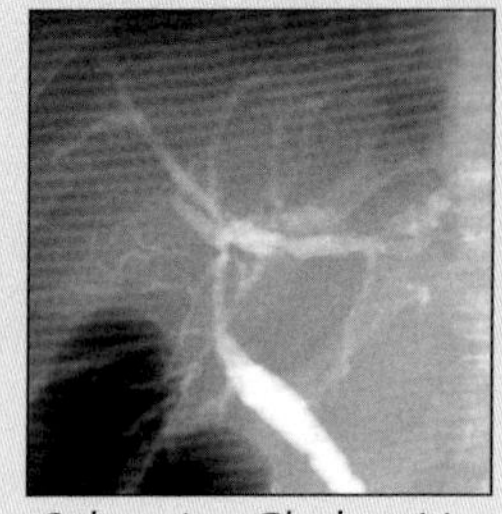

Sclerosing Cholangitis

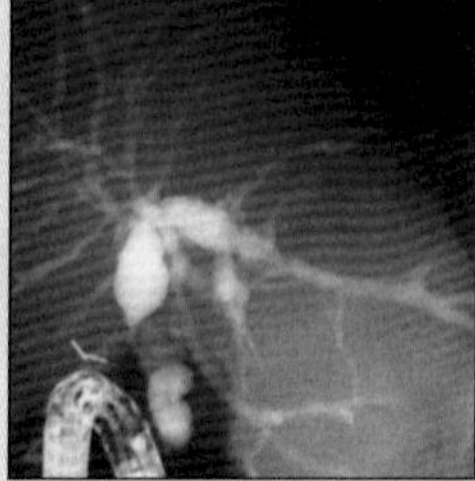

Ascending Cholangitis

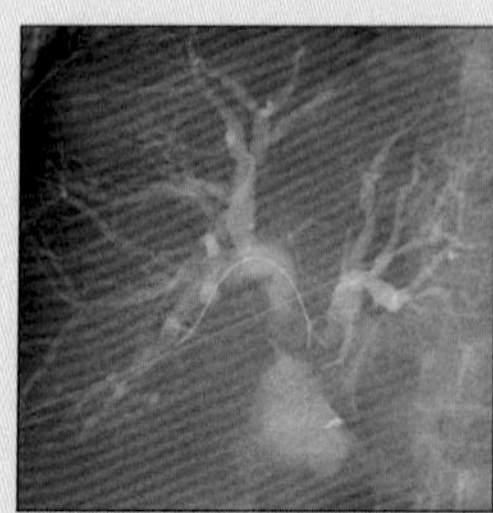

Iatrogenic Biliary Injury

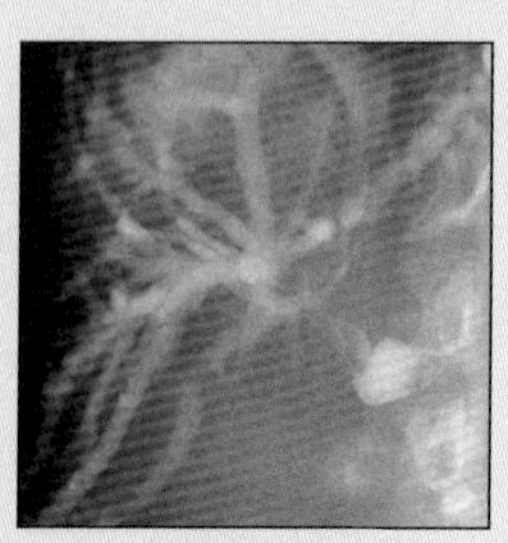

Metastases

CHEMOTHERAPY CHOLANGITIS

Key Facts

Terminology
- Iatrogenic cholangitis following intra-arterial chemotherapy for treatment of primary hepatic tumors and metastases

Imaging Findings
- Strictures of common hepatic duct are characteristic
- Strictures frequently involve biliary bifurcation
- Sparing of distal common bile duct; hallmark

Top Differential Diagnoses
- Primary sclerosing cholangitis (PSC)
- Ascending cholangitis
- Iatrogenic biliary injury
- Extrinsic compression (liver masses)

Pathology
- Ischemic cholangiopathy
- Floxuridine (FUDR): Offending agent in most cases

- Gallbladder & cystic duct are usually more severely involved in chemotherapy cholangitis than in PSC

Ascending cholangitis
- Presence of bile duct stones; prior sphincterotomy

Iatrogenic biliary injury
- Such as common hepatic duct injury during laparoscopic cholecystectomy

Extrinsic compression (liver masses)
- May compress common hepatic duct

PATHOLOGY

General Features
- Etiology
  - Ischemic cholangiopathy
  - Direct chemical toxic effects of drug on duct &/or occlusion of peribiliary vascular plexus with resultant biliary fibrosis
  - Floxuridine (FUDR): Offending agent in most cases
- Epidemiology: Cholangiographic abnormalities reported in 7-30% of patients undergoing intra-arterial chemotherapy
- Associated abnormalities: Aneurysm or pseudoaneurysm of hepatic artery at infusion site, pancreatitis, cholecystitis, gallbladder perforation

CLINICAL ISSUES

Presentation
- Obstructive, severe, persistent jaundice
- Lab: Hyperbilirubinemia with progressively elevated serum glutamic-pyruvic transaminase & alkaline phosphatase

Natural History & Prognosis
- Extrahepatic biliary stenosis may represent primary event leading to secondary intrahepatic biliary damage that does not correlate with specific floxuridine toxicity but results from bile stasis & infection, recurrent cholangitis & eventually biliary sclerosis
- Evolution to biliary cirrhosis is sometimes possible
- Complications: Ischemic cholecystitis, if chemoembolization particles injected "upstream" from origin of cystic artery
- Severe complications: Acute hepatic failure or death
- Mortality rate: 1%

Treatment
- Immediate cessation of intra-arterial FUDR; surgical or percutaneous drainage of biliary tree; balloon dilation of stricture
- In rare cases liver transplantation may be indicated despite a history of metastasizing carcinoma

DIAGNOSTIC CHECKLIST

Consider
- When clinical signs of hepatic dysfunction occur in absence of tumor progression, biliary sclerosis from chemotherapy must be suspected

SELECTED REFERENCES

1. Aldrighetti L et al: Extrahepatic biliary stenoses after hepatic arterial infusion (HAI) of floxuridine (FUdR) for liver metastases from colorectal cancer. Hepatogastroenterology. 48(41):1302-7, 2001
2. Shea WJ Jr et al: Sclerosing cholangitis associated with hepatic arterial FUDR chemotherapy: radiographic-histologic correlation. AJR Am J Roentgenol. 146(4):717-21, 1986
3. Botet JF et al: Cholangitis complicating intraarterial chemotherapy in liver metastasis. Radiology. 156(2):335-7, 1985

IMAGE GALLERY

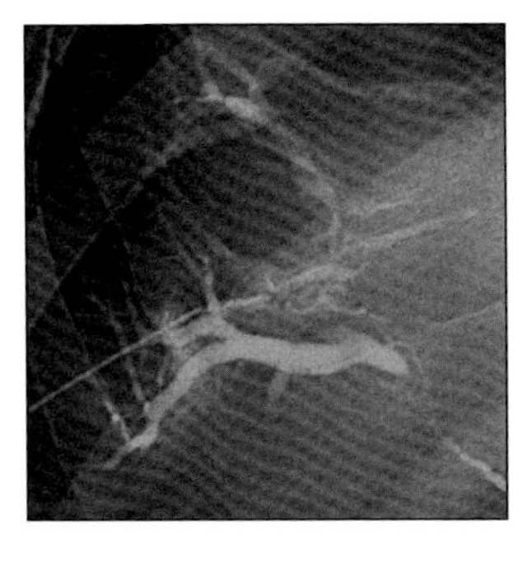

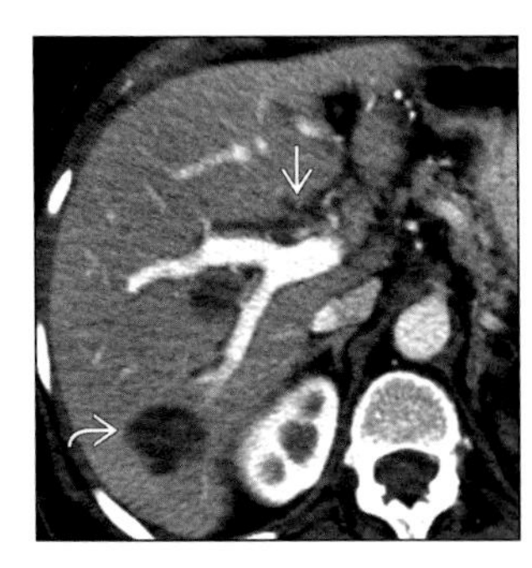

(Left) *Right transhepatic cholangiogram shows dilated IHBD. High grade central biliary strictures developed after intra-arterial chemotherapy with FUDR for metastatic colon carcinoma to the liver.* ***(Right)*** *Axial CECT shows centrally dilated IHBD (arrow) in patient with metastatic colon carcinoma (curved arrow) who had undergone intra-arterial chemotherapy.*

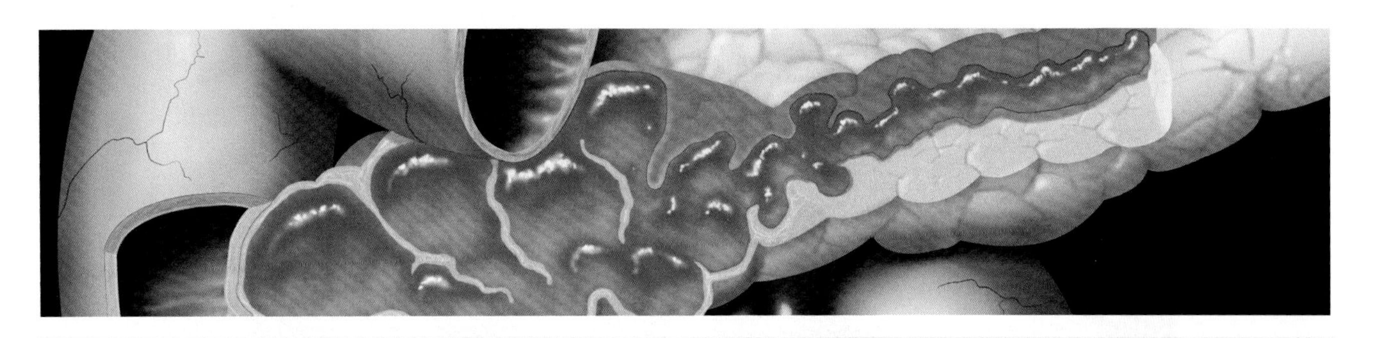

SECTION 3: Pancreas

Introduction and Overview

Normal Variants and Congenital

Inflammation

Trauma

Neoplasm, Benign

Neoplasm, Malignant

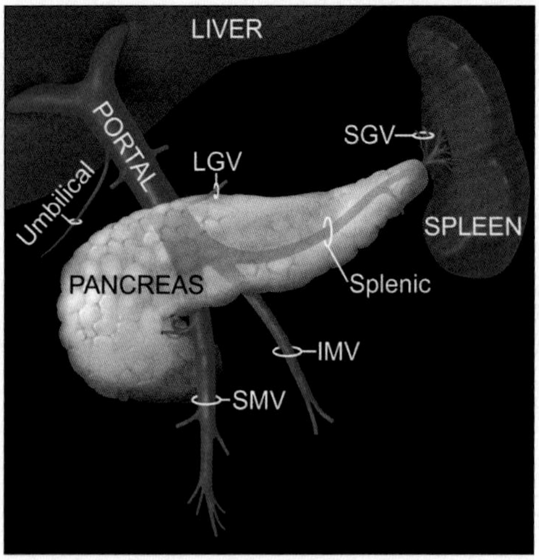

PANCREAS ANATOMY AND IMAGING ISSUES

Graphic shows the major venous anatomy relevant to the pancreas. The splenic vein lies in a groove on the dorsal surface of the pancreas. LGV + SGV = left and short gastric veins.

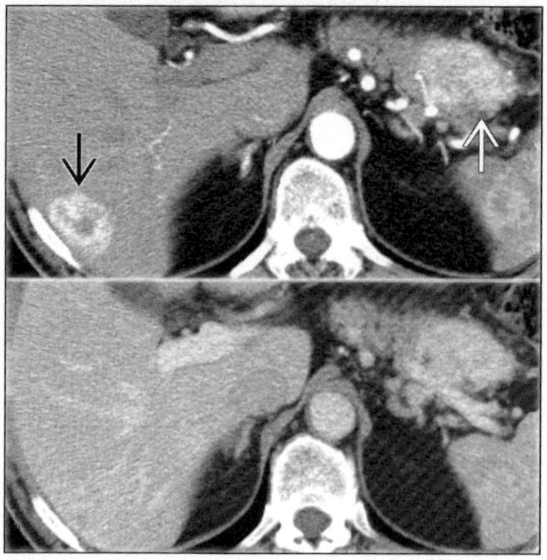

Axial pancreatic (top) and hepatic (bottom) phase CECT shows hypervascular islet cell tumor with liver metastases (arrows), much less obvious on hepatic phase image.

IMAGING ANATOMY

Location

- Pancreas lies in the anterior pararenal space (APRS)
- Size of pancreas varies among population and in individuals
  - Tends to atrophy with age > 70, and pancreatic duct dilates
  - Pancreatic head is widest part; neck narrowest (ventral to mesenteric vessels); gentle taper to tail

ANATOMY-BASED IMAGING ISSUES

Key Concepts or Questions

- Advantages and disadvantages of CT vs. MR for pancreatic evaluation
  - MR and CT have similarly high accuracy in experienced hands
  - MR does not require iodinated contrast administration (but should be performed with bolus injection of gadolinium contrast with multiphasic imaging of liver and pancreas
  - MR allows easy evaluation of common bile duct and pancreatic duct (using MRCP sequences)
  - CT shows calcifications, generally less prone to technical and interpretive errors, faster, more available, more practical for acutely ill patients
- Which MR sequences are most useful for evaluation of a known or suspected pancreatic tumor?
  - Volume-acquired 3D fat-suppressed gradient echo sequences; with and without bolus dynamic Gd-enhancement
- What CT protocol should be used for known/suspected pancreatic tumor?
  - Nonenhanced images through liver and pancreas: Can aid detection of some hypervascular tumors and calcifications
  - Pancreatic phase images (35-45 sec delay after initiation of bolus of 125 ml contrast at 4 ml/sec
    - Through liver and pancreas: Allows detection of hypervascular (e.g., islet cell) and hypovascular (e.g., pancreatic ductal) carcinomas
    - With multidetector-row CT, acquire 1 to 2 mm thick sections to facilitate CT angiography and 3D displays of ductal anatomy: Display axial section as 5 mm thick
  - Hepatic parenchymal phase images (70 sec delay)
    - 5 mm thick images; best for hypovascular liver metastases, venous and nodal anatomy
  - Water as oral contrast medium
    - Facilitates detection of hypervascular tumor in or near duodenum (e.g., gastrinoma), and improves quality of 3D reformations, because dense oral contrast medium would interfere with vessel display
- What CT protocol should be used for "acute pancreatitis"?
  - The same "pancreatic tumor protocol" can be used, but it is usually sufficient to use an "acute abdomen protocol" (iodinated oral contrast medium; all scans obtained after IV administration of 125 ml of contrast at 3 ml/sec; 5 mm thick contiguous sections, 70 sec delay)
- Is there a role for ultrasonography in patient suspected of having pancreatic tumor?
  - Ultrasonography, especially endoscopic US, can reveal some small tumors that are not apparent on CT or MR
  - Intra-operative US is often useful to detect small islet cell tumors and to minimize surgical manipulation of the pancreas
  - Ultrasonography is often used to guide percutaneous (or endoscopic) biopsy of pancreatic tumors
- What criteria can be used to distinguish pancreatitis from carcinoma?

PANCREAS ANATOMY AND IMAGING ISSUES

DIFFERENTIAL DIAGNOSIS

Tumors of exocrine pancreas

Common
- Ductal adenocarcinoma

Uncommon
- Mucinous cystic neoplasm
- Intraductal papillary mucinous tumor (IPMT)
- Serous (microcystic) adenoma

Rare
- Cystic and solid (pseudopapillary) neoplasm
- Acinar cell carcinoma
- Pancreatoblastoma
- Mature cystic teratoma

Tumors of endocrine pancreas
- Islet cell (neuroendocrine) tumor
- Small cell carcinoma

Nonepithelial tumors
- Soft tissue tumors (neurofibroma, etc.)
- Lymphoma

Hypervascular pancreatic mass(es)
- Islet cell (neuroendocrine) tumor
- Metastases
- ⇒ Especially renal cell carcinoma, melanoma

  - Pancreatic ductal carcinoma is a scirrhous hypovascular mass best detected on bolus enhanced CT (or MR) and typically causes narrowing or obstruction of vessels and ducts, and extends dorsally to the celiac and superior mesenteric arterial origins
  - Acute pancreatitis causes fluid exudation and fat infiltration, extending ventrally and laterally into the mesentery and anterior pararenal space
  - Necrotizing pancreatitis may result in a nonenhancing pancreatic "mass", but there are multiple other radiographic and clinical signs of pancreatitis in these patients
  - Chronic pancreatitis, like carcinoma, may result in pancreatic head enlargement, ductal dilation, and (rarely) infiltration of perivascular planes: The presence of lymph node or liver metastases or biopsy may be necessary for diagnosis of carcinoma
- How vigorously and by what means should a cystic pancreatic mass be investigated?
  - If clinical signs, symptoms, and lab tests support a diagnosis of pseudocyst, merely monitor (CT or sonography) for stability and resolution
  - If incidental, small (< 2 cm) lesion in elderly patient or patient with limited (< 10 years) life expectancy, can usually ignore as insignificant
  - If lesion has characteristic features of serous cystadenoma and causes few symptoms, continued surveillance at 6 to 12 month intervals (CT, MR or US)
  - If lesion has characteristic features of side-branch IPMT (intraductal papillary mucinous tumor), confirm with ERCP and surveillance with CT or MR
  - If pancreatic cysts are asymptomatic and are discovered as part of a multisystem syndrome disorder (e.g., polycystic disease , von Hippel Lindau), no further evaluation unless cyst is complex or growing
    - Consider possibility of pseudoaneurysm and obtain color Doppler US, contrast-enhanced CT or MR, or angiography
    - Consider the possibility of metastases in patient with a known tumor
  - Most other pancreatic cystic masses should be regarded as malignant or premalignant and surgical exploration should be considered

CLINICAL IMPLICATIONS

Clinical Importance
- Location in the anterior pararenal space
  - Duodenum also located n APRS: Pancreatitis and pancreatic carcinoma may result in duodenal wall thickening, luminal narrowing or obstruction, or an intramural mass (tumor or pseudocyst)
  - Ascending and descending colon also located in APRS: Lateral spread of inflammation from pancreatitis may result in "colon cut-off sign" due to spasm of the proximal descending colon: Rightward spread of inflammation can affect the ascending colon and clinically simulates acute appendicitis or colitis
- Location adjacent to lesser sac: Pancreas is separated from the lesser sac (omental bursa) by only the posterior parietal peritoneum
  - Acute pancreatitis often results in fluid exudation into lesser sac which may be mistaken for a pseudocyst: These acute fluid collection usually resolve quickly, while pseudocysts take longer to develop and resolve, and have a fibro-inflammatory wall
- Position of pancreatic tail within the splenorenal ligament
  - The tail of the pancreas constitutes only the distal few centimeters of gland and lies intraperitoneally: Acute inflammation of the pancreatic tail may result in an intrasplenic pseudocyst and pancreatic ascites
- Islet cell tumors may be "functioning " (excess hormone-secreting) or not
  - Nonfunctioning tumors remain asymptomatic until large, and are frequently invasive and metastatic
  - Functioning tumors result in symptoms and permit earlier diagnosis
  - Insulinoma (solitary, small, symptomatic hypoglycemia)

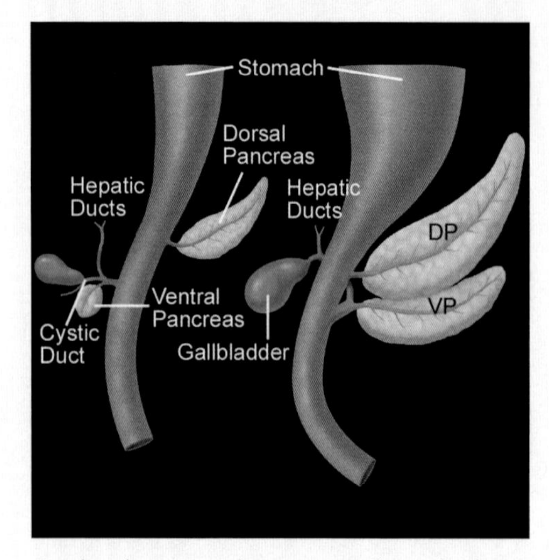

Graphic shows ventral pancreatic anlagen developing as an outpouching of the hepatic-biliary diverticulum. As the stomach and duodenum elongate, the ventral pancreas and bile ducts rotate clockwise and posteriorly to fuse with the dorsal pancreas.

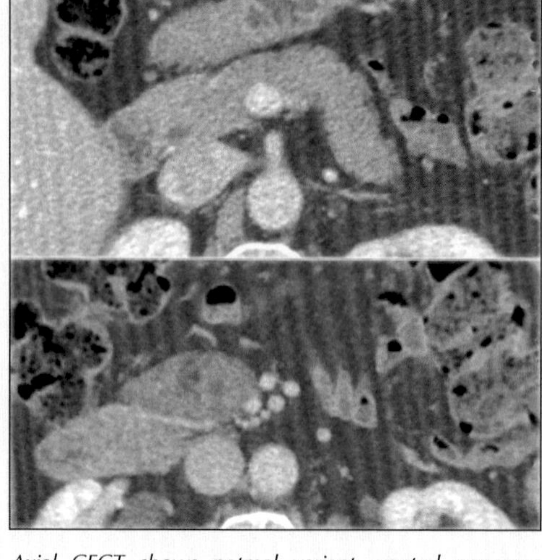

Axial CECT shows notmal variant, ventral pancreas (head/uncinate) with greater fatty lobulation than dorsal pancreas (body/tail), simulating a pancreatic head mass.

- Gastrinoma (Zollinger-Ellison syndrome)

EMBRYOLOGY

Embryologic Events

- The body-tail segment of the pancreas developed from the embryologic dorsal pancreatic bud, while the head-uncinate segment develops from the embryologic ventral bud which also gives rise to the biliary tree
- During normal development, the ventral bud migrates clockwise around the fetal duodenum and eventually merges with the dorsal bud to form the pancreas with the branching pancreatic and bile ducts

Practical Implications

- Failure or anomalies of rotation and fusion may result in annular pancreas, pancreas divisum, agenesis of dorsal pancreas
- The ventral (head-uncinate) and dorsal (body-tail) segments may have a different "texture" in adults that may be misinterpreted as pancreatic pathology

CUSTOM DIFFERENTIAL DIAGNOSIS

Dilated pancreatic duct

- Chronic pancreatitis
- Pancreatic ductal carcinoma
- Ampullary tumor
- Intraductal papillary mucinous tumor (IPMT)
- Distal common duct stone
- Advanced age

Cystic pancreatic mass

- Common
  - Pseudocyst
  - Mucinous cystic tumor
  - IPMT
  - Serous cystadenoma
  - Necrotic pancreatic ductal carcinoma
  - Duodenal diverticulum
- Uncommon
  - Simple cyst
  - Congenital polycystic conditions
    - von Hippel Lindau, polycystic disease, CF
  - Cystic and solid (pseudopapillary) tumor
  - Lymphangioma
  - Metastases (cystic, squamous or sarcoma)
  - Choledochal cyst

Hypovascular pancreatic mass

- Common
  - Pancreatic ductal carcinoma
- Uncommon
  - Metastases/lymphoma
  - Chronic pancreatitis
  - Islet cell tumor (usually hypervascular)
- Rare
  - Pancreaticoblastoma
  - Acinar cell carcinoma
  - Small cell carcinoma
  - Stromal tumors (neurofibroma, etc.)

SELECTED REFERENCES

1. Koeller KK, et al (eds): Radiologic Pathology (2nd ed). Washington , D.C., Armed Forces Institute of Pathology, 2003
2. McNulty NJ et al: Multi--detector row helical CT of the pancreas: effect of contrast-enhanced multiphasic imaging on enhancement of the pancreas, peripancreatic vasculature, and pancreatic adenocarcinoma. Radiology. 220(1):97-102, 2001
3. Dähnert W: Radiologic Review Manual (4th ed). Philadelphia. Lippincott, Williams and Wilkins, 2000
4. Fukuoka K et al: Complete agenesis of the dorsal pancreas. J Hepatobiliary Pancreat Surg. 6(1):94-7, 1999;6(1):94-7
5. Gore RM: Pancreas: Differential Diagnosis. In Gore RM, Levine MS (eds) Textbook of Gastrointestinal Radiology (2nd ed.) Philadelphia, WB Saunders, 2000, pp 1836-41

IMAGE GALLERY

Typical

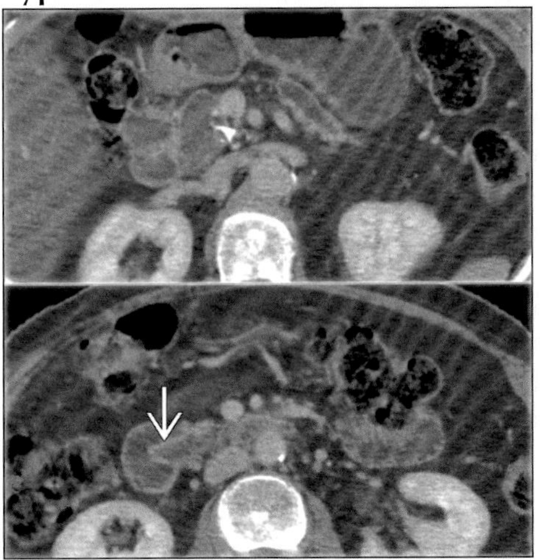

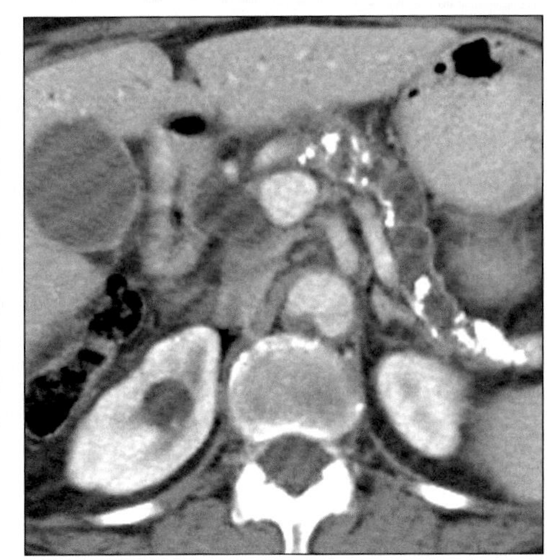

(Left) *Axial CECT shows dilated pancreatic duct and glandular atrophy due to IPMT. Note bulging papilla (arrow).* ***(Right)*** *Axial CECT shows dilated PD, glandular atrophy, and ductal calculi, due to chronic pancreatitis.*

Typical

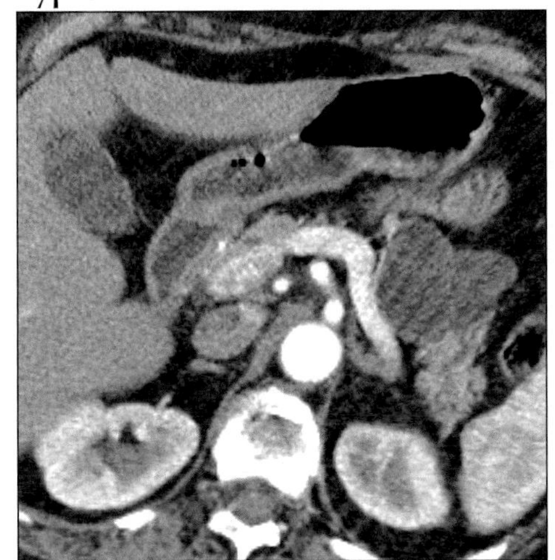

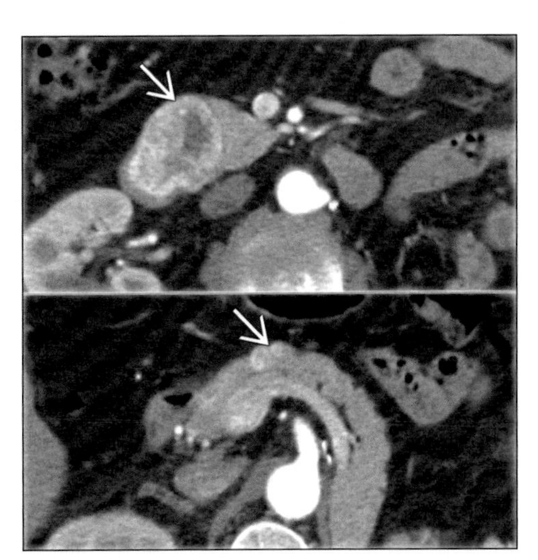

(Left) *Axial CECT shows lobulated, septated cystic mass in pancreatic body due to mucinous cystic tumor.* ***(Right)*** *Axial CECT shows hypervascular pancreatic masses (arrows) due to metastatic renal cell carcinoma.*

Typical

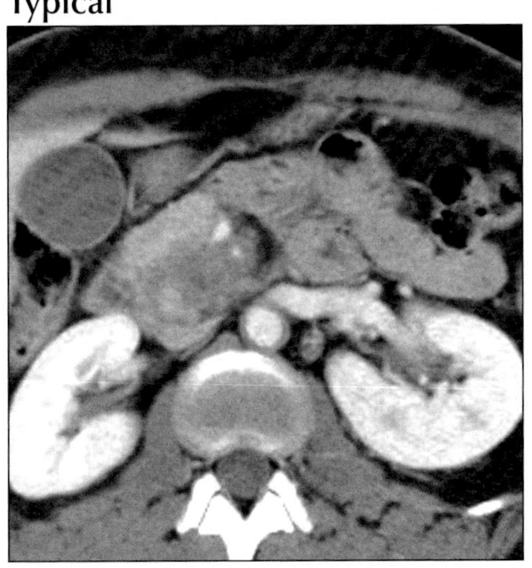

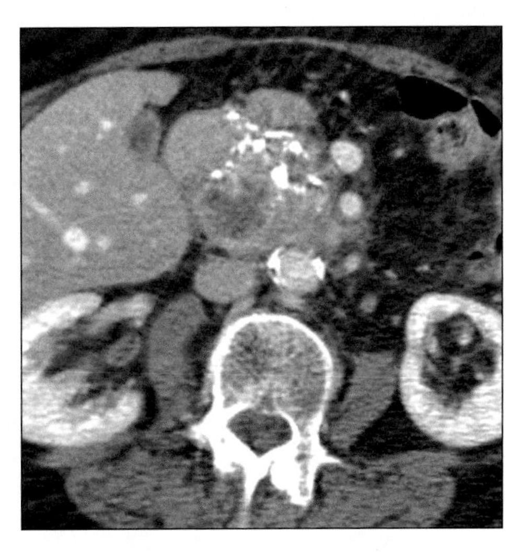

(Left) *Axial CECT shows heterogeneous hypovascular mass in pancreatic head and perivascular invasion; pancreatic carcinoma.* ***(Right)*** *Axial CECT shows heterogeneous hypovascular mass in pancreatic head and calcifications; chronic pancreatitis.*

AGENESIS OF DORSAL PANCREAS

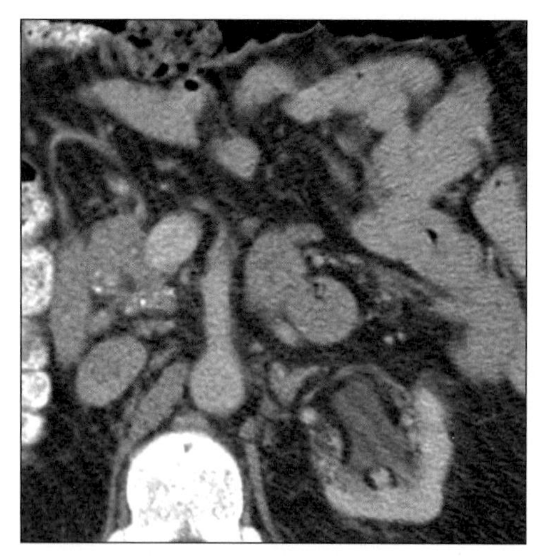

Axial CECT shows pancreatic head with focal calcifications indicative of chronic pancreatitis. Glandular tissue and duct were completely absent in the body-tail segment.

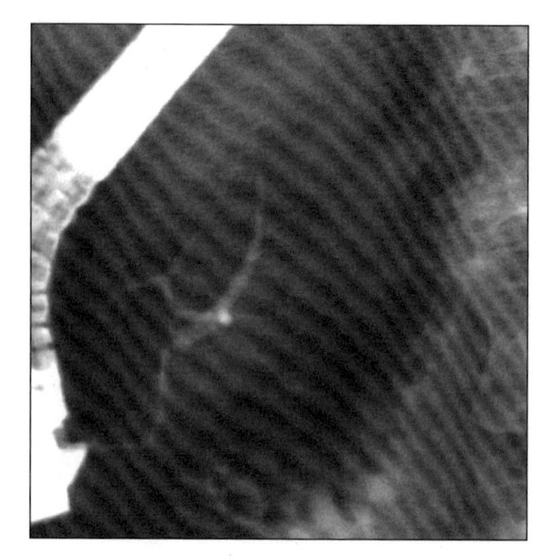

ERCP shows filling of ventral pancreatic duct within the head and uncinate process only. Note normal tapering of the pancreatic duct and its side branches.

TERMINOLOGY

Abbreviations and Synonyms

- Complete or partial pancreatic agenesis; pancreatic aplasia/hypoplasia

Definitions

- Agenesis of dorsal pancreas is a rare congenital anomaly which results from defective pancreas formation

IMAGING FINDINGS

General Features

- Best diagnostic clue: Absence of pancreatic tissue in expected location of neck, body & tail

Radiographic Findings

- ERCP: Filling of ventral duct without identifiable dorsal ducts or minor papilla in agenesis of entire dorsal pancreas
- Hypoplasia: Main pancreatic duct is shortened, some dorsal ducts remain, as evidenced by filling of accessory duct of Santorini
  - Minor papilla may be present in these cases; indicates only portions of dorsal pancreas failed to develop
- Visualized ducts: Taper normally, small in caliber, confined to head of pancreas

CT Findings

- Short pancreas; normal pancreatic head; absence of tail with complete or partial absence of body

MR Findings

- MRCP: In complete agenesis of dorsal pancreas, accessory & dorsal duct system are not observed

Ultrasonographic Findings

- Pancreatic head in normal location with nonvisualization of body & tail
- Limited value: Pancreatic nonvisualization may be due to overlying bowel gas or other technical factors

Angiographic Findings

- No feeding arteries for pancreatic body or tail

Imaging Recommendations

- Best imaging tool: CT & MRCP or ERCP

DIFFERENTIAL DIAGNOSIS

Pancreatic carcinoma

- Abrupt obstruction of pancreatic duct
- Irregular, nodular, rat-tail eccentric obstruction
- Localized encasement with prestenotic dilatation

DDx: Obstruction of Pancreatic Duct

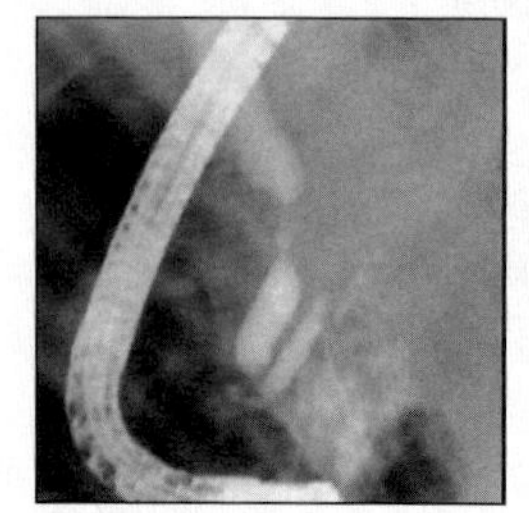

Pancreatic Carcinoma

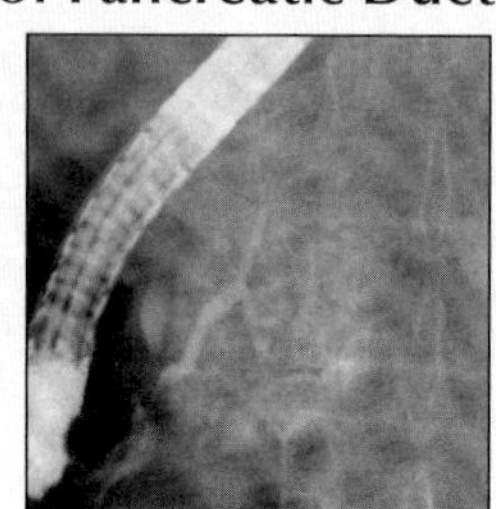

Pancreas Divisum

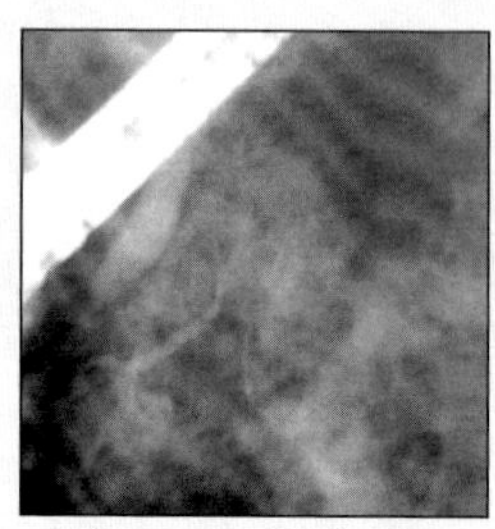

Pancreas Divisum

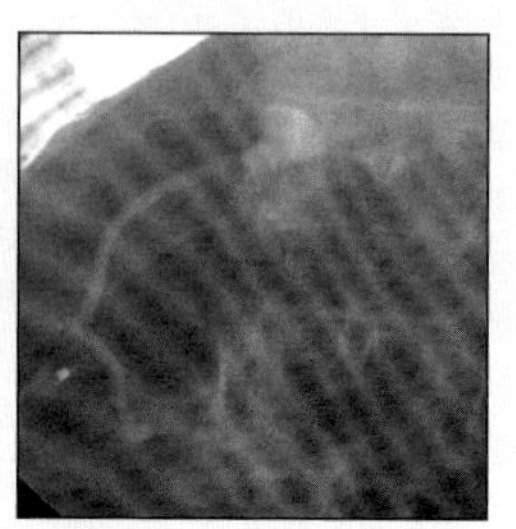

Chronic Pancreatitis

AGENESIS OF DORSAL PANCREAS

Key Facts

Terminology
- Complete or partial pancreatic agenesis; pancreatic aplasia/hypoplasia

Imaging Findings
- Best diagnostic clue: Absence of pancreatic tissue in expected location of neck, body & tail
- Short pancreas; normal pancreatic head; absence of tail with complete or partial absence of body

Top Differential Diagnoses
- Pancreatic carcinoma
- Pancreas divisum
- Chronic pancreatitis

Diagnostic Checklist
- May radiologically mimic acquired atrophy of pancreatic body & tail (pseudo-agenesis)

Pancreas divisum
- Contrast injection into major papilla demonstrates only short ventral duct with normal tapering
- Contrast injection into minor papilla fills dorsal duct
- No communication between ventral + dorsal ducts

Chronic pancreatitis
- Dilated & beaded main pancreatic duct + radicles; intraductal calculi

PATHOLOGY

General Features
- General path comments: Familial occurrence of agenesis of dorsal pancreas suggests hereditary mechanisms may play a role in pathogenesis
- Etiology
  - Defect of dorsal pancreatic bud (anlage)
  - Dorsal bud develops from foregut & extends into dorsal mesentery; forms body, tail, anterior head
    - Duct of dorsal pancreas becomes main pancreatic duct & distal to site of fusion, it becomes accessory duct & drains through minor papilla
- Epidemiology
  - Complete agenesis of dorsal pancreas is rare
  - Few reported cases of dorsal agenesis or hypoplasia
- Associated abnormalities
  - Complete pancreatic agenesis (very rare); related with stillbirth or early neonatal death
  - Polysplenia syndrome (7 cases reported)
  - Associated absence of uncinate process

Microscopic Features
- Biopsy specimens from head show normal pancreatic tissue but only fat from expected position of body
- Absence of recognizable islets of Langerhans

CLINICAL ISSUES

Presentation
- May be asymptomatic, incidental finding
- Diabetes mellitus & recurrent abdominal pain
- May present with jaundice/steatorrhea

Demographics
- Age: Mean: 31.6; range: 0-56 years

Natural History & Prognosis
- Complicated by pancreatitis, insulin-requiring diabetes mellitus, duodenal papillary dysfunction

DIAGNOSTIC CHECKLIST

Consider
- May radiologically mimic acquired atrophy of pancreatic body & tail (pseudo-agenesis)
  - In pancreatic atrophy due to pancreatic carcinoma or chronic pancreatitis, there may be a mass with area of architectural distortion; some portion of body or tail should remain, often containing dilated duct
  - Traumatic transection of pancreatic duct may also cause proximal atrophy, but there should be some remaining pancreas & history of trauma

SELECTED REFERENCES

1. Fukuoka K et al: Complete agenesis of the dorsal pancreas. J Hepatobiliary Pancreat Surg. 6(1):94-7, 1999
2. Oldenburg B et al: Pancreatitis and agenesis of the dorsal pancreas. Eur J Gastroenterol Hepatol. 10(10):887-9, 1998
3. Wildling R et al: Agenesis of the dorsal pancreas in a woman with diabetes mellitus and in both of her sons. Gastroenterology. 104(4):1182-6, 1993
4. Shah KK et al: CT diagnosis of dorsal pancreas agenesis. J Comput Assist Tomogr. 11(1):170-1, 1987

IMAGE GALLERY

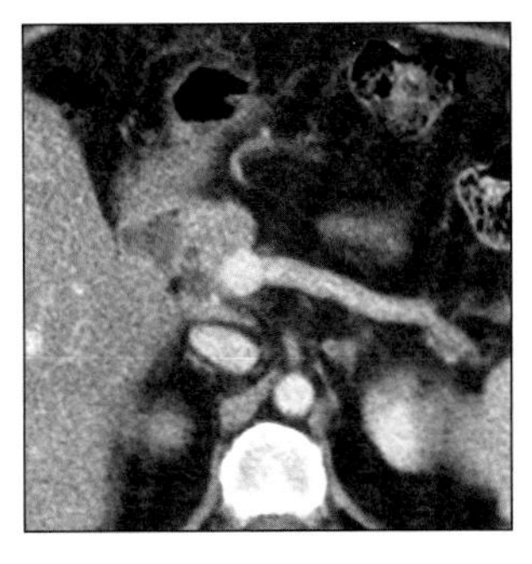

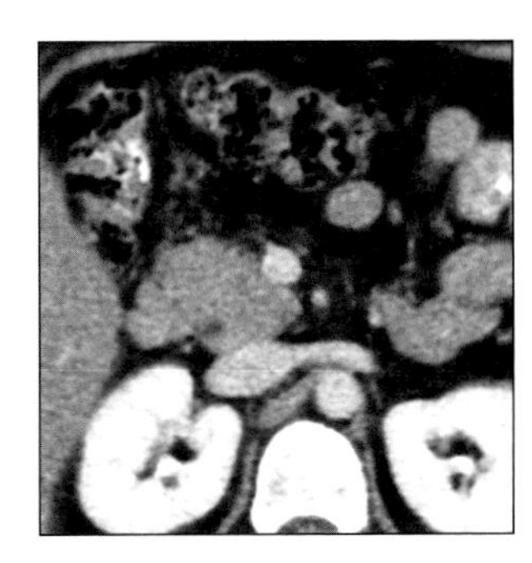

*(**Left**) Axial CECT shows no pancreatic tissue or duct in the expected position of the body-tail segments. (**Right**) Axial CECT shows normal appearance of pancreatic head and uncinate in a patient with dorsal agenesis; initially misdiagnosed as a pancreatic head mass which prompted an unnecessary biopsy.*

ANNULAR PANCREAS

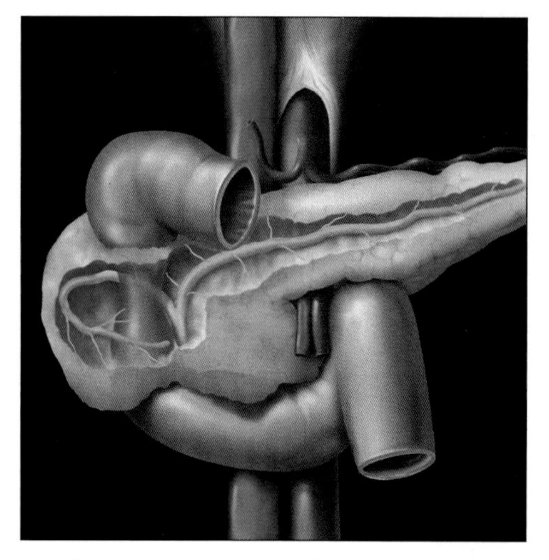

Graphic shows concentric duodenal luminal narrowing by encircling pancreatic tissue. The small pancreatic head duct also encircles the descending duodenum. Note proximal duodenal dilatation.

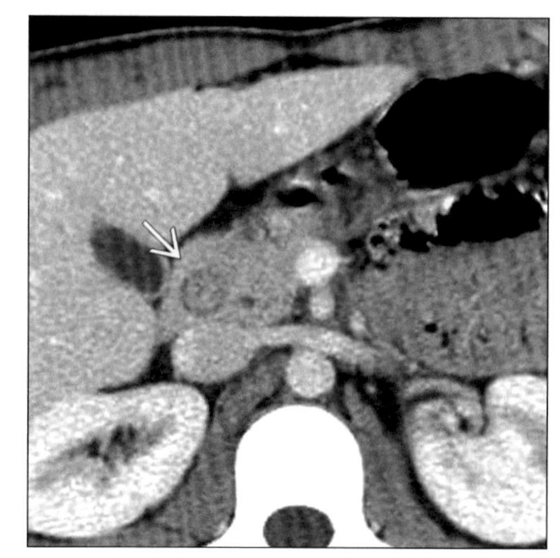

Axial CECT shows pancreatic tissue completely encircling the second part of the duodenum (arrow).

TERMINOLOGY

Definitions

- Ring of pancreatic tissue that encircles second part of duodenum

IMAGING FINDINGS

General Features

- Best diagnostic clue: Pancreatic duct encircles the endoscope or duodenum on ERCP
- Location: 2nd part of duodenum (85%); 1st or 3rd parts (15%)
- Morphology
  - Uncommon congenital anomaly of pancreas
  - Usually manifests in neonates (52% cases) or asymptomatic until adulthood (48% cases)
  - Three theories in development of annular pancreas
    - Persistence or hypertrophy plus abnormal migration of ventral left pancreatic bud to right of duodenum rather to left (more common)
    - Hypertrophy of both dorsal & ventral ducts leads to complete ring
    - Adherence of ventral duct to duodenum before rotation
  - Acquired associated pathology
    - Gastric & duodenal ulcers in 26-48% of cases
    - Pancreatitis seen in 15-30% of patients

Radiographic Findings

- Plain x-ray abdomen
  - Narrowing of descending duodenum
  - Classic "double-bubble" sign
    - Large proximal bubble: Dilated stomach
    - Small distal bubble: Dilated duodenal bulb
- Fluoroscopic guided Barium Study (UGI)
  - Extrinsic, eccentric defect on medial margin of second part of duodenum
  - Concentric narrowing of second part of duodenum & dilated proximal duodenum
  - Reverse peristalsis; duodenal ulcer (2nd part) may be seen
  - Periampullary duodenal ulcer in an adult
    - Diagnosis: Annular pancreas or Zollinger-Ellison syndrome (ZES)
- ERCP
  - Normal main pancreatic duct in body & tail communicating with small duct of pancreatic head encircling duodenum
  - Small duct of pancreatic head seen originating on right anterior surface of duodenum, passing posteriorly around duodenum & entering main duct

CT Findings

- Often nonspecific

DDx: Narrowed Duodenal Lumen

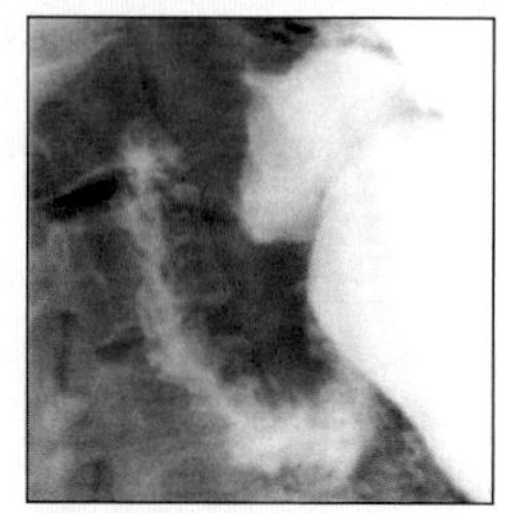

Duodenal Carcinoma

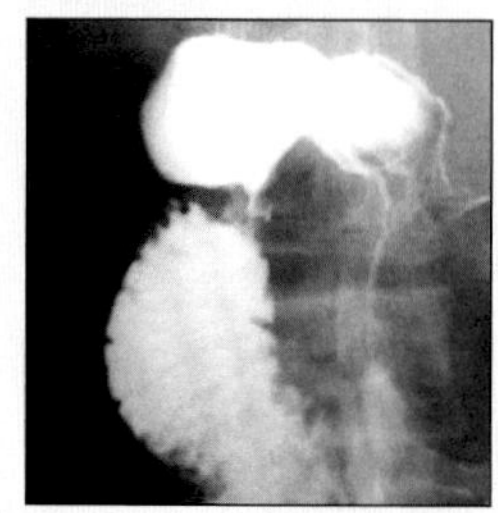

Postbulbar Ulcer

Pancreatic Carcinoma

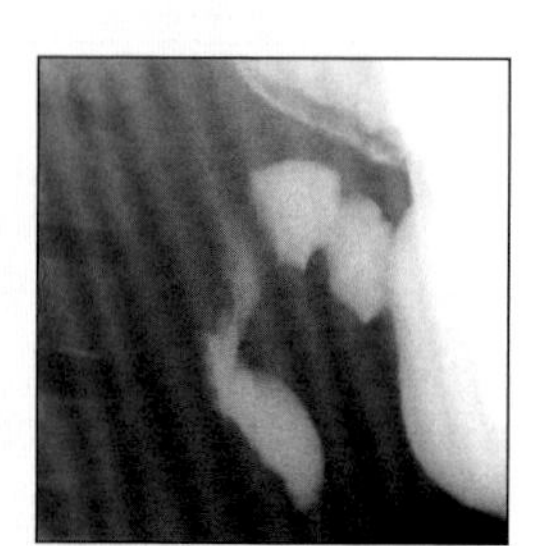

Duodenal Hematoma

ANNULAR PANCREAS

Key Facts

Terminology
- Ring of pancreatic tissue that encircles second part of duodenum

Imaging Findings
- Best diagnostic clue: Pancreatic duct encircles the endoscope or duodenum on ERCP
- Location: 2nd part of duodenum (85%); 1st or 3rd parts (15%)
- Gastric & duodenal ulcers in 26-48% of cases
- Pancreatitis seen in 15-30% of patients
- Classic "double-bubble" sign

Top Differential Diagnoses
- Duodenal carcinoma
- Postbulbar peptic ulcer
- Pancreatic carcinoma

Pathology
- Congenital anomaly
- Persistence of left ventral pancreatic bud & abnormal migration to right of duodenum than to left
- Associated congenital anomalies in up to 75% cases
- Intestinal malrotation; imperforate anus
- Esophageal atresia; duodenal atresia/stenosis

Clinical Issues
- Nausea, vomiting, epigastric pain, jaundice

Diagnostic Checklist
- Differentiate other causes from annular pancreas which can produce "narrow duodenal lumen" (descending part)
- Periampullary ulcer in an adult suggests diagnosis of annular pancreas or Zollinger-Ellison syndrome (ZES)

- Large pancreatic head & central area of increased attenuation representing contrast within narrowed duodenal segment
- Gastric & duodenal dilatation plus circumferential thickening of wall
- Obstructing band of pancreatic tissue & dilated main pancreatic duct
- Dilated CBD & intrahepatic bile ducts (IHBD)

MR Findings
- Fat-suppressed T1WI
  - Normal pancreatic tissue encircling duodenum
- MRCP
  - Depicts course & drainage pattern of pancreatic duct

Ultrasonographic Findings
- Real time
  - Nonspecific enlargement of pancreatic head

Angiographic Findings
- Celiac angiography
  - Anomalous branch from posterior pancreaticoduodenal artery that courses in a right & inferior direction supplying annular moiety

Imaging Recommendations
- MR + MRCP; ERCP; barium study

DIFFERENTIAL DIAGNOSIS

Duodenal carcinoma
- Rare malignant tumor
- Accounts for 1% of all GIT neoplasms
- Location: Postbulbar portion of duodenum
  - At or below level of ampulla of Vater
- Increased incidence reported in patients with Gardener syndrome & celiac disease
- Occasionally associated with Crohn disease & neurofibromatosis
- On barium studies
  - Polypoid, ulcerated or annular lesions
  - Narrowed duodenal lumen
  - Annular lesion at second part of duodenum: Indistinguishable from annular pancreas
    - MRCP or ERCP: Depicts course & drainage pattern of pancreatic duct
- On CT: Discrete mass or thickening of duodenal wall
- Hypotonic duodenography differentiates periampullary & ampullary tumors causing narrow duodenal lumen

Postbulbar peptic ulcer
- Location: Usually medial wall of proximal descending duodenum above ampulla of Vater
- On barium studies
  - Ulcer on medial wall of second part of duodenum
  - Folds radiate toward ulcer crater
  - Edema & spasm: Result in smooth, rounded indentation on lateral wall
  - "Ring stricture" with eccentric narrowing of postbulbar duodenum due to scarring & fibrosis of ulcer
  - Edema with spasm & ring stricture causes narrow duodenal lumen
    - Mimics narrow duodenal lumen of annular pancreas

Pancreatic carcinoma
- Irregular, heterogeneous, poorly enhancing mass
- Location: Head (60% of cases)
- Pancreatic duct and/or CBD show abrupt obstruction & dilatation
- Obliteration of retropancreatic fat
- Extensive local invasion & regional metastases
- Invasion: Medial part of duodenal sweep, narrowing lumen may be seen
- Barium UGI finding
  - "Inverted 3" contour: Medial part of duodenal sweep
- ERCP
  - Irregular eccentric obstruction of CBD & pancreatic duct
- Trousseau sign: Migratory thrombophlebitis
  - Characteristic of cancer
- 65% of patients present with advanced local disease & distant metastases

ANNULAR PANCREAS

Duodenal hematoma
- Results from trauma or anticoagulation
- Narrowed lumen, folds often thickened in "picket fence" pattern

PATHOLOGY

General Features
- General path comments
  - Embryology-anatomy: Normal pancreas
    - Normal development: From ventral anlage
    - Hepatic diverticulum: Forms right & left ventral pancreatic buds
    - Right ventral bud persists to form head & uncinate process
    - Left ventral bud: Atrophies
- Etiology
  - Congenital anomaly
  - Most common pathogenesis
    - Persistence of left ventral pancreatic bud & abnormal migration to right of duodenum than to left
    - Ring of pancreatic tissue encircles 2nd part
    - Duodenal narrowing leads to vomiting
  - Ventral anlage undergoes 180° counter clockwise rotation while duodenum undergoes 90° clockwise rotation, so that ventral anlage contiguous with dorsal anlage, medial to duodenum
- Epidemiology: Uncommon in adults
- Associated abnormalities
  - Associated congenital anomalies in up to 75% cases
    - Intestinal malrotation; imperforate anus
    - Esophageal atresia; duodenal atresia/stenosis
    - Down syndrome & cardiac anomalies

Gross Pathologic & Surgical Features
- Concentric narrowing of second part of duodenum encircled by band of tissue

Microscopic Features
- Normal pancreatic tissue & ductal epithelium
- No inflammatory cells

CLINICAL ISSUES

Presentation
- Most common signs/symptoms
  - Children & adults
    - Nausea, vomiting, epigastric pain, jaundice
  - Neonates: Persistent vomiting since first day of life
    - History of polyhydramnios in utero
    - Other manifestations of GIT obstruction

Demographics
- Age: Neonates (52%); children & adults (48%)
- Gender: Males more than females

Natural History & Prognosis
- Complications
  - Gastric & duodenal ulcers (26-48%)
  - Pancreatitis (15-30%)
- Prognosis
  - Good: After surgical correction

Treatment
- Pediatric population
  - Retrocolic duodenojejunostomy/gastrojejunostomy
- Adult population
  - Surgical & interventional endoscopic procedures

DIAGNOSTIC CHECKLIST

Consider
- Differentiate other causes from annular pancreas which can produce "narrow duodenal lumen" (descending part)
- Periampullary ulcer in an adult suggests diagnosis of annular pancreas or Zollinger-Ellison syndrome (ZES)

Image Interpretation Pearls
- ERCP: Small pancreatic duct encircling second part of duodenum with narrow lumen
- Barium UGI: Concentric narrowing of second part of duodenum with proximal dilatation & reverse peristalsis

SELECTED REFERENCES

1. Yamaguchi Y et al: Annular pancreas complicated by carcinoma of the bile duct: diagnosis by MR cholangiopancreatography and endoscopic ultrasonography. Abdom Imaging. 28(3):381-3, 2003
2. Harthun NL et al: Duodenal obstruction caused by intraluminal duodenal diverticulum and annular pancreas in an adult. Gastrointest Endosc. 55(7):940-3, 2002
3. Shan YS et al: Annular pancreas with obstructive jaundice: beware of underlying neoplasm. Pancreas. 25(3):314-6, 2002
4. Benya EC: Pancreas and biliary system: imaging of developmental anomalies and diseases unique to children. Radiol Clin North Am. 40(6):1355-62, 2002
5. McCollum MO et al: Annular pancreas and duodenal stenosis. J Pediatr Surg. 37(12):1776-7, 2002
6. Kamisawa T et al: A new embryologic hypothesis of annular pancreas. Hepatogastroenterology. 48(37):277-8, 2001
7. Jayaraman MV et al: CT of the duodenum: an overlooked segment gets its due. Radiographics. 21 Spec No:S147-60, 2001
8. Fulcher AS et al: MR pancreatography: a useful tool for evaluating pancreatic disorders. Radiographics. 19(1):5-24; discussion 41-4; quiz 148-9, 1999
9. Jadvar H et al: Annular pancreas in adults: imaging features in seven patients. Abdom Imaging. 24(2):174-7, 1999
10. Berrocal T et al: Congenital anomalies of the upper gastrointestinal tract. Radiographics. 19(4):855-72, 1999
11. Weiss H et al: Ultrasonography of fetal annular pancreas. Obstet Gynecol. 94(5 Pt 2):852, 1999
12. Lecesne R et al: MR cholangiopancreatography of annular pancreas. J Comput Assist Tomogr. 22(1):85-6, 1998
13. Kallen B et al: Major congenital malformations in Down syndrome. Am J Med Genet. 65(2):160-6, 1996
14. Rizzo RJ et al: Congenital abnormalities of the pancreas and biliary tree in adults. Radiographics. 15(1):49-68; quiz 147-8, 1995
15. Reinhart RD et al: MR imaging of annular pancreas. Abdom Imaging. 19(4):301-3, 1994

IMAGE GALLERY

Typical

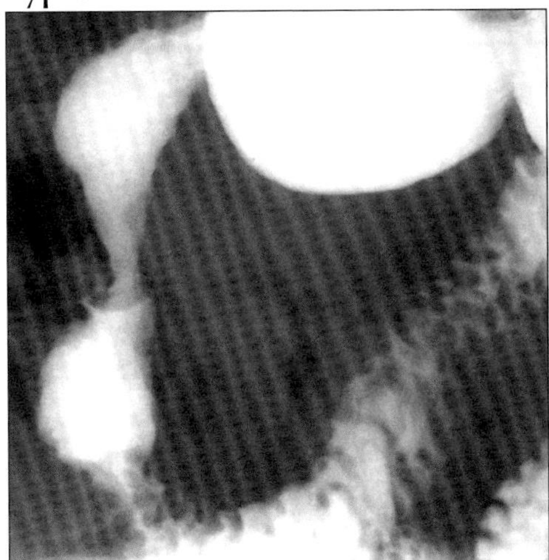

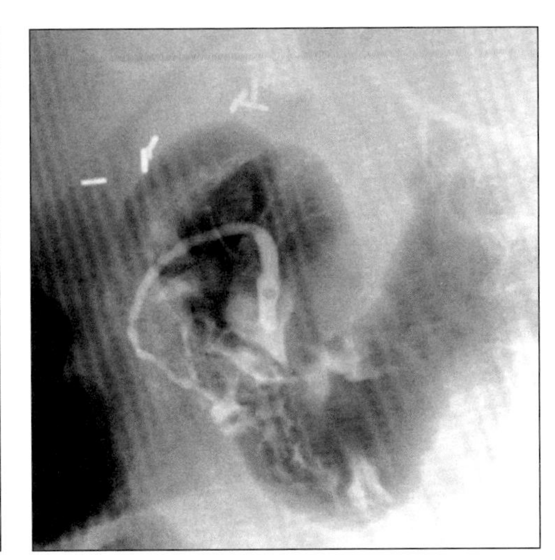

(Left) *Upper GI series shows circumferential narrowing of the lumen of the duodenum by annular pancreas.* ***(Right)*** *ERCP shows small pancreatic head duct originating on right anterior surface of duodenum, encircling the duodenum and emptying into the main pancreatic duct near the ampulla.*

Typical

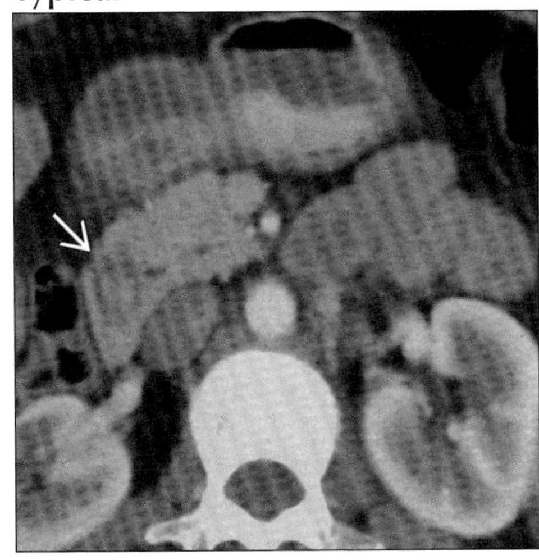

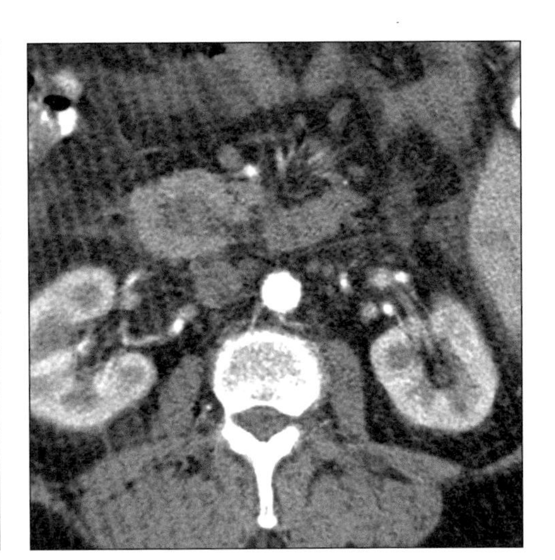

(Left) *Axial CECT shows pancreatic tissue completely encircling the second part of the duodenum (arrow).* ***(Right)*** *Axial CECT shows pancreatic head tissue encircling duodenum. Infiltrated mesentery due to pancreatitis related to annular pancreas.*

Typical

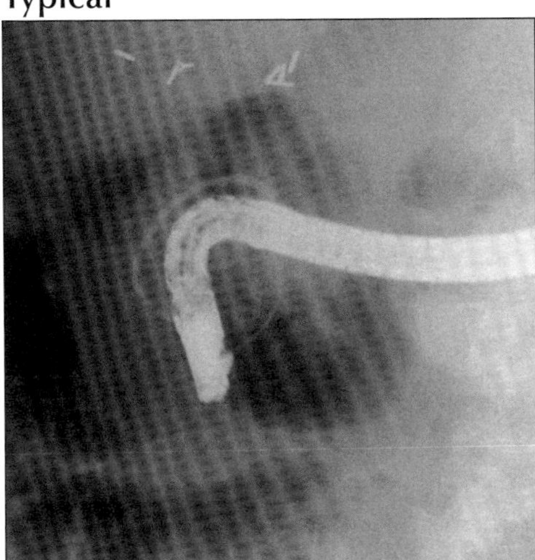

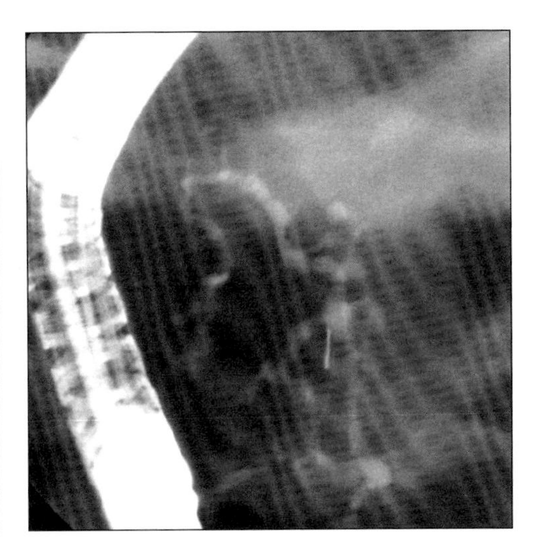

(Left) *ERCP shows pancreatic duct branch encircling endoscope within duodenum.* ***(Right)*** *ERCP shows a portion of annular pancreatic duct that has a beaded appearance due to chronic pancreatitis.*

PANCREAS DIVISUM

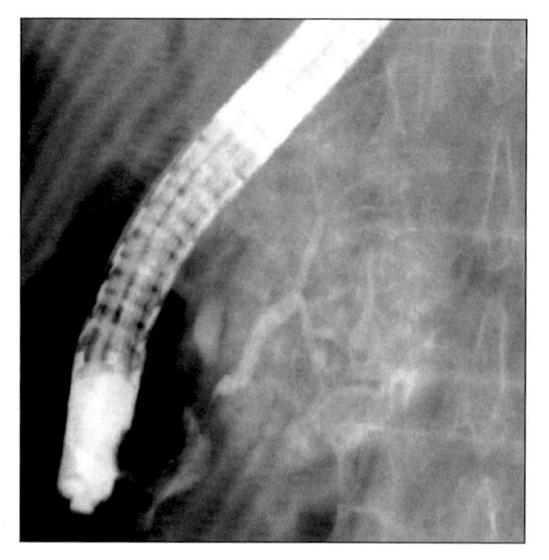

ERCP with cannulation of major papilla shows opacification of the ventral pancreatic duct. The duct is small and short with normal tapering. Note adjacent acinarization of pancreatic head.

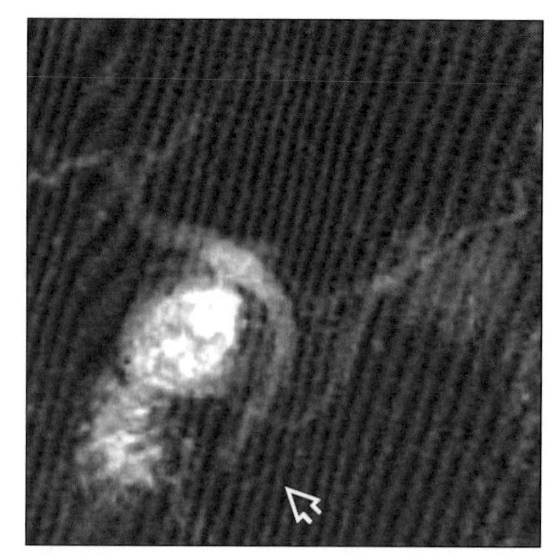

MRCP in a patient with recurrent episodes of pancreatitis shows faintly the ventral duct (open arrow) with a pancreas divisum configuration of the dorsal pancreatic duct.

TERMINOLOGY

Definitions

- Failure of fusion of ventral & dorsal pancreatic buds

IMAGING FINDINGS

General Features

- Best diagnostic clue: Normal branching of short ventral duct (head), not communicating with long dorsal pancreatic duct (body & tail) on ERCP
- Morphology
  - Most common congenital anatomic variant of pancreas
  - Most common variant of pancreatic ductal fusion & drainage anomalies
  - Seen in 12-26% of patients with idiopathic recurrent pancreatitis
  - Drainage consequences of anatomical variant
    - Head & uncinate process of pancreas: Drained by ventral pancreatic duct of Wirsung via major papilla
    - Body & tail of pancreas: Drained by dorsal pancreatic duct of Santorini via minor papilla
  - Clinically: Most cases asymptomatic or may contribute to pancreatitis

Radiographic Findings

- ERCP
  - Cannulation of major papilla
    - Opacification of short, tapered ventral pancreatic (Wirsung) duct
  - Cannulation of minor or accessory papilla (technically difficult)
    - Opacification of long ± dilated dorsal pancreatic (Santorini) duct
  - No communication between dorsal (long) & ventral (short) pancreatic ducts

CT Findings

- Abnormal contour of pancreatic head & neck
- Large pancreatic head
- Two distinct pancreatic moieties separated by a fat cleft
- Unfused ductal system on thin collimation scans
- Changes of pancreatitis (acute or chronic)
  - Focal or diffuse enlargement of gland
  - Fluid collections or pseudocyst
  - Infiltration or obliteration of peripancreatic fat
  - Intra-/peripancreatic cysts or atrophy of gland

MR Findings

- T2* GRE: Depicts pancreatic ductal anomaly
- MRCP

DDx: Obstruction of Pancreatic Duct

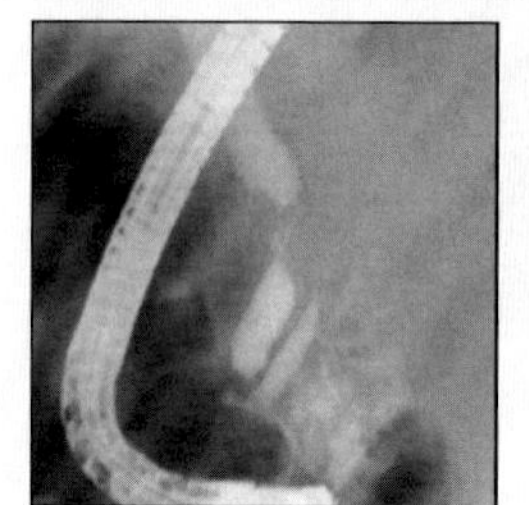

Pancreatic Carcinoma

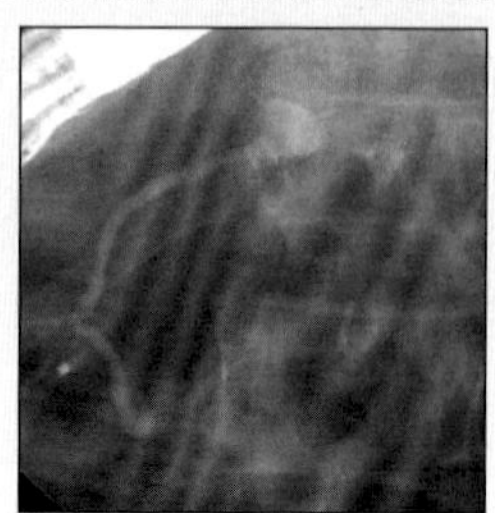

Chronic Pancreatitis

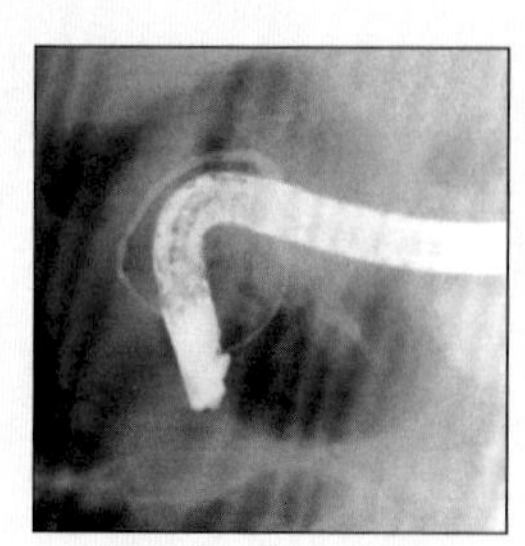

Annular Pancreas

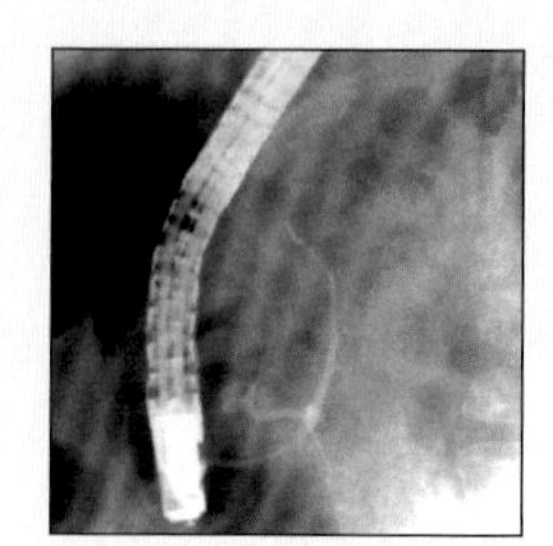

Dorsal Agenesis

PANCREAS DIVISUM

Key Facts

Terminology

- Failure of fusion of ventral & dorsal pancreatic buds

Imaging Findings

- Best diagnostic clue: Normal branching of short ventral duct (head), not communicating with long dorsal pancreatic duct (body & tail) on ERCP
- Two distinct pancreatic moieties separated by a fat cleft
- Unfused ductal system on thin collimation scans
- Changes of pancreatitis (acute or chronic)
- Best imaging tool: MRCP or ERCP

Top Differential Diagnoses

- Pancreatic carcinoma
- Chronic pancreatitis
- Annular pancreas

Pathology

- On day 37 ventral pancreas rotates posterior to duodenum & comes in contact with dorsal pancreas, failure of fusion results in pancreas divisum
- Ventral pancreas: Head & uncinate process (short duct of Wirsung)
- Dorsal pancreas: Body & tail (long & narrow duct of Santorini)

Clinical Issues

- Most cases are asymptomatic

Diagnostic Checklist

- Rule out other causes of "pancreatic duct obstruction"
- MRCP & ERCP: Demonstrate short ventral & long dorsal pancreatic ducts with lack of communication between two ducts

  - Shows course & drainage pattern of dorsal & ventral pancreatic ducts
    - Dorsal duct: Long & narrow entering minor papilla
    - Ventral duct: Short; entering major papilla
    - No communication between dorsal & ventral ducts

Ultrasonographic Findings

- Secretin Test
  - Performed to identify patients who will benefit from surgical sphincterotomy
  - Secretin mechanism increases HCO3 secretion which overloads a functionally inadequate papilla
  - Secretin-induced ductal dilatation occurs in 72% of symptomatic patients due to stenotic minor or accessory papilla in pancreas divisum anomaly
  - Normal result: No change in size of duct before & 20 minutes after secretin administration
  - Grade I response
    - 1 mm dilated duct in only one segment of pancreas
    - Result is equivocal & patient probably will not benefit from surgery
  - Grade II response
    - More than 2 mm dilated duct in 2 segments of pancreas
  - Grade III response
    - More than 2 mm dilated duct in 3 segments of pancreas
  - Grade II & III: Patients benefit from surgery

Imaging Recommendations

- Best imaging tool: MRCP or ERCP

DIFFERENTIAL DIAGNOSIS

Pancreatic carcinoma

- Heterogeneous, poorly-enhancing mass
- Location: Head of pancreas (more common)
- Abrupt obstruction & dilatation of main pancreatic duct
- Obliteration of retropancreatic fat
- Extensive local invasion & regional metastases
- Contiguous organ invasion seen:
  - Duodenum, stomach & mesenteric root
- ERCP & MRCP
  - Main pancreatic duct: Irregular, nodular, rat-tailed eccentric obstruction & dilatation

Chronic pancreatitis

- Focal or diffuse atrophy of gland showing heterogeneous enhancement
- Obstruction & dilatation of pancreatic +/- bile ducts
- Intraductal calculi & areas of calcification
- Focal masses (areas of fibrosis & fat necrosis)
- Intra or peripancreatic cysts
- Thickening of peripancreatic fascia
- ERCP & MRCP: Obstruction & dilatation of pancreatic duct/radicles

Annular pancreas

- Congenital anomaly of pancreas
- Ring of pancreatic tissue that encircles second part of duodenum
- Barium study
  - Concentric narrowing of second part with dilated proximal duodenum
  - Reverse peristalsis & duodenal ulcer may be seen
- CT
  - Gastric & duodenal dilatation with circumferential thickening of wall
  - Obstructing band of pancreatic tissue & dilated main pancreatic/bile ducts may be seen
- ERCP & MRCP
  - Depicts course & drainage pattern of pancreatic duct
  - Pancreatic duct encircling endoscope or second part of duodenum with narrow lumen
  - Obstruction & dilatation of pancreatic/bile ducts

Agenesis of dorsal pancreas

- ERCP appearance may be identical to pancreas divisum
- Diagnostic key: Absence of pancreatic body-tail on CT or MR

PANCREAS DIVISUM

PATHOLOGY

General Features

- General path comments
  - Embryology
    - On day 37 ventral pancreas rotates posterior to duodenum & comes in contact with dorsal pancreas, failure of fusion results in pancreas divisum
  - Anatomy of pancreas divisum
    - Ventral pancreas: Head & uncinate process (short duct of Wirsung)
    - Dorsal pancreas: Body & tail (long & narrow duct of Santorini)
  - Pathogenesis: Embryological anomaly
  - Dorsal pancreas (body & tail)
    - Long & narrow duct of Santorini & minor papilla
    - Poor drainage of secretions from body & tail
    - Increased stasis & ductal pressure: Pancreatitis
    - Head (ventral pancreas) is spared
  - Ventral pancreas (head & uncinate process)
    - Alcoholics: Pancreatitis due to reflux of bile via short duct of Wirsung
    - Body & tail (dorsal pancreas) is spared due to pancreas divisum
- Etiology: Congenital anatomical variant: Failure of fusion of dorsal & ventral pancreatic buds
- Epidemiology
  - Prevalence
    - Seen in 3-6% of general population
    - 4-11% of autopsy series & 3-4% of ERCP series
- May be associated with other congenital anomalies

Gross Pathologic & Surgical Features

- Dorsal & ventral pancreatic tissues
- Two separate duct systems

Microscopic Features

- Normal pancreatic tissue & ductal epithelium
- With or without inflammatory cells

CLINICAL ISSUES

Presentation

- Most common signs/symptoms
  - Most cases are asymptomatic
  - Young patients: Epigastric pain, nausea & vomiting due to pancreatitis
  - May be seen in multiple family members
- Clinical profile: Young patient with history of episodes of idiopathic recurrent pancreatitis

Demographics

- Age
  - Varies widely at diagnosis
  - Common between 30 & 50 yrs
- Gender: Males > females

Natural History & Prognosis

- Complications
  - Recurrent pancreatitis (mostly in children)
  - Pancreaticolithiasis, serous cystadenoma (speculative)
- Prognosis
  - Good: After medical & surgical correction in symptomatic patients

Treatment

- Asymptomatic
  - No specific treatment
- Symptomatic
  - Medical
    - Conservative measures/pancreatic enzyme therapy
  - Surgical or endoscopic correction
    - Sphincteroplasty of minor or accessory papilla

DIAGNOSTIC CHECKLIST

Consider

- Rule out other causes of "pancreatic duct obstruction"

Image Interpretation Pearls

- MRCP & ERCP: Demonstrate short ventral & long dorsal pancreatic ducts with lack of communication between two ducts

SELECTED REFERENCES

1. Khalid A et al: Secretin-stimulated magnetic resonance pancreaticogram to assess pancreatic duct outflow obstruction in evaluation of idiopathic acute recurrent pancreatitis: a pilot study. Dig Dis Sci. 48(8):1475-81, 2003
2. Mishra D et al: Pancreas divisum: an uncommon cause of acute pancreatitis. Indian J Pediatr. 70(7):593-5, 2003
3. Benya EC: Pancreas and biliary system: imaging of developmental anomalies and diseases unique to children. Radiol Clin North Am. 40(6):1355-62, 2002
4. Manfredi R et al: Idiopathic chronic pancreatitis in children: MR cholangiopancreatography after secretin administration. Radiology. 224(3):675-82, 2002
5. Masatsugu T et al: Serous cystadenoma of the pancreas associated with pancreas divisum. J Gastroenterol. 37(8):669-73, 2002
6. Vitellas KM et al: MR cholangiopancreatography of bile and pancreatic duct abnormalities with emphasis on the single-shot fast spin-echo technique. Radiographics. 20(4):939-57; quiz 1107-8, 1112, 2000
7. Morgan DE et al: Pancreas divisum: implications for diagnostic and therapeutic pancreatography. AJR Am J Roentgenol. 173(1):193-8, 1999
8. Fulcher AS et al: MR cholangiography: technical advances and clinical applications. Radiographics. 19(1):25-41; discussion 41-4, 1999
9. Bret PM et al: Pancreas divisum: evaluation with MR cholangiopancreatography. Radiology. 199(1):99-103, 1996
10. Rizzo RJ et al: Congenital abnormalities of the pancreas and biliary tree in adults. Radiographics. 15(1):49-68; quiz 147-8, 1995

IMAGE GALLERY

Typical

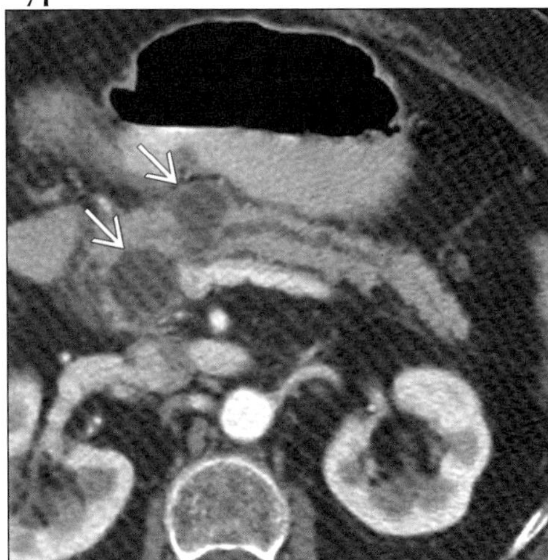

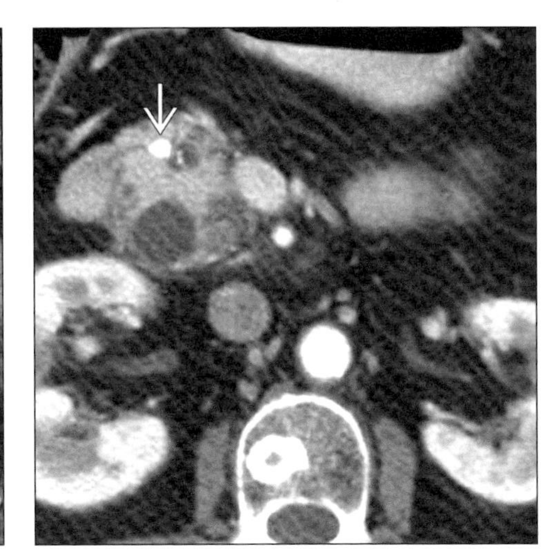

(Left) *Axial CECT shows dilated pancreatic duct and small pseudocysts (arrows) in patient with pancreas divisum complicated with recurrent episodes of pancreatitis.* ***(Right)*** *Axial CECT shows calculus (arrow) within duct of Santorini, which is of normal caliber "downstream" as it enters the minor papilla.*

Typical

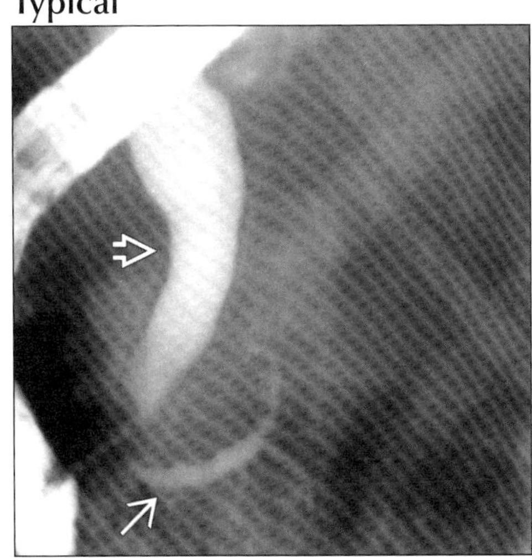

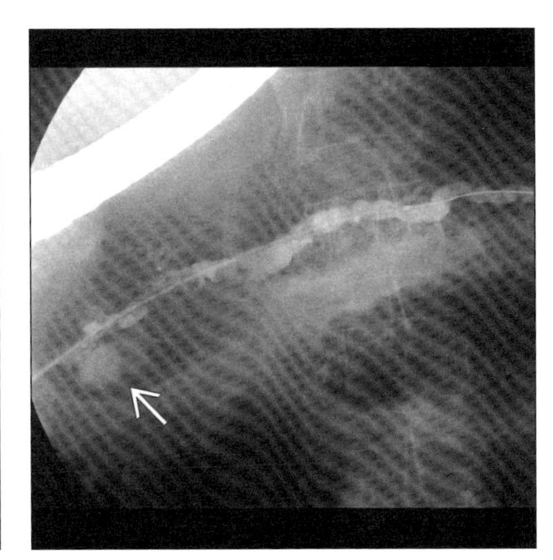

(Left) *ERCP in a patient with chronic pancreatis and pancreas divisum shows filling of ventral pancreatic duct (arrow) and a dilated common bile duct (open arrow), following cannulation of major papilla.* ***(Right)*** *ERCP shows dilatation and irregularity of dorsal pancreatic duct and small pseudocysts (arrow), following cannulation of minor papilla. Pancreas divisum and chronic pancreatitis.*

Typical

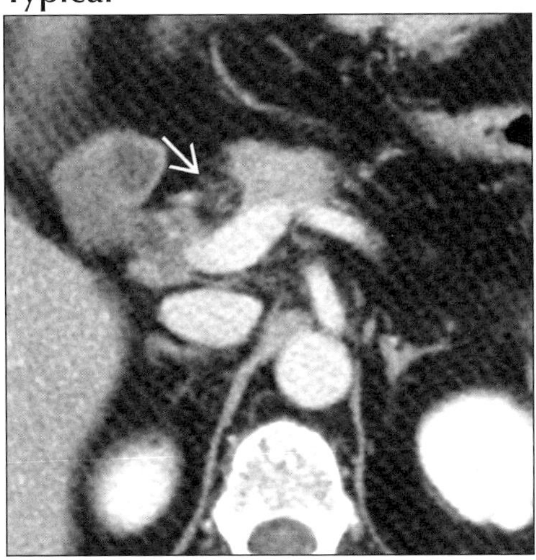

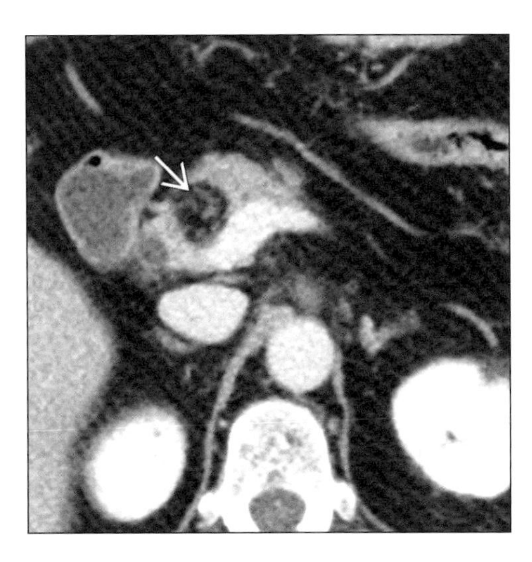

(Left) *Axial CECT shows a fatty cleft (arrow) separating the ventral and dorsal pancreatic segments.* ***(Right)*** *Axial CECT shows a fatty cleft (arrow) separating the ventral (head-uncinate process) and dorsal (body-tail) segments.*

ECTOPIC PANCREATIC TISSUE

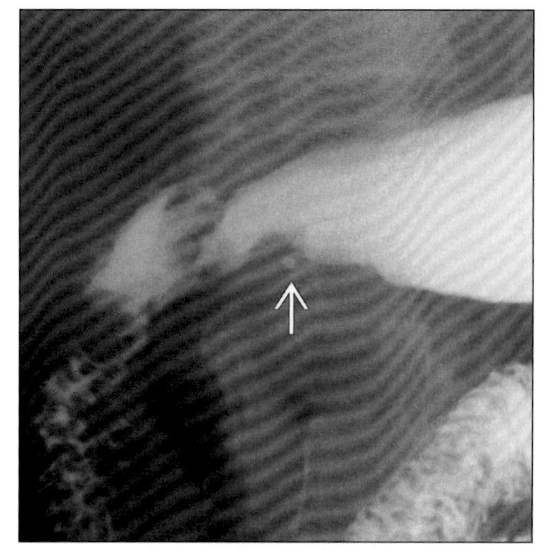

Upper GI series shows small intramural antral mass with a collection of barium (arrow) marking the rudimentary duct.

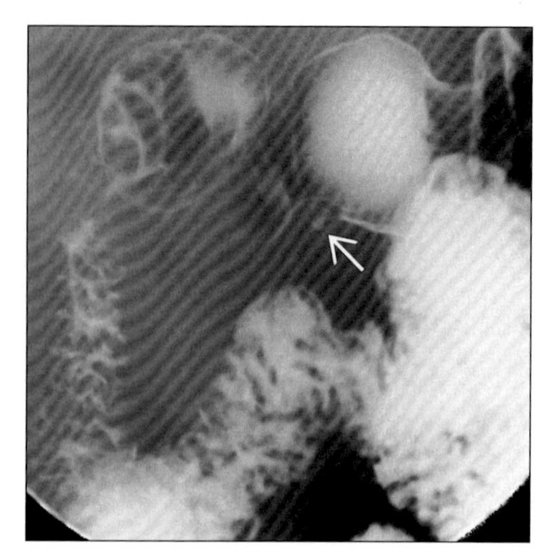

Upper GI series shows a small antral mass with intact mucosa. Central "dot" of barium (arrow) fills rudimentary duct.

TERMINOLOGY

Abbreviations and Synonyms

- Eptopic pancreatic tissue (EPT)
- Pancreatic rests; heterotopic, aberrant, accessory pancreas

Definitions

- Pancreatic tissue located outside normal confines of pancreas & lacking any anatomic or vascular connection with it

IMAGING FINDINGS

General Features

- Best diagnostic clue
  - Small intramural gastric mass with central umbilication is diagnostic (45%)
    - Central umbilication: Orifice of rudimentary duct into which EPT opens & empties into gut lumen near summit of nodule
- Location
  - Those identifiable at imaging are nearly always in stomach or duodenum
  - Remaining sites: Ileum, Meckel diverticulum, liver, biliary tract, spleen, omentum, mesentery, lung, mediastinum, fallopian tube, esophagus, colon
  - Mainly lies submucosally (73%); can be located in muscular layer (17%), or in subserosa (10%)
- Size
  - Nodule: 0.5-2 cm; may be up to 5 cm in diameter
  - Pit: May be 5 mm in diameter & 10 mm in length
- Morphology
  - In submucosal layer, appears as well-defined flat or nodular projection into gut lumen, with intact overlying mucosa
  - In muscularis or subserosal layers, produces smooth bulge or area of wall thickening

Radiographic Findings

- Ability of radiographic contrast studies to visualize EPT depends on size & location of deposit
- Barium study may show narrowed pyloric channel with or without a polypoid or sessile mass
- Characteristic appearance: Well-defined, smoothly marginated, round or oval broad-based mass
  - In stomach: Typically 1 to 2 cm in diameter, along greater curvature or posterior aspect of antrum, within 6 cm of pylorus
  - Nodule may be larger, have a narrow base & appear polypoid, or located in more proximal antrum
- Streaking or central depression; with contrast filling pit in center of mound; specific feature
  - This may be mistaken for an ulcerative lesion
- Reflux of contrast into rudimentary duct-like structure may extend below central pit

DDx: Ulcerated Gastric Polypoid Lesion

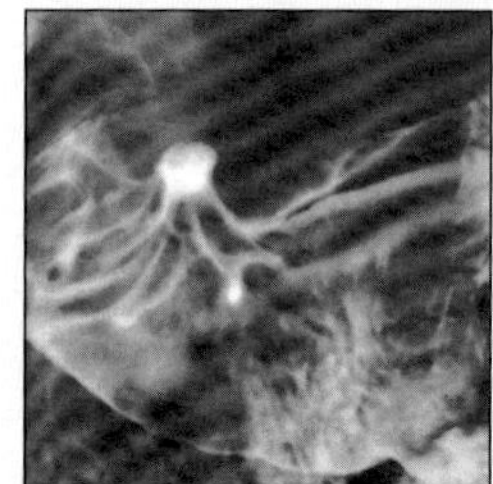

Gastric Ulcer

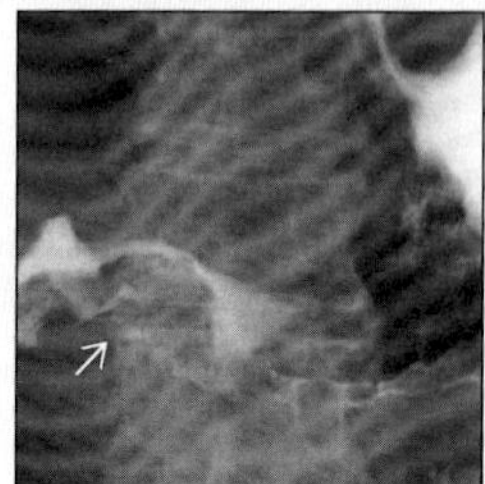

Gastric Carcinoma

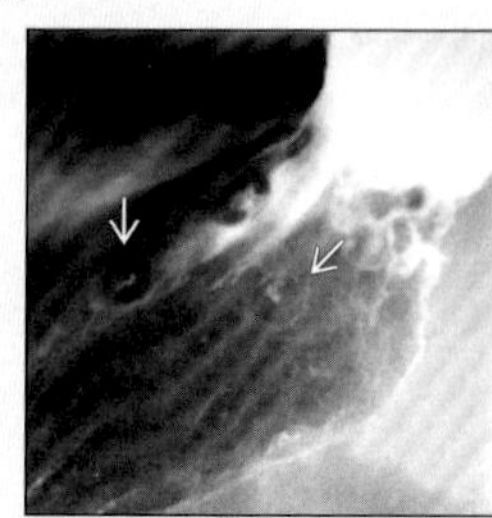

Gastric Metastases

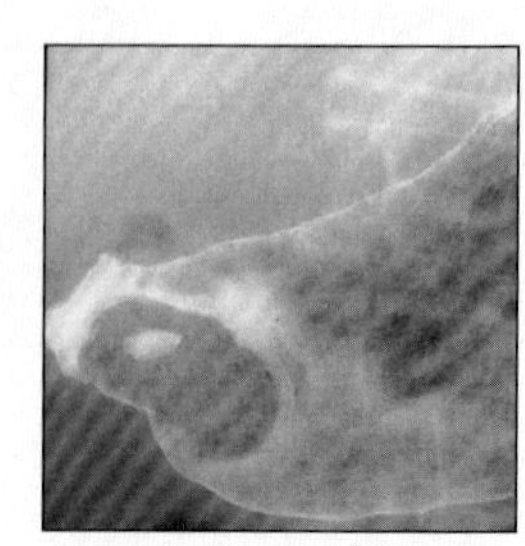

Gastric Stromal Tumor

ECTOPIC PANCREATIC TISSUE

Key Facts

Terminology
- Pancreatic tissue located outside normal confines of pancreas & lacking any anatomic or vascular connection with it

Imaging Findings
- Small intramural gastric mass with central umbilication is diagnostic (45%)
- Characteristic appearance: Well-defined, smoothly marginated, round or oval broad-based mass

Top Differential Diagnoses
- Gastric ulcer
- Gastric carcinoma
- Gastric metastases
- Gastric stromal tumor

CT Findings
- Usually too small to be detected
- Rarely intramural cystic collections in stomach & duodenum

Other Modality Findings
- Endoscopy: More capable of identifying EPT when nodule is small & located in duodenum
  - Often nonspecific due to submucosal location
  - Central umbilication may be visualized, & if injected, rudimentary duct system may be seen

DIFFERENTIAL DIAGNOSIS

Gastric ulcer
- Round ulcer, smooth mound of edema, radiating folds to ulcer edge, Hampton line, ulcer collar

Gastric carcinoma
- Polypoid or circumferential mass, ± ulceration, focal wall thickening with mucosal irregularity, focal infiltration of wall

Gastric metastases
- "Bull's eye" sign: Ulceration in center of lesion
- Melanoma, Kaposi sarcoma

Gastric stromal tumor
- Large, lobulated, submucosal mass with ulceration; requires biopsy, histologic diagnosis

PATHOLOGY

General Features
- Epidemiology: Incidence of autopsy series: 2-14%
- Seen in organs, like pancreas, derived from endoderm as a result of heteroplastic differentiation of parts of embryonic endoderm that do not normally produce pancreatic tissue

Microscopic Features
- May contain all or only some elements of normal pancreas; including acini, ducts & islet cells

CLINICAL ISSUES

Presentation
- Asymptomatic: Incidental finding
- Symptomatic: May simulate duodenal ulcer, gallbladder disease, or appendicitis
- In stomach, symptoms of pyloric obstruction
- Periampullary, rare biliary obstruction

Natural History & Prognosis
- Complicated by bleeding or mucosal ulceration
- Acute pancreatitis +/- hemorrhage and necrosis
- Chronic pancreatitis with pseudocyst formation
- Malignancy: Ductal adenocarcinoma

Treatment
- Surgical intervention for obstruction or hemorrhage
- Endoscopic resection if lesion confined to submucosa
- When asymptomatic may be treated expectantly

DIAGNOSTIC CHECKLIST

Consider
- If central umbilication is absent, lesion may not be differentiated from other submucosal tumors

SELECTED REFERENCES

1. Jeong HY et al: Adenocarcinoma arising from an ectopic pancreas in the stomach. Endoscopy. 34(12):1014-7, 2002
2. Hayes-Jordan A et al: Ectopic pancreas as the cause of gastric outlet obstruction in a newborn. Pediatr Radiol. 28(11):868-70, 1998
3. Kaneda M et al: Ectopic pancreas in the stomach presenting as an inflammatory abdominal mass. Am J Gastroenterol. 84(6):663-6, 1989

IMAGE GALLERY

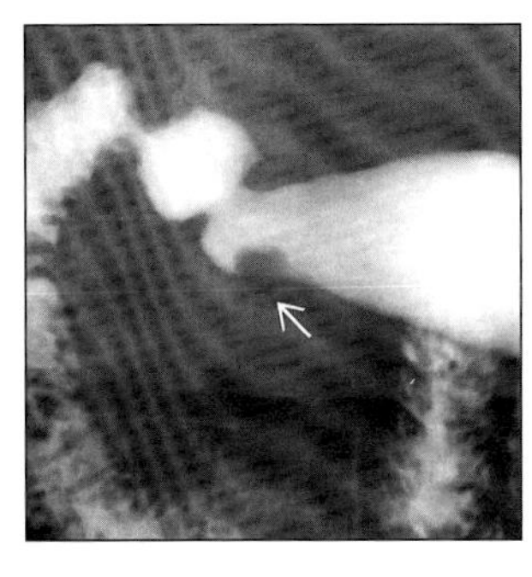

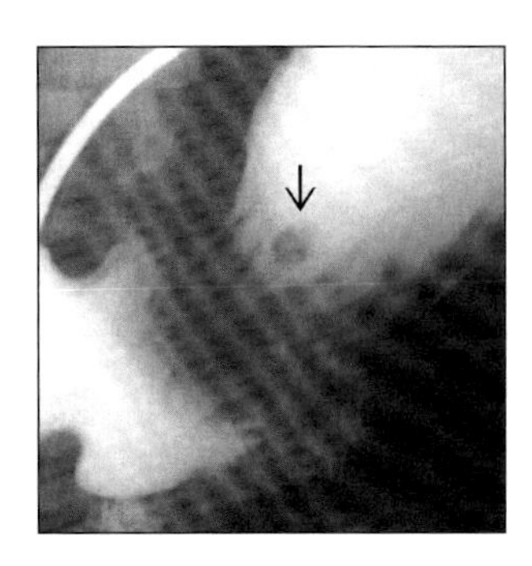

*(**Left**) Upper GI series show small intramural mass (arrow) along greater curvature of antrum with intact mucosa. (**Right**) Upper GI series show small mass with central umbilication (arrow).*

CYSTIC FIBROSIS, PANCREAS

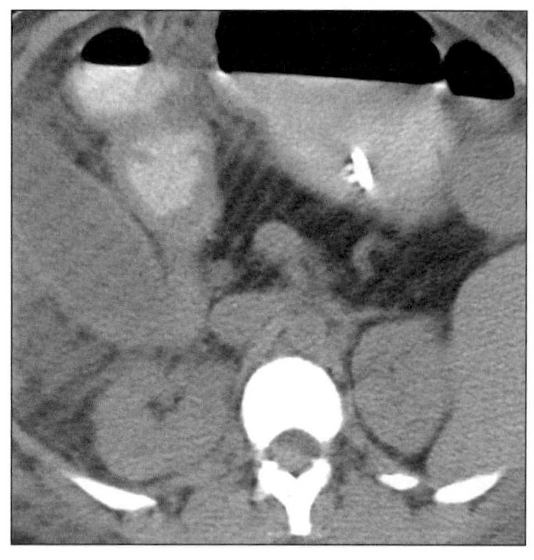

Axial NECT shows lipomatous replacement of entire pancreas.

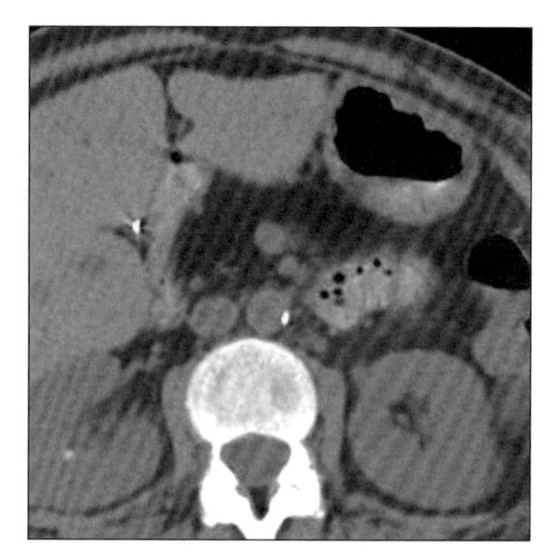

Axial NECT shows lipomatous replacement and pseudohypertrophy of entire pancreas.

TERMINOLOGY

Abbreviations and Synonyms

- Cystic fibrosis (CF)
- Mucoviscidosis/fibrocystic disease

Definitions

- Cystic fibrosis is a recessively inherited disorder of epithelial chloride transport & characterized by abnormality of exocrine gland function

IMAGING FINDINGS

General Features

- Best diagnostic clue: Lipomatous pseudohypertrophy; fatty replacement of pancreas (-90 to -120 HU) on CT
- Location: Multisystem disease; affecting primarily pancreas, lungs, gut, liver & exocrine glands

CT Findings

- Early: Inhomogeneous attenuation
- Later: Low attenuation with complete fatty infiltration & replacement
- Microcysts may develop (multiple microscopic); some may become small macroscopic cysts demonstrable with CT
- ± Scattered foci of calcifications
- Dilatation of pancreatic duct (uncommon)
- Pancreatic dysfunction in advanced cases; on CT demonstrated as fibrosis & marked atrophy
- Pancreatic cysts: Late manifestation; diffuse replacement of pancreas by multiple fluid-filled cysts
  - Related to inspissation of tenacious secretions, leading to ductal ectasia

MR Findings

- Fat deposition: Hyperintense on T1WI
- Fibrosis: Hypointense on T1 & T2WI
- Pancreatic cysts are relatively common finding; these true cysts are typically quite small but are well demonstrated at MR imaging & MRCP
- Pancreatic duct abnormalities also occasionally seen

Ultrasonographic Findings

- Diffuse increased echogenicity (fatty replacement & fibrosis)
- Hypoechoic enlargement of pancreas noted in some patients who exhibit clinical signs of pancreatitis
- Chronic calcific pancreatitis
- Macroscopic multiple cysts (dilated acini & ducts)
- Pancreas may be barely identifiable or not demonstrated at all (advanced disease; atrophy)

DDx: Fatty Replacement of Pancreas

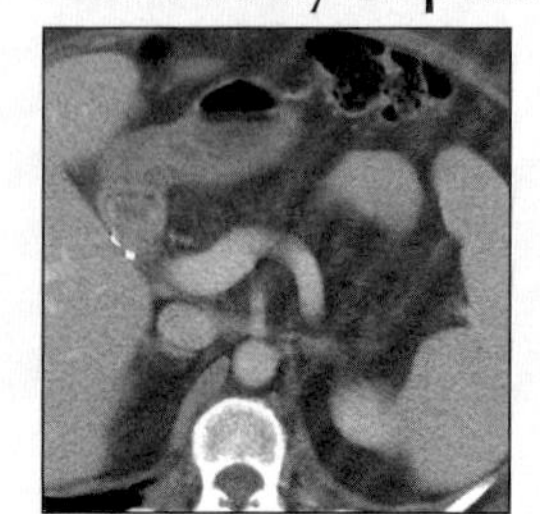

Fatty Lobulation

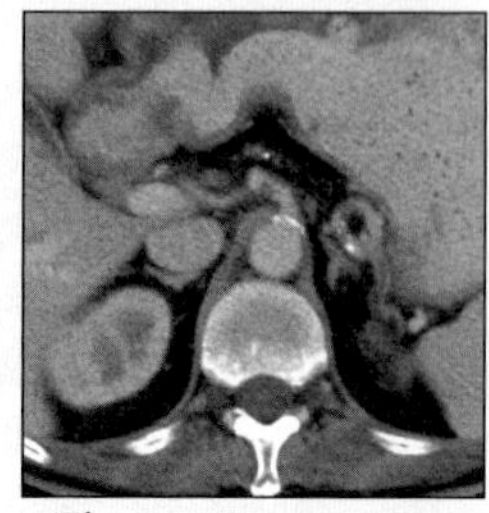

Chronic Pancreatitis

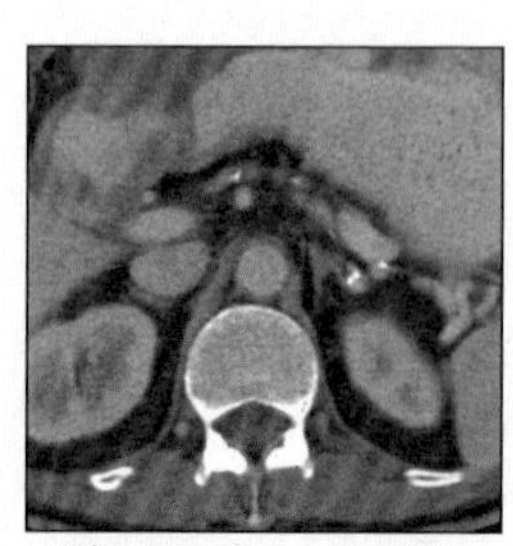

Chronic Pancreatitis

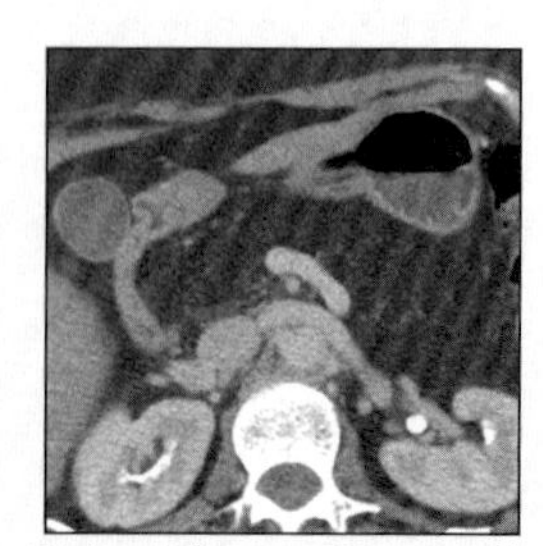

Schwachman-Diamond

CYSTIC FIBROSIS, PANCREAS

Key Facts

Terminology

- Cystic fibrosis is a recessively inherited disorder of epithelial chloride transport & characterized by abnormality of exocrine gland function

Imaging Findings

- Best diagnostic clue: Lipomatous pseudohypertrophy; fatty replacement of pancreas (-90 to -120 HU) on CT
- Pancreatic dysfunction in advanced cases; on CT demonstrated as fibrosis & marked atrophy
- Pancreatic cysts: Late manifestation; diffuse replacement of pancreas by multiple fluid-filled cysts

Pathology

- Mutations in transmembrane conductance regulator (CFTR) gene

DIFFERENTIAL DIAGNOSIS

Normal fatty lobulation

- Profuse fat with glandular atrophy seen occasionally in elderly, obese and diabetic

Chronic pancreatitis

- Focal/diffuse enlargement or atrophy; dilated pancreatic duct; ductal calculi
- Intra/peripancreatic cysts; thickening of peripancreatic fascia; heterogeneous enhancement
- Small, hypodense, focal masses (fibrosis & fat necrosis)

Shwachman-Diamond syndrome

- Pancreatic lipomatosis: Pancreas is completely replaced by fat; has low attenuation value on CT scans

PATHOLOGY

General Features

- General path comments
  - Cystic fibrosis (CF) is major cause of pancreatic exocrine failure in childhood
  - Primary ductal cell chloride channel abnormality results in dehydrated protein-rich secretions obstructing proximal ducts, leading to acinar cell destruction, fibrosis & exocrine insufficiency
- Genetics
  - Autosomal recessive
  - CF gene located on long arm of chromosome 7
  - Mutations in transmembrane conductance regulator (CFTR) gene
- Etiology
  - Mutations of CFTR gene leads to pancreatic pathology
  - Disruption of chloride ion, bicarbonate & water transport in duct cells
- Epidemiology
  - Prevalence: 1 per 2,000
  - Pancreatic abnormalities in 85-90% of CF patients

CLINICAL ISSUES

Presentation

- Steatorrhea, malabsorption, fat intolerance
- Diabetes mellitus
- Pancreatitis: Susceptivity to infection by Staphylococcus aureus & Pseudomonas aeruginosa
- Chronic pulmonary disease, hepatic fibrosis, intestinal obstruction, infertility in males
- Lab: Positive sweat test
- Genotyping has proved useful in identifying gene carriers; antenatal diagnosis & treatment

Demographics

- Age: Children; patients reaching adulthood represent a rapidly growing percentage of CF population
- Ethnicity: More common in Caucasians

Natural History & Prognosis

- Pancreatic involvement varies depending on degree of ductal obstruction by mucus; leading to exocrine gland atrophy, progressive fibrosis & cyst formation
- Pancreatitis in less than 1% of CF patients
- Predisposes to pancreatic cancer

Treatment

- Aggressive nutritional & pancreatic enzyme therapy

SELECTED REFERENCES

1. Taylor CJ et al: The pancreas in cystic fibrosis. Paediatr Respir Rev. 3(1):77-81, 2002
2. King LJ et al: Hepatobiliary and pancreatic manifestations of cystic fibrosis: MR imaging appearances. Radiographics. 20(3):767-77, 2000
3. Soyer P et al: Cystic fibrosis in adolescents and adults: fatty replacement of the pancreas--CT evaluation and functional correlation. Radiology. 210(3):611-5, 1999

IMAGE GALLERY

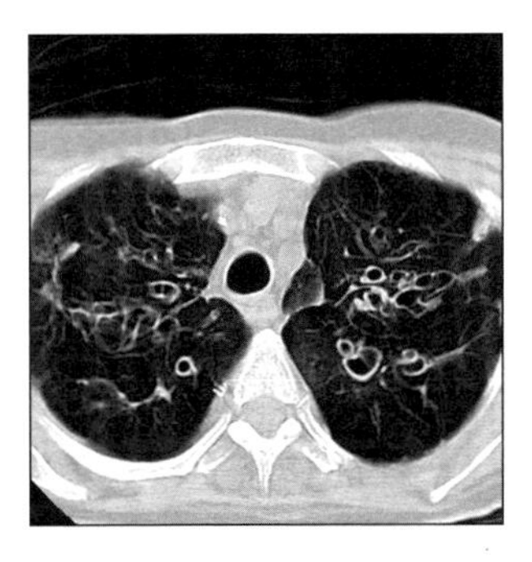

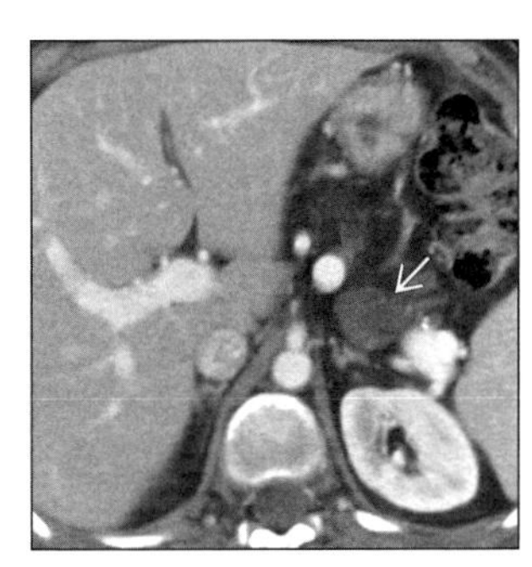

*(**Left**) Axial NECT shows hyperaeration of the upper lobes with cystic bronchiectasis in a patient with cystic fibrosis. (**Right**) Axial CECT shows lipomatous replacement of the pancreas and one of several small cysts (arrow).*

ACUTE PANCREATITIS

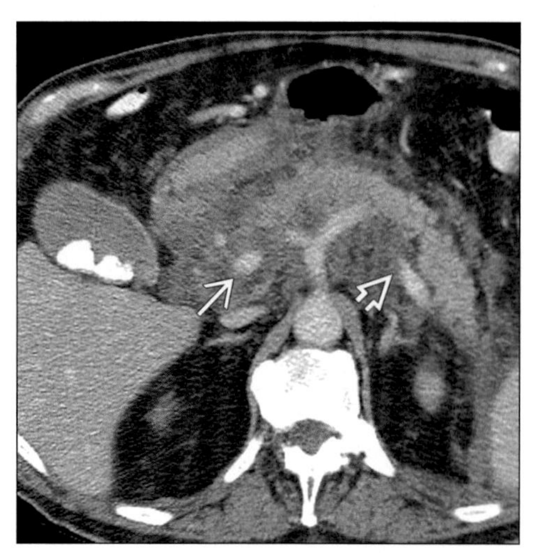

Axial CECT. Extensive infiltration of the peripancreatic fat planes. The celiac axis and portal vein (arrow) are surrounded, splenic vein (open arrow) is occluded, gastric wall is thickened

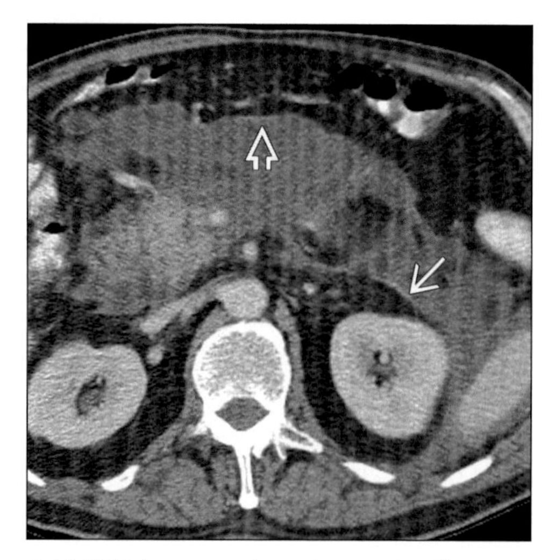

Axial CECT shows extensive peripancreatic infiltration to the perirenal fascia (arrow) and ventrally into the mesentery (open arrow).

TERMINOLOGY

Definitions

- Acute inflammatory process of pancreas with variable involvement of other regional tissues or remote organ systems

IMAGING FINDINGS

General Features

- Best diagnostic clue: Enlarged pancreas, fluid collections & obliteration of fat planes
- Location: Pancreatic & peripancreatic
- Size: Pancreas increased in size (focal or diffuse)
- Morphology: Inflammatory disease of pancreas producing temporary changes with restoration of normal anatomy and function following resolution

Radiographic Findings

- Radiography
  - Duodenal ileus
  - Sentinel loop: Mildly dilated, gas-filled segment of small bowel with or without air-fluid levels
  - "Colon cutoff" sign
    - Markedly distended transverse colon with air
    - Absence of gas distal to splenic flexure caused by functional colonic spasm due to spread of pancreatic inflammation to proximal descending colon
- ERCP
  - Dilated or normal main pancreatic duct (MPD)
  - Communication of pseudocyst with MPD (acutely)
  - May show narrowed & tapered distal common bile duct (CBD) with prestenotic biliary dilatation

CT Findings

- Focal or diffuse enlargement of pancreas
- Heterogeneous enhancement; nonenhancing necrotic areas
- Rim-enhancement of acute fluid collections, abscesses and pseudocysts
- Infiltration of peripancreatic fat; gall stones
- Pseudoaneurysm: May simulate pseudo cyst; on CECT enhances like adjacent blood vessels
- Chest: Pleural effusions & basal atelectasis

MR Findings

- T1WI
  - T1WI gradient-echo image
    - Variable decreased signal intensity & enlarged gland
- T2WI
  - Fat-suppressed T2WI

DDx: Peripancreatic Infiltration

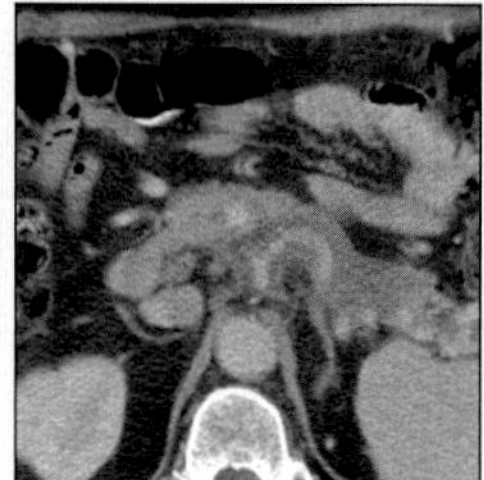

Pancreatic Carcinoma

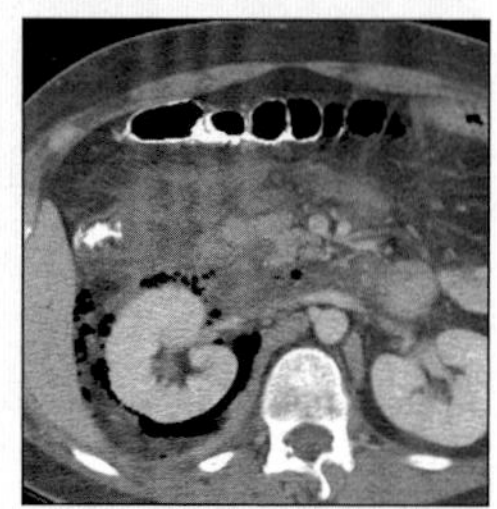

Perf. Ulcer (Duodenal)

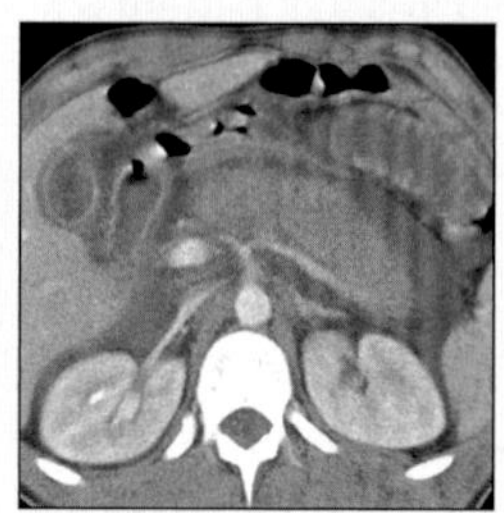

"Shock" Pancreas

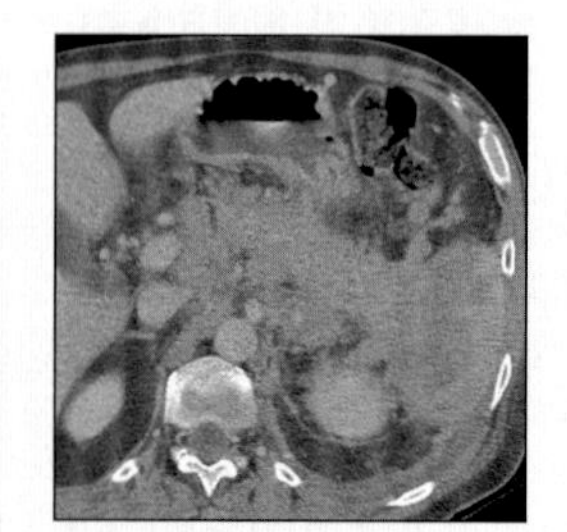

Lymphoma

ACUTE PANCREATITIS

Key Facts

Terminology
- Acute inflammatory process of pancreas with variable involvement of other regional tissues or remote organ systems

Imaging Findings
- Best diagnostic clue: Enlarged pancreas, fluid collections & obliteration of fat planes
- Communication of pseudocyst with MPD (acutely)
- Chest: Pleural effusions & basal atelectasis

Top Differential Diagnoses
- Infiltrating pancreatic carcinoma
- Perforated duodenal ulcer
- "Shock" pancreas
- Lymphoma & metastases

Pathology
- Alcohol/gallstones/metabolic/infection/trauma/drugs
- Pathogenesis: Due to reflux of pancreatic enzymes, bile, duodenal contents & increased ductal pressure

Clinical Issues
- Clinical profile: Patient with history of alcoholism, fever & severe mid-epigastric pain radiating to back
- Increased serum amylase & lipase
- Leukocytosis, hypocalcemia (poor prognostic sign)

Diagnostic Checklist
- Rule out other pathologies which can cause "peripancreatic infiltration"
- Bulky, irregularly enlarged pancreas with obliteration of peripancreatic fat planes, fluid collections, pseudocyst or abscess formation

    - Fluid collections, pseudocyst, necrotic areas: Hyperintense
    - Gallstones or intraductal calculi: Hypointense
- T1 C+
  - Heterogeneous enhancement pattern
    - Nonenhancing decreased signal areas (necrosis/fluid collection/pseudocyst)
    - Pancreatic pseudocyst contiguous with MPD
    - Vascular occlusions can be easily demonstrated
- MRCP
  - All fluid-containing structures: Hyperintense
  - Dilated or normal main pancreatic duct (MPD)
  - Pseudocyst contiguous with MPD

Ultrasonographic Findings
- Real Time: Enlarged hypoechoic gland/fluid collection/abscess/pseudocyst

Angiographic Findings
- Conventional
  - Performed when a pseudoaneurysm is suspected
  - Useful when pancreatitis due to a vascular cause
    - Vasculitis, polyarteritis nodosum, lupus
    - Postaortic aneurysm resection

Imaging Recommendations
- NE + CECT; MR (fat-suppressed images); MRCP; ERCP

DIFFERENTIAL DIAGNOSIS

Infiltrating pancreatic carcinoma
- Irregular, heterogeneous, poorly-enhancing mass
- Abrupt obstruction & dilatation of pancreatic duct
- Obliteration of retropancreatic fat
- No hemorrhage, calcium very rare
- Local tumor extension: Splenic hilum & porta hepatis
- Contiguous organ invasion
  - Duodenum, stomach & mesenteric root
- ERCP: Main pancreatic duct
  - Irregular, nodular, rat-tailed eccentric obstruction
  - Prestenotic dilatation
- Angiography: Hypovascular mass encasing vessels

Perforated duodenal ulcer
- Penetrating ulcers may infiltrate anterior pararenal space, simulating pancreatitis
- Less than 50% of cases have evidence of extraluminal gas or contrast medium collections
- Pancreatic head may be involved

"Shock" pancreas
- Infiltration of peripancreatic & mesenteric fat planes following hypotensive episode (e.g., blunt trauma)
- Pancreas itself looks normal or diffusely enlarged
- Usually presents with "shock bowel" appearance with mucosal enhancement and submucosal edema
- Quickly resolves following resuscitation

Lymphoma & metastases
- Nodular, bulky, enlarged pancreas due to infiltration
- Retroperitoneal adenopathy
- MPD & side branches
  - Show extrinsic mass effect & ductal draping
    - Smooth ductal splaying or some narrowing
  - Lack of communication with tumor
- Peripancreatic infiltration (obliteration of fat planes)
- Primary may be seen in case of metastatic infiltration

PATHOLOGY

General Features
- General path comments
  - Embryology-anatomy
    - Congenital anomalies may cause pancreatitis
    - Annular pancreas: Failure of migration of ventral bud to contact dorsal
    - Pancreas divisum: Ventral & dorsal pancreatic buds fail to fuse; relative block at minor papilla
- Genetics
  - Hereditary pancreatitis
    - Autosomal dominant & incomplete penetrance
- Etiology
  - Alcohol/gallstones/metabolic/infection/trauma/drugs
  - Pathogenesis: Due to reflux of pancreatic enzymes, bile, duodenal contents & increased ductal pressure

- MPD or terminal duct blockage
- Edema; spasm; incompetence of sphincter of Oddi
- Periduodenal diverticulum or tumor

- Epidemiology
  - In USA: Urban & VA hospitals (alcohol); suburban & rural (gallstones)
  - Incidence in USA
    - 0.005 to 0.01% in general population

Gross Pathologic & Surgical Features
- Bulky pancreas, necrosis, fluid collection & pseudocyst

Microscopic Features
- Acute edematous pancreatitis
  - Edema, congestion, leukocytic infiltrates
- Acute hemorrhagic pancreatitis
  - Tissue destruction, fat necrosis & hemorrhage
- Dilated ducts & protein plugs may be seen

Staging, Grading or Classification Criteria
- CT classification: Five grades based on severity
  - Grade A: Normal pancreas
  - Grade B
    - Focal or diffuse enlargement of gland
    - Contour irregularities & heterogeneous attenuation
    - No peripancreatic inflammation
  - Grade C: Intrinsic pancreatic abnormalities & associated inflammatory changes in peripancreatic fat
  - Grade D
    - Small & usually single, ill-defined fluid collection
  - Grade E
    - Two or more large fluid collections
    - Presence of gas in pancreas or retroperitoneum
- Most important criterion: Presence & extent of necrotizing pancreatitis (nonenhancing parenchyma)

CLINICAL ISSUES

Presentation
- Most common signs/symptoms
  - Epigastric pain, often radiating to back
  - Tenderness, fever, nausea, vomiting
  - Grey Turner sign: Bluish discoloration of flanks
  - Cullen sign: Periumbilical discoloration
- Clinical profile: Patient with history of alcoholism, fever & severe mid-epigastric pain radiating to back
- Lab data
  - Increased serum amylase & lipase
  - Hyperglycemia, increased lactate dehydrogenase
  - Leukocytosis, hypocalcemia (poor prognostic sign)
  - Fall in hematocrit, rise in blood urea nitrogen (BUN)

Demographics
- Age
  - Usually young & middle age group
  - Can be seen in any age group
- Gender: Males more than females

Natural History & Prognosis
- Complications
  - Pancreatic
    - Fluid collections, pseudocyst, necrosis, abscess
  - Gastrointestinal
    - Hemorrhage, infarction, obstruction, ileus
  - Biliary: Obstructive jaundice
  - Vascular: Pseudoaneurysm, porto-splenic vein thrombosis, hemorrhage
  - Disseminated intravascular coagulation (DIC)
  - Shock due to pulmonary & renal failure
  - Cardiac, CNS & metabolic complications
- Prognosis
  - Early detection with minor complications: Good
  - Late detection with major complications: Poor
  - Infected pancreatic necrosis - almost 50% mortality even with surgical debridement

Treatment
- Conservative
  - NPO, gastric tube, atropine, analgesics, antibiotics
- Treat complications of acute pancreatitis
  - Infected or obstructing pseudocysts require drainage
  - Most resolve spontaneously
  - Infected necrosis requires surgery, not catheter drainage

DIAGNOSTIC CHECKLIST

Consider
- Rule out other pathologies which can cause "peripancreatic infiltration"

Image Interpretation Pearls
- Bulky, irregularly enlarged pancreas with obliteration of peripancreatic fat planes, fluid collections, pseudocyst or abscess formation

SELECTED REFERENCES

1. Balthazar EJ: Acute pancreatitis: assessment of severity with clinical and CT evaluation. Radiology. 223(3):603-13, 2002
2. Balthazar EJ: Staging of acute pancreatitis. Radiol Clin North Am. 40(6):1199-209, 2002
3. Balthazar EJ: Complications of acute pancreatitis: clinical and CT evaluation. Radiol Clin North Am. 40(6):1211-27, 2002
4. Piironen A: Severe acute pancreatitis: contrast-enhanced CT and MRI features. Abdom Imaging. 26(3):225-33, 2001
5. Lecesne R et al: Acute pancreatitis: interobserver agreement and correlation of CT and MR cholangiopancreatography with outcome. Radiology. 211(3):727-35, 1999
6. Vitellas KM et al: Pancreatitis complicated by gland necrosis: evolution of findings on contrast-enhanced CT. J Comput Assist Tomogr. 23(6):898-905, 1999
7. Sica GT et al: Comparison of endoscopic retrograde cholangiopancreatography with MR cholangiopancreatography in patients with pancreatitis. Radiology. 210: 605-10, 1999
8. Balthazar EJ et al: Imaging and intervention in acute pancreatitis. Radiology. 193(2):297-306, 1994

IMAGE GALLERY

Typical

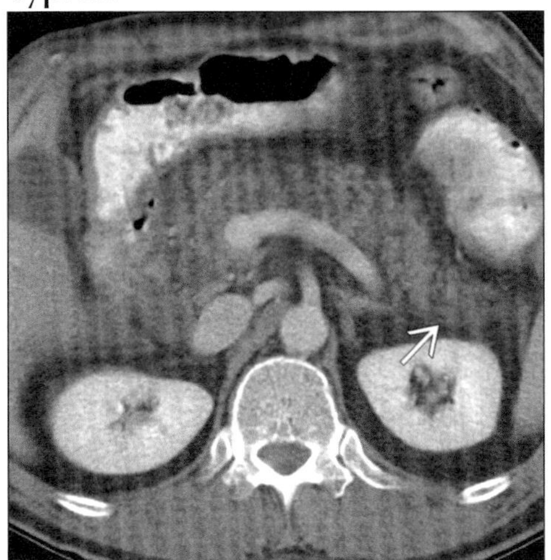

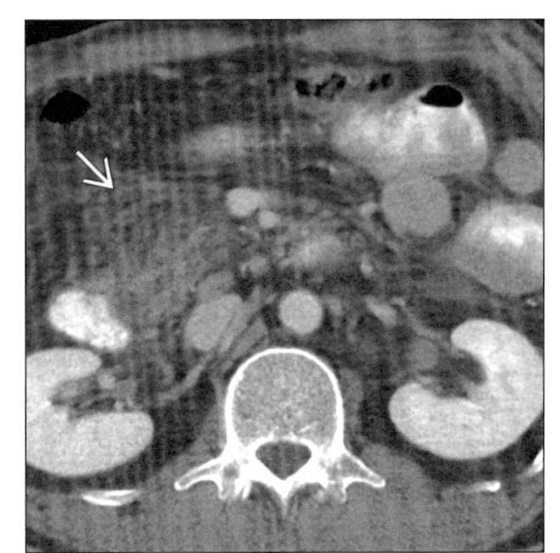

(Left) *Mild acute pancreatitis. Gland is diffusely enlarged with minimal peripancreatic infiltration (arrow).* ***(Right)*** *Mild acute pancreatitis. Axial CECT shows enlarged pancreatic head with infiltration of mesenteric fat (arrow).*

Typical

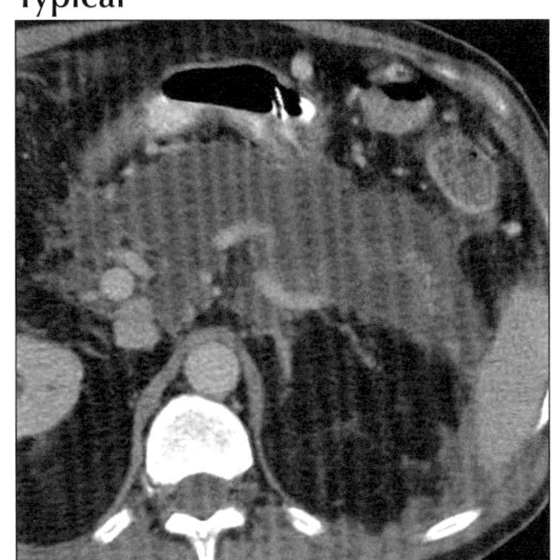

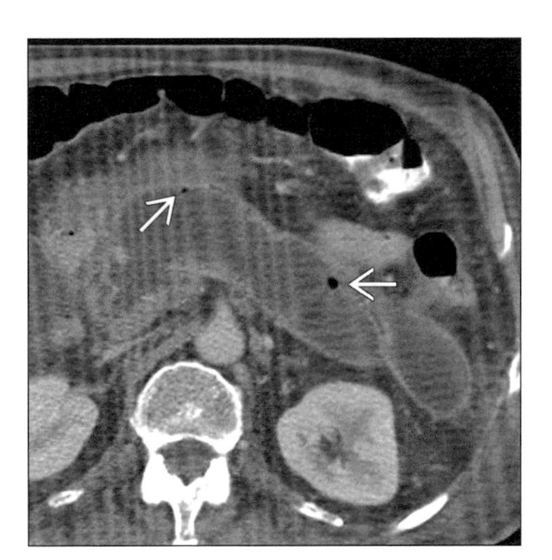

(Left) *Necrotizing pancreatitis. Axial CECT shows almost no enhancing viable pancreatic tissue, only fluid and necrotic tissue.* ***(Right)*** *Infected pancreatic necrosis. Axial CECT shows no enhancing parenchyma. The necrotic tissue contains gas bubbles (arrows) indicating infection.*

Typical

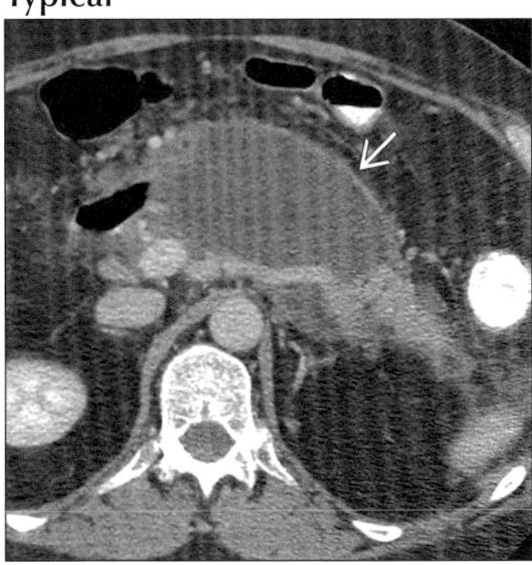

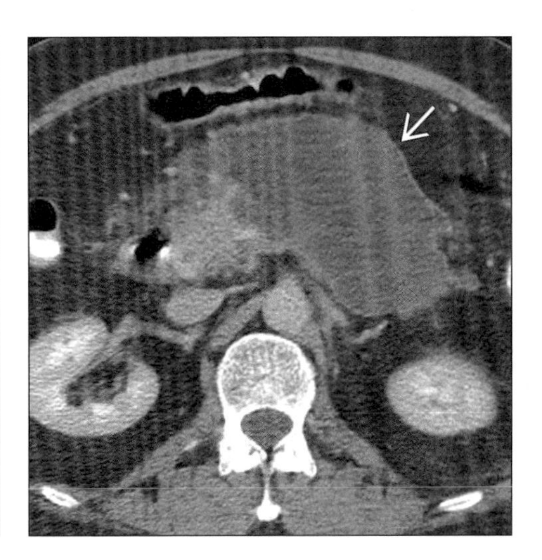

(Left) *Central pancreatic necrosis. Axial CECT shows enhancing viable tissue in pancreatic tail. The pancreatic body is necrotic and a large pseudocyst has formed (arrow).* ***(Right)*** *Axial CECT shows viable enhancing pancreatic head and pseudocyst (arrow).*

PANCREATIC PSEUDOCYST

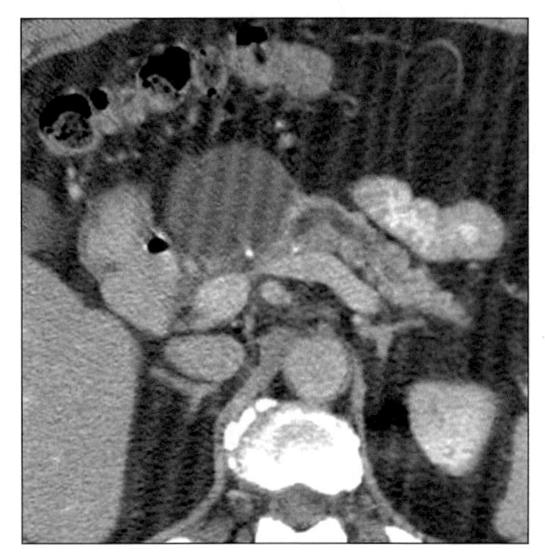

Axial CECT shows dilated main pancreatic duct and pseudocyst in and adjacent to pancreatic head.

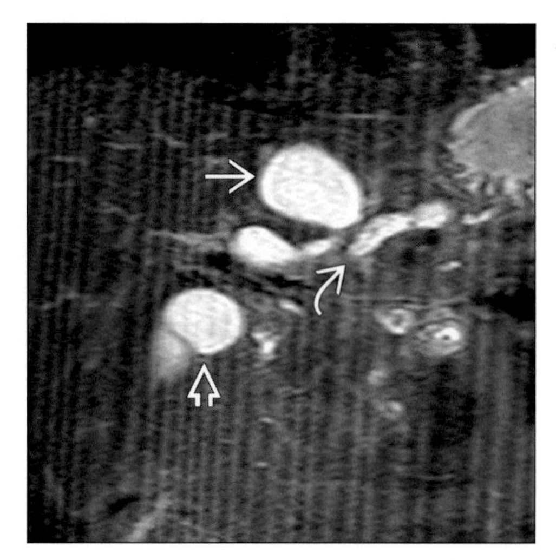

Coronal T2WI MR shows dilated main pancreatic duct (curved arrow) and contiguous pseudocyst (arrow). Gallbladder (open arrow).

TERMINOLOGY

Definitions

- Collection of pancreatic fluid & inflammatory exudate encapsulated by fibrous tissue

IMAGING FINDINGS

General Features

- Best diagnostic clue: Cystic mass with infiltration of peripancreatic fat planes
- Location
  - Two thirds within pancreas: Usually in lesser sac
    - Body & tail (85%); head (15%)
  - One third
    - Juxtasplenic, retroperitoneum & mediastinum
    - Pararenal, left lobe of liver
- Size: Varies from 2-10 cm
- Morphology
  - One of the complications of acute pancreatitis
    - Other pancreatic complications: Fluid collections, abscess, infected necrosis, pseudoaneurysm
  - Seen in approximately 15% of patients with acute pancreatitis
  - Can also be seen with chronic pancreatitis
  - Develop over a period of 4-6 weeks after onset of acute pancreatitis
  - In contrast to true cysts, pseudocysts lack a true epithelial lining

Radiographic Findings

- ERCP
  - Communication of pseudocyst with pancreatic duct seen in 70% of cases (decreases over time)

CT Findings

- NECT
  - Round or oval, homogeneous, hypodense lesion with a near water density ("mature" pseudocyst)
  - Hemorrhagic, infected pseudocyst: Lobulated, heterogeneous, mixed density lesion
  - ± Pancreatic calcification; main pancreatic duct (MPD) & common bile duct (CBD) dilatation
- CECT
  - Enhancement of thin rim of fibrous capsule
  - No enhancement of pseudocyst contents
  - Gas within pseudocyst suggests superimposed infection
  - Decompression of pseudocyst into pancreatic duct, stomach or bowel may result in gas within cyst
  - Pseudoaneurysms can be caused by or simulate a pseudocyst
    - CECT shows enhancement like adjacent blood vessels

DDx: Cystic Pancreatic Mass

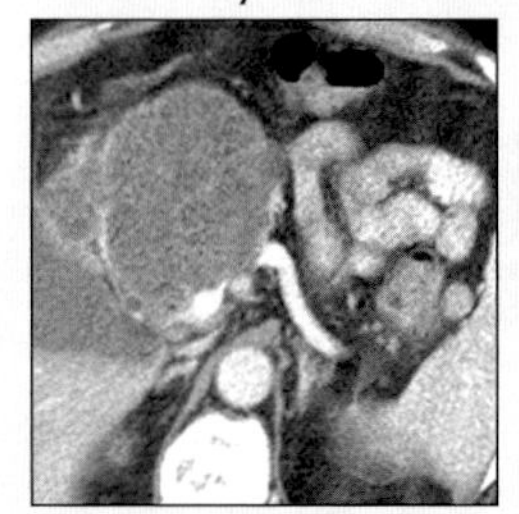

Serous Cystadenoma

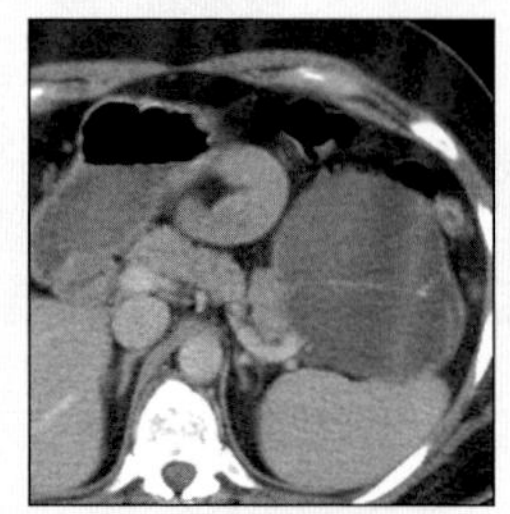

Mucinous Cystic tumor

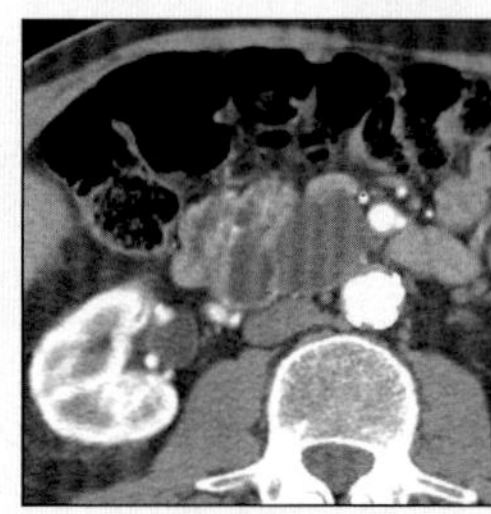

IPMT of Pancreas

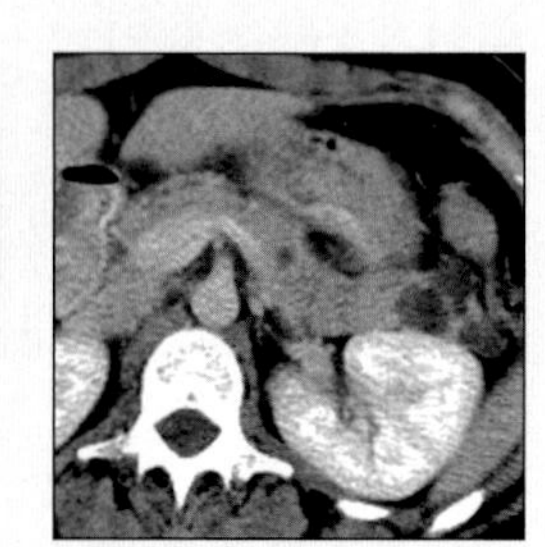

Congenital Cysts (VHL)

PANCREATIC PSEUDOCYST

Key Facts

Terminology
- Collection of pancreatic fluid & inflammatory exudate encapsulated by fibrous tissue

Imaging Findings
- Best diagnostic clue: Cystic mass with infiltration of peripancreatic fat planes
- Communication of pseudocyst with pancreatic duct seen in 70% of cases (decreases over time)
- Enhancement of thin rim of fibrous capsule

Top Differential Diagnoses
- Serous (microcystic) cystadenoma
- Mucinous cystic tumor of pancreas
- IPMT of pancreas
- Cystic islet cell tumor
- Congenital cysts

Pathology
- Unabsorbed fluid collections organize & within 4-6 weeks develop a fibrous capsule
- Pseudocyst: Major complication of acute (more common) & chronic pancreatitis

Clinical Issues
- Clinical profile: Patient with history of chronic alcoholism, abdominal pain & palpable tender mass
- May persist, resolve or can even continue to grow
- Complications: Pseudocysts larger than 4-5 cm in size

Diagnostic Checklist
- Rule out other "cystic lesions of pancreas"
- Consider possibility of pseudoaneurysm, especially if drainage is contemplated

MR Findings
- T1WI: Hypointense
- T2WI
  - Hyperintense (fluid)
  - Mixed intensity (fluid + debris)
- T1 C+: May show enhancement of fibrous capsule
- MRCP
  - Hyperintense cyst contiguous with dilated pancreatic duct

Ultrasonographic Findings
- Real Time
  - Usually solitary unilocular cyst (body or tail)
  - Multilocular in 6% of cases
  - Fluid-debris level & internal echoes due to autolysis (blood clot/cellular debris)
  - Septations (rare; sign of infection or hemorrhage)
  - Dilated pancreatic duct & CBD may be seen
  - Calcification of pancreas (chronic pancreatitis)

Angiographic Findings
- Conventional
  - To confirm diagnosis of pseudoaneurysm
    - Splenic artery is most frequently involved, followed by inferior & superior pancreatico-duodenal arteries

Imaging Recommendations
- NE + CECT, MRCP, ERCP, US

DIFFERENTIAL DIAGNOSIS

Serous (microcystic) cystadenoma
- Synonym: Glycogen-rich cystadenoma of pancreas
- Benign pancreatic tumor (arises from acinar cells)
- Slowly growing tumor & may become quite large
- Most frequently seen in middle-aged females
- CECT
  - Honeycomb or sponge appearance
  - Enhancement of septa delineating small cysts
  - Enhancement of cyst wall seen
  - May have calcification in central scar
- Angiography: Highly vascular tumor due to extensive capillary network within septa

Mucinous cystic tumor of pancreas
- CT: Multiloculated hypodense mass
- CT: Enhancement of thin internal septa & wall
- T2WI MR
  - Cysts: Hyperintense
  - Internal septations: Hypointense
- Location: Tail of pancreas (more common)
- Sonography: Multiloculated cystic mass with echogenic internal septa
- Angiography: Predominantly avascular
- Multilocularity or mural nodules favor tumor
- Often indistinguishable from pseudocyst by imaging alone
- More common in middle-aged females
- Most consider this tumor as premalignant

IPMT of pancreas
- IPMT: Intraductal papillary mucinous tumor
- Cystic lesion contiguous with dilated MPD sometimes indistinguishable from pseudocyst
- Low grade malignancy arises from main pancreatic duct (MPD) or branch pancreatic duct (BPD)
- Side branch type usually arises in BPD of pancreatic head/uncinate, resembling cluster of grapes or small tubular cysts
- Main duct type causes gross dilatation of MPD +/- cystic spaces
- Maybe be indistinguishable from chronic pancreatitis & pseudocyst

Cystic islet cell tumor
- Usually non-insulin producing & nonfunctioning
- Tumor: Cystic on NECT & nonenhancing on CECT
  - No pancreatic ductal dilatation
- Angiography: Hypervascular primary & secondary

Congenital cysts
- Associated with von Hippel-Lindau (VHL) & ADPKD
- Rare, usually small & multiple nonenhancing cysts
- No pancreatic ductal dilatation

PANCREATIC PSEUDOCYST

PATHOLOGY

General Features

- General path comments
  - Fluid collection
    - Rupture of pancreatic duct
    - Release of enzymes & pancreatic juice
    - Exudation of fluid from surface of pancreas due to activation of enzymes within gland
    - Usually absorbed within 2-3 weeks
    - Seen in up to 50% of patients with acute pancreatitis; does not constitute pseudocyst
  - Pseudocyst
    - Unabsorbed fluid collections organize & within 4-6 weeks develop a fibrous capsule
- Etiology
  - Pseudocyst: Major complication of acute (more common) & chronic pancreatitis
    - Chronic alcoholism (75%)
    - Abdominal trauma (13%): Major cause in children
    - Cholelithiasis, pancreatic carcinoma, idiopathic
- Epidemiology
  - Pseudocysts form during initial attack of pancreatitis in 1-3% of patients
  - Pseudocysts develop after several episodes of alcoholic pancreatitis in 12% of patients
- Associated abnormalities: Acute or chronic pancreatitis

Gross Pathologic & Surgical Features

- Collection of fluid, tissue, debris, pancreatic enzymes & blood covered by a thin rim of fibrous capsule

Microscopic Features

- Inflammatory cells, necrosis, hemorrhage
- Absence of epithelial lining
- Walls consist of necrotic, granulation or fibrous tissue

CLINICAL ISSUES

Presentation

- Most common signs/symptoms
  - Clinical significance is related to its size & complications
    - Abdominal pain with or without radiation to back (common complaint)
    - Palpable, tender mass in middle or left upper abdomen
    - Signs of hemorrhage: Increase in size, bruit over mass, decrease in hemoglobin level & hematocrit
- Clinical profile: Patient with history of chronic alcoholism, abdominal pain & palpable tender mass
- Lab data
  - Acute pancreatitis
    - Increased serum amylase & lipase
  - Chronic pancreatitis
    - Secretin test: Decreased amylase & bicarbonate

Demographics

- Age: More common in young & middle age group
- Gender: Males > females

Natural History & Prognosis

- Natural history: Difficult to predict
  - May persist, resolve or can even continue to grow
  - Spontaneous resolution of pseudocyst can occur by
    - Drainage into pancreatic duct
    - Erosion into adjacent hollow organ (stomach, small bowel, colon)
    - Rupture with spillage into peritoneal cavity
- Complications: Pseudocysts larger than 4-5 cm in size
  - Compression of adjacent bowel or bile duct
    - Obstruction, severe pain, jaundice
  - Spontaneous rupture into peritoneal cavity
    - Ascites, peritonitis
  - Secondary infection: Infected pseudocyst not as lethal as infected pancreatic necrosis
  - Erosion into adjacent vessel
    - Hemorrhage or pseudoaneurysm
  - Rupture & hemorrhage are prime causes of death from pseudocyst
- Prognosis
  - Spontaneous resolution in 25-40% of patients
  - Percutaneous drainage cure in 90% of cases
    - Complications in 5-10% of cases

Treatment

- Conservative therapy
  - Infected pseudocyst
  - Asymptomatic or decrease in size on serial scans
- Percutaneous drainage
  - Size more than 4-5 cm
  - Symptomatic or increase in size
  - Requires long-term catheter if pseudocyst still communicates with pancreatic duct
  - Drainage routes
    - Retroperitoneal, transperitoneal
    - Transgastric, transhepatic or duodenal
- Surgical therapy: Internal (usually into stomach) or external drainage of cyst

DIAGNOSTIC CHECKLIST

Consider

- Rule out other "cystic lesions of pancreas"
- Consider possibility of pseudoaneurysm, especially if drainage is contemplated

Image Interpretation Pearls

- Correlate with ancillary imaging findings and clinical setting of prior pancreatitis to confirm diagnosis and avoid mismanagement

SELECTED REFERENCES

1. Morgan DE et al: Pancreatic fluid collections prior to intervention: Evaluation with MR imaging compared with CT and US. Radiology. 203: 773-8, 1997
2. Lee MJ et al: Acute complicated pancreatitis: Redefining the role of interventional radiology. Radiology. 183: 171-4, 1992
3. Sonnenberg EV et al: Percutaneous drainage of infected and noninfected pancreatic pseudocysts: Experience in 101 cases. Radiology. 170: 757-61, 1989

IMAGE GALLERY

Typical

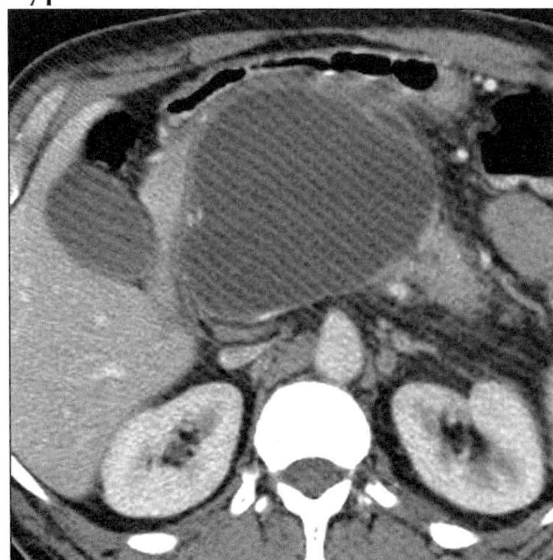

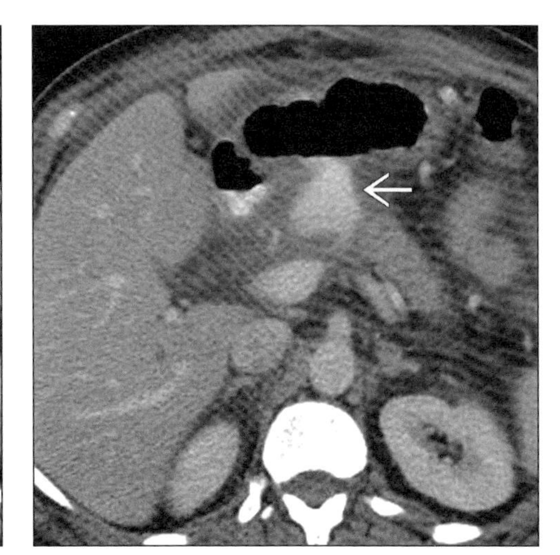

(Left) *Axial CECT shows large pseudocyst displacing stomach anteriorly.* ***(Right)*** *Axial CECT in patient who had large retrogastric pseudocyst. High density material within small pseudocyst (arrow) is oral contrast medium, indicating spontaneous rupture of pseudocyst into stomach.*

Typical

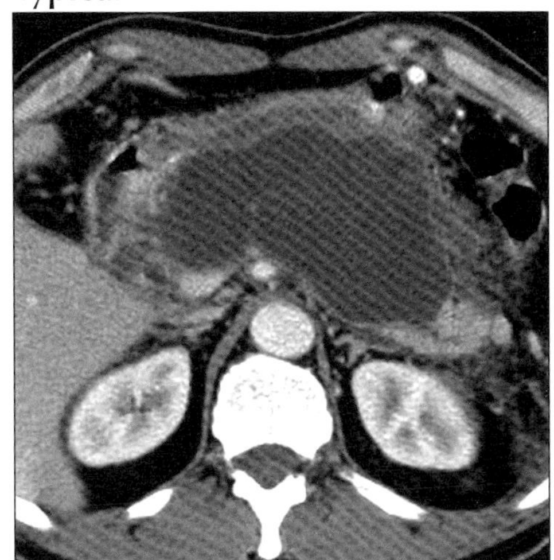

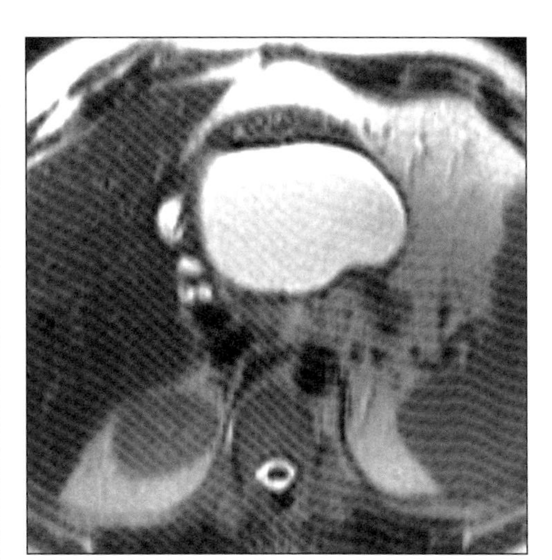

(Left) *Axial CECT shows large pseudocyst resulting from central pancreatic necrosis. Pancreatic tail is intact; pancreatic duct is disrupted and empties into pseudocyst.* ***(Right)*** *Axial T2WI MR shows hyperintense pseudocyst within lesser sac.*

Typical

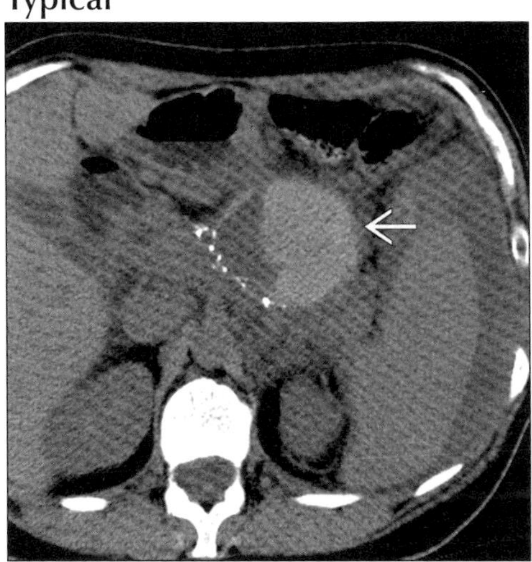

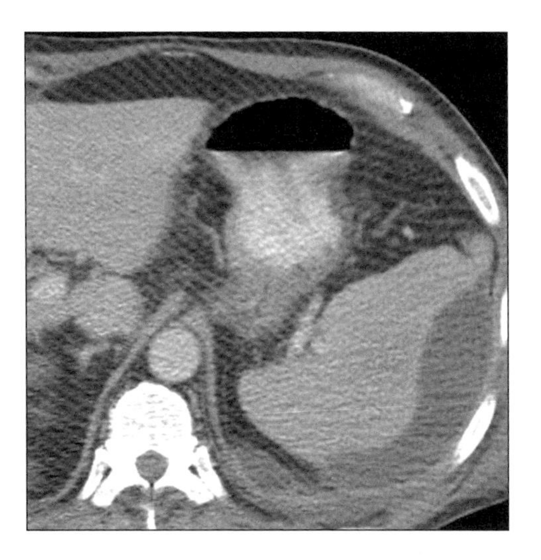

(Left) *Axial NECT shows hyperdense clotted blood within pseudocyst (arrow); parenchymal calcifications from chronic pancreatitis. Ascites around spleen.* ***(Right)*** *Axial CECT shows intrasplenic subcapsular pseudocyst resulting from retroperitoneal extension of inflammation from pancreas into splenic hilum.*

CHRONIC PANCREATITIS

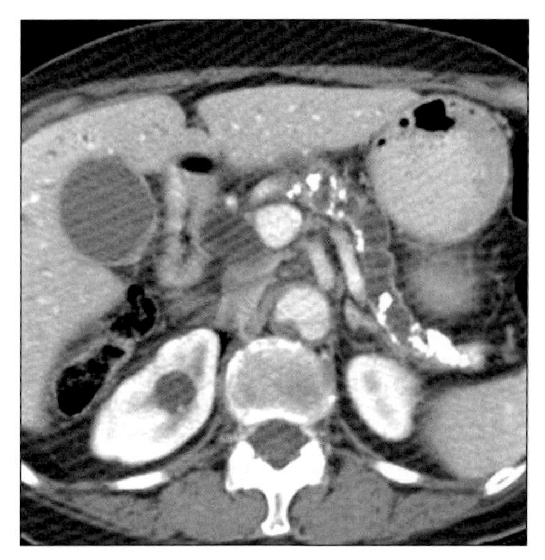

Axial CECT shows glandular atrophy, dilated main pancreatic duct and intraductal calculi.

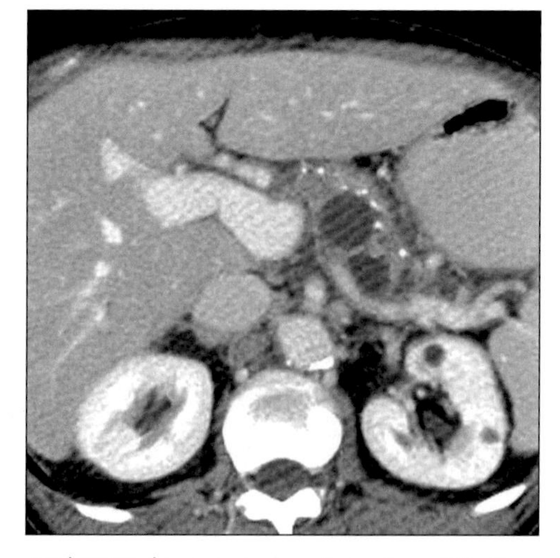

Axial CECT shows parenchymal atrophy, calcifications and small pseudocysts.

TERMINOLOGY

Definitions

- Irreversible inflammatory damage of pancreas usually evident on imaging or functional testing

IMAGING FINDINGS

General Features

- Best diagnostic clue: Atrophy of gland, dilated main pancreatic duct (MPD), intraductal calculi
- Size: Pancreas usually decreased in size (atrophy)
- Morphology
  - Inflammatory disease of pancreas characterized by irreversible damage to morphology & function
  - Pancreatic calcification
    - Almost diagnostic of chronic pancreatitis
    - In 40-60% of patients with alcoholic pancreatitis
    - Approximately 90% of calcific pancreatitis is caused by alcoholism
- Other features
  - In USA 75% of cases are due to alcoholism
  - Developing countries: Malnutrition & alcoholism

Radiographic Findings

- Radiography
  - Plain x-ray abdomen
    - Pancreatic calcification
    - Small, irregular calcifications (local or diffuse)
- Barium (UGI series)
  - Changes seen in second part of duodenum
    - Varying degrees of atony
    - Thickened, irregular & spiculated mucosal folds
    - Stricture & proximal dilatation
    - Enlarged papilla of Vater (Poppel papillary sign)
  - Frostberg sign: Inverted-3 configuration of duodenal loop (seen occasionally)
- ERCP
  - Dilated & beaded MPD plus radicles
  - MPD filling defects: Intraductal calculi
  - CBD may appear dilated with distal narrowing

CT Findings

- NECT
  - Gland
    - Atrophy (more common)
    - Focal or diffuse enlargement (occasionally)
  - Dilated MPD with ductal calculi
  - Intra & peripancreatic cysts
  - Thickening of peripancreatic fascia
  - Splenic vein thrombosis, splenomegaly, varices
  - Hypodense focal mass (fibrosis & fat necrosis)
    - Usually in pancreatic head
    - May simulate neoplasm
- CECT
  - Heterogeneous enhancement of pancreas

DDx: Dilated Main Pancreatic Duct and Gland Atrophy

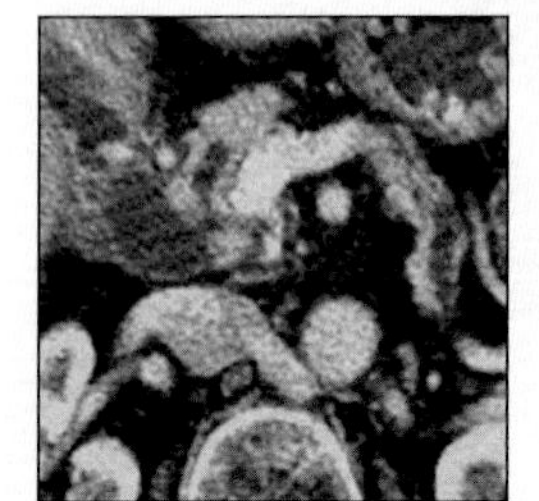

Pancreatic Carcinoma

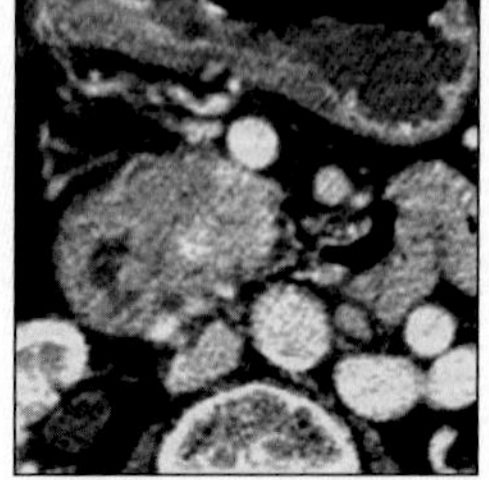

Pancreatic Carcinoma

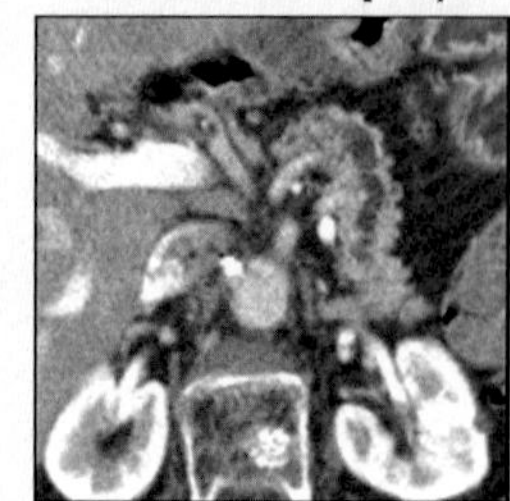

IPMT of Pancreas

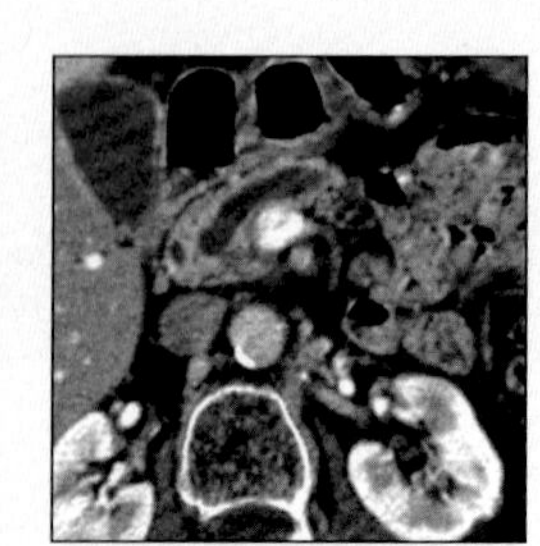

IPMT of Pancreas

CHRONIC PANCREATITIS

Key Facts

Imaging Findings
- Best diagnostic clue: Atrophy of gland, dilated main pancreatic duct (MPD), intraductal calculi
- Pancreatic calcification
- Dilated MPD with ductal calculi
- Thickening of peripancreatic fascia
- Splenic vein thrombosis, splenomegaly, varices
- Hypodense focal mass (fibrosis & fat necrosis)
- Heterogeneous enhancement of pancreas
- Pseudocyst contiguous with MPD
- CBD may be dilated with smooth distal tapering

Top Differential Diagnoses
- Pancreatic carcinoma
- IPMT of pancreas

Pathology
- Chronic pancreatitis usually caused by alcohol abuse

Clinical Issues
- Clinical profile: Patient with history of chronic alcoholism, recurrent attacks of mid-epigastric pain radiating to back, jaundice, steatorrhea & diabetes
- Elevated serum amylase & lipase

Diagnostic Checklist
- Differentiate from other conditions which can cause "MPD dilatation & glandular atrophy"
- Glandular atrophy with dilated MPD, ductal calculi, thickened peripancreatic fascia ± pseudocyst are best signs for chronic pancreatitis

- Mass due to chronic pancreatitis: Varied enhancement due to presence or absence of fibrosis
  - Fibroinflammatory mass: Common in pancreatic head
  - Hypoenhanced mass: Due to fibrosis
  - Isodense enhancing mass: Lack of fibrosis

MR Findings
- T1WI GEI
  - Decreased or loss of signal intensity
- Fat-suppressed T2WI
  - Pseudocyst, necrotic areas: Hyperintense
  - Gallstones, intraductal calculi: Hypointense
- T1 C+ GEI
  - Heterogeneous enhancement pattern
    - Nonenhancing decreased signal areas: Necrosis, pseudocyst
    - Pancreatic pseudocyst contiguous with dilated MPD is well depicted
    - Vascular occlusions can be demonstrated
- MRCP
  - Fluid-containing structures are well depicted
    - Dilated MPD plus radicles
    - Pseudocyst contiguous with MPD
    - CBD may be dilated with smooth distal tapering

Ultrasonographic Findings
- Real Time
  - Atrophic gland
  - Dilated MPD
    - Tubular anechoic structure
    - Seen in up to 90% of cases
  - Echogenic foci (calcifications) with posterior acoustic shadowing
  - Pseudocyst
    - Unilocular, anechoic & sharply defined

Imaging Recommendations
- NE + CECT; MR (fat-suppressed images); MRCP; ERCP

DIFFERENTIAL DIAGNOSIS

Pancreatic carcinoma
- Irregular, heterogeneous, poorly enhancing mass
- Location: Head (60% of cases); body (20%); tail (15%)
- Atrophy of parenchyma may be seen
- Main pancreatic duct and/or common bile duct
  - Obstruction & dilatation
- Obliteration of retropancreatic fat
- Extensive local invasion & regional metastases
  - Local invasion to medial wall of duodenal sweep; narrows lumen
- ERCP
  - Irregular, nodular, rat-tailed eccentric obstruction
- 65% of patients present with advanced local disease & distant metastases
- Some cases of chronic pancreatitis & pancreatic cancer are impossible to differentiate without surgical excision & histology

IPMT of pancreas
- IPMT: Intraductal papillary mucinous tumor
- Low grade malignancy arises from main pancreatic duct (MPD) or branch pancreatic duct (BPD)
- Involvement of main pancreatic duct may simulate chronic pancreatitis clinically & on CT/MR
- Dilated MPD and parenchymal atrophy
- ERCP: Best diagnostic tool
  - Visualizes mucus ± polypoid lesions within MPD

PATHOLOGY

General Features
- General path comments
  - Embryology-anatomy
    - Congenital anomalies may cause pancreatitis
    - Pancreas divisum: Ducts too small to adequately drain pancreatic secretions leading to chronic stasis
    - Annular pancreas: Pancreatic ductal obstruction and stasis of secretions
  - Chronic calcifying pancreatitis (alcoholism)

- Lesions are diffuse
- Chronic obstructive pancreatitis (gallstones)
  - Lesions are more prominent in head of pancreas
  - Pattern does not have a lobular distribution
- Genetics
  - Hereditary pancreatitis
    - Autosomal dominant & incomplete penetrance
- Etiology
  - Chronic pancreatitis usually caused by alcohol abuse
  - Gallstones, hyperlipidemia, trauma, drugs often cause acute but rarely chronic pancreatitis
  - Pathogenesis: Due to chronic reflux of pancreatic enzymes, bile, duodenal contents & increased ductal pressure
    - MPD or terminal duct blockage
    - Edema, spasm or incompetent sphincter of Oddi
    - Periduodenal diverticulum or tumor
- Epidemiology: More common in developing countries

Gross Pathologic & Surgical Features

- Hard atrophic pancreas with intraductal calculi & dilated MPD
- Areas of multiple parenchymal calcifications
- Pseudocysts may be seen

Microscopic Features

- Atrophy & fibrosis of acini with dilated ducts
- Mononuclear inflammatory reaction
- Occasionally squamous metaplasia of ductal epithelium

CLINICAL ISSUES

Presentation

- Most common signs/symptoms
  - Recurrent attacks of mid-epigastric pain, typically radiates to back
  - Jaundice, steatorrhea & diabetes mellitus
  - Endocrine & exocrine deficiencies due to progressive destruction of gland & stricture of common bile duct (CBD)
  - Mid-epigastric pain & weight loss
  - Continued consumption of alcohol for a period of 3-12 years is usually required to develop manifestations of chronic pancreatitis
- Clinical profile: Patient with history of chronic alcoholism, recurrent attacks of mid-epigastric pain radiating to back, jaundice, steatorrhea & diabetes
- Lab data
  - Elevated serum amylase & lipase
  - Increased blood glucose levels & fat in stool
  - Secretin test: Decreased amylase & bicarbonate

Demographics

- Age: Usually middle age group
- Gender: Males more than females

Natural History & Prognosis

- Complications
  - Diabetes
  - Malabsorption
  - Biliary obstruction; jaundice
  - GI bleeding & splenic vein thrombosis
  - Significant increase in pancreatic cancer incidence
- Prognosis
  - Poor

Treatment

- Surgical or endoscopic intervention
  - Ductal & GI obstruction; GI bleeding
  - Large pseudocyst or persistently symptomatic

DIAGNOSTIC CHECKLIST

Consider

- Differentiate from other conditions which can cause "MPD dilatation & glandular atrophy"
- May be very difficult to distinguish chronic pancreatitis with a focal fibrotic mass (in head) from pancreatic carcinoma

Image Interpretation Pearls

- Glandular atrophy with dilated MPD, ductal calculi, thickened peripancreatic fascia ± pseudocyst are best signs for chronic pancreatitis

SELECTED REFERENCES

1. Remer EM et al: Imaging of chronic pancreatitis. Radiol Clin North Am. 40(6):1229-42, v, 2002
2. Manfredi R et al: Idiopathic chronic pancreatitis in children: MR cholangiopancreatography after secretin administration. Radiology. 224(3):675-82, 2002
3. Varghese JC et al: Value of MR pancreatography in the evaluation of patients with chronic pancreatitis. Clin Radiol. 57(5):393-401, 2002
4. Matos C et al: MR imaging of the pancreas: a pictorial tour. Radiographics. 22(1):e2, 2002
5. Remer EM et al: Imaging of chronic pancreatitis. Radiol Clin North Am. 40(6):1229-42, v, 2002
6. Varghese JC et al: Value of MR pancreatography in the evaluation of patients with chronic pancreatitis. Clin Radiol. 57(5):393-401, 2002
7. Matos C et al: MR imaging of the pancreas: a pictorial tour. Radiographics. 22(1):e2, 2002
8. Ichikawa T et al: Duct-penetrating sign at MRCP: usefulness for differentiating inflammatory pancreatic mass from pancreatic carcinomas. Radiology. 221(1):107-16, 2001
9. Kim T et al: Pancreatic mass due to chronic pancreatitis: correlation of CT and MR imaging features with pathologic findings. AJR Am J Roentgenol. 177(2):367-71, 2001
10. Manfredi R et al: Severe chronic pancreatitis versus suspected pancreatic disease: dynamic MR cholangiopancreatography after secretin stimulation. Radiology. 214(3):849-55, 2000
11. Johnson PT et al: Pancreatic carcinoma versus chronic pancreatitis: dynamic MR imaging. Radiology. 212(1):213-8, 1999
12. Sica GT et al: Comparison of endoscopic retrograde cholangiopancreatography with MR cholangiopancreatography in patients with pancreatitis. Radiology. 210(3):605-10, 1999
13. Fulcher AS et al: MR pancreatography: a useful tool for evaluating pancreatic disorders. Radiographics. 19(1):5-24; discussion 41-4; quiz 148-9, 1999
14. Sica GT et al: Comparison of endoscopic retrograde cholangiopancreatography with MR cholangiopancreatography in patients with pancreatitis. Radiology. 210(3):605-10, 1999

IMAGE GALLERY

Typical

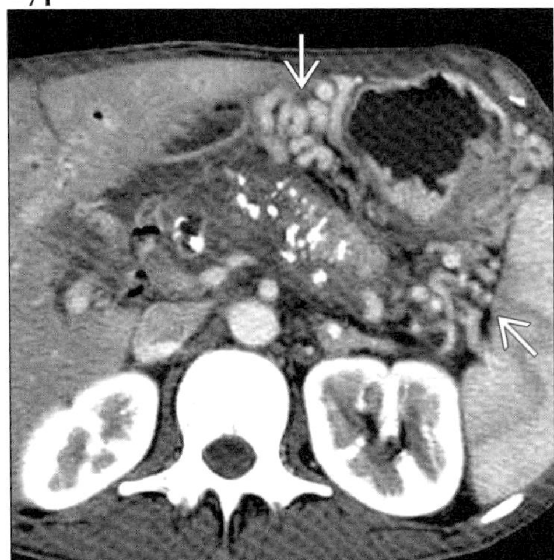

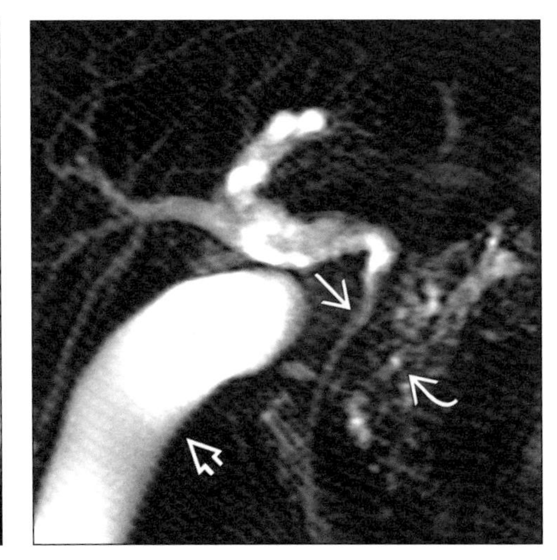

(Left) *Axial CECT shows pancreatic calculi, perisplenic and perigastric varices (arrows) due to obstruction of splenic vein.* ***(Right)*** *MRCP shows long tapered narrowing of CBD (arrow) as it passes through head of pancreas. Irregular strictures and dilatation of pancreatic duct (curved arrow). Gallbladder (open arrow).*

Typical

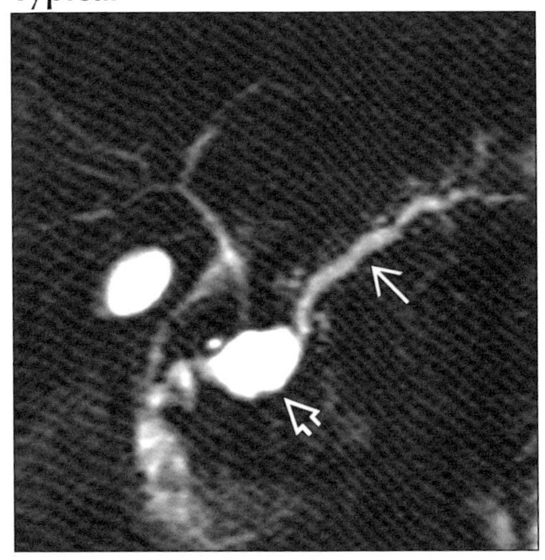

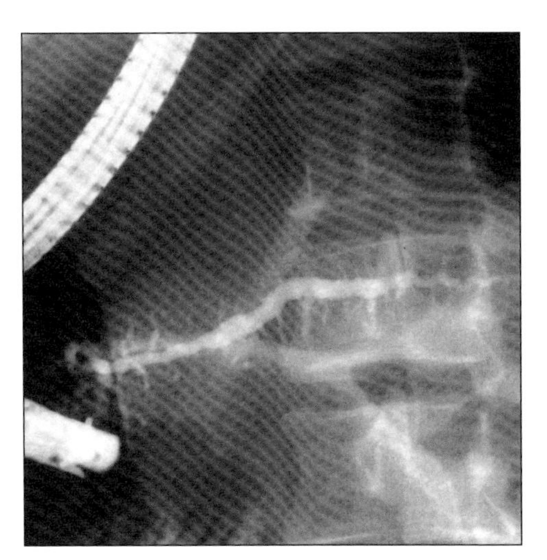

(Left) *MRCP shows dilated main pancreatic duct (arrow) and adjacent pseudocyst (open arrow).* ***(Right)*** *ERCP shows irregular dilatation of main pancreatic duct and side branches.*

Typical

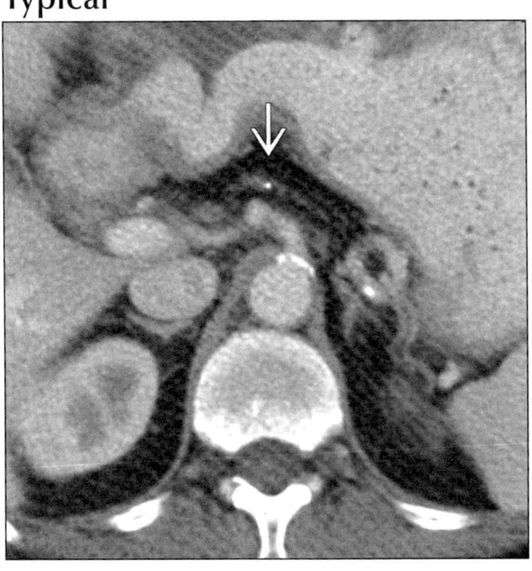

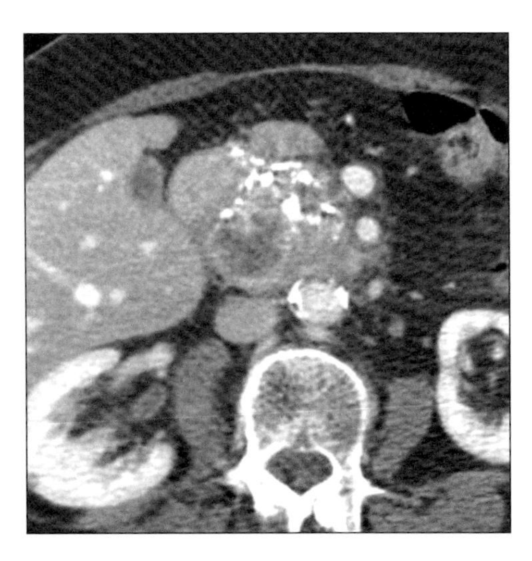

(Left) *Axial CECT shows marked parenchymal atrophy with fatty replacement and a dilated main pancreatic duct containing calculi (arrow).* ***(Right)*** *Chronic pancreatitis. Axial CECT shows a large heterogeneous mass in the pancreatic head with parenchymal calculi. No tumor was found after surgical resection.*

TRAUMATIC PANCREATITIS

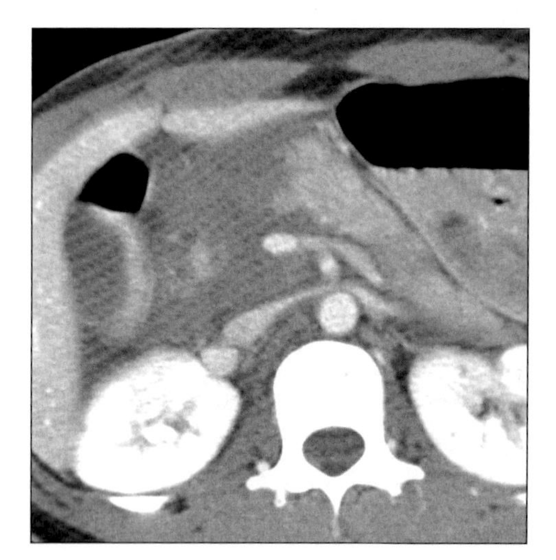

Pancreatic transection. Axial CECT shows fracture plane through neck of pancreas and peripancreatic edema and hemorrhage.

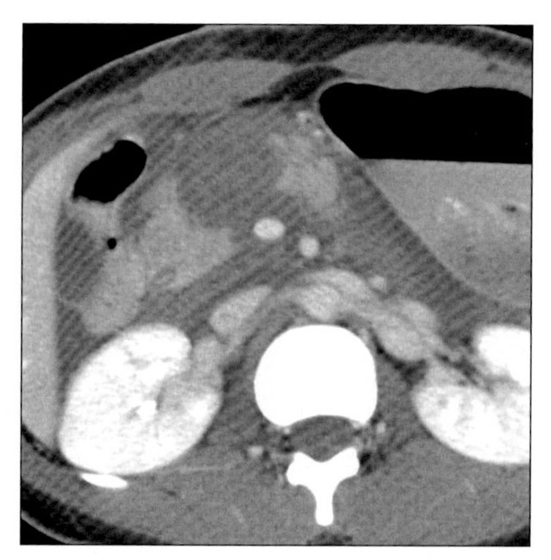

Pancreatic transection. Axial CECT shows fracture plane through neck of pancreas. Pancreatic duct was disrupted and body-tail of pancreas was resected at surgery.

TERMINOLOGY

Abbreviations and Synonyms

- Traumatic pancreatic injury

Definitions

- Inflammatory disease of pancreas secondary to trauma

IMAGING FINDINGS

General Features

- Best diagnostic clue: Enlargement of gland, heterogeneous parenchyma, peripancreatic fluid collections & history of trauma
- Morphology: Spectrum of pancreatic injuries ranges from acute pancreatitis, contusions, deep lacerations & fractures with ductal disruption

Radiographic Findings

- ERCP
  - Normal in cases of pancreatic "contusion"
  - Transection of pancreatic duct: Abrupt duct termination or contrast extravasation
  - Communication of pseudocyst with pancreatic duct
  - May cause pancreatitis

CT Findings

- Focal/diffuse pancreatic enlargement
- Irregularity of pancreatic contour
- Edema/fluid in peripancreatic fat
  - Loss of normal fat plane: Peripancreatic infiltration
- Heterogeneous parenchymal attenuation
- Thickening of anterior renal fascia
- Peripancreatic soft tissue changes of traumatic pancreatitis are often subtle; becoming more evident within 24-48 hours
- Laceration
  - Area of low attenuation (actual size of laceration difficult to visualize)
  - Linear cleft, usually oriented anteroposteriorly
- Pancreatic fracture or transection
  - Ill-defined low density area
  - Results in clear separation of two ends of gland
  - Nearly always extends through pancreatic neck
- Lacerations/fractures may produce subtle changes in parenchymal density; may be undetectable on CT
- Extrapancreatic fluid collections: Perivascular, transverse mesocolon, pararenal space, lesser sac, juxtasplenic, root of mesentery
- Pancreatic contusion/hematoma: Range from apparent contour deformity of pancreas to rounded-mass several centimeters in diameter
  - Peripancreatic hematoma may mimic fluid-filled proximal small bowel loop
  - Often have concomitant injury to liver (left lobe) and bowel

DDx: Peripancreatic Infiltration, Following Trauma

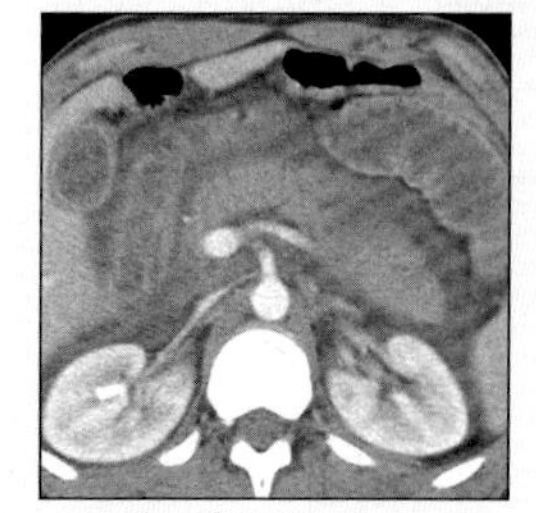

"Shock" Pancreas

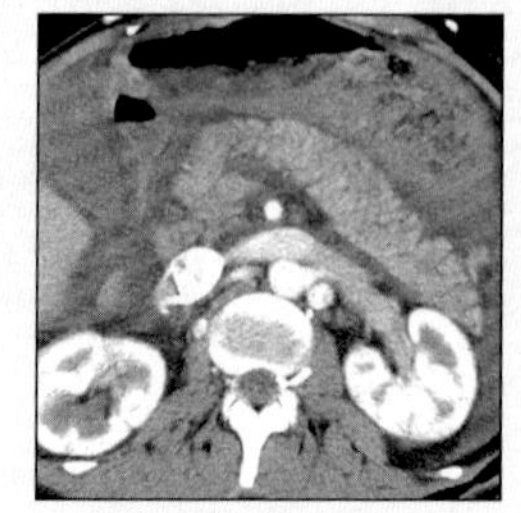

"Shock" Pancreas

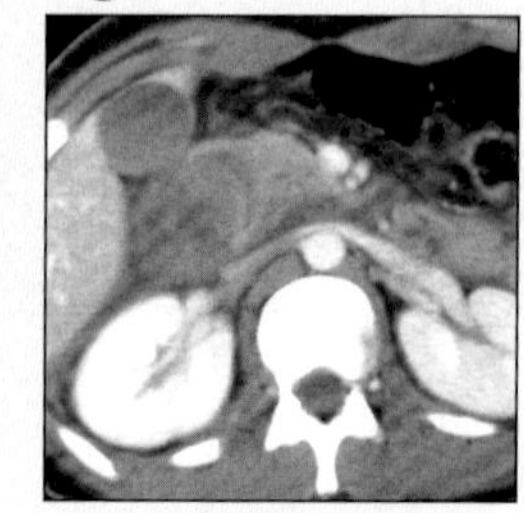

Duodenal Hematoma

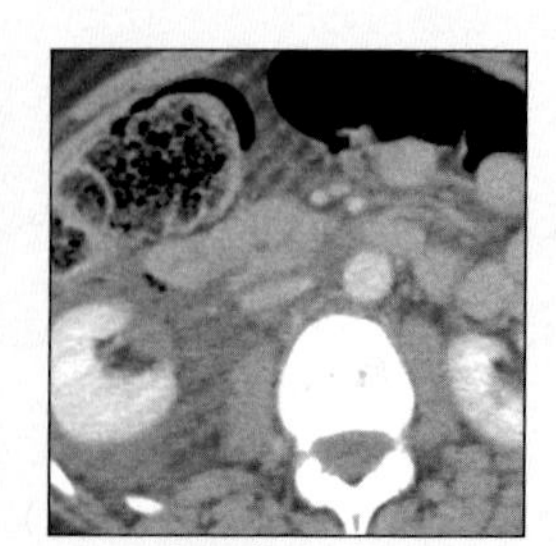

Duodenal Perforation

TRAUMATIC PANCREATITIS

Key Facts

Imaging Findings

- Best diagnostic clue: Enlargement of gland, heterogeneous parenchyma, peripancreatic fluid collections & history of trauma
- Morphology: Spectrum of pancreatic injuries ranges from acute pancreatitis, contusions, deep lacerations & fractures with ductal disruption
- Irregularity of pancreatic contour
- Edema/fluid in peripancreatic fat
- Rupture of main pancreatic duct (MPD) (23%)
- Protocol advice: 24-48 hours delayed scans may uncover findings not present earlier

Top Differential Diagnoses

- "Shock" pancreas
- Intramural duodenal hematoma ± duodenal rupture

Pathology

- Penetrating/blunt trauma
- Trauma to pancreas is uncommon, accounts for 3-12% of all abdominal injuries

Clinical Issues

- Complications: Recurrent pancreatitis, pseudocyst, hemorrhage, pseudoaneurysm, fistula & abscess
- Mortality from pancreatic injuries is nearly 20%

Diagnostic Checklist

- CT diagnosis of pancreatic trauma may be difficult in selected patients who are scanned soon after injury
- CT signs of traumatic pancreatitis become more evident after 24-48 hours

- Pancreatitis caused by ERCP (+/- papillotomy, etc.) usually more severe in & around pancreatic head

MR Findings

- T1WI: Variable decreased signal intensity
- T2WI: Fluid collections & pseudocyst: Hyperintense on fat suppressed T2WI
- T1 C+
  - Heterogeneous enhancement pattern
  - Nonenhancing hypointense areas (fluid collection, pseudocyst, necrosis)
- MRCP
  - Rupture of main pancreatic duct (MPD) (23%)
  - Pseudocyst contiguous with MPD

Ultrasonographic Findings

- US (& CT) findings in traumatic pancreatitis may be similar to those of nontraumatic pancreatitis
- Not as sensitive as CT in diagnosing acute injury

Imaging Recommendations

- Best imaging tool
  - CT: More accurate method of detecting extrapancreatic fluid collections, pancreatic lacerations or fractures
  - Emergency ERCP to investigate pancreatic injuries when CT shows injury, but status of pancreatic duct is uncertain
- Protocol advice: 24-48 hours delayed scans may uncover findings not present earlier

DIFFERENTIAL DIAGNOSIS

"Shock" pancreas

- In severe injury; "hypoperfusion complex"
- Abnormally intense contrast-enhancement of pancreas, bowel wall & kidneys
- Moderate to large peritoneal fluid collections
- Decreased caliber of aorta & inferior vena cava
- Diffuse dilatation of intestine with fluid
- Mesenteric and peripancreatic fat planes are infiltrated
- Findings resolve spontaneously within 24 hours of fluid resuscitation

Intramural duodenal hematoma ± duodenal rupture

- Hematoma
  - Focal high attenuation thickening of duodenal wall
  - Hemorrhage can cause a picket-fence appearance
  - Smooth intramural mass causing incomplete bowel obstruction
- Rupture
  - Air/fluid level in adjacent extraperitoneal space
  - Gas or fluid tracking into anterior pararenal space
  - Extravasation of oral contrast into anterior pararenal space
- May simulate or coexist with pancreatic injury

PATHOLOGY

General Features

- Etiology
  - Penetrating/blunt trauma
    - Gunshot (45%), blunt (37%), stab wound (18%)
  - Mechanism in blunt trauma
    - Compression against vertebral column with shear across pancreatic neck
    - Relatively fixed position of pancreas anterior to spine
  - In children: Trauma from a bicycle handlebar, motor vehicle accident, child abuse
  - Lacerations usually accompany midline compression injury, which may also involve left hepatic lobe, duodenum, central renal vascular pedicle
  - Endoscopic procedures:
    - ERCP, especially with papillotomy, stone extraction, stent placement
  - Surgery: Billroth II resections, splenectomy, biliary surgery, aortic graft surgery
- Epidemiology
  - Trauma to pancreas is uncommon, accounts for 3-12% of all abdominal injuries
  - Acute post traumatic pancreatitis is an infrequent disease; representing 0.4% of acute pancreatitis with pseudocyst formation

TRAUMATIC PANCREATITIS

- Pancreatic penetrating injuries more common than blunt trauma
- Combined injury of other organs seen in 80% of patients
- Accounts for about 5% of all abdominal injuries in childhood
- Trauma is most common cause of pseudocyst in children (often related to child abuse)

Staging, Grading or Classification Criteria

- Grade 1: Contusion/hematoma; pancreatic duct intact
- Grade 2: Parenchymal injury; pancreatic duct intact
- Grade 3: Major ductal injury
- Grade 4: Severe crush injury
- Grade of pancreatic injury is an independent predictor of both pancreatic complications & mortality
- Grade 1 & 2: Conservative management
- Grade 3 & 4: Require surgery within 24 hours

CLINICAL ISSUES

Presentation

- Most common signs/symptoms
  - History of traumatic injury
  - Upper abdominal pain
  - Postprandial vomiting, abdominal distention
- Clinical profile
  - Serum amylase/lipase levels
    - Elevated in 90% of patients
    - May be normal immediately after trauma
  - Leukocytosis, hyperamylasemia
  - Diagnosis: Exploratory laparotomy
  - Patients with penetrating trauma generally undergo immediate laparotomy
  - Blunt injuries to pancreas may be clinically occult & may go unrecognized on initial evaluation

Demographics

- Gender: More frequently seen in males (69%)

Natural History & Prognosis

- Complications: Recurrent pancreatitis, pseudocyst, hemorrhage, pseudoaneurysm, fistula & abscess
- Subcutaneous fat necrosis & polyarthritis secondary to post-traumatic pancreatitis is reported in less than 1% of patients with pancreatic disease
- Cerebral fat embolism is a rare possible complication of traumatic pancreatitis
- Mortality from pancreatic injuries is nearly 20%
- Morbidity (42%)
  - Pancreatic fistula (11%)
  - Pancreatitis (7%)
  - Pancreatic pseudocyst (3%)
  - Intra-abdominal abscesses (8%)
  - Associated liver or intestinal injuries (> 80%)
- Morbidity is higher with external drainage compared to exploration without drainage

Treatment

- Conservative
  - Total parenteral nutrition
  - Somatostatin or octreotide
  - More recently, endoscopic with pancreatic stenting
  - Superficial lesions not affecting major pancreatic duct can be managed nonoperatively
- Surgical: For lacerations or fractures
  - Surgical drainage
  - Partial pancreatectomy
  - Persistently elevated serum amylase levels & increasing cyst size are indications for surgical intervention

DIAGNOSTIC CHECKLIST

Consider

- CT diagnosis of pancreatic trauma may be difficult in selected patients who are scanned soon after injury
- Thickening of anterior renal fascia on CT of trauma patient should prompt critical evaluation of the pancreas
- CT signs of traumatic pancreatitis become more evident after 24-48 hours

SELECTED REFERENCES

1. Kao LS et al: Predictors of morbidity after traumatic pancreatic injury. J Trauma. 55(5):898-905, 2003
2. Mayer JM et al: Pancreatic injury in severe trauma: early diagnosis and therapy improve the outcome. Dig Surg. 19(4):291-7; discussion 297-9, 2002
3. Akhrass R et al: Pancreatic trauma: a ten-year multi-institutional experience. Am Surg. 63(7):598-604, 1997
4. Portis M et al: Traumatic pancreatitis in a patient with pancreas divisum: clinical and radiographic features. Abdom Imaging. 19(2):162-4, 1994
5. Lewis G et al: Traumatic pancreatic pseudocysts. Br J Surg. 80(1):89-93, 1993
6. Jeffrey RB Jr et al: Computed tomography of pancreatic trauma. Radiology. 147(2):491-4, 1983
7. Ivancev K et al: Value of computed tomography in traumatic pancreatitis in children. Acta Radiol Diagn (Stockh). 24(6):441-4, 1983
8. Chintapalli K et al: Renal fascial thickening in pancreatitis. J Comput Assist Tomogr. 6(5):983-6, 1982

IMAGE GALLERY

Typical

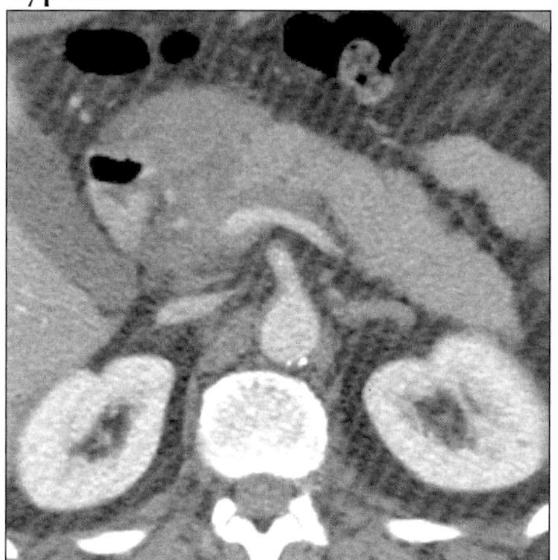

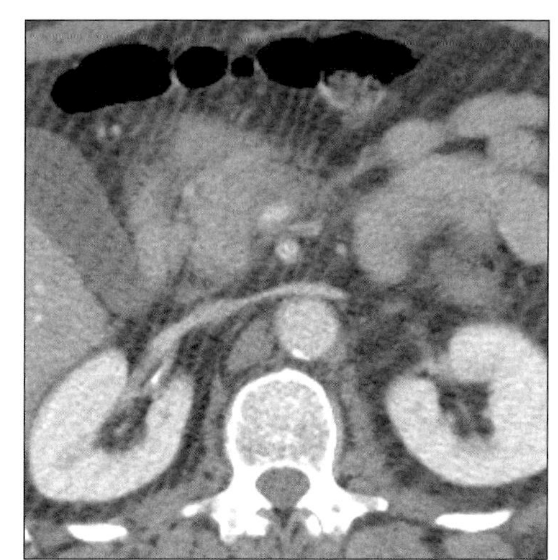

(Left) *Pancreatic contusion. Axial CECT shows heterogeneous mass effect at pancreatic neck, fluid between splenic vein and pancreatic body.* ***(Right)*** *Pancreatic contusion. Axial CECT shows infiltration of fat planes anterior and posterior to pancreatic neck.*

Typical

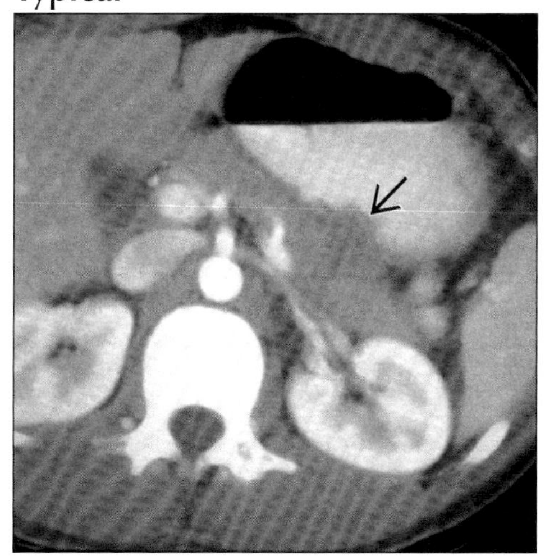

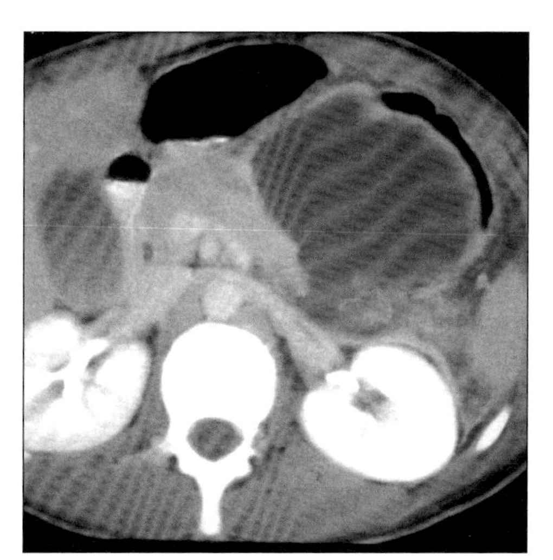

(Left) *Pancreatic transection. Initial axial CECT shows subtle fracture plane and hematoma in body of pancreas (arrow).* ***(Right)*** *Pancreatic transection. Axial CECT 48 hours after trauma shows "pseudocyst" in lesser sac.*

Typical

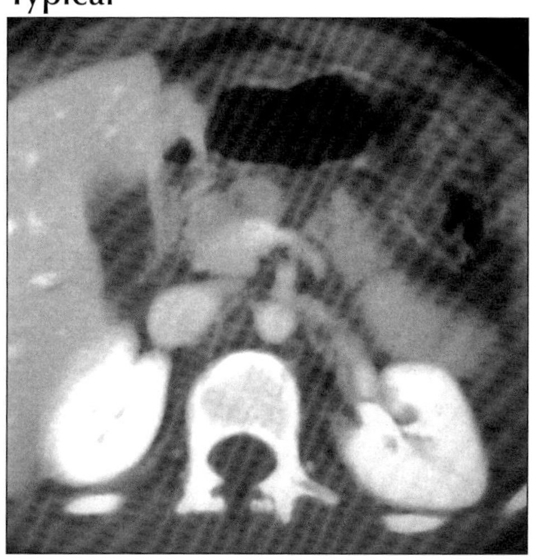

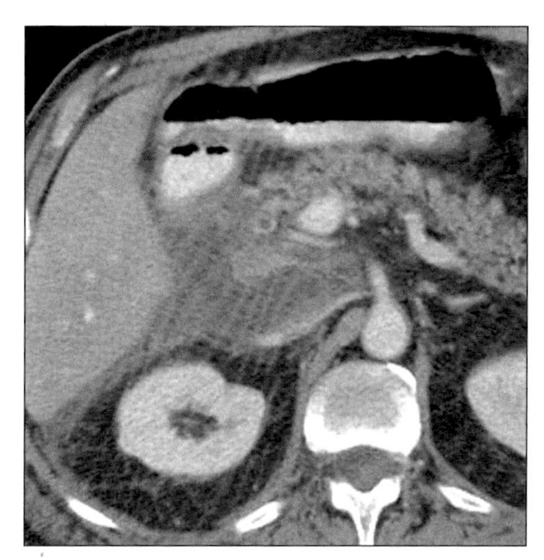

(Left) *Pancreatic transection. Axial CECT shows fracture plane completely through pancreatic neck.* ***(Right)*** *Iatrogenic (post ERCP) pancreatitis. Axial CECT shows extensive infiltration of fat planes and spaces around pancreatic head, while pancreatic body and tail are uninvolved.*

SEROUS CYSTADENOMA OF PANCREAS

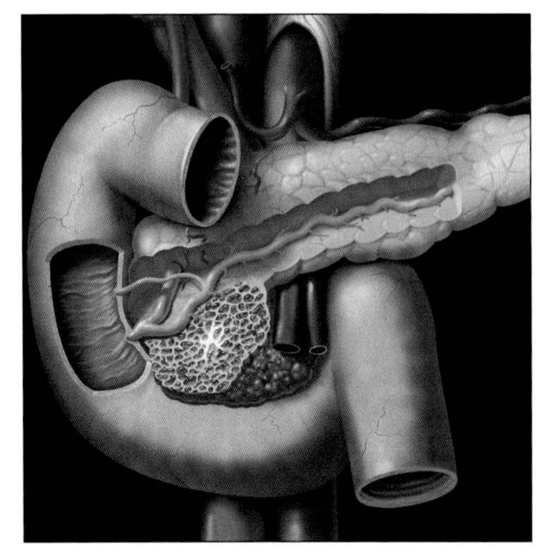

Graphic shows mass with a sponge or honeycomb appearance in the pancreatic head. Innumerable small cysts, a central scar and no obstruction of the pancreatic or bile duct.

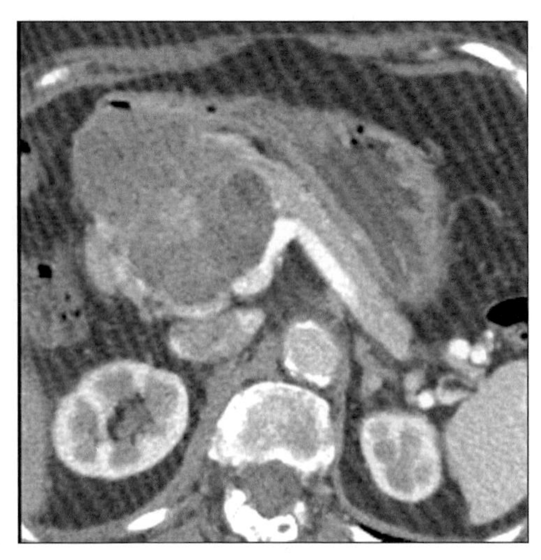

Serous cystadenoma. Axial CECT shows a large mass in the pancreatic head having a sponge appearance with a central scar. No pancreatic or biliary obstruction was present.

TERMINOLOGY

Abbreviations and Synonyms

- Glycogen-rich or micro-/macrocystic serous adenoma

Definitions

- Benign pancreatic tumor that arises from acinar cells

IMAGING FINDINGS

General Features

- Best diagnostic clue: Honeycomb or sponge-like mass in pancreatic head (microcystic serous cystadenoma)
- Location
  - Head of pancreas (more common)
  - Can be seen in any part of pancreas
- Size
  - Large cystic lesion varies from 5-10 cm
  - Innumerable small cysts (1-20 mm) within large cystic lesion
- Morphology
  - Found incidentally in 10-30% of cases
  - Slowly growing tumors & may become quite large masses
  - Calcification is more common in serous than mucinous tumor (38:16%)
  - Most frequently seen in middle-aged/elderly women
  - Tumors show increased frequency in patients with von Hippel-Lindau disease
  - Based on WHO subclassification: Two types
    - Serous microcystic adenomas (more common)
    - Serous oligocystic ("macrocystic" variant) adenoma

Radiographic Findings

- ERCP
  - Displacement, narrowing & dilatation of adjacent MPD and/or CBD

CT Findings

- CECT
  - Microcystic adenoma: Honeycomb pattern
    - Enhancement of septa delineating small cysts
  - Capsular enhancement is noted
  - Calcification within central scar may be seen
  - CBD and/or pancreatic duct dilatation may be seen
  - Atrophy of pancreas distal to tumor (rarely)
- Macrocystic serous cystadenoma (usually unilocular)
  - Location: Usually in pancreatic head
  - One or few cystic components (locules)
  - Thin nonenhancing imperceptible wall
  - Otherwise indistinguishable from mucinous cystic tumor

MR Findings

- T1WI

DDx: Cystic Pancreatic Mass

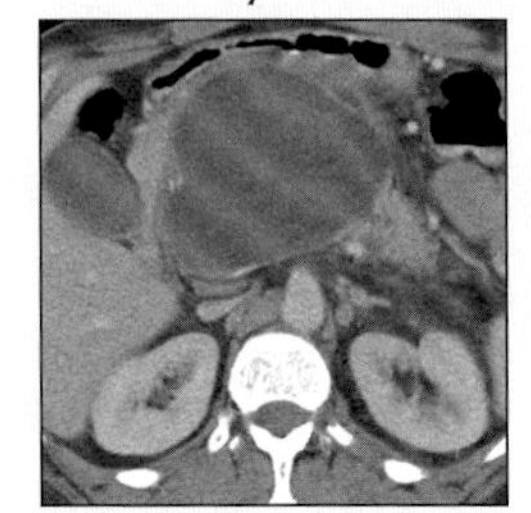

Pseudocyst

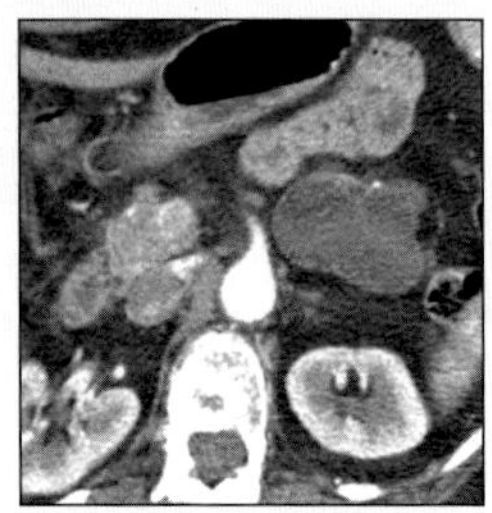

Mucinous Cystic

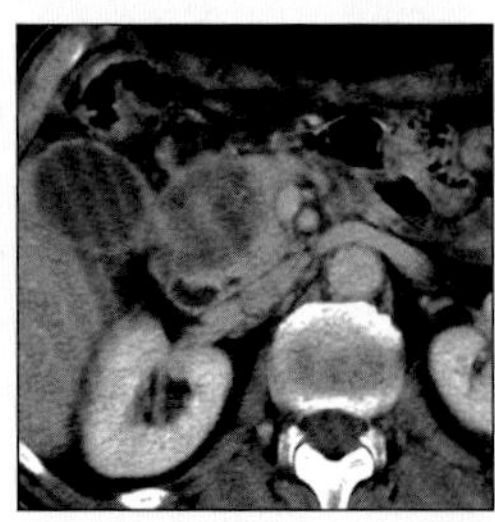

Pancreatic Carcinoma

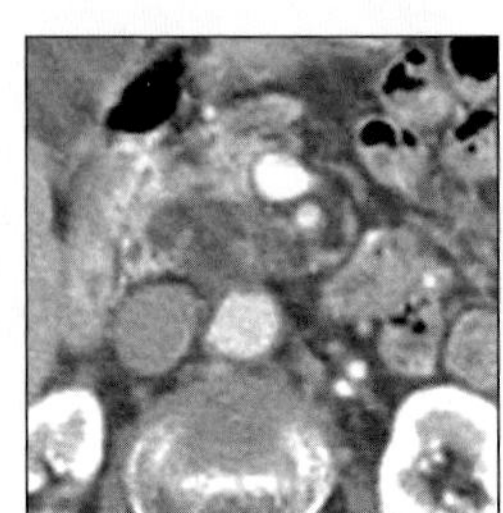

IPMT Pancreas

SEROUS CYSTADENOMA OF PANCREAS

Key Facts

Terminology

- Glycogen-rich or micro-/macrocystic serous adenoma
- Benign pancreatic tumor that arises from acinar cells

Imaging Findings

- Best diagnostic clue: Honeycomb or sponge-like mass in pancreatic head (microcystic serous cystadenoma)
- Enhancement of septa delineating small cysts
- Capsular enhancement is noted
- Macrocystic serous cystadenoma (usually unilocular)
- Thin nonenhancing imperceptible wall

Top Differential Diagnoses

- Pseudocyst
- Mucinous cystadenoma of pancreas
- Pancreatic carcinoma
- Intraductal papillary mucinous tumor (IPMT)
- Congenital pancreatic cysts

Pathology

- Etiology: Uncertain
- Associated abnormalities: Seen with increased frequency in Von Hippel-Lindau disease patients

Clinical Issues

- Asymptomatic, epigastric pain, palpable mass
- Carcinoembryonic antigen level (CEA): Negative
- Gender: M < F (M:F = 1:4)

Diagnostic Checklist

- Rule out other "cystic pancreatic masses"
- Large, well-demarcated, lobulated cystic lesion composed of innumerable small cysts (1-20 mm) separated by thin septa located in head of pancreas

  - Tumor: Hypointense
  - Blood within cysts: Varied intensity
  - Central scar & calcification: Hypointense
- T2WI
  - Tumor: Hyperintense
  - Central scar & calcification: Hypointense
- T1 C+
  - Capsular enhancement
  - Enhancement of septa delineating small cysts
  - Central scar: Enhancement on delayed scan
- MRCP
  - Depict pancreatic duct & CBD dilatation

Ultrasonographic Findings

- Real Time
  - Tumor with tiny cysts
    - Hyperechoic mass but with through transmission
  - Tumor with large cysts
    - Discrete anechoic areas with thin walls
  - Calcification
    - Highly reflective echoes with acoustic shadowing

Angiographic Findings

- Conventional
  - Highly vascular tumor due to extensive capillary network within septa
  - Neovascularity & dense tumor blush
  - Large cysts: Produce lucent regions
  - Dilated feeding arteries
  - Prominent draining veins

Imaging Recommendations

- NE + CECT, MR + CEMR, US

DIFFERENTIAL DIAGNOSIS

Pseudocyst

- Collection of pancreatic fluid encapsulated by fibrous tissue
- Location: More common in body or tail
- Usually unilocular, key is history of pancreatitis
- Cystic mass with infiltration of peripancreatic fat
- NECT
  - Round/oval, hypodense (near water HU)
  - Lobulated, mixed density lesion
    - Hemorrhagic or infected
  - Acute pancreatitis: Enlarged pancreas
  - Chronic pancreatitis: Gland atrophy, dilated MPD & intraductal calculi
- CECT: Rim- or capsule-enhancement
- T1WI: Hypointense
- T2WI
  - Hyperintense (fluid); mixed intensity (fluid/debris)
- MRCP: Hyperintense pseudocyst contiguous with MPD

Mucinous cystadenoma of pancreas

- Most consider this tumor as premalignant
- Location: Tail of pancreas (more common)
- CECT: Multiloculated cystic mass
  - Enhancement of thin internal septa & wall
- T2WI
  - Cysts: Hyperintense
  - Internal septations: Hypointense
- Sonography: Multiloculated cystic mass with echogenic internal septa
- Angiography: Predominantly avascular
- Multilocularity or mural nodules favor tumor
- May be indistinguishable from macrocystic serous cystadenoma of pancreas by imaging alone
- Gross pathology: Multiloculated cystic mass with septa
  - Cysts
    - Fewer than 6 in number, larger than 2 cm
    - Peripheral calcification may be seen

Pancreatic carcinoma

- May appear cystic due to necrosis/fibrosis

Intraductal papillary mucinous tumor (IPMT)

- Low grade malignancy, arises from
  - Main pancreatic duct (MPD)
  - Branch pancreatic duct (BPD) or combined
- CECT
  - BPD lesion: Lobulated "multicystic" lesion
    - May show ring enhancement
  - MPD lesion: Markedly dilated & tortuous MPD
  - Combined type
    - Cystic lesion in head & grossly dilated MPD

- MR
  - T1WI: Hypointense BPD ± dilated, tortuous MPD
  - T2WI: Hyperintense BPD cysts ± dilated MPD
  - MRCP: Hyperintense cysts, dilated MPD
- BPD & combined type lesions of IPMT may simulate serous microcystic adenoma due to presence of dilated small branch ducts in pancreatic head
  - Appear as "grape-like" clusters or small cysts

Congenital pancreatic cysts

- Examples: von Hippel-Lindau & ADPKD
- Rare, usually small & multiple nonenhancing cysts
- No pancreatic ductal dilatation

Cystic islet cell tumor

- Usually non-insulin producing & nonfunctioning
- Tumor: Nonenhancing cyst contents
  - No pancreatic ductal dilatation
- Angiography: Hypervascular primary & secondary

PATHOLOGY

General Features

- General path comments
  - Cell of origin: Centroacinar cell
  - Positive staining for tumor cells is found with epithelial membrane antigen & cytokeratins of low and high molecular weights
  - Tumor shares both clinical & pathologic characteristics of biliary & ovarian tumors
  - Tumors: Composed of smaller cysts (1-20 mm)
  - Serous cystadenomas: No malignant potential
- Etiology: Uncertain
- Epidemiology
  - Cystic pancreatic neoplasms are rare
  - Accounts 10-15% of all pancreatic cysts
  - Accounts only 1% of all pancreatic neoplasms
- Associated abnormalities: Seen with increased frequency in von Hippel-Lindau disease patients

Gross Pathologic & Surgical Features

- Well-circumscribed, round/ovoid, cystic, multilocular
- Lobulated edges secondary to bulging cysts
- On cut section
  - Honeycombed or spongy appearance (due to small, innumerable cysts)
  - Fluid in cysts
    - Typically clear with no mucoid plugs
    - Hemorrhagic (rarely)
  - Thin fibrous septa radiating from central scar
  - Dystrophic calcification within central scar

Microscopic Features

- Cysts are lined by cuboidal/flat epithelial cells separated by fibrous septa
- Cells are glycogen-rich
- No cytologic atypia nor mitotic figures
- Areas of calcium; cholesterol clefts
- Hemosiderin-laden macrophages
- Pancreatic tissue adjacent to tumor is normal or focally atrophic

CLINICAL ISSUES

Presentation

- Most common signs/symptoms
  - Asymptomatic, epigastric pain, palpable mass
  - Weight loss, jaundice, diabetes mellitus
  - Other signs/symptoms of mass effect on adjacent structures (stomach & bowel)
- Lab data
  - Carcinoembryonic antigen level (CEA): Negative
- Diagnosis
  - Endoscopic US with cyst aspiration & cytology

Demographics

- Age
  - Middle & elderly age group (more common)
  - Mean age 65 years
- Gender: M < F (M:F = 1:4)

Natural History & Prognosis

- Complications
  - Due to mass effect
    - Bowel (second part of duodenum) obstruction
    - CBD obstruction (jaundice)
    - Atrophy of pancreatic gland distal to tumor
- Prognosis
  - Completely excised: Good prognosis
  - No malignant potential

Treatment

- Asymptomatic & small tumors
  - No surgical excision if confidently diagnosed
- Symptomatic & large tumors
  - Complete surgical excision & follow-up

DIAGNOSTIC CHECKLIST

Consider

- Rule out other "cystic pancreatic masses"

Image Interpretation Pearls

- Large, well-demarcated, lobulated cystic lesion composed of innumerable small cysts (1-20 mm) separated by thin septa located in head of pancreas

SELECTED REFERENCES

1. Goldsmith JD: Cystic neoplasms of the pancreas. Am J Clin Pathol. 119 Suppl:S3-16, 2003
2. Sheth S et al: Imaging of uncommon tumors of the pancreas. Radiol Clin North Am. 40(6):1273-87, vi, 2002
3. Yeh HC et al: Microcystic features at US: a nonspecific sign for microcystic adenomas of the pancreas. Radiographics. 21(6):1455-61, 2001
4. Curry CA et al: CT of primary cystic pancreatic neoplasms. AJR. 175: 99-103, 2000
5. Procacci C et al: Characterization of cystic tumors of the pancreas: CT accuracy. Journal of Computer Assisted Tomography. 23(6): 906-12, 1999
6. Healy JC et al: CT of microcystic (serous) pancreatic adenoma. J Comput Assist Tomogr. 18(1):146-8, 1994
7. Buck JL et al: From the Archives of the AFIP. Microcystic adenoma of the pancreas. Radiographics. 10(2):313-22, 1990

IMAGE GALLERY

Typical

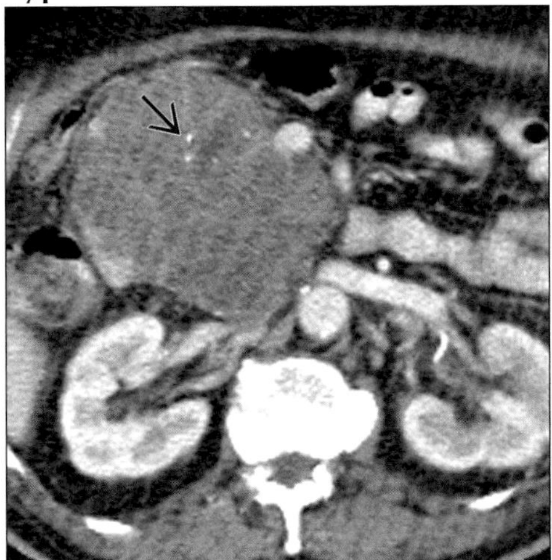

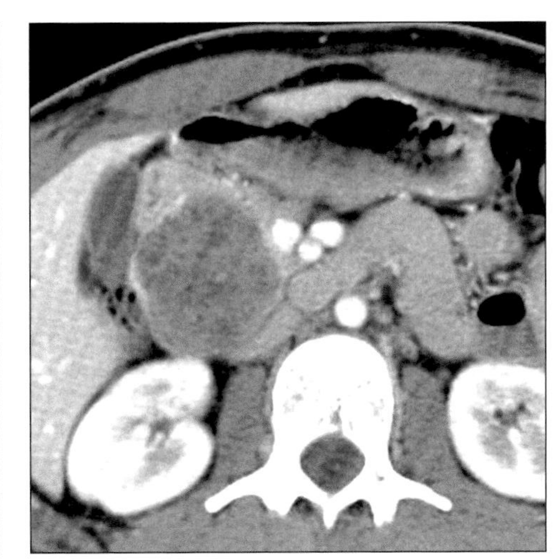

(Left) *Axial CECT shows large sponge-like microcystic serous cystadenoma in pancreatic head. Small calcifications are present in the central scar and septa (arrow).* ***(Right)*** *Axial CECT shows a sponge-like microcystic serous cystadenoma in the pancreatic head.*

Variant

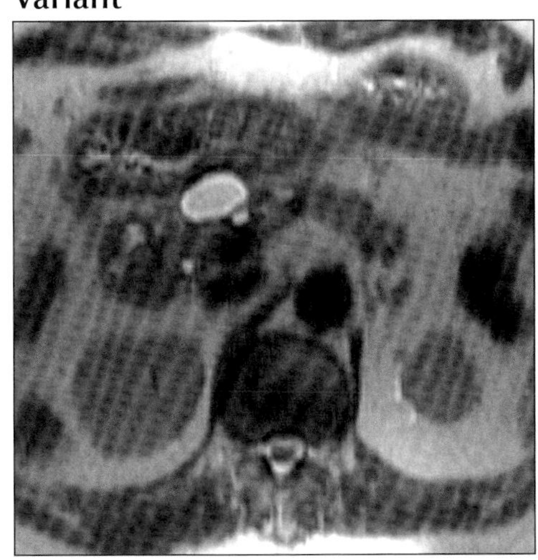

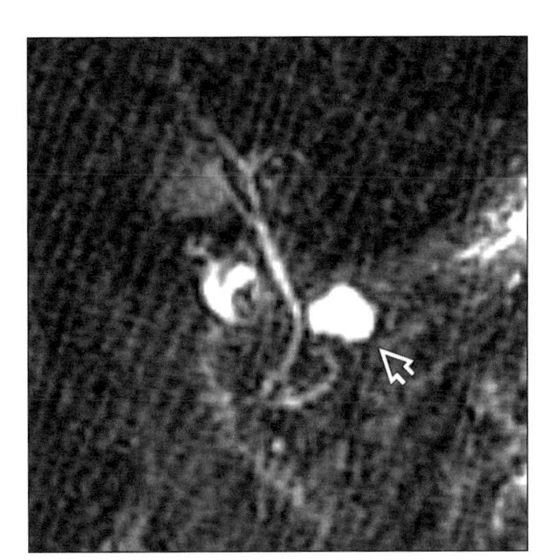

(Left) *Axial T2WI MR shows a unilocular macrocystic serous cystadenoma in the pancreatic head (Courtesy V. Vilgrain, MD).* ***(Right)*** *MRCP shows unilocular macrocystic serous cystadenoma in pancreatic head (open arrow).*

Typical

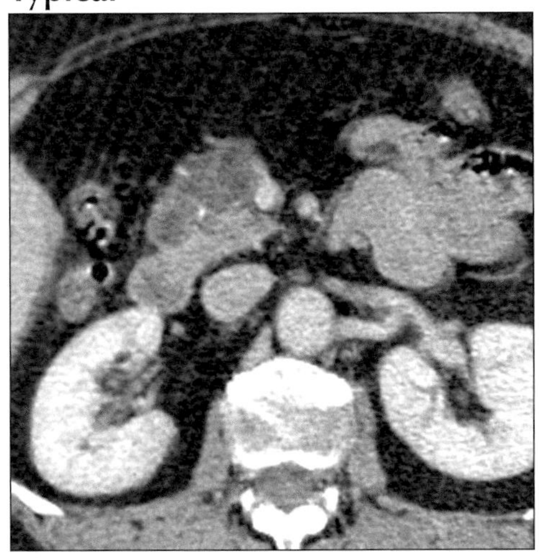

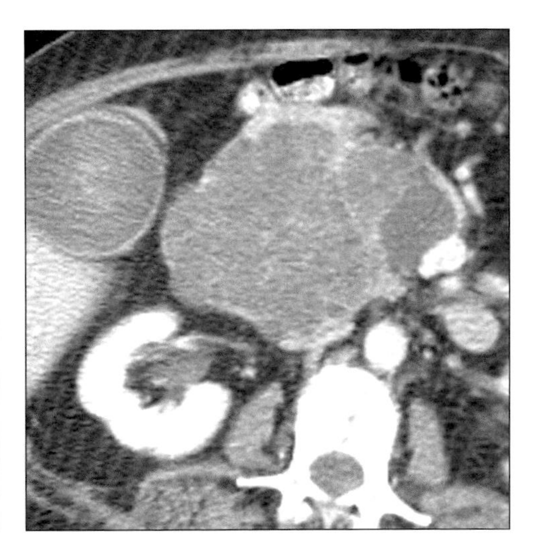

(Left) *Axial CECT shows multiloculated tumor in the region of the pancreatic head. Note fibrous septa surrounding cystic structures within tumor.* ***(Right)*** *Axial CECT shows both microcystic and macrocystic components within this serous cystadenoma.*

PANCREATIC CYSTS

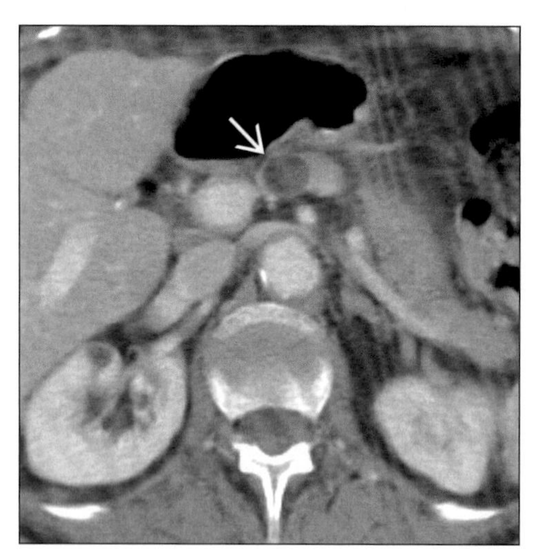

Axial CECT shows a water density thin-walled cyst (arrow) in the pancreatic neck.

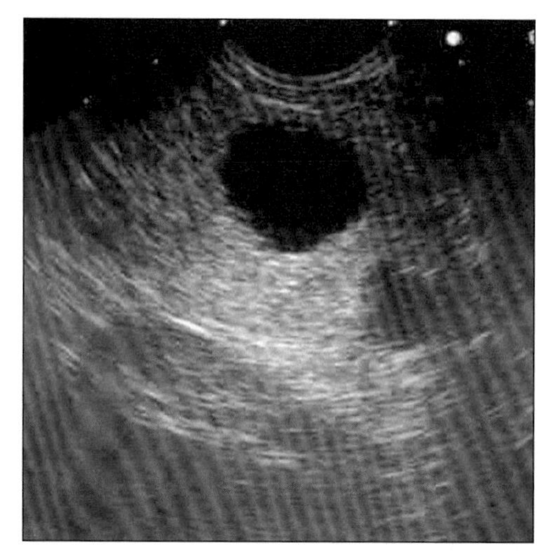

Endoscopic sonography shows a 2 cm diameter simple cyst in the neck of the pancreas.

TERMINOLOGY

Abbreviations and Synonyms

- Congenital/true cysts

Definitions

- Congenital true pancreatic cyst is a very rare cause of cystic lesion of pancreas
- Refers to non-neoplastic, non-inflammatory cysts

IMAGING FINDINGS

General Features

- Best diagnostic clue: Differential diagnosis of true cyst from other cystic lesions of pancreas is usually based on histology
- Size: Usually quite small; giant cyst as large as 15 cm in diameter reported
- Morphology
  - Round or oval shape, smooth thin wall, absence of internal structures
  - Usually unilocular
  - Solitary or multiple (associated with cystic syndromes)

Radiographic Findings

- ERCP: No connection between cyst & pancreatic ductal system

CT Findings

- Imaging modalities show cystic nature of lesion
- Round/oval homogeneous hypodense lesion with near water density
  - Thin imperceptible wall

MR Findings

- T1WI: Low signal intensity
- T2WI: Very high signal intensity

Ultrasonographic Findings

- Anechoic; usually devoid of internal echoes
- Trauma & internal hemorrhage can cause a more complex appearance

Imaging Recommendations

- Best imaging tool: Endoscopic US is diagnostic procedure of choice

DIFFERENTIAL DIAGNOSIS

Pseudocyst

- Usually more complex; history of pancreatitis

Serous cystadenoma

- Honeycomb-like microcysts; sponge-like mass with innumerable small cysts in pancreatic head
- Unilocular or macrocystic serous adenoma

DDx: "Pseudocyst"

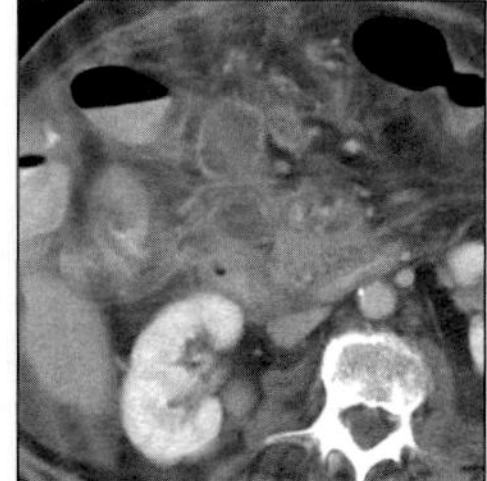

Pseudocyst

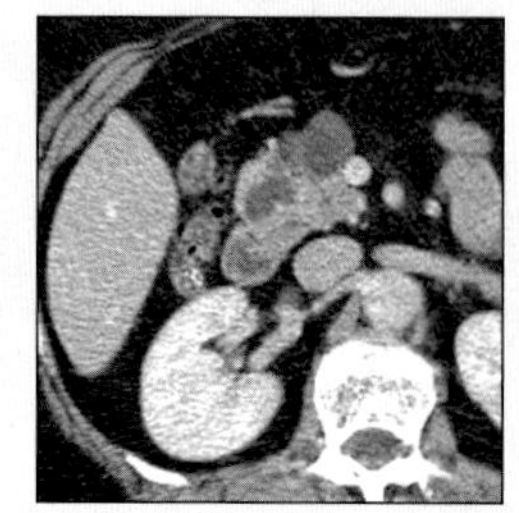

Serous Cystadenoma

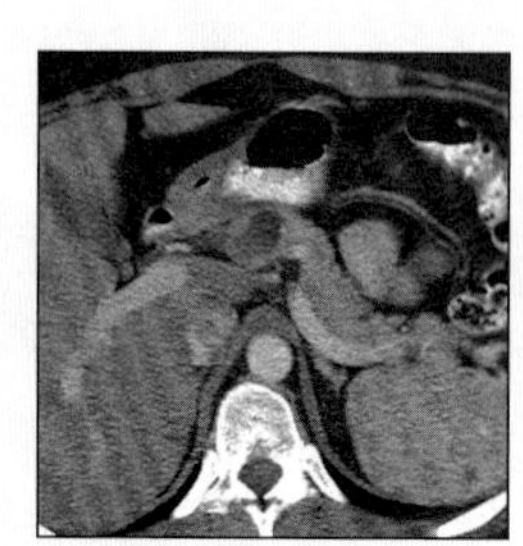

Mucinous Tumor

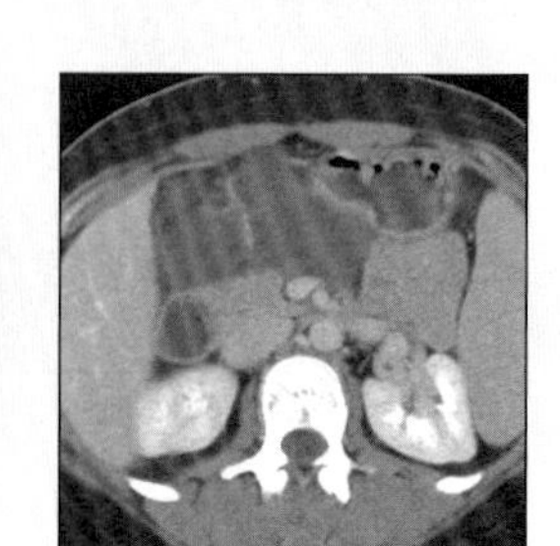

Lymphangioma

PANCREATIC CYSTS

Key Facts

Terminology
- Refers to non-neoplastic, non-inflammatory cysts

Imaging Findings
- Usually unilocular
- Round/oval homogeneous hypodense lesion with near water density
- Thin imperceptible wall
- Anechoic; usually devoid of internal echoes

Top Differential Diagnoses
- Pseudocyst
- Serous cystadenoma
- Mucinous cystic neoplasm
- Intraductal papillary mucinous tumor (IPMT)

Diagnostic Checklist
- May be impossible to distinguish from macrocystic serous cystadenoma

- May be indistinguishable from true cyst except by histology of wall

Mucinous cystic neoplasm
- Septated mass, usually in body/tail of pancreas
- Mutilocularity or mural nodules favor tumor

Intraductal papillary mucinous tumor (IPMT)
- Dilated main pancreatic duct & adjacent cystic lesions

Lymphangioma
- Multiseptated mesenteric cystic mass

PATHOLOGY

General Features
- Epidemiology
  - Isolated cysts are very rare
  - Comprise less than 1% of all pancreatic cysts
- Syndromes account for most non-neoplastic cysts
  - von Hippel Lindau disease
  - Autosomal dominant polycystic kidney (ADPKD)
  - Beckwith-Wiedemann syndrome

Gross Pathologic & Surgical Features
- True epithelial lining (absent in pseudocysts)

Microscopic Features
- Cyst wall: Cuboidal epithelium
- High amylase & lipase contents of cyst
- Fluid of cyst does not contain any mucus

CLINICAL ISSUES

Presentation
- Most common signs/symptoms
  - Asymptomatic, painless, epigastric mass
  - Incidental finding; no history of pancreatic disease

Natural History & Prognosis
- "Simple" cyst ≤ 2 cm in asymptomatic adult is rarely of any clinical significance
- Can follow with imaging, especially in elderly

Treatment
- Complete excision if symptomatic
- Laparotomy may be performed with presumptive diagnosis of cystic tumor of pancreas

DIAGNOSTIC CHECKLIST

Consider
- Possibility of pseudocyst if history of pancreatitis
- Excision for symptomatic or "complex" cystic neoplasms
- May be impossible to distinguish from macrocystic serous cystadenoma
  - Both are benign & of minimal clinical significance

Image Interpretation Pearls
- Imaging characteristics of simple pancreatic cysts on CT & endosonography are uncommonly similar to those of benign cystic neoplasms

SELECTED REFERENCES

1. Cohen-Scali F et al: Discrimination of unilocular macrocystic serous cystadenoma from pancreatic pseudocyst and mucinous cystadenoma with CT: initial observations. Radiology. 228(3):727-33, 2003
2. Bergin D et al: Simple pancreatic cysts: CT and endosonographic appearances. AJR Am J Roentgenol. 178(4):837-40, 2002
3. Takahashi O et al: Solitary true cyst of the pancreas in an adult: report of a case. Int J Gastrointest Cancer. 30(3):165-70, 2001
4. Mao C et al: Solitary true cyst of the pancreas in an adult. Int J Pancreatol. 12(2):181-6, 1992

IMAGE GALLERY

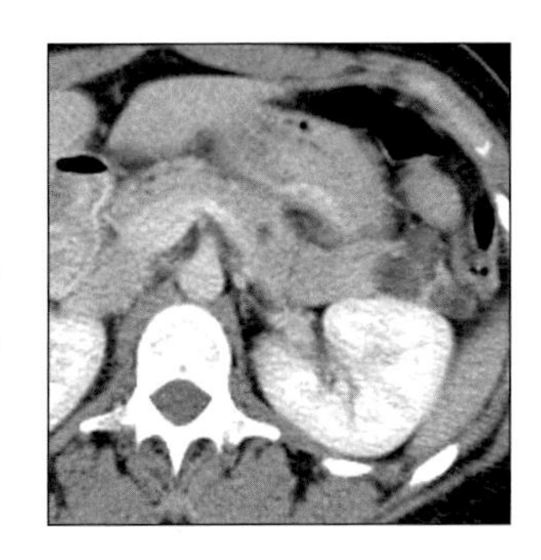

(Left) *Axial NECT in patient with ADPKD shows multiple renal cysts, some with calcified walls. Also present are several pancreatic cysts (arrow).* ***(Right)*** *Axial CECT in a patient with von Hippel Lindau syndrome shows several small cysts in pancreatic tail.*

MUCINOUS CYSTIC PANCREATIC TUMOR

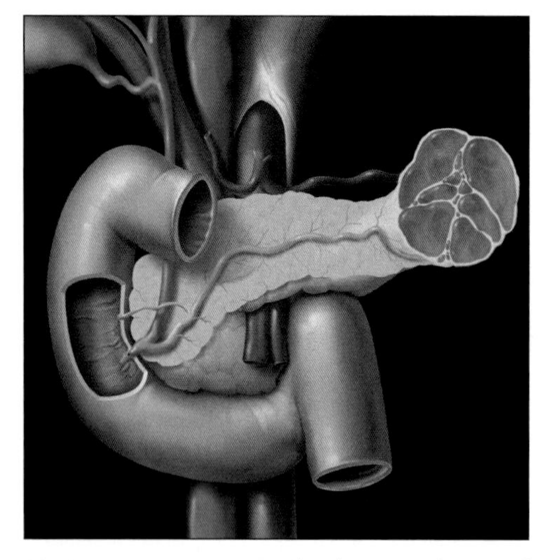

Mucinous cystic tumor. Graphic show a multiseptated mucin-filled cystic mass in the pancreatic tail that displaces the pancreatic duct.

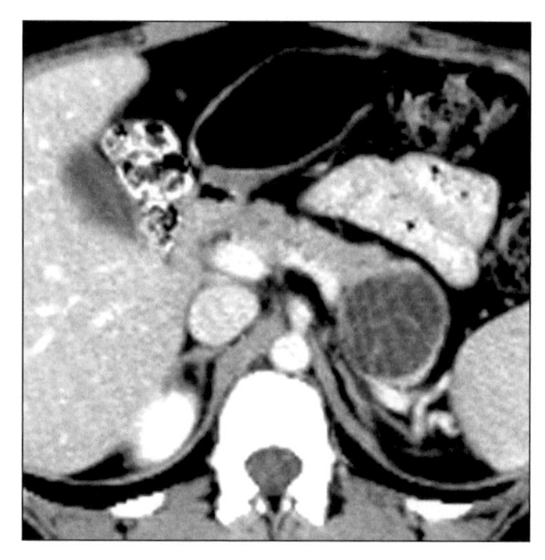

Axial CECT shows a mucinous cystic tumor in pancreatic tail with multiple septations and displacement of the pancreatic duct.

TERMINOLOGY

Abbreviations and Synonyms

- Mucinous macrocystic neoplasm, macrocystic adenoma, mucinous cystadenoma or cystadenocarcinoma

Definitions

- Thick-walled, uni-/multilocular low grade malignant tumor composed of large, mucin-containing cysts

IMAGING FINDINGS

General Features

- Best diagnostic clue: Enhancing multiseptated mass in body or tail of pancreas
- Location: Tail of pancreas (more common)
- Size: Varies from 2-12 cm in diameter
- Morphology
  - Classified under pancreatic mucinous tumors along with intraductal papillary mucinous tumor (IPMT) of pancreas
  - Mucin producing tumors must be considered when cystic lesions of pancreas are found
  - Most consider this tumor as premalignant, if not low grade malignancy

Radiographic Findings

- ERCP
  - Displacement & narrowing of main pancreatic duct adjacent to tumor

CT Findings

- NECT
  - Hypodense unilocular or multilocular cyst
  - Focal calcifications may be seen (16% of cases)
    - Location: Wall, septum or peripheral
- CECT
  - Multilocular cystic lesion
    - Enhancement of thin internal septa & cyst wall
  - Unilocular cystic lesion
    - Enhancement of cyst wall

MR Findings

- T1WI
  - Variable signal intensity based on cyst content
    - Fluid-like material: Hypointense
    - Proteinaceous or hemorrhagic: Hyperintense
  - Focal calcifications: Hypointense
- T2WI
  - Cysts: Hyperintense
  - Internal septations: Hypointense
  - Focal calcifications: Hypointense
- T1 C+
  - Fat suppression sequence

DDx: Cystic Pancreatic Mass

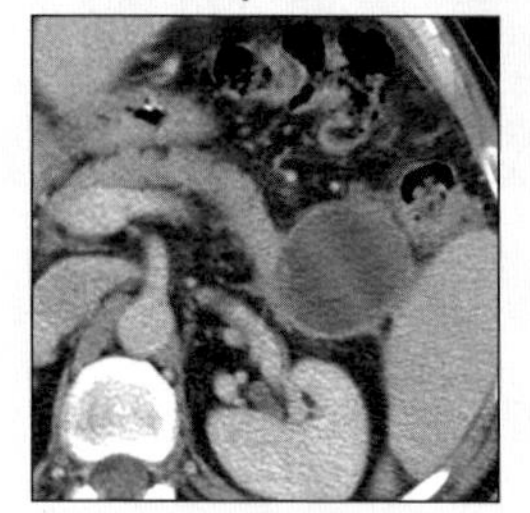

Pseudocyst

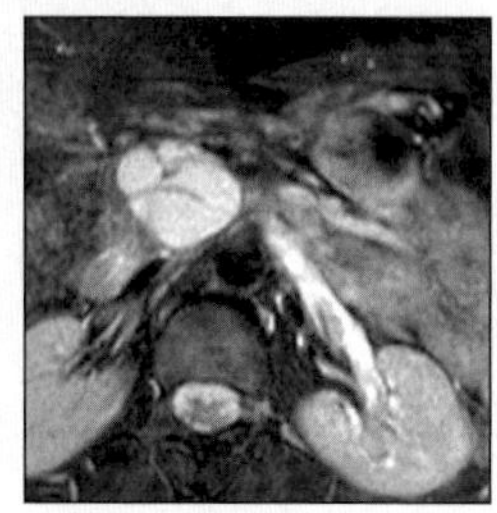

Serous Cystadenoma

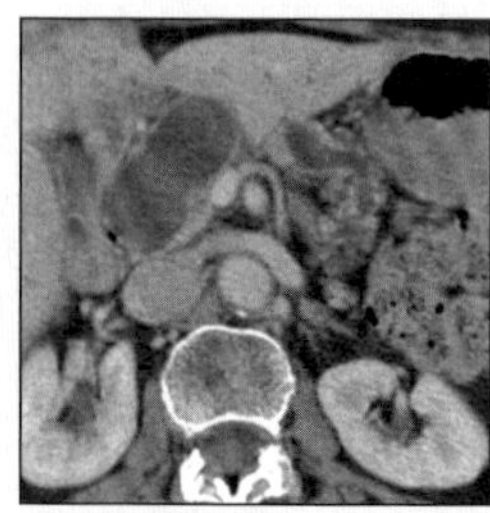

IPMT

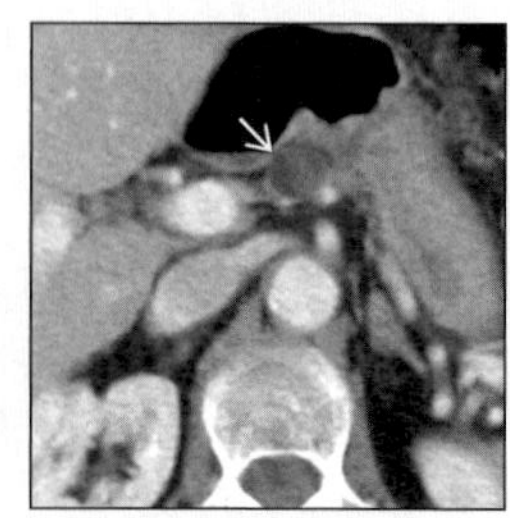

Simple Cyst

MUCINOUS CYSTIC PANCREATIC TUMOR

Key Facts

Terminology

- Thick-walled, uni-/multilocular low grade malignant tumor composed of large, mucin-containing cysts

Imaging Findings

- Best diagnostic clue: Enhancing multiseptated mass in body or tail of pancreas
- Variable signal intensity based on cyst content
- Predominantly avascular mass

Top Differential Diagnoses

- Pseudocyst
- Serous cystadenoma of pancreas
- IPMT of pancreas
- Cystic islet cell tumor
- Congenital pancreatic cysts

Pathology

- May be related to germ cell migration during 1st 8 weeks of gestation
- Etiology: Is uncertain

Clinical Issues

- Asymptomatic, epigastric pain, palpable mass
- Cyst fluid levels of CA 72-4 (more than 4 U/ml)

Diagnostic Checklist

- Differentiate from other "cystic pancreatic lesions"
- Cyst aspiration & check for mucin/tumor markers
- Large, multiloculated cystic mass with enhancing septa & cyst wall in pancreatic body or tail

        - Enhancement of septations & cyst wall
- MRCP
    - Depicts displacement, narrowing & prestenotic dilatation of pancreatic duct

Ultrasonographic Findings

- Real Time
    - Multiloculated cystic mass with echogenic internal septa
    - Unilocular anechoic mass

Angiographic Findings

- Conventional
    - Predominantly avascular mass
    - Cyst wall & solid component
        - Show small areas of vascular blush & neovascularity
    - Displacement of surrounding arteries & veins by cysts

Imaging Recommendations

- MR + T1 C+; CECT

DIFFERENTIAL DIAGNOSIS

Pseudocyst

- Inflammatory changes in peripancreatic fat
- Pancreatic calcifications & temporal evolution of lesion
- Communicate with pancreatic duct (70% of cases)
- Clinical history of pancreatitis or alcoholism
- Lab data: Increased levels of amylase
- Simulates unilocular mucinous cystic tumor

Serous cystadenoma of pancreas

- Large, well-defined, encapsulated, sponge-like mass in pancreatic head
- Innumerable small cysts separated by thin septa
- Central scar with calcification
- Calcification more common in serous than mucinous pancreatic neoplasms (38:16%)
- Macrocystic variant of serous cystadenoma
    - Difficult to distinguish from mucinous tumor
    - Serous lesion usually has thinner wall

IPMT of pancreas

- IPMT: Intraductal papillary mucinous tumor
- Low-grade malignancy arises from
    - Main pancreatic duct (MPD)
    - Branch pancreatic duct (BPD) or combined
- BPD or combined type of IPMT may simulate mucinous cystic neoplasm due to presence of dilated cystic branch ducts in pancreatic tail

Cystic islet cell tumor

- Usually non-insulin producing & nonfunctioning
- Tumor: Cystic on NECT & nonenhancing cyst contents
    - Cyst wall shows enhancement
- No pancreatic ductal dilatation
- Angiography: Hypervascular primary & secondary

Congenital pancreatic cysts

- Examples: von Hippel-Lindau disease & autosomal dominant polycystic kidney disease (ADPKD)
- Rare, usually small & multiple nonenhancing cysts
- No pancreatic ductal dilatation

Variant of ductal adenocarcinoma

- Mucinous colloid adenocarcinoma or mucin-hypersecreting cancer
- Pancreatic ductal obstruction & dilatation
- Local invasion & regional metastases

Lymphangioma

- Often extends from or into retroperitoneal soft tissues
- Water density; imperceptible wall; thin septations

PATHOLOGY

General Features

- General path comments
    - Embryology-anatomy
        - May be related to germ cell migration during 1st 8 weeks of gestation

- A neoplasm with number of cysts less than 6 & more than 2 cm in diameter seen in 95% cases
- Stromal component is must for diagnosis of mucinous cystic neoplasm
- Tumor shares both clinical & pathologic characteristics of biliary & ovarian tumors
- Great propensity for invasion of adjacent organs
- Hypovascular mass with sparse neovascularity

- Etiology: Is uncertain
- Epidemiology
  - Uncommon primary tumor of pancreas
  - Frequency: 10% of pancreatic cysts & 1% of pancreatic neoplasms

Gross Pathologic & Surgical Features

- Large encapsulated mass by thick fibrous capsule (2-12 cm in diameter)
- Smooth & round; a lobulated surface may be seen
- Cut section: Multi-/unilocular large cysts
  - More than 2 cm; thin septa less than 2 mm thick
- Cystic cavity may be filled with thick mucoid material/clear/green/blood-tinged fluid
- Solid papillary projections protrude into interior of tumor (sign of cancer)

Microscopic Features

- Tall, mucin-producing columnar cells
- Subtended by a densely cellular mesenchymal stroma
- Characteristic ovarian-type stroma with spindle cells

CLINICAL ISSUES

Presentation

- Most common signs/symptoms
  - Asymptomatic, epigastric pain, palpable mass
  - Symptoms of mass effect on adjacent structures (stomach/bowel)
  - Rarely tumor may manifest with local invasion/distant metastases
  - Very rarely present with systemic manifestations caused by tumor production of gastrin/VIP
- Laboratory data
  - Increased levels of serum CEA
  - Increased cyst fluid levels of CA 19-9 (80% of cases)
  - Cyst fluid levels of CA 72-4 (more than 4 U/ml)
    - 80% sensitivity & 95% specificity for tumor
- Diagnosis
  - Endoscopic ultrasound with cyst aspiration/cytology
  - Tumor markers
  - Surgical resection

Demographics

- Age
  - Mean age: 50 years (range of 20-95 years)
  - 50% between 40-60 years
- Gender: Females more than males (M:F = 1:9)

Natural History & Prognosis

- Complications
  - Due to mass effect
    - Bowel obstruction
    - Pancreatic duct narrowing or extrinsic obstruction
- Prognosis
  - Completely excised: Good prognosis
  - Incompletely excised, marsupialized or drained
    - Poor prognosis
  - 5 year survival rate with malignancy regardless of surgery (74.3%)

Treatment

- Complete surgical excision

DIAGNOSTIC CHECKLIST

Consider

- Differentiate from other "cystic pancreatic lesions"
- Cyst aspiration & check for mucin/tumor markers

Image Interpretation Pearls

- Large, multiloculated cystic mass with enhancing septa & cyst wall in pancreatic body or tail

SELECTED REFERENCES

1. Maire F et al: Benign inflammatory pancreatic mucinous cystadenomas mimicking locally advanced cystadenocarcinomas. Presentation of 3 cases. Pancreatology. 2(1):74-8, 2002
2. Hara T et al: Mucinous cystic tumors of the pancreas. Surg Today. 32(11):965-9, 2002
3. Oshikawa O et al: Dynamic sonography of pancreatic tumors: comparison with dynamic CT. AJR Am J Roentgenol. 178(5):1133-7, 2002
4. Yamaguchi K et al: Radiologic imagings of cystic neoplasms of the pancreas. Pancreatology. 1(6):633-6, 2001
5. Balci NC et al: Radiologic features of cystic, endocrine and other pancreatic neoplasms. Eur J Radiol. 38(2):113-9, 2001
6. Friedman AC et al: CT of primary cystic pancreatic neoplasms: nihilism may be unwarranted. AJR Am J Roentgenol. 177(2):469-70, 2001
7. Megibow AJ et al: Cystic pancreatic masses: cross-sectional imaging observations and serial follow-up. Abdom Imaging. 26(6):640-7, 2001
8. Grogan J et al: Making sense of mucin-producing pancreatic tumors. AJR. 176: 921-9, 2001
9. Sarr MG et al: Clinical and pathologic correlation of 84 mucinous cystic neoplasms of the pancreas: can one reliably differentiate benign from malignant (or premalignant) neoplasms? Ann Surg. 231(2):205-12, 2000
10. Lundstedt C et al: Serous and mucinous cystadenoma/cystadenocarcinoma of the pancreas. Abdom Imaging. 25(2):201-6, 2000
11. Taouli B et al: Intraductal papillary mucinous tumors of the pancreas: helical CT with histopathologic correlation. Radiology. 217(3):757-64, 2000
12. de Lima JE Jr et al: Mucinous cystic neoplasm of the pancreas. Radiographics. 19(3):807-11, 1999
13. Le Borgne J et al: Cystadenomas and cystadenocarcinomas of the pancreas: a multiinstitutional retrospective study of 398 cases. French Surgical Association. Ann Surg. 230(2):152-61, 1999
14. Buetow PC et al: From the Archives of the AFIP. Mucinous cystic neoplasms of the pancreas: radiologic-pathologic correlation. Radiographics. 18(2):433-49, 1998

IMAGE GALLERY

Typical

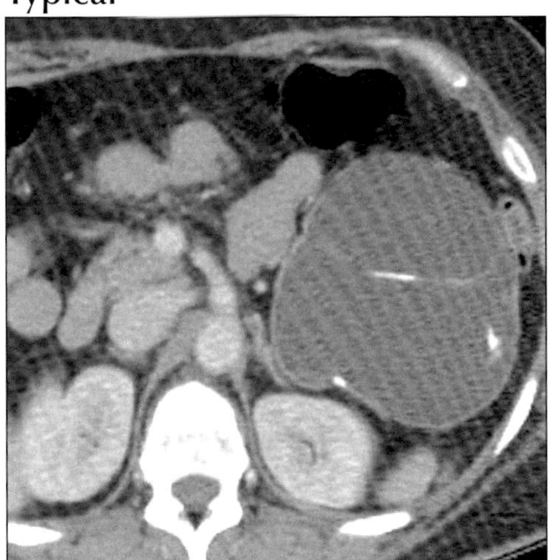

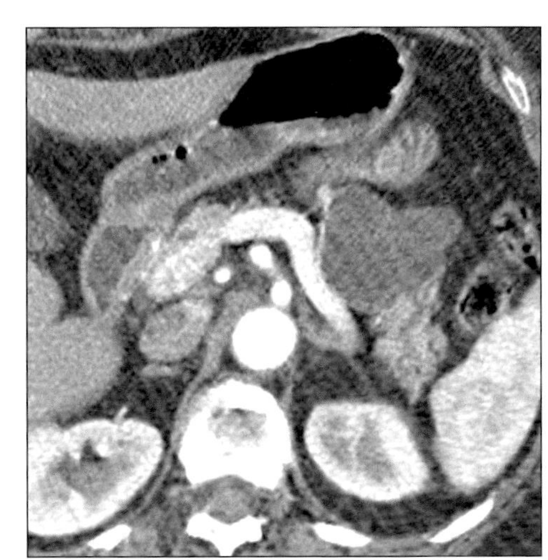

(Left) *Axial CECT shows a cystic mucinous tumor in the pancreatic tail, with a few large cystic spaces separated by visible septa and focal calcifications.* ***(Right)*** *Axial CECT shows a cystic mucinous tumor in the pancreatic body containing a few cystic spaces and septa. Pancreatic duct is compressed and dilated.*

Typical

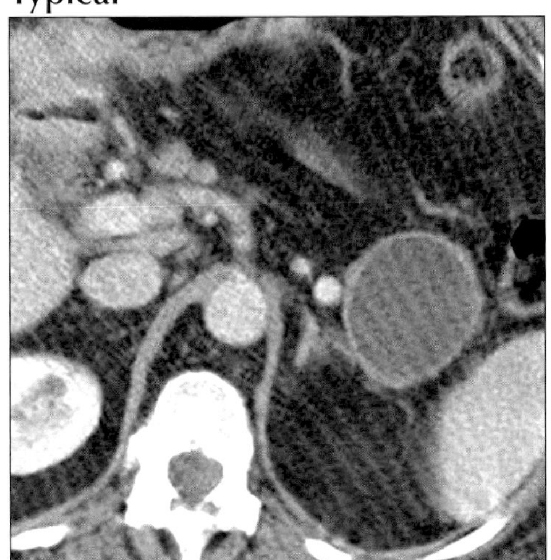

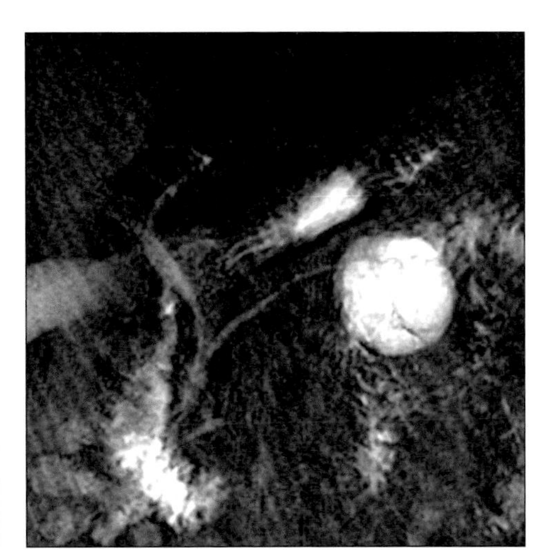

(Left) *Axial CECT shows a unilocular mucinous cystic tumor in the pancreatic tail. Visible non-calcified wall.* ***(Right)*** *Coronal MRCP shows cystic mucinous tumor in pancreatic tail with a few cystic spaces and septa. Pancreatic duct is deviated but otherwise normal.*

Typical

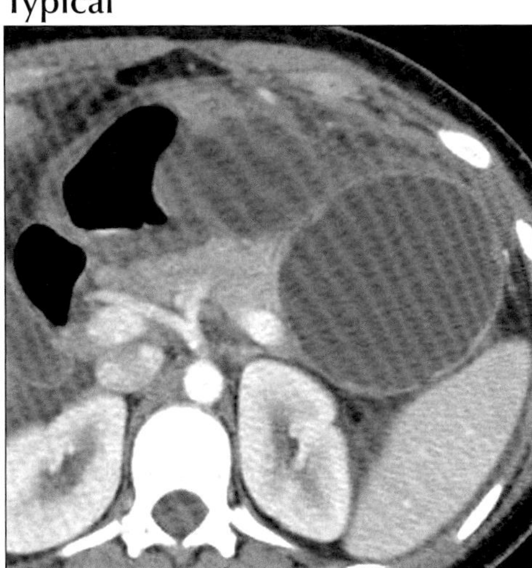

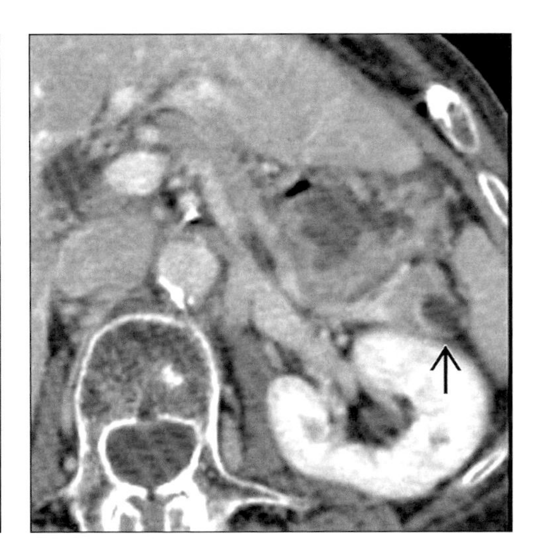

(Left) *Mucinous cystic tumor. CECT in 20 year old woman thought to have pseudocysts. Lack of resolution led to cyst aspiration yielding thick mucinous material with a low amylase and high CA 19-9 levels.* ***(Right)*** *Axial CECT shows a 1.5 cm cystic mass (arrow) in the pancreatic tail. Confirmed mucinous cystic tumor at surgical resection.*

IPMT, PANCREAS

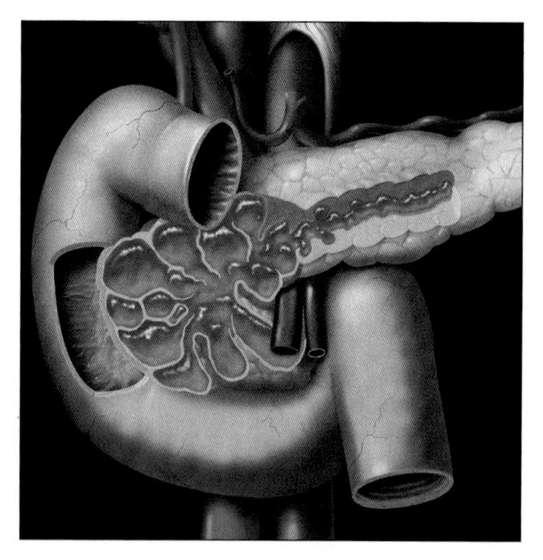

Graphic shows combined main and branch type IPMT with gross dilatation of all ducts by mucin, which pours out of a bulging papilla into the duodenum. Parenchyma in head is atrophic.

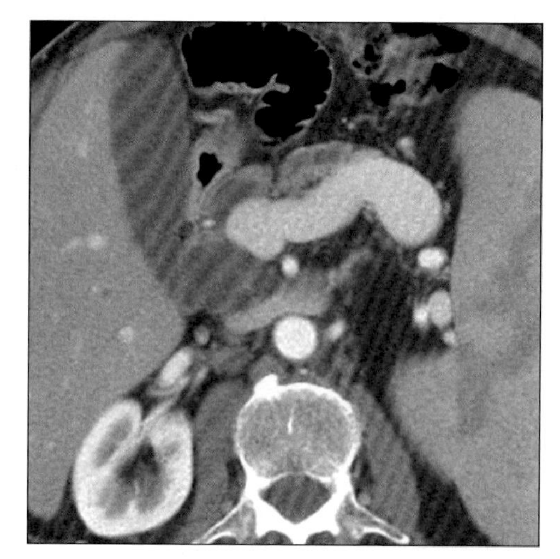

Axial CECT shows combined main and branch type IPMT, with dilatation of all pancreatic ducts and pancreatic parenchymal atrophy.

TERMINOLOGY

Abbreviations and Synonyms

- Intraductal papillary mucinous tumor (IMPT)
- Intraductal mucin-hypersecreting neoplasm, ductectatic mucinous cystadenoma/carcinoma

Definitions

- Low grade malignancy that arises from epithelial lining of main pancreatic duct (MPD) and/or branch pancreatic ducts (BPD) with excessive mucin production

IMAGING FINDINGS

General Features

- Best diagnostic clue: "Multicystic" lesion in uncinate process/head contiguous with dilated MPD on CECT
- Location
  - BPD lesion: Uncinate process & head
  - MPD lesion: Usually body or tail
- Size: BPD cysts: 5-20 mm
- Morphology
  - IPMT is a subdivision of mucin producing tumors along with mucinous macrocystic neoplasm
  - Classified into three types
    - Branch pancreatic duct (BPD) type: Focal lobulated "multicystic" dilatation of branch ducts
    - Main pancreatic duct (MPD) type: Diffuse dilatation of main pancreatic duct
    - Combined type: Dilatation of both BPD & MPD
  - Most common type is combined variety
    - BPD plus MPD

Radiographic Findings

- ERCP
  - Cystic BPDs, dilated MPD
  - Thick intraductal mucinous secretions
    - Elongated or band-like filling defects in MPD
  - Mucin & papillary tumors clearly detected as nodular filling defects
  - Diagnostic of IPMT of pancreas
    - Real-time visualization of patulous ampulla & increased mucus production

CT Findings

- BPD type: Lobulated "multicystic" lesion ("grape-like" clusters or tubes & arcs)
  - Thin, irregular, peripheral ring-enhancing "multicystic" lesion
  - Bulging ampulla at duodenal sweep with thin rim enhancement
- MPD type: Markedly dilated tortuous MPD
  - May visualize polypoid lesions lining MPD
  - Punctate calcifications may be seen

DDx: Dilated Main Pancreatic Duct with or without Cyst

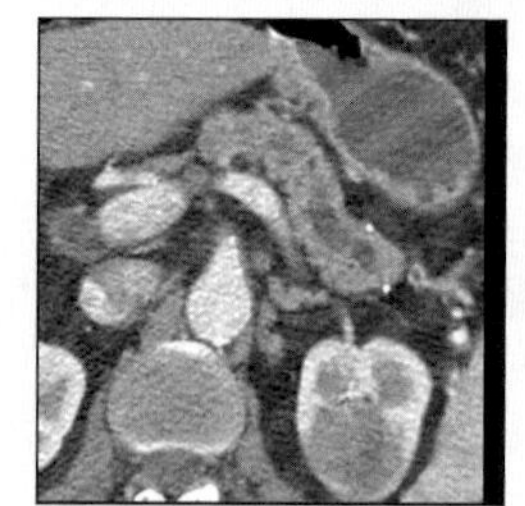

Chronic Pancreatitis

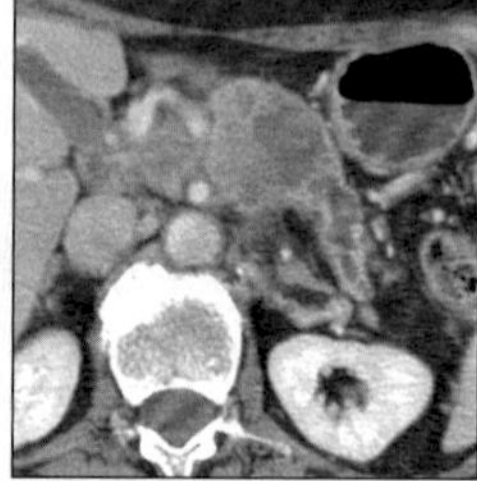

Pancreatic Carcinoma

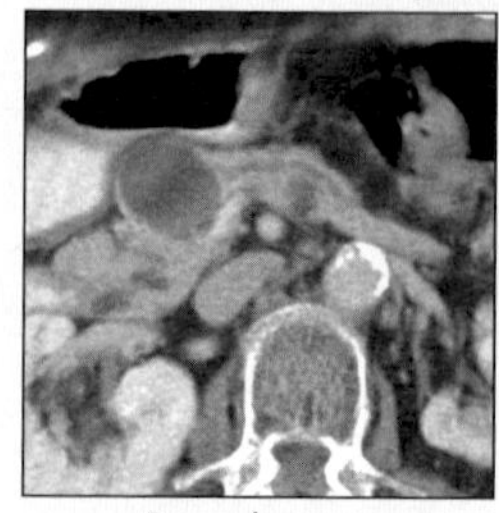

Pseudocyst

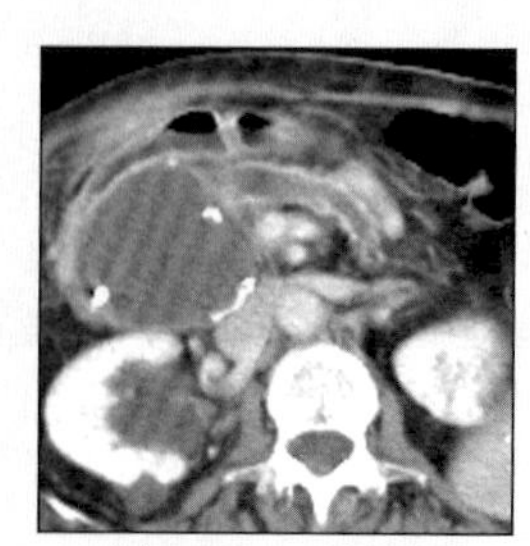

Mucinous Cystic

IPMT, PANCREAS

Key Facts

Terminology

- Intraductal papillary mucinous tumor (IMPT)
- Low grade malignancy that arises from epithelial lining of main pancreatic duct (MPD) and/or branch pancreatic ducts (BPD) with excessive mucin production

Imaging Findings

- Best diagnostic clue: "Multicystic" lesion in uncinate process/head contiguous with dilated MPD on CECT
- BPD type: Lobulated "multicystic" lesion ("grape-like" clusters or tubes & arcs)
- MPD type: Markedly dilated tortuous MPD
- May visualize polypoid lesions lining MPD
- Combined type: "Multicystic" lesion in uncinate process contiguous with grossly dilated MPD
- Reveals communication between cystic lesions/ducts

Top Differential Diagnoses

- Chronic pancreatitis
- Pancreatic pseudocyst
- Mucinous cystic neoplasm
- Serous cystadenoma

Diagnostic Checklist

- Rule out other cystic pancreatic pathologies associated with dilated main pancreatic duct (MPD)
- Markedly dilated MPD & adjacent cystic lesions in uncinate process or head of pancreas
- MRCP shows communication between cystic lesions & dilated ductal system

- Combined type: "Multicystic" lesion in uncinate process contiguous with grossly dilated MPD
- Atrophy of gland may be seen distal to tumor

MR Findings

- T1WI
  - Axial: Hypointense branch duct cysts ± dilated MPD
  - Coronal: Clustered cystic lesion with thin septa
- T2WI
  - Hyperintense dilated branch duct cysts ± dilated MPD
  - May show papillary excrescence in cystic lesion of pancreatic head or along MPD
- MRCP
  - Lobulated clustered cysts, dilated MPD
  - Reveals communication between cystic lesions/ducts
  - Intraductal mucin & papillary tumors: May be detected as nodular filling defects

Ultrasonographic Findings

- Real Time
  - Septated cystic lesion
  - Dilated MPD
  - Punctate calcifications (occasionally)
  - Intra-operative or endoscopic sonography affords superior visualization

Imaging Recommendations

- MR, MRCP, ERCP & CT coronal & oblique reconstructed images
- Thin-sections facilitate quality reformations

DIFFERENTIAL DIAGNOSIS

Chronic pancreatitis

- Focal or diffuse atrophy of gland
- Obstruction & dilatation of pancreatic/bile ducts
- Intraductal calculi & areas of calcification
- Thickening of peripancreatic fascia
- Pseudocyst: Round, homogeneous, hypodense cystic lesion
- ERCP & MRCP
  - Obstruction & dilatation of pancreatic duct/radicles
  - Pseudocyst contiguous with pancreatic duct

Pancreatic carcinoma

- Hypovascular mass with abrupt obstruction of ducts

Pancreatic pseudocyst

- Collection of pancreatic fluid encapsulated by fibrous capsule
- Round/oval, homogeneous, hypodense cystic lesion
- Inflammatory changes in peripancreatic fat
- MRCP: Communicate with dilated pancreatic duct (70% of cases)
- Clinical history of pancreatitis or alcoholism
- Sometimes mimics combined type of IPMT lesion

Mucinous cystic neoplasm

- Most consider this tumor as premalignant
- Septated globular mass
- Location: Tail of pancreas (more common)
- Enhancement of thin septa & wall
- Multilocularity or mural nodules favor tumor
- ERCP or MRCP: Displacement, narrowing & dilatation of pancreatic duct adjacent to tumor
- Angiography: Predominantly avascular
- Gross pathology: Multiloculated cystic mass with septa
  - Cysts
    - Fewer than 6 in number, larger than 2 cm
    - Peripheral calcification may be seen

Serous cystadenoma

- Glycogen-rich cystadenoma of pancreas
- Benign pancreatic tumor (arises from acinar cells)
- Slowly growing tumor & may become quite large
- Most frequently seen in middle-aged females
- Sponge-like mass of innumerable small cysts in head
  - May simulate BPD type of IPMT in head
- ERCP or MRCP
  - Displacement, narrowing & dilatation of pancreatic duct adjacent to tumor
  - No communication of lesion to pancreatic duct
- Calcification within central scar may be seen
- Enhancement of septa & cyst wall
- Angiography: Highly vascular tumor

IPMT, PANCREAS

PATHOLOGY

General Features

- General path comments
  - Tumors are typically papillary lesions
  - Ranging in size from a few millimeters to panductal
  - Varying proportions of dysplasia & in situ cancer
- Etiology
  - Uncertain
  - Pathogenesis: Sequence of events in IPMT
    - Hyperplasia of columnar epithelial cells lining ducts
    - Dysplasia & proliferation to form papillary projections
    - Papillary projections protrude into & expand BPD & MPD
    - Excessive mucin production, obstruction & dilatation of BPD/MPD
    - Malignant transformation over many years
- Epidemiology: Rare pancreatic cystic neoplasm

Gross Pathologic & Surgical Features

- BPD
  - Cystically dilated branch ducts with rough internal surface
  - Few/multiple intraductal elevated papillary tumors (adenomas)
  - Communicating channels between cystic spaces
- MPD
  - Dilated MPD filled with mucin
  - Flat elongated tumor (hyperplasia/malignancy) with a rough surface

Microscopic Features

- Simple hyperplasia, papillary adenomas, dysplasia, in situ carcinoma
- Innumerable papillary projections covered with columnar epithelial cells
- Intervening septa that separate mucin-filled lacuna

CLINICAL ISSUES

Presentation

- Most common signs/symptoms
  - BPD & MPD types
    - Pain, weight loss, diarrhea, attacks of pancreatitis & diabetes
  - Most specific predictive signs of malignancy
    - Diabetes mellitus
    - Solid mass on imaging
    - Dilated MPD: More than 10 mm
    - Diffuse/multifocal involvement
    - Attenuating/calcified intraluminal content
- Laboratory data
  - Increased serum & urinary amylase
  - Altered pancreatic function tests

Demographics

- Age: Onset between 60 & 80 years
- Gender: M > F

Natural History & Prognosis

- Complications
  - Recurrent attacks of acute & chronic pancreatitis
  - Biliary disease
- Prognosis
  - Localized lesion
    - After resection have better prognosis than ductal adenocarcinoma & mucinous cystadenocarcinoma
  - Invasive carcinoma: Poor prognosis

Treatment

- In older & less symptomatic cases
  - Periodic monitoring of head/uncinate BPD variant
- In younger & more symptomatic cases
  - Complete surgical excision & frozen section analysis

DIAGNOSTIC CHECKLIST

Consider

- Rule out other cystic pancreatic pathologies associated with dilated main pancreatic duct (MPD)

Image Interpretation Pearls

- Markedly dilated MPD & adjacent cystic lesions in uncinate process or head of pancreas
- MRCP shows communication between cystic lesions & dilated ductal system

SELECTED REFERENCES

1. Sugiyama M et al: Predictive factors for malignancy in intraductal papillary-mucinous tumours of the pancreas. Br J Surg. 90(10):1244-9, 2003
2. Sai JK et al: Management of branch duct-type intraductal papillary mucinous tumor of the pancreas based on magnetic resonance imaging. Abdom Imaging. 28(5):694-9, 2003
3. Prasad SR et al: Intraductal papillary mucinous tumors of the pancreas. Abdom Imaging. 28(3):357-65, 2003
4. Taouli B et al: Intraductal papillary mucinous tumors of the pancreas: features with multimodality imaging. J Comput Assist Tomogr. 26(2):223-31, 2002
5. Peters HE et al: Magnetic resonance cholangiopancreatography (MRCP) of intraductal papillary-mucinous neoplasm (IPMN) of the pancreas: case report. Magn Reson Imaging. 19(8):1139-43, 2001
6. Lim JH et al: Radiologic spectrum of intraductal papillary mucinous tumor of the pancreas. Radiographics. 21(2):323-37; discussion 337-40, 2001
7. Silas AM et al: Intraductal papillary mucinous tumors of the pancreas. AJR Am J Roentgenol. 176(1):179-85, 2001
8. Taouli B et al: Intraductal papillary mucinous tumors of the pancreas: helical CT with histopathologic correlation. Radiology. 217(3):757-64, 2000
9. Fukukura Y et al: Intraductal papillary mucinous tumors of the pancreas: thin-section helical CT findings. AJR Am J Roentgenol. 174(2):441-7, 2000
10. Procacci C et al: Intraductal papillary mucinous tumor of the pancreas: a pictorial essay. Radiographics. 19(6):1447-63, 1999
11. Ariyama J et al: Endoscopic ultrasound and intraductal ultrasound in the diagnosis of small pancreatic tumors. Abdom Imaging. 23(4):380-6, 1998
12. Koito K et al: Mucin-producing pancreatic tumors: comparison of MR cholangiopancreatography with endoscopic retrograde cholangiopancreatography. Radiology. 208(1):231-7, 1998

IMAGE GALLERY

Typical

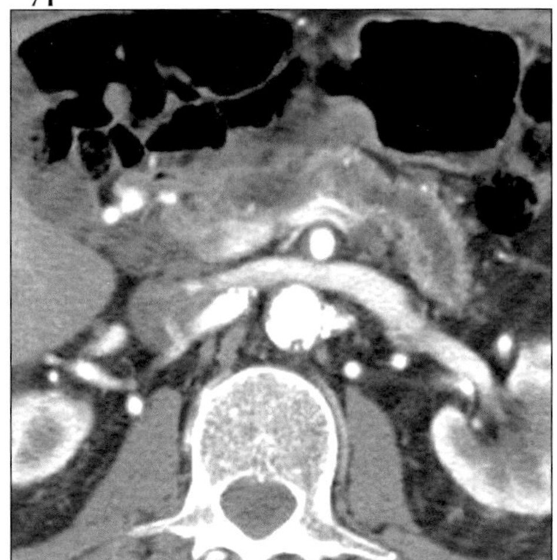

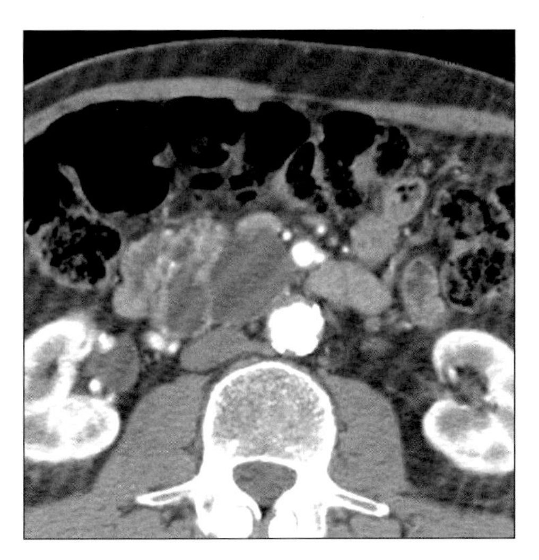

(Left) *Axial CECT shows combined branch pancreatic duct (BPD) and main pancreatic duct (MPD) IPMT. Dilated main pancreatic duct, glandular atrophy, no mass.* ***(Right)*** *Combined branch pancreatic duct (BPD) and main pancreatic duct (MPD) IPMT. "Cystic" dilatation of branch ducts in pancreatic head and uncinate.*

Typical

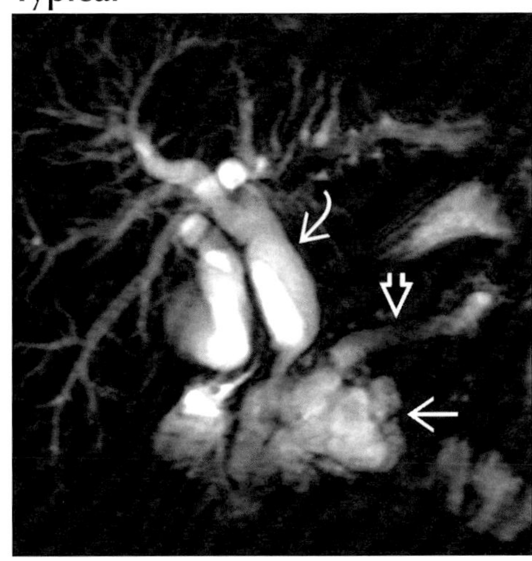

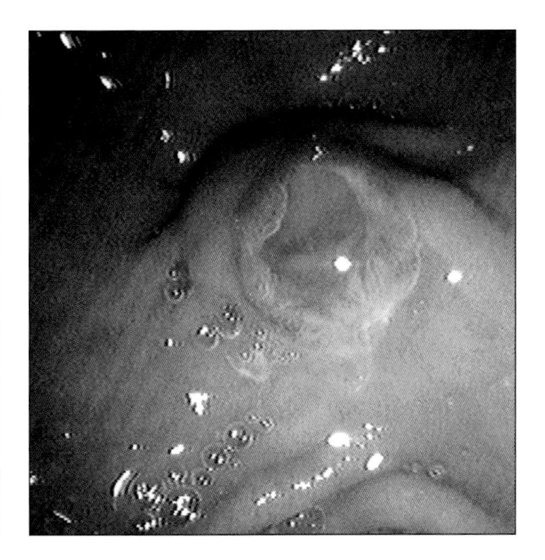

(Left) *Coronal MRCP in combined BPD and MPD IPMT. Pancreatic duct (open arrow) and common bile duct (curved arrow) are dilated. Cluster of dilated mucin-filled branch ducts in pancreatic head (arrow).* ***(Right)*** *Endoscopic view showing mucin pouring out of a patulous papilla.*

Typical

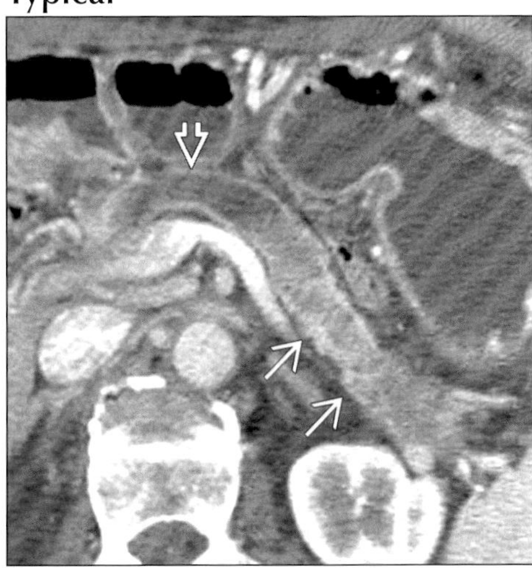

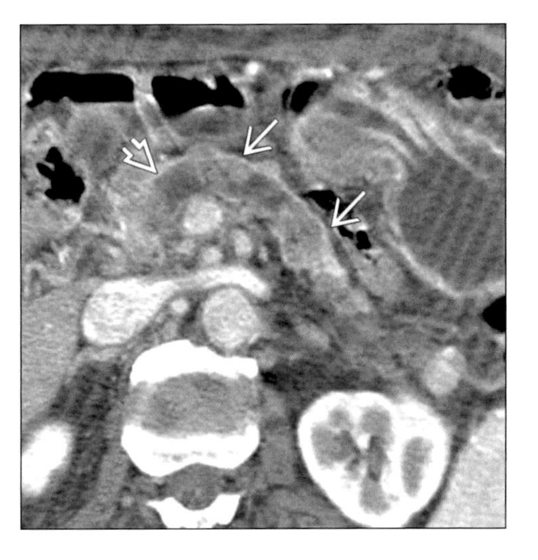

(Left) *Malignant main pancreatic duct (MPD) IPMT. The main pancreatic duct (open arrow) is grossly distended. Solid papillary tumor nodules are evident (arrows).* ***(Right)*** *Malignant main pancreatic duct (MPD) IPMT. The MPD (open arrow) is grossly dilated. Solid papillary tumor nodules are evident (arrows).*

PANCREATIC DUCTAL CARCINOMA

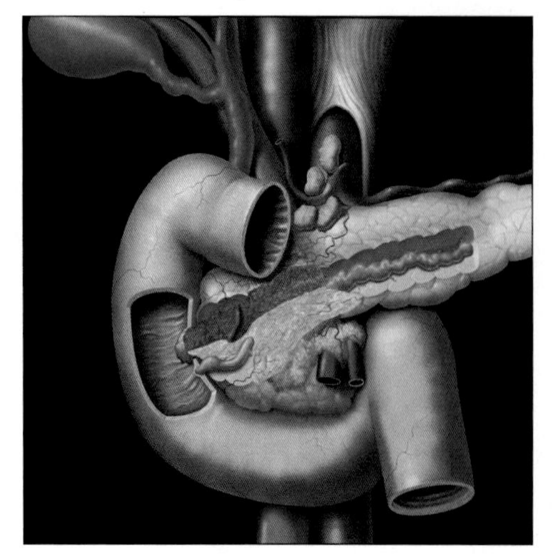

Graphic shows scirrhous mass in pancreatic head that partially obstructs the common bile and pancreatic ducts. The mesenteric vessels are encased by tumor; celiac lymphadenopathy is present.

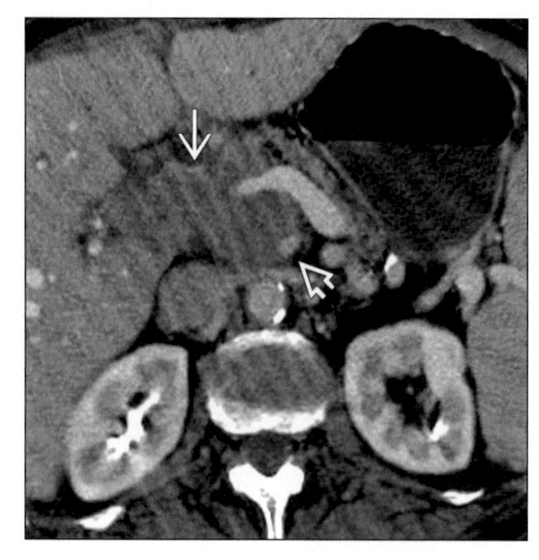

Axial CECT (venous phase) shows hypodense mass (arrow) in pancreatic head that encases splenoportal confluence & SMA (open arrow); occludes SMV. Body & tail are atrophic; duct is dilated.

TERMINOLOGY

Abbreviations and Synonyms

- Pancreatic adenocarcinoma, pancreatic cancer

Definitions

- Malignancy that arises from ductal epithelium of exocrine pancreas

IMAGING FINDINGS

General Features

- Best diagnostic clue: Irregular, heterogeneous, poorly-enhancing mass with abrupt obstruction of pancreatic and/or common bile duct ("double duct sign")
- Location: Head (60%), body (20%), diffuse (15%), tail (5%)
- Size
  - Varies; average diameter is 2-3 cm
  - Large tumor can be up to 8-10 cm
- Morphology
  - Most common primary malignant tumor of exocrine pancreas
  - Accounts more than 75% of pancreatic tumors
  - Rarely resectable for cure
  - Small & ill-defined or large tumor, with extensive local invasion & regional metastasis

Radiographic Findings

- Barium (UGI) study
  - "Frostberg 3" sign
    - "Inverted 3" contour to medial part of duodenal sweep
  - Spiculated duodenal wall, traction & fixation
  - "Antral padding"
    - Extrinsic indentation of posteroinferior margin of antrum
- ERCP
  - Irregular, nodular, rat-tailed eccentric obstruction
  - Localized encasement with prestenotic dilatation
  - "Double duct" sign: Obstruction of pancreatic and common bile duct at same level

CT Findings

- NECT
  - Isodense mass, no hemorrhage, Ca++ very rare
  - Dilated pancreatic duct & obliteration of retropancreatic fat
- CECT
  - Heterogeneous, poorly-enhancing mass
  - Parenchymal atrophy distal to tumor may be seen
  - Pancreatic ductal dilatation distal to tumor
  - Lesion in head may also cause CBD obstruction & dilatation of bile ducts

DDx: Pancreatic Mass with or without Dilated Main Pancreatic Duct

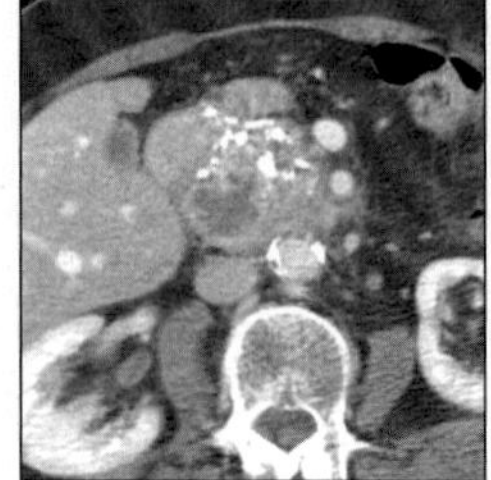

Chronic Pancreatitis

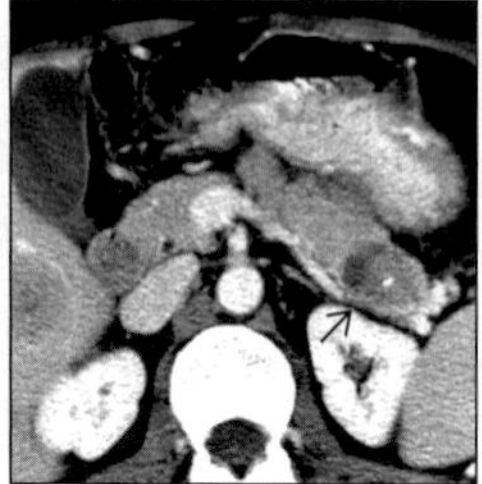

Islet Cell Tumor

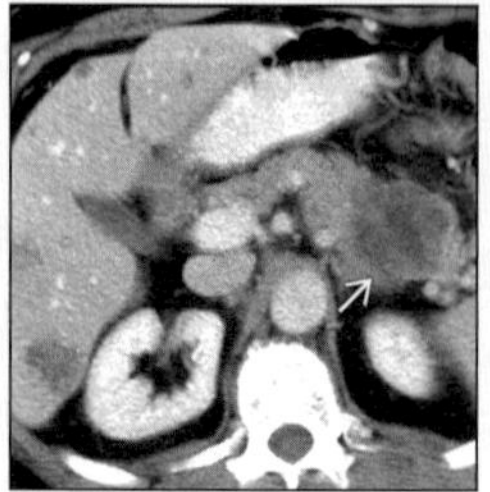

Metastases

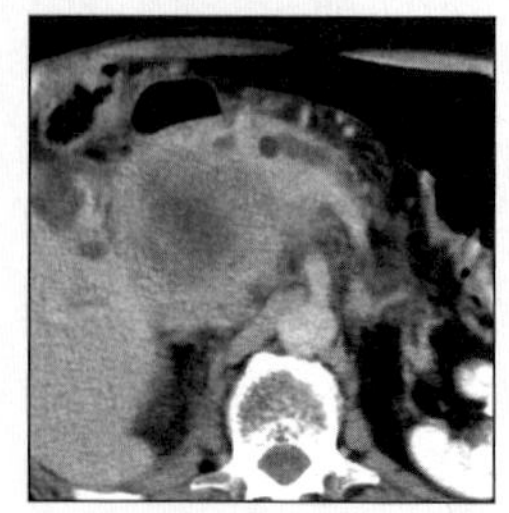

Lymphoma

PANCREATIC DUCTAL CARCINOMA

Key Facts

Terminology
- Malignancy that arises from ductal epithelium of exocrine pancreas

Imaging Findings
- Best diagnostic clue: Irregular, heterogeneous, poorly-enhancing mass with abrupt obstruction of pancreatic and/or common bile duct ("double duct sign")
- Location: Head (60%), body (20%), diffuse (15%), tail (5%)
- Small & ill-defined or large tumor, with extensive local invasion & regional metastasis
- Parenchymal atrophy distal to tumor may be seen
- Hypovascular tumor

Top Differential Diagnoses
- Chronic pancreatitis
- Islet cell carcinoma
- Metastases
- Lymphoma

Clinical Issues
- Usually asymptomatic until late in its course
- Obstructive jaundice, pain & weight loss
- Elevated levels of tumor markers: CEA, CA19-9

Diagnostic Checklist
- Differentiate from other solid pancreatic masses with or without main pancreatic duct dilatation
- Irregular heterogeneous mass in head of pancreas with eccentric ductal obstruction/dilatation & extensive local invasion & regional metastases

  - Local tumor extension into splenic hilum & porta hepatis
  - Contiguous organ invasion
    - Duodenum, stomach & mesenteric root
  - Vascular invasion: "Tear drop" shaped SMV
    - Encasement of more than half circumference of vessel, narrowing or occlusion
    - Mesenteric collateral veins may be seen
  - Distant metastases
    - Liver, peritoneum & regional nodes (common)
    - Adrenals, bones, lungs & pleura (rare)

MR Findings
- T1WI
  - Low signal intensity relative to normal parenchyma due to fibrous nature of tumor
  - Fat-suppressed T1WI
    - Hypointense lesion compared to high signal intensity of normal pancreatic parenchyma
- T2WI: Variable signal intensity
- T1 C+
  - Poor or no enhancement on dynamic study
  - No significant diagnostic improvement over CT
- T2 GRE & T1WI spin-echo sequences
  - Detects vascular invasion
- MRCP: Show level & degree of ductal obstruction

Ultrasonographic Findings
- Real Time
  - Hypoechoic mass & contour deformity of gland
  - Pancreatic ductal dilatation distal to tumor

Angiographic Findings
- Conventional
  - Hypovascular tumor
  - Effective in detecting carcinoma of body & tail
    - Demonstrates vascular narrowing
    - Displacement or occlusion by tumor

Imaging Recommendations
- Helical CT
  - With thin collimation (3-5 mm)
  - Rapid IV bolus contrast injection
- CT & MR
  - High predictive value (near 100%) for tumor unresectability
  - Less predictive value (75-85%): Resectable tumor
- ERCP; PET combined CT; endoscopic US

DIFFERENTIAL DIAGNOSIS

Chronic pancreatitis
- Focal or diffuse atrophy of gland, fibrotic mass in head
- Dilated main pancreatic duct with ductal calculi
- Parenchymal calcification is also seen
- Distal CBD long stricture causes prestenotic dilatation
- Thickening of peripancreatic fascia & fat necrosis
- May be indistinguishable from cancer on imaging

Islet cell carcinoma
- Hypervascular primary & secondary tumors
- Ring enhancement seen in insulinoma
- No pancreatic ductal dilatation
- Usually functioning tumors are small in size & non-functioning tumors are large in size

Metastases
- Hypovascular metastases (e.g., lung, colon)
- Hypervascular metastases (e.g., renal, melanoma)
- Rarely obstruct pancreatic and biliary ducts

Lymphoma
- May show focal or diffuse glandular enlargement
- Rarely obstructs ducts & widely disseminated in nodes

PATHOLOGY

General Features
- General path comments
  - Produces mucin & dense, collagenous desmoplastic stroma
  - Spread: Local, peripancreatic, perivascular, perineural & lymphatic invasion
- Genetics
  - Mutations in K-ras genes & p16INK4 gene on chromosome 9p21

- Abnormal high levels of p53 gene
- Etiology: Increased risk factors: Cigarette smoking, diabetes mellitus, chronic pancreatitis, high-fat diet
- Epidemiology
  - Fourth leading cause of cancer deaths in US
  - 11th most common cancer in US
  - Accounts 2-3% of all cancers
- Associated abnormalities
  - Heritable syndromes
    - Hereditary pancreatitis, ataxia telangiectasia
    - Familial colon cancer, Gardner syndrome
    - Familial aggregation of pancreatic cancer

Gross Pathologic & Surgical Features

- Hard nodular mass obstructing pancreatic duct/CBD
- Hypovascular, locally invasive, desmoplastic response

Microscopic Features

- White fibrous lesion, dense cellularity, nuclear atypia
- Most ductal cancers are mucinous adenocarcinomas

Staging, Grading or Classification Criteria

- Stage I
  - Confined to pancreas (or)
  - Extension into peripancreatic tissues
- Stage II: Stage I plus regional lymph node metastases
- Stage III: Stage I & II plus distant metastases

CLINICAL ISSUES

Presentation

- Most common signs/symptoms
  - Usually asymptomatic until late in its course
  - Head of pancreas
    - Obstructive jaundice, pain & weight loss
  - Body & tail
    - Weight loss & massive metastases to liver
  - Courvoisier law suggests pancreatic cancer
    - Painless jaundice with a palpable gallbladder
  - At presentation
    - 65% patients: Advanced local disease/metastases
    - 21%: Localized disease with spread to regional lymph nodes
    - 14%: Tumor confined to pancreas
  - Lab data
    - Elevated levels of tumor markers: CEA, CA19-9

Demographics

- Age
  - Mean age at onset: 55 years
  - Peak age: 7th decade
- Gender: M:F = 2:1
- Ethnicity: Blacks more than whites

Natural History & Prognosis

- Complications: Venous thrombosis, GI hemorrhage
- Prognosis
  - 1 & 5 year survival (poor, even with surgery)
  - With surgery (pancreaticoduodenectomy)
    - 5-year survival rate is approximately 20%
  - Without surgery
    - 5-year survival rate is less than 5%
  - Tumor markers used for prognosis are
    - CEA, CA19-9 & CA 242

Treatment

- Complete surgical resection for potentially curative tumor (< 15%)
- Pancreaticoduodenectomy ("Whipple resection")
- Radiotherapy: External beam radiation
- Chemotherapy
- Endoscopic stenting-palliates obstructive jaundice
- Gastric bypass-palliates duodenal obstruction
- Chemical splanchnicectomy or celiac nerve block to palliate abdominal pain

DIAGNOSTIC CHECKLIST

Consider

- Differentiate from other solid pancreatic masses with or without main pancreatic duct dilatation

Image Interpretation Pearls

- Irregular heterogeneous mass in head of pancreas with eccentric ductal obstruction/dilatation & extensive local invasion & regional metastases

SELECTED REFERENCES

1. Yusoff IF et al: Preoperative assessment of pancreatic malignancy using endoscopic ultrasound. Abdom Imaging. 28(4):556-62, 2003
2. Roche CJ et al: CT and pathologic assessment of prospective nodal staging in patients with ductal adenocarcinoma of the head of the pancreas. AJR Am J Roentgenol. 180(2):475-80, 2003
3. Ly JN et al: MR imaging of the pancreas: a practical approach. Radiol Clin North Am. 40(6):1289-306, 2002
4. Rodallec M et al: Helical CT of pancreatic endocrine tumors. J Comput Assist Tomogr. 26(5):728-33, 2002
5. McNulty N et al: Multi-detector row helical CT of pancreas: Effect of contrast enhanced multiphasic imaging on enhancement of pancreas,peripancreatic vasculature, and pancreatic adenocarcinoma. Radiology. 220: 97-102, 2001
6. Johnson D: Pancreatic carcinoma: Developing a protocol for multi-detector row CT. Radiology. 220: 3-4, 2001
7. Jadvar H et al: Evaluation of pancreatic carcinoma with FDG PET. Abdom Imaging. 26(3):254-9, 2001
8. Brizi MG et al: Staging of pancreatic ductal adenocarcinoma with spiral CT and MRI. Rays. 26(2):151-9, 2001
9. Ros PR et al: Imaging features of pancreatic neoplasms. JBR-BTR. 84(6):239-49, 2001
10. Nishiharu T et al: Local extension of pancreatic carcinoma: Assessment with thin-section helical CT versus with breath-hold fast MR imaging-ROC analysis. Radiology. 212: 445-52, 1999
11. Demachi H et al: Histological influence on contrast-enhanced CT of pancreatic ductal adenocarcinoma. J Comput Assist Tomogr. 21(6):980-5, 1997
12. Ichikawa T et al: Pancreatic ductal adenocarcinoma: preoperative assessment with helical CT versus dynamic MR imaging. Radiology. 202(3):655-62, 1997
13. Soyer P et al: Involvement of superior mesenteric vessels and portal vein in pancreatic adenocarcinoma: detection with CT during arterial portography. Abdom Imaging. 19(5):413-6, 1994

IMAGE GALLERY

Typical

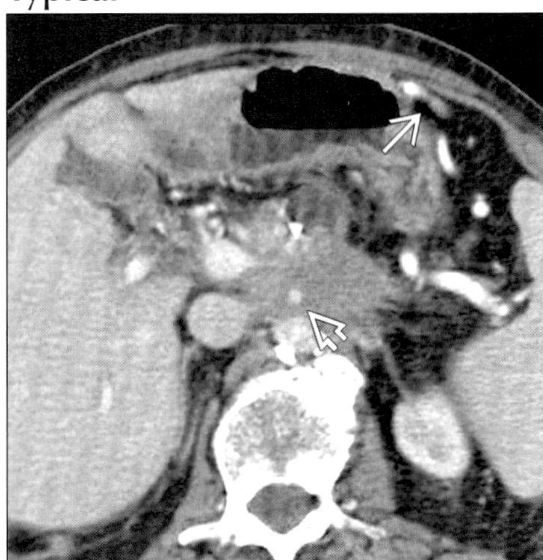

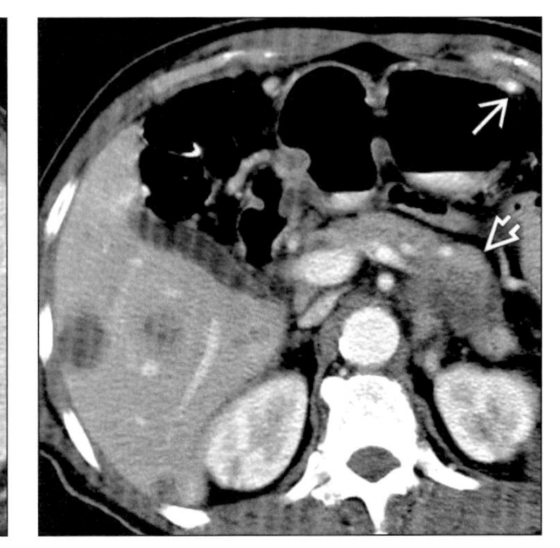

(Left) *CECT shows cancer arising from pancreatic head/uncinate with extensive encasement of SMA (open arrow) and splenoportal confluence. Perigastric collaterals (arrow) indicate splenic vein occlusion.* ***(Right)*** *Axial CECT shows hypovascular mass in pancreatic body (open arrow) that occludes the splenic artery and vein, with perigastric varices (arrow). Note liver metastases.*

Typical

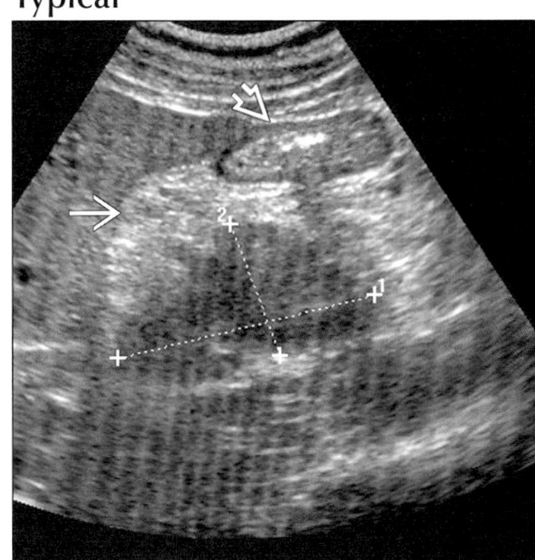

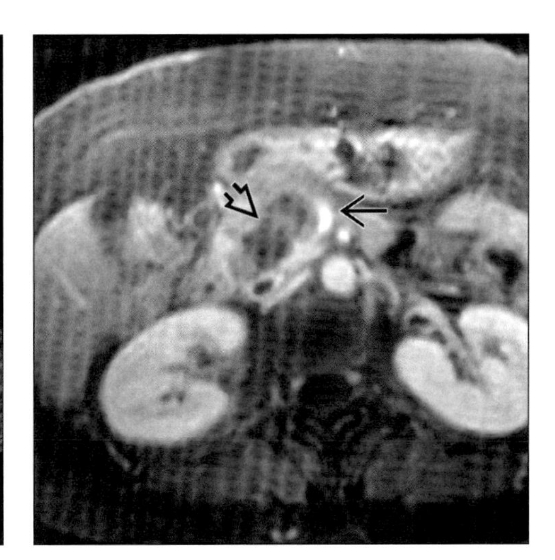

(Left) *Transverse abdominal sonogram shows hypoechoic mass (cursors) within pancreatic head (arrow). Gastric antrum (open arrow).* ***(Right)*** *Axial T1 C+ MR shows hypointense mass (open arrow) within pancreatic head. Note teardrop shape of SMV (arrow) indicating tumor invasion.*

Typical

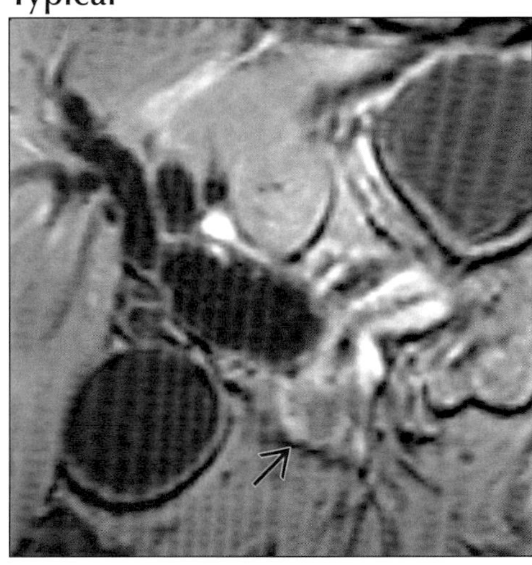

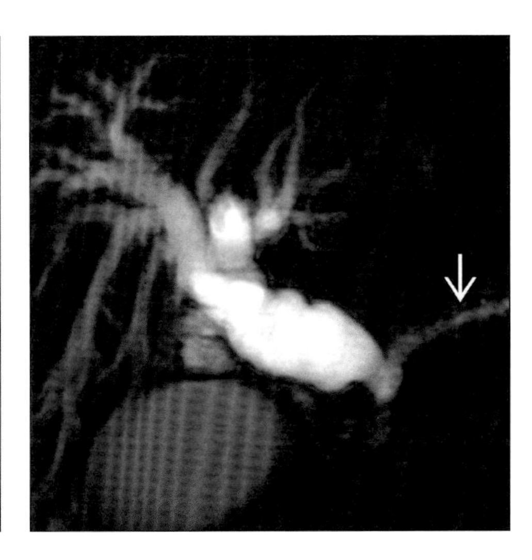

(Left) *Coronal T1WI MR shows dilated biliary tree obstructed by hypointense mass (arrow) within pancreatic head (Courtesy V. Vilgrain, MD).* ***(Right)*** *Coronal MRCP shows "double duct" sign: Common bile duct and pancreatic duct (arrow) are obstructed at the same point by invasive pancreatic ductal carcinoma.*

PANCREATIC ISLET CELL TUMORS

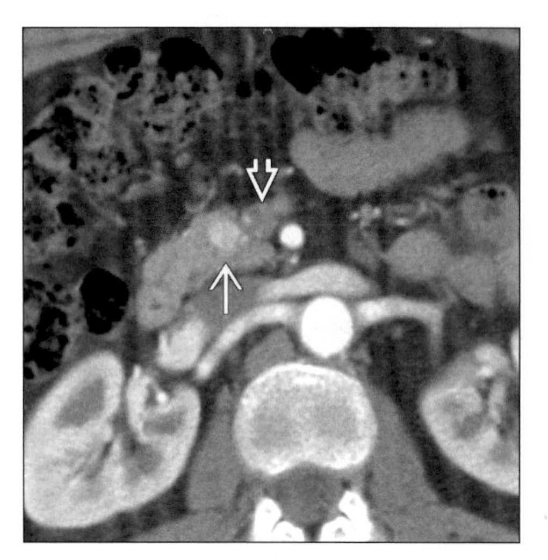

Axial CECT in arterial phase shows an 8 mm hypervascular insulinoma (arrow) in pancreatic head that was not detected on portal venous phase CT. Opacified SMA & unopacified SMV (open arrow).

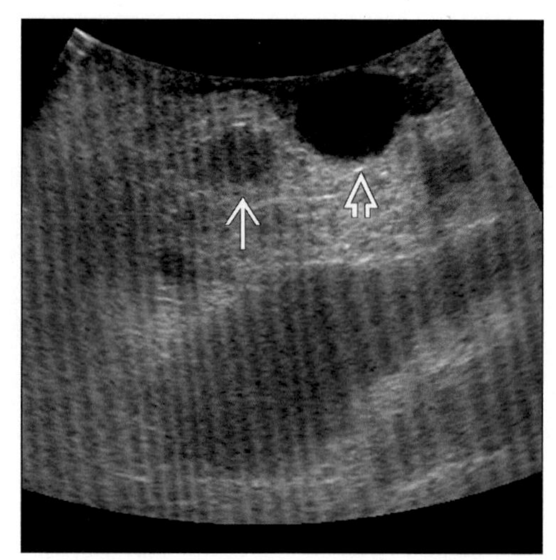

Insulinoma. Intra-operative sonography shows 8 mm hypoechoic mass (arrow) in pancreatic head, just lateral to superior mesenteric vein (SMV) (open arrow).

TERMINOLOGY

Abbreviations and Synonyms

- Pancreatic or gastroenteropancreatic neuroendocrine tumor (NET)

Definitions

- Tumors arising from pancreatic endocrine cells (islets of Langerhans)

IMAGING FINDINGS

General Features

- Best diagnostic clue: Hypervascular mass(es) in pancreas (primary) & liver (metastases)
- Location
  - Pancreas (85%); ectopic (15%)
    - Ectopic: Duodenum, stomach, nodes, ovary
  - Gastrinoma: Gastrinoma triangle
    - Superiorly: Cystic & common bile duct (CBD)
    - Inferiorly: 2nd & 3rd parts of duodenum
    - Medially: Pancreatic neck & body
- Size: Varies from few millimeters to 10 centimeters
- Morphology
  - Rare compared to tumors of exocrine pancreas
  - Benign or malignant
  - Single or multiple (with different cell types)
  - May be hormonally functional (85%) or nonfunctional
  - Functioning tumors: Often secrete multiple pancreatic hormones, with dominant single defining clinical presentation
    - Insulinoma, glucagonoma, gastrinoma, somatostatinoma, VIPoma (vasoactive intestinal polypeptide), PPoma (pancreatic polypeptide), APUDoma (carcinoid clinical syndromes)
  - Nonfunctioning tumors
    - Hypofunctioning or clinically silent large tumors
    - Larger than functioning tumors at diagnosis
    - Cystic islet cell tumor: Usually non-insulin producing & nonfunctioning

CT Findings

- Functioning tumors
  - NECT
    - Small or large in size; calcification may be seen
    - Small lesions: Usually undetectable
    - Cystic & necrotic areas (usually non-insulin tumors)
  - CECT: Arterial phase (AP) & portal venous phase
    - Most are hypervascular (hyperdense on AP)
    - Delayed scan: Solid/ring-enhancement (insulinoma)
    - Enhancing metastases (AP) in liver & nodes
- Nonfunctioning tumors
  - NECT

DDx: Pancreatic Mass

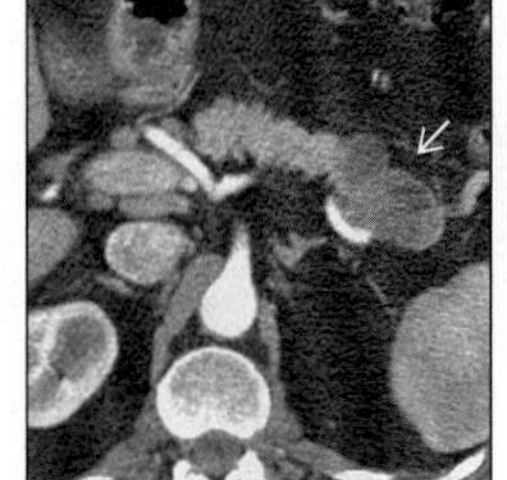

Pancreatic Carcinoma

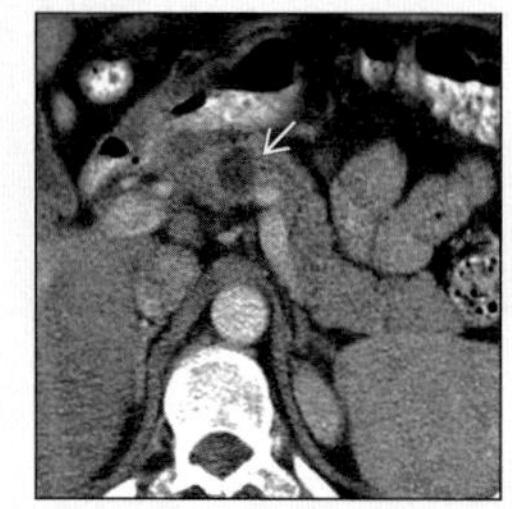

Mucinous Cystic tumor

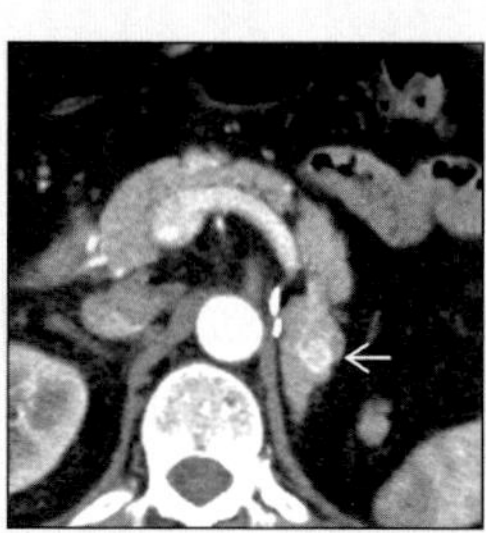

Metastases (Renal)

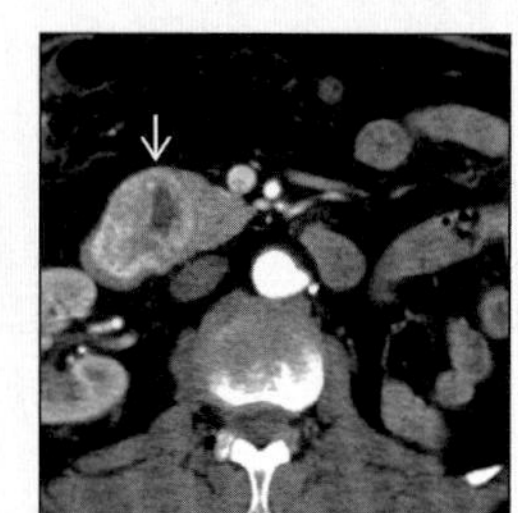

Metastases (Renal)

PANCREATIC ISLET CELL TUMORS

Key Facts

Terminology
- Pancreatic or gastroenteropancreatic neuroendocrine tumor (NET)
- Tumors arising from pancreatic endocrine cells (islets of Langerhans)

Imaging Findings
- Best diagnostic clue: Hypervascular mass(es) in pancreas (primary) & liver (metastases)
- Pancreas (85%); ectopic (15%)
- Cystic & necrotic areas (usually non-insulin tumors)
- Liver metastases often extensive even in relatively healthy patient

Top Differential Diagnoses
- Pancreatic ductal adenocarcinoma
- Mucinous cystic tumor of pancreas
- Metastases
- Serous cystadenoma of pancreas

Pathology
- Insulinoma: Most common islet cell tumor

Clinical Issues
- Palpitations, sweating, tremors, headache, coma
- Can live with metastases for many years

Diagnostic Checklist
- Differentiate from other solid, cystic, vascular tumors
- Correlate with clinical & biochemical information
- Hypervascular pancreatic tumor & liver metastases suggests islet cell/neuroendocrine tumor
- Large functioning & nonfunctioning tumors: Hypervascular, complex & highly malignant

        - Mixed density; usually large & complex
        - Cystic & necrotic areas (seen in large tumors)
        - Calcification
    - CECT
        - Usually hypervascular
        - Nonenhancing cystic or necrotic areas
        - Enhancing viable tumor
        - Enhancing metastases (AP)
        - Liver metastases often extensive even in relatively healthy patient
- Large functional & nonfunctional tumors: Highly malignant
    - Calcification; local invasion
    - Early invasion of portal vein leads to liver metastases

MR Findings
- Functional tumors
    - Fat-saturated T1WI: Hypointense
    - T2WI SE & STIR sequences: Hyperintense (both primary & secondaries)
    - T1 C+
        - T1WI: Solid or ring-enhancement (insulinoma)
        - Fat-saturated delayed enhanced T1WI SE: Hyperintense (solid enhancing lesions)
- Nonfunctioning tumors
    - T1WI SE image
        - Small tumors: Isointense
        - Large tumors: Heterogeneous (cystic & necrotic)
    - T2WI SE image
        - Small tumors: Isointense
        - Large tumors: Hyperintense (cystic & necrotic)
    - T1 C+
        - Fat-saturated delayed enhanced T1WI SE: Hyperintense (small)
        - Nonenhancing (cystic + necrotic areas) & increased enhancing viable tumor

Ultrasonographic Findings
- Real Time
    - Endoscopic ultrasound (EUS)
        - Detects small islet cell tumors
        - Homogeneously hypoechoic mass
    - Intra-operative US
        - Detects very small lesions; sensitivity (75-100%)

Angiographic Findings
- Conventional
    - Functioning & nonfunctioning tumors
        - Hypervascular (primary & secondary)
    - Hepatic venous sampling after intra-arterial stimulation of pancreas
        - Functioning tumors: Elevated levels of hormones
        - Nonfunctioning: Decreased levels or absent

Imaging Recommendations
- NE + CECT
- MR & T1 C+ (including fat-suppressed delayed images)
- Endoscopic ultrasound (EUS)
- PET study with 68Ga(DFO)-octreotide
    - Complimentary in cases with disseminated disease

DIFFERENTIAL DIAGNOSIS

Pancreatic ductal adenocarcinoma
- Hypovascular tumor; pancreatic ductal obstruction
- Location: Head (60%)
- Obliteration of retropancreatic fat
- Extensive local invasion & regional metastases
- ERCP
    - Irregular, nodular, rat-tailed eccentric obstruction

Mucinous cystic tumor of pancreas
- Can be similar to cystic/necrotic islet cell tumor
- Location: Tail of pancreas (more common)
- NECT: Multiloculated hypodense mass
- CECT: Enhancement of thin internal septa & wall
- T2WI: Cysts (hyperintense); septations (hypointense)
- Angiography: Predominantly avascular

Metastases
- Examples: Renal cell carcinoma & melanoma
- Small, well-defined, round hypervascular lesions
- Indistinguishable from islet cell tumor metastases

Serous cystadenoma of pancreas
- Benign, glycogen-rich cystadenoma of pancreas

PANCREATIC ISLET CELL TUMORS

- Honeycomb or sponge appearance
- Location: Head of pancreas (more common)
- Enhancement of septa delineating small cysts
- Cyst wall enhances; angiography (highly vascular)
- Macrocystic type: Thin wall/septa than cystic islet cell

PATHOLOGY

General Features

- General path comments
  - Embryology-anatomy
    - Originate from embryonic neuroectoderm
- Etiology
  - Arise from APUD cells
    - APUD: Amine precursor uptake & decarboxylation
  - Pathogenesis
    - Insulinoma: β-cell tumor → hyperinsulinemia → hypoglycemia
    - Gastrinoma: Islet cell tumor → increased gastrin → increased gastric acid → peptic ulcer
    - Glucagonoma: α-cell tumor → increased glucagon→ erythema migrans & diabetes mellitus
    - Nonfunctioning: Derived from α & β cells
- Epidemiology
  - Insulinoma: Most common islet cell tumor
    - Solitary benign (90%); malignant (10%)
  - Gastrinoma: 2nd common
    - Multiple & malignant (60%); MEN I (20-60%)
  - Nonfunctioning: 3rd common
    - Accounts 20-45% of all islet cell tumors
    - Malignant (80-100%)
- Associated abnormalities
  - Gastrinoma (Zollinger-Ellison syndrome)
    - Associated with MEN type I

Gross Pathologic & Surgical Features

- Small tumor: Encapsulated & firm
- Large tumor: ± Cystic, necrotic, calcified

Microscopic Features

- Sheets of small round cells, uniform nuclei/cytoplasm
  - Electron microscopy: Neuron specific enolase ("neuro-endocrine")

CLINICAL ISSUES

Presentation

- Most common signs/symptoms
  - Insulinoma: Whipple triad (hypoglycemia + low fasting glucose + relief by IV glucose)
    - Palpitations, sweating, tremors, headache, coma
  - Gastrinoma (Zollinger-Ellison syndrome)
    - Peptic ulcer, increased acidity & diarrhea
  - Glucagonoma: Necrolytic erythema migrans, diarrhea, diabetes, weight loss
  - Nonfunctional
    - Mostly asymptomatic
    - Pain, jaundice, variceal bleeding

Demographics

- Age
  - Insulinoma: 4th-6th decade
  - Gastrinoma: 4th-5th decade
- Gender
  - Insulinoma: M < F
  - Gastrinoma: M > F

Natural History & Prognosis

- Complications
  - Glucagonoma: Deep venous thrombosis (DVT) & pulmonary embolism
- Prognosis
  - Insulinoma (good); gastrinoma (poor)
  - Nonfunctional
    - 3 year survival (60%), 5 year survival (44%)
    - Can live with metastases for many years

Treatment

- Acute phase: Octreotide (potent hormonal inhibitor)
- Insulinoma: Surgery curative
- Gastrinoma
  - Medical: Omeprazole, 5-fluorouracil
  - Surgery curative in 30% cases
- Nonfunctional: Resection/embolization
- Transarterial chemoembolization: Liver metastases

DIAGNOSTIC CHECKLIST

Consider

- Differentiate from other solid, cystic, vascular tumors
- Correlate with clinical & biochemical information

Image Interpretation Pearls

- Hypervascular pancreatic tumor & liver metastases suggests islet cell/neuroendocrine tumor
- Solid/ring-enhancement (insulinoma): Delayed scans
- Large functioning & nonfunctioning tumors: Hypervascular, complex & highly malignant

SELECTED REFERENCES

1. Dromain C et al: MR imaging of hepatic metastases caused by neuroendocrine tumors: comparing four techniques. AJR Am J Roentgenol. 180(1):121-8, 2003
2. Marcos HB et al: Neuroendocrine tumors of the pancreas in von Hippel-Lindau disease: spectrum of appearances at CT and MR imaging with histopathologic comparison. Radiology. 225(3):751-8, 2002
3. Ichikawa T et al: Islet cell tumor of the pancreas: biphasic CT versus MR imaging in tumor detection. Radiology. 216(1):163-71, 2000
4. Thoeni RF et al: Detection of small, functional islet cell tumors in the pancreas: selection of MR imaging sequences for optimal sensitivity. Radiology. 214(2):483-90, 2000
5. Stafford-Johnson DB et al: Dual-phase helical CT of nonfunctioning islet cell tumors. J Comput Assist Tomogr. 22(2):335-9, 1998
6. Sohaib SA et al: Cystic islet cell tumors of the pancreas. AJR Am J Roentgenol. 170(1):217, 1998
7. Stafford Johnson DB et al: Dual-phase helical CT of nonfunctioning islet cell tumors. J Comput Assist Tomogr. 22(1):59-63, 1998
8. Buetow PC et al: Islet cell tumors of the pancreas: clinical, radiologic, and pathologic correlation in diagnosis and localization. Radiographics. 17(2):453-72; quiz 472A-472B, 1997

IMAGE GALLERY

Typical

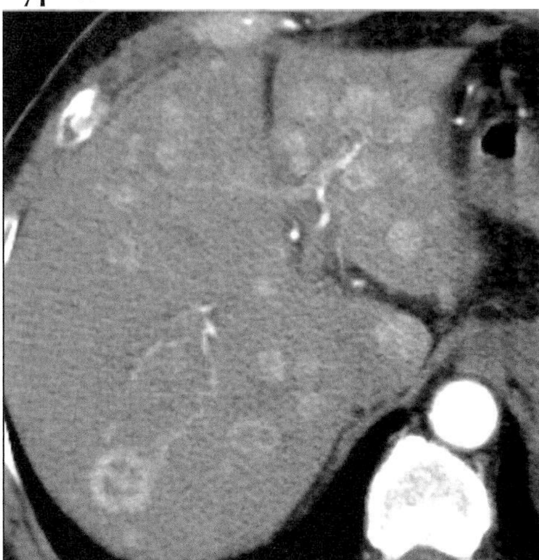

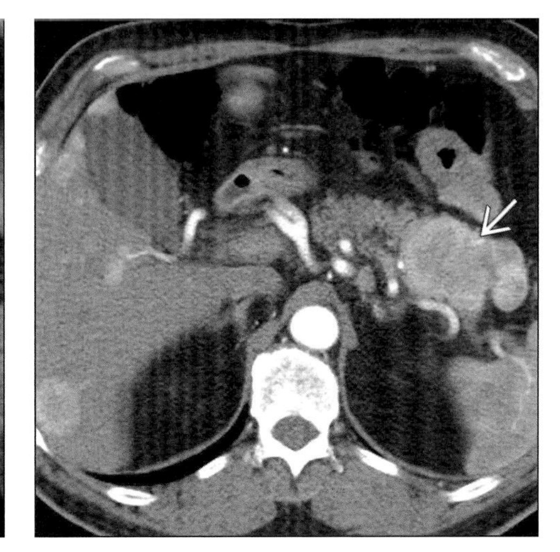

(Left) *Axial CECT during arterial phase, in patient with metastatic glucagonoma. Solid and ring-enhancing hypervascular liver metastases.* ***(Right)*** *Malignant glucagonoma. Axial CECT (arterial phase) shows hypervascular mass in pancreatic tail (arrow) with hypervascular liver metastases.*

Typical

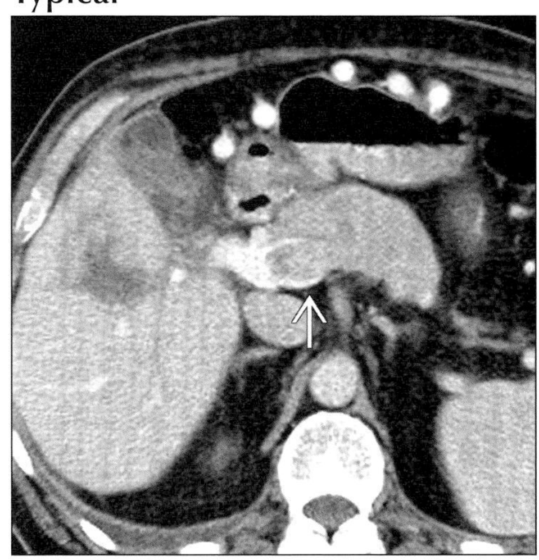

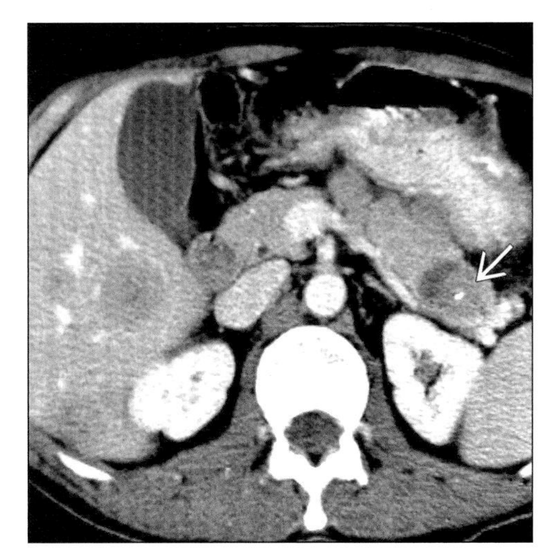

(Left) *Malignant non-functional NET. Axial CECT (portal venous phase) shows hypovascular mass in pancreatic body with direct invasion of splenic vein (arrow). Hypodense liver metastases.* ***(Right)*** *Malignant non-functional NET. Axial CECT (portal venous phase) shows hypodense mass in pancreatic tail with focal calcification (arrow). Liver metastases.*

Typical

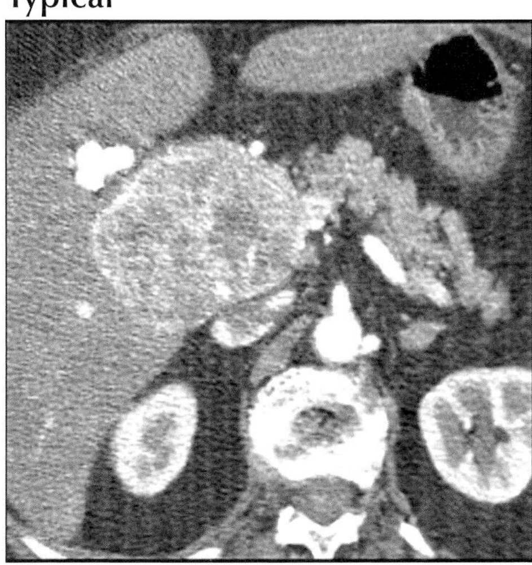

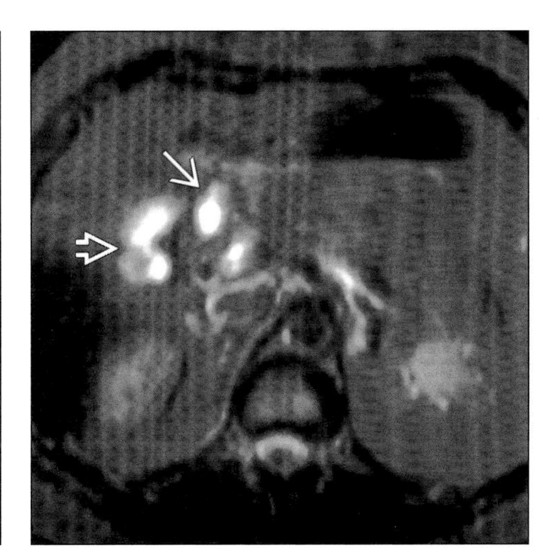

(Left) *Malignant non-functioning NET. Axial CECT (arterial phase) shows large hypervascular mass in pancreatic head. Note absence of pancreatic ductal dilatation.* ***(Right)*** *Benign gastrinoma. Axial T2 WI MR shows 1 cm hyperintense mass (arrow) in pancreatic head, just medial to duodenum (open arrow).*

SOLID AND PAPILLARY NEOPLASM

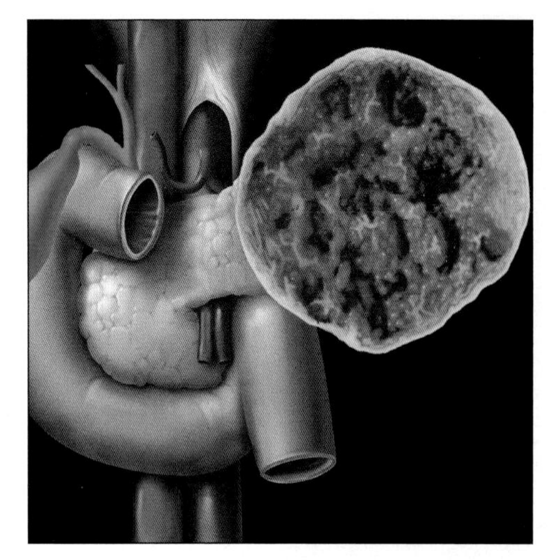

Graphic shows large mass arising from pancreatic tail, having prominent solid and cystic/hemorrhagic components.

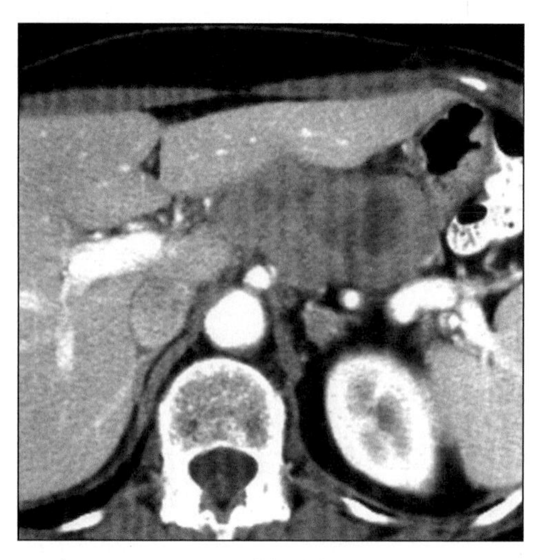

Axial CECT in a 15 year old girl shows a large solid and cystic mass in the pancreatic body/tail (Courtesy B. Jeffrey, MD).

TERMINOLOGY

Abbreviations and Synonyms

- Solid and papillary epithelial neoplasm; papillary epithelial neoplasm; papillary cystic carcinoma; solid and cystic tumor of the pancreas

Definitions

- Pancreatic mass of low malignant potential with solid and cystic features

IMAGING FINDINGS

General Features

- Best diagnostic clue: Well-demarcated large mass with solid and cystic areas in pancreatic tail region in CT
- Location: Commonly in body and/or tail of pancreas
- Size: Average 10 cm, range of 2.5-20 cm

CT Findings

- NECT
  - Encapsulated, hypodense mass
  - ± Calcification (rare)
  - ± Metastases to liver, lymph nodes
- CECT
  - Well-defined heterogenous large mass
  - Low density areas of variable size within the lesion; depends on degree of hemorrhage and necrosis
  - Thick capsule enhancement

MR Findings

- T1WI
  - Large well-demarcated mass with central areas of low and high signal intensity
  - High signal intensity secondary to hemorrhage
  - Capsule appears as rim of low intensity

Ultrasonographic Findings

- Fluid-debris levels; posterior enhancement

Angiographic Findings

- Avascular/hypovascular; depends on degree of necrosis

Imaging Recommendations

- Best imaging tool: Helical CT
- Protocol advice
  - Helical CT: Pancreatic mass protocol
    - 125 ml IV at 4 ml per second; water for oral contrast

DIFFERENTIAL DIAGNOSIS

Mucinous cystic pancreatic tumor

- Most common in middle age to elderly women
- Usually a spherical mass in body or tail of the pancreas with several cystic spaces separated by thin septa

DDx: Cystic Pancreatic Mass

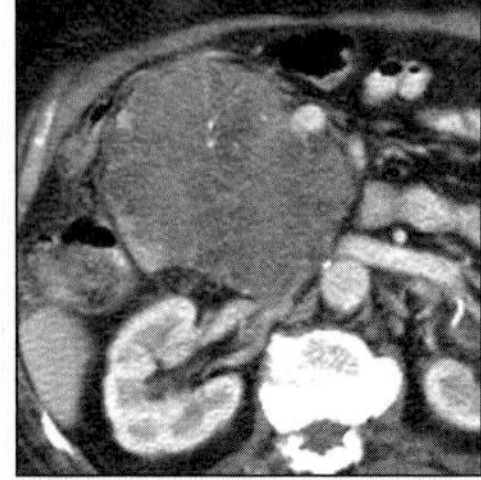

Serous Cystadenoma

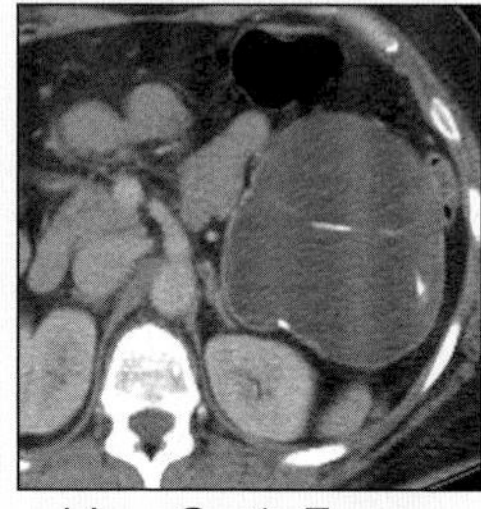

Muc. Cystic Tumor

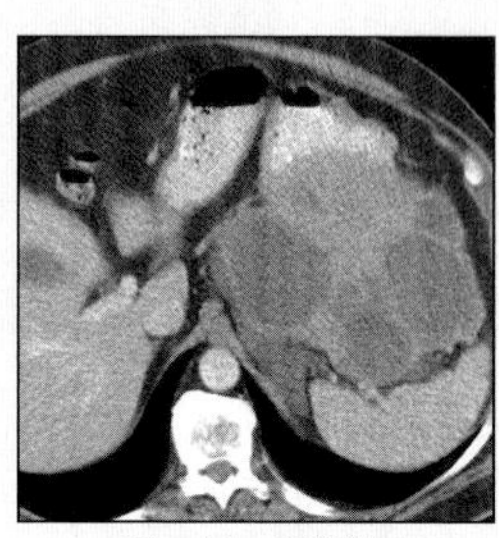

Gastric GIST

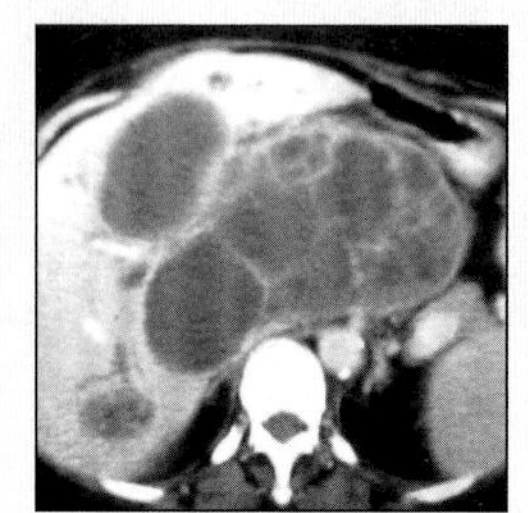

Ovarian Cancer

SOLID AND PAPILLARY NEOPLASM

Key Facts

Imaging Findings
- Best diagnostic clue: Well-demarcated large mass with solid and cystic areas in pancreatic tail region in CT
- Location: Commonly in body and/or tail of pancreas
- Low density areas of variable size within the lesion; depends on degree of hemorrhage and necrosis

Top Differential Diagnoses
- Mucinous cystic pancreatic tumor
- Serous cystadenoma of pancreas
- Exophytic gastric mass
- Pancreatic pseudocyst
- Pancreatic metastases

Diagnostic Checklist
- Palpable pancreatic mass in a young African-American female
- Encapsulated mass with prominent cystic and solid components

Serous cystadenoma of pancreas
- CT
  - Usually located in head of pancreas
  - Typically has "sponge" appearance with innumerable small cysts
  - Never has thick soft tissue component like solid and papillary neoplasm

Exophytic gastric mass
- GIST may closely simulate solid and papillary neoplasm
- Necrotic mass arising from gastric wall
- Usually in older adult

Pancreatic pseudocyst
- CT: Cystic mass in or around pancreas; usually no mural nodularity
- History or signs of pancreatitis

Pancreatic metastases
- Example: Breast, lung, melanoma and ovarian cancer
- Usually clinically silent; in older population

PATHOLOGY

General Features
- General path comments
  - 0.13-2.7% of all pancreatic tumors
  - Low malignant potential

Gross Pathologic & Surgical Features
- Areas of central hemorrhage and necrosis surrounded by solid and pseudopapillary structure
- Thick, fibrous, hypervascular capsule surrounding a mixture of solid and cystic areas

Microscopic Features
- Homogeneous, small epithelioid cells present singly, in aggregates, small sheets or papillary structures

CLINICAL ISSUES

Presentation
- Most common signs/symptoms
  - Asymptomatic or abdominal pain
  - Palpable abdominal mass
- Lab-Data: Normal

Demographics
- Age: < 35 years of age
- Gender: M:F = 1:9.5
- Ethnicity: African-Americans or other non-Caucasian groups

Natural History & Prognosis
- Complications: Hemorrhage, pseudocyst, sepsis, shock
- Prognosis: Good, after surgical resection; rarely recurs

Treatment
- Complete surgical excision

DIAGNOSTIC CHECKLIST

Consider
- Palpable pancreatic mass in a young African-American female

Image Interpretation Pearls
- Encapsulated mass with prominent cystic and solid components

SELECTED REFERENCES

1. Madan AK et al: Solid and papillary epithelial neoplasm of the pancreas. J Surg Oncol. 85(4):193-8, 2004
2. Buetow PC et al: Solid and papillary epithelial neoplasm of the pancreas: imaging-pathologic correlation on 56 cases. Radiology. 199(3):707-11, 1996
3. Choi BI et al: Solid and papillary epithelial neoplasms of the pancreas: CT findings. Radiology. 166(2):413-6, 1988

IMAGE GALLERY

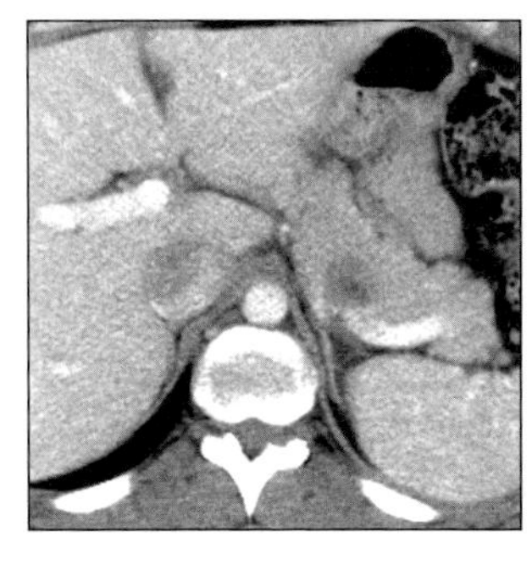

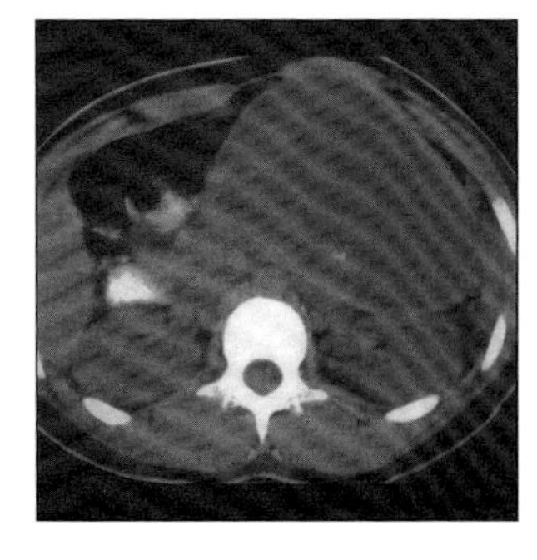

(Left) *Axial CECT shows poorly defined solid and cystic mass in pancreatic body.* ***(Right)*** *Axial NECT shows a solid and cystic pancreatic tumor in an adolescent girl.*

PANCREATIC METASTASES AND LYMPHOMA

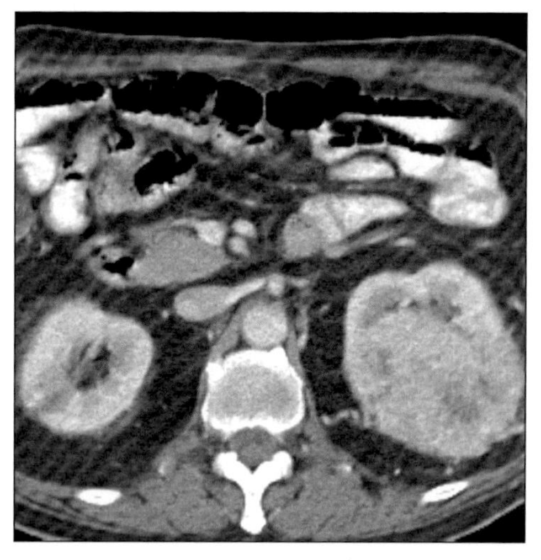

Axial CECT shows primary malignancy in left kidney, renal cell carcinoma.

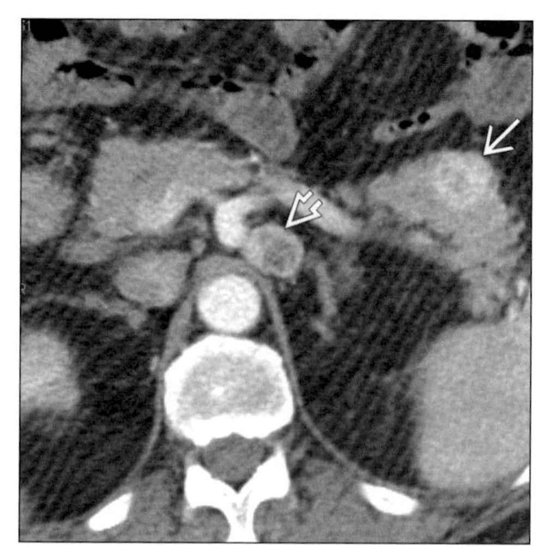

Axial CECT in patient with new diagnosis of renal cell carcinoma shows hypervascular metastases in pancreatic tail (arrow) and a retroperitoneal lymph node (open arrow).

TERMINOLOGY

Definitions

- Metastases from primary cancer of other sites
- Lymphoma: Malignant tumor of B lymphocytes

IMAGING FINDINGS

General Features

- Best diagnostic clue: Mass(es) in pancreas without pancreatic or biliary ductal obstruction
- Other general features
  - Pancreatic metastases
    - Types of spread: Hematogenous, lymphatic, direct
    - Hematogenous spread: Most common
  - Pancreatic lymphoma
    - Secondary: Direct extension from peripancreatic lymphadenopathy
    - May be only apparent site of involvement

CT Findings

- Pancreatic metastases
  - Solitary (78%) or multiple (17%)
  - May cause diffuse infiltration
  - Discrete mass(es) more common
  - Enhancement pattern is variable; mimics primary tumor
    - Hyperattenuation: Heterogenous (60%) or homogenous (15%)
    - Hypoattenuation (20%)
    - Isoattenuation (5%)
  - Concomitant intraabdominal metastases
    - Liver (36%)
    - Lymph nodes (30%)
    - Adrenal glands (30%)
  - Dilatation of the pancreatic duct or biliary tree not common (33%)
  - Encasement of the major peripancreatic vascular structure (rare)
- Pancreatic lymphoma
  - Homogeneous soft tissue mass
  - Minimal enhancement
  - Focal and circumscribed single or multiple masses
  - Diffuse enlargement of pancreas with infiltrating tumor ± peripancreatic fat involvement; may simulate acute pancreatitis
  - Peripancreatic lymph node enlargement
  - ± Disseminated lymph nodes
  - Lymphadenopathy below level of renal veins; feature of lymphoma
  - Dilatation of the pancreatic duct (uncommon); distinguish feature from adenocarcinoma
  - Encasement of the peripancreatic vessels

MR Findings

- Pancreatic metastases

DDx: Solid Pancreatic Mass

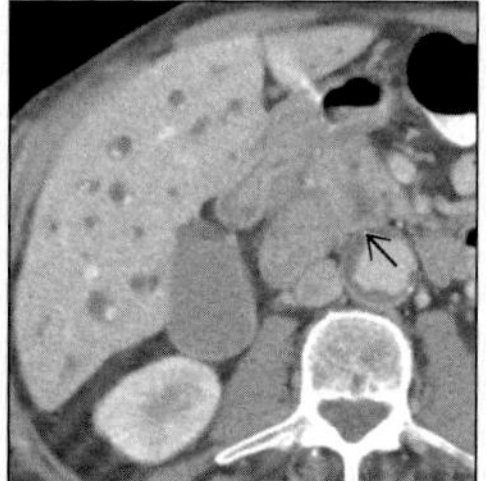

Pancreatic Cancer

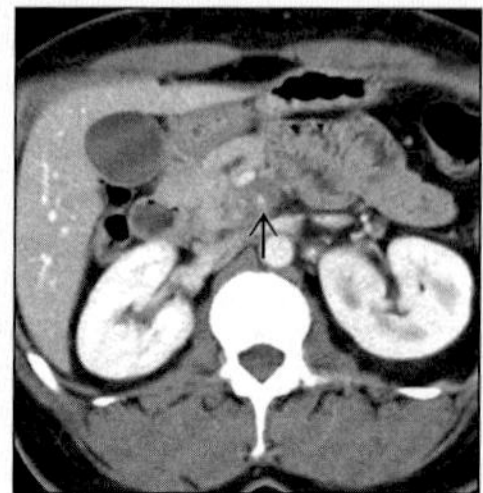

Ductal Carcinoma

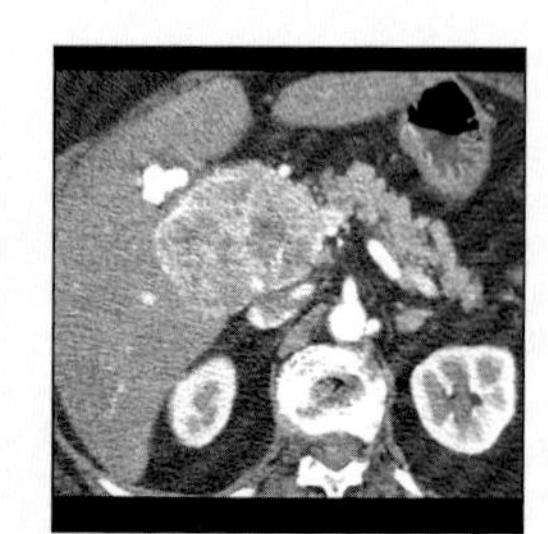

Islet Cell Tumor

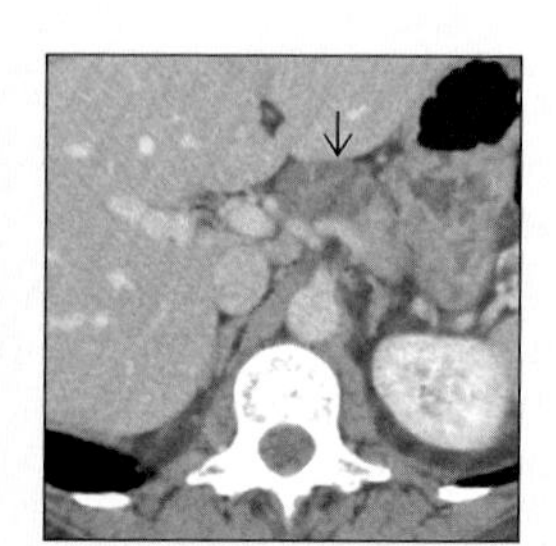

Mucinous Tumor

PANCREATIC METASTASES AND LYMPHOMA

Key Facts

Imaging Findings

- Best diagnostic clue: Mass(es) in pancreas without pancreatic or biliary ductal obstruction
- May cause diffuse infiltration
- Focal and circumscribed single or multiple masses
- Diffuse enlargement of pancreas with infiltrating tumor ± peripancreatic fat involvement; may simulate acute pancreatitis
- Peripancreatic lymph node enlargement
- Encasement of the peripancreatic vessels
- Best imaging tool: Helical CT

Top Differential Diagnoses

- Pancreatic ductal carcinoma
- Pancreatic Islet cell tumors
- Mucinous cystic pancreatic tumor

Pathology

- Most common lymphoma is non-Hodgkin B cell subtype
- Epidemiology: Increased diagnosis by incidental findings by CT

Clinical Issues

- Asymptomatic or abdominal pain, jaundice, weight loss
- Age: Usually in middle age and elderly

Diagnostic Checklist

- Check for history of primary cancer
- Overlapping radiographic features of pancreatic metastases, lymphoma and primary carcinoma
- Consider renal cell metastases to pancreas even years after resection of primary tumor

- T1WI
  - Hypointense ± fat-saturation
- T1 C+
  - Rim of enhancement in larger tumors
  - Homogenous enhancement in smaller tumors
- T2WI
  - Slightly heterogenous and moderately hyperintense
  - Diffuse, enlarged metastases may appear as hypointense nodules

Imaging Recommendations

- Best imaging tool: Helical CT

DIFFERENTIAL DIAGNOSIS

Pancreatic ductal carcinoma

- Location: Head (60%)
- CT
  - Heterogenous, poorly-enhancing mass
  - Abrupt pancreatic and/or common bile duct obstruction
  - Encasement of vessels
  - Obliteration of retropancreatic fat
  - Extensive local invasion & regional metastases

Pancreatic Islet cell tumors

- CT
  - Ring-enhancement seen in insulinoma
  - No pancreatic ductal dilatation
  - Functioning tumors usually small in size
  - Non-functioning tumors usually large in size
- Hypervascular primary and secondary tumors

Mucinous cystic pancreatic tumor

- Location: Body/tail of pancreas (more common)
- CT
  - Multiloculated hypodense mass
  - Enhancement of thin internal septa & wall
- MR: T2WI
  - Hyperintense cysts
  - Hypointense internal septations
- Treated as malignant or premalignant
- Most common in middle age to elderly women

PATHOLOGY

General Features

- General path comments
  - Pancreatic metastases
    - 3-10% at autopsy, less common "clinically"
  - Pancreatic lymphoma
    - Most common lymphoma is non-Hodgkin B cell subtype
    - Primary: < 1% of pancreatic neoplasms
- Etiology
  - Pancreatic metastases
    - Renal cell carcinoma (30%)
    - Bronchogenic carcinoma (23%)
    - Breast carcinoma (12%)
    - Soft tissue sarcoma (8%)
    - Colonic carcinoma (6%)
    - Melanoma (6%)
    - Prostate carcinoma
    - Ovarian carcinoma
    - Other gastrointestinal tumors
  - Pancreatic lymphoma
    - Primary (rare)
    - Secondary (30% of patients with widespread lymphoma)
- Epidemiology: Increased diagnosis by incidental findings by CT
- Associated abnormalities
  - Pancreatic lymphoma
    - Immunocompromised patients, particularly HIV

Gross Pathologic & Surgical Features

- Solitary or multiple; polypoid masses

Microscopic Features

- Pancreatic metastases: Varies based on primary cancer
- Pancreatic lymphoma: Lymphoepithelial lesions

PANCREATIC METASTASES AND LYMPHOMA

CLINICAL ISSUES

Presentation

- Most common signs/symptoms
  - Pancreatic metastases
    - Asymptomatic or abdominal pain, jaundice, weight loss
    - Acute pancreatitis (uncommon)
  - Pancreatic lymphoma
    - Abdominal pain
    - Obstructive jaundice & hyperbilirubinemia (uncommon)
- Diagnosis
  - Pancreatic lymphoma
    - Percutaneous or endoscopic biopsy

Demographics

- Age: Usually in middle age and elderly
- Gender
  - Pancreatic metastases
    - M:F = 1:1
  - Pancreatic lymphoma
    - M:F = 1.4:1

Natural History & Prognosis

- Pancreatic metastases
  - Few months to several years after primary tumor; usually widespread
  - Renal cell carcinoma
    - Occasionally presents 5-10 years after diagnosis of primary tumor
    - May be isolated only to pancreas
- Prognosis
  - Pancreatic metastases
    - Very poor
  - Pancreatic lymphoma
    - Poor, 30% cure rate after treatment

Treatment

- Pancreatic metastases
  - Mostly palliative treatment only
  - Surgical resection if metastases is only isolated to pancreas
- Pancreatic lymphoma
  - Chemotherapy

DIAGNOSTIC CHECKLIST

Consider

- Check for history of primary cancer

Image Interpretation Pearls

- Overlapping radiographic features of pancreatic metastases, lymphoma and primary carcinoma
- Consider renal cell metastases to pancreas even years after resection of primary tumor

SELECTED REFERENCES

1. Fenchel S et al: Multislice helical CT of the pancreas and spleen. Eur J Radiol. 45 Suppl 1:S59-72, 2003
2. Merkle EM et al: Helical computed tomography of the pancreas: potential impact of higher concentrated contrast agents and multidetector technology. J Comput Assist Tomogr. 27 Suppl 1:S17-22, 2003
3. Schima W et al: Evaluation of focal pancreatic masses: comparison of mangafodipir-enhanced MR imaging and contrast-enhanced helical CT. Eur Radiol. 12(12):2998-3008, 2002
4. Sheth S et al: Imaging of uncommon tumors of the pancreas. Radiol Clin North Am. 40(6):1273-87, vi, 2002
5. Hanbidge AE: Cancer of the pancreas: the best image for early detection--CT, MRI, PET or US? Can J Gastroenterol. 16(2):101-5, 2002
6. Scatarige JC et al: Pancreatic parenchymal metastases: observations on helical CT. AJR Am J Roentgenol. 176(3):695-9, 2001
7. Tamm E et al: Pancreatic cancer: current concepts in imaging for diagnosis and staging. Cancer J. 7(4):298-311, 2001
8. Salvatore JR et al: Primary pancreatic lymphoma: a case report, literature review, and proposal for nomenclature. Med Oncol. 17(3):237-47, 2000
9. Merkle EM et al: Imaging findings in pancreatic lymphoma: differential aspects. AJR Am J Roentgenol. 174(3):671-5, 2000
10. Kassabian A et al: Renal cell carcinoma metastatic to the pancreas: a single-institution series and review of the literature. Urology. 56(2):211-5, 2000
11. Ng CS et al: Metastases to the pancreas from renal cell carcinoma: findings on three-phase contrast-enhanced helical CT. AJR Am J Roentgenol. 172(6):1555-9, 1999
12. Bouvet M et al: Primary pancreatic lymphoma. Surgery. 123(4):382-90, 1998
13. Klein KA et al: CT characteristics of metastatic disease of the pancreas. Radiographics. 18(2):369-78, 1998
14. Merkle EM et al: Metastases to the pancreas. Br J Radiol. 71(851):1208-14, 1998
15. Cario E et al: Diagnostic dilemma in pancreatic lymphoma. Case report and review. Int J Pancreatol. 22(1):67-71, 1997
16. Ferrozzi F et al: Pancreatic metastases: CT assessment. Eur Radiol. 7(2):241-5, 1997
17. Jones WF et al: AIDS-related non-Hodgkin's lymphoma of the pancreas. Am J Gastroenterol. 92(2):335-8, 1997
18. Miller FH et al: Pancreaticobiliary manifestations of AIDS. AJR Am J Roentgenol. 166(6):1269-74, 1996
19. Das DK et al: Ultrasound guided percutaneous fine needle aspiration cytology of pancreas: a review of 61 cases. Trop Gastroenterol. 16(2):101-9, 1995
20. Keogan MT et al: Computed tomography and magnetic resonance imaging in the assessment of pancreatic disease. Gastrointest Endosc Clin N Am. 5(1):31-59, 1995
21. Friedman AC et al: Rare pancreatic malignancies. Radiol Clin North Am. 27(1):177-90, 1989

IMAGE GALLERY

Typical

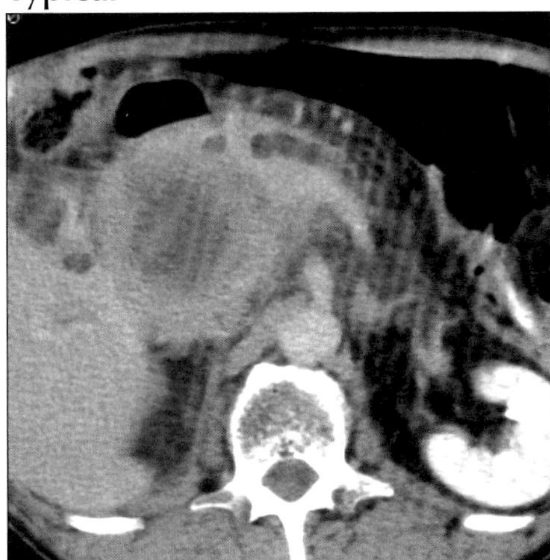

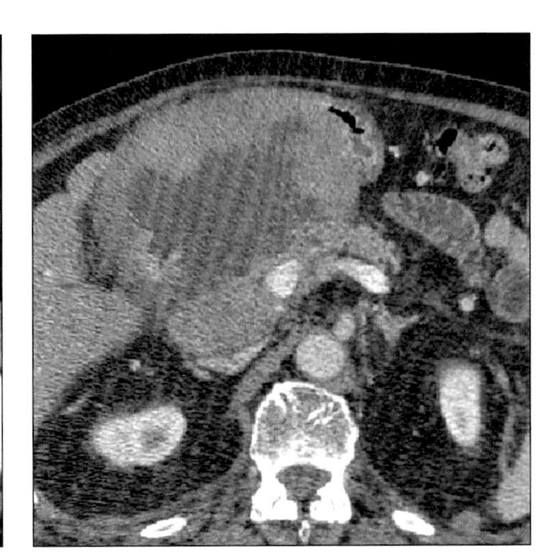

*(**Left**) Axial CECT shows large heterogenous mass in pancreatic head and dilation of the pancreatic duct due to lymphoma. (**Right**) Axial CECT in a patient with NH lymphoma shows a large heterogeneous pancreatic head mass without pancreatic or biliary ductal dilation. The vessels are also encased but not obstructed.*

Typical

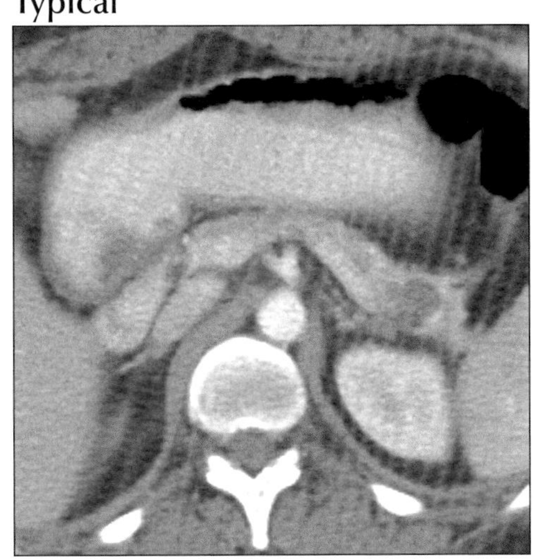

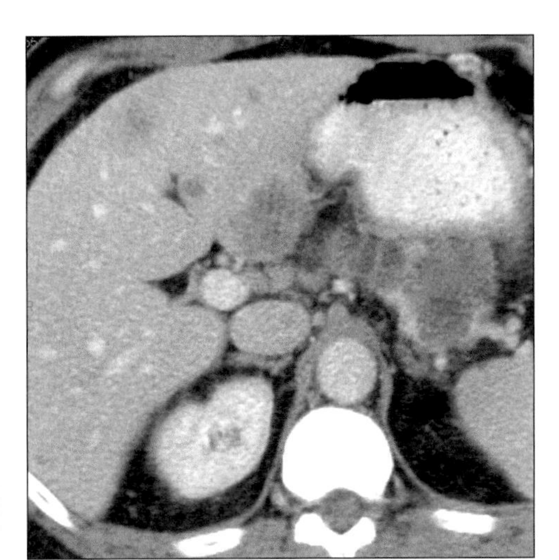

*(**Left**) Axial CECT shows a hypodense mass in pancreatic tail due to metastatic sarcoma. (**Right**) Axial CECT shows multiple hypodense masses in the liver and pancreas due to lung cancer.*

Typical

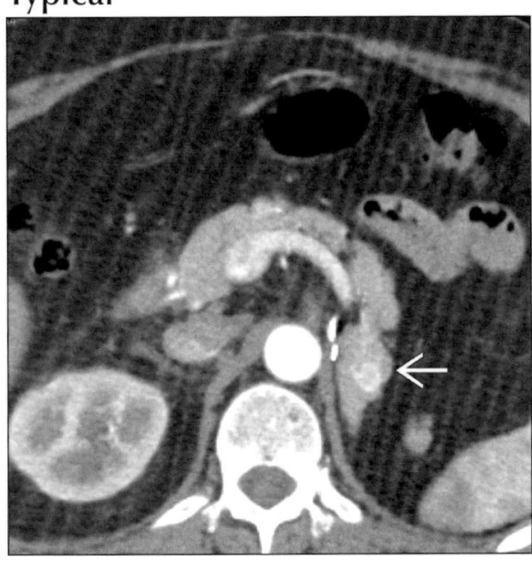

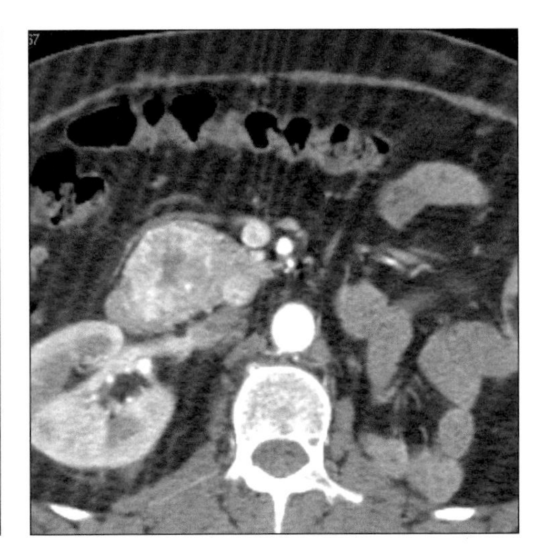

*(**Left**) Axial CECT in a patient 12 years post nephrectomy for renal cell carcinoma (RCC). Multiple pancreatic hypervascular masses (arrow) are proven RCC metastases. (**Right**) Axial CECT shows hypervascular metastases to head of pancreas from renal cell carcinoma, 12 years after nephrectomy.*

PART III

Genitourinary and Retroperitoneum

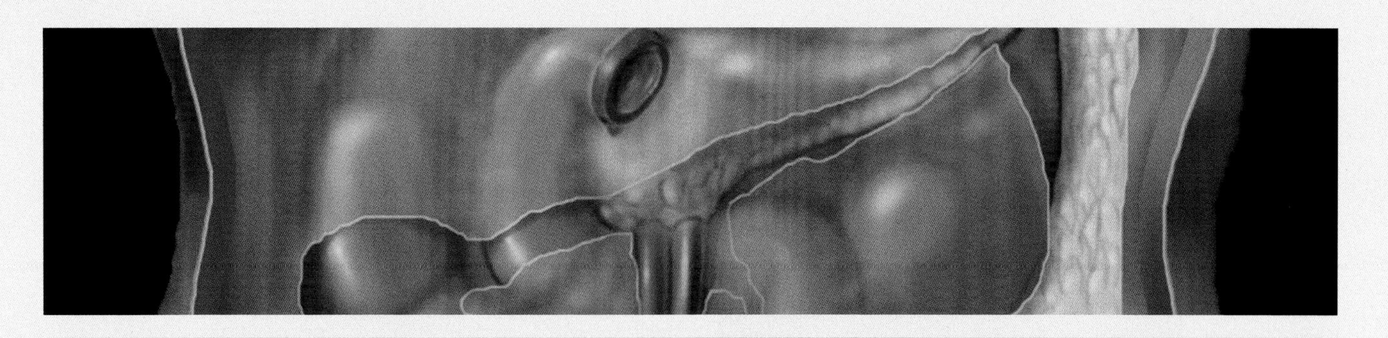

SECTION 1: Retroperitoneum

Introduction and Overview

Normal Variants

Inflammation

Trauma

Neoplasm

Treatment Related

RETROPERITONEUM ANATOMY AND IMAGING ISSUES

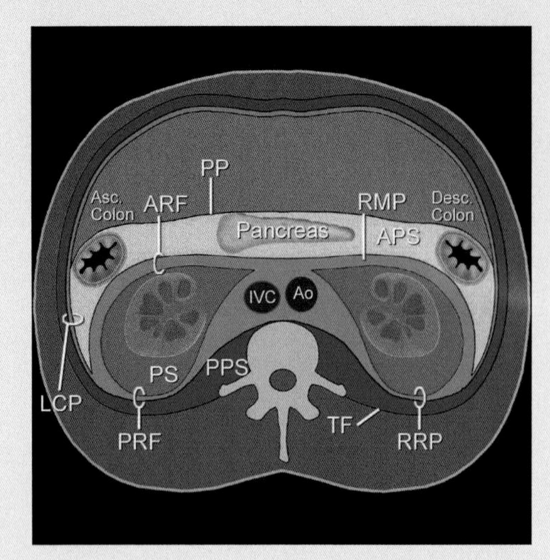

Graphic shows 3 main divisions of retroperitoneum, the anterior pararenal (yellow), perirenal (purple), and posterior pararenal space (blue). Duodenum (not shown) lies in APS.

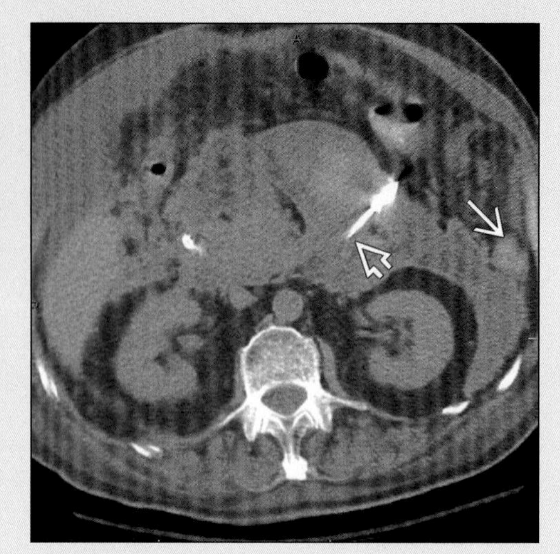

Axial NECT in pancreatitis shows extensive infiltration of APS, limited laterally by the lateroconal fascia, and posteriorly by the renal fascia. Duodenum (open arrow); colon (arrow).

TERMINOLOGY

Abbreviations and Synonyms

- Parietal peritoneum (PP)
- Anterior renal (Gerota) fascia (ARF)
- Posterior renal fascia (PRF)
- Lateroconal plane (fascia) (LCP)
- Retromesenteric plane (RMP)
- Retrorenal plane (RRP)

IMAGING ANATOMY

Location

- Basic compartments
  - Anterior pararenal space (APS)
  - Perirenal space (PS)
  - Posterior pararenal space (PPS)
- Compartments communicate inferiorly
- APS contains
  - Colon (ascending and descending)
  - Pancreas
  - Duodenum (2nd and 3rd portions)
- Perirenal space contains
  - Kidneys
  - Adrenals
  - Proximal ureters
- PPS contains
  - No solid organs

Anatomic Relationships

- Parietal peritoneum separates peritoneal cavity from APS
- Anterior renal fascia (Gerota) separates perirenal from APS
- Posterior renal fascia (Zuckerkandl) separates perirenal from PPS
- Lateroconal fascia separates from APS from PPS and marks the lateral extent of the APS
- Renal and lateroconal fascia are laminated planes and can form spaces as pathways of spread for rapidly expanding fluid collections or inflammatory processes (e.g., hemorrhage, or pancreatitis)
- Anterior renal fascia can "split" into a retromesenteric plane, which is continuous across the midline abdomen (green in graphic above)
- Posterior renal fascia splits into retrorenal plane
- Lateroconal fascia splits into lateroconal plane
- All three of these planes communicate at the junction of the lateroconal and renal fascia

Internal Structures-Critical Contents

- Sympathetic nerves and ganglia
  - Extend in parallel paraspinal chains along spine
  - Potential site of neural tumors
- Inferior vena cava (IVC)
  - Major conduit for thrombi from legs and pelvis to lungs (pulmonary emboli)
  - Major conduit for tumor emboli to lungs (pulmonary metastases)
    - Kidneys, liver and adrenals are major organs of origin
  - Uncommon site of primary tumor
    - Sarcoma of IVC
- Aorta
  - Atherosclerotic occlusive disease and aneurysm are major diseases
    - Rupture of aneurysm can cause hemorrhage into any or all of retroperitoneal compartments
    - Extension along renal hilum into perirenal space can simulate primary renal pathology
  - Dissection
    - Rarely starts in abdomen, unless iatrogenic (e.g., angiography, placement of stent graft)
    - Often spreads into abdominal aorta and its branches from thoracic aortic dissection
  - Inflammation
    - Aortitis ("peri-aneurysmal fibrosis") can simulate retroperitoneal fibrosis

RETROPERITONEUM ANATOMY AND IMAGING ISSUES

DIFFERENTIAL DIAGNOSIS

Retroperitoneal neoplastic masses
- Lymphoma
- Metastatic lymphadenopathy
- Primary (benign and malignant)
- ⇒ Mesenchymal (e.g., liposarcoma)
- ⇒ Neurogenic (e.g., paraganglioma)
- ⇒ Germ cell (e.g., teratoma)

Retroperitoneal non-neoplastic masses
- Hematoma
- Pancreatitis
- ⇒ (E.g., pseudocyst)
- Urinoma
- ⇒ (E.g., post obstructive, trauma)
- Retroperitoneal fibrosis
- Extramedullary hematopoiesis

Fat-containing retroperitoneal mass
- Liposarcoma
- Teratoma
- Primary or metastatic (testicular or ovarian)
- Myelolipoma
- ⇒ Adrenal
- Angiomyolipoma
- ⇒ Kidney

Retroperitoneal neurogenic tumors
- Usually paraspinal (sometimes intra-psoas)
- Often elongated, smooth, encapsulated
- Usually benign
- May enlarge neural foramina, extend intraspinally
- May be multiple (neurofibromatosis)

CLINICAL IMPLICATIONS

Clinical Importance
- Disease within APS is common
  - Pancreatic disease > duodenal > colonic
  - Pancreatitis often spreads throughout APS to affect duodenum and colon
    - Duodenal effects: Spasm, fold thickening, stricture, intramural pseudocyst, fistula
    - Colonic effects: Same processes, affecting ascending or descending colon; "colon cut off" sign due to pancreatitis causing spasm or stricture of anatomic splenic flexure
  - Perforated duodenal ulcer
    - May result in fluid and gas in APS plus intraperitoneal
  - Colonic inflammation
    - Colitis, diverticulitis, epiploic appendagitis
- Disease within PS is common
  - Any inflammatory or neoplastic process of kidney or adrenal
  - Renal fascia is very strong
    - Effective in containing most primary renal pathology within PS
    - Effective in excluding most other pathology from PS
  - Perirenal space can be breached through renal hilum or through opening in inferior cone of renal fascia
- Perirenal fluid
  - Blood
    - Trauma (including iatrogenic), tumor, anticoagulation, abdominal aortic aneurysm, vasculitis
    - Attenuation: 50 to 80 HU acutely
  - Pus or inflammation
    - Renal and perirenal abscess (or less commonly, abscess originating outside kidney, such as diverticulitis)
    - Pancreatitis (may result in perirenal pseudocyst; must breach renal fascia or enter through renal hilum)
    - Attenuation: Approximately 10 to 20 HU
  - Urine
    - Trauma with parenchymal laceration into collecting system or uretero-pelvic disruption
    - Acute extravasation (common with ureteral calculus; results in perirenal "stranding", not much fluid)
    - Chronic extravasation (results in urinoma, can wall off into a "uriniferous pseudocyst")
    - Attenuation: Varies from water to density of contrast-opacified urine
- Retroperitoneal tumors
  - Liposarcoma: Most common mesenchymal tumor
    - Women slightly more than men
    - Peak incidence between fourth and sixth decades
    - Does not arise from a lipoma or normal retroperitoneal fat
    - Insidious symptoms, large size at diagnosis
    - 90% have enough fat to be recognizable on CT
    - Poorly differentiated or myxoid tumors (or parts of a larger tumor) may have soft tissue or water attenuation
    - Complete removal often impossible; recurrence is common
- Leiomyosarcoma
  - Second most common primary mesenchymal tumor
  - Most common in women, fifth or sixth decade
  - May arise from within IVC or retroperitoneal smooth muscle
  - Symptoms may reflect obstruction of IVC (e.g., Budd-Chiari syndrome, lower extremity edema)
  - Like other sarcomas, often has a vascular periphery and a necrotic core
- Malignant fibrous histiocytoma
  - Does not contain fat or involve IVC
  - Otherwise indistinguishable from other retroperitoneal tumors
- Nerve sheath tumors
  - Occur in younger patients
  - May be isolated mass or extensive (neurofibromatosis)
  - May be benign or malignant
    - Indistinguishable by imaging

RETROPERITONEUM ANATOMY AND IMAGING ISSUES

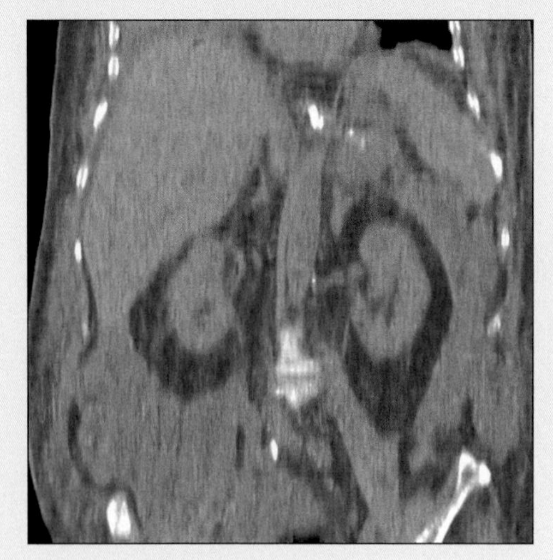

Coronal reformation of NECT in acute pancreatitis shows infiltration of fat in the APS outlining the fat in the perirenal space, separated by the renal fascia.

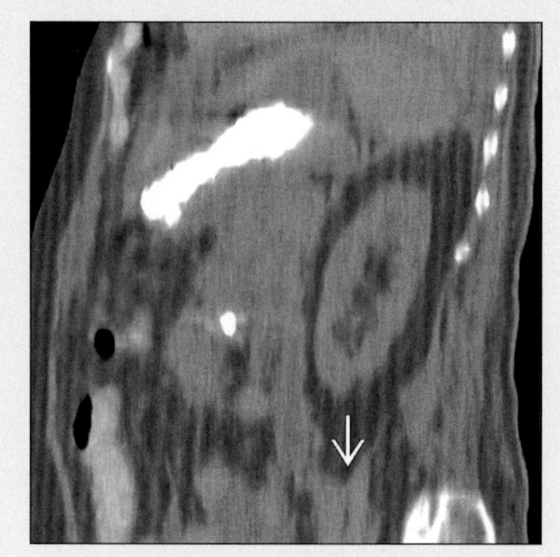

Sagittal reformation of NECT shows fat in perirenal space outlined by extensive infiltration of the APS. Below the perirenal space, the anterior and posterior pararenal spaces join (arrow).

- Key findings: Bilobed, dumbbell-shaped mass with one component extending into neural foramen or eroding spine or rib
- Neuroblastoma, ganglioneuroblastoma, ganglioneuroma
  - Derived from sympathetic ganglion cells
  - Can arise anywhere along chain of sympathetic ganglia
    - Paraspinal or pre-sacral in location
  - Neuroblastoma: Malignant tumor; children
  - Ganglioneuroblastoma: Variable age and biologic behavior
  - Ganglioneuroma: Benign
- Paraganglioma
  - Derived from paraganglionic cells nears sympathetic chain
  - Diagnosed at age 35-45, typically
  - Most are hormonally active (identical to pheochromocytoma; catecholamines) with same symptoms (headache, hypertension, palpitations, sweating)
  - Key findings: Very bright on T2WI MR; brightly enhancing; well-defined mass < 7cm
- Teratoma
  - Most common primary retroperitoneal tumor arising from an embryonic rest
  - Diagnosed in infancy, with second peak in early adulthood
  - Are mature, benign: Displace, rather than invade
  - Key findings: Fat, calcification and soft tissue components
  - Imaging may resemble liposarcoma, but teratoma is more common in childhood
- Disease within the PPS is uncommon
  - Due to absence of viscera
  - Common site of hemorrhage due to coagulopathy
  - May be site of primary retroperitoneal sarcoma
  - Disease originating in other space can "invade" PPS (e.g., diverticulitis)

CUSTOM DIFFERENTIAL DIAGNOSIS

Retroperitoneal mesenchymal neoplasms (benign and malignant)

- Common
  - Lipoma, liposarcoma (most common)
- Uncommon or rare
  - Leiomyoma, sarcoma
  - Malignant fibrous histiocytoma
  - Lymphangioma
  - Hemangioma
  - Hemangiopericytoma
  - Angiosarcoma

SELECTED REFERENCES

1. Nishino M et al: Primary retroperitoneal neoplasms: CT and MR imaging findings with anatomic and pathologic diagnostic clues. Radiographics. 23(1):45-57, 2003
2. Koeller KK et al: Radiologic pathology. 2nd ed. Washington, DC, Armed Forces Institute of Pathology, 531-7, 2003
3. Vivas I et al: Retroperitoneal fibrosis: typical and atypical manifestations. Br J Radiol. 73(866):214-22, 2000
4. Bass JE et al: Spectrum of congenital anomalies of the inferior vena cava: cross-sectional imaging findings. Radiographics. 20(3):639-52, 2000
5. Heiken JP et al: Textbook of gastrointestinal radiology: Peritoneal cavity and retroperitoneum: normal anatomy and examination techniques. 2nd ed. Philadelphia, WB Saunders. 39-57, 2000
6. Aizenstein RI et al: Interfascial and perinephric pathways in the spread of retroperitoneal disease: refined concepts based on CT observations. AJR Am J Roentgenol. 168(3):639-43, 1997
7. Meyers MA: Dynamic radiology of the abdomen: normal and pathologic anatomy. 3rd ed. New York, Springer -Verlag, 1-100, 1988

IMAGE GALLERY

Typical

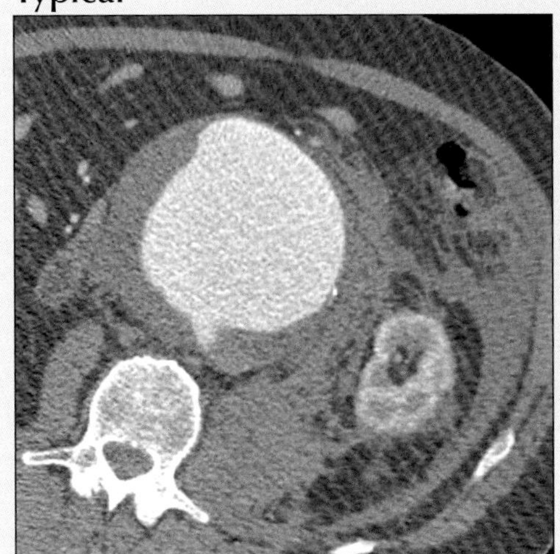

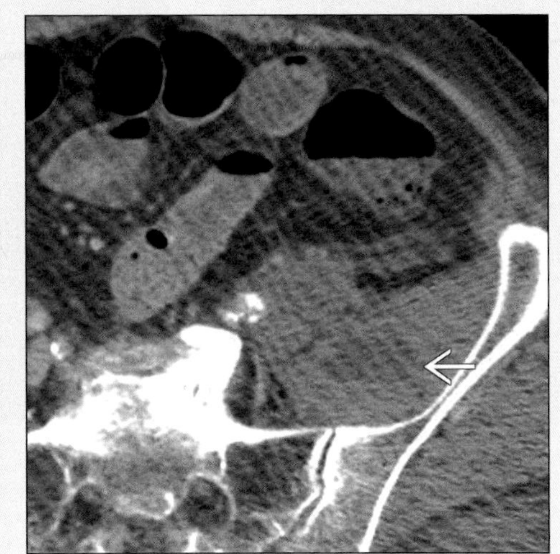

*(**Left**) Axial CECT shows large aortic aneurysm with bleeding into perirenal and anterior pararenal spaces, distending the retromesenteric plane. (**Right**) Axial CECT shows spontaneous "retroperitoneal" hemorrhage (heparin) into iliopsoas compartment. Hematocrit effect (arrow).*

Typical

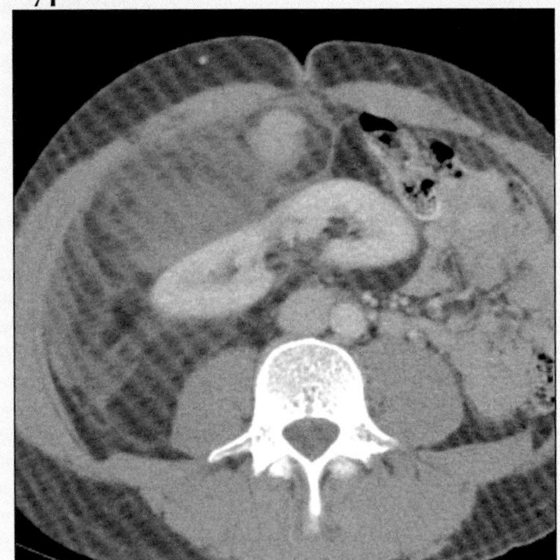

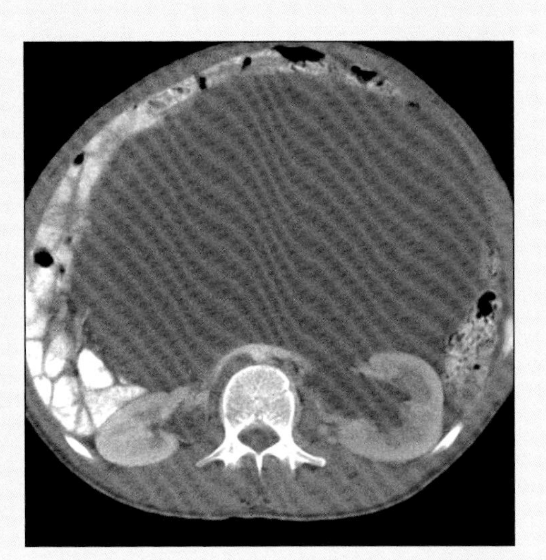

*(**Left**) Axial CECT shows shows a liposarcoma with fatty and myxoid elements displacing kidney and bowel. (**Right**) Axial CECT shows huge liposarcoma, mostly myxoid, displacing bowel and obstructing ureters.*

Typical

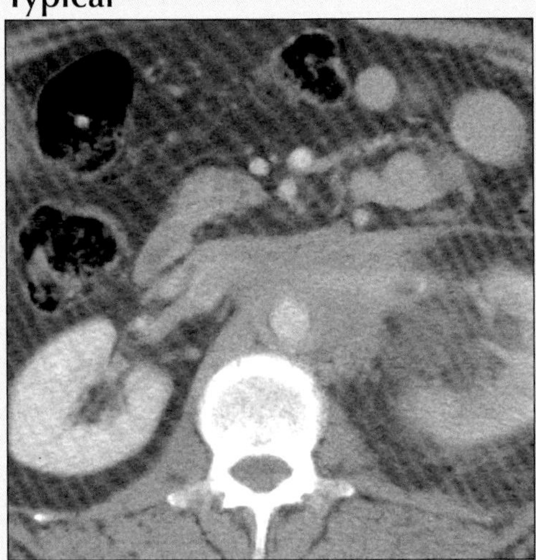

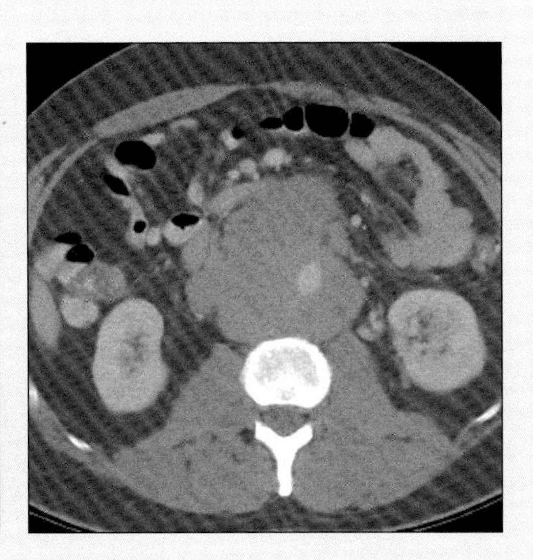

*(**Left**) Axial CECT shows retroperitoneal fibrosis as a mantle of soft tissue surrounding aorta and IVC, and obstructing left ureter. (**Right**) Axial CECT shows large soft tissue mass surrounding aorta + IVC but causing no obstruction; metastatic testicular carcinoma.*

DUPLICATIONS AND ANOMALIES OF IVC

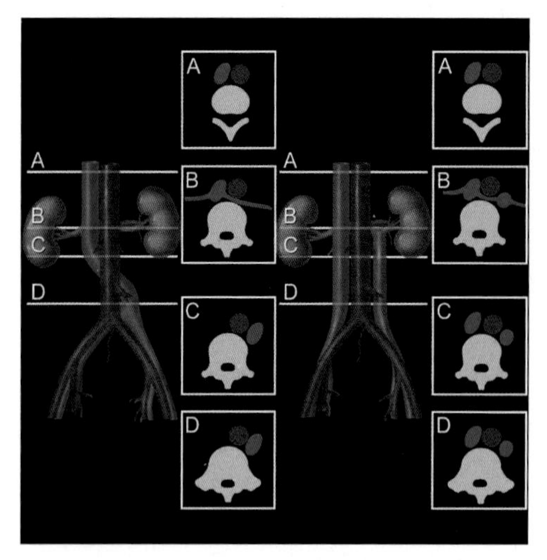

Graphic shows transposition of IVC (on left picture); duplication of IVC (right). Note that duplicated IVC continues as left iliac vein and empties into left renal vein.

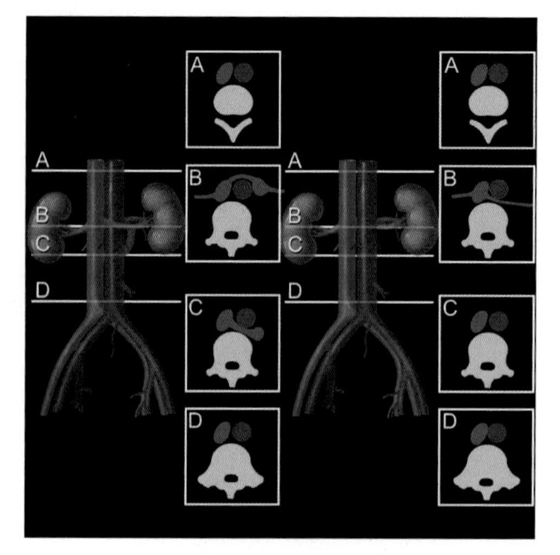

Graphic shows circumaortic left renal vein, with smaller ventral vein crossing cephalad to dorsal vein (left). Right graphic shows completely retroaortic renal vein.

TERMINOLOGY

Definitions

- Congenital anomalies of inferior vena cava (IVC)

IMAGING FINDINGS

General Features

- Best diagnostic clue: Malposition or duplication of IVC inferior to renal vein
- Other general features
  - Types of IVC anomalies
    - Duplication of or double IVC
    - Left IVC
    - Azygos continuation of the IVC
    - Circumaortic left renal vein
    - Retroaortic left renal vein
    - Duplication of IVC with retroaortic right renal vein and hemiazygos continuation of the IVC
    - Duplication of IVC with retroaortic left renal vein and azygos continuation of the IVC
    - Circumcaval or retrocaval ureter
    - Absence of infrarenal or entire IVC

CT Findings

- Duplication of IVC
  - Left and right IVC inferior to renal vein
  - Usually, left IVC ends at left renal vein, which crosses anterior to aorta in normal fashion to join right IVC
  - Left and right IVC may have significant asymmetry in size
- Left IVC
  - Left IVC ends at left renal vein, which crosses anterior to aorta in normal fashion, uniting with right renal vein to form normal right suprarenal IVC
  - ↑ Enhancement of right renal vein relative to left renal vein (dilution from unenhanced venous return from lower extremities)
- Azygos continuation of the IVC
  - IVC passes posterior to diaphragmatic crus to enter thorax as azygos vein
  - Azygos vein joins superior vena cava at normal location in right peribronchial location
  - Hepatic veins drain directly into right atrium
  - Enlarged azygos vein is similar in attenuation to superior vena cava
  - Gonadal veins drain to ipsilateral renal veins
- Duplication of IVC with retroaortic right renal vein and hemiazygos continuation of the IVC
  - Left and right IVC inferior to renal vein
  - Right IVC ends at right renal vein, which crosses posterior to aorta to join left IVC
  - Suprarenal IVC passes posterior to diaphragmatic crus to enter thorax as hemiazygos vein

DDx: Paraaortic Soft Tissue "Mass" Simulating IVC

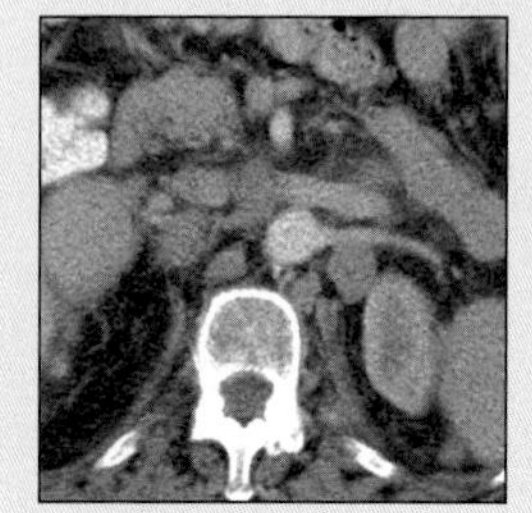

Lymphadenopathy

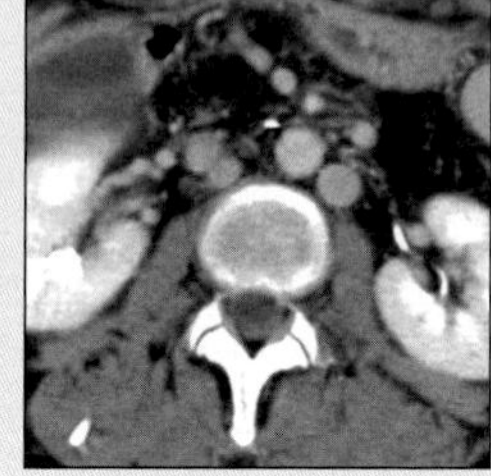

IVC Stenosis, Collaterals

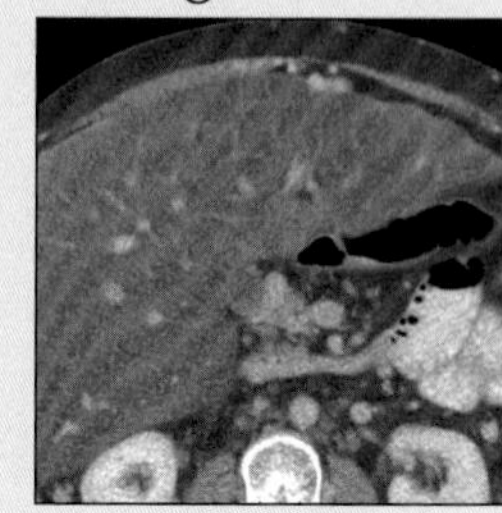

Collaterals

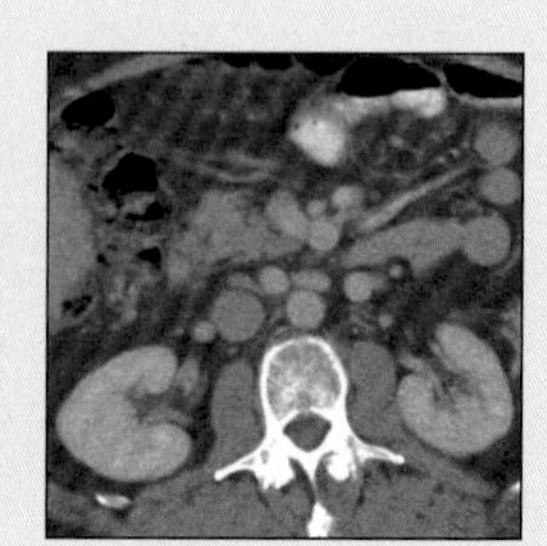

Cirrhosis, Varices

DUPLICATIONS AND ANOMALIES OF IVC

Key Facts

Terminology
- Congenital anomalies of inferior vena cava (IVC)

Imaging Findings
- Best diagnostic clue: Malposition or duplication of IVC inferior to renal vein
- Usually, left IVC ends at left renal vein, which crosses anterior to aorta in normal fashion to join right IVC
- Left and right IVC may have significant asymmetry in size
- Best imaging tool: CT; consider multiplanar reformations

Top Differential Diagnoses
- Retroperitoneal lymphadenopathy
- Varices/Collaterals
- Gonadal vein

Pathology
- Duplication of IVC: 0.2-3% of general population

Clinical Issues
- Asymptomatic
- Usually diagnosed incidentally by imaging
- Prognosis: Very good
- Usually no treatment

Diagnostic Checklist
- Duplication of IVC should be suspected in recurrent pulmonary embolism following IVC filter placement
- Pre-operative imaging may be important in planning abdominal surgery, liver or kidney transplantation or interventional procedures (e.g., IVC filters, varicocele sclerotherapy, venous renal sampling)

    - In thorax, collateral pathways for hemiazygos vein include
        - Crosses posterior to aorta at T8-9 to join azygos vein
        - Continues superiorly to join coronary vein of heart via persistent left superior vena cava
        - Accessory hemiazygos continuation to left brachiocephalic vein
- Duplication of IVC with retroaortic left renal vein and azygos continuation of the IVC
    - Mixture of findings previously mentioned
- Circumaortic left renal vein (common variant)
    - 2 left renal veins
        - Superior renal vein joined by left adrenal vein and crosses aorta anteriorly
        - Inferior renal vein, 1-2 cm below to superior renal vein, joins by left gonadal vein and crosses aorta posteriorly
- Retroaortic left renal vein
    - 1 left renal vein, crosses aorta posteriorly
- Circumcaval ureter
    - Proximal ureter courses posterior IVC, emerges to right of aorta and lies anterior to right iliac vessels
- Absence of infrarenal or entire IVC
    - External and internal iliac veins join to form enlarged ascending lumbar veins
    - Venous return from lower extremities to azygos and hemiazygos vein via anterior paravertebral collateral veins
    - ± Suprarenal IVC formed by left and right renal veins
    - May be acquired abnormality following thrombosis of IVC

MR Findings
- Flow voids or flow-related enhancement; may distinguish aberrant vessels from masses

Ultrasonographic Findings
- Infrahepatic IVC ends with azygos or hemiazygos continuation
- Hepatic veins drain directly into right atrium
- Right renal artery crossing anteriorly to azygos vein; may demonstrate azygos continuation of the IVC

Angiographic Findings
- Most accurate diagnostic method

Imaging Recommendations
- Best imaging tool: CT; consider multiplanar reformations

DIFFERENTIAL DIAGNOSIS

Retroperitoneal lymphadenopathy
- E.g., metastases and lymphoma, granulomatous disease
- Left sided paraaortic adenopathy; may mimic duplication of or left IVC
    - Differentiate by renal vein drainage or contrast-enhancement of IVC
- Retrocrural adenopathy; may mimic enlarged azygos vein in retrocrural space
    - Differentiate by tubular structure of azygos vein extending from diaphragm to azygos arch
    - Retrocrural adenopathy lacks enhancement
- Retroperitoneal adenopathy; may mimic circumaortic left renal vein

Varices/Collaterals
- E.g., cirrhosis, IVC obstruction

Gonadal vein
- May appear as paraaortic soft tissue “mass" or mimic left sided IVC
- Follow inferiorly; does not "join" left iliac vein

PATHOLOGY

General Features
- General path comments
    - Embryology

DUPLICATIONS AND ANOMALIES OF IVC

- 6-8th gestational weeks: Infrahepatic IVC develops from appearance and regression of three paired embryonic veins; postcardinal, subcardinal and supracardinal veins
- Normal IVC comprised of hepatic, suprarenal, renal and infrarenal segments
- Hepatic segment develops from vitelline vein
- Suprarenal segment develops from subcardinal-hepatic anastomosis
- Renal segment develops from right supra-subcardinal and post-subcardinal anastomoses
- Infrarenal segment develops from right supracardinal vein
- In thorax, supracardinal veins form to azygos and hemiazygos veins
- In abdomen, subcardinal and supracardinal veins progressively replace postcardinal veins
- In pelvis, postcardinal veins form common iliac veins

- Etiology
  - Congenital
  - Risk factor: First degree relatives
  - Pathogenesis
    - Duplication of IVC: Persistence of both supracardinal veins
    - Left IVC: Regression of right supracardinal vein with persistence of left supracardinal vein
    - Azygos continuation of the IVC: Failure to form right subcardinal-hepatic anastomosis, resulting atrophy of right subcardinal vein
    - Circumaortic left renal vein: Persistence of dorsal limb of embryonic left renal vein and of dorsal arch of renal collar (intersupracardinal anastomosis)
    - Retroaortic left renal vein: Persistence of dorsal arch of renal collar and regression of ventral arch (intersubcardinal anastomosis)
    - Duplication of IVC with retroaortic right renal vein and hemiazygos continuation of the IVC: Persistence of left lumbar and thoracic supracardinal vein and left suprasubcardinal anastomosis, with failure to form right subcardinal-hepatic anastomosis
    - Duplication of IVC with retroaortic left renal vein and azygos continuation of the IVC: Persistence of left supracardinal vein and dorsal limb of the renal collar with regression of ventral arch, with failure to form subcardinal-hepatic anastomosis
- Epidemiology
  - Prevalence
    - Duplication of IVC: 0.2-3% of general population
    - Left IVC: 0.2-0.5%
    - Azygos continuation of the IVC: 0.6%
    - Circumaortic left renal vein: 8.7%
    - Retroaortic left renal vein: 2.1%

CLINICAL ISSUES

Presentation

- Most common signs/symptoms
  - Asymptomatic
  - Circumcaval ureter: Partial right ureteral obstruction or recurrent urinary tract infections
  - Absence of infrarenal or entire IVC: Venous insufficiency of lower extremities or idiopathic deep venous thrombosis
- Diagnosis
  - Usually diagnosed incidentally by imaging

Demographics

- Age: Any age

Natural History & Prognosis

- Prognosis: Very good

Treatment

- Usually no treatment
- Circumcaval ureter: Surgical relocation of ureter anterior to IVC

DIAGNOSTIC CHECKLIST

Consider

- Duplication of IVC should be suspected in recurrent pulmonary embolism following IVC filter placement

Image Interpretation Pearls

- Pre-operative imaging may be important in planning abdominal surgery, liver or kidney transplantation or interventional procedures (e.g., IVC filters, varicocele sclerotherapy, venous renal sampling)

SELECTED REFERENCES

1. Yilmaz E et al: Interruption of the inferior vena cava with azygos/hemiazygos continuation accompanied by distinct renal vein anomalies: MRA and CT assessment. Abdom Imaging. 28(3):392-4, 2003
2. Basile A et al: Embryologic and acquired anomalies of the inferior vena cava with recurrent deep vein thrombosis. Abdom Imaging. 28(3):400-3, 2003
3. Brochert A et al: Unusual duplication anomaly of the inferior vena cava with normal drainage of the right IVC and hemiazygous continuation of the left IVC. J Vasc Interv Radiol. 12(12):1453-5, 2001
4. Bass JE et al: Spectrum of congenital anomalies of the inferior vena cava: cross-sectional imaging findings. Radiographics. 20(3):639-52, 2000
5. Bass JE et al: Absence of the infrarenal inferior vena cava with preservation of the suprarenal segment as revealed by CT and MR venography. AJR Am J Roentgenol. 172(6):1610-2, 1999
6. Mayo J et al: Anomalies of the inferior vena cava. AJR Am J Roentgenol. 140(2):339-45, 1983
7. Breckenridge JW et al: Azygos continuation of inferior vena cava: CT appearance. J Comput Assist Tomogr. 4(3):392-7, 1980
8. Garris JB et al: Ultrasonic diagnosis of infrahepatic interruption of the inferior vena cava with azygos (hemiazygos) continuation. Radiology. 134(1):179-83, 1980
9. Royal SA et al: CT evaluation of anomalies of the inferior vena cava and left renal vein. AJR Am J Roentgenol. 132(5):759-63, 1979
10. Faer MJ et al: Inferior vena cava duplication: demonstration by computed tomography. Radiology. 130(3):707-9, 1979

IMAGE GALLERY

Typical

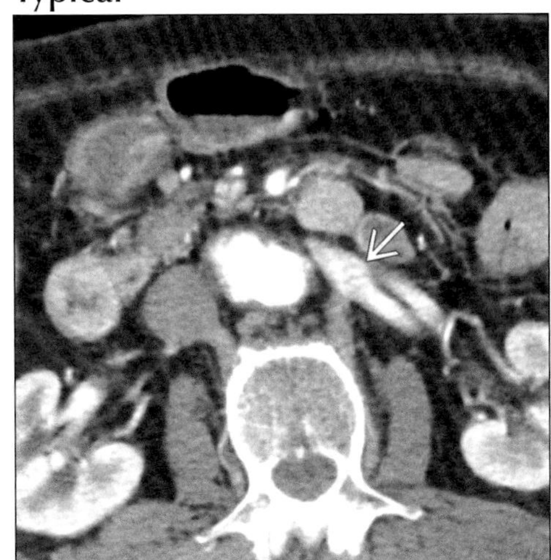

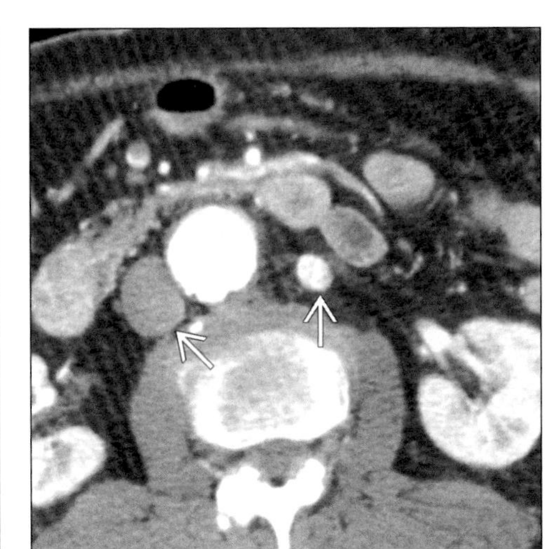

(Left) *Axial CECT shows duplicated IVC at level of renal vein which is dilated (arrow) as it receives the left-sided IVC.* ***(Right)*** *Axial CECT shows duplicated IVC (arrows) with a smaller left-sided vessel.*

Typical

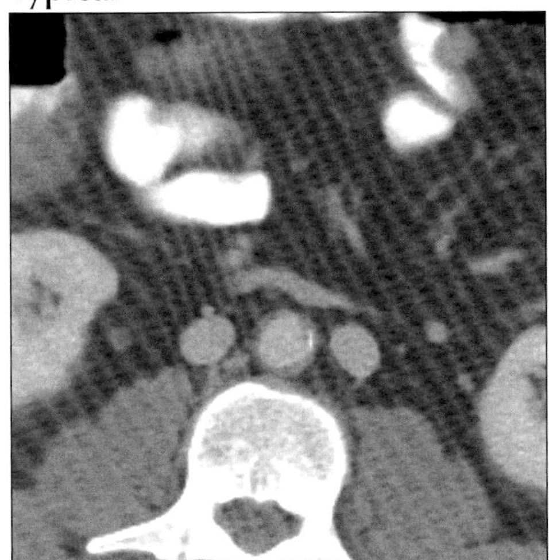

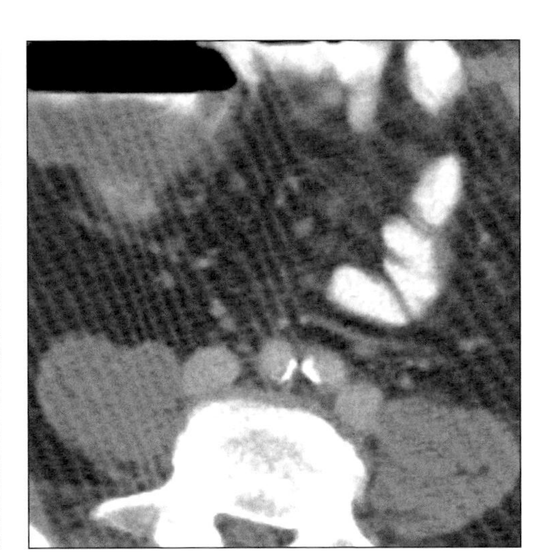

(Left) *Axial CECT shows duplicated IVC with equal size vessels.* ***(Right)*** *Axial CECT shows duplicated IVC. The left common iliac vein appears normal, but does not cross to join the right iliac vein.*

Typical

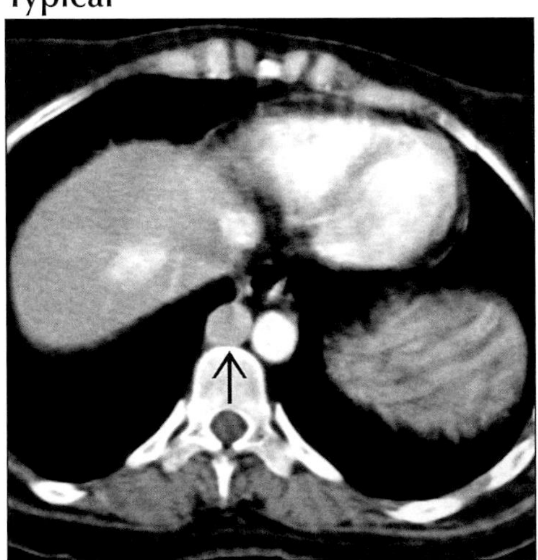

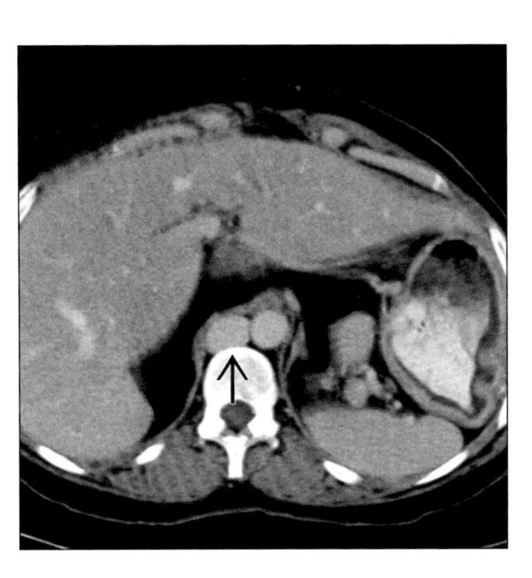

(Left) *Axial CECT shows polysplenia syndrome with absent IVC and continuation of dilated azygous vein (arrow).* ***(Right)*** *Axial CECT shows polysplenia syndrome with absent IVC and continuation of dilated azygous vein (arrow).*

RETROPERITONEAL FIBROSIS

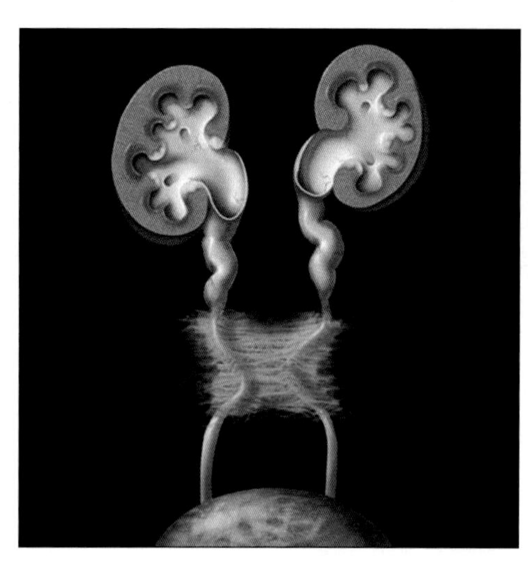

Graphic shows encasement and displacement of mid-ureters by a band of fibrous tissue. Hydronephrosis.

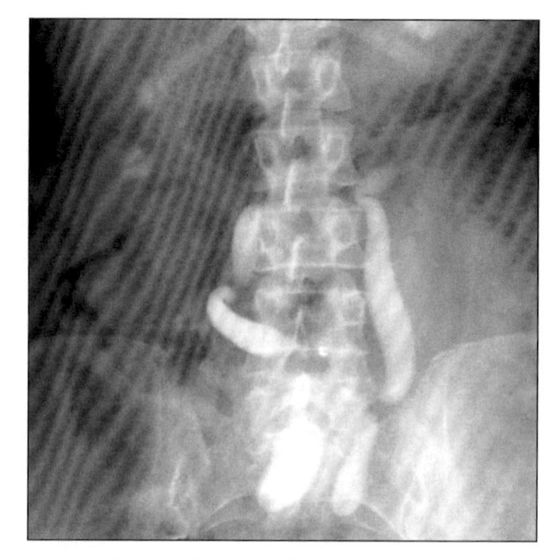

Retrograde pyelogram shows encasement and displacement of mid-ureters by retroperitoneal fibrosis.

TERMINOLOGY

Definitions

- Chronic inflammatory process in lumbar retroperitoneum

IMAGING FINDINGS

General Features

- Best diagnostic clue: Mantle of soft tissue encasing aorta, inferior vena cava (IVC) & ureters
- Location: Lower lumbar region & extends into pelvis
- Morphology
  - Exuberant mass of woody fibrous tissue that usually encases abdominal aorta, IVC & ureters
  - Pinkish or glistening
- Other general features
  - Two types based on etiology
    - Primary or idiopathic (common): 2/3 cases
    - Secondary: 1/3 cases
  - Classification based on pathology & radiology
    - Limited (common) & extensive (rare)
    - Malignant & nonmalignant
  - Limited type (common form)
    - Common manifestation: Isolated plaque over lower lumbar spine
  - Extensive type (rare form)
    - May involve root of mesentery & adjacent organs
    - Extend via crus of diaphragm → thorax as fibrous mediastinitis
    - Fibrotic mass in true pelvis involving iliac vessels + lower ureters
  - 15% cases associated with fibrotic process elsewhere

Radiographic Findings

- Excretory urography (IVU)
  - Classic triad
    - Upper ureteral hydronephrosis (above L4/5)
    - Medial deviation of ureters in middle third, typically bilateral
    - Gradual tapering of ureters, extrinsic compression
- Retrograde pyelography (RGP)
  - Pyelocaliectasis & ureterectasis to the level of L4-5
  - Medial deviation of ureters
  - Valuable study to assess location, extent & severity of ureteral obstruction
- Fluoroscopic guided barium enema
  - Retroperitoneal fibrosis involving pelvic cavity may show extrinsic compression, displacement of rectum or rectosigmoid loop

CT Findings

- Fibrous plaque
  - Soft tissue density fibrotic plaque or mantle of variable thickness

DDx: Mantle of Tissue Around Aorta

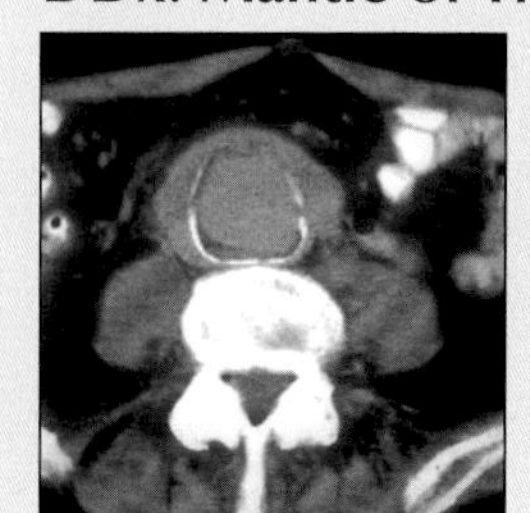

Aortitis

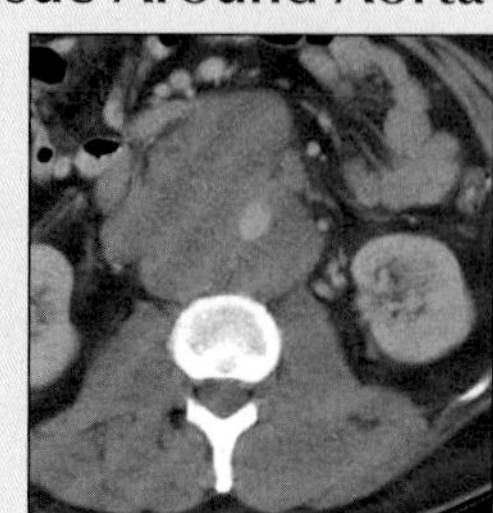

Testicular Mets.

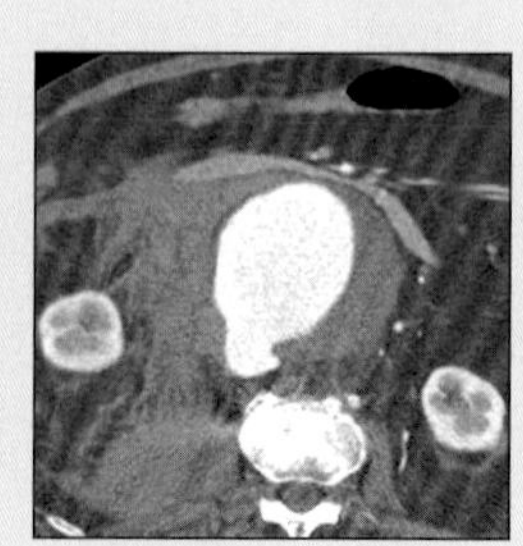

Leaking AAA

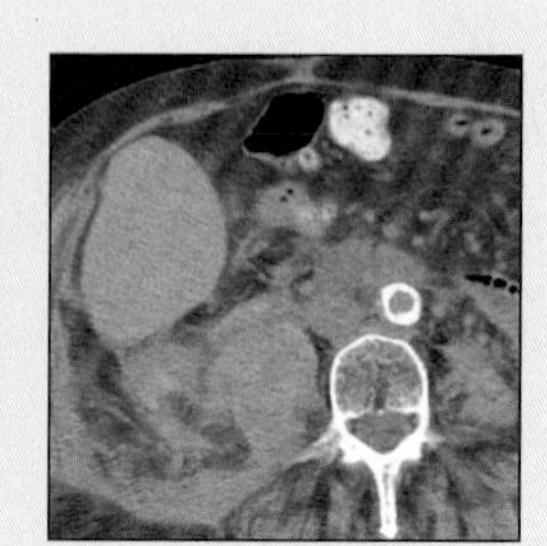

Retroperitoneal Bleed

RETROPERITONEAL FIBROSIS

Key Facts

Terminology

- Chronic inflammatory process in lumbar retroperitoneum

Imaging Findings

- Best diagnostic clue: Mantle of soft tissue encasing aorta, inferior vena cava (IVC) & ureters
- ± Displacing structures anteromedially/anterolaterally
- Rarely invades aorta, IVC & ureters
- ↑ Enhancement: Active inflammatory process
- ↓ Enhancement: Well-organized fibrous tissue

Top Differential Diagnoses

- Aortitis
- Retroperitoneal metastases & lymphoma
- Ruptured abdominal aortic aneurysm (AAA)
- Retroperitoneal hemorrhage

Pathology

- Primary (idiopathic): Probably autoimmune disease with antibodies → stimulate desmoplastic reaction
- Drugs: Methysergide, β blocker, hydralazine, ergotamine, LSD
- Epidemiology: Prevalence, 1 in 200,000 population
- Other associated inflammatory fibrotic processes
- Mass of woody fibrous tissue; pinkish & glistening

Clinical Issues

- Extrinsic ureteral obstruction → renal failure
- Great vessels (aorta & IVC) obstruction

Diagnostic Checklist

- Check for underlying malignancy, AAA
- Mantle or rind of soft tissue density encircling great vessels & ureters in retroperitoneum

---

- Plaque may extend from crus of diaphragm to common iliac vessels
    - Most commonly extends from renal hilum to pelvic brim
    - Rarely extends to involve kidneys, pancreas, spleen, mediastinum
    - Plaque may be asymmetrical, sharply localized or very extensive
- Usually surrounds aorta, IVC & ureters; ± caudal extension to iliac vessels
    - ± Displacing structures anteromedially/anterolaterally
    - ± Compression or narrowing, proximal dilatation
    - Rarely invades aorta, IVC & ureters
    - No tissue plane visible between fibrosis & muscles
- Contrast enhancement varies depending on maturity of fibrous tissue
    - ↑ Enhancement: Active inflammatory process
    - ↓ Enhancement: Well-organized fibrous tissue
- Difficult to differentiate a confluent malignant retroperitoneal adenopathy & fibrosis

MR Findings

- T1WI: Low-medium homogeneous signal intensity
- T2WI
    - Low-moderate heterogeneous signal intensity
    - Signal intensity, less than fat but more than muscle

Ultrasonographic Findings

- Real Time: Hypoechoic "halo": Irregularly contoured periaortic soft tissue mass

Imaging Recommendations

- NE + CECT; IVU or RGP

DIFFERENTIAL DIAGNOSIS

Aortitis

- Severe degree is known as inflammatory abdominal aortic aneurysm or perianeurysmal fibrosis
- Cause: Unknown, may be due to hypersensitivity to antigens in atheromatous plaques
- Gross pathologic finding
    - Thick cuff of firm, fibrous granulation tissue around aorta encompassing adjacent structures
- Morphology & histology are identical to idiopathic retroperitoneal fibrosis
- Imaging
    - NECT
        - Perivascular, irregular mantle of tissue with variable attenuation
        - Isodense or hypodense to aorta
    - CECT
        - Enhancing fibrotic rind of variable attenuation
        - Ureters: Normal; displaced or obstructed
    - Multiplanar CT & MR imaging: Demonstrate relationship of perianeurysmal mass to aneurysm, ureters & psoas muscle
    - MRA: Evaluates blood flow in affected vessels
    - Aortography: Determines extent of aneurysmal disease & its relationship to peripheral vessels

Retroperitoneal metastases & lymphoma

- Metastases
    - E.g., prostate, cervix, breast & lung carcinoma
    - May cause mantle of tissue + desmoplastic response similar to retroperitoneal fibrosis
    - Additional pelvic + retroperitoneal nodes usually seen
    - Usually more discrete or asymmetrical
- Lymphoma
    - May simulate retroperitoneal fibrosis, especially after treatment
    - Rarely obstructs ureters
    - Usually involves higher nodes as well

Ruptured abdominal aortic aneurysm (AAA)

- Location of AAA
    - Usually infrarenal with extension to iliac arteries
- NECT
    - Thickened diaphragmatic crura
    - "Enlarged" left psoas muscle
    - Aortic outline is obscured by soft tissue density
- CECT
    - Saccular or fusiform dilatation of aorta
    - Indistinct focal area of aortic wall

RETROPERITONEAL FIBROSIS

- Anterior displacement of kidney
- Extravasation of contrast material
- Fluid collection (hematoma) within posterior pararenal + perirenal spaces

Retroperitoneal hemorrhage

- Most cases are iatrogenic
  - Usually due to over anticoagulation
- Rupture of abdominal aortic aneurysm
  - Second most common cause of retroperitoneal bleed
- Spontaneous tumor bleed is 3rd most common cause
  - Kidney: Renal cell carcinoma & angiomyolipoma
  - Adrenal: Carcinoma or myelolipoma
- CT findings
  - Acute: High density fluid collection or hematoma
  - Chronic: Low density (organized clot)
  - Associated renal, adrenal tumors or aortic aneurysm
- MR findings
  - Varied signal intensity (evolution of blood products)
  - Hyperacute phase: Due to oxyhemoglobin
    - T1WI; slightly hypointense; T2WI; hyperintense
  - Acute phase: Iso-/hypointense on T1WI & markedly ↓ signal on T2WI due to deoxyhemoglobin
  - Chronic phase: ↓ Signal (T1WI); ↑ signal (T2WI)
- Diagnosis: History of anticoagulation; imaging evidence of tumor or aneurysm

PATHOLOGY

General Features

- Etiology
  - Primary (idiopathic): Probably autoimmune disease with antibodies → stimulate desmoplastic reaction
    - Ormond speculation: Retroperitoneal fibrosis is similar to collagen vascular disease, supported by coexistence with other inflammatory processes
  - Secondary
    - Drugs: Methysergide, β blocker, hydralazine, ergotamine, LSD
    - Diseases that stimulate desmoplastic reaction: Malignant tumors, metastases, Hodgkin, carcinoid tumor, hematoma, radiation, retroperitoneal injury, surgery, infection, urinary extravasation
- Epidemiology: Prevalence, 1 in 200,000 population
- Associated abnormalities
  - Other associated inflammatory fibrotic processes
    - Pseudotumor of orbit; Reidel thyroiditis
    - Sclerosing cholangitis; chronic fibrosing mediastinitis

Gross Pathologic & Surgical Features

- Mass of woody fibrous tissue; pinkish & glistening
- Encases vessels & ureters

Microscopic Features

- Early: Collagen, plasma cells, histiocytes, giant cells
- Late: Acellular fibrosis

CLINICAL ISSUES

Presentation

- Most common signs/symptoms
  - Pain: Back, flank, abdomen
  - Renal insufficiency, HTN, leg edema, anemia
- Lab data: ↑ ESR; ↓ hematocrit; ± azotemia

Demographics

- Age: Usually above 40 years
- Gender: M:F = 2:1

Natural History & Prognosis

- Complications
  - Extrinsic ureteral obstruction → renal failure
  - Great vessels (aorta & IVC) obstruction
- Prognosis
  - Good; may require surgery

Treatment

- Withdrawal of possible causative agent
- Corticosteroids; ureteral stent; ureterolysis

DIAGNOSTIC CHECKLIST

Consider

- Check for underlying malignancy, AAA

Image Interpretation Pearls

- Mantle or rind of soft tissue density encircling great vessels & ureters in retroperitoneum

SELECTED REFERENCES

1. Fukukura Y et al: Autoimmune pancreatitis associated with idiopathic retroperitoneal fibrosis. AJR Am J Roentgenol. 181(4):993-5, 2003
2. Hamano H et al: Hydronephrosis associated with retroperitoneal fibrosis and sclerosing pancreatitis. Lancet. 359(9315):1403-4, 2002
3. Nishimura H et al: MR imaging of soft-tissue masses of the extraperitoneal spaces. Radiographics. 21(5):1141-54, 2001
4. Sung MS et al: Myxoid liposarcoma: appearance at MR imaging with histologic correlation. Radiographics. 20(4):1007-19, 2000
5. Vivas I et al: Retroperitoneal fibrosis: typical and atypical manifestations. Br J Radiol. 73(866):214-22, 2000
6. Rominger MB et al: Perirenal involvement by retroperitoneal fibrosis: the usefulness of MRI to establish diagnosis. Urol Radiol. 13(3):173-6, 1992
7. Amis Jr ES: Retroperitoneal fibrosis. AJR 157: 321-9, 1991
8. Arrive L et al: Malignant versus nonmalignant retroperitoneal fibrosis: differentiation with MR imaging. Radiology. 172(1):139-43, 1989
9. Mulligan SA et al: CT and MR imaging in the evaluation of retroperitoneal fibrosis. J Comput Assist Tomogr. 13(2):277-81, 1989
10. Arger PH et al: Retroperitoneal fibrosis: An analysis of the clinical spectrum and roentgenographic signs. AJR 119: 812, 1973

RETROPERITONEAL FIBROSIS

IMAGE GALLERY

Typical

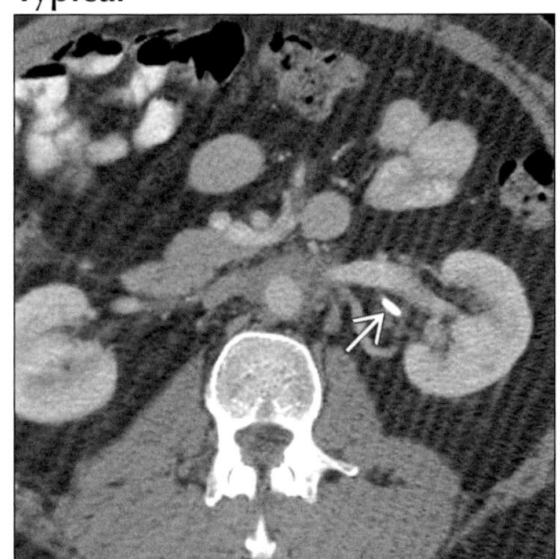

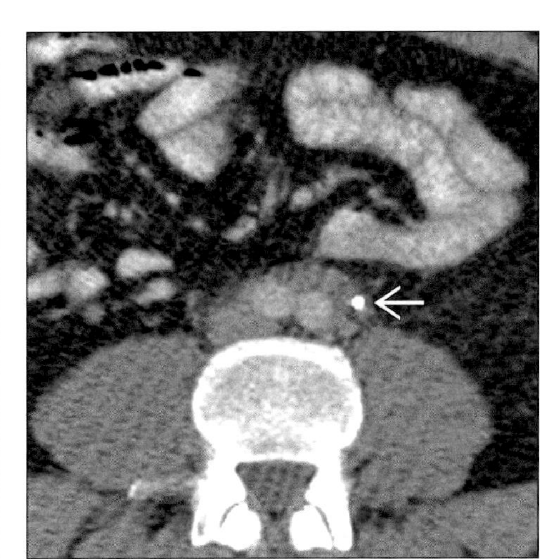

(Left) *Axial CECT shows mantle of soft tissue around aorta. Stent (arrow) in left ureter to bypass obstruction.* ***(Right)*** *Axial CECT shows mantle of soft tissue surrounding the aortic bifurcation and left ureteral stent (arrow).*

Typical

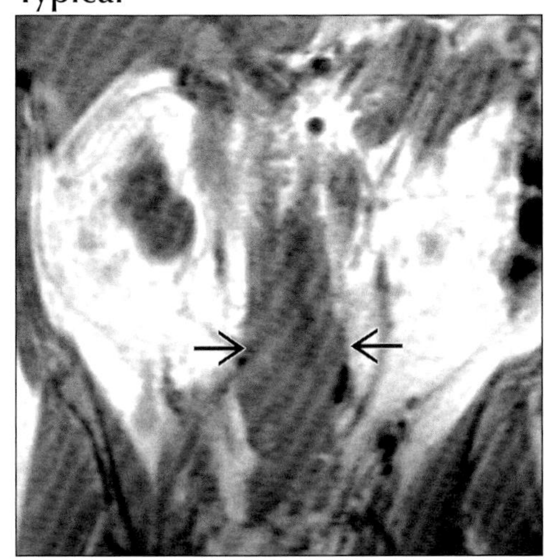

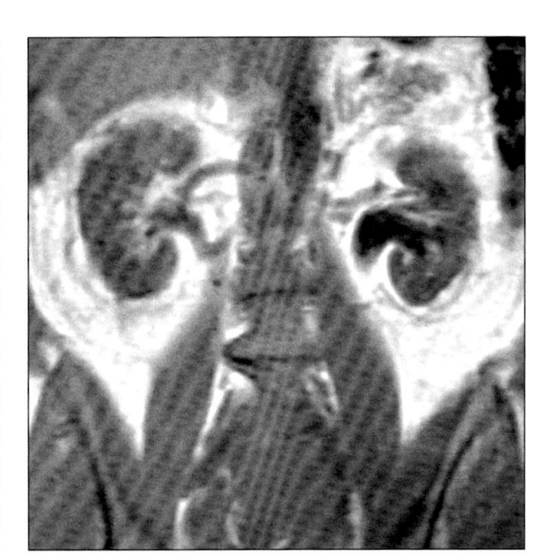

(Left) *Coronal T2WI MR shows mantle of soft tissue (arrows) surrounding aorta and IVC.* ***(Right)*** *Coronal T2WI MR shows hydronephrosis, worse on left.*

Variant

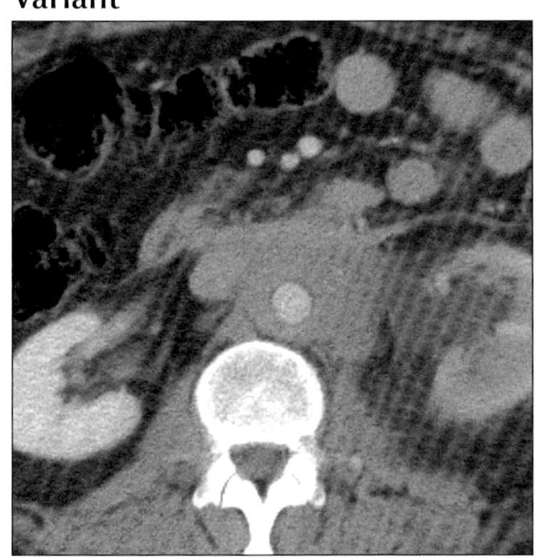

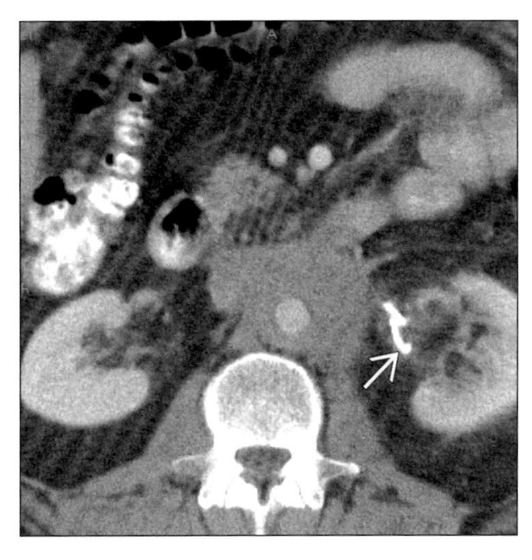

(Left) *Axial CECT shows unusually massive retroperitoneal fibrosis in September, 2003. Obstructed left kidney.* ***(Right)*** *Axial CECT in November, 2003, shows slight decrease in fibrotic mass following steroid therapy, and relief of obstruction by stent (arrow).*

PELVIC LIPOMATOSIS

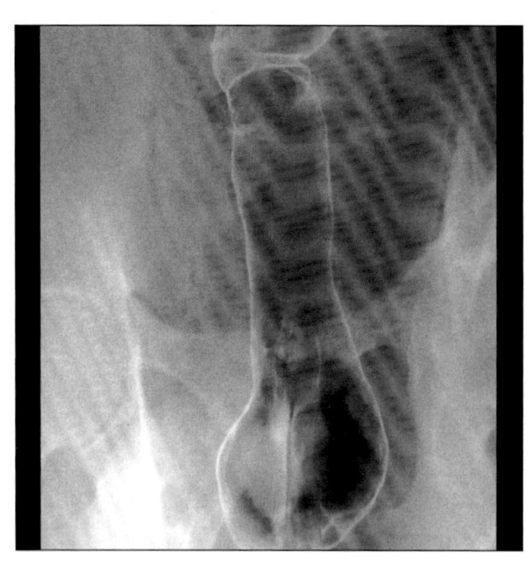

Air-contrast barium enema (BE) shows smooth, long stricture of rectum due to pelvic lipomatosis.

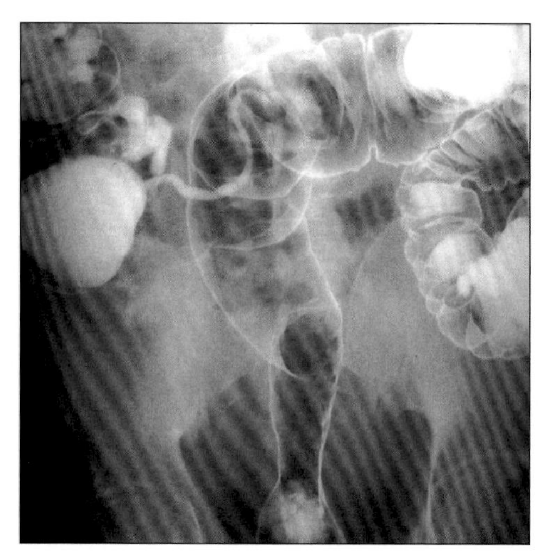

Air-contrast BE shows extrinsic compression of rectum by pelvic lipomatosis.

TERMINOLOGY

Abbreviations and Synonyms

- Pelvic lipomatosis (PL)

Definitions

- Nonmalignant overgrowth of nonencapsulated fatty tissue in perirectal and perivesical spaces of pelvis

IMAGING FINDINGS

General Features

- Best diagnostic clue: Classic triad 1) pelvic radiolucency, 2) elevation of an intact rectosigmoid and 3) elevation of urinary bladder with symmetric inverted pear shape

Radiographic Findings

- Radiography
  - ↑ Radiolucency of the perivesical area
  - High positioned, pear-shaped or inverted teardrop bladder
  - Tubular narrowing of rectosigmoid colon
- IVP: Dilated, tortuous and medially displaced ureters
- Fluoroscopic-guided barium enema
  - "Lower rectum": Elongation & symmetrical extrinsic compression of rectum and cephalad displacement of sigmoid colon

CT Findings

- Nonencapsulated fatty mass surrounding pelvic organs symmetrically
- Fat density attenuation (-80 to -120 HU); may contain strands with higher attenuation than fat

MR Findings

- Cephalad displacement of bladder base
- Elevation of prostate gland
- Elongation of bladder neck and posterior urethra
- Medial and superior displacement of seminal vesicles
- Medial or lateral displacement of ureters
- ↑ Distance between prostate gland & rectum by fat

Ultrasonographic Findings

- Real Time
  - Tubular or cigar shaped urinary bladder compressed by extensive echogenic perivesical tissue
  - Bladder "floats" in pelvic fat and unable to distend normally to the pelvic side walls

Imaging Recommendations

- Best imaging tool: CT: Definitive diagnosis

DIFFERENTIAL DIAGNOSIS

Proctitis

- E.g., radiation, lymphogranuloma venereum

DDx: Narrowed Rectum +/or Urinary Bladder

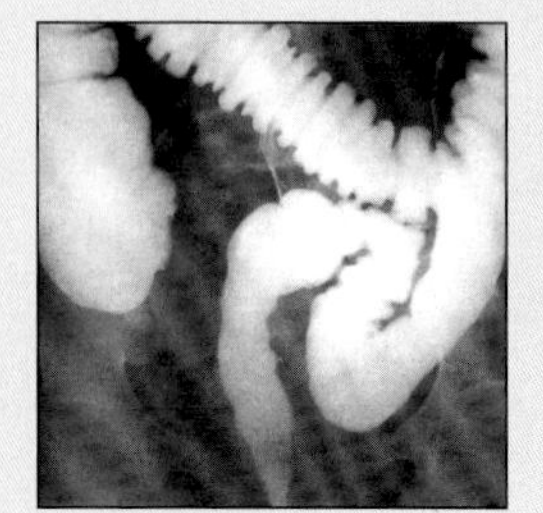

Proctitis

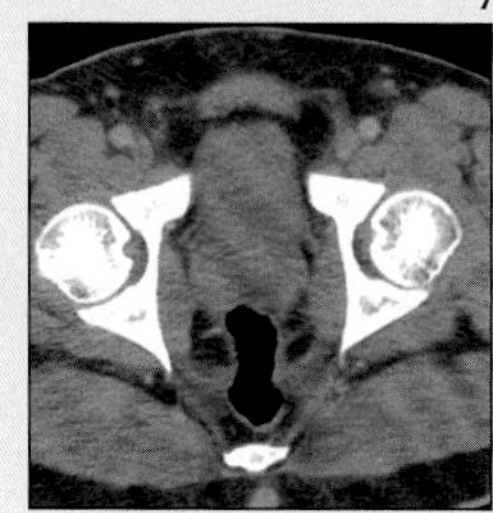

Narrow Pelvis

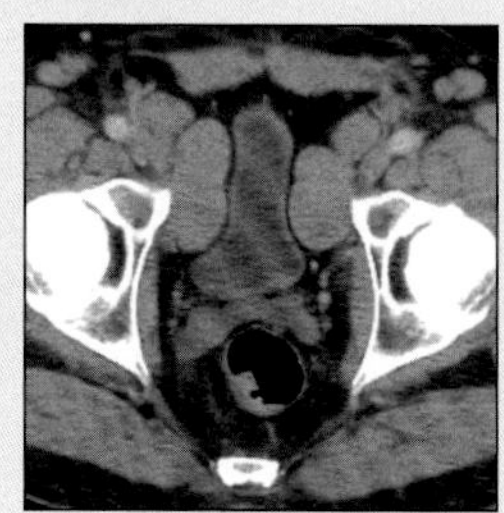

Lymphadenopathy

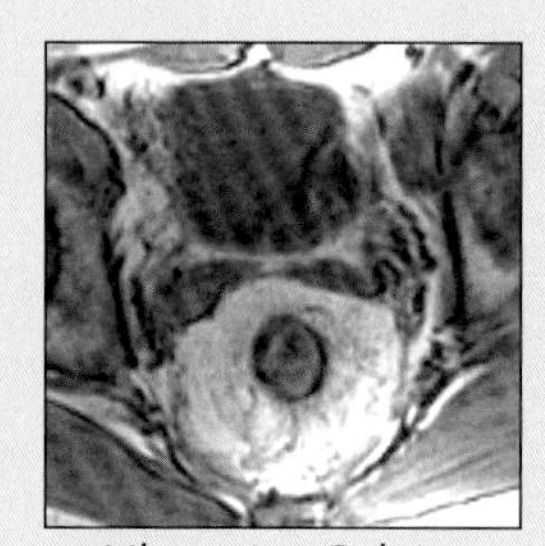

Ulcerative Colitis

PELVIC LIPOMATOSIS

Key Facts

Imaging Findings
- Best diagnostic clue: Classic triad 1) pelvic radiolucency, 2) elevation of an intact rectosigmoid and 3) elevation of urinary bladder with symmetric inverted pear shape

Top Differential Diagnoses
- Proctitis
- Normal variant
- Ulcerative colitis
- Post operation

Pathology
- General path comments: Nonmalignant condition

Diagnostic Checklist
- Close follow-up to monitor ureteral obstruction
- Use CT to differentiate fatty infiltration from other conditions

Normal variant
- Large pelvic muscles with narrow bony pelvis
- Pear-shaped bladder caused by large iliopsoas muscles

Ulcerative colitis
- Pancolitis with decreased haustration & multiple ulcerations
- Diffuse & symmetric wall thickening of colon
- Chronic phase → lead pipe colon

Post operation
- E.g., proctosigmoid resection/re-anastomosis

PATHOLOGY

General Features
- General path comments: Nonmalignant condition
- Etiology: Unknown
- Epidemiology: 0.6-1.7 per 100,000 hospital admissions
- Associated abnormalities: Cystitis glandularis

Gross Pathologic & Surgical Features
- Abnormal deposition of mature adipose tissue

Microscopic Features
- Fat overgrowth without fibrosis or inflammation

CLINICAL ISSUES

Presentation
- Most common signs/symptoms
  - Asymptomatic
  - Compressed urinary system (i.e., ↑ frequency, dysuria, nocturia, and hematuria)
  - Compressed intestinal tract (i.e., constipation, rectal bleeding, tenesmus, ribbon-like stools with mucus)
  - Compressed vasculature (i.e., edema of lower extremities)

Demographics
- Age: 9-80 years old, peak 25-60 years old
- Gender: M:F = 10:1
- Ethnicity: African-Americans

Natural History & Prognosis
- Complications: Hydroureteronephrosis, renal failure, ureteral obstruction, obstruction of inferior vena cava or pelvic veins

Treatment
- Urinary diversion to relieve obstruction
- Role of surgical resection is unclear

DIAGNOSTIC CHECKLIST

Consider
- Close follow-up to monitor ureteral obstruction

Image Interpretation Pearls
- Use CT to differentiate fatty infiltration from other conditions

SELECTED REFERENCES

1. Heyns CF: Pelvic lipomatosis: a review of its diagnosis and management. J Urol. 146(2):267-73, 1991
2. Demas BE et al: Pelvic lipomatosis: diagnosis and characterization by magnetic resonance imaging. Urol Radiol. 10(4):198-202, 1988
3. Klein FA et al: Pelvic lipomatosis: 35-year experience. J Urol. 139(5):998-1001, 1988
4. Clark WM et al: Ultrasonographic features of pelvic lipomatosis. Urol Radiol. 1(3):183-6, 1980
5. Chang SF: Pear-shaped bladder caused by large iliopsoas muscles. Radiology. 128(2):349-50, 1978
6. Ambos MA et al: The pear-shaped bladder. Radiology. 122(1):85-8, 1977
7. Moss AA et al: Pelvic lipomatosis: a roentgenographic diagnosis. Am J Roentgenol Radium Ther Nucl Med. 115(2):411-9, 1972

IMAGE GALLERY

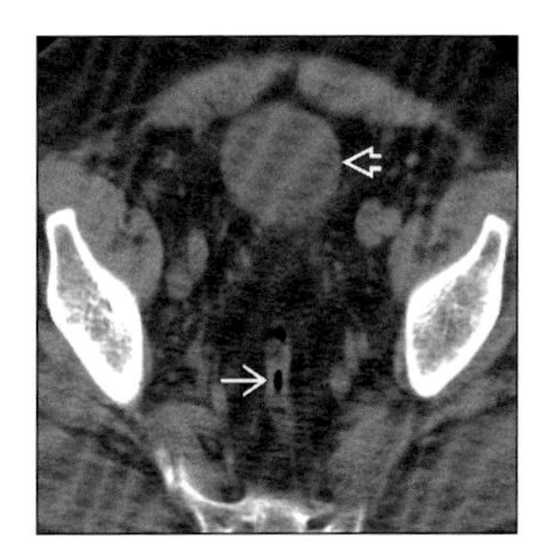
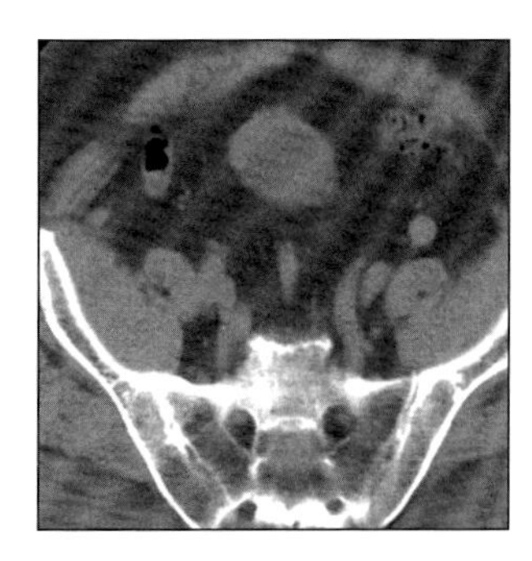

(Left) *Axial NECT shows markedly narrowed recto-sigmoid colon (arrow) + displacement of bladder (open arrow).* ***(Right)*** *Axial NECT shows displacement + narrowing of colon and bladder by pelvic lipomatosis.*

RETROPERITONEAL HEMORRHAGE

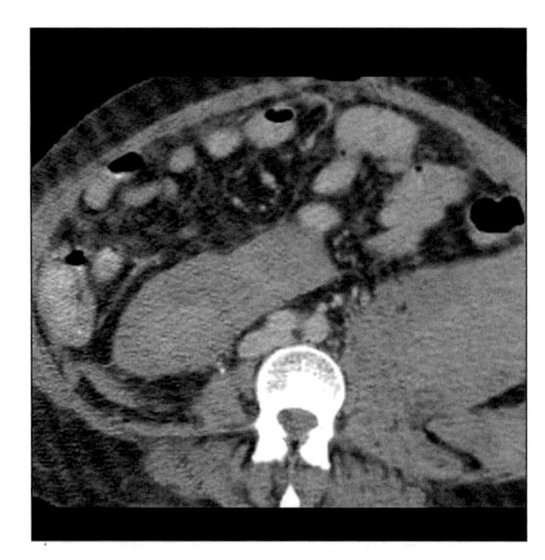

Axial CECT shows spontaneous hemorrhage into multiple sites in the retroperitoneum and body wall in an anticoagulated patient.

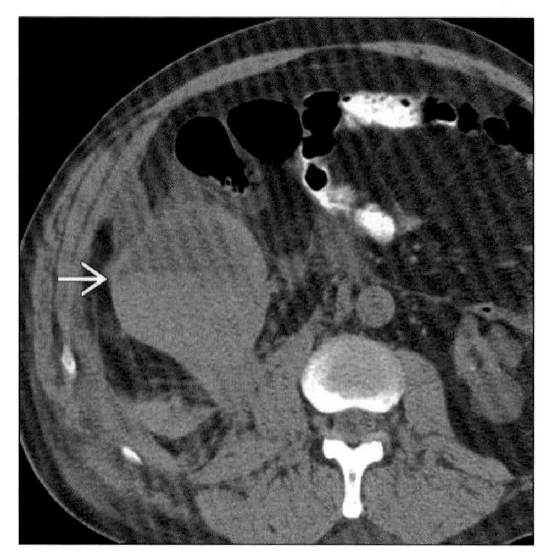

Axial NECT shows multifocal collections in the right retroperitoneum and body wall, plus a fluid-fluid level ("hematocrit effect") (arrow) in an anticoagulated patient.

TERMINOLOGY

Abbreviations and Synonyms

- Hemoretroperitoneum

Definitions

- Bleeding in the retroperitoneum or posterior abdominal wall muscles

IMAGING FINDINGS

General Features

- Best diagnostic clue: High density collection in retroperitoneal space with fluid-fluid level
- Location: Localized to area of bleeding; blood vessel or specific organ
- Other general features
  - Associated findings depend on etiology

CT Findings

- Active bleeding
  - Linear or flame-like appearance isodense to enhanced vessels
  - Extravasation of vascular contrast (80-370 HU)
- Acute (60-80 HU): High attenuating fluid collection or hematoma
- Chronic (20-40 HU): Low density (organized clot)
- Mixed-density mass (acute and chronic)
- ± Hematoma within perihepatic, perisplenic, perirenal, pararenal and/or pelvic spaces
- ± Mass effect: Displacement of spleen anteriorly, kidney anteriorly and/or midline
- ± Extension of hematoma: Superiorly to diaphragm, inferiorly to pelvis
  - Extensive swelling of psoas & iliac muscle
- Signs of bleeding from coagulopathy or anticoagulation
  - Hematocrit effect
    - Fluid-fluid level within "mass"
  - Bleeding into several anatomic spaces (e.g., retroperitoneum, body wall muscles)
  - Bleeding out of proportion to injury
- Associated abdominal aortic aneurysm
  - NECT
    - Located infrarenal; extension to iliac arteries
    - Thickened diaphragmatic crura
    - "Enlarged" left psoas muscle
    - Aortic outline is obscured by soft tissue density
  - CECT
    - Saccular or fusiform dilatation of aorta
    - Indistinct focal area of aortic wall
    - Blood tracks along perirenal fascia → can follow renal hilum into perirenal space; simulates renal source of bleeding
    - Anterior displacement of kidney
    - Extravasation of contrast

DDx: Retroperitoneal Mass or Fluid

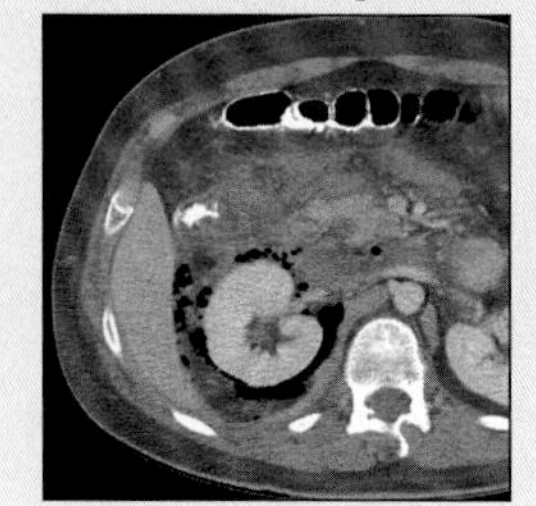

Perforated Duod. Ulcer

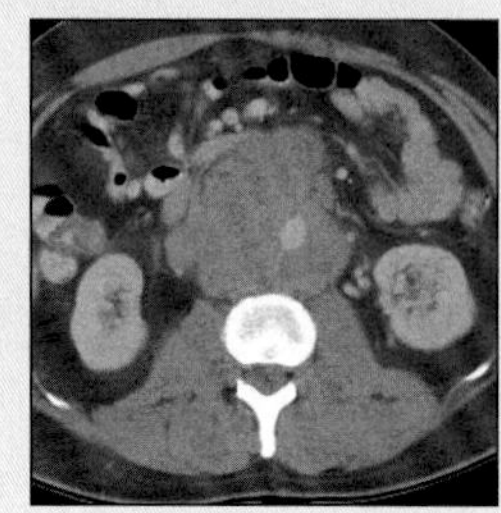

Testicular Ca Mets

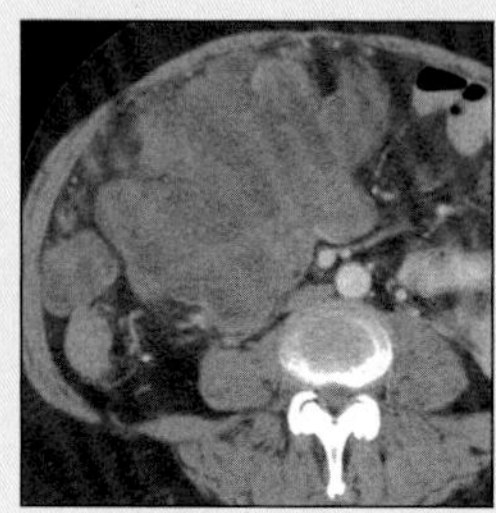

Liposarcoma

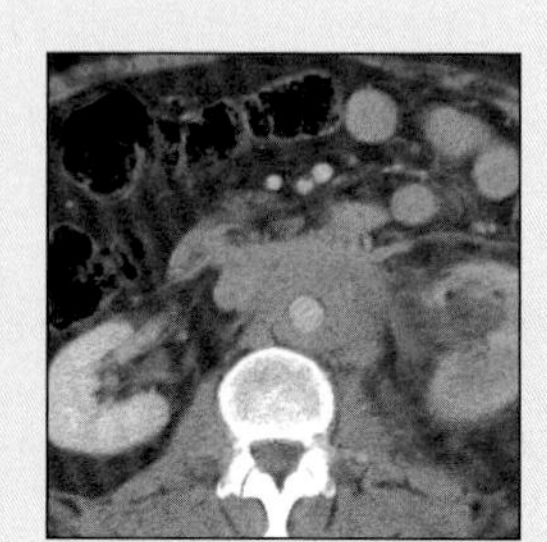

Retroperit. Fibrosis

RETROPERITONEAL HEMORRHAGE

Key Facts

Terminology
- Bleeding in the retroperitoneum or posterior abdominal wall muscles

Imaging Findings
- Best diagnostic clue: High density collection in retroperitoneal space with fluid-fluid level
- ± Hematoma within perihepatic, perisplenic, perirenal, pararenal and/or pelvic spaces
- ± Mass effect: Displacement of spleen anteriorly, kidney anteriorly and/or midline

Top Differential Diagnoses
- Retroperitoneal abscess
- Retroperitoneal tumor
- Asymmetrical muscles

Pathology
- Anticoagulation (most common): Warfarin, low-molecular weight heparin
- Abdominal aortic aneurysm (2nd most common)
- Spontaneous tumor rupture (3rd most common)

Clinical Issues
- Abdominal, back or flank pain; pain radiating to groin and anteromedial thigh

Diagnostic Checklist
- Most common cause is iatrogenic; discuss with referring physician
- Hematocrit effect and bleeding into several spaces indicates anticoagulation
- Spontaneous perirenal hemorrhage: Look for underlying tumor

- Associated renal or adrenal tumors
  - Spherical heterogeneous mass replacing part of kidney
  - Renal cell carcinoma
    - Most common neoplastic cause of spontaneous hemorrhage
  - Angiomyolipoma
    - Contains visible fat in most cases
    - May be obscured by hemorrhage or distorted anatomy
  - Adrenal carcinoma, pheochromocytoma or metastases
    - Distorted anatomy due to bleeding may obscure tumor
- Trauma
  - Significant retroperitoneal bleeding usually due to renal traumatic injury

MR Findings
- Varied signal intensity (evolution of blood products)
- Hyperacute phase (due to oxyhemoglobin)
  - T1WI: Slightly hypointense
  - T2WI: Hyperintense
- Acute phase (due to deoxyhemoglobin)
  - T1WI: Isointense or slightly hypointense
  - T2WI: Markedly hypointense
- Chronic phase
  - T1WI: Hypointense
  - T2WI: Hyperintense

Imaging Recommendations
- Best imaging tool: Helical CT
- Protocol advice
  - Helical CT
    - Injection rate ≥ 3 ml per second; 70 seconds delay
    - Diaphragm to symphysis using 5 mm collimation

DIFFERENTIAL DIAGNOSIS

Retroperitoneal abscess
- Less likely to extend far from site of origin versus peritoneal abscesses
- Etiology
  - Affected adjacent segment of bowel
    - Example: Retrocecal appendicitis
  - Penetrating trauma
  - Inflammatory bowel disease
  - Perforated duodenal ulcer
  - Perforated colonic carcinoma
- Radiography
  - Mottled gas collection
  - Increased density at area of abscess
  - Obliteration of psoas outline
  - Obliteration of renal outline and fat within the perirenal space (perirenal abscess)
  - Renal "halo" sign: Fluid in pararenal space and outside the renal fascia outlines perirenal fat
- Fluoroscopic-guided barium studies
  - Displacement of organs due to abscess (e.g., medial displacement of kidney due to perirenal abscess)
  - ± Obstruction (distal colon)
- CT
  - Gas and fluid within enclosed space
  - ± Extravasation of enteric contrast

Retroperitoneal tumor
- Example: Retroperitoneal sarcoma
  - CT
    - Usually located peri-/pararenal region; > 10 cm
    - Large heterogeneous mass of fat and soft tissue attenuation displacing retroperitoneal structures or viscera
    - Poorly or sharply marginated and encapsulated mass; ± calcification
    - Displacement, compression, distortion of adjacent structures (kidneys, bowel, colon)
    - Liver metastases (necrotic or cystic)
  - MRI: Variable signal intensities depending on fat, solid, cystic, necrotic, hemorrhagic components

Asymmetrical muscles
- Examples: Polio or amputation
- CT: Asymmetrical, enlarged iliopsoas muscle → indentation or compression of bladder dome (asymmetrical pear-shaped bladder)

RETROPERITONEAL HEMORRHAGE

PATHOLOGY

General Features

- Etiology
  - Iatrogenic
    - Anticoagulation (most common): Warfarin, low-molecular weight heparin
    - Femoral vein cannulization or arteriography
    - Percutaneous nephrostomy or renal biopsy
    - Translumbar aortography
  - Arterial aneurysms
    - Abdominal aortic aneurysm (2nd most common)
    - Renal artery aneurysm
    - Inferior adrenal artery aneurysm
    - Lumbar artery aneurysm
    - Iliac artery aneurysm
    - Ovarian artery aneurysm
    - Uterine artery aneurysm
    - Inferior pancreaticoduodenal artery aneurysm
  - Spontaneous tumor rupture (3rd most common)
    - Renal cell carcinoma
    - Angiomyolipoma
    - Adrenal carcinoma
    - Adrenal myelolipoma
    - Adrenal pheochromocytoma
    - Renal or adrenal cyst
  - Trauma
    - Abdominal wall, renal or adrenal trauma
    - Trauma to postero-superior region of segment VII (liver bare area)
  - Other disease
    - Pancreatitis

Gross Pathologic & Surgical Features

- Contusion, laceration and/or ischemia of organ

Microscopic Features

- Contusion, laceration and/or ischemia of tissue

CLINICAL ISSUES

Presentation

- Most common signs/symptoms
  - Abdominal, back or flank pain; pain radiating to groin and anteromedial thigh
  - Hypotension, nausea, vomiting
  - Abdominal distention, peritoneal signs
  - Weakness and ↓ sensation of lower extremities

Demographics

- Age: Any age, increase likelihood with age
- Gender: M > F

Natural History & Prognosis

- Complications: Paresis, shock and death
- Prognosis: Poor, but dependent on cause

Treatment

- Stop medication ± protamine
- Infusion of blood products (i.e., fresh frozen plasma, packed red blood cells)
- Transcatheter or angiographic embolization for active bleeding
- Surgical repair and resection (i.e., tumor, organ injury)

DIAGNOSTIC CHECKLIST

Consider

- Most common cause is iatrogenic; discuss with referring physician

Image Interpretation Pearls

- Hematocrit effect and bleeding into several spaces indicates anticoagulation
- Spontaneous perirenal hemorrhage: Look for underlying tumor

SELECTED REFERENCES

1. Harris AC et al: Ct findings in blunt renal trauma. Radiographics. 21 Spec No:S201-14, 2001
2. Lindner A et al: Images in clinical medicine. Retroperitoneal hemorrhage. N Engl J Med. 344(5):348, 2001
3. Amano T et al: Retroperitoneal hemorrhage due to spontaneous rupture of adrenal myelolipoma. Int J Urol. 6(11):585-8, 1999
4. Patten RM et al: Traumatic laceration of the liver limited to the bare area: CT findings in 25 patients. AJR Am J Roentgenol. 160(5):1019-22, 1993
5. Trerotola SO et al: Bleeding complications of femoral catheterization: CT evaluation. Radiology. 174(1):37-40, 1990
6. Wilms G et al: Embolization of iatrogenic pelvic and retroperitoneal hemorrhage. J Belge Radiol. 72(4):279-82, 1989
7. Sclafani SJ et al: Lumbar arterial injury: radiologic diagnosis and management. Radiology. 165(3):709-14, 1987
8. Weinbaum FI et al: The accuracy of computed tomography in the diagnosis of retroperitoneal blood in the presence of abdominal aortic aneurysm. J Vasc Surg. 6(1):11-6, 1987
9. Ralls PW et al: Renal biopsy-related hemorrhage: frequency and comparison of CT and sonography. J Comput Assist Tomogr. 11(6):1031-4, 1987
10. Illescas FF et al: CT evaluation of retroperitoneal hemorrhage associated with femoral arteriography. AJR Am J Roentgenol. 146(6):1289-92, 1986
11. Cronan JJ et al: Retroperitoneal hemorrhage after percutaneous nephrostomy. AJR Am J Roentgenol. 144(4):801-3, 1985
12. Rosen A et al: CT diagnosis of ruptured abdominal aortic aneurysm. AJR Am J Roentgenol. 143(2):265-8, 1984
13. Federle MP: Computed tomography of blunt abdominal trauma. Radiol Clin North Am. 21(3):461-75, 1983
14. Samuelsson L et al: Ruptured aneurysm of the internal iliac artery. J Comput Assist Tomogr. 6(4):842-4, 1982
15. Jeffrey RB et al: Computed tomography of splenic trauma. Radiology. 141(3):729-32, 1981
16. Amendola MA et al: Evaluation of retroperitoneal hemorrhage by computed tomography before and after translumbar aortography. Radiology. 133(2):401-4, 1979
17. Cisternino SJ et al: Diagnosis of retroperitoneal hemorrhage by serial computed tomography. J Comput Assist Tomogr. 3(5):686-8, 1979
18. Sagel SS et al: Detection of retroperitoneal hemorrhage by computed tomography. AJR Am J Roentgenol. 129(3):403-7, 1977
19. Stephens DH et al: Computed tomography of the retroperitoneal space. Radiol Clin North Am. 15(3):377-90, 1977
20. Einarsson GV: Spontaneous retroperitoneal hemorrhage as first presentation of carcinoma of the kidney. A case report. Scand J Urol Nephrol. 20(3):235-6, 1986

IMAGE GALLERY

Typical

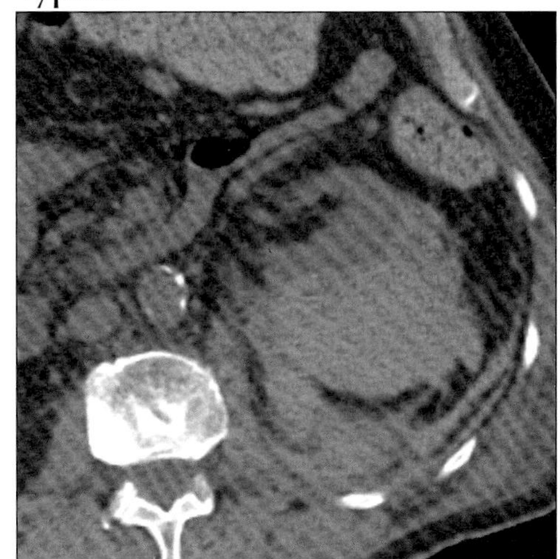

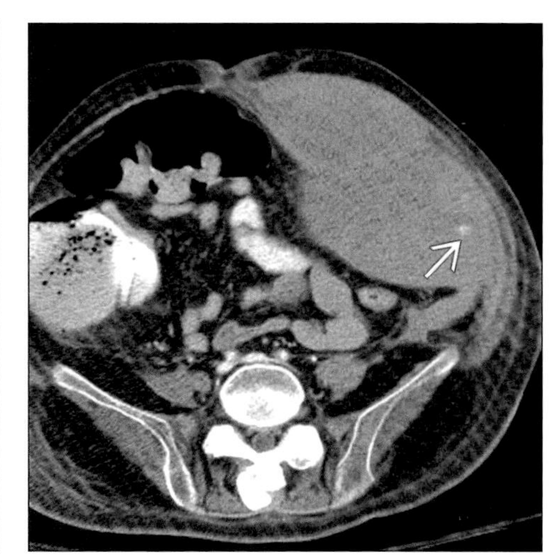

*(**Left**) Axial CECT shows peri- and pararenal hemorrhage following renal biopsy. (**Right**) Axial CECT shows a large rectus sheath hematoma with foci of active bleeding (arrow) in an anticoagulated patient.*

Typical

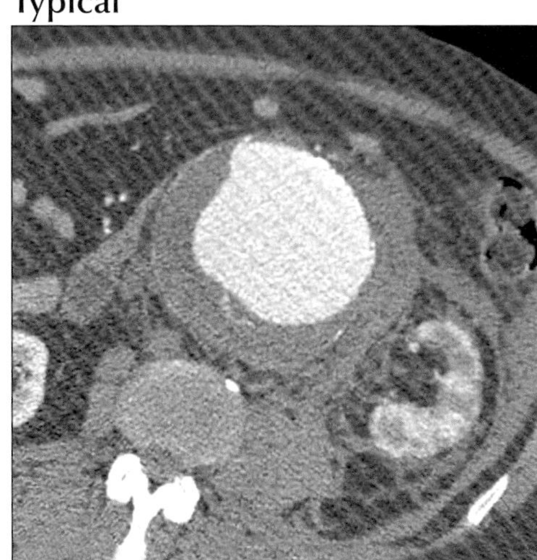

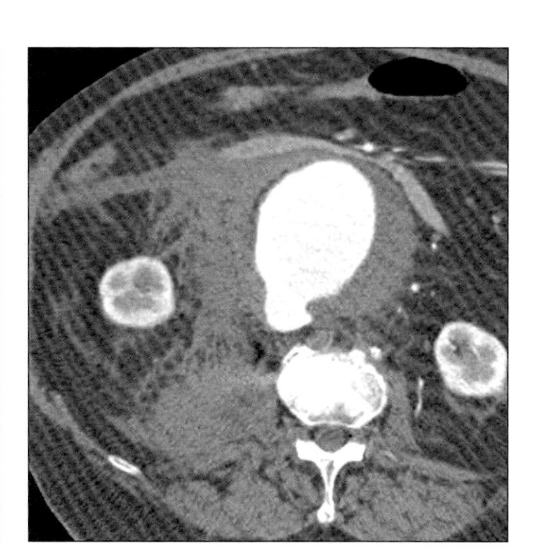

*(**Left**) Axial CECT shows a ruptured abdominal aortic aneurysm (AAA) with blood tracking throughout all three retroperitoneal compartments. (**Right**) Axial CECT shows a ruptured AAA with blood tracking throughout the right retroperitoneal spaces.*

Typical

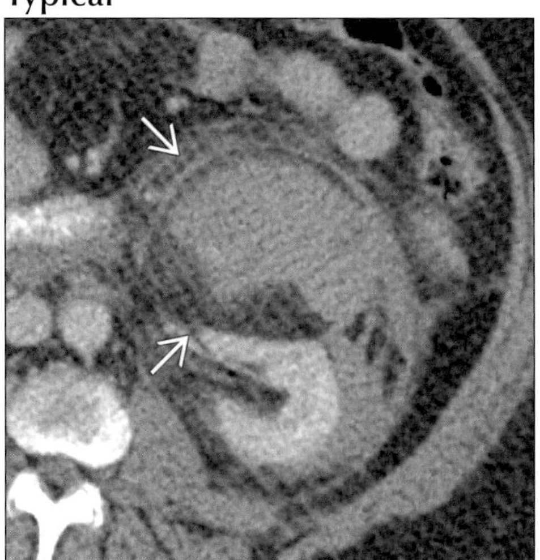

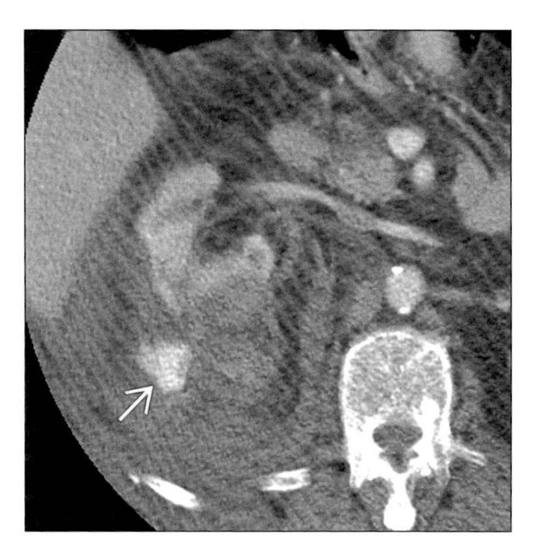

*(**Left**) Axial CECT shows perirenal hemorrhage that occurred spontaneously from underlying angiomyolipoma (arrows). (**Right**) Axial CECT shows peri- and pararenal hemorrhage following blunt abdominal trauma. Note active arterial bleeding (arrow).*

RETROPERITONEAL SARCOMA

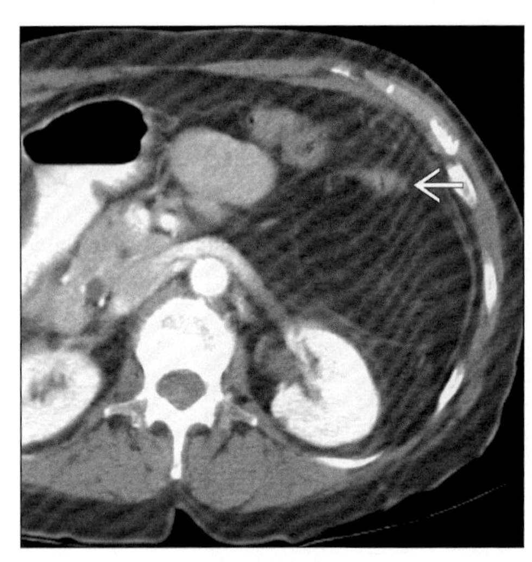

Axial CECT shows predominantly fatty mass that displaces bowel, including descending colon (arrow); liposarcoma.

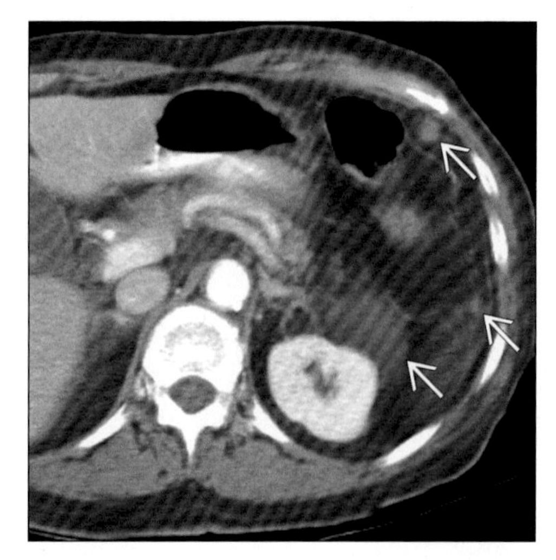

Axial CECT shows liposarcoma with heterogeneous areas whose attenuation is similar to fat or soft tissue (arrows), probably representing myxoid stroma.

TERMINOLOGY

Definitions

- Malignant primary retroperitoneal tumor arising from various elements of primitive mesenchyme, urogenital ridge, embryonic remnants

IMAGING FINDINGS

General Features

- Best diagnostic clue: Large heterogeneous mass of fat + soft tissue attenuation displacing retroperitoneal structures or viscera
- Location: Usually peri-/paranephric region
- Size: Usually large at diagnosis (> 10 cm)
- Key concepts
  - 80% of primary retroperitoneal tumors: Malignant
    - Lymphomas (most common); retroperitoneal sarcomas
  - Most malignant retroperitoneal tumors are of mesodermal origin
  - Classified into four types based on tissue of origin
    - Liposarcoma
    - Leiomyosarcoma
    - Fibrosarcoma
    - Rhabdomyosarcoma
  - Liposarcoma: Adipose tissue tumor
    - Most common primary retroperitoneal malignant tumor
    - 2nd Most common adult soft tissue sarcoma after malignant fibrous histiocytoma
    - 15% of all soft tissue tumors, 10-20% of them in retroperitoneum
    - Malignant from inception; rarely arises from lipomas
    - Usually grow slowly & attain a large size before they are detected
    - Histologically: 5 types (well-differentiated, myxoid 40-50% (most common), round cell, pleomorphic, mixed)
  - Leiomyosarcoma: Smooth muscle tumor (vessels & embryonic remnants)
    - 2nd Most common primary retroperitoneal tumor after liposarcoma
    - Accounts for 11% of all retroperitoneal malignant tumors
    - Leiomyosarcomas are more common than benign leiomyomas
  - Fibrosarcoma & malignant fibrous histiocytoma: Connective tissue tumors
    - Fibrosarcoma: 3% of all retroperitoneal malignant tumors
    - Malignant fibrous histiocytoma: Most common soft tissue sarcoma in adults, 15% of these arise in abdominal cavity & retroperitoneum

DDx: Large Mass in the Retroperitoneum

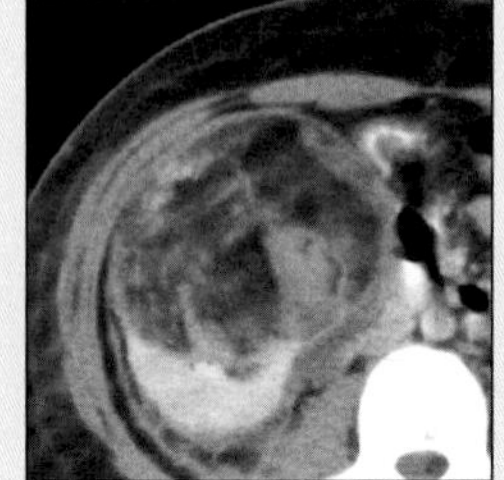

Renal AML

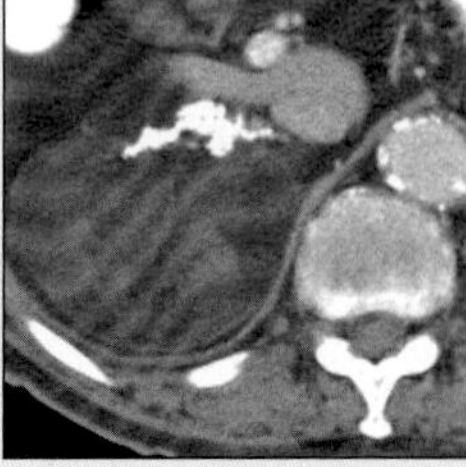

Adrenal Myeolipoma

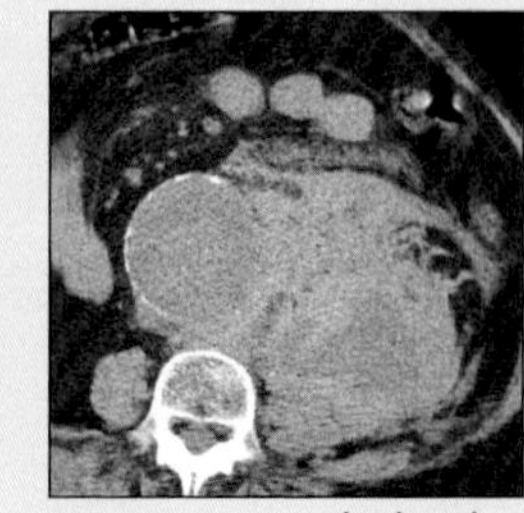

Retroperineal Bleed

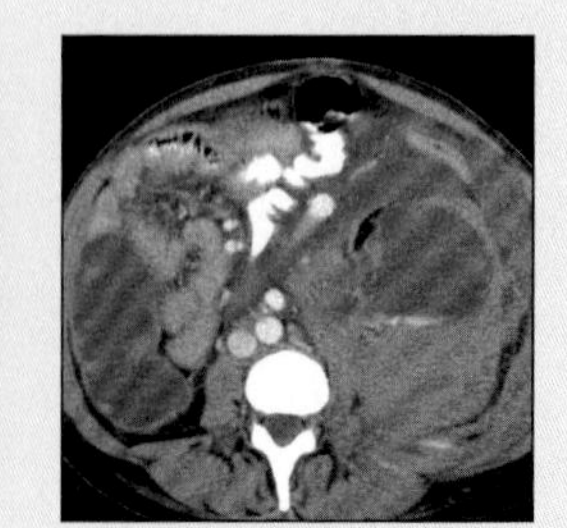

Retroperitoneal Bleed

RETROPERITONEAL SARCOMA

Key Facts

Terminology
- Malignant primary retroperitoneal tumor arising from various elements of primitive mesenchyme, urogenital ridge, embryonic remnants

Imaging Findings
- Best diagnostic clue: Large heterogeneous mass of fat + soft tissue attenuation displacing retroperitoneal structures or viscera
- Location: Usually peri-/paranephric region
- Size: Usually large at diagnosis (> 10 cm)
- Poorly or sharply marginated + encapsulated mass ± calcification
- Displacement, compression, distortion of adjacent structures (kidneys, bowel, colon)
- Liver metastases (necrotic or cystic)

Top Differential Diagnoses
- Renal angiomyolipoma (AML)
- Retroperitoneal hemorrhage

Pathology
- Mesodermal origin: Most adult malignant retroperitoneal tumors
- Large, encapsulated mass

Clinical Issues
- Abdominal, back, flank, radicular pain
- Palpable mass; GI & urinary tract symptoms

Diagnostic Checklist
- Differentiate from adrenal & renal tumors
- Large mixed attenuation mass of fat & soft tissue density displacing retroperitoneal structures

---

- Rhabdomyosarcoma: Striated muscle tumor
  - Embryonal rhabdomyosarcomas: 60% of striated muscle tumors
- Angiosarcoma: Malignant tumor of vascular endothelium (rare)
- Lymphangiosarcoma (lymph vessels) & myxosarcoma (mesenchyme)

Radiographic Findings
- Fluoroscopic guided barium study
  - Show displacement of stomach, small bowel, colon
- Excretory urography findings
  - Demonstrate displacement of kidney, ureter, bladder ± hydronephrosis
  - Usually medial displacement of ureters (most tumors arise laterally)

CT Findings
- Liposarcoma: 3 CT patterns based on amount & distribution of fat in tumor (often coexist)
  - Solid pattern: Attenuation values > +20 HU
  - Mixed pattern: Discrete fatty areas with HU < -20 & areas > +20 HU
  - Pseudocystic pattern: Homogeneous density between +20 & -20 HU
  - Poorly or sharply marginated + encapsulated mass ± calcification
  - Displacement, compression, distortion of adjacent structures (kidneys, bowel, colon)
  - ± Invasion of adjacent structures
  - CECT: Hetero-/homogeneous enhancement; usually lack prominent vessels
- Leiomyosarcoma
  - Extravascular (62%)
    - Large retroperitoneal mass ± necrotic or cystic degeneration
    - Liver metastases (necrotic or cystic)
  - Intravascular mass (6%)
    - Solid mass within IVC + dilatation or obstruction
    - Proximal IVC mass: Show dilated hepatic veins + portal radicles
    - CECT: Heterogeneous enhancement
  - Extra + intravascular mass (33%)
    - Solid & necrotic extraluminal tumor
    - Enhancing intravascular component
  - Intramural tumor: Extremely rare

MR Findings
- Variable signal intensities depending on amount of fat, solid, cystic, necrotic, hemorrhagic components

Ultrasonographic Findings
- Real Time
  - Liposarcoma: Large well-defined solid mass with internal echoes (fat)
  - Leiomyosarcoma: Large solid mass + hypoechoic cystic & necrotic content

Angiographic Findings
- Conventional
  - Show hypo-/hypervascularity; displacement of great vessels
  - Liposarcoma: Hypovascular tumor
  - Leiomyosarcoma: Hypervascular + feeding vessels
- Inferior venacavography
  - For primary or secondary involvement of IVC

Imaging Recommendations
- NE + CECT; MR; angiography

DIFFERENTIAL DIAGNOSIS

Renal angiomyolipoma (AML)
- Composed of blood vessels, muscle & fat tissue
- Hamartomatous lesion (indicates a benign tumor)
- 20% of renal angiomyolipoma patients have tuberous sclerosis
- 80% of tuberous sclerosis patients have renal angiomyolipomas
- Large angiomyolipoma may simulate retroperitoneal liposarcoma (both contain fat)
  - Renal parenchymal defect & enlarged vessels favor angiomyolipoma (AML)
  - Smooth compression of kidney & extension beyond perirenal space favor liposarcoma

RETROPERITONEAL SARCOMA

Retroperitoneal hemorrhage

- Most are iatrogenic (over anticoagulation)
- Abdominal aortic aneurysm rupture is 2nd most common cause of retroperitoneal bleed
- Spontaneous tumoral bleed: 3rd most common cause
  - Kidney: Renal cell carcinoma, angiomyolipoma
  - Adrenal: Carcinoma or myelolipoma
- CT findings
  - Acute: High density fluid collection or hematoma
  - Chronic: Low density (organized clot)
  - Associated renal, adrenal tumors or aortic aneurysm
- MR findings
  - Varied signal intensity (evolution of blood products)
  - Hyperacute phase: Due to oxyhemoglobin
    - T1WI: Slightly hypointense
    - T2WI: Hyperintense
  - Acute phase: Due to deoxyhemoglobin
    - T1WI: Isointense or slightly hypointense
    - T2WI: Markedly hypointense
  - Chronic phase: ↓ Signal (T1WI); ↑ signal (T2WI)
- Diagnosis
  - History of anticoagulation
  - Imaging evidence of tumor or aneurysm

PATHOLOGY

General Features

- General path comments
  - Embryology/anatomy
    - Mesodermal origin: Most adult malignant retroperitoneal tumors
- Etiology: Unknown
- Epidemiology: 1 in 11,800 admissions

Gross Pathologic & Surgical Features

- Liposarcoma
  - Large, encapsulated mass
  - White or yellow glistening, brain-like
- Leiomyosarcoma
  - Lobulated, encapsulated tumor
  - Cystic, necrotic, blood & Ca++ components

Microscopic Features

- Liposarcoma
  - Myxoid (mucinous + fibrous tissue + fat < 10%)
- Leiomyosarcoma
  - Smooth muscle + atypical giant cells
  - Invasion of vessels

CLINICAL ISSUES

Presentation

- Most common signs/symptoms
  - Abdominal, back, flank, radicular pain
  - Palpable mass; GI & urinary tract symptoms
  - Leg edema, varicosities, hypoglycemia

Demographics

- Age: 40-60 y
- Gender
  - Liposarcoma: M > F
  - Leiomyosarcoma: M < F

Natural History & Prognosis

- Liposarcoma
  - 5 year survival rate of 32%
- Leiomyosarcoma
  - High mortality within 5 years
  - Local recurrence: 40-70% cases

Treatment

- Complete resection, radiotherapy, chemotherapy
- Follow-up imaging & excision of local recurrence

DIAGNOSTIC CHECKLIST

Consider

- Differentiate from adrenal & renal tumors

Image Interpretation Pearls

- Large mixed attenuation mass of fat & soft tissue density displacing retroperitoneal structures

SELECTED REFERENCES

1. Nishino M et al: Primary retroperitoneal neoplasms: CT and MR imaging findings with anatomic and pathologic diagnostic clues. Radiographics. 23(1):45-57, 2003
2. Tateishi U et al: Primary dedifferentiated liposarcoma of the retroperitoneum. Prognostic significance of computed tomography and magnetic resonance imaging features. J Comput Assist Tomogr. 27(5):799-804, 2003
3. Bellin MF et al: Evaluation of retroperitoneal and pelvic lymph node metastases with MRI and MR lymphangiography. Abdom Imaging. 28(2):155-63, 2003
4. Israel GM et al: CT differentiation of large exophytic renal angiomyolipomas and perirenal liposarcomas. AJR. 179: 769-73, 2002
5. Grubnic S et al: MR evaluation of normal retroperitoneal and pelvic lymph nodes. Clin Radiol. 57(3):193-200; discussion 201-4, 2002
6. Gupta AK et al: CT of recurrent retroperitoneal sarcomas. AJR Am J Roentgenol. 174(4):1025-30, 2000
7. Heslin MJ et al: Imaging of soft tissue sarcomas. Surg Oncol Clin N Am. 8(1):91-107, 1999
8. Kurosaki Y et al: Well-differentiated liposarcoma of the retroperitoneum with a fat-fluid level: US, CT, and MR appearance. Eur Radiol. 8(3):474-5, 1998
9. Radin R et al: Adrenal and extra-adrenal retroperitoneal ganglioneuroma: imaging findings in 13 adults. Radiology. 202(3):703-7, 1997
10. Engelken JD et al: Retroperitoneal MR imaging. Magn Reson Imaging Clin N Am. 5(1):165-78, 1997
11. Kim T et al: CT and MR imaging of abdominal liposarcoma. AJR. 166: 829-33, 1996
12. Mesurolle B et al: Retroperitoneal extramedullary hematopoiesis: sonographic, CT, and MR imaging appearance. AJR Am J Roentgenol. 167(5):1139-40, 1996
13. Kim T et al: CT and MR imaging of abdominal liposarcoma. AJR Am J Roentgenol. 166(4):829-33, 1996
14. Miyazaki T et al: Retroperitoneal leiomyosarcoma. Its MR manifestations. Clin Imaging. 17(3):207-9, 1993
15. Lane RH et al: Primary retroperitoneal neoplasms: CT findings in 90 cases with clinical & pathologic correlation. AJR. 152: 83-9, 1989

IMAGE GALLERY

Typical

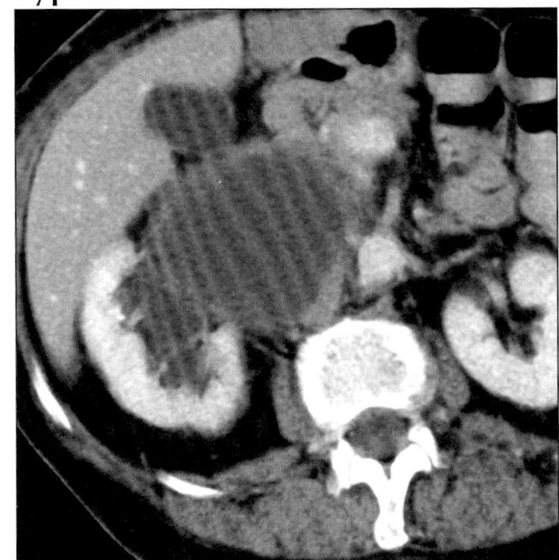

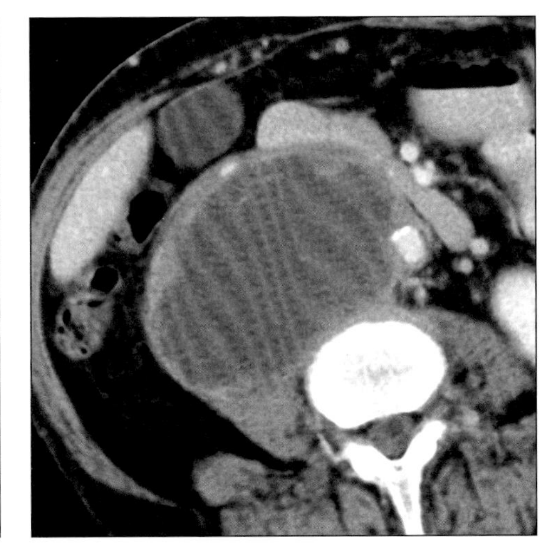

(Left) *Axial CECT shows a low density mass that displaces right kidney, pancreas, etc. Leiomyosarcoma.* ***(Right)*** *Axial CECT shows leiomyosarcoma that invades or displaces right psoas, aorta, and other structures.*

Typical

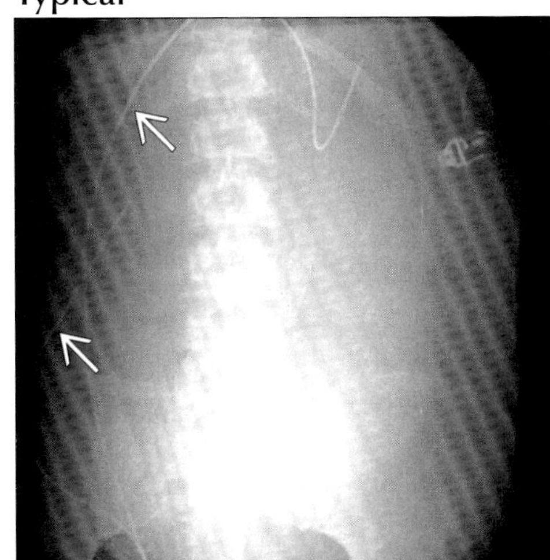

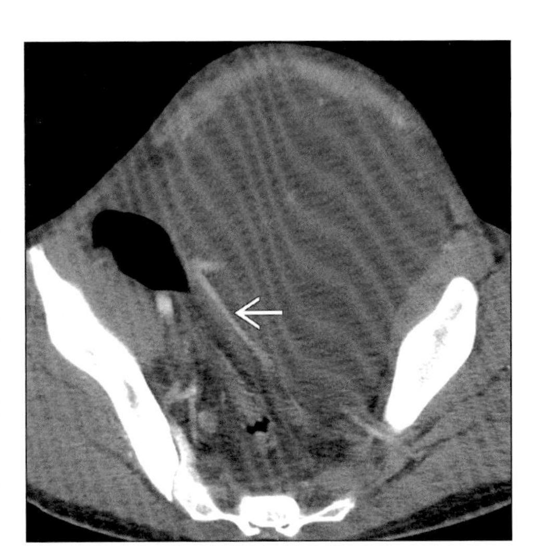

(Left) *Radiograph shows massive displacement of kidneys and ureters, marked by ureteral stents (arrows); liposarcoma in 30 year-old man.* ***(Right)*** *Axial CECT in a 30 year-old man with liposarcoma shows displacement of all pelvic (and abdominal) contents, including left iliac vessels (arrow).*

Typical

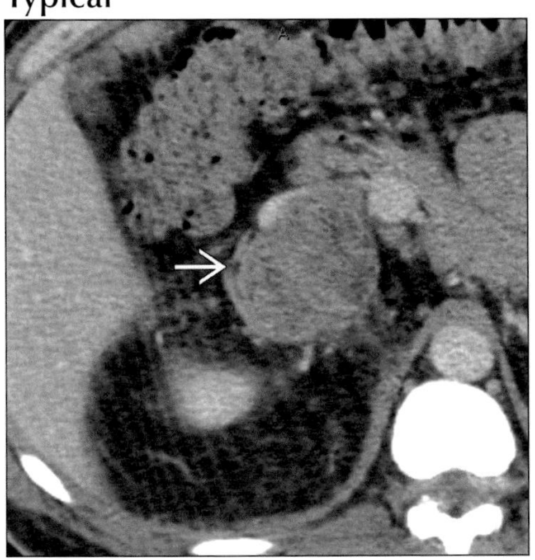

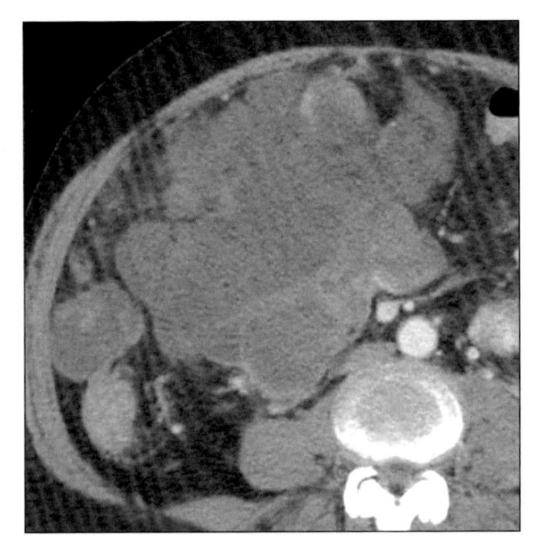

(Left) *Axial CECT shows a primary sarcoma of the IVC (arrow). Note collateral vessels.* ***(Right)*** *Axial CECT shows liposarcoma arising in the mesentery, which bridges the intraperitoneal and retroperitoneal spaces.*

RETROPERITONEAL LYMPHOMA

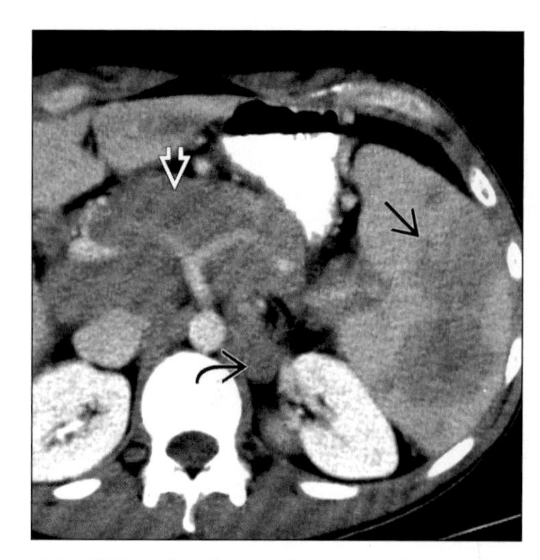

Axial CECT with splenic and nodal involvement from Hodgkin disease. Note multiple low-density lesions (arrow) with celiac (open arrow) and retroperitoneal (curved arrow) involvement.

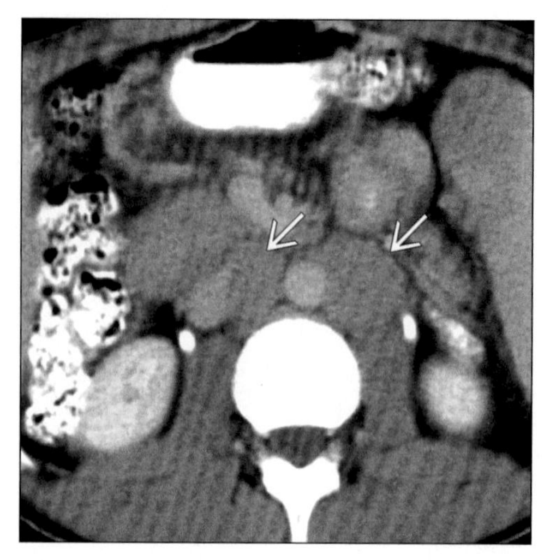

Axial CECT of retroperitoneal nodes from Hodgkin lymphoma. Note enlarged para-aortic and aortocaval nodes (arrows).

TERMINOLOGY

Definitions

- Hodgkin (HD) and non-Hodgkin (NHL) lymphoma involving retroperitoneal nodes

IMAGING FINDINGS

General Features

- Best diagnostic clue: Mantle of soft tissue adenopathy surrounding aorta, inferior vena cava (IVC)
- Location: Para-aortic, aortocaval and retrocaval nodal groups
- Size: Nodes > 1.5 cm in short axis
- Morphology: Confluent soft tissue mantle of adenopathy surrounding aorta and IVC

Radiographic Findings

- Radiography
  - Lymphangiography (LAG)
    - Used for HD to detect involvement in normal-sized nodes
    - Positive nodes have "foamy appearance" or focal filling defects

CT Findings

- CECT
  - Enlarged nodes (> 1.5 cm in short axis) involving bilateral retroperitoneal nodal chains
  - Confluent soft tissue mantle of nodes surrounding aorta and IVC
  - Nodes may displace aorta from spine, unusual for other types of nodal mets
  - Attenuation of nodes similar to muscle
  - In untreated patients nodes are rarely calcified (< 1%) or cystic
  - Nodes involved with NHL typically greater in size than HD and demonstrate skip areas
  - 25% of newly-diagnosed HD patients have positive para-aortic nodes at presentation, compared to 50% with NHL
  - HD nodes typically involve upper para-aortic nodes first without skip areas of involvement
  - HD nodes may be normal in size but contain microscopic tumor leading to false-negative CT
  - Involvement of peripancreatic nodes from NHL results in large peripancreatic mass (pancreatic lymphoma)
  - Involvement of mesenteric nodes more commmon with NHL (> 50%) than HD (< 5%)

MR Findings

- T1WI: Low signal para-aortic lymphadenopathy
- T2WI: High signal para-aortic lymphadenopathy
- T1 C+: Nodes enhance similar to muscle without necrosis

DDx: Retroperitoneal Disorders Mimicking Lymphoma

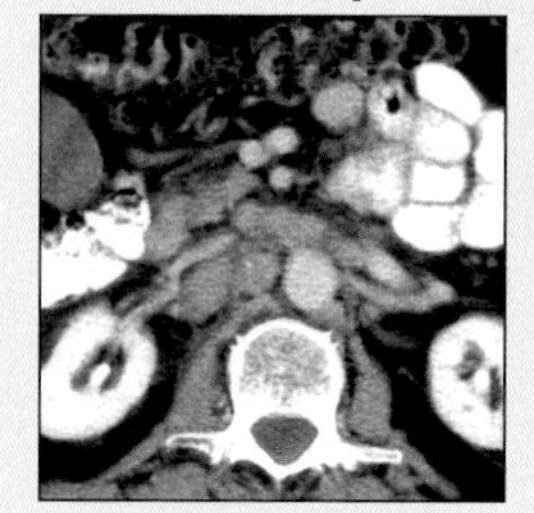

Carcinoma

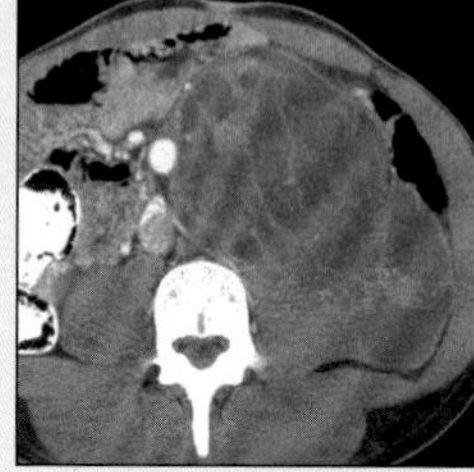

Testicular CA

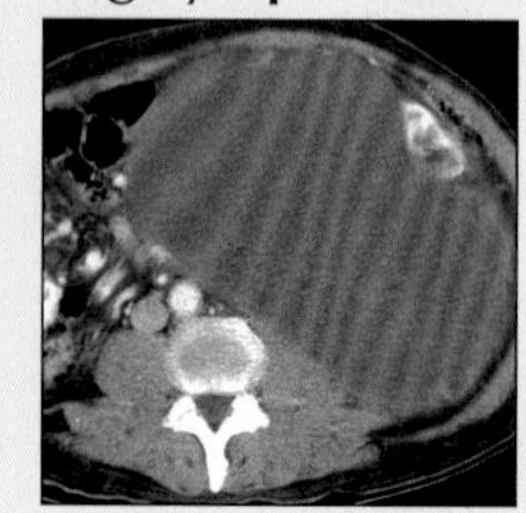

Liposarcoma

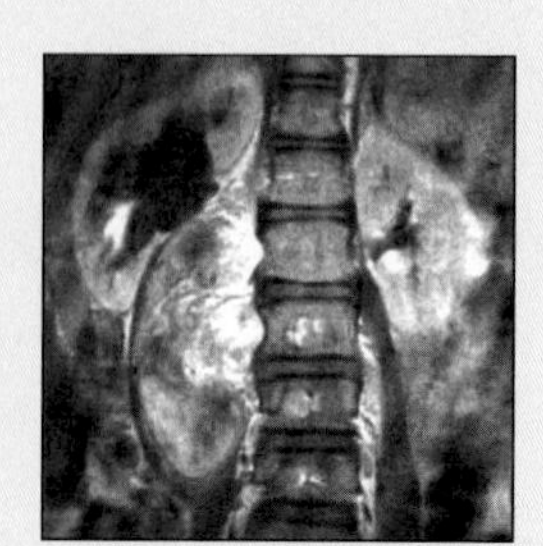

Neurogenic mass

RETROPERITONEAL LYMPHOMA

Key Facts

Terminology
- Hodgkin (HD) and non-Hodgkin (NHL) lymphoma involving retroperitoneal nodes

Imaging Findings
- Best diagnostic clue: Mantle of soft tissue adenopathy surrounding aorta, inferior vena cava (IVC)
- Location: Para-aortic, aortocaval and retrocaval nodal groups
- Best imaging tool: PET-CT with FDG; CECT
- Protocol advice: Combining PET with CT reduces false-positive and false-negative rates

Top Differential Diagnoses
- Carcinoma
- Sarcoma
- Testicular metastases
- Neurogenic mass
- Tuberculosis (TB)
- Retroperitoneal fibrosis

Clinical Issues
- Nodal mass; fever; night sweats; weight loss
- Clinical profile: Young patient with palpable nodal mass
- Age: HD bimodal; 2 peak incidences: 20-24, 80-84
- HD: 80% cure with chemotherapy and/or radiation, or stem cell transplant
- NHL: Depends on grade and stage; in general, worse than HD, worst in AIDS-related NHL; poor prognosis if patients don't respond to initial therapy

Diagnostic Checklist
- Metastatic nodes from carcinoma

Ultrasonographic Findings
- Real Time: Hypoechoic nodes without enhanced through sound transmission

Nuclear Medicine Findings
- PET
  - Hypermetabolic nodes with FDG
  - Sensitivity 90-95%, compared to 80-85% for CECT
  - More sensitive for thoracic lymphoma (> 90%) than abdominal or pelvic disease (> 75%)
  - Detects more lesions than CT, may result in changing stage of patients in 10-40% of cases
  - Modifies therapy in 25% of cases
  - Useful to determine early response to chemotherapy
  - Helpful to distinguish residual lymphoma vs. fibrosis post-radiation
  - Not useful for B-cell lymphoma or for mucosa-associated lymphoid tissue (MALT)

Imaging Recommendations
- Best imaging tool: PET-CT with FDG; CECT
- Protocol advice: Combining PET with CT reduces false-positive and false-negative rates

DIFFERENTIAL DIAGNOSIS

Carcinoma
- Enlarged nodes that are discrete but not "confluent" adenopathy, or "mantle" appearance of soft tissue retroperitoneal mass
- Calcification for mucinous adenocarcinoma or serous carcinoma of the ovary

Sarcoma
- Typically unilateral large retroperitoneal mass
- Displaces organs such as colon or kidney
- Often heterogeneous enhancement on CECT
- Fat attenuation in liposarcoma

Testicular metastases
- Nodes enlarged along gonadal vein as it drains into IVC; left renal hilus and retrocaval on right
- Low attenuation myxomatous element

Neurogenic mass
- Paraganglioma, ganglioneuroma typically paraspinous in location
- Widening of neural foramen
- Heterogeneous enhancement on CECT and MRI

Tuberculosis (TB)
- Low attenuation nodes
- Associated TB peritonitis includes mesenteric or omental nodules, ascites & ileo-colic mural thickening

Retroperitoneal fibrosis
- Usually limited to mantle of tissue surrounding aorta and IVC from mid-lumbar level through common iliac

Miscellaneous
- Kaposi sarcoma: Hypervascular retroperitoneal nodes in AIDS patient
- Castleman disease (angiofollicular lymphoid hyperplasia): Hypervascular retroperitoneal nodes
- Growing teratoma syndrome: Water- or fat-density nodal mass due to development of mature teratoma following chemotherapy for testicular nonseminomatous tumor

PATHOLOGY

General Features
- General path comments: Rubbery enlarged nodes without necrosis or calcification
- Genetics: Several types of NHL associated with oncogenes bcl 1, 2, 3, and 6
- Etiology: Unknown
- Epidemiology
  - 5% of newly-diagnosed cancer in USA
    - HD: 8,000 cases/year in USA
    - NHL: 50,000 cases/year in USA
  - Incidence has increased due to AIDS-related NHL and organ transplantation
  - Third most common cancer death

RETROPERITONEAL LYMPHOMA

Gross Pathologic & Surgical Features

- Excisional biopsy or core biopsy preferred for HD subtyping
- Fine needle aspiration (FNA) adequate for recurrent or NH lymphoma

Microscopic Features

- HD
  - Few malignant cells within background of normal T-lymphocytes and inflammatory cells
  - Malignant cells: Reed-Sternberg cells and variants
    - Lacunar cells, mononuclear and pleomorphic variants
  - Distorted nodal architecture
  - Histologic subtypes: Nodular sclerosing, lymphocyte predominant, mixed cellularity, lymphocyte depletic
- NHL
  - Classification currently based on World Health Organization criteria
  - Cell type origin determines subtype (B-cell, T-cell or natural killer cell) as well as immune phenotype

Staging, Grading or Classification Criteria

- Ann Arbor staging classification
  - Stage I: Single lymph node region/extralymphatic site
  - Stage II: Two or more nodal regions on same side of diaphragm (including single extralymphatic site)
  - Stage III: Nodal group involvement above and below diaphragm
  - Stage IV: Diffuse involvement of one or more extralymphatic sites (liver, lungs, marrow)

CLINICAL ISSUES

Presentation

- Most common signs/symptoms
  - Nodal mass; fever; night sweats; weight loss
  - Other signs/symptoms: Axillary, neck or groin mass
- Clinical profile: Young patient with palpable nodal mass

Demographics

- Age: HD bimodal; 2 peak incidences: 20-24, 80-84
- Gender: Slight increase in males for HD
- Ethnicity: HD: Higher in Caucasians

Natural History & Prognosis

- HD: 80% cure with chemotherapy and/or radiation, or stem cell transplant
- NHL: Depends on grade and stage; in general, worse than HD, worst in AIDS-related NHL; poor prognosis if patients don't respond to initial therapy

Treatment

- Options, risks, complications
  - HD: Initial chemotherapy or radiation; stem cell transplant for treatment failure or relapse
  - NHL: Initial treatment with chemotherapy; stem cell transplant for failure or recurrence

DIAGNOSTIC CHECKLIST

Consider

- Metastatic nodes from carcinoma

Image Interpretation Pearls

- Confluent soft tissue mass surrounding aorta and IVC

SELECTED REFERENCES

1. Morgan PB et al: Uncommon presentations of Hodgkin's disease. Case 1. Hodgkin's disease of the jejunum. J Clin Oncol. 22(1):193-5, 2004
2. O'Malley ME et al: US of gastrointestinal tract abnormalities with CT correlation. Radiographics. 23(1):59-72, 2003
3. Sheth S et al: Mesenteric neoplasms: CT appearances of primary and secondary tumors and differential diagnosis. Radiographics. 23(2):457-73; quiz 535-6, 2003
4. Ahmad A et al: Gastric mucosa-associated lymphoid tissue lymphoma. Am J Gastroenterol. 98(5):975-86, 2003
5. Vinnicombe SJ et al: Computerised tomography in the staging of Hodgkin's disease and non-Hodgkin's lymphoma. Eur J Nucl Med Mol Imaging. 30 Suppl 1:S42-55, 2003
6. Elis A et al: Detection of relapse in non-Hodgkin's lymphoma: role of routine follow-up studies. Am J Hematol. 69(1):41-4, 2002
7. Apter S et al: Calcification in lymphoma occurring before therapy: CT features and clinical correlation. AJR Am J Roentgenol. 178(4):935-8, 2002
8. Kurosawa H et al: Burkitt lymphoma associated with large gastric folds, pancreatic involvement, and biliary tract obstruction. J Pediatr Hematol Oncol. 24(4):310-2, 2002
9. Boni L et al: Primary pancreatic lymphoma. Surg Endosc. 16(7):1107-8, 2002
10. Jerusalem G et al: Whole-body positron emission tomography using 18F-fluorodeoxyglucose compared to standard procedures for staging patients with Hodgkin's disease. Haematologica. 86(3):266-73, 2001
11. Daskalogiannaki M et al: Splenic involvement in lymphomas. Evaluation on serial CT examinations. Acta Radiol. 42(3):326-32, 2001
12. Sheth S et al: Non-Hodgkin lymphoma: pattern of disease at spiral CT. Crit Rev Diagn Imaging. 42(6):307-56, 2001
13. Fields S et al: CT-guided aspiration core needle biopsy of gastrointestinal wall lesions. J Comput Assist Tomogr. 24(2):224-8, 2000
14. Jung G et al: Abdominal lymphoma staging: is MR imaging with T2-weighted turbo-spin-echo sequence a diagnostic alternative to contrast-enhanced spiral CT? J Comput Assist Tomogr. 24(5):783-7, 2000
15. Hwang K et al: Imaging of malignant lymphomas with F-18 FDG coincidence detection positron emission tomography. Clin Nucl Med. 25(10):789-95, 2000
16. Oh YK et al: Stages I-III follicular lymphoma: role of CT of the abdomen and pelvis in follow-up studies. Radiology. 210(2):483-6, 1999
17. Chang DK et al: Lymph node involvement rate in low-grade gastric mucosa-associated lymphoid tissue lymphoma--too high to be neglected. Hepatogastroenterology. 46(28):2694-700, 1999
18. Zinzani PL et al: The role of positron emission tomography (PET) in the management of lymphoma patients. Ann Oncol. 10(10):1181-4, 1999
19. Kessar P et al: CT appearances of mucosa-associated lymphoid tissue (MALT) lymphoma. Eur Radiol. 9(4):693-6, 1999

IMAGE GALLERY

Typical

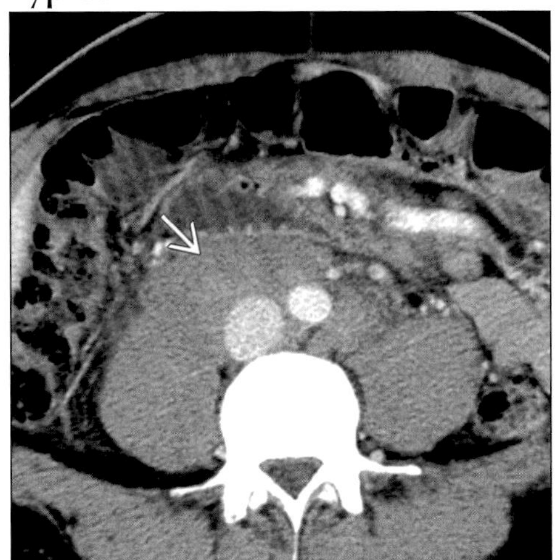

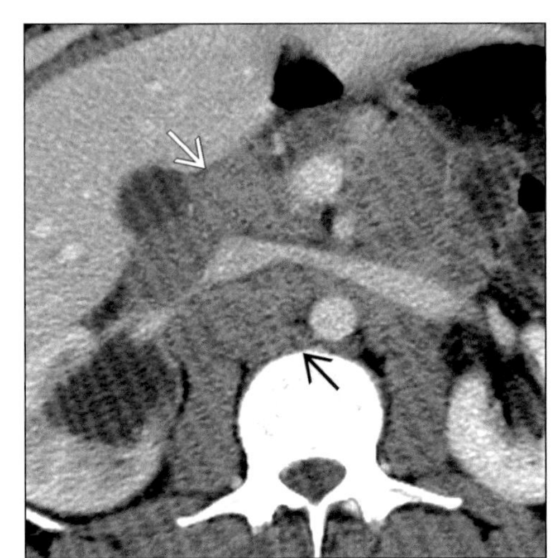

(Left) *Axial CECT of typical confluent "mantle" of nodal disease in non-Hodgkin lymphoma. Note soft tissue mass (arrow) surrounding aorta and inferior vena cava.* ***(Right)*** *Axial CECT of nodal disease in non-Hodgkin lymphoma. Note nodal mass surrounding inferior vena cava (arrows).*

Typical

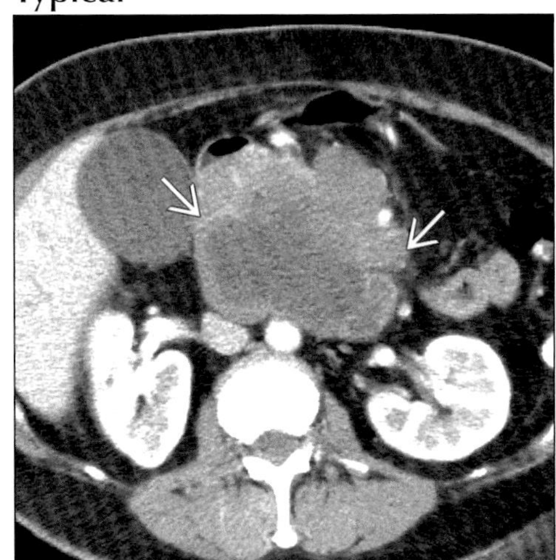

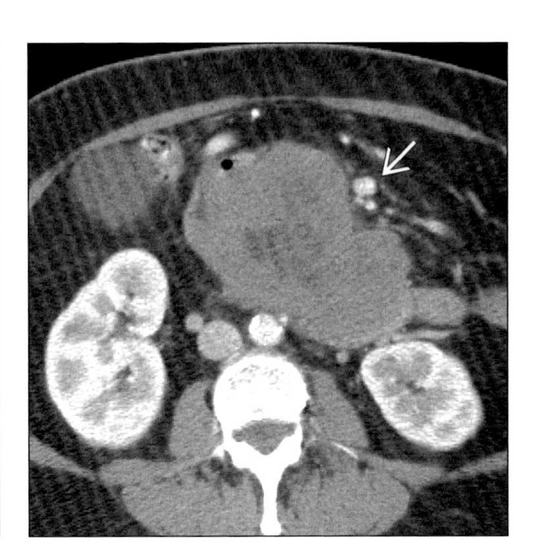

(Left) *Pancreatic non-Hodgkin lymphoma on axial CECT. Note bulky mass in region of head of pancreas (arrow).* ***(Right)*** *Pancreatic non-Hodgkin lymphoma on axial CECT. Note lack of encasement of mesenteric vessels (arrow).*

Typical

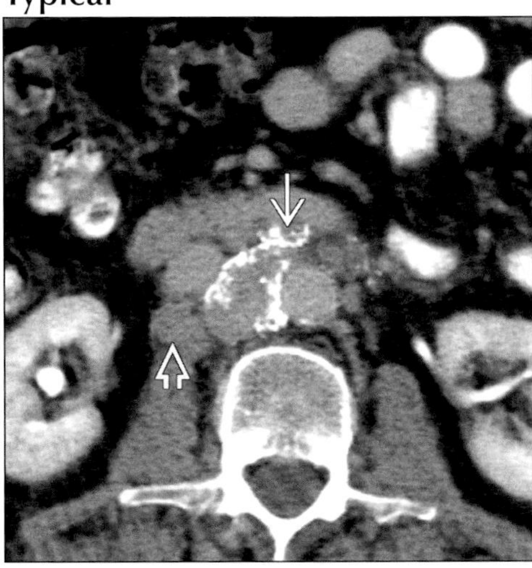

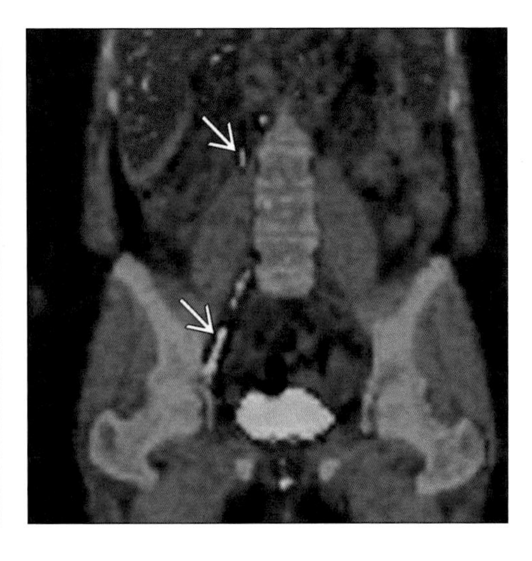

(Left) *Axial CECT after lymphangiogram (LAG) for Hodgkin disease. Note high attenuation LAG contrast in some nodes (arrow) but not others (open arrow).* ***(Right)*** *Fused coronal PET-CT image in patient with Hodgkin disease. Note FDG avid right para-aortic and iliac nodes (arrows).*

RETROPERITONEAL METASTASES

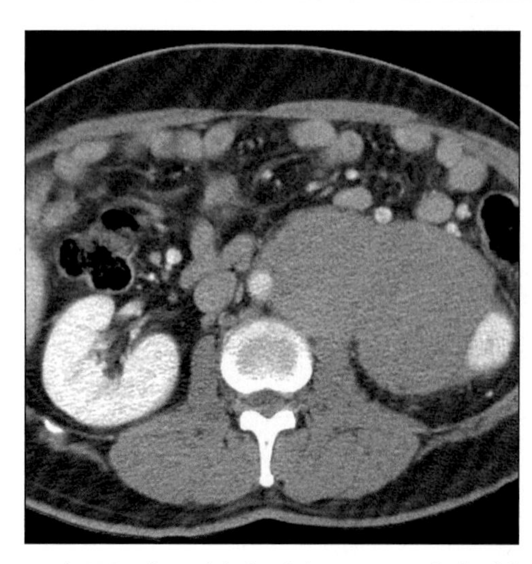
Axial CECT shows lobulated, homogenous bulky left para-aortic lymphadenopathy in patient with left testicular seminoma.

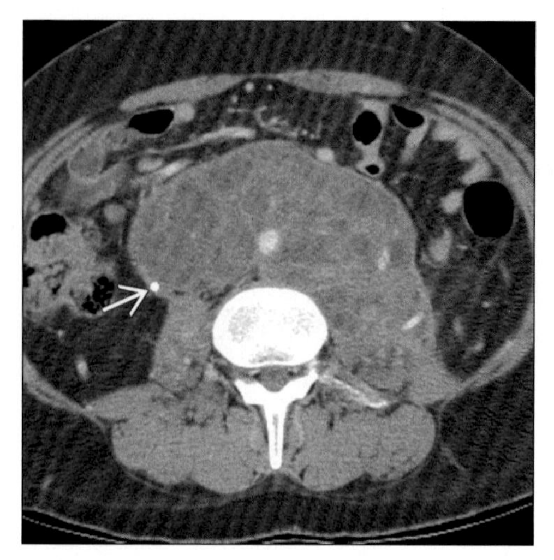
Axial CECT shows heterogeneous retroperitoneal nodal mass encasing vessels in patient with non-seminomatous germ cell tumor. A stent has been placed in the right ureter (arrow).

TERMINOLOGY

Abbreviations and Synonyms

- Retroperitoneal (RP) metastases (mets)

Definitions

- Spread of malignant cells to retroperitoneal lymph nodes and surrounding tissues

IMAGING FINDINGS

General Features

- Best diagnostic clue
  - Enlarged RP lymph nodes in patient with known malignancy
  - Cluster of small RP nodes in patient with nonseminomatous testicular cancer
- Location
  - RP lymphadenopathy: Periaortic, interaortocaval, pericaval
    - Testicular cancers: Lymphatic drainage follows gonadal vessels; nodal mets at ipsilateral renal hilum
    - Pelvic cancers: Obturator nodes may be first site of involvement
  - Perinephric and properitoneal fat
  - Iliopsoas muscle compartment
- Size
  - Para-aortic, aortocaval, pelvic lymphadenopathy: > 10 mm short axis
  - Retrocrural nodes: > 6 mm short axis diameter
- Morphology
  - Focal or diffusely infiltrating masses
  - Round nodes (max diameter < 1.5x min diameter) suggest malignancy

Radiographic Findings

- Radiography: Lymphangiography no longer indicated for detection of retroperitoneal adenopathy

CT Findings

- NECT: Calcification in metastatic nodes from ovarian cancer, mucin-producing tumors
- CECT
  - Testicular cancer: Bulky RP nodes in seminomas, small nodes in nonseminomatous germ cell tumors (GCT)
  - Discrete or confluent nodal masses
    - Low density nodes can be seen in untreated testicular CA
    - Growing low density RP mass in treated nonseminomatous GCTs: Residual teratoma
  - Discrete masses in retroperitoneal fat
  - Malignant retroperitoneal fibrosis: Confluent soft tissue mass; anterior displacement of aorta

DDx: Retroperitoneal Masses

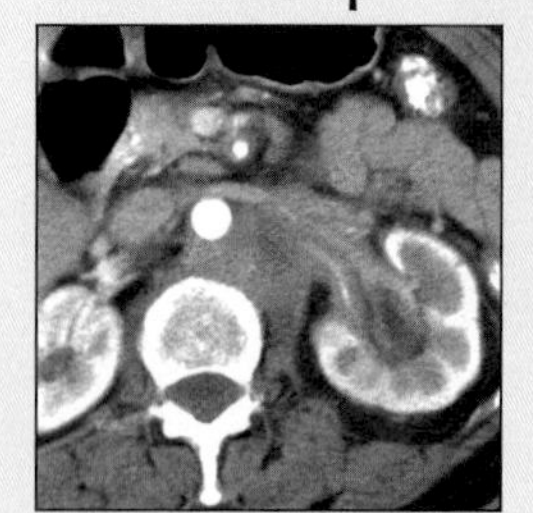
Lymphoma

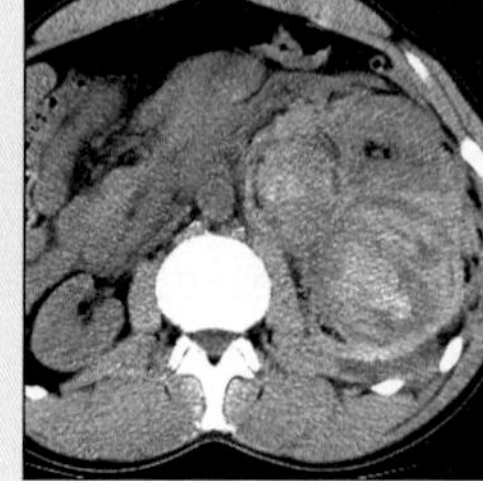
Hemorrhage

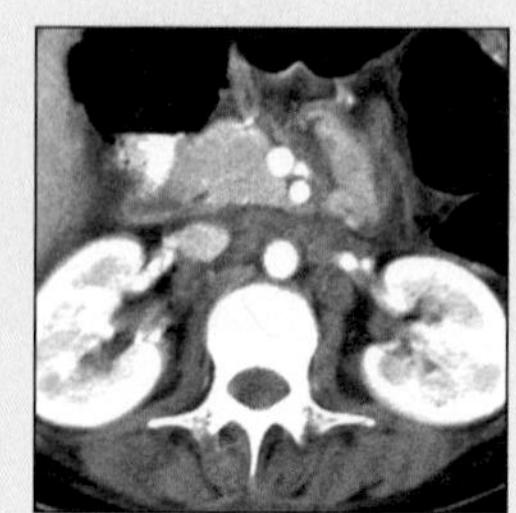
Tuberculosis

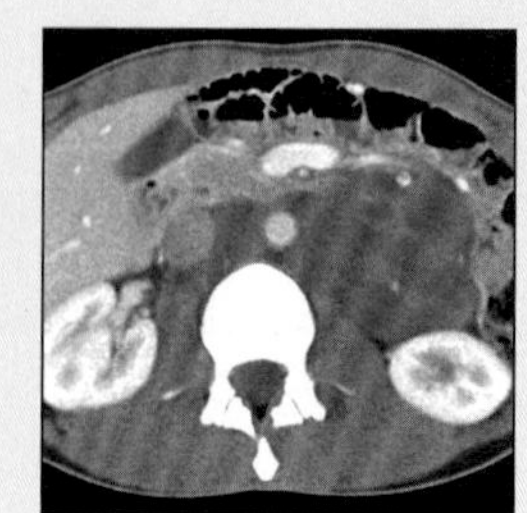
LAM

RETROPERITONEAL METASTASES

Key Facts

Imaging Findings
- Testicular cancers: Lymphatic drainage follows gonadal vessels; nodal mets at ipsilateral renal hilum
- Para-aortic, aortocaval, pelvic lymphadenopathy: > 10 mm short axis
- Retrocrural nodes: > 6 mm short axis diameter
- Round nodes (max diameter < 1.5x min diameter) suggest malignancy
- Testicular cancer: Bulky RP nodes in seminomas, small nodes in nonseminomatous germ cell tumors (GCT)
- Low density nodes can be seen in untreated testicular CA
- Growing low density RP mass in treated nonseminomatous GCTs: Residual teratoma

Top Differential Diagnoses
- Retroperitoneal lymphoma
- Primary retroperitoneal tumors
- Infection: Tuberculosis, AIDS
- Retroperitoneal fibrosis
- Lymphangioleiomyomatosis (LAM)
- Retroperitoneal hemorrhage

Pathology
- Lymphatic or hematogenous spread: Testicular cancer, melanoma, ovarian, prostate, lung, breast
- Direct extension from primary intra-abdominal neoplasms: Pancreas, GI cancers
- Testicular germ cell tumors: Nodes at level of ipsilateral renal hilum

MR Findings
- T1WI
  - RP lymphadenopathy: Low-intermediate SI nodes
  - Metastatic melanoma may have high signal intensity
- T2WI
  - Tumor usually high signal intensity
  - Low signal intensity masses after treatment may be fibrosis
- T1 C+: Metastatic disease in nodes may enhance like primary tumor

Ultrasonographic Findings
- Real Time
  - Round or oval lymph nodes with ratio longitudinal/transverse diameter < 2
  - Narrow or absent echogenic fatty hilum
  - Echogenic and heterogeneous; more echogenic than lymphoma
- Color Doppler
  - Malignant nodes may have increased vascularity
  - Displacement of normal nodal hilar vessels

Nuclear Medicine Findings
- PET
  - May be more sensitive and specific than CT for certain RP mets depending on primary
  - Non-seminomatous GCTs with well-differentiated components not very FDG-avid

Imaging Recommendations
- Best imaging tool: CT
- Protocol advice
  - Oral contrast may be helpful in thin patients to distinguish bowel from RP nodes
  - Obtain non-contrast scans if hemorrhage suspected

DIFFERENTIAL DIAGNOSIS

Retroperitoneal lymphoma
- Confluent soft tissue mantle of adenopathy or enlarged bilateral nodes (> 1.5 cm short axis)
- Nodes may displace aorta anteriorly from spine

Primary retroperitoneal tumors
- Mesenchymal tumors: Liposarcoma, leiomyosarcoma, malignant fibrous histiocytoma
- Usually large size
  - Malignant: Average size 11-20 cm
  - Benign: Average size 4-7 cm

Infection: Tuberculosis, AIDS
- TB: Often have low density adenopathy
- AIDS: Adenopathy common but enlargement usually not massive

Retroperitoneal fibrosis
- 70% idiopathic; may be immune response to atherosclerotic disease
- 8-10% of cases are due to metastatic foci inciting desmoplastic response

Lymphangioleiomyomatosis (LAM)
- Proliferation of smooth muscle cells in lymph vessels causes obstruction
- Low density masses surround and displace retroperitoneal vessels

Retroperitoneal hemorrhage
- High attenuation on noncontrast CT
- Clinical setting of anticoagulation, trauma and/or dropping hematocrit

Cirrhosis/hepatitis
- Upper abdominal lymphadenopathy common with biliary and chronic viral etiologies

Sarcoidosis
- Enlarged abdominal lymph nodes in 30%

PATHOLOGY

General Features
- General path comments
  - Lymphatic or hematogenous spread: Testicular cancer, melanoma, ovarian, prostate, lung, breast

RETROPERITONEAL METASTASES

  - Direct extension from primary intra-abdominal neoplasms: Pancreas, GI cancers
- Etiology
  - Malignant RP lymphadenopathy
    - Testicular germ cell tumors: Nodes at level of ipsilateral renal hilum
    - Pelvic malignancies: Ovarian, bladder, prostate, uterus, colorectal
    - Patients with poorly differentiated cancer of unknown primary affecting primarily retroperitoneal-mediastinal nodes considered to have extragonadal germ-cell syndrome
  - Malignant retroperitoneal fibrosis: Desmoplastic reaction to cancer cells
    - Carcinomas: Breast, stomach, colon, lung
    - Lymphoma, melanoma, carcinoid, sarcomas
- Associated abnormalities: Ureteral obstruction with bulky retroperitoneal disease

Gross Pathologic & Surgical Features

- Seminoma: May have bulky retroperitoneal masses
- Non-seminomatous GCTs: Often have metastases in normal sized nodes

Microscopic Features

- Testicular GCTs: Metastases may have different histology from primary testicular lesion
  - Embryonal CA: May show teratomatous histology in nodes
  - Teratoma may show choriocarcinoma in the lymph nodes

Staging, Grading or Classification Criteria

- Retroperitoneal lymphadenopathy makes testicular cancer stage II

CLINICAL ISSUES

Presentation

- Most common signs/symptoms
  - Retroperitoneal disease often asymptomatic
  - Other signs/symptoms
    - Back pain, hematuria, obstructive uropathy
- Clinical profile
  - RP lymphadenopathy: Known lymphoma, testicular, renal cell, cervical, prostate, or ovarian cancer
  - RP fat metastases: Melanoma, small cell lung cancer

Demographics

- Age: Any; depends on primary cancer
- Gender: Both; depends on primary cancer

Natural History & Prognosis

- Testicular cancer: Excellent prognosis even with RP lymphadenopathy

Treatment

- Options, risks, complications
  - Testicular germ-cell tumors: 3 possibilities for residual masses
    - Residual cancer: Rx with chemotherapy
    - Residual benign teratoma following chemotherapy ("growing teratoma syndrome"): Rx with surgery
    - Fibrosis: No Rx needed

DIAGNOSTIC CHECKLIST

Consider

- Lymph node enlargement > 1.5 cm is unusual in the AIDS complex alone; should prompt CT-guided biopsy to rule out lymphoma, Kaposi sarcoma or MAI

Image Interpretation Pearls

- Nonseminomatous GCTs: Any lymph nodes at ipsilateral hilum should be considered suspicious regardless of size
- RP nodes > 4 mm located anterior to mid-portion of aorta suspicious for metastatic nonseminomatous GCT

SELECTED REFERENCES

1. Haaga JR et al: CT and MR imaging of the whole body, 4th ed. St. Louis, Mosby, 1684-1714, 2003
2. Nishino M et al: Primary retroperitoneal neoplasms: CT and MR imaging findings with anatomic and pathologic diagnostic clues. Radiographics. 23(1):45-57, 2003
3. Pallisa E et al: Lymphangioleiomyomatosis: pulmonary and abdominal findings with pathologic correlation. Radiographics. 22 Spec No:S185-98, 2002
4. Avila NA et al: Lymphangioleiomyomatosis: abdominopelvic CT and US findings. Radiology. 216(1):147-53, 2000
5. Vesselle HJ et al: FDG PET of the retroperitoneum: normal anatomy, variants, pathologic conditions, and strategies to avoid diagnostic pitfalls. Radiographics. 18(4):805-23; discussion 823-4, 1998
6. Hilton S et al: CT detection of retroperitoneal lymph node metastases in patients with clinical stage I testicular nonseminomatous germ cell cancer: assessment of size and distribution criteria. AJR Am J Roentgenol. 169(2):521-5, 1997
7. Jeffrey RB Jr et al: Abdominal CT in acquired immunodeficiency syndrome. AJR Am J Roentgenol. 146(1):7-13, 1986

IMAGE GALLERY

Typical

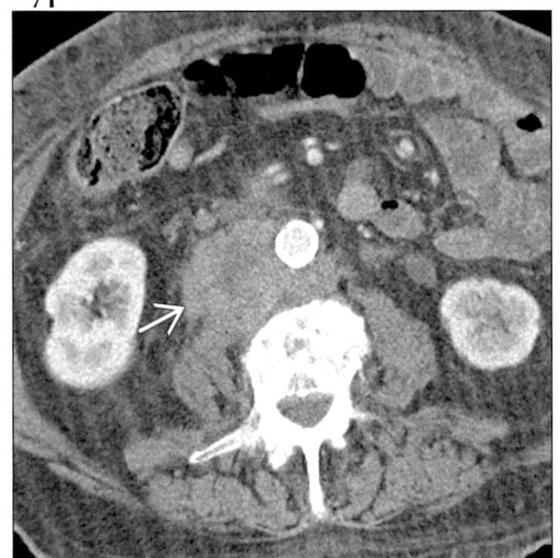

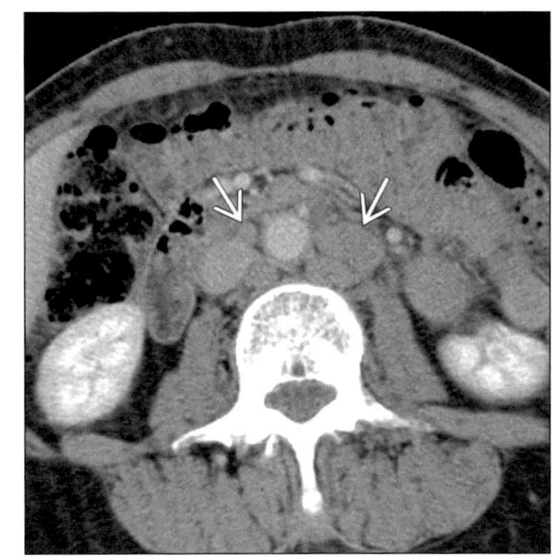

(Left) *Axial CECT shows heterogeneous pericaval nodal mass (arrow) representing metastatic prostate CA. Sclerotic bony metastases are also present.* ***(Right)*** *Axial CECT shows left para-aortic and pericaval nodes (arrows) in patient with metastatic rectal cancer.*

Variant

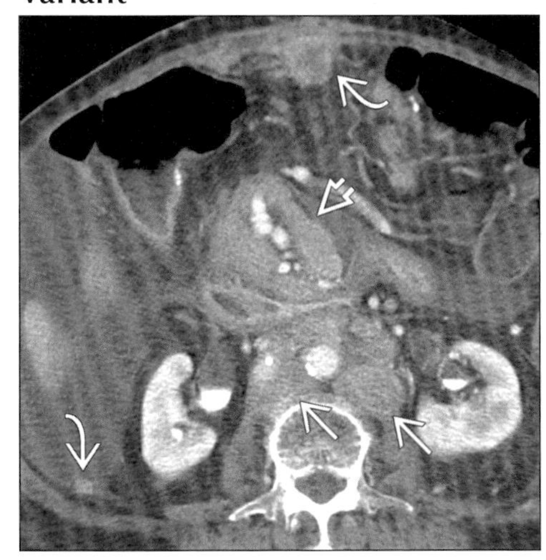

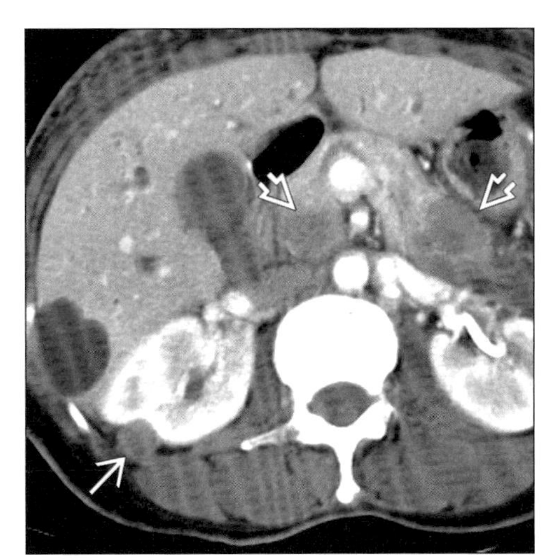

(Left) *Axial CECT shows para-aortic and paracaval nodes (arrows) in patient with ovarian CA. Mesenteric nodes (open arrows), ascites, and peritoneal implants (curved arrows) are also present.* ***(Right)*** *Axial CECT shows tumor implant in perinephric fat (arrow) in patient with metastatic neuroendocrine tumor of the cervix. Pancreatic metastases are also present (open arrows).*

Variant

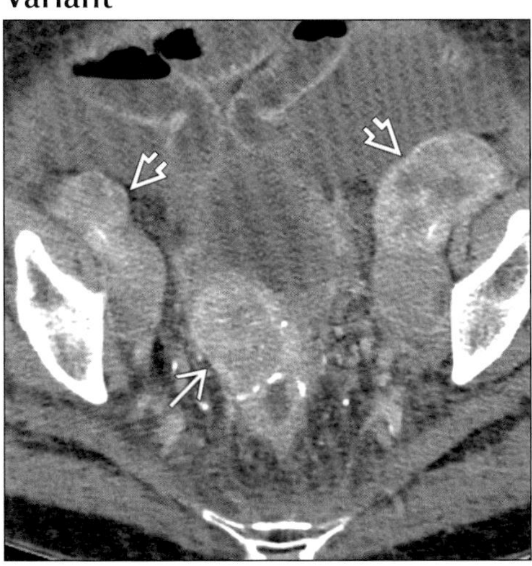

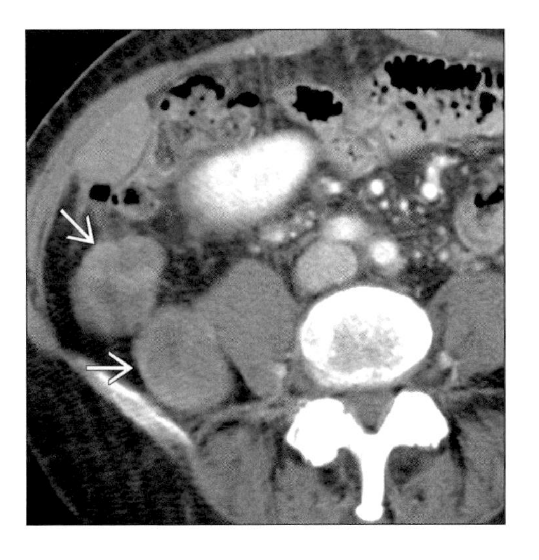

(Left) *Axial CECT shows heterogeneously enhancing enlarged pelvic lymph nodes (open arrows) in patient with metastatic ovarian cancer. A peritoneal drop metastasis is also noted (arrow).* ***(Right)*** *Axial CECT shows masses in the right retroperitoneal fat (arrows) representing metastatic leiomyosarcoma.*

RETROPERITONEAL LYMPHOCELE

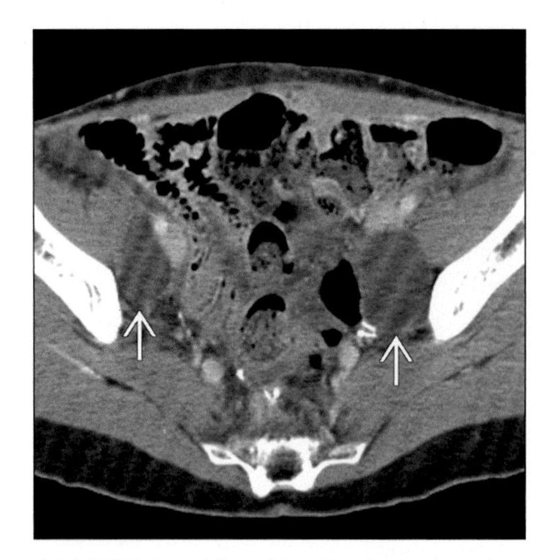

Axial CECT shows bilateral lymphoceles (arrows) along the iliac vessels in a woman following bilateral salpingo-oophorectomy and pelvic lymph node dissection.

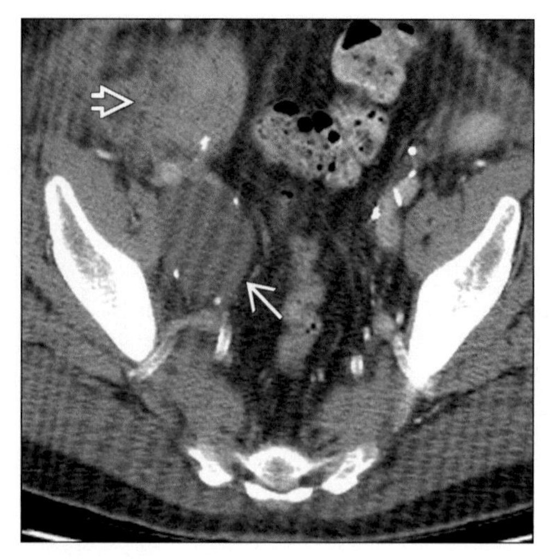

Axial CECT shows lymphocele (arrow) along right iliac vessels in a patient following pancreatic transplantation (open arrow). Note high attenuation clips along periphery of lymphocele.

TERMINOLOGY

Abbreviations and Synonyms

- Retroperitoneal (RP) lymphocele, lymph cyst

Definitions

- Pseudocyst formed when lymph leaks from disrupted lymphatics

IMAGING FINDINGS

General Features

- Best diagnostic clue: Low density retroperitoneal fluid collection in post-surgical patient
- Location: Along iliac vessels, para-aortic retroperitoneum, inguinal area
- Size: Variable
- Morphology
  - Unilocular or multilocular
  - Round or oval

CT Findings

- NECT
  - Low density fluid collection when uncomplicated
  - Calcification is rare
- CECT: Imperceptible wall; does not enhance with contrast

MR Findings

- T1WI: Round or ovoid mass with low signal intensity
- T2WI: Round or ovoid mass with high signal intensity
- T1 C+: Low signal masses without enhancement

Ultrasonographic Findings

- Real Time
  - Anechoic or hypoechoic mass
  - Increased through transmission
  - May have dependent debris, septations

Imaging Recommendations

- Best imaging tool: CECT
- Protocol advice: Delayed imaging (> 15 minutes) useful to detect filling of urinoma or bladder diverticulum

DIFFERENTIAL DIAGNOSIS

Other pelvic cystic masses

- Urinoma: Should fill with contrast on delayed imaging
- Abscess: Rim-enhancement
- Hematoma, endometrioma: High attenuation; fluid-fluid levels

Bladder (Bl.) diverticulum

- Typically near ureteral insertions
- Fill with contrast on delayed imaging

DDx: Cystic Pelvic Masses

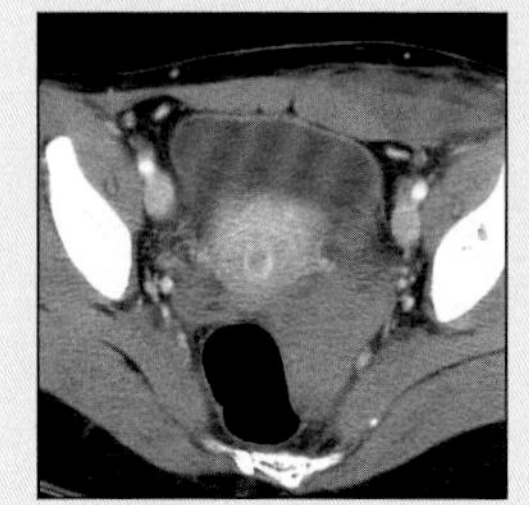

Pelvic Hematoma

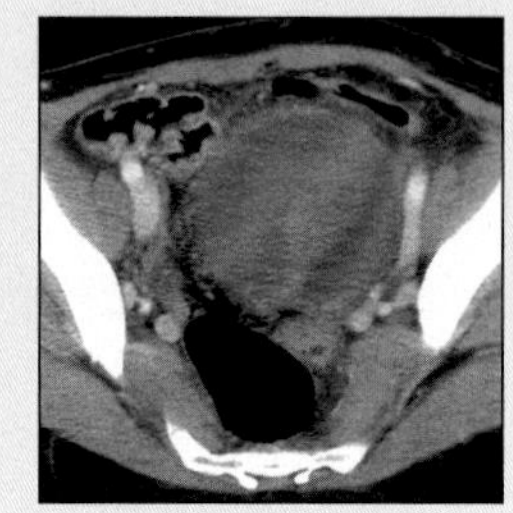

Endometrioma

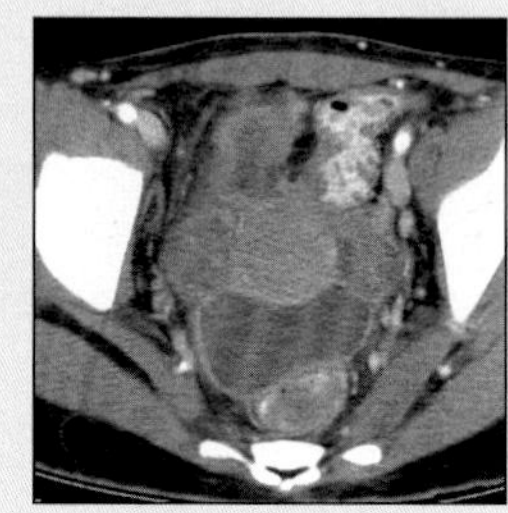

Pelvic Abscess

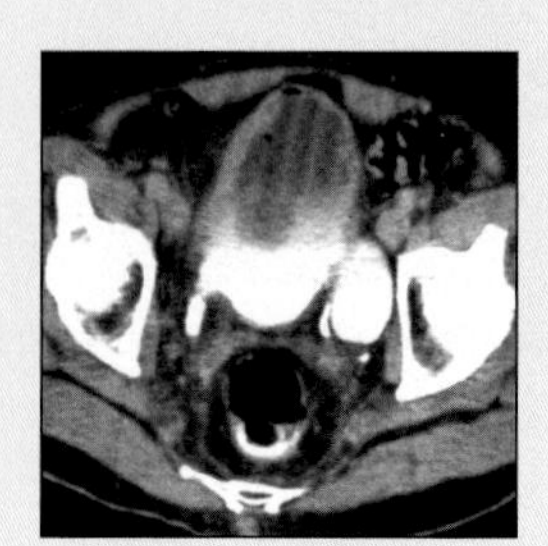

Bl. Diverticulum

RETROPERITONEAL LYMPHOCELE

Key Facts

Terminology
- Pseudocyst formed when lymph leaks from disrupted lymphatics

Imaging Findings
- Best diagnostic clue: Low density retroperitoneal fluid collection in post-surgical patient
- Location: Along iliac vessels, para-aortic retroperitoneum, inguinal area
- CECT: Imperceptible wall; does not enhance with contrast
- Anechoic or hypoechoic mass

Clinical Issues
- Most lymphoceles are asymptomatic
- May compromise function of renal transplants by obstructing blood supply or ureter

---

- May see narrow neck connecting to bladder

Lymphangioma
- More often multiloculated
- May have calcification

PATHOLOGY

General Features
- General path comments: Common after surgery near large lymphatic trunks
- Etiology
  - Lymphatics have slow flow and normally collapse after being disrupted
  - Persistent lymphatic leakage from divided lymphatic channels → lymph pseudocyst
- Epidemiology
  - Up to 30% incidence post radical pelvic lymphadenectomy
  - 0.6-20% incidence post renal transplant
    - 3-4% symptomatic

Gross Pathologic & Surgical Features
- Tan, dark yellow or brown fluid

Microscopic Features
- Fat globules, lymphocytes, few red blood cells
- Fluid with high protein content

CLINICAL ISSUES

Presentation
- Most common signs/symptoms
  - Only 5-7% of patients symptomatic
  - Renal transplant patients
    - Iliac vein compression: Unilateral leg edema
    - Swelling over transplant
    - Elevated creatinine due to ureteral compression
  - Abdominal distention, pain, secondary infection
- Clinical profile
  - Post-operative patient: RP node dissection; groin surgery
  - Renal transplant patient
  - Typical presentation is several weeks after surgery

Demographics
- Age: Any
- Gender: Both

Natural History & Prognosis
- Most lymphoceles are asymptomatic
- May compromise function of renal transplants by obstructing blood supply or ureter
- May become secondarily infected

Treatment
- Options, risks, complications
  - Small lymphoceles: Often resolve spontaneously
  - Percutaneous aspiration: Not definitive
  - Long-term catheter drainage: Success in 50-87%
  - Sclerotherapy: Success in 79-94%
  - Transplants: Surgical lymphocele fenestration

DIAGNOSTIC CHECKLIST

Image Interpretation Pearls
- Enhancing thick wall suggests superinfection

SELECTED REFERENCES

1. Fuller TF et al: Management of lymphoceles after renal transplantation: laparoscopic versus open drainage. J Urol. 169(6):2022-5, 2003
2. Chow CC et al: Complications after laparoscopic pelvic lymphadenectomy: CT diagnosis. AJR Am J Roentgenol. 163(2):353-6, 1994
3. vanSonnenberg E et al: Lymphoceles: imaging characteristics and percutaneous management. Radiology. 161(3):593-6, 1986

IMAGE GALLERY

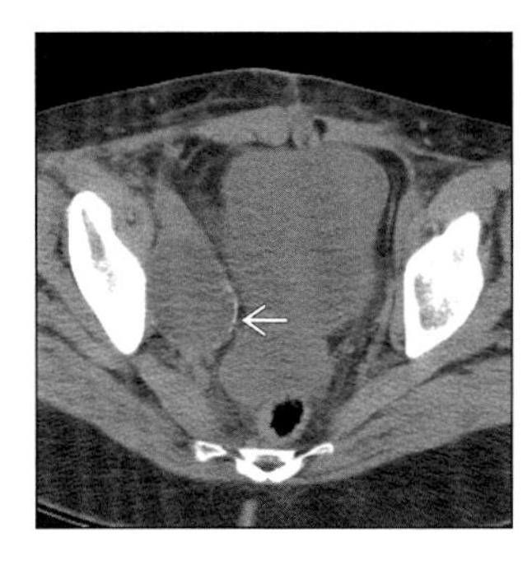
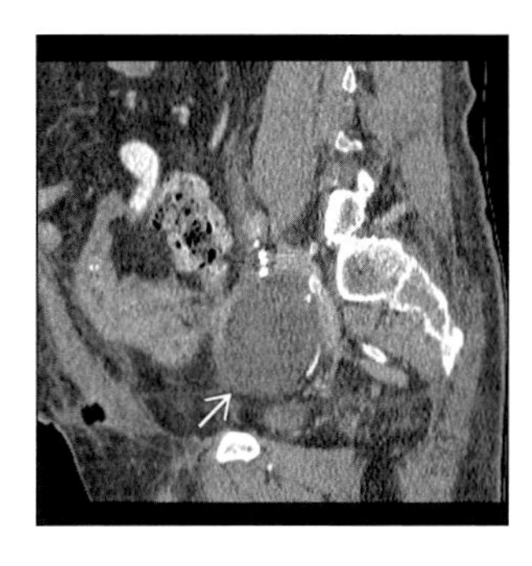

*(**Left**) Axial NECT shows right sided lymphocele with thin rim of calcification (arrow). (**Right**) Sagittal CECT shows lymphocele (arrow) anterior to internal iliac vessels. Note adjacent surgical clips.*

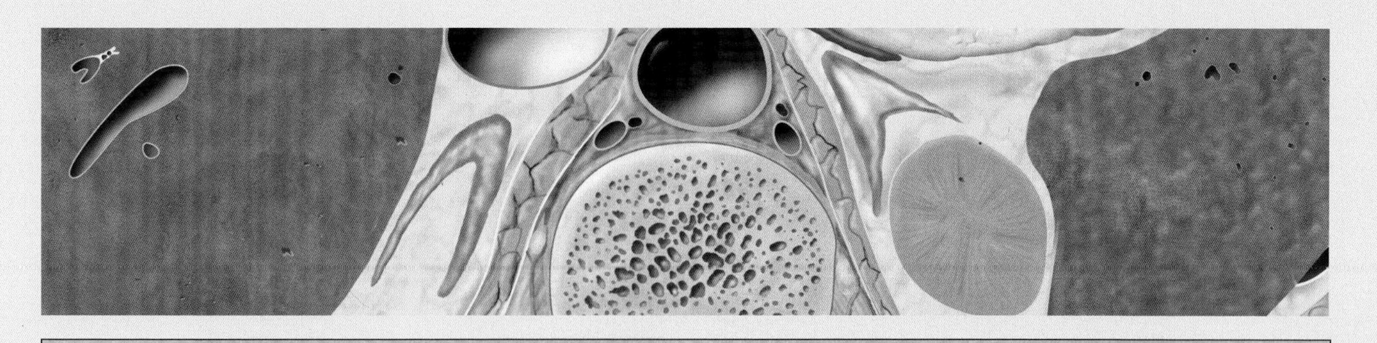

SECTION 2: Adrenal

Introduction and Overview

Infection

Vascular - Traumatic

Metabolic

Neoplasm, Benign

Neoplasm, Malignant

ADRENAL ANATOMY AND IMAGING ISSUES

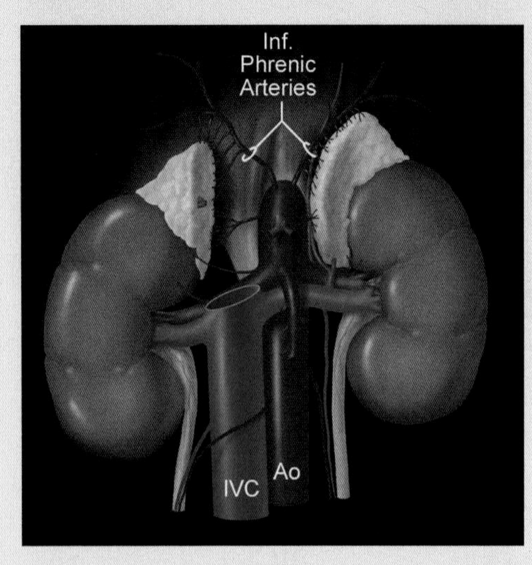

Graphic shows adrenal anatomy. Note multiple arterial sources. Right adrenal vein drains directly into the IVC, while the left enters the renal vein.

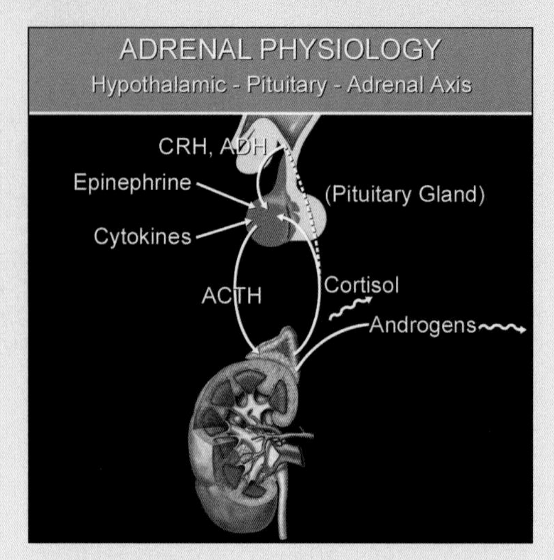

Graphic shows stimulation of the anterior pituitary by epinephrine and cytokines, or by hypothalamic release of CRH or ADH, resulting in release of ACTH causing adrenal to secrete cortisol.

TERMINOLOGY

Abbreviations and Synonyms

- Tuberculosis (TB)
- Adrenal corticotrophic hormone (ACTH)
- Corticotropin releasing hormone (CRH)
- Antidiuretic hormone (ADH)

IMAGING ANATOMY

- Location
  - Adrenal glands lie in the perirenal space (within the perirenal fascia), usually cephalic to the kidneys
  - Right adrenal is suprarenal, lies lateral to diaphragmatic crus, medial to liver, and touches the back of the IVC
  - Left adrenal is usually ventral to upper pole of left kidney, dorsal to the splenic vein and pancreas

ANATOMY-BASED IMAGING ISSUES

Key Concepts or Questions

- Adrenal cortex and medulla are essentially different organs within the same structure
  - Cortex is an endocrine gland secreting primarily cortisol, aldosterone, and androgenic steroids (all derived from cholesterol, which contributes to the high lipid content characteristic of adrenals)
  - Medulla is derived from neural crest and secretes epinephrine and norepinephrine
- Adrenal physiology is controlled by elaborate interaction between the hypothalamus, pituitary, and adrenals
  - Stress results in release of epinephrine (and cytokines with fever, trauma, etc.) which cause the pituitary to secrete ACTH, which stimulates adrenal release of cortisol
  - Elevated serum cortisol has a suppressing effect on the hypothalamus and pituitary, reducing further release of ACTH
- Cushing syndrome (due to excess cortisol)
  - Characterized by truncal obesity, hirsutism, amenorrhea, hypertension, weakness, abdominal striae
  - Causes
    - Adrenal hyperplasia (75-80%)
    - Adrenal adenoma (20-25%)
    - Adrenal carcinoma (< 5%)
    - Exogenous corticosteroid medication
  - Diagnostic tests for Cushing syndrome
    - Serum: High cortisol level
    - Urine: High 17-hydroxycorticoid level
    - Serum: Elevated ACTH = hyperplasia; low ACTH = autonomous adenoma
    - Dexamethasone suppression test: To distinguish pituitary from "ectopic" source of excess ACTH
- Conn syndrome (excess aldosterone)
  - Characterized by hypertension, hypokalemic alkalosis, muscle weakness, cardiac dysfunction
  - Causes
    - Adrenal adenoma (65-70%)
    - Adrenal hyperplasia (25-30%)
    - Adrenal carcinoma (< 1%)
- Addison syndrome (disease) (adrenal insufficiency)
  - Causes of slow onset insufficiency
    - Autoimmune (most common in developed countries)
    - TB or systemic fungal or granulomatous disease
    - Metastases
    - AIDS (acquired immunodeficiency syndrome)
  - Causes of abrupt onset insufficiency
    - Adrenal hemorrhage (sepsis, shock, septicemia, anticoagulation, vasculitis)
    - Post-partum pituitary necrosis (Sheehan syndrome)
    - Withdrawal of long term steroid medication
- What is the likelihood of a small (< 1.5 cm) adrenal mass to be an adenoma?

ADRENAL ANATOMY AND IMAGING ISSUES

DIFFERENTIAL DIAGNOSIS

Adrenal neoplasms

- Adenoma
- Metastases
- Lymphoma
- Pheochromocytoma
- Carcinoma
- Myelolipoma
- Hemangioma

Non-neoplastic adrenal masses

- Hemorrhage
- Infection
- Hyperplasia
- Cyst
- Pseudocyst
- (Gastric diverticulum; vessels)

Incidental adrenal mass

- Adenoma
- Nodular hyperplasia
- Myelolipoma
- (Pseudomass)
- Hematoma, hemorrhage
- Metastasis
- Ganglioneuroma
- Pheochromocytoma
- Carcinoma

Neural crest tumors

- Neuroblastoma (malignant, children)
- Ganglioneuroblastoma
- Ganglioneuroma (benign, adults)
- Pheochromocytoma (usually benign, adults)

    - In the absence of a known malignancy, close to 100%
        - (At least 2% of the general population have a non-hyperfunctioning adrenal adenoma)
    - In a patient with cancer, most small adrenal masses are still adenomas
- If an adrenal mass is of indeterminate attenuation (> 15 HU) on nonenhanced CT, will a subsequent MR evaluation be of value in distinguishing adenoma from metastasis?
    - Probably not because both nonenhanced CT and MR rely on imaging evidence of high lipid content to diagnose adenoma
        - "Lipid-poor" adenomas are best evaluated by "adrenal protocol CT" (including delayed phase)
- What is a practical imaging approach for diagnosing pheochromocytoma?
    - Diagnosis should be established with reasonable certainty prior to imaging (symptoms of headache, palpitations and excessive sweating; urine assay for catecholamine metabolites, metanephrine and vanillylmandelic acid)
    - For CT, include lower thorax to symphysis
        - 98% of pheochromocytomas arise in the abdomen
        - 90% arise in the adrenals
        - Pericardium is rare site, but easily included on CT
    - For MR, include at least adrenals through aortic bifurcation
        - T2WI and gadolinium-enhanced sequences are best
- Is there an important distinction between the lipid content, characteristic of adenoma, and fat content, characteristic of myelolipoma?
    - The lipid in adenoma is intra- and intercellular and not in macroscopic deposits
        - NECT detects this usually as near-water density (-5 to +15 HU)
        - MR detects this best with a combination of in-phase and opposed-phase sequences; the latter designed to suppress signal from tissue with an even admixture of lipid and water protons
    - The fat in myelolipomas is usually macroscopic, mature
        - NECT detects this usually as fat-density (-80 to -100 HU) deposits: Thin sections and pixel analysis may be necessary to detect small foci of fat
        - Opposed-phase MR will not suppress signal from large foci of fat
- Is it necessary and safe to give I.V. contrast material for CT or MR evaluation of pheochromocytoma?
    - Because most sporadic pheochromocytomas are large (3 to 5 cm) and are easily recognized in an adrenal site, contrast administration is usually unnecessary
    - In the setting of "syndromic " pheochromocytoma (e.g., von Hippel Lindau, multiple endocrine neoplasia), pheochromocytomas are often multiple, small, and extra-adrenal, and I.V. contrast may be necessary
    - It is extremely rare for I.V. administration of nonionic iodinated or gadolinium-based contrast media to induce a hypertensive crisis in patients with pheochromocytoma
    - Most authorities consider it unnecessary to give prophylactic α and β-adrenergic blockade prior to a contrast-enhanced CT or MR exam
- What is the role of MIBG in diagnosis of pheochromocytoma?
    - MIBG (metaiodobenzylguanidine) has a sensitivity of nearly 90% and a specificity of 99%
    - It is generally available only in large academic referral centers and is very expensive
    - Main roles are in patients with familial pheochromocytoma syndromes, and in patients with strong clinical evidence of pheochromocytoma in whom CT or MR have failed to identify the tumor
- What is the role of adrenal biopsy?
    - Rarely necessary with the proper use and interpretation of CT or MR
    - Dangerous and usually contraindicated for pheochromocytoma
- What is the role of PET or PET-CT for evaluation of adrenal cancer (primary or metastatic)?
    - Usually unnecessary
    - Unproven in large clinical trials to date

ADRENAL ANATOMY AND IMAGING ISSUES

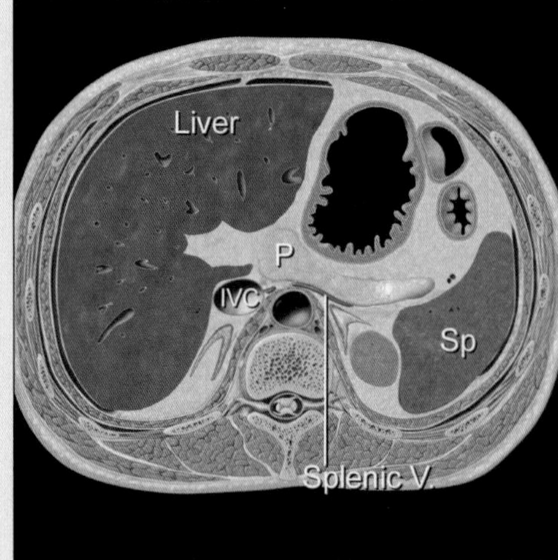

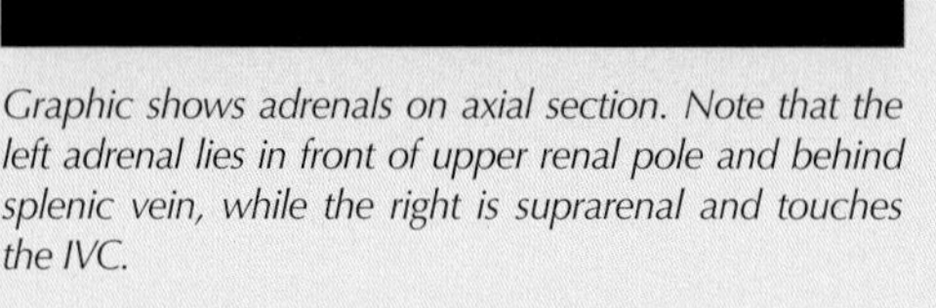

Graphic shows adrenals on axial section. Note that the left adrenal lies in front of upper renal pole and behind splenic vein, while the right is suprarenal and touches the IVC.

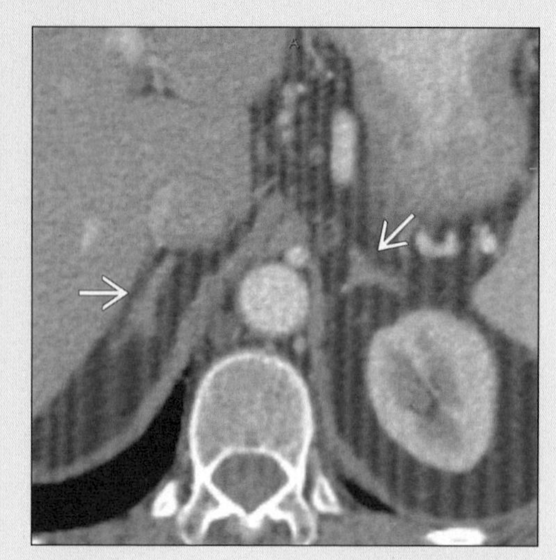

Axial CECT shows one of the typical appearances of normal adrenal glands (arrows).

- Probably accurate and useful for diagnosing adrenal metastases as part of a "whole body" PET-CT exam for recognized indications (e.g., lymphoma, melanoma)

CUSTOM DIFFERENTIAL DIAGNOSIS

Calcified adrenal lesion

- Infection
  - TB, histoplasmosis
- Hemorrhage ("old")
- Carcinoma (small foci)
- Pheochromocytoma
  - Plus heterogeneous hypervascularity
- Myelolipoma
  - Plus foci of fat density
- Metastases
  - Rare

Pheochromocytoma associated diseases

- Multiple endocrine neoplasia (type II and III)
  - Plus medullary thyroid carcinoma and parathyroid adenoma (Sipple syndrome)
  - Mucosal neuroma syndrome (type III)
    - Medullary thyroid cancer and intestinal ganglioneuromatosis
- Neurofibromatosis
- von Hippel Lindau
  - In some families, the predominant finding
  - Usually with CNS hemangioblastomas, renal and pancreatic cysts and tumors
- Carney syndrome
  - Plus pulmonary chondromas, GI stromal tumors
- Tuberous sclerosis
  - Plus neurological lesions, renal cysts and angiomyolipomas

Adrenal hemorrhage in adults: Etiology

- Anticoagulant therapy
- Stress
- Sepsis (Waterhouse-Friderichsen)
- Surgery
- Tumor
- Blunt abdominal trauma

Ectopic (nonpituitary) sources of ACTH

- Oat cell carcinoma of lung
- Liver cancer
- Bronchial or thymic carcinoid
- Bronchia adenoma
- Pancreatic islet cell tumor
- Medullary carcinoma (thyroid)
- Thymoma
- Pheochromocytoma

SELECTED REFERENCES

1. Caoili EM et al: Adrenal masses: characterization with combined unenhanced and delayed enhanced CT. Radiology. 222(3):629-33, 2002
2. Dunnick NR et al: Imaging of adrenal incidentalomas: current status. AJR Am J Roentgenol. 179(3):559-68, 2002
3. Mayo-Smith WW et al: State-of-the-art adrenal imaging. Radiographics. 21(4):995-1012, 2001
4. Krebs TL et al: MR imaging of the adrenal gland: radiologic-pathologic correlation. Radiographics. 18(6):1425-40, 1998
5. Cirillo RL Jr et al: Pathology of the adrenal gland: imaging features. AJR Am J Roentgenol. 170(2):429-35, 1998
6. Dunnick NR et al: Adrenal radiology: distinguishing benign from malignant adrenal masses. AJR Am J Roentgenol. 167(4):861-7, 1996
7. Korobkin M et al: Delayed enhanced CT for differentiation of benign from malignant adrenal masses. Radiology. 200(3):737-42, 1996

IMAGE GALLERY

Typical

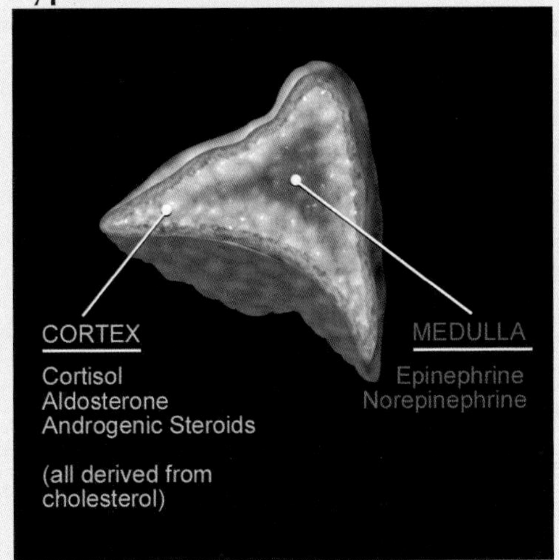

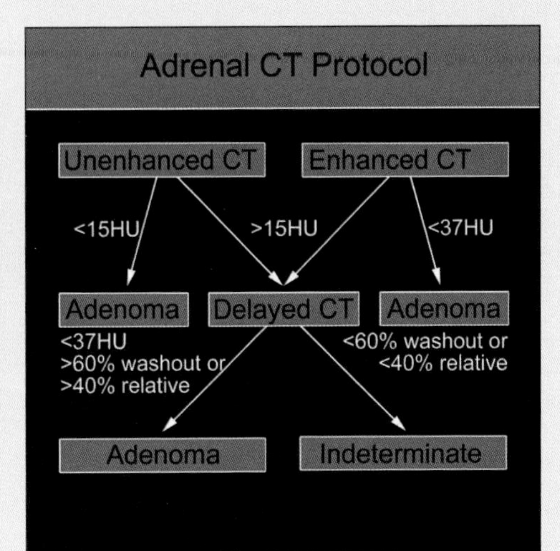

*(**Left**) Graphic shows location and function of adrenal cortex and medulla. (**Right**) Graphic shows adrenal CT protocol and criteria for diagnosing adenoma.*

Typical

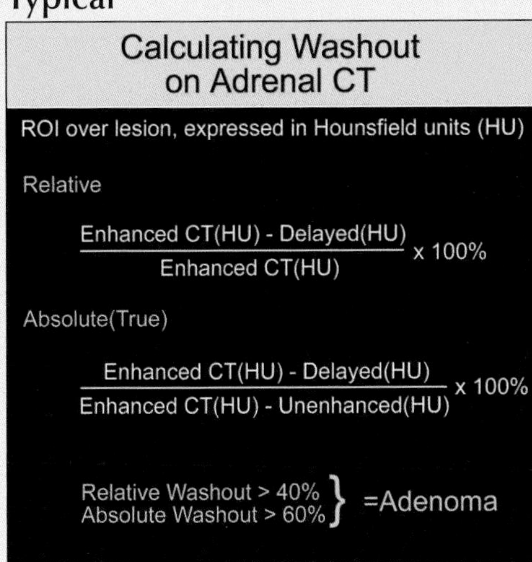

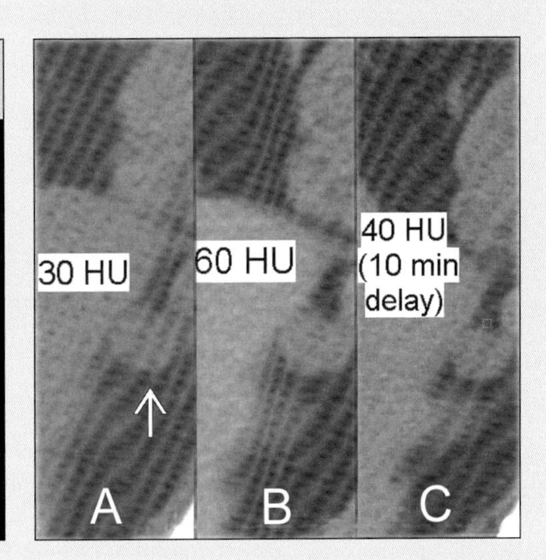

*(**Left**) Graphic shows the formula for calculating adrenal washout on CT. (**Right**) Adrenal protocol CT. NECT (A) shows adrenal mass (arrow) measuring 30 HU. The mass measures 60 HU on CECT (B), and 40 HU on delayed CT (C). Lipid-poor adenoma.*

Typical

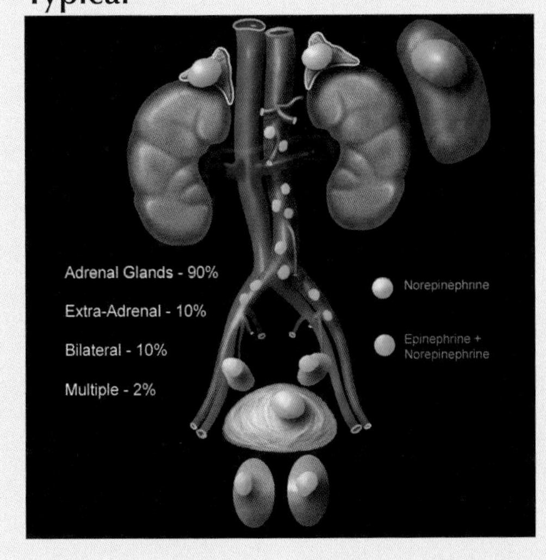

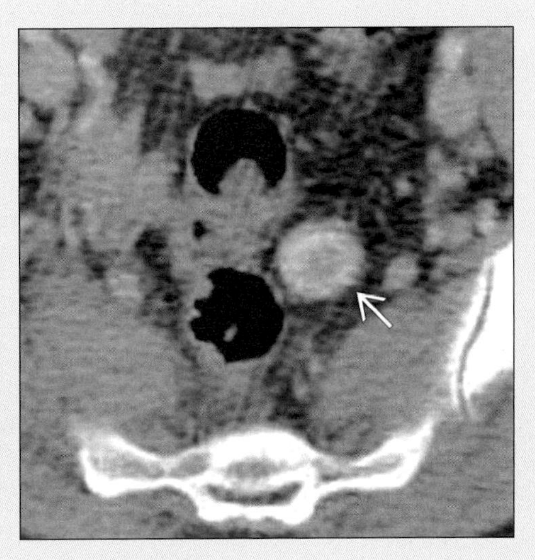

*(**Left**) Graphic indicates potential sites for pheochromocytoma (blue) and paraganglioma (green) organ of Zuckerkandl at aortic bifurcation. (**Right**) Axial CECT shows hypervascular mass (arrow) in pelvis; paraganglioma with clinical symptoms identical to those of pheochromocytoma.*

ADRENAL TB AND FUNGAL INFECTION

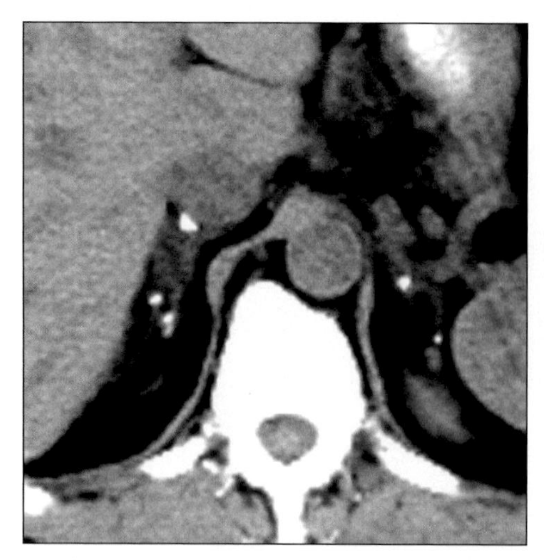

Axial NECT shows nodular calcification and atrophy of the adrenal glands.

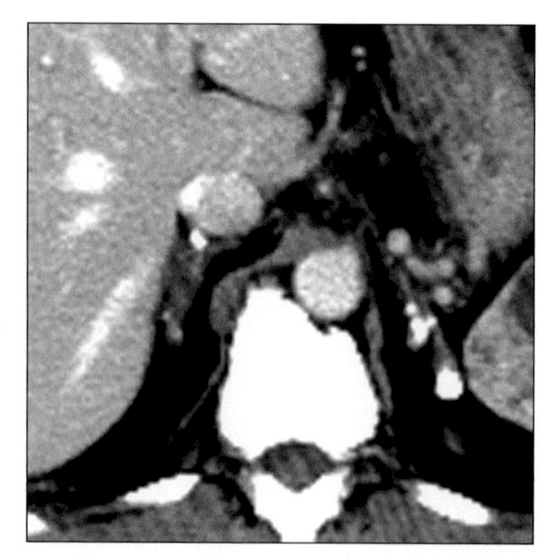

Axial NECT shows nodular calcification and atrophy of the adrenal glands.

TERMINOLOGY

Definitions

- Tuberculosis and fungal infections in adrenal glands

IMAGING FINDINGS

General Features

- Best diagnostic clue: Calcifications of adrenal glands
- Location: Bilateral > unilateral
- Other general features
  - Stages of infection: Acute, chronic

Radiographic Findings

- Radiography: Suprarenal calcifications

CT Findings

- Acute
  - Homo-/heterogenous masses
  - Mild to marked symmetrical enlargement of adrenal glands with preserved contour
  - Central areas of low attenuation (caseous necrosis)
  - Enhancement in the peripheral rim
  - Calcifications
    - Faint flecks of calcium
    - Focal low attenuation nodules
    - Large areas of necrosis or dense calcification
- Enlargement of lymph nodes
- Chronic: Atrophic glands with calcifications

MR Findings

- T1WI: Signal intensity similar to spleen
- T2WI: Signal intensity similar to or higher than fat
- DWI: Granulomas show little enhancement

Ultrasonographic Findings

- Real Time: Homo-/heterogeneous, hypoechoic masses

Imaging Recommendations

- Best imaging tool: CT

DIFFERENTIAL DIAGNOSIS

Adrenal adenoma

- Mass with low attenuation (< 10 HU)
- Focal mass without diffuse enlargement
- Enhancing mass that "de-enhances" rapidly

Pheochromocytoma

- > 3 cm, hypervascular mass
- Bilateral in multiple endocrine neoplasia (MEN)

Adrenal metastases and lymphoma

- Metastases
  - Unilateral or bilateral, invasive, enhancing masses
  - Central necrosis ± hemorrhage
  - Focal masses without diffuse enlargement

DDx: Adrenal Mass +/- Calcification

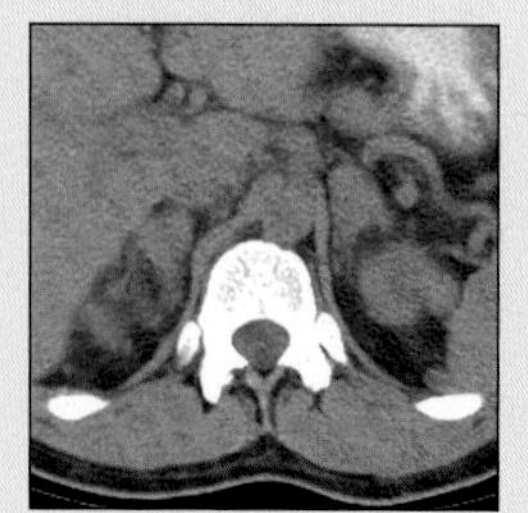

Lymphoma

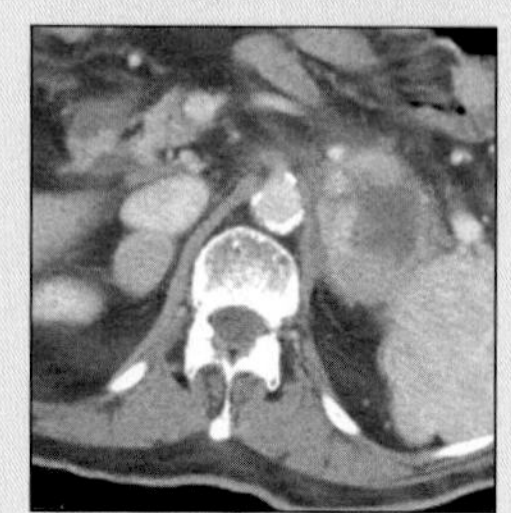

Metastases

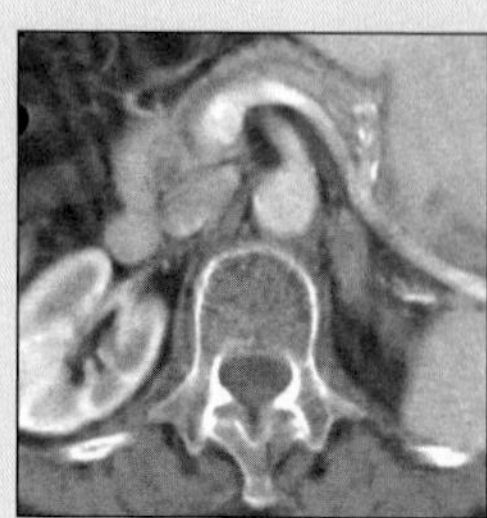

Hyperplasia

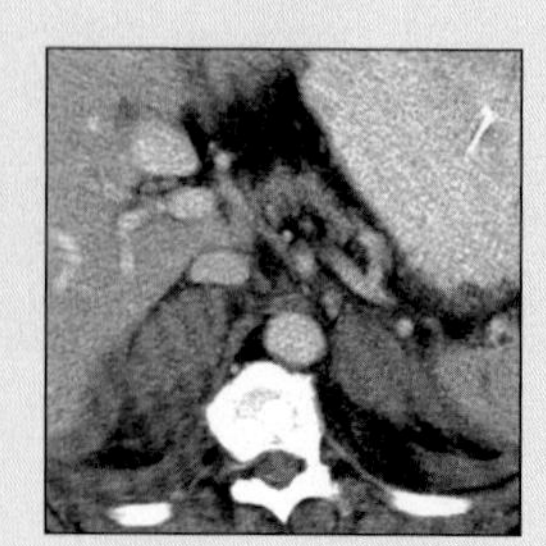

Hemorrhage

ADRENAL TB AND FUNGAL INFECTION

Key Facts

Imaging Findings
- Best diagnostic clue: Calcifications of adrenal glands
- Mild to marked symmetrical enlargement of adrenal glands with preserved contour
- Enlargement of lymph nodes
- Best imaging tool: CT

Top Differential Diagnoses
- Adrenal adenoma
- Pheochromocytoma
- Adrenal metastases and lymphoma
- Adrenal hemorrhage
- Adrenal hyperplasia
- Adrenal myelolipoma

Diagnostic Checklist
- Infection in patients with acute adrenal insufficiency
- Bilateral enlargement with preserved contour

- Lymphoma
  - Usually bilateral, triangular shaped masses

Adrenal hemorrhage
- Heterogenous hyperdense mass (50-90 HU)

Adrenal hyperplasia
- Width of adrenal gland limb > 10 mm
- Contour of adrenal gland preserved
- No discrete mass or nodule

Adrenal myelolipoma
- Small, asymptomatic, well-defined suprarenal mass
- Mass with fat attenuation (-30 to -115 HU)

PATHOLOGY

General Features
- Etiology
  - Tuberculosis (most common in underdeveloped countries)
  - Histoplasmosis (most common in Southeastern, South Central U.S.)
  - Blastomycosis (South American, North American)
  - Coccidioidomycosis
  - Cryptococcosis
  - Paracoccidioidomycosis
  - Risk factors: Immunocompromised patients
- Associated abnormalities: Pulmonary or other disseminated infections

Gross Pathologic & Surgical Features
- Calcifications; caseous necrosis

Microscopic Features
- Granulomatous or fungal organisms

CLINICAL ISSUES

Presentation
- Most common signs/symptoms: Fever, night sweats, weight loss, lethargy, weakness
- Diagnosis
  - Tuberculosis: Positive purified protein derivative (PPD) skin test
  - Percutaneous aspiration with cytology (e.g., acid fast bacilli (AFB) staining) for definitive diagnosis

Demographics
- Age: Any age; 20-40 years; most common
- Gender: M = F

Natural History & Prognosis
- Complications: Adrenal insufficiency (Addison disease), sepsis
- Prognosis: Good, if recognized and treated

Treatment
- Tuberculosis: Antituberculosis therapy
- Fungal infections: Antifungal therapy

DIAGNOSTIC CHECKLIST

Consider
- Infection in patients with acute adrenal insufficiency

Image Interpretation Pearls
- Bilateral enlargement with preserved contour

SELECTED REFERENCES

1. Kumar N et al: Adrenal histoplasmosis: clinical presentation and imaging features in nine cases. Abdom Imaging. 28(5):703-8, 2003
2. Wilson DA et al: Histoplasmosis of the adrenal glands studied by CT. Radiology. 150(3):779-83, 1984
3. Wilms GE et al: Computed tomographic findings in bilateral adrenal tuberculosis. Radiology. 146(3):729-30, 1983
4. Morgan HE et al: Bilateral adrenal enlargement in Addison's disease caused by tuberculosis. Nephrotomographic demonstration. Radiology. 115(2):357-8, 1975

IMAGE GALLERY

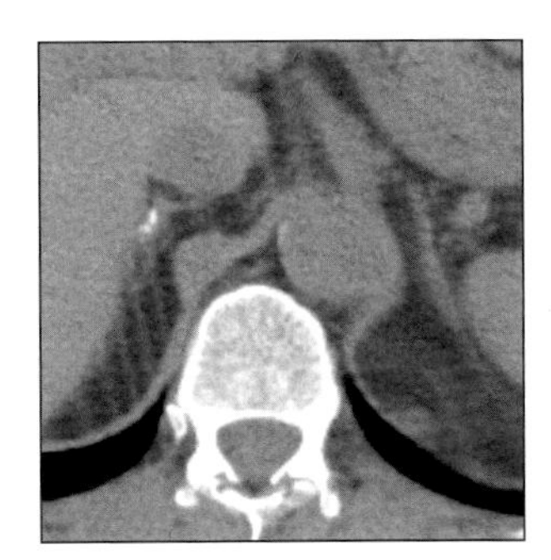

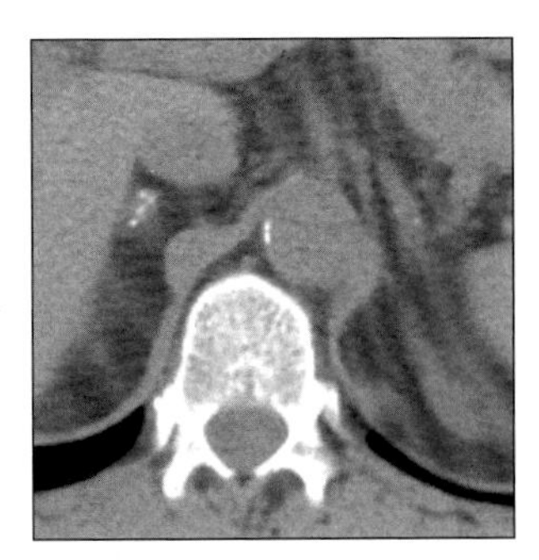

*(**Left**) Axial NECT shows focal calcifications and atrophy of the adrenal glands. (**Right**) Axial NECT shows shows focal calcifications and atrophy of the adrenal glands.*

ADRENAL HEMORRHAGE

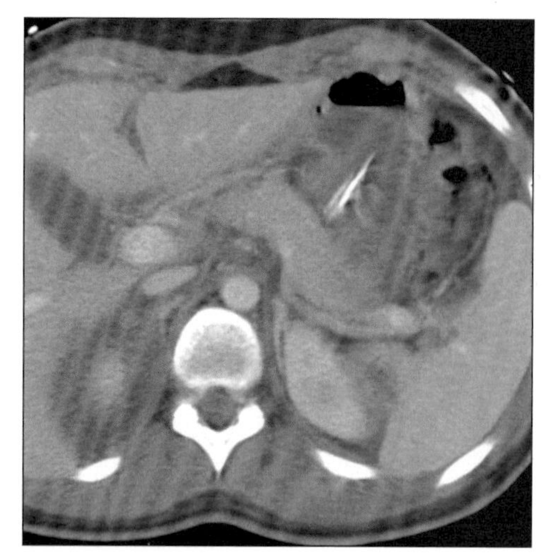

Axial CECT shows hemorrhage around right adrenal following motor vehicle crash.

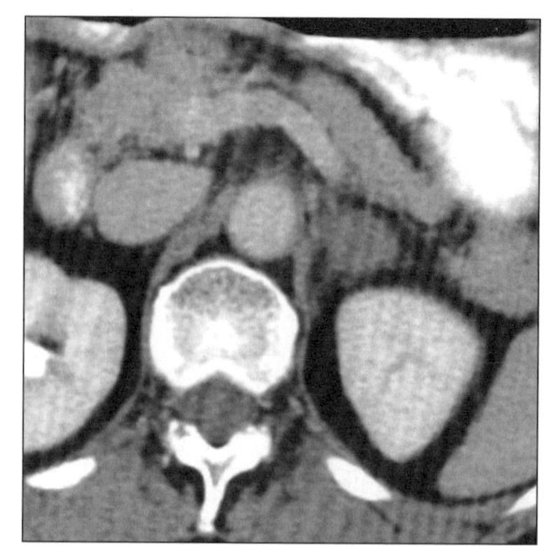

Axial CECT shows nonspecific left adrenal mass; hematoma following motor vehicle crash.

TERMINOLOGY

Abbreviations and Synonyms

- Adrenal hemorrhage (AH)

Definitions

- Hemorrhage within adrenal gland or tumor

IMAGING FINDINGS

General Features

- Best diagnostic clue: Hyperdense lesion within adrenal gland on NECT
- Key concepts
  - Relatively uncommon condition but potentially catastrophic event seen in patients of all ages
    - More common in neonates than children & adults
  - Secondary to traumatic & nontraumatic causes
    - Traumatic more common than nontraumatic
    - May be unilateral or bilateral
  - Traumatic hemorrhage: Blunt abdominal trauma
    - Unilateral in 80% of cases: Right (85%), left (15%)
    - Bilateral in 20% of cases
  - Nontraumatic hemorrhage (often bilateral): Causes are classified into 5 categories
    - Stress, hemorrhagic diathesis or coagulopathy
    - Neonatal stress, adrenal tumors, idiopathic
  - Bilateral AH in 15% of individuals who die of shock
  - Manifest with adrenal insufficiency when 90% of adrenal tissue is destroyed
  - Neonatal adrenal hemorrhage
    - Most common cause of adrenal mass in infancy
    - Usually seen during first week of life
    - Incidence ranges from 1.7-3% per 1000 births
    - Gland, hypervascular & weighs twice of adults

CT Findings

- Unilateral or bilateral adrenal hematomas
- ± Associated adrenal or renal vein thrombosis
- Acute or subacute hematoma
  - Round or oval mass of high attenuation (50-90 HU)
  - Asymmetric enlargement of adrenal glands
  - Homogeneous & no enhancement with contrast
  - Distortion of normal adrenal gland shape
  - Inflammatory stranding of periadrenal fat
  - Thickening of adjacent diaphragmatic crura
  - ± Periadrenal hemorrhage + perinephric extension
  - ± Upper abdominal traumatic findings
    - Pneumothorax, hydropneumothorax, rib fracture
    - Contusion of lung, liver, spleen or pancreas
- Chronic hematoma
  - Mass with a hypoattenuating center (pseudocyst)
  - Lack of enhancement confirms cystic nature of mass
  - Calcifications (usually seen after 1 year in adults)
    - Neonates: Seen within 1-2 weeks after trauma
- Hematomas ↓ in size, attenuation over a period of time

DDx: Adrenal Mass

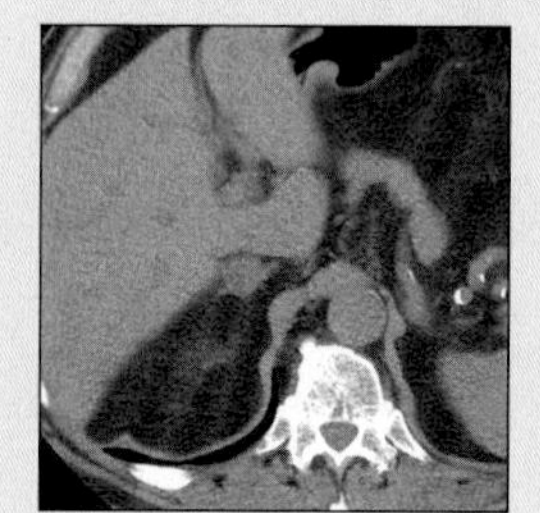

Bilateral Adenomas

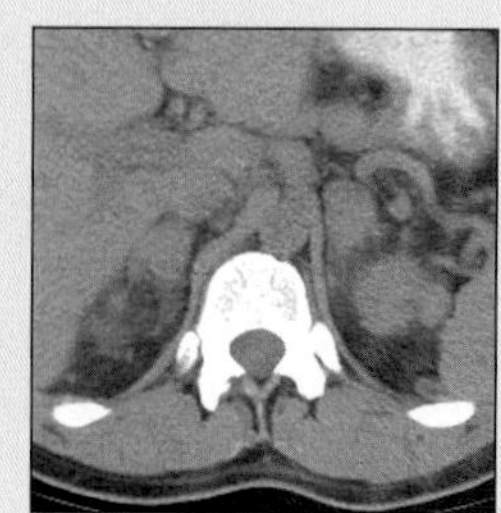

Bilateral Lymphoma

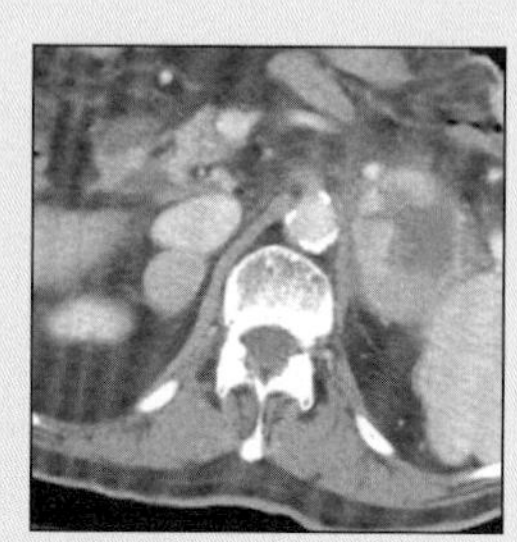

Bilateral Metastases

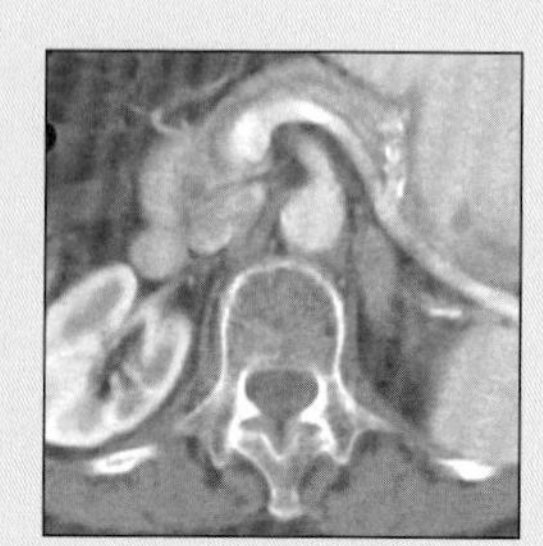

Hyperplasia

ADRENAL HEMORRHAGE

Key Facts

Imaging Findings

- Best diagnostic clue: Hyperdense lesion within adrenal gland on NECT
- ± Associated adrenal or renal vein thrombosis
- Round or oval mass of high attenuation (50-90 HU)
- Homogeneous & no enhancement with contrast
- Distortion of normal adrenal gland shape
- ± Underlying large adrenal mass (cyst, myelolipoma)
- T1 & T2WI: Varied signal based on age of hematoma

Top Differential Diagnoses

- Adrenal adenoma
- Adrenal metastases & lymphoma
- Adrenal hyperplasia
- Adrenal infection
- Adrenal carcinoma

Pathology

- Anticoagulation therapy (most common)
- Antiphospholipid antibody syndrome & disseminated intravascular coagulopathy
- Stress: Surgery, sepsis, burns, hypotension, steroids
- Metastases: Lung cancer & malignant melanoma
- Blunt abdominal trauma (Rt gland > Lt gland)
- Adrenal vein thrombosis; adrenal tumor
- Meningococcal septicemia (Waterhouse-Friderichsen syndrome)
- Stress or adrenal tumor → ↑ ACTH → ↑ arterial blood flow + limited venous drainage → hemorrhage

Diagnostic Checklist

- Check for history of trauma, anticoagulant therapy, coagulopathies, malignancies, stress, adrenal tumor

- ± Underlying large adrenal mass (cyst, myelolipoma)
  - Intracystic or intratumoral hemorrhage
    - Adrenal cyst: Mass of water density
    - Adrenal myelolipoma: Heterogeneous fatty mass

MR Findings

- T1 & T2WI: Varied signal based on age of hematoma
- Acute hematoma (less than 7 days after onset)
  - T1WI: Isointense or slightly hypointense
  - T2WI: Markedly hypointense
  - Due to high concentration of intracellular deoxyhemoglobin → T2 proton relaxation
- Subacute hematoma (7 days to 7 weeks after onset)
  - T1WI: Hyperintense
    - Due to free methemoglobin (Fe^{3+}), produced by oxidation of hemoglobin (Fe^{2+}) as hematoma ages
  - T2WI: Markedly hyperintense
    - Due to serum & clot lysis products
  - Large hematoma: Varied signal (irregular clot lysis)
    - Multilocular, fluid-fluid levels
- Chronic hematoma (beyond 7 weeks after onset)
  - T1 & T2WI: Hyperintense hematoma
    - Due to persistence of free methemoglobin
  - T1WI & T2WI: Hypointense rim
    - Due to hemosiderin deposition in fibrous capsule
- Adrenal or renal vein thrombosis
  - Clot: ↑ Signal on both T1 & T2WI
- Extension of thrombus into IVC can be seen by MR
- Underlying large adrenal mass (cyst or myelolipoma)
  - Varied signal based on content of mass lesion
- Gradient-echo imaging
  - Demonstrate "blooming" effect (magnetic susceptibility) due to hemosiderin deposition
    - Detect blood & monitor hemorrhage as it progresses from methemoglobin to hemosiderin
    - Useful in large hematomas which exhibit slower clot evolution

Ultrasonographic Findings

- Real Time
  - Acute hematoma: Hyperechoic mass-like lesion
  - Subacute hematoma
    - Mixed echogenicity + central hypoechoic area
  - Chronic hematoma
    - Anechoic & cyst-like lesion; wall calcification
- Color Doppler: Shows avascular nature of mass

Angiographic Findings

- Conventional
  - Usually not recommended in adrenal hemorrhage
  - Adrenal hemorrhage & pseudocyst: Avascular
  - Adrenal mass: Neovascularity seen

Nuclear Medicine Findings

- Tc99m dimercaptosuccinic acid study (DMSA)
  - Adrenal hematoma: Photopenic suprarenal mass with inferior displacement of kidney

Imaging Recommendations

- Best imaging tool: Helical CT & MR
- Protocol advice
  - CT: 3 mm thick section at 3 mm intervals or less
  - MR: Spin-echo & gradient-echo imaging

DIFFERENTIAL DIAGNOSIS

Adrenal adenoma

- NECT: Lipid rich adenoma (< 10 HU)
- CECT: Washout of adenoma 15 min post I.V. > 40% washout
- T1W out-of-phase
  - Marked signal "drop-out" (lipid rich adenoma)

Adrenal metastases & lymphoma

- Adrenal metastases
  - Lung cancer: Hemorrhagic; enhancing adrenal mass
  - Malignant melanoma: Hypervascular metastases
- Adrenal lymphoma
  - Primary (rare); secondary (non-Hodgkin, common)
  - Often bilateral; retroperitoneal disease usually seen
  - Discrete or diffuse mass, shape is maintained

Adrenal hyperplasia

- Adrenal glands are often symmetrically enlarged
- Width of adrenal gland limbs > 10 mm (diagnostic)
- No discrete mass or nodule seen as a rule

Adrenal infection
- Granulomatous infection: Tuberculosis, histoplasmosis
- Adrenal calcification seen on CT imaging

Adrenal carcinoma
- Rare, unilateral, invasive & enhancing mass
- More than 6 cm when initially diagnosed

PATHOLOGY

General Features
- Etiology
  - Bilateral adrenal hemorrhage
    - Anticoagulation therapy (most common)
    - Antiphospholipid antibody syndrome & disseminated intravascular coagulopathy
    - Stress: Surgery, sepsis, burns, hypotension, steroids
    - Pheochromocytoma; rarely trauma
    - Metastases: Lung cancer & malignant melanoma
  - Unilateral adrenal hemorrhage
    - Blunt abdominal trauma (Rt gland > Lt gland)
    - Adrenal vein thrombosis; adrenal tumor
  - Neonates
    - Difficult labor or delivery; renal vein thrombosis
    - Asphyxia or hypoxia; hemorrhagic disorders
    - Meningococcal septicemia (Waterhouse-Friderichsen syndrome)
  - Pathogenesis (non-traumatic)
    - Stress or adrenal tumor → ↑ ACTH → ↑ arterial blood flow + limited venous drainage → hemorrhage
    - Stress or tumor → ↑ catecholamines → adrenal vein spasm → stasis → thrombosis → hemorrhage
    - Coagulopathies → ↑ venous stasis → thrombosis → hemorrhage
- Epidemiology
  - Autopsy studies: 0.3-1.8% of unselected cases
  - 15% of individuals who die of shock
  - 2% of orthotopic liver transplantation cases
- Associated abnormalities
  - Adrenal or renal vein thrombosis
  - Adrenal tumor, hemorrhagic disorders

Gross Pathologic & Surgical Features
- Hematoma, enlarged gland, periadrenal stranding

Microscopic Features
- Necrosis of all 3 cortical layers + medullary cells

CLINICAL ISSUES

Presentation
- Most common signs/symptoms
  - Nonspecific: Abdominal, lumbar, thoracic pain
  - Fever, tachycardia, hypotension
  - Acute adrenal insufficiency
    - Fatigue, anorexia, nausea & vomiting
  - Acute abdomen
    - Guarding, rigidity, rebound tenderness
  - Confusion, disorientation, shock in late phase
  - Symptoms of associated underlying conditions
  - Rarely, asymptomatic; incidental finding (imaging)
  - Waterhouse-Friderichsen syndrome: Skin rash, cough, headache, dizziness, arthralgias & myalgias
- Lab-data
  - ↓ Hematocrit or hemoglobin; ↑ WBC
  - Hyponatremia, hyperkalemia & prerenal azotemia
  - ↓ Serum cortisol, aldosterone, androgens & ↑ ACTH

Demographics
- Age
  - Any age group
    - More common in neonates than children & adults
  - Nontraumatic (40-80 years); traumatic (20-30 years)
- Gender: M:F=2:1

Natural History & Prognosis
- Complications
  - Prerenal azotemia, adrenal abscess, shock
- Prognosis
  - Prognosis depends on etiology rather than extent of adrenal hemorrhage
  - Unilateral adrenal hemorrhage (e.g., from blunt trauma or liver transplantation)
    - Rarely of clinical concern, resolve on its own & adrenal functions normally
  - Overall, AH is associated with a 15% mortality rate
    - Waterhouse-Friderichsen syndrome: 55-60%
  - Adults-adrenal crisis; neonate- death (> blood loss)

Treatment
- Medical
  - Correct fluid, electrolytes & treat underlying cause
- Surgical: Adrenalectomy (open or laparoscopic)
  - Surgery not required, except in adrenal tumors

DIAGNOSTIC CHECKLIST

Consider
- Check for history of trauma, anticoagulant therapy, coagulopathies, malignancies, stress, adrenal tumor

Image Interpretation Pearls
- NECT: Hyperdense lesion within adrenal gland
- MR: Varied signal intensity based on age of hematoma

SELECTED REFERENCES

1. Dunnick NR et al: Imaging of adrenal incidentalomas: Current status. AJR. 179:559-68, 2002
2. Caoili EM et al: Adrenal masses: characterization with combined unenhanced and delayed enhanced CT. Radiology. 222(3):629-33, 2002
3. Vella A et al: Adrenal hemorrhage: a 25-year experience at the Mayo Clinic. Mayo Clin Proc. 76(2):161-8, 2001
4. Mayo-Smith WW et al: State-of-the-art adrenal imaging. Radiographics. 21(4):995-1012, 2001
5. Khati NJ et al: Adrenal adenoma and hematoma mimicking a collision tumor at MR imaging. Radiographics. 19(1):235-9, 1999
6. Kawashima A et al: Imaging of nontraumatic hemorrhage of the adrenal gland. Radiographics. 19(4):949-63, 1999
7. Krebs TL et al: MR imaging of the adrenal gland: radiologic-pathologic correlation. Radiographics. 18(6):1425-40, 1998

ADRENAL HEMORRHAGE

IMAGE GALLERY

Typical

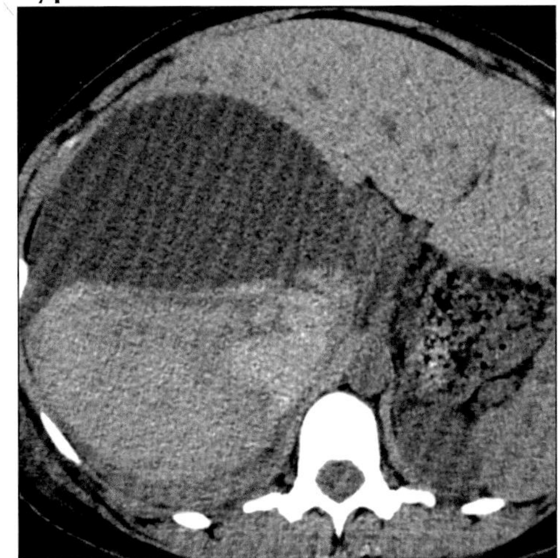

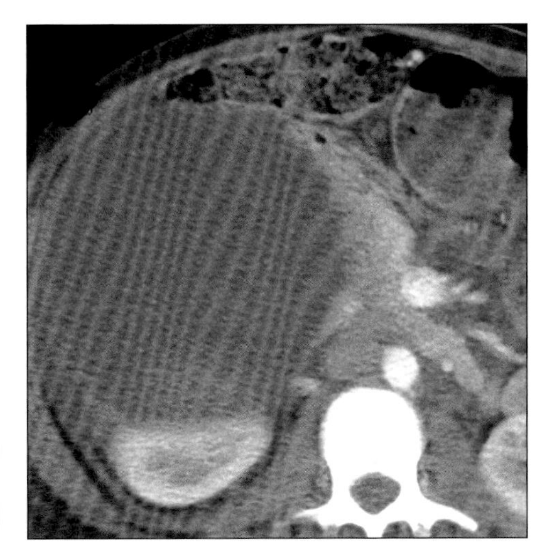

*(**Left**) Axial NECT shows high density hemorrhage in the dependent part of a large cystic mass; proven adrenal hemorrhage in a post-partum 31 year old woman. (**Right**) Axial CECT shows heterogeneous right adrenal mass; postpartum hemorrhage.*

Typical

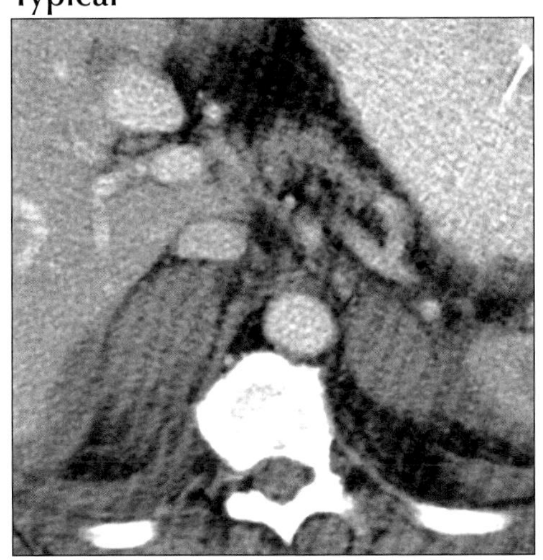

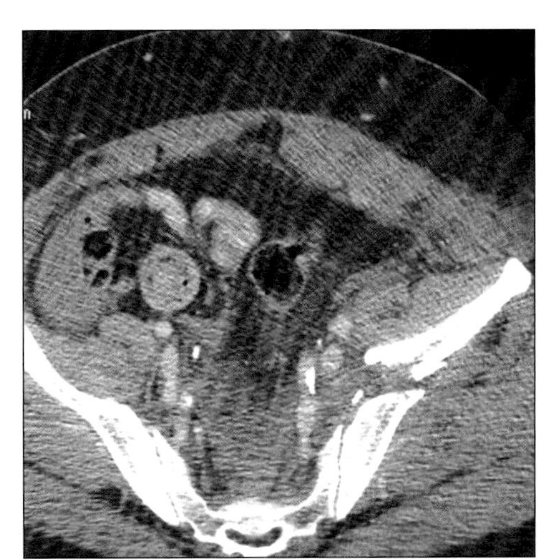

*(**Left**) Axial CECT shows bilateral adrenal hemorrhage due to shock from pelvic fractures. (**Right**) Axial CECT shows pelvic fractures in a patient with bilateral adrenal hemorrhage.*

Typical

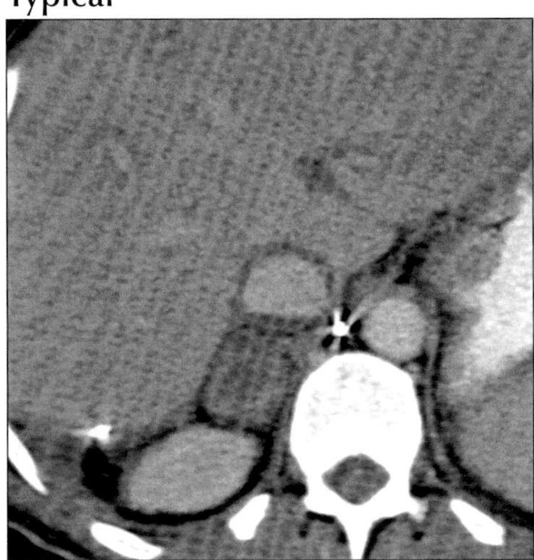

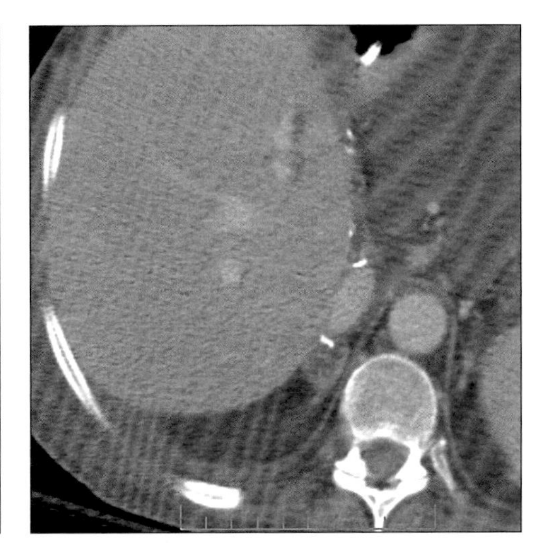

*(**Left**) Axial CECT shows a right adrenal hematoma due to trauma from recent liver transplantation. (**Right**) Axial CECT shows a right adrenal hematoma due to recent liver transplantation.*

ADRENAL HYPERPLASIA

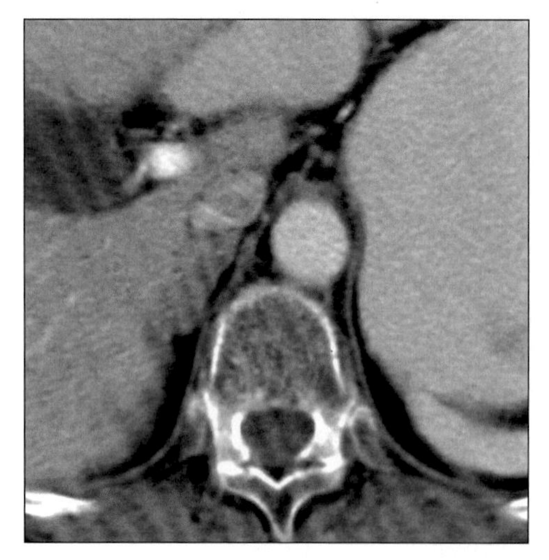

Axial CECT shows enlargement of both adrenal glands but preservation of normal shape.

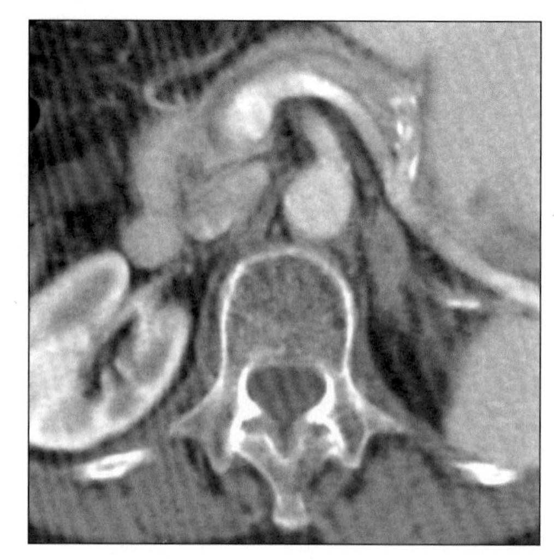

Axial CECT shows enlargement of both adrenal glands but preservation of normal shape.

TERMINOLOGY

Definitions

- Bilateral adrenal gland hyperfunction ± enlargement

IMAGING FINDINGS

General Features

- Best diagnostic clue: Enlarged limbs of one or both adrenal glands > 10 mm width on CT
- Location: Suprarenal
- Key concepts
  - Width of adrenal gland limbs > 10 mm is consistent with hyperplasia
  - Glands may appear normal size or multinodular
  - Adrenal gland shape is maintained
  - Clinical syndromes caused by adrenal hyperplasia
    - Cushing syndrome (hypercortisolism)
    - Conn syndrome (hyperaldosteronism)
  - Bilateral adrenocortical hyperplasia
    - ACTH dependent Cushing syndrome
    - Conn syndrome (primary hyperaldosteronism)
    - PPNAH or PPNAD: Primary pigmented nodular adrenocortical hyperplasia or dysplasia
    - AIMAH: ACTH-independent macronodular adrenocortical hyperplasia
    - Congenital adrenal hyperplasia (CAH)
  - Cushing syndrome (hypercortisolism)
    - Adrenocorticotrophic hormone (ACTH) dependent (80-85%)
    - Adrenocorticotrophic hormone (ACTH) independent (15-20%)
  - ACTH dependent Cushing syndrome: Bilateral adrenal hyperplasia in 80-85% of cases
    - ACTH-secreting anterior pituitary adenoma: Cushing disease (75-85% of cases)
    - Ectopic ACTH-secreting tumors (15%)
    - Hypothalamic tumor → ↑ corticotropin releasing factor → ↑ pituitary ACTH (rare)
  - ACTH independent Cushing syndrome (15-20%)
    - Usually due to adrenal adenoma > carcinoma
    - Rarely caused by PPNAH or PPNAD & AIMAH
  - Conn syndrome (primary hyperaldosteronism)
    - 20% of cases are due to adrenal gland hyperplasia
    - 80% of cases are due to adrenal adenoma
  - Congenital adrenal hyperplasia (CAH)
    - Autosomal recessive; due to enzyme deficiencies
    - Decreased secretion of cortisol, aldosterone or both with compensatory increase in ACTH
    - Cause most cases of adrenogenital syndrome
    - Usually present in children, rarely in adults
    - Associated with testicular & ovarian tumors that arise from ectopic adrenal cortical rests
  - Bilateral adrenal medulla hyperplasia
    - Nodular or diffuse

DDx: Bilateral Adrenal Enlargement

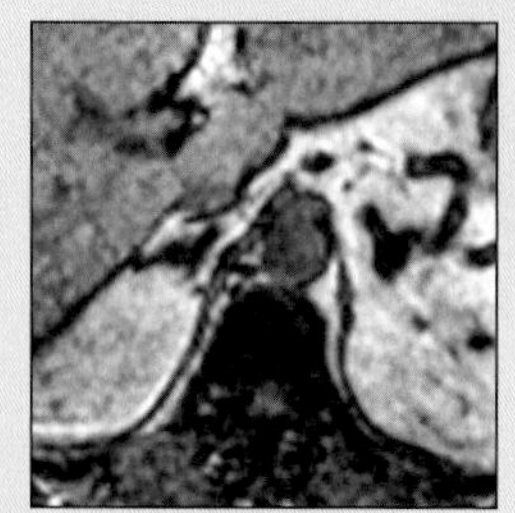

Adenomas

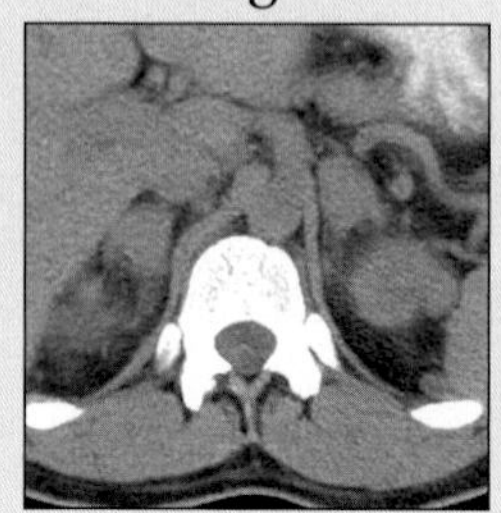

Lymphoma

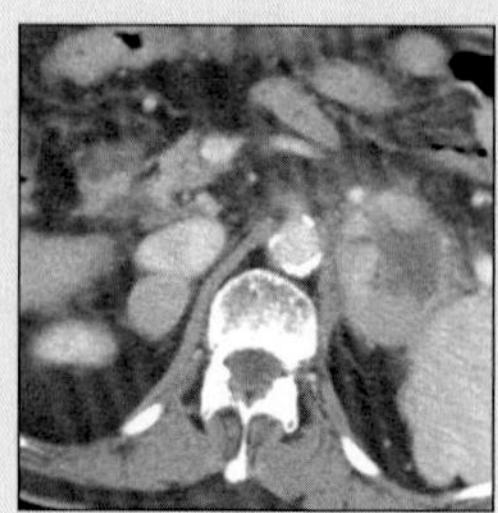

Metastases

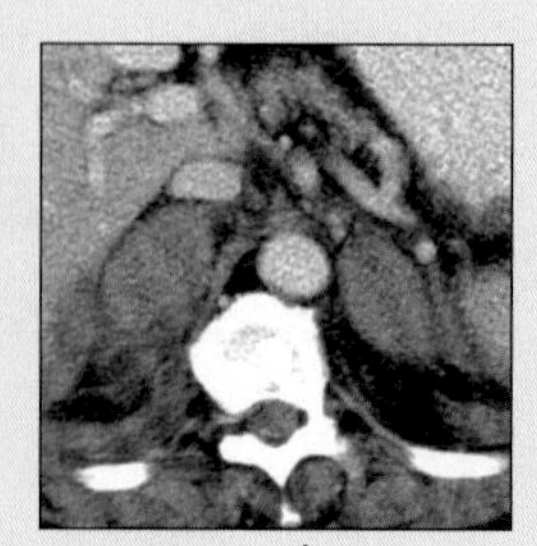

Hemorrhage

ADRENAL HYPERPLASIA

Key Facts

Imaging Findings
- Best diagnostic clue: Enlarged limbs of one or both adrenal glands > 10 mm width on CT
- Adrenal glands are often symmetrically enlarged
- Usually both adrenal glands involved
- No discrete mass or nodule seen as a rule
- Cushing syndrome can show nodular hyperplasia
- Up to 30% of cases may show normal glands

Top Differential Diagnoses
- Adrenal adenoma
- Adrenal metastases & lymphoma
- Adrenal hemorrhage
- Pheochromocytoma

Pathology
- Congenital (due to enzyme deficiencies)
- Cushing disease: Pituitary adenoma → ↑ ACTH → adrenal gland hyperplasia
- Cushing syndrome: Adrenal adenoma/carcinoma, iatrogenic, ectopic ACTH production
- Conn syndrome (primary hyperaldosteronism)

Clinical Issues
- Moon facies, truncal obesity, buffalo hump
- Hypertension & hypokalemia
- Lab data: ↑ Levels of ACTH, cortisol, aldosterone
- Diagnosis: Clinical, biochemical, imaging, histology

Diagnostic Checklist
- Correlate clinical, biochemical & imaging findings
- Enlarged limbs of adrenal glands (> 10 mm) on CT
- Multinodular hyperplasia difficult to distinguish from adenoma by CT

    - May cause hypertensive symptoms similar to pheochromocytoma
    - Associated with multiple endocrine neoplasia (MEN 2b/3), duodenal carcinoid

CT Findings
- ACTH dependent Cushing syndrome (80-85% of cases)
  - Adrenal glands are often symmetrically enlarged
  - Limbs of adrenal gland > 10 mm makes this diagnosis
  - Usually both adrenal glands involved
  - No discrete mass or nodule seen as a rule
  - Cushing syndrome can show nodular hyperplasia
  - Up to 30% of cases may show normal glands
- Conn syndrome (primary hyperaldosteronism)
  - Findings similar to Cushing syndrome
  - Adrenal glands may show nodular hyperplasia
- PPNAH or PPNAD
  - Normal sized adrenal glands
  - Small discrete nodules between 2-5 mm in size
  - Moderate enhancement of nodular glands
- AIMAH
  - Massively enlarged bilateral adrenal glands
  - Normal adrenal shape is retained
  - Adrenal limb width & nodule size 30 mm
  - Marked enhancement of hyperplastic nodular glands, predominantly at periphery

MR Findings
- Adrenal hyperplasia due to Cushing, Conn syndromes
  - T1WI define both adrenals within retroperitoneal fat
  - Increase in width of adrenal gland limbs > 10 mm
  - Adds no new information compared to CT findings
- T1WI
  - PPNAH or PPNAD
    - Hyperintense relative to muscle & spleen
    - Isointense relative to liver
  - AIMAH
    - Isointense relative to spleen & hypointense relative to liver
  - ACTH-dependent (hyperplasia) Cushing syndrome
    - Out of phase sequence: 35-40% signal dropout within glands indicating intracellular lipid
- T2WI
  - PPNAH or PPNAD
    - Hyperintense relative to muscle & liver
    - Isointense relative to spleen
  - AIMAH
    - Hyperintense relative to both liver & spleen
- T1 C+
  - PPNAH or PPNAD: Moderate nodular enhancement
  - AIMAH: Marked homogeneous enhancement of hyperplastic nodular glands

Nuclear Medicine Findings
- Adrenocortical scintigraphy
  - NP-59 is a cholesterol analog that binds to low density lipoprotein receptors of adrenal cortex
  - Normal NP-59: When both adrenal glands are seen on day 5 after injection or thereafter
  - Adrenal gland hyperplasia: Bilateral early adrenal visualization before day 5 after injection
  - Adenoma (adrenal): Unilateral early adrenal visualization before day 5 after injection

Imaging Recommendations
- Abdominal CT best imaging tool for this diagnosis
- Contiguous ≤ 3 mm CT images best technique

DIFFERENTIAL DIAGNOSIS

Adrenal adenoma
- Intracellular lipid makes CT density -20 to +10 HU
- Focal mass, not diffuse enlargement
- Over 40% washout of contrast on a 15 minute delayed CT scan is diagnostic of an adenoma

Adrenal metastases & lymphoma
- Adrenal metastases
  - Invasive, enhancing mass in adrenal gland
  - Focal mass, loss of adrenal shape
- Adrenal lymphoma
  - Usually spread to adrenal gland of retroperitoneal
  - Bilateral primary lymphoma (non-Hodgkin) can simulate hyperplasia; usually more mass-like
  - Hypovascular; moderate enhancement with contrast

ADRENAL HYPERPLASIA

Adrenal hemorrhage

- Adults
  - Septicemia, burns, trauma, severe stress or hypotension
  - Excessive anticoagulation, thrombocytopenia
  - Disseminated intravascular coagulation, antiphospholipid antibody syndrome
- Neonates: Meningococci, pneumococci & gonococci
- Adrenal hematomas usually appear round in shape
- Some poorly marginated & show periadrenal stranding
- CT
  - Acute: High density fluid collection (40-60 HU)
  - Chronic: Low density (clot of 20-30 HU)

Pheochromocytoma

- Tumor of adrenal medulla & usually more than 3 cm
- Highly vascular; prone to hemorrhage & necrosis
- Tumors are very hyperintense on T2WI
- Bilateral adrenal tumors in MEN IIA & IIB syndromes

PATHOLOGY

General Features

- General path comments
  - Normal adrenal anatomy
    - Classically described as inverted Y, V or T-shaped
    - Width of limb (4-9 mm); gland weight (4-6 grams)
- Etiology
  - Congenital (due to enzyme deficiencies)
  - Cushing disease: Pituitary adenoma → ↑ ACTH → adrenal gland hyperplasia
  - Cushing syndrome: Adrenal adenoma/carcinoma, iatrogenic, ectopic ACTH production
  - Conn syndrome (primary hyperaldosteronism)
    - ↑ Aldosterone → ↑ preservation of sodium & loss of potassium at renal tubular level
    - ↑ Sodium → water retention → ↑ extracellular volume → hypertension → ↓ renin production
    - Hypokalemia causes hypokalemic alkalosis
- Associated abnormalities
  - Pituitary adenoma in ACTH-dependent hyperplasia
  - PPNAH associated with Carney complex
    - Cushing disease, spotty skin pigmentation
    - Cutaneous & cardiac myxomas
    - Sertoli cell tumors of testis in males
    - Multiple myxoid fibroadenomas of breast, females
  - Adrenal medulla hyperplasia: MEN 2b/3, duodenal carcinoid

Gross Pathologic & Surgical Features

- Bilateral diffusely enlarged adrenal glands
- Nodular, macronodular or pigmented nodular glands

Microscopic Features

- Thickened zona glomerulosa, fasciculata, reticularis
- Fasciculata may show excess lipid, lipid depleted or atypical cells with hyperchromatic nuclei in nodules

CLINICAL ISSUES

Presentation

- Most common signs/symptoms
  - Asymptomatic incidental CT finding
  - Cushing syndrome (hypercortisolism)
    - Moon facies, truncal obesity, buffalo hump
  - Conn syndrome (hyperaldosteronism)
    - Hypertension & hypokalemia
- Lab data: ↑ Levels of ACTH, cortisol, aldosterone
- Diagnosis: Clinical, biochemical, imaging, histology

Demographics

- Age
  - Adults (70-80%); children (15-20%)
    - Cushing syndrome: 25-40 years
    - Conn syndrome: 30-50 years
- Gender: Cushing syndrome (M:F = 1:5); Conn syndrome (M:F = 1:2)

Natural History & Prognosis

- Complications: Untreated adrenal crisis → death
- Prognosis: Usually good with treatment

Treatment

- Adrenal hyperplasia without symptoms: Follow-up
- Adrenal hyperplasia with symptoms: Surgical resection of pituitary adenoma or source of hormone

DIAGNOSTIC CHECKLIST

Consider

- Correlate clinical, biochemical & imaging findings

Image Interpretation Pearls

- Enlarged limbs of adrenal glands (> 10 mm) on CT
- Multinodular hyperplasia difficult to distinguish from adenoma by CT

SELECTED REFERENCES

1. Rockall AG et al: CT and MR imaging of the adrenal glands in ACTH-independent cushing syndrome. Radiographics. 24(2):435-52, 2004
2. Imaki T et al: Adrenocortical hyperplasia associated with ACTH-dependent Cushing's syndrome: comparison of the size of adrenal glands with clinical and endocrinological data. Endocr J. 51(1):89-95, 2004
3. Caoili EM et al: Adrenal masses: characterization with combined unenhanced and delayed enhanced CT. Radiology. 222(3):629-33, 2002
4. Mayo-Smith WW et al: State-of-the-art adrenal imaging. Radiographics. 21(4):995-1012, 2001
5. Doppman JL et al: Adrenocorticotropin-independent macronodular adrenal hyperplasia: an uncommon cause of primary adrenal hypercortisolism. Radiology. 216(3):797-802, 2000
6. Sohaib SA et al: CT appearance of the adrenal glands in adrenocorticotrophic hormone-dependent Cushing's syndrome. AJR Am J Roentgenol. 172(4):997-1002, 1999
7. Kawashima A et al: Spectrum of CT findings in nonmalignant disease of the adrenal gland. Radiographics. 18(2):393-412, 1998
8. Krebs TL et al: MR imaging of the adrenal gland: radiologic-pathologic correlation. Radiographics. 18(6):1425-40, 1998
9. Doppman JL et al: Cushing syndrome due to primary pigmented nodular adrenocortical disease: findings at CT and MR imaging. Radiology. 172(2):415-20, 1989

IMAGE GALLERY

Typical

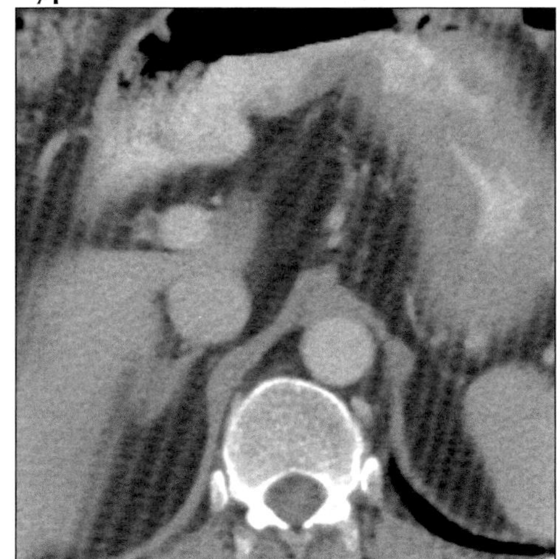

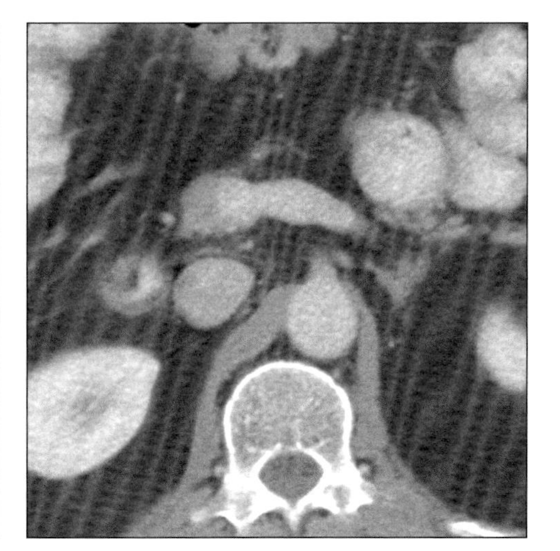

(Left) *Axial CECT shows multiple small nodules in both adrenal glands, one manifestation of adrenal hyperplasia.* ***(Right)*** *Axial CECT shows multiple small nodules in both adrenal glands, one manifestation of adrenal hyperplasia.*

Typical

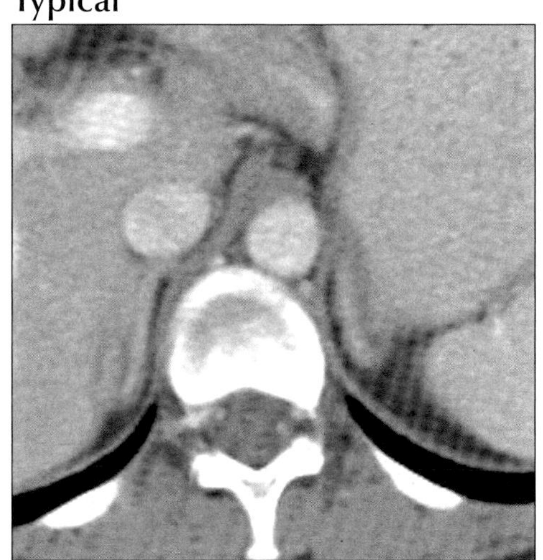

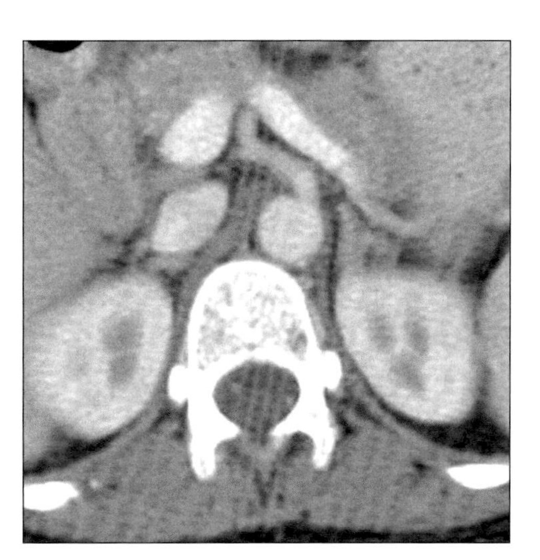

(Left) *Axial CECT shows prominent adrenal glands without obvious hypertrophy or nodularity.* ***(Right)*** *Axial CECT shows prominent adrenal glands without obvious hypertrophy or nodularity.*

Typical

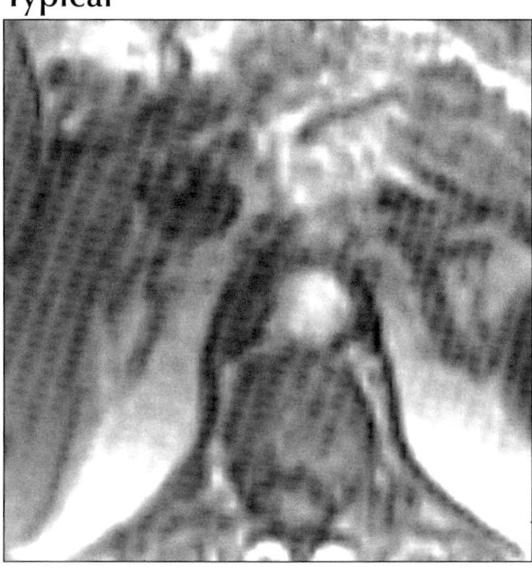

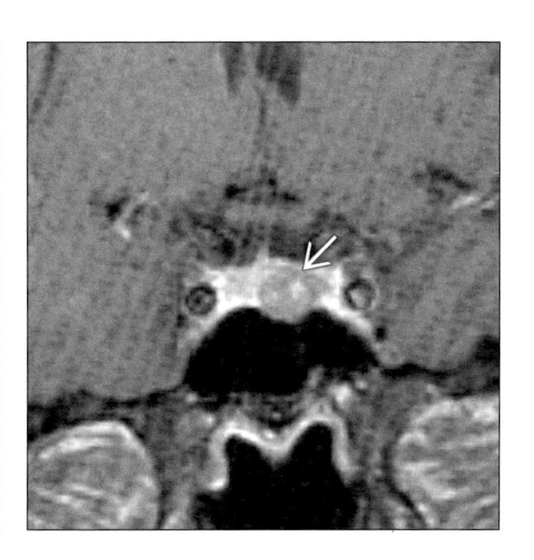

(Left) *Axial T2WI MR shows prominent adrenal glands.* ***(Right)*** *Coronal T2WI MR shows microadenoma (arrow) in pituitary; one potential manifestation of excess ACTH production and adrenal hyperplasia.*

ADRENAL INSUFFICIENCY

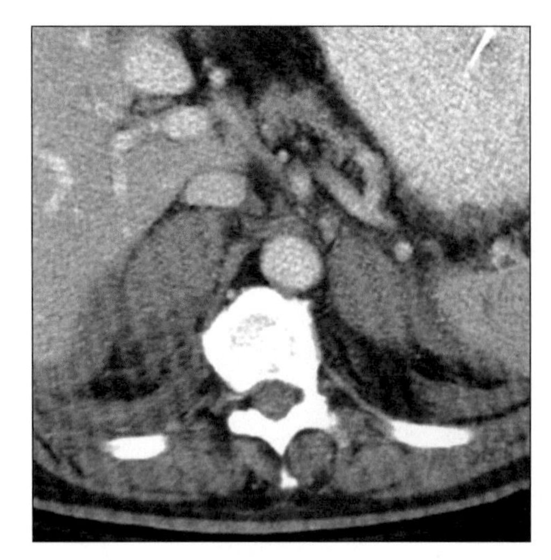

Axial CECT shows bilateral adrenal hemorrhage following shock due to motor vehicle crash + pelvic fractures.

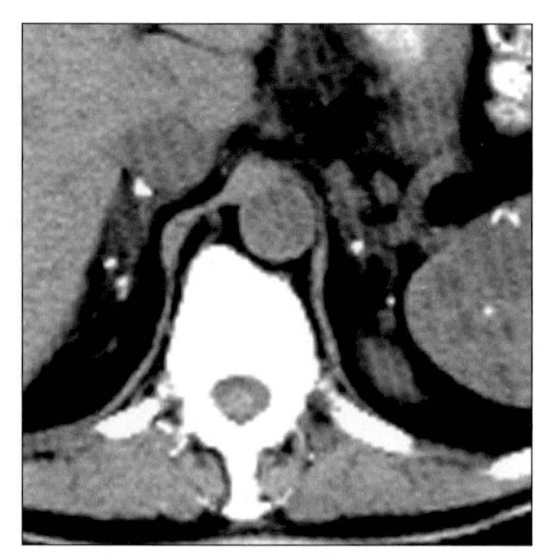

Axial NECT shows splenic and adrenal calcifications indicating prior fungal or tuberculous infection.

TERMINOLOGY

Abbreviations and Synonyms

- Primary adrenal insufficiency, Addison disease

Definitions

- Inadequate secretion of corticosteroids resulting from partial or complete destruction of adrenal glands

IMAGING FINDINGS

General Features

- Location: Bilateral
- Other general features
  - Imaging findings depend on course
    - Acute, subacute (< 2 years), chronic

Radiographic Findings

- Radiography: Bilateral calcifications above the kidneys

CT Findings

- Acute: Addisonian crisis, adrenal apoplexy
  - Enlarged adrenal glands
  - Heterogenous enhancement (hemorrhage)
- Subacute
  - Enlarged glands with normal adrenal contours
  - Reduced density due to caseation or necrosis
- Chronic
  - Small, atrophic glands (autoimmune)
  - Dense, chunky calcifications (infection)

Imaging Recommendations

- Best imaging tool: CT

DIFFERENTIAL DIAGNOSIS

Adrenal metastases and lymphoma

- Metastases
  - Unilateral or bilateral, invasive, enhancing masses
  - Central necrosis ± hemorrhage
  - Focal masses without diffuse enlargement
  - May not cause adrenal insufficiency
- Lymphoma
  - Usually bilateral, triangular shaped masses
  - Rarely causes adrenal insufficiency

Adrenal hemorrhage

- Heterogenous enhancing mass (40-80 HU)

Adrenal infection

- Unilateral or bilateral enlarged adrenal glands

Adrenal adenoma

- May be bilateral

DDx: Adrenal Masses or Atrophy

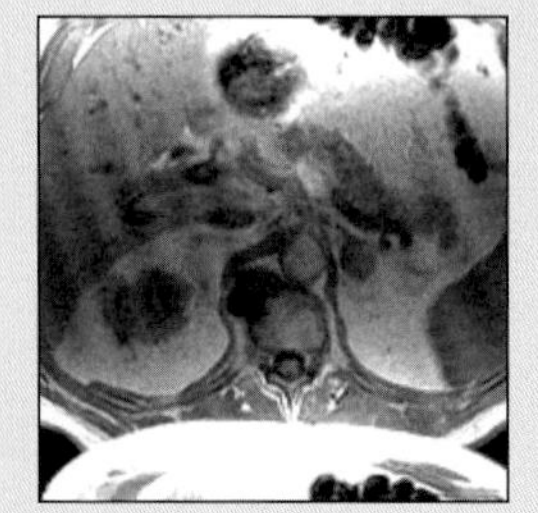

Adrenal Adenomas

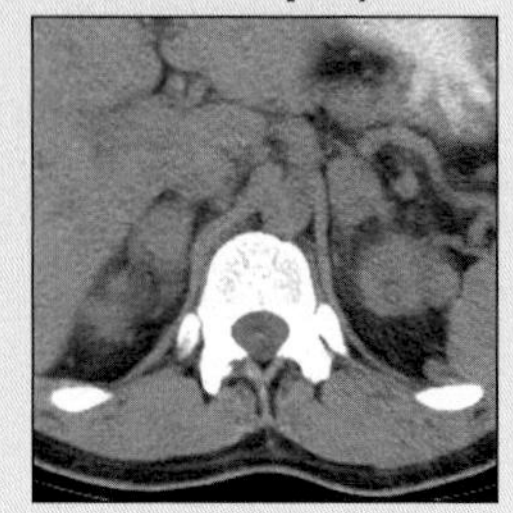

Lymphoma

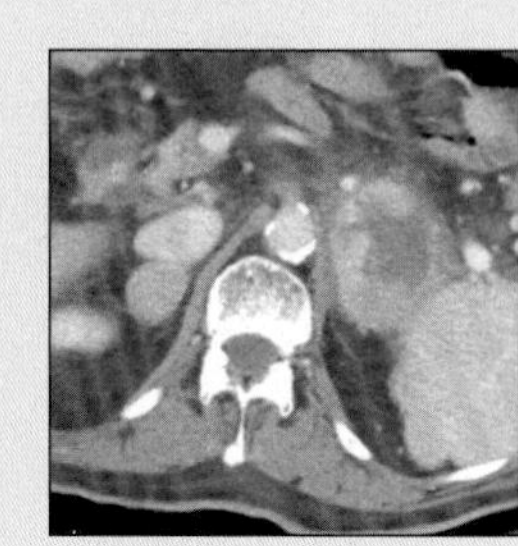

Metastases

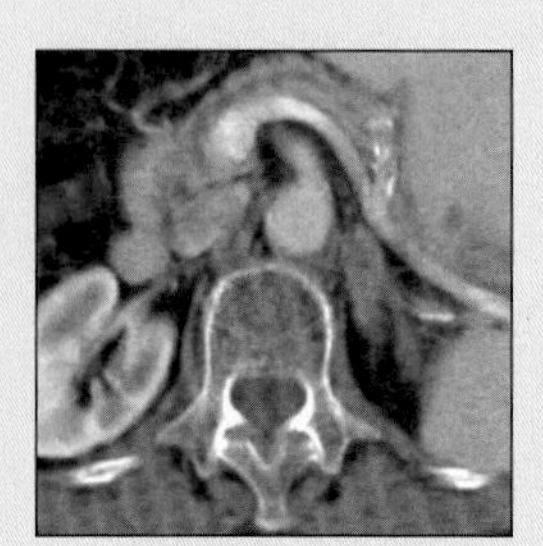

Hyperplasia

ADRENAL INSUFFICIENCY

Key Facts

Imaging Findings

- Location: Bilateral
- Imaging findings depend on course
- Heterogenous enhancement (hemorrhage)
- Enlarged glands with normal adrenal contours
- Small, atrophic glands (autoimmune)
- Dense, chunky calcifications (infection)
- Best imaging tool: CT

Top Differential Diagnoses

- Adrenal metastases and lymphoma
- Adrenal hemorrhage
- Adrenal infection

Diagnostic Checklist

- Aggressive diagnosis & treatment to avoid complications

PATHOLOGY

General Features

- Etiology
  - Idiopathic autoimmune disorders (80% of cases)
  - Granulomatous diseases: Tuberculosis (most common cause in underdeveloped nations), sarcoidosis
  - Systemic fungal infections: Histoplasmosis (most common infection in southeastern, south central U.S.), cryptococcosis, blastomycosis
  - Adrenal hemorrhage, necrosis or thrombosis: Stress after surgery, shock, anticoagulation (e.g., warfarin), sepsis (e.g., Waterhouse-Friderichsen syndrome), coagulation disorders, antiphospholipid syndrome
  - Neoplasms: Metastases (e.g., lung, ovary, breast, kidney), lymphoma, leukemia
  - Acquired immune deficiency syndrome (AIDS): Opportunistic infections (e.g., cytomegalovirus), neoplasms (e.g., Kaposi sarcoma)
  - Other causes: Adrenomyeloneuropathy, familial glucocorticoid deficiency, amyloidosis
- Epidemiology: 50 per 1,000,000 persons

Gross Pathologic & Surgical Features

- Enlarged glands with hemorrhagic and necrotic tissues
- Small, atrophic glands or dense, chunky calcifications

Microscopic Features

- Hemorrhagic destruction; infectious organisms
- Lymphocytic, granulomatous or neoplastic infiltration

CLINICAL ISSUES

Presentation

- Most common signs/symptoms
  - Acute: Fever, abdominal or back pain, hypotension, weakness, nausea, vomiting, diarrhea
  - Chronic: Progressive lethargy, weakness, cutaneous pigmentation, weight loss
- Lab data
  - Chemistry: Hyponatremia, hyperkalemia, azotemia, hypercalcemia, hypoglycemia
  - Adrenocorticotrophic hormone (ACTH) stimulation test: Cortisol level fail to rise
  - 24-hour urine cortisol: Decreased

Demographics

- Gender: M:F = 1:1; autoimmune: 1:2-3

Natural History & Prognosis

- Complications: Acute: Shock, death
- Prognosis: Good, if recognized and treated

Treatment

- Acute
  - Glucocorticoid therapy
  - Volume and electrolytes replacement
  - Correct any etiology (e.g., antibiotics for infections)
- Chronic
  - Glucocorticoid + mineralocortoid replacement

DIAGNOSTIC CHECKLIST

Consider

- Aggressive diagnosis & treatment to avoid complications

Image Interpretation Pearls

- Can not diagnose adrenal hyper- or hypofunction by imaging alone

SELECTED REFERENCES

1. Kawashima A et al: Imaging of nontraumatic hemorrhage of the adrenal gland. Radiographics. 19(4):949-63, 1999
2. Kawashima A et al: Spectrum of CT findings in nonmalignant disease of the adrenal gland. Radiographics. 18(2):393-412, 1998
3. Doppman JL et al: CT findings in Addison's disease. J Comput Assist Tomogr. 6(4):757-61, 1982

IMAGE GALLERY

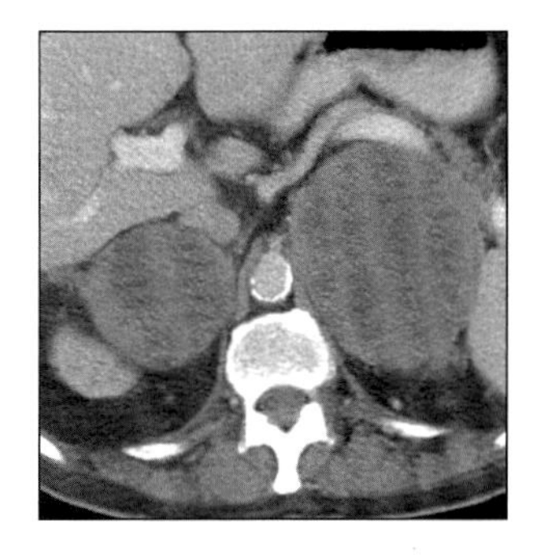

(Left) *Axial CECT shows massive adrenal metastases from lung cancer that resulted in adrenal insufficiency.* ***(Right)*** *Axial CECT shows massive adrenal metastases from malignant melanoma.*

ADRENAL CYST

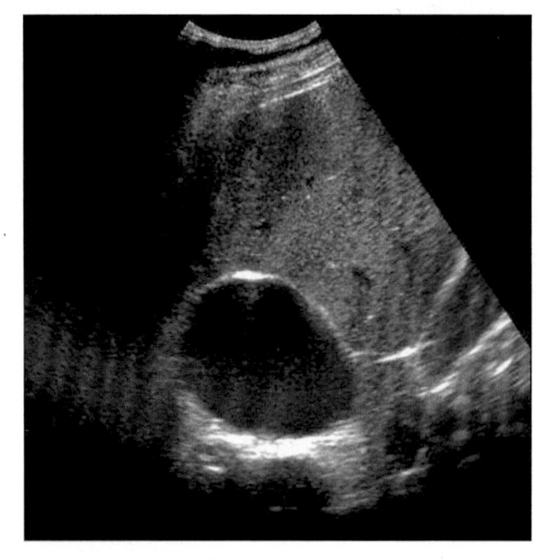

Sagittal sonogram shows sonolucent mass above right kidney; adrenal cyst.

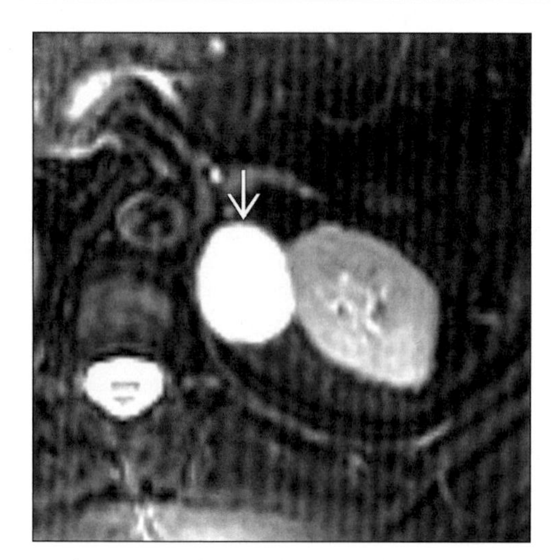

Axial T2WI MR shows high intensity left adrenal cyst (arrow).

TERMINOLOGY

Definitions

- A cystic mass arising in adrenal gland

IMAGING FINDINGS

General Features

- Best diagnostic clue: Well-defined nonenhancing water-density adrenal mass ± calcifications
- Location
  - Suprarenal
  - Right as common as left
  - Unilateral > bilateral (8-10% of cases)
- Size: < 5 cm (50%), up to 20 cm

CT Findings

- NECT
  - Unilocular or multilocular mass
  - Well-defined, round to oval, homogeneous mass with water (0 HU) or near water density
  - Higher or mixed attenuation mass (hemorrhage, intracystic debris, crystals)
  - ↑ Wall thickness, up to 3 mm
  - Calcifications
    - Rim-like or nodular (51-69%)
    - Centrally in intracystic septation (19%)
    - Punctate within intracystic hemorrhage (5%)
- CECT: No central enhancement ± wall enhancement

MR Findings

- T1WI
  - Homogeneous, hypointense mass
  - Hyperintense mass (hemorrhage)
- T2WI: Hyperintense mass

Imaging Recommendations

- Best imaging tool: CT or MR

DIFFERENTIAL DIAGNOSIS

Adrenal adenoma

- NECT: Homogeneous, well-defined, ovoid mass with ≤ 30 HU density
- CECT: Enhancing mass without visible wall or peripheral calcifications

Gastric diverticulum

- Abnormal rounded soft tissue shadow, often lies in suprarenal location
- Air-filled, fluid-filled or contrast-filled mass with no enhancement of contents

Adrenal myelolipoma

- Usually has foci of fat density

DDx: "Cystic"- Appearing Suprarenal Mass

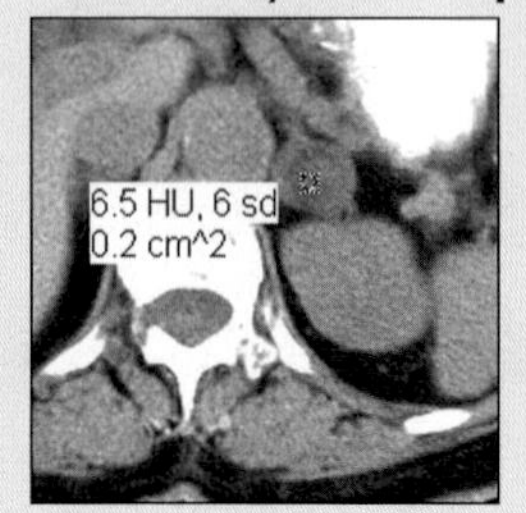

Adenoma

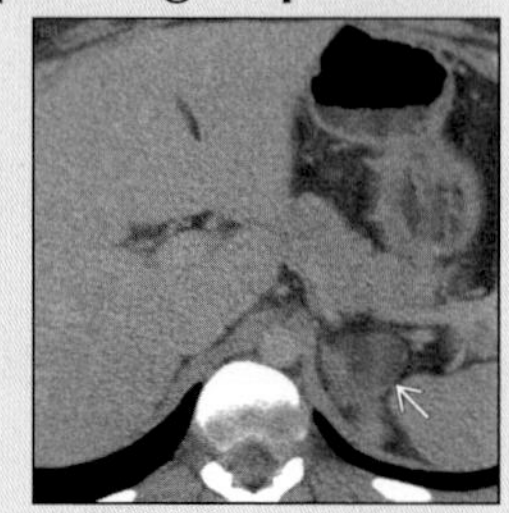

Gastric Diverticulum

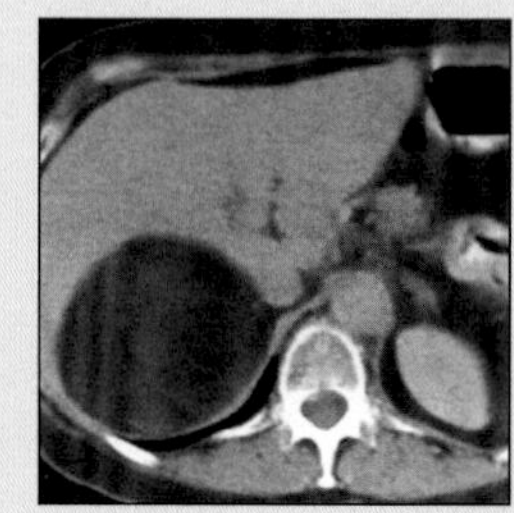

Myelolipoma

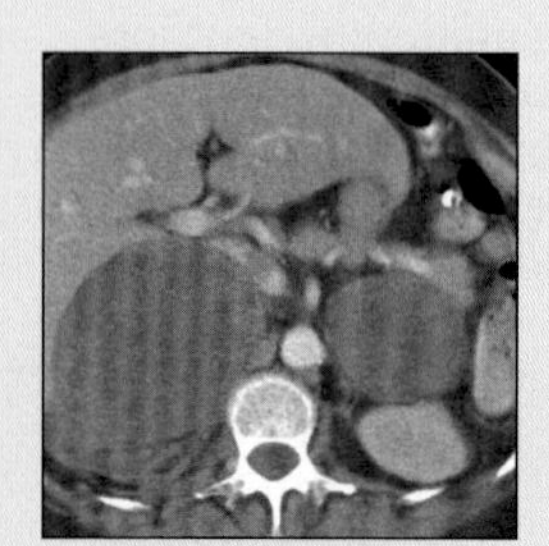

Metastases

ADRENAL CYST

Key Facts

Imaging Findings

- Best diagnostic clue: Well-defined nonenhancing water-density adrenal mass ± calcifications
- Unilocular or multilocular mass
- ↑ Wall thickness, up to 3 mm

Top Differential Diagnoses

- Adrenal adenoma
- Gastric diverticulum
- Adrenal myelolipoma
- Necrotic adrenal tumor

Clinical Issues

- Asymptomatic

Diagnostic Checklist

- Complicated cyst has ↑ malignant potential
- Complicated cyst has high attenuation, thick enhancing wall and/or septations

Necrotic adrenal tumor

- Primary or metastatic

PATHOLOGY

General Features

- Etiology
  - Endothelial lining (45-48%)
    - Lymphangioma
    - Hemangioma
  - Pseudocyst (39-42%)
    - Prior hemorrhage (e.g., vascular neoplasm, primary adrenal mass) or infarction
    - Cystic degeneration
  - Epithelial lining: True cyst (9-10%)
    - Glandular or retention cyst
    - Embryonal cyst
    - Cystic adenoma
    - Mesothelial inclusion cyst
  - Parasitic cyst (7%)
    - Hydatid or echinococcal cyst
- Epidemiology: 1-2 per 10,000 persons

CLINICAL ISSUES

Presentation

- Most common signs/symptoms
  - Asymptomatic
  - Abdominal pain, gastrointestinal symptoms, palpable mass
- Diagnosis
  - Usually found incidentally on imaging

Demographics

- Age: Any age; 20-50 years of age most common
- Gender: M:F = 1:3

Natural History & Prognosis

- Complications
  - Hypertension, infection, rupture, hemorrhage
- Prognosis
  - Good

Treatment

- Fine-needle aspiration
  - Cyst with low malignant potential
- Surgical resection
  - Usually by laparoscopic approach
  - Cyst with high malignant potential, > 5 cm
  - Patients with symptoms, endocrine abnormalities, complications

DIAGNOSTIC CHECKLIST

Consider

- Complicated cyst has ↑ malignant potential

Image Interpretation Pearls

- Complicated cyst has high attenuation, thick enhancing wall and/or septations

SELECTED REFERENCES

1. Otal P et al: Imaging features of uncommon adrenal masses with histopathologic correlation. Radiographics. 19(3):569-81, 1999
2. Neri LM et al: Management of adrenal cysts. Am Surg. 65(2):151-63, 1999
3. Kawashima A et al: Imaging of nontraumatic hemorrhage of the adrenal gland. Radiographics. 19(4):949-63, 1999
4. Kawashima A et al: Spectrum of CT findings in nonmalignant disease of the adrenal gland. Radiographics. 18(2):393-412, 1998
5. Rozenblit A et al: Cystic adrenal lesions: CT features. Radiology. 201(2):541-8, 1996
6. Tung GA et al: Adrenal cysts: imaging and percutaneous aspiration. Radiology. 173(1):107-10, 1989
7. Johnson CD et al: CT demonstration of an adrenal pseudocyst. J Comput Assist Tomogr. 9(4):817-9, 1985

IMAGE GALLERY

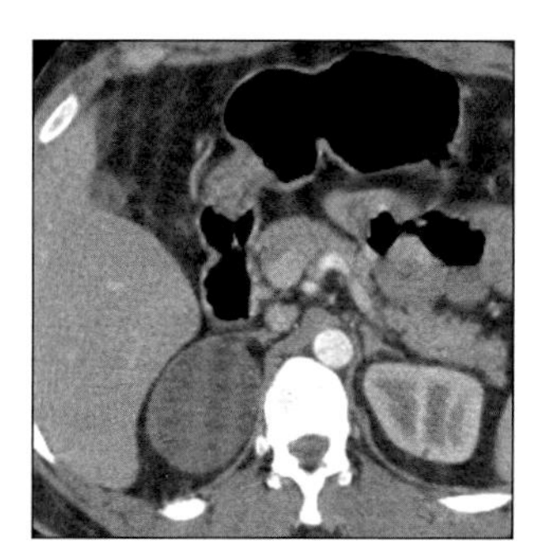

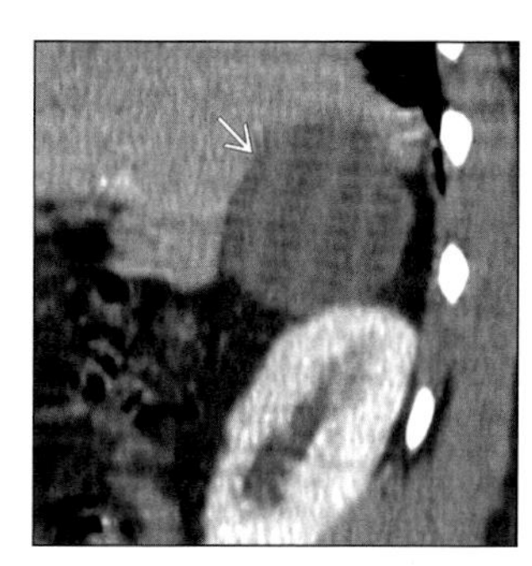

(Left) *Axial CECT shows nonenhancing water density right adrenal cyst.* ***(Right)*** *Sagittal reformation of CECT shows nonenhancing right adrenal cyst (arrow).*

ADRENAL ADENOMA

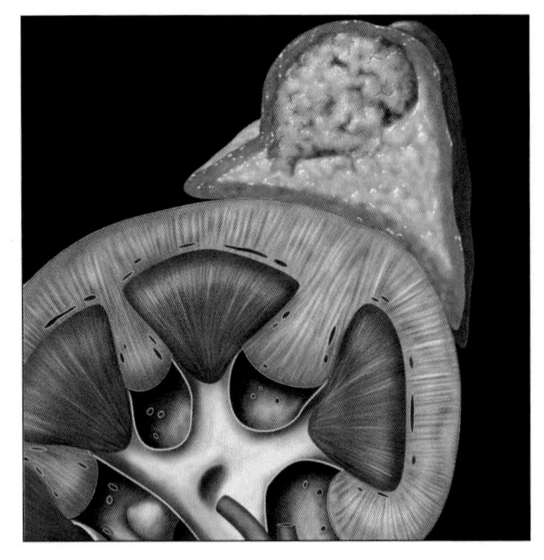

Graphic shows small lipid-rich mass within the adrenal gland.

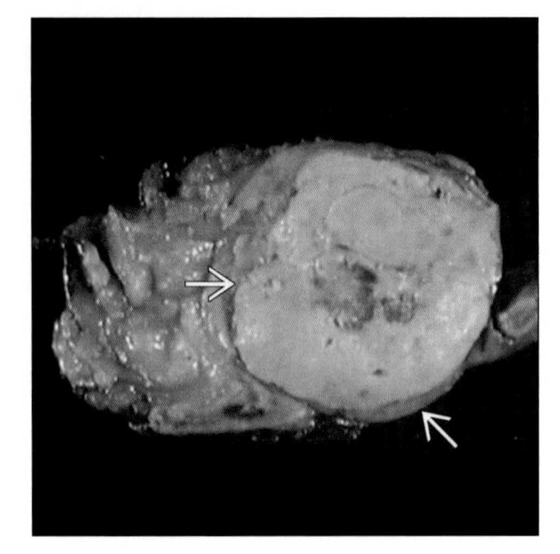

Surgical photograph of resected adenoma shows a 2 cm lipid-rich nodule (arrows) within the adrenal gland.

TERMINOLOGY

Definitions

- Benign tumor of adrenal gland cell origin

IMAGING FINDINGS

General Features

- Best diagnostic clue: Well-circumscribed, low density, small adrenal mass on CT
- Size
  - Adenoma of Cushing syndrome: Varies from 2-5 cm
  - Adenoma of Conn syndrome: < 2 cm (20% < 1 cm)
- Morphology: Usually round to oval suprarenal mass
- Key concepts
  - Most common adrenal cortex tumor (10% bilateral)
    - Accounts for 90% of all "incidentalomas"
  - May occur in up to 9% of general population
  - Lipid rich adrenal adenoma: 70% of all adenomas
  - Lipid poor adrenal adenoma: 30% of adenomas
  - ↑ Incidence in patients with diabetes & HTN
  - CT: Study of choice to diagnose adrenal incidentalomas
  - Classified into two types based on function
    - Nonhyperfunctioning: Normal hormone levels
    - Hyperfunctioning: Primary hyperaldosteronism; Cushing syndrome; hyperandrogenism
  - Primary hyperaldosteronism (Conn syndrome)
    - 80% of cases are due to adrenal adenoma
    - 20% of cases are due to adrenal hyperplasia
    - Adenomas are often small & difficult to detect
  - Cushing syndrome
    - 15-25% of cases are due to adrenal adenoma
    - 80-85% of cases are due to adrenal hyperplasia
    - Adenomas are usually greater than 2.0 cm

CT Findings

- NECT
  - Smooth, well-defined, round or oval in shape
  - Homogeneous soft tissue mass of 0-20 HU
  - Lipid rich adrenal adenoma (70% of cases)
    - Low-attenuation (less than 10 HU)
    - Characteristic & diagnostic of adenoma
  - Lipid poor adrenal adenoma (30% of cases)
    - Attenuation varies from 10-30 HU
    - Difficult to differentiate from metastases on NECT
  - Cushing syndrome due to adrenal adenoma
    - Remainder of ipsilateral gland & contralateral adrenal gland are atrophic due to ↓ ACTH levels
    - ↑ Cortisol: Feedback inhibition on pituitary ACTH
  - Conn syndrome due to adrenal adenoma
    - Remainder of ipsilateral gland & contralateral adrenal gland appear normal
  - Large adenomas
    - More heterogeneous than small adenomas
    - ± Hemorrhage, cystic degeneration, calcification

DDx: Adrenal Mass

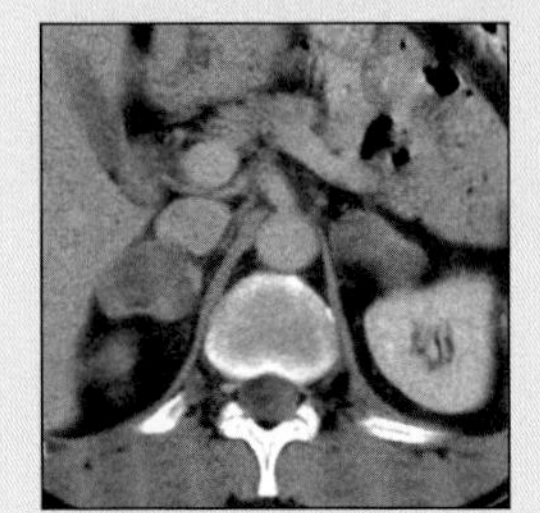

Metastatic Melanoma

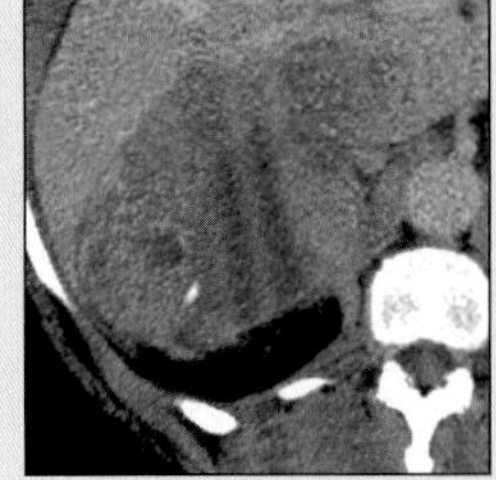

Adrenal Carcinoma

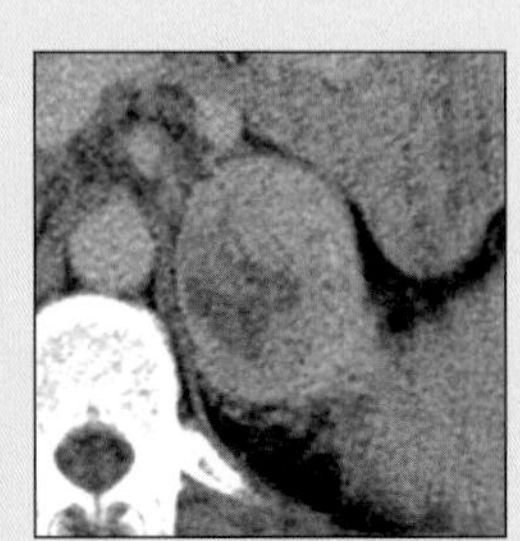

Pheochromocytoma

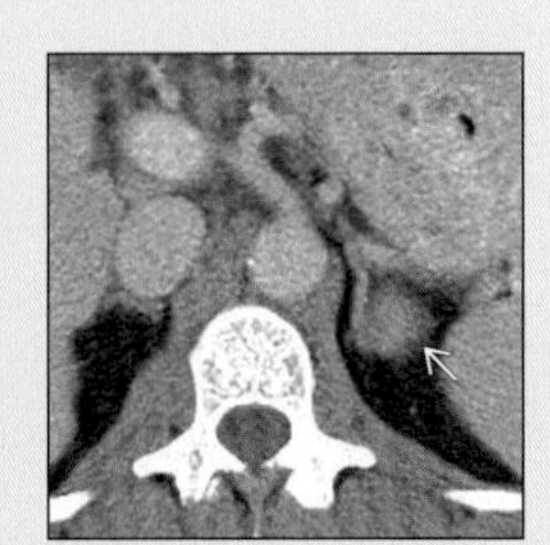

Gastric Diverticulum

ADRENAL ADENOMA

Key Facts

Imaging Findings

- Best diagnostic clue: Well-circumscribed, low density, small adrenal mass on CT
- Smooth, well-defined, round or oval in shape
- Homogeneous soft tissue mass of 0-20 HU
- Enhancing adrenal mass that "de-enhances" rapidly
- Washout of adenoma: 10 min. post injection > 50%
- Enhanced-delayed x 100/enhanced-unenhanced
- T1WI out of phase: ↑ Signal "drop-out" (lipid-rich)
- Adrenal adenoma: No increased uptake of FDG
- Adrenal adenoma: Unilateral early adrenal visualization before day 5 after NP-59 injection
- Washout value of > 50%: Sensitivity (96%), specificity (near 100%) for adrenal adenoma
- Washout value of < 50%: Indicative of either metastases or an atypical adenoma

Top Differential Diagnoses

- Adrenal metastases & lymphoma
- Adrenal carcinoma
- Pheochromocytoma
- Adrenal hyperplasia
- Gastric diverticulum

Clinical Issues

- Asymptomatic incidental CT imaging finding
- Conn syndrome: Hypertension & weakness
- Cushing syndrome: Moon facies, truncal obesity, purple striae & buffalo hump
- Diagnosis: Clinical, biochemical, imaging, histology

Diagnostic Checklist

- Asymptomatic mass: Usually a non-hyperfunctioning adenoma, even in a patient with a known cancer

- CECT
    - Enhancing adrenal mass that "de-enhances" rapidly
        - Enhanced phase: Attenuation varies, 40-50 HU
        - 10 min. delayed phase: Attenuation ↓ 20-25 HU
    - Washout of adenoma: 10 min. post injection > 50%
        - Enhanced-delayed x 100/enhanced-unenhanced
    - Washout pattern is diagnostic for adenoma
        - No follow-up is required if seen
    - Lipid poor adenoma have similar washout pattern to lipid rich adenoma

MR Findings

- T1WI & T2WI
    - Varied signal intensity
- Lipid rich adrenal adenoma
    - T1WI out of phase: ↑ Signal "drop-out" (lipid-rich)
    - T1WI in phase: Hyperintense
- T1 C+
    - Early phase: Adenoma shows enhancement
    - Delayed phase: > 50% washout is seen

Ultrasonographic Findings

- Real Time
    - May show a mass lesion in suprarenal area
        - Right suprarenal mass seen more clearly than left side due to acoustic window provided by liver

Angiographic Findings

- Conventional
    - Adrenal arteriography
        - Catheterization of renal or inferior adrenal arteries show vascular supply of adrenal tumors
        - Adenomas are usually hypovascular to moderately vascular
        - No arterial encasement, venous laking or puddling which are of malignant vascular features
    - Adrenal venography
        - Most commonly to obtain adrenal vein samples
        - Adrenal adenoma is seen as a filling defect within adrenal gland displacing adjacent vessels
        - Circumferential vein frequently seen around adrenal adenoma

Nuclear Medicine Findings

- FDG-PET
    - Adrenal adenoma: No increased uptake of FDG
        - Differentiates from malignant lesion (↑ uptake)
- Adrenocortical scintigraphy by using NP-59
    - NP-59 is a cholesterol analog that binds to low-density lipoprotein receptors of adrenal cortex
    - NP-59 used + dexamethasone: Accentuate uptake in non-ACTH-dependent adrenal tissues (adenoma)
    - Normal NP-59: When both adrenal glands are seen on day 5 after injection or thereafter
    - Adrenal adenoma: Unilateral early adrenal visualization before day 5 after NP-59 injection
    - Adrenal hyperplasia: Bilateral early adrenal visualization before day 5 after NP-59 injection

Imaging Recommendations

- CT is study of choice to confirm the diagnosis of adrenal adenoma
    - CT technique: 3 mm thick section at 3 mm intervals
- If suspect adrenal adenoma, NECT alone sufficient
    - Attenuation value < 10 HU is diagnostic
- If CECT done, assess the following
    - If lesion < 37 HU on CECT, call it adenoma
    - If lesion > 37 HU, on CECT, get 10 min delayed scan to determine washout
    - Washout value of > 50%: Sensitivity (96%), specificity (near 100%) for adrenal adenoma
    - Washout value of < 50%: Indicative of either metastases or an atypical adenoma
- MR with in and out of phase imaging
    - Diagnostic for lipid-rich adenomas

DIFFERENTIAL DIAGNOSIS

Adrenal metastases & lymphoma

- Adrenal metastases
    - Unilateral or bilateral masses ± central necrosis, hemorrhage
    - Usually known to have malignancy elsewhere
    - NECT: Metastases mimic lipid poor adenoma

- CECT: Hypervascular & prolonged washout pattern
- Adrenal lymphoma
  - Usually spread to adrenal gland of retroperitoneal
  - Unilateral primary lymphoma (non-Hodgkin) can mimic adenoma
  - Hypovascular; moderate enhancement with contrast

Adrenal carcinoma

- Rare, unilateral invasive & enhancing mass
- More than 6 cm when initially diagnosed

Pheochromocytoma

- Tumor > 3 cm in most cases, T2WI very hyperintense
- Highly vascular tumor prone to hemorrhage, necrosis
- Bilateral adrenal tumors in MEN syndromes

Adrenal hyperplasia

- Adrenal glands are often symmetrically enlarged
- Width of adrenal gland limbs > 10 mm (diagnostic)
- No discrete mass or nodule seen as a rule
- Dominant macronodule of macronodular hyperplasia mimic small adrenal adenoma
  - Cortisol-secreting adenoma: Remainder of ipsilateral & contralateral glands, atrophic (↓ ACTH)
  - Macronodular hyperplasia: Both glands are enlarged (due to elevated ACTH levels)
- No obvious enhancement & washout pattern seen

Gastric diverticulum

- Abnormal rounded soft tissue shadow in left suprarenal area, mimics adrenal mass
- Diverticular contents do not enhance; adenomas do

Adrenal myelolipoma

- Small or large, asymptomatic adrenal mass
- Intramural fatty elements recognized on imaging

Ganglioneuroma

- Younger patients, mean age = 27 y
- Larger mass with average tumor size of 8 cm

Unilateral adrenal hemorrhage

- Chronic hematoma: Well-defined, round, low density, mass-like lesion simulating adenoma

PATHOLOGY

General Features

- General path comments
  - Most adrenals with adenoma: Normal function
  - Occasionally adenoma causes adrenal hyperfunction
  - Normal adrenocortical secretory hormones
    - Adrenal cortex: Cortisol, aldosterone, androgens
- Etiology: Exact etiology unknown
- Epidemiology
  - Most common adrenal tumor of all incidentalomas
  - ↑ Incidence in patients with diabetes or HTN
  - Occur in up to 9% of population (postmortem data)
- Associated abnormalities: Multiple endocrine neoplasia syndromes (MEN)

Gross Pathologic & Surgical Features

- Well-delineated, tan-yellow, ovoid mass

Microscopic Features

- 70% of adenomas: High % of intracytoplasmic lipid
- 30% of adenomas: Less % of intracytoplasmic lipid

CLINICAL ISSUES

Presentation

- Most common signs/symptoms
  - Asymptomatic incidental CT imaging finding
  - Conn syndrome: Hypertension & weakness
  - Cushing syndrome: Moon facies, truncal obesity, purple striae & buffalo hump
  - Virilization in women
  - Lab data: ↑ Aldosterone, cortisol & androgens
  - Diagnosis: Clinical, biochemical, imaging, histology

Demographics

- Age
  - Prevalence of adenoma increases with age
    - Peak at 60-69 y and decreasing thereafter

Natural History & Prognosis

- Prognosis: Excellent when incidental & nonsecretory

Treatment

- No treatment when asymptomatic incidental finding
- Laparoscopic removal of gland if hyperfunctioning

DIAGNOSTIC CHECKLIST

Consider

- Asymptomatic mass: Usually a non-hyperfunctioning adenoma, even in a patient with a known cancer

Image Interpretation Pearls

- Well-defined, low density (< 10 HU) suprarenal mass
- Enhances + wash out pattern of > 50% within 10 min.
- T1W out of phase: ↑ Signal "drop-out" lipid-rich mass

SELECTED REFERENCES

1. Kebapci M et al: Differentiation of adrenal adenomas (lipid rich and lipid poor) from nonadenomas by use of washout characteristics on delayed enhanced CT. Abdom Imaging. 28(5):709-15, 2003
2. Caoili EM et al: Adrenal masses: characterization with combined unenhanced and delayed enhanced CT. Radiology. 222(3):629-33, 2002
3. Dunnick NR et al: Imaging of adrenal incidentalomas: Current status. AJR. 179:559-68, 2002
4. Mayo-Smith WW et al: State-of-the-art adrenal imaging. Radiographics. 21(4):995-1012, 2001
5. Pena CS et al: Characterization of indeterminate (lipid-poor) adrenal masses: use of washout characteristics at contrast-enhanced CT. Radiology. 217(3):798-802, 2000
6. Caoili EM et al: Delayed enhanced CT of lipid-poor adrenal adenomas. AJR Am J Roentgenol. 175(5):1411-5, 2000
7. Kawashima A et al: Spectrum of CT findings in nonmalignant disease of the adrenal gland. Radiographics. 18(2):393-412, 1998
8. Korobkin M et al: Adrenal adenomas: relationship between histologic lipid and CT and MR findings. Radiology. 200(3):743-7, 1996

IMAGE GALLERY

Typical

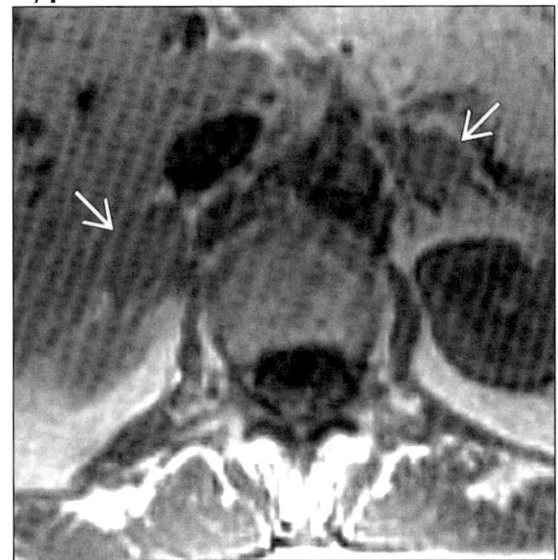

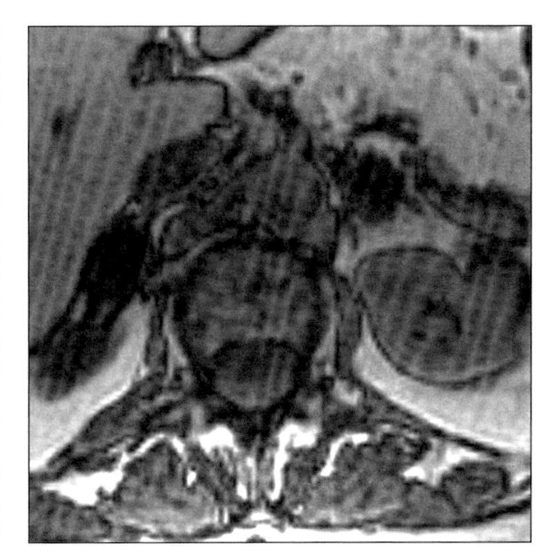

(Left) *Axial T1WI MR in phase GRE sequence shows bilateral small adrenal masses (arrows).* ***(Right)*** *Axial T1WI MR opposed phase GRE sequence shows loss of signal in both adenomas, confirming presence of lipid.*

Typical

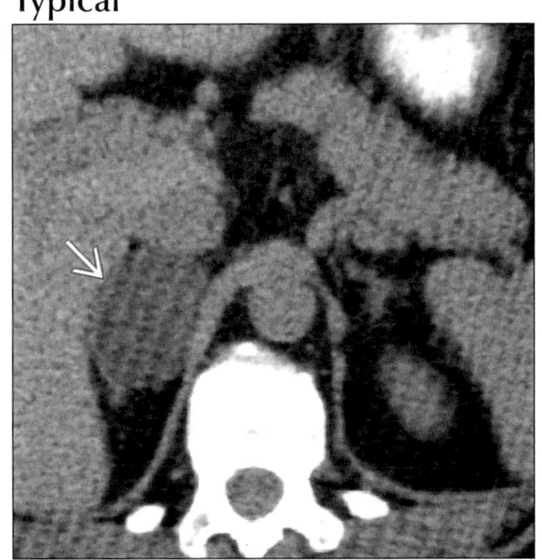

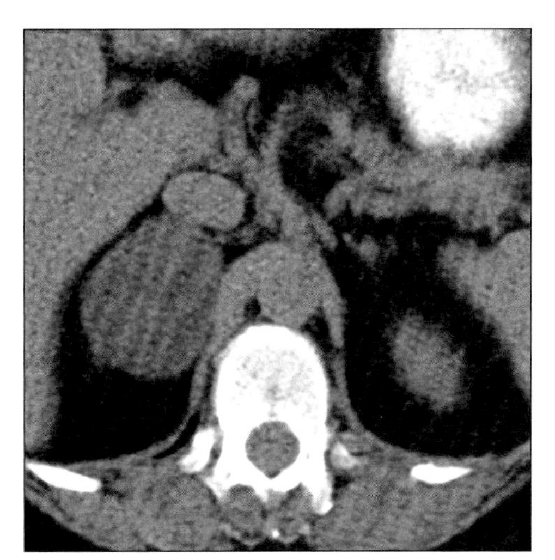

(Left) *Axial NECT shows homogeneous low density right adrenal adenoma (arrow), normal left adrenal.* ***(Right)*** *Axial NECT shows homogeneous low density adenoma that is larger than typical and which had grown slowly over 10 years.*

Typical

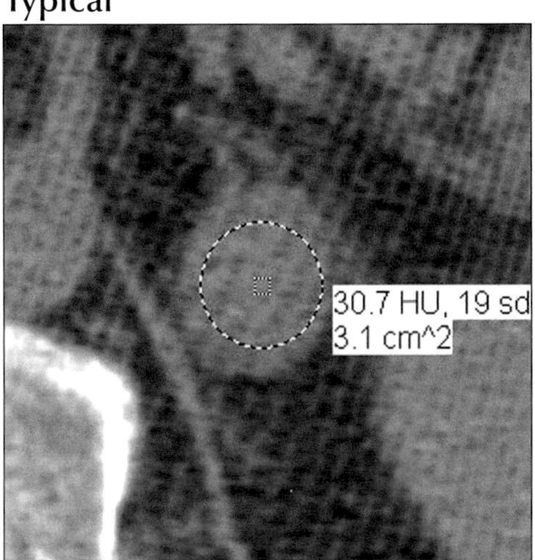

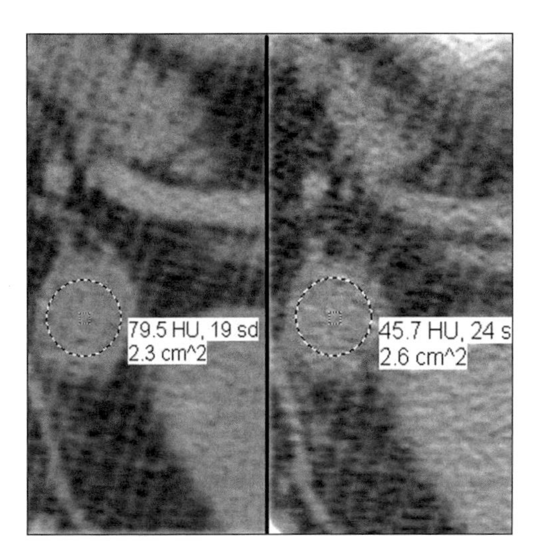

(Left) *Axial NECT shows an indeterminate left adrenal mass, proved to be a lipid-poor adenoma.* ***(Right)*** *Axial CECT in parenchymal + delayed phases shows significant enhancement and rapid washout. Absolute washout is 69%, relative is 42%, indicating lipid-poor adenoma.*

ADRENAL MYELOLIPOMA

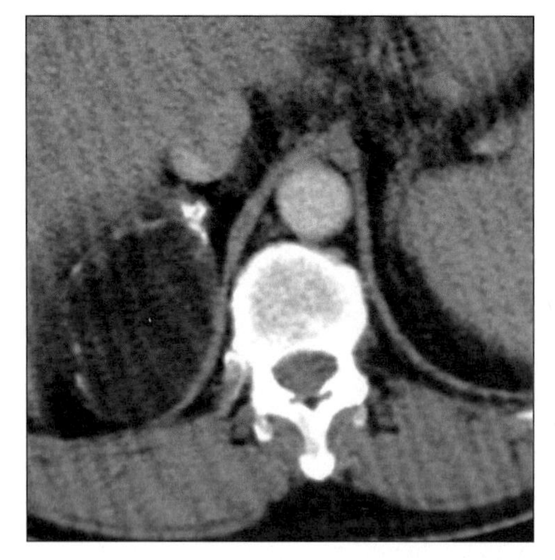

Axial CECT shows heterogeneous, predominantly fatty, right adrenal mass with calcified foci.

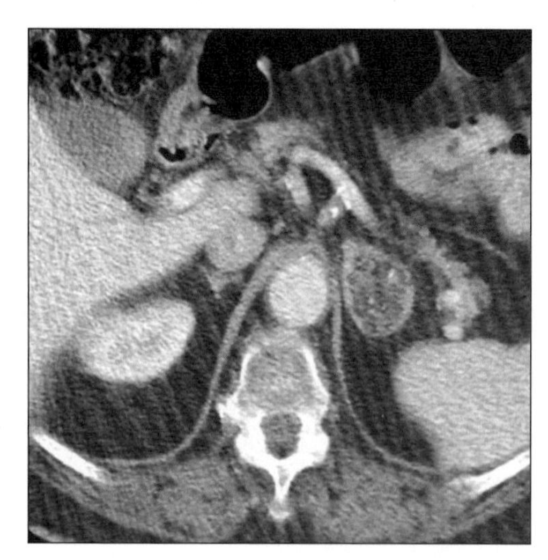

Axial CECT shows predominantly fatty left adrenal mass.

TERMINOLOGY

Definitions

- Rare benign tumor composed of mature fat tissue & hematopoietic elements (myeloid & erythroid cells)

IMAGING FINDINGS

General Features

- Best diagnostic clue: Heterogeneous fatty adrenal mass
- Location: Suprarenal
- Size: Usually 2-10 cm, rarely 10-20 cm
- Key concepts
  - Rare, benign neoplasm of adrenal gland
  - Seen in 0.2-0.4% of cases based on autopsy series
  - Frequency among all incidentalomas, 7-15%
  - Incidental finding on CT in older people
  - Typically unilateral & very rarely bilateral
  - Large tumor can bleed spontaneously or necrose
  - Nonfunctioning (do not secrete any hormones)
  - Large myelolipoma can mimic retroperitoneal lipoma or liposarcoma

CT Findings

- CT appearance depends on histologic composition
  - Most tumors are heterogeneous fatty adrenal masses
    - Low-attenuation of fat density (-30 to -90 HU)
    - Presence of pure fat within tumor is diagnostic
- Usually well-defined mass with recognizable capsule
- Punctate calcifications seen (20% of cases)

MR Findings

- MR appearance depends on histologic composition
  - Tumor with major fat component
    - T1WI in phase: Typically hyperintense
    - Fat suppression sequences: Loss of signal
  - Bone marrow elements (myeloid & erythroid cells)
    - Low signal on T1WI; moderate signal on T2WI

Ultrasonographic Findings

- Real Time
  - Well-defined, echogenic mass (↑ fat tissue)
  - Heterogeneous mass (↑ myeloid cells)

Angiographic Findings

- Conventional: Differentiate myelolipoma from retroperitoneal liposarcoma by determining origin of blood supply & vascularity of tumors

Imaging Recommendations

- Helical NECT or MR with fat suppression sequence

DIFFERENTIAL DIAGNOSIS

Adrenal adenoma

- Lipid rich adenoma: ↓ Attenuation (less than 10 HU)

DDx: Adrenal Mass

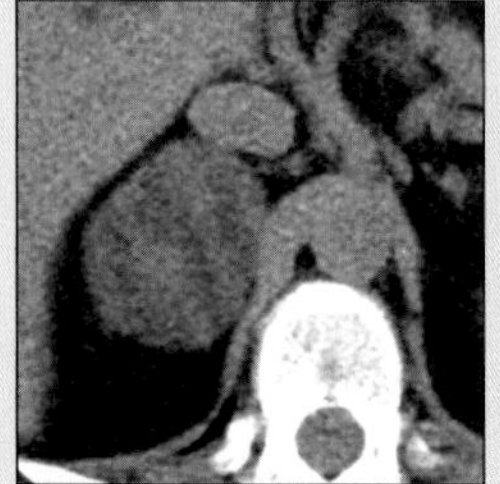

Adenoma

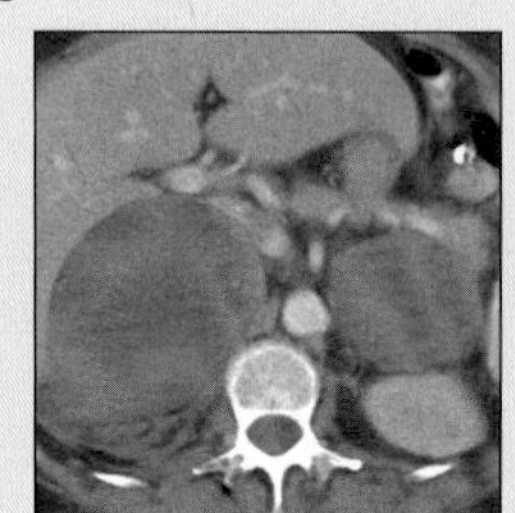

Metastases

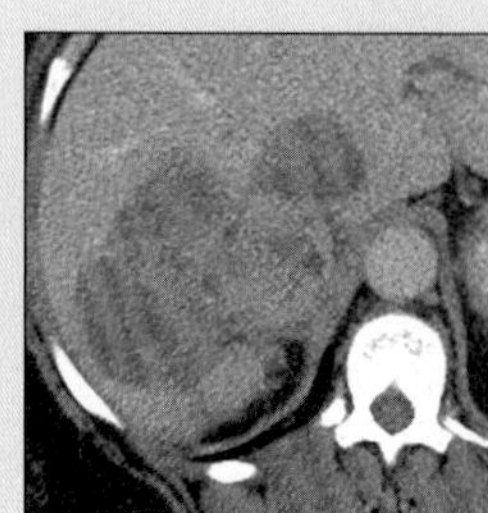

Metastases

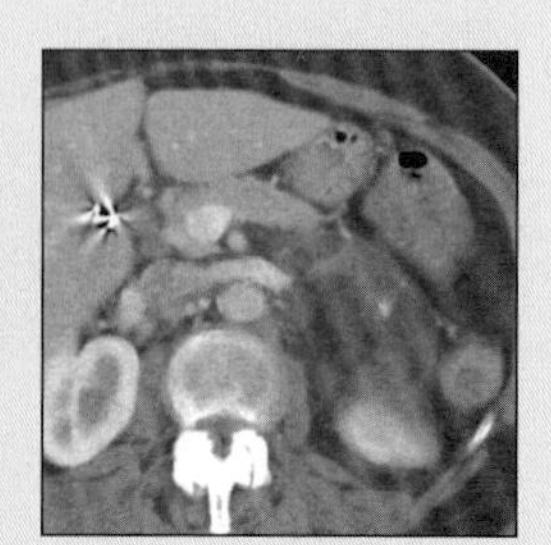

Liposarcoma

ADRENAL MYELOLIPOMA

Key Facts

Terminology
- Rare benign tumor composed of mature fat tissue & hematopoietic elements (myeloid & erythroid cells)

Imaging Findings
- Most tumors are heterogeneous fatty adrenal masses
- Low-attenuation of fat density (-30 to -90 HU)
- T1WI in phase: Typically hyperintense
- Fat suppression sequences: Loss of signal

Top Differential Diagnoses
- Adrenal adenoma
- Adrenal metastases & lymphoma
- Pheochromocytoma

Diagnostic Checklist
- Differentiate from other tumors (lipid-rich adenoma)
- Key is presence of tumoral fat & benign nature, avoid further workup for incidental mass

- No true fat density, unlike adrenal myelolipoma
- CECT: Wash out 10 min. post injection > 50%

Adrenal metastases & lymphoma
- Adrenal metastases
  - Invasive, enhancing mass in adrenal gland
  - Higher density on CT than myelolipoma
- Adrenal lymphoma
  - Primary (rare); secondary (non-Hodgkin common)
  - Often bilateral; retroperitoneal disease usually seen
  - Discrete or diffuse mass, shape is maintained

Liposarcoma
- Retroperitoneal primary sarcoma involving perirenal space may simulate adrenal (or renal) fatty tumor

Pheochromocytoma
- Highly vascular; prone to hemorrhage & necrosis
- Hyperintense on T2WI; bilateral in MEN syndromes

Adrenal carcinoma
- Rare, unilateral, invasive & enhancing mass

PATHOLOGY

General Features
- Etiology
  - Unknown
  - Best hypothesis: Reticuloendothelial cell metaplasia of capillaries in adrenal (stress/infection/necrosis)
  - Secondary hypothesis: Myelolipoma represents a site of extramedullary hematopoiesis
- Epidemiology: Autopsy incidence 0.2-0.4%
- Associated abnormalities: Adrenal collision tumors

Gross Pathologic & Surgical Features
- Cut section: Fat & soft tissue components

Microscopic Features
- Mature fat cells & megakaryocytes; no malignant cells

CLINICAL ISSUES

Presentation
- Most common signs/symptoms
  - Asymptomatic
  - Usually an incidental finding on CT, MR or US
  - "Acute abdomen": Rupture with hemorrhage (rare)
- Diagnosis: CT or MR; biopsy prone to sampling error

Demographics
- Age: Usually elderly age group: 50-70 years

Natural History & Prognosis
- Complication: Rupture with hemorrhage (rare)
- Prognosis: Excellent

Treatment
- When diagnosis is certain, surgery not needed

DIAGNOSTIC CHECKLIST

Consider
- Differentiate from other tumors (lipid-rich adenoma)
- Key is presence of tumoral fat & benign nature, avoid further workup for incidental mass

Image Interpretation Pearls
- Well-defined heterogeneous fat density tumor on CT
- T1WI hyperintense, signal loss with fat suppression

SELECTED REFERENCES

1. Dunnick NR et al: Imaging of adrenal incidentalomas: current status. AJR Am J Roentgenol. 179(3): 559-68, 2002
2. Mayo-Smith WW et al: State-of-the-Art adrenal imaging. Radiographics. 21(4): 995-1012, 2001
3. Rao P et al: Imaging and pathologic features of myelolipoma. Radiographics. 17: 1373-85, 1997
4. Cyran KM et al: Adrenal myelolipoma. AJR 166: 395-400, 1996

IMAGE GALLERY

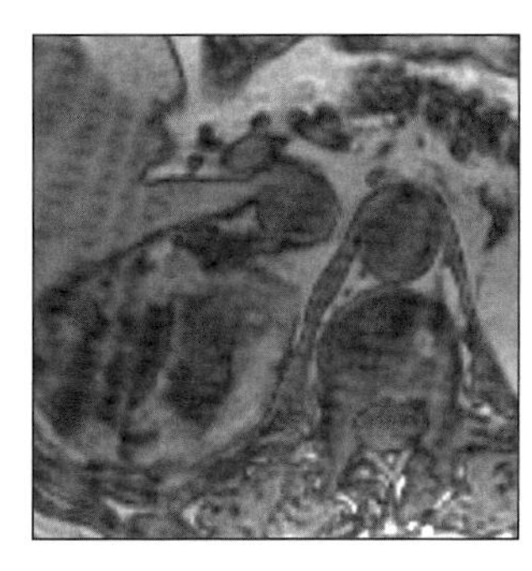

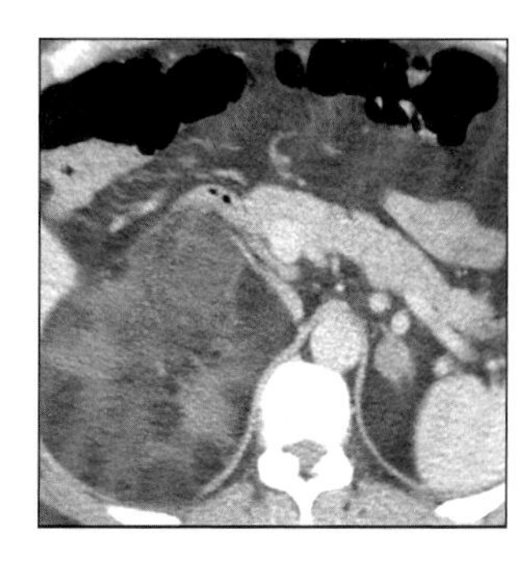

*(**Left**) Chemical shift (opposed phase) axial MR shows heterogeneous right adrenal mass with signal loss at fat/soft tissue interfaces. (**Right**) Axial CECT shows a large heterogeneous fatty myelolipoma in right adrenal (and an adenoma in left gland).*

PHEOCHROMOCYTOMA

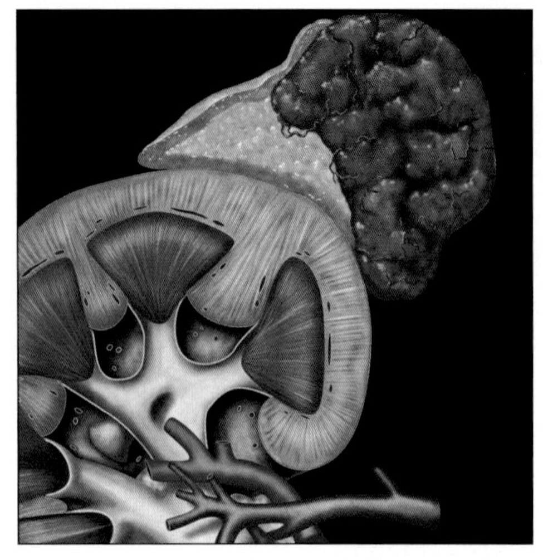

Graphic shows heterogeneous hypervascular adrenal pheochromocytoma.

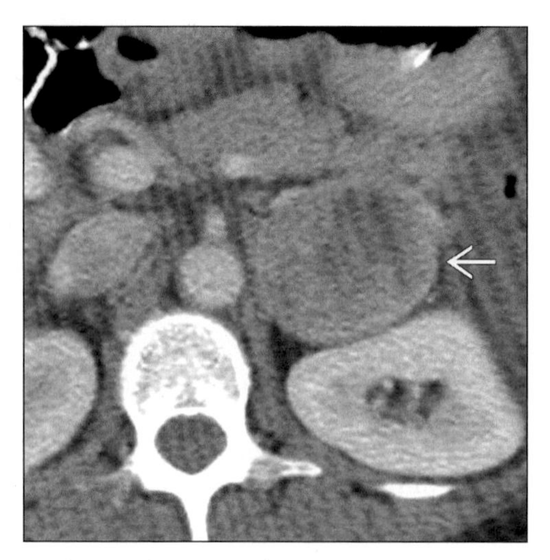

Axial CECT shows 5 cm heterogeneous hypervascular left adrenal pheochromocytoma (arrow).

TERMINOLOGY

Abbreviations and Synonyms

- Paraganglioma or ganglioneuroma (extra-adrenal)

Definitions

- Tumor arising from chromaffin cells of adrenal medulla or extra-adrenal ectopic tissue

IMAGING FINDINGS

General Features

- Best diagnostic clue: Very hyperintense 3-5 cm mass on T2WI with bright heterogeneous enhancement
- Location
  - Along sympathetic chain: Neck to urinary bladder
  - Subdiaphragmatic (98%); thorax (1-2%)
    - Adrenal medulla (90%); extra-adrenal (10%)
- Size
  - Usually more than 3 cm in most cases
  - Weight ranging from 1 gm to over 4 kg
- Morphology
  - Well-circumscribed, encapsulated tumor
    - Solitary (sporadic); multiple (familial)
- Key concepts
  - Also called tumor with "rule of 10s" or 10% tumors
  - 10% extra-adrenal: Paragangliomas/chemodectomas
  - 10% bilateral, malignant & extra-abdominal
  - 10% familial, pediatric, silent
  - 10% have autosomal dominant transmission & associated with various other dominant conditions
  - Extra-adrenal tumors arise from sympathetic ganglia
    - Neck, mediastinum, pelvis or urinary bladder
    - Aortic bifurcation (organ of Zuckerkandl): Ganglia at origin of inferior mesenteric artery
  - 90% patients present with HTN secondary to release of catecholamines
  - Term pheochromocytoma refers to dusky color
    - Tumor "stains" this color when treated with chromium salts
  - Imaging: Difficult to distinguish benign, malignant
    - Benign lesions can be locally invasive of IVC & renal capsule
    - Distant metastases indicate malignancy

CT Findings

- NECT
  - Well-defined, round, homogeneous (muscle density)
  - ± Areas of ↑ density (hemorrhage)
  - ± Areas of ↓ density (cystic, necrotic, septate)
  - ± Areas of curvilinear or mural calcification
- CECT
  - Shows marked homogeneous enhancement
  - Heterogeneous enhancement
    - Due to tissue necrosis & hemorrhage
  - Peripheral enhancement with fluid-levels

DDx: Adrenal Mass

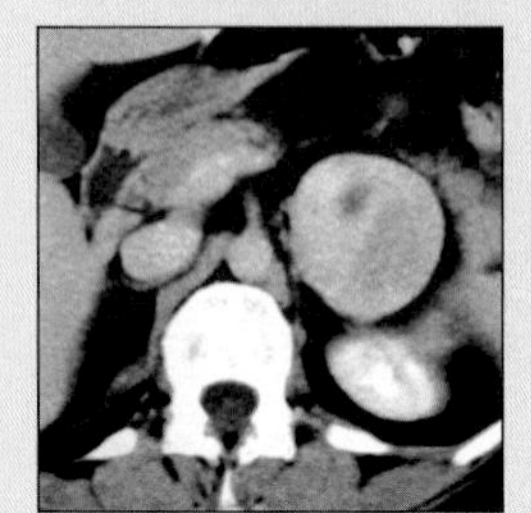

Adenoma with Bleed

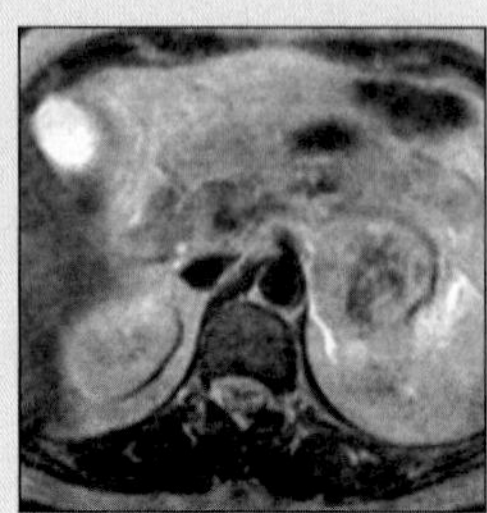

Adenoma with Bleed

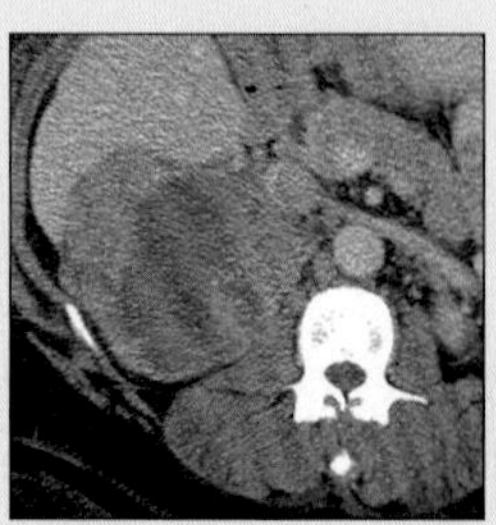

Adrenal Metastasis

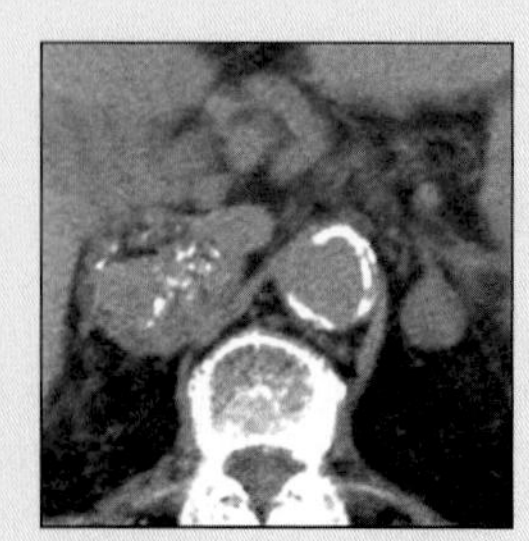

Myelolop. + Adenoma

PHEOCHROMOCYTOMA

Key Facts

Terminology
- Paraganglioma or ganglioneuroma (extra-adrenal)
- Tumor arising from chromaffin cells of adrenal medulla or extra-adrenal ectopic tissue

Imaging Findings
- Best diagnostic clue: Very hyperintense 3-5 cm mass on T2WI with bright heterogeneous enhancement
- Adrenal medulla (90%); extra-adrenal (10%)
- 10% bilateral, malignant & extra-abdominal
- Well-defined, round, homogeneous (muscle density)
- ± Areas of ↑ density (hemorrhage)
- ± Areas of ↓ density (cystic, necrotic, septate)
- ± Areas of curvilinear or mural calcification
- Markedly hyperintense on T2WI (characteristic)
- After 24-72 hrs.: ↑ Uptake of I-131 MIBG in tumor

Top Differential Diagnoses
- Adrenal adenoma
- Adrenocortical carcinoma
- Adrenal metastases & lymphoma
- Adrenal myelolipoma (myelolop.)
- Adrenal hemorrhage
- Granulomatous infection

Clinical Issues
- Clinical profile: A young patient with paroxysmal attacks of headache, palpitations, sweating & tremors
- ↑ Levels of vanillylmandelic acid (VMA) 24-hr. urine

Diagnostic Checklist
- Imaging findings + history & labs (usually diagnostic)
- Spherical suprarenal mass, 3-5 cm, very hyperintense on T2WI MR & brightly enhancing with contrast

MR Findings
- T1WI
  - Isointense to muscle & hypointense to liver
  - Heterogeneous signal intensity
    - Due to areas of hemorrhage & necrosis
  - ± Areas of increased signal intensity
    - Due to acute or subacute hemorrhage
- T2WI
  - Markedly hyperintense on T2WI (characteristic)
  - Long T2 relaxation time
    - Due to ↑ water content as a result of necrosis
  - Heterogeneous signal intensity (in 33% of cases)
    - Due to hemorrhage & necrosis with fluid levels
- T1 C+
  - Characteristic salt & pepper pattern (due to increased tumor vascularity)
    - Salt: Represents enhancing parenchyma
    - Pepper: Represents flow void of vessels
  - Can show marked early as well as prolonged contrast-enhancement

Ultrasonographic Findings
- Real Time
  - Iso-/hypoechoic (77%) & hyperechoic (23%) in contrast to normal renal parenchyma
  - Round & well-circumscribed mass

Angiographic Findings
- Conventional: Hypervascular tumor

Nuclear Medicine Findings
- I-131 or 123 Metaiodobenzylguanidine (MIBG)
  - After 24-72 hrs.: ↑ Uptake of I-131 MIBG in tumor
  - Particularly useful for detecting extra-adrenal tumors
    - Metastatic disease in malignant condition
    - Recurrent & extra-abdominal tumors
  - Sensitivity (80-90%); specificity (90-100%)

Imaging Recommendations
- Helical NE + CECT
  - Hypertensive crisis is rare or nonexistent with I.V. administration of nonionic contrast material
    - Routine premedication (α and β blockade) is not recommended by most authorities
- MR & T1 C+
- MIBG: For ectopic, recurrent & metastatic tumors

DIFFERENTIAL DIAGNOSIS

Adrenal adenoma
- Most common benign tumor of adrenal gland (cortex)
- Histopathologically: Rich in lipid
- Imaging
  - NECT
    - Well-defined mass of < 10 HU (lipid rich)
    - Well-defined mass of > 10 HU (lipid poor)
  - CECT: Enhancing mass that "de-enhances" rapidly
    - Washout adenoma: 10 min. post injection (> 50%)
    - Washout pattern is diagnostic for adenoma
  - MR: Signal suppression on out of phase T1WI
- Hyperfunctioning: Clinical symptoms, signs; lab data

Adrenocortical carcinoma
- Rare; usually unilateral; rarely bilateral (up to 10%)
- Functioning tumors (small); nonfunctioning (large)
- Imaging
  - Large, unilateral adrenal mass with invasive margins
  - Large solid mass, areas of necrosis, hemorrhage
  - ± Calcification (30% cases); variable enhancement
  - T1WI hypointense; T2WI hyperintense
  - Local spread: Renal vein or IVC extension
  - Metastatic tumor spread: Lungs, liver, nodes & bone
- Diagnosis: Biopsy & histology

Adrenal metastases & lymphoma
- Adrenal metastases
  - E.g., lung, breast, renal cell carcinoma & melanoma
  - Unilateral or bilateral; central necrosis ± hemorrhage
  - History: Patient usually known to have malignancy
  - Lung cancer
    - Adrenal metastases (35-38% of cases); usually solid
  - Breast carcinoma: Adrenal metastases (50% of cases)
  - Renal cell carcinoma (RCC)
    - Adrenal metastases (seen in 18-25% of cases)

PHEOCHROMOCYTOMA

- Usually ipsilateral & hypervascular
  - Malignant melanoma
    - Adrenal metastases (50% cases); usually bilateral
    - CT: Solid or cystic; rim calcification may be seen
    - MR: Hyperintense on T1WI (melanin pigment)
- Adrenal lymphoma
  - 25% cases of secondary lymphoma; primary (rare)
  - Non-Hodgkin most common; usually bilateral
  - Histologically: Diffuse cell type > nodular type
  - CECT: Mild enhancement (hypovascular)
  - Diagnosis: Percutaneous aspiration biopsy

Adrenal myelolipoma (myelolop.)

- Rare benign tumor (fat + hematopoietic elements)
- Unilateral fatty adrenal tumor (-100 to -30 HU)
- T1WI: Typically hyperintense; size varies (2-10 cm)
- Out of phase T1WI: Focal areas of signal loss
- US: Echogenic mass in adrenal bed

Adrenal hemorrhage

- Etiology: Septicemia, burns, trauma, stress, hypotension & hematological abnormalities
- CT findings
  - Usually bilateral
  - Old hemorrhage: Soft tissue attenuation (20-35 HU)
  - Recent hemorrhage: ↑ Attenuation values
- MR findings: T1 & T2WI
  - Varied signal depending on hematoma age
  - Subacute phase: Usually ↑ signal (methemoglobin)
  - Perilesional dark ring (hemosiderin or ferritin)

Granulomatous infection

- E.g., tuberculosis, histoplasmosis, other fungal diseases
- Usually bilateral, heterogeneous, poorly enhancing (acute)
- Chronic: Small & calcified adrenals
- Diagnosis: Clinical history & lab data

PATHOLOGY

General Features

- General path comments
  - Embryology/anatomy
    - Neoplasm of chromaffin cells derived from neural crest or neuroectoderm
- Etiology
  - Chromaffin cells of sympathetic nervous system
    - Adrenal medulla: Pheochromocytoma
    - Extra-adrenal: Paraganglioma
- Epidemiology
  - Incidence
    - 0.13% in autopsy series; 0.1-0.5% of HTN cases
- Associated abnormalities
  - With 10% autosomal dominant variety
    - von Hippel-Lindau syndrome
    - Type 1 neurofibromatosis
    - Multiple endocrine neoplasia syndromes (MEN) type IIA & type IIB
    - Tuberous sclerosis; Sturge-Weber syndrome
    - Carney syndrome: Pulmonary chondroma, gastric leiomyosarcoma, pheochromocytoma

Gross Pathologic & Surgical Features

- Round, tan-pink to violaceous, encapsulated mass
- ± Cystic, mucoid, serosanguineous, hemorrhage

Microscopic Features

- Large cells: Granular cytoplasm & pleomorphic nuclei
- Chromaffin reaction: Cells stained with chromium salt

CLINICAL ISSUES

Presentation

- Most common signs/symptoms
  - Symptoms may be episodic or paroxysmal
  - Crisis: Headaches, HTN, palpitations, sweating, tremors, arrhythmias, pain
  - Classic: Paroxysmal HTN ± visual changes
  - Atypical: Labile HTN, myocardial infarction, CVA
- Clinical profile: A young patient with paroxysmal attacks of headache, palpitations, sweating & tremors
- Lab data
  - ↑ Levels of vanillylmandelic acid (VMA) 24-hr. urine
    - Normal range of VMA levels: 1.8-6.7 mg/24 hours

Demographics

- Age: 3rd & 4th decades; ↑ familial incidence
- Gender: M = F

Natural History & Prognosis

- Complications: During hypertensive crisis
  - Cerebrovascular accidents (CVA)
  - Pregnancy + pheochromocytoma: Mortality (48%)
  - Malignancy in 2-14% cases; distant metastases
- Prognosis
  - Noninvasive & nonmetastatic: Good prognosis
  - Malignant & metastatic: Poor prognosis
    - 5 year survival rate is < 50%

Treatment

- Medical therapy: Before, during, after surgery
  - Alpha-adrenergic blockers
    - Phenoxybenzamine, phentolamine
  - Beta-adrenergic blocker: Propranolol
- Surgical resection: Benign & malignant
- Chemotherapy
  - Cyclophosphamide + vincristine + dacarbazine

DIAGNOSTIC CHECKLIST

Consider

- Imaging findings + history & labs (usually diagnostic)

Image Interpretation Pearls

- Spherical suprarenal mass, 3-5 cm, very hyperintense on T2WI MR & brightly enhancing with contrast

SELECTED REFERENCES

1. Dunnick NR et al: Imaging of adrenal incidentalomas: current status. AJR Am J Roentgenol. 179(3):559-68, 2002
2. Mayo-Smith WW et al: State-of-the-art adrenal imaging. Radiographics. 21(4):995-1012, 2001
3. Kawashima A et al: Spectrum of CT findings in nonmalignant disease of the adrenal gland. Radiographics. 18(2):393-412, 1998

IMAGE GALLERY

Typical

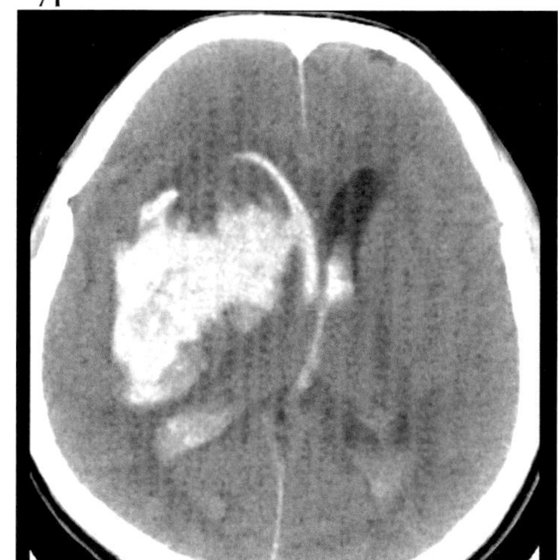

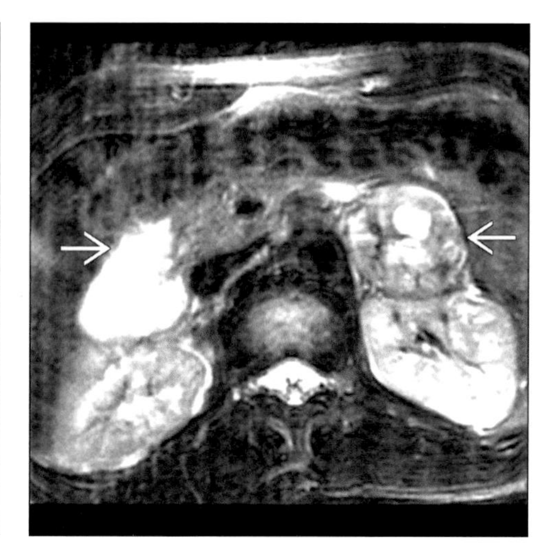

*(**Left**) Axial NECT of brain in 30 year old man with severe hypertension shows massive hemorrhage. (**Right**) Axial T2WI MR in 30 year old man shows bilateral hyperintense heterogeneous adrenal pheochromocytomas (arrows).*

Typical

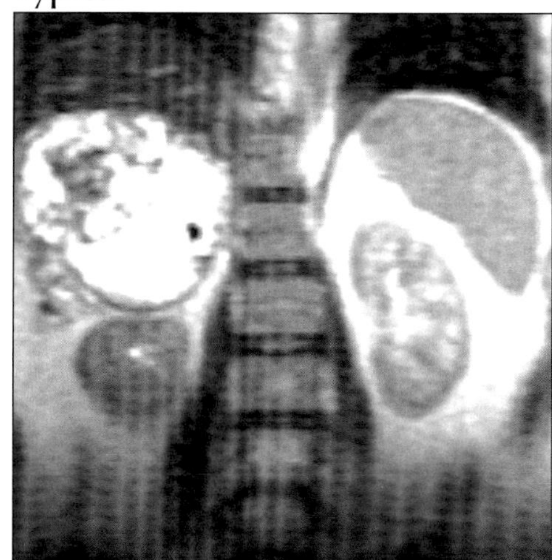

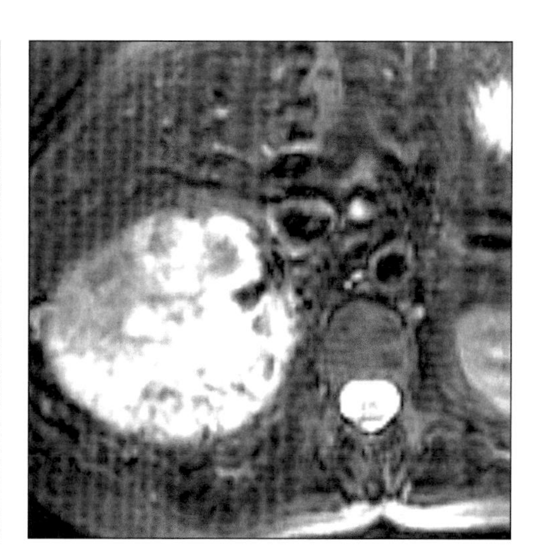

*(**Left**) Coronal T2WI MR shows large right adrenal mass that is heterogeneous and hyperintense. (**Right**) Axial T2WI NEMR shows heterogeneous hyperintense right adrenal pheochromocytoma.*

Typical

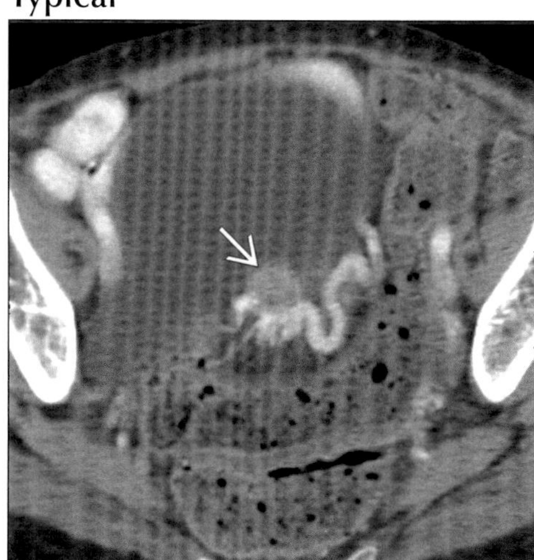

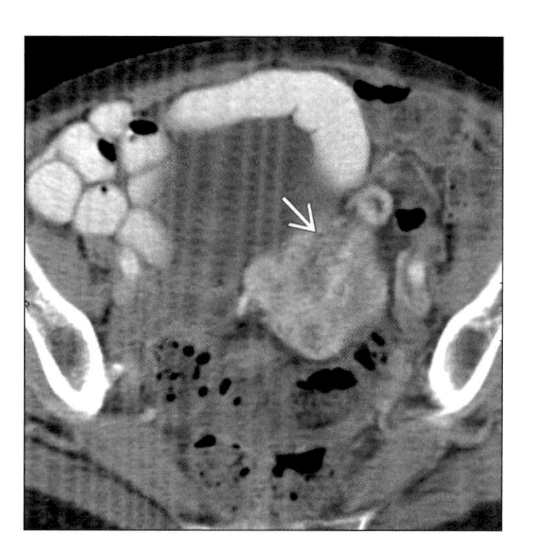

*(**Left**) Axial CECT shows hypervascular mass (arrow) adjacent to bladder; pelvic pheochromocytoma (paraganglioma). (**Right**) Axial CECT shows hypervascular paraganglioma in a patient with headache and palpitation associated with voiding.*

ADRENAL CARCINOMA

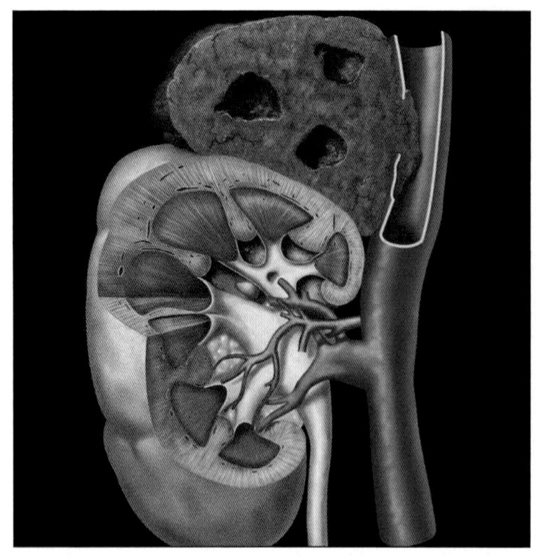

Graphic shows large, hypervascular adrenal mass directly invading the inferior vena cava.

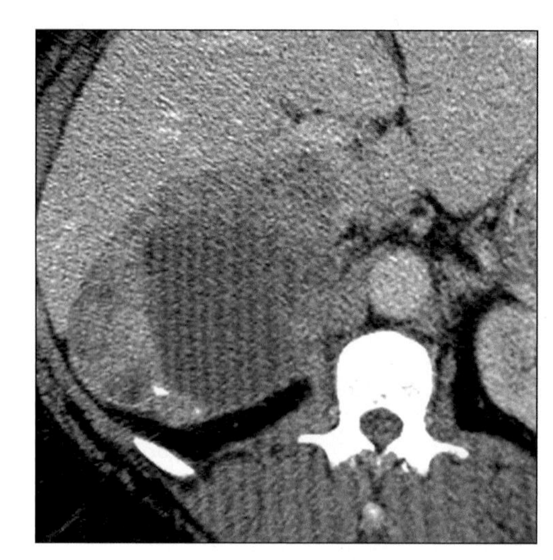

Axial CECT shows large heterogeneous adrenal carcinoma with direct invasion of the inferior vena cava.

TERMINOLOGY

Abbreviations and Synonyms

- Adrenocortical carcinoma, adrenal cancer

Definitions

- Malignant growth from one of the adrenal cell lines

IMAGING FINDINGS

General Features

- Best diagnostic clue: Large, solid, unilateral adrenal mass with invasive margins (bilateral in 10%)
- Location: Suprarenal, usually unilateral (left > right)
- Size
  - Functioning tumors: Usually 5 cm at presentation
  - Nonfunctioning tumors: 10 cm or more
- Morphology
  - Large suprarenal invasive lesion
  - Usually contain hemorrhagic, cystic & calcific areas
- Key concepts
  - Rare & highly malignant neoplasm of adrenal cortex
  - Accounts for 0.05-0.2% of all cancers in U.S.A
  - Mostly unilateral, but bilateral in up to 10% of cases
  - Rarely, etiology of asymptomatic incidentaloma
  - Local spread: Renal vein, inferior vena cava (IVC)
  - Metastatic spread: Lungs, liver, nodes, bone
    - 20% of cases metastatic at presentation
  - Most patients are at stage 3 or 4 at time of diagnosis
  - Functioning (< 50%); nonfunctioning (> 50%)
  - Clinical syndromes with functioning tumors
    - Cushing syndrome most common (30-40%)
    - Virilization in females (20-30%)
    - Conn syndrome, feminization (males) < common
  - Accounts for only 0.002% of all childhood cancers
    - Mostly functional, virilization seen in 95% cases

CT Findings

- Solid, well-defined suprarenal mass + invasive margins
- Usually unilateral; may be bilateral in 10% of cases
- ± Areas of necrosis, hemorrhage, calcification, fat
  - Calcification within tumor seen in 30% of cases
- Variable enhancement (necrosis & hemorrhage)
- ± Renal vein, IVC, adjacent renal extension
- Metastases to lung bases, liver or nodes

MR Findings

- T1WI: Hypointense adrenal mass compared to liver
- T2WI: Hyperintense adrenal mass compared to liver
- T1 C+: Heterogeneous enhancement (tumor necrosis)
- Multiplanar contrast enhanced imaging
  - Renal vein, IVC & adjacent renal parenchymal invasion well-depicted on MR
  - Sagittal imaging helps to evaluate IVC invasion
  - Delineate tumor-liver interface if tumor is on right

DDx: Adrenal Mass

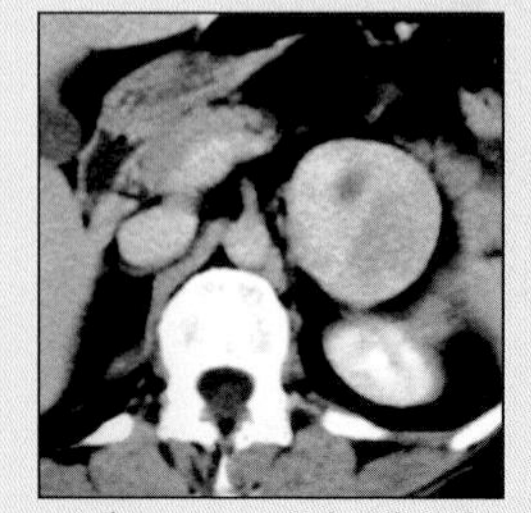

Adenoma with Bleed

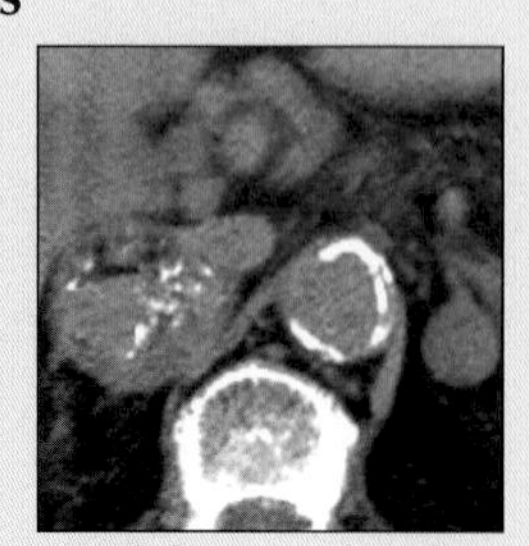

Myelolop. + Adenoma

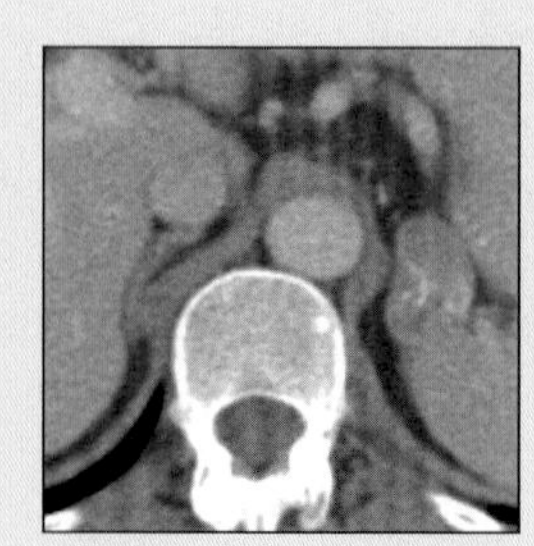

Gastric Diverticulum

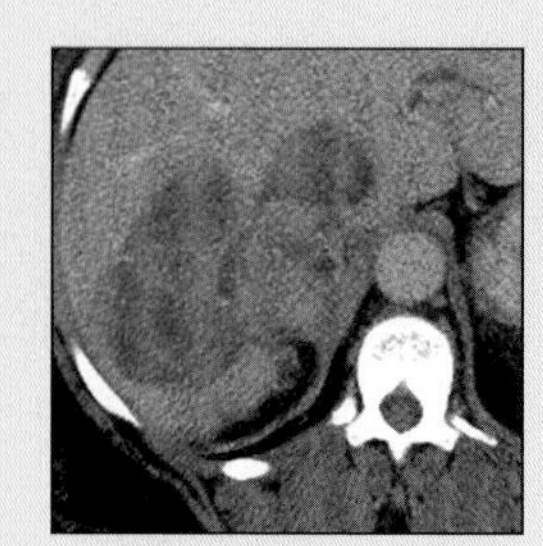

Adrenal Metastasis

ADRENAL CARCINOMA

Key Facts

Imaging Findings

- Best diagnostic clue: Large, solid, unilateral adrenal mass with invasive margins (bilateral in 10%)
- Functioning tumors: Usually 5 cm at presentation
- Nonfunctioning tumors: 10 cm or more
- ± Areas of necrosis, hemorrhage, calcification, fat
- Variable enhancement (necrosis & hemorrhage)
- ± Renal vein, IVC, adjacent renal extension
- Metastases to lung bases, liver or nodes

Top Differential Diagnoses

- Adrenal adenoma
- Adrenal Metastases & lymphoma
- Adrenal myelolipoma (myelolop.)
- Pheochromocytoma
- Renal cell carcinoma (RCC) upper pole

Pathology

- May be associated with genetic syndromes
- Beckwith-Wiedemann, Li-Fraumeni, Carney & MEN type 1

Clinical Issues

- Cushing syndrome (30-40%): ↑ Cortisol
- Virilization in females (20-30%): ↑ Androgens
- Conn syndrome (primary hyperaldosteronism)
- Feminization in males: ↑ Androgens

Diagnostic Checklist

- Rule out other adrenal tumors especially adenoma
- Large, unilateral adrenal mass with invasive margins + venous, nodal invasion + distant metastases
- Precise definition of cephalad extension of tumor venous thrombus is essential for surgical resection

Ultrasonographic Findings

- Real Time
  - Variable appearance depending on size & contents
  - Small tumors: Echo pattern similar to renal cortex
  - Large tumors: Mixed heterogeneous echo pattern (due to areas of necrosis & hemorrhage)

Angiographic Findings

- Conventional
  - Selective catheterization
    - Inferior phrenic artery opacifies superior adrenal artery, which is often predominant arterial supply
    - Renal artery opacifies inferior adrenal artery
    - Middle adrenal artery arise from aorta
  - Enlarged adrenal arteries; minimal neovascularity
  - Inferior venacavography: Confirms tumor invasion

Nuclear Medicine Findings

- FDG-PET
  - Adrenal carcinoma: Increased uptake of FDG
    - Differentiates from adenoma by lack of ↑ uptake
- Adrenocortical scintigraphy by using NP-59
  - No uptake in either gland with large tumor
    - Carcinoma side: Most of the gland is destroyed
    - Contralateral side: Carcinoma ↑ hormone release → pituitary feedback shutdown of normal gland

Imaging Recommendations

- NE + CECT: Study of choice to exclude adenoma
  - 3 mm slice sections with 3 mm increments for precise cephalad extension of tumor thrombus

DIFFERENTIAL DIAGNOSIS

Adrenal adenoma

- Well-defined, soft tissue adrenal mass of 0-20 HU
  - Lipid rich adenoma: Low density (< 10 HU) mass
  - Lipid poor adenoma: Varied density of 10-20 HU
- Enhancing adrenal mass that "de-enhances" rapidly
- Washout of adenoma: 10 min. post injection > 50%
  - Characteristic & diagnostic of adrenal adenoma
  - Similar washout pattern: Lipid rich & poor adenoma

Adrenal Metastases & lymphoma

- Adrenal metastases
  - Unilateral or bilateral masses ± necrosis, hemorrhage
  - Size: Usually < 5 cm, may be larger in melanoma
  - Mostly hypervascular & prolonged washout pattern
  - Usually known to have malignancy elsewhere
- Adrenal lymphoma
  - Primary (rare); secondary (non-Hodgkin common)
  - Often bilateral; retroperitoneal disease usually seen
  - Discrete or diffuse mass, shape is maintained
  - Extensive retroperitoneal tumor engulfing adrenal
  - Hypovascular, moderate enhancement with contrast

Adrenal myelolipoma (myelolop.)

- Small or large, asymptomatic suprarenal mass
- Intramural fatty elements seen on imaging

Pheochromocytoma

- Tumor > 3 cm in most cases
- Highly vascular tumor prone to hemorrhage, necrosis
- Tumors are very hyperintense on T2WI
- Bilateral adrenal tumors in multiple endocrine neoplasia (MEN) IIA & IIB syndromes
- Clinical presentation & lab data may be helpful

Renal cell carcinoma (RCC) upper pole

- Large upper pole RCC mimics large adrenal carcinoma
- Angiography
  - Selective injection of renal artery: Downward displacement of kidney with tumor vascularity
  - Selective injection of superior or middle adrenal arteries: Opacifies tumor & confirm adrenal etiology
  - Hypervascular (RCC); hypovascular (adrenal cancer)
  - Neovascularity: Predominant with RCC than adrenal

Adrenal hemorrhage

- Adrenal hematomas usually appear round in shape
- Some poorly marginated & show periadrenal stranding
- Acute: High density fluid collection (40-60 HU)
- Chronic: Large, well-defined, low density hematoma
- No enhancement of hematoma on CECT

ADRENAL CARCINOMA

Ganglioneuroma
- Younger patients, mean age of 27 years
- Larger mass with average tumor size of 8 cm
- Well-circumscribed mass simulating adrenal cancer

PATHOLOGY

General Features
- Genetics
  - Usually shows loss of heterozygosity at some loci
  - More likely to be aneuploid or tetraploid
  - Genetic syndromes may ↑ incidence of tumor
- Etiology: Unknown for sporadic adrenal carcinoma
- Epidemiology
  - 0.05-0.2% of all cancers
  - 2 new cases per 1 million population
  - 1 per 1500 adrenal tumors are malignant
  - 20% have metastatic disease at presentation
- Associated abnormalities
  - May be associated with genetic syndromes
    - Beckwith-Wiedemann, Li-Fraumeni, Carney & MEN type 1

Gross Pathologic & Surgical Features
- Usually large & predominantly yellow on cut surface
- Necrotic, hemorrhagic, calcific, lipoid & cystic areas

Microscopic Features
- Well differentiated to markedly anaplastic cells

Staging, Grading or Classification Criteria
- Staging of adrenal carcinoma
  - T1: Diameter ≤ 5 cm without local invasion
  - T2: Diameter > 5 cm without local invasion
  - T3: Any size tumor with local invasion but not involving adjacent organs
  - T4: Any size tumor with local invasion & extension into adjacent organs, nodes & distant metastases

CLINICAL ISSUES

Presentation
- Most common signs/symptoms
  - Presentation with non-hormonally active malignancy
    - Abdominal pain, fullness or palpable mass
    - Incidentally discovered mass on imaging exam
    - Metastatic disease in lung, liver ± bone (20% at presentation)
    - 54% of cases nonfunctioning at presentation
- Presentation with hormonally active malignancy
  - Cushing syndrome (30-40%): ↑ Cortisol
    - Moon facies, truncal obesity, purple striae & buffalo hump
  - Virilization in females (20-30%): ↑ Androgens
    - 95% of children with functioning adrenal carcinoma present with virilization
  - Conn syndrome (primary hyperaldosteronism)
    - Hypertension & weakness
  - Feminization in males: ↑ Androgens
- Other clinical syndromes at presentation
  - Hypoglycemia, polycythemia & nonglucocorticoid-related insulin resistance

Demographics
- Age
  - Bimodal distribution
    - 1st peak below age 5
    - 2nd peak in 4-5th decades of life
- Gender
  - Overall, females more than males
    - Females account for 65-90% of all cases
    - Functioning tumors: More common in females
    - Nonfunctioning tumors: More common in males

Natural History & Prognosis
- Rapid growth with local invasion & distant metastases
- Tumor thrombus: IVC & renal vein
- Mean survival 18 months; children better than adults
- 5 year survival for stage 3 disease is under 30%
- Stage 1 & 2: Good prognosis after surgical removal
- Stage 3 & 4: Poor prognosis with or without treatment

Treatment
- Small lesions: Laparoscopic adrenalectomy
- Large lesions with extension: Radical resection of ipsilateral kidney, adrenal gland, adjacent structures
- Metastatic sites also resected as possible
- Chemotherapy: Mitotane, cisplatin, 5-FU & suramin

DIAGNOSTIC CHECKLIST

Consider
- Rule out other adrenal tumors especially adenoma

Image Interpretation Pearls
- Large, unilateral adrenal mass with invasive margins + venous, nodal invasion + distant metastases
- Precise definition of cephalad extension of tumor venous thrombus is essential for surgical resection

SELECTED REFERENCES

1. Caoili EM et al: Adrenal masses: characterization with combined unenhanced and delayed enhanced CT. Radiology. 222(3):629-33, 2002
2. Dunnick NR et al: Imaging of adrenal incidentalomas: current status. AJR Am J Roentgenol. 179(3):559-68, 2002
3. Dunnick NR: Adrenal masses. AJR Am J Roentgenol. 179(5):1344, 2002
4. Mayo-Smith WW et al: State of the art adrenal imaging. RadioGraphics. 21:995-1012, 2001
5. Siegelman ES: MR imaging of the adrenal neoplasms. Magn Reson Imaging Clin N Am. 8(4):769-86, 2000
6. Otal P et al: Imaging features of uncommon adrenal masses with histopathologic correlation. Radiographics. 19(3):569-81, 1999
7. Dunnick NR et al: Adrenal radiology: distinguishing benign from malignant adrenal masses. AJR Am J Roentgenol. 167(4):861-7, 1996
8. McLoughlin RF et al: Tumors of the adrenal gland: findings on CT and MR imaging. AJR Am J Roentgenol. 163(6):1413-8, 1994
9. Fishman EK et al: Primary adrenocortical carcinoma. CT evaluation with clinical correlation. AJR Am J Roentgenol. 148: 531-5, 1987

IMAGE GALLERY

Typical

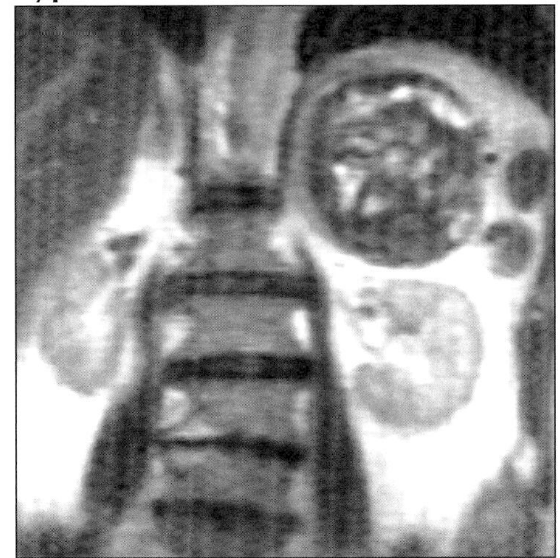

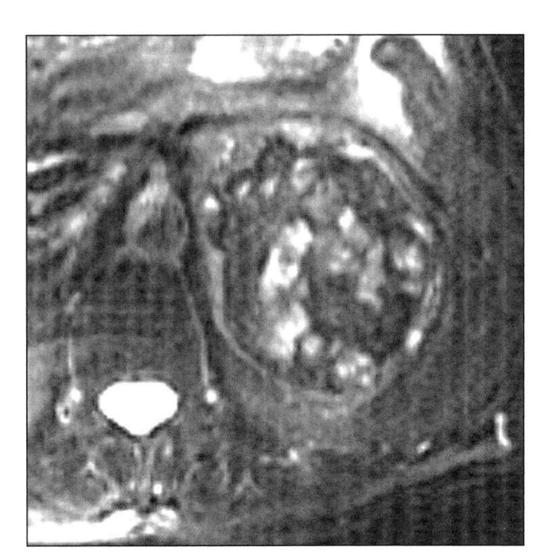

(Left) *Coronal T2WI MR shows heterogeneous, hypointense, large left adrenal carcinoma.* ***(Right)*** *Axial T2WI MR shows heterogeneous, large left adrenal carcinoma.*

Typical

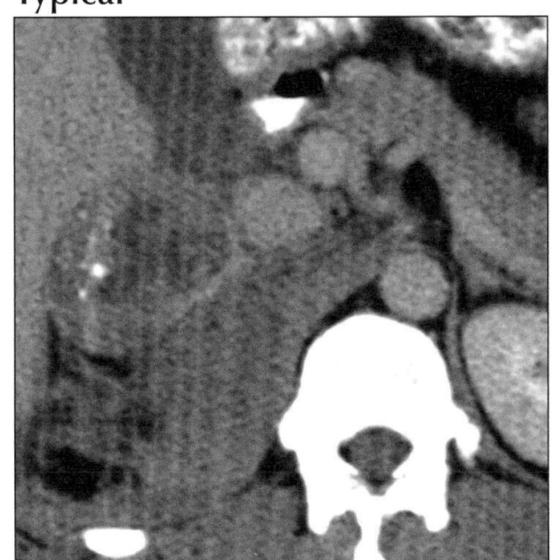

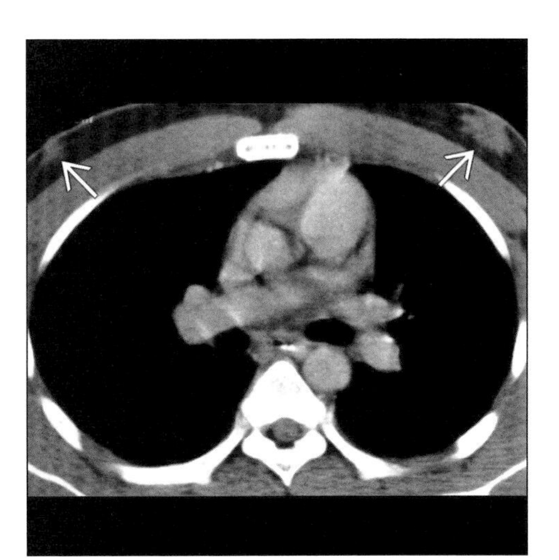

(Left) *Axial CECT shows large, heterogeneous, partially calcified right adrenal carcinoma with spontaneous retroperitoneal bleed.* ***(Right)*** *Axial CECT of 30 year old man shows gynecomastia (arrows), his presenting complaint, due to adrenal carcinoma.*

Typical

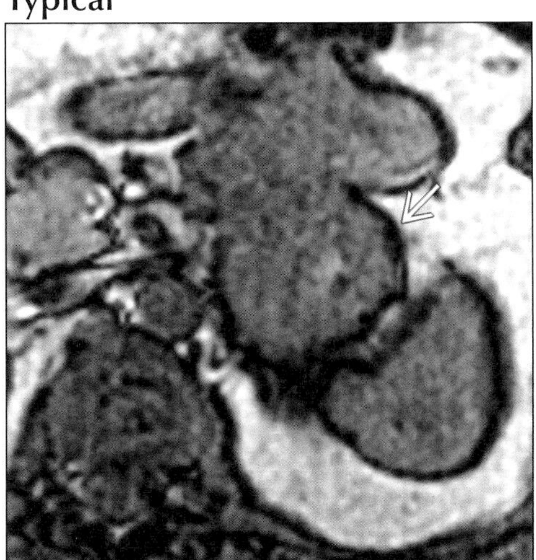

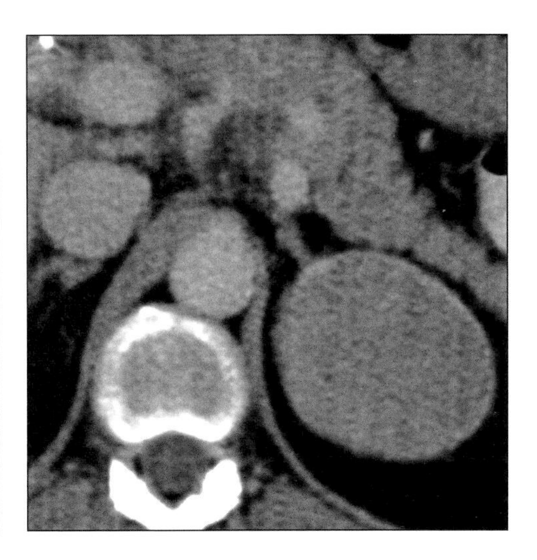

(Left) *Axial NEMR, opposed phase GRE, shows no signal dropout from adrenal mass (arrow); carcinoma.* ***(Right)*** *Axial CECT shows an unusually homogeneous left adrenal carcinoma.*

ADRENAL METASTASES AND LYMPHOMA

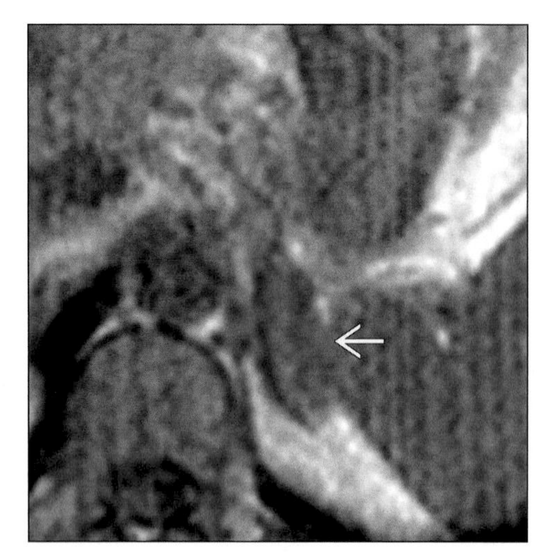

Axial T1WI MR in phase shows left adrenal mass (arrow); metastatic lung cancer.

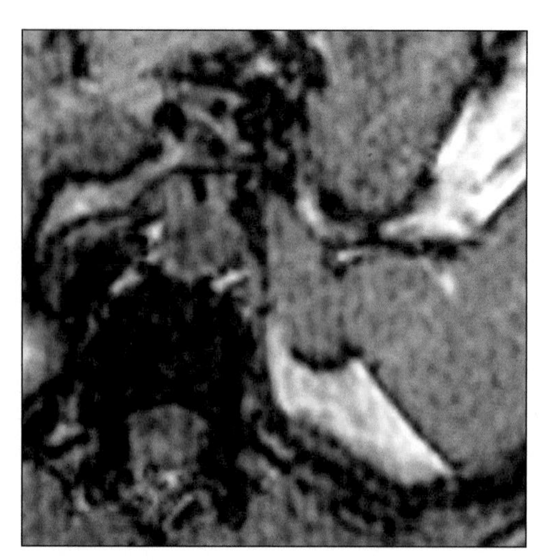

Axial T1WI MR opposed phase shows no signal dropout; metastatic lung cancer.

TERMINOLOGY

Definitions

- Adrenal metastases from other primary cancer sites
- Adrenal lymphoma: Malignant tumor of B-lymphocytes

IMAGING FINDINGS

General Features

- Best diagnostic clue: Discrete or diffuse suprarenal masses of soft tissue density
- Key concepts
  - Adrenal metastases
    - Adrenals are common site of metastatic disease
    - Adrenal glands are 4th most common site of metastases after lungs, liver & bone
    - Most common primary sites: Lung, breast, skin (melanoma), kidney, thyroid & colon cancers
    - Indicates stage IV of distant metastatic disease
    - Seen in 27% of autopsy cases of epithelial origin
    - May be unilateral or bilateral; small or large
    - Mostly, discrete intraparenchymal masses
    - Direct contiguous extension into adrenals may occur due to malignancy in surrounding organs
    - Malignant melanoma: 50% metastasize to adrenal glands
    - Lung & breast cancers: 30-40% metastasize to adrenal glands
    - Renal & GI tract malignancies: 10-20% metastasize to adrenal glands
    - Most often, adrenal metastases are clinically silent
    - Occasionally, present with adrenal insufficiency
  - Adrenal lymphoma
    - Lymphoma of adrenal gland is unusual
    - Autopsy: 25% cases of lymphoma involve adrenals
    - Primary lymphoma (rare); secondary (common)
    - Non-Hodgkin (common); Hodgkin (rare)
    - Non-Hodgkin: 4% cases have adrenal involvement
    - Diffuse cell type predominate over nodular type
    - Bilateral (50%); usually associated with retroperitoneal adenopathy

CT Findings

- Adrenal metastases
  - Small metastases
    - Well-defined, round or oval in shape
    - Homogeneous soft tissue density masses
    - Unilateral or bilateral
    - Adrenal gland contour is maintained
    - Necrosis, hemorrhage & calcification (rare)
    - May mimic lipid poor adrenal adenoma on NECT
  - Large metastases
    - Unilateral or bilateral enlarged adrenal glands
    - Lobulated or irregular in shape
    - Heterogeneous density (necrosis, hemorrhage)

DDx: Adrenal Mass

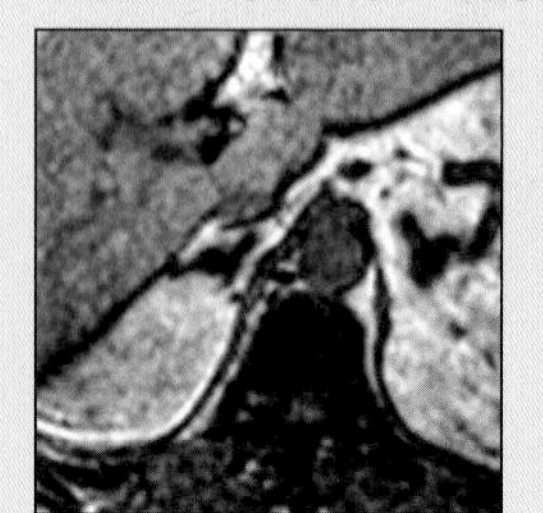

Adrenal Adenomas

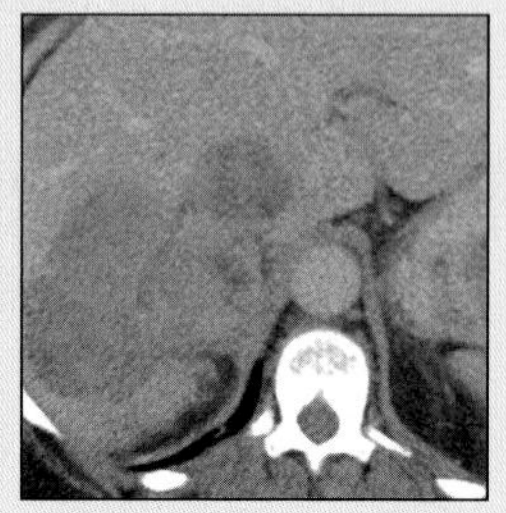

Adrenal Carcinoma

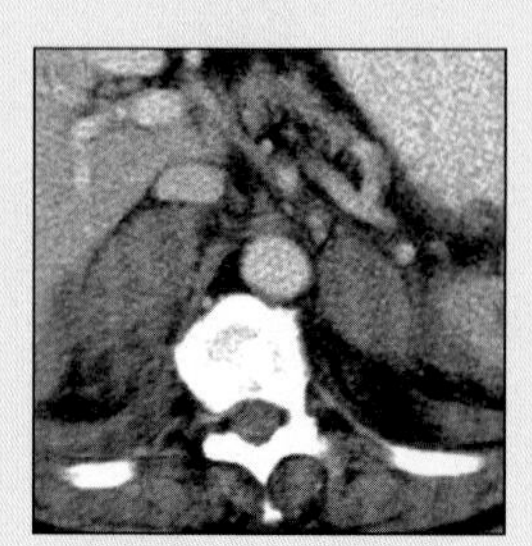

Adrenal Hemorrhage

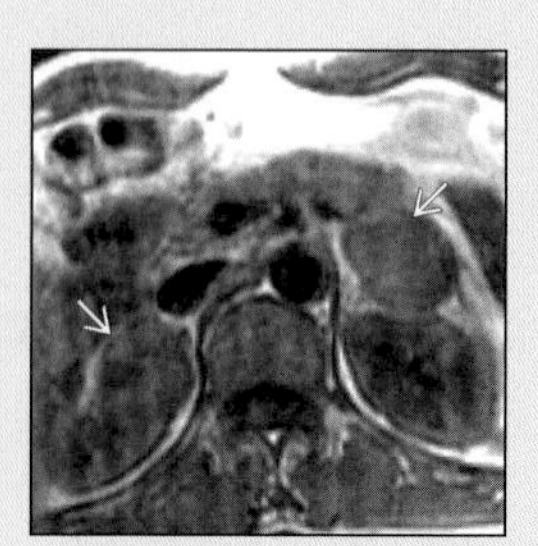

Pheochromocytomas

ADRENAL METASTASES AND LYMPHOMA

Key Facts

Imaging Findings

- Best diagnostic clue: Discrete or diffuse suprarenal masses of soft tissue density
- ± Central necrosis, hemorrhage, calcification
- May have thick enhancing rims (metastases)
- Hypervascular or hypovascular
- Prolonged washout pattern on CECT (metastases)
- Washout value of < 50% after 10-15 min.: Indicates either metastases or atypical adenoma

Top Differential Diagnoses

- Adrenal adenoma
- Adrenal carcinoma
- Adrenal hemorrhage
- Adrenal pheochromocytoma
- Adrenal myelolipoma

Pathology

- Primary malignant tumors of lung, breast, skin (melanoma), kidney, thyroid & GI tract
- Adrenal lymphoma: Often secondary, non-Hodgkin

Clinical Issues

- Almost always clinically silent
- Addison disease: Weakness, weight loss, anorexia, nausea, vomiting, hypotension, skin pigmentation
- Clinical profile: Patient with history of malignancy elsewhere & manifestations of Addison disease
- Diagnosis: Imaging & percutaneous needle biopsy

Diagnostic Checklist

- Check for history of primary cancer or lymphoma
- Overlapping findings: Adrenal metastases, lymphoma
- Imaging important to suggest & stage malignancy

        - Distortion of normal contour of adrenal gland
        - ± Central necrosis, hemorrhage, calcification
        - May have thick enhancing rims (metastases)
        - Hypervascular or hypovascular
        - ± Invasion of contiguous organs such as kidneys
        - Prolonged washout pattern on CECT (metastases)
        - Washout value of < 50% after 10-15 min.: Indicates either metastases or atypical adenoma
    - Direct contiguous adrenal invasion
        - Tumor growth of kidney, pancreas, stomach, liver & retroperitoneal sarcoma may be seen
- Adrenal lymphoma
    - Discrete or diffuse mass, shape is maintained
    - Configuration of adrenal limbs is preserved
    - Homogeneous soft tissue density on NECT
    - Hypovascular, moderate enhancement with contrast
        - Attenuation values vary between 40-60 HU
    - Unilateral or bilateral (in 50% of cases)
    - Usually associated with retroperitoneal adenopathy
    - Extensive retroperitoneal tumor engulfing adrenal gland
    - Necrosis is uncommon without prior therapy
        - Rapidly growing lymphoma may show necrosis
    - Adrenal lymphoma may mimic lipid poor adenoma

MR Findings

- Adrenal metastases
    - Without necrosis & hemorrhage
        - T1WI: Usually homogeneous & hypointense
        - T2WI: Relatively hyperintense (due to fluid content)
    - With necrosis & hemorrhage
        - T1 & T2WI: Heterogeneous signal intensity
    - Exception with metastatic malignant melanoma
        - Hyperintense on T1WI
        - Occasionally remain hyperintense on T2WI mimicking pheochromocytoma
        - Hyperintense on T1WI out of phase & T1WI fat-saturated sequences exclude adenoma
- Adrenal lymphoma
    - Nonspecific MR findings

Ultrasonographic Findings

- Real Time
    - Adrenal metastases
        - Solid lesions with heterogeneous echogenicity
        - Usually echogenicity is less than surrounding fat (hypoechoic)
    - Adrenal lymphoma
        - Relatively homogeneous, hypoechoic lesions
        - ± Areas of echogenicity within mass lesion
        - Enlarged suprarenal glands

Angiographic Findings

- Conventional
    - Adrenal metastases
        - Hypervascular: RCC or sarcoma metastases
        - Hypovascular: Squamous cell cancer metastases
    - Adrenal lymphoma
        - Hypovascular adrenal mass
        - Palisading of vessels may be seen
        - Infiltration: Encasement & amputation of vessels

Nuclear Medicine Findings

- FDG-PET
    - Malignant lesions (adrenal metastases & lymphoma)
        - Increased uptake
    - Benign lesion (adrenal adenoma)
        - No increased uptake
- [Iodine-131] 6-iodomethyl-19-norcholesterol (NP-59)
    - Malignant lesions (adrenal metastases & lymphoma)
        - Lack of uptake
    - Benign lesion (adrenal adenoma)
        - Increased uptake
        - Unilateral early adrenal visualization before day 5 after NP-59 injection
    - Normal adrenal glands or normal NP-59
        - When both adrenal glands are seen on day 5 after injection or thereafter

Imaging Recommendations

- Best imaging tool: Helical NE + CECT; MR
- Protocol advice
    - CT: 3 mm thick section at 3 mm intervals or less
    - MR T1WI in & out of phases to exclude adenoma

ADRENAL METASTASES AND LYMPHOMA

DIFFERENTIAL DIAGNOSIS

Adrenal adenoma

- NECT
  - Lipid rich adenoma: Low attenuation (< 10 HU)
    - Characteristic & diagnostic of adenoma
  - Lipid poor adenoma: Attenuation varies, 10-30 HU
    - Simulate metastases on NECT
  - Large adenoma: Heterogeneous density
    - Due to hemorrhage, cystic degeneration, Ca++
- CECT
  - Lipid rich & poor adenomas
    - Washout of adenoma: 10 min. post I.V. > 50%
    - Washout pattern: Diagnostic for adenoma
- T1WI out of phase
  - Lipid rich adenoma: Marked signal "drop-out"
  - Lipid poor adenoma: Minimal signal loss

Adrenal carcinoma

- Rare, unilateral, invasive & enhancing mass
- More than 6 cm when initially diagnosed

Adrenal hemorrhage

- Acute: High density fluid collection (40-60 HU)
- Chronic: Low density collection (clot of 20-30 HU)
  - May mimic as metastases or lymphoma

Adrenal pheochromocytoma

- Highly vascular; prone to hemorrhage & necrosis
- Tumors are very hyperintense on T2WI
- Bilateral adrenal tumors in MEN IIA & IIB syndromes

Adrenal myelolipoma

- Tumor composed of fat & hematopoietic elements
- Usually are heterogeneous fatty adrenal masses
- Tumor with ↑ fat content
  - Fat suppression sequence: Hypointense
  - Marked signal drop-out at fat-soft tissue interfaces

PATHOLOGY

General Features

- General path comments
  - Normal anatomy of adrenal glands
    - Located between vertebrae T11 & L2, lateral to body of L1
    - Gland has an anteromedial ridge & 2 limbs
    - Maximum body width: 0.79 mm (Rt); 0.6 mm (Lt)
    - Length of adrenal limbs varies: May be up to 4 cm
    - Width of limbs: < 1 cm (right thinner than left)
- Etiology
  - Adrenal metastases
    - Primary malignant tumors of lung, breast, skin (melanoma), kidney, thyroid & GI tract
  - Adrenal lymphoma: Often secondary, non-Hodgkin
- Epidemiology
  - Adrenal metastases
    - Autopsy series: 27% of cases of epithelial origin
  - Adrenal lymphoma
    - Incidence: 4% of cases with non-Hodgkin
    - Autopsy series: 25% of cases with lymphoma
- Associated abnormalities
  - Primary malignant tumor in case of metastases
  - Generalized adenopathy in case of lymphoma
  - Adrenal collision tumor
    - Metastases & adenoma in same adrenal gland

Gross Pathologic & Surgical Features

- Discrete or diffuse; unilateral or bilateral
- ± Cystic, necrotic, hemorrhagic, calcific areas

Microscopic Features

- Metastases: Varies based on etiology
- Lymphoma: Lymphoepithelial, Reed-Sternberg cells

CLINICAL ISSUES

Presentation

- Most common signs/symptoms
  - Almost always clinically silent
  - Extensive masses: Adrenocortical insufficiency (Addison disease) when 90% of tissue is damaged
  - Addison disease: Weakness, weight loss, anorexia, nausea, vomiting, hypotension, skin pigmentation
- Clinical profile: Patient with history of malignancy elsewhere & manifestations of Addison disease
- Lab data
  - Adrenocortical insufficiency
    - ↓ Cortisol, aldosterone, androgens & ↑ ACTH
    - ↓ Na+, Cl & ↑ K+ levels
- Diagnosis: Imaging & percutaneous needle biopsy

Demographics

- Age: Any age group
- Gender: M = F

Natural History & Prognosis

- Complications: Adrenocortical insufficiency
- Prognosis: Usually poor

Treatment

- Chemotherapy & follow-up

DIAGNOSTIC CHECKLIST

Consider

- Check for history of primary cancer or lymphoma

Image Interpretation Pearls

- Overlapping findings: Adrenal metastases, lymphoma
- Imaging important to suggest & stage malignancy

SELECTED REFERENCES

1. Dunnick NR et al: Imaging of adrenal incidentalomas: Current status. AJR. 179:559-68, 2002
2. Mayo-Smith WW et al: State-of-the-art adrenal imaging. Radiographics. 21(4):995-1012, 2001
3. Dunnick NR et al: Adrenal radiology: Distinguishing benign from malignant adrenal masses. AJR. 167:861-867, 1996
4. Alvarez-Castells A et al: CT of primary bilateral adrenal lymphoma. J Comput Assist Tomogr .17:408-409, 1993
5. Glazer HS et al: Non-Hodgkin lymphoma: Computed tomographic demonstration of unusual extranodal involvement. Radiology. 149:211-217, 1983

IMAGE GALLERY

Typical

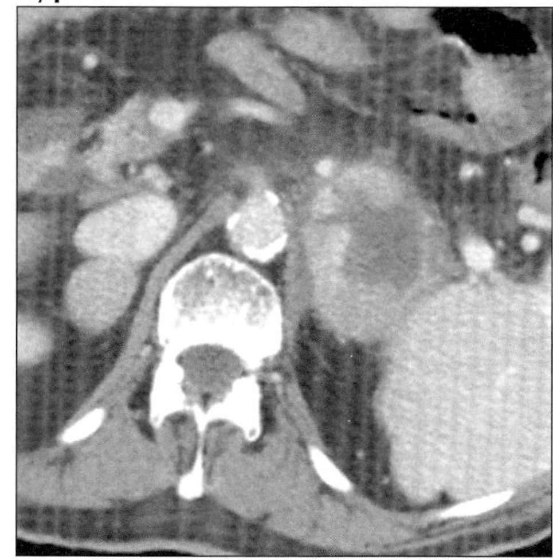

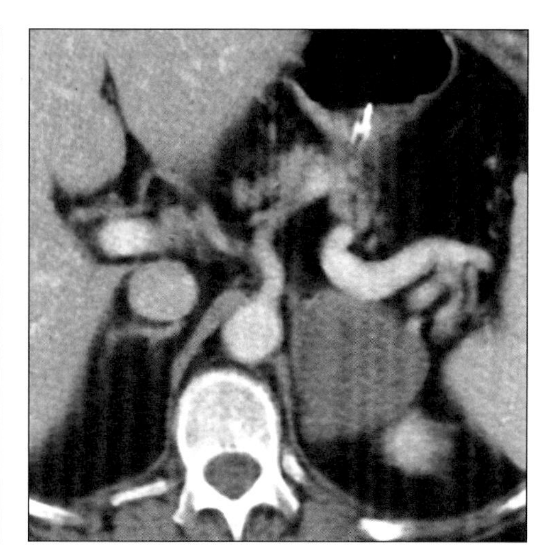

(Left) *Axial CECT shows heterogeneous, bilateral adrenal masses; metastatic breast cancer.* ***(Right)*** *Axial CECT shows an unusually hypodense left adrenal metastasis from breast cancer.*

Typical

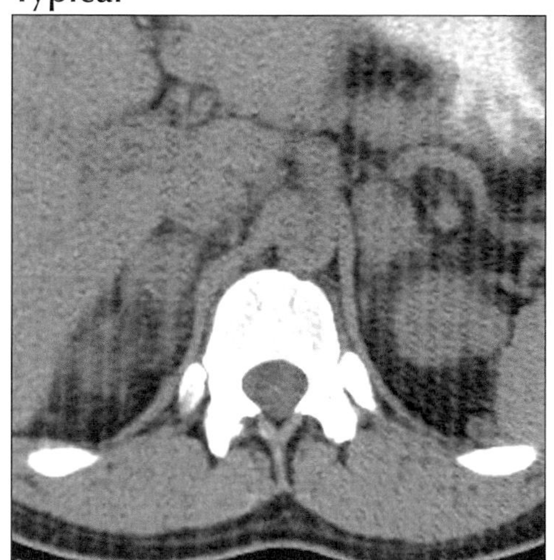

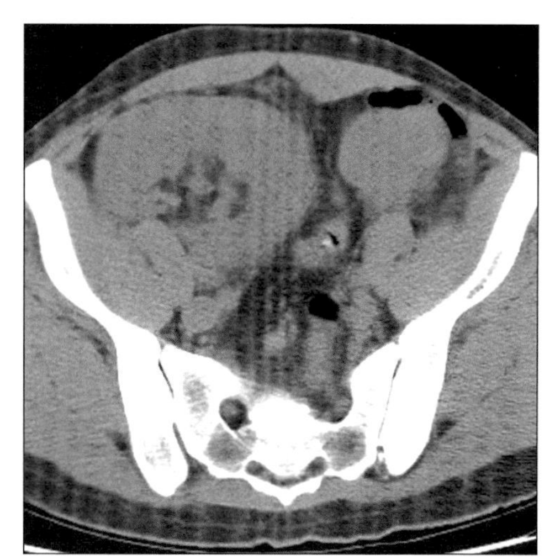

(Left) *Axial NECT in renal transplant patient shows bilateral adrenal masses of lymphoma (post-transplant lymphoproliferative disorder).* ***(Right)*** *Axial NECT shows pelvis in a renal transplant patient with bilateral adrenal lymphoma (PTLD).*

Typical

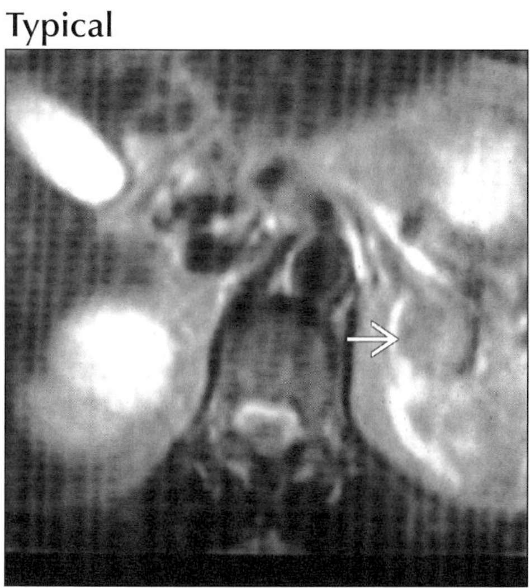

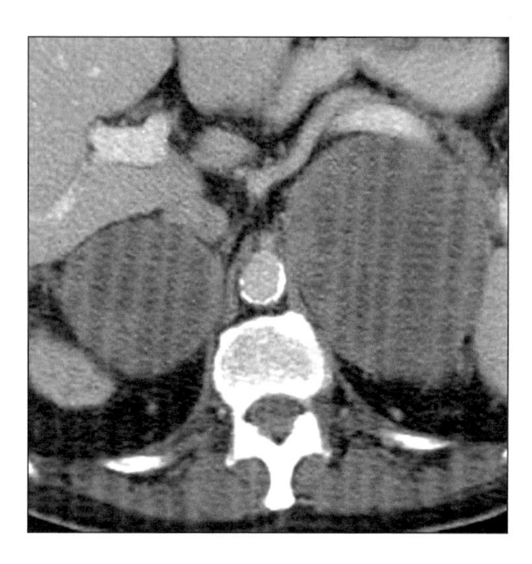

(Left) *Axial T2WI MR (nonenhanced) shows a left adrenal metastasis (arrow) and a right renal cyst.* ***(Right)*** *Axial CECT shows large adrenal masses due to metastatic melanoma.*

ADRENAL COLLISION TUMOR

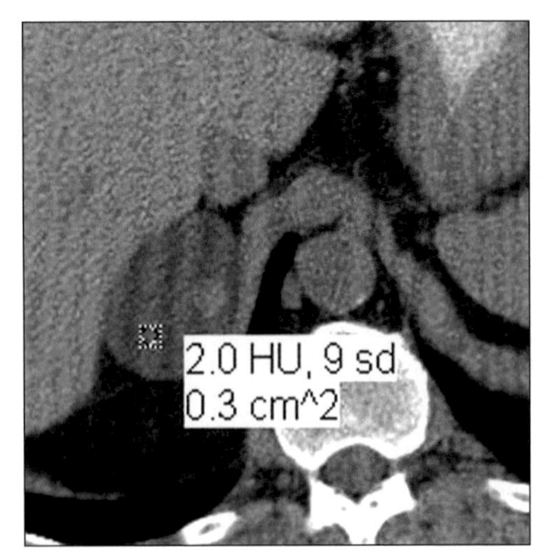

Axial NECT shows near-water density adenoma + a small focus of higher density.

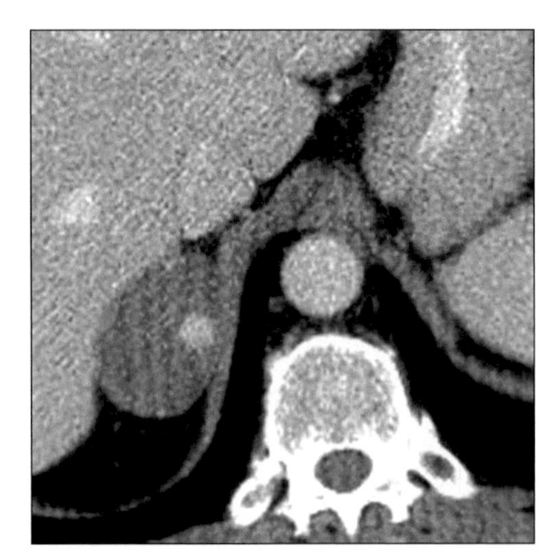

Axial CECT shows bright enhancement of small nodule (melanoma) within larger mass (adenoma) = collision tumor.

TERMINOLOGY

Abbreviations and Synonyms

- Adrenal collision tumor (ACT)

Definitions

- Coexistence of two contiguous but histologically different tumors within same adrenal gland

IMAGING FINDINGS

General Features

- Best diagnostic clue
  - Histological verification by percutaneous biopsy with imaging guidance
    - Two histologically distinct masses; without significant tissue admixture
- Location: Unilateral; both within same adrenal gland
- Both tumors may be malignant, or one may be benign & other malignant; or both benign
  - Adrenal masses may be adenomas, metastases, cysts, myelolipomas, hemangiomas, carcinomas etc.
    - Some have characteristic imaging features

CT Findings

- Nature of mass: Cystic; solid; solid fatty lesions
  - Macroscopic fat within mass; may be association of solid soft tissue lesion with myelolipoma
  - May see sharp demarcation between myelolipomatous & adenomatous components
- Differentiate benign from malignant lesions
  - CT attenuation values; threshold of 10 HU on NECT
    - Sensitivity 71% and specificity 98%
  - Discriminate adenomas from nonadenomas
    - Attenuation values on CECT after 15 minute delay
    - Adenomas diagnosed on delayed CECT at thresholds of 37 HU (specificity, sensitivity; 96%)

MR Findings

- Demonstrates & enables characterization of 2 masses
- Chemical shift MR: Demonstrates lipid; ↑ specificity
  - In-phase & opposed-phase gradient-echo technique
    - Indeterminate masses at CT or conventional MR
    - ACT suspected when there is only focal ↓ in signal intensity of mass on opposed-phase images
    - Opposed-phase image: Adenoma, (quantitative) ↓ signal; metastatic component ↑ signal intensity
    - Myelolipoma, benign cortical mass; show loss of signal intensity; not metastases, hemorrhage, cyst
  - Adrenal-spleen ratio (ASR); quantitative analysis
    - Identifies all adenomas with ratio of less than 70
    - ASR lower for adenoma than for metastases

Nuclear Medicine Findings

- Adrenocortical scintigraphy with NP-59 (131I-6b-iodomethyl-norcholesterol)
  - Differentiate adenomas from nonadenomas

DDx: Adrenal Mass

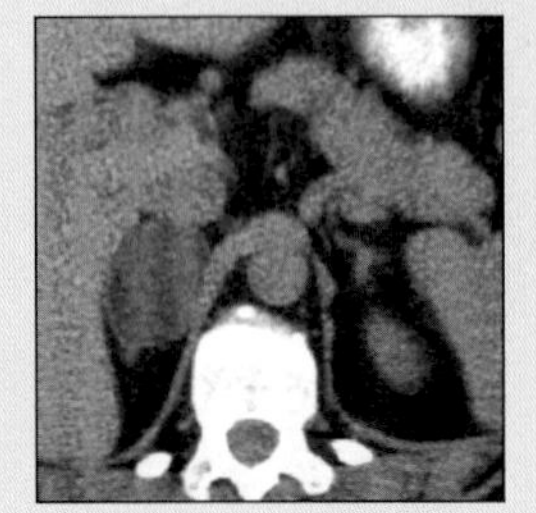

Adenoma

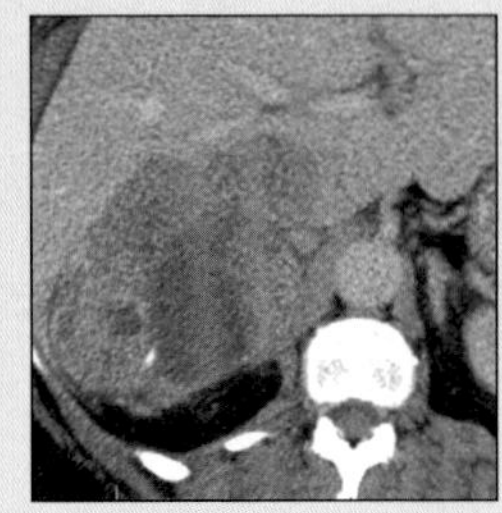

Adrenal Carcinoma

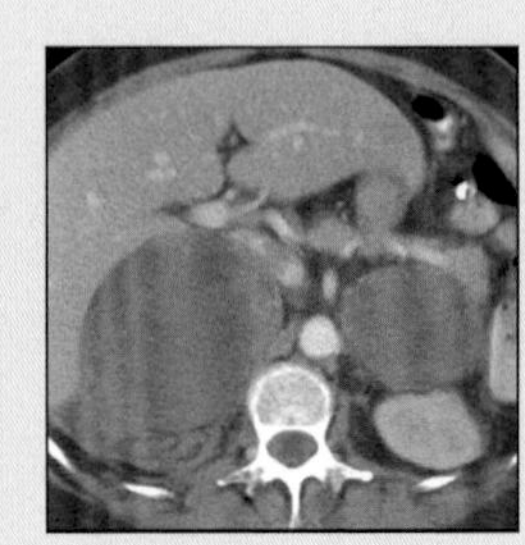

Adrenal Metastases

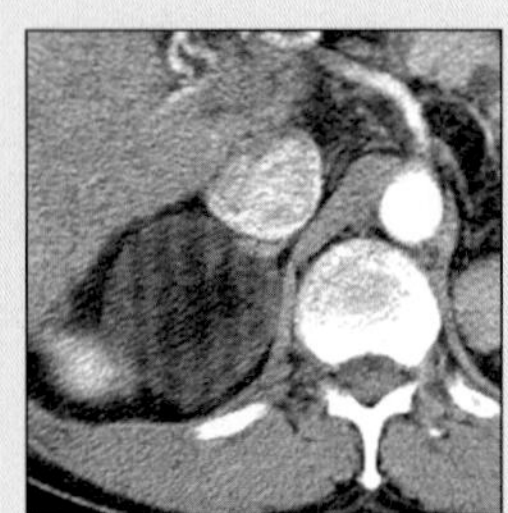

Myelolipoma

ADRENAL COLLISION TUMOR

Key Facts

Terminology

- Coexistence of two contiguous but histologically different tumors within same adrenal gland

Imaging Findings

- Histological verification by percutaneous biopsy with imaging guidance
- Both tumors may be malignant, or one may be benign & other malignant; or both benign
- ACT suspected when there is only focal ↓ in signal intensity of mass on opposed-phase images

Top Differential Diagnoses

- Adenoma
- Carcinoma
- Lymphoma; metastasis
- Myelolipoma
- Hemorrhage; infection

  - Limited in masses < 2 cm in diameter with indeterminate uptake

Imaging Recommendations

- Best imaging tool
  - NECT, CECT or MR
    - Characterization of separate components of ACT
    - Planning & guiding percutaneous needle biopsy
- Protocol advice
  - Lesions ≤ 10 HU; likely benign; no further work-up
  - Lesions ≥ 20 HU; likely malignant; biopsied when result may influence management
  - For indeterminate NECT: Consider CECT w/delayed scans for washout calculation or chemical shift MR
    - ASR threshold of 70; benign lesion; no work-up
    - ASR > 70; biopsy depending on clinical situation

DIFFERENTIAL DIAGNOSIS

Adenoma

- Hypointense to liver on T2WI; < 10 HU on NECT
- Rapid washout pattern of contrast material

Carcinoma

- Large (usually ≥ 5 cm); central necrosis & hemorrhage
- Lung metastases, lymphadenopathy, venous invasion

Lymphoma; metastasis

- Lymphoma: Usually bilateral; hypovascular
- Metastases: Unilateral or bilateral; known malignancy
  - Hyperintense on T2WI; delayed contrast washout

Myelolipoma

- Intramural fatty elements recognized on imaging

Hemorrhage; infection

- Hemorrhage: High density on NECT
- Infection: Granulomatous; AIDS

PATHOLOGY

General Features

- Epidemiology: Extremely rare; prevalence unknown
- Associated abnormalities: Unusual variant of multiple endocrine neoplasia syndrome; ACT as component

Gross Pathologic & Surgical Features

- Two neoplastic processes; may replace adrenal gland

Microscopic Features

- 2 neoplasms; without significant tissue admixture

CLINICAL ISSUES

Presentation

- May be asymptomatic; incidental finding at imaging
- Diagnosis: Percutaneous needle biopsy
  - Noninvasive differential diagnosis using CT, MR

Natural History & Prognosis

- Variable; depends on histology of each component

DIAGNOSTIC CHECKLIST

Consider

- ACT suspected on chemical shift MR; only focal ↓ in signal intensity of mass on opposed-phase images

SELECTED REFERENCES

1. Khati NJ et al: Adrenal adenoma and hematoma mimicking a collision tumor at MR imaging. Radiographics. 19(1):235-9, 1999
2. Otal P et al: Imaging features of uncommon adrenal masses with histopathologic correlation. Radiographics. 19(3):569-81, 1999
3. Schwartz LH et al: Collision tumors of the adrenal gland: demonstration and characterization at MR imaging. Radiology. 201(3):757-60, 1996

IMAGE GALLERY

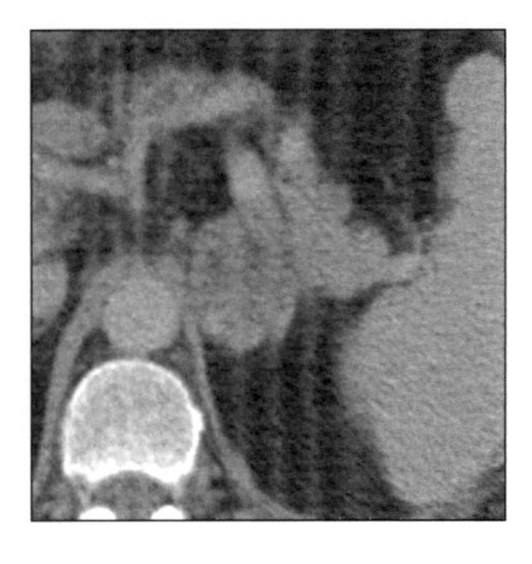

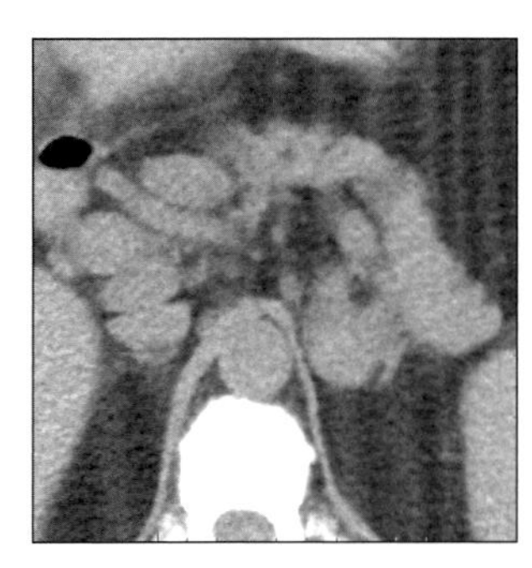

(Left) *Axial NECT shows heterogeneous left adrenal mass, myelolipoma + adenoma = collision tumor* ***(Right)*** *Axial NECT shows heterogeneous left adrenal mass, myelolipoma + adenoma = collision tumor.*

SECTION 3: Kidney and Urinary Tract

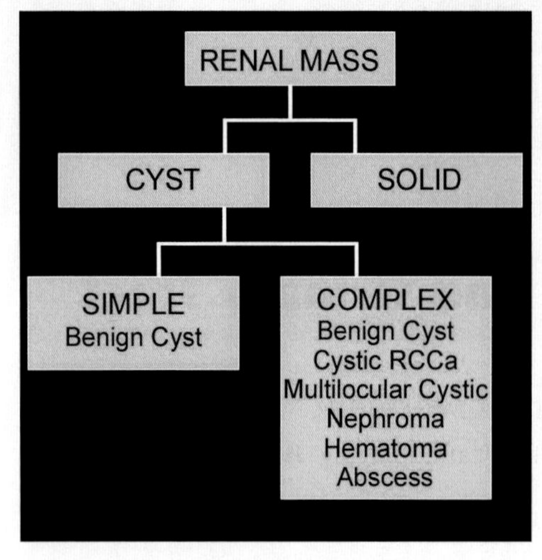

Graphic shows algorithm for analyzing imaging features of a renal cystic mass, listing the most common causes.

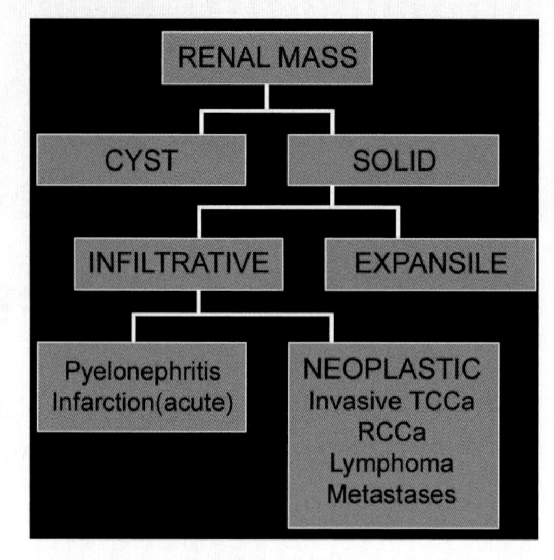

Graphic shows algorithm for analyzing a renal mass, showing main differential for an infiltrative mass.

TERMINOLOGY

Abbreviations and Synonyms

- Renal cell carcinoma (RCC)
- Transitional cell carcinoma (TCC)
- Intravenous pyelogram (excretory urogram) (IVP)
- Xanthogranulomatous pyelonephritis (XGP)
- Tuberculosis (TB)

IMAGING ANATOMY

Location

- Kidneys lie within the perirenal space (with adrenals)
- Congenital anomalies of location (ectopic, rotational), fusion, and number (congenital absence) are relatively common

ANATOMY-BASED IMAGING ISSUES

Key Concepts or Questions

- Current role of IVP
  - Detection of anomalies such as ureteral duplication, ureterocele
  - Complementary to retrograde pyelography and CT/MR for transitional cell carcinoma
  - Alternative (suboptimal) to CT for evaluation of neatly all etiologies of hematuria ± flank pain
    - Calculi, tumor, infection, trauma, vascular
- How can we replace the IVP (urography) with CT completely?
  - By performing CT urography
- How do you perform CT urography?
  - 1000 cc water orally 15 to 20 min prior to CT (to produce diuresis)
  - Nonenhanced scans through kidneys (diaphragm to iliac crest)
  - 100 to 125 ml nonionic contrast IV at 3 ml/sec
  - After 3 min , roll patient 360° (to opacify all segments of bladder and collecting system)
  - Scan diaphragm to pubis with 4 min delay
  - May add earlier phase (40 sec delay) if vascular anatomy is important
  - Rarely add 10 min delay in event of high grade ureteral obstruction
- How can we minimize or deal with the hypodense renal mass "too small to characterize" on CT?
  - Minimize occurrence by
    - Obtaining CT sections less than ½ the diameter of the lesion
    - Using adequate volume of IV contrast (≥ 100 ml in average adult)
    - Avoid scanning in corticomedullary phase
  - Check nonenhanced or delayed CT, if available
    - No visible enhancement or de-enhancement, almost always benign
    - If lesion less dense than blood on NECT, almost always benign
    - Consider clinical setting (e.g., septic; oncology patient; elderly): ≤ 2 cm hypodense mass in very elderly patient usually requires no further evaluation
    - Sonography good at resolving cyst versus solid
- What are the key morphologic features used to distinguish a benign (Bosniak class I or II) form a "surgical" (Bosniak III or IV) cyst?
  - Bosniak II
    - Calcification: Thin, peripheral, milk of calcium
    - Hyperdense: Homogeneous, no enhancement
    - Septations: < 2 mm thick, no nodularity, no enhancement
  - Bosniak III
    - Calcification: Thick (≥ 2 mm), central, irregular
    - Hyperdense: Heterogeneous, enhancement
    - Septations: ≥ 2mm, nodularity, enhancement
- Renal calcifications
  - Dystrophic
  - Nephrocalcinosis
    - Cortical
    - Medullary
  - Calculi (nephrolithiasis)

DIFFERENTIAL DIAGNOSIS

Expansile renal mass (benign)
- Cyst
- Angiomyolipoma
- Oncocytoma
- Multilocular cystic nephroma
- Focal severe pyelonephritis)
- Other rare tumors
- ⇒ (Leiomyoma, fibroma, etc.)

Expansile renal mass (malignant)
- Renal cell carcinoma
- Metastases
- Lymphoma

Infiltrative renal process/mass (benign)
- Pyelonephritis (including XGP, TB)
- Infarction (acute)

Infiltrative renal process/mass (malignant)
- Transitional cell carcinoma
- Lymphoma
- Renal cell carcinoma
- Medullary carcinoma
- Squamous cell carcinoma

Multiple renal cysts
- Multiple simple cysts
- Autosomal dominant polycystic disease
- Multicystic dysplastic kidney
- Acquired cystic kidney disease (uremia)*
- Tuberous sclerosis complex*
- von Hippel Lindau*
- ⇒ *Associated with multiple renal neoplasms

  - Vascular, usually arterial
- Dystrophic calcification
  - Calcification in abnormal, damaged tissue
    - Tumor, infection, infarction
- Cortical nephrocalcinosis
  - Often appears as "egg-shell" calcification
  - Kidneys usually small with diminished or absent function
  - Typical causes
    - Chronic glomerulonephritis
    - Acute cortical necrosis (shock, sepsis, toxins)
    - Transplant rejection
- Medullary nephrocalcinosis
  - Calcification in normal tissue
  - Renal size and function are often normal
  - Pattern: May conform to renal pyramids
  - Often associated with nephrolithiasis (calculi)
    - E.g., medullary sponge kidney, hypercalcemic condition
- Distinguishing among causes of nephrocalcinosis
  - Medullary sponge kidney: Cystic dilation of renal tubules that contain stones +/or fill with contrast on urography; clustered at papillary tips: Calculi become obscured by contrast-opacified urine (unique)
    - Papillary blush" is a normal finding; does not constitute medullary sponge
  - Renal tubular (transient dense opacification of tubules in papilla; in dehydration) acidosis: Dense and extensive calcification of the medullary portions of the renal lobes; electrolyte abnormalities
    - Calcification is not obscured by contrast-opacified urine
  - Renal papillary necrosis
    - Sloughed papilla that may calcify, often as a ring shape
    - May have amorphous calcification of papilla
    - Calcifications are not obscured by opacified urine

CLINICAL IMPLICATIONS

Clinical Importance
- Histologic variations and cell of origin of primary renal carcinomas affect imaging appearance and approach, as well as prognosis
  - Clear cell RCC (hypernephroma)
    - Constitutes 70 to 85% of renal carcinomas
    - Refers to lipid-rich cytoplasm
    - Hypervascular, heterogeneous
    - Often aggressive growth and widely metastatic
  - Papillary RCC
    - Constitutes 10 to 15% of RCC
    - Hypovascular, homogeneous, often encapsulated
    - Easily mistaken for a cyst on imaging studies
    - Slow to metastasize, better prognosis
  - Medullary carcinoma
    - Arises from epithelium of papilla or distal collecting duct
    - Affects young men with sickle cell anemia
    - Very aggressive, infiltrative, metastatic
    - Survival < 4 months average
  - Transitional cell carcinoma
    - Arises from uroepithelium
    - Appears as obstruction or filling defect within collecting system of kidney, ureter or bladder
    - May diffusely infiltrate the kidney
  - Squamous cell carcinoma
    - Due to squamous metaplasia of uroepithelium caused by chronic irritation (stones, infection)
    - Infiltrative, aggressive; survival < 1 year average
    - Imaging resembles XGP
    - Survival < 1 year average

CUSTOM DIFFERENTIAL DIAGNOSIS

Focal or global small, scarred kidney
- Reflux nephropathy (chronic atrophic pyelonephritis)
- Ischemia/infarction

Renal and ischemic infarction
- Embolic more common cause than thrombotic

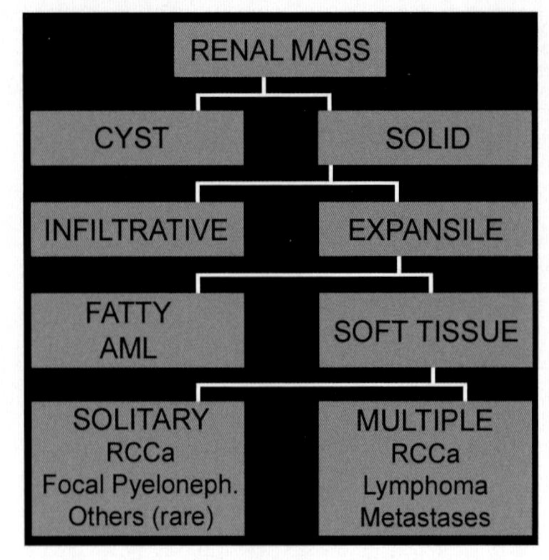

Graphic shows algorithm for analyzing imaging features of a renal mass, showing main differential for a solid expansile mass.

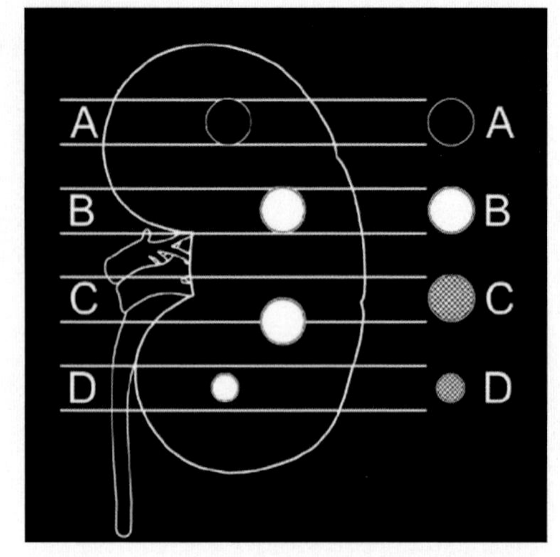

Graphic shows effect of partial volume averaging on appearance of small masses. If a cystic mass (B) fills the CT section, it will appear of water attenuation. If it fills only part of the section (C, D), it will appear of higher density.

- Usually a cardiac source
  - Prosthetic cardiac valves
  - Endocarditis
  - Myocardial infarction
  - Atrial fibrillation
  - Intracardiac catheters/pacemaker wires
  - Cardiac tumors
  - Vasculitis
  - Aortic dissection
  - Atheromatous plaque
- May be asymptomatic; usually causes flank pain, fever, nausea
- Key finding: Wedge-shaped or global nonenhancement of kidney on CECT (acute)
  - Cortical rim sign (enhancement of capsule and peripheral cortex)
  - (Subacute infarction collateral vessels)
  - Focal or global atrophy without caliceal dilation (chronic)

Reflux nephropathy

- Previously known as chronic atrophic pyelonephritis
- Usually due to reflux and episodic urinary infection beginning in infancy or childhood
  - Often present in young adult with hypertension and renal insufficiency
- Can be unilateral or bilateral
- Leads to focal or global decrease in renal size
- Key finding: Cortical loss (scar) over dilated calyx

Unilateral delayed or persistent nephrogram

- Slow blood inflow
  - Renal artery stenosis
- Slow blood outflow
  - Renal vein stenosis/thrombosis
- Slow urine outflow
  - Ureteral obstruction
- Decreased nephron function

Striated nephrogram

- Pyelonephritis
- Contusion
- Vasculitis
- Renal vein thrombosis

Bilateral persistent nephrograms

- Hypotension/shock
- Obstruction of renal tubules
  - Myoglobin, urate, protein, acute tubular necrosis
- Bilateral ureteral, arterial, or venous obstruction

Medullary nephrocalcinosis causes

- Hypercalcemia
  - Hyperparathyroidism, milk-alkali, vitamin D toxicity
- Medullary sponge kidney (renal tubular ectasia)
- Renal tubular acidosis
- Chronic infection
  - TB, AIDS organisms, chronic pyelonephritis

SELECTED REFERENCES

1. Koeller KK et al: Radiologic pathology. 2nd ed. Washington, DC, Armed Forces Institute of Pathology. 487-594, 2003
2. Urban BA et al: Three-dimensional volume-rendered CT angiography of the renal arteries and veins: normal anatomy, variants, and clinical applications. Radiographics. 21(2):373-86; questionnaire 549-55, 2001
3. Pollack HM et al: Clinical Urography. 2nd ed. Philadelphia, WB Saunders, 2000
4. Dähnert W: Radiology review manual. 4th ed. Philadelphia, Lippincott, Williams and Wilkins. 723-56, 2000
5. Nicolau C et al: Autosomal dominant polycystic kidney disease types 1 and 2: assessment of US sensitivity for diagnosis. Radiology. 213(1):273-6, 1999
6. Bosniak MA: Diagnosis and management of patients with complicated cystic lesions of the kidney. AJR Am J Roentgenol. 169(3):819-21, 1997
7. Kawashima A et al: CT of renal inflammatory disease. Radiographics. 17(4):851-66; discussion 867-8, 1997

IMAGE GALLERY

Typical

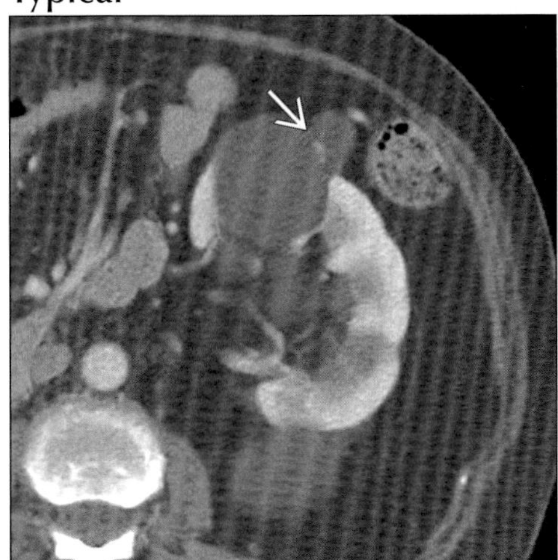

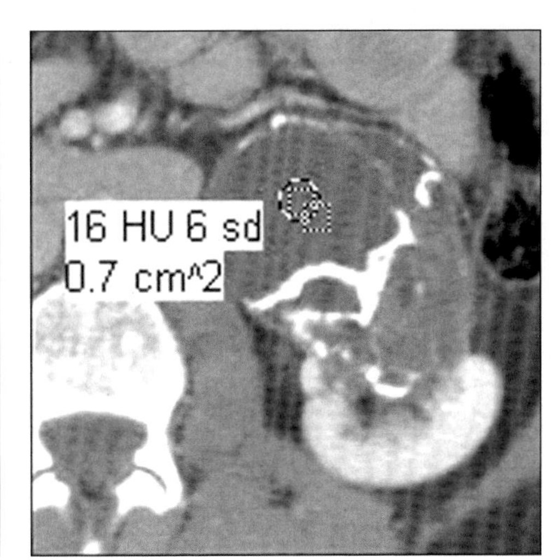

(Left) *Axial CECT shows Bosniak II cyst (arrow) with thin calcified septum, water density.* ***(Right)*** *Axial CECT shows Bosniak III cystic mass (multilocular cystic nephroma); thick, calcified septa.*

Typical

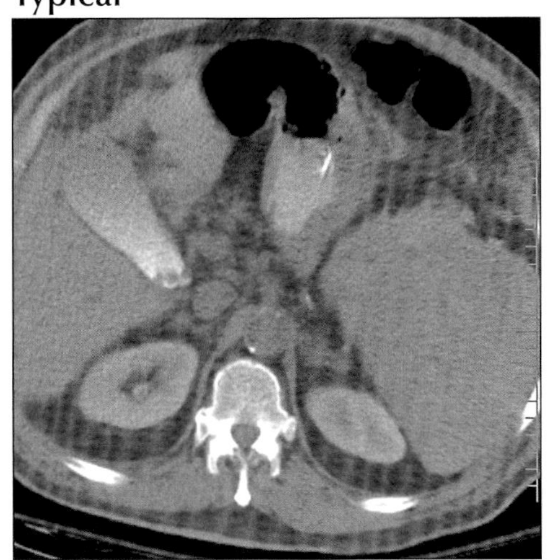

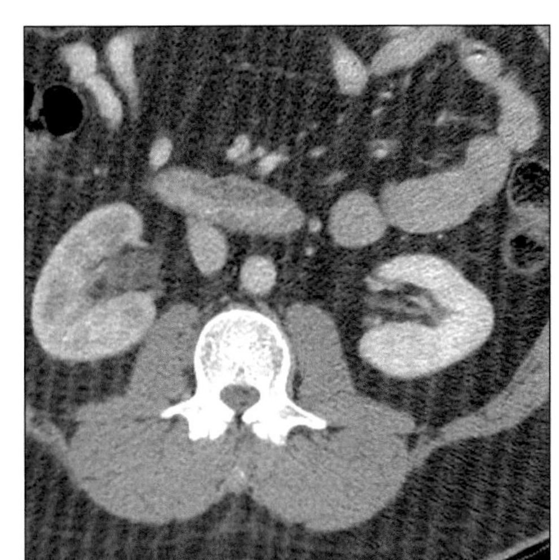

(Left) *Nonenhanced, axial CT shows vicarious excretion (dense bile) and persistent nephrograms due to acute tubular necrosis; shock following splenic laceration.* ***(Right)*** *Axial CECT shows delayed right nephrogram due to acute ureteral obstruction (ureteral calculus).*

Typical

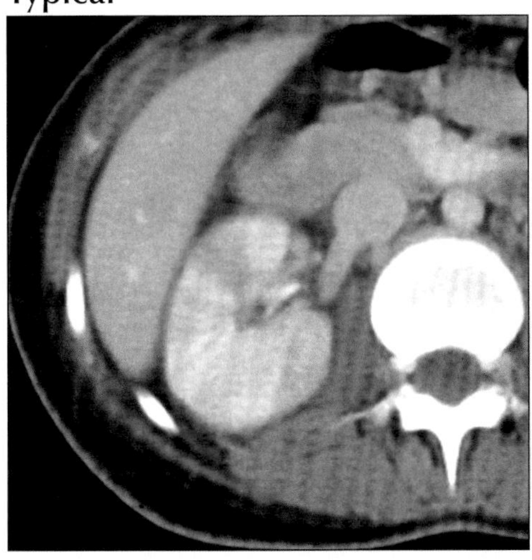

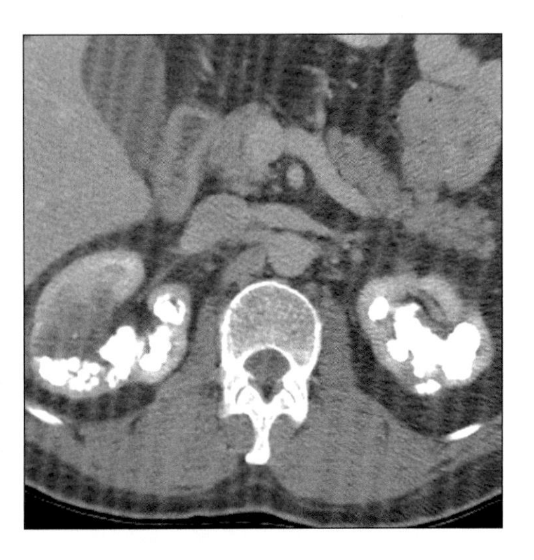

(Left) *Axial CECT shows striated nephrogram; acute pyelonephritis.* ***(Right)*** *Axial CECT shows extensive medullary nephrocalcinosis; primary oxalosis.*

RENAL ECTOPIA

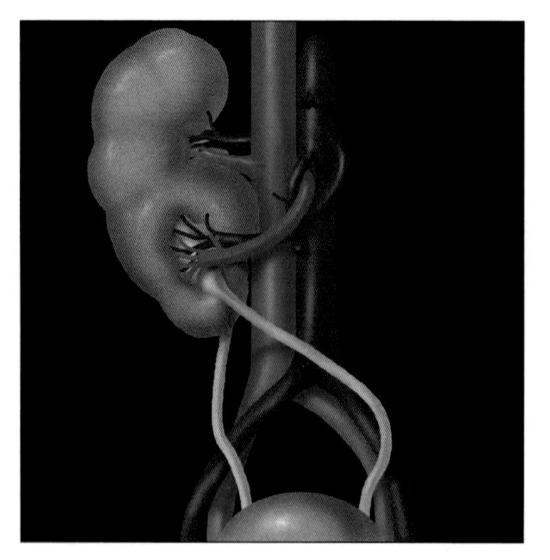

Graphic shows crossed inferior fused renal ectopia.

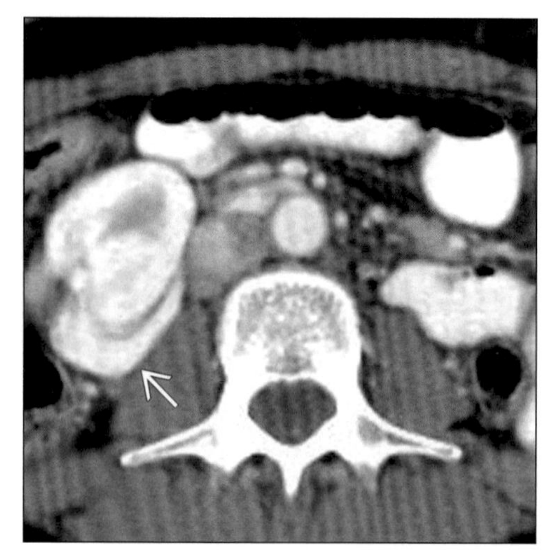

Axial CECT shows embryologic left kidney in an inferior crossed ectopic location. Note dilated left renal artery (arrow) and renal pelvis opening toward the right.

TERMINOLOGY

Abbreviations and Synonyms

- Renal ectopia (RE)

Definitions

- Abnormal location of kidney due to developmental anomaly

IMAGING FINDINGS

General Features

- Best diagnostic clue: Abnormal location of kidney
- Location
  - Kidneys normal location: 1st-3rd lumbar vertebrae
  - Ipsilateral RE: Kidney on same side of body as orifice of its attendant ureter
    - Cranial (superior RE): Above normal position; intrathoracic; or below eventrated diaphragm
    - Caudal (simple RE): Below normal position; abdominal, iliac or pelvic
    - Abdominal: Kidney lies above iliac crest, below L2
    - Iliac: Kidney located opposite iliac crest or in iliac fossa
    - Pelvic (sacral): Kidney located in true pelvis; below iliopectineal line
  - Crossed RE: Kidney located on opposite side of midline from its ureteral orifice
- Size: Ectopic kidneys vary in size
- Other general features
  - Caudal RE
    - Unilateral (more common)
    - Involvement of both kidneys (rare)
    - In solitary kidney (least common)
  - Crossed RE: With fusion (most common); without fusion (10-15%); in solitary kidney (least common)
    - With fusion (most common)
    - Without fusion (10-15%)
    - In solitary kidney (least common)
  - Classification of unilateral fused kidney or crossed fused RE
    - Superior: Kidney crosses over midline; lies superior to resident kidney
    - Sigmoid (S-shaped): Crossed kidney lies inferiorly
    - Unilateral lump kidney: Both kidneys completely fused; large irregular lump
    - Unilateral L-shaped: Crossed kidney inferior & transverse; resident kidney normally oriented
    - Unilateral disc: Each kidney fused to other along medial concave border
    - Inferior RE: Crossed kidney inferior to resident; its upper pole fused to lower pole of resident kidney

Radiographic Findings

- Radiography

DDx: Misplaced Kidney

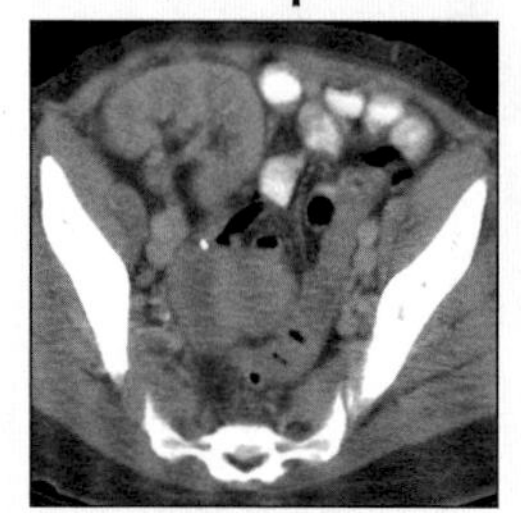

Renal Transplant

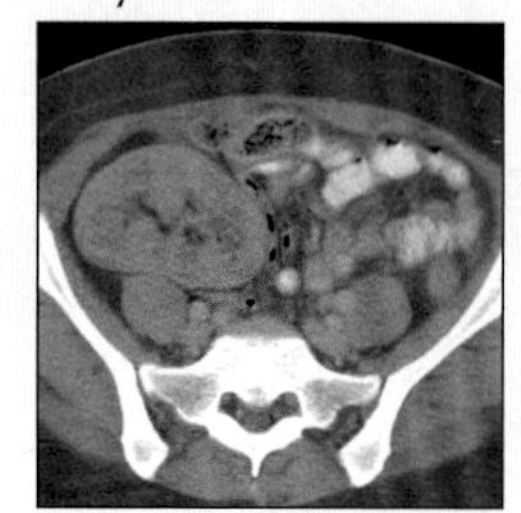

Renal Transplant

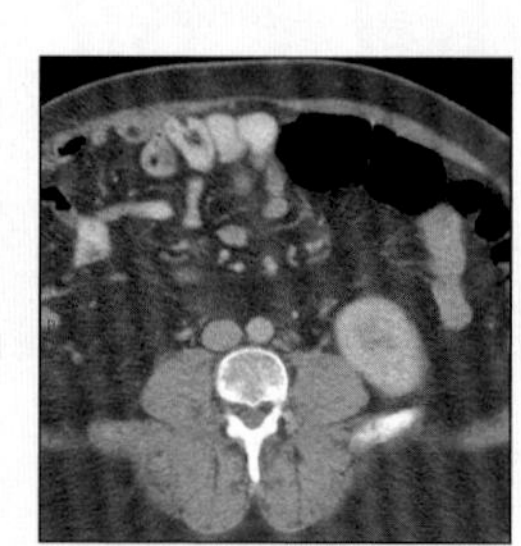

Displaced Kidney

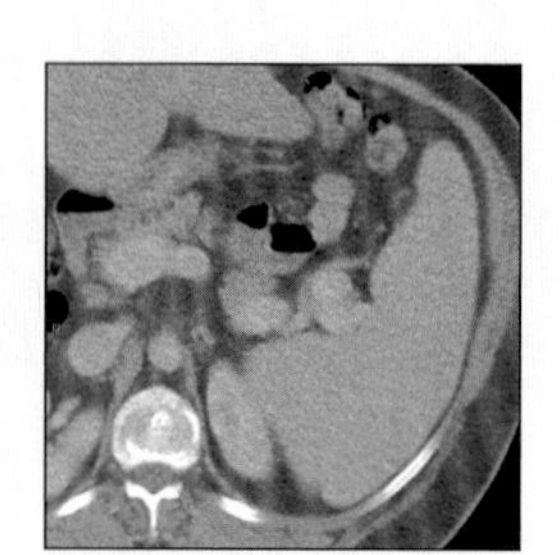

Displaced Kidney

RENAL ECTOPIA

Key Facts

Terminology

- Renal ectopia (RE)
- Abnormal location of kidney due to developmental anomaly

Imaging Findings

- Best diagnostic clue: Abnormal location of kidney
- Renal outline not visible in expected position
- Soft-tissue outline of kidney may be visible in ectopic position
- Cranial RE: Kidney residing in thorax; differentiate from a mediastinal mass
- Abdominal or iliac RE
- Pelvic RE: Differentiate RE from various pelvic masses
- Crossed RE: CT with thin (4-5 mm) slices may show degree of separation of kidneys

Top Differential Diagnoses

- Renal allograft
- Renal autotransplantation
- Horseshoe kidney
- Acquired renal displacement

Diagnostic Checklist

- Retroperitoneal mass, huge renal cyst, gigantic renal pelvis secondary to UPJ obstruction can force kidney to opposite side simulating crossed unfused RE (IVP)
- CECT helps in detecting RE & cause of displacement
- On IVP, pelvic kidney may be difficult to locate (due to sacral superimposition & bowel gas)
- Clue to pelvic kidney is ureter, which can be visible even if collecting system & kidney are not visible
- Important not to confuse RE with renal ptosis

    - Chest x-ray (in cranial RE)
        - Well-defined posteroinferior mediastinal mass
    - Abdominal x-ray (in caudal RE)
        - Renal outline not visible in expected position
        - Soft-tissue outline of kidney may be visible in ectopic position
        - Malposition of colon
    - Abdominal x-ray (in crossed RE)
        - Soft-tissue outline of kidney may be visible on opposite side
        - Malposition of colon
- IVP
    - Cranial RE
        - Kidney lies partially or completely in thorax
        - Length of attendant ureter longer than normal
    - Abdominal or iliac RE: Kidney in either abdominal or iliac area
        - Can be simple (unilateral, bilateral,or solitary) or crossed
        - Kidney usually smaller & ureter shorter than normal
        - Bizarre pattern of calyces; extrarenal calyces (common)
        - Rotational anomaly: Incomplete, reverse or nonrotation
    - Pelvic kidney
        - Left (70%) > right; if bilateral, left usually lower than right kidney & generally fused
        - May see anomalies of rotation
        - Ureter is frequently too high as it exits renal pelvis ("high insertion")
        - May see ectopic ureter, extrarenal calyces, calyceal diverticula
    - Crossed RE: Distal ureter inserts into trigone on side of origin
        - Both pelvises rotated anteriorly in superior, unilateral lump & inferior RE
        - Sigmoid (S-shaped): Resident kidney pelvis is medial; lateral in crossed kidney
        - Unilateral disc: Resident kidney pelvis is anteromedial, pelvis of other is anterolateral
    - Bilateral crossed RE: Both kidneys on wrong side but their attendant ureters arise normally

CT Findings

- Cranial RE: Kidney residing in thorax; differentiate from a mediastinal mass
    - Adrenal gland may lie above, behind or below ectopic kidney
- Abdominal or iliac RE
    - Adrenal gland in normal place; appears linear on CT
    - Colonic flexures, duodenum, loops of small bowel, spleen, tail of pancreas in abnormal position
- Pelvic RE: Differentiate RE from various pelvic masses
- Crossed RE: CT with thin (4-5 mm) slices may show degree of separation of kidneys

Ultrasonographic Findings

- Real Time
    - Cranial RE
        - Kidney lies just below an eventrated diaphragm
        - Passing through defect in diaphragm
    - Caudal RE (abdominal, iliac, or pelvic)
        - Renal sinus echo complex: Eccentric or absent
    - Crossed RE
        - Separation of kidneys can be demonstrated
        - Kidneys move separately from each other during respiration
    - Crossed fused RE
        - One or two anterior or posterior notches in renal parenchyma
        - Renal sinuses lie in different planes, run in different directions & echoes reflect differently

Angiographic Findings

- Conventional
    - Cranial RE
        - Elongated renal artery from aorta at normal level
        - Occasionally, accessory RA from thoracic aorta
    - Abdominal or iliac RE
        - Renal arteries arise lower in aorta than normal
        - Multiple renal arteries (common)
    - Pelvic RE
        - Renal arteries are often multiple
        - Usually arise from distal end or aortic bifurcation
    - Crossed RE: Anomalous blood supply arising from vessels in vicinity

RENAL ECTOPIA

Nuclear Medicine Findings

- Tc99m-DMSA or Tc99m-glucohepatanate scan
  - Detects ectopic kidney by outlining kidney shape
  - Crossed fused renal ectopia: Isotope excretion or localization by a kidney, with no contralateral isotope excretion or localization

Imaging Recommendations

- Helical CECT; IVP

DIFFERENTIAL DIAGNOSIS

Renal allograft

- Transplanted kidney in iliac fossa
- Renal vessels anastomosed to external iliac artery, vein
- Ureter reimplanted into bladder via submucosal tunnel; variable axis of pelvis

Renal autotransplantation

- Surgically repositioning patient's own kidney

Horseshoe kidney

- Fusion of lower poles of kidneys in low mid-abdomen

Acquired renal displacement

- Due to large liver, splenic or any retroperitoneal tumor

PATHOLOGY

General Features

- Etiology
  - Cranial RE: Kidney herniated into thorax through lumbocostal triangle or foramen of Bochdalek
  - Caudal RE: Diminished ureteral growth; umbilical arteries block cranial ascent of kidney; asymmetry in level of development of 2 kidneys
  - Crossed RE: Mesonephric ducts & ureteral buds may stray from normal course
  - RE inherited as autosomal recessive trait; reported in monozygotic twins
- Epidemiology
  - Cranial RE: 1 in 15,000 autopsies
  - Abdominal or iliac RE: 1 in 600 on IVP
  - Pelvic kidney: 1 in 725 live births
  - Unilateral crossed fused RE: 1 in 1,300 to 1 in 7,600
- Associated abnormalities
  - Genitourinary (50%): Malrotation, hypospadias, high insertion of ureter into renal pelvis, ectopic ureter, extrarenal calyces, calyceal diverticula, bladder extrophy
  - Skeletal (40%): Anomalies of ribs, vertebral bodies; skull asymmetry & absence of radius
  - Cardiovascular (40%): Valvular & septal defects
  - Gastrointestinal (33%): Anorectal malformations, malrotation.
  - Ears, lips, palate (33%): Low-set or absent ears; hare lip; cleft palate
  - Hematopoietic (7%): Fanconi anemia
  - Cranial RE: Omphalocele
  - Pelvic kidney: Vesicoureteral reflux, contralateral renal agenesis, absent or hypoplastic vagina
  - Crossed ectopia: Megaureter, cryptorchidism, urethral valves, multicystic dysplasia

CLINICAL ISSUES

Presentation

- Most common signs/symptoms
  - May be asymptomatic, incidental finding
  - May present with signs & symptoms of obstruction, urolithiasis, reflux & infection

Demographics

- Gender: Cranial RE (M > F); crossed fused RE (M < F)

Natural History & Prognosis

- Complications
  - Obstruction, urolithiasis, reflux, infection
  - Pelvic kidneys: ↓ Function & may obstruct labor
  - Aberrant arteries may cross & obstruct ureter
  - Abdominal & iliac ectopic kidneys more injury prone; prone to vascular injury during aortic surgery
- Prognosis
  - Recurrent obstruction, reflux, infection: Poor

Treatment

- Treat complications of renal ectopia

DIAGNOSTIC CHECKLIST

Image Interpretation Pearls

- Retroperitoneal mass, huge renal cyst, gigantic renal pelvis secondary to UPJ obstruction can force kidney to opposite side simulating crossed unfused RE (IVP)
- CECT helps in detecting RE & cause of displacement
- On IVP, pelvic kidney may be difficult to locate (due to sacral superimposition & bowel gas)
  - Clue to pelvic kidney is ureter, which can be visible even if collecting system & kidney are not visible
- Important not to confuse RE with renal ptosis
  - Kidney drops further down in abdomen from its normal position, but attendant ureter of normal length & renal arteries arise from normal site

SELECTED REFERENCES

1. Li J et al: Single ureteral ectopia with congenital renal dysplasia. J Urol. 170(2 Pt 1):558-9, 2003
2. Gu LL et al: Crossed solitary renal ectopia. Urology. 38(6):556-8, 1991
3. Hawass ND et al: Intrathoracic kidneys: report of 6 cases and a review of the literature. Eur Urol.14(1):83-7, 1988
4. Goodman JD et al: Crossed fused renal ectopia: sonographic diagnosis. Urol Radiol. 8(1):13-6, 1986
5. McCarthy S et al: Ultrasonography in crossed renal ectopia. J Ultrasound Med. 3(3):107-12, 1984
6. Hertz M et al: Crossed renal ectopia: clinical and radiological findings in 22 cases. Clin Radiol. 28(3):339-44, 1977
7. Rubinstein ZJ et al: Crossed renal ectopia: angiographic findings in six cases. Am J Roentgenol. 126(5):1035-8, 1976

RENAL ECTOPIA

IMAGE GALLERY

Typical

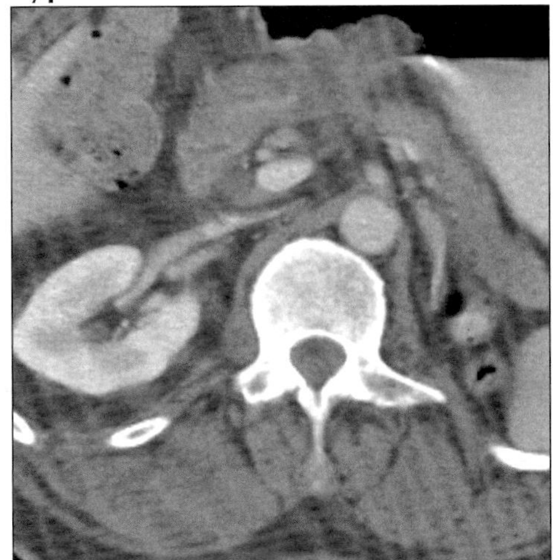

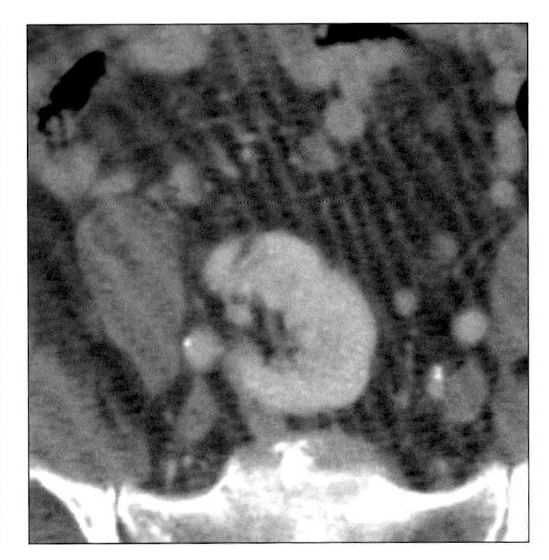

(Left) *Axial CECT shows right kidney in normal position, no kidney in left renal fossa.* ***(Right)*** *Axial CECT shows left kidney in midline, non-fused, pelvic location.*

Typical

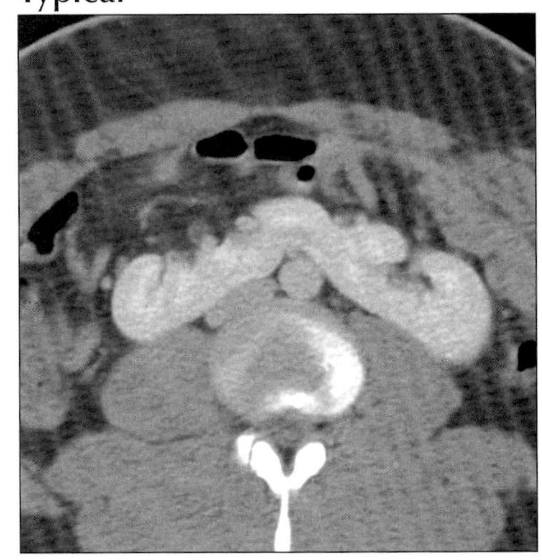

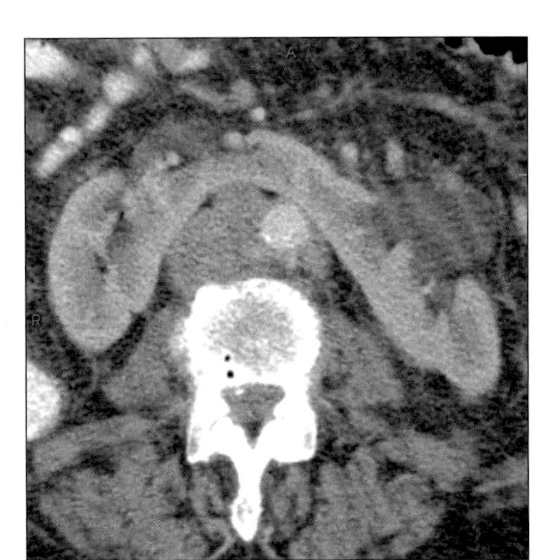

(Left) *Axial CECT shows horseshoe kidney with fusion of the lower poles across the midline and low in position.* ***(Right)*** *Axial CECT shows a horseshoe kidney with hydronephrosis. Retro-renal lymphadenopathy from prostate metastases.*

Typical

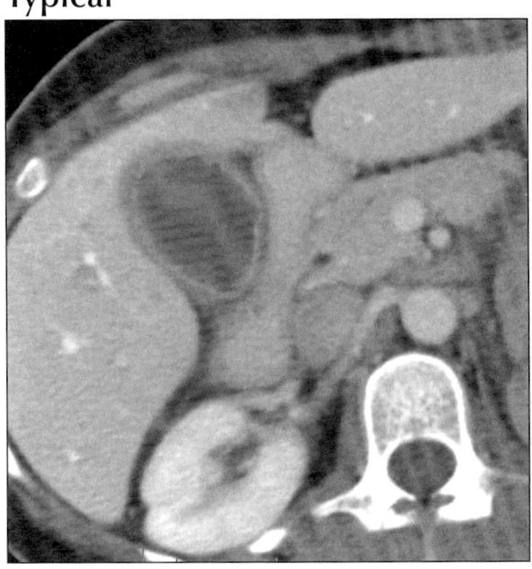

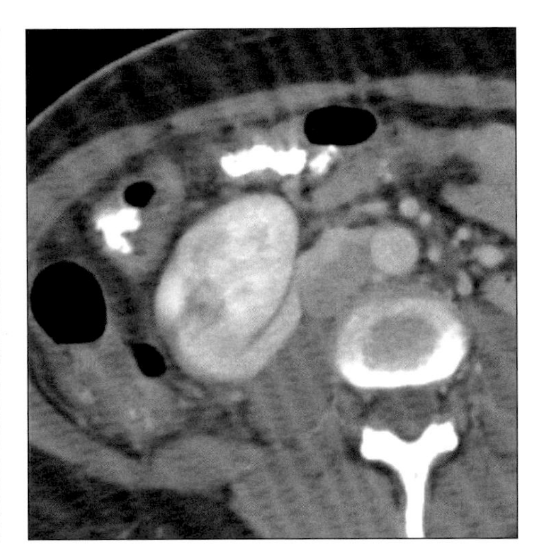

(Left) *Axial CECT shows normal right kidney, but no left.* ***(Right)*** *Axial CECT shows the left kidney in a crossed-fused inferior ectopic location with its pelvis opening postero-laterally.*

HORSESHOE KIDNEY

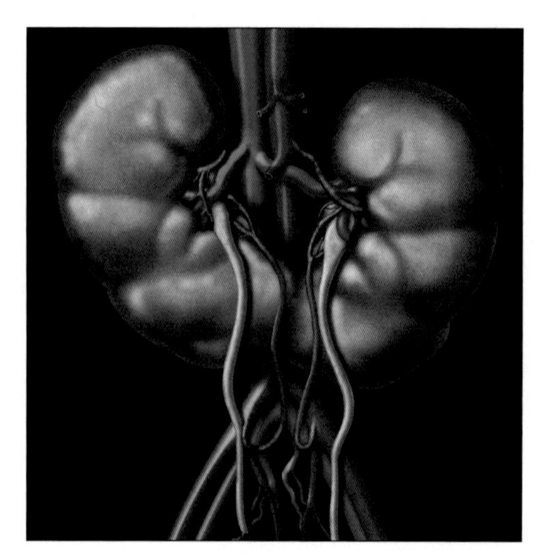

Graphic shows horseshoe kidney. Note multiple renal arteries.

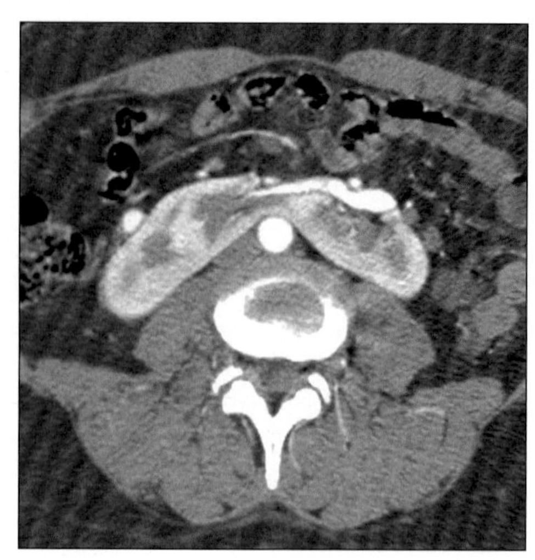

Axial CECT shows fusion of the lower poles of the kidneys across the midline and low position of kidneys.

TERMINOLOGY

Definitions

- A congenital anomaly of the kidney where 2 kidneys fused by isthmus at the lower poles

IMAGING FINDINGS

General Features

- Best diagnostic clue: 2 kidneys on opposite sides of the body with the lower poles fused in midline
- Location
  - Ectopic, lies lower than normal kidney
  - Isthmus usually anterior to aorta and inferior vena cava (IVC)
  - Rarely, isthmus is posterior or in between aorta (posterior) and IVC (anterior)
- Morphology
  - 2 types of fusion
    - Midline or symmetrical fusion (90% of cases)
    - Lateral or asymmetrical fusion

Radiographic Findings

- Radiography
  - Kidney appears too close to the spine
  - Vertical long axis of kidney may be seen, lower poles lie closer to spine
  - Visualize the Isthmus of the 2 kidneys
- IVP
  - Midline fusion
    - Hand holding calyces: Lower calyces descend toward midline near isthmus
    - Nephrogram is U-shaped
    - If width of isthmus is < 1/3 length of kidney, renal pelvis lies between normal 30° anteromedial and 90° direct anterior angle
    - If width of isthmus is ≥ 1/3 length of kidney, renal pelvis lies between 90° direct anterior and lateral angle
    - Rarely, calyces from opposing kidneys joined to form a common renal pelvis with 1 ureter
  - Lateral fusion
    - Lower calyces crosses midline and drain part of renal parenchyma on opposite kidney
    - Nephrogram is L-shaped
    - One part crosses midline and lies in transverse position, renal pelvis lies anteriorly or laterally
    - Remaining part lies in vertical position, renal pelvis lies anteriorly or medially
  - Large and extrarenal renal pelvis
  - Renal pelvis often large and flabby; ureter inserts abnormally high in renal pelvis
  - Rarely, kidney is fused at the upper poles (5%)
  - Ipsilateral lower calyces medial to ureter; may simulate renal malrotation without fusion

DDx: Abnormal Position and Rotation of Kidney

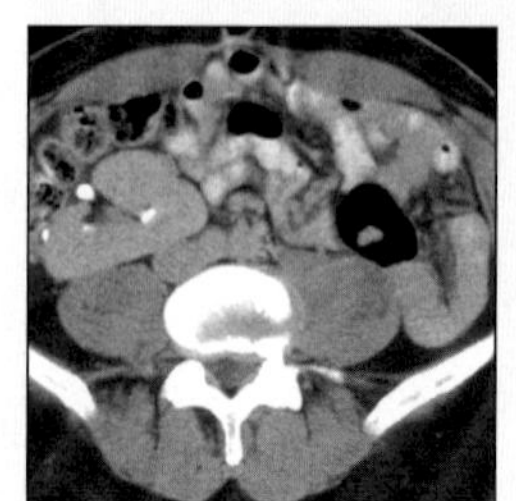

Crossed Ectopia

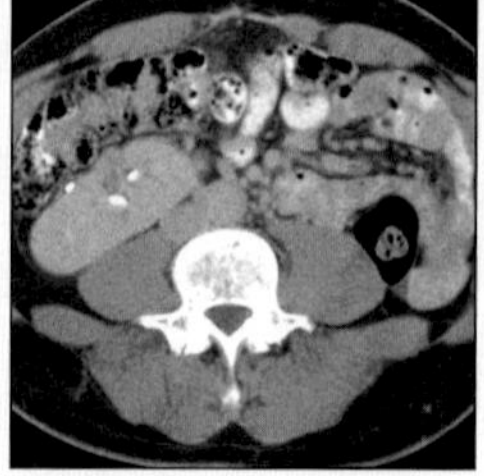

Crossed Ectopia

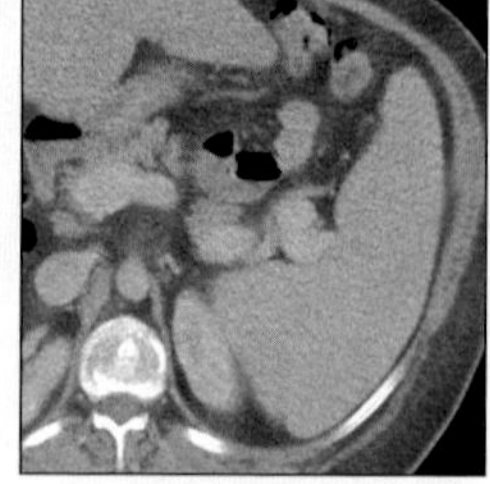

Displacement

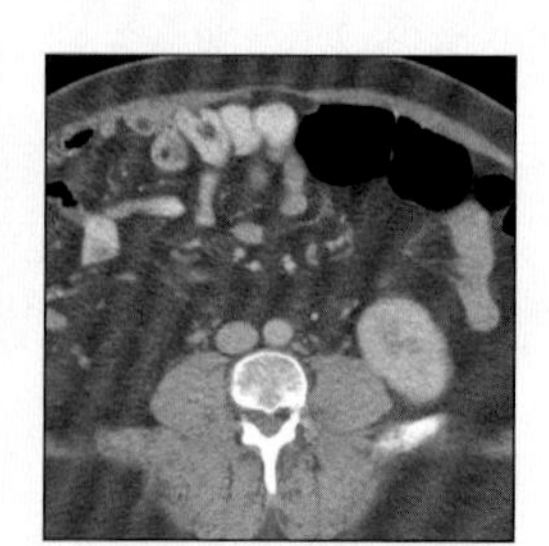

Displacement

HORSESHOE KIDNEY

Key Facts

Terminology

- A congenital anomaly of the kidney where 2 kidneys fused by isthmus at the lower poles

Imaging Findings

- Best diagnostic clue: 2 kidneys on opposite sides of the body with the lower poles fused in midline
- Ectopic, lies lower than normal kidney
- Isthmus usually anterior to aorta and inferior vena cava (IVC)
- Midline or symmetrical fusion (90% of cases)
- Kidney appears too close to the spine
- Large and extrarenal renal pelvis
- Multiple, bilateral renal arteries
- IVP followed by CT or scintigraphy for pre-operative assessment

Top Differential Diagnoses

- Renal ectopia
- Renal displacement

Pathology

- Epidemiology: 1 in 400 people

Clinical Issues

- Asymptomatic or associated abnormalities
- Any age
- Gender: M:F = 2:1

Diagnostic Checklist

- Associated abnormalities and other complications in imaging, treatment and prognosis
- Kidney appears U-shaped with isthmus in midline

 - Ureteropelvic (more common) or ureterovesical junction obstruction with delayed clearing of contrast
 - "Flower-vase" appearance: Each ureter crosses isthmus and curves laterally and continues medially, assuming a normal course distally
 - Bifid or double ureters drain the kidneys
- Voiding Cystourethrography
 - Vesicoureteral reflux

CT Findings

- CECT
 - Define structural abnormalities
   - Degree and site of fusion: Midline or lateral fusion
   - Degree of renal malrotation
   - Renal parenchymal changes (e.g., scarring, cystic disease)
   - Collecting system abnormalities (e.g., duplex system, hydronephrosis)
 - Differentiate composition of isthmus between fibrous or normal parenchymal tissue
- CTA
 - Variant arterial supply
   - Multiple, bilateral renal arteries
   - Inferior mesenteric artery always crosses the isthmus
   - Arteries arising from aorta or common iliac, internal iliac, external iliac or inferior mesenteric arteries

Ultrasonographic Findings

- Real Time
 - Isthmus lies anterior to spine and continues with lower poles of opposite kidneys
 - Curved configuration, elongation and poorly defined lower poles
 - Inverted triangular or pyriform shape (longitudinal scan)

Angiographic Findings

- Conventional: Variant arterial supply

Nuclear Medicine Findings

- Demonstrate fusion with functional parenchymal tissue

Imaging Recommendations

- Best imaging tool
 - IVP followed by CT or scintigraphy for pre-operative assessment
 - US for diagnosis in utero
- Protocol advice: CTA: Use 3-D volume-rendered CT to better define the vessels

DIFFERENTIAL DIAGNOSIS

Renal ectopia

- Kidney congenitally in abnormal position
- Ipsilateral or simple ectopia: Kidney on proper side of body as its ureter
 - Abdominal: Kidney lies above iliac crest but below L2
 - Iliac: Kidney is located opposite iliac crest or in iliac fossa
 - Pelvic (sacral): Kidney in true pelvis
- Crossed renal ectopia: 2 kidneys are on the same side of the body (right side > left)
 - With fusion (90%): 2 fused kidneys lie on the same side of spine; ureter of crossed kidney crosses midline to insert into bladder
 - Without fusion: 2 kidneys lie on the same side of spine without fusion; ureter of crossed kidney crosses midline to insert into bladder
 - Solitary: 1 kidney arises on the wrong side, ureter crosses midline to insert into bladder
 - Bilateral: Left and right kidneys arise on the wrong side, both ureters crosses midline to insert into bladder

Renal displacement

- Thoracolumbar gibbous deformity: Alters the renal axis → pseudohorseshoe kidney
- Displacement by mass: Forcing one kidney to opposite side of body or into pelvis

HORSESHOE KIDNEY

- Retroperitoneal masses
- Giant hydronephrosis secondary to ureteropelvic junction (UPJ) obstruction
- Giant renal cyst (e.g., autosomal dominant polycystic kidney disease)

PATHOLOGY

General Features

- General path comments: Most common renal fusion anomaly
- Genetics: Reported in identical twins, but no clear evidence
- Epidemiology: 1 in 400 people
- Associated abnormalities
  - Congenital disorders
    - Chromosomal abnormalities: Turner syndrome, trisomy 18
    - Hematological abnormalities: Fanconi anemia, dyskeratosis congenita with pancytopenia
    - Laurence-Biedl-Moon syndrome
    - Thalidomide embryopathy
  - Anomalies (most common to least common)
    - UPJ obstruction
    - Vesicoureteral reflux
    - Unilateral or bilateral duplication
    - Megaureter
    - Ectopic ureter
    - Unilateral triplication
    - Renal dysplasia
    - Retrocaval ureter
    - Supernumerary kidney
    - Anorectal malformation
    - Esophageal atresia
    - Rectovaginal fistula
    - Omphalocele
    - Cardiovascular, vertebral, neurological, peripheral skeletal or facial anomalies

Gross Pathologic & Surgical Features

- Isthmus is composed of normal parenchyma or connective tissue

CLINICAL ISSUES

Presentation

- Most common signs/symptoms
  - Asymptomatic or associated abnormalities
  - Vague abdominal pain, radiating to the back
  - Nausea and vomiting
  - Rovsing sign, palpable abdominal mass

Demographics

- Age
  - Any age
  - Still births > infants > children > adults; ↓ with age because many diagnosed based on associated abnormalities
- Gender: M:F = 2:1

Natural History & Prognosis

- Complications
  - Trauma injury: Isthmus lies anteriorly without protection by ribs → split by hard blow to abdomen
  - UPJ obstruction: High "insertion" of ureter
  - Recurrent infections: Vesicoureteral reflux and UPJ obstruction
  - Urolithiasis: 75% metabolic calculi, 25% struvite calculi
  - Wilms tumors in children: 2-8 times more common
  - Primary renal carcinoid tumor: ↑ Prevalence
- Prognosis
  - Poor, with associated abnormalities causing significant morbidity and mortality
  - Good, without other abnormalities

Treatment

- Surgical separation in symptomatic patients

DIAGNOSTIC CHECKLIST

Consider

- Associated abnormalities and other complications in imaging, treatment and prognosis

Image Interpretation Pearls

- Kidney appears U-shaped with isthmus in midline

SELECTED REFERENCES

1. Strauss S et al: Sonographic features of horseshoe kidney: review of 34 patients. J Ultrasound Med. 19(1):27-31, 2000
2. Pozniac MA et al: Three-dimensional computed tomographic angiography of a horseshoe kidney with ureteropelvic junction obstruction. Urology 49:267-268, 1997
3. Banerjee B et al: Ultrasound diagnosis of horseshoe kidney. Br J Radiol. 64(766):898-900, 1991
4. Mesrobian HG et al: Wilms tumor in horseshoe kidneys: a report from the National Wilms Tumor Study. J Urol. 133(6):1002-3, 1985
5. Grainger R et al: Horseshoe kidney--a review of the presentation, associated congenital anomalies and complications in 73 patients. Ir Med J. 76(7):315-7, 1983
6. Evans WP et al: Horseshoe kidney and urolithiasis. J Urol. 125(5):620-1, 1981
7. Pitts WR Jr et al: Horseshoe kidneys: a 40-year experience. J Urol. 113(6):743-6, 1975
8. Whitehouse GH: Some urographic aspects of the horseshoe kidney anomaly-a review of 59 cases. Clin Radiol. 26(1):107-14, 1975
9. Boatman DL et al: Congenital anomalies associated with horseshoe kidney. J Urol. 107:205-7, 1973
10. Kolln CP et al: Horseshoe kidney: a review of 105 patients. J Urol. 107(2):203-4, 1972
11. Segura JW et al: Horseshoe kidney in children. J Urol. 108:333-6, 1972
12. Boatman DL et al: The arterial supply of horseshoe kidneys. Am J Roentgenol Radium Ther Nucl Med. 113(3):447-51, 1971
13. Zondek LH et al: Horseshoe kidney and associated congenital malformations. Urol Int. 18:347-56, 1964

IMAGE GALLERY

Typical

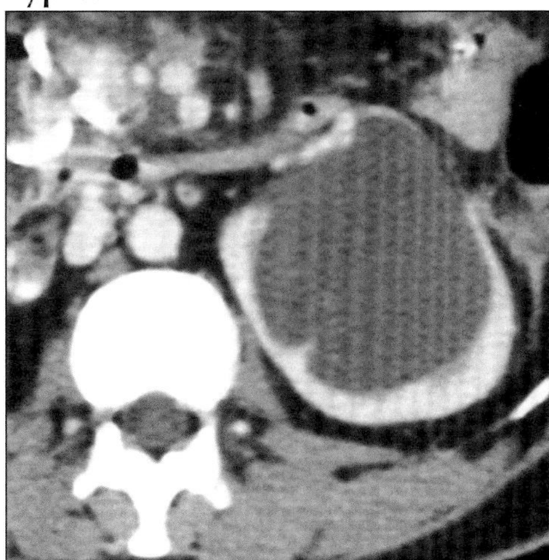
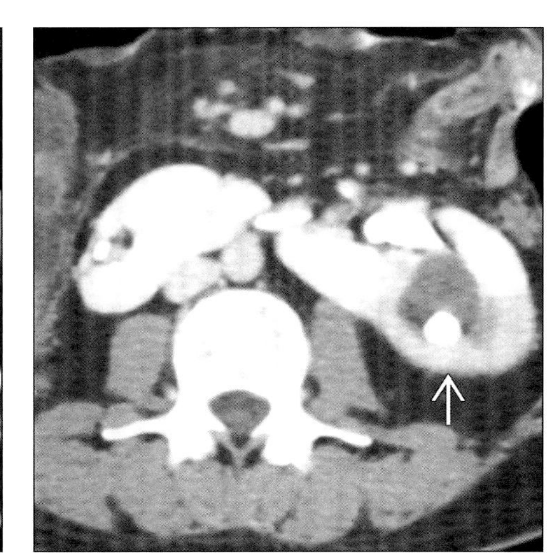

(Left) *Axial CECT shows hydronephrosis of left side of a horseshoe kidney due to stones and UPJ obstruction.* ***(Right)*** *Axial CECT shows a horseshoe kidney with hydronephrosis, UPJ obstruction and multiple calculi (arrow).*

Typical

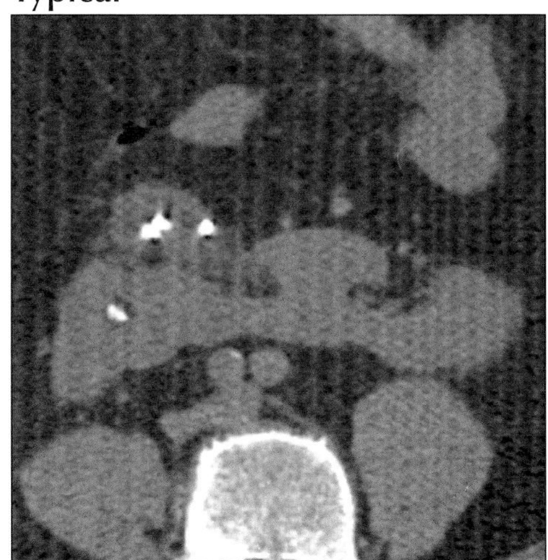
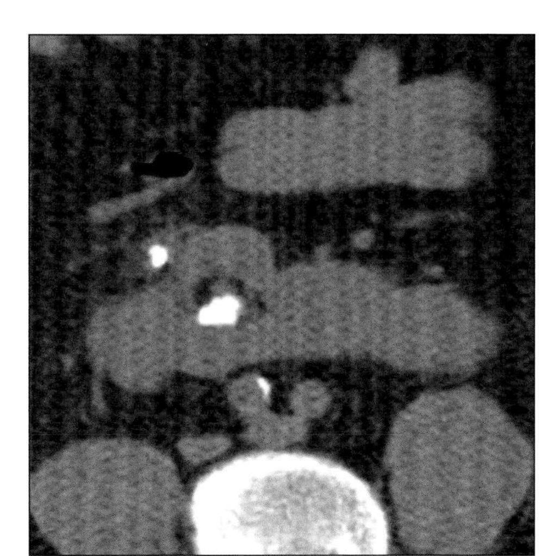

(Left) *Axial NECT shows horseshoe kidney, multiple calculi, and renal scarring.* ***(Right)*** *Axial NECT shows horseshoe kidney, calculi, and renal scarring.*

Typical

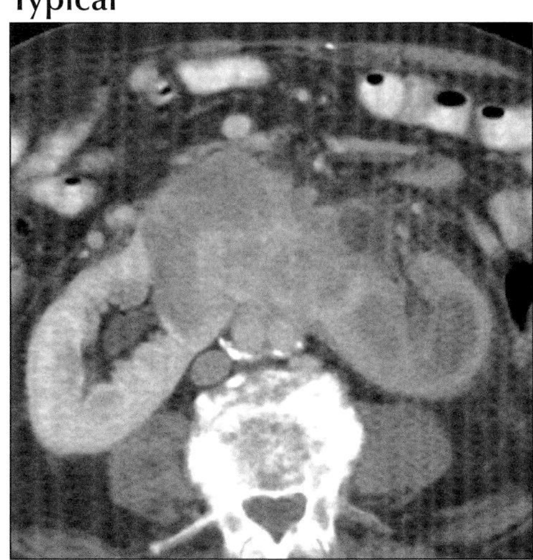
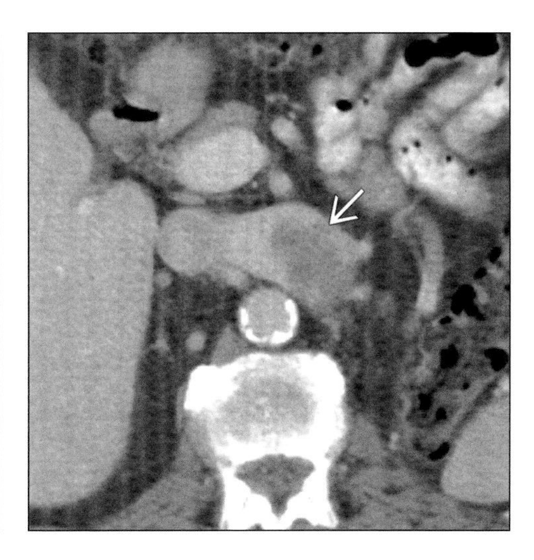

(Left) *Axial CECT shows a horseshoe kidney with a large mass (RCC) replacing most of the isthmus and left side.* ***(Right)*** *Axial CECT shows tumor thrombus (arrow) within the left renal vein of a patient with horseshoe kidney and RCC.*

COLUMN OF BERTIN

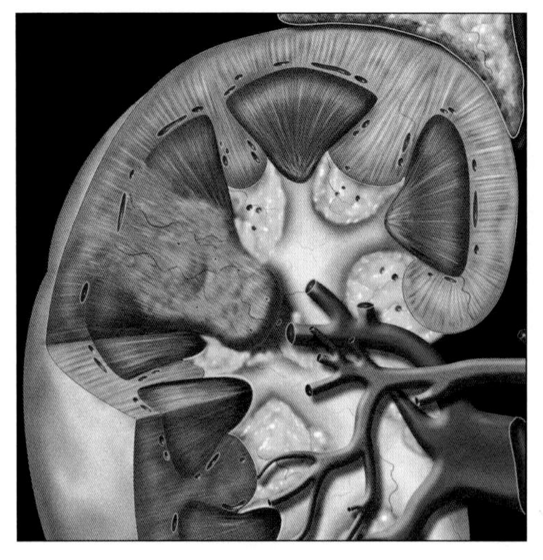

Graphic shows mass-like extension of renal cortex between upper and middle calices.

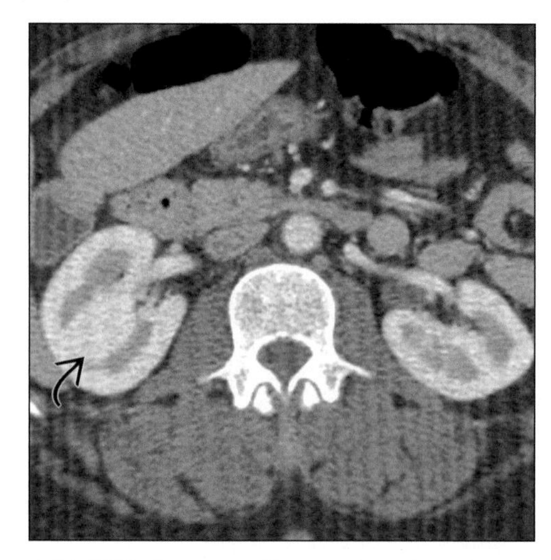

Axial CECT shows mass-like extension of renal cortex (curved arrow) between upper and middle calices.

TERMINOLOGY

Abbreviations and Synonyms

- Septal cortex, hypertrophied or enlarged column of Bertin, focal cortical hyperplasia, benign cortical rest, cortical island, focal renal hypertrophy, junctional parenchyma

Definitions

- Hypertrophic medial bands of cortical tissue that separate the pyramids of the renal medulla

IMAGING FINDINGS

General Features

- Best diagnostic clue: Normally enhancing renal cortex protruding into renal sinus
- Location
  - Between upper and middle calyces
  - Left side > right side
  - Unilateral > bilateral (18% of cases)

Radiographic Findings

- IVP
  - Mass effect on pelvicaliceal system, always at level of emerging renal vein
  - Splaying and abnormal separation of upper and lower pole of collecting system

CT Findings

- NECT
  - Mass continuous with renal cortex
  - Lateral indentation of renal sinus
  - Deformed adjacent calices and infundibula
- CECT
  - Absence of a mass
  - Similar enhancement as normal renal cortex

Ultrasonographic Findings

- Real Time
  - Normal renal outline
  - Isoechoic with renal cortex; hypoechoic to medulla
  - Mass with an echogenic linear rim of renal sinus fat
  - Splaying of central sinus echoes

Nuclear Medicine Findings

- In equivocal cases, renal scintigraphy can differentiate normal renal parenchyma from pathologic mass

Imaging Recommendations

- Best imaging tool: Multiphasic CT
- Protocol advice: CT: Corticomedullary phase best

DDx: Renal Mass or Pseudomass

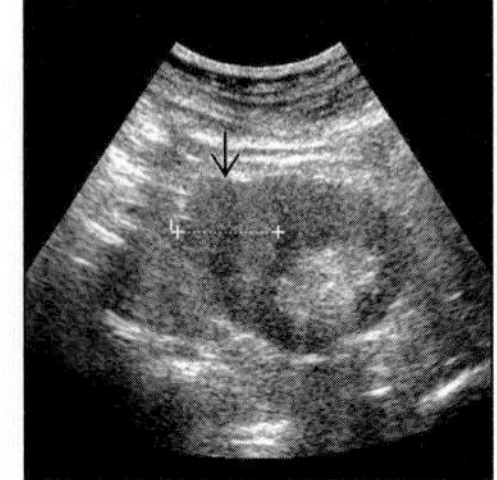

Renal Cell Carcinoma

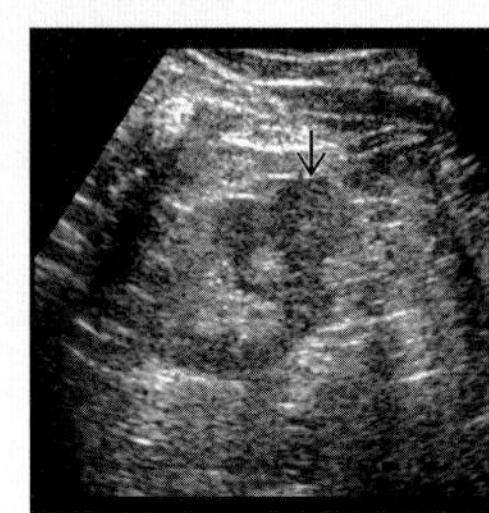

Fetal Lobulation

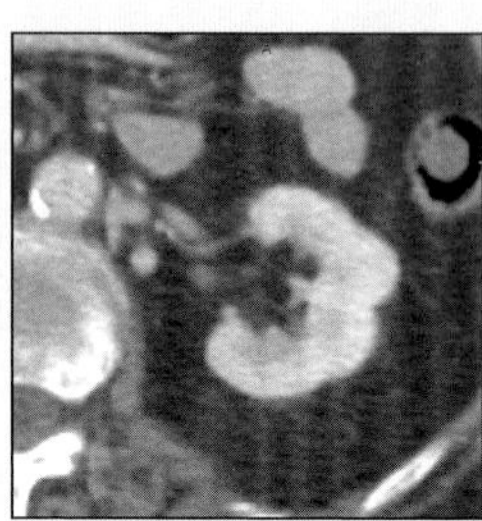

Fetal Lobulation

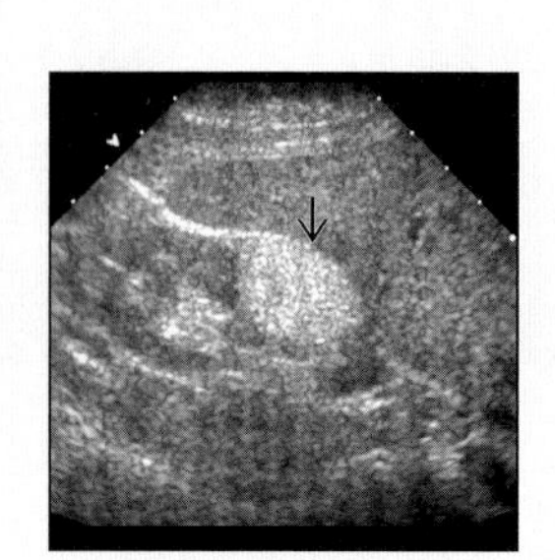

Angiomyolipoma

COLUMN OF BERTIN

Key Facts

Terminology
- Hypertrophic medial bands of cortical tissue that separate the pyramids of the renal medulla

Imaging Findings
- Best diagnostic clue: Normally enhancing renal cortex protruding into renal sinus
- Isoechoic with renal cortex; hypoechoic to medulla

Top Differential Diagnoses
- Renal cell carcinoma
- Fetal lobulation
- Renal scarring
- Other renal masses

Diagnostic Checklist
- Pseudotumor, not pathological disease
- Absence of a mass on CECT

DIFFERENTIAL DIAGNOSIS

Renal cell carcinoma
- Solitary renal mass with central necrosis, heterogeneous
- Mass is usually rounded or oval in cortical location and grows by expansion
- Multiple septations, septal thickening and nodularity
- Hypervascular mass

Fetal lobulation
- Persistent cortical lobulation
- Fourteen individual lobes with centrilobar cortex located around calices

Renal scarring
- Focal scarring from chronic reflux nephropathy, surgery, trauma or infarction
- Nodular compensatory hypertrophy of areas of unaffected tissue

Other renal masses
- E.g., renal metastases and lymphoma, renal oncocytoma, renal angiomyolipoma

PATHOLOGY

General Features
- General path comments: Embryology: Unresorbed polar parenchyma of one or both of two sub-kidneys that fuse to form a normal kidney

CLINICAL ISSUES

Presentation
- Most common signs/symptoms: Asymptomatic
- Diagnosis
  - Usually found incidentally on imaging
  - Most likely to simulate a mass on sonography

Natural History & Prognosis
- Normal variant
- Complications: None
- Prognosis: Very good

Treatment
- None

DIAGNOSTIC CHECKLIST

Consider
- Pseudotumor, not pathological disease

Image Interpretation Pearls
- Absence of a mass on CECT

SELECTED REFERENCES

1. Yeh HC et al: Junctional parenchyma: revised definition of hypertrophic column of Bertin. Radiology. 185(3):725-32, 1992
2. Seppala RE et al: Sonography of the hypertrophied column of Bertin. AJR Am J Roentgenol. 148(6):1277-8, 1987
3. Lafortune M et al: Sonography of the hypertrophied column of Bertin. AJR Am J Roentgenol. 146(1):53-6, 1986
4. Leekam RN et al: The sonography of renal columnar hypertrophy. J Clin Ultrasound. 11(9):491-4, 1983
5. Mahony BS et al: Septa of Bertin: a sonographic pseudotumor. J Clin Ultrasound. 11(6):317-9, 1983
6. Williams ED et al: Kidney pseudotumour diagnosed by emission computed tomography. Br Med J (Clin Res Ed). 285(6352):1379-80, 1982
7. Green WM et al: "Column of Bertin": diagnosis by nephrotomography. Am J Roentgenol Radium Ther Nucl Med. 116(4):714-23, 1972

IMAGE GALLERY

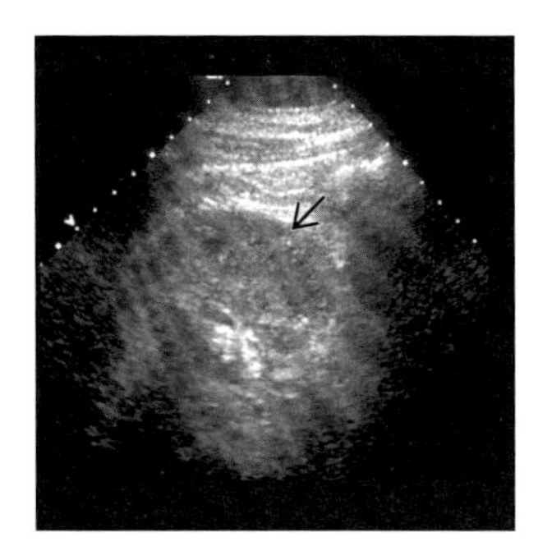

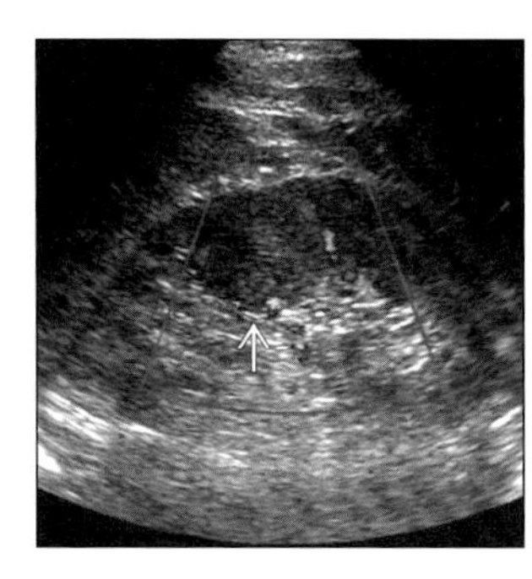

(Left) *Sonogram shows pseudomass (arrow) in upper pole of kidney isoechoic to renal cortex.* ***(Right)*** *Sonogram shows pseudomass (arrow) in upper pole of kidney; column of Bertin.*

AD POLYCYSTIC DISEASE, KIDNEY

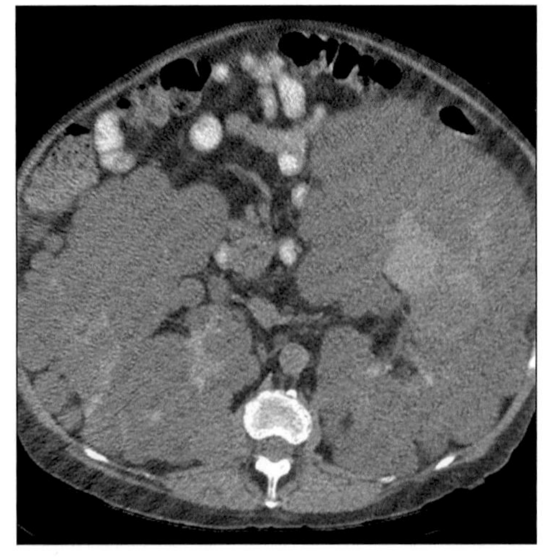

Axial NECT shows massively enlarged kidneys with innumerable cysts.

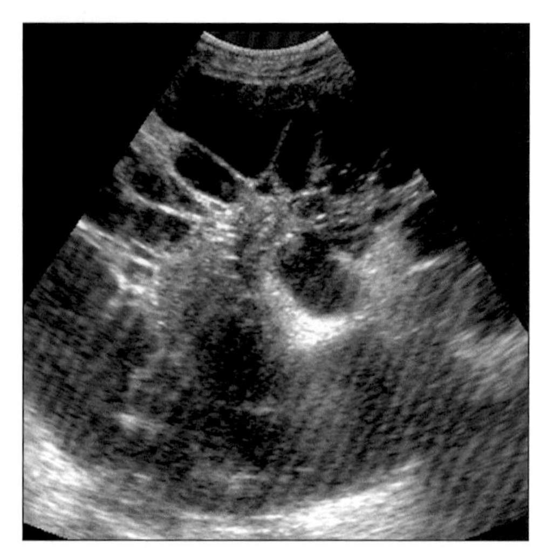

Sagittal sonogram shows massively enlarged kidney with innumerable cysts.

TERMINOLOGY

Abbreviations and Synonyms

- Autosomal dominant polycystic kidney disease (ADPKD) or adult PKD
- Autosomal recessive polycystic kidney disease (ARPKD)
  - Infantile, newborn, childhood, hamartomatous PKD

Definitions

- Hereditary disorder characterized by multiple renal cysts & various systemic manifestations

IMAGING FINDINGS

General Features

- Best diagnostic clue: Massively enlarged kidneys with innumerable cysts
- Location: Renal cortex, medulla & subcapsular areas
- Size: Variable in size
- Morphology: Well-defined, round or oval cysts + thin imperceptible or calcified wall
- Other general features
  - PKD: One of the classifications of renal cystic disease
  - PKD: Two major genetically inherited types (ADPKD & ARPKD)
  - ADPKD or adult PKD
    - One of the most common monogenetic disorders
    - Fourth leading cause of chronic renal failure in the world
    - 50% chance of inheriting mutant gene from ADPKD parent → child
    - Multisystemic disorder with cystic & non-cystic manifestations
  - Cystic manifestations of ADPKD
    - Kidneys (100%); liver (75%); pancreas (10%); ovaries & testis
  - Noncystic manifestations of ADPKD
    - Cardiac valvular disorders (26%); hernias (25%); colonic diverticula
    - Aneurysms: Cerebral "berry" aneurysms (5-10%); aorta & coronary arteries
  - ARPKD or infantile PKD
    - Very rare cystic renal disease in infancy/childhood
    - Enlarged polycystic kidneys + hepatic cysts
    - Usually bilateral, often present with renal failure & may develop hepatic fibrosis & portal HTN
    - Poor prognosis due to early onset of renal failure

Radiographic Findings

- IVP
  - Usually not recommended
  - Plain film
    - ± Curvilinear, dystrophic cyst wall calcification
    - ± Renal calculi
  - Mild or markedly enlarged kidneys

DDx: Multiple Cysts in Kidney

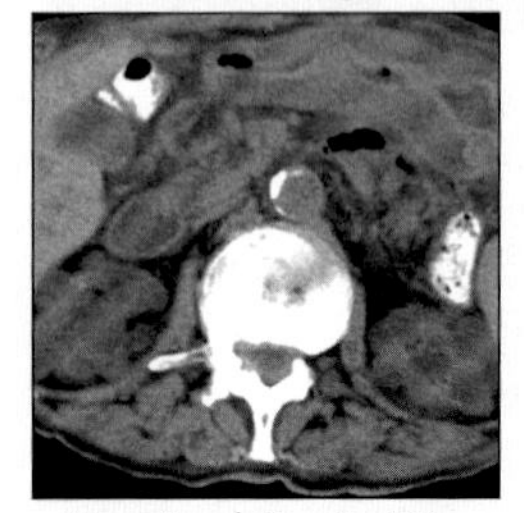

Acquired Cystic Dis.

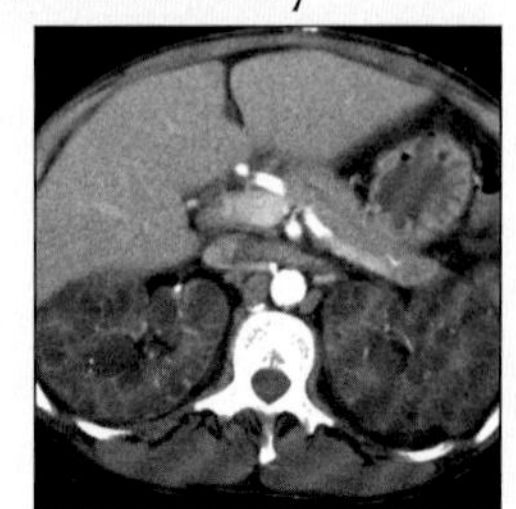

Acquired Cystic Dis.

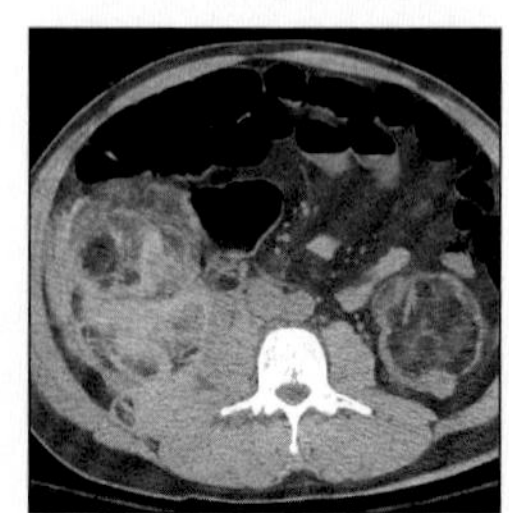

Tuberous Sclerosis

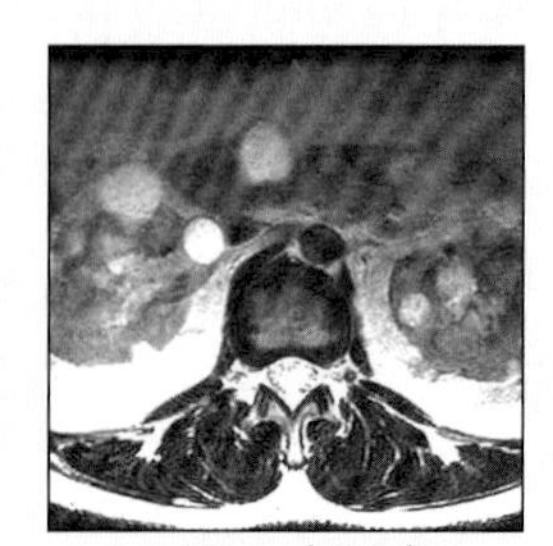

von Hippel Lindau

AD POLYCYSTIC DISEASE, KIDNEY

Key Facts

Terminology
- Autosomal dominant polycystic kidney disease (ADPKD) or adult PKD
- Hereditary disorder characterized by multiple renal cysts & various systemic manifestations

Imaging Findings
- Best diagnostic clue: Massively enlarged kidneys with innumerable cysts
- Cysts: Multiple, well-defined, round or oval in shape; variable in size; bilateral; ± calculi
- Bosselated kidneys: Multiple cysts projecting beyond renal contours
- Hypodense cysts (water attenuation) + thin wall
- CECT: No enhancement of cysts + normal renal tissue enhancement

Top Differential Diagnoses
- Multiple simple cysts
- Acquired cystic disease of uremia
- von Hippel-Lindau disease
- Tuberous sclerosis
- Medullary cystic disease

Clinical Issues
- Asymptomatic, flank pain
- Hematuria, HTN, renal failure (depends on age)

Diagnostic Checklist
- Check for genetic or family history of PKD
- Check for associated cystic lesions of liver, pancreas, spleen, ovaries & intracranial aneurysms + cardiac valve defects
- Check for history of dialysis (acquired cystic disease)

- "Swiss cheese" pattern: Smoothly marginated radiolucencies in cortex & medulla seen on nephrographic phase
- Smooth, bosselated renal contour
- Normal or effaced collecting system

CT Findings
- Cysts: Multiple, well-defined, round or oval in shape; variable in size; bilateral; ± calculi
- Early stage: Kidneys are normal in size & contour
- Later stage: ↑ In size, number of cysts, renal volume; ± asymmetrical kidneys
- Bosselated kidneys: Multiple cysts projecting beyond renal contours
- Uncomplicated cysts
  - Hypodense cysts (water attenuation) + thin wall
  - CECT: No enhancement of cysts + normal renal tissue enhancement
- Complicated (hemorrhagic cysts)
  - Hyperdense cysts (60-90 HU); usually subcapsular
  - ± Perinephric large hematomas (due to rupture)
  - ± Curvilinear mural calcification within cysts
  - CECT: Hypodense relative to normal renal tissue
- Complicated (infected cysts)
  - Hypodense; ± gas within infected cyst
  - Thick irregular cyst wall & adjacent renal fascia
  - CECT: ± Wall enhancement

MR Findings
- T1WI
  - Uncomplicated & infected cysts: Hypointense
  - Complicated (hemorrhagic cysts)
    - Varied signal (depending on age of hemorrhage)
    - Hyperintense (methemoglobin-, paramagnetic + short T1 relaxation time)
    - ± Fluid-iron levels (> intense settling layer posteriorly-, methemoglobin)
- T2WI
  - Uncomplicated: Hyperintense (thin wall)
  - Complicated (infected cysts): Hyperintense (marked mural thickening)
  - Complicated (hemorrhagic cysts): Varied signal or markedly hyperintense

Ultrasonographic Findings
- Real Time: Multiple well-defined round anechoic areas in both enlarged kidneys

Angiographic Findings
- Conventional
  - Arterial phase: Stretching & displacement of intrarenal vessels around cysts
  - Nephrographic phase: Multiple avascular masses (Swiss cheese pattern) in renal cortex & medulla

Nuclear Medicine Findings
- Technetium-99m (Tc99m) mercaptoacetylinetriglycerine (MAG3)
  - Multiple photopenic masses
  - Delayed clearance of radionuclide
    - Obstruction of collecting system due to cysts
    - Decreased renal function
- Indium-111 labeled WBC scan
  - Detects infected renal cysts
  - Increased tracer uptake, markedly at cyst periphery

Imaging Recommendations
- U.S.: Sensitivity 97%; specificity 100%; accuracy 98%
  - Recommended for screening family members
- NE + CECT; MR

DIFFERENTIAL DIAGNOSIS

Multiple simple cysts
- Rarely as numerous as in ADPKD
- Usually not associated with nephromegaly
- Normal renal function
- Differential features favoring ADPKD
  - Family history; presence of renal failure
  - Cysts (other organs): Liver, pancreas, spleen, ovaries
  - Intracranial aneurysms

Acquired cystic disease of uremia
- History: Chronic renal failure with long term dialysis
- Early stage: Small kidneys with few cysts
- Advanced stage: Large kidneys + multiple small cysts

von Hippel-Lindau disease
- Autosomal dominant; multiple renal cysts & cysts in other organs
- Renal cysts are usually less numerous than in ADPKD
- Hemangioblastomas: Cerebellar, spinal & retinal
- Multifocal renal cell carcinomas, pheochromocytomas

Tuberous sclerosis
- Multiple bilateral renal cysts
- Small fat-containing renal angiomyolipomas (AMLS)
- Cerebral paraventricular calcifications
- Renal cysts have typical histological features in tuberous sclerosis

Medullary cystic disease
- Nephronophthisis or salt wasting nephropathy
- Two types based on age related & inherited patterns
  - Childhood nephronophthisis: Autosomal recessive + associated eye, CNS, hepatic, skeletal abnormalities
  - Adult form: Autosomal dominant + none associated
- Kidneys are almost invariably small in size
- Clinically, progressive renal failure in young patients
- Imaging
  - Renal cysts may be too small to be seen
  - Visible cysts occur only in renal medulla
    - Cysts in ADPKD involve both cortex & medulla

PATHOLOGY

General Features
- Genetics
  - Autosomal dominant (90%); spontaneous mutations (10%)
  - 50% Chance of inheriting mutant gene from ADPKD parent to child
  - Three types of ADPKD based on gene location
    - ADPKD1: Short arm of chromosome 16 (90%)
    - ADPKD2: Long arm of chromosome 4 (10%), type 3 gene unknown
- Etiology
  - Hereditary: Autosomal dominant
  - Pathogenesis: Abnormal gene → proliferation of renal tubular cells→ diverticula of nephrons (collecting ducts) → cystogenesis
- Epidemiology
  - Incidence
    - 1 in 400-1,000 persons in U.S.
  - Prevalence in U.S.: Higher than cystic fibrosis, hemophilia, sickle cell disease & muscular dystrophy
- Associated abnormalities
  - Spectrum of hepatic cystic disease
    - Polycystic liver disease; congenital hepatic fibrosis
    - Biliary hamartomas; Caroli disease
  - Pancreatic & splenic cysts
  - Intracranial aneurysms & cardiac valve defects

Gross Pathologic & Surgical Features
- Enlarged kidneys
- Cysts with clear, serous, turbid or hemorrhagic fluid

Microscopic Features
- Cysts lined by simple flattened or cuboidal epithelium
- ± Wall calcification

CLINICAL ISSUES

Presentation
- Most common signs/symptoms
  - Asymptomatic, flank pain
  - Hematuria, HTN, renal failure (depends on age)

Demographics
- Age: Childhood to 8th decade
- Gender: M = F

Natural History & Prognosis
- Complications
  - Hemorrhage; infection; rupture
  - Renal failure; malignancy (rare)
- Prognosis: Fair following renal transplantation

Treatment
- Treat symptoms & complications
  - Hypertension, pain, renal infection
- Renal transplantation

DIAGNOSTIC CHECKLIST

Consider
- Differentiate from other multiple renal cystic diseases
- Check for genetic or family history of PKD
- Check for associated cystic lesions of liver, pancreas, spleen, ovaries & intracranial aneurysms + cardiac valve defects
- Check for history of dialysis (acquired cystic disease)

Image Interpretation Pearls
- Bilateral, multiple cysts + enlarged kidneys
- ± Cystic lesions: Liver, pancreas, spleen, ovaries
- ± Cerebral "berry" aneurysms & cardiac valve defects

SELECTED REFERENCES

1. Chatha RK et al: Von Hippel-Lindau disease masquerading as autosomal dominant polycystic kidney disease. Am J Kidney Dis. 37(4):852-8, 2001
2. Slywotzky CM et al: Localized cystic disease of the kidney. AJR Am J Roentgenol. 176(4):843-9, 2001
3. Nascimento AB et al: Rapid MR imaging detection of renal cysts: age-based standards. Radiology. 221(3):628-32, 2001
4. Nicolau C et al: Autosomal dominant polycystic kidney disease types 1 and 2: assessment of US sensitivity for diagnosis. Radiology. 213(1):273-6, 1999
5. Perrone RD: Extrarenal manifestations of ADPKD. Kidney Int. 51(6):2022-36, 1997
6. Fick GM et al: Natural history of autosomal dominant polycystic kidney disease. Annual Review of Medicine. 45: 23-9, 1994
7. Gabow PA: Autosomal dominant polycystic kidney disease N. Engl. J. Med., 329(5): 332 - 342, July 29, 1993
8. Parfrey PS et al: The diagnosis and prognosis of autosomal dominant polycystic kidney disease. N Engl J Med. 323(16):1085-90, 1990
9. Walker FC Jr et al: Diagnostic evaluation of adult polycystic kidney disease in childhood. AJR Am J Roentgenol. 142(6):1273-7, 1984

IMAGE GALLERY

Typical

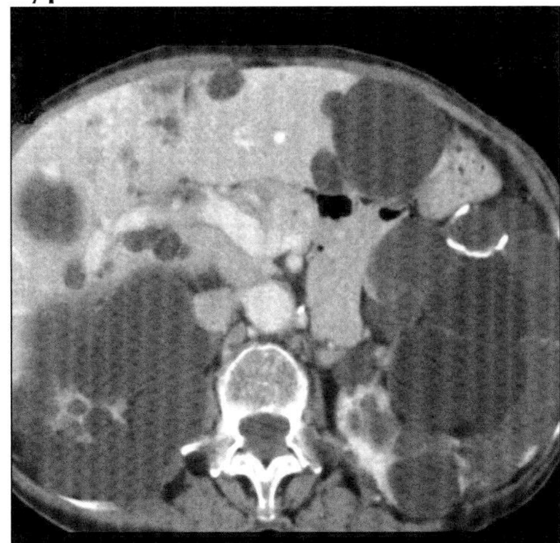

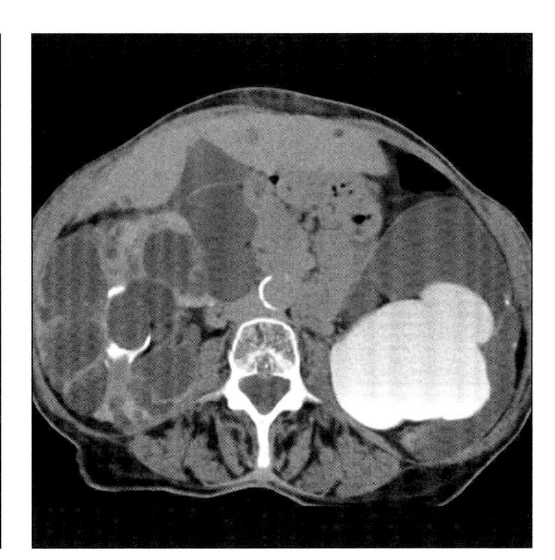

(Left) *Axial CECT shows polycystic involvement of kidneys and liver, with varying density of cysts + mural calcification due to prior hemorrhage.* ***(Right)*** *Axial CECT (delayed) shows bilateral polycystic kidneys and dilated left collecting system due to obstructing calculus.*

Typical

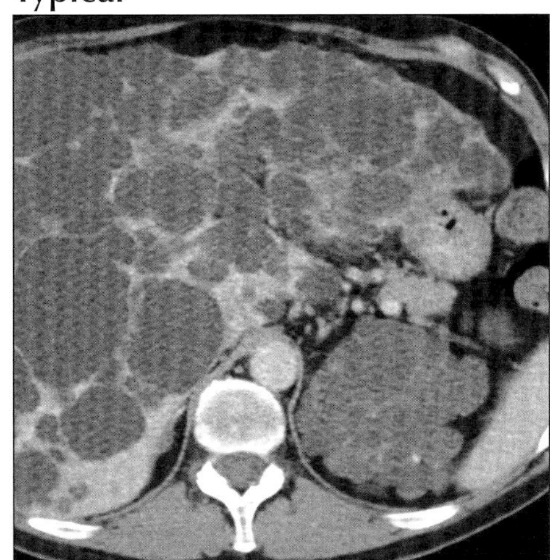

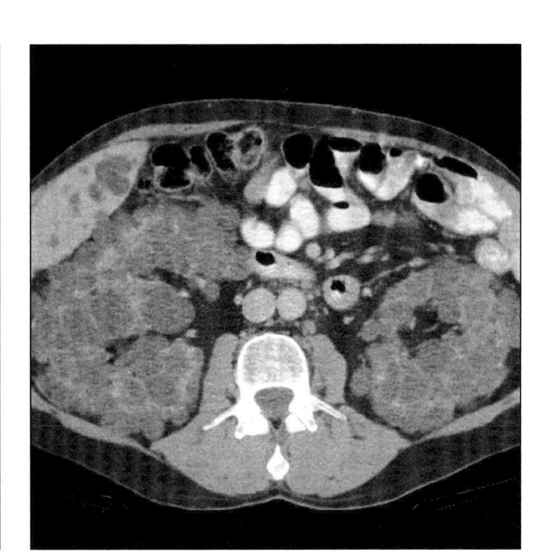

(Left) *Axial CECT shows massive cystic enlargement of kidneys and liver.* ***(Right)*** *Axial CECT shows extensive renal and liver involvement.*

Variant

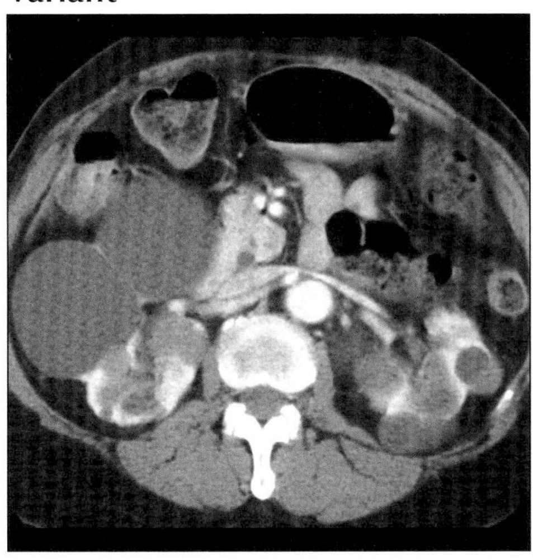

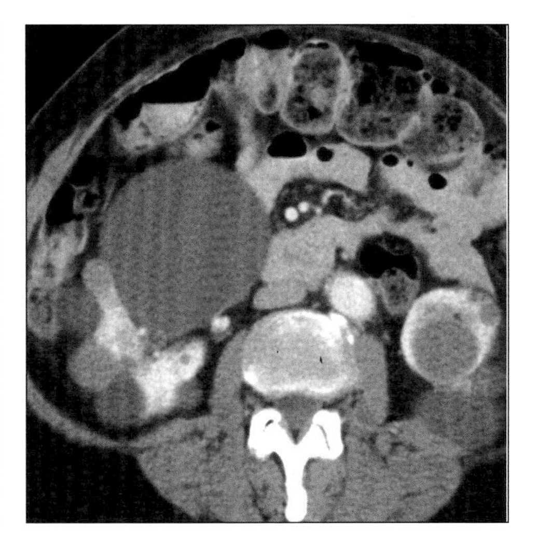

(Left) *Axial CECT in a 79 year old woman with ADPKD shows functioning kidneys with numerous cysts.* ***(Right)*** *Axial CECT in a 79 year old woman with ADPKD but preserved renal function.*

VON HIPPEL LINDAU DISEASE

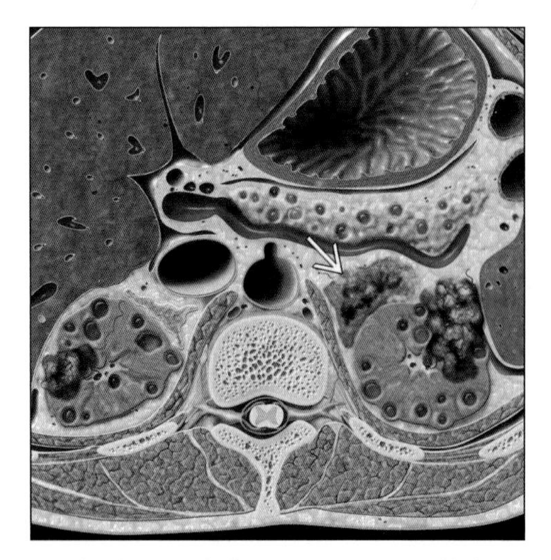

Graphic shows multiple cysts in pancreas and kidneys, solid masses (renal cell carcinoma) in both kidneys and a left adrenal pheochromocytoma (arrow).

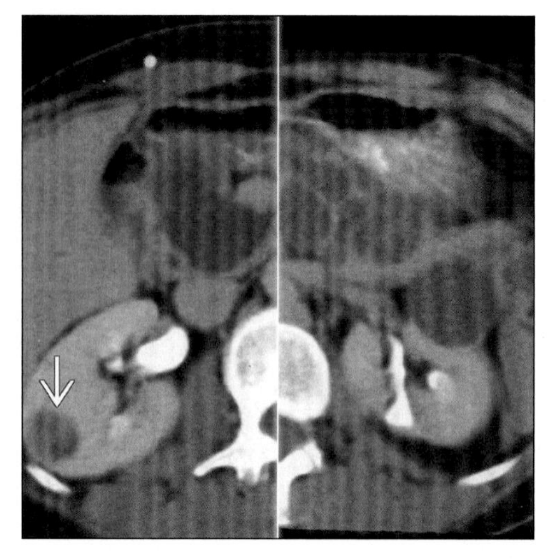

Axial CECT shows multiple pancreatic and renal cysts and a renal cell carcinoma (arrow) in a patient with VHL.

TERMINOLOGY

Abbreviations and Synonyms

- von Hippel-Lindau (VHL) syndrome

Definitions

- Rare, autosomal dominant multi-systemic disorder characterized by abnormal growth of tumors

IMAGING FINDINGS

General Features

- Best diagnostic clue: Retinal or cerebellar tumor (hemangioblastoma) + multiple renal cysts or adrenal tumor (pheochromocytoma)
- Other general features
  - Autosomal dominant (AD) familial tumor syndrome
  - Sporadic mutations occur rarely resulting in a syndrome indistinguishable from hereditary VHL
  - Manifestations of VHL
    - Retinal & CNS (cerebellar, medullary & spinal cord) hemangioblastomas
    - Endolymphatic sac tumor: Papillary cystadenoma
    - Renal: Cysts & renal cell carcinoma (RCC)
    - Adrenal: Pheochromocytoma
    - Pancreas: Cysts, serous cystadenoma, islet cell tumor, adenocarcinoma, hemangioblastoma
    - Liver cysts, papillary cystadenoma of epididymis & broad ligament (rare)
  - Eyes: Retinal hemangioblastomas (RHb)
    - Most frequent & earliest detected lesions of VHL
    - Also called "retinal angiomas" or "hemangiomas"
    - Rarely, optic nerve & chiasm hemangioblastomas
    - 45-59% of patients with VHL exhibit RHb
    - About 50% of cases have bilateral RHb
    - Diagnosed in one & two year olds; 5% of cases detected before age 10
  - CNS: Cerebellar hemangioblastomas (CHb)
    - One of the most common manifestations of VHL
    - CHb occurs in 44-72% of all VHL cases
    - Represent 2% of all brain & 7-10% of posterior fossa tumors (only 5-30% of these are due to VHL)
    - Usually occurs at a young age & often multiple
    - Solitary CHb, considered potentially a VHL case
    - Medullary Hb (5%); spinal Hb (13%) of VHL cases
  - Endolymphatic sac tumors
    - Relatively newly recognized complication of VHL
    - Inner ear: End of endolymphatic duct (labyrinth)
  - Renal cysts & tumors
    - Renal cysts: 50-75% of VHL cases; simple & complex; usually bilateral & multiple
    - RCC: 28-45% of VHL cases; multicentric & bilateral (75% of cases); seen in young age
    - Hereditary cancer syndromes: Bilateral solid clear cell renal cancers without other VHL features

DDx: Multiple Renal Cysts/Masses

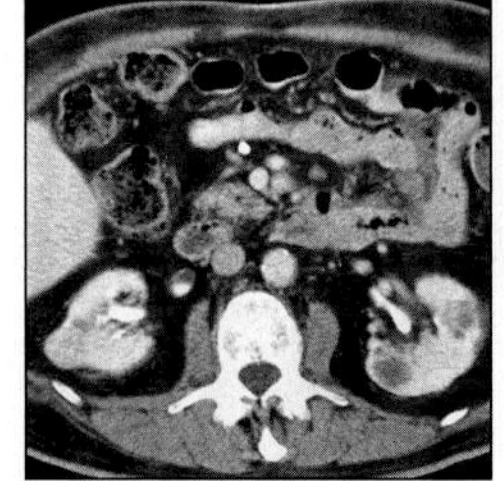

Cysts + RCC

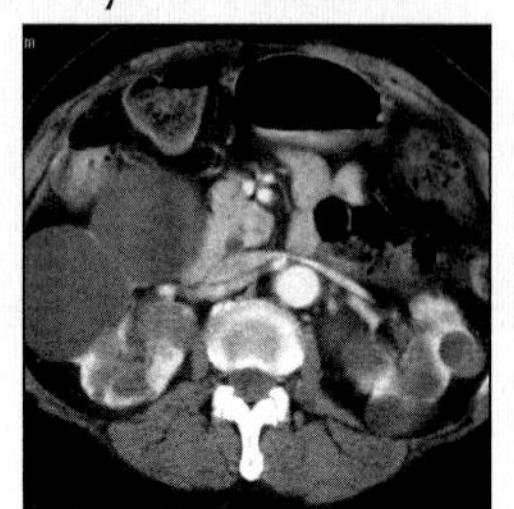

AD Polycystic

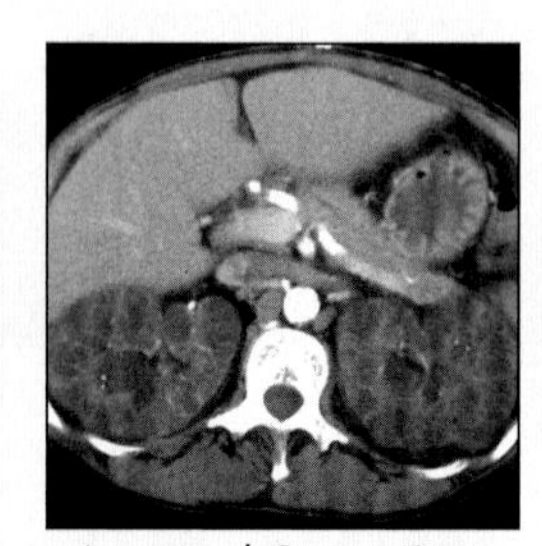

Acquired Cystic Dis.

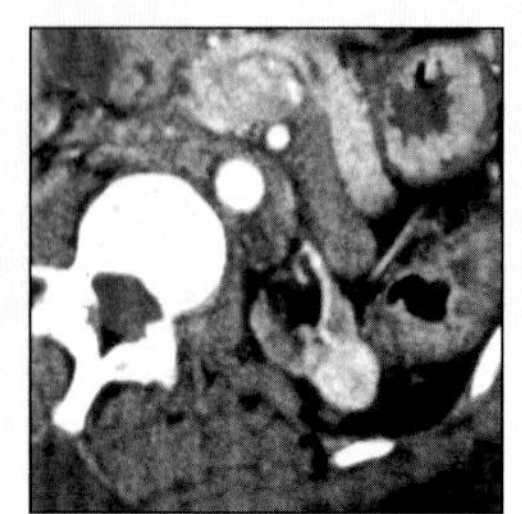

ACKD + RCC

VON HIPPEL LINDAU DISEASE

Key Facts

Terminology

- von Hippel-Lindau (VHL) syndrome
- Rare, autosomal dominant multi-systemic disorder characterized by abnormal growth of tumors

Imaging Findings

- Best diagnostic clue: Retinal or cerebellar tumor (hemangioblastoma) + multiple renal cysts or adrenal tumor (pheochromocytoma)
- CHb occurs in 44-72% of all VHL cases
- Simple cysts: Well-defined, rounded, thin-walled, nonenhancing, near water density lesions
- Multiple, bilateral solid hypovascular or complex cystic masses + mural nodules, septa (RCC)
- Well-defined, round, suprarenal mass (muscle HU)
- Honeycomb or sponge-like pancreatic head mass

Top Differential Diagnoses

- AD polycystic kidney disease (ADPKD)
- Acquired cystic kidney disease of uremia (ACKD)
- Medullary cystic disease

Clinical Issues

- More than one hemangioblastoma in CNS or retina
- One hemangioblastoma + visceral manifestation
- Known family history + one of above manifestation

Diagnostic Checklist

- Annual clinical, biochemical & imaging screening of VHL patients & at risk family members, relatives
- Genetic testing in family members, define their status
- Retinal & CNS Hbs; renal & pancreatic cysts
- Bilateral solid renal masses-CT (RCCs); hyperintense adrenal mass T2WI (pheochromocytoma)

- Adrenal tumors: Pheochromocytomas
    - Occur in 7-18% of all patients with VHL
    - Some families have ↑ pheochromocytomas frequency + low frequency of CHb & renal cancers
    - 50-80% are bilateral; often multiple & benign
    - Malignancy rate, 10-15% with ↑ incidence of metachronous tumors developing after surgery
    - Mostly localized, but 15-18% are extraadrenal
    - 20% of VHL families with high prevalence of pheochromocytoma will develop islet cell tumors
- Pancreatic cysts & tumors
    - Frequency of pancreatic involvement, 15-77%
    - Simple cysts, 80-90%; serous cystadenoma, 8-12%
    - Islet cell (neuroendocrine) tumors (7-12%): Usually nonfunctional & often multiple
    - Combined lesions (11%); ± pheochromocytoma

CT Findings

- Renal masses
    - Renal cysts
        - Usually bilateral & multiple
        - Simple cysts: Well-defined, rounded, thin-walled, nonenhancing, near water density lesions
        - Complex cysts: Irregular, septate, minimally calcified, slightly ↑ attenuation, no enhancement
    - Renal cell carcinoma
        - Multiple, bilateral solid hypovascular or complex cystic masses + mural nodules, septa (RCC)
- Adrenal tumors: Pheochromocytomas
    - Well-defined, round, suprarenal mass (muscle HU)
    - ± Areas of necrosis, hemorrhage, calcification
    - Homogeneous or heterogeneous enhancement
- Pancreatic cysts & tumors
    - Pancreatic cysts
        - Round or oval shaped water density lesions
        - Usually multiple; thin imperceptible wall
    - Serous (microcystic) cystadenoma
        - Honeycomb or sponge-like pancreatic head mass
        - Enhancement of septa delineating small cysts
        - Size varies from 5-10 cm + capsular enhancement
        - ± Calcification; ± CBD, pancreatic duct dilatation
    - Islet cell (neuroendocrine) tumor
        - Usually large (nonfunctional) mixed density
        - ± Calcification, cystic & necrotic areas
        - Hypervascular, enhancing metastases

MR Findings

- Renal masses
    - Renal cysts
        - Simple cysts: Hypointense (T1WI); hyperintense (T2WI); no enhancement (T1 C+)
        - Complex cysts: Varied signal (T1 & T2WI), ± enhancement (T1 C+)
    - Renal cell carcinoma
        - Multiple solid lesions of varied intensity (T1 & T2WI); heterogeneous enhancement (T1 C+)
- Adrenal tumors: Pheochromocytomas
    - T1WI
        - Isointense to muscle & hypointense to liver
        - Heterogeneous signal (necrosis & hemorrhage)
    - T2WI
        - Markedly hyperintense (characteristic)
        - Long T2 relaxation time (due to increase water content as a result of necrosis)
    - T1 C+: Brightly enhancing with contrast
- Pancreatic cysts & tumors
    - Pancreatic cysts
        - Hypointense (T1WI); hyperintense (T2WI)
    - Serous (microcystic) cystadenoma
        - Hypointense (T1WI); hyperintense (T2WI)
        - T1 C+: Capsular enhancement; enhancement of septa delineating small cysts
        - MRCP: Depict pancreatic duct & CBD dilatation
    - Islet cell (neuroendocrine) tumor
        - T1WI: Isointense (small); heterogeneous (large)
        - T2WI: Isointense (small); hyperintense (large)
        - T1 C+: Hyperintense (small); heterogeneous (large)

Ultrasonographic Findings

- Real Time
    - Renal & pancreatic cysts: Well-defined & anechoic
    - RCC: Hyperechoic, isoechoic or hypoechoic
    - Serous cystadenoma of pancreas
        - Tumor with tiny cysts: Hyperechoic mass but with through transmission
    - Pancreatic islet cell tumor: Hypoechoic mass

- Pheochromocytoma
  - Iso-/hypoechoic (77%); hyperechoic (23%)

Nuclear Medicine Findings

- I-131 or 123 Metaiodobenzylguanidine (MIBG)
  - After 24-72 hrs: ↑ Uptake in pheochromocytoma

Imaging Recommendations

- Helical NE + CECT & MR, T1 C+

DIFFERENTIAL DIAGNOSIS

Renal cyst

- May be present in kidney with RCC without VHL
- Cyst and RCC are both common

AD polycystic kidney disease (ADPKD)

- Multiple, bilateral renal cysts of VHL mimic ADPKD
- Massively enlarged kidneys with innumerable cysts
- Cystic manifestations of ADPKD
  - Kidneys (100%); liver (75%); pancreas; ovaries; testes

Acquired cystic kidney disease of uremia (ACKD)

- Bilateral small kidneys with multiple small cysts
- History of long term dialysis in endstage renal disease
- Increased prevalence of renal carcinoma

Medullary cystic disease

- Nephronophthisis or salt wasting nephropathy
- Adult form (AD); childhood form (AR)
- Small sized kidneys with cysts in medulla
- Clinically, progressive renal failure in young patients

PATHOLOGY

General Features

- Genetics
  - Gene is located on short arm of chromosome 3
    - Due to inactivation of tumor suppressor gene
    - 30% of patients have no exact mutation identified
- Etiology: Autosomal dominant; very rarely sporadic
- Epidemiology: Prevalence, one in 39,000-53,000

Gross Pathologic & Surgical Features

- Varies depending on underlying lesion of VHL

Microscopic Features

- Varies depending on tumor of origin

Staging, Grading or Classification Criteria

- Proposed national cancer institute (NCI) classification
  - I. VHL without pheochromocytoma (most common)
    - Retinal & CNS Hb, renal cysts & cancers, pancreatic cysts, no pheochromocytoma
  - II. VHL with pheochromocytoma
    - A. Retinal & CNS Hbs, islet cell tumors of pancreas, pheochromocytomas
    - B. Retinal & CNS Hbs, renal cell carcinomas, pancreatic lesions, pheochromocytomas

CLINICAL ISSUES

Presentation

- Most common signs/symptoms: Symptoms & signs varies based on tumor of origin
- Lab data
  - CBC to look for polycythemia vera
  - Urine: Vanillylmandelic acid (VMA), & hematuria
- Diagnostic criteria
  - More than one hemangioblastoma in CNS or retina
  - One hemangioblastoma + visceral manifestation
  - Known family history + one of above manifestation

Demographics

- Age: Infancy to 7th decade (average age 26 years)

Natural History & Prognosis

- Complications: Varies depending on underlying tumor
- Prognosis
  - RCC is the leading cause of death in VHL patients
    - 35-75% prevalence in one autopsy series
  - 2nd most common cause of morbidity & mortality in VHL disease is CNS hemangioblastomas
    - CHb: Neurologic compromise & death
    - Retinal Hb: Retinal detachment or visual loss

Treatment

- Medical: Symptomatic treatment
- Surgical: Organ-sparing surgery

DIAGNOSTIC CHECKLIST

Consider

- Annual clinical, biochemical & imaging screening of VHL patients & at risk family members, relatives
- Genetic testing in family members, define their status
- Consider VHL disease in whom retinal or cerebellar Hb, renal tumors, or pheochromocytoma is diagnosed

Image Interpretation Pearls

- Retinal & CNS Hbs; renal & pancreatic cysts
- Bilateral solid renal masses-CT (RCCs); hyperintense adrenal mass T2WI (pheochromocytoma)

SELECTED REFERENCES

1. Taouli B et al: Spectrum of abdominal imaging findings in von Hippel-Lindau disease. AJR Am J Roentgenol. 181(4):1049-54, 2003
2. Marcos HB et al: Neuroendocrine tumors of the pancreas in von Hippel-Lindau disease: spectrum of appearances at CT and MR imaging with histopathologic comparison. Radiology. 225(3):751-8, 2002
3. Couch V et al: von Hippel-Lindau disease. Mayo Clin Proc. 75(3):265-72, 2000
4. Choyke PL et al: von Hippel-Lindau disease: genetic, clinical, and imaging features. Radiology. 194(3):629-42, 1995
5. Hough DM et al: Pancreatic lesions in von Hippel-Lindau disease: prevalence, clinical significance, and CT findings. AJR Am J Roentgenol. 162(5):1091-4, 1994
6. Choyke PL et al: von Hippel-Lindau disease: radiologic screening for visceral manifestations. Radiology. 174(3 Pt 1):815-20, 1990

VON HIPPEL LINDAU DISEASE

IMAGE GALLERY

Typical

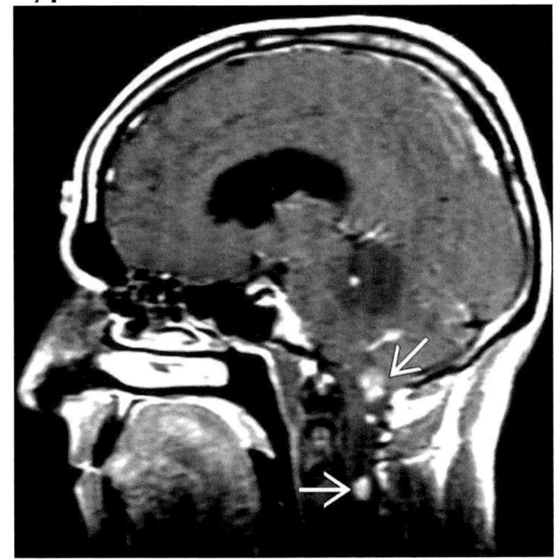

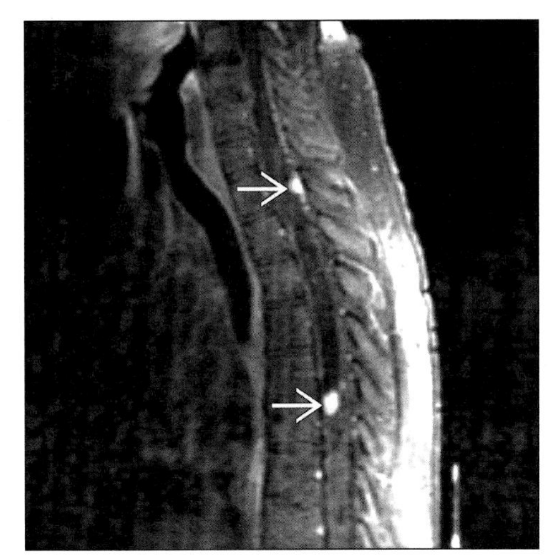

(Left) *Sagittal MR shows cerebellar and spinal cord hemangioblastomas (arrows).* ***(Right)*** *Sagittal MR shows multiple spinal hemangioblastomas (arrows).*

Typical

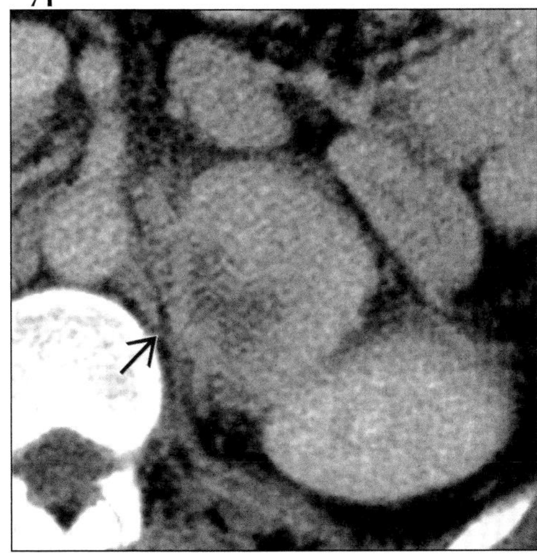

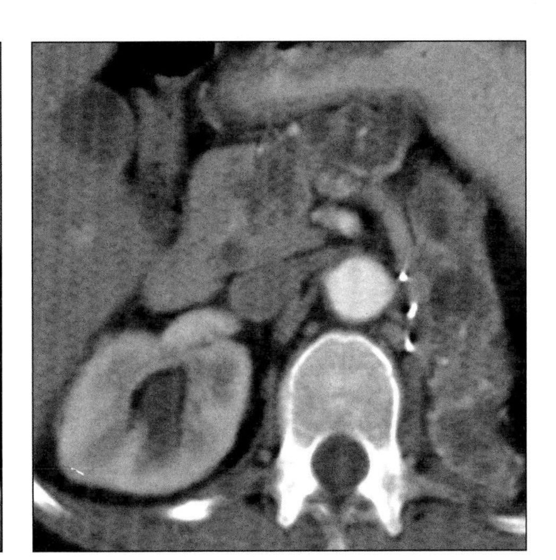

(Left) *Axial CECT shows a heterogeneous left adrenal mass (arrow), a pheochromocytoma in a patient with VHL.* ***(Right)*** *Axial CECT shows numerous pancreatic cysts. Left nephrectomy for renal cell carcinoma.*

Typical

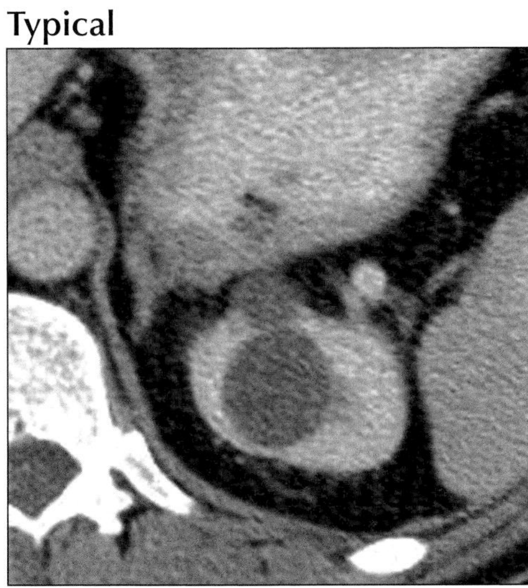

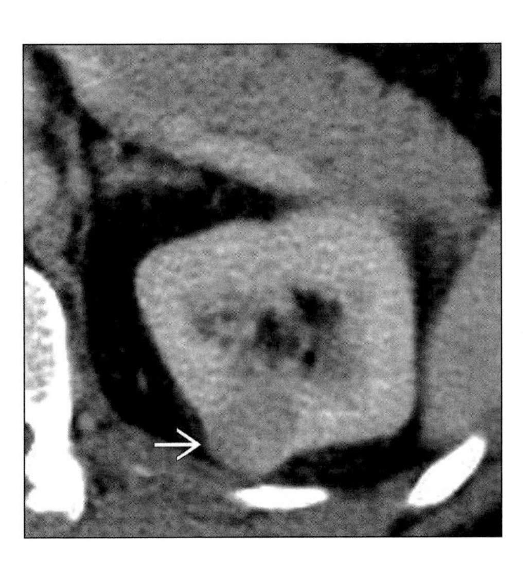

(Left) *Axial CECT shows two of multiple renal cysts in a patient with VHL.* ***(Right)*** *Axial CECT shows a small renal cell carcinoma (arrow) in a patient with VHL.*

URETEROPELVIC JUNCTION OBSTRUCTION

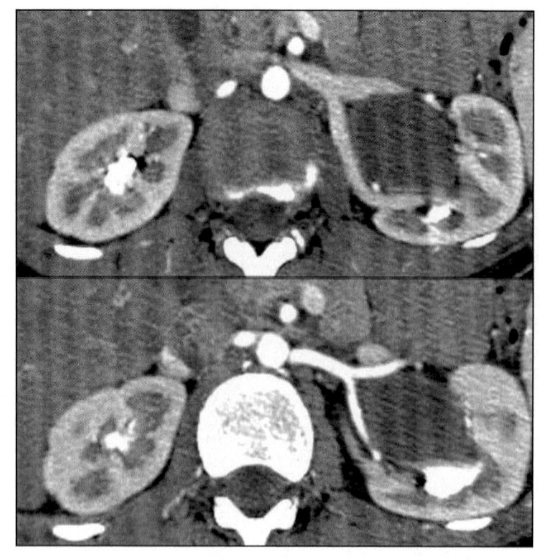
Axial CECT in arterial phase shows relationship of dilated renal pelvis to arteries and veins.

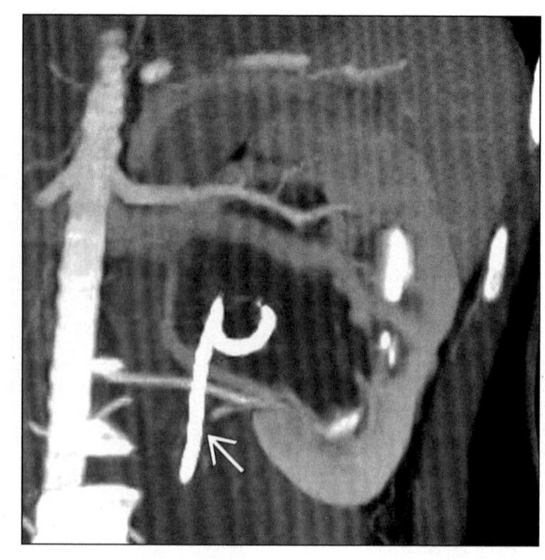
Coronal MIP CT image shows two left renal arteries and the renal vein branches relative to dilated renal pelvis. Pigtail ureteral stent (arrow).

TERMINOLOGY

Abbreviations and Synonyms

- Ureteropelvic junction (UPJ) obstruction, pelviureteric junction obstruction, idiopathic, pelvic or congenital hydronephrosis

Definitions

- Obstructed urine flow from renal pelvis to proximal ureter → pressure increase in renal pelvis

IMAGING FINDINGS

General Features

- Best diagnostic clue: Pyelocaliectasis with UPJ narrowing
- Location
  - Left kidney (2 times) > right kidney
  - Unilateral > bilateral obstruction (10-30% of cases)

Radiographic Findings

- Radiography
  - Calcification in wall of dilated renal pelvis
  - Staghorn calculus, calculus fragments or "explosion"
- IVP
  - Marked pyelocaliectasis; UPJ narrowing
  - Giant hydronephrosis; may displace, rotate and obstruct contralateral kidney and ureter
  - Incomplete visualization of normal caliber ureter
  - Increased or decreased renal size due to partial or complete obstruction, respectively
  - "Negative pyelogram" appearance in early films: Unilateral delayed opacification of collecting system
  - "Calyceal crescents": Thin, semilunar, opacified collecting tubules in the periphery of dilated calyces
  - Dilated papillary (Bellini) ducts: Small dots of contrast at medial edge of dilated calyces
  - "Puddles" or "Ball" pyelogram: Contrast settles in dependent portions of dilated calyces
  - "Soap bubble" nephrogram: Parenchymal thinning → opacified, thin, overlapping rim of parenchyma
  - "Shell" or "rim" nephrogram: Opacify only the peripheral portion of thin rim
  - Delayed clearing of contrast from collecting system
  - "Linear band" sign: Linear oblique crossing defect in proximal end of ureter
  - "Short segment" sign: Small amount of contrast trapped in segment just below the UPJ
  - Enlarged extrarenal pelvis out of proportion to caliectasis; may simulate normal variant
    - Usually flabby or distensible
    - Relatively narrow or "closed" outlet
    - Turgid in some phases of IVP
  - Diuresis IVP as adjunct

DDx: Mimics of UPJ Obstruction

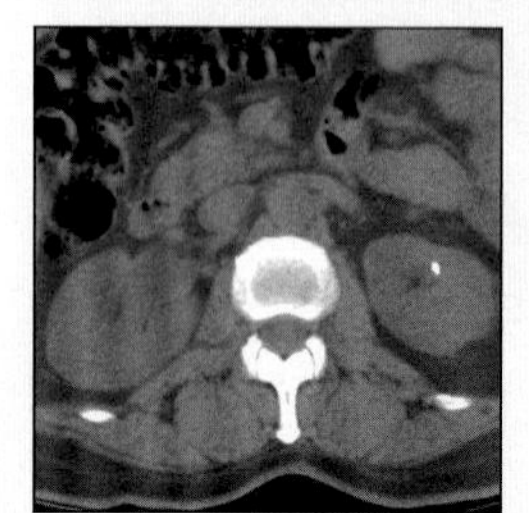
Urolithiasis

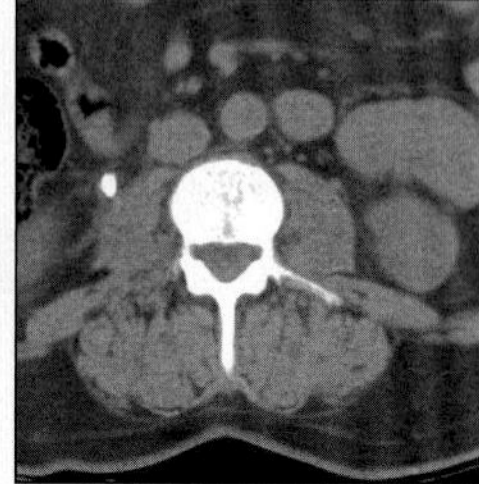
Urolithiasis

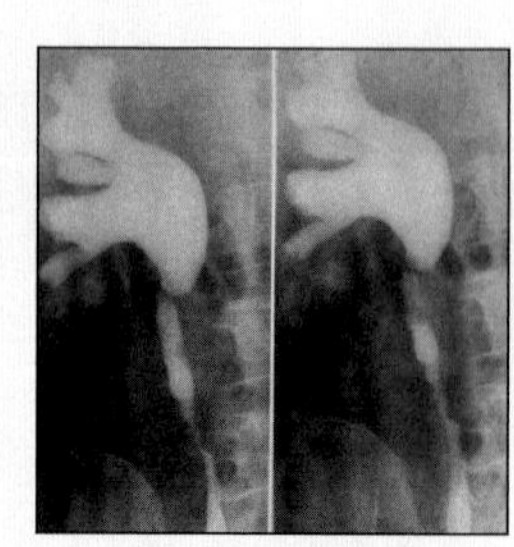
TCC at UPJ

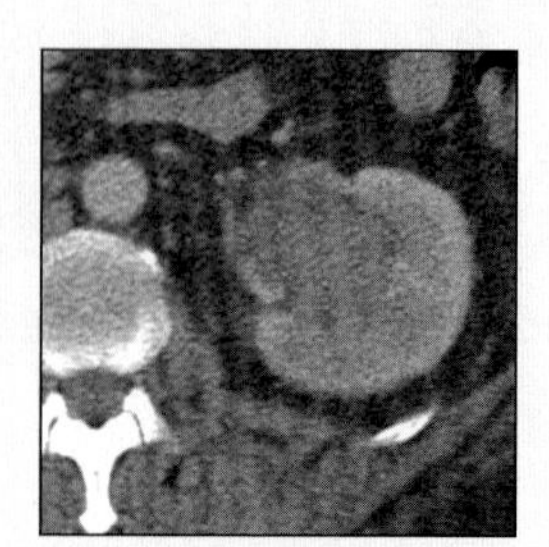
Renal Pelvic Blood Clot

URETEROPELVIC JUNCTION OBSTRUCTION

Key Facts

Terminology

- Obstructed urine flow from renal pelvis to proximal ureter → pressure increase in renal pelvis

Imaging Findings

- Best diagnostic clue: Pyelocaliectasis with UPJ narrowing
- Left kidney (2 times) > right kidney
- Incomplete visualization of normal caliber ureter
- Increased or decreased renal size due to partial or complete obstruction, respectively
- "Negative pyelogram" appearance in early films: Unilateral delayed opacification of collecting system
- Delayed clearing of contrast from collecting system
- Best imaging tool: IVP: Adult; US: Neonates and children

Top Differential Diagnoses

- Urolithiasis
- Tumor
- Extrinsic indentation
- Retrocaval ureter
- Renal pelvis blood clot

Clinical Issues

- Indicated when patient has symptoms, stones, infection or renal function is impaired or at risk

Diagnostic Checklist

- Use CT or MR to evaluate potential acquired etiologies of UPJ obstruction
- Chronic obstruction → progressive changes to renal pelvis and calyces → distinctive urographic signs

    - Delayed clearing (> 10 min.) of contrast, pyelocaliectasis and flank pain; suggest intermittent UPJ obstruction
- Voiding cystourethrography
  - Exclude severe vesicoureteral reflux in infants
- Retrograde ureteropyelography
  - Assess ureter if not visualized in other studies
  - "Jet" of contrast squirting through narrow UPJ into non-opacified, urine-filled renal pelvis
  - Angulated UPJ by crossing bands or vessels

CT Findings

- NECT
  - Hydronephrosis ± ureterectasis
  - Level of obstruction
  - ± Acquired etiologies (e.g., crossing vessels, neoplasm, retroperitoneal inflammatory conditions) and associated abnormalities (e.g., renal malformation)
- CTA
  - Use 3D reconstruction to better define vessels prior to endoscopic pyelotomy
  - 3D and multiplanar reformations
    - Useful in planning endopyelotomy
    - Show relation of pelvis to adjacent vessels

MR Findings

- MRA: Detect crossing vessels
- MR Urography: ± Acquired etiologies

Ultrasonographic Findings

- Real Time
  - Use US in both prenatal and postnatal evaluation
  - Prenatal findings
    - Anteroposterior (AP) pelvic diameter > 10 mm, pelvic to AP renal diameter ratio > 0.5, caliectasis: Fetal hydronephrosis
    - Mild pyelectasis (pelvic diameter 4-10 mm in < 20 weeks of gestation and 5-10 mm in 20-24 weeks): 10-15% obstructed
    - Pelvic or pyelocalyceal dilatation
    - Oligo-, poly- or euhydroamnios
    - ± Urine ascites
  - Postnatal findings
    - Hydronephrosis ± ureterectasis
    - Large, medial sonolucent area (renal pelvis) surrounded by smaller, rounded sonolucent areas (dilated calyces)
    - Assess severity and level of obstruction
    - Advanced parenchymal atrophy
    - Hypertrophy of normal kidney contralateral to hydronephrotic kidney

Nuclear Medicine Findings

- Diuresis renography
  - Separates obstructive from nonobstructive dilatation
  - Localize level of obstruction
  - Assess renal function, often pre-operatively
  - "Homsy" sign: Delayed double-peak pattern; suggests intermittent UPJ obstruction

Imaging Recommendations

- Best imaging tool: IVP: Adult; US: Neonates and children
- Protocol advice
  - IVP
    - Visualize UPJ with prone oblique view; left/right anterior oblique for left/right UPJ, respectively
    - Diuresis IVP: Furosemide I.V. 0.5 mg/kg 15-20 min. into IVP; film at 5, 10, 15 min. after injection
  - CTA: Use 3D reconstruction to better define vessels and their relation to UPJ
  - US: Serial US should be done several days postnatally due to relative neonatal oliguria
  - Diuresis Renography: Tc99m labeled mercaptoacetyltriglycine (MAG3) is preferred due to lower radiation burden

DIFFERENTIAL DIAGNOSIS

Urolithiasis

- Ureteral filling defects
- Most calculi are markedly hyperdense (> 200 HU)
- Usually acute; rarely cause prolong dilatation and obliteration of renal pelvis and calyces

URETEROPELVIC JUNCTION OBSTRUCTION

Tumor

- Benign or malignant retroperitoneal mass may cause ureteral obstruction; use CT to evaluate
- Primary transitional cell carcinoma (TCC) near UPJ may cause obstruction

Extrinsic indentation

- E.g., crossing bands, crossing vessels
- "Linear band" sign

Retrocaval ureter

- Ureter courses behind and medial to inferior vena cava
- Obstruction at mid-ureter → dilatation proximally from mid-ureter, not proximal

Renal pelvis blood clot

- E.g., anticoagulated patient, trauma
- May obstruct UPJ

PATHOLOGY

General Features

- General path comments: Obstruction caused by spectrum of pathophysiological processes of varying etiologies
- Genetics: Familial occurrences in some cases
- Etiology
  - Congenital (most common)
    - Intrinsic stenosis
    - Adynamic segment
    - Valves and folds
    - Kinks or angulations
    - Adhesions or bands
    - Crossing vessels near UPJ
    - High "insertion" of ureter
  - Acquired
    - Scarring: Inflammation, surgery, trauma
    - Vesicoureteral reflux
    - Malignant neoplasm: Transitional cell carcinoma, squamous cell carcinoma, metastasis
    - Benign neoplasm: Polyp, mesodermal tumor
    - Intraluminal lesion: Stone, clot, papilla, fungus ball, cholesteatoma, bullet, miscellaneous
    - Herniation of kidney: Through lumbodorsal fascia defect, into Bochdalek hernia
- Epidemiology: Neonates: 40% of all significant neonatal hydronephrosis (1/500 pregnancies)
- Associated abnormalities
  - Cystic renal dysplasia, primary megaureter
  - Lower or upper segment of duplex kidney
  - Ectopic, malrotated, pelvic and horseshoe kidneys
  - Complex congenital anomaly: VATER (vertebral, anus, tracheoesophageal, renal and radial)

CLINICAL ISSUES

Presentation

- Most common signs/symptoms
  - Neonates
    - Asymptomatic, diagnosed by prenatal screening
    - Palpable, sometimes visible abdominal mass
  - Children and adults
    - Intermittent abdominal or flank pain, nausea, vomiting
    - Hematuria, renovascular hypertension (rare)
- Lab data: Microhematuria, pyuria, urinary tract infection

Demographics

- Age: Any age; less common in adults
- Gender
  - Overall, M:F = 2:1
  - In infants, M:F = 5:1

Natural History & Prognosis

- Complications: Failure to thrive, renal insufficiency, urinary tract infection, urolithiasis, gastroduodenal obstruction, traumatic or spontaneous kidney rupture
- Prognosis: Good, after treating unilateral obstruction

Treatment

- Indicated when patient has symptoms, stones, infection or renal function is impaired or at risk
  - Infants and children: Open pyeloplasty
  - Adults: Endopyelotomy
- Follow-up: 3-6 months with diuresis renography

DIAGNOSTIC CHECKLIST

Consider

- Use CT or MR to evaluate potential acquired etiologies of UPJ obstruction

Image Interpretation Pearls

- Chronic obstruction → progressive changes to renal pelvis and calyces → distinctive urographic signs

SELECTED REFERENCES

1. Khaira HS et al: Helical computed tomography for identification of crossing vessels in ureteropelvic junction obstruction-comparison with operative findings. Urology. 62(1):35-9, 2003
2. Mitsumori A et al: Evaluation of crossing vessels in patients with ureteropelvic junction obstruction by means of helical CT. Radiographics. 20(5):1383-93; discussion 1393-5, 2000
3. Rouviere O et al: Ureteropelvic junction obstruction: use of helical CT for preoperative assessment--comparison with intraarterial angiography. Radiology. 213(3):668-73, 1999
4. Farres MT et al: Helical CT and 3D reconstruction of ureteropelvic junction obstruction: accuracy in detection of crossing vessels. J Comput Assist Tomogr. 22(2):300-3, 1998
5. Siegel CL et al: Preoperative assessment of ureteropelvic junction obstruction with endoluminal sonography and helical CT. AJR Am J Roentgenol. 168(3):623-6, 1997
6. Quillin SP et al: Helical (spiral) CT angiography for identification of crossing vessels at the ureteropelvic junction. AJR Am J Roentgenol. 166(5):1125-30, 1996
7. Wolf JS Jr et al: Imaging for ureteropelvic junction obstruction in adults. J Endourol. 10(2):93-104, 1996
8. Bagley DH et al: Endoluminal sonography in evaluation of the obstructed ureteropelvic junction. J Endourol. 8(4):287-92, 1994
9. Bush WH et al: Ureteropelvic junction obstruction: treatment with percutaneous endopyelotomy. Radiology. 171(2):535-8, 1989

URETEROPELVIC JUNCTION OBSTRUCTION

IMAGE GALLERY

Typical

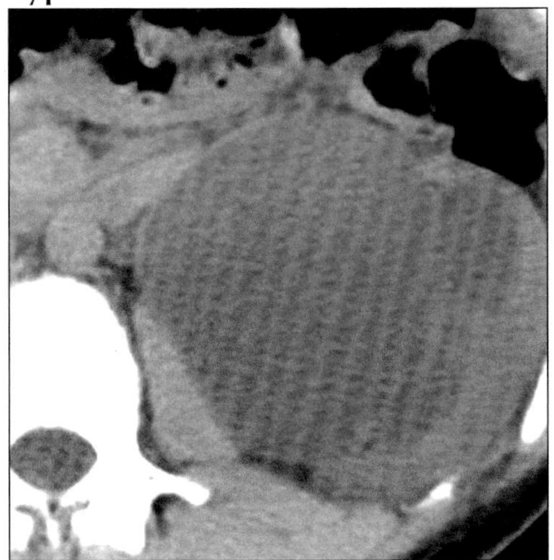

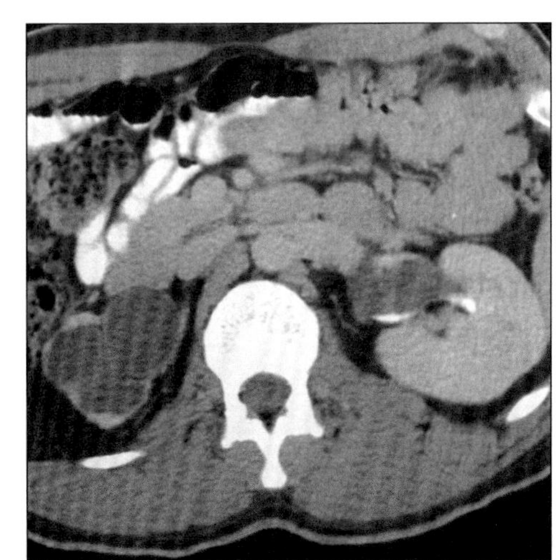

(Left) *Axial NECT shows massive dilation of pelvis + calices in patient with congenital UPJ obstruction.* ***(Right)*** *Axial CECT shows dilated renal pelvis, caliectasis and parenchymal atrophy of the right kidney.*

Typical

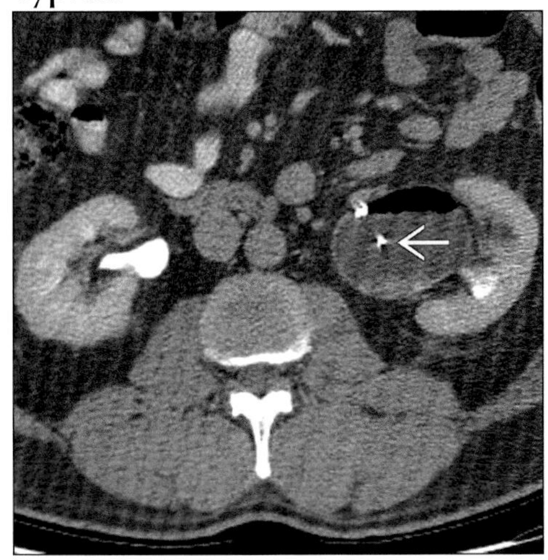

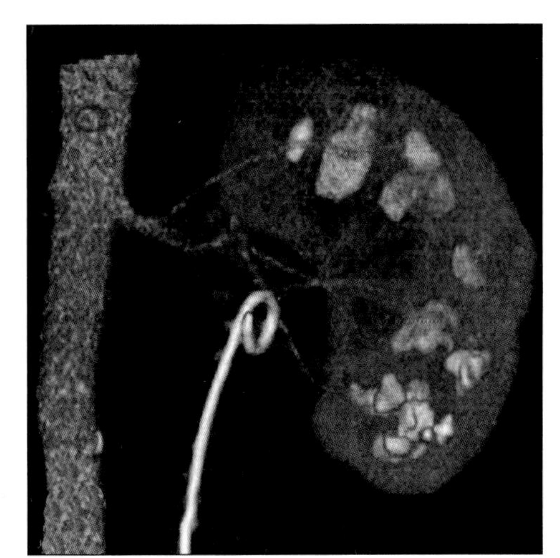

(Left) *Axial CECT shows left side pelvocaliectasis due to congenital UPJ obstruction. Gas due to ureteral stent (arrow).* ***(Right)*** *Volume-rendered 3D CT image shows pigtail ureteral stent marking the UPJ plus the adjacent arteries.*

Typical

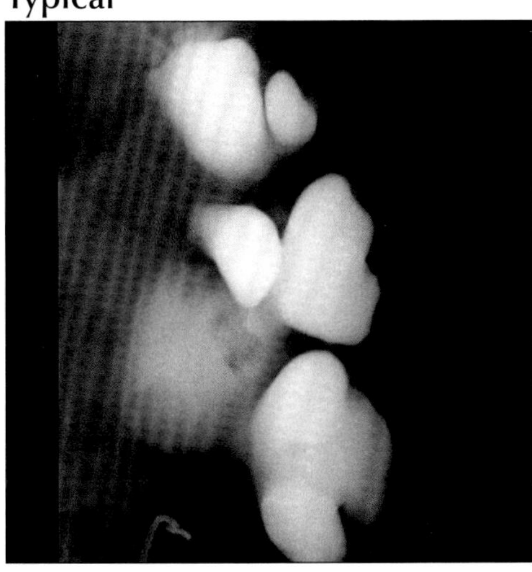

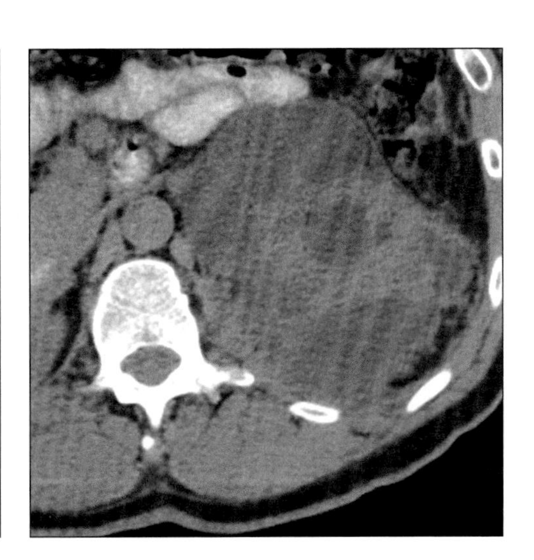

(Left) *Retrograde pyelogram shows severe pelvocaliectasis.* ***(Right)*** *Axial NECT shows perirenal urinoma due to rupture of a chronically dilated left renal pelvis.*

PYELONEPHRITIS

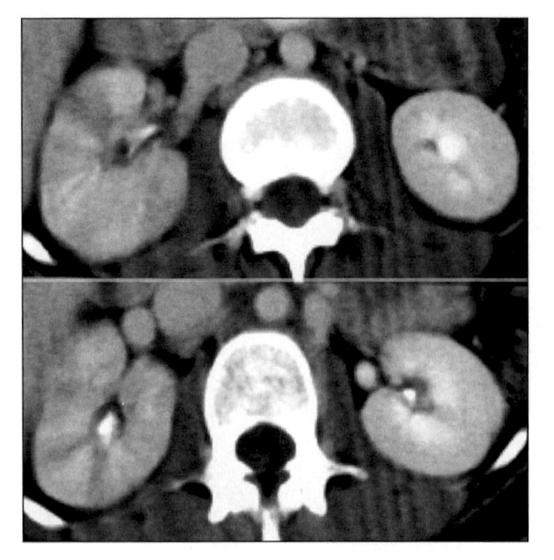

Axial CECT shows a swollen right kidney with a striated nephrogram.

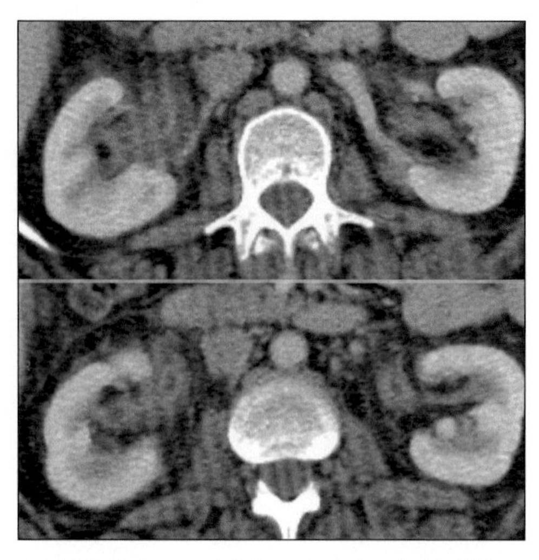

Axial CECT shows subtle heterogeneity of right kidney, + thick enhancing uroepithelium indicating pyelitis.

TERMINOLOGY

Definitions

- Infection of renal pelvis, tubules & interstitium (not glomerulus)

IMAGING FINDINGS

General Features

- Best diagnostic clue: Wedge-shaped + striated areas of ↓ enhancement & renal swelling on CECT (acute)
- Location: Usually multifocal
- Morphology
  - Acute pyelonephritis: Enlarged kidney
  - Chronic pyelonephritis: Scarred contracted kidney
- Other general features
  - Tubulointerstitial disease of kidney (upper UTI)
  - Route of infection
    - Ascending infection (> common); hematogenous
  - Classification based on clinical onset & pathology
    - Acute pyelonephritis
    - Chronic pyelonephritis
    - Emphysematous pyelonephritis
    - Xanthogranulomatous pyelonephritis

Radiographic Findings

- IVP
  - Acute pyelonephritis
    - Global or focal enlargement
    - Impaired excretion: Delayed appearance, ↓ density, ↓ nephrogram
    - Striated nephrogram or lucent areas (↓ filling); streaking & blushing
    - Calyceal compression, pelvicaliceal or ureteral dilatation, ± calculi
  - Chronic pyelonephritis
    - Contracted small sized kidney, ↓ & delayed excretion, dilated ureter
    - Focal or diffuse calyceal clubbing or blunting + cortical scarring
    - Contralateral diffuse or focal compensatory hypertrophy

CT Findings

- Acute pyelonephritis
  - Renal enlargement, focal swelling, sinus obliteration
  - Thickening of Gerota fascia + perinephric stranding
  - ± Areas of ↑ HU (hemorrhagic bacterial nephritis)
  - Nephrographic phase: "Patchy" nephrogram
    - Cortical wedge-shaped areas of decreased density (hypoperfusion, edema, hypoconcentration)
    - Striated nephrogram
    - Loss of normal corticomedullary differentiation
    - Focal severe pyelonephritis: Mimics renal neoplasm
  - Excretory phase

DDx: Striated or Wedge Defects on CT Nephrogram

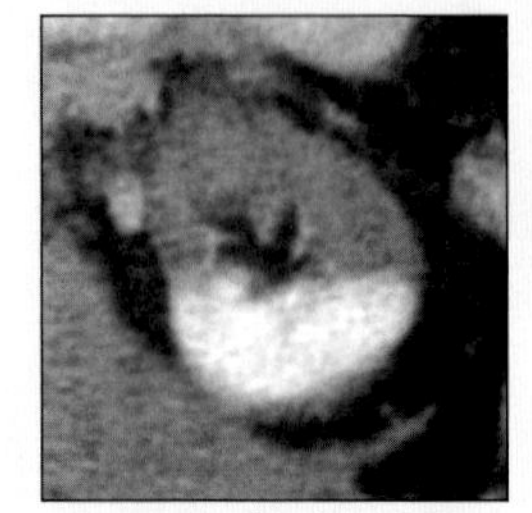

Renal Infarct

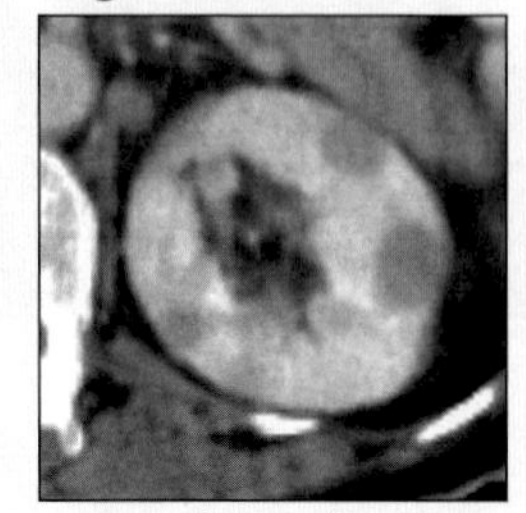

Renal Lymphoma

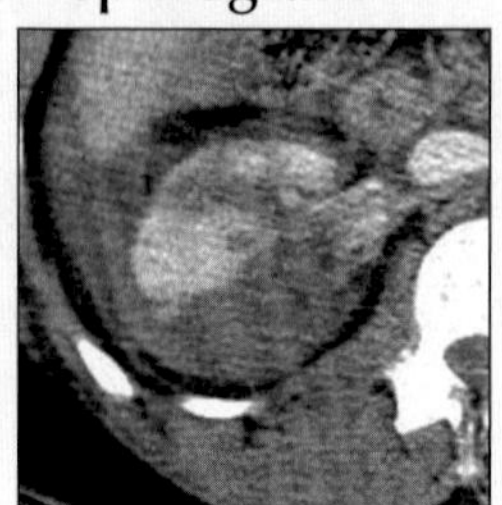

Renal Trauma

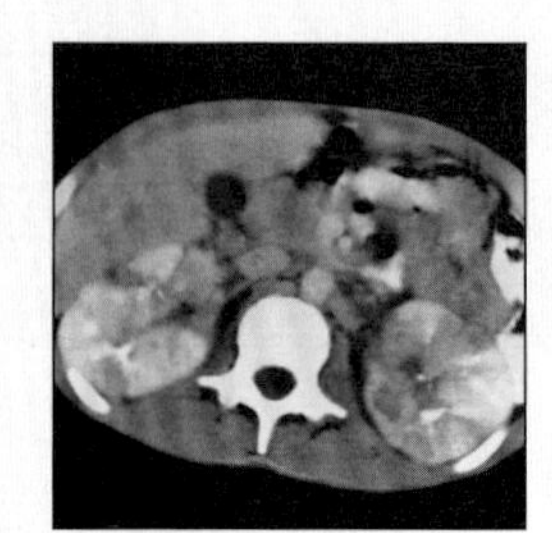

Polyarteritis Nodosa

PYELONEPHRITIS

Key Facts

Terminology
- Infection of renal pelvis, tubules & interstitium (not glomerulus)

Imaging Findings
- Best diagnostic clue: Wedge-shaped + striated areas of ↓ enhancement & renal swelling on CECT (acute)
- Characteristic appearance: Small kidney with cortical scarring over dilated calices (chronic)
- Nonfunctional kidney or part of kidney with obstructing calculi (XGPN)
- Necrosis + gas replacement of renal parenchyma with little or no pus (emphysematous type)

Top Differential Diagnoses
- Renal infarction
- Renal trauma
- Vasculitis

Pathology
- Gram negative: E. coli, proteus, klebsiella, Enterobacter (from fecal flora)
- Vesicoureteral reflux (VUR); obstructive uropathy
- Risk factor → ascending infection (most common)
- ↑ Incidence: Women under age 40, men above 65
- Associated abnormalities: BPH; VUR; UT obstruction

Diagnostic Checklist
- Distinction of pyelonephritis from vasculitis or renal infarction often requires clinical correlation
- Wedge-shaped parenchymal defects with renal swelling, usually acute pyelonephritis
- Defects with cortical scarring usually chronic pyelonephritis, vasculitis or infarction

        - Streaky linear bands: Alternating ↑ + ↓ attenuation
        - Due to ↓ concentration of contrast in tubules from ischemia + tubular obstruction by inflammatory cells + debris
        - Calyceal effacement, dilated renal pelvis & ureter
        - Thickening of walls of renal pelvis, calyces, ureter
    - Delayed phase (3-4 hrs)
        - ↑ Enhancement in previously low density wedge-shaped zones
        - Due to eventual filling of tubules that are partially obstructed by surrounding interstitial inflammatory edema
- Chronic pyelonephritis
    - Deep cortical scarring: Focal, segmental, diffuse; unilateral or bilateral
    - Atrophy: Focal (> in upper pole) or diffuse
    - Unilateral with compensatory hypertrophy of contralateral kidney
    - Characteristic appearance: Small kidney with cortical scarring over dilated calices (chronic)
    - Loss of corticomedullary differentiation
- Xanthogranulomatous pyelonephritis (XGPN)
    - Nonfunctional kidney or part of kidney with obstructing calculi (XGPN)
    - Low attenuation collections: Foot print of a bear paw (markedly dilated collecting system filled with pus, xanthoma cells + mildly dilated pelvis)
    - Lack of contrast excretion + bright rim enhancement (due to capillary proliferation in granulation tissue)
    - Thickened Gerota fascia, perinephric soft tissue stranding, abscess, extension into abdominal wall
- Emphysematous pyelonephritis
    - Necrosis + gas replacement of renal parenchyma with little or no pus (emphysematous type)
        - Occurs almost exclusively in diabetics
        - Due to infarction & infection with gas forming organisms
        - Usually requires urgent nephrectomy

Ultrasonographic Findings
- Real Time
    - Acute pyelonephritis
        - Normal or swollen kidney with ↓ echogenicity; loss of sinus echoes
        - Wedge-shaped hypo-/isoechoic zones; hyperechoic (hemorrhage)
        - Blurred corticomedullary junctions; ± anechoic areas (abscesses)

Nuclear Medicine Findings
- Tc99m-DMSA or glucoheptonate cortical imaging
    - Decreased uptake in foci of inflammation
    - Used in follow-up of renal scarring
    - Highly sensitive in diagnosing acute pyelonephritis
    - Dimercaptosuccinic acid (DMSA) renal SPECT
        - > Sensitive than planar scintigraphy in children
- Indium-111 labeled leukocytes
    - Normally do not accumulate in kidneys
    - More specific than gallium (detecting inflammation)
    - Increased uptake
        - Acute focal or diffuse pyelonephritis; renal abscess

Imaging Recommendations
- Helical NE + CECT
- Indium-111 labeled leukocytes
    - For early detection of renal & perinephric infection
    - When renal ultrasonography is normal in a clinically positive patient

DIFFERENTIAL DIAGNOSIS

Renal infarction
- Focal segmental infarction
    - Sharply demarcated, nonenhancing wedge-shaped
- Global infarction
    - Total absence of renal enhancement, no excretion
    - ± Medullary striations: "Spoke wheel" enhancement
        - Due to collateral circulation
- Acute infarction
    - Normal or large kidney with smooth contour
    - Absent or ↓ nephrogram + cortical enhancement
    - "Cortical rim" sign on CECT (6-8 hrs after infarction)
        - Preserved capsular & subcapsular enhancement

PYELONEPHRITIS

Lymphoma/metastases

- Multifocal tumor may simulate multifocal pyelonephritis
- Unifocal tumor may simulate focal severe pyelonephritis
- Clinical presentation usually characteristic

Renal trauma

- Best imaging clue
  - Irregular linear or segmental nonenhancing tissue & subcapsular or perinephric hematoma
- Subcapsular hematoma
  - Round or elliptic fluid collection (40-70 HU clot)
- Lacerations: Irregular or linear hypodense areas
- Segmental infarction: Nonenhancing wedge-shaped
- "Shattered kidney"
  - Renal artery avulsion
    - Global infarction + perinephric hematoma
  - Renal artery thrombosis
    - Global infarction + no perinephric hematoma

Vasculitis

- E.g., polyarteritis nodosa; SLE; scleroderma, drug abuse
- Wedge-shaped or striated nephrogram
- Key differential
  - Vasculitis: Capsular retraction over parenchymal lesions
  - Acute pyelonephritis: Capsular bulge over parenchymal lesions
- Microaneurysms of small vessels are commonly seen

PATHOLOGY

General Features

- Etiology
  - Gram negative: E. coli, proteus, klebsiella, Enterobacter (from fecal flora)
  - Predisposing or increased risk factors
    - Vesicoureteral reflux (VUR); obstructive uropathy
    - Pregnancy, benign prostatic hypertrophy (BPH)
    - Urethral instrumentation
    - Diabetes mellitus & other renal pathology
  - Pathogenesis
    - Risk factor → ascending infection (most common)
    - Hematogenous infection (less common)
- Epidemiology
  - Incidence of urinary tract infection (UTI)
    - ↑ Incidence: Women under age 40, men above 65
- Associated abnormalities: BPH; VUR; UT obstruction

Gross Pathologic & Surgical Features

- Acute pyelonephritis
  - "Polar abscesses": Microabscesses on renal surface
    - Lower & upper poles are most common
  - Narrowed calyces, enlarged kidney
- Chronic pyelonephritis
  - Blunted calyces + scarred shrunken kidney

Microscopic Features

- Acute pyelonephritis
  - Interstitial or tubular necrosis
  - Mononuclear cell infiltrate + fibrosis
- Chronic pyelonephritis
  - Chronic inflammation, atrophy, interstitial fibrosis
- Xanthogranulomatous pyelonephritis
  - Foamy, lipid-laden histiocytes; pus cells & necrosis

CLINICAL ISSUES

Presentation

- Most common signs/symptoms
  - Acute pyelonephritis
    - Fever, malaise, dysuria, flank pain & tenderness
- Lab data
  - ↑ ESR; ↑ WBC; ↑ proteinuria
  - Positive urine culture for bacilli; impaired RFTs
- Diagnosis: Clinical findings, imaging & biopsy

Demographics

- Age: Common in adults (also seen in children)
- Gender: Females under age 40 & males above age 65

Natural History & Prognosis

- Complications
  - Renal abscess, perinephric abscess, pyonephrosis
  - Renal papillary necrosis (RPN)
  - Renal atrophy (focal or global), renal failure
- Prognosis
  - Acute pyelonephritis: Good
  - Chronic, XGPN, emphysematous types: Poor

Treatment

- Acute: Antibiotic therapy
- Chronic: Treat reflux & obstruction; nephrectomy

DIAGNOSTIC CHECKLIST

Consider

- Distinction of pyelonephritis from vasculitis or renal infarction often requires clinical correlation

Image Interpretation Pearls

- Wedge-shaped parenchymal defects with renal swelling, usually acute pyelonephritis
- Defects with cortical scarring usually chronic pyelonephritis, vasculitis or infarction

SELECTED REFERENCES

1. Kawashima A et al: Radiologic evaluation of patients with renal infections. Infect Dis Clin North Am. 17(2):433-56, 2003
2. Kawashima A et al: CT of renal inflammatory disease. Radiographics. 17(4):851-66; discussion 867-8, 1997
3. Saunders HS et al: The CT nephrogram: implications for evaluation of urinary tract disease. Radiographics. 15(5):1069-85; discussion 1086-8, 1995
4. Talner LB et al: Acute pyelonephritis: Can we agree on terminology? Radiology 192: 297-305, 1994
5. Soulen MC et al: Bacterial renal infection: Role of CT. Radiology 171: 703-707, 1989
6. Morehouse HT et al: Imaging in inflammatory disease of the kidney. AJR Am J Roentgenol. 143(1):135-41, 1984

IMAGE GALLERY

Typical

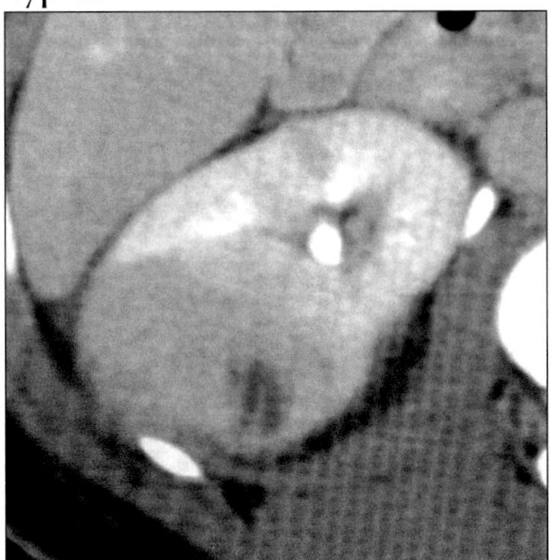

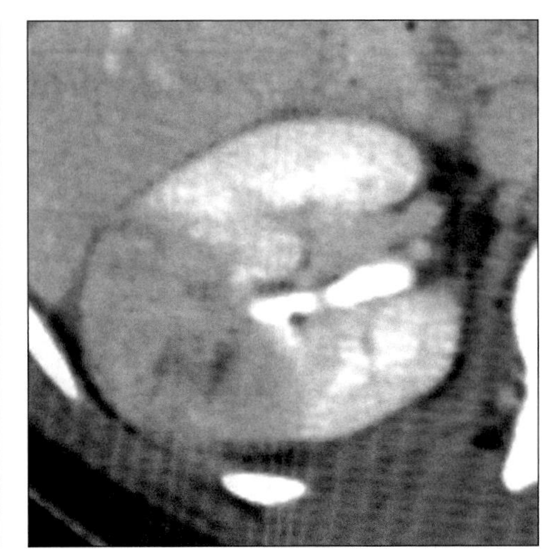

(Left) *Axial CECT shows focal severe pyelonephritis with mass-like swelling and a striated nephrogram.* ***(Right)*** *Axial CECT shows severe focal pyelonephritis with decreased enhancement and focal swelling of right kidney.*

Typical

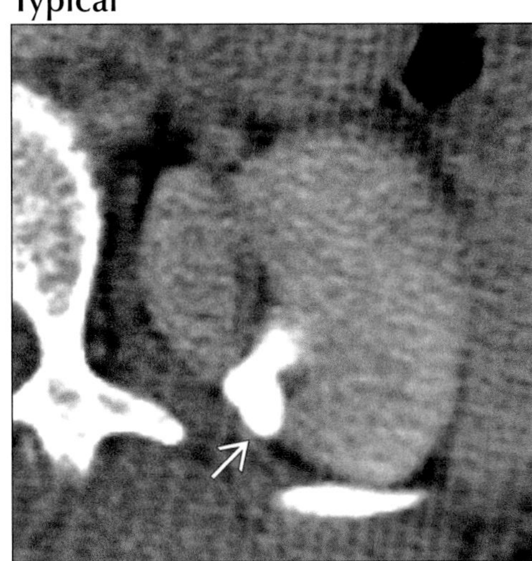

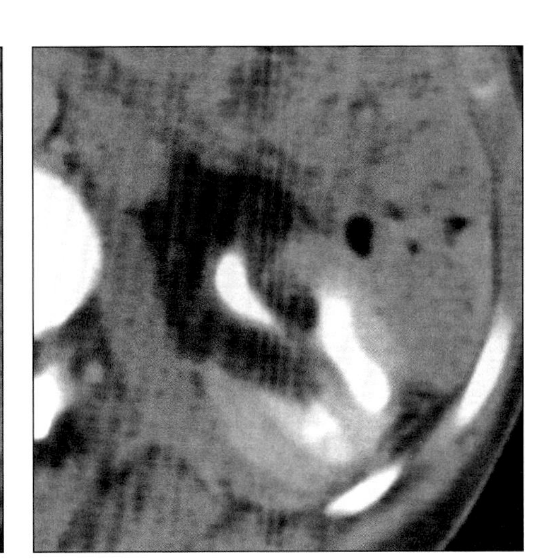

(Left) *Axial CECT shows result of chronic pyelonephritis with cortical loss over a dilated calyx in the left kidney (arrow).* ***(Right)*** *Axial CECT shows effects of chronic pyelonephritis with global atrophy of left kidney and proliferation of perirenal fat.*

Typical

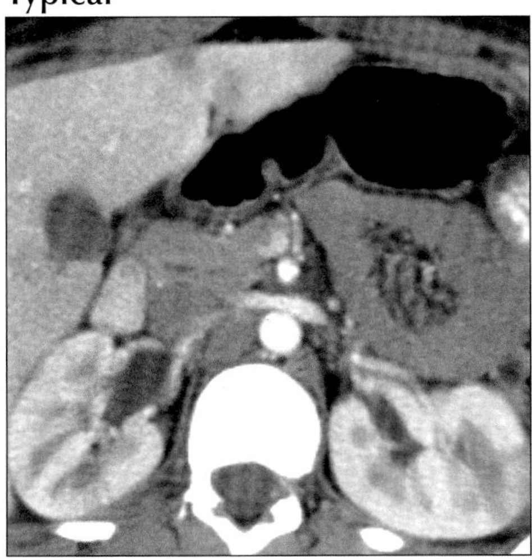

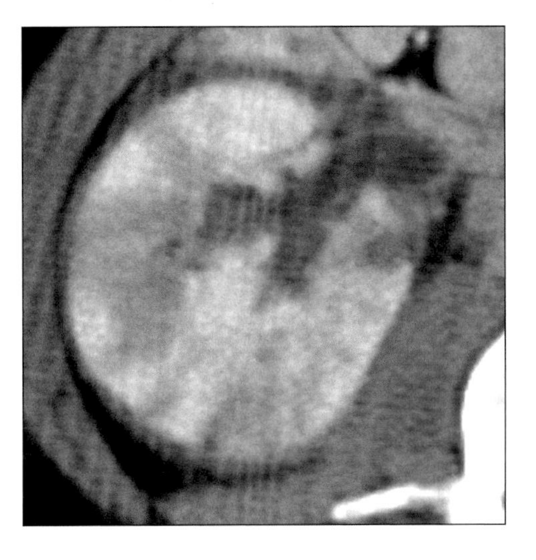

(Left) *Axial CECT shows heterogeneous parenchymograms bilaterally, more subtle due to corticomedullary phase of scan.* ***(Right)*** *Axial CECT shows wedge-shaped and striated nephrogram, plus inflamed uroepithelium and perirenal infiltration.*

RENAL ABSCESS

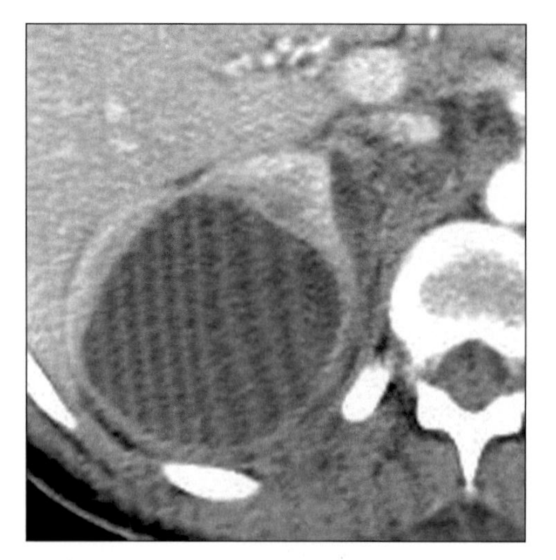

Axial CECT shows an encapsulated nonenhancing renal mass with an attenuation of 20 HU; abscess.

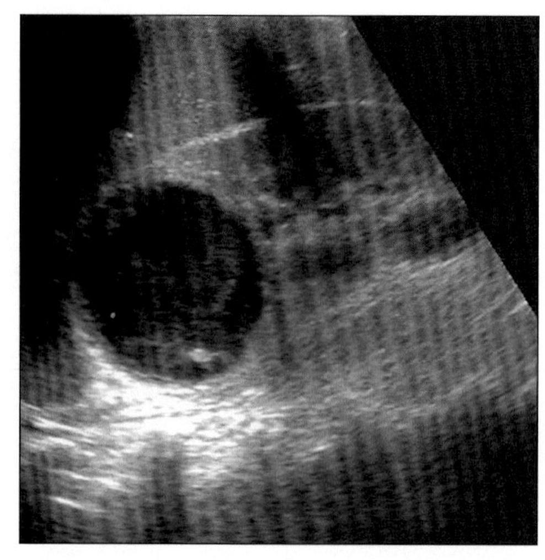

Sagittal sonogram shows spherical mass with low level echogenicity and acoustic enhancement.

TERMINOLOGY

Definitions

- Localized collection of pus caused by suppurative necrosis in kidney

IMAGING FINDINGS

General Features

- Best diagnostic clue: Spherical nonenhancing renal mass with perinephric stranding on CECT
- Other general features
  - Progress to perinephric abscess: Extension of renal abscess through capsule

Radiographic Findings

- IVP
  - Impaired excretion
    - Delayed appearance time, decreased contrast density or decreased nephrogram
    - ± Absence of nephrogram and calyceal opacification
  - Heterogeneous nephrogram
  - Single or multiple, well-defined, round or irregular lucent mass
  - Calyceal or pelvic effacement
  - ± Calyceal, pelvic or ureteral dilatation

CT Findings

- NECT
  - Single (more common) or multiple; unilateral or bilateral
  - Round, well-marginated, low-attenuation masses
  - ± Gas within collection
- CECT
  - Enlarged kidney with focal areas of hypoattenuation (acute)
  - "Rim or ring" sign: Enhancement of abscess wall (subacute or chronic)
  - No central enhancement of lesion; enhancement of normal renal tissue
  - Obliteration of renal sinus or calyceal effacement
  - Thickened walls and mild dilatation of renal pelvis and ureter
  - Perinephric reaction or extension
    - Altered renal contour, indistinct renal outline or renal displacement
    - Edema or obliteration of perinephric fat
    - Thickened Gerota fascia and perinephric septa

MR Findings

- T1WI: Hypointense mass
- T2WI: Hyperintense mass and increased signal surrounding the mass (perilesional edema)
- T1 C+: Shows rim-enhancement (lesion < 1 cm enhances homogeneously)

DDx: Thick-walled Cystic or Necrotic Lesion

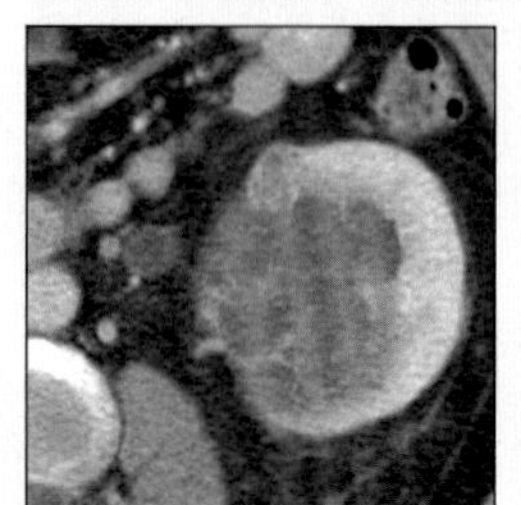

Cystic RCC

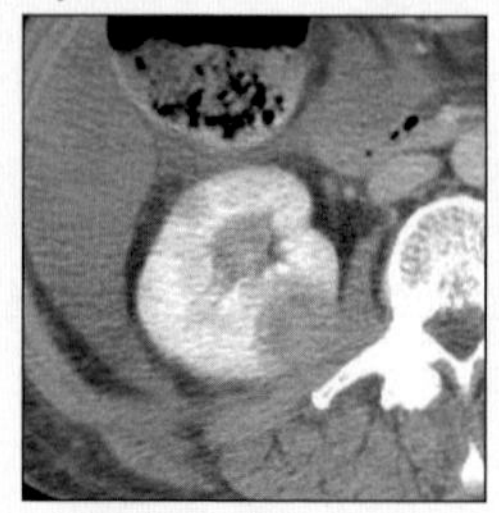

Lymphoma

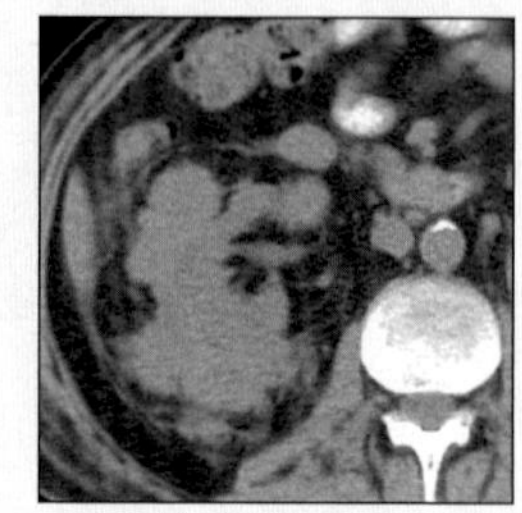

Multiple Myeloma

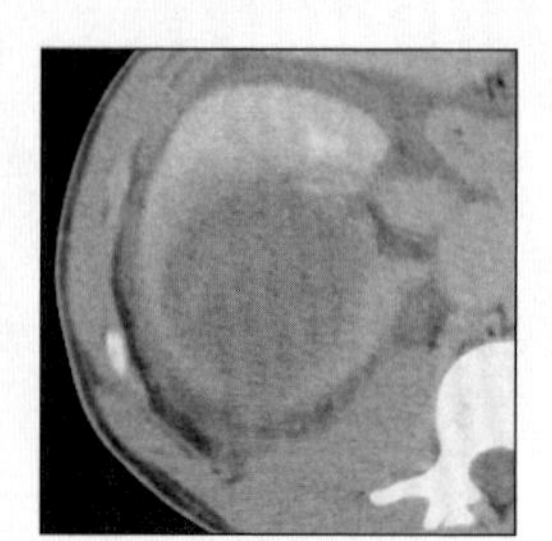

Hemorrhagic Cyst

RENAL ABSCESS

Key Facts

Terminology
- Localized collection of pus caused by suppurative necrosis in kidney

Imaging Findings
- Best diagnostic clue: Spherical nonenhancing renal mass with perinephric stranding on CECT
- ± Gas within collection
- "Rim or ring" sign: Enhancement of abscess wall (subacute or chronic)
- No central enhancement of lesion; enhancement of normal renal tissue

Top Differential Diagnoses
- Renal carcinoma (RCC)
- Metastases and lymphoma
- Infected or hemorrhagic cyst

Pathology
- Accounts for 2% of all renal masses
- Sequelae of acute renal infections
- Urinary tract infection → ascends to kidney → acute pyelonephritis or acute focal bacterial nephritis → liquefaction → sequestration → renal abscess

Clinical Issues
- Fever, flank or abdominal pain, chills and dysuria
- Antibiotic therapy

Diagnostic Checklist
- "Rim" sign; Absence of lesion enhancement; perinephric stranding

Ultrasonographic Findings
- Real Time
  - Anechoic or hypoechoic to echogenic fluid collection that blends with the normal echogenic fat within Gerota fascia
  - Mass within or displacing kidney
  - Round, thickened, or smooth-walled complex mass
  - Low level internal echoes move with change of position (internal debris)
  - "Comet sign": Internal echogenic foci (gas within abscess)
    - Associated posterior "dirty" shadowing
  - ± Internal septations or loculations

Angiographic Findings
- Conventional: Peripheral distribution and fine neovascular pattern

Nuclear Medicine Findings
- WBC Scan
  - ↑ Uptake of indium-111 labeled leukocytes noted within renal ± perinephric abscess
  - Possible false negative leukocyte scans: Prior antibiotic therapy, walled-off abscesses or poorly developed inflammatory responses
  - More sensitive or specific in early detection of renal ± perinephric infection

Imaging Recommendations
- NECT and CECT
  - To distinguish abscess from tumor

DIFFERENTIAL DIAGNOSIS

Renal carcinoma (RCC)
- CT: Enhancement of solitary mass
- Angiography: Hypervascularity
- Symptoms are rare, usually asymptomatic
- 25-40% diagnosed by incidental findings on CT or US
- 25-30% present with metastases (e.g., lung, mediastinum, bone, liver)
- Clinical history and urinalysis can differentiate

Metastases and lymphoma
- Metastases (e.g., lung cancer, breast cancer, gastrointestinal cancer, malignant melanoma)
  - Almost all develop via hematogenous spread
  - Lung, breast and colon carcinoma can occasionally be large and solitary; difficult to differentiate from renal carcinoma
  - CT: Multifocal, small, enhancing (5-30 HU) nodules; widespread nonrenal metastases
  - Angiography: Hypovascular pattern
  - Asymptomatic (most common) or flank pain and hematuria
  - CT or US guided biopsy for pathologic confirmation
- Lymphoma
  - CT: Multiple distinct masses (45%); direct invasion from enlarged retroperitoneal nodes (25%); solitary mass (15%); diffuse infiltration (10%); predominantly perinephric involvement (5%)
  - Asymptomatic (most common), fever, weight loss, flank pain, hematuria and renal failure

Infected or hemorrhagic cyst
- May be indistinguishable on imaging; solitary, nonenhancing lesion
- CT: Absence of perinephric stranding, "rim" sign, shaggy wall, hyperintense mass

PATHOLOGY

General Features
- General path comments
  - Accounts for 2% of all renal masses
  - Sequelae of acute renal infections
    - Acute pyelonephritis or focal bacterial nephritis
    - Usually seen 1 to 2 weeks after infection
- Etiology
  - Ascending urinary tract infections (80%)
    - Calculi, obstruction, renal anomalies and urinary reflux (diabetes or pregnancy)
    - Iatrogenic intervention (catheterization)

- Gram negative organisms (E. coli, Proteus species or Klebsiella species)
- Abscesses likely form at corticomedullary junction
- Hematogenous spread (20%)
  - I.V. drug users and skin infection
  - Hematogenous seeding from other infected sites (e.g., valvular heart disease, prosthesis)
  - Iatrogenic intervention (cyst aspiration, embolization of kidney vessels)
  - Gram positive and negative organisms (Staphylococcus aureus, Streptococcus species or species within Enterobacteriaceae family)
  - Abscesses likely form at renal cortex
- Risk factors: Diabetes Mellitus, long term hemodialysis and intravenous drug users
- Pathogenesis
  - Urinary tract infection → ascends to kidney → acute pyelonephritis or acute focal bacterial nephritis → liquefaction → sequestration → renal abscess

- Epidemiology: Incidence: Renal abscess (0.2%); perinephric abscess (0.02%)
- Associated abnormalities: 20-60% of patients with renal or perinephric abscess have urolithiasis

Gross Pathologic & Surgical Features

- Well-defined, round, thickened or smooth-walled mass

Microscopic Features

- Infected and necrotic tissue; ± gas

CLINICAL ISSUES

Presentation

- Most common signs/symptoms
  - Fever, flank or abdominal pain, chills and dysuria
  - Symptoms are longer than 2 weeks
  - Costovertebral angle tenderness and palpable flank mass
- Lab data
  - Urinalysis: Elevated white blood cells (WBC) (75%), positive bacterial culture (33%)
  - Blood tests: Elevated erythrocyte sedimentation rate (ESR), positive bacterial culture (50%)

Demographics

- Age: All ages
- Gender: M = F

Natural History & Prognosis

- Complications
  - Rupture → into perinephric space (perinephric abscess) → beyond Gerota fascia (paranephric abscess) → psoas or transversalis muscles → anterior peritoneal cavity → subdiaphragmatic or pelvic abscess
  - Rupture → renal collecting system → pyonephrosis
  - Compression or obstruction → hydronephrosis → renal atrophy
  - Necrosis and cavitation
- Prognosis
  - Good, in early diagnosis and treatment
  - Poor, in delayed diagnosis and treatment

Treatment

- Antibiotic therapy
  - If causative organisms are known, use specific antibiotics
  - If unknown, treat empirically with broad-spectrum antibiotics (ampicillin or vancomycin with aminoglycoside or third-generation cephalosporin)
- If abscess not resolved within 48 hours after treatment with antibiotics, do percutaneous aspiration & drainage under CT or US
  - Well-defined mass on CT or fluid-filled mass on US indicate a "ripe" abscess for drainage
- If abscess still not resolved, open surgical drainage or nephrectomy
- Follow-up
  - Imaging to confirm resolution of abscess
  - Evaluate for underlying urinary tract abnormalities

DIAGNOSTIC CHECKLIST

Consider

- Clinical history and urinalysis to diagnose and differentiate from malignancy

Image Interpretation Pearls

- "Rim" sign; Absence of lesion enhancement; perinephric stranding

SELECTED REFERENCES

1. Dalla Palma L et al: Medical treatment of renal and perirenal abscesses: CT evaluation. Clin Radiol. 54(12):792-7, 1999
2. Yen DH et al: Renal abscess: early diagnosis and treatment. Am J Emerg Med. 17(2):192-7, 1999
3. Kawashima A et al: CT of renal inflammatory disease. Radiographics. 17: 851-866, 1997
4. Davidson AJ et al: Radiologic assessment of renal masses: Implications for patient care. Radiology 202: 297-305, 1997
5. Brown ED et al: Renal abscesses: appearance on gadolinium-enhanced magnetic resonance images. Abdom Imaging. 21(2):172-6, 1996
6. Siegel JF et al: Minimally invasive treatment of renal abscess. J Urol. 155(1):52-5, 1996
7. Rinder MR: Renal abscess: an illustrative case and review of the literature. Md Med J. 45(10):839-43, 1996
8. Talner LB et al: Acute pyelonephritis: can we agree on terminology? Radiology 192: 297-305, 1994
9. Goldman SM et al: Upper urinary tract infection: the current role of CT, ultrasound, and MRI. Semin Ultrasound CT MR. 12(4):335-60, 1991
10. Soulen MC et al: Sequelae of acute renal infections: CT evaluation. Radiology. 173(2):423-6, 1989
11. Soulen MC et al: Bacterial renal infection: role of CT. Radiology. 171(3):703-7, 1989
12. Edelstein H et al: Perinephric abscess. Modern diagnosis and treatment in 47 cases. Medicine (Baltimore). 67(2):118-31, 1988
13. Jeffrey RB et al: CT and ultrasonography of acute renal abnormalities. Radiol Clin North Am. 21(3):515-25, 1983
14. Hoddick W et al: CT and sonography of severe renal and perirenal infections. AJR Am J Roentgenol. 140(3):517-20, 1983
15. Thornbury JR: Acute renal infections. Urol Radiol. 12(4):209-13, 1991

IMAGE GALLERY

Typical

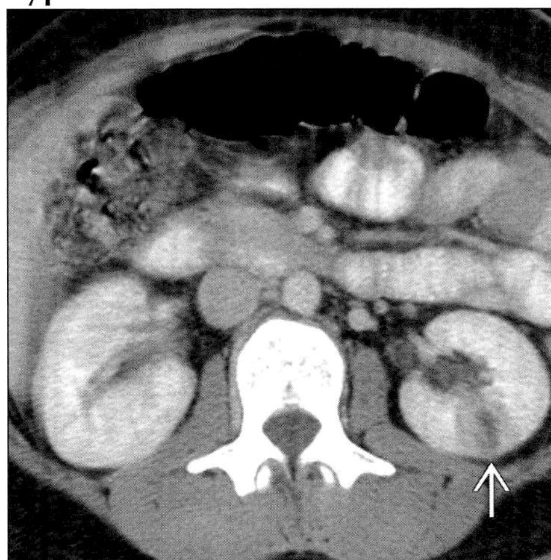

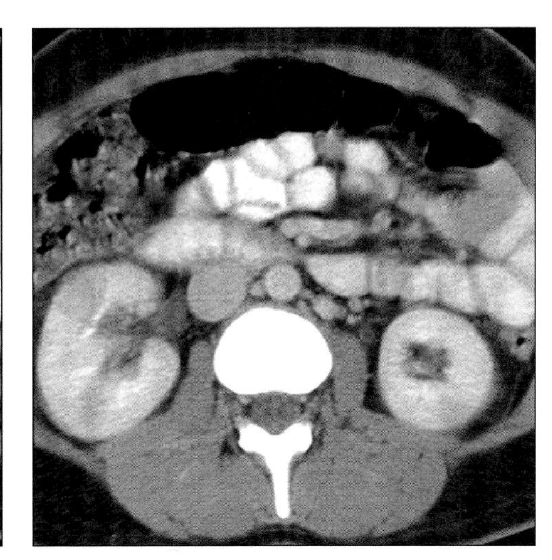

(Left) *Axial CECT shows heterogeneous mass (arrow) simulating tumor; early abscess + pyelonephritis.* ***(Right)*** *Axial CECT in a patient with pyelonephritis + early abscess (left kidney) shows striated and wedge-shaped defects in right kidney.*

Typical

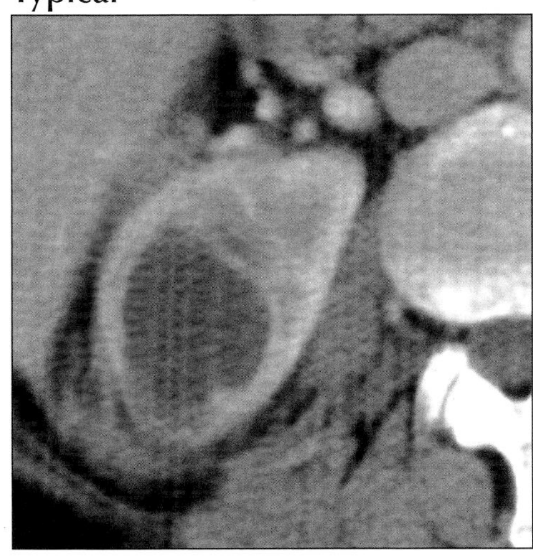

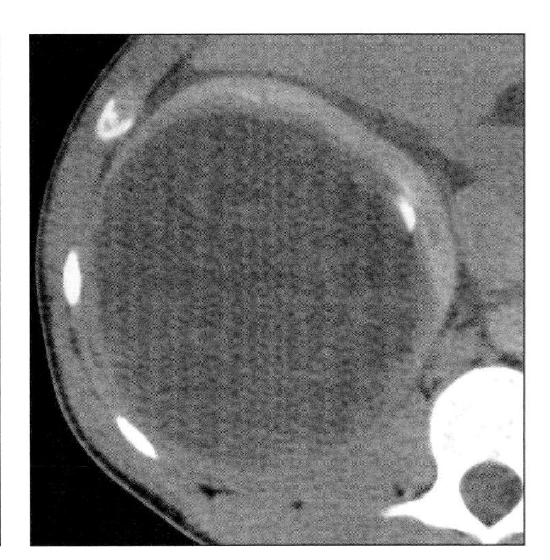

(Left) *Axial CECT shows low density mass in right kidney plus perirenal infiltration due to abscess.* ***(Right)*** *Axial CECT in a young woman (one week postpartum) shows large low density, nonenhancing renal abscess.*

Typical

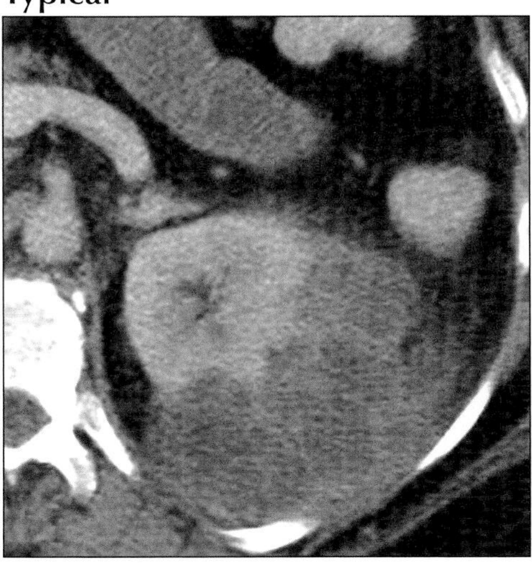

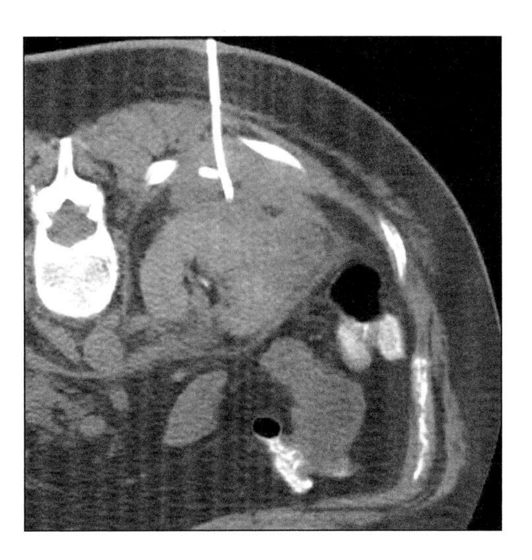

(Left) *Axial CECT shows renal and perirenal abscess.* ***(Right)*** *Axial NECT shows placement of a pigtail catheter for abscess drainage, obviating surgery.*

XANTHOGRANULOMATOUS PYELONEPHRITIS

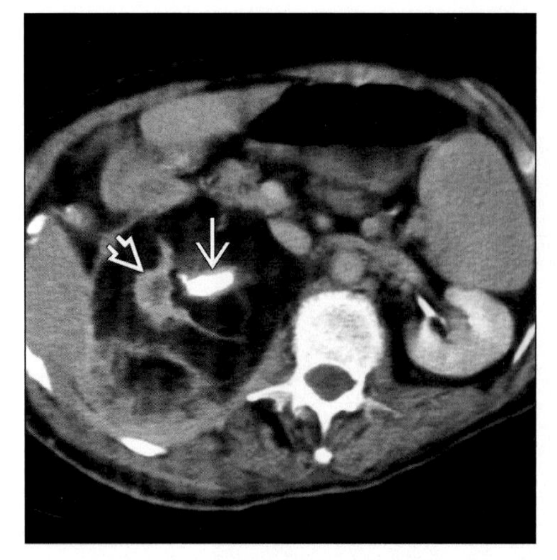

Axial CECT shows large calculus in renal pelvis (arrow), atrophic kidney (open arrow), and proliferation of perirenal fibrofatty tissue extending into back musculature.

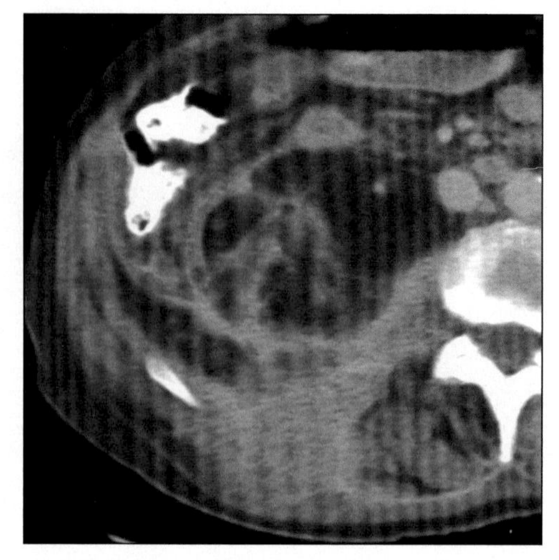

Axial CECT shows proliferated fibrofatty tissue in perirenal space with infection and inflammation extending into back musculature.

TERMINOLOGY

Abbreviations and Synonyms

- Xanthogranulomatous pyelonephritis (XGP) or (XGPN)

Definitions

- Chronic infection of kidney and surrounding tissues characterized by destruction and replacement of renal parenchyma by lipid-laden macrophages

IMAGING FINDINGS

General Features

- Best diagnostic clue: Obstructing calculus with atrophic or nonfunctioning kidney and perirenal fibrofatty proliferation
- Location
  - Unilateral (most cases) > bilateral
  - 2 forms of XGP
    - Diffuse (83-90%): Involves entire kidney
    - Segmental or focal (10-17%): Tumefactive due to obstructed single infundibulum; one moiety of duplex system
- Size: 2.5-5.8 cm; mean is 3.8 cm
- Morphology: Well-circumscribed mass with global or focal renal enlargement
- Other general features
  - 3 stages of XGP
    - Stage 1: Lesion confined to kidney
    - Stage 2: Lesion extends to perirenal space
    - Stage 3: Lesion spreads to pararenal spaces ± abdominal wall

Radiographic Findings

- IVP
  - Diffusely or focally absent nephrogram
  - Contracted pelvis; dilated calices
  - Centrally obstructing calculus; staghorn (75%)
- Retrograde pyelography
  - Complete obstruction at ureteropelvic junction; infundibulum; proximal ureter
  - Contracted renal pelvis; dilated deformed calices with nodular filling defects
  - Irregular parenchymal masses with cavitation

CT Findings

- Multiple, focal low attenuating (-10 to +30 HU) masses scattered throughout involved portions of kidney; represent dilated, debris-filled calyces & xanthoma collections
- Bright enhancement of rims of xanthoma collections due to inflammatory hypervascularity; no enhancement of collections
- Poor or no excretion of contrast into collecting system (nonfunctioning kidney)

DDx: Decreased Function and Mass

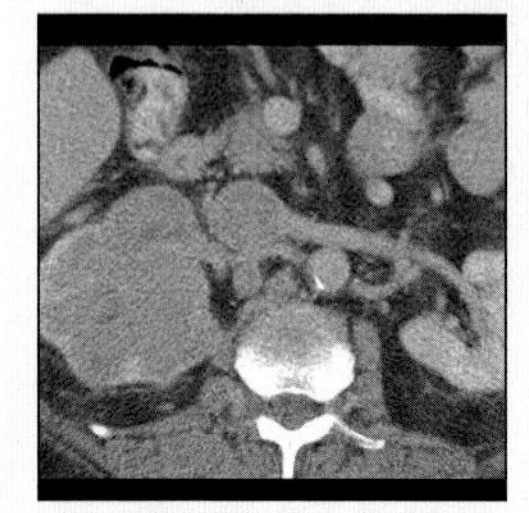

Renal Cell Carcinoma

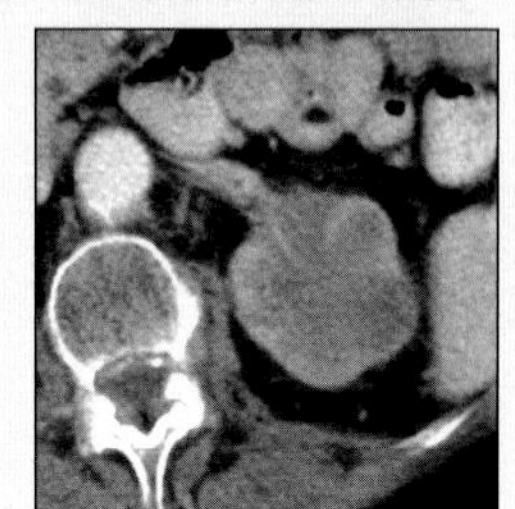

TCC

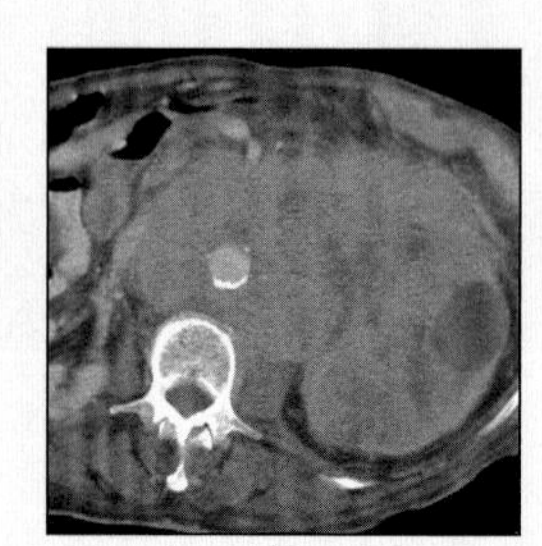

Lymphoma

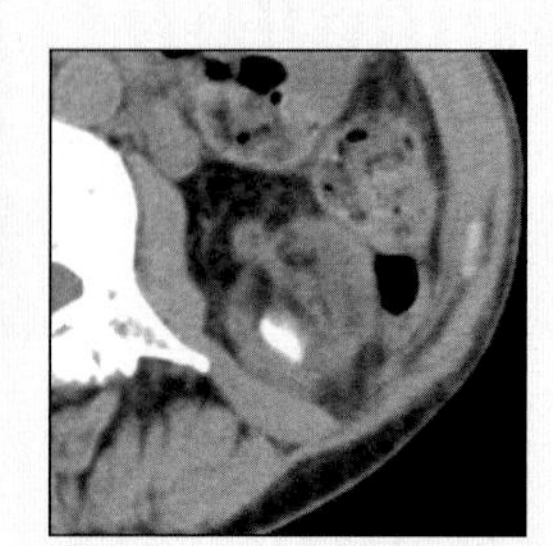

Chronic Pyelonephritis

XANTHOGRANULOMATOUS PYELONEPHRITIS

Key Facts

Terminology
- Chronic infection of kidney and surrounding tissues characterized by destruction and replacement of renal parenchyma by lipid-laden macrophages

Imaging Findings
- Best diagnostic clue: Obstructing calculus with atrophic or nonfunctioning kidney and perirenal fibrofatty proliferation
- Contracted pelvis; dilated calices
- Centrally obstructing calculus; staghorn (75%)

Top Differential Diagnoses
- Renal cell carcinoma
- Transitional cell carcinoma
- Renal metastases and lymphoma
- Renal abscess

Pathology
- Accumulation of lipid-laden "foamy" macrophages (xanthoma cells) and granulomatous infiltrate

Clinical Issues
- Nephrectomy usually required

Diagnostic Checklist
- Some variations may occur (i.e., small kidney, lack of calculi); difficult to distinguish XGP from other infections or neoplasms
- Histologic diagnosis must be made; cannot base solely on radiographic imaging studies
- Poor or no excretion of contrast from kidney; staghorn calculi

- Renal sinus fat obliterated (inflammation)
- Perinephric extension (14%): Extension of inflammation into perirenal space; pararenal space, ipsilateral psoas muscle, colon, spleen, diaphragm, posterior abdominal wall and skin
- Large central calculus, often staghorn
- Gas may rarely be seen

MR Findings
- Thin rim of renal parenchyma
- Loss of corticomedullary differentiation
- Dilated collecting system and surrounding renal parenchyma are of intermediate to high signal intensity on both T1 & T2WI (↑ fat content in macrophages)
- Renal calculi: Negative defects within dilated collecting system on T2WI

Ultrasonographic Findings
- Real Time
  - Multiple anechoic or hypoechoic masses replacing normal corticomedullary differentiation and contracted pelvis
  - Peripelvic fibrosis may obscure acoustic shadowing from central staghorn calculus
  - Parenchymal thinning and hydronephrosis
  - Sonographic signs of chronic obstructive uropathy caused by stones; echoes in the dilated collecting system
  - Perinephric fluid collection

Angiographic Findings
- Conventional
  - Stretching of segmental/interlobar arteries around large avascular masses
  - Hypervascularity/blush around periphery of masses in late arterial phase (granulation tissue)
  - Venous encasement with occlusion

Imaging Recommendations
- Best imaging tool: CT

DIFFERENTIAL DIAGNOSIS

Renal cell carcinoma
- Solitary, soft tissue density mass (30-50 HU range) with with central necrosis
- Hypervascular mass at renal cortex

Transitional cell carcinoma
- Renal infiltration → calyceal invasion, renal enlargement and poorly defined margins without change in shape
- Renal pelvic filling defect, irregular narrowing of collecting system
- Encasing pelvicaliceal system → hypovascular tumor

Renal metastases and lymphoma
- Metastases (e.g., lung cancer, breast cancer, colon cancer, malignant melanoma)
  - Usually hypovascular with infiltrative growth
- Lymphoma
  - Usually multiple or bilateral with infiltrative growth
  - Hypoechoic, hypovascular, solitary, intrarenal mass ± adenopathy

Renal abscess
- Solitary or multiple, round, well-marginated, low attenuation mass
- Enhancement of walls without central enhancement
- Perinephric stranding
- ± Presence of gas

Hydronephrosis/pyonephrosis
- Pus-filled collecting system may simulate XGP

PATHOLOGY

General Features
- General path comments
  - Usually found in setting of chronic obstruction (e.g., calculus, stricture, carcinoma)
  - Often renal pelvis less dilated than expected for high grade chronic obstruction
- Etiology

- Escherichia coli
- Proteus mirabilis
- Staphylococcus aureus
- Klebsiella species
- Pseudomonas species
- Enterobacter species
- Risk factors
  - Recurrent or chronic urinary tract infections
  - Immunocompromised patients: Diabetes mellitus
  - Abnormal lipid metabolism
- Pathogenesis
  - Chronic renal obstruction and infection with failure of local host immunity
- Epidemiology: Incidence: 1% all renal infections
- Associated abnormalities: Urolithiasis

Gross Pathologic & Surgical Features

- Diffuse form
  - Renal enlargement with indurated or thickened perinephric fat
  - Dilated renal pelvis with staghorn calculus
  - Replacement of corticomedullary junction with soft yellow nodules; calices filled with pus and debris
- Focal form
  - Yellowish white, solid or semisolid renal mass

Microscopic Features

- Accumulation of lipid-laden "foamy" macrophages (xanthoma cells) and granulomatous infiltrate
- Foam cells contain neutral fat and cholesterol (ester granules); positive for periodic acid-Schiff (PAS) stain
- Diffuse infiltration by plasma cells and histiocytes

CLINICAL ISSUES

Presentation

- Most common signs/symptoms
  - Dull or persistent flank pain, fever
  - Palpable mass, weight loss
- Lab data
  - Urinalysis: Microscopic hematuria, proteinuria, pyuria
  - Urine culture: Specific bacterial species
  - Liver function tests: Elevated
  - Erythrocyte sedimentation rate: Elevated

Demographics

- Age
  - Any age: 45-65 years of age most common
  - Very rare in childhood; focal form is more common in children
- Gender: M:F = 1:3-4

Natural History & Prognosis

- Symptomatic for 6 months prior to diagnosis (40% of cases)
- Complications: Hepatic dysfunction (reversible), extrarenal extension, fistulas (e.g., pyelocutaneous, ureterocutaneous), hemorrhage
- Prognosis: Good, mortality is rare, morbidity can be substantial

Treatment

- Antibiotics: Treatment prior to surgical intervention
- Nephrectomy usually required
  - Radical nephrectomy in complete XGP
  - Partial nephrectomy in focal XGP

DIAGNOSTIC CHECKLIST

Consider

- Some variations may occur (i.e., small kidney, lack of calculi); difficult to distinguish XGP from other infections or neoplasms
- Histologic diagnosis must be made; cannot base solely on radiographic imaging studies

Image Interpretation Pearls

- Poor or no excretion of contrast from kidney; staghorn calculi

SELECTED REFERENCES

1. Kim J: Ultrasonographic features of focal xanthogranulomatous pyelonephritis. J Ultrasound Med. 23(3):409-16, 2004
2. Cakmakci H et al: Pediatric focal xanthogranulomatous pyelonephritis: dynamic contrast-enhanced MRI findings. Clin Imaging. 26(3):183-6, 2002
3. Tiu CM et al: Sonographic features of xanthogranulomatous pyelonephritis. J Clin Ultrasound. 29(5):279-85, 2001
4. Kim JC: US and CT findings of xanthogranulomatous pyelonephritis. Clin Imaging. 25(2):118-21, 2001
5. Verswijvel G et al: Xanthogranulomatous pyelonephritis: MRI findings in the diffuse and the focal type. Eur Radiol. 10(4):586-9, 2000
6. Fan CM et al: Xanthogranulomatous pyelonephritis. AJR Am J Roentgenol. 165(4):1008, 1995
7. Eastham J et al: Xanthogranulomatous pyelonephritis: clinical findings and surgical considerations. Urology. 43(3):295-9, 1994
8. Rabushka LS et al: Pictorial review: computed tomography of renal inflammatory disease. Urology. 44(4):473-80, 1994
9. Chuang CK et al: Xanthogranulomatous pyelonephritis: experience in 36 cases. J Urol. 147(2):333-6, 1992
10. Hayes WS et al: From the Archives of the AFIP. Xanthogranulomatous pyelonephritis. Radiographics. 11(3):485-98, 1991
11. Mulopulos GP et al: MR imaging of xanthogranulomatous pyelonephritis. J Comput Assist Tomogr. 10(1):154-6, 1986
12. Goldman SM et al: CT of xanthogranulomatous pyelonephritis: radiologic-pathologic correlation. AJR Am J Roentgenol. 142(5):963-9, 1984
13. Bissada NK et al: Preoperative diagnosis of xanthogranulomatous pyelonephritis. Urology. 7(2):228-30, 1976
14. Cha EM et al: Xanthogranulomatous pyelonephritis. Angiographic evaluation. Urology. 3(2):159-62, 1974
15. Vinik M et al: Xanthogranulomatous pyelonephritis: angiographic considerations. Radiology. 92(3):537-40, 1969

XANTHOGRANULOMATOUS PYELONEPHRITIS

IMAGE GALLERY

Typical

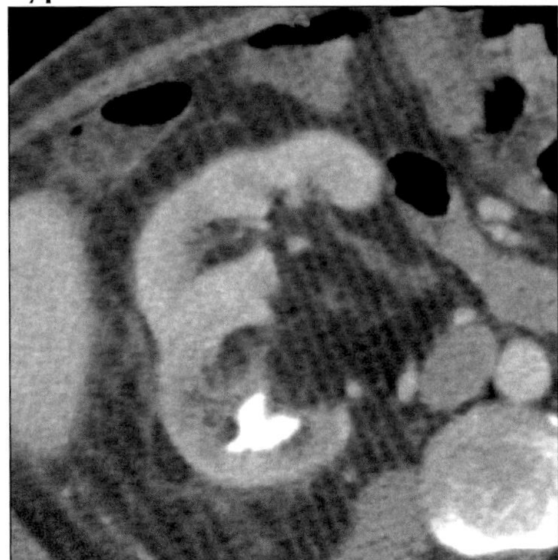

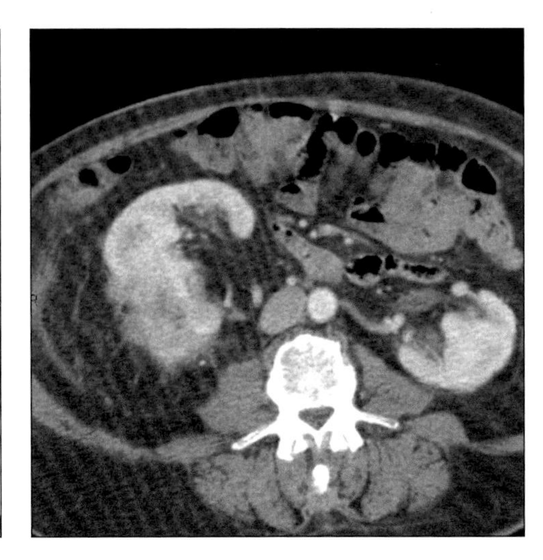

(Left) *Axial CECT shows enlarged right kidney with minimal function, large calculus and perirenal fibrofatty proliferation.* ***(Right)*** *Axial CECT shows enlarged right kidney with infiltrative "mass" in dorsal segment and extensive perirenal inflammation.*

Typical

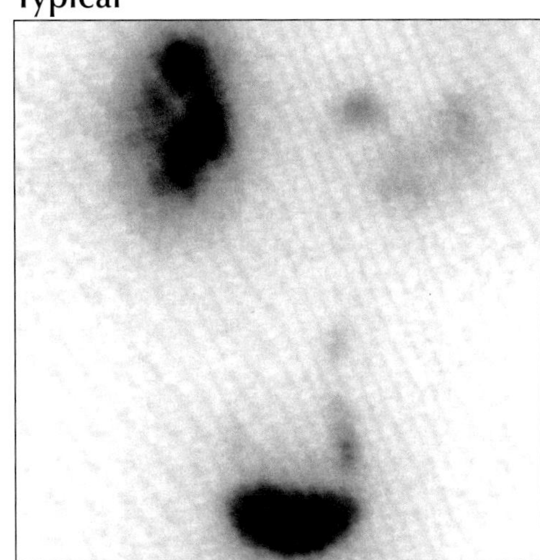

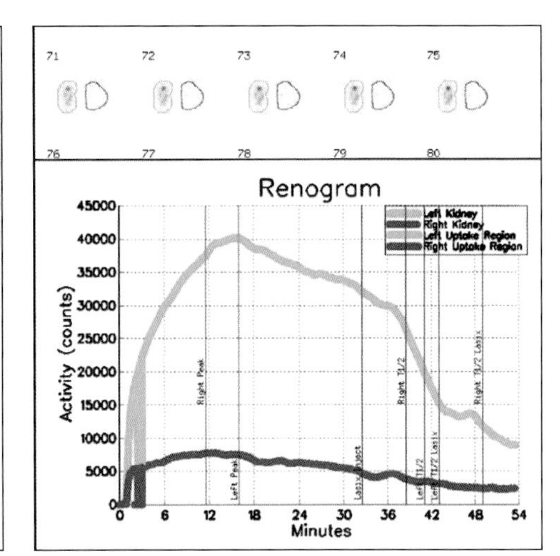

(Left) *Renal scintigram (posterior view) shows minimal function of right kidney.* ***(Right)*** *Tc MDP renogram shows minimal perfusion or function of right kidney due to XGP.*

Typical

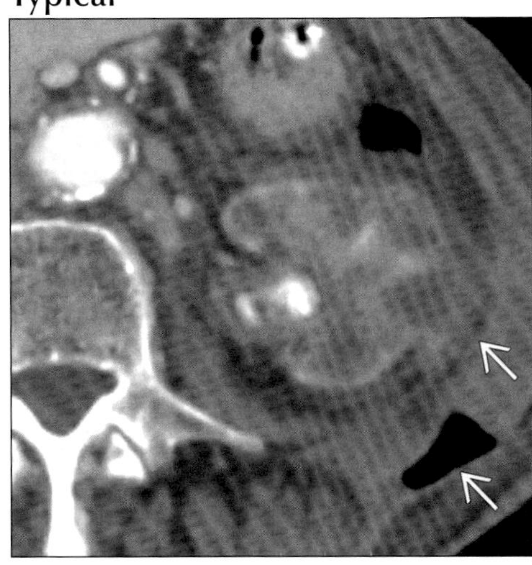

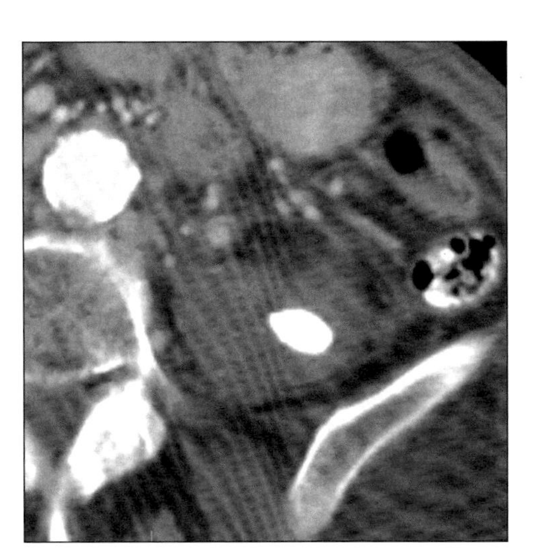

(Left) *Axial CECT shows staghorn calculus and nonfunctional left kidney. Extensive retroperitoneal and body wall inflammation/infection (arrows).* ***(Right)*** *Axial CECT shows calculus lying outside kidney in pocket of chronic perirenal inflammation due to XGP.*

EMPHYSEMATOUS PYELONEPHRITIS

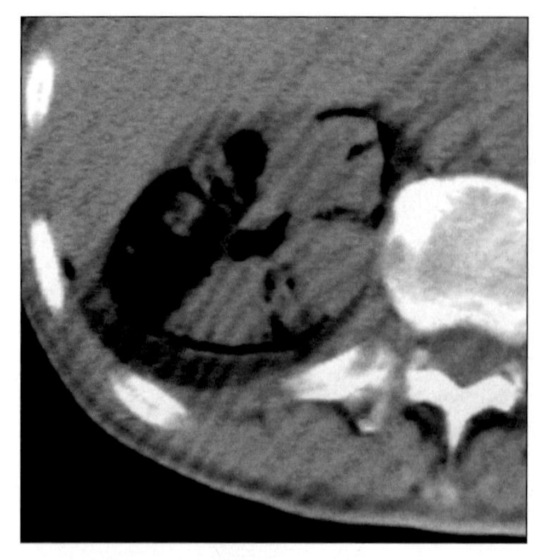

Axial NECT shows gas replacement of infarcted, infected renal parenchyma.

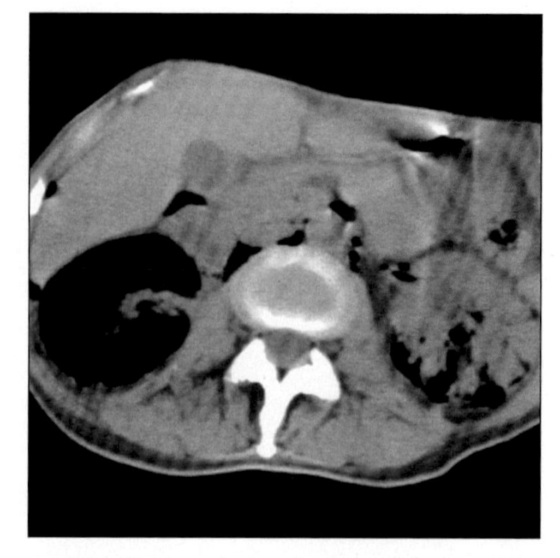

Axial NECT shows almost complete replacement of the right kidney by gas, segmental gas in the left kidney, and gas dissecting through retroperitoneum. Fatal EPN in diabetic.

TERMINOLOGY

Abbreviations and Synonyms

- Emphysematous pyelonephritis (EPN)

Definitions

- Life-threatening, fulminant, necrotizing upper urinary tract infection (UTI) associated with gas within kidney

IMAGING FINDINGS

General Features

- Best diagnostic clue: Gas in renal parenchyma on CT
- Location
  - Unilateral > bilateral (5-7% of cases)
  - Left (52%) > right (43%)

Radiographic Findings

- Radiography: Gas in parenchyma ± paranephric space
- IVP (generally contraindicated)
  - Renal enlargement with delayed or absent excretion
  - Renal edema with destruction of renal parenchyma

CT Findings

- 2 types of EPN
  - Type I (33%) (true EPN)
    - Parenchymal destruction without fluid; streaky or mottled gas radiating from medulla to cortex
    - ± Crescent of subcapsular or perinephric gas
  - Type II (66%)
    - Renal or perirenal fluid abscesses with bubbly gas pattern ± gas within renal pelvis
- Mottled ↓ attenuation areas extend radially along pyramids
- Intraparenchymal, intracaliceal and intrapelvic gas
- Gas often extends into subcapsular, perinephric, pararenal, contralateral retroperitoneal spaces
- ± Gas extends into renal veins and inferior vena cava

MR Findings

- T1WI, T2WI: Void of signal

Ultrasonographic Findings

- Real Time
  - Highly echogenic areas within renal sinus and parenchyma with "dirty" shadowing
  - Ring-down artifacts: Air bubbles trapped in fluid
  - Gas in perinephric space; may obscure kidney

Imaging Recommendations

- Best imaging tool: CT

DIFFERENTIAL DIAGNOSIS

Emphysematous pyelitis

- Gas in renal pelvis & calices, not parenchyma

DDx: Gas in or Around Kidney

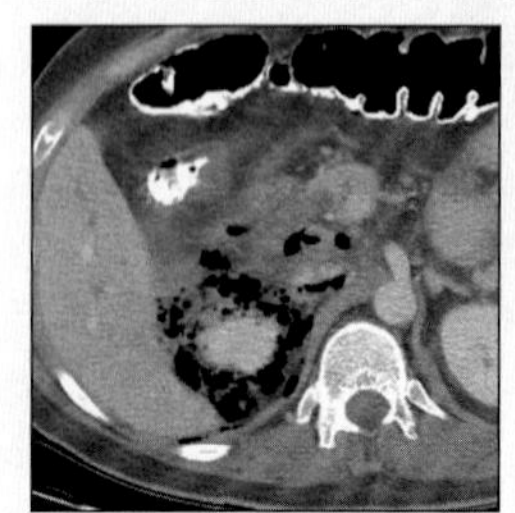

Perforated Ulcer

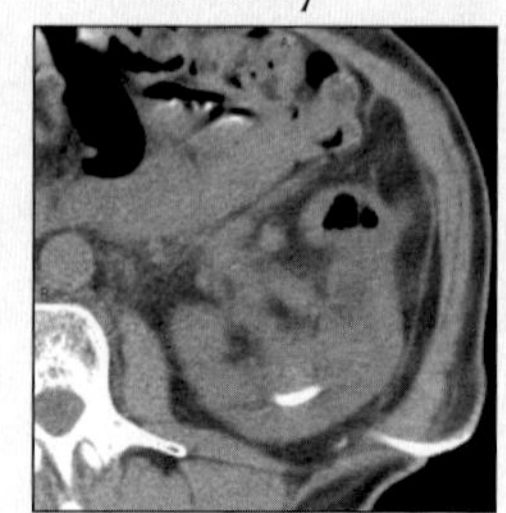

Nephrostomy

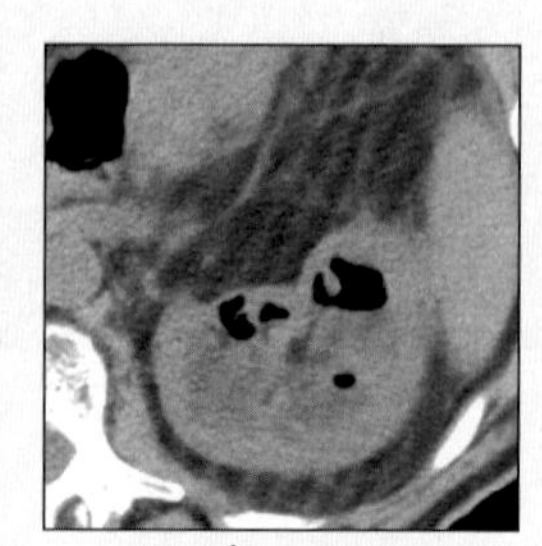

Nephrostomy

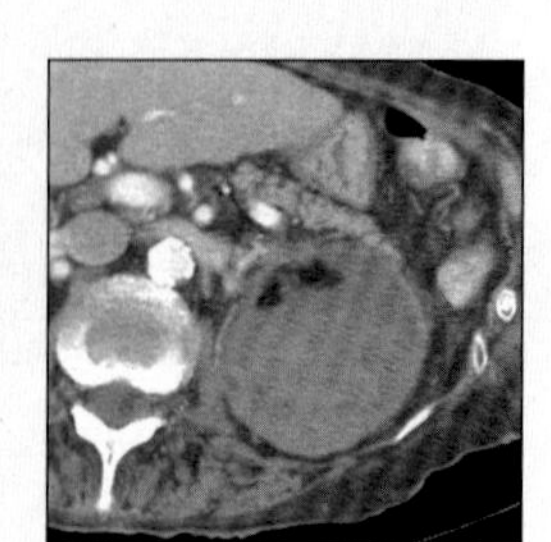

Acute Infarction

EMPHYSEMATOUS PYELONEPHRITIS

Key Facts

Imaging Findings
- Best diagnostic clue: Gas in renal parenchyma on CT
- Parenchymal destruction without fluid; streaky or mottled gas radiating from medulla to cortex
- Renal or perirenal fluid abscesses with bubbly gas pattern ± gas within renal pelvis
- Best imaging tool: CT

Top Differential Diagnoses
- Emphysematous pyelitis
- Perforated duodenal ulcer
- Iatrogenic

Diagnostic Checklist
- Exclude EPN in diabetics with pyelonephritis
- Type I (true EPN) is an medical and surgical emergency

- 50% diabetics; less grave prognosis than EPN

Perforated duodenal ulcer
- Occasionally, gas "outlines" the kidney

Iatrogenic
- E.g., retrograde pyelography, nephrostomy
- Chemoembolization or ablation of renal tumor with sudden infarction of tissue

PATHOLOGY

General Features
- Etiology
  - Single or mixed organism(s) infection
    - E. coli (68%), Klebsiella pneumoniae (9%)
    - Proteus mirabilis, Pseudomonas, Enterobacter, Candida, Clostridia species
  - Risk factor
    - Recurrent or chronic UTIs
    - Immunocompromised: Diabetes mellitus (87-97%)
    - Ureteral obstruction (20-40%): Calculi, stenosis
    - Renal failure: Polycystic kidney, end-stage
  - Pathogenesis
    - Pyelonephritis → ischemia and low oxygen tension → facultative anaerobes proliferation in an anaerobic environment → CO_2 production
- Epidemiology: Very rare

Gross Pathologic & Surgical Features
- Suppurative necrotizing infection of renal parenchyma and perirenal tissue with multiple cortical abscesses

CLINICAL ISSUES

Presentation
- Most common signs/symptoms
  - Fever, chills, flank pain, lethargy, confusion
  - Nausea, vomiting, dyspnea, crepitant mass
- Lab data
  - Hyperglycemia, acidosis, electrolyte imbalance, thrombocytopenia
  - Blood, urine, aspiration culture: Positive

Demographics
- Age: 19-81 years of age; mean 54 years of age
- Gender: M:F = 1:2-6

Natural History & Prognosis
- Complications: Sepsis
- Prognosis: Poor
  - Mortality
    - 66% with type I, 18% with type II
    - 60-75% with antibiotic therapy
    - 21-29% after antibiotic therapy and nephrectomy
    - 80% if extends into perinephric space

Treatment
- Antibiotic therapy; nephrectomy for type I
- CT-guided drainage procedures: Safe, quick and life-saving with 70% success rate, only for type II

DIAGNOSTIC CHECKLIST

Consider
- Exclude EPN in diabetics with pyelonephritis

Image Interpretation Pearls
- Type I (true EPN) is an medical and surgical emergency

SELECTED REFERENCES

1. Grayson DE et al: Emphysematous infections of the abdomen and pelvis: a pictorial review. Radiographics. 22(3):543-61, 2002
2. Wan YL et al: Acute gas-producing bacterial renal infection: correlation between imaging findings and clinical outcome. Radiology. 198(2):433-8, 1996
3. Joseph RC et al: Genitourinary tract gas: imaging evaluation. Radiographics. 16(2):295-308, 1996
4. Rodriguez-de-Velasquez A et al: Imaging the effects of diabetes on the genitourinary system. Radiographics. 15(5):1051-68, 1995

IMAGE GALLERY

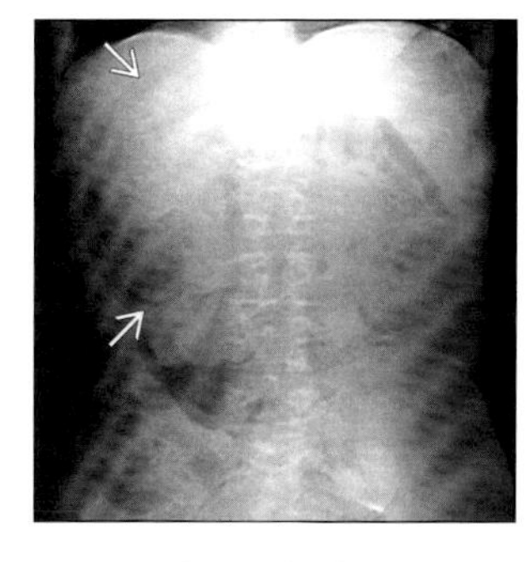

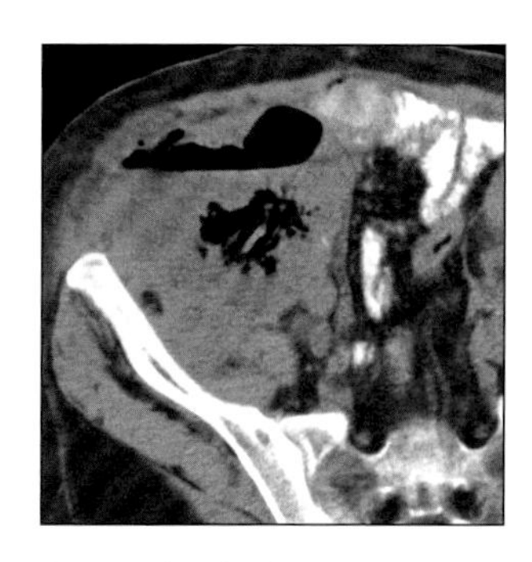

(Left) *Radiograph shows gas in and around right kidney (arrows).* ***(Right)*** *Axial NECT shows gas in and around a transplanted kidney in the iliac fossa, due to emphysematous pyelonephritis.*

HIV NEPHROPATHY

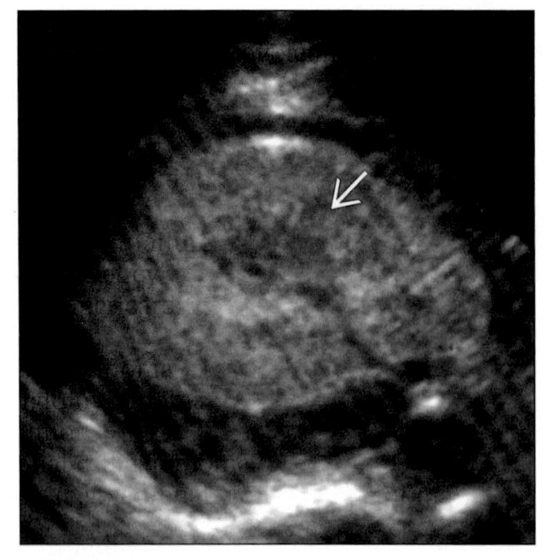

Sonography of HIV nephropathy. Transverse scan demonstrates markedly echogenic renal parenchyma (arrow).

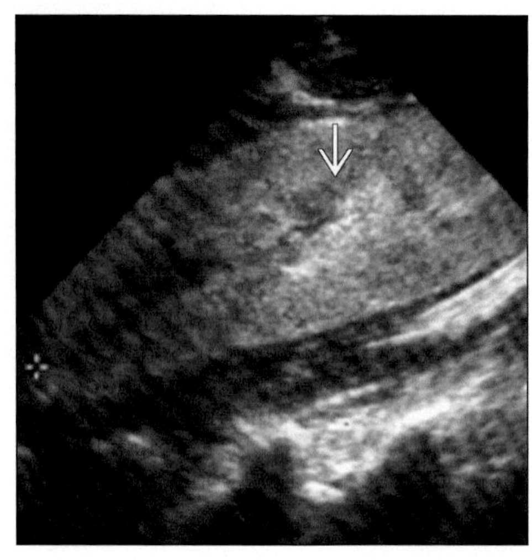

Sagittal sonogram of HIV nephropathy demonstrates enlarged kidney with loss of cortico-medullary interface (arrow).

TERMINOLOGY

Abbreviations and Synonyms

- HIV-associated nephropathy (HIVAN), focal segmental glomerulosclerosis (FSGS)

Definitions

- HIV-1 infection associated with focal segmental glomerulosclerosis

IMAGING FINDINGS

General Features

- Best diagnostic clue: Enlarged echogenic kidneys on sonography
- Size: Kidneys > 11 cm in length
- Morphology: Kidneys may be globular (53%)

CT Findings

- NECT: Hyperattenuation of medulla (14%)
- CECT: Striated nephrogram

MR Findings

- T2WI: Loss of corticomedullary differentiation

Ultrasonographic Findings

- Real Time
    - Abnormally increased echogenicity (89%), globular renal morphology (53%), decreased corticomedullary definition (38%)
    - Heterogeneous parenchyma (43%), renal enlargement (20%), pelvocalyceal thickening
    - No hydronephrosis

Imaging Recommendations

- Best imaging tool: US

DIFFERENTIAL DIAGNOSIS

Medical renal disease

- Acute interstitial nephritis results in enlarged echogenic kidneys
- Renal US in glomerulonephritis is variable, may demonstrate normal size & echogeneity
- Kidneys > 11 cm; increased echogenicity; prominent renal pyramids

Acute tubular necrosis (ATN)

- Most common cause of reversible acute renal failure; hypotension most common cause
- Other etiologies include drugs (e.g., contrast reaction), heavy metals or solvent exposure
- US of hypotension-related ATN is often normal
- US of ATN caused by drugs, heavy metals or solvent exposure results in enlarged echogenic kidneys

DDx: Renal Lesions Mimicking HIVAN

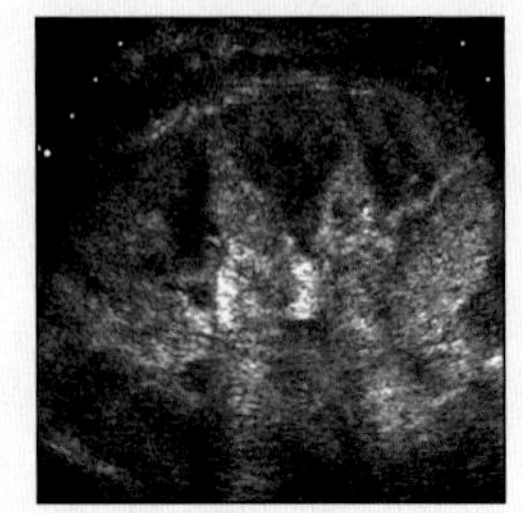

Medical Renal Disease

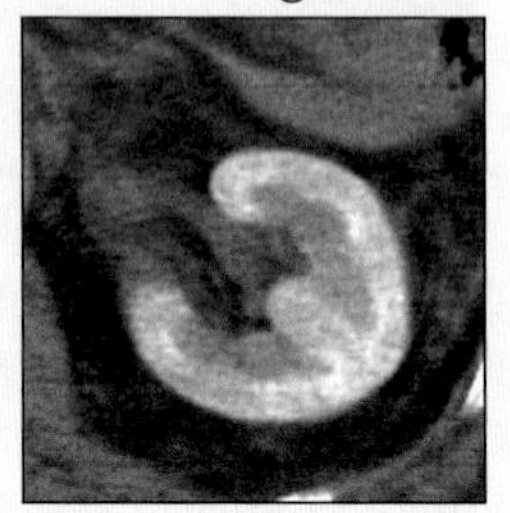

ATN

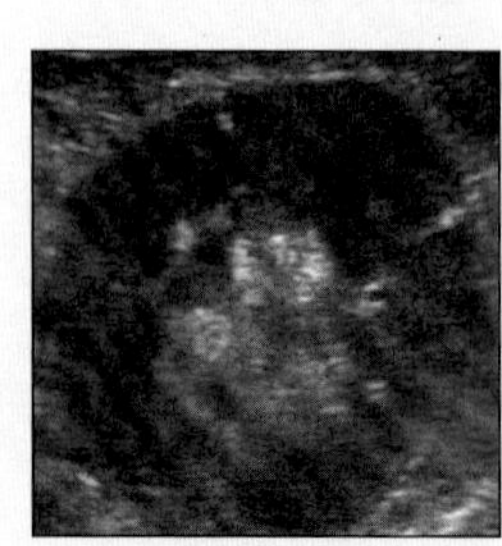

Renal PC

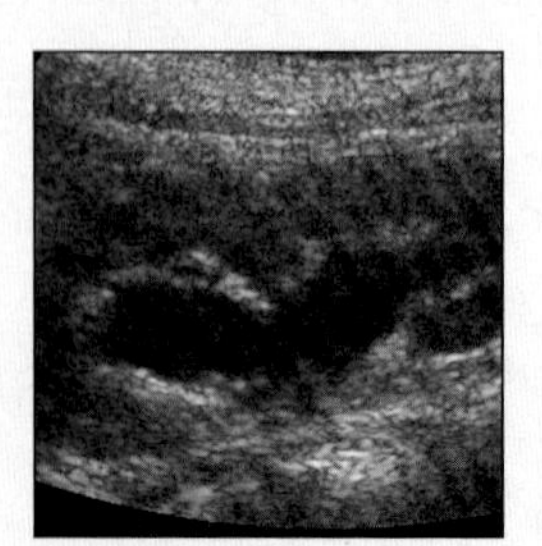

Obstructive Uropathy

HIV NEPHROPATHY

Key Facts

Terminology
- HIV-1 infection associated with focal segmental glomerulosclerosis

Imaging Findings
- Best diagnostic clue: Enlarged echogenic kidneys on sonography

Top Differential Diagnoses
- Medical renal disease
- Acute tubular necrosis (ATN)
- Renal Pneumocystis carinii (PC)

Clinical Issues
- Clinical profile: African-American, HIV-positive patient with renal failure

- Elevated resistive index (RI > 0.7)
- Kidneys after contrast reaction demonstrate delayed cortical transit & excretion

Renal Pneumocystis carinii (PC)
- Punctate calcifications of renal cortex in AIDS patients; associated calcifications in liver, spleen, lymph nodes

Acute obstructive uropathy
- Bilateral hydronephrosis
- Non-dilated obstructive uropathy may lead to false-negative US; causes: Prostate and bladder cancers, retroperitoneal fibrosis

PATHOLOGY

General Features
- General path comments
  - Focal and segmental glomerulosclerosis; interstitial infiltrates
  - Degenerative changes within tubules; interstitial tubular microcystic ectasia with protein casts
- Etiology: HIV-1 infection
- Epidemiology
  - HIV infected individuals
  - Substantially higher incidence in African-Americans - 90% of reported cases

Microscopic Features
- Focal segmental glomerulosclerosis with collapse of glomerular tuft associated with dilated tubules with microcysts

CLINICAL ISSUES

Presentation
- Most common signs/symptoms
  - Azotemia, hypoalbuminemia, proteinuria
  - Other signs/symptoms: Normotensive
- Clinical profile: African-American, HIV-positive patient with renal failure

Demographics
- Age: 20-64 yrs
- Gender: M > F
- Ethnicity: African-American: HIVAN is third leading cause of renal failure in African-Americans age 20-64 yrs

Natural History & Prognosis
- 2 yr survival 36%

Treatment
- Triple drug anti-retroviral therapy reduces incidence, stabilizes patient
- Dialysis for end-stage renal disease

DIAGNOSTIC CHECKLIST

Consider
- Acute medical renal disease

Image Interpretation Pearls
- Enlarged echogenic kidneys

SELECTED REFERENCES

1. Kopp JB et al: HIV-associated nephropathy in African Americans. Kidney Int Suppl. (83):S43-9, 2003
2. Herman ES et al: HIV-associated nephropathy: Epidemiology, pathogenesis, and treatment. Semin Nephrol. 23(2):200-8, 2003
3. Rao TK: Human immunodeficiency virus infection in end-stage renal disease patients. Semin Dial. 16(3):233-44, 2003

IMAGE GALLERY

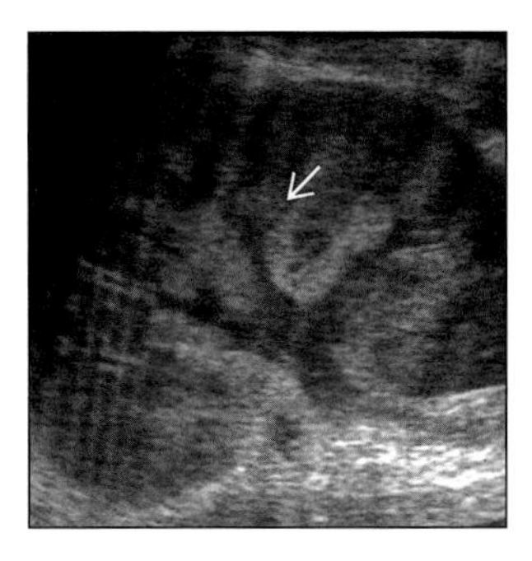
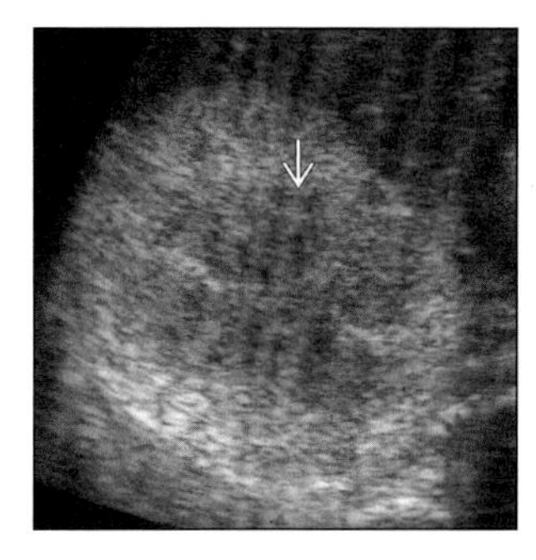

(Left) *Sonography of HIVAN. Longitudinal scan shows echogenic enlarged kidney with minimal pelvo-calyceal dilation (arrow) (Courtesy Philip W. Ralls, MD).* ***(Right)*** *Sonography of HIVAN shows striking increase in renal echogenicity and poor definition of renal pyramids (arrow) (Courtesy Philip W. Ralls, MD).*

GLOMERULONEPHRITIS

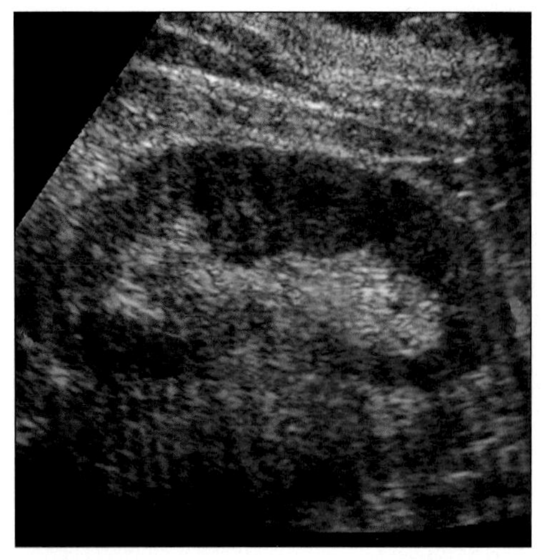

Coronal oblique US shows normal appearing kidney in patient with biopsy-proven acute glomerulonephritis.

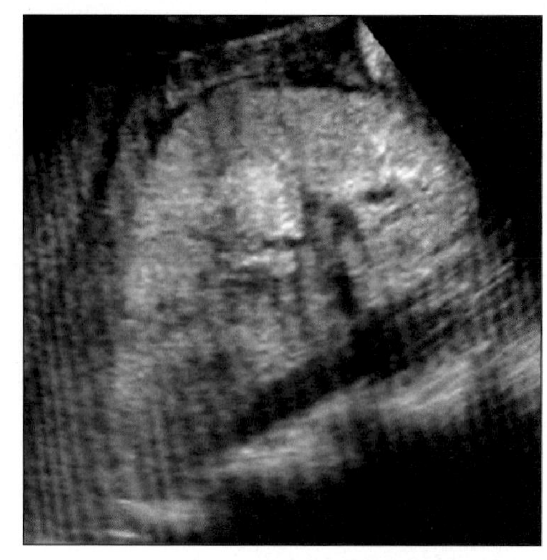

Coronal oblique US shows markedly echogenic kidney in patient with acute renal failure due to glomerulosclerosis (Courtesy H. Harvin, MD).

TERMINOLOGY

Abbreviations and Synonyms

- Glomerulonephritis (GN)

Definitions

- Inflammation and proliferation of glomerular tissue

IMAGING FINDINGS

General Features

- Best diagnostic clue: Nonspecific imaging findings
- Size
  - Enlarged kidneys: Acute GN
  - Small kidneys: Chronic GN
- Morphology: Smooth renal contour

CT Findings

- NECT
  - Acute GN: Normal or bilateral renal enlargement
  - Chronic GN: May have cortical calcification

Ultrasonographic Findings

- Real Time
  - Acute GN
    - Normal to increased renal echogenicity
    - Bilateral normal or enlarged kidneys
    - No hydronephrosis to explain renal failure
  - Chronic GN
    - Increased echogenicity
    - Small, smooth kidneys
    - Proliferation of sinus fat
- Color Doppler: May see renal vein thrombosis

Imaging Recommendations

- Best imaging tool: US to rule out hydronephrosis
- Protocol advice: US useful to document signs of chronic vs. acute renal failure (ARF)

DIFFERENTIAL DIAGNOSIS

Bilateral smooth renal enlargement

- Amyloidosis, multiple myeloma
- Acute tubular necrosis (ATN)
- Acute interstitial nephritis (AIN)

Bilateral small kidneys with smooth contour

- Arteriosclerosis, nephrosclerosis
- Renal embolic disease

PATHOLOGY

General Features

- General path comments: GN is one of several intrinsic causes of acute renal failure

DDx: Acute Renal Failure

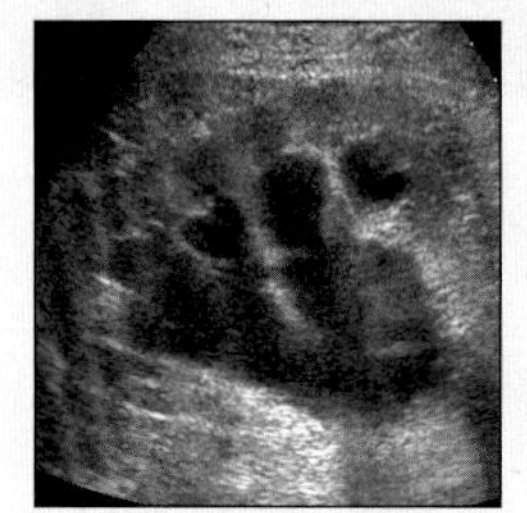

Hydronephrosis

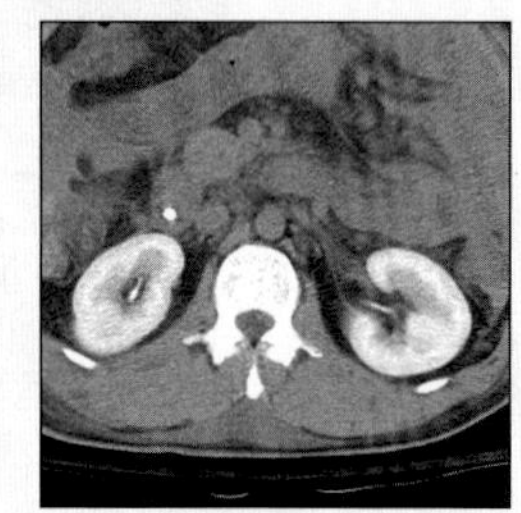

ATN

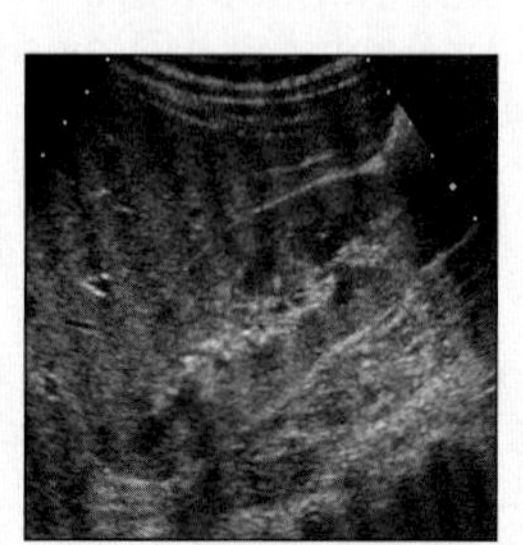

AIN

GLOMERULONEPHRITIS

Key Facts

Imaging Findings
- Best diagnostic clue: Non-specific imaging findings
- Chronic GN: May have cortical calcification
- Normal to increased renal echogenicity
- No hydronephrosis to explain renal failure
- Proliferation of sinus fat
- Best imaging tool: US to r/o hydronephrosis

Clinical Issues
- Post-streptococcal: Follows pharyngitis or impetigo
- May progress to end-stage renal disease

Diagnostic Checklist
- Look for associated renal vein thrombosis
- Try to distinguish acute from chronic and pre/post-renal causes of renal failure from intrinsic causes

- Etiology
  - Primary GN: Confined to kidney
    - Systemic features secondary to renal dysfunction
    - Most are immune-mediated
    - Common: Post-streptococcal
  - Secondary: Part of multisystem disorder; many immune-mediated
- Epidemiology
  - Causes of acute renal failure
    - Pre-renal (renal hypoperfusion): 55%
    - Intrinsic renal disease ≈ 40%
    - Post-renal (obstruction) ≈ 5%
- Associated abnormalities: Renal vein thrombosis

Gross Pathologic & Surgical Features
- Kidneys grossly normal or enlarged
- May see renal pallor or petechial hemorrhages

Microscopic Features
- Hypercellularity
- Glomerular basement membrane (GBM) thickening
- Hyalinization: Deposition of amorphous proteinaceous material
- Sclerosis: Obliteration of glomerular tuft

Staging, Grading or Classification Criteria
- Morphologic criteria
  - Proliferative: Increased glomerular cell number
  - Crescentic: Half-moon collection of cells in Bowman space
  - Membranous: Expansion of GBM with immune deposits
- Extent within kidney
  - Focal: Involves < 50% of glomeruli
  - Diffuse: Involves majority of glomeruli
- Glomerular involvement
  - Global: Involves whole of glomerular tuft
  - Segmental: Involves part of glomerular tuft

CLINICAL ISSUES

Presentation
- Most common signs/symptoms
  - Major presentations of glomerular disease
    - Acute nephritic or nephrotic syndrome
    - Asymptomatic urinalysis abnormalities
  - Acute GN: Develops over days
    - Proteinuria, hematuria, red cell casts
    - Hypertension, edema, oliguria
  - Rapidly progressive GN: Weeks → months
  - Chronic GN: Develops over months-years
- Clinical profile
  - Post-streptococcal: Follows pharyngitis or impetigo
  - Secondary: Autoimmune diseases, other entities

Natural History & Prognosis
- May progress to end-stage renal disease

Treatment
- Options, risks, complications
  - Immunosuppressive therapy
  - Treatment of underlying systemic disorder

DIAGNOSTIC CHECKLIST

Consider
- Look for associated renal vein thrombosis
- Try to distinguish acute from chronic and pre/post-renal causes of renal failure from intrinsic causes

SELECTED REFERENCES

1. Braunwald E et al: Harrison's principles of internal medicine. 15th ed. New York, McGraw-Hill. 1541-98, 2001
2. Davidson AJ et al: Radiology of the kidney and genitourinary tract. 3rd ed. Philadelphia, W.B. Saunders, 143-83, 1999

IMAGE GALLERY

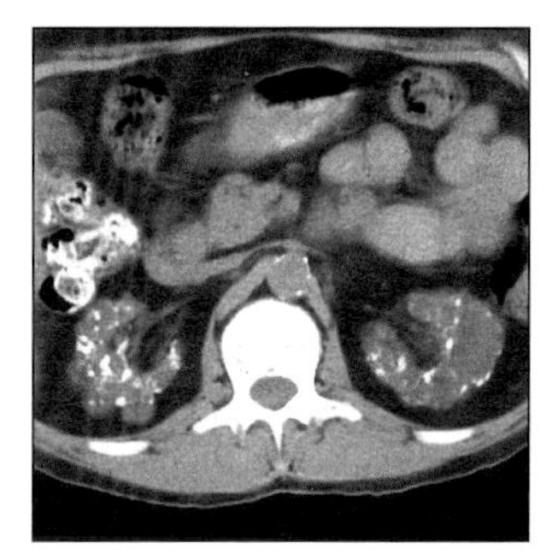
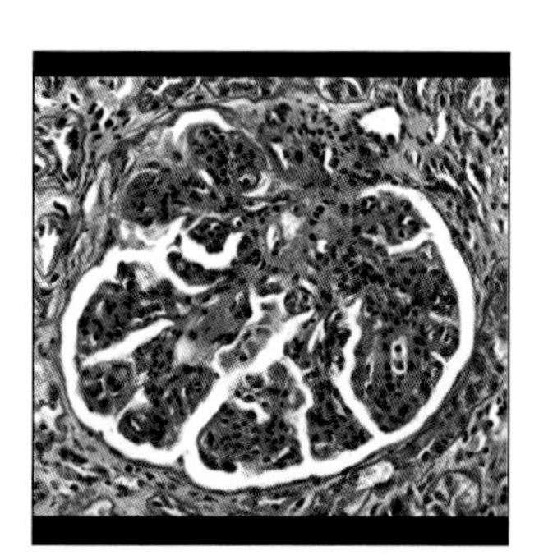

(Left) *Axial NECT shows punctate parenchymal calcification in dialysis patient with chronic glomerulonephritis and secondary hyperparathyroidism.* ***(Right)*** *Micropathology shows diffusely enlarged glomerulus with increased cellularity and thickening of the glomerular capillary walls in Hepatitis C virus-associated MPGN (Courtesy T. Morgan, MD).*

RENAL PAPILLARY NECROSIS

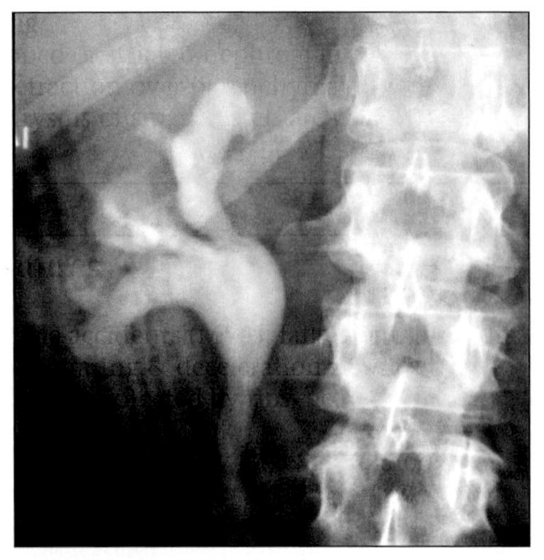

Urogram shows club-shaped calices bilaterally due to papillary necrosis attributed to chronic analgesic abuse.

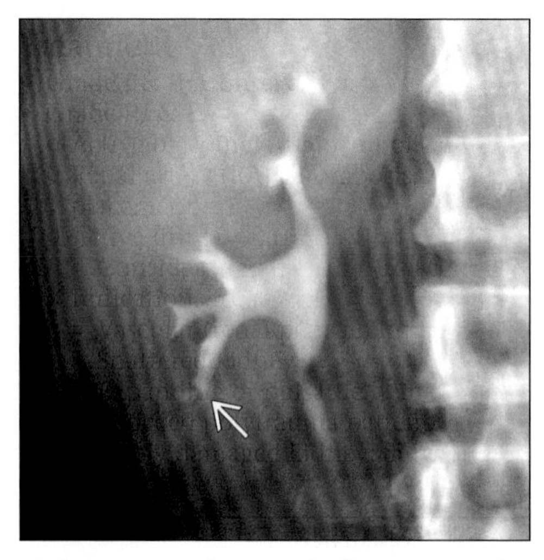

Nephrotomogram shows streak of contrast extending from fornix parallel to papilla (arrow).

TERMINOLOGY

Definitions

- Necrosis of renal papilla within medulla secondary to interstitial nephritis or ischemia

IMAGING FINDINGS

General Features

- Best diagnostic clue: Triangular or bulbous cavitation adjacent to calyx on IVP or retrograde pyelogram
- Location
  - Bilateral: Analgesics & diabetes
  - Unilateral: Obstruction; infection; venous thrombus
- Morphology: Papillary swelling, tract formation, cavitation & sloughing

Radiographic Findings

- IVP
  - Subtle streak of contrast extending from fornix parallel to long axis of papilla
  - Triangular or bulbous cavitation of papilla
  - Widened fornix (necrotic shrinkage of papilla)
  - Calyx: Club-shaped or saccular (sloughed papilla)
  - Sloughed papilla: Triangular filling defect in calyx, pelvis, ureter or ring-shaped peripheral calcification
  - "Ring shadow": Outlining detached papilla
  - ↓ (Rarely ↑) density of contrast in nephrogram
  - ± Displaced collecting system (due to edema)
  - ± Calcification or ossification of necrotic papillae
- Retrograde pyelography: Rapid filling of cystic cavities

CT Findings

- Normal, enlarged or small contracted kidneys
- Ring shaped medullary calcification
- Hematoma, lobar infarct, scarring (sickle cell)
- Contrast filled clefts in renal parenchyma
- Filling defects: Renal pelvis/ureter (sloughed papillae)
- ± Hydronephrotic changes (due to obstruction)
- ± Renal vein thrombosis
- ± Heterogeneous parenchymal enhancement due to pyelonephritis

Ultrasonographic Findings

- Real Time
  - Early stage
    - Necrotic renal papillae: Seen as echogenic foci
    - Sonolucent rims: Fluid around necrotic papillae
  - Advanced stage
    - Single or multiple cystic cavities in medullary pyramids continuous with calices ± calcification

Imaging Recommendations

- IVP or RGP; helical NE + CECT
  - Nonionic contrast agents recommended due to ↑ incidence of contrast associated nephropathy

DDx: Blunted or Dilated Calices

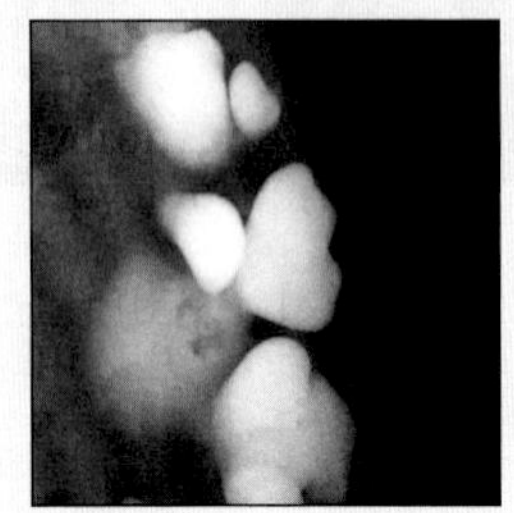

Hydronephrosis

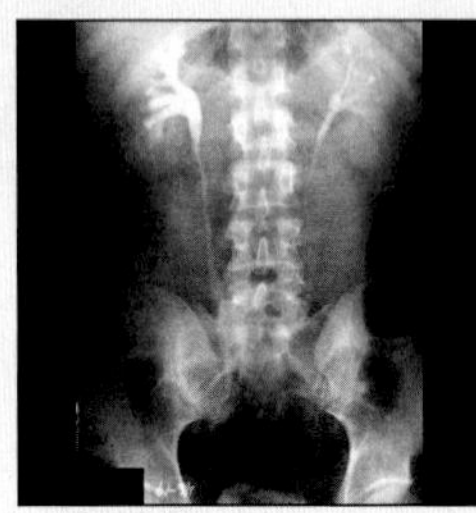

Hydronephrosis

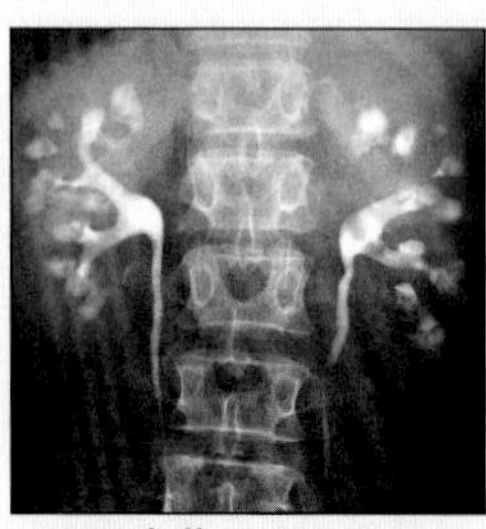

Medullary Sponge

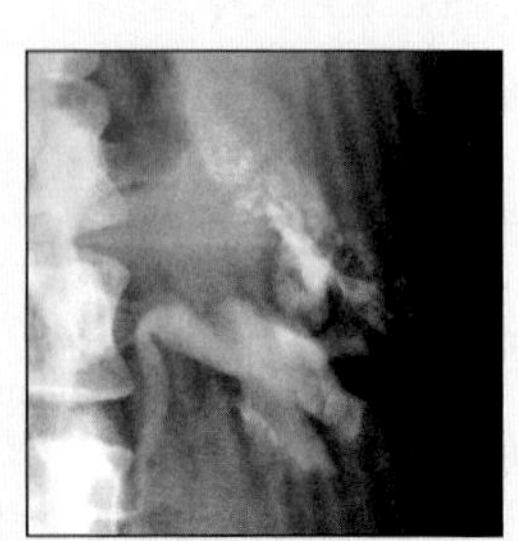

Medullary Sponge

RENAL PAPILLARY NECROSIS

Key Facts

Terminology
- Necrosis of renal papilla within medulla secondary to interstitial nephritis or ischemia

Imaging Findings
- Best diagnostic clue: Triangular or bulbous cavitation adjacent to calyx on IVP or retrograde pyelogram
- Calyx: Club-shaped or saccular (sloughed papilla)
- "Ring shadow": Outlining detached papilla

Top Differential Diagnoses
- Hydronephrosis
- Medullary sponge kidney

Pathology
- Analgesic nephropathy, DM, S-hemoglobinopathy

Diagnostic Checklist
- Correlate imaging with history of analgesics, DM, SC

DIFFERENTIAL DIAGNOSIS

Hydronephrosis
- Dilated pelvicaliceal system & ureter; ↓ renal function

Medullary sponge kidney
- Multiple small cystic cavities or tubular ectasia; medullary nephrocalcinosis
- Excretory urography
  - Linear densities ("brush" appearance) in pyramids

Congenital megacalices
- Large & wide calyceal system; normal renal function

PATHOLOGY

General Features
- General path comments
  - Three types of parenchymal involvement in RPN
    - Medullary type: Central cavitation of papillae extending from fornix
    - Papillary type: Necrosis & cavitation at periphery of papillae → sloughing of papillae
    - Necrosis in situ of papillae ± Ca++ or ossification
  - Ca++ within necrotic papillae: Analgesics (common)
- Etiology
  - Analgesic nephropathy, DM, S-hemoglobinopathy
  - UTI & obstruction, TB, kidney transplant, alcoholics

Gross Pathologic & Surgical Features
- Mild cases: Normal or enlarged kidneys
- Severe cases: Small, smooth or scarred

Microscopic Features
- Analgesics: RPN, centrilobar cortex atrophy, septal cortex hypertrophy
- Sickle disease: Vessel occlusion, lobar infarcts, tubular obliteration, fibrosis

CLINICAL ISSUES

Presentation
- Most common signs/symptoms
  - Flank pain, dysuria, fever, ureteral colic, HTN
  - Pyuria, hematuria, acute oliguric renal failure
- Lab: ± ↑ WBC or ↓ RBC; proteinuria; impaired RFT

Demographics
- Age: Middle-aged women (analgesic nephropathy)

Natural History & Prognosis
- Complications: Obstruction, infection, renal failure, transitional cell carcinoma
- Prognosis: Early stage (good); advanced stage (poor)

Treatment
- Early stage: Symptomatic treatment
- Advanced stage (obstruction, infection, failure)
  - Ureteral stent, surgical treatment

DIAGNOSTIC CHECKLIST

Consider
- Correlate imaging with history of analgesics, DM, SC

Image Interpretation Pearls
- Club-shaped calyx, triangular cavitation of papillae or filling defect of sloughed papillae with ring shadow

SELECTED REFERENCES

1. Hoffman JC et al: Demonstration of renal papillary necrosis by sonography. Radiology 145: 785-787, 1982
2. McCall IW et al: Urographic findings in homozygous sickle cell disease. Radiology 126: 99-104, 1978
3. Poynter JD et al: Necrosis in situ: A form of renal papillary necrosis seen in analgesic nephropathy. Radiology 111: 69-76, 1974

IMAGE GALLERY

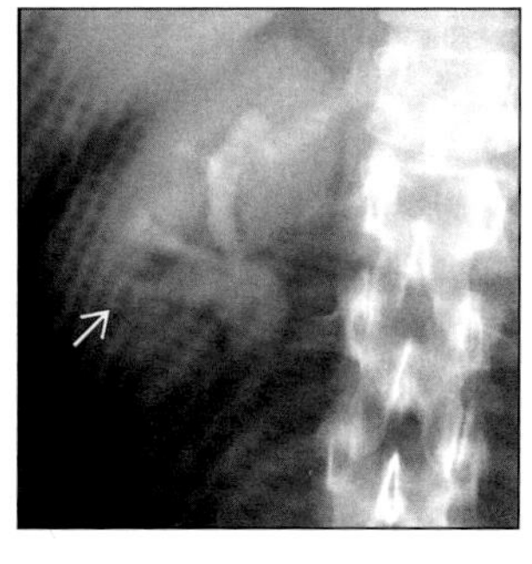
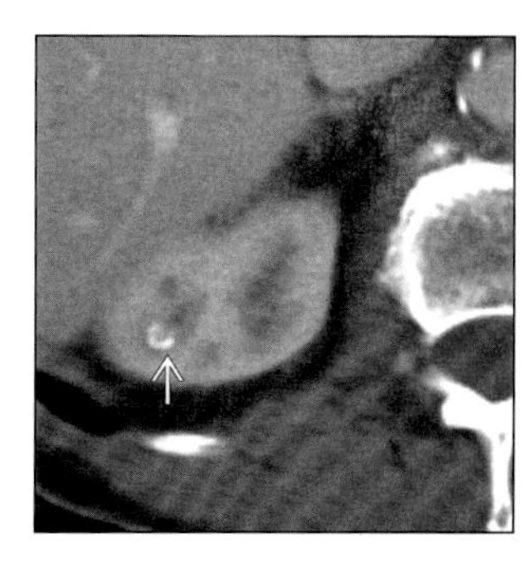

(Left) *Nephrotomogram shows sloughed papilla (arrow) as a filling defect in calyx; clubbed calices.* ***(Right)*** *Axial CECT shows ring-shaped medullary calcification (arrow) on the surface of a necrotic papilla.*

UROLITHIASIS

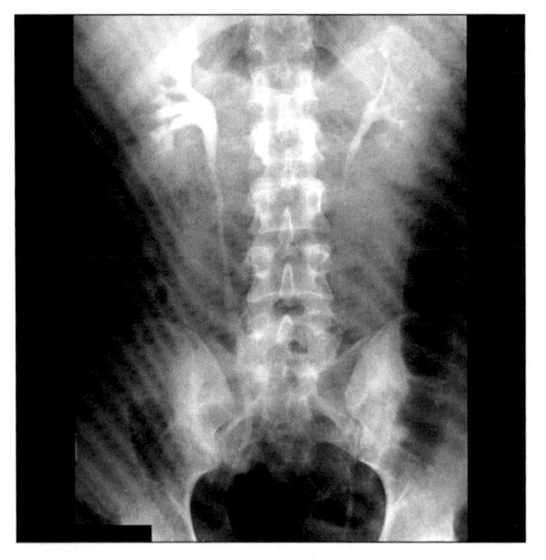

Urogram shows mild right hydronephrosis with complete opacification of ureter to UV junction; no stone seen.

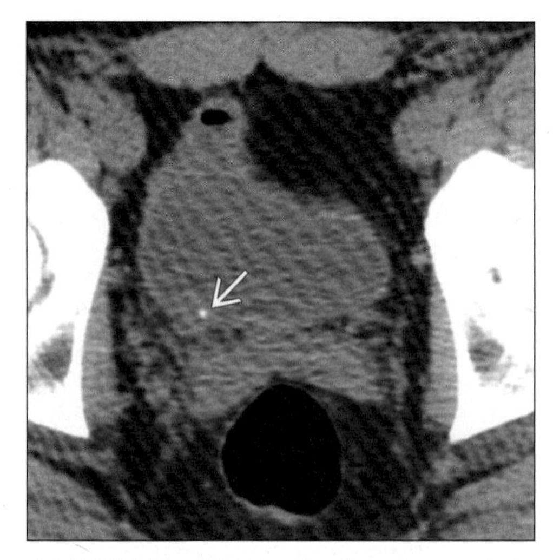

Axial NECT through UV junction shows 2 mm calculus (arrow) responsible for obstruction and pain.

TERMINOLOGY

Abbreviations and Synonyms

- Calculous disease; nephrolithiasis; kidney, renal or urinary stones

Definitions

- Concretions within the urinary system

IMAGING FINDINGS

General Features

- Best diagnostic clue: Small dense focus at ureterovesical junction (UVJ) with proximal hydronephrosis and perinephric stranding
- Location
  - Upper urinary tract (UT): Calyceal, renal pelvis or ureteropelvic junction (UPJ)
  - Ureteral calculi: Ureter or UVJ
  - Lower UT: Bladder, urethral, prostatic
- Other general features
  - Types of stones
    - Calcium stones (75-80%): Calcium oxalate and/or calcium phosphate
    - Struvite stones (15-20%): Magnesium ammonium phosphate (struvite), magnesium ammonium phosphate + calcium phosphate (triple phosphate)
    - Uric acid stones (5-10%)
    - Cystine stones (1-3%)
    - Matrix stones (rare): Mucoproteins
    - Xanthine stones (extremely rare)
    - Milk of calcium: Calcium carbonate + calcium phosphate (carbonate apatite)
    - Protease inhibitor stones: Indinavir-induced

Radiographic Findings

- Radiography
  - Pre-CT belief
    - Radiography detects 90% of calcium stones, some struvite and "misses" uric acid stones
  - Based on CT correlation: Radiography "misses" majority of calculi
    - Due to small size, insufficient radiopacity, overlying bones, bowel, etc.
  - Calcium oxalate or phosphate stones
    - Usually very opaque, visible if large
    - Can be smooth, irregular or jagged
  - Struvite and cystine stones
    - Staghorn calculi: Shape may conform to pelvocaliceal system
    - Usually opaque, detectible if large
  - Uric acid and xanthine stones
    - Usually small, smooth, disc-shaped
    - Rarely opaque or detectible (when mixed with calcium salts)
  - Milk of calcium

DDx: High Density Focus in Kidney or Ureter

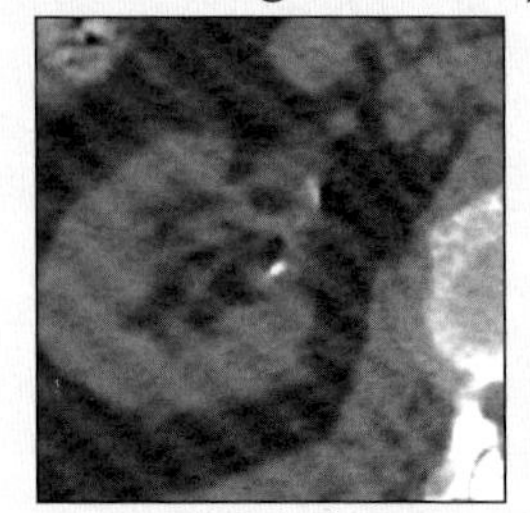

Arterial Calcification

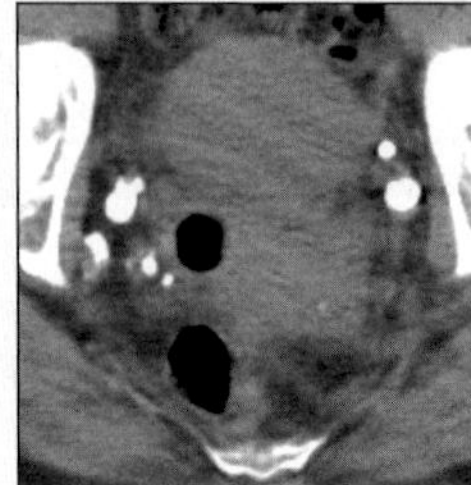

Phleboliths

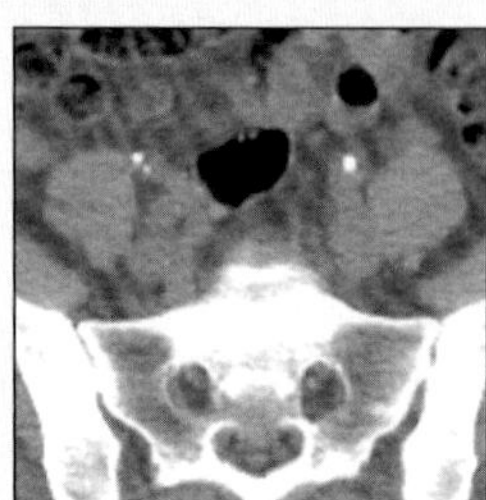

Phleboliths

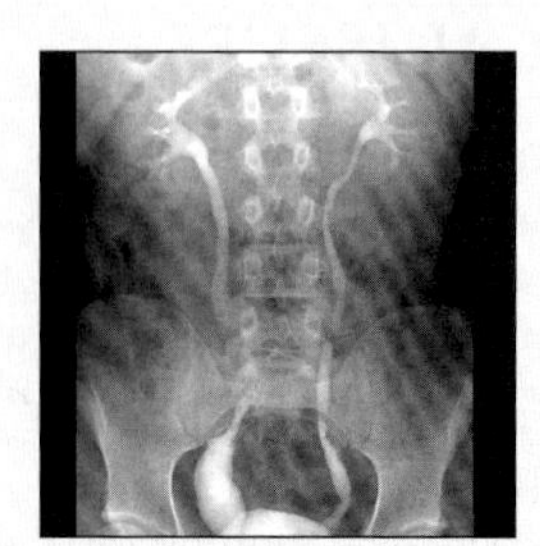

Primary Megaureter

UROLITHIASIS

Key Facts

Terminology
- Calculous disease; nephrolithiasis; kidney, renal or urinary stones

Imaging Findings
- Best diagnostic clue: Small dense focus at ureterovesical junction (UVJ) with proximal hydronephrosis and perinephric stranding
- Based on CT correlation: Radiography "misses" majority of calculi
- Radiopacity (most to least): Calcium oxalate and/or phosphate > cystine > struvite > uric acid

Top Differential Diagnoses
- Renal artery calcification
- Nephrocalcinosis
- Phlebolith

Pathology
- Majority are mixed composition; > 50% contain calcium salts

Clinical Issues
- Acute colicky flank pain radiating to groin (60-95% with these symptoms have stones)
- Age: 1 in 8 have stones by 70 years of age
- Spontaneous passage through ureter: 80% (< 4 mm), 50% (4-6 mm), 20% (> 8 mm)

Diagnostic Checklist
- Avoid repeat CT in young patients with known or clinically evident stones (excessive radiation)
- Small dense focus in course of ureter with perinephric stranding ± hydronephrosis

    - Liquid suspension of calcium salts that are "trapped" in a calyceal diverticulum or ureterocele
    - Moderately opaque
  - Protease inhibitor stones
    - Nonopaque
- IVP
  - Lucent (uric acid, cystine, matrix): Filling defects
  - Opaque (calcium, milk of calcium): Obscured by contrast-opacified urine
  - ± PC diverticula, UPJ obstruction, tubular ectasia (medullary sponge kidney), urinary diversion
  - Ureteral calculi: Nephrographic phase
    - Enlarged kidney
    - Delayed ("obstructive", peak at 6 hrs.); prolonged
    - Dense; striated; absent ("negative")
  - Ureteral calculi: Pyelographic phase
    - Delayed opacification (≥ 24 hrs)
    - Hydronephrosis; stone in ureter
    - ↓ Contrast density in collecting system
    - Contrast extravasation; ± forniceal rupture
    - Asymmetry of ureteral caliber to obstructed level
    - "Standing column" of contrast to obstructed level
    - Interureteric ridge or edema (pseudoureterocele)
  - Ureteral calculi: Late phase
    - Vicarious excretion of contrast (to gallbladder)

CT Findings
- NECT
  - Stones are uniformly dense except matrix & indinavir stones
  - Radiopacity (most to least): Calcium oxalate and/or phosphate > cystine > struvite > uric acid
  - Calcium stones: 400-600 HU
  - Uric acid & cystine stones: 100-300 HU
  - Matrix stones
    - Soft tissue attenuation (pure)
    - Laminated peripheral calcification, diffuse ↑ density or round faintly opaque nodules with densely calcified center (when mixed with calcium salts)
  - Milk of calcium: Layered opaque suspension; stone movement
  - Indinavir stones: Not or faintly opaque; deduced from secondary findings (obstruction)
  - Ureteral calculi: Visualize stone and secondary signs
    - "Soft tissue rim" sign: Ureteral wall edema at stone
    - Pseudoureterocele: UVJ edema around calculus
    - Hydronephrosis; hydroureter; perinephric or periureteral stranding
- CECT: Lucent (matrix and indinavir stones): Filling defects

MR Findings
- No signal (no mobile protons); large: Signal voids
- Ureteral calculi: Abrupt change in ureteral caliber indicates obstruction level; secondary signs

Ultrasonographic Findings
- Real Time
  - Visualize calculi as bright echogenic foci with acoustical shadowing
    - Not dependent on chemical consistency
  - Best visualized in kidney and at UVJ; poor visualization of ureteral stones
- Color Doppler: Ureteral jet: Visualization of "jet" of urine from UVJ into bladder excludes obstructing distal stone

DIFFERENTIAL DIAGNOSIS

Renal artery calcification
- Radiography: Curvilinear parallel lines of calcification in extrarenal, intrarenal or pelvic arteries
- US: Echogenic foci within renal sinus with acoustical shadowing; indistinguishable from < 1 cm calculi
- US: Curvilinear appearance; distinguish > 1 cm calculi
- Differentiate by correlation with radiography

Nephrocalcinosis
- Calcification within parenchyma: Cortex & medulla (most common); can erode & evolve into urolithiasis
- CT & US: Indistinguishable except by location

Phlebolith
- Stone within vein (pelvic, gonadal)

- Radiography: Multiple, round, central lucency
- CT: ≤ 4 mm calcification without "tissue rim" sign
- CT: "Comet" or "tail" sign: Adjacent eccentric, tapering soft tissue mass (noncalcified portion of pelvic vein) & a central lucency within calcification
- Distinguishing ovarian vein phlebolith from ureteral calculus is difficult
  - Follow course of vessel and look for signs of ureteral obstruction

Primary megaureter
- Rare congenital anomaly causes dilated ureters

PATHOLOGY

General Features
- General path comments: Majority are mixed composition; > 50% contain calcium salts
- Etiology
  - Calcium stones
    - Idiopathic (85%): Idiopathic hypercalciuria
    - Acquired (15%): Hyperparathyroidism, sarcoidosis, renal tubular acidosis, hyperoxaluria, steroids, Cushing syndrome, immobilization, ↑ vitamin D
  - Struvite stones: Urinary tract infections (UTI) (Proteus, Klebsiella, Pseudomonas; urea-splitting)
  - Uric acid stones: Hyperuricosuria (25% with gout), ileostomy, chemotherapy, acidic & concentrated urine, adenine phosphoribosyltransferase deficiency
  - Cystine stones: Cystinuria (autosomal recessive)
  - Matrix stones: Chronic UTI, urine stasis, obstruction
  - Xanthine stones: Xanthine oxidase deficiency
  - Milk of calcium: PC diverticula, ureteroceles
  - Risk factors
    - Environment: Warm climates, summer
    - Medications: Acetazolamide, indinavir
    - Anatomical abnormalities: UPJ obstruction (horseshoe or ectopic kidney), PC diverticula, tubular ectasia, urinary diversion
  - Pathogenesis
    - Supersaturated solution → crystal formation in urine (excessive excretion & precipitation theory)
    - Lack of substances that inhibit crystal deposition, stone formation & growth (inhibitor theory)
    - Presence of specific macromolecules that are essential for stone formation (matrix theory)
- Epidemiology
  - Prevalence: 2-3%; 40-60 years of age (in Caucasians)
  - Incidence: 1-2 per 1,000; peak at 20-40 years of age

Gross Pathologic & Surgical Features
- Matrix stones: Gelatinous or soft putty texture; tan to red-brown

Microscopic Features
- Crystals dependent on type of stones

CLINICAL ISSUES

Presentation
- Most common signs/symptoms
  - Upper UT: Asymptomatic, flank pain, fever
  - Ureteral calculi
    - Acute colicky flank pain radiating to groin (60-95% with these symptoms have stones)
  - Lower UT: Asymptomatic, dysuria, dull/sharp pain radiating to penis, buttocks, perineum or scrotum
- Lab data
  - Urinalysis: Hematuria, crystals ± bacteruria or pyuria
- Diagnosis: Ureteral calculi by CT
  - ≤ 4 mm with "tissue rim" sign (91% sensitive)
  - Hydroureteronephrosis & perinephric stranding (97% positive & 93% negative predictor value)

Demographics
- Age: 1 in 8 have stones by 70 years of age
- Gender
  - M:F = 3:1
  - Struvite & matrix stone: M:F = 1:2-3

Natural History & Prognosis
- Spontaneous passage through ureter: 80% (< 4 mm), 50% (4-6 mm), 20% (> 8 mm)
- Complications: Obstruction, infection, abscess and renal insufficiency
- Prognosis: Recurrence without treatment: 10% at 1 year, 35% at 5 years, 50% at 10 years

Treatment
- ↑ Hydration (2L urine/day), restrict diet (protein, sodium, calcium) & drugs (thiazides or allopurinol)
- Extracorporeal shock wave lithotripsy (ESWL), percutaneous nephrostolithotomy, endoscopic retrieval or suprapubic cystolithotomy
- Follow-up recurrence only: 4-6 weeks after treatment, 24-hour urine (volume, calcium, phosphorus, uric acid, creatine, oxalate, citrate, cystine screen)

DIAGNOSTIC CHECKLIST

Consider
- Avoid repeat CT in young patients with known or clinically evident stones (excessive radiation)
  - Consider US, especially in young women

Image Interpretation Pearls
- Small dense focus in course of ureter with perinephric stranding ± hydronephrosis

SELECTED REFERENCES

1. Sheafor DH et al: Nonenhanced helical CT and US in the emergency evaluation of patients with renal colic: Prospective comparison. Radiology. 217: 792-7, 2000
2. Smith RC et al: Diagnosis of acute flank pain: Value of unenhanced helical CT. AJR. 166: 97-101, 1996
3. Katz DS et al: Unenhanced helical CT of ureteral stones: Incidence of associated urinary tract findings. AJR. 166: 1319-22, 1996
4. Sommer FG et al: Detection of ureteral calculi in patients with suspected renal colic: value of reformatted noncontrast helical CT. AJR Am J Roentgenol. 165(3):509-13, 1995
5. Jeffrey RB et al: CT and ultrasonography of acute renal abnormalities. Radiol Clin North Am. 21(3):515-25, 1983

UROLITHIASIS

IMAGE GALLERY

Typical

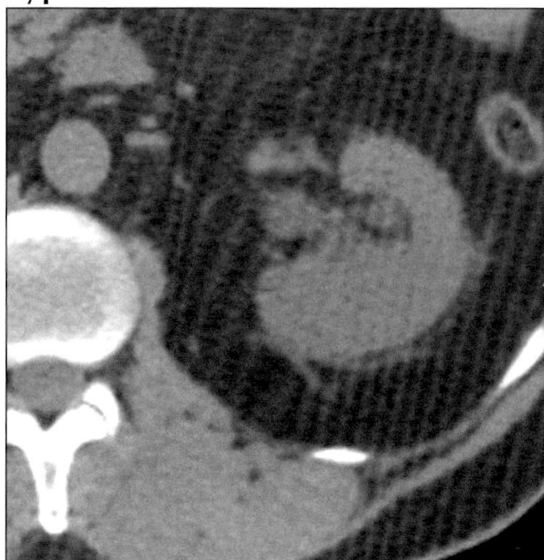

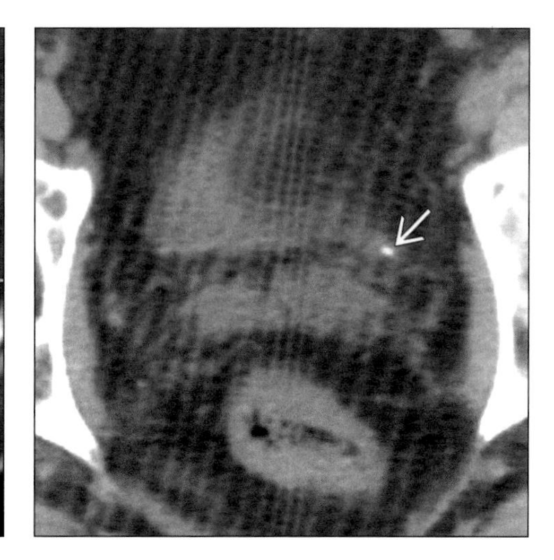

(Left) *Axial NECT shows mild left hydronephrosis and perirenal stranding.* ***(Right)*** *Axial NECT shows small stone (arrow) at ureterovesical junction.*

Typical

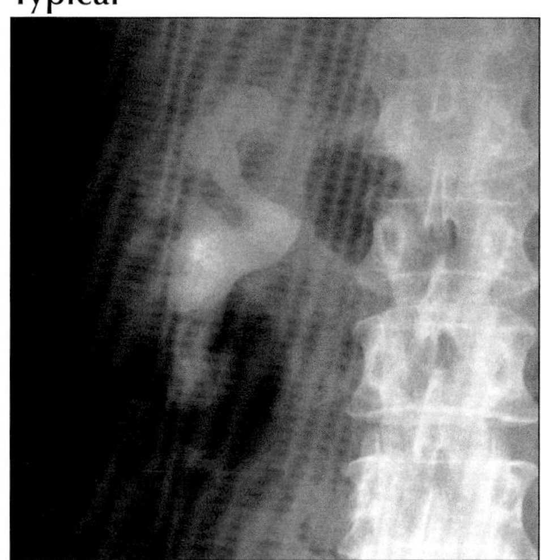

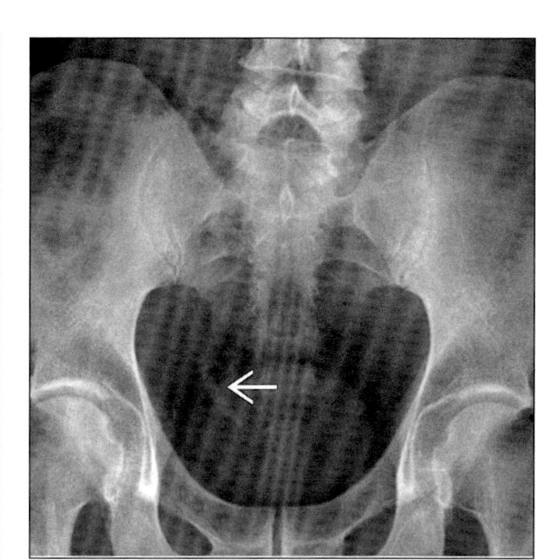

(Left) *Radiograph shows large staghorn calculus.* ***(Right)*** *Radiograph shows multiple stone fragments filling distal ureter (arrow) following lithotripsy.*

Typical

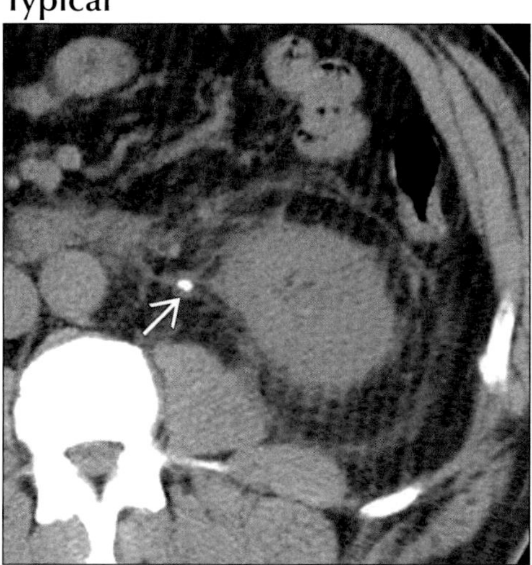

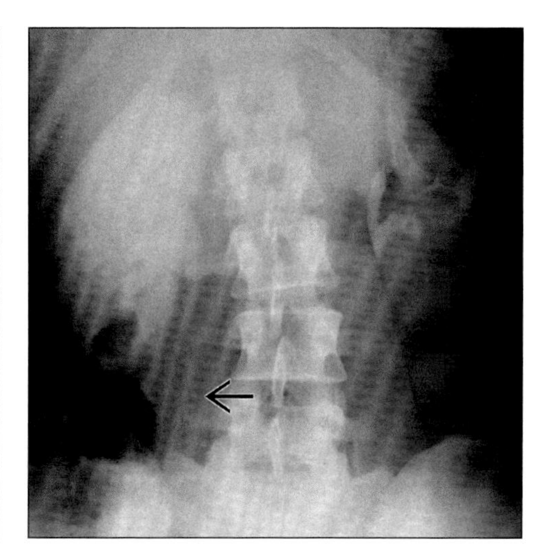

(Left) *Axial NECT shows proximal ureteral stone (arrow) with a "rim" of soft tissue representing the ureteral wall.* ***(Right)*** *Excretory urogram shows persistent right nephrogram and delayed excretion due to proximal ureteral stone (arrow).*

NEPHROCALCINOSIS

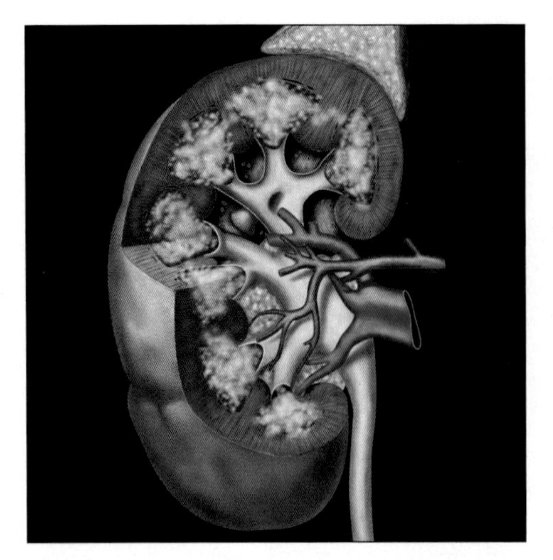

Drawing shows calcification in renal pyramids, representing nephrocalcinosis.

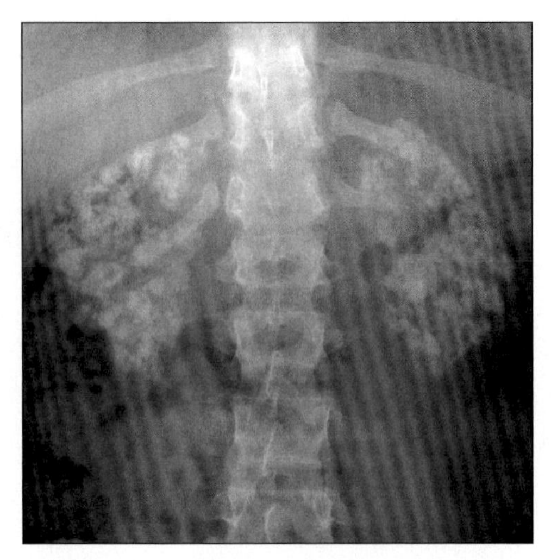

Anteroposterior radiography shows dense calcification within the renal pyramids in a patient with sarcoid and medullary nephrocalcinosis.

TERMINOLOGY

Abbreviations and Synonyms

- Medullary nephrocalcinosis, cortical nephrocalcinosis

Definitions

- Radiologically detectable diffuse calcium deposition within the renal substance

IMAGING FINDINGS

General Features

- Best diagnostic clue: Calcification within renal parenchyma
- Location
  - Renal parenchyma
    - Medullary nephrocalcinosis: 95%
    - Cortical nephrocalcinosis: 5%
    - Both cortical and medullary: Rare
- Size: Kidneys often normal size and contour
- Morphology
  - Variable patterns of calcification
    - Scattered punctate calcifications in renal medullae
    - Dense, confluent medullary calcification: Common in renal tubular acidosis
    - "Tram line" calcification or punctate calcifications in renal cortex

Radiographic Findings

- Radiography
  - Fine stippled calcification in renal pyramids
  - Coarse, confluent calcification
  - Punctate or "tramline" cortical calcification
- IVP
  - Medullary: Calcification in renal pyramids on preliminary films
  - May see linear striations and/or cystic spaces in papillae in patients with underlying medullary sponge kidney

CT Findings

- NECT
  - Stippled or confluent calcifications in renal parencyma
  - May see ringlike pattern due to relatively increased calcification at corticomedullary junction

Ultrasonographic Findings

- Real Time
  - Medullary: Echogenic renal pyramids
    - May have ring-like pattern with increased echogenicity along periphery of pyramid
  - Acoustic shadowing may be absent

Imaging Recommendations

- Best imaging tool: Noncontrast CT

DDx: Mimics of Nephrocalcinosis

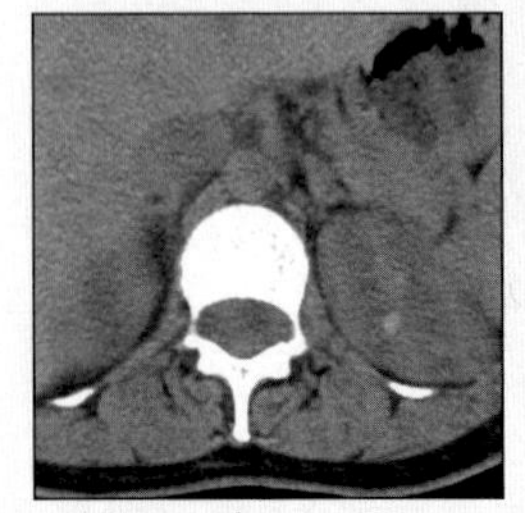

Hemorrhagic Cysts

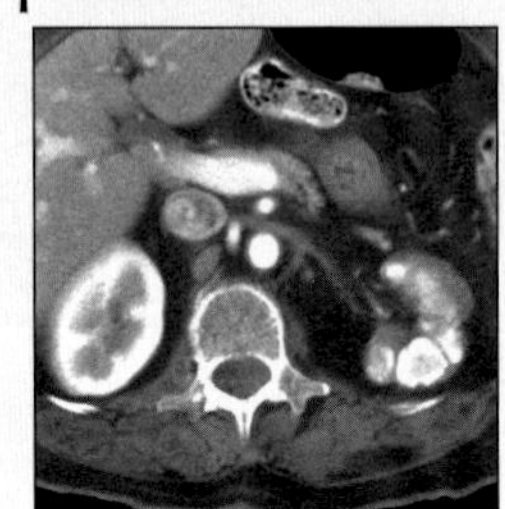

Renal TB

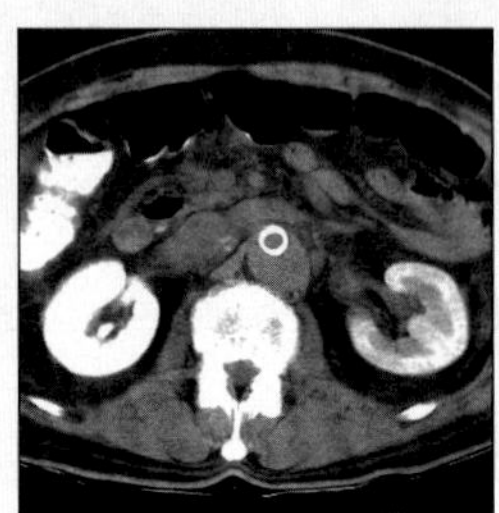

Renal Failure

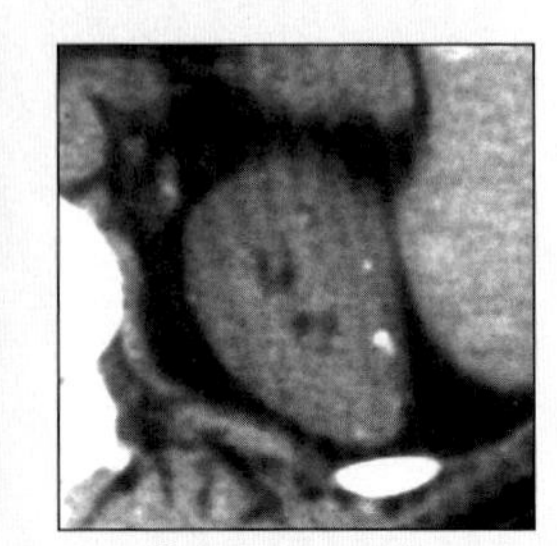

Pneumocystis

NEPHROCALCINOSIS

Key Facts

Terminology
- Radiologically detectable diffuse calcium deposition within the renal substance

Imaging Findings
- Medullary nephrocalcinosis: 95%
- Cortical nephrocalcinosis: 5%
- Fine stippled calcification in renal pyramids
- Coarse, confluent calcification
- May see ringlike pattern due to relatively increased calcification at corticomedullary junction
- Medullary: Echogenic renal pyramids
- Best imaging tool: Noncontrast CT

Top Differential Diagnoses
- Papillary necrosis
- Renal tuberculosis

Pathology
- Medullary: Renal tubular acidosis type 1 (distal)
- Medullary: Hyperparathyroidism
- Medullary sponge kidney (MSK): Cystic or fusiform dilation of collecting ducts in renal pyramids
- Acute cortical necrosis secondary to shock, placental abruption, nephrotoxins
- Cortical: Chronic glomerulonephritis
- Associated abnormalities: Urolithiasis when calculi formed in renal medulla erode into collecting system

Diagnostic Checklist
- Massive, dense medullary nephrocalcinosis usually due to RTA type I
- Unilateral or segmental medullary nephrocalcinosis → medullary sponge kidney

- Protocol advice: Detection of nephrocalcinosis on plain films is improved by low kV technique

DIFFERENTIAL DIAGNOSIS

Technical
- NECT: Residual contrast material

Papillary necrosis
- Common in analgesic nephropathy
- Curvilinear, ring-shaped, or triangular calcification in sloughed papillae
- Calcifications arranged in arc conforming to positions of papillary tips
- May be indistinguishable from nephrocalcinosis

Renal tuberculosis
- Focal amorphous calcification within kidney at sites of caseating infection
- Diffuse: Dense calcification throughout autonephrectomized kidney ("putty kidney")

Extrapulmonary pneumocystis carinii or MAI infection in AIDS
- May have both cortical and medullary punctate calcification

PATHOLOGY

General Features
- General path comments
  - Calcium stones grow on papillae
    - Most break loose and enter collecting system → urolithiasis
    - If calcium stones remain in place → medullary nephrocalcinosis
- Genetics
  - Type I renal tubular acidosis (RTA): Familial form
    - Autosomal dominant inheritance pattern most common
    - May be due to defect in chloride-bicarbonate exchange gene AE1
  - Hyperoxaluria: Familial form
    - Autosomal recessive
- Etiology
  - Medullary nephrocalcinosis
    - 40%: Hyperparathyroidism
    - 20%: Renal tubular acidosis type I
    - 20%: Medullary sponge kidney
  - Cortical nephrocalcinosis
    - Chronic glomerulonephritis
    - Renal cortical necrosis
    - Transplant kidney: Chronic rejection
  - Three primary mechanisms for calcium deposition
    - Metastatic: Metabolic abnormality leads to calcium deposition in the medullae of morphologically normal kidneys
    - Urinary stasis: Calcium salts precipitate in dilated collecting ducts containing static urine
    - Dystrophic: Calcium deposition in damaged renal tissue
  - Entities causing metastatic calcification
    - Medullary: Renal tubular acidosis type 1 (distal)
    - Medullary: Hyperparathyroidism
    - Medullary: Hypercalcuria
    - Medullary: Hyperoxaluria
  - Nephrocalcinosis due to urinary stasis
    - Medullary sponge kidney (MSK): Cystic or fusiform dilation of collecting ducts in renal pyramids
  - Nephrocalcinosis due to dystrophic calcification
    - Acute cortical necrosis secondary to shock, placental abruption, nephrotoxins
    - Cortical: Chronic glomerulonephritis
- Epidemiology
  - Incidence 0.1-6%
  - Medullary sponge kidney: Seen in 0.5% of excretory urograms
- Associated abnormalities: Urolithiasis when calculi formed in renal medulla erode into collecting system

Gross Pathologic & Surgical Features
- Depends on underlying etiology of nephrocalcinosis

NEPHROCALCINOSIS

Microscopic Features

- Calcium deposition in the interstitium, tubule epithelial cells, along basement membranes
- Calcium deposition within lumina of tubules

CLINICAL ISSUES

Presentation

- Most common signs/symptoms
  - Most often asymptomatic
  - Other signs/symptoms
    - Flank pain, hematuria if associated with urolithiasis
- Clinical profile: Cortical nephrocalcinosis
  - Acute cortical necrosis
    - Nephrotoxic drugs (ethylene glycol, methoxyflurane anesthesia, amphotericin B)
    - Acute vascular insult (shock, placental abruption)
  - Chronic glomerulonephritis
  - Alport syndrome: Hereditary nephritis and nerve deafness
- Clinical profile: Medullary nephrocalcinosis
  - Skeletal deossification
    - Primary and secondary hyperparathyroidism
    - Bony metastases
    - Prolonged immobilization
  - Increased intestinal absorption of calcium
    - Sarcoidosis
    - Milk-alkali syndrome
  - Medullary sponge kidney
  - Hyperoxaluria
    - Hereditary type
    - Acquired: Secondary to small bowel disease or bariatric surgery
  - Renal tubular acidosis (RTA) type 1 (distal RTA)
    - May be primary or secondary to other systemic disease (Sjogren, lupus, others)
    - Distal tubule unable to secrete hydrogen ions
    - Metabolic acidosis with urinary pH > 5.5
    - Type II (proximal) RTA never causes nephrocalcinosis

Demographics

- Age: Any
- Gender: M > F

Natural History & Prognosis

- Depends on underlying cause of nephrocalcinosis

Treatment

- Options, risks, complications: Medullary nephrocalcinosis often complicated by urolithiasis

DIAGNOSTIC CHECKLIST

Consider

- Focal areas of dystrophic calcification in masses or infection are not considered nephrocalcinosis

Image Interpretation Pearls

- Massive, dense medullary nephrocalcinosis usually due to RTA type I
- Unilateral or segmental medullary nephrocalcinosis → medullary sponge kidney

SELECTED REFERENCES

1. Hoppe B et al: Diagnostic and therapeutic approaches in patients with secondary hyperoxaluria. Front Biosci. 8:e437-43, 2003
2. Peacock M: Primary hyperparathyroidism and the kidney: biochemical and clinical spectrum. J Bone Miner Res. 17 Suppl 2:N87-94, 2002
3. Gobel U et al: The protean face of renal sarcoidosis. J Am Soc Nephrol. 12(3):616-23, 2001
4. Unwin RJ et al: The renal tubular acidoses. J R Soc Med. 94(5):221-5, 2001
5. Sayer JA et al: Diagnosis and clinical biochemistry of inherited tubulopathies. Ann Clin Biochem. 38(Pt 5):459-70, 2001
6. Boonen S et al: Primary hyperparathyroidism: pathophysiology, diagnosis and indications for surgery. Acta Otorhinolaryngol Belg. 55(2):119-27, 2001
7. Sharma OP: Hypercalcemia in granulomatous disorders: a clinical review. Curr Opin Pulm Med. 6(5):442-7, 2000
8. Audran M et al: Hypercalciuria. Joint Bone Spine. 67(6):509-15, 2000
9. Davidson AJ et al: Radiology of the kidney and genitourinary tract. 3rd ed. Philadelphia, W.B. Saunders, 213-216, 1999
10. Chen MY et al: Abnormal calcification on plain radiographs of the abdomen. Crit Rev Diagn Imaging. 40(2-3):63-202, 1999
11. Dyer RB et al: Abnormal calcifications in the urinary tract. Radiographics. 18(6):1405-24, 1998
12. Miller FH et al: Renal manifestations of AIDS. Radiographics. 13(3):587-96, 1993
13. Cockerell CJ et al: Widespread cutaneous and systemic calcification (calciphylaxis) in patients with the acquired immunodeficiency syndrome and renal disease. J Am Acad Dermatol. 26(4):559-62, 1992
14. Dunnick NR et al: Textbook of uroradiology. Baltimore, Williams and Wilkins. 189-190, 1991
15. Amis ES et al: Essentials of uroradiology. Boston, Little, Brown and company. 213-216, 1991
16. Falkoff GE et al: Partial, combined cortical and medullary nephrocalcinosis: US and CT patterns in AIDS-associated MAI infection. Radiology. 162(2):343-4, 1987
17. Hoffbrand BI et al: Nodular cortical nephrocalcinosis: a benign and hitherto undescribed form of renal calcification. Nephron. 46(4):370-2, 1987
18. Martijn A et al: Radiologic findings in primary hyperoxaluria. Skeletal Radiol. 8(1):21-4, 1982

IMAGE GALLERY

Typical

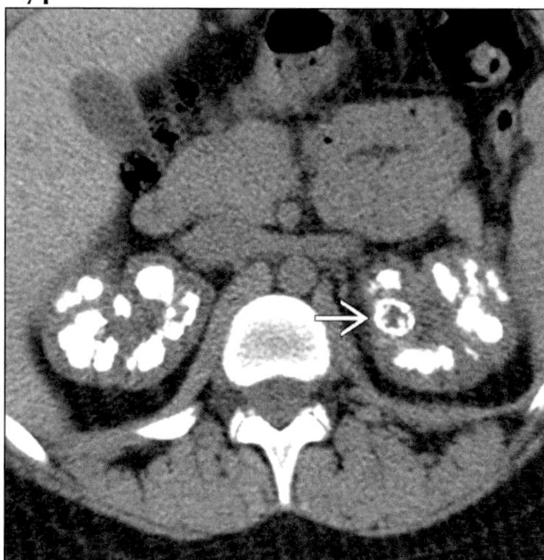

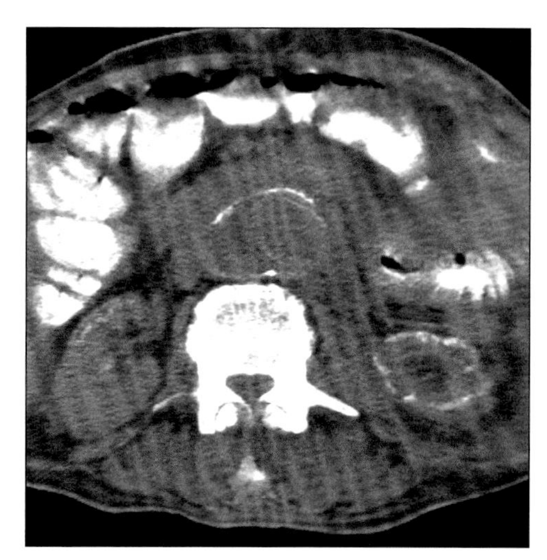

(Left) *Axial NECT shows densely calcified renal pyramids in a patient with sarcoid and medullary nephrocalcinosis. Note ring-like pattern (arrow) in left kidney at cortico-medullary junction.* ***(Right)*** *Axial NECT shows thin rim of calcification in renal cortices bilaterally, greater on the left than right in a vascular surgery patient with renal cortical necrosis.*

Typical

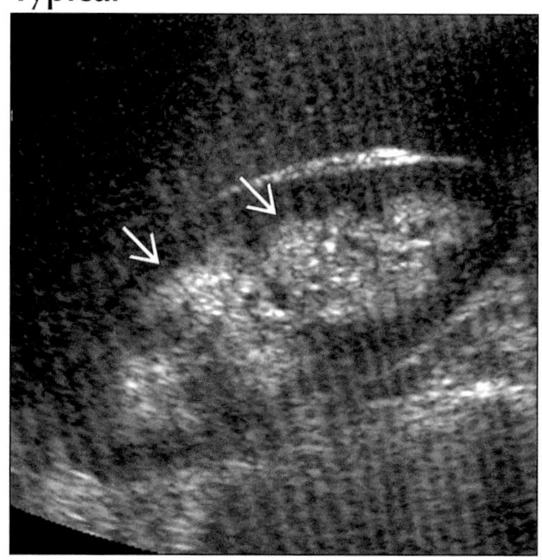

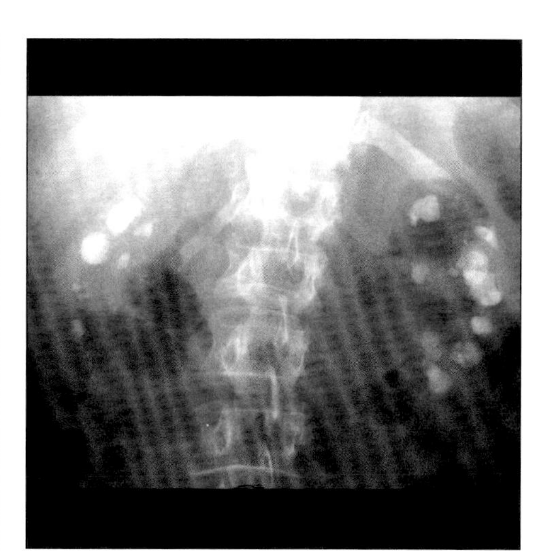

(Left) *Sagittal US shows echogenic renal pyramids (arrows) in patient with medullary nephrocalcinosis.* ***(Right)*** *Anteroposterior radiography shows densely calcified pyramids in patient with renal tubular acidosis.*

Variant

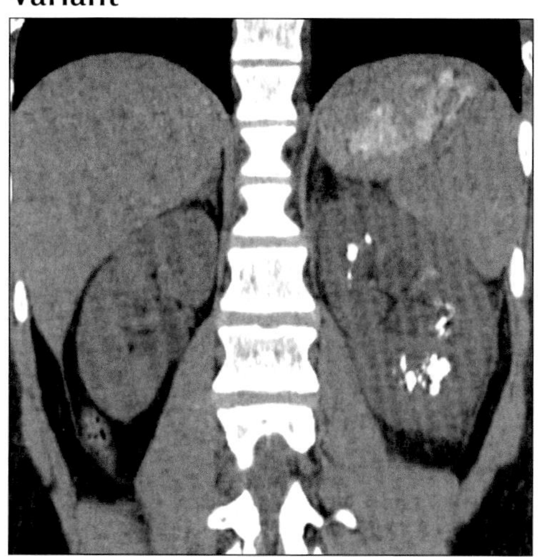

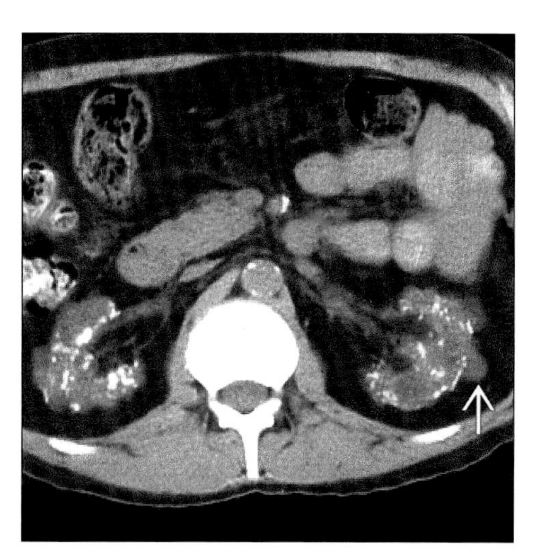

(Left) *Coronal NECT shows unilateral nephrocalcinosis in a patient with medullary sponge kidney.* ***(Right)*** *Axial NECT shows coarse calcification in renal cortical regions in patient with chronic glomerulonephritis. Exophytic cyst in left kidney (arrow) is secondary to chronic dialysis.*

ACQUIRED CYSTIC DISEASE OF UREMIA

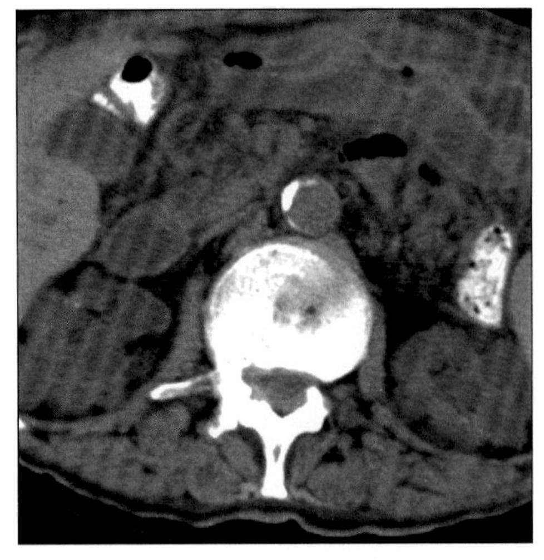

Axial NECT shows small multicystic kidneys.

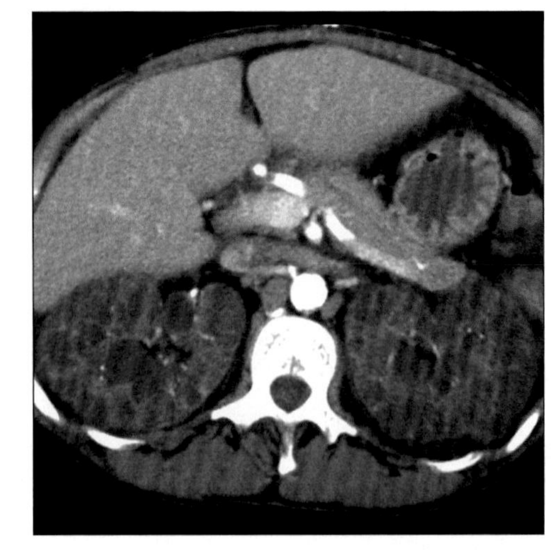

Axial CECT shows innumerable cysts in bilaterally enlarged kidneys in a patient who has been on dialysis for many years.

TERMINOLOGY

Abbreviations and Synonyms

- Acquired cystic kidney disease (ACKD) of uremia

Definitions

- Development of multiple bilateral renal cysts with end-stage renal disease (ESRD) on long term dialysis

IMAGING FINDINGS

General Features

- Best diagnostic clue: Small kidneys with multiple small cysts bilaterally
- Location: Renal cortex (predominantly) & medulla
- Size
  - Cysts: Usually smaller than 0.5 cm
  - Kidneys: Small to normal size, may enlarge as cysts expand
- Morphology: Multiple small bilateral renal cysts in shrunken kidneys
- Other general features
  - ACKD develops in patients with ESRD undergoing long term dialysis
  - Prevalence of ACKD in dialysis patients
    - After 1-3 years: 10-20%
    - After 3-5 years: 40-60%
    - After 5-10 years: More than 90%
  - Prevalence, size & number of cysts correlate with duration of dialysis, but not with renal disease
  - ACKD has similar prevalence in patients treated with hemodialysis & peritoneal dialysis
  - Several types of renal neoplasms occur in ESRD
    - Adenomas or carcinomas (extensive papillary epithelial hyperplasia in renal cysts of ACKD)
    - Histologically 50% nonpapillary & 50% papillary
  - Approximately 7% of ACKD patients develop renal cell carcinoma of low grade malignancy
  - 86% of RCCs occurring in patients undergoing dialysis affect those with ACKD

CT Findings

- Bilateral small renal cysts in small to normal kidneys
- ± Hemorrhage within cysts may be seen
- Larger cysts may occur
  - When present, mimics polycystic kidney disease
- Calcification of cyst walls may occur
- Focal hypervascular areas suggest associated renal cell carcinoma
  - Small tumors are the rule

MR Findings

- Bilateral multiple small cysts in small-normal kidneys
  - T1WI: Hypointense
  - T2WI: Hyperintense
  - T1 C+: Nonenhancing

DDx: Multiple Renal "Cysts"

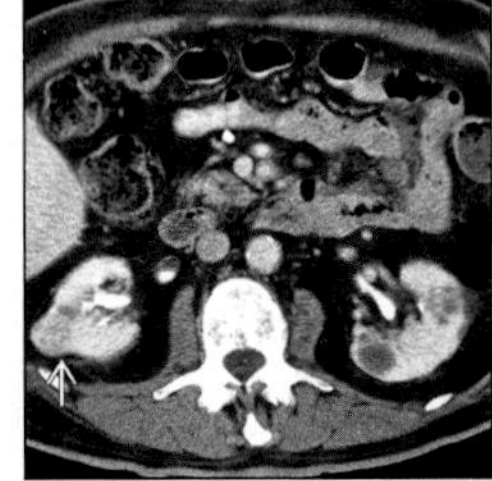

Renal Cysts + RCC

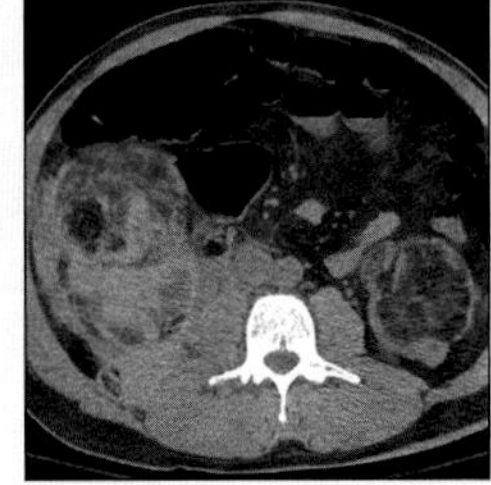

Tuberous Sclerosis

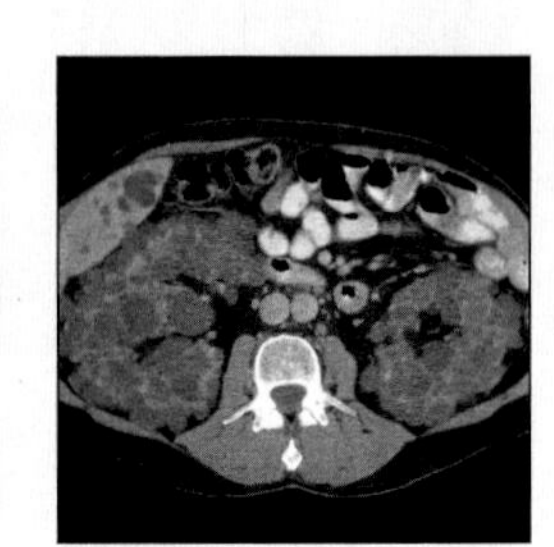

AD Polycystic

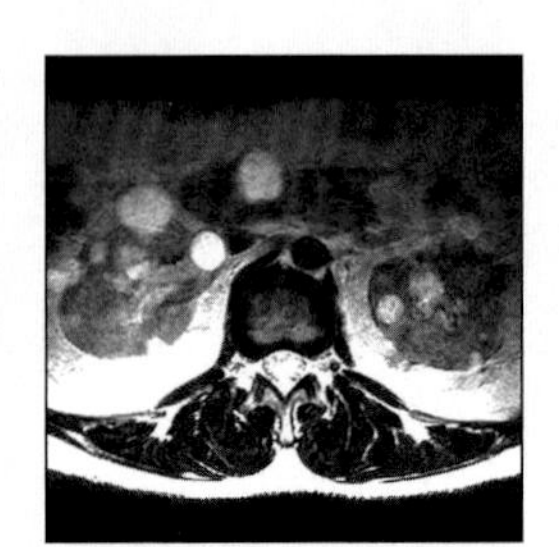

von Hippel Lindau

ACQUIRED CYSTIC DISEASE OF UREMIA

Key Facts

Terminology
- Acquired cystic kidney disease (ACKD) of uremia
- Development of multiple bilateral renal cysts with end-stage renal disease (ESRD) on long term dialysis

Imaging Findings
- Best diagnostic clue: Small kidneys with multiple small cysts bilaterally
- Location: Renal cortex (predominantly) & medulla
- Cysts: Usually smaller than 0.5 cm
- ± Hemorrhage within cysts may be seen
- Calcification of cyst walls may occur
- Focal hypervascular areas suggest associated renal cell carcinoma
- RCC distinguished from cysts by their lower signal intensity on T2WI images

Top Differential Diagnoses
- Multiple simple renal cysts
- Tuberous sclerosis (TS)
- Adult polycystic kidney disease (ADPKD)
- Medullary cystic disease
- von Hippel-Lindau disease

Diagnostic Checklist
- Check history & rule out hereditary cystic diseases
- CT or US screening of asymptomatic patients for RCC & ACKD at the end of 3rd year of dialysis
- Annual screening of native kidneys for RCC in transplant recipients
- Bilateral small cysts within small to normal kidneys
- Differentiate hemorrhagic cyst from a small RCC
- Enhancement: Hemorrhagic cyst (no) ; RCC (yes)

- Hemorrhagic cysts
  - T1 & T2WI: Variable signal (age of hemorrhage)
  - T1 C+: No enhancement
- Renal cell carcinoma (RCC)
  - RCC distinguished from cysts by their lower signal intensity on T2WI images
  - T1 C+: Significant enhancement
- Enhancement (T1 C+) is helpful in distinguishing hemorrhagic cysts from small renal cell carcinomas

Ultrasonographic Findings
- Real Time
  - Small to normal size kidneys with ↑ echogenicity
  - Corticomedullary differentiation is lost
  - Cysts are seen as sonolucent lesions
    - 3-5 cysts in patients with ESRD is diagnostic of ACKD
  - Hemorrhagic cyst
    - Anechoic with a smooth wall
    - Prominent posterior acoustic enhancement
    - Internal echoes may be seen due to prior hemorrhage
    - Differentiation from neoplasm often not possible
  - RCC
    - Solid or with mural nodules

Imaging Recommendations
- Ultrasound establishes initial diagnosis
- Enhanced CT or MR used to look for RCC
- Early phase enhanced CT best for detecting RCC
  - In this setting minimal renal enhancement seen

DIFFERENTIAL DIAGNOSIS

Multiple simple renal cysts
- Patients with normal renal function
- No history of dialysis present
- Kidneys are normal size

Tuberous sclerosis (TS)
- Multiple bilateral renal cysts
  - Cysts have typical histological features in TS
- Often associated with renal angiomyolipomas
  - Solid lesion, usually with fat
- Brain: Paraventricular calcifications

Adult polycystic kidney disease (ADPKD)
- Autosomal dominant polycystic kidney disease
- Cysts & kidneys are large in size
- Liver cysts may also be present

Medullary cystic disease
- Rare disease associated with progressive salt wasting nephropathy with renal insufficiency
- Cysts are predominantly in medulla
- Kidneys are almost invariably small in size
- Clinically, progressive renal failure in young patients
- Imaging
  - Renal cysts may be too small to be seen
  - Visible cysts occur only in renal medulla
    - Cysts in ACKD involve both cortex & medulla

von Hippel-Lindau disease
- Autosomal dominant multisystemic disorder
- Multiple renal cysts & cysts in other organs (e.g., pancreas)
- Kidneys are normal in size & functioning
- Hemangioblastomas: Retinal, cerebellar, spinal, medulla
- Bilateral renal cell carcinomas; pheochromocytomas

PATHOLOGY

General Features
- Etiology
  - Seen in patients with all etiologies of ESRD, particularly in patients who are dialysis dependent
  - Three most common causes of ESRD in USA
    - Nephrosclerosis
    - Diabetic nephropathy
    - Chronic glomerulonephritis
  - Incidence, number & size of cysts increase in proportion to duration of dialysis
  - Pathophysiology: Theories of ACKD
    - Secondary to obstruction of tubules by fibrosis or oxalate crystals

- Altered compliance of tubular basement membrane
- Toxicity from circulating metabolites (endogenous/exogenous toxins, growth factors)
- Vascular insufficiency
- Pathogenesis of cyst formation
  - Most cysts derive from proximal renal tubules
  - Begins with proliferation of epithelial cells
  - Marked ↑ in surface area of affected tubule wall
  - Originates as mural diverticula of nephrons
  - Early stage, cyst retains its connection with tubule & glomerular filtrate enters the enlarging cyst
  - Later stage, cyst loses its connection with parent tubule & becomes an isolated fluid-filled sac
- Epidemiology
  - More than 90% of cases after 5-10 years of dialysis
  - Up to 7% of ACKD patients develop RCC
- Associated abnormalities
  - Renal cell carcinoma in 7% of ACKD
  - Small papillary, tubular or solid clear cell adenomas

Gross Pathologic & Surgical Features

- Contracted small size kidneys + multiple small cysts
  - Cysts: Renal cortex more than medulla
- ± Associated renal tumors: Renal cell carcinoma

Microscopic Features

- Typical Cysts: Lined with a single layer of flattened or cuboidal epithelial cells
- Atypical cysts: Multilayered epithelial lining
- Papillary projections from cyst walls often seen
- Papillary lining cells may exhibit nuclear atypia
- Cyst content: Usually clear fluid; few hemorrhagic
- Crystal deposition: Prominent feature in cyst walls, cyst lumina & renal interstitium

CLINICAL ISSUES

Presentation

- Most common signs/symptoms
  - Most patients with ACKD are asymptomatic
  - Symptomatic, if they develop complications
    - Flank pain, fever, hematuria
    - Oliguria or anuria
- Clinical profile: Patient with history of chronic renal failure undergoing long term dialysis
- Lab data: ↑ Blood urea nitrogen & creatinine; ↓ GFR
- Diagnosis: Bilateral, with at least 4-5 cysts per kidney in ESRD patient on long term dialysis

Demographics

- Age: Any age group, but more common in older age
- Gender: M > F
- Ethnicity: African Americans > Caucasians

Natural History & Prognosis

- Complications
  - Intra- and pericystic bleeding
  - Microhematuria, cyst infection, renal calculi
  - Rupture with retroperitoneal hemorrhage
  - ↑ Risk of renal neoplasms particularly RCC
    - 7 times more common in men than in women
- Prognosis
  - In absence of RCC, prognosis relates to course of their renal failure
  - Mortality rate in dialysis-treated population with ESRD in US is high relative to general population
    - Average life expectancy in 30-49 yrs: 5-10 yrs
    - Average life expectancy in > 50 yrs: 1.5-6 yrs

Treatment

- If renal cell carcinoma detected, nephrectomy is done using laparoscopic flank approach
- Renal transplantation

DIAGNOSTIC CHECKLIST

Consider

- Check history & rule out hereditary cystic diseases
- CT or US screening of asymptomatic patients for RCC & ACKD at the end of 3rd year of dialysis
- Annual screening of native kidneys for RCC in transplant recipients

Image Interpretation Pearls

- Bilateral small cysts within small to normal kidneys
- Differentiate hemorrhagic cyst from a small RCC
- Enhancement: Hemorrhagic cyst (no); RCC (yes)

SELECTED REFERENCES

1. Takebayashi S et al: Using helical CT to evaluate renal cell carcinoma in patients undergoing hemodialysis: value of early enhanced images. AJR Am J Roentgenol. 172(2):429-33, 1999
2. Levine E: Acquired cystic kidney disease. Radiol Clin North Am. 34(5):947-64, 1996
3. Levine LA et al: Acquired cystic disease and renal adenocarcinoma following renal transplantation. J Urol. 151(1):129-32, 1994
4. Levine E et al: Natural history of acquired renal cystic disease in dialysis patients: a prospective longitudinal CT study. AJR Am J Roentgenol. 156(3):501-6, 1991
5. Ishikawa I et al: Ten-year prospective study on the development of renal cell carcinoma in dialysis patients. Am J Kidney Dis. 16(5):452-8, 1990
6. Matson MA et al: Acquired cystic kidney disease: occurrence, prevalence, and renal cancers. Medicine (Baltimore). 69(4):217-26, 1990
7. Miller LR et al: Acquired renal cystic disease in end-stage renal disease: an autopsy study of 155 cases. Am J Nephrol. 9(4):322-8, 1989
8. Taylor AJ et al: Renal imaging in long-term dialysis patients: a comparison of CT and sonography. AJR Am J Roentgenol. 153(4):765-7, 1989
9. Soffer O et al: CT findings in complications of acquired renal cystic disease. J Comput Assist Tomogr. 11(5):905-8, 1987
10. Ishikawa I: Uremic acquired cystic disease of kidney. Urology. 26(2):101-8, 1985
11. Levine E et al: CT of acquired cystic kidney disease and renal tumors in long-term dialysis patients. AJR Am J Roentgenol. 142(1):125-31, 1984

IMAGE GALLERY

Typical

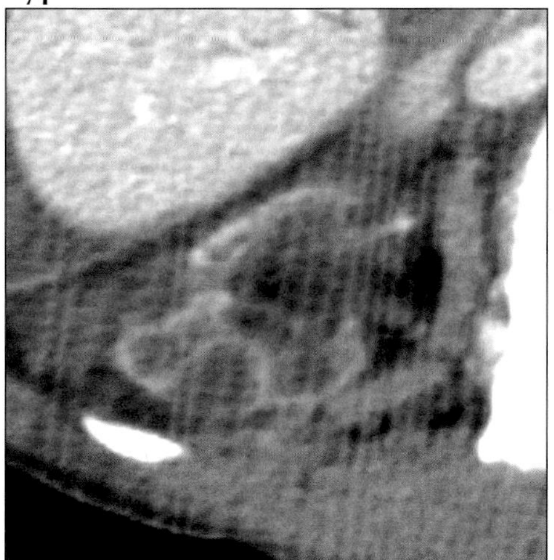

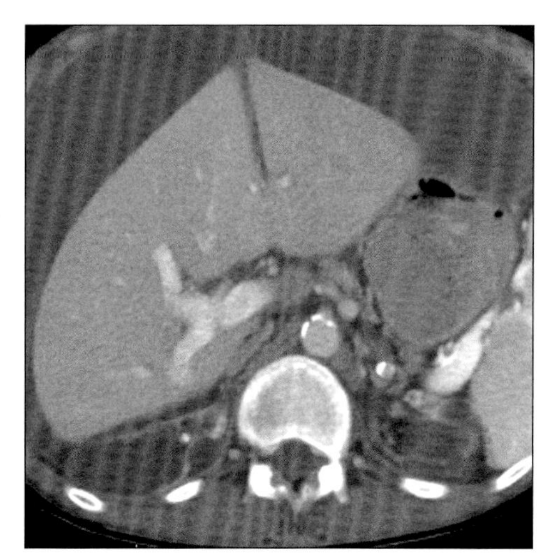

(Left) *Axial CECT shows a small, perfused but nonfunctional kidney with innumerable cysts.* ***(Right)*** *Axial CECT shows bilateral small, multicystic kidneys and ascites due to peritoneal dialysis-induced peritonitis.*

Typical

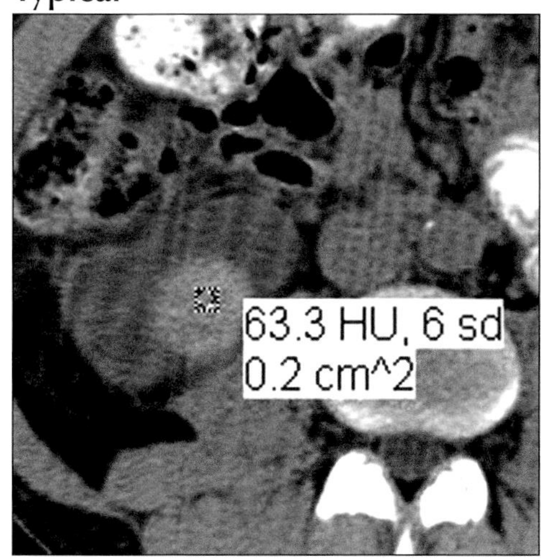

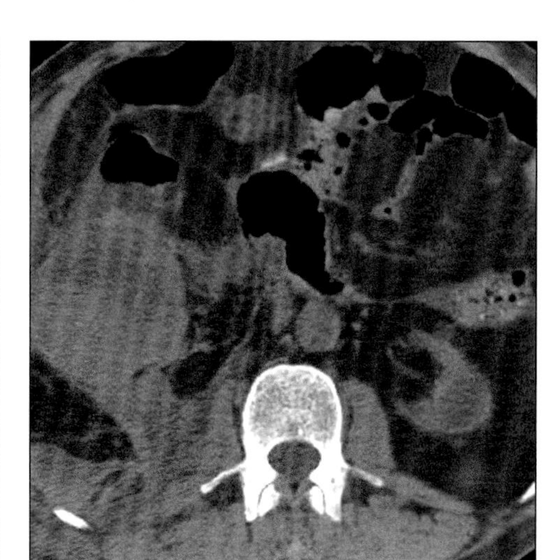

(Left) *Axial NECT shows high density, representing spontaneous hemorrhage within one of many renal cysts due to ACKD.* ***(Right)*** *Axial NECT shows bilateral ACKD with spontaneous retroperitoneal hemorrhage from right kidney.*

Typical

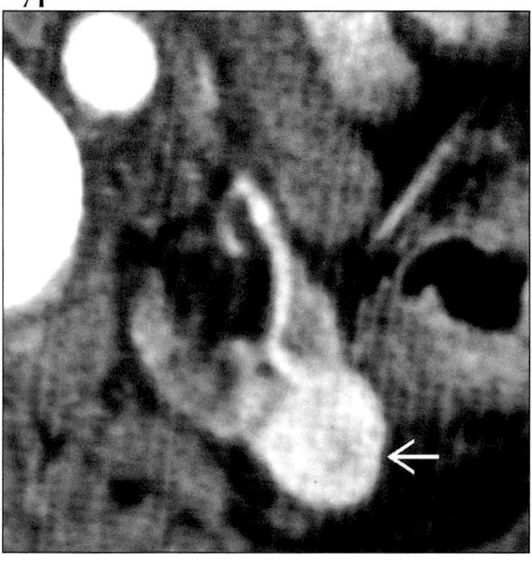

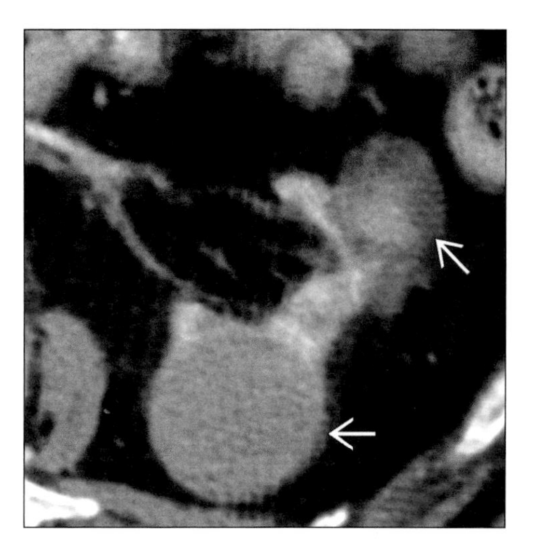

(Left) *Axial CECT shows an enhancing mass (arrow) representing renal cell carcinoma in a patient with ACKD.* ***(Right)*** *Axial CECT shows two enhancing masses (arrows), renal cell carcinomas in a patient with ACKD.*

MEDULLARY SPONGE KIDNEY

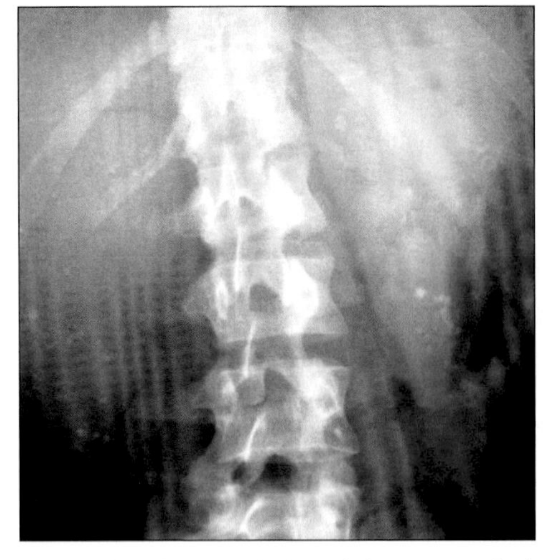

Radiograph shows multiple calcifications in both kidneys.

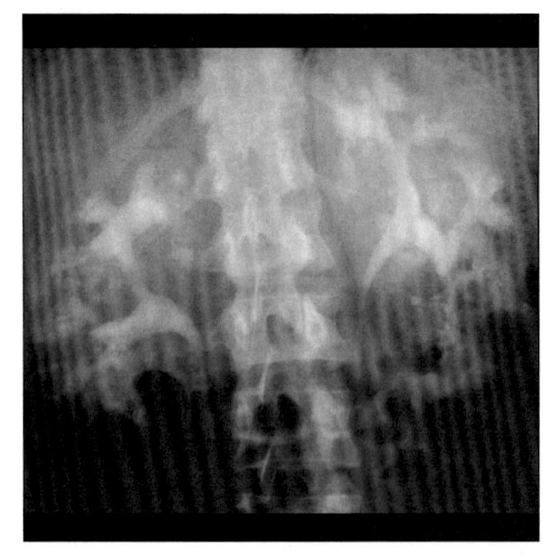

Urogram shows dilated collecting ducts within papillae bilaterally and a "bouquet of flowers" appearance caused by ectatic ducts and calculi.

TERMINOLOGY

Abbreviations and Synonyms

- Medullary sponge kidney (MSK); renal tubular ectasia; cystic disease of the renal pyramids; precalyceal canalicular ectasia; cystic dilatation of the renal tubules

Definitions

- Multiple cystic dilations in collecting ducts of medulla

IMAGING FINDINGS

General Features

- Best diagnostic clue: "Paint brush" appearance of dilated opacified pyramid tubules on IVP

Radiographic Findings

- Radiography: ± Medullary nephrocalcinosis (calcifications within medulla) or urolithiasis
- IVP
  - Bilateral (75%); unilateral (25%); can be in single pyramid or segment
  - Mild ductal ectasia
    - "Paint brush" appearance: Discrete linear densities 3-4 in ≥ 1 papillae (benign tubular ectasia)
    - Collecting ducts wider than 200-300 μm in diameter (normal)
  - Moderate ductal ectasia
    - More prominent linear radiodense striae
    - Clusters of small rounded opacities in the papillae (cystic dilatation of the collecting ducts)
    - ± Enlarged papillae and splayed calyceal cups
    - "Bouquet of flowers" appearance: Ectatic collecting ducts filled with calcification
    - Medullary calcifications: Large & dense
  - Advanced ductal ectasia
    - Gross deformity of the papillae with beaded or striated cavities
    - Distortion of calyces: Broad, shallow or widely cupped
    - Numerous and large calcifications
  - ± Nephrolithiasis
  - Obstruction, calyceal distortion or destruction, urinary tract infection due to calculi complications
    - Early films: No visualization of affected kidney
    - Delayed films: Hydronephrosis or contrast-filled areas of destruction
  - Evaluation of adjacent calyx for tumor (e.g., transitional cell carcinoma) or for obstructing calculi
  - Complications ↓ visualization (↓ renal function)
- Retrograde Pyelography
  - No filling or poor filling of dilated collecting ducts; in contrast injection that is not forceful
  - Tips of papillary ducts are not prone to reflux, regardless of disease severity; same as normal

DDx: Nephrocalcinosis+/- Pelvo-Calyceal Abnormality

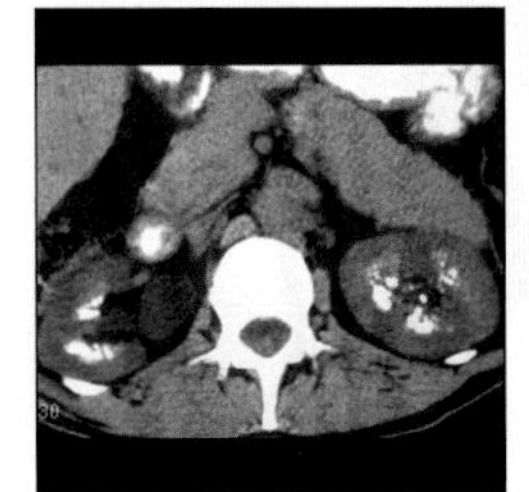

Med. Nephrocalcinosis

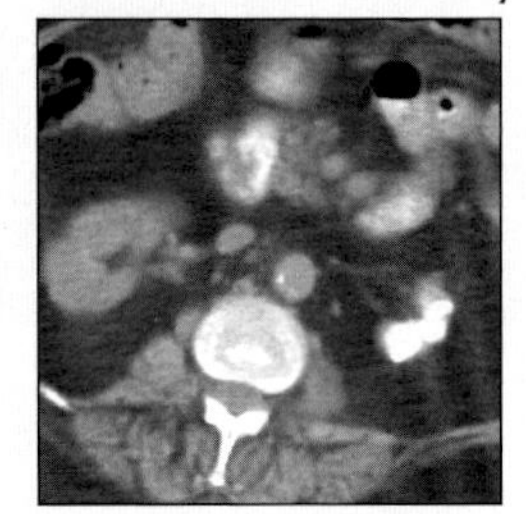

Renal Tuberculosis

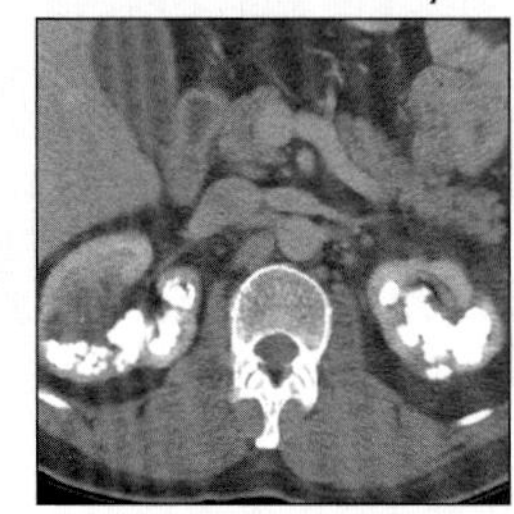

Oxalosis (NECT)

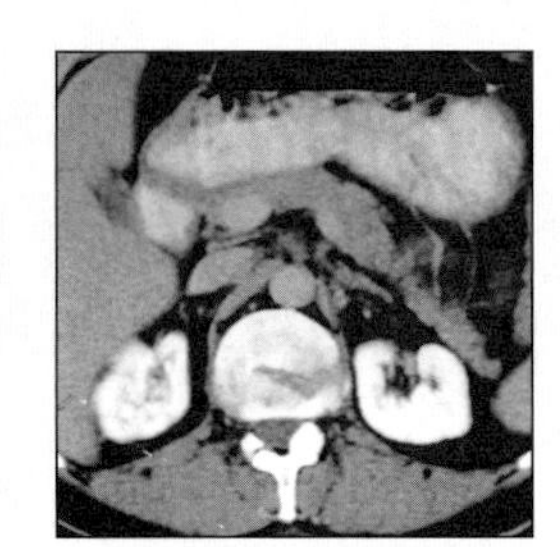

Oxalosis (NECT)

MEDULLARY SPONGE KIDNEY

Key Facts

Terminology
- Multiple cystic dilations in collecting ducts of medulla

Imaging Findings
- Best diagnostic clue: "Paint brush" appearance of dilated opacified pyramid tubules on IVP
- Bilateral (75%); unilateral (25%); can be in single pyramid or segment
- "Paint brush" appearance: Discrete linear densities 3-4 in ≥ 1 papillae (benign tubular ectasia)
- ± Medullary nephrocalcinosis and urolithiasis

Top Differential Diagnoses
- Hypercalcemia
- Acute cortical necrosis
- Renal papillary necrosis
- Renal tubular acidosis
- Tuberculosis
- Oxalosis
- Papillary Blush

Pathology
- Predisposing to urolithiasis (urinary stasis)

Clinical Issues
- Asymptomatic

Diagnostic Checklist
- Tubular ectasia can be caused by a adjacent tumor; need to evaluate
- Bilateral, "paint brush" appearance; medullary calcifications

 - Differentiate cysts of medullary sponge kidney from cysts of other medullary cavities (e.g., papillary necrosis or tuberculosis)
- Nephrotomography
  - Before contrast: Multiple areas of calcification
  - After contrast: Rounded opacities in medulla

CT Findings
- NECT
  - ± Medullary nephrocalcinosis and urolithiasis
  - ± Hydronephrosis and hydroureter (obstruction)
- CECT
  - "Paint" brush appearance: Retention of contrast within dilated tubules in pyramids
  - ± Medullary nephrocalcinosis and urolithiasis: Localizes the calcifications
  - Severe disease: ± Extracaliceal contrast accumulation within papillae or abscess

MR Findings
- Insensitive in detecting calcium
- Decreased sensitivity in detecting tubular ectasia

Ultrasonographic Findings
- Real Time
  - Nonspecific; can demonstrate medullary cavities
  - ± Medullary nephrocalcinosis
    - Discrete hyperechoic foci with acoustic shadowing within the pyramids
    - Increased echogenicity between interlobar cortices, surrounding the tips and sides of the pyramids (tiny calcifications)

Imaging Recommendations
- Best imaging tool
  - IVP: Discrete linear papillary densities
  - CT: Medullary nephrocalcinosis and dilated tubules ("paint brush")

DIFFERENTIAL DIAGNOSIS

Hypercalcemia
- Examples: Milk-alkali syndrome, sarcoidosis, hyperparathyroidism
- ↑ Calcium uptake by gut or ↑ bone destruction
- Associated with nephrocalcinosis (calcification within the parenchyma, including the medulla and cortex)
- Calcifications of medulla without ectatic ducts or cysts
- Calcifications more diffuse

Acute cortical necrosis
- Acute renal failure secondary to ischemic necrosis of renal cortex (prolonged vascular spasm, microvascular injury or intravascular coagulation)
- Radiography: Thin cortical shells & tram lines caused by calcification
- CT: Absent opacification of the renal cortex
- Anuria with poor prognosis

Renal papillary necrosis
- Destruction of the apex of the pyramid → irregular cavitation & sinus formation between papilla & calyx
- Ring or triangular calcification of sloughed papilla
- Retrograde pyelography: Fill 1 or 2 irregular cavities

Renal tubular acidosis
- Failure to secrete hydrogen ion & form ammonium ion → "fixed base" sodium, potassium & calcium
- 50% have medullary nephrocalcinosis
- Calcifications more diffuse
- Vomiting, poor growth, weakness & bone discomfort
- Metabolic acidosis with urinary anion gap

Tuberculosis
- Reactive or secondary tuberculosis of the kidney
- Unilateral ulceration of papilla in areas that are initially small and irregular
- Tendency toward coalescence into large cavities, stricture formation & calcification in caseous necrosis

Oxalosis
- Secondary to ↑ urinary oxalate (> 40 mg/24 hrs)
- Three types

- Primary (hereditary, rare) → ↑ hepatic production
- Acquired (↑ oxalate ingestion)
- Absorptive or enteric (i.e., malabsorption, inflammatory bowel disease) → dehydration & ↑ uptake by gut
- ↑ Calcium oxalate calculi → medullary and cortical (5%) nephrocalcinosis and nephrolithiasis

Papillary blush

- Homogeneous with no individual tubules
- IVP: Cone-shaped density fades into cortex at base
- Dose related, less prominent on delayed film, accentuated by ureteral compression and no nephrocalcinosis

PATHOLOGY

General Features

- General path comments: No cysts in other organs including liver (unlike autosomal dominant or autosomal recessive polycystic kidney disease
- Genetics: Sporadic; occasionally hereditary (i.e., Caroli disease, hemihypertrophy, Beckwith-Wiedemann syndrome, Ehlers-Danlos syndrome, autosomal recessive polycystic kidney disease)
- Etiology
  - Unknown
  - Pathogenesis: Unknown; developmental/progressive degeneration of collecting tubules by ↑ in age
- Epidemiology
  - Prevalence estimated at 1 in 5,000-20,000
  - Incidence by urography: 0.5%
  - Incidence in nephrolithiasis patients: 2.6-21%
- Associated abnormalities
  - Predisposing to urolithiasis (urinary stasis)
  - 33-50% hypercalcemia (i.e., hyperparathyroidism)

Gross Pathologic & Surgical Features

- Cross-section of kidney appears as a sponge
- Multiple cystic cavities ranging from 1-7.5 mm in diameter in one or more pyramids
- Dilatation of the terminal portion of the collecting tubules and papillary Bellini's ducts
- Cysts confined to papillary portion of the pyramids
- Cysts contain yellow-brown fluid and desquamated cells or calcified material
- Cysts communicate proximally with collecting tubules and distally with papillary ducts or the calyx
- Occasionally have inter- or noncommunication
- Calculi within cysts and their walls; calcium oxalate ± calcium phosphate

Microscopic Features

- Minimally ectatic collecting tubules
  - Normal columnar epithelium in interpapillary & high cuboidal epithelium in proximal intramedullary regions
  - Normal appearing medullary interstitium
- Large or prominent cysts
  - Cell lining: Transitional, columnar, or stratified (rare) squamous epithelium
  - Interstitial fibrosis and inflammation

CLINICAL ISSUES

Presentation

- Most common signs/symptoms
  - Asymptomatic
  - Ureteral colic, fever, dysuria and flank pain
- Lab data
  - Renal function: ± Impaired acidification or concentration of urine, absorptive and renal leak hypercalciuria and distal renal tubular acidosis
  - Urinalysis: ± Hematuria, ± positive culture
  - Blood tests: ± Positive culture
- Diagnosis: Incidental findings by urography

Demographics

- Gender
  - Unknown
  - Males < females in disease morbidity

Natural History & Prognosis

- Number and size of renal calcifications ↑ with time
- Previously uninvolved papillae develops new cavities (less common)
- Complications: Infection, extensive calculus formation, renal insufficiency and failure
- Prognosis: Very good, patients have normal life span if without complications

Treatment

- Only treat complications, otherwise, no treatment
  - Infection: Antibiotics
  - ↓ Stone formation: Thiazide & alkali therapy
  - Stone removal: Extracorporeal lithotripsy and percutaneous nephrolithotomy

DIAGNOSTIC CHECKLIST

Consider

- Tubular ectasia can be caused by a adjacent tumor; need to evaluate

Image Interpretation Pearls

- Bilateral, "paint brush" appearance; medullary calcifications

SELECTED REFERENCES

1. Levine E et al: Current concepts and controversies in imaging of renal cystic diseases. Urol Clin North Am. 24(3):523-43, 1997
2. Ginalski JM et al: Medullary sponge kidney on axial computed tomography: comparison with excretory urography. Eur J Radiol. 12(2):104-7, 1991
3. Ginalski JM et al: Does medullary sponge kidney cause nephrolithiasis? AJR Am J Roentgenol. 155(2):299-302, 1990
4. Boag GS et al: CT visualization of medullary sponge kidney. Urol Radiol. 9(4):220-1, 1988
5. Parks JH et al: Calcium nephrolithiasis and medullary sponge kidney in women. N Engl J Med. 306(18):1088-91, 1982
6. Madewell JE et al: Radiologic-pathologic correlations in cystic disease of the kidney. Radiol Clin North Am. 17(2):261-79, 1979

IMAGE GALLERY

Typical

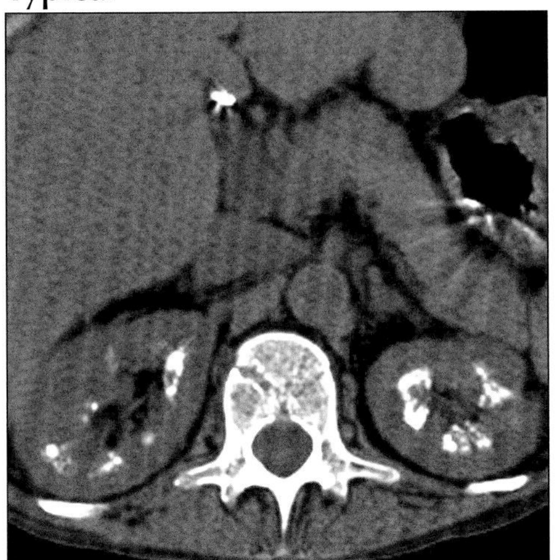

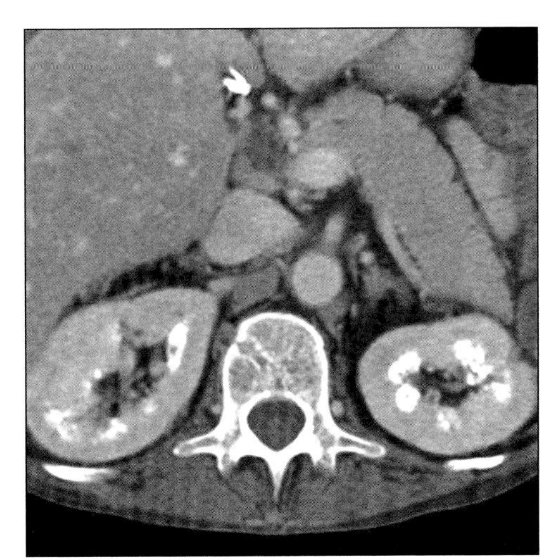

*(**Left**) Axial NECT shows medullary nephrocalcinosis and calculi. (**Right**) Axial CECT shows paintbrush appearance of dilated tubules and nephrocalcinosis.*

Typical

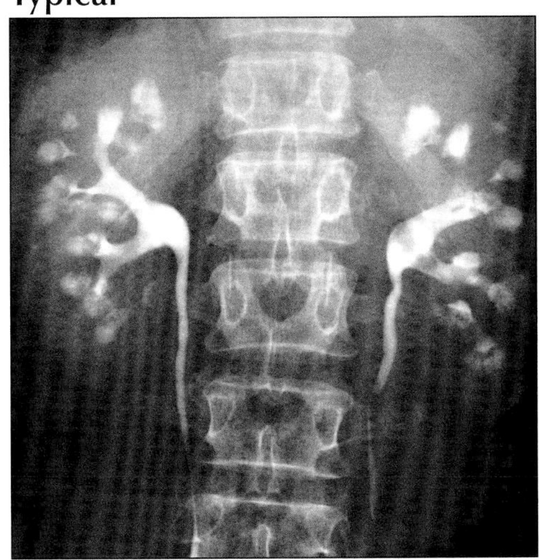

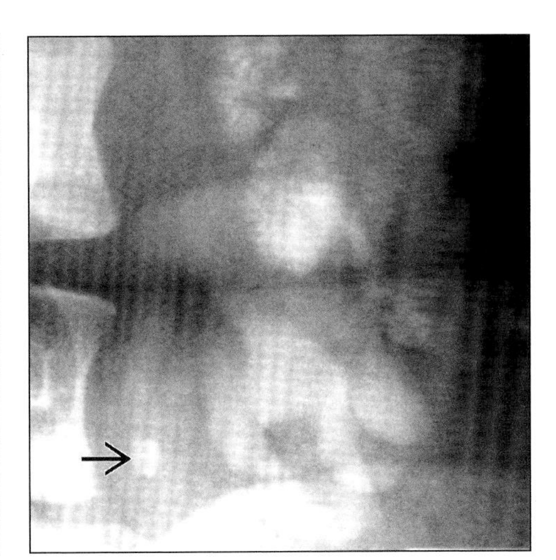

*(**Left**) Excretory urogram shows classic "paintbrush" appearance of tubular ectasia bilaterally. (**Right**) Oblique film from urogram shows severe ductal ectasia with beaded cavities replacing the renal papillae and wide calices. Note obstructing calculus (arrow).*

Typical

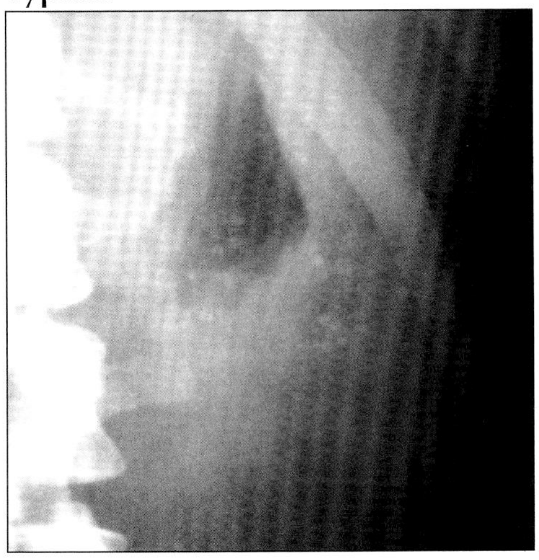

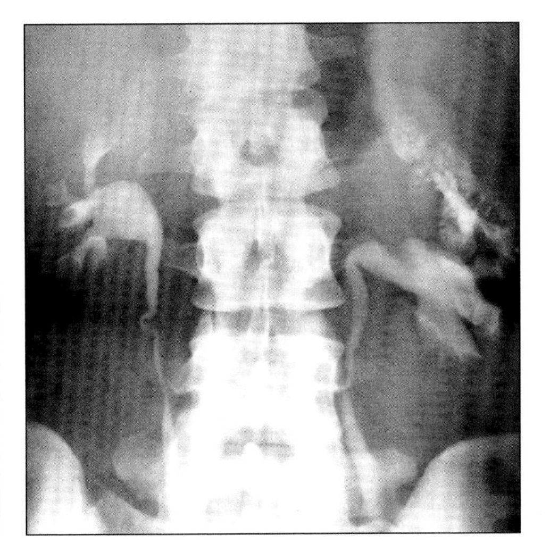

*(**Left**) Radiograph shows unilateral medullary sponge kidney with multiple medullary calculi. (**Right**) Urogram shows unilateral (left) medullary sponge kidney with dilated tubules containing innumerable calculi.*

RENAL ARTERY STENOSIS

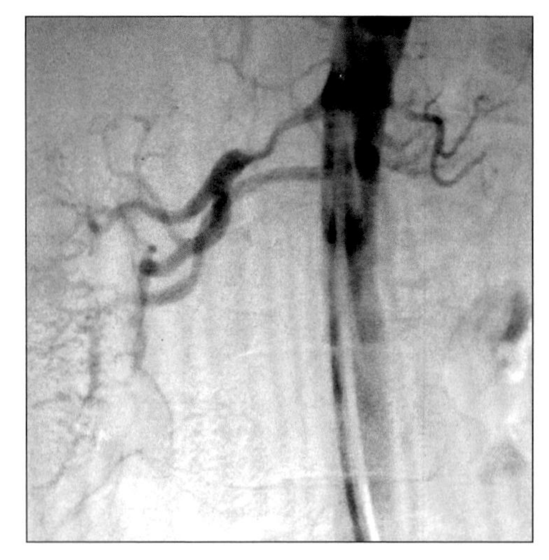

Angiography shows small left renal artery (and kidney). Right renal artery is concentrically stenotic just beyond its origin from atherosclerosis.

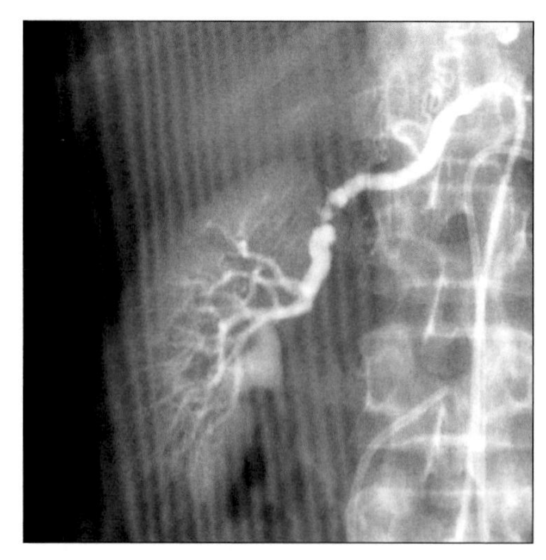

Angiogram shows beaded appearance of mid portion of renal artery due to fibromuscular dysplasia.

TERMINOLOGY

Abbreviations and Synonyms

- Renal artery stenosis (RAS)

Definitions

- Narrowing of renal arterial lumen

IMAGING FINDINGS

General Features

- Best diagnostic clue: Focal or segmental luminal narrowing of renal artery
- Location
  - RAS (atherosclerosis)
    - Renal artery ostium or within proximal 2 cm
  - RAS (fibromuscular dysplasia): Right > left
    - Mid or distal main RA ± intrarenal arteries
- Other general features
  - Most common cause of secondary hypertension
  - Renovascular hypertension (RVH)
    - Accounts 1-4% of all patients with hypertension
  - Atherosclerotic disease
    - Most common cause of renal artery stenosis (RAS)
    - Accounts for two-thirds of cases of RAS
    - Stenosis: Due to atherosclerotic plaque & Ca++
    - Location: Ostium or proximal 2 cm of renal artery
    - Majority of affected: Above 50 years old (M > F)
    - Bilateral: In 30-40% of atherosclerotic RAS cases
  - Fibromuscular dysplasia (FMD)
    - Second most common cause of RAS
    - Most common cause of RVH in children & young adults
    - Account for one-third of cases of RAS
    - Stenosis: Due to medial fibroplasia (> common)
    - Location: Mid or distal main renal artery (R > L)
    - Majority of affected: Young & middle-aged (F > M)
    - Bilateral: In 2/3 cases of fibromuscular dysplasia
  - Classification of fibromuscular dysplasia: Based on site of involvement & pathology within vessel wall
    - Medial fibroplasia (70-80%): Most common type
    - Medial or fibromuscular hyperplasia (2-3%)
    - Intimal hyperplasia (10-15%)
    - Perimedial or subadventitial fibroplasia (15%)

Ultrasonographic Findings

- Color Doppler
  - Direct signs: Criteria to diagnose RAS or occlusion
    - ↑ In peak systolic velocity 100-200 cm/sec (mild with < 50% stenosis)
    - ↑ In peak systolic velocity > 200 cm/sec (severe with 50-99% stenosis)
    - Renal-to-aortic ratio of peak systolic velocity > 3.5
    - Poststenotic turbulent flow + spectral broadening ± flow reversal

DDx: Focal Narrowing of Renal Artery

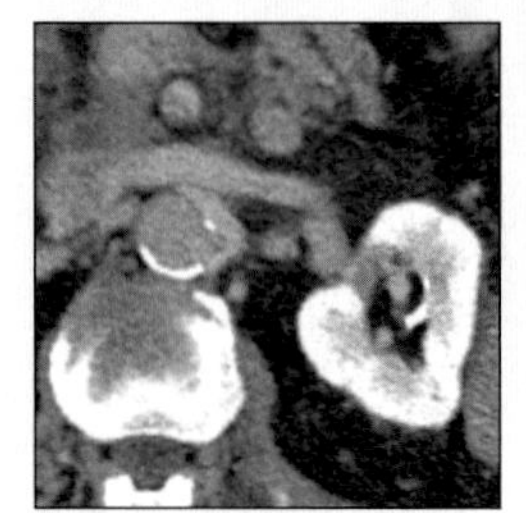

Aortic Dissection

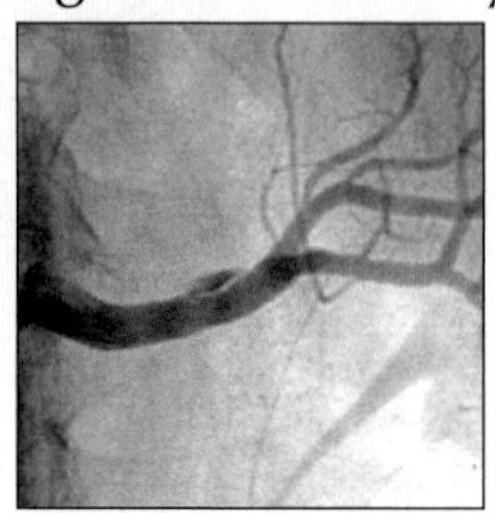

Traumatic RA Dissect.

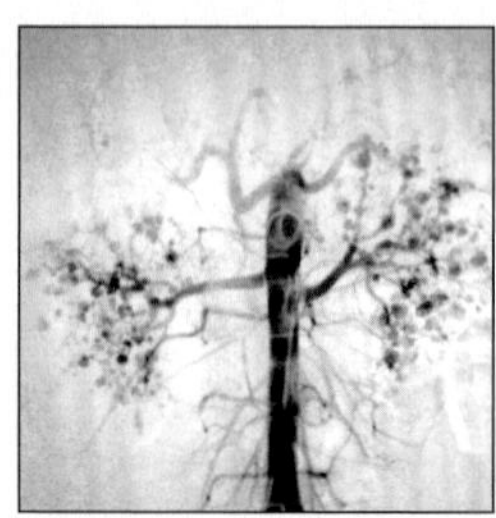

Polyarteritis Nodosa

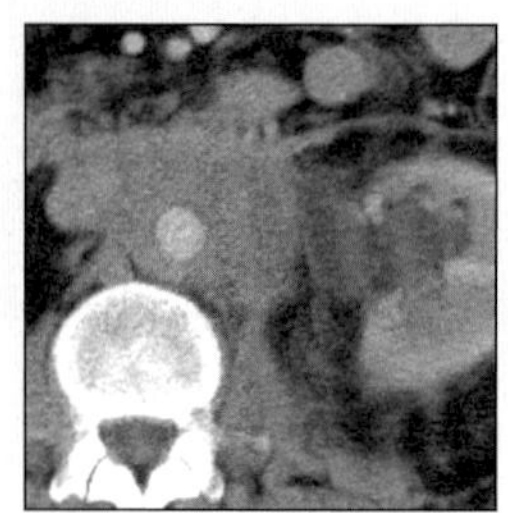

Lymphoma

RENAL ARTERY STENOSIS

Key Facts

Terminology

- Renal artery stenosis (RAS)

Imaging Findings

- ↑ In peak systolic velocity 100-200 cm/sec (mild with < 50% stenosis)
- ↑ In peak systolic velocity > 200 cm/sec (severe with 50-99% stenosis)
- Renal-to-aortic ratio of peak systolic velocity > 3.5
- Poststenotic turbulent flow + spectral broadening ± flow reversal
- Color imaging: Focal color aliasing (in systolic turbulence phase)
- Focal/segmental; eccentric/concentric stenosis
- Truncated (ostial stenosis); ± Ca++; uni-/bilateral
- Stenosis; "string-of-beads" pattern; focal aneurysms

Top Differential Diagnoses

- Arterial dissection
- Vasculitis
- Extrinsic compression

Pathology

- Congenital RAS
- Atherosclerosis (60-90%): > Common in elderly
- Fibromuscular dysplasia (10-30%): Young females

Diagnostic Checklist

- RAS in a patient with high or sudden ↑ in blood pressure + unexplained impairment of renal function
- Atherosclerotic RAS: Ostial or proximal 2 cm narrowing of renal artery (RA)
- FMD-RAS: Mid or distal main RA narrowing ± intrarenal arteries ("string-of-beads" appearance)

        - Color imaging: Focal color aliasing (in systolic turbulence phase)
    - Indirect signs: Distal manifestations of proximal RAS
        - Pulsus tardus-parvus: Dampened Doppler waveform during systole
        - Early systolic acceleration: < 20-30 cm/sec (best predictor)
        - Time of acceleration: > 0.05-0.08 seconds
        - Resistive index: < 5% on affected side compared to contralateral normal kidney
    - Administration of captopril
        - ↑ Detection when percentage of RAS is > 50%

Angiographic Findings

- CTA & MRA following bolus contrast injection
- Angiography: Catheter injection + digital subtraction angiography (DSA)
    - Atherosclerotic lesions
        - Focal/segmental; eccentric/concentric stenosis
        - Location: Ostium or proximal 2 cm of renal artery
        - Truncated (ostial stenosis); ± Ca++; uni-/bilateral
    - Medial fibroplasia
        - Stenosis; "string-of-beads" pattern; focal aneurysms
        - Location: Mid or distal long segment of main RA ± intrarenal arteries
        - Uni-/bilateral RAS
    - Medial hyperplasia or fibromuscular hyperplasia
        - Focal concentric narrowing of distal main RA + intrarenal branches
    - Secondary signs of RAS
        - Post-stenotic dilatation ± collateral vessels
        - Decreased or delayed or absent nephrogram
        - Normal or atrophic kidney

Imaging Recommendations

- Digital subtraction angiography (DSA)
- MR angiography (MRA)
    - Multiplanar, contrast-enhanced, 3D, maximum intensity projection, fast gradient echo sequences
- Color Doppler US (3.5-MHz curvilinear probe)
- CT angiography (CTA)
    - MIP; volume rendering techniques (complimentary)

DIFFERENTIAL DIAGNOSIS

Arterial dissection

- E.g., aortic dissection extending into renal artery
- Classification of aortic dissection based on DeBakey system
    - Type I: Extend from aortic root into descending aorta
    - Type II: Involve only ascending aorta
    - Type III: Extend from distal to left subclavian artery into abdomen
- Stanford classification of aortic dissection
    - Type A: Involves ascending aorta ± distal extent
        - Surgical treatment
    - Type B: Distal to origin of left subclavian artery
        - Medical treatment
- Aortic dissection + extension into renal artery (type III)
- Most frequently seen in elderly people (M > F)
- Pathogenesis: Aging → aortic media degeneration
- Predisposing conditions
    - Atherosclerosis, hypertension, pregnancy
    - Aortic stenosis, bicuspid aortic valve, coarctation, trauma
    - Marfan syndrome & other connective tissue disorders
- Digital subtraction angiography (DSA) or CT angiography
    - Irregular caliber of aortic lumen
    - False or occluded lumen + intimal flap
    - Thickened aortic wall
    - Narrowing or occlusion of renal artery due to a false lumen of dissection
        - May occlude or narrow renal artery at its origin
        - May extend into renal artery producing more distal narrowing
- MR imaging
    - Spin echo sequences: Demonstrate aortic dissection with intimal flap + extension up to renal arteries

Vasculitis

- E.g., polyarteritis nodosa; Takayasu arteritis
- Takayasu: Inflammation of medium-large arteries
- Most common in Asian women 15-45 y

- Etiology: Unknown
- Pathology: Fibrous thickening of wall of aorta narrowing orifices of major branches
- Common location: Orifices of aortic arch vessels & pulmonary arteries
  - Occasionally distal part of aorta & its branches
    - Renal artery involvement is particularly common
- May also cause atypical coarctation of distal aorta with narrowing of renal arteries
- Clinical presentation
  - Aortic arch vessels
    - Visual & neurological abnormalities
    - Weak pulses in upper extremities
  - Renal artery involved: Hypertension

Extrinsic compression

- Causes of extrinsic compression of renal artery
  - Aortic aneurysm: Suprarenal & infrarenal
  - Retroperitoneal tumors
  - Retroperitoneal fibrosis

PATHOLOGY

General Features

- Etiology
  - Congenital RAS
    - Due to intimal fibroplasia of distal two-third renal artery + branches
  - Atherosclerosis (60-90%): > Common in elderly
  - Fibromuscular dysplasia (10-30%): Young females
  - Other causes of RAS
    - Thromboembolic; arterial dissection; infrarenal aortic aneurysm; post radiation
    - Buerger & Takayasu disease; polyarteritis nodosa
    - Neurofibromatosis; retroperitoneal fibrosis
    - Pheochromocytoma; diaphragmatic crus
- Epidemiology: Seen in adults < 5% of HTN patients

Gross Pathologic & Surgical Features

- RAS (atherosclerosis)
  - Eccentric plaque in ostium or proximal RA
- RAS (fibromuscular dysplasia): Medial fibroplasia
  - "String-of-beads" appearance in mid or distal RA
- Mild RAS: Normal size kidney
- Moderate-severe RAS: Atrophic kidney
- Post-stenotic dilatation of main RA ± collateral vessels

Microscopic Features

- RAS (atherosclerosis)
  - White-yellow intimal plaque of vessel wall
  - Mixture of fatty, fibrous & calcium components
- RAS (fibromuscular dysplasia)
  - Focal muscular thinning of media + fibrous ridges

CLINICAL ISSUES

Presentation

- Most common signs/symptoms
  - Very high or sudden ↑ BP in a child, young adult or after age 50 years; epigastric or flank bruit
  - ↑ BP + unexplained impairment of renal function
  - Symptoms of increased intracranial pressure
    - Papilledema, headache, vomiting
- Lab data
  - Positive captopril test
    - Exaggerated ↑ in plasma renin activity (PRA)
  - ± Impaired renal function tests (based on stenosis)

Demographics

- Age
  - RAS (atherosclerosis): Adult age group
  - RAS (fibromuscular dysplasia): Young & middle-age
- Gender
  - RAS (atherosclerosis): M = F
  - RAS (fibromuscular dysplasia): M < F (1:3)

Natural History & Prognosis

- Complications
  - Severe HTN; CVA; azotemia; occlusion
  - Infarction; dissection; thrombosis
- Prognosis
  - Transluminal angioplasty
    - Success rate: Nonostial 80%; ostial lesions 25-30%
  - Surgical revascularization
    - Success rate: 80-90% for any lesion & location
    - FMD: ↓ Incidence of restenosis & improved GFR

Treatment

- Angiotensin converting enzyme (ACE) inhibitors
- Transluminal angioplasty
- Surgical revascularization (80-90% success rate)

DIAGNOSTIC CHECKLIST

Consider

- Rule out other causes of RAS especially dissection
- RAS in a patient with high or sudden ↑ in blood pressure + unexplained impairment of renal function

Image Interpretation Pearls

- Atherosclerotic RAS: Ostial or proximal 2 cm narrowing of renal artery (RA)
- FMD-RAS: Mid or distal main RA narrowing ± intrarenal arteries ("string-of-beads" appearance)

SELECTED REFERENCES

1. Sheth S et al: Multi-detector row CT of the kidneys and urinary tract: techniques and applications in the diagnosis of benign diseases. Radiographics. 24(2):e20, 2004
2. Urban BA et al: Three-dimensional volume-rendered CT angiography of the renal arteries and veins: normal anatomy, variants, and clinical applications. Radiographics. 21(2):373-86; questionnaire 549-55, 2001
3. Kawashima A et al: CT evaluation of renovascular disease. Radiographics. 20(5):1321-40, 2000
4. Johnson PT et al: Renal artery stenosis: CT angiography--comparison of real-time volume-rendering and maximum intensity projection algorithms. Radiology. 211(2):337-43, 1999
5. Gilfeather M et al: Renal artery stenosis: evaluation with conventional angiography versus gadolinium-enhanced MR angiography. Radiology. 210(2):367-72, 1999
6. Rubin GD et al: Helical CT angiography of renal artery stenosis. AJR Am J Roentgenol. 168(4):1109-11, 1997

IMAGE GALLERY

Typical

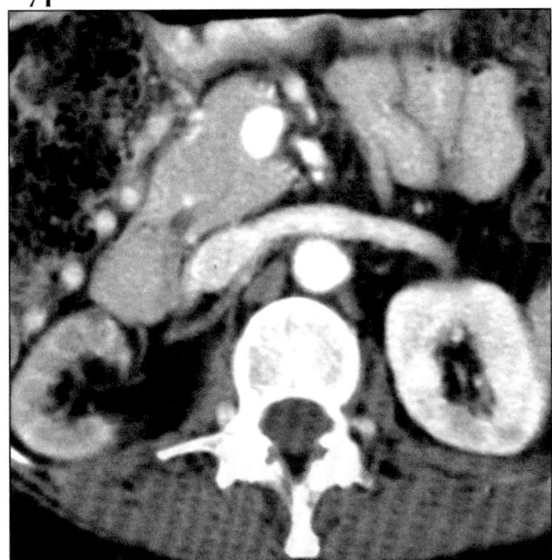

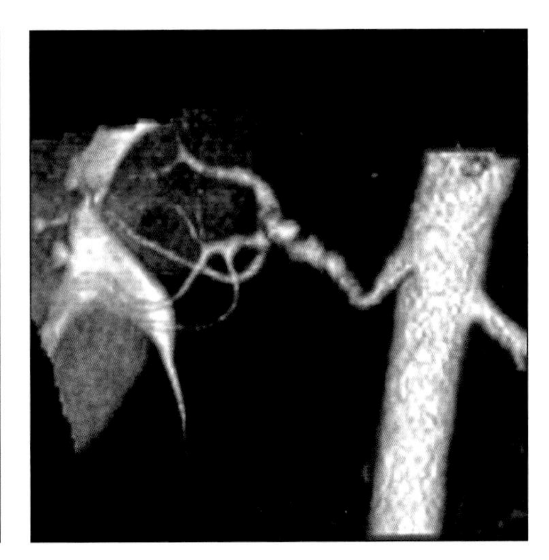

(Left) *Axial CECT in a 41 year old woman with worsening hypertension shows global atrophy of right kidney and a beaded renal artery due to fibromuscular dysplasia (FMD).* ***(Right)*** *CT Angiography shows beaded appearance of renal artery due to fibromuscular dysplasia.*

Typical

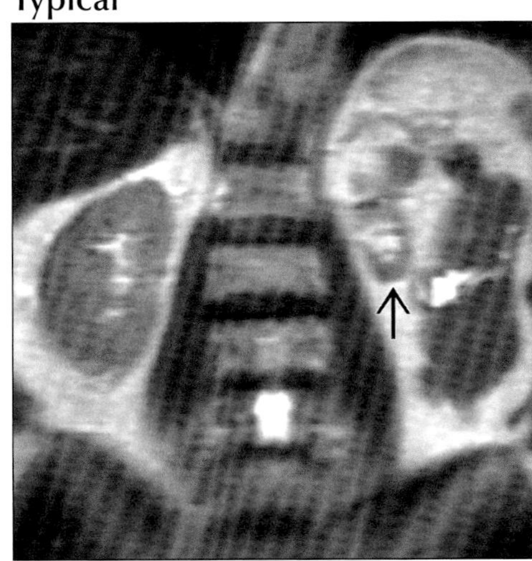

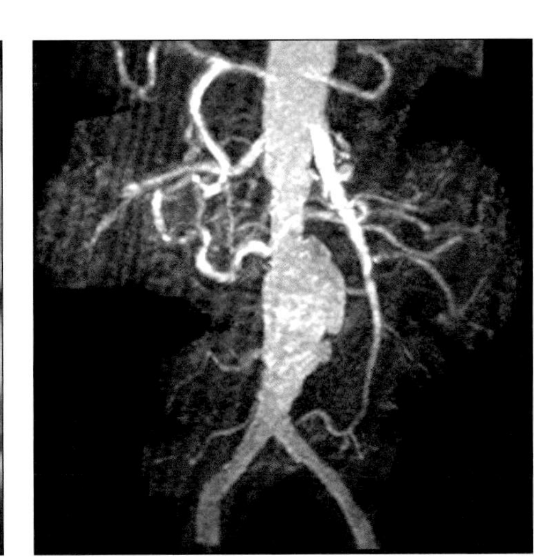

(Left) *Coronal T1WI MR (nonenhanced) in 70 year old man shows atrophic left kidney (arrow).* ***(Right)*** *MR Angiography shows abdominal aortic aneurysm, tiny left and stenotic right renal artery.*

Typical

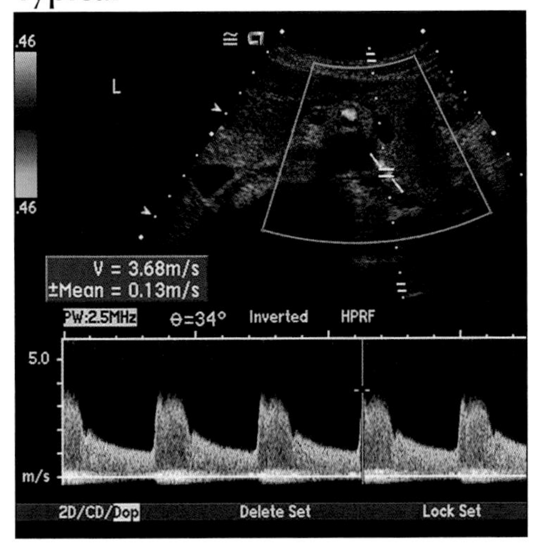

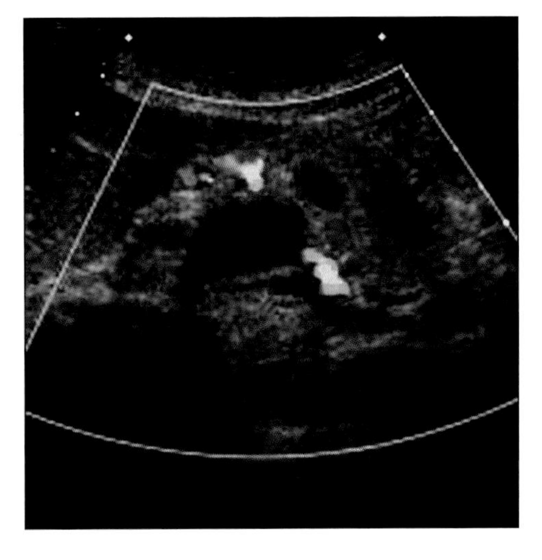

(Left) *Color Doppler sonography with gate over left renal artery shows high velocity flow (3.7 m/sec) indicating renal artery stenosis.* ***(Right)*** *Color Doppler sonography shows color aliasing within the left renal artery indicating turbulent, rapid flow, signs of renal artery stenosis.*

RENAL INFARCTION

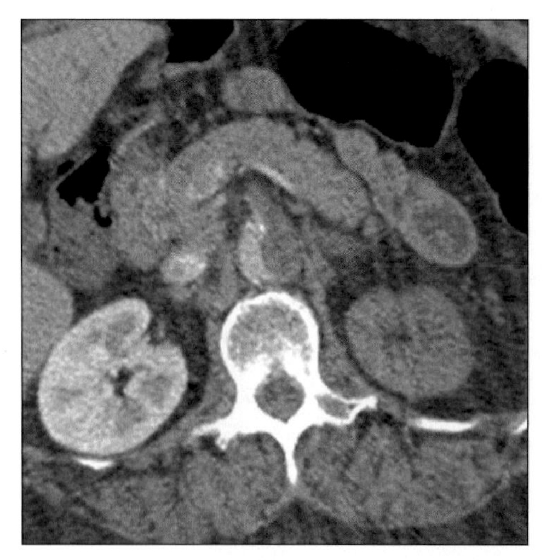

Axial CECT shows nonenhancement of left kidney due to aortic dissection occluding renal artery.

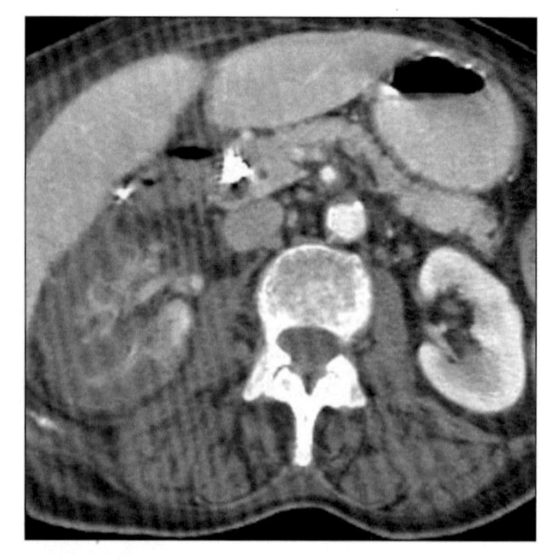

Axial CECT shows global nonenhancement of right kidney due to embolic occlusion of artery. Atrial fibrillation.

TERMINOLOGY

Definitions

- Localized or global area of ischemic necrosis in kidney, resulting most often from sudden occlusion of its arterial supply

IMAGING FINDINGS

General Features

- Best diagnostic clue: Nonenhancing wedge-shaped area + enhancing cortical rim
- Other general features
  - Major causes of renal infarction
    - Thrombosis, embolism, trauma
  - Classification based on onset
    - Acute, subacute & chronic renal infarction
  - Classification based on anatomy & vascular distribution
    - Focal: Segmental or subsegmental infarction (cortex ± medulla)
    - Global infarction
  - Focal traumatic segmental or subsegmental renal infarct (more common)
    - Categorized as grade I & II renal injuries
    - Thrombosis or laceration of a segmental renal arterial (RA) branch
    - Solitary or multiple, frequently associated with other renal injuries
    - Finally results in a renal scar
  - Global traumatic renal infarct (less common)
    - Categorized as grade III renal injury
    - Thrombosis, transection or avulsion of main RA

Radiographic Findings

- IVP
  - Segmental or subsegmental infarction
    - Focal absent or decreased nephrogram
  - Global infarction
    - Complete absence of nephrogram + excretion

CT Findings

- CT findings may be variable based on etiology of infarction
  - Embolic infarction: Multifocal & bilateral
  - Traumatic & thrombotic infarction: Segmental or global & unilateral
- Focal subsegmental infarction
  - Small, sharply demarcated, wedge-shaped area of decreased or poor contrast-enhancement
  - Base of wedge pointed towards renal capsule & apex towards hilum
- Focal segmental infarction
  - Sharply demarcated, dorsal or ventral segmental decreased enhancement area

DDx: Striated, Wedge or Global Nonenhancement

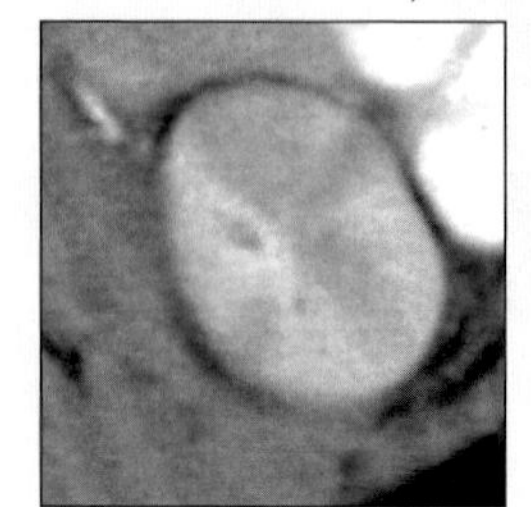

Pyelonephritis

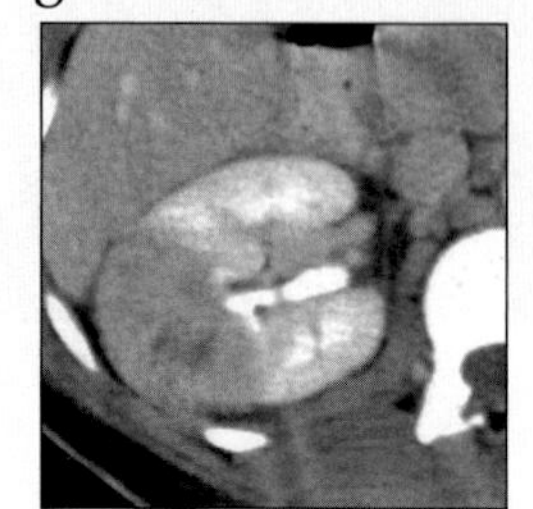

Pyelonephritis

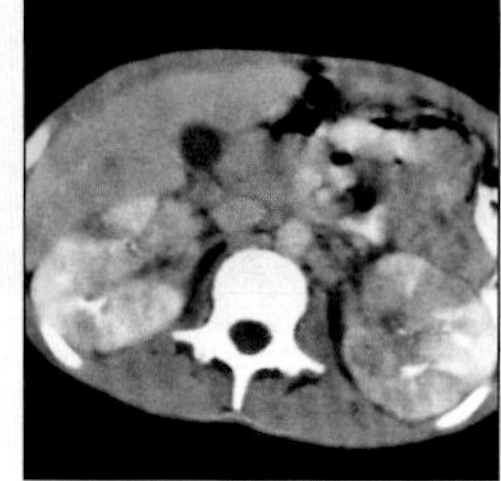

Polyarteritis Nodosa

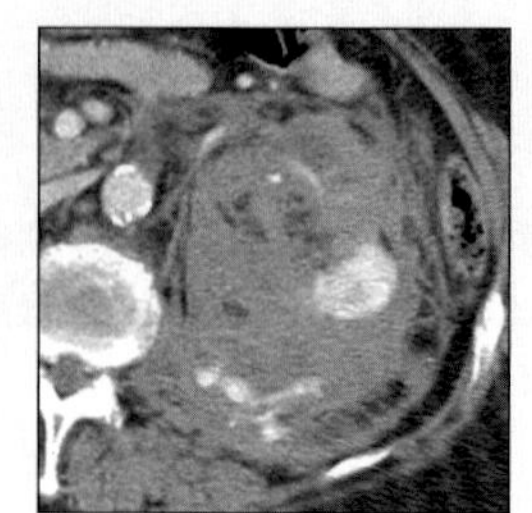

Renal Laceration

RENAL INFARCTION

Key Facts

Terminology

- Localized or global area of ischemic necrosis in kidney, resulting most often from sudden occlusion of its arterial supply

Imaging Findings

- Best diagnostic clue: Nonenhancing wedge-shaped area + enhancing cortical rim
- Total absence of renal enhancement + no excretion + no perinephric hematoma (renal artery thrombosis)
- Total absence of enhancement + large perinephric hematoma (renal artery avulsion)
- "Cortical rim" sign: Reliable sign of subacute infarction
- Small-sized kidney with smooth or irregular contour (chronic infarction)

Top Differential Diagnoses

- Pyelonephritis
- Vasculitis
- Renal trauma

Pathology

- Embolism: Cardiac (most common)
- Thrombosis
- Trauma

Diagnostic Checklist

- Correlate with history: Trauma, cardiac, aortic disease
- Pyelonephritis + acute infarction may have similar appearance
- Straight line demarcation + "cortical rim" sign favor infarction
- Perinephric stranding favors pyelonephritis

- A straight-line demarcation between normal enhancing & abnormal nonenhancing parenchyma
  - Strongly suggestive of ischemia
- Global infarction
  - Total absence of renal enhancement + no excretion + no perinephric hematoma (renal artery thrombosis)
    - Renal outline is preserved
  - Total absence of enhancement + large perinephric hematoma (renal artery avulsion)
  - ± Medullary striations: "Spoke-wheel" pattern enhancement (due to collateral circulation)
- Acute renal infarction
  - Normal or large-sized kidney; smooth contour; ± subcapsular fluid collection
  - Absent or decreased nephrogram
  - "Cortical rim" sign: Reliable sign of subacute infarction
    - Preserved capsular or subcapsular enhancement
    - Usually seen 6-8 hrs after infarction
    - Seen in 50% cases of infarction (due to intact collateral circulation)
- Chronic renal infarction
  - Small-sized kidney with smooth or irregular contour (chronic infarction)
  - Absent or diminished enhancement; no "cortical rim" sign

MR Findings

- T1WI: Low signal intensity lesion
- T2WI: Low signal intensity lesion
- T1 C+: Sharply demarcates nonenhanced infarcted area from densely enhancing noninfarcted portion

Ultrasonographic Findings

- Real Time: Findings of renal infarction are nonspecific
- Color Doppler: May show focal or global absence of blood flow in involved kidney

Angiographic Findings

- Conventional
  - Diagnosis of renal infarction: Confirmed at selective renal arteriography
    - Focal segmental infarction: Nonopacification of focal cortical area
    - Global infarction: Total absence of renal enhancement

Nuclear Medicine Findings

- SPECT imaging with Tc99m DMSA
  - Acute infarction: Photon deficient area

Imaging Recommendations

- Helical CECT; color Doppler sonography
- Selective renal angiography

DIFFERENTIAL DIAGNOSIS

Pyelonephritis

- Acute pyelonephritis
  - Cortical wedge-shaped or striated nephrogram
    - May simulate focal segmental or subsegmental infarction
    - "Cortical rim" sign favors infarction
  - Loss of normal corticomedullary differentiation
    - Corticomedullary differentiation is intact in focal infarction & is lost in global infarction
  - Renal enlargement, focal swelling
  - Thickening of Gerota fascia + perinephric stranding
  - Calyceal effacement, dilated renal pelvis & ureter which are not seen in acute infarction
- Chronic pyelonephritis
  - Small kidney with cortical scarring over dilated calices
  - Unilateral with compensatory hypertrophy of contralateral kidney
  - May simulate chronic global renal infarction

Vasculitis

- eg; Polyarteritis nodosa; SLE; scleroderma; drug abuse
- Wedge-shaped or striated nephrogram (usually bilateral & diffuse)
- Both vasculitis & infarction cause parenchymal scarring + capsular retraction (may be same pathology)
- Microaneurysmal dilatation of small vessels

RENAL INFARCTION

Renal trauma

- Best imaging clue
  - Irregular linear or segmental nonenhancing tissue & subcapsular or perinephric hematoma
- Lacerations: Irregular or linear hypodense areas
- Segmental infarction
  - Nonenhancing wedge-shaped areas
- "Shattered kidney"
  - Renal artery avulsion
    - Global infarction + perinephric hematoma
  - Renal artery thrombosis
    - Global infarction + no perinephric hematoma

PATHOLOGY

General Features

- Etiology
  - Embolism: Cardiac (most common)
    - Rheumatic, arrhythmias, MI, prosthetic valve, SBE
  - Thrombosis
    - Atherosclerosis, polyarteritis nodosa
    - Aneurysm or dissection (aorta/renal artery)
    - Sickle cell disease, thrombotic thrombocytopenic purpura, thromboangiitis obliterans
  - Trauma
    - Blunt or penetrating
    - Surgery, interventional procedures
- Associated abnormalities: Cardiac abnormalities, hypercoagulable state, aortic aneurysm or dissection

Gross Pathologic & Surgical Features

- Wedge shaped infarct (white or pale in color)
- Renal artery thrombus or traumatic avulsion
- Large or small sized kidney + smooth or irregular contour

Microscopic Features

- Focal or global renal ischemic changes
- Necrosis & scarring

CLINICAL ISSUES

Presentation

- Most common signs/symptoms
  - Asymptomatic, flank pain, tenderness (traumatic), hematuria
  - Hypertension in chronic infarction

Demographics

- Age: Any age group
- Gender: Equal in both males & females (M = F)

Natural History & Prognosis

- Complications
  - Necrosis, infection, abscess formation
- Prognosis
  - Focal infarction: Good
  - Global infarction: Poor

Treatment

- Medical
  - Antithrombolytics, anticoagulants, antihypertensives
- Surgery or angioplasty
  - Atherosclerotic renal artery stenosis
- Nephrectomy
  - Irreversible traumatic global infarction

DIAGNOSTIC CHECKLIST

Consider

- Correlate with history: Trauma, cardiac, aortic disease

Image Interpretation Pearls

- Pyelonephritis + acute infarction may have similar appearance
- Straight line demarcation + "cortical rim" sign favor infarction
- Perinephric stranding favors pyelonephritis

SELECTED REFERENCES

1. Suzer O et al: CT features of renal infarction. Eur J Radiol. 44(1):59-64, 2002
2. Kawashima A et al: Imaging evaluation of posttraumatic renal injuries. Abdom Imaging. 27(2):199-213, 2002
3. Kawashima A et al: Imaging of renal trauma: a comprehensive review. Radiographics. 21(3):557-74, 2001
4. Kawashima A et al: CT evaluation of renovascular disease. Radiographics. 20(5):1321-40, 2000
5. Carey HB et al: Bilateral renal infarction secondary to paradoxical embolism. Am J Kidney Dis. 34(4):752-5, 1999
6. Kawashima A et al: CT of renal inflammatory disease. Radiographics. 17(4):851-66; discussion 867-8, 1997
7. Dalla-Palma L et al: Delayed CT in acute renal infection. Semin Ultrasound CT MR. 18(2):122-8, 1997
8. Nunez D Jr et al: Traumatic occlusion of the renal artery: helical CT diagnosis. AJR Am J Roentgenol. 167(3):777-80, 1996
9. Kamel IR et al: Assessment of the cortical rim sign in posttraumatic renal infarction. J Comput Assist Tomogr. 20(5):803-6, 1996
10. Krinsky G: Unenhanced helical CT in patients with acute flank pain and renal infarction: the need for contrast material in selected cases. AJR Am J Roentgenol. 167(1):282-3, 1996
11. Saunders HS et al: The CT nephrogram: implications for evaluation of urinary tract disease. Radiographics. 15(5):1069-85; discussion 1086-8, 1995
12. Fanney DR et al: CT in the diagnosis of renal trauma. Radiographics. 10(1):29-40, 1990
13. Bankoff MS et al: Computed tomography differentiation of pyelonephritis and renal infarction. J Comput Tomogr. 8(3):239-43, 1984
14. Wong WS et al: Renal infarction: CT diagnosis and correlation between CT findings and etiologies. Radiology. 150(1):201-5, 1984
15. Haaga JR et al: CT appearance of renal infarct. J Comput Assist Tomogr. 4(2):246-7, 1980
16. Sant GR et al: Computed tomography in evaluation of blunt renal trauma. Potential for misdiagnosis of renal infarction. Urol Int. 43(6):321-3, 1988

IMAGE GALLERY

Typical

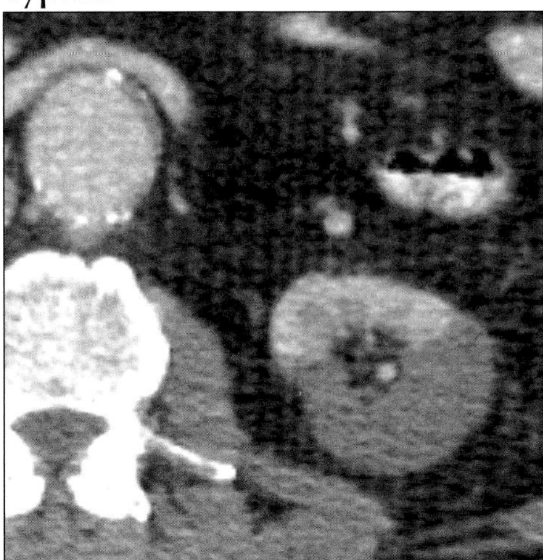

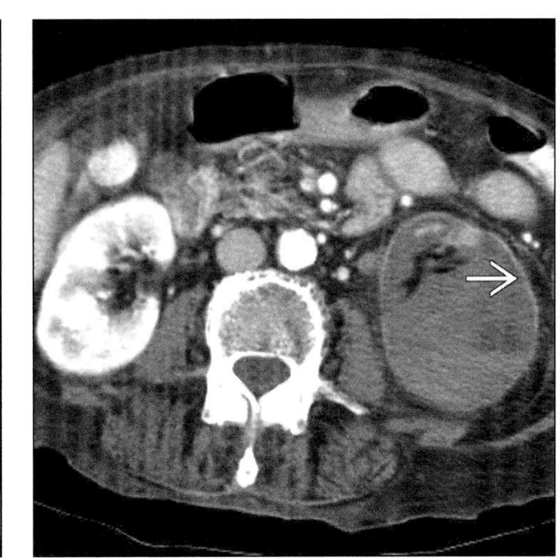

(Left) *Axial CECT shows straight line demarcation between normal ventral kidney and infarcted dorsal half.* ***(Right)*** *Axial CECT shows global infarction of left kidney due to emboli from prosthetic mitral valve. Note "cortical rim" sign (arrow).*

Typical

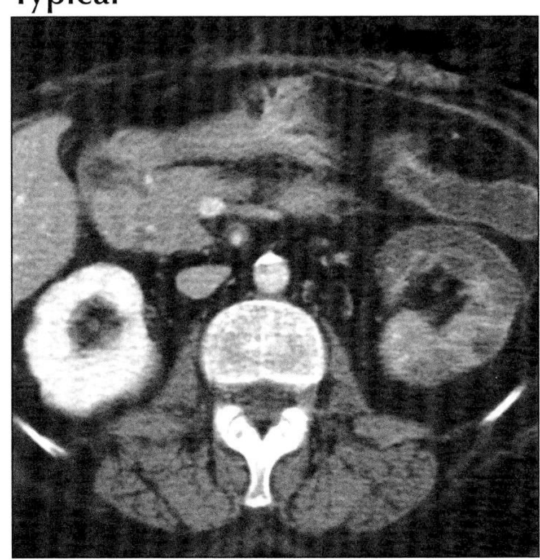

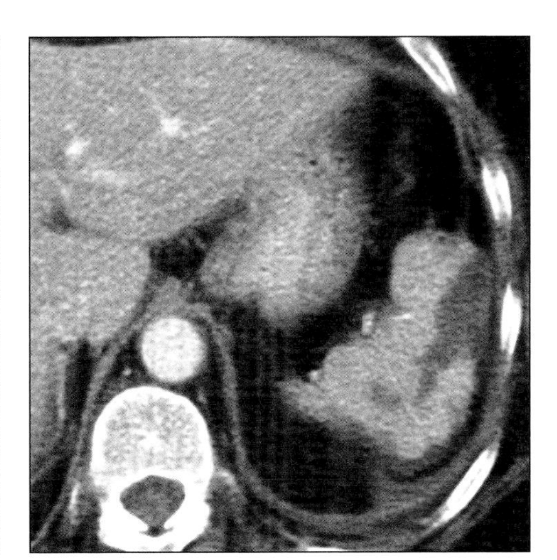

(Left) *Axial CECT in a patient with recent myocardial infarct shows global infarction of left kidney.* ***(Right)*** *Axial CECT in a patient with recent myocardial infarction shows multiple splenic infarctions as well.*

Typical

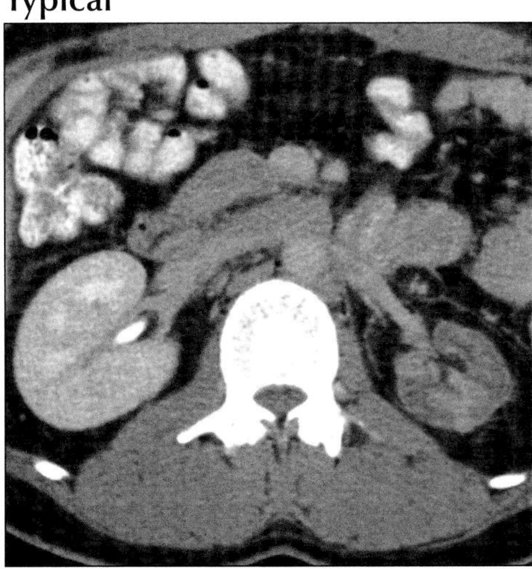

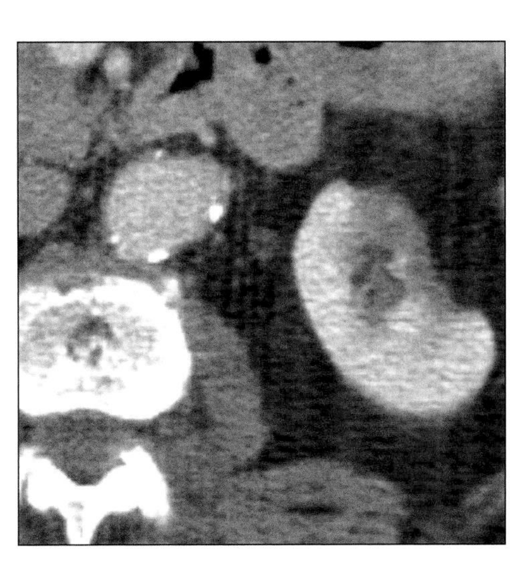

(Left) *Axial CECT shows global atrophy and minimal function of left kidney due to chronic ischemia.* ***(Right)*** *Axial CECT shows renal cortical loss due to chronic focal infarction.*

RENAL VEIN THROMBOSIS

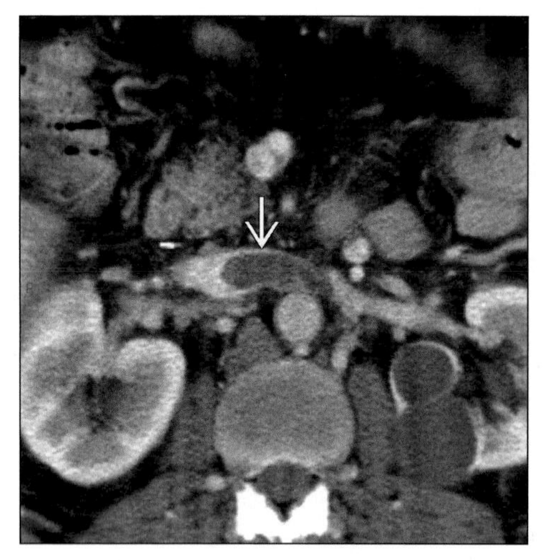

Axial CECT shows thrombus as a filling defect (arrow) in the left renal vein.

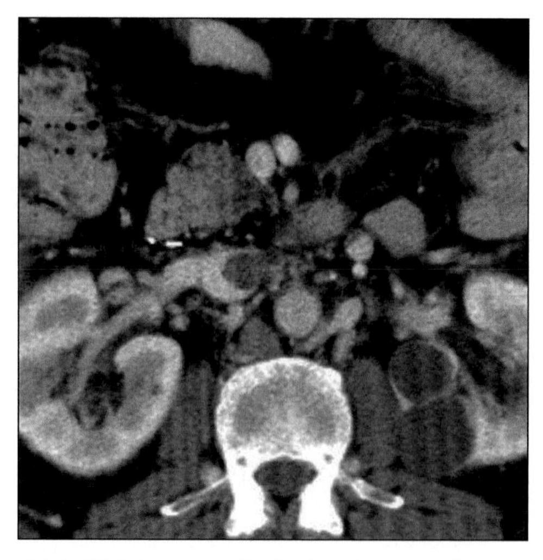

Axial CECT shows renal vein thrombus extending into the inferior vena cava.

TERMINOLOGY

Abbreviations and Synonyms

- Renal vein thrombosis (RVT)

Definitions

- Obstruction of renal vein by a thrombus

IMAGING FINDINGS

General Features

- Best diagnostic clue: Mass in renal vein with renal enlargement and delayed renal function
- Location
  - Unilateral > bilateral (more common in children)
  - Left renal vein > right renal vein
- Size
  - Renal enlargement (75% of cases)
  - Small shrunken kidney (rare)
- Morphology: Mass in renal vein ± extension to inferior vena cava (IVC) ± right atrium
- Other general features
  - Features may vary depending on type of obstruction
    - Acute or chronic (more common)
    - Partial or complete

Radiographic Findings

- IVP
  - Dense, prolonged nephrogram (partial)
  - Little or no nephrographic opacification (complete)
  - ↓ Amount of opaque urine in renal calyces, infundibula and pelvis
  - Opacified veins in perinephric space
  - Venous notching or indentation of ureter by tortuous collateral veins
  - Rarely, alternating radiopaque and radiolucent striations in renal cortex

CT Findings

- CECT
  - Low-attenuating filling defect within renal vein
  - ↓ Nephrographic attenuation
  - Persistent parenchymal opacification
  - No corticomedullary differentiation
  - Delayed corticomedullary junction time and contrast excretion into renal calyces and pelvis
  - Renal vein attenuation and enlargement; easier seen on left side
  - Thickening of the Gerota fascia and perinephric "whiskering" (edema or hemorrhage)
  - Opacified periureteral and perinephric ("cobwebs") collaterals
- CTA
  - Tortuous and dilated collateral veins close to ureters
  - Venous collateralization
  - Retrograde flow in dilated superficial epigastric veins

DDx: Enlarged Kidney Delayed Function

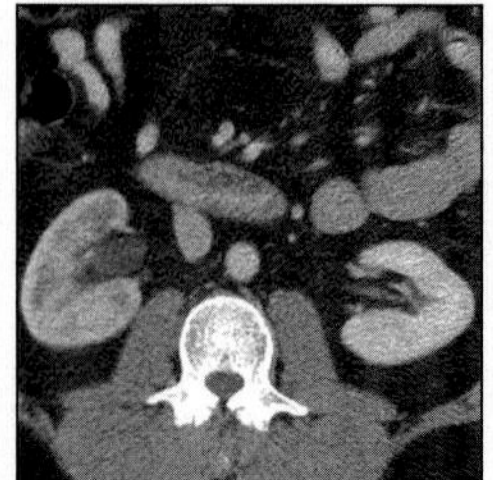

Ureteral Calculus

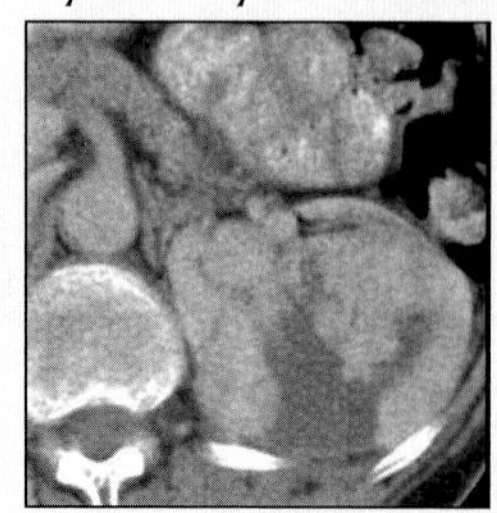

Renal Cell Cancer

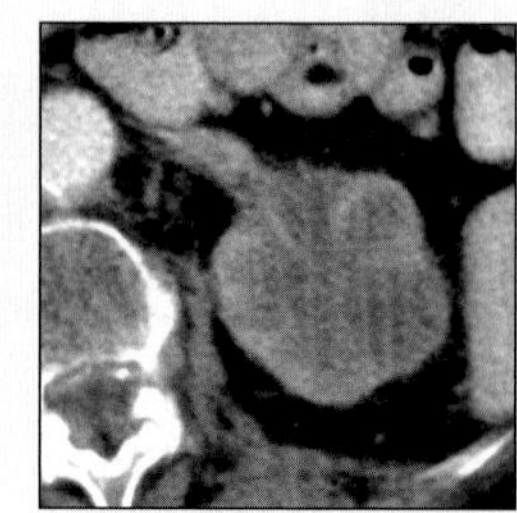

Transitional Cell Ca

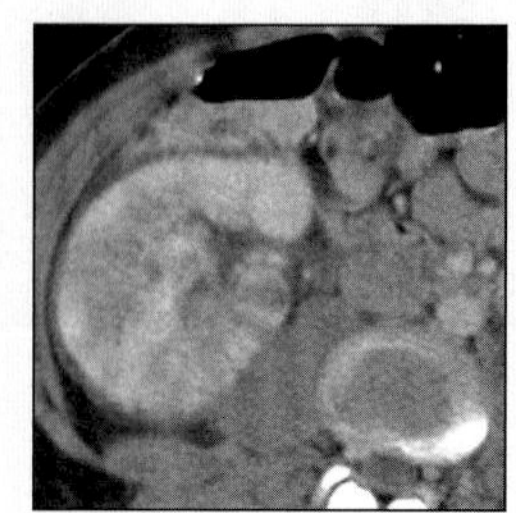

Pyelonephritis

RENAL VEIN THROMBOSIS

Key Facts

Terminology
- Obstruction of renal vein by a thrombus

Imaging Findings
- Best diagnostic clue: Mass in renal vein with renal enlargement and delayed renal function
- Renal enlargement (75% of cases)
- Morphology: Mass in renal vein ± extension to inferior vena cava (IVC) ± right atrium
- Low-attenuating filling defect within renal vein
- No corticomedullary differentiation
- Venous collateralization
- Best imaging tool: US followed by CT or MR

Top Differential Diagnoses
- Ureteral obstruction
- Infiltrating tumor
- Pyelonephritis

Pathology
- Most commonly associated with nephrotic syndrome in adults
- Most commonly associated with dehydration and sepsis in children

Clinical Issues
- Asymptomatic or thromboembolic disease
- Age: Adults (more common) or < 2 years of age
- Anticoagulation therapy

Diagnostic Checklist
- Adequate return of renal circulation may prevent renal function deterioration
- Filling defect within renal vein; venous collaterals

MR Findings
- T1WI: No corticomedullary differentiation
- MRA: High contrast between flowing blood, vascular walls and surrounding tissues
- T1WI, T2WI
  - Filling defect in renal vein
  - Prolongation of renal cortex and medulla relaxation times → low signal intensity
  - Low signal intensity band in outer part of medulla
  - Obliteration of renal sinus fat and compression of renal collecting systems
  - ↑ Attenuation of renal veins
  - Multiple perinephric collateral veins
  - Gonadal vein dilatation

Ultrasonographic Findings
- Real Time
  - Renal edema
  - ↓ Echogenicity (acute), then ↑ (after 10-14 days)
  - ± Corticomedullary differentiation
  - Thrombus or tumor in IVC (≤ 20%)
  - Intraluminal echoes with renal vein dilatation proximal to occlusion; easier seen on left > right side
- Color Doppler
  - Renal artery and proximal branches
    - Peaked, sudden ↓ systolic-frequency shifts
    - Retrograde plateau-like shifts during diastole
  - Absent venous signal
  - ↑ Blood velocity and turbulence (partial)
  - No blood flow (complete)
  - Anechoic or hypoechoic clot (acute)
  - Echogenic thrombus (chronic)

Angiographic Findings
- Conventional
  - Reflux into renal vein
  - Venous collateralization

Nuclear Medicine Findings
- Delayed or absent renal perfusion

Imaging Recommendations
- Best imaging tool: US followed by CT or MR
- Protocol advice: CTA: Corticomedullary phase best; second helical acquisition performed 90-120 seconds

DIFFERENTIAL DIAGNOSIS

Ureteral obstruction
- Filling defect and irregular narrowing of ureter
- ↓ Contrast density in collecting system
- Hydronephrosis; hydroureter; perinephric or periureteral stranding

Infiltrating tumor
- E.g., transitional cell carcinoma, renal metastases and lymphoma
- Poorly defined margins without change in shape
- Heterogenous mass ± calcification
- Enhancing mass with soft tissue attenuation

Pyelonephritis
- Multiple necrotic masses with renal enlargement
- Differentiate by clinical history and urinalysis

PATHOLOGY

General Features
- General path comments
  - Most commonly associated with nephrotic syndrome in adults
  - Most commonly associated with dehydration and sepsis in children
- Genetics: Inherited hypercoagulable states (e.g., antithrombin III, protein S, protein C deficiency)
- Etiology
  - Nephrotic syndrome
    - Membranous glomerulonephritis
    - Membranoproliferative glomerulonephritis
    - Focal sclerosis
    - Amyloidosis
    - Lipoid nephrosis
  - Other primary renal disease
    - Other glomerulonephritis
    - Pyelonephritis

        - Vasculitis
    - Renal hypoperfusion by hypovolemia or vascular stasis
        - Dehydration
        - Sepsis
        - Gastrointestinal fluid loss
        - Hemorrhage
        - Congestive heart failure
        - Aortic insufficiency
        - Constrictive pericarditis
    - Tumor extension
        - Renal cell carcinoma
        - Renal angiomyolipoma
        - Wilms tumor
        - Transitional cell carcinoma
        - Metastasis (e.g., adrenal, gonadal carcinoma)
    - Iatrogenic
        - Drugs (e.g., oral contraceptive pills, estrogens)
        - Abdominal surgery
        - Renal transplant rejection
    - Other hypercoagulable states
        - Pregnancy
        - Disseminated malignancy
        - Septic abortion
        - Genetics
    - Abdominal trauma
    - Thrombus extension
        - Left ovarian vein thrombosis
        - Deep vein thrombosis (e.g., leg, pelvic)
        - Retrograde IVC extension
    - Other systemic disease
        - Polyarteritis nodosa
        - Diabetes mellitus (e.g., maternal diabetes, glomerulosclerosis)
        - Sickle cell anemia
        - Systemic lupus erythematosus
    - Mechanical compression
        - Pregnancy
        - Retroperitoneal fibrosis
        - Tumor (e.g., lymphoma, carcinoma of tail of pancreas)
        - Abscess
        - Hematoma (e.g., adrenal)
        - Urinoma
        - Lymphocele
        - Aberrant arteries
        - Arterial aneurysms
- Epidemiology
    - Incidence
        - Unknown in asymptomatic patients
        - Nephrotic syndrome: 16-42% of patients

Gross Pathologic & Surgical Features

- Renal fibrosis, hemorrhage, necrosis, calcification

CLINICAL ISSUES

Presentation

- Most common signs/symptoms
    - Acute (more common in children)
        - Flank pain, nausea, vomiting
        - Palpable kidney, hypertension
    - Chronic
        - Asymptomatic or thromboembolic disease
        - Fever, edema
- Lab data
    - Acute: Urinalysis: Gross or microscopic hematuria
    - Chronic: Urinalysis: Proteinuria

Demographics

- Age: Adults (more common) or < 2 years of age

Natural History & Prognosis

- Effects of RVT depend on site of origin, time to occlusion, collateral veins and extent of recanalization
- Complications: Pulmonary embolism (most common), widespread thrombosis, renal hemorrhage, atrophy and failure
- Prognosis: Good, with frequent spontaneous recovery

Treatment

- Anticoagulation therapy
    - Intravenous heparin, then oral warfarin
- Thrombolytic therapy: Bilateral RVT, extension of RVT into IVC, massive clot, pulmonary emboli, severe flank pain or failed anticoagulation therapy
- Steroids or other immunosuppressive medications: Autoimmune disease
- Suprarenal vena cava filter: Thrombus extending into IVC
- Surgical thrombectomy or nephrectomy: Failed medical treatment, tumor thrombus

DIAGNOSTIC CHECKLIST

Consider

- Adequate return of renal circulation may prevent renal function deterioration

Image Interpretation Pearls

- Filling defect within renal vein; venous collaterals

SELECTED REFERENCES

1. Urban BA et al: Three-dimensional volume-rendered CT angiography of the renal arteries and veins: normal anatomy, variants, and clinical applications. Radiographics. 21(2):373-86; questionnaire 549-55, 2001
2. Kawashima A et al: CT evaluation of renovascular disease. Radiographics. 20(5):1321-40, 2000
3. Tempany CM et al: MRI of the renal veins: assessment of nonneoplastic venous thrombosis. J Comput Assist Tomogr. 16(6):929-34, 1992
4. Gatewood OM et al: Renal vein thrombosis in patients with nephrotic syndrome: CT diagnosis. Radiology. 159(1):117-22, 1986
5. Glazer GM et al: Computed tomography of renal vein thrombosis. J Comput Assist Tomogr. 8(2):288-93, 1984
6. Jeffrey RB et al: CT and ultrasonography of acute renal abnormalities. Radiol Clin North Am. 21(3):515-25, 1983
7. Bradley WG Jr et al: Renal vein thrombosis: occurrence in membranous glomerulonephropathy and lupus nephritis. Radiology. 139(3):571-6, 1981
8. Chait A et al: Renal vein thrombosis. Radiology. 90(5):886-96, 1968

IMAGE GALLERY

Typical

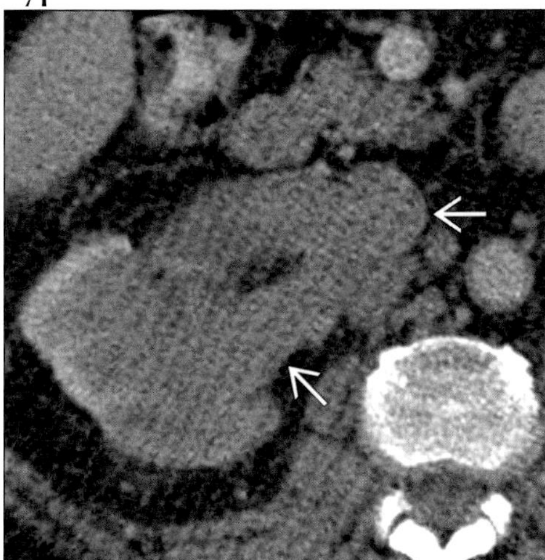

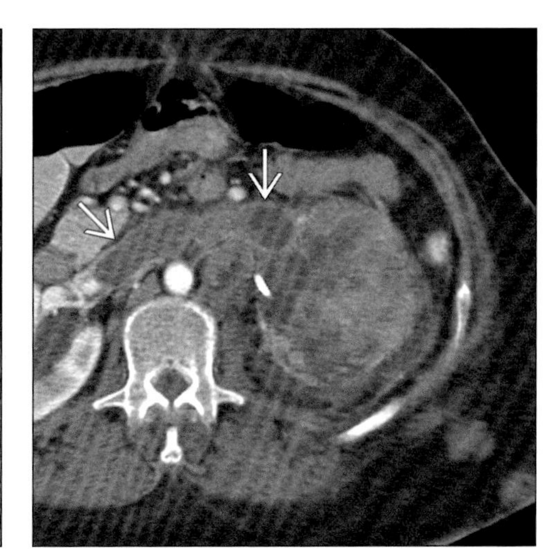

*(**Left**) Axial CECT shows renal cell carcinoma replacing most of kidney and invading renal vein and IVC (arrows). (**Right**) Axial CECT shows transitional cell carcinoma infiltrating left kidney and extending into renal vein and IVC (arrows).*

Typical

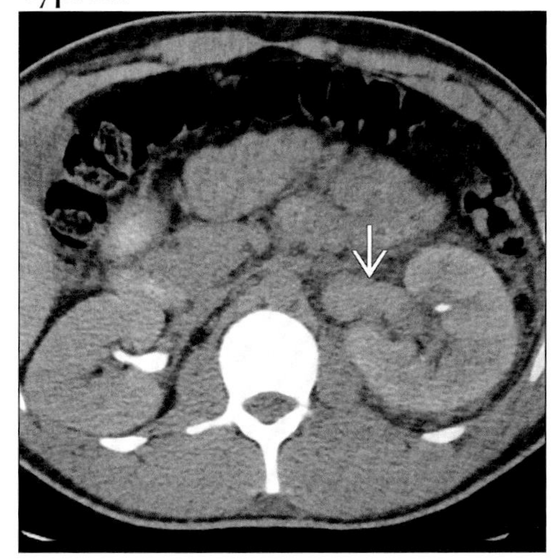

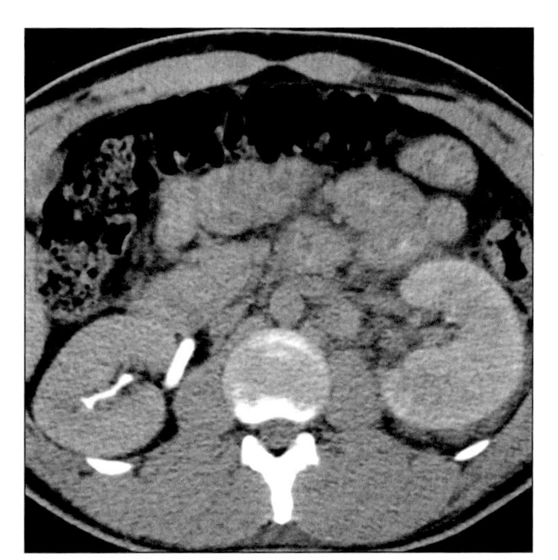

*(**Left**) Axial CECT (delayed excretory phase) in a 19 year old boy shows prolonged nephrogram in left kidney, hyperdense clot in renal vein (arrow). (**Right**) Axial CECT in a 19 year old boy shows left persistent nephrogram, decreased function, perirenal standing. Acute RVT due to glomerulonephritis.*

Typical

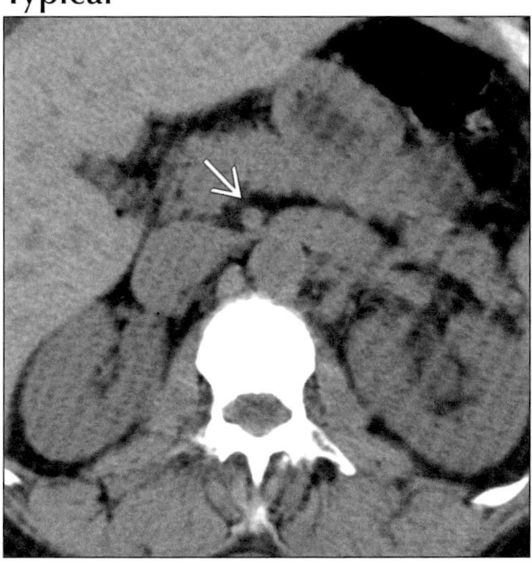

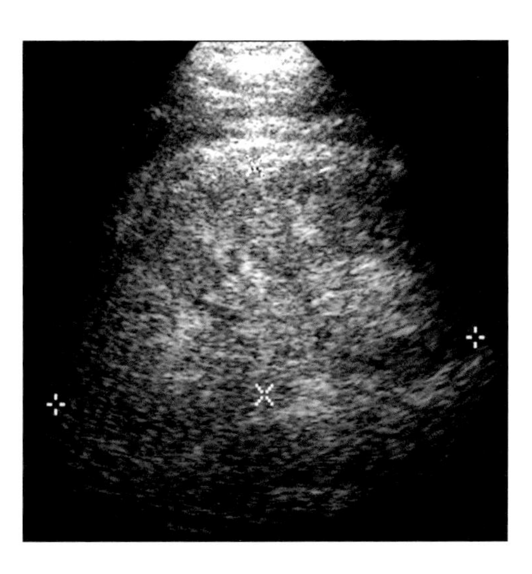

*(**Left**) Axial NECT in a girl with left flank pain and chronic hematuria shows compression of left renal vein by superior mesenteric artery; "nutcracker" (arrow). (**Right**) Sagittal sonogram of patient with two-week history of flank pain and hematuria. Kidney is swollen, echogenic with loss of corticomedullary differentiation due to RVT.*

RENAL TRAUMA

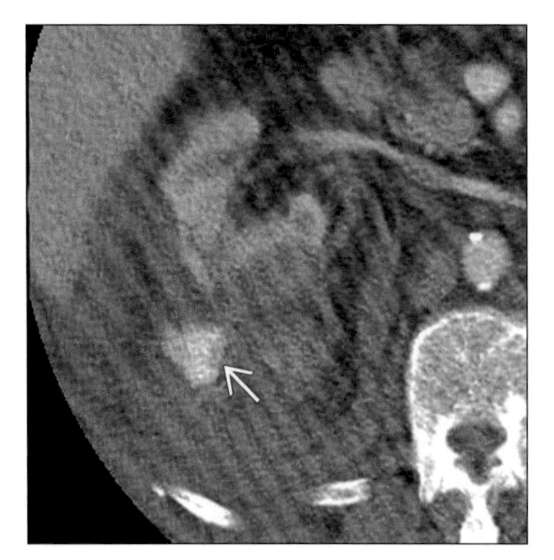

Axial CECT shows deep laceration of right kidney with perirenal blood and active hemorrhage (arrow).

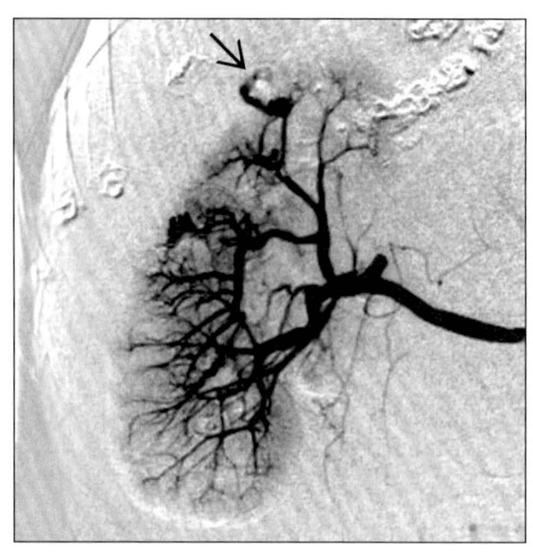

Renal angiogram shows active bleeding (arrow) that was controlled by coil embolization of the bleeding vessel.

TERMINOLOGY

Definitions

- Injury to the kidney

IMAGING FINDINGS

General Features

- Best diagnostic clue: Renal parenchymal defect with perirenal hemorrhage ± extravasation of blood/urine
- Other general features
  - Seen in 8-10% of patients with blunt or penetrating abdominal injuries
  - 80-90% of cases involve blunt rather than penetrating injury
  - Serious renal injuries usually associated with multi-organ involvement
    - Due to penetrating trauma: 80% of cases
    - Due to blunt trauma: 75% of cases
  - 98% of isolated renal injuries are minor & require no specific therapy
  - Radiologic classification of renal injuries
    - Grade I to IV
  - Grade I: 75-85% of all renal injuries
    - Minor injury (contusion; intrarenal or subcapsular hematoma)
    - Minor laceration + limited perinephric hematoma
    - No extension to collecting system or medulla
    - Small subsegmental cortical infarct
  - Grade II: 10% of all renal injuries
    - Major injury (major cortical laceration + extension to medulla & collecting system)
    - With or without urine extravasation or segmental renal infarct
  - Grade III: 5% of cases
    - Catastrophic injury (multiple renal lacerations & vascular injury involving renal pedicle)
  - Grade IV: Rare consequence
    - Ureteropelvic junction injury: Complete transection or laceration

Radiographic Findings

- IVP
  - Grade I: Normal
  - Grade II-IV
    - Delayed, absent excretion or extravasation

CT Findings

- Grade I lesions
  - Intrarenal hematoma or contusion
    - Ill-defined, round or ovoid lesion
    - Parenchymal phase: ↓ Enhancement relative to normal kidney
    - Delayed phase: Hyperdense due to urine stasis + clot filled tubules
  - Subcapsular hematoma

DDx: Parenchymal Lesion & Perirenal Fluid

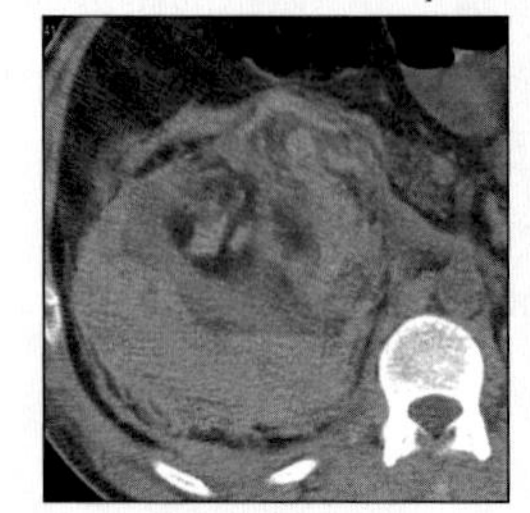

Bleeding AML

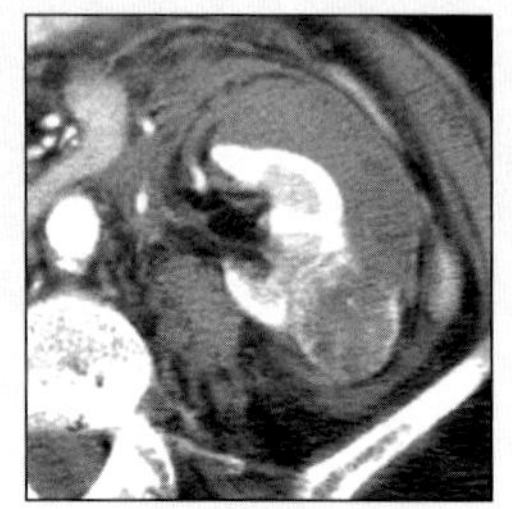

Bleeding RCC

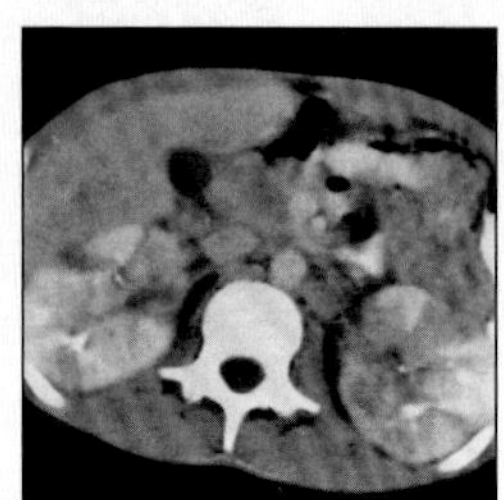

Polyarteritis Nodosa

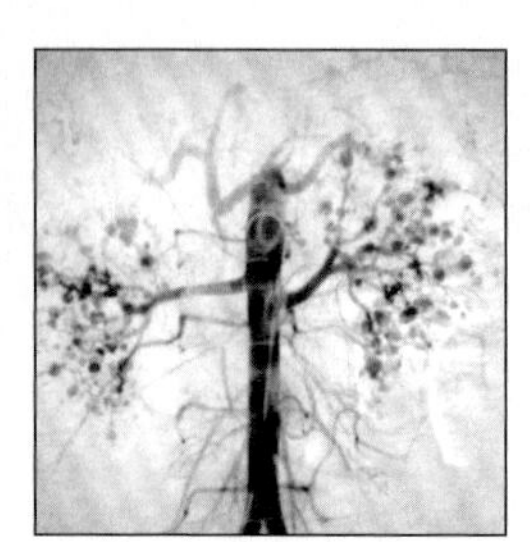

Polyarteritis Nodosa

RENAL TRAUMA

Key Facts

Imaging Findings
- Best diagnostic clue: Renal parenchymal defect with perirenal hemorrhage ± extravasation of blood/urine
- Segmental renal infarct: Sharply demarcated, wedge-shaped area of decreased enhancement
- Global infarction (nonenhancement) + no perinephric hematoma (renal artery thrombosis)
- Global infarction (nonenhancement) + perinephric hematoma (renal artery avulsion)
- Protocol advice: For any renal laceration evident on CT, must obtain 8-10 min. delayed scans to evaluate for urinary extravasation

Top Differential Diagnoses
- Renal tumor
- Vasculitis

Pathology
- Blunt, penetrating & deceleration injuries
- Other organ injuries in 75% of cases

Clinical Issues
- Flank pain, tenderness or ecchymosis
- Hematuria, anuria, uremia, shock
- Poor correlation between degree of hematuria & severity of renal injury

Diagnostic Checklist
- Possibility of underlying renal tumor if bleeding seems out of proportion to degree of trauma
- Arterial extravasation usually requires catheter embolization to control bleeding
- Urinary extravasation often requires ureteral stent ± catheter drainage of urinoma

        - Round or elliptic fluid collection (40-70 HU clotted blood)
    - Minor lacerations: Small linear hypodense areas in periphery
    - Limited perinephric hematoma: Adjacent to laceration
    - Subsegmental cortical infarct
        - Small, sharply demarcated, wedge-shaped decreased attenuation area → scar
- Grade II lesions
    - Major laceration through cortex extending to medulla
        - Long irregular or linear hypodense area
    - When laceration extends into collecting system
        - Nephrographic phase: Large, distracted renal fracture (hypodense)
        - Excretory phase: Contrast extravasation into perinephric space
        - ± Antegrade filling of ureter
    - Segmental renal infarct: Sharply demarcated, wedge-shaped area of decreased enhancement
- Grade III lesions
    - Multiple renal lacerations & vascular injury
        - Nephrographic phase: Several irregular, linear or band like interpolar hypodense areas ± areas of active arterial contrast extravasation
    - Subacute infarction
        - "Cortical rim" sign: Preserved capsular or subcapsular enhancement (reliable sign)
        - Seen 6-8 hrs after infarction
    - "Shattered kidney"
        - Segmental infarction: Nonenhancing wedge-shaped area (devitalized upper or lower renal pole branch)
        - Global infarction (nonenhancement) + no perinephric hematoma (renal artery thrombosis)
        - Global infarction (nonenhancement) + perinephric hematoma (renal artery avulsion)
- Grade IV lesions
    - Ureteropelvic junction: Complete transection (avulsion) or laceration
        - Good excretion of contrast + medial perinephric extravasation
        - A circumferential urinoma may be seen around affected kidney

Imaging Recommendations
- Protocol advice: For any renal laceration evident on CT, must obtain 8-10 min. delayed scans to evaluate for urinary extravasation
- Helical CECT: Gold standard imaging
- IVU: Limited urography (to evaluate hemodynamically unstable patient)
    - Take a plain film abdomen & administer 100-150 ml of 60% contrast I.V. obtain immediate "cone down" nephrogram film + full film after 8 min.
    - "One-shot IVU": Not to evaluate injured kidney but to asses normal kidney
- Retrograde pyelography
    - To assess ureteral & renal pelvic injuries
- US: To assess hemoperitoneum in a hemodynamically unstable patient

DIFFERENTIAL DIAGNOSIS

Renal tumor
- Spontaneous bleed or rupture may be seen in renal tumors
- Perinephric fluid collection of blood density
- Look for underlying renal mass lesion
    - Renal cell carcinoma (RCC)
    - Angiomyolipoma
- Renal cell carcinoma
    - Solid renal mass & is usually hypervascular
    - Renal vein & IVC extension may be seen
    - Hypervascular metastatic foci are often seen
- Angiomyolipoma (AML)
    - Composed of vascular, smooth muscle & fat components
    - Renal mass with variable amounts of fat is diagnostic
    - May enhance significantly with contrast infusion

Vasculitis
- E.g., polyarteritis nodosa; SLE; scleroderma, drug abuse

RENAL TRAUMA

- Renal imaging
  - Wedge-shaped or striated nephrogram
  - Capsular retraction over parenchymal lesions
  - Microaneurysms of small vessels are usually seen

PATHOLOGY

General Features

- Etiology
  - Motor vehicle accidents (MVA), falls, fights, assaults
    - Blunt, penetrating & deceleration injuries
  - Adults: Kidneys protected by ribs, heavy musculature of back & flank
  - Children: Kidneys relatively large, more mobile & more vulnerable to trauma
- Epidemiology
  - Renal trauma incidence
    - 8-10% of injuries in emergency department
- Associated abnormalities
  - Other organ injuries in 75% of cases
    - Liver, spleen, bowel, pancreas

Gross Pathologic & Surgical Features

- Contusion, laceration, hematoma, infarction, vascular or ureteropelvic injury

Microscopic Features

- Contusion, laceration, ischemia of cortico-medullary or collecting system

CLINICAL ISSUES

Presentation

- Most common signs/symptoms
  - Flank pain, tenderness or ecchymosis
  - Hematuria, anuria, uremia, shock
  - Poor correlation between degree of hematuria & severity of renal injury
    - 14% of major & 10% of minor injuries may not have hematuria
- Clinical profile: Patient with history of MVA, flank pain, hematuria or anuria
- Lab data
  - Blood in urine (> 5 RBCs per high power field)
- Diagnosis: Clinical & classic imaging features are diagnostic of renal trauma

Demographics

- Age: Any age group (children > vulnerable than adults)
- Gender: M = F

Natural History & Prognosis

- Complications
  - Early: Urinoma, perinephric abscess, sepsis, AV fistula, pseudoaneurysm
  - Late: Hydronephrosis, HTN, calculus formation, chronic pyelonephritis, renal failure & atrophy
- Prognosis
  - Grade I & II: Good
  - Grade III & IV
    - Unilateral after treatment: Good
    - Bilateral: Poor

Treatment

- Grade I & II: Conservative therapy
- Grade III & IV
  - Active bleeding: Angioembolization
  - Renal artery (RA) thrombosis: Anticoagulants; stent placement
  - Active urinary extravasation
    - Consider ureteral stent & catheter drainage
  - Indications for surgery
    - Vascular (renal pedicle) injury
    - Shattered kidney
    - Expanding or pulsatile hematoma
    - Shocked polytrauma patient
  - Severely damaged kidney: Surgical nephrectomy

DIAGNOSTIC CHECKLIST

Consider

- Possibility of underlying renal tumor if bleeding seems out of proportion to degree of trauma

Image Interpretation Pearls

- Arterial extravasation usually requires catheter embolization to control bleeding
- Urinary extravasation often requires ureteral stent ± catheter drainage of urinoma

SELECTED REFERENCES

1. Smith JK et al: Imaging of renal trauma. Radiol Clin North Am. 41(5):1019-35, 2003
2. Kawashima A et al: Imaging evaluation of posttraumatic renal injuries. Abdom Imaging. 27(2):199-213, 2002
3. Yao DC et al: Using contrast-enhanced helical CT to visualize arterial extravasation after blunt abdominal trauma: incidence and organ distribution. AJR Am J Roentgenol. 178(1):17-20, 2002
4. Harris AC et al: Ct findings in blunt renal trauma. Radiographics. 21 Spec No:S201-14, 2001
5. Mizobata Y et al: Successful evaluation of pseudoaneurysm formation after blunt renal injury with dual-phase contrast-enhanced helical CT. AJR Am J Roentgenol. 177(1):136-8, 2001
6. Kawashima A et al: Imaging of renal trauma: a comprehensive review. Radiographics. 21(3):557-74, 2001
7. Morey AF et al: Single shot intraoperative excretory urography for the immediate evaluation of renal trauma. J Urol. 161(4):1088-92, 1999
8. Kawashima A et al: Ureteropelvic junction injuries secondary to blunt abdominal trauma. Radiology. 205(2):487-92, 1997
9. Mirvis SE: Trauma. Radiol Clin North Am. 34(6):1225-57, 1996
10. Pollack HM et al: Imaging of renal trauma. Radiology. 172(2):297-308, 1989
11. Federle MP et al: Penetrating renal trauma: CT evaluation. J Comput Assist Tomogr. 11(6):1026-30, 1987
12. Lang EK et al: Renal trauma: radiological studies. Comparison of urography, computed tomography, angiography, and radionuclide studies. Radiology. 154(1):1-6, 1985
13. Federle MP et al: The role of computed tomography in renal trauma. Radiology. 141(2):455-60, 1981

IMAGE GALLERY

Typical

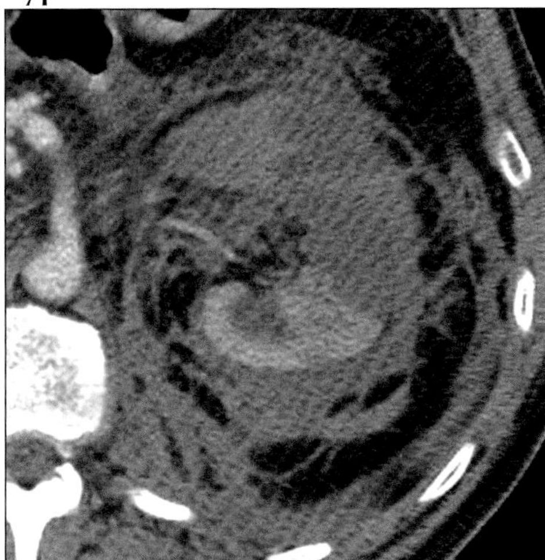

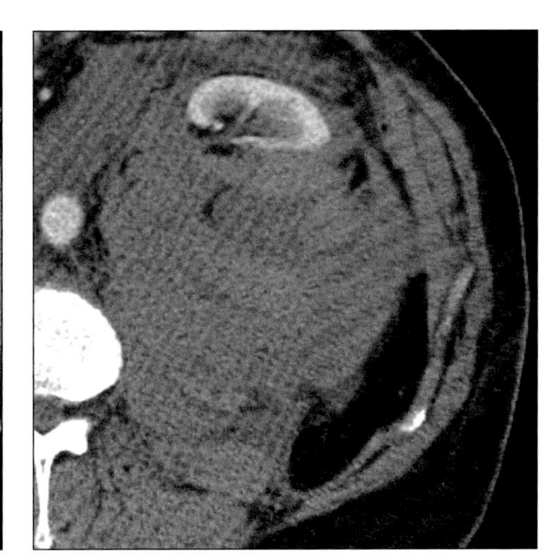

(Left) *Axial CECT shows deep laceration through the kidney and a large perirenal hemorrhage.* ***(Right)*** *Axial NECT shows a deep laceration through the kidney and large perirenal hemorrhage.*

Typical

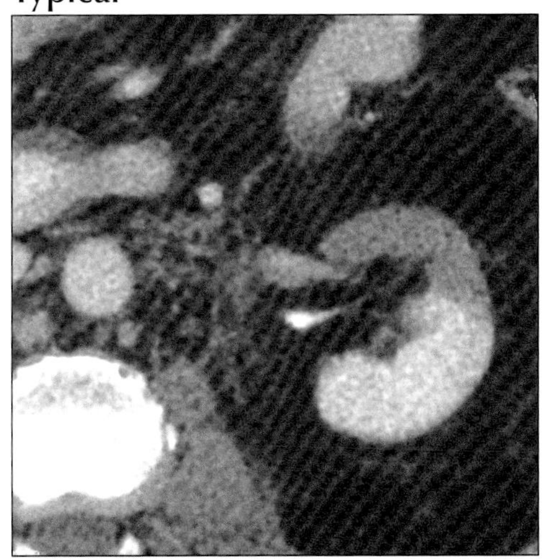

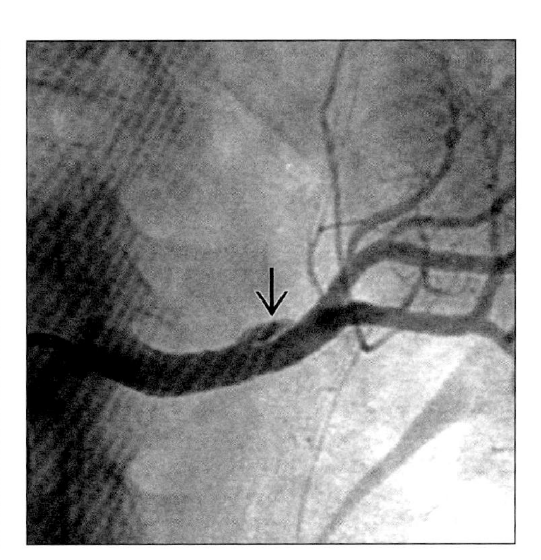

(Left) *Axial CECT shows nonenhancement of the ventral half of the kidney but minimal bleeding, suggesting traumatic infarction due to renal arterial branch occlusion.* ***(Right)*** *Renal arteriogram shows traumatic dissection (arrow) that resulted in segmental renal infarction.*

Typical

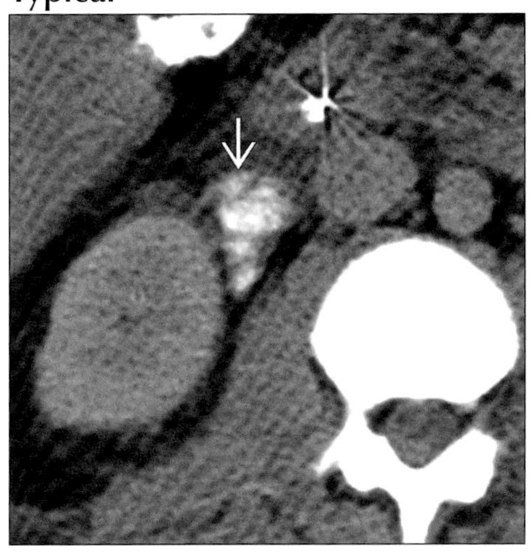

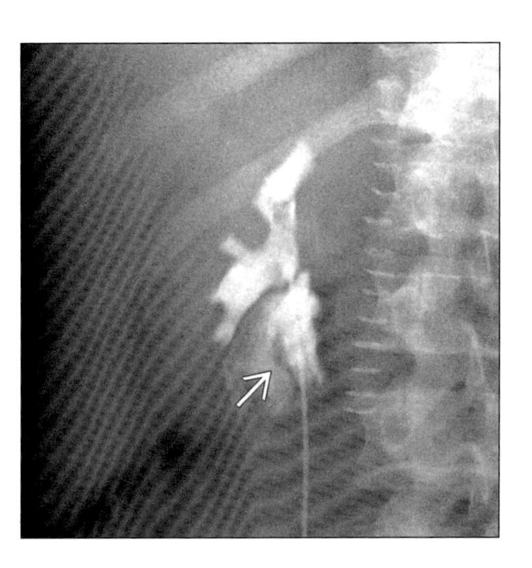

(Left) *Axial NECT (delayed) shows opacified urine (arrow) without a parenchymal renal injury, suggesting ureteral injury.* ***(Right)*** *Retrograde pyelogram shows urinary extravasation (arrow) due to traumatic uretero-pelvic laceration.*

RENAL CYST

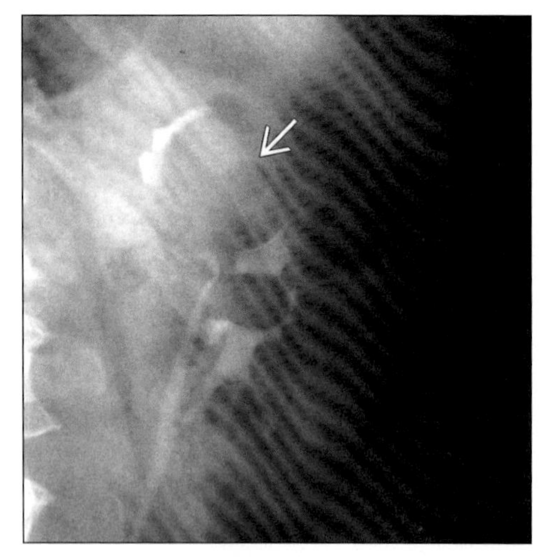

Excretory urogram shows spherical absence of nephrogram and splaying of calices in upper pole (arrow).

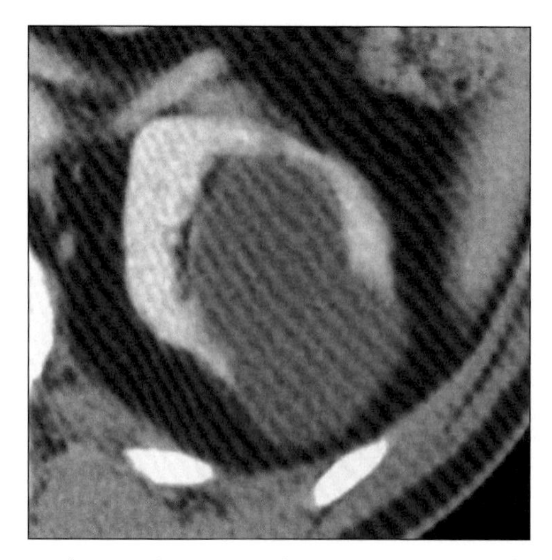

Axial CECT shows water density mass in upper pole with no visible wall; Bosniak I cyst.

TERMINOLOGY

Definitions

- A benign, fluid-filled, nonneoplastic renal lesion

IMAGING FINDINGS

General Features

- Best diagnostic clue
  - Water density, nonenhancing, spherical lesion with no visible wall on CT
  - Lucent mass with no visible wall and acoustic enhancement on US
- Size: Usually 2-5 cm diameter (up to 10 cm)
- Other general features
  - Simple renal cyst classification
    - Typical or uncomplicated
    - Complicated: Hemorrhagic, infected, ruptured, neoplasm from cystic wall
    - Atypical: Calcified, hyperdense, septated, multiple simple, localized cystic disease, milk of calcium

Radiographic Findings

- IVP
  - Large cyst → distorts contour & splays, elongates, displaces, obstructs or totally obliterates calyces
  - Uncomplicated cyst: Parenchymal deformity at edges of cyst ("beak or claw" sign)
  - Neoplastic wall: Rare calcification of mural nodule

CT Findings

- Bosniak classification for renal cysts is standard
  - Class I: Benign cysts
    - Well-defined, rounded, homogeneous, lucent (< 20 HU, near water density) mass with a thin or imperceptible, nonenhancing wall
  - Class II: Minimally complicated cysts; benign
    - Minimally irregular, calcified, septated
    - Hyperdense cyst: Contents 60 to 90 HU, no enhancement, spherical, partially exophytic, usually ≤ 3 cm diameter
    - Septa are thin (< 1 mm), smooth, and attached to wall without thickened elements or enhancement
  - Class IIF: Require follow-up (3, 6, 12-month interval)
    - Some hyperdense cysts, lesions with more calcium in wall or slightly more complicated cysts
  - Class III: More complicated cysts; biopsy (controversial) or excision
    - Irregular and thickened septa ± enhancement
    - Thickened and irregular calcification
    - Irregular margination
    - Uniform wall thickening
    - Multilocular; small, nonenhancing nodular mass
    - Usually simulate cystic or necrotic renal cell carcinoma or multilocular cystic nephroma

DDx: Hypodense Renal Mass

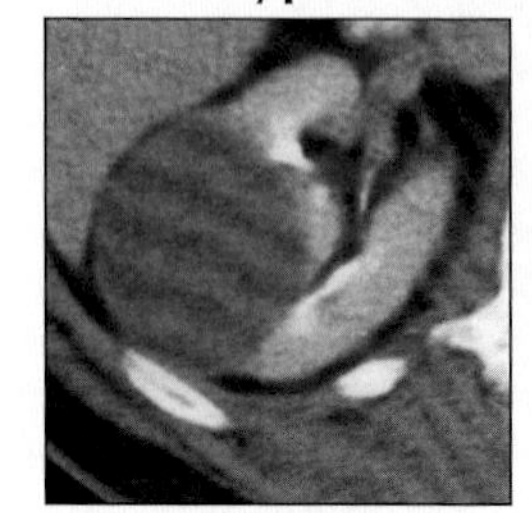

Papillary RCC

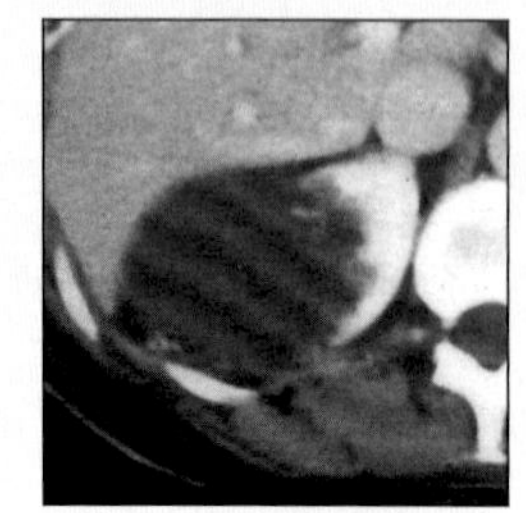

Cystic RCC

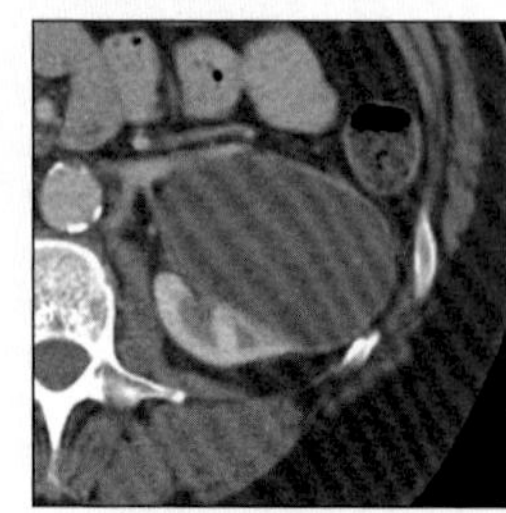

MC Nephroma

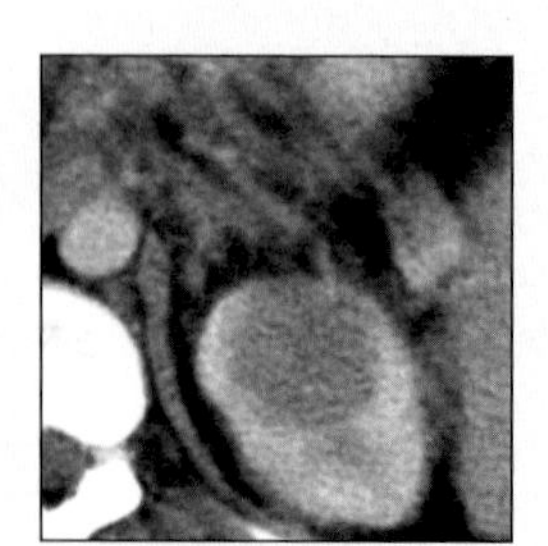

Lymphoma

RENAL CYST

Key Facts

Terminology
- A benign, fluid-filled, nonneoplastic renal lesion

Imaging Findings
- Water density, nonenhancing, spherical lesion with no visible wall on CT
- Lucent mass with no visible wall and acoustic enhancement on US
- Bosniak classification for renal cysts is standard
- Best imaging tool: CT (100% sensitivity); US (> 98%)

Top Differential Diagnoses
- Renal carcinoma
- Multilocular cystic nephroma (MCN)
- Renal abscess
- Hydronephrosis
- Metastases and lymphoma

Pathology
- Most common renal mass in adults (62%)

Clinical Issues
- Asymptomatic or palpable mass and flank pain
- Renal cyst aspiration: Used only to culture infected cyst and diagnose indeterminate mass by cytology
- Age: 50% > 50 years of age; rare in < 30 years of age
- Bosniak class I, II and IIF: No treatment ± follow-up
- Bosniak class III and IV: Surgical excision

Diagnostic Checklist
- Image findings generally more reliable than "clinical correlation"
- Image evaluation and classification of cystic masses is key to management

- Class IV: Malignant lesions with large cystic or necrotic components; require surgery
    - Irregular wall thickening or enhancing mass
- Uncomplicated cyst
    - Sharply marginated, round, smooth, homogeneous, hypodense (-10 - +20 HU) mass
    - Thin, imperceptible, nonenhancing wall
    - Small (< 1 cm): Cannot measure region of interest; if less than blood density on NECT, probably cyst
- Infected cyst: Thick wall, septated, heterogeneous enhancing fluid, debris- or gas-fluid level; ± calcification (chronic)
- Hemorrhagic cyst
    - NECT: Hyperdense & CECT: Hypodense (flip-flop phenomenon), homogeneous (70-90 HU) (acute)
    - No contrast-enhancement
    - Heterogeneous (clot or debris), ↑ wall thickness & ↓ attenuation ± calcification (chronic)
- Ruptured cyst: Retroperitoneal or perinephric collection of fluid, blood (varied density) or pus
- Neoplastic wall: Focal thickening or enhancing nodule
- Calcified cyst: Calcification, usually in thickened wall
- Hyperdense cyst: Homogeneous, hyperdense, nonenhancing mass
- Septated cyst: ≥ 1; partial or complete thin septa
- Multiple simple cysts: Bilateral > unilateral
- Localized cystic disease
    - 1 portion of unilateral kidney; ill-defined multiloculated mass and/or cluster of cysts
    - No capsule; small cysts not within main cluster
- Milk of calcium cyst: Dependent, fluid-calcium layer

MR Findings
- Uncomplicated cyst
    - T1WI: Round/oval, homogenous, hypointense mass
    - T2WI: Homogeneous, hyperintense mass with imperceptible wall; smooth & distinct inner margin
    - CEMR: No enhancement
- Infected cyst: T1WI: ↑ Intensity, less homogenous than uncomplicated cyst; ↓ intensity than subacute hemorrhage (similar to chronic); ± thickened wall
- Hemorrhagic cyst
    - Variable signal intensity due to age of hemorrhage
    - T1WI: Highest intensity in subacute (< 72 hours)
    - T2WI: High intensity (less than uncomplicated cyst); fluid-debris level (intensity less in dependent); ± heterogeneous mass and lobulation of contour
- Neoplastic wall: Focal mass or wall thickening; fluid simulates uncomplicated or hemorrhagic cyst
- Hyperdense cyst: ↑ Protein simulates hemorrhage

Ultrasonographic Findings
- Real Time
    - Uncomplicated cyst: Spherical or ovoid; sharply defined distant wall with smooth, distinct margins; no internal echoes; acoustical enhancement
    - Infected cyst: Thick wall with scattered internal echoes ± debris-fluid level
    - Hemorrhagic cyst: Internal echoes (clot); thick calcified wall ± multiloculated (chronic)
    - Neoplastic wall: Tumor nodule
    - Hyperdense cyst: Few scattered internal echoes
    - Localized cystic disease: 1 large multiloculated mass
    - Milk of calcium cyst: Line of calcified debris

Angiographic Findings
- Conventional
    - Vascular displacement; hypo- or avascular
    - Infected cyst: Neovascularity in granulation tissue
    - Neoplastic wall: Focal blush of nodule

Imaging Recommendations
- Best imaging tool: CT (100% sensitivity); US (> 98%)
- Protocol advice: CT: Noncontrast and nephrographic phase (90 seconds after contrast)
    - Same day, scanner, CT technique and collimation

DIFFERENTIAL DIAGNOSIS

Renal carcinoma
- Angiography: Neovascularity (papillary renal cell carcinoma may be "avascular")
- Features to distinguish from calcified cyst
    - CT: Hyperintense (> water), discrete enhancing mass
- Features to distinguish from localized cystic disease
    - Capsule between cluster of cysts and parenchyma

- No cysts other than within main cluster of cysts

Multilocular cystic nephroma (MCN)
- Septated, encapsulated cystic mass
- Herniation into renal sinus

Renal abscess
- "Rim" sign; ± perinephric stranding
- Shaggy wall; hyperdense (> water)

Hydronephrosis
- No distinct mass; dense "obstructive" nephrogram with enlarged kidney & dilated collecting structures

Metastases and lymphoma
- Metastases (e.g., melanoma & lung, breast & GI cancer)
  - CT: Multifocal, small, Isodense, enhancing (5-30 HU) nodules; widespread nonrenal metastases
- Lymphoma (non-Hodgkin > Hodgkin)
  - CT: Multiple (45%); invasion from retroperitoneal nodes (25%); solitary (15%)

PATHOLOGY

General Features
- General path comments
  - Most common renal mass in adults (62%)
  - Hemorrhagic cyst: 6% all cysts; calcified cyst: 1-3%
- Etiology
  - Uncomplicated cyst: Unknown, ischemia, tubular obstruction by solid tumor (sentinel cyst) or medullary interstitial fibrosis
  - Infected cyst: Hematogenous spread, vesicoureteric reflux, surgery or cyst puncture
  - Hemorrhagic cyst: Unknown, trauma, bleeding diathesis or varicosities in simple cyst
  - Ruptured cyst: Spontaneous
  - Calcified cyst: Hemorrhage, infection or ischemia
  - Hyperdense cyst: Hemorrhage or tumor
  - Septated cyst: 2 adjacent cysts or healing and organization of hemorrhagic or infected cyst
  - Localized cystic disease: No family history

Gross Pathologic & Surgical Features
- Uncomplicated cyst
  - Unilocular; arise in cortex (superficial) and bulge from renal surface; less common from medulla
  - Clear or straw-colored fluid; up to several liters
  - Smooth, yellow-white, thin and translucent wall
  - Rarely calcified; no communication to renal pelvis
- Infected cyst: Markedly thickened wall ± calcification; varying pus, fluid and calcified or noncalcified debris
- Hemorrhagic cyst: Rust-colored putty-like material surrounded by thick fibrosis and plates of calcification
- Neoplastic wall: Discrete nodule at base of cyst
- Hyperdense cyst: Rusty-colored material, ↑ protein content, ↑ viscosity, solidified colloid
- Localized cystic disease: Nonencapsulated, numerous simple cysts in 1 portion of kidney
- Milk of calcium cyst: Small calcified debris (calcium carbonate) in cystic fluid

Microscopic Features
- Uncomplicated cyst
  - Cystic wall: Cuboidal or flattened, discontinuous epithelium; Fibrosis & hyalinization; ± calcification
  - 1-2 mm thickness of wall; rare thin septa divide cysts ± communicate with each other
  - Adjacent tissue: Compression and fibrosis
- Hemorrhagic cyst: Uni- or multilocular, thickened wall
- Neoplastic wall: Well differentiated clear/granular cell
- Septated cyst: Compressed normal parenchyma or nonneoplastic connective tissue
- Localized cystic disease: Dilatation of ducts & tubules (mm to cm); no neoplastic or dysgenetic stroma

CLINICAL ISSUES

Presentation
- Most common signs/symptoms
  - Asymptomatic or palpable mass and flank pain
  - Infected cyst: Pain in flank, malaise and fever
  - Hemorrhagic cyst: Abrupt and severe pain
  - Ruptured cyst: Severe abdominal pain, hematuria
- Lab data
  - Cystic fluid: Similar to plasma transudate
- Renal cyst aspiration: Used only to culture infected cyst and diagnose indeterminate mass by cytology

Demographics
- Age: 50% > 50 years of age; rare in < 30 years of age
- Gender: M > F

Natural History & Prognosis
- Complications: Hydronephrosis, hemorrhage, infection, cyst rupture or carcinoma
- Prognosis: Very good

Treatment
- Bosniak class I, II and IIF: No treatment ± follow-up
- Bosniak class III and IV: Surgical excision
- Follow-up: Changes size, configuration & internal consistency; excision if changes suggest carcinoma

DIAGNOSTIC CHECKLIST

Consider
- Image findings generally more reliable than "clinical correlation"

Image Interpretation Pearls
- Image evaluation and classification of cystic masses is key to management

SELECTED REFERENCES

1. Bosniak MA: Diagnosis and management of patients with complicated cystic lesions of the kidney. AJR. 169: 819, 1997
2. Davidson AJ et al: Radiologic assessment of renal masses: Implication for patient care. Radiology. 202: 297, 1997
3. Siegel CL et al: CT of cystic renal masses: Analysis of diagnostic performance and interobserver variation. AJR. 169: 813, 1997

IMAGE GALLERY

Typical

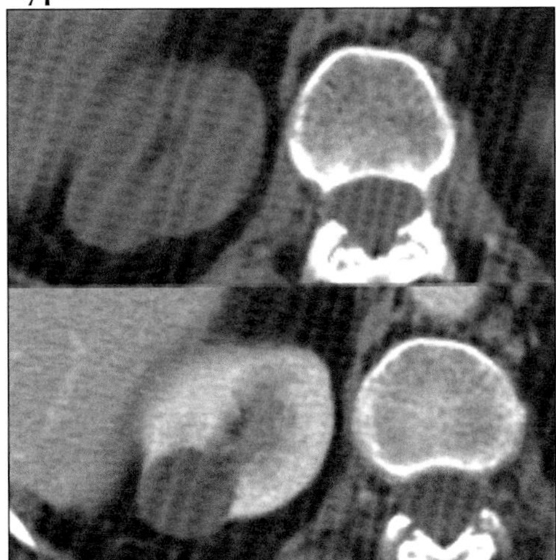

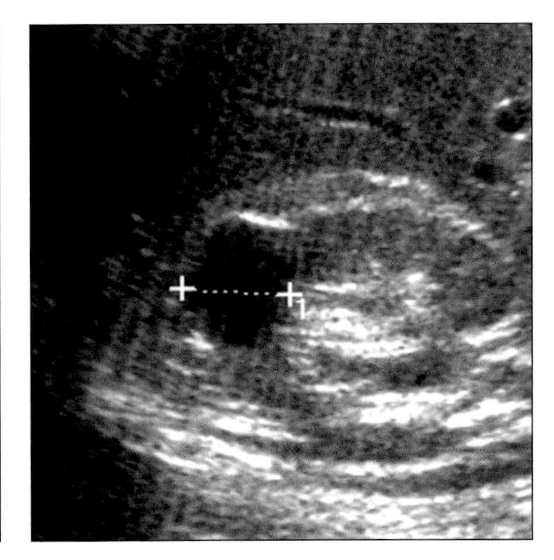

(Left) *Axial NECT and CECT show a water density mass with no enhancement and no definable wall; Bosniak I cyst.* ***(Right)*** *Sagittal sonogram shows anechoic mass in upper pole with posterior acoustic enhancement.*

Typical

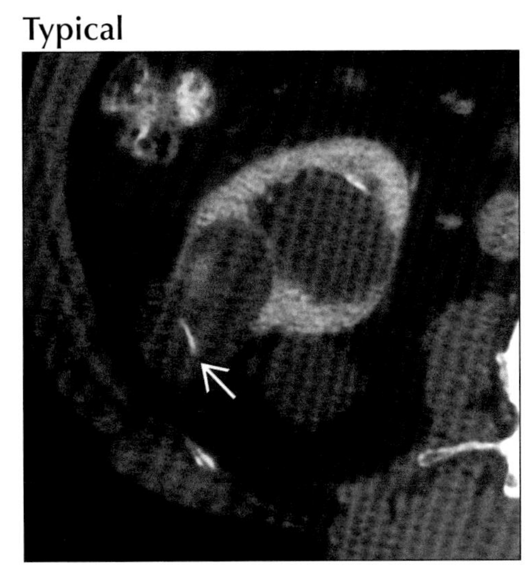

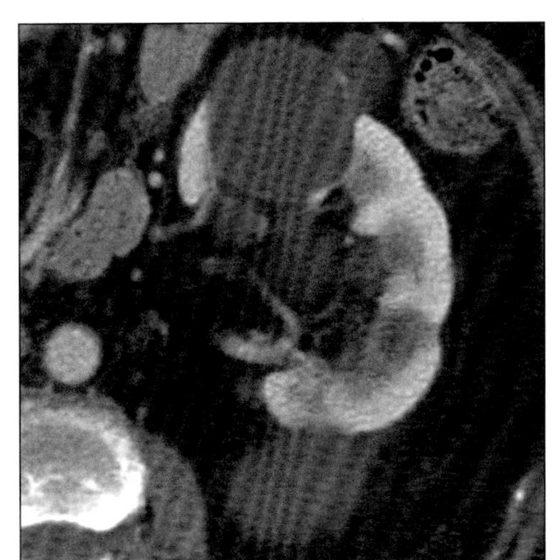

(Left) *Axial CECT shows multiple water density cystic lesions, including calcified thin septum (arrow); Bosniak II cyst.* ***(Right)*** *Axial CECT shows multiple cortical and parapelvic cysts. Thin septal calcification indicates benign Bosniak II classification.*

Typical

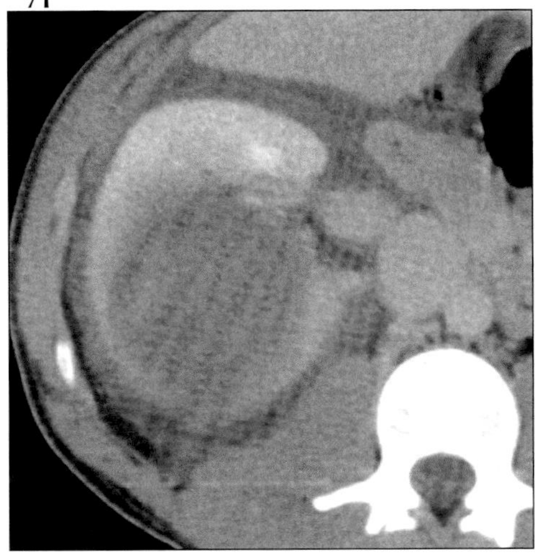

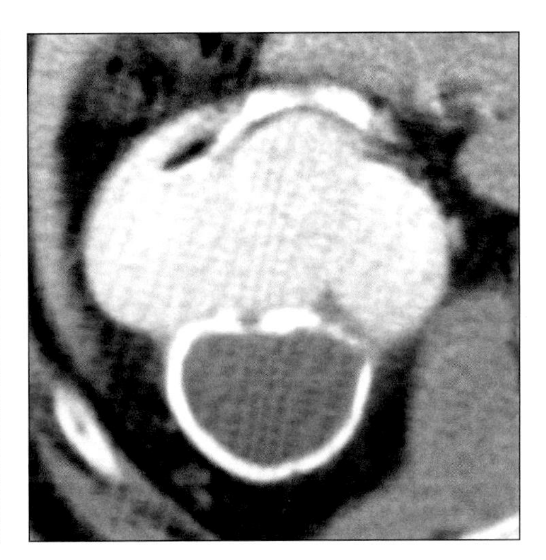

(Left) *Axial CECT shows heterogeneous nonenhancing mass and perirenal stranding. Hemorrhagic cyst.* ***(Right)*** *Axial CECT shows water density mass with relatively thick, calcified wall; was resected and was a benign cyst.*

RENAL ONCOCYTOMA

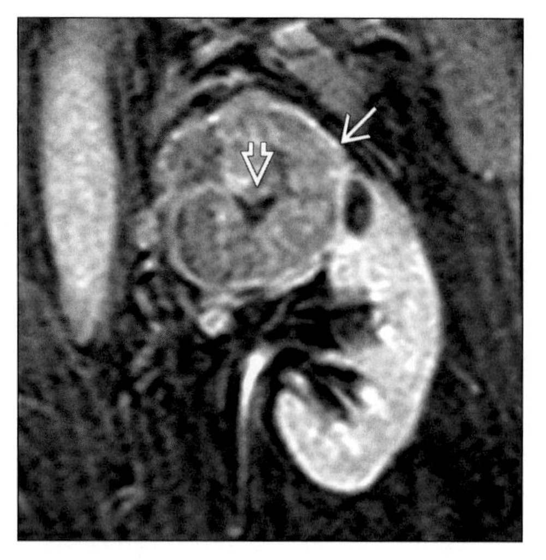

Coronal T1WI MR (enhanced) shows well-defined spherical renal mass (arrow) with a central scar (open arrow) and a pseudocapsule characteristic of an oncocytoma.

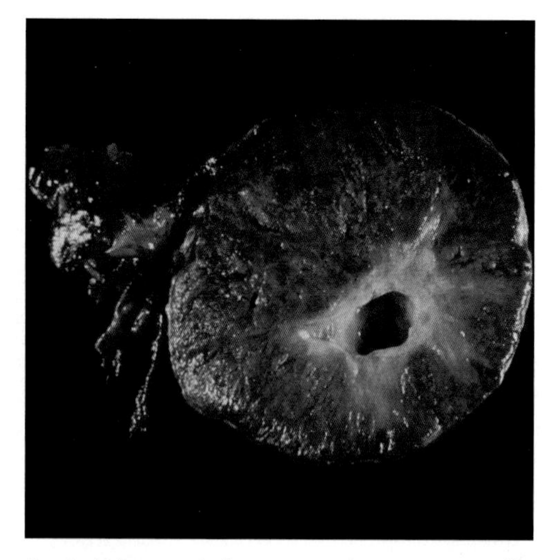

Surgical photograph shows resected oncocytoma with a central scar and thin septa within a homogeneous solid mass.

TERMINOLOGY

Abbreviations and Synonyms

- Synonyms
  - Proximal tubular adenoma with oncocytic features
  - Benign oxyphilic adenoma

Definitions

- Epithelial tumor composed of eosinophilic epithelial cells, arising from intercalated cells of collecting ducts

IMAGING FINDINGS

General Features

- Best diagnostic clue: Solid renal cortical mass lesion with central stellate scar
- Location: Usually renal cortex
- Size: Varies from 3 to 10 cm (average of 7 cm)
- Morphology: Encapsulated, well-marginated, smooth contour, solid mass
- Other general features
  - 2nd Most common benign renal tumor after AML
  - Accounts for approximately 3-7% of all renal neoplasms
  - About 2-12% are multifocal & 4-14% are bilateral
  - 50-90% of tumors are incidental on imaging
  - 17-21% of cases present with clinical manifestations
  - Tumors composed of oncocytes occur in thyroid, parathyroid, adrenal & salivary glands
  - In less than 10% of cases, oncocytoma & chromophobe renal cell carcinoma (RCC) may coexist

Radiographic Findings

- Radiography
  - Plain radiographic findings are nonspecific
    - Large, soft tissue mass in renal area with displacement of fat planes
    - Calcification is rare
- IVP: Large mass with a renal contour abnormality & compression of collecting system

CT Findings

- NECT
  - Well-defined, homogeneous, solid density mass
  - Isodense or slightly hyperdense relative to kidney
  - Central hypoattenuating area suggestive of scar
    - Seen in 33-54% of cases
    - Difficult to differentiate from central necrosis of renal cell carcinoma
  - Calcification, necrosis & hemorrhage are rare
  - Occasionally multifocal & bilateral
- CECT
  - Nephrographic phase
    - Tumor appears less dense than renal parenchyma

DDx: Expansile Renal Mass

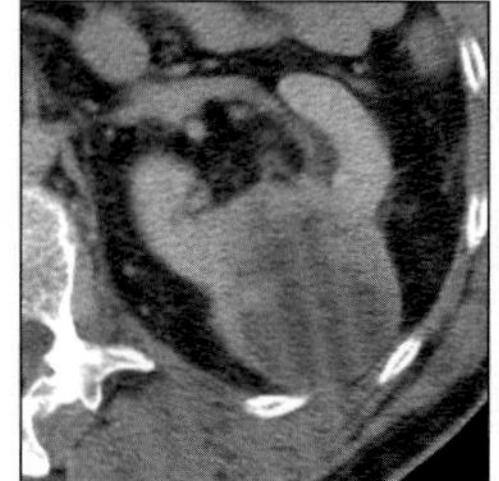

Renal Cell Carcinoma

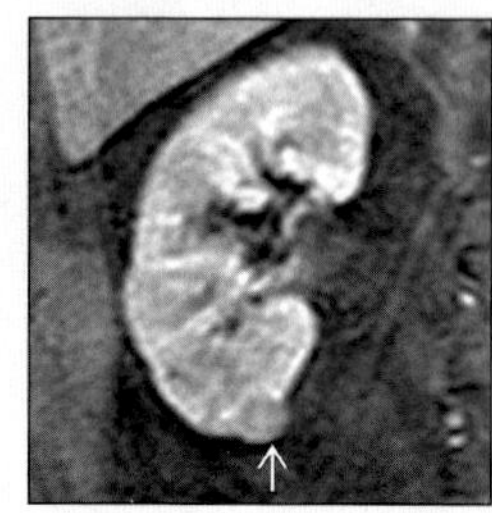

Renal Cell Carcinoma

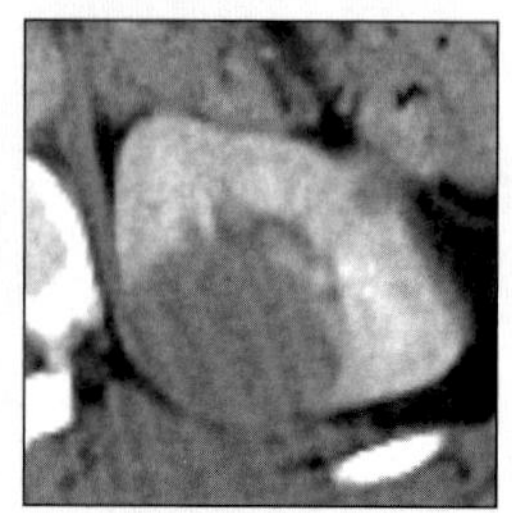

Lymphoma

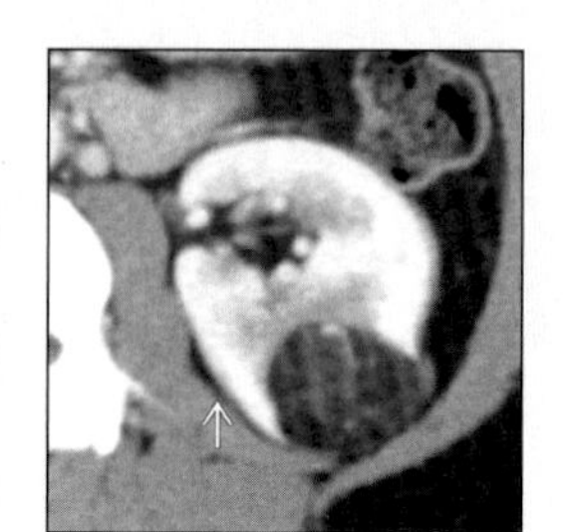

Angiomyolipoma

RENAL ONCOCYTOMA

Key Facts

Terminology
- Proximal tubular adenoma with oncocytic features
- Benign oxyphilic adenoma
- Epithelial tumor composed of eosinophilic epithelial cells, arising from intercalated cells of collecting ducts

Imaging Findings
- Best diagnostic clue: Solid renal cortical mass lesion with central stellate scar
- Calcification, necrosis & hemorrhage are rare
- Tumor: Homogeneous enhancement
- Central stellate scar: Nonenhancing
- "Spoke-wheel" pattern of tumor vascularity
- Absence of bizarre neoplastic vessels
- No AV shunting, vascular puddling, renal vein invasion

Top Differential Diagnoses
- Renal cell carcinoma (RCC)
- Renal angiomyolipoma (AML)
- Renal metastases & lymphoma

Pathology
- Deep-brown or mahogany color in contrast to yellowish-orange RCC
- Fleshy central scar: Characteristic of oncocytoma

Diagnostic Checklist
- Not possible to distinguish from RCC on imaging
- May suggest diagnosis → nephron-sparing surgery
- Pathologic diagnosis requires entire tumor due to oncocytic features of RCC
- Well-defined, solid cortical mass lesion with central scar & "spoke-wheel" pattern of vessels on angiogram

    - Central scar with radiating low-density areas (seen in large tumors)
    - ± Displacement of renal collecting system
    - Absence of malignant features: Invasion or infiltration into perinephric fat, collecting system or vessels
    - No regional lymphadenopathy & metastases

MR Findings
- T1WI
  - Well-defined, homogeneous mass
  - Iso-/hypointense relative to renal cortex
  - Internal radiating architecture may be seen
- T2WI
  - Slightly hyperintense relative to renal cortex
  - "Spoke-wheel" pattern
    - Internal radiating low signal areas
    - Rarely may appear bright on T2WI
  - Tumor necrosis: Common feature of malignant mass
    - Hypointense on T1WI
    - Hyperintense on T2WI
- T1 C+
  - Tumor: Homogeneous enhancement
  - Central stellate scar: Nonenhancing
  - Malignant features of an invasive tumor are absent

Ultrasonographic Findings
- Real Time
  - Well-defined, homogeneous, hypo-/isoechoic mass
  - Central scar: Usually echogenic
- Color Doppler: Tumor with central radiating vessels

Angiographic Findings
- Conventional
  - "Spoke-wheel" pattern of tumor vascularity
  - Homogeneous dense tumoral contrast during capillary phase
  - Sharp demarcation from kidney and surrounding areas
  - Peritumoral halo (lucent-rim sign)
  - In contrast to RCC
    - Absence of bizarre neoplastic vessels
    - No AV shunting, vascular puddling, renal vein invasion

Nuclear Medicine Findings
- FDG-PET
  - Less FDG uptake than RCCs
  - Amount of uptake, usually isointense to renal tissue
  - Occasionally have uptake in the range of RCC
- Tc99m dimercaptosuccinic acid (DMSA) scans
  - Photopenic area displacing cortex & collecting system

Imaging Recommendations
- Helical NE + CECT; MR

DIFFERENTIAL DIAGNOSIS

Renal cell carcinoma (RCC)
- Most common primary renal malignant neoplasm
- NECT
  - Solid mass lesion of density, 30-50 HU range
  - Necrotic, calcific, hemorrhagic, cystic components often seen in RCC, rare or absent in oncocytoma
    - Central necrosis of RCC may mimic central scar of oncocytoma
- CECT
  - Highly enhancing mass with thick septa & nodularity
  - ± Invasion or infiltration into perinephric fat, collecting system
  - ± Tumor extension into renal vein & IVC
  - ± Regional lymphadenopathy & hypervascular metastases
- Bilateral RCC
  - von Hippel-Lindau disease, tuberous sclerosis
- Cut section: Yellowish-orange appearance
- Histology: Granular eosinophilic cytoplasm with perinuclear clearing
- Immunohistochemical tests: Vimentin positivity

Renal angiomyolipoma (AML)
- Tumor composed of abnormal blood vessels, smooth muscle & fatty components
- NECT

- Well-defined, cortical heterogeneous tumor, predominantly of fat density (-30 to -100 HU)
  - Renal mass with fat is almost diagnostic of AML
- Calcification rarely seen (if present, suspect RCC)
- Hemorrhage may be seen in large tumors > 4 cm
- When multiple AML seen, suspect tuberous sclerosis

- CECT
  - Varied enhancement based on amount of fat & vascular components
- CTA: Aneurysmal renal vessels may be seen
- MR
  - Tumor with increased fat content
    - T1WI: Hyperintense
    - Fat-suppression sequences: Show signal "drop-out"

Renal metastases & lymphoma

- Renal metastases
  - Multiple renal masses of low density
  - When seen as a solitary dense mass, may mimic oncocytoma
  - Usually known primary tumor & metastases in other sites strongly suggests renal metastases
- Renal lymphoma
  - Primary very rare; secondary from generalized spread (more common)
  - Bilateral involvement seen in 40-60% of cases
  - Non-Hodgkin lymphoma > common than Hodgkin
  - Hypovascular; moderate enhancement with contrast

PATHOLOGY

General Features

- Genetics
  - Deletion of chromosome 1 & sex chromosomes
  - Balanced translocation involving 11q13
- Etiology: Exact etiology unknown
- Epidemiology: Usually incidental on imaging or autopsy
- Associated abnormalities: Oncocytoma & RCC may coexist

Gross Pathologic & Surgical Features

- Large & spherical tumor with a pseudocapsule
- Cut section
  - Deep-brown or mahogany color in contrast to yellowish-orange RCC
  - Fleshy central scar: Characteristic of oncocytoma
  - Rarely necrosis, hemorrhage, calcification

Microscopic Features

- Composed of oncocytes
  - Large cells with granular eosinophilic cytoplasm
  - Cells arranged in sheets, tubulocystic or combined
  - Nuclei appear smooth, round, with minimal atypia
- Electron microscopy: Cells packed with mitochondria

CLINICAL ISSUES

Presentation

- Most common signs/symptoms
  - Mostly asymptomatic
  - Occasionally flank pain, hematuria
  - Hypertension or palpable mass on physical exam
- Lab-data
  - Immunohistochemical tests
    - Oncocytoma: Cathepsin H positivity
    - Chromophobe RCC: Vimentin positivity
- Diagnosis
  - Percutaneous needle biopsy unreliable
  - Whole tumor resection & histopathology

Demographics

- Age
  - Usually seen in older age group
  - Peak incidence occurring in sixth & seventh decades
- Gender: M:F = 2-3:1

Natural History & Prognosis

- Oncocytomas are benign tumors
- Complications (rare)
  - Hematuria
  - Obstructive nephropathy (extrinsic compression of renal pelvis)
- Prognosis
  - Excellent after partial or total nephrectomy

Treatment

- Surgical resection: Partial or total nephrectomy

DIAGNOSTIC CHECKLIST

Consider

- Not possible to distinguish from RCC on imaging
- May suggest diagnosis → nephron-sparing surgery
- Pathologic diagnosis requires entire tumor due to oncocytic features of RCC

Image Interpretation Pearls

- Well-defined, solid cortical mass lesion with central scar & "spoke-wheel" pattern of vessels on angiogram

SELECTED REFERENCES

1. Landau A et al: Adrenocortical oncocytoma: benign or malignant? ANZ J Surg. 74(5):390, 2004
2. Bandhu S et al: Spoke-wheel pattern in renal oncocytoma seen on double-phase helical CT. Australas Radiol. 47(3):298-301, 2003
3. Katz DS et al: Renal oncocytomatosis. Am J Kidney Dis. 27(4):579-82, 1996
4. Licht MR: Renal adenoma and oncocytoma. Semin Urol Oncol. 13(4):262-6, 1995
5. Davidson AJ et al: Renal oncocytoma and carcinoma: failure of differentiation with CT. Radiology. 186(3):693-6, 1993
6. Remark RR et al: Magnetic resonance imaging of renal oncocytoma. Urology. 31(2):176-9, 1988
7. Curry NS et al: Small renal neoplasms: diagnostic imaging, pathologic features, and clinical course. Radiology. 158(1):113-7, 1986
8. Levine E: Small renal neoplasms: diagnostic imaging, pathologic features, and clinical course. Radiology. 159(3):817-8, 1986
9. Levine E et al: Computed tomography of renal oncocytoma. AJR Am J Roentgenol. 141(4):741-6, 1983

IMAGE GALLERY

Typical

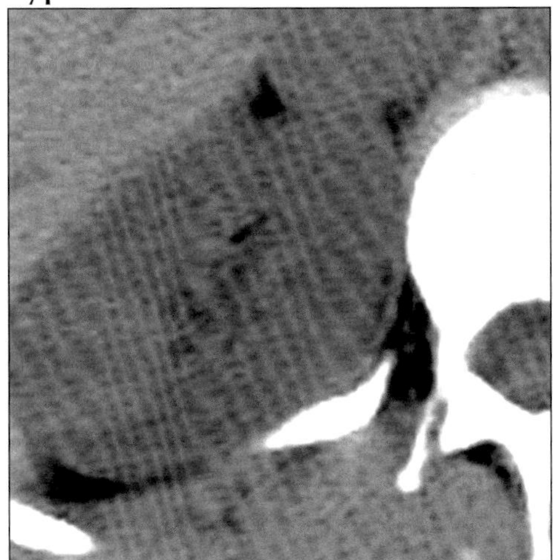

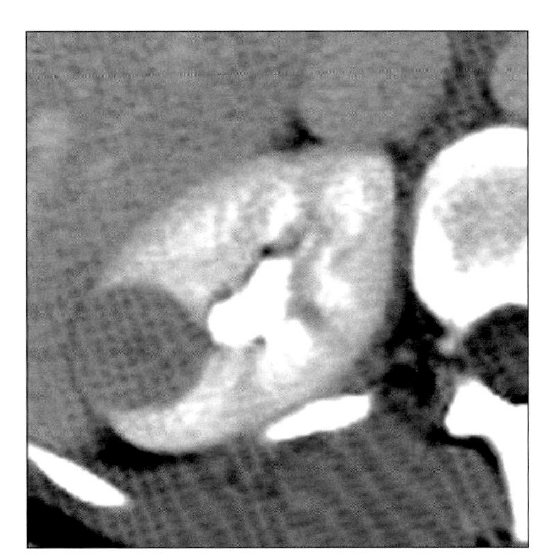

(Left) *Axial NECT shows spherical mass that is slightly hyperdense to the kidney.* ***(Right)*** *Axial CECT shows only slight enhancement of the mass and a small central scar.*

Typical

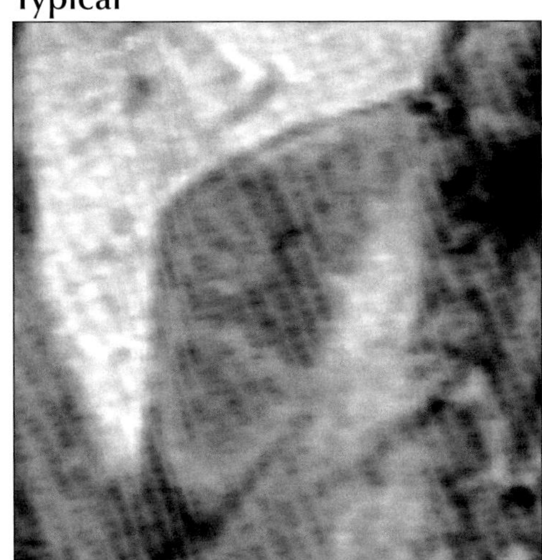

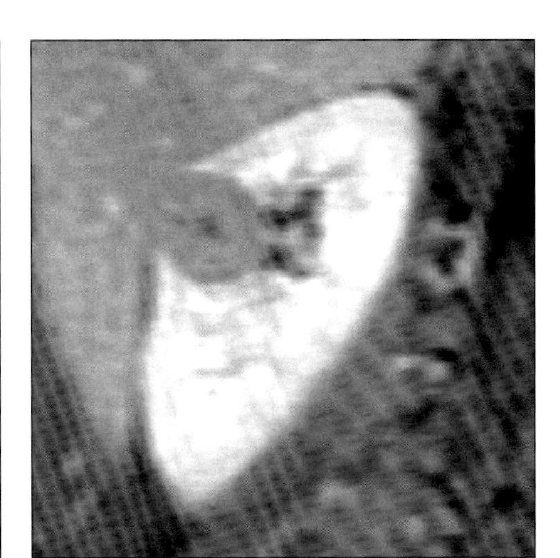

(Left) *Coronal T1WI MR (nonenhanced) shows an encapsulated spherical mass that is slightly hyperintense to kidney.* ***(Right)*** *Coronal T1WI MR (enhanced) shows slight enhancement of the mass and a small central scar.*

Variant

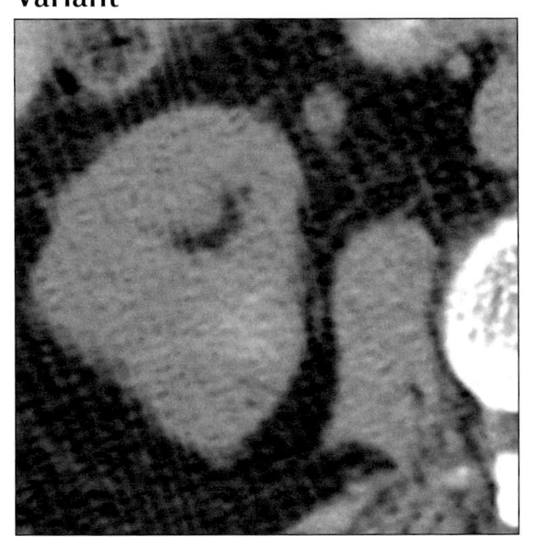

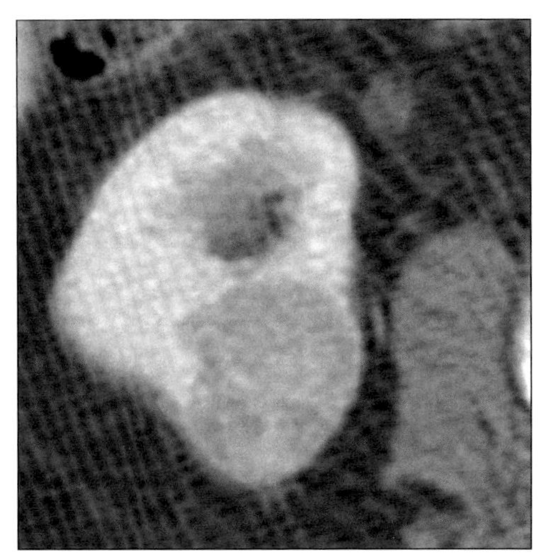

(Left) *Axial NECT shows an exophytic spherical mass that is moderately heterogeneous with hyperdense focus.* ***(Right)*** *Axial CECT shows slight enhancement of the mass, including the hyperdense focus, indicating that this does not represent hemorrhage.*

RENAL ANGIOMYOLIPOMA

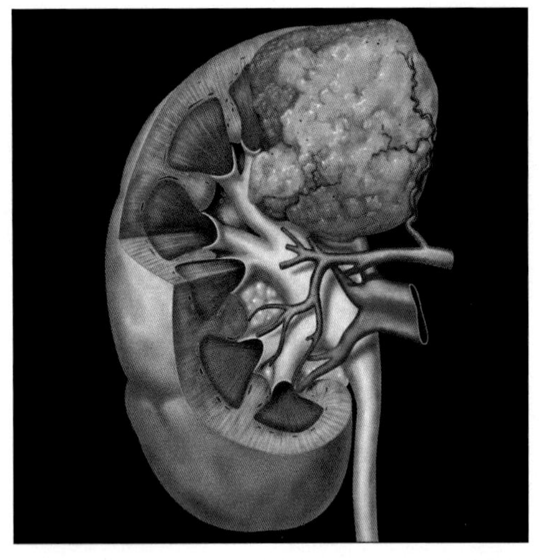

Graphic shows vascular renal mass with fatty and soft tissue components.

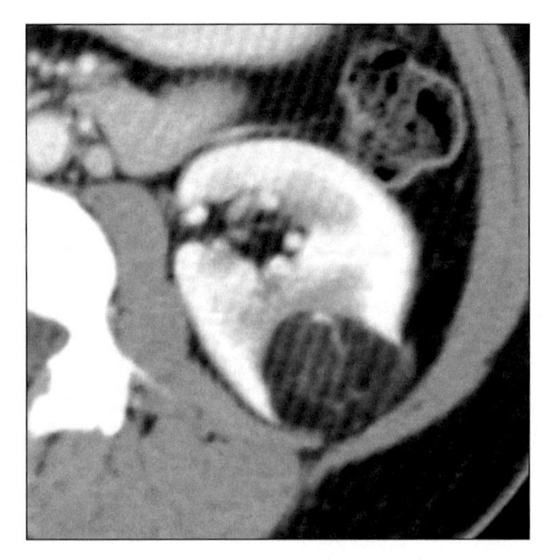

Axial CECT shows spherical renal mass with fat component and prominent vessels.

TERMINOLOGY

Abbreviations and Synonyms

- Angiomyolipoma (AML)
- Synonyms: Renal hamartoma or choristoma

Definitions

- Benign renal tumor composed of abnormal blood vessels, smooth muscle & fatty components

IMAGING FINDINGS

General Features

- Best diagnostic clue: Intrarenal fatty mass
- Location: Intrarenal (cortex) or exophytic in location
- Size
  - Varies in size
    - May range from few mm to 25 cm or more
- Morphology
  - Lobulated mass lesion of varied size + number
  - Usually discrete, rarely diffuse parenchymal mass
  - Tumor components
    - Abnormal blood vessels
    - Smooth muscle
    - Adipose tissue
  - One or two of the components may predominate
- Other general features
  - Most common benign tumor of kidney
  - 80% of cases are incidental findings on imaging
  - Usually solitary & unilateral; occasionally multiple & bilateral
  - 90% unilateral: Usually not associated with tuberous sclerosis
  - 10% multiple & bilateral: Usually due to tuberous sclerosis
  - 20% of renal AML cases have tuberous sclerosis
  - 80% of tuberous sclerosis patients have renal AMLs
  - Tuberous sclerosis (Bourneville disease)
    - Facial angiofibroma, ungual fibroma
    - Retinal hamartoma
    - Cortical tubers & subependymal nodules
    - Multiple renal angiomyolipomas
  - AML occurs in 15% of patients with lymphangiomyomatosis
  - Lymphangiomyomatosis
    - Hamartomatous proliferation of smooth muscle along lymphatic system, most commonly in lung
    - Exclusively affects women

Radiographic Findings

- Radiography
  - Tomographic sections: Radiolucent mass
    - When lesion contains large amount of fat
- IVP
  - Small tumor: Well defined radiolucent defect
  - Large tumor: Distortion of collecting system

DDx: Renal Mass

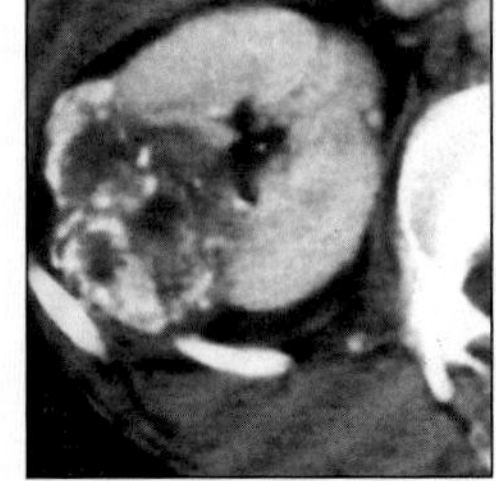

Renal Carcinoma

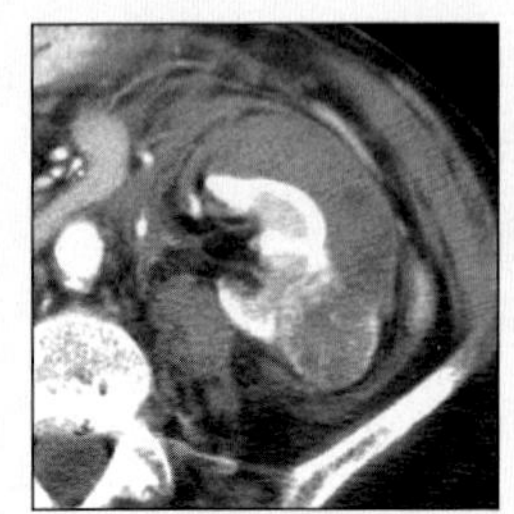

RCC with Bleed

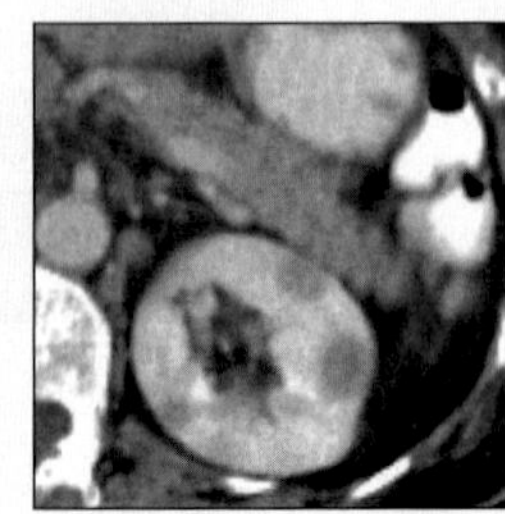

Renal Lymphoma

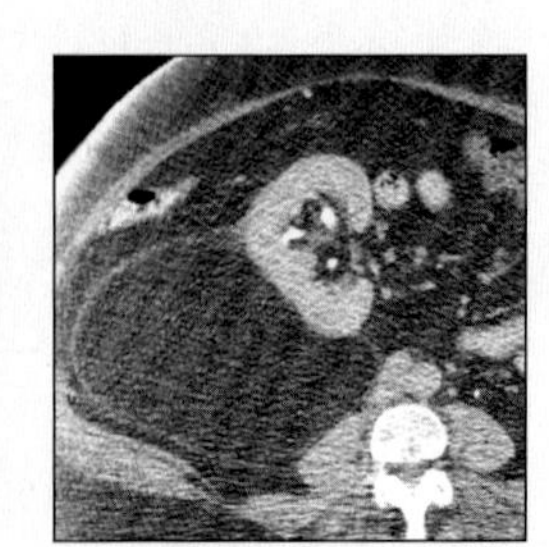

Liposarcoma

RENAL ANGIOMYOLIPOMA

Key Facts

Terminology

- Benign renal tumor composed of abnormal blood vessels, smooth muscle & fatty components

Imaging Findings

- Best diagnostic clue: Intrarenal fatty mass
- May range from few mm to 25 cm or more
- Well-marginated cortical heterogeneous tumor, predominantly of fat density (-30 to -100 HU)
- When multiple AML seen, suspect tuberous sclerosis
- ~ 5% have no detectable fat on CT, such AML cannot be diagnosed by CT or other imaging modalities
- Hemorrhage may be seen in large tumors > 4 cm
- Calcification rarely seen; if present suspect RCC
- RCC may "de-differentiate", form bone + fat
- CTA: Aneurysmal renal vessels may be seen

Top Differential Diagnoses

- Renal cell carcinoma (RCC)
- Renal metastases & lymphoma
- Renal oncocytoma
- Perirenal liposarcoma

Pathology

- Benign mesenchymal tumor of kidney
- 80% isolated (sporadic) AML
- 20% AML associated with tuberous sclerosis

Diagnostic Checklist

- Use CT pixel analysis to detect small foci of fat on NECT
- Calcification or ossification within tumor highly suggestive of renal cell carcinoma, not AML

  - Multiple lucent defects when associated with tuberous sclerosis

CT Findings

- NECT
  - Well-marginated cortical heterogeneous tumor, predominantly of fat density (-30 to -100 HU)
  - Renal mass with fat is almost diagnostic of AML
    - Variable amounts of fat present
  - When multiple AML seen, suspect tuberous sclerosis
  - ~ 5% have no detectable fat on CT, such AML cannot be diagnosed by CT or other imaging modalities
  - Hemorrhage may be seen in large tumors > 4 cm
  - Calcification rarely seen; if present suspect RCC
    - RCC may "de-differentiate", form bone + fat
- CECT
  - Varied enhancement pattern based on amount of fat & vascular components
  - AML is not known to undergo malignant change, but benign satellite deposits may be seen in regional lymph nodes, liver & spleen
  - Rarely extension of tumor into IVC seen
- CTA: Aneurysmal renal vessels may be seen

MR Findings

- Varied signal intensity due to vessels, muscle & fat
- Tumor with increased fat content
  - T1WI: Hyperintense
  - Fat suppression sequences: Signal loss
- T1 C+
  - Tumor with increased fat content
    - Show minimal enhancement
  - Tumor with high vascular component
    - Show significant enhancement

Ultrasonographic Findings

- Real Time
  - Markedly hyperechoic mass relative to normal renal tissue
  - May look similar to small renal cell carcinoma which may also be hyperechoic

Angiographic Findings

- Tumor with increased vascular component
  - Highly vascular mass lesion
  - Multisacculated pseudoaneurysms
  - Absence of arteriovenous shunts
  - "Sunburst" appearance of capillary nephrogram
  - "Onion peel" appearance of peripheral vessels in venous phase

Imaging Recommendations

- Best imaging approach is thin-section (≤ 3 mm) NECT and CECT

DIFFERENTIAL DIAGNOSIS

Renal cell carcinoma (RCC)

- Rarely reported to contain fat (engulfed renal sinus fat)
- If AML is fat deficient, may mimic renal cell carcinoma
- Calcification or ossification within tumor highly suggestive of RCC (calcification very rare in AML)

Renal metastases & lymphoma

- Renal metastases
  - Multiple renal masses of low density easily differentiated from multiple fatty AML
  - Usually known primary tumor & metastases in other sites strongly suggests renal metastases
  - Occasionally, present as a large solitary mass, but devoid of fat
- Renal lymphoma
  - Primary very rare; secondary from generalized spread (more common)
  - Bilateral involvement is seen in 40-60% of cases
  - Tuberous sclerosis patients with bilateral AML can simulate renal lymphoma if AML lesions lack fat
  - Non-Hodgkin lymphoma > common than Hodgkin
  - Hypovascular; moderate enhancement with contrast

Renal oncocytoma

- Rare benign renal tumor; rarely contains fat
- Proximal tubular adenoma with oncocytic features
- Well-defined mass of homogeneous density

RENAL ANGIOMYOLIPOMA

Perirenal liposarcoma

- Large exophytic AML may simulate retroperitoneal liposarcoma (both contain fat)
  - Renal parenchymal defect & enlarged vessels favor angiomyolipoma (AML)
  - Smooth compression of kidney & extension beyond perirenal space favor liposarcoma
- Very rarely liposarcoma originate within kidney
  - Typically subcapsular in location, lenticular in shape
- Diagnosis: Require entire tumoral resection

Wilms tumor

- Pediatric renal tumor that may contain fat
- Age difference at presentation avoids confusion

PATHOLOGY

General Features

- Etiology
  - Benign mesenchymal tumor of kidney
  - Hamartoma: Benign tumor consisting of tissues that normally occur in organ of origin
  - Choristoma: Benign tumor composed of tissues not normally occurring within organ of origin
- Epidemiology
  - 0.3-3% in autopsy series
  - 80% isolated (sporadic) AML
  - 20% AML associated with tuberous sclerosis
- Associated abnormalities: Tuberous sclerosis; lymphangiomyomatosis

Gross Pathologic & Surgical Features

- Round, lobulated, yellow-to-gray color secondary to fat content

Microscopic Features

- Variable amounts of angioid (vascular), myoid (smooth muscle), & lipoid (fatty) components

CLINICAL ISSUES

Presentation

- Most common signs/symptoms
  - Often asymptomatic, incidental CT finding
  - Hematuria, flank pain or palpable flank mass
  - Acute abdomen (spontaneous hemorrhage, rupture)
  - Occasionally, hypertension & chronic renal failure
  - Detected while screening tuberous sclerosis patients for renal cell carcinoma
  - Tuberous sclerosis
    - Mental retardation, epilepsy
    - Sebaceous adenomas of face
- Diagnosis: Imaging & biopsy

Demographics

- Age: Usually beyond 40 years old age group
- Gender
  - More common in females than males (M:F = 1:4)
  - AML associated with tuberous sclerosis (M:F = 1:1)

Natural History & Prognosis

- Complications
  - No malignant potential
  - Hemorrhage & rupture
- Prognosis
  - AML usually a slowly growing tumor
  - Usually good: After partial or complete nephrectomy
  - Poor: With hemorrhage, rupture, no treatment

Treatment

- Debate exists as to how to manage asymptomatic patients with AML
- Tumor size less than 4 cm
  - Conservative management with follow-up
- Tumor size more than 4 cm
  - Partial nephrectomy often recommended
- Those patients presenting with spontaneous bleeding treated with embolization initially
  - Surgery postponed until patient stabilized

DIAGNOSTIC CHECKLIST

Consider

- Use CT pixel analysis to detect small foci of fat on NECT

Image Interpretation Pearls

- Well-marginated, discrete fatty renal mass
- Calcification or ossification within tumor highly suggestive of renal cell carcinoma, not AML

SELECTED REFERENCES

1. Kim JK et al: Angiomyolipoma with minimal fat: differentiation from renal cell carcinoma at biphasic helical CT. Radiology. 230(3):677-84, 2004
2. Israel GM et al: CT differentiation of large exophytic renal angiomyolipomas and perirenal liposarcomas. AJR Am J Roentgenol. 179(3):769-73, 2002
3. Wilson SS et al: Angiomyolipoma with vena caval extension. Urology. 60(4):695-6, 2002
4. Yamakado K et al: Renal angiomyolipoma: relationships between tumor size, aneurysm formation, and rupture. Radiology. 225(1):78-82, 2002
5. Lemaitre L et al: Imaging of angiomyolipomas. Semin Ultrasound CT MR. 18(2):100-14, 1997
6. Katz DS et al: Massive renal angiomyolipoma in tuberous sclerosis. Clin Imaging. 21(3):200-2, 1997
7. Siegel CL et al: Angiomyolipoma and renal cell carcinoma: US differentiation. Radiology. 198(3):789-93, 1996
8. Curry NS: Small renal masses (lesions smaller than 3 cm): imaging evaluation and management. AJR Am J Roentgenol. 164(2):355-62, 1995
9. Curry NS et al: Intratumoral fat in a renal oncocytoma mimicking angiomyolipoma. AJR Am J Roentgenol. 154(2):307-8, 1990
10. Bosniak MA et al: CT diagnosis of renal angiomyolipoma: the importance of detecting small amounts of fat. AJR Am J Roentgenol. 151(3):497-501, 1988

IMAGE GALLERY

Typical

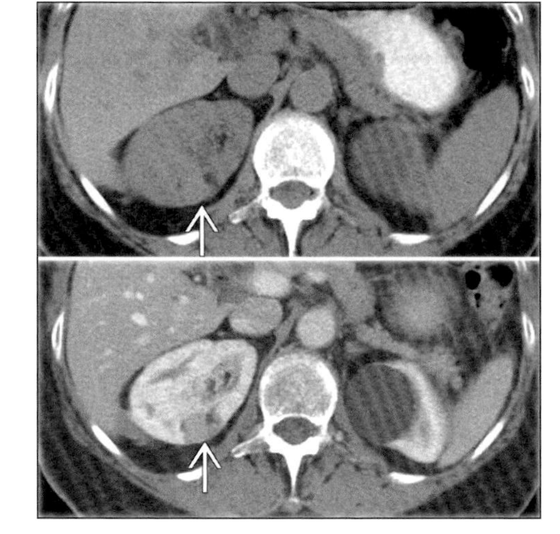

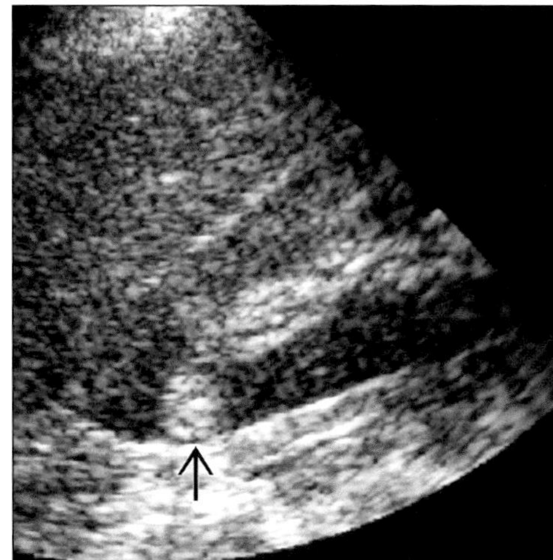

(Left) *Axial NECT and CECT show subtle foci of fat in multiple AMLs in right kidney. Simple cyst in left kidney (arrows).* ***(Right)*** *Sagittal sonogram shows hyperechoic mass (arrow) in upper pole right kidney.*

Typical

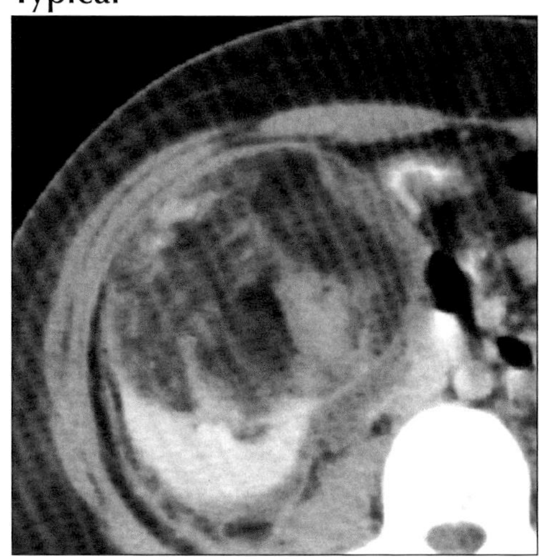

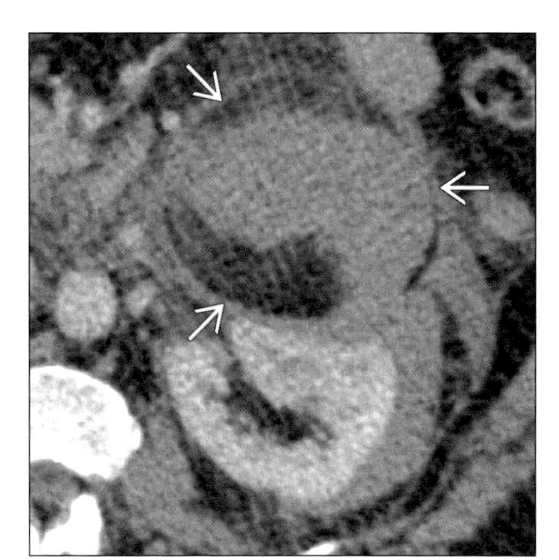

(Left) *Axial CECT shows large heterogeneous fat-containing mass in right kidney.* ***(Right)*** *Axial CECT shows spontaneous perirenal bleed from left renal AML (arrows).*

Variant

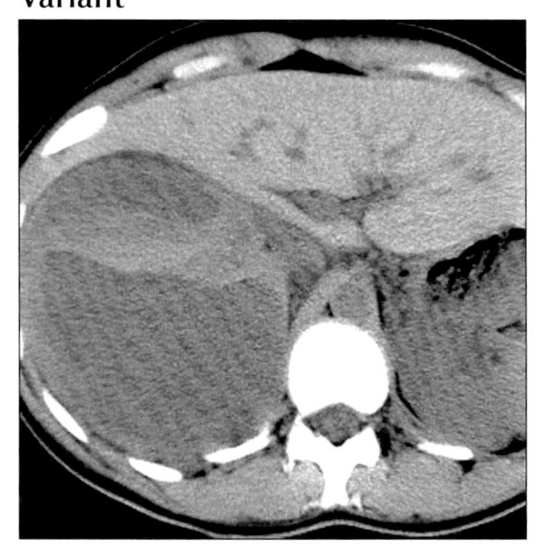

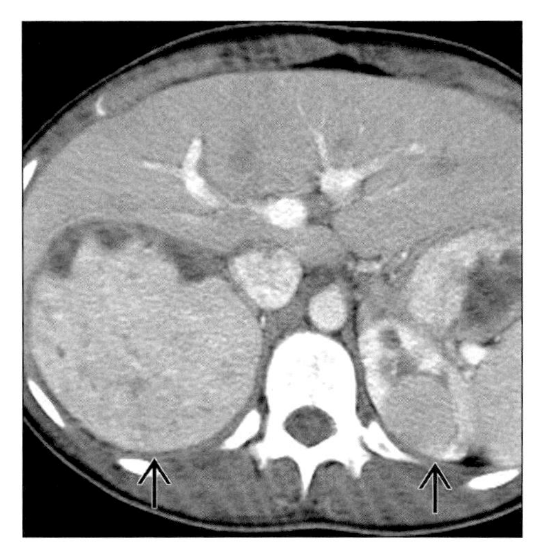

(Left) *Axial NECT shows spontaneous bleed from right renal mass (AML) without an obvious fat component.* ***(Right)*** *Axial CECT shows bilateral enhancing renal masses (arrows) in a patient with tuberous sclerosis; AMLs without obvious fat component.*

MULTILOCULAR CYSTIC NEPHROMA

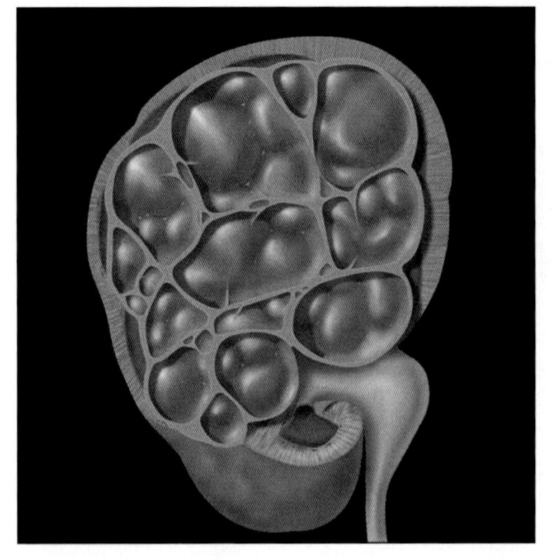

Graphic shows multilocular cystic mass that herniates into the renal hilum.

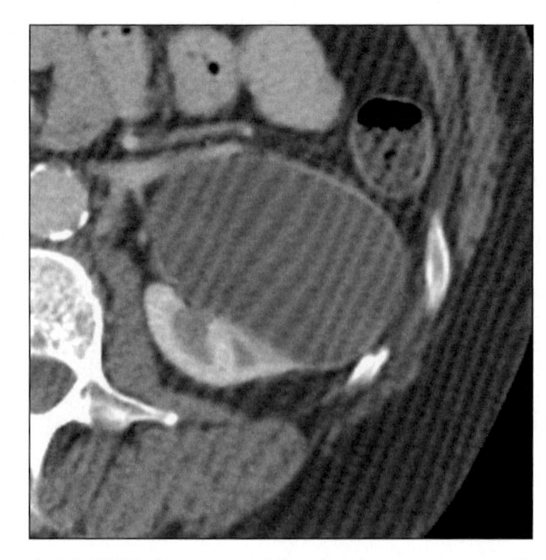

Axial CECT shows a multiloculated cystic mass with enhancing septa.

TERMINOLOGY

Abbreviations and Synonyms

- Multilocular cystic nephroma (MLCN)
- Synonym(s)
  - Multilocular cystic renal tumor; benign cystic nephroma; cyst adenoma
  - Cystic nephroblastoma; cystic partially differentiated nephroblastoma (CPDN)

Definitions

- Rare nonhereditary benign renal neoplasm arising from metanephric blastema

IMAGING FINDINGS

General Features

- Best diagnostic clue: Multilocular cystic mass herniating into renal hilum on CECT
- Location
  - Typically unilateral (usually lower pole)
  - Occasionally mass completely replaces kidney
- Size
  - Mass: Few cm to > 30 cm (average size 10 cm)
  - Locules: Few millimeters to 2.5 cm
- Morphology: Solitary, unilateral, well-circumscribed cystic mass with a thick fibrous capsule ± herniation into renal pelvis
- Other general features
  - Usually solitary (rarely multiple) & unilateral
  - Usually symptomatic in adults & asymptomatic palpable mass in children
  - Tumor may grow slowly over years or rapidly within months
  - Histological classification of multilocular cystic renal tumors
    - Multilocular cystic nephroma
    - Cystic partially differentiated nephroblastoma
  - MLCN & CPDN indistinguishable by imaging alone

Radiographic Findings

- Radiography
  - Plain x-ray abdomen
    - Soft tissue mass (particularly if large & displaces adjacent structures)
    - ± Curvilinear or amorphous calcification
- IVP
  - Depending on size & location of mass
    - ± Obstructed collecting system/pyelocaliectasis
    - ± Tumor herniates/protrudes into renal pelvis (mimicking a primary tumor of collecting system)
  - Tomogram of nephrographic phase show septations

DDx: Unilateral Multicystic Kidney

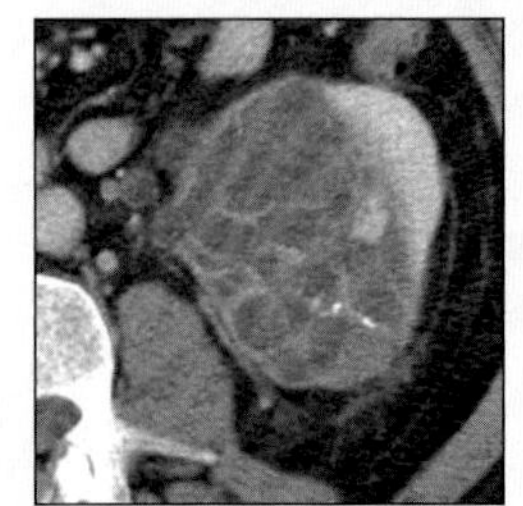

Renal Cell Carcinoma

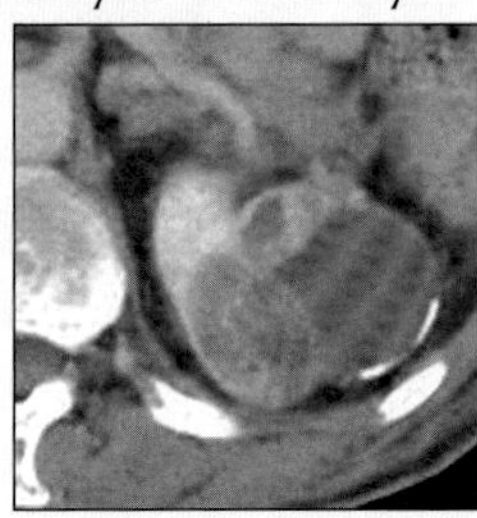

Renal Cell Carcinoma

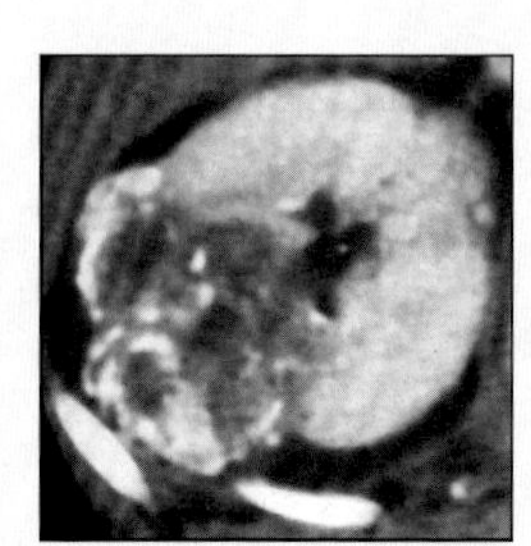

Renal Cell Carcinoma

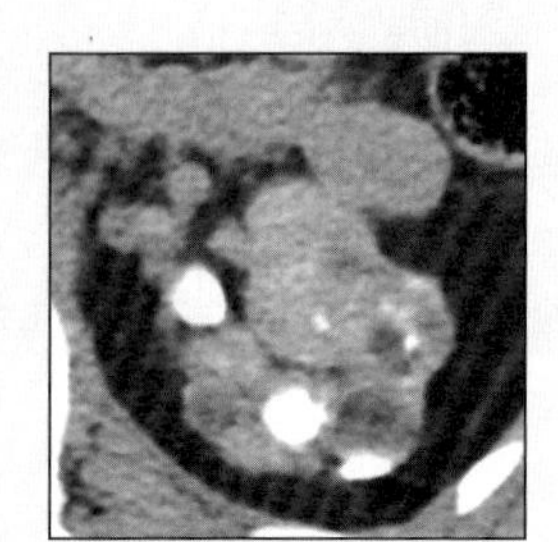

Segmental MCDK

MULTILOCULAR CYSTIC NEPHROMA

Key Facts

Terminology
- Multilocular cystic nephroma (MLCN)
- Rare nonhereditary benign renal neoplasm arising from metanephric blastema

Imaging Findings
- Best diagnostic clue: Multilocular cystic mass herniating into renal hilum on CECT
- Typically unilateral (usually lower pole)
- CT density: Equal to water or higher than water (proteinaceous fluid)
- Capsule: ± Enhancement
- Septa: Moderate enhancement (thin & linear)
- Cystic component: No enhancement
- No communication with collecting system
- Varied signal intensity (protein or blood product)

Top Differential Diagnoses
- Multilocular cystic renal cell carcinoma
- Cystic Wilms tumor
- Multicystic dysplastic kidney (MCDK)
- Cortical (simple) cysts

Clinical Issues
- Boys between 3 months & 2 years old (90% tumors)
- Females: < 5 yrs & peak in 5th & 6th decades

Diagnostic Checklist
- MLCN appears as a Bosniak class III or IV complex cystic mass (surgical excision is treatment of choice)
- Difficult to distinguish from multilocular cystic RCC
- MLCN: Unilateral, solitary, multiloculated, cystic mass of near water HU ± herniation into renal pelvis

CT Findings
- NECT
  - Large, well-defined, encapsulated, multiloculated cystic mass
  - CT density: Equal to water or higher than water (proteinaceous fluid)
  - Small locules (< 1 cm) & myxomatous material within cysts + fibrous septa may appear as solid mass
  - ± Curvilinear or amorphous calcification
- CECT
  - Capsule: ± Enhancement
  - Septa: Moderate enhancement (thin & linear)
  - Cystic component: No enhancement
    - No communication with collecting system
  - ± Distortion of collecting system or herniation of mass into renal hilum + obstructive changes

MR Findings
- T1WI
  - Multiloculated hypointense mass (clear fluid)
  - Varied signal intensity (protein or blood product)
- T2WI
  - Multiloculated hyperintense mass (clear fluid)
  - Varied signal intensity (protein or blood product)
  - Capsule & septa: Hypointense (fibrous content)
- T1 C+
  - Enhancement of thin septa
  - Herniation of mass into renal pelvis well depicted

Ultrasonographic Findings
- Real Time
  - Large, well-defined multiloculated cystic mass
  - Innumerable anechoic cysts + hyperechoic septa
  - Thick hyperechoic fibrous capsule
  - ± Hyperechoic areas within cystic tumor mimicking a solid component (due to numerous tiny cysts causing acoustic interfaces)

Angiographic Findings
- Conventional: Hypovascular mass (rarely avascular or hypervascular)

Nuclear Medicine Findings
- Technetium-99m diethylenetriaminepentaacetic acid
  - Photon defect corresponding to renal mass
  - Occasionally faint uptake by septa noted

Imaging Recommendations
- Helical NE + CECT; MR + T1 C+; US

DIFFERENTIAL DIAGNOSIS

Multilocular cystic renal cell carcinoma
- On imaging it mimics multilocular cystic nephroma
- MLCN may herniate into renal pelvis causing filling defect, obstructive uropathy & hematuria which is not seen in multilocular cystic RCC
- Occasionally focal areas of solid component may be seen to differentiate from MLCN
- Usually seen in adult men more than women
  - MLCN seen in boys & adult women
- Gross pathology: Simulates MLCN
- Diagnosis: Histology (clear cell type, ↓ nuclear grade)

Cystic Wilms tumor
- Grows entirely by expansion of large cystic spaces within stroma
- Septa: Numerous/thick where as thin/linear in MLCN

Multicystic dysplastic kidney (MCDK)
- Usually involves entire kidney
- Rarely focal or segmental
- Segmental MCDK: Mostly associated with ureteral duplication
- Segmental MCDK + absence of ureteral duplication may simulate multilocular cystic tumor
  - Both are histologically distinctive
  - Segmental MCDK have no capsule where as MLCN have a capsule

Cortical (simple) cysts
- Solitary or multiple; unilateral or bilateral
- Most cysts are unilocular & arise in cortex; less commonly from medulla

MULTILOCULAR CYSTIC NEPHROMA

- Sharply marginated, round & smooth in contour
- Thin imperceptible wall & no septations
- Low density masses of near water HU (-10 to + 20)
- Unilateral, solitary, multiloculated cyst with thin septa may simulate MLCN

PATHOLOGY

General Features

- General path comments
  - Embryology/anatomy
    - Tumor arises from metanephric blastema
- Etiology
  - Unknown
  - Pathogenesis: Benign equivalents of nephroblastoma, hamartoma
- Epidemiology: Rare tumor
- Associated abnormalities
  - Sporadic association with other anomalies
  - Coexist with angiomyolipoma; adrenal adenoma

Gross Pathologic & Surgical Features

- "Honeycombed" cystic areas of varied sizes
- Encapsulated by a thick fibrous capsule
- Noncommunicating locules separated by thin septa
  - CPDN locules separated by thick or irregular septa
- Mostly intraparenchymal tumors

Microscopic Features

- Locules: Lined by flattened or cuboidal epithelium
- Septa contains fibrous tissue + tubular elements
- Between cysts, loose stroma with spindle-shaped fibroblasts & cells resembling smooth muscle cells
- Tumor extended into renal pelvis may show foci of necrosis & hemorrhage
- No capsular, lymphatic or vascular invasion
- No mitotic figures
- MLCN: Septa have no undifferentiated elements
- Cystic partially differentiated nephroblastoma (CPDN)
  - Septa contain blastemal ± other embryonal elements

CLINICAL ISSUES

Presentation

- Most common signs/symptoms
  - Adults
    - Abdominal or flank pain; ± palpable mass (adults)
    - Hematuria; urinary tract infection (UTI); HTN
  - Children
    - Nonpainful; palpable abdominal or flank mass
    - ± Hematuria & urinary tract infection (UTI)
- Lab data
  - Urine analysis may show RBC ± WBC
- Diagnosis: Biopsy & histology

Demographics

- Age
  - Biphasic age distribution
  - Boys between 3 months & 2 years old (90% tumors)
  - Females: < 5 yrs & peak in 5th & 6th decades
- Gender
  - Biphasic sex distribution
  - Predominantly boys (2/3rd) in childhood & women (1/3rd) in adulthood

Natural History & Prognosis

- Complications
  - Local recurrence; obstructive uropathy; infection
  - Hemorrhagic when extended into renal pelvis
  - Very rarely malignant transformation
- Prognosis
  - Good: After nephrectomy
  - Few cases: Local recurrence

Treatment

- Surgical excision: Nephrectomy
  - Complete or partial with tumor-free margins

DIAGNOSTIC CHECKLIST

Consider

- MLCN appears as a Bosniak class III or IV complex cystic mass (surgical excision is treatment of choice)
- Difficult to distinguish from multilocular cystic RCC

Image Interpretation Pearls

- MLCN: Unilateral, solitary, multiloculated, cystic mass of near water HU ± herniation into renal pelvis

SELECTED REFERENCES

1. Hopkins JK et al: Best cases from the AFIP: cystic nephroma. Radiographics. 24(2):589-93, 2004
2. Fujimoto K et al: Spontaneously ruptured multilocular cystic nephroma. Int J Urol. 9(3):183-6, 2002
3. Kural AR et al: Multilocular cystic nephroma: an unusual localization. Urology. 52(5):897-9, 1998
4. Eble JN et al: Extensively cystic renal neoplasms: cystic nephroma, cystic partially differentiated nephroblastoma, multilocular cystic renal cell carcinoma, and cystic hamartoma of renal pelvis. Semin Diagn Pathol. 15(1):2-20, 1998
5. Levine E et al: Current concepts and controversies in imaging of renal cystic diseases. Urol Clin North Am. 24(3):523-43, 1997
6. Kettritz U et al: Multilocular cystic nephroma: MR imaging appearance with current techniques, including gadolinium enhancement. J Magn Reson Imaging. 6(1):145-8, 1996
7. Agrons GA et al: Multilocular cystic renal tumor in children: radiologic-pathologic correlation. Radiographics. 15(3):653-69, 1995
8. Yamashita Y et al: Multilocular cystic renal cell carcinoma presenting as a solid mass: radiologic evaluation. Abdom Imaging. 20(2):164-8, 1995
9. Castillo OA et al: Multilocular cysts of kidney. A study of 29 patients and review of literature. Urology. 37(2):156-62, 1991
10. Joshi VV et al: Multilocular cyst of the kidney (cystic nephroma) and cystic, partially differentiated nephroblastoma. Terminology and criteria for diagnosis. Cancer. 64(2):466-79, 1989
11. Madewell JE et al: Multilocular cystic nephroma: a radiographic-pathologic correlation of 58 patients. Radiology. 146(2):309-21, 1983

IMAGE GALLERY

Typical

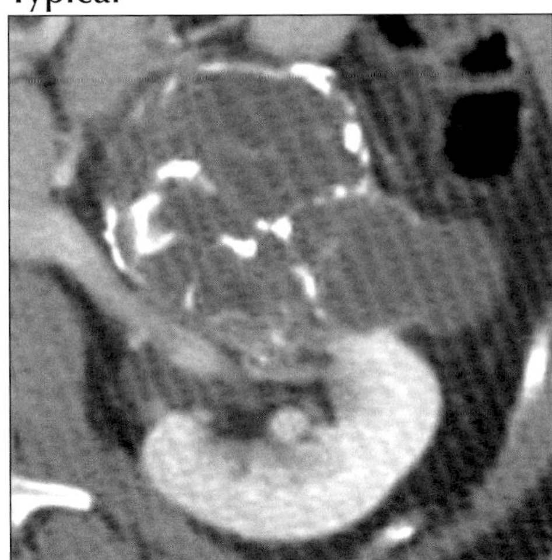
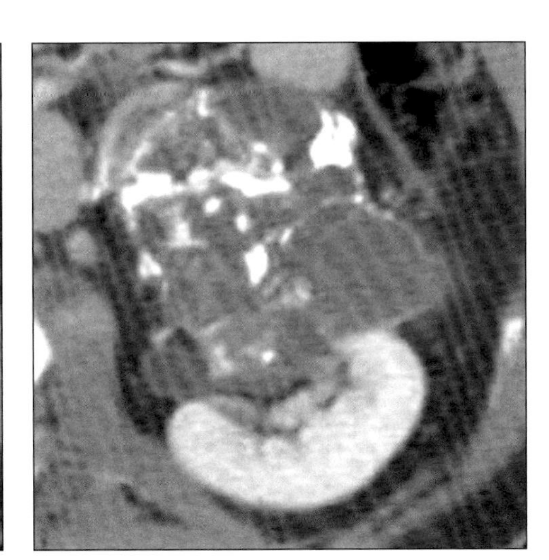

(Left) *Axial CECT shows multiloculated cystic mass with calcification of many of the septa.* ***(Right)*** *Axial CECT shows multicystic mass with calcified septa, herniating into the renal hilum.*

Typical

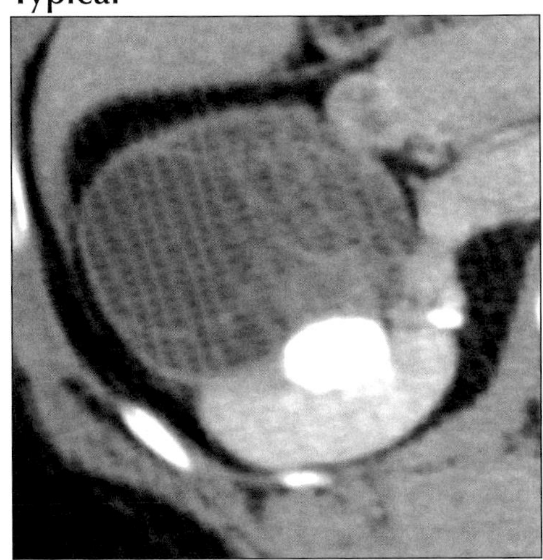

(Left) *Axial CECT shows multiloculated cystic mass with enhancing septa, herniating into the renal hilum.* ***(Right)*** *Surgical photograph of cross section of excised kidney shows multiloculated cystic mass extending into renal hilum.*

Typical

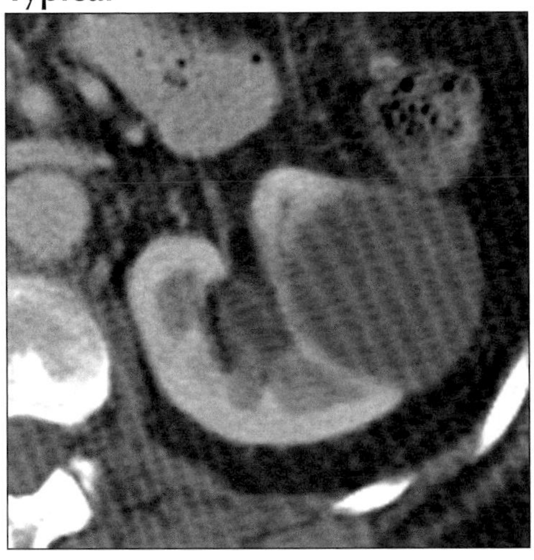
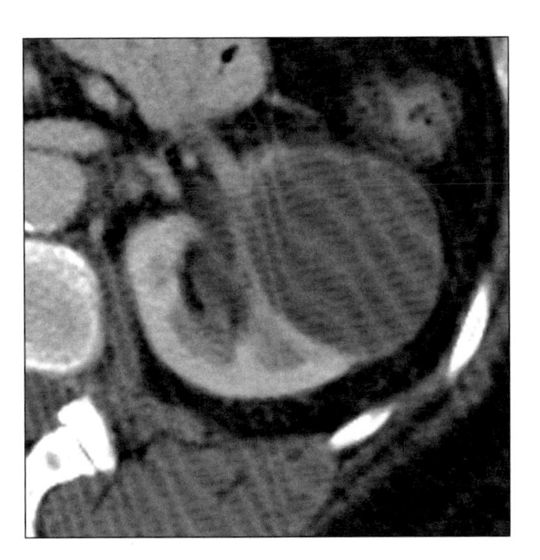

(Left) *Axial CECT shows multiloculated cystic mass with enhancing septa.* ***(Right)*** *Axial CECT shows multiloculated cystic mass with enhancing septa.*

RENAL CELL CARCINOMA

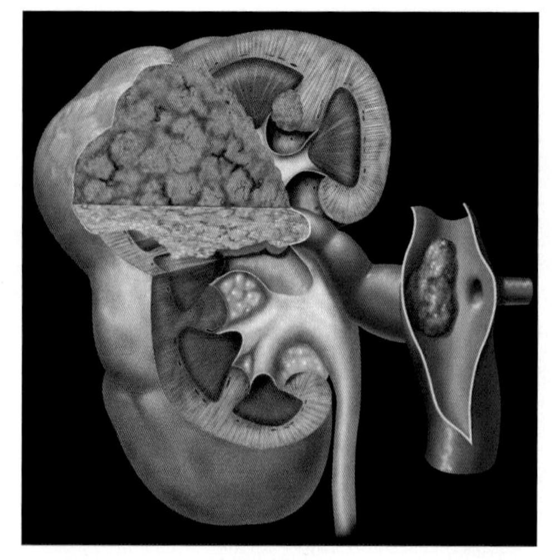

Graphic shows heterogeneous, vascular, expansile mass arising from renal cortex, invading renal vein and IVC.

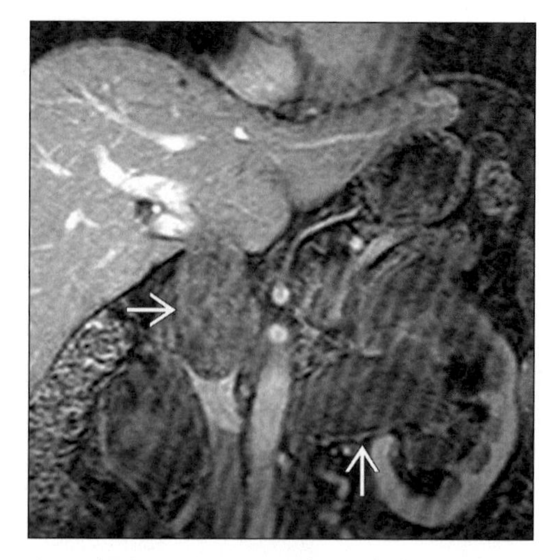

Coronal T1WI MR (contrast enhanced) shows expansile mass in upper pole of left kidney extending into renal vein and IVC (arrows).

TERMINOLOGY

Abbreviations and Synonyms

- Renal cell carcinoma (RCC)
- Hypernephroma, renal carcinoma

Definitions

- Malignant renal tumor arises from tubular epithelium

IMAGING FINDINGS

General Features

- Best diagnostic clue: Hypervascular cortical renal mass
- Location
  - 2% of sporadic RCC are bilateral and 16-25% of sporadic RCC are multicentric in the same kidney
  - Renal cortex (most common)
- Morphology: Usually solid mass; occasionally cystic
- Other general features
  - 25-40% found incidentally on abdominal CT or US

CT Findings

- NECT
  - Hyperdense, isodense or hypodense mass as compared with normal renal tissue
  - High density rim may separate mass from adjacent renal tissue (pseudocapsule)
  - Heterogenous mass (hemorrhage and necrosis); high (acute hemorrhage) or low attenuation (chronic)
  - Alteration of renal contour
  - ± Calcifications (10% of cases); amorphous internal (most common), curvilinear, (peripheral or central), dense or diffuse calcification
  - Small areas of fat attenuation (-80 to -120 HU) (rare)
  - Combination of fat and calcification suggests RCC, not renal angiomyolipoma
  - Cystic RCC
    - Uni- or multilocular cystic mass with a thick wall
    - Calcification of septa or tumor capsule
- CECT
  - Enhancement (attenuation value ↑ ≥ 20 HU) of solid mass, usually less than normal renal tissue
  - Small (≤ 3 cm), hypervascular mass better seen on nephrographic phase than arterial or corticomedullary phase
  - Heterogenous enhancement (hemorrhage and necrosis)
  - Rounded, lobulated margin demarcating from adjacent renal tissue
  - Lucent zone surrounding the mass (pseudocapsule)
  - ± Infiltration of calyces or renal pelvis; may simulate transitional cell carcinoma
  - Subcapsular hemorrhage, perinephric hemorrhage or both; hyperdense fluid collection (hematoma)
  - Exophytic growth pattern

DDx: Renal Mass

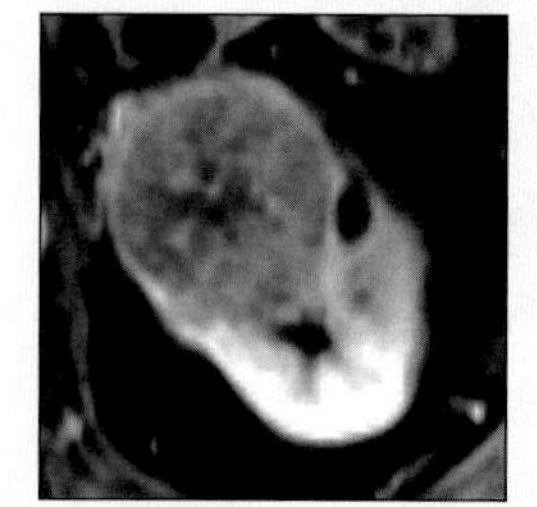

Oncocytoma

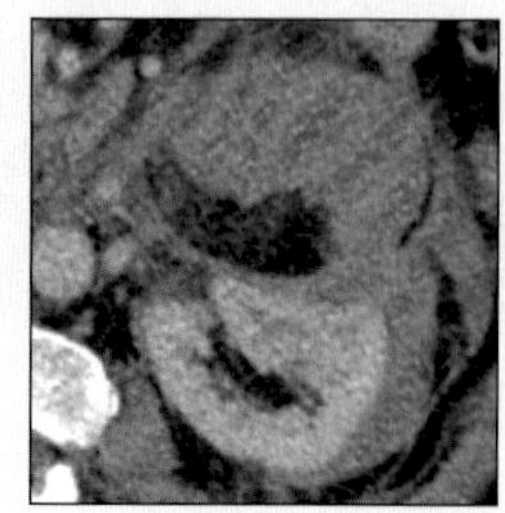

Angiomyolipoma

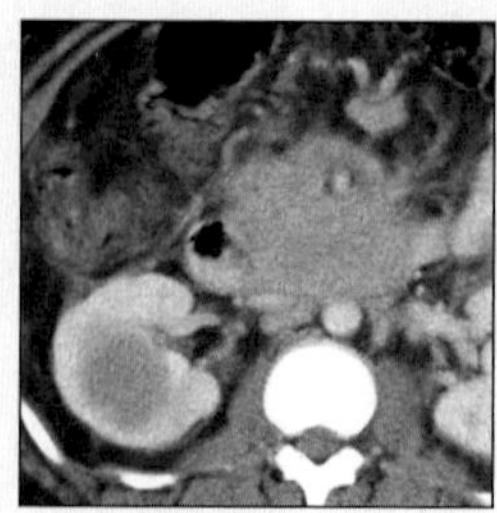

Lymphoma

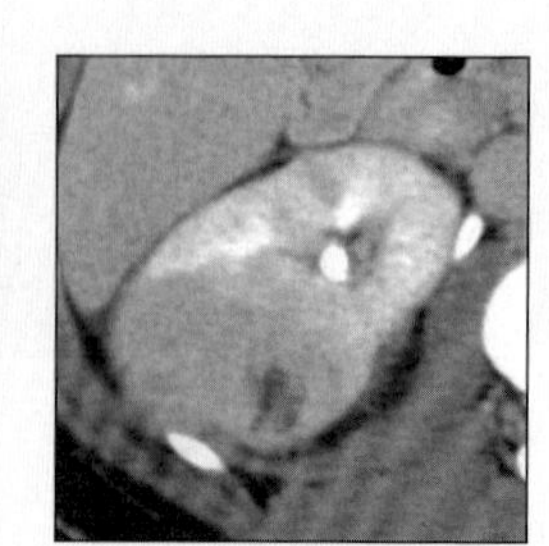

Pyelonephritis

RENAL CELL CARCINOMA

Key Facts

Imaging Findings

- Best diagnostic clue: Hypervascular cortical renal mass
- Morphology: Usually solid mass; occasionally cystic
- Rounded, lobulated margin demarcating from adjacent renal tissue
- Lucent zone surrounding the mass (pseudocapsule)
- Metastases (most to least common) in lung, liver, bone, adrenal, opposite kidney and brain; often hypervascular
- Mandatory: Nonenhanced and parenchymal phase (≥ 80 sec delay); optional corticomedullary (60 sec), excretory (2-5 min delay)

Top Differential Diagnoses

- Renal oncocytoma
- Renal angiomyolipoma
- Transitional cell carcinoma
- Renal metastases and lymphoma
- Renal infection
- Hemorrhagic renal cyst

Clinical Issues

- Gross hematuria (60%), flank pain (40%), palpable flank mass (30-40%) (classical triad < 10%)
- Age: 50-70 years of age
- Gender: M:F = 1.5-2:1

Diagnostic Checklist

- RCC frequently found "by accident" on CT
- Papillary type RCC is hypovascular, easily mistaken for cyst on all imaging modalities

- Renal venous (23%) and inferior vena cava (IVC) tumor extension (7%)
- Direct extension to adjacent muscles (i.e., diaphragm, psoas, quadratus lumborum, erector spinae) & viscera (i.e., colon, liver, pancreas, spleen)
- Metastases in local lymph nodes (≥ 1 cm)
- Metastases (most to least common) in lung, liver, bone, adrenal, opposite kidney and brain; often hypervascular
- Cystic RCC: Enhancing, smooth or nodular septa

MR Findings

- Isointense (60%) on T1 & T2WI or hyperintense (40%) on T2WI
- Hypointense band/rim on T1WI (25%) & T2WI (60%)
- T1 C+: Enhances, usually less than renal tissue
- Multiplanar ideal for renal venous & IVC extension

Ultrasonographic Findings

- Real Time
  - Liver overlies right kidney, a good acoustic window
  - Detect 85% of mass > 3 cm, ≤ 60% < 2 cm
  - Hyperechoic (48%), isoechoic (42%) or hypoechoic (10%) renal mass
  - Poor acoustic transmission, lobulated contour
  - Hypoechoic rim, irregular or poorly defined interface with adjacent renal tissue
  - Cystic RCC
    - Multiple septations with septal thickening nodules and/or calcification
    - Mural thickening, nodules and calcification
    - Diffuse, hypoechoic mass (hemorrhage and necrosis) with fluid-debris levels
- Color Doppler
  - Assess vascularity of renal mass; most prominent color flow around the periphery of the mass
  - High velocity signal from arteriovenous shunting

Imaging Recommendations

- Best imaging tool
  - Multiphase CT: Diagnosis and staging
    - Arterial phase: Early corticomedullary phase
    - Corticomedullary phase: Enhancing renal cortex, limited medullary
    - Nephrographic phase: Full capillary level renal enhancement
    - Excretory phase: Contrast in collecting system
  - MR: Staging is equal or better than CT
- Protocol advice
  - Multiphase CT
    - Mandatory: Nonenhanced and parenchymal phase (≥ 80 sec delay); optional corticomedullary (60 sec), excretory (2-5 min delay)
    - 3D mapping with volume rendering ± maximum intensity projection technique for pre-operative staging

DIFFERENTIAL DIAGNOSIS

Renal oncocytoma

- Sharply marginated, homogeneous mass with smooth, rounded margins
- Stellate area of low attenuation representing scar; may simulate hemorrhagic or necrotic RCC
- Imaging cannot differentiate; surgical resection needed

Renal angiomyolipoma

- Intratumoral, fat attenuating mass without calcification; very rare in RCC
- Indistinguishable from RCC if no visible fat on CT

Transitional cell carcinoma

- Renal infiltration → kidney enlargement and poorly defined margins without change in shape
- Renal pelvic filling defect, irregular narrowing of collecting system
- Encasing pelvicaliceal system → hypovascular tumor
- Rarely simulates RCC

Renal metastases and lymphoma

- Metastases (e.g., lung cancer, breast cancer, colon cancer, malignant melanoma)
  - Usually hypovascular with infiltrative growth
  - If hypervascular, may simulate RCC
- Lymphoma
  - Usually multiple or bilateral with infiltrative growth

- Hypoechoic, hypovascular, solitary, intrarenal mass ± adenopathy may simulate RCC
- Biopsy or chemotherapy may be used for diagnosis

Renal infection

- E.g., focal pyelonephritis, renal abscess
- Necrotic mass with renal enlargement
- Differentiate by clinical history and urinalysis

Hemorrhagic renal cyst

- Avascular RCC appears similar; use multiphase CT

PATHOLOGY

General Features

- Genetics: Mutation of von Hippel-Lindau (VHL) gene (3p25); autosomal dominant
- Etiology
  - Risk factors
    - Genetics: VHL
    - Environment: Smoking
    - Chemical: Diethylstilbestrol, fluoroacetamide, dimethylnitrosamine, lead, cadmium
- Epidemiology
  - Accounts for 2% of all cancers
  - Incidence: 30,000 per year in U.S.
  - Mortality: 12,000 per year in U.S.
- Associated abnormalities: Hereditary RCC, VHL, tuberous sclerosis, acquired cystic renal disease

Gross Pathologic & Surgical Features

- Completely solid to cystic mass with irregular lobulated margins
- Heterogenous appearance with hemorrhage and necrosis (cut section)

Microscopic Features

- Clear cell (70%), papillary (10-15%), granular cell (7%), chromophobe cell (5%), sarcomatoid (1.5%), collecting duct (< 1%)

Staging, Grading or Classification Criteria

- Robson classification of RCC with TNM correlation
  - Stage I (T1 or 2, N0M0): Tumor confined to kidney
  - Stage II (T3a, N0M0): Tumor spread to perinephric fat, but confined within renal fascia; possible ipsilateral adrenal gland
  - Stage IIIA (T3b-3c, N0M0): Tumor spread to renal vein, IVC or both
  - Stage IIIB (T1-3a, N1-3, M0): Tumor spread to local lymph nodes
  - Stage IIIC (T3b-3c, N1-3, M0): Tumor spread to local vessels and lymph nodes
  - Stage IVA (T4, any N, M0): Tumor spread to adjacent organs (except ipsilateral adrenal gland)
  - Stage IVB (any T & N, M1): Distant metastases

CLINICAL ISSUES

Presentation

- Most common signs/symptoms
  - Gross hematuria (60%), flank pain (40%), palpable flank mass (30-40%) (classical triad < 10%)
  - Fever, anorexia, weight loss, malaise, nausea, vomiting, constipation
  - Systemic manifestations by paraneoplastic (e.g., erythropoietin secretion, renin secretion), paraendocrine (e.g., hepatopathy) and serological (e.g., amyloidosis) syndromes
  - Distant metastases may cause symptoms of cough, hemoptysis, bone pain

Demographics

- Age: 50-70 years of age
- Gender: M:F = 1.5-2:1

Natural History & Prognosis

- Prognosis
  - 5, 10 year survival rate
    - Stage I: 67%, 56%
    - Stage II: 51%, 28%
    - Stage III: 33.5%, 20%
    - Stage IV: 13.5%, 3%
  - 5, 10 year survival rate in patients with multiple distant metastases: 5-10%, < 5%
  - Bilateral or multiple RCC have poorer survival rate; solitary RCC or metastasis has better survival rate
  - 20-30% with localized tumor who underwent surgical resection relapse in 15-18 months

Treatment

- Radical nephrectomy is the standard treatment
- Partial nephrectomy is a common alternative
  - Preoperatively, do CT angiography & 3D mapping
  - Requires ≤ 5 cm tumor size, peripheral location, exophytic extension and no invasion of vessels or lymph nodes
- Chemotherapy and hormone therapy: Limited
- Immunotherapy: 15% complete/partial response

DIAGNOSTIC CHECKLIST

Consider

- RCC frequently found "by accident" on CT

Image Interpretation Pearls

- Papillary type RCC is hypovascular, easily mistaken for cyst on all imaging modalities

SELECTED REFERENCES

1. Catalano C et al: High-resolution multidetector CT in the preoperative evaluation of patients with renal cell carcinoma. AJR Am J Roentgenol. 180(5):1271-7, 2003
2. Sheth S et al: Current concepts in the diagnosis and management of renal cell carcinoma: Role of multidetector CT & three-dimensional CT. Radiographics. 21:S237-54, 2001
3. Herts BR et al: Triphasic helical CT of the kidneys: Contribution of vascular phase scanning in patients before urologic surgery. AJR. 173: 1273-77, 1999
4. Birnbaum BA et al: Multiphasic renal CT: comparison of renal mass enhancement during the corticomedullary and nephrographic phases. Radiology. 200(3):753-8, 1996
5. Motzer RJ et al: Renal-cell carcinoma. N Engl J Med. 335(12):865-75, 1996

IMAGE GALLERY

Typical

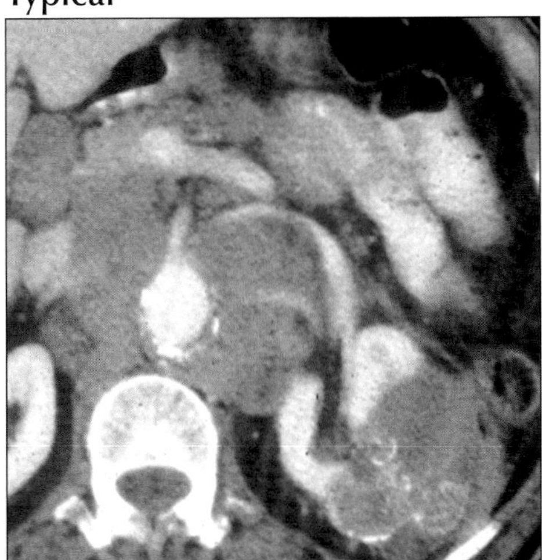

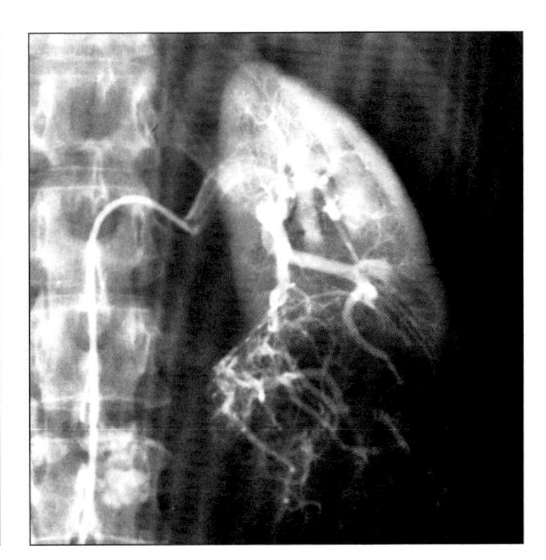

(Left) *Axial CECT shows an expansile renal mass with calcification and extensive regional lymphadenopathy.* ***(Right)*** *Selective angiogram shows hypervascular lower pole renal mass.*

Typical

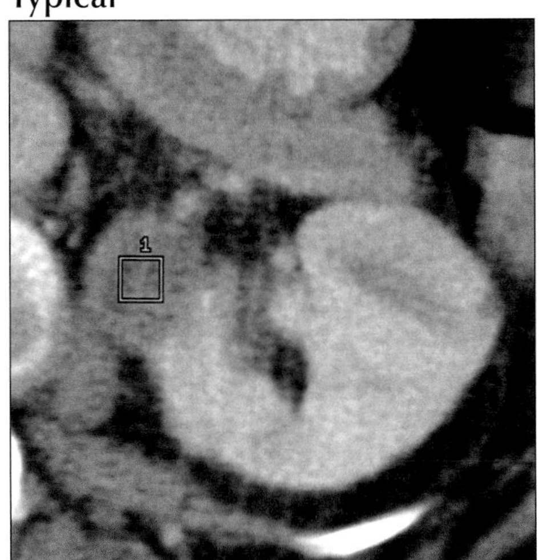

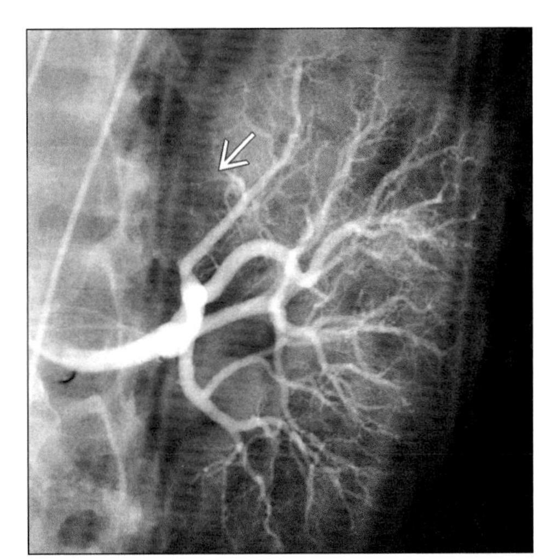

(Left) *Axial CECT shows a homogeneous renal mass that enhanced only 20 HU over the unenhanced scan; papillary renal cell carcinoma.* ***(Right)*** *Selective angiography in a patient with papillary RCC shows no obvious vascularity to this malignant mass (arrow).*

Typical

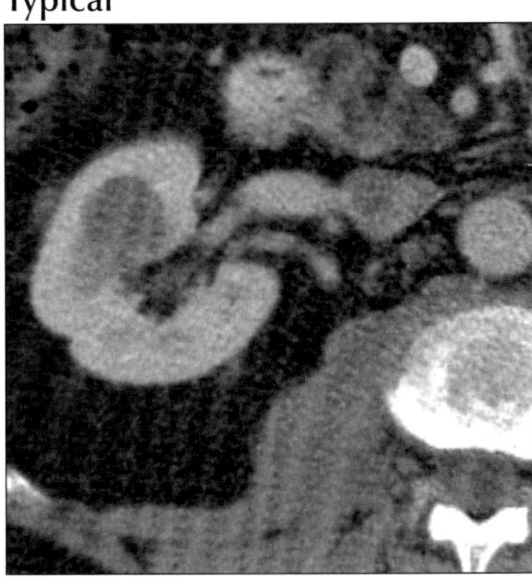

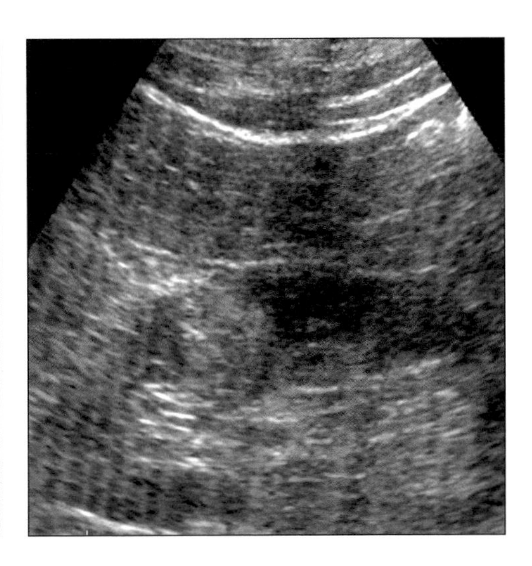

(Left) *Axial CECT shows a spherical hypodense right renal mass (ROI=60 HU): RCC.* ***(Right)*** *Sagittal sonogram shows echogenic mass in right kidney; RCC.*

RENAL MEDULLARY CARCINOMA

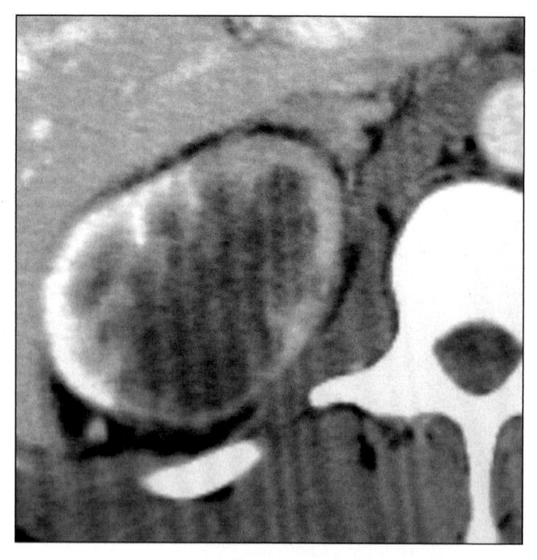

Axial CECT in a 32 year old African-American man with sickle cell trait shows an infiltrative tumor with the preservation of renal outline due to sickle cell disease.

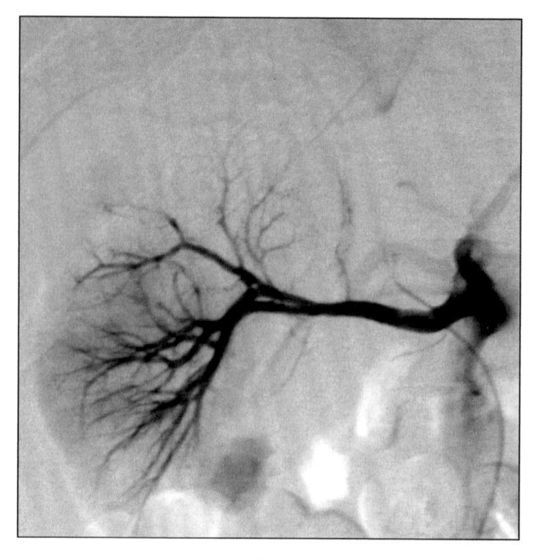

Arteriogram in a 32 year old man with sickle cell anemia shows a hypovascular mass in the center of the kidney.

TERMINOLOGY

Definitions

- Highly aggressive, very rare tumor of renal medulla arises in calyceal transitional epithelium

IMAGING FINDINGS

General Features

- Best diagnostic clue: Infiltrative, centrally located mass with preserved renal shape in a sickle cell patient
- Size: 4-12 cm; mean is 7 cm

CT Findings

- CECT
  - Central infiltrative tumor growth → caliectasis without pelviectasis
  - Fistula between necrotic cavity & collecting system
  - Infiltration from renal medulla to renal cortex → renal enlargement & preservation of reniform shape
  - Heterogeneous enhancement (necrosis)
  - Infiltration into peripelvic or perinephric fat
  - Renal venous & inferior vena cava tumor extension
  - Direct extension to retroperitoneal structures and ipsilateral adrenal gland
  - Metastases in lung, liver, local & retroperitoneal lymph nodes

Angiographic Findings

- Hypovascular mass

Imaging Recommendations

- Best imaging tool: CECT

DIFFERENTIAL DIAGNOSIS

Renal cell carcinoma

- Usually rounded or oval mass in cortical location and grow by expansion
- Collecting duct variant: Tumor location and pattern of growth may simulate renal medullary carcinoma

Transitional cell carcinoma

- Renal infiltration → kidney enlargement and poorly defined margins without change in shape
- Renal pelvic filling defect, irregular narrowing of collecting system
- Encasing pelvicaliceal system → hypovascular tumor

Renal angiomyolipoma

- Intratumoral fat attenuating mass

Renal metastases and lymphoma

- Metastases (e.g., lung cancer, breast cancer, colon cancer, malignant melanoma)
  - Usually hypovascular with infiltrative growth

DDx: Renal Mass

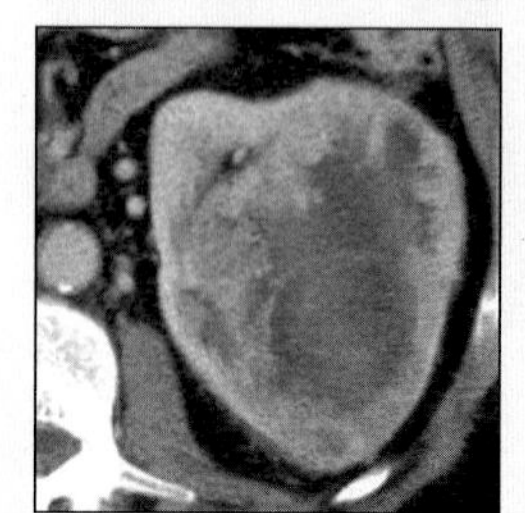

Renal Cell Carcinoma

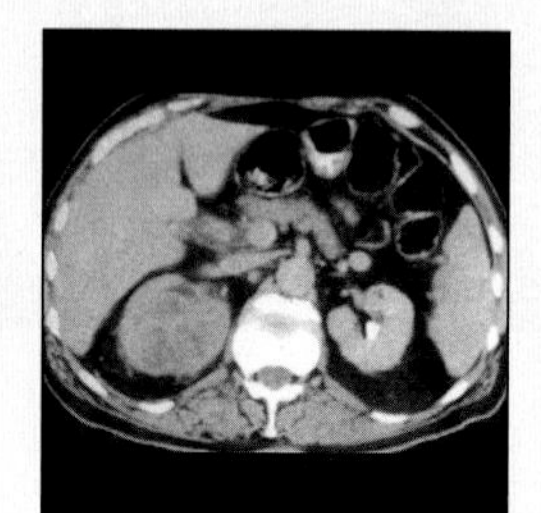

Transitional Cell Ca.

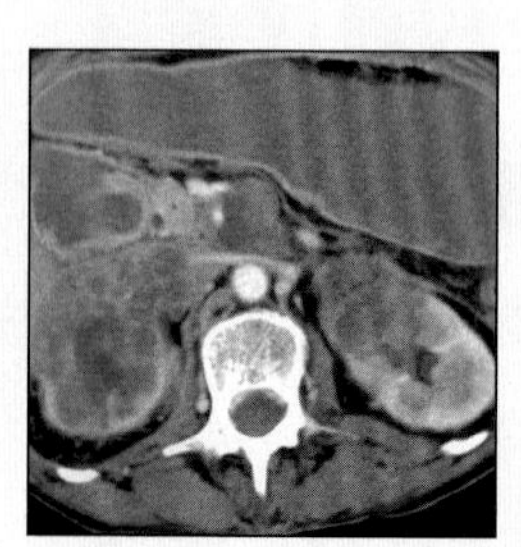

Renal Lymphoma

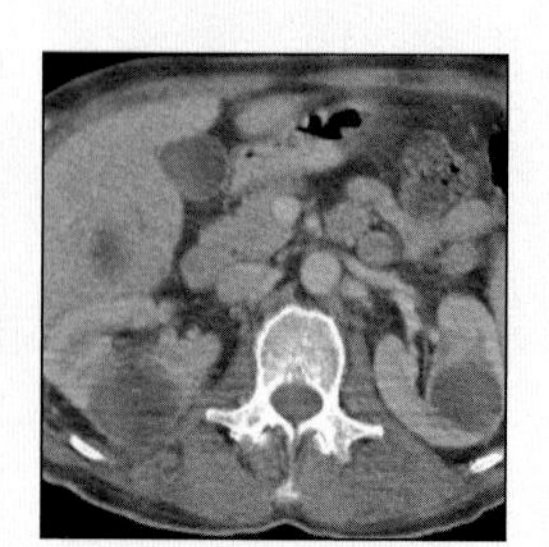

Renal Metastases

RENAL MEDULLARY CARCINOMA

Key Facts

Imaging Findings
- Best diagnostic clue: Infiltrative, centrally located mass with preserved renal shape in a sickle cell patient
- Best imaging tool: CECT

Top Differential Diagnoses
- Renal cell carcinoma
- Transitional cell carcinoma
- Renal angiomyolipoma
- Renal metastases and lymphoma

Clinical Issues
- Hematuria, abdominal or flank pain
- Prognosis: Very poor

Diagnostic Checklist
- Young, African-American patient with sickle cell trait
- Centrally located renal mass with Infiltrative pattern

- Lymphoma
  - Usually multiple or bilateral with infiltrative growth

Focal pyelonephritis
- Differentiate by clinical history and urinalysis

PATHOLOGY

General Features
- Genetics: Hemoglobin (Hb) beta globin gene mutation
- Etiology: Genetics: Sickle cell trait or Hb SC disease
- Epidemiology: Very rare; ~ 50 cases reported in U.S.

Gross Pathologic & Surgical Features
- Infiltration of renal cortex
- Nephromegaly with preservation of renal shape
- Poorly defined lobulated mass in renal medulla
- Tan to gray with variable hemorrhage and necrosis
- Usually with edematous or myxoid appearance, sometimes collagenous

Microscopic Features
- Reticular pattern with the cell aggregates forming spaces of varying size (most common)
- Microcystic pattern with micropapillae projecting into the cyst lumina (less common)
- Associated with areas of compact adenoid cystic growth pattern and poorly differentiated areas as solid sheets of cells
- Cells show a dark cytoplasm, clear nuclei and prominent nucleoli
- ± Cytoplasmic mucin droplets
- ± Cells display squamoid appearance
- Polymorphonuclear leukocytes within tumor
- Sickled erythrocytes in focal areas of tumor or adjacent renal tissue
- Tumor invasions of lymphatics and veins

CLINICAL ISSUES

Presentation
- Most common signs/symptoms
  - Hematuria, abdominal or flank pain
  - Fever, weight loss, palpable abdominal mass

Demographics
- Age: 11-39 years of age, average is 22 years of age
- Gender: M:F = 2:1
- Ethnicity: African-American patients

Natural History & Prognosis
- Prognosis: Very poor
  - Mean survival is 15 weeks, rarely survive > 1 year

Treatment
- Nephrectomy for disease without metastases
- Chemotherapy and radiation therapy: Limited

DIAGNOSTIC CHECKLIST

Consider
- Young, African-American patient with sickle cell trait

Image Interpretation Pearls
- Centrally located renal mass with Infiltrative pattern

SELECTED REFERENCES

1. Choyke PL et al: Hereditary renal cancers. Radiology. 226(1):33-46, 2003
2. Khan A et al: Renal medullary carcinoma: sonographic, computed tomography, magnetic resonance and angiographic findings. Eur J Radiol. 35(1):1-7, 2000
3. Avery RA et al: Renal medullary carcinoma: clinical and therapeutic aspects of a newly described tumor. Cancer. 78(1):128-32, 1996
4. Davidson AJ et al: Renal medullary carcinoma associated with sickle cell trait: radiologic findings. Radiology. 195(1):83-5, 1995

IMAGE GALLERY

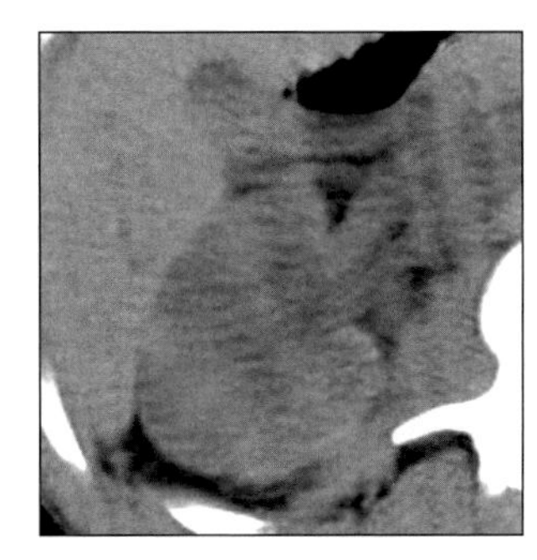
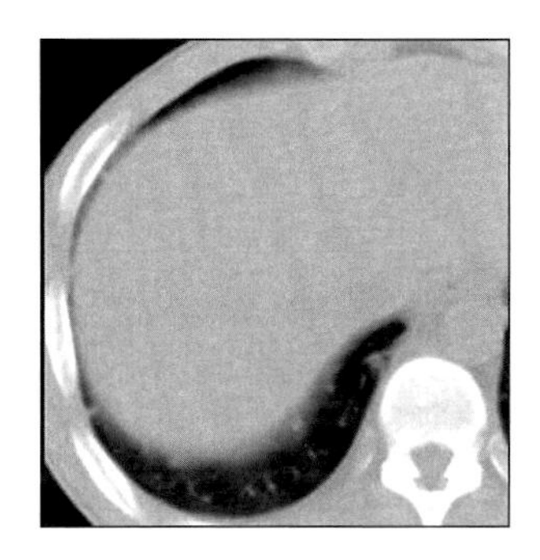

(Left) *Axial NECT in a 35 year old man with sickle cell trait shows a heterogeneous, infiltrative mass with invasion of perirenal space. Sclerotic bones due to sickle cell.* ***(Right)*** *Axial NECT in 35 year old man with renal medullary carcinoma shows multiple lung and pleural metastases.*

TRANSITIONAL CELL CARCINOMA

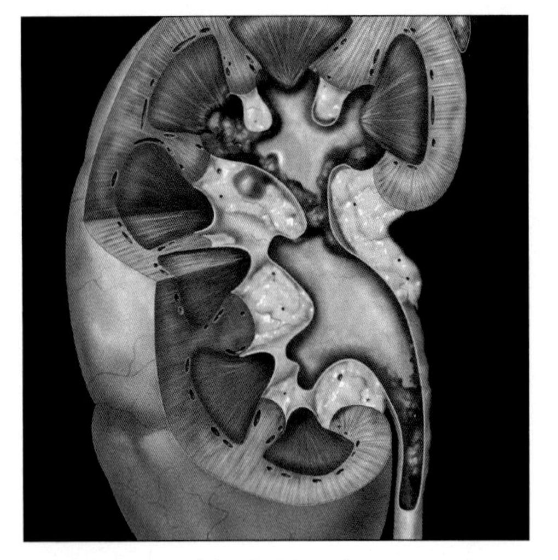

Graphic shows multifocal TCC involving proximal ureter & upper pole calices. Partial obstruction of infundibulum results in upper pole hydronephrosis with filling defects ("oncocalices").

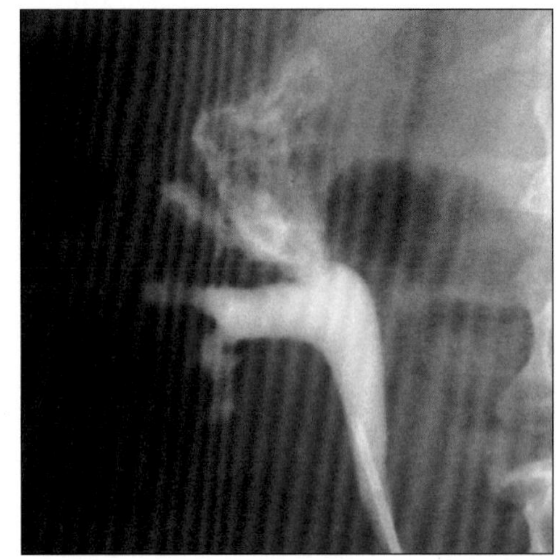

Retrograde pyelogram shows "oncocalices" in upper pole, distended with tumor.

TERMINOLOGY

Abbreviations and Synonyms

- Transitional cell carcinoma (TCC)

IMAGING FINDINGS

General Features

- Best diagnostic clue: Dilated lumen upstream and downstream ureteral mass on retrograde pyelography
- Location
  - Bladder (90%): Lateral wall, anterior superior surface
  - Kidney (8%): Extrarenal part of renal pelvis more common than infundibulocalyceal
  - Ureter & proximal two thirds urethra (2%)

Radiographic Findings

- IVP
  - Nonfunctioning parenchymal segments secondary to collecting duct obstruction; global nonfunction if obstructed at ureteropelvic junction or ureter
  - Single or multiple discrete filling defects; surface is usually irregular, stippled, serrated or frond-like
    - Can also be seen in retrograde pyelography
  - Mass may be flat with minimal or no intraluminal intrusion (nonpapillary); or have a pedicle
  - Renal pelvis
    - "Stipple sign": Contrast within interstices of tumor
    - "Oncocalyx": Ballooned tumor-filled calyx
    - "Phantom calyx": Unopacified calyx from obstruction of calyceal infundibulum
  - Ureter
    - Normal or delayed excretion (partial obstruction)
    - Hydronephrosis ± hydroureter
    - Fixation of ureter with irregular narrowing of lumen and non-tapering margins
  - Bladder
    - Irregular filling defect with broad base and frondlike projections
- Retrograde pyelography
  - Renal pelvis
    - Pyelotumoral backflow: Contrast in interstices
    - Opacification of phantom calyces; irregular papillary or nodular mucosa
  - Ureter
    - Goblet or champagne glass sign: Cup-shaped contrast collection distal to an intraluminal filling defect of ureter (intraluminal growth of tumor)
    - Ovoid filling defect from wall: Short stalk
    - Smooth or frondlike surface; brush border or stippled appearance
    - Irregular decrease in luminal diameter over variable lengths (varying obstruction)

DDx: Filling Defect or Infiltrative Mass

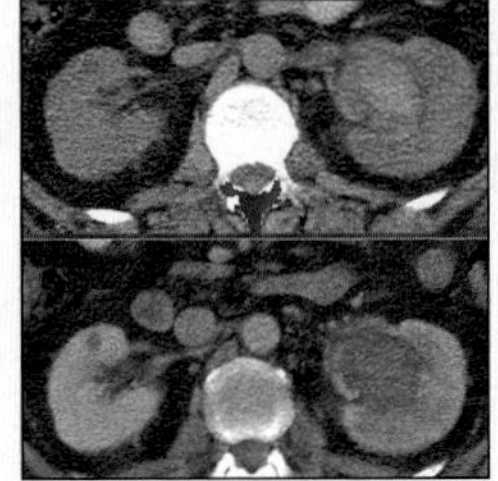

Renal Pelvic Clot

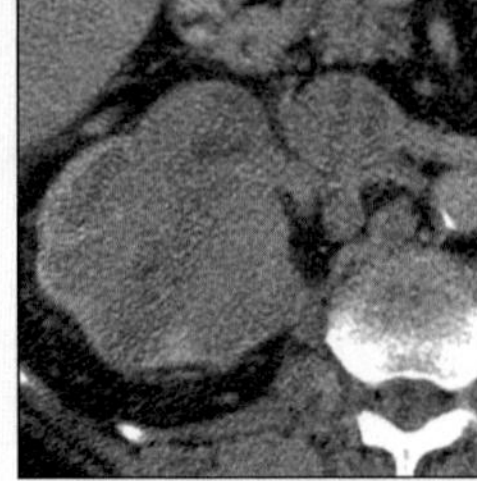

Renal Cell Cancer

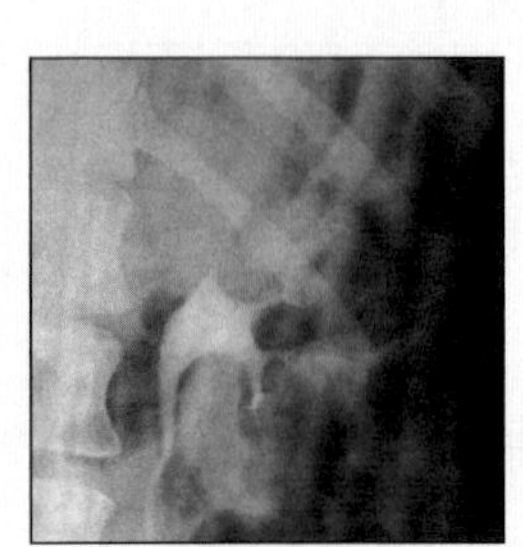

Renal Tuberculosis

TRANSITIONAL CELL CARCINOMA

Key Facts

Imaging Findings

- Best diagnostic clue: Dilated lumen upstream and downstream ureteral mass on retrograde pyelography
- Bladder (90%): Lateral wall, anterior superior surface
- Kidney (8%): Extrarenal part of renal pelvis more common than infundibulocalyceal
- Ureter & proximal two thirds urethra (2%)
- Single or multiple discrete filling defects; surface is usually irregular, stippled, serrated or frond-like
- Goblet or champagne glass sign: Cup-shaped contrast collection distal to an intraluminal filling defect of ureter (intraluminal growth of tumor)
- Best imaging tool: Retrograde pyelography and/or CT urography

Top Differential Diagnoses

- Calculus
- Blood clot
- Infection (e.g., tuberculosis)
- Papillary necrosis
- Renal cell carcinoma (RCC)

Clinical Issues

- Gross hematuria (70-80%), dull or colicky pain (50%)
- Age: > 60 years of age
- Gender: M:F = 2-3:1
- Follow-up: Lifetime increased risk of recurrent TCC; require yearly surveillance

Diagnostic Checklist

- Synchronous or metachronous TCC
- Filling defects; "stipple" sign; goblet sign

CT Findings

- NECT: Isodense to slightly hyperdense (31-48 HU) mass relative to urine
- CECT: Hypovascular infiltrative tumor with minimal enhancement (43-82 HU); preserved renal shape
- Renal pelvis
  - Sessile, flat or polypoid solid mass
  - ± Calcifications (2% of renal pelvis TCC)
  - Focal renal pelvic wall thickening
  - Identify intracaliceal or anterior/posterior wall tumor; obscured in IVP
  - Compression or invasion of renal sinus fat and parenchyma
  - Crust-like rims: Contrast in curvilinear calyceal spaces around periphery of the tumor
  - Coarse striations or diminished nephrogram (partial obstruction)
  - Oncocalyx: Ballooned tumor-filled calyx
  - ± Hydronephrosis
  - Variable enhancement: Based on growth pattern, vascularity and obstruction
  - ± Metastases
- Ureter
  - Dilated, urine-filled collecting system
  - Demonstrate level of obstruction
  - Intraluminal soft tissue mass (30-60 HU)
  - ± Extraureteral extension; ± enhancement
  - Eccentric or circumferential wall thickening
- Bladder
  - Focal wall thickening and mass protruding into lumen; ± enhancement

MR Findings

- Renal pelvis & ureter
  - T1WI: Same or slightly ↓ versus normal parenchyma
  - T2WI: Same or slightly ↑ versus normal parenchyma
  - T1 C+: ↓ or ↑ Enhancement (hypo-/hypervascularity)
- Bladder: MR is staging modality of choice
  - T1WI: TCC isointense to bladder muscle
  - T2WI: Hyperintense than normal bladder wall; ± perivesical invasion
  - T1 C+: Mild enhancement (primary, perivesical, nodal or bone invasion)

Imaging Recommendations

- Best imaging tool: Retrograde pyelography and/or CT urography
  - CT urography
    - Combines axial CECT with subsequent radiography or multiplanar reconstruction
    - Provides "IVP"-type image plus conventional CT views
    - Most informative, accurate single imaging modality for renal or ureteral TCC

DIFFERENTIAL DIAGNOSIS

Calculus

- IVP: Ureteral filling defects
- Retrograde pyelography: No goblet sign
- CT: Most calculi are markedly hyperdense (> 200 HU); differentiate from tumors
- US: Echogenic foci with posterior shadowing

Blood clot

- IVP: "Stipple sign" simulates TCC
- CT: Denser than renal parenchyma (50-90 HU), no enhancement; disappear with time
- US: Same echogenicity as tumors; mobile and disappear with time

Infection (e.g., tuberculosis)

- IVP: Amputated calyx (TB)
- CT: Parenchymal mass ± calcifications (TB)
  - Filling defect in pelvis; nonenhancing (fungus ball)

Papillary necrosis

- Destruction of the apex of the pyramid → irregular cavitation & sinus formation between papilla & calyx
- Retrograde pyelography: Fill 1 or 2 irregular cavities
- CT: Ring or triangular calcification of sloughed papilla

Renal cell carcinoma (RCC)
- May also cause infiltrative mass, extend into renal pelvis

PATHOLOGY

General Features
- General path comments
  - Uroepithelial cancers
    - TCC: 90% of renal pelvic & 97% of ureteral cancers
    - Squamous cell carcinoma: 5-10% of renal pelvic & rare in ureteral cancers
    - Adenocarcinoma: Very rare in both
  - Synchronous or metachronous (multicentricity)
    - Metachronous upper tract TCC: 11-13%
    - Metachronous ipsilateral upper tract TCC: 14-30%
    - Synchronous bilateral renal pelvic TCC: 1-2%; ureteral: 2-9%
    - Synchronous bladder TCC & renal pelvic TCC: 24%; ureteral: 39%
    - Upper tract & bladder TCC at some time: 30-75%
    - Upper tract TCC in 10%, 26% & 34% at 5, 5-10 & > 10 years of follow-up of bladder TCC
- Etiology
  - Environment: Smoking (≥ 3-folds), Balkan countries
  - Occupation: Dyes, rubber, cable, plastics
  - Iatrogenic: Analgesic, cyclophosphamide, Thorotrast
  - Infections: Chronic bacterial urinary tract infections, Schistosoma haematobium (squamous cell carcinoma of bladder and ureter)
- Epidemiology
  - Urothelial tumors: 3% of mortality in U.S.
  - TCC is 5% of all urothelial tumors
    - Most TCC are bladder cancers
  - Ureteral: Incidence is ¼ of renal pelvis
  - Bladder: Cancer mortality: Males: 4th; females: 10th

Gross Pathologic & Surgical Features
- ≥ 85% papillary (low grade); ± infiltrative (high)
- 4 types of features
  - Nonpapillary, noninfiltrative (carcinoma in situ): Slight planar thickening ± hyperemia or hemorrhage
  - Papillary, noninfiltrative: Long, cylindrical villi; few mm to several cm
  - Papillary infiltrative: Large, thick villi; broad base
  - Nonpapillary, infiltrative: Thickening of renal pelvic or ureteral walls

Microscopic Features
- Transitional epithelium: In renal pelvis, ureter, bladder and proximal two thirds urethra
- Epithelial atypia or dysplasia; abnormal fibrovascular core of lamina propria

Staging, Grading or Classification Criteria
- T0: No tumor; Tis: Carcinoma in situ; Ta: Tumor confined to mucosa
- T1: Invades lamina propria
- T2a: Invades inner muscularis; T2b: Outer muscularis
- T3a: Microscopic perivesical fat invasion; T3b: Macroscopic perivesical fat invasion
- T4a: Invades adjacent organs; T4b Invades pelvic or abdominal wall
- N1-3: Pelvic nodes; N4: Above renal bifurcation
- M1 Distant metastases

CLINICAL ISSUES

Presentation
- Most common signs/symptoms: Gross hematuria (70-80%), dull or colicky pain (50%)
- Lab data
  - Urinalysis: Hematuria, ± pyuria and bacteriuria
  - Urine cytology or ureteroscopy with biopsy: Positive

Demographics
- Age: > 60 years of age
- Gender: M:F = 2-3:1

Natural History & Prognosis
- Complications: Hydronephrosis, infections, fistulas
- Renal pelvis & ureteral
  - ≤ T1: 77-80% 5 year survival rate
  - T2: 44% 5 year survival rate
  - ≥ T3: 0-20% 5 year survival rate
- Bladder: Overall 5 year survival rate is 30%

Treatment
- Renal pelvis and ureteral
  - Total nephroureterectomy and bladder cuff excision
  - Metastases: Chemotherapy and/or radiation
- Bladder
  - Superficial: Transurethral resection + Bacillus Calmette-Guérin (BCG) (↓ 40-45% recurrence rate)
  - Deep: Partial/radical cystectomy (with "neobladder" or ileal conduit) or radiation ± chemotherapy
  - Metastases: Surgery or radiation ± chemotherapy
- Follow-up: Lifetime increased risk of recurrent TCC; require yearly surveillance

DIAGNOSTIC CHECKLIST

Consider
- Synchronous or metachronous TCC
- Cystoscopy still necessary to diagnose bladder cancer

Image Interpretation Pearls
- Filling defects; "stipple" sign; goblet sign

SELECTED REFERENCES

1. Akbar SA et al: Multidetector CT urography: techniques, clinical applications, and pitfalls. Semin Ultrasound CT MR. 25(1):41-54, 2004
2. Caoili EM et al: Urinary tract abnormalities: initial experience with multi-detector row CT urography. Radiology. 222(2):353-60, 2002
3. Wong-You-Cheong JJ et al: Transitional cell carcinoma of the urinary tract: Radiologic-pathologic correlation. Radiographics 18: 123-42, 1998
4. Urban BA et al: CT appearance of transitional cell carcinoma of the renal pelvis: Part 2. advanced-stage disease. AJR 169: 163-8, 1997

IMAGE GALLERY

Typical

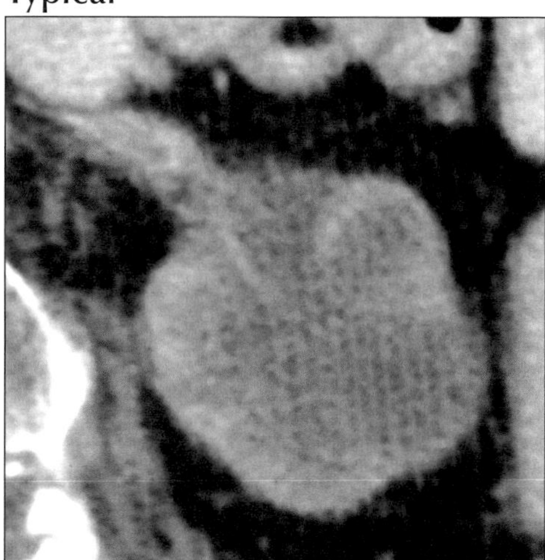

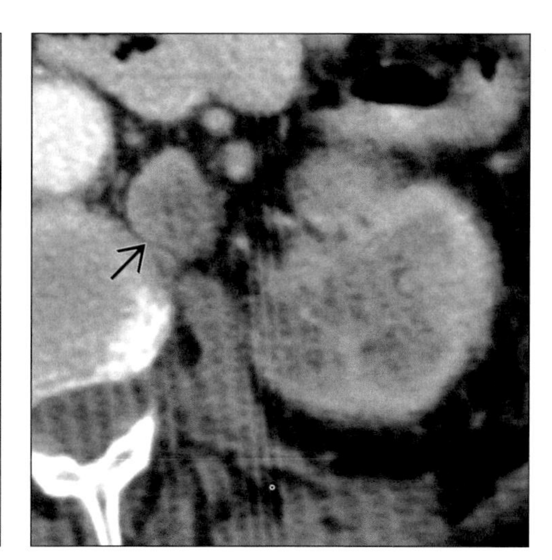

*(**Left**) Axial CECT shows infiltrative hypovascular tumor "replacing" renal parenchyma. (**Right**) Axial CECT shows infiltrative hypovascular mass with preserved renal outline; enlarged node (arrow).*

Typical

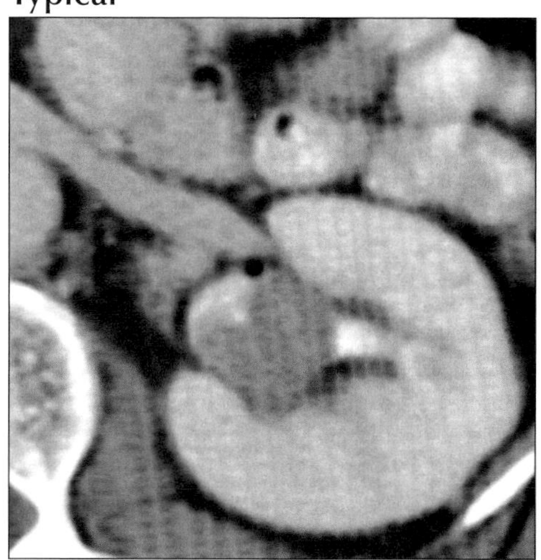

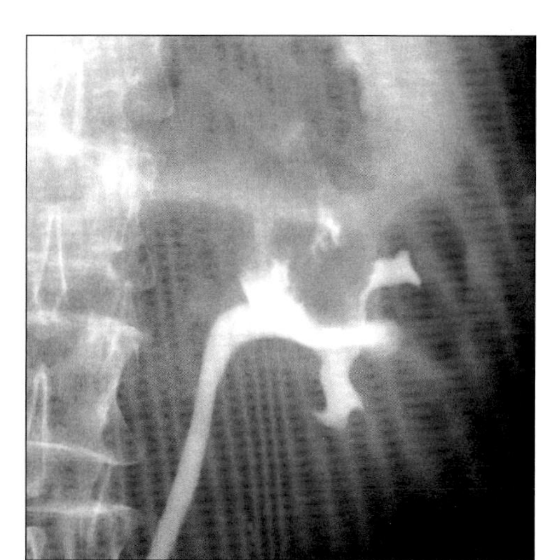

*(**Left**) Axial CECT shows hypovascular mass in renal pelvis. (**Right**) Retrograde pyelogram shows amputation of upper pole infundibulum and mass effect on mid and lower pole calices.*

Typical

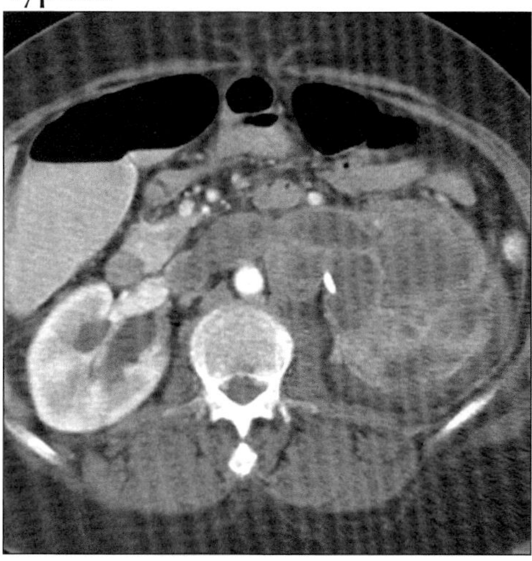

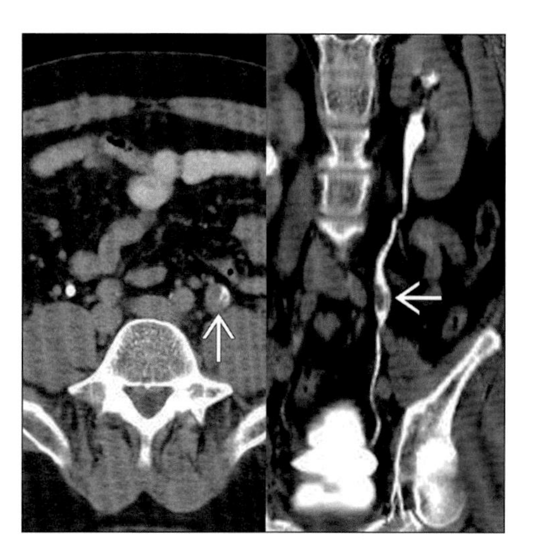

*(**Left**) Axial CECT shows infiltrative heterogeneous hypovascular mass with invasion of left renal vein and nodes. (**Right**) Axial CECT with coronal reformation shows irregular mass (arrows) with dilation of ureteral lumen ("goblet" sign).*

RENAL METASTASES AND LYMPHOMA

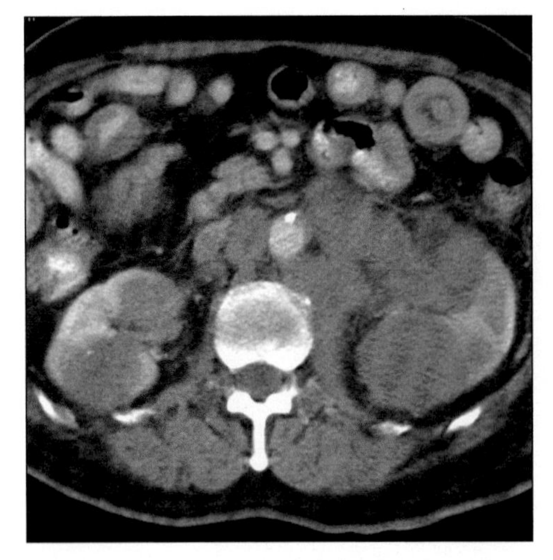

Axial CECT shows multiple homogeneous hypodense masses in the kidneys and retroperitoneal nodes; non-Hodgkin lymphoma.

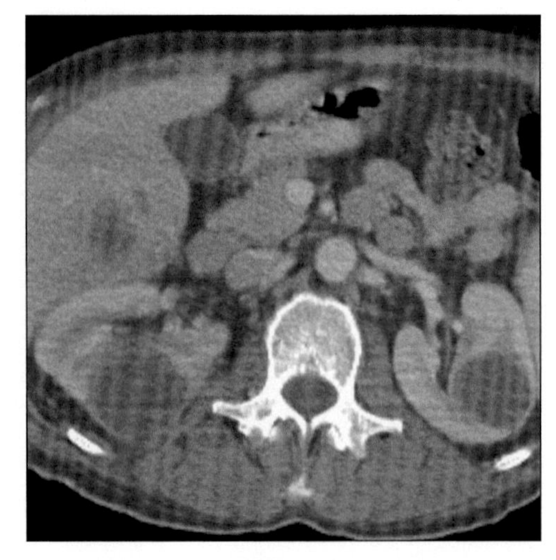

Axial CECT shows multiple masses in the kidneys and liver; metastatic squamous cell cancer (from paranasal sinuses).

TERMINOLOGY

Definitions

- Metastases from primary cancer of other sites
- Lymphoma: Malignant tumor of B lymphocytes

IMAGING FINDINGS

General Features

- Best diagnostic clue: Multiple masses in bilateral kidneys
- Other general features
  - Renal metastases
    - Types of spread: Hematogenous, direct
    - Hematogenous spread: Most common
  - Renal lymphoma
    - Primary: Rare, 3% of all cases of renal lymphoma
    - Secondary: Hematogenous spread (90%) or direct extension via retroperitoneal lymphatic channels

CT Findings

- Renal metastases
  - Multiple, small, low density renal masses
  - Mass occasionally is large and solitary; may simulate renal cell carcinoma
  - Usually bilateral renal involvement; rarely unilateral
  - Wedge-shaped appearance occasionally
  - Usually located within renal parenchyma; rarely cause deformity of renal contour or capsule
  - Iso- to hypodense (10-40 HU) masses on NECT
  - Limited enhancement after contrast (10-20 HU), except highly vascular tumors (i.e., melanoma)
  - Extension through renal capsule into perinephric space
  - Widespread extrarenal metastases usually present
- Renal lymphoma
  - Renal masses are homogeneous with minimal contrast-enhancement (10-20 HU)
  - Multiple, bilateral renal masses (40-60%)
  - Retroperitoneal adenopathy (25%), splenomegaly or lymphadenopathy at other sites
  - Solitary renal mass in one kidney (15%)
  - Infiltration of renal parenchyma or sinus, maintaining reniform shape; may simulate transitional cell carcinoma
  - Direct extension into renal hilum or pericapsular region from retroperitoneal lymph nodes (25%)
  - Diffuse parenchymal involvement (10%) of one or both kidneys → diffuse enlargement
  - Attenuated and poorly opacified collecting systems with variable degree of ↓ enhancement on CECT
  - Renal vein & inferior vena cava tumor thrombus
  - Extranodal involvement of gastrointestinal tract, brain, liver and bone marrow (especially in AIDS)

DDx: Renal Masses

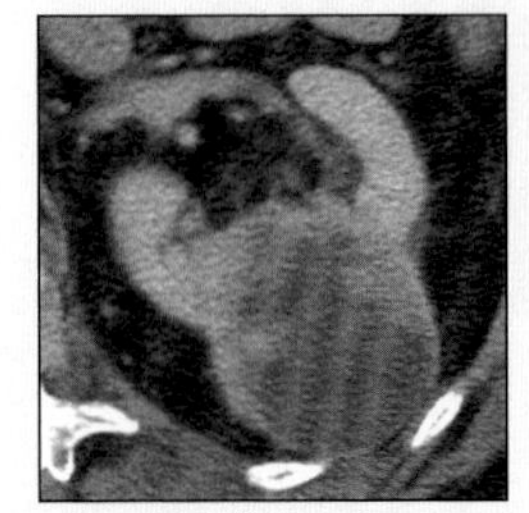

Renal Cell Carcinoma

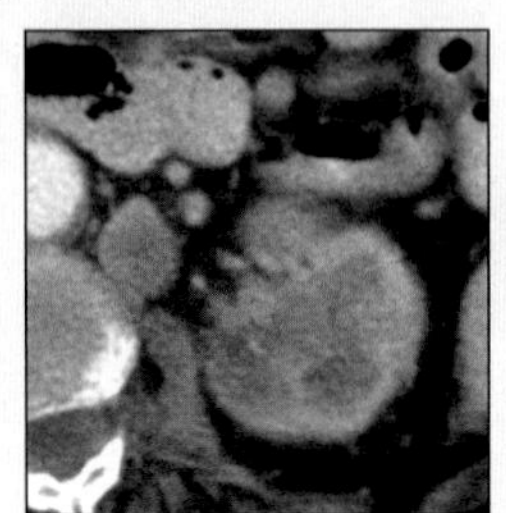

Transitional Cell

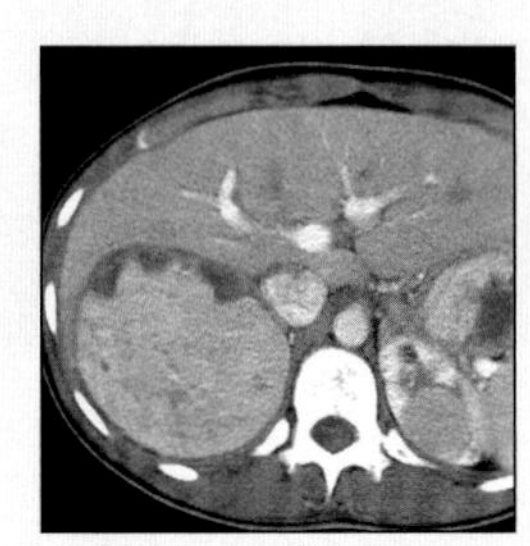

Angiomyolipomas

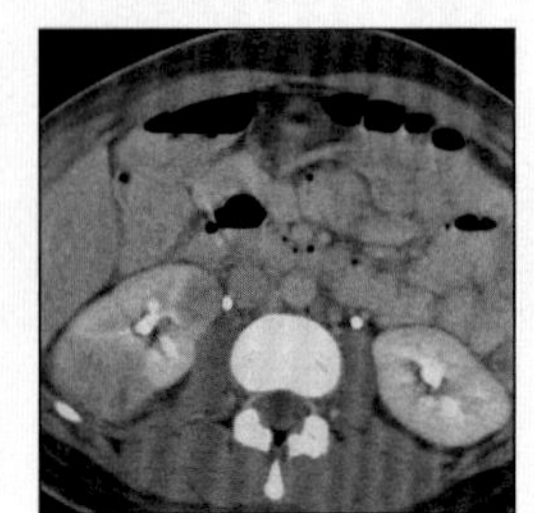

Renal Infection

RENAL METASTASES AND LYMPHOMA

Key Facts

Imaging Findings
- Best diagnostic clue: Multiple masses in bilateral kidneys
- Usually located within renal parenchyma; rarely cause deformity of renal contour or capsule
- Limited enhancement after contrast (10-20 HU), except highly vascular tumors (i.e., melanoma)
- Infiltration of renal parenchyma or sinus, maintaining reniform shape; may simulate transitional cell carcinoma
- Best imaging tool: Helical CT
- Protocol advice: CT: Renal lymphoma: Nephrographic phase imaging is ideal

Top Differential Diagnoses
- Renal cell carcinoma
- Transitional cell carcinoma
- Renal angiomyolipoma
- Renal infection

Clinical Issues
- Usually asymptomatic
- Renal lymphoma: 57% have complete remission after treatment

Diagnostic Checklist
- Check for history of primary cancer
- Overlapping radiographic features of renal metastases, lymphoma and primary carcinoma
- Suspected metastases or lymphoma is one of the rare indications for percutaneous biopsy of a renal mass

MR Findings
- Renal lymphoma
  - T1WI: Iso- to slightly hypointense
  - T2WI: Hypointense
  - T2* GRE, T1 C+: Minimal enhancement

Ultrasonographic Findings
- Real Time
  - Renal lymphoma
    - Hypoechoic relative to renal parenchyma; may simulate renal cyst
    - Limited posterior acoustic enhancement
    - Rarely shows a sonographic halo or hypoechoic mass in perinephric regions

Angiographic Findings
- Conventional
  - Renal metastases
    - Usually masses are hypovascular with infiltrative renal pattern with little or no vascularity
    - Hypervascularity (e.g., melanoma); may simulate renal cell carcinoma
    - Perinephric hemorrhage (e.g., melanoma)
  - Renal lymphoma
    - Marked attenuation of segmental and interlobar arteries
    - Masses are usually hypovascular or avascular
    - Rarely, masses demonstrate neovascularity or hypervascularity with arterial venous shunting
    - Multiple low density cortical defects (nephrogram phase)

Nuclear Medicine Findings
- Bone Scan
  - Renal metastases
    - Increased activity (e.g., breast carcinoma, sarcoma)

Imaging Recommendations
- Best imaging tool: Helical CT
- Protocol advice: CT: Renal lymphoma: Nephrographic phase imaging is ideal

DIFFERENTIAL DIAGNOSIS

Renal cell carcinoma
- Solitary renal mass with central necrosis
- Mass is usually rounded or oval in cortical location and grows by expansion
- Hypervascular mass

Transitional cell carcinoma
- Renal infiltration → kidney enlargement and poorly defined margins without change in shape
- Renal pelvic filling defect, irregular narrowing of collecting system
- Encasing pelvicaliceal system → hypovascular tumor

Renal angiomyolipoma
- Mass with foci of fat

Renal infection
- E.g., focal pyelonephritis, renal abscess
- Multiple necrotic masses with renal enlargement
- Differentiate by clinical history and urinalysis

PATHOLOGY

General Features
- General path comments
  - Renal metastases
    - In autopsy, 2-3 times more common than primary renal neoplasms
    - Fifth most common site of metastases in the body after lung, liver, bone and adrenals
    - Usually too small to be found by imaging due to brief survival of patient
  - Renal lymphoma
    - 6% of patients with lymphoma at presentation
    - 33% of patients dying from malignant lymphoma
    - Renal involvement of non-Hodgkin lymphoma to Hodgkin lymphoma ratio: 10:1
- Etiology
  - Renal metastases
    - Bronchogenic carcinoma (19.8-23.3%)

- Breast carcinoma (12.3%)
- Gastric carcinoma (11.1-15.1%)
- Malignant melanoma
- Renal cell carcinoma of contralateral kidney
- Colonic carcinoma
- Cervical carcinoma
- Ovarian carcinoma
- Pancreatic carcinoma
- Prostate carcinoma
- Thyroid carcinoma
- Testicular carcinoma
- Various sarcomas (i.e., osteogenic sarcoma)

- Renal lymphoma
  - Immunosuppression: Iatrogenic (i.e., organ transplantation) or acquired (i.e., acquired immunodeficiency syndrome (AIDS), autoimmune diseases, Burkitt lymphoma)

Microscopic Features

- Renal metastases: Varies based on primary cancer
- Renal lymphoma
  - Lymphoepithelial lesions
  - Diffuse (more common) or nodular
  - Non-Hodgkin lymphoma: Large cell lymphoma is most common
  - AIDS: Small cell lymphoma is most common

CLINICAL ISSUES

Presentation

- Most common signs/symptoms
  - Renal metastases
    - Usually asymptomatic
    - Hematuria or flank pain; rarely as the initial symptoms of the primary tumor elsewhere
  - Renal lymphoma
    - Usually asymptomatic
    - Hematuria, flank pain, palpable mass or renal insufficiency
  - Other signs/symptoms
    - Fever, weight loss
- Diagnosis
  - CT or US-guided percutaneous biopsy

Demographics

- Age: Any age; middle-age to elderly more common
- Gender: Unknown

Natural History & Prognosis

- Complications
  - Renal or perinephric hemorrhage, renal obstruction, renovascular hypertension, acute renal failure
- Prognosis
  - Renal metastases: Very poor
  - Renal lymphoma: 57% have complete remission after treatment
    - 4 year survival rate: ~ 40%

Treatment

- Renal metastases: Systemic therapy or palliative treatment only
- Renal lymphoma: Chemotherapy ± radiation therapy
- Nephrectomy if metastases and lymphoma is small and isolated to one kidney or have other extenuating circumstances (i.e., severe renal hemorrhage)

DIAGNOSTIC CHECKLIST

Consider

- Check for history of primary cancer
- Patient with multiple renal masses and widespread systemic metastases or a history of extrarenal primary cancer should suspect renal metastases

Image Interpretation Pearls

- Overlapping radiographic features of renal metastases, lymphoma and primary carcinoma
- Suspected metastases or lymphoma is one of the rare indications for percutaneous biopsy of a renal mass

SELECTED REFERENCES

1. Rendon RA et al: The natural history of small renal masses. J Urol. 164(4):1143-7, 2000
2. Urban BA et al: Renal lymphoma: CT patterns with emphasis on helical CT. Radiographics. 20(1):197-212, 2000
3. Sheeran SR et al: Renal lymphoma: spectrum of CT findings and potential mimics. AJR Am J Roentgenol. 171(4):1067-72, 1998
4. Smith PA et al: Spiral computed tomography evaluation of the kidneys: state of the art. Urology. 51(1):3-11, 1998
5. Ferrozzi F et al: Computed tomography of renal metastases. Semin Ultrasound CT MR. 18(2):115-21, 1997
6. Wyatt SH et al: Spiral CT of the kidneys: role in characterization of renal disease. Part II: Neoplastic disease. Crit Rev Diagn Imaging. 36(1):39-72, 1995
7. Honda H et al: CT analysis of metastatic neoplasms of the kidney. Comparison with primary renal cell carcinoma. Acta Radiol. 33(1):39-44, 1992
8. Volpe JP et al: The radiologic evaluation of renal metastases. Crit Rev Diagn Imaging. 30(3):219-46, 1990
9. Levine E et al: Small renal neoplasms: clinical, pathologic, and imaging features. AJR Am J Roentgenol. 153(1):69-73, 1989
10. Pollack HM et al: Other malignant neoplasms of the renal parenchyma. Semin Roentgenol. 22(4):260-74, 1987
11. Curry NS et al: Small renal neoplasms: diagnostic imaging, pathologic features, and clinical course. Radiology. 158(1):113-7, 1986
12. Mitnick JS et al: Metastatic neoplasm to the kidney studied by computed tomography and sonography. J Comput Assist Tomogr. 9(1):43-9, 1985
13. Nishitani H et al: Computed tomography of renal metastases. J Comput Assist Tomogr. 8(4):727-30, 1984
14. Heiken JP et al: Computed tomography of renal lymphoma with ultrasound correlation. J Comput Assist Tomogr. 7(2):245-50, 1983
15. Bhatt GM et al: CT diagnosis of renal metastases. J Comput Assist Tomogr. 7(6):1032-4, 1983
16. Hartman DS et al: Renal lymphoma: radiologic-pathologic correlation of 21 cases. Radiology. 144(4):759-66, 1982
17. Jafri SZ et al: CT of renal and perirenal non-Hodgkin lymphoma. AJR Am J Roentgenol. 138(6):1101-5, 1982
18. Rubin BE: Computed tomography in the evaluation of renal lymphoma. J Comput Assist Tomogr. 3(6):759-64, 1979

IMAGE GALLERY

Typical

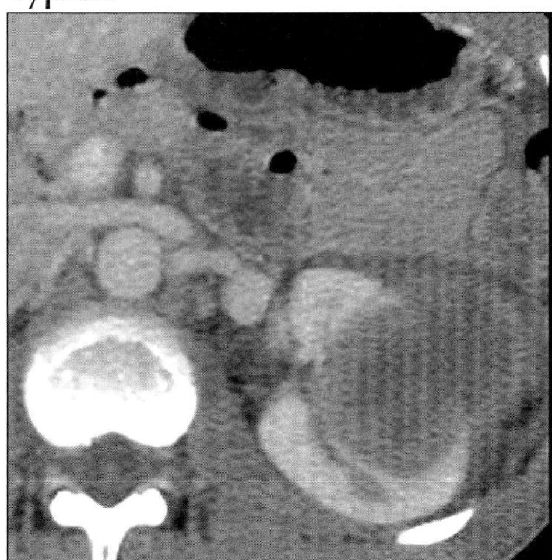

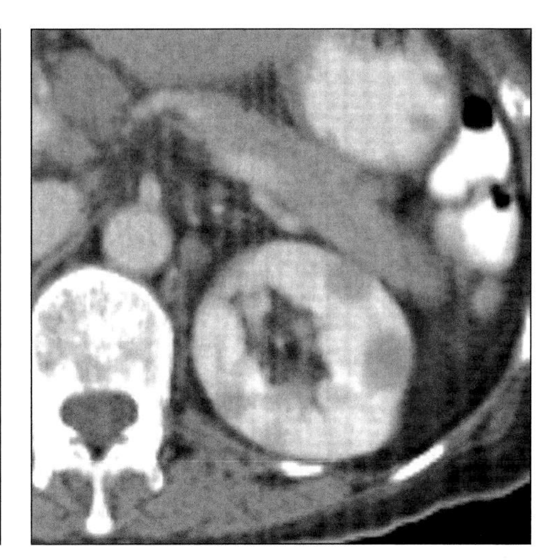

*(**Left**) Axial CECT shows a lobulated renal mass in a patient with AIDS; isolated renal lymphoma. (**Right**) Axial CECT shows multiple small homogeneous hypodense masses; lymphoma.*

Typical

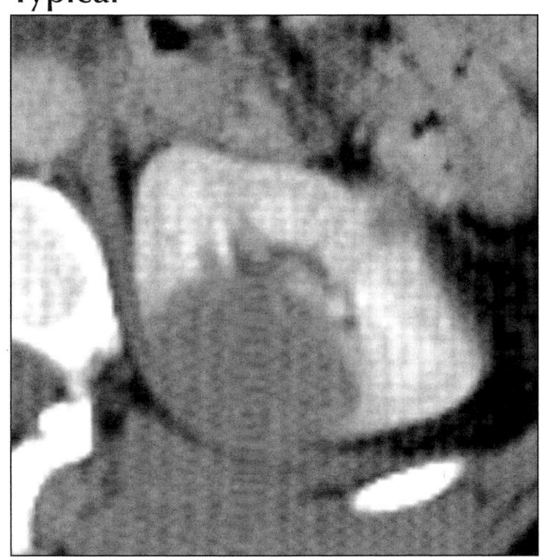

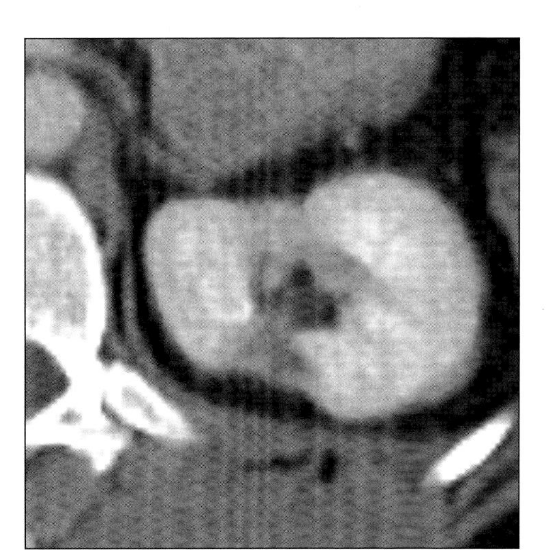

*(**Left**) Axial CECT in May shows two spherical hypodense renal masses due to lymphoma. (**Right**) Axial CECT in July shows no residual mass, only scarring at site of larger tumor, following chemotherapy.*

Typical

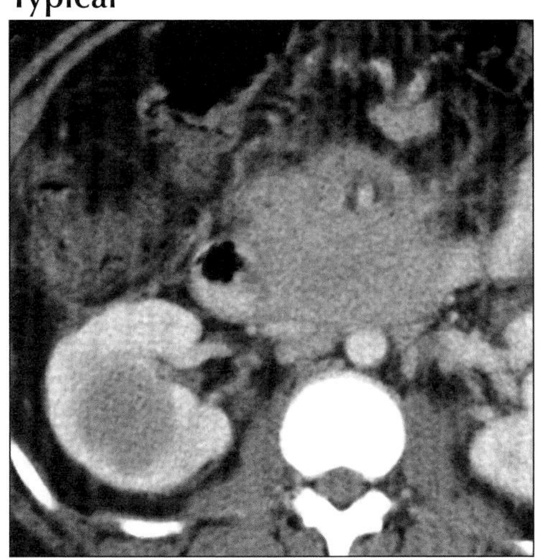

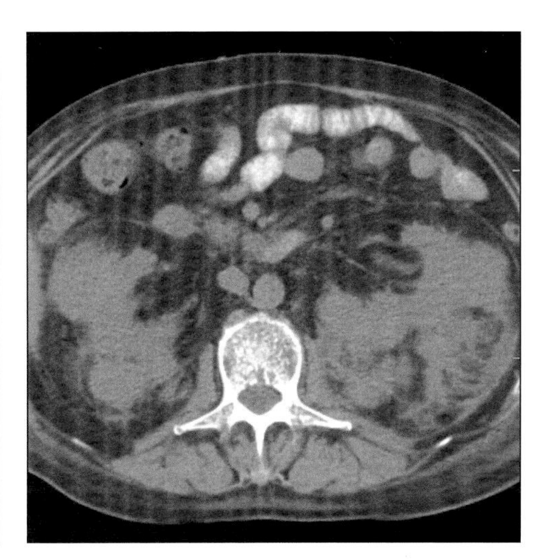

*(**Left**) Axial CECT shows renal and nodal masses due to leukemia. (**Right**) Axial NECT shows bilateral renal and perirenal masses due to multiple myeloma.*

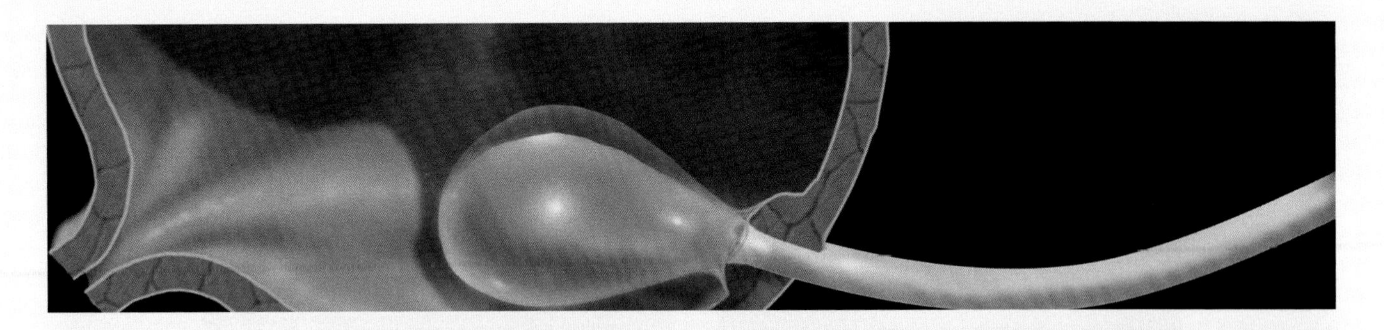

SECTION 4: Ureter

Congenital

Infection

URETERAL DUPLICATION

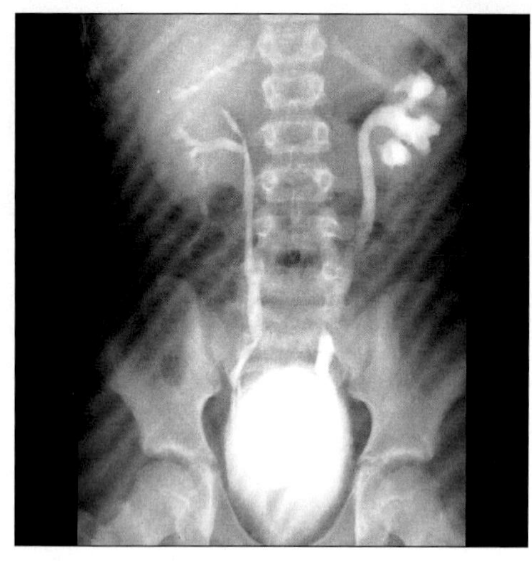

Excretory urogram shows "drooping lily" sign, with downward displacement of left lower pole calices by a hydronephrotic, nonfunctioning upper pole of a duplicated collecting system.

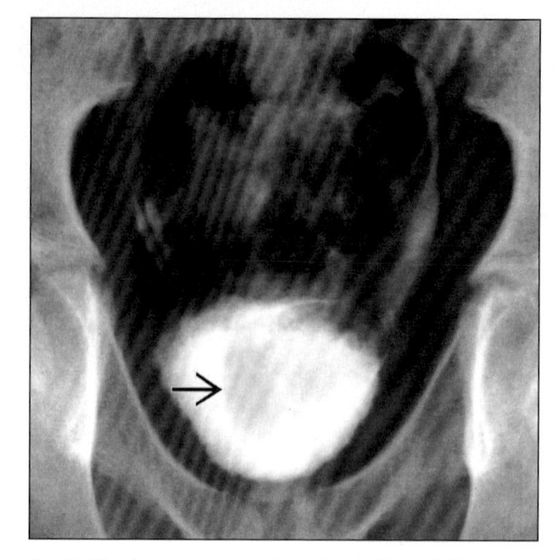

Pelvic film from urogram shows large filling defect in the bladder (arrow) representing a ureterocele from the ectopic upper pole ureter that inserted into the prostatic urethra.

TERMINOLOGY

Abbreviations and Synonyms

- Double ureters

Definitions

- 2 ureters drain a duplex kidney and remain separate to bladder or beyond

IMAGING FINDINGS

General Features

- Best diagnostic clue: Double collecting system ipsilaterally on IVP or CT urography
- Other general features
  - 85% obey Weigert Meyer rule: Upper pole ureter inserts medial & caudal to lower pole ureter
  - 15% upper pole ureter inserts anywhere along ectopic pathway
  - Most commonly, upper pole ureter is ectopic & obstructed and lower pole ureter refluxes
  - Kidney & ureter may be normal, except duplicated
  - 20% of contralateral ureter is also duplicated

Radiographic Findings

- IVP (or CT urography)
  - Duplex kidney with double ureters; 2 jets of contrast
  - Poor or no excretion by upper pole of duplex kidney
  - "Drooping lily" sign: Hydronephrosis and ↓ function of obstructed upper pole → downward displacement of lower pole calyces
  - "Nubbin" sign: Scarring, atrophy and ↓ function of lower pole moiety; may simulate renal mass
  - Fewer calyces & infundibula of lower pole collecting system; shortened upper pole infundibulum
  - Single or diffuse calyceal clubbing, thin overlying parenchyma ± scarring in lower pole
  - ± Ureteropelvic junction obstruction of lower pole
- Voiding cystourethrogram
  - ± Reflux, ureterocele, diverticulum of urethra

CT Findings

- "Faceless kidney": No renal sinus or collecting system at junction of upper & lower pole of a duplex kidney
- ± Obstruction in either pole of a duplex kidney

Ultrasonographic Findings

- 2 central echo complexes with intervening renal parenchyma
- Will show upper pole hydronephrosis

Nuclear Medicine Findings

- ± Reflux up one or both ureters
- Estimate degree of renal function

Imaging Recommendations

- Best imaging tool: IVP or CT urography

DDx: Variations or Mimics of Duplication

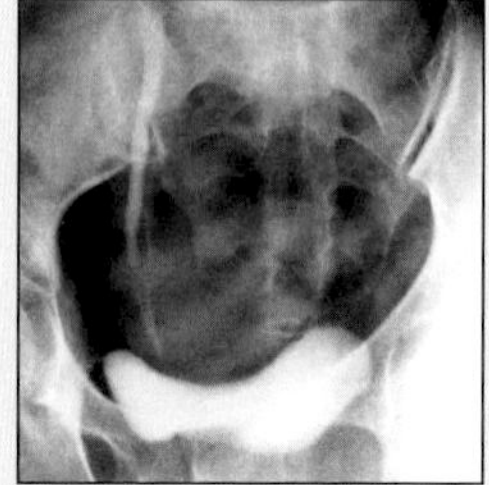

Bifid Ureter

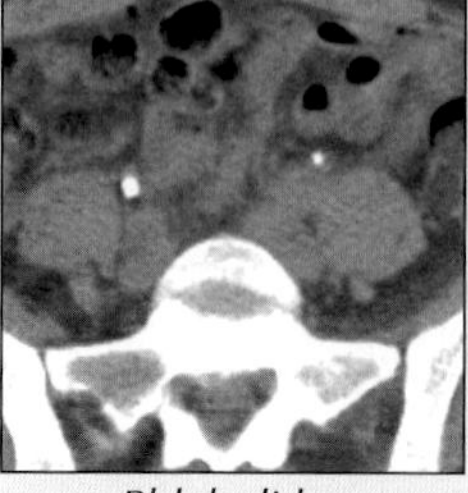

Phleboliths

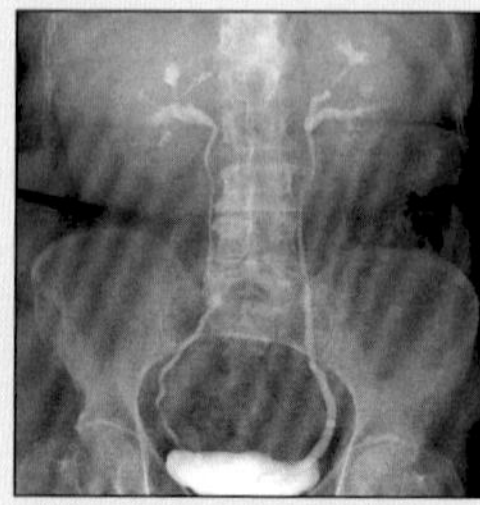

Ureterocele

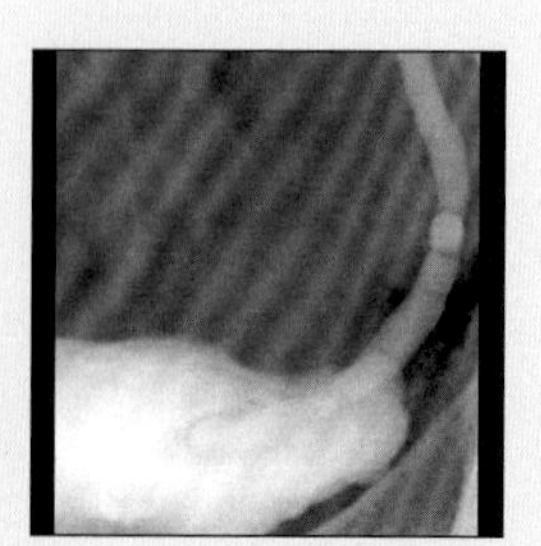

Ureterocele

URETERAL DUPLICATION

Key Facts

Terminology
- Double ureters
- 2 ureters drain a duplex kidney and remain separate to bladder or beyond

Imaging Findings
- Best imaging tool: IVP or CT urography

Top Differential Diagnoses
- Bifid ureters
- Gonadal vein phlebolith
- Ureterocele

Diagnostic Checklist
- Young females with recurrent urinary tract infections
- 2 distinct ureters; "drooping lily" sign

DIFFERENTIAL DIAGNOSIS

Bifid ureters
- 2 ureters drain a duplex kidney, but join before emptying into bladder; can be ectopic
- Ureters may join at intravesical junction: V-shaped; extravesical (most common) junction: Y junction
- IVP
  - Ureteroureteral reflux or yo-yo phenomenon: Peristalsis down 1 limb of ureter forcing urine via reflux up the other
  - Asymmetric dilatation of 1 ureteral segment

Gonadal vein phlebolith
- May simulate stone in ureter (single or duplicated)

Ureterocele
- Can be isolated finding

PATHOLOGY

General Features
- General path comments: Both ureters pass through bladder wall through a common tunnel
- Etiology
  - Genetics: Autosomal dominant with low penetrance
  - Environment: Geographic areas → ↑ prevalence
- Epidemiology: 1 per 500 persons
- Associated abnormalities
  - Solitary or dysplastic kidney, hypoplastic kidneys, all types of fused kidneys or posterior urethral valves
  - Complex congenital anomalies: VATER, VACTERL (vertebral, anal, cardiovascular, tracheoesophageal, renal and limb)

CLINICAL ISSUES

Presentation
- Most common signs/symptoms
  - Asymptomatic, nocturnal incontinence
  - Incontinence, only in females due to insertion of upper pole ureteral orifice below bladder sphincter
  - Intermittent or persistent urinary tract infections ± acute pyelonephritis, frequently in females

Demographics
- Gender: M:F = 1:10

Natural History & Prognosis
- Complications: Urolithiasis, abscess, renal failure

Treatment
- Lower grades of reflux: Medical treatment
- Higher grades of reflux, upper pole obstruction, ectopy, poor renal function: Surgical treatment

DIAGNOSTIC CHECKLIST

Consider
- Young females with recurrent urinary tract infections

Image Interpretation Pearls
- 2 distinct ureters; "drooping lily" sign

SELECTED REFERENCES

1. Zissin R et al: Renal duplication with associated complications in adults: CT findings in 26 cases. Clin Radiol. 56(1):58-63, 2001
2. Fernbach SK et al: Ureteral duplication and its complications. Radiographics. 17(1):109-27, 1997
3. Share JC et al: The unsuspected double collecting system on imaging studies and at cystoscopy. AJR Am J Roentgenol. 155(3):561-4, 1990

IMAGE GALLERY

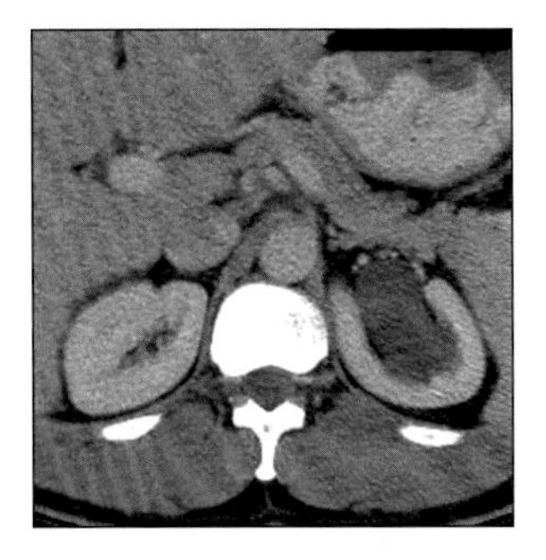

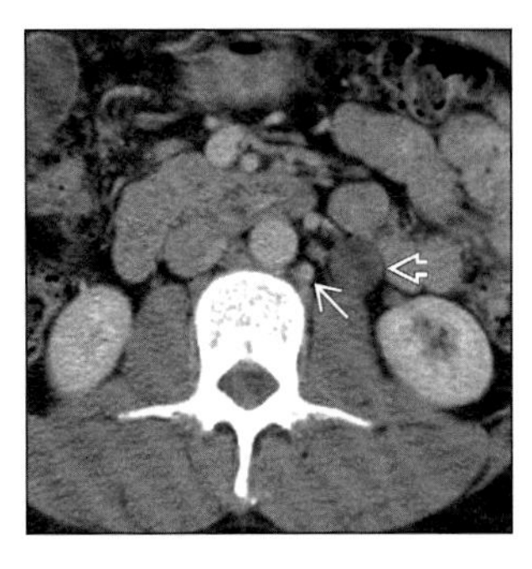

*(**Left**) Axial CECT shows hydronephrotic upper pole collecting system. (**Right**) Axial CECT shows normal lower pole calices + ureter (arrow). Ureter draining upper pole (open arrow) is dilated.*

ECTOPIC URETER

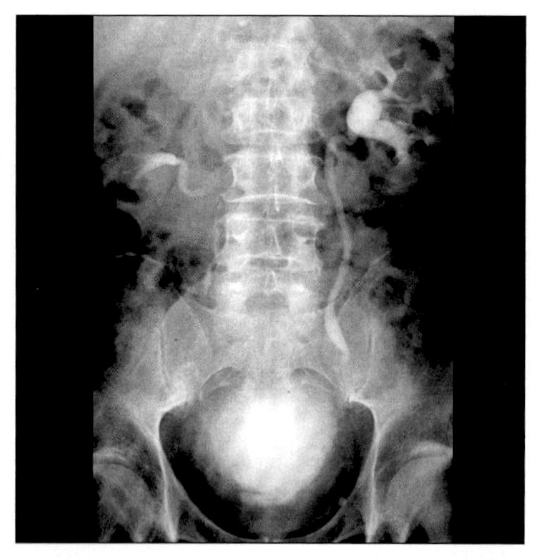

IVP shows "drooping lily" sign, with lower pole moiety displaced inferiorly and laterally by obstructed upper pole moiety and dilated ectopic ureter (Courtesy FG Sommer, MD).

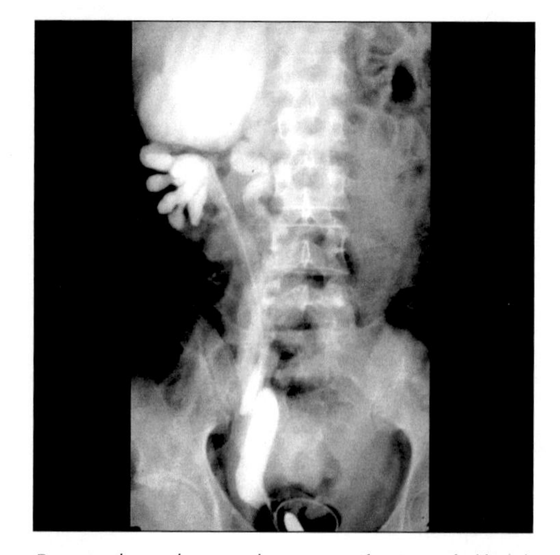

Retrograde pyelogram shows opacification of dilated, tortuous ectopic ureter and severely hydronephrotic upper pole moiety. Note displaced lower pole moiety (Courtesy FG Sommer, MD).

TERMINOLOGY

Abbreviations and Synonyms

- Ectopic ureter (EU), ureteral ectopia

Definitions

- Ureter that does not terminate at bladder trigone
  - Ectopic insertion within bladder: Usually no significant pathology
- Common usage: Ureter that terminates outside bladder

IMAGING FINDINGS

General Features

- Best diagnostic clue: 70-80% associated with complete ureteral duplication
- Location
  - Usually extravesicular insertion; males always above external sphincter
  - Males: Vas deferens 10%, seminal vesicle 28%, prostatic urethra 54%, ejaculatory duct 8%
    - Prostatic urethra most common insertion site in male
  - Females: Uterus or cervix 3%, vagina 27%, urethra 32%, vestibule 38%
    - Urethra or vestibule most common insertion site in female
  - 5-17% of ectopic ureters are bilateral
- Morphology
  - Complete duplication: Ectopic ureter drains upper moiety
  - Orifice commonly stenotic, leading to obstruction of upper pole moiety

Radiographic Findings

- Fluoroscopy
  - Voiding cystourethrogram (VCUG): Reflux into either moiety
  - VCUG useful to locate insertion of ectopic ureter if within urinary tract
    - Will not visualize ectopic insertion if outside urinary tract
- IVP
  - Dilated upper pole collecting system
  - Non-visualization of upper pole moiety with severe obstruction/dysplasia
    - Visualized lower pole moiety: Fewer calyces than normal for whole kidney
    - Lower pole displaced infero-laterally ("drooping-lily" sign)
  - Ectopic insertion of single system ureter: Involved kidney usually small, dysplastic and nonfuntional

DDx: Mimics of Ectopic Ureter

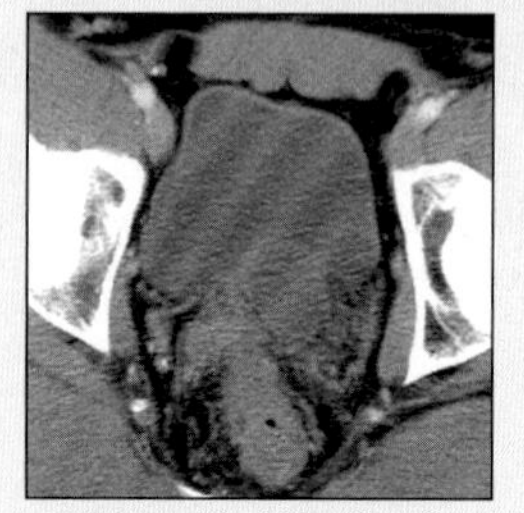

Sem. Vesicle Cyst

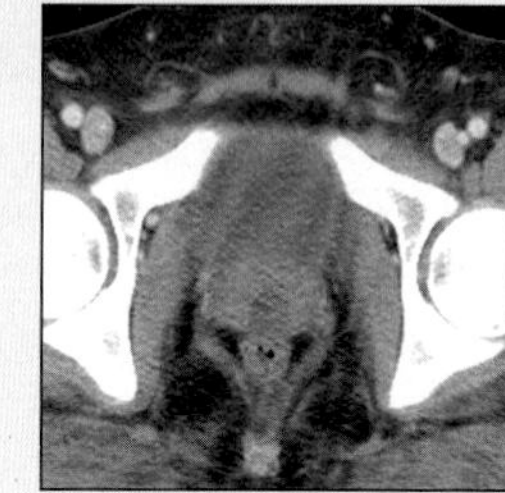

Prostatic Cyst

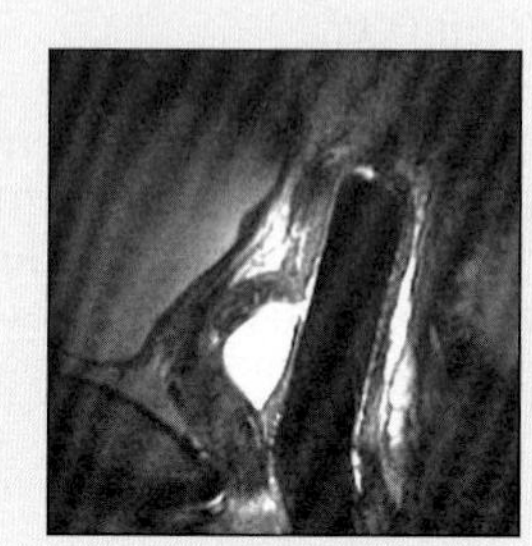

Gartner Duct Cyst

ECTOPIC URETER

Key Facts

Terminology
- Common usage: Ureter that terminates outside bladder

Imaging Findings
- Best diagnostic clue: 70-80% associated with complete ureteral duplication
- Prostatic urethra most common insertion site in male
- Urethra or vestibule most common insertion site in female
- Complete duplication: Ectopic ureter drains upper moiety
- Orifice commonly stenotic, leading to obstruction of upper pole moiety
- Lower pole displaced infero-laterally ("drooping-lily" sign)

Pathology
- Hypoplasia or dysplasia of renal moiety drained by ectopic ureter
- Distance from trigone correlates with degree of ipsilateral renal dysplasia
- More distal the ureter, the greater the dysplasia

Clinical Issues
- Females: Continual dribbling urinary incontinence (50%)
- Males: No incontinence because ectopic ureteral orifice always above external sphincter

Diagnostic Checklist
- Weigert-Meyer law: Upper moiety ureter inserts inferior and medial to lower moiety ureter

CT Findings
- CECT
  - Hydronephrotic upper pole moiety with variable function
  - Dilated, tortuous ureter to level of insertion
  - Males with single ectopic ureters: Non-functional kidney and dilated ipsilateral seminal vesicle

MR Findings
- T1WI
  - Tortuous low signal intensity ureter dilated to level of ectopic insertion
  - Severe hydronephrosis of upper pole moiety
- T2WI
  - Tortuous high signal intensity ectopic ureter dilated to level of insertion
  - High signal cystic dysplasia of ipsilateral upper pole moiety
- T1 C+: Variable degree of function in obstructed upper pole moiety

Ultrasonographic Findings
- Real Time
  - Cystic area in medial upper pole of kidney = non-functional obstructed upper moiety
  - Dilated ureter behing otherwise normal bladder

Nuclear Medicine Findings
- Renal scintigraphy
  - Variable function of moiety drained by ectopic ureter

Imaging Recommendations
- Best imaging tool
  - Ultrasound
  - CT may be useful to locate small poorly functioning dysplastic kidney with single ectopic ureter
  - MR urography can display ectopic ureteral insertions even if outside urinary tract
- Protocol advice: Trace dilated ureter on US to its terminus below bladder

DIFFERENTIAL DIAGNOSIS

Seminal vesicle cyst
- Associated with renal agenesis

Gartner duct cyst
- Inclusion cyst in lateral wall of vagina, remnant of mesonephric duct

PATHOLOGY

General Features
- General path comments
  - Ectopic ureters opening to bladder neck or posterior urethra may reflux
  - Ectopic ureters terminating outside urinary tract: Usually obstructed
- Etiology
  - Congenital: Abnormal ureteral bud migration
    - Failure of separation of ureteral bud from Wolffian duct results in caudal ectopia
- Epidemiology
  - Incidence: At least 1 in 1,900
  - True incidence uncertain since many cases asymptomatic
- Associated abnormalities
  - Hypoplasia or dysplasia of renal moiety drained by ectopic ureter
  - Degree of ureteral ectopia correlates with degree of renal abnormality
  - Imperforate anus, tracheo-esophageal fistula

Gross Pathologic & Surgical Features
- Single system ectopic ureter: Absent ipsilateral hemitrigone
- Distance from trigone correlates with degree of ipsilateral renal dysplasia
  - More distal the ureter, the greater the dysplasia
  - Very distal insertions ⇒ usually very poor renal function

ECTOPIC URETER

Microscopic Features

- Muscularis of ectopic ureteral wall may have ultrastructural abnormalities

CLINICAL ISSUES

Presentation

- Most common signs/symptoms
  - Recurrent or chronic urinary tract infections (UTIs)
  - Females: Continual dribbling urinary incontinence (50%)
    - Males: No incontinence because ectopic ureteral orifice always above external sphincter
  - Males: Chronic or recurrent epididymitis
- Clinical profile
  - Girl with continuous dribbling urinary incontinence
  - Prepubertal boy with epididymitis or UTI

Demographics

- Age
  - Age at diagnosis varies widely; some cases not detected during life
  - Many cases diagnosed with prenatal ultrasound
- Gender
  - M:F = 1:6
  - Single system ectopic ureters more common in males
  - Ectopic ureters in males usually drain single systems
  - Females: 80% of ectopic ureters are duplicated systems

Natural History & Prognosis

- Most ectopic ureters drain single kidneys or upper pole moieties with minimal function

Treatment

- Options, risks, complications
  - Ectopic ureter with duplicated system: Surgical upper pole nephrectomy
  - Single system: Nephrectomy if minimal function
  - If renal function preserved or dx made prenatally: Ureteropyelostomy or common sheath ureteral implantation

DIAGNOSTIC CHECKLIST

Image Interpretation Pearls

- Weigert-Meyer law: Upper moiety ureter inserts inferior and medial to lower moiety ureter

SELECTED REFERENCES

1. Wille S et al: Magnetic resonance urography in pediatric urology. Scand J Urol Nephrol. 37(1):16-21, 2003
2. Berrocal T et al: Anomalies of the distal ureter, bladder, and urethra in children: embryologic, radiologic, and pathologic features. Radiographics. 22(5):1139-64, 2002
3. Riccabona M et al: Feasibility of MR urography in neonates and infants with anomalies of the upper urinary tract. Eur Radiol. 12(6):1442-50, 2002
4. Oge O et al: Nephron-sparing surgery in a duplex system associated with a vaginal ectopic ureter. Pediatr Nephrol. 16(12):1135-6, 2001
5. Staatz G et al: Magnetic resonance urography in children: evaluation of suspected ureteral ectopia in duplex systems. J Urol. 166(6):2346-50, 2001
6. Patel PM et al: Ureteral triplication with ectopic upper pole moiety. Urology. 58(2):279-80, 2001
7. Damry N et al: Ectopic vaginal insertion of a duplicated ureter: demonstration by magnetic resonance imaging (MRI). JBR-BTR. 84(6):270, 2001
8. Lee SS et al: Giant hydronephrosis of a duplex system associated with ureteral ectopia: a cause of retrograde ejaculation. Arch Androl. 45(1):19-23, 2000
9. Engin G et al: MR urography findings of a duplicated ectopic ureter in an adult man. Eur Radiol. 10(8):1253-6, 2000
10. Bourian M et al: Clinical quiz. Ectopic ureter. Pediatr Nephrol. 14(8-9):879-81, 2000
11. Cabay JE et al: Ectopic ureter associated with renal dysplasia. JBR-BTR. 82(5):228-30, 1999
12. Komatsu K et al: Single ectopic vaginal ureter diagnosed by computed tomography. Urol Int. 63(2):147-50, 1999
13. De Vuyst D et al: Ectopic ureter. J Belge Radiol. 81(5):239, 1998
14. Carrico C et al: Incontinence due to an infrasphincteric ectopic ureter: why the delay in diagnosis and what the radiologist can do about it. Pediatr Radiol. 28(12):942-9, 1998
15. Dunnick NR et al: Textbook of uroradiology. 2nd ed. Baltimore, Williams and Wilkins, 29-33, 1997
16. Plaire JC et al: Management of ectopic ureters: experience with the upper tract approach. J Urol. 158(3 Pt 2):1245-7, 1997
17. Yanagisawa N et al: Diagnostic magnetic resonance-urography in an infant girl with an ectopic ureter associated with a poorly functioning segment of a duplicated collecting system. Int J Urol. 4(3):314-7, 1997
18. Amatulle P et al: Ureteral duplication anomaly with ectopic intraprostatic insertion. J Ultrasound Med. 16(3):231-3, 1997
19. el Ghoneimi A et al: Ectopic ureter with complete ureteric duplication: conservative surgical management. J Pediatr Surg. 31(4):467-72, 1996
20. Rothpearl A et al: MR urography: technique and application. Radiology. 194(1):125-30, 1995
21. Dudak SD et al: Transitional cell carcinoma in a duplicated ectopic ureter. Urology. 46(2):251-3, 1995
22. Jelen Z: The value of ultrasonography as a screening procedure of the neonatal urinary tract: a survey of 1021 infants. Int Urol Nephrol. 25(1):3-10, 1993
23. Herman TE et al: Radiographic manifestations of congenital anomalies of the lower urinary tract. Radiol Clin North Am. 29(2):365-82, 1991

ECTOPIC URETER

IMAGE GALLERY

Typical

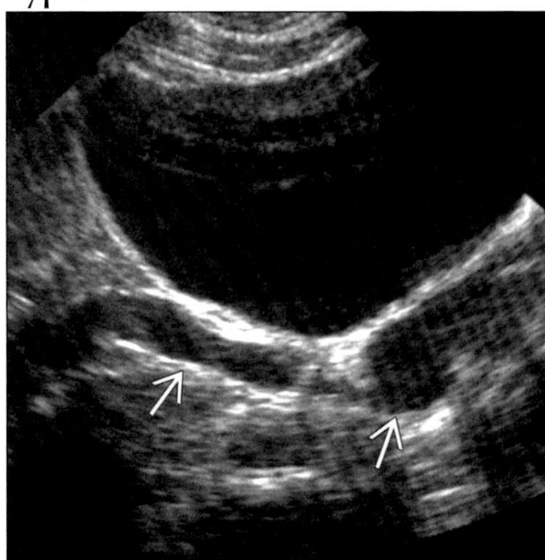

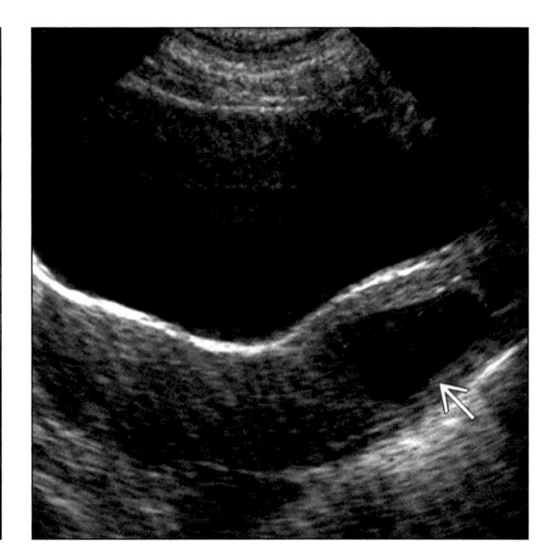

(Left) *Sagittal US shows tortuous dilated ureter (arrows) continuing beyond the expected location for the UVJ.* ***(Right)*** *Sagittal US shows cystic mass in vagina (arrow), representing insertion of dilated ectopic ureter (Courtesy FG Sommer, MD).*

Variant

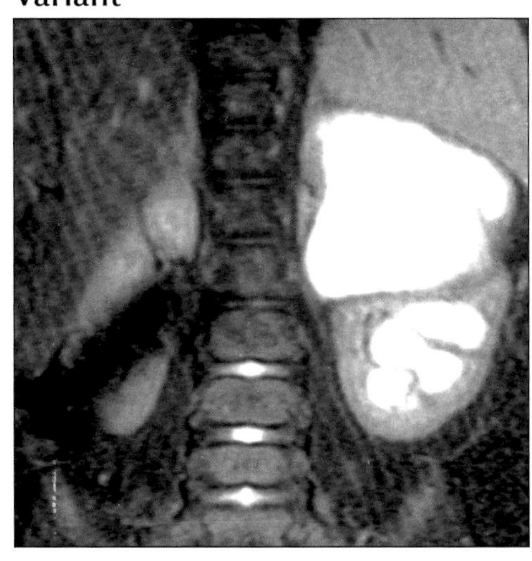

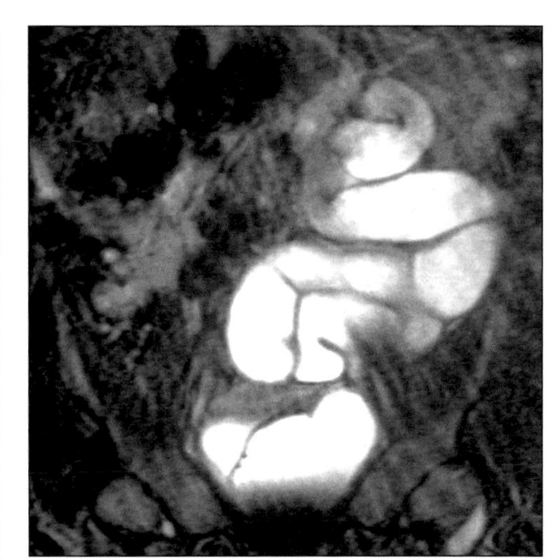

(Left) *Coronal T2WI MR shows duplicated left collecting system with hydronephrotic upper pole and mild lower pole dilatation (Courtesy FG Sommer, MD).* ***(Right)*** *Coronal T2WI MR slightly more anterior to previous image shows markedly dilated, tortuous ectopic ureter (Courtesy FG Sommer, MD).*

Typical

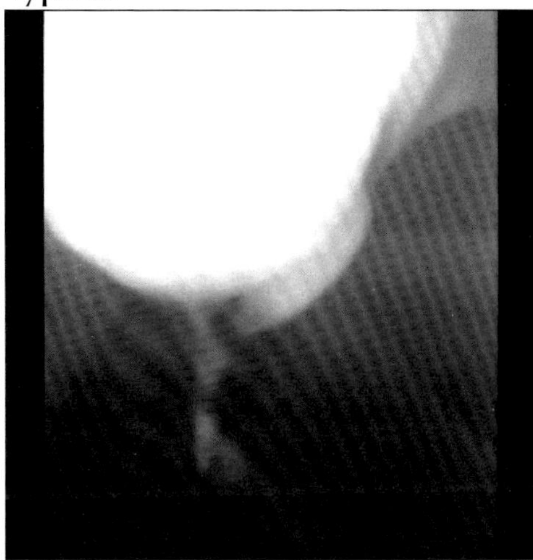

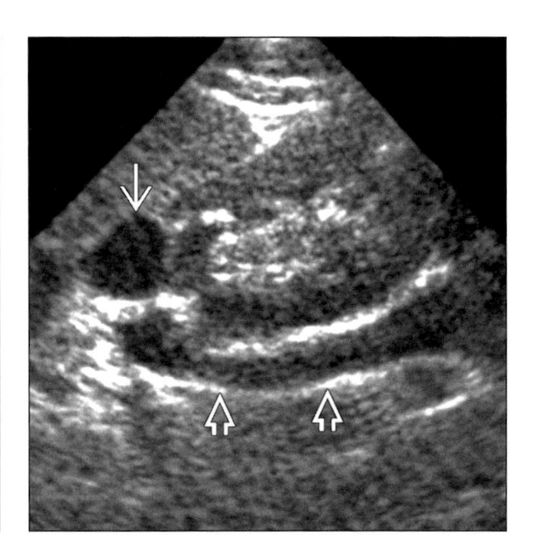

(Left) *Voiding cystourethrogram shows ectopic left ureter draining into urethra (Courtesy G. Friedland, MD).* ***(Right)*** *Sagittal US shows dilated ectopic ureter (open arrows) deep to kidney and cystic dysplasia of upper pole moiety (arrow) (Courtesy FG Sommer, MD).*

URETERO CELE

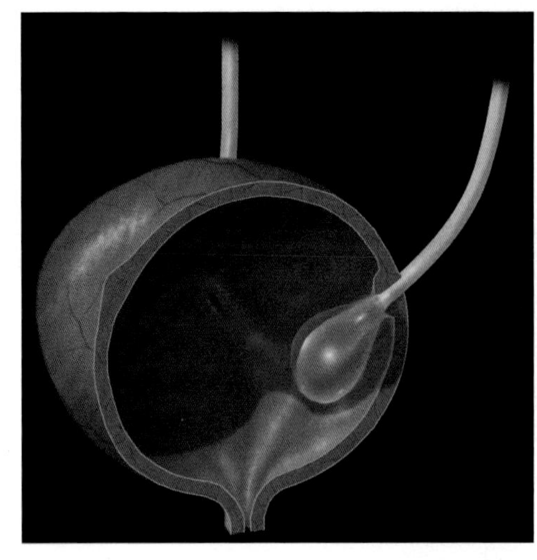

Drawing shows simple ureterocele, with prolapsing, dilated end of orthotopic ureter.

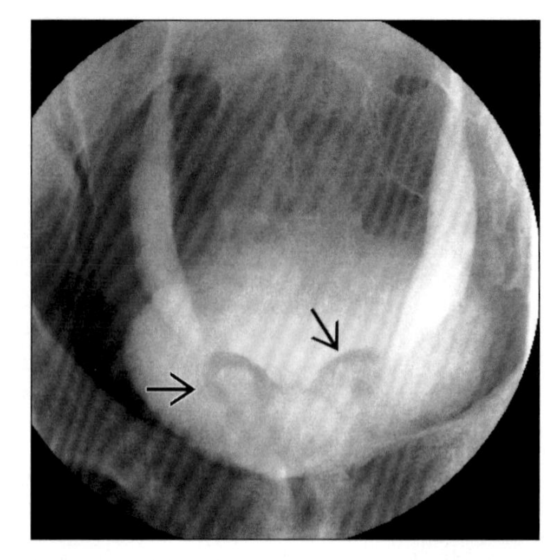

IVP shows bilateral simple ureteroceles, with cystic dilatation of intramural portions of orthotopic ureters and surrounding radiolucent halos (arrows) (Courtesy M. Nino-Murcia, MD).

TERMINOLOGY

Abbreviations and Synonyms

- Simple = orthotopic = adult-type ureterocele

Definitions

- Cystic dilatation of distal ureter with prolapse into bladder lumen
- Orthotopic ureterocele: Normal insertion at trigone
  - Orthotopic: Single ureter systems
- Ectopic ureterocele: Inserts below trigone
  - Ectopic: Duplicated systems in 80%

IMAGING FINDINGS

General Features

- Best diagnostic clue: Orthotopic: "Cobra-head" or "spring onion" deformity of distal ureter with surrounding radiolucent halo
- Location
  - Ectopic: ½ in bladder and ½ in posterior urethra; 10% bilateral
    - Males: Insertion always above external sphincter
- Size: Up to several cm in diameter
- Morphology: Smooth, round or ovoid

Radiographic Findings

- IVP/cystography
  - Orthotopic ureterocele: Cobrahead deformity
    - Dilated distal ureter projecting into lumen of bladder with surrounding radiolucent halo
  - Ectopic ureterocele: Smooth, radiolucent intravesicular mass near bladder base
    - May evert during voiding and mimic diverticulum
    - Lumen opacification depends on function of upper pole moiety
    - Drooping lily sign: Displacement of lower pole collecting system by obstructed upper pole moiety

CT Findings

- CECT: Intravesicular mass at UVJ

MR Findings

- T2WI
  - Intravesicular mass at UVJ
    - Ectopic: May see ectopic insertion into urethra, vagina, etc.

Ultrasonographic Findings

- Real Time: Thin walled, cystic intravesical mass near ipsilateral ureter

Imaging Recommendations

- Best imaging tool: Excretory urography
- Protocol advice: Cystogram: Get early images of bladder filling; overfilling may collapse low-pressure ureterocele

DDx: Ureterocele Mimics

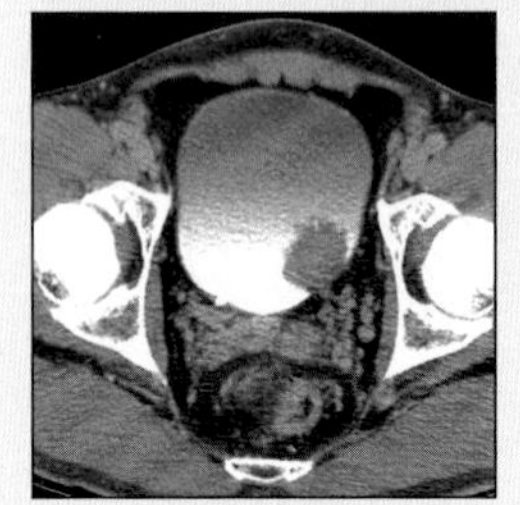

Bladder Cancer

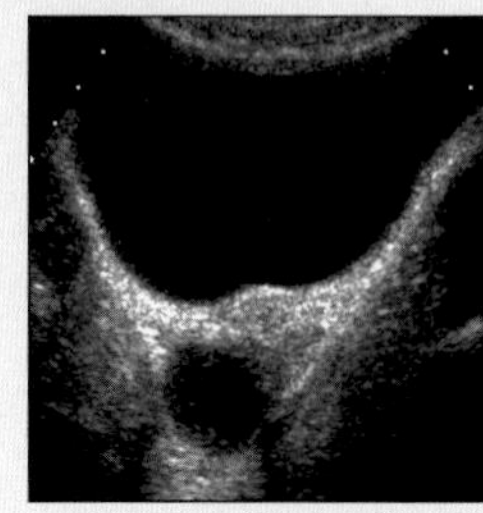

Gartner Duct Cyst

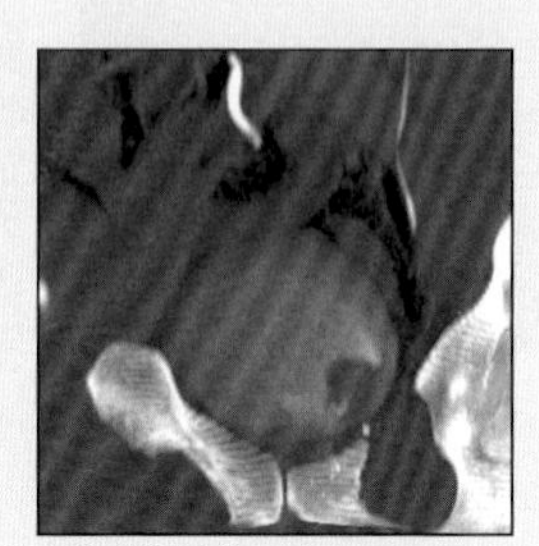

Bladder Cancer

URETERERCELE

Key Facts

Terminology

- Cystic dilatation of distal ureter with prolapse into bladder lumen
- Orthotopic ureterocele: Normal insertion at trigone
- Orthotopic: Single ureter systems
- Ectopic ureterocele: Inserts below trigone
- Ectopic: Duplicated systems in 80%

Imaging Findings

- Best diagnostic clue: Orthotopic: "Cobra-head" or "spring onion" deformity of distal ureter with surrounding radiolucent halo
- Ectopic ureterocele: Smooth, radiolucent intravesicular mass near bladder base
- May evert during voiding and mimic diverticulum
- Real Time: Thin walled, cystic intravesical mass near ipsilateral ureter

DIFFERENTIAL DIAGNOSIS

Pseudoureteroceles

- Bladder or cervical cancer invading UVJ
- Radiation cystitis
- Distal UVJ stone or edema 2° to recent stone passage

Prolapsing ureterocele: Vaginal mass in girls

- Rhabdomyosarcoma: Grape-like clustered mass
- Gartner duct cyst: Cyst in vaginal wall

PATHOLOGY

General Features

- Etiology: Congenital anomaly
- Epidemiology: US: 1 in 12,000 to 1 in 5,000
- Associated abnormalities: Single system ectopic ureteroceles: Cardiac and genital anomalies

Gross Pathologic & Surgical Features

- Simple ureteroceles: Pin-point orifices but no significant obstruction
- Ectopic ureteroceles: Often obstructed, with dysplasia of upper pole kidney

Microscopic Features

- Wall: Covered by bladder mucosa and lined by ureteral mucosa

Staging, Grading or Classification Criteria

- Orifice type: Stenotic, sphincteric, sphincterostenotic, cecoureterocele
  - Sphincteric: Orifice distal to bladder neck
  - Cecoureterocele: Intravesical orifice; submucosal extension to urethra

CLINICAL ISSUES

Presentation

- Most common signs/symptoms
  - Orthotopic: Usually asymptomatic; incidental finding
  - Ectopic ureteroceles: Urinary tract infection (UTI), incontinence, vaginal mass
- Clinical profile: Ectopic: Infant or child with UTI or sepsis

Demographics

- Age
  - Ectopic: Median age 3 months at diagnosis
    - Often diagnosed with prenatal ultrasound
- Gender: Ectopic ureterocele with duplicated system: M:F = 1:4
- Ethnicity: Almost exclusively Caucasians

Natural History & Prognosis

- Severe obstruction: Primarily ectopic ureteroceles
  - Dysplasia of obstructed upper pole moiety

Treatment

- Options, risks, complications: Obstructed ureteroceles may cause stasis and stone formation

DIAGNOSTIC CHECKLIST

Consider

- Check obliques for mucosal irregularity to rule out tumor

Image Interpretation Pearls

- Long axis of ectopic ureterocele points to side of origin

SELECTED REFERENCES

1. Walsh PC et al: Campbell's Urology. 8th ed. Philadelphia, Saunders. 2007-52, 2002
2. Davidson AJ et al: Radiology of the kidney and genitourinary tract. 3rd ed. Philadelphia, W.B. Saunders. 213-6, 1999

IMAGE GALLERY

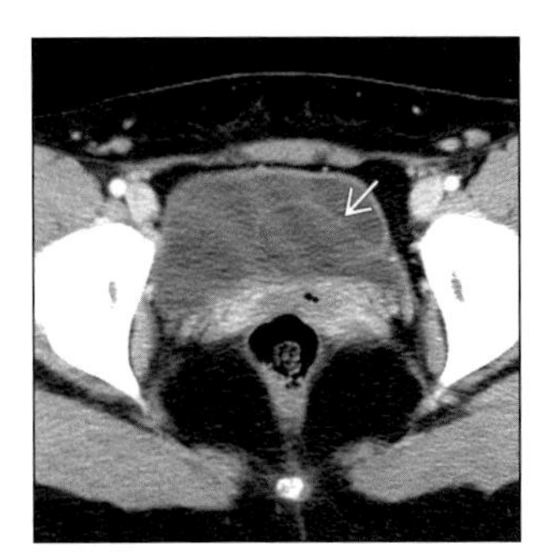

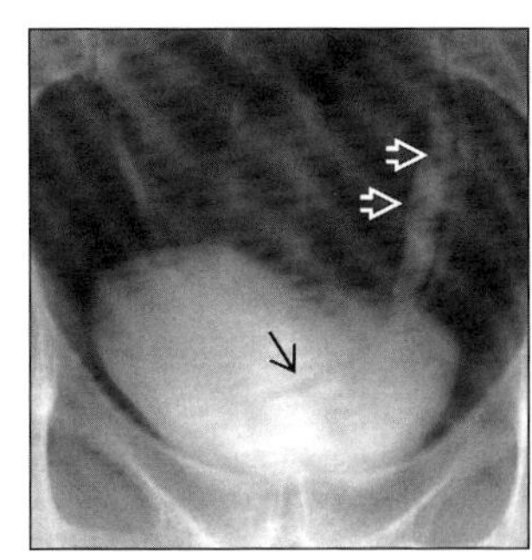

*(**Left**) Axial CECT shows large left orthotopic ureterocele (arrow). (**Right**) IVP shows complete left system duplication and ectopic ureterocele at bladder base (arrow). Filling defects in dilated ureter (open arrows) represent ureteritis cystica (Courtesy M. Nino-Murcia, MD).*

URETERITIS CYSTICA

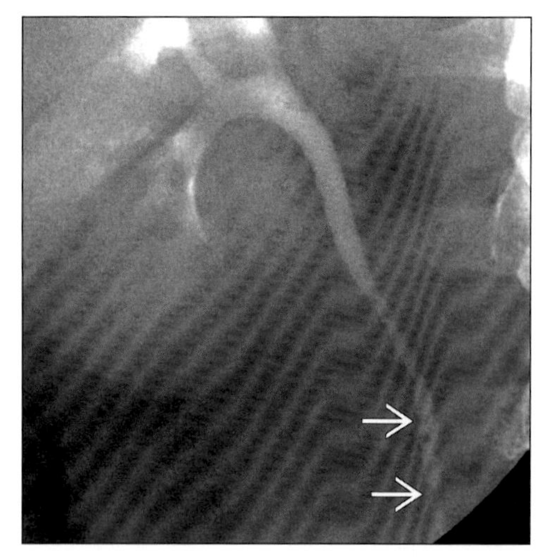

IVP shows tiny smooth, round filling defects (arrow) in right ureter in patient with ureteritis cystica (Courtesy M. Nino-Murcia, MD).

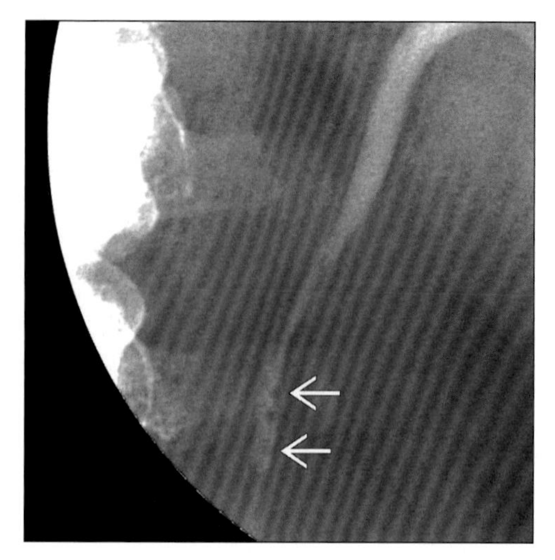

IVP shows image of left ureter in the same patient, with several tiny nodular filling defects (arrows) representing submucosal cysts of ureteritis cystica (Courtesy M. Nino-Murcia, MD).

TERMINOLOGY

Abbreviations and Synonyms

- Ureteritis cystica
- Pyeloureteritis cystica if renal pelvis involved

Definitions

- Multiple small subepithelial cysts in ureter wall

IMAGING FINDINGS

General Features

- Best diagnostic clue: Multiple small mural-based filling defects in ureter lumen
- Location
  - Proximal 1/3 of ureter
  - May also occur in renal pelvis (pyelitis cystica) and urinary bladder (cystitis cystica)
  - May be unilateral or bilateral, symmetric or asymmetric
- Size
  - Cysts 2-3 mm in size; usually < 5 mm
  - Occasionally may be as large as 2 cm
- Morphology
  - Smooth, spherical radiolucent filling defects
  - May have few cysts or dozens
  - Lesions may be isolated or clustered in groups

Radiographic Findings

- IVP
  - Nodular radiolucent small filling defects in ureter lumen
  - In profile: Scalloping of ureteral margins
  - May have ragged appearance of ureter wall from confluent groups of tiny cysts
- Retrograde pyelogram: Tiny radiolucent filling defects

MR Findings

- T2WI: Beading of ureter wall on MR urography

Imaging Recommendations

- Best imaging tool: IVP or retrograde pyelography
- Protocol advice: Oblique films helpful to show mural location and smooth surface of tiny cysts

DIFFERENTIAL DIAGNOSIS

Air bubbles

- Should move; do not elevate uroepithelium

Multifocal transitional cell carcinoma (TCCA)

- Usually have more irregular surface
- Not as numerous as ureteritis cystica

Metastatic melanoma

- Usually larger, fewer nodules

DDx: Ureteral Filling Defects

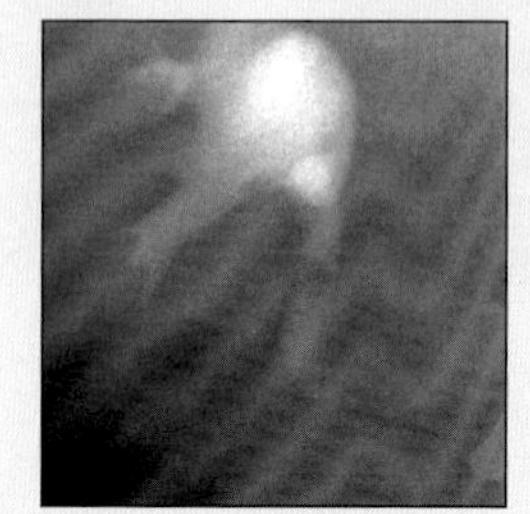

Melanoma

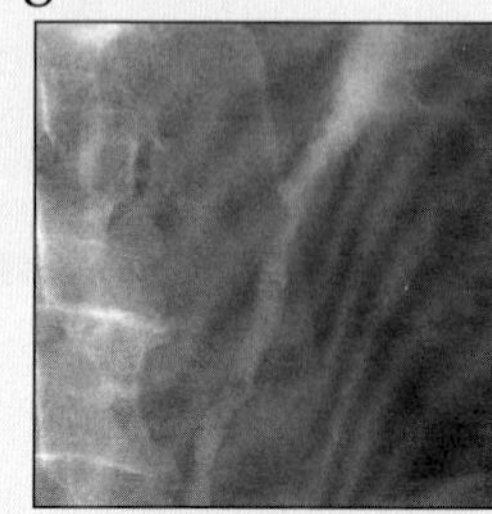

Multifocal TCCA

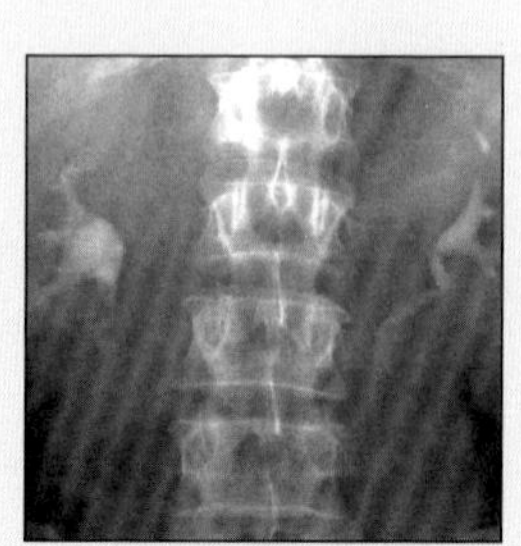

Melanoma

URETERITIS CYSTICA

Key Facts

Terminology
- Multiple small subepithelial cysts in ureter wall

Imaging Findings
- May also occur in renal pelvis (pyelitis cystica) and urinary bladder (cystitis cystica)
- May be unilateral or bilateral, symmetric or asymmetric
- Cysts 2-3 mm in size; usually < 5 mm

Top Differential Diagnoses
- Air bubbles
- Multifocal transitional cell carcinoma (TCCA)
- Metastatic melanoma

Pathology
- Relatively rare reactive hyperplastic lesion of uroepithelium
- Not premalignant or neoplastic

Crossing vessels
- Usually not spherical; filling defects not as numerous

Blood clots, radiolucent stones
- Fewer, larger, mobile

PATHOLOGY

General Features
- General path comments
  - Relatively rare reactive hyperplastic lesion of uroepithelium
  - Not premalignant or neoplastic
- Etiology
  - May be response to chronic urinary tract infection (UTI)
  - May persist even after infection resolves; can occur without UTI
  - Theory of pathogenesis: Von Brunn cell nest
    - Inflammatory stimulation → proliferation of urothelium → "cell nest" shifting mucosa inward
    - Cells become isolated within tunica propria → metaplastic transformation into glandular structures → fluid-filled cysts
  - Can be seen with bacterial infection and schistosomiasis
    - Ureteritis cystica calcinosa: Punctate calcifications
- Associated abnormalities: Calculi, urinary tract infections (UTIs)

Gross Pathologic & Surgical Features
- Smooth, round, grayish subepithelial nodules containing translucent or yellowish fluid
- Cyst fluid may be thin or thick
- Fluid may be hemorrhagic

Microscopic Features
- Tiny cysts in ureter wall lined with epithelial cells

CLINICAL ISSUES

Presentation
- Most common signs/symptoms
  - Usually asymptomatic
  - Other signs/symptoms
    - Hematuria from UTI, cyst rupture or coexistent stones
    - UTI symptoms
- Clinical profile: Older patient with chronic UTI

Demographics
- Age: All ages, most common in ages 50-60
- Gender: M < F

Natural History & Prognosis
- Usually inconsequential; should rule out UTI
- No risk of malignant degeneration

Treatment
- Options, risks, complications: Treat any concomitant pathology such as stones or UTI

DIAGNOSTIC CHECKLIST

Image Interpretation Pearls
- Ureteritis cystica does not cause obstruction

SELECTED REFERENCES

1. Wang IK et al: Pyeloureteritis cystica: case report. Chang Gung Med J. 24(1):62-7, 2001
2. Menendez V et al: Cystic pyeloureteritis: review of 34 cases. Radiologic aspects and differential diagnosis. Urology. 50(1):31-7, 1997
3. Dunnick NR et al: Textbook of uroradiology. 2nd ed. Baltimore, Williams and Wilkins, 381-2, 1997
4. Noda S et al: Histopathological studies on the cystic formation of the human urothelium. Kurume Med J. 37(2):55-65, 1990

IMAGE GALLERY

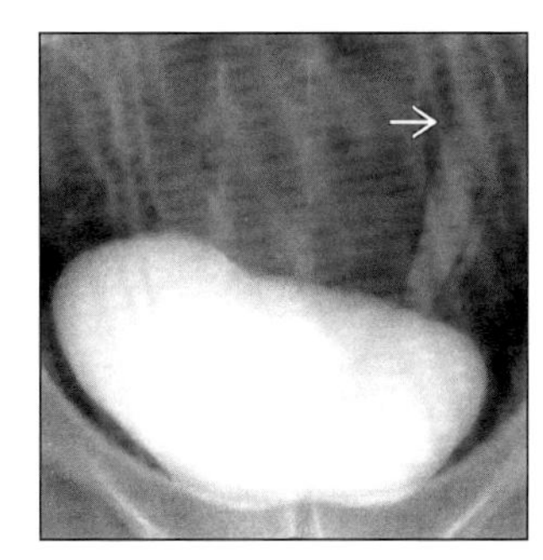

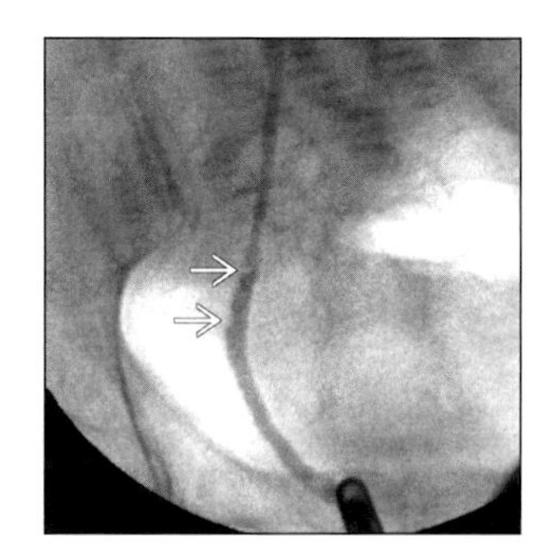

(Left) *IVP shows duplicated collecting system with ureteritis cystica (arrows) (Courtesy M. Nino-Murcia, MD).* ***(Right)*** *Retrograde pyelogram shows mural-based nodular filling defects (arrows) in distal ureter.*

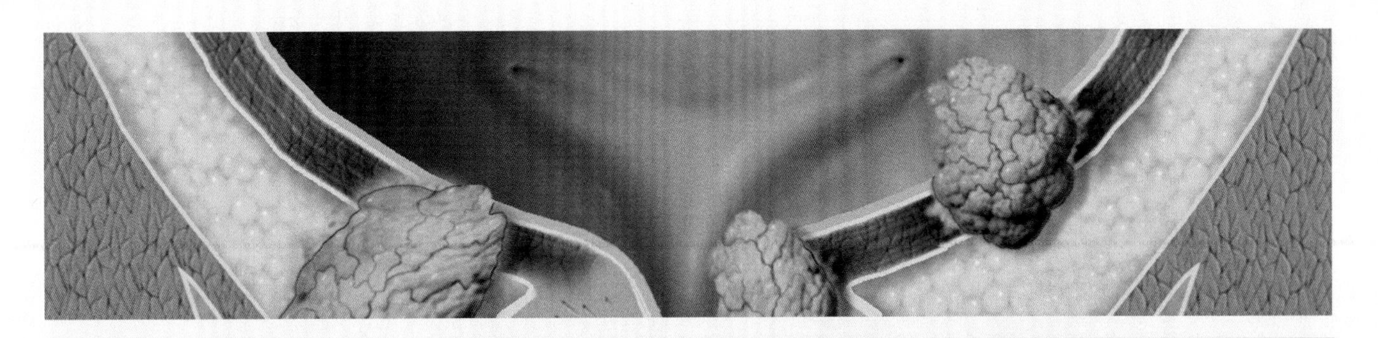

SECTION 5: Bladder

Introduction and Overview

Congenital

Infection

Inflammation

Trauma

Neoplasm

BLADDER ANATOMY AND IMAGING ISSUES

Graphic shows staging system for bladder carcinoma, based on depth of invasion of bladder wall. T3 tumor involves the perivesicle fat.

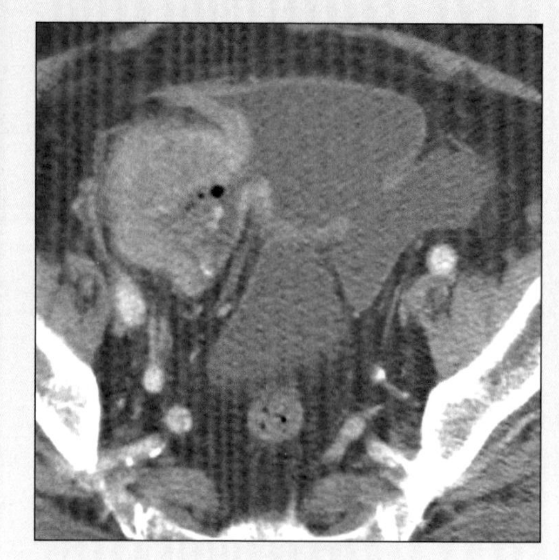

Axial CECT shows multiple bladder diverticula, one of which contains a solid mass; transitional cell carcinoma.

IMAGING ANATOMY

- Bladder and ureters are lined by transitional epithelium
  - Can undergo metaplasia to squamous epithelium in response to chronic inflammation
- Bladder is predominately a pelvic, extraperitoneal organ
  - Dome is covered by peritoneum, prone to intraperitoneal rupture when distended and subjected to blunt trauma
  - Extraperitoneal space between anterior bladder wall and the symphysis pubis is the anterior perivesical space (of Retzius)

ANATOMY-BASED IMAGING ISSUES

Key Concepts or Questions

- Ureteral duplication
  - Can be complete (due to second ureteric bud from mesonephric duct)
  - Can be incomplete (branching of a single ureteric bud)
  - Weigert-Meyer rule
    - Upper moiety ureter inserts inferior and medial to the ureter draining lower moiety
    - Upper pole ureter inserts below the trigone into any Wolffian duct derivative (e.g., vagina, prostate)
  - Complications
    - Vesicoureteral reflux (most common)
    - Ureteral obstruction
    - Enuresis (only in females; insertion always above external sphincter in males)
    - Obstruction: Upper pole ureter often obstructed (ectopic ureterocele, etc.); often non-functional upper renal pole at time of diagnosis
  - Diagnosis
    - "Drooping lily sign" on urography
    - Hydronephrosis and decreased function of upper pole displace the lower pole calices
    - Both ureters evident on CT
    - Upper pole ureter often dilated and unopacified on CECT
- Retrocaval ureter
  - Congenital anomaly of IVC
    - Persistent subcardinal vein "traps" ureter
  - Appearance: Proximal ureter swings medially over the pedicle of L3 or 4 vertebra, behind IVC, then returns to normal position
  - Causes variable degree of hydronephrosis
- Primary megaureter (megaloureter)
  - "Achalasia" of ureter
  - Congential dilation of lower ureter (occasionally, entire collecting system)
  - Due to faulty development of muscle layers of ureter
    - (Not obstruction)
  - Can be associated with other renal and ureteral anomalies
    - (Duplication, ectopia, ureterocele)
  - May be asymptomatic or associated with pain, infection
- Ureteropelvic junction obstruction
  - Most common cause of neonatal hydronephrosis
  - Commonly encountered in adults also
  - Sharp transition from dilated pelvis to normal ureter
  - May be congenital or acquired
- Hydronephrosis (obstructive uropathy)
  - Dilation of collecting structures without functional impairment
  - Obstructive nephropathy = dilation + impairment
  - Degree of dilation depends on
    - Duration of obstruction
    - Renal output, hydration
    - Presence of spontaneous decompression
  - Acute causes
    - Ureteral calculus
    - Passage of blood clot (e.g., tumor, trauma)
    - Surgical mishap (e.g., ligation)
    - Ureteral edema following instrumentation

BLADDER ANATOMY AND IMAGING ISSUES

DIFFERENTIAL DIAGNOSIS

Hydronephrosis on Sonography

False negatives
- Staghorn calculus filling lumen
- Very acute obstruction
- Spontaneous decompression (e.g., forniceal rupture)
- Dehydration

False positives
- Distended bladder
- Overhydration
- Prior obstruction
- Vesico-ureteral reflux
- (Parapelvic cyst)
- (Renal sinus vessels)
- (Prominent extrarenal pelvis)

Ureteropelvic junction obstruction (UPJ)
- Congenital
- Eosinophilic ureteritis
- Ischemia
- Extrinsic
- ⇒ Crossing vessels
- ⇒ Adventitial bands
- ⇒ Renal mass
- ⇒ Aortic aneurysm

Urinary bladder tumors (malignant)
- Transitional cell carcinoma (90%)
- Squamous cell cancer (chronic inflammation)
- Adenocarcinoma (urachal)
- Rhabdomyosarcoma (childhood)
- Lymphoma

        - Sulfonamide crystallization in urine
        - Pregnancy
- Emphysematous pyelitis
    - Gas is confined to collecting system (not renal parenchyma)
    - Usually in diabetes with coliform bacterial infection
    - Better prognosis than emphysematous pyelonephritis
- Pyeloureteritis cystica
    - Hyperplastic proliferation of transitional epithelium projecting into ureteral lumen
    - Caused by chronic urinary infection +/or inflammation
    - More common in diabetes, women
    - Affects bladder more than ureter
    - Appearance: Multiple small (1-3 mm) lucent detects in ureteral or bladder lumen
    - Can simulate or lead to transitional cell carcinoma
- Ureteral calculi (urolithiasis)
    - Affects 12% of population by age 70
        - 2 to 3% will have an attack of renal colic
        - Peak age: 20-30
    - Most patients have a specific metabolic disorder
        - Hypercalcemia (e.g., sarcoidosis, Cushing syndrome, hyperparathyroidism)
        - Excess dietary calcium or sodium
        - Bowel disease
        - Gout
        - Oxalosis
    - Patients with Crohn disease or other cause of ileal resection have excess oxalate resorption
    - Patients with urinary tract infection with urea-splitting organisms can get "triple phosphate" (struvite) stones
        - Especially common in patients with neurogenic bladder, indwelling catheter
        - Can lead to staghorn calculi
    - Diagnosis
        - CT is the most accurate and simple
        - No need to document every instance of renal colic with CT (many patients are young, will have many stones and will form more; danger of excessive radiation)
- Neurogenic bladder
    - Denervation or neurological disorder of the bladder
    - "Spastic bladder"
        - Lesion of "upper motor neuron" (above conus)
    - Atonic bladder
        - "Lower motor neuron" lesion (below conus)
        - Resembles a pine cone or Christmas tree on cystography
    - Etiology
        - Congenital (e.g., myelomeningocele)
        - Trauma
        - Neoplasm (spinal or brain)
        - Infection (polio)
        - Inflammation (multiple sclerosis, syrinx)
        - Systemic disorder (e.g., diabetes)
- Urachal Anomalies
    - Urachus is the remnant of the allantois
    - Should regress to a cord (median umbilical ligament) in fetal stage
    - Variations
        - Patent urachus - fistula from bladder to umbilicus (urine draining from umbilicus)
        - Urachal sinus - patent only at umbilicus
        - Urachal diverticulum - communicates only with bladder dome
        - Urachal cyst - closed at both ends; most common in adults, symptomatic midline cystic mass between umbilicus and bladder; symptoms due to enlargement or infection; can undergo malignant degeneration (usually adenocarcinoma)
- Schistosomiasis (Bilharziasis)
    - S. hematobium affects the urinary system
    - Parasitic infection transmitted through fresh water snails
        - Usually invades humans through feet in contact with contaminated water
        - Passes through lymphatics to vessels
    - Endemic in Africa and Puerto Rico
    - Appearance
        - Bladder wall calcification
        - Ureteral, seminal vesicle calcification
        - Ureteral strictures
        - Bladder calculi

BLADDER ANATOMY AND IMAGING ISSUES

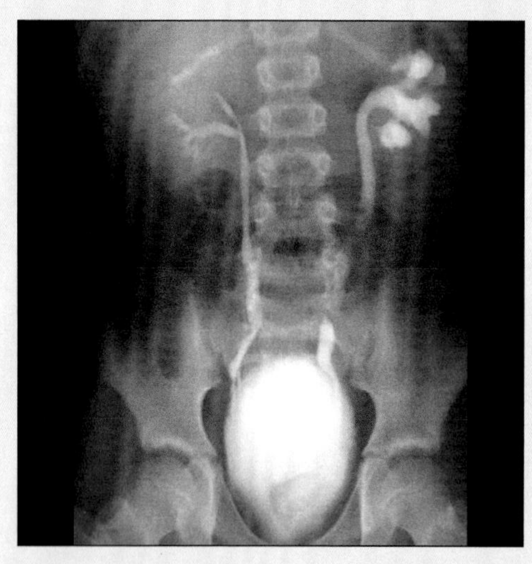

Urogram shows "drooping lily" sign; left renal calices displaced downward by non-functioning hydronephrotic duplicated upper pole.

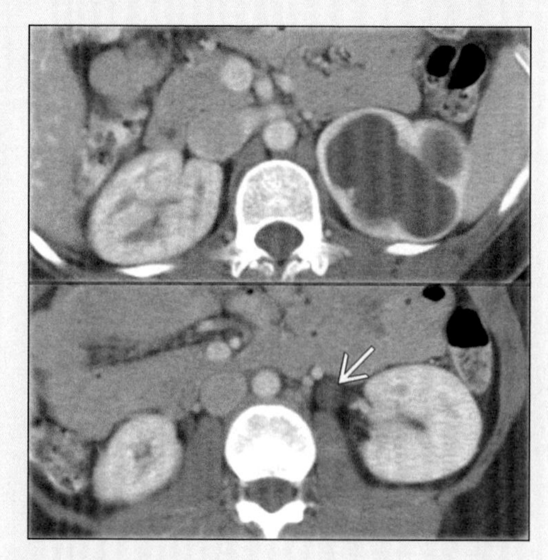

Axial CECT shows hydronephrotic upper pole calices and ureter (arrow), normal lower pole and ureter; ureteral duplication with obstructed upper ureter.

- Bladder diverticulum
  - Usually associated with bladder outlet obstruction
    - E.g., prostate enlargement; neurogenic bladder
    - Often multiple
    - Can be source of urinary stasis, calculi, infection, tumor
  - Can be congenital
    - Hutch diverticulum (above and lateral to ureteral orifice)
  - Can be simulated by surgical augmentation of bladder (with bowel) or by herniation of bladder into inguinal hernia ("bladder ears", most common in male infants)
- Bladder cancer
  - Transitional cell cancer = 85 to 90%
  - Epidemiology: M:F = 3:1; mean age 64
  - Risk factors
    - Chemical exposure (e.g., aniline dye)
    - Tobacco
    - Cyclophosphamide therapy
    - Analgesic abuse
    - Recurrent or chronic infection
  - Staging accuracy
    - Clinical (50%)
    - CT (30 to 80%)
    - MR (70%)
  - Cytoscopy essential for diagnosis, staging
  - T1: Limited to submucosa
  - T2 : Into superficial muscle
  - T3: Into deep muscle
  - T3b: Into perivesicle fat
  - T4: Into adjacent organs or distant metastases
- Bladder fistulas
  - Can be caused by trauma, tumor, infection, surgery or instrumentation of bladder or any adjacent organ
  - Colovesicle fistula is most common
    - Sigmoid diverticulitis most common etiology; tumor second
    - Symptoms: Dysuria, urinary infection, fecaluria
    - Key findings: Imaging evidence of gas, stool, enteric contrast in bladder (best shown by CT; also shows associated findings of diverticulitis, tumor, etc.)

CUSTOM DIFFERENTIAL DIAGNOSIS

Focal hydronephrosis

- Congenital: Duplicated system with obstruction
  - Usually upper pole
- Infection
  - Usually TB; stricture of infundibulum
  - Infundibular calculus
  - Tumor
    - Transitional cell carcinoma

Radiolucent filling defects in renal pelvis

- Calculus
  - Many are nonopaque on radiography
  - Only matrix and indinavir stones are nonopaque on CT
- Blood clot
- Fungus ball or pus
- Tumor (TCC or RCC)
- Sloughed tissue (papillary necrosis)
- Gas (e.g., iatrogenic; pyelitis cystica)
- Foreign body

SELECTED REFERENCES

1. Yu JS et al: Urachal remnant diseases: spectrum of CT and US findings. Radiographics. 21(2):451-61, 2001
2. Morgan DE et al: CT cystography: radiographic and clinical predictors of bladder rupture. AJR Am J Roentgenol. 174(1):89-95, 2000
3. Kim B et al: Bladder tumor staging: comparison of contrast-enhanced CT, T1- and T2-weighted MR imaging, dynamic gadolinium-enhanced imaging, and late gadolinium-enhanced imaging. Radiology. 193(1):239-45, 1994

IMAGE GALLERY

Typical

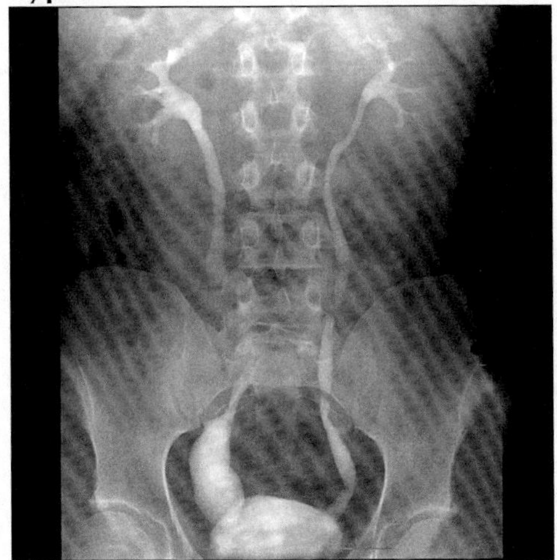

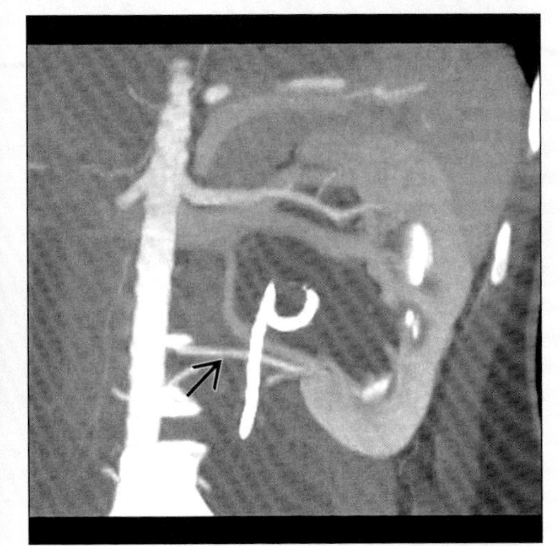

(Left) *Urogram shows congenital primary megaureter, simulating obstruction.* ***(Right)*** *CECT, MIP coronal reformation shows dilated renal pelvis, ureteral stent, and accessory renal artery (arrow) crossing the UPJ.*

Typical

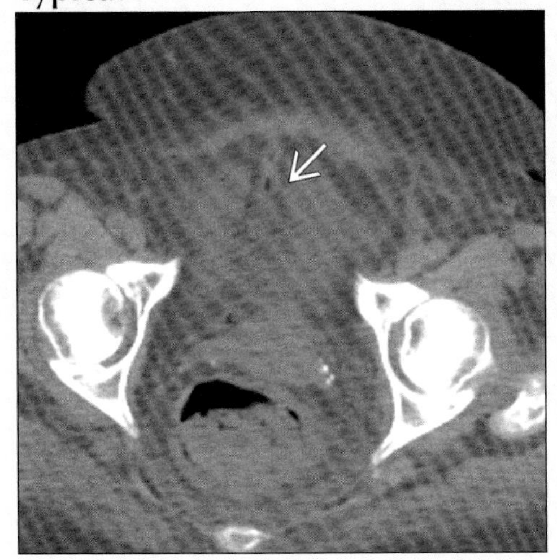

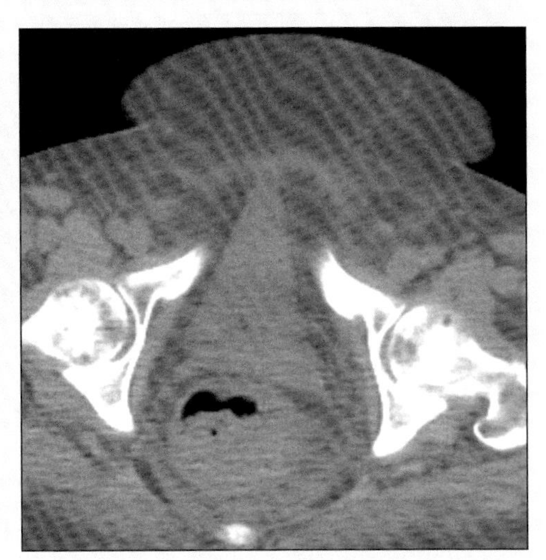

(Left) *Axial NECT shows median umbilical ligament containing gas bubble (arrow), with ascites on both sides.* ***(Right)*** *Axial NECT shows triangular shaped bladder due to urachal diverticulum.*

Typical

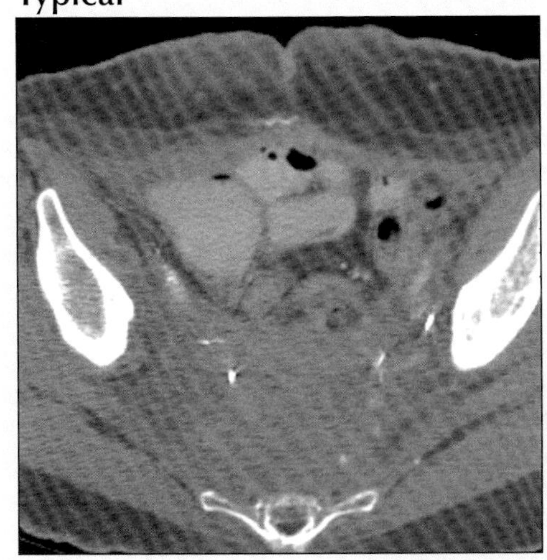

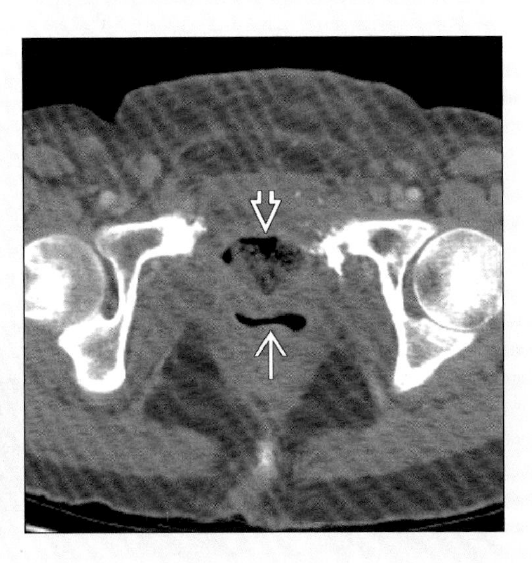

(Left) *Axial CECT shows extensive pre-sacral soft tissue following surgery and radiation therapy for cervical carcinoma.* ***(Right)*** *Axial CECT shows gas and feces in vagina (arrow), and bladder (open arrow), due to fistulas to rectum.*

URACHAL REMNANT

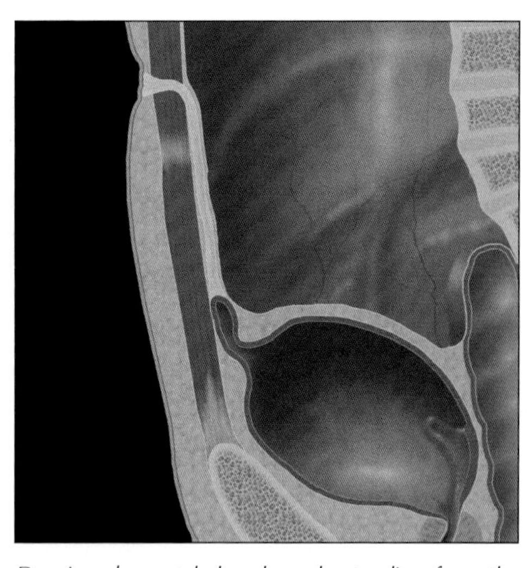

Drawing shows tubular channel extending from the dome of bladder along the anterior abdominal wall in the midline toward the umbilicus, representing a urachal diverticulum.

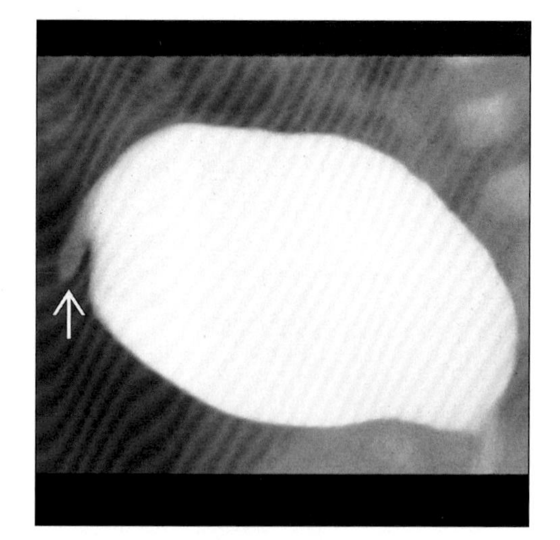

Cystogram shows tubular structure (arrow) arising from bladder dome and extending to umbilicus, representing patent urachus (Courtesy G. Friedland, MD).

TERMINOLOGY

Abbreviations and Synonyms

- Vestigial, obliterated urachus = median umbilical ligament

Definitions

- Urachal abnormalities: Incomplete obliteration of embryonic connection between bladder dome and allantoic duct
- Patent urachus: Entire urachal channel fails to close
- Urachal cyst: Umbilical and bladder openings close; channel in between remains open and fluid-filled
- Urachal sinus: Dilatation of urachus at umbilical end; no communication with bladder
- Urachal diverticulum: Dilatation of urachus at vesical end; no communication with umbilicus
- Urachal carcinoma (Ca): Adenocarcinoma (90%) arising from urachus

IMAGING FINDINGS

General Features

- Best diagnostic clue: Cyst or mass between bladder and umbilicus
- Location: Between peritoneum and transversalis fascia (space of Retzius)
- Size
  - Urachus: 5-6 cm in length
  - Urachal diverticulum: Small or large
- Morphology: Urachal cyst: May be single or multiple

Radiographic Findings

- Radiography: Urachal Ca: Stippled, granular or curvilinear calcifications above bladder outline
- Fluoroscopy
  - Sinogram: Tract extending inferiorly from umbilicus
  - Voiding cystourethrogram: Tubular channel arising from bladder dome

CT Findings

- CECT
  - Urachal diverticulum: Midline cystic lesion at anterosuperior aspect of bladder
  - Urachal cyst: Midline cyst above bladder dome
    - May have rim calcification
    - Low density contents due to mucin production
    - No soft tissue component unless complicated
  - Urachal carcinoma: Midline supravesical soft tissue mass; calcification in 70%

Ultrasonographic Findings

- Real Time
  - Urachal cyst: Midline cyst above bladder
  - Complicated urachal cyst: Heterogeneous echogenicity

DDx: Mimics of Urachal Carcinoma

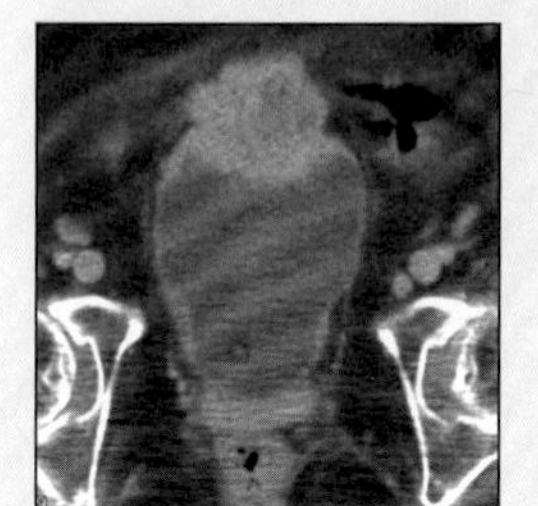

Bladder Ca

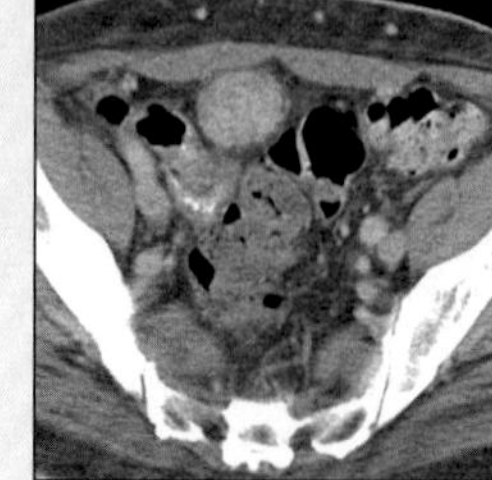

Gastric Ca Met

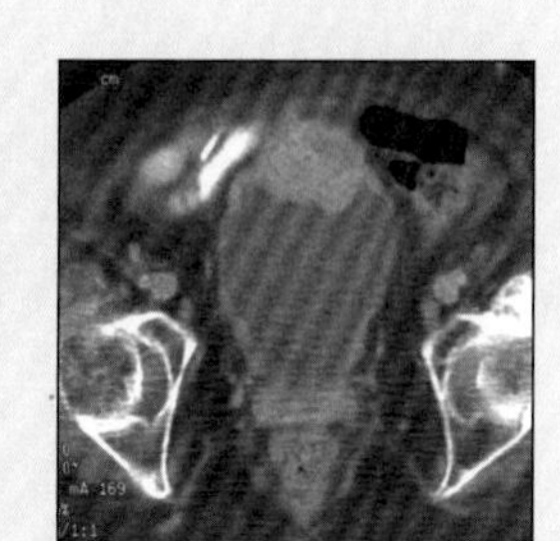

Bladder Ca

URACHAL REMNANT

Key Facts

Terminology

- Urachal abnormalities: Incomplete obliteration of embryonic connection between bladder dome and allantoic duct
- Patent urachus: Entire urachal channel fails to close
- Urachal cyst: Umbilical and bladder openings close; channel in between remains open and fluid-filled
- Urachal sinus: Dilatation of urachus at umbilical end; no communication with bladder
- Urachal diverticulum: Dilatation of urachus at vesical end; no communication with umbilicus

Imaging Findings

- Urachal carcinoma: Midline supravesical soft tissue mass; calcification in 70%

Clinical Issues

- Urachal carcinoma: 5 year survival ≤ 15%

Imaging Recommendations

- Best imaging tool: CT

DIFFERENTIAL DIAGNOSIS

Cystic lesions near bladder/umbilicus

- Vitelline cyst, mesenteric cyst

Urachal carcinoma: Differential diagnosis

- Primary bladder carcinoma at dome
- Metastasis (met) to bladder from rectal, stomach, ovarian or other adenoca

PATHOLOGY

General Features

- Etiology: Congenital anomaly
- Epidemiology
  - Patent urachus: Rare; prevalence 0.25-15 per 10,000
  - Urachal carcinoma: 0.2-0.3% of all bladder cancers; 20-40% of primary bladder adenocarcinomas
- Associated abnormalities: Patent urachus: Congenital lower urinary tract obstruction: Posterior urethral values, prune belly syndrome, urethral atresia

Gross Pathologic & Surgical Features

- Urachal carcinoma: Frequently contains calcium; invades bladder dome

Microscopic Features

- Urachal anomalies: Mucosal lining of transitional epithelium
- Urachal carcinoma: 85-90% are adenocarcinomas resulting from metaplasia of transitional epithelium

CLINICAL ISSUES

Presentation

- Most common signs/symptoms
  - Patent urachus: Urine leakage from umbilicus
  - Complicated urachal cyst: Suprapubic pain, fever
  - Urachal carcinoma: Clinically silent until bladder dome invaded; then get hematuria and/or mucoid discharge

Demographics

- Age
  - Patent urachus: Diagnosed at birth
  - Urachal carcinoma: 40-70
- Gender: M:F = 3:1

Natural History & Prognosis

- Urachal carcinoma: 5 year survival ≤ 15%
- Poor prognosis related to late presentation

Treatment

- Options, risks, complications
  - Complications: Adenocarcinoma in urachal remnant
  - Stone formation in urachal diverticulum

DIAGNOSTIC CHECKLIST

Image Interpretation Pearls

- Look for center of mass; if outside bladder, consider urachal CA with bladder invasion

SELECTED REFERENCES

1. Yu JS et al: Urachal remnant diseases: spectrum of CT and US findings. Radiographics. 21(2):451-61, 2001
2. Davidson AJ et al: Radiology of the kidney and genitourinary tract. 3rd ed. Philadelphia, W.B. Saunders. 213-16, 1999
3. Dunnick NR et al: Textbook of uroradiology. 2nd ed. Baltimore, Williams and Wilkins. 29-33, 1997

IMAGE GALLERY

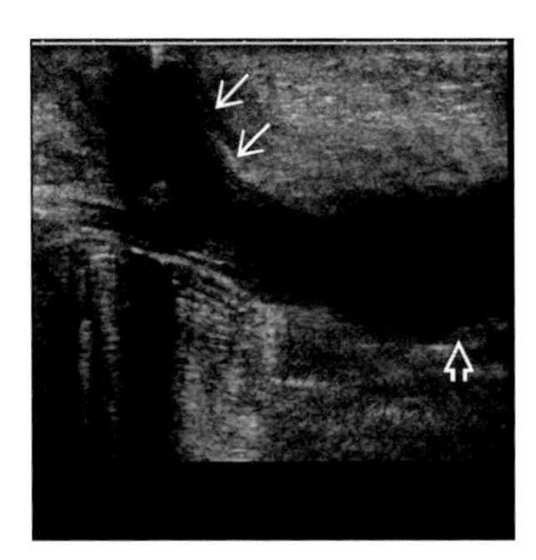

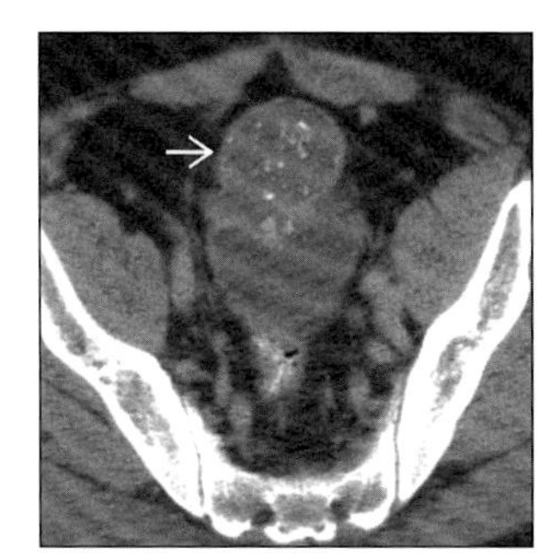

(Left) *Sagittal US shows tubular anechoic channel (arrows) connecting umbilicus to bladder (open arrow) representing patent urachus.* ***(Right)*** *Axial CECT shows exophytic mass (arrow) with coarse internal calcification invading bladder dome, representing urachal adenocarcinoma.*

CYSTITIS

Axial CECT shows thick-walled, irregular bladder representing radiation cystitis in patient who underwent radiation therapy for rectal cancer.

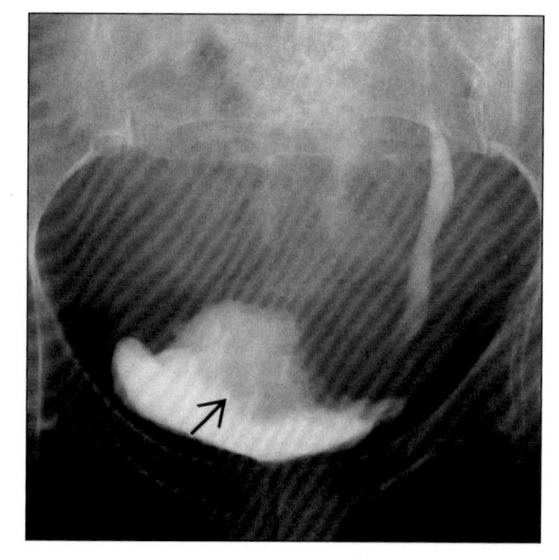

IVP shows "cobblestone" appearance of bladder mucosa in patient with cystitis secondary to perforated sigmoid diverticulitis (Courtesy M. Nino-Murcia, MD).

TERMINOLOGY

Definitions

- Inflammation of the urinary bladder wall
- Acute cystitis: Recent onset, short duration
- Chronic cystitis: Longer duration (months-years)
  - Variants: Cystitis cystica, cystitis glandularis

IMAGING FINDINGS

General Features

- Location: Usually involves entire bladder wall, but can be focal
- Morphology: Chronic cystitis: Contracted, nodular bladder wall

Radiographic Findings

- Radiography: TB, schistosomiasis: Bladder wall calcification
- IVP
  - Bladder often normal in acute cystitis
  - Acute: Thickened, coarse mucosal folds with cobblestone appearance
  - Acute: Bullous edema of bladder mucosa
  - Chronic: Contracted, irregular thick-walled bladder

CT Findings

- CECT
  - Bladder wall thickening +/- hypodense wall
  - Emphysematous cystitis: Gas in bladder wall and/or lumen

Ultrasonographic Findings

- Real Time: Hypoechoic edematous bladder wall

DIFFERENTIAL DIAGNOSIS

Thickened bladder wall: Nondistention, trabeculation, infiltrating Ca

- Indistinguishable by imaging alone

PATHOLOGY

General Features

- General path comments: Schistosomiasis is most frequent cause of bladder wall calcification
- Etiology
  - Infectious cystitis
    - Bacterial: Transurethral invasion of bladder by perineal flora in sexually active women
    - Bacterial: Bladder outlet obstruction and urinary stasis in men
    - Schistosomiasis: Inflammatory response to ova deposited in bladder submucosa

DDx: Bladder Wall Thickening

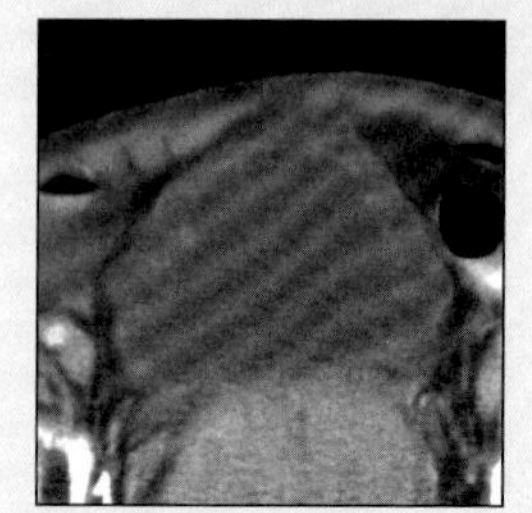

Trabeculation

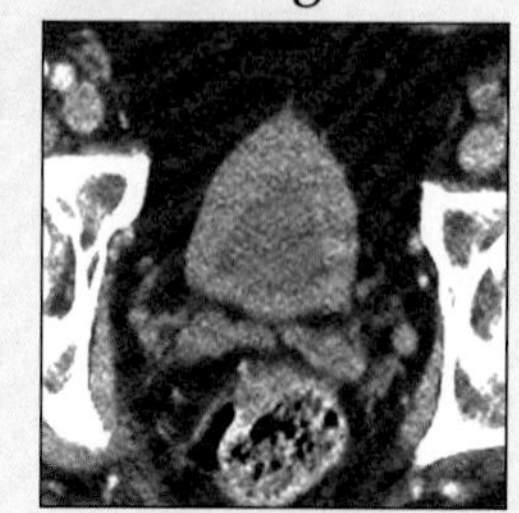

Non-Distention

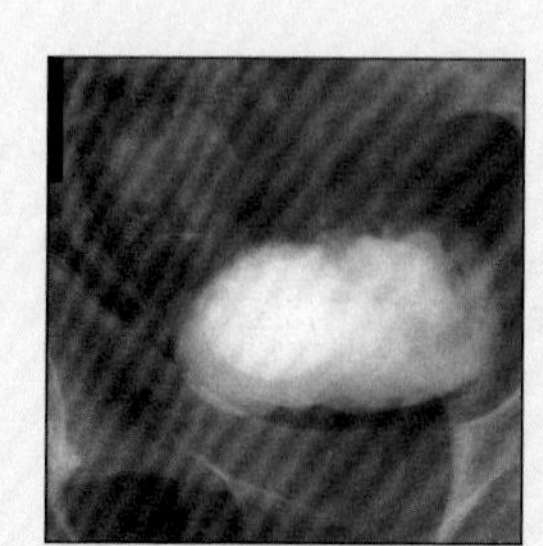

Melanoma

Key Facts

Imaging Findings
- Acute: Bullous edema of bladder mucosa
- Chronic: Contracted, irregular thick-walled bladder
- Bladder wall thickening +/- hypodense wall
- Emphysematous cystitis: Gas in bladder wall and/or lumen

Top Differential Diagnoses
- Thickened bladder wall: Nondistention, trabeculation, infiltrating Ca

Pathology
- Bacterial: Transurethral invasion of bladder by perineal flora in sexually active women
- Bacterial: Bladder outlet obstruction and urinary stasis in men

        - Tuberculosis: Descending infection from kidney
        - Emphysematous: E. coli, Aerobacter aerogenes or Candida infection
        - Alkaline encrustation: Infection by urea-splitting organism → alkaline urine + focal necrosis → dystrophic calcification
        - Others: Viral, fungal (Candida)
    - Noninfectious cystitis
        - Mechanical: Local irritation from Foley catheter, stone, foreign body, etc
        - Drug-related: Cyclophosphamide (hemorrhagic cystitis)
        - Radiation-induced
        - Idiopathic: Interstitial cystitis = pancystitis causing severe urgency and frequency
- Associated abnormalities
    - Chronic cystitis: Decreased bladder capacity and vesicoureteral reflux
    - Recurrent bacterial infection: Malakoplakia
        - Associated with E. coli infection
        - Granulomatous inflammatory process
        - Caused by deficient function of lysosomes in macrophages

Gross Pathologic & Surgical Features
- Ulceration, petechiae, erythema of bladder mucosa
- Chronic: Fibrosis and contraction of bladder wall

CLINICAL ISSUES

Presentation
- Most common signs/symptoms
    - Dysuria, frequency, urgency
    - Other signs/symptoms
        - Gross hematuria, pyuria, bacteriuria

Demographics
- Gender: Bacterial cystitis: M < F due to short urethra

Natural History & Prognosis
- Emphysematous cystitis: Responds well to antibiotics; not life-threatening

Treatment
- Options, risks, complications: Acute: Usually responds to antibiotics
- Complications of chronic cystitis
    - Hyperplastic uroepithelial cell clusters (Brunn nests) form in bladder submucosa
        - Fluid accumulation → pseudocysts = cystitis cystica
        - Transformation into glands → cystitis glandularis
    - Squamous metaplasia: Transformation of urothelium into keratin producing squamous cells
        - May see white patches (leukoplakia) on foci of squamous metaplasia
    - Schistosomiasis: Bladder carcinoma, especially squamous

DIAGNOSTIC CHECKLIST

Consider
- Nodular bladder wall thickening in cystitis may mimic bladder carcinoma

Image Interpretation Pearls
- Schistosomiasis progresses proximally; TB progresses distally
- Gas in bladder wall always secondary to infection

SELECTED REFERENCES

1. Ravakhah K et al: Images in clinical medicine. Unusual cystitis. N Engl J Med. 350(22):2280, 2004
2. Zagoria R: Genitourinary radiology, 2nd ed. Philadelphia, Mosby Inc. 211-29, 2004
3. Thoumas D et al: Imaging characteristics of alkaline-encrusted cystitis and pyelitis. AJR Am J Roentgenol. 178(2):389-92, 2002

IMAGE GALLERY

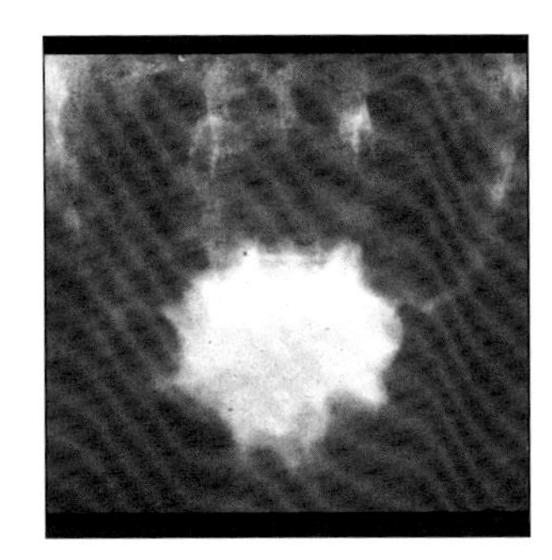

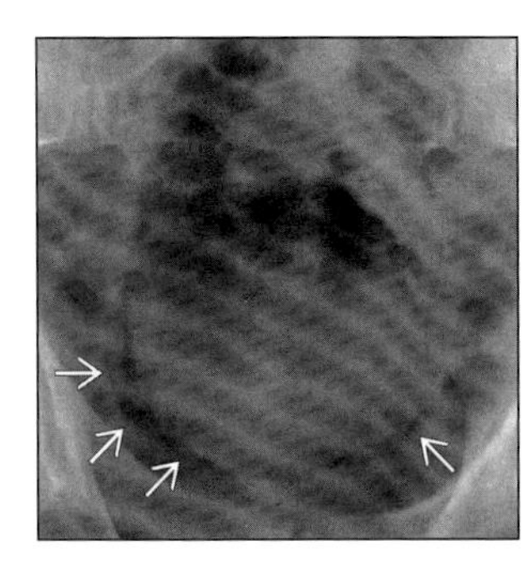

(Left) *IVP shows small capacity bladder with nodular wall in patient with cystitis cystica (Courtesy G. Friedland, MD).* ***(Right)*** *Radiography shows crescent of gas in suprapubic location representing gas in the bladder wall in a patient with emphysematous cystitis (Courtesy M. Nino-Murcia, MD).*

BLADDER CALCULI

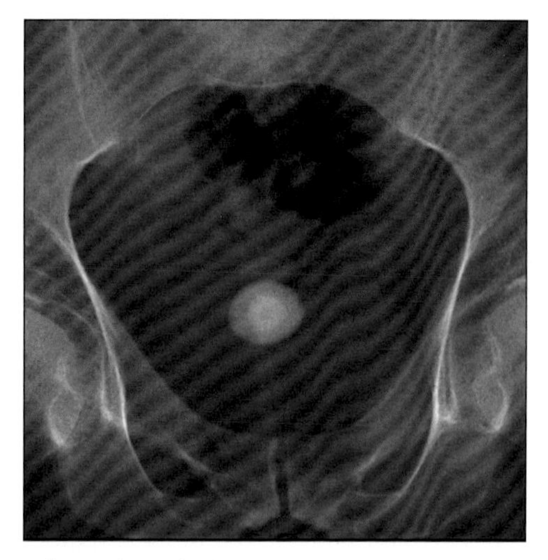

Pelvic radiograph shows smooth, round, lamillated calcification overlying bladder typical of a bladder calculus.

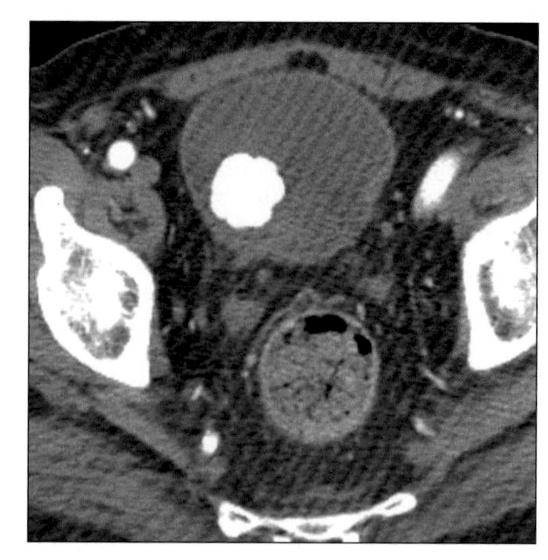

Axial CECT shows lobulated large stone in bladder in a male patient with a history of prostatic enlargement.

TERMINOLOGY

Abbreviations and Synonyms

- Bladder stones, vesical calculi, cystolithiasis

Definitions

- Concretions of mineral salts within bladder lumen

IMAGING FINDINGS

General Features

- Best diagnostic clue: Smooth round or ovoid laminated calcification in bladder
- Location
  - Bladder lumen: Usually midline with patient supine
    - Eccentric if within bladder augmentation or diverticulum
- Size: Variable
- Morphology
  - Round, oval, spiculated, laminated, faceted
    - Stones with multiple spicules: "Jackstones"
    - Stones with lobulated margins: "Mulberry stones"

Radiographic Findings

- Radiography
  - Solitary or multiple calcifications overlying bladder
  - Most are radiopaque but opacity variable
- IVP: Filling defect or radiopacity, depending on relative density of stone versus contrast material

CT Findings

- NECT: All bladder calculi radiopaque on CT

MR Findings

- All pulse sequences: Signal void(s) in bladder

Ultrasonographic Findings

- Real Time: Mobile echogenic foci in bladder with acoustic shadowing

Imaging Recommendations

- Best imaging tool: CT
- Protocol advice: Low kilovoltage (60-70 kV) best for detecting stones on plain film

DIFFERENTIAL DIAGNOSIS

Other bladder calcifications

- Urachal carcinoma: Punctate or coarse calcification in mass at bladder dome
- Schistosomiasis: Calcification in bladder wall

Other pelvic calcifications

- Prostate calcification, calcified fibroids

DDx: Calcified Pelvic Masses

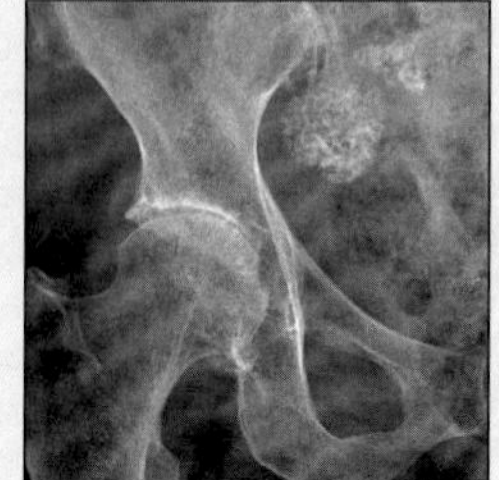

Calcified Fibroid

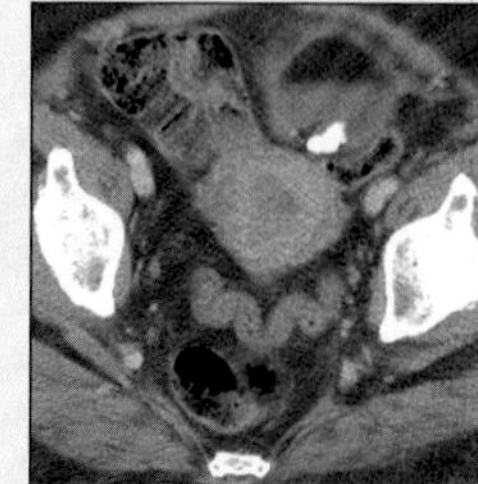

Dermoid

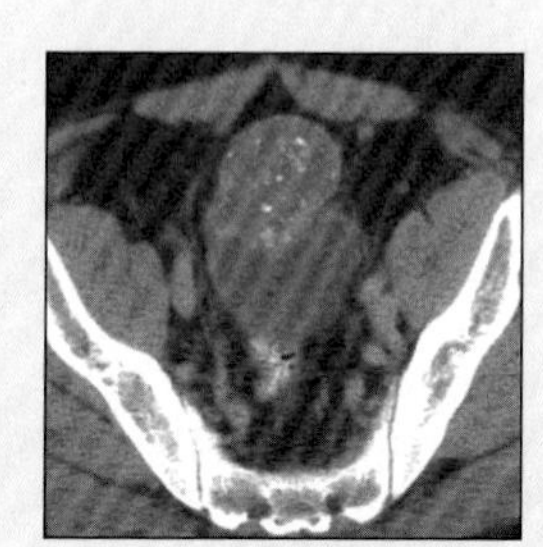

Urachal Carcinoma

BLADDER CALCULI

Key Facts

Imaging Findings
- Round, oval, spiculated, laminated, faceted
- Most are radiopaque but opacity variable
- NECT: All bladder calculi radiopaque on CT

Top Differential Diagnoses
- Other bladder calcifications
- Other pelvic calcifications

Pathology
- Stasis: Bladder outlet obstruction, neurogenic bladder, bladder diverticula
- Infection, especially Proteus mirabilis
- Renal calculus passing into bladder
- Suture material, migrated IUDs or other gyn applicances
- Bladder augmentation: Local metabolic derangement
- Nutritional deficiencies in third world children

Filling defect in bladder
- Blood clot, fungus ball, ureterocele, bladder neoplasm

PATHOLOGY

General Features
- General path comments
  - Most are mixture of calcium oxalate and calcium phosphate
  - Infection stones: Magnesium ammonium phosphate ("struvite")
- Etiology
  - Stasis: Bladder outlet obstruction, neurogenic bladder, bladder diverticula
  - Infection, especially Proteus mirabilis
  - Foreign bodies: Nidus for crystal growth
    - Renal calculus passing into bladder
    - Suture material, migrated IUDs or other gyn applicances
    - Pubic hairs introduced by catheterization
  - Bladder augmentation: Local metabolic derangement
  - Nutritional deficiencies in third world children
- Epidemiology
  - 5% of all urinary tract calculi in Western world
  - Incidence declining in West over past 100 years
    - Upper tract > lower tract stones in developed countries
  - Idiopathic: Endemic in underdeveloped areas, especially in children

CLINICAL ISSUES

Presentation
- Most common signs/symptoms
  - Most asymptomatic
  - Other signs/symptoms
    - Suprapubic pain, microhematuria
    - Gross hematuria is rare
- Clinical profile
  - Spinal cord injury patient with indwelling Foley
  - Elderly male with bladder outlet obstruction
  - Older woman with pelvic prolapse
  - Bladder augmentation with ileal or colonic segments

Demographics
- Gender: M > F

Natural History & Prognosis
- Complication: Malignant bladder tumors in patients with stones from indwelling Foley catheters

Treatment
- Options, risks, complications
  - Prevention and eradication of underlying cause
  - Cystoscopic fragmentation and removal
  - Open cystolithotomy only for very large or very hard stones

DIAGNOSTIC CHECKLIST

Consider
- Obstruction accounts for 70% of adult bladder stones

Image Interpretation Pearls
- Carcinoma resulting from chronic bladder irritation may co-exist with bladder stone

SELECTED REFERENCES

1. Ord J et al: Bladder management and risk of bladder stone formation in spinal cord injured patients. J Urol. 170(5):1734-7, 2003
2. Schwartz BF et al: The vesical calculus. Urol Clin North Am. 27(2):333-46, 2000
3. Dyer RB et al: Abnormal calcifications in the urinary tract. Radiographics. 18(6):1405-24, 1998

IMAGE GALLERY

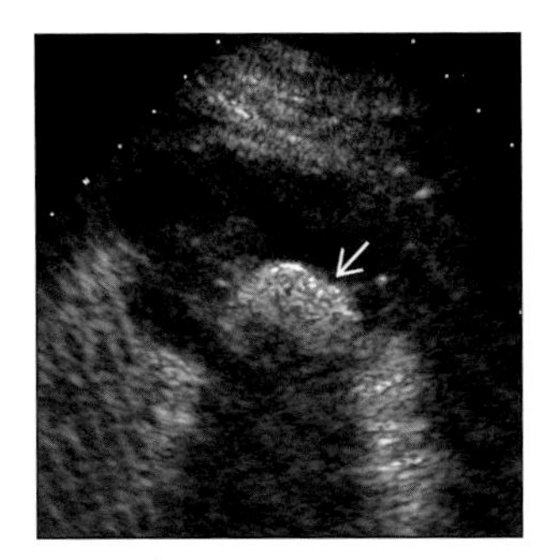
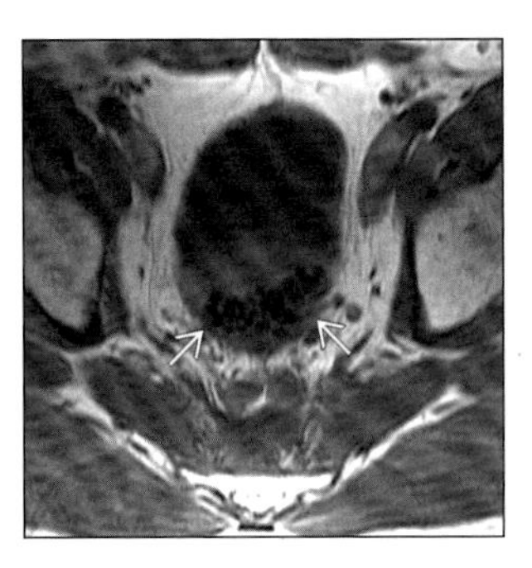

(Left) *Axial oblique US shows large echogenic focus (arrow) with posterior acoustic shadowing representing a large bladder calculus.* ***(Right)*** *Axial T1WI MR image shows innumerable round foci of low signal (arrows) in the dependent portion of the bladder representing calculi in a man with marked prostatic enlargement.*

BLADDER DIVERTICULUM

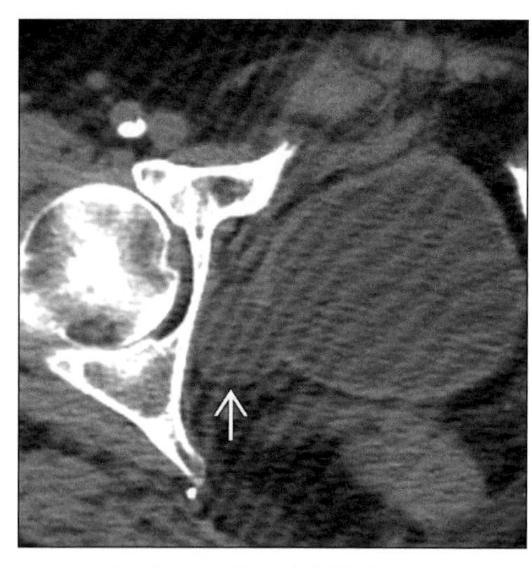

Axial NECT shows right sided bladder diverticulum (arrow) protruding through the muscle of the bladder wall.

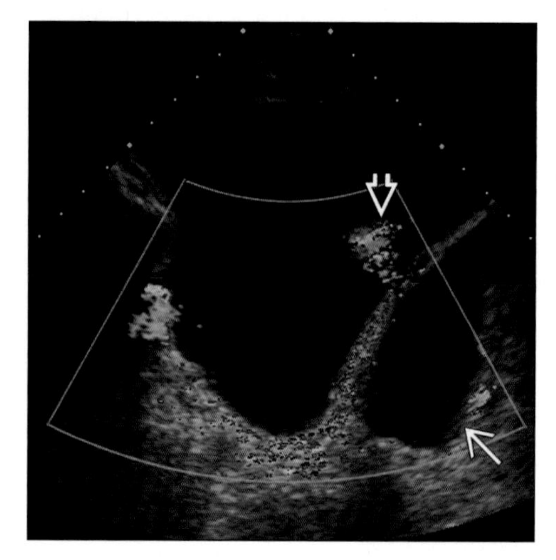

Axial color Doppler US shows a large bladder diverticulum posterior to the bladder (arrow). Note color jet traversing neck of diverticulum (open arrow) and connection to bladder lumen.

TERMINOLOGY

Abbreviations and Synonyms

- Bladder diverticulum (tic)

Definitions

- Sac formed by herniation of bladder mucosa and submucosa through muscular wall

IMAGING FINDINGS

General Features

- Best diagnostic clue: Perivesical cystic mass with connection to bladder lumen
- Location
  - Near ureterovesical junction (UVJ)
  - Bladder dome: Likely urachal if solitary
- Size: Small to very large; can exceed size of bladder
- Morphology: Single or multiple; smooth wall

Radiographic Findings

- IVP
  - Medial deviation of ipsilateral ureter
  - Usually fills with contrast unless obstructed
  - Diverticulum may contain stones, debris, or tumor
- Cystogram: Oblique films may show configuration of diverticulum neck

CT Findings

- CECT
  - Fluid attenuation outpouching from bladder
  - Usually fills with contrast on delayed images

MR Findings

- T1WI: Low signal mass contiguous with bladder
- T2WI
  - High signal mass contiguous with bladder
  - May see dephasing with motion of urine between tic and bladder lumen

Ultrasonographic Findings

- Real Time
  - Anechoic outpouching from bladder
  - Narrow or wide neck
  - May contain stones, debris, or tumor
- Color Doppler: Color jet connecting to bladder very useful to distinguish diverticulum from other paravesical masses

Imaging Recommendations

- Best imaging tool: Ultrasound, cystogram
- Protocol advice: Check emptying of diverticulum on post-void studies

DDx: Mimics of Bladder Diverticula

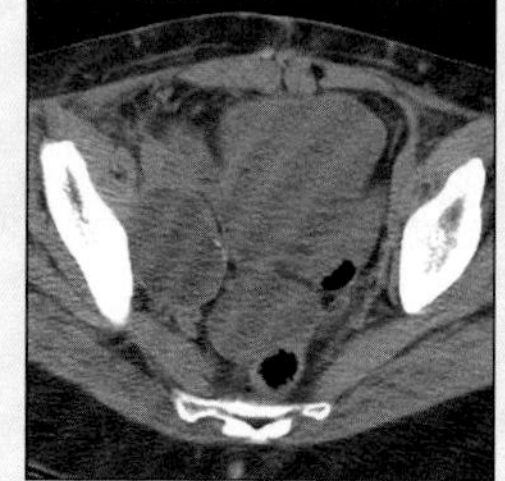

Lymphocele

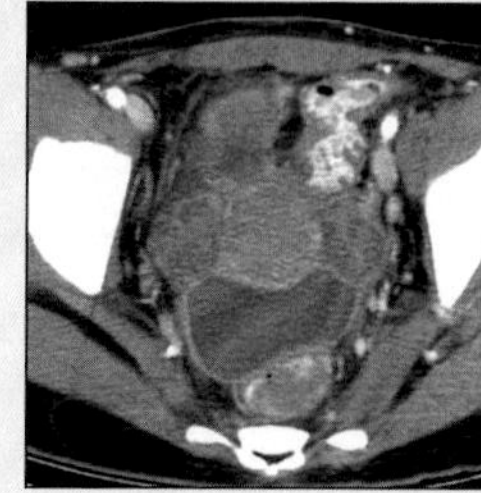

Pelvic Abscess

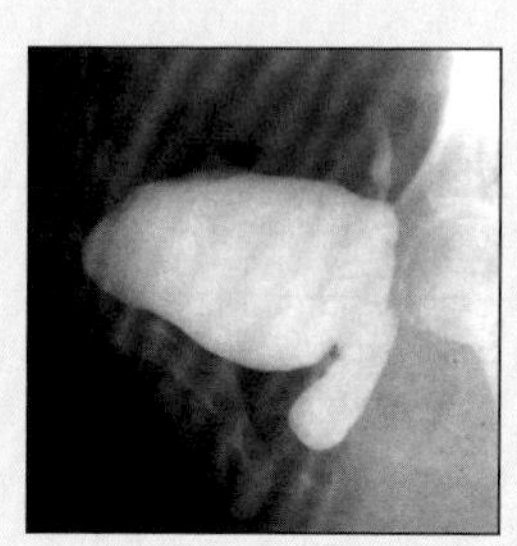

Ev. Ureterocele

BLADDER DIVERTICULUM

Key Facts

Imaging Findings
- Best diagnostic clue: Perivesical cystic mass with connection to bladder lumen
- Near ureterovesical junction (UVJ)
- Size: Small to very large; can exceed size of bladder
- Color Doppler: Color jet connecting to bladder very useful to distinguish diverticulum from other paravesical masses

Clinical Issues
- Narrow-neck diverticula often drain poorly → chronic infection and stone formation
- Secondary inflammation predisposes to development of carcinoma within diverticulum

Diagnostic Checklist
- Continuity with urethra distinguishes bladder from diverticulum

DIFFERENTIAL DIAGNOSIS

Lymphocele
- Adjacent to extraperitoneal vessels; doesn't fill with contrast

Everted (Ev) ureterocele
- Continuous with ureter

Cystic masses: Abscess, ovarian cystic lesions
- No communication with bladder

PATHOLOGY

General Features
- Etiology
  - Most common: Acquired
    - Secondary to bladder outlet obstruction: Prostatic enlargement, neurogenic bladder
    - Children: Posterior urethral valves, prune-belly syndrome, Ehlers-Danlos
  - Congenital
    - Weakness in detrusor muscle anterolateral to ureteral orifice (Hutch diverticulum)
- Epidemiology: Prevalence 1.7% in children

Microscopic Features
- Uroepithelial lining

CLINICAL ISSUES

Presentation
- Most common signs/symptoms: Usually asymptomatic
- Clinical profile: Older male with BPH; spinal cord injury patient

Demographics
- Age: 6th and 7th decade
- Gender: M:F = 9:1

Natural History & Prognosis
- Wide-neck diverticula: Empty readily with the bladder
- Narrow-neck diverticula often drain poorly → chronic infection and stone formation
- Secondary inflammation predisposes to development of carcinoma within diverticulum
- Tumors in diverticula have worse prognosis; poorly formed wall leads to more rapid local spread

Treatment
- Options, risks, complications
  - Complications: Carcinoma, vesico-ureteral reflux, ureteral obstruction
  - Surgery may be indicated for persistent infection, stone formation, or ureteral obstruction

DIAGNOSTIC CHECKLIST

Consider
- Large diverticulum may be confused with bladder especially if bladder is contracted

Image Interpretation Pearls
- Continuity with urethra distinguishes bladder from diverticulum

SELECTED REFERENCES

1. Maynor CH et al: Urinary bladder diverticula: sonographic diagnosis and interpretive pitfalls. J Ultrasound Med. 15(3):189-94, 1996
2. Bellah RD et al: Ureterocele eversion with vesicoureteral reflux in duplex kidneys: findings at voiding cystourethrography. AJR Am J Roentgenol. 165(2):409-13, 1995
3. Dondalski M et al: Carcinoma arising in urinary bladder diverticula: imaging findings in six patients. AJR Am J Roentgenol. 161(4):817-20, 1993

IMAGE GALLERY

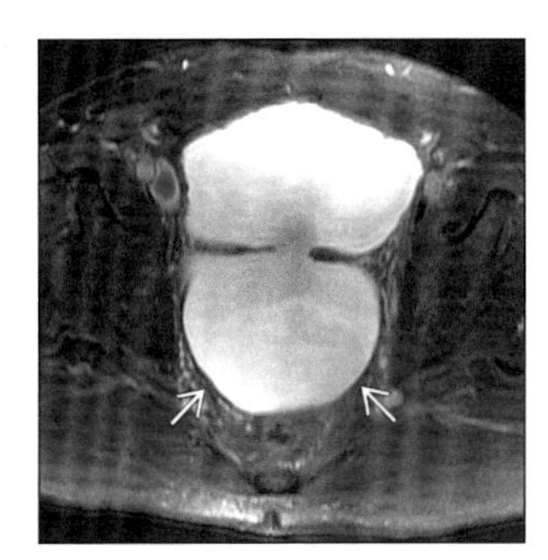

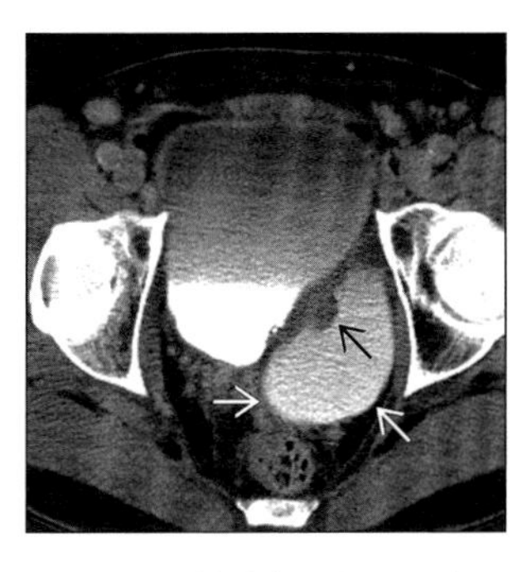

(Left) *Axial T2WI MR shows large posterior bladder diverticulum (arrows) connected to bladder via a broad neck. Signal loss in connecting channel is due to motion of urine and resultant dephasing.* ***(Right)*** *Axial CECT shows large bladder diverticulum (white arrows). Filling defect along medial wall of diverticulum (black arrow) is transitional cell carcinoma (Courtesy M. Nino-Murcia, MD).*

BLADDER FISTULAS

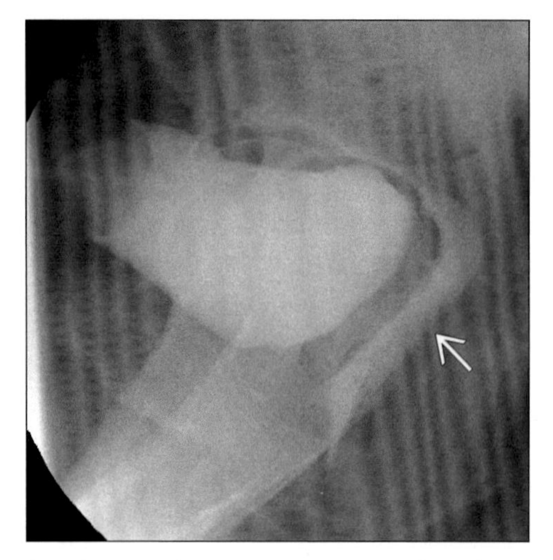

Lateral view from a cystogram performed via suprapubic catheter shows filling of vagina posteriorly (arrow) in patient with vesicovaginal fistula following hysterectomy.

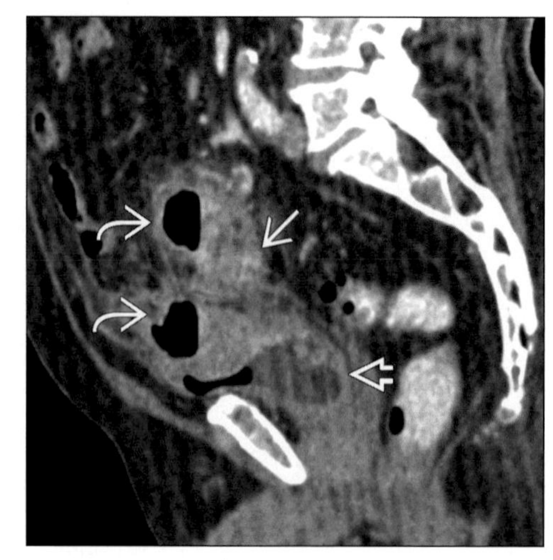

Sagittal CECT shows fistula from sigmoid colon (arrow) to bladder (open arrow) in patient with diverticulitis. Note associated abscesses (curved arrows) (Courtesy H. Harvin, MD).

TERMINOLOGY

Abbreviations and Synonyms

- Enterovesical, colovesical, vesicovaginal, vesicocutaneous, or vesicouterine fistula

Definitions

- Abnormal communication between bladder and adjacent viscus

IMAGING FINDINGS

General Features

- Best diagnostic clue: Gas within bladder lumen

Radiographic Findings

- Radiography: Gas in bladder
- IVP
  - Bladder wall irregularity +/- mass effect
  - Fistula rarely seen
- Cystography
  - Enterovesical: Filling of tract with contrast material seen in < 50% of cases; may see only bladder wall irregularity
  - Enterovesical: Tract may be unidirectional
  - Vesicovaginal: Contrast opacifies vagina posteriorly
    - Post-void and lateral films can be helpful to demonstrate vaginal contrast

CT Findings

- CECT
  - Enterovesical: Gas in bladder (90%), bladder and/or adjacent bowel wall thickening (90%)
  - Extravesical soft tissue mass (75% of cases)
  - Enterovesical: Direct visualization of fistulous tract in < 50% of cases
  - Vesicovaginal: Contrast in vagina on delayed imaging

MR Findings

- T1 C+
  - High signal enhancing fistula wall and low signal intensity tract on early post-contrast images
  - Delayed imaging may show high signal fluid in tract
  - Fat-saturation improves contrast between enhancing tract and background

Imaging Recommendations

- Best imaging tool: CT
- Protocol advice
  - CT-cystogram technique or rectal contrast may improve ability to visualize tract
  - Multiplanar reformations may be helpful to demonstrate course of fistula
  - Post-void or post-evacuation scans may improve fistula detection

DDx: Mimics of Bladder Fistulae

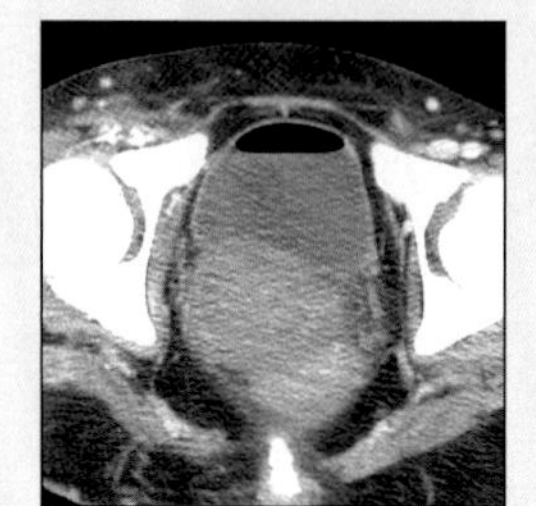

Catheterization

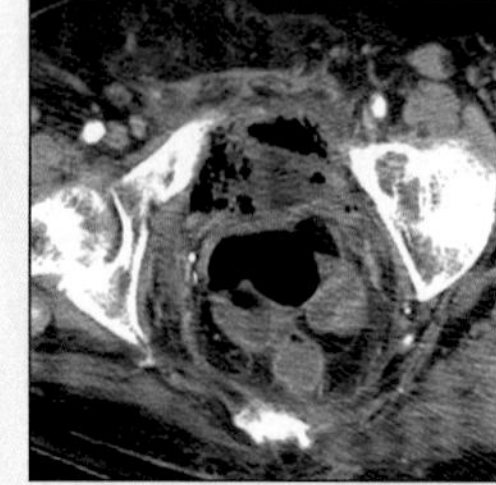

Emph. Cystitis

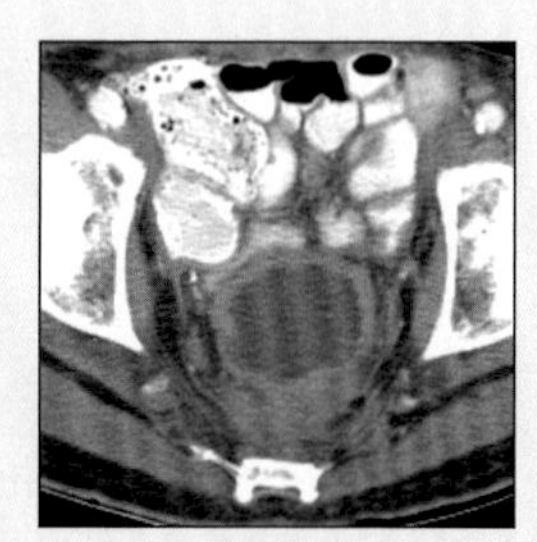

Cystitis

BLADDER FISTULAS

Key Facts

Imaging Findings

- Vesicovaginal: Contrast opacifies vagina posteriorly
- Enterovesical: Gas in bladder (90%), bladder and/or adjacent bowel wall thickening (90%)
- Extravesical soft tissue mass (75% of cases)
- Enterovesical: Direct visualization of fistulous tract in < 50% of cases
- Best imaging tool: CT

Pathology

- Inflammatory: Diverticulitis, Crohn disease, appendicitis, other pelvic abscesses
- Inflammatory: Granulomatous infection of bladder with TB or schistosomiasis
- Neoplastic: Colon, bladder, gynecologic cancers
- Penetrating trauma and iatrogenic injury

DIFFERENTIAL DIAGNOSIS

Gas in bladder lumen: Gas-forming infection; catheterization

- Emphysematous (Emph.) cystitis: May have gas in both wall and lumen

Thick walled bladder: Cystitis

- Clinical diagnosis

PATHOLOGY

General Features

- General path comments
  - Diverticulitis is most common cause of colovesical fistula, followed by colon CA
  - Crohn disease is most common cause of enterovesical fistula
- Etiology
  - Inflammatory: Diverticulitis, Crohn disease, appendicitis, other pelvic abscesses
  - Inflammatory: Granulomatous infection of bladder with TB or schistosomiasis
  - Neoplastic: Colon, bladder, gynecologic cancers
  - Gynecologic surgery: Pelvic hematoma → infection → erosion into bladder/vaginal wall
  - Penetrating trauma and iatrogenic injury
  - Radiation: Usually 12-18 months after therapy
  - Congenital: Rare; associated with other GU malformations
- Epidemiology
  - 0.5-1% frequency of bladder injury with total abdominal hysterectomy
    - 75% of vesicovaginal fistulae occur following hysterectomy for benign disease
  - Obstetrical trauma most common cause of vesicovaginal fistula in developing countries

Gross Pathologic & Surgical Features

- Rectosigmoid colon is bowel segment most frequently involved
- Cystoscopy shows inflammatory changes but fistula usually not seen

CLINICAL ISSUES

Presentation

- Most common signs/symptoms
  - Most common: Nonspecific cystitis symptoms
  - Fecaluria or pneumaturia present in 40-70% of cases
  - Persistent urinary tract infections (UTIs)
  - Vesicovaginal fistula: Painless dribbling incontinence of urine through vagina
- Clinical profile
  - Vesicovaginal fistula: Post gynecologic surgery
    - Pre-operative irradiation predisposes to fisula
  - Enterovesical fistula: Patient with Crohn disease
  - Colovesical fistula: Patient with diverticulitis or colon CA

DIAGNOSTIC CHECKLIST

Consider

- Bowel fistulae harder to detect than vaginal

Image Interpretation Pearls

- Bladder wall irregularity may be sole clue to presence of fistula

SELECTED REFERENCES

1. Pickhardt PJ et al: Acquired gastrointestinal fistulas: classification, etiologies, and imaging evaluation. Radiology. 224(1):9-23, 2002
2. Pollack HM et al: Clinical urography. 2nd ed. Philadelphia, WB Saunders, 3001-6, 2000
3. Semelka RC et al: Pelvic fistulas: appearances on MR images. Abdom Imaging. 22(1):91-5, 1997

IMAGE GALLERY

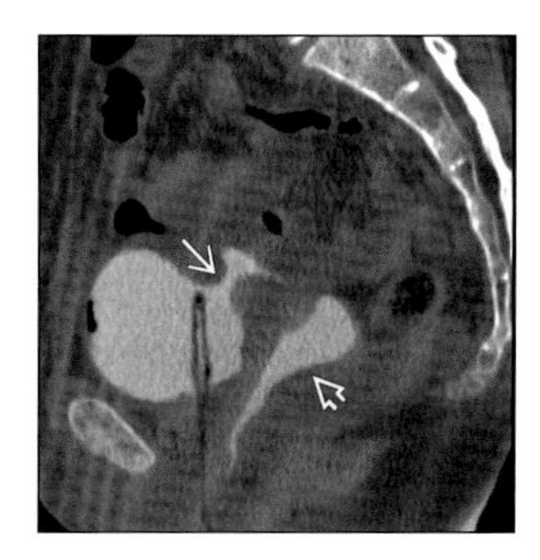

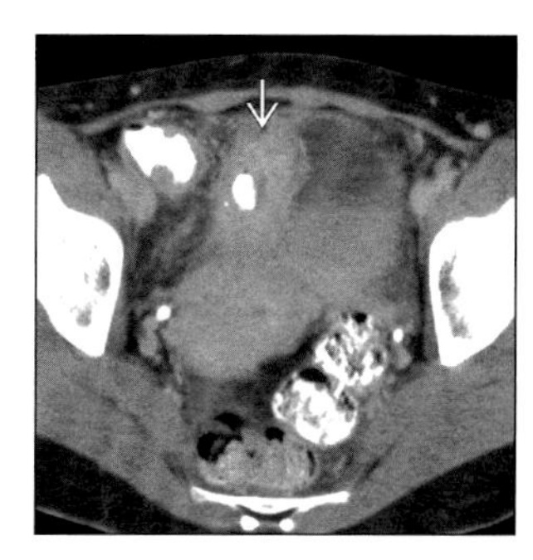

__(Left)__ Sagittal CT-cystogram shows fistulous connection (arrow) between bladder & uterus, with filling of vagina (open arrow). Patient was post Cesarian section for placenta accreta. __(Right)__ Axial CECT shows thickened loop of small bowel (arrow) adjacent to thick-walled bladder in patient with Crohn disease & bladder fistula.

NEUROGENIC BLADDER

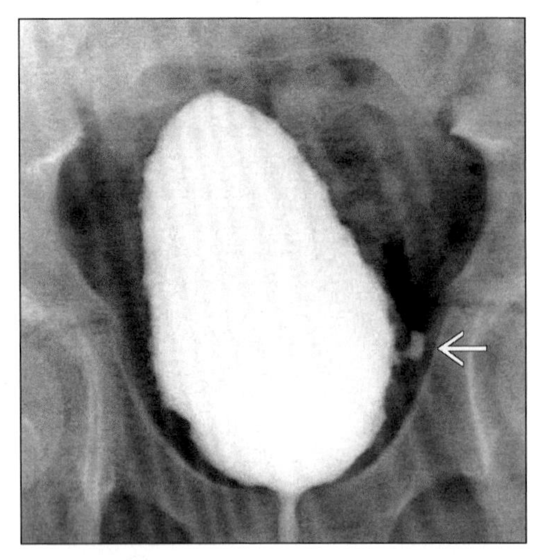

Cystogram shows "pine cone" bladder, elongated and pointed bladder with pseudodiverticula (arrow).

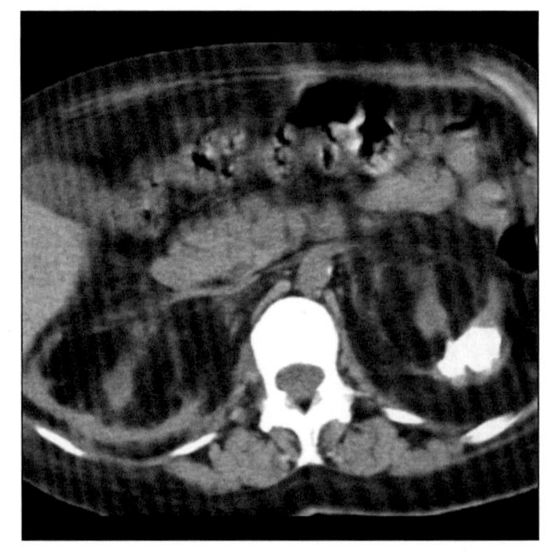

Axial NECT in a patient with neurogenic bladder shows bilateral renal cortical atrophy and parenchymal calcification due to chronic reflux.

TERMINOLOGY

Definitions

- Lesion of nervous system causing bladder dysfunction

IMAGING FINDINGS

General Features

- Best diagnostic clue: Contracted or distended bladder
- Other general features
  - Types of lesions: Suprapontine; suprasacral; epiconal; peripheral; sensory

Radiographic Findings

- IVP, cystourethrography or cystography findings
  - Suprapontine (above pontine center) lesions → detrusor hyperreflexia
    - Rounding of the bladder shape & serration of mucosa above the trigone (detrusor contractions)
    - Prominent interureteric ridge apparent only during voiding (detrusor contractions)
    - ± Trabeculations & ↑ wall thickness
  - Suprasacral (above S2-4) lesions → detrusor hyperreflexia with detrusor-sphincter dyssynergia
    - Beak of contrast (slightly open posterior urethra)
    - Contrast extends to external sphincter/urogenital diaphragm (distended posterior urethra)
    - "Christmas tree" or "pine cone" shape (severe): Elongated & pointed with pseudodiverticula
  - Epiconal (in and around S2-4) lesions → weak detrusor hyperreflexia (storing), detrusor areflexia (voiding), detrusor-sphincter dyssynergia
    - Upper & lower motor neuron features; "pine cone"
  - Peripheral (below S2-4) lesions → detrusor areflexia
    - Large atonic bladder: No signs of detrusor contraction during voiding and poor emptying
    - Downward displacement of bladder base & vagina → functional infravesical obstruction (females)
  - Sensory lesions → unable to sense bladder fullness
    - Voiding is preserved (early)
    - Rounding of bladder shape & serration of mucosa
    - Large atonic & smooth bladder: Indistinguishable from peripheral nerve lesions (chronic)
  - Secondary bladder & upper tract abnormalities
    - Trabeculation (i.e., hypertrophy), dilated upper tracts, pseudodiverticula, vesicoureteral reflux
    - Indistinguishable between detrusor-sphincter dyssynergia & functional infravesical obstruction

Ultrasonographic Findings

- Real Time: Small contracted/large atonic bladder; wall thickening; vesicoureteric reflux during voiding; increased post-void residual urine

Imaging Recommendations

- IVP or cystourethrography: At rest and voiding

DDx: Dilated Bladder or Mimics

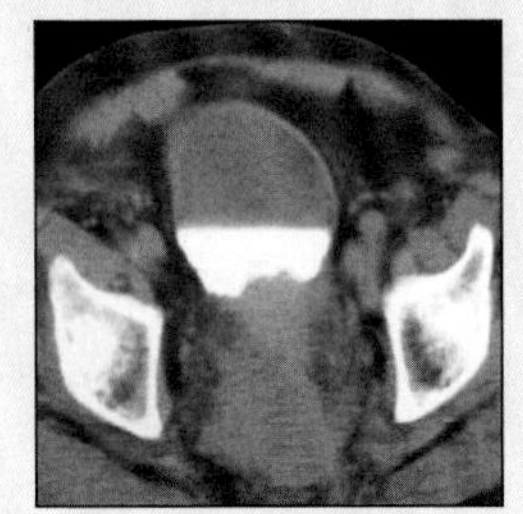

Prostate Carcinoma

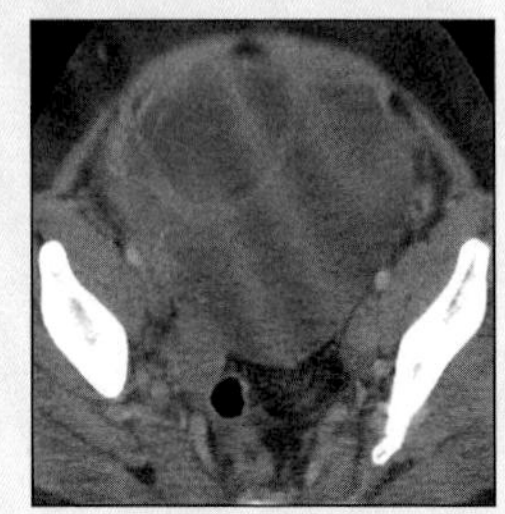

Ovarian Tumor

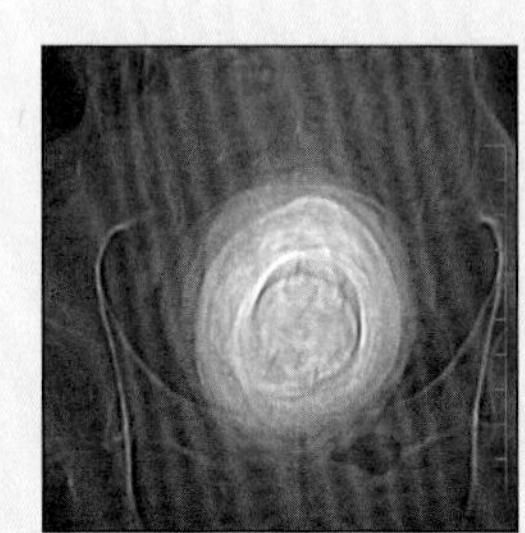

Foreign Body

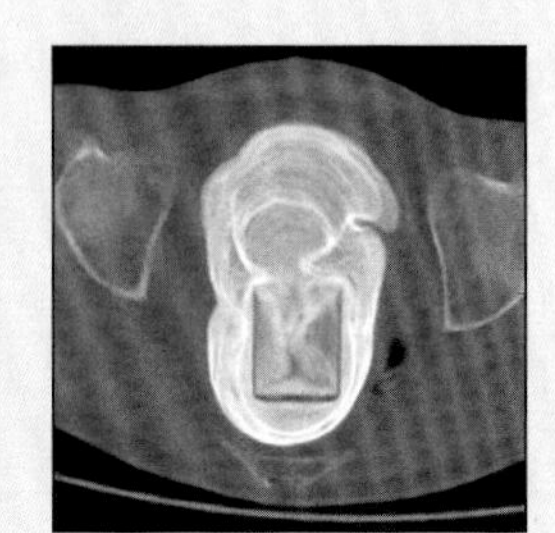

Foreign Body

NEUROGENIC BLADDER

Key Facts

Terminology
- Lesion of nervous system causing bladder dysfunction

Imaging Findings
- Best diagnostic clue: Contracted or distended bladder

Top Differential Diagnoses
- Bladder outlet obstruction
- Pelvic cystic mass
- Bladder foreign body

Clinical Issues
- Frequency, nocturia, urgency, retention, incontinence

Diagnostic Checklist
- Multiple sites of injury → mixed features
- ± Trabeculations; "pine tree" appearance

DIFFERENTIAL DIAGNOSIS

Bladder outlet obstruction
- Examples: Urethral strictures, benign or malignant enlargement of prostate, congenital urethral valves
- Trabeculation of detrusor and urine retention
- Detrusor hyperreflexia (early) → areflexia (chronic)
- Relieve obstruction → ↑ function, but never normal

Pelvic cystic mass
- Example: Ovarian cyst/carcinoma, tuboovarian abscess
- No trabeculation; features do not change with voiding

Bladder foreign body
- Can distort lumen, calcify, cause dysuria

PATHOLOGY

General Features
- Etiology
  - Suprapontine: Stroke, arteriosclerosis or multiple sclerosis
  - Suprasacral: Trauma, tumor or multiple sclerosis
  - Epiconal: Myelomeningocele, tumor or trauma
  - Peripheral (pelvic ± pudendal nerve): Pelvic surgery, cauda equina
  - Sensory: Diabetes mellitus, pernicious anemia or tabes dorsalis

Gross Pathologic & Surgical Features
- Small (hyperreflexic)/large capacity (areflexic) bladder

CLINICAL ISSUES

Presentation
- Most common signs/symptoms
  - Frequency, nocturia, urgency, retention, incontinence
  - Hyper- or hyporeflexia; impaired or no sensation

Natural History & Prognosis
- Complications: Infection, hydronephrosis, urolithiasis, sexual dysfunction, autonomic dysreflexia
- Prognosis: Good in early stage (treatment); poor in late stage (renal damage → renal failure)

Treatment
- Hyperreflexia
  - ↑ Volume: Cystoplasty, muscular or fascial slings, parasympatholytic drugs, botulinum-a toxin
  - ↑ Voiding: Catheter, transurethral sphincterotomy
- Hyporeflexia
  - Bladder training, catheter, bladder neck resection/denervation, parasympathomimetic drugs

DIAGNOSTIC CHECKLIST

Consider
- Multiple sites of injury → mixed features

Image Interpretation Pearls
- ± Trabeculations; "pine tree" appearance

SELECTED REFERENCES

1. Zawin JK et al: Neurogenic dysfunction of the bladder in infants and children: Recent advances and the role of radiology. Radiology. 182: 297, 1992
2. Ruutu M et al: The value of urethrocystography in the investigation of patients with spinal cord injury. Clinical Radiology. 35: 485-9, 1984
3. Shapeero G et al: Transrectal sonographic voiding cystourethrography: Studies in neuromuscular dysfunction. AJR. 141: 83-90, 1983

IMAGE GALLERY

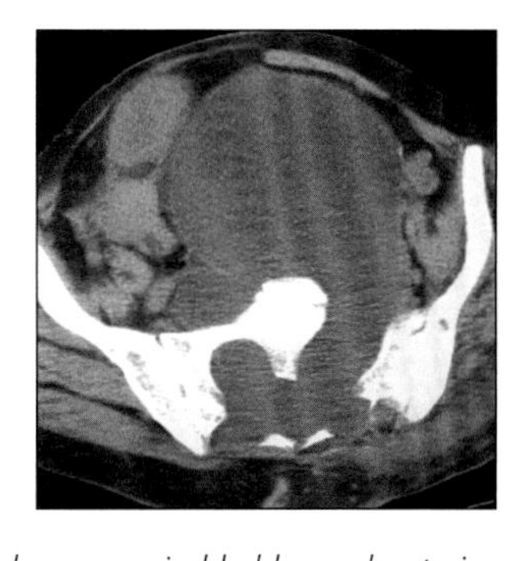

(Left) *Axial CECT shows distended neurogenic bladder and anterior meningocele in a patient with spina bifida.* ***(Right)*** *Axial CECT in a patient with spina bifida shows a distended neurogenic bladder and anterior meningocele.*

BLADDER TRAUMA

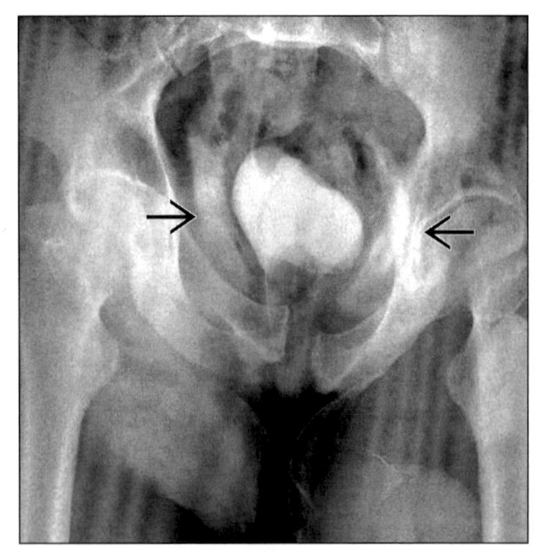

Cystogram shows displaced pelvic fractures and extraperitoneal bladder rupture (arrows).

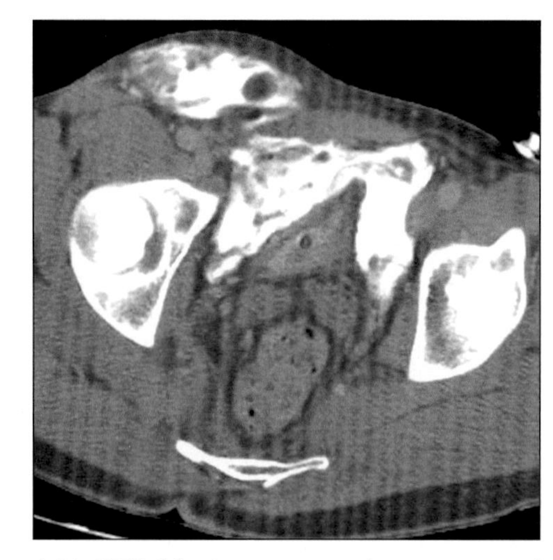

Axial CECT following cystogram shows extravasated urine in extraperitoneal pelvic spaces and proximal right thigh; type 4B.

TERMINOLOGY

Definitions

- Bladder injury due to blunt, penetrating or iatrogenic trauma

IMAGING FINDINGS

General Features

- Best diagnostic clue: Extravasation of contrast in cystography
- Other general features
  - Classification of bladder injury after blunt trauma
    - Type 1: Bladder contusion
    - Type 2: Intraperitoneal rupture
    - Type 3: Interstitial injury
    - Type 4A: Simple extraperitoneal rupture
    - Type 4B: Complex extraperitoneal rupture
    - Type 5: Combined injury

Radiographic Findings

- Cystography
  - Bladder contusion (type 1)
    - Teardrop bladder: Extrinsic compression of bladder (symmetrical extraperitoneal hematoma)
  - Intraperitoneal rupture (type 2)
    - Contrast in paracolic gutters, around bowel loops, pouch of Douglas and intraperitoneal viscera
    - ± Pelvic fractures
  - Interstitial injury (type 3)
    - Focal mural defect along wall of bladder
  - Extraperitoneal rupture (type 4)
    - Simple (type 4A): Flame-shaped extravasation around bladder
    - Complex (type 4B): Extravasation extends beyond the pelvis
    - Extravasation best seen on post-drainage films
  - Combined rupture (type 5)
    - Features of intra- and extraperitoneal ruptures
  - Penetrating injury
    - Foreign bodies (i.e., metallic fragments from a bullet) along course of penetration
    - Extravasation ± foreign body (i.e., bullet) best seen on post-drainage films

CT Findings

- CT cystography
  - Bladder contusion (type 1)
    - Normal findings
  - Intraperitoneal rupture (type 2)
    - Contrast in paracolic gutters, around bowel loops, pouch of Douglas and intraperitoneal viscera
  - Interstitial injury (type 3)
    - Intramural and submucosal extravasation of contrast without transmural extension

DDx: Pelvic Trauma with Hematuria

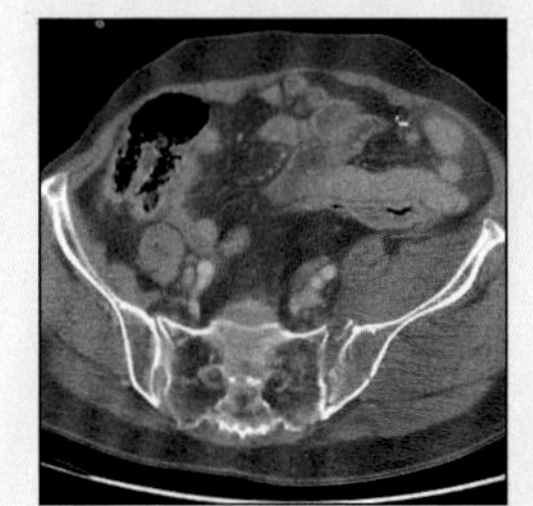

Hemoperitoneum

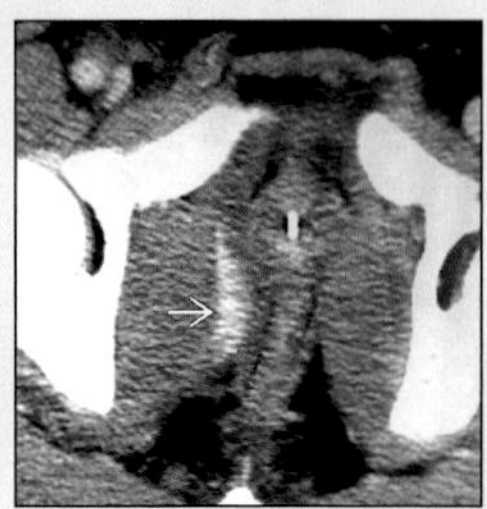

Active Bleeding

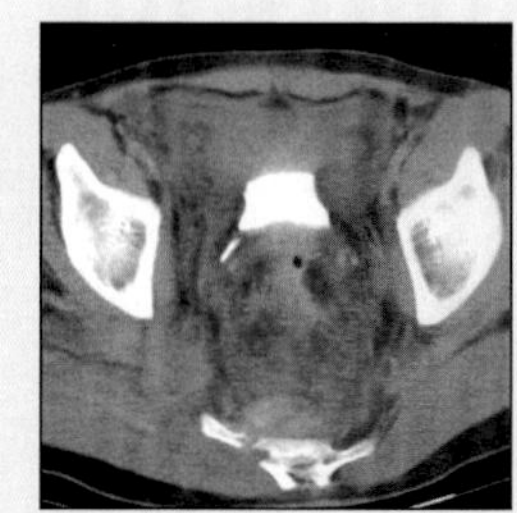

Pelvic Bleeding

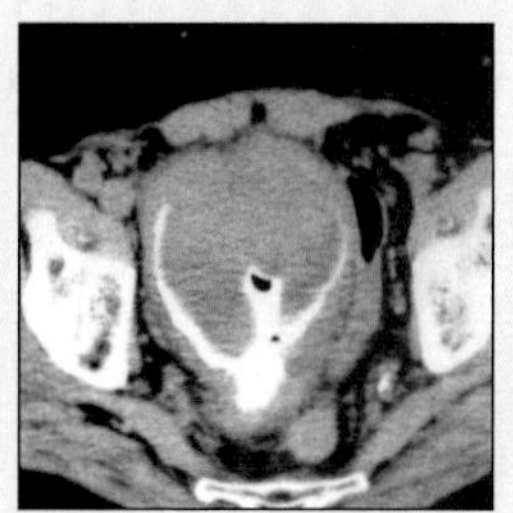

Blood Clot in Bladder

BLADDER TRAUMA

Key Facts

Terminology
- Bladder injury due to blunt, penetrating or iatrogenic trauma

Imaging Findings
- Best diagnostic clue: Extravasation of contrast in cystography
- Focal mural defect along wall of bladder
- Extravasation best seen on post-drainage films
- Best imaging tool: Cystography or CT cystography: 85-100% sensitivity

Top Differential Diagnoses
- Simple male urethral injury
- Hemoperitoneum
- Pelvic bleeding

Pathology
- Propensity for bladder injury is related to degree of bladder distention; more distended, more injury
- Blunt trauma: Seat belt or steering wheel injury
- 84-100% of bladder trauma have pelvic fractures
- 10-29% of bladder trauma have urethral injuries

Clinical Issues
- Gross hematuria (84-95%)
- Bladder trauma including penetrating injury: 12-22% mortality

Diagnostic Checklist
- Cystography ± CT still the procedure of choice
- Bloody urethral discharge requires urethrogram prior to catheterization
- CT after only IV contrast may miss bladder rupture

    - Extraperitoneal rupture (type 4)
        - Perforation by bony spicules
        - "Knuckle" of bladder: Trapped bladder by displaced fracture of anterior pelvic arch
        - Simple (type 4A): Extravasation is confined to perivesical space
        - Complex (type 4B): Extravasation extends beyond perivesical space; thigh, scrotum, penis, perineum, anterior abdominal wall, retroperitoneum (paranephric or perinephric spaces) or hip joint
        - "Molar tooth sign": Rounded cephalic contour (due to vertical perivesicle components of extraperitoneal fluid)
    - Combined rupture (type 5)
        - Features of intra- and extraperitoneal ruptures

Ultrasonographic Findings
- Real Time
    - "Bladder within a bladder": Fluid around bladder
    - Pseudomass appearance: Intravesical hematoma
    - Fluid in abdomen and pelvis (i.e., pouch of Douglas)

Imaging Recommendations
- Best imaging tool: Cystography or CT cystography: 85-100% sensitivity
- Protocol advice
    - Cystography
        - Exclude urethral injury in males prior to cystography; if positive, use suprapubic cystostomy tube
        - Post-drainage film to check for extravasation hidden by contrast
        - CT cystograph: Instill diluted contrast (10 ml contrast, 300 ml saline

DIFFERENTIAL DIAGNOSIS

Simple male urethral injury
- Contrast extravasation into base of penis or thigh may result form urethral or extraperitoneal bladder
- Blood in urethral meatus is indicative of urethral injury; avoid "blind" catheterization to bladder

Hemoperitoneum
- Clotted blood/active extravasation of enhanced blood may simulate extravesicle urine from bladder rupture
- Compare with density of blood and urine
- CT cystography usually diagnostic

Pelvic bleeding
- Bleeding from other pelvic trauma can appear similar to contrast extravasation from bladder
- Synchronous occurrence with bladder trauma is likely; complete pelvic assessment is needed

Blood clot in bladder
- Blood from renal or bladder injury may cause irregular filling defect within bladder

PATHOLOGY

General Features
- General path comments
    - Propensity for bladder injury is related to degree of bladder distention; more distended, more injury
    - Location of extravasation depends on anatomy
        - Anterosuperior rupture: Extravasation to intraperitoneal, Retzius space or both
        - Posterosuperior rupture: Extravasation to intraperitoneal, extraperitoneal or both
        - Intact superior fascia of urogenital diaphragm prohibits extravasation from escaping the pelvis
        - Intact inferior fascia of urogenital diaphragm prohibits extravasation into perineum
    - Bladder contusion (type 1)
        - Incomplete or partial tear of bladder mucosa; ecchymosis of a localized segment of bladder wall
        - Most common minor bladder injuries
        - Diagnosed by exclusion
    - Intraperitoneal rupture (type 2)
        - Direct blow to lower abdomen with a distended bladder
        - ↑ Bladder pressure → horizontal tear along bladder wall; at dome of bladder covered by peritoneum
        - 25% of major bladder injuries

- more common in < 20 years of age due to intra-abdominal bladder
- Interstitial injury (type 3)
  - Intramural or partial-thickness laceration with intact serosa
  - Incomplete perforation; seen on either intra- or extraperitoneal portion of bladder
  - Rare bladder injuries
- Extraperitoneal rupture (type 4)
  - Classic mechanism: Anterolateral laceration at base of bladder by bony spicules (anterior pelvic arch fractures)
  - Other mechanism: Stress applied to hypogastric wing or puboprostatic ligaments → laceration; direct blow to distended bladder
  - 62% (most common) of major bladder injuries
- Combined rupture (type 5)
  - Both intra- and extraperitoneal bladder ruptures
  - 12% bladder ruptures, include both blunt and penetrating trauma
- Penetrating injury
  - Results in intraperitoneal, extraperitoneal or combined bladder injury
  - ↑ Incidence of vascular injuries → high mortality
- Etiology
  - Blunt trauma: Seat belt or steering wheel injury
  - Penetrating Injury: Gunshot or knife wounds
  - Iatrogenic injury: Pelvic, urological, obstetrical and gynecological procedures
- Associated abnormalities
  - 6-10% of pelvic fractures have bladder trauma
  - 84-100% of bladder trauma have pelvic fractures
  - 10-17% of urethral injuries have bladder trauma
  - 10-29% of bladder trauma have urethral injuries
  - Penetrating injury
    - Gunshot wounds: 83% have bowel injuries
    - Knife wounds: 33% have colon injuries

Gross Pathologic & Surgical Features

- Asymmetrical compression, congestion, laceration or fill-thickness tear

Microscopic Features

- Mural defect; contusion or clot within bladder mucosa

CLINICAL ISSUES

Presentation

- Most common signs/symptoms
  - Gross hematuria (84-95%)
  - Suprapubic pain and tenderness, anuria, fever
  - Bladder contusion (type 1): Hematuria
  - Intraperitoneal rupture (type 2)
    - No bowel sounds; acute abdomen
- Lab data
  - Urinalysis: Hematuria
  - Intraperitoneal rupture (type 2)
    - Chemistry: Hypernatremia, hyperkalemia, uremia and acidosis (reabsorption of urine)
    - Peritoneal lavage: Urinary ascitic fluid
    - Catheterization: No urine

Natural History & Prognosis

- Complications: Fistulas, sinuses, abscess, sepsis, bladder calculi, hematoma, hemorrhage or shock
- Minor injuries: Good, spontaneous healing within 10-14 days
- Major injuries: Poor, if not corrected surgically and with complications
  - Bladder trauma including penetrating injury: 12-22% mortality
  - Penetrating injury: 12% mortality

Treatment

- Bladder contusion (type 1) and interstitial injury (type 3): Self limiting, no treatment
- Intraperitoneal rupture (type 2): Open surgical correction via transperitoneal approach; postoperatively, use suprapubic and urethral catheters
- Extraperitoneal rupture (type 4): Antibiotics and urethral catheter drainage (10 days); large blood clots or bladder neck injuries require surgery
- Penetrating injury: Exploratory laparotomy with surgical correction
- Follow-up: Cystography 10 days after conservative or surgical treatment

DIAGNOSTIC CHECKLIST

Consider

- Cystography ± CT still the procedure of choice
- Bloody urethral discharge requires urethrogram prior to catheterization

Image Interpretation Pearls

- CT after only IV contrast may miss bladder rupture
- CT or conventional cystogram

SELECTED REFERENCES

1. Morgan DE et al: CT cystography: Radiographic and clinical predictors of bladder rupture. AJR. 174: 89-95, 2000
2. Vaccaro JP et al: CT cystography in the evaluation of major bladder trauma. RadioGraphics. 20:1373-81, 2000
3. Bigongiari LR et al: Trauma to the bladder and urethra. American College of Radiology. ACR Appropriateness Criteria. Radiology. 215 Suppl:733-40, 2000
4. Deck AJ et al: Computerized tomography cystography for the diagnosis of traumatic bladder rupture. J Urol. 164(1):43-6, 2000
5. Peng MY et al: CT cystography versus conventional cystography in evaluation of bladder injury. AJR Am J Roentgenol. 173(5):1269-72, 1999
6. Kane NM et al: The value of CT in the detection of bladder and posterior urethral injuries. AJR Am J Roentgenol. 153(6):1243-6, 1989
7. Corriere JN Jr et al: Mechanisms of injury, patterns of extravasation and management of extraperitoneal bladder rupture due to blunt trauma. J Urol. 139(1):43-4, 1988
8. Sandler CM et al: Bladder injury in blunt pelvic trauma. Radiology. 158: 633-8, 1986

IMAGE GALLERY

Typical

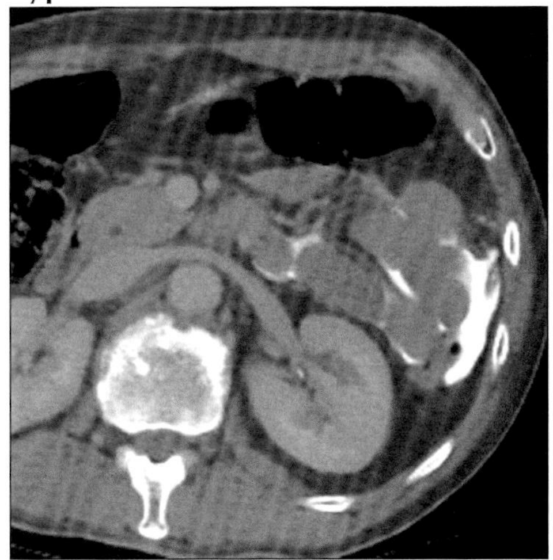

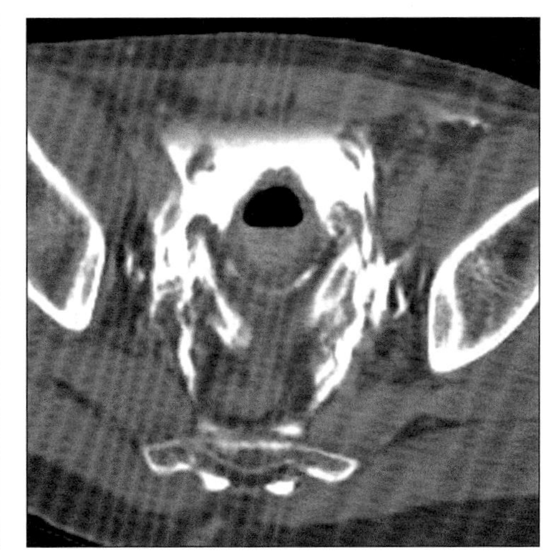

*(**Left**) Axial CECT following cystogram shows opacified urine in left paracolic gutter and between bowel loops (intraperitoneal). (**Right**) Axial CECT following cystogram shows extraperitoneal extravasated urine with "molar tooth" appearance. Type 5 (combined) rupture.*

Typical

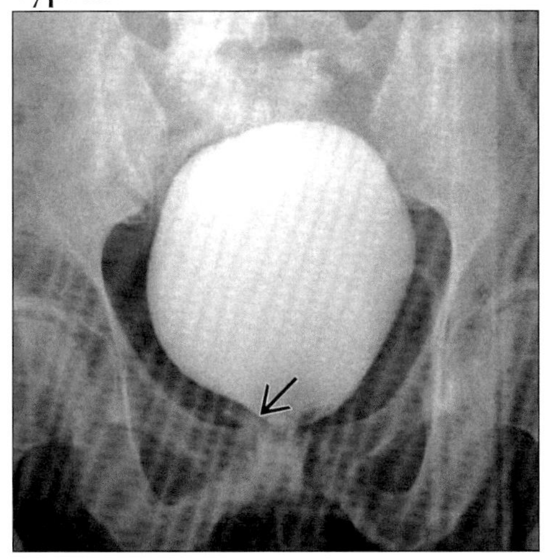

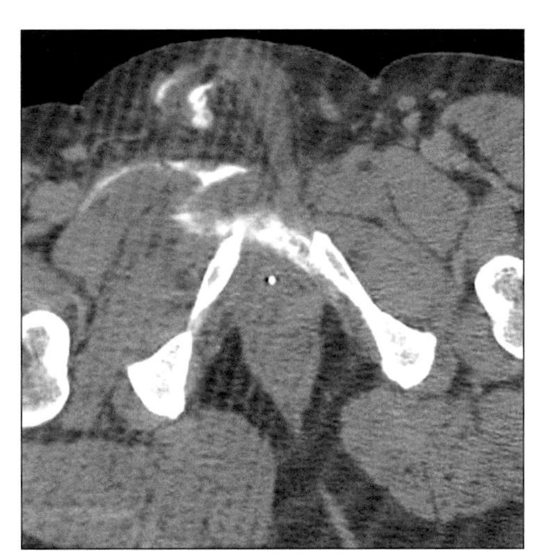

*(**Left**) Cystogram shows extravasation from base of bladder (arrow). (**Right**) Axial CECT following cystogram shows extravasation of urine into thigh and base of penis.*

Typical

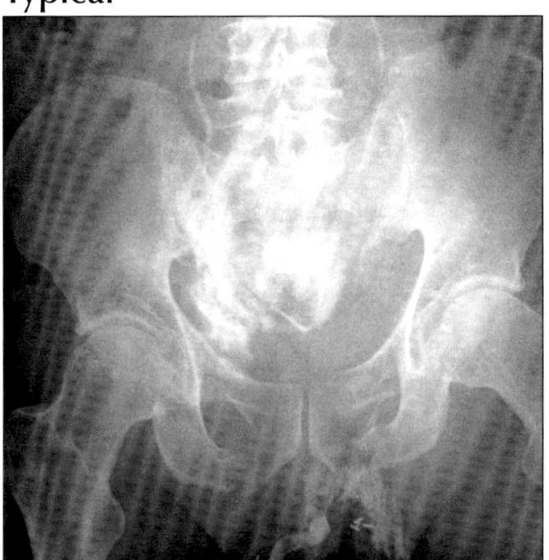

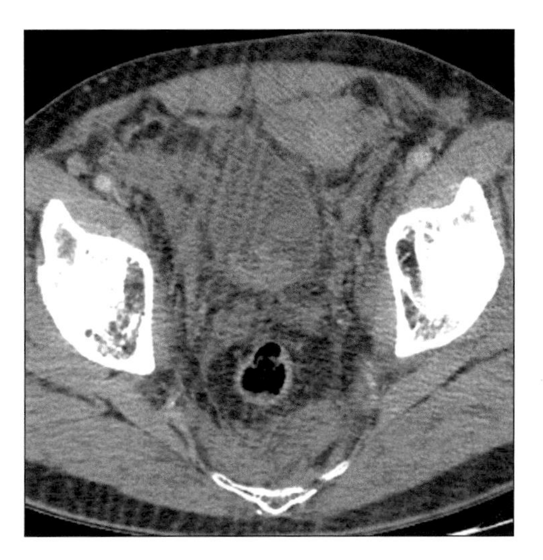

*(**Left**) Retrograde urethrogram shows extravasation into scrotum and pelvis from tear of base of bladder; pelvic fractures. (**Right**) Axial CECT prior to cystogram shows extraperitoneal blood. Difficult to recognize extravasated urine.*

BLADDER CARCINOMA

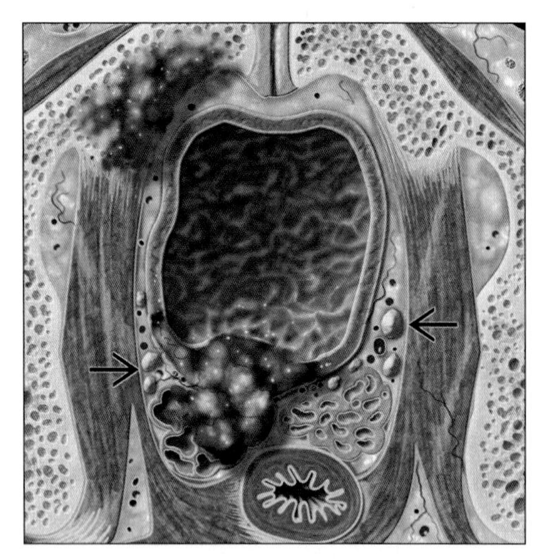

Graphic shows tumor arising from posterior wall of bladder, invading through the right seminal vesicle with hematogenous metastases to the right pubic ramus; lymph nodes metastases (arrows).

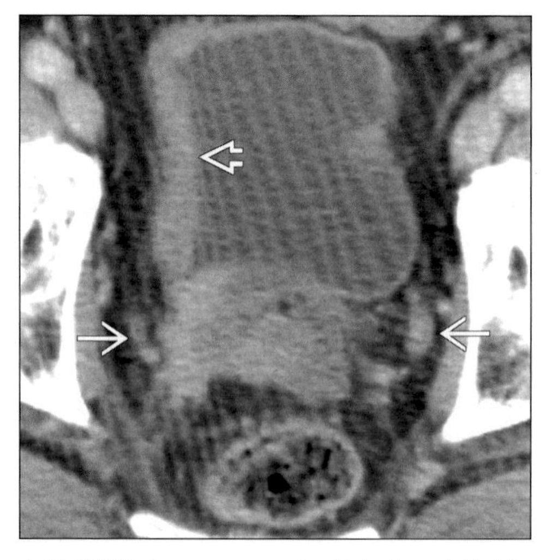

Axial CECT shows asymmetric thickening of bladder wall (open arrow) due to carcinoma. Lymph node metastases (arrows).

TERMINOLOGY

Definitions

- Malignant tumor growth within the bladder

IMAGING FINDINGS

General Features

- Best diagnostic clue: Bladder wall invasion by intraluminal soft tissue mass on CT or MRI

Radiographic Findings

- IVP
  - Multifocal (2-3% of urothelial cancer)
  - Punctate or speckled calcification on fronds of villous, papillary tumors (en face view)
  - Linear or curvilinear calcification on the surface of sessile tumors
  - Central calcification (necrosis)
  - ± Lytic bony metastasis or sclerotic
  - ± Urinary tract obstruction
  - ± Deviation of bladder from pelvic or retroperitoneal adenopathy
  - ± Luminal protrusions or invaginations
  - Nonspecific filling defects within bladder
- Cystography
  - ± Bladder diverticulum (2-10% contain neoplasm)

CT Findings

- Sessile or pedunculated soft tissue mass projecting into the lumen; similar density to bladder wall
- ± Enlarged (> 10 mm) metastatic lymph nodes; extravesical tumor extension
- Fine punctate calcifications with tumor; may suggest mucinous adenocarcinoma
- Ring pattern of calcification; may suggest pheochromocytoma
- Inability to distinguish tumors from bladder wall hypertrophy, local inflammation and fibrosis
- Unable to differentiate Ta-T3a, invasion of dome/base of bladder or local organ (due to partial volume effect), nonenlarged lymph nodes
- Urachal adenocarcinoma
  - Midline abdominal mass ± calcification
  - Solitary lobulated tumor arising from dome of bladder on ventral surface

MR Findings

- T1WI
  - Tumor has intermediate signal intensity, equal to muscle layer of bladder wall
  - Infiltration of perivesical fat (high signal intensity)
  - Endoluminal tumor in urine filled bladder (low signal intensity)
  - Bone marrow metastases; similar signal intensity as primary tumor

DDx: Bladder Wall Thickening with Hematuria

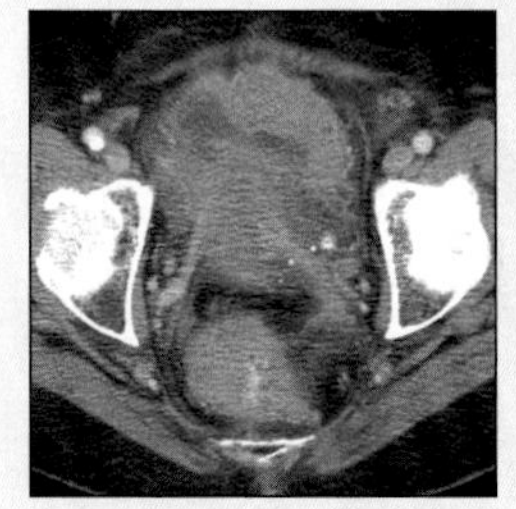

Lymphoma of Bladder

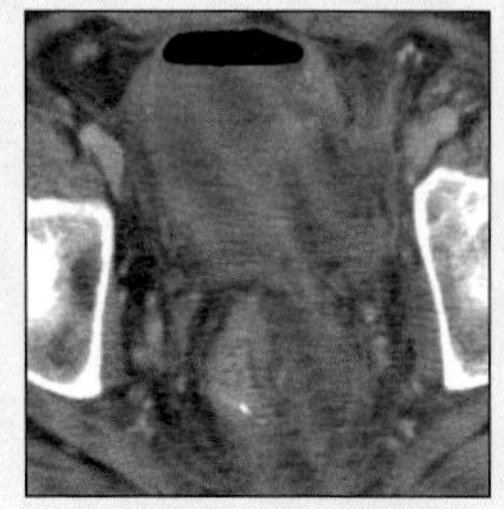

Colovesicle Fistula

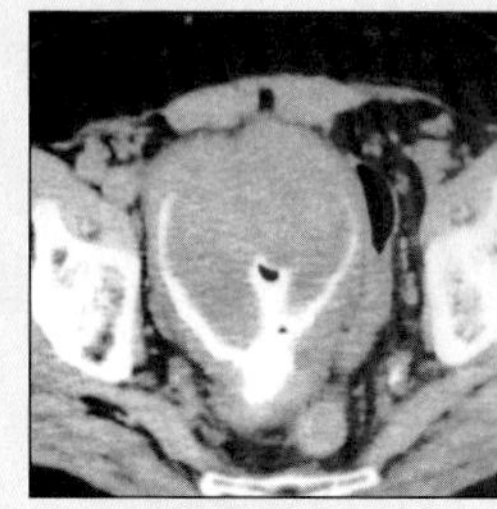

Blood Clot

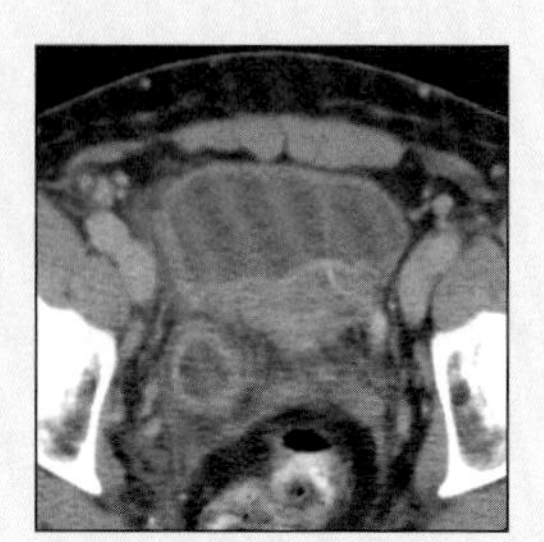

Hemorrhagic Cystitis

BLADDER CARCINOMA

Key Facts

Terminology
- Malignant tumor growth within the bladder

Imaging Findings
- Best diagnostic clue: Bladder wall invasion by intraluminal soft tissue mass on CT or MRI
- Punctate or speckled calcification on fronds of villous, papillary tumors (en face view)
- Sessile or pedunculated soft tissue mass projecting into the lumen; similar density to bladder wall
- ± Enlarged (> 10 mm) metastatic lymph nodes; extravesical tumor extension

Top Differential Diagnoses
- Extrinsic tumor
- Extrinsic inflammation
- Blood clot
- Bladder inflammation
- Trabeculation

Clinical Issues
- Most common signs/symptoms: Painless hematuria
- 50-60 years of age
- Gender: M:F = 4:1
- 5 year survival rate: 82% in all stages combined

Diagnostic Checklist
- Detection by cystoscopy; diagnosis by transurethral resection ± biopsy
- Staging is needed for treatment and prognosis
- Check kidneys, ureters for synchronous and metachronous tumors
- MRI is superior in staging and used in patients with high grade stage T1 or > stage T2

- T2WI
  - Tumor has intermediate signal intensity, higher than bladder wall or fibrosis, lower than urine
  - Determine tumor infiltration of perivesical fat (either low or high signal intensity)
  - Invasion of prostate, rectum, uterus, vagina → ↑ signal intensity
  - Direct invasion of seminal vesicles (sagittal plane) → ↑ size, ↓ signal intensity & obliteration of angle between seminal vesicle & posterior bladder wall
  - Confirm bone marrow metastases
- T1 C+
  - Mild enhancement in primary, perivesical, nodal or bone invasion
  - Tumor shows early & ↑ enhancement than bladder wall or other benign tissues; assess infiltration
  - Earlier enhancement than edema and granulation tissue
- ± Enlarged (> 10 mm) metastatic lymph nodes
- Unable to differentiate stage T1 from stage T2, acute edema or hyperemia from first week post-biopsy or nonenlarged lymph nodes
- Urachal adenocarcinoma
  - Varied appearance
  - T2WI: Increased signal intensity

Ultrasonographic Findings
- Fixed echogenicities not casting an acoustical shadow
- Inability to distinguish tumors from chronic cystitis, local bladder wall hypertrophy, blood clots, bladder wall invasion, abnormal lymph nodes
- Transabdominal: Unsuitable images caused by obesity, scars on wall and poor bladder distention
- Intravesical: Invasive and requires anesthesia

Imaging Recommendations
- Best imaging tool
  - IVP: Screening upper urinary tract
  - MRI: Staging bladder carcinoma
    - Accuracy: 73-96%; 10-33% greater than CT
- Protocol advice: MRI: Multiplanar imaging including sagittal plane, gadolinium contrast, fast dynamic first-pass MRI with acquisition every 2 seconds

DIFFERENTIAL DIAGNOSIS

Extrinsic tumor
- Rectal, ovarian, vaginal tumor or fibroids overlying bladder; may simulate bladder carcinoma

Extrinsic inflammation
- Diverticulitis often inflames and thickens bladder wall; may cause fistula

Blood clot
- IVP: "Stipple sign": Contrast trapped within interstices of tumor
- CT: Disappear with time, no enhancement
- MR: ↓ Signal intensity, no infiltration
- US: Mobile mass, does not cast an acoustical shadow

Bladder inflammation
- Cystitis may cause mural thickening and hemorrhage

Trabeculation
- Secondary sign of outlet obstruction
- Associated with increased bladder wall thickness
- May vary inversely with amount of urine in bladder; 2 cm thick bladder wall and crinkled mucosa are normal for an empty bladder
- Caused by tumor or benign prostate hypertrophy

PATHOLOGY

General Features
- General path comments
  - 95% of bladder neoplasms are malignant
  - Types of epithelial bladder carcinoma
    - Transitional cell carcinoma (90-95%)
    - Squamous cell carcinoma (5%)
    - Adenocarcinoma (2%): Urachal origin, secondary to cystitis glandularis, secondary to extrophy
    - Carcinosarcoma
    - Other rare tumors: Carcinoid, rhabdoid, villous, small cell
    - Metastasis: Gastrointestinal tract, melanoma
  - Types of nonepithelial bladder carcinoma

    - Pheochromocytoma
    - Leiomyosarcoma
    - Embryonal rhabdomyosarcoma (most common bladder neoplasm in children)
    - Lymphoma
    - Plasmacytoma
- Genetics: Lack of GSTM1 (1.8-fold ↑ risk in smokers)
- Etiology
  - Aromatic amines, nitrosamines, aldehydes (e.g., acrolein)
  - Risk factors
    - Environment: Smoking
    - Infection: Schistosomiasis, chronic cystitis
    - Iatrogenic: Cyclophosphamide, radiation therapy
    - Occupation: Chemical, dye (e.g., aniline dye), rubber and textile industries
- Epidemiology
  - Mortality: 12,710 in U.S.
  - Incidence: 60,240 in U.S.
  - Fourth leading cause of cancer incidence in males
  - Tenth leading cause of cancer incidence in females

Gross Pathologic & Surgical Features

- Superficial and are usually papillary (66%)
- Infiltrating in/beyond muscular layer of wall (33%)

Staging, Grading or Classification Criteria

- TNM classification of bladder carcinoma
  - T0: No tumor
  - Tis: Carcinoma in situ
  - Ta: Papillary tumor confined to mucosa (epithelium)
  - T1: Invasion of lamina propria (subepithelial connective tissue)
  - T2: Invasion of inner half of muscle (detrusor)
  - T2b: Invasion of outer half of muscle
  - T3a: Microscopic invasion of perivesical fat
  - T3b: Macroscopic invasion of perivesical fat
  - T4a: Invasion of surrounding organs
  - T4b: Invasion of pelvic or abdominal wall
  - N1-3: Pelvic lymph node metastases
  - N4: Lymph node metastases above bifurcation
  - M1: Distant metastases

CLINICAL ISSUES

Presentation

- Most common signs/symptoms: Painless hematuria
- Lab data
  - Positive urine dip stick
  - ± Micro- to normocytic anemia
- Diagnosis: Transurethral resection ± biopsy

Demographics

- Age
  - 50-60 years of age
  - Increasing incidence in patients < 30 years of age
- Gender: M:F = 4:1
- Ethnicity: Caucasian-to-African-American ratio: 1.5:1

Natural History & Prognosis

- Complications
  - Hydronephrosis, incontinence & urethral stricture
- Prognosis
  - Multifocal, high grade or infiltrating tumor → ↑ risk of metastases
  - 5 year survival rate: 82% in all stages combined
    - 94% in localized stages
    - 48% in regional stages
    - 6% in distant stages

Treatment

- < T2: Local endoscopic resection ± intravesical instillation or bacille Calmette-Guérin therapy
- T2 to T4a: Radical cystectomy or radiotherapy (cure)
- > T4b: Chemotherapy or radiotherapy ± adjuvant surgery (palliative)
- Follow-up
  - < T2: Repeated cystoscopy every 3-6 months
  - T2 to T4a: Clinical; MRI if indicated
  - > T4b: MRI

DIAGNOSTIC CHECKLIST

Consider

- Detection by cystoscopy; diagnosis by transurethral resection ± biopsy
- Staging is needed for treatment and prognosis
- Check kidneys, ureters for synchronous and metachronous tumors

Image Interpretation Pearls

- MRI is superior in staging and used in patients with high grade stage T1 or > stage T2

SELECTED REFERENCES

1. Barentsz JO et al: Staging urinary bladder cancer after transurethral biopsy: value of fast dynamic contrast-enhanced MR imaging. Radiology. 201(1):185-93, 1996
2. Kim B et al: Bladder tumor staging: comparison of contrast-enhanced CT, T1- and T2-weighted MR imaging, dynamic gadolinium-enhanced imaging, and late gadolinium-enhanced imaging. Radiology. 193(1):239-45, 1994
3. Narumi Y et al: Bladder tumors: staging with gadolinium-enhanced oblique MR imaging. Radiology. 187(1):145-50, 1993
4. Tanimoto A et al: Bladder tumor staging: comparison of conventional and gadolinium-enhanced dynamic MR imaging and CT. Radiology. 185(3):741-7, 1992
5. Rholl KS et al: Primary bladder carcinoma: evaluation with MR imaging. Radiology. 163(1):117-21, 1987
6. Amendola MA et al: Staging of bladder carcinoma: MRI-CT-surgical correlation. AJR Am J Roentgenol. 146(6):1179-83, 1986
7. Hillman BJ et al: Recognition of bladder tumors by excretory urography. Radiology. 138(2):319-23, 1981
8. Morgan CL et al: Computed tomography in the evaluation, staging, and therapy of carcinoma of the bladder and prostate. Radiology. 140(3):751-61, 1981

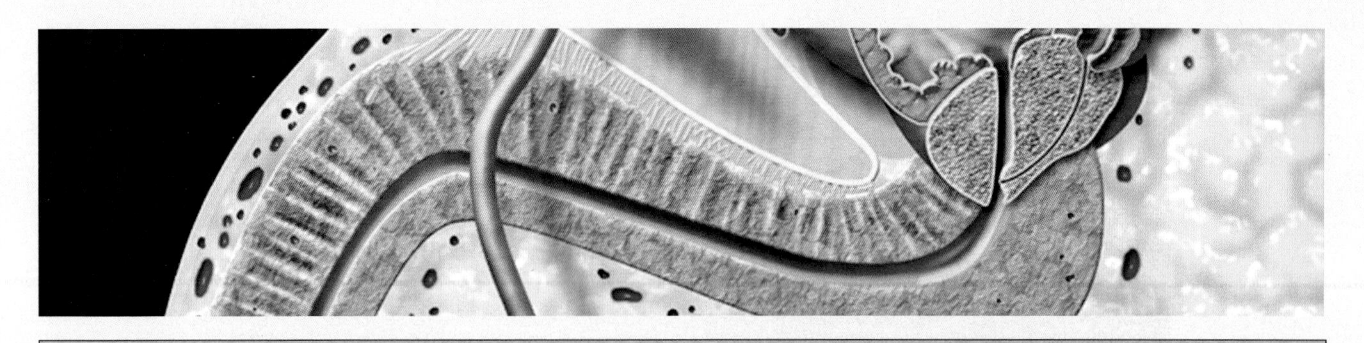

SECTION 6: Genital Tract (Male)

Introduction and Overview

Congenital

Infection

Inflammation

Trauma

Neoplasm

GENITAL TRACT ANATOMY AND IMAGING ISSUES

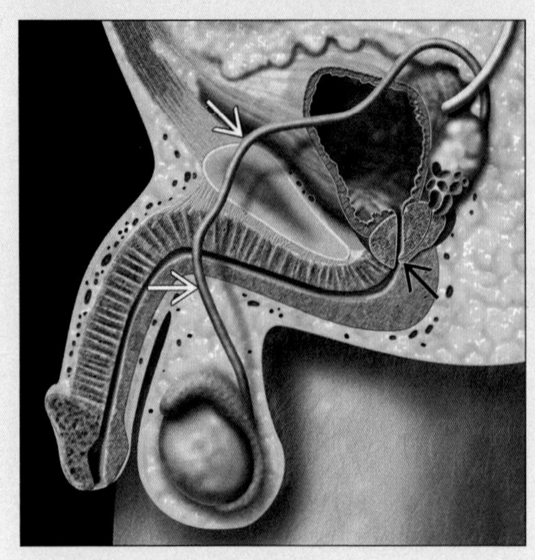

Sagittal graphic of male genital tract anatomy. Vas deferens (white arrows) passes from scrotum through inguinal ring to unite with ejaculatory ducts. Note verumontanum (black arrow).

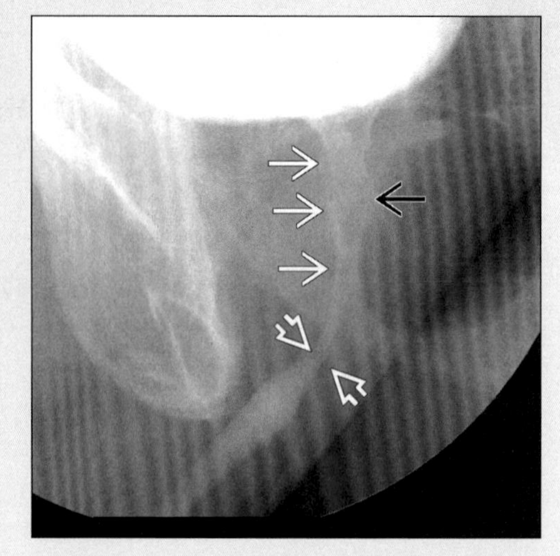

Voiding cystourethrogram shows normal male posterior urethral anatomy. Note verumontanum (black arrow), prostatic urethra (white arrows) and membranous urethra (open arrows).

IMAGING ANATOMY

- Prostatic capsule = blending of prostatic fibromuscular stroma with endopelvic fascia
  - Absent anteriorly and at apex
- Prostate
  - Ovoid gland; tapers as it extends from bladder base to urogenital diaphragm
  - Base of prostate: Abuts base of bladder
  - Apex of prostate: Inferior to base of prostate; abuts urogenital diaphragm
  - Glandular prostate surrounded by smooth muscle and connective tissue stroma
    - Anterior fibromuscular stroma blends with periurethral muscle fibers at bladder neck
  - Peripheral zone
    - Acinar glandular tissue
    - Accounts for 70% of prostatic volume
    - Site of origin of > 70% of prostate cancers
    - Extends from prostate base to apex
  - Central zone
    - Deep to peripheral zone
    - Primarily stromal cells
    - Site of origin for 10% of prostate cancers
    - Surrounds ejaculatory ducts from insertion at prostatic base to verumontanum
  - Transition zone
    - Bilobed; surround prostatic urethra
    - Site of hyperplastic nodules of BPH
    - Site of origin for 20% of prostate cancers
    - Represents 5% of gland volume in youth versus up to 90% of volume in old age
  - Periurethral zone
    - Short ducts lining prostatic urethra
  - Neurovascular bundle
    - Prostatic arteries and veins; cavernosal nerves
    - Common site of extracapsular extension of prostate CA
    - At 5 and 7 o'clock positions posterolateral to prostate
  - Periprostatic venous plexus
  - Lymphatic drainage of prostate
    - Obturator, external iliac, internal iliac nodes
- Seminal vesicles
  - Extend laterally from base of prostate
  - Outpouchings of vas deferens
  - Each is a highly convoluted tubule that produces protein in seminal fluid; contracts during ejaculation
- Ejaculatory ducts
  - At junction of seminal vesicles and vas deferens
  - Cross peripheral and central zones of prostate to end at verumontanum in prostatic urethra
- Testis
  - Fibrous capsule: Tunica albuginea
  - Mediastinum: Invagination of tunica into testis; site of spermatic cord insertion
  - Divided into cone-shaped lobules that converge to mediastinum
  - Lobules contain branching seminiferous tubules that coalesce into 12-20 efferent ductules at mediastinum
  - Efferent ductules unite and continue as epididymis
- Vas deferens
  - Distal continuation of tail of epididymis
  - Ascends posteriorly in spermatic cord to deep inguinal ring
  - Crosses anterior to external iliac artery then curves downward into pelvis
  - Joins seminal vesicle at base of prostate to form ejaculatory duct
- Spermatic cord
  - Contains vas deferens, lymphatics, blood vessels and nerves
- Scrotum
  - Outpouching of abdominal cavity
  - Concentric layers reflect layers of abdominal wall
    - Skin→dartos→external spermatic fascia (ESF)→cremaster muscle→Internal spermatic fascia (ISF)→parietal tunica vaginalis→potential space→visceral tunica vaginalis→tunica albuginea
- Male urethra

GENITAL TRACT ANATOMY AND IMAGING ISSUES

DIFFERENTIAL DIAGNOSIS

Intratesticular mass
- Germ cell tumors, stromal tumors
- Leukemia/lymphoma
- Metastases
- Orchitis/abscess
- Hematoma/contusion
- Focal infarct

Extratesticular mass
- Cysts: Spermatoceles, epididymal cyst, tunica albuginea cyst
- Calcifications: Scrotal pearl, tunica plaques
- Non-neoplastic masses: Hernia, varicocele, abscess, sperm granuloma
- Tumors: Adenomatoid tumor, lipoma, sarcoma, lymphoma, leiomyoma

Urethral stricture
- Infectious/inflammatory: Gonococcus, nongonococcal urethritis, TB
- Bulbar urethra
- Multiple and serial strictures
- Iatrogenic: Surgery, instrumentation
- Membranous urethra or penile-scrotal junction
- Traumatic: Following complete transection
- Solitary, short segment

Urethral filling defects
- Common: Stone, polyp, carcinoma
- Unusual: Condyloma acuminata, malacoplakia, urethritis cystica, metastases

  - Posterior urethra: From bladder neck to urogenital diaphragm
    - Prostatic urethra: Traverses transitional zone of prostate
    - Verumontanum: Site of insertion of prostate glands
    - Membranous: At external urethral sphincter
  - Anterior urethra: Extends through corpus spongiosum of penis
    - Bulbar urethra
    - Penile urethra
  - Cowper glands
    - Glands surround membranous urethra
    - Drain into proximal bulbar urethra

Anatomic Relationships
- Arteries
  - In spermatic cord: Testicular artery (from aorta), cremasteric artery (from inferior epigastric), artery to vas deferens (from vesical artery)
    - Testicular artery: Arises from aorta; supplies testis and epididymis
    - Cremasteric artery: Arises from inferior epigastric; supplies cremaster muscle and other layers of spermatic cord
    - Artery to vas deferens: Arises from superior vesical artery; supplies vas
    - Cremasteric and vas deferens arteries anastamose with testicular artery
- Veins
  - From scrotum: Pampiniform plexus in spermatic cord→ipsilateral gonadal vein
  - Right gonadal vein→directly into IVC
  - Left gonadal vein→left renal vein
- Nerves: Cremasteric nerve, genital branch of genitofemoral nerve, testicular sympathetic plexus
- Lymphatics
  - Ascend in spermatic cord with testicular vessels
  - Drain to lateral aortic and preaortic nodes

Fascia
- Scrotal wall layers: Derived from layers of abdominal wall
  - Tunica vaginalis: Continuation of peritoneal processus vaginalis
- Dartos: Outer layer of the scrotum

ANATOMY-BASED IMAGING ISSUES

Normal Measurements
- Testis: 3.5-4 cm length, 2-3 cm width
- Prostate: 4.8 cm width x 2.8 cm x 2.8 cm
  - Normal weight: 20 ± 6 grams
- Epididymis: Head normally 7-8 mm in diameter
- Seminal vesicles: Each 3 cm length x 1.5 cm width but size and shape variable
- Ejaculatory duct: 2 cm in length
- Male urethra: 20 cm in length

EMBRYOLOGY

Embryologic Events
- Primitive sex cords form as coelomic epithelium of gonadal ridge proliferates and penetrates mesenchyme
- Gonad differentiates into testis under influence of testes determining factor
- Two pairs of genital ducts in early embryo: Mesonephric and paramesonephric ducts
- Mesonephric (Wolffian) duct persists in male: Forms vas deferens, epididymis and ejaculatory ducts
- Mullerian ducts regress in male
- Testis descends from posterior abdominal wall into scrotal sac following ligamentous band known as gubernaculum
- Outpouching of peritoneal cavity follows migrating gubernaculum to form processus vaginalis
- Evagination of peritoneum through abdominal wall produces inguinal canal; fascial layers of scrotum reflect abdominal wall layers
- Regression of gubernaculum fixes testis to scrotal wall

Practical Implications
- Incomplete testicular fixation predisposes to torsion
- Embryonic remnants

GENITAL TRACT ANATOMY AND IMAGING ISSUES

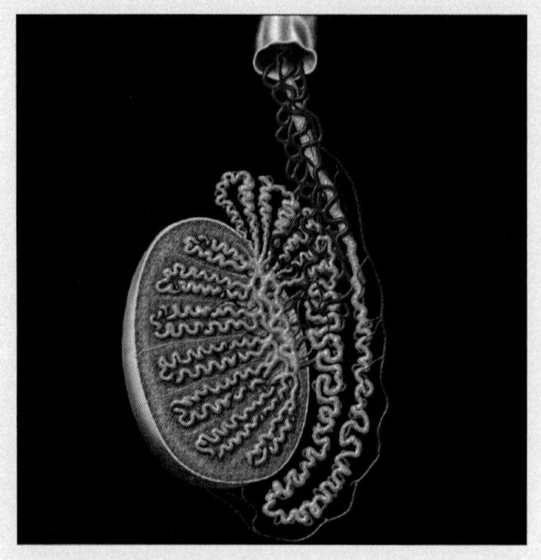

Normal testis, epididymis, and spermatic cord. Note lobular architechture of testis.

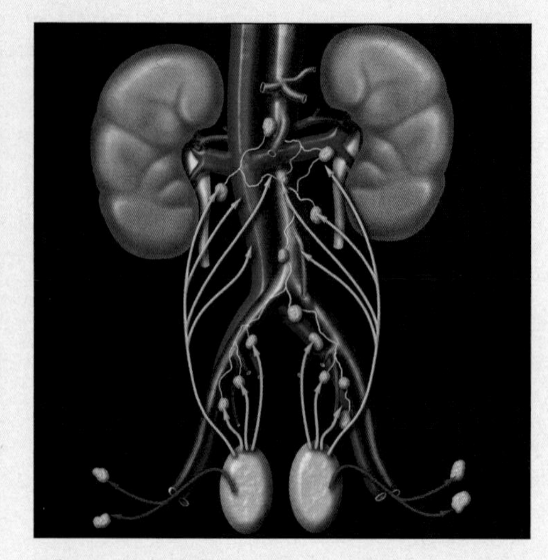

Schematic drawing of testicular lymphatic drainage. Primary pathways are shown in yellow.

- Appendix epididymis: Wolffian duct remnant; from cranial end
- Appendix testis (Hydatid of Morgagni): Mullerian duct remnant
- Inferior end of Mullerian duct persists as prostatic utricle at distal end of verumontanum

CUSTOM DIFFERENTIAL DIAGNOSIS

Intratesticular mass

- Germ cell tumor
- Stromal tumor
- Leukemia/lymphoma
- Metastases
- Orchitis/abscess
- Hematoma/contusion
- Focal infarct

Extratesticular masses

- Cysts: Spermatoceles, epididymal cyst, tunica albuginea cyst
- Calcifictions
  - Epididymis: Prior infection or trauma; age-related
  - Scrotal pearl: Within layers of tunica vaginalis
    - Adherent or mobile
    - Detached and calcified testicular appendages secondary to prior torsion; or post-inflammatory
  - Calcified tunica plaques
- Varicocele
- Hernia
- Abscess
- Non-neoplastic masses
  - Sperm granulomas: Iso-hypoechoic avascular mass; may be painful
  - Fibrous pseudotumor
- Tumors
  - Spermatic cord tumors: Rare
    - 70% benign: Lipomas, mixed, fibromas, leiomyomas
    - Malignant: Sarcomas (rhabdomyosarcoma, liposarcoma, leiomyosarcoma, others), lymphoma
  - Epididymal tumors: Rare
    - Adenomatoid tumor: Most common; represents 30% of extratesticular masses
    - Leiomyomas
    - Metastatic prostate CA, lymphoma
    - Papillary cystadenoma: 65% of patients with von Hippel-Lindau

Urethral stricture

- Infectious/inflammatory: Gonococcus, nongonococcal urethritis, TB
  - Bulbar urethra
  - Multiple and serial strictures
- Iatrogenic: Surgery, instrumentation
  - Membranous urethra or penile-scrotal junction
- Traumatic: Following complete transection
  - Solitary, short segment

Urethral filling defects

- Stone
- Polyp
- Carcinoma
- Unusual: Condyloma acuminata, malacoplakia, urethritis cystica, metastases

SELECTED REFERENCES

1. Zagoria R: Genitourinary radiology, 2nd ed. Philadelphia, Mosby Inc. 312-51, 2004
2. Pavlica P et al: Imaging of male urethra. Eur Radiol. 13(7):1583-96, 2003
3. Ali M et al: CT signs of urethral injury. Radiographics. 23(4):951-63; discussion 963-6, 2003
4. Sudakoff GS et al: Scrotal ultrasonography with emphasis on the extratesticular space: anatomy, embryology, and pathology. Ultrasound Q. 18(4):255-73, 2002
5. Kubik-Huch RA et al: CT and MRI of the male genital tract: radiologic-pathologic correlation. Eur Radiol. 9(1):16-28, 1999

IMAGE GALLERY

Typical

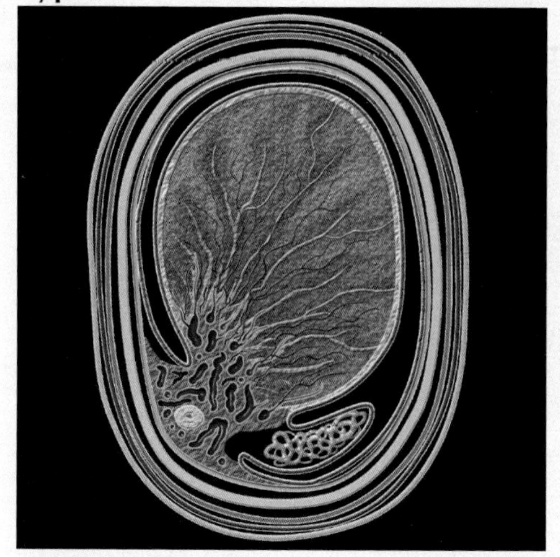

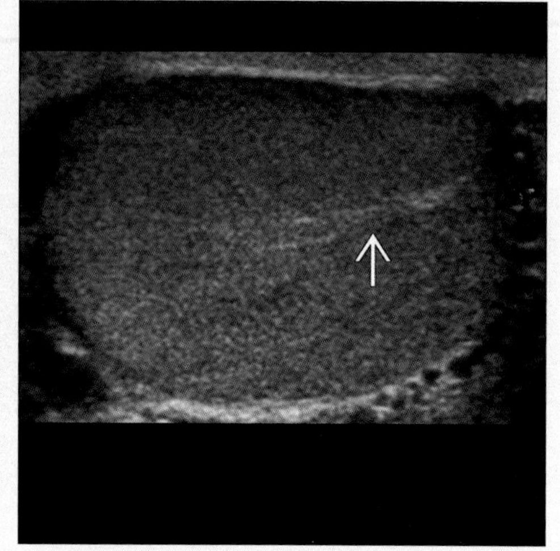

(Left) *Cross-sectional graphic shows concentric layers surrounding testis: scrotal skin (beige), Dartos (red), ESF (orange), cremaster m. (yellow), ISF (green), t. vaginalis (blue), and t. albuginea (gray).* ***(Right)*** *Axial US shows normal testis with echogenic mediastinum testis (arrow). Vessels enter the testis through the mediastinum.*

Typical

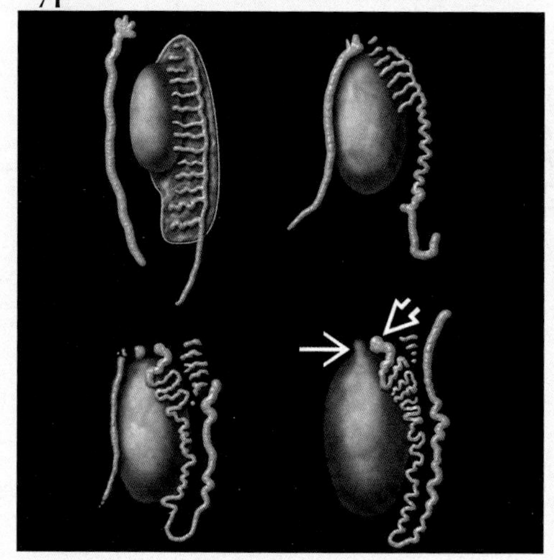

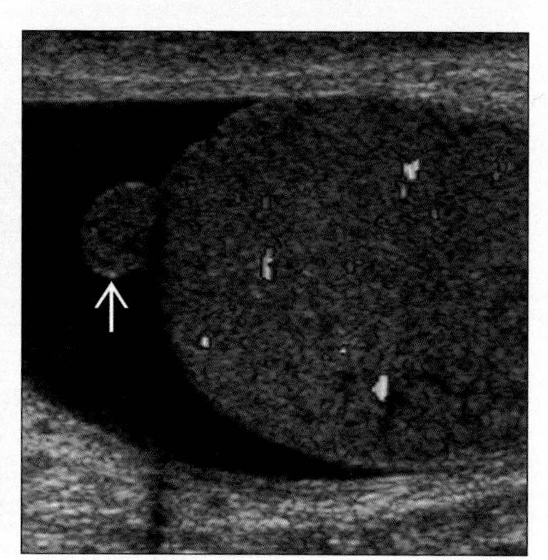

(Left) *Evolution of efferent duct system. Wolffian ducts (yellow) are precursors of epididymis, vas deferens and appendix epididymis (open arrow). Appendix testis (arrow) is Mullerian duct (green) remnant.* ***(Right)*** *Sagittal color Doppler US shows appendix testis (arrow) arising from upper pole of testis.*

Typical

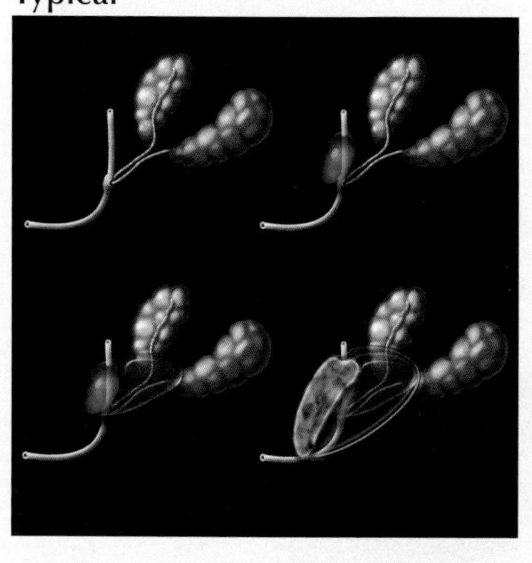

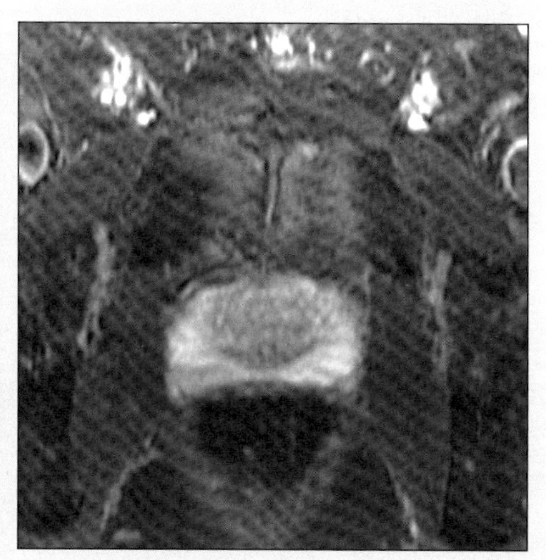

(Left) *Drawing shows prostate zonal anatomy relative to urethra and ejaculatory ducts: Transition zone (green), central zone (red), peripheral zone (transparent) and anterior fibromuscular stroma (yellow).* ***(Right)*** *Axial T2WI MR shows U-shaped high signal peripheral zone surrounding lower signal intensity central gland (transitional and central zones).*

CRYPTORCHIDISM

Axial CECT shows ovoid mass (arrow) anterior to left iliac vessels representing cryptorchid testis.

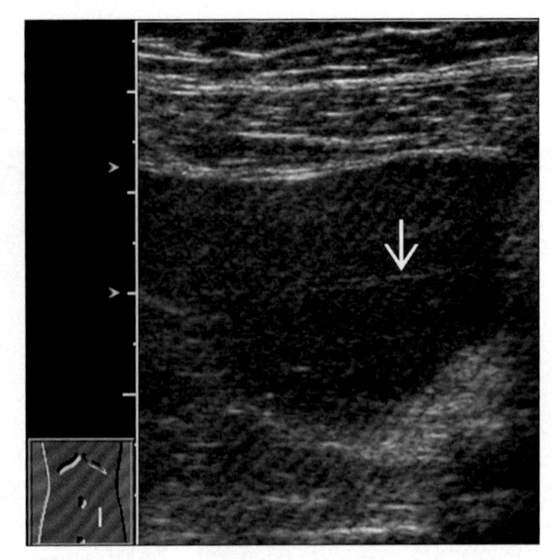

Sagittal US at level just above left inguinal crease shows hypoechoic ovoid mass representing cryptorchid testis. Note mediastinum testis (arrow),

TERMINOLOGY

Abbreviations and Synonyms

- Cryptorchidism = cryptorchism = undescended testis

Definitions

- Incomplete descent of testis into base of scrotum

IMAGING FINDINGS

General Features

- Best diagnostic clue: Unilateral absence of spermatic cord
- Location
  - Anywhere from kidney to inguinal canal
  - Inguinal canal most common (80%)
  - Bilateral in 10%
- Size: Cryptorchid testis smaller than normal testis
- Morphology: Ovoid well-circumscribed mass

CT Findings

- CECT: Low attenuation mass above scrotum in path of testicular descent

MR Findings

- T1WI: Low signal intensity ovoid mass
- T2WI: High signal intensity ovoid mass

Ultrasonographic Findings

- Real Time
  - Ovoid homogeneous well-circumscribed structure smaller than normal descended testis
  - Echogenic line of mediastinum testis

Angiographic Findings

- Gonadal venography: Visualization of pampiniform plexus used to locate undescended testis
  - Testicular agenesis: Absent plexus and blind-ending vein

Imaging Recommendations

- Best imaging tool
  - Ultrasound for inguinal testis
    - MRI and CT: 90-95% sensitive for intra-abdominal testis
- Protocol advice: Oral contrast may be helpful to distinguish bowel from small undescended testis

DIFFERENTIAL DIAGNOSIS

Absent testis: Anorchia

- Congenital or prior resection

Inguinal masses

- Lymphadenopathy: Hypoechoic mass
- Hernia: Bowel signature; real-time motion

DDx: Inguinal Masses

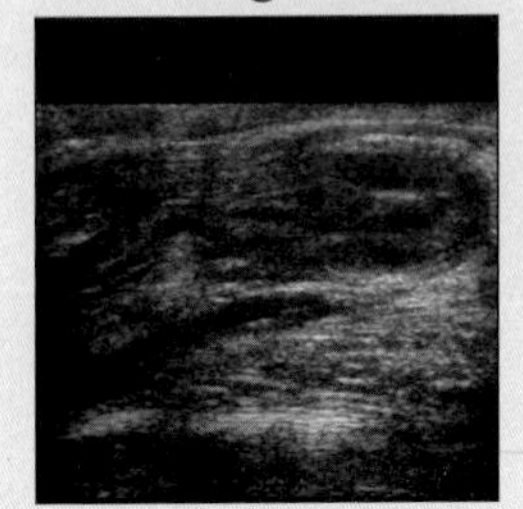

Inguinal Hernia

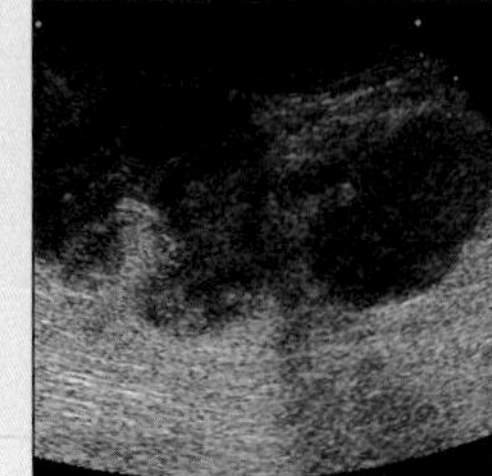

Lymphadenopathy

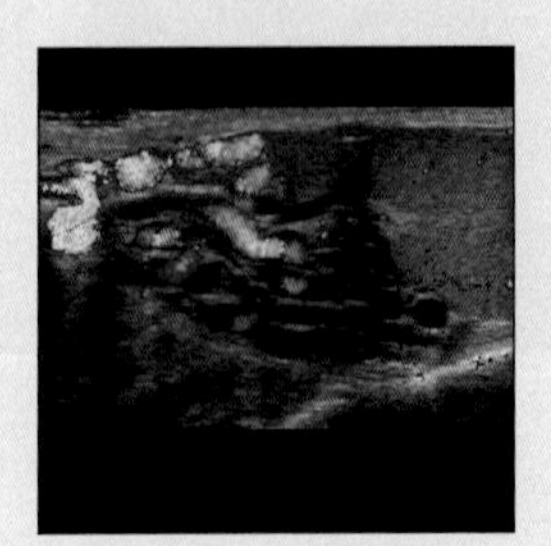

Varicocele

CRYPTORCHIDISM

Key Facts

Terminology
- Incomplete descent of testis into base of scrotum

Imaging Findings
- Best diagnostic clue: Unilateral absence of spermatic cord
- Anywhere from kidney to inguinal canal
- Echogenic line of mediastinum testis

Clinical Issues
- Infertility: Impaired function of cryptorchid testis
- ↑ Risk of malignant neoplasm in cryptorchid testis; also ↑ risk in contralateral testis

Diagnostic Checklist
- Identify mediastinum to distinguish cryptorchid testis from other inguinal masses on US

- Varicocele: Distended veins; ↑ with Valsalva

PATHOLOGY

General Features
- General path comments: Testis may be absent in 5%
- Etiology: Interruption of embryologic process of testicular descent from abdomen into scrotal sac
- Epidemiology
  - 4% of newborns; 0.8% by age 3 months (get spontaneous descent after birth)
  - Incidence of testis cancer in men with undescended testis is 1:1,000-1:2,500
- Associated abnormalities
  - Renal agenesis/ectopia, prune belly syndrome, epispadias
  - Wolffian duct anomalies: Seminal vesical cyst, ectopic ureter, etc.

Gross Pathologic & Surgical Features
- Pre-operative imaging usually not necessary; surgical exploration required
- Most common tumors in cryptorchid testis: Seminoma, embryonal cell CA

Microscopic Features
- Histologic changes in cryptorchid testis irreversible after age 2.5 years

Staging, Grading or Classification Criteria
- Types: Retractile, canalicular, abdominal, ectopic
  - Retractile: Moves intermittantly between groin and scrotal base; not truly undescended
  - Canalicular: Between internal and external inguinal rings
  - Abdominal: Most immediately proximal to internal ring
  - Ectopic: Remote from normal pathway of descent

CLINICAL ISSUES

Presentation
- Most common signs/symptoms
  - Absent testis in scrotal sac
  - Infertility: Impaired function of cryptorchid testis

Demographics
- Age: Usually diagnosed in newborns

Natural History & Prognosis
- ↑ Risk of malignant neoplasm in cryptorchid testis; also ↑ risk in contralateral testis

Treatment
- Orchiopexy before age 2 to preserve fertility; surgical removal post-puberty
- Complications: Torsion, infertility, malignant change, trauma

DIAGNOSTIC CHECKLIST

Consider
- Absent spermatic cord on one side suggests cryptorchidism

Image Interpretation Pearls
- Identify mediastinum to distinguish cryptorchid testis from other inguinal masses on US

SELECTED REFERENCES

1. Zagoria RJ. Genitourinary Radiology 2nd ed. Philadelphia, Mosby. 327-329, 2004
2. Shadbolt CL et al: Imaging of groin masses: inguinal anatomy and pathologic conditions revisited. Radiographics. 21 Spec No:S261-71, 2001
3. Nguyen HT et al: Cryptorchidism: strategies in detection. Eur Radiol. 9(2):336-43, 1999

IMAGE GALLERY

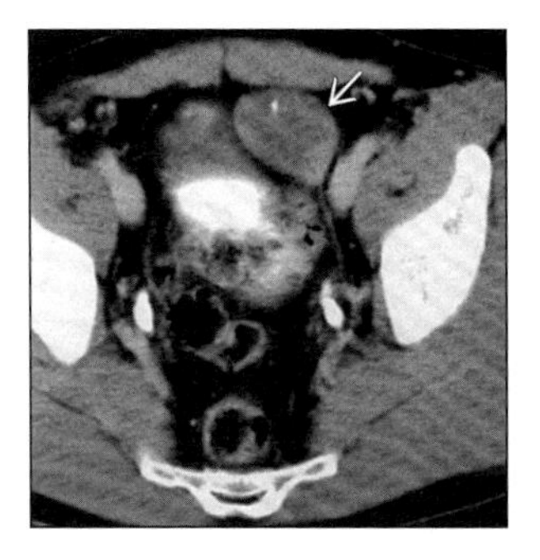

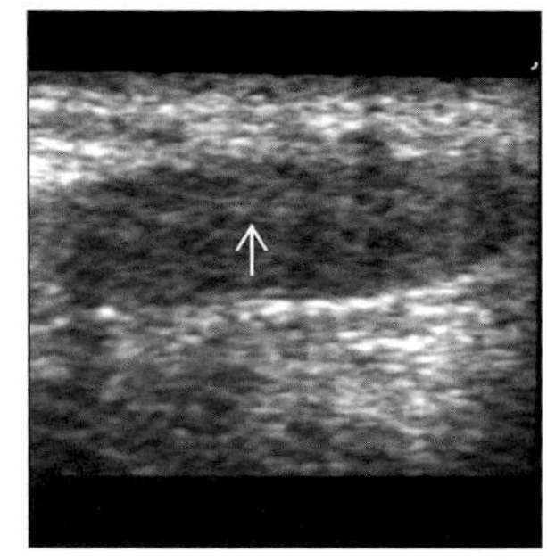

(Left) *Axial CECT shows heterogeneous ovoid mass (arrow) adjacent to bladder and iliac vessels. Mass proved to be a seminoma arising in a cryptorchid testis (Courtesy H. Harvin, MD).* ***(Right)*** *Sagittal US shows hypoechoic mass in inguinal area representing atrophic, cryptorchid testis. Note echogenic mediastinum testis (arrow).*

EPIDIDYMITIS

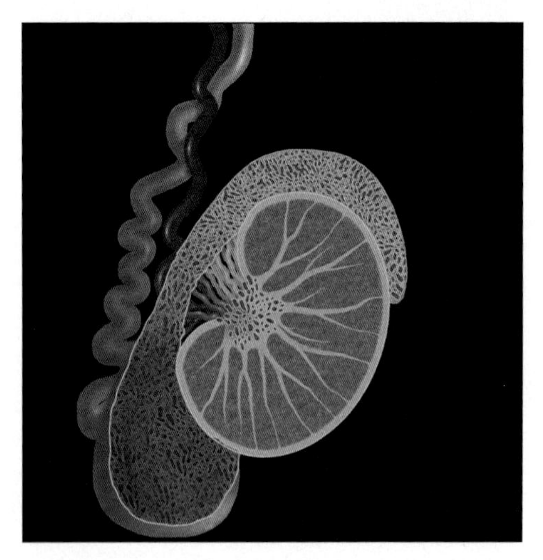

Schematic depiction of acute epididymitis. Note swollen and inflammed tail of the epididymis with adjacent normal testis.

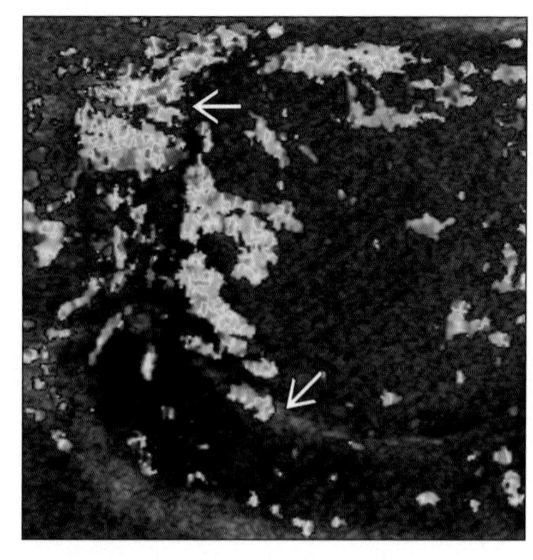

Sagittal color Doppler US demonstrates hyperemia of enlarged epididymis (arrows), consistent with epididymitis.

TERMINOLOGY

Abbreviations and Synonyms

- Acute scrotum, orchitis, epididymo-orchitis

Definitions

- Infectious inflammation of epididymis and/or testicle

IMAGING FINDINGS

General Features

- Best diagnostic clue: Enlarged, hyperemic epididymis and/or testicle on color Doppler US
- Location: Early epididymitis often involves tail of epididymis
- Size: Epididymis typically 2-3 times larger than normal
- Morphology: Focal enlargement of tail or diffuse enlargement of entire epididymis

Ultrasonographic Findings

- Color Doppler
  - Enlarged epididymis
    - Diffusely or focally in tail of epididymis; increased vascularity of epididymis, testicle; reactive hydrocele with internal low level echoes
    - Echotexture may be decreased or increased from normal; Doppler flow dramatically increased
    - Scrotal wall inflammation with skin thickening

Nuclear Medicine Findings

- Technetium 99m
  - 90% accurate in differentiating torsion from epididymitis
  - Increased flow within testicular vessels and vas deferens on flow study

Imaging Recommendations

- Best imaging tool: Color Doppler US; high frequency transducers optimal (≥ 10 mHz)
- Protocol advice: Comparison with contralateral testicle useful in subtle cases of increased vascularity

DIFFERENTIAL DIAGNOSIS

Testicular torsion

- Absent or diminished color Doppler flow, "twist" of spermatic cord in inguinal region
- Epididymis may be enlarged but not hyperemic on color Doppler US

Testicular tumor

- Focal mass on US with abnormal vessels on color Doppler

DDx: Testicular Lesions Mimicking Epididymitis

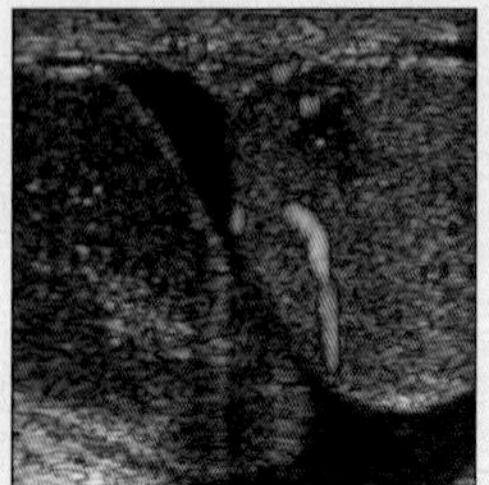

Testicular Torsion

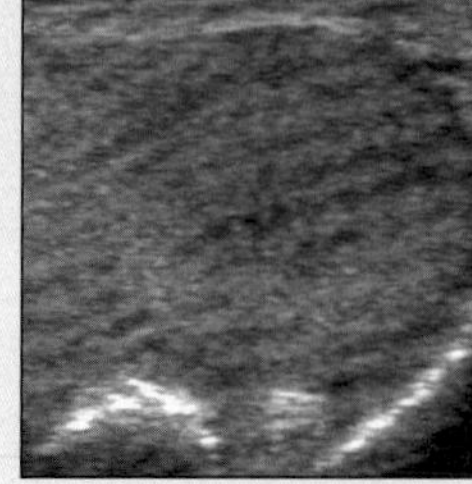

Testicular Tumor

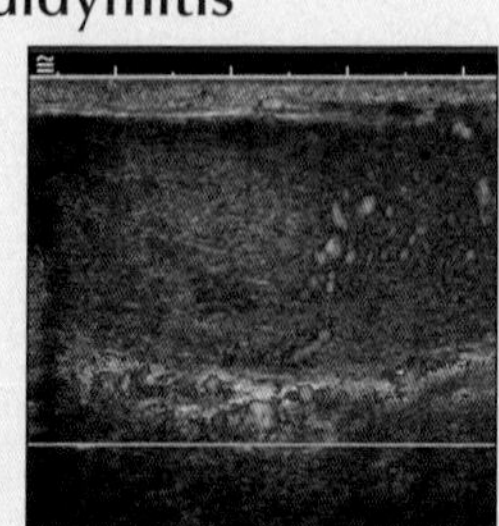

Segmental Infarct

EPIDIDYMITIS

Key Facts

Terminology
- Infectious inflammation of epididymis and/or testicle

Imaging Findings
- Best diagnostic clue: Enlarged, hyperemic epididymis and/or testicle on color Doppler US
- Protocol advice: Comparison with contralateral testicle useful in subtle cases of increased vascularity

Top Differential Diagnoses
- Testicular torsion
- Testicular tumor
- Segmental infarct

Clinical Issues
- Prognosis excellent if treated early with antibiotics
- Recurrent cases can lead to fertility problems long term

Segmental infarct
- Acute pain, no history of trauma; focal hypoechoic area avascular on color Doppler; increased incidence in patients with hypercoagulable states or advanced atherosclerosis, such as diabetics

PATHOLOGY

General Features
- General path comments: Infectious inflammatory response, can lead to abscess if not treated (6%)
- Etiology
  - Bacterial seeding occurs directly in cases with genitourinary (GU) anomalies, and presumably hematogenously in cases without demonstrable anomaly
  - Sexually transmitted ascending GU tract infection: Chlamydia, E. coli, Staphylococcus aurea, mumps virus
- Epidemiology: Most frequently seen in sexually active young men; also seen in infants and boys

Gross Pathologic & Surgical Features
- Only treated surgically if abscess forms despite antibiotic treatment

Microscopic Features
- Inflammatory infiltrate of testis and epididymis

Staging, Grading or Classification Criteria
- Isolated epididymitis; isolated orchitis (may be seen in boys with mumps); combined epididymitis & orchitis

CLINICAL ISSUES

Presentation
- Most common signs/symptoms: Acutely painful scrotum; scrotal swelling, erythema; fever; dysuria
- Clinical profile: Positive urinalysis for WBC and bacteria; may have elevated WBC

Demographics
- Age: Most commonly 15-35 yrs
- Gender: Male

Natural History & Prognosis
- Prognosis excellent if treated early with antibiotics
- Complications
  - Abscess formation
  - Testicular ischemia if entrapment of testicular artery with chronic inflammation
- Recurrent cases can lead to fertility problems long term

Treatment
- Antibiotic therapy; follow-up scans to exclude abscess if not improved
- Work-up for GU anomalies in younger children and recurrent cases

DIAGNOSTIC CHECKLIST

Consider
- Torsion if low flow to testicle

Image Interpretation Pearls
- Hyperemic and enlarged epididymis and/or testis

SELECTED REFERENCES

1. Kraus SJ et al: Genitourinary imaging in children. Pediatr Clin North Am. 48:1381-424, 2001
2. Munden MM et al: Scrotal pathology in pediatrics with sonographic imaging. Curr Probl Diagn Radiol. 29(6):185-205, 2000
3. Bukowski TP et al: Epididymitis in older boys: Dysfunctional voiding as an etiology. J Urol. 154:762-5, 1995

IMAGE GALLERY

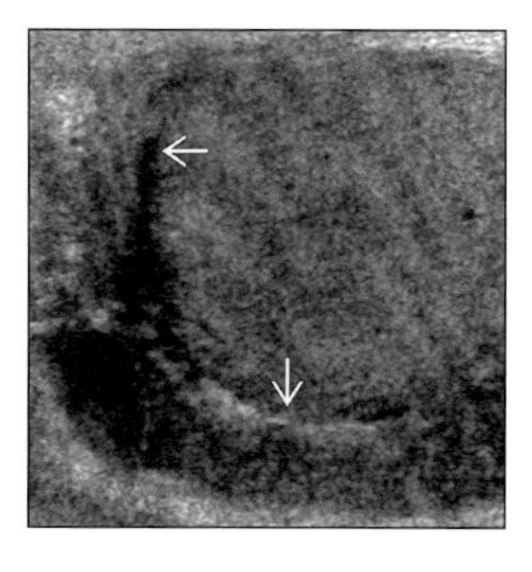

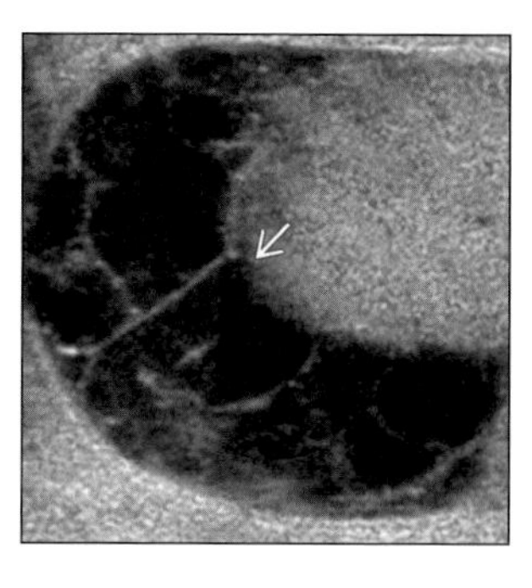

(Left) *Grayscale imaging of epididymitis. Sagittal US of testis demonstrates diffuse hypoechoic enlargement of epididymis (arrows).* ***(Right)*** *Sagittal US demonstrates reactive hydrocele with multiple septations (arrow).*

PROSTATITIS

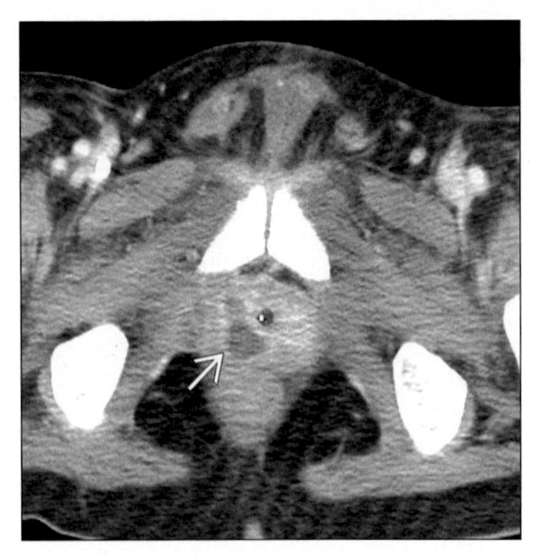

Axial CECT shows rim-enhancing collection (arrow) in right lateral aspect of prostate gland representing prostatic abscess in a patient with leukemia.

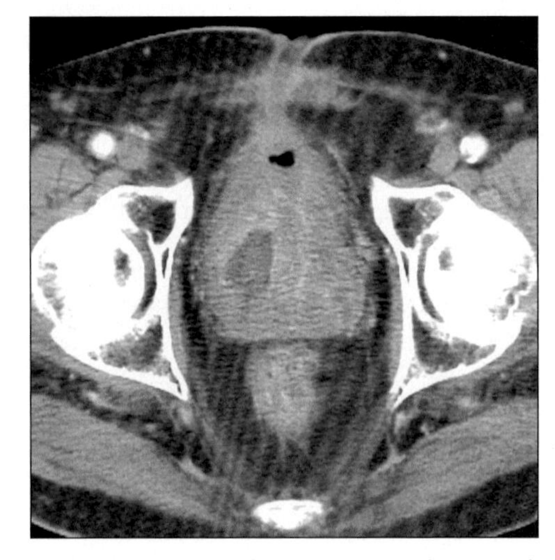

Axial CECT shows low attenuation lesion with rim-enhancement in central portion of right prostate lobe, consistent with abscess.

TERMINOLOGY

Abbreviations and Synonyms

- Acute/chronic/nonbacterial bacterial prostatitis (ABP/CBP/NBP)

Definitions

- Clinical syndrome of pelvic/perineal pain often caused by inflammation of the prostate gland
- Prostatodynia: Prostatitis symptoms without histologic evidence of prostatic inflammation
  - Prostatodynia = chronic prostatitis-chronic pelvic pain syndrome (CPPS)

IMAGING FINDINGS

General Features

- Best diagnostic clue
  - Acute: Swollen, warm and tender prostate on digital rectal exam
  - Chronic bacterial: Greater than 10-fold bacterial growth in the expressed prostatic secretions
- Location
  - Prostatitis: Peripheral gland
  - Abscess: Usually peripheral; can be central following transurethral prostatectomy (TURP)
- Morphology: ABP: Diffusely enlarged, edematous gland

Radiographic Findings

- Fluoroscopy
  - Urethrography: Compression of prostatic urethra
    - Chronic: May see reflux into prostate gland
- IVP: Normal, or edema of interureteric ridge

CT Findings

- CECT
  - Abscess: Rim-enhancing unilocular or multilocular low attenuation mass in prostate
    - May extend through capsule into periprostatic tissues, seminal vesicles, peritoneum

MR Findings

- T2WI
  - Low signal areas indistinguishable from prostate CA
  - Abscess: Focal area of high signal intensity

Ultrasonographic Findings

- Real Time
  - Transrectal ultrasound (TRUS)
    - Focal hypoechoic areas indistinguishable from prostate CA
    - Hypoechoic rim surrounding gland
    - Hypoechoic periurethral region
  - Discrete hypoechoic mass suggests abscess formation

DDx: Prostatic Pathology

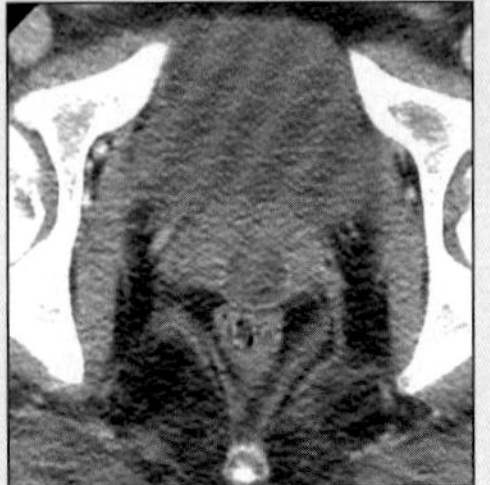

Prostatic Cyst

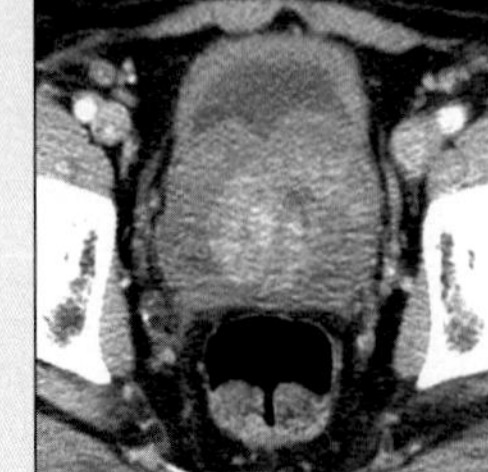

BPH

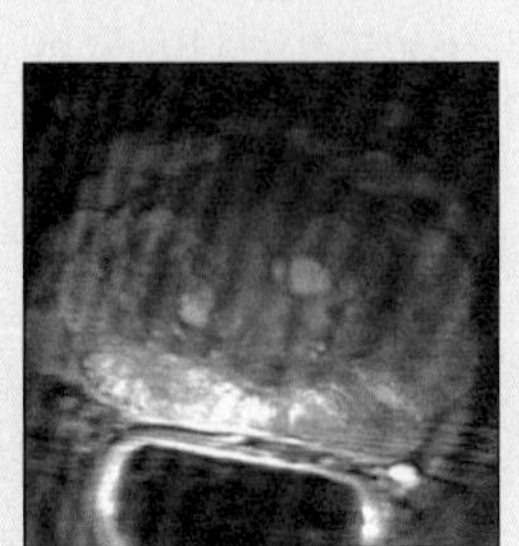

Prostate Cancer

PROSTATITIS

Key Facts

Terminology
- Clinical syndrome of pelvic/perineal pain often caused by inflammation of the prostate gland

Imaging Findings
- Abscess: Usually peripheral; can be central following transurethral prostatectomy (TURP)
- Focal hypoechoic areas indistinguishable from prostate CA

Pathology
- Ascending colonization of urinary tract, or post-biopsy
- Granulomatous: BCG therapy used for treatment of superficial bladder CA

Clinical Issues
- Acute: Fever, chills, dysuria, frequency and pain
- Chronic: Dysuria, urgency, pain

- Color Doppler: Increased flow in area of abscess

Imaging Recommendations
- Best imaging tool: CT if large abscess suspected

DIFFERENTIAL DIAGNOSIS

Prostate carcinoma
- Low signal intensity in peripheral zone on T2WI

Congenital prostatic cysts
- Usually in or near midline, or along course of ejaculatory duct

Cystic degeneration of BPH
- Not "pure" water intensity

PATHOLOGY

General Features
- Etiology
  - Ascending colonization of urinary tract, or post-biopsy
  - Bacterial prostatitis: E. coli, pseudomonas, enterococci most common
  - Chronic non-bacterial: Chlamydia, others
  - Granulomatous: BCG therapy used for treatment of superficial bladder CA
  - Eosinophilic prostatitis, malacoplakia: Stages in resolution of bacterial prostatitis
- Epidemiology: Worldwide prevalence: 2-10%

Microscopic Features
- Acute: Glands and ducts of the prostate contain exudate of polymorphonuclear leukocytes

Staging, Grading or Classification Criteria
- NIH classification: Based on leukocyte (WBC) count and culture of midstream urine specimen (VB2) and expressed prostate secretions (EPS)
- NIH category I: + VB2 and EPS culture and WBCs; bacteremia = ABP
- NIH category II: + VB2 and EPS culture and WBCs; recurrent bacteruria = CBP
- NIH category IIIa: - VB2 culture and WBCs; + EPS WBCs = NBP = inflammatory CPPS
- NIH category IIIb: - VB2 and EPS culture and WBCs = Prostatodynia = noninflammatory CPPS
- NIH category IV: +/- VB2 and EPS cultures, + VB2 and EPS WBC but without symptoms

CLINICAL ISSUES

Presentation
- Most common signs/symptoms
  - Acute: Fever, chills, dysuria, frequency and pain
  - Chronic: Dysuria, urgency, pain

Demographics
- Age: CBP: Men aged 40-70 years

Treatment
- Antibiotics; abscess drainage

DIAGNOSTIC CHECKLIST

Consider
- Prostatitis is not an imaging diagnosis

SELECTED REFERENCES

1. Hua VN et al: Acute and chronic prostatitis. Med Clin North Am. 88(2):483-94, 2004
2. Zagoria R: Genitourinary radiology, 2nd ed. Philadelphia, Mosby, Inc. 211-229, 2004
3. Pollack H et al: Clinical urography. 2nd ed. Philadelphia, WB Saunders. 1058-1069, 2000
4. Nghiem HT et al: Cystic lesions of the prostate. Radiographics. 10(4):635-50, 1990

IMAGE GALLERY

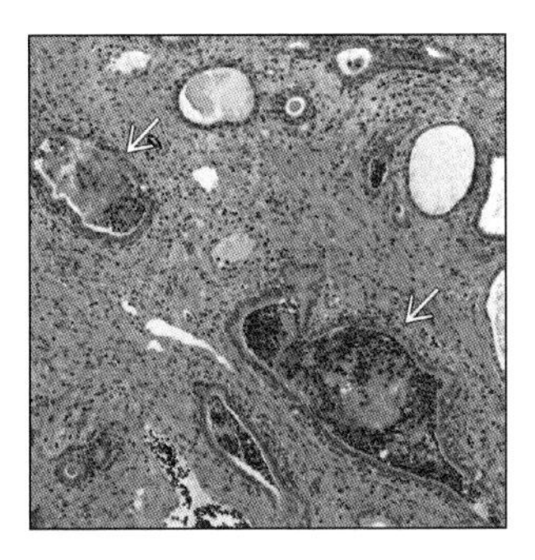

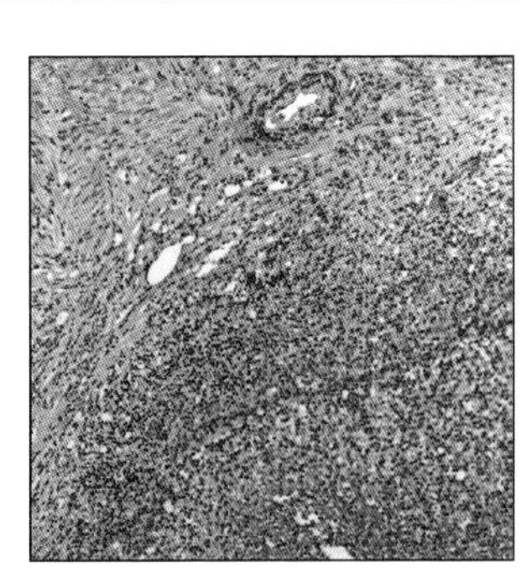

(Left) *Micropathology shows polymorphonuclear leukocytes in glandular lumens (arrows) and infiltrating epithelium of glands in a patient with acute prostatitis (Courtesy T. Morgan, MD).* ***(Right)*** *Micropathology shows necrotizing granuloma in lower right of prostate biopsy specimen from a patient with chronic prostatitis (Courtesy T. Morgan, MD).*

URETHRAL STRICTURE

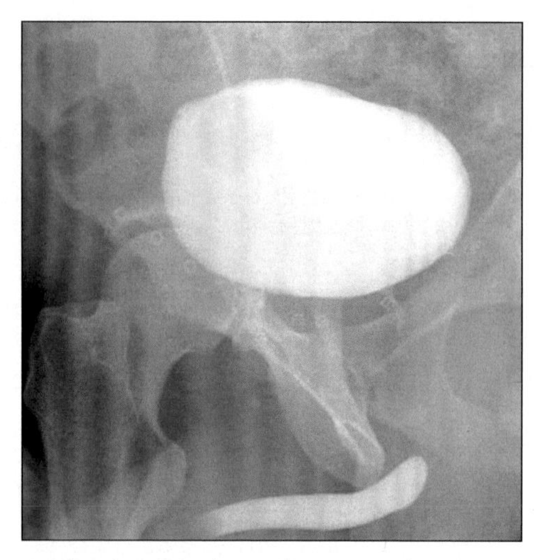

Combined retrograde urethrogram and cystogram (suprapubic catheter) seems to demonstrate long urethral stricture one month following pelvic trauma.

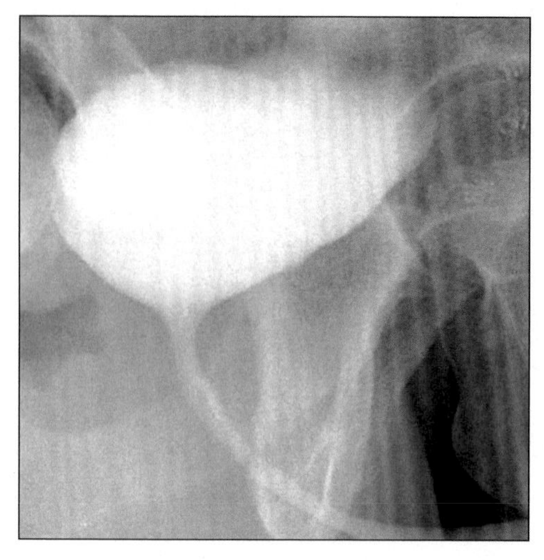

Post-operative cystogram following repair of 2 cm long stricture shows normal urethra.

TERMINOLOGY

Definitions

- A narrowing of urethra due to scar tissue or tumor

IMAGING FINDINGS

General Features

- Best diagnostic clue: Focal narrowing of urethra with proximal dilation

Radiographic Findings

- Retrograde urethrography
  - Focal, asymmetrical indentations
  - Long segmental, tubular or beaded constrictions
  - Marked narrowing to slight irregularity of urethral wall closely proximal and/or distal to main stricture
  - Proximal & distal to stricture is normal (acute)
  - Dilatation proximal to stricture, not distal (chronic)
  - Normal cone shape of proximal bulb
  - Small amount of reflux into Cowper gland duct
  - Thick-walled bladder (obstruction)
  - Iatrogenic
    - Located in penoscrotal junction and membranous portion of urethra (most common)
    - Other narrow sites (meatus and bladder neck)
    - Occasionally entire urethra; diffuse or multifocal
  - Gonorrhea: Located in bulbous portion of urethra or immediately proximal to meatus; multifocal
  - Reiter syndrome: Involves entire penile urethra
- Voiding cystourethrography
  - Evaluate anterior urethra proximal to severe stricture

Ultrasonographic Findings

- Sonourethrography
  - Used mainly in evaluating anterior strictures
  - Greater accuracy in predicting stricture length
  - Staging of complex or reoperative strictures: Calculi, urethral hair, false passage & stent encrustation

Imaging Recommendations

- Best imaging tool
  - Retrograde urethrogram and voiding urethrography
    - Combined to evaluate proximal & distal stricture

DIFFERENTIAL DIAGNOSIS

Urethral trauma

- Stretching or narrowing; ± pelvic fractures, fragments
- ± Extravasation from a partial or complete tear
- Blunt trauma → partial or complete urethral tears → post-traumatic stricture

Urethral tumor

- Long, irregular narrowing ± fistula

DDx: Mimics of Urethral Stricture

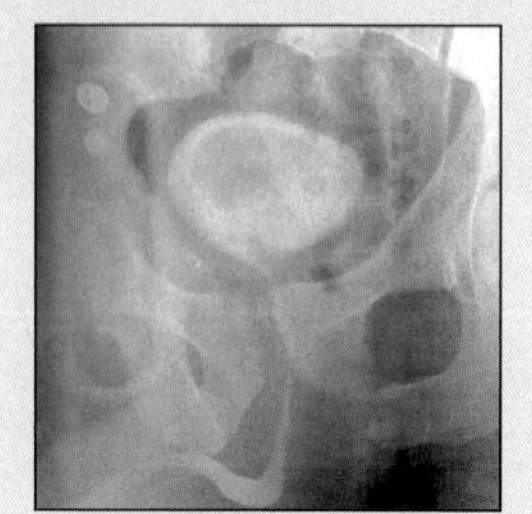

Trauma, Stretching

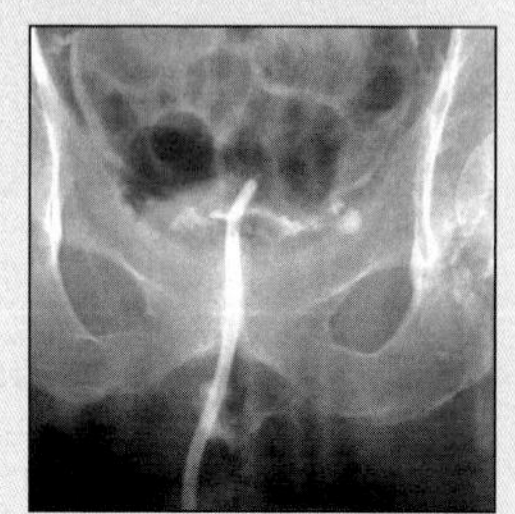

Urethral Tear

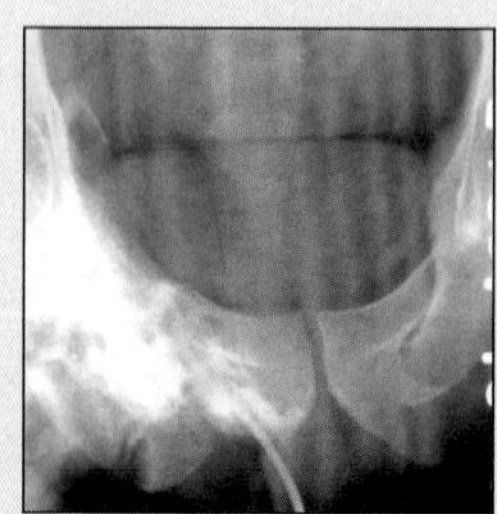

Urethral Tear

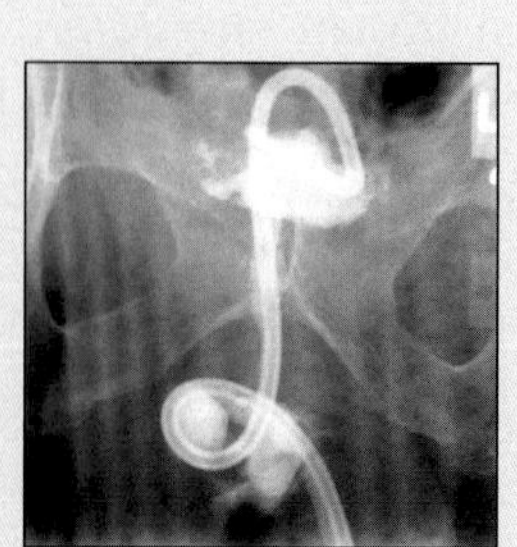

Urethral Tear

URETHRAL STRICTURE

Key Facts

Terminology
- A narrowing of urethra due to scar tissue or tumor

Imaging Findings
- Best diagnostic clue: Focal narrowing of urethra with proximal dilation
- Focal, asymmetrical indentations
- Long segmental, tubular or beaded constrictions
- Normal cone shape of proximal bulb
- Retrograde urethrogram and voiding urethrography

Top Differential Diagnoses
- Urethral trauma
- Urethral tumor

Diagnostic Checklist
- Clinical presentation (e.g., trauma, infection)
- Evaluate proximal and distal to stricture

- Intraluminal filling defect (rare)
- Primary neoplasm (rare): Squamous cell carcinoma from bulbar or bulbomembranous portions
- Secondary neoplasm, in or near urethra
  - Transitional cell carcinoma, melanoma, sarcoma, adenocarcinoma, Cowper gland adenocarcinoma
  - Metastases: Prostate, bladder, colon, rectum

PATHOLOGY

General Features
- Etiology
  - Idiopathic
  - Iatrogenic trauma (most common)
    - Instrumentation: Cystoscopy
    - Urethral catheters: Straight, indwelling, rubber
    - Surgery: Transurethral resection of prostate, urethroplasty
    - Irradiation: Treat uterine cervical cancer
  - Chemical trauma: Injection of corrosive substances
  - Infection: Gonorrhea, tuberculosis, syphilis
  - Inflammatory disease: Reiter syndrome, amyloidosis
  - Pathogenesis
    - Urethritis or periurethritis → fibrotic changes → stricture

CLINICAL ISSUES

Presentation
- Most common signs/symptoms: Obstructive symptoms: Diminished force of stream, ↓ urine flow rate, prolonged voiding, frequency, nocturia

Demographics
- Age: Any age, but incidence ↑ with age
- Gender: Male > female

Natural History & Prognosis
- Complications: Obstruction, abscess, fistula
- Prognosis: Very good

Treatment
- Use one or more procedures
  - Dilatation with catheters, sounds or balloons
  - Internal or external urethrotomy
  - Excision of stricture with reanastomosis
  - Marsupialization of urethra with urethroplasty
  - Pedicle flap repair
  - Full-thickness skin graft urethroplasty
  - UroLume endourethral prosthesis

DIAGNOSTIC CHECKLIST

Consider
- Clinical presentation (e.g., trauma, infection)

Image Interpretation Pearls
- Evaluate proximal and distal to stricture

SELECTED REFERENCES

1. Gallentine ML et al: Imaging of the male urethra for stricture disease. Urol Clin North Am. 29(2):361-72, 2002
2. Morey AF et al: Sonographic staging of anterior urethral strictures. J Urol. 163(4):1070-5, 2000
3. Gupta S et al: Sonourethrography in the evaluation of anterior urethral strictures: correlation with radiographic urethrography. J Clin Ultrasound. 21(4):231-9, 1993
4. McCallum RW et al: The radiologic assessment of iatrogenic urethral injury. J Can Assoc Radiol. 36(2):122-6, 1985
5. Cass AS: Urethral injury in the multiple-injured patient. J Trauma. 24(10):901-6, 1984
6. Singh M et al: The pathology of urethral stricture. J Urol. 115(6):673-6, 1976
7. Shaver WA et al: Changes in the male urethra produced by instrumentation for transurethral resection of the prostate. Radiology. 116(3):623-6, 1975
8. Lapides J et al: Usefulness of retrograde urethrography in diagnosing strictures of the anterior urethra. J Urol. 100(6):747-50, 1968

IMAGE GALLERY

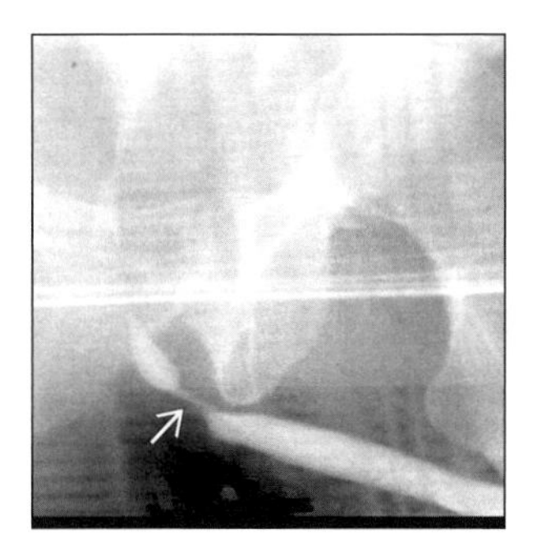

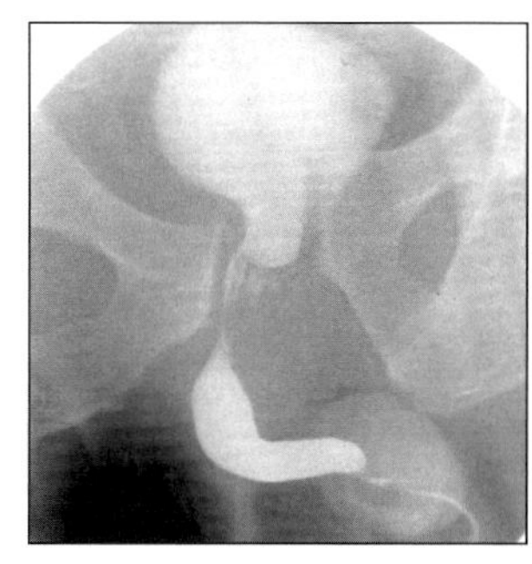

(Left) *Retrograde urethrogram shows stricture (arrow) of membranous urethra.* ***(Right)*** *Retrograde urethrogram shows stricture of prostatic urethra following pelvic fracture.*

HYDROCELE

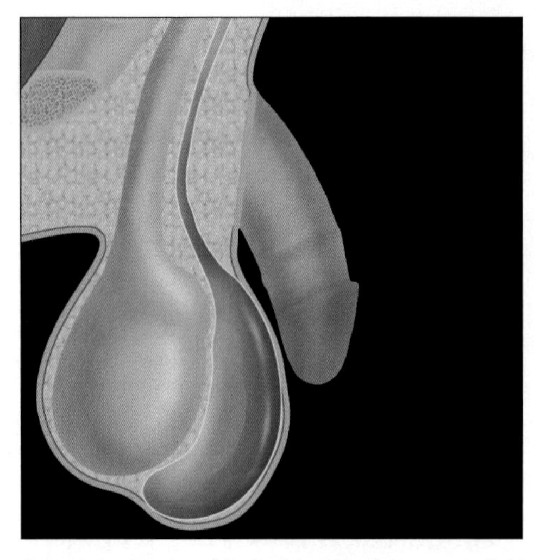

Schematic drawing of hydrocele. Note large serous fluid collection anterior to testis in tunica vaginalis.

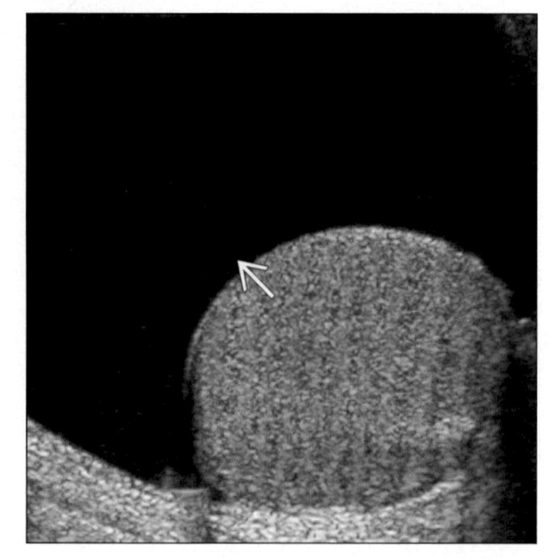

Transverse grayscale scan of large hydrocele. Note anechoic collection anterior and lateral to normal testis (arrow).

TERMINOLOGY

Definitions

- Congenital or acquired serous fluid contained within layers of tunica vaginalis

IMAGING FINDINGS

General Features

- Best diagnostic clue: Scrotal fluid collection surrounding testis on US except for "bare area" where tunica vaginalis does not cover testis & is attached to epididymis
- Location: Tunica vaginalis, located anterolateral aspect of testis
- Morphology: Simple fluid collection

MR Findings

- T1WI: Low signal fluid collection
- T2WI: High signal consistent with serous fluid

Ultrasonographic Findings

- Real Time: Crescentic anechoic fluid collection surrounding testis
- Color Doppler: Avascular on color Doppler

Imaging Recommendations

- US is modality of choice

DIFFERENTIAL DIAGNOSIS

Pyocele

- Septated fluid with low-level echoes
- Associated with epididymitis, intrascrotal abscess and clinical signs of inflammation

Spermatocele

- Retention cyst of tubules connecting rete testis to head of epididymis
- Cystic mass in head of epididymis
- Large spermatoceles have low level echoes and septations
- Displacement of testis inferiorly

Hematocele

- Complex echogenic fluid in tunica vaginalis on ultrasound
- Associated with trauma, torsion and infarct

Scrotal Hernia

- Bowel or echogenic omentum seen within scrotum due to indirect inguinal hernia

PATHOLOGY

General Features

- General path comments

DDx: Scrotal Lesions Mimicking Hydrocele

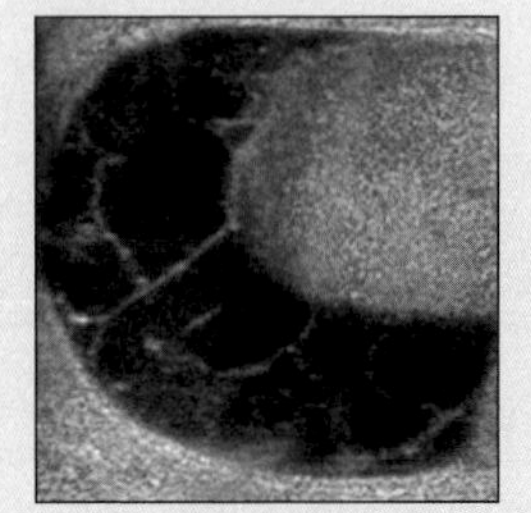

Pyocele

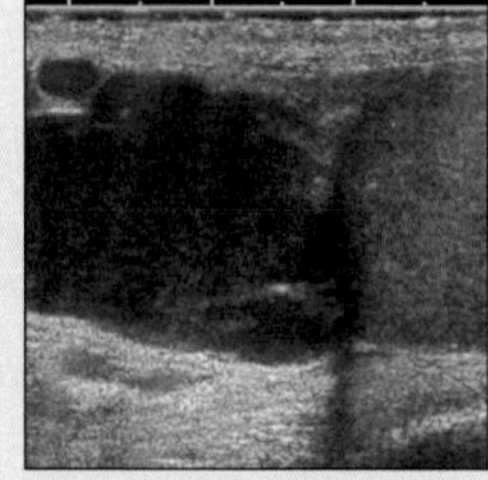

Spermatocele

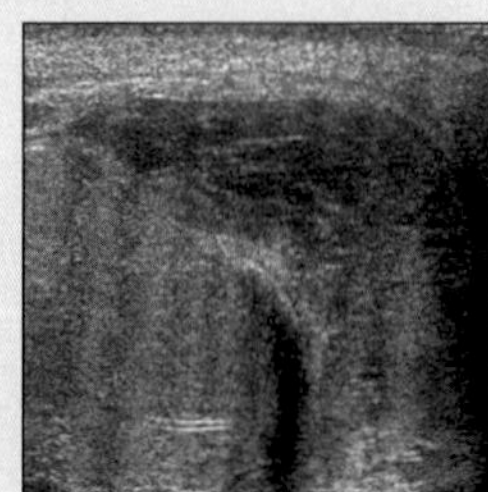

Hematocele

HYDROCELE

Key Facts

Imaging Findings
- Best diagnostic clue: Scrotal fluid collection surrounding testis on US except for "bare area" where tunica vaginalis does not cover testis & is attached to epididymis
- US is modality of choice

Top Differential Diagnoses
- Pyocele
- Spermatocele
- Hematocele
- Scrotal Hernia

Clinical Issues
- Excellent with surgical repair

Diagnostic Checklist
- Anechoic fluid collection along anterolateral aspect of testis

  - Simple serous fluid collection within tunica vaginalis
  - Chronic cases show thickened layers of tunica and septation
- Etiology
  - Embryology-anatomy
    - Congenital or communicating hydrocele due to failure of processus vaginalis to close
    - Congenital occurrance in 6% of male term infants
    - Secondary occurrence in adults due to epididymitis or surgery for varicocele
- Epidemiology: 10% of testicular tumors have associated hydrocele

Gross Pathologic & Surgical Features
- Serous fluid collection

Staging, Grading or Classification Criteria
- Congenital: Due to incomplete closure of tunica vaginalis
- Acquired: Epididymitis, torsion, trauma; rarely tumor

CLINICAL ISSUES

Presentation
- Most common signs/symptoms
  - Asymptomatic scrotal mass
  - Other signs/symptoms
    - Mass "transilluminates" when evaluated with light source
- Clinical profile
  - Painless scrotal mass
  - Transilluminating fluid collection
  - Chronic hydrocele with thick walls
    - May not transilluminate

Demographics
- Age
  - All ages
    - Pediatric patients invariably have congenital form of hydrocele with persistent communication with peritoneal cavity
- Gender: Male

Natural History & Prognosis
- Excellent with surgical repair

Treatment
- Surgical resection of hydrocele sac with oversewing of edges

DIAGNOSTIC CHECKLIST

Consider
- Spermatocele

Image Interpretation Pearls
- Anechoic fluid collection along anterolateral aspect of testis

SELECTED REFERENCES

1. Belman AB: Abdominoscrotal hydrocele in infancy: A review and presentation of the scrotal approach for correction. J Urol. 165(1): 225-7, 2001
2. Kapur P et al: Pediatric hernias and hydroceles. Pediatr Clin North Am. 45(4): 773-9, 1998
3. Rathour DP et al: Sonography of abdomino-scrotal hydrocele. Australas Radiol. 38(4): 282-3, 1994
4. Fowler RC et al: Scrotal ultrasonography: a clinical evaluation. Br J Radiol. 60(715):649-54, 1987
5. Krone KD et al: Scrotal ultrasound. Radiol Clin North Am. 23(1):121-39, 1985

IMAGE GALLERY

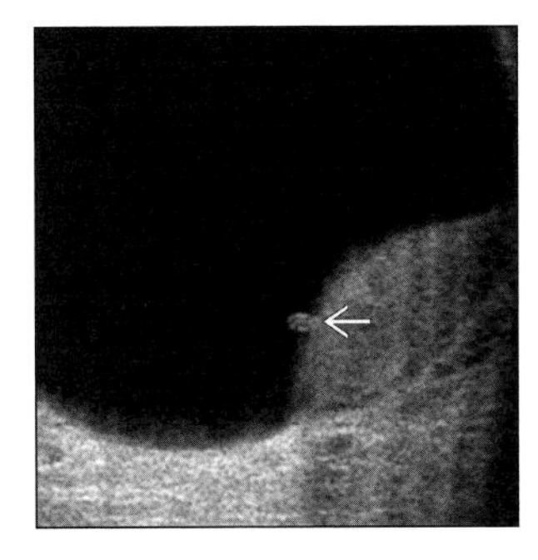
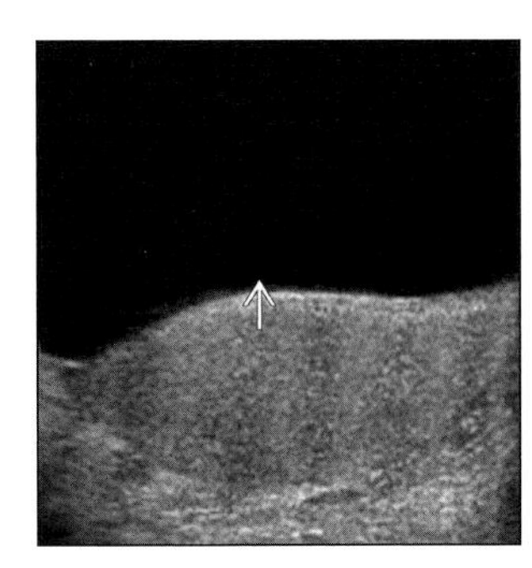

*(**Left**) Sagittal grayscale sonogram of large hydrocele in pediatric patient. Note fluid surrounding appendix testis (arrow). (**Right**) Transverse sonogram of large hydrocele in pediatric patient. Note siimple fluid collection anterior to normal testis (arrow).*

VARICOCELE

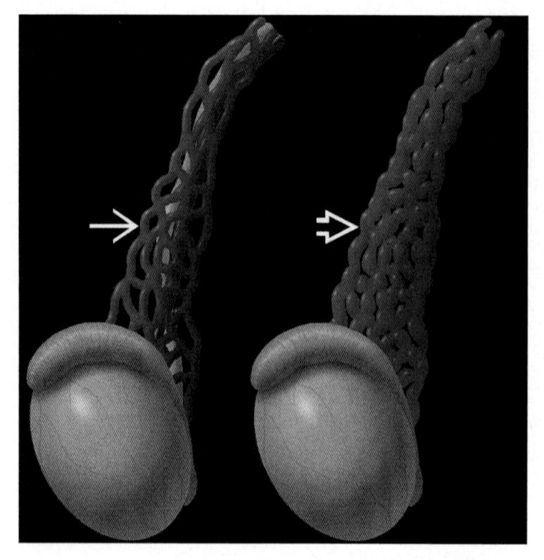

Anatomic drawing of varicocele. On left side, note normal veins of pampiniform plexus (arrow). Right side of image demonstrates dilated veins representing varicocele (open arrow).

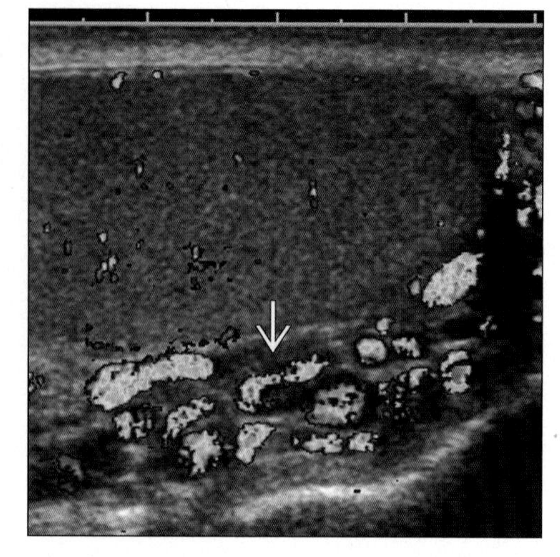

Sagittal color Doppler sonogram of left varicocele. Note multiple color-coded venous channels within epididymis (arrow).

TERMINOLOGY

Definitions

- Dilatation of veins of pampiniform plexus > 2-3 mm in diameter due to retrograde flow in internal spermatic vein

IMAGING FINDINGS

General Features

- Best diagnostic clue
  - Dilated serpiginous veins behind superior pole of testis on color Doppler US
  - Distention due to retrograde flow with Valsalva
- Location: Dilated veins in cremasteric plexus, vein of vas deferens and internal spermatic vein
- Size: Normally veins in pampiniform plexus are ≤ 2 mm; varicocele diagnosed when multiple veins are > 2-3 mm and increased in size with Valsalva
- Morphology: Tortuous vascular channels representing dilated veins

Ultrasonographic Findings

- Real Time: Tubular serpiginous vessels posterior to testis on grayscale
- Color Doppler: Prominent color flow within vessels with Valsalva due to retrograde flow

Other Modality Findings

- Catheter venography via retrograde injection in testicular vein demonstrates dilated venous channels

Imaging Recommendations

- Best imaging tool: US with color Doppler
- Protocol advice: Resting and Valsalva color Doppler images of epididymis

DIFFERENTIAL DIAGNOSIS

Tubular ectasia/rete testis

- Normal variant of dilated seminiferous tubules in mediastinum of testis
- No flow on color Doppler
- May be associated with spermatocele

Tumor

- Focal intratesticular mass on grayscale images
- Abnormal pattern of vasculature on color Doppler

Epididymitis

- Enlarged hypoechoic epididymis with increased flow on color Doppler

Testicular torsion

- Absent or decreased flow to testis on color Doppler
- Hypoechoic testis in late stage

DDx: Testicular Lesions Mimicking Varicocele

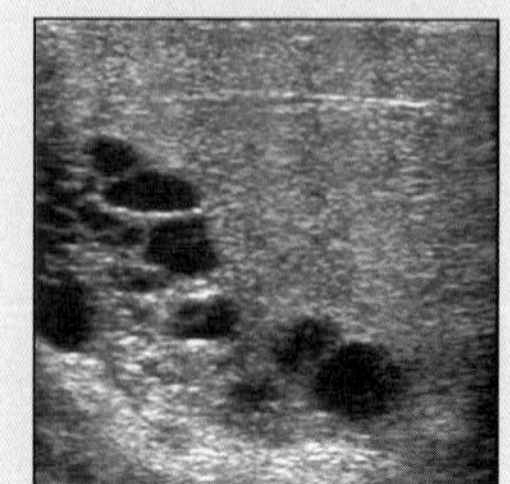

Rete Testis

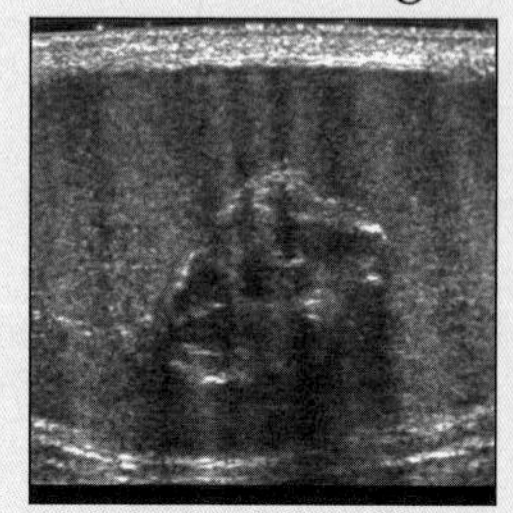

Testicular Tumor

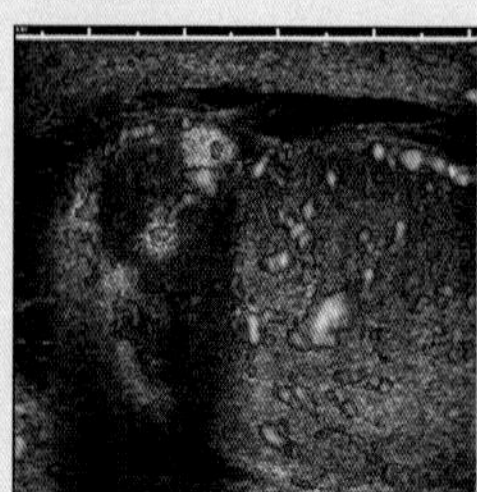

Epididymitis

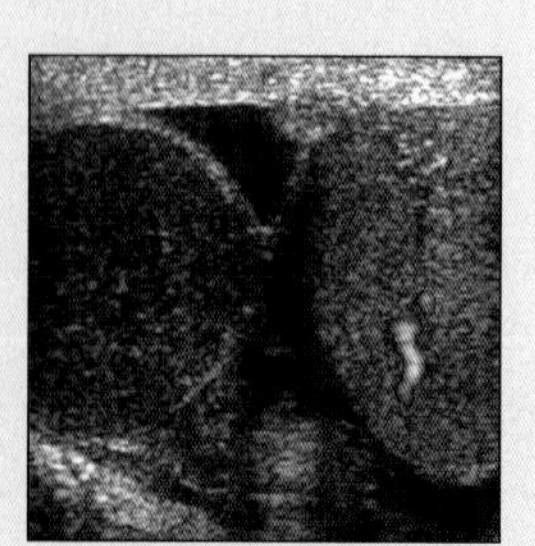

Torsion

VARICOCELE

Key Facts

Terminology
- Dilatation of veins of pampiniform plexus > 2-3 mm in diameter due to retrograde flow in internal spermatic vein

Imaging Findings
- Best imaging tool: US with color Doppler
- Protocol advice: Resting and Valsalva color Doppler images of epididymis

Pathology
- Most frequent cause of male infertility

Clinical Issues
- Clinical profile: Majority (80%) are left-sided; bilateral in 15% of patients

Diagnostic Checklist
- Left renal vein occlusion by tumor in elderly male

PATHOLOGY

General Features
- General path comments: Dilated veins within pampiniform plexus
- Etiology
  - Primary: Incompetent venous valve
  - Secondary: Venous obstruction or invasion of left renal vein by renal mass, nodes or adrenal tumor
- Epidemiology
  - 10-15% of men in US have varicoceles
  - Subclinical varicocele in 40-75% of infertile men
  - Most frequent cause of male infertility
- Associated abnormalities
  - Low sperm count
  - Secondary (reactive) varicoceles caused by retroperitoneal mass obstructing testicular veins

Gross Pathologic & Surgical Features
- Dilated veins within pampiniform plexus

Staging, Grading or Classification Criteria
- Primary
  - Idiopathic (incompetent valves) - 98% on left
  - Most common cause of correctable infertility
- Secondary
  - Obstruction of left renal vein by mass or renal tumor

CLINICAL ISSUES

Presentation
- Most common signs/symptoms
  - Infertility
  - Vague scrotal discomfort or pressure, primarily when standing
- Clinical profile: Majority (80%) are left-sided; bilateral in 15% of patients

Demographics
- Age: > 15 yrs
- Gender: Male

Natural History & Prognosis
- Excellent prognosis in treated cases
- Results for increased fertility have been mixed, may not be as effective as previously thought

Treatment
- Catheter embolization if symptomatic or causing low sperm count
- Surgical ligation

DIAGNOSTIC CHECKLIST

Consider
- Left renal vein occlusion by tumor in elderly male

Image Interpretation Pearls
- Valsalva essential for diagnosis of small varicoceles

SELECTED REFERENCES

1. Forti G et al: Varicocele and infertility. J Endocrinol Invest. 26(6):564-9, 2003
2. Evers JL et al: Assessment of efficacy of varicocele repair for male subfertility: a systematic review. Lancet. 361(9372):1849-52, 2003
3. Naughton CK et al: Pathophysiology of varicoceles in male infertility. Hum Reprod Update. 7(5): 473-81, 2001
4. Munden MM et al: Scrotal pathology in pediatrics with sonographic imaging. Curr Probl Diagn Radiol. 29(6): 185-205, 2000
5. Cornud F et al: Varicocele: Strategies in diagnosis and treatment. Eur Radiol. 9(3): 536-45, 1999

IMAGE GALLERY

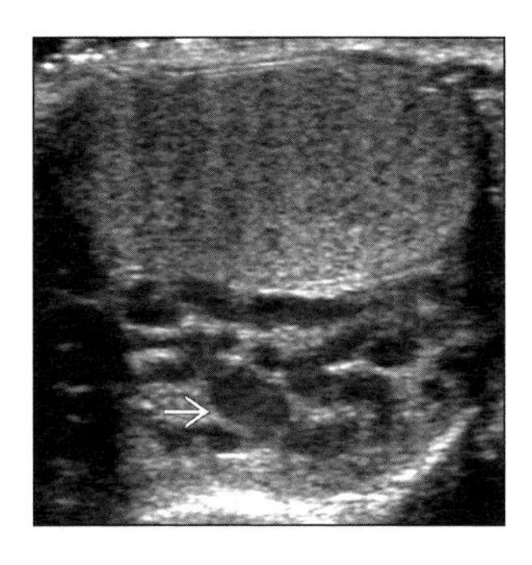

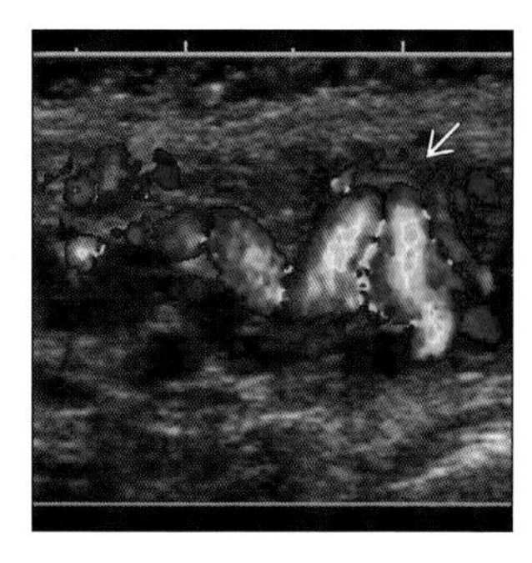

*(**Left**) Grayscale sagittal sonogram of varicocele. Note hypoechoic dilated venous channels within epididymis (arrow). (**Right**) Sagittal color Doppler sonogram of varicocele. Note numerous color-coded veins along spermatic cord (arrow).*

TESTICULAR TORSION

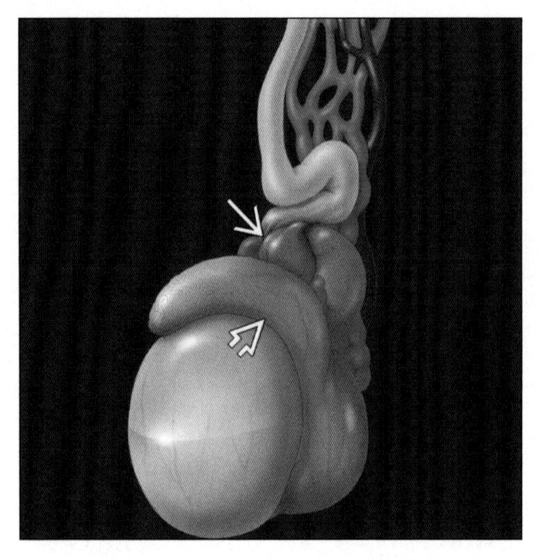

Anatomic drawing of testicular torsion. Note twisted cord (arrow) and enlarged epididymis (open arrow).

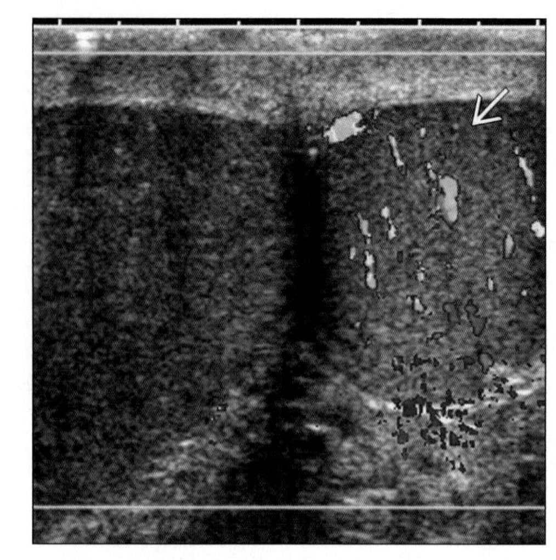

Transverse color Doppler sonogram of testicular torsion. Note normal flow to left testis (arrow) and absence of flow to the ischemic right testis.

TERMINOLOGY

Abbreviations and Synonyms

- Torsion, late or "missed" torsion, acute scrotum

Definitions

- Spontaneous or traumatic twisting of testis & spermatic cord within scrotum, resulting in vascular occlusion/infarction

IMAGING FINDINGS

General Features

- Best diagnostic clue: Decreased or absent blood flow to testicle on color Doppler US
- Location: Unilateral in 95% of patients

Ultrasonographic Findings

- Real Time
  - May be entirely normal
  - Enlarged testicle and epididymis, heterogeneous echotexture, most often decreased echogenicity; secondary hydrocele
  - Intratesticular necrosis, hemorrhage or fragmentation if delayed diagnosis
  - "Spiral" twist of spermatic cord from inguinal canal to testis
- Color Doppler: Absent or decreased flow; sensitivity 80-90%; 10% of patients with early or partial torsion have normal exam

Nuclear Medicine Findings

- Tc99m pertechnetate: Dynamic flow imaging at 2-5 second intervals for 1 minute (vascular phase); 5 minute intervals for tissue phase; sensitivity 80-90%

Imaging Recommendations

- Best imaging tool: US with high-frequency linear transducer & color Doppler; power Doppler preferable
- Protocol advice: Power Doppler with comparison to contralateral normal testis

DIFFERENTIAL DIAGNOSIS

Epididymo-orchitis

- Enlarged hypoechoic epididymis with increased flow on color Doppler

Testicular tumor

- Focal mass on US; abnormal flow within tumor

Testicular trauma

- Hematocele, irregular contours, heterogeneous parenchymal echogenicity

DDx: Testicular Lesions Mimicking Torsion

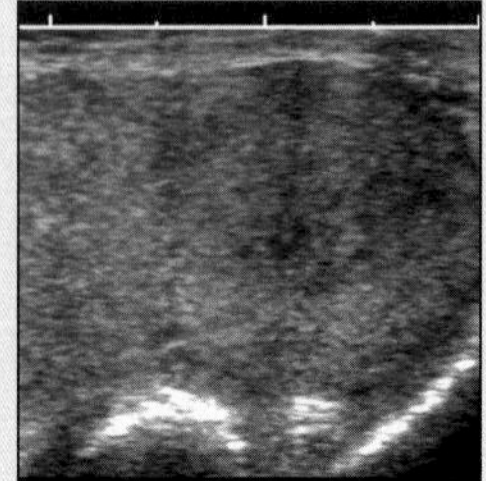

Carcinoma

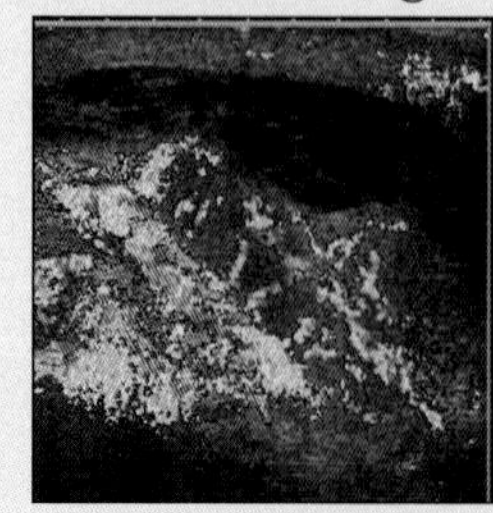

Epididymo-Orchitis

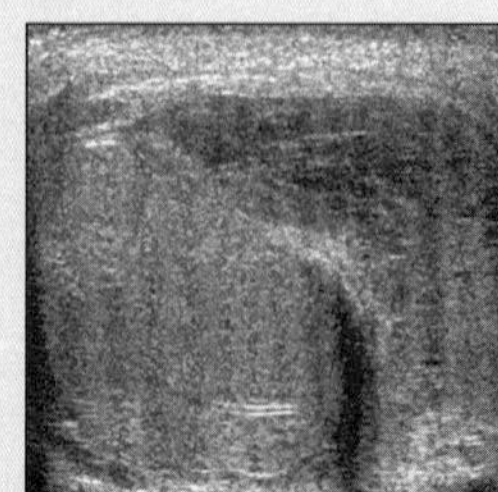

Testicular Trauma

TESTICULAR TORSION

Key Facts

Imaging Findings
- Best diagnostic clue: Decreased or absent blood flow to testicle on color Doppler US
- Best imaging tool: US with high-frequency linear transducer & color Doppler; power Doppler preferable

Top Differential Diagnoses
- Epididymo-orchitis
- Testicular tumor
- Testicular trauma

Clinical Issues
- Clinical profile: Male child with acute scrotal pain

Diagnostic Checklist
- Pitfall of normal color Doppler flow in early or partial torsion; normal US doesn't exclude early torsion

PATHOLOGY

General Features
- General path comments
  - Varying degrees of ischemic necrosis & fibrosis depending on duration of symptoms
  - Embryology-anatomy: Deficient testicular fixation related to tunica vaginalis & gubernaculum ("bell clapper" deformity); testicle rotates within scrotum and twists spermatic cord
- Etiology: Most occur spontaneously; rarely occurs traumatically
- Epidemiology: Infant & adolescent boys most often affected

Gross Pathologic & Surgical Features
- Purple, edematous, ischemic testicle, may rapidly re-perfuse when manually de-torsed

Microscopic Features
- Hemorrhagic, interstitial edema; necrosis

Staging, Grading or Classification Criteria
- Previously classified acute, subacute, or delayed based on duration of symptoms

CLINICAL ISSUES

Presentation
- Most common signs/symptoms
  - Acute scrotal/inguinal pain; swollen, erythematous hemiscrotum without recognized trauma
  - Other signs/symptoms: Low grade torsion may be tolerated for long periods
- Clinical profile: Male child with acute scrotal pain

Demographics
- Age: < 15 yrs
- Gender: Male

Natural History & Prognosis
- Surgical emergency: Testicular infarction if not treated promptly
- Testicular viability depends on degree of torsion (> 540° worse), duration of symptoms & rapid surgical intervention
- Unilateral testicular loss typically does not lead to infertility problems

Treatment
- Surgical exploration; de-torsion; bilateral orchidopexy if viable testicle
  - Non-viable testicle usually removed; higher risk of subsequent torsion on contralateral side

DIAGNOSTIC CHECKLIST

Consider
- Pitfall of normal color Doppler flow in early or partial torsion; normal US doesn't exclude early torsion

Image Interpretation Pearls
- Decreased or absent flow on power Doppler

SELECTED REFERENCES

1. Mernagh JR et al: Testicular torsion revisited. Curr Probl Diagn Radiol. 33(2):60-73, 2004
2. Dogra VS et al: Sonography of the scrotum. Radiology. 227(1):18-36, 2003
3. Kravchick S et al: Color Doppler sonography: its real role in the evaluation of children with highly suspected testicular torsion. Eur Radiol. 11(6):1000-5, 2001

IMAGE GALLERY

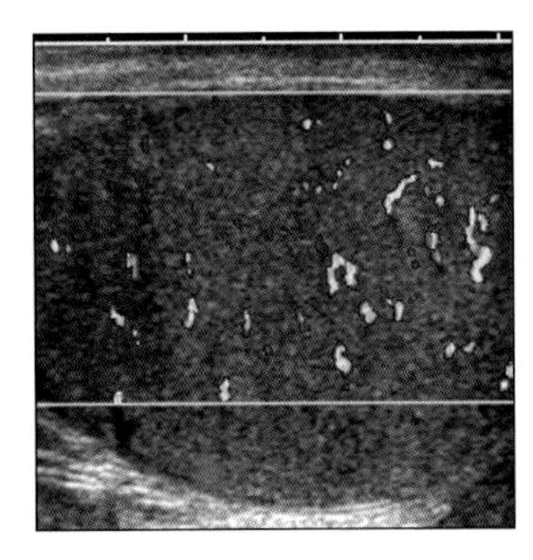

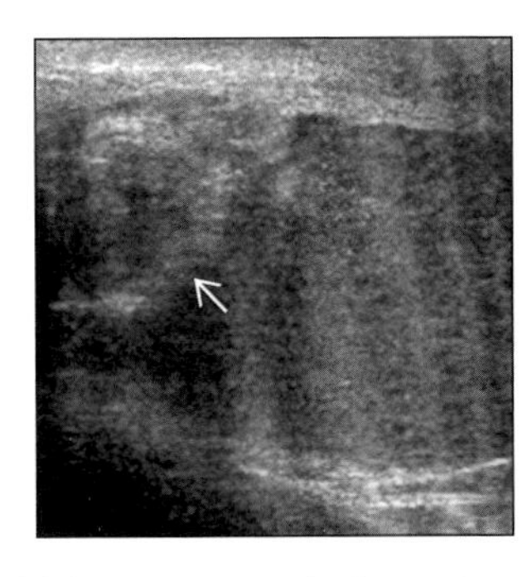

*(**Left**) Color Doppler sonogram of false-negative study for torsion. Note preserved flow to the testis on longitudinal scan. At surgery, testicle was twisted 180° and was only minimally ischemic. (**Right**) Twisted spermatic cord on sagittal grayscale US image in testicular torsion. Note "mass" (arrow) near upper pole of testis representing twisted spermatic cord and thickened epididymis due to torsion.*

BENIGN PROSTATIC HYPERTROPHY

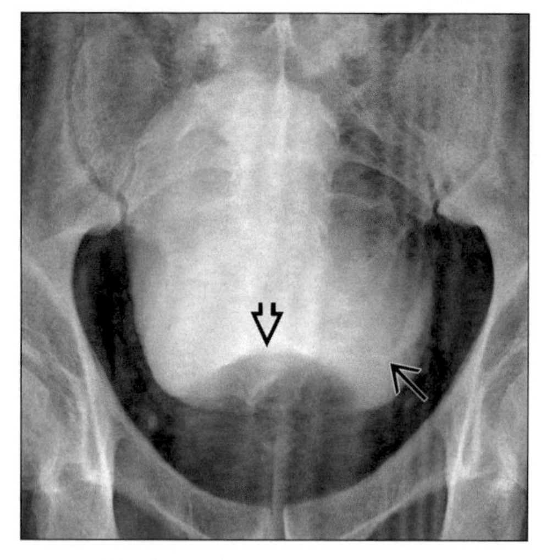

BPH on IVP. Note extrinsic impression (open arrow) on base of bladder with "fish hooking" of left ureter (arrow).

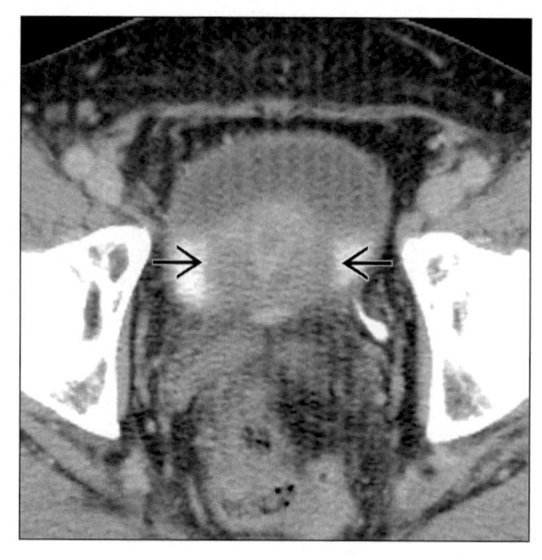

BPH on axial CECT. Note medial lobe hypertrophy of prostate pressing on base of bladder (arrows).

TERMINOLOGY

Abbreviations and Synonyms

- Benign prostatic hypertrophy (BPH), nodular hyperplasia

Definitions

- Enlargement of prostate from benign hyperplasic nodule (fibromyoadenomatous nodule)

IMAGING FINDINGS

General Features

- Best diagnostic clue: Enlarged prostate on CT, US or MRI with nodular hypertrophy in transitional or periurethral zone
- Location: Transition zone and periurethral zone proximal to verumontaneum; "lateral lobe" = 82%, median lobe = 12%
- Size: Variable; may be up to 10-12 cm
- Morphology: Rounded or lobulated soft tissue hypertrophy; nodules typically 60-100 gm

Radiographic Findings

- IVP: Extrinsic impression on base of bladder with "J hooking" or "fish hooking" of distal ureters

CT Findings

- NECT: Enlarged prostate; calcifications when prostate enlarged
- CECT: Enlarged prostate with extrinsic impression on base of bladder

MR Findings

- T1WI: Enlarged prostate
- T2WI: Low or heterogeneous signal nodular adenoma involving transition or periurethral zone; cannot distinguish BPH from carcinoma
- T1 C+
  - Data mixed on whether dynamic contrast enhancement may aid in differentiating BPH from carcinoma
  - Carcinoma in general has more rapid uptake of gadolinium on dynamic MRI

Ultrasonographic Findings

- Real Time
  - 80% demonstrate hypoechoic nodular enlargement of transition and periurethral zones; 10-20% of hyperplastic nodules are isoechoic or echogenic
  - Large residual urine volume post-voiding

Imaging Recommendations

- Best imaging tool: MRI, endorectal US
- Protocol advice: MRI: T2 axial and coronal images

DDx: Prostatic Lesions Mimicking BPH

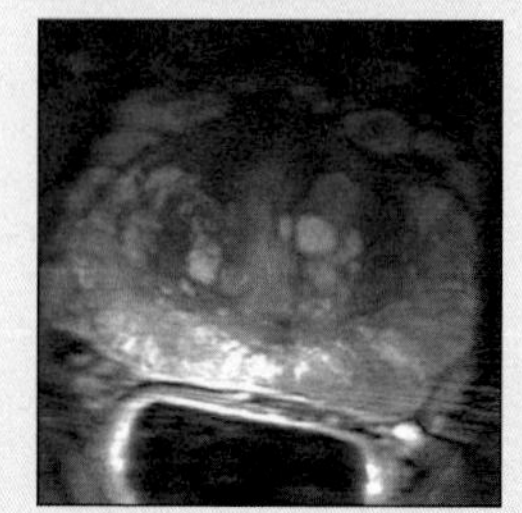

Prostate Carcinoma

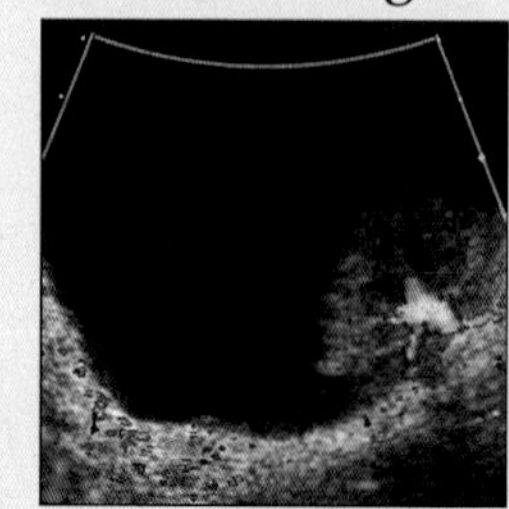

Bladder Carcinoma

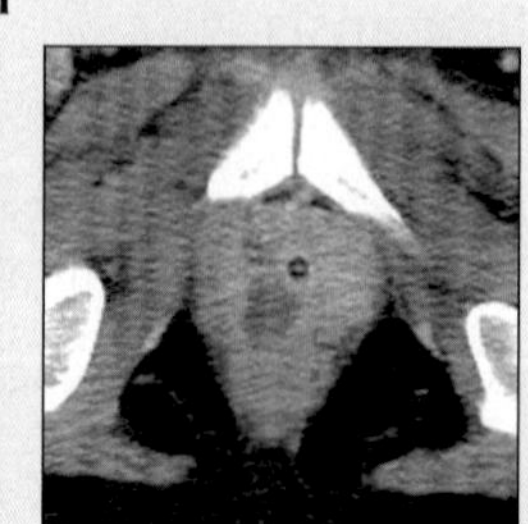

Prostate Abscess

BENIGN PROSTATIC HYPERTROPHY

Key Facts

Terminology
- Enlargement of prostate from benign hyperplasic nodule (fibromyoadenomatous nodule)

Imaging Findings
- Best diagnostic clue: Enlarged prostate on CT, US or MRI with nodular hypertrophy in transitional or periurethral zone

Top Differential Diagnoses
- Prostate carcinoma
- Prostatic abscess
- Bladder carcinoma

Pathology
- Epidemiology: 70% of men have BPH by age 70; 80% of men have BPH by age 80

DIFFERENTIAL DIAGNOSIS

Prostate carcinoma
- Typically involves peripheral zone (70%); hypoechoic on endorectal US with increased color Doppler flow; low signal mass on T2 MRI

Prostatic abscess
- Low attenuation fluid collection bulging capsule

Bladder carcinoma
- Posterior epithelial mass may simulate median lobe as polypoid mass

PATHOLOGY

General Features
- General path comments: Firm hypertrophied tissue
- Etiology: Stromal hyperplasia stimulated by normal action of dehydrotestosterone and growth factors (fibroblast growth factor, insulin-like growth factor)
- Epidemiology: 70% of men have BPH by age 70; 80% of men have BPH by age 80
- Associated abnormalities: Bladder wall hypertrophy with trabeculation and diverticula

Gross Pathologic & Surgical Features
- Enlarged, firm gland at prostatectomy

Microscopic Features
- Hyperplastic nodules due to glandular proliferation and/or fibrous or muscular proliferation of stroma

CLINICAL ISSUES

Presentation
- Most common signs/symptoms
  - Symptoms of retention and distended bladder; nocturnal dribbling, poor urethral stream
  - Other signs/symptoms: Symptom severity doesn't correlate strongly with size of gland on imaging
- Clinical profile: Acute retention with bladder outlet obstruction (BOO); enlarged prostate on rectal exam; may have elevated prostate-specific antigen

Demographics
- Gender: Male

Natural History & Prognosis
- May progress to BOO, hydronephrosis if severe
- May lead to urinary infection, gross hematuria

Treatment
- Options, risks, complications: Open surgery for gland > 80 gm; transurethral resection for smaller glands; medical therapy with alpha-adrengeric blockers, finasteride for mild symptoms

DIAGNOSTIC CHECKLIST

Consider
- Prostate carcinoma

Image Interpretation Pearls
- Median lobe hypertrophy simulating bladder mass

SELECTED REFERENCES

1. Tubaro A et al: Investigation of benign prostatic hyperplasia. Curr Opin Urol. 13(1):17-22, 2003
2. Grossfeld GD et al: Benign prostatic hyperplasia: clinical overview and value of diagnostic imaging. Radiol Clin North Am. 38(1):31-47, 2000
3. Aarnink RG et al: Aspects of imaging in the assessment and follow up of benign prostatic hyperplasia. Curr Opin Urol. 9(1):21-9, 1999

IMAGE GALLERY

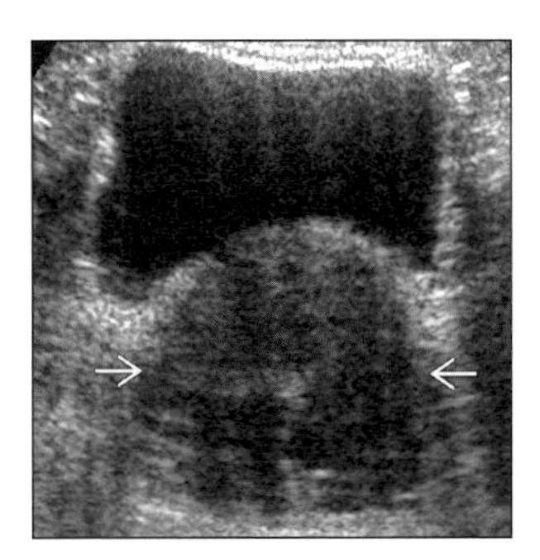
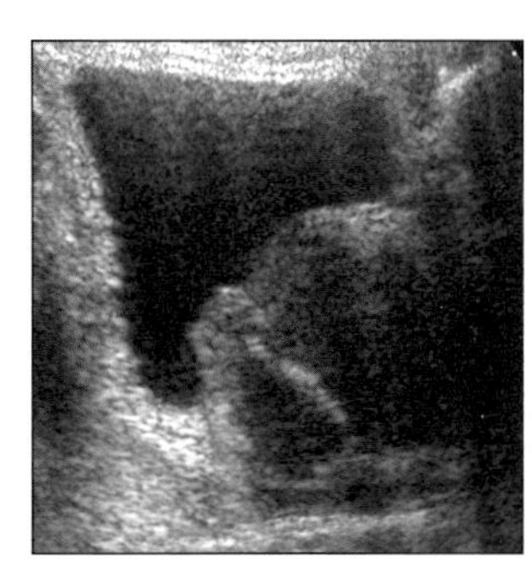

(Left) *Pre-void transverse sonogram of BPH. Note enlarged hypoechoic prostate gland (arrows).* ***(Right)*** *Sagittal post-void sonogram demonstrates no significant change in bladder volume, consistent with bladder outlet obstruction from BPH.*

SCROTAL TRAUMA

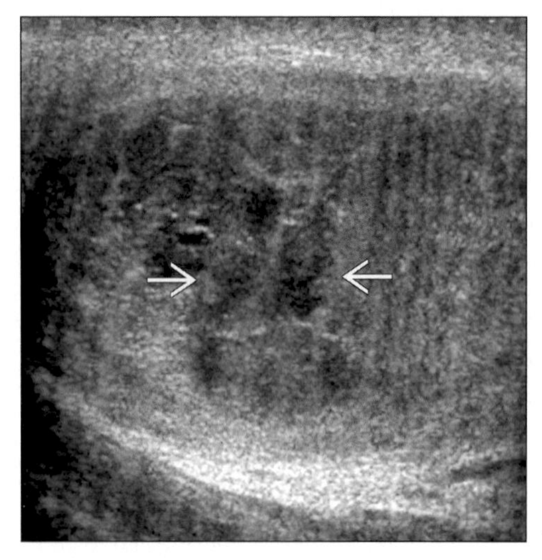

Grayscale sagittal sonogram of testicular fracture. Note hypoechoic fracture traversing testis (arrows).

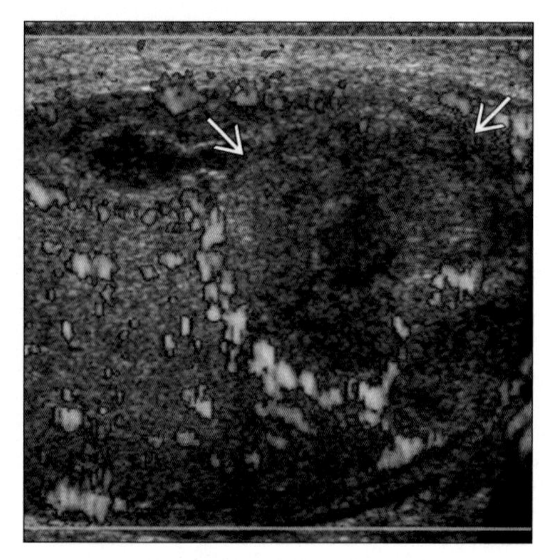

Sagittal power Doppler sonogram of testicular rupture with intraparenchymal hematoma. Note rounded hypoechoic hematoma (arrows) that is avascular.

TERMINOLOGY

Abbreviations and Synonyms

- Testicular rupture, fracture of testis

Definitions

- Laceration of tunica albugines, extrusion of testicular tissue into scrotal sac

IMAGING FINDINGS

General Features

- Best diagnostic clue: Heterogeneous parenchymal echogenicity of testis on sonography
- Morphology: Irregularity of testicular contour

Radiographic Findings

- Radiography: May have associated pelvic fractures

MR Findings

- T1WI: High signal acute hematoma
- T2WI: High signal hematoma
- T1 C+: Areas of injury are avascular post-contrast

Ultrasonographic Findings

- Real Time
  - Hematocele with echogenic fluid, linear stranding in tunica vaginalis
  - Abnormal echogenicity of testicular parenchyma
  - Epididymal enlargement
  - Discrete linear or irregular fracture plane within testis (17%)
  - Focal intraparenchymal testicular hematoma
- Color Doppler
  - Distorted intraparenchymal vascularity with interruption of vessels in area of injury
  - Avascular intraparenchymal hematoma
  - Post-traumatic epididymitis may demonstrate enlarged epididymis with increased flow

Imaging Recommendations

- Best imaging tool: High-resolution US (≥ 7.5 mHz)

DIFFERENTIAL DIAGNOSIS

Testicular tumor

- Mild to moderate pain with palpable testicular mass but no history of trauma
- Focal grayscale abnormality with abnormal interval vascularity, vascular displacement on color Doppler

Epididymo-orchitis

- Acute or chronic pain without history of trauma
- Enlarged hypoechoic epididymis with increased vascularity to epididymis and testis on color Doppler

DDx: Testicular Lesions Mimicking Scrotal Trauma

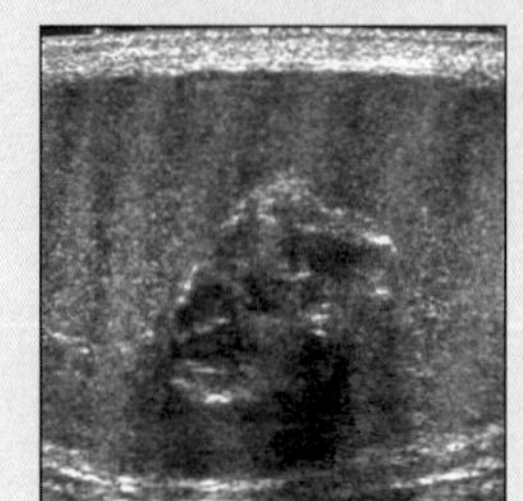

Testicular Tumor

Epididymo-orchitis

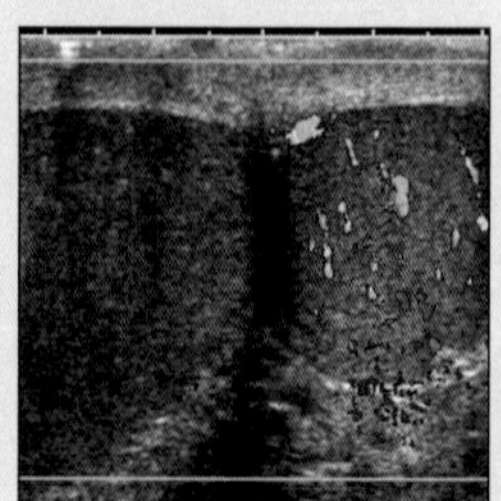

Testicular Torsion

Segmental Infarct

SCROTAL TRAUMA

Key Facts

Imaging Findings

- Best diagnostic clue: Heterogeneous parenchymal echogenicity of testis on sonography
- Best imaging tool: High resolution US (≥ 7.5 mHz)

Clinical Issues

- Most common signs/symptoms: Acute scrotal hematoma following blunt trauma
- Unless repaired within 72 hours, salvage rate only 45%

Diagnostic Checklist

- Consider isolated hematocele without rupture
- Irregularity of testicular contour; heterogeneous testicular parenchyma

Testicular torsion

- Decreased vascularity compared to normal testis

Segmental infarct

- Acute pain, no history of trauma
- Focal hypoechoic area avascular on color Doppler
- Increased incidence in patients with hypercoagulable states or advanced atherosclerosis, such as diabetics

PATHOLOGY

General Features

- General path comments
  - Capsular disruption; hematocele
  - Necrotic testicular parenchyma extrudes into tunica vaginalis
- Etiology: Blunt trauma impales scrotal contents to symphysis pubis or pubic rami
- Associated abnormalities: Pelvic fracture

Microscopic Features

- Intraparenchymal hemorrhage and necrotic lacerated tissue

Staging, Grading or Classification Criteria

- Extratesticular: Injury to scrotal wall, tunica vaginalis or tunical & testicular parenchyma
- Intratesticular: Injury to tunical and testicular parenchyma

CLINICAL ISSUES

Presentation

- Most common signs/symptoms: Acute scrotal hematoma following blunt trauma

Demographics

- Age: < 50 yrs
- Gender: Male

Natural History & Prognosis

- Unless repaired within 72 hours, salvage rate only 45%

Treatment

- Options, risks, complications: Surgery mandatory

DIAGNOSTIC CHECKLIST

Consider

- Consider isolated hematocele without rupture

Image Interpretation Pearls

- Irregularity of testicular contour; heterogeneous testicular parenchyma

SELECTED REFERENCES

1. Micallef M et al: Ultrasound features of blunt testicular injury. Injury. 32(1):23-6, 2001
2. Wessells H et al: Testicular trauma. Urology. 47(5):750, 1996
3. Simmons MZ: Re: Accuracy of ultrasound diagnosis after blunt testicular trauma. J Urol. 152(3):968-9, 1994
4. Patil MG et al: The value of ultrasound in the evaluation of patients with blunt scrotal trauma. Injury. 25(3):177-8, 1994
5. Altarac S: Management of 53 cases of testicular trauma. Eur Urol. 25(2):119-23, 1994

IMAGE GALLERY

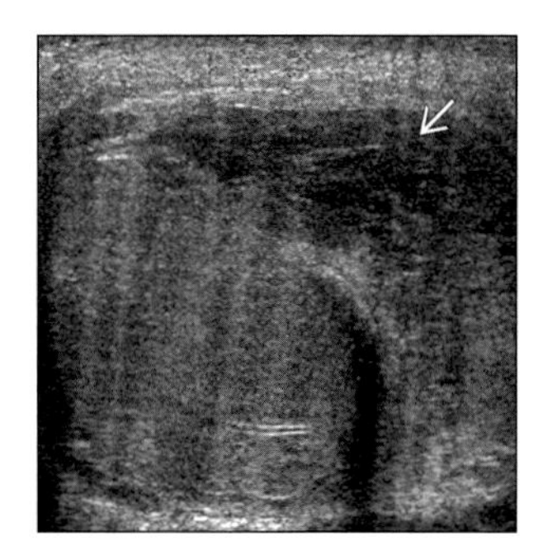

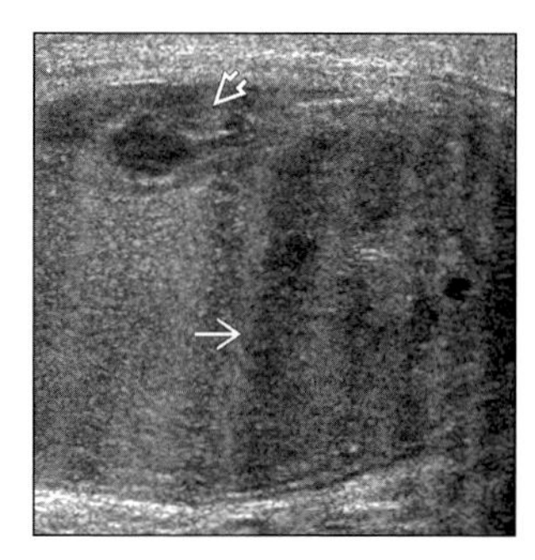

(Left) *Testicular fracture with large hematoma. Transverse grayscale image of scrotum demonstrates large complex fluid collection with septations representing hematocele (arrow).* ***(Right)*** *Testicular fracture. Sagittal grayscale sonogram demonstrates large hypoechoic area representing intraparenchymal hematoma (arrow) with adjacent hematocele (open arrow).*

GONADAL STROMAL TUMORS

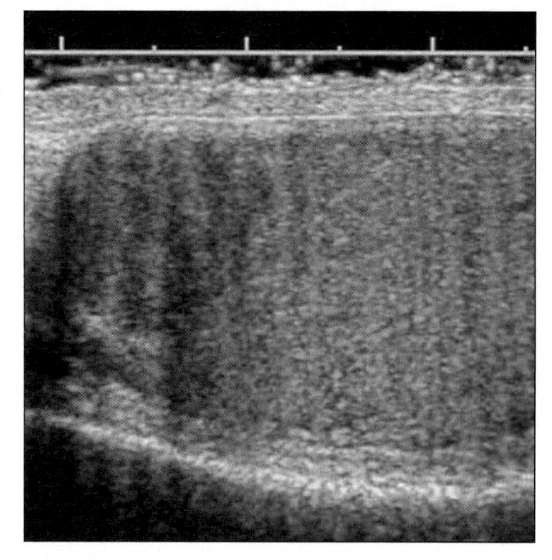

Sagittal US shows small, round, well circumscribed hypoechoic mass at upper pole of testis representing Leydig cell tumor (Courtesy B.F. King, MD).

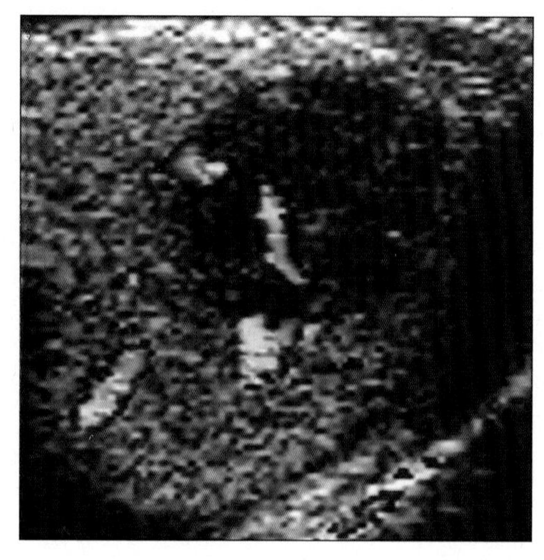

Axial color Doppler US in a middle-aged man shows a hypoechoic mass with internal color flow, which proved to be Leydig cell tumor at pathology.

TERMINOLOGY

Abbreviations and Synonyms

- Gonadal stromal tumors (GSTs) = nongerm cell tumors = interstitial cell tumors = sex cord tumors

Definitions

- Neoplasms arising from non-germ cell elements
- Leydig cell tumors (LCT): Arise from interstitial cells
- Sertoli cell tumors (SCT): Arise from sustentacular cells lining seminiferous tubules
- Granulosa cell tumor: Rare benign tumor
- Gonadoblastoma: Contains both stromal and germ cell elements

IMAGING FINDINGS

General Features

- Location: Bilateral in 3%
- Size
  - Usually < 3 cm
  - Malignant tumors: Usually > 5 cm
- Morphology: Well-circumscribed, round-lobulated

MR Findings

- T2WI
  - Low signal intratesticular mass +/- high signal fibrous capsule rim
  - High signal intensity foci secondary to central scars

Ultrasonographic Findings

- Real Time
  - Indistinguishable from germ cell tumors
  - LCT: Small solid hypoechoic intratesticular mass
    - Larger tumors: Hemorrhage or necrosis
  - SCT: Solid and cystic components, +/- calcification
- Color Doppler: Internal/perinodular flow

Imaging Recommendations

- Best imaging tool: Ultrasound

DIFFERENTIAL DIAGNOSIS

Testicular germ cell tumors

- Indistinguishable from stromal tumors on US

Testicular metastases, lymphoma, leukemia

- Often multiple; otherwise indistinguishable

Focal orchitis

- Symptomatic with pain, tenderness

Intratesticular hematoma

- No internal color flow

DDx: Intratesticular Masses

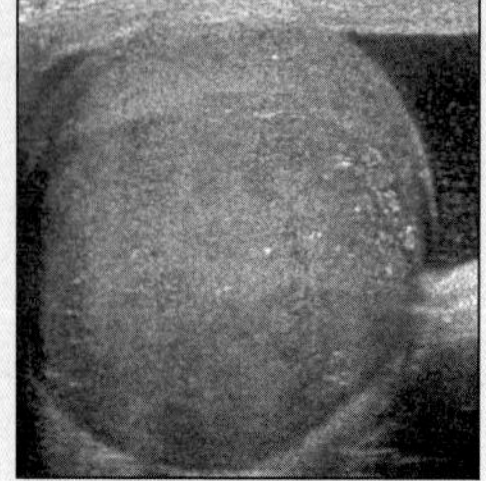

Germ Cell Tumor

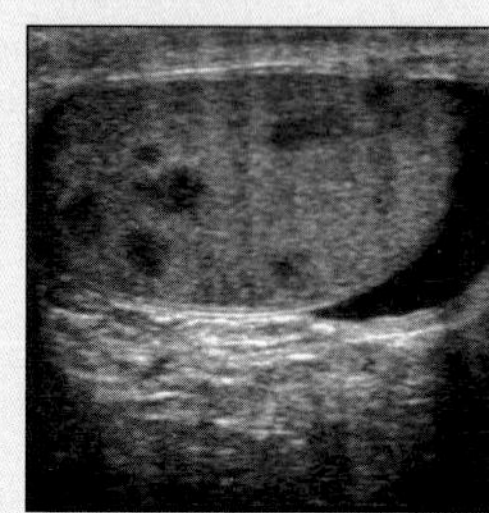

Lymphoma

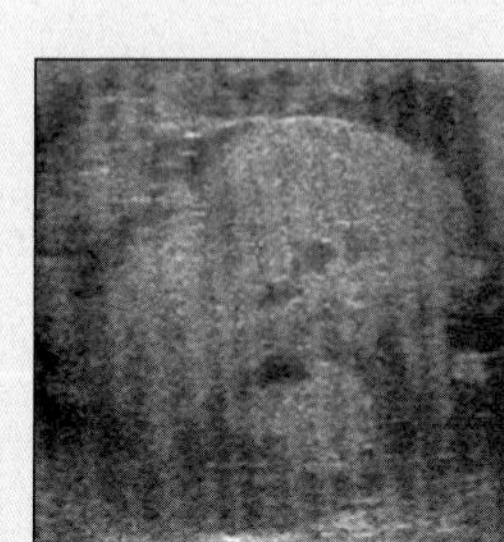

Hematoma

GONADAL STROMAL TUMORS

Key Facts

Terminology
- Leydig cell tumors (LCT): Arise from interstitial cells
- Sertoli cell tumors (SCT): Arise from sustentacular cells lining seminiferous tubules

Imaging Findings
- Indistinguishable from germ cell tumors

Pathology
- Represent 4-5% of all testis tumors; 10-30% occur in childhood
- Leydig cell tumors: 90% benign
- Sertoli cell tumors (SCT): 85-90% benign

Clinical Issues
- Precocious virilization in children
- Gynecomastia, impotence, ↓ libido in adults

PATHOLOGY

General Features
- General path comments
  - Represent 4-5% of all testis tumors; 10-30% occur in childhood
    - Leydig cell tumors: 3% of all testis tumors; most common stromal tumor
    - Sertoli cell tumors: 1% of all testis tumors
  - Leydig cell tumors: 90% benign
    - May produce testosterone
    - Malignant potential difficult to assess histologically
    - Best criterion for malignancy: Presence of metastases
  - Sertoli cell tumors (SCT): 85-90% benign
    - May produce estrogen or Mullerian inhibiting factor
- Epidemiology: Represent 10-13% of testicular neoplasms in pediatrics
- Associated abnormalities
  - Leydig cell tumor: Kleinfelter syndrome
  - SCT: Large cell calcifying subtype associated with Peutz-Jegher syndrome and Carney syndrome

Gross Pathologic & Surgical Features
- Well-circumscribed, solid, lobulated intratesticular mass

Microscopic Features
- Tumors may have single or multiple cell types
- LCT: Large cells with central round nuclei, small nucleoli and eosinophilic cytoplasm
  - Crystals of Reinke in 30-40% = rod-shaped, intracytoplasmic inclusion bodies
- SCT: Sheets of uniform cells with areas of tubule formation

CLINICAL ISSUES

Presentation
- Most common signs/symptoms
  - Painless testicular enlargement
  - 30% of patients have endocrinopathy secondary to testosterone or estrogen production by tumor
    - Precocious virilization in children
    - Gynecomastia, impotence, ↓ libido in adults
- Clinical profile: Malignant Leydig cell tumors: Elderly men

Demographics
- Age
  - LCT: Age 30-60; 25% occur before puberty
  - Malignant Leydig cell tumors only occur in adults
  - SCT: All age groups; 1/3 < age 12

Natural History & Prognosis
- Malignant tumors metastasize in same pattern as testicular germ cell tumors

Treatment
- Orchiectomy or testis-sparing surgery

DIAGNOSTIC CHECKLIST

Image Interpretation Pearls
- Consider stromal tumor in ddx of patient with endocrinopathy and intratesticular mass

SELECTED REFERENCES

1. Fernandez GC et al: MRI in the diagnosis of testicular Leydig cell tumour. Br J Radiol. 77(918):521-4, 2004
2. Dogra VS et al: Sonography of the scrotum. Radiology. 227(1):18-36, 2003
3. Woodward PJ et al: From the archives of the AFIP: tumors and tumorlike lesions of the testis: radiologic-pathologic correlation. Radiographics. 22(1):189-216, 2002

IMAGE GALLERY

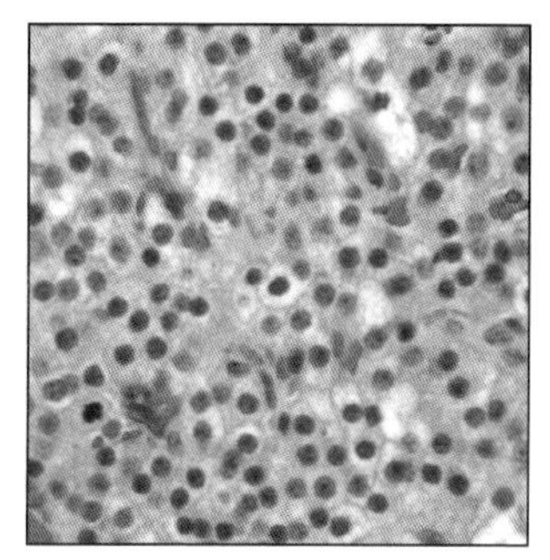

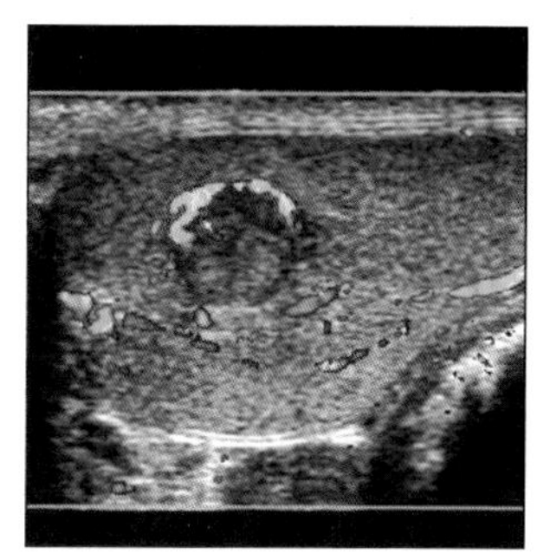

(Left) *Pathology shows sheet-like nested pattern of Leydig cell tumor. Cells are polygonal with eosinophilic cytoplasm and round nuclei. Crystals of Reinke are absent in this case (Courtesy T. Morgan, MD).* ***(Right)*** *Sagittal color Doppler US shows prominent vessel and internal color flow in a small, round, hypoechoic intratesticular mass. Pathology confirmed Leydig cell tumor (Courtesy B.F. King, MD).*

TESTICULAR CARCINOMA

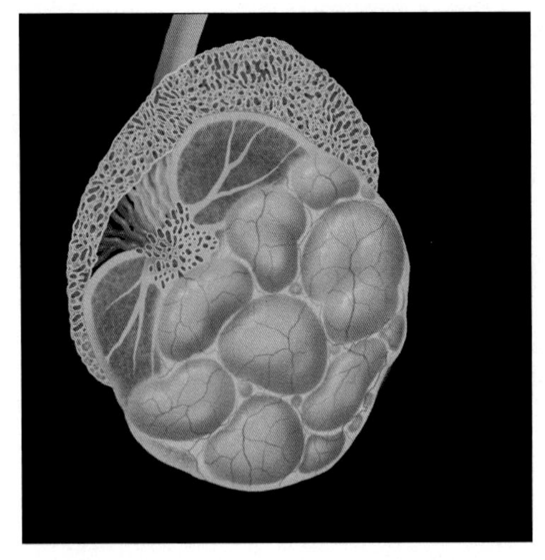

Schematic drawing of testicular carcinoma. Note heterogenous, solid mass arising from testicular parenchyma and distending tunica albuginea.

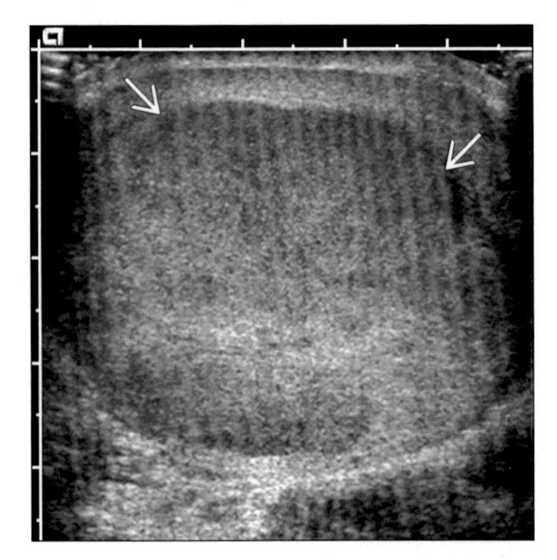

Sagittal grayscale sonogram of pure seminoma. Note uniform hypoechoic solid mass (arrows).

TERMINOLOGY

Abbreviations and Synonyms

- Germ cell tumor of testis

Definitions

- Malignant germ cell tumor of testis

IMAGING FINDINGS

General Features

- Best diagnostic clue: Discrete hypoechoic or mixed echogenic mass on US

CT Findings

- CECT
  - Low attenuation nodes
  - Even nodes < 1 cm suspicious if located in typical drainage areas; left renal hilus and retrocaval location on right
  - Residual low attenuation masses common in bulky disease
  - Useful for staging retroperitoneal, nodal and pulmonary metastases
  - Helpful to identify retroperitoneal recurrence and/or "growing teratoma" syndrome

MR Findings

- T2WI: Moderate high signal intensity lymphadenopathy in retroperitoneum

Ultrasonographic Findings

- Real Time
  - Seminoma
    - Hypoechoic, well-defined solid mass without cystic areas, calcification or invasion
  - Embryonal cell carcinoma
    - Heterogeneous predominantly solid mass of mixed echogenicity
    - Poorly marginated
    - 1/3 have cystic areas
    - Invasion of tunica may distort contour of testis
  - Teratoma/teratocarcinoma
    - Heterogenous with mixed echogenicity
    - Cystic areas and calcification (cartilage, immature bone, scarring or fibrosis)
  - Choriocarcinoma
    - Heterogeneous with mixed echogenicity
    - Cystic areas with calcification common
- Color Doppler
  - Variable vascularity; disorganized flow is typical
  - Cystic areas are avascular

Nuclear Medicine Findings

- PET

DDx: Testicular Lesions Mimicking Carcinoma

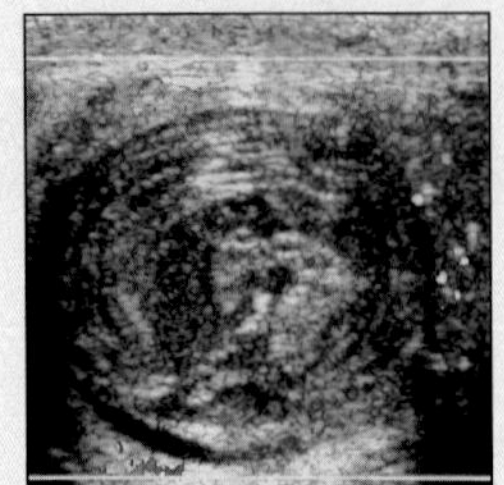

Epidermoid Cyst

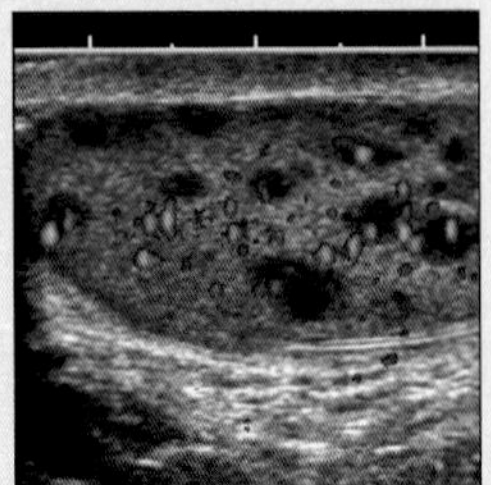

Lymphoma

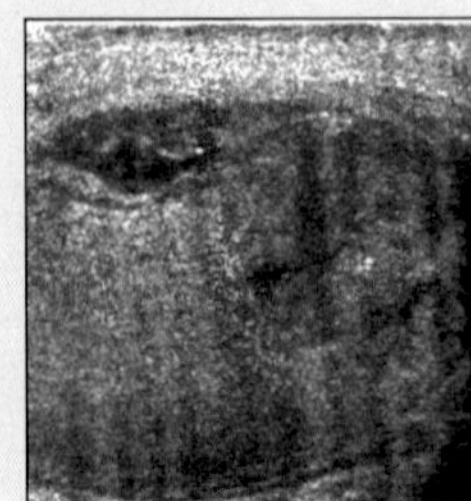

Testicular Hematoma

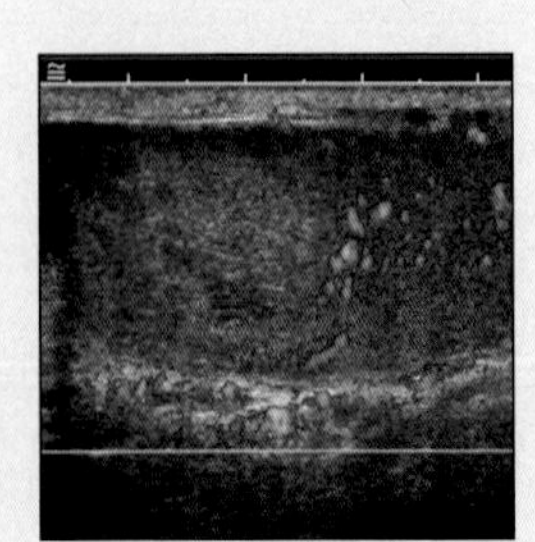

Segmental Infarct

TESTICULAR CARCINOMA

Key Facts

Terminology

- Malignant germ cell tumor of testis

Imaging Findings

- Best imaging tool: US to identify and characterize scrotal mass; CT or MR for initial staging; PET to evaluate post-treatment residual masses
- Protocol advice: High frequency (≥ 10 mHz) linear array transducer

Top Differential Diagnoses

- Epidermoid cyst
- Lymphoma
- Subacute hematoma
- Segmental Infarct
- Focal orchitis

Pathology

- Etiology-pathogenesis: Associated with cryptorchidism, previous contralateral cancer; possible association with mumps orchitis, microlithiasis and family history of tumor
- Most common cancer in men aged 15-34

Clinical Issues

- Clinical profile: Young male with palpable testicular mass, elevated tumor markers such as beta-HCG, alfa-feto-protein

Diagnostic Checklist

- Consider testicular lymphoma if bilateral lesions identified, particularly if patient is > 50 yrs
- Discrete mass on grayscale with abnormal intrinsic vessels on color Doppler

---

- Helpful to reduce false-negative CT
  - May aid in differentiating residual tumor from scar in treated patients

Imaging Recommendations

- Best imaging tool: US to identify and characterize scrotal mass; CT or MR for initial staging; PET to evaluate post-treatment residual masses
- Protocol advice: High frequency (≥ 10 mHz) linear array transducer

DIFFERENTIAL DIAGNOSIS

Epidermoid cyst

- Cystic cavity lined by stratified squamous epithelium
- "Onion skin" appearance on US due to alternating layers of keratin and desquamated squamous cells
- May have peripheral calcified rim
- No enhancement on MRI

Lymphoma

- Older age group
- Most common tumor of testes in men > 60 years
- 50% of cases bilateral
- Often multiple lesions
- Hypoechoic and hypervascular on color Doppler

Subacute hematoma

- History of trauma
- Associated hematocele
- Hypoechoic on US

Segmental Infarct

- Acute pain most common symptom
- Focal avascular area on color Doppler
- Infarct is typically hypoechoic
- No palpable mass

Focal orchitis

- Irregular hypoechoic area on US
- Increased vascularity on color Doppler without displacement of vessels
- Signs of systemic infection
- Reactive hydrocele with low level echoes
- Scrotal wall thickening
- Enlarged epididymis

PATHOLOGY

General Features

- General path comments
  - 95% of testicular tumors are malignant germ cell tumors
  - Single histologic subtype in 65% of tumors (seminoma is the most common)
  - Multiple subtypes in 35%
    - Teratoma and embryonal cell (teratocarcinoma)
    - Seminoma and embryonal cell
    - Seminoma and teratoma
  - 40-50% are seminoma
  - 25% have embryonal subtype (often mixed with other subtypes)
  - 5-10% are teratomas
  - Etiology-pathogenesis: Associated with cryptorchidism, previous contralateral cancer; possible association with mumps orchitis, microlithiasis and family history of tumor
- Genetics: Family history increases risk
- Epidemiology
  - Most common cancer in men aged 15-34
  - 1% of all cancers in men
  - Seminomas most common in men 35-39 years old
  - Seminomas rare before 10 years and after 60 years
  - Lymphoma most common tumor over 50 years
- Associated abnormalities: Gynecomastia, pre-pubescent virilization

Gross Pathologic & Surgical Features

- Solid or solid/cystic intratesticular mass
- 10-15% have epididymis or spermatic cord involvement
- Bilateral in 2-3% of cases

Microscopic Features

- Malignant germ cell tumors with numerous histologic subtypes

Staging, Grading or Classification Criteria

- Stage I (A): Tumor confined to testis
- Stage II (B): Tumor metastatic to nodes below diaphragm
- Stage IIA (B1): Retroperitoneal node enlargement < 2 cm (5 cm^3)
- Stage IIB (B2): Retroperitoneal node enlargement < 2 cm x < 5 cm (10 cm^3)
- Stage IIC (B3): Retroperitoneal node enlargement > 5 cm
- Stage III (C): Tumor metastatic to lymph nodes above diaphragm
- Stage IIIA (C1): Metastases confined to lymphatic
- Stage IIIB or IV: Extranodal metastases

CLINICAL ISSUES

Presentation

- Most common signs/symptoms
  - Palpable mass in testis
  - Dull pain (27%)
  - Acute pain (10%)
- Clinical profile: Young male with palpable testicular mass, elevated tumor markers such as beta-HCG, alfa-feto-protein

Demographics

- Age: 15-40 yrs
- Gender: Male
- Ethnicity: Increased incidence in Caucasian and Jewish males

Natural History & Prognosis

- 95% 5-year survival rate overall

Treatment

- Radical orchiectomy; retroperitoneal node dissection for non-seminomatous tumor
- XRT or chemotherapy for metastatic disease

DIAGNOSTIC CHECKLIST

Consider

- Consider testicular lymphoma if bilateral lesions identified, particularly if patient is > 50 yrs

Image Interpretation Pearls

- Discrete mass on grayscale with abnormal intrinsic vessels on color Doppler

SELECTED REFERENCES

1. Boyle P: Testicular cancer: the challenge for cancer control. Lancet Oncol. 5(1):56-61, 2004
2. Owens EJ et al: Imaging of testicular tumours. Hosp Med. 65(1):18-21, 2004
3. MacVicar GR et al: Testicular cancer. Curr Opin Oncol. 16(3):253-6, 2004
4. Jones RH et al: New directions in testicular cancer; molecular determinants of oncogenesis and treatment success. Eur J Cancer. 39(2):147-56, 2003
5. Deshpande MS et al: Metastatic adenocarcinoma in testis presenting as a testicular mass--a case report and review of literature. Can J Urol. 10(1):1770-1, 2003
6. Hussain A et al: The unsuspected nonpalpable testicular mass detected by ultrasound: a management problem. Can J Urol. 10(1):1764-6, 2003
7. Bach AM et al: Is there an increased incidence of contralateral testicular cancer in patients with intratesticular microlithiasis? AJR Am J Roentgenol. 180(2):497-500, 2003
8. Hellerstedt BA et al: Germ cell tumors: review of selected studies from 2002. Curr Opin Oncol. 15(3):234-8, 2003
9. Rajpert-De Meyts E et al: Testicular neoplasia in childhood and adolescence. Endocr Dev. 5:110-23, 2003
10. Huyghe E et al: Increasing incidence of testicular cancer worldwide: a review. J Urol. 170(1):5-11, 2003
11. Hain SF et al: Positron emission tomography for urological tumours. BJU Int. 92(2):159-64, 2003
12. Patel MI et al: Management of recurrence and follow-up strategies for patients with seminoma and selected high-risk groups. Urol Clin North Am. 30(4):803-17, 2003
13. Jewett MA et al: Management of recurrence and follow-up strategies for patients with nonseminoma testis cancer. Urol Clin North Am. 30(4):819-30, 2003
14. Scholz M et al: Stage I testicular cancer. Curr Opin Urol. 13(6):473-6, 2003
15. Vergouwe Y et al: Predictors of occult metastasis in clinical stage I nonseminoma: a systematic review. J Clin Oncol. 21(22):4092-9, 2003
16. Jones RH et al: Part II: testicular cancer--management of advanced disease. Lancet Oncol. 4(12):738-47, 2003
17. Jones RH et al: Part I: testicular cancer--management of early disease. Lancet Oncol. 4(12):730-7, 2003
18. Woodward PJ et al: From the archives of the AFIP: tumors and tumorlike lesions of the testis: radiologic-pathologic correlation. Radiographics. 22(1):189-216, 2002
19. Albers P et al: Positron emission tomography in the clinical staging of patients with Stage I and II testicular germ cell tumors. Urology. 53(4):808-11, 1999
20. Lawton AJ et al: Staging and prognostic factors in testicular cancer. Semin Surg Oncol. 17(4):223-9, 1999
21. Foster RS et al: Testicular cancer: what's new in staging, prognosis, and therapy. Oncology (Huntingt). 13(12):1689-94; discussion 1697-700, 1703, 1999
22. Bender H et al: Possible role of FDG-PET in the evaluation of urologic malignancies. Anticancer Res. 17(3B):1655-60, 1997
23. Leibovitch L et al. Improved accuracy of computerized tomography based clinical staging in low stage nonseminomatous germ cell cancer using size criteria of retroperitoneal lymph nodes. J Urol. 154(4):1759-63, 1995
24. Newhouse JH: Clinical use of urinary tract magnetic resonance imaging. Radiol Clin North Am. 29(3):455-74, 1991
25. Heiken JP et al: Testicular tumors: oncologic imaging and diagnosis. Int J Radiat Oncol Biol Phys. 10(2):275-87, 1984

TESTICULAR CARCINOMA

IMAGE GALLERY

Typical

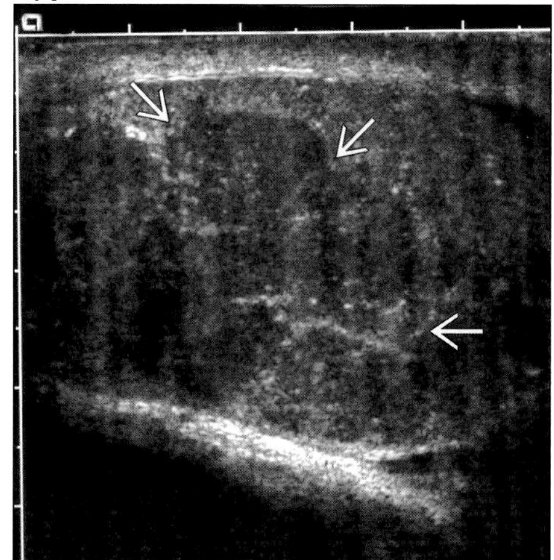

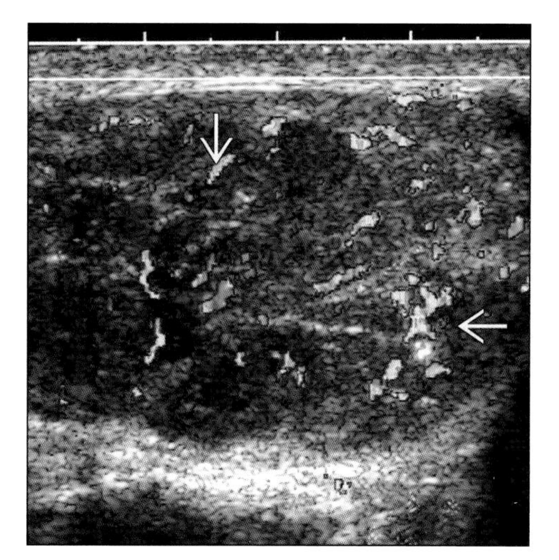

*(**Left**) Sagittal grayscale sonogram of testicular seminoma. Note well-defined hypoechoic mass (arrows). (**Right**) Sagittal color Doppler sonogram of testicular seminoma. Note internal flow within mass and disorganized pattern of vascularity (arrows).*

Typical

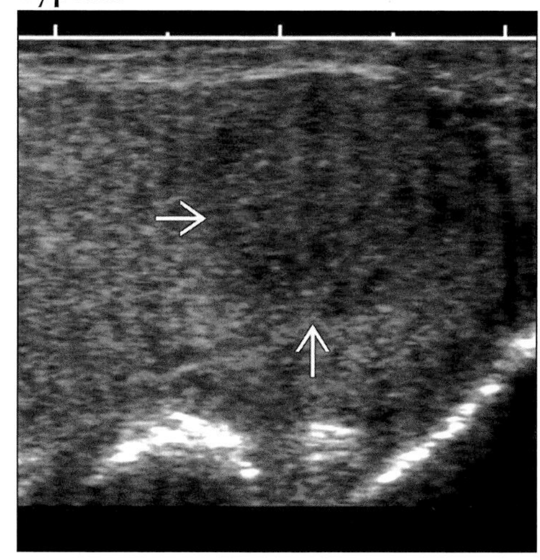

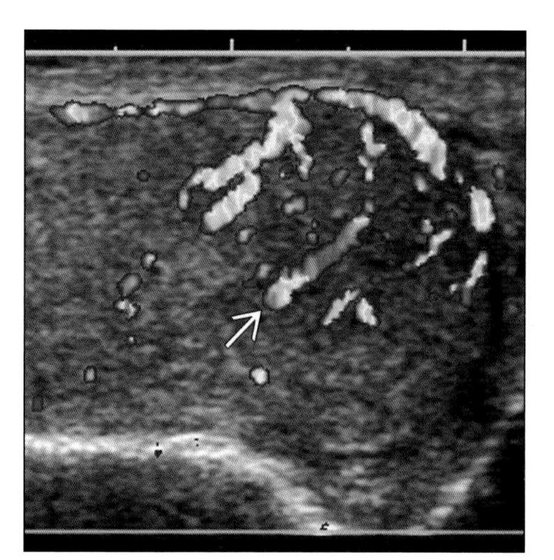

*(**Left**) Embryonal cell carcinoma of the testis on sagittal sonogram. Note ill-defined hypoechoic mass (arrows). (**Right**) Sagittal color Doppler sonogram demonstrates increased flow within mass (arrow).*

Typical

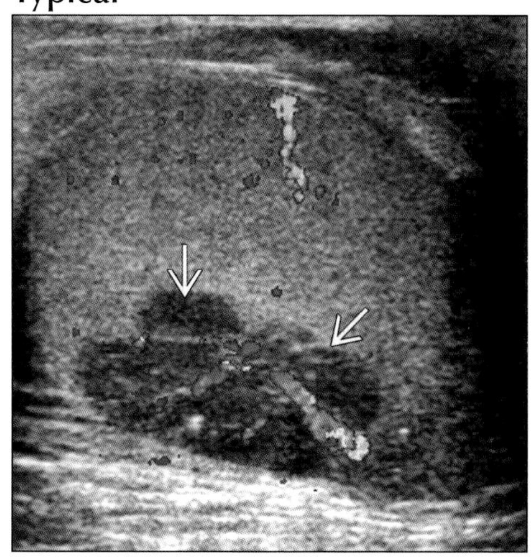

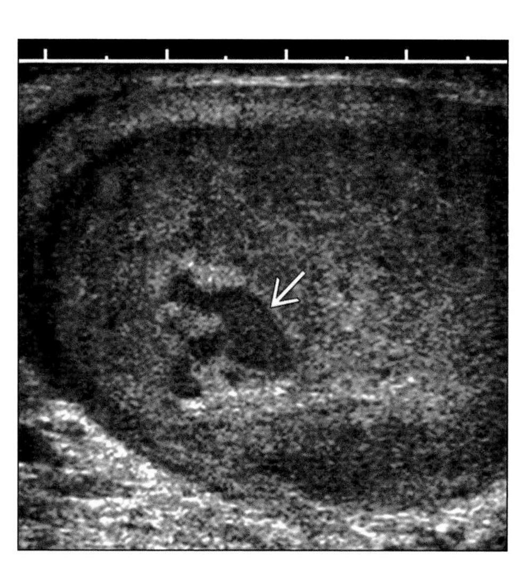

*(**Left**) Choriocarcinoma on sagittal color Doppler sonogram. Note hypoechoic mass (arrows) with internal flow. (**Right**) Mixed-cell type testicular cancer on sagittal sonogram. Note cystic component (arrow) to mixed seminoma and embryonal cell carcinoma.*

PROSTATE CARCINOMA

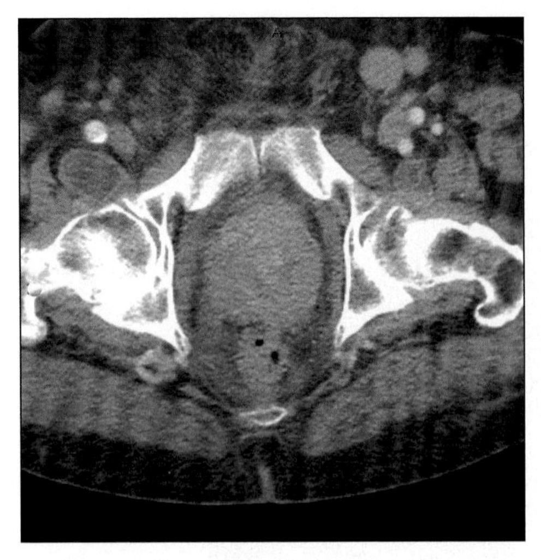

Axial CECT shows enlarged prostate with extensive malignant lymphadenopathy.

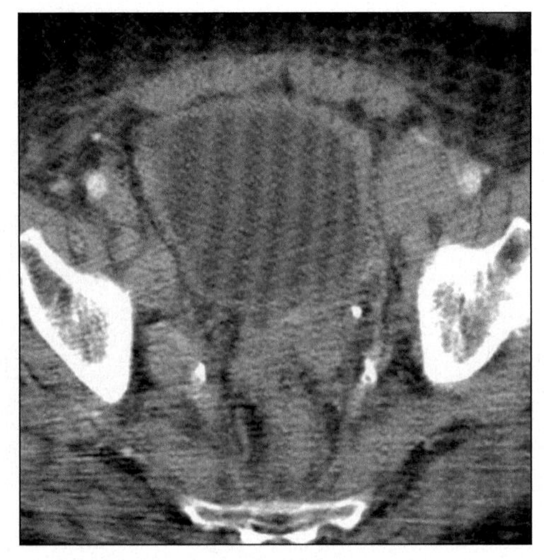

Axial CECT shows extensive pelvic lymphadenopathy from prostate cancer.

TERMINOLOGY

Definitions

- Malignancy of prostate gland

IMAGING FINDINGS

General Features

- Best diagnostic clue: ↓ Signal in a normally high signal peripheral zone on T2WI
- Location
  - Peripheral zone (70%): Posterior region (> common)
  - Transition zone (20%), central (10%)
  - Typical bony metastases (pelvis & lower vertebrae)
- Other general features
  - Most common noncutaneous cancer in American men
  - 2nd most common cause of male cancer deaths in US (after lung cancer)
  - Lifetime risk of developing prostate cancer is 18-19% & ↑ with age
  - Diagnosis of prostate cancer is suggested on basis of
    - Abnormal digital exam, transrectal US, MR findings
    - Confirmed at biopsy
    - ↑ Prostate specific antigen (PSA) level
  - Most important factor affecting prognosis & choice of treatment is presence or absence of extracapsular extension (ECE)
  - Role of diagnostic imaging in prostate cancer remains unclear
  - Imaging accuracy for local staging suboptimal
    - Better for advanced disease & metastases
  - 3D MR spectroscopy (↑ choline & ↓ citrate levels) + endorectal MR imaging ↑ accuracy in detecting & staging of local + ECE of prostate cancer
  - Histopathological types of prostate carcinoma
    - Adults: 95% adenocarcinoma
    - Children: Rhabdomyosarcoma
  - Spread
    - Hematogenous: Prostatic venous plexus drains into internal iliac veins & communicates with vertebral venous plexus (osteoblastic metastases)
    - Lymphatic: Drains mostly to internal iliac nodes, then to retroperitoneal nodes

CT Findings

- Not accurate in detection of cancer within prostate
- Signs of extracapsular extension (ECE)
  - Obliteration of periprostatic fat plane
  - Abnormal enhancement of contiguous neurovascular bundle
  - Urinary bladder, rectal invasion
  - Lymphadenopathy

DDx: Prostate Mass or Pelvic Malignancy

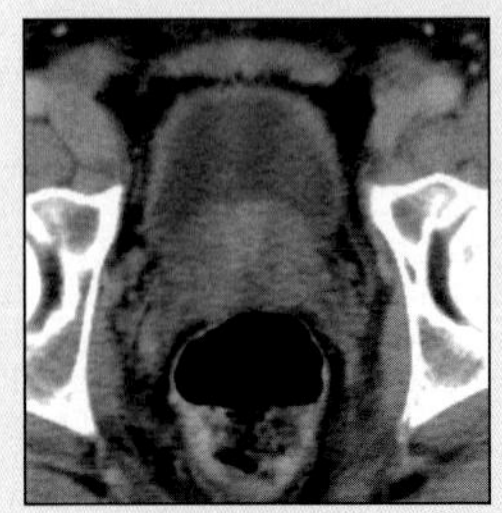

BPH

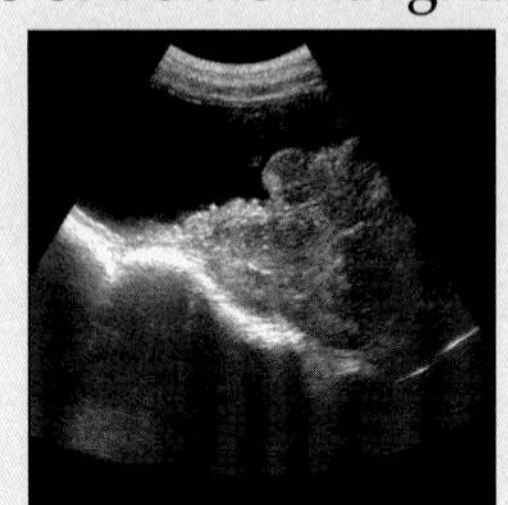

BPH

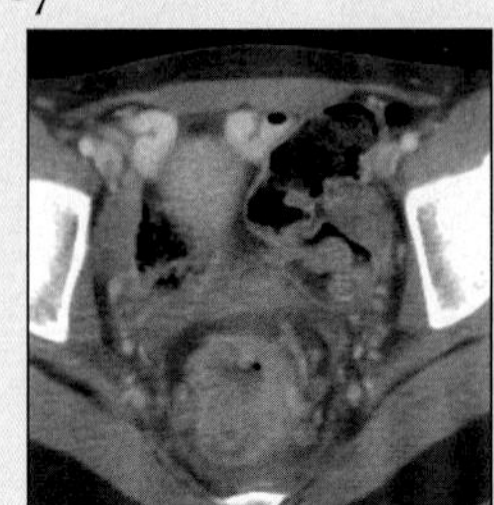

Rectal Cancer

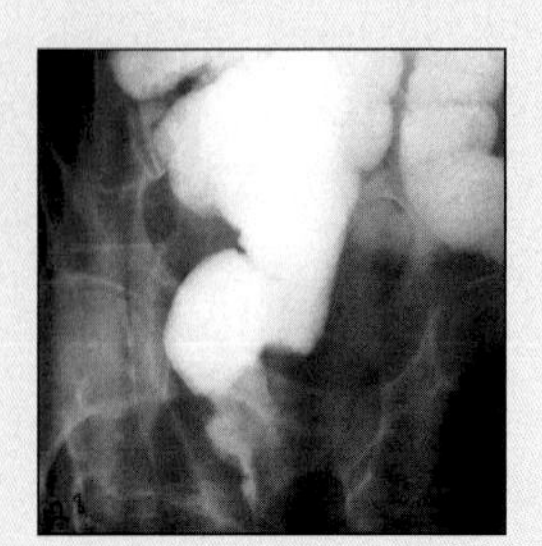

Rectal Cancer

PROSTATE CARCINOMA

Key Facts

Imaging Findings

- Best diagnostic clue: ↓ Signal in a normally high signal peripheral zone on T2WI
- Peripheral zone (70%): Posterior region (> common)
- Imaging accuracy for local staging suboptimal
- Signs of extracapsular extension (ECE)
- Obliteration of periprostatic fat plane
- Obliteration of rectoprostatic angle & neurovascular bundle
- Osteoblastic bone metastases
- Hypoechoic (60-70%) lesion in peripheral zone, abutting the adjacent normal tissue

Top Differential Diagnoses

- Benign prostatic hypertrophy (BPH)
- Rectal carcinoma

Pathology

- Advancing age, hormonal influence, environmental & genetic factors play a role in development
- 1 in 11 males will develop prostate cancer
- More common in Western world & rare in Asians
- Approximately 300,000 new cases detected annually
- 95% of tumors are adenocarcinoma

Diagnostic Checklist

- Prostate cancer until proven otherwise in an elderly man with osteoblastic metastases on a plain x-ray
- Probable cancer in a man with abnormally ↑ PSA levels + enlarged prostate on digital rectal exam
- Annual screening: PSA levels + digital rectal exam
- Area of abnormal low signal in a normal high signal peripheral zone on T2WI ± ECE & bony metastases

MR Findings

- Prostate cancer is best seen on T2WI
  - Abnormal low signal in a normally high signal peripheral zone
  - Extracapsular extension (ECE)
    - Obliteration of rectoprostatic angle & neurovascular bundle
    - Urinary bladder, rectal invasion well depicted
  - Osteoblastic bone metastases
    - Low signal intensity on both T1WI & T2WI
    - T1WI: Normal ↑ signal in fatty marrow is replaced by low-intermediate signal
- T1 C+
  - Tumor foci & ECE are well depicted

Ultrasonographic Findings

- Transrectal ultrasonography (TRUS)
  - Hypoechoic (60-70%) lesion in peripheral zone, abutting the adjacent normal tissue
  - Hyperechoic (1-5%)
  - Isoechoic (30-40% of lesions)
  - Most peripheral isoechoic lesions close to capsule
    - Asymmetrical contour + bulging along lateral aspect of prostate
  - ± ECE: Lesion extending beyond prostatic margin into periprostatic fat
  - Size versus rate of detection
    - < 5 mm (36%); 6-10 mm (53%)
    - 11-15 mm (53%); 16-20 mm (84%)
    - 21-25 mm (75%); > 26 mm (92%)
  - TRUS in local staging: Most widely used but still remains controversial

Nuclear Medicine Findings

- PET
  - Increased uptake of FDG
    - Early detection of metastatic foci
- Tc99m methylene diphosphonate (MDP) bone scan
  - Detects osteoblastic metastases by ↑ uptake

Imaging Recommendations

- Transrectal US; MRI (T2WI multiplanar images)
- Tc MDP scan for bone metastases

DIFFERENTIAL DIAGNOSIS

Benign prostatic hypertrophy (BPH)

- MR: T2WI
  - Normal prostate: ↑ Peripheral zone signal & ↓ transition zone signal
  - BPH: Enlarged gland nodular heterogeneous signal intensity
    - Glandular hyperplasia: ↑ Signal intensity
    - Fibrous & muscular hyperplasia: Intermediate-low signal intensity
    - Compressed peripheral zone into a thin rind by enlarged gland
    - Fibrous capsule: Low signal intensity

Rectal carcinoma

- Barium enema: Advanced cancer (apple-core lesion)
  - Circumferential narrowing of bowel
  - Shelf-like overhanging borders (mucosal destruction)
- Imaging
  - Mass with circumferential wall thickening
  - Extracolonic tumor extension
    - Loss of fat planes between rectum & adjacent structures
    - MR shows obliteration of rectoprostatic angle & neurovascular bundle with invasion of prostate
  - Metastases to lung, bone, liver & lymph nodes

PATHOLOGY

General Features

- General path comments
  - Embryology/anatomy
    - Base (cranially); apex (caudally)
    - Zones (peripheral, central & transition)
- Genetics
  - Familial in 10% of caucasian American men
  - Cancer susceptibility gene mapped to 1q24-25
- Etiology
  - Unknown
  - Advancing age, hormonal influence, environmental & genetic factors play a role in development

PROSTATE CARCINOMA

- Epidemiology
  - Incidence
    - 1 in 11 males will develop prostate cancer
    - Mortality rate: 13%
  - More common in Western world & rare in Asians
  - Approximately 300,000 new cases detected annually

Gross Pathologic & Surgical Features

- Growth: Usually more common in peripheral zone
  - Localized, diffuse or extracapsular extension
  - Firm or "gritty" as a result of fibrosis

Microscopic Features

- 95% of tumors are adenocarcinoma
- Well-differentiated glandular pattern
  - Uniform epithelium, oval nuclei
  - Pale cytoplasm & rare mitotic figures
- Poorly differentiated glandular pattern
  - Sheets of tumor cells, mitoses, cellular atypia

Staging, Grading or Classification Criteria

- Jewett-Whitmore & TNM staging
  - A & T1: Clinically localized (tumor not palpable on digital rectal exam)
  - A1 & T1a: Focal tumor or low grade
  - A2 & T1b: Diffuse tumor or high grade
  - B & T2: Clinically localized (tumor palpable)
  - B1 T2a: Tumor involves < ½ lobe; B2 & T2b - tumor involves > ½ lobe
  - C & T3: Locally invasive beyond prostatic capsule (tumor palpable)
  - C1 & T3a: Unilateral extracapsular extension
  - C1 & T3b: Bilateral extracapsular extension
  - C1 & T3c: Seminal vesicle invasion
  - C2 & T4: Invades adjacent tissues
    - E.g, bladder, rectum, levator ani
  - D & N/M: Lymph node & distant metastases (bones, lung, liver & brain)

CLINICAL ISSUES

Presentation

- Most common signs/symptoms
  - Asymptomatic, or bone pain
  - Urination: Hesitancy, urgency, increased frequency
- Lab data
  - Increased prostate specific antigen (PSA) level
    - Normal levels: 0-4.0 ng/mL
    - Values 4-10 ng/ml also seen in BPH
  - Ratio of free PSA - bound to α1-antichymotrypsin distinguishes prostate cancer from BPH
    - Percent of free PSA is lower in cancer compared to BPH
  - ↑ Prostatic acid phosphatase levels (non-specific)
    - Seen only with metastases; prostatic manipulation
- Diagnosis: Imaging findings & transperineal biopsy

Demographics

- Age: Adults above 40 years (risk ↑ with age)
- Ethnicity: African-Americans > Caucasians

Natural History & Prognosis

- Complications
  - Bladder outlet & rarely rectal obstruction
  - Obstructive uropathy, uremia, pathological fractures
- Prognosis
  - After radical prostatectomy (for local cancer)
    - Life expectancy more than 15 years
  - Radiation & chemotherapy without surgery
    - Life expectancy less than 5 years

Treatment

- Radical resection (for cancer confined to capsule)
- Radiation therapy (for cancer confined to capsule + outside capsule & no spread)
- Hormonal therapy for spread
  - Diethylstilbestrol & leuprolide; surgical orchiectomy
- Chemotherapy & cryosurgery

DIAGNOSTIC CHECKLIST

Consider

- Prostate cancer until proven otherwise in an elderly man with osteoblastic metastases on a plain x-ray
- Probable cancer in a man with abnormally ↑ PSA levels + enlarged prostate on digital rectal exam
- Annual screening: PSA levels + digital rectal exam

Image Interpretation Pearls

- Area of abnormal low signal in a normal high signal peripheral zone on T2WI ± ECE & bony metastases

SELECTED REFERENCES

1. Dhingsa R et al: Prostate cancer localization with endorectal MR imaging and MR spectroscopic imaging: effect of clinical data on reader accuracy. Radiology. 230(1):215-20, 2004
2. Beyersdorff D et al: Patients with a history of elevated prostate-specific antigen levels and negative transrectal US-guided quadrant or sextant biopsy results: value of MR imaging. Radiology. 224(3):701-6, 2002
3. Kurhanewicz J et al: The prostate: MR imaging and spectroscopy. Present and future. Radiol Clin North Am. 38(1):115-38, viii-ix, 2000
4. Yu KK et al: Prostate cancer: prediction of extracapsular extension with endorectal MR imaging and three-dimensional proton MR spectroscopic imaging. Radiology. 213(2):481-8, 1999
5. Scheidler J et al: Prostate cancer: localization with three-dimensional proton MR spectroscopic imaging--clinicopathologic study. Radiology. 213(2):473-80, 1999
6. Presti JC Jr et al: Local staging of prostatic carcinoma: comparison of transrectal sonography and endorectal MR imaging. AJR Am J Roentgenol. 166(1):103-8, 1996
7. Harris RD et al: Staging of prostate cancer with endorectal MR imaging: Lessons from a learning curve. Radiographics. 15: 813-829, 1995
8. Ellis JH et al: MR imaging and sonography of early prostatic cancer: pathologic and imaging features that influence identification and diagnosis. AJR Am J Roentgenol. 162(4):865-72, 1994
9. Mirowitz SA et al: Evaluation of the prostate and prostatic carcinoma with gadolinium-enhanced endorectal coil MR imaging. Radiology. 186(1):153-7, 1993
10. Hricak H et al: Prostatic carcinoma: staging by clinical assessment, CT, and MR imaging. Radiology. 162(2):331-6, 1987

IMAGE GALLERY

Typical

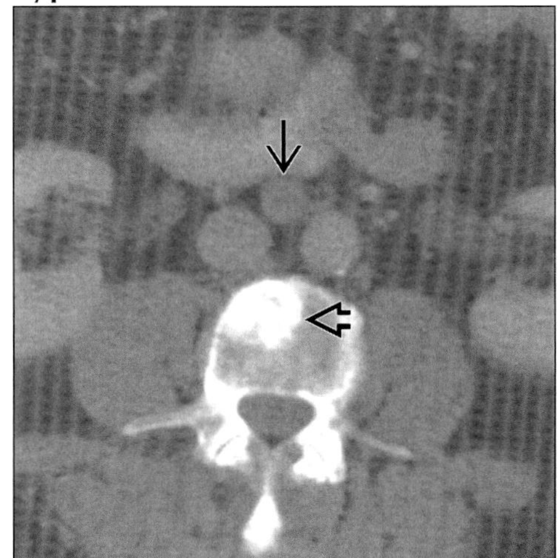

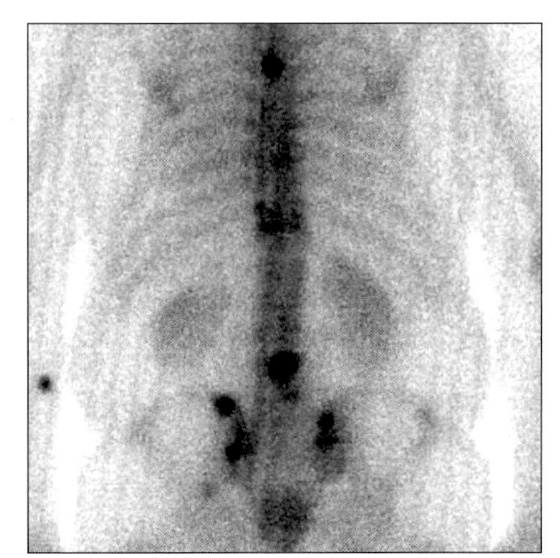

(Left) *Axial CECT shows retroperitoneal lymphadenopathy (arrow) and sclerotic vertebral metastasis (open arrow).* ***(Right)*** *Radionuclide bone scan shows multiple bone metastases, especially in the spine and pelvis.*

Typical

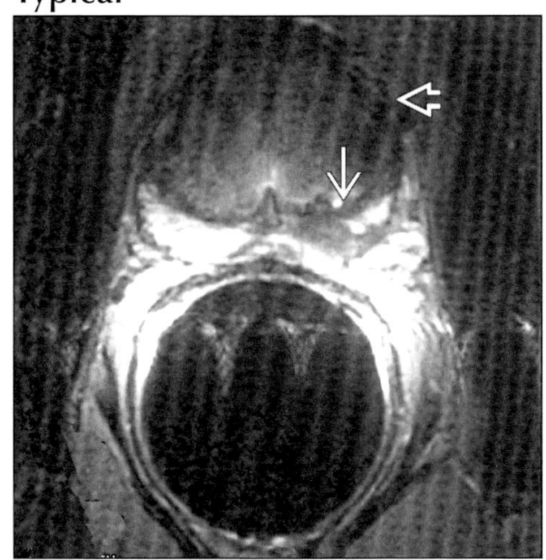

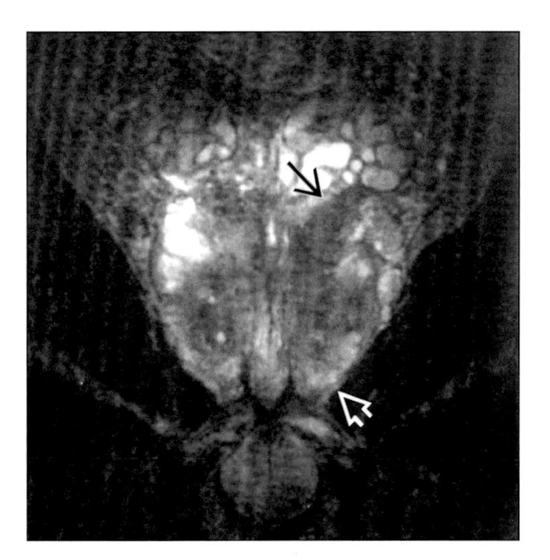

(Left) *Axial T2WI MR shows enlarged prostate (open arrow) with loss of normal high signal peripheral zone at site of carcinoma (arrow).* ***(Right)*** *Coronal T2WI MR shows enlarged prostate (open arrow) and loss of normal high signal in peripheral zone (arrow) at site of carcinoma.*

Typical

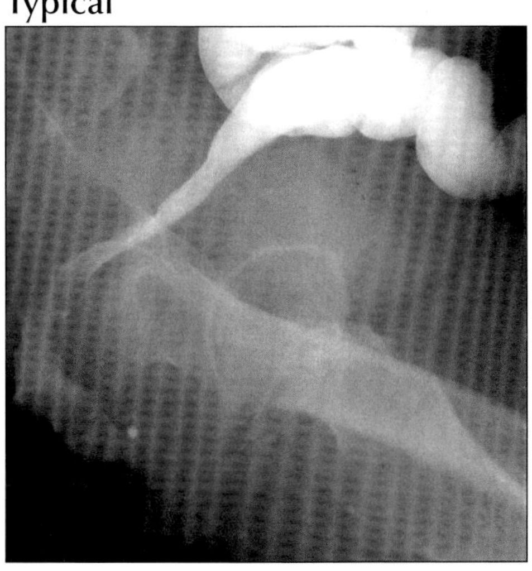

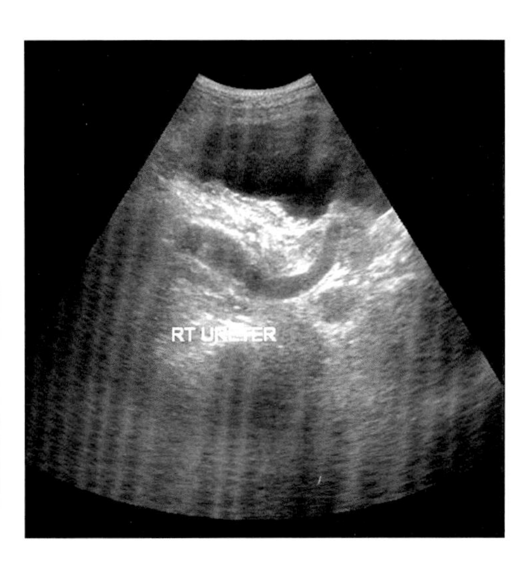

(Left) *Barium enema shows encasement and narrowing of rectum by large prostate carcinoma.* ***(Right)*** *Sagittal sonogram shows encasement and partial obstruction of distal right ureter by prostate cancer.*

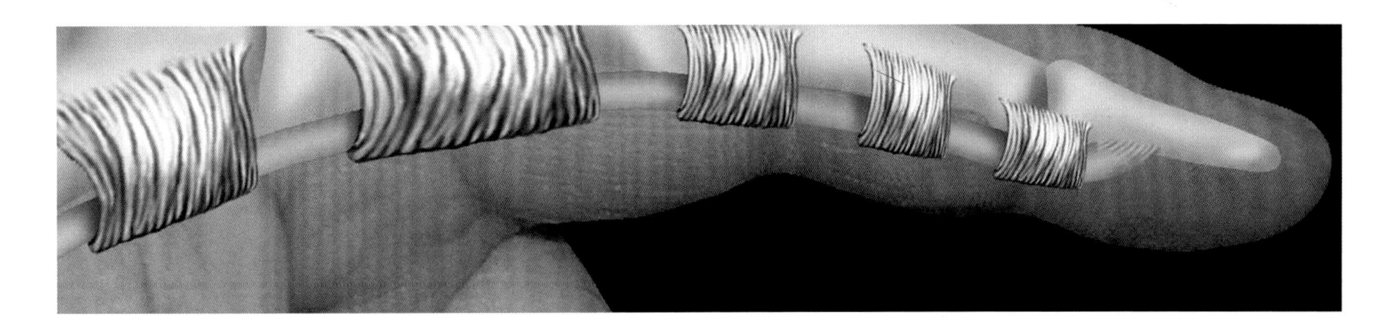

INDEX

A

INDEX

B

INDEX

C

INDEX

INDEX

i

INDEX

D

E

INDEX

F

i

INDEX

G

INDEX

H

INDEX

INDEX

I

J

K

L

INDEX

M

i

INDEX

N

INDEX

O

P

INDEX

INDEX

R

INDEX

S

INDEX

T

INDEX

INDEX

W

Y

Z